2007 7th International Conference on Power Electronics

Daegu, South Korea
22 – 26 October 2007

Pages 1-618

IEEE Catalog Number: CFP07CPB-PRT
ISBN: 978-1-4244-1871-8

Copyright © 2007 by the Institute of Electrical and Electronic Engineers, Inc
All Rights Reserved

Copyright and Reprint Permissions: Abstracting is permitted with credit to the source. Libraries are permitted to photocopy beyond the limit of U.S. copyright law for private use of patrons those articles in this volume that carry a code at the bottom of the first page, provided the per-copy fee indicated in the code is paid through Copyright Clearance Center, 222 Rosewood Drive, Danvers, MA 01923.

For other copying, reprint or republication permission, write to IEEE Copyrights Manager, IEEE Service Center, 445 Hoes Lane, Piscataway, NJ 08854. All rights reserved.

***This publication is a representation of what appears in the IEEE Digital Libraries. Some format issues inherent in the e-media version may also appear in this print version.

IEEE Catalog Number: CFP07CPB-PRT
ISBN 13:   978-1-4244-1871-8

Additional Copies of This Publication Are Available From:

Curran Associates, Inc
57 Morehouse Lane
Red Hook, NY 12571 USA
Phone: (845) 758-0400
Fax: (845) 758-2633
E-mail: curran@proceedings.com

TABLE OF CONTENTS

A Transformerless Hybrid Active Filter for Integration Into a Medium-Voltage Motor Drive with a Passive Front End..1
    H. Akagi

Digital Current Mode Control Approach for the Parallel Module DC-DC Converters............................9
    B. Cho, H. Bae

Energy Storage Technology Markets and Applications: Ultracapacitors in Combination with Lithium-ion...16
    J. Miller

Impact of Power Density Maximization on Efficiency of DC-DC Converter Systems.........................23
    J. Kolar, J. Biela, U. Badstuebner

On the Control of Active Power Filters..33
    J. Liu, X. Wang, C. Yuan, Z. Wang

Adjustable Speed Generator Systems - An Emerging Technology for Efficient Electrical Energy Generation..43
    W. Koczara

Intelligent High-voltage High-power Ballast IC..53
    J. Huang, D. Kim, M. Jung, G. Cho

Thermal Analysis of PT IGBT by Using ANSYS...59
    S. Ryu, H. Ahn, D. Han, M. Nokali

The Effect of a Shielding Layer on Breakdown Voltage in a Trench Gate...62
    J-S. Lee, H-H. Shin, H-S. Lee, E-G. Kang

Utility Interactive PV System with Improved Peak-cut Characteristics..66
    H-S. Heo, G-H. Choe, Y-H. Choi, H-S. Kim, J-C. Kim

High Voltage AlGaN/GaN Schottky Barrier Diode Employing the Inductively Coupled Plasma-chemical Vapor Deposition SiO2...71
    Y-H. Choi, J. Lim, K-H. Cho, M-K. Han

Optimized Operation and Stabilization of Microgrids with Multiple Energy Resources.....................74
    T. Tanabe, Y. Ueda, S. Suzuki, T. Ito, N. Sasaki, R. Yokoyama

A New Converter-Inverter-Brake (CIB) Module with Shoot-through Immunity..................................79
    S. Park, J-S. Kim, S-J. Kim, B-H. Kwon

Analysis of Sensitivity of the Performance of Interleaved Flyback Converter to the Principal Design Parameters...85
    Y. Wang, S. De Haan, A. van Zwam

Formulation, Measurement and Analysis for the Thrust Force of HB-Type Linear Pulse Motor..90
    D-H. Kim, J-Y. Ahn, G-I. Kang, K-H. Kim, Y-C. Lim

Application of Fast Lifting Wavelet Transform to the Estimation of Harmonic for Shunt Active Power Filters..96
    H. Liu, G. Liu, Y. Shen

Small-Signal Analysis of the Lower-frequency Power Transfer Model..102
    R. Wang, J. Liu

Generation of Differentiable Homomorphism Function Based on Self-Organizing Polynomial Data-mining Algorithm 107
Y-W. Kim, T. Narikiyo

Development of the Low-Cost Impedance Spectroscopy System for Modeling the Electrochemical Power Sources 113
W. Choi, J. Lee

3-d Education System of DCC Motor Drive Using RecurDyn and CoLink 119
H-S. Mok, J-H. Lee, G-H. Choe, S-H. Kim, D-J. Yun

A Study on the Modeling of Piezoelectric Transformer for CCFL Using a PSPICE 122
L-H. Hwang, J-H. Yoo, H. Xiao, Y-M. Zhi, H-S. Kim, H-S. Oh, M-T. Cho, G-S. Choi

An Active Gate Drive Circuit for High Power Inverter System to Reduce Turn-off Spike Voltage of Igbt 127
J-H. Kim, D-H. Park, J-B. Kim, B-H. Kwon

An Improved Strategy to Detect the Switching Device Fault in NPC Inverter System 132
J-D. Lee, T-J. Kim, J-C. Lee, D-S. Hyun

Overmodulation Strategy of NPC Type 3-level Inverter for Traction Drives 137
J. Lee, J. Choi, Y. Nishida

Discrete-time Model Following Control for PWM Inverter with Electric Double Layer Capacitor 143
S. Suzuki, T. Haneyoshi

Synchronized Pwm Control of Symmetrical Six-phase Drives 147
V. Oleschuk, G. Griva, F. Profumo, A. Tenconi

Optimized State Controller on DC-DC Converter 153
M. Poodeh, S. Esdtehardiha, M. Namnabat

A Study on Rotor Foc Method Using Matrix Converter Fed Induction Motor with Common-Mode Voltage Reduction 159
H-H. Lee, H. Nguyen, T-W. Chun

Robust Digital Controller Design for Three-phase Inverter Using CRA Method 164
J. Lee, G. Park, J. Choi, G-B. Chung

Analysis and Design of Pwm Inverter System for Flywheel Energy Storage System 169
J-C. Park, G-H. Choe, Y-S. Kim, B. Dugarjav, Z. Baljinnyam, J-P. Lee

Novel Single-state Pwm Technique in Multilevel Inverter for Unbalanced DC Sources 174
N. Nho, H. Lee, N. Tuyen

A Novel Optimization Method for Solving Harmonic Elimination Equations 180
A. Samadi, S. Farhangi

Verification of Autonomous Decentralized Control UPS Systemusing Fpga Based Hardware Controller 186
T. Saito, N. Doi, T. Yokoyama

Novel Dual-band Controller for Regulating DC-Bus Voltage That Contains a Large Voltage Ripple 192
Y. Wang, S. De Haan, A. van Zwam

A Study of the Train Running Simulation for Train Propulsion System Performance Analysis 197
Y-C. Kim, Y-G. Seo, S-C. Hong, J-S. Ko, B-S. Lee

Development of DC Line Voltage Simulator for Control of Regenerative 203
K-H. Cho, S-J. Jang, B-K. Lee, C-Y. Won, G-D. Kim

The Design and Analysis of the Piezoelectric Inverter Too the EEFL Driving of LCD Backlight 208
H-S. Park, S-H. Yang, Y-C. Lim

Direct Torque Control for Induction Machine Based on AC-AC Matrix Converter 213
G. Yougui, Z. Jianlin, C. Caixue

A Development of the Solar Position Tracker on the Program Method for the Small Typed Stand-alone PV System 217
L-H. Hwang, S-K. Na, H-S. Oh, Y-S. Kim, H. Xiao, Y-M. Zhi, M-T. Cho, S-C. Chang, G-S. Choi

A Study on Driving of 35W(T5) Fluorescent Lamp by the Electronic Ballast Using Piezoelectric Transformer 221
L-H. Hwang, J-H. Yoo, H-S. Song, S-K. Na, H-S. Kim, H-S. Oh, S-H. Lee, K-H. Choi, M-T. Cho

An Internet Based Embedded Network Monitoring System for Renewable Energy Systems 225
P-Y. Chen, S-K. Ho, W-J. Lee, C-C. Chu, C-T. Pan

Fabrication of 4H-SiC Schottky Barrier Diodes with the Epilayer Grown by Bis-trimethylsilylmethane Precursor 229
M. Oh, H. Song, J. Moon, J. Yim, J. Lee, H. Seo, Y. Choi, J. Choi, H. Kim

High Accuracy and High Stability Magnet Power Supply by Four-phase Buck Type DC/DC Converter 232
S-C. Kim, K-M. Ha, J-Y. Huang

Automatic Interleaving Control for Paralleled Converter System and Its Ripple Estimation with Simplified Circuit Model 238
T. Kohama, T. Ninomiya

Start-up Inrush Current Reduction Technique of Asymmetrical Half-bridge DC/DC Converter for PC Power Supply 243
J-K. Kim, S. Lee, W-S. Oh, C-H. Gil, J-R. Cho

Comparison the Control Methods in Improvement the Performance of the DC-DC 246
M. Namnabat, M. Poodeh, S. Eshtehardiha

Optimal Design of Bus Converter in On-board Distributed Power Architecture 252
S. Abe, M. Hirokawa, T. Ninomiya

A New Topology for Photovoltaic 4 Series DC/DC Converter with High Efficiency Under Wide Load Range 257
J. Lee, B. Min, J. Kim, K. Ryul, T. Kim, D. Yoo, B. Song, J. Yoo

Static and Dynamic Characteristics of DC-DC Converter Using Digital Filter 261
F. Kurokawa M. Okamatsu

A Study on Novel Buck-boost AC-DC Converter of High Performance by Partial Resonance Technique 266
D-K. Kwak, B-S. Lee, C-S. Kim, J-S. Shim

Hybrid Ballast for Field Emission Lamp with CNT Emitter 271
J-H. Kim, M-H. Yoo, D-W. Yoo, H-M. Jung, S-J. Ha

Band-gap Reference Voltage Control Strategy of Power Conditioning System for Fuel Cell Hybrid Vehicle 274
Y-D. Kim, K-B. Park, C-E. Kim, G-W. Moon

Development of a 72V IPMSM Drive System for Golf Carts 278
H-S. Mok, B-C. Kim, S-H. Kim, S-L. Ryu, Y-J. Oh

Development Operation Algorithm for a 2.5-Ton Electric Forklift Using an Induction Motor 283
N-C. Park, J-H. Kim, H-S. Mok, S-H. Kim

A Monitoring System with Ubiquitous Sensors for Passenger Safety in Railway Platform 289
S. Oh, G. Kim, H. Lee

Development of Water-cooled Heat Sink for High-power IGBT Inverter 295
M. Han, S-D. Lee, C. Hong, C-S. Yang, K-S. Kim

The Research of Model Reference Adaptive Control of Static Var Compensator (SVC) 300
Z. Guang, Z. Lijuan, W. Quanhai, S. Yanmin

The Study on the Protection Circuit of the Inverter for LCD Backlight Using Digital Control Method 305
J-W. Lee, Y-C. Lim, S-H. Yang

A Simple Partial Discharge Detector for Low-Voltage Rotating Electrical Machines 309
C-C. Tai, C-C. Tu, J-C. Hsieh, C-T. Wang, T-C. Huang, C-Y. Chen, Y-S. Lin, J-H. Lai

Validation of Electrical Power System Using Electrical Test Bed of GEO Satellite 315
J-D. Choi, J-C. Koo, C-H. Koo, E-C. Kim

Implement of Power Line Communication Module for an Electric Power Energy Monitoring System 319
Y-C. Jung, H-S. Jung, J-U. Kim

Electrical Interfaces Compatibility Analysis for the COMS EPS 323
J-C. Koo, E-C. Kim

Analysis of Effects of Inductance Component in Electrodeless Lamp on Ballast Performances 329
S-B. Han, S. Park, E. Song, H-G. Jeong, B-M. Jung

A Study on the Resonant Inverter for Corona Generators 333
C-L. Choi, S. Kwak, D-S. Lee

Reduced Voltage Drop Characteristics of the Series Transformer in a Voltage Disturbance Generator 337
E-C. Nho, I-D. Kim, S-D. Park, T-W. Chun, H-G. Kim, N-S. Choi

A Study on Optimal Braking Control Using Adhesion Coefficient 342
H. Lee, G. Kim, S. Park

Stress Grading in Intergrated Power Modules 346
C. Duchesne, M. Mermet-Guyennet, E. Dutarde, T. Lebey, S. Dagdag

Modelling of Iron Losses in Salient Pole Permanent Magnet Synchronous Motors 351
Y. Navrapescu, M. Popescu, D. Kisck, G. Andronescu, M. Kisck

Low Cost Fault Detection System for Inverter Driven Induction Motor Using Currents Signal 357
N-H. Kim, O. Yabg, M-H. Kim, H-A. Toliyar, Y. Oh, C-H. Choi

Control of Induction Motor Using IMC Approach 361
D. Nghia; N. Nho; H-H. Lee

Development of HBPI Controller for High Performance Control of IPMSM Drive. 367
J-S. Ko, K-T. Park, D-H. Chung, J-S. Choi, H-S. Park

Efficiency Optimization Control of SynRM Drive by LM-FNN Controller 372
J-S. Choi, K-T. Park, D-H. Chung, J. Ko, B-S. Park

A Study on Control Strategy of Electric Power Steering Based on Fuzzy Control 377
H. Zhang, J. Ren, Y. Zhong, J. Chen

A Four-Switch and Half-Bridge Boost Converter Based BLDC Drive System for DC-Link Voltage Unbalancing Compensation 381
S-H. Park, Y-H. Ryu, W-C. Lee, J-H. Lee, J-S. Yu, Y-R. Kim, C-Y. Won

Sensorless Speed Control and Initial Rotor Position Estimation of an IPMSM.........387
W-T. Joung, J-H. Lee, S-H. Moon, H-J. Wang, T-S. Hwang, C-H. Choi, Y-S. Kim

Rotor Position Detection Method of a Single Phase Switched Reluctance Motor Using Back EMF.........393
H-G. Sun, D-S. Shin, H-Y. Yang, Y-C. Lim

Modified Double-Modulation Signal PWMm Control for Large Motor Drive Using Five-Level Converter.........398
N. Kimura, T. Hamada, T. Morizane, K. Taniguchi

Improved Initial Pole-Position Estimation of SPMSM Sensorless Servo Drive by Absolute Integrals of Torque-Component Currents.........404
T-W. Kim, P. Wheeler, C-I. Jeong, J. Choi

Design and Development of Distorted Source Device for Circuit Breakers Failure Analysis.........410
S-I. Lee, J-G. Yoo, H-J. Jeon, G-H. Choe

Analysis and Simulation of Matrix Converter Using PSIM.........413
A. Sahoo, J. Meenakshi, S. Dash, T. Thyagarajan

Hybrid Three-Phase HPF Rectifier with Programmable Input Current THD, Using FPGA Device and VHDL Language.........419
J. Soares, C. Canesin, L. de Freitas

A Novel Soft Switched Topology for Power Factor Correction with Load Isolation Capability.........425
M. Yazdanian, S. Farbangi

A Grid Connected Photovoltaic System with Irradiation Injected Current Control.........430
J. Wang, J. Chen, R. Li

Performance Analysis of Soft-switching Inverter for the Photovoltaic Power System.........435
H-J. Kim, Y-H. Chung, K-S. Lee, Y-S. Ion, K-S. Kim

Development of the Power Conditioning System for High Power Fuel Cell System.........439
J-H. Lee, S-T. Baek, H-J. Jung, H-H. Kang, J-M. Chung, I-Y. Suh

A Coordinated Control Method for Leveling Output Power Fluctuations.........444
T. Senjyu, M. Datta, A. Yona, H. Sekine, T. Funabashi

Performance Monitoring and Aanlysis of Middle Scale Grid-Connected PV System.........450
J-H. So. B-G. Yu, H-M. Hwang, G-J. Yu, J-Y. Choi, I. Choi

Development of On-Line Type Dynamic Voltage Compensation System.........454
J-H. Han, J-G. Shon, I-D. Seo, H-J. Jeon

Development of Grid-connected 200kW Photovoltaic Inverter.........460
Y-C. Shin, Y-R. Kim, B-H. Ra, J-S. Moon, S-K. Kim

Development of 10kw Grid-Connected Photovoltaic Inverter with High Frequency Transformer.........463
Y-H. Son, Y-R. Kim, B-H. Ra, S-H. Kim, S-K. Kim

Z-Source Active Power Filter with a Fuel Cells Source.........466
J-H. Oum, Y-C. Lim, Y-G. Jung

A Stand Alone Type Fuel Cells Micro-Source System with a Voltage Sag.........471
J-H. Kim, Y-C. Lim, Y-G. Jung

A Novel Grid-connected PV PCS with New High Efficiency Converter.........477
B-D. Min, J-P. Lee, J-H. Kim, T-J. Kim, D-W. Yoo, K. Ryu, J. Kim, E-H. Song

Low-Cost Converters for Micro Wind Turbine Systems Using PMSG.........482
H-G. Park, S-H. Jang, D-C. Lee, H-G. Kim

Study on Voltage Regulation Method in the Power Distribution System 487
K. Yamaguchi, K. Lee, K. Kurokawa

High Efficiency Photovoltaic Pump System with Double Pumps 492
F. Qing, C. Xiaogao, X. Lei

A Variable Structure Approach to Control the Active and Reactive Power for Doubly Fed Induction Generator 497
W-S. Kim, K-B. Lee, B-C. Jeong, S-H. Song

Improved RPV (Reactive-Power-Variation) Anti-islanding Method for Grid-connected Three-Phase PVPCS 502
K-O. Lee, J-Y. Choi, I. Cho, S-H. Song, G-J. Yu, J-H. So, Y-S. Jung, B-G. Yu

Power Control of a Grid-Connected Hybrid Generation System with Photovoltaic/Wind Turbine/Battery Sources 505
J. Jeon, S. Kim, C. Cho, J. Ahn, J. Kim

Analysis and Proposition of a PV Module Integrated Converter with High Voltage Gain Capability in a Non-Isolated Topology 510
S. Araujo, P. Zacharias, T. Torrical Bascope, F. Antunes

Modeling and Analysis of Jeju Power System Operation Connected with Wind Turbine System and HVDC in Jeju Island Using PSCAD/EMTDC 517
B-H. Kim, S-B. Oh, K-B. Song, S-H. Song

An Analysis of Pemfc & Photovoltaic 500W Hybrid System 521
H-J. Choi, S-J. Park, J-S. Choi, I-S. Cha, J-P. Yoon, J-S. Suh, S-D. Gun

The Study of Operating Characteristics 1KW Households PEMFC System 524
H-J. Choi, S-J. Park, J-S. Choi, J-S. Cha, J-P. Yoon, J-S. Suh, J-I. Lee

Performance Evaluation and Analysis of 50KW Grid-connected PV System 527
J-M. Park, Z-G. Piao, Y-O. Cho, H-L. Baek

A Study on Performance Improvement of DVR System Using EDLC 530
W-R. Kim, C-J. Lim, B-K. Kwon, H-J. Jeon, J-G. Shon

Three-level VSI Based Low Switching Frequency 10 MVA STATCOM in Reactive Power and Harmonics Compensation 535
R. Lauttamus, H. Tuusa

10MVA STATCOM Installation and Commissioning 541
Y. Han, C. Chung, J. Choi, D. Kim, J. Yoon

A Test Facility for Large Scale Inverter Valve and Pole Using Resonant Circuit 547
Y-S. Han, C. Chung, H. Yoo, J. Kim, S. Kim

An Improved Synchronization Control Scheme of a Low Cost 400Hz Power Supply for No-Break Power Transfer 553
S-E. Joung, B-G. Park, D-S. Hyun, I-H. Cha

Economic Dispatch for Power Generation Using Artificial Neural Network 557
S. Pauta, S. Premrudeepreechacham

ISC-STS Development for Power Supply to Improve Reliability in Communication System 562
H-C. Jung, D-S. Hyun

Optimal Current Controller in a Three-phase Grid Connected Inverter with an LCL Filter 567
K-J. Lee, N-J. Park, D-S. Hyun

A Solar Battery Charging Module by Means of Limit-Cycle MPPT Control 571
M. Matsui, B. Yu, K. Koh, T. Kitano

Frequency Regulation for Parallel-connected UPS System Under Independent Control 575
A. Kawamura, R. Oikawa, Y. Yoshioka, Y. Matsumoto

Carrier Based Two-state Pwm Method for Optimising Voltage Error in Multilevel Inverters 580
N. Nho, T. Cong, H-H. Lee

Comparison of OMTHD and OHSW Harmonic Optimization Techniques in Multi-level Voltage-Source Inverter with Non-Equal DC Sources 586
M. Hosseiniaghdam S. Fathi, G. Gharehpetian

A Modified Space Vector PWM for Common Mode Voltage Mitigation 591
Q-V. Tran, I-R. Ahn, T-W. Chun, H-G. Kim, E-C. Nho

Development of a Current-Sensorless Multi-Loop Control for Standalone PWM Inverters 597
Y. Yisheng, L. Chang, S. Pinggang

A Novel Soft-switching Two-Stage Step-Up DC/DC Converter 601
S. Du, Z. Chen, L. Chang

A New Technique to Achieve Zero Voltage Switching in a Resonant Reset Single Switch Forward Converter 607
N. Vangala, R. Mannam

Design of Bidirectional Pwm Sepic/Zeta DC-DC Converter 613
I-D. Kim, Y-H. Lee, B-H. Min, E-C. Nho, J-W. Ahn

Complete Analysis of Steady-state and Efficiency Considerations in a Forward-Flyback Mixed Converter 619
Y. Kusuhara, T. Ninomiya, A. Nakayama, S. Nakagawa

Effect of Sampling Frequency of A/D Converter on Controller Stability and Bandwidth of Digital-Controlled Power Converter 624
Y-T. Chang, Y-S. Lai

Power Spectra of the Single Phase 6/6 SRM Drives by the Separately Randomized Pulse Position (SRP) PWM Method 629
N. Khai, D-S. Shin, Y-G. Jung, Y-C. Lim

Inductance Vector Angle Based Sensorless Speed Estimation in Switched Reluctance Motor Drive 634
F. Kucuk, H. Goto, H-J. Guo, O. Ichinokura

A Position Sensorless Control System of SRM Over Wide Speed Range 639
M-H. Kim, W-S. Baik, D-H. Kim, N-H. Kim, K-H. Choi

Efficiency Map of a Switched Reluctance Motor Using Finite Element Method in Vehicular Applications 643
H. Mokhtari, E. Tara

A Novel Direct Instantaneous Pressure Control of Hydraulic Pump System with SRM Drive 649
J. Liang, D-H. Lee, J-W. Ahn, Y-J. An

Multiloop Control Strategy for Four-wire Current Source Active Power 654
S. Pettersson, M. Salo, H. Tuusa

A Novel Online Controller for Tuning Shunt Active Power Filters Based Upon Switched-Capacitors 659
O. Elgendy, A. Abass, A. Mahmoud, A. Alkoshairy

A Multiband Shunt Hybrid Active Filter with Sensorless Control 665
S. Kumar, P. Sensama

A Shunt Active Filter for Reactive Power Compensation and Harmonic Mitigation 671
N. Apte, V. Bapat, A. Jog

A Novel Control Method of Papf for Resonance Damping and Harmonics Compensation in Power System 676
L. Wu, F. Zhuo, J. Liu, Z. Wang

Power Semiconductors State-of-the-Art and Future Development Trends 682
L. Lorenz

Optimization of Power MOSFETs for Low Power Motor Drive Applications 686
J-H. Song, S-T. Han, T-S. Kwon, S-I. Yong

High-Power and High-Speed Semiconductor Switch RSD Applied in Pulsed Power System 690
L. Liang, Y. Yu, L. Deng, Y. Peng

New Smart Power Module for Low Power Motor Drives 694
T-S. Kwon, J-H. Song, J-B. Lee, S-H. Paek, S-I. Yong

Cost-effective Driving System for Plasma Displays by Replacing and Combining Functions of Power Switches 699
D-M. Lee, D-S. Hyun

A Cold Cathode Fluorescent Lamp Driving Circuit Without a Transformer for Liquid Crystal Display Backlight Unit 705
E-S. Choi, J-H. Cho, H-K. Yoon, G-W. Moon, M-J. Youn

A Novel Adaptive Dimming Technique with X-Y Channels for LED Backlight System of LCD TVs 709
W-S. Oh, K-M. Cho, D. Cho, G-W. Moon, B. Yang, T. Jang

Cim Based Soft Switching of Energy Recovery Sustain Driver for AC PDPs 713
S-B. Lim, J-Y. Lee, J-S. Ko, S-C. Hong

Automated Diagnosis of Rolling Bearing Faults in Electrical Drives 717
H. Zoubek

Rotor Fault Detection System for Inverter Driven Induction Motor Using 723
N-H. Kim, M-H. Kim, H. Toliyar, S-H. Lee, C-H. Choi, W-S. Baik

Fault Diagnosis of Induction Motor Using Decision Tree with an Optimal 728
N-T. Nguyen, J-M. Kwon, H-H. Lee

A Simple Fault Detection of the Open-Switch Damage in BLDC Motor Drive Systems 732
J-D. Lee, B-G. Park, T-S. Kim, J-S. Ryu, D-S. Hyun

Interleaved Or Sequential Switching - for Increasing the Switching 737
A. Cabral Ferreira, R. Kennel

Space Vector Modulation Method for Unidirectional Four-Wire Three-Phase/Level/Switch (Vienna) Rectifier 741
J. Alahuhtala

Voltage Balancing Technique with Low Switching Frequency for Cascade Multilevel Active Front-End 748
O. Alizadeh, S. Farhangi

Diode Rectifier Circuits with Commutation Capacitors for Three-phase 753
T. Takeshita

Two-phase Interleaved Buck Converter with a New Digital Self-Oscillating Modulator 759
L. Jakobsen, M. Andersen

Feedback Linearization Control of Three-phase AC/DC PWM Converters 765
D-E. Kim, D-C. Lee

Digital Control of Phase-shifted Full Bridge PWM Converter 771
J-G. Lim, S-H. Lim, S-K. Chang

Autonomous Variable Speed Power Generating System with Five-level Cascaded Converter 777
L. Grzesiak, J. Tomasik

Modularized Charge Equalization Converter with High Power Density and Low Voltage Stress for HEV Lithium-Ion Battery String 783
H-S. Park, C-E. Kim, C-H. Kim, J-H. Lee

Design and Control Algorithm Research of Active Regenerative Bidirectional DC/DC Converter Used in Electric Railway 789
C-H. Park, S-J. Jang, B-K. Lee, C-Y. Won, H-M. Lee

Charge Equalization Converter with Parallel Primary Winding for Series Connected Lithium-Ion Battery Strings in HEV 794
C-H. Kim, H-S. Park, C-E. Kim, J-H. Lee, J. Oh

Calculation of Regenerative Energy in DC 1500V Electric Railway Substations 800
C-H. Bae, D-U. Jang, Y-G. Kim, S-K. Chang, J-K. Mok

Robust Speed Sensorless Induction Motor Drives 805
E. Geetha, T. Thyagarajan, V. Subrahmanyam

Improved Rotor Position Estimation Employing Voltage Distortion Compensation for Sensorless PMSM Drives at Low Speed 810
S-M. Jung, J-S. Park, H-W. Kim, M-J. Youn

Indirect Position and Speed Sensing for PMSM Sensorless Control 816
D. Yousfi, M. Adnani

Hybrid Sensor-less Control of Permanent Magnet Synchronous Motor in Low-Speed Region 822
Y. Yamamoto, S. Ogasawa, H. Funato

A New Method for Smoothing Output Power Fluctuations of PV System Connected to Small Power Utility 828
T. Senjyu, M. Datta, A. Yona, H. Sekine, T. Funabashi

Case Study of Distrubution-unified Power Flow Controller (D-UPFC) in the Clustered PV System 834
K. Lee, K. Yamaguchi, K. Kurokawa

A Novel Islanding Detection Method Based on Minute Asymmetrical Current Injection for Three-Phase Grid-connected PV Inverters 840
X-D. Sun, M. Matsui, B-G. Yu

A Novel Single-voltage-sensor-based Maximum Power Point Tracking Method 846
X-D. Sun, M. Matsui, K. Yanagimura

Frequency-Domain Transformation Approaches to Develop the D-Q Synchronous-Frame Models of Three-phase Symmetrical Networks and Applications in Modeling of PWM Power Converters 850
X. Wang, J. Liu, Y. Meng, J. Hu

A Gate Drive Circuit of Power MOSFETs and IGBTs for Low Switching Loss 856
T. Shimizu, K. Wada

Load Modeling for the Drum Washing Machine System Simulation 860
J-H. Lee, T-W. Kong, W-C. Lee, J-S. Yu, C-Y. Won

Advanced Simulation Concept for the Power Train of an AC Locomotive and Its Verification 865
V. Staudt, C. Heising, A. Steimel

Modeling Diode Reverse Recovery and Corresponding Implementation in Fast Time-Domain Simulation 870
H. Wang, J. Liu, R. Wang, Q. Hou

A Half Bridge Flyback Converter with ZVS and ZCS Operations 875
L-M. Wu, C-Y. Pong

A New Half Bridge Converter Without DC Offset of Magnetizing Current 882
K-M. Cho, W-S. Oh, K-W. Lee, G-W. Moon

A Novel Low-loss Modulation Strategy for High-Power Bi-Directional Buck+Boost Converters 888
S. Waffler, J. Kolar

Comparison of Pwm Strategies for Three-phase Current-fed DC/DC Converter 894
H. Cha, S. Choi

Effect of AC Inductance in a Phase Shifted DC-DC Bridge Converter 899
P. Freere, W. Kong, D. Holmes

Improved Low Speed Performance of a PI-DTC Interior Pm Machine with Compensations of Dead-time Effects and Forward Voltage Drops 905
S. Sayeef, M. Rahman

Study on Efficiency Optimizing of PMSM for Pump Applications 911
G-X. Zhou, H-J. Wang, D-H. Lee, J-W. Ahn

Wide-Speed-Range Optimal PAM Control for Permanent Magnet Synchronous Motors 915
T. Senjyu, Y. Noguchi, N. Yrasaki, A. Yona, H. Sekine, T. Funabashi

Development Issues of an ISG PM Machines and Control System 921
X. Chen, M. Edington, R. Thornton, Y. Fang, Q. Peng

Control of IPMSM Drive System for Drum Washing Machine 929
W-C. Lee, S-H. Park, J-H. Lee, Y-R. Kim, C-Y. Won

Fuzzy Logic Controller for Boost Converter with Active Power Factor Correction 935
F. Martinez, D. Gomez

A Low Cost and High Reliability Control Scheme in Parallel Operation for 400Hz Power Supply System 940
S-E. Joung, D-S. Hyun

Frequency Characteristics of the D-Q Synchronous-Frame Current Reference Generation Methods for Active Power Filter 944
X. Wang, J. Liu, J. Hu, Y. Meng, C. Yuan

IHTSs Development for Uninterruptible Power Supply at UPS Fault 950
H-C. Jung, D-S. Hyun

A Grid Current-Controlling Shunt Active Power Filter 955
H. Tumbelaka, L. Borle, C. Nayar, S. Lee

A New Battery Equalizer Based on Buck-Boost Topology 961
S-H. Park, T-S. Kim, J-S. Park, G-W. Moon, M-J. Yoon

Design and Test of Controller in Power Conditioning System for Superconducting Magnetic Energy Storage 965
H. Zhang, J. Ren, Y. Zhong, J. Chen

Research of Fuzzy Logic-Controlled Smes for Power System Transient Voltage Stability......971
  X-H. Huang, L-Y. Xiao

A Calculation of Predicting the Expected Life of Super Capacitor Following Current Pattern of Railway Vehicles......976
  J-Y. Kim, S-J. Jang, B-K. Lee, C-Y. Won, C-M. Lee

Comparative Study on a Single Energy Recovery Circuit with Dividing Energy Recovery Path for Plasma Display Panels (PDPs)......982
  K-H. Yi, S-W. Choi, G-W. Moon

MFFL Driving System with Current Feedback Maintaining Glow Discharge Mode......988
  J. Baek, J. Park, B. Cho

A New PWM-controlled Quasi-resonant Converter for High Efficiency PDP Sustaining Power Module......994
  W-J. Lee, D-Y. Kim, C-E. Kim, G-W. Moon

Current Stress Minimizing Control Scheme for Power Factor Correction(PFC) Boost Pre-regulator......1001
  H-C. Lee, C-E. Kim, H-S. Park, K-B. Park, G-W. Moon

Novel Three Phase Flux Reversal Machine with Full Pitch Winding......1005
  D. Moore, B. Fernandes

Comparison of Winding Concepts for Bearingless Pumps......1011
  K. Raggl, J. Kolar, T. Nussbaumer

A Multi-Winding Transformer Model for Predictive Smps Design and Analysis......1019
  P. Okyere

Comparison of Topologies for Linear Drives in Industrial Material Handling and Processing Applications.......1025
  P. Mutschler

AC-SPWM-cycloconverter Based on an Extended Chopper Scheme......1031
  Q. Zhou, W. Wu, J. Lin

Matrix Converter with a Novel General Commutation Strategy......1036
  N-S. Choi, Y. Li, B-M. Han, J-S. Ko, E-C. Nho

Novel DC Link Balancing Scheme in Generic N-Level Back-to-Back Converter System......1042
  L. Grzesiak, J. Tomasik

Matrix Converter As UPFC for Transmission Line Compensation......1048
  A. Dasgupta, P. Tripathy, P. Sensarma

Novel Carrier PWM Technique with Extension Range for 4-Switch Inverter......1054
  N. Nho, T. Phuc, N. Bac

Current Control of PMSM in Overmodulation Region......1060
  J-S. Park, S-M. Jung, H-W. Kim, M-J. Youn

Characteristics of a Pulse-Link Inverter for Fuel Cells......1064
  K. Fukushima, I. Norigoe, K. Tsukakoshi, T. Ninomiya, Y. Harada, Z. Dai

Analysis of Conduction and Switching Losses of a Matrix-Z-Source Converter......1069
  K. You, M. Rahman

Advanced Development of High Frequency Transformer Parasitic Inductive Components and Lossless Inductive Snubber-assisted Series......1075
  B. Saha, M. Ishirobi, H. Sugimura, S. Mun, M. Nakaoka

Common-mode Noise Reduction in the Watkins-Johnson Converter......1081
  M. Shoyama, T. Ninomiya

Current Ripple Reduction Technique of DC/DC Converter 1085
  Y-T. Chang, C-A. Yeh, M. Hamaogi, F. Takahashi

An Improved Energy Recovery Clamp Circuit for PWM Converters with a Wide Range of Input Voltage 1090
  H. Cha, L. Chen, F. Peng, Q. Tang

Power Saving of PWM Rectifier-VSI Fed Induction Machines 1096
  S. Halasz, I. Kadar

A New Multi-machine Control System Based on Direct Torque Control Algorithm 1101
  H. Mahkhtari, A. Alizadeh

New Current Controller for Inverter Fed Medium Voltage Drives with LC Filter 1107
  T. Laczynski, A. Martens

Low Frequency Stability Study of a Three-Phase Induction Motor 1113
  M. Uddin, M. Pramanik, S. Reza

Modeling of Battery Charging Wind Turbines 1119
  J. Reed, G. Venkataramanan, J. Rose

Development of Simulator for DFIG-Based Wind Turbine 1125
  Y-G. Seo, Y-C. Kim, J-S. Ko, S-C. Hong, B-M. Han

Lcl Filter Design for Grid-connected NPC Inverters in Offshore Wind Turbines 1131
  S. Araujo, A. Engler, F. Antunes

Decoupled Voltage and Frequency Controller for an Isolated Pico Hydro System Feeding Dynamic Loads 1137
  B. Singh, G. Kasal

Power Electronic Conversion Systems for Deepwater Applications 1143
  J. Aguillon-Garcia, A. Kristoffersen, T. Undeland

Accurate Mixed Electrical and Electromagnetic Model of a 6,5kV Igbt Module 1149
  E. Batista, J. Dienot, M. Mermet-Guyennet, A. Castellazi, M. Ciappa, W. Fitchner

Research on a Flexible Waveform Power Amplifier Adopting Switch-Linear Hybrid(SLH) Scheme 1154
  X. Liu, S. Liu, L. Ge

Switching Transient Shaping of Rf Power MOSFETs for a 2.5 MHz, Three-Phase PFC 1158
  M. Hartmann, A. Muesing, J. Kolar

Controlling Voltage Profile in Loop Distribution System with Distributed Generation Using Series Type BTB Converter 1165
  R. Simanjorang, Y. Miura, T. Ise

A New Direct Current Internal Resistance and State of Charge Relationship for the Li-Ion Battery Pulse Power Estimation 1171
  J-H. Kim, S-J. Lee, J-M. Lee, B-H. Cho

A New Single-stage PFC AC/DC Converter with Voltage-Doubler Rectified Asymmetric Half-Bridge Converter 1177
  B-H. Lee, C-E. Kim, K-B. Park, G-W. Moon

Direct High Frequency Soft Switching PWM Cyclo-Converter-Fed AC-DC Converter Without DC Link for Consumer Magnetron Drive 1183
  H. Sugimura, B. Saha, S. Mun, E. Hiraki, H. Omori, M. Nakaoka

An Estimation Method on the Initial Pole Position of a Z-Axis PMLSM 1189
  J-W. Lee

Direct Torque Control Permanent Magnet Synchronous Motor Drive with Asymmetrical Multilevel Inverter Supply 1194
M. Kadir, S. Mekhilef, H. Ping

Robust Measurement Disturbance Observer Design for AC Motor Drive Systems with Current Measurement Errors 1200
K-R. Cho, J-K. Seok

New Flux Weakening Control for Surface Mounted Permanent Magnet Synchronous Machine Using Gradient Descent Method 1206
Y-D. Yoon, S-K. Sul

HVDC Control Development for Isolated Small Ac System with Real Time Digital Simulator 1211
T. Yoshino, J. Lee, C. Lee, S. Park

Using a Cascaded H-Bridge STATCOM for Rebalancing Unbalanced Voltages 1217
R. Betz, T. Summers

Distributed Power Supply with Power Factor Correction: a Solution to Feed All Modules of Power Electronic Transformers 1223
M. Saghaleini, S. Farhangi

Development on 31.5MVA STATCOM and Digital Evaluation Tool for Voltage Flicker Compensation 1228
K. Hidesi, T. Ishizuka, F. Aoyama, S. Ota, K. Ogushi

Author Index

A Transformerless Hybrid Active Filter for Integration into a Medium-Voltage Motor Drive with a Passive Front End

Hirofumi Akagi, *Fellow, IEEE*
Department of Electrical and Electronic Engineering
Tokyo Institute of Technology
S3-17, 2-12-1, O-okayama, Meguro-ku, Tokyo, 152-8552, JAPAN
E-mail: akagi@ee.titech.ac.jp

Abstract—This paper discusses a transformerless hybrid active filter integrated into the 6.6-kV, 1-MW adjustable-speed motor drive having a three-phase diode rectifier at the front end. The hybrid filter consists of an active filter using a three-level diode-clamped PWM converter rated at 60 kVA, and a 250-kVA passive filter tuned to the 7th-harmonic frequency. They are directly connected in series without transformer. This circuit configuration enables to use 1.2-kV IGBTs because the dc voltage of the three-level converter is 1.32 kV (20% of 6.6 kV). Voltage-balancing control characterized by superimposing a 6th-harmonic zero-sequence voltage on the active-filter voltage reference in each phase is introduced to the three-level converter with triangle-carrier modulation. Experimental waveforms obtained from a 400-V 15-kW down-scaled system verify the viability and effectiveness of the proposed hybrid filter, keeping the two dc-capacitor voltages balanced well.

Index Terms—Active filters, diode rectifiers, harmonics, hybrid filters, passive filters, three-level converters.

I. INTRODUCTION

A. Background

With the emergence of high-voltage IGBTs rated at 3.3 kV, 4.5 kV and 6.5 kV, attention has been paid to medium-voltage adjustable-speed motor drives without transformers. Generally, their nominal motor voltages are in a range of 2.3 kV, 3.3 kV, 4.16 kV and 6.6 kV [1]. Medium-voltage adjustable-speed motor drives for energy savings require neither fast speed response nor regenerative braking, particularly in applications to fans, blowers and pumps. As a result, a three-phase diode rectifier can be used as the front-end converter of such a motor drive, instead of a three-phase PWM rectifier. The diode rectifier is much more efficient and reliable as well as less expensive than the PWM rectifier. However, the diode rectifier produces a large amount of harmonic current, and therefore it does not comply with the harmonic guidelines.

Hybrid active filters consist of single or multiple voltage-source PWM converters and passive components such as capacitors, inductors, and/or resistors. They are more attractive in harmonic filtering than pure active filters from both viability and economical points of view, particularly for medium-voltage applications [2]-[12].

The authors have proposed a transformerless hybrid active filter for harmonic compensation of a three-phase diode

rectifier with a capacitive load [13]-[16]. This hybrid filter is formed by a three-phase passive filter tuned to the 7th-harmonic frequency and a small-rated active filter using a three-phase two-level PWM converter. They are directly connected in series without transformer. This paper follows the previouly-published papers [13]-[16], with focus on a hybrid active filter using a three-level diode-clamped or neutral-point-clamped PWM converter. This hybrid filter is well suited to a 6.6-kV motor drive having a three-phase diode rectifier at the front end, because the hybrid filter can use 1.2-kV IGBTs that are currently available from the market at reasonable cost. A concern resulting from introducing the three-level converter into the hybrid filter is voltage imbalance of the two split dc capacitors.

B. The Three-Level Diode-Clamped Converter

Since the three-level PWM inverter was invented in 1980 [17], comprehensive research has been achieved on voltage-balancing control of the two split dc capacitors [18]-[26]. However, the research has been confined to the three-level converters and inverters with sinusoidal current inputs or outputs because they have been applied to STATCOMs (STATic synchronous COMpensators) and motor drives.

A three-phase pure active filter has the same power circuit as a three-phase PWM rectifier. However, the active filter is controlled to draw a three-phase non-sinusoidal current from the point of installation, unlike the PWM rectifier drawing a three-phase sinusoidal current. This makes more complicated and/or difficult voltage-balancing control for the three-level converter used as the active filter than that for the three-level converter used as the PWM rectifier. As a result, no literature has addressed voltage-balancing control for the three-level converter in pure and hybrid active filters, including experimental verification of its validity and effectiveness, although a few papers have dealt with a pure active filter using the three-level converter [27]-[28].

C. Triangle-Carrier and Space-Vector Modulations

The three-level PWM converter can be classified into triangle-carrier modulation and space-vector modulation by pulsewidth modulation. The authors of this paper prefer

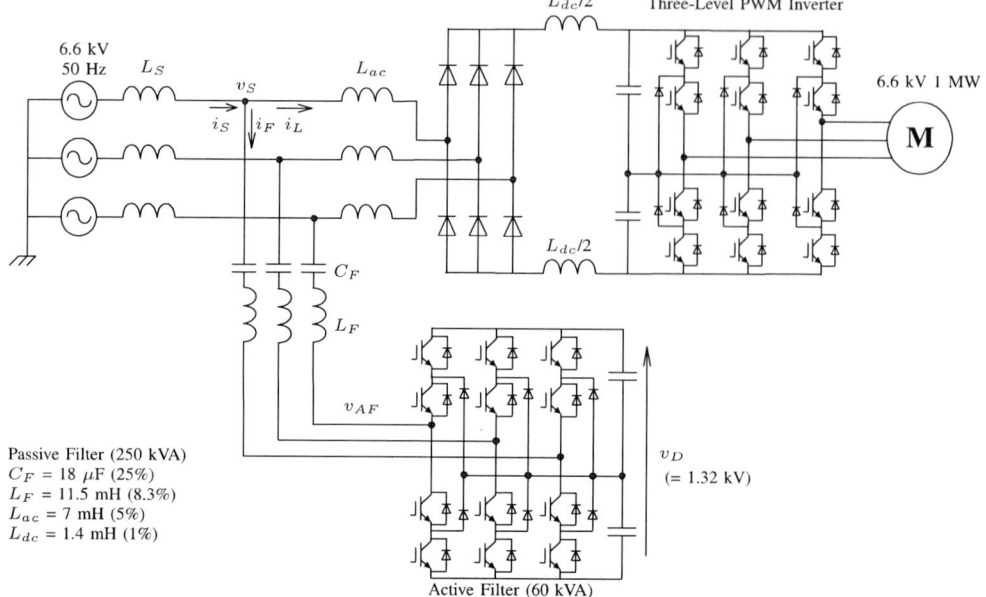

Fig. 1. The 6.6-kV, 1-MW transformerless adjustable-speed drive equipped with the hybrid filter.

triangle-carrier modulation to space-vector modulation in terms of simple and easy implementation of voltage-balancing control, when the three-level converter is used as an active filter. The main reason is that the three-level converter in active filters requires a higher switching frequency than that in motor drives, which lies usually around 10 kHz to provide satisfactory filtering performance.

This paper lays emphasis on voltage-balancing control for the three-level converter with triangle-carrier modulation. A down-scaled laboratory system rated at 400 V and 15 kW is designed, constructed and tested to verify the viability and effectiveness of the hybrid filter. Experimental waveforms obtained from the laboratory system show that the hybrid filter provides satisfactory filtering performance in steady and transient states, maintaining the two dc-capacitor voltages balanced well, even in transient states.

II. SYSTEM CONFIGURATION

Fig. 1 shows the circuit configuration of a 6.6-kV, 1-MW transformerless adjustable-speed motor drive integrated with a hybrid active filter. The hybrid filter consists of the 250-kVA passive filter tuned to the 7th-harmonic frequency and the 60-kVA active filter using a three-phase three-level diode-clamped PWM converter. Note that each IGBT symbol in the 6.6-kV, 1-MW three-level inverter represents either a string of two 4.5-kV IGBTs connected in series or a string of three 3.3-kV IGBTs connected in series. The dc voltage of the 60-kVA active filter is designed as 1.32 kV (= 6.6 kV × 0.2), so that a dc voltage of 660 V is applied across each split dc capacitor. Therefore, each IGBT symbol in the three-level converter of the 60-kVA active filter represents a single 1.2-kV IGBT. This makes the active filter inexpensive because the 1.2-kV IGBT is available on the market at reasonable cost.

Although the active filter has a carrier frequency of 10 kHz, the actual switching frequency of each IGBT is 5 kHz, that is,

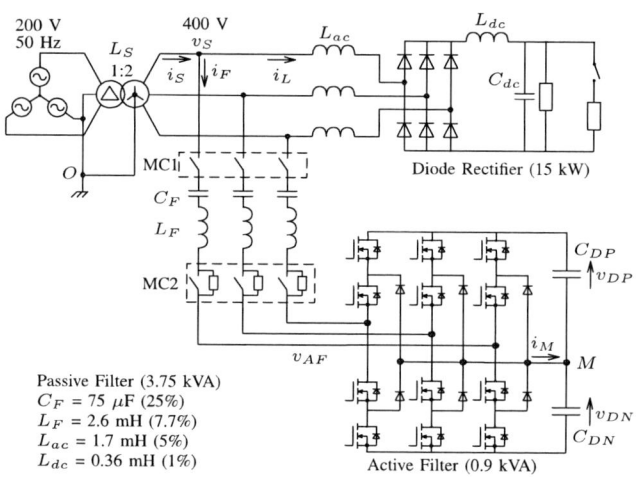

Fig. 2. The 400-V, 15-kW down-scaled system.

half of the carrier frequency. This leads to less switching loss. Generally, the ac inductor L_{ac} of the 6.6-kV diode rectifier ranges from 5 to 10%. Fig. 1 assigned it to 5%, thus resulting in cost and size reductions. Note that L_S (= 1.8%) is not an intentionally-connected inductor for the hybrid filter, but an equivalent background system inductor seen upstream of the point of installation of the hybrid filter.

Fig. 2 shows the circuit configuration of a 400-V, 15-kW down-scaled system that was designed, constructed and tested to confirm the validity of Fig. 1. Table I summarizes the circuit and control parameters of the down-scaled system. The active filter in Fig. 2 has a carrier frequency of 10 kHz, that is the same as that in Fig. 1. The passive filter is tuned, not to the most dominant 5th-harmonic frequency, but to the second most dominant 7th-harmonic frequency, thus bringing cost and size reductions to the passive filter. As a result, the passive filter

TABLE I

SPECIFICATIONS AND PARAMETERS OF THE 400-V, 15-KW
DOWN-SCALED SYSTEM.

Diode rectifier rating	15 kW
Nominal line-to-line rms voltage	400 V
Line frequency	50 Hz
Background system inductance: L_S	0.6 mH (1.8%)
AC inductor: L_{ac}	1.7 mH (5%)
DC inductor: L_{dc}	0.36 mH (1%)
DC capacitor of rectifier: C_{dc}	2 mF
Unit capacitance constant: H_{DR} [29]	18.8 ms
Filter capacitor: C_F	75 μF (25%)
Filter inductor: L_F	2.6 mH (7.8%)
Resonant frequency between C_F and L_F	360 Hz
Quality factor of L_F: Q	33
Active filter rating	0.9 kVA
DC capacitor of active filter: $C_{DP} = C_{DN}$	10 mF
DC voltage of active filter: V_D	80 V
Unit capacitance constant: H_{HF} [29]	1.1 ms
Cut-off frequency of HPF	50 Hz
Cut-off frequency of LPF	16 Hz
Feedback gain: K	21 Ω (2 p.u.)

on a three-phase, 400-V, 50-Hz, 15-kVA base

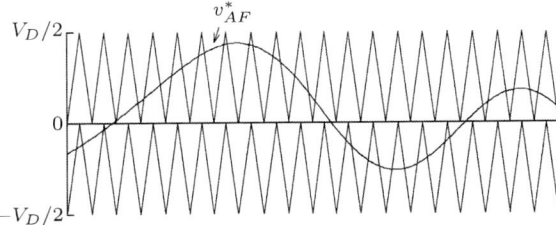

Fig. 4. Common triangle-carrier signals, and the active-filter voltage reference v_{AF}^* in one phase.

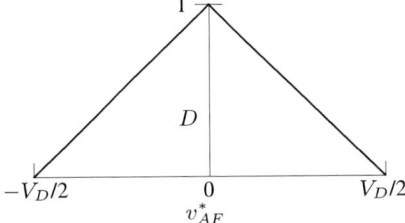

Fig. 5. The active-filter voltage reference v_{AF}^* on the horizontal axis and the duty factor D on the vertical axis.

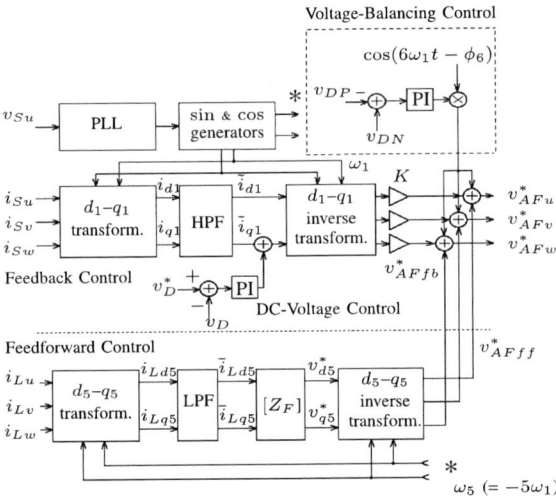

Fig. 3. Control block diagram of the active filter, where ϕ_6 is a constant value of 1.4 rad in the voltage-balancing control.

sinks the 7th-harmonic current from the diode rectifier, while the active filter compensates for the other harmonic currents produced by the diode rectifier [13].

The hybrid filter is directly connected to the 400-V system. The active filter consists of a three-phase three-level diode-clamped PWM converter using twelve 100-V MOSFETs, and two split dc capacitors C_{DP} and C_{DN}. The active filter controller described in the next section regulates the total dc voltage v_D (= $v_{DP} + v_{DN}$) at 80 V (= 400 V × 0.2), and keeps the two dc capacitor voltages v_{DP} and v_{DN} balanced well. Neither auxiliary circuit or component exists on the dc side except for two voltage sensors.

The ac inductor L_{ac} (= 5%) should be designed to be larger than the background system inductance L_S (= 1.8%).

Note that L_S includes the leakage inductance of the 15-kW transformer with a primary voltage of 200 V and a secondary voltage of 400 V in Fig. 2. The dc inductor L_{dc} (= 1%) is connected to the dc side of the diode rectifier.

Reference [16] described a starting procedure of the hybrid filter replacing the three-level converter with a conventional two-level converter, along with experimental waveforms during starting. This starting procedure using two three-phase magnetic contactors MC1 and MC2, and a resistor in each phase is also applicable to Figs. 1 and 2.

III. CONTROL SYSTEM OF THE ACTIVE FILTER

Fig. 3 shows the control block diagram of the active filter. It can be divided into feedback control, feedforward control, dc-voltage control and voltage-balancing control. All the signal processing is achieved by a fully-digital controller based on a DSP and FPGAs. References [13]-[16] have already described the whole control system of the hybrid active filter using a traditional two-level converter with triangle-carrier modulation.

This section gives an overview of the control system, except for voltage-balancing control for a three-level diode-clamped converter with triangle-carrier modulation.

A. Feedback Control

The feedback control plays an important role in reducing supply harmonic currents. The instantaneous supply currents i_{Su}, i_{Sv} and i_{Sw} are detected and converted into two-phase currents i_{d_1} and i_{q_1} on the reference frames rotating at the fundamental frequency ω_1 [2][30][31]. Two first-order high-pass filters with the same cut-off frequency as 50 Hz [16] extract ac components from i_{d_1} and i_{q_1}. The inverse transformation produces their supply harmonic currents, i_{Shu}, i_{Shv}, and i_{Shw}.

Each harmonic current i_{Sh} is amplified by a gain of K. The feedback voltage reference is given as follows [13]:

$$v^*_{AFfb} = K \cdot i_{Sh}. \tag{1}$$

B. Feedforward Control

The task of the feedforward control is active compensation for the most dominant 5th-harmonic load current. The measured load currents i_{Lu}, i_{Lv} and i_{Lw} are transformed into two-phase currents i_{Ld5} and i_{Lq5} on the reference frame rotating at the 5th-harmonic frequency ω_5. The dc currents $\bar{i}_{Ld5}$ and $\bar{i}_{Lq5}$ are extracted by two first-order low-pass filters with the same cut-off frequency as 16 Hz.

In Fig. 3, an impedance matrix $[Z_F]$ has a dimension of 2×2, corresponding to equation (12) in [13]. Note that the four elements in the matrix are related to the circuit parameters of the passive filter. The inverse transformation produces each feedforward voltage reference v^*_{AFff}.

C. DC-Voltage Control

The dc-voltage control is based on a proportional-plus-integral (PI) regulator. The active filter can build up and regulate the dc voltage by itself without any external power supply. The dc voltage v_D is detected and compared to the dc voltage reference v^*_D ($= 80$ V). If the active filter is assumed to produce no loss inside, it is delivered to the dc capacitor. Therefore, the electrical quantity adjusted by the dc-voltage control is not $\bar{i}_{d1}$ but $\bar{i}_{q1}$, as shown in Fig. 3. This dc-voltage control with a proportional gain of 0.6 A/V and an integral time constant of 20 ms is designed to present an underdamped characteristic against a step response of the voltage reference. Reference [16] made a detailed description of the dc-voltage control.

Fig. 4 shows two triangle-carrier signals with the same frequency as 10 kHz, and the active-filter voltage reference v^*_{AF} in one phase. The sampling time for digital signal processing is 50 μs. Note that the actual switching frequency of the three-level converter is 5 kHz, that is, a half of the carrier frequency.

IV. VOLTAGE-BALANCING CONTROL OF THE TWO SPLIT DC CAPACITORS

This section discusses voltage-balancing control for the active filter playing an important part of the hybrid filter, making a graphic description of the operating principle.

A. Duty Factor

A duty factor in one phase is defined by the ratio of a time interval, during which the filter current i_F flows into, or out of, the mid-point M through a clamping diode, with respect to the line cycle (20 ms at 50 Hz). The product of the duty factor D and the filter current i_F yields the mid-point current i_M, that is, $i_M = D \cdot i_F$ in one phase. Fig. 5 shows a relation between the voltage reference v^*_{AF} and the duty factor D, where the duty factor is a function of v^*_{AF} as follows [18]:

$$D = \begin{cases} 1 + 2v^*_{AF}/V_D & (-V_D/2 \le v^*_{AF} < 0) \\ 1 - 2v^*_{AF}/V_D & (0 \le v^*_{AF} \le V_D/2) \end{cases} \tag{2}$$

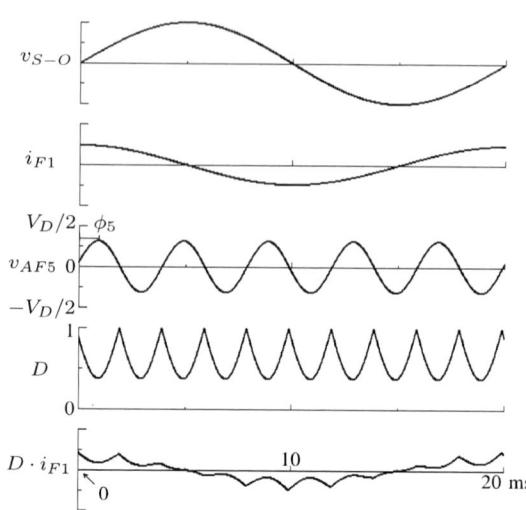

Fig. 6. The waveforms of v_{AF5}, D and i_M ($= D \cdot i_{F1}$) when no voltage-balancing control is applied in an ideal condition.

B. Mid-Point Current When No Voltage-Balancing Control is Applied

The following assumptions in one phase are made for the sake of simplicity:

- The line-to-neutral voltage v_{S-O} is purely sinusoidal: $v_{S-O} = V_{S-O} \sin(\omega_1 t)$.
- The filter current i_F consists of only the fundamental component: $i_{F1} = I_{F1} \cos(\omega_1 t)$.
- The active-filter voltage v_{AF} consists of only the 5th-harmonic component: $v_{AF5} = V_{AF5} \cos(5\omega_1 t - \phi_5)$.

Fig. 6 shows some waveforms to draw the waveform of i_M when no voltage-balancing control is applied in an ideal condition, where ϕ_5 is the initial phase difference at $t = 0$ between the filter current i_{F1} and the active-filter voltage v_{AF}. Experimental waveforms in Fig. 9 gives a relation of $\phi_5 = 1.40$ rad ($= 80°$). Note that the waveforms in Fig. 6 are not described by using circuit simulators such as a software of PSCAD/EMTDC but calculated by using a software of Excel.

The dc mean current of i_M over a period of 20 ms becomes zero, so that no voltage imbalance occurs in ideal systems. However, component tolerances including unequal conducting and switching losses produced by power switching devices, along with signal imbalance and resolution issues inherent in the digital control circuit including voltage/current sensors, may bring a voltage imbalance to the dc capacitors in actual systems. In fact, it was observed in Fig. 2 that a voltage imbalance of 4 V (10% of 40 V) occurred when no voltage-balancing control was applied. In other words, the positive dc voltage v_{DP} was higher by 4 V than the negative dc voltage v_{DP}.

C. Operating Principle of Voltage-Balancing Control

The basic idea of the voltage-balancing control proposed in this paper is to superimpose a 6th-harmonic zero-sequence voltage with an appropriate amplitude and initial phase on the

978-1-4244-1871-8/07 $25.00 © 2007 IEEE

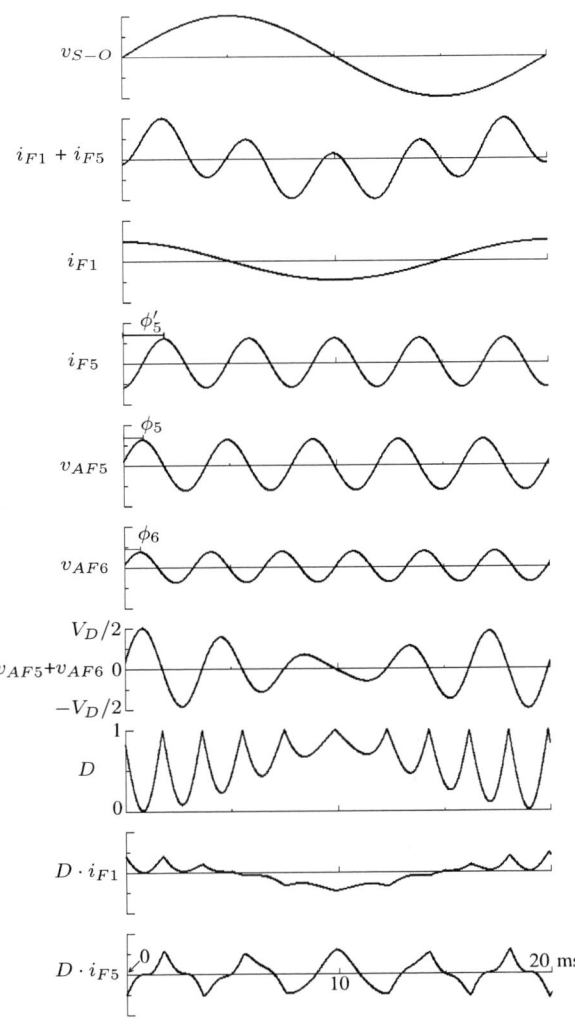

Fig. 7. The waveform of $v_{AF5} + v_{AF6}$, and the mid-point current i_M ($= D \cdot i_{F1} + D \cdot i_{F5}$), when the 6th-harmonic zero-sequence voltage with an initial phase of $\phi_6 = \phi_5 = 1.4$ rad is superimposed on each voltage reference.

active-filter voltage reference in each phase. Fig. 7 depicts waveforms in one phase when the voltage-balancing control is applied. These waveforms are drawn by using a software of Excel, like those in Fig. 6.

Fig. 7 makes the following assumptions:

- The line-to-neutral voltage v_{S-O} is purely sinusoidal: $v_{S-O} = V_{S-O} \sin(\omega_1 t)$.
- The filter current consists of the fundamental and 5th-harmonic components: $i_F = I_{F1} \cos(\omega_1 t) + I_{F5} \cos(\omega_5 t - \phi_5')$.
- The 5th-harmonic active-filter voltage is given by $v_{AF5} = V_{AF5} \cos(5\omega_1 t - \phi_5)$, where $\phi_5 = 1.40$ rad.
- The 6th-harmonic superimposed voltage[1] is given as: $v_{AF6} = V_{AF6} \cos(6\omega_1 t - \phi_6)$.

[1] The value of V_{AF6}/V_{AF5} is assumed as 0.6 in Fig. 7 to explain the operating principle of the voltage-balancing control. However, it is as small as 0.02 in the experimental result of Fig. 9.

TABLE II

MEASURED HARMONIC COMPONENTS OF v_{AF} AND i_F, AND CALCULATED DC MEAN CURRENT OF $D \cdot i_{Fn}$ ($n = 1, 5, 7, 11$ AND 13), EXPRESSED IN [%].

Harmonic order : n	1st	5th	7th	11th	13th
Harmonic component of v_{AF}	30	100	6	27	30
Harmonic component of i_F	100	125	31	16	10
DC mean current of $D \cdot i_{Fn}$	100	20	8	4	1

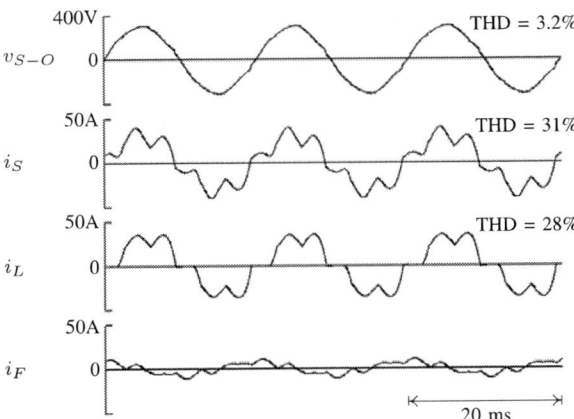

Fig. 8. Experimental waveforms with a dc load rated at 15 kW, when the stand-alone passive filter is connected.

The amplitude I_{F5} and the initial phase ϕ_5' of i_{F5} are obtained from the experimental waveform of i_F in Fig. 9. The waveforms of $D \cdot i_{F1}$ and $D \cdot i_{F5}$ are described under an assumption of $v_{AF} = v_{AF5} + v_{AF6}$. As a result, the dc component of $D \cdot i_{F1}$ over a period of 20 ms in Fig. 7 becomes negative. Therefore, the positive dc capacitor C_{DP} is charged, whereas the negative dc capacitor C_{DN} is discharged. In the other words, the mid-point voltage falls down. Hence, Fig. 7 means that the voltage-balancing control is effective.

Numerical analysis concludes that the dc component of i_M gets maximal when $\phi_5 = \phi_6$. The reason can be explained in the following. The 5th- and 6th-harmonic voltages have the same phase at $t = 0$, and they have the opposite phase at $t = 10$ ms, when $\phi_5 = \phi_6$. The reason is that the waveform of the 5th-harmonic voltage has two and a half periods for 10 ms, whereas that of the 6th-harmonic voltage has three periods for 10 ms. As a result, the sum of the 5th- and 6th-harmonic voltages ($v_{AF5} + v_{AF6}$) gets small, so that the duty factor D gets large around $t = 10$ ms. The filter current i_{F1} in Fig. 7 reaches the negative maximum of $-I_F$ at $t = 10$ ms. As a result, the dc mean current of $D \cdot i_{F1}$ over a period of 20 ms becomes negative.

The amplitude of the 5th-harmonic filter current I_{F5} is larger by 25% than that of the fundamental filter current I_{F1}, as shown in Table II. However, the dc mean current of $D \cdot i_{F5}$ is much smaller than that of $D \cdot i_{F1}$, because some differences exists between the waveforms of $D \cdot i_{F1}$ and $D \cdot i_{F5}$ in Fig. 7. The average value of D over a period of $i_{F1} > 0$ is different from that of $i_{F1} < 0$. This means that a dc mean current of

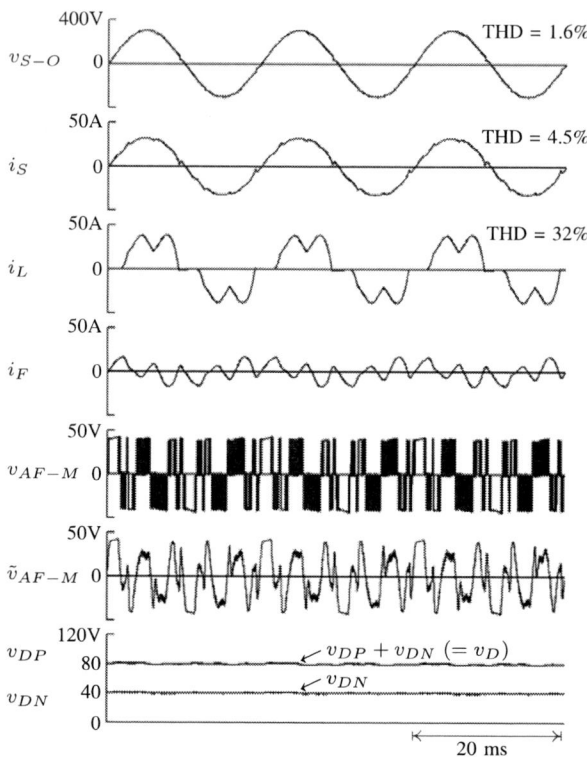

Fig. 9. Experimental waveforms with a dc load rated at 15 kW, where $K = 2.0$ p.u. and $V_D^* = 80$ V.

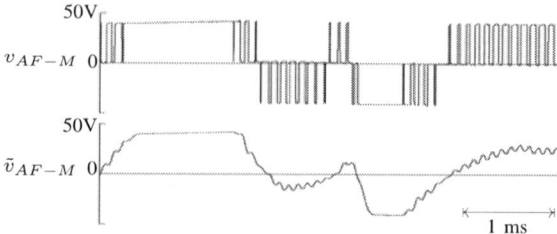

Fig. 10. Time-expanded waveforms of v_{AF-M} and $\tilde{v}_{AF-M}$ in Fig. 9.

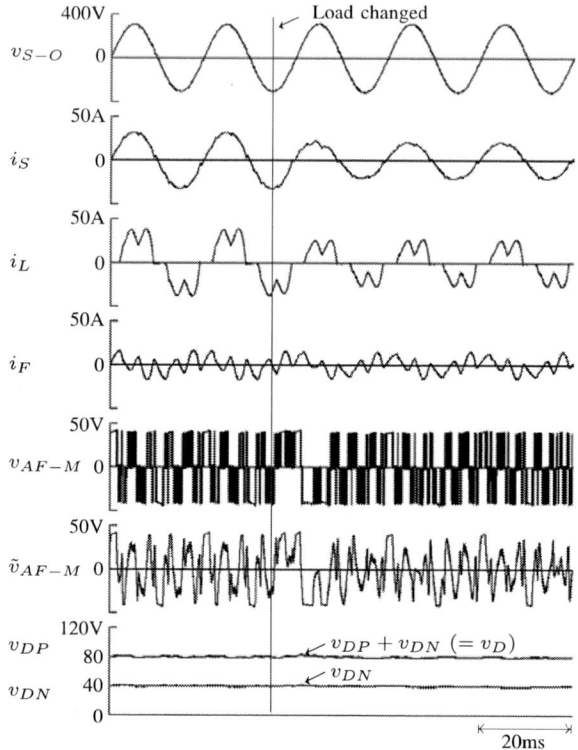

Fig. 11. Experimental waveforms when a step change occurred in a dc load from 15 kW to 10 kW, where $K = 2.0$ p.u. and $V_D^* = 80$ V.

$D \cdot i_{F1}$ over a period of 20 ms is not equal to zero. On the other hand, when attention is paid to the waveforms of i_{F5} and D, i_{F5} has a period of 4 ms, whereas D has an average period of 2 ms. Roughly speaking, the average value of D over the period of 2 ms is independent of whether i_{F5} is positive or negative. Thus, such a small dc component appears in $D \cdot i_{F5}$.

D. Effects of Harmonic Components Contained in v_{AF} and i_F on Voltage-Balancing-Control Performance

Table II summarizes the fundamental, 5th-, 7th-, 11th- and 13th-harmonic components of v_{AF} and i_F, along with the dc mean currents of $D \cdot i_{Fn}$. Note that V_{AF5}, I_{F1} and the dc mean current of $D \cdot i_{F1}$ are set to 100%, respectively. Individual harmonic components of v_{AF} and i_F are calculated from the experimental waveforms of Fig. 9. The dc mean currents of $D \cdot i_{Fn}$ are obtained by numerical analysis, where the value of V_{AF6}/V_{AF5} is assumed as 0.02.

Table II concludes that the fundamental filter current i_{F1} produces the most dominant effect (100%) on the performance of the voltage-balancing control. The 5th-harmonic filter current i_{F5} produces an effect as low as 20%. The 7th-harmonic filter current i_{F7}, 11th-harmonic filter current i_{F11}, and 13th-harmonic filter current i_{F13} are negligible, as shown in Table II.

V. EXPERIMENTAL RESULTS

A. Performance of the Stand-alone Passive Filter

Fig. 8 shows experimental waveforms with a dc load rated at 15 kW when the stand-alone passive filter was connected.

The total harmonic distortion (THD) of i_L was 28%, whereas that of i_S became 32%. The reason why the THD of i_S is higher than that of i_L is that a weak resonance at 325 Hz ($= 1/2\pi\sqrt{(L_F + L_S)C_F}$) occurred between the background system inductance and the passive filter tuned to the 7th-harmonic frequency.

B. Performance of the Hybrid Filter

Fig. 9 shows experimental waveforms when the diode rectifier had a dc load rated at 15 kW. The proportional gain and integral time constant of the voltage-balancing control in Fig. 3 was assigned as 1 V/V and 20 ms, and the initial phase of the 6th-harmonic zero-sequence voltage superimposed on the active-filter voltage reference was set as a constant value of $\phi_6 = 1.4$ rad in the following experiments.

The waveform of v_{AF-M} was observed as the ac voltage of the active filter with respect to the mid-point M, while the waveform of $\tilde{v}_{AF-M}$ was observed through a first-order

978-1-4244-1871-8/07 $25.00 © 2007 IEEE

TABLE III
CURRENT THD AND HARMONICS OF THE EXPERIMENTAL RESULTS, EXPRESSED AS THE HARMONIC-TO-FUNDAMENTAL CURRENT RATIO[%].

Hybrid Filter	3rd	5th	7th	11th	13th	17th	19th	23rd	25th	29th	31st	35th	37th	THD
i_S (15 kW)	1.2	1.8	0.9	1.9	1.9	1.2	0.9	0.9	0.7	0.6	0.6	0.5	0.4	4.2
i_L (15 kW)	1.4	31.1	7.4	4.6	3.1	2.0	1.6	0.9	0.8	0.6	0.5	0.5	0.4	32
Passive Filter	3rd	5th	7th	11th	13th	17th	19th	23rd	25th	29th	31st	35th	37th	THD
i_S (15 kW)	0.7	30.6	2.5	3.1	2.8	1.0	1.2	0.7	0.6	0.5	0.4	0.3	0.4	31
i_L (15 kW)	0.8	26.6	5.8	4.3	3.2	1.3	1.4	0.8	0.7	0.6	0.5	0.3	0.4	28

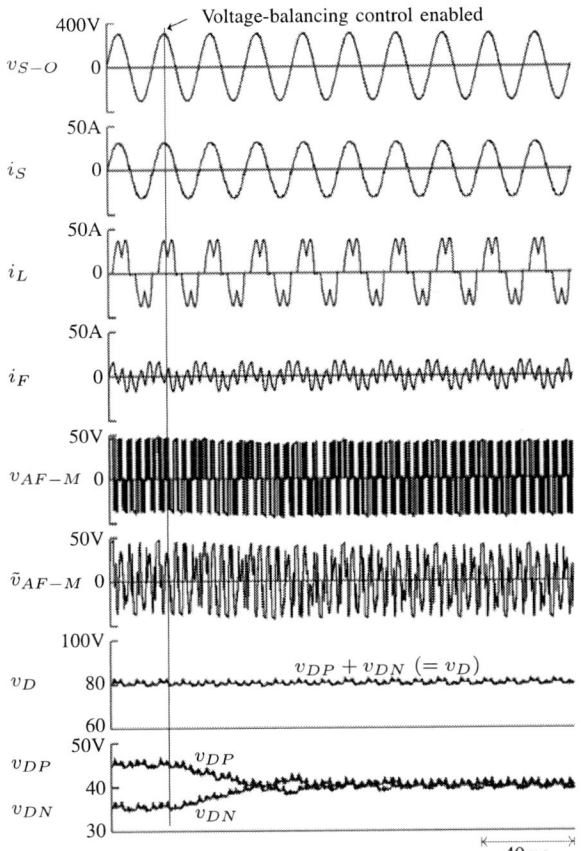

Fig. 12. Experimental waveforms with a dc load rated at 15 kW before and after the voltage-balancing control was enabled, where a 250-Ω resistor was intentionally connected across C_{DN}.

low-pass filter with a cut-off frequency of 4 kHz, to make the waveform of v_{AF-M} clear. No switching ripple appeared in the waveforms of v_{S-O} and i_S because a voltage step change in the active filter is as low as 40 V, as shown in the waveform of v_{AF-M}, which is only 10% of 400 V.

Fig. 10 shows time-expanded waveforms[2] of v_{AF-M} and $\tilde{v}_{AF-M}$ in Fig. 9. The waveform of v_{AF-M} verified that the three-level converter worked properly as expected, because PWM switching was achieved either between 40 V and 0 or between 0 and −40 V, and no switching occurred between 40 V and −40 V. This PWM switching voltage as low as

[2]A digital recorder WE7000, manufactured by Yokogawa, was used for observing the waveform of the active-filter voltage v_{AF-M}. The sampling frequency was 1 MS/s.

40 V results in a good "side effect" that neither switching-ripple voltage nor current appeared in v_{S-O} and i_S although no switching-ripple filter was installed on the 400-V line.

Table III summarizes the THD and each harmonic current of i_S. The THD was calculated up to the 37th-harmonic current[3]. The passive filter absorbed the 7th-harmonic current, and the active filter compensated for the 5th-, 11th-, and higher-other harmonic currents. When the 15-kW load was applied, each harmonic current contained in i_S was reduced below 2%. The THD of i_S was as low as 4.5%, whereas that of i_L was as high as 32%. Fig. 9 also shows that the dc-voltage of v_D was regulated at 80 V, and that the two dc-capacitor voltages v_{DP} and v_{DN} were well balanced.

Fig. 11 shows experimental waveforms in a step load change from 15 kW to 10 kW. After the load change occurred, the supply current i_S was slightly distorted for about half a cycle (about 10 ms). The dc capacitor voltage reached 82 V during the step load change, and this overvoltage was only 2.8% of 80 V. These waveforms verified that the hybrid filter provided much better transient performance than stand-alone passive filters suffering from undesirable harmonic resonances.

C. Effectiveness of the Voltage-Balancing Control

Fig. 12 shows experimental waveforms to verify the effectiveness of the voltage-balancing control. A 250-Ω resistor was intentionally connected across the negative dc capacitor C_{DN}. This brought a "forced voltage imbalance" to the two split dc capacitors. Before the voltage-balancing control was applied, the positive dc voltage v_{DP} (= 46 V) got higher by 12 V than the negative dc voltage v_{DN} (= 34 V). However, the total dc voltage v_D (= $v_{DP} + v_{DN}$) was kept to be a constant voltage of 80 V because the dc-voltage control was enabled through this experiment. Note that the negative dc capacitor was producing a power loss of 4.6 W (= $34^2/250$), that is, only 0.5% of 900 VA.

As soon as the voltage-balancing control was enabled, the two dc capacitor voltages started converging, and finally reached the same value as 40 V in 80 ms. An amplitude ratio of the 6th-harmonic voltage with respect to the 5th-harmonic voltage contained in v_{AF-M} was 8% when the two dc capacitor voltages were balanced. On the other hand, the ratio was as small as 2% when the 250-Ω resistor was eliminated from Fig. 12. This means that the voltage-balancing

[3]The Japanese harmonic guideline prescribes that the THD of supply current must take into account up to the 40th-harmonic current. Even 38th- and 40th-harmonic, and triple 39th-harmonic currents were low enough to be eliminated from Table III.

control has no bad effect on the filtering performance of the hybrid filter because the 6th-harmonic zero-sequence voltage superimposed by the voltage-balancing control is much smaller than the most dominant 5th-harmonic voltage produced by the active filter.

D. Discussions on the Initial Phase

It is verified by careful experiments that the initial phase of the 6th-harmonic zero-sequence voltage can be assigned as an appropriate and constant value for all the dc-load conditions of the three-phase diode rectifier. This concludes that neither fine tuning nor adjustment is required for the initial phase ϕ_6, thus being of great advantage to simple and easy implementation of the voltage-balancing control proposed in this paper. Note that the hybrid filter is installed for harmonic compensation of the three-phase diode rectifier or the so-called "passive frond end" in the motor drive, as shown in Fig. 1.

When the hybrid filter is intended for harmonic compensation of a three-phase thyristor rectifier with an inductive dc load, the initial phase should be adjusted with load change in such a way as to constitute an additional feedback or feedforward control loop because the firing angle of the thyristor rectifier is controllable.

VI. CONCLUSIONS

This paper has addressed voltage-balancing control for a three-level diode-clamped converter with triangle-carrier modulation in a transformerless hybrid active filter. This hybrid filter is intended for integration into the 6.6-kV motor drive having a three-phase diode rectifier at a frond end.

The 400-V, 15-kW down-scaled system has been designed, constructed, and tested, with focus on voltage balancing of the two split dc capacitors. A 6th-harmonic zero-sequence voltage with an appropriate and constant initial phase is superimposed on the active-filter voltage reference in each phase. Experimental results in steady and transient states have verified the effectiveness and viability of the voltage-balancing control.

REFERENCES

[1] B. Wu, *High-Power Converters and AC Drives*: IEEE Press, 2006.
[2] H. Akagi, E. H. Watanabe, and M Aredes, *Instantaneous Power Theory and Applications to Power Conditioning*: IEEE Press, 2007.
[3] F. Z. Peng, H. Akagi, and A. Nabae, "A new approach to harmonic compensation in power systems–A combined system of shunt passive and series active filters," *IEEE Trans. Ind. Appl.*, vol. 26, no. 6, pp. 983-990, Nov./Dec. 1990.
[4] H. Fujita and H. Akagi, "A practical approach to harmonic compensation in power systems: Series connection of passive and active filters," *IEEE Trans. Ind. Appl.*, vol. 27, no. 6, pp. 1020-1025, Nov./Dec. 1991.
[5] F. Z. Peng, H. Akagi, and A. Nabae, "Compensation characteristics of the combined system of shunt passive and series active filters," *IEEE Trans. Ind. Appl.*, vol. 29, no. 1, pp. 144-152, Jan./Feb. 1993.
[6] H. Akagi, "Trends in active power line conditioners," *IEEE Trans. Power Electron.*, vol. 9, no. 3, May/Jun. pp. 263-268, 1994.
[7] M. Rastogi, N. Mohan, and A. A. Edris, "Filtering of harmonic currents and damping of resonances in power systems with a hybrid-active filter," in *Conf. Rec. IEEE-APEC*, 1995, pp. 607-612.
[8] H. Akagi, "New trends in active filters for power conditioning," *IEEE Trans. Ind. Appl.*, vol. 32, no. 6, pp. 1312-1322, Nov./Dec. 1996.
[9] S. Bhattacharya, P. Cheng, and D. M. Divan, "Control of square-wave inverters in high power hybrid active filter systems," *IEEE Trans. Ind. Appl.*, vol. 34, no. 3, pp. 458-472, May/Jun. 1998.

[10] B. N. Singh, B. Singh, A. Chanda, and K. Al-Haddad, "Digital implementation of a new type of hybrid filter with simplified control strategy," in *Conf. Rec. IEEE-APEC*, 1999, pp. 642-648.
[11] D. Basic, V. S. Ramsden, and P. K. Muttik, "Harmonic filtering of high-power 12-pulse rectifier loads with a selective hybrid filter system," *IEEE Trans. Ind. Electron.*, vol. 48, no. 6, pp. 1118-1127, Dec. 2001.
[12] D. Detjen, J. Jacobs, R. W. De Doncker, and H. G. Mall, "A new hybrid filter to dampen resonances and compensate harmonic currents in industrial power systems with power factor correction equipment," *IEEE Trans. Power Electron.*, vol. 16, no. 6, pp. 821-827, Nov. 2001.
[13] S. Sriangthumrong and H. Akagi, "A medium-voltage transformerless ac/dc power conversion system consisting of a diode rectifier and a shunt hybrid filter," *IEEE Trans. Ind. Appl.*, vol. 39, no. 3, pp. 874-882, May/Jun. 2003.
[14] H. Akagi, S. Srianthumrong, and Y. Tamai, "Comparisons in circuit configuration and filtering performance between hybrid and pure shunt active filters," in *Conf. Rec. IEEE-IAS Annual Meeting*, 2003, pp. 1195-1202.
[15] H. Akagi, "Active harmonic filters," *Proceedings of the IEEE*, vol. 93, no. 12, pp. 2128-2141, Dec. 2005.
[16] W. Tangtheerajaroonwong, T. Hatada, K. Wada, and H. Akagi, "Design and performance of a transformerless shunt hybrid filter integrated into a three-phase diode rectifier," *IEEE Trans. Power Electron.*, vol. 22, no. 5, pp. 1882-1889, Sep. 2007.
[17] A. Nabae, I. Takahashi, and H. Akagi, "A new neutral-point-clamped PWM inverter," *IEEE Trans. Ind. Appl.*, vol. 17, no. 5, pp. 518-523, Sep/Oct. 1981.
[18] S. Ogasawara and H. Akagi, "Analysis of variation of neutral point potential in neutral-point-clamped voltage source PWM inverters," *Conf. Rec. IEEE-IAS Annual Meeting*, 1993, pp. 965-970.
[19] M. Matsui, "Static var compensator using neutral-point-clamped PWM inverter and its control scheme," *Conf. Rec. IPEC-Yokohama*, 1995, pp. 488-493.
[20] N. Celanovic and D. Boroyevich, "A comprehensive study of neutral-point voltage balancing problem in three-level neutral-point-clamped voltage source PWM inverters," *IEEE Trans. Power Electron.*, vol. 15, no. 2, pp. 242-249, Mar. 2000.
[21] H. D. T. Mouton, "Natural balancing of three-level neutral-point-clamped PWM inverters," *IEEE Trans. Ind. Electron.*, vol. 49, no. 5, pp. 1017-1024, Oct. 2002.
[22] K. Yamanaka, A. M. Have, H. Kirino, Y. Tanaka, N. Koga, and T. Kume, "A novel neutral point potential stabilization technique using the information of output current polarities and voltage vector," *IEEE Trans. Ind. Appl.*, vol. 38, no. 6, pp. 1572-1580, Nov./Dec. 2002.
[23] B. P. McGrath, D. G. Holmes, and T. A. Lipo, "Optimized space vector switching sequences for multilevel inverters," *IEEE Trans. Power Electron.*, Vol. 18, no. 6, pp. 1293-1301, Nov. 2003.
[24] T. Brückner and D. G. Holmes, "Optimal pulse width modulation for three-level inverters," *IEEE Trans. Power Electron.*, vol. 20, no. 1, pp. 82-89, Jan. 2005.
[25] A. K. Gupta and A. M. Khambadkone, "A simple space vector PWM scheme to operate a three-level NPC inverter at high modulation index including overmodulation region, with neutral point balancing," *IEEE Trans. Ind. Appl.*, vol. 43, no. 3, pp. 751-760, May/Jun. 2007.
[26] J. Holtz and N. Oikonomous, "Neutral point potential balancing algorithm at low modulation index for three-level inverter medium-voltage drives," *IEEE Trans. Ind. Appl.*, vol. 43, no. 3, pp. 761-768, May/Jun. 2007.
[27] V. Aburto, M. Schenider, L. Moran, and J. Dixon, "An active power filter implemented with a three-level NPC voltage source inverter," *Conf. Rec. IEEE-PESC*, 1997, pp. 1121-1126.
[28] T. Jin, J. Wen, and K. Smedley, "Control and topologies for three-phase three-level active power filters," *Conf. Rec. IEEE-APEC*, 2005, pp. 655-664.
[29] H. Fujita, S. Tominaga, and H. Akagi, "Analysis and design of a dc voltage-controlled static var compensator using quad-series voltage-source inverters," *IEEE Trans. Ind. Appl.*, vol. 32, no. 4, pp. 970-978, Jul./Aug. 1996.
[30] H. Akagi, Y. Kanazawa, and A. Nabae, "Generalized theory of the instantaneous reactive power in three-phase circuits," in *Conf. Rec. IEEJ-IPEC*, 1983, pp. 1375-1386.
[31] H. Akagi, Y. Kanazawa, and A. Nabae, "Instantaneous reactive power compensators comprising switching devices without energy storage components," *IEEE Trans. Ind. Appl.*, vol. 20, no. 3, pp. 625-630, May/Jun. 1984.

Digital Current Mode Control Approach for the Parallel Module DC-DC Converters

Bohyung Cho
Seoul National University
Department of Electrical Engineering
San 56-1, Sillim-dong, Gwanak-gu, Seoul, 151-742
Email: bhcho@snu.ac.kr

Hyunsu Bae
Seoul National University
Department of Electrical Engineering
San 56-1, Sillim-dong, Gwanak-gu, Seoul, 151-742
Email: bhsue@pesl.snu.ac.kr

Abstract—This paper presents the digital current mode control employing nonlinear resistive current mode control and digital state feedback control using the pole placement technique for parallel interleaved DC-DC converters. The former method achieves current sharing by converting the effective input characteristic of the converters seen by the source to a resistive load sink. The latter method derives the discrete state feedback controller structure for robust tracking control using the error state. For the design example of the two control scheme, a prototype hardware system with two 100W parallel module buck converters with a TMS320F2812 DSP has been built and tested.

I. INTRODUCTION

With recent advances in digital systems, digital control has become increasingly feasible even for high frequency, low to medium power switching converters. Digital control offers the potential advantages of immunity to analog component variations, programmability and possibilities to improve performance using more advanced and sophisticated algorithms [1-10].

Ever increasing demands for the development of compact, lightweight power supplies with more power density, higher efficiency and fast dynamics, often require power conversion through parallel connected converters [11-12]. In order to achieve the current sharing among the converter modules, various analog current mode control methods such as peak current mode control, charge current mode control, average current mode control, etc [13-18] are used.

However, a direct implementation of a digital control scheme for an analog current mode control method is not easy. In analog current mode control, the switch current or the inductor current is sensed and the switch duty cycles of each converter module are generated by comparing the sensed current to a reference. Because the switch or inductor current is a fast changing waveform and the switching frequency is high, the need for a very fast analog to digital converter (ADC) to produce multiple samples of the sensed current per switching period, which may require excessively complex hardware. Thus, a digital current mode control method that can match or exceed the performance of the analog current control method has been of much interest. In [4, 5], current estimation algorithms are proposed. These algorithms focus on the current mode control to improve the dynamics. Even though the dynamics is improved, the current sharing among the parallel converter modules may

not be implemented. Thus, a parallel interleaved converter control method using the sliding mode control was proposed [3]. However, this controller design technique is very complex and the expansion of the modules is not convenient. A simplified digital peak current mode controller whose operation is similar to analog peak current program mode was proposed in [8, 9]. However, this approach still used an analog current loop.

An all state feedback control for the DC-DC converters has been proposed for the systematic controller design [3, 19-24]. These algorithms used a LQR approach [19, 24], sliding mode control [3, 21] for the robust control, and an (observed based) integral control scheme for reference voltage tracking [20, 22-23]. However, most of these algorithms are based on the continuous time domain, which needs the approximated transformation to the discrete time domain for a realization of digital controller, and single module converter control. In [23], the discrete time domain design is mentioned but the basic controller structure is similar to the continuous integral control. In [3, 22], the parallel module converter control issue is discussed. However, this controller design technique is also very complex and the expansion of the converter module is not convenient because the algorithms use a high order state and a multi-input system.

In this paper, two kinds of a simple digital control scheme are presented for the parallel interleaved DC-DC converter. Of the two proposed schemes, one is a constant resistive load sink control scheme. This algorithm is very simple to implement with a digital controller and there is no power stage parameter dependency in the controller design. By converting the effective input characteristic of the converters seen by the power source to a resistive load sink, the proposed control scheme achieves the current sharing by using a nonlinear transformation. The second scheme is a digital state feedback control method using the pole placement technique. By changing a multi-input/multi-output system into a single-input/single-output system using a two loop control scheme, the proposed control scheme can precisely achieve the interleaved current sharing and can achieve the systematical controller design for digital implementation due to the fact that the controller design and analysis are performed in the discrete time domain. Thus, these simple algorithms fully utilizing the advantages of the digital controller are convenient for the expansion of the converter modules. For the verification of the proposed digital control schemes, a parallel module Buck converter with a TMS320F 2812 DSP has been built and tested.

978-1-4244-1871-8/07 $25.00 © 2007 IEEE

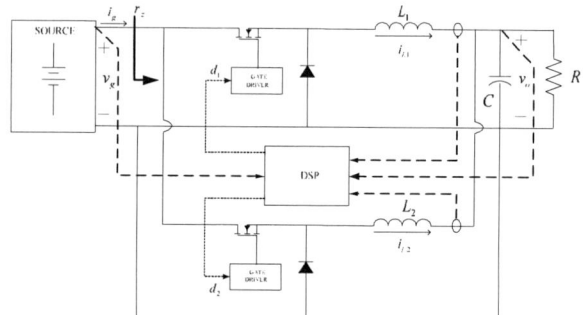

Fig. 1: The parallel interleaved buck converter

II. DIGITAL RESISTIVE CURRENT CONTROL USING THE EFFECTIVE INPUT CHARACTERISTIC OF THE CONVERTERS SEEN BY THE SOURCE

A. Digital Resistive Current (DRC) Controller for the current mode control

The proposed DRC control scheme is illustrated in Fig. 1. If the effective input characteristic of the converter seen by the source is a resistive load sink, the converter input current, i_g, and the source voltage, v_g, have the relation of

$$v_g = r_z \cdot i_g \quad (1)$$

where, r_z is the equivalent input resistance of the converter and is set to the given reference value of the DRC controller. If the buck converter is operated in the continuous conduction mode (CCM), the control duty ratio, d, required to realize the current mode control can be derived based on the DC gain of the converter as follows;

$$v_g = r_z \cdot d \cdot i_L$$

$$d_i = \frac{v_g}{r_{ze} \cdot i_{L,i}} \quad (i = 1, 2, \Lambda, N \text{ and } r_{ze} = N \cdot r_z) \quad (2)$$

where, i_L is the output inductor current and N is the number of converter modules. For other types of converters, the control duty ratio can be derived by a similar procedure.

If the parallel module converter is controlled using equation (2), the effective input characteristic seen by the source is a constant resistive load. Also, the current sharing among the converter modules is automatically achieved because the resistive load seen by the source is the same. The proposed DRC control method is quite simple and easy to realize in a digital controller. Furthermore, there is no power stage parameter dependency in the controller design. The control duty ratio is determined by the sensed voltage and current.

B. Small signal modeling and stability analysis

Small signal modeling was performed to analyze the dynamics and the stability for the proposed control system. Since the controller is realized in the digital domain, the discrete time domain analysis is useful for design guidelines.

Using the discrete time modeling of the DC-DC converters [25 -26], the open loop transfer functions of the buck converter are as follows;

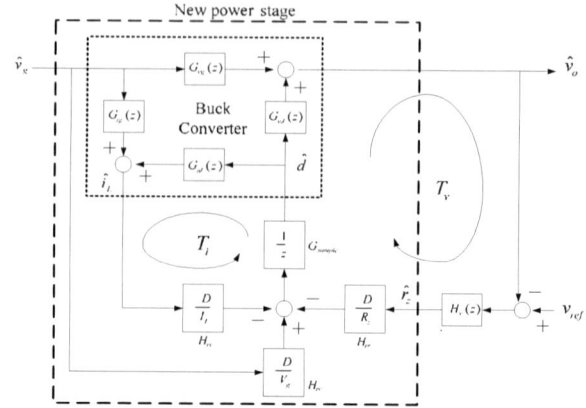

Fig. 2: The small signal block diagram in the discrete time domain

$$G_{id}(z) = \frac{V_g T_s}{L} \frac{z - (1 - a + ab)}{z^2 + (a - 2)z + (1 - a + ab)}$$

$$G_{vd}(z) = \frac{V_g T_s^2}{LC} \frac{z}{z^2 + (a - 2)z + (1 - a + ab)} \quad (3)$$

where $a = \frac{T_s}{RC}, b = \frac{RT_s}{L}, L = \frac{L_{nor}}{N}, L_i \cong L_{nor}$

where, L is an equivalent inductance for the equivalent single module model of the parallel module converter [11], and T_s is the sampling time, which is the same as the switching period. For the small signal modeling of the controller, the perturbation and linearization of equation (2) gives

$$\hat{d}(k+1) = \frac{D}{V_g}\hat{v}_g(k) - \frac{D}{I_L}\hat{i}_L(k) - \frac{D}{R_z}\hat{r}_z \text{ where } D = \frac{V_g}{R_z I_L} \quad (4)$$

Fig. 2 shows the small signal block diagram of the proposed control system. From the small signal block diagram, the control-to-output transfer function of the new power stage is given as follows;

$$G_{vr}(z) = -\frac{V_o T_s^2}{LCR_z} \frac{z}{z^3 + (a+2)z^2 + (1-a+b+ab)z - b(1-a+ab)} \quad (5)$$

Fig. 3 shows the root locus of the control-to-output transfer function with respect to the load range in the CCM operation. It can be seen that the poles are placed outside of the unit circle, which means that sub-harmonic oscillation occurs at the light load condition.

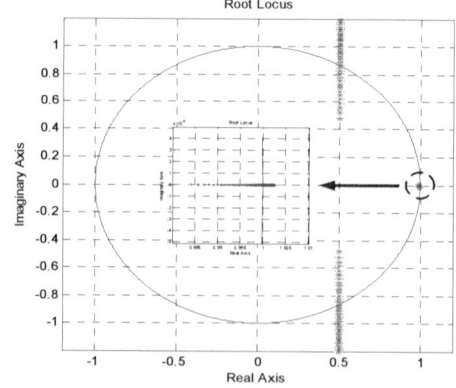

Fig. 3: The root locus of control-to-output transfer function (sub-harmonic)

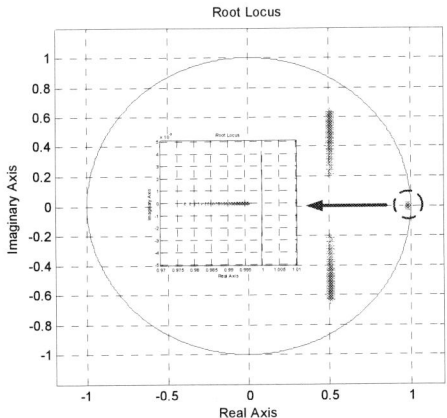

Fig. 4: The root locus of the control-to-output transfer function with the offset inductor current

To prevent this sub-harmonic oscillation, an offset current value, I_{Lk}, is added to the proposed algorithm. The small signal model shown in Fig. 2 is still valid as follows;

$$v_g = r_z \cdot d \cdot (i_L + I_{Lk}) \Leftrightarrow d = \frac{v_g}{r_z \cdot (i_L + I_{Lk})} \quad (6)$$

where, I_{Lk} is the offset inductor current value, which is programmed in the digital processor.

$$G_{vr}(z) = -\frac{V_o T_s^2}{LCR_z} \frac{z}{z^3 + (a-2)z^2 + (1-a+\alpha b+ab)z - \alpha b(1-a+ab)}$$

$$\text{where } \alpha = \frac{I_L}{I_L + I_{Lk}} \quad (7)$$

By selecting the proper $I_{Lk}(>0)$ value, the sub-harmonic oscillation can be suppressed as shown in Fig. 4. The poles of the control-to-output transfer function are placed inside of the unit circle for the given load range.

C. Origin shift method of the resistance and Simple PI voltage loop compensation

It is observed from equation (7) that the DC gain of the control-to-output transfer function is affected by the equivalent resistance of the converter. Thus, the system has poor dynamics at light load and it is difficult to design a voltage compensator for obtaining the desired loop gain. To solve these problems, the origin of the equivalent resistance can be shifted as shown in Fig. 5. Therefore, the DRC control algorithm is modified and the complete small signal model is expressed as follows;

$$v_g = V_1 + r_z \cdot (i_g - I_1) = V_1 + r_z \cdot d \cdot (i_L + I_{Lk}) - r_z \cdot I_1$$

$$d = \frac{v_g - V_1 + r_z \cdot I_1}{r_z \cdot (i_L + I_{Lk})}$$

$$G_{vr}(z) = -\frac{V_o T_s^2}{LCR_z} \left\{ 1 - \frac{I_1}{D(I_L + I_{Lk})} \right\} \quad (8)$$

$$\frac{z}{z^3 + (a-2)z^2 + (1-a+\alpha b+ab)z - \alpha b(1-a+ab)}$$

where, V_1, I_1 are programmed in the digital processor as well.

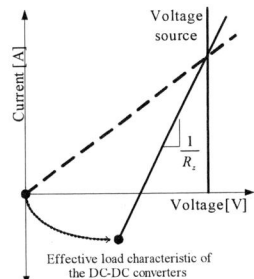

Fig. 5: The origin shift of the equivalent resistance

After the origin shift, the DC gain of the control-to-output transfer function can be similar at both the heavy and light load conditions, which enables the voltage loop compensation for the fast dynamics. Since there is only one low-frequency pole below the switching frequency in the control-to-output transfer function, a simple discrete PI voltage controller with an anti-windup loop is applied to the proposed DRC control. When the voltage compensator is designed, the positive feedback loop is needed because the control-to-output transfer function has a negative DC gain.

D. Experimental verification

The prototype hardware system, which consists of two 100W parallel module buck converters, was built as shown in Fig. 1. The converter parameters are: $V_g = 60[V]$, $V_o = 31[V]$, $L_1 = L_2 \equiv 200[\mu H]$, $C = 100[\mu F]$ and $F_s = 100[kHz]$. The DSP control strategy using the synchronous PWM is used, which can sample the average inductor current and can easily generate two phase-shifted duty cycles for the interleaving control. Fig. 6 and 7 show the transient performance of the DRC control without the voltage loop at the resistive reference variation. Fig. 8 and 9 show the converter behavior, which closes the PI voltage control loop, in the presence of stepped load variation from 60[W] to 120[W] and vice versa. It is observed that the voltage regulation can be achieved and the interleaved current sharing is also possible during the transient period, as well as in steady state.

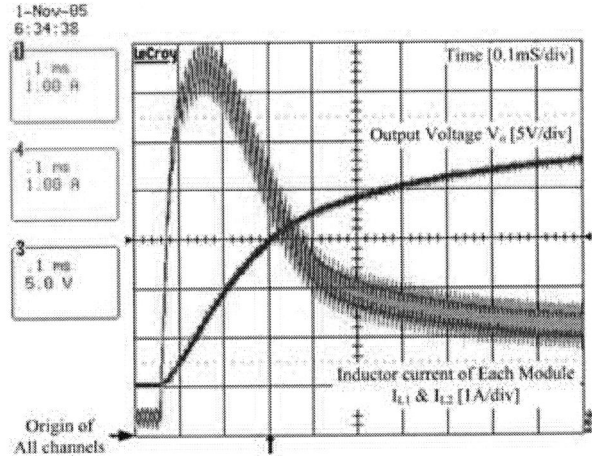

Fig. 6: The resistive reference step variation (3W to 120W)

978-1-4244-1871-8/07 $25.00 © 2007 IEEE

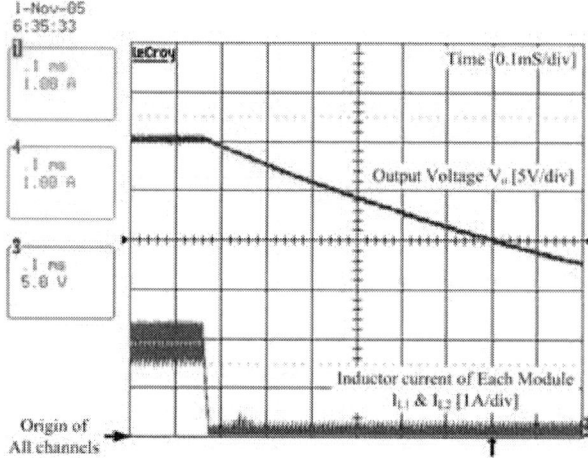

Fig. 7: The resistive reference step variation (120W to 3W)

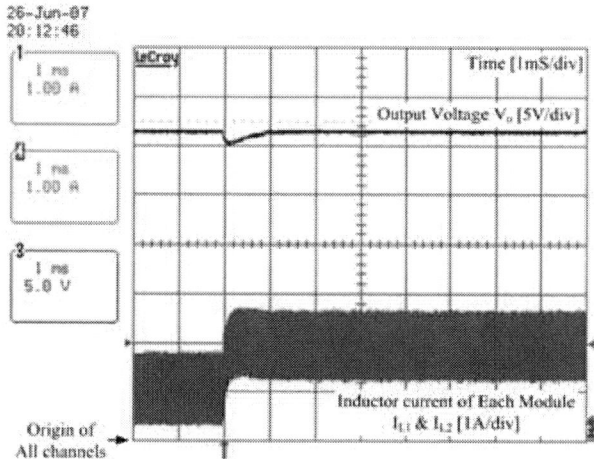

Fig. 8: The load step variation (60W to 120W)

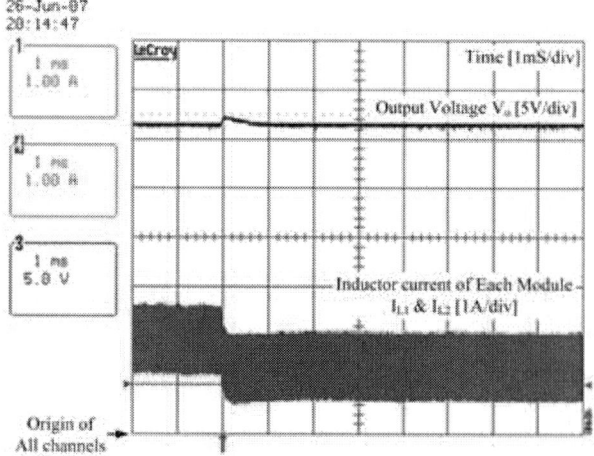

Fig. 9: The load step variation (120W to 60W)

III. DISCRETE STATE FEEDBACK CONTROL – AVERAGE AND PROPORTIONAL TYPE CURRENT CONTROL

A. State feedback controller structure in the discrete time domain for robust tracking control

Using the discrete time modeling method [25 -26], the average discrete time state equation of the DC-DC converters can be derived for the discrete state feedback controller.

$$x(k+1) = A_z x(k) + B_z d(k) + B_{vgz} v_g(k) + B_{ioz} i_o(k)$$
$$y(k) = C_z x(k)$$
(9)

The objective is to design an overall system such that the output $y(k)$ will track asymptotically any step reference input, $r(k) = R(\text{constant})$, even with the presence of an input disturbance and with plant parameter variations. The error state, $e(k)$, and augmented state variable, $z(k)$, $u(k)$, $w(k)$, are defined, and the input voltage, v_g, and output current, i_o, are assumed to be sustained with one switching period.

$$e(k) = r(k) - y(k) = R - C_z x(k), z(k) = \&(k) \cong \frac{x(k+1) - x(k)}{T_s}$$

$$u(k) = \&(k) \cong \frac{d(k+1) - d(k)}{T_s}, \quad \&(k) \cong -C_z z(k)$$
(10)

$w_1(k) = \&_g(k) \cong 0$, $w_2(k) = \&_o(k) \cong 0$ using Euler's method

Then, the system can be expressed by the augmented state vector $[e(k) \quad z(k)]^T$.

$$\begin{bmatrix} e(k+1) \\ z(k+1) \end{bmatrix} = \begin{bmatrix} 1 & -T_s C_z \\ 0 & A_z \end{bmatrix} \begin{bmatrix} e(k) \\ z(k) \end{bmatrix} + \begin{bmatrix} 0 \\ B_z \end{bmatrix} u(k)$$

$$u(k) = -\begin{bmatrix} K_1 & K_2 \end{bmatrix} \begin{bmatrix} e(k) \\ z(k) \end{bmatrix}$$
(11)

If the expanded system is controllable, then there exists a state feedback gain such that the expanded system is stable. That means the output of the converters tracks the reference value. Furthermore, using the state feedback gain, the system's eigenvalues can be placed to the desired poles, which determine the performance of the feedback control system. The duty ratio for the converters can be derived from equation (10) and (11) as follows;

$$d(k) = d(k-1) - K_1 T_s e(k-1) - K_2 [x(k) - x(k-1)] \quad (12)$$

However, the system has a high order state and multi input when the converter module is expanded. For example, the system shown in Fig. 1 has three state variables and two independent input duty ratios. Furthermore, whenever the converter module is expanded, the redesign of entire controllers is always required. Thus, in order to perform the systematical controller design for the parallel module system, it is necessary to make a simple state feedback algorithm.

B. Average type current control scheme

First, the inner current loop controller is designed for the each converter module without the output capacitor, which has only one state variable, the inductor current, as shown in Fig. 1. Using the un-terminated modeling method, the state equation is derived and the discrete state equation employing equation (9) becomes

SW ON) $\dot{i}_{L,i}(t) = \frac{1}{L_i}\left(v_g(t) - v_o(t)\right)$, SW OFF) $\dot{i}_{L,i} = -\frac{1}{L}v_o(t)$ (13)

$$\hat{i}_{L,i}(k+1) = \hat{i}_{L,i}(k) + \frac{V_g T_s}{L_i}\hat{d}_i(k), \quad i = 1,2,\Lambda, N$$

where, the input voltage and output voltage are assumed to be the sustained during one switching period. From the same procedure in Section III.A, the state equation of the overall closed loop system is given as follows;

$$\begin{bmatrix} e(k+1) \\ z(k+1) \end{bmatrix} = \begin{bmatrix} 1 & -T_s \\ -\dfrac{K_1 V_g}{L}T_s & 1 - \dfrac{K_2 V_g}{L}T_s \end{bmatrix} \begin{bmatrix} e(k) \\ z(k) \end{bmatrix}, \quad L_i \cong L \quad (14)$$

The augmented system is a second order system. Therefore, the desired pole location can be easily obtained from the given converter specification, as well as the general second order system's response using the continuous and discrete time domain relationship. The given converter specifications are defined as follows; Settling time $\leq 100[\mu S]$, P.O. $\leq 1[\%]$. From the required system dynamics, the desired pole locations are given by

$$\text{Settling time} \cong \frac{4T_s}{\ln(r)} \leq 100[\mu S], \ \text{P.O.} \cong 100 e^{-\ln(r)\pi/\theta} \leq 1[\%] \quad (15)$$

$$z_{desired} = r\angle\pm\theta = r\cos(\theta)\pm jr\sin(\theta)$$

Fig. 10 shows the state feedback current loop control system for the two module buck converter using average type current control. The state equation of the output capacitor is derived to design the outer voltage loop from Fig. 1.

$$\dot{v}_o(t) = \frac{1}{C}\left(i_L(t) - i_o(t)\right) \Leftrightarrow \hat{v}_o(k+1) = \hat{v}_o(k) + \frac{T_s}{C}\hat{i}_L(k) \quad (16)$$

$$\text{where } i_L = i_{L,1} + i_{L,2} + \Lambda + i_{L,N}$$

where, output current is assumed to be the sustained during one switching period. After closing the current loop, the parallel module converter becomes a SISO system, which has one control input, $I_{L,i,ref}$, and one output, i_L. Thus, the small signal model of the new power stage equation, which is terminated to the output capacitor, can be derived as follows;

$$\begin{bmatrix} i_L(k) \\ i_L(k+1) \\ d(k) \\ v_o(k+1) \end{bmatrix} = \begin{bmatrix} 0 & 1 & 0 & 0 \\ \dfrac{V_g(K_1 T_s + K_2)T_s}{L_{eq}} & 1 - \dfrac{V_g K_2 T_s}{L_{eq}} & \dfrac{V_g T_s}{L_{eq}} & 0 \\ K_1 T_s + K_2 & -K_2 & 1 & 0 \\ 0 & \dfrac{T_s}{C} & 0 & 1 \end{bmatrix} \begin{bmatrix} i_L(k-1) \\ i_L(k) \\ d(k-1) \\ v_o(k) \end{bmatrix}$$

$$+ \begin{bmatrix} 0 & -\dfrac{V_g K_1 T_s^2}{L_{eq}} & -K_1 T_s & 0 \end{bmatrix}^T I_{ref}(k), \ L_{eq} = \frac{L}{N}$$

(17)

It is difficult to design the outer voltage controller with the same procedure as in Section III.A because the state equation has four state variables due to the current loop states. Furthermore, the augmented system using equation (17) is uncontrollable because the duty ratio is determined by the internal state. Thus, another approach for designing the outer voltage loop is needed. However, the average type current mode control is still useful for the current control in applications such as the battery charger, the bi-directional converter in the hybrid automobile system and so on.

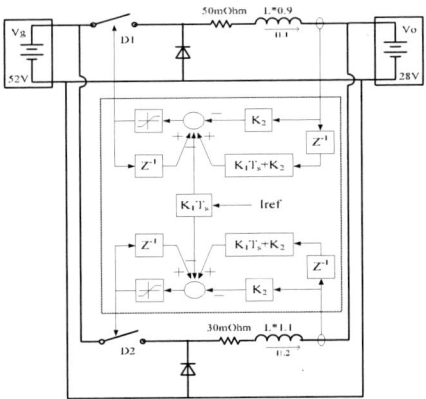

Fig. 10: The average type state feedback current loop

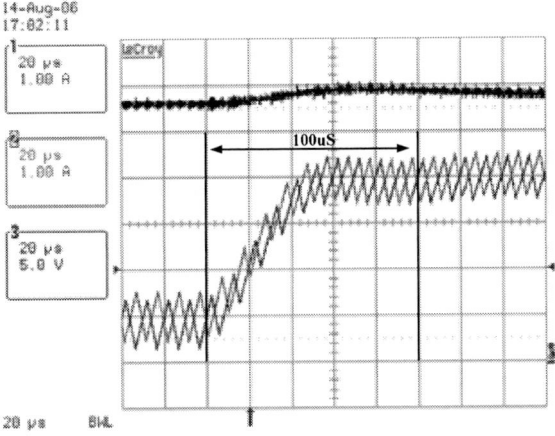

Fig. 11: Inductor current response of average type current control

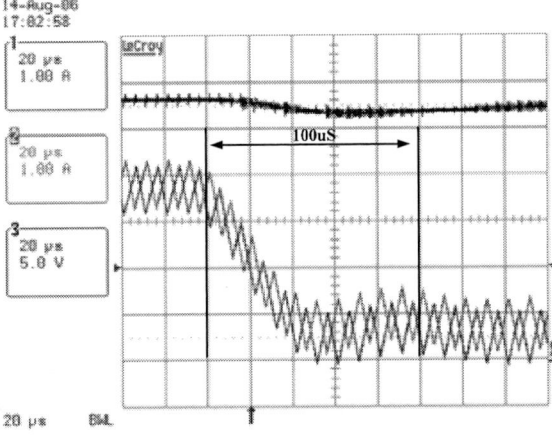

Fig. 12: Inductor current response of average type current control

To verify the theoretical analysis, the prototype hardware was built and tested as shown in Fig. 1. The converter parameters are: $V_g = 52[V]$, $V_o = 28[V]$, $L_1 = L_2 \cong 100[\mu H]$, $C = 100[\mu F]$ and $F_s = 100[kHz]$. In this implementation, the output is set to 28[V] using a constant voltage mode using electric load equipment. Fig. 11 and 12 show the inductor current of each module, which meets the required dynamics for the given specification.

C. Proportional Type current control scheme

The average type current mode control scheme is proposed for the current sharing. Even though the exact current control is possible, this approach poses a difficulty in designing the output voltage regulation loop. Thus, the proportional type current control scheme is proposed for the systematical output voltage control loop design.

The control duty ratio is defined as $d_i(k) = K_c \lfloor I_{ref} - i_{L,i}(k) \rfloor$, which is a proportional type state feedback controller in order to achieve the current sharing. Thus, the state equation of the current loop closed system employing equation (13) is a first order system as follows;

$$\hat{i}_{L,i}(k+1) = \left(1 - \frac{V_g T_s K_c}{L_i}\right)\hat{i}_{L,i}(k), \quad \hat{d}_i(k) = -K_c \hat{i}_{L,i}(k) \quad L_i \cong L$$

$$1 - \frac{V_g T_s K_c}{L} = \exp\left(-\frac{4T_s}{\tau_{spec}}\right) \Leftrightarrow K_c = \frac{L}{V_g T_s}\left(1 - \exp\left(-\frac{4T_s}{\tau_{spec}}\right)\right) \quad (18)$$

where, τ_{spec} is a settling time specification and K_c is a feedback gain using the general 1st order system's response. In this case, although a steady state error for a given reference value exists, the current sharing can be achieved. For this analysis, the ESR of the inductor, $r_{L,i}$, is considered and the final value theorem of the average large signal converter model from equation (13) is used.

$$\overset{\&}{i}_{L,i}(t) = -\frac{r_{L,i}}{L_i} i_{L,i}(t) + \frac{1}{L_i} d_i(t) v_g(t) - \frac{1}{L_i} v_o(t), \quad v_g \cong V_g, \quad v_o \equiv V_o$$

$$i_{L,i}(\infty) = \lim_{s\to 0} sI_{L,i}(s) = \frac{V_g K_c I_{ref} - V_o}{r_{L,i} + V_g K_c} \cong I_{ref} - \frac{D}{K_c}\left(r_{L,i} \ll V_g K_c\right) \quad (19)$$

It is confirmed that the steady state inductor current is not related to the inductance, and the effect of ESR can be ignored. Thus, the proposed proportional current loop can achieve the current sharing.

After closing the current loop, the parallel module converter becomes a SISO system, which has one control input, $I_{L,i,ref}$, and one output, i_L. Thus, the small signal model of the new power stage equation, which is terminated to the output capacitor, can be derived as follows;

$$\begin{bmatrix} \hat{i}_L(k+1) \\ \hat{v}_o(k+1) \end{bmatrix} = \begin{bmatrix} 1 - \dfrac{V_g T_s K_c}{L_{eq}} & 0 \\ \dfrac{T_s}{C} & 1 \end{bmatrix} \begin{bmatrix} \hat{i}_L(k) \\ \hat{v}_o(k) \end{bmatrix} + \begin{bmatrix} \dfrac{V_g T_s K_c}{L_{eq}} \\ 0 \end{bmatrix} \hat{I}_{ref}(k) \quad (20)$$

From the same procedure in Section III.A, the state feedback controller can be designed to regulate the output voltage of the parallel module converter. The augmented discrete state equation becomes a 3rd order system. Thus, using the dominant pole approach for the pole placement technique, the desired pole locations can be obtained relating the given specification and the general 2nd order system's response as equation (15). The given converter's specifications are defined as follows; Settling time $\tau_{spec} \le 300[\mu S]$, P.O. $\le 1[\%]$. From the required system dynamics, the desired pole locations are given by

$$z_{desired} = r\cos(\theta) \pm jr\sin(\theta) \text{ and } \left[r\cos(\theta)\right]^{10} \quad (21)$$

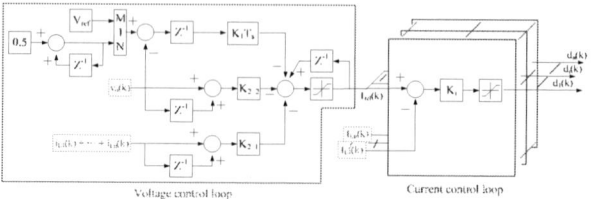

Fig. 13: The complete state feedback control structure with the proportional type current control

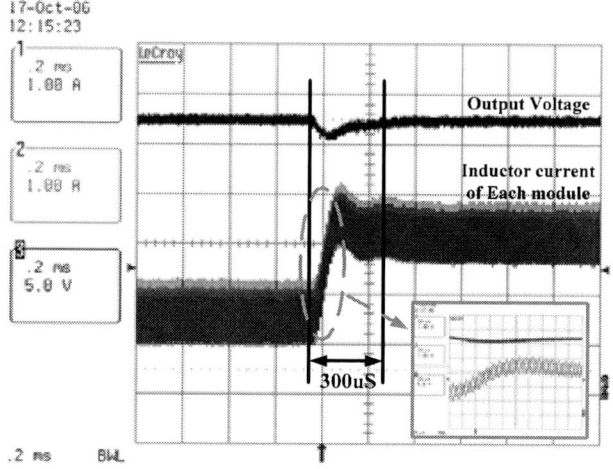

Fig. 14: The proposed control system response during step load variation

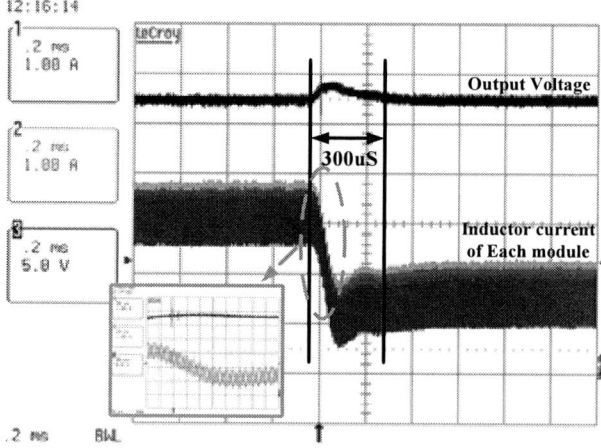

Fig. 15: The proposed control system response during step load variation

Fig. 13 shows the complete state feedback controller with the proportional type current control for the parallel module buck converter. Fig. 14 and 15 show the converter behavior, which closes the overall state feedback control loop in the presence of a stepped load variation from 85[W] to 170[W] and vice versa. The same converter parameters in the average type current control implementation are used. It is observed that the voltage regulation as well as the interleaved current sharing is achieved during the transient period, as well as in steady state. The response time is in good agreement with the required converter specification and thus confirms the theoretical analysis.

IV. CONCLUSIONS

This paper presents the digital current mode control employing nonlinear resistive current control and digital state feedback control using the pole placement technique for the parallel interleaved DC-DC converters. The digital resistive current control converts the effective input characteristic of the converters seen by the source to a resistive load. Thus, this control scheme can achieve current mode control like the existing analog current mode control, while also achieving current sharing among the converter modules. Also, the sub-harmonic oscillation at the light load condition is defined and the method of avoiding this oscillation is presented by adding an offset inductor current value. For the design and stability analysis, the complete small signal analysis in the discrete time domain is performed. To regulate the output voltage, a simple discrete PI controller with an anti-windup loop and an origin shift of the resistance is designed for the purpose of fast dynamic response, as well as controller design convenience. Also, the discrete state feedback controller structure for robust tracking control is derived to regulate the converter output. Using the un-terminated modeling method, two types of the current control loops are designed to achieve current sharing; namely, the proportional and average current mode control methods. For the output voltage regulation, the outer voltage controller is designed using the new power stage, which is the current loop closed converter system without an output capacitor in the proportional type current control. For the design example of the proposed control schemes, two 100W parallel module buck converter with a TMS320F2812 DSP has been built and tested. Since these algorithms are quite simple and are designed in the discrete time domain, it is convenient for the digital realization and the expansion of the converter modules, as well as for the dynamics and stability analysis.

ACKNOWLEDGMENT

This study is partly supported by Korea Aerospace Research Institute (KARI).

REFERENCES

[1] Y. Duan and H. Jin, "Digital Controller Design for Switch Mode Power Converters," *IEEE Annual Applied Power Electronics Conference*, 1999, Vol. 2, pp. 967-973.

[2] Angel V. Perterchev and Seth R. Sanders, "Quantization Resolution and Limit Cycling in Digitally Controlled PWM Converters," *IEEE Transactions on Power Electronics*, Vol. 18, No. 1, Jan. 2003.

[3] C. Sudhakarababu and Mummadi Veerachary, "DSP Based Control of Interleaved Boost Converter," *Journal of Power Electronics*, Vol. 5, No. 3, pp. 180-189, July 2005..

[4] Jingquan Chen, Alecksandar Prodic, Robert W. Ericson and Dragan Maksimovic, "Predictive Digital Current Programmed Control," *IEEE Transactions on Power Electronics*, Vol. 18, No. 1, Jan. 2003..

[5] P. Matttavelli, "Digital Control of dc-dc Boost Converters with Inductor Current Estimation," *IEEE Annual Applied Power Electronics Conference and Exposition*, 2004..

[6] J. H. Lee, S. H. Park, H. S. Bae and B. H. Cho, "Constant Resistance Control of Solar Array Regulator using Average Current Mode

Control," *IEEE Annual Applied Power Electronics Conference and Exposition*, Mar. 2006, Vol. 1, pp. 1544-1549.

[7] H. S. Bae, S. H. Park, J. H. Lee and B. H. Cho, "Digital Control of the Parallel Interleaved Solar Array Regulator using the Digital Signal Processor," *IEEE 37th Power Electronics Specialists Conference*, June 2006, pp. 2828-2832.

[8] Stefano Saggini, Massimo Ghioni and Angelo Geraci, "An Innovative Digital Control Architecture for Low-Voltage, High-Current DC-DC Converters With Tight Voltage Regulation," *IEEE Transactions on Power Electronics*, Vol. 19, No. 1, pp. 210-218, Jan. 2004.

[9] Olivier Trescases, Zdravko Lukic, Wai Tung Ng and Alecksandar Prodic, "A Low-Power Mixed-Signal Current-Mode DC-DC Converter Using a One-Bit delta-sigma DAC," *IEEE Annual Applied Power Electronics Conference and Exposition*, pp. 700-704, Mar. 2006.

[10] Hyun-Su Bae, Jeong-Hwan Yang, Jae-Ho Lee and Bo-Hyung Cho, "Digital State Feedback Current Control using the Pole Placement Technique," *Journal of Power Electronics*, Vol. 7, No. 3, July 2007.

[11] B. Choi, B. H. Cho, R. B. Ridley and F. C. Lee, "Control Strategy for Multi-Module Parallel Converter System, " *IEEE Proc. On Power Electronics Specialists Conference*,1990, pp. 225-234.

[12] K. Siri and C. Q. Lee, "Current Distribution Control for Parallel connected Converters," *IEEE Transactions on Aerospace and Electronic Systems*, Vol. 28(3), pp. 829-840, 1992.

[13] Wei Tang, Fred C. Lee and Raymond B. Ridley, "Charge Control: Modeling, Analysis, and Design," *IEEE Transactions on Power Electronics*, Vol. 8, No. 4, Oct. 1993.

[14] Raymond B. Ridley, "A New, Continuous-Time Model For Current-Mode Control," *IEEE Transactions on Power Electronics*, Vol. 6, No. 2, pp. 271-280, April 1991.

[15] Wei Tang, Fred C. Lee and Raymond B. Ridley, "Small-Signal Modeling of Average Current-Mode Control," *IEEE Transactions on Power Electronics*, Vol. 8, No. 2, pp. 112-119, April 1993.

[16] M. M. Jovanovic, D. E. Crow and Fang-Yi, "A novel, low-cost implementation of "democratic" load-current sharing of paralleled converter modules," *IEEE Transaction on Power Electronics*, Vol. 11, No. 4, pp. 604-611, July 1996.

[17] Xunwei Zhou, Peng Xu and Fred C. Lee, "A Novel Current-Sharing Control Technique for Low-Voltage High-Current Voltage Regulator Module Applications," *IEEE Transactions on Power Electronics*, Vol. 15, No. 6, pp. 1153-1162, Nov. 2000.

[18] Jung-Won Kim, Hang-Seok Choi and Bo Hyung Cho, "A Novel Droop Method for Converter Parallel Operation," *IEEE Transactions on Power Electronics*, Vol. 17, No. 1, pp. 25-32, Jan. 2002.

[19] Frank H. F. Leung, Peter K. S. Tam and C. K. Li, "The Control of Switching dc-dc Converters-A General LQR Problem," *IEEE Transactions on Industrial Electronics*, Vol. 38, No. 1, Feb. 1991.

[20] Pompeo Marino and Francesco Vasca, "A New Nonlinear Feedforward Compensation for Feedback Controlled DC-DC Converters," *IEEE Power Electronics Specialists Conference*, June 1993, pp. 728-734.

[21] Vietson M. Nguyen and C. Q. Lee, « Indirect Implementations of Sliding-Mode Control Law in Buck-Type Conveters, » *IEEE Applied Power Electronics Conference and Exposition*, Mar. 1996, Vol. 1, pp. 111-115.

[22] R. W. Ashton, J. G. Ciezki and M. G. Badorf, "The Synthesis and Haardware Validation of DC-to-DC Converter Feedback Controls," *IEEE Power Electronics Specialists Conference*, May 1998, Vol. 1, pp. 65-71.

[23] Chung-Chieh Fang and Eyad H. Abed, "Output Regulation of DC-DC Switching Converters Using Discrete-Time Integral Control," *American Control Conference*, Jun. 1999, Vol. 2, pp. 1052-1056.

[24] Fei-Hu Hsieh, Nie-Zen Yen and Yau-Tarng Juang, "Optimal Controller of a Buck DC-DC Converter Using the Uncertain Load as Stochastic Noise," *IEEE Transactions on Circuit and Systems – II: Express Briefs*, Vol. 52, No. 2, Feb. 2005.

[25] Dennis John Packard, "Discrete Modeling and Analysis of Switching Regulators," *Ph. D. Dissertation*, California Institute of Technology, 1976.

[26] Dragan Maksimovic and Regan Zane, "Small-signal Discrete-time Modeling of Digitally Controlled DC-DC Converters," *IEEE COMPEL Workshop*, July 2006, pp. 231-235.

978-1-4244-1871-8/07 $25.00 © 2007 IEEE

Energy Storage Technology Markets and Application's: Ultracapacitors in Combination with Lithium-ion

John M. Miller PE, PhD
Maxwell Technologies, Inc
9244 Balboa Avenue
San Diego, CA. 92123
Email: jmiller@maxwell.com

Abstract— **Ultracapacitors are becoming widely accepted in the energy storage industry in both standalone and in combination with batteries. Standalone applications, once niche, are now rapidly expanding as the technical and economic benefits of this power dense component become more widely understood. Combination examples continue to proliferate, primarily in battery electric and hybrid electric commercial transportation segments such as transit buses and trains. The ultracapacitor offers a fast energy buffer to advanced chemistry energy reservoirs such as nickel metal hydride and lithium batteries and offer an opportunity for truly energy optimized battery systems. This presentation will discuss trends in energy storage systems, advanced chemistry batteries such as nickel-hydrogen and lithium-ion and why such components would benefit from working in combination with carbon capacitors. The presentation will focus on the strengths and weakness of each technology and why the active parallel combination of ultracapacitor plus electronic converter with the lithium-ion battery is a complimentary and synergistic alternative.**

I. INTRODUCTION

Interest in lithium-ion batteries is now very high because of announced activities in plug-in hybrids by the major automakers, especially the recent news from General Motors about their Chevy Volt plug-in and eFlex battery electric vehicles [1]. Toyota has announced plans for a major innovation in their hybrid propulsion systems [2] for 3rd generation hardware that will reduce electric drive component sizes by half and will introduce lithium-ion battery technology, an industry first, that will appear in the fall of 2008 (now delayed 2 years). Nissan Motor Co. has announced similar plans for lithium beginning 2010 in a hybrid vehicle that will use NEC lithium-ion packs [3]. Mitsubushi has announced plans to work collaboratively with GS Yuasa Corp. [4] for lithium-ion battery development to produce in-house hybrid vehicle lithium-ion designs for release in 2009 as MY2010 hybrid electrics.

Energy storage system attributes can be distinguished by power and energy capability as well as the type of vehicle application. Table 1 summarizes these points for conventional low energy storage hybrid electric vehicles such as the Saturn Vue Greenline hybrid, a more battery dominant Saturn Vue plug-in hybrid and battery dominant plug-in hybrid exemplified by the GM E-Flex range extended battery electric vehicle.

TABLE 1
ENERGY STORAGE NEEDS AND VEHICLE TYPES

Attribute	Hybrid Electric Vehicle	Plug-in Hybrid Electric	E-Flex Range Extended
Energy Storage System	Low storage needed, ultracap possible	Engine dominant PHEV, power optimized lithium	Battery dominant PHEV, energy optimized lithium
All Electric Range, AER	Minimal	10+ mile in-city drive	40+ mile in-city drive
Recharge Strategy	During driving only	During drive and when plugged-in	During driving and when plugged-in
Power	For low speed city drive	EV mode	Full vehicle performance available in EV mode
Life	10y, 150k miles	10y, 150k miles	10y, 150k miles

Ultracapacitors are envisioned as enablers for plug-in hybrid electric vehicles (PHEV) where electric energy storage system (ESS) ON-cost and reliability are strong functions of power to energy ratio P/E. ON-cost is lower when P/E is low, a consequence of electrode design for energy optimized lithium-ion. Also, ON-cost is minimized when state-of-charge (SOC) loading is low and cycling is modest. The challenge therefore is how to combine ultracapacitor technology with low P/E lithium-ion to expand useable SOC window (loading) from today's 50% to 70% and wider in the future. A major European automaker puts this into perspective [5] by stating that battery technology is needed that is both weight and package efficient while providing an AER of 500 km. This demands loading factors of 80% to 90% without loss of cycle life and reliability (or safety). The challenge then is for ultracapacitors combined with low P/E lithium-ion in a battery dominant PHEV that is designed to deliver 40% to 60% charge depleting (CD) performance and not incur reliability deterioration during its remaining charge sustaining (CS) operation before a plug-in opportunity. This means that a lithium-ion pack designed today for sustained operation at 60% SOC will be reduced to 30% SOC and still be expected to deliver full CS mode functionality. This has not been studied to date and remains a wide open opportunity for ultracapacitors. This paper addresses that opportunity.

978-1-4244-1871-8/07 $25.00 © 2007 IEEE

II. LITHIUM TECHNOLOGY CHALLENGES THAT ULTRACAPACITORS BENEFIT

There remain many issues facing lithium-ion battery implementation into vehicle propulsion energy storage system packs [6]. Chief among these issues are the following:

- Cost remains a factor of three too high on a power basis ($/kW)
- Performance at cold temperatures, especially for discharge power remains poor and fades with ageing. Temperature goals for lithium-ion are -30°C to +65°C.
- Cycle life is close to meeting target expectations but calendar life remains elusive, especially for 15y life in power assist and 10y in electric mode energy batteries.
- SOC window. Lithium batteries today are limited to 20% to 30% loading factor, or SOC window, because deep cycling significantly reduces life. For instance, a lithium-ion battery capable of 2,600 cycles to 30% depth of discharge (DOD) would only be capable of 1,000 cycles at 80% DOD. Loading factors must increase to 80% to 90% for Plug-in hybrids to be feasible.
- Tolerance to abuse: short circuit, over voltage, over temperature, and rapid discharge and charge, and crush tolerance remain particularly problematic in the lithium-ion cell. This means strong reliance on proper charge/discharge controls and energy management are necessary.

Conventional electrochemical storage technologies are simply inadequate to match the needs of plug-in hybrid electric vehicles. Demands for 20 to 40 miles of all electric range (AER), regardless of pure charge depleting or blended mode, requires that the energy storage system (ESS) exceed a specific energy of 200 Wh/kg at a cost of <$20/Wh. The best of today's high cycling advanced batteries such as nickel metal hydride (NiMH) and lithium-ion fall far short of these goals as shown in Table I. Red text in Table 2 highlights the energy throughput (i.e., Wh-cycles) of these technologies.

an energy optimized lithium-ion, nickel metal hydride or even advanced lead acid batteries be?

According to data in Table 2, lithium excels in cost of energy, while ultracapacitors have the best cost of power. Moreover, ultracapacitors excel on total energy throughput over their operational life, with lithium-titanate coming in a fairly strong second because of its very high deep cycling capability (15,000 cycles to 25,000 cycles at 80% DOD).

III. ULTRACAPACITOR CHALLENGES THAT LITHIUM TECHNOLOGY BENEFITS

Ultracapacitors offer the following advantages that compliment the lithium-ion weaknesses:

- Cost on a power basis is low – ultracapacitors are power dense components
- Performance at cold temperatures, in both discharge and charge, is strong and very efficient
- Cycle life is high, in the +1M cycles at 75% to 80% of SOC. Calendar life is high because the ultracapacitor does not wear out – it stores energy in the same form as it will be used, as electrons, there is no electrical to chemical energy conversion as takes place in the battery.
- SOC window of 75% to 80% is typical of ultracapacitor applications without significant loss of cycle or calendar life.
- Ultracapacitors have very fast response and can deliver very high pulse power virtually instantaneously. Therefore, the ultracapacitor must be handled carefully when charged because of its exceptional power density. Proper controls and energy management are essential.

For the combination of lithium cells plus ultracapacitor and dc-dc converter to generate a positive value proposition the dialogue must now address the combined efficiency of ultracapacitor (95% to 98% depending on loading) plus dc-dc converter efficiency (~97%) and contrast this with the efficiency of a standalone lithium cell string. This is the R&D gap that needs further investigation [7].

TABLE 2
ENERGY STORAGE NEEDS AND VEHICLE TYPES

ESS Component	Specific Energy (Wh/kg)	Energy specific cost ($/Wh)	Power specific cost ($/kW)	Cycle Capability at 80% DOD # Cycles (Wh-cycles)		
Ultracap	5	16	12	>10⁶ 4x10⁶		
VRLA	30	0.12	80	3*10² 7x10⁰		
NiMH	44	0.65	75	4*10³ 1.5x10⁵		
Lithium	70	0.50	75	JCS 2.5*10³ 1.4x10⁵ A123 5*10³ 2.5x10⁶ AltairNano 15*10³ 8.4x10⁶		

The question that arises is what benefit will the combination of a power dense component such as an ultracapacitor with

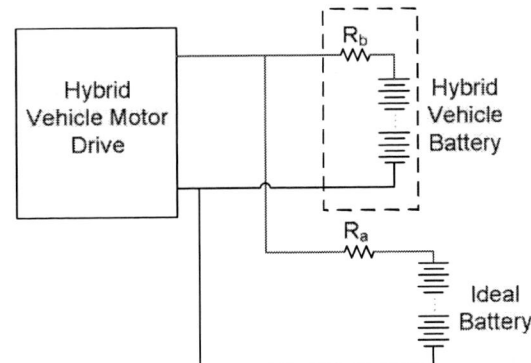

Fig. 1 Illustration of active parallel hybrid energy storage system (Excerpted from [7])

978-1-4244-1871-8/07 $25.00 © 2007 IEEE

Ultracapacitors are power dense components capable of absorbing and releasing their energy instantaneously because their charge is stored in the electrochemical double layer capacitor. Electrochemical energy storage relies on electron transfer in the redox reactions taking place in the electrodes which take time to complete. Response is therefore slower.

Energy storage components are commonly compared based on their power and energy capabilities, a plot known as the Ragone relationship, where energy and power are charted on log-log axes. The Ragone relationship reveals the straight line character of energy (E) as a function of power (P) in the ideal case (i.e., when internal parameters are linear and not a function of potential or current).

$$E_{avail} = E_{stored} - T_{char}P_0 \qquad (1)$$

The characteristic time (T_{char}) is particularly interesting and can be viewed as the intersection of a diagonal line with the Ragone trace and a horizontal line representing half the stored energy. At this triple intersection on the Ragone chart the energy storage component is loaded at matched load power (P_{ML}) and its terminal efficiency is 50%. The characteristic time is therefore the average time constant of the energy storage medium be it electrochemical (Faradaic) or ultracapacitor (non-Faradaic). Load power in (1) is the terminal power applied during discharge. Taking the case of an ultracapacitor having an average time constant, $\tau =$ ESR*C, and noting that matched load power is defined as: $P_{ML} = U_{mx}^2/(4*ESR)$ then (1) can be restated as:

$$E_0 = 2\tau P_{ML} - \tau P_0$$
$$\therefore \tau = \frac{E_{stored}}{2P_{ML}} = T_{char} \qquad (2)$$

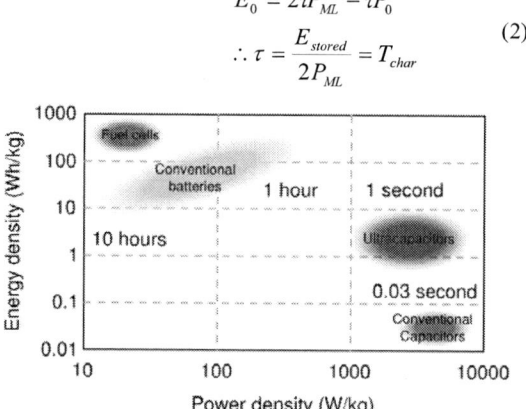

Fig. 2 Ragone relationship chart for energy storage components

In order to improve the performance of batteries and ultracapacitors manufacturers must take the following actions:

- Increase thickness of current collector foils. Ultracapacitor products use 20 μm aluminum foil current collectors. Electrochemical cells use similar thickness foils. Lithium cells for example are constructed of copper foil anodes (graphite based) and aluminum foil cathodes (lithium mix).

- Use of more conductive electrolytes. Higher molarity electrolytes help but are limited by the particular salts available. Electrode film (coated, extruded or dry cast) conductivity improvements through application of additives such as carbon black and better binders are needed.

- Increase the available reaction surface area by making electrode films thinner and longer. Thinner electrodes mean the electrolyte ions can fully permeate the electrode thereby increasing power capability.

IV.   ULTRACAPACITOR – BATTERY COMBINATION

Ultracapacitors can be combined with batteries, including lithium-ion, either directly as in passive parallel or via an electronic converter as in the active parallel configuration. Passive parallel has the benefit of not needing additional electronics, but also makes a limited contribution to battery wear and life. Active parallel provides the most benefit in terms of extracting a full 75% to 80% state-of-charge (SOC) from the ultracapacitor while protecting the battery from undue current, voltage and power stress. Fig. 3 illustrates the active parallel configuration of battery with ultracapacitor. This architecture is also the most adaptable to energy management strategies via the power electronic converter controller for power level and direction of power flow.

Active Parallel HESS

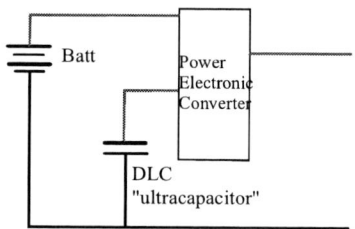

Fig. 3 Active Parallel Combination of Ultracapacitors with Battery

A.   Dc-dc Converter for Energy Storage Buffering

Advances in minimized switching loss converters open the door to high dynamic power flow buffers for power dense storage components. Fast dynamic response to bidirectional power flow facilitates turn around efficiencies in the 96% to 98% range over a large fraction of operating power levels.

Hard switched dc-dc converters operating up to 70 kHz suffer significant efficiency loss so that nowadays soft switching techniques are replacing them. The Honda Motor Co. Civic hybrid for example has transitioned to soft switching [8]. Fig. 4 shows the Honda Civic hybrid battery tray that houses the NiMH 144V, 5.5Ah pack plus air conditioning inverter and dc-dc inverter. All these components are in a package of 59 liters and 55 kg.

Fig. 4 Integrated NiMH pack with dc-dc converter and air conditioning inverter in Honda Civic Hybrid (from [8])

Zero voltage transition (ZVT) switching is used in inverter to implement soft switching at variable frequency between 90 kHz to 115 kHz with overall improved efficiency. Supporting technologies include use of film capacitors in place of aluminum electrolytics, thinner dielectric on the transistor stacks, and improved thermal management. Fahimi and Wang [9] give an excellent presentation on ZVT for dc-dc converters for fuel cell propulsion systems. The circuit configuration for their work is repeated here as Fig. 5 to highlight the major switching elements. This architecture is highlighted because it represents recent progress in dc-dc converter technologies having power processing efficiencies greater than 96%.

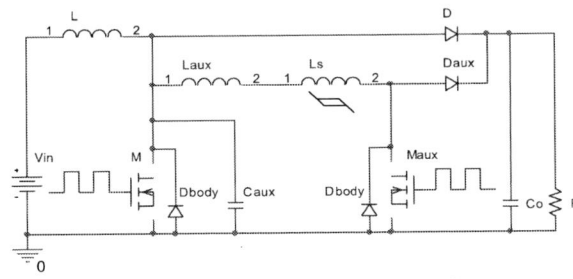

Fig. 5 ZVT dc-dc converter schematic (from [9])
Used with permission.

Hard switched dc-dc converters are simply impractical for today's energy storage system interface because power levels are too high and losses would seriously complicate the thermal management budgets. Soft switching as used by Honda Motor Co and many others and as illustrated in Fig. 6 below is superior to hard switched converters. In fact, the differences are very significant above 100 kHz where many designs tend to be for package volume and mass reduction.

Soft switching is appearing in very high power converters, for interfacing to energy storage and generation components. Brusa [10] for example offers a 97% to 98% efficiency 100V to 425V converter rated 150A low side input. Switching frequency is 22 kHz and the unit is liquid cooled.

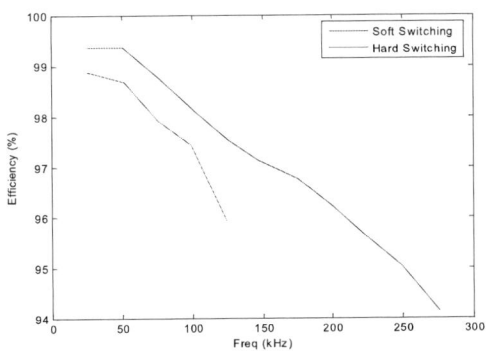

Fig. 6 ZVT dc-dc converter efficiency versus frequency.
(From [7] with permission)

Converter efficiencies in the high 90% range are necessary for buffering energy storage components and for executing advanced energy storage strategies on combination technologies.

B. Ultracapacitor Efficiency Under Constant Power

The active parallel combination is receiving renewed interest because the battery does not recharge the ultracapacitor following a high power pulse as it will in passive parallel connection since it is series connected with a dc-dc converter that manages the power flows into and out of the ultracapacitor. Investigations into ultracapacitor-battery hybrid energy storage systems is growing [11-15] and shows considerable potential to realize energy and power optimized combinations.

Fig. 7 illustrates the essential nature of electric propulsion events as constant power pulses having various dwell times and flow directions. The ultracapacitor is unique in its application to such profiles because its efficiency is very close to being the same regardless of power flow direction. This is a strong benefit of the combination. Analytical modeling of the ultracapacitor under constant power conditions has been performed by various researchers and shows that charge or discharge efficiency can be >90% for a significant fraction of matched load power.

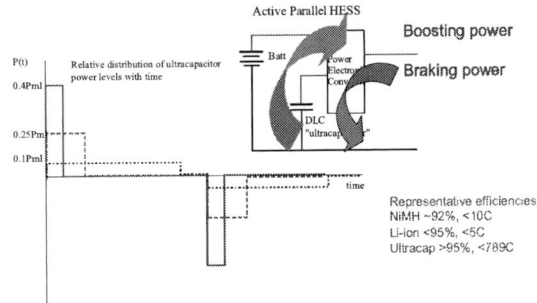

Fig. 7 Constant Power Profiles Applied to the Ultracapacitor.

978-1-4244-1871-8/07 $25.00 © 2007 IEEE

Under constant power operation the ultracapacitor efficiency is high initially and drops with increasing pulse dwell time. This is because the accumulated charge depletes during discharge resulting in diminished cell potential and the consequent demand for higher current (i.e., faster rate of change of stored charge). Fig. 8 captures the main attributes of ultracapacitor efficiency during discharge under constant power conditions at 10%, 25% and 40% of matched load.

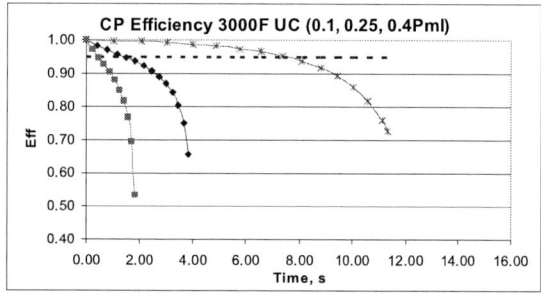

Fig. 8 Ultracapacitor Efficiency Under Constant Power Discharge Conditions: Analytical (solid trace) and Measured (symbols).

C. Lithium-ion Efficiency Under Constant Power Loading.

The most basic equivalent circuit model for lithium battery consists of an internal open circuit potential source, a series resistance corresponding to the electronic components of Joule losses and an ionic contribution. The ionic component models the electrolyte and electrode kinetics and losses and is approximated as a parallel RC element shown in Fig. 9.

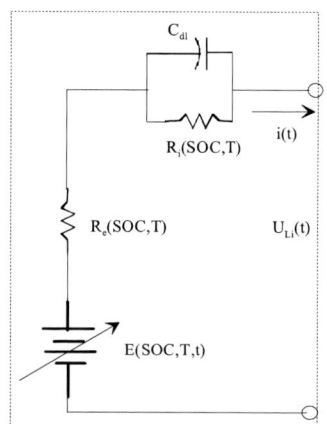

Fig. 9 Lithium Battery Equivalent Circuit Model
(300V, ~35Ah, 10 kWh, 112 mΩ at 10s)

Component values for Fig. 9 are approximated based on commercially available cells and scaled to a 10 kWh, 300V pack suitable for plug-in hybrid electric vehicle propulsion. The model is generic and not intentionally linked to any vendor of lithium technology. The internal EMF source represents the electrochemical potential that is SOC, temperature and time dependent. Circuit values are representative of an energy optimized pack of the type promoted in this paper. In fact, a very high quality lithium.

V. PERFORMANCE OF THE ULTRACAP-LITHIUM COMBINATION

The equivalent circuit model in Fig. 9 is evaluated in simulation (Ansoft Simplorer) for various constant current pulses with dwell times of 1s and 10s. The goal of this exercise is to quantify the power dissipation of the lithium pack, the ultracapacitor and the dc-dc converter under the specific power pulse profiles illustrated in Fig. 7.

A. Response of the Lithium Pack Model

For this analysis the generic lithium pack suitable for a PHEV has the parameter values listed in Table 3. Dynamic performance is specified at 1s and 10s pulse widths.

TABLE 3
GENERIC LITHIUM PACK PARAMETERS

Parameter	EMF	Relectronic	Rionic	Cdl
	296V	80mΩ	32mΩ	26F
Capacity	126k Coul			

Fig. 10 exhibits the lithium pack performance to a 15C discharge and charge pulse with dwell of 1s and 10s. The 1s impedance is 80mΩ increasing to 112mΩ for the 10s pulse.

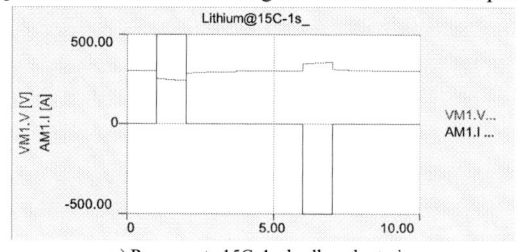

a) Response to 15C, 1s dwell, pulse train

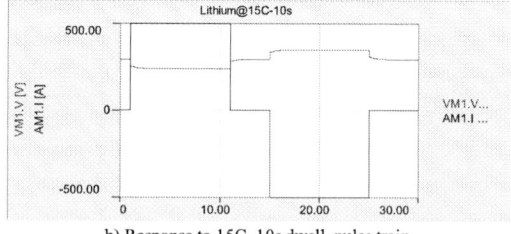

b) Response to 15C, 10s dwell, pulse train

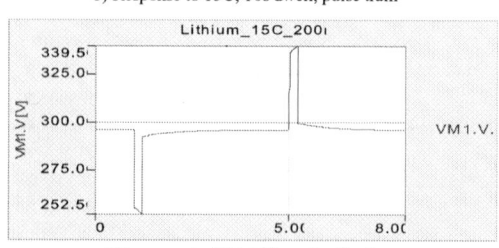

c) Response to 15C, 200 ms dwell pulse
Fig. 10 Response to 15C, 200ms pulse loading

To more clearly illustrate the dynamic response of the lithium pack a 15C pulse for 200ms is applied in Fig. 10c. This pulse duration more closely resembles the demands on the energy storage system in actual vehicle propulsion applications. A 200ms power pulse is also representative of hybrid electric engine starting under the warm restart conditions of idle-stop (i.e., a strategy start).

In Fig. 10b for the 10s pulse dwell the individual cell potential is 3.7V open circuit with min and max of 3.15V/cell and 4.24V/cell respectively and within nominal cell voltage ranges for discharge and charge conditions.

B. Response of the Ultracapacitor Pack Model

Maxwell Technologies has developed an ultracapacitor cell/module equivalent circuit model that is validated for large and small signal conditions. The model captures the nonlinear SOC characteristics of carbon-carbon ultracapacitors while preserving charge and energy storage characteristics of the production parts. The ultracapacitor, a BCAP0650P cell in a 112s x 1p pack is shown as Fig. 11. The inner salmon colored inset represents the cell model and the outer cyan colored box represents the overall module after scaling.

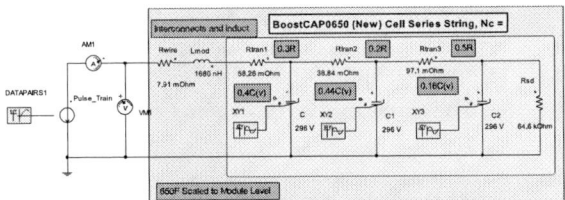

Fig. 11 Ultracapacitor model, BCAP0650P, power cell.

TABLE 4

ULTRACAPACITOR MODEL PARAMETERS

Rwire	7.91mΩ	Lmod	1680 nH
Rtran1	58.26mΩ	C(246V)	2.15F
Rtran2	38.84mΩ	C1(246V)	2.37F
Rtran3	97.1mΩ	C2(246V)	0.86F
Rsd	64.6kΩ	Sum(C's)=	5.38F

The ultracapacitor pack of Fig. 11 is pre-charged to 296V open circuit for compatibility with the generic lithium pack discussed previously. In this configuration the individual ultracapacitor cell potential is 2.64V. During the high current pulse test of 150kW the ultracapacitor pack exhibits an ESR of 60 mΩ, a voltage decay of 92.3V and SOC window of 53 %. Fig. 12 shows the response of the ultracapacitor pack to the current loading used in Fig. 10a.

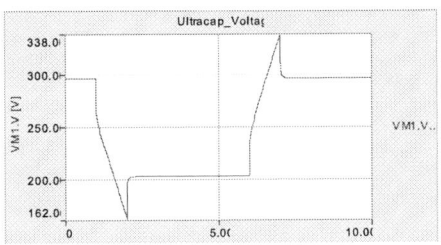

Fig. 12 Ultracapacitor voltage response to 500A pulse for 1s

The efficiency of the ultracapacitor can be summarized as the loss fraction on an energy basis using total stored energy and dissipated energy in the calculation of efficiency (relationships shown in the tables). The result is that the ultracapacitor operating under extreme loading (650F cell at 500A, or 770mA/F or 1136C) results in efficiency of 84% to 86% versus the very large 10 kWh lithium pack simulated efficiency of 83% at 15C loading ignoring the dynamics present. The ultracapacitor is therefore at 41% of P_{ML}.

The higher efficiency of the small ultracapacitor pack (0.44Ah, 131Wh available and 98 Wh useable) is very respectable in contrast to the very large capacity 10,000 Wh lithium pack assumed.

C. Comparison of Models

In practice a buck-boost dc-dc converter will interface the converter. These buck/boost converters account for converter losses and will be used later to approximate real product performance when connected as the buffer between the ultracapacitor pack and the lithium pack and load.

Module power dissipation and full energy survey results are shown in Fig. 13 for the ultracapacitor pack. Table 5 summarizes the efficiency for the ultracapacitor.

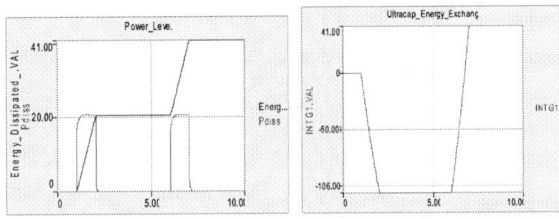

a) Power dissipation and energy b) Discharge and re-charge energy

Fig. 13 Power dissipation and energy survey for ultracapacitor @ 500A

TABLE 5

EFFICIENCY SUMMARY FOR ULTRACAPACITOR

Module	Unit	Discharge	Charge	Total
Pdiss	W	20,729	20,581.4	
Ediss	J	20,479.5	20,429.5	40,909
Eo	J	-105,503	+146,414	40,911
Eff	#	0.837	0.86	0.0.721
		$\eta_d = \dfrac{1}{1+\dfrac{E_{diss}}{E_0}}$	$\eta_c = 1-\dfrac{E_{diss}}{E_{in}}$	$\eta_{ta} = \dfrac{E_0}{E_{in}} = \eta_d \eta_c$ Turn-around eff.

The same procedure is repeated for the lithium pack given the 500A for 1s discharge followed by charge at constant current. Input power during charge is 174 kW for 15C.

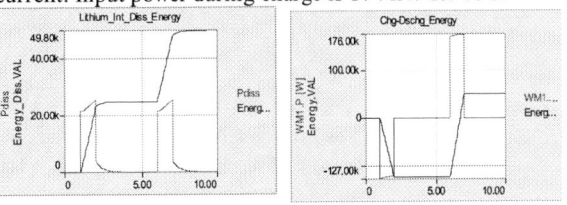

a) Internal dissipation and energy b) Discharge and charge energy

Fig. 14 Lithium pack power dissipation and energy survey @ 15C

TABLE 6
EFFICIENCY SUMMARY FOR LITHIUM

Module	Unit	Discharge	Charge	Total
Pdiss	W	25,200	25,250	
Ediss	J	24,838	24,859	49,697
Eo	J	-123,905	+173,590	49,685
Eff	#	0.83	0.857	0.714
		$\eta_d = \dfrac{1}{1 + \dfrac{E_{diss}}{E_0}}$	$\eta_c = 1 - \dfrac{E_{diss}}{E_{in}}$	$\eta_{ta} = \dfrac{E_0}{E_{in}} = \eta_d \eta_c$ Turn around eff.

Deeper insights can be gained from closer observation of the dynamic response of the ultracapacitor model in Fig. 13 and the large lithium pack model in Fig. 14. For example, the leading edge character of the internal power dissipation of both is characteristically different. The ultracapacitor internal power dissipation exhibits a parabolic down trend as the pulse time runs out while the lithium model exhibits a parabolic up trend with pulse time.

A plausible explanation for this is that in ultracapacitors the terminal resistance changes from ESRdc to ESRac where the ratio: ESRac/ESRdc ~ 0.7, whereas for the lithium model the 10s impedance is generally 1.4 times the 1s impedance. This observation would account for the trends in internal power dissipation noted in Fig. 15.

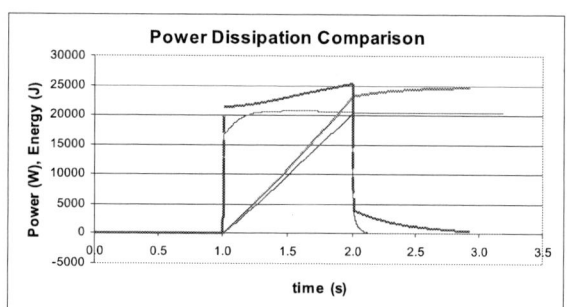

Fig. 15 Comparison of Ultracapacitor (light traces) with Lithium pack (heavy traces) 1s high rate pulse internal power dissipation characteristic.

In summary the ultracapacitor pack consisting of 112 small cells will have an approximate mass of approximately 26 kg including packaging and volume of roughly 30 dm^3. The lithium pack will have a mass of 143 kg.

Influence of the dc-dc converter on overall efficiency, package and cost will be the subject of future work.

VI. CONCLUSIONS

There is growing recognition that lithium technology has shortcomings when long term charge depleting (PHEV) operation is demanded that has superimposed the shallow power cycling characteristic of charge sustaining (HEV mode). Issues with SOC window, charge acceptance above 80% SOC and discharge performance below 30% SOC along with poor pulse power performance at cold temperature are prompting intense research into optimized lithium chemistries such as the 1/3-1/3-1/3 lithium-ion cathode and non-graphitic, lithium-titanate anode designs.

Ultracapacitors offer new opportunities when combined with fast dynamic dc-dc converters and the high efficiency of ultracapacitors under constant power cycling (>95% for loading <25% of P_{ML}) in combination with efficient half-bridge dc-dc converters (~96% to 98%) efficient. The combination shows promise of the composite energy storage system delivering power and energy at greater than 90% efficiency overall and over the high dynamic working range of application power levels.

REFERENCES

[1] GM's View on Battery Technologies for Hybrids, Plug-Ins and E-Flex, Green Car Congress, Energy, Technologies, Issues and Policies for Sustainability Mobility, 13 MARCH 2007

[2] Toyota's Bid for a Better Battery , MARCH 5, 2007 THE CORPORATION/Online Extra BusinessWeek Online

[3] Nissan, NEC say mulling joint car battery company Reuters / AutoNews.com, December 21, 2006 - 4:49 am

[4] Mitsubishi, GS Yuasa to form lithium-ion battery joint venture, Reuters, May 8, 2007 - 9:00 am

[5] SAE International, "Searching for Fossil Alternatives," Automotive Electronics International, AEI, March 2007.

[6] U.S. Department of Energy 2006 Annual Progress on Energy Storage Research and Development, Office of FreedomCAR and Vehicle Technologies, January 2007

[7] T. Bohn, "Plug-in Hybrid Vehicles: Decoupling Battery Load Transients with Ultracapacitor Storage," Advanced Capacitor World Summit, San Diego, CA., 25 July 2007

[8] T. Hasebe, "Honda Hybrids Overview – Innovation of IMA (Integrated Motor Assist) Technologies," SAE Hybrid Vehicle Technology Symposium, San Diego, CA, 7 Feb 2007

[9] B. Fahimi, S. Wang, "High Efficiency and Compact DC/DC Converter for High Power Fuel Cells Systems," IEEE Power Electronics Society newsletter, Vol. 19, Nr. 3, July 2007.

[10] Brusa BDC414 DC/DC Converter. www.brusa.biz

[11] John M. Miller, "Ultracapacitor Efficiency under Constant Power Conditions: Prospects for Lithium Battery plus Ultracapacitor Hybrid Energy Storage Systems," 3rd International Symposium on Large Ultracapacitor Technology and Application (UCAP), Advanced Automotive Battery Conference, Hyatt Regency hotel, Long Beach, CA, 15-17 May 2007

[12] James Marco, Nicholas D. Vaughan, "Defining Performance Metrics for Hybrid Electric Vehicles," SAE technical paper 2007-01-0287, Society of Automotive Engineers International World Congress, Cobo Center, Detroit, MI. 16-19 April 2007

[13] Juan Dixon, Ian Nakashima, Fabian Arcos, Micah Ortuzar, "Test Results in an Electric Vehicle using a combination of Ultracapacitors and Zebra Battery," 22nd International Battery, Hybrid and Fuel Cell Electric Vehicle Symposium and Exposition, Yokohama, Japan, 23-25 Oct. 2006

[14] Ahmad Pesaran, Tony Markel, Matthew Zolot, Sam Sprik, "Ultracapacitors and Batteries in Hybrid Electric Vehicles," Advanced Capacitor World Summit, Hilton San Diego Resort, 11-13 July 2005

[15] Roberto M. Schupbach, Juan C. Balda, "The Role of Ultracapacitors in an Energy Storage Unit for Vehicle Power Management," IEEE Vehicular Technology Conference VTC2005, DisneyWorld Hotel, Florida, Sept. 2005.

Impact of Power Density Maximization on Efficiency of DC-DC Converter Systems

J.W. Kolar, J. Biela and U. Badstübner
Power Electronic Systems Laboratory, ETH Zurich
ETH-Zentrum, ETL H23, CH-8092 Zurich
Email: kolar@lem.ee.ethz.ch

Abstract—**The demand for decreasing costs and volume leads to a constantly increasing power density of industrial converter systems. In order to improve the power density further different aspects, like thermal management and electromagnetic effects must be considered in conjunction with the electrical design. Therefore, a comprehensive optimization procedure based on analytical models for minimizing volume of dc-dc converter systems has been developed at the Power Electronic Systems Laboratory of the ETH Zurich.**

Based on this procedure three converter topologies – a phase shift converter with current doubler and with capacitive output filter and a series-parallel resonant converter – are optimized with respect to power density for a telecom supply (400 V/48 V).

There, the characteristic of the power density, the efficiency and the volume distribution between the components as function of frequency is discussed. For the operating points with maximal power density also the loss distribution is presented. Furthermore, the sensitivity of the optimum with respect to junction temperature, cooling and core material is investigated.

The highest power density is achieved by the series-parallel resonant converter. For a 5 kW supply a density of approximately 12 kW/ltr. and a switching frequency of ca. 130 kHz results.

I. INTRODUCTION

The power density of power electronic converters has roughly doubled every 10 years since 1970. Propelling this trajectory has been the increase of converter switching frequencies, by a factor of 10 every decade, due to the continuous advancement of power semiconductor device technology. This increase in power density has been especially important in the design of telecom power supplies which have to operate in a limited space and have maximum weight requirements.

In the near future, the short term operating costs of telecom power supplies will outweigh their capital cost. Along with the high operating cost, due to rising energy prices, the negative environmental effects of increasing energy consumption will demand power supplies with the highest possible efficiency. Therefore, an optimization of power supplies with respect to power density and efficiency for future "Green Data Centers" [1], which enables cost and cooling effort reduction, is required. The main question that arises is: "To what extent does a power density optimization influence the efficiency of the converter system"? In order to address this question an optimization of the converter design is required.

Modern power supply design must consider the thermal issues (thermal interfaces, heat distribution and fluid dynamics) and electromagnetic effects (parasitic elements, electromagnetic coupling, HF-losses and EMI filtering) in conjunction

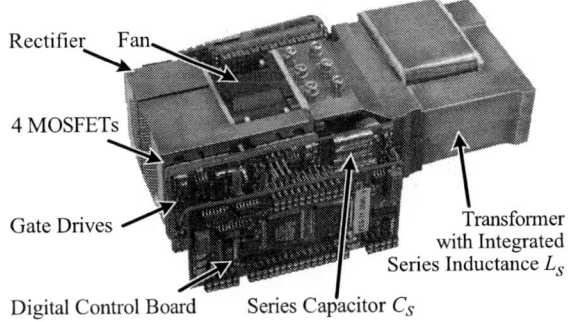

Figure 1. Prototype of an optimized series-parallel resonant converter for telecom applications with a power density of 10 kW/ltr. and the specification given in table I.

with the electrical design since all these areas significantly influence the size and the efficiency of the system. Therefore, an automatic optimization procedure is applied in this paper to maximize efficiency and/or power density. With this procedure the efficiency, the power density and their mutual influence on two, widely used, telecom power supplies concepts, i.e. a resonant converter and a phase shift converter with capacitive/inductive output filtering, are investigated.

The optimization procedure is based on analytic approaches with sufficient accuracy but limited calculation effort instead of general FEM/CFD simulations in order to limit the calculation time. Consequently, analytical models and equations, which include the magnetic devices, ZVS/ZCS switching losses, and HF-losses in the integrated transformer, have been derived and validated for the two converter types. Moreover, the thermal models for the transformer/inductor with integrated cooling system and models for the volume of the required cooling system including the fan have been developed [2], [3]. The optimization procedure also includes methods for calculating

TABLE I
Specifications of the telecom power supplies considered in the optimization presented in this paper.

Input voltage	400 V
Input current	13.0 A
Output voltage	54 V
V_{Ripple} at Output	300 mV$_{pp}$
Output current	92.6 A
Output power	5 kW
Maximum ambient temp.	45 °C

the volume of the resonant and output capacitors.

Based on this procedure, the power supplies are optimized with respect to the power density for the parameters given in table I. There, the characteristic of the power density, the efficiency and the volume distribution between the components as function of frequency is discussed. For the operating points with maximal power density also the loss distribution is presented. Furthermore, the sensitivity of the optimum with respect to junction temperature, cooling and core material is investigated.

Before the optimization procedure is presented in **Section III** the current and voltage waveforms of the three topologies and the main differences are explained in **Section II**. In **Section IV** the models applied in the optimization procedure are briefly described. Thereafter, the results of the optimization and the comparison of the topologies are presented in **Section V** and the two converter concepts are compared with respect to the maximal achievable power density and efficiency.

II. CONVERTER TOPOLOGIES

In order to find the limits of the achievable power density and efficiency with an optimization procedure, first the topologies must be identified which show the best potential for volume minimization while maintaining high efficiency. In literature many different topologies have been proposed for telecom applications [4] - [11], which could be basically divided into hard switched, soft switched and resonant converters. Due to the high switching losses, the hard switched topologies do not allow to reduce the volume of the passive components by increasing the switching frequency and simultaneously having a high efficiency.

In the area of soft switched converters the phase shift converter with current doubler or with capacitive output filter [12] (cf. Fig. 2) are promising representatives, which show low switching losses/high efficiency, a simple control, a low number of components and the potential for high power density. Therefore, this concept and a series-parallel resonant converter are optimized in this paper. The series-parallel resonant converter with capacitive or LC output filter as shown in Fig. 4 is a promising converter structure since it combines the advantages of the series resonant converter and the parallel resonant converter. On the one hand the resonant current decreases with the decrease of the load and the converter can be regulated at no load and on the other hand good part-load efficiency can be achieved [13], [14]. Furthermore, the converter is naturally short circuit proof.

A. Phase Shift Converter

In Fig. 2 a phase shift converter with the two considered rectifier structures – a current doubler (CDR) and a center tapped transformer with capacitive output filter (CTC) – are shown. The primary side of these converters and also the control of the switches is the same for both rectifier topologies. However, the current waveforms (cf. Fig. 3) and the related switching and conduction losses, as well as the transformer design are significantly influenced by the rectifier stage.

TABLE II
Main difference between the three considered topologies. For the current doubler output the turns ratio is half the turns ratio of the phase shift full bridge with capacitive output.

	Phase Shift		Resonant
	Current-D.	Cap.-Filter	Cap.-Filter
Turns Ratio	2.75:1	5.5:1:1	7:1:1
Rectifier Volt.	$V_{IN}/2.75$	$2 \times V_{OUT}$	$2 \times V_{OUT}$
I_{Off} Leg B	medium	high	low/medium
I_{Off} Leg A	medium	low	zero
Series Inductance	low	medium	high
Reactive Energy	low	medium	zero/low

With a current doubler on the secondary side, a transformer with two standard windings can be applied. There, the turns ratio is $N_P/(2N_S)$ – in the considered case 2.75:1 – what leads to a high secondary voltage, which must be blocked by the rectifier diodes (here: worst case 400 V/2.75≈145 V). The average current in the output inductors is half of the output current, what is roughly also true for the secondary winding of the transformer. During states 1 & 3 (cf. Fig. 3) the current in the secondary winding is equal to the output inductor current, which rises from $I_{1,CDR}$ to $I_{Off,CDR,B}$, which is turned off by a MOSFET of the leg B. During states 2 & 4 the current is determined by the leakage inductance of the transformer. There, the current is decreasing relatively slowly down to $I_{Off,CDR,A}$, since the voltage across the leakage inductance is approximately equal to the forward voltage drop of the conducting power transistors. The value of the leakage inductance and the current at the switching instant must not be too small in order to guarantee ZVS conditions for all four switches. For determining the power density/efficiency limit only the operating point with nominal output power is considered. Part load efficiency is neglected, since the aim is to determine the upper achievable limits. Consequently, a relatively low leakage inductance value (here ≈ 2μH) is sufficient for charging/discharging the parasitic output capacitors of the MOSFETs. This inductance is realized by spatial separation of the transformer windings.

In case the full bridge is combined with a center tapped transformer and a capacitive output filter the series induc-

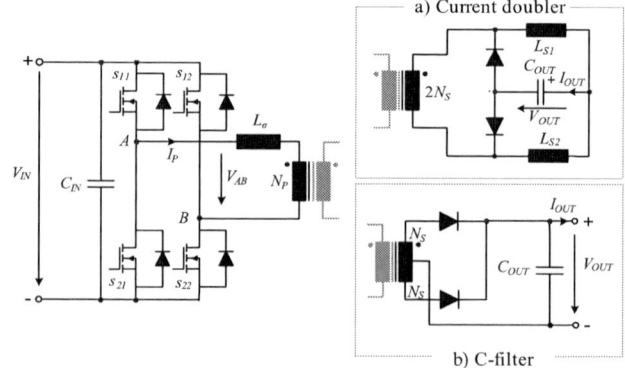

Figure 2. Schematic of the phase-shift converter with (a) current doubler and (b) capacitive output

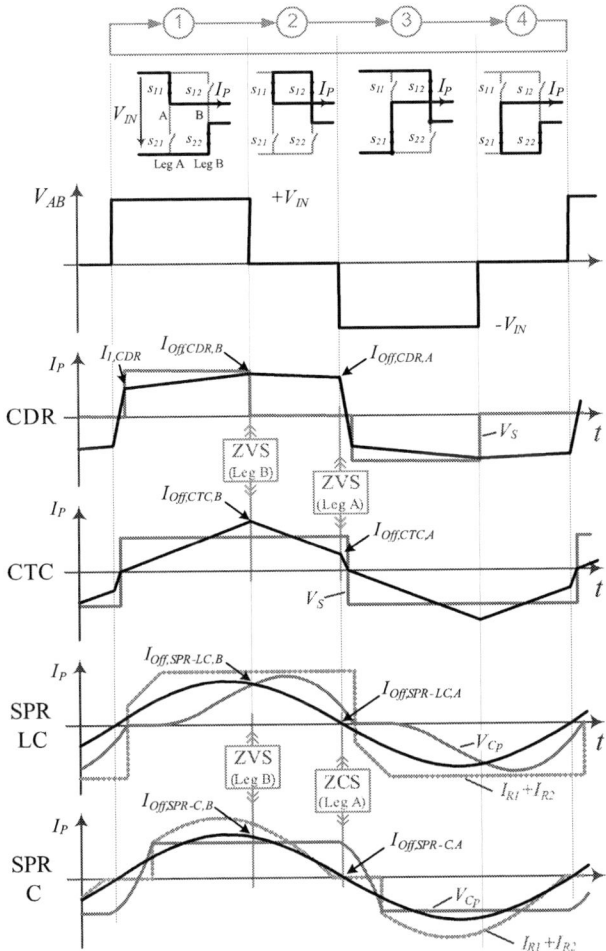

Figure 3. Switching states and primary currents of the phase-shift converter with (a) current doubler (CDR), (b) capacitive output (CTC) and series-parallel resonant converter (SPR) with (c) LC-filter and (d) C-Filter. There, a smaller duty cycle has been used than in nominal, in order to improve the readability of the figure.

tance, which is integrated as leakage inductance L_σ into the transformer, must be larger in order to limit the rise of the current I_P (cf. Fig. 3), since two voltage sources are directly connected via the transformer. During the states 1 & 3 the voltage across L_σ is $V_{IN} - N_P/N_S \cdot V_{OUT}$ and the current rises up to $I_{Off,CTC,B}$, which must be turned off by a MOSFET of the leg B. In states 2 & 4 the voltage $-N_P/N_S \cdot V_{OUT}$ lies across L_σ and the current decreases down to $I_{Off,CTC,A}$, which is turned off by a MOSFET of the leg A. At the beginning of the following state 3 or 1, the current in the leakage inductance first must be decreased down to zero before it could rise again. During this period the energy stored in the leakage inductance is fed back to the DC link, what results in a reactive power flow. This reactive power flow is required for obtaining ZVS condition in the leg A. With the CDR analogue behavior could be observed, but the energy is lower since the inductance is smaller.

The required turns ratio of the CTC-transformer is $N_P:N_S:N_S$ – in the considered case 5.5:1:1 (assumed max.

duty cycle < 0.85, cf. table V) – resulting in $2 \times V_{OUT}$ across the rectifier diodes during the blocking state.

B. Series-Parallel Resonant Converter

For the series-parallel resonant converter two different rectifier circuits are considered – a center tapped transformer with LC-output filter and one with capacitive filter as shown in Fig. 4. There, the output filter topology influences the behavior of the dc-dc converter significantly, since the LC-filter acts more like a current source at the output and the capacitive filter as voltage source. The current and voltage waveforms are shown in Fig. 3, where besides the resonant current I_P also the voltage across the parallel capacitor V_{C_P} and the current $I_{R1} + I_{R2}$ in the rectifier diodes is shown.

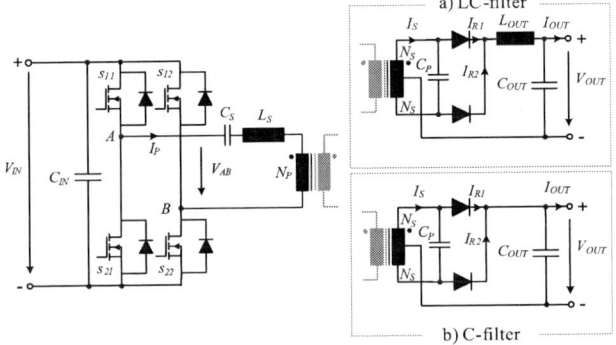

Figure 4. Schematic of the series-parallel resonant converter with a) LC-filter and b) capacitive output

In case of the LC output filter the continuous (CCV) and the discontinuous (DCV) capacitor voltage mode must be distinguished [15]. Since the DCV-mode usually occurs at heavy load conditions, in Fig. 3 the waveforms for this mode are shown. There, the voltage waveform of the parallel capacitor V_{C_P} is clamped by the output current to zero for a certain period of time, which leads to a discontinuous parallel capacitor voltage V_{C_P} showing a large deviation from the purely sinusoidal shape. At the time when V_{C_P} falls to zero the resonant current I_S is smaller than I_{OUT}. As long as I_S is smaller the capacitor voltage V_{C_P} is clamped to zero by the negative difference between I_S and I_{OUT} [$-(I_S - I_{OUT})$] which flows through the rectifier diodes. From the angle on when I_S is larger than I_{OUT} the positive difference between the currents ($I_S - I_{OUT}$) charges the capacitor C_P.

Due to the sinusoidal resonant current and the output inductor L_{OUT} the current in the rectifier diodes starts more smoothly resulting in a lower diode turn on stress. In the secondary winding, however, flows a constant DC current plus AC component causing higher transformer losses.

With the capacitive output filter the current in the rectifier steps from zero to the value of the output current. This could result in higher diode forward recovery losses depending on the diode semiconductor technology and in an increased EMI noise level. The transformer secondary RMS current, however, is lower what results in lower losses and a compacter transformer design.

978-1-4244-1871-8/07 $25.00 © 2007 IEEE

TABLE III
Component values of the LC and the capacitive output filter shown in Fig. 4 for an output voltage ripple of $300\,\mathrm{mV}_{pp}$.

LC-Filter		Capacitive Filter	
C	$30\,\mu\mathrm{F}$	C	$470\,\mu\mathrm{F}$
-		R_{ESR}	$50\,\mu\Omega$
L	$5\,\mu\mathrm{H}$	-	
$I_{C,Ripple,RMS}$	$4.6\,\mathrm{A}$	$I_{C,Ripple,RMS}$	$52\,\mathrm{A}$

Based on the required output voltage ripple of maximum $300\,\mathrm{mV}_{pp}$ and a maximum inductor ripple current of $\pm 7.5\,\%$ for the LC-filter, the component values for the two topologies can be determined (cf. table III). The ripple current in the filter capacitor in the topology with capacitive filter is much higher than the one for the LC-filter. Applying electrolytic capacitors this high ripple current results in a large filter volume due to the relatively high ESR/low current carrying capability of electrolytic capacitors. This is shown in the second line of row "LC-Filter Size" in table IV. The first line represents the volume of the capacitor if just the capacitance value is considered and the ripple current is neglected. With the ripple current the capacitor volume increases more than by a factor of ten. Both values are based on the cuboid volume of cylindrical electrolytic capacitors.

TABLE IV
Comparison of LC and capacitive output filter with electrolytic or ceramic capacitors. For electrolytic capacitors two volumes are given: The first value gives the volume if just the capacitance value is realized - neglecting the ripple current I_{AC} and the current carrying capability of the capacitors. With the second value also the ripple current is accounted for. The inductor of the LC-filter has a volume of $40.9\,\mathrm{cm}^3$. for the ceramic capacitors also two volumes are considered: The volume "SMD-Device" is the pure volume of the ceramic capacitor and the volume "Mounted" also accounts for the volume of the PCB where the capacitors are mounted. There, a double sided, $1.5\,\mathrm{mm}$ thick board is assumed.

	Electrolytic		Ceramic	
	Cylindrical	Cuboid	SMD-Device	Mounted
$\mu\mathrm{F/cm}^3$	170	134	111	90
I_{AC}/cm^3	$0.25\,\mathrm{A}$	$0.19\,\mathrm{A}$	$41.6\,\mathrm{A}$	$35.1\,\mathrm{A}$
LC-Filter Size in [cm³]	$\mu\mathrm{F}$ only: $\mathbf{0.22} + 40.9\ (L)$		$\mathbf{0.32} + 40.9$ (Inductor)	
	I_{AC}: $\mathbf{24.2} + 40.9\ (L)$		(Max. $I_{AC} < 11.3\,\mathrm{A}$)	
C-Filter Size in [cm³]	$\mu\mathrm{F}$ only: $\mathbf{3.7}$, $I_{AC} < 1.5\,A$		$\mathbf{5.4}$	
	I_{AC}: $\mathbf{273} \Rightarrow C = 36\,\mathrm{mF}$		(Max. $I_{AC} < 233\,\mathrm{A}$)	

If the capacitance is realized by ceramic capacitors the volume decreases down to $0.32\,\mathrm{cm}^3$ including ripple current considerations. In both cases the volume of the inductor ($40.9\,\mathrm{cm}^3$) is based on very compact commercially available inductors e.g. [16], [17]. Due to the large inductor volume the size of the LC-filter is relatively large and would consume approximately $10\,\%$ (including interconnection) of the volume if a power density of $10\,\mathrm{kW/ltr}$. is assumed.

With a capacitive filter and ceramic capacitors the volume of the filter elements decreases down to $5.4\,\mathrm{cm}^3$ including the volume of the PCB for mounting and the maximum thermally possible ripple current increases to $233\,\mathrm{A}$. A capacitive filter with electrolytic capacitors would result in a filter volume of $273\,\mathrm{cm}^3$ and a capacitance value of $36\,\mathrm{mF}$ if the ripple current

is considered. Taking just the required capacitance value into account this volume decreases to $3.7\,\mathrm{cm}^3$.

Since with the capacitive output filter the volume of the filter is much smaller and the current in the secondary winding is lower, in the following only the resonant converter with capacitive output filter is considered in the optimization performed with the optimization procedure presented in the following section.

III. OPTIMIZATION PROCEDURE

After the topologies with the best potential for a high power density and a high efficiency have been identified, the components values must be chosen, so that the system volume becomes minimal and/or the efficiency maximal. Since the volume of the single components, which are mainly limited by the respective maximum operation temperature, interdepend to some extent on each other, the optimization of the overall volume/efficiency is a quite involved task with many degrees of freedom.

Therefore, an automatic optimization procedure is applied for determining the optimal component values of the telecom supply.

In Fig. 5 a possible flow chart of such a procedure is given, where the specification of the design parameters like the input and output voltage, the output power, temperature limits, material characteristic, etc. is the starting point of the procedure. These parameters as well as starting values for the

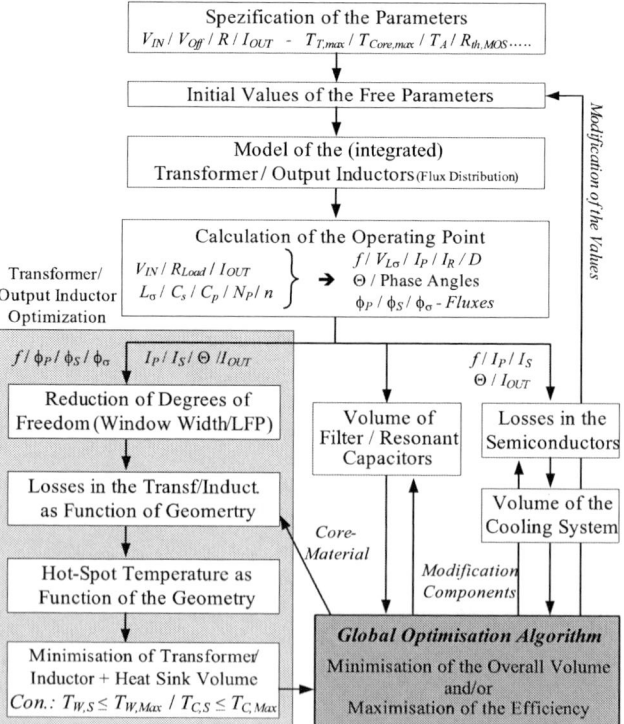

Figure 5. Automatic procedure for optimizing the volume/efficiency of a series-parallel resonant converter while keeping the device temperatures below given limits.

free parameters like N_P, N_S or C_S, C_P, L_S/L_{OUT} must be specified by the user.

Based on the values for the series/leakage respectively the output inductance and the turn numbers the magnetic components are modeled. In case of the phase shift converter the models mainly consist of analytic expressions for the flux distribution and the optimal thickness/diameter of the foil/wires for the windings [18]. For the resonant converter a reluctance model of the transformer with integrated series inductance is calculated. This model is combined with the converter model for calculating the flux distribution in the core [15], [19].

The converter model for the phase shift converters is based on analytic expressions with which the currents and the voltages as well as the duty cycle and the constraints for ZVS conditions are calculated. For the resonant converter first, the operating point of the dc-dc converter is estimated by an approximated fundamental frequency analysis [20]. With the estimated values the solution space for the analytic converter model [19] is restricted and the calculation time is reduced. The converter model is based on a set of equations which are derived with the extended fundamental frequency analysis [21] and solved numerically. The solution are the operating frequency, the voltages and currents as well as the flux distribution of the integrated transformer including phase information.

With the currents in the converter the switching and conduction losses of the MOSFETs and the rectifier diodes are determined. These losses, the ambient temperature and the maximum junction temperatures of the semiconductor devices are used for calculating the volume of the semiconductor heat sink including the fan based on the $CSPI$ (Cooling System Performance Index), which is defined by

$$CSPI\left[\tfrac{\mathrm{W}}{\mathrm{K\,ltr.}}\right] = \frac{G_{th,S-A}\left[\tfrac{\mathrm{W}}{\mathrm{K}}\right]}{Vol_{HS,Magn.}\,[\mathrm{ltr.}]} \qquad (1)$$

and has been introduced in [3] ($G_{th,S-A}$ is the thermal conductivity of the heat sink). There, it is important to check the resulting volume of the cooling system, since with the scaling factor $CSPI$ quite small volume for the heat sink/fan can result, which is difficult to manufacture. A possible solution is to combine heat sinks which are on comparable temperature levels, so that one larger fan could be used for both heat sinks.

Besides the losses in the semiconductors also the volume and the losses in the resonant tank capacitors and/or the output capacitors are calculated with the voltages/currents. There, the losses in the capacitors must be limited to the maximum admissible values.

The volume and the shape of the transformer/inductor core and the two windings is determined in a second, inner optimization procedure (light gray shaded in Fig. 5). There, the volume of the transformer/inductor is minimized while keeping the temperatures below the allowed limits. For this purpose, first the geometrical degrees of freedom are reduced to 3 by determining the core window width using the optimal winding thickness and the turns number [20]. In case of transformers with integrated leakage inductance also the width of the leakage path (LFP) is fixed by setting the flux density in the leakage path to the same value as in the middle leg conducting the main flux.

Thereafter, the core and winding losses are calculated as function of the three remaining geometrical variables (a, b, d cf. Fig. 7(a)). With the losses and the thermal model of the transformer/inductor the temperature distribution in the core and the winding also could be calculated as function of the variables a, b and d. The peak temperatures in the windings and the core are together with the maximum allowed temperatures the constraints for the following minimization of the volume including the volume of cooling system for the magnetic device (cf. Fig. 7(b)). Furthermore, the variables $a - e$, defining the transformer/inductor geometry, can be restricted in order to preserve certain limitations resulting from the manufacturing process.

In the inner optimization loop it is also possible to maximize the efficiency of the transformer/inductor, if an upper limit for the volume is given.

Together with the volumes of the capacitors/heat sink the minimized transformer/inductor volumes are passed to the global optimization algorithm. This algorithm systematically varies the values of the free parameters until a minimal system volume or a maximal efficiency is obtained. This procedure is relatively fast/simple for the phase shift converters since the number of interdependencies is small, but in case of the resonant converter the models are complex and the calculation/computation effort is huge.

IV. MODELS

In the subsequent paragraphs the different models of the optimization procedure are explained shortly. First, the analytical converter model is derived, then the equations for the semiconductor losses and the model for the resonant tank capacitor volumes. Finally, the loss equation and the thermal model of the transformer are presented.

A. Analytical Converter Model

With the analytical converter models the currents and the voltages as well as the operating point (duty cycle, frequency, phase shift, etc.) for the phase shift full bridge with current doubler or center tapped transformer and the series-parallel resonant converter with capacitive output filter are calculated.

In case of the series-parallel resonant converter (cf. Fig. 4(b)) the models are partly based on the extended fundamental frequency analysis (E-FFA) proposed in [15], [19], [21], where the currents and voltages are represented by their fundamentals as shown in Fig. 6. In the model also the control method described in [15] with ZVS condition in one leg and ZCS condition in the other leg as well as control by frequency and duty cycle is considered in the equations. This control method reduces the switching losses significantly.

For the resonant converter with purely capacitive output filter the E-FFA has been improved so that not only the fundamental component is considered but also the third harmonic,

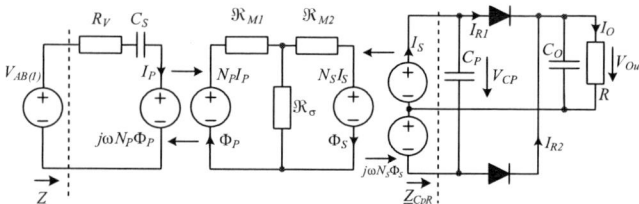

Figure 6. Equivalent circuit of the series-parallel resonant converter with capacitive output filter.

since it influences the behavior of the converter significantly. Thus, both the primary resonant current I_P and the secondary resonant current I_S are sinusoidal with a superimposed third harmonic component. Furthermore, it is assumed that the output voltage is constant and that the components are ideal. The major procedure of the analysis is to determine the impedance Z_{CpR} of the parallel connection of C_P and the rectifier at first. With this impedance the input impedance Z of the resonant circuit, seen by the H-bridge / voltage source V_{AB} (cf. figure 6), could be calculated. In the impedance Z also the reluctance model of the transformer is included.

In the next step two equations for each harmonic can be set up. The first one describes the relation between the phase shift of the primary current I_P and the fundamental component of the H-bridge voltage $V_{AB(1)}$, which is determined on the one hand by the duty cycle D and on the other hand by the impedance Z. The second expression relates the input impedance of the resonant tank to the amplitude of the resonant current. These equations are derived in [19] and are solved numerically in the optimization procedure.

B. Semiconductor Losses

For calculating the volume of the heat sink for the semi-conductors with (1) the maximum operating temperature and the thermal resistance between junction and heat sink of the semiconductors are required. These values can be derived from the data sheets of the applied semiconductors. Furthermore, the losses in the 4 MOSFETs including antiparallel diode and the 2 rectifier diodes must be calculated. Based on the currents calculated with the converter models the RMS currents in the MOSFETs and the resulting conduction losses can be calculated. There, it is assumed that always one MOSFET per leg is turned on, so that the current does not flow via the antiparallel diode but in reverse direction through the MOSFET.

For the rectifier diodes an approximately constant forward voltage drop is assumed, so that the conduction losses can be calculated with the average currents. The switching losses of the diodes are neglected since it is assumed that Schottky diodes are used.

Due to the ZVS condition the switching losses can not be calculated based on data sheet information. Instead measurements, which have been performed with the applied APT50M75 MOSFETs from Microsemi (former Advanced Power Semiconductors) [20], are used in the optimization procedure. Based on these measurements the losses per

MOSFET can be determined by

$$P_{ZVS}[\text{W}] = \left(1.9^{e-7} I_{Off}^2[\text{A}] - 3.8^{e-6} I_{Off}[\text{A}] + 1.4^{e-5}\right) f[\text{Hz}]$$

in case the current turned off by the MOSFET is

$$I_{Off} \geq 15\,\text{A}$$

and they are negligible if the current I_{Off} is below 15 A.

With the applied control method one leg of the resonant converter would switch at ZCS condition. However, if the ZCS leg, which should switch at the zero crossing of the resonant current, is switched slightly before the zero crossing, the MOSFET has to turn off a small current. Because of the fast switching and the large output capacitance of the MOSFET this current does not cause relevant turn off losses. In case the turned off current is large enough, so that it charges the MOSFET capacitances during the interlocking delay [15], the opposite MOSFET is turning on at zero voltage. Consequently, the switching losses in the ZCS leg are negligible [20].

With the explained approaches the semiconductor losses can be calculated and the heat sink temperature and volume (cf. (1)) can be determined so that the maximal junction temperature is not exceeded. For the efficiency calculation also the losses in the gate drive circuits, which can be calculated with the gate-charge/capacitance and which increase linearly with frequency, must be considered.

C. Resonant Tank Capacitors

The capacitors of the resonant tank and the capacitive output filter are carrying high frequent currents with a relatively high amplitude. In order to limit the losses and the temperature rise dielectrics with a low loss factor $\tan \delta$ are required. There are basically two good choices: either foil capacitors with polypropylene or ceramic capacitors with COG/NPO material. Since the resulting volume with foil capacitors is significantly larger than for ceramics as could be derived from data sheets, the latter are chosen for the considered telecom power supply.

For calculating the volume required for realizing the series and parallel capacitor a commercial 3.9 nF/800 V COG ceramic capacitor in a 1210 SMD housing from Novacap [22] has been chosen as reference component, since it offers the highest capacitance per volume ratings at AC voltages with a high frequency and amplitude. Based on this capacitor the resulting volume could be calculated by scaling. There, also the volume for mounting the components on a double sided PCB is considered in the optimization procedure.

In case of the filter capacitor a 2.2 μF/100 V X7R ceramic capacitor in a 1210 housing manufactured by muRata [23] is used as reference element. The capacitance value is calculated with the currents based on the maximum allowed ripple voltage of 300 mV$_{pp}$ at the output.

With the currents also the losses in the capacitors can be determined with the loss factor. These are compared with the maximal allowed values, which can be derived by the loss limit of 0.35 W per 1210 housing at 40 °C ambient temperature and 125 °C maximal dielectric temperature. There, also the decrease of the capacitance with temperature and DC voltage is considered in the optimization procedure.

978-1-4244-1871-8/07 $25.00 © 2007 IEEE

D. Transformer Model

In the optimization procedure shown in Fig. 5 also the shape of the transformer is optimized for minimal volume in the inner loop while the hot spot temperatures are kept below the limits. For this inner optimization loop the volume, the losses and the temperature distribution in the transformer are needed as function of the geometry. The geometry could be described by 5 variables as shown in Fig. 7(a), where the construction of the transformer and the definition of the variables are given. There, it is assumed that copper foil is used for realizing the primary and secondary winding, since the thermal resistance between the winding layers is lower. Furthermore, per layer only one turn is realized.

In Fig. 7(a) a transformer with integrated leakage inductance is shown [24], where the secondary winding encloses the middle leg and the primary winding the middle and one outer leg, which conducts the leakage flux. This type of transformer is used in the resonant converter and in the phase shift converter with capacitive output filter. In case of the phase shift converter with CDR a transformer where both windings just enclose the middle leg is assumed. There, the required series inductance is integrated by spatial separation of the windings.

In order to maximize the power density of the transformer an advanced cooling method as described in [2] has been

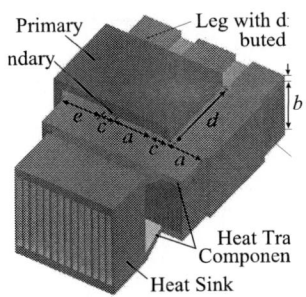

(a) Transformer with integrated cooling

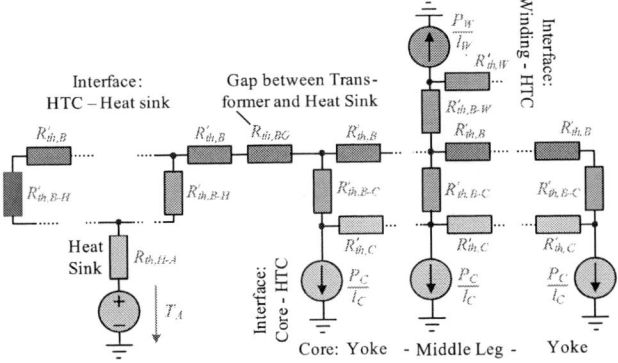

(b) Thermal model of transformer with HTC

Figure 7. (a) Geometry of transformer with integrated series inductance and heat transfer component (HTC) / heat sink for cooling. Parameters a, b and d are the degrees of freedom in the optimization. (b) Distributed thermal model of the cooling system shown above with the heat sink on the left hand side, a gap between heat sink bridged by the HTC and transformer and the transformer with winding on the right hand side.

applied. With this method the losses in the windings and the core are transferred via a heat transfer component (HTC) to an additional heat sink for the transformer.

For calculating the temperature rise the losses in the windings and the core are required. There, the winding losses are calculated by a 1D approach, which includes skin- and proximity effect losses, and the thickness of the foil is optimized as described in [18]. The core losses are calculated by the approach presented in [25], which is based on Steinmetz parameter [26] and on the rate of magnetization (dB/dt).

With the losses the temperature distribution in the transformer could be calculated based on the thermal model shown in Fig. 7. This model describes the heat flow from the winding/core via the thermal interfaces and HTC to the heat sink/ambient by distributed thermal resistances (R_{th} per length). The calculation of the temperature profile is based on transmission line equations, what is described in detail and validated in [2].

For improving the heat flow within the windings made of foil and also from the winding to the HTC a thermally conductive insulating material is used [27]. Moreover, thermal grease between the core and the HTC and a cover pressing the winding on the HTC are used. This cover is not shown in Fig. 1 and Fig. 7(a).

After the volume has been minimized it is passed to the global optimization algorithm, where it is used for calculating the system volume and for varying the parameters systematically.

V. CALCULATION RESULTS

Based in the presented procedure shown in Fig. 5 the three considered topologies have been optimized for a telecom supply with the specification given in table I. The results presented below are based on the following components/limitations if not stated differently:

- Core material: N87 from Epcos ($T_{Max} \leq 115\,°C$)
- Windings: Foil windings ($T_{Max} \leq 125\,°C$)
- Center tapped secondary winding
- MOSFETs: APT50M75 from Microsemi (former APT)
- Rectifier diode: APT100S20 from Microsemi
- Capacitors C_S and C_P: Reference 3.9 nF 800 V COG series from Novacap
- Cooling system performance index: 23 (for transformer and semiconductor heat sink)
- Maximal junction temperature $T_{j,max} \leq 140\,°C$

A further requirement is the overall height of the supply which should be below 1 U ($\approx$44 mm). That significantly influences the design of the transformer/inductors (especially the height b - cf. Fig. 7(a)) as well as the cooling system.

In Fig. 8 the characteristic of the power density and the efficiency as function of frequency are shown. The maximal achievable power density is 19.1 kW/ltr. for the resonant converter and 15 kW/ltr. / 11.7 kW/ltr. for the phase shift converter with capacitive output filter/current doubler (1 kW/ltr. = 16.4 W/inch³). There, only the net volume of the components including PCBs/housings is considered. The volume between

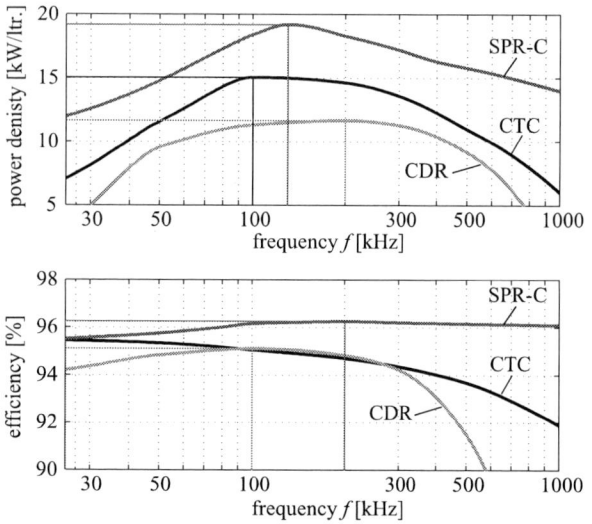

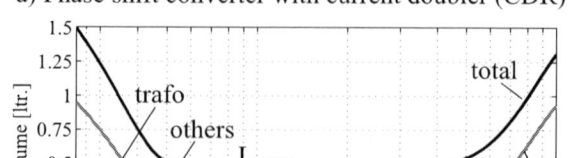

a) Phase shift converter with current doubler (CDR)

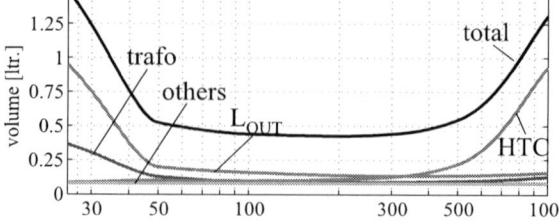

b) Phase shift converter with capacitive output (CTC)

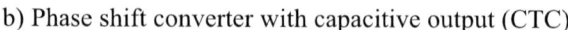

c) Resonant converter with capacitive output (SPR-C)

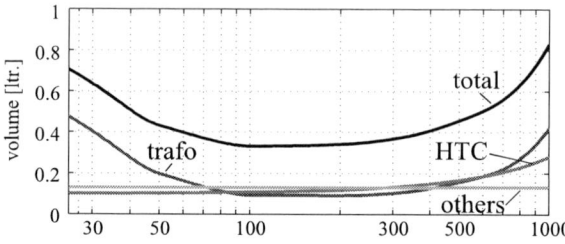

Figure 8.   a) Power density and b) efficiency of the phase shift converter with capacitive filter (CTC) and current doubler (CDR) and series-parallel resonant converter with C-Filter (SPR-C) in dependency of switching frequency.

Figure 9.   Volume vs. frequency for the phase shift converter with a) current doubler (CDR), b) capacitive filter (CTC) and c) series-parallel resonant converter with C-Filter (SPR-C), split in transformer, heat sink and components volume including output capacitor.

the components required for mounting/insulation and due to not fitting housings (e.g. cube type and cylindrical shapes) is not considered, since it depends significantly on the 3D arrangement of the components and the design of the supply. This volume adds significantly to the total converter volume, so that the resulting power density is lower than calculated here. In case of the resonant converter shown in Fig. 1 the calculated power density is 15 kW/ltr. and the power density of the final system is 10 kW/ltr. If a similar scaling factor is assumed for the three optimized converter topologies, 12.7 kW/ltr. (SPR-C), 10 kW/ltr. (CTC) and 7.8 kW/ltr. (CDR) would result for the power density.

The efficiencies at the operating point with maximal power density are 96.2 % for the resonant converter and 95 % / 94.8 % for the CTC / CDR as shown in Fig. 8(b). At the operating point with minimal losses an efficiency of 96.3 % at a switching frequency of approximately 220 kHz could be reached with the resonant converter. There, the switching frequency is higher than at the operating point for maximal power density, since the losses of the passive components reach a minimum at this frequency. Due to the increased volume for the semiconductor heat sink this operating point results only in a power density of 17.1 kW/ltr.

In case of the phase shift converter with capacitive output filter the maximal efficiency is achieved at the lowest considered operating frequency of 25 kHz. There, the CTC reaches 95.4 %. The maximal efficiency for the CDR is 95.1 %, which is achieved at 100 kHz.

In Fig. 9 the distributions of the volumes on the magnetic devices (transformer and inductor), the capacitors (resonant tank and output), the heat sink for the semiconductors and the remaining components like housings, control board or gate drive for the three topologies are shown. There, it could be seen that with rising switching frequency the volume of the semiconductors' heat sink is increasing due to rising switching

losses. The shape of the volume distribution as function of frequency is strongly determined by the passive components,

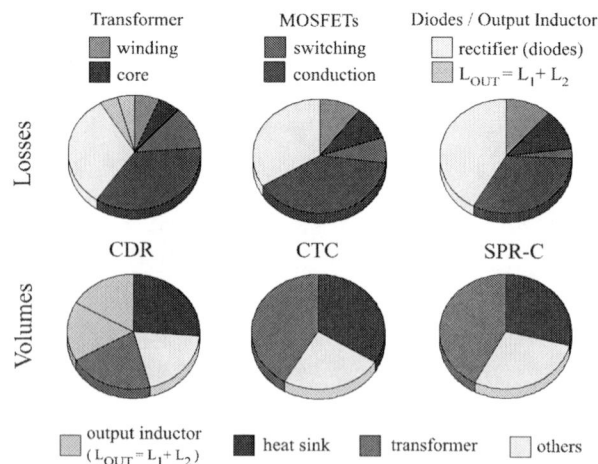

Figure 10.   Distribution of the losses and volumes of the phase shift converter with capacitive filter (CTC) / current doubler (CDR) and series-parallel resonant converter with C-Filter (SPR-C).

TABLE V
Resulting specification of the optimized 5 kW telecom supplies. There, the losses in the two inductors of the CDR are 2×10.9 W, the AC flux density is 77 mT and the inductance is 6.7 μH. (In brackets: simulated values including more parasitic elements than considered in the analytic model.)

| | Phase Shift | | Resonant |
	Current-D.	Cap.-Filter	Cap.-Filter
Power Density	11.7 kW/ltr.	14.7 kW/ltr.	19.1 kW/ltr.
Operating Point			
• Frequency	200 kHz	100 kHz	135(129) kHz
• Duty Cycle	0.81(0.86)	0.82(0.88)	0.78(0.83)
• Efficiency	94.8 %	95 %	96.2 %
Transformer			
• Winding Losses	16 W	24.6 W	21.9 W
• Core Losses	13.8 W	23.7 W	23.5 W
• Turns Ratio	11:4	11:2	14:2
• Winding Temp.	125 °C	124 °C	125 °C
• Core Temp.	107 °C	115 °C	115 °C
• Flux Density	175 mT	240 mT	300 mT
Semiconductors			
• $P_{V,SW,MOSFET}$	31.2 W	17.8 W	5 W
• $P_{V,Cond,MOSFET}$	93.7 W	96.8 W	64 W
• Rectifier Losses	83.3 W	83.3 W	83.3 W
• Heat Sink Temp.	125.1 °C	125 °C	130.8 °C

which define a frequency range from app. 100 kHz. . .300 kHz of high power density. Outside this range the volume of the magnetic devices and at higher frequencies also the volume of the semiconductor heat sink rises significantly, what limits the achievable power density.

A more detailed distribution of the volume and also of the losses of the three considered topologies is given in Fig. 10. There, the volume and the losses of the magnetic devices, the capacitors, the semiconductors/heat sink and the remaining components for the operating point with maximal power density is presented. Further details are shown in table V.

The relatively low optimal operating frequency for maximal power density results, since the whole system is considered in the optimization. In case only the transformer for a fixed input voltage/current level would be considered the optimal operating frequencies for maximal power density of the transformer would be (significantly) higher.

A. Power Density Barriers

In the preceding paragraph the power density values for available components/technologies have been presented. Now, the influence of different parameters on the achievable power density is investigated. In table VI the achievable power density and efficiency for the phase shift converter with capacitive output filter is shown for different parameter variations. First, it is assumed that the maximal allowed junction temperature is increased from 150 °C to 200 °C (e.g. by applying SiC switches and appropriate packaging) – the remaining parameters are not modified. By this means the volume of the semiconductor heat sink significantly decreases, so that a power density of 17.1 kW/ltr. (before 14.7 kW/ltr.) could be achieved. There, it is important to note that with the heat sink

also the size of the area for mounting the semiconductors is shrinking. Consequently, the thermal resistance between the semiconductors and the heat sink increases, so that the gain in power density is very limited. This effect has not been considered in table VI.

TABLE VI
Influence of device parameters/max. allowed temperatures on maximal achievable power density for the phase shift converter with capacitive output filter. The maximum remains at a switching frequency of app. 100 kHz, except for case 5 where all improvements are considered at the same time. There the optimal switching frequency rises to approximately 200 kHz.

	Parameter	Old Value	New value	New Density/Eff.
1	Junction Temperature	150 °C	200 °C	17.1 $\frac{kW}{ltr.}$ / 95 %
3	$R_{th,MOSFET}$	0.22 $\frac{K}{mW}$	0.11 $\frac{K}{mW}$	16.1 $\frac{kW}{ltr.}$ / 95 %
	$R_{th,Rectifier}$	0.2 $\frac{K}{mW}$	0.1 $\frac{K}{mW}$	
2	Winding / Core Temp.	125 °C / 115 °C	150 °C	16.8 $\frac{kW}{ltr.}$ / 94.9 %
4	$P_{Switching}$	P_{SW}	$\frac{1}{2}P_{SW}$	15.9 $\frac{kW}{ltr.}$ / 95.2 %
5	$V_{F,Rectifier}$	0.9 V	0.45 V	17 $\frac{kW}{ltr.}$ / 95.8 %
6	1+2+3+4+5	-	-	21.6 $\frac{kW}{ltr.}$ / 96.3 %

In the second row of table VI it has been assumed that the thermal resistance between the chip and the heat sink is decreased. Again, this measure allows to shrink the volume of the semiconductor heat sink, since its temperature could be increased. There, the power density rises up to 16.1 kW/ltr. By increasing the operating temperature of the transformer a value of 16.8 kW/ltr. could be reached. There, the rising copper losses due to the rising resistivity limit the gain of power density.

Another option would be to decrease the switching losses by 50 %. Due to the ZVS condition this measure only results in small increase of the power density to 15.9 kW/ltr. A bigger step could be achieved by reducing the forward voltage drop of the rectifier diodes by a factor of two, what results in a smaller volume of the heat sink. This would lead to a power density of 17 kW/ltr. and an efficiency of 95.8 %. A reduction of the conduction losses of the rectifier could be achieved by synchronous rectification. The $R_{DS,on}$ of the applied MOSFETs, however, must be very low – in the considered case ≈ 8 mΩ@125 °C for cutting the losses in half. Since the rectifier diodes and their heat sink are already quite small, a synchronous rectifier would not help to improve the power density but only the efficiency.

In case all measures are combined a power density of 21.6 kW/ltr. and an efficiency of 96.3 % can be achieved. Similar improvements can be expected for the other systems in case the same measures are taken.

Summing up, this leads to two possible ways to improve the power density:

- For components users / supply manufacturers
  - Improve the thermal coupling between the semiconductors and the heat sink (e.g. by low temperature soldering) and

- apply advanced cooling methods for the passive components [2].
- For device manufacturers
  - Further reduce the conduction losses of the power semiconductors,
  - reduce the magnetic materials HF losses,
  - decrease the thermal resistance between chip and housing/base plate,
  - increase operating temperatures.

VI. CONCLUSION

In this paper three topologies for telecom supplies – a phase shift converter with capacitive output filter and with current doubler and a series-parallel resonant converter with capacitive output filter – have been optimized with respect to power density. There, maximal 12 kW/ltr. (19 kW/ltr. pure component volume) are obtained for the series-parallel resonant converter (SPR). The optimal operating frequency with respect to the power density is approximately 135 kHz. For the phase shift converter 10 kW/ltr. result for a capacitive output filter (CTC) and 7.8 kW/ltr. for a current doubler (CDR). Again, the optimal operating frequencies are relatively low – approximately 100 kHz for the CTC and 200 kHZ for the CDR. There, the efficiencies are 96.2 % for the SPR, 95 % for the CTC and 94.8 % for the CDR. These values slightly improve ($\approx 0.8\,\%$) if the converter is optimized for efficiency, but there the power density decreases significantly.

The presented optimizations have been performed for operation at nominal output power – part load efficiency, soft switching range, costs or EMI issues, etc. have been neglected. In case also these constraints are also considered the achievable power density will decrease to approximately 6-8 kW/ltr. for the resonant converter. In combination with a PFC converter with a power density of also 6-8 kW/ltr. a system power density of 3-4 kW/ltr. for an air cooled supply would result.

For increasing the power density especially the thermal management is decisive. A direct cooling of the magnetic components as presented in [2] or improving the thermal resistance between the chip and the heat sink by low temperature solders, which could replace the thermal grease, are effective measures to reduce the system volume. Semiconductors with reduced losses or improved core materials are other possibilities to increase the power density and efficiency. These approaches, however, can only be followed by device/material manufacturers and are not directly accessible to supply manufacturers.

The largest gain results from optimizing the system parameters for the given specifications. The question which topology should be applied is important but does not influence the achievable power density as much as the optimization.

REFERENCES

[1] Homepage of IBM: www.ibm.com, "Project Big Green".

[2] J. Biela and J. W. Kolar, "**Cooling Concepts for High Power Density Magnetic Devices**," Nagoya, Japan, 2-5 April, 2007, pp 1-8.

[3] U. Drofenik, G. Laimer und J.W. Kolar, "**Theoretical Converter Power Density Limits for Forced Convection Cooling**," Proceedings of the International PCIM Europe 2005 Conference, Nuremberg, Germany, June 7-9, pp. 608-619.

[4] T.F. Vescovi and N.C.H. Vun, "**A switched-mode 200 A 48 V rectifier/battery charger for telecommunications applications**," 12th International Telecommunications Energy Conference (INTELEC'90), 1990, pp. 112-118.

[5] S. Moisseev, S. Hamada and M. Nakaoka, "**Double two-switch forward transformer linked soft-switching PWM DC-DC power converter using IGBTs**," Electric Power Applications, IEE Proceedings, Vol.150., January 2003, pp 31-38.

[6] R. Chen, J.T. Strydom and J.D. van Wyk, "**Design of Planar Integrated Passive Module for Zero-Voltage-Switched Asymmetrical Half-Bridge PWM Converter**," IEEE transactions on industry applications, Vol.39, No.6, Nov./Dec. 2003, pp. 1648-1655.

[7] B. Yang, F.C. Lee, A.J. Zhang and G. Huang, "**LLC resonant converter for front end DC/DC conversion**," Applied Power Electronics Conference and Exposition (APEC), Vol.2, March 2002, pp. 1108-1112.

[8] J. Jacobs, A. Averberg, S. Schröder and R. De Doncker, "**Multi-phase series resonant dc-to-dc converters: Transient investigations**," 36th Annual Power Electronics Specialists Conference, 2005, pp. 1972-1978.

[9] A.K.S. Bhat, "**A Resonant Converter Suitable for 650 V dc Bus Operation**," IEEE Transaction on Power Electronics, Vol.6, Oktober 1991, pp. 739-748.

[10] J. Elek and D. Knurek, "**Design of a 200 amp telecom rectifier family using 50 amp dc-dc converters**," 21st International Telecommunications Energy Conference (INTELEC), June 1999.

[11] J.A. Sabate, V. Vlatkovic, R.B. Ridley, F.C. Lee and B.H. Cho, "**Design considerations for high-voltage high-power full-bridge zero-voltage-switched PWM converter**," Fifth Annual Applied Power Electronics Conference and Exposition (APEC), March 1990, pp. 275-284.

[12] I.D. Jitaru, "**High efficiency converter using current shaping and synchronous rectification**," 24th Annual International Telecommunications Energy Conference, Intelec02, 29. Sept.-3 Oct. 2002 pp. 48-54.

[13] A. K. S. Bhat, S. B. Dewan, "**Analysis and design of a high-frequency resonant converter using LCC-type commutation**," IEEE Transaction on Power Electronics, Vol.2, No.4, 1987, pp. 291-300.

[14] R. L. Steigerwald, "**A comparison of half-bridge resonant converter topologies**," IEEE Transactions on Power Electronics, Vol. 3, Issue: 2, April, 1988, pp. 174-182.

[15] J. Biela and J.W. Kolar, "**Analytic Model inclusive Transformer for Resonant Converters based on Extended Fundamental Frequency Analysis for Resonant Converter-Design and Optimization**," Published in IEEJ Transactions of Institute of Electrical Engineers of Japan, Vol. 126, No. 5, May 2006, pp. 568-577.

[16] Homepage of Vishay: http://www.vishay.com - SMD inductors.

[17] Homepage of Payton: http://www.paytongroup.com

[18] W.G. Hurley, E. Gath and J.G. Breslin, "**Optimizing the AC Resistance of Multilayer Transformer Windings with Arbitrary Current Waveforms**," IEEE Transaction on Power Electronics, Vol. 15, No. 2, March 2000, pp. 369-376.

[19] J. Biela, U. Badstübner and J.W. Kolar, "**Design of a 5 kW, 1 U, 10 kW/ltr. Resonant DC-DC Converter for Telecom Applications**," Intelec Conference, Rome, Italy, 1-4 October, 2007.

[20] J. Biela, "**Optimierung des elektromagnetisch integrierten Serien-Parallel-Resonanzkonverters mit eingeprägtem Ausgangsstrom**," Ph.D. Thesis, Eidgenösische Technische Hochschule – ETH Zürich, Switzerland, 2005.

[21] A.J. Forsyth, G.A. Ward und S.V. Mollov, "**Extended fundamental frequency analysis of the LCC resonant converter**," IEEE Transactions on Power Electronics, Vol. 18, Issue 6, Nov., 2003, pp. 1286-1292.

[22] Homepage of Novacap: http://www.novacap.com/

[23] Homepage of muRata: http://www.murata.com

[24] J. Biela and J.W. Kolar, "**Electromagnetic Integration of High Power Resonant Circuits Comprising High Leakage Inductance Transformers**," Power Electronic Specialist Conference PESC'04, Aachen, June 2004, pp. 4537 - 4545.

[25] K. Venkatachalam, C.R. Sullivan, T. Abdallah, H. Tacca, "**Accurate prediction of ferrite core loss with nonsinusoidal waveforms using only Steinmetz parameters**," IEEE Workshop on Computers in Power Electronics, 3-4 June 2002, pp. 36 - 41.

[26] C.P. Steinmetz, "**On the law of hysteresis**," Proc. IEEE, Vol. 72, February 1984, pp. 197-221.

[27] Homepage of Bergquist: http://www.bergquistcompany.com

On the Control of Active Power Filters

LIU Jinjun, WANG Xiaoyu, YUAN Chang, WANG Zhaoan
Power Electronics and Renewable Energy Center
School of Electrical Engineering
Xi'an Jiaotong University
28 West Xianning Road, Xi'an, Shaanxi 710049 China
Email: jjliu@mail.xjtu.edu.cn

Abstract—This paper compares the popular control methods for two dominant types of active power filters (APFs). With the equivalent transformation of their control diagrams, the recently proposed current-source control method for series active power filter (SAPF) is revealed to have the same control structure as the traditional voltage-source control method therefore have same compensation performances. Similarly, two kinds of popular control methods for parallel active power filter (PAPF) are also compared and found to have the similar control structure. The recently proposed PAPF control scheme without harmonics detection is revealed as a reduced version of the traditional control scheme with harmonics detection. Therefore both have same basic control principles regarding load current harmonics compensation. A general expression of the control methods for PAPF is then proposed in the form of both a control diagram and a set of equations based on current detection. With a particular set of filter gains of the three detected currents, each particular control scheme can be specifically identified and described by the general expression. Subsequently the 64 possible sets of filter gains are examined by the proposed necessary conditions to locate practically valid control schemes. Finally 4 valid control schemes come out and among which some are schemes that have never been reported before. With the comprehensive comparison of the control methods and the proposed general expression in this paper, the essential relationships of the control methods are revealed, some misinterpretations from previous publications are clarified, and some new control schemes are obtained.

I. INTRODUCTION

With the development of the modern electronics industry and the continuous proliferation of nonlinear electric load, power quality problems have been drawing increased attention [1]. In the beginning, passive power filters were used to solve the power quality problems. Since the 1970s, active power filter (APF) has been one of the most competitive solutions to suppress power system harmonics and enhance power quality. Active power filters have been designed, used, and improved upon for many years and is now in the commercial stage [2]. According to its connection to the power system, there are two most dominant types of APF as series active power filter (SAPF) and parallel active power filter (PAPF), shown as Fig. 1 [2]. Researches and application show that the SAPF is preferable to compensate the voltage-source-type harmonic loads, while PAPF is more suitable to compensate the current-source-type harmonic loads [3].

Control strategies play a key role in active power filters,

and now and again "novel" control schemes are proposed with advantages claimed. Yet most methods are reported separately, focusing only on the advantages of certain features. Hence these claims may not be well proven or verified strictly. This paper compares popular reported control methods for these two dominant types of active power filters. Some essential relationships of the control methods are reveals and some misinterpretations from previous publications are also clarified.

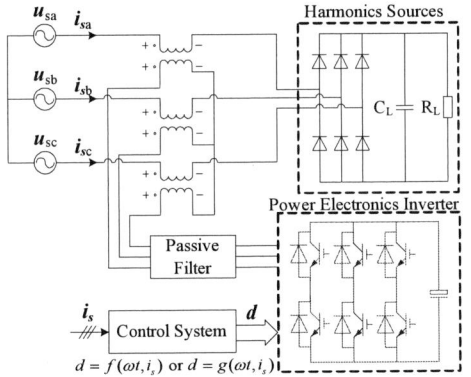

a). System configuration of series active power filter (SAPF).

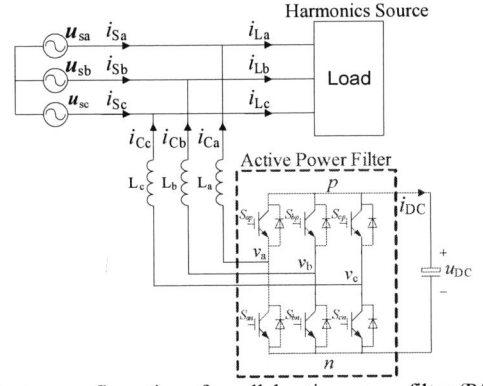

b). System configuration of parallel active power filter (PAPF).
Fig. 1. System configuration of two dominant types of active power filter.

Firstly in this paper, the control methods for series active power filters are reviewed and compared with equivalent transformation of their control diagrams. Among the control methods for SAPF, voltage-source control is the traditional scheme, in which the SAPF is controlled as a voltage source shown as Fig. 2(a). With the amplitude controlled directly proportional to the current harmonics, the SAPF works like frequency-selective impedance. Yet in practical applications,

978-1-4244-1871-8/07 $25.00 © 2007 IEEE

large impedance will direct the system to unstable. So it has been a puzzling issue so far to compromise between the compensation characteristics and system stability. As a solution, some publications reported a new control strategy called current-source control in late 1990s [7-9], which is also called fundamental magnetic flux control [10-11]. Current-source control is based on the alternative principle that the SAPF behaves like a current regulator rather than frequency-selective impedance [2, 6, 7-11]. And it has been claimed that current-source control is prior to voltage-source control in many aspects [2, 7-11]. However it is quite confusing that how current-source control can achieve all these better performances with just the same power stage as voltage-source control. This paper compares the traditional voltage-source control and the recent proposed current-source control with equivalent transformation of their control diagrams.

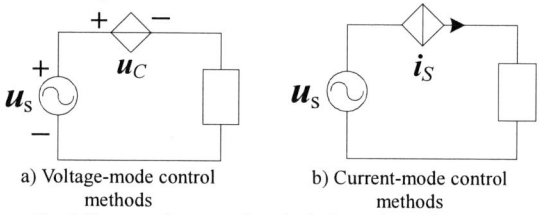

a) Voltage-mode control methods   b) Current-mode control methods

Fig. 2 Two popular control methods for series active power filter

Then, the popular control methods for PAPF are compared. According to the harmonics detection, the control methods for PAPF can be generally categorized as two types: one is the control methods with harmonics detection which are proposed with the concept of active power filtering techniques, the other kind is the methods without harmonics detection which are proposed in late 1990s by many researchers from different point of view. Fig. 3 shows the categories of popular control schemes for PAPF.

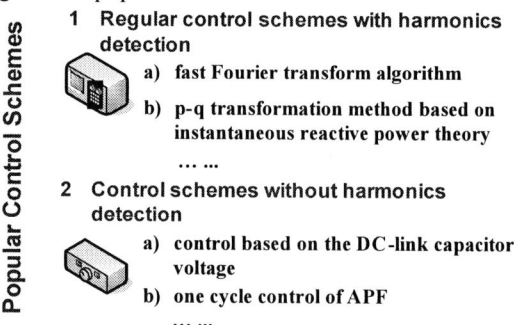

Fig. 3 Categories of popular control schemes for parallel active power filter

This paper compares these control schemes in the viewpoint of whole system. Similarities are found by the equivalent transformation and reduction of their control diagrams. The recently proposed control scheme without harmonics detection is revealed as just a reduced version of the traditional one with harmonics detection. Focusing on the control loops, the control without harmonics detection can be explained as a simple feedback control, while the control with load harmonics detection is a composite control of both feedback of DC-link voltage and feedforward of the

fundamental load current which results in better dynamic performance.

The comparison of the control methods stimulated further study of the generalized voltage-mode control, which leads to a systematic analysis and design method for PAPF. In view of the whole system, a general expression is proposed in the form of both a control diagram and equations to describe any linear control method based on current detection. With particular gains of the three detected currents, different control schemes can be specifically identified. To advance the analysis, the gains of each detected current are simplified and limited to four typical equivalent filters in the general control diagram: high-pass filter (HPF), low-pass filter (LPF), zero-pass filter, and full pass filter. If the four kinds of gains are applied to the three detected currents, a total of 64 possible control schemes can be obtained from the proposed general expression for the control of PAPF. To verify the effectiveness of these control schemes, the necessary conditions are given, by which several new practical voltage-mode control schemes, including the previously proposed scheme, are identified in a systematic approach.

II. COMPARISON OF THE CONTROL METHODS FOR SERIES ACTIVE POWER FILTER (SAPF)

A. Review of Voltage-Source Control for SAPF

The original control methods for SAPF is voltage-source control, in which the SAPF is controlled as a voltage source with amplitude directly proportional to the current harmonics:

$$u_c = K \cdot i_{1h} \qquad (1)$$

where i_{1h} is the harmonic current, u_c is the compensation voltage and K is the control regulator. In this way, the APF works just like frequency-selective impedance. It presents zero impedance for the fundamental frequency and acts as a resistor with high resistance of K [Ω] for harmonics frequencies. If the ratio K [Ω] is large enough, the line current will be close to sinusoidal [3].

Fig. 4 shows the equivalent circuit of SAPF, and Fig. 4b) illustrates the basic control diagram.

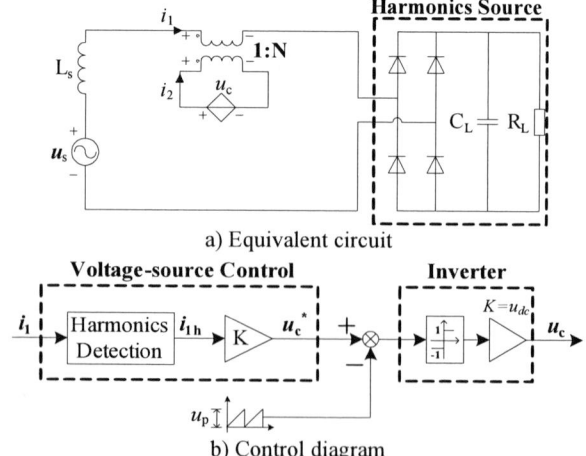

a) Equivalent circuit

b) Control diagram

Fig. 4. Voltage-mode control of series active power filter

978-1-4244-1871-8/07 $25.00 © 2007 IEEE   34

The controller detects the line current $\mathbf{i}_1$, calculates its harmonics $\mathbf{i}_{1h}$, and then controls the SAPF as selective-frequency impedance. The algorithm of harmonics detection can be chosen freely from lots of algorithm proposed, such as fast Fourier transform algorithm (FFT), p-q method based on instantaneous reactive power theory, etc.

The weakness of voltage-source control is the stability problem, because the ratio K [Ω] could not be too large in the close-loop control [3, 4]. It has been a puzzling issue so far to compromise between the compensation characteristics and system stability.

B. Review of Current-Source Control for SAPF

In late 1990s, some publications reported a new control strategy called current-source control [5][6]. This control strategy is also called fundamental magnetic flux control [6][7]. Current-source control is based on the alternative principle that the SAPF behaves like a current regulator rather than frequency-selective impedance. It has been claimed that current-source control is prior to voltage-source control in many aspects [5]-[7]. In current-source control, the SAPF acts as a sinusoidal current source in phase with the line voltage supply, presenting very high impedance for all the current harmonics.

Fig. 4a) shows the equivalent circuit of SAPF with current-source control, and Fig. 4b) illustrates the corresponding control diagram. The controller detects the line current $\mathbf{i}_1$ and calculates the fundamental current $\mathbf{i}_{1F}$, and controls the current flew series active power filter follow $\mathbf{i}_2$. A compensator is used in current-tracking modulation to keep a good close-loop characteristic.

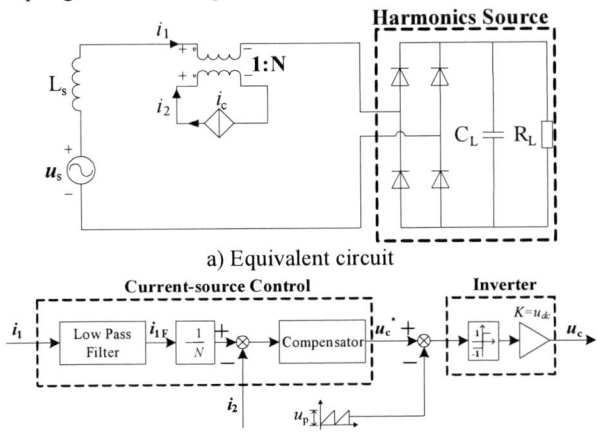

a) Equivalent circuit

b) Control diagram

Fig. 5. Current-mode control of series active power filter

The current-source control provides a fresh implementation of the operation principle and control of SAPF. However, in comparison with the traditional voltage-source control, how does current-source control achieve all these better performances with just the same power stage as voltage-source control. And what is the fundamental relationship between these two control strategies? This section compares these two control methods by investigating their equivalent control diagrams.

C. Comparison of the Control Methods for SAPF

In view of the whole system, the SAPF consists of two basic parts: power stage and control algorithm, shown as Fig. 6. The function of either control algorithm is to calculate the duty cycle d from the detected source current. Either voltage-source control or current-source control deals with absolutely the same power stage and hardware configuration. So it becomes vital to compare these two control methods.

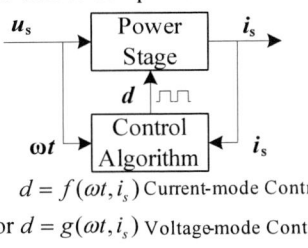

$d = f(\omega t, i_s)$ Current-mode Control

or $d = g(\omega t, i_s)$ Voltage-mode Control

Fig. 6. System diagram of SAPF.

Before comparative study on the control strategies it is necessary to build the model of the power stage. Thanks to the research to model the uncontrolled rectifier, the nonlinear load could be modeled as harmonic voltage sources in frequency domain. Then the power stage model can be obtained by solving the system equations, which reveals the relationship between the line current and the three independent sources including the utility source $\mathbf{u}_s$, the load harmonics source $\mathbf{u}_L$ and the controlled voltage source $\mathbf{u}_c$. Fig. 7. illustrates the derived power stage model of single-phase SAPF in terms of Laplace transformation. The power stage model of three-phase SAPF may be obtained in the same way.

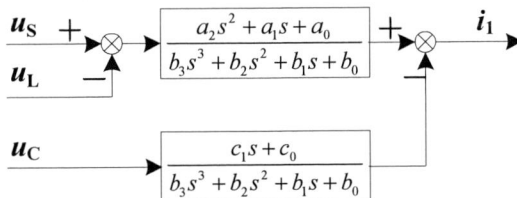

Fig. 7 Block diagram of power stage.

Combination of the power stage model and the control diagrams yields to the system control diagrams. Fig. 8 and Fig. 9 show the system diagram with voltage-source control and current-source control separately.

To clearly reveal their relationship, the system control diagram of current-source control as Fig. 8 can be equivalently transformed to Fig. 10, which is much like the control diagram of voltage-source control as Fig. 9. With careful comparison of Fig. 9 and Fig. 10, it can be concluded that although these two control strategies seem entirely different in appearance, they have identical control structure and operation principles by nature. In the view of the whole system, the compensator in the current loop of current-source control is actually the corresponding element to the ratio K [Ω] in voltage-source control. This opens our view to possibly upgrade the algorithm of voltage-source control by replacing the ratio K with a regulator. This conclusion also

978-1-4244-1871-8/07 $25.00 © 2007 IEEE

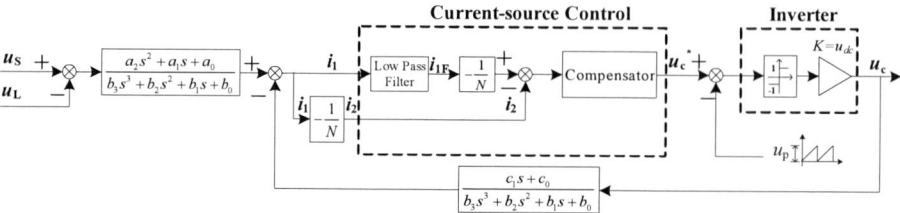

Fig. 8. System control diagram of SAPF with current-source control method.

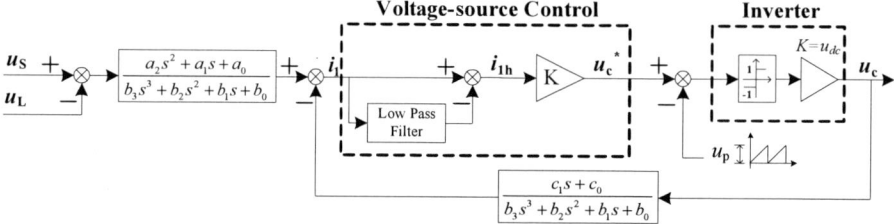

Fig. 9. System control diagram of SAPF with voltage-source control method.

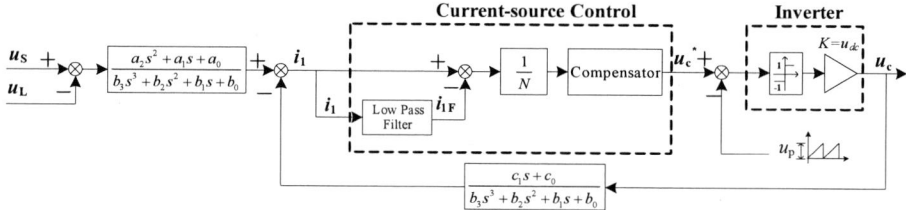

Fig. 10. Equivalent control diagram of SAPF with current-source control method.

helps explaining some previously confusing issues between these two control strategies.

III. COMPARISON OF THE CONTROL METHODS FOR PARALLEL ACTIVE POWER FILTER (PAPF)

After the comparison of the popular control methods for SAPF, the control methods for PAPF are comparatively investigated in the same way.

Parallel active power filters have been used and improved upon for many years and is now in the commercial stage. The original control idea and basic operating principles of parallel active power filter are generally illustrated with current injection. When the PAPF injects harmonic current into the system with the same amplitude and reverse phase to load current harmonics, the source current harmonics are compensated [2]. This idea is straightforward, but it is quite challengeable to detect the instantaneous harmonics from load current quickly and precisely. Lots of harmonics detection algorithms have been proposed, but most of cost a lot.

As a simple solution to avoid the complicated algorithm to detect harmonics, the control schemes without harmonics detection are propose in late 1990s from different point of view, and have been drew a lot of research attention these years[8]. Many advantages were reported such as simplicity to implement with low cost and comparable compensation characteristics with the traditional control schemes [9]. Among these control schemes without harmonics detection, two typical representatives are the control based on the DC-link capacitor voltage [10] and one-cycle control [11][12].

In our practice of power quality control with active power filters, the authors feel puzzled to choose the best one from so many control schemes, which motivates us to research on the internal relationship and essential compensating mechanism of them, and consider the question as: what's the difference between the two main kinds, what are their internal relationships, and what's the actual function of harmonics detection algorithm.

A. Review of Control Methods with Harmonics Detection

To control the PAPF, the straightforward control idea is based on harmonics detection-and-compensation strategy, and the basic operating principles are generally illustrated with current injection. The PAPF injects harmonic current into the system with the same amplitude and reverse phase to load current harmonics. Fig. 11a) shows the system equivalent circuit and Fig. 11b) shows its corresponding control scheme. The controller detects the load current i_L and calculates its harmonics i_{Lh} as the reference to PAPF. Following this reference, the PAPF is controlled as a harmonic current source shown as Fig. 11a).

The performances of the harmonics detection algorithms are very important but it is quite challengeable to detect the instantaneous harmonics from load current quickly and precisely. Lots of harmonics detection algorithms have been proposed, such as fast Fourier transform algorithm (FFT), p-q method based on instantaneous reactive power theory, etc. Most of them need large computation workload, and commonly implemented with high-cost DSP. Fig. 11c) shows the control diagram with the d-q method employed in synchronous frame to detect the current harmonics, which is very popular in industry application due to its quick dynamic characteristics and simple implement.

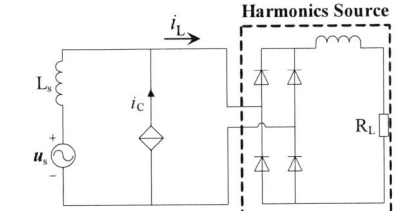

a) Equivalent circuit of the system with PAPF

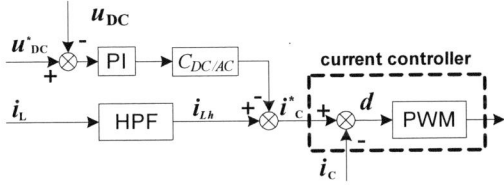

b) The traditional control scheme with harmonics detection.

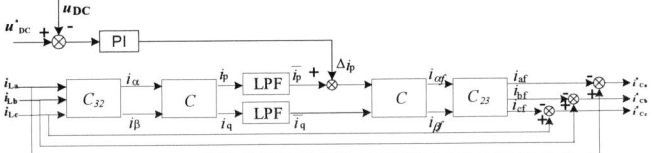

c) Harmonics detection algorithm based on instantaneous reactive power theory or *p-q* transformation

Fig. 11. The traditional control scheme with harmonics detection.

To maintain a constant DC-link voltage of the power converter, the DC capacitor voltage is fed back to control the active power flow of PAPF shown as Fig. 11b) and Fig. 11c).

B. Review of the Control Methods without Harmonics Detection

As a solution to simply the control system, the control schemes without harmonics detection are proposed in late 1990s from different point of view, and have been drew a lot of research attention these years. Some well-marked advantages are proclaimed as:

- The compensation is achieved without harmonics extraction, hence simplifying the control and reducing the cost.
- The control logic and the associated hardware are simple, thereby enhancing the system reliability.
- It requires minimum measurement with only source currents and DC-link capacitor voltage.

There are two key point of operating principle for these kinds of control schemes. First, the source current is fed back to a controller to shape the waveform. The controller ensures the source current track a sinusoidal waveform reference whatever the load current is. As the result, the load harmonic current is forced to the shunt branch of PAPF. The other key operating principle is the power balance control. If the sinusoidal source current is not equal to the fundamental component of load current, the PAPF will either consume or deliver active power, and can't work for a long period. With the feedback control of DC voltage, a const DC-link voltage is obtained; hence the active power flow through shunt

branch of PAPF is eliminated. These two key points keep the source current not only sinusoidal but also has equal amplitude to fundamental component of load current -.

To implement these two key control principles, Fig. 12 shows the control diagram based on DC-link capacitor voltage control. It has two feedback control loops. The inner loop is current waveshaping controller, which shapes the source current to sinusoidal. While the outer loop is active power control which keeps the DC-link capacitor voltage const and generates the reference to source current with equal amplitude to fundamental component of load current. Unlike the traditional control schemes, the reference to the inner current loop comes directly from the DC-link capacitor voltage. To maintain a constant DC-link capacitor voltage, the PAPF consumes no average power from the source; hence in steady state the current of PAPF is purely harmonics, which could exactly compensate the harmonics part of load current.

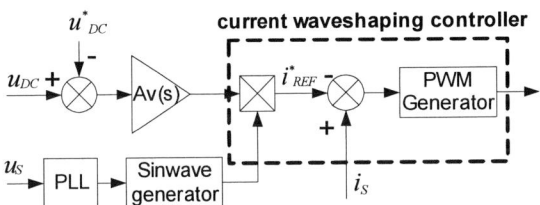

Fig. 12 Control schemes of PAPF based on DC-link capacitor voltage control.

Another popular control scheme without harmonics detection is based on the theory of one cycle control (OCC). In 1990s, one cycle control is proposed as a nonlinear method to modulate the pulse width. From the year of 2000, the idea of one cycle control is applied into the control of APF. Fig. 13 illustrates the basic control scheme. It contains a regulator of $A_v(s)$, which may general be proportional-integration (PI) regulator, a special integrator with a reset switch, a comparator, and a flip-flop. The voltage of DC-link capacitor is sensed and its error is fed to a controller of $Av(s)$ which usually contains a PI controller. This reference leads the source current i_S, which is sensed with a resistor R_S at the source side, to sinusoidal waveform to achieve the goals of active power filter.

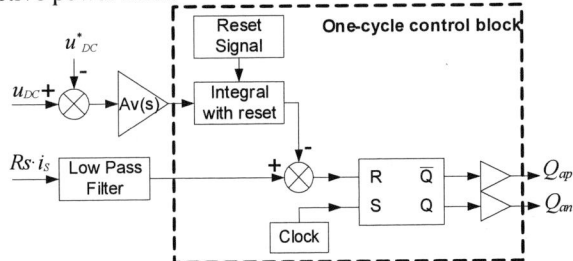

Fig. 13. Control scheme of PAPF based on one cycle control (OCC).

In this control scheme, one cycle control modal is used to generator PWM signal instead of regular PWM techniques, which makes it look different at first appearance. Yet actually it has the same control loop and hence essential principle with previously discussed control scheme without harmonics extraction based on DC-link capacitor voltage.

C. Equivalent Transformation and Analysis of the Schemes with Harmonics Detection

The essential relationship among these control schemes can be revealed by the equivalent transformation of their control diagrams. Starting from the control diagram with harmonics detection shown in Fig. 11, Fig. 14 exhibits a series of its equivalent control diagram transformation. In the figures, the complicated harmonics detection algorithm is represented with a filter in order to simplify the analysis. The harmonics detection component (shown as HPF) in Fig. 11b) is substituted by the subtraction of a full pass filter and a fundamental detection component (shown as LPF) as Fig. 14a) if the amplitude is unified. Then, the input node of load current i_L splits as Fig. 14a), and shifts backward as fig. 14b). Considering the currents around PCC (Point of Common Coupling), governed by Kirchoff's Current Law

$$i_S = i_L - i_C \qquad (2)$$

the input node of load current i_L and compensated current i_C in Fig. 14b) could be combined together as source current i_S, then another equivalent diagram is obtained as Fig. 14c). We redraw Fig. 14c) as Fig. 14e).

D. Comparisons of the Control Schemes

The control diagram with harmonics detection shown as Fig. 4e) has two parts: the highlight part with a rectangle and a feedforward path of fundamental load current. The highlight part in this control diagram looks much similar with the diagram of the control scheme without harmonics detection shown as Fig. 12 and Fig. 13. Both of them detect the DC-link capacitor voltage; adjust the voltage error with a PI compensator, and feedback the source current to generate PWM signals. It can be proved that the voltage control loops are absolutely the same between the control methods with and without harmonics detection. The only slight difference between these two kinds of methods is the feedforward path.

Take the three-phase PAPF for example. In three-phase case, the function block of $C_{DC/AC}$ is realized with

synchronous frame transformation based on instantaneous reactive power theory shown as Fig. 15a). The output of PI compensator is fed to synchronous frame transformation block as d channel, which controls the active power of PAPF. While in the control without harmonics detection, there-phase source current reference is generated with phase shift of single phase reference shown as Fig. 15b). The reference from DC voltage control in the scheme with harmonics detection can be proven as the same with the reference in the control scheme without harmonics detection.

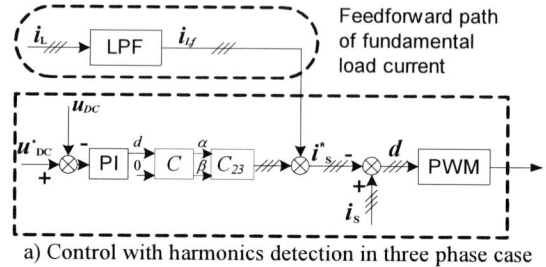

a) Control with harmonics detection in three phase case

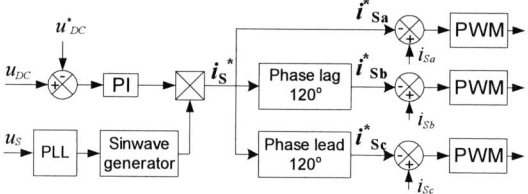

b) Control without harmonics detection in three phase case

Fig. 15. Comparison of the control diagrams with and without harmonics detection in three-phase PAPF

The essential relationship can be concluded as:

1) The control scheme without harmonics detection is just a reduced version of the broadly used regular control with load current harmonics detection, and by nature has the similar essential control principles in the view of system point. It has a simple feedback, while the control with load harmonics detection is a composite control, with both feedback of DC-link voltage and feedforward of the fundamental load current which results in better dynamic

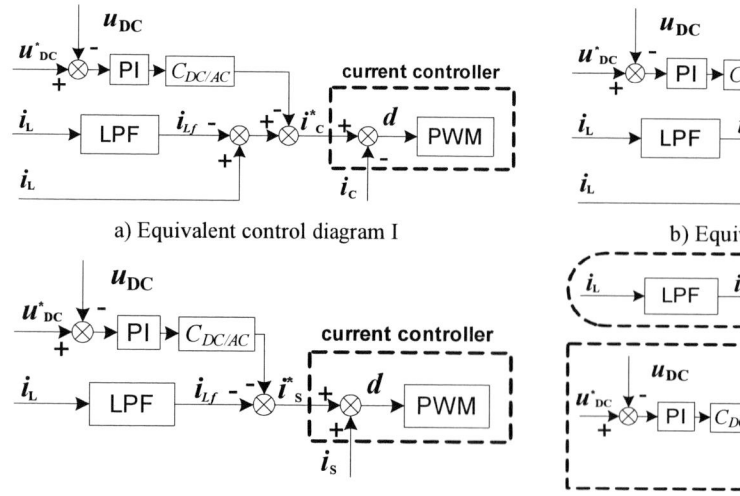

c) Equivalent control diagram III

d) Equivalent control diagram IV

Fig. 14. Equivalent transformation and reduction of the control diagram derivation with harmonics detection.

performance.

2) The control without harmonics detection is simpler and easy to implement, hence is suitable with low cost application. While the regular control is more complicated, and should be chosen when dynamic performance is concerned more.

3) One cycle control is proposed in a fresh view and seems quite different from normal control schemes, yet it is actually one kind of implementation of the control scheme without harmonics detection. Based on similar main control loop, one cycle control generates PWM signal with nonlinear modulation method to achieve a better dynamic response.

IV. GENERAL EXPRESSION OF THE CONTROL METHODS FOR PARALLEL ACTIVE POWER FILTER

The previous comparison of control diagrams stimulated further investigation of generalized voltage-mode control and a systematic approach to analyze and design the control system for PAPF. In view of the whole system, PAPF consists of two basic parts: power stage and control algorithm. Fig. 16a) shows the whole system of PAPF[15][16]. Whatever the control algorithm, it calculates duty ratio d or output compensation voltage u_C from detected currents. So it is straightforward and general to investigate the output voltage u_C or duty ratio d during the analysis and design of the control system.

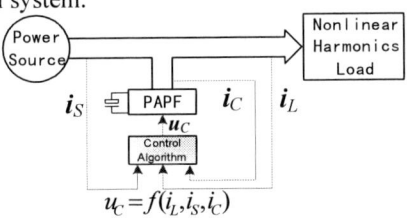

$$u_C = f(i_L, i_S, i_C)$$

a) Whole system of PAPF

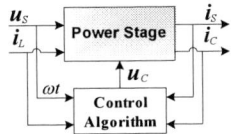

b) PAPF abstractly represented with block diagram

Fig. 16. General system diagram of PAPF.

The system can be described with block diagram as Fig. 8b). Because the load harmonic currents and characteristics are less dependent up the AC side, the harmonic load behaves like an independent current source i_L. So in the black-box of the whole system, the inputs are two independent sources u_S, i_L. While the outputs are the concerned source current i_S and compensation current i_C. The aim of the control is to keep the source current i_S to be pure sinusoidal and the compensation current i_C to be pure harmonic as large as load harmonic current.

A. Two-Port Representation of Power Stage

Many efforts have been done to model the uncontrolled rectifier in both time-domain and frequency-domain. With frequency-domain modeling techniques, the harmonic load can be modeled as independent harmonic current source. Then the system becomes linear. Fig. 17 shows the equivalent circuit, where u_S represents the source, u_C represents the voltage of power electronics inverter in PAPF, and i_{Lh} represents the harmonic current source of the load.

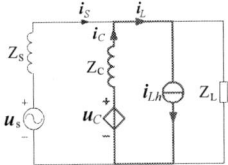

Fig. 17. Equivalent circuit of whole system with PAPF.

By solving these equations that govern the power stage, the relationship between the output currents (source current and compensated current) and the inputs (two independent sources u_S, i_{Lh}, plus one controllable voltage source u_C) can be obtained. Assume the load impedance Z_L is much larger than source impedance Z_S, as is generally true in industrial applications, the results can be simplified to (3).

$$\begin{bmatrix} I_S \\ I_C \end{bmatrix} = \begin{bmatrix} \dfrac{1}{Z_S+Z_C} & \dfrac{Z_C}{Z_S+Z_C} & -\dfrac{1}{Z_S+Z_C} \\ -\dfrac{1}{Z_S+Z_C} & \dfrac{Z_S}{Z_S+Z_C} & \dfrac{1}{Z_S+Z_C} \end{bmatrix} \cdot \begin{bmatrix} U_S \\ I_{Lh} \\ U_C \end{bmatrix} \quad (3)$$

Fig. 18 represents the power stage in a black-box. It should be point out that although the interaction between the PAPF and the source impedance and load variation are within the scope of this paper for clarity; their interaction could be included into this power stage model, and be analyzed in the same systematical method.

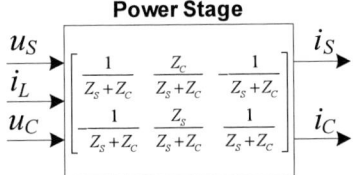

Fig. 18. Simplified linear model of power stage.

B. A General Expression of the Control Methods for PAPF

The control algorithm calculates the output voltage u_C^* from the detection currents.

$$u_C^* = f(i_L, i_S, i_C) \quad (4)$$

Taken linear control method, the algorithm can be represented as

$$U_C^*(s) = k_L(s) \cdot I_L(s) + k_S(s) \cdot I_S(s) - k_C(s) \cdot I_C(s) \quad (5)$$

This general control equation can be illustrated with the control diagrams as Fig. 11. The detected currents are adjusted by the corresponding gains of K_L, K_C, and K_S, which may be complex compensator or equivalent filters, such as high pass filter (HPF) or low pass filter (LPF) on rotational coordinate based instantaneous reactive power theory or p-q transformation. The compensator $H(s)$ is often chosen as proportional integral (PI) regulator.

978-1-4244-1871-8/07 $25.00 © 2007 IEEE

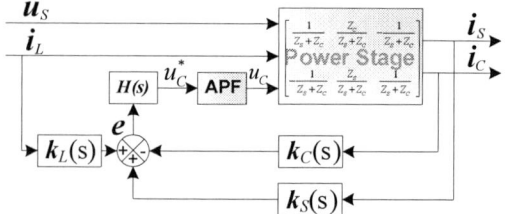

Fig. 19. General control diagram of PAPF

With particular gains of K_L, K_C, and K_S, different control scheme can be specifically identified from this general control diagram. For example, when K_L=HPF, K_C=1 and K_S =0, the general control diagram yields the traditional current-mode control shown as Fig. 20. The controller detects the harmonics i_{Lh} from load current i_L, and the compensated current i_C tracks this harmonic reference to guarantee that all harmonic current flow through the PAPF as shunt branch. Actually this is the regular current-mode control because when compensated current i_C tracks quickly enough with little error, the PAPF could be viewed as current source. Fig. 12b) abstractly represents this control scheme in frequency domain. A high pass filter represents the harmonics detection algorithm, whether it is fast Fourier transform algorithm (FFT), *p-q* method based on instantaneous reactive power theory, or the other. Here, the feedback gain of compensated current i_C is full pass, and the gain of source current i_S is zero pass because the source current is not detected.

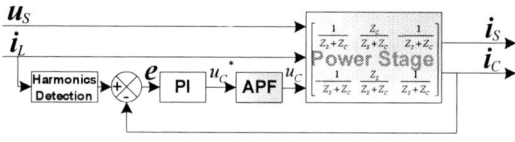

a) Control diagram

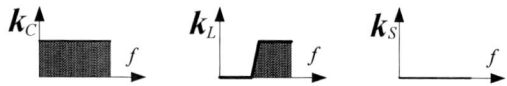

b) Equivalent current gains in frequency domain

Fig. 20. Representation of popular current-mode control with general voltage-mode control diagram.

C. Necessary Conditions for Valid Control Scheme

Although the gain of each detected current has many options, this paper only discusses four kinds of gains, including zero pass, full pass, high-frequency pass and low-frequency pass for simplicity. Depending on these four types of current gains with unity amplitude, 64 possible control schemes have been obtained. Because they are not all valid, it becomes vital to examine the necessary conditions according to several basic constrains.

The aim of control in PAPF is

$$\begin{cases} i_{Sh}(t) = 0 \\ i_{Cf}(t) = 0 \end{cases} \tag{6}$$

which means that both the harmonic source current and fundamental compensation current should tend to zero in PAPF. Here and in the following paper, the subscript letters $_L$, $_S$, and $_C$ indicate the load, source, and compensation respectively, and the subscript $_f$ and $_h$ indicate the fundamental component and the harmonics respectively.

The currents are governed by Kirchoff's Current Law (KCL)

$$i_L(t) = i_S(t) + i_C(t) \tag{7}$$

With well-designed feedback control, the error tends to zero:

$$i_L(s)K_L(s) - i_C(s)K_C(s) + i_S(s)K_S(s) \to 0 \tag{8}$$

Substitution of (7) to (8) yields

$$i_C(s)\big(K_L(s) - K_C(s)\big) + i_S(s)\big(K_L(s) + K_S(s)\big) = 0 \tag{9}$$

Because different frequency components are independent, separating the fundamental and harmonic frequency parts of (9) gives

$$\begin{cases} i_{Cf}(s)\big(K_{Lf}(s) - K_{Cf}(s)\big) + i_{Sf}(s)\big(K_{Lf}(s) + K_{Sf}(s)\big) = 0 \\ i_{Ch}(s)\big(K_{Lh}(s) - K_{Ch}(s)\big) + i_{Sh}(s)\big(K_{Lh}(s) + K_{Sh}(s)\big) = 0 \end{cases} \tag{10}$$

Considering the control aims of PAPF as (6), $i_{Cf}(s)$ and $i_{Sh}(s)$ is expected to be zero. Then the necessary conditions for any valid control scheme are obtained as

$$\begin{cases} K_{Lf} + K_{Sf} = 0 \\ K_{Lh} - K_{Ch} = 0 \end{cases} \Leftrightarrow \begin{cases} K_{Lf} = -K_{Sf} \\ K_{Lh} = K_{Ch} \end{cases} \tag{11}$$

These conditions are satisfied in only 16 possible control schemes. However, some of them are invalid. For example, the control method with zero gains ($K_L = K_S = K_C = 0$) generated constant zero voltage, and the PAPF is uncontrollable.

So the condition of controllability is added. In neither fundamental nor harmonic frequency, the three current gains are allowed to be the same. In fact equal current gains, whatever they are, result in constant zero output voltage. Then the necessary conditions are modified to be

$$\begin{cases} K_{Lf} = -K_{Sf} \ne K_{Cf} \\ K_{Lh} = K_{Ch} \ne -K_{Sh} \end{cases} \tag{12}$$

D. Systematic Derivation of Practical Control Scheme

For simplicity, assume the current gains have unity amplitude. Then according to necessary conditions of (12), there are totally 4 valid control schemes among all 64 options. Fig. 21 abstractly represents these control schemes with the current gains shown in frequency domain.

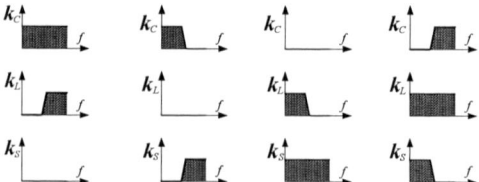

a) Regular current b) New control c) New control d) New control
mode control.   I   II   IIII

Figure 21. Practical control schemes in a systematic procedure.

The control scheme shown as Fig. 21a) is actually the widely-used current mode control. Detail discussions of the new control schemes are given in [16]. Here gives the illustrative example of the first new control scheme shown as Fig. 21. The basic operating principle is illustrated with virtual impedance in [14], here another analysis of control loops is given. There are two independent controls in the control diagram as Fig. 22b). One is fundamental frequency control, and the other is harmonics frequency. The fundamental-frequency loop controls the fundamental compensation current i_{cf} to track zero command, and the harmonic-frequency control loop suppresses the harmonic source current i_{sh} tend to zero. With well-designed current compensators $H(s)$, the harmonics of source current tends to zero, and less current flows through the shunt branch of APF.

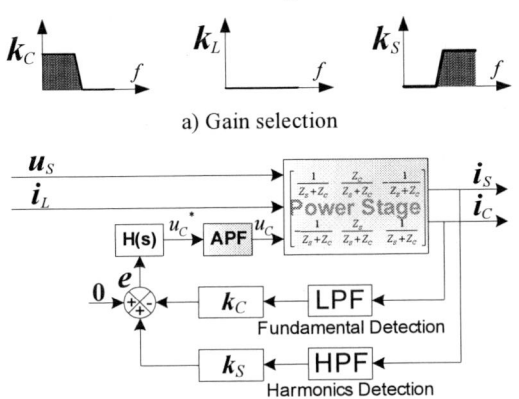

a) Gain selection

b) Control diagram

Figure 22. Gain selection and control diagram of one new control scheme.

V. SIMULATION AND EXPERIMENTAL RESULTS

Saber was taken to run the simulations to verify the pervious analysis and conclusion. Fig. 23 and Fig.24 shows the simulation to verify the comparison of PAPF. From the simulation result, it can be verified that the control method either with or without harmonics detection is valid and has the similar performance with the other in steady state. But when there is any disturbance at the load, the control scheme with harmonics detection has quicker response than that without harmonics detection, shown by the waveform of U_{DC}. Fig. 25 shows the simulation result of the new derived control methods.

A 3kVA three-phase experimental prototype was built and comparative experimental was conducted to verify the previous study. Fig. 25 and Fig. 26 show some snapshots of the experimental results with different control schemes. For comparison, the same compensators are used in all control schemes, with the same power stage.

VI. CONCLUSION

This paper compared the popular control methods for series and parallel active power filters. With the equivalent transformation of their control diagrams, the voltage-source control and current-source control for SAPF were proven to have the same control structure, and hence they could achieve the same compensation performance. In a same way, two kinds of control schemes for PAPF were compared, and the

control without harmonics detection can be explained as a simple feedback control, while the control with load harmonics detection is a composite control of both feedback of DC-link voltage and feedforward of the fundamental load current which results in better dynamic performance.

After the comprehensive comparison of the popular control methods for two dominant types APFs, a general expression of the control methods for PAPF was proposed in the form of both a control diagram and equations based on current detection. A systematic approach was subsequently presented to the analysis of the different control schemes for PAPF, and a family of new control methods was derived with this systematic approach.

REFERENCES

[1] R. D. Henderson and P. J. Rose, "Harmonics: The effects on power quality and transformers," *IEEE Trans. Industrial Applications*, pp. 528–532, 1994.

[2] M. EI-Habrouk, M. K. Darwish, and P. Mehta, "Active power filters: A review," *in Proc. Inst. Elect. Eng. Electric Power Applications*, vol. 147, 2000, pp. 403–412.

[3] F. Z. Peng, H. Akagi, and A. Nabae, "Compensation characteristics of the combined system of shunt passive and series active filters," *IEEE Trans. Industry Applications*, Vol. 29, No. 1, pp. 144-152, 1993

[4] J. W. Dixon, G. Venegas, and L. A. Morán, "A series active power filter based on a sinusoidal current-controlled voltage-source inverter," *IEEE Trans. Ind. Electron.*, vol. 44, pp. 612-620, Oct. 1997.

[5] Guozhu Chen, Zhengyu Lu, Zhaoming Quan, and Fang Zheng Peng. "A new sieial hybrid active power filter using controllable current source" *in PESC 2002*, pp364-368.

[6] Chen Qiaofu, LiDayi, and Wei Yanan, "A series active power filter based on fundamental magnetic flux compensation", *Huazhong Univ. of Sci. & Tech.* pp. 27-40, 2000 (in Chinese).

[7] Li Dayi, Chen Qiaofu, Xue Jianke, Zhang Saijun and Jia Zhengchun, "Three-phase active power filter based on fundamental magnetic flux compensation", *Automation of Electric Power System*, pp. 48-52, 2003(in Chinese).

[8] Wu, J.-C.; Jou, H.-L., "Simplified control method for the single-phase active power filter ," *Electric Power Applications, IEE Proceedings-*, vol.143, no.3pp.219-224, May 1996.

[9] Xiaoyu Wang, Jinjun Liu, Chang Yuan, and Zhaoan Wang, " Comparative analysis of popular control schemes for parallel active power filter and experimental verification," *Power Electronics Specialist, PESC. 2006 IEEE 37th Annual Conferece on*: 2006.

[10] Casadei, D.; Grandi, G.; Reggiani, U.; Rossi, C., "Control methods for active power filters with minimum measurement requirements," *Applied Power Electronics Conference and Exposition, 1999. APEC '99.* Fourteenth Annual , vol.2, no.pp.1153-1158 vol.2, 14-18 Mar 1999.

[11] Zhou, L.; Smedley, K.M., "Unified constant-frequency integration control of active power filters," *Applied Power Electronics Conference and Exposition, 2000. APEC 2000.* Fifteenth Annual IEEE , vol.1, no.pp.406-412 vol.1, 2000

[12] K. M. Smedley, L. Zhou, and C. Qiao, "Unified constant-frequency integration control of active power filters-steady-state and dynamics" *Power Electronics, IEEE Transactions on*, vol. 16, no. 3, pp. 428-436, 2001.

[13] WANG Xiaoyu, LIU Jinjun, YUAN Chang, WANG Zhaoan. "A Comparative Study on Voltage-Source Control and Current-Source Control of Series Active Power Filter". *Proceedings of IEEE 21st Annual Applied Power Electronics Conference and Exposition (APEC 2006)*, Dallas, Texas, USA, Mar. 19-23, 2006, pp. 1570~1575.

[14] Xiaoyu Wang, Jinjun Liu, Chang Yuan, Zhaoan Wang, "A Novel Voltage-Mode Control of Parallel Active Power Filter," *IAS Annual Meeting (IEEE Industry Applications Society)*, United States. pp. 169-174, 2006.

[15] Xiaoyu Wang, Jinjun Liu, Chang Yuan, Zhaoan Wang, "Generalized Control Approach for Active Power Filters," in Power Electronics and

978-1-4244-1871-8/07 $25.00 © 2007 IEEE

Motion Control Conference, 2006.IPEMC '06.CES/IEEE 5th International, Shanghai, China, August 13-16, 2006, pp. 714-718.

[16] Xiaoyu Wang, Jinjun Liu, Chang Yuan, Zhaoan Wang. "A Family of Control Methods for Parallel Active Power Filters Based on Current

Detection," in *Applied Power Electronics Conference, APEC 2007 - Twenty Second Annual IEEE, Anaheim, California, USA*, Feb. 25 – Mar. 1, 2007, pp. 675-681.

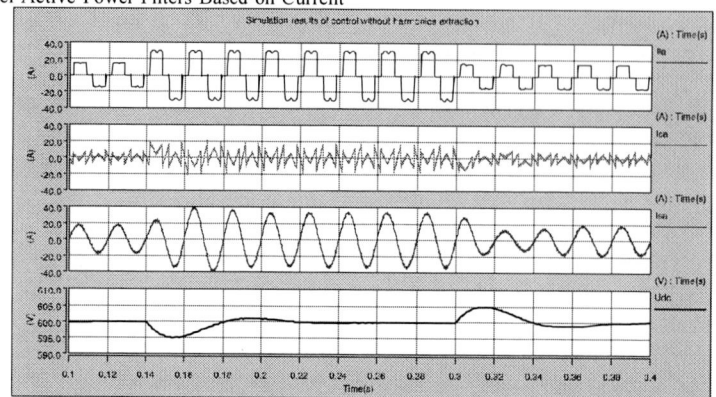

i_{La}: Load current; i_{Ca}: Compensated current; i_{Sa}: Source current; u_{DC}: DC-link capacitor voltage

Fig. 23. Simulation results of three-phase PAPF controlled without harmonics extraction.

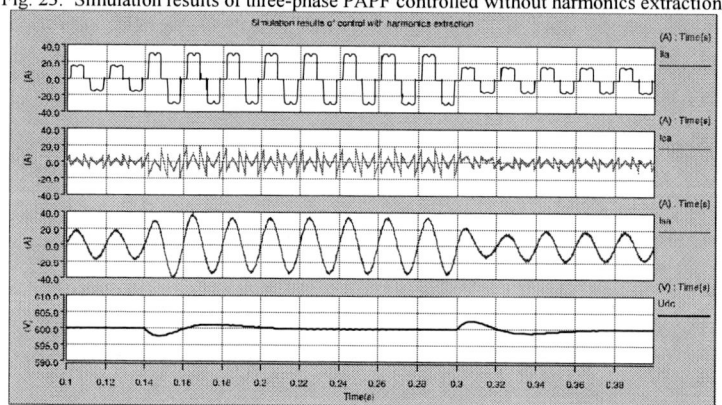

i_{La}: Load current; i_{Ca}: Compensated current; i_{Sa}: Source current; u_{DC}: DC-link capacitor voltage

Fig. 24. Simulation results of three-phase PAPF controlled with harmonics extraction.

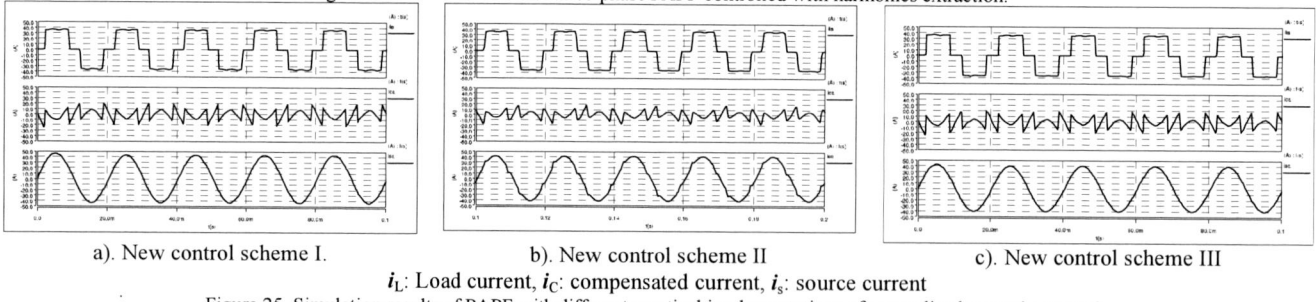

a). New control scheme I.　　b). New control scheme II　　c). New control scheme III

i_L: Load current, i_C: compensated current, i_s: source current

Figure 25. Simulation results of PAPF with different practical implementations of generalized control approach.

a). New control scheme I.　　b). New control scheme II　　c). New control scheme III.

①i_L: Load current, ②i_C: compensated current, ③i_s: source current

Figure 26.Experimental results of PAPF with different practical implementations of generalized control approach.

978-1-4244-1871-8/07 $25.00 © 2007 IEEE

Adjustable Speed Generator Systems – an Emerging Technology for Efficient Electrical Energy Generation

Wlodzimierz Koczara
Warsaw University of Technology
Institute of Control and Industrial Electronics
Pl. Politechniki 1, 00-661 Warszawa, Poland
Email: koczara@isep.pw.edu.pl

Abstract— **The paper presents adjustable speed generation systems providing high quality output AC standard voltage independently to type of load and saving primary energy. The energy saving is result of speed control in a range of high efficiency of the engine. The new adjustable speed generation systems can be driven by any internal or external combustion engine but presented examples are related mostly to Diesel engine. The prime mover is driving a newly designed and built axial flux shaftless and bearingless permanent magnet generator, which supplies power electronic converter acting as universal AC power source with single or three phase output. Speed of the prime mover is adjusted to meet load demand by real time load measurement. Thus the power electronics converter decouples the generator and load and then speed of the generator is not related to frequency of the output voltage. Therefore such a system has degree of freedom related to speed. There are two basic topologies providing decoupled generation first called "decoupled by topology" and second called "decoupled by control and topology". Systems are fully controlled and provide high quality output power. Theory and results of tests of both decupled systems confirm their performances.**

I. INTRODUCTION

A. Primary energy and environment

The world existence is now dependent on coal, oil and natural gas for its energy. Unfortunately fossil fuels are not renewable and they draw on finite resources which will be eventually shortly run out. Moreover, they are becoming too expensive or environmentally damaging . We noticed that since last 10 years the average price of the fuels is growing continuously. Fig. 1 shows history of oil prices during last 6 decades [47]. In 2006 the average price of crude oil passed US $60 and in September 2007 prices jumped over US $74. The coal is following the oil and its price between year 2005 and 2006 did increase by 20% [46]. Hence, the growing world economy will notice problems of both high primary energy prices and lack the energy. Moreover, today electricity production, based on fossil fuels, is significant source of earth pollution. Our responsibility is to save environment by developing all possible technologies of energy savings and particularly in process of electricity

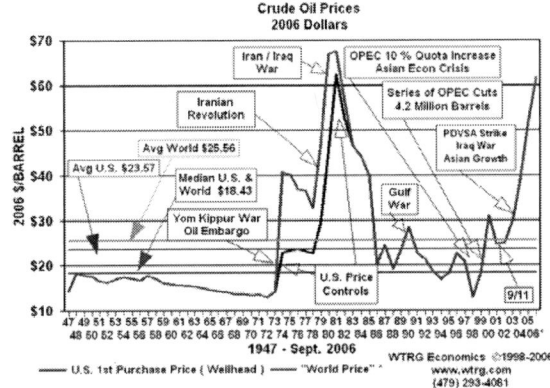

Fig. 1: Crude oil prices [47]

production and use. The way to produce enough electrical energy from unstable and not reliable ("wild") renewable sources as wind and sun is still very long and needs several decades to get demanded high quality and quantity. Moreover, these renewable sources need complementary fully controllable energy sources assuring demanded high quality.

B. Examples of energy saving equipment

Drives are main user of electricity. The power electronic controlled variable/adjustable speed drives application resulted in enormous electrical energy savings and they are still in progress. The drive speed control and precise adjustment provides quality of products, increases productivity of the machine tools and saves energy. Automotive industry just started to develop hybrid vehicle with fully controllable and efficient electrical powertrain. This is very promising filed of enormous saving of energy.

C. Typical demand of power and today electrical power sources technology

The demand of the electrical power varies in time in wide range. Fig. 2 shows an example of load profile. The load power varies depending on hour, day of week or season of the year. Usually the average power differs considerably from minimum and peak power. However, today technology of the electricity production using fuels and synchronous generator is based on fixed and very precise maintained speed. Operation, leading to energy saving, is limited mainly to control of number of active power sources, which are

978-1-4244-1871-8/07 $25.00 © 2007 IEEE

optimized to only one fixed speed of the synchronous generator.

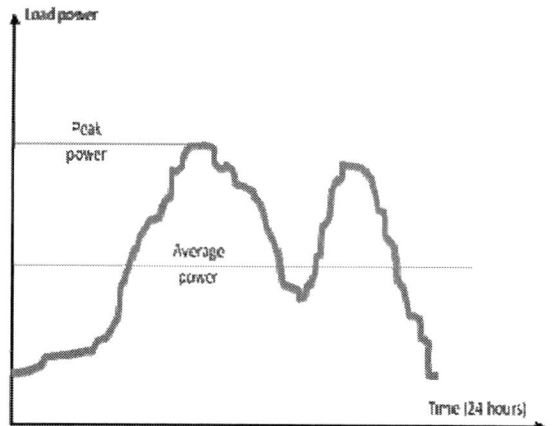

Fig. 2: An example of variable power demand

There is significant progress in exhaust filtration but still huge energy (fuel) is lost and passes to earth atmosphere contributing to green house effect. A quantity of gases exhaust is proportional to total use of primary energy.

The conventional wound rotor synchronous generator, which is in practice only source of electricity in power stations, is genius invention of XIX century, which is improved during electromagnetic era of the all XX century. The wound rotor fixed speed synchronous generator (Fig. 3 and Fig. 4) combines two functions: the conversion of the mechanical power to the electrical one and delivering given quality of power.

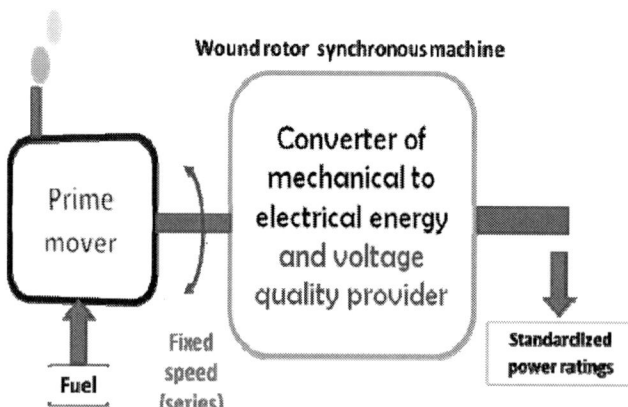

Fig. 3. Fixed speed generation system based on wound synchronous generator.

The results of this "two in one" concept are following: the lack of speed flexibility requiring constant speed operation, the tendency to swing, easily going out of synchronism, limiting the speed to 3000rpm (1500 rpm) or 3600 rpm (1800

rpm), an enormous weight and no prospects for the weight reduction, a slow response in voltage control or reactive power control, no prospects for instantaneous voltage control and distorted voltage production in case of nonlinear loads.

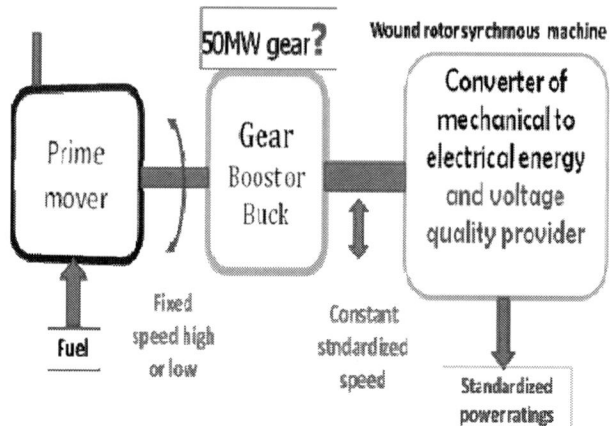

Fig. 4 Fixed speed generation based on wound synchronous generator driven via gearbox.

In case of grid disturbances a healthy generating set has to be disconnected and the following resynchronization takes a very long time. Brushless construction has poor dynamics, whereas direct excitation current control requires slip-rings application.

The synchronous generator is successful in very high power units reaching 1000 MW. However an optimization of the efficiency of the total generating set (including prime mover) is limited only to given fixed speed (Fig. 5)

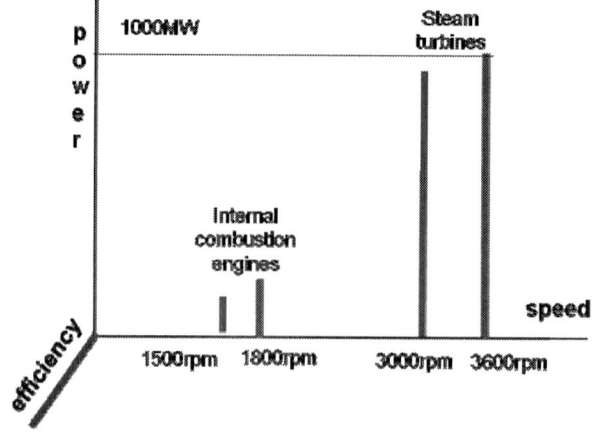

Fig. 5. The constant speed generation system has limited area of the engine efficiency improvement to given fixed speed.

Now the choice of best operation of the synchronous generator is limited to only 2 speeds one for high speed steam engines (3000 rpm or 36000 rpm) and one for low speed Diesel engines (1500 rpm or 1800 rpm).The synchronous generator operation requires fixed speed prime movers. The energy, delivered to the generator, comes mostly from fossil fuels, bio-fuels, gas, nuclear fuels, water potential energy and

978-1-4244-1871-8/07 $25.00 © 2007 IEEE   44

wind kinetic energy. The use of fuels, that are still the main source of energy, requires application of internal or external combustion engines (ICE or ECE) that convert burned fuel kinetic energy to mechanical rotational movement energy. Another method, based on fuel burning requires intermediate carries as steam, which kinetic energy is converted to mechanical rotational energy. All the processes related to conversion of fuel to mechanical energy, delivered to the generator, have a very low energy efficiency. Moreover, as the rotational speed of the driving engine has to be constant and fixed on specific series of speeds, then the possibility to increase the efficiency of the conversion is extremely limited.

Hence, several serious questions arise: Is it possible to improve the existing generators or to replace them by other technology during the next 10 or even 20 years? Is it feasible any challenge resulting in invention of competitive technology?

II. DECOUPLED GENERATION AS AN EMERGING TECHNOLOGY

Application of power electronic converters in the generation opens new way for the generation system design and building. The power electronic converter can be supplied from any voltage source and produces standardized AC or DC voltage. Therefore, the variable speed generators producing variable frequency and amplitude are going to be source of power. There are several types of generators which can be applied for such a systems. There are synchronous wound rotor machines, permanent magnet machines, induction machines. The speed free prime movers can be redesign according to additional degree of freedom related to speed. There are two basic systems representing decoupled generation shown in Fig. 6 and 7.

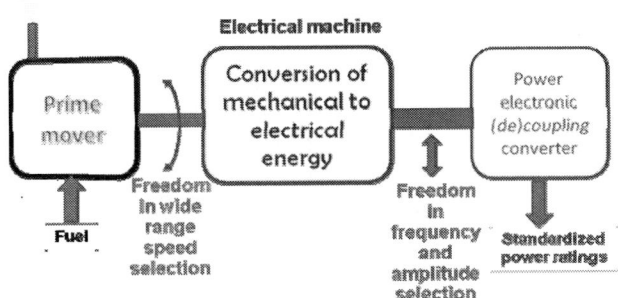

Fig. 6. Variable speed generation system – an additional degree of freedom for engine and generator design – "decoupling by topology".

The system "decoupled by topology" (Fig. 6) [22], consists of prime mower, generator and power electronic converter. The prime mover operates in wide speed range driving the generator producing voltage having variable frequency and amplitude. This variable voltage is converted by power electronics converter to demanded (e.g. standardized)

voltage. This ability of the power electronic converter to control power flow with high efficiency, results in additional degree of freedom in speed. This open the road to redesign the prime mover and generator according higher efficiency and weight reduction.

The concept of "decoupling by control and topology" shown in Fig. 7 is based on existing slipring generator and power electronic converter connected to the rotor [33].

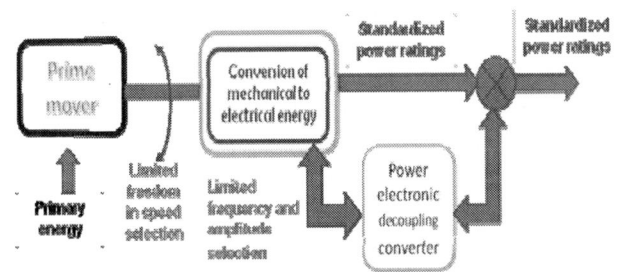

Fig. 7. Variable speed generation – an additional degree of freedom for engine design – "decoupling by control and topology".

This is sort of "synchronous" generator excited from the rotor side by current controlled from the power electronic converter connecting the rotor and stator of the generator. The difference between well known double fed induction generator DFIG is autonomous feature and is called ADFIG The power electronics produces variable frequency of the excitation current in the way that a sum of the frequency of the rotor speed and frequency of the excitation current is fixed (e.g. standard frequency). This system does not permit to redesign the generator but its main advantage is low power electronic converter.

The variable speed generator operation results in more freedom in design of engine generator of the generation system resulting in optimization of energy use.

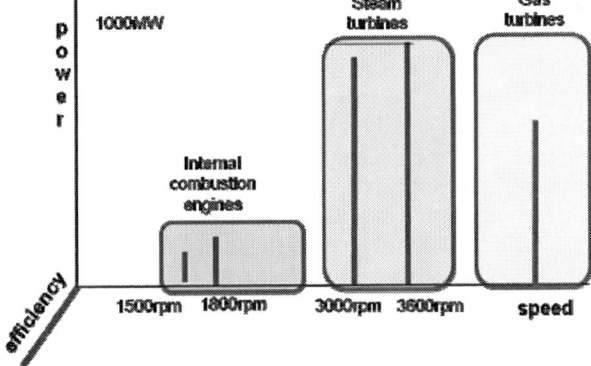

Fig. 8. Variable speed generation systems permits to optimisation of the engine efficiency in wide range of speed.

978-1-4244-1871-8/07 $25.00 © 2007 IEEE   45

III. ADJUSTABLE SPEED GENERATING SYSTEMS DECOUPLED BY TOPOLOGY

A Power Delivered by the Engine in Fixed and Variable Speed Operation

The efficiency of internal combustion engine (ICE) is low. Current achievement in Diesel engine places the efficiency in range below 40%. Therefore the important issue is an adjustment of the speed to regions of high efficiency. An example of simplified dependence of efficiency from power and speed of typical Diesel engine is shown in Fig.9. There are: narrow region of constant speed and wide region of variable speed. The variable speed operation may be adjusted along high efficiency range of the ICE.

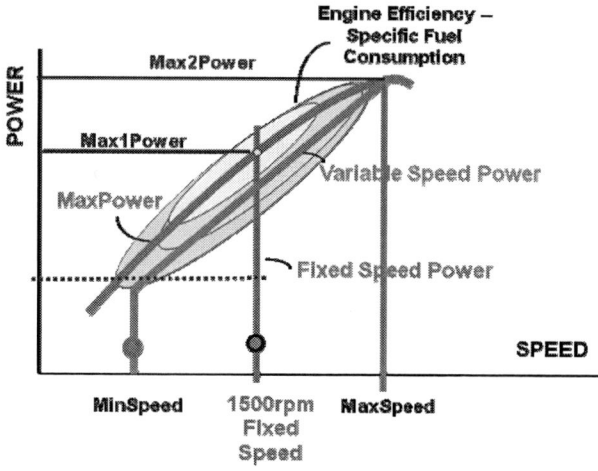

Fig. 9. Simplified presentation of power and efficiency (specific fuel consumption) of the (ICE) Diesel engine for fixed speed and variable speed operation..

The region of the high efficiency (i.e. low specific fuel consumption as kWh/g) is close to maximum available power. Therefore, the most economic operation is to adjust speed along curve of maximum power. If the engine operates with fixed speed, which is determined by generator output fixed frequency, then most of operation is far from the area of high efficiency. Moreover, the freedom in the speed permits operation with speed higher than one coming from standard frequency of the generator. Hence, higher speed responds higher available power. The maximum speed is determined only by efficiency and engine ability to produce demanded torque. Fig. 10 shows an example of application of lower size of the engine producing the same power as the bigger size one operating with synchronous speed. The increase of output power is result of higher maximum speed.

Reduction of speed at low loads results in extension of the engine life cycle and reduction of fuel consumption. Moreover during low speed operation the engine produces less noise. When lower power engine is used then the noise level is also reduced.

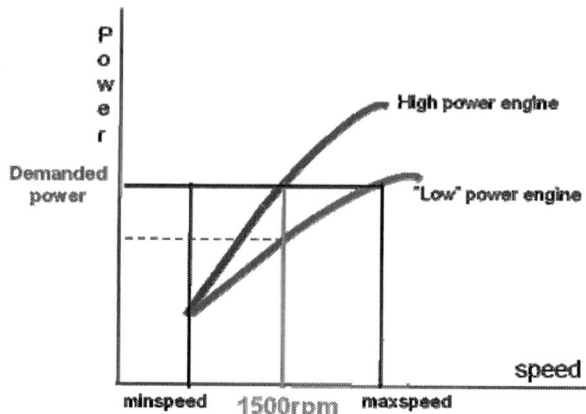

Fig. 10. Increase of speed results in higher power delivered by the engine

B An Example of Modern Permanent Magnet Generator

Another key features of the system "decoupled by topology" are: freedom in speed as an input and in frequency and amplitude of the generator output voltage. Therefore, the additional degrees of freedom opens the way to the new era of the electrical generator construction and optimization. Especially there is an opportunity to simplify the generator design by application of permanent magnets. Another advantage is free choice of number of poles and then induced stator frequency. One of the first industrialized design of the generator is an axial flux permanent magnet generator (Fig. 11). In the axial flux generator the poles are built by pasting permanent magnet plates on the side of the rotor. The shape and size of the poles are specially designed [24].

Fig. 11. The 40 kW bearingless and shaftless permanent magnet axial flux generator produced by Cummins Generator Technologies (former NEWAGE). in WUT Laboratory.

The increase of frequency and axial flux construction results in very short length of the generator. This feature permits to build shaftless and bearingless generator. The very light (1,5kg/kW) and very short generator is just mounted on the shaft of the driving engine.

C Power Electronics Converters

Main stream of power electronic converters manufacturing is oriented to dedicated drives. This means that the converter is sized to symmetrical well known load. The active and reactive power is limited to the values coming from the electrical motor. The converters, produced for drives, have usually not true sinusoidal voltage but pulse nature i.e. they are PWM sources. Another group of the power electronic converters with usually sinusoidal voltage is for power quality application as UPS. Although the UPS technique delivers sinusoidal voltage (mostly low power single phase) the converters dedicated to this application are usually designed to be supplied from batteries with rather constant DC voltage. A time of the UPS operation is limited to quantity of energy stored in battery. The UPS load is well known.

The variable speed generation systems (Fig. 12) have to be dedicated to general application as main source of power. Hence, besides of rated power any load from zero to short circuit may be applied. A reliable operation, in case of short circuit events and high current, activating, in short time, protection, is required. Such loads and requirement of high reliability needs special design of the PTG converter.

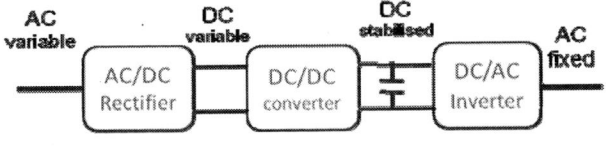

Fig. 12. The power electronic converter PTG as an universal source of voltage and power

The power electronic converter topology depends on voltage source i.e. generator voltage characteristic features. Usually the power electronic converter is built as voltage source converter with intermediate DC link voltage. Figures 13, 14, 15 and 16 show block diagrams of typical AC/AC converters with DC link.

The variable speed operation of the generator results in variable AC voltage. Therefore after rectification (Fig. 13) a DC/DC converter is used. When DC voltage is well stabilized

Fig. 13. Basic topology of the AC/AC converter for wide speed range generation set.

then the DC/AC converter can produce high quality of the AC single or multiphase voltage. In high power systems the rectifier and DC/DC converter is replaced by bidirectional AC/DC converter (Fig. 14) as three phase classical two level bridge topology or multilevel topology.

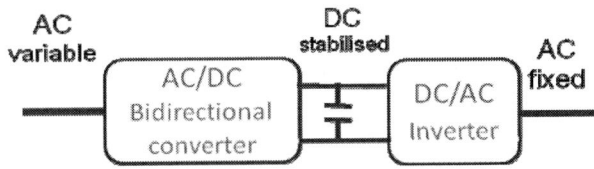

Fig. 14. AC/AC application of fully controlled reversible converter

The DC link of the converter can be supported by supercapacitor energy storage (Fig. 15) or/and by hybrid storage made from supercapacitor and battery banks (Fig.16)

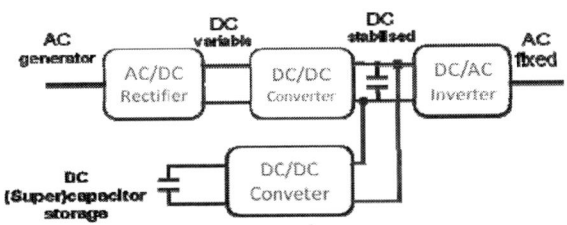

Fig.15. AC/AC converter with capacitor energy storage

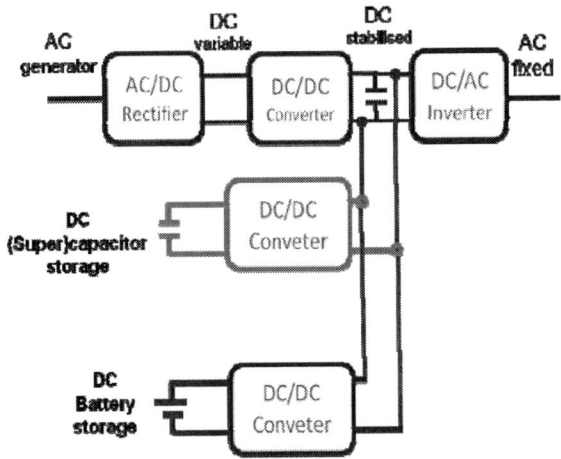

Fig.16. AC/AC converter with capacitor and battery energy storages

The system with battery storage can operate as conventional UPS too [13].

D Adjustable Speed Generation System

In the adjustable speed generation systems the speed follows the demand of output power. There are different strategies of the speed control. One them is shown in Fig. 17. The "Variable Speed Controller" senses demanded power and adjust a reference speed of the engine.

978-1-4244-1871-8/07 $25.00 © 2007 IEEE

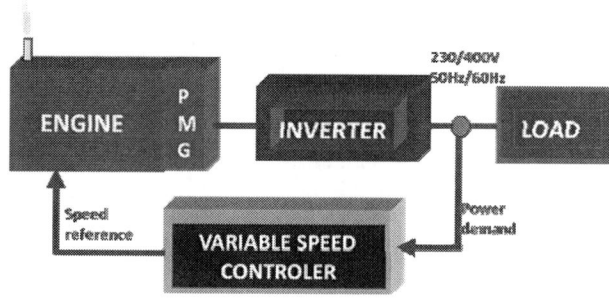

Fig. 17. Adjustable speed generation system control

The 40 kW generation system is designed and built as is shown in Fig. 18 and Fig. 19.

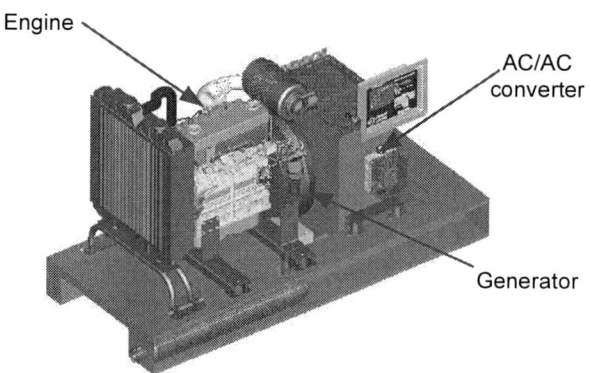

Fig. 18. Design of the adjustable speed generation system

Fig. 19. Tested adjustable speed generation system

Fig. 20 shows oscillogram of the high frequency input voltage and 50 Hz output high quality sinusoidal phase voltage. Case of step of RL load is presented in Fig. 21. The transient state has no any negative impact on the amplitude of the AC voltage.

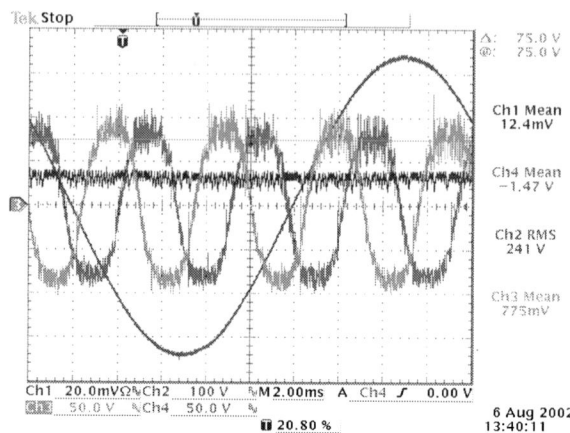

Fig. 20. Oscillogram of the generator phase voltage and output voltage – steady state operation

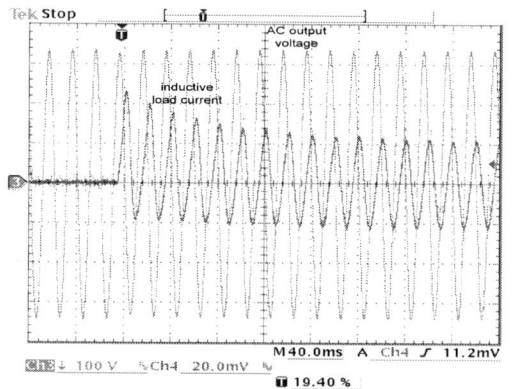

Fig. 21. Case of turn-on resistive-inductive load – Oscillogram of the output voltage and load current

The method "decoupled by topology' is applied also to high speed turbines systems. However, a wide range of speed is needed but direct drive and freedom in speed selection makes for the feasibility of quick applications.

V – ADJUSTABLE SPEED GENERATING SYSTEMS DECOUPLED BY CONTROL AND BY TOPOLOGY

A – Case of Autonomous Operation

The doubly-fed induction generator has popular application in wind and water turbines. In these cases the induction generator is magnetized (excited) by grid voltage. Therefore in case of grid failure the generator is useless. However, an autonomous operation is possible. Concept of self excitation is shown in Fig. 22 [31], [32], [33]. In this case the generator is excited mainly by rotor current from stator side controlled by PWM of the power electronic converter PEC. One part of the excitation current can be delivered by the capacitor placed in the stator side. However, this capacitor has to be limited. In case of high its capacitance will appear noncontrolled oscillations caused by resonance between the capacitor and the generator inductance. The power electronic converter PEC produces reactive current component which is needed for the excitation (magnetization) and active current component according to active power flow of the generator operation. Principle of the doubly fed induction generator operating in autonomous mode and grid connected mode is shown in Fig. 23.

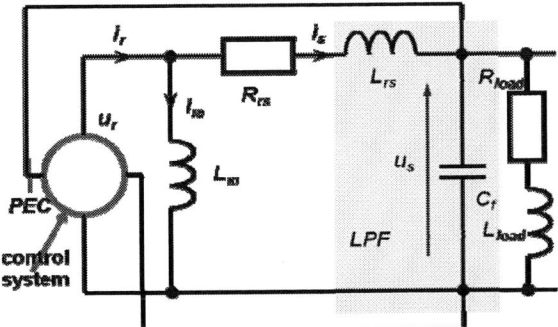

Fig. 22. Equivalent diagram of the autonomous operation of the doubly-fed induction generator

The control system provides sensorless "Direct Voltage Control" **DVC** which fulfill the principle of the decoupled operation. The sensorless means not use any sensor providing signal of speed and rotor position. The DVC method keeps the stator voltage according to reference amplitude and frequency(phase) of the stator voltage by control rotor current components. The reference voltage vector is rotating synchronously in *dq* frame. The system is called "**decoupled by topology**" because there is part of power electronic converter which is able to produce variable frequency and amplitude rotor current. However, the rotor current must be very precise controlled in range of phase angle, amplitude and frequency in relation to rotor speed and required stator voltage frequency. Therefore the system is also called "**decoupled by control**". The control system DVC is operating in autonomous mode providing fixed frequency sinusoidal voltage. When the system is additionally equipped in controller fulfilling grid connected mode (marked by red color blocks) then in case of grid failure the autonomous operation is possible [37], [39]. The DVC control system is based on precise control of the rotor current. Fig .24a shows an oscillogram of rotor current in transient state of passing by rotor synchronous speed. The amplitude, phase and frequency of the rotor current varies according to speed and load. The load being a three phase diode rectifier is very nonlinear as is shown in Fig. 24b [41]. Small pulsation of the rotor current i_{ra} is coming from tendency to compensate voltage deformation produced by commutation process of the load rectifier connected to the stator. In spite of very nonlinear load the AC output voltage distortion is not significant. The ability of the generator, controlled by the DVC method, to cope with sudden change of nonlinear load is presented in Fig. 25. The Fig. 25a shows case of sudden increase of the load whereas the Fig. 25b illustrates the case of turn-of transients. Instant reaction of the rotor (excitation) current results in very short stator voltage distortion.

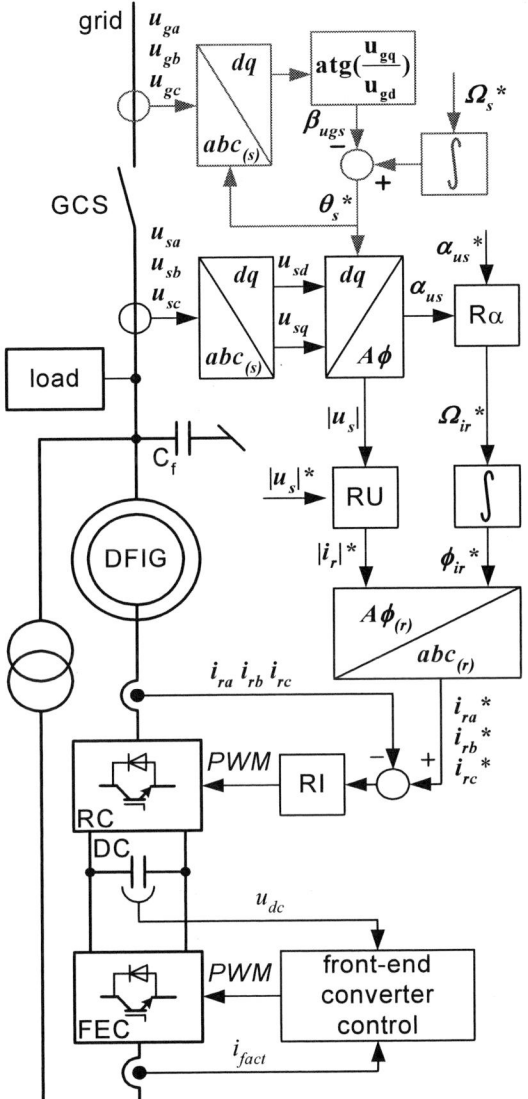

Fig. 23. Topology and control system of the doubly fed autonomous and grid connected operation

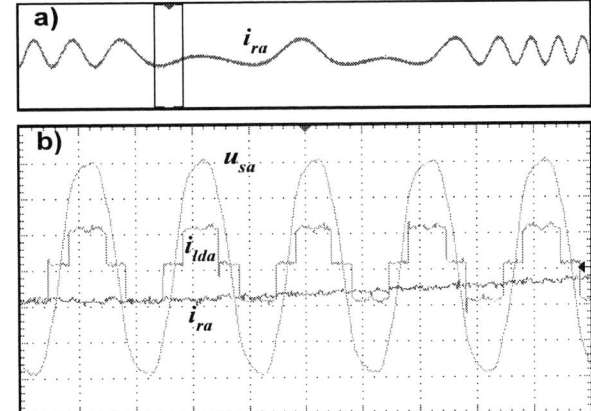

Fig. 24. Oscillogram presenting the autonomous operation of the double fed induction generator, a) rotor current i_{ra}, b) stator voltage u_{sa} (100V/div), rotor current i_{ra} (20A/div), load current i_{lda} (2A/div) of 2.2kW machine.

978-1-4244-1871-8/07 $25.00 © 2007 IEEE

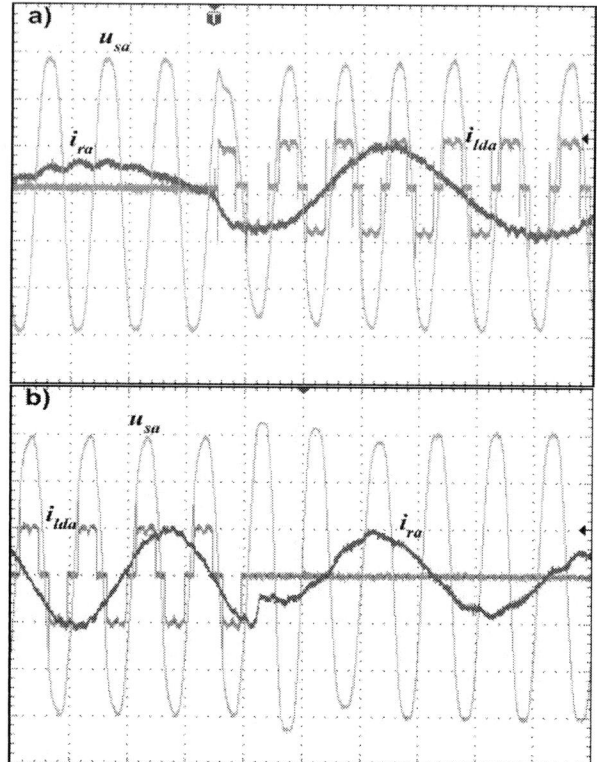

Fig. 25. Oscillogram presenting the autonomous operation of the double fed induction generator a) step of nonlinear load, b) transient in case of turn-of the nonlinear load

The doubly fed induction generator system is only economical when it has limited speed range. Therefore, the range of "degree of freedom" related to speed is not so wide like in system using permanent magnet generator. Moreover, the double fed generator speed series is close to synchronous speeds. Furthermore the slip-ring induction generator has brushes, is heavy and so more there is no any prospects to reduce its weight and size. They are unfortunately significant limitations reducing application of this technology. The main advantage of the DFIG and the ADFIG systems is lower power electronic converter. This advantage makes the system applicable in wind energy harvesting and in water turbine multimegawat hydro stations. The ADFIG system can be useful especially in generator driven by Diesel engines in gensets producing 60 Hz where synchronous speed is 1800rpm. An increase of supersynchonous speed to 2200-2330 rpm is very suitable for the Diesel engine high efficiency operation.

VI SUMMARY

The today generation systems are based on classical synchronous generator invented in IXX century and they have no prospects to further development or even improvement. The power electronic technology application to power systems opens the way to the new techniques of the power generation. The new techniques breaks stiff limitations in the driving engine and generator design and development.

The paper has presented two basic theories of power generation:

- Concept of the decoupled generation by topology
- Concept of the decoupled generation by control and topology.

The decoupled generation by topology is an emerging technique which can be used in autonomous and grid connected mode. This topology uses permanent magnet generator and can operate in wide speed range. However, this technique is not limited only to such type of the electrical machine. The great number of advantages of system "decoupled by topology" as:

- reduction of weight of the engine
- reduction of weight of the generator
- freedom in rated speed selection
- freedom in speed range
- real prospects of specific fuel consumption reduction

make it very feasible in further development.

The system "decoupled by control and topology" in autonomous operation (ADFIG) is also new technique however is based on known and already applied DFIG operating only in grid connected mode. The ADFIG application make the induction generator useful in grid failure event whereas in such a case the classical DFIG system is useless.

The further progress of the decoupled by topology generation systems requires development of matured, high power converters and generators. It is related to low speed system based on ICE (internal combustion engines) and high power external combustion engines (ECE) as high speed high power turbines powered by steam or by gas.

VII REFERENCES

[1] Koczara W., L. Grzesiak, M. da Ponte.: Hybrid Load-Adaptive Variable-Speed Generating Set: New System Topology and Control Strategy. Proceedings of International Conference on Power Generation, Powergen, Orlando, Florida 7-9 December 1998, USA.

[2] . Al-Khayat, R. Seliga, W. Koczara, A. Krasnodebski, B. Kaminski: DSP Control of Variable Speed Integrated Generator". IEEE ISIE'02, 8-11 July 2002, L'Aquila, Italy.

[3] E. Clark, N. Sidell, Jewell, D. Hove.: High Temperature Electromagnetic Devices. TRW Automotive Technical Centre. The U.K. Magnetic Society UK. 5 October 2000. Solihull.

[4] N. Brown, L. Haydock, E. Spooner.: 3 Dimensional Finite Element Analysis of a Toroidal Wound Axial Flux Permanent Magnet Generato". UPEC'99, Leicester. UK 1999.

[5] Naser M. Abdel-Rahim, John E. Quaicoe, Analysis and Design of a Multiple Feedback Loop Control Strategy for Single-Phase Voltage-Source UPS Inverters" , IEEE Transactions on power electronics, Vol. 11, NO.4, 1996.

[6] Tzou Ying-Yu: DSP-Based Fully Digital Control of a PWM DC-AC Converter for AC Voltage Regulation", IEEE Transactions on power electronics 1995.

[7] Osman Kukrer: Deadbeat Control of a Three-Phase Inverter with an Output LC filter, IEEE Transactions on power electronics, Vol. 11, NO.1, 1996.

[8] Von Jouanne, Parasad N. E., Lucas J.L.: DSP Control of High Power UPS Systems Feeding Nonlinear Loads. IEEE Transaction on Industrial Electronics Vol. 43, No. 1 February 1996. pp. 121 – 124.

[9] J Loghead: A New Electrical Industry. Editorial Power Engineering Journal, February 1998, pp. 2-3.

[10] Time for the Tine Turbines. Electrical Review. Vol. 231 No. 12 pp. 28-30.

[11] L Grzesiak, W Koczara, M Da Ponte: Novel Hybrid Load-Adaptive Variable-speed Generating System. Proceedings IEEE International Symposium on Industrial Electronics ISIE'98, Pretoria, South Africa 7-10 July 1998, pp. 271.276.

[12] Hygen Variable-speed Generating System for Quality Electric Power Supply, Technical overview, Volt Ampere, Pretoria, South Africa 1997.

[13] M Da Ponte, L Grzesiak, W Koczara, A Niedzialkowski, P Pospiech: Hybrid Generator Apparatus, South Africa Patent No. 97/11503.

[14] Boldea I., Tutlea L., Serban I.: Variable Speed Electric Generators and Their Control: an emerging technology. Journal of Electrical Engineering, Vol. 2 / 2002

[15] Bolognani S., Venturo A., Zigliotto M.: Novel Control Technique for High-Performance Diesel-Driven Generator-Sets. Conference Proceedings on Power Electronics and Variable Speed Drives, 18-19 September 2000. Conference Publication No. 475 © IEE 2000, pp. 18-523.

[16] Clark W.Gellings CEIDS Consortium for Electric Infrastructure to Support A Digital Society. Proceedings of 6th International Conference on Electrical Power Quality and Utilisation, September 19-21, 2001 Cracow, Poland pp. 25-34.

[17] L. Grzesiak, Koczara W., M. Da Ponte: Application of Permanent Magnet Machine in the Novel Hygen Adjustable-Speed Load-Adaptive Electricity Generating System. Proceedings of the 1999 IEEE International Electric machines and Drives Conference IEMDC'99, Seattle, Washington USA, May 9-12, 1990, pp. 398-400.

[18] Grzesiak L., Koczara W. M. da Ponte. Power Quality of the Hygen Autonomous Load –Adaptive Adjustable Speed Generating System,. Applied Power Electronics Conf. APEC'99. Dallas, Texas, March 1999.

[19] Jawad Al-Tayie R. Seliga, Nazar Al-Khayat, W.Koczara: Steady State and Transient Performances of New Variable Speed Generating Set. CD Proceedings of 10th European Conference on Power Electronics and Application, 2-4 September 2003 Tolouse, France ISBN 90-75815-07-7.

[20] Nazar Al-Khayat, Robert Seliga, Wlodzimierz Koczara, Artur Krasnodebski, Bartlomiej Kaminski: DSP Control of Variable Speed Integrated Generator. 2002 IEEE International Symposium on Industrial Electronics IEEE-ISIE 2002 July 8-11, 2002 L'Aquila, Italy CD proceedings IEEE catalog number 02TH8606C.

[21] Tolbert L.M., Peterson W.A., Scudiere M.B White C.P., Theiss T. J., Andriulli J.B., Ayers C.W., Farquharson G., Ott G.W., Seiber L.E.: Electronic Power Conversion System for an Advanced Mobile Generator Set. IEEE Trans vol. IA – 38, 2002M. Young, *The Techincal Writers Handbook*. Mill Valley, CA: University Science, 1989.

[22] W. Koczara, E. Ernest, N. A. Khayat, J. A.Tayie: Smart and Decoupled Power Electronic Generation System. Conference Proceedings IEEE -PESC'2004 Aachen 21-15 June 2004.

[23] L.H. Lasseter, L.Paigi: Microgrid – A Conceputal Solution. Proceedings of the 35 Annual Power Electronics Specialists Conference PESC'2004, Aachen, June 2004, Germany, pp. 4285-4290.

[24] W.Koczara, J. Leonarski, R.Dziuba, N. Al.-Khayat, N, Jakeman: Variable Speed Three Phase Power Generation Set. Proceedings of 9th European Conference on Power Electronics and Application EPE2001, 27-29 August 2001, Graz, Austria

[25] W.Koczara, J. Leonarski, R.Dziuba, ˙N. Al.-Khayat, N, Jakeman: Variable Speed Set for Embedded Power Generation. Proceedings of 6th International Conference on Electrical Power Quality and Utilisation EPQU2001, 19-21 September 2001. Cracow, Poland.

[26] I, Von Jouszzo, N. Parasad, J. Lucas: DSP Control of Hig Power UPS Systems Feeding Non-linear Loads. IEEE Transaction on Industrial Electronics vol. 43, no.1 Feb. 1996, pp. 121-124.

[27] G. Iwanski and W. Koczara, "Sensorless Stand Alone Variable Speed System for Distributed Generation", Proceedings of the 35th Annual IEEE Power Electronics Specialist Conference – PESC'04, Aachen, Germany, vol. 3, pp. 1915-1921.

[28] G. Iwanski and W. Koczara, "Sensorless Direct Voltage Control Method for Stand-Alone Slip-Ring Induction Generator" Proceedings

of 11th European Conference on Power Electronics and Applications – EPE'05, Dresden, Germany.

[29] D. Forchetti, G. Garcia and M.I. Valla, "Vector Control Strategy for a Doubly-Fed Stand-Alone Induction Generator". Proceedings of Industrial Electronics Conference - IECON 02 Vol 2, 5-8 Nov. 2002 pp. 991 – 995

[30] R. Cardenas, R. Pena, J. Proboste, G. Asher and J. Clare, "Sensorless Control of a Doubly-Fed Induction Generator for Stand Alone Operation" in Proceedings the 35th IEEE Power Electronics Specialist Conference – PESC'04, Aachen, Germany.

[31] G. Iwański, W. Koczara: „Simple autonomous sensorless generation system with wound induction machine", Proc. of IEEE International Symposium on Industrial Electronics – ISIE'04, Ajaccio, France, 4-7 May 2004, vol. 2, pp. 929- 934.

[32] G. Iwański, W. Koczara: „Island operation of the variable speed induction generator", Proc. of 4th International Power Electronics and Motion Control Conference – IPEMC'04, Xi'an, China, 14-16 Aug. 2004, vol. 2, pp. 896 – 901.

[33] G. Iwański, W. Koczara: „Sensorless stand alone variable speed system for distributed generation", Proc. of 35th Annual IEEE Power Electronics Specialists Conference – PESC'04, Aachen, Germany, 20-25 June 2004, vol. 3, pp. 1915 – 1921.

[34] G. Iwański, W. Koczara: „Control system of the variable speed autonomous doubly fed induction generator", CD Proc. of Power Electronics and Motion Control Conference – EPE-PEMC'04, Riga, Latvia, 2-4 Sept., 2004.

[35] G. Iwański, W. Koczara: „Method of Voltage Control of Variable Speed Sensorless Three Phase Four Wires Autonomous Generation System Operating in Unbalanced Load Conditions", Proc. on Intelligent Power Conversion and Intelligent Motion Conference – PCIM'06, Nurnberg, Germany, 30 May – 1 June, 2006, pp. 461-466.

[36] W. Koczara, B. Kamiński, G. Iwański: „Prototyping of Double Fed Induction Machine based Power Generation Systems Using High Performance DSP/FPGA Controller", Proc. on Intelligent Power Conversion and Intelligent Motion Conference – PCIM'06, Nurnberg, Germany, 30 May – 1 June, 2006, pp. 421-426.

[37] G. Iwański, W. Koczara: „Grid Connection to Stand Alone Transitions of Slip Ring Induction Generator During Grid Faults", CD Proc. of International Power Electronics and Motion Control Conference – IPEMC'06, Shanghai, China, 13-16 Aug., 2006.

[38] G. Iwański, W. Koczara: „Positive and negative sequence based sensorless control for stand alone slip ring generator", CD Proc. of Power Electronics and Motion Control Conference – PEMC'06, Portoroz, Slovenia, 30 Aug. – 1 Sept., 2006.

[39] G. Iwański, W. Koczara: „Synchronization and Mains Outage Detection for Controlled Grid Connection Processes of the Wind Driven Variable Speed Power Generation System", ICCEP'07, Capri, Italy.

[40] G. Iwański, W. Koczara: „Extended Direct Voltage Control of the Stand-Alone Double Fed Induction Generator", POWERENG'07, Setubal, Portugal.

[41] Iwański G., W.Koczara: "Sensorless Direct Voltage Control of the Stand-Alone Slip-Ring Induction Generator" IEEE Transactions on Industrial Electronics.

[42] Tomasik J.: Autonomous Power Generating System with Multi-level Converters. PhD thesis, Warsaw University of Technology 2007.

[43] Kaminski B.: Bidirectional switch neutral point clamped three-level inverter with sinusoidal voltage output. PhD thesis, Warsaw University of Technology 2006.

[44] Wejrzanowski K.: Active rectifier providing two symmetrical DC voltages. PhD thesis, Warsaw University of Technology 2007.

[45] Quality of Power Produced by Variable Speed Generation System. Włodzimierz Koczara, Emil Ernest, Nazar Al-Khayat Proceedings of PEMC 2004 Conference.

[46] Angela Macdonald-Smith, Bloomberg News July 9, 2006.

[47] Oil Proce History and Analysis. http://www.wtrg.com/prices.html

THE AUTHOR

Wlodzimierz Koczara, PhD, Habil. Professor of Warsaw University of Technology (WUT), Institute of Control and Industrial Electronic. Since 1981 head

of Electrical Drive Division WUT Specialization: Permanent magnet machine sensorless control, energy saving drives, variable/adjustable speed electrical power generation, renewable energy sources and power conditioning, DFIG & ADFIG.

Member of the EPEA, Institutions of Polish Electrical Engineers and IEEE. His works were awarded by European Union Program in range of Excellence Centre and by European Power Electronic Center awarding his achievements as Centre of Competence. He promoted 16 PhD and currently supervises 10 PhD thesis.

Intelligent High-Voltage High-Power Ballast IC

Jong Tae Hwang*, Dae Ho Kim, Moon Sang Jung, Gye Hyun Cho
Power Conversion Product Line
Fairchild Semiconductor
82-3, Dodang-dong, Wonmi-gu, Bucheon-si, Gyeonggi-do, Korea
Email*: jthwang@fairchildsemi.co.kr

Abstract— The ballast IC which includes 625V high-side gate driver, lamp control function, protection circuitry and power MOSFETs for the half-bridge is proposed using a cost-effective 9SIP package. This IC detects output swing of the inverter and automatically controls the switching time to meet ZVS operation. The additional benefit of active ZVS control, the IC can protect the system from an open load or abnormal load conditions. When it drives two 35W lamps, the efficiency is 89.3% including the power consumption of a PFC circuit.

I. INTRODUCTION

Electronic ballasts show better performance in luminous efficiency, efficiency, size and audible noise reduction comparing with conventional electromagnetic ballasts. Most popular electronic ballast is based on self-oscillation. It is possible to high frequency operation and moreover lowest cost approach is possible. However, it is hard to control the oscillation frequency according to the lamp condition. Such disadvantages can be solved using ballast control IC. However, conventional IC requires external high-side switch driver and control circuitry. Increased total build-cost has been made barrier for the IC to penetrate in ballast system.

Considering cost-conscious ballast market, highly integrated system-in-package smart ballast IC is proposed. It contains 600V/11A MOSFETs for half-bridge inverter and 600V high voltage control IC including smart functions such as filament preheating, lamp ignition, active ZVS control, open lamp detection, ignition fail detection and thermal shutdown function in 9SIP(single inline package) as shown in Fig. 1.

This chip has only two control pins: one is for resistor RT to control the oscillation frequency and the other is for capacitor CPH to adjust preheating, ignition time and ZVS control loop. Owing to the minimized control pins, it is easy to use and can reduce total cost.

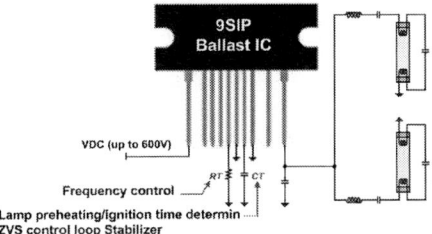

Fig. 1. 9SIP ballast IC.

II. LAMP CONTROL

A. Lamp Preheating & Ignition

Ls, Cs, Cp and the lamp configure the resonant tank as in Fig. 2 and this configuration is called LCC resonant tank [1]. The LCC is possible to get the high gain to low gain according to lamp's equivalent impedance, RL. Also, resonant frequency is dependent on RL. To ignite the fluorescent lamp, sufficient high voltage must be applied to the lamp. After the ignition, the lamp must be driven lower voltage than ignition voltage. These requirements can be easily obtained using LCC network and adjusting operating frequency as well. Thus, the LCC is one of the best choices for the fluorescent lamp drive.

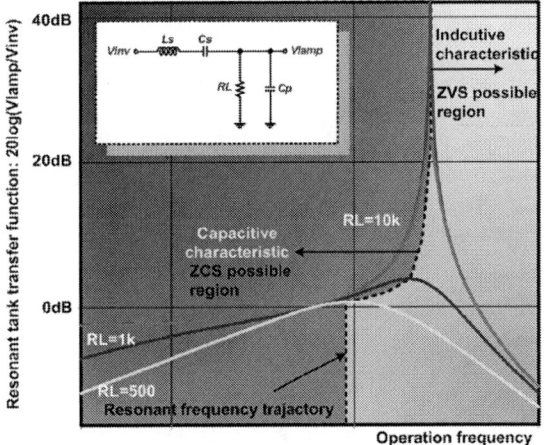

Fig. 2. LCC resonant tank.

The lamp impedance shows quite nonlinear characteristic. To effectively drive the lamp, it is required to understand the impedance change before and after ignition. Fig. 3 shows how to ignite lamp using ballast IC. When the IC starts to drive the lamp, the lamp impedance, RL, is very high. Besides, the switching frequency is higher than normal running frequency. Consequently, almost current flows through the filaments of the lamps and Cp as depicted in the figure. During the preheating period, the ballast IC drives the constant preheating frequency and warms up the filament for easy turning on of the lamp. Also, warm-up time helps to increase the lifetime of the lamp. After the preheating period, IC decreases the oscillation frequency. At this time, the lamp voltage, *Vlamp*, is increased since RL is still high. Normally during the ignition period, the lamp is turned on. Once the

978-1-4244-1871-8/07 $25.00 © 2007 IEEE

lamp is ignited, RL decreases rapidly. Therefore, almost current out of the IC flows through the tube of the lamp. A small portion of residual current flows through Cp.

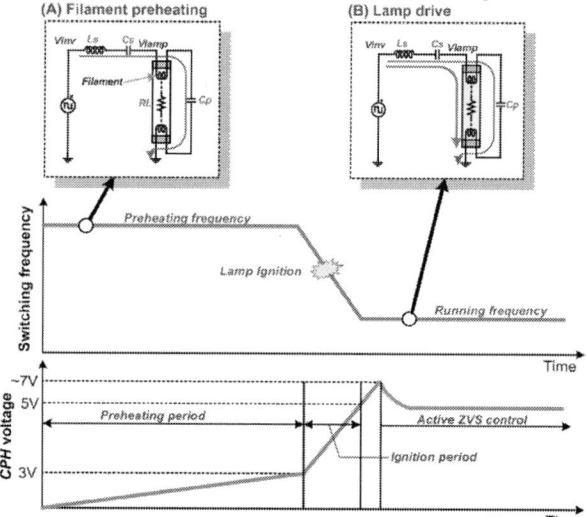

Fig.3 Filament preheating & Lamp ignition.

The preheating and ignition periods are determined by where the CPH voltage is located in. Until CPH voltage is below 3V, the oscillation frequency is fixed to higher frequency than the running frequency. When CPH voltage lies in 3V to 5V, the oscillation frequency is changed like the figure in accordance with CPH voltage variation. Once CPH is reaches about 7V, the oscillation frequency is fixed to running frequency rather than affected by CPH voltage. After that voltage, the IC operates in active ZVS mode and CPH voltage level determines the dead-time.

B. On-chip Lamp Protection

Since the ballast system is based on the resonant inverter technique, tank characteristic and operating frequency heavily govern the performance of the ballast system. Those are also affected by the functions of many factors and uncertainties. One of them is the temperature. At high temperature, the inductor wounded on ferrite core has the tendency to easy to saturate at low current level. Let see Fig. 4. To understand the inductance variation over the temperature, the inductor is driven by the pulse gated MOSFET. Therefore, the inductor current was built like the figure when the MOSFET was turned on. Since the inductor is designed not to fall into saturation until the current reaches 1A, the slope of the current must be constant below 1A level. However, as show in the figure, the slope became steeper below 1A level at the high temperature. The point of inflection becomes lower as the temperature goes up. Considering the high temperature circumstance, which the ballast circuit operates, it is undesirable.

The inductance degradation can change the resonant peak frequency like Fig. 5. In the simulation, we assumed that the inductance changed from 2.5mH to 1.4mH. If the inductance changes to 1.4mH, the ballast drives the capacitive load and the temperature of the half-bridge MOSFET increases

rapidly. It could also cause the thermal burnt of the MOSFETs.

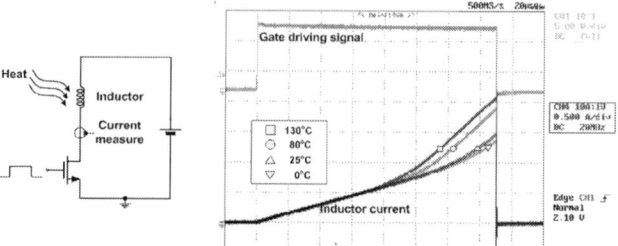

Fig.4. Example of inductor saturation problem in high temperature.

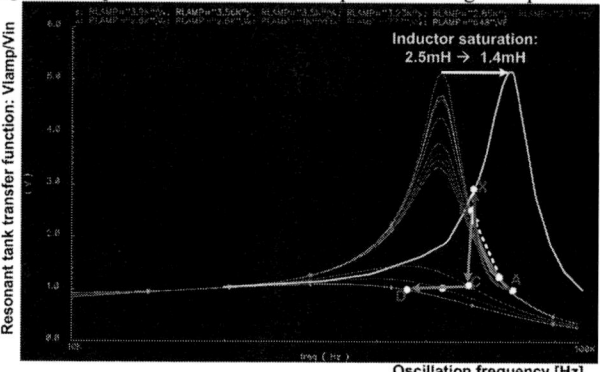

Fig.5. Inductor saturation problem in high temperature.

Fig. 6(a) shows normal ignition process. The preheating time is enough to warm to the filament to ignite the lamp during ignition period. In this case operating points moves from A to C. In all cases, the operating frequency is always higher than the resonant frequency. Therefore, the inverter can meet ZVS or Quasi-ZVS operation at least. But, at the high ambient temperature, the capacitance of CPH, which is connected to the IC and determines preheating period, can be decreased because of the temperature dependency of the given capacitor. It decreases the preheating time like shown in Fig. 6(b).

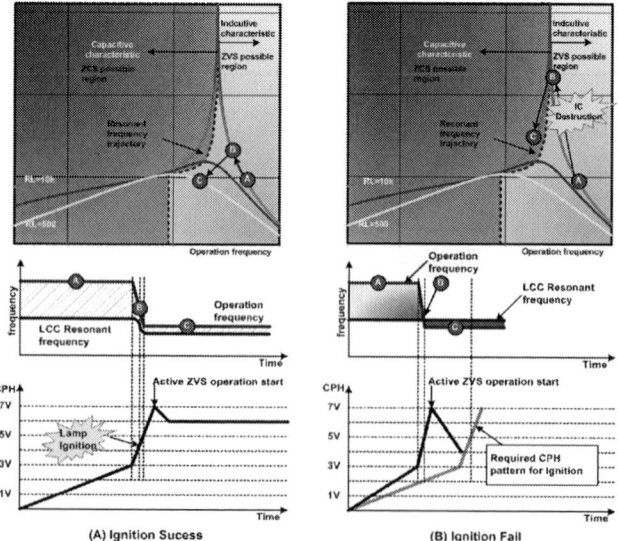

Fig. 6 (a) Normal ignition process, and (b) Ignition fail.

978-1-4244-1871-8/07 $25.00 © 2007 IEEE   54

Thus, the lamp could not turn on during ignition period. It is very dangerous for the switches because the quality factor of the LCC is still high due to the high lamp impedance. At B point, the large inductor current can destroy MOSFET switches.

In these two cases, switching devices can be permanently damaged since temperature of switches increases rapidly. To prevent damage of the switching devices, on-chip hard switching protection circuit is implemented. When ZVS is accomplished, inverter output transits during dead-time. However, when hard-switching is occurred, output changes according to the driving signals of power devices. To detect the switching condition, D-flipflop is used as shown in Fig. 7. When hard-switching continues, the IC ceases operation to protect the switching devices.

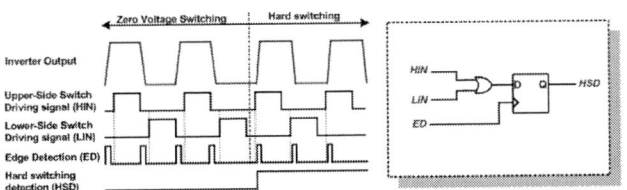

Fig. 7. Hard switching detector: (left) waveforms of ZVS and hard-switching and (right) detection circuit.

III. ACTIVE ZVS CONTROL LOOP

A. Dead-time control loop

Fig.8 shows the schematic of the intelligent ZVS control loop, which drives LCC resonant circuit. The edge detection circuit detects the output transition. REDGE and FEDGE signals represent the duration of the rising and falling edges of the output, respectively. Using high and low side driver's inputs, HSDRV and LSDRV, and those edge detection signals, the DFFs determine whether the dead-time is enough or not to meet the ZVS condition. When hard-switching or quasi-ZVS conditions are detected, the one of the DFFs' outputs become high and have CPH discharged. Thus, VC voltage decreases. The dead-time, which is made by the dead-time controller from the oscillator's output, is inversely proportional to VC. Thus, If the determined dead-time by the dead-time control circuit is too short to meet ZVS condition, CPH is discharged and the dead-time is increased.

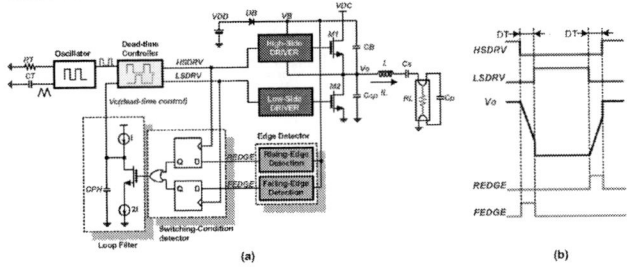

Fig. 8. Active ZVS control circuit.

B. Feedback Loop Modeling

To build an equivalent circuit, the target of the feedback control must be defined. The objective of the control is to meet ZVS condition. Therefore, Ccp, which is connected on the inverter's output, must be charged from ground level to VDC during the dead-time or vice versa. Consequently, it is reasonable to select the charge, Qref, as a target value, which satisfies following equation:

$$Qref = VDC \cdot Ccp \qquad [1]$$

The feedback loop compares the total charge (Qsense) of Ccp by the inductor current during the dead-time. The latched comparator compares Qref and Qsense at every rising instance of HSDRV and LSDRV. If Qsense is smaller than Qref, the half-bridge does not meet ZVS. Therefore, the comparator makes the switch of the loop filter turned on and decreases Vc to increase the dead-time. Otherwise, Vc is increased by the constant current source and dead-time is decreased.

To finish the modeling, the transfer function between Qsense and DT must be found including the resonant tank characteristic. Fig.9 shows waveforms of inductor current and inverter's output voltage. In Fig. 9(a), the area of S is the total charge of Ccp, which is supplied by the inductor current during the dead-time. For easy calculation, the inductor current is represented as Fig. 9(b). Assuming the dead-time is shorter than the oscillation period and the current variation around ϕ is almost linear, the total charge, Qsense, is approximately given as

$$Qsense = \int_{\phi/2\pi f - DT/2}^{\phi/2\pi f + DT/2} IL(t)dt = \frac{Io}{\pi f}\sin\phi \cdot \sin(\pi f \cdot DT) \approx DT \cdot Io \cdot \sin\phi$$

$$[2]$$

, where Io is the peak of the inductor current, which is determined by the dc-link voltage, VDC, and the impedance of the LCC resonant tank including the equivalent lamp's impedance, and ϕ is the phase delay between Vo and the inductor current.

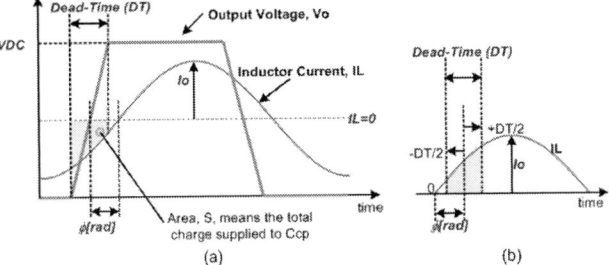

Fig. 9. inverter voltage and inductor current.

Io and ϕ are almost fixed since the tank impedance can be regarded as a constant when the lamp is operated in the steady state. Therefore, if VDC and driving frequency sustain fixed values, those are also constants. Consequently, as depicted in Fig. 10, the poles of the LCC resonant tank do not affect the loop stability. Moreover, the completed loop has the scheme of the Bang-bang controller. Therefore, the loop meets stability unless the lamp impedance is totally different from a desired value due to filament-open, an old lamp closed to its intrinsic lifetime and lamp-broken.

978-1-4244-1871-8/07 $25.00 © 2007 IEEE

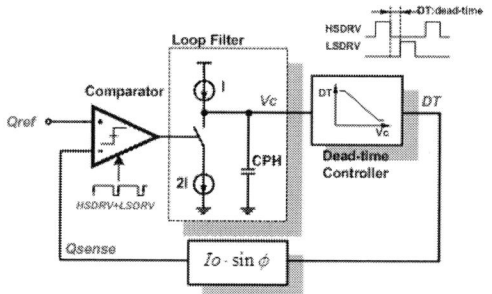

Fig. 10. Equivalent circuit of the active ZVS control loop.

IV. IMPLEMENTATION

A. High Voltage Process

625V BCDMOS process shown in Fig. 11 can be divided into low and high voltage circuit regions. The devices such as MOSFETs, BJT, capacitors and resistors in each region guarantees 25V breakdown voltage. However, the high voltage circuit region can be isolated up to 600V offset voltage. To satisfy 600V offset voltage constraint, the high-side circuit is conceptually incorporated in the cathode area of the high voltage diode, which is comprised by highly resistive P-substrate (PSUB), N-buried layer (NBL) and N-epitaxial layer (NEPI). To horizontally isolate this diode from other devices, a high voltage isolation (HVI) structure based on the RESURF technique is developed using selective NBL in HVI.

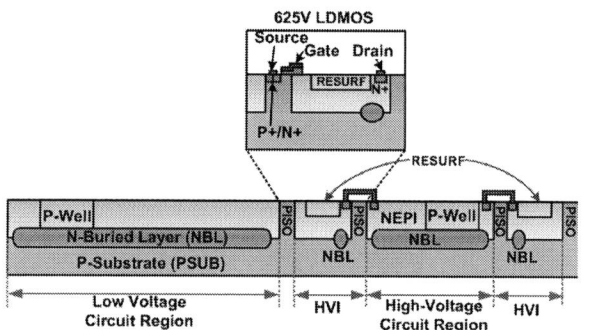

Fig. 11. 625V High Voltage Junction Isolation Technique and High Voltage LDMOS using selective buried layer technique.

Owing to the isolation technique using PISO, it is possible to use a part of HVI as a drain of LDMOS [2]. Since a conventional process has been used additional LDMOS at the outside of high voltage area, a high voltage connection method between LDMOS and the high voltage circuit area such as field plating technique [3] is needed. This process is area effective and frees from high voltage connection problem as well. Another advantage of NBL is that it gives robustness to the latch-up event. Consequently, this process shows stable operation against a capacitive coupled noise and the displacement current due to the high dv/dt noise generated by the half-bridge output.

B. Circuit Configuration

Fig. 12 shows the block diagram of the IC. The oscillation frequency of the oscillator is controlled by RT. From the oscillator ouput, the dead-time control circuit makes the dead-time varies from 1us to 3us. The ratio between dead-time and CPH voltage is about 1us/1V. If CPH voltage is below 2V, the IC falls into shutdown mode.

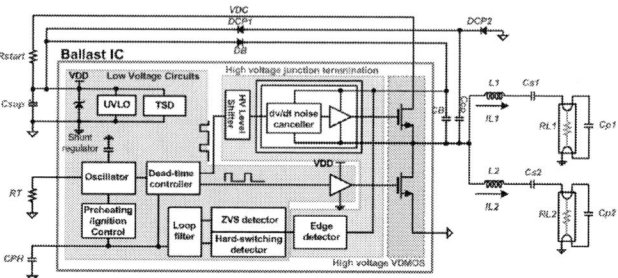

Fig. 12. Block diagram of the IC.

The CPH capacitor is also used to determine the preheating and ignition times. Until the voltage of CPH is below 3V, CPH is charged by 2uA current source. Between 3V to 5V, CPH is charged by 10uA. At preheating period, the oscillation frequency is fixed to 1.3 times of a normal running frequency. When CPH voltage is between 3V to 5V, the oscillation frequency is inversely proportional to the CPH voltage.

This IC senses output transition without an external device and determines the switching conditions such as ZVS, quasi-ZVS and hard-switching. In both quasi-ZVS and hard-switching conditions, CPH voltage decreases to meet ZVS condition.

The outputs of dead-time controller are used as the inputs of high-side and low-side gate driver. In high-side driver, it has dv/dt noise cancellation circuit [4]. Owing to this circuit, it can guarantee +/-50V/ns of dv/dt noise. The micro-photo of the designed controller is depicted in Fig. 13.

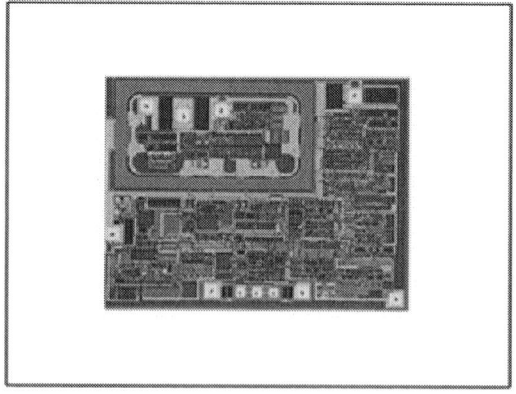

Fig. 13. Micro-photo of the controller.

978-1-4244-1871-8/07 $25.00 © 2007 IEEE

C. Package

Fig. 14 presents the 9SIP package. It has three separate islands for 2 MOSFETs and a control IC. The dimension of the 9SIP is only 25.5x10x3mm³ excluding leads. This package is fully packed by molding compound. It is easy to attach to heat-sink and can reduce fabrication cost.

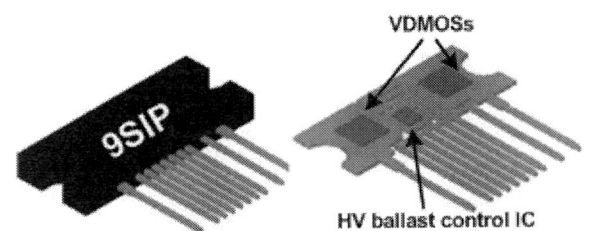

Fig. 14. 9SIP package.

V.   TEST RESULTS

The IC is tested with PFC circuits as shown in Fig. 16 test board. Thus, applied voltage to the IC is set to 400V from 210Vac input. To check the efficiency and temperature of the IC, two 35W fluorescent lamps were used. When input power is 68.26W, lamp power is 60.96W. Thus, the efficiency of the ballast system is 89.3%. Since the supplied power includes the dissipation of PFC circuitry, the efficiency of ballast IC itself exceeds 90%.

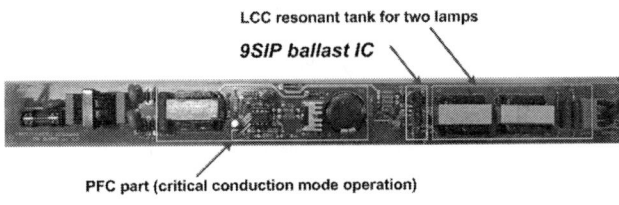

Fig. 15. Ballast IC test board.

The temperature of the package ranked 51.5°C at 25°C room temperature at above test condition. Fig. 14 shows infrared image of the package. When discrete MOSFET has same current rating is used with control IC, its temperature was 41 °C. But, considering 9SIP contains two MOSFET, thermal performance of the package is good enough to use in ballast system. In table I, performance of the IC is summarized.

Fig. 17 shows lamp removal test. When the IC starts, two lamp are connected to the ballast system. Thus, two inductor currents charges and discharges Ccp during the dead-time. Few seconds later, one lamp is removed. From this time, CPH voltage decreased to meet ZVS condition. When both lamps are removed, hard-switching is occurred, thus, peak pulsating current makes noise in waveform. Since the IC detected hard-switching condition, it automatically shuts down whole system and protects switching devices.

Fig. 18 shows dimming control of the lamp. To change the brightness of the lamp, the oscillation frequency is changed. As increasing the oscillation frequency, the lamp current is decreased as the figure. If the oscillation frequency

is changed, the required dead-time for ZVS is slightly changed. The IC tracks such variation and changes the dead-time.

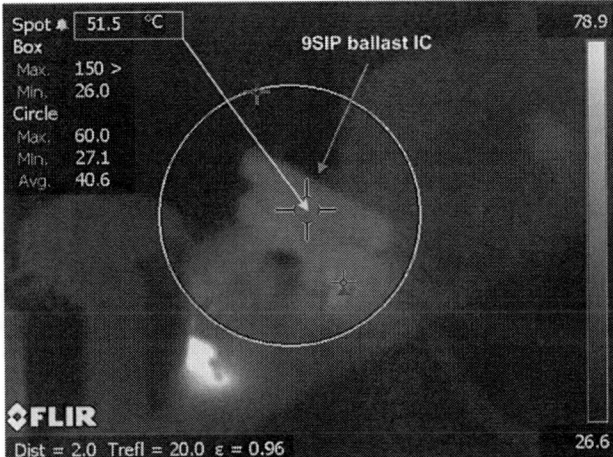

Fig. 16. Thermo image of 9SIP when delivers 61W to two 35W fluorescent lamps.

TABLE I. Performance Summary

		Test board
Steady state	Input power [W]	68.26
	Output power [W]	60.96
	Lamp voltage [V]	133
	Lamp current [A]	0.229
Preheat state	Filament voltage [V]	1.08
	Filament current [A]	0.197
	Consumption power [W]	0.87
Preheating time	Time[s]	0.8
Glow state	Lamp voltage [V]	136.6
	Lamp current [mA]	5
Strike state	Ignition Voltage (pk)	562
	Ignition Voltage (rms)	391

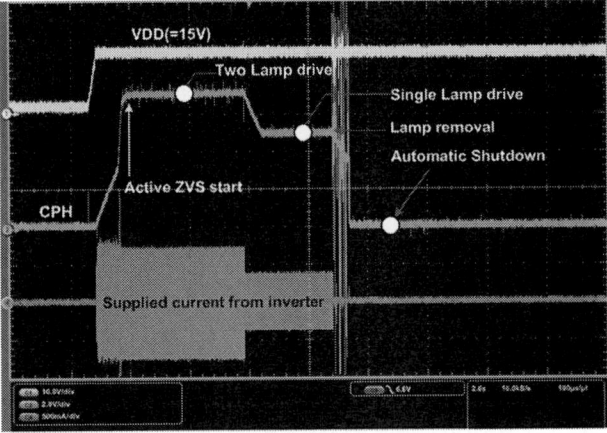

Fig. 17. Lamp removal test.

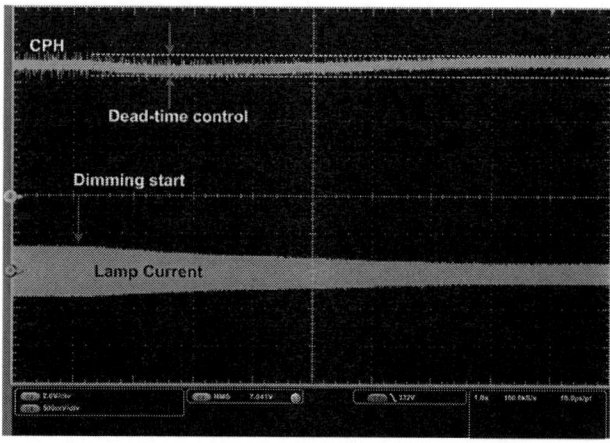

Fig. 18. Lamp Dimming and dead-time control.

VI. CONCLUSION

Based on 9SIP package, highly integrated system-in-package concept ballast IC is proposed. It includes 600V rate high-side gate drive, control functions and 600V/11A rate MOSFETs. To reduce the system build-cost, additional smart functions are implemented without requiring external components such as: (1) preheating control, (2) ignition control, (3) active ZVS control, (4) open lamp protection, (5) ignition fail protection and (6) thermal protection. The proposed IC successfully drives two 35W fluorescent lamps without a heat-sink. Even though the test vehicle does not use the heat-sink, the package temperature settles down 51°C at the room temperature.

REFERENCES

[1] Melvin C. Cosby, Jr, R.M. Nelms, "A Resonant Inverter for Electronic Ballast Applications," IEEE Trans. Industrial Electronics, VOL.41, No. 41, pp.418-425, Aug. 1994

[2] S.L.Kim, C.K. Jeon, M.H. Kim, and J.J. Kim, "Realization of Robust 600V High-side Gate Driver IC with a New Isolated Self-Shielding Structure," Proc. ISPSD 2005.

[3] Anthony F.J. Murray, William A. Lane, "Optimization of interconnection-induced breakdown voltage in junction isolated IC's using biased polysilicon plates," IEEE Trans. Electron Devices, Vol. 44, No.1, pp. 185-189, Jan. 1997.

[4] Jong Tae Hwang, "High Noise Immunity High-Side Gate Driver ," Power System Design Europe, pp. 24-28, May 2005.

Thermal analysis of PT IGBT by using ANSYS

Sehwan Ryu
Konkuk University
Department of Electrical Engineering
1 Hwayang-dong, Gwangjin-gu, Seoul, 143-701
Email: ryu382@konkuk.ac.kr

Hyungkeun Ahn
Konkuk University
Department of Electrical Engineering
1 Hwayang-dong, Gwangjin-gu, Seoul, 143-701
Email: hkahn@konkuk.ac.kr

Deukyoung Han
Konkuk University
Department of Electrical Engineering
1 Hwayang-dong, Gwangjin-gu, Seoul, 143-701
Email: dyhan@konkuk.ac.kr

Mahmoud El. Nokali
University of Pittsburge
Department of Electrical Engineering
Pittsburgh,PA 15260, USA
Email: elnokali@ee.pitt.edu

Abstract—As the power density and switching frequency increase, thermal analysis of power electronics system becomes imperative. The analysis provides valuable information on the semiconductor rating, long-term reliability and efficient heat-sink design. In this paper, the thermal model and thermal distribution of discrete Insulated Gate Bipolar Transistor with non heat-sink and heat-sink has been studied. For analysis of thermal distribution, we obtained results by using finite element simulator, ANSYS and compared with experimental data.

I. INTRODUCTION

Nowadays, the insulated gate bipolar transistors (IGBTs) are widely used in power systems for high switching frequency and medium power ranges [1]

Electro-thermal effect for IGBT is becoming more and more important. Therefore the demand for faster and more accurate thermal models is increasing [2].

When turn-on or turn-off the switch, the transient process takes time and the device consumes power while both the voltage and current are non-zero. Power losses inside an IGBT can restrict its use and therefore it becomes an important factor. The power losses work as a heat source inside the semiconductor device and this heat will elevate the junction temperature and change the temperature profile inside the device.

Thermal resistor-capacitor networks are widely used for the thermal analysis because they are easy to integrate into existing circuit simulators. The RC thermal model is flexible and can be used describe one-dimensional [3], two-dimensional [4], or three-dimensional [5] problems. The model can be built through the discretization of the thermal conduction equation by using either finite difference or finite element method (FEM)[6].

In this work, we used 3-D FEM software ANSYS [7] to implement the thermal analysis of discrete IGBT. The 3-D simulation resulted in the thermal distribution of IGBT with non heat-sink and heat-sink. To prove simulation result, we compared with experimental data obtained by using thermo-CAM.

II. THERMAL MODEL

Figure 1 shows the schematic structure of the discrete IGBT. The silicon layer is the active device part, while the substrate is Cu and the package material is EMC.

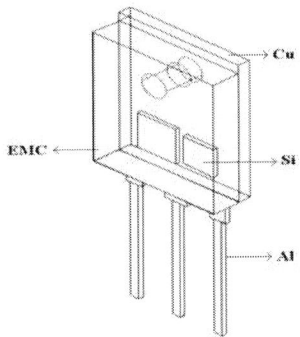

Fig. 1: Schematic of the discrete IGBT

The material properties used for simulation are listed in table 1.

TABLE 1
Material properties used for simulation

Material	K[W/m-K]	C[J/Kg-K]	ρ[Kg/m³]
Si	153	703	2340
Cu	395	385	8960
EMC	0.73	273	2270
Al	237	910	2702

The model uses 8-nodeed 3-D thermal solid elements (SOLID 60). The finite element mesh is constructed to obtain 7 mesh size. The final structure mesh of the discrete IGBT with non heat-sink is shown in figure 2. The final structured mesh has 53299 elements and 10765 nodes. Inside IGBT, the dominant heat transfer mechanism is thermal conduction

978-1-4244-1871-8/07 $25.00 © 2007 IEEE

between different layers, convection and radiation of the different materials to ambient is ignored.

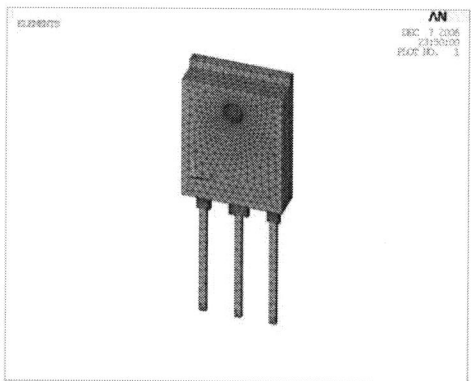

Fig. 2: Mesh structure of discrete IGBT

Although the properties of the all materials are temperature –dependent, only the thermal conductivity of silicon varies sufficiently over the temperature range. The temperature dependent thermal conductivity of silicon is listed in table 2.

Since the device active region results in most of the heat generation, and because the active region is small compared to the device dimension, the electrical power losses will be treated as surface load which is applied uniformly on the silicon chip surface.

TABLE 2

Temperature dependent thermal conductivity of silicon

Temperature[℃]	Conductivity, K[W/m-K]
25	153
77	119
127	98.9
227	76.2

III. SIMULATION RESULTS

The temperature distribution of IGBT with non heat-sink is shown in figure 3, 4. The temperature ranges are about 28℃ to 105℃. In the part of silicon chip, maximum temperature is obtained.

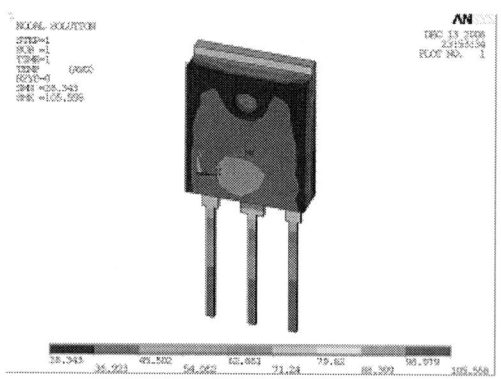

Fig. 3: the Front side of thermal distribution of IGBT

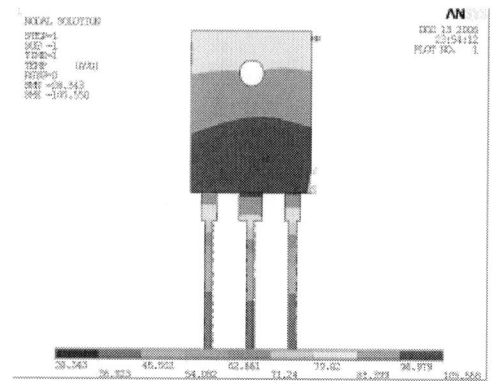

Fig. 4 the back side of thermal distribution of IGBT

Fig. 5: the front side of thermal distribution of IGBT obtained by Thermo-CAM

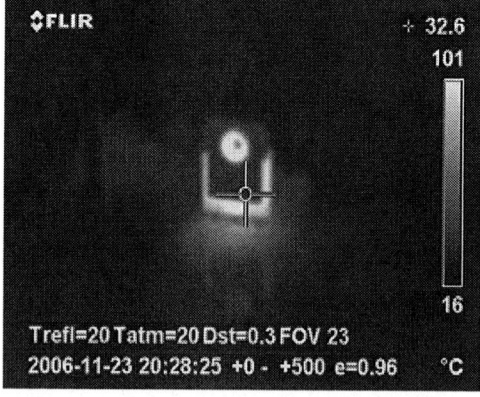

Fig. 6: the back side of thermal distribution of IGBT obtained by Thermo-CAM

Figure 5, 6 show thermal distribution of IGBT obtained by Thermo-CAM. The temperature ranges are about 17℃ to 122℃ and maximum spot point temperature is 122℃.

Figure 7 shows thermal distribution of IGBT connected to heat-sink which surface area is 113.58 [cm²]. The temperature ranges are about 31℃ to 41℃.

The thermal distribution of IGBT through experiment is

978-1-4244-1871-8/07 $25.00 © 2007 IEEE   60

shown by figure 8, 9. The temperature rages are about 29℃ to 44℃. When heat-sink is connected and is not connected to IGBT,

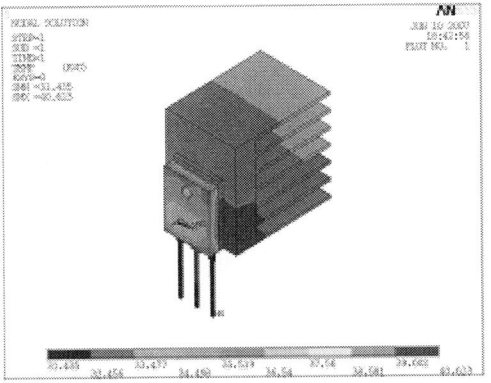

Fig. 7: Thermal distribution of IGBT connected to heat-sink

Fig. 8: : the front side of thermal distribution of IGBT connected to heat sink obtained by Thermo-CAM

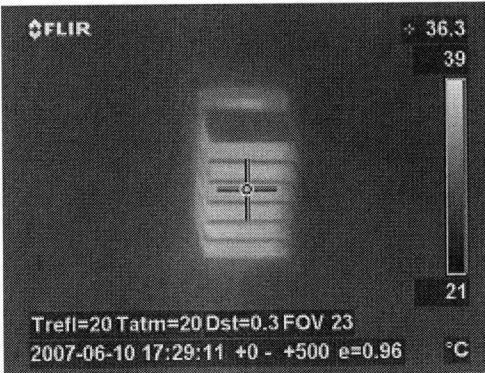

Fig. 9: The back side of thermal distribution of IGBT connected to heat-sink obtained by Thermo-CAM

IV.   CONCLUSION

In this paper, the thermal model and thermal distribution of discrete Insulated Gate Bipolar Transistor with non heat-sink and heat-sink has been studied. For analysis of thermal distribution, we obtained results by using finite element simulator, ANSYS and compared with experimental data.

We will try to study RC thermal network model for discrete IGBT and will propose optimum heat-sink model.

REFERENCES

[1] B. Jayant Baliga, "Trends in power semiconductor devices", IEEE trans. on Electron Devices, vol. 43, pp. 1717-1731, Oct. 1996.

[2] Kuang Shen, Stephen J. Finney, and Barry W, Williams, "Thermal stability of IGBT High- frequency operation", IEEE Trans. on Industrial Electronics, vol. 47, pp. 9-16, Feb. 2000.

[3] A. R. Hefner, "A dynamic electro-thermal model for the IGBT", IEEE Trans. Ind. Applicat., vol. 30, pp. 394-405, Mar./Apr.1994.

[4] Anis Ammous, Kaiçar Ammous, Hervé Morel, Bruno Allard, Dominique Bergogne, Fayçal Sellami, and Jean Pierre Chante, "Electrothermal modeling of IGBT's: application to short- circuit conditions", IEEE Trans. On Power Electronics, Vol.,15, No. 4, July 2000.

[5] T. Kikunaga and T. Ohi, "Analysis and simulation technologies for high reliability design of power modules", R&D Progress Rep., Mitsubishi, 2003.

[6] Jia Tzer and loc Vu-Quoc, "A rational formulation of thermal circuit models for electro-thermal simulation-Part I: Finite element method", IEEE Trans. on Circuit and Systems I : Fundamental and Applications, vol. 43, pp. 721-732, Sep. 1996.

[7] ANSYS Thermal Transfer Analysis, TAE SUNG S&E Inc., 2002.

The Effect of a Shielding Layer on Breakdown Voltage in a Trench Gate IGBT

Jong-Seok Lee, Ho-Hyun Shin, Han-Sin Lee,
Man young Sung[*]
Korea University
Department of Electrical Engineering
Anam-Dong, Seongbuk-Gu, Seoul, 136-701
[*]Email: semicad@korea.ac.kr

Ey-Goo Kang
Far East University
Department of Information Technology
Uemsung-gun, Chung-buk 369-851
Email: keg@kdu.ac.kr

Abstract—In this paper we introduced the shielding layer concept in order to alleviate the electric field of concentrated on the trench bottom corner. The shielded trench gate IGBT is a trench gate IGBT with a P+ Shielding layer bottom of a trench gate. By simulation results, we verified that a shielding layer reduced the electric fields not only in the gate oxide but in the p-base region. Compared with conventional trench gate IGBT, about 33% increment of forward breakdown voltages are achieved.

I. INTRODUCTION

Insulated Gate Bipolar Transistors (IGBTs) are widely used in high voltage applications such as inverters and motor driver circuits[1]. Especially, after the arrival of the trench IGBT, several research groups have studied in order to reach the demands of high breakdown voltage, low on-state voltage drop and improved switching characteristics[2, 3]. Generally, IGBTs are made with DMOS process which make the device easily influenced by diffusion process and restricted by parasitic JFET element[4-6]. However, Trench IGBTs have improved on-state characteristics compared to DMOS-IGBTs since there are no JFET elements in the device and the current flow is stable through the vertical channel, whereas the breakdown voltage of a trench IGBT is highly influenced by the concentrated electric field near the bottom of the trench gate[7-9]. For the enhancement of breakdown voltage, we proposed a shielding layer in trench bottom corner and it was verified that the shielding layer are effective by 2D simulator, MEDICI.

II. DEVICE STRUCTURE AND PROCESS

The structure of shielded trench gate IGBT is shown in Figure 1. The shielding layer consists of a heavily doped P-type(P+) region located at the bottom of the trench. The shielded trench gate IGBT structure can be fabricated by using the same process of the conventional trench gate IGBT structure with the addition of an ion implantation step to form the P+ shielding layer.

The ion implant used to form the P+ shielding layer must be performed after etching the trenches. This implantation step should not dope the sidewalls of the trenches because of the variation of threshold voltage. So ion implantation must be accomplished by doing the ion implant orthogonal to the wafer surface. Alternately, a conformal oxide can be deposited on the trench sidewalls and removed from the trench bottom using anisotropic reactive ion etching to selectively expose the trench bottom to the P-type dopant during the ion implantation.

III. BREAKDOWN CHARACTERISTIC CONSIDERATIONS

The shielded trench gate IGBT operates in the forward blocking mode when the gate electrode is shorted to the emitter by the external gate drive circuit. At low collector bias voltages, the voltage is supported by a depletion region formed on both sides of the P-base/N-drift junction. Consequently, the collector potential appears across the IGBT located at the top of the structure. This produces a positive potential at location 'A' in Figure 1, which reverse biases the junction between the P+ shielding layer and the N-drift region. The depletion region that extends from the P+/N junction pinches off the JFET region producing a potential barrier at location 'A'. The potential barrier tends to isolate the P-base region form any additional bias applied to the collector electrode. Consequently, a high electric field can develop in the N-drift region below the P+ shielding layer while the electric field at the P-base region remains low. This has the beneficial effects of mitigating the reach-through of the depletion region within the P-base region and in keeping the electric field in the gate oxide low at location 'B' where it is exposed to the N-drift region. The maximum blocking voltage of the shielded trench gate IGBT is determined by the properties of the drift region. This allows reduction of the drift region resistance close to that of the ideal case.

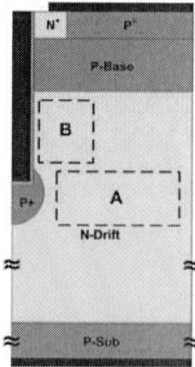

Fig. 1. Shielded trench gate IGBT structure

In addition, the reduction of the electric field in the vicinity of the P-base region allows reduction of the channel length as well as the gate oxide thickness. This is beneficial for further reduction of the device on-state resistance.

IV. DEVICE SIMULATION RESULTS

In order to gain insight into the operation of the shielded trench gate IGBT structure, two-dimensional numerical simulations were performed with a drift region doping concentration of $1 \times 10^{14} cm^{-3}$ and thickness 100 microns. The baseline device had a gate oxide thickness of 0.1 microns. The N+ emitter region and P-base region had a depth of 1 micron and 3 microns, respectively. The channel length is 2 microns. A trench width (W_T) of 1 micron was chosen with a mesa width (W_M) of 9 microns, resulting in a cell pitch of 10 microns. The P+ region located at the bottom of the 7 micron deep trench had a junction depth of 1.5 microns, resulting in a P+ shielding layer thickness of 3 microns. The doping concentration of the P+ region was $1 \times 10^{19} cm^{-3}$.

The blocking capability of the shielded trench gate IGBT structure was investigated by maintaining zero gate bias while increasing the collector voltage. It was found that the device could sustain a collector bias of upto 563 volts as shown in Figure 2.

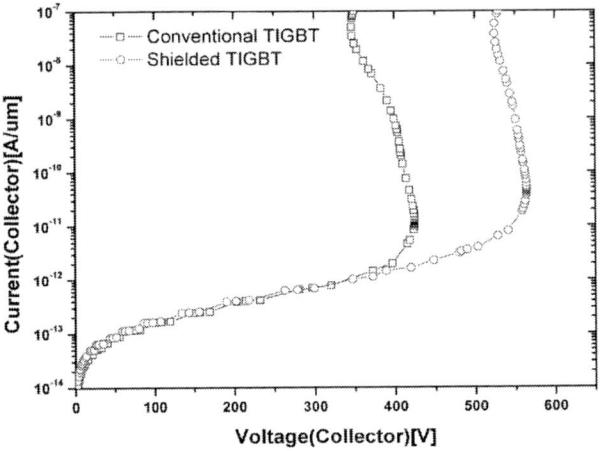

Fig. 2. Comparison of breakdown voltage characteristics of a conventional and a shielded trench gate IGBT

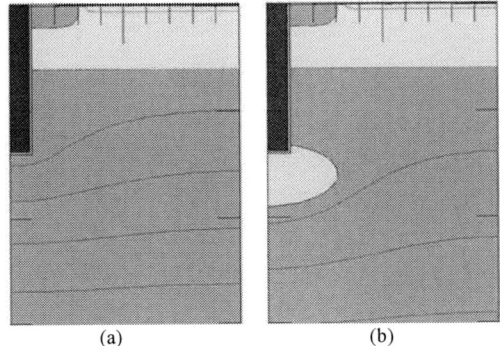

Fig. 3. Potential contour (a)conventional (b)shielded trench IGBT

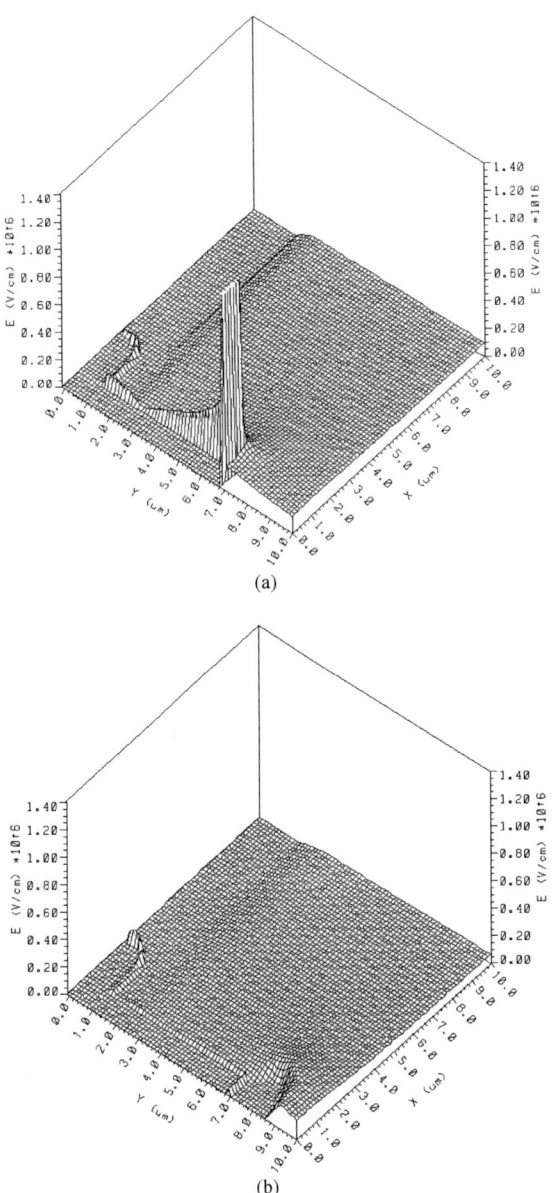

Fig. 4. 3D E-Field distribution (a)conventional (b)shielded trench IGBT

The potential distribution within the upper portion of the shielded trench gate IGBT structure is shown in Figure 3 at a collector bias of 563 volts with the gate held at zero volts. It can be seen that there is an electric field enhancement at the corner of the P+ shielding layer. This demonstrates that the presence of the P+ shielding layer under the N+ emitter region together with the formation of the JFET region suppresses the depletion of the relatively thin P-base region without the occurrence of reach-through. In addition, the potential contours are widely separated in the vicinity of the gate oxide indicating a low electric field within the gate insulator.

A serious problem with the conventional trench gate IGBT structure is associated with the high electric field generated at the gate oxide interface. In order for the channel to span the

P-base region, it is necessary for the trench to penetrate through the P-base region into the N-drift region. This exposes the gate oxide located at the bottom of the trench to the high electric field developed in the drift region. This problem is overcome in the shielded trench gate structure. Three-dimensional view of the electric field distributions for the conventional and the shielded trench gate IGBT structure is shown in Figure 4 for a collector bias of breakdown with zero bias applied to the gate electrode. The maximum electric field is generated at the junction between the P+ shielding layer and the N-drift region. The field distribution shown in this figure indicates that the electric field at the junction is close to the breakdown field strength for silicon(about 3×10^5V/cm). These results demonstrate the ability to fully utilize the breakdown field strength of the semiconductor without problems of rupture or reliability for the gate oxide.

From Figure 4, it can be seen that the electric field at the P-base region is significantly reduced when compared with the maximum electric field generated in the semiconductor.

The electric field strength of p-base region bottom is decreased about 38%.

The behavior of the electric field at x=10 microns in the cell structure with increasing collector bias is shown in Figure 5 for a conventional and a shielded trench gate IGBT.

In case of the shielded trench gate IGBT, it can be observed that the formation of a potential barrier due to the presence of the JFET region suppresses the electric field at the P-base region to about on-half of the electric field generated in the drift region. This is beneficial for preventing reach-through even with a narrow width for the P-base region. The reduction of the electric field in the gate oxide by the shielding provided by the JFET region can be observed in the electric field profile shown in Figure 6 for various collector bias voltages. This profile was taken at a depth of 4 microns from the surface at a location below the P-base region. It can be seen that the electric field in the oxide is only 6.5×10^4V/cm even when the collector bias reaches 563 volts.

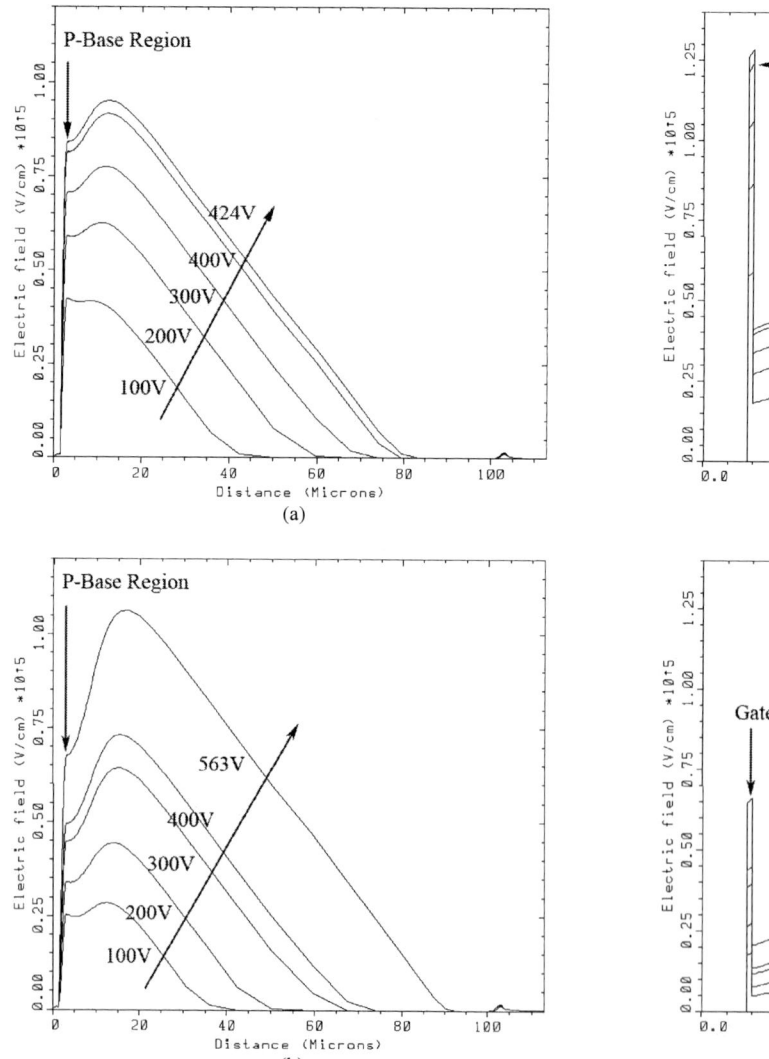

Fig. 5. 1D E-Field at x=10μm (a)conventional (b)shielded trench IGBT

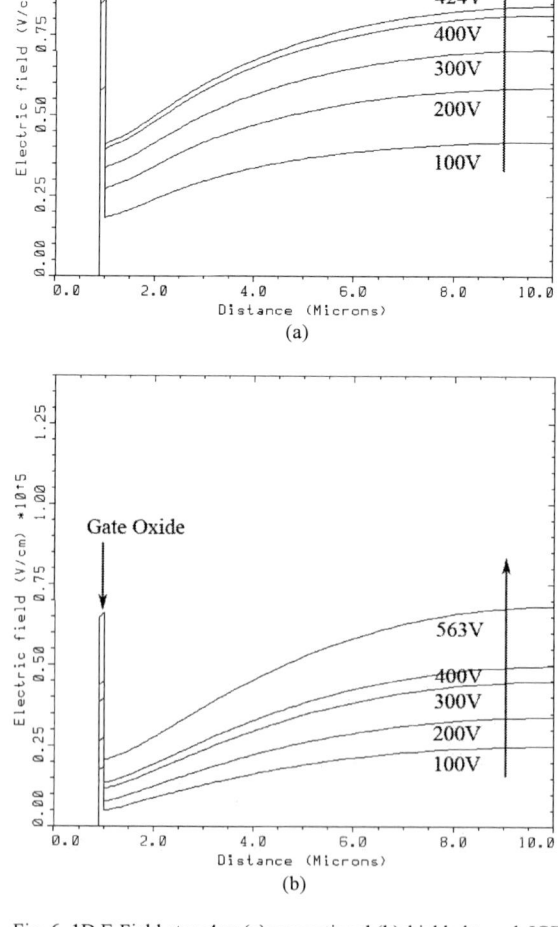

Fig. 6. 1D E-Field at y=4μm (a)conventional (b)shielded trench IGBT

V. CONCLUSIONS

The shielded trench gate IGBT structure was proposed in order to shield the P-base region and the gate oxide from the high electric field generated in the drift region during the blocking mode. The shielding is provided by the addition of a P-type layer located at the bottom of the trenches. Maximum electric fields shifted from gate oxide of trench bottom to p+/N-drift junction by the presence of shielding layer. So the electric field at the junction is close to the breakdown field strength for silicon and the problems of rupture or reliability for the gate oxide are not occur.

This P-type shielding layer presence creates two JFET regions in the trench gate structure, which can increase the on-state voltage drop unless the doping concentration in the vicinity of the trenches is enhanced. However with the enhanced doping concentration, it will be expected that the lower on-state voltage drop and the shorter channel length than that of a conventional trench IGBT can be obtained.

This shielded trench gate IGBT structure provides a very attractive for power devices. Furthermore detailed study about shielded layer(For example, doping concentration, thickness, width, shape, etc.) will give us a more useful information to increase breakdown voltage and decrease on-state voltage drop.

ACKNOWLEDGEMENT

The authors would like to thank the Ministry of Commerce, Industry and Energy in the process of Power IT "Development on Power Semiconductor for Electric Power Generation and Industrial Inverter Applications."

REFERENCES

[1] B. J. Baliga, *Power Semiconductor Devices*, 1st ed. PWS, Boston: 1996

[2] T. Laska, J. Fugger, F. Hirler, W. Scholz, "Optimizing the vertical IGBT Structure - The NPT concept as the most economic and electrically ideal solution for a 1200V-IGBT," in *Proc. 8th International Symposium on Power Semiconductor*, Hawaii, 1996, pp.169-172.

[3] F. Udrea, S. S. M. Chan, J. Thomson, T. Trajkovic, P. R. Waind, G. A. J. Amaratunga et al, "1.2kV Trench insulated gate bipolar transistors with ultralow on-resistance," *IEEE Elec. Device Letters*, vol. 20, No. 8, p. 428-430, 1999.

[4] E. G. Kang, D. S. Oh, D. W. Kim, D. J. Kim, M. Y. Sung, "A Novel Lateral Trench electrode IGBT for Superior Electrical Characteristics," *J. KIEEME*, vol. 15, no. 9, pp.758-763, Sep. 2002.

[5] E. G. Kang, S. H. Moon, M. Y. Sung, "A new trench electrode IGBT having superior electrical characteristics for power IC systems," *Microelectronics J.*, vol. 32, no. 8, pp.641-647, Aug. 2001.

[6] E. G. Kang, M. Y. Sung, "A small sized lateral trench electrode IGBT for improving latch-up and breakdown characteristics," *Solid State Electronics*, vol. 46, pp. 295-300, Feb. 2002.

[7] X. Yuan, T. Trajkovic, F. Udreak, J. Thomson, P. R. Waind, P, Taylor et al, "Suppression of parasitic JFET effect in trench IGBTs by using a self-aligned p base process," *Solid-State Electronics*, vol. 46, pp. 1907-1912, Nov. 2002.

[8] J. S. Lee, E. G. Kang, M. Y. Sung, "Improvement of Electrical Characteristics of Vertical NPT Trench gate IGBT Using Trench Emitter Electrode," *J. KIEEME*, vol. 19, no. 10, pp. 912-917, Oct. 2006.

[9] H. H. Shin, H. S. Lee and M. Y. Sung, "A Study on the Breakdown Voltage Characteristics with Process and Design Parameters in Trench Gate IGBT," *J. KIEEME*, vol. 20, no. 5, pp. 403-409, May 2007.

978-1-4244-1871-8/07 $25.00 © 2007 IEEE

Utility Interactive PV System with Improved Peak-Cut Characteristics

H.S Heo, G.H Choe, Y.H Choi
Konkuk University
Department of Electrical Engineering
1, HwaYang-Dong, Gwangjin-Gu,
Seoul
Email: ghchoe@konkuk.ac.kr

H.S Kim
Hanbit EDS Corporation
Department of Research &Development
48-28, Munpyeong-dong, daedeok-gu,
Daejeon
Email: pvkim@hanmail.net

J.C Kim
Soong-Sil University
Department of Electrical Engineering
Sangdo 5-dong, Dongjak-gu,
Seoul
Email: jckim@ssu.ac.kr

Abstract—This paper describes Utility Interactive PV(UIPV) system that is able to improve peak-cut characteristic without major alteration of power circuit structure which adds energy storage device like batteries and alters software application algorithm. A new power circuit and application algorithm has been simulated to prove this.

I. INTRODUCTION

It has been being difficult to secure energy in summer because of rapid increasing of demands of energy, even though, there has been being increased amounts of the energy in the world. In addition, it is expected that the amounts of the produced energy would not be able to catch up the demands due to the serious climate fluctuations. For this reason, we have to find a solution of this shortage of the energy.

The daily peak of power consumption during summer occurs at 2:00 pm. Electrical power supply facilities should have been enlarged to cover this peak electrical demand. Hence, reducing electrical power demand in summer is intensively required to improve the efficient utilization of facilities. The characteristic of PV power generation is suitable to reduce the peak electrical power consumption in summer. However, the time of peak PV power generation does not necessarily cover the peak of electrical power demand [1][2]. So this paper introduces a new system that is able to improve peak-cut characteristic without major alteration of power circuit structure which adds energy storage device like battery and alters software application algorithm

II. BACKGROUND OF SYSTEM PROPOSED

Table.1 shows the peak time of power consumption in recent 11 years, which indicates that most of the maximum demand occurred p between 12 and 15 o'clock. This gives us the information that the peak time of power consumption did not correspond to the time of peak power generation of PV system around noon like fig.1. The utilization factor of system is defined as follows:

$$\text{Utilization factor of system}$$
$$= \frac{\text{annual generated energy } (kWh)}{\text{Rated capacity of system } (kWp) \times 24 \times 365} \times 100 \, [\%] \quad (1)$$

TABLE 1
Peak time and date of power consumption

Production Date	Peak Time
Aug. 13th,Tues, 1996	15:00
Aug. 20th,Wed, 1997	12:00
Sep.10th,Thurs, 1998	15:00
Aug. 17th,Tues, 1999	17:00
Aug. 8th,Fri, 2000	12:00
Jul. 26th,Tues, 2001	15:00
Aug. 29th,Thurs, 2002	15:00
Aug. 22th,Fri, 2003	12:00
Jul. 29th,Thurs, 2004	15:00
Aug. 17th,Wed, 2005	12:00

According to this equation, the utilization factor of system is 12[%] on the annual average, which shows that Utility Interactive PV system is converted into about 3-hour rated operation time. By chance, this result corresponded to 3-hour of between 12 and 15 o'clock at which peak power demand occurs.

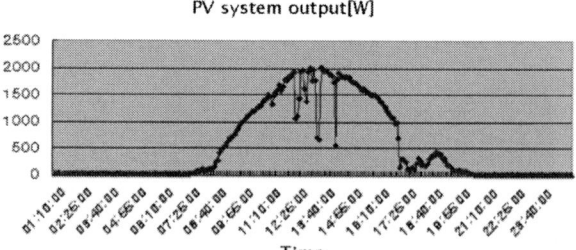

Fig 1 : Output characteristic of PV system in summer

Therefore, this study suggest proposed system which able to improve peak-cut characteristic with control PV array output using discharge and charge of energy storage system like battery.

III. CONFIGURATION OF PROPOSED SYSTEM

A. Design of power circuit

Fig. 2 shows the composition method of a proposed system which is just adding the energy storage system such as flywheel, batteries comparison to existing system.

978-1-4244-1871-8/07 $25.00 © 2007 IEEE

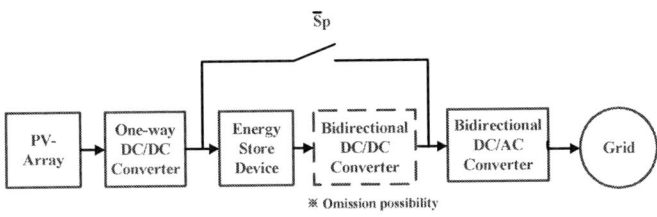

Fig. 2: Composition method of proposed system

It shows a power circuit in detail and with fig. 3 it is same. The one-way DC-DC converter that plays a role in maximum power tracking delivers the PV-array output to the energy storage system, the reason why we use a boost converter is the operating point of PV system for home is lower than the DC link voltage. And the reason which uses the high-transformer, it is a possibility of making a continuous current. If an incoming current i_{dsol} is discontinuous from DC-DC converter, this fact causes to fall-off in a life span of energy storage like batteries.

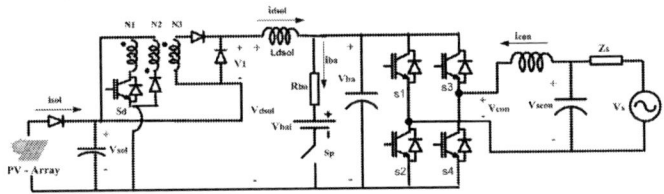

Fig.3 Proposed System which it omits bidirectional DC/DC converter

When like fig.3 it composes, current i_{dsol} will be able to control continuously, on this condition voltage v_1 can be expressed as follows:

$$V_1 = \frac{N_3}{N_1} \cdot D \cdot V_{sol} \qquad (2)$$

Where

D: Duty ratio of switch S_d

When it uses the duty ratio and turn ratio of a insulating transformer, v_{dsol} will be able to control more highly than v_{sol} as follows:

$$v_{dsol} = v_{sol} + v_1 \qquad (3)$$

The case which composes the boost converter in this way without respect to a switching loss, we obtain the power balancing equation in steady state as follows:

$$P_{sol} = V_{sol}I_{sol} = V_{dsol}I_{dsol} = V_{sol}I_{dsol} + V_1I_{dsol}$$
$$(4)$$

Where

V_{sol} : PV-array output voltage without respect to ripple in steady state

I_{sol} : PV-array output current without respect to ripple in

steady state

V_{dsol} : Average output voltage of boost converter in steady state

I_{dsol} : Average output current of boost converter in steady state

V_1 : Average voltage of v_1 in steady state

The delivered power from boost converter corresponds to V_1I_{dsol} (4), we could know that only the process which delivers this power occurs switching loss. Therefore proposed boost converter has advantage of high efficiency compare to existing method. Due to the current i_{dsol} is controlled continuously when composing the system which uses the energy storage system like batteries, it is a good composition method from life span of batteries. And the winding N2 of high-frequency transformer is for demagnetizing winding to reset the magnetizing current of high-frequency transformer during one switching cycle.

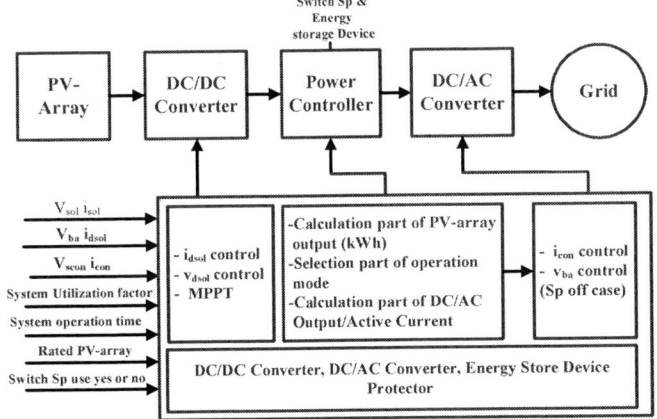

Fig.4 Overall proposed control system

Fig.4 shows the overall proposed control system of power circuit.

B. System operating modes

We can know the daily average generation time by using utilization factor of system, rated pv-array and operation time of system determined. With this, we have three operation mode depend on generation quantity of pv-array which could be calculated by (5) ; base on annual average output, in case of the same with daily pv-array output, lower case(ex, if it rains) and higher case(ex, in a clear summer day)

$$PV - array\ output = \int_0^{generation\ time} v_{sol}i_{sol}dt\ [kWh] \qquad (5)$$

① Annual average output (standard)
= daily pv-array output[kWh]

978-1-4244-1871-8/07 $25.00 © 2007 IEEE

In fact, this mode is ideal but unlikely to happen. However, it was mentioned because it is a important point to explain the operation mode of system. In this mode, daily pv-array output is the same with the standard. Therefore, DC/AC converter works during operation time which set up manually. It is described in fig. 5

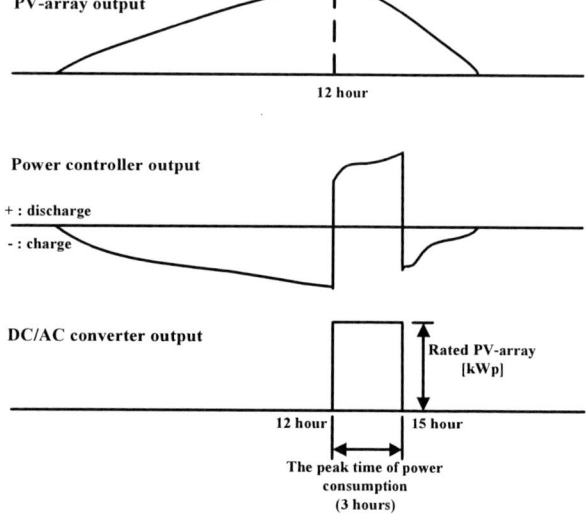

Fig. 5 : Daily PV-array output = Annual average output
(Utilization factor 12%)

② Annual average output (standard)
                    < daily pv-array output[kWh]
This mode is a case that the daily pv-array output is above a standard and operation mode is described in fig. 6.

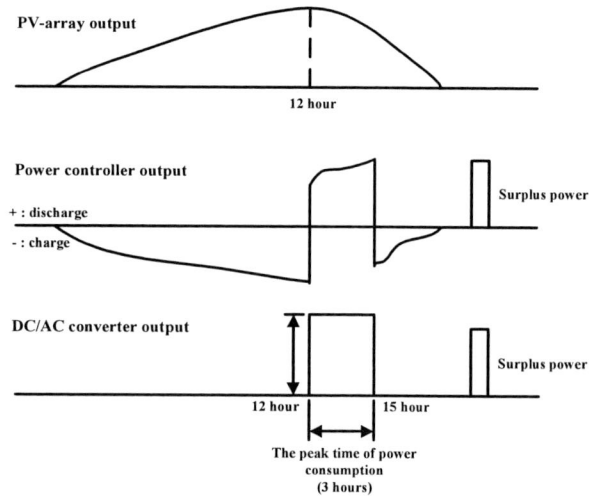

Fig. 6 : Daily PV-array output > Annual average output
(Utilization factor 12%)

Repeating this mode, the battery is likely to reach to over discharge mode when using energy storage device such ad a battery. This may make power generated from pv-array the following day unable to charge, which discharges power

quantity [kWh] as follows;

$Surplus\ power\ quantity[kWh]$

$$= \int_0^{generation\ time} v_{sol} i_{sol} dt - daily\ standard[kWh]\ (>0) \quad (6)$$

And by using the discharge time of the rest of peak time set up manually, the effective of current i_{con} is calculated as follows:

$$I_{s.dis} = \frac{surplus\ power\ quantity[kWh]}{V_{scon} \cdot discharge\ time\ of\ the\ rest\ of\ peak\ time} \quad (7)$$

③ Annual average output (standard)
                    > daily pv-array output[kWh]
This mode occurs when the weather is not good. Repeating this mode, it is unable to obtain enough power quantity needed in peak time of power consumption. Therefore, in this case by using the power control characteristic of bidirectional DC/AC converter from time of the rest of peak time (ex, at midnight that has lower utilization factor of generation facility), the following power quantity can be charged, operation mode is in fig. 7

$charge\ power\ quantity[kWh]$

$$= \int_0^{generation\ time} v_{sol} i_{sol} dt - daily\ standarad\ [kWh]\ (<0) \quad (8)$$

And by using the charge time of the rest of peak time set up manually, the effective of current i_{con} is calculated as follows:

$$I_{s.dis} = \frac{charge\ power\ quantity[kWh]}{V_{scon} \cdot charge\ time\ of\ the\ rest\ of\ peak\ time} \quad (9)$$

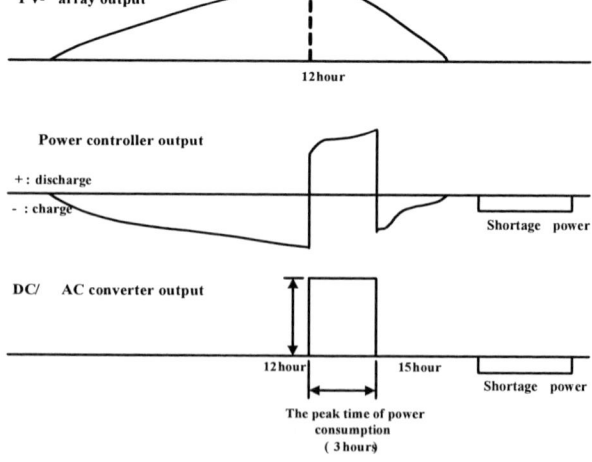

Fig. 7 : Daily PV-array output < Annual average output
(Utilization factor 12%)

④ The mode not using power controller

As mentioned above, this is a switch Sp opened mode and the same with the existing utility Interactive PV System. The peak demand of power is usually occurred in the middle of the summer. Therefore, except this time, we can assume that power reserve ratio is enough. This mode is opening the switch Sp and operating like the existing system. In this mode, operation time of DC/AC converter is the same with the generation time of pv-array.

II.   RESULTS

The characteristic of proposed system was analyzed by simulation to verify the validity. Table 2. indicates operation conditions, the battery is assumed as energy storage system.

TABLE 2
Operation condition of system

Rated PV-array	1[kWp]
Rated PV-array Voltage	340[V]
Battery rated Voltage	380[V]
Battery Voltage Change Range	357~462[V]
Battery internal resistance	0.66[Ω]
Grid Voltage	220[V]

System simulation was carried out according to the following 3-way scenario.

Scenario 1 - The charge characteristic of the case which does not generate power on the AC side

- PV-array output : 0.9 [kW]
- Battery voltage : 357 [V] (10.8*33)
- Generated power : 0 [kW]

Scenario 2 - The discharge characteristic

- PV-array output : 0 [kW]
- Battery voltage : 363 [V] (10.8*33)
- Generated power : 1 [kW] (DC/AC converter output)

Scenario 3 – The characteristic which operates simultaneously.

- PV-array output : 0.5 [kW]
- Battery voltage : 363 [V] (10.8*33)
- Generated power : 1 [kW] (DC/AC converter output)

Fig. 8 shows the charge characteristic of the case which does not generate power on the AC side (scenario 1), and the assumed initial value of pv-array voltage v_{sol} was 300 [v].

We can see pv-array voltage v_{sol} showed overshoot of about 10 [V], and then reached to steady state (output 0.9[kW], 340[V]). Working in the charging mode, it is stabilized at more than battery open voltage (357[V]) due to battery internal resistance.

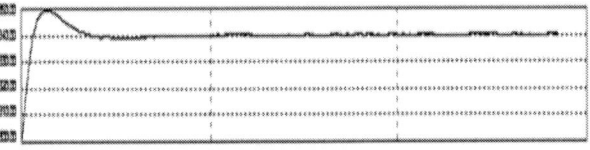

(a) PV-array output voltage (Vsol)

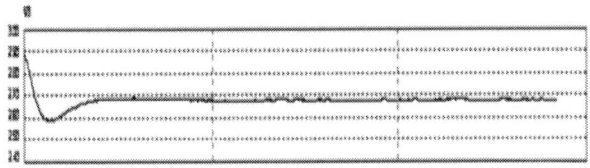

(b) PV-array output current (isol)

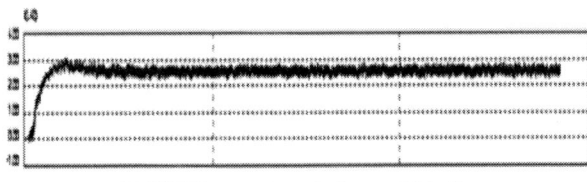

(c) DC/DC converter output current (idsol)

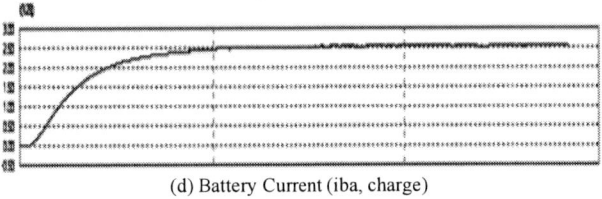

(d) Battery Current (iba, charge)

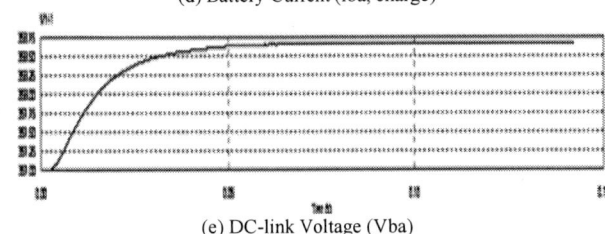

(e) DC-link Voltage (Vba)

Fig. 8: Senario-1 system waveform

Fig.9 shows The discharge characteristic when the power is not generated from pv-array (scenario 2). Because it is discharge mode, we can find battery voltage decreased in comparison whit battery open voltage (363[V]). And it has a built-in soft start function making the charge current of DC/AC converter reach to the steady state after 5 cycles. Therefore, the output increased slowly and got to the steady state after 5 cycles.

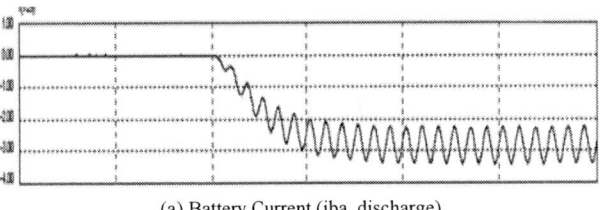

(a) Battery Current (iba, discharge)

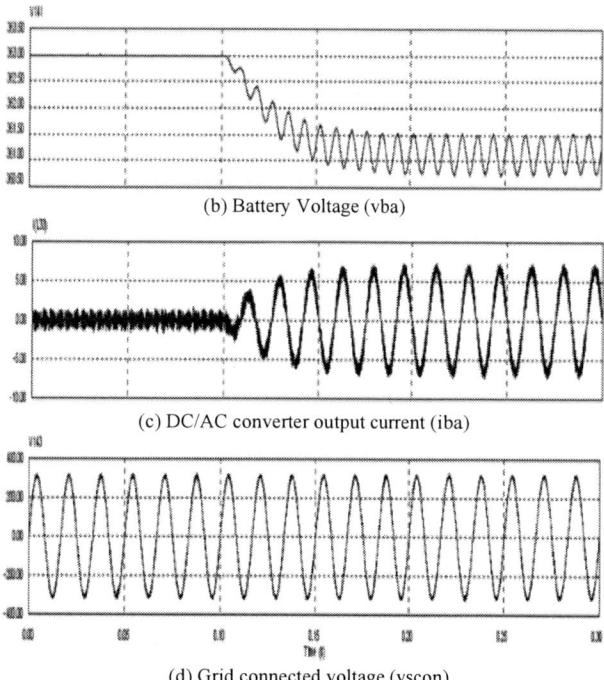

(b) Battery Voltage (vba)

(c) DC/AC converter output current (iba)

(d) Grid connected voltage (vscon)

Fig. 9: Senario-2 system waveform

Fig. 10 shows the battery of pv-array in the middle of generating power changed its mode from charge mode to discharge mode (scenario 3), and the assumed initial value of pv-array voltage was 300[V]. After pv-array voltage approached to 340[V], the charge current starts to flow and the battery voltage increases. Also DC/AC converter starts to generate power at X=0.1[sec], the battery converted its mode into discharge mode, which led to decrease battery voltage. These results prove that the proposed control system and power circuit are valid.

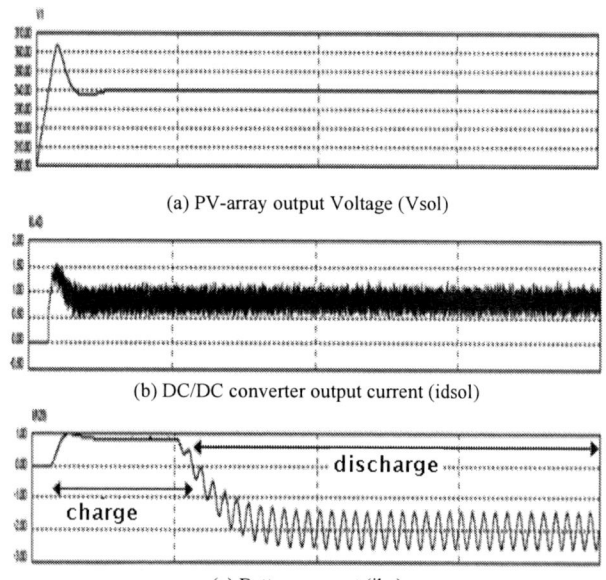

(a) PV-array output Voltage (Vsol)

(b) DC/DC converter output current (idsol)

(c) Battery current (iba)

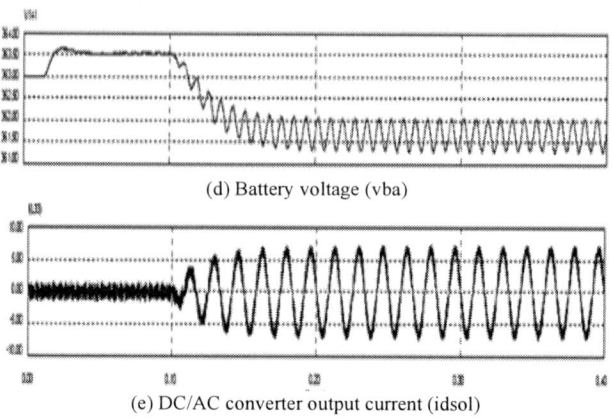

(d) Battery voltage (vba)

(e) DC/AC converter output current (idsol)

Fig. 10: Senario-3 system waveform

III. CONCLUSION

In this paper, the proposed concept has been confirmed with simulation by designing the power circuit and applying new operation algorithm which sends out stored power in battery and pv-array power together at the peak time of power consumption. What we wish to show in this paper, proposed system is able to improve the peak-cut characteristic by controlling the DC/AC converter output.

ACKNOWLEDGMENT

The authors would like to thank the financial support by Korea Electrical Engineering & Science Research Institute.

REFERENCES

[1] H. Nagayoshi et al, "Peak-power reduction with 100kW PV and battery combined system at Shonan Institute of Technology" Solar Energy Materials & Solar Cell 67 (2001) 601-609

[2] Hong-Sung Kim et al., "Advanced grid connected PV system with functions to suppress disturbance by PV output variation and customer load change" Solar Energy Materials & Solar Cell 67 (2001) 559-569

High Voltage AlGaN/GaN Schottky Barrier Diode Employing the Inductively Coupled Plasma-Chemical Vapor Deposition SiO₂ Passivation

Young-Hwan Choi, Jiyong Lim, Kyu-Heon Cho and Min-Koo Han
School of Electrical Eng. & Computer Science #50, Seoul National University,
Shillim-dong, Gwanak-gu, Seoul, 151-742, KOREA
Phone: +82-2-880-7254 Fax. : +82-2-875-7254 E-mail: wink7@emlab.snu.ac.kr

Abstract—The SiO₂ passivation using the inductively coupled plasma-chemical vapor deposition(ICP-CVD) was proposed for the high voltage AlGaN/GaN Schottky barrier diode(SBD). The ICP-CVD is well known for the high-density remote plasma, so this method reduces the plasma damage on the surface of semiconductor. Experimental results showed that the SiO₂ passivation improved the electrical characteristics of AlGaN/GaN SBDs. The specific on-resistance was decreased from 15.1 mΩ cm² to 13.2 mΩ cm², and the reverse breakdown voltage increased from 87.5 V to 497.0 V. Schottky barrier height was also increased from 0.64 to 0.92. The figure-of-merit (VB²/R_{ON}) of the SiO₂ passivated device was 18.71 MW/cm². These improvements of the proposed device were attributed to the suppression of electron trapping at the surface states.

I.   INTRODUCTION

For the high operating frequency in power electronic applications, the device which have high speed are required. The speed of diodes is strongly dependent on the reverse recovery time. Silicon P-i-N diodes and SBDs are commonly used, but Silicon P-i-N diodes have a large reverse recovery time and Silicon SBDs which have fast switching speed show a low breakdown voltage.

In these aspects, AlGaN/GaN Schottky barrier diodes (SBDs) have attracted a considerable attention, because the material property of GaN is superior to that of Silicon. GaN has high critical breakdown field (> 3MV/cm) due to wide band-gap, so that high voltage SBDs can be fabricated. Also, two-dimensional electron gas (2DEG) formed in the hetero-interface of AlGaN and GaN offers a large advantage to forward characteristics of SBDs due to large electron concentration and high electron mobility [1].

The surface passivation is critical for GaN-based devices because the electron trapping at surface states depletes 2DEG and induces the surface leakage current. Various passivation materials such as Si₃N₄ and SiO₂ have been applied to solve the problems related with surface states [2-4]. The plasma-enhanced chemical vapor deposition (PECVD) is widely used for the SiO₂ passivation. The plasma made by PECVD is adjacent to the surface of semiconductor, so that the device may be damaged during passivation process. The inductively coupled plasma-chemical vapor deposition (ICP-CVD) is well known for the high-density remote plasma and reduces the plasma damage on the surface of semiconductor compared with PECVD [5].

The purpose of our work is to propose the SiO₂ passivation employing an ICP-CVD process for AlGaN/GaN SBDs to suppress the electron trapping at surface states. AlGaN/GaN SBDs employing the ICP-CVD SiO₂ passivation were successfully fabricated. Experimental results showed that the SiO₂ passivation improved the specific on-resistance from 13.2 mΩ cm² to 15.1 mΩ cm², increased the reverse breakdown voltage from 87.5 V to 497.0 V.

II.   DEVICE STRUCTURE AND FABRICATION

AlGaN/GaN heterostructure was grown on semi-insulating 4H-SiC substrate by metal-organic chemical vapor deposition. The epi structure used in this experiment is as follows. Firstly, the nucleation layer was grown, followed by the 3 μm-thick Fe doped GaN layer as a buffer. The 30 nm-thick unintentionally doped (UID) Al₀.₂₆Ga₀.₇₄N layer and the UID GaN capping layer were grown in a sequence. The UID capping layer was for the protecting the UID AlGaN layer during the fabrication. The device isolation was performed by 200 nm-thick mesa structure by dry etching. Ohmic contacts were formed with Ti/Al/Ni/Au (20/80/20/100 nm) and annealed at 860 °C for 30 s in N₂ ambient. Schottky contact, Ni/Au (50/250 nm), was deposited. The e-gun evaporation and the lift-off method were used to deposit and define the metal contact. The mobility and the sheet charge concentration was 1590 x 10³ cm²/Vs and 8.27 x 10¹² cm⁻². After Schottky metallization, 350 nm-thick SiO₂ layer was deposited by an ICP-CVD process with He/N₂O/SiH₄ (100/30/5 sccm). The process temperature, RF power and pressure were 300 °C, 400 W and 50 mTorr, respectively. The cross-sectional view and the photograph of the proposed device are showed in Fig. 1 and Fig. 2.

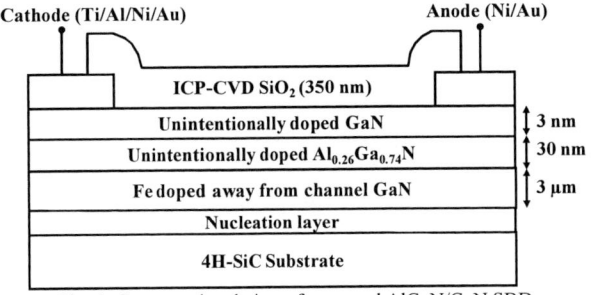

Fig. 1. Cross-sectional view of proposed AlGaN/GaN SBD

978-1-4244-1871-8/07 $25.00 © 2007 IEEE

Fig. 2. Photograph of fabricated AlGaN/GaN SBD

III. EXPERIMENTAL RESULTS

The electron tapping at the surface state causes the leakage current of AlGaN/GaN devices. When the reverse bias is applied to AlGaN/GaN devices, electrons flow through the surface state of semiconductor and the leakage current of AlGaN/GaN devices is increased. To investigate the effect of SiO_2 passivation on the surface leakage current, the buffer leakage before and after the SiO_2 passivation was measured. Fig. 3 shows the cross-sectional view of the buffer leakage measurement pattern, which was fabricated simultaneously with the proposed device. The distance between ohmic patterns is 20 μm and the width of ohmic pattern is 100 μm. The voltage was applied up to 100 V and the buffer leakage current was measured. The leakage current through the surface of buffer is dominant in this measurement, because GaN buffer layer has semi-insulating characteristics for the device isolation and the etched buffer region may have a lot of surface states due to the plasma damage during dry etching process.

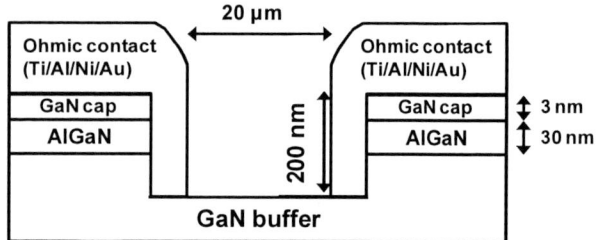

Fig. 3. Cross-sectional view of buffer leakage measurement pattern

Before the SiO_2 passivation, the leakage current of the buffer leakage pattern at $V_{applied} = 100$ V was 974.06 μA/mm. However, the leakage current of the SiO_2 passivated buffer leakage pattern at the same voltage was 12.66 nA/mm. The buffer leakage current was decreased by 4-orders after the SiO_2 passivation. These results shown in Fig. 4 indicate that the buffer leakage current is dominantly dependent on the current through the surface of buffer layer and the SiO_2 passivation successfully prevents electrons from trapping at the surface state.

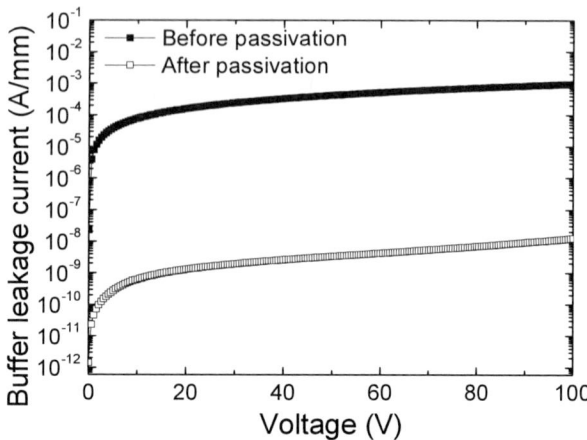

Fig. 4. Measured leakage current of the buffer leakage pattern before and after the SiO_2 passivation

The measured I-V characteristics of the unpassivated device and the SiO_2 passivated device are shown in Fig. 5. After the SiO_2 passivation, the turn-on voltage was shifted to the negative direction, which is due to the suppression of electron trapping at the anode region. The trapped electrons near the anode region deplete 2DEG and extend depletion region. Hence, the device has a negatively biased region with zero bias, which was known as the virtual gate [2]. For this reason, the higher anode voltage is required to turn on the unpassivated device. In Fig. 4, the SiO_2 passivated buffer leakage measurement pattern showed the remarkable reduction of buffer leakage current and the suppression effect on the electron trapping at surface states. The channel of proposed device was not depleted and the negatively biased region was not created at the surface of the proposed device. Therefore, the turn-on voltage of the proposed device was decreased. The specific on-resistance of proposed device was decreased from 15.1 mΩcm^2 to 13.2 mΩcm^2 due to the decrease of the turn-on voltage.

The leakage current of the unpassivated device and the SiO_2 passivated device were measured at $V_{Anode} = -10$ V. The leakage current of unpassivated device was 683.30 mA/cm^2, while that of SiO_2 passivated device was 1.94 mA/cm^2. The SiO_2 passivation suppresses electrons to be trapped at surface states, and these results are consistent with the reduction of the buffer leakage current.

The ideality factor (n) and the barrier height (Φ_{BN}) were calculated from Eq. 1 and Eq. 2 [7].

$$n = \frac{qV}{kT \ln(J/J_S)} \tag{1}$$

$$\Phi_{BN} = \frac{kT}{q} \ln(\frac{A^{**}T^2}{J_S}) \tag{2}$$

J_s is the saturation current density obtained by the log-linear plot of I-V characteristics at V = 0, A^{**} is the effective Richardson constant assuming 26.64 $Acm^{-2}K^{-2}$ [6], and the temperature is 300 K.

The ideality factor and the barrier height were calculated from the measured I-V characteristics and summarized in Table. 1. The ideality factor at $V_{Anode} = 0.5$ V was improved from 3.09 to 1.04 and the barrier height was increased from 0.64 eV to 0.92 eV. The SiO_2 passivation suppresses electrons to flow through the surface states of active region so that the reverse characteristics of proposed device are improved by the reduction of the surface leakage current.

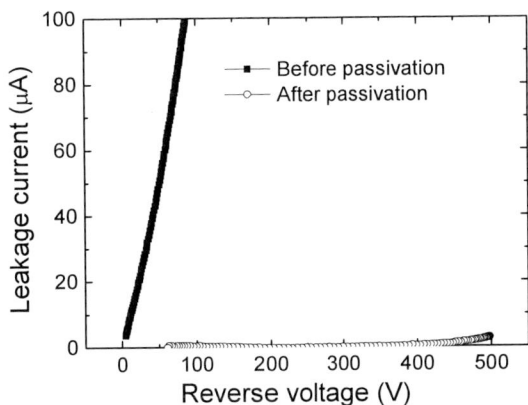

Fig. 6. Measured reverse characteristics of the unpassivated device and the SiO_2 passivated one

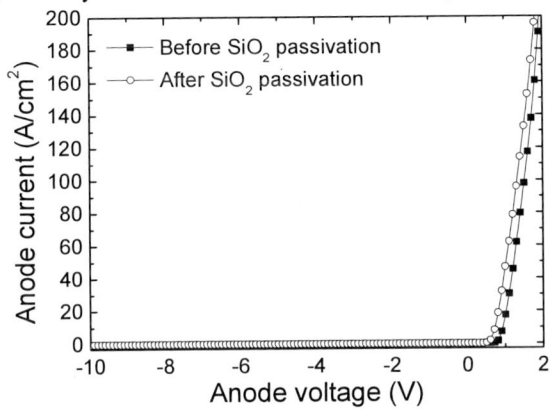

Fig. 5. Measured I-V characteristics of the unpassivated device and the SiO_2 passivated one

TABLE 1. IDEALITY FACTORS AND BARRIER HEIGHTS OF THE UNPASSIVATED DEVICE AND THE SiO_2 PASSIVATED ONE

	Ideality factor (at $V_{Anode} = 0.5$ V)	Schottky barrier height
Unpassivated device	3.09	0.64
SiO_2 passivated device	1.04	0.92

The reverse characteristics of both devices were measured. The reverse breakdown voltage of unpassivated device was 87.5 V at $I_{Anode} = 100.0$ μA. The SiO_2 passivated device showed the reverse breakdown voltage of 497.0 V at $I_{Anode} = 2.7$ μA and the proposed device was burned-out at this voltage. The leakage current of the proposed device is considerably decreased due to the suppression of surface leakage current. The figure-of-merit (VB^2/R_{ON}) of the SiO_2 passivated device was 18.71 MW/cm^2.

IV. CONCLUSION

We have fabricated the AlGaN/GaN Schottky barrier diodes employing the SiO_2 passivation layer by using an ICP-CVD process. The ICP-CVD reduces the plasma damage compared with PECVD. The SiO_2 passivation suppressed the electron trapping at surface states so that the SiO_2 passivated device showed the superior characteristics compared with that of unpassivated device. The specific on-resistance and the breakdown voltage of SiO_2 passivated device were 15.1 $m\Omega cm^2$ and 497.0 V, while those of unpassivated device were 13.2 $m\Omega cm^2$ and 87.5 V.

ACKNOWLEDGMENT

This work was supported by the power IT project of the Ministry of Commerce, Industry and Energy in the Republic of Korea.

REFERENCES

[1] S. J. Pearton, J. C. Zolper, R. J. Shul, and F. Ren J., "GaN: Processing, defects, and devices", *J. Appl. Phys.*, Vol. 86, pp 1-78, July, 1999.

[2] R. Vetury, N. Q. Zhang, S. Keller, and U. K. Mishra, "The impact of surface states on the DC and RF characteristics of AlGaN/GaN HFETs", *IEEE Trans. Electron Devices*, vol. 48, no. 3, pp. 560-566, March, 2001.

[3] B. M. Green, K. K. Chu, E. M. Chumbes, J. A. Smart, J. R. Shealy, and L. F. Eastman, "The effect of surface passivation on the microwave characteristics of undoped AlGaN/GaN HEMT's", *IEEE Electron Device Lett.*, vol. 21, no. 6, pp. 268-270, June, 2000.

[4] M.-W. Ha, S.-C. Lee, J.-H. Park, J.-C. Her, K.-S. Seo, and M.-K. Han, "Silicon dioxide passivation of AlGaN/GaN HEMTs for high breakdown voltage", *Int. Symp. Power Semiconductor Device and ICs*, pp. 169-172, 2006.

[5] M. Goto, H. Toyota, M.Kitagawa, T. Hirao, and H. Sugai, "Low temperature growth of amorphous and polycrystalline silicon films from a modified inductively coupled plasma", *Jpn. J. Appl. Phys.*, vol. 36, no. 6A, pp. 3714-3720, June, 1997.

[6] A. R. Arehart, B. Moran, J. S. Speck, U. K. Mishra, S. P. DenBaars, and S. A. Ringel, "Effect of threading dislocation density on Ni/n-GaN Schottky diode I-V characteristics", *J. Appl. Phys.*, vol. 100, 023709, July, 2006.

[7] S.M. Sze, *Physics of Semiconductor Devices*, pp. 256-280, Wiley, New York, 1981.

Optimized Operation and Stabilization of Microgrids with Multiple Energy Resources

T.Tanabe, Y.Ueda, S.Suzuki, T.Ito, N.Sasaki,
T.Tanaka and T.Funabashi
Meidensha Corporation
ThinkPark Tower, 2-1-1, Osaki, Shinagawa-Ku,
Tokyo, 141-6029 Japan
Email: tanabe-t@mb.meidensha.co.jp

R.Yokoyama
Waseda University
Graduate School of Environment and Energy Engineering
1-1-7 Nishi-Waseda Shinjuku Tokyo
EMAIL: Yokoyama-ryuichi@waseda.jp

Abstract— A microgrid supply/demand control system is proposed. In this control system, operation planning is realized based on generation and load forecasting by using artificial neural network and fuzzy systems. Unit commitment of generations includes start/stop of power generations and energy storages. Load following function is accomplished based on PI control scheme. To compensate for the power fluctuation in microgrid, power system stabilizer with EDLC is proposed..

I.   INTRODUCTION

A concept of microgrid was proposed and is being developed in many laboratories and organizations [1-4]. Microgrid is a small grid in which distributed generations and electric loads are placed together and controlled efficiently in an integrated manner. It contributes to utility grid's load leveling by controlling power flow between utility grid and Microgrid according to a predetermined power flow pattern. Also, it will be able to contribute to an efficient operation of distributed generations and suppression of CO2 gas emission by operation planning considering grid economics and energy efficiency. In addition, Microgrid is useful for the area with no power system or weak power system in rural area. It can be operated in an islanded manner with appropriate control scheme.

Shimizu Corporation and Meidensha Corporation recently developed a Microgrid control system and Power Systems Stabilizer (PSS) with Electric Double Layer Capacitors (EDLCs). In this paper, a microgrid control system is proposed. In this control system, operation planning is realized based on generation and load forecasting by using neural network and fuzzy systems. It includes multi-objective evaluation of generation cost and CO2 gas emission with some constraints. Unit commitment of generations includes start/stop of power generations and storages. Load following function is accomplished based on PI control scheme. Power Systems Stabilizer (PSS) for microgrids has been developed with an Electric Double Layer Capacitor (EDLC). Rapid fluctuations of renewable generations are compensated with PSS. It enables rapid and frequent charge/discharge and realizes a long lifetime. PSS compensates both active and reactive power and it can also realize constant reactive power supply and consequently it can decrease reactive power of generators in microgrids.

According to a study for microgrid written in this paper,

Shimizu Corporation and Meidensha Corporation are jointly executing a field test.

II.   OUTOLINE OF FIELD TESTS

A.   Outline of Microgrid field test in Japan

The NEDO (New Energy and Industrial Technology Development Organization) started three research projects, which deal with new energy integration to local power system field test in 2004. The sites are in Aomori, Aichi and Kyoto [1-3].

In Aomori project, field tests were started to develop a distributed energy supply system, in which some loads in special districts are supplied by this supply system with private power lines and makes no influence to utility power system with which the energy supply system is connected at one point. In this project gas engine generator and battery system are used to control variable power output of photovoltaic and wind turbine generators. Electric power is supplied with private power lines to four school and some buildings. Heats are supplied to water treatment works. Total DER capacity is 710kW, in which PV: 80kW, Wind: 20kW, Biomass gas engine: 510kW and Battery: 100kW.

In Aichi project, fuel cells are used as main generations. The site was World Expo 2005. Fuel cell systems are MCFC: 350kW, 370kW, PAFC: 800kW, and SOFC: 50kW. Total DER capacity is 2400kW. PV: 330kW and NAS battery: 500kW are also used. In Kyoto project, capacity of each resources are wind: 50kW, Biogas: 400kW, Fuel cell: 250kW, and PV: 50kW.

The Shimizu Corporation has built a Microgrid at its research labs in Tokyo, Japan [3-4]. It includes two gas engines (90kW and 350kW) as well as extra storage in form of a 400kWh NiMH battery (50kW×8h, with 200kW inverter) and an 100kW EDLC.

B.   Outline of Mass PV field test in Japan

A field test of voltage control for a distribution feeder connected with multiple PV systems was started in 2003. In this project about 500 PV systems are connected to the power system. Power quality, such as voltage fluctuation and harmonic distortion, are recorded and performance of voltage controller and anti-islanding detector is being tested. Field test site of this project is in Ohta city, Gunma prefecture. At September 2007, more than 500 houses are installed with

978-1-4244-1871-8/07 $25.00 © 2007 IEEE

photovoltaic modules. Voltage and current are measured every one second. Using GPS clock synchronizes these data. Voltage and current waveform can be record and are used to analyze harmonics.

III. OBJECTIVES

A. Supply and demand control

By making the supply power from the utility grid to the microgrid constant or a specified pattern, the SDC system contributes to the load leveling of the utility power system. Fig.1 shows an example of microgrid and Fig.2 shows an example of supply and demand control.

B. Efficient operation of the distributed generations

By the control based on the operation plan of the distributed generations considering their economics and energy efficiency, the SDC system realizes the efficient operation of the distributed generations in the microgrid.

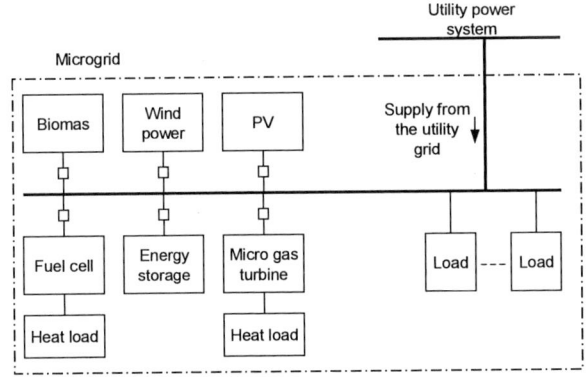

Fig.1. An example of microgrid.

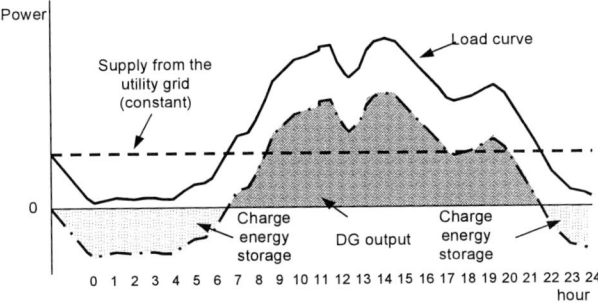

FIG.2. AN EXAMPLE OF SUPPLY AND DEMAND CONTROL.

IV. CONFIGURATION OF SUPPLY AND DEMAND CONTROL SYSTEM

The SDC system consists of the control system that executes the operation planning and supervisory control of the total microgrid and the supervisory control terminals that executes supervisory control of distributed generations and loads. Fig.3 shows a block diagram of the SDC system.

A. Control system

This is the computer system that executes supply and demand control keeping the quality and reliability of the electric power and heat in an affordable level.

B. Supervisory and control terminals (SC terminals)

The SC terminal collects measurement items and fault/state signals, and sends them to the main control system. These terminals also receive signals from the main control system for breakers on/off, DG operation start/stop and generator output control. Fig.4 shows a block diagram of a supervisory control terminal.

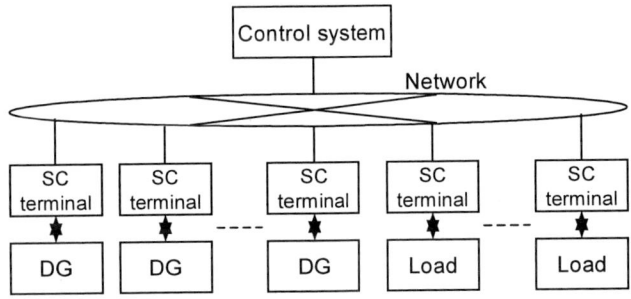

Fig.3. A block diagram of SDC system.

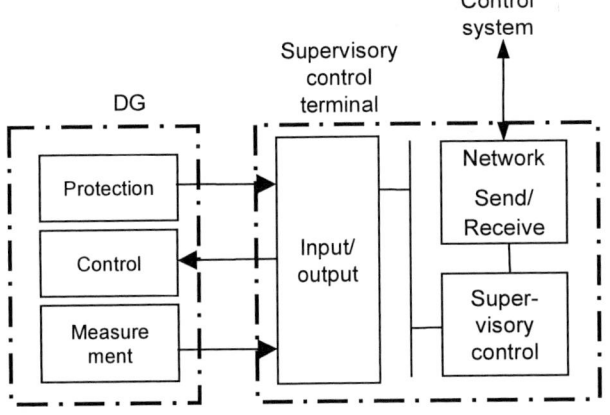

Fig.4. A block diagram of SC terminal

V. SUPPLY AND DEMAND CONTROL SYSTEM

Generator output control is executed according to the operation plan that is made based on the pattern of supply power from the grid (In this system it is constant as shown in Fig.1) and the load demand pattern forecasted in advance. Fig.5 shows an outline of the control method.

A. Operation plan

The SDC system forecasts a next day's load pattern and makes a next day's operation plan taking into account their economics and efficient operation of the distributed generations and reserve power (supply reliability). Then, the SDC system determines operating time of DGs without and

978-1-4244-1871-8/07 $25.00 © 2007 IEEE   75

DGs with load following and its operating pattern. Next day generation forecasting is done for wind generation and photovoltaic generations based on weather forecast data, wind data and photovoltaic data. Heat demand is also included in generation planning. Next day's load forecasting is done using neural network [5], which have been learned by past load data. From temperature and load data for a day of forecasting and temperature and average load data for a past day of which these data are resembled to forecasting day's data, a correction factor is calculated. Correction factor is added to past day's average load data. Then, if needed, some more fuzzy estimated correction is done for a special day such as a public holiday and sudden climate change. Fig.6 shows a flow chart of load forecast..

B. Modification of operation plan using short time load forecast

By forecasting load at next fifteen minutes to three hours based on the demand at present, accuracy of forecast for supply power that is used for load following, is improved [6-7]. If the supply power reserve for load change is insufficient, the operation plan is modified.

C. Data acquisition for supply/load power

By collecting real power, imaginary power, voltage, current data at each DG and load, the total supply/demand power at present is calculated.

D. Load following control

For making the supply power coincident to the planned pattern at high accuracy, the output control of DG dispatched to load following is done. Control is carried out to realize both control modes of the constant power flow control (to assure kW following accuracy) by which the power flow at the interconnected point is maintained as constant and the power balancing control (to assure simultaneity in terms of kWh). The output control value is determined from the planned output value and real-time collected total supply and total demand. Output control signal is determined considering DG's response time and data transmission delay time in the information network [8].

Constant power flow control is used to define a generator output command by observing the active power (W) at the interconnected point and computing the component of generator adjustment so that the observed active power (W) can be maintained as the reference input The generator output command is defined in the power balancing control block following the steps that the differential component between the active power (W) at the interconnected point and the reference input is integrated (converted into Wh) and the component of generator adjustment is computed to maintain the reference input (Wh). In the developed supply and demand control system, the control block is designed to realize the simultaneous support of the above-mentioned two control blocks.

In this case, the power balancing control block does not

control the generator, but it simply generates an output of the auto-control result. This auto-control output and the reference input for constant power flow control are added to make up a new reference input, which value is then applied to the constant power flow control block. As a result, the computed generator command value is effective to realize the simultaneous support of both blocks [9].

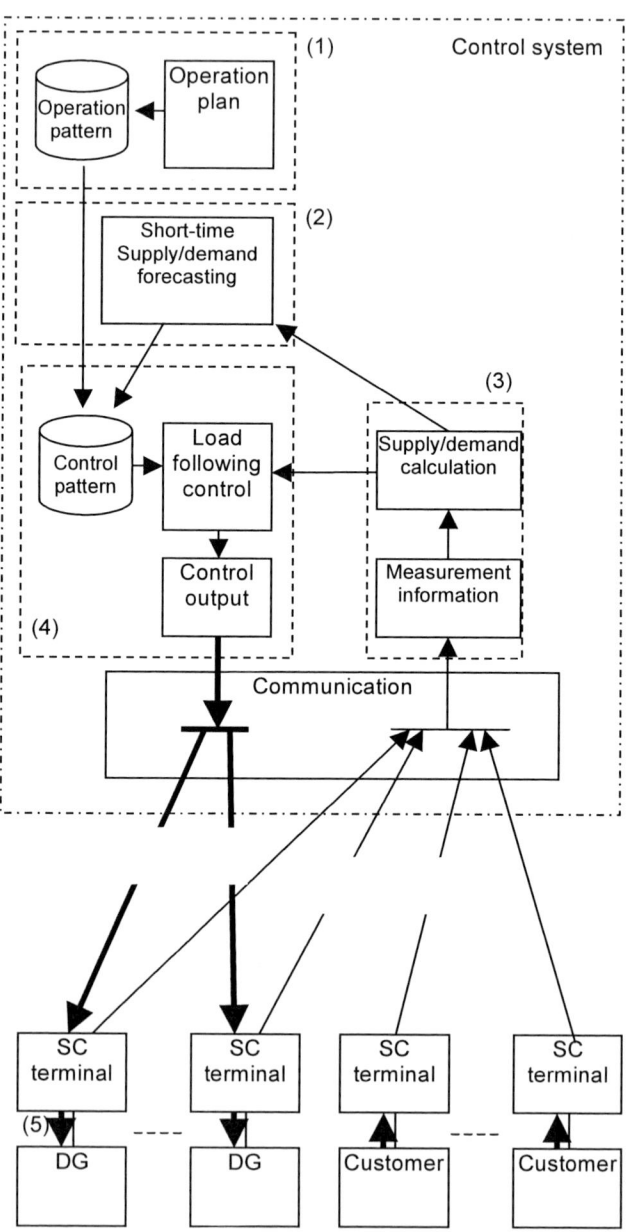

Fig.5 An outline of control method

E. Control of DGs

According to the control signal received from the SDC system, supervisory control terminals send to the DGs such signals as operation start/stop or changing output.

978-1-4244-1871-8/07 $25.00 © 2007 IEEE

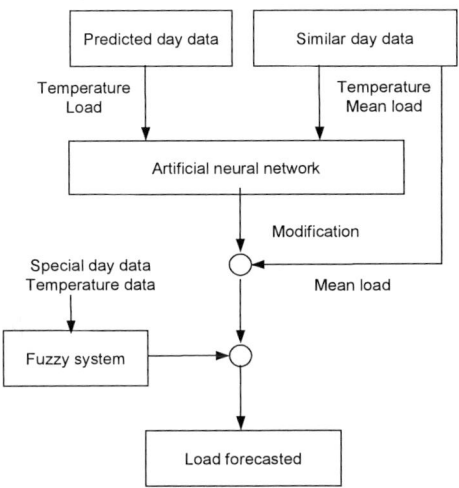

Fig.6 A flowchart of load forecast [6]

VI. POWER SYSTEM STABILIZER WITH EDLC

A. Features

In order to compensate for sudden load fluctuations occurring in the microgrid, the EDLCs are used as the charge/discharge elements because they are capable of high-speed charging and discharging.

The features of equipment are described below.

1) High-speed response

Compared with storage batteries, the EDLCs are capable of high-speed charging and discharging.

2) Long operational life after frequent charge/discharge actions

Compared with storage batteries, the EDLCs can withstand frequent charge/discharge duty cycles. In addition, the operational life is long.

3) Generation of active power in high preference

When fluctuations occur in active power reactive power at the same time, active power is output in higher performance.

4) Generation of a constant reactive power

According to the reactive power command value sent from the supply and demand control system, it is possible to generate a constant level of reactive power burden to be shouldered by the generator.

B. Equipment Configuration

Fig.7 shows the equipment configuration. The power is 100kVA. The EDLC manufactured by Meidensha Corporation comes in a module of 45F in capacitance. Six units of such modules are connected into a capacitor bank consisting of three-serial and two parallel modules. Thus, the resultant capacitor bank has a total capacitance of 30F. In order to cope with load switching (closure and tripping), the operating DC voltage is defined based on the rated voltage (middle voltage) of 336 volts and this voltage is practically used within a voltage range of 240 – 400 volts according to the charging and discharging cycles. The time needed for

charging and discharging is 2 seconds for charging and 2 seconds for discharging, respectively. This is based on the assumption that this time is consumed from the middle voltage to the maximum or minimum voltage with a +/-100kW output in the charging or discharging phase.

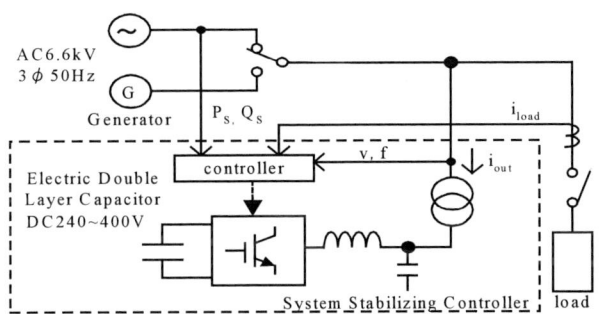

Fig.7 A configuration of power system stabilizer [11]

C. Operation of Power System Stabilizer

Fig.2 shows the operation of the power system stabilizer. During interconnected operation with the power system, the power system stabilizer performs power flow control to compensate for the power flow fluctuation at the point of common coupling (PCC). By this control active or reactive power at the PCC is used to determine the power flow fluctuation at the PCC. The Power system stabilizer then generates an output to compensate for the determined power flow fluctuation.

When the microgrid stays in its islanding operation independent of the utility power network system, the power system stabilizer continues its control after selecting either power flow control to compensate for the load power fluctuation or a CVCF control to maintain the same levels of voltage and frequency in the microgrid system.

In the power flow control for islanding operation, the load current is used to determine the load power fluctuation and the power system stabilizer generates an output to compensate for this power fluctuation. In the CVCF control mode, the fluctuation components of system voltage and frequency ar determined. Based on these components, frequency fluctuations are controlled by the active power and voltage fluctuations are controlled by the reactive power.

D. Evaluation Test

The evaluation test has been carried out in the factory [11]. Test circuit is shown in Fig.7. Fig.8 shows an example, the result of the active power fluctuation control under the interconnected condition. Due to the resistor load insertion to the load insertion point, power at the PCC fluctuates rapidly. But the power system stabilizer compensates for this in 0.1 sec or so.

The evaluation test has been carried out also in Shimizu Corporation's research labs (real–world). Fig.9 shows an example, the result of the active power fluctuation control under the interconnected condition. A test system includes gas-engine generator, loads and power system stabilizer with EDLC.

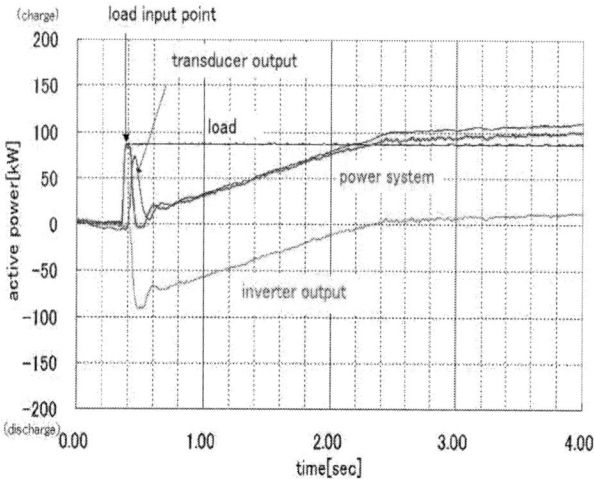

Fig.8 Power flow control of interconnected operation (laboratory) [11]

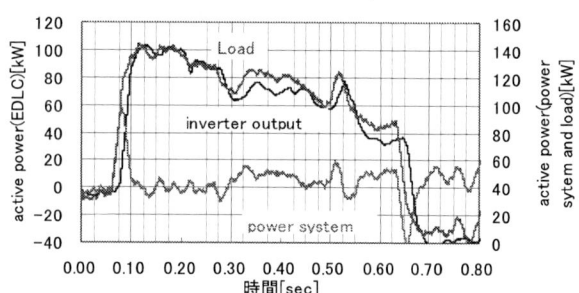

Fig.9 Power flow control of interconnected operation (real world)

VII. CONCLUSIONS

This paper introduced a microgrid supply/demand control system. Some features of the system were outlined. And also, power system stabilizer to compensate for the fluctuation of active power in microgrid is proposed. The microgrid with these new techniques can contribute to efficient operation of DGs and stable power supply for customers.

For improving the practical system, some challenges exist.

A. Improving demand and generation forecast accuracy

Improving accuracy for electricity load forecast and heat demand forecast is needed. And also is needed output power forecast for wind generations and photovoltaic generations.

B. Optimized operation of DGs

Optimization method for DGs operation by taking into account economics and energy efficiency of heat and power is needed [12].

C. Load following

Adaptive control to automatically change the control gain according to the power system parameters change is needed.

D. Application of multi agent system

For flexibly adding new DGs or loads to the supply and demand control system, multi agent technology might be required [13-14].

ACKNOWLEDGEMENT

The authors give a special thank to Shimizu Corporation for a cooperated study of designing a Microgrid and executing some tests in a facility of Shimizu Corporation.

REFERENCES

[1] T.Funabashi and R.Yokoyama, "Microgrid Field Test Experiences in Japan," IEEE PES General Meeting, 1996

[2] S.Morozumi and K.Nara, "Recent Trend of New Type Power Delivery System and its Demonstrative Project in Japan," IEEJ Trans. PE, Vol.127, No.7, 2007, pp.770-775

[3] Mike Barnes, Giri Ventakaramanan, Junji Kondoh, Robert Lasseter, Hiroshi Asano , Nikos Hatziagyriou, Jose Oyarzabal, and Tim Green, "Real-World MicroGrids- An Overview," 2007 IEEE International Conference on System of Systems Engineering, San Antonio, USA., April 16-18, 2007

[4] A.Denda, "Shimizu's microgrid research activities," Symposium on Microgrids, Montreal, June 2006.

[5] T.Senjyu, P.Mandel, K.Uezato and T.Funabashi, "Next day load curve forecasting using neural network structure," IEE Proc.-Gener. Transm. Distrib., Vol.151, No.3, May 2004

[6] T.Senjyu, H.Takara, K.Uezato and T.Funabashi, "One-Hour-Ahead Load Forecasting Using Neural Network", IEEE Trans. Power Systems, Vol.17, No.1, February 2002, pp.113-118.

[7] P.Mandel, T.Senjyu, K.Uezato and T.Funabashi, "Forecasting Several-Hours-Ahead Electricity Demand Using Neural Network," IEEE DRPT 2004 (The 2nd International Conference on Electric Utility Dereguration, Restructuring and Power Technologies), 5-8 April 2004, Hong Kong

[8] T.Senjyu, H.Takara, K.Uezato, T.Funabashi and T.Ito, "Load Following Control for Dispersed Generators Using PI Controller," T.IEE Japan, Vol.122-C, No.8, 2002, pp.1333-1340 (in Japanese)

[9] T.Ito, T.Tanabe and N.Sasaki, "Development of Microgrid Control Systems," Meiden Review, No.140, No.2 2007, pp.4-7

[10] S.Suzuki and Y.Ueda, "System Stabilizing Controller by Electric Double Layer Capacitor," Meiden Review, No.140, No.2 2007, pp.8-11

[11] S.Suzuki, Y.Ueda, S.Numata and A.Denda, "Combined Power Supply Control of Distributed Generators in Micro Grid (Part...2) – Development of System Stabilizing Controller by Electric Double Layer Capacitor -," IEE Japan Power and Energy Society annual meeting, September 2006 (in Japanese)

[12] H.Otoguro, T.Funabashi, Y.Okuno and Y.Hoshi, "Power-Supply Facilities Optimization Plan for Microgrid," Meiden Review, No.140, No.2 2007, pp.16-18

[13] T.Hiyama and T.Funabashi, "Multi-Agent Based Operation and Control of Isolated Power System with Dispersed Power Sources including New Energy Storage device," ICREPQ 2004

[14] A. Dimeas and N. Hatziargyriou, "A Multi Agent System for Microgrids," , IEEE Trans. on Power Systems, Vol. 20, No. 3, pp. 1447-1455, 2005.

A New Converter-Inverter-Brake (CIB) Module with Shoot-Through Immunity

Shihong Park*, Jun-Sik Kim

School of Electronics and Computer Engineering
Dankook University
126, Jugjeon-dong, Suji-gu, Yongin-si,
Gyounggi-do, 448-701, Korea
* Corresponding author

Sun-Ja Kim, Bong-Hyun Kwon

LS Industrial Systems Co., Ltd
533, Hogye-dong, Dongan-gu, Anyang-si,
Gyeonggi-do 431-080, Korea

Abstract— This paper introduces a new 25-pin CIB power module with inherent immunity to shoot-through failures and the need for control dead-time. As a result, the need for dead-time compensation in pulse-width-modulated (PWM) inverters that often requires additional current sensors and fast dedicated logic is also eliminated in many applications that now require it. The 25-pin CIB power module was manufactured and tested in 300W inverter system. The experimental results of proposed and conventional system are provided to verify its performance. The parasitic impedance due to interconnections in the power module is also investigated using Q3D Extractor and PSPICE.

I. INTRODUCTION

Modern power semiconductor modules have been widely used from household appliances to industrial applications owing to technology innovation and evolution of power devices such as IGBTs and MOSFETs. Standardization of power modules in invert systems is relatively simple because of their similar topology compare to other power electronic systems. Many kinds of invert power module classified by voltage and current ratings have been manufactured for a variety of applications in these days.

Fig. 1 (a) shows the basic power circuit building block for inverters, consisting of a high- and low-side switch spanning the dc voltage source V_{dc}. One of the most serious faults that can occur in this circuit is the shoot-through condition that occurs when both switches are inadvertently turned on at the same time, shorting the supply voltage. In order to avoid short-through failure, a dead-time is required in control signals. A wide variety of protection techniques have been developed to remove the shoot-through fault if it occurs by rapidly turning off one or both of the switches, and the majority of these techniques are based on detecting overcurrent conditions [1-7].

An enhanced phase-leg configuration shown in Fig. 2 overcomes the limitations of conventional phase-legs, providing inherent shoot-through immunity. The biggest difference embodied in the new configuration is the presence of series diode D_S between the two switches. This diode eliminates the shoot-through current path E

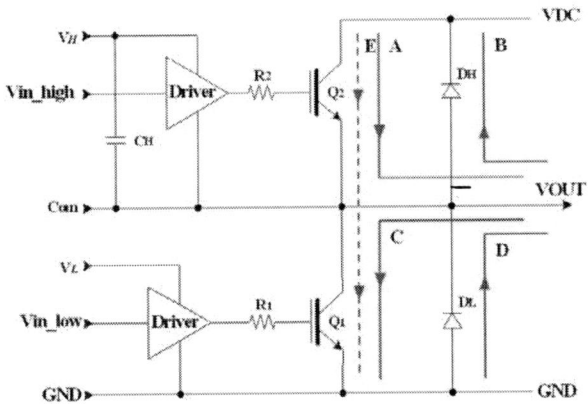

Fig. 1. Basic schematic and available current paths for a conventional inverter configuration.

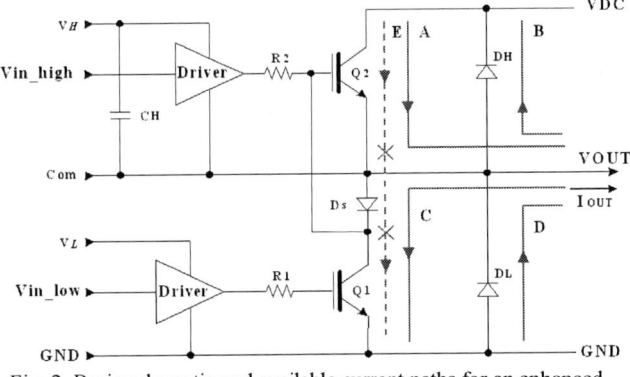

Fig. 2. Basic schematic and available current paths for an enhanced inverter configuration.

because D_S and the gate-emitter (-source) junction of the high-side switch cannot be simultaneously forward-biased. As a result, the phase-leg shoot-through condition is topologically prevented. [8]

In this paper 25-pin CIB power module adapting enhanced inverter scheme realized shoot-through protection was manufactured and investigated in a 300W inverter system. In addition, the effects of module interconnection impedance are analyzed using Q3D Extractor and PSPICE tools in order to confirm the switching characteristics.

978-1-4244-1871-8/07 $25.00 © 2007 IEEE

II. STRUCTURE OF 25-PIN CIB POWER MODULE

Fig. 3 shows the circuit diagram, package outline, internal power device layout of proposed 25-pin CIB power module. The package dimension and pin layout are exactly the same as the conventional one. It includes 600V-25A IGBTs for 3-phase inverter, bridge diodes for 3-phase rectifier and IGBT/diode for boost converter. The power devices were manufactured by Fairchild Korea and the new prototype power module was manufactured by LS industry systems (LSIS). The additional diodes for shoot-through protection are added in each phase-leg as shown in Fig.3 (a).

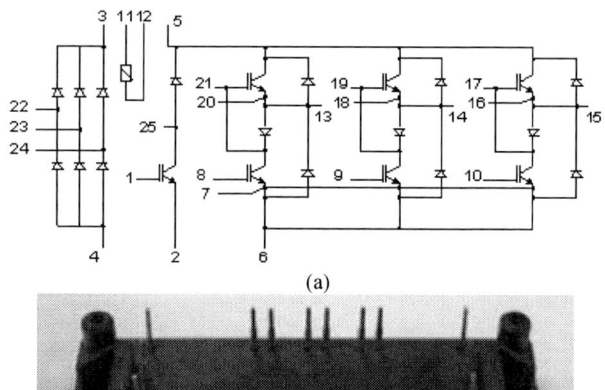

(a)

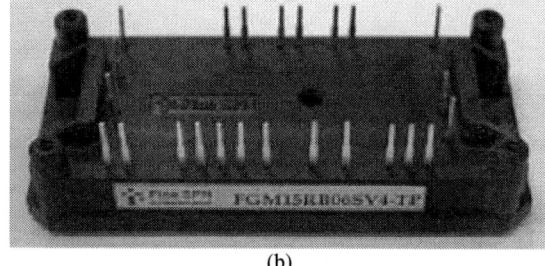

(b)

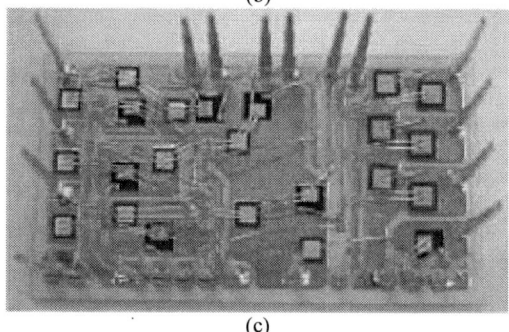

(c)

Fig. 3. A new 25-pin CIB power module (a) circuit diagram (b) package outline (c) power device layout

The effect of interconnection impedance in new inverter topology plays an important role in high current operation especially when discrete power devices are occupied, as discussed in [9]. Although the proto-type power module is designed for low power applications, the effect of interconnection impedance in switching characteristics was investigated using commercially available software Q3D Extractor and PSPICE. Fig.4

shows a depicted 3-Dimensional layout of a phase-leg in the CIB power module including all the information about materials and physical dimension of power devices, bonding wires and DBC. Q3D Extractor program generates the interconnection impedances and their net list from 3-D drawings. It provides resistance, capacitance and inductance as well as mutual inductance between interconnections. Fig. 4 shows the equivalent circuit of interconnection impedance between two wires extracted by Q3D program.

The simulation circuit of a phase-leg and gate drive including interconnection impedance is shown in Fig. 5. Each box has the circuit level information for its interconnection impedance.

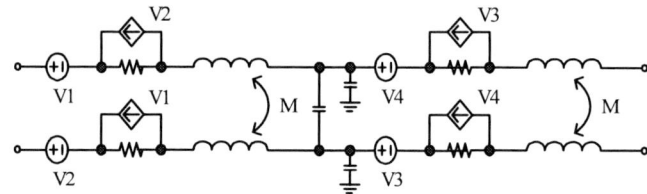

Fig. 4. Equivalent circuit of interconnection impedance of two wires.

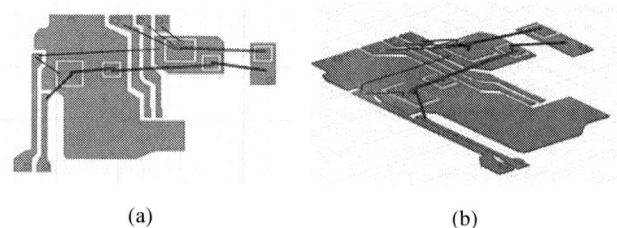

(a)   (b)

Fig. 5. 3-Dimensional layout of an inverter phase-leg in CIB module with all material information (a) plane figure (b) bird's-eye-view

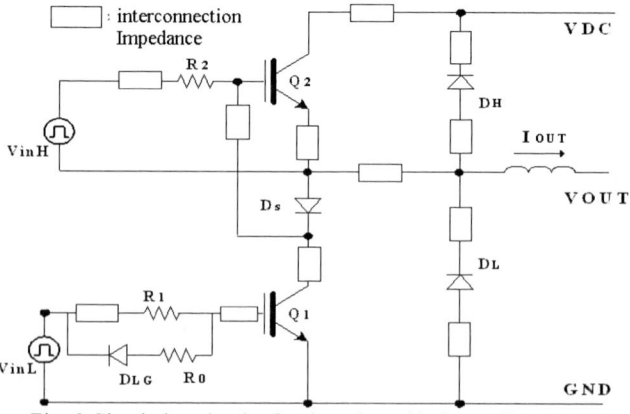

Fig. 6. Simulation circuit of a phase-leg with the net list generated from the interconnection impedance

III. SIMULATION

Simulation of the phase-leg configuration with interconnection impedance was conducted using PSPICE with the parameters for a 600V, 25A IGBT. The device model used in the 25-pin CIB power module was not available so instead, the model parameters of IRG4PC30U manufactured by International Rectifier are applied.

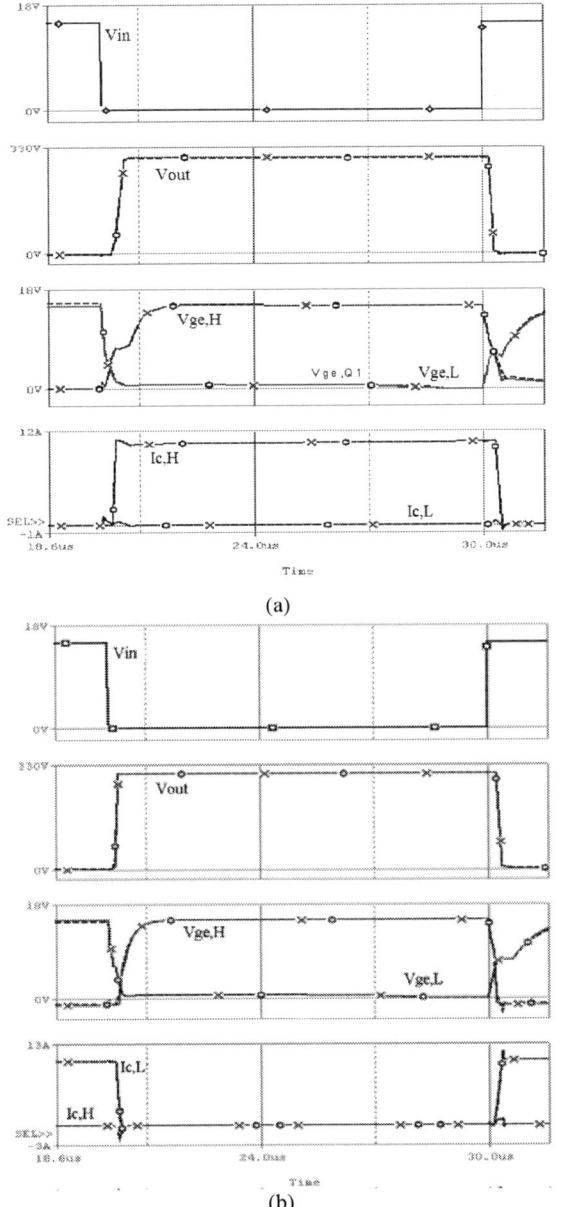

(a)

(b)

Fig. 7. Simulated circuit waveforms for (a) $i_{out} > 0$ and (b) $i_{out} < 0$. Waveforms (top to bottom): input voltage command V_{in}; phase-leg output voltage V_{out}; gate-emitter voltages $V_{ge.L}$, $V_{ge.H}$; collector currents $I_{c.L}$, $I_{c.H}$

For these PSPICE simulations, the following parameter values and operating conditions are used:

$V_{dc} = 300\ V$, $i_{out} = 10\ A$, $L_{load} = 300\ \mu H$,
$f_{in} = 250\ kHz\ (for\ V_{in})$, $V_{cc} = 18\ V$
$R_B = 5\ \Omega$, $R_1 = 100\ \Omega$, $R_2 = 5\ \Omega$
$R_3 = 25\ \Omega$, $R_{p1} = R_{p2} = 1\ k\Omega$

Figure 7 shows the simulated phase-leg voltage and current waveforms for both polarities of output current. The waveforms for $i_{out} > 0$ and $i_{out} < 0$ are shown in Fig. 7(a) and (b) respectively. Zero dead-time is applied in the gate command signals. Closer examination of Fig. 7 reveals that there is no harmful shoot-through current in both the high-side switch ($I_{c.H}$) and the low-side switch ($I_{c.L}$) even after the effect of parasitic interconnection impedance is considered. As discussed in [8], it is apparent in the two simulated gate voltages $V_{ge.H}$ and $V_{ge.L}$ which cross each other without any delay times during both switching events. However, there is no shoot-through current due to the presence of series diode D_S despite the absence of any dead-time delay intervals in the controls.

The simulation results show that the parasitic impedance of interconnection in this size of module is so small that it cannot affect the switching performance.

IV. EXPERIMENTAL RESULTS

A. Experimental System

Experimental testing of the new 25-pin CIB power module was conducted in the drive system that was a three-phase PWM inverter iG5 manufactured by LSIS. Since the new CIB module has the same pin configuration as the conventional module, two identical inverter systems with different power modules were tested for performance comparison. Fig. 8 shows the photo of two drive systems with different power modules located at the bottom of the inverter.

The specifications of tested system are summarized in Table 1.

Fig. 8. Three-phase PWM inverter iG5 manufactured by LSIS

Table 1. Specifications of Tested System

Rated capacity	0.75 kW
Rated torque	4.14 Nm
Rated frequency	60 Hz
Rated phase current	3.5 A
DC link voltage	270 V
Number of pole pairs	2
PWM frequency	3 kHz
Dead-time	3 us

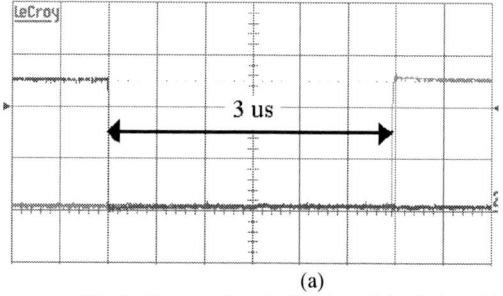 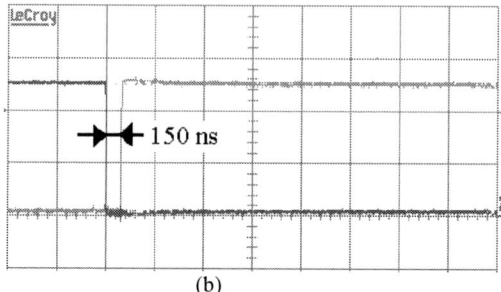

(a)　　　　　　　　　　　　　　(b)

Fig. 9. Measured controller-based dead-time (a) conventional CIB module (b) proposed CIB module

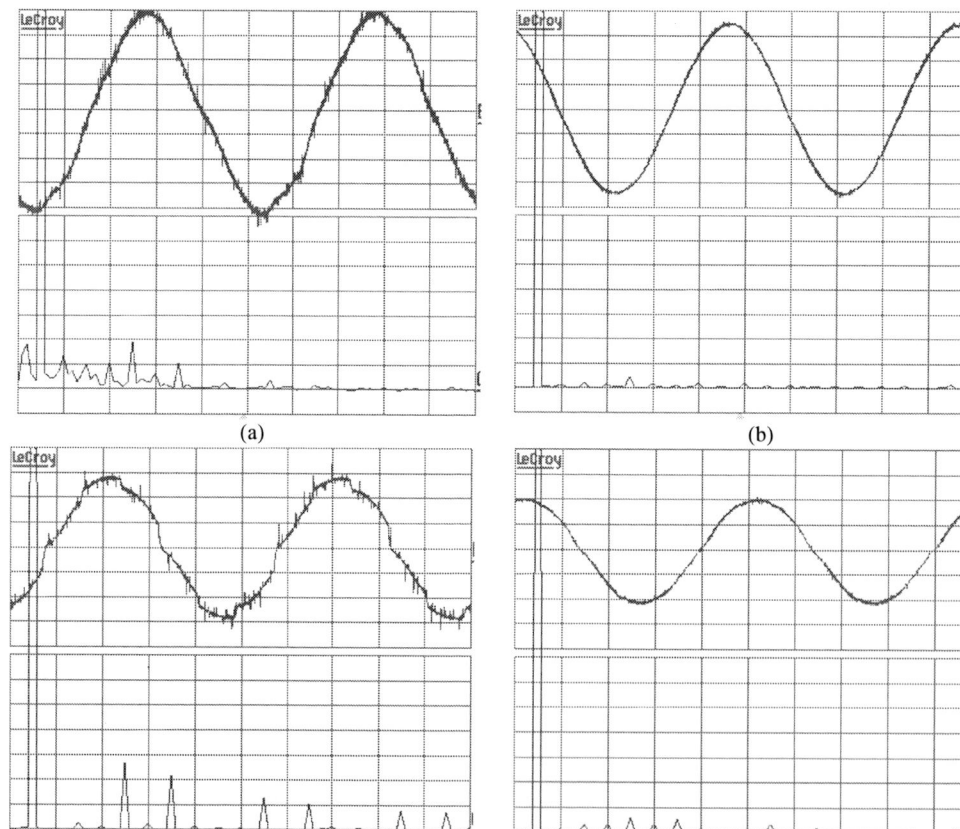

(a)　　　　　　　　　　　　　　(b)

(c)　　　　　　　　　　　　　　(d)

Fig. 10. Measured output phase current waveforms and harmonics (a) Conventional CIB module under 10 Hz condition (b) Proposed CIB module under 10 Hz condition (c) Conventional CIB module under 1 Hz condition (d) Proposed CIB module under 1 Hz condition

Table 2. Harmonics at 10 Hz operation

10Hz	Conventional CIB module		Proposed CIB Module	
	Value	HF	Value	HF
1	778		684	
2	12	0.015	0	0.000
3	10	0.013	3	0.004
4	10	0.013	0	0.000
5	19	0.024	5	0.007
6	7	0.009	2	0.003
7	11	0.014	0	0.000
8	0	0.000	2	0.003
9	3	0.004	0	0.000
10	0	0.000	2	0.003
11	3	0.004	0	0.000
12	0	0.000	0	0.000
13	1	0.001	0	0.000
THD	3.8%		1.0%	

Table 3. Harmonics at 1 Hz operation

1Hz	Conventional CIB module		Proposed CIB Module	
	Value	HF	Value	HF
1	552		412	
2	0	0.000	0	0.000
3	3	0.005	4	0.010
4	0	0.000	4	0.010
5	27	0.049	7	0.017
6	3	0.005	4	0.010
7	22	0.040	5	0.012
8	2	0.004	1	0.002
9	0	0.000	0	0.000
10	0	0.000	0	0.000
11	13	0.024	4	0.010
12	0	0.000	0	0.000
13	10	0.018	0	0.000
THD	7%		2.9%	

The drive system uses a general V/F control method with 3kHz PWM frequency. Two systems have different dead-time as shown in Fig. 9. The controller-based dead-time of the conventional power module is 3 us, while the new power module uses 150 ns dead-time which is the minimum dead-time of a given Microcontroller in drive system. It is possible to apply such a small dead-time since new topology itself has the immunity to shoot-through failure.

B. Experimental Results

For a direct comparison between the conventional and new CIB module, the same frequency command is applied to the same motor load. Fig. 9 (a) and (b) show the phase current at 10 Hz operation of inverter. The red waveforms represent the output phase currents and the blue waveforms show the harmonics of the measured output phase current. It is evident that the output phase current with the new CIB module has much less current distortion and noise without additional dead-time compensation scheme. Fig. 9 (c) and (d) show the phase currents at 1 Hz operation of inverter. It also shows that the new CIB module has much better performance.

In order to compare THD (Total Harmonic Distortion) of output phase currents of two modules, the amplitude of harmonic components were measured and summarized in Table 2 and 3. THD of drive system with the conventional CIB module are 3.8 % and 7 % at 10 Hz and 1Hz operation frequency, respectively. On the other hand, THDs with the new CIB module are 1.0 % and 2.9 % at 10 Hz and 1Hz operation frequency. It shows that THD of new CIB module is about 3 times better than the conventional module.

Since the distortion of output phase current (voltage) heavily depends on the ratio between dead-time and switching period, the higher value of dead-time generates the higher THD in PWM operations.

In the conventional drive system, the minimum dead-time has to be long enough to minimize the shoot-through current under PWM operations. Generally, the minimum dead-time is higher than 500 ns. The dead-time generated by microcontroller, controller-based dead-time, is different with the actual dead-time. The actual dead-time includes the turn-on/turn-off delay of a gate drive IC and the rising/falling time of power devices. The output voltage and current distortion are determined by the actual dead-time.

Although the new CIB module with an enhance phase-leg topology does not require any dead-time, the minimum 150 ns dead-time of given microcontroller was applied. Therefore, it is expected that the negative effects of 150ns dead-time in the enhance phase-leg topology are considerably larger than when the dead-time is zero. And also, the output phase current waveforms and THD will be better if zero controller-based dead-time is applied.

V. CONCLUSION

This paper has presented a new 25-pin CIB power module with shoot-through protection that provides the attractive output voltage and current waveforms with much less dead-time nonlinearities compared to conventional power modules. In addition, it provides inherent simple and robust protection scheme for shoot-through failure at any circumstances.

The drive system with new CIB power module provides the minimization of dead-time nonlinearities and improved output phase voltage waveforms with minimum distortion and nonlinearity compared to the conventional PWM inverters.

Experimental results have been presented in this paper to confirm the desired operating characteristics of CIB power module with an enhanced phase-leg configuration. Work is continuing to explore how the enhance phase-leg topology can be best utilized in PWM inverters and future generations of power module designs.

ACKNOWLEDGMENT

This work was supported by the Power IT project of the Ministry of Commerce, Industry and Energy in the Republic of Korea

REFERENCES

[1] N. Mohan, T.M. Undeland and W.P. Robbins, *Power Electronics: Converters, Applications and Design*, John Wiley & Sons, New York, 1995.

[2] Sasagawa, K., Miki, H., "A new driving and protective circuit of IGBT for motor drive application", in *Proc. of 1993 IEEE Appl. Power Elec. Conf. (APEC)*, pp. 402-407, 1993.

[3] S. Musumeci, R. Pagano, A. Raciti, G. Belverde, M. Melito, "A new gate circuit performing fault protections of IGBTs during short circuit transients", in *Rec. of the 2002 IEEE Ind. Appl. Society Ann. Meeting*, vol. 4, pp. 2614–2621, Oct. 2002.

[4] A.R. Munoz, T.A. Lipo, "On-line dead-time compensation echnique for open-loop PWM-VSI drives", *IEEE Transactions on Power Electronics*, vol. 14, pp. 683-689, Jul. 1999.

[5] Y. Murai, T. Watanabe, and H. Iwasaki, "Waveform distortion and correction circuit for PWM inverters with switching lag-times," *IEEE Trans. Ind. Applications*, vol. 23, pp. 881–886, Sept./Oct. 1987.

[7] J.W. Choi and S.K. Sul, "Inverter output voltage synthesis using novel dead-time compensation," *IEEE Trans. Power Electronics*, vol. 11, pp. 221–227, Mar. 1996.

[8] Park, S. and T.M. Jahns, "An Augmented Phase-Leg Configuration with Shoot-Through Immunity and Improvements for High-Current Operation", IEEE APEC Conference, Anaheim, CA, February 22-26, 2004.

[9] Park, S. and Jahns, T.M.; "A novel dead-time elimination method using single-input enhanced phase-Leg configuration", Industry Applications Conference, 2003. 38th IAS Annual Meeting. Conference Record of the , Volume: 3 , 12-16 Oct. 2003 , pp. 2033 – 2040, vol.3

Analysis of sensitivity of the performance of interleaved flyback converter to the principal design parameters

Y.Wang, S.W.H. de Haan
Delft University of Technology
Department of Electrical Engineering, EPP Group
Mekelweg 4, 2628CD, Delft, the Netherlands
Email: yi.wang@tudelft.nl
s.w.h.dehaan@tudelft.nl

A. van Zwam
Mastervolt BV international
Snijdersbergweg 93, Amsterdam, the Netherlands
Email: avanzwam@mastervolt.com

Abstract—Design for optimizing the efficiency of interleaved power converters requires considering many design parameters, such as the magnetizing inductance of the transformer, switching frequency, number of parallel units and parallel semiconductors. Because of the number of parameters and the non-linear relationships among the parameters, a multidimensional design optimization is not straightforward.

In this paper the sensitivity of the efficiency in dependency of the value of principal design parameters is investigated by a loss model built in Mathcad file. The results are used to find a set of parameters for optimal efficiency. In addition, the results can be utilized to make trade-offs between some conflicting requirements such as efficiency versus cost.

I. INTRODUCTION

Optimizing the design of power converters from efficiency point of view means that many parameters have to be taken into account the operating parameters, such as the magnetizing inductance of the transformer L_{mag} and switching frequency f_{sw}. In interleaved converters the number of parameters that determine the converter performance is even larger than in non-interleaved converters, since the number of parallel units (N_m) and parallel semiconductors (N_{sw}) come out as additional parameters. Because of the number of parameters and the non-linear relationship between the performance and the parameter values it is cumbersome to do a multi-dimensional design optimization. In addition such a method would give little insight in interrelationship between efficiency and design parameters. In this paper the sensitivity of the efficiency in dependency of the value of design parameters is investigated by building a loss model in Mathcad file which is used to predict the losses and efficiency with different parameters. This analysis can be used to find an optimized design.

The method is applied to optimize the prototype design of a 12V/360 V, 3600 W interleaved flyback converter with bidirectional power transfer ability. The prototype in Fig. 1 shows the unidirectional version of battery-discharging mode. Interleaving techniques are applied to increase the power density. Six flyback units are connected in parallel and three MOSFETs are paralleled at the switch position in each unit. In order to investigate the sensitivity of the efficiency to the principal design parameters including L_{mag}, f_{sw}, N_m and N_{sw}, a Mathcad loss model is developed that calculates losses as function of these principal parameters and operating points (V_{in}, I_{in}, V_{out}, I_{out} and power level P). Based on this loss model,

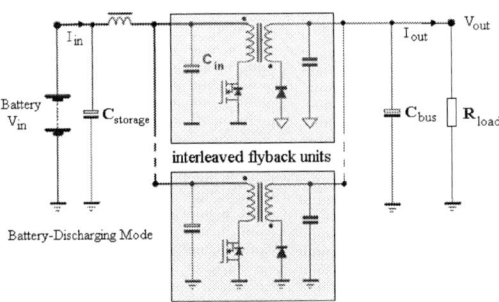

Fig. 1: the circuit of proposed system consisting of an interleaved flyback converter

the parameters are toned to investigate the sensitivity of the efficiency to their variations.

Modelling the voltage and current waveforms on main components accurately at different operating conditions is the first critical step for the precise loss prediction. The loss calculations of all main components in the circuit are based on these modelled voltage and current waveforms.

Because of the high power level, the relatively low input voltage (12 V) and the existence of the parasitic elements, the estimated efficiency by this method is not accurate at all if only the ideal voltage and current waveforms are considered. In this paper, by including most influential the parasitics and realistic parameters of components in the converter, the loss model gives a satisfactory result quite close to the actual efficiency on the prototype and it is justified to be used to predict the system loss and efficiency with different operating conditions.

II. LOSS MODEL DESCRIPTION

A. Parameter Extraction

Most of the component parameters which can be related to prediction of loss can be found in datasheet provided by manufacturer, but they deviate somewhat from the real values in our application. This is because the parameters in datasheet are calibrated under certain standard conditions which are quite different from those in our case. Therefore, to get an accurate enough loss model, several crucial component parameters in the varied operating points and some other practical aspects are also taken into account which influences converter operating behaviours. These realistic factors include:

– The temperature of the transformer winding conductor at

different input powers.

- The on-resistance of MOSFET at different input powers.
- The inductor voltage drop at switch current peak deviated from the input voltage with varied input powers.
- The temperature of inductor core at different input power.
- The ambient temperature at different input powers.
- The influence of the loss in switch on the duty ratio.
- Equivalent series resistance (ESR) of input and output capacitors.

Waveform shapes and most of components' parameters have certain relations with the varied operating condition, for instance, the on-resistance of MOSFET (R_{on}) changes with the junction temperature (T_{junc}) and the switch duty ratio varies with the power level. If all the ever-changing parameters are used as variables in mathematical function, this function will become awkward and it is inconvenient if you want to obtain waveforms under different conditions, you need to check all parameters all your desired conditions and insert them into your function parenthesis. Fortunately, all the variables have certain relations with input power. Defining them beforehand will simplify the function quite a lot. Equation (1) shows this designing principle.

$$P_{loss}(P, D, R_{on}(T_{junc}), v_L) \xrightarrow{D(P), R_{on}(P), v_L(P)} P_{loss}(P) \quad (1)$$

where the D is the switch duty ratio and v_L is the voltage across the magnetizing inductance of the transformer, they all have relations with power P. It is difficult to define these relations precisely. In this paper, the linearized relations approximated from the measurement are utilized. For example, the on-resistances of MOSEFT calculated (R_{low} and R_{high}) from measured voltage and current is obtained on the prototype at input power P_{low} and P_{high}, respectively, then a linear function between these two values is defined as:

$$R_{on}(P) = \frac{P_{high}R_{low} + P_{low}R_{high}}{P_{high} - P_{low}} + \frac{R_{high} - R_{low}}{P_{high} - P_{low}} \cdot P \quad (2)$$

This function is defined at nominal input 12 V.
Similar methods are used on the voltage across the transformer, the ambient temperature and the conductor temperature, etc. Although this method is not very accurate, it is enough and handy to get an acceptable indication of the system losses and efficiency.

B. Loss Calculations

Main loss sources are MOSFET losses and transformer losses. MOSFET loss includes the conduction loss and the switching loss. Transformer losses consist of the core loss and the copper loss.

MOSFET losses

The conduction loss in the MOSFET P_{sw_cond} is modeled as $N_{sw} \cdot R_{on}(P) \cdot I_{sw}^2(P)$, where the N_{sw} is the number of the MOSFETs in parallel and $I_{sw}(P)$ is the current flowing through one single MOSFET. The switching loss calculation should not only involve the current and voltage at the switching transition but also the turn-off interval (t_{off}) and

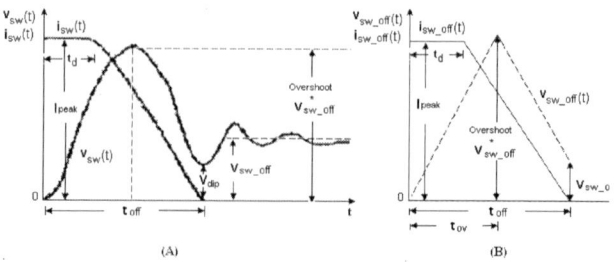

Fig. 3 Measured (A) and modeled (B) switch waveforms at turn-off transitions

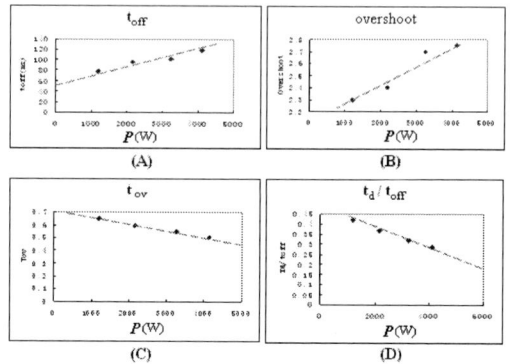

Fig. 2 Measured (in dots) and modeled (in dash lines) t_{off}, overshoot, t_{ov} and t_d/t_{off} with varying power

turn-on interval (t_{on}) of the current and voltage, etc, which are all changes with processed power P. The typical waveforms of the switched current and voltage at turn-off are shown in Fig. 3(A). Due to the resonance between the output capacitance of the MOSFET and the inductance in the circuit, oscillation happens in voltage waveform at turn-off (Fig. 3(A)). To facilitate the waveform modeling, this kind of oscillating waveform is piece-wise linearized as shown in Fig. 3(B), where the modeled current and voltage waveforms are in solid and dash lines, respectively. Parameters like the t_{off}, overshoot, t_d and t_{ov} defining the switched waveforms vary with P. These parameters are verified by the measurement on the prototype to be almost linear with P, shown in Fig. 2, where the measured and modeled values are in dots and dash lines. Therefore, the switched waveforms with these parameters are justified to be modeled as linear with power P and the switching losses can be calculated by the integration of the current and voltage product within the turn-off interval and turn-on intervals. The expression of the switching losses is defined in (3).

$$P_{sw}(P) = P_{sw\_off}(P) + P_{sw\_on}(P)$$
$$= f_{sw} \cdot n_{sw} \cdot \int_0^{t_{off}(P)} \left(i_{sw\_off}(P) \cdot v_{sw\_off}(P) \cdot dt \right) \quad (3)$$
$$+ f_{sw} \cdot n_{sw} \cdot \int_0^{t_{on}(P)} \left(i_{sw\_on}(P) \cdot v_{sw\_on}(P) \cdot dt \right)$$

What should be noticed here is that the turn-on switching loss can be regarded as zero in DCM thanks to the zero initial current.

Transformer losses

978-1-4244-1871-8/07 $25.00 © 2007 IEEE

The major loss sources in transformer include the core loss and copper loss. The hysteresis loss of ferrites was modeled by Mulder [1] using Steinmetz-equation.

$$P_{core} = C_m \cdot f_{sw}^x \cdot B^y \cdot (ct_2 \cdot T_{core}^2 - ct_1 T_{core} + ct) \qquad (4)$$

where

C_m, x, y: loss coefficient;

ct_2, ct_1, ct: temperature coefficient

T_{core}: core temperature (°C);

f_{sw}: switching frequency (Hz);

B: magnitude of flux density in the core (T)

The loss coefficients are usually provided by the core manufacturer for sinusoidal excitation only. The current exciting the flyback transformer core is in triangular shape, so (4) does not apply to this case. Alback et al. [2] extended the applicability of loss coefficients to non-sinusoidal magnetizing currents by the concept equivalent sinusoidal frequency f_{eq}.

$$P_{core} = f_{sw} \cdot C_m \cdot f_{eq}^{x-1} \cdot B^y \cdot (ct_2 \cdot T_{core}^2 - ct_1 T_{core} + ct) \qquad (5)$$

For flyback converter, this f_{eq} can be calculated as

$$f_{eq}(P) = \frac{2}{\pi^2} f_{sw} \frac{1}{D(P)\left(1 - D(P)\right)} \qquad (6)$$

The inductor core has an ambient temperature 27°C at very low input power and 80°C at nominal power. To take into account temperature effect on the core loss, the temperature of the core is linearized with input power between these two power levels and the equation is given as

$$T_{core}(P) = (27 + \frac{53 \cdot P}{600W}) \cdot {}^oC \qquad (7)$$

The flux density in the core is also a function of the power P as other parameters are and should be defined as B (P). The final loss equation of the core is calculated as (8)

$$P_{core}(P) = f_{sw} \cdot C_m \cdot f_{eq}^{x-1}(P) \cdot B(P)^y \cdot [ct_2 \cdot T_{core}(P)^2 - ct_1 T_{core}(P) + ct] \qquad (8)$$

The primary and secondary windings of the flyback transformer are interleaved to decrease the magnetic field strength between the wires and reduce the losses. The configuration of the windings and the current flowing through them basically determine the copper losses. The calculation method can be found in [4]. Since the resistance of the winding is influenced by the copper temperature, temperature relation with power are also linearized and embedded into the loss calculation.

Other losses

Other losses in the loss model involve the diode reverse-recovery loss, input and output capacitors losses and control circuit loss. The calculations of these losses are neglected here due to their relatively small amounts in

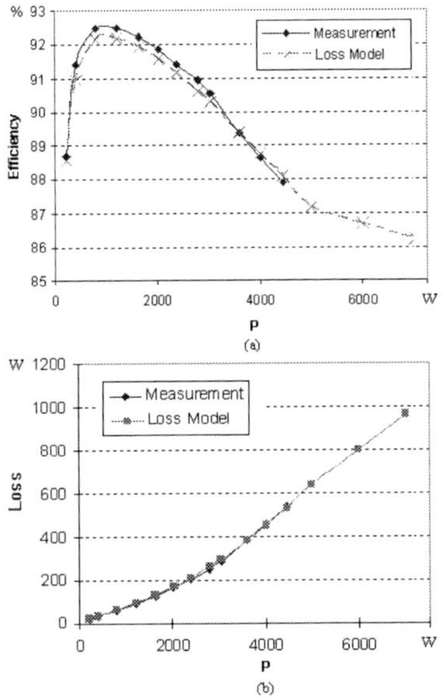

Fig. 4 Measured and modeled efficiency curves (a) and loss curves (b) of the prototype systems with 12V input voltage and different power levels

comparison with the losses in the transformer and the MOSFETs.

III. LOSS MODEL VERIFICATION

The loss model should be verified to be with enough accuracy before it is used to test the sensitivity of the performance to several main parameters and optimize the final design based on the prototype system.

Fig. 4 shows the measured and modeled efficiency and loss curves of the prototype system with 12 V input voltage and varying power level. Since this is a prototype without any protection mechanism, in order to operate it in the safe range, the measurements were done only up to 4.5 kW input power. However, the calculations can be done within a full power range. As you can see in Fig. 4, the efficiency predicted by the loss model is only underestimated a little bit less than half percent for the power below 3.5 kW and overestimated slightly above 4 kW. Measured and modeled losses are almost overlapped with each other owing to the graph resolution. When the converter steps into the CCM (at about 4800 W power level), the calculated efficiency curve has a flatter slope. This is because the peak current of the MOSFET increases slower than that in DCM. From these two figures we can only initially say the loss model is satisfactory. To further investigate the accuracy, the loss in the individual component should be verified. Since the loss in MOSFETs calculated by the loss model takes up more than 50% of the overall losses, the loss model can be verified with full confidence if the MOSFETs losses are proved to be close to the measured values.

The method used to measure the loss in MOSFETs is quite simple. A ceramic layer with known thermal resistance (R_{th}) was placed between the MOSFETs and the heatsink when the prototype system is running. The circumstance has a stable

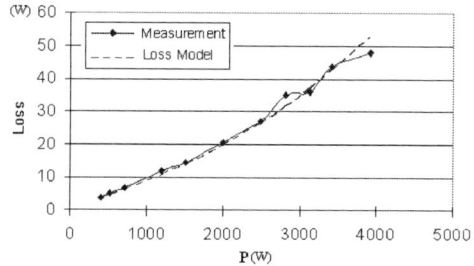

Fig. 5 Measured and modeled MOSFET losses in the prototype system

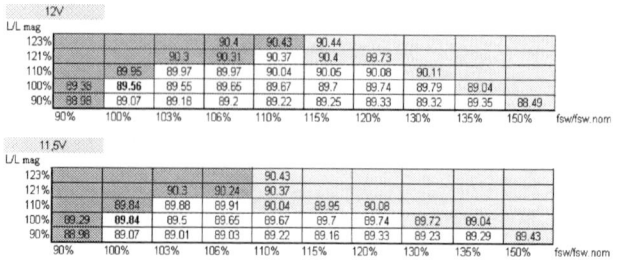

Fig. 6 Efficiency matrices with combination of switching frequency and magnetizing inductance. Upper graph shows the efficiencies with 12 V input and the low one shows with 11.5 V input.

temperature but the air flow is not strictly controlled for this quick and simple measuring method. The temperatures of the MOSFET base (T_b) and the heatsink (T_h) were measured respectively. Assuming all the loss power is transferred by conduction mechanism, the MOSFET losses can then be calculated as

$$P_{MOSFET} = \frac{T_b - T_h}{R_{th}} \qquad (9)$$

Fig. 5 shows the results of the measured and calculated MOSFET losses in the prototype system. In the low power level, these two losses are quite close but at the power higher than 2.5 kW, large discrepancy happens. This is probably due to the unstable air flowing environment in which the measurement was conducted. However, the MOSFET loss calculation in the loss model can be proved to be fairly precise and the overall loss calculation can be deemed as a feasible tool to optimize the converter design.

IV. SENSITIVITY ANALYSIS AND DESIGN OPTIMIZATION

The principal design parameters of the interleaved flyback converters involve f_{sw}, L_{mag}, N_m and N_{sw}. How does the efficiency of the system vary with all these parameters is a complicated question and it is difficult and impractical to try all combinations of these parameters on the hardware. The idea of this paper is to utilize the above verified loss model to predict the sensitivity of the system performance to these

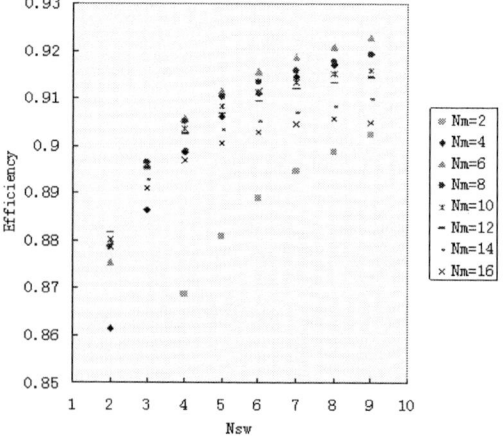

Fig. 7 System efficiency as a function of number of parallel switches N_{sw} in one module and number of parallel power module N_m at nominal load

main design parameters. Based on the sensitivity analysis, the prototype design is optimized to achieve the highest efficiency at the nominal power 3.5 kW.

First, the system efficiency with different number of parallel units and optimum number of MOSFETs in parallel in one unit are investigated by the loss model. Fig. 7 shows the system efficiency as a function of N_{sw} and N_m at nominal load calculated by the loss model. It can be observed that the efficiency increase is the largest if three MOSFETs are used instead of two and when more than 8 MOSFETs the system efficiency does not increase any more but decrease. More than five MOSFETs used will not boost the efficiency distinctly. From Fig. 7, 6 parallel modules and 3 parallel switches in one module can be chosen as the optimum options from performance and cost point of view.

Second, the optimum magnetizing inductance of the flyback transformer and the switching frequency are determined. The combining effect of inductance and switching frequency is considered. Efficiency matrices predicted from the loss model shown in Fig. 6 is made at 12 V input and 11.5 V input voltage by the loss model. The matrices give the efficiencies with different combination of the inductances and switching frequencies. The varying normalized switching frequency is used as x-axis and varying normalized magnetizing inductance as it y-axis. The switching frequency and the magnetizing inductance of the transformer used in the prototype system are the base values.

It is found that the efficiency is the highest when L_{mag} is fixed and the switching frequency is $1.1f_{sw.nom}$. Hence, a switching frequency of $1.1f_{sw.nom}$ can be seen as a trade-off point between the high switching loss and the high conduction loss in MOSFET. The relation between the efficiency and the magnetizing inductance is more straightforward: the efficiency increases as the L_{mag} is increasing. This is because larger L_{mag} will result in a small current peak in the circuit.

The area in purple colour represents the design point causing the transformer core saturated at peak power. To guarantee certain margin for the system working in DCM at the nominal power, the power level of 105% nominal power should be designed in DCM. The matrix area in yellow colour stands for the design point resulting in a CCM operation at the 105% nominal power. Hence, the un-colored area represents the points satisfying an acceptable design requirement at 12 V. In order to guarantee the operation performance when the battery voltage drops to 11.5V, the design requirement for 11.5 V input voltage should be implemented as well. The lower map in Fig. 6 illustrates the efficiency matrix of the system with 11.5 V input voltage. Similar to the upper map, the un-colored area represents the design point which fulfils the design requirement at 11.5 V.

The optimum design point should be around the highest efficiency point existing in both un-colored areas, which results in the decision that the design point with $1.06f_{sw}$ and $1.1L_{mag}$ is selected. Due to the limited resolution of the efficiency matrix, the design around this point should be checked to obtain the highest efficiency under the design requirement.

V. CONCLUSIONS

This paper presents a method for analyzing sensitivity of efficiency to principal design parameters in the interleaved flyback converter system. The method is based on a verified Mathcad loss model which uses all principal design parameters, including the magnetizing inductance of the transformer, switching frequency, number of parallel units and parallel semiconductors, as variables. This loss model gives a convenient and powerful way to investigate the influence of the variation in a single design parameter or a set of parameters on the overall efficiency.

It is found from the efficiency sensitivity analysis that when the converter has six parallel modules and three parallel switches in one unit have the best efficiency performance from efficiency and cost point of view. The optimum magnetizing inductance and switching frequency has been determined by considering the efficiency, core saturation level and operation mode at the nominal condition.

REFERENCES

[1] S. Mulder, "Loss formulas for power ferrites and their use in transformer design," Eindhoven, the Netherlands, Philips, 1994.

[2] M. Albach, T. Duerbaum, and A. Brockmeyer, "Calculating core losses in transformers for arbitrary magnetizing currents, a comparison of different approaches," *Proc. 27th Annu. IEEE Appl. Power Electron. Conf.*, vol. 2, pp. 1463-1468, 1996.

[3] L. Dixon, *"Eddy current losses in transformer windings and circuit wiring"*, Unitrode Seminar Manual SEM600, 1988.

[4] L. Dixon, "Transformer and inductor design for optimum circuit performance", *Unitrode Power Supply Design Seminar*, 2003.

978-1-4244-1871-8/07 $25.00 © 2007 IEEE

Formulation, Measurement and Analysis
for the Thrust Force of HB-type Linear Pulse Motor

Dong-Hee Kim, Jae-Young Ahn, Geon-Il Kang
Chonnam National University
Department of Electrical Engineering
300 Yongbong-dong, Buk-gu, Gwangju, 500-757
Email: kimdh630@hanmail.net

Kwang-Heon Kim, Young-Cheol Lim
Chonnam National University
Department of Electrical Engineering
300 Yongbong-dong, Buk-gu, Gwangju, 500-757
Email: khk@chonnam.ac.kr

Abstract— The linear motor is available for linear transition motion, because of its advantages, the motor design and its application have gradually increased, but the quantitative measurement system of thrust force has not been generalized. Need analysis of correct thrust for control performance improvement of HB-LPM (HB-type Linear Pulse Motor). It is difficult to analyze HB-LPM's thrust. In this paper, HB-LPM's thrust is expressed to mathematical expression. And it is proved validity of this numerical formula by thrust measurement system. Two phase driver is composed. It is verified validity of numerical formula that measure waveform of electric current and voltage that is supplied in each phase.

I. INTRODUCTION

Need analysis of correct thrust for control improvement of performance of HB-LPM(HB-type Linear Pulse Motor). To be original and unfolded static thrust by the most suitable method to HB-LPM that is analysis target, and verifies validity of this numerical formula, developing supplemented static thrust measurement system newly, measured static thrust. Also, compose two phase driver by IGBT inverter, LabVolt's power electronic practice equipment, run HB-LPM, and measured voltage and electric current of A and B phase. Hereupon, in waveform of electric current and voltage that is measured when two phase control run, thrust's calculation data and experimental data, I wish to analyze about thrust.

II. FORMULATION OF HB-LPM'S THRUST

A. Static Thrust when magnetize two phases

$$F_T = -\frac{\sqrt{2}}{2} \frac{2\pi N_r}{\tau} \left(\frac{F F_m P_m P_{(1)}}{P_{(0)}} \right)$$

$$\times \left[\sin(\Theta - 45°) - \frac{F P_{(1)}}{F_m P_m} \sqrt{2} \sin 2(\Theta - 45°) \right] \quad (1)$$

$$+ \triangle F_T$$

B. Thrust when magnetize two phases
Thrust formula is as following.

$$F_T = K_t [-(I_1 - I_3)\sin\Theta + (I_2 - I_4)\cos\Theta]$$

$$\text{(Here, } K_t = \frac{\frac{2\pi N_r}{\tau}}{2} N B_0 A_m \frac{P_{(1)}}{P_{(\theta)}} \quad (2)$$

K_t : Thrust Constant)

C. Total excited thrust

Found out that permeance distribution is near in sine wave according to 1 [pu] displacement about HB-LPM that have rectangular tooth's structure. But, higher harmonics of total excited thrust are existed because distortion happens why with figure 1 if compose thrust that happen according to magnet and coil. Stoppage thrust (Detent Force) of permanent magnet can display High thrust special quality, and know that compares with fundamental frequency and acts by the fourth higher harmonics in figure 2.

III. MEASUREMENT OF STATIC THRUST

To prove whether HB-LPM's thrust form, numerical formula (1), (2) that present in front, are reasonable, need static thrust measurement system. So, developed static thrust measurement system improved newly for this study. Devise with figure 3 so that might measure static thrust precisely by inexpensive price and manufactured system. Table 1 displays HB-LPM's, analysis target, specification.

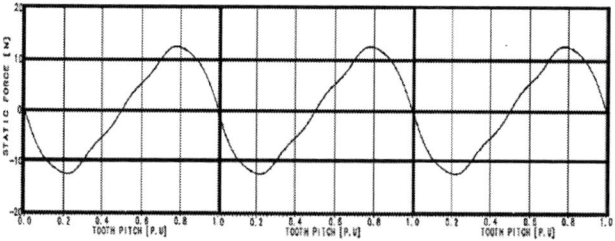

Fig. 1: Total excited thrust of 3 cycle

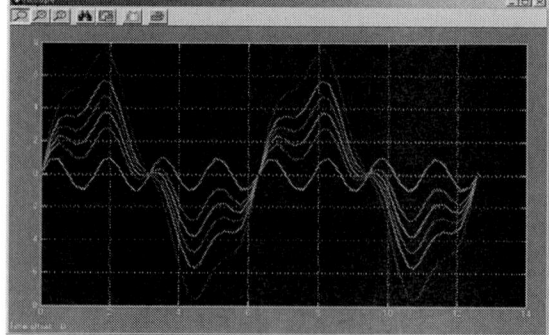

Fig. 2: Thrust feature be followed by exciting current

978-1-4244-1871-8/07 $25.00 © 2007 IEEE

TABLE 1: Spec. of HB-LPM

Item	Unit	LPM
Exit distance	mm	1,000
Resistance/coil	ohm	6
Inductance/coil	mH	3.3
Current/phase	A	6
Series thrust	kgf	25
Static thrust	kgf	28
Air pressure	kg/(cm*cm)	6
Coil phase	phase	2
Loading capacity	kg	10

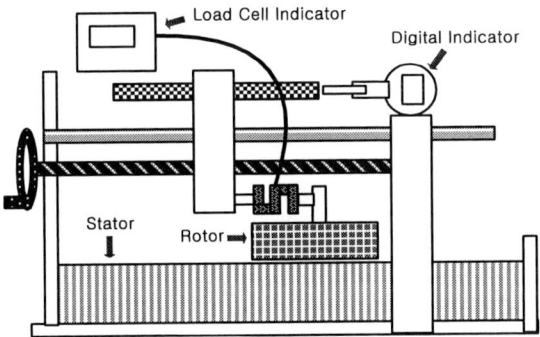

Fig. 3: Measurement system for the static thrust of HB-LPM

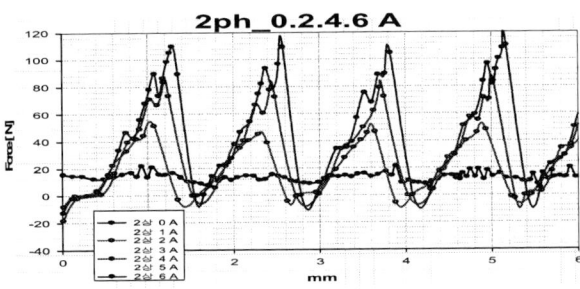

Fig. 4: Four cycle static thrust of two phase exciting

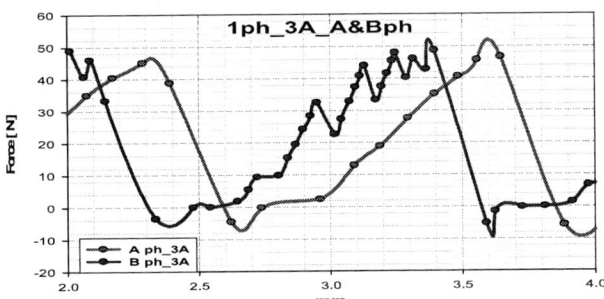

Fig. 5: Static thrust comparison of A phase and B phase

Got following waves and conclusion according as have new static thrust measurement system and measures HB-LPM's static thrust.

First, operation mover exists 12 modules, so each module causes of extreme value, and exists mechanically deflection. Also, could know that 1 mechanical interval of tooth is 1.28mm. (reference Figure 4)

Second, could know that thrust value that calculate mathematically and actual measurement value of static thrust conform. Also, could know that actual measurement value agrees with mathematical expansion of one phase and two phase exciting thrust form.

Third, we can know Kt value is 20, and permanent magnet has 15 ~ 20N's stoppage thrust. Also, when magnetize winding, A and B phase have difference of 90°. (reference Figure 5)

About analysis target HB-LPM, validity of presented thrust formula (1), (2) was proved as measuring thrust.

IV. MEASURED CURRENT AND VOLTAGE (I)

Compose two phase driver by IGBT inverter, LabVolt's power electronic practice equipment, run HB-LPM, and measured voltage and current of A and B phase.

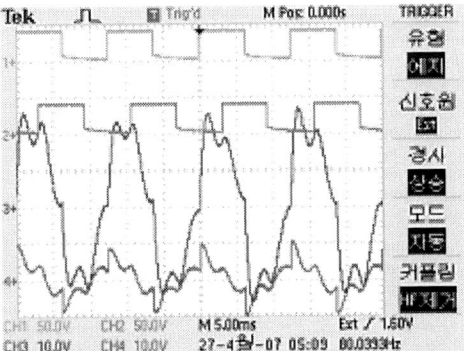

(a) Two phase driver control signal

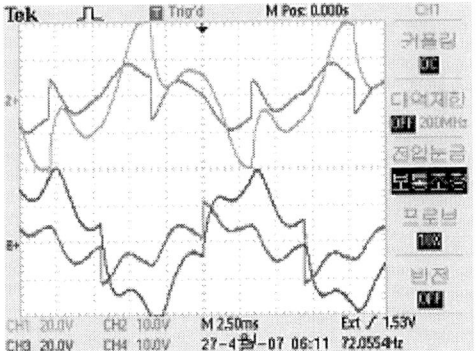

(b) Waveform when forward, control frequency is 72Hz

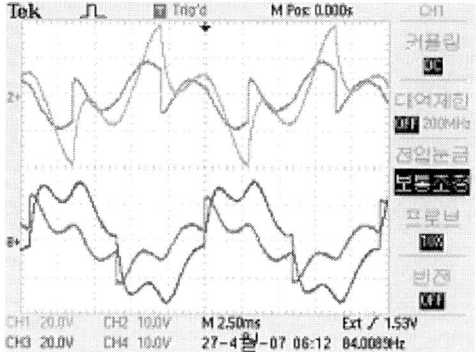

(c) Waveform when forward, control frequency is 84Hz

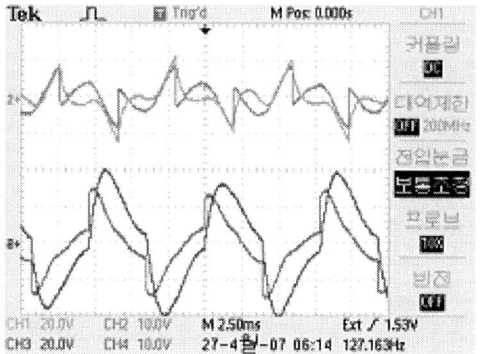

(d) Waveform when forward, control frequency is 127Hz

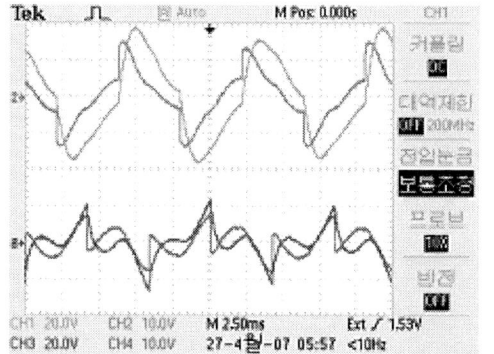

(e) Waveform when backward, control frequency is 127Hz

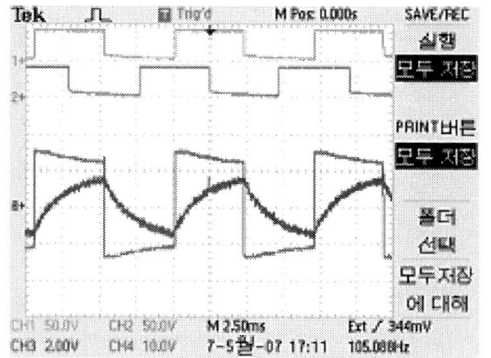

(f) Waveform when increase R, L value 100 times

Fig. 6: Wave of current and voltage at moving (at C=50uF)

Waveform of 1 and 2 in figure 6 (a), is two phase 180 control signal of driver. This time, current of A phase is 3, and voltage of A phase is 4. If compares with figure 1, 2, 4, current of 3 is waveform when run 80Hz, can know that waveform is alike.

Figure 6 (b), (c), (d) are current and voltage waveform of A and B phase, when control frequency is 72Hz, 84Hz, 127Hz. As control frequency increases, as can know in upside three waveform, current distortion (harmonic component) is decrescent.

If compare figure 6 (d) and (e), when control frequency is 127Hz, waveform (d) is forward running, and waveform (e) is backward running. Waveform of 1 and 2 in (d) and (e) is current and voltage waveform of B phase, and Waveform of 3 and 4 is current and voltage waveform of A phase. When run forward direction, as can know in this, current and voltage real value(value that is measured by meter of current and voltage) of B phase could knew that are small than value of A phase, and when run backward, that value of B is bigger than value of A. That is, according to transfer direction of HB-LPM's Mover, could know that value of current and voltage that is approved to each phase is different.

Figure 6 (f) is ideal waveform of current and voltage that is approved in load, when run two phase driver. Because the value of resistance(R) and inductance (L) is very small as can know in Table 1, when increase load R, L value 100 times, current and voltage that is approved to A phase, is tri-angle and square waveform that does not exist distortion.

V. MEASURED CURRENT AND VOLTAGE (II)

Figure 7 is waveform when capacitance is 15uF, run by using LabVolt's IGBT Driver. We can know that distortion of waveform is very serious.

Figure 8 is waveform when use AVR chip, and runs using IPM (IRAMY20UP60B, 20A, 600V) when capacitance is 50uF.

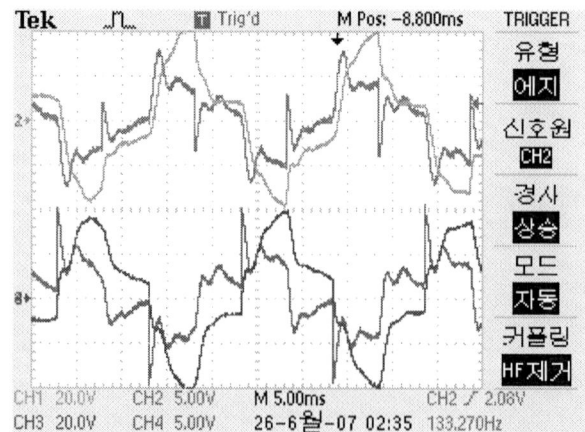

Fig. 7: Wave of current and voltage at moving (at C=15uF)

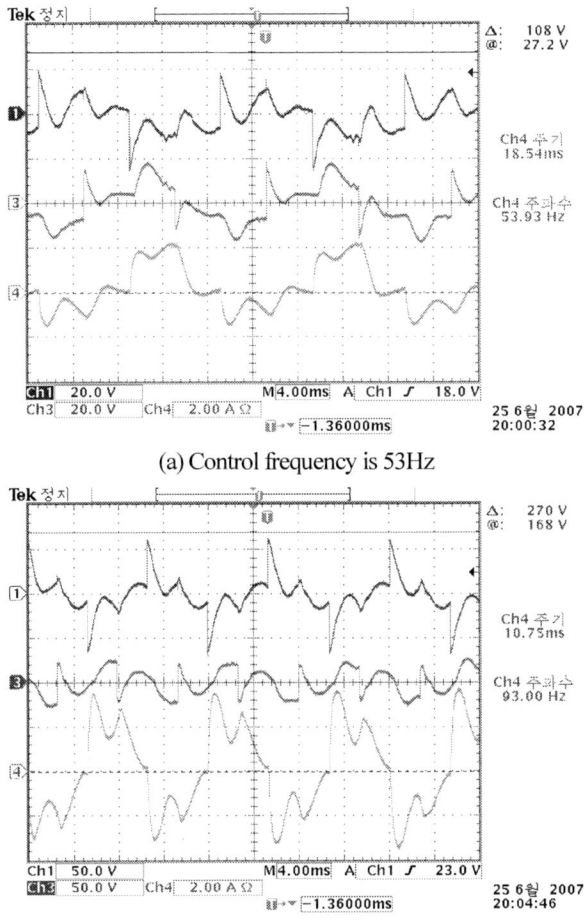

(a) Control frequency is 53Hz

(b) Control frequency is 93Hz

Fig. 8: Wave at moving by AVR driver (at C=50uF)

Figure 9 is waveform when capacitance is 22000uF, run by using LabVolt's IGBT Driver. (d) Waveform is breakaway image of HB-LPM. While frequency is distorted ripple from 72Hz, can know that become breakaway crossing 80Hz.

Figure 10 is waveform when capacitance is 5000uF, run by using LabVolt's IGBT Driver. (a) Waveform is compulsion breakaway image indeed when stop because mover is bumped against a wall. While frequency is distorted ripple from 78Hz, can know that become breakaway crossing 80Hz.

Figure 11 is waveform when use AVR chip, and runs using IPM when capacitance is 5000uF. When experiment, run by Open Loop without breakaway more than 300Hz.

In this study, composed two phase driver, advantage of this, IPM module 1 by accumulated IGBT element 6 is possible in Driver composition. That is, can economize 1 module. In other words, Driver composition is available by IGBT or MOSFET (metal oxide semiconductor field effect transistor) element 4. This is economical big gains.

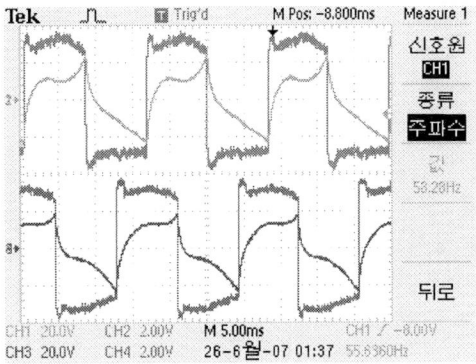

(a) Waveform when backward

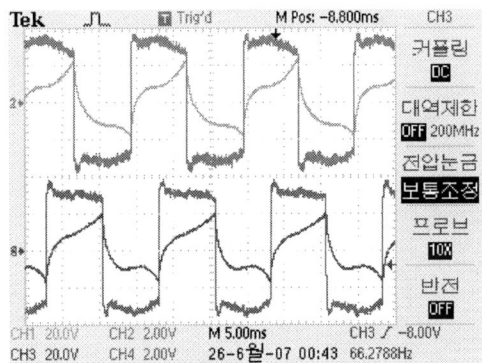

(b) Waveform when forward

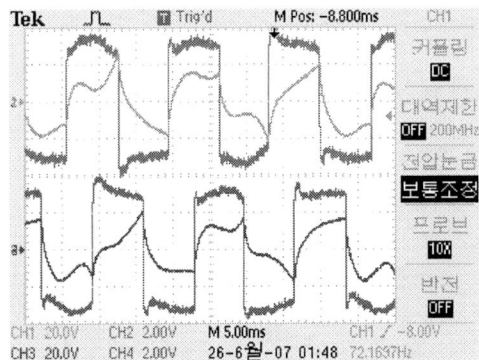

(c) Waveform when critical breakaway

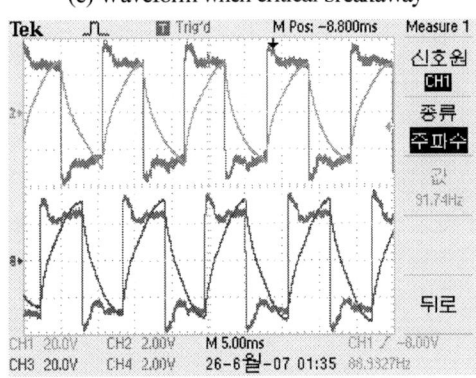

(d) Waveform when breakaway

Fig. 9: Wave of current and voltage at moving (at C=22000uF)

978-1-4244-1871-8/07 $25.00 © 2007 IEEE   93

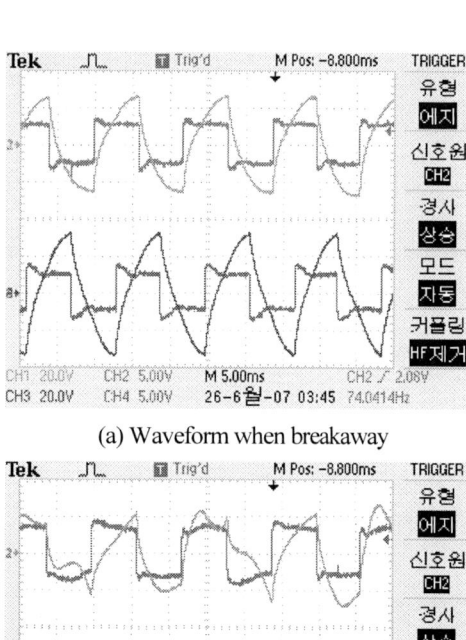

(a) Waveform when breakaway

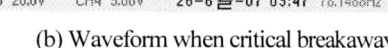

(b) Waveform when critical breakaway

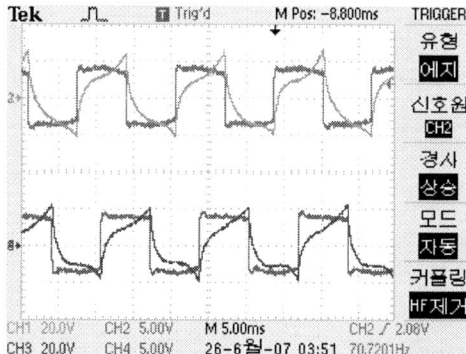

(c) Waveform when forward

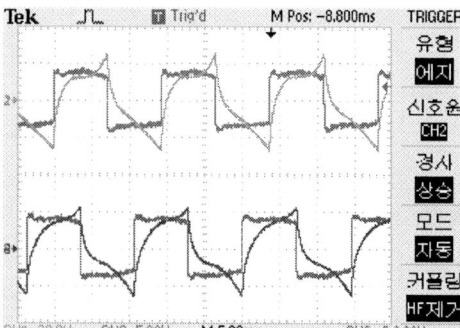

(d) Waveform when backward

Fig. 10: Wave of current and voltage at moving (at C=5000uF)

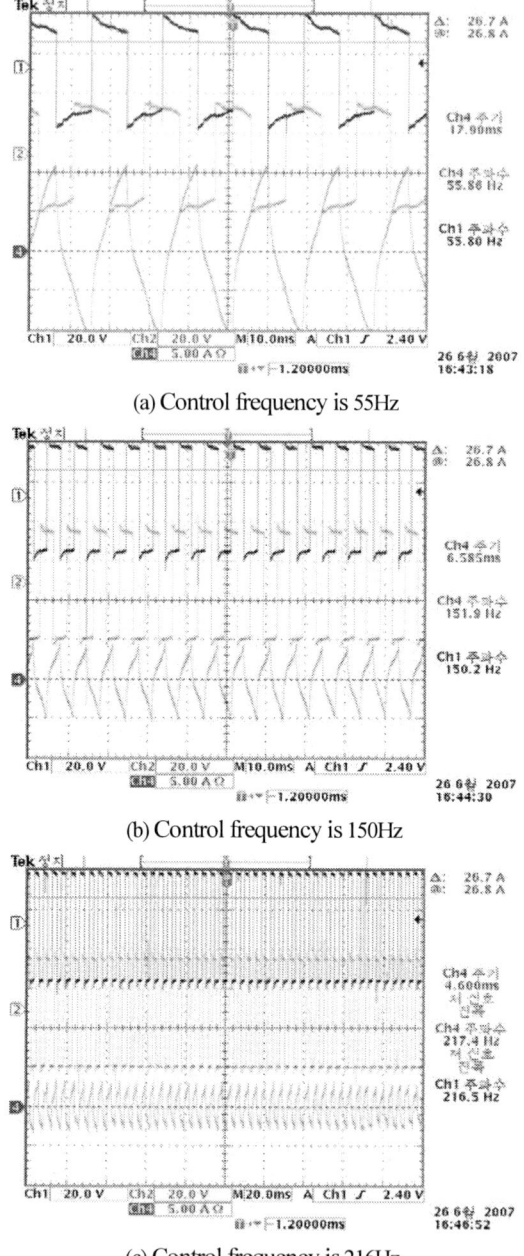

(a) Control frequency is 55Hz

(b) Control frequency is 150Hz

(c) Control frequency is 216Hz

Fig. 11: Wave at moving by AVR driver (at C=5000uF)

Figure 12 is PISM Modeling and simulation result value of voltage and current. Can know that is very similar with figure 10 that is result that measure reality.

Figure 13 does modeling, and shows simulation result value of voltage and current equally by PSCAD. Can know correctness much more in actuality measured value than value of upside figure 12.

Figure 14 (a) is picture of HB-LPM, this study analysis target, (b) is AVR Driver and, (c) is actuality picture of LabVolt's Driver.

978-1-4244-1871-8/07 $25.00 © 2007 IEEE

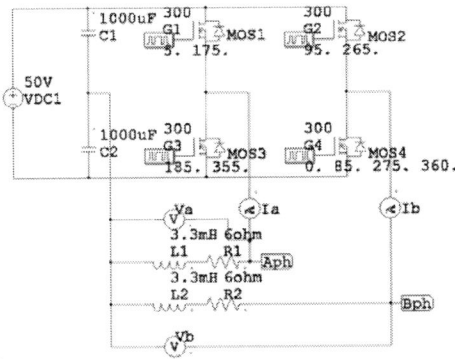

(a) PSIM Modeling (MOSFET)

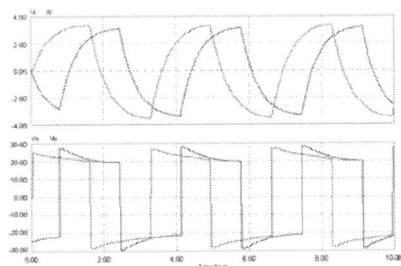

(b) PSIM Result

Fig. 12: PSIM Simulation of HB-LPM

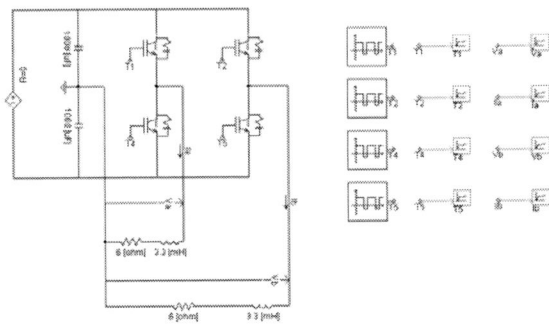

(a) PSCAD Modeling

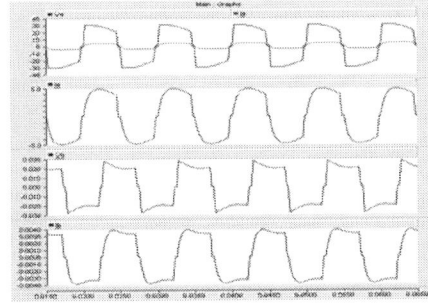

(b) PSCAD Result

Fig. 13: PSCAD Simulation of HB-LPM

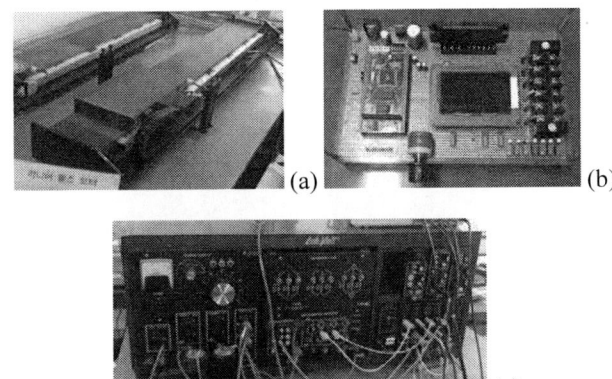

Fig. 14: HB-LPM, AVR driver and LabVolt's driver

VI. CONCLUSION

It is not easy to analyze HB-LPM's thrust. Verified validity of upside numerical formula once more that measure waveform of current and voltage that is been expressed HB-LPM's static thrust and thrust to mathematical expression, and also proved validity of this numerical formula by static thrust measurement system, and composes two phase driver in actuality and magnetizes each phase A, B. Through current waveform of figure 6 (b), (c), (d) that it is measured when two phase run, figure 4 and figure 5, measuring by static thrust measurement system, figure 1 and figure 2, that is waveform of mathematical modeling, can know that each waveforms are each other very alike. This is proving accuracy of mathematical formula. Also, Because waveform that is gotten by an upside experiment is serious of distortion degree, can know why is difficult of HB-LPM control, and know that become cause of vibration and noise, when runs HB-LPM.

ACKNOWLEDGMENT

This paper was studied by research expense support of Chonnam National University RIC (Regional Innovation Center).

REFERENCES

[1] S.H. Kim, Ph.D. Thesis, Chungnam National University, Daejeon, 1999.
[2] H.S. Oh, Ph.D. Thesis, Yeungnam University, Daegu, 2000.
[3] S.J. Kim, M.S. Thesis, Chungnam National University, Daejeon, 2001.
[4] K.P. Park, M.S. Thesis, Chonnam National University, Kwangju, 2002.
[5] I.J. Kim, Ph.D. Thesis, Chungnam National University, Daejeon, 1995.
[6] Syed A. Nasar and I. Boldea, "Linear Electric Motors: Theory, Design, and Practical Applications", pp. 159-195, Prentice-Hall, New Jersey, 1987.
[7] Syed A. Nasar and I. Boldea, "Linear Motion Electric Machines", pp. 1-52, John Wiley&Sons, Toronto, 1976.
[8] Benjamin C. Kuo(Ed.), "Step Motors and Control Systems", pp. 1-113, SRL, Illinois, 1979.
[9] Benjamin C. Kuo, "Theory and Applications of Step Motors", pp. 33-153, West, St. Paul, 1974.

978-1-4244-1871-8/07 $25.00 © 2007 IEEE

Application of Fast Lifting Wavelet Transform to the Estimation of Harmonic for Shunt Active Power Filters

Hui Liu, Guohai Liu and Yue Shen
JiangSu University
School of Electrical and Information Engineering
Zhen Jiang, Jiang Su Province, China
Email: amity@ujs.edu.cn

Abstract—This paper presents a new harmonic measurement scheme based on lifting wavelet transform (LWT) for shunt active power filters (APFs). As one of methods to build second-generation wavelets, LWT can be easy to achieve in time domain for its original bit location calculation. And it could lessen calculation complexity and depress course due to its particular structure, compared with first generation wavelet algorithm (WT). The theoretical and simulative results have verified that the novel LWT approach can abstract harmonic components currents effective with short transient time even when source voltages are asymmetrical distorted. It can accurately detect harmonic components more quickly than classical WT algorithm.

Index Terms—active power filters (APFs), harmonic detection, lifting wavelet transform (LWT)

I.   INTRODUCTION

The compensation for harmonic currents becomes increasingly important owing to the wide use of power electronic equipment. In recent years active power filters (APFs) [1-4] have been widely studied to eliminate the unwanted harmonic and compensate power factor by injecting equal but opposite compensation currents. The operation principle of the shunt active power filter is shown in Fig.1.

Harmonic detection is one of the important steps for APFs in order to generate reference compensation currents. Many methods have been developed in the field of generating reference current [1-7] such as the fast Fourier transform (FFT) methods, the instantaneous reactive power algorithm, the artificial neural networks (ANNs) and Wavelet transform (WT) [4-7], etc. FFT method needs a large amount of calculation and the response time is too long. ANN method is accurate and quick in response, but the network designing and the selection of cut-off harmonics order lack theory foundation. And the instantaneous reactive power algorithms are all time-domain solutions.

Classical WT algorithm, usually named one-generation wavelet, is a powerful signal-processing tool for computing time-frequency representation of signals. However, high computation complexity and long delay of detection result of classical WT algorithm baffles its active application. And one-generation wavelets are all translates and dilates of one or a few basic shapes; the Fourier transform is the crucial tool for wavelet construction. On this account, lifting wavelet

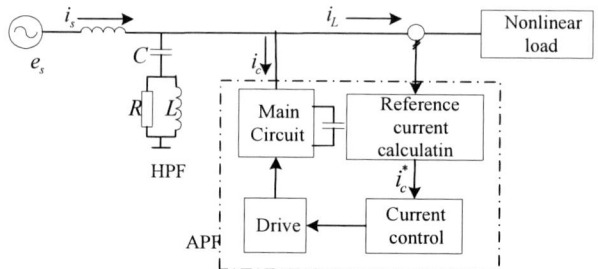

Fig. 1: The principle diagram of shunt active power filter

transform (LWT)[8-10], whose original motivation was to build second-generation wavelets, is proposed. It could lessen calculation complexity and depress course due to its particular structure, adapted to situations that do not allow transform and dilation like non-Euclidean spaces. Now more and more research for LWT are circumfused about two dimension and three dimension signal process domain, such as image compression and image de-noising .

This paper presents a new harmonic measurement scheme based on LWT for shunt active power filters. LWT is used to analysis the high frequency harmonic and results express that LWT, compare to WT, could lessen calculation complexity and depress course due to its particular structure. The theoretical and simulative results have verified that the novel approach can abstract fundamental components currents effective with short transient time even when source voltages are asymmetrical distorted.

II.   WAVELET AND THE LIFTING SCHEME

LWT algorithm is a new method of biorthogonal wavelet construction, by which analysis of frequency domain characteristic of signal can be completed rightly after signal transformation at time domain. Lifting scheme was introduced by Sweldens in 1995 [8-9] had simplified the mechanism in constructing wavelets and this approach becomes more practical to be realized for real-time applications. The scheme does not require the information in Fourier transform because the wavelet transform can be implemented in spatial or time domain.

Fig.2 shows the general block scheme of a wavelet transform. The forward transform uses two analysis filters $\widetilde{h}$ (low-pass) and $\widetilde{g}$ (high-pass) followed by sub sampling, while the inverse transform first up samples and then uses

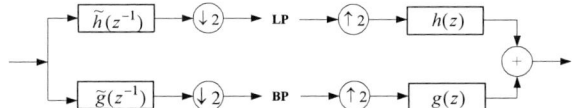

Fig. 2: Decomposition and Reconstruction Chart of Mallet Discrete Wavelet Transform Algorithm

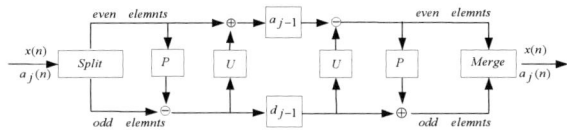

Fig. 3: Principle Chart of Lifting wavelet algorithm

two synthesis filters h (low-pass) and g (high-pass).The condition for perfect reconstructions are given by

$$\begin{cases} h(z)\widetilde{h}(z^{-1}) + g(z)\widetilde{g}(z^{-1}) = 2 \\ h(z)\widetilde{h}(-z^{-1}) + g(z)\widetilde{g}(-z^{-1}) = 0 \end{cases} \quad (1)$$

The lifting scheme is an easy relationship between reconstruction filter pairs(h,g).Split, Predict and Merge Principle Chart of Lifting wavelet algorithm is shown in Fig.3.

The basic steps in lifting operations are[11]:

1) Split: The original signal of length n where $n = 2^j$ is separated into two disjoint sets of even and odd samples.

2) Predict: The predict step replaces the odd element with this difference as in (2) and can be considered as high frequency or detail components. Therefore the predict step can be viewed as high-pass filter. This is done by the following equation:

$$d_{j-1}[n] = x_0[n] - P(x_e[n]) \quad (2)$$

where P is the predict operator.

3) Update: This step replaces the even element with an approximation that is the signal becomes smoother compare to the previous scale. Hence this operation is viewed as low-pass filtering since the smoother signal contains fewer high frequency components. The update equation is as follows:

$$a_{j-1}[n] = x_e[n] + U(d_{j-1}[n]) \quad (3)$$

4) Normalization: The approximation and details coefficients must be normalized in the final step of the transformation. The lifting step is depicted in Fig. 3 for the decomposition or analysis of the forward wavelet transform. The update and predict stages can become a pair but sometimes they may not be together in a lifting step.

In the inverse step, the update step is followed by predict step and finally the odd and even components are merged which interleaves the odd and even elements back into one data stream. The equations for the inverse lifting steps are given by:

$$x_e[n] = a_{j-1}[n] - U(d_{j-1}[n]) \quad (4)$$
$$x_0[n] = d_{j-1}[n] + P(x_e[n]) \quad (5)$$

III. COMPUTATIONAL COMPLEXITY

The filter h is a Finite Impulse Response (FIR) in case only a finite number of filter coefficients h_k are non-zero, as showed in Fig.2. The z-transform of FIR filter h is a Laurent polynomial $h(z)$ given by [8]

$$h(z) = \sum_{k=k_b}^{k_e} h_k z^{-k} \quad (6)$$

The degree of a Laurent polynomial h is defined as

$$|h| = k_e - k_b \quad (7)$$

So the length of the filter is the degree of the associated polynomial plus one. We use the standard wavelet algorithm comparing with lifting scheme. This already takes advantage of the fact that the filters will be sub sampled and thus avoids computing samples that will be sub sampled immediately. The unit we use is the cost, measured in number of multiplications and additions, of computing one sample pair (s_l, d_l).The cost of applying a filter h is |h|+1 multiplications and |h| additions. The cost of the standard wavelet algorithm is 2(|h|+|g|)+2.

Consider a general case not involving symmetry [8]. Take |h|=2N,|g|=2M,and assume M>=N. The cost of standard algorithm now is 4(N+M)+2.And the cost of the lifting algorithm is 2(N+M)+2.In particular case the numbers can differ slightly. Table 1 gives the cost S of the standard WT algorithm, the cost of the LWT algorithm, and the relative speedup (S/L-1) for several examples. Asymptotically, for long filters, the cost of the lifting algorithm for computing the wavelet transform is one half of the cost of the standard algorithm, as showed in Table 1.

IV. SIMULATION RESULTS

A. Common Simulation Conditions

The simulation conditions are described as follows. The three-phase sinusoidal voltages are balanced, and the source frequency is 50Hz. Three-phase controlled bridge rectifier is selected as the circuit load. The line voltage of power source is 100V in RMS value.

Two conditions are considered, ideal mains voltages and asymmetric distorted mains voltages. Remind that all simulation figures below are based on phase-a, and both methods aim at harmonic compensating. In addition every scope waveform ranges from 0.08s to 0.18s.

TABLE 1
Computational cost of LWT versus the WT algorithm

Wavelet	WT	LWT	Speedup
Haar	6	6	0%
Db2	14	9	56%
Db3	22	14	57%
Db8	62	38	63%
Bior6.8	56	30	87%
|h|=2N,|g|=2M	4(N+M)+2	2(N+M)+2	≈100%

978-1-4244-1871-8/07 $25.00 © 2007 IEEE

B. Stable Harmonic detection with Ideal Mains Voltages

The load constants of the three-phase bridge rectifier are load resistance $R = 10\Omega$; trigger angle $\alpha = \pi/6$; Two conditions about direct current (DC) load inductance L are considered, L is considered ideal infinite and $L = 10mH$. A sample rate of 6400 per second is considered for sampling of the simulation signals, and Bior6.8 wavelet is selected. Five-level (M=5) transforms will meet requirement, for the load currents mainly contain $6k \pm 1$ order harmonic, where k is positive.

Fig.4 and Fig.6 show estimated phase–a fundamental component, respectively using WT and LWT with idea mains voltages when DC load inductance is infinite. Fig.5 and Fig.7 show corresponding harmonic currents.

Total Harmonic Distortion (THD) of load currents before detection is 28.78%. It reduces to 3.34% and 3.66% respectively with WT and the proposed LWT. And both detected fundamental current magnitude are close to expectation amplitude, as showed in Table 2.

Another condition when $L = 10mH$ is also analyzed. For limited paper length, Figures about it are omitted. But several parameters result of two methods can be seen clearly in Table 3. THD of load currents before detection is 29.33%. It reduces to 3.21% and 3.38% respectively with WT and the proposed LWT, as showed in Table 3.

Both methods can abstract fundamental current accurately whether DC load inductance is infinite or equals to 10Mh with ideal voltages, except that the proposed lifting scheme cost less calculation.

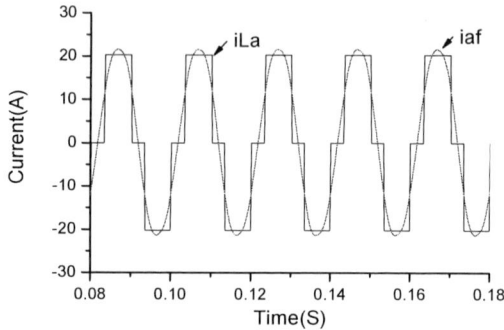

Fig. 6: Estimated fundamental current iaf using Bior6.8 LWT with idea mains voltages when DC load inductance is infinite

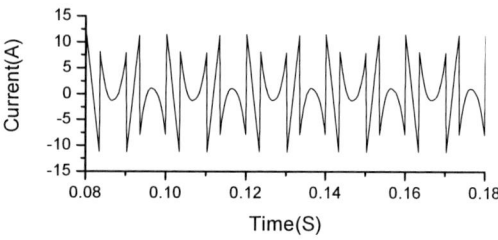

Fig. 7: Estimated harmonic current using Bior6.8 LWT with idea mains voltages when DC load inductance is infinite

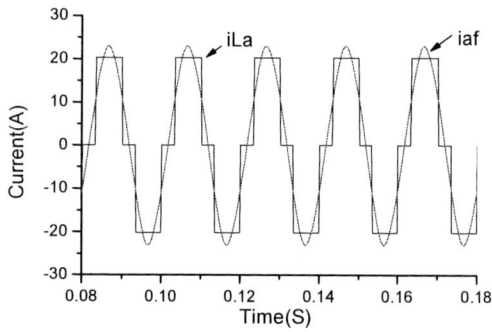

Fig. 4: Estimated fundamental current iaf using Bior6.8 WT with idea mains voltages When DC Load Inductance is infinite

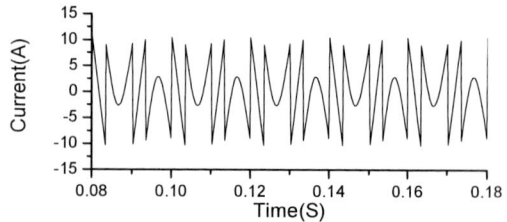

Fig. 5: Estimated harmonic current using Bior6.8 WT with idea mains voltages when DC load inductance is infinite

TABLE 2
Several Parameters Contrast between LWT and WT Methods With Ideal Mains Voltages and Infinite DC Load Inductance

Detection objection	Harmonic currents	
Sample Frequency	6400Hz	
Mains voltages condition	Ideal mains voltages	
THD (%) of mains voltages	0	
THD (%)of load currents before algorithms	28.78%	
Load inductance	Infinite	
Wavelet Base	Bior6.8	
Wavelet level	5	
Wavelet Methods	WT	LWT
Estimated amplitude (A)	23.0931	22.2874
Expectation amplitude (A)	22.2913	
Amplitude Relative Error(%)	3.60%	0.02%
THD (%)of iaf after algorithms	3.34%	3.66%
Performance estimation	good	good
Algorithmic Complexity	complex	simple

TABLE 3
Several Parameters Contrast between LWT and WT Methods With Ideal Mains Voltages and Finite Load Inductance

Detection objection	Harmonic currents	
Sample Frequency	6400Hz	
Mains voltages condition	Ideal mains voltages	
THD (%) of mains voltages	0	
THD (%)of load currents before algorithms	29. 33%	
Load Inductance(mH)	10	
Wavelet Base	Bior6. 8	
Wavelet level	5	
Wavelet Methods	WT	LWT
Estimated amplitude (A)	22. 3678	22. 4527
Expectation amplitude (A)	22. 2635	
Relative Error (%)	0.47%	0.85%
THD (%)of iaf after algorithms	3.21%	3.38%
Performance estimation	good	good
Algorithmic Complexity	complex	simple

C. Stable Harmonic Detection with Distorted Mains Voltages

The line voltage is also 100V in RMS value. But two harmonic are imposed on the fundamental with positive sequence and negative sequence respectively. And load inductance value is not infinite but $L = 10mH$. Else parameters are same as before. Due to the faulty voltages, load currents become distorted too, with THD 35. 92%. .

Fig. 8 shows estimated phase-a fundamental current iaf using Bior6.8 WT with distorted voltage in stable state. Fig.9 shows its corresponding harmonic component. Fig. 10 and Fig. 11 are contrasted with Fig.8 and Fig.9 using the proposed LWT method. Detailed contrast results can be seen in Table 4.

The proposed LWT algorithm behaves good Performance in speed-up computation. And distorted currents can be both decreased to allowable THD level, separately 3.21% using WT and 3.37% using LWT.

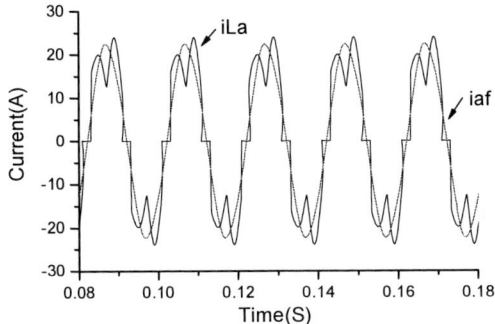

Fig. 8: Estimated fundamental current iaf using Bior6.8 WT with distorted mains voltages

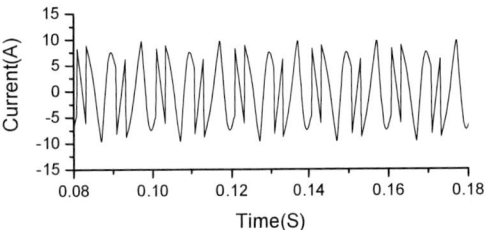

Fig. 9: Estimated harmonic current using Bior6.8 WT with distorted mains voltages

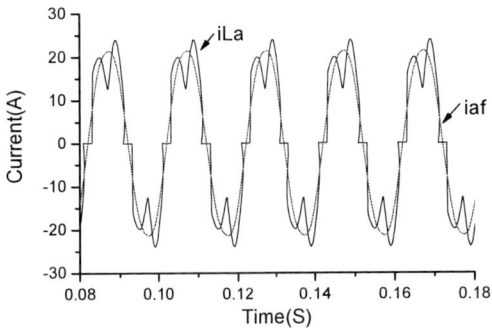

Fig. 10: Estimated fundamental current iaf using Bior6.8 LWT with distorted mains voltages

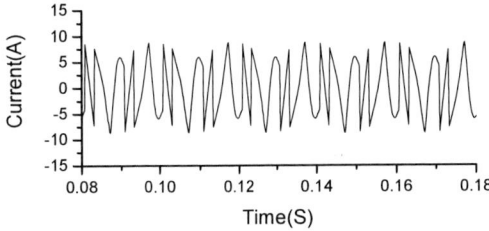

Fig. 11: Estimated harmonic current using Bior6.8 LWT with distorted mains voltages

D. Suspend Harmonic Detection with Ideal Mains Voltages

Besides forenamed simulation, the classical WT algorithm is presented to compare with the novel LWT method in suspend state. Figures of WT algorithm results are omitted for limited paper length.

The line voltage of power source is 100V in RMS value. The load constants of the three-phase bridge rectifier are load resistance $R = 20\Omega$; trigger angle $\alpha = \pi / 6$ and load inductance L is considered ideal infinite. Load currents with resistance load variation take place at 0.1s from $R = 20\Omega$ to $R = 10\Omega$.

Fig.12 shows the proposed LWT method to abstract phase–a fundamental component. Fig.13 shows phase-a harmonic current by LWT algorithm, i.e. reference current, which to generate by subtracting abstracted fundamental components current iaf from practical load current iLa. The transient time is less than one cycle because of the fast characteristic of LWT. And THD level of iaf after detection, 3.21%, is in the allowed range.

TABLE 4
Several Parameters Contrast between LWT and WT Methods
With Distorted Voltages and Finite Load Inductance

Detection objection	Harmonic currents	
Sample Frequency	6400Hz	
Mains voltages condition	Asymmetrical and distorted voltages	
THD (%) of mains voltages		
THD (%)of load currents before algorithms	35. 92%	
Load Inductance(mH)	10	
Wavelet Base	Bior6. 8	
Wavelet level	5	
Wavelet Methods	WT	LWT
Estimated amplitude (A)	21. 9033	22. 2199
Expectation amplitude (A)	21. 2884	
Relative Error (%)	2.89%	4.38%
THD (%)of iaf after algorithms	3.21%	3.37%
Performance estimation	good	good
Algorithmic Complexity	complex	simple

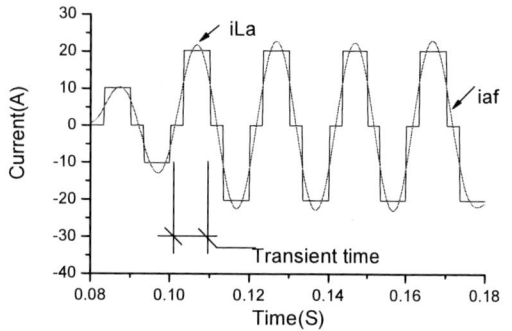

Fig. 12: Estimated fundamental current iaf using Bior6.8 LWT method with idea mains voltages in suspend status

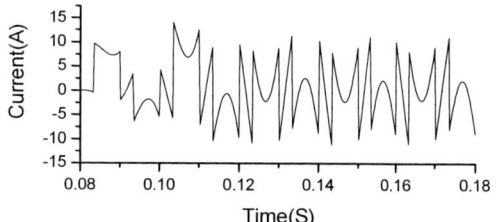

Fig. 13: Estimated harmonic current using Bior6.8 LWT method with idea mains voltages in suspend status

E. Suspend Harmonic Detection with Distorted Mains Voltages

Conditions are almost same as stable harmonic detection with distorted mains voltages, except that load currents with resistance load variation take place at 0.1s , where $R = 20\Omega$ turns into $R = 10\Omega$. Else parameters are same as before.

Fig.14 shows the proposed LWT method to abstract phase–a fundamental component when source voltages are distorted and symmetrical, and LWT algorithm parameter selection is same as those of ideal mains voltages. Fig.15 shows corresponding phase-a harmonic current. The

transient time is less than one cycle And THD level of iaf, 3.34% in the allowed range.

Simulation results show that the two approaches are both effective on compensating harmonic components of load currents under ideal and destroyed source voltages. However, LWT has good performance of low computation complex compared with WT algorithm.

V. CONCLUSIONS

LWT[12] is one of methods to build second-generation wavelets. It has some properties which are not found in many other transforms. One of the advantages of LWT is that it allows for an introduction into wavelet theory without the use of Fourier theory. And LWT is immediately clear .It is easily implemented and can be done in-place. The most important property is the calculation complexity. It is proven that for long filters the lifting scheme cuts computation complexity in half, compared to the standard FIR filter bank algorithm for WT. It is a fast wavelet transform and therefore we will refer to it as the fast lifting wavelet transform (FLWT).

A new harmonic detection method based on LWT has been proposed to improve the performance of APF. The authors provide information about principle of lifting algorithm and procedure of its implementation; and apply lifting scheme of bior6. 8 wavelet filter to the design of active power filter. The computer simulations have verified the effectiveness of lifting schemes under ideal and asymmetrical distorted mains voltages conditions. From the simulation results, the proposed LWT approach has an improved real time character and a better practical performance. It's a kind of fast method to implement wavelet transform and of good value on the measurement of harmonic swiftly and truly.

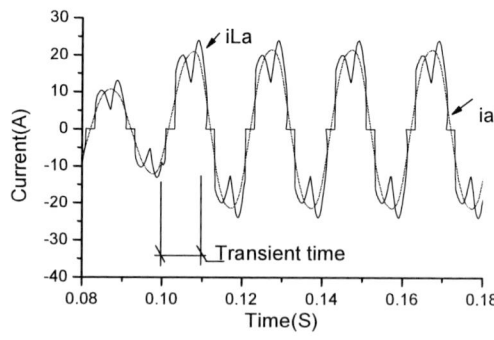

Fig. 14: Estimated fundamental current iaf using Bior6.8 LWT method with distorted mains voltages in suspend status

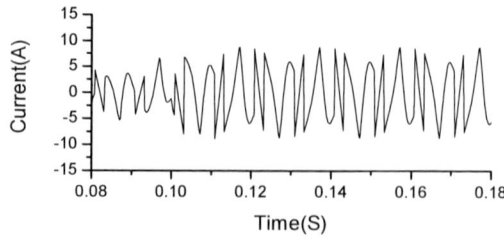

Fig. 15: Estimated harmonic current iaf using Bior6.8 LWT method with distorted mains voltages in suspend status

REFERENCES

[1] Hurng-Liahng Jou, "A Novel Active Power Filter for Harmonic Suppression," *IEEE Transactions on Power Delivery,* Vol. 20, NO. 2, pp.1507-1513,APRIL 2005

[2] Jaime prieto, "A Series-Parallel Configuration of Active Power Filters for VAR And Harmonic Compensation", *28th Annual Conference of the IEEE Industrial Electronics Society,* pp.2945 vol.4,2002

[3] Gary W.Chang, "A comparative study of Active Power Filter Reference Compensation Approaches," *2002 IEEE Power Engineering Society Summer Meeting,* pt. 2, pp.1017-21 vol.2 ,2002

[4] H.L.Jou, " Performance comparison of the three-phase active-power -filter algorithms, " *IEE Proc.Gener.Transm.Distib.* , vol.142, pp.646 -652, 1995

[5] Satish and B. I. Gururaj, "Wavelet Analysis for Estimation of Mean-Curve of Impulse Waveforms Superimposed by Noise, Oscillations and Overshoot," *IEEE Transactions on Power Delivery,* Vol. 16, No. 1,pp.117, January 2001

[6] Ren Zhen, Wavelet Analysis and Applications to Power Systems, *Chinese Electric Power Press,* p.206

[7] Wu Jun Ji, "Application of wavelet filter in harmonic wave detection of power system," *Proceedings o f the EPSA,* Vol.11 No.5-6 pp.51-52, Dec 1999

[8] I. Daubechies and W. Sweldens, "Factoring wavelet transforms into lifting steps," *Journal Fourier and Applications.* Vol. 4, no. 3, pp. 247–269, 1998.

[9] W. Sweldens, "The Lifting Scheme: A Construction of Second Generation Wavelets," SIAM Journal in Math. Analysis, vol. 29, no. 2, pp. 511-546., 1998.

[10] Xu Chuanzhong ,Yang Guanlu , "A Wavelet Lifting Based Active Power Filter," *Journal of Huaqiao University (Natural Science),China* Vol. 25,No. 4,pp.366-370, Oct.2004

[11] Salwani, M.D., Jasmy, Y, "Comparison of few wavelets to filter ocular artifacts in EEG using lifting wavelet transform," *Digital Object Identifier* Nov. 2005 pp:1 - 6

[12] C.Valens,The Fast Lifting Wavelet Transform,1999
c.valens@mindless.com

Small-Signal Analysis of the Lower-Frequency Power Transfer Model

Runxin Wang, *Student Member, IEEE*, Jinjun Liu, *Member, IEEE*,
Hao Wang, *Student Member, IEEE*

School of Electrical Engineering,
Xi'an Jiaotong University, Xi'an, CHINA
E-mail: rxwang@ieee.org

Abstract—This paper presents the most recent progress in deep research to the large-signal modeling technique for current-mode-controlled SMPS modules. The small-signal analysis to the lower-frequency power transfer model is carried out and the conclusion is obtained that the power stage model does not lose any small-signal information so the proposed large-signal simulation scheme is able to be used in small-signal analysis when the system under study reaches its steady state point. This conclusion is quite valuable in both theory and engineering practices.

Index Terms-SMPS, computer simulation, small-signal analysis, lower-frequency power transfer model

I. INTRODUCTION

The lower-frequency power transfer model was proposed to treat each of the current-mode-controlled power stages in a complex dc power supply system, which was formed by inter-connecting different converter modules, to find a way for efficiently implementing the large-signal computer simulations of the whole system[1, 2, 3]. Having known its capability in reflecting the large-signal properties on both the input and output ports of the power stage under investigation in a dynamic process, one may ask such questions very naturally: How about the small-signal properties? Can one derive out the small-signal characteristics from the proposed model? If so, what relationship is kept between the derived results and those obtained from the conventional averaging method?

To answer above questions, this paper will take the lower-frequency power transfer model for buck-boost power stage as an example and analyze its small-signal behaviors near the equilibrium point. It will be shown that the small-signal characteristics obtained from the proposed model are consistent to those from the traditional averaging method, although the results came from different ways and even with different appearances. This conclusion is quite valuable in revealing that the previously-proposed large-signal simulation scheme does not lose any small-signal information and it is able to be used in small-signal analysis when the system under study reaches its steady state point.

The rest of this paper is organized as follows. Section 2 briefly summarizes the conventional averaging method in treating the current-mode-controlled power stages for small-signal analysis. Section 3 derives the small-signal characteristics of buck-boost power stage based on the large-signal lower-frequency power transfer model and compares the results with those obtained in Section 2. Section 4 gives the conclusion.

This work was supported by the National Nature Science Foundation of China (NSFC) under Grant Number 50677053.

II. SUMMARIZING CONVENTIONAL AVERAGING METHODS IN TREATING CURRENT-MODE-CONTROLLED POWER STAGES

Let us consider the buck-boost power stage in Fig. 1. Two approaches exist in deriving the small-signal ac model of the current-mode-controlled power stages:

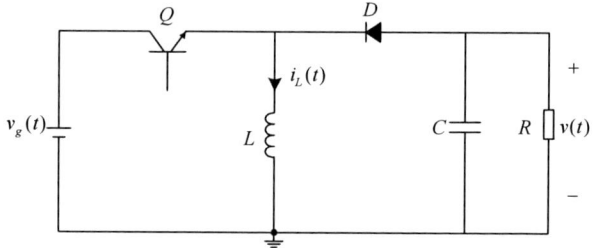

Figure 1 Buck-boost power stage.

A. Algebraic Approach[4]

Under duty cycle control, the small-signal averaged equations for the converter in Fig.1 is written as:

$$L\frac{d\hat{i}_L(t)}{dt} = D'\hat{v}(t) + D\hat{v}_g(t) + (V_g - V)\hat{d}(t)$$

$$C\frac{d\hat{v}(t)}{dt} = -D'\hat{i}_L(t) - \frac{1}{R}\hat{v}(t) + I_L\hat{d}(t) \quad (1)$$

$$\hat{i}_g(t) = D\hat{i}_L(t) + I_L\hat{d}(t)$$

The Laplace transforms of these equations, with initial conditions set to zero, are

$$sL\hat{i}_L(s) = D'\hat{v}(s) + D\hat{v}_g(s) + (V_g - V)\hat{d}(s)$$

$$sC\hat{v}(s) = -D'\hat{i}_L(s) - \frac{1}{R}\hat{v}(s) + I_L\hat{d}(s) \quad (2)$$

$$\hat{i}_g(s) = D\hat{i}_L(s) + I_L\hat{d}(s)$$

Under current-mode control, one of the independent control variable $\hat{d}(s)$ becomes controlled by another control variable, $\hat{i}_{ctrl}(s)$, and

$$\hat{i}_L(s) = \hat{i}_{ctrl}(s) \quad (3)$$

holds well with the assumption that the current-mode-controlled power stages operate ideally. (Actually a constant coefficient K should be inserted but in our discussion at present we can consider K=1 without losing

978-1-4244-1871-8/07 $25.00 © 2007 IEEE

generality[1,2,3,5,6].) To eliminate the old independent control variable $\hat{d}(s)$, one can solute it from the first equation in Eq. (2) and substitute the solution into all other equations which yields

$$sC\hat{v}(s) = -D'\hat{i}_{ctrl}(s) - \frac{1}{R}\hat{v}(s)$$
$$+ I_L \frac{sL\hat{i}_{ctrl}(s) - D\hat{v}_g(s) - D'\hat{v}(s)}{V_g - V} \qquad (4)$$

$$\hat{i}_g(s) = D\hat{i}_{ctrl}(s)$$
$$+ I_L \frac{sL\hat{i}_{ctrl}(s) - D\hat{v}_g(s) - D'\hat{v}(s)}{V_g - V}$$

By collecting terms and using the steady-state relationships, Eq. (4) becomes

$$sC\hat{v}(s) = \left(\frac{sLD}{D'R} - D'\right)\hat{i}_{ctrl}(s)$$
$$- \left(\frac{D}{R} + \frac{1}{R}\right)\hat{v}(s) + \frac{D^2}{D'R}\hat{v}_g(s) \qquad (5)$$

$$\hat{i}_g(s) = \left(\frac{sLD}{D'R} + D'\right)\hat{i}_{ctrl}(s)$$
$$- \frac{D}{R}\hat{v}(s) - \frac{D^2}{D'R}\hat{v}_g(s)$$

Eq. 5 is the final small-signal ac model, from which one can derive transfer functions.

B. Averaged Switch Modeling

Averaged switch modeling method, which brings a more physical insight to the operation properties, can also be used for small-signal model derivation for current-mode-controlled power stages. By eliminating the duty cycle $d(t)$ in the averaged circuit model of the power stage, one can obtain a modified model in which the averaged switch network becomes to reflect the power transfer relationship between the input and output ports of the switch network only. Perturbation and linearization to nonlinear elements in this modified model will lead to the small-signal ac model[4].

Literature [4] has shown that the model obtained from the averaged switch modeling approach is equivalent to that obtained from the algebraic approach by an example of buck converter.

III. SMALL-SIGNAL ANALYSIS OF THE LOWER-FREQUENCY POWER TRANSFER MODEL

Fig. 2 shows the new buck-boost power stage definition in deriving the large-signal lower-frequency power transfer model and Fig. 3 shows the corresponding derived model. Attentions should be paid to that the reference direction of the output port in Fig. 2 is reversed comparing to that in Fig. 1. This section will derive the small-signal model based on Fig. 3 and analyze its consistency with the results obtained from the traditional averaging methods.

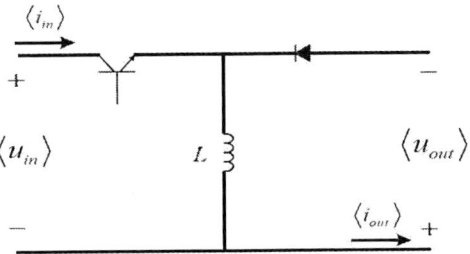

Figure 2 Buck-boost power stage definition.

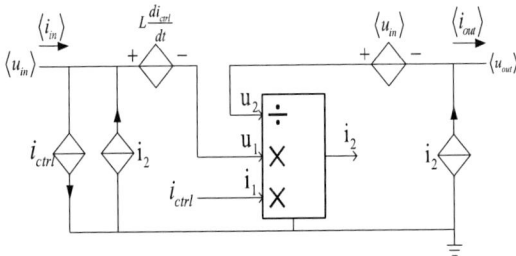

Figure 3 The larger-signal lower-frequency power transfer model for the buck-boost power stage.

In the lower-frequency power transfer model shown in Fig. 3, the algebraic operation block is the unique nonlinear component, which constraints the following relationship among the four terminal variables:

$$\langle u_1 \rangle \langle i_{ctrl} \rangle = \langle u_2 \rangle \langle i_2 \rangle \qquad (6)$$

To construct the small-signal models by perturbation and linearization method, let

$$\langle u_1 \rangle = U_1 + \hat{v}_1$$
$$\langle u_2 \rangle = U_2 + \hat{v}_2$$
$$\langle i_{ctrl} \rangle = I_{ctrl} + \hat{i}_{ctrl} \qquad (7)$$
$$\langle i_2 \rangle = I_2 + \hat{i}_2$$

Eq. (6) will become

$$\left(U_1 + \hat{v}_1\right)\left(I_{ctrl} + \hat{i}_{ctrl}\right) = \left(U_2 + \hat{v}_2\right)\left(I_2 + \hat{i}_2\right) \qquad (8)$$

Upon equating and eliminating the dc terms, the linear relation among small-signal terms of Eq. (8) can be written as

$$\hat{v}_1 I_{ctrl} + U_1\hat{i}_{ctrl} = \hat{i}_2 U_2 + I_2\hat{v}_2 \qquad (9)$$

Solution for the small signal output current of the algebraic operation block yields

$$\hat{i}_2 = \hat{v}_1 \frac{I_{ctrl}}{U_2} + \hat{i}_{ctrl}\frac{U_1}{U_2} - \hat{v}_2\frac{I_2}{U_2} \qquad (10)$$

Embedding Eq. (10) into Fig. 3 and eliminating all the dc terms will lead to the small-signal model near the equilibrium point.

To make it clear for comparison with results obtained from traditional averaging method, equivalent circuits for large signal, dc signal and small-signal are constructed as those respectively shown in Fig. 4, Fig. 5 and Fig. 6. The Laplace transforms are performed in Fig. 6 with the assumption that initial conditions set to zero.

978-1-4244-1871-8/07 $25.00 © 2007 IEEE

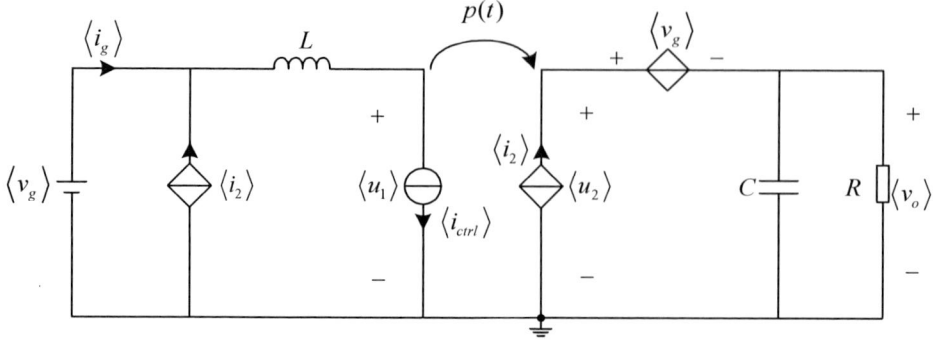

Figure 4　The equivalent circuit of the large-signal lower-frequency power transfer model.

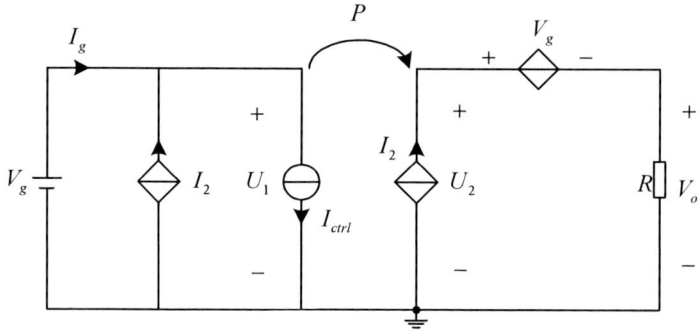

Figure 5　The equivalent circuit for dc signal.

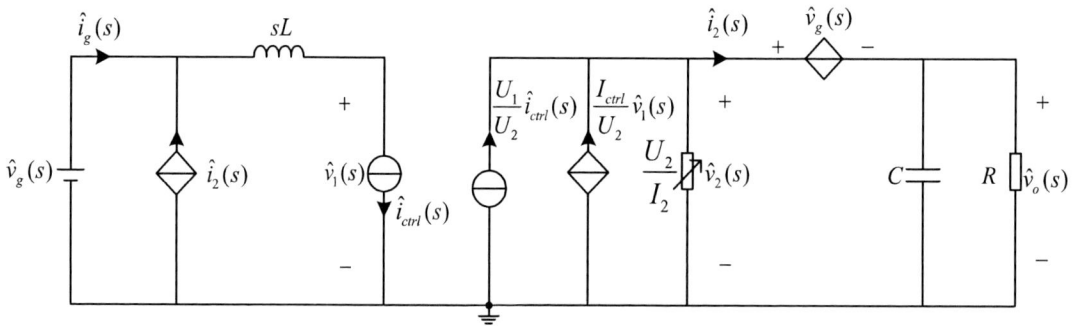

Figure 6　The derived linear small-signal ac model.

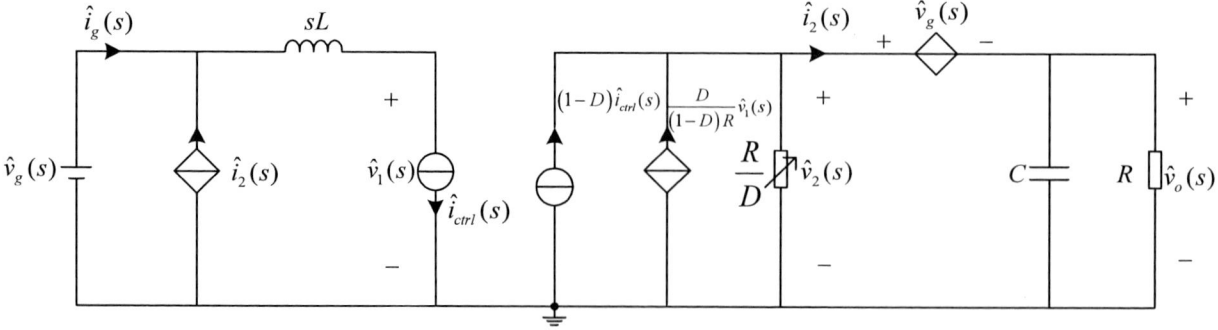

Figure 7　The modified linear small-signal ac model, in which the dc relationships are represented by duty cycle D.

978-1-4244-1871-8/07 $25.00 © 2007 IEEE　　　　104

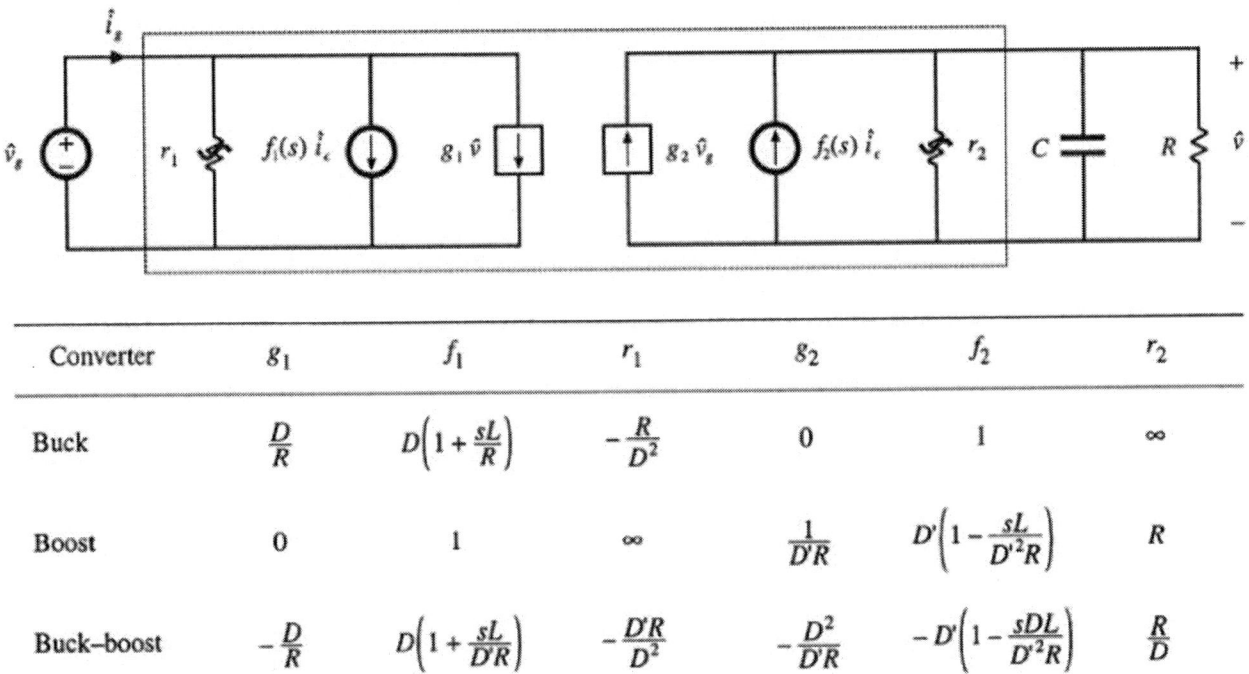

Converter	g_1	f_1	r_1	g_2	f_2	r_2
Buck	$\dfrac{D}{R}$	$D\left(1+\dfrac{sL}{R}\right)$	$-\dfrac{R}{D^2}$	0	1	∞
Boost	0	1	∞	$\dfrac{1}{D'R}$	$D'\left(1-\dfrac{sL}{D'^2R}\right)$	R
Buck–boost	$-\dfrac{D}{R}$	$D\left(1+\dfrac{sL}{D'R}\right)$	$-\dfrac{D'R}{D^2}$	$-\dfrac{D^2}{D'R}$	$-D'\left(1-\dfrac{sDL}{D'^2R}\right)$	$\dfrac{R}{D}$

Figure 8 The equivalent small-signal circuit and parameters obtained by traditional averaging method[4].
(Upper: equivalent circuit; Lower: parameters)

Considering the following dc relationships: $U_1=V_g$, $U_2=V_g+V_o$, $I_2=I_o$, $I_{ctrl}=I_g+I_2$, $V_o=RI_o$ and $V_o/V_g=I_g/I_o=D/(1-D)$ in the equilibrium point under the output voltage/current definitions in Fig. 2, Eq. (10) can be simplified by eliminating dc quantities and representing them by the duty cycle D in steady state,

$$\hat{i}_2 = \hat{v}_1 \frac{D}{(1-D)R} + \hat{i}_{ctrl}(1-D) - \hat{v}_2 \frac{D}{R} \quad (11)$$

By using Eq. (11), Fig. 6 is redrawn as Fig. 7, from which the following transfer functions can easily be derived:

A. *The line-to-output transfer function*

$$G_{vg}(s) = \left.\frac{\hat{v}_o(s)}{\hat{v}_g(s)}\right|_{\hat{i}_c=0} = \frac{D^2}{(1-D)R}\left(\frac{R}{D}//R//\frac{1}{sC}\right) \quad (12)$$

B. *The control-to-output transfer function*

$$G_{vg}(s) = \left.\frac{\hat{v}_o(s)}{\hat{i}_{ctrl}(s)}\right|_{\hat{v}_g=0}$$

$$= (1-D)\left(1-\frac{sDL}{(1-D)^2 R}\right)\left(\frac{R}{D}//R//\frac{1}{sC}\right) \quad (13)$$

C. *The output impedance*

$$Z_o(s) = \frac{R}{D}//R//\frac{1}{sC} \quad (14)$$

D. *The input impedance*

$$Z_{in}(s) = \left.\frac{\hat{v}_g(s)}{\hat{i}_g(s)}\right|_{\hat{i}_c=0} = -\frac{1-D}{D}\left(\frac{R}{D}+R//\frac{1}{sC}\right) \quad (15)$$

Considering the reversed output port reference direction in Fig. 2 compared with Fig. 1, one can verify that all above transfer functions are consistent to those derived from Fig. 8, which shows the equivalent circuits constructed on the base of the algebraic approach and the corresponding parameters.

Actually, similar verification process can be carried out to other basic power stages such as buck and boost converters. We have the same conclusion that the small-signal characteristics obtained from the proposed model are consistent to those from the traditional averaging method, although the results came from different ways.

IV. CONCLUSION

This paper presents the most recent progress in deep research to the large-signal modeling technique for current-mode-controlled SMPS modules. The small-signal analysis to the lower-frequency power transfer model is carried out and we get the conclusion that the power stage model does not lose any small-signal information so that the proposed large-signal simulation scheme is able to be used in small-signal analysis when the system under study reaches its steady state point. This conclusion is quite valuable in both theory and engineering practices.

REFERENCES

[1] Runxin Wang, Jinjun Liu, Hao Wang and Qinsan Hou, "Multi-level modeling of PFC rectifier and dc-dc converter modules in inter-connected switching power supply systems," *IEEE APEC 2007*, pp. 1475-1479.

[2] Runxin Wang, Jinjun Liu and Hao Wang, "Universal approach to modeling current mode controlled converters in distributed power systems for large-signal subsystem interactions investigation," *IEEE APEC 2007*, pp. 442-448.

[3] Runxin Wang, Jinjun Liu and Hao Wang, "Modeling Current-Mode-Controlled Power Stages for Simulating Multiple-Module Inter-Connected Power Supply Systems,"*IEEE PESC 2007*, pp. 975~979.

[4] R. W. Erickson and D. Maksimovic, Fundamentals of Power Electronics 2nd Edition, Kluwer Academic Publishers, MA, 2001.

[5] L. Dixon, "Average current mode control of switching power supplies," *Unitrode Application Note*, U-140.

[6] Venable Industries, "New Techniques for Measuring Feedback Loop Transfer Functions in Current Mode Converters," available on the Venable Industries web site. at: http://www.venable.biz/

Generation of Differentiable Homomorphism Function Based on Self-Organizing Polynomial Data-mining Algorithm

YoungWoo Kim
Nagoya University
EcoTopia Science Institute
Furo-cho, Chikusa-ku, Nagoya, 464-8603, Japan
Email: ywkim@esi.nagoya-u.ac.jp

Tatsuo Narikiyo
Toyota Technological Institute
Department of Electrical Engineering
2-12-1 Hisagata Tenpaku, Nagoya 468-8511, Japan
Email: n-tatsuo@toyota-ti.ac.jp

Abstract— This paper presents a new controller design method based on the data-mining polynomial algorithm. We show application of a polynomial data-mining algorithm to controller design, where an input-state linearized polynomial vehicle model is developed for a very low speed operation and without introducing any process with fudge factor, control inputs of nonlinear system are obtained in the original coordinate. We verify the developed modeling method and controller design method through some numerical experiments.

I. INTRODUCTION

This paper proposes a new controller design method based on application of the data mining polynomial algorithm. Generally the model obtained by applying data mining polynomial algorithm is high order nonlinear dynamics, and constrained by the permissible sets in consideration for reproducibility of the developed model.

It is not easy to apply model predictive policy to the systems with the developed polynomial models. Strict Linearization method [1] is an approach to nonlinear controller design, where the central idea is to algebraically transform nonlinear system dynamics into linear system dynamics. Although the usefulness of this method, a very critical point is as follows: there is no general method to find a differentiable homeomorphism function which is used for the dynamics transformation.

After the strict linearization, the newly developed input variables and state variables generally do not have any physical meaning. Therefore when we construct controller for the strictly linearized model, the newly developed input must have been transformed to the values in the space made by the original input variables. Since this process requires inverse of the matrix consists of high order polynomial element, it contains large fudge factor.

The proposed controller design method involves strict linearization process and the original input is obtained using the original nonlinear state equation together with the strictly linearized state equation. The proposed method does not require any inverse process of the matrix which may provokes large error. The proposed method is verified through some numerical experiments.

II. INPUT-STATE LINEARIZED MODEL

In the previous applications of the data-mining algorithm,

model in a fixed form. In [2], the relationship between the base angular distance of planar space robot and the size of patterned movement of both arms is successfully developed in a single variable system form. In [3], nonlinear dynamics of an unmanned vehicle system is modeled in an affine system form. If the system is modeled in linear form, various control theories such as model predictive policy, multi-parametric programming and so on, will be applied easily to the system. The proposed modeling method is a statistical input-state linearization procedure, where affine systems of the form (1) are input-state linearized.

$$\dot{x} = f(x) + \sum_i g_i(x) u_i$$

$$x = \begin{bmatrix} x_1 & \cdots & x_J \end{bmatrix}^T \in R^J, \quad u = \begin{bmatrix} u_1 & \cdots & u_I \end{bmatrix}^T \in R^I$$

$$f(x) = \begin{bmatrix} f_1(x) \\ \vdots \\ f_J(x) \end{bmatrix}, \quad g_1(x) = \begin{bmatrix} g_{11}(x) \\ \vdots \\ g_{J1}(x) \end{bmatrix}, \quad g_I(x) = \begin{bmatrix} g_{1I}(x) \\ \vdots \\ g_{JI}(x) \end{bmatrix}$$

$$x_{j,\min} \le x_j \le x_{j,\max}, \quad u_{\min} \le u \le u_{\max} \quad (1)$$

From the viewpoint of the differentiable homeomorphism, following theorem must be satisfied for the input-state linearization.

Theorem 1: The system (1) is input-state linearized if we find V around origin and positive integer k_1, k_2, ... k_I as follows,

(a) $k_1 + k_2 + ... + k_I = J$
(b) All the element x of the set of vector field S is linearly independent with each other.
(c) For arbitrary $k \ge 0$ on V,
$$C^\infty \text{span}(D_k) = C^\infty \text{span}(S_k) \quad (2)$$
(d) $S_{k_i - 2}$ is involutive on V

, where

$$D_k = \left\{ ad_f^l g_i(x) \middle| 0 \le l \le k, 1 \le i \le I \right\} \quad (3)$$

$$S = \left\{ ad_f^l g_i(x) \middle| 0 \le l \le k_i - 1, 1 \le i \le I \right\} \quad (4)$$

$$S_k = \left\{ ad_f^l g_i(x) \middle| 0 \le l \le \min(k_i - 1, k), 1 \le i \le I \right\} \quad (5)$$

Here, the involutiveness of the distribution

$$span(\Delta(x)): \quad \Delta(x)=\left(t_1(x),t_2(x),\cdots,t_p(x)\right) \quad (6)$$

implies for a given vector field $t_i(x)$ $(1 \le i \le l)$ to satisfy Lie bracket

$$\left[t_i,t_j\right] \in span(\Delta) \quad (7)$$

, where the pair $(^\forall i, j)$ is arbitrary combination. ■

The system models in industrial world often contain high order nonlinear dynamics and it is almost impossible to show the involutivenss of S_{k_j-2} in a general manner. In the proposed method, the differentiable homomorphism function T that is used for the dynamics transformation is obtained satisfying the condition (4) as follows.

Polynomial Input-State Linearization Method

Step1: Initialization: Set all the observational data sets as a set of ζ_T as follows.

$$\zeta_T = \{x_1,\cdots x_J\}$$

Set the iteration number $l_{no}=0$, number of the previously developed variable $n_N=J$, the first iteration optimality level $z_{0,best}=\infty$ and the threshold value J_{th}.

Step2: Acquisition of Data Sets: The observational data sets are subdivided into the training set and the checking set.

Step3: Propagation of Variables: All n independent intermediate variables in the training set are combined by the partial polynomial as follows.

$$z_{l_{no},k}(x_{l_{no},p},x_{l_{no},q}) = c_{1,k}x_{l_{no},p} + c_{2,k}x_{l_{no},q} \\ + c_{3,k}x^2_{l_{no},p} + c_{4,k}x^2_{l_{no},q} + c_{5,k}x_{l_{no},p}x_{l_{no},q} \quad (8)$$

, where $x_{l_{no},p}$ is the variable in the training set at the iteration l_{no}, i denotes variable index, and $1 \le p,q \le n_N(n_N-1)/2$.

Step4: Minimization of Performance Criterion: In order to minimize the performance criterion

$$J_p\left(z_{l_{no},k}\right) = \sum_{i=1}^{l} W_i\left(L_g z_{l_{no},k}\right)^2 \quad (9)$$

, J_p is recast to a general form of the nonlinear programming problem as follows.

$$J_p\left(z_{l_{no},k}\right) =$$

$$\frac{1}{2}\begin{bmatrix}c_1\\c_2\\c_3\\c_4\\c_5\end{bmatrix}^T \begin{bmatrix}2h_{11}(x)&2h_{12}(x)&2h_{13}(x)&2h_{14}(x)&2h_{15}(x)\\0&2h_{22}(x)&2h_{23}(x)&2h_{24}(x)&2h_{25}(x)\\0&0&2h_{33}(x)&2h_{34}(x)&2h_{35}(x)\\0&0&0&2h_{44}(x)&2h_{45}(x)\\0&0&0&0&2h_{55}(x)\end{bmatrix}\begin{bmatrix}c_1\\c_2\\c_3\\c_4\\c_5\end{bmatrix}$$

$$+\begin{bmatrix}h_1(x)\\h_2(x)\\h_3(x)\\h_4(x)\\h_5(x)\end{bmatrix}\begin{bmatrix}c_1\\c_2\\c_3\\c_4\\c_5\end{bmatrix}^T$$

subject to

$$1-\varepsilon \le \sum_{k=1}^{5} c_k^2 \le 1+\varepsilon \quad (11)$$

, where W_i is weighing parameter, ε is small tolerance and function h is easily obtained by partially differentiating the performance criterion.

Step 5: Representativeness Assessment: Reorder $z_{l_{no},k}$ in the order of the performance criterion, where $(1 \le k \le n_N(n_N-1)/2)$.
Screen out $z_{l_{no},k}$ which satisfies $J_p\left(z_{l_{no},k}\right) \ge J_{th}$. Set $z_{l_{no},best}$ among $z_{l_{no},k}$, $1 \le k \le n_N(n_N-1)/2$. For each $z_{no,k}$, do the following: $\zeta_T = \zeta_T \cup z_{l_{no},k}$.

Step 6: Model Optimality Test: If $z_{l_{no},best} < z_{l_{no}-1,best}$, set $l_{no} = l_{no}+1$, and go to Step 2. Otherwise terminate with end.

In step 2, the percentage of the training set among overall observational data set is selected empirically. While the training set is used to propagate each variable in order to minimize the performance criterion, the checking set is used to verify the generation of the developed model. In step 3, n_N is initially set to the number of observational variables J, and at the next iteration, this value will be $n_N \equiv n_N(n_N-1)/2$ by the combination of n_N variables at this iteration. $z_{l_{no},k}$ is the newly generated variable at the iteration of l_{no}. This variable is redefined as new x in step 5. In step 4 $L_{g_i} z_{l_{no},k}$ is Lie differe--ntiation, equivalently

$$L_{g_i}z_{l_{no},k} = \frac{\partial z_{l_{no},k}}{\partial x}g_i = \left(\frac{\partial z_{l_{no},k}}{\partial x_1}g_{1i}+\cdots+\frac{\partial z_{l_{no},k}}{\partial x_J}g_{Ji}\right). \quad (12)$$

In the nonlinear programming problem of the step 4, new planning variables are $C(x) = [c_1\ c_2\ c_3\ c_4\ c_5]^T$, $h_i(x)$ $(1 \le i \le 5)$ is the coefficient polynomial in x of the variable c_i, and $h_{ij}(x)$ $(1 \le i, j \le 5, i \ne j)$ is the coefficient polynomial in x of the variable c_i^2, respectively.

Note that the square sum of the planning variables $[c_1 c_2 c_3 c_4 c_5]^T$ (of the nonlinear programming problem) is obtained at every iteration, l_{no}. Since these variables are also coefficient of T, they are quantified with the constraint of (11). Without this constraint, the obtained coefficient of c_i would be always zero to minimize (9).

Using the differentiable homeomorphism function T, the nonlinear model of (1) is linearized as follows.

978-1-4244-1871-8/07 $25.00 © 2007 IEEE

$$\frac{d}{dt}\begin{bmatrix}\xi_1\\\xi_2\\\xi_3\end{bmatrix}=\begin{bmatrix}0&1&0\\0&0&0\\0&0&0\end{bmatrix}\begin{bmatrix}\xi_1\\\xi_2\\\xi_3\end{bmatrix}+\begin{bmatrix}0&0\\1&0\\0&1\end{bmatrix}\begin{bmatrix}\upsilon_1\\\upsilon_2\end{bmatrix}\quad(13)$$

$$\xi_{min}\leq\xi\leq\xi_{max},\ \upsilon_{min}\leq\upsilon\leq\upsilon_{max}$$

, where ξ and υ are new state variable and input variable defined as follows.

$$\begin{bmatrix}\xi_1\\\xi_2\\\xi_3\end{bmatrix}=\begin{bmatrix}T_1(x)\\T_2(x)\\T_3(x)\end{bmatrix}=\begin{bmatrix}T_1(x)\\L_fT_1(x)\\T_3(x)\end{bmatrix}\quad(14)$$

$$\begin{bmatrix}\upsilon_1\\\upsilon_2\end{bmatrix}=\begin{bmatrix}L_fT_2(x)\\L_fT_3(x)\end{bmatrix}+\begin{bmatrix}L_{g_1}T_2(x)&L_{g_2}T_2(x)\\L_{g_1}T_3(x)&L_{g_2}T_3(x)\end{bmatrix}\begin{bmatrix}u_1\\u_2\end{bmatrix}\quad(15)$$

The vehicle system with a polynomial model is considered in order to show the usefulness of the proposed algorithm, where the dynamics with three states and two inputs is shown as follows.

$$\frac{d}{dt}\begin{bmatrix}x_1\\x_2\\x_3\end{bmatrix}=\begin{bmatrix}f_1(x)\\f_2(x)\\f_3(x)\end{bmatrix}+\begin{bmatrix}g_{11}(x)&g_{12}(x)\\g_{21}(x)&g_{22}(x)\\g_{31}(x)&g_{32}(x)\end{bmatrix}\begin{bmatrix}u_1\\u_2\end{bmatrix}\quad(16)$$

$$x=\begin{bmatrix}x_1\\x_2\\x_3\end{bmatrix}=\begin{bmatrix}v_x\\v_y\\\theta\end{bmatrix},\ u=\begin{bmatrix}u_1\\u_2\end{bmatrix}=\begin{bmatrix}\dot\delta_{SW}\\\dot v\end{bmatrix}$$

, where v_x, v_y and θ are longitudinal speed, lateral speed and yaw angle respectively, $\dot\delta_{SW}$ and $\dot v$ are steering wheel angular speed of the vehicle and acceleration of the vehicle, respectively, and the functions f and g are obtained in a polynomial form.

In order to take observational data sets of the vehicle dynamics operating at a low speed, we used CarSim. CarSim is the driving simulation system developed by the University of Michigan Transportation Research Institute which simulates various operating conditions related with accelerating, braking, wheel angle steering and so on, and provides the variables in consideration to plot and analyze.

Since the proposed data-mining based algorithm tries to represent the prominent behavior of the target variable, some idiosyncratic deviations may not always be identified through the developed model. Therefore the operating condition applied to the simulator should be fully considered that for the task to be realized, operating conditions such as input pattern, hardware specification to be adopted, and so on, should be used for construction of the model.

The control objective of this work is to realize an exact behavior of the vehicle dynamics. The driving conditions used in this simulation are as follows: The vehicle is 3800 mm long and 1527 kg weigh with the distance from gravity center to front wheel and rear wheel of 1014 mm and 1676 mm respectively, and radii of front and rear wheel is 277.566

mm and 286.387 mm respectively, where roll, pitch and yaw inertias are 606.1 kg m^2, 2741.9 kg m^2, and 2741.9 kg m^2, respectively.

Since the system of (14) is three state two input system, the only combination to satisfy **Theorem 1** is $k_1=2$ and $k_2=1$. We show the developed linearized model obtained by applying the proposed method as follows.

$$\dot\xi=\overline A_c\xi+\overline B_c\upsilon$$

$$\overline A_c=\begin{bmatrix}0&1&0\\0&0&0\\0&0&0\end{bmatrix}\quad\overline B_c=\begin{bmatrix}0&0\\1&0\\0&1\end{bmatrix}\quad(17)$$

$$\xi_{min}\leq\xi\leq\xi_{max},\ \upsilon_{min}\leq\upsilon\leq\upsilon_{max}$$

$$\xi=\begin{bmatrix}\xi_1\\\xi_2\\\xi_3\end{bmatrix}=\begin{bmatrix}T_1(x)\\T_2(x)\\T_3(x)\end{bmatrix}=\begin{bmatrix}T_1(x)\\L_fT_1(x)\\T_3(x)\end{bmatrix}\quad(18)$$

$$\upsilon=\begin{bmatrix}\upsilon_1\\\upsilon_2\end{bmatrix}=\begin{bmatrix}K_{11}(x)\\K_{21}(x)\end{bmatrix}+\begin{bmatrix}Y_{11}(x)&Y_{12}(x)\\Y_{21}(x)&Y_{22}(x)\end{bmatrix}\begin{bmatrix}u_1\\u_2\end{bmatrix}\quad(19)$$

$$K(x)=\begin{bmatrix}K_{11}(x)\\K_{21}(x)\end{bmatrix}=\begin{bmatrix}L_fT_2\\L_fT_3\end{bmatrix}\quad(20)$$

$$Y(x)=\begin{bmatrix}Y_{11}(x)&Y_{12}(x)\\Y_{21}(x)&Y_{22}(x)\end{bmatrix}=\begin{bmatrix}L_{g_1}T_2&L_{g_2}T_2\\L_{g_1}T_3&L_{g_2}T_3\end{bmatrix}\quad(21)$$

, where for example T_1 are obtained as follows.

$$\begin{aligned}T_1(x)=&-2.4744\times10^{-4}x_1+2.8434\times10^{-5}x_2+9.9115\times10^{-5}x_1^2\\
&+7.8909\times10^{-6}x_2^2+1.0347\times10^{-5}x_1x_2-1.8061\times10^{-9}x_3\\
&+6.0813\times10^{-13}x_3^2+2.0834\times10^{-12}x_1x_3-1.2974\times10^{-12}x_1^3\\
&+7.8782\times10^{-15}x_1x_2^2-8.9715\times10^{-16}x_1^2x_2+2.4462\times10^{-17}x_2^3\\
&+2.5650\times10^{-13}x_1^4-3.0983\times10^{-15}x_1^2x_2^2-4.0705\times10^{-15}x_1^3x_2\\
&+3.3944\times10^{-18}x_2^4+8.9021\times10^{-18}x_2x_3^2-1.7294\times10^{-17}x_1x_3^2\\
&-8.0266\times10^{-13}x_1^2x_3+1.2262\times10^{-17}x_3^3-7.8624\times10^{-18}x_1^2x_3^2\\
&+2.3840\times10^{-17}x_1^3x_3+1.6710\times10^{-14}x_2x_3+1.6370\times10^{-19}x_2x_3^2\\
&+6.0805\times10^{-15}x_1x_2x_3+4.6374\times10^{-15}x_2^2x_3+4.5431\times10^{-20}x_2^2x_3^2\\
&-1.3774\times10^{-19}x_1x_2^2x_3+5.9573\times10^{-20}x_1x_2x_3^2\\
&-1.8062\times10^{-19}x_1^2x_2x_3\end{aligned}$$

Note that we can take an arbitrary linearly independent scalar function T_3. In this work, we used

$$T_3(x)=100+x_3.$$

We show the equations (13), (14) and (15) are true as follows.

978-1-4244-1871-8/07 $25.00 © 2007 IEEE

Proof 1:

The relation $\dot{T}_1(x) = \dot{\xi}_1$ is transformed as follows

$$\dot{T}_1(x) = \frac{\partial T_1}{\partial x}\dot{x}$$

$$= \frac{\partial T_1}{\partial x}\left(f + g_1 u_1 + g_2 u_2\right) \qquad (22)$$

$$= \frac{\partial T_1}{\partial x}f + \frac{\partial T_1}{\partial x}g_1 u_1 + \frac{\partial T_1}{\partial x}g_2 u_2$$

$$= L_f T_1 + \left(L_{g_1}T_1\right)u_1 + \left(L_{g_2}T_1\right)u_2$$

. Since $L_{g_1}T(x) \approx 0$ and $L_{g_2}T(x) \approx 0$, we have $\dot{T}_1(x) = L_f T_1$. We introduce the new variable ξ_2 which satisfies the foll--owing equation.

$$\dot{\xi}_1 = \xi_2 = T_2(x) = L_f T_1 \qquad (23)$$

In a similar way, we differentiate $T_2(x)$ as follows.

$$\dot{T}_2(x) = \frac{\partial T_2}{\partial x}\dot{x}$$

$$= \frac{\partial T_2}{\partial x}\left(f + g_1 u_1 + g_2 u_2\right) \qquad (24)$$

$$= \frac{\partial T_2}{\partial x}f + \frac{\partial T_2}{\partial x}g_1 u_1 + \frac{\partial T_2}{\partial x}g_2 u_2$$

$$= L_f T_2 + \left(L_{g_1}T_2\right)u_1 + \left(L_{g_2}T_2\right)u_2$$

By selecting the new input υ_1 which satisfies

$$\upsilon_1 = L_f T_2 + \left(L_{g_1}T_2\right)u_1 + \left(L_{g_2}T_2\right)u_2 \qquad (25)$$

, we have

$$\dot{\xi}_2 = T_2(x) = \upsilon_1 \qquad (26)$$

, since

$$\xi_2 = T_2(x) \qquad (27)$$

. We also have

$$\dot{T}_3(x) = L_f T_3 + \left(L_{g_1}T_3\right)u_1 + \left(L_{g_2}T_3\right)u_2 \qquad (28)$$

, which derives

$$\dot{\xi}_3 = T_3(x) = \upsilon_2 \qquad (29)$$

by selecting

$$\upsilon_2 = L_f T_3 + \left(L_{g_1}T_3\right)u_1 + \left(L_{g_2}T_3\right)u_2 \qquad (30)$$

s

III. Optimal Controller Design

The obtained linearized model (17) to (21) is discretized using the zero order holders as follows.

$$\xi_{k+1} = \overline{A}\xi_k + \overline{B}\upsilon_k \qquad (31)$$

$$\overline{A} = e^{\overline{A}_c T_s} = I + \overline{A}_c T_s + \frac{\overline{A}_c^2}{2!}T_s^2 + \cdots + \frac{\overline{A}_c^k}{k!}T_s^k + \cdots = I + \overline{A}_c T_s$$

$$= \begin{bmatrix} 1 & T_s & 0 \\ 0 & 1 & 0 \\ 0 & 0 & 1 \end{bmatrix} \qquad (32)$$

$$\overline{B} = \int_0^{T_s} e^{\overline{A}_c \tau}d\tau \overline{B}_c$$

$$= \begin{bmatrix} T_s & 0.5T_s^2 & 0 \\ 0 & T_s & 0 \\ 0 & 0 & T_s \end{bmatrix}\begin{bmatrix} 0 & 0 \\ 1 & 0 \\ 0 & 1 \end{bmatrix} = \begin{bmatrix} 0.5T_s^2 & 0 \\ T_s & 0 \\ 0 & T_s \end{bmatrix} \qquad (33)$$

Note that since the developed model is a constrained polynomial dynamics, we have following constrains.

$$\forall k, \quad \xi_{min} \le \xi_k \le \xi_{max}, \quad \upsilon_{min} \le \upsilon_k \le \upsilon_{max} \qquad (34)$$

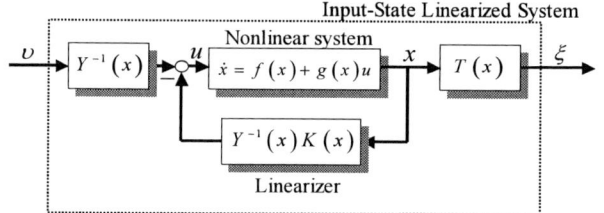

Figure 1 Structure of Input-State Linearized System

The constraint of (34) is for guaranteeing reproducibility of the data-mining based statistical model. If the variables are in the ranges specified in (34), they are fully reproduced since they were included in the pattern of the learning data that we used to construct the model. Actually the values of ξ_{min}, ξ_{max} and υ_{max}, υ_{min} are determined by selecting the minimum and maximum value of the observational data.

Note also that newly introduced variables ξ and υ do not have any physical meaning. When the input-state linearized models are used in the conventional methods for control purpose, the control input υ should be transformed to u. Since the matrices $K(x)$ and $Y(x)$ are generally high-order polynomials in our polynomial data-mining algorithm based approach, the transformation of υ to u comes to have big fudge factor. See Figure 1, where linearizer contains inverse of matrix $Y(x)$. Since mean values of the state and input variables are sometimes very different, this transformation comes to have big error.

In the proposed method, the original nonlinear state equation together with the linearized state equation is used for obtaining the original input u without using transformation from υ.

Based on the model developed in the previous chapter, the control problem for attitude control of the unmanned vehicle system is stated as follows.

Find u which minimizes the following performance criteria

$$\min_{\xi_{k+1},\upsilon_k,u_k} J = w_1\left(\xi_{1,k+1}-\xi_{1d,k}\right)^2 + w_2\left(\xi_{2,k+1}-\xi_{2d,k}\right)^2$$
$$+ w_3\left(\xi_{3,k+1}-\xi_{3d,k}\right)^2 + w_4\left(\upsilon_{1,k}-\upsilon_{1d,k}\right)^2 \quad (35)$$
$$+ w_5\left(\upsilon_{2,k}-\upsilon_{2d,k}\right)^2 + w_6\left(u_{1,k}-u_{1d,k}\right)^2$$
$$+ w_7\left(u_{2,k}-u_{2d,k}\right)^2$$

subject to

$$\begin{bmatrix} I & -\bar{B} & 0 \\ 0 & I & Y(x_k) \end{bmatrix}\begin{bmatrix} \xi_{k+1} \\ \upsilon_k \\ u_k \end{bmatrix} = \begin{bmatrix} \bar{A}x_k \\ K(x_k) \end{bmatrix} \quad (36)$$

$$\begin{bmatrix} \xi_{\min} \\ \upsilon_{\min} \\ u_{\min} \end{bmatrix} \leq \begin{bmatrix} \xi_{k+1} \\ \upsilon_k \\ u_k \end{bmatrix} \leq \begin{bmatrix} \xi_{\max} \\ \upsilon_{\max} \\ u_{\max} \end{bmatrix} \quad (37)$$

, where $\xi_{\max}$ and $\xi_{\min}$ are obtained from $x_{\max}$ and $x_{\min}$.
The structure of the proposed controller is illustrated in Figure 2.

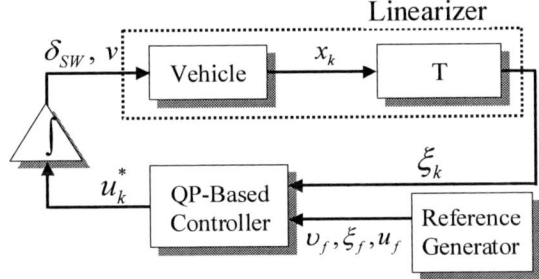

Figure 2 Structure of the proposed controller

Note that although the vehicle system is originally represented with an extremely constrained nonlinear model, it is easily optimized in the proposed method solving a linear problem coupled with input-state linearization.

IV.   EXPERIMENTAL RESULTS

A.   Model Verification

In this chapter, we show some results to verify the usefulness of the proposed modeling method. We used the vehicle shown in the previous chapter. First, we show the model constructed by applying the presented method. In Figure 3 to Figure 5, the trajectories of the state variables obtained from both an original model and a newly developed linearized model are compared. These behaviors are obtained in full consideration of a variety of driving situations. The correlation values between both trajectories in Figure 3 to Figure 5 are in the cases of all three state variables over 0.999 respectively.

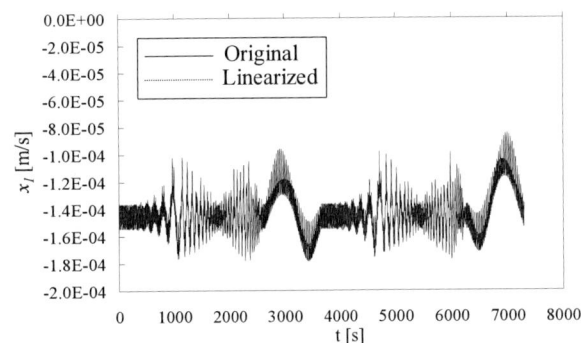

Figure 3 Behavior of x_1

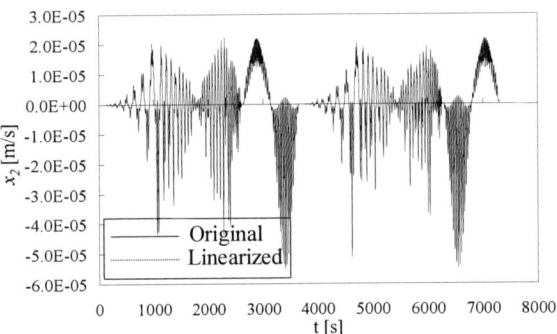

Figure 4 Behavior of x_2

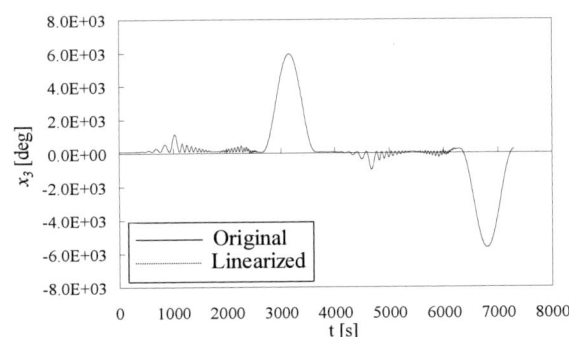

Figure 5 Behavior of x_3

B.   Verification of Controller Design Method

In this subchapter, the proposed controller design method is verified through some numerical experiments. We show in Figure 6 to Figure 8, behaviors of steering angle, driving speed and driving trajectory to track circle with radius of 10 *m*. As we see in these figures circular movement is well achieved. During first several seconds in these simulations, controller does not operate since the developed model is only valid among the speed region of 3 km/h $\leq v \leq$ 6 km/h. After attaining the driving speed of 3 km/h with steering wheel angle unchanged, the controller start to operate.

978-1-4244-1871-8/07 $25.00 © 2007 IEEE   111

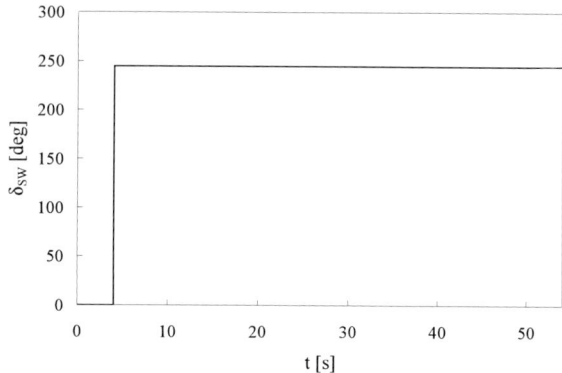

Figure 6 Circle Tracking Behavior of δ_{SW}

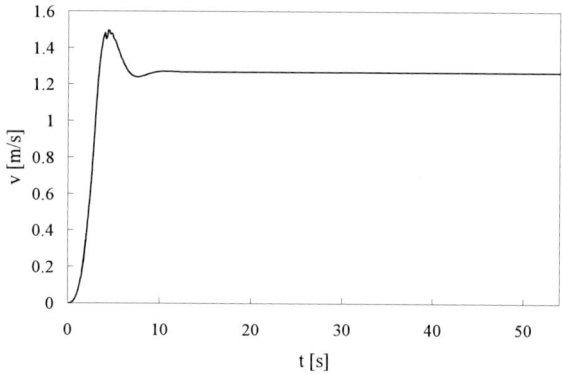

Figure 7 Behavior of Driving Speed In Case Of Tracking Circle

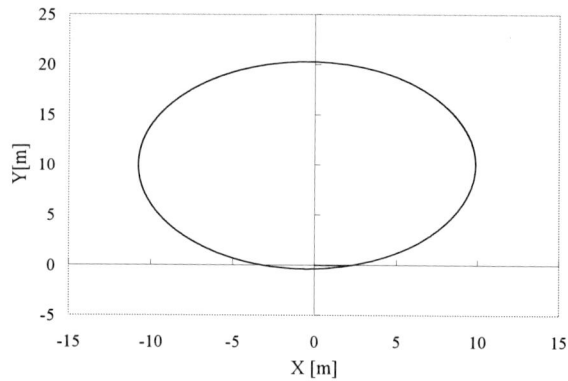

Figure 8 Circle Tracking Behavior of Vehicle

V. CONCLUSION

In this paper we have presented a controller design method based on application of the data-mining polynomial algorithm. We showed an input-state linearization method using the proposed polynomial data-mining algorithm, where a polynomial vehicle model operating at a very low speed was linearized. Application of the developed model to controller design was shown where without introducing any process, which may invoke fudge factor, control inputs are

obtained in the original coordinate. We verified the developed modeling method and controller design method through some numerical experiments.

REFERENCES

[1] Slotine, J. J, Li, W., *Applied Nonlinear Control*, Prentice-Hall, Inc. A Division of Simon & Schuster, Englewood Cliffs, New Jersey 07632, ISBN:0-13-040890-5, 1991.

[2] Kim, Y., Narikiyo, T, Kim, J.H., Attitude Control of Planar Space Robot Based on Self-Organizing Polynomial Data Mining Algorithm, Proddedings of the *8th IASTED International Conference Intelligent Systems and Control*, 303- 308, October 31 - November 2, 2005, Cambridge, USA, 2005.

[3] Kim, Y., Matsuzaki, S., and Narikiyo, T., Efficient Modeling of Vehicle Dynamics Operating at Low Speed and Its Application to Non-Linear Optimal Controller Design, *JSME International Journal* **49,4**, 1040-1047,2006.

978-1-4244-1871-8/07 $25.00 © 2007 IEEE

Development of the Low-Cost Impedance Spectroscopy System for Modeling the Electrochemical Power Sources

Woojin Choi
Soongsil University
Department of Electrical Engineering
San 1-1, Sangdo-dong, Dongjak-gu, Seoul, 156-743
E-mail: cwj777@ssu.ac.kr

Juhyung Lee
Soongsil University
Department of Electrical Engineering
San 1-1, Sangdo-dong, Dongjak-gu, Seoul, 156-743
E-mail: t_tcrying@hotmail.com

Abstract—This paper presents a low-cost impedance spectroscopy system (LCISS) suitable for modeling the electrochemical power sources such as fuel cells, batteries and supercapacitors. Since the developed LCISS is composed of simple sensor circuits, commercial data acquisition board and software, it provides a low cost solution for modeling and diagnosing the electrochemical power sources compared with the commercial electrochemical impedance spectroscopy instruments. The system is capable of automatically measuring the impedance of the electrochemical power sources over the specified frequency range and from the measured impedance data an equivalent circuit can be derived by the developed software. In the proposed system, the digital lock-in amplification technique is adopted to achieve the accurate measurements even in the presence of the high level of noises. The validity of the developed system is proven by comparing the measured frequency response of a simple R-C filter with those obtained by using the commercial frequency response analyzer (NF FRA 5097). The developed hardware and software is applied to measure the impedance spectrum of the 1.2kW proton exchange membrane fuel cell and the results are detailed.

I. INTRODUCTION

In recent years, Electrochemical Impedance Spectroscopy (EIS) has seen a tremendous increase in popularity [1-10]. EIS is a well-known technique in electrochemistry field to characterize electrode processes and complex interfaces [1]. Since the analysis of the system response contains information about the interface, its structure, and the reactions taking place there, it has become a valuable research tool in many areas including applied chemistry and engineering in medicine and biology [2]. Many of the research have been conducted to utilize this useful tool for modeling and diagnosing the electrochemical power sources such as fuel cell, battery and supercapacitor [3-5]. However, most of the commercially available EIS instruments are developed suitable for measuring the impedance of the single cell or small module and very high in cost [11].

In this paper a low-cost impedance spectroscopy system is introduced. The system is composed of simple sensor circuits, commercial data acquisition board and software, it provides a low cost solution for modeling and diagnosing the electrochemical power sources compared with the

commercial electrochemical impedance spectroscopy instruments. This paper details the structure of the software including digital lock-in amplification and curve-fitting technique. The validity of the system will be proven by comparing the measurement results obtained by the developed low-cost impedance spectroscopy system (LCISS) with those by the commercial instrument, NF FRA 5097.

II. SYSTEM CONSTRUCTION

A. Construction of Hardware

Fig 1 shows the system configuration of the proposed LCISS. The electrochemical power sources such as fuel cell, battery and supercapacitor are directly connected to the programmable electronic load, and small sensing & signal conditioning circuit board for measuring the current and voltage is connected in between them. The swept sine waveform superimposed on a certain DC level is generated by the software and transferred to the external control port of the electronic load to draw the same form of the current from the electrochemical power sources. The measured current and voltage signals are amplified and transferred to the data acquisition board and converted into the digital format.

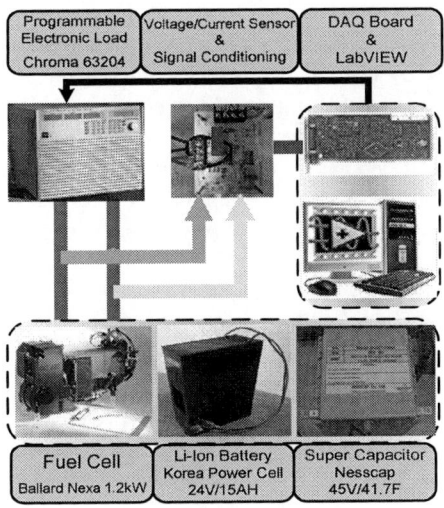

Fig. 1: Configuration of the proposed LCISS

In the system an isolated multifunction data acquisition (DAQ) board (National Instruments PCI-6154) is employed to simultaneously sample the current and voltage signal. The DAQ PCI-6154 has four differential simultaneously sampled 16-bit analog inputs with channel-to-channel isolation and 250 kS/s sampling rate. This feature makes it possible to sample the current and voltage data simultaneously with a high resolution of 305[μV/Bit] and significantly reduce the problems caused by common-mode voltage and ground loop in the measurements. It also has 4 channels of analog outputs with 16-bit resolution and 250 kS/s sampling rate, only one of them is used to generate the high resolution sinusoidal waveform for the electronic load control.

In the experiments, Chroma 63204 programmable electronic load is used and its range of dynamic loading is from 0.1Hz to 10kHz. For the current measurements LEM LA55-P is used and the resistance divider circuit for the voltage measurements. From the sampled data, only the desired frequency component is extracted and the impedance at a certain frequency is calculated. The process is detailed in the next section.

B. Implementation of the Digital Lock-in Amplifier

Fast Fourier Transform (FFT) is an efficient algorithm to compute the discrete Fourier transform (DFT) and widely used in variety of applications in digital signal processing. However, FFT has a drawback of low noise immunity and resolution in the low frequency region, which is critical in the impedance measurements [12]. Further, FFT requires more computation time than the Lock-in Amplification technique.

Digital Lock-in Amplifier (DLA) was proposed and now commercially available [13-16]. Further, thanks to the highly developed PC based hardware, it has become possible to implement the DLA with commercial DAQ board such as NI PCI-6154. Usually ALA(Analog Lock-in Amplifier) generates its internal reference signal by the PLL(Phase-Locked Loop) locked to the external reference. Unlike the ALA, however, DLA can generate the high purity sine wave without external VCO(Voltage Controlled Oscillator) and PLL, and provide the perfect synchronization between signal generation and data acquisition. The generated reference signal is injected to the objective and from the response the in-phase and in-quadrature components are calculated only at the reference frequency.

Thus the proposed LCISS employs the DLA which allows higher accuracy for the AC measurement in the low frequency range and less computation time. Two DLAs are implemented to extract the Fourier components of the reference frequency from the current and voltage measurements and the impedance at the reference frequency can be calculated from the results. Though the code for the DLA can be written in any high level language, in the developed system the LabVIEW, a powerful graphical language, has been chosen.

The reference signal generated by the Labview software for the DLA can be written as (1). Then, the response by the applied reference signal, including parasitic noise, can be expressed as (2).

$$S_{ref} = \cos(\omega_{ref} \cdot t) \tag{1}$$

$$S_{res} = A_{res}\cos(\omega_{ref} \cdot t + \delta_{res}) + \sum A_{noise}\cos(\omega_{noise} \cdot t + \delta_{noise}) \tag{2}$$

If the twice reference signal is multiplied with the response signal, the resulting equation can be easily simplified by using some trigonometry identity as (3).

$$
\begin{aligned}
S_{mult} &= 2 \cdot S_{ref} S_{res} \\
&= 2\cos(\omega_{ref} \cdot t) \cdot A_{res}\cos(\omega_{ref} \cdot t + \delta_{res}) \\
&\quad + 2 \cdot \cos(\omega_{ref} \cdot t)\sum A_{noise}\cos(\omega_{noise} \cdot t + \delta_{noise}) \\
&= A_{in}\cos(\delta_{res}) + A_{in}\cos(2 \cdot \omega_{ref} \cdot t + \delta_{res}) \\
&\quad + \sum A_{noise}\cos((\omega_{ref} + \omega_{noise}) \cdot t + \delta_{noise}) \\
&\quad + \sum A_{noise}\cos((\omega_{ref} - \omega_{noise}) \cdot t - \delta_{noise})
\end{aligned} \tag{3}
$$

If the multiplied signal is low pass filtered, then only two terms remain as (4), the DC term due to the output of the system, and the component of noise with frequency near the reference signal.

$$S_{filtered} = A_{res}\cos(\delta_{res}) + A_{noise@ref}\cos(\delta_{noise@ref}) \tag{4}$$

If the reference is adjusted such that $\delta res = 0$ and the noise is ignored, then the amplitude of the filtered signal is the amplitude of the system response as (5).

$$S_{filtered} = A_{res} \tag{5}$$

The above process is performed for both in phase and in quadrature components of the response signal, respectively. The magnitude and phase of the response at each frequency can be calculated as (6) and (7) for both current and voltage.

$$\text{Magnitude} = \sqrt{\text{Real}^2 + \text{Imaginary}^2} \tag{6}$$

$$\text{Phase} = \arctan\left(\frac{\text{Imaginary}}{\text{Real}}\right) \tag{7}$$

Fig. 2 shows the block diagram of the digital lock-in amplification process explained above and the interconnection between hardware and software.

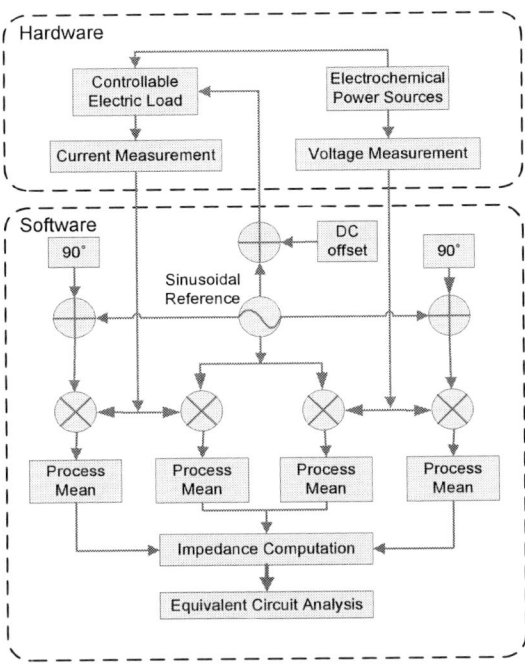

Fig. 2: Block Diagram of the LCEISS

C. Construction of Software

The developed software is composed of two different modes. One is the data acquisition mode and the other is the analysis mode. Fig. 3 shows the data acquisition mode which has the embedded digital lock-in amplifier. In this mode, parameters for the measurements such as DC offset, amplitude of the current perturbation and sweeping frequency range are specified with desired values. When the start button is pushed, the sweeping sinusoidal waveforms are generated and the data acquisition is performed. At each frequency, the DLAs extract only the reference frequency component from the measured current and voltage waveforms, and calculate the impedance.

Fig. 4 shows the analysis mode of the developed software. After the data acquisition and corresponding impedance calculation is completed, the mode is automatically transferred into the analysis mode to develop the suitable equivalent circuit with the measurement results. At the beginning of this mode, a white curve shows up in the screen, which represents the measured impedance spectrum in the data acquisition mode. Then the suitable equivalent circuit model for electrochemical power sources is selected and the manual curve fitting is started to calculate the parameters for the equivalent circuit. After selecting the suitable equivalent circuit, the user types in the initial value into the small window below the slide bar in the left side of the screen. Then the corresponding impedance spectrum of the equivalent circuit (red curve in Fig. 4) shows up in the screen. When the parameters vary with the slide bar and the impedance spectrum of the selected equivalent circuit also

varies simultaneously. This helps the user understand the relationship between the variation in the parameter value and the variation in the shape of the impedance spectrum. This is a very unique and powerful feature that the commercially available software does not provide.

During the course of manual curve fitting the accuracy of the fitting is continuously evaluated by calculating the chi-square value which represents the sum of the least squared error between the original data and calculated fit as (8) [17-18]. Also the calculated chi-square value is displayed in the screen. The manual curve fitting continues until the chi-square value becomes small enough, which is usually less than 1%. Fig. 5 shows the flow chart of the developed software.

$$\chi^2 = \sum_{i=1}^{n} \left[\frac{y_i - f(x_i)}{\sigma_i} \right]^2 \tag{8}$$

(y_i:Measured Data, $f(x_i)$:Fitted Data, σ_i: Standard Deviation)

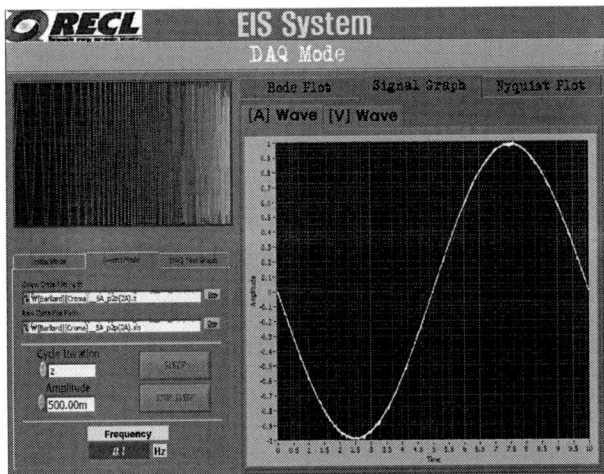

Fig. 3: Data acquisition mode of the proposed LCEISS

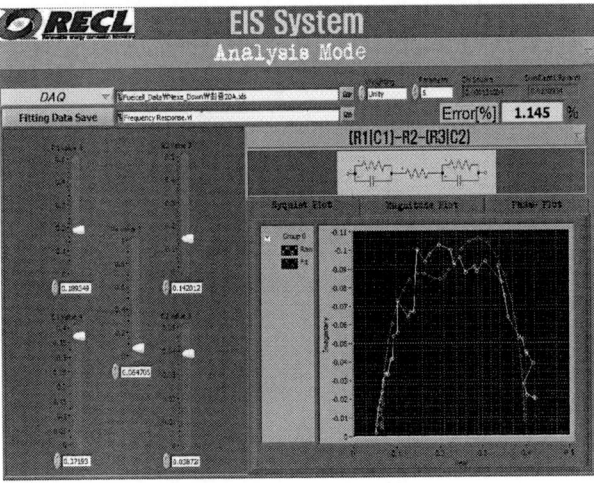

Fig. 4: Analysis mode of the proposed LCISS

978-1-4244-1871-8/07 $25.00 © 2007 IEEE

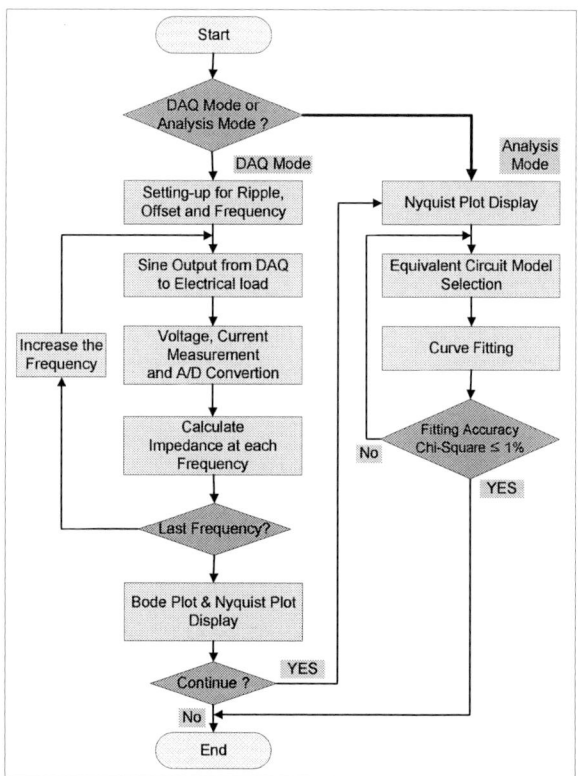

Fig. 5: Flow chart of the developed software

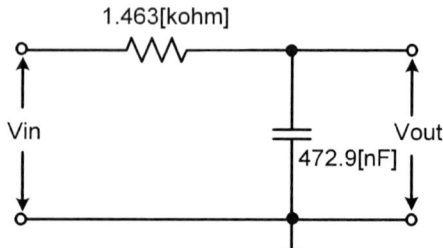

Fig. 6: Simple R-C filter for accuracy comparison

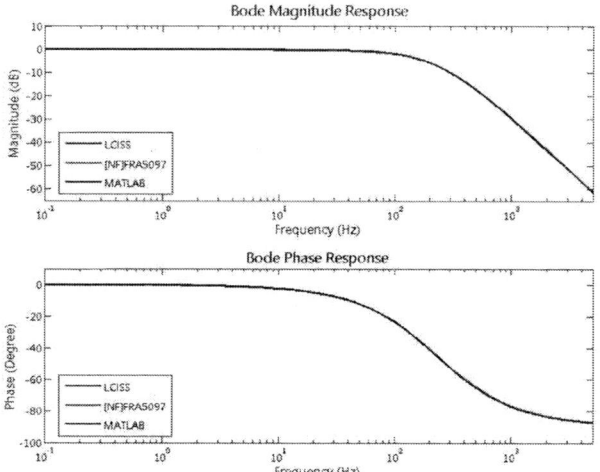

Fig. 7: Bode plots of the R-C filter

III. EXPERIMENTATION

A. Comparison with Commercial Instrument

In order to verify the validity and the accuracy of the proposed system, measurement results were compared with those of commercial Frequency Response Analyzer, NF FRA 5097.

The selected test circuit is a simple low pass filter composed of a resistor (1.463[kΩ]) and a ceramic capacitor (472.9[nF]) connected in parallel, of which cut-off frequency is 230[Hz] (Fig 6). The parameter values of the passive components were measured by the Agilent 4263B LCR meter.

The frequency response of the LPF was measured over the frequency range from 0.1Hz to 5kHz by the LCISS and FRA 5097, respectively and the results are compared with the simulated data by the Matlab software. Bode plots are drawn with the results and it is found that all the plots are almost matched perfect as seen in Fig 7. Some of the results at specific frequencies are also shown in the Table 1 and the calculated Mean Absolute Percentage Error (MAPE) by (9) shows less than 1% relative error in both magnitude and phase, which is acceptable.

$$\text{MAPE} = \frac{1}{n} \sum_{t=1}^{n} \left| \frac{A_t - F_t}{A_t} \right| \tag{9}$$

TABLE 1

Measured frequency response of the R-C filter

	MATLAB		NF-FRA5097		LCISS	
[Hz]	Magnitude	Phase	Magnitude	Phase	Magnitude	Phase
0.1	1	−0.02	0.997	−0.02	1	−0.01
1	1	−0.25	0.997	−0.25	1	−0.24
10	0.999	−2.49	0.996	−2.52	0.998	−2.49
100	0.917	−23.5	0.913	−23.6	0.911	−24.1
1000	0.224	−77	0.225	−76.8	0.219	−76.4
5000	0.046	−87.4	0.046	−87.1	0.0447	−87

B. Application of the LCISS to the Fuel Cell Impedance Measurements

During the impedance measurements, a small ac perturbation is applied to the system. The impedance derived is valid provided that the four criteria of linearity, causality, stability, and finiteness are met. The details of the premise for the measurements can be found in the reference [19]. Thus the voltage ripple caused by the current perturbation is limited by the thermal voltage as shown in (10) [8].

$$V_T = \frac{RT}{F} = 26\,mV \;@\; 25^o C \tag{10}$$

(R=8.314[J/mole-K], T=Temp[K], F=96485[C/equiv])

In the experiments, hydrogen (>99.99% at 5 psig) and air are supplied to the 1.2kW PEM Fuel Cell Stack (PEMFCS). It usually takes several minutes for a PEMFCS to become stable because the membrane needs to be humidified for optimum performance. Thus, before the test, the PEMFCS was started after a short period of warm-up and then run for an hour at the full-load to ensure a thermally stable operation. It should be noticed that the entire test was performed repeatedly and consistent data have been obtained.

The developed system is used to measure the impedance of a Ballard Nexa 1.2kW PEMFC system over the specified frequency range. The experiments were performed by perturbing the PEMFCS at a certain DC operating point. This process can be simply done in the data acquisition mode as described in the section II-C. Fig. 8 shows the experimental setup for the measurements.

The measured response of the PEMFCS is used to compute the AC impedance of the FC equivalent circuit. The impedance of the PEMFC were measured over the specified frequency range (0.1Hz ~2kHz) at seven different operating points from 10[A] to 40[A] at every 5[A]. Then the curve fitting was performed with the results by the equivalent circuit model shown in the Fig. 9. This model is derived considering the internal structure of the PEMFCS and the electrochemical reaction taking place inside of it [1].

Fig. 10 shows the resulting curve-fitted impedance plots drawn at each DC operating points. It can be noted that the diameter of the impedance plot varies significantly at different operating points. This matches with the slope of the polarization curve at each DC operating points since the diameter of the impedance plot represents the sum of the resistance values of the equivalent circuit.

Fig. 11 shows the variation of the each equivalent circuit parameters. These results are almost similar with those obtained with same kind of PEMFCS [6]. Though many of the interesting facts can be found from the plot, more research is required to properly explain the relationship between the load variations and parameter variations of the equivalent circuit. However, it can be easily inferred from the results that the impedance variation of the PEMFCS has a certain trend and it is possible to utilize it for diagnosing or prognosing the lifetime.

IV. CONCLUSION

In this paper a low-cost impedance spectroscopy system was presented. The proposed system employs digital lock-in amplifiers to detect the reference frequency components of the current and voltage to calculate the impedance at a certain frequency. It has been proven by the performance comparison with commercial instruments that the developed instrument provides reliable results with required accuracy at a lower cost. An equivalent circuit model of the Ballard Nexa 1.2kW PEMFCS was derived by the LCISS and the successful results were obtained. Also, since the developed

LabVIEW-based software is user friendly and provides powerful analysis functions, it can be a useful tool for the research of modeling, diagnosing and prognosing the electrochemical power sources.

Fig. 8: Ballard Nexa 1.2kW PEMFCS and the sensing & signal conditioning circuit board

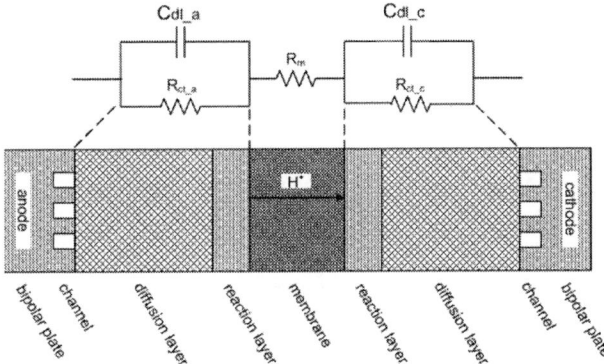

Fig. 9: Applied equivalent circuit model of the PEMFCS

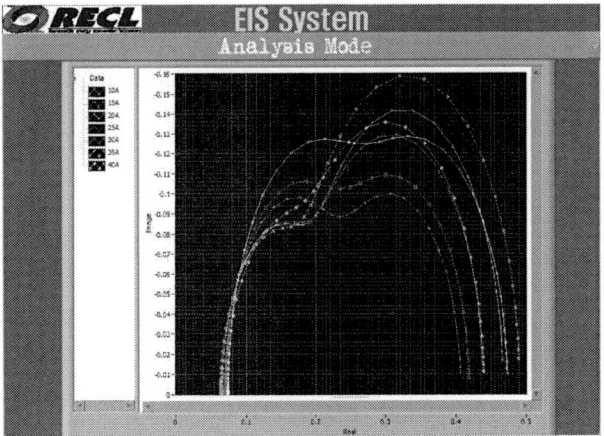

Fig. 10: Curve-fitted impedance plots of the Ballard Nexa 1.2kW PEMFCS at each DC operating points

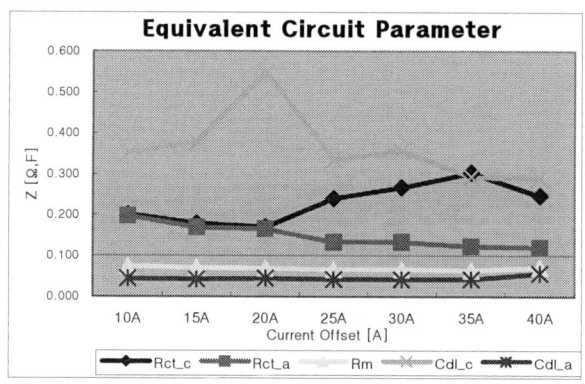

Fig. 11: Variation of the equivalent circuit parameters of the PEMFCS at each DC operating points

REFERENCES

[1] J.R. Macdonald, *Impedance Spectroscopy: Emphasizing Solid Materials and Systems*, John Wiley & Sons, 1987

[2] B.E. Conway, J.O'M. Bockris, Ralph E. White, *Modern Aspects of Electrochemistry*, Kluwer Academic Publishers, 2002

[3] Eckhard Karden, Stephan Buller, Rik W. De Doncker, "A frequency-dpmain approach to dynamical modeling of electrochemical power sources," Electrochemica Acta, 2002, pp. 2347-2356

[4] Eckhard Karden, Stephan Buller, Rik W. De Doncker, "A Method for measurement and interpretation of impedance spectra for industrial batteries," Journal of Power Sources 85, 2002, pp. 72-28

[5] Stephan Buller, Rik W. De Doncker, Eckhard Karden, "Impedance-Based Simulation Models of Supercapacitors and Li-Ion Batteries for Power Electronic Applications," IEEE Transactions on Industry Applications, Vol. 41, No. 3, 2005, pp. 742-747

[6] W. Choi, J. Howze, P. Enjeti, "Development of an equivalent circuit of the a fuel cell to evaluate the effects of inverter ripple current," Journal of Power Source 158, 2006. pp. 1324–1332.

[7] Xiaozi Yuan, Jian Colin Suna, Mauricio Blanco, Haijiang Wang, Jiujun Zhang, David P. Wilkinson, "AC impedance diagnosis of a 500W PEM fuel cell stack Part I: Stack impedance," Journal of Power Source 161, Issue 2, Oct. 2006, pp. 920-928

[8] Kevin Cooper, Impedance Spectroscopy for Fuel Cell Diagnostics, The 2006 Fuel Cell Short Course, February 5-8, 2006 Cocoa, Florida

[9] J.R. Sleman, Y.P. Lin, "Application of AC impedance in fuel cell research and development," Electrochem. Acta 38 (14), 1993, pp. 2063–2073.

[10] J.M. Correa, F.A. Farret, L.N. Canha, "An Analysis of the Dynamic Performance of Proton Exchange Membrane Fuel Cells Using an Electrochemical Model," The 27th Annual Conference of the IEEE IECON, vol. 1, 2001, pp. 141–146.

[11] Solartron Analytical, "1255A HF Frequency Response Analyzer," Avaliable: http://www.solartronanalytical.com

[12] M. Ordonez, M. T. Iqbal, J. E. Quaicoe, M. O. Sonnaillon and F. J. Bonneto, "An Embedded DSP-based Frequency Response Analyzer for Fuel Cells Monitoring and Characterization," 37th IEEE Power Electronics Specialists Conference, June 2006, pp. 2197-2203

[13] A.L. S. Cova and I. Freias, " Versatile digital lock-in detection technique: Applicaion to spectroflourometry and other fields," Rev. Sci. Intrum. 50. 296, 1979, pp. 296-301

[14] Pierrea-Alain Probst and B.Collet, "Low-frequency digital lock-in amplifier," Rev. Sci. Intrum. 56. 466, 1985, pp. 466-470

[15] Pierrea-Alain Probst and Alain Jaqier, "Multi-channel digital lock-in amplifier with PPM resolution," Rev. Sci. Instrum. 56. 466, 1985 pp.747-750

[16] Stanford Research System, About Lock-in Amplifier, Application Note #3. Available: http://www.thinkSRS.com

[17] Resources for Electrochemistry, Available: http://www.consultrsr.com/

[18] W.H. Press, S.A. Teukolsky, W.T. Vetterling and B.P. Flannery, *Numerical Recipes in C: The Art of Scientific Computing, Second Edition*, Cambridge University Press, 1992, ch15.

[19] Andrzej Lasia, B. Conway, J. Bockris and Ralph White, *Modern Aspects of Electrochemistry*, Springer US, 2002, Volume 32

3D Education System of DC Motor Drive using RecurDyn and CoLink

H.S. Mok, J.H. LEE, G.H. CHOE
Konkuk University
Department of Electrical
Engineering
1, HwaYang-Dong, Gwangjin-Gu,
Seoul
Email: hsmok@konkuk.ac.kr

S.H. KIM
Kangwon National University
Department of Electrical and
Electronics Engineering
192-1, Hyoja2-Dong, Chunchon,
Kangwon-Do
Email: kshoon@kangwon.ac.kr

D.J. YUN
FunctionBay,Inc.
Development Department Ⅲ
R&D Center
246-5, Seohyeon-Dong, Bundang-Gu,
Seongmam-Si, Gyeonggi-Do
Email: djyun@functionbay.co.kr

Abstract—An increasing number of schools and industrial workplaces has recently applied theory-based computer simulation to their education system for electric drives, making computer simulation an essential element of the program. This paper introduces an education system for motor drive using the dynamic 3D analysis simulation tools RecurDyn and Colink to provide more accurate analyses and realistic training.

I. INTRODUCTION

The general task of power electronics, which includes motor control, involves supplying voltage and current to process or control electric energy as shown in Fig. 1[1]. Moreover, performing computer simulation using programs such as Matlab/Simulink, PSIM, Spice and Simplorer prior to a series of experiments has become an essential part of the process. The benefits of computer simulation include easy inquiry of results according to changes in numerous design variables, reduction of time and cost required for solving problems, and consideration of production and operation from the design stage. Accordingly, computer simulation has become an essential part of the design process[2].

In order to apply simulation techniques to motor control, models for the motor, power converter, control unit and load are required. There are accurate models readily available for the motor, power converter and control unit, and most are provided as libraries within the simulation programs mentioned above, making it easy for the user to access the proven models. On the other hand, the load model requires pre-calculation using principles of dynamics such as the equivalent inertia coefficient, friction coefficient and load torque according to the subject of application[3].

From the perspective of education, computer simulation has become an indispensable element of education programs since it provides students with an opportunity to indirectly experience the actual industrial worksites. However, most conventional computer simulation programs are based on mathematical equations and 2D plane viewpoints, making it difficult to provide realistic results and draw interest from the students.

RecuDyn is a dynamics simulation program for modeling

and simulating mechanical systems capable of 3D analyses. Once the mechanical 3D configuration is drawn and physical properties are input, RecurDyn provides outputs for the load torque, equivalent inertia coefficient and friction coefficient, allowing easy access to the parameters of the mechanical system model. Furthermore, linking the motor drive program with Colink allows simulation of the overall drive system without load modeling.

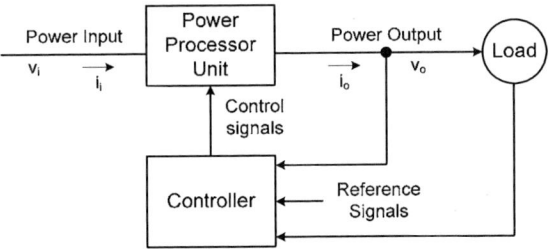

Fig. 1: Block diagram of Power Electronics System.

This paper introduces applications of RecurDyn and Colink for the PMDC motor drive system and examines whether they can be applied in education programs.

II. PMDC DRIVE SYSTEM USING RECURDYN AND COLINK

Fig. 2 displays an example of using RecurDyn and Colink for a PMDC drive system.

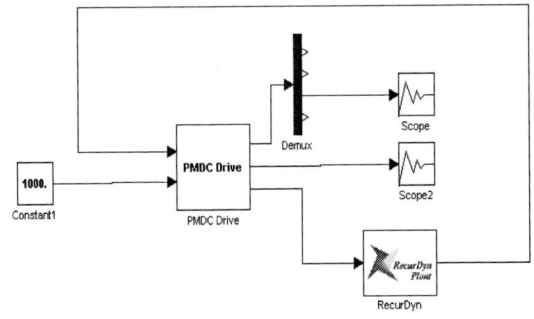

Fig. 2(a). : PMDC and Speed Controller Modeling using Colink

978-1-4244-1871-8/07 $25.00 © 2007 IEEE

Colink receives as inputs the load torque and inertia coefficient calculated from RecurDyn. Colink then updates the received input parameters to the motor and the controller to allow joint analysis. Fig. 3 exhibits an inertia load model created using RecurDyn.

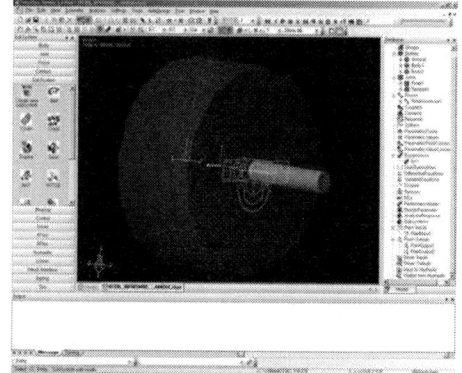

Fig. 3. : Inertia Load Modeling using RecurDyn

The results of joint analysis can be verified through 2D graphs and 3D animation. The graph in Fig. 4 represents the resultant waveform of the PMDC speed control system.

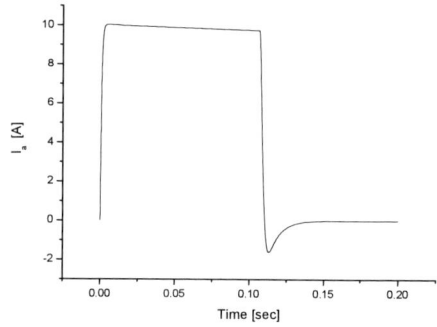

Fig. 4(a) : Resultant Waveform of Amateur Current

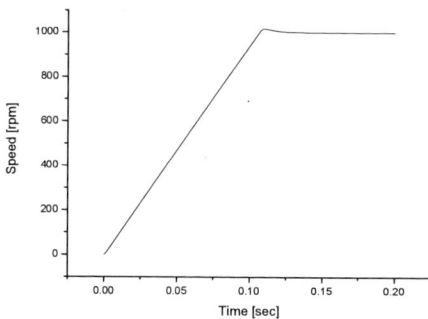

Fig. 4(b). : Resultant Waveform of Speed (RPM)

Furthermore, RecurDyn provides animation of the load dynamics that allows the user to view the rotation mechanism. Fig. 5 is a model created to check RecurDyn's animation functions.

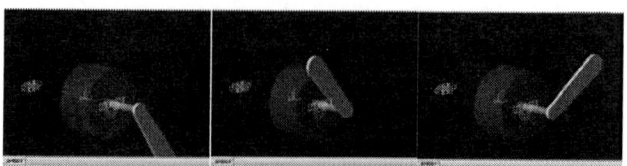

Fig. 5. : Animation using RecurDyn

III. MOTOR MODELING USING COLINK

Fig. 6 represents the PMDC motor circuit diagram applied to Colink. 6. Equations (1)-(4) were applied to Colink using the C language along with the numerical analysis algorithm to perform modeling. The user can eliminate this process by providing motor parameter inputs as shown in Fig. 7.

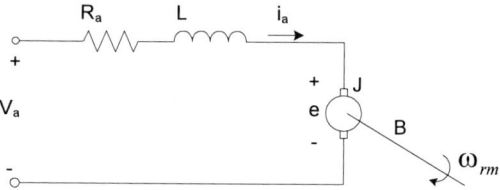

Fig. 6. : Motor Model Circuit Diagram applied to Colink

$$V_a = R_a i_a + L\frac{di_a}{dt} + e \qquad (1)$$

$$e = K_e \lambda_f \omega_m \qquad (2)$$

$$T_e = K_e \lambda_f i_a \qquad (3)$$

$$T_e = J\frac{d\omega_{rm}}{dt} + B\omega_{rm} + T_L \qquad (4)$$

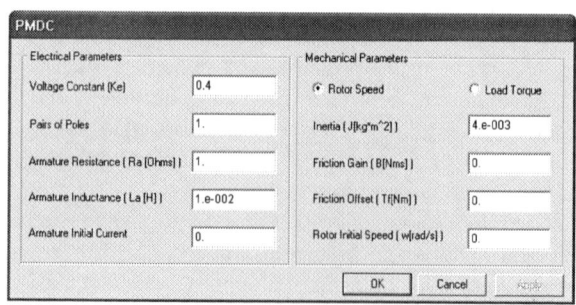

Fig. 7. : Motor Parameter Inputs

IV. CONTROLLER MODELING USING COLINK

Fig. 8 outlines the algorithm applied to Colink. The algorithm consists of six modes, which can easily be modeled by providing each gain as shown in Fig. 9.

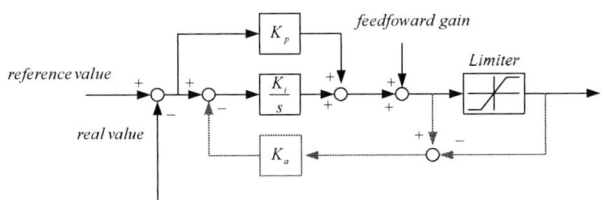

Fig. 8. : The control algorithm for PMDC controller

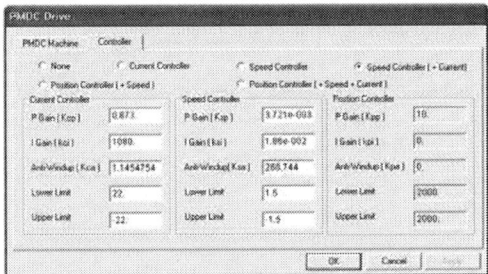

Fig. 9. : Control System using CoLink

V. CREATING LOAD MODEL USING RECURDYN

As shown in Fig. 10, RecuDyn is a dynamics simulation program for modeling and simulating mechanical systems capable of 3D analyses. Once the mechanical 3D configuration is drawn with detailed CAD data and inputs of dimensions and physical properties are provided, RecuDyn yields outputs for the load torque, equivalent inertia coefficient and friction coefficient. For a more complex load modeling, types of joint, force and contact as well as points and parameters can be provided to create a load model that closely resembles performance and characteristics of the actual mechanism.

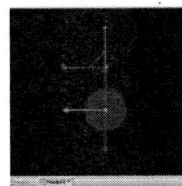 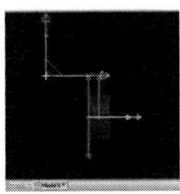 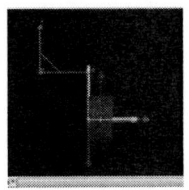

Fig. 10(a). : Creating External Configuration using RecurDyn

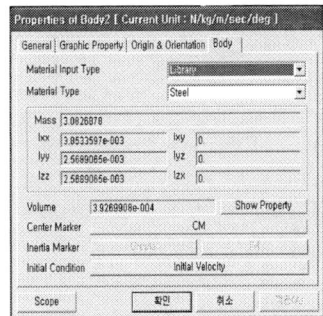

Fig. 10(b). : Automatic Calculation of Inertia Coefficient according to Motor's Physical Properties

Fig. 10. Inertia Load Modeling using RecurDyn

TABLE 1
PMDC Motor Parameters

Amateur Resistance	$1\,[\Omega]$
Amateur Inductance	$0.01\,[H]$
Back EMF Constant	$0.4\,[\text{Volt}/(\text{Weber-turn-}rad/\sec]$
Poles	2
Inertia	$0.004\,[Kg-m^2]$

VI. CONCLUSION

This study examined an education approach using RecurDyn and Colink for the PMDC motor drive system to provide more accurate and visually attractive outcomes. Applying this approach to the motor drive education system is expected to provide a more effective and interesting learning environment for the students.

REFERENCES

[1] Chunting Mi, Z. John Shen and Theresa Ceccarelli, "Continuing Education in Pwer Electronics", *IEEE Trans. Educ.*, vol. 48, no. 1, Feb. 2005.

[2] P.J van Duijsen, P. Bauer, B.Davat, "Simulation andAnimation of Power Electronics and Drives, Requirements for Education" ,www. **Simulation**research.com/sr/home/**SimulationRequirementsforEducation**.pdf

[3] S.H. Kim, H.S. Mok, D.J. Yun, "Co-Simulation for ElecticForklift System Using RecurDyn and Matlabwith Simulink", Power Electronics Annual Conference, Jun. 2006, jeju, Korea, pp.324-344,

[4] S.K. Sul, *Theory of Electric Machines Control* : DC Machines, seoul : hongpub, 2005.

[5] Functionbay, "RecurDyn v6.21 Solver Theoritical Manual", 2005.

[6] Functionbay, " RecurDyn v6.21 Basic Traning Guide", 2005

A study on the Modeling of Piezoelectric Transformer for CCFL using a PSPICE

L H Hwang *, J H Yoo *, HanXiao*,YuMingZhi*, H S Kim *, H S Oh*, M T Cho** and G S Choi ***
*Dept. of Electircal Engineering, Semyung University, Jechon, Chungbuk, 390-711, Korea
**Dept. of Electrical Engineering, Daewon Science College, Jechon, Chungbuk, 390-702 Korea
*** InTech-FA.Co . Boondanggu Sungnam Kyunggido, Korea
Email: lhhwang@semyung.ac.kr

Abstract— In this paper, a cold cathode Fluorescent(CCFL) model for high frequency dimming electric ballast simulation is presented. The model can be utilized for an electronic ballast simulation with continues dimming and transient mode simulation such as step dimming. The model is implemented using a PSPICE. The piezoelectric transformer is evaluated using an AC equivalent analysis. We are proposes an electronic ballast utilizing piezoelectric transformer for fluorescent lamps. The electronic ballast composed of piezoelectric transformers and Zero Voltage Switching(ZVS) inverter was implemented. It enables a flurorescent lamp to be turned on stably. Accordingly, simulation of a high frequency electronic ballast which operates a fluorescent lamp at high frequency is proposed. Simulation is carried out using PSPICE program to illustrate the performance for the circuit.

I. INTRODUCTION

The Resent Characteristics of a piezoelectric transformer were analyzed by combining an equivalent circuit with piezo-electric equations, and the relation between the characteristics and the dimension of the PT was studied.
Piezoelectric transformer were shown to be advantageous in DC-DC converter application and in Cold Cathode Fluorescent Lamps(CCFL) drivers[5],[6].
Simulation of a high frequency electronic ballast which operates a fluorescent lamp at high frequency is proposed. Simulation is carried out using PSPICE program to illustrate the performance of the circuit.

The steady-state characteristics of the push-pull inverter with a piezoelectric transformer are analyzed. The piezoelectric transformer operating in the 3rd-order longitudinal vibration mode is used in place of a conventional magnetic transformer to produce a high output voltage to light up a cold cathode fluorescent lamp, The circuit operation, the load characteristics, the efficiency and the ZVS conditions are analyzed using equivalent circuits. These analytical results are confirmed by experiments. An example of the output current control is also shown.

II. PIEZOELECTRIC TRANSFORMER MODELING

PT Modeling is express in equation (1) and (2)

$$V_{LR}(t) = V_{LR} \sin \omega_s t \tag{1}$$

$$V_{Lr} = \frac{2}{\pi} \frac{V_{in}}{(1-D)} \sin D\pi \tag{2}$$

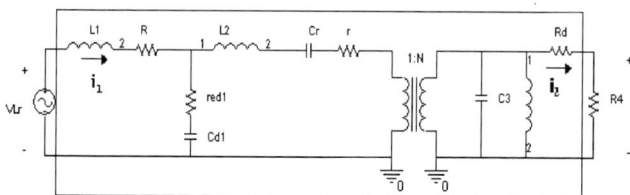

Fig.1. AC equivalent circuit of PT.

An equivalent impedance seen output side in expressed in equation (3)

$$R_{EQ} = \frac{\pi^2}{2} R_L \tag{3}$$

which is the average of half a cycle is given by equation (4)

$$V_0 = \frac{1}{2\pi} \int_0^\pi V_2 \sin \omega t d\omega t = \frac{v_2}{\pi} \tag{4}$$

Consequently, voltage gain G of the AC equivalent circuit is expressed as follows

$$G = \frac{V_2}{V_{Lr}} = \frac{(1-D)\pi^2}{2 \sin D\pi} \frac{V_0}{V_{in}} \tag{5}$$

As shown in the Fig.2., F-matrix configuration of Fig.1 which uses those symbols. For this figure, the following expressions are given

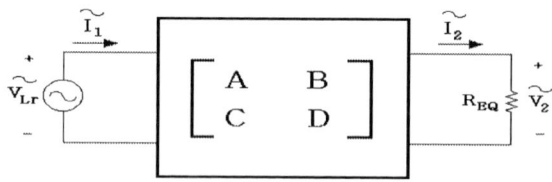

Fig.2. F-Matrix

$$\begin{bmatrix} \tilde{V}_{Lr} \\ \tilde{I}_1 \end{bmatrix} = \begin{bmatrix} A & B \\ C & D \end{bmatrix} \begin{bmatrix} \tilde{V}_2 \\ \tilde{I}_2 \end{bmatrix} \tag{6}$$

$$F = \begin{bmatrix} A & B \\ C & D \end{bmatrix}$$

$$= \begin{bmatrix} 1 & j\omega L_s \\ 0 & 1 \end{bmatrix} \begin{bmatrix} 1 & R_{Ls} \\ 0 & 1 \end{bmatrix}$$

$$\begin{bmatrix} \dfrac{1}{j\omega C_{rn}} & 0 \\ 1+j\omega C_{rn}r_{rn} & 1 \end{bmatrix}\begin{bmatrix} 1 & r \\ 0 & 1 \end{bmatrix}\begin{bmatrix} 1 & j\omega L_{r} \\ 0 & 1 \end{bmatrix}$$

$$\begin{bmatrix} 1 & \dfrac{1}{j\omega C_r} \\ 0 & 1 \end{bmatrix}\begin{bmatrix} \dfrac{1}{N} & 0 \\ 0 & N \end{bmatrix}\begin{bmatrix} 1 & 0 \\ j\omega C_{d2} & 1 \end{bmatrix}$$

$$\begin{bmatrix} 1 & 0 \\ \dfrac{1}{j\omega L_p} & 1 \end{bmatrix}\begin{bmatrix} 1 & R_d \\ 0 & 1 \end{bmatrix} \qquad (7)$$

Voltage gain G of The AC equivalent circuit is expressed in equation (8) and output voltage V_0 of the converter is derived from eq.(4) and eq.(8) as follows

$$G = \frac{V_2}{V_{Lr}} = \frac{\tilde{V}_2}{\tilde{V}_{Lr}} = \left| \frac{R_{EQ}}{A \cdot R_{EQ} + B} \right| \qquad (8)$$

$$V_0 = \frac{2 \sin D\pi}{(1-D)\pi^2} \left| \frac{R_{EQ}}{A \cdot R_{EQ} + B} \right| V_{in} \qquad (9)$$

Given from equation (6) as follows

$$P_{in} = \mathrm{Re}\, al \left(\frac{C \cdot R_{EQ} + D}{A \cdot R_{EQ} + B} \right) \left| \tilde{V}_{Lr} \right|^2 \qquad (10)$$

$$P_0 = \frac{R_{EQ}}{\left| A \cdot R_{EQ} + B \right|^2} \left| \tilde{V}_{LR} \right|^2 \qquad (11)$$

$$\eta = \frac{p_0}{P_{in}} \qquad (12)$$

III. CCFL MODELING

The dimming characteristic of shown in the CCFL of negative impedance characteristics of a fluorescent lamp in higher dimming levels and these two different dimming characteristics of a lamp, employed the following equation

$$V_{rms} = 60.966 + 110.45 \cdot e^{-1.9404 \times I_{rms}}$$
$$-48.578 \times e^{-60.182 I_{rms}} \qquad (13)$$

In equation (13), the first term shows the base value of the curve, the second term shows the negative impedance characteristics at higher dimming levels, and the third term shows the positive impedance characteristics at the lower

dimming level area. The equation parameters are derived easily by using the method of least squares.

$$R_{LAMP} = \frac{V_{rms}}{I_{rms}} \qquad (14)$$

From equation (13) and (14) can obtain the equation (15)

$$R_{LAMP} = \frac{60.966 + 110.45 \cdot e^{-1.904 \times I_{rms}} - 48.578 \times e^{-60.182 \times I_{rms}}}{I_{rms}}$$
$$(15)$$

Then we can obtain the equation for the instantaneous lamp voltage using the instantaneous lamp current as in the following

$$V_t = R_{LAMP} \times I_t$$
$$= \frac{60.966 + 110.45 \cdot e^{-1.904 \times I_{rms}} - 48.578 \times e^{-60.182 \times I_{rms}}}{I_{rms}} \times I_t$$
$$(16)$$

is calculated as a function of the simulation time T as in the following equation (17)

$$V_A = \int_0^T I_t^2 \frac{dt}{T} = I_{rms}^2 \qquad (17)$$

By using this RC integrating circuit to calculate the RMS lamp current, the lamp model equation (16) can be applied to the PSPICE model circuit as shown in Fig.3.

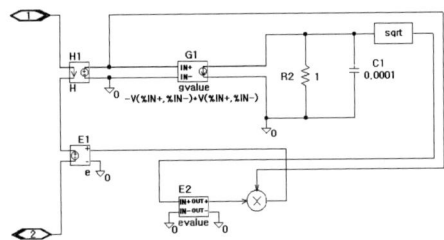

Fig.3. Simulation Model of CCFL.

IV. SIMULATION RESULT

As shown in the Fig.4., the push-pull inverter circuit with PT. The CCFL is connected between the secondary electrode strips of PT and the ground. MOSFETs Q1 and Q2 are driven alternately with 50% duty ratio, which is not critical for overlap nor gap under the ZVS conditions.

There are four operation states in one switching cycle as shown in Fig.5.

In state1, Q1 is on and Q2 is off. Both of the upper primary electrodes are connected to the ground. During this state, Vd2(=Vin) shows a resonant waveform due to L2, Cds2 in parallel with C1 and the input impedance of PT.

In state2, Vds2 is grounded by conducting the body diode of Q2. During this state, no voltage is applied to the PT.

In state3, in turn, Q2 is on and Q1 is off. Both of the lower primary electrodes are connected to the ground. During this state, Vds1(=Vin) shows a resonant waveform due to L1, Cds1 in parallel with C1 and the input impedance of PT.

In state4, Vds1 is grounded by conducting the body diode of Q1. During this state, no voltage is applied to the PT.

By repeating these four operation states in turn, the waveform of Vin makes a bipolar quasi-sinusoidal voltage, which is fed to the input terminal of the PT. The input voltage is boosted up and changed to be almost sinusoidal at the secondary side by the filtering effect of the PT.

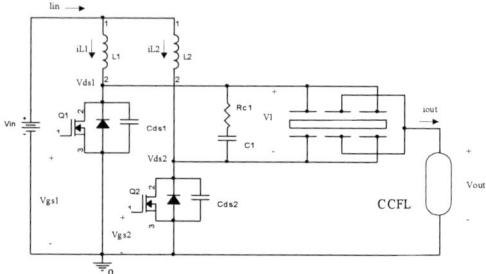

Fig.4. Push-poll Inverter with Piezoelectric

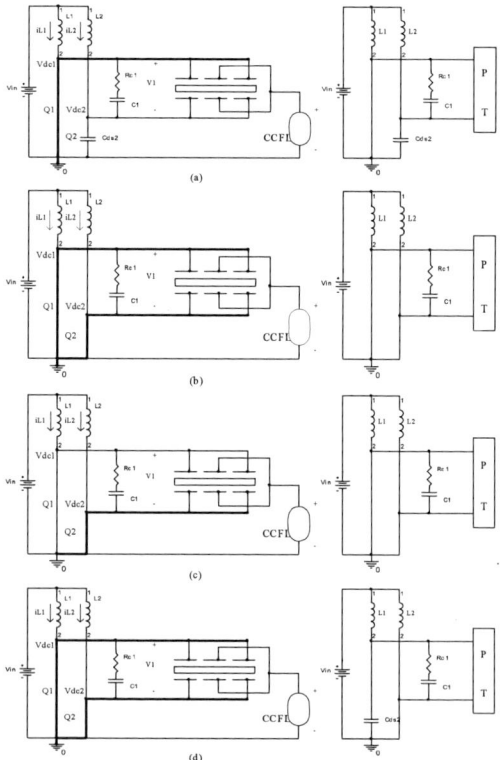

Fig.5. Operation states of the Push-pull Inverter

As shown in the Fig.6., the Push-pull Inverter PSPICE Model using Piezoelectric Transformer, As shown in Fig.7., the simulated results of the frequency characteristics of output current with resistor variation. Fig.8. shows the simulated

waveform of PT output voltage as a function of driving frequency. Fig.9. shows the simulated waveforms Vds1 taking L(=L1=L2) as a parameter using the equivalent impedance of Rs and Cs.

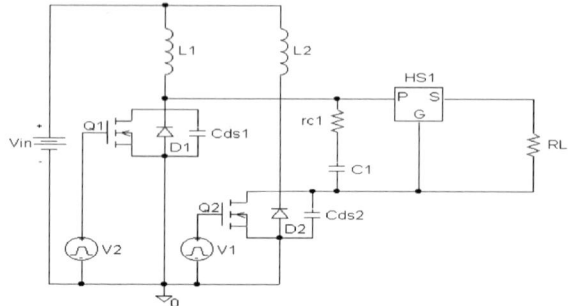

Fig.6. Push-pull Inverter PSPICE Model using Piezoelectric Transformer

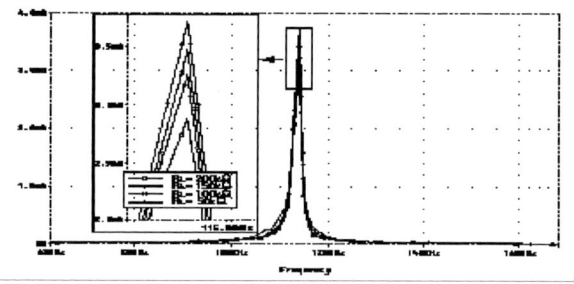

Fig.7. Frequency Characteristics of Output Current with Resistor Variation

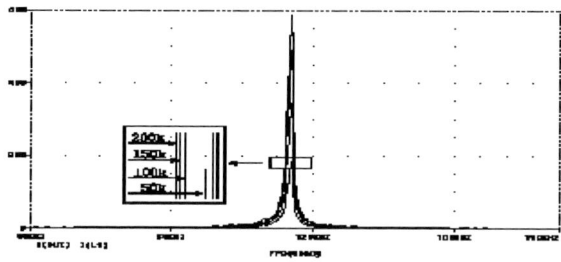

Fig.8. Output Voltage of Piezoelectric Transformer as a function of driving frequency.

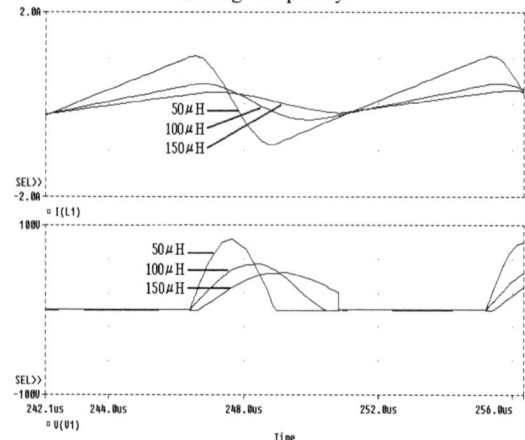

978-1-4244-1871-8/07 $25.00 © 2007 IEEE   124

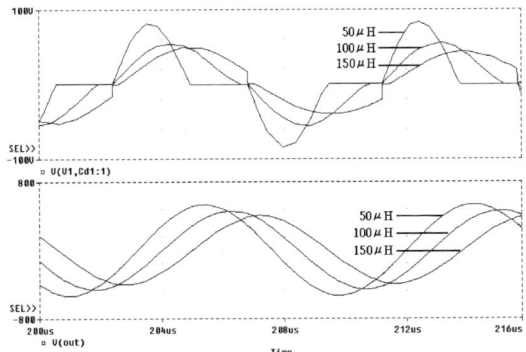

Fig.9. Simulation Waveforms with Inductance Variation

The results as shown in Fig.10. are for a frequency step from 75kHz to 30kHz. The simulation results of shown in Fig.10., this clearly show that is proposed of model scopes very well with the step produces the correct current and voltage waveforms before and after the transient.

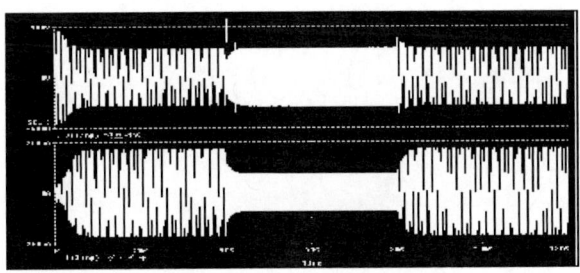

Fig.10. Simulation Waveforms of Lamp Voltage and Current with Input Frequency

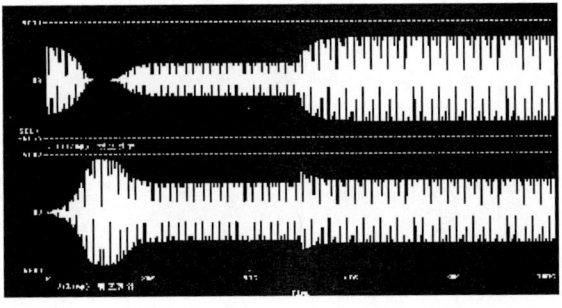

Fig.11. Simulation Waveforms of Lamp Voltage and Current with Resistor Variation

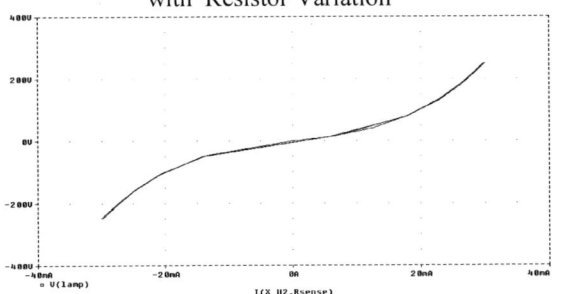

Fig.12. V-I Characteristics of CCFL

The experiment results can be used as data for the system that in order to solve the problem is proposed using a new type of electronic ballast that the traditional magnetic ballasts operated at 50-60Hz have been suffered from noticeable flicker, high loss, large crest factor and heavy weight which is composed of rectifier, active power corrector, series resonant half bridge inverter, micro-controller and piezoelectric transformer for driving for driving T5 fluorescent lamp were manufactured.

As shown in Fig.13., this realation of output voltage and frequency characteristics in a detecting by fluorescent lamp.

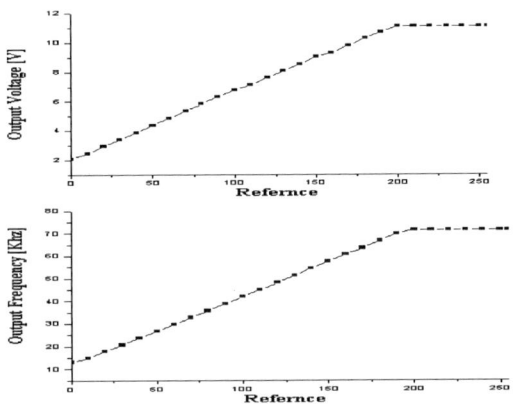

Fig.13. Output voltage and frequency fluorescent lamp detecting

As shown in Fig.14. and Fig.15., for a voltage and current characteristics of fluorescent lamp at ignition. Also, the system operating shown in Fig. 16.

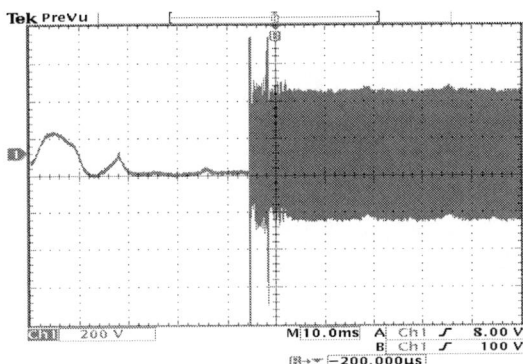

Fig.14. The voltage waveform of fluorescent lamp at ignition (voltage)

These characteristics have low ignition voltage, low crest factor and soft-start. Therefore, proposed ballast was convinced longer lamp life and stable driving than typical electronic ballast.

978-1-4244-1871-8/07 $25.00 © 2007 IEEE

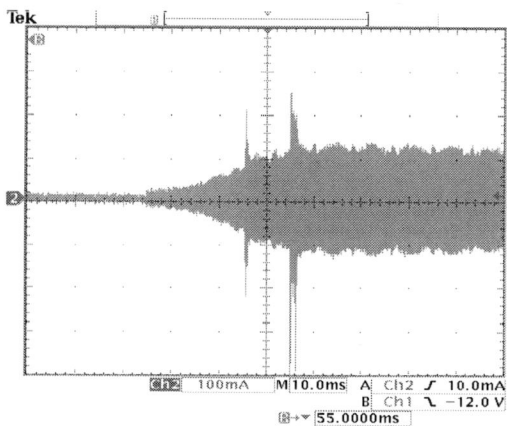

Fig.15. Current waveform of fluorescent lamp at ignition(current)

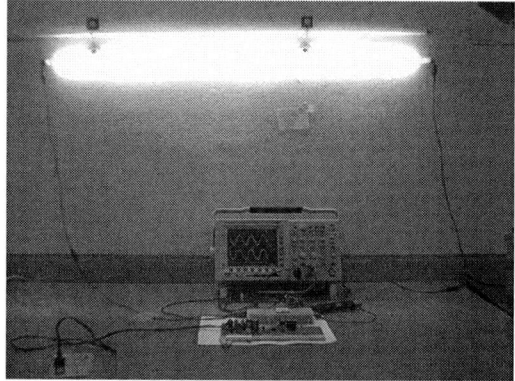

Fig. 16. The photograph of fluorescent lamp drive

These characteristics have been low ignition voltage, low crest factor and soft-start. Therefore, proposed ballast was convinced longer lamp life and stable driving than typical electronic ballast.

V. CONCLUSION

A fluorescent lamp model for high frequency dimming electronic ballasts is proposed and verified in this paper. This lamp model is derived by using the method of least squares and this model has a possibility of being applied for transient mode simulations such as step dimming mode or start-up mode

The steady-state characteristics of the push-pull piezoelectric inverter has been analyzed using an equivalent circuit of the piezoelectric transformer. The circuit operation, the load characteristics, the efficiency and the ZVS condition have been analyzed. Accordingly, the experiment results were also proposed electronic ballast operated at high frequency(about 75kHz) shows a input power factor of more than 0.995, total harmonic distortion of less than 12% and lamp current crest factor of less than 1.5, respectively, Output power and efficiency showed 28W and 85%, respectively, Accordingly, it is considered that the ballast using

piezoelectric transformer can replace the typical electronic ballast.

REFERENCE

[1] U.Mader and P.Hom,"A Dynamic Model for the Electrical Characteristics of Fluorescent Lamps" IEEE IAS Annual Meeting Proceeding, pp.1928-1934, 1992.

[2] N.Sun and B.Hesterman,"Pspice High Frequency Dynamic Fluorescent Lamps Model" IEEE APEC Conference Record, pp.641-647, 1997.

[3] T.Zaitsu, T.Shigehisa, M.Shoyama, and T.Ninomiya, "piezoelectric Transformer Converter with PWM Control, "IEEE APEC'96 Proc., pp.279-283, March. 1996.

[4] T.Zaitsu, T.Shigehisa, T.Inoue, M.Shoyama, and T. Ninomiya, "Piezoelectric Transformer Converter with Frequency Control," IEEE INTELEC' 95 Proc., pp. 175-180, Oct.1995.

[5] C.Y.Lin and F.C.Lee, "Design of a Piezoelectric Transformer Converter and its matching networks." IEEE PESC'94 Record . pp.607-612, June. 1994.

[6] T.E.Wu, J.C.Hung and T.H.Yu, "A Pspice Model for Fluorescent Lamps Operated at High Frequencies" IECON Proceeding. pp.

An Active Gate Drive Circuit for High Power Inverter System to Reduce Turn-off Spike Voltage of IGBT

Jin-Hong, Kim   Dong-Hyun, Park   Jeong-Bin, Kim   Bong-Hyun, Kwon

LS Industrial Systems Co., Ltd.

Central Research & Development Center

533, Hogye-dong, Dongan-gu, Anyang-si, Gyeonggi-do, 431-080, Korea

Email : jhkim6@lsis.biz , dhparkb@lsis.biz , jeongbink@lsis.biz , bonkwon@lsis.biz

Abstract — This paper presents a new gate drive method for high power IGBT to reduce the turn-off spike voltage which is appeared between collector and emitter. In the power circuit of inverter, there is a commutation loop stray inductance between the dc-link capacitors and IGBTs. It affects the turn-off spike voltage of IGBT. In general, snubber circuits are used to control the spike voltage but they are consists of passive components result in the limitation to reduce the spike voltage. In this paper, a new active method is proposed and based on the saturation voltage characteristics of IGBT which is related with collector current and collector-emitter voltage. In order to reduce the spike voltage, gate current is controlled in the active region during the turn-off. The proposed method is implemented with 220kW inverter and verified by the experimental results.

I. INTRODUCTION

IGBT(Insulated Gate Bipolar Transistors), a voltage driving semiconductor with high input impedance has the advantage that the gate control is easy with low gate driving power and the fast switching. In addition, as the voltage and current ratings and chip characteristics have been constantly improved, it has expanded widely in power circuit of inverters.

In the power circuit of inverter, there is a commutation loop stray inductance between the dc-link capacitors and IGBTs. During turn-off switching transient, spike voltage occurs by commutation loop stray inductances, and stresses are added to IGBT due to the addition of dc-link voltage and spike voltage. The spike voltage not only causes EMI problem by noise but also significantly affects the inverter reliability. Therefore, many research for reducing such a spike voltage has been studied. In order to reduce commutation loop stray inductance, use of Kapton film between bus plates is proposed.[1] However, it also has a weak point, that the surface of both plates has to be manufactured very precisely, and very expensive insulated film should be used. Conventional method to reduce spike voltage is to make snubber circuit.[2][3] However, for inverter, snubber circuit has limits in the size, heat emission and the stray inductance of snubber capacitor itself to reduce spike. Therefore, many studies on making optimal snubber circuit and snuberless system through the control of gate voltage and current of IGBT has been done by researcher in order to resolve the problem like spike voltage.[4][5][6][7][8][9] Because spike voltage increases with increased load

current, turn-off of large currents like short circuit of a leg significantly affects inverter reliability. Therefore, gate drive ICs embedded with soft turn-off function has been used widely.[10] However, in case of output short circuit with inductance within short circuit loop, collector-emitter saturation voltage($V_{CE(SAT)}$) can not be reached to the detection voltage(7~9[V]). In such case, the gate to the protection circuit may be turned off by the current information entered with CT(Current Transducer), and soft turn-off does not operate, it can lead to the serious physical problem of inverter.

In this paper, an active turn-off circuit is presented for driving high power inverter. The proposed active gate drive circuit can detect over-current by using IGBT's saturation voltage characteristic and change turn-off gate current only when the current exceeds over the over-current level, so spike voltage can be reduced during operation. Also, in case of output short fault where soft turn-off circuit in gate drive IC cannot operate, the proposed gate drive circuit can also reduce spike voltage efficiently. The proposed circuit is implemented with 220kW inverter and verified by the experimental results.

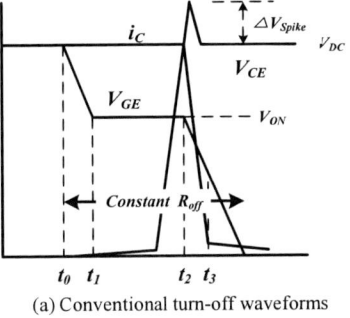

(a) Conventional turn-off waveforms

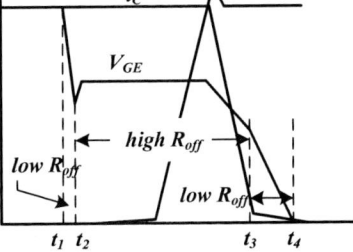

(b) Proposed turn-off waveforms

Fig. 1 : Waveforms of IGBT at turn-off

978-1-4244-1871-8/07 $25.00 © 2007 IEEE

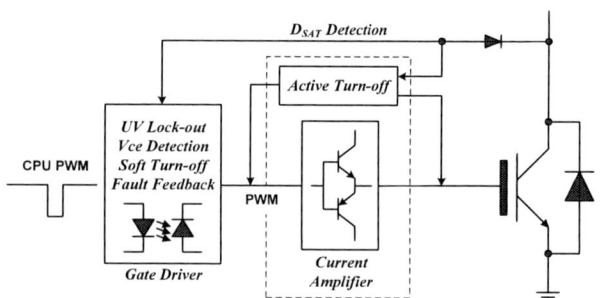

Fig. 2 : Entire Block Diagram of the Proposed Low Side Gate Drive Circuit

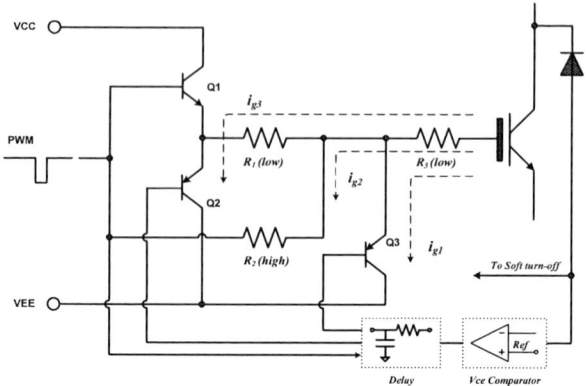

Fig. 3 : Circuit Diagram of the Proposed Gate Drive

II.   TURN-OFF CHARACTERISTICS

A.   The Turn-Off Operation of Conventional Gate Drive Circuit

Fig. 1(a) shows the turn-off process of IGBT with inductive load. In t0~t1, the gate-emitter voltage(V_{GE}) decreases exponentially with time due to discharging of the gate-emitter capacitance(C_{GE}) via the gate resistor. In this time interval, collector-emitter voltage(V_{CE}) and collector current(I_C) remain unchanged. In t1~t2, after V_{GE} decreases to saturated I_C equal to load current, V_{CE} slowly increases. Due to the Miller effect, the gate voltage will remain at the constant value because of the modulation of the collector-gate capacitor(C_{CG}). If V_{CE} increases near to V_{GE}, C_{GC} decreases to a much smaller value and V_{CE} increases rapidly towards the dc-link voltage. In t2~t3, after V_{CE} exceeds the dc-link voltage, the upper Free Wheeling Diode(FWD) turns on and V_{CE} is clamped and the load current begins to divert from the IGBT to the upper FWD.

When the load current is transferred to upper FWD from IGBT, in the commutation loop stray inductance existing within the closed circuit between dc-link capacitor and IGBT, spike voltage occurs as in equation (1), and in IGBT, voltage stress increases added by the dc link voltage and spike voltage.[11]

$$\Delta V_{Spike} = L_{LK} \cdot di_c / dt \ [V] \qquad (1)$$

Here, L_{LK} shows the stray inductance between dc-link capacitor and IGBT.

B.   The Turn-Off Operation of the Proposed Gate Drive Circuit

Fig. 1(b) shows the turn-off waveform of the proposed gate drive circuit. In t1~t2 which has the constant value of V_{CE} and I_C, V_{GE} is quickly discharged by small turn-off resistance to reduce total turn-off time. In t2~t3, switching transient is generated during this time interval, V_{GE} is discharged by large turn-off resistance and di_c/dt slop is reduced. After the switching transient period is over, rest of the V_{GE} is discharged through small turn-off resistance and turn-off operation is completed. The spike voltage generated during turn-off can be reduced by decreasing di_c/dt. Though the

spike voltage occurring during turn-off can be decreased by reducing di_c/dt, the turn-off time increase. Therefore, the proposed gate drive circuit is chosen for the completion of turn-off operation within the deadtime.

III.   PROPOSED ACTIVE GATE DRIVE CIRCUIT

A.   Block Diagram

Fig. 2 shows the entire block diagram of the proposed active gate drive circuit. The proposed circuit has the gate drive IC(Integrated circuit) which has these functions such as under voltage lockout, V_{CE}(sat) detection, soft turn-off , the push-pull circuit which can amplify the gate current. In addition the active turn-off block for limiting turn-off spike voltage is inserted.

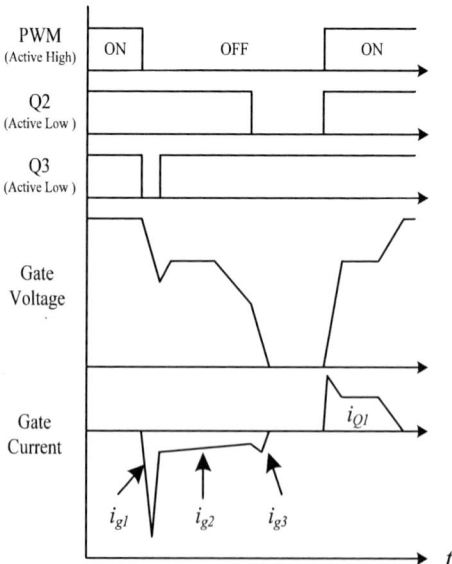

Fig. 4 : Timing Diagram of the Proposed Gate Drive

978-1-4244-1871-8/07 $25.00 © 2007 IEEE   128

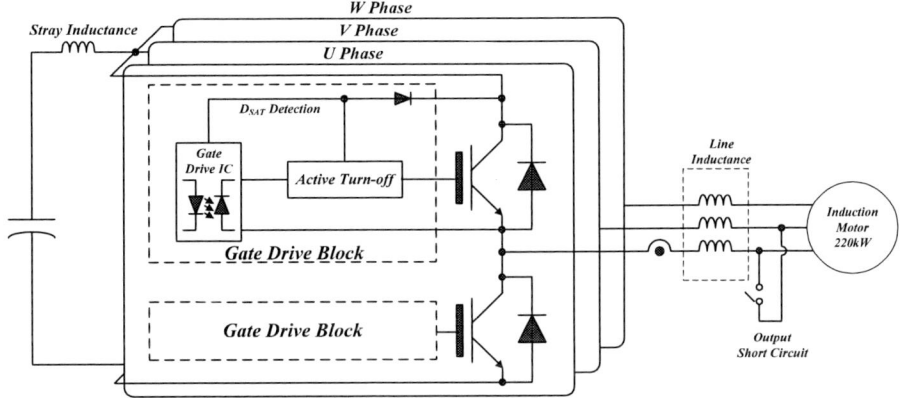

Fig. 5 : Experiment setup for the proposed circuit

B. Circuit Configuration

The proposed circuit has the characteristics that reduces di_c/dt slope by operating the active turn-off block in case of inverter over-current, and limits the turn-off spike voltage during inverter operation by optimizing turn-off time. Fig. 3 is the basic structure of the proposed active gate drive circuit. Upon a conventional gate drive circuit, this circuit is added with gate resistance R2 and turn-off transistor Q3 for active turn-off, V_{CE} comparator that compares collector emitter voltage with reference voltage and RC-delay block that adjusts the turn-on time of Q2 and Q3.

Q3 does not turn-on until V_{CE} comparator operates, therefore the active gate drive circuit does not affect the turn-off characteristics of IGBT. As I_C increase, V_{CE} also increases during IGBT's turn-on. This phenomenon can be explained by the saturation voltage characteristic. Using such characteristics, when over-current flows through IGBT, the V_{CE} comparator generates transistor turn-on signals by comparing V_{CE} and V_{REF}. When V_{CE} comparator signal occurs, base signals of Q2 and Q3 are generated with the time delay block composed of R-C components. In Fig. 4, waveforms of each part in the proposed active gate drive circuit are shown. When V_{CE} comparator operates, Q3 is turned on, synchronized with the falling edge of PWM. V_{GE} is quickly discharged by small turn-off resistance R3 and Q3. When Q3 is turned off, V_{GE} is discharged only with large resistance R2, and at this time, V_{CE} increases. After V_{CE} increase, Q2 is turned on, and discharges the remaining V_{GE}. Saturation voltage of IGBT is changed by T_j(Junction Temperature). Therefore, in this paper, the proposed circuit is operated only over the rated inverter current, V_{REF} is selected based on the 80% the rated current of IGBT at 25℃. The turn-on time of Q2 and Q3 are previously determined by the time delay block of Fig. 3 composed of R-C.

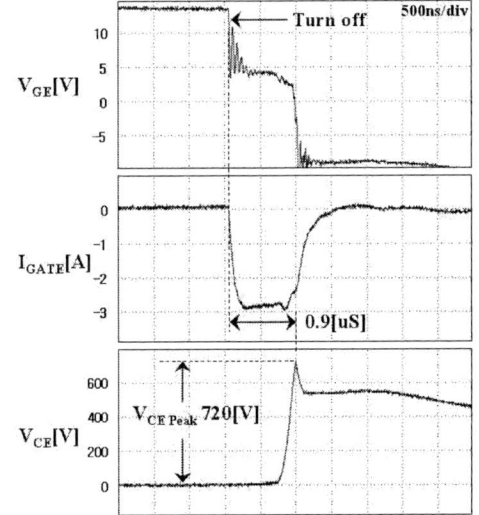

Fig. 6 : Experimental results at turn-off with the conventional circuit
(Load Current 1,000A)

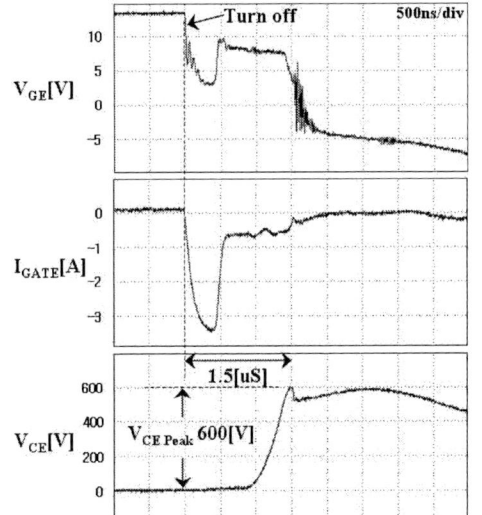

Fig. 7 : Experimental results at turn-off with the proposed circuit
(Load Current 1,000A)

978-1-4244-1871-8/07 $25.00 © 2007 IEEE   129

IV. EXPERIMENTAL RESULTS OF *di/dt* CONTROL WITH 220kW INVERTER

Active gate drive circuit was confirmed experimentally by applying to 220kW inverter, and the experiment structure is shown in Fig. 5. This experiment has two types of spike voltage waveform. One was measured when the current reached to the maximum current value at 1000[A]. The other was measured at output phase short circuit. The bus-bar between dc-link and IGBT was made with the planar type to reduce stray inductances. And 2[uF] film capacitor is inserted in parallel. 1200[V], 1000[A] rating of the IGBT was used for the experiment, and turn-off resistances are R1 = 1.5[Ω], R2 = 25[Ω], R3 = 1.5[Ω]. Rated current of motor is 416[A$_{rms}$] and the maximum output peak current of inverter is 1000[A]. The detection level of over-current is set to operate over or same as 800[A], and the on-time of Q3 is set as 500[ns] by experiment. The time constant of delay block is chosen for the completion of the proposed turn-off operation within the inverter dead-time. The carrier frequency is 3[kHz].

A. Over Current

Fig. 6 and 7 shows the experimental waveforms of the conventional gate drive circuit and the proposed active gate drive circuit at dc-link voltage of 500[V] and the IGBT rated current 1000[A]. The V$_{CE}$ peak voltage of conventional gate drive circuit is 720[V]; turn-off time is 0.9[us]; the V$_{CE}$ peak voltage of the proposed circuit is 600[V]; and the turn-off time is 1.5[us]. Though the turn-off time of the proposed circuit increased 0.6[us], the turn-off operation completes within the dead-time of the inverter. It is shown that the spike voltage can be reduced by limiting the gate current at the time of turn-off.

B. Output Short Circuit

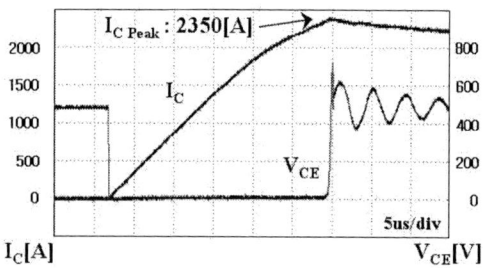

Fig.8: Waveform of output short circuit

The conventional protection of short circuit is to detect the V$_{CE}$ voltage of IGBT. If there is a short circuit current which is multiple times of IGBT's rating, gate voltage is turned off softly to reduce spike voltage by detecting V$_{CE}$ which has increased rapidly. However, if there is a inductance in short circuit path, the short circuit current is going up more slowly at that moment. In this case, IGBT might be normally turned off by other current detector such as CT. Finally, it may lead to IGBT's destruction. Fig. 8 is experiment waveform of the output short circuit that has long short circuit wire when dc link voltage is 500[V]. The length of the short circuit wire

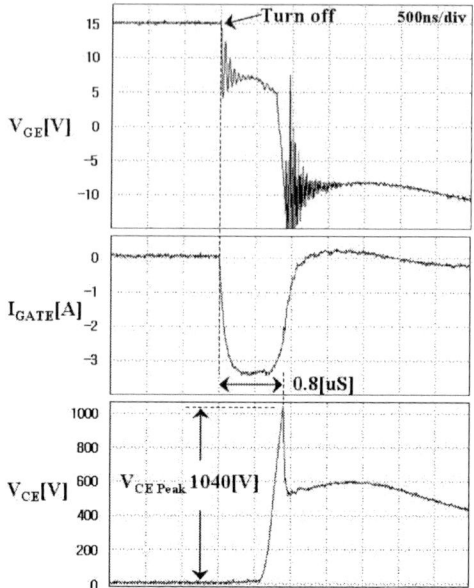

Fig.9: Experimental results at turn-off with the conventional circuit

(Output phase short circuit current : 2,350A)

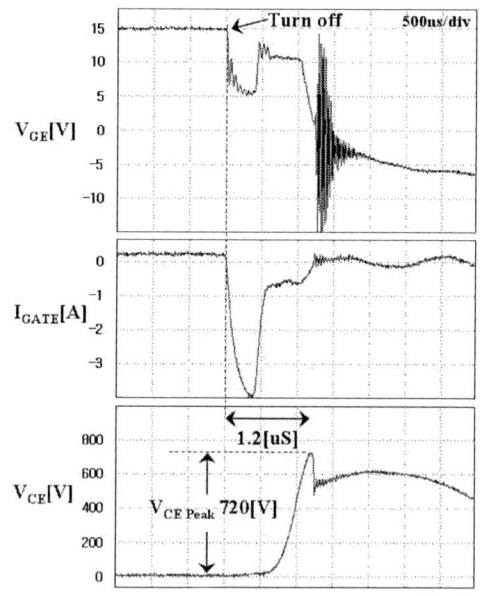

Fig.10 Experimental results at turn-off with the proposed circuit

(Output phase short circuit current : 2,350A)

used in the experiment is 10[m]. The short circuit current is 2350[A]. Fig. 9 and Fig. 10 show the waveforms of the gate voltages, gate currents and V$_{CE}$ for conventional gate drive circuit and the proposed gate drive circuit in case of output short circuit. In the case of conventional gate drive circuit, high peak voltage of 1,040[V] occurs. Supposing the inverter dc-link voltage is allowed up to 800[V], such peak voltage can even break the IGBT. In the case of the proposed circuit,

978-1-4244-1871-8/07 $25.00 © 2007 IEEE

it has such an excellent voltage limiting characteristics even at output short circuit with V_{CE} at 720[V].

V. CONCLUSION

In this study the proposed gate drive circuit that can operate even during high power inverter operation and in case of output short circuit. It was confirmed that the gate drive circuit can reduce the spike voltage that occurs at turn-off through the detection of over-current by using IGBT saturation voltage characteristics, and by limiting the turn-off gate current to reduce di_c/dt. The proposed gate drive circuit can make it easy to design bus-bar. Also, the proposed circuit can improve the system reliability in designing high power inverter.

REFERENCES

[1] Caponet. M.C, Profumo. F, De Doncker. R.W, Tenconi.A, "Low Stray Inductance Bus Bar Design and Construction for Good EMC Performance in Power Electronic Circuits," in *Rec. of 2000 IEEE Power Electronics Specialists Conf.(PESC)*,June 2000, pp. 916-921.

[2] Hughes.K.B, Schauder.C.D, Gernhardt.M.G, Stacey.E.J, "Effect of Snubber Circuit Parameters on the Turn off Voltage Spike of Large GTO Thyristors" in *Rec. of 1993 IEEE Industry Applications Society Annual Meeting , Oct.* 1993 pp. 1280-1285.

[3] Harada. K, Ninomiya. T, "Optimum Design of a Nondissipative Snubber," in *Rec. of 1994 IEEE Power Electronics Specialists Conf.(PESC)*, June 1994, pp. 1188-1195.

[4] Weis.B, Bruckmann.M, "A New Gate Driver Circuit for Improved Turn-off Characteristics of High Current IGBT Modules" in *Rec. of 1998 IEEE Industry Applications Society Annual Meeting*, Oct. 1998 pp. 1073-1077.

[5] John.V, Bum-Seok Suh, Lipo.T.A, "High-Performance Active Gate Drive for High-Power IGBTs" in *Rec. of 1998 IEEE Industry Applications Society Annual Meeting , Oct.* 1998 pp.1519-1529.

[6] Kimata. M, Chikai. S, Tanaka. T, Ishii.K, "High Performance Gate Drive Circuit of High Voltage IPMs(HVIPMs)" in *Proc. of 1998 IEEE Power Electronics Specialists Conf.(PESC)*, May. 1998 pp.1196-1200.

[7] Shihong Park. Jahn. T.M, "Flexible dv/dt and di/dt Control Method for Insulated Gate Power Switch" *in Rec. of 2001 IEEE Industry Applications Society Annual Meeting*, Sept. 2001 pp. 1038-1045.

[8] Idir. N, Bausiere. R, Franchaud.J .J, "Active Gate Voltage Control of Turn-on didt and Turn-off dvdt in Insulated Gate Transistors" *in Rec. of 2006 IEEE Power electronics*, July 2006 pp. 849-855.

[9] Chokhawala.R, Catt. J, Pelly. B, "Gate Drive Consideration for IGBT Modules" in *Rec. of 1992 IEEE Industry Applications Society Annual Meeting , Oct.* 1992 pp. 1186-1195.

[10] Chokhawala.R.S, Sobhani.S, "Switching Voltage Transient Protection Schemes for High-Current IGBT modules" in *Rec. of 1997 IEEE Industry Applications Society Annual Meeting , Nov/Dec.* 1997 pp. 1601-1610.

[11] Dulau.L Pontarollo.S, Boimond.A, Garnier. J-F, Giraudo.N, Terrasse. O, "A new Gate Driver integrated Circuit for IGBT devices with advanced protections" *in Rec. of 2006 IEEE Power electronics*, Jan 2006 pp 38-44.

An Improved Strategy to Detect the Switching Device Fault in NPC Inverter System

Jung-Dae Lee, Tae-Jin Kim, Jae-Chul Lee, and Dong-Seok Hyun

Hanyang University
Dept. of Electrical Engineering
17 Haengdang-dong Seongdong-gu, Seoul, 133-791
Korea
E-mail: skyljd78@hanyang.ac.kr

Abstract—In this paper, a simple fault detection scheme is proposed for improving the reliability under the open-switch fault of a three-level neutral point clamped (NPC) inverter. The fault of switching device is detected by checking the change of a pole-voltage in a NPC inverter. This method has the advantages of fast detection ability and a simple realization for fault detection, compared with existing methods. Reconfiguration is also performed by two-phase control method that used to supply the balancing three-phase power to the load continuously. This proposed method minimizes a bad influence on the load caused by fault occurrence. This method can also be embedded into the existing NPC inverter system as a subroutine without excessive computational effort. The proposed scheme has been verified by simulation and experimental results.

I. INTRODUCTION

A multilevel inverter system is promising technology in high voltage and high power applications because it overcomes the limitation of the maximum power rating of the solid-state device and it does not require output bulky from transformers. Moreover, as the number of levels increases, the output waveform generates a lower harmonic distortion, compared with the counterparts of conventional 2-level inverters [1]-[3]. With the help of those characteristics, multilevel inverter topologies have been widely studied by many researchers. They are classified into Diode Clamped Multilevel Inverters (DCMIs), Clamping Capacitor Multilevel Inverters (CCMIs), and Isolates H-bridge Multilevel Inverters (IHMIs) [4], [5].

Among them, the 3-level diode clamped inverter has been regarded as the most popular inverter. It is also known as the neutral-point clamped (NPC) inverter and does not have clamping capacitors and isolation transformers, resulting in the possibility of hardware simplification. Due to these advantages, compared with the other multilevel circuits, it has been widely used in high power industrial applications, such as voltage source converter based HVDC transmissions, static var compensators, high-power adjustable-frequency motor drives and so on.

The specific application among widely industrial utilities, the NPC inverter is needed to operate continuously. When the faults in a NPC inverter occur in industrial field such as aerospace, military, electrical machine, and transmission system, it can lead to serious consequences. For example, when one of the faults does occur in the practical applications

above, the drive operation has to be stopped for a nonprogrammed maintenance schedule. The cost of this schedule can be high. Also, the situation changes considerably if a fault arises under these high-speed operating conditions which cause the NPC inverter to "shot down" at high speed.

Therefore, these justify the development of fault tolerant control strategies for improving reliability. In recent years, several papers for faults in an inverter system have been published using the following schemes:
fault detection and identification methods [6], [7], and reconfiguration schemes for isolating faulty power devices [8], [9]. However, most of the diagnosis methods suggested above take either at least one forth of fundamental period or general one fundamental period between the fault occurrence and the fault detection.

This paper proposes a fault tolerant control system that can quickly recover the control performance by fast fault detection time and reconfiguration of system topology. The fault detection and identification are achieved by the measurement of pole-voltage. And this detection method requires detection time within maximum two sampling ($2Ts$). Fast fault detection ability in proposed method minimizes a bad influence on the load caused by a delay of fault detection. The novel proposed method has been verified by simulation and experimental results.

II. NEUTRAL-POINT-CLAMPED INVERTER SYSTEM WITH BRIEF DESCRIPTION

Fig. 1 presents structure of conventional three-level NPC inverter. The switching states and the output pole-voltages of the three-level NPC inverter are listed in Table I.

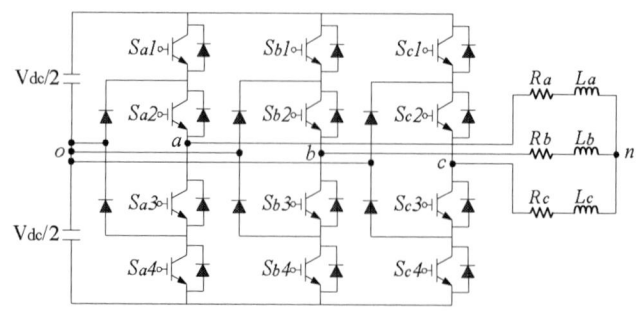

Fig. 1. Structure of conventional three-level NPC inverter.

978-1-4244-1871-8/07 $25.00 © 2007 IEEE

TABLE I
THE SWITCHING STATE AND POLE-VOLTAGE
FOR A THREE-LEVEL NPC INVERTER

Switching sequence / Switching states	S_1	S_2	S_3	S_4	V_{xo} (x=a,b,c)
P	ON	ON	OFF	OFF	$+V_{dc}/2$
O	OFF	ON	ON	OFF	0
N	OFF	OFF	ON	ON	$-V_{dc}/2$

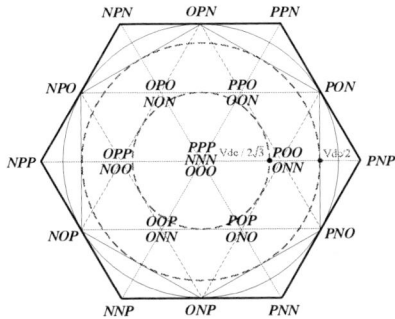

Fig. 2. Space voltage vector diagram for a three-level NPC inverter.

All available voltage space vectors for a three-level NPC inverter are shown in Fig. 2 (27 vectors). These vectors, called switching-state vectors, represent inverter output pole-voltage V_{xo} (x=a, b, c) and are produced by switching different states of the inverter, also shown in Fig.2. Each phase a, b, or c can be connected to either the positive (P), negative (N), or neutral (O) point of the DC-link.

Shown in Fig. 2, switching state P, for example, is producing pole-voltage $+V_{dc}/2$. In this case, the switch (S_1 and S_2) turn on.

Incidentally, pole-voltage has 0 as state O when the switch (S_2 and S_3) turn on. In the same way, the following switching states can be explained.

III. AN OPEN-SWITCH FAULT OF NPC INVERTER

When the open fault of switching device occurs, voltage waveform has to be analyzed in order for fault detection.

Fig. 3(a) shows waveform of pole-voltage under normal condition. Fig. 3(b) and (c) show a waveform of pole-voltage at switch (S_1 and S_2) fault respectively. Whatever condition in switch fault, the pole-voltage is different from reference value. Therefore, unbalancing power appears, when open fault is generated at switching devices of NPC inverter. These results make distorted phase currents and voltages.

Fig. 4 shows conduction path in each switching state when phase current (i_a) is positive ($i>0$). Fig. 4 (a), (b), and (c) show current paths in P, O and N of normal condition.

When the open fault of the switch (S_1 and S_2) in phase a occurs as shown in Fig.4 (d) and (e), current flows normally in state O and N. In conclusion, if the open fault of the switching devise happens, this switching devise influences partial conduction path in each switching state.

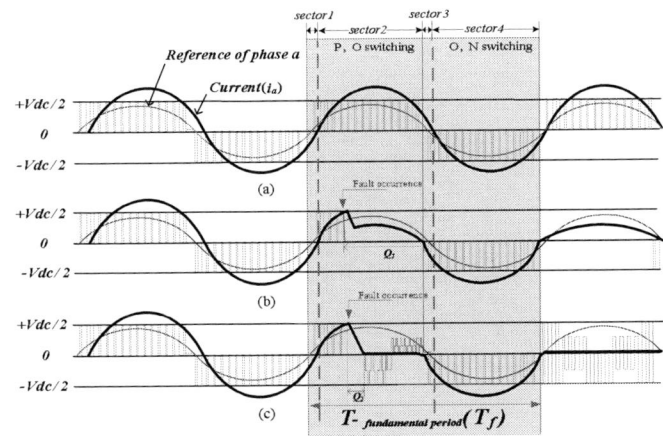

Fig. 3. Pole-voltage and current under the condition of the fault of switching device. (a) Normal condition. (b) Switch S_1 fault. (c) Switch S_2 fault.

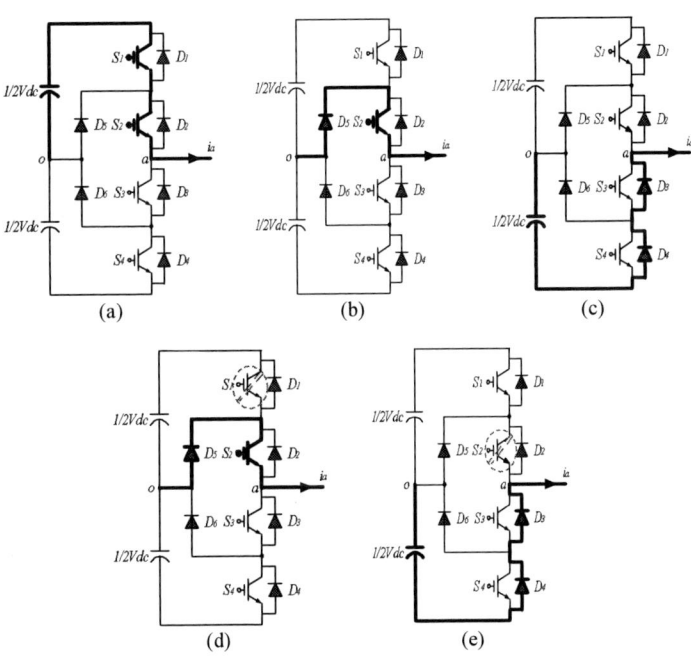

Fig. 4. Conduction path under switching states ($i_a > 0$). (a) P switching state. (b) O switching state. (c) N switching state. (d) Switch S_1 fault. (e) Switch S_2 fault.

IV. PRINCIPLE OF PROPOSED FAULT DIAGNOSIS

Pole-voltage, machine phase-voltage, and system line-voltage are commonly utilized in fault analysis. Among them, machine phase voltage and machine neutral voltage depend on the machine balance and an accessible neural terminal, while pole-voltage and system line-voltage do not depend on the machine model because they refer to the mid-point 'O' of the capacitor bank at the DC-link. Also, analysis of System line-voltage is complex in comparison with analysis of pole-voltage. Therefore, in this work the analysis of open-switch fault effect in NPC inverter has been used by the pole-voltage V_{xo} (x=a, b and c). According to switching state, pole-voltage is presented as $\pm V_{dc}/2$ or 0 in normal condition. However, in fault condition, pole-voltage is presented as $\pm V_{dc}/4$. Otherwise, $\pm V_{dc}/2$ or 0 presents continuously more than regular time ($Q_{fault\ time}$).

978-1-4244-1871-8/07 $25.00 © 2007 IEEE   133

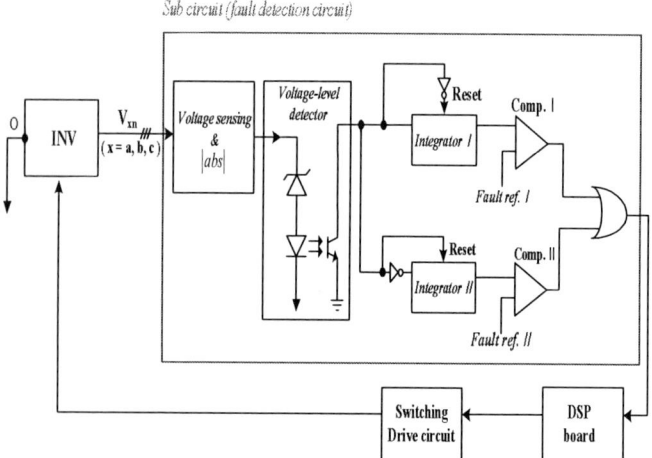

Fig. 5. Equivalent circuit of a fault detection system.

The pole-voltage magnitude in open-switch fault is distorted or disappears in comparison with pole-voltage magnitude of normal condition. In order to detect open-switch fault condition, the proposed fault detector uses information of measured pole-voltage magnitude and maintenance time in each switching state. This information of pole-voltage that has constant magnitude and fundamental period is becoming the singular of fault diagnosis. This method employs a direct comparison of the measured pole-voltage to Fault reference signal.

As shown in Fig. 5, the proposed fault detection circuit consists of *voltage sensor*, *absolute value circuit*, *voltage-level detector*, *integrator I*, *integrator II* and *comparator*. This circuit performs fault detection and identification through the measured pole-voltage of a NPC inverter.

The fault diagnosis system is accomplished in four steps as follows:

A. *First step*: Pole-voltage is measured by voltage sensor.

B. *Second step*: Output of voltage-level detector is integrated through *integrator*.

C. *Third step*: Output of *integrator* compares with fault reference during the T_s.

IF $V_{xo} \neq V_{dc}/2$ *or* 0 **THEN** *Fault detection*

ELSE *normal*

IF $V_{xo} = \pm V_{dc}/2$ *or* 0

ELSE IF $V_{xo} = \pm V_{dc}/2$ *or* 0 after $Q_{fault\ time}$

THEN *Fault detection*

ELSE *normal*

D. *Fourth step*: Reconfiguration of a NPC inverter system.

V. ISOLATION AND RECONFIGURATION

This paper proposes a novel fault diagnosis scheme. To validate the reliability of this algorithm, existing isolation and reconfiguration method is introduced as shown in Fig. 6.

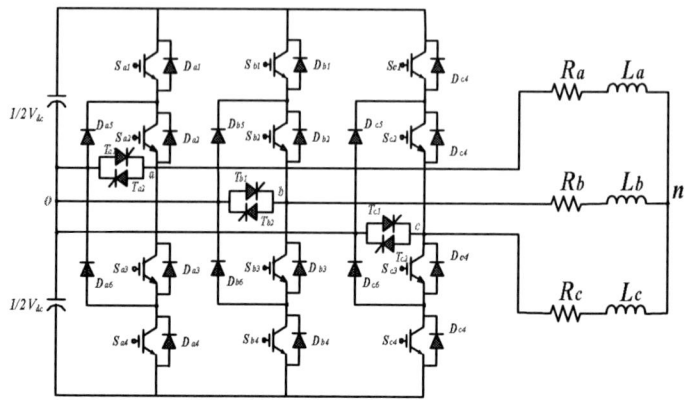

Fig. 6. Schematic of the NPC inverter system for reconfiguration.

After the fault identification, the reconfiguration scheme is achieved by two-phase control topology using bidirectional switches. When the fault of a switch is identified, the bidirectional switch of the faulty phase is fired in order to connect the faulty phase to the midpoint of the DC-link and the switching signals of the faulty phase are removed at the same time. The system after isolating the faulty phase operating by eight-switch two-phase drives has been proposed in [10]. This current magnitude that is decreased by two-phase drive can lead to serious consequences in the NPC inverter system, but it can be controllable by adjusting modulation ration size. It, therefore, is possible that the NPC inverter system operates continuously like a normal condition.

VI. SIMULATION RESULTS

Simulations are performed for the NPC inverter with ratings of parameters in Table II.

Fault condition of switching device is made by off-signal of gate drive. In this work, only the open fault of switch (S_2) will be considered. Also, the effect of this fault type in any other device is assumed to be symmetrical.

TABLE II
SIMULATION CONDITIONS

DC-link voltage	V_{dc}		200 [V]
Modulation ratio	m_i	Before fault	0.5333
		After fault	0.9237
Load resistance	R		9 [Ω]
Load inductance	L		5 [mH]
DC-link capacitance	C		5000 [μF]
Control period	T_s		250 [μsec]

978-1-4244-1871-8/07 $25.00 © 2007 IEEE

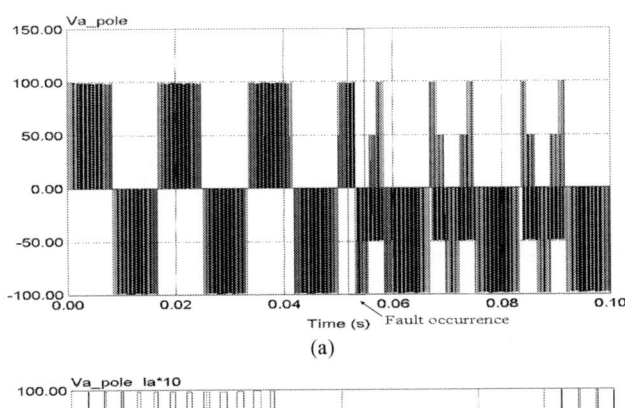

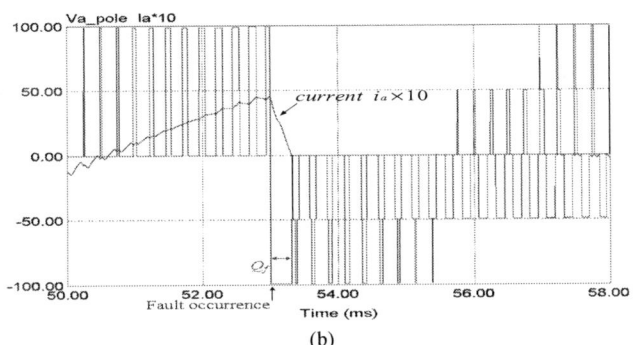

Fig. 7. Pole-voltage under fault occurrence. (a) Pole-voltage in faulty phase. (b) Output of the phase current.

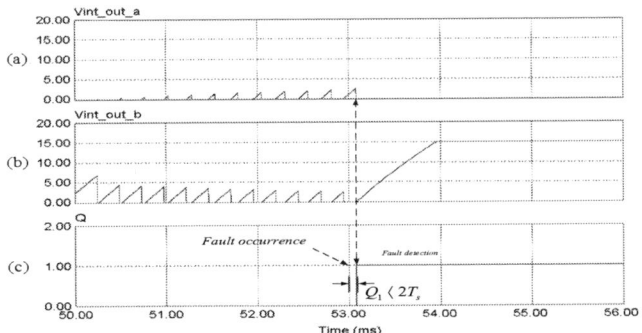

Fig. 8. Output of *integrator I, II* under switch S_2 is fault. (a) Output of *integrator I*. (b) Output of *integrator II*. (c) Fault detection signal.

Fig. 7(a) shows pole-voltage of the NPC inverter in switch (S_2) fault condition and Fig. 7(b) shows magnified current of faulty phase *a* in time base.

When the open fault of the switch (S_2) occurs, even if the switch (S_2) operates abnormally, phase currents of phase *a* flow through freewheeling diode in state O, P. Accordingly, the pole-voltage in Q_2 presents $-V_{dc}/2$.

Fig. 8 presents output of *integrator I, integrator II* and fault signal. The output of *integrator I* increase at $Q_f \le 2T_s$ after fault occurrence (53ms). As shown in Fig. 8(a), when the output of *integrator I* go over the fixed threshold value, the fault signal represents within T_s. The output of *integrator II* in Fig. 8(b) continuously increases up to saturation region because pole-voltage magnitude remains at 0.

The NPC inverter system also is reconfigured after fault detection. After this situation, the pole-voltage of faulty phase maintains 0 due to isolation of faulty phase.

Fig. 9 shows pole-voltage waveform of isolated phase *a* and output of controlled phase current.

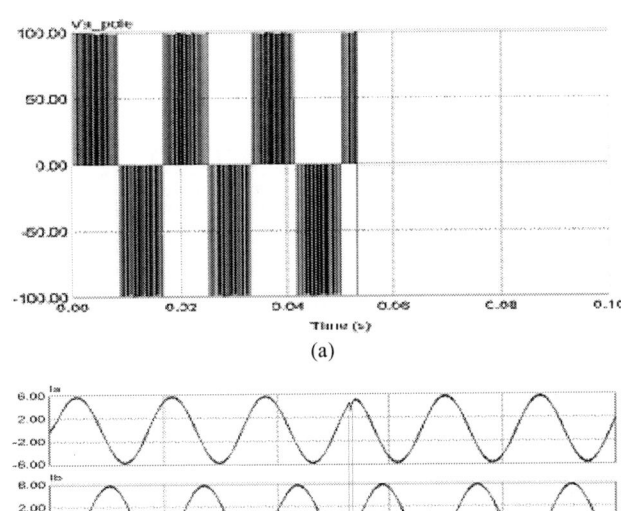

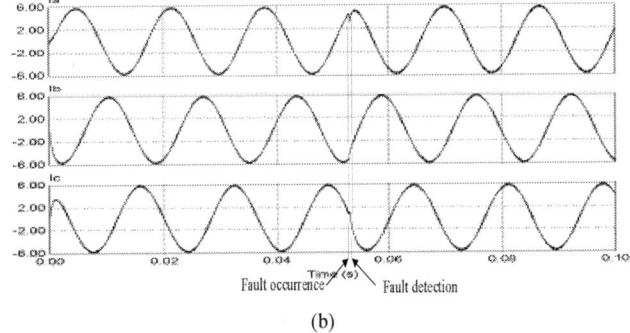

Fig. 9. Pole-voltage and phase currents under the controlled system. (a) Pole-voltage in faulty phase. (b) Output of the phase current.

VII. EXPERIMENTAL RESULTS

The proposed fault detection method has been tested experimentally with a NPC inverter with NF34AB50N6S2D IGBT at the laboratory. In order to control a NPC inverter system and generate the PWM pulse, we employed a digital signal processor (DSP) TMS320C31 and EPLD. The experimental condition is same as simulation condition.

Fig.10 shows experimental results for output waveform of *integrator I, II* and fault detection signal. Fault condition is made by off-signal of gate drive and phase current presents positive value. $-V_{dc}/2$ of the pole-voltage occurs until the phase current becomes from positive value to 0. Therefore the output of *integrator I* rises and exceeds threshold value because of this fault phase. Then the *integrator I* detects the fault condition at detection circuit.

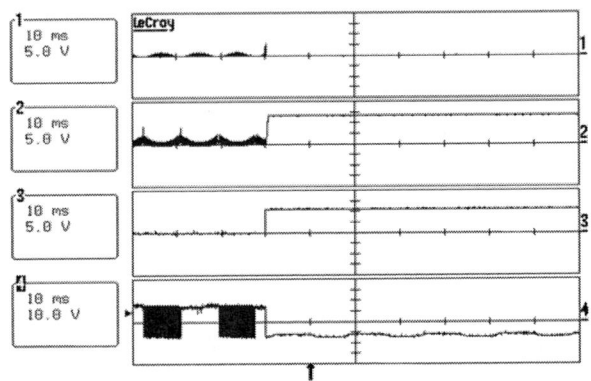

Fig. 10. Experimental results for output waveform of *integrator I, II* and fault detection signal. (ch.1: *integrator I* (5V/div), ch.2: *integrator II* (5V/div), ch.3: detection signal (5V/div), ch.4: gate pulse (10V/div)).

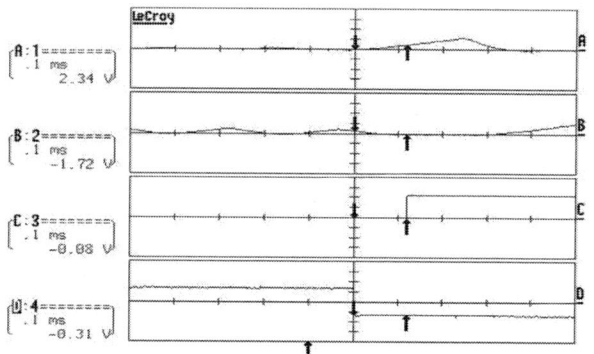

Fig. 11. Experimental results for amplifier of the fault point. (ch.1: *integrator I* (5V/div), ch.2: *integrator II* (5V/div), ch.3: detection signal (5V/div), ch.4: gate pulse (10V/div)).

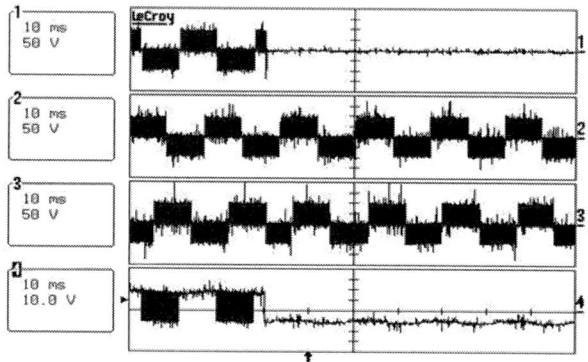

Fig. 12. Experimental results for pole-voltage under the open-switch (S_2) fault signal. (ch.1: pole-voltage in V_a (50V/div), ch.2: pole-voltage in V_b (50V/div), ch.3: detection signal (50V/div), ch.4: gate pulse (10V/div)).

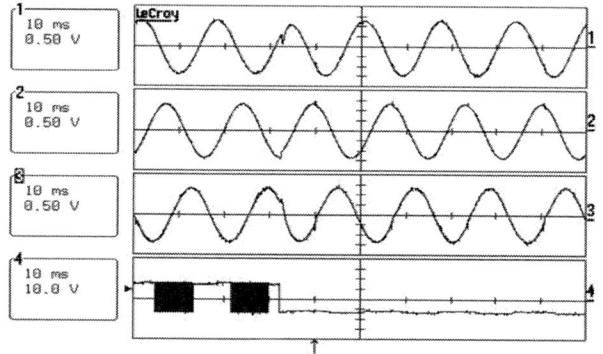

Fig. 13. Experiment results under the controlled system signal. (ch.1: out-current of V_a (5A/div), ch.2: out-current of V_b (50V/div), ch.3: output current of V_c (5A/div), ch.4: gate pulse (10V/div)).

As shown in Fig. 11, fault detection time of 118.6μsec is shorter than 250μsec of one sampling because of 0.533 modulation ratio. Output of the NPC inverter has balancing power after the fault tolerant control.

Fig. 12 represents pole-voltage under the open-switch (S_2) fault signal in the NPC inverter. After fault tolerant control, pole-voltage in fault phase from the DC-link neutral point remains at 0.

Fig. 13 shows the output current of the NPC inverter under the fault control. Outputs of the NPC inverter have balancing three phase current size in the normal condition after fault tolerant control.

VIII. CONCLUSION

This paper proposes the novel fault detection method at the case of open fault of switching devices. The proposed method reduces fault detection time within maximum $2T_s$.

In the case of the low modulation ratio, this method has possibility that fault condition is detected within minimum T_s. There is also no decline of output power magnitude because modulation ratio can be again mediated.

Besides, the fault of switching device is detected by observing the change of pole-voltage in a three-level NPC inverter. This voltage doesn't respond sensitively to the variable machine balance. These merits of proposed scheme can improve reliability of a three-level NPC inverter system.

In comparison to the existing detection methods, the proposed method allows for easier implementation and can be introduced into commercial a three-level NPC inverter system without any complexity. This is justified when it is necessary to improve the reliability of those systems.

Experimental and simulated results confirm the validity of the proposed method, which can be embedded into the existing ac drive software without an excessive computational effort.

REFERENCE

[1] C. Hochgraf, R. Lasseter, D. Divan, and T. Lipo, "Comparison of multi-level inverters for static Var compensation", *in Proc. IEEE Ind. Applicat. Soc. Annu Conf, Vol.* 2, pp. 921–928, 1994.

[2] R. W. Menzies, P. Steimer, and J. K. Steinke, "Five level GTO inverters for large induction motor drives", *in Conf. Rec. IEEE IAS Annu. Meeting.*, pp. 593–601, 1993.

[3] H. Stemmler, "Power electronics in electric traction applications" *in Proc. IEEE IECON'93*, pp. 707–713, 1993.

[4] A. Steimel, "Electric railway traction in Europe", *IEEE Ind. Applicat. Mag.*, Vol. 2, pp. 7–17, Nov./Dec. 1996.

[5] Jih Sheng Lai and Fang Zheng Peng, "Multilevel Converters-A New Breed of Power Converters", *IEEE Trans. on Industry Application*, Vol. 32, pp. 509–517, 1996.

[6] R. L. A. Ribeiro, C. B. Jacobina, E. R. C. daSilva and A. M. N. Lima, "Fault Detection of Open-Switch Damage in Voltage-Fed PWM Motor Drive Systems", *IEEE Trans. Power Electronics*, Vol. 18, No. 2, pp. 587–593, Mar. 2003.

[7] N. Bianchi, S. Bolognani, M. Zigliotto and M. Zordan, "Innovative Remedial Strategies for Inverter Faults in IPM Synchronous Motor Drives", *IEEE Trans. Energy Conversion*, Vol. 18, No. 2, pp. 306–312, June 2003.

[8] J. R. Fu and T. Lipo, "A Strategy to Isolate the Switching Device Fault of a Current Regulated Motor Drive," *in Conf. Rec. IEEE-IAS Annu.* Meeting, Vol. 1, pp. 1015–1020, 1993.

[9] R. Peuget, S. Courtine, J. Rognon, "Fault detection and isolation on a PWM inverter by knowledge-based model", *IEEE Trans.Ind. Applications*, Vol. 34, pp. 1318–1326, Nov./Dec. 1998.

[10] Gun-Tae Park, Tae-Jin Kim, Dae-Wook Kang, Dong-Seok Hyun, "Control method of NPC inverter for continuous operation under one phase fault condition", *IAS annual meeting, Conf. Rec.*,Vol. 4, pp. 2188–2193, Oct./2004.

978-1-4244-1871-8/07 $25.00 © 2007 IEEE

Overmodulation Strategy of NPC Type 3-Level Inverter for Traction Drives

Jaemoon Lee,   Jaeho Choi
Chungbuk National University
School of Electrical & Computer Engineering
410 Sungbong-ro, Heungduk-gu, Cheongju
Chungbuk, Korea   361-763
choi@chungbuk.ac.kr

Yasuyuki Nishida
Nihon University
Department of Electric & Electronic Engineering
Tokusada, Tamura-machi
Kouriyama, 963-8642   Japan
nishida@ee.ce.nihon-u.ac.jp

Abstract—This paper proposes an overmodulation method for improving the voltage utilization of NPC type 3-level inverter for traction drives.

The conventional railway vehicle has used a vector control to MI=0.907 and a slip-frequency control from MI=0.907 to the six-step mode. The slip-frequency control has less output torque dynamics than the vector control. In this paper, the linear region of NPC type 3-level inverter output voltage is extended to the six-step mode by using overmodulation strategy. As a result, vector control can be adapted in the whole regions and it leads to better output torque dynamics comparing with the conventional slip frequency control in overmodulation region.

The overmodulation range is divided into two modes depending on the modulation index (MI), In the overmodulation region I, the reference angles are derived from the Fourier series expansion of the reference voltage corresponds to the MI. In the overmodulation region II, the holding angles are also derived in the same way. Therefore, it is possible to obtain the linear control and the maximized utilization of PWM inverter output voltage.

I. INTRODUCTION

This paper presents an overmodulation strategy and indirect vector control of NPC type 3-level inverter. Recently, the railway vehicle needs to operate in higher speed within the limited plant for the higher efficiency. The conventional railway vehicle has used the vector control to MI=0.907 and used the slip-frequency control from MI=0.907 to six-step mode. The slip-frequency control is suitable for traction drives because the drive patterns of electric railway system do not request a rapid change. However that control can not realize a quick torque response. In high-speed trains, the slip between train wheels and rails are likely to occur, and the fast torque control is crucial to overcome this slip problems. The vector control provides an instantaneous torque response because it is possible to control the flux and torque of induction motor independently.

NPC type 3-level inverter has three output voltage levels. With this circuit configuration, the voltage stress on its power switching devices is half that of the conventional two-level inverter. Because of this nature, it has been applied to the medium and high voltage drives. In addition to the capability to handle the high voltage, the NPC inverter has favorable features; lower line to line and common-mode voltage steps, more frequent voltage steps in one carrier cycle, and lower ripple components in the output current at the same carrier frequency. These features lead to significant advantages for motor drives over the conventional two level inverters in the form of lower stresses to the motor windings and bearings, less influence of noise to the adjacent equipment, etc [1]. Early applications included the railroad traction areas in Japan. Recently, most of Japanese electrical train companies have developed a three level PWM inverter drive system. It is said that the three-level configuration has greatly reduced the size of main transformer and traction-motor electrical, the current harmonics in the signaling band, the acoustic noise, the volume and weight of the equipment [2,3]. And also, the power rating of the system can be increased.

This paper presents the vector control of induction motor of the NPC type inverter. Especially, the overmodulation mode control method is well described to be selected as the change of speed to meet the wide speed range of the traction drives [4,5,6]. Asynchronous space vector PWM mode and synchronous one pulse mode are basically used under the full speed range with the overmodulation PWM mode for the soft transfer between them. The validity of this proposed control scheme is verified with the simulation results.

II. NPC TYPE 3-LEVEL INVERTER

Figure 1 shows the circuit diagram of the NPC type 3-level inverter. Each phase has four switching devices (IGBTs) connected in series. For example taking phase U, when IGBTs S_1 and S_2 are turned on, the output U is connected to the positive rail of the dc bus. When S_2 and S_3 are on, it is

TABLE 1 Output voltage according to switching state

SW	S_1	S_2	S_3	S_4	V_{Uo}
	On	On	Off	Off	$+V_{dc}/2$
State	Off	Off	On	On	$-V_{dc}/2$
	Off	On	On	Off	0

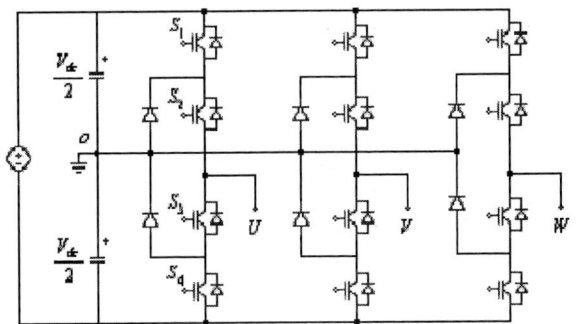

Fig. 1. Power circuit of NPC type 3-level inverter

978-1-4244-1871-8/07 $25.00 © 2007 IEEE

connected to the mid-point of the dc bus, and when S_3 and S_4 are on, it is connected to the negative rail. Thus, the output can take three voltage levels. Relations between the switching states of IGBTs and the resulting output voltage with respect to the dc mid-point are summarized in Table 1. The total switching states are consisted of 27 states and can be shown as the space voltage vectors as in Fig. 2. The corresponding 27 switching states of NPC three-level inverter are classified by 4 voltage vectors according to the voltage vector value. Four voltage vectors are Zero Vector (ZV), Small Vector (SV), Middle Vector (MV), and Large Vector (LV). These 4 voltage vectors are summarized in Table 2.

NPC type inverter has a merit to direct connect to the existing system. But has a demerit to occur unbalance of dc-link capacitor voltages. The problem of unbalanced dc-link capacitor voltage can be mitigated with the following technique.

The mid-point of dc bus capacitors is connected to the inverter bridge circuit through diodes as shown in Fig. 1. The current flow in and out of this point causes the voltage imbalance between the upper and lower capacitors. For example, in the state of SV, the mid-point current flows differently in positive and negative direction and makes the capacitor voltage imbalanced. This can be controlled by choosing the appropriate switching combinations from those which produce the same motor voltage but cause the

capacitor current to flow in the opposite direction. SV is classified by USV and LSV. USV and LSV provide the motor with the same voltage and current under a given operation. On the other hand, the mid-point currents have opposite directions to each other. Therefore, the capacitor voltage balancing can be realized by choosing appropriately from USV and LSV.

III. OVERMODULATION STRATEGY

An over-modulation strategy for higher voltage utilization is driven from developing Fourier series expansion of the reference phase voltage waveform which generates the desired fundamental component. According to the modulation index (MI), the PWM range is divided into three regions as linear region ($0 \le MI \le 0.906$), over-modulation region I ($0.906 \le MI \le 0.952$), over-modulation region II ($0.952 \le MI \le 1$) as shown in Fig. 3.

The linear region is located in the inscribed circle of an outer hexagon. It consists of linear region mode I, hybrid region, and linear region mode II. Linear region mode I is located in the inscribed circle of inner hexagon. Hybrid region is located between the inscribed circle of inner hexagon and the circumscribed circle of inner hexagon. Linear region mode II is located between the circumscribed circle of inner hexagon and the inscribed circle of outer hexagon. Linear region mode I has 3 steps output line-to-line voltage like as 2-level inverter. Linear region mode II has 5 steps output line-to-line voltage and hybrid region have characteristics of linear region mode I and linear region mode II [7].

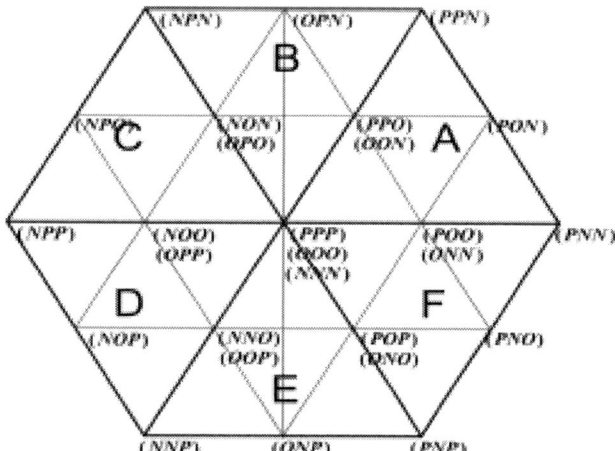

Fig. 2. Space voltage vector diagram of NPC type 3-level inverter

TABLE 2   Vector classification, magnitude and capacitor voltage variation according to switching state

Vector		Switching Vector			CP	CN	Value
ZV		PPP	OOO	NNN	0	0	0
SV	USV	POO	PPO	OPO	−	+	$\dfrac{V_{dc}}{3}$
		OPP	OOP	POP			
	LSV	ONN	OON	NON	+	−	
		NOO	NNO	ONO			
MV		PON	OPN	NPO	+/−	+/−	$\dfrac{V_{dc}}{\sqrt{3}}$
		NOP	ONP	PNO			
LV		PNN	PPN	NPN	−	−	$\dfrac{2V_{dc}}{3}$
		NPP	NNP	PNP			

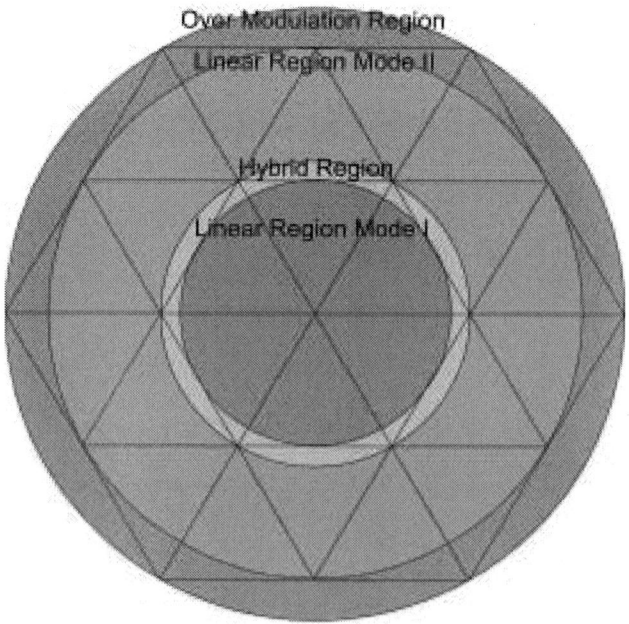

Fig. 3. NPC type 3-level inverter operation diagram

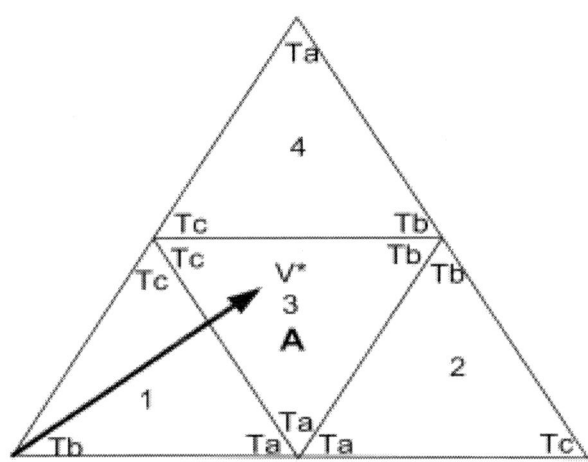

Fig. 4 Voltage vector :Sector A

TABLE 3. Durations of the voltage vectors in each region

Region	T_a	T_b	T_c
1	$2kT_s \sin(\frac{\pi}{3}-\theta)$	$T_s[1-2k\sin(\theta+\frac{\pi}{3})]$	$2kT_s \sin\theta$
2	$2T_s[1-k\sin(\theta+\frac{\pi}{3})]$	$2kT_s \sin\theta$	$T_s[2k\sin(\frac{\pi}{3}-\theta)-1]$
3	$T_s[1-2k\sin\theta]$	$T_s[2k\sin(\theta+\frac{\pi}{3})-1]$	$T_s[2k\sin(\theta-\frac{\pi}{3})+1]$
4	$T_s[2k\sin\theta-1]$	$2kT_s \sin(\frac{\pi}{3}-\theta)$	$2T_s[1-k\sin(\theta+\frac{\pi}{3})]$

A. Linear region ($0 \le MI \le 0.906$)

Figure 4 shows one of six sectors in linear region. This triangle can be divided into four smaller regions 1, 2, 3, and 4. In the space vector PWM method, generally, the reference voltage vector can be generated by its nearest three voltage vectors in order to minimize the harmonic components of the output line-to-line voltage. The durations of each voltage vector can be calculated in the same way. The results are presented in Table 3.

B. Overmodulation region I ($0.906 \le MI \le 0.952$)

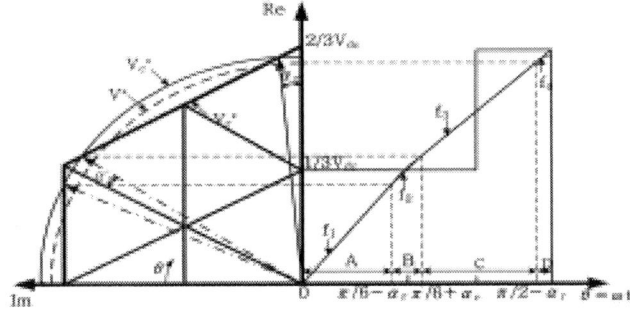

Fig. 5. Trajectory of reference voltage vector and phase voltage waveform in mode I.

In the overmodulation region I, the reference vector V^* exceeds the outer hexagon which is the boundary of making a maximum voltage value. So V^* is boosted up to V_c^* for the compensation of voltage loss due to the excess region. Figure 5 presents the trajectory of V^*, V_c^*, and V_r^* which is the actual reference vector applying to control in time domain. The angle α_r is the reference angle which is an intersection of V_c^* and boundary of hexagon. The regions of V_r^* are presented four kinds of equations per $\pi/2$ as the angle of reference voltage vector, α_r, as shown in Eqs. from (1) to (4). In both side regions of each triangle sector to α_r, V_c^* is usedto compensate the voltage loss due to the excess of to theouter hexagon. And it generates the maximum voltage value to follow the outer hexagon between those two regions. The value of the fundamental component of V_r^* to take is direct proportion to MI. the V_r^* is presented in Fig. 5.

$$f_1 = \frac{V_{dc}}{\sqrt{3}}\tan\theta , \qquad 0 \le \theta \le \left(\frac{\pi}{6}-\alpha_r\right) \quad (1)$$

$$f_2 = \frac{V_{dc}}{\sqrt{3}\cos\left(\frac{\pi}{6}-\alpha_r\right)}\sin\theta , \qquad \left(\frac{\pi}{6}-\alpha_r\right) \le \theta \le \left(\frac{\pi}{6}+\alpha_r\right) \quad (2)$$

$$f_3 = \frac{V_{dc}}{\sqrt{3}\cos\left(\frac{\pi}{3}-\theta\right)}\sin\theta , \qquad \left(\frac{\pi}{6}+\alpha_r\right) \le \theta \le \left(\frac{\pi}{2}-\alpha_r\right) \quad (3)$$

$$f_4 = \frac{V_{dc}}{\sqrt{3}\cos\left(\frac{\pi}{6}-\alpha_r\right)}\sin\theta , \qquad \left(\frac{\pi}{2}-\alpha_r\right) \le \theta \le \frac{\pi}{2} \quad (4)$$

$$\alpha_r = -30.23 \times MI + 27.04 (0.9068 \le MI \le 0.9095) \quad (5)$$

$$\alpha_r = -8.58 \times MI + 8.23 (0.9095 \le MI \le 0.9485) \quad (6)$$

$$\alpha_r = -26.43 \times MI + 25.15 (0.9485 \le MI \le 0.9517) \quad (7)$$

C. Overmodulation region II ($0.952 \le MI \le 1$)

Over the compensation limit by using the technique in overmodulation region I, V_r^* is held during α_h for the compensation of voltage loss when V^* is rotating according to time. And V_r^* is rotating according to the exterior of hexagon during the extra time. It is called an overmodulation region II. The regions of V_r^* are presented four kinds of equations per $\pi/2$ as the angle of reference voltage vector, α_h, as shown in Eqs. from (8) to (11). The value of the fundamental component of V_r^* is directly proportional to MI. Figure 6 shows the trajectory of reference voltage vector and phase voltage waveform in the overmodulation region II.

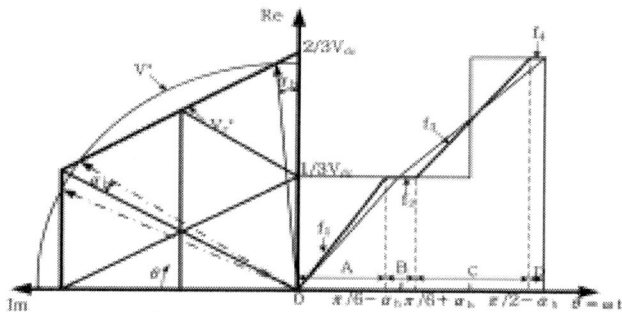

Fig. 6. Trajectory of reference voltage vector and phase voltage waveform in mode I I.

$$f_1 = \frac{V_{dc}}{\sqrt{3}} \tan \alpha_p, \qquad 0 \le \theta \le \left(\frac{\pi}{6} - \alpha_h\right) \qquad (8)$$

$$f_2 = \frac{V_{dc}}{3}, \qquad \left(\frac{\pi}{6} - \alpha_h\right) \le \theta \le \left(\frac{\pi}{6} + \alpha_h\right) \qquad (9)$$

$$f_3 = \frac{V_{dc}}{\sqrt{3} \cos\left(\frac{\pi}{3} - \alpha_p'\right)} \sin \alpha_p', \quad \left(\frac{\pi}{6} + \alpha_h\right) \le \theta \le \left(\frac{\pi}{2} - \alpha_h\right) \qquad (10)$$

$$f_4 = \frac{2V_{dc}}{3}, \qquad \left(\frac{\pi}{2} - \alpha_h\right) \le \theta \le \frac{\pi}{2} \qquad (11)$$

$$\alpha_h = 6.40 \times MI - 6.09 \left(0.9517 \le MI \le 0.9800\right) \qquad (12)$$

$$\alpha_h = 11.75 \times MI - 11.34 \left(0.9800 \le MI \le 0.9975\right) \qquad (13)$$

$$\alpha_h = 48.96 \times MI - 48.43 \left(0.9975 \le MI \le 1.0000\right) \qquad (14)$$

$$\alpha_p = \frac{\pi}{\pi - 6\alpha_h} \theta \qquad (15)$$

$$\alpha_p' = (1 - \frac{6}{\pi} \alpha_h)\theta \qquad (16)$$

IV. SIMULATION AND EXPERIMENTAL RESULTS

The validity of the proposed algorithm is verified through the simulation and experiment for three-level inverter with RL load. The simulation results are shown in Figs. 7 to 10. Figure 7 shows the results of phase current and line-to-line voltage at MI=0.4 in the linear region mode I. The line-to-line voltage has the same characteristic as that of 2-level inverter. Figure 8 shows the results of phase current and line-to-line voltage at MI=0.8 in the linear region mode II. The line-to-line voltage has 5 steps so the harmonic level is much less than that of 2-level inverter. Figure 9 shows the results of phase current and line-to-line voltage at MI=0.95 in the over-modulation region I and Figure 10 shows those at MI=0.99 in the over-modulation region II. The line-to-line voltage waveform is almost same as that in six-step mode at MI=1.0. From those figures, we know that the voltage

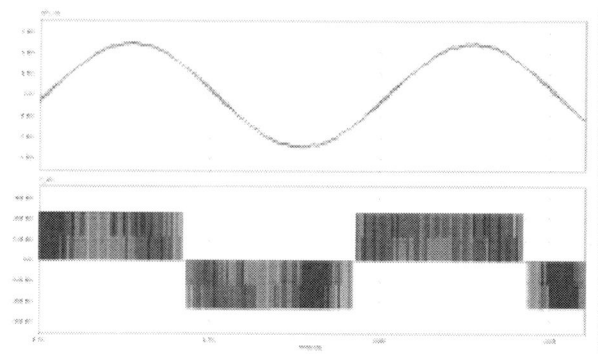

Fig. 7. Simulation results of phase current and line voltage waveforms at MI=0.4 in linear region I.

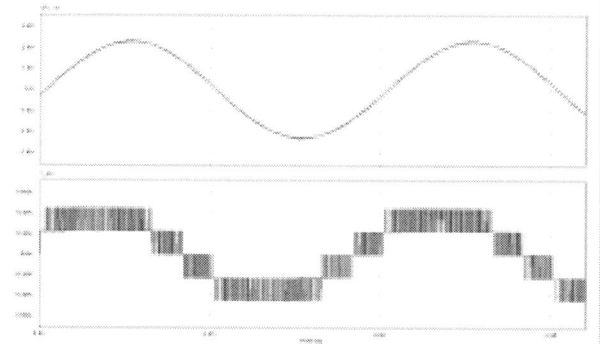

Fig. 8. Simulation results of phase current and line voltage waveforms at MI=0.8 in linear region II.

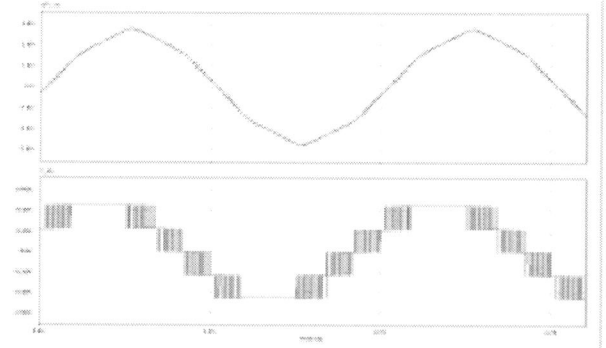

Fig. 9. Simulation results of phase current and line voltage waveforms at MI=0.95 in over-modulation region I.

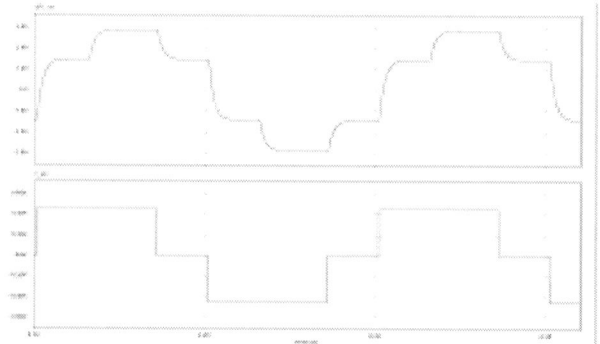

Fig. 10. Simulation results of phase current and line voltage waveforms at MI=0.99 in over-modulation region II.

978-1-4244-1871-8/07 $25.00 © 2007 IEEE

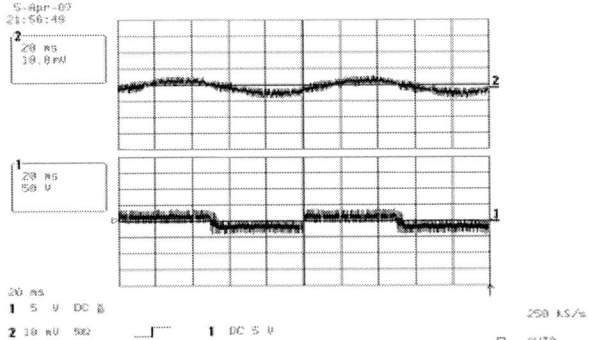

Fig. 11. Phase current and line voltage experimental waveform at MI=0.4, Linear region model.

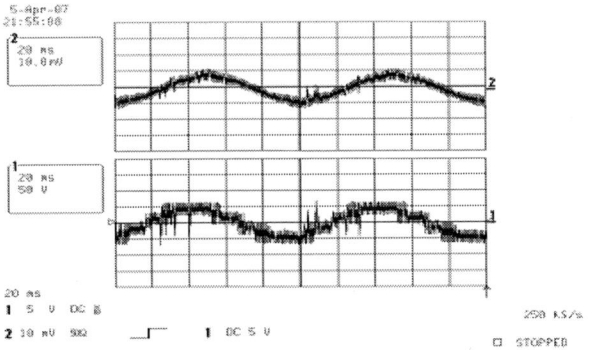

Fig. 12. Phase current and line voltage experimental waveform at MI=0.8, Linear region modelI.

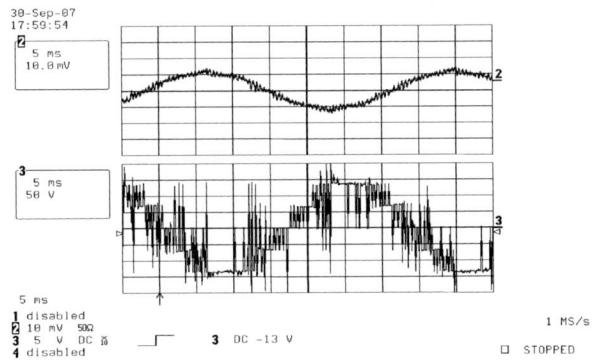

Fig. 13. Phase current and line voltage experimental waveform at MI=0.95, Overmodulation region model.

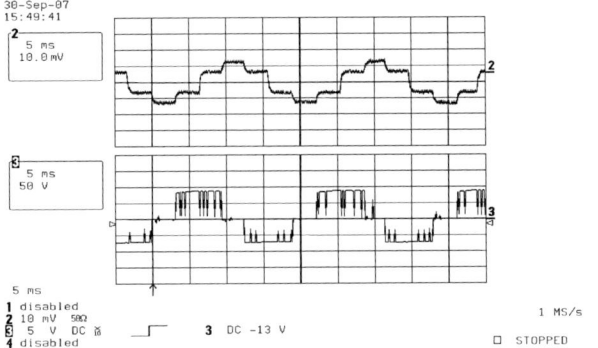

Fig. 14. Phase current and line voltage experimental waveform at MI=0.99, Overmodulation region modelI.

magnitude can be controlled by changing the modulation index from 0 to 1 and the current waveform is almost sinusoidal in the linear region but distorted as the increase of modulation index in over- modulation region

The experimental results are shown from Fig. 11 to Fig. 14 as the same values of modulation index as the above simulations. These results are matches well with the simulation results.

For the application of the over-modulation techniques for traction drives, it is simulated with the induction motor vector drives. Figure 15 shows the system block diagram of the indirect vector control of induction motor drives. The output of the speed controller determines the reference value of the torque component of current i_{qs}^*. The flux component of current i_{ds}^* is given to regulated the rotor flux to the desired value, Then the slip frequency is generated from in feedforward manner from Eq. (17) to satisfy the vector diagram [7].

$$\omega_{sl}^* = \frac{R_r}{L_r} \frac{i_{qs}^*}{i_{ds}^*} \tag{17}$$

The rated parameters of the induction motor are shown at Table.4.

TABLE 4. Motor Parameters.

Rated Power	0.75[KW]	Rs	1.6282[Ω]
Rated Speed	1440[rpm]	Rr	1.5042[Ω]
Rated Voltage	220[V]	Ls	4.4[mH]
Rated Current	2.1[A]	Lr	4.4[mH]
Poles	4	Lm	158[mH]

Figure 16 shows the simulation waveforms of the speed, the torque current component i_{qs}^* and the flux current component i_{ds}^* when the reference motor speed set to 700[rpm] from 0[sec] to 1[sec] and set to 1400[rpm] from 1[sec] to 2[sec] and set to 700[rpm] from 2[sec] to 3[sec] in no load condition. As the change of the speed reference value, the real speed is well regulated to follow the reference value. The torque current component is regulated to generate the positive and negative value in acceleration and deceleration region and also the flux current component is well regulated to follow the reference value.

Figure 17 shows the simulation waveforms under the load changing condition. The reference motor load torque set to no load condition during the acceleration region from 0[sec] to 1[sec], and set to 50% rated motor load torque from 1[sec] to 2[sec], and set to no load condition again from 2[sec] to 3[sec]. From the figure, we know that as the change of the load condition, the torque current component is well regulated to follow the reference value and the speed is well regulated at same speed even though the torque changes.

978-1-4244-1871-8/07 $25.00 © 2007 IEEE

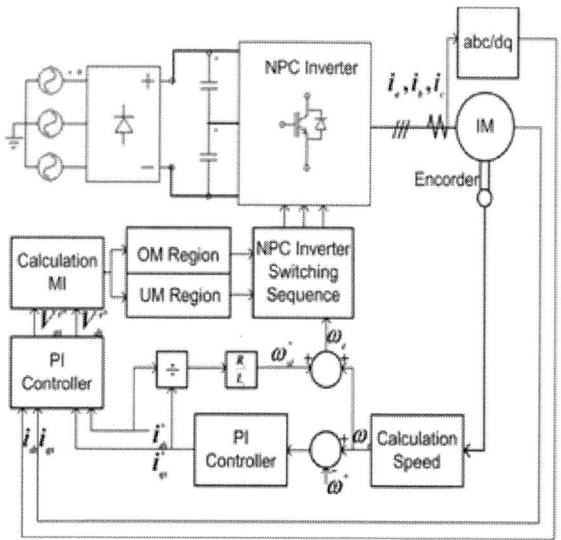

Fig. 15. Indirect vector control block diagram of NPC type inverter

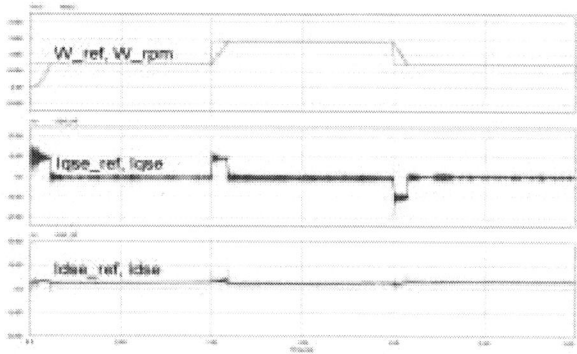

Fig. 16. Simulation results under the speed change condition.

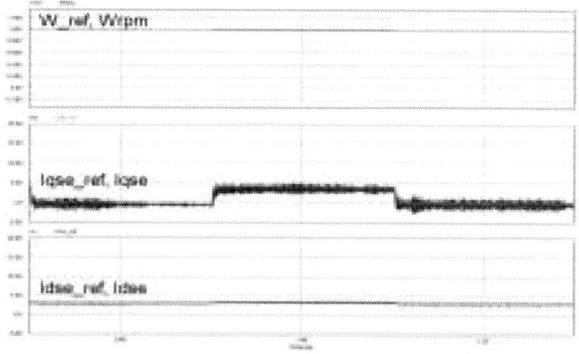

Fig. 17. Simulation results under the load change condition.

V. CONCLUSION

This paper proposed the over-modulation technique for NPC type 3-level inverter. With this proposed over-modulation strategy, the output voltage of 3-level inverter can be controlled from the linear SVM region to the six-step region. The proposed algorithm is verified through the simulation and experiment with the phase current and line-to-line voltage waveforms as the typical values of the modulation index in linear and over-modulation regions. And also, it is simulated with the 3-level inverter fed induction motor drives and well verified with the regulation performance under the speed change and load change conditions.

ACKNOWLEDGEMENT

This research was supported by the Program for the Training of Graduate Students in Regional Innovation which was conducted by the Ministry of Commerce Industry and Energy of the Korean Goverment.

REFERENCES

[1] A. Nabae, I. Takahashi, and H. Akagi, "A new neutral point clamped PWM inverter," *IEEE Trans. Ind. Appl.*, IA-17, (5), pp. 518-523, 1981.

[2] A. Horie, S. Saito, S. Ito, T. Takasaki, and H. Ozawa "Development of a three-level converter-inverter system with IGBT's for AC electric cars," *IEE Japan IAS Annu. Meet. Rec.*, pp. 75-78, 1995.

[3] E. Akagawa, S. Kawamoto, S. Tamai, H. Okayama, and T. Uemura, "Three-level PWM converter-inverter system for next-generation Shinkansen," *IEE Japan IAS Annu. Meet. Rec.*, pp. 81-82, 1995.

[4] J.Holtz, W.Lotzkat, and A. M. Khambadkone, " On continuous control of PWM inverters in the overmodulation range including the six-step mode," *IEEE Trans. Power Electron.*, vol. 8, no. 4, pp. 546-553, 1993.

[5] Dong-Choon Lee and G-Myoung Lee, "A Novel Overmodulation Technique for Space-Vector PWM Inverters," *IEEE Trans. Power Electron.* vol. 13, no. 6, pp. 1144-1151, 1998.

[6] Subrata K. Mondal, João O. P. Pinto, and Bimal, K. Bose, "A Neural-Network-Based Space-Vector PWM Controller for a Three-Level Voltage-Fed Inverter Induction Motor Drive", IEEE *Trans. Ind. Appl.*, vol. 38, no. 3, pp. 660-669, 2002.

[7] Bimal K. Bose, *Modern Power Electronics and AC Drives*, pp. 368-387, Prentice-Hall, Inc., 2002

Discrete-time Model Following Control for PWM Inverter with Electric Double Layer Capacitor

Shota Suzuki
Tokyo Denki University
Department of Computer and System Engineering
Ishizaka, Hatoyama-cho, Hiki-gun, Saitama-ken, 350-0394
Email: suzuki@hanelab.k.dendai.ac.jp

Toshimasa Haneyoshi
Tokyo Denki University
Department of Computer and System Engineering
Ishizaka, Hatoyama-cho, Hiki-gun, Saitama-ken, 350-0394
Email: hane@k.dendai.ac.jp

Abstract— In a single phase inverter, momentary power pulsate with double frequency of the output. In the case of UPS with single phase inverter, it causes the fact that DC input voltage to inverter is pulsated and output voltage is distorted.

In this paper, to control this pulsating input voltage, the single phase PWM inverter is applied to a discrete-time model following control. The output characteristics of the inverter changed with of the voltage of Electric Double Layer Capacitor(EDLC) are described.

The results are shown that this control method can amend for the change of the input voltage.

I. INTRODUCTION

In the highly-networked information society, a steady power supply is not only important for UPS to improve the reliability in the power supply system but in great demand recently. Additionally, in the global environment feared, a new storage battery in UPS system is proposed as a substitute for conventional one, such as lead acid battery.

As for EDLC, there is little environmental impact and it has a long life than the rechargeable battery. However, EDLC causes voltage form to distort itself, which is different from conventional one. A single and three-phases inverter are predominantly employed in UPS. In a single phase inverter, differently from that in another, momentary power pulsates with double frequency. It causes an output voltage wave pattern to be distorted. The crooked voltage waveform causes an electricity loss and damage of an informative appliance.

Two or more methods such as deadbeat control[5][6] and repetitive control[7][8] are proposed as control techniques of the inverter that amends for the distortion of the output voltage waveform. There is model following control as one of the methods. This is a control method to synchronize dynamic characteristics to coincide with given dynamic response. Inverter is assumed a control object. This method can identify the output of inverter as a sine wave in a steady state.

In this paper, the single phase PWM inverter with EDLC is applied to a discrete-time model following control. We consider whether it can amend for the change of the input voltage and whether it controls the output voltage to be distorted. In addition, the case where the rectifier load is connected to this system is examined. In the final, setting up experimental equipment, all the results are verified through

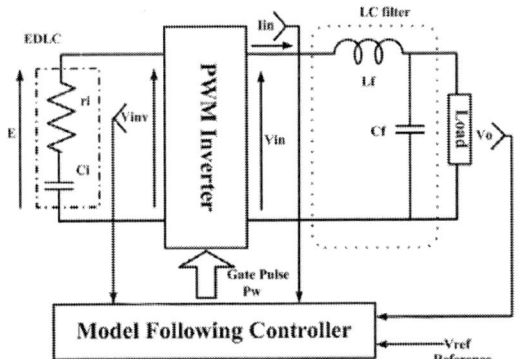

Fig.1 Single phase PWM inverter with EDLC

comparing with the simulation results.

II. DISCRETE-TIME MODEL FOLLOWING CONTROL

The model following control indicates one of the feedback control which makes the output follow to the reference as shown in Fig.1.

The voltage E of EDLC decreases with time. Supposing the input voltage to V_{invk} to be uniformity during sampling time, the control method of state feedback is derived by [1] and [4].

From Fig.1, the error ε_k between the output voltage V_{ok} and the reference voltage V_{refk} at a sampling time is given by (1).

$$\varepsilon_k = V_{ok} - V_{refk} . \qquad (1)$$

The equation in an enlarged control system, which includes quantity of control object x_k, quantity of r_k and error ε_k, is shown as follows:

$$\begin{aligned} x_{k+1} &= \Phi_s x_{sk} + \Gamma_s u_k \\ \varepsilon_{k-1} &= C_s x_{sk} \end{aligned} \qquad (2)$$

where

$$x_{sk} = \begin{pmatrix} x_k - x_{k-1} \\ \varepsilon_{k-1} \\ r_k - r_{k-1} \end{pmatrix} \quad \Phi_s = \begin{pmatrix} \Phi & 0 & 0 \\ C & 1 & -C_r \\ 0 & 0 & \Phi_r \end{pmatrix} \quad \Gamma_s = \begin{pmatrix} \Gamma \\ 0 \\ 0 \end{pmatrix}$$

$$C_s = \begin{pmatrix} 0 & 1 & 0 \end{pmatrix} \quad u_k = V_{ink} - V_{ink-1} . \qquad (3)$$

978-1-4244-1871-8/07 $25.00 © 2007 IEEE

From (3), the conditional feedback control method to minimize linear second evaluation is derived. An evaluation function, which sets heaviness at q and r, is shown as (4),

$$J = \sum_{k=0}^{\infty} (\varepsilon_{k-1}^2 q + u_k^2 r) \ . \tag{4}$$

where F_D is defined as a feedback gain. Operation to satisfy (4) is as follows

$$u_k = F_D x_{xk} = -(r + \Gamma_s^T P \Gamma_s)^{-1} \Gamma_s^T P \Phi_s x_{sk} \ . \tag{5}$$

Here, P is a solution of a Riccati algebraic equation in discrete time system as following

$$P = \Phi_s^T P \Phi_s + C_s^T Q_D C_s$$
$$- \Phi_s^T P \Gamma_s (R_D + \Gamma_s^T P \Gamma_s)^{-1} \Gamma_s^T P \Phi_s \ . \tag{6}$$

Equation (6) brings ε_k close to zero asymptotically and allows the output to follow the reference voltage.

In this control method, it determines quantity of operation u_k with a value of the input voltage V_{invk} at a sampling point. Where $F_D = \begin{pmatrix} f_1 & f_2 & f_3 \end{pmatrix}$, and substituting (3) into (5),

$$u_k = V_{ink} - V_{ink-1} = F_D x_{sk}$$
$$= \begin{pmatrix} f_1 & f_2 & f_3 \end{pmatrix} \begin{pmatrix} x_k - x_{k-1} \\ \varepsilon_{k-1} \\ r_k - r_{k-1} \end{pmatrix} . \tag{7}$$

where a total of an error ε_k at a sampling point in k time is e_k and $e_k - e_{k-1} = \varepsilon_k$.
From (7),

$$u_k = V_{ink} - V_{ink-1} = F_D x_{sk}$$
$$= (x_k - x_{k-1})f_1 + (e_{k-1} - e_{k-2})f_2 + (r_k - r_{k-1})f_3 . \tag{8}$$

From (8), the output voltage of inverter V_{ink} is

$$V_{ink} = x_k f_1 + e_{k-1} f_2 + r_k f_3 . \tag{9}$$

On the basis of (9), this control system is constructed.

III. SINGLE PHASE PWM INVERTER WITH EDLC

The basic idea of an input and output wave pattern to a filter for a single phase PWM inverter with EDLC resumes in Fig.2.

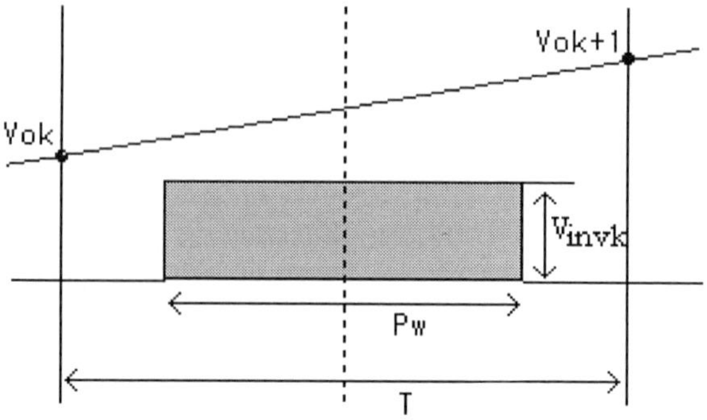

Fig.2. Output voltage of inverter and load

The adaptive control is derived from (4) in which q = 10.0 is arriving time and r = 1.0 is heaviness of operation quantity. The output voltage V_{ok} and the input voltage to the filter I_{ink} are fed back at every sampling time. The amount of the operation V_{ink} is formulated by multiplying the feedback gain through the above. The input voltage pulse waveform from the inverter is obtained by the pulse width from (10) [2][3] and the output voltage waveform is shown in Fig.2.

$$P_W = \frac{T}{E} V_{ink} . \tag{10}$$

where T is sampling period and E is EDLC voltage at the time of each sampling.

Based on the above, it is simulated to two kinds of loads of rated resistive load and rectifier load considering voltage charge changed with EDLC. In the simulation, the frequency is assumed to be 50[Hz], the reference execution voltage to be 100[Vrms] assumed that capacitance and the initial voltage of EDLC is 170[V]. In the case of rated resistive load, the output voltage and the reference voltage waveforms are examined in Fig.3.and the input voltage waveform to inverter is shown in Fig.4. The simulation result for rectifier load is shown in Fig.5 and Fig.6. In addition, error rate and Total Harmonic Distortion(THD) are evaluated in table 1 and table 2.

TABLE 1   ERROR AND THD(RATED RESISTIVE LOAD)

Reference[Vrms]	100
Output[Vrms] (Error[%])	99.636 (-0.364)
THD[%]	0.182

TABLE 2   ERROR AND THD(RECTIFIER LOAD)

Reference[Vrms]	100
Output[Vrms] (Error[%])	97.269 (-2.731)
THD[%]	2.944

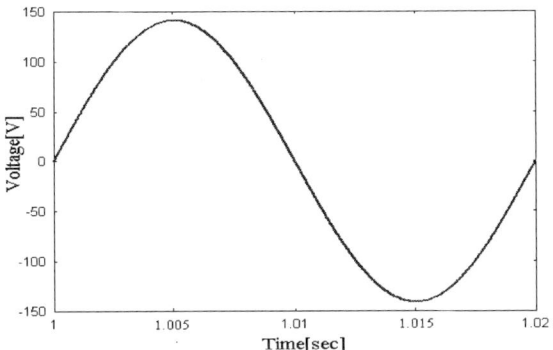

Fig.3 Output voltage waveforms(RATED RESISTIVE LOAD)

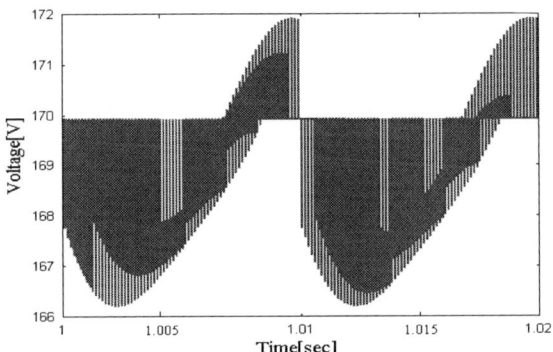

Fig.4 Input voltage to inverter waveform(RATED RESISTIVE LOAD)

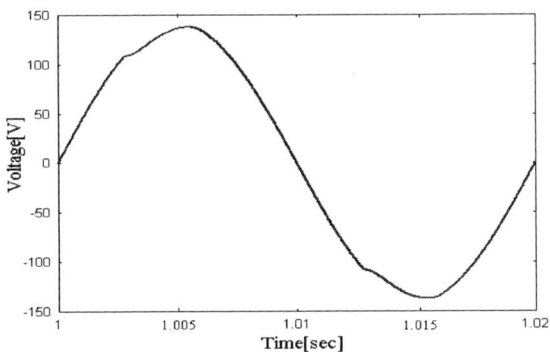

Fig.5 Output voltage waveforms(RECTIFIER LOAD)

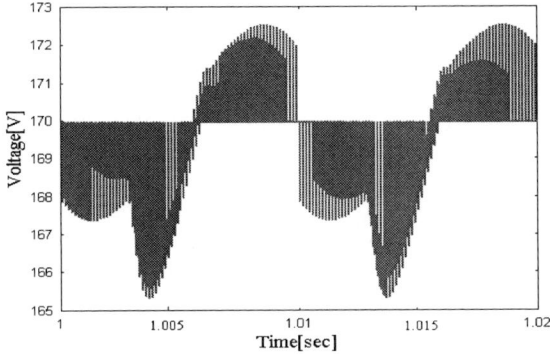

Fig.6 Input voltage to inverter waveform(RECTIFIER LOAD)

These results show that even if input voltage to inverter E changes, the output voltage can follow the reference voltage. However, table 1 and table 2 indicate that while the charge voltage of EDLC is higher enough than the reference value, an error of the effectiveness voltage and THD is controlled with a low value comparatively.

IV. EXPERIMENT

As displayed in Figs. 7 and 8, we carried out experimental investigation employing the discrete-time model following control. Experimental conditions are shown in table 3. In this experiment, 30 EDLC of 2200[F] is connected with the series, charged up to 56.5[V] in total.

Fig.7 The experimental equipment

Fig.8 The experimental equipment (EDLC BANK)

TABLE3   EXPERIMENTAL CONDITIONS

Switching Frequency f_s [kHz]		7.5
Reference	Output voltage [Vrms]	21.5
	Frequency [Hz]	50
EDLC	E [V]	56.5
	C_i [F]	73
	r_i [mΩ]	150
LC-filter	C_f [μF]	300
	L_f [μH]	500
Rated resistive load [Ω]		30

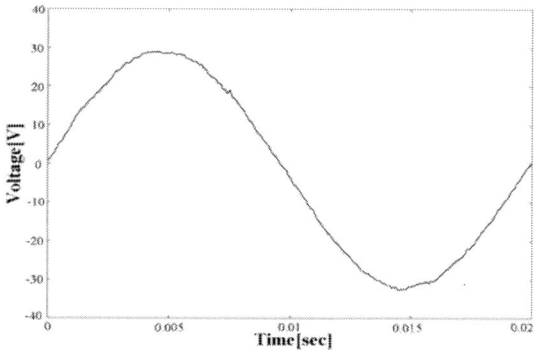

Fig.9 Output voltage waveforms

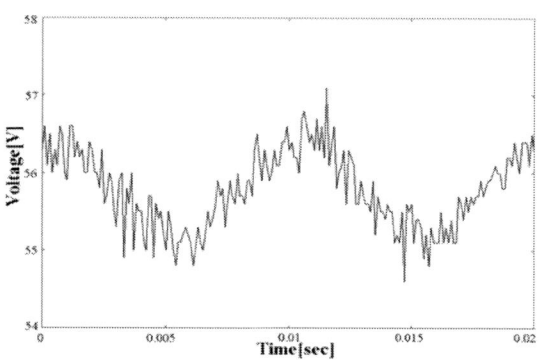

Fig.10 Input voltage to inverter waveform

The output voltage waveform is shown in Fig.9 and the inverter input voltage waveform is shown in Fig.10. In Fig.10, input voltage to inverter (EDLC voltage) is pulsated since momentary power pulsates with double frequency of the output. However, Fig.9 shows that the output voltage can follow the reference voltage in this system.

V. CONCLUSION

Model following control was applied to single phase PWM inverter. It was confirmed that this method enable to amends for the change of the power supply voltage in EDLC. Even if the big pulse of input voltage was generated at that time, it was cleared that the method amended for it. Moreover, it was inspected that this system was controlled similarly when the rectifier load was connected. Next, our method applied to the single PWM inverter was examined in an experimental stage. We finally demonstrated consistent operation and enhanced effectiveness of this system.

REFERENCES

[1] Atushi Umemura, Toshimasa Haneyoshi, Yukio Saito and Fumio Harashim: "Discrete-time Model Following Control of Inverter with Rectifer Load",The Institute of Electrical Engineers of Japan Trans. Industry Applications Society, Vol 125,No 7, 2005

[2] Atushi Umemura and Toshimasa Haneyoshi: "Pulse Width Calculation Method for Model Following Controlled PWM Inverter", Institute of Electrical Engineers of Japan Conference, Vol3, No73, pp.104-105,2003

[3] Atushi Umemura and Toshimasa Haneyoshi: "Pulse Raise & Lower Point Calculation Method for Model Following Controlled PWM Inverter", The Society of Instrument and Control Engineers Annual Conference pp.2367-2372, 2003-8.

[4] Katuhisa Furuta and Katumi Komiya : "Design of Model Following Servo Controller", The Institute of Electrical and Electronics Engineers Trans. Automatic Control, Vol.AC-27, No3, pp.725-727, 1982

[5] Tomoki Yokoyama, Yoshiki Igarashi, Toshimasa, Haneyoshi, and Tomoki Izumi: "A Study of Digital Instantaneous Value Control with Filter Capacitor Current Compensation for PWM Inverter", Institute of Electrical Engineers of Japan Trans. Industry Applications Society, Vol123, No5, pp.500-506, 2003-5C. J. Kaufman, Rocky Mountain Research Lab., Boulder, CO, private communication, May 1995.

[6] Tomoki Yokoyama and Atuio Kawamura : "Digital Control of Three Phase PWM Inverter for UPS Application Based on Disturbance Observer and Deadbeat Control", Institute of Electrical Engineers of Japan Trans. Industry Applications Society, Vol113, No5, pp617-624, 1993-5M. Young, *The Techincal Writers Handbook.* Mill Valley, CA: University Science, 1989.

[7] Toshimasa Haneyoshi and Atuo Kawamura, and Richard G. Hoft: "Waveform Compensation of PWM Inverter with Cyclic Fluctuation Loads" The Institute of Electrical and Electronics Engineers Trans. Industry Applications, Vol24, No4, pp.582-589, 1988 S. Chen, B. Mulgrew, and P. M. Grant, "A clustering technique for digital communications channel equalization using radial basis function networks," *IEEE Trans. Neural Networks*, vol. 4, pp. 570–578, July 1993.

[8] Ying-Yu Tzou, Shih-Liang Jung, and Hsin-ChungYeh : "Adaptive Repetitive Control of PWM Inverter for Very Low THD AC-Voltage Regulator with Unkown Loads", The Institute of Electrical and Electronics Engineers Trans. Power Electronics, Vol14, No5, 1999-9S. P. Bingulac, "On the compatibility of adaptive controllers (Published Conference Proceedings style)," in *Proc. 4th Annu. Allerton Conf. Circuits and Systems Theory*, New York, 1994, pp. 8–16.

Synchronized PWM Control of Symmetrical Six-Phase Drives

V. Oleschuk[*,**]

[*]Academy of Sciences of Moldova
Institute of Power Engineering
5 Academy Str., Kishinau, MD-2028, MOLDOVA
Email: oleschukv@hotmail.com

G. Griva, F. Profumo, A. Tenconi

[**]Politecnico di Torino
Department of Electrical Engineering
Corso Duca degli Abruzzi, 24, Turin, 10129, ITALY
Email: giovanni.griva@polito.it

Abstract — In order to avoid asynchronism of standard schemes of space-vector modulation, new method of synchronized pulsewidth modulation (PWM) has been applied for scalar control of symmetrical six-phase (dual three-phase) induction motor drives fed by two three-phase inverters. Basic schemes of synchronized continuous and discontinuous PWM, and also its combinations, disseminated for control of symmetrical dual three-phase systems, have been analyzed. Simulations give the behavior of symmetrical dual three-phase system with low switching frequency, controlled by algorithms of synchronized PWM. The spectra of the phase voltage of six-phase drives with synchronized pulsewidth modulation do not contain even harmonics and sub-harmonics during the whole control range, which is especially important for the high power/high current applications.

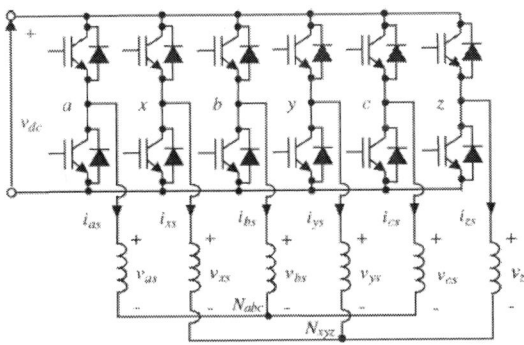

Fig. 1: Dual three-phase drive system with two neutral points

I. INTRODUCTION

Principle of modulation of pulse signals is the basic for control of power converters. Last decades have been marked by fast application of methods of space-vector modulation, which are ones of the most suitable for adjustable speed drives [1], including the developed methods of space-vector PWM, applied for control of multiphase drives [2].

Multiphase induction motor drives are a subject of increasing interest in the last years due to some advantages compared with conventional three-phase adjustable speed drives. Ones of the interesting and perspective topologies of multiphase drives are now six-phase (split-phase or dual three-phase) induction motor drives [3]-[9].

Dual three-phase motor drives have several advantages over their three-phase counterparts, such as: reduction of torque pulsations, of the rotor harmonic losses, and of the rated current of power switches; improved reliability at system level, etc. [5].

There are two basic versions of dual three-phase induction motor drives. The most famous asymmetrical systems are based on asymmetrical six-phase machine which has two sets of windings spatially shifted by 30 electrical degrees [3]-[6]. Less known symmetrical six-phase drive systems are based on symmetrical induction motor, which has two sets of windings spatially shifted by 60 electrical degrees [7]-[9]. In particular, Fig. 1 presents basic topology of dual three-phase drive system with two isolated neutral points [5]. To provide increased efficiency of six-phase drives fed by power converters, novel space-vector-based control and modulation strategies have been proposed for both asymmetrical and symmetrical systems [3]-[9].

One of the most perspective applications of six-phase induction motor drives lies in the field of high power/high current systems (ship propulsion, locomotive, electrical vehicles, etc.), which are characterized by low switching frequency of power switches.

It is known, that for the drive systems with increased power rating it is necessary to synchronize the output voltage of power converters for elimination of undesirable sub-harmonics of voltage and current [1],[10]. It is known also, that reliability of operation of drive systems with PWM can be increased strongly after minimization or elimination of the common-mode voltages and currents [7].

So, this paper presents results of dissemination of novel method (methodology) of synchronized space-vector-based PWM for control of symmetrical dual three-phase drives, providing continuous synchronization of the phase voltages in these systems. Both combined and non-combined PWM algorithms for control of two inverters have been analyzed. In particular, some of these schemes provide cancellation of undesirable common-mode voltages in systems during the whole control range, including the zone of overmodulation.

II. PRINCIPLES OF CONTINUOUS OUTPUT VOLTAGE SYNCHRONIZATION

In order to provide synchronization of the process of space-vector PWM, a novel approach has been proposed for synthesis of the output voltage of standard three-phase inverters [11]. As an illustration of continuous scheme of modulation, Fig. 2 presents basic control and output signals of standard three-phase voltage source inverter with synchronized PWM inside the interval 0^0-90^0.

978-1-4244-1871-8/07 $25.00 © 2007 IEEE

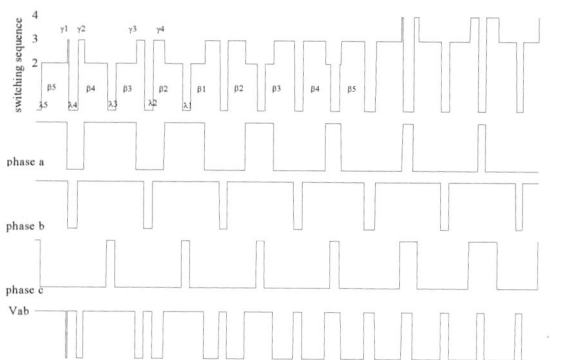

Fig. 2: Control and output signals for quarter-period of three-phase inverter with continuous PWM (CPWM)

The upper traces in Fig. 2 are switching state sequence (in accordance with conventional designation [1],[11]), then - control signals for the cathode switches of the phases *a, b, c* of the inverter. The lower trace in Fig. 2 shows the corresponding quarter-wave of the line-to-line output voltage of the inverter. Signals βj represent total switch-on durations during switching sub-interval τ, signals γ_k are generated on the borders or in the centers of the corresponding β-signal. Widths of notches λ_k represent duration of zero sequences.

In accordance with principle of voltage space-vector modulation, for symmetrical PWM versions, applied for control of three-phase systems, during a sub-cycle the ratio of the total durations of the corresponding active switching state signals β_j to the total duration of the central active switching state signal β_1, which is formed in the centre of the 60^0-clock-intervals, can be written as

$$\frac{\beta_j}{\beta_1} = \frac{\sin(60^0 - \alpha_j) + \sin\alpha_j}{\sin(60^0 - \alpha_1) + \sin\alpha_1} = \quad (j = 2,...i) \quad (1)$$
$$= [\sin(60^0 - \alpha_j) + \sin\alpha_j],$$

where α_j - angle position of the centre of the β_j-signal from the beginning of the 60^0-clock-interval; $\alpha_1 = 30^0$ - angle position of the central β_1-signal.

In particular, in accordance with (1):

$$\beta_j = \beta_1 \cos(30^0 - \alpha_j) \qquad (2)$$

At the same time, the angle position α_j of the centre of the corresponding β-signal is equal to:

$$\alpha_j = 30^0 - (j-1)\tau \quad (j=1,...i), \qquad (3)$$

where τ is width of sub-cycle. And for $j=2,...i$:

$$\beta_j = \beta_1 \cos[(j-1)\tau] \qquad (4)$$

So, one of the basic ideas of the method of synchronized PWM is in continuous synchronization of the positions of all

central signals β_1 in the centres of the 60^0-clock-intervals, and then – in symmetrical generation of all other active β-signals around the centres of the 60^0-clock-intervals.

Special signals λ' (λ_5 in Fig. 2) with the neighboring β'' (β_5 in Fig. 2) are formed in the clock-points (0^0, 60^0, 120^0..) of the output curve of inverters with synchronized PWM. They are reduced simultaneously until close to zero value at the boundary frequencies F_i and F_{i-1} between control sub-zones [11]:

$$F_i = \frac{1}{6(2i-1)\tau} \qquad (5)$$

$$F_{i-1} = \frac{1}{6(2i-3)\tau}, \qquad (6)$$

where index i is equal to the number of notches inside a half of the 60^0-clock-intervals, including the notch on the border of the clock-intervals.

So, this control principle provides continuous adjustment of the voltage waveforms of inverters, with smooth shock-less pulses-ratio changing during the whole control range.

Equations (7)-(12) present set of control functions for determination of parameters of signals of inverters with synchronized PWM in absolute values (seconds) for scalar *V/F* control mode of the system during the whole control diapason including the zone of overmodulation [11].

For $j=2,...i-1$:

$$\beta_j = \beta_1 \cos[(j-1-K_3)\tau K_{ov1}] \qquad (7)$$

$$\gamma_j = \beta_{i-j+1}\{0.5 - 0.87\tan[(i-j-K_3)\tau]\}K_{ov2} \qquad (8)$$

$$\beta_i = \beta'' = \beta_1 \cos[(i-K_3-1)\tau K_{ov1}]K_s \qquad (9)$$

$$\gamma_1 = \beta''\{0.5 - 0.87\tan[(i-K_3-2)\tau + \\ (\beta_{i-1} + \beta_i + \lambda_{i-1})/2]\}K_s K_{ov2} \qquad (10)$$

$$\lambda_j = \tau - (\beta_j + \beta_{j+1})/2 \qquad (11)$$

$$\lambda_i = \lambda' = (\tau - \beta'')K_{ov1}K_s \qquad (12)$$

where: β - total switch-on duration inside switching interval; γ - minor parts of the total switch-on durations; λ - duration of notches; $m = F/F_m$ – modulation index; $\beta_1 = 1.1\tau m$ until $F_{ov1} = 0.907F_m$, and $\beta_1 = \tau$ if frequency is more than F_{ov1}; $K_s = [1-(F-F_i)/(F_{i-1}-F_i)]$ - coefficient of synchronization; the first coefficient of overmodulation $K_{ov1} = 1$ until F_{ov1}, and $K_{ov1} = [1-(F-F_{ov1})/(F_{ov2}-F_{ov1})]$ between F_{ov1} and $F_{ov2} = 0.952F_m$; the second coefficient of overmodulation $K_{ov2} = 1$ until F_{ov2}, and $K_{ov2} = [1-(F-F_{ov2})/(F_m-F_{ov2})]$ in the zone between F_{ov2} and F_m; $K_3=0$ for continuous PWM, and $K_3=0.25$ for the discontinuous PWM schemes.

III. SYMMETRICAL SIX-PHASE DRIVES WITH COMMON-MODE VOLTAGE ELIMINATION

Control of symmetrical six-phase induction machine drives is based on the 60^0-phase-shifting of the corresponding signals of two inverters [7]-[9]. In accordance with the theory of vector space decomposition, the basic six-dimensional space (*as, bs, cs, xs, ys, zs*) of a dual-three phase induction machine with isolated neutral points can be transformed into two orthogonal two-dimensional subspaces (*sa, sb*) and (*m1, m2*) [4]. Voltage components V_{sa} and V_{m1} in these subspaces, and also the phase voltage $V_{as} = V_{sa} + V_{m1}$ (and, by analogy, V_{xs}), are calculated for symmetrical six-phase drives with two isolated neutrals as [8]:

$$V_{sa} = 0.333(V_a - 0.5V_b - 0.5V_c + 0.5V_x - V_y + 0.5V_z) \quad (9)$$

$$V_{m1} = 0.333(V_a - 0.5V_b - 0.5V_c - 0.5V_x + V_y - 0.5V_z) \quad (10)$$

$$V_{as} = V_{sa} + V_{m1} = V_a - 0.333(V_a + V_b + V_c), \quad (11)$$

where V_a, V_b, V_c, V_x, V_y, V_z are the corresponding pole voltages of each inverter.

In this case, the V_{sa} component, which produces rotating MMF k-th order voltage harmonics ($k = 12m \pm 1$, $m=1,2,3,.$), is the useful component. But the V_{m1} component, which generates loss-producing harmonics ($k = 6m \pm 1$, $m=1,3,5,.$), is the undesirable voltage component.

The common-mode voltage V_0 in six-phase systems is calculated in accordance with (12):

$$V_0 = (1/6)*(V_a + V_b + V_c + V_x + V_y + V_z) \quad (12)$$

Elimination of undesirable common-mode voltages in symmetrical dual three-phase systems can be provided by special order of switching sequences of two three-phase inverters [8]. In particular, during one time-interval zero sequences of two inverters must be different (**07** or **70**), and active switching states of two inverters for one time-interval must be different too [7]. If, as an example, zero sequence of the first inverter is **0**, the corresponding zero sequence of the second inverter at the same time must be **7**, and quite the reverse.

Algorithms of synchronized PWM allow providing these conditions during the whole control range. As an illustration of this control principle, Figs. 3 – 6 show the corresponding switching sequences and pole voltages of two inverters during 120^0 for four basic schemes of synchronized PWM, providing cancellation of the common-mode voltages.

So, basic switching sequence for the dual three-phase system with continuous PWM for the first 60^0-clock-interval (Fig. 3) is: **70-21-32-07-32-21-70-.**, where the first number corresponds to the switching sequence of the inverter with the **abc** phases (Fig. 1), and the second number corresponds to the switching sequence of the second (**xyz**) inverter.

For the system with two inverters, controlled by algorithms of discontinuous PWM with the 30^0-nonswitching intervals (DPWM30, Fig. 4) basic switching sequence is: **70-21-32-21-70-...-07-32-21-32-07-...**

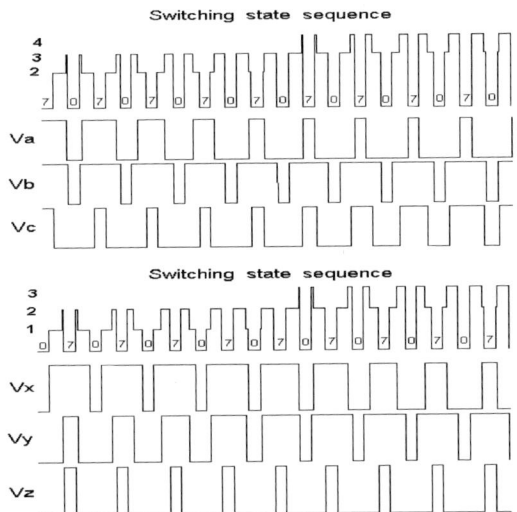

Fig. 3: Switching state sequences and pole voltages of two inverters of the six-phase system with continuous synchronized PWM (CPWM)

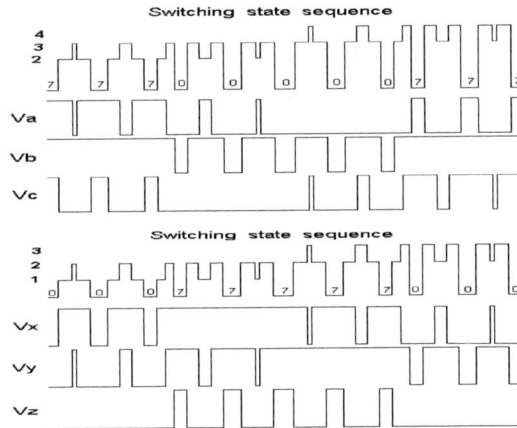

Fig. 4: Switching sequences and pole voltages of two inverters of the system with discontinuous PWM with the 30^0-nonswitching intervals (DPWM30)

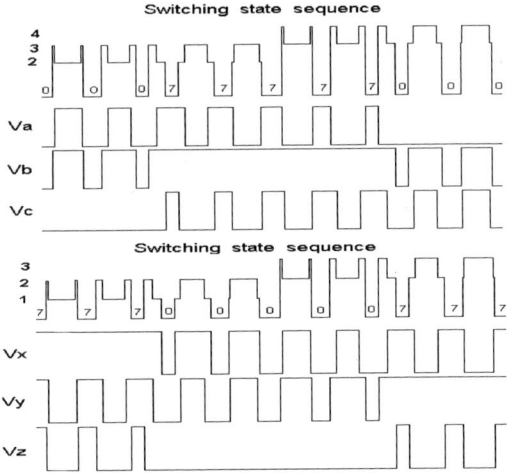

Fig. 5: Switching sequences and pole voltages of two inverters of the system with discontinuous PWM with the 60^0-nonswitching intervals (DPWM60)

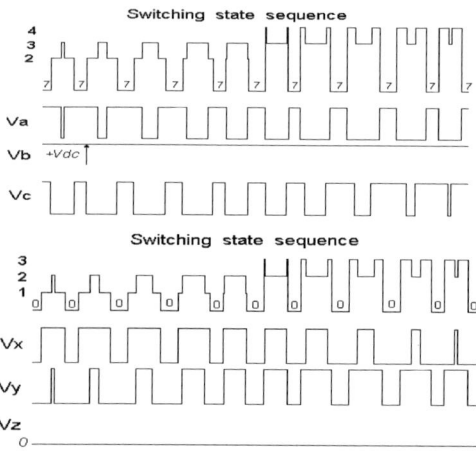

Fig. 6: Switching sequences and pole voltages of two inverters of the system with discontinuous PWM with the 120^0-nonswitching intervals (DPWM120)

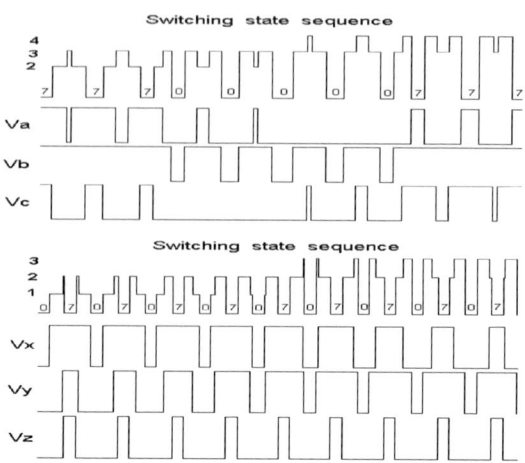

Fig. 7: Switching sequences and pole voltages of two inverters of dual three-phase system with combined (DPWM30+CPWM) control

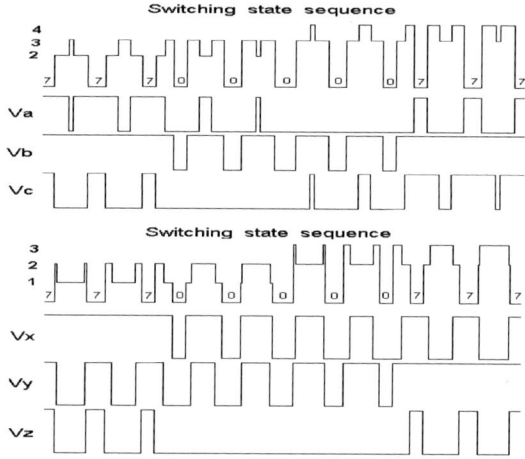

Fig. 8: Switching sequences and pole voltages of two inverters of dual three-phase system with combined (DPWM30+DPWM60) control

For six-phase system with two inverters, controlled in accordance with algorithms of discontinuous PWM with the 60^0-nonswitching intervals (DPWM60, Fig. 5) basic switching sequence is: **07-32-21-32-07-...70-21-32-21-70...**

For the dual three-phase system, controlled by algorithms of discontinuous PWM with the 120^0-nonswitching intervals (DPWM120, Fig. 6), switching sequence for elimination of the common-mode voltages is **70-21-32-21-70-** during the whole 60^0-clock-interval, and this scheme has specific peculiarities in comparison with other schemes of discontinuous synchronized PWM.

IV. COMBINED PWM CONTROL OF TWO INVERTERS IN SYMMETRICAL DUAL THREE-PHASE SYSTEM

Dual three-phase power conversion systems have additional degrees of freedom for PWM control of two inverters. In particular, different modulation schemes can be used for control of each inverter in dual three-phase systems with combined PWM algorithms.

As an example, Fig. 7 illustrates combined PWM control and shows switching sequences and pole voltages of two inverters, where the first one (with the **abc** phases, see Fig. 1) is controlled in accordance with algorithms of discontinuous PWM with the 30^0-nonswitching intervals (DPWM30), and the second inverter (with the **xyz** phases) is controlled by algorithms of continuous PWM (CPWM). Average switching frequency of two voltage source inverters is equal in this case.

As an another example, Fig. 8 shows combined PWM control of two inverters, based on the schemes of discontinuous pulsewidth modulation with the 30^0-nonswitching intervals (DPWM30, the first inverter), and with the 60^0-nonswitching intervals (DPWM60, the second inverter).

It is necessary to mention, that the presented (and others) combined schemes of PWM do not provide elimination of the common-mode voltages in symmetrical dual three-phase systems.

V. OPERATION OF SYMMETRICAL DUAL THREE-PHASE SYSTEMS WITH SYNCHRONIZED PWM

Fig. 9 – Fig. 14 present basic voltage waveforms (with its spectra) during period of the fundamental frequency of the symmetrical six-phase system controlled in accordance with three basic schemes of synchronized PWM: 1) continuous synchronized PWM (CPWM, Figs. 9–10); 2) discontinuous PWM with the 30^0-nonswitching intervals (DPWM30, Figs. 11-12), and 3) discontinuous PWM with the 120^0-nonswitching intervals (DPWM120, Figs. 13-14).

The average switching and fundamental frequencies of each inverter of dual three-phase system are, respectively, equal to *900 Hz* and *35 Hz* (modulation index *m=0.7*). Figs. 9, 11 and 13 present pole voltages in the system, and the corresponding useful V_{sa} (9) and loss-producing V_{ml} (10) components of the phase voltage. The V_{ml} voltage is equal to zero in these cases. The common-mode voltage V_0 (12) relatively mid-point of the DC-source (the signal $V_0-V_{dc}/2$) is equal to zero too. Figs. 10, 12 and 14 show spectra of the V_{sa} voltage, which include only odd non-triplen harmonics.

978-1-4244-1871-8/07 $25.00 © 2007 IEEE

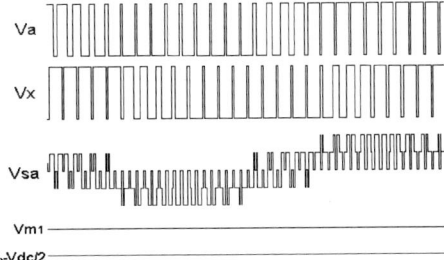

Fig. 9: Pole voltages V_a and V_x, useful V_{sa} and loss-producing V_{m1} components of the phase voltage, and common-mode voltage V_0, for the system with continuous synchronized PWM (CPWM)

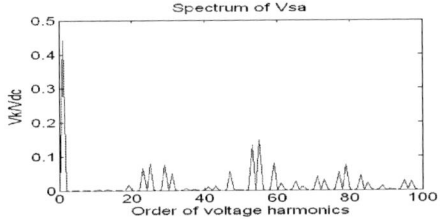

Fig. 10: Spectrum of the V_{sa} voltage for the system with CPWM

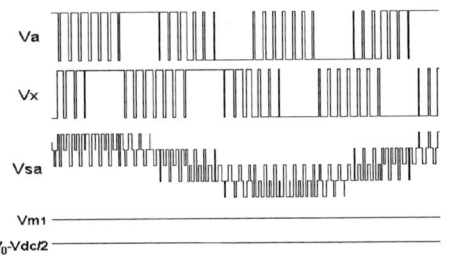

Fig. 11: Pole voltages V_a and V_x, useful V_{sa} and loss-producing V_{m1} components of the phase voltage, and common-mode voltage V_0, for the system with discontinuous synchronized PWM (DPWM30)

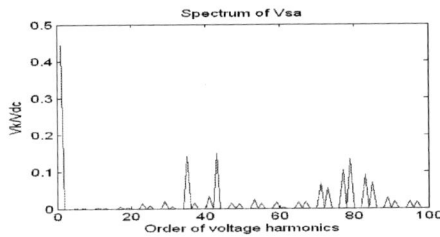

Fig. 12: Spectrum of the V_{sa} voltage for the system with DPWM30

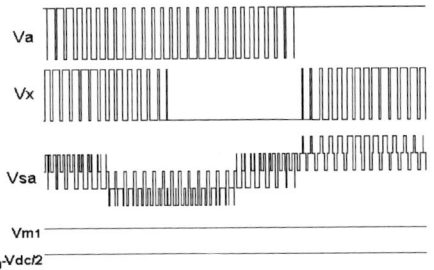

Fig. 13: Pole voltages V_a and V_x, useful V_{sa} and loss-producing V_{m1} components of the phase voltage, and common-mode voltage V_0, for the system with discontinuous synchronized PWM (DPWM120)

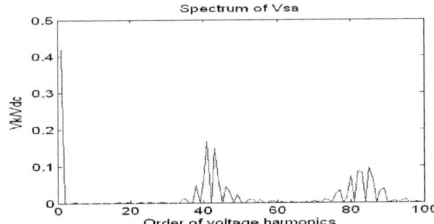

Fig. 14: Spectrum of the V_{sa} voltage for the system with DPWM120

Fig. 15 – Fig. 20 show basic voltage waveforms (with its spectra) during period of the fundamental frequency of the symmetrical six-phase system with combined DPWM30+ CPWM synchronized control (Figs. 15-17), and with combined DPWM30+DPWM60 control (Figs. 18-20). Figs. 16 and 19 present spectra of the V_{xs} and V_{as} phase voltages. Figs. 17 and 20 show spectra of the V_{sa} voltage. The average switching and fundamental frequencies of each inverter are, respectively, equal to *900 Hz* and *35 Hz*.

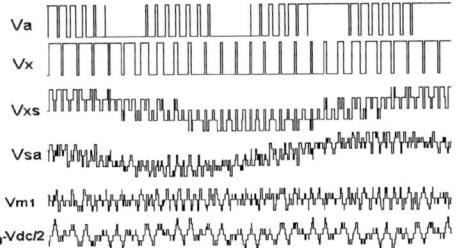

Fig. 15: Pole voltages V_a and V_x, phase voltage V_{xs}, useful V_{sa} and loss-producing V_{m1} components of the phase voltage, and common-mode voltage V_0, for the system with combined DPWM30+CPWM control

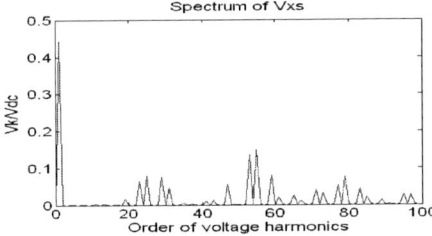

Fig. 16: Spectrum of the V_{xs} voltage (DPWM30+CPWM control)

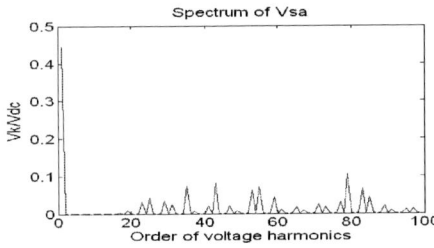

Fig. 17: Spectrum of the V_{sa} voltage (DPWM30+CPWM control)

Algorithms of synchronized PWM, including combined PWM control of two inverters, provide continuous voltage synchronization in dual three-phase systems. The spectra of the V_{sa}, V_{xs} and V_{as} voltages (Figs. 10, 12, 14, 16-17, 19-20) do not contain even harmonics and sub-harmonics, which is especially important for high power/high current systems.

978-1-4244-1871-8/07 $25.00 © 2007 IEEE

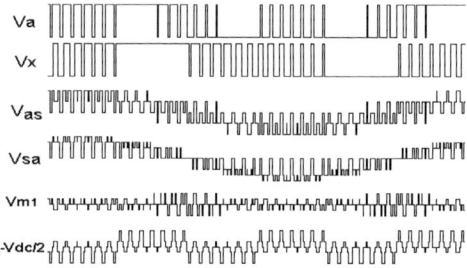

Fig. 18: Pole voltages V_a and V_x, phase voltage V_{as}, useful V_{sa} and loss-producing V_{m1} components of the phase voltage, and common-mode voltage V_0, for the system with combined DPWM30+DPWM60 control

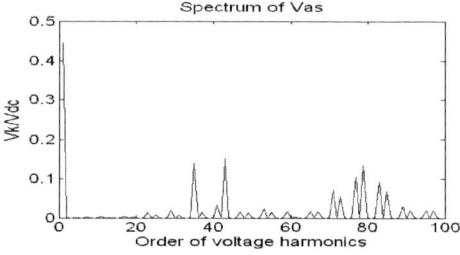

Fig. 19: Spectrum of the V_{as} voltage (DPWM30+DPWM60 control)

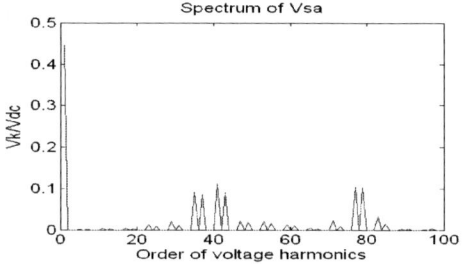

Fig. 20: Spectrum of the V_{sa} voltage (DPWM30+DPWM60 control)

Fig. 21 presents calculation results of Weighted Total Harmonic Distortion factor (*WTHD*) for the V_{sa} voltage (averaged values of $WTHD = (1/V_{sa1})\sqrt{\sum_{i=2}^{1000}(V_{sai}/i)^2}$) of symmetrical dual three-phase system with non-combined control algorithms, providing common-mode voltages elimination, which are based on continuous (CPWM) and three discontinuous (DPWM30, DPWM60 and DPWM120), versions of synchronized space-vector PWM. The average switching frequency of inverters is equal to *900 Hz* during standard scalar *V/F* control.

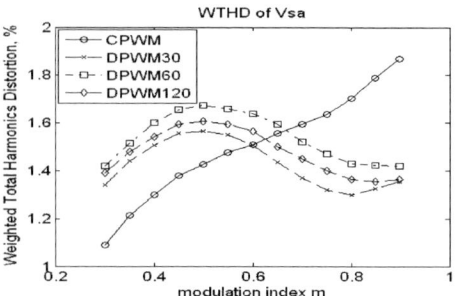

Fig. 21: Averaged *WTHD* factor of the V_{sa} voltage versus modulation index

Presented in Fig. 21 characteristics show some advantage of continuous synchronized PWM at low modulation indices. At higher modulation indices, the discontinuous scheme with the 30^0-nonswitching intervals provides slightly better *WTHD* factor of the V_{sa} voltage in comparison with other versions of synchronized PWM.

VI. CONCLUSION

New method of synchronized PWM has been applied for control of symmetrical six-phase (dual three-phase) induction motor drives. Both combined and non-combined schemes of synchronized PWM provide continuous phase voltage synchronization in these systems. All these schemes, based on space-vector approach for determination of the pulse patterns, are characterized by minimum number of switchings in power circuits, providing the minimization of switching losses in inverters. And, in particular, non-combined versions of synchronized PWM allow elimination of undesirable common-mode voltages in symmetrical six-phase systems during the whole control range.

The spectra of the phase voltage of symmetrical dual three-phase drives with synchronized PWM do not contain even harmonics and sub-harmonics for any (integral or fractional) ratios between the switching and fundamental frequencies, which is especially important for high power/high current applications.

ACKNOWLEDGMENT

This research has been supported in part by Marie Curie International Fellowships Award of the FP6 Program of the EC, and by the NATO Collaborative Linkage Grant.

REFERENCES

[1] J. Holtz, "Pulsewidth modulation for electronic power conversion," *Proc. of IEEE*, vol. 82, no. 8, pp. 1194-1213, 1994.

[2] M. Jones, and E. Levi, "A literature survey of state-of-the art in multiphase ac drives," in *2002 Proc. Universities Power Eng. Conf.*, pp. 505-510.

[3] R. Bojoi, A. Tenconi, F. Profumo, G. Griva, and D. Martinello, "Complete analysis and comparative study of digital modulation techniques for dual three-phase AC motor drives", in *2002 Proc. IEEE PESC Conf.*, pp. 851-857.

[4] Y. Zhao, and T.A. Lipo, "Space vector PWM control of dual three-phase induction machine using vector decomposition," *IEEE Trans. Ind. Appl.*, vol. 31, no. 5, pp. 1100-1109, 1995.

[5] R. Bojoi, F. Farina, F. Profumo, and A. Tenconi, "Dual-three phase induction machine drives control – a survey," in *2005 CD-ROM Proc. IEEE IPEC Conf.*, 10 p.

[6] V. Oleschuk, R. Bojoi, F. Profumo, A. Tenconi, and A.M. Stankovic, "Multifunctional six-phase motor drives with algorithms of synchronized PWM," in *2006 Proc. IEEE IECON Conf.*, pp. 1852-1859.

[7] M.B.R. Correa, C.B. Jacobina, C.R. da Silva, A.M.N. Lima, and E.R.C. da Silva, "Six-phase AC drive system with reduced common-mode voltage," in *2003 Proc. IEEE ICEMD Conf.*, pp. 1852-1858.

[8] M.B.R. Correa, C.B. Jacobina, C.R. da Silva, A.M.N. Lima, and E.R.C. da Silva, "Vector modulation for six-phase voltage source inverters," in *2003 CD-ROM Proc. EPE Conf.*, 10 p.

[9] D. Dujic, A. Iqbal, and E. Levi, "A space vector PWM techniques for symmetrical six-phase voltage source inverters," *EPE Journal*, vol. 17, no. 1, pp. 24-32, 2007.

[10] N. Mohan, T.M. Undeland, and W.P. Robbins, *Power Electronics*, 3rd ed., John Wiley & Sons, 2003.

[11] V. Oleschuk, and F. Blaabjerg, "Direct synchronized PWM techniques with linear control functions for adjustable speed drives," in *2002 Proc. IEEE APEC Conf.*, pp. 76-82.

978-1-4244-1871-8/07 $25.00 © 2007 IEEE

Optimized State Controller on DC-DC Converter

M. Bayati Poodeh
Islamic Azad University
Najafabad Branch
Department of Electrical Engineering
Email: bayati@iaun.ac.ir

S. Eshtehardiha
Islamic Azad University
Najafabad Branch
Department of Electrical Engineering
Email: eshtehardiha@gmail.com

M. Namnabat
Islamic Azad University
Majlesi Branch
Department of Electrical Engineering
Email: namnabat@iaumajlesi.ac.ir

Abstract -Due to their switching property included in their structure, DC-DC converters have a non-linear behavior and their controlling design is accompanied with complexities. But by employing the average method it is possible to approximate the system by linear system and exploiting linear control methods. In this paper the central method of Linear Quadratic Regulator (LQR) **is employed for controlling Buck converter and improving its dynamic functioning, but determining the LQR controlling matrices is very difficult. So, genetic algorithm is employed to eliminate this obstacle. The simulation of DC-DC converter with LQR based on genetic algorithm is shown the improvement in voltage control response and the converter output current.**

Keywords: DC-DC converter, State controller, LQR optimum controller, Genetic algorithm

I. INTRODUCTION

Buck converter has a wide application in electrical industry and power systems and specifically it can supply the voltage for a direct current consumer. Buck is a one-input and multiple-output in structure with a non-linear property due to its switching behavior, but at the same time when the switch is on and off its behavior is linear. Therefore, by employing the averaging method, it is possible to exchange a non-linear system with a linear one. Many of the methods are centered on the isolation of the system variables and PI controller design [1]. Some control methods have defined the subject of control based on pole placement. One of the other control methods is the use of feedback stage in DC converter control [2]. Other control method is the feedback loop control [3]. Another one is the use of Linear Quadratic Regulator (LQR) method in the control of Buck converter [4], [5]. To administer different control methods, varied models of DC-DC converters are presented which have satisfactory responses [6]. In this article an optimum LQR is designed that can improve the Buck converter response. There is no specific method in LQR design which is based on trial and error. The best constant values for state feedback matrix are laboriously obtained through trial and error, although time consuming. Genetic algorithm is employed to find the best values for LQR controller in a very short time. Therefore, a new method is presented for optimizing the systems with two factors of the least response time and the highest precision.

This paper is organized as follows: Introduction, that contains the identification of DC-DC converter and digest of control methods to improve its performance. DC-DC converter circuit model is described in section II. Then in section III, Linear Quadratic Regulator design is explained. In section IV, the genetic algorithm method for optimal performance of DC-DC converter is discussed in detail. The simulation results are given in section V. Finally, a brief conclusion in section VI.

II. BUCK CONVERTER CIRCUIT MODEL

The Buck converter circuit model is depicted in Fig. 1.

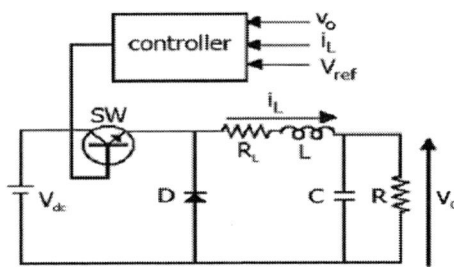

Fig. 1 : Buck converter

In this model, V_o is the system output voltage and V_{ref}, is the converter voltage. To obtain the converter state equations in low-frequency state, it is required that the system state be studied in two states of on and off as shown in Fig. 2, and Fig. 3.

Switch ON:

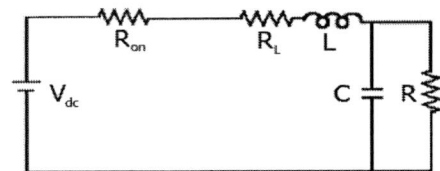

Fig. 2 : Circuit topology during t_{on}.

$$V_{dc} - V_o = (R_s + R_L)i_L + L(di_L / dt) \ ,$$

$$i_L = C(dV_o / dt) + V_o / R \ , \qquad\qquad (1)$$

$$X_1 = i_L \ , \qquad X_2 = V_o \ , \qquad X = \begin{bmatrix} X_1 \\ X_2 \end{bmatrix} ,$$

978-1-4244-1871-8/07 $25.00 © 2007 IEEE   153

$$\dot{X} = A_1 X + B_1 V_{dc} ,$$

$$A_1 = \begin{bmatrix} \dfrac{-(R_L + R_S)}{L} & -\dfrac{1}{L} \\ \dfrac{1}{C} & -\dfrac{1}{RC} \end{bmatrix} , \quad B_1 = \begin{bmatrix} \dfrac{1}{L} \\ 0 \end{bmatrix} .$$

Switch OFF:

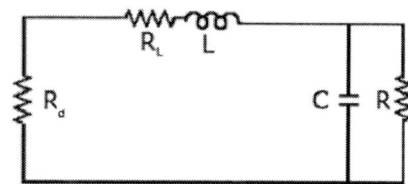

Fig. 3 : Circuit topology during t_{off}.

$$(R_d + R_L)i_L + L(di_L / dt) + V_O = 0 ,$$

$$i_L = C(dV_O / dt) + V_O / R ,$$

$$\dot{X} = A_2 X + B_2 V_{dc} , \qquad (2)$$

$$A_2 = \begin{bmatrix} \dfrac{-(R_L + R_d)}{L} & -\dfrac{1}{L} \\ \dfrac{1}{C} & -\dfrac{1}{RC} \end{bmatrix} , \quad B2 = \begin{bmatrix} 0 \\ 0 \end{bmatrix} .$$

Now it is required to show the effect of on and off durations of switch in (1) and (2) to obtain the mean values of state equations.

$$\dot{X} = AX + BV_{dc} ,$$

$$A = mA_1 + (1 - m)A_2 ,$$

$$B = mB_1 + (1 - m)B_2 , \qquad (3)$$

$$m = \frac{t_{on}}{T_s} .$$

III. LQR DESIGN

It is possible to improve the converter response by employing the LQR control method.

The LQR design problem has been extensively investigated for the past four decades [7-11]. Consequently, the theory of linear optimal control has been very well developed. Central to the optimal stabilizer design problem is the formulation of the cost function and the choice of the state and control weighting matrices. Although the need for the design of a cost function that reflects the physical characteristics and practicalities of the studied system is reasonably understood, the choice of the elements of the weighting matrices that would translate this understanding still remains a difficulty. This is mainly due to the interactive nature of the states and controls of linear multivariable dynamic systems. In the following we present a novel systematic approach to this problem which overcomes the shortcomings of existing approaches.

In this method, the feedback gain matrix is determined if J energy function is optimized. To achieve equilibrium among range control parameters, response speed, settling time, and proper overshoot rate, all of which guarantee the system stability, the LQR is employed.

A. LQR algorithm

For a system in the form of $\dot{X} = AX + BU$ the LQR method determines the K matrix of the equation $U(t) = -kX(t)$ to minimize the $J = \int_0^\infty (X^T QX + U^T RU)dt$ function.

R and Q matrices express the relation between error and energy expense rate. R and Q are also the definite positive matrices. From the above equation, we have:

$$J = \int_0^\infty (X^T QX + X^T K^T RKX)dt$$
$$= \int_0^\infty (X^T (Q + K^T RK)X)dt \qquad (4)$$

Subsequent to the solution stages of the equation and optimization of the parameters of the following equation:

$$X^T (Q + K^T RK)X = \frac{d}{dt}(X^T PX) \qquad (5)$$

The following equation is presented:

$$X^T (Q + K^T RK)X = -X^T PX - X^T PX$$
$$= -X^T [(A - BK)^T P + P(A - BK)]X \qquad (6)$$

With regard to the values on the both sides of the above equation and with regard to this fact that these equations are true for every X, then the following equation is obtained:

$$(A - BK)^T P + P(A - BK) = -(Q + K^T RK) \qquad (7)$$

If $R = T^T$. T matrix is positive and definite and T matrix is also gross:

$$A^T P + PA + [(TK - T^T)^{-1} B^T P]^T * [TK - (T^T)^{-1} B^T P]$$
$$- PBR^{-1}B^T P + Q = 0 \qquad (8)$$

For quantifying J in relation to K we have:

$$X^T [TK - T^{-T} B^T P]^T TK - T^{-T} B^T P]x \qquad (9)$$

This equation is non-negative and the minimum amount takes place when it is zero or when:

$$TK = T^- TB^T P \qquad (10)$$

So

$$K = T^{-1}T^{-T} B^T P = R^{-1}B^T P \qquad (11)$$

We can also obtain the control matrix for U input.

$$U(t) = -KX(t) = -R^{-1}B^T PX(t) \qquad (12)$$

And P should be true in the following Reecati equation.

978-1-4244-1871-8/07 $25.00 © 2007 IEEE

$$A^T P + PA - PBR^{-1} B^T P + Q = 0 \qquad (13)$$

In LQR design, R and Q weight matrix which determines the quotient related to the closed loop feedback system within the least time is determined. The selection of R and Q has the least dependence on the specification of system administration and requires a long range of trial and error.

B. Flow chart of LQR controller

Statement of the Problem	
Given the plant as x(t) = Ax(t) + Bu(t) the performance index as $$J = \int_0^\infty (X^T Q X + X^T K^T R K X)\,dt = \int_0^\infty (X^T(Q + K^T R K)X)\,dt$$ and the its conditions $x(t_0) = x_0 ; x(\infty) = 0,$ Find the optimal control, index.	
Solution of the Problem	
Step 1	Solve the matrix algebraic Riccati equation $$A^T P + PA - PBR^{-1} B^T P + Q = 0$$
Step 2	Solve the optimal state x* (t) from $$X^*(t) = (A - BR^{-1} B^T \bar{P}) X^*(t)$$ with initial condition x(t$_0$) = x$_0$.
Step 3	Obtain the optimal control u* (t) from $$u^*(t) = -R^{-1} B^T P X^*(t)$$
Step 4	Obtain the optimal performance index from $$J^* = \frac{1}{2} e^{2at_0} x^{*'}(t_0) \bar{P} x^*(t_0)$$

IV. GENETIC ALGORITHM

In 1960 the first serious investigation into Genetic Algorithms (GAs) was undertaken by John Holland. Genetic algorithms have become popular due to several factors. They have been successfully applied to self-adaptive control systems and to function optimization problems. They are a powerful search technique, yet are computationally simple. The search method they use is robust since it is not limited like other search methods with regard to assumptions about the search space.

The genetic algorithm is an algorithm which is based on natural evolution and the survival of the best chromosome. There are three basic differences between genetic algorithm and optimization classical methods. Firstly, the genetic algorithm works on the encoded strings of the problem parameters. Each string is the representative of one answer to the problem, and the real quantities of the parameters are obtained from the decoding of these strings. Secondly, the genetic algorithm is a search algorithm which works on a population of search spaces. This quality causes the genetic algorithm to search different response spaces simultaneously reducing the possibility of being entrapped at local optimized points. Thirdly, the genetic algorithm does not need previous data from the problem response space such as convexity and derivable. It is only necessary to calculate a response function named fitness function. This function expresses the rate of response proximity to the goal function of the intended algorithm. Different methods are used for encoding the parameters in genetic algorithm.

One of the common methods is the binary encoding in which for each parameter depending on the precision required, a few bits are allocated. The problem with this method of encoding is the encoding of the continuous parameters. In order to provide the required precision for the continuous parameters in this method of encoding we have to increase the number of the allocated bits for each parameter which results in longer strings and an increase in the volume of calculations resulting in longer algorithm time.

Another method used in genetic algorithm for the encoding of data is the real encoding. For every parameter in this method one real number is taken into consideration and in addition to the above methods, one can use encoding methods of the genetic functioning defined for these algorithms for real numbers (conversion, value, and tree). The following stages are carried out in the genetic algorithm:

i. The formation of initial population
A definite number of chromosomes are randomly selected with regard to the type of the problem.

ii. Evaluation
Each chromosome from the initial population is processed on the basis of the initial goal of the problem.

iii. Production of new population
In this stage a new population is selected on the basis of the previous one. The stages for the manufacture of this population are:

Transmit: in this stage the chromosomes with high efficiency are directly transmitted to the new population.

Selection: Two pairs of the remaining chromosomes from the previous population are selected according to their rate of efficiency (each with higher efficiency has in fact higher chance for being selected). Different methods can be selected for the selection of the chromosomes according to their efficiency values.

Crossover: By selecting two chromosomes from the present population, it is tried to improve the efficiency rate of one of the produced chromosomes by employing crossover method. These operations are accompanied with a reduction probability and can happen in one or several points of chromosomes. Fig. 4, depicts the functions of crossover.

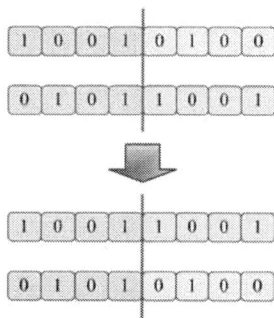

Fig. 4 : Single point crossover

Mutation: in this stage it is tried to correct the gene in a chromosome which has caused a reduction in the competence rate of the chromosome. With regard to the kind of the employed encoding the type of this functioning is determined. This functioning is also accompanied with a reduction probability. Fig. 5, depicts the functions of mutation.

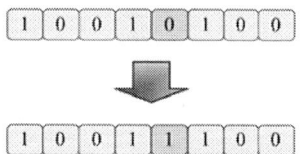

Fig. 5 : Binary mutation operator

Reception: the obtained chromosome is placed in the new population according to the previous states.

iv. Replacement
The newly obtained population is replaced with the previous population and then we return to stage *ii*.

v. Stoppage
This factor determines the stoppage method of the algorithm, which can be based on the rate of competence proximity of the chromosomes to each other in the new population or is determined on the basis of the number of the produced population in the algorithm. After specifying the critical sections in this article we embark on the formation of chromosomes by binary encoding.

After the determination of the amount of the processing function related to each of the chromosomes, we get on the production of a new population on the basis of the previous population.

Then produce a new population and determine the optimization function rate of the produced chromosomes in the new population. Then, with regard to the amount of optimization function we continue to produce a new population until we reach to the desired point.

B. R and Q design by genetic algorithm

The target function is as follows:

$$J = (1-\exp(-1))\times(Ess+Mp)+(\exp(-1)\times(ts-tr)) \quad (14)$$

That *tr* is rise time, *ts* is settling time, *Mp* is overshoot and *Ess* is steady state error.

The genetic algorithm flowchart is presented as follows to optimize the target function in shown Fig. 6.

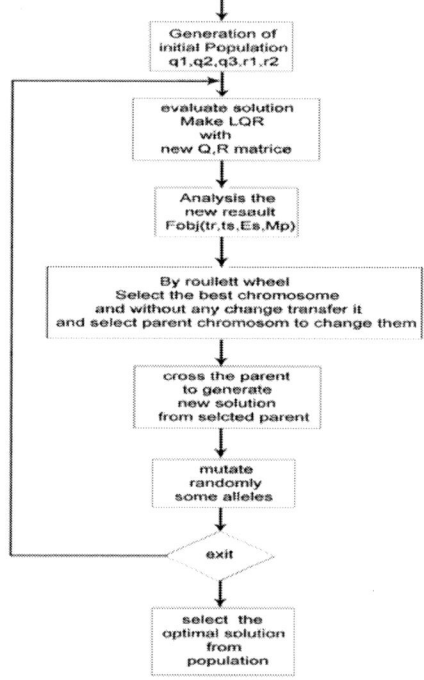

Fig. 6 : Genetic algorithm flowchart to optimize the target function

Through this method the time-consuming stage of determining Q and R matrices is performed with great precision and the system is optimized to reach the intended specifications in closed loop automatically and at the end, the output of the system is optimized with less overshoot, less oscillation, low settling time, and little steady error stage.

V. THE RESULT OF SIMULATION

With regard to the state equations for the converter and taking into consideration $r_L = 0.7\Omega$, $r_d = 0.7\Omega$, $R_L = 118\Omega$, $r_c = 1.18\Omega$, $r_l = 0.2\Omega$, $L = 0.42mH$, $C = 1450\mu F$, $V_{dc} = 50v$, $d = 0.5$ the Buck converter open loop response will be in the form of Fig. 7.

978-1-4244-1871-8/07 $25.00 © 2007 IEEE   156

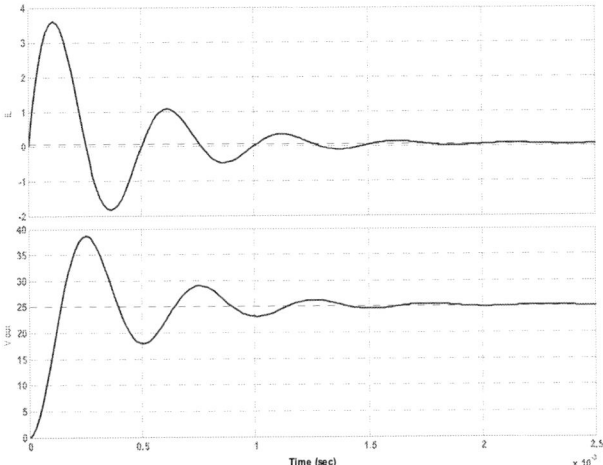

Fig. 7 : Open Loop response of Buck converter

As it is evident from the study of outputs, the open loop system cannot hold the output in constant and ideal conditions. Of course, the outputs will eventually approximate the ideal rate with regard to the stability of the system. In the design of LQR weight matrices, R and Q are the determining elements for the quotient related to closed loop feedback system. Paying close attention to Fig. 8, and Fig. 9. This result is obtained that with fluctuation of R and Q matrix amounts, the following results are obtained:

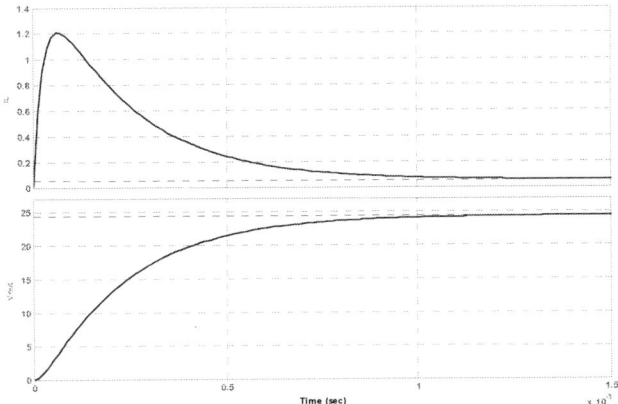

Fig. 9 : Time response of system with LQR

With using genetic algorithm the following results are obtained in Fig. 10.

With paying close attention to the system responses in Fig. 8, and Fig. 9, it is observed that the LQR controller has optimized the time needed to reach the desired system response *i.e.*, to the possible minimum amount.

For this purpose the genetic algorithm on the system, within several successive seasons, has improved the convergence of the system and has brought the best chromosome of that generation into optimum amount from generation to generation. This trend is observed in the evolution and optimization of the chromosomes within several successive generations in Fig. 11.

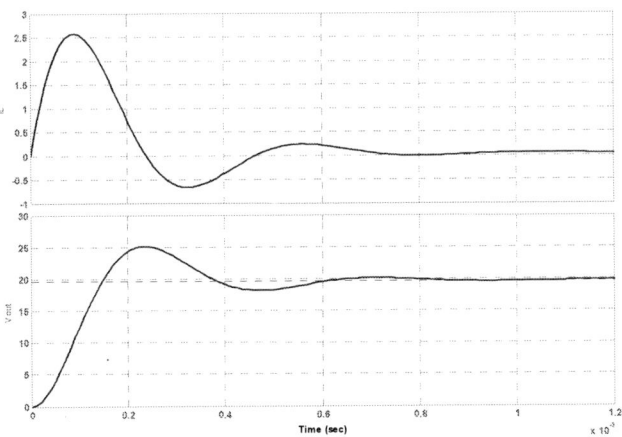

Fig. 8 : Time response of system with LQR

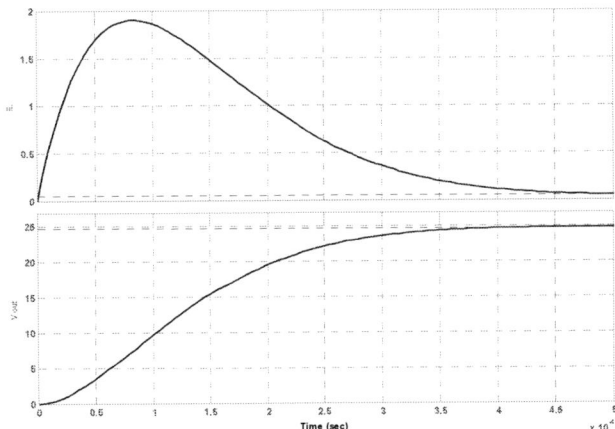

Fig. 10 : System response with LQR controller based on genetic algorithm

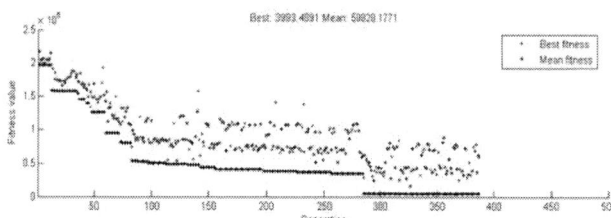

Fig. 11 : Optimizing of the chromosomes within several successive generations

Table. 1, shows the LQR parameters in manual and genetic algorithm.

In order to compare the results of the administration of LQR-GA and LQR methods Table. 2, is presented.

TABLE 1
LQR parameters

LQR1	$Q = \begin{bmatrix} 10 & 0 \\ 0 & 1 \end{bmatrix}, R = \begin{bmatrix} 1000 \end{bmatrix}$
LQR2	$Q = \begin{bmatrix} 460 & 0 \\ 0 & 0.09 \end{bmatrix}, R = \begin{bmatrix} 966 \end{bmatrix}$
LQR-GA	$Q = \begin{bmatrix} 213 & 0 \\ 0 & 0.097 \end{bmatrix}, R = \begin{bmatrix} 1658 \end{bmatrix}$

TABLE 2
Result of simulation

	Ess	Mp	tr	ts
Open loop	0.1	%53.5	94.9 µs	1.56 ms
LQR1	5.3	%27.7	99.7 µs	0.735 ms
LQR2	0.6	% 0	4.95 µs	0. 9 ms
LQR-GA	0.03	% 0.049	2.11 µs	0.335 ms

VI. CONCLUSION

The reduction of output voltage ripple of the reduction converter is very important. The optimum design method for linear controller is able to control the dynamic behavior of the converter. The using of genetic algorithm for the calculation of optimum coefficients of the matrices in the design of LQR controllers can bring about optimum dynamic response. In this paper, a LQR controller is designed to improve the Buck converter performance, in this way, genetic algorithm is used to optimized the LQR matrices. The results of simulation prove the improvement of the functioning of this converter compared with simple LQR method.

REFERENCES

[1] K. S. Kostov, J. Kyyra, and T. Suntio, "The input impedance of a buck converter," *European power Electronics Conf.*, Sept. 2003.

[2] I. G. Keller, D. Lasea, J. M. A.Myrzik, "State- space control structures for buck converters with / without Input Filter," *IEEE. 2005., Power Electron, 2005 European Conf.*, 10 pp, 2005.

[3] S. Uran and M. Milanovic, "State controller for buck converter," *EUROCON, 2003*, Vol. 1, pp. 381 – 385, 2003.

[4] J. M. Dores Costa, "Design of linear quadratic regulators for quasi-resonant DC-DC converter," *IEEE. 2001. Power Electron Conf., PESC. 2001.*, Vol. 1, pp. 422 – 426, 2001.

[5] F. H. F. Leung, P. K. S. Tam and C. K. Li, "The control of switching DC-DC converter S-A general LQR problem," *IEEE Trans., Ind Electron.*, Vol. 38, pp. 65-71, 1991.

[6] Qi. Feng, J. Y. Hung and R. M. Nelms, "The application of posicast control to DC-DC converters," *2002. 37th Intersociety Energy Converesion Engineering Conference (IECEC)*.

[7] M. ATHANS, and P. FALB, *Optimal control: an introduction to the theory and its application*, McGraw-Hill, 1966.

[8] F. L. LEWIS, *Optimal control*, John Wiley, 1986.

[9] B. D. O. ANDERSON, and J. B. MOORE, *Optimal control: linear quadratic methods*, Prentice-Hall, 1989.

[10] W. S. LEVINE, and M. ATHANS, "On the determination of the optimal constant output feedback gains for linear multivariable systems," *IEEE Trans., 1970, AC-15, (I)*, pp. 44-48, 1970.

[11] D. D. MOERDER, and A. J. CALISE, "Convergence of a numerical algorithm for calculating optimal output feedback gains," *IEEE Trans., 1985, AC-30, (9)*, pp. 900-903, 1985.

[12] R. L. Haupt and S.E. Haupt, *Practical genetic algorithms*, John wiley & Sons Inc., New york, 2004.

978-1-4244-1871-8/07 $25.00 © 2007 IEEE

A Study on Rotor FOC Method Using Matrix Converter Fed Induction Motor With Common-Mode Voltage Reduction

Lee, Hong-Hee
University of Ulsan
School of Electrical Engineering
Muger-2-Dong Nam-Gu, Ulsan, 680-749
Email: hhlee@ulsan.ac.kr

Hoang M. Nguyen
University of Ulsan
School of Electrical Engineering
Muger-2-Dong Nam-Gu, Ulsan, 680-749
Email: nmhoang@ulsan.ac.kr

Chun, Tae-Won
University of Ulsan
School of Electrical Engineering
Muger-2-Dong Nam-Gu, Ulsan, 680-749
Email: twchun100@ulsan.ac.kr

Abstract—Currently, most of researches on matrix converter concerns about indirect rotor flux oriented control (FOC) method use MC indirect space vector modulation (SVM) technique. This paper introduces a direct rotor flux oriented control method using MC fed induction motor with the direct space vector modulation technique. Furthermore, the proposed technique of common-mode voltage reduction in direct space vector has been firstly applied to reduce the bad effect of common-mode voltage on induction motor. The under-rated speed and field-weakening regions has been examined to verify the operation of matrix converter under different load conditions. The experiments are carried out with 5HP induction motor fed by matrix converter to show the high dynamic performance for the proposed control system.

Keyword: **Matrix Converter, common-mode voltage, flux oriented control**

I. INTRODUCTION

In past two decades, the progress of power device technology and the development of large power integrated circuits have reviewed direct AC-AC power conversion technologies. This converter fulfills all the requirements of the conventionally used rectifier/ dc link/ inverter structures and provides an efficient way to convert electric power for motor drives, UPS, VF generators and reactive energy control, [1] [4] [5].

In general, the desirable characteristics of AC to AC converters are
- Sinusoidal input and output wave forms with minimal higher order harmonics and no sub-harmonics
- Bidirectional energy flow capability,
- Minimal energy storage requirements (minimal size reactive component),
- A controllable power factor.
Furthermore, according to [2], the matrix converter has more advanced potential as compared with conventional voltage source inverters. Advantages include:
- Compact design and long life due to the absence of a bulky electrolytic capacitor,
- Unity input power factor at the power supply side,
- Availability of the continuous zero speed operation because no current concentrates in any of the switches.

In the past, the direct rotor FOC method using voltage source inverters are widely employed in field-oriented induction motor drive systems [3] [5] because of its fast current response and can be extremely easily implemented with a current controlled PWM inverter.

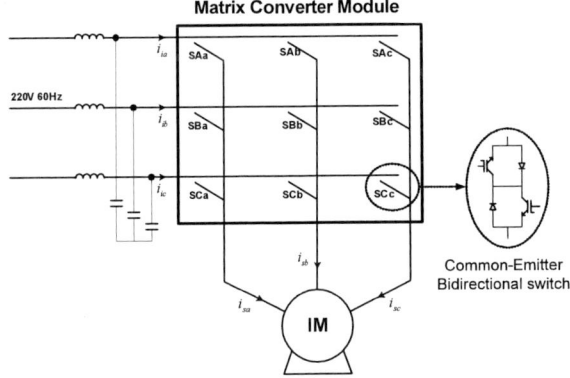

Fig. 1: Matrix Converter schematic block diagram

Due to the advantages of matrix converter and the rotor FOC method, this paper presents the combined method using matrix converter fed induction motor which utilizes the advantages of both methods. The direct SVM method with common-mode voltage reduction has been firstly applied to MC control system. Experimental results are shown to validate the proposed control method.

II. DIRECT ROTOR FOC METHOD USING MC FED IM

A. Matrix Converter Theory

Three-phase matrix converter module includes nine bidirectional switches as shown in figure 1. There are 27 switching configuration states, which mean 27 possible space vectors can be used to control IM and can be split respectively into 3 groups as shown in Table I; group I: two output lines are connected to one of the other input lines, group II: all output lines are connected to a common input line, group III: each output line is connected to a different input line. And, the corresponding output line-to-neutral voltage vector and input line current vector have fixed directions as represented in figure 2. However, group III is not useful, only 18 non-zero space vectors in group I (±1, ±2, …, ±9) and 3 zero space vectors in group II (0a, 0b, 0c) can be usually employed in the modern control techniques for matrix converter such as Space Vector Modulation, DTC methods and etc..[6] [7].

For simplicity of the SVM technique explanation, according to figure 2, assuming the desired output space vector is located in sector 1 and input current space vector is

978-1-4244-1871-8/07 $25.00 © 2007 IEEE   159

also located in sector 1 without missing the generality of the analysis.

TABLE 1
Possible Switching Configuration of MC

Group	Vector	A	B	C	v_s	α_0	i_i	β_i
I	$+1_{MC}$	a	b	b	$2/3v_{ab}$	0	$2/\sqrt{3}i_{sa}$	$-\pi/6$
	-1_{MC}	b	a	a	$-2/3v_{ab}$	0	$-2/\sqrt{3}i_{sa}$	$-\pi/6$
	$+2_{MC}$	b	c	c	$2/3v_{bc}$	0	$2/\sqrt{3}i_{sa}$	$\pi/2$
	-2_{MC}	c	b	b	$-2/3v_{bc}$	0	$-2/\sqrt{3}i_{sa}$	$\pi/2$
	$+3_{MC}$	c	a	a	$2/3v_{ca}$	0	$2/\sqrt{3}i_{sa}$	$7\pi/6$
	-3_{MC}	a	c	c	$-2/3v_{ca}$	0	$-2/\sqrt{3}i_{sa}$	$7\pi/6$
	$+4_{MC}$	b	a	b	$2/3\,v_{ab}$	$2\pi/3$	$2/\sqrt{3}i_{sb}$	$-\pi/6$
	-4_{MC}	a	b	a	$-2/3v_{ab}$	$2\pi/3$	$-2/\sqrt{3}i_{sb}$	$-\pi/6$
	$+5_{MC}$	c	b	c	$2/3v_{bc}$	$2\pi/3$	$2/\sqrt{3}i_{sb}$	$\pi/2$
	-5_{MC}	b	c	b	$-2/3v_{bc}$	$2\pi/3$	$-2/\sqrt{3}i_{sb}$	$\pi/2$
	$+6_{MC}$	a	c	a	$2/3v_{ca}$	$2\pi/3$	$2/\sqrt{3}i_{sb}$	$7\pi/6$
	-6_{MC}	c	a	c	$-2/3v_{ca}$	$2\pi/3$	$-2/\sqrt{3}i_{sb}$	$7\pi/6$
	$+7_{MC}$	b	b	a	$2/3v_{ab}$	$4\pi/3$	$2/\sqrt{3}i_{sc}$	$-\pi/6$
	-7_{MC}	a	a	b	$-2/3v_{ab}$	$4\pi/3$	$-2/\sqrt{3}i_{sc}$	$-\pi/6$
	$+8_{MC}$	c	c	b	$2/3v_{bc}$	$4\pi/3$	$2/\sqrt{3}i_{sc}$	$\pi/2$
	-8_{MC}	b	b	c	$-2/3v_{bc}$	$4\pi/3$	$-2/\sqrt{3}i_{sc}$	$\pi/2$
	-9_{MC}	a	a	c	$2/3v_{ca}$	$4\pi/3$	$2/\sqrt{3}i_{sc}$	$7\pi/6$
	$+9_{MC}$	c	c	a	$-2/3v_{ca}$	$4\pi/3$	$2/\sqrt{3}i_{sc}$	$7\pi/6$
II	0_a	a	a	a	0	-	0	-
	0_b	b	b	b	0	-	0	-
	0_c	c	c	c	0	-	0	-
III	x	a	b	c	x	x	x	x
	x	a	c	b	x	x	x	x
	x	b	c	a	x	x	x	x
	x	b	a	c	x	x	x	x
	x	c	a	b	x	x	x	x
	x	c	b	a	x	x	x	x

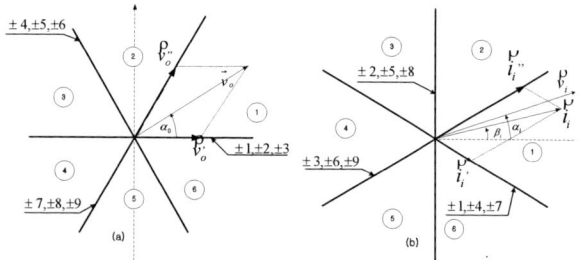

Fig. 2: (a) The output line-to-neutral voltage vectors

(b) The input line current vectors

The suitable space vectors used to generate the desired voltage vector and the unity power factor are -7, +9, -3, +1 and zero switching state with the following duty ratios:

$$\delta_1 = \frac{2}{\sqrt{3}}q\sin\left[\alpha_o - (k_v - 1)\frac{\pi}{3}\right]\sin\left[\frac{\pi}{6} - \left(\alpha_i - (k_i - 1)\frac{\pi}{3}\right)\right] \quad (1)$$

$$\delta_2 = \frac{2}{\sqrt{3}}q\sin\left[\alpha_o - (k_v - 1)\frac{\pi}{3}\right]\sin\left[\frac{\pi}{6} + \left(\alpha_i - (k_i - 1)\frac{\pi}{3}\right)\right] \quad (2)$$

$$\delta_3 = \frac{2}{\sqrt{3}}q\sin\left[k_v\frac{\pi}{3} - \alpha_o\right]\sin\left[\frac{\pi}{6} - \left(\alpha_i - (k_i - 1)\frac{\pi}{3}\right)\right] \quad (3)$$

$$\delta_4 = \frac{2}{\sqrt{3}}q\sin\left[k_v\frac{\pi}{3} - \alpha_o\right]\sin\left[\frac{\pi}{6} + \left(\alpha_i - (k_i - 1)\frac{\pi}{3}\right)\right] \quad (4)$$

$$\delta_5 = 1 - \frac{2}{\sqrt{3}}q\cos\left[\alpha_o - (2k_v - 1)\frac{\pi}{6}\right]\cos\left[\alpha_i - (k_i - 1)\frac{\pi}{3}\right] \quad (5)$$

where k_v, k_i are the output voltage vector sector and input current vector sector.

B. Rotor FOC Method using MC Fed IM

The basic problem involved in the implementation of current control PWM inverters is the choice of a suitable current control method which affects both the parameters obtained and the final configuration of the entire system. In figure 3, the PWM current control loop operates in synchronous field-oriented co-ordinates x-y. The feedback stator currents i_{sd}, i_{sq} are obtained from the measured values i_a, i_b after Clarke transformation:

$$i_{s\alpha} = i_a$$
$$i_{s\beta} = (1/\sqrt{3})(i_a + 2i_b) \quad (6)$$

And Park^{-1} transformation

$$i_{sd} = i_{s\alpha}\cos\theta_{cm} + i_{s\alpha}\sin\theta_{cm}$$
$$i_{sq} = -i_{s\alpha}\sin\theta_{cm} + i_{s\alpha}\cos\theta_{cm} \quad (7)$$

Where the rotor flux vector position θ_{cm} is calculated from the stator currents i_{sd}, i_{sq} and the measured speed, ω_m as follows:

$$\theta_{cm} = \int(\omega_m + \omega_{sl})dt$$
$$= \int\left(\omega_m + \frac{T_r i_{sd}}{i_{sq}(1 + pT_r)}\right)dt \quad (8)$$

Where $p = d/dt$ and $T_r = L_r/R_r$ is rotor-time constant.

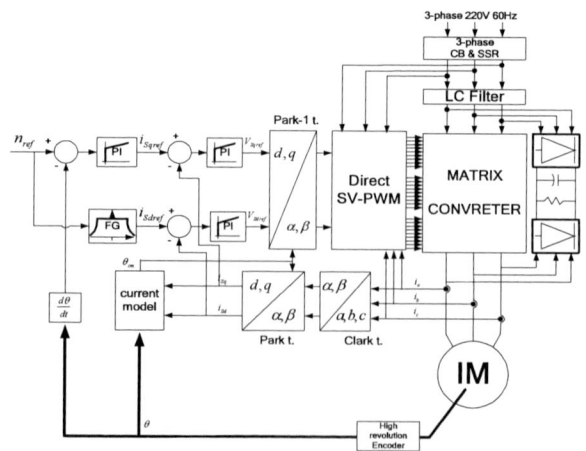

Fig. 3: Rotor FOC method block using MC fed IM

978-1-4244-1871-8/07 $25.00 © 2007 IEEE

III. SVM METHOD WITH COMMON-MODE VOLTAGE REDUCTION

A. Common-Mode Voltage Effects

Figure 4 shows a matrix converter connected to an induction motor. The leakage impedance between the motor neutral point and ground is represented by impedance Z_L. It is common to connect the induction motor to neutral point of the secondary side of the input transformer and the motor frame to the same ground. The common-mode voltage v_{sg} can be expressed as follows

$$v_{sg} = \begin{cases} v_{phase}/\sqrt{3} & , Group\ I \\ v_{phase} & , Group\ II \\ 0 & , Group\ III \quad (not\ used) \end{cases} \tag{9}$$

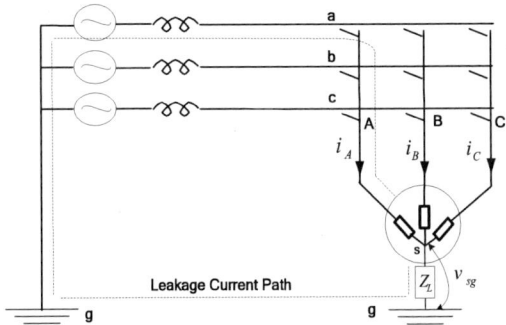

Fig. 4: Path of motor leakage current

B. Common-Voltage Reduction

According to [4], , instead of using zero SCs to complete the sampling period, T_s, two opposite SCs are used with the same duty ratio. They will not change the synthesized output voltage reference vector or the direction of the input current reference vector. However, there are nine pairs of the opposite SCs. The criteria to choose suitable SCs pairs are as follows.

1. The smallest output current and voltage ripple and the low harmonics distortion of the MC input current.
2. The smallest switching commutation within one sampling period and only one switch change at any instant in time.

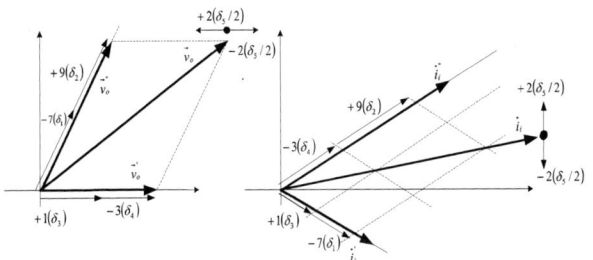

Fig. 5: Rotor FOC method block using MC fed IM

Figure 5 shows the suitable space configurations to achieve the common-mode voltage reduction in case both input current space vector and output space voltage are located in sector 1. The switching configuration ±2 are used to complete the rest of the redundant status instead of zero switching configurations. Table 2 shows the entire space vectors for proposed direct SVM method.

TABLE 2
Switching Table for Common-Mode Voltage Reduction

Input Current Sector / Output Voltage Sector	Sector 1 or Sector 4						Sector 2 or Sector 5						Sector 3 or Sector 6					
	$\delta_0/4$					$\delta_0/4$	$\delta_0/4$					$\delta_0/4$	$\delta_0/4$					$\delta_0/4$
Sector 1 or Sector 4	+2	-3	+9	-7	+1	-2	+7	-8	+2	-3	+9	-7	+3	-1	+7	-8	+2	-3
Sector 2 or Sector 5	-5	+4	-7	+9	-6	+5	-7	+9	-6	+5	-8	+7	-6	+5	-8	+7	-4	+6
Sector 3 or Sector 6	+5	-6	+3	-1	+4	-5	+1	-2	+5	-6	+3	-1	+6	-4	+1	-2	+5	-6
Sector 4 or Sector 1	-8	+7	-1	+3	-9	+8	-1	+3	-9	+8	-2	+1	-9	+8	-2	+1	-7	+9
Sector 5 or Sector 2	+8	-9	+6	-4	+7	-8	+4	-5	+8	-9	+6	-4	+9	-7	+4	-5	+8	-9
Sector 6 or Sector 3	-2	+1	-4	+6	-3	+2	-4	+6	-3	+2	-5	+4	-3	+2	-5	+4	-1	+3

IV. EXPERIMENTAL RESULTS

A. Experimental Setups

The experiment has been carried out on 7.5kW Eupec MC module (FR35R12KE3V1) and 5HP Induction Motor with the following parameters

TABLE 3
Induction Motor Parameters

Parameters	Value
Rated Power	5[Hp]
Rated Voltage	220/380[V], 60[Hz]
Number of Poles	4
Stator Resistance	1.4587 [Ω]
Rotor Resistance	0.9058 [Ω]
Stator Inductance	0.1511 [H]
Rotor Inductance	0.1639 [H]
Mutual Inductance	0.1483 [H]

As shown in Fig. 6, the input voltages and output currents are measured by a 12-bit A/D converter (AD7891). The calculation of rotor FOC method is executed with a floating-point 32-bit DSP TMS320C32 by Texas Instrument,

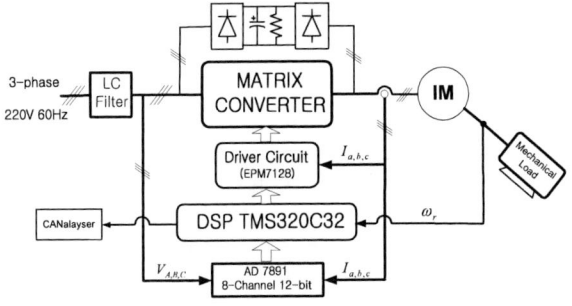

Fig. 6: Block diagram of matrix converter hardware experiment

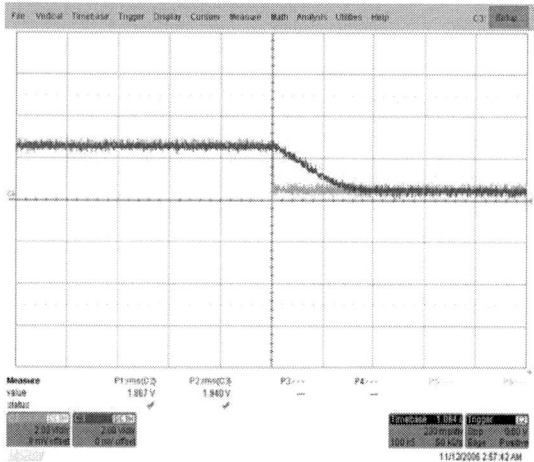

Fig. 8: Speed response as the speed step changes from 500 rpm – 100 rpm, under load torque 2N.m

running at 60MHz. The DSP board includes CAN (Control Area Network) function which is connected to PC through a CAN card to monitor the entire system. The sampling period

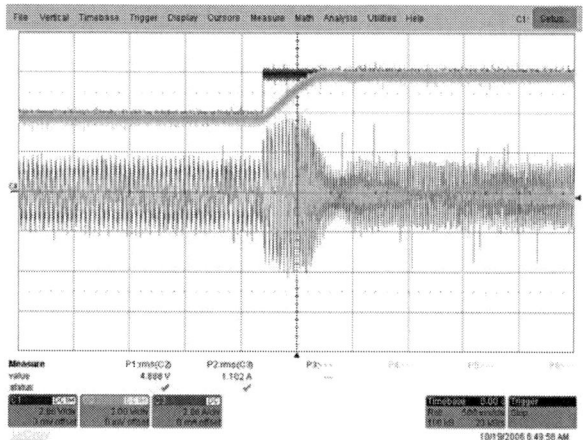

Fig.7: Speed response and stator current as the speed step changes from 720 rpm – 1145 rpm, under load torque 2N.m

is 333 µs, corresponding to a switching frequency of 3.3 kHz.

The control of the commutation process is performed by a high-CMOS (electrically erasable) programmable logic device (CPLD) board using EPM7128, manufactured by Altera, running at 40MHz. The 4-step commutation is carried out within 2 µs. A gate driver board (TC4420) with six isolated power supplies is sufficient to drive the MC power modules

B. Experimental Results

Figure 7 shows the dynamic response at the high speed of induction motor as the speed change from 720 rpm – 1145 rpm. Figure 8 shows the low speed operation as motor speed change from 500 rpm – 100 rpm. Furthermore, the entire speed range operation is shown in Fig. 9.

Figure 11 shows that the common-mode voltage peak value

is greatly reduced to 42% and contains much smaller harmonic components as compared to the conventional direct SVM method shown in figure 10.

Figure 12 shows the input phase current and input phase voltage waveforms with the unity power factor at the power supply side.

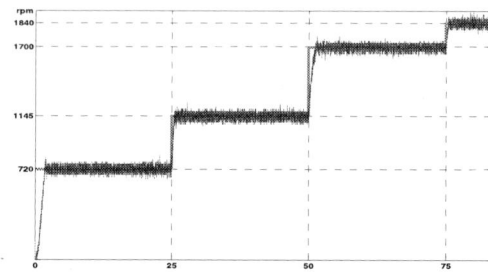

Fig. 9: IM speed responses with under rated speed (<1800rpm) and field weakening region (>1800 rpm)

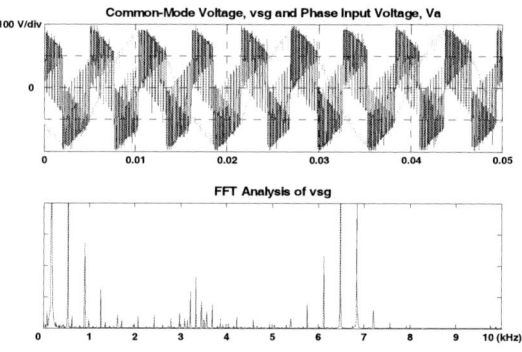

Fig. 10: Common-mode voltage and spectrum analysis

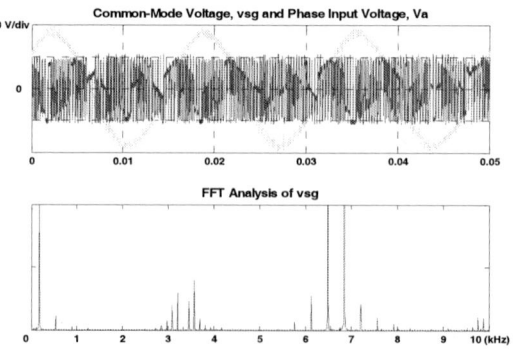

Fig. 11: Common-mode voltage reduction and spectrum analysis

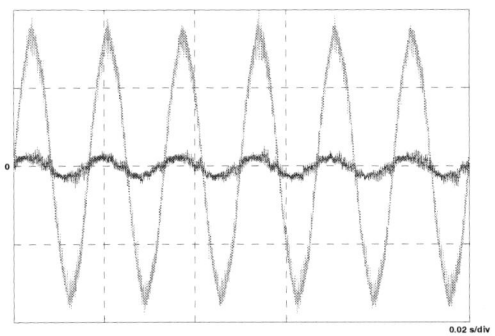

0.02 s/div

Fig. 12: Input phase current (10 A/div) and input phase voltage (100V/div)

[6] D. Casadei, G. Serra, A. Tani, L. Zarri, "Matrix converter modulation strategies: a new general approach based on space-vector representation of the switch state", IEEE Transactions on Industrial Electronics, Volume 49, Issue 2, April 2002, Page(s): 370– 381.

[7] Casadei, D.; Grandi, G.; Serra, G.; Tani, A. "Space vector control of matrix converters with unity input power factor and sinusoidal input/output waveforms" Fifth European Conference on Power Electronics and Applications, 13-16 Sep 1993 Page(s):170 - 175 vol.7

V. CONCLUSION

This paper describes the combination between the rotor FOC method and the MC technique. This method inherits the advantages of the rotor FOC with the dynamic responses and easy implementation. The unity power factor at the power supply side and the compact design due to the absence of the bulky capacitor make this proposed method more suitable and necessary for the induction motor control filed, especially with the high IM power capacity. Furthermore, the common-mode voltage reduction has been achieved by adding only some C-code program and without any extra hardware.

ACKNOWLEDGMENT

The authors would like to thank Ulsan Metropolitan city and MOCIE and MOE of Korean Government which partly supported this research through the Network-based Automation Research Center (NARC) and the Post BK 21 Program at University of Ulsan.

REFERENCES

[1] Hong-Hee Lee, Hoang M Nguyen "A New Study on Indirect AC Current Control Method Using a Matrix Converter Fed Induction Motor", Journal of Power Electronics, KIPE, Vol. 1, pp. 67-72, 2006.

[2] Watanabe, E.; Ishii, S.; Yamamoto, E.; Hara, H.; Jun-Koo Kang; Hava, A.M., "High performance motor drive using matrix converter", Advances in Induction Motor Control (Ref. No. 2000/072), IEE Seminar , 23rd , May 2000.

[3] Bojoi, R.; Lazzari, M.; Profumo, F.; Tenconi, A, "Digital field oriented control for dual three-phase induction motor drives", 37th IAS Annual Conference of Industry Applications. Record of the IEEE 2, vol. 2, Pages:818 - 825, October 2003.

[4] Hong-Hee Lee; Nguyen, H.M.; Eui-Heon Jung, "A study on reduction of common-mode voltage in matrix converter with unity input power factor and sinusoidal input/output waveforms", 32nd Annual Conference of Industrial Electronics Society, IECON 2005, 6-10 Nov. 2005 Page(s):7 pp.

[5] M.P. Kazmierkowski, R. Krishman, F. Blaabjerg, Control in Power Electronics–Selected Problems, Academics press, ISBN 0-12-402772-5, ch. 3, 2002.

Robust Digital Controller Design for Three-Phase Inverter Using CRA Method

Jinmok Lee, Gawoo Park, Jaeho Choi
Chungbuk National University
School of Electrical and Computer Engineering
410 Sungbong-ro, Heungduk-gu
Chungbuk, Korea, 361-763
Email: choi@chungbuk.ac.kr

Gyo-Bum Chung
Hongik University
School of Electrical, Electronic & Computer Eng.
Jochiwon, Chungnam, 339-701, Korea
Email: gbchung@wow.hongik.ac.kr

Abstract— This paper proposes a CRA based robust digital controller for three-phase PWM VSI inverter. To regulate the inverter output voltage with constant voltage and constant frequency, the usual inverter controller consists of double looped PI controllers. Those are an outer voltage controller and an inner current controller, then the order of system characteristic polynomial is high and so the gain tuning of the controllers is difficult. Considering the limited switching frequency of the devices and sampling frequency of the digital controller, the gain tuning is practically dependent on the engineering experiences with the try and error method. In this paper, the error-space approach is used to get the system model including the controller with low order, and the characteristic ratio assignment (CRA) method is proposed for the design of robust controller which has the advantage to design the optimal gain to meet the referenced response and overshoot within the limit range. The validity of the proposed controller is verified through the PSiM simulation and experimental results.

I. INTRODUCTION

The PWM voltage source inverter has the operation of constant voltage and constant frequency for the application of UPS or renewable energy sourced power supply. It has been generally used for the control of PWM inverter with double feedback control loops such as the outer voltage controller and the inner current controller as shown in Fig. 1. The voltage and current controllers usually consist of PI controller but it makes the order of closed loop transfer function high and so it is very difficult to tune the gains of PI controllers [1, 2]. And what is more, if it is considered the switching frequency of the switching devices and the sampling frequency of microprocessor, it is more complicate and practically dependent on the engineering experiences with the try and error method. As one of the efficient methods for the gain tuning of PI controllers, the CDM controller was proposed [3]. It described the relation between the parameter values of the output filter and the PI controller gains with the closed form. By using this technique, it is possible to get the optimal parameter values of output filter but it was still based on the engineering experiences to design the controller with CDM controller.

As the developed method of CDM controller, the CRA (characteristic ratio assignment) method was proposed and applied for the control of power electronics converter [4, 5, 6]. It is one of the model matching techniques, and the coefficients of characteristic polynomial of the transfer function are determined by comparing the coefficients of that

of predesigned system based on CRA method. By using this method, the characteristic ratio and generalized time constant can be assigned independently, and then the coefficients of characteristic polynomial to meet the referenced damping characteristic and response time can be calculated with only two parameters of the principal characteristic ratio and the generalized time constant.

In this paper, the CRA method is introduced firstly. And then, the system model of three phase PWM inverter for the control of constant voltage and constant frequency described by using the error-space feedback control. Finally, the PSiM simulation and experimental results will be shown to verify the validity of the proposed CRA based voltage controller.

II. CHARACTERISTIC RATIO ASSIGNMENT

The characteristic ratio assignment (CRA) method is one of the model matching methods. The reference model is described not from the poles and zeros but from the characteristic ratio (CR) and generalized time constant. By using this technique, the control performance, such as the system response speed and overshoot, can be considered.

Consider a linear all pole system whose transfer function is given as Eq. (1):

$$G(s) = \frac{a_0}{\delta(s)} = \frac{a_0}{a_n s^n + a_{n-1} s^{n-1} + \cdots + a_1 s + a_0} \tag{1}$$

where all coefficients are positive.

Then it has been verified that the coefficients of characteristic polynomial of all pole system described as in Eq. (2) have the close relation with the overshoot and the

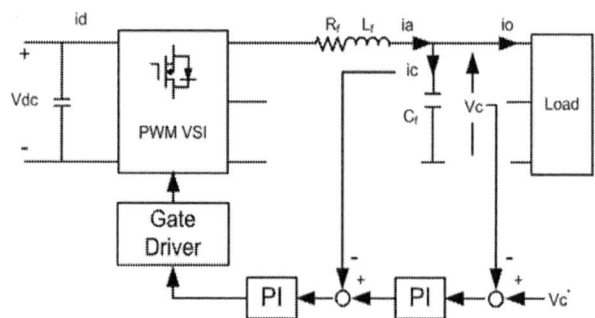

Fig. 1. Three phase inverter circuit and double looped controller.

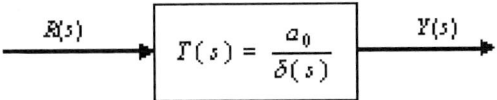

Fig. 2. All-pole systems

response speed to the step input [7].

$$\delta(s) = a_n s^n + a_{n-1} s^{n-1} + \cdots + a_1 s + a_0, \quad (a_i > 0) \tag{2}$$

Then the characteristic ratios, α_i, and the generalized time constant, τ, are defined as in Eqs. (3) and (4):

$$\alpha_1 = \frac{a_1^2}{a_0 a_2}, \alpha_2 = \frac{a_2^2}{a_1 a_3}, \cdots, \alpha_{n-1} = \frac{a_{n-1}^2}{a_{n-2} a_n} \tag{3}$$

$$\tau = \frac{a_1}{a_0}$$

(4)

where $\alpha_1 = a$ is so called as the principal characteristic ratio. And then all the coefficients of characteristic polynomial, $\delta(s)$, can be described with the characteristic ratios and the generalized time constant as shown in Eqs. (5) and (6):

$$a_1 = a_0 \tau \tag{5}$$

$$a_i = \frac{a_0 \tau^i}{\alpha_{i-1} \alpha_{i-2}^2 \alpha_{i-3}^3 \cdots \alpha_2^{i-2} \alpha_1^{i-1}} \tag{6}$$

where $i = 2, \cdots, n$.

Let's define a so-called K-polynomial whose coefficients are generated by using Eqs. (5) and (6) with arbitrarily chosen positive parameters and the characteristic ratios, obeying the following formula as shown in Eqs. (7) and (8):

(i) $\alpha_1 > 2$

(7)

(ii) $\alpha_k = \Gamma_k \cdot \alpha_1 = \dfrac{\sin\left(\dfrac{k\pi}{n}\right) + \sin\left(\dfrac{k\pi}{n}\right)}{2\sin\left(\dfrac{k\pi}{n}\right)} \cdot \alpha_1 \tag{8}$

where $k = 2, 3, \cdots, n-1$.

Then the characteristic polynomial can be described as shown in Eq. (9):

$$\delta(s, \alpha_1, \tau) = a_0 \begin{pmatrix} 1 + \tau s + \dfrac{\tau^2}{\alpha_1} s^2 + \cdots \\ + \left(\displaystyle\prod_{i=1}^{n-1} \dfrac{1}{\Gamma_{n-1}^i \alpha_1^i} \right) \tau^n s^n \end{pmatrix} \tag{9}$$

$$= a_0 + a_1 s + \cdots + a_n s^n$$

These conditions allow us to characterize the reference

all-pole system by adjusting the design parameters and to reduce the overshoot and response speed, respectively. These are so called the method of characteristic ratio assignment (CRA) [7]. The frequency magnitude response $|G(j\omega)|$ of the transfer function of all pole system with K-polynomial is gradually reduced from $|G(0)| = 1$ and always stable. From the study of Chestnut, for the transfer function which has the characteristic polynomial as shown in Eq. (9), the attenuation slope of $|G(j\omega)|$ in high frequency band is reduced as the increase of α_1, so the overshoot of step response can be reduced. The minimum damping ratio of the characteristic polynomial, $\zeta_{\min}$ is higher as the increase of α_1. These results mean that an all pole system can be developed which is stable and meets the damping characteristic and response speed by the selection of α_1 and τ with the arbitrary coefficient of a_0.

III. CONTROLLER DESIGN

A. Controller design using error-space approach for single phase inverter

If we consider the single phase equivalent circuit for the output of three phase inverter, the system equation can be described as Eqs. (10) and (11).

$$i_c' + \frac{R_f}{L_f} i_c = \frac{1}{L_f} v_a - \frac{1}{L_f} v_c - (i_o' + \frac{R_f}{L_f} i_o) \tag{10}$$

$$v_c' = \frac{1}{C_f} i_c \tag{11}$$

where R_f is the resistance value of the output filter inductor. Let $x_1 = i_c$, $x_2 = v_c$, then Eqs. (10) and (11) can be represented by using the state variables as shown in Eqs. (12) and (13).

$$x_1' = -\frac{R_f}{L_f} x_1 - \frac{1}{L_f} x_2 + \frac{1}{L_f} v_a - (i_o' + \frac{R_f}{L_f} i_o) \tag{12}$$

$$x_2' = \frac{1}{C_f} x_1 \tag{13}$$

And also, it can be described as the space-state model can be described as shown Eqs. (14) and (15).

$$\dot{x} = \begin{bmatrix} -\dfrac{R_f}{L_f} & -\dfrac{1}{L_f} \\ \dfrac{1}{C_f} & 0 \end{bmatrix} \begin{bmatrix} x_1 \\ x_2 \end{bmatrix} + \begin{bmatrix} \dfrac{1}{L_f} \\ 0 \end{bmatrix} v_a$$

$$- \begin{bmatrix} 1 \\ 0 \end{bmatrix} (i_o' + \frac{R_f}{L_f} i_o) \tag{14}$$

$$= Fx + Gu + G_1 w$$

$$y = v_c = Hx \tag{15}$$

In the above state equation, the system input, r is given as the referenced sinusoidal voltage waveform, so it satisfies the Eq. (16).

$$r^{*"} + \omega_0^2 r^* = 0 \tag{16}$$

If we define a tracking error as Eq. (17)

$$e = r - y = v_c^* - v_c \tag{17}$$

Then the error terms of the state is given as shown in Eq. (18).

$$e^{"} + \omega_0^2 e = -x^{"} - \omega_0^2 x \tag{18}$$

From the above equations, the control input error space is expressed after eliminating $r^{*"}$ by using the new state vectors, ξ and μ, in error space as followings:

$$\xi^{'} = -\frac{R_f}{L_f}\xi - \frac{1}{L_s}\mu \tag{19}$$

where $\xi = x^{"} + \omega_0^2 x$ and $\mu = u^{"} + \omega_0^2 u$ $\quad$ (20)

And then, the error-space state equation can be defined as followings by using the new state vectors, ξ and μ for the design of controller which makes the tracking error be zero to the input, r.

$$\dot{z} = Az + B\mu \tag{21}$$

where $z = \begin{bmatrix} e & e' & \xi_1 & \xi_2 \end{bmatrix}^T$ $\quad$ (22)

$$A = \begin{bmatrix} 0 & 1 & 0 & 0 \\ -\omega_o^2 & 0 & 0 & -1 \\ 0 & 0 & -\dfrac{R_f}{L_f} & -\dfrac{1}{L_f} \\ 0 & 0 & \dfrac{1}{C_f} & 0 \end{bmatrix}, \quad B = \begin{bmatrix} 0 \\ 0 \\ \dfrac{1}{L_f} \\ 0 \end{bmatrix} \tag{23}$$

If we assume that the error system {A, B} is controllable, there exists a control law of the form:

$$\mu = -\begin{bmatrix} k_1 & k_2 & k_3 & k_4 \end{bmatrix} z = -K z \tag{24}$$

It is noted that if Eq. (24) stabilized the closed-loop system, all states z becomes zero as time goes to infinity. In other words, the controller achieves perfect tracking. Substituting Eq. (24) into Eq. (21), the characteristic polynomial of the error state equation can be given as followings:

$$\Delta(s) = |sI - (A - BK)|$$

$$= s^4 + \left(\frac{R_f}{L_f} + \frac{k_3}{L_f}\right)s^3 + \frac{1}{C_f}\left(\frac{1}{L_f} + \frac{k_4}{L_f}\right) + \omega_0^2 s^2$$

$$+ \left[-\frac{k_2}{L_f C_f} + \omega_0^2\left(\frac{R_f}{L_f} + \frac{k_3}{L_f}\right)\right]s$$

$$+ \left[-\frac{k_1}{L_f C_f} + \frac{\omega_0^2}{L_f C_f} + \frac{\omega_0^2 k_4}{L_f C_f}\right] \tag{25}$$

B. Implementation of voltage and current Controller

Let us set the target polynomial for CRA based controller design.

$$\Delta^*(s) = s^4 + \delta_3 s^3 + \delta_2 s^2 + \delta_1 s + \delta_0 \tag{26}$$

For the implementation of the control law given as Eq. (24), we have to design the controller $K = \begin{bmatrix} k_1 & k_2 & k_3 & k_4 \end{bmatrix}$ to satisfy $\Delta(s) \equiv \Delta^*(s)$. Then it needs to be expressed μ in terms of e and u. Using Eqs. (20) and (24), it is described as Eq. (27) and the relation between the input and output error and state variables is given as Eq. (28).

$$\mu = u^{"} + \omega_o^2 u = -k_1 e - k_2 e^{'} - k_3 \xi_1 - k_4 \xi_2$$

$$= -k_1 e - k_2 \dot{e} - k_3(x_1^{"} + \omega_o^2 x_1) - k_4(x_2^{"} + \omega_o^2 x_2) \tag{27}$$

$$(u + k_3 x_1 + k_4 x_2)^{"} + \omega_0^2(u + k_3 x_1 + k_4 x_2) = -k_1 e - k_2 e^{'} \tag{28}$$

It can be described simply after substituting $u + k_3 x_1 + k_4 x_2$ in Eq. (28) with η to design the controller in observer form and taking a Laplace transform as Eq. (30).

$$\eta^{"} + \omega_0^2 \eta = -k_1 e - k_2 e^{'} \quad \text{where} \quad \eta = u + k_3 x \tag{29}$$

$$s^2\eta(s) + \omega_0^2\eta(s) = -(k_1 + sk_2)e(s) \tag{30}$$

and it can be simplified as Eqs. (31) and (32),

$$s\eta_2(s) = \eta_1(s) - k_2 e(s) \quad \text{where} \quad \eta_2(s) = \eta(s) \tag{31}$$

$$s\eta_1(s) = -\omega_0^2\eta_2(s) - k_1 e(s) \tag{32}$$

Taking an inverse Laplace transform to Eqs. (31) and (32), we have a state equations as followings:

$$\begin{bmatrix} \eta_1' \\ \eta_2' \end{bmatrix} = \begin{bmatrix} 0 & -\omega_0^2 \\ 1 & 0 \end{bmatrix}\begin{bmatrix} \eta_1 \\ \eta_2 \end{bmatrix} - \begin{bmatrix} k_1 \\ k_2 \end{bmatrix} e \tag{33}$$

$$\eta = \begin{bmatrix} 0 & 1 \end{bmatrix}\begin{bmatrix} \eta_1 \\ \eta_2 \end{bmatrix} \tag{34}$$

From Eqs. (33) and (34), the block diagram of designed current controller for PWM-VSI inverter using error state equation is shown in Fig. 3.

978-1-4244-1871-8/07 $25.00 © 2007 IEEE

C. Definition of controller parameter K using CRA

From Eqs. (14) and (15), the control input can be shown as Eq. (35).

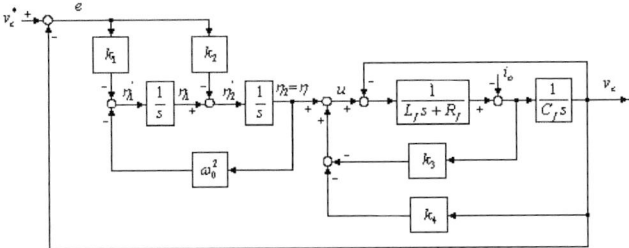

Fig. 3. Error-state controller for single phase inverter.

$$\begin{aligned}
\dot{x} &= Fx + G(\eta_2 + u_2) + G_1 w \\
&= Fx + Gu_2 + G\eta_2 + G_1 w \\
&= [F - GK_o]x + G\eta_2 + G_1 w \\
y &= Hx
\end{aligned} \tag{35}$$

Transfer function of Eq. (35) with $w = 0$ is given by Eq. (36). Consider the target polynomial for inner loop as shown Eq. (37).

$$\begin{aligned}
G_{pc} &= \frac{y(s)}{\eta_2(s)} = H[sI - (F - GK_o)]^{-1} G \\
&= \frac{\dfrac{1}{L_f C_f}}{s^2 + (\dfrac{R_f}{L_f} + \dfrac{k_3}{L_f})s + \dfrac{k_4 + 1}{L_f C_f}}
\end{aligned} \tag{36}$$

So the transfer function determined by the inner look control block is given as second order as shown in Eq. (36) and designed by two control coefficients of k_3 and k_4. Then the target polynomial to determine k_3 and k_4 is given as in Eq. (37).

$$\Delta_i^*(s) = s^2 + \delta_{i1}s + \delta_{i0} \tag{37}$$

From Eqs. (36) and (37), the relation of the characteristic ratio. α and general time constant, τ which are important to CRA based controller design is given as Eq. (38).

$$\alpha_{i1} = \frac{C_f(R_f + k_3)^2}{L_f(k_4 + 1)} \quad \text{and} \quad \tau_i = \frac{C_f(R_f + k_3)}{(k_4 + 1)} \tag{38}$$

And also if we describe the coefficients of $\Delta_i^*(s)$ in Eq. (37) with only α and τ, then

$$\delta_{i1} = \frac{\alpha_1}{\tau} \quad \text{and} \quad \delta_{i0} = \frac{\delta_{i1}}{\tau} \tag{39}$$

With given value of α_{i1}^* and τ_i^*, the controller gains of k_3 and k_4 can be determined using Eqs. (38) and (39).

$$\delta_{i1} = \frac{\alpha_{i1}^*}{\tau_i^*} = \frac{1}{L_f}(R_f + k_3) \quad \text{and} \quad k_3 = L_f \delta_{i1} - R_f \tag{40}$$

$$\delta_{i0} = \frac{\delta_{i1}}{\tau_i^*} = \frac{k_4 + 1}{L_f C_f} \quad \text{and} \quad k_4 = L_f C_f \delta_{i0} - 1 \tag{41}$$

After design the inner controller, the control gains of k_1 and k_2 also can be determined by comparing the characteristic polynomial of total closed transfer function in Eq. (25) to meet the target polynomial to meet the CRA controller in Eq. (26).

IV. SIMULATION AND EXPERIMENTAL RESULTS

The proposed CRA based digital controller for three-phase PWM VSI inverter is simulated with PSiM. Table I shows the specification of inverter system for simulation and experiment. The resistance load is used.

TABLE 1
Specification of a inverter system for simulation and experiment.

System parameter		Value
Filter		Inductance (L_f)
Reference voltage	Resistance (R_f)	3 mH
	Capacitance (C_f)	0.01Ω
	80Vpeak	100uF
DC-link voltage		300V
Resistance load		15Ω
Switching and Sampling frequency		5 kHz

Figure 4 (a) and (b) show the waveforms of a simulation and Fig. 4 (c) and (d) show those of experiment. Capacitor current and inductor current are shown in Fig. 4 (a) when load was changed from half load to full load at 0.25s. It shows that inverter system is controlled robustly without overshoot. Three-phase output voltage and capacitor current waveforms are shown in Fig. 4 (b). Output voltages and capacitor voltage waveforms are controlled well without any transient though load is changed.

Experimental voltage waveforms of inductor and capacitor are shown in Fig. 4 (c). When load is changed, the capacitor voltage is controlled well and the inductor current is changed smoothly without any overshoot. Output voltage waveforms and capacitor current waveform are shown in Fig. 4 (d). It takes 3 periods for output voltage to be regulated. The output voltage of the inverter system with LC filter is well regulated to follow the reference voltage even though the load is changed from half load to full load. The proposed algorithm is verified through the experiment. The control board has

978-1-4244-1871-8/07 $25.00 © 2007 IEEE

main DSP of TMS320VC33 and 2 isolated A/D channels and 6 non-isolated A/D channels and 7 8-bit D/A channels.

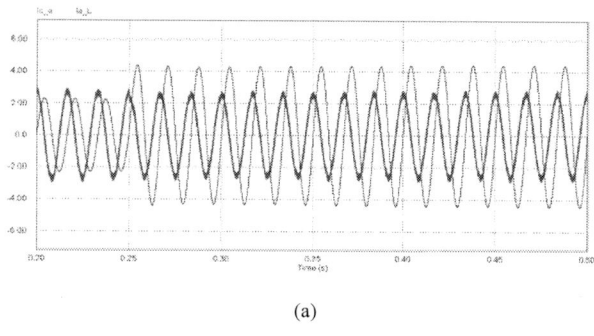

(a)

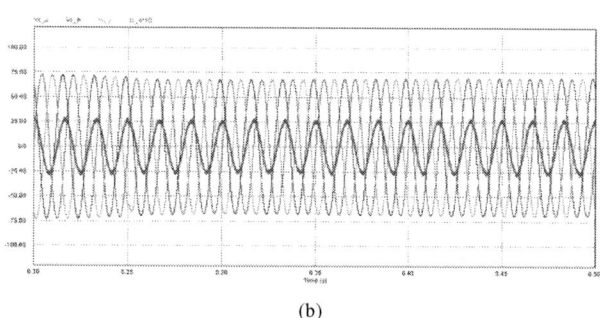

(b)

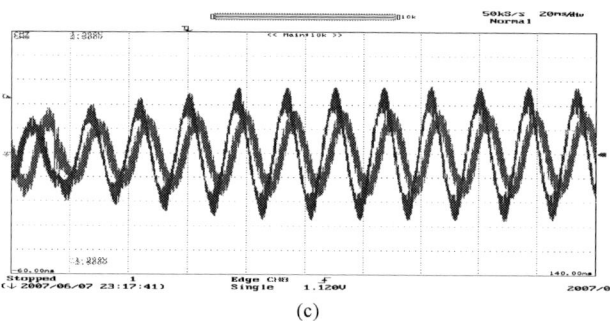

(c)

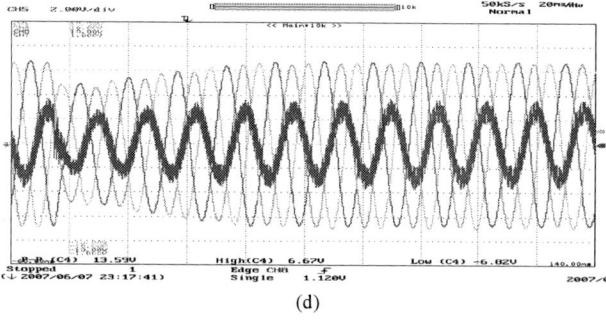

(d)

Fig. 4 (a) Simulation waveforms of capacitor current and inductor current, (b) Simulation waveforms of three-phase voltage and capacitor current, (c) Experimental waveforms of capacitor current and inductor current, (d) Experimental waveforms of three-phase voltage and capacitor current.

V. CONCLUSION

In this paper, a robust digital controller for three phase PWM voltage source inverter using CRA method is proposed. The error-space approach is used to get the system model including the

controller with low order, and CRA method is proposed for the design of robust controller which has the advantage to design the optimal gain to meet the referenced response and overshoot within the limit range. The validity of the proposed controller is well verified through the PSiM simulation and experimental results.

In the modeling of the system, the disturbance due to the load change is neglected. So the response characteristics to the capacitor input type rectifier load is not good enough and there is a little bit of voltage depression at the top of the voltage waveform. It can be improved with the feedforward compensation of the load current interruption or the conventional repetitive controller.

ACKNOWLEDGMENT

This work was supported by New & Renewable Energy RD&D Program from National RD&D Organization for Hydrogen and Fuel Cell, Korea Energy Management Corporation, and MOCIE (Ministry of Commerce, Industry and Energy).

REFERENCES

[1] M. Carpita, M. Mazzucchelli, S. Savio, and G. Sciutto, "A New Control System for UPS Using Hysteresis Comparator," *in Conf. Rec, IEEE IAS Annu, Mee.t*, pp.749-754, 1987.

[2] M. J. Ryan and R. D. Lorenz, "A high Performance Sine Wave Inverter Controller with Capacitor Current Feedback and "Back-EMF" Decoupling," *in Conf. Rec. IEEE PESC'95*, pp. 507-513, 1995.

[3] J. Kim, *Design of Output Filter and Controller for UPS Inverter*, Ph. D. Dissertation in Chungbuk National University, 2001.

[4] Y. T. Woo and Y. C. Kim, "Digital Control of a Single-Phase UPS Inverter for Robust AC-Voltage Tracking," *in Conf. Rec. IEEE IECON'2004*, vol. 2, pp. 1623-1628, 2004

[5] S. Kim, H. Kim, and J. Choi, "CRA based Robust Digital Current Controller for AC/DC PWM Converter," *in Conf. Rec. PCC-Nagoya' 2007*, DS4-2-2, 2007.

[6] J. Lee, G., Park, and J. Choi, "CRA Based Robust Digital Controller for a Single Phase UPS Inverter," *in Conf. Rec. EPE'07*, p0945, 2007

[7] P. Naslin, *Essentials of Optimal Control*, Boston Technical Publishers, Inc. 1969.

[8] Y. C. Kim, L. H. Keel, and S. P. Bhattacharyya, "Transient response control via characteristic ratio assignment," *IEEE Trans. on Automatic Control*, vol. AC-48, no. 12, pp. 2238-2244, 2003.

978-1-4244-1871-8/07 $25.00 © 2007 IEEE

Analysis and Design of PWM Inverter System
for Flywheel Energy Storage System

Jong-Chan Park, Guy-Ha Choe, Young-Sik Kim, Bayasgalan Dugarjav, Zayabaatar Baljinnyam, Jae-Pil Lee
Konkuk University
Department of Electrical Engineering,
1 Hwayang-dong, Gwangjin-gu, Seoul 143-701, Korea
Email: pjc1230@konkuk.ac.kr

Abstract— This paper presents energy input and output modeling for a flywheel energy storage system that can store and supply mechanical energy, which is emerging as one of clean energy sources, and the analysis and control of a PWM inverter system. For this, we discuss the characteristics of flywheel storage system and system modeling through PI controllers. The variations of voltage and current according to the velocity characteristic of flywheel energy storage system were simplified into numerical expressions and analyzed, and flywheel energy storage was analyzed through simulation.

I. INTRODUCTION

Recently, the remarkable increase in the use of heaters and air conditioners is raising the peak load, and to cope with this, power generation capacity is being expanded continuously for a higher power reserve ratio. However, the expansion of power generation capacity has a limitation because of the extremely high construction cost. As a effort to ease the concentration of power consumption, energy storage system based on auxiliary power systems are utilized. Particularly with increasing interest in technologies for the efficient use of energy, EMB (electro-mechanical battery), a mechanical energy storage system that converts electric energy into rotating kinetic energy and stores it using the inertia of flywheel instead of chemical storages such as storage batteries and produces electric energy if necessary, is spotlighted due to its good dynamic characteristics such as energy storage efficiency, charging time and the depth of discharge (DOD) although its initial installation cost is higher than chemical batteries [1],[2],[3].

The present study analyzed the variations of voltage and current according to the characteristic of flywheel and the velocity characteristic of flywheel energy storage system by simplifying them into numerical expressions, and analyzed the state of energy storage of flywheel through simulation. In addition, we modelled an inverter of single-phase PWM using PI controllers in order to use as an online UPS that can supply energy regardless of the condition of load-side power supply.

II. FLYWHEEL STORAGE SYSTEM

For the numerical expression of the relation between the amount of energy storage in flywheel and its velocity, we need
to divide situation into operation under steady state and operation under power source trouble. Figure 1 shows a flywheel operation circuit diagram under steady state, and Figure 2 shows a flywheel operation circuit diagram under power source trouble.

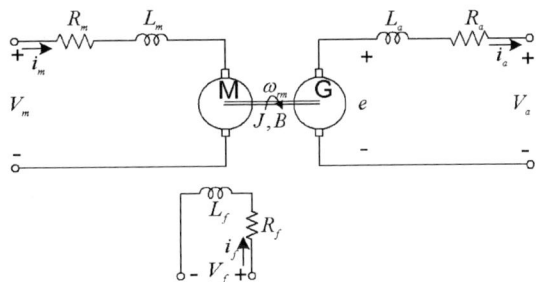

Fig. 1: Flywheel operation circuit diagram under steady state

① When the amount of energy storage under steady state is considered,

$$E_n = \frac{1}{2} j J \omega_0^2.$$ (1)

$$J = \frac{2 \cdot E_n}{\omega_0^2}$$ (2)

Considering these equations, the initial state is calculated, and assuming that output energy is constant over time and that the energy loss rate of the flywheel energy storage system is α [%] and the energy loss rate of the output device is β [%], we can derive Equation (3).

$$B_m = \frac{100 \cdot P_0}{\beta \cdot \omega_0^2} \left(\frac{100}{\alpha} - 1 \right)$$ (3)

where, J : Rotational inertia moment $[kg \cdot m^2]$
P_0 : Pure load-side output energy

978-1-4244-1871-8/07 $25.00 © 2007 IEEE

Flywheel under steady state is supplied with energy from the power source, and energy is continuously in the state of storage except energy loss in the flywheel[4]. At that time, the equivalent equations can be expressed as Equation (4)~(8).

$$T_m = J\frac{d\omega_{rm}}{dt} + B\omega_0 + T_L \tag{4}$$

$$T_m = \frac{P_{rm}}{\omega_0} = \frac{10000 \cdot P_0}{\alpha \cdot \beta \cdot \omega_0} \tag{5}$$

$$T_L = \frac{100 \cdot P_0}{\beta \cdot \omega_0} = K \cdot \Phi \cdot i_a = K_T \cdot i_a \tag{6}$$

$$i_a = \frac{P_0}{V_{out}},\ K_T = \frac{100 \cdot V_{out}}{\beta \cdot \omega_0} \tag{7}$$

$$E_a = K \cdot \Phi \cdot \omega_{rm} = K_E \cdot \omega_{rm},\ K_E\frac{E_a}{\omega_{rm}} \tag{8}$$

② Considering the amount of energy storage under power source trouble

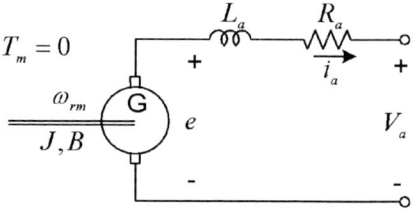

Fig. 2 : Flywheel operation circuit diagram under power source trouble

When the power source is abnormal, the power source is blocked and energy stored in the flywheel is supplied. Assuming that output energy is constant over time, the equivalent circuit can be expressed numerically as Equation (9).

$$T_m = J\frac{d\omega_{rm}}{dt} + B\omega_{rm} + T_L\ ,\quad T_{rm} = 0 \tag{9}$$

The amount of energy storage according to velocity and time can be obtained by Equation (10) and (11).

$$w_{rm}(t) = \sqrt{\frac{2C}{J \cdot \beta} \cdot e^{-\frac{2B}{J}t} - \frac{100 \cdot P_0}{B \cdot \beta}} \tag{10}$$

$$E(t) = \frac{1}{2} \cdot J \cdot w(t)^2 = \frac{C}{\beta} \cdot e^{-\frac{2B}{J}t} - \frac{50 \cdot J \cdot P_0}{B \cdot \beta} \tag{11}$$

III. PWM INVERTER SYSTEM CONTROLLER SYSTEM

As mentioned in the introduction, this study modelled a PI control system method. In general, the controllers of single-phase PWM inverters are divided into voltage controllers and current controllers[5][6]. Most inverters uses the double loop control system, which has a voltage control loop outside and a current control loop inside the voltage control loop. Figure 3 shows an inverter control system that adopts double loop control.

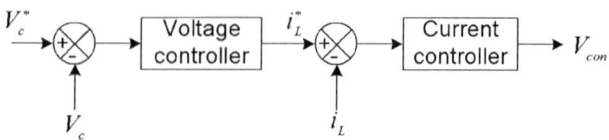

Fig. 3 : Block diagram of the inverter system with double control loop

A. Voltage Controller

The voltage controller controls using the difference between the output voltage and the reference voltage, and its output is the reference current of the current controller. In the PI controller applied to the voltage controller, the control response has phase delay due to the effect of the zero point at AC reference input, so it is difficult to calculate the optimal controller gain. In addition, P controller requires a relatively large gain in order to reduce the error under steady state, does not impair system stability, and does not have phase delay. Equation (14) shows the system model for the composition of PI voltage controller, and Equation (15) is the control rule for the voltage controller. Figure 4 is a control block diagram.

$$V_O = (V_O^* - V_O)(K_P + \frac{K_i}{s})\frac{1}{Cs} \tag{12}$$

$$\frac{V_O}{V_O^*} = \frac{K_P}{C} \cdot \frac{s + \frac{K_i}{K_P}}{s^2 + \frac{K_P}{C}s + \frac{K_i}{C}} \tag{13}$$

$$K_P = 2\zeta \cdot w_n \cdot C \tag{14}$$

$$K_i = w_n^2 \cdot C \tag{15}$$

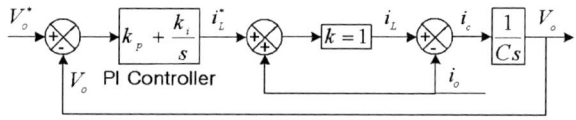

Fig. 4 : Block diagram of voltage control loop

If current control works ideally, the commanded current is the same as actual current flowing in the inductance, and from Equation (14) is derived Equation (15), a voltage controller transfer function in the form of closed loop that adopts PI controller. Accordingly, with the exclusion of the zero point, controller gain is calculated as follows by the pole placement method, a circular secondary system design method by poles. Here, ξ is the damping ratio of voltage controller, and ω_n is the non-damping sinusoidal frequency and is in the $\omega_c = \omega_n \cdot \sqrt{1-\xi^2}$ relation with the cut-off frequency(ω_c) of the voltage control system. In designing a PI controller, if the zero point of the closed loop voltage controller is sufficiently far from the origin, it is possible to approximate it to a circular secondary system.

Figure 5 is a bode diagram through voltage controller gain, in which ω_n is 1885[rad/s] and ξ is 0.707.

978-1-4244-1871-8/07 $25.00 © 2007 IEEE

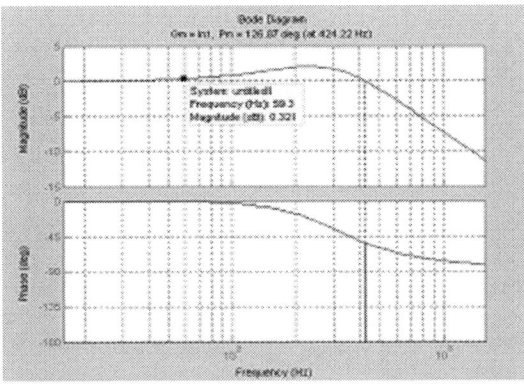

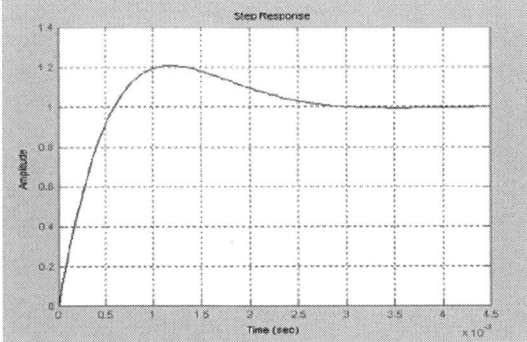

Fig. 5 : Frequency response and step response of voltage controller

B. Current Controller

In an inverter system, the current controller is positioned in the inner loop of the voltage controller and its function is to compensate the error between the reference current and feedback sensing current through the current control loop. Here, the commanded output value becomes the commanded modulated voltage of PWM inverter. When composing a double loop controller, the frequency bandwidth of the current control loop is designed over 5 times larger than that of the voltage controller so that the inner loop controller can respond more quickly than the external loop controller. Equation (16) is the system model of PI current controller, and Equation (17) is the control rule applied to the current controller. Figure 6 shows a control block diagram.

$$I_L = ([(I_L^* - I_L)(K_P + \frac{K_i}{s}) + K_{mc} \cdot V_O]\frac{V_{dc}}{V_{tri}} - V_O)\frac{1}{LS} \quad (16)$$

$$\frac{I}{I^*} = \frac{K_P}{L} \cdot \frac{s + \frac{K_i}{K_P}}{s^2 + \frac{K_P}{L}s + \frac{K_i}{L}} \quad (17)$$

By rewriting Equation (16), we can obtain a current controller transfer function as in Equation (17) in the form of closed loop that adopts PI controller. With the exclusion of the zero point, controller gain is calculated as in Equation (18~19) by the pole placement method, a circular secondary system design method by poles. Figure 7 is a bode diagram through current controller gain, in which ω_n is 7539[rad/s] and ξ is 0.707.

$$K_P = 2\zeta \cdot w_n \cdot L \quad (18)$$

$$K_i = w_n^2 \cdot L \quad (19)$$

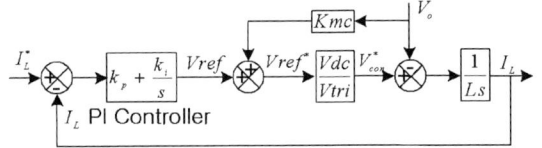

Fig. 6 : Block diagram of current control loop

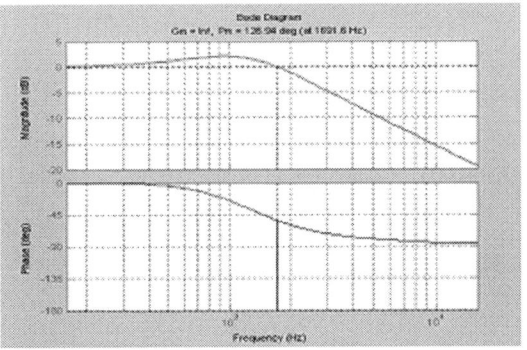

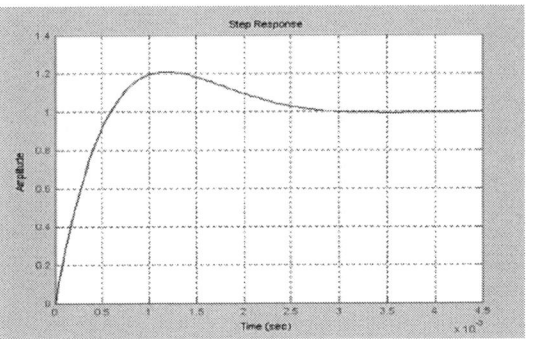

Fig. 7 : Frequency response and step response of current control loop

IV. SIMULATION

We simulated and experimented upon the steady state and overload stat of single-phase PWM inverter in order to supply AC power using a flywheel energy storage system. The system implemented in this study has switching frequency of 10[KHz]. Accordingly, the simulation was implemented and analyzed at switching frequency of 10[KHz], and the results were compared with the results of the present system.

The simulation modelled a single-phase PWM inverter using Psim, a simulation tool developed by Powersim Company. Table 1 shows parameters related to the simulation. Using these parameters, we implemented a PI controller and simulated it. Figure 8 shows the circuit diagram of the simulated single-phase PWM inverter.

The single-phase PWM inverter was simulated using the unipolar modulation method, and resistance load was used under the condition that 1[p.u] is 3[kVA] and the load varied from 0.1[p.u] to 1[p.u]. In addition, other variables such as temperature and switch characteristic were excluded from

978-1-4244-1871-8/07 $25.00 © 2007 IEEE

Flywheel under steady state is supplied with energy from the power source, and energy is continuously in the state of storage except energy loss in the flywheel[4]. At that time, the equivalent equations can be expressed as Equation (4)~(8).

$$T_m = J\frac{d\omega_{rm}}{dt} + B\omega_0 + T_L \tag{4}$$

$$T_m = \frac{P_{rm}}{\omega_0} = \frac{10000 \cdot P_0}{\alpha \cdot \beta \cdot \omega_0} \tag{5}$$

$$T_L = \frac{100 \cdot P_0}{\beta \cdot \omega_0} = K \cdot \Phi \cdot i_a = K_T \cdot i_a \tag{6}$$

$$i_a = \frac{P_0}{V_{out}}, \quad K_T = \frac{100 \cdot V_{out}}{\beta \cdot \omega_0} \tag{7}$$

$$E_a = K \cdot \Phi \cdot \omega_{rm} = K_E \cdot \omega_{rm}, K_E \frac{E_a}{\omega_{rm}} \tag{8}$$

② Considering the amount of energy storage under power source trouble

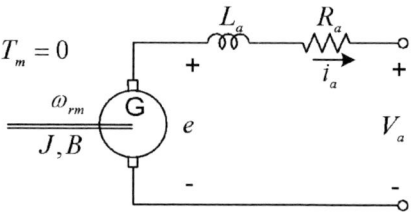

Fig. 2 : Flywheel operation circuit diagram under power source trouble

When the power source is abnormal, the power source is blocked and energy stored in the flywheel is supplied. Assuming that output energy is constant over time, the equivalent circuit can be expressed numerically as Equation (9).

$$T_m = J\frac{d\omega_{rm}}{dt} + B\omega_{rm} + T_L, \quad T_{rm} = 0 \tag{9}$$

The amount of energy storage according to velocity and time can be obtained by Equation (10) and (11).

$$w_{rm}(t) = \sqrt{\frac{2C}{J \cdot \beta} \cdot e^{-\frac{2B}{J}t} - \frac{100 \cdot P_0}{B \cdot \beta}} \tag{10}$$

$$E(t) = \frac{1}{2} \cdot J \cdot w(t)^2 = \frac{C}{\beta} \cdot e^{-\frac{2B}{J}t} - \frac{50 \cdot J \cdot P_0}{B \cdot \beta} \tag{11}$$

III. PWM INVERTER SYSTEM CONTROLLER SYSTEM

As mentioned in the introduction, this study modelled a PI control system method. In general, the controllers of single-phase PWM inverters are divided into voltage controllers and current controllers[5][6]. Most inverters uses the double loop control system, which has a voltage control loop outside and a current control loop inside the voltage control loop. Figure 3 shows an inverter control system that adopts double loop control.

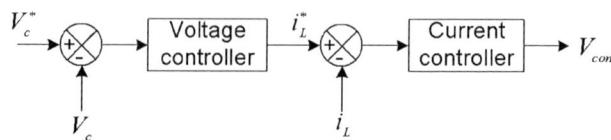

Fig. 3 : Block diagram of the inverter system with double control loop

A. Voltage Controller

The voltage controller controls using the difference between the output voltage and the reference voltage, and its output is the reference current of the current controller. In the PI controller applied to the voltage controller, the control response has phase delay due to the effect of the zero point at AC reference input, so it is difficult to calculate the optimal controller gain. In addition, P controller requires a relatively large gain in order to reduce the error under steady state, does not impair system stability, and does not have phase delay. Equation (14) shows the system model for the composition of PI voltage controller, and Equation (15) is the control rule for the voltage controller. Figure 4 is a control block diagram.

$$V_O = (V_O^* - V_O)(K_P + \frac{K_i}{s})\frac{1}{Cs} \tag{12}$$

$$\frac{V_O}{V_O^*} = \frac{K_P}{C} \cdot \frac{s + \frac{K_i}{K_P}}{s^2 + \frac{K_P}{C}s + \frac{K_i}{C}} \tag{13}$$

$$K_P = 2\zeta \cdot w_n \cdot C \tag{14}$$

$$K_i = w_n^2 \cdot C \tag{15}$$

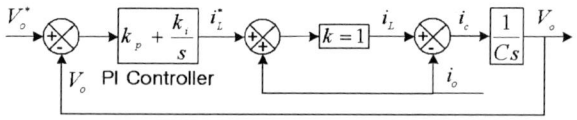

Fig. 4 : Block diagram of voltage control loop

If current control works ideally, the commanded current is the same as actual current flowing in the inductance, and from Equation (14) is derived Equation (15), a voltage controller transfer function in the form of closed loop that adopts PI controller. Accordingly, with the exclusion of the zero point, controller gain is calculated as follows by the pole placement method, a circular secondary system design method by poles. Here, ξ is the damping ratio of voltage controller, and ω_n is the non-damping sinusoidal frequency and is in the $\omega_c = \omega_n \cdot \sqrt{1-\xi^2}$ relation with the cut-off frequency(ω_c) of the voltage control system. In designing a PI controller, if the zero point of the closed loop voltage controller is sufficiently far from the origin, it is possible to approximate it to a circular secondary system.

Figure 5 is a bode diagram through voltage controller gain, in which ω_n is 1885[rad/s] and ξ is 0.707.

978-1-4244-1871-8/07 $25.00 © 2007 IEEE

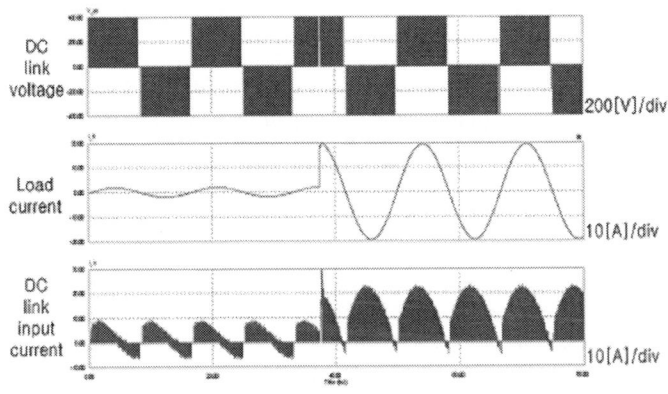

Fig. 12 : Output characteristics of 0.1~1[p.u] resistive load.

V. CONCLUSIONS

The present study analyzed the variations of voltage and current according to the velocity characteristic of flywheel energy storage system by simplifying them into numerical expressions, and analyzed the state of energy storage in flywheel through simulation. In addition, we modelled a single-phase PWM inverter to be used as an online UPS that can supply energy regardless of the condition of the load-side power source, and simulated it using the Psim tool of Powersim in order to examine its load characteristic. Based on established theories and designs, we need to make further experimental research on the operation of flywheel storage system, load output, DC link voltage variation, velocity under different load, the variation of voltage when disconnected from the system power source, etc. through linked operation of single-phase UPS inverter system and flywheel energy storage system.

REFERENCES

[1] J.G.Bitterly, "Flywheel technology past, present, and 21st Century projections", Energy Conversion Engineering Conference, Aug. 1997, pp. 2312-2315.

[2] R.S.Weissbach, G.G.Karady, R.G.Farmer, "A combined uninterruptible power supply and dynamic voltage compensator using a flywheel energy storage system", Power Delivery, IEEE Trans. Vol. 16, No.2, 2001, pp.265-270.

[3] R.G.Lawrence, K.L.Craven, G.D.Nichols, "Flywheel UPS", Industry Applications Magazine, IEEE, Vol. 9, Issue. 3, May-June. 2003, pp.44-50.

[4] Zhang Jiancheng, Huang Lipei, Chen Zhiye, "Research on flywheel energy storage system for power quality", PowerCon Proc., 2002, pp.496-499.

[5] O. Kukrer, H. Komurcugil, N.S.Bayindir, "Control strategy for single-phase UPS inverters", Electric Power Applications, IEEE Proc. Vol. 150, Issue. 6, Nov. 2003, pp.743-746.

[6] M.J.Ryan, W.E.Brumsickle, R.D.Lorenz, "Control topology option for single-phase UPS inverters" IEEE Transsaction on Industry Application, Vol. 33, Issue. 2, March-April. 1997, pp.493-50.

[7] Abdel-Rahim, N.M. "Analysis and design of a multiple feedback loop control strategy for single-phase voltage-source UPS inverters" pp.532-541, July. 1996

[8] Abdel-Rahim, N. "Multiple feedback loop control strategy for single-phase voltage-source UPS inverter", pp.958-964, June. 1994

Novel Single-state PWM technique in Multilevel Inverter for unbalanced dc sources

Nguyen Van Nho[1],
HochiminhCity University Of Technology
Department of Electrical Engineering
268 LyThuongKiet, District 10, Hochiminh City
Email: nvnho@hcmut.edu.vn

Hong Hee Lee[2], Nguyen Dinh Tuyen[3]
University Of Ulsan
Department of Electrical Engineering
680-749 San 29, Muger 2-dong, Ulsan, Korea
Email: hhlee@mail.ulsan.ac.kr

Abstract— Single-state PWM technique (SS PWM0 presents an optimised switching losses PWM scheme to be applied for multilevel inverter for possible reducing number of switchings in a sampling period. Nonlinear control characteristics and harmonic distortion factor can be significantly improved if the number of levels is high. To implement a single- state PWM control with minimised voltage error on condition of unbalance dc sources, a compensating technique is required. The paper will propose a possible solution for implementing the SS PWM with minimised voltage error, using carrier based approach. This again demonstrates the utilisation of carrier based and space vector PWM relationship in solving various optimised PWM problems in multilevel inverters.

I. INTRODUCTION

Recently, multilevel inverters have been developing very intensively and they have become a significant part in power applications. There are several basic topologies as diode clamped multilevel inverters, cascaded multilevel inverter and capacitor clamped multilevel inverters. Normally, for attaining required voltages, these power converters can be controlled using conventional carrier based PWM or space PWM methods [1]-[4]. It appears a question, if it is possible to effectively implement a PWM method with reduced number of switchings in sampling period. In practice, there exist several PWM methods, which can produce approximately the reference. The simplest case as single-state PWM, which normally requires only one commutation for three-phase in half carrier period, can obtain the output with acceptable error if the number of levels is high [5]. In another case, the switching voltage vector will be selected for purpose of satisfying several optimal conditions. The Direct Torque Control and hysteresis current controller in control of AC motor drives can be seen as different applications of single-state PWM techniques.

This paper presents a carrier based approach to implement single-state PWM method with minimised error. Comparing with recent solution [5], several new characteristics will be considered: a) solution is of carrier based PWM approach with possible common mode control, b) variation of dc sources and c) solution in whole modulation range.

For a given PWM scheme, dc voltage unbalancing or oscilating can reduce the output quality [6]. It is needed to be eliminated or compensated, particulaly for diode-clamped inverters, capacitor clamped inverters and related hybrid inverters. Recent developing of hybrid inverters enables to build very high level inverters of reduced hardware components [7]. In this case, the SS PWM method can be a prospective solution for hybrid topologies. In some case, diode clamped inverter can be used in thie hybrid inverters and compensating the influence of unbalanced dc sources by a proper SS PWM will improve the voltage output.

In the paper, the principle of proposed single state PWM can be deduced from the generalised PWM theory for multileg multilevel inverter on unbalanced dc sources [8]-[9]. The control characteristics and harmonic distortion will be calculated for several multilevel inverters. The method is explained for diode-clamped multilevel inverters (Fig.1) but its principle can be properly applied to other inverter topologies.

II. CIRCUIT ANALYSIS

The circuit model and equations of a generalized l-leg n-level inverter can be applied for three-phase n-level inverter, in which each voltage vector can be described in 3-coordinates as follows:

$$\vec{V} = [V_1, V_2, V_3]^T \qquad (1)$$

Reference voltage in single-leg inverter: $V_{ref(x)}$, $x = 1,2,3$ between inverter outputs and dc-neutral point "0", analysed as a sum of active voltage and the relevant offset as follows [8-9]:

$$\vec{V}_{ref} = \vec{V}_{12} + V_{0ref}\vec{I} ; \qquad (2)$$

where $\vec{V}_{ref} = [V_{ref(1)}, V_{ref(2)}, V_{ref(3)}]^T$; $\vec{I} = [1,1,1]^T$
and $\vec{V}_{12} = [V_{12(1)}, V_{12(2)}, V_{12(3)}]^T$. $\qquad (3)$

Define two active high $V_{H(x)}$ and low $V_{L(x)}$ voltage levels, which are closest to the reference $V_{ref(x)}$ and corresponding to levels of $L_{(x)}$ and $H_{(x)}$ on dc side (Fig.1b) as:

$$0 \le V_{L(x)} \le V_{ref(x)} \le V_{H(x)} \le V_{(n-1)} \qquad (4)$$

978-1-4244-1871-8/07 $25.00 © 2007 IEEE

The reference modulating vector $\vec{v}_{ref} = [v_{ref(1)}, v_{ref(2)}, v_{ref(3)}]^T$ can be determined as follows [3]:

$$\vec{v}_{ref} = \vec{L} + \vec{\xi}_{ref} \tag{5}$$

where $\vec{\xi}_{ref} = [\xi_{ref(1)}, \xi_{ref(2)}, \xi_{ref(3)}]^T$ is determined as

$$\vec{\xi}_{ref} = [V_{Ad}]^{-1}(\vec{V}_{ref} - \vec{V}_L) \tag{6}$$

Corresponding matrix of active dc voltage sources $V_{Ad(x)}$ is of 3x3-dimension and described as:

$$[V_{Ad}] = \begin{bmatrix} V_{Ad(1)} & 0 & 0 \\ 0 & V_{Ad(2)} & 0 \\ 0 & 0 & V_{Ad(3)} \end{bmatrix}$$

(7)

and $V_{Ad(x)}$ - active dc voltage source:

$$V_{Ad(x)} = V_{H(x)} - V_{L(x)} \; ; \; x = 1,2,3 \tag{8}$$

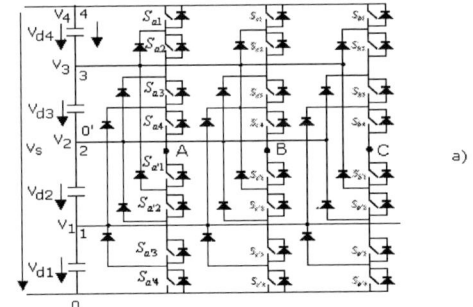

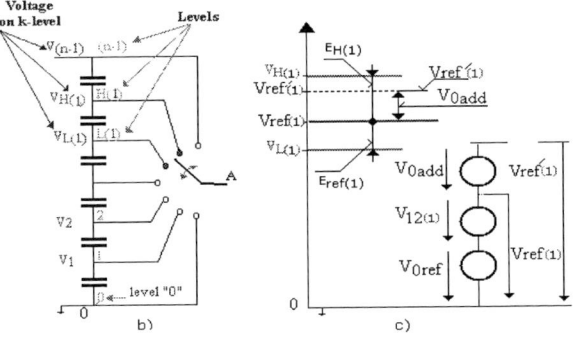

Figure 1: a) 5-level NPC inverter; b) c) and d) Analysis of inverter leg voltages

Reference vector can be implemented by 4 subsequent switching voltage vectors as follows:

$$\vec{V}_{ref} = K_1\vec{V}_1 + K_2\vec{V}_2 + K_3\vec{V}_3 + K_4\vec{V}_4$$
$$K_1 + K_2 + K_3 + K_4 = 1$$

(9)

where

$$\vec{V}_j = [V_{j(1)}, V_{j(2)}, V_{j(3)}]^T . \tag{10}$$

Switching states $\vec{S}_j$ corresponding switching voltage vectors $\vec{V}_j$ can be described as:

$$\vec{S}_j = \vec{L} + \vec{s}_j \; ; \; j = 1,2,3,4 \tag{11}$$

$$\vec{V}_j = \vec{V}_L + [V_{Ad}]\vec{s}_j \; ; \; j = 1,2,3,4 . \tag{12}$$

where $\vec{s}_j$ is nominal switching state and can be determined as below [9].

Nominal switching states $\vec{s}_j$: can be deduced as below. First, let's re-arrange nominal modulating signals $\xi_{ref(x)}$, $x = 1,2,3$ from (6) in decreasing order as:

$$1 \ge \xi_1 \ge \xi_2 \ge \xi_3 \ge 0 = \xi_4 \tag{13}$$

where $\xi_1 = Max\{\xi_{ref(1)}, \xi_{ref(2)}, \xi_{ref(3)}\}$
, $\xi_3 = Min\{\xi_{ref(1)}, \xi_{ref(2)}, \xi_{ref(3)}\}$ $j = 1,2,3$. (14)

The nominal switching state $\vec{s}_j$; $j = 1,2,3,4$ can be determined as :

$$s_{j(x)} = \begin{cases} 1 & if & \xi_{ref(x)} > \xi_j \\ 0 & else. \end{cases} \tag{15}$$

Depending on the vector diference $(\vec{V}_{ref} - \vec{V}_L)$ and active voltage $[V_{ad}]$, the states $\vec{s}_2$ can be any from vectors $[0,0,1]^T$, $[0,1,0]^T$, and $[1,0,0]^T$ and $\vec{s}_3$ can be any from vectors $[0,1,1]^T$, $[1,0,1]^T$ and $[1,1,0]^T$. It can be deduced that:

- voltage vectors to appear in (9) are depending on the reference common mode voltage

- adding an additional common mode voltage V_{0add} to reference in modified PWM techniques can change the order of commutations between phases. As result, combination $(\vec{s}_2, \vec{s}_3)$ can be some from 6 available variants described above.

III. PROPOSED SINGLE STATE PWM METHOD

To determine the vector in SS PWM method with absolutely minimum active error from the reference, it would require to consider all vectors to be involved when the offset voltage goes through entire operating range (i.e.

$V_{0Min} \leq V_0 \leq V_{0Max}$). However, if the the offset voltage is modified very large, some extra switchings may be required and they can cause more switching losses.

An appropriate modified PWM can be proposed so that the modified leg voltages will be limited in corresponding active voltage levels, i.e. $V_{L(x)} \leq V'_{xref} \leq V_{H(x)}$. There can be several approaches as:

    a. The selected vector will be considered as one from 4 related vectors in (9). This approach does not need much calculating.

    b. A more generalised criterium than case a) is that the implemented vector will be proposed as one from 8 vectors deduced from equation as

$$\vec{V}_k = \vec{V}_L + [V_{Ad}]\vec{s}_k; \ k = 0,1,2,...,7. \quad (16)$$

where

$$\vec{s}_k \in \left\{ \begin{matrix} [0,0,0]^T, [0,0,1]^T, [0,1,0]^T, [0,1,1]^T \\ [1,0,1]^T, [1,1,0]^T, [1,1,1]^T \end{matrix} \right\}$$

This approach enables to locally select vector with minimum active error while reducing the number of extra switchings to minimum.

Discussion: For balanced dc sources, the vector with minimum error in single-state PWM method can be determined from 4 vectors $\vec{V}_j$ involved in (9) and changing common mode voltage has no effect on the location of selected vector and the resulted error. This will be different for unbalanced dc sources, the vector with minimum error is needed to be determined from all 8 variants described above and the selected vector and voltage error will depend on the reference common mode voltage.

Algorithm of proposed single-state PWM method:

1. For given reference active voltage $\vec{V}_{12}$ and measuried dc sources, to select reference common mode voltage V_{0ref} within (V_{0Min}, V_{0Max}) range

2. To determine active low voltage level $\vec{V}_L$, active dc sources $V_{Ad(1)}, V_{Ad(2)}, V_{Ad(3)}$ and vector errors of reference $\vec{V}_{ref}$ from switching vector $\vec{V}_j$ as:

$$\vec{E}_j = \vec{V}_{ref} - \vec{V}_j = \vec{V}_{12} + V_{0ref}\vec{I} - \vec{V}_L - [V_{Ad}]\vec{s}'_j \ ;$$
where j=0,1,2,3..,7.   (17)

$$\vec{s}'_0 = [0,0,0]^T; \vec{s}'_1 = [0,0,1]^T;$$
$$\vec{s}'_2 = [0,1,0]^T; \vec{s}'_3 = [0,1,1]^T; ...; \vec{s}'_7 = [1,1,1]^T \quad (18)$$

3. To calculate error Ej. The selected vector is one, whose error attains minimum value:

$$E_j = |\vec{E}_j|$$
$$= k[(E_{j(1)} - 0.5E_{j(2)} - 0.5E_{j(3)})^2 + 0.75(E_{j(2)} - E_{j(3)})^2] \ ;$$
j=0,1,2,3,...,7 (16)
$$\vec{S}_{ref} = \vec{S}_j = \vec{L} + \vec{s} \ \text{ for}$$
$$E_j = Min(E_0, E_2, ..., E_7) \ ; \text{k=const.} \quad (19)$$

IV. OVERMODULATION

Let's suppose that:
- the value of reference modulation index is defined the same as for case of standard balanced dc sources.
- the voltage values are described without dimension as ratio (V / V_{dcN}), where the normal dc voltage V_{dcN} can selected as voltage value on each dc source in case of dc voltage balancing.

Because of producing output voltage with nonzero error, the single-state modulation has a non-linear control characteristic and generates low-order harmonic voltages for the whole modulation range. Compared to conventional PWM methods, overmodulation in single-state PWM method losses its original meaning. However, overmodulation can be supposed as an approach to extend the reference fundamental voltage $V_{(1)mref}$ to a maximum value of six-step mode, ie, attaining a value of $2\dfrac{V_{SMin}}{\pi}$, where V_{sMin} is a selected value not to exceed the minimum total instant dc voltage. That is, overmodulation happens if reference fundamental voltage $V_{(1)mref}$ exceeds the value of $V_{sMin}/\sqrt{3}$. The active voltages in single-state overmodulation can be deduced from principle control between two-limit trajectories. As a result, three following active voltages, corresponding to 3 limit modulation indexes of m_{ref}=1,1.05 and 1.1 can be deduced for two-mode overmodulation.

For m_{ref}=1, the active voltages are defined as

$$V_{x12} = \frac{V_{SMin}}{\sqrt{3}} \cos(\theta - \frac{2\pi k_x}{3} + \frac{2\pi}{3}); k_x = 1,2,3 \ (20).$$

For m_{ref}=1.05, the leg voltage can be described in the form as:

$$V_{x0} = \begin{cases} V_{SMin} & for & f_x > V_{SMin} \\ f_x & for & 0 \leq f_x < V_{SMin} \\ 0 & for & f_x < 0 \end{cases} \quad (21)$$

where $f_x = 0.5V_{SMin} + V_{SMin} \cos(\theta - \dfrac{2\pi x}{3} + \dfrac{2\pi}{3})$ (22);

$x = 1,2,3$.

For m_{ref}=1.1, the leg voltage is of square waveform with peak-to-peak value equal to V_{sMin}. For the last two cases,

978-1-4244-1871-8/07 $25.00 © 2007 IEEE

the active voltages can be deduced from the defined leg voltages after rejecting the zero sequence component. Principle control between two limit trajectories: for a given modulation index m_{ref}, active voltages can be deduced from the active voltages $V_{x12,mA}, V_{x12,mB}$, corresponding two defined limit modulation indexes of m_A and m_B:

$$V_{x12,m} = (1-\eta)V_{x12,mA} + \eta V_{x12,mB} \qquad (23)$$

$$\eta = \frac{m - m_A}{m_B - m_A} \qquad (24)$$

After having the active voltages, the further step similarly as the above described algorithm can be implemented for overmodulation.

V. SIMULATION AND CHARACTERISTICS

A generalised conclusion for PWM method for various changes of dc voltage sources, would be impossible. The results would be deduced case by case. Because the method is proposed for possible minimised voltage error, it can expect an improvement of output voltage compared with another conventional SS PWM method for similar number of switchings.

The control characteristics expressed as the function of amplitude of the fundamental voltage for variable reference modulation index. In the diagrams, for several discrete reference modulation indexes selected as 0.1,0.2,..,1 the fundamental voltage were calculated for 5-, 7- ang 11- level inverters and drawn as shown in Fig.2,4 and 6. For demonstration, the offset was proposed with medium common mode voltage $V_{0ref} = (V_{0Max} + V_{0Min})/2$ [2]. In the diagrams, FkCOM, presents the amplitude of fundamental voltage obtained from the proposed method, FkUNCOM presents the amplitude of fundamental voltage obtained from conventional single-state PWM without consideration of dc-source unbalance. FkIdeal presents the control characteristic of SS PWM for dc balance condition. The LinMod presents a linear control characteristic of three-state and four-state PWM methods on condition of dc source balance.

A significant advantage of the proposed method is that approximate linearity of control characteristic can be obtained in a large range of modulation index (Fig. 2,4,6,8 and 9). As a result, the reducing of harmonic content in the proposed method compared to conventional single-state PWM would be expected in most voltage range. For demonstration, the diagrams of harmonic analysis for m_{ref}=0.6 for 5-,7- and 11-level inverters were drawn for two cases of : a) SS PWM method without unbalance consideration and b) proposed SS PWM method with dc unbalance consideration. If the difference between dc sources are held low (compare Fig. 4 and 9), and particularly if the number of level is high (Fig.7), then the obtained control characteristics are nearly linear. The advantageous

control characteristic of proposed method compared to others can be obviously followed for case of small number of levels (Fig.2,4 and 8) and large difference of dc source values. For dc source unbalance, the implementing of SSPWM method can produce some even harmonics of low orders. In the studies, the proposed SS PWM can favorably contribute to reducing their amplitudes (Fig.3,5 and 7).

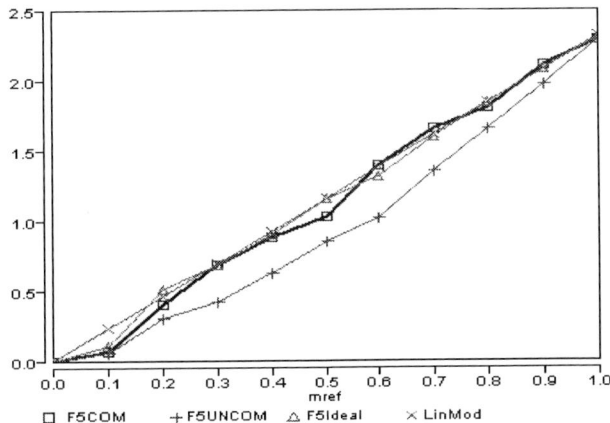

Figure 2: Five-level inverter. Diagrams of Control characteristics of Single-state PWM methods for unbalanced dc sources : V_{d1}=1.2; V_{d2}=0.6; V_{d3}=0.9; V_{d4}=1.4.

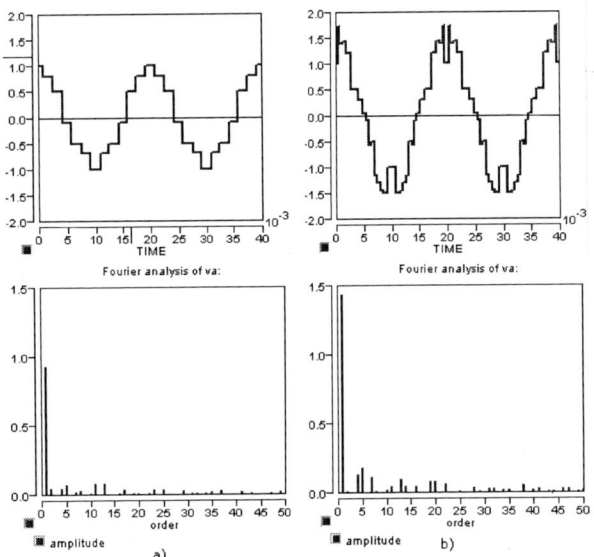

Figure 3: 5-level inverter. Influence of dc voltage unbalance on the output. Diagrams of phase load voltage and harmonic analysis in SS PWM for case a) conventional SS PWM and b) proposed SS PWM. Dc sources: V_{d1}=1.2; V_{d2}=0.6; V_{d3}=0.9; V_{d4}=1.4 ; m_{ref}=0.6.

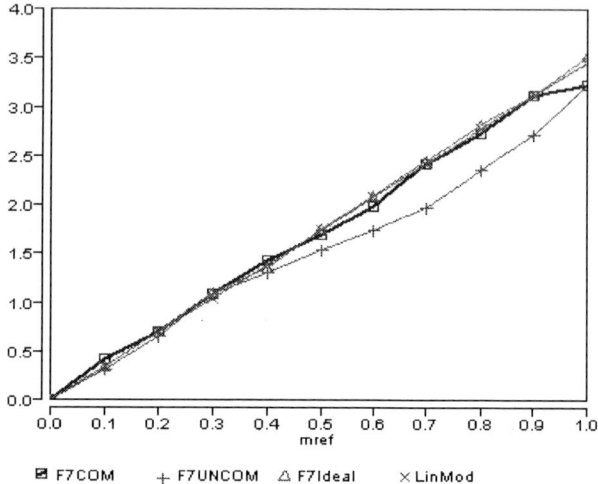

Figure 4: Seven-level inverter. Diagrams of Single-state PWM methods for unbalanced dc sources : V_{d1}=1.2; V_{d2}=0.6; V_{d3}=0.9; V_{d4}=1.1; V_{d5}=0.5; V_{d6}=1.3;

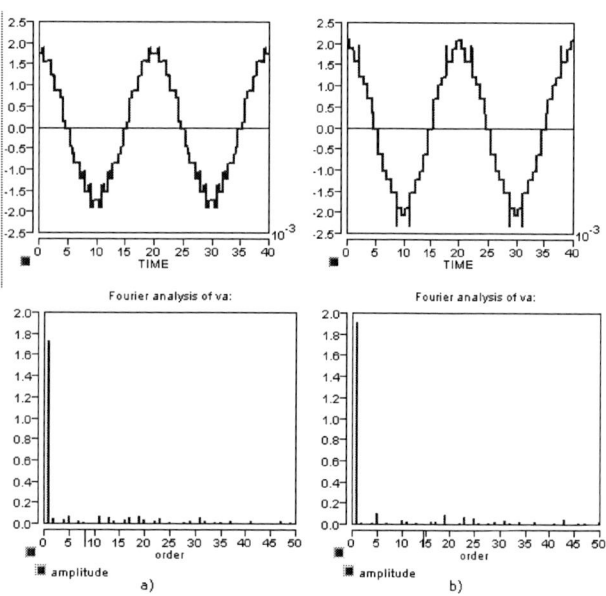

Figure 5: Seven-level inverter. Single-state PWM method for unbalanced dc sources : V_{d1}=1.2; V_{d2}=0.6; V_{d3}=0.9; V_{d4}=1.1; V_{d5}=0.5; V_{d6}=1.3. Diagrams of phase load voltage and harmonic analysis of phase load voltage for m_{ref}=0.6 for a) without dc source unbalance consideration and b) with dc source unbalance consideration.

VI. CONCLUSIONS

The paper has proposed a single-state PWM technique with minimised voltage error for multilevel inverter on condition of unbalanced dc sources. The flexible behavior of the method is presented by possible control of common mode, whose proper selection can be efficient for optimising the switching losses. The algorithm has shown the

advantages of proposed PWM method for improving the control characteristic and possible reducing the harmonics of low orders. Since the described features of output voltages were studied for open loop control, it is expected to obtain even better output performances in closed loop control.

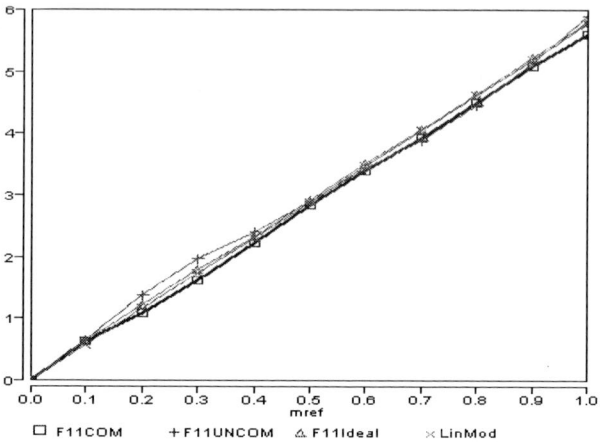

Figure 6: 11-level inverter. Diagrams of Control characteristics of Single-state PWM methods for unbalanced dc sources : V_{d1}=1.2; V_{d2}=0.6; V_{d3}=0.9; V_{d4}=1.1; V_{d5}=1.0; V_{d6}=1.3; V_{d7}=0.6; V_{d8}=1; V_{d9}=1.2; V_{d10}=1.5.

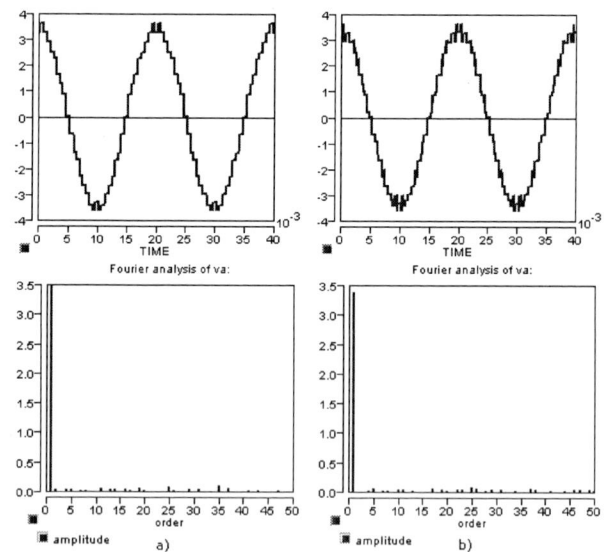

Figure 7: 11-level inverters . Single-state PWM method for unbalanced dc sources : V_{d1}=1.2; V_{d2}=0.6; V_{d3}=0.9; V_{d4}=1.1; V_{d5}=1.0; V_{d6}=1.3; V_{d7}=0.6; V_{d8}=1; V_{d9}=1.2; V_{d10}=1.5. Diagrams of phase load voltage and harmonic analysis of phase load voltage for m_{ref}=0.6 for a) without dc source unbalance consideration and b) with dc source unbalance consideration.

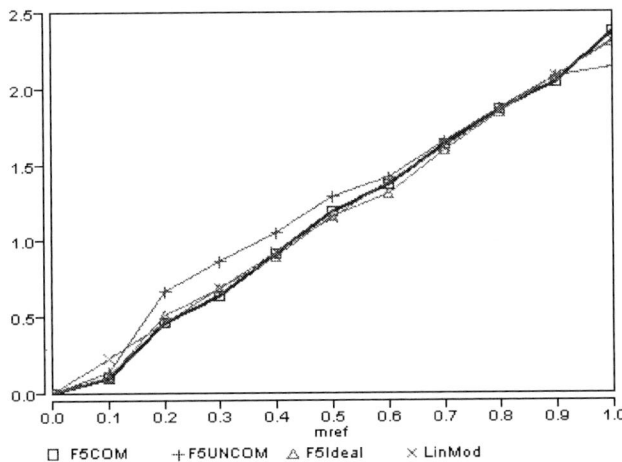

Figure 8: Diagrams of Control characteristics of Single-state PWM methods in 5-level inverters for unbalanced dc sources V_{d1}=0.8; V_{d2}=1.3; V_{d3}=0.9; V_{d4}=1.2.

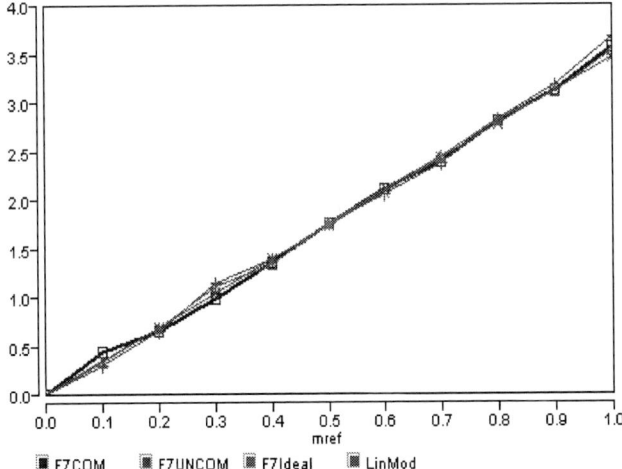

Figure 9: Diagrams of Control characteristics of Single-state PWM methods in 7-level inverters for unbalanced dc sources :V_{d1}=1.3; V_{d2}=0.9; V_{d3}=0.9; V_{d4}=1.2; V_{d5}=0.8; V_{d6}=1.2.

ACKNOWLEDGMENT

The authors would like to thank Vietnam National University-HCM for partly supported. The authors would like to also thank to Korea Ministry of Commerce, Industry and Energy and Ulsan Metropolitan City which partly supported this research through the Network-based Automation Research Center (NARC) at University of Ulsan.

REFERENCES

[1] G. Carrara, S.Gardella,M. Marchesoni,R. Salutari, and G.Sciutto, "A new multilevel PWM method- A theoretical analysis," IEEE Trans. Power Electronics, vol.7, pp.497-505 1992

[2] B. P. McGrath, D.G.Holmes," Multi-carrier PWM strategies for multilevel inverters," IEEE Trans. Industrial Electronics, vol. 49, pp.858-867, August 2002

[3] Wei, S., Wu, B., Li, F., and Liu, C.: 'A general space vector PWM control algorithm for multilevel inverters'. Proc. IEEE Conf. APEC, 2003, pp. 562–568

[4] Celanovic, N., and Boroyevich, D.: 'A fast space vector modulation algorithm for multilevel three phase converters', IEEE Trans., 2001,IA-37, pp. 637–641

[5]Jose Rodríguez, Luis Morán, Pablo Correa and Cesar Silva,"A Vector Control Technique for Medium-Voltage Multilevel Inverters', IEEE TRANSACTIONS ON INDUSTRIAL ELECTRONICS, VOL. 49, NO. 4, AUGUST 2002

[6] N. Celanovic, D. Borojevic," A comprehensive study of neutral-point voltage balancing problem in the three-level neutral-point-clamped voltage source PWM inverters" APEC '99 Proceedings , Volume: 1 , Page(s): 535 -541

[7] M.D. Manjrekar, P.K. Steimer, and T. A. Lipo," Hybrid Multilevel Power Conversion System:
A Competitive Solution for High-Power Applications", IEEE TRANSACTIONS ON INDUSTRY APPLICATIONS, VOL. 36, NO. 3, MAY/JUNE 2000,pp. 834

[8] N.V.Nho,H.H. Lee,Generalized Carrier PWM Algorithms For Multilevel Inverters With Unbalanced DC Voltages,Proceeding of the 37th IEEE Power Electronics Specialists Conference PESC 18-22nd June 2006, Jeju , Korea

[9] N.V.Nho,H.H.Lee,"Carrier PWM Algorithm For Multi-leg Multilevel Inverters', EPE 2007 - 12th European Conference on Power Electronics and Applications 2 - 5 September 2007, Aalborg, Denmark

A Novel Optimization Method for Solving Harmonic Elimination Equations

A. Samadi
School of Electrical and computer Eng.
University of Tehran
North Kargar Ave, 14395/515, Tehran, Iran
Email: a.samadi@ece.ut.ac.ir

S. Farhangi
School of Electrical and computer Eng.
University of Tehran
North Kargar Ave, 14395/515, Tehran, Iran
Email: farhangi@ut.ac.ir

Abstract—**This paper proposes an optimization method for solving nonlinear equations of both selective harmonic elimination PWM and staircase fundamental switching waveforms. The proposed method finds all set of solutions irrespective of the switching angle numbers. The proposed optimization method is based on *Mathmatica Optimization Toolbox*. The suggested method can overcome the drawbacks associated with harmonic elimination problems. In contrast to the results reported in previous works, the approach here produces all possible solutions without computational and human efforts.**

I. INTRODUCTION

The Harmonic Elimination (HE) technique, which eliminates the specified lower order harmonics, can produce higher quality output waveforms with the lower switching frequency, compared with the traditional carrier modulated PWM schemes. This technique has been extensively used for two and three level, and recently for multilevel converters [1]-[10]. The main challenge associated with the HE method is to obtain the arithmetical solutions of nonlinear transcendental equations, which naturally exhibit multiple solutions. There have been several approaches to the HE problem reported in the technical literature, including: numerical method [1], sequential homotopy-based computation [2], resultants theory [3]-[4], optimization search [5]-[6], Walsh functions [7]-[8] and optimal methods [9].

The numerical process is one of the approaches that was employed to solve HE problem. This approach relies on the proper starting points to converge. In [1] an algorithm has been proposed to approximate exact solution. The starting points with the aid of numerical method can find exact solution, even for the large number of switching angels. However, the paper reports the algorithm and discusses multiple solutions for a three phase case; but, it does not report the performance of the mentioned technique against its capability to find all possible solutions.

A systematic approach to solve the HE problem is proposed in [2]. It is based on a homotopy method, which finds the multiple solutions for a specific degree of freedom $M = m$ from those existing for $M = m-1$ sequentially, by the mathematical induction of varying a fundamental component

value as the homotopy parameter. However, this method is long and bulky.

Another approach based on the resultants theory is proposed in [3]. Specifically, the transcendental equations which specify the harmonic elimination problem are converted into polynomial equations using trigonometric identities. When the number of harmonics to be eliminated increases, the degree of polynomials becomes large, and the use of resultants theory to solve the system of polynomial equations is limited. In [4] it was proposed to apply the symmetric polynomials to reduce the degree of the polynomial equations that must be solved. However, this method finds all set of solutions, when the number of switching angels increases over than seven (finding seven switching angels has only been claimed in [4] but not presented), even with the use of the symmetric polynomials solving the polynomial equations is limited. Furthermore, the paper discusses the bipolar waveform briefly, reporting only the angles that minimize the fifth and seventh harmonics.

A minimization technique combined with random search that directly applied to transcendental equations was proposed in [5]. However, this technique finds all possible solutions, it does not seek the values for the angles that make the coefficients zero. This method tries to find those that minimize function, and with employing an iterative algorithm such as Newton-Raphson finds exact solutions, i.e. it is two-stage approach. Moreover, the paper does not treat the staircase waveform.

In this paper a new optimization method, with the aid of *Mathematica Optimization Toolbox* [10], is presented. This is a one-stage approach that can be directly applied to the transcendental equations, results in finding all possible solutions. This method finds all local minimums even those which make the coefficients zero. The new proposed method can be applied on both staircase waveform for the multilevel inverter and bipolar waveform for the traditional inverter, which is extendable to the multilevel selective harmonic elimination PWM. Results are reported for the staircase and bipolar waveform to confirm the robustness of the proposed method.

II. HARMONIC ELIMINATION WAVEFORMS

A. Fundamental Switching Staircase Waveform

978-1-4244-1871-8/07 $25.00 © 2007 IEEE

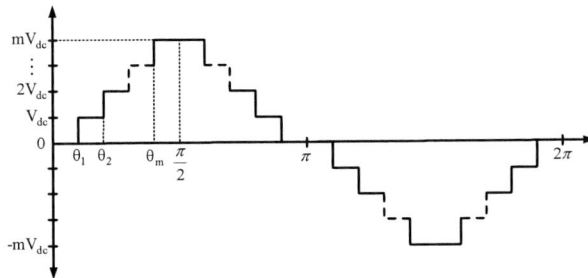

Fig. 1: Staircase waveform of the $2m+1$ level cascade multilevel.

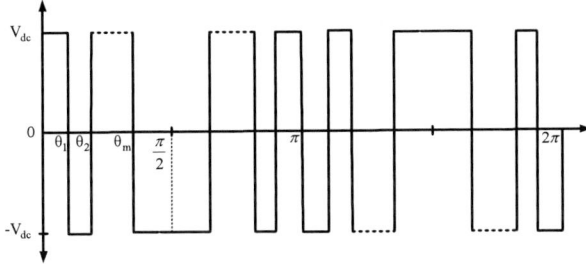

Fig. 2: Output voltage waveform of bipolar PWM.

The $2m+1$ level output voltage waveform can be generated by the fundamental frequency switching scheme as Fig. 1. Where m is the number of DC sources in a cascaded H-bridge multilevel converter. The Fourier series expansion of the output voltage waveform using fundamental frequency switching scheme is as follows

$$V(\omega t) = \sum_{n=1,3,5,\ldots}^{\infty} \frac{4V_{dc}}{n\pi}\left(\cos(n\theta_1)+\cos(n\theta_2)+\right.$$
$$\left. \ldots+\cos(n\theta_m)\right)\sin(n\omega t) \quad (1)$$

According to Fig. 1, θ_1 to θ_m must satisfy the following condition, in order to eliminate a certain number of harmonics

$$0<\theta_1<\theta_2<\ldots<\theta_m<\frac{\pi}{2} \quad (2)$$

To minimize the harmonic distortion and achieve an adjustable amplitude for the fundamental component, up to $m-1$ harmonic components can be removed from the voltage waveform by solving the following system of equations.

$$\left(\cos(\theta_1)+\cos(\theta_1)+\ldots+\cos(\theta_m)\right)-m\cdot\frac{\pi}{4}\cdot M_a=0$$
$$\left(\cos(5\theta_1)+\cos(5\theta_1)+\ldots+\cos(5\theta_m)\right)=0$$
$$\vdots \quad (3)$$
$$\left(\cos(n\theta_1)+\cos(n\theta_1)+\ldots+\cos(n\theta_m)\right)=0$$

Where M_a is defined as follows

$$M_a=\frac{V_1}{m\cdot V_{dc}} \quad (4)$$

And V_1 is the fundamental voltage harmonic.

For a three-phase application, the triplen harmonics in each phase need not to be cancelled, as they automatically cancel in the line-to-line voltages. Then n, harmonic order, will be

$$n=\begin{cases}3m-1 & m=even\\3m-2 & m=odd\end{cases} \quad (5)$$

B. Bipolar PWM Waveform

The bipolar PWM technique can be used for both single phase and three phase converters. But here we consider only three phase case. Fig .2 shows the bipolar PWM waveform. The generalized idea is to find the proper angels, θ_1 to θ_m, to eliminate $m-1$ nontriplen odd harmonics and control the fundamental harmonic component.

The Fourier series expansion of the output voltage waveform using bipolar scheme is as follows

$$V(\omega t) = \sum_{n=1,3,5,\ldots}^{\infty} \frac{4V_{dc}}{n\pi}\left[1+2\left(-\cos(n\theta_1)+\cos(n\theta_2)-\right.\right.$$
$$\left.\left.\ldots+(-1)^m\cos(n\theta_m)\right)\right]\sin(n\omega t) \quad (6)$$

Where, m is the number of switching angles and

$$0<\theta_1<\theta_2<\ldots<\theta_m<\pi/2. \quad (7)$$

In order to eliminate up to $m-1$ non triplen harmonics, and have an adjustable amplitude for fundamental component, the following system of equations must be solved.

$$\left(1+2\sum_{i=1}^{m}(-1)^i\cos(\theta_i)\right)-\frac{\pi}{4}\cdot m_a=0$$
$$1+2\sum_{i=1}^{m}(-1)^i\cos(5\theta_i)=0$$
$$\vdots$$
$$1+2\sum_{i=1}^{m}(-1)^i\cos(n\theta_i)=0 \quad (8)$$

Where n is odd integer as (5), and m_a is modulation index and defined as follows

$$m_a=\frac{V_1}{V_{dc}} \quad (9)$$

In bipolar HE case when the $-1<m_a<0$, the angels which are found cause a PWM waveform with a phase shift of 180°

against square wave.

III. PROPOSED OPTIMIZATION METHOD

As described earlier, the main problem of harmonic elimination methods is to find a procedure to solve the set of transcendental equations. Many researches have been done in this field and a series of methods has been proposed. These methods are time consuming and sometimes hard to work with them. So, presenting an approach that can abolish these disadvantages is desirable. The proposed method is an optimization one, which is based on the *Mathematica Optimization Toolbox*.

To find the desire switching angels for any value of the modulation index, the system of equations must be solved. For this purpose a cost function is defined. As example for staircase waveform the cost function is defined as follows

$$f = \left(\left(\left[\cos(\theta_1) + \cos(\theta_2) + ... + \cos(\theta_m) \right] - \pi/4 \cdot m \cdot M_a \right)^2 \right.$$
$$+ \left(\cos(5\theta_1) + \cos(5\theta_2) + ... + \cos(5\theta_m) \right)^2$$
$$+ ... \qquad (10)$$
$$+ \left. \left(\cos(n\theta_1) + \cos(n\theta_2) + ... + \cos(n\theta_m) \right)^2 \right)$$
$$\Rightarrow \quad Min(f)$$

The optimization routine is to minimize the proposed cost function and find the exact set of solutions for any value of modulation index, employing the *Mathematica Optimization Toolbox*. In the proposed method, with applying minimization command of the *Mathematica Optimization Toolbox* "*Nminimize*" the set of solutions will be achieved. This command provides four degree of freedoms, which improves the flexibility of finding all possible solutions. The first degree of freedom is the ability to select different algorithms. The second defines constraints which may be a combination of equalities, inequalities, and domain specifications. The third describes a rectangular initial region for starting points. And finally the fourth specifies the number of starting points. In this paper we do not get involve in analytical exploration of algorithms. The probation shows the effectiveness of *NelderMead* algorithm for the staircase waveform and *DifferentialEvolution* algorithm for the bipolar PWM waveform. After selecting the algorithms, the constraints should be specified as (2). The switching angels θ_1 to θ_m are the variables. The initial region for each variable is specified as follows

$$0 < \theta_i < \pi/2 \qquad (11)$$

In the following we show that the initial region can change. By specifying the number of starting points, which usually for the first run is about thirty, the process of optimization begins. The answer contains a set of global minimums of *f*. Some of these global minimums which make *f* function zero, are the exact solution of transcendental equations. To clarify how the procedure is done in *Mathematica*, the case for elimination of three harmonics is presented.

```
(teta = Table[
    NMinimize[{(Cos[x1] + Cos[x2] + Cos[x3] - 3*π/4*Ma)^2
        +(Cos[5*x1] + Cos[5*x2] + Cos[5*x3] )^2
        +(Cos[7*x1] + Cos[7*x2] + Cos[7*x3] )^2,
        0 < x1 < x2 < x3 < π/2}, {{x1, 0, π/2},
        {x2, 0, π/2}, {x3, 0, π/2}},
        Method -> {"NelderMead", "RandomSeed" -> i}],
        {i, 30}]) // MatrixForm
```

When the number of switching angels increases, applying optimization with large number of starting points is time consuming. The mentioned four degree of freedoms combined with the specific characteristics of each waveform incredibly increase the speed of solving. In the following the procedure for each waveform is described.

A. Bipolar PWM Waveform

Initial region defined in (11) is not efficient. In bipolar waveform, on entire range of modulation indices, sets of solutions numbers do not change. This is the major characteristics of bipolar waveform. Then, for one modulation index all sets of solutions with large number of starting points are calculated. Now for entire range of modulation indices, redefine initial region of each variables around the calculated solution with the respect to each set of solutions. This interval of initial region does not need to be large, for example $\pi/9$ would be sufficient. So the number of starting points is reduced to one or two.

B. Staircase Waveform

The problem in staircase waveform is more complicated as we know that the set of solutions numbers are irregularly distributed over the entire range of modulation indices. However, the same procedure as bipolar waveform can be used, but we need the sets of solutions in more than one modulation index with large number of starting points.

The staircase waveform which has the Fourier series expansion as (1), has only the positive sign in its coefficient equations. This characteristic lets us to redefine the constraint equation (2) as follows

$$0 < Min\left(\{\theta_1, \theta_2, ..., \theta_m\} \right)$$
$$Max\left(\{\theta_1, \theta_2, ..., \theta_m\} \right) < \frac{\pi}{2} \qquad (12)$$

With this technique the number of starting points is reduced. After solving the equations the switching angels are rearranged from down to up. This technique makes the random initial starting points more powerful and finds all sets of equations with less starting points.

IV. RESULTS

The optimization method has been applied successfully to both bipolar and staircase waveform described earlier in order to illustrate its robustness. It is confirmed that the

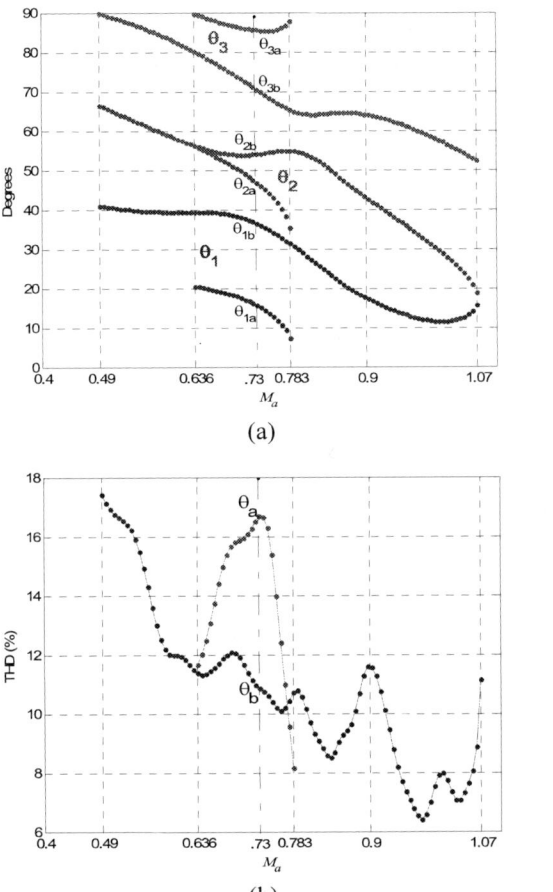

(a)

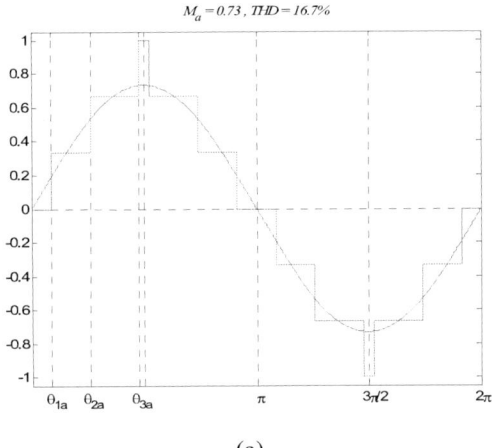

(a)

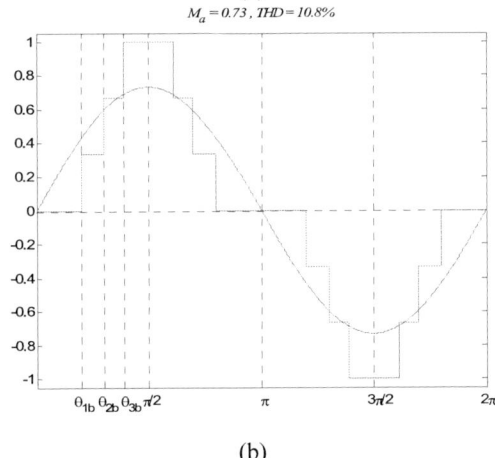

(b)

Fig. 3: Staircase waveform, 5th and 7th harmonics are eliminated. (a) Switching angels. (b) THD.

(b)

Fig. 4: Staircase waveform with three switching angels and corresponding fundamental harmonic for $m_a = 0.73$ (a) THD = 16.7%, θ_{1a}=15.6°, θ_{2a}=46.82° and θ_{3a}=85.55° (b) THD = 10.8%, θ_{1b}=36.42°, θ_{2b}=54.1° and θ_{3b}=70.5°

proposed optimization method finds all sets of solutions for each waveform.

To indicate the quality of the output waveform, the line voltage THD in per unit is defined as follows

$$THD(\%) = \frac{\sqrt{\sum_{i=1}^{7} V_{6i-1}^2 + \sum_{i=2}^{7} V_{6i-5}^2}}{V_1} \times 100 \qquad (13)$$

A. Staircase

1) Staircase waveform with 3 switching angels, 5th and 7th harmonics are eliminated:

The switching angels for the specified harmonics to be eliminated are plotted in Fig. 3(a). As can be seen the proposed method finds all sets of solutions. It is important to recognize optimum set of solutions base on their quality. For this purpose their THD are plotted in Fig. 3(b).

Fig. 4 is used to visualize how the output staircase waveform can be formed by two sets of switching angels for m_a= 0.73, marked in Fig. 3. As can be seen the both set of

switching angels (θ_{1a}, θ_{2a}, θ_{3a}) and (θ_{1b}, θ_{2b}, θ_{3b}) provide the same fundamental amplitude. But, the set (θ_{1b}, θ_{2b}, θ_{3b}) leads to the lower THD.

2) Staircase waveform with 5 switching angels, 5th, 7th, 11th and 13th harmonics are eliminated:

The switching angels which eliminate the above mentioned harmonics are plotted in Fig. 5(a). It is interesting to note that the proposed method finds all sets of solutions. To identify the best quality of solutions, the THD of solutions are plotted in Fig. 5(b).

B. Bipolar PWM waveform

1) Bipolar PWM waveform with 5 switching angels, 5th, 7th, 11th and 13th harmonics are eliminated:

The switching angels for the bipolar waveform which eliminates the above mentioned harmonics are plotted in Fig. 6(a) and 6(b), respectively. Again the proposed method finds all sets of solutions. There is two set of solutions for $0 < m_a < 1.16$ and two more for $-1.16 < m_a < 0$. To recognize optimum set of solutions with respect to harmonic contents, the THD of solutions are plotted in Fig. 6(c)

978-1-4244-1871-8/07 $25.00 © 2007 IEEE

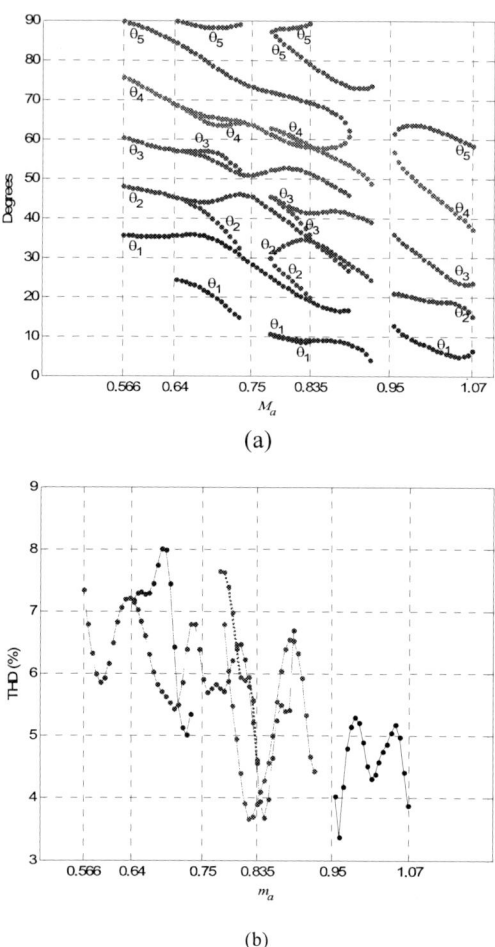

(a)

(b)

Fig. 5: Staircase waveform, 5th, 7th, 11th and 13th harmonics are eliminated. (a) Switching angles. (b) THD.

2) Bipolar PWM waveform with 7 switching angles, 5th, 7th, 11th, 13th, 17th and 19th harmonics are eliminated:

Fig. 7(a) and 7(b) illustrate the seven switching angles of mentioned bipolar waveform, respectively. There are four sets of solutions for negative values of m_a. On the other hand, there are no solutions for positive values of m_a. To specify the quality of solutions, their THD graph is plotted in Fig. 7(c).

CONCLUSION

In this paper an optimization technique for solving the equations of harmonic elimination method has been proposed, and applied successfully on both staircase and bipolar waveforms. The proposed technique rather than other optimization methods which need an extra iterative algorithm such as Netwon-Raphson, finds all sets in the one-stage. The benefit of this technique is that no information about the algorithms is needed, and further reduces the computational and human effort. The simulation results on both bipolar PWM waveform and staircase waveform, confirm the robustness of the proposed method. Results prove that there exist a set of solutions that have better quality with respect to THD.

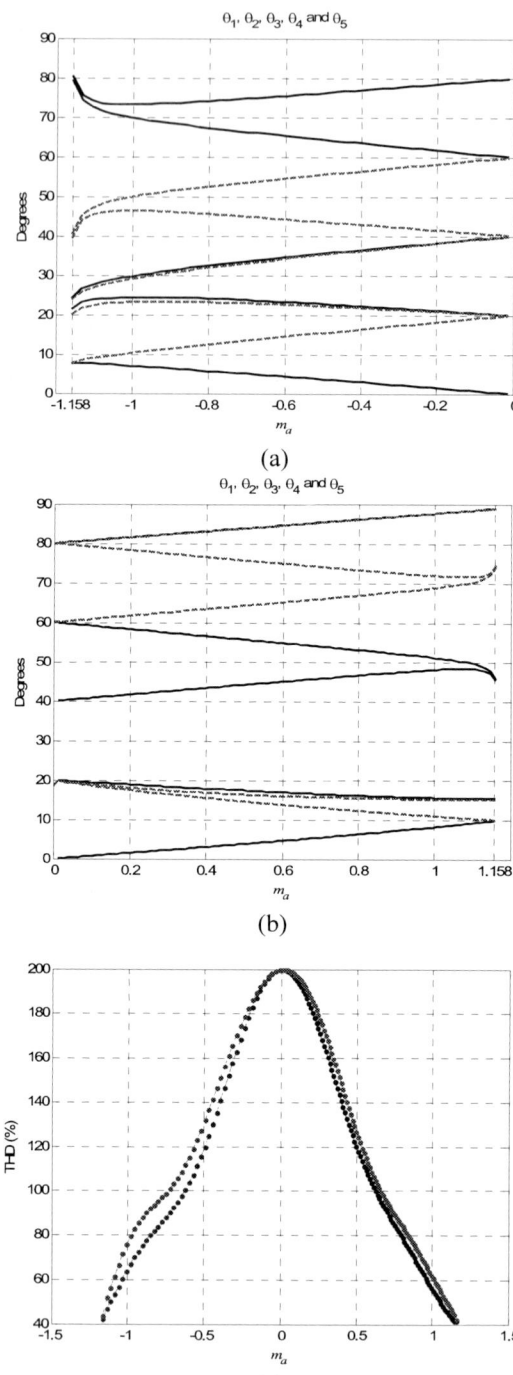

(a)

(b)

(c)

Fig. 6: Bipolar waveform, 5th, 7th, 11th and 13th harmonics are eliminated. (a) Switching angels for negative m_a. (b) Switching angels for positive m_a. (c) THD for both negative and positive m_a

V.   REFERENCES

[1]   P. Enjeti and J. F. Lindsay, "Solving nonlinear equations of harmonic elimination PWM in power control," *Electron. Lett.*, vol. 23, no. 12, pp. 656–657, June 1987.

[2]   T. Kato, "Sequential homotopy-based computation of multiple solutions for selected harmonic elimination in PWM inverters", in *IEEE Trans. Circuits and Systems* – I: Fundamental Theory and Applications, Vol. 46, No. 5, pp. 586-593, May 1999.

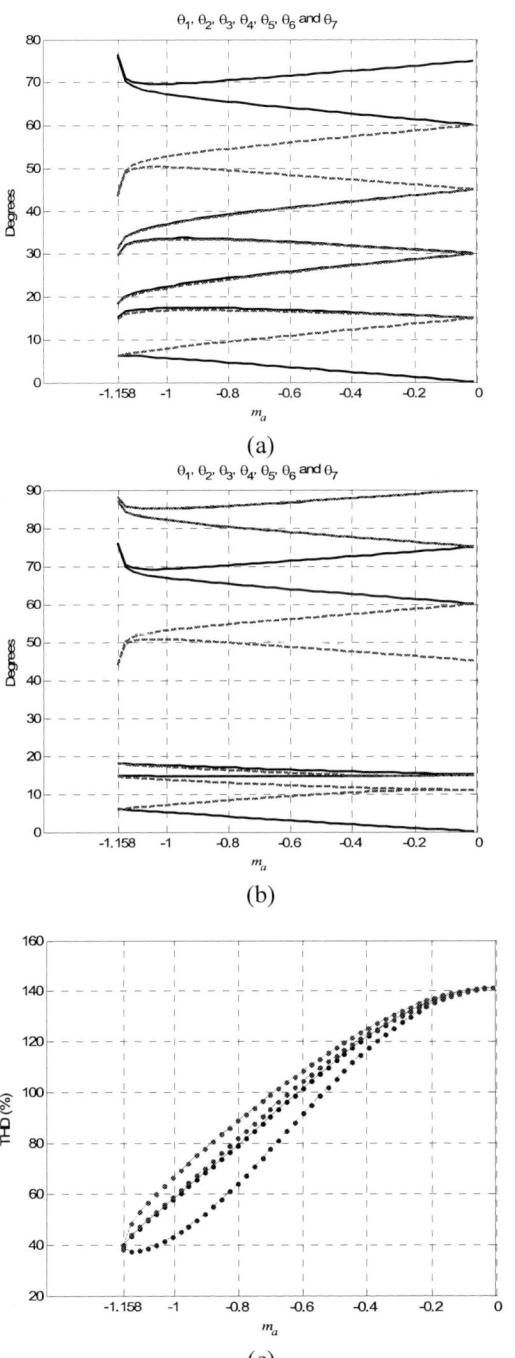

(a)

(b)

(c)

Fig. 7: Bipolar waveform, 5th, 7th, 11th, 13th, 17th and 19th harmonics are eliminated. (a) Two sets of switching angels for negative m_a. (b) Another two sets of switching angels for negative m_a. (c) THD

[3] J. Chiasson, L. M. Tolbert, K. McKenzie, and Z. Du, "A complete solution to the harmonic elimination problem," *IEEE Trans. Power Electron*, vol. 19, no. 2, pp. 491–499, Mar. 2004.

[4] J.N. Chiasson, L.M. Tolbert, K.J. McKenzie, and D. Zhong, "Elimination of harmonics in a multilevel converter using the theory of symmetric polynomials and resultants", in *IEEE Transactions on Control Systems Technology*, Vol. 13, No. 2, March 2005, pp. 216-223.

[5] V. G. Agelidis, A. Balouktsis, and I. Balouktsis, "On applying a minimization technique to the harmonic elimination PWM control: The bipolar waveform," IEEE Power Electron. Lett, vol. 2, no. 2, pp. 41–44, Jun. 2004.

[6] J. Vassallo, J. C. Clare, and P. W. Wheeler, " A Power-Equalized Harmonic-Elimination Scheme for Utility Connected Cascaded H-Bridge Multilevel Converters," *Industrial Electronics Society, 2003. IECON . 29th Annu Conf of the IEEE*, Vol.2 pp:1185- 1190.

[7] F. Swift and A. Kamberis, "A new Walsh domain technique of harmonic elimination and voltage control in pulse-width modulated inverters", in IEEE Trans. Power Electronics, Vol. 8, No. 2, pp. 170-185, April 1993

[8] T.J. Liang, R.M. O'Connell and R.G. Hoft, "Inverter harmonic reduction using Walsh function harmonic elimination method", in *IEEE Transactions on Power Electronics*, Vol. 12, No. 6, November 1997, pp. 971-982.

[9] J.R. Wells, B.M. Nee, P.L. Chapman, and P.T. Krein, "Optimal harmonic elimination control", in Conf. Proc. of *IEEE Power Electronics Specialists Conference* 2004, pp. 4214-4219.

[10] *Mathematica* 5.0, Wolfram Research Inc

Verification of Autonomous Decentralized Control UPS system using FPGA based Hardware Controller

Tsuyoshi Saito and Nobuaki Doi
TOKYO DENKI UNIVERSITY
Isizaka, Hiki-gun, Saitama, Japan, 350-0394
Email:saito@yokolab.k.dendai.ac.jp, doi@yokolab.k.dendai.ac.jp

Tomoki Yokoyama
TOKYO DENKI UNIVERSITY
Isizaka, Hiki-gun, Saitama, Japan, 350-0394
Email:yoko@k.dendai.ac.jp

Abstract— Design concept of FPGA based hardware controller with HW/SW codesign for autonomous decentralized control UPS system is proposed. Progress of FPGA technology makes it possible to include the software macro CPU core into the FPGA chip, a high flexibility can be realized for the construction of the control processor in power electronics application.

In the proposed method, all the control system is implemented in one FPGA chip. Complicated calculations are assigned to hardware calculation logic, and the parallel processing circuit makes it possible to realize minimizing the calculation time. Also Nios II CPU core is implemented in the same FPGA chip, and the software development can be applied for non-time critical calculations.The advantages of the proposed system is discussed through simulations and experiments.

I. INTRODUCTION

Various kinds of digital control methods were proposed for power electronics applications with DSP based software controller. Generally in the DSP based software controller, DSP treats the digital control schemes and ASIC or PLD treat the PWM logic control. Nowadays DSP has many features to treat power electronics applications such as PWM generation logic and other gate control logic. On the other hand, FPGA has the hardware configurable features using Hardware Description Language (HDL). The required circuit only can be built into the chip, the optimized control circuit can be realized. But if all logics with complicated control schemes are made into hardware, the total circuit will become huge, and the gate size of FPGA is limited in some cases. So it is important to balance the process sharing with software and hardware. In this paper, an autonomous decentralized control UPS system is constructed with Hardware/Software (HW/SW) codesign procedure using FPGA based hardware controller.

In the proposed system, the calculation logic for a quasi dq transformation, PLL control and gate pulse generation block were implemented in one FPGA chip. The calculation for a droop control is implemented in software for Nios II CPU. Therefore each operation are processed by hardware and software, the optimal control system can be realized. The design concept of the HW/SW codesign for the proposed method is discussed and verified through simulations and experiments.

II. SYSTEM MODELING AND CONTROL METHOD

Fig. 1 shows a system configuration of parallel-connected UPS system , and Fig. 2 shows a basic parallel-connected UPS system where the voltage v_i is the output voltage for the ith UPS unit as $v_i = E_i \cos(\omega t + \phi_i)$. Each UPS is controlled independently with the same control law, and it is assumed that each UPS can observe only its output voltage and current.

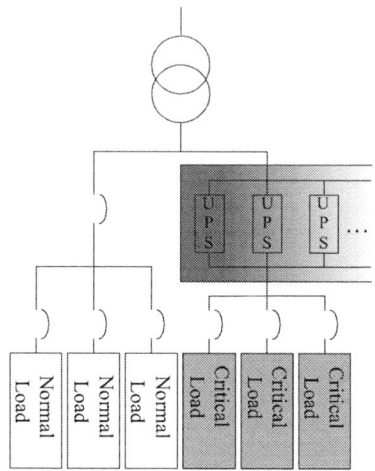

Fig. 1. Autonomous Decentralized Control UPS system

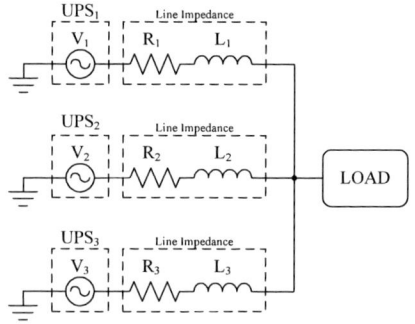

Fig. 2. Model of parallel-connected single-phase UPS system

A. Droop Control

The phase and the voltage amplitude are controlled based on a droop characteristics in this system. Control factor for the output voltage are the phase ϕ_i and the voltage amplitude E_i. The reference of the phase $\phi_i{}^*$ and the reference of the voltage amplitude $E_i{}^*$ are determined according to the active power of ith UPS P_i and the reactive power of ith UPS Q_i. Fig.3 shows how to decide the phase $\phi_i{}^*$ and the voltage amplitude $E_i{}^*$ based on the droop characteristics. The active element and the reactive element of the output voltage, and the active element and the reactive element of the current output can be detected by quasi dq transformation [1].

$$\phi_i^* = \phi_0 - m_i (P_{0i} - P_i) \tag{1}$$

$$E_i^* = E_0 - n_i (Q_{0i} - Q_i) \tag{2}$$

978-1-4244-1871-8/07 $25.00 © 2007 IEEE   186

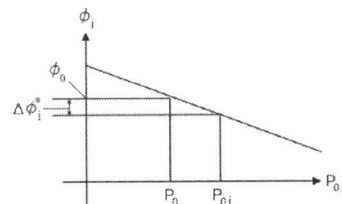

Fig. 3.   Droop characteristics

E_i^* : reference of voltage amplitude
E_0 : nominal voltage amplitude
ϕ_i^* : reference of phase
ϕ_0 : nominal phase
P_{0i} : rated active power for ith UPS
Q_{0i} : rated reactive power for ith UPS
n_i, m_i : droop characteristics gain
Vd_{0i} : d phase voltage for ith UPS
Id_{0i} : d phase current for ith UPS
Iq_{0i} : q phase current for ith UPS

Equation (1) and (2) indicate how to decide the phase and the voltage amplitude from the active power and the reactive power. Differentiate the rated active power of the ith UPS P_{0i} and the active power of the ith UPS P_i, then multiplying by the droop characteristics gain m_i, subtract from the nominal phase ϕ_0, desired phase reference ϕ_i^* is derived. Also differentiate the rated reactive power of the ith UPS Q_{0i} and the reactive power of the ith UPS Q_i, then multiplying by the droop characteristics gain n_i, subtract from the nominal voltage amplitude E_0, desired voltage amplitude reference E_i^* is derived.

B. Quasi dq Transformation

Quasi dq transformation was used to detect an instantaneous phase and an amplitude of a single phase voltage. Target waveform V_s is assumed as the normal line voltage, and is sampled in every sampling period T_s, so the present sampled data $Vs(k)$ can be described as equation (3). Also pre-sampled data $Vs(k-1)$ and $Vs(k-2)$ can be expressed as (4) and (5). Here, V_L is the maximum peak value of Vs, and $V_s(k-1)$ is defined as V_{re} [8].

$$V_s(k) = V_L \cos(\omega t) \qquad (3)$$
$$V_s(k-1) = V_L \cos\{\omega(t - T_s)\} = V_{re} \qquad (4)$$
$$V_s(k-2) = V_L \cos\{\omega(t - 2T_s)\} \qquad (5)$$

Differentiate (3) and (5), adapting the formula of trigonometric function, (6) can be derived.

$$
\begin{aligned}
\frac{V_s(k-2) - V_s(k)}{2T_s} &= \frac{V_L}{2T_s}[\cos\{\omega(t - 2T_s)\} - \cos\omega t] \\
&= \frac{V_L}{2T_s}[-2\sin(-T_s)\sin\{\omega(t - T_s)\}] \\
&\cong V_L \sin\{\omega(t - T_s)\} = V_{im} \qquad (6)
\end{aligned}
$$

Here, it is obvious that the phase of V_{im} delays $90°$ than the phase of V_{re}, so using only three sampled data of the single phase waveform, V_{re} and V_{im} form the rectangular coordinates system, and dq transformation can be adopted to the single phase system. Therefore, the phase and the voltage amplitude can be detected from three sampling data using quasi dq transformation. In order to reduce the influence of

the noise, three kind of sampling frequencies are adopted for the implementation, which are 20[kHz], 10[kHz] and 6.7[kHz] respectively. Three parallel calculation circuit blocks are prepared, each block is driven by three different sampling frequencies, averaging the calculated values of every circuit to derive dq elements of the target waveform.

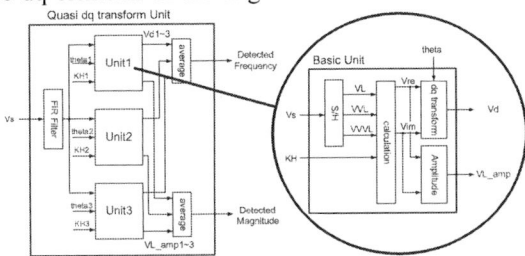

Fig. 4.   Quasi dq Transformation

C. Control Block Diagram for Single Phase UPS system

Fig.5 shows the proposed control block diagram of the single phase UPS system. ϕ_{V_i} is the phase of the output voltage for the ith UPS. Single phase UPS is controlled by the droop control and the PLL control based on the quasi dq transformation [1]. It is assumed that each UPS can observe only its output voltage and current. The phase and the voltage amplitude can be detected by the quasi dq transformation. The operation frequency is detected by the PLL control using the phase ϕ_{V_i} of the ith UPS output voltage.

Also, the active power of the ith UPS can be obtained by multiplying the voltage amplitude and the current amplitude, and the reactive power of the ith UPS can be obtained by multiplying the voltage amplitude and the phase of the output current, which is nominalized to calculate the reactive power. The phase reference ϕ_i^* and the voltage amplitude reference E_i^* can be decided by applying the phase and the voltage amplitude control based on the droop characteristics using the active power P_{0i} and the reactive power Q_{0i}. Therefore the voltage of each UPS can be controlled and each UPS supplies a balanced power to the load.

III. SIMULATION

Simulations were carried out for three kinds of load conditions, which were R–L load, rectifier load and R–L load with parallel-connected capacitor. The parameters for the simulations are listed in Table.I.

TABLE I

SIMULATION CONDITIONS

rated active power of ith UPS P_{0i}	2000.0 [W]
rated reactive power of ith UPS Q_{0i}	1000.0 [Var]
nominal frequency f	50.0 [Hz]
sampling frequency	20.0 [kHz]
reference of voltage amplitude E_{0i}	$100\sqrt{2}$ [V]
reference of phase difference ϕ_{0i}	$2\pi f$ [rad]
maximum impedance R	0.03+j0.004 [Ω]
droop characteristic gain m_{0i}	-0.000525
droop characteristic gain n_{0i}	-0.0025

TABLE II

LOAD PARAMETERS

rectifier load R	5.0 [Ω]
rectifier load L	0.5 [mH]
rectifier load C	1.0 [mH]
variation range for voltage amplitude	±3 [%]
variation range for phase angle	±1 [%]

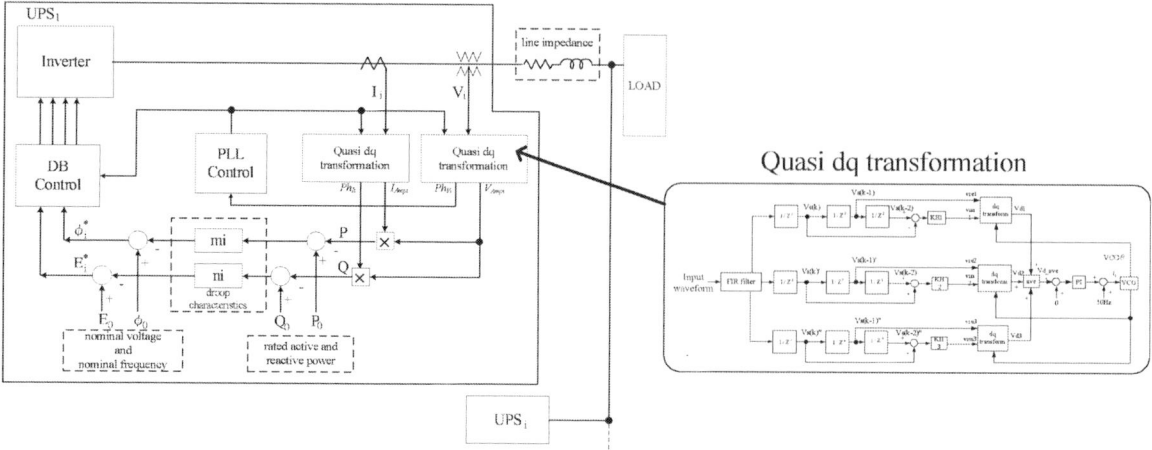

Fig. 5. Control block diagram for one single phase UPS system

1) Load Variation: The capacity of the rectifier load is changed at 0.5 seconds with three parallel-connected UPSs, each UPS is defined as an equal rating of the output power. Fig.6 shows the active power and the reactive power for the rectifier load of each UPS. From this figure, it is confirmed that the active power and the reactive power of the load is supplied equally balanced from the three UPSs while the load is changed.

2) Variation of connected UPSs: The stability of the single phase UPS system is verified when the number of connected UPS with an equal rating of the output power is changed from 3 to 2 in 0.2 seconds and 2 to 3 in 0.4 seconds.

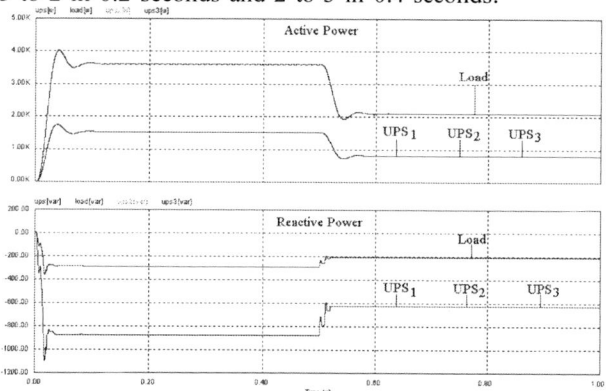

Fig. 6. Active power and reactive power of the rectifier load and UPSs

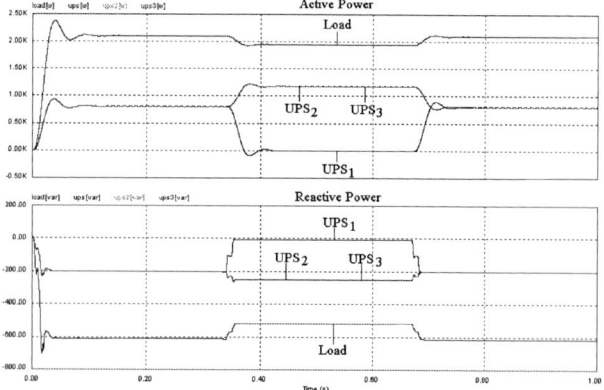

Fig. 7. Active power and reactive power of the rectifier load and UPSs

Fig.7 shows the simulation result. From this result, the active power and the reactive power of the load is supplied equally from the connected UPSs when the number of connected UPS is changed.

IV. EXPERIMENT

A. System Design

The parameters for the experimental setup are listed in Table.III. The proposed single phase UPS system was implemented using FPGA with a 32bit Nios II processor[11] as shown in Fig.8. StratixII EP2S60(ALTERA Corp.) is applied in this system. The Nios II CPU has a pipelined RISC architecture and reconfigurable features. In the experiments of the proposed single phase UPS system, a quasi dq transformation, PLL control and gate pulse generation block were implemented in FPGA logic. The operations such as droop control were implemented in Nios II CPU software.

TABLE III
EXPERIMENT CONDITIONS

Controller Board	Altera NiosII Development Kit
FPGA Device	Altera Stratix2 2S60
CPU Core	Altera NiosII CPU
rated active power of ith UPS P_{0i}	125.0 $[W]$
rated reactive power of ith UPS Q_{0i}	62.5 $[Var]$
nominal frequency f	50.0 $[Hz]$
sampling frequency	20.0 $[kHz]$
reference of voltage amplitude E_{0i}	$25\sqrt{2}$ $[V]$
reference of phase difference ϕ_{0i}	$2\pi f$ $[rad]$

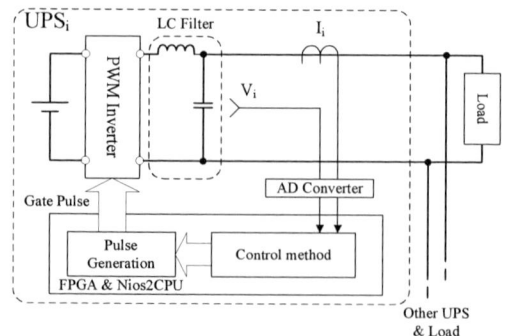

Fig. 8. FPGA Based Hardware Controller System Configuration

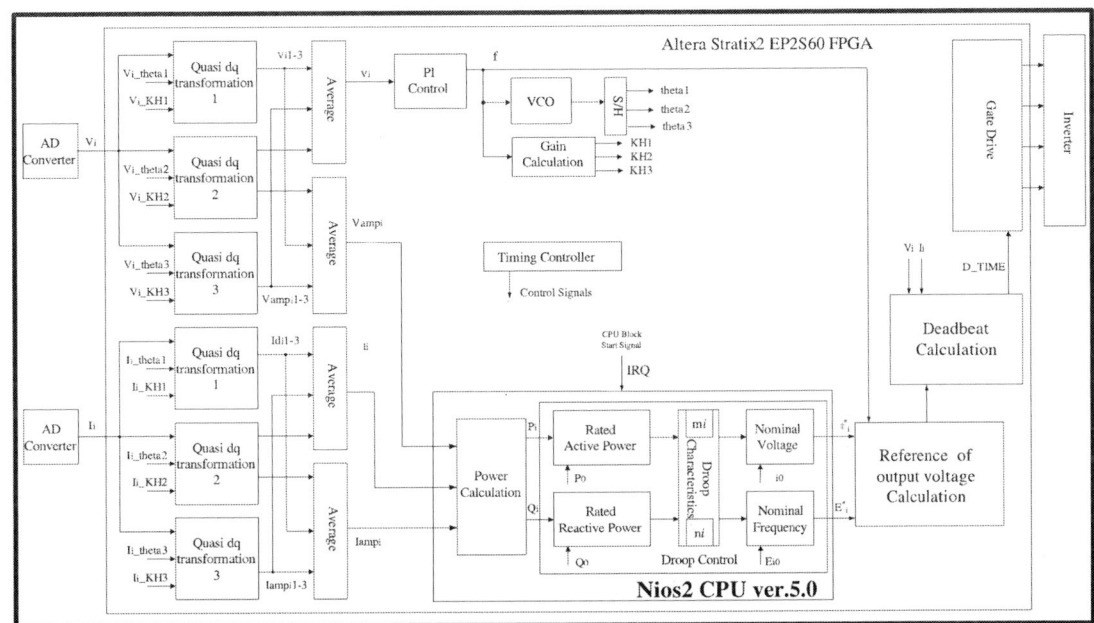

Fig. 9. Module block diagram of FPGA controller

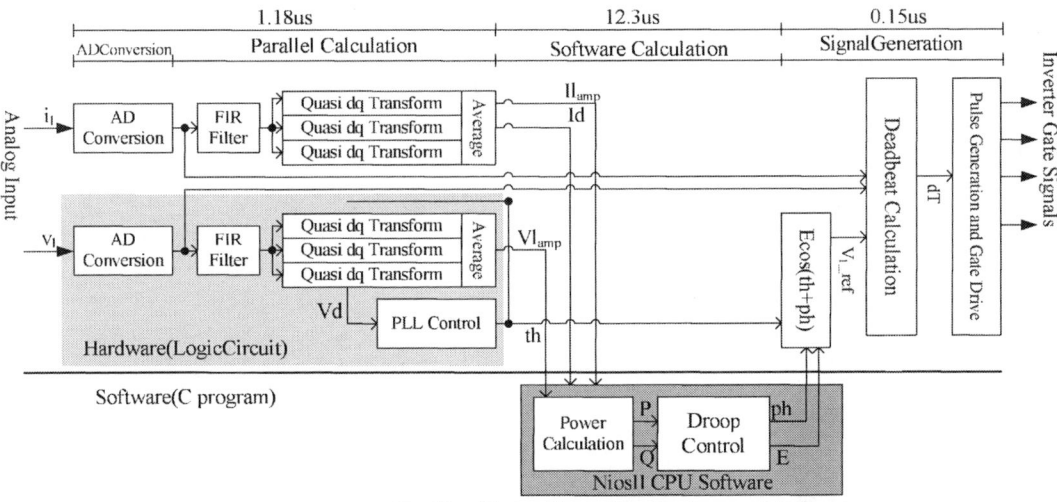

Fig. 10. Timing Chart of the Hardware and Software

B. Experimental System

Fig.9 shows the module block diagram of the proposed single phase UPS system in the FPGA with a 32bit Nios II processor. The FPGA blocks are constructed with quasi dq transformation block, PI control block and VCO block. In the Nios II CPU blocks, power calculation block, droop control block and reference of output voltage calculation block are implemented. Two quasi dq transformation blocks are constructed in parallel, which calculates in the different sampling frequency, and the phase difference and the amplitude are calculated for every control stage. In FPGA, the phase difference ϕ_{V_i}, ϕ_{I_i} and the voltage amplitude V_{amp_i}, I_{amp_i} can be calculated by the quasi dq transformation using the ith UPS output voltage and the ith UPS output current. Also, the operation frequency f is detected by the PLL control using the phase difference of the ith UPS output voltage ϕ_{V_i}. These values (V_{amp_i}, I_{amp_i}, ϕ_{I_i}, f) are sent to Nios II CPU. In Nios

II CPU, the reference of the output voltage is decided by the droop control.

Fig.10 shows the timing chart of the FPGA based hardware controller. AD conversion and hardware calculation is finished in 1.18us(73clocks), the software calculation is finished in 12.3us, and all the calculation is finished in 13.6us.

Stratix II (EP2S60) is used for the FPGA controller, which has 60,440LE (logic elements), and the 20% of LE is used for the proposed control logic including Nios II CPU core.

Fig.11 shows the RTL circuit diagram of the proposed method which indicates the circuit configuration of the FPGA chip. Fig.12 shows the FPGA based hardware controller, and Fig.13 shows the experimental system for parallel connected UPS. Fig.14 shows the output voltage and current for one UPS unit. Fig.15 shows the output voltage and current waveforms when the 2 UPSs are connected in parallel. UPS2 is connected in time 3.2 second, then the current is balanced between 2 UPSs. Also it is verified that the output power of the 2 UPSs are also balanced.

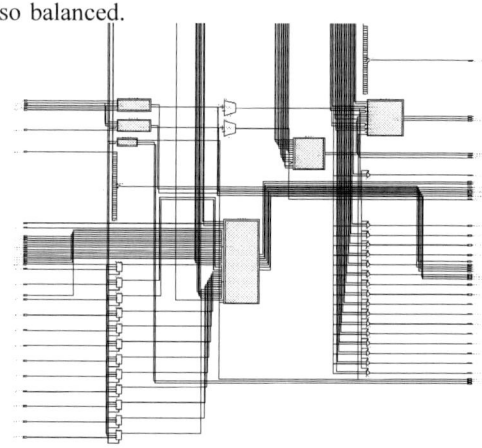

Fig. 11. RTL view

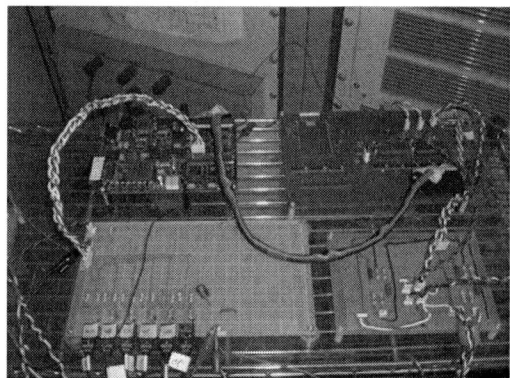

Fig. 12. Controller circuit

Fig. 13. Load system

V. CONCLUSION

In this paper, an autonomous decentralized control system for single phase UPS inverter with FPGA based hardware controller using software CPU core is proposed. By applying the proposed method in parallel-connected single phase UPS system, each UPS shares the power of the load equally when the power of the load changed or the number of the connected UPS is changed, an autonomous decentralized control for single phase UPS system can be realized. Constructing the control system using FPGA, and HW/SW codesign procedure is applied to realize the optimal control system.

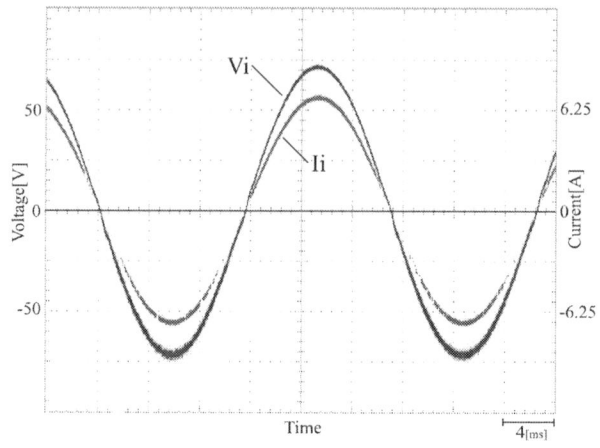

Fig. 14. Experimental result (1UPS)

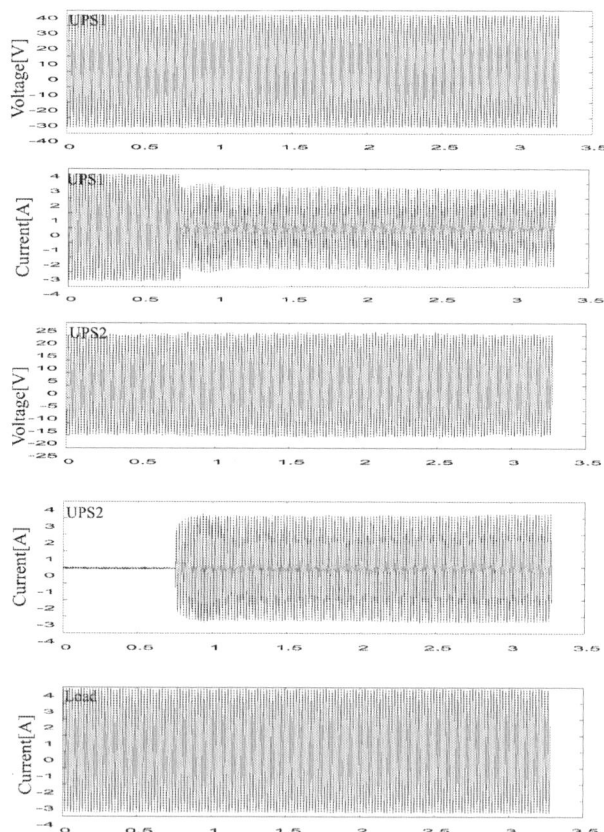

Fig. 15. Waveforms for parallel connected operation for two UPSs

REFERENCES

[1] T.Komiyama, T.Kojima, T.Yokoyama : "Implementation of High-speed Frequency Detection for Single Phase Utility Interactive Inverter System" EPE-PEMC' 04

[2] T.Yokoyama, S.Simogata, M.Horiuchi, T.Ide : "Instantaneous Deadbeat Control for Single Phase and Three Phase PWM Inverter Using FPGA Based Hardware Controller" EPE-PEMC' 04

[3] H.Takeda, T.Yokoyama : "A Study of Autonomous Decentralized Control for Single Phase UPS System based on quasi dq transformation" PESC' 04

[4] M.Horiuchi, T.Fujii, T.Yokoyama : "Comparison of Multirate Deadbeat Control for Single Phase PWM Inverter" IPEC' 05

[5] E.Shimada, K.Aoki, T.Komiyama, T.Yokoyama : "Implementation of Deadbeat Control for Single Phase Utility Interactive Inverter Using FPGA" IPEC' 05

978-1-4244-1871-8/07 $25.00 © 2007 IEEE

[6] T.Fujii, T.Yokoyama : "FPGA based Multi-Sampling Pulse Compensation Method with Deadbeat Control for Single Phase PWM Inverter" EPE-PEMC' 05

[7] E.Shimada, K.Aoki, T.Komiyama, T.Yokoyama : "Implementation of Deadbeat Control for Single Phase Utility Interactive Inverter Using FPGA based Hardware Controller" EPE-PEMC' 05

[8] A.Takeuchi, S.Kondo : "Quick Voltage Amplitude Control on Parallel Operation of Single Phase Inverters" 2002 National Convention Record I.E.E of Japan p122 (in Japanese)

[9] W.Fujii, T.Yokoyama : "Construction of FPGA based hardware controller for autonomous decentralized control for UPS application" EPE-PEMC 2006

[10] Myway labs -PSIM- http://www.myway-labs.co.jp/psim/

[11] ALTERA http://www.altera.com/

Novel Dual-Band Controller for Regulating DC-Bus Voltage That Contains A Large Voltage Ripple

Y.Wang, S.W.H. de Haan
Delft University of Technology
Department of Electrical Engineering, EPP Group
Mekelweg 4, 2628CD, Delft, the Netherlands
Email: yi.wang@tudelft.nl
s.w.h.dehaan@tudelft.nl

A. van Zwam
Mastervolt BV international
Snijdersbergweg 93, Amsterdam, the Netherlands
Email: avanzwam@mastervolt.com

Abstract—Autonomous system with DC-AC converter as an interface from the battery source to the inverter load easily suffers from voltage ripples on the DC-bus at the twice load frequency (100 Hz). A fast voltage controller for tightly regulating the DC-bus voltage with the inevitable 100 Hz ripples bears a rapid transient response, however always contaminating battery current with large ripples at 100 Hz, which do harm to the battery life span.

In the paper, an average current mode controller with a novel dual-band selector is introduced. This controller has two voltage-loop branches with 5 Hz and 641 Hz bandwidths, for smoothing the battery current in the steady state and regulating the deviating bus voltage in the transients, respectively. Dual-band selector determines how these two voltage-loop branches are switched to each other, aimed at attenuating transients as soon as possible.

This control methodology is demonstrated by simulation for an interleaved bi-directional flyback converter that powers the DC-bus from a battery.

I. INTRODUCTION

In the marine applications where the utility grid is not permanently available, the storage battery is quite frequently utilized as the back-up power source when sailing. To bi-directionally interface the AC-fed domestic appliances on board and utility grid with the storage battery, an interleaved flyback converter cascaded by a dc-ac inverter both with bi-directional power transfer ability are designed. Six flyback units are evenly phase-shifted in each switching cycle to boost the battery voltage for the AC loads with galvanic isolation guaranteed, meanwhile decreasing the passive component volume and increasing the power efficiency. The desired nominal power for each unit is 600 W at which all the flyback units are designed to work in discontinuous conduction mode (DCM). Fig. 1 illustrates such a system structure that includes the storage battery, the interleaved flyback converter and the inverter.

It is well-known that the inverter load will inherently generate the current ripple of double grid frequency (100 Hz) through the DC-bus capacitors at the input of the inverter, resulting in 100 Hz voltage ripple riding on the DC-bus. If the DC-bus voltage is tightly regulated, a large current ripple will consequentially propagate into the storage battery [1]. This undesired current ripple causes the extra temperature rise in the battery and therefore, the shorter battery life span. Additionally, the ripple current probably accelerates the positive grid corrosion leading to premature failures [2]. More bus capacitors [3] or an active filter [4] could be used to

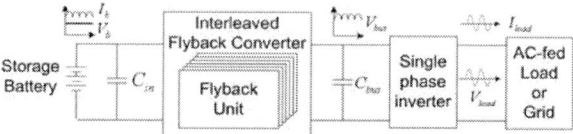

Fig. 1: Autonomous system structure with bidirectional interleaved flyback converter and inverter

suppress the voltage ripple on the DC-bus, however, the resultant more cost and larger component volume are not preferable in the present power electronics trend. Reference [1] proposed an active control method to reduce the low frequency current ripple in the power source (fuel cell) without more unnecessary passive storage components. Its closed voltage control loop of the DC-DC converter only possesses a bandwidth of 2 Hz such that the voltage ripple of 100 Hz on the DC-bus in the steady state could be ignored and nearly decoupled with the current in the fuel cell source, therefore, the fuel current ripple is significantly reduced. However, the transient response will obviously be suffered. Williams [5] proposed a nonlinear regulation-band method in which the closed voltage control loop gain K is flexible such that the DC-bus voltage is controlled in a given range in steady state and regulated back into this range and stabilized quickly when load transients occur. This paper presents a novel dual-band controller using average current mode control (ACMC) scheme. Besides the ACMC current loop for controlling the average current of the power stage, two voltage-loop branches with two different bandwidths (2 Hz and 641 Hz) are implemented for steady state and load transients. The slow loop branch significantly reduces the battery current ripple in the steady state by decoupling the 100 Hz DC-bus voltage ripples in the allowed range with the battery current. The fast loop branch pulls the DC-bus voltage back into the allowed voltage range rapidly and the temporarily huge current peaks in the battery can be smoothed rapidly with benefit of negligible harm to battery life. The core part of this controller is the dual-band selector. The bus voltage is divided with two bands: outer band and inner band. Outer band determines when the fast voltage loop branch is activated and inner band decides when the slow branch is switched on without pushing the bus voltage out of the outer band. Simulations demonstrate this new control method. The current ripple in the battery is only 5 % in the steady state and the response time in the transients is less than one grid cycle.

978-1-4244-1871-8/07 $25.00 © 2007 IEEE

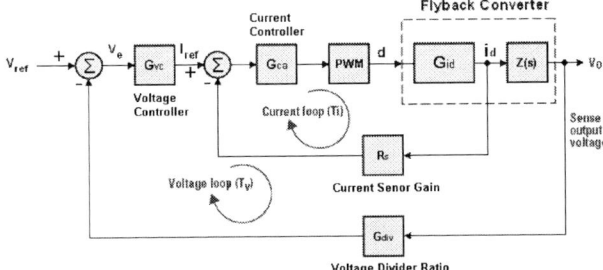

Fig. 2 Control block diagram of average current mode control

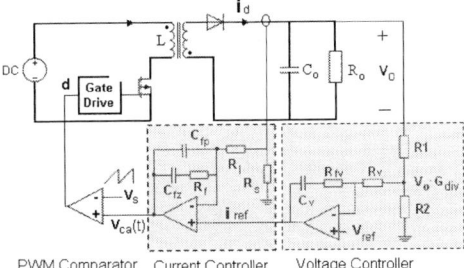

PWM Comparator   Current Controller   Voltage Controller

Fig. 3 Op amp circuit of average current mode control

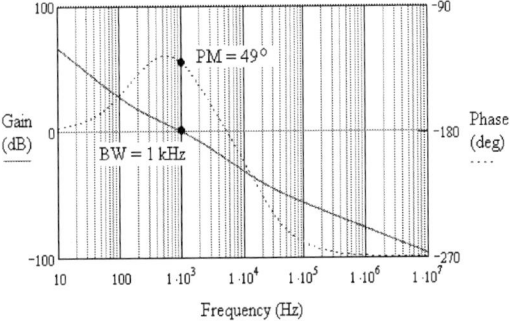

Fig. 4 Bode plot of the close current loop (phase margin 49° and 1 kHz bandwidth)

II. AVERAGE CURRENT MODE CONTROL STRUCTURE

As shown in Fig. 1, six flyback units connected in parallel are switched in such a way that the current flowing through them are evenly phase-shifted in one switching period. By doing this, the current ripples in the capacitors are dramatically reduced compared to the non-interleaved structure at the same power level, resulting in less capacitors volume and losses. However, since the six flyback units are interleaved to share the load current, one of the primary control requirement is to guarantee an equal current distribution among them. The well-know peak current control scheme, which limits the maximum current in one power unit, can only make the units look like voltage sources and the equal current share among all units cannot be realized even tiny differences in tolerance of components happens existing. Average current mode control [6] can properly carry out this task.

ACMC system has two control loops: One is the current loop which is laid inside in the outer voltage loop. The inner current loop is to keep the average current of the power stage in one switching cycle at certain value and the outer voltage loop is to regulate the output voltage of the power stage meanwhile feed the current loop with reference value. The control block diagram for the interleaved DC-DC converter under study is illustrated in Fig. 2. It should be noticed that the output impedance transfer function of the flyback converter is not located in the current loop but in the voltage loop. All flyback units have their own current loops and share one common voltage loop, fed together with the same current reference. Under such a control scheme, all flyback units can be regarded as identical current sources and the case that only one unit is overburdened can be prevented.

III. CURRENT CONTROL LOOP DESIGN

Because of the functional identity and the common voltage reference of all flyback units, the controller design can focus on one single unit. The capacitance of DC-bus should be seen as equally distributed on each flyback unit, so one sixth of the DC-bus capacitance is actually utilized by each flyback unit. The diode current on the high voltage side is sensed and controlled for both power flow directions. The controller design for the battery-discharging mode is presented here only. The flyback converter is designed to work in DCM under nominal power and in continuous conduction mode

(CCM) at higher power level. Basically, the flyback converter working in CCM has the same gain characteristics to that working in DCM except an annoying right-half-plane (RHP) zero at high frequency which places the difficulty in the controller design [7]. To handle this extra RHP zero in CCM, the current loop is designed for CCM.

The transfer function of the current loop controller (PI type) with op amp circuit parameters shown in Fig. 4 is given in (1):

$$G_{ca}(s) = \frac{C_{fz} + C_{fp}}{R_i(C_{fp} + C_{fz})} \cdot \frac{(1 + s \cdot R_f C_{fz})}{s \cdot (C_{fz} + C_{fp} + s \cdot R_f C_{fz} C_{fp})} \quad (1)$$

Two poles and one zero are given in this controller.

The closed current loop gain is derived as:

$$T_i(s) = G_{ca}(s) \cdot K_m \cdot R_s \cdot G_{id}(s) \quad (2)$$

where K_m is the PWM compensator gain, R_s is the current sensing resistor and $G_{id}(s)$ is the open loop transfer function from control to the diode current. $G_{id}(s)$ can be derived as

$$G_{id}(s) = \frac{\hat{i}_d(s)}{\hat{d}(s)} = \frac{\dfrac{V_o}{R_o DD'}(\dfrac{1}{N} - \dfrac{s \cdot LND}{R_o D'^2})(sR_o C_o + 1)}{s^2 \dfrac{N^2 LC_o}{D'^2} + s \cdot \dfrac{N^2 L}{R_o D'^2} + 1} \quad (3)$$

where D is the duty ratio which is 0.5 in CCM and D', N is the transformer turn ratio. As can be seen here, RHP exists in $G_{id}(s)$ which is

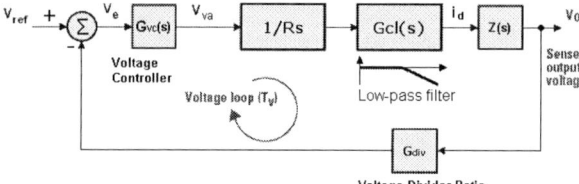

Fig. 5 The closed current loop can be considered as a low-pass filter $G_{cl}(s)$ seen from voltage loop

$$Z_{RHP} = \frac{R_o D'^2}{N^2 LD} \qquad (4)$$

As you can see in equation (4), the RHP zero is proportional to the load resistance. To get rid of the negative influence of this RHP zero in the loop, the current controller should be designed to crossover the gain characteristics of the current loop about one decade before this RHP zero, which means the smaller load resistance results in the lower limit of the closed loop bandwidth. In the event of the worst case, the load resistance is assumed to be zero and the current loop is designed with the lowest RHP zero.

The op amp circuit of PI current controller is shown in Fig. 3. The pole at origin is used to boost DC and low-frequency gain of the current loop and the high-frequency pole is placed at half of switching frequency to filter the switching ripple of the sensed current signal and increase noise immunity. The zero at about one decade before the high-frequency pole is required for crossing over the closed loop gain. The crossover frequency should be at least one decade before the RHP zero to obtain enough phase margin for the stability concerns. The bode plot of the closed current loop is illustrated in Fig. 4, where 49° phase margin and 1 kHz bandwidth can be observed. 1 kHz bandwidth for this current loop is high enough that the dynamic response and ripple performance are no longer sensitive to the current loop bandwidth changes [1].

Viewed from outer voltage loop, the closed current loop behaves like a low-pass filter $G_{cl}(s)$ shown in Fig. 5 which rolls off so distant before the RHP zero that RHP zero does not have any influence on the phase margin. This low-pass filter characteristic is the same to the closed current loop working in DCM. The RHP zero in CCM just looks like 'buried' in the current loop and doesn't appear in the voltage loop [6]. Therefore, the boundary cross between DCM and CCM is invisible to the outer loop, which means average current mode control works well in both operating mode.

IV. Voltage Hoop and Dual-Band Selector Design

A. Voltage Loop with Two Branches

ACMC is a dual-loop system, in which the current loop limits the average current value at the given level and the voltage loop regulates the output voltage. As discussed above, if a voltage loop with low bandwidth compared with 100 Hz, as done in [1], the battery current will be quite smooth in the steady state as if it could not feel the voltage ripple on the DC-bus. However, the bus voltage regulation is sadly poor

when the load transients happen. If the bandwidth of the voltage loop is much larger than the bus voltage ripple frequency 100 Hz, the battery current will tremendously fluctuate to try to correct the inevitable bus voltage ripples in the steady state, even this ripple is small. Although the dynamic response is excellent, the battery life will definitely be shortened.

The task of the appropriate voltage loop design is clearly seen as two-folded. One is to attenuate the battery current ripple by ignoring the DC-bus voltage ripple in an acceptable band under the steady state and the other is to regulate the DC-bus voltage back into the desired band and attenuate the disturbed battery current as soon as possible when the load transients occur. From these two requirements, a thought of designing a voltage loop of two branches can be easily settled: a voltage-loop branch which has a bandwidth much lower than 100 Hz should be used to fulfill the first requirement and a voltage-loop branch with a bandwidth much larger than 100 Hz should be active when transients take place.

Because the closed current loop observed from the outer voltage loop has a simple low-pass filter characteristic as mentioned in the previous section, the voltage loop can easily be shaped by a PI compensator $G_{vc}(s)$ for the desired closed loop bandwidth.

The closed current loop transfer function $G_{cl}(s)$ can be expressed as

$$G_{cl}(s) = \frac{T_i(s)}{1 + T_i(s)} \qquad (5)$$

The gain of the closed voltage loop is

$$T_v(s) = \frac{G_{cl}(s) \cdot Z(s) \cdot G_{div} \cdot G_{vc}(s)}{R_s} \qquad (6)$$

where G_{div} is the voltage divider gain, $Z(s)$ is the output impedance transfer function which is

$$Z(s) = \frac{1}{sC_o} // R_o = \frac{R_o}{sR_oC_o + 1} \qquad (7)$$

and $G_{vc}(s)$ is the voltage controller transfer function

$$G_{va}(s) = \frac{R_{fv}}{R_v} \cdot \frac{s + R_{fv}C_v}{s} \qquad (8)$$

Due to the simple low-pass characteristic of the closed loop, the op amp parameters in the voltage controller shown in Fig. 3 can be easily selected such that two different closed voltage loop bandwidth of 5 Hz and 641 Hz can be realized, with bandwidths both higher than 55°.

B. Dual-Band Selector Design

The slow and fast voltage-loop branches are designed aimed at the steady state and transient state respectively, but how to switch between two branches is critical: If these two branches are designed to switch from one to the other in an

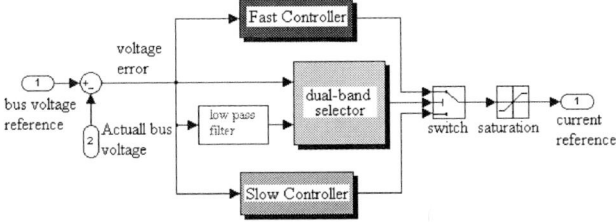

Fig. 6 Dual-band selector used to decide when and which voltage loop branch works

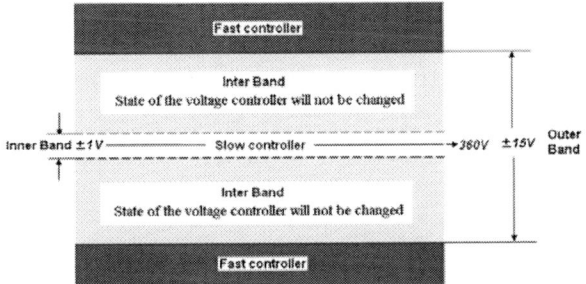

Fig. 7 The dual-band state map of the voltage loop

imprudent way, continual jumps between two branches will probably take happen due to the bus voltage transients caused by the loop-switching and battery will suffer from the current peaks in series. This loop-switching transient should be solved properly.

An originally-designed dual-band selector for managing the states of two voltage-loop branches presented below can be used to perfectly accomplish this kind of design task. This dual-band means that there are two DC-bus voltage bands determining which voltage-loop branch (with low or high bandwidth) will govern the bus voltage control without loop-switching transients. One voltage band is set for steady state control, whose range is decided by the cascading inverter and the allowed current ripple of the DC-bus capacitor. The bus voltage ripple should be in such a range that the inverter's output voltage quality is guaranteed and current ripples produced by the voltage ripples in the bus capacitors are confined under the capacitor ripple limits [1]. In our case, this voltage band is set within ± 15 V around the nominal bus voltage 360 V. When the bus voltage stays in this band in the steady state, the voltage-loop branch with 5 Hz bandwidth works and the battery current is controlled to be almost constant. If the bus voltage jumps out of this range, the voltage loop branch with 641 Hz bandwidth will be active immediately, trying to drive the deviating voltage back into this range as soon as possible.

The other voltage band is for determining when to switch from the fast voltage loop branch to the slow one with assuring that the voltage will not jump out of ± 15 V band once the slow loop works again. Settling this band range is tricky. Since the slow branch can only effectively control the low frequency voltage components less than 5 Hz, if the slow branch is activated once these low frequency components are regulated very close to the reference by the fast branch, a large current fluctuation is not needed to correct this slight

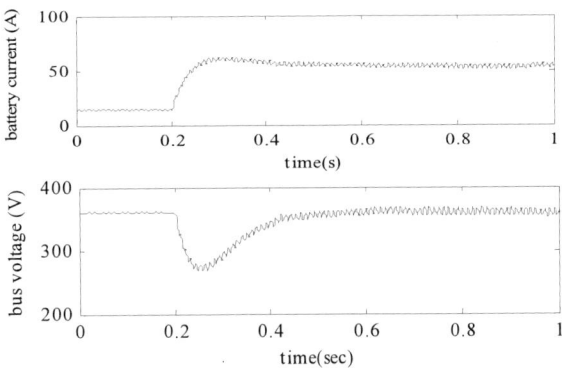

Fig. 8 Step response of the battery current and bus voltage when only 5 Hz bandwidth voltage-loop is used.

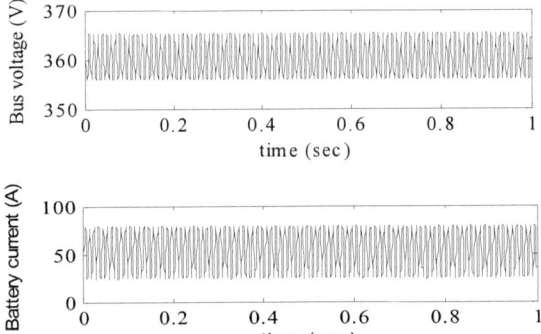

Fig. 9 Voltage and current behaviors when only 641 Hz bandwidth voltage-loop is used, at the nominal load in the steady state.

voltage error and the bus voltage will stay into the ± 15 V band. Transients will be removed thoroughly and the steady state recovers. However, if this band is too narrow close to zero, the low frequency voltage will take too much time to get into this band and the fast loop will generate large battery current oscillations for longer time, which does harm to the battery life. The band range of ± 1 V has been tested as a good choice. A low-pass filter with 5 Hz crossover frequency, as shown in Fig. 6, can be used to extract the low frequency components out from the bus voltage error signal. If the filtered bus voltage is located in the ± 1 V range, the slow voltage-loop branch takes charge of the voltage loop. The state map of the voltage loop is shown in Fig. 7. The selecting principle which is implemented in the dual-band selector is summarized as follows:

- Fast voltage-loop branch stays online when the voltage error jumps out of 360 ± 15 V band.

- Slow voltage-loop branch is working when the filtered voltage error lies in the 360 ± 1 V band.

- The state of the voltage-loop will not be changed as long as the un-filtered and filter voltage error stay between these two bands.

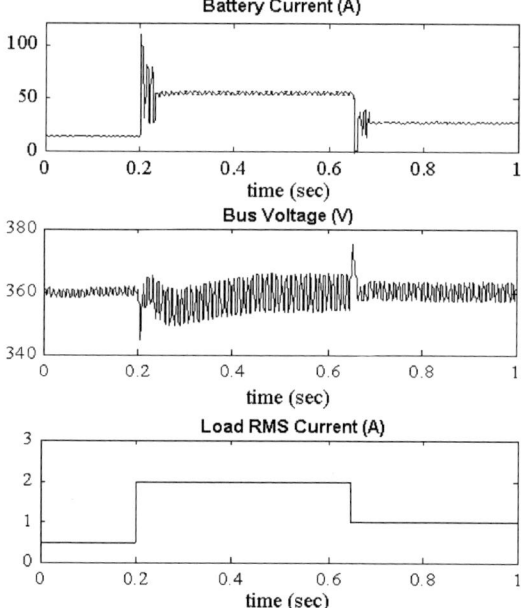

Fig. 10 Simulated bus voltage and battery current behaviors under load step transients when the dual-band controller is used

V. SIMULATION RESULTS

Simulations have been done to evaluate the voltage and current behaviors when only fast or only slow voltage loops is implemented. As you can see in Fig. 8 , when the voltage loop with 5 Hz bandwidth is implemented, the bus voltage has a fierce drop from 360 V to 280 V due to a load step from 0.5 A to 2 A, but in the steady state the battery current is quite smooth, with only 5% ripple. Voltage-loop branch with 641 Hz bandwidth gives out a 48% ripple in battery current at the nominal load in the steady state (Fig. 9). From these figures, it could be understood that a voltage-loop with single bandwidth is not a good design with consideration of all load states.

An ACMC controller with double voltage-loop branches and dual-band selector has been proposed in the paper. Simulation results are presented in Fig. 10, where you can see that when the DC-bus voltage runs out of the 360 ± 15 V band due to the output RMS current of the flyback jumping from 0.5 A to 2 A or from 2 A to 1 A, it is regulated rapidly back into this band again with less than one grid cycle, by fast voltage-loop branch. Before the unfiltered bus voltage error goes back into the ± 1 V band, the fast branch is still effective which cause a large battery current oscillation in less than 100 ms. Once the slow voltage-loop is activated, bus voltage stays in the 360 ± 15 V and the steady state recovers.

VI. CONCLUSIONS

In the paper, a novel ACMC controller with dual-band selector is introduced. This controller has two voltage-loop branches, one with 5 Hz bandwidth and other one with 641 Hz, for significantly reducing the battery current ripple in the

steady state and rapidly regulating the bus voltage in the transients, respectively. An originally-designed dual-band selector in the voltage-loop determines when and which voltage-loop branch is switched on by checking the filtered and unfiltered bus voltage's status in a dual-band state map. Switching actions between two branches only result in small and short disturbances on voltage and current behaviors. This method protects the battery from large current ripples in the steady state and possesses a very short DC-bus voltage transient response time, within less than one grid cycle.

This control methodology has been demonstrated by simulations on an interleaved bi-directional flyback converter that powers the DC-bus from a battery. The experimental verification will be done in the future.

REFERENCES

[1] Changrong Liu, and Jih-Sheng Lai, " Low frequency current ripple reduction technique with active control in a fuel cell power system with inverter load," *IEEE Trans on Power Electronics* , vol. 22, no. 4, pp. 1429–1436, July 2007 .

[2] N. Kutkut, "Output ac ripple effect," Power Designer, available: http://www.powerdesigners.com/pdf/AC%20Voltage%20Ripple%20 Effect.pdf

[3] M. Schenck and K. Stanton and J. S. Lai, "Fuel cell and power conditioning system interactions," *IEEE Applied Power Electronics Conf.*, Austin, TX, Mar. 2005, pp. 114-120.

[4] A. Monti, E. Santi, D. Franzoni, D. Patterson, and N. Barry "Fuel cell and power conditioning system interactions," *IEEE Applied Power Electronics Conf.*, Austin, TX, Mar. 2005, pp. 114-120.

[5] J. B. Williams, "Design of feedback loop in unity power factor ac to dc converter," *Power Electron. Spec. Conf. Rec.*, 1989, pp. 959-967.

[6] L. Dixon, "Average current mode control of switching power supplies," presented at the *Unitrode Power Supply Design Seminar Merrimack*, NH, 1990.

[7] J. Lempinen and T. Sunito, "Small-signal modeling for design of robust variable-frequency flyback battery chargers for portable device applications," *IEEE Applied Power Electronics Conf.*, vol. 1, 1987, pp. 548–554.

978-1-4244-1871-8/07 $25.00 © 2007 IEEE

A Study of the Train Running Simulation for Train Propulsion System Performance Analysis

Young-Chan Kim, Young-Ger Seo,
Soon-Chan Hong, Jong-Sun Ko
Dankook University
Department of Electrical Engineering
San 44-1, Jukjeon-dong, Suji-gu, Youngin-si
Gyeonggi-do, 448-160
Email: weist14g@nate.com, syger01@hanmail.net
schong@dku.edu, jsko@dku.edu

Byung-Sung Lee
Korean Railroad Research Institute
Signaling & Electrical Enginerring Research Department
Power supply, Collection & Conversion Team
Woram-dong, Uiwng-City, Gyeonggi-do, Korea, 437-757
Email: bslee@krri.re.kr

Abstract— The aim of the paper is to analysis on train propulsion system and to study for energy saving. For this study, we make the program that simulates actual operation of the train route to get same experiment result. The train running simulation is performed from starting station to 4th station by using the route data of Deajeon city Metro Subway. The study for control method of electrical motor and energy recovery to save energy is selected. The train propulsion system is constituted as a M-G Set, which is realized via Space Vector Modulation(SVM) - Direct Torque Control(DTC), the energy consumption during train operation and energy recovery during breaking is simulated by Simplorer program. From this result, the energy consumption and recovery of train with SVM-DTC is studied.

I. INTRODUCTION

To meet the transportation needs and increase the competency in nationwide logistic, many attentions and investments are drawn to the construction of express railway system. However, cost for oil energy is increasing as time goes. Therefore, high performance and reliability in electrical systems of railway infrastructures and cars should be attained due to the increased role of electric railway system. Major studies on this subject, such as quality compensation device, electricity collection and operation tester, lightning protection relay system for DC electric supply, performance testing on propulsion system, etc are currently undergone.

In this paper, the performance testing of propulsion system and simulation of propulsion system tester is studied. Through this study, accident prevention, securing safety in performance testing and safety and reliability of propulsion system tester is guaranteed for maintenance cost savings[1]. Induction motor in propulsion system has several benefits like easy manufacturing and stability than PMSM, but disadvantages of difficulty in rotor flux measurement and complexity in precise control. Robustness of controller that is required for parameter change and precise current control becomes a necessary property. By simulating the operation of train in actual line, effects of load and driving unit are studied. Induction motor in this study is composed in complication, and this complexity is also applied to inverter via vector controller. Therefore, program that features simple schematic is required. For this purpose, Simplorer program is used to construct the simulation of central parts in train.

This program is applied to running simulation of metropolitan railcar and express way railcar on actual line, and its energy consumption and recovery is studied.

II. TRAIN RUNNING

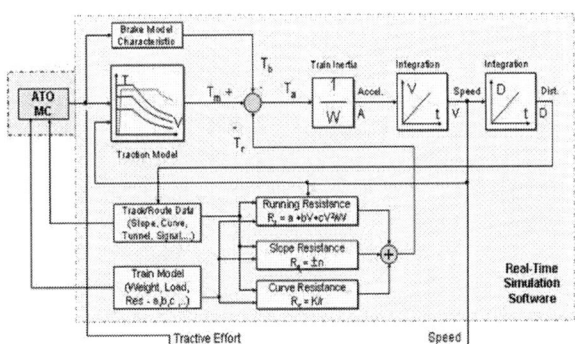

(a) Block diagram for actual train running System

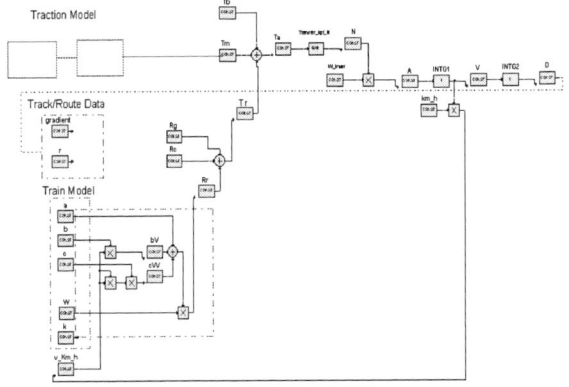

(b) Block diagram for Simplorer train running

Fig. 1: Block diagram for Simplorer train running system

Train running is consisted of speed increase up to limiting speed after starting, inertial movement, braking, stop[2][3]. In Fig. 1 (a), block diagram for actual train running is shown. Fig. 1 (b) shows the recreation of actual train running through Simplorer program for better understanding of program and modification by user.

978-1-4244-1871-8/07 $25.00 © 2007 IEEE

This system can be divided into four parts, traction (T_m), train resistance (T_r), air resistance (T_b), train inertia (T_a), and following equations explain the relationship[4][5].

$$T_m - T_r - T_b = T_a \ (kgf). \tag{1}$$
$$Accel = (T_a \times 9.8)/W \ (m/s^2) \tag{2}$$
$$Velocity = \int_t Accel \ dt \ (m/s) \tag{3}$$

Where, W is weight of train and 113 ton at fully loaded. If traction (T_m) is given, through air resistance and train resistance, train inertia can be determined from Eq. (1). By this, acceleration of train can be calculated through Eq. (2). Then, train speed can be determined via Eq. (3).

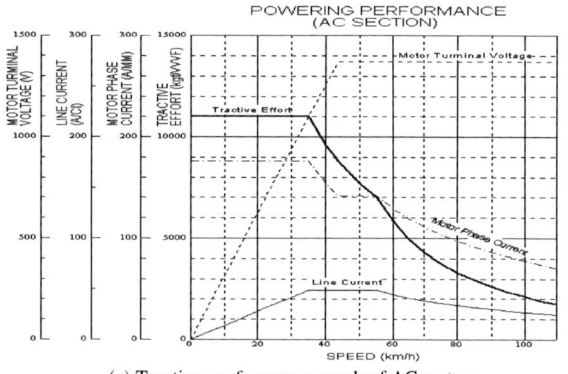

(a) Traction performance graph of AC motor

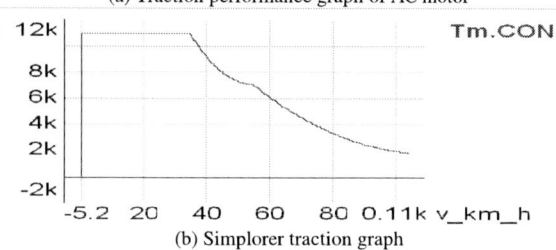

(b) Simplorer traction graph

Fig. 2: Train Traction - T_m

Traction of train in program can simulate traction graph by actual speed of train (km/h) as in Fig. 2 (a) through unit conversion. Fig. 2 (b) shows the simulated traction for same condition using Eq. (3). This value will be inputted for torque reference of motor for enabling the study on energy consumption during acceleration.

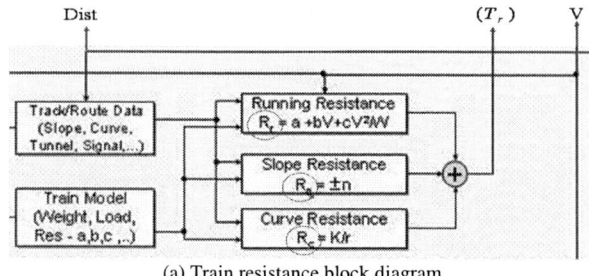

(a) Train resistance block diagram

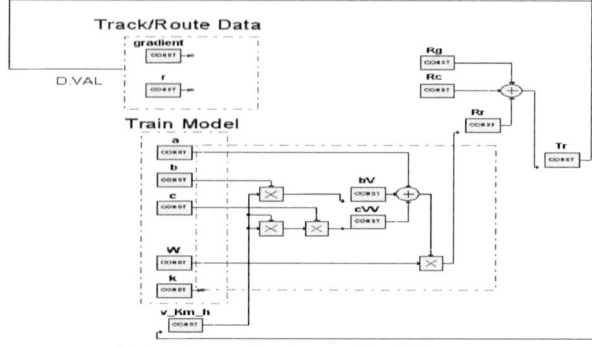

(b) Simplorer train resistance block diagram

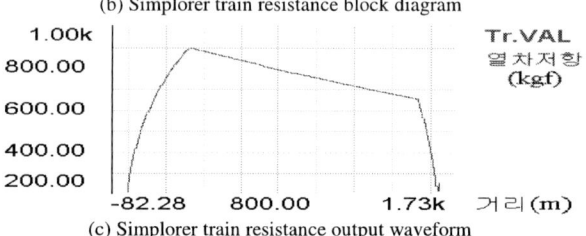

(c) Simplorer train resistance output waveform

Fig. 3: Train resistance

Fig. 3 shows train resistance (T_r), and is composed of three sub items, running resistance (R_r), gradient resistance (R_g), curved resistance (R_c). It can be denoted as Eq. (4). Via feedback of train speed, distance value, gradient and curvaed is determined from line data. These values are put into each equations to complete following equations[4].

$$T_r = R_r + R_g + R_c \ (kgf) \tag{4}$$
$$R_r = (1.867 + 0.0359V + 0.000745V^2) \times W(kgf) \tag{5}$$
$$R_g = W \times (i \times m) \ (kgf) \tag{6}$$
$$R_c = 600/r(kgf) \tag{7}$$
$$Distance = \int Velocity \ dt \ (m) \tag{8}$$

From Eq. (5) for running resistance, 1.867 is mechanical friction resistance, 0.0359 is friction resistance between vehicle axis and rail, 0.000745 indicates air resistance. V is train speed (Km/h). Eq. (6) denotes gradient resistance, and m is gradient distance, i is gradient height. Eq. (7) is curved resistance, and 600(m) is the summation of distance between rails and fixed axis distance. r(m) is curved radius. These values are programmed to be changed by Eq. (8) for running distance from line data.

$$Brake = 28.34 \times (113 + 40 \times 0.14 + 33 \times 0.66) \times 3.5 \tag{9}$$
$$= 11960 \ kgf$$

By Eq. (9) for regenerative braking (reverse torque) of train, calculated value is 11960kgf. This value is summation of air braking (T_b) and vehicle braking ($T_{m\_}brake$). As described before, corresponding reverse torque is generated for braking action like traction. This value generates reverse torque from braking moment to under 5km/h of speed near stopping area. After that, only air braking is used for stopping. Fig. 4 shows

978-1-4244-1871-8/07 $25.00 © 2007 IEEE   198

electric braking performance graph, and it is programmed to generate corresponding output to speed in Simplorer. Graph value for electric braking performance is different to electric braking performance in train, but traction change by speed is same.

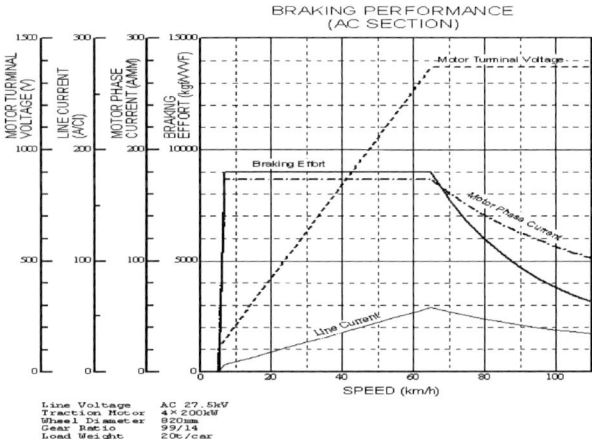

Fig. 4: Braking performance graph

Air braking is generated from braking moment till complete stopping for precise control of stop line. This value is 1262kgf, and can be calculated from the deduction of total braking power with required braking power from Eq. (10) for M car. Eq. (11) describes general equation for air braking[2][4].

$$60 \times 1000 \times 0.1783 = 10698kgf \qquad (10)$$
$$11960 - 10698 = 1262kgf. \qquad (11)$$

During braking, reverse torque is applied in motor. At the same time, connected generator is turned in right direction, and drives as generator. Through this, the ultimate goal of this study, relationship between consumption energy and regenerative energy can be explained.

III. M-G SET SVM CONTROL ALGORITHM FOR TRAIN PROPULSION SYSTEM

In case of driven by AC motor, the maximum AC voltage should be achieved from constant and given DC voltage to make the maximum torque from whole operation range. Therefore, linear composition ability for output phase voltage from given DC voltage, V_{dc} in inverter can be important aspect to evaluate the performance of voltage modulation methods. From this point of view, space vector voltage modulation (SVM) can generate the largest AC voltage from given DC voltage. In addition, harmonic content in output phase current is recognized to show far superior property to other types of modulation. Considering the relationship between phase voltages in load (V_{as}, V_{bs}, V_{cs}) and existence function, 8 different voltages can be output by existence function, and shown in 3 phase frame (d-q frame by stator) for a, b, c as shown in Fig. 5. Effective voltage vectors, $V_1 \sim V_6$, has $60°$ phase difference, and size is fixed to ($2\ V_{dc}/3$). Zero voltage vectors, V_0, V_7, do not generate actual voltage[6].

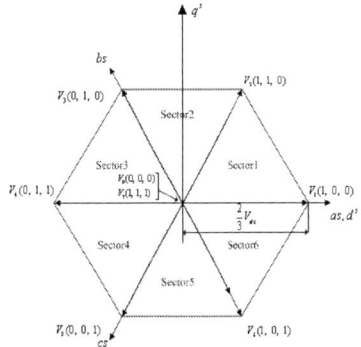

Fig. 5: Space voltage vector by existence function

Basic principle for SVM is composition of same voltage to standard voltage vector in average at one cycle using effective vectors of inverter. In other word, if reference output voltage vector, V^*, is given, the same voltage to voltage reference vector should be generated in average via the closest two effective vector to the reference (V_n, V_{n+1}) and zero vector (V_0, V_7) for control cycle, T_s. If T_1, T_2, T_0 is applied time for V_n, V_{n+1}, zero vector, respectively, the following equation can be made,

$$\int_0^{T_s} V^* dt = \int_0^{T_1} V_n dt + \int_{T_1}^{T_1+T_2} V_{n+1} dt + \int_{T_1+T_2}^{T_s} V_{0.7} dt \quad (12)$$

If V^* and V_{dc} is same for switching cycle, T_s, Eq. (12) can be expressed as follows.

$$V^* T_s = V_n T_1 + V_{n+1} T_2 \qquad (13)$$

As shown Fig. 6, if voltage reference is given at Sector 1, Eq. (13) can be denoted as Eq. (14).

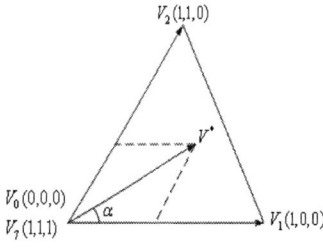

Fig. 6: Reference voltage vector

$$T_s |V^*| \begin{bmatrix} \cos(\alpha) \\ \sin(\alpha) \end{bmatrix} = T_1 \frac{2}{3} V_{dc} \begin{bmatrix} 1 \\ 0 \end{bmatrix} + T_1 \frac{2}{3} V_{dc} \begin{bmatrix} \cos(\pi/3) \\ \sin(\pi/3) \end{bmatrix} (14)$$

978-1-4244-1871-8/07 $25.00 © 2007 IEEE   199

Where, size α of is $0 \le \alpha \le \pi/3$. Therefore, if reference voltage vector is located in sector 1, applied time for finally calculated effective vectors and zero vector can be expressed as general equations, (15), (16), (17)[3][4].

$$T_1 = \frac{\sqrt{3}T_s}{V_{dc}}\left[\sin(\frac{\pi}{3}m)V_{ds}^* - \cos(\frac{\pi}{3}m)V_{qs}^*\right] \quad (15)$$

$$T_2 = \frac{\sqrt{3}T_s}{V_{dc}}\left[-\sin(\frac{\pi}{3}(m-1))V_{ds}^* + \cos(\frac{\pi}{3}(m-1))V_{qs}^*\right] \quad (16)$$

$$T_0 = T_s - (T_1 + T_2) \quad (17)$$

Where, m=sector number (1,2,.....,6).

In SVM method, various voltage modulation methods can be realized by changing k value within 0 to 1. If k is 0<k<1, status of every switch in one cycle can be diverted, and this is called continuous voltage modulation method or 3 phase voltage modulation method. If k is 0 or 1, only two phase switches (3 switches) can be diverted among 3 phase. This is called as discontinuous voltage modulation or 2 phase voltage modulation method. For SVM, various continuous or discontinuous voltage modulations are possible[7][8][9][10].

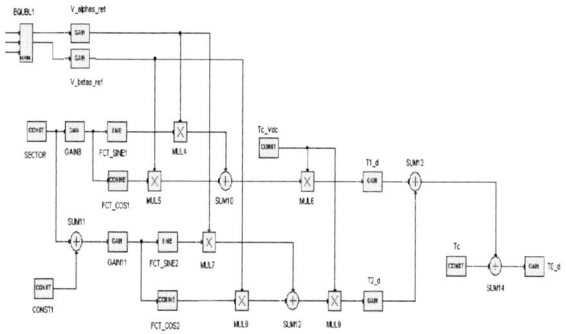

Fig. 7: Simplorer SVM block diagram

The block diagram for SVM in Simplorer program for easier modification and realization in Fig. 7. Inputs to upper left, V_{ds}^*, V_{qs}^*, θ_e can realize overall SVM by this value. A general type for train propulsion system as block diagram in Fig. 8(a). From this, Simplorer can makes similar type, and it is shown in Fig. 8 (b)[11]. Power system in motor can use either AC or DC, but only AC is realized for this study.

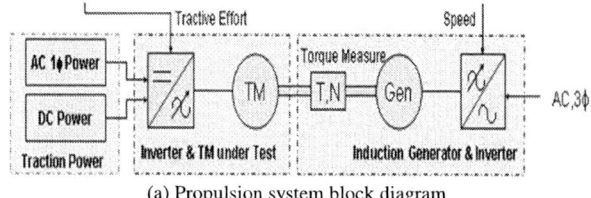

(a) Propulsion system block diagram

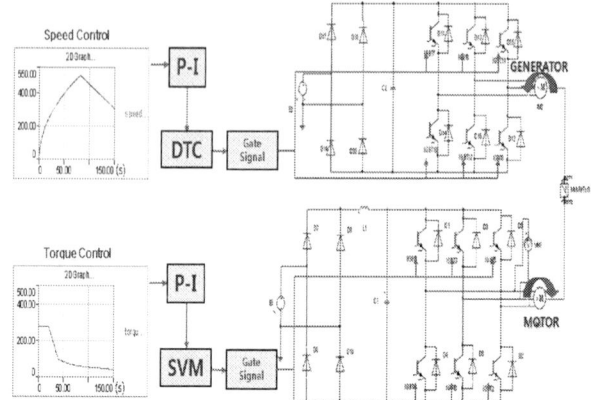

(b) Simplorer propulsion system block diagram
Fig. 8: Train propulsion system block diagram

IV. SIMULATION

TABLE 1 : Motor and Generator parameter

	MOTOR	GENERATOR		MOTOR	GENERATOR
Voltage	11000[V]	11000[V]	Stator inductance	0.0482[H]	0.0454[H]
Torque	99.082[Nm]	3524[Nm]	Rotor inductance	0.0483[H]	0.00457[H]
Rotor resistance	0.13488[Ω]	0.00509[Ω]	Mutual inductance	0.047[H]	0.00439[H]
Stator resistance	0.0777[Ω]	0.00388[Ω]	Moment of inertia	42.62[kg·m²]	42.62[kg·m²]

Parameters of induction motor and generator for simulation are shown in Table 1. As shown in Fig. 9, overall train running system is completed from train running simulation and propulsion system simulation. M-G set of actual traction train needs 4 unit, but it complicates program and increases data calculations leading the unable to run. However, overall system is linear, and one M-G set is used for experiment. For the study of recovery energy, which is the goal of the research, train traction is converted and applied to torque reference of motor. In addition, speed is converted to rpm, and applied to generator reference.

Traction conversion equation is shown in Eq. (18), and Eq. (19) is for converting speed (km/h, m/s) to rpm. In Eq. (18), (19), 0.82 stands for wheel diameter in meter. 7.07 is gear ratio. These values are divided by four before entered as reference of propulsion system. Simulation can realize up to 4th station of actual line for simulation, but it is too big to run the simulation for propulsion system. Therefore, it is programmed for section from starting first section, stopping at next station to restarting.

$$Nm = kgf/\{(9.8 \times 2)/0.82\} \quad (18)$$

$$rpm = Velocity \times \frac{7.07}{\pi \times 0.82} \times 60 \quad (19)$$

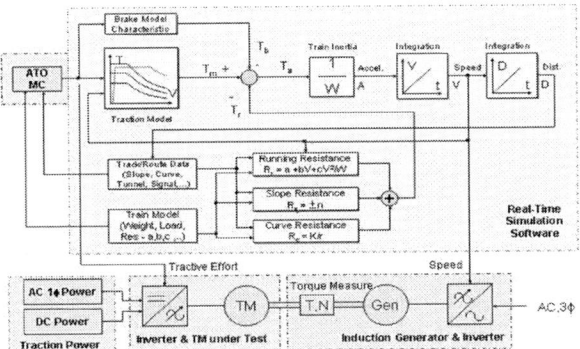

(a) Train operation system block diagram

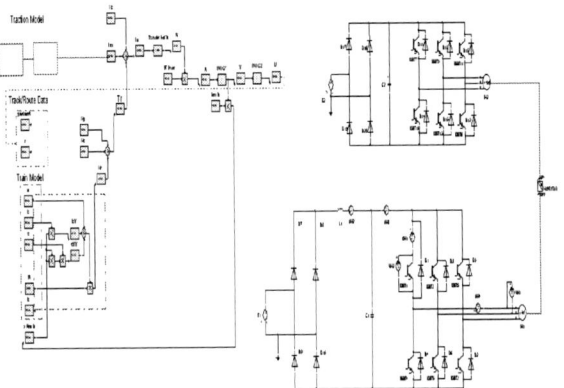

(b) Simplorer train operation system block diagram

Fig. 9: Train operation system

The state diagram for sector determination in SVM, as show in Fig. 10. It can be divided into two cases (sector 1-3 and 4-6) by that $V_{\beta s}^*$ is larger than 0 or smaller than 0. Each sector can be divided by the size of $V_{\beta s}^* / V_{\alpha s}^*$.

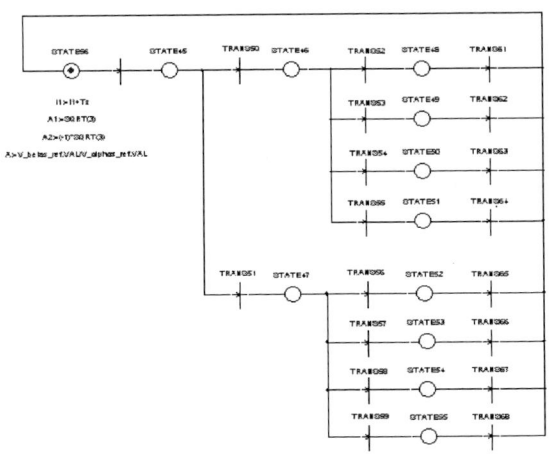

Fig. 10: The switching table for section determination in SVM

The result waveforms for motor and generator by torque and speed reference are compared in Fig. 11. The output

keeps up corresponding reference well. Eq. (20) and Eq. (21) is for calculation of power of motor and generator using torque and rotational angular velocity. The values from these equations are consumption energy and recovery energy.

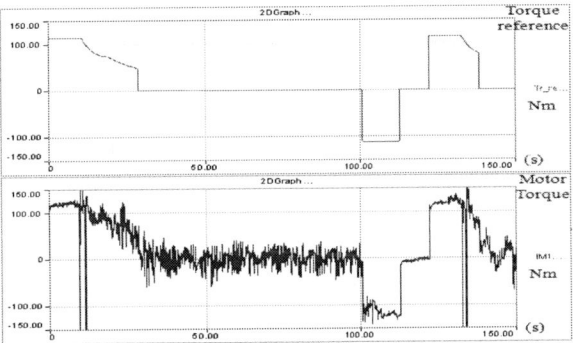

(a) Torque reference and result waveform

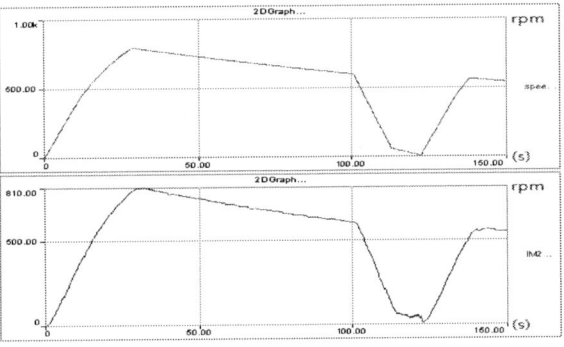

(b) Speed reference and result waveform

Fig. 11 Simplorer simulation result waveform

$$P_{m\_out} = IM1.MI \times IM1.rad/s \qquad (20)$$
$$P_{g\_out} = IM2.MI \times IM2.rad/s \qquad (21)$$

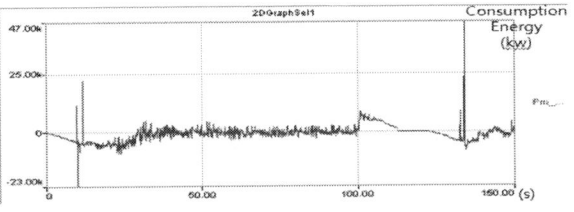

(a) Consumption energy waveform

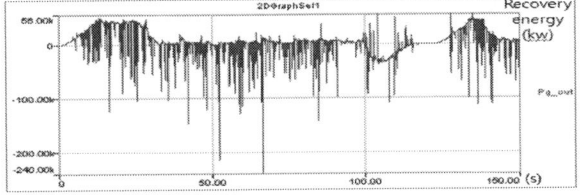

(b) Recovery energy waveform

Fig. 12 Simplorer simulation result waveform

The simulation result for consumption and recovery energy in generator during running first section, stopping, and restarting is shown in Fig. 12. Fig. 12 (a) shows the energy consumption during train running, and Fig. 12 (b) shows the energy recovery during braking.

V. CONCLUSION

For the study of train operation, the program that can simulate the actual operation from every data without running train in actual track is completed. Using this program, energy consumption at motor and regeneration in generator can be studied at first section running from the data based on Daejeon Metro Subway. In addition, efficiency of device, loss measurement test, temperature rise test, traction/braking performance test, efficiency measurement test, parallel operation test by input of torque and speed reference from actual testing equipment can be conducted. If these testing are completed, the propulsion system with less energy consumption can be selected via comparing consumed energy in current vehicle. This can contribute energy saving and research works for train vehicle by conducting various testing for composition and control method in propulsion system.

ACKNOWLEDGMENT

The financial support of Korea Railroad Research Institute in 2006 is gratefully acknowledged.

REFERENCES

[1] The practical technical development for tilting system, Korea Railroad Research Institute, 2004.07.

[2] Korea Railroad Research Institute, "Propulsion performance execution plan technical consultation final report, Facility technical examination report", pp14-15, Experiment simulation, 2000. 10.

[3] The Ministry of commerce, "Electric vehicle VVVF propulsion development" chapter 1~2 Propulsion control system, 2000. 12.

[4] Fei Lu, Mumin Song, Guohui Tian, Xiaolei Li, "The Application of Robot Formation Approach in the Control of Subway Train", *IEEE International Conf. 06*, pp5937-5941. 2006.

[5] Lee-Gunther, J., Bolduc, M., Butler, S., "Vista rail network simulation" *IEEE Railroad conf.* pp93-98, 1995.

[6] D.Rathnakumar. J.KakshmanaPerumal and T.Srinivasan "A New Software Implementation of Space Vector PWM" *IEEE Southeast conf 2005*, pp131-136, April 2005.

[7] Soon-Chan Hong, Duck-Yong Yoon, Min Cha, Young-Suk You, "The Transition character analysis and Control algorithms study of the 2.2kw induction motor-inverter", Experiment report, 1997.

[8] Jae-Do Lee, Jae-Ho Hwang, Hak-Ju Lee, Young-Nok Kim, Ki-Hong Lee, Se-jin Seong, "Space Voltage Vector PWM Modeling and Simulation Using Simulink", *KIPE Conf* pp15-18, 2003.

[9] Moon-Goo Jung, Byung-San Baek, Tae-Wan Ki, Seung-Pyo Ryu, Nam-Hae Kim, "Development of Regenerative Inverter for Electric Railway Using Space Vector PWM", *KIPE Conf* pp413-416, 2000.

[10] Ma Hao, Lang Yunping, and Chen Huiming, "A Simplified Algorithm for Space Vector Modulation of Three-phase Voltage Source PWM Rectifier" *IEEE PECS 04*, vol. 5, pp3665-3670, June 2004.

[11] Sam-Yong Jeong, Jin-Seop Lee, Ji-Yeun Seo, Dong-Hui Kim, Youn-Ok Choi, Guem-Bae Cho, "Indirect Vector Control of Induction Motor using Nonlinear Observer" *KIPE Conf* July 98, pp366-370.

978-1-4244-1871-8/07 $25.00 © 2007 IEEE

Development of DC Line Voltage Simulator for Control of Regenerative Energy Storage Device in DC Traction System

Kee-Hyun Cho[1], Su-Jin Jang[1], Byoung-Kuk Lee[1], Chung-Yuen Won[1] and Gil-Dong kim[2]

[1] Department of Information and Communication Engineering
Sungkyunkwan University, Su-won, Korea
[2] Korea Railroad Research Institute, Uiwang, Korea
Email:polo30@skku.edu

Abstract—In this paper, a regenerative energy conversion system for subway is proposed. In order to simulate the actual voltage and current fluctuation in subway system, The DC line voltage simulator, which is based on PWM ac-dc converter, is developed and implemented. The theoretical explanation is carried out and the validity of the proposed system is verified by simulation and experimental results.

Index Terms –DC line voltage simulator, regenerative energy, regenerative energy storage system.

I. INTRODUCTION

The DC subway system consumed a lot of electric energy and according to movement of electric car in moving field, there is generated regenerative energy. In other word, if the movement range is download stream or braking range, there is changed in 45% of input energy to regenerative energy. Fig. 1 is a distribution chart of regenerative energy [1].

The 55% of input energy consumed in electric car. All of the 45% of energy generated in form of regenerative energy, there is used 25% to the adjacency vehicles start and accelerating and remaining 20% of regenerative energy, there is pretended could've recycled energy [2]-[3].

In DC subway system, this type of regenerative energy raised the DC line voltage so that became a main cause the problem making such as malfunction of protecting equipment and coil burning. But in this time, there is no other equipment to store or to use in this regenerative energy in DC traction system, so the DC line voltage rose over determined level, it is consumed by using resistor load.

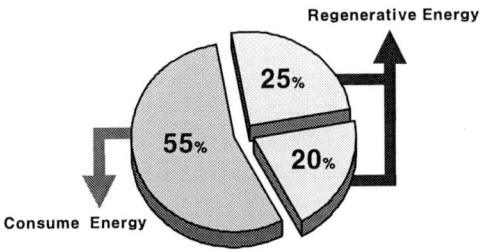

Fig. 1 Distribution chart of regenerative energy.

Therefore, to solve the problem cause by raised the DC line voltage and raise the using method of regenerative energy; a lot of research such as energy storage equipment has been studied.

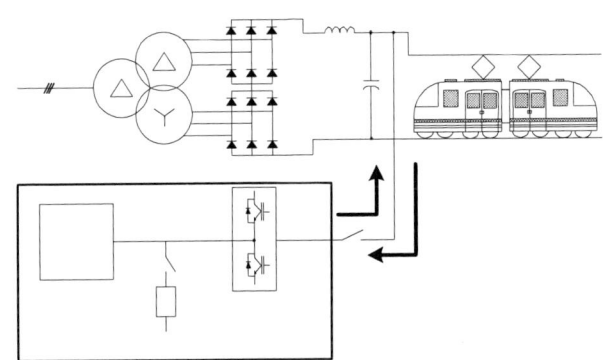

Fig. 2 Regeneration energy storage equipment.

Fig. 2 is block diagram for regenerative energy storage equipment in subway system. In this state, it is not possible to simulate the real size scale to test DC line voltage which has got enormous rate in regenerative energy using equipment. So it could not efficiently control the real system. Therefore in this study has been adapted to solve the problem cause by risen the DC line voltage and propose the DC line voltage simulator to efficiently control the energy storage equipment.

II. THE DC VOLTAGE ACCORDING TO CHARACTERISTIC OF ELECTRIC CAR IN DC TRACTION SYSTEM

In DC traction system, the electric car operated in combination of 4ea of operating mode such as start-move-brake-stop when it moved in station to station. The combination method of this operation mode affected to the electric car's run time and also affected to energy consumption and amount of revived energy. We could divide the characteristic of electric car's operation mode to energy consumption mode and energy regenerative mode. Fig. 3 indicated the standard operation chart in electric car.

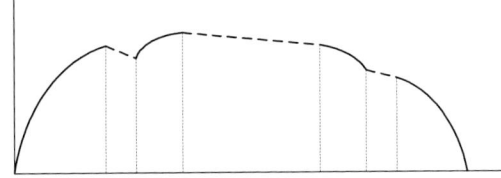

Fig. 3 Standard operation chart in electric car.

978-1-4244-1871-8/07 $25.00 © 2007 IEEE

The constantly maintained DC line voltage in the level of 1500V by 3-phase rectifier, affected by regenerative energy which was cause of standard operation characteristic pattern, it is changed like Fig. 4 which is included in regenerative energy, so it has got non-linear wire voltage pattern. Like this, In DC line voltage which was regenerative energy, if there is non-existence of adjacency vehicles or there is no energy consumption, the maximum value of DC line voltage positioned more than 1800V, so it produce a cause of important system's problem such as burning of coil or malfunction of protecting equipment.[4][5]

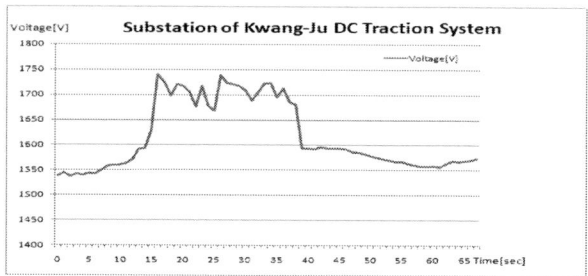

Fig. 4 DC line voltage affected by regeneration energy.

III. ENERGY STORAGE SYSTEM UNSING DC LINE VOLTAGE SIMULATOR

Fig. 5 Show the overall energy storage system. The DC line simulator has been connected with output of rectifier to the type of DC, so it could control the DC line voltage in 1500V type of included regenerative energy. In this time, the reference of simulator taken the sensing of voltage and current by actually measured DC line voltage and compared then compensate the tolerance.

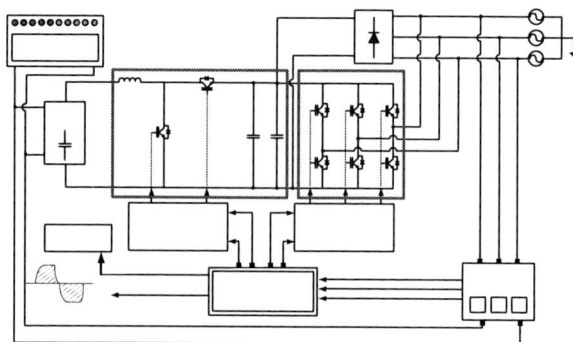

Fig. 5 Regeneration energy simulator system.

A. Analysis of operating mode

The simulator's energy flow which was cause of occurred by revived energy could've displayed like as Fig. 6 In the movement range of electric car was downstream range or braking range, the DC line voltage shall be risen by means of occurring regenerative energy. Fig. 6 (a) showed up the

electric current which was raised of DC line voltage. In reverse side, such as the case of start or acceleration of electric car, the down stream of DC line voltage could've displayed like as Fig. 6 (b).

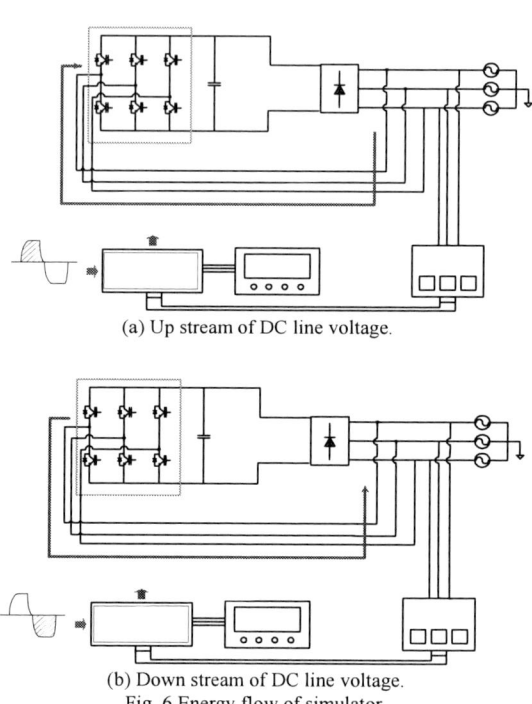

(a) Up stream of DC line voltage.

(b) Down stream of DC line voltage.
Fig. 6 Energy flow of simulator.

B. Control Algorithm of Simulator

The proposed DC line voltage simulator is bidirectional AC/DC converter type. Fig. 7 indicates the AC/DC converter which was used of DC line voltage simulator.

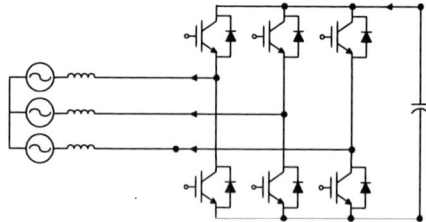

Fig. 7 DC line voltage simulator.

The regeneration power flows from traction to the ac source. For this mode, voltage equations are below.

$$e_a = L_i \frac{di_a}{dt} + V_a$$
$$e_b = L_i \frac{di_b}{dt} + V_b \qquad (1)$$
$$e_c = L_i \frac{di_c}{dt} + V_c$$

978-1-4244-1871-8/07 $25.00 © 2007 IEEE

d-q voltage equations are shown in (2) in the stationary coordinate system.

$$e_d^s = L_i \frac{di_d^s}{dt} + V_d^s$$

$$e_q^s = L_i \frac{di_q^s}{dt} + V_q^s \qquad (2)$$

The synchronous d-q voltage equations can be shown using (3).

$$e_d^e = L_i \frac{di_d^e}{dt} + \omega L i_q^e + V_d^e$$

$$e_q^e = L_i \frac{di_q^e}{dt} - \omega L i_d^e + V_q^e \qquad (3)$$

If the ac source is a three-phase sinusoidal voltage with an angular frequency, the three-phase line voltage on basis of a-phase is presented by (4).

$$e_a = -E \sin \omega t$$

$$e_b = -E \sin\left(\omega t - \frac{2\pi}{3}\right)$$

$$e_a = -E \sin\left(\omega t + \frac{2\pi}{3}\right) \qquad (4)$$

In (4), *E* is the maximum phase voltage and the ac source voltage in d-q stationary and synchronous coordinate is (5), (6).

$$e_d^e = E \sin \omega t$$

$$e_q^e = -E \cos \omega t \qquad (5)$$

$$e_d^e = E$$

$$e_q^e = 0 \qquad (6)$$

By (3), Equation (6) can be expressed by using (7).

$$E = L_i \frac{di_d^e}{dt} + \omega L i_q^e + V_d^e$$

$$0 = L_i \frac{di_q^e}{dt} - \omega L i_d^e + V_q^e \qquad (7)$$

The error between the transformed three-phase current to synchronous coordinates and the DC line voltage is the input value of the *PI* controller and makes *d*-axis current reference. This *d*-axis current reference compensates *q*-axis interference and makes *d*-axis voltage reference. [6]

Therefore, the notion of the proposed scheme in this paper uses a three-phase SVPWM inverter for regeneration of surplus power. Fig. 8 show the overall control block diagram of DC line voltage simulator.

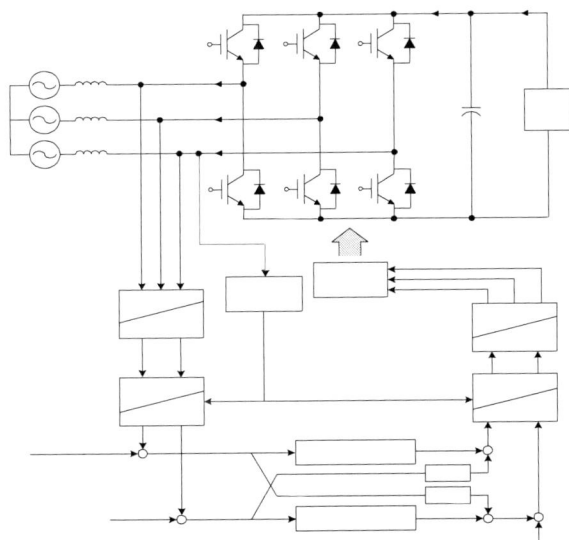

Fig. 8 Overall control block diagram.

IV. SIMULATION RESULTS

The DC line voltage simulator which was simulate the change of DC line voltage, bidirectional DC-DC converter which was changed the regenerative energy and capacitor which has got function of energy storage, had been adapted real Traction system and simulated. The Fig. 9 indicated the circuit diagram which was proposed in this paper

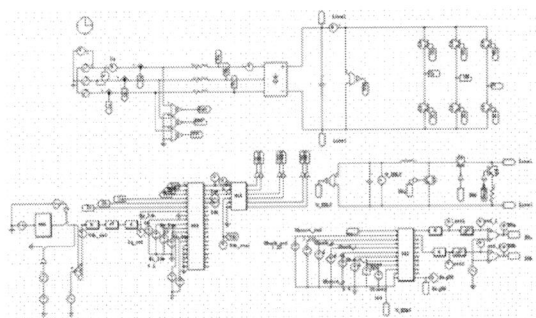

Fig. 9 The proposed regenerative system.

Fig. 10 was a displayed waveform which was viewed by using program in DC line voltage which was measured in substation of Kwang-Ju DC traction system. We could realize the fact that the displayed DC line voltage was not uniform because of the regenerative energy.

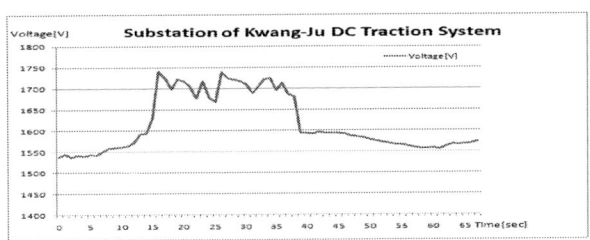

Fig. 10 Measured in substation Kwang-ju DC Traction System.

978-1-4244-1871-8/07 $25.00 © 2007 IEEE

The DC line voltage data which were gained using special program has been simulated by 1 second interval used of DLL of PSIM. Fig. 11 was displayed DLL of really measured DC voltage datum.

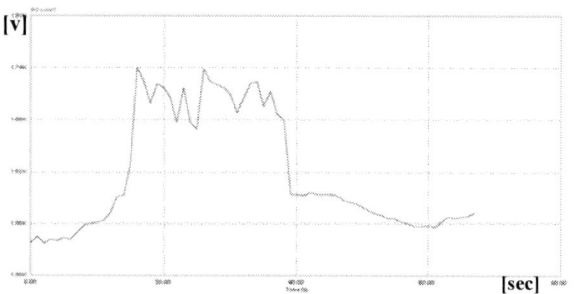

Fig. 11 Simulation output of really measured DC voltage data.

In DC traction system, no-load input was DC 1500V, but for the purpose of experiment, it is scaled for 350V, and in the substation of Kwang-Ju DC traction system picked up real data and simulated based of these.

A something amount of DC line voltage has been affected by regenerative energy which were caused from electric car. Fig. 12 showed up the DC line voltage which was affected from regenerative energy of simulator.

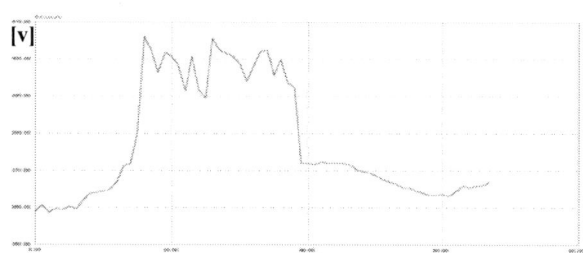

Fig. 12 Affected from regenerative energy of simulator. [sec]

Fig. 13 is waveform of I_{q_ref}, I_q and I_d current which has been flown to simulator of regenerative energy. Simulator could control the bidirectional current, therefore to made energy balance.

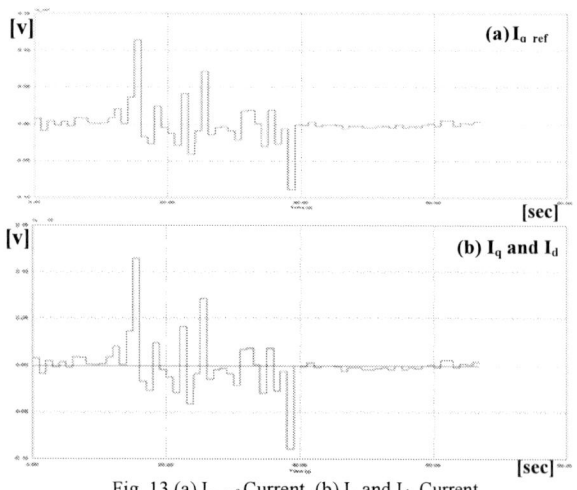

Fig. 13 (a) I_{q_ref} Current (b) I_q and I_d. Current

Fig. 14 shows the stabilization of DC line voltage. While the DC line voltage above 370[V], The DC line voltage is not up stream because energy charging of capacitor. In case of below the 350[V], the energy is discharged by capacitor. So the voltage ripple is decrease.

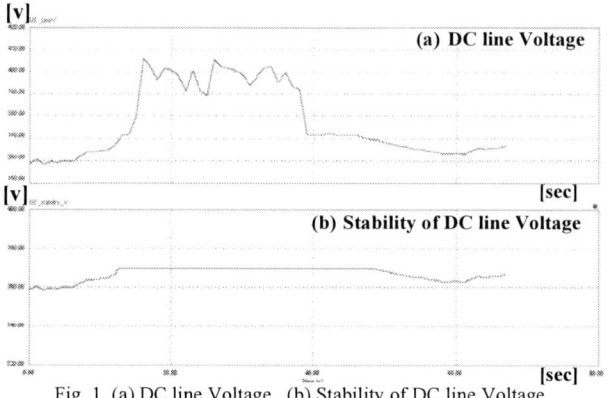

Fig. 1. (a) DC line Voltage, (b) Stability of DC line Voltage.

V. EXPERIMENT RESULTS

The real experimental test bed consists of a DC line voltage simulator and control DSP board. Fig. 15 shows experiment set.

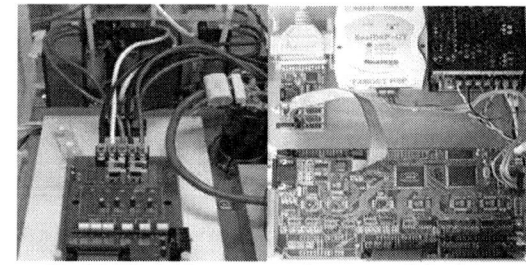

Fig. 15 Affected from regenerative energy of simulator.

The DC line voltage has been affected by regenerative energy which was caused from electric car. Fig. 16 showed up the DC line voltage which was affected from DC line voltage simulator.

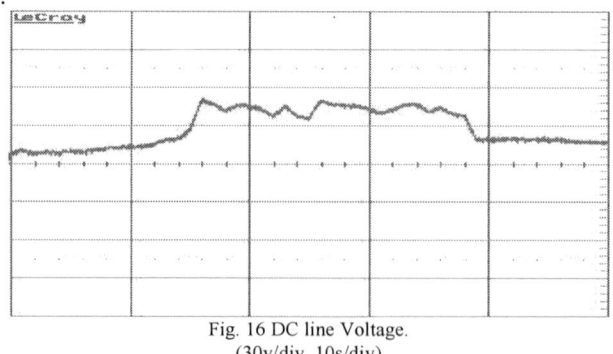

Fig. 16 DC line Voltage.
(30v/div, 10s/div)

q axis current is controlled to zero because it is reactive power current. The error between voltage reference and dc line voltage is the input value of the PI controller and makes d-axis current reference. Fig. 17 shows I_{q_ref}, I_q and I_d which is reactive current.

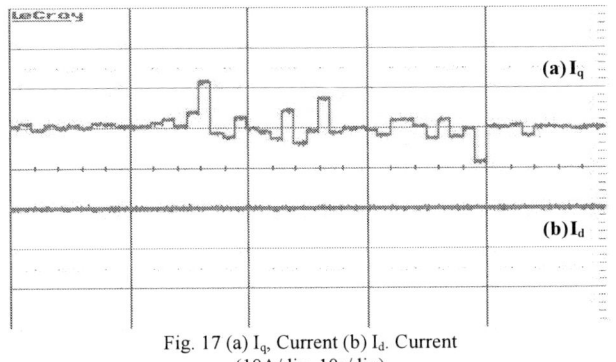

Fig. 17 (a) I_q, Current (b) I_d. Current
(10A/div, 10s/div)

We expressed the voltage waveform that is integrated DC line voltage simulator and energy storage device. Fig. 18 shows waveform that is decreased DC line voltage ripple by charge/discharge control of energy storage device.

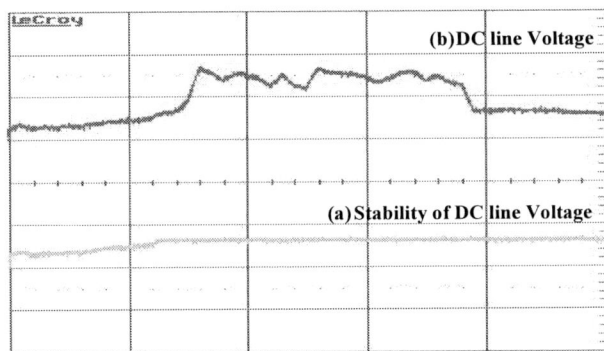

Fig. 18 (a) DC line Voltage (b) Stability DC line Voltage.
(30v/div, 10s/div)

VI. CONCLUSION

In this paper, based on the actual simulator, which can simulate DC line voltage fluctuation of subway, regenerative energy storage system has been designed and implemented. Using super-capacitors as the energy storage system, a test-bed has been built and the static and dynamic performance of the proposed system has been analyzed. From the experimental results, it can be expected that the proposed system can be utilized for energy efficient subway systems.

ACKNOWLEDGMENT

The preferred spelling of the word "acknowledgment" in America is without an "e" after the "g." Try to avoid the stilted expression, "One of us (K. B. G.) thanks ..." Instead, try "K.B.G. thanks ..."

REFERENCES

[1] K. H Cho, S. J Jang, J. Y Kim, C. Y. Won and Y. K. Kim, "Regeneration inverter system for DC traction system," KIEE Annual Spring Conference, 2007.

[2] K.W. Kim, I. S. Yoon, Y. M. Seo, D. Y. Yoon, and S. C. Hong, "A study on the power factor improvement of DC power regenerating systems," KIPE, vol. 6, no. 5, pp. 410-415, 2001.

[3] K. W. Kim, Y. M. Seo, and S. C. Hong, "A study on the implementation of inverter systems for regenerated power control," KIPE, vol. 7, no. 2, pp. 205-213, 2002.

[4] A. Horn, R. H. Wilkinson, and T. H. R Enslin, "Evaluation of converter topologies for improved power quality in DC traction substations," IEEE-ISIE'96, vol.2, pp.802-807, 1996.

[5] P. J. Randewijk and J. H. R Enslin, "Inverting DC traction substation with active power filtering incorporated," IEEE-PESC'95, Vol.1, pp. 360-366, 1995.

[6] H. I. Han, N. K. Hahm, N. K Sung, G.. D. Kim, D. K Kim, and K. H. Han, "A study on current controller comparision for three phase PWM converter," KIEE, pp.1248-1250, 2000.

The Design and Analysis of the Piezoelectric Inverter to the EEFL Driving of LCD Backlight

Hong-Sun Park, Seung-Hak Yang*, and Young-Cheol Lim

Chonnam National University
Department of Electricalti Engineering
300 Youngbong-dong, Buk-Gu, Gwangju,
Korea, 500-757
Email: hspak1749@paran.com

*Honam University
Department of Electrical Engineering
59-1 Seobong-Dong, Gwangsan-Gu, Gwangju,
Korea, 506-714
Email: yangsh@honam.ac.kr

Abstract—Cold Cathode Fluorescent Lamp(CCFL) has been used for a source of light on LCDs, and External Electrode Fluorescent Lamp(EEEFL), Flat Fluorescent Lamp (FFL), Field Emission Lamp (FEL) including LED is actively progressing application for other source of light. This paper used EEFL that has advantage for decrease inverter round figures, because is advantageous to multi lamp drive, transformer decrease loss in itself, miniaturization is possible, connect piezoelectric transformer that have high step-up ratio by multiple and did so that multi lamp drive may be possible.

To compose EEFL driver circuit of ideal has been designed Push-Pull type of piezoelectric inverter. As achieve simulation analysis for designed inverter circuit, and hereafter will apply drive methods of several form, presented that manufacture of inverter for wide screen multi lamp drive is available by piezoelectric transformer.

I. INTRODUCTION

Generally, LCD (Liquid Crystal Display) is widely used for electronic information machine such as TV, Laptop PC, Monitor, Video Cam, Mobile Phone, navigation, etc. However, because it has no ability of emitting light, it needs a luminous source for backlight. And usually, an electric discharge lamp having a Cold Cathode Fluorescent Lamp play a role as a luminous source.

And also, recently, as LCD Monitor screen grows larger, several electric discharge lamp attached in the back of panel are used for backlight and EEFL (External Electrode Fluorescent Lamp) is used instead of the established Cold Cathode Fluorescent Lamp. While the existing Cold Cathode Fluorescent Lamps need several inverters in order to supply distributed electric current at the beginning of multi-lamp operation, EEFL doesn't need many inverters for it.

Generally, a coiled-current transformer is used in operation inverters of electric discharge lamp. It has low-efficiency of electric energy because of lose of coils and core, and it has a larger and heavier than the piezoelectric transformer. Also, it is likely to make a fire by overheat from continual operation. For these shortcomings of the coiled-current transformer, the piezoelectric transformer has the following advantage: high-efficiency, smaller size, light weight, incombustibility, low-EMI, etc. However, because

the material of it is a kind of ceramic that has a resonance frequency, the reliance of the resonance frequency is high and it can be substituted for vibrating frequency of narrow maximum electric current conveyance. And also, due to capacitance of input, it is difficult to plan a matching circuit and the ceramic is brittle.

This piezoelectric transformer measured the resonance point of the piezoelectric transformer by the admittance orbit-tracing way of impedance analyzer. On the base of the measure, we made up an electric equivalence circuit and to compose the best EEFL operation circuit, we made a piezoelectric-inverter taking advantage of Push-Pull.

II. THE STRUCTURES AND CHARACTERISTICS OF THE EEFL AND THE PIEZOELECTRIC TRANSFORMER

A. The structure and operation-principle of the EEFL

The EEFL doesn't have any electrode in the inside of emission room, and it is operated by the combination of capacity of External Electrode attached to the outer part of both the ends of glass tube. While in the CCFL, metal electrode attached to both the ends of the inside of emission tube is exposed to emission plasma and the electric current flows in the metal electrode directly, in the EEFL the charged particles are piled up in both the ends of the emission tube and the electric current of the plasma flows in the inside of the emission tube alternately. Therefore, in the EEFL because electrode can avoid interaction with the plasma, the life of the lamp grows longer and as the electrode is established in the outer part of both the ends of glass tube, it is very easy to make the lamp.

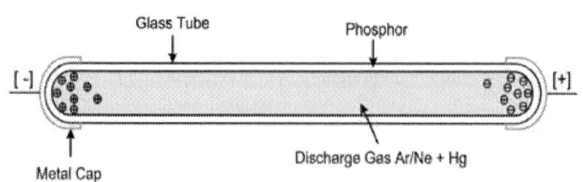

Fig. 1: Construction of EEFL

The figure 1 indicates the lamp's structure of the EEFL and the operation of the charged particles. In the inside wall of glass tube, the luminous and fluorescent material of three original colors, RGB is coated. In the inside room of glass tube, the mixed gases of neon and argon is put with a small amount of mercury. After that, both the ends of glass tube are closed and the External Electrode made of copper is established in both the ends as a cap-shape.

As the EEFL uses a rectangular wave rather than a sine wave, it has a high luminescence and high energy efficiency, and also it is an example of self-emission motivation way by wall-electric charge.

B. The Structures of the piezoelectric transformer and an electric equivalence circuit

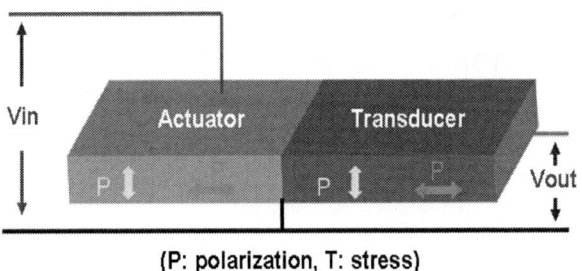

(P: polarization, T: stress)

Fig. 2: Structure of Piezoelectric Transformer

The size of the piezoelectric transformer used on the study is 7.4(W) x 35.0(L) x 3.1(T) [mm]. The resonance of these current transformers is usually one of $\lambda/2$, λ, or $3\lambda/2$. The resonance used on the study is $\lambda/2$. The equivalence of the piezoelectric transformer by the admittance orbit-tracing way is done as the following : after the input terminal and the output terminal are short-circuited alternately, we can measure each variable with an impedance analyzer and make a equivalence circuit with the parameter from the admittance orbit.

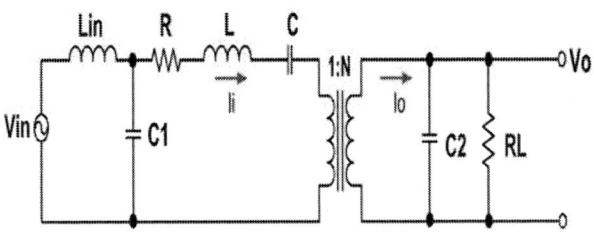

Fig. 3: Equivalent Circuit of Piezoelectric Transformer

The figure 3 shows that, when we plan a circuit, important factors to be considered are capacitance (C1, C2) and rate of boosting voltage (1:N). Usually, C1 is scores and hundreds of nano, and C2 is lots of pico.

With the impedance analyzer, we could get the measures from the equivalence circuit of the piezoelectric transformer as the following : C1=155Nf, C2= 18.9pF, N=50, R=1.75Ω, L=1.13mH, C=7.99nF.

III. THE TOPOLOGY OF THE PIEZOELECTRIC INVERTER APPLICATION

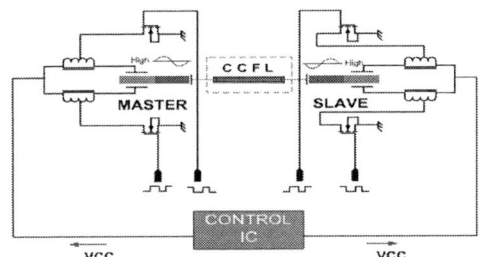

Fig. 4: Piezoelectric Inverter of Push-Pull Type

The inverter for the operation of backlight must have the following three essential factors. First, it must have a DC/DC converter to cope with a variety of DC input. Especially, in general laptop computers, it is very necessary because the outputs of battery and adaptor are different each other. Second, it must have a DC/AC converter to operate the lamp keeping the highest energy efficiency under the condition of a sine wave. Third, it must have the control of radiation to tune the light of monitor screen.

In consideration of these factors, Full-Bridge has a good efficiency of power conversion, and an advantage in operation under the low voltage. But because it has many switching elements, it also has complicated circuits. Push-Pull and Half-Bridge have an easy establishment of ZVS (Zero Voltage Switching), a good conversion efficiency, and an easy control. But in the Push-Pull, the same volume of resonance inductor is needed and in the Half-Bridge, because the upper side resonance voltage is used, it can cause the aging of current transformer.

The figure 4 shows Push-Pull typed inverter applying the High-High operation way. On the study, the EEFL made up of the tube having diameter 4.0mm and length 740mm that is used in 32inch large LCD backlight inverters is composed in order to control the inverters attached to both sides, Master and Slave through one control IC (BD9825). And Push-Pull typed topology applying High-High lamp operation way is made. ZVS (Zero Voltage Switching) is operated by resonance wave and ZVS happens by the alternate operations at the rate, 50:50(duty rates) of MOSFET (Q1&Q2), Push-Pull output. The electric power is supplied through the Buck Regular (Q3) of Push-Pull terminal. The control circuit is operated by the mixed rates of average voltage, standard electric current-sensing resistance, the standard voltage of rectifier. And also, a various and balanced sine wave more than 300-1500[V] occurs in the second current transformer and condenser, C2 makes balance of load voltage and electric current. The Unlike the coiled- current transformer, the piezoelectric transformer must be operated near the resonance wave and according to the loads, the resonance wave is changed. In addition, the equivalence impedance is changed and so, operation wave must be changed, too according to the conditions of turning on EEFL connected with the load of the piezoelectric transformer.

As you see in the figure 5, the piezoelectric inverter for operation is composed of three main parts: one is PWM PULSE circuit, another is the piezoelectric transformer, and the other is EEFL. PWM PULSE circuit gives an input signal to the piezoelectric transformer. The piezoelectric transformer raises the voltage for the input signal. The EEFL is operated by a high AC voltage.[1]

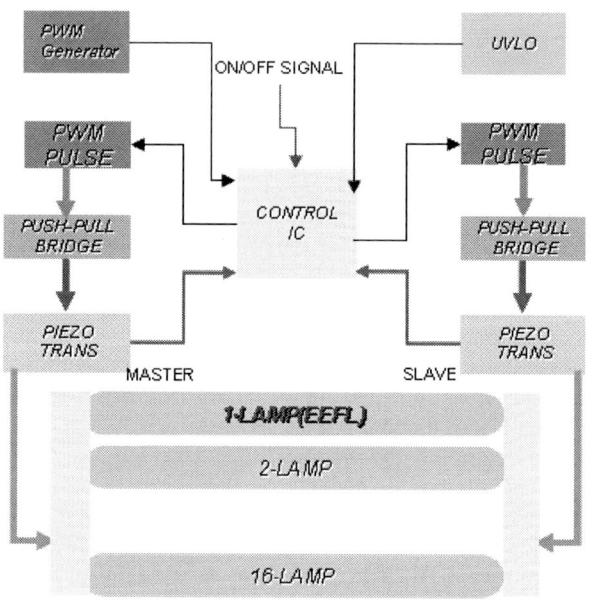

Fig. 5: Block Diagram of Piezoelectric Inverter

On the study, the switching circuit of the inverter is planned in order to control the wave by changing the input voltage with Voltage Controlled Oscillator that outputs the wave in accordance with the input voltage. And also, it is designed in order to decide the wave for operation according to the conditions of turning on EEFL when the input voltage of VCO (Voltage Controlled Oscillator) feedbacks the output electric current of the EEFL, the last load. The beginning output of VCO is set up in order to be equal to the non-loaded resonance wave of the piezoelectric transformer and also, it is fixed in order to return to the input of the VCO.[2][3]

IV.   SIMULATION AND TEST RESULT

A. *Simulation*

We have done the simulation of the loaded lamp having the modeled piezoelectric transformer with PSpice, circuit-interpreting program. As you see in the figure 6, the simulation has been done under the same conditions, EEFL as a real-made piezoelectric inverter.[5]

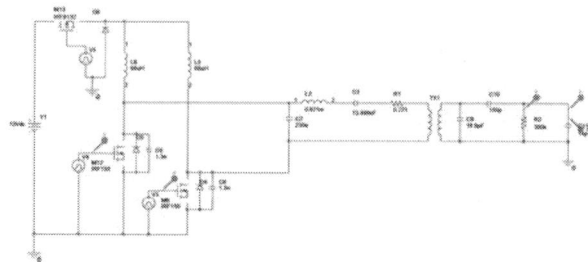

Fig. 6: PSpice Model of the Push-Pull Inverter Using Piezoelectric Transformer

In the simulation test, we applied the Push-Pull referred on the study. The wave of switching element is the same 48[kHz] as the switching wave for resonance to the piezoelectric transformer, having the delay time at the duty rate, 0.5. The figure 7(a) indicates the wave form of MOSFET that is switching while the simulation is going on.

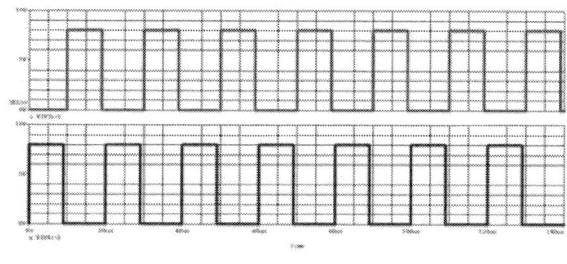

(a) Switching Waveforms

The EEFL of the figure 7 is not made up of only the resistance of the existing CCFL because of capacitor by external electrode. And after the capacitor was connected with the EEFL in series and parallel circuit, the simulation has done.

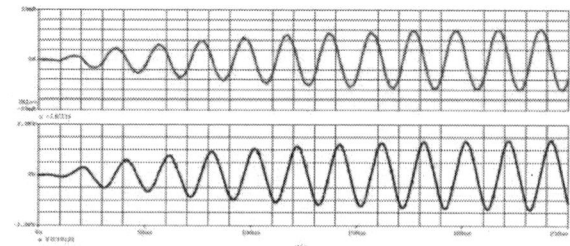

(b) Output Current (Top) and Output Voltage (Low)

Fig. 7: Simulation Waveforms of Push-Pull Inverter

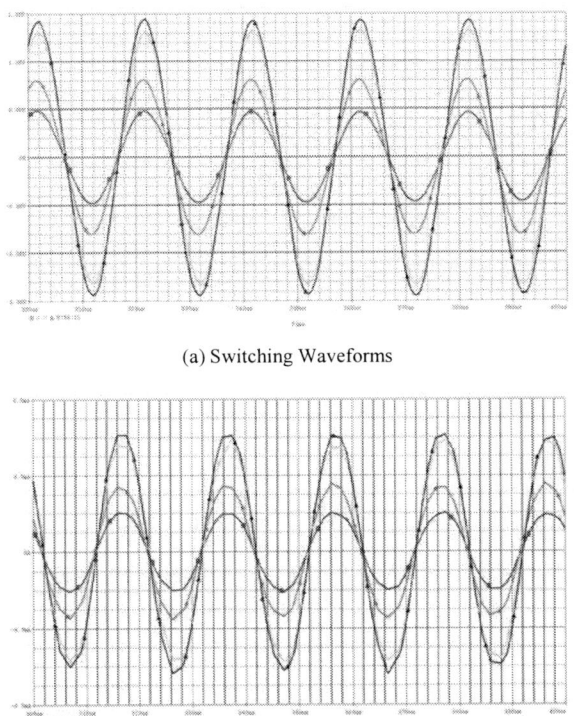

(a) Switching Waveforms

(b) Waveforms of Output Current

Fig. 8: Output Voltage·Current Waveforms with Load Resister Variation

The figure 8 shows the output wave form of the EEFL when loads are changed. The sine wave made from switching by the rectangular wave of the figure 8(a) operates the EEFL and as a result, we can know the wave forms of the lamp, 8(b) voltage and electric current. And In the EEFL, we can also know the fact that the high voltage-low electric current operation is done-- one of the characteristics of EEFL. And also, we can know that the electric current flows faster than the voltage-- 90 degrees.

B. Dimming Control

The dimming control of the screen is one of the essential functions that inverter for LCD Backlight must make. The way of the dimming control is the following: the brightness of the lamp is controlled by amount of electric current that flows in the lamp. And sensing the change of the electric current is done to keep the high efficiency of inverter. Generally, there are two ways in the control of brightness: one is Burst control way that, the other is analogue control way. The former takes advantage of PWM method, the later applies a variable resistance. PWM method is the way to control the degree of brightness according to the duty rate, width of pulse, comparing PWM signal with standard voltage.

The variable resistance method is the way to control the degree of brightness by tuning the amount of standard DC. It is a simple way, but it is uneasy because of using a manual element. On the study, as you see in the figure 9, Burst control way is adopted.

The Burst way to control the On/Off of inverter, using a variable pulse width, is generalized.[4]

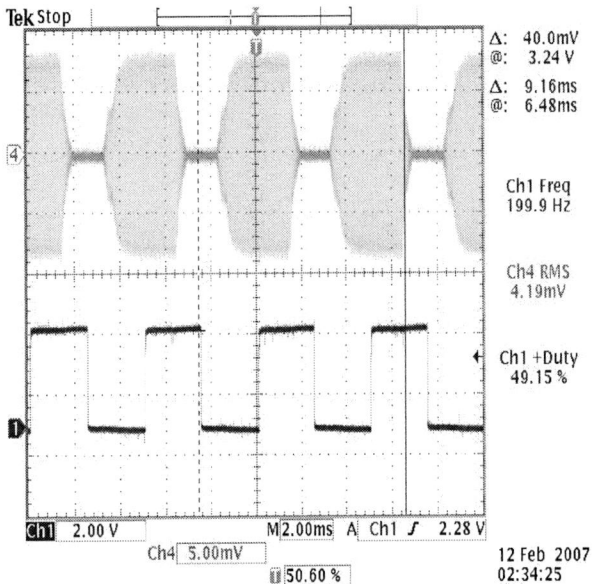

Fig. 9: Burst Dimming Control (Dim 50%)

C. The Test Result

In the planned inverter, when the load of the EEFL is tube diameter 4.0mm and tube length 740mm, the operation voltage is about 1700-1790[v] and the electric current of the lamp is about 7.40-7.51[mA], and the resistance of the emitting tube is about 100K[Ω].

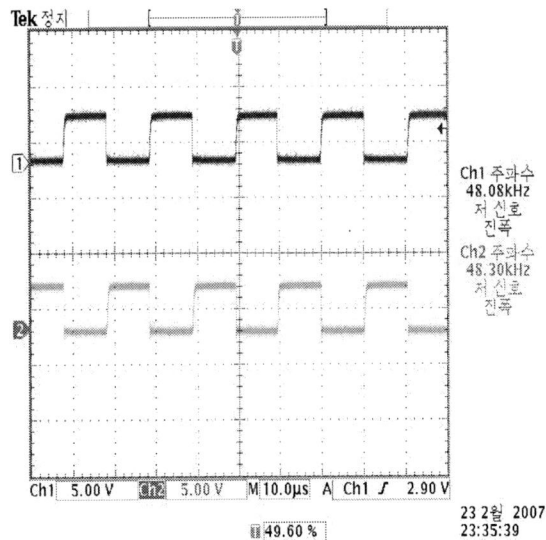

Fig. 10: Switching Voltage Waveform

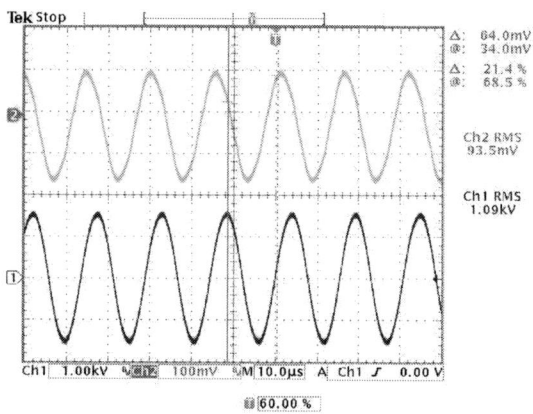

(a) Master Voltage (High) and Current (Low)

The figure 10 shows the wave form of input switching voltage and the figure 11 indicates the wave forms of the output voltage and output electric current of the inverter to

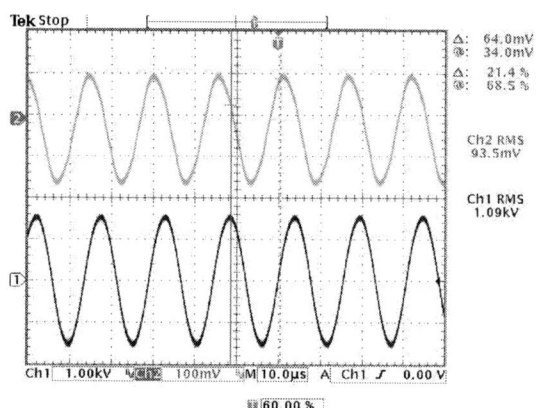

(b) Slaves Voltage (High) and Current (Low)

Fig. 11: Output Waveforms of EEFL

apply High-High lamp operation method, one of the operation ways of EEFL. And in the figure 10, we can know that the wave forms are steady sine wave ones. We can find out that High-High lamp operation method adopted in the test study makes each lamp radiate well. And also, we can measure the resonance wave 48[kHz] for the operation of the inverter. We can know that the wave 48[kHz] matches well with the piezoelectric transformer.

V. CONCLUSION

The study analyzed the electrical characteristics of EEFL and made the inverter for operation of EEFL to backlight 32inch large LCD screen, using the piezoelectric transformer. We could get the above test results through the simulation of the operation- piezoelectric inverter of EEFL, applying

PSpice, electronic circuit interpretation program. The inverter of the test was made by Push-Pull way with ZVS to get high energy efficiency. The use of the piezoelectric transformer removed unnecessary intervention (EMI) between the inverter and LCD panel. We found out that the piezoelectric inverter usually applied to a small LCD Backlight can be used for more than 32inch large LCD Backlight. At the present, the increase of the capacity of piezoelectric transformer has emerged as an urgent problem. If the piezoelectric transformer having a large capacity is made, it can replace a coil transformer in the future. And if various topologies such as Full-Bridge are applied, we expect that we can get lower operation voltage, lower electric current, and higher energy efficiency in the future.

ACKNOWLEDGMENT

This study was supported by the Research Center for Stabilization of Power Systems with Large Distributed Generations at Chonnam National University in KOREA.

REFERENCES

[1] Doo-Oh Hur, Tae-Ku Kang, Cheol-Hwan Cho, Hae-min Lee, Hyung-Keun Ahn, Deuk-Young Han, Design and Fabrication of Piezoelectric ceramic Transformer for the LCD Backlight, Properties and Application of Dielectric Materials, *Proceedings of the 5th International Conference*, 1999: p843-846.

[2] J. H. Han, Y. C. Lim, S. H. Yang, G. H. Kweon, Design of the Backlight Inverter for Multi-lamp Driving, ICPE-2001, 2001: p80-83.

[3] Y. Takeda, M. Takegi, Y. Watanaba, M. Amano, H. Nakano, Thigh Reliability External Electrode Mercury Fluorescent Lamp for a LCD TV's Backlight, HSID 2002 DIGEST, 2002: p364-349.

[4] H. L. Chang, L. Ying, J. P. Pai, Digital-Dimming Controller with Current Spikes Elimination Technique for LCD Backlight Electronic Ballast, HIEEE APEC 2004, 2004: p153-158.

[5] G. H. Kweon, Y. C. Lim, S. H. Yang. " An analysis of the Backlight Inverter by Topologies", IEEE, ISIE-2001, pp 896-900.

978-1-4244-1871-8/07 $25.00 © 2007 IEEE

Direct Torque Control for Induction Machine based on AC-AC Matrix Converter

Guo Yougui
College of Information and Engineering of
Xiangtan University, Hunan, China, 411105
Email: guoyougui@sina.com

Zhu Jianlin, Chen Caixue
College of Information and Engineering of
Xiangtan University, Hunan, China, 411105

Abstract—The combination of matrix converter with direct torque control is a way towards green energy-saving of the world. It has great sense for theory research and good developing future. First, it is analyzed about working principle of matrix converter, the control principle and its realization process of direct torque control for induction machine inverter-based. Then the comprehensive switching control table is deduced according to the special properties of matrix converter together with direct torque control for induction machine. And the comprehensive modulation strategy can be concluded by looking up the table. Third, the simulation model is set up. The simulation results have verified the feasibilities and advantages of this novel alternative current adjustable speed system. Finally, an adaptive fuzzy PI controller was designed to improve the electric magnetic torque of induction machine.

I. INTRODUCTION

AC-AC matrix converter (named matrix converter later) is a matrix array consisting of a number of bidirectional switching devices, which is used to converter the input voltages into the output voltages with expected amplitude and frequency. Its topology is represented as fig.1. It has a series of advantages and is a green converter [1-2]. With its ripeness of development, several countries have begun its application research.

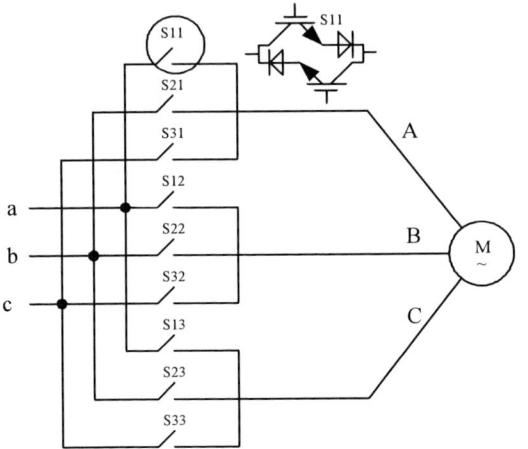

Fig.1 The simplified topology of matrix converter

Direct torque control [3-4] is a high performance AC adjustable speed technology. It was put forward by M. Depenbrock for the first time in 1985. Afterwards, Japanese scholar, I. Takahashi also did a similar control strategy and got an exciting control results. It is an internal control with feed back of torque and can overcome the effect of torque on flux variation, and can be decoupled between speed and flux.

The combination of matrix converter (MC) and direct torque control (DTC) for induction machine is a new AC adjustable speed technology with a very bright future. The research has just started [5-6]. Three-phase to three-phase matrix converter is selected as a supply in this control system. The simulation model is setup for this novel system and simulation results are given in this paper. Then the electronic magnetic torque of induction machine is improved by designing an adaptive PI controller.

II. BASIC PRINCIPLE

A. The working principle of matrix converter

The working principle is that the switching on or off is effectively controlled for nine directional switches according to defined rule, and the expected output voltages are gotten by input voltages multiplied by a switching matrix; similarly, the expected input currents are also gotten. The mathematical models are represented as follows.

$$\vec{V}o = M \times \vec{V}i \qquad (1)$$

$$\vec{I}i = M^T \times \vec{I}o \qquad (2)$$

Where M is a 3×3 modulation matrix, which consists of working states of 9 bidirectional switches; $\vec{V}_o$ is three-phase output voltage vector gotten by modulation, $\vec{V}_i$ is three-phase input voltage vector, $\vec{I}_o$ is three-phase output current vector, $\vec{I}_i$ is three-phase input current vector gotten by modulation.

B. The basic principle of DTC for induction machine

Here take two-level inverter driving three-phase induction machine as main parts of AC adjustable speed system in order to exemplify [3-4]. Three-phase output voltages are work out respectively under the 8 kinds of safe switching modes of inverter. Then 8 output voltage vectors are gotten according to the expression of PARK transformation definition. Among these 8 vectors, 6 are working ones and 2 are zero ones. Thus a voltage vector hexagon can be made out. The flux is controlled by two-level hysteresis. As is represented in Fig.2. The estimation value of flux and its phase angle is calculated in expression (3)-(6).

$$\psi_{s\alpha} = \int \left(u_{s\alpha} - i_{s\alpha} R_s \right) dt \qquad (3)$$

$$\psi_{s\beta} = \int \left(u_{s\beta} - i_{s\beta} R_s \right) dt \qquad (4)$$

$$|\psi_s| = \sqrt{\psi_{s\alpha}^2 + \psi_{s\beta}^2} \qquad (5)$$

$$\theta_{\psi_s} = tg^{-1} \frac{\psi_{s\beta}}{\psi_{s\alpha}} \qquad (6)$$

978-1-4244-1871-8/07 $25.00 © 2007 IEEE

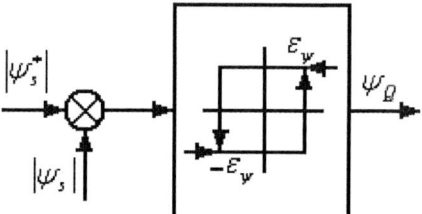

Fig.2 The stator flux hysterisis regulator

And the electronic torque is controlled by three-level Hysteresis. As is represented in Fig.3. Its estimation value is calculated in expression (7). The given torque is gotten through the given speed and measured one by PI regulator. Then the switching combination table of stator flux can be gotten according to all kinds of output combinations of flux hysteresis and torque hysteresis, between the changes of flux and torque and the corresponding relationship of each sector of output voltage vector hexagon. Direct torque control can be concluded that the table is looked up for various combinations of the outputs of flux hysteresis and torque hysteresis. Thus the control signal of switching are decided to reach the aim.

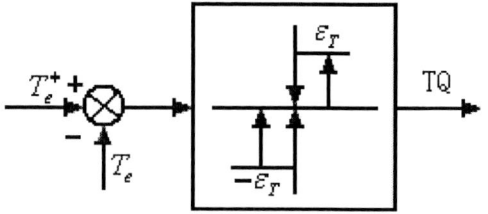

Fig.3 The torque hysterisis regulator

$$T_e = \frac{3}{2} P \left(\psi_{s\alpha} i_{s\beta} - \psi_{s\beta} i_{s\alpha} \right) \tag{7}$$

III. THE COMBINATION OF MATRIX CONVERTER AND DIRECT TORQUE CONTROL OF INDUCTION MACHINE

A. The basic switching combinations of matrix converter

For a 3 to 3 matrix converter, such conditions must be satisfied as follows: input port must not be shortened off and output port must not be cut off. So there are only 27 switching combinations to be permitted. Except for 6 kinds with variable direction of voltage vectors, only 21 switching combinations with still voltage vectors are convenient for switching control [5]. As is shown in table 1.

Table 1
The basic switching combinations of matrix converter

+1	-1	+2	-2	+3	-3	+4	-4
a b b	b a a	b c c	c b b	c a a	a c c	b a b	a b a
+5	-5	+6	-6	+7	-7	+8	-8
c b c	b c b	a c a	c a c	b b a	a a b	c c b	b b c
+9	-9	0	0	0			

a a c	c c a	a a a	b b b	c c c			

Where a, b, c are correspond to three-phase inputs respectively, while each switching combination correspond to three-phase outputs A, B, C respectively. Such as switching combination abb correspond to three-phase outputs A, B, C respectively. $\pm 1 - \pm 9$ represent 18 kinds of effective switching combinations. For example +1 corresponds to abb, 0 corresponds to zero switching combination.

B. Combination control of matrix converter and DTC

To begin with, the eight output voltage space vectors are gotten by the basic principle of direct torque control. Second, the switching combinations in 6 input current sectors are achieved according to the corresponding relationship among each output voltage space vector, the changes of stator flux and input power factor, as well as the output values of hysteresis control for torque and the basic switching combinations, then the total switching control table is designed out, which is very important for the control of matrix converter – direct torque control system. As is represented in table 2. In table 2, I~VI represents 6 sectors of input current in ascending order respectively, pf represents the input power factor, V1~V6 represent switching optimally six output voltage effective space vectors of direct torque control based on inverter, $\pm 1 - \pm 9$ correspond to table 1. The control of this novel AC adjustable speed system can be comprehended to look up the total switching-control table according to various control demands. Thus it is realized to effectively control this novel AC adjustable speed system. Its realization process is represented in Fig.4.

Table 2
The switching table of MC together with DTC

	I		II		III		IV		V		VI	
pf	+1	-1	+1	-1	+1	-1	+1	-1	+1	-1	+1	-1
V1	-3	1	2	-3	-1	2	3	-1	-2	3	1	-2
V2	9	-7	-8	9	7	-8	-9	7	8	-9	-7	8
V3	-6	4	5	-6	-4	5	6	-4	-5	6	4	-5
V4	3	-1	-2	3	1	-2	-3	1	2	-3	-1	2
V5	-9	7	8	-9	-7	8	9	-7	-8	9	7	-8
V6	6	-4	-5	6	4	-5	-6	4	5	-6	-4	5

In this system, the most important part is S-function. It flexibly completes the comprehensive control strategy of matrix converter together with DTC of induction machine. It works in such a way: when flag is 0, it callbacks the initialized module to initialize the system; when fag is 1 or 2 or 9, it does nothing; when flag is 4, it callbacks the sampling time module to calculate the next sampling time; when flag is 3, it callbacks the output module to decide the switching control signal matrix 3×3, and outputs a vector with 9 elements, which is the control signals to control the switching on or off of 9 bidirectional switches. This output module is very complex. Its working process is represented in Fig.5.

978-1-4244-1871-8/07 $25.00 © 2007 IEEE

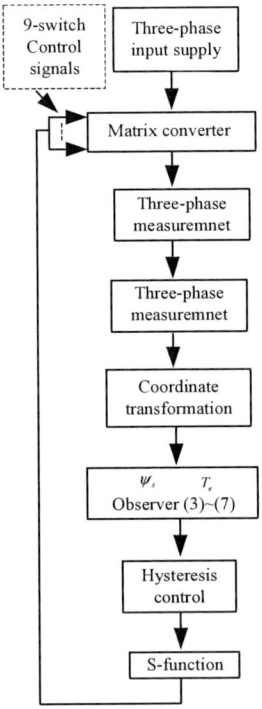

Fig.4 The brief flow chart of simulation

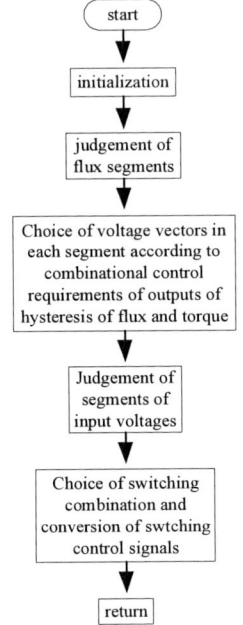

Fig.5 The flow chart of producing switching control signals

IV. INSTANCES RESEARCH

Basic parameters set for three-phase induction machine: P_N= 3kVA, rated line to line voltage U_N=380V, f_N= 50Hz, Rs=0.435 Ω , Ls=2.0mH, Rr=0.816 Ω , Lr=2.0mH, Lm=69.31mH, the pairs of magnet poles=2. Three-phase input line to neutral voltages 220V, input frequency=50Hz.

The simulation model of this novel AC adjustable speed system is set up with MATLAB/SIMULINK power system blockset, etc. Above all, the control strategy is realized by S-function.

A. Induction machine operation with constant load

Speed of induction machine is 600rpm, load torque is 25N.m. The simulation waveforms are as follows (Fig.6-7). Seen from the waveforms, the novel control strategy is feasible which can use advantages of MC and DTC.

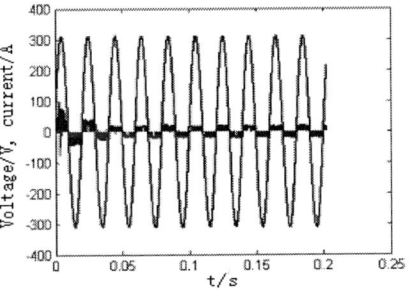

Fig.6 Input voltage and current of MC

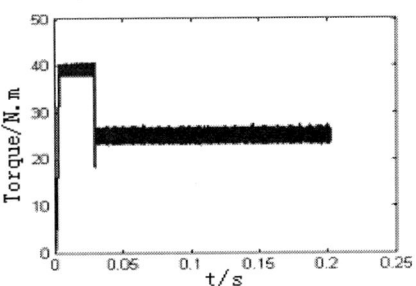

Fig.7 Waveform of electric magnetic torque

B. Improvement of control strategy of the system

Fuzzy controller has its own features. For example, it doesn't require the precise mathematical model of controlled object, it only demands to master experiences of operators or knowledge of experts in this field; it has a strong robustness, especially fitting for control of nonlinear, variable-time, and lagging system.

As for fuzzy PI, that is to say, on the basis of PI algorithm, it uses fuzzy rules to carry out fuzzy inference by calculating the error e and its variable rate etc. of the present system. Then look up the fuzzy matrix table to adjust parameters. It is shown in Fig. 8. The kernel of fuzzy PI control design is to summarize technological knowledge and actual experience of engineering technicians. A fitting fuzzy rule table is set up to get a fuzzy control table for respectively setting two parameters-K_P and K_I. It is shown in table 3.

An adaptive PI fuzzy controller is designed for the case 4.1 with proper fuzzy control rules [7-8]. The original PI regulator deciding the given torque is replaced by this new controller in the simulation model. The simulation waveform shows it has improved the torque effectively. As is represented by Fig.9.

Table 3 The fuzzy rule table of fuzzy PI two parameters

e	Ec (K$_P$/K$_I$)						
	NB	NM	NS	ZO	PS	PM	PB
NB	PB/NB/PS	PB/NB/NS	PM/NM/NB	PM/NM/NB	PS/NS/NB	ZO/ZO/NM	ZO/ZO/PS
NM	PB/NB/PS	PB/NB/NS	PM/NM/NB	PS/NS/NM	PS/NS/NM	ZO/ZO/NS	NS/ZO/ZO
NS	PM/NB/ZP	PM/NM/NS	PM/NS/NM	PS/NS/NM	ZO/ZO/NS	NS/PS/NS	NS/PS/PZ
ZO	PM/NM/ZP	PM/NM/NS	PS/NS/NS	ZO/ZO/NS	NS/PS/NS	NM/PM/NS	NM/PM/ZO
PS	PS/NM/ZO	PS/NS/ZO	ZO/ZO/ZO	NS/PS/ZO	NS/PS/ZO	NM/PM/ZO	NM/PB/ZO
PM	PS/ZO/PB	ZO/ZO/NS	NS/PS/PS	NM/PS/PS	NM/PM/PS	NM/PB/PS	NB/PB/PB
PB	ZO/ZO/PB	ZO/ZO/PM	NM/PS/PM	NM/PM/PM	NM/PM/PS	NB/PB/PS	NB/PB/PB

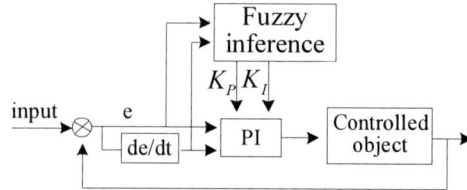

Fig.8 The controller structure of adaptive fuzzy PI

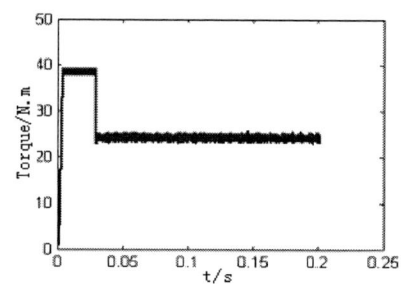

Fig.9 Waveform of electric magnetic torque

V.   CONCLUSION

The combination of matrix converter with direct torque control for induction machine is feasible and a development direction in the future AC adjustable speed system. The responses of electric magnetic torque and flux is fast. Further, the torque can be improved by adaptive fuzzy PI regulator. Application of intelligent control algorithm will be a good improvement tendency to this novel system.

REFERENCES

[1] L.Huber, D.Borojevic. "Space vector modulated three phase to three phase matrix converter with input power factor correction," IEEE Trans. on Ind. Appl., 1995, vol. 31 no. 6, pp 1234-1246

[2] Zuang Xinfu. The control principle and experimental study for AC-AC matrix converter [J]. Power electronics, 1994, (2): 1-6.

[3] Li Su. "Direct torque control for induction machine". Beijing: mechanical industry press, 2001.

[4] Xiao Anwen, She Zhiting. "The simulation study of direct torque control for induction machine". The world of inverters, 2004, pp125-128.

[5] Casadei, D. Serra, G. Tani, A. "The use of matrix converters in direct torque control of induction machines," IEEE 24th Annual Conference of Industrial Electronics Society, vol. 2 pp744-749, 1998.

[6] Casadei, D. Serra, G. Tani, A. "The use of matrix converters in direct torque control of induction machines," IEEE Transactions on industry electronics, 2001, vol. 48, no. 6, pp. 1057-1064.

[7] Zhang Weiguo, Yang Xiangzhong. Fuzzy control theory and its application. Shanxi: Xibei industry university press, 2000.

[8] Liu Jinkun. MATLAB simulation of advanced PID control, Beijing: electronics industry press, 2004.

A Development of the Solar Position Tracker on the Program Method for the Small Typed Stand-alone PV System

L H Hwang*, S K Na*,H S Kim*, H S Oh*, Y S Kim*, HanXiao*, YuMingZhi*,
M T Cho **, S C Chang *** and G S Choi ****

*Dept. of Electircal Engineering, Semyung University, Jechon, Chungbuk, 390-711, Korea
**Dept. of Electrical Engineering, Daewon   Science College, Jechon, Chungbuk, 390-702 Korea
***KEPCO, Daejeon, Korea
**** InTech-FA.Co . Boondanggu Sungnam Kyunggido,Korea
Email: lhhwang@semyung.ac.kr

Abstract— In this paper, I used to microprocessor and sensor and designed to improve the efficiency of the photovoltaic system the photovoltaic position tracker device, and compared the normal photovoltaic system of fixed form with the photovoltaic system of solar position tracked form. Moreover, compared the catalogue of solar cell module and the simulation through a mathematics modeling with the solar cell's characteristic interpreting and, composed an power conversion system with boost converter and voltage source inverter. Used the constant voltage control method for maximum power point tracking in boost converter control and, used the SPWM(Sinusoidal Pulse Width Modulation) control method in inverter control. The result was less then 5% when compared the catalogue of solar cell module and the simulation through a mathematics modeling.

I. INTRODUCTION

The main drawbacks of PV systems are high fabrication cost and low energy-conversion efficiency, which are partly caused by their nonlinear and temperature-dependent V-I and P-I characteristics. To overcome these drawbacks, three essential approaches can be Improving manufacturing processes of solar arrays [1], controlling the insolation input to PV arrays [2] and Utilization of output electric power of solar arrays : the main reasons for the low electrical efficiency are the nonlinear variations of output voltage and current with solar-radiation levels, operating temperature, and load current. overcome these problems, the maximum power operating point of the PV system is tracked using online or offline algorithms and the system operating point is forced toward this optimal condition [3]-[4].

Maximum power point tracking(MPPT) is performed by some battery charge controllers and by most grid connected PV inverters. The principle is to adjust the actual operating voltage V (or current I) of the PV array so that the actual power P approaches the optimum value Pmax as closely as possible.

Many different ways exist to track the MPP which can be classified as either direct or indirect methods

Direct methods include algorithms that use measured DC input current and voltage or AC output power values, and, by varying the PV array operational points, determine the actual MPP. Adjustment of MPP may well or not include artificial MPP search movements.

Indirect methods are those which use an outside signal to estimate the MPP. Such outside signals may be given by measuring the irradiance, the module temperature, the short circuit current, or the open circuit voltage of a reference solar cell.

In this paper, two simple and powerful maximum power-point tracking indirect methods (based on "computational" methods) known as voltage-based VMPPT [3] and current-based CMPPT [4] are simulated, and compared.

The model must be valid under different isolation, temperature, and degradation conditions. Based on the modeled V-I characteristics, the corresponding maximum power point are computed for different load conditions as a function of cell open-circuit voltages or cell short-circuit currents.

Comparison and analyses results illustrate the advantages and shortcomings of both techniques. Finally, the optimal applications of each tracker are classified.

II. RADIATION- AND TEMPERATURE DEPENDENT SOLAR CELL CHARACTERISTICS

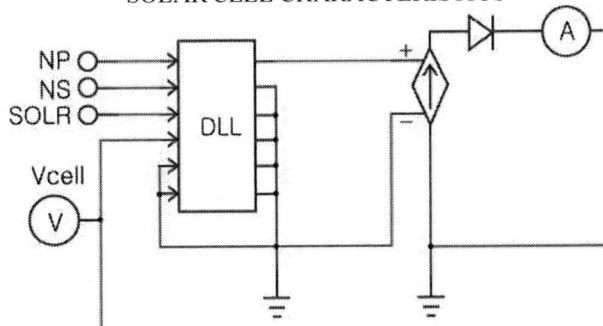

Fig1. Schematic circuit of PV array with PSIM

The PV generator is formed by the combination of many PV cells connected in series(Ns) and parallel(Np) to provide the desired output voltage and current. This PV generator exhibits a nonlinear insolation dependent V-I characteristic. Equation (1) is result modeling of solar cell and fig.1 is the schematic circuit of PV array with PSIM using (1).

$$V_{PV} = -I_{PV}R_S(\frac{N_S}{N_P}) + (\frac{N_S}{\lambda})\ln\left[1 + \frac{N_P I_{SC} - I_{PV}}{N_P I_O}\right] \qquad (1)$$

where, I_{sc} is the cell short-circuit current (representing insolation level), I_o is the reverse saturation current, R_s is the series cell resistance, and λ is a constant coefficient and depends upon the cell material. I_{PV}, V_{PV} are the solar cell array current and voltage, respectively. For given values of solar cell array parameters, the V-I characteristic depends on the solar insolation and the MPP varies with the solar insolation. Rewriting (1) as

$$V_{PV} = -I_{PV} R_S (\frac{N_s}{N_P}) + (\frac{N_s}{\lambda}) \times \left[\ln(\frac{I_{pha}}{I_o}) + \ln(1 - \frac{I_{PV}}{N_P I_{pha}}) \right] \quad (2)$$

where, $I_{pha} = I_{SC} + I_o$ Expanding the term $\ln(1 - (I_{PV}/N_P I_{pha}))$ into Taylor series and neglecting higher orthe terms[5] results in the following equation.

$$V_{PV} = -I_{PV} (R_{sg} + (\frac{2N_s}{\lambda N_P I_{pha}}) + (\frac{N_s}{\lambda} \left[\ln(\frac{I_{pha}}{I_o}) \right] \quad (3)$$

Simplifying the above equation for the solar cell array current results in the following equation.

$$I_{PV} = \frac{1}{(R_{sg} + \frac{2N_s}{\lambda N_P I_{pha}})} \left[\frac{N_s}{\lambda} \ln(\frac{I_{pha}}{I_o}) - V_{PV} \right] \quad (4)$$

where, $R_{sg} = N_S R_S / N_P$. The equations (3) and (4) are used in the simulation studies.

Computed (2) (3) V-I as well as P-V characteristics for the PSE panel are shown in Fig. 2 for two insolation levels. This figure illustrates the variations of the cell maximum power point (e.g., the maximum of the P-V curves) with rspect to insolation levels. The key specifications are shown in table 1.

Table 1. Specifications of silicon solar panels(PSEM50)

Open –circuit cell voltage	Voc=21	[V]
Short-circuit cell current	Isc =3.17	[A]
Reverse saturation current	Io = 0.5×10^{-4}	[A]
Voltage, max power	Vpv =17.1	[V]
Current, max power	Ipv =2.92	[A]
Maximum power	Ppv =49.9	[W]
Cell resistance	Rs =0.0277	[Ω]
Cell material coefficient	λ =0.049	[1/V]

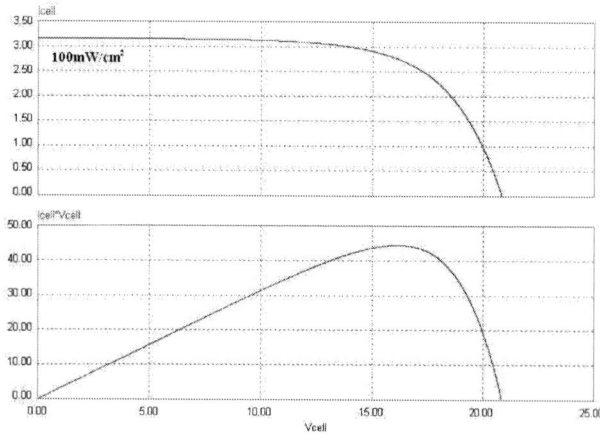

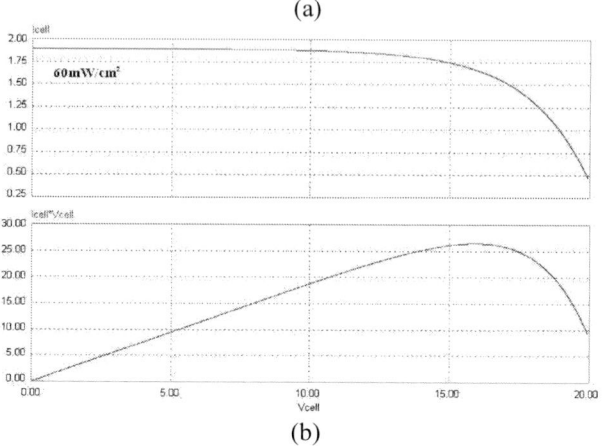

(a)

(b)

Fig 2. PSIM model V-I and P-V curves for different insolation levels. ((a):1000w/cm2, (b):600w/cm2)

III. VOLTAGE- AND CURRENT-BASED MPPT TECHINQUES.

To determine the operation point corresponding to maximum power for different insolation levels, (2) is used to compute the partial derivative of power with respect to cell current. Reference [4] employs numerical methods to show a linear dependence between the "cell currents corresponding to maximum power" and the "cell-short circuit currents"

$$I_{MP} = M_c I_{sc} \quad (5)$$

This equation characterizes the main idea of the current-based maximum power-point tracking (CMPPT) technique. M_c is cellde the "current factor" and is equal to 0.86 for the PSE silicon panel (Table I). Equation (5) is plotted in Fig. 3 together with the computed (almost linear) dependence of I_{MP} with respenct to I_{sc} (indicated by "+" signs).

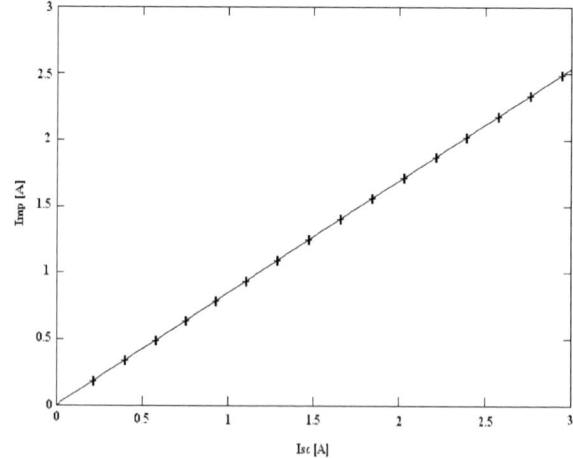

Fig. 3. Computed ("+" signs) and linear (5) dependence of "cell current corresponding to maximum power" versus "cell short-circuit current" for one PSE panel (T=25°C at varying insolation levels).

978-1-4244-1871-8/07 $25.00 © 2007 IEEE

A similar approach is taken in [3]. It is shown that "cell voltages corresponding to maximum power" exhibit a linear dependence, independent of panel configuration, with respect to cell "open-circuit voltages" for different insolation and temperature levels

$$V_{MP} = M_V V_{oc} \qquad (6)$$

This equation represents the concept of the voltage-based maximum power-point tracking (VMPPT) technique and M_V, which is called the "voltage factor" is equal to 0.71 for the PSE silicon panel. Equation (6) is plotted in Fig. 4 along with the computed (almost linear) dependence.

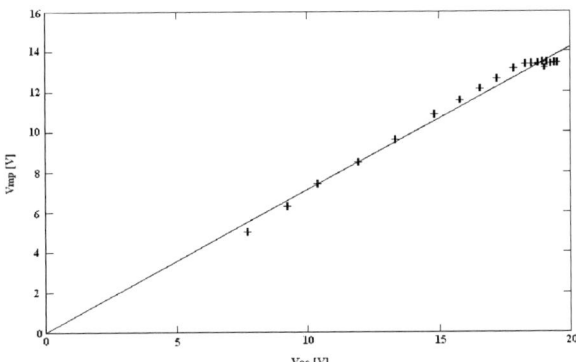

Fig. 4. Computed ("+" signs) and linear (6) dependence of "cell voltage corresponding to maximum power" versus "cell open-circuit voltage" for one PSE panel (T=25° C at varying insolation levels)

To a first order approximation, the Mpp current is dependent solely on the in-plane irradiance. If the MPPT works correctly, the input current I should be close to MPP current. Since the proportionality between irradiance and MPP current is a linear relation, it is valid not only for instantaneous values, but also for integrals as sums or average data. According to Fig. 3 and 4, CMPPT and VMPPT techniques are simple and fast methods for maximum power-point estimation.

3-1. SIMULATION OF VMPPT AND CMPPT TECHNIQUES

Simulink software and its facilities are used to model a resistive-load solar system with a VMPPT (in boost mode) tracker as shown in Fig. 5.
For the solar cell equivalent circuit, we have created a block called "DLL(Dynamic Link Library)" as shown in Fig. 5, which simulates the nonlinear V-I characteristics of one PSE panel (1), employing the cell short circuit (Isc) as a measure of insolation level.
We have introduced a delay function to limit the fast current response of the "controlled voltage source" and to improve the convergence of solution.
For the voltage-based MPPT equivalent circuit, we have used

a block call "VMPPT" as shown in Fig. 5. This block computes cell open-circuit voltage [using Isc and (2), compares it with the PV output voltage using (6) and calculates the firing commands for the pulse-width modulation (PWM) block].

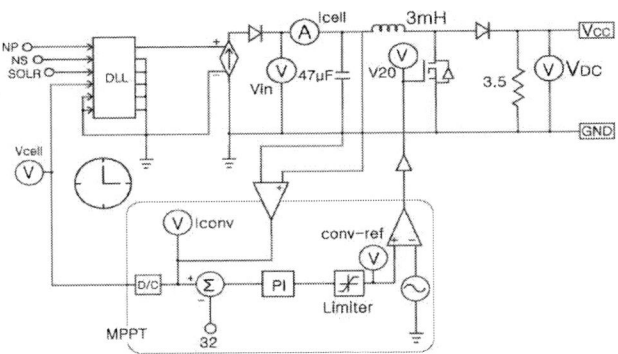

Fig. 5. Simulation of a resistive-load PV system with VMPPT.

Fig. 6 shows computed voltage, current, and power characteristics of Fig. 5 at the output of the "DLL" block (top graph) and across the 3.5Ω resistive load (bottom graph). The captured maximum power is approximately 32 W.

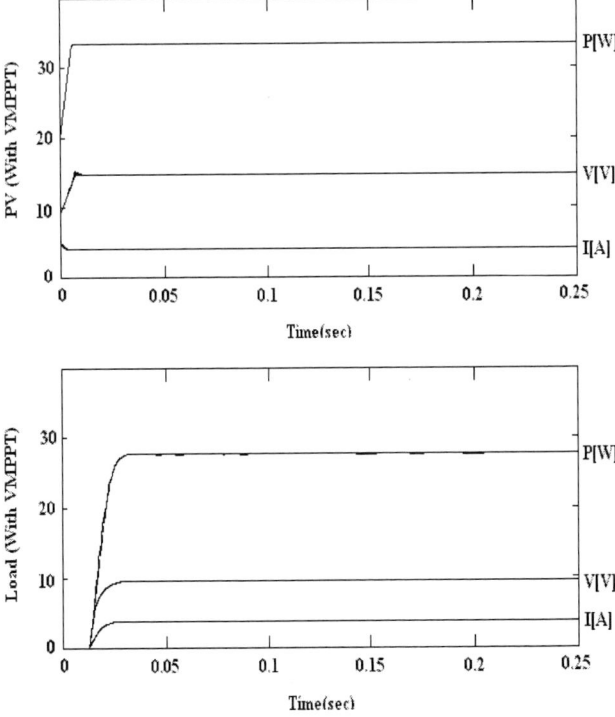

Fig. 6. Computed (using Fig. 5) voltage, current, and power characteristics of the PV panel (top graph) and resistive-load (bottom graph) for the PV system with a boost-mode VMPPT.

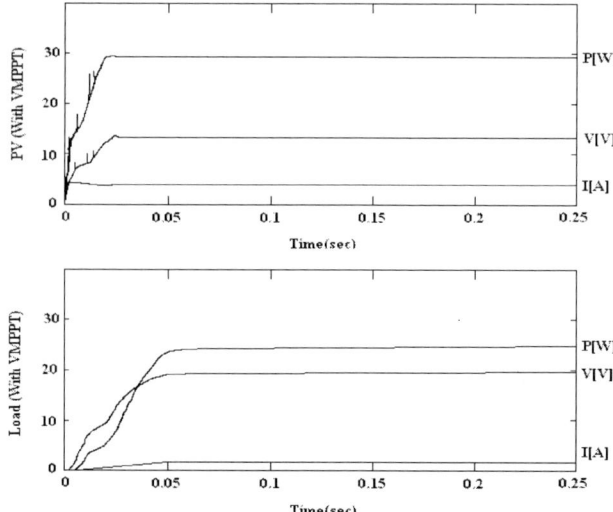

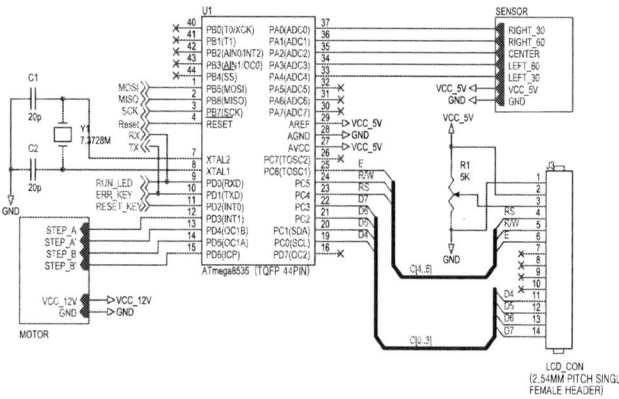

Fig.7. Computed voltage, current, and power characteristics of the PV panel (top graph) and at the 15.5Ω resistive-load (bottom graph) for the PV system with a boost-mode VMPPT

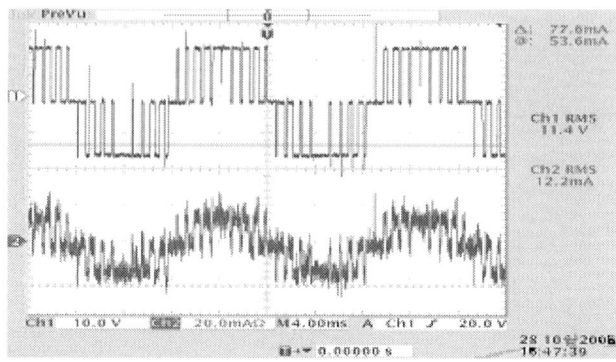

Fig. 8. Circuits of microprocessor part.

IV. COMPARISON OF VMPPT AND CMPPT TECHNIQUES

Based on the theoretical investigations of this paper, the following results may be stated.

(1) Both VMPPT and CMPPT techniques are fast, practical, and powerful methods for MPP estimation of PV generators under all isolation and temperature conditions. The increase in output power depends on load characteristics, environmental factors (isolation and temperature), and the type of tracker used.

(2) Online measurements of PV short-circuit and output currents make CMPPT hardware more complicated and expensive compared with (same rating) VMPPT circuitry, requiring voltage measurements only. Fig. 9 shown the output voltage and current waveforms of inverter is operated in the Small Typed Stand-alone PV System.

Fig. 9. Output voltage and current waveforms of inverter.

V. CONCLUSIONS

Two powerful and practical methods for maximum power-point tracking of PV systems are investigated and compared. For the theoretical analysis, "PSIM" and "Simulink" facilities are employed. Based on the results presented before, the following conclusions may be stated.

(1) The linear current function used by the CMPPT technique is a more accurate approximation of the actual nonlinear PV characteristics compared with the linear voltage function of the VMPPT technique.

(2) VMPPT technique is naturally more efficient and has less circuit losses .

As a result, the optimal MPPT methodology strongly depends on matching load and tracker characteristics. Considering the natural behavior, advantage, and limitations of CMPPT and VMPPT techniques, the following suggestions are made for some typical PV loads;

(3) PV loads with high voltage and low current (e.g., motors and high resistive loads) could be matched with either VMPPT or CMPPT systems, but the VMPPT technique will result in simple hardware with higher efficiency and lower noise and cost.

REFERENCES

[1] H. S. Rauschenbach, Solar Cell Array Design Handbook; The Principles and Technology of Photovoltaic Energy Conversion. New York; Van Nostrans, 1980.

[2] K. E. Teager, " Electric vehicles and solar power: Enhancing the advantages of electricity," *IEEE Power Eng. Rev.*, vol. 12, Oct. 1992.

[3] "Design, construction and testing of a voltage-based maximum power point tracker(VMPPT) for small satellite power supply," in 13th Annu. AIAA/USU Conf. small Satellite, Aug. 23-26, 1999.

[4] S. M. Alghu wainem, "Matching of a dc motor to a photovoltaic generator using a step up coverter with a current-locked loop,*" IEEE Trans. Energy Conversion* , vol. 9,pp. 192-198, Mar. 1994

978-1-4244-1871-8/07 $25.00 © 2007 IEEE   220

A study on Driving of 35W(T5) fluorescent lamp by the electronic ballast using piezoelectric transformer

L H Hwang*, J H Yoo*, H S Song*, S K Na*, H S Kim*, H S Oh*, S H Lee*, K H Choi*
And M T Cho **

*Dept. of Electircal Engineering, Semyung University, Jechon, Chungbuk, 390-711, Korea
**Dept. of Electrical Engineering, Daewon Science College, Jechon, Chungbuk, 390-702 Korea
Email: lhhwang@semyung.ac.kr

Abstract—Recently, 35W class Fluorescent lamps with 32mm tube diameter are replaced with 32W one with 26 mm in diameter to conserve lamp materials and to increase luminance efficiency. Moreover, 35, 28 and 14 W fluorescent lamps with 16mm (T5) in diameter, which are nowadays developed, also may replace 32 W lamps again. Application of slim lamps, however, requires small sized electronic ballast to full fill the design philosophy of miniaturizing. However, the traditional magnetic ballasts operated at 50-60Hz have been suffered from noticeable flicker, high loss, large crest factor and heavy weight. In this study, in order to solve these problems, a new type of electronic ballast, which is composed of rectifier, active power factor corrector, series resonant half bridge inverter and piezoelectric transformer, was proposed for driving T5 fluorescent lamp. Driving of piezoelectric transformer was carried out with input region for the ring electrode and output region for the dot electrode. A 35W (T5) fluorescent lamp is successfully driven by the fabricated ballast with piezoelectric transformer. After driving the lamp using the proposed electronic ballast for 20 min, the input power factor and efficiency of ballast shown 0.95 and 86%, respectively, at operating frequency of 81 kHz. And also, the output power and temperature rise of piezoelectric transformer showed 35.07W and 20.5℃, respectively.

I. INTRODUCTION

As PDA (personal digital assistants) has become widely used for wireless mobile communications, their important components have also actively developed. PDA is composed of LCD (liquid crystal display), CCFL (cold cathode fluorescent lamp), and inverter for driving CCFL (LCD backlight), etc. Recently, the piezoelectric transformer has been widely applied to the LCD backlight inverter for notebook, camcorder and PDA, etc because it has favorable characteristics such as electromagnetic-noise free, compact size, higher efficiency and superior power density, etc, compared with electromagnetic one[1][2]. Accordingly, various attempts to improve piezoelectric and structural characteristics of piezoelectric transformer have been intensively performed [3]. Especially, Rosen-type piezoelectric transformers can be effectively used for driving LCD backlight for PDA, because their electrical characteristics accord with required ones of LCD backlight that needs high input voltage at ignition state and low input voltage in static one. Piezoelectric transformer is a device to convert electrical energy to mechanical vibration. Accordingly, mechanical strength is important requirements

for supporting mechanical vibration. Therefore, the dense composition ceramics with fine-grain size is necessary for inhibiting its crack caused by the rise of vibration velocity [4]. Temperature rise of the piezoelectric transformer is also inevitable when it is operated with high vibration velocity under high power [4][5]. Accordingly, its mechanical quality factor (Qm) must be as high as possible in order to inhibit temperature rise caused by generation of heat[6]. In this study, $Pb(Mn_{1/3}Nb_{2/3})O_3$-$Pb(Zr,Ti)O_3$(PMN-PZT)composition ceramics that have been well known for high mechanical quality factor was chosen and $Pb(Ni_{1/2}W_{1/2})O_3$ was added to PMN-PZT ceramics to increase dielectric constant of PMN-PZT system[4]. Moreover, to manufacture the dense composition ceramics with fine-grain size 0.3wt% Nb_2O_5 was added into $Pb(Ni_{1/2}W_{1/2})O_3$-$Pb(Mn_{1/3}Nb_{2/3})O_3$-$Pb(Zr,Ti)O_3$(PNW-PMN-PZT) ceramics. Piezoelectric transformer using the composition was fabricated as Rosen-type one with the size of $1\times16\times5mm^3$. And their electrical characteristics were investigated with the variations of load resistance and driving frequency. The following problems should be solved in designing PDA LCD backlight driving circuits using piezoelectric transformer; 1) developing a driving circuit with high efficiency, 2) controlling input voltage and output voltage tolerances for static operation, 3) detecting current of lamp to maintain a constant output current, etc. To solve above problems, a new driving circuit which was composed of the two MOSFET, connected in series with the output part of the VCO (voltage control oscillator) through PFC (power factor corrector) circuits and feedback circuit using one-chip microprocessor was designed and manufactured. Especially, one-chip microprocessor was originally used in order to finely control the driving frequency of piezoelectric transformer. At 0.6W PDA LCD backlight (1.8mmφ × 67mm) driving circuits, electrical characteristics of piezoelectric transformer were also investigated.

II. Experiments

2- 1. Fabrication of piezoelectric transformer.

The specimens were manufactured using conventional mixed oxide process. The composition used in this study is as follows.

978-1-4244-1871-8/07 $25.00 © 2007 IEEE

$(Pb_{0.94}Sr_{0.06})[(Ni_{1/2}W_{1/2})_{0.02}(Mn_{1/3}Nb_{2/3})_{0.07}(Zr_{0.51}Ti_{0.49})_{0.91}]O_3$
$+0.5wt\%PbO+0.3wt\%Fe_2O_3+0.25wt\%CeO_2+0.3wt\%Nb_2O_5$.
The powders were ball-milled for 24hrs and calcined at
850°C for 2 h after drying. The calcined material was ground
in a ball mill again for 24 h and mixed with 5wt % polyvinyl
alcohol (PVA). The powders were consolidated into the discs
with diameter of 21mm at $1000kg/cm^2$. The green bodies
were sintered at temperature of 1230°C for 2 h. The sintered
material was fabricated as Rosen-type transformer with the
size of $1\times16\times5mm^3$, pasted with Ag, and finally heat treated
at 600°C for 10 minutes. The poling treatment was followed
in the 120°C silicon oil bath at 30kV/Cm. The dielectric
constant was calculated from the 1kHz capacitance data
measured using LCR meter, Ando AG-4304. The
piezoelectric characteristics were evaluated based on
electro-mechanical coupling coefficient and mechanical
quality factor calculated from the resonance and
anti-resonance frequencies measured by impedance analyzer,
Agilent 4294. And with the variations of load resistance and
driving frequency, electrical characteristics of piezoelectric
transformer were investigated.

2.2 Design of PDA LCD backlight driving circuits

2.2.1 Half bridge inverter circuit for driving piezoelectric transformer

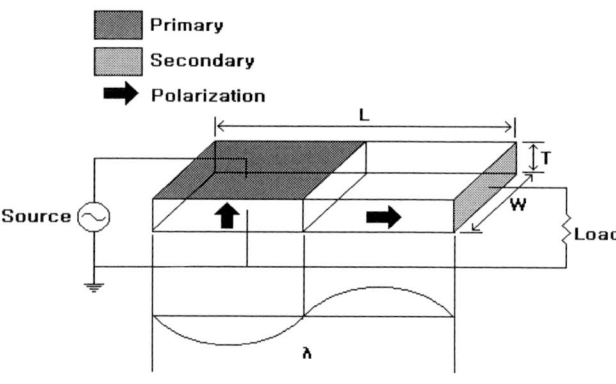

Fig. 1. Rosen-type piezoelectric transformer.

Figure 1 shows PDA LCD backlight driving circuit using
Rosen-type piezoelectric transformer. The driving inverter
that composed of a half-bridge type with two switches is
fabricated. Figure 2 shows that the two MOSFET, which
have no charging time at turn on and the fast switching time,
were used as half-bridge inverter [7]. During their operation,
the dead time setting is necessary for avoiding the
cross-condition of MOSFET power devices. The dead time
provides the MOSFET for arriving at enough SOA(Safety
Operating Area)[8].
Therefore, two MOSFET can be turn on and off, alternatively.
The driving frequency of half-bridge was provided by means
of VOC(voltage controlled oscillator). In this study,
inductance of 1.64mH in input part of piezoelectric
transformer was inserted for L-C resonant circuit [9].

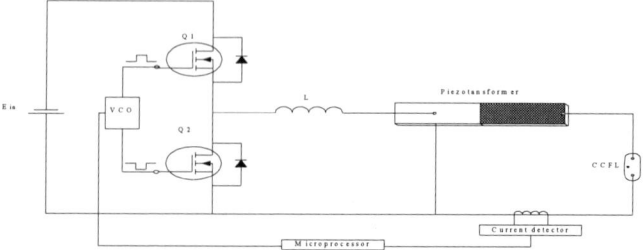

Fig. 2. Half-bridge type inverter.

2.2.2 Design of feedback circuit

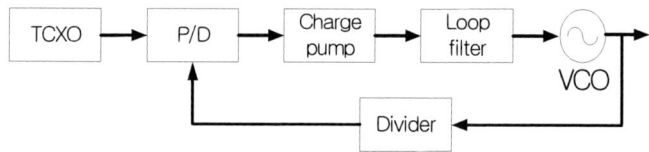

Fig. 3. Feedback circuit diagram.

Figure 3 shows that the voltage step up ratio of piezoelectric
transformer is remarkably varied with the variations of
driving frequency and load resistance. Accordingly,
feedback circuit is necessary for adjusting the driving
frequency, detecting output current of LCD backlight.
The fine control of driving frequency is required for driving
piezoelectric transformer because the output current of
piezoelectric transformer is very sensible for its input
frequency. Therefore, a micro-controller is used in order to
finely control the driving frequency. The micro-controller
used for oscillating was a one chip micro controller
(AT90S4433) with a timer/counter and A/D conversion.
The voltage changed into the analog signal by D/A converter
can be reduced, step-by-step, by setting the time intervals of
internal timer interrupt. Accordingly, the frequency
generated by VCO (Voltage to Frequency) IC becomes low
too and the LCD backlight (CCFL) is slowly changed from
ignition state to the steady state. It can be illustrated as the
soft starting effect to protect the piezoelectric transformer.

III. Results and discussion

Voltage step-up ratio of Rosen-type piezoelectric transformer
at no load is generally expressed as follows [10].

$$\frac{V_o}{V_i} = \frac{4}{\pi^2} K_{31}K_{33}Q_m \frac{l}{t} \qquad (1)$$

As can be seen in equation (1), voltage step-up ratio depends
on not only electromechanical coupling factor (K_{31},K_{33}) and
mechanical quality factor (Q_m) but also the ratio of length($2l$)
and thickness(t) of piezoelectric transformer. Also, fine
grained and non-porous ceramics were required for high
output power. For the piezoelectric transformer with high
voltage step-up ratio and high output power, the high values
of the dielectric constant, mechanical quality factor (Q_m) and

electromechanical coupling factor and the small size of grain are necessary. Physical properties of manufactured specimen are shown in Table 1.

Table 1. Physical and piezoelectric properties of PNW-PMN-PZT ceramics

Dielectric constant	**1704**
Grain size[μm]	**2.50**
kp	**0.55**
Qm	**2041**
Density[g/cm³]	**7.71**
Tetragonality (c/a)	**1.0161**

Taking into consideration mechanical quality factor (Q_m) of 2,041, electromechanical coupling factor (K_p) of 0.55, grain size of 2.5μm and dielectric constant of 1,704, this composition can be effectively utilized for piezoelectric transformer application.

Voltage step-up ratio with the variations of load resistance and driving frequency was shown in Fig. 4.

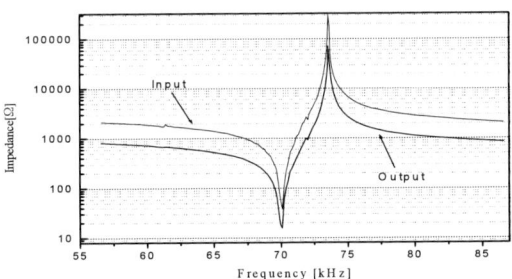

Fig. 4. Impedance curve of piezoelectric transformer.

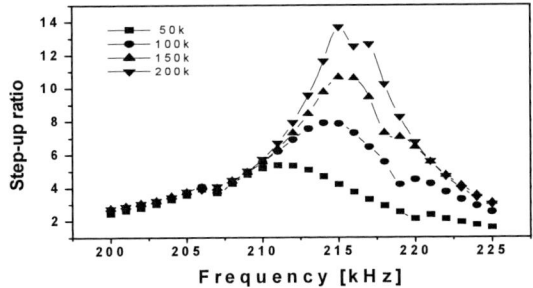

Fig.5. Step-up ratio with the variations of load resistance and driving frequency

With the increase of load resistance, voltage step-up ratio increased gradually and maximum step-up ratio of 13.68 was

shown at the load resistance of 200 kΩ. Above those results coincided with general electrical characteristics of piezoelectric transformer and it was confirmed that applying this piezoelectric transformer for PDA LCD backlight-driving circuit would be possible. That is, since the starting voltage of 0.6W LCD backlight (CCFL) was above 600V$_{rms}$ and the steady state voltage and current were 300V$_{rms}$ and 2mA, respectively, the step up ratios of the transformer before and after lighting the lamp must be maintained higher than 19.3 and 9.6 under the input source of about 31V$_{rms}$.

As can be seen in Figure 5, the transformer can be used for driving 0.6W LCD backlight because the step-up ratios measured at the driving frequency were more than 20 at no load. And also, the ratios at load resistance of 150kΩ were measured to be nearly same value required for driving 0.6W LCD backlight.

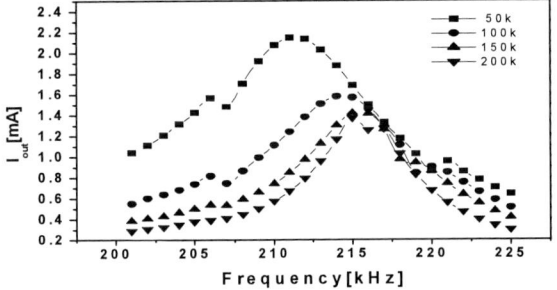

Fig. 6. Output current with the variations of load resistance and driving frequency.

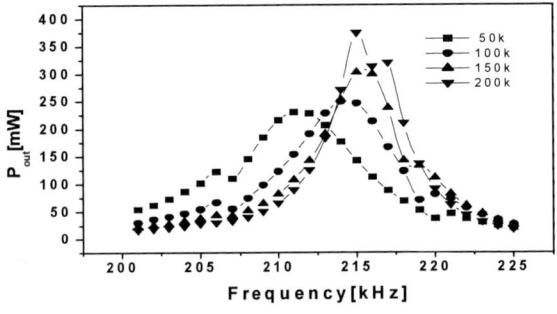

Fig. 7. Output power with the variations of load resistance and driving frequency

Also, in Fig. 6 and Fig .7 output current and power with the variations of load resistance and driving frequency were shown, respectively. The frequency that shows the maximum output current and power of piezoelectric transformer was moved to rightward with the increase of load resistance of piezoelectric transformer.

Figure 8and 9 show input and output voltage after driving LCD backlight for 25 min. Those waveforms show the perfect sinusoidal one. Driving frequency of 214.4 kHz, input voltage of 31.78 V and input current of 21.1mA were measured at the input part of piezoelectric transformer. And

978-1-4244-1871-8/07 $25.00 © 2007 IEEE

also, output voltage of 293.2 V and output current of 2.2mA were measured at the output part of piezoelectric transformer. At the same time, efficiency of 96.2% and temperature rise of 3.6°C were appeared at the piezoelectric transformer.

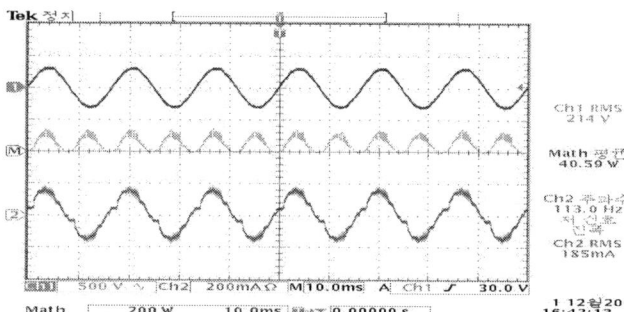

Fig. 8. Characteristic of ballast Input waveform.

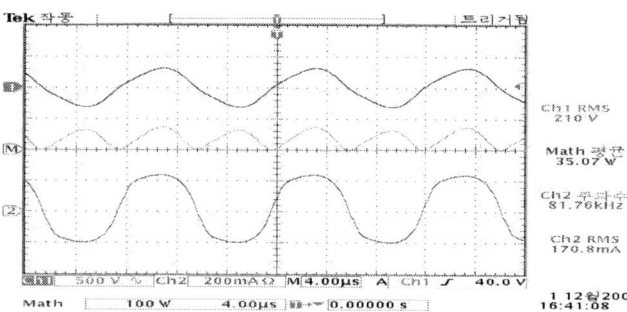

Fig. 9. Characteristic of ballast output waveform.

Figure 10 shows LCD backlight (CCFL) lit by inverter manufactured using piezoelectric transformer.

Fig.10. Photograph of driving fluorescent lamp by electronic ballast

IV. Conclusion

In this paper, to apply piezoelectric transformer for PDA LCD backlight driving circuits, piezoelectric transformer using PNW-PMN-PZT composition was fabricated as Rosen-type one with the size of $1\times16\times5mm^3$. And their electrical characteristics were investigated with the variations of load resistance and driving frequency. And then, new PDA LCD backlight driving circuits was manufactured using piezoelectric transformer, VCO and one-chip microprocessor. Following conclusion was obtained through the experiments.

1. With the increase of load resistance of piezoelectric transformer, voltage step-up ratio increased gradually and maximum voltage step-up ratio of 13.68 was shown at the load resistance of 200 kΩ.

2. The frequency that shows the maximum output power of piezoelectric transformer was moved to rightward with the increase of load resistance of piezoelectric transformer

3. After driving PDA LCD backlight for 25 min using the proposed circuit, driving frequency of 214.4 kHz, input voltage of 31.78 V and input current of 21.1mA were measured at the input part of piezoelectric transformer. And also, output voltage of 293.2 V and output current of 2.2mA were measured at the output part of piezoelectric transformer. At the same time, efficiency of 96.2% and temperature rise of 3.6°C were appeared at the piezoelectric transformer.

References

[1] S. Kawashima, O. Ohnishi. S. Tagami, A. Fukuoka, T.Inoue and S. Hirose, Proc.1. Int. IEEE Ultrasonic Symposium, (1994) 525.

[2] Y. Sasaki, M. Yamamoto, A .Ochi,t .Inoue and S. Takahashi ,Jpn. J. Appl. Phys. 38(1999) 5598.

[3] J .H. Yoo, Y .W. Lee, K. H. Yoon, H. S. Jung, Y .H. Jeong, C.Y. Park, KIEEME 11 (1998) No.10, 849.

[4] J .H. Yoo, K .H. Yoon, S.M Hwang, Y. W. Lee, S. J. Suh, J. S. Kim and C. S. Yoo, Jpn. J. Appl. Phys. 40(2001) 3256.

[5] O. Ise, K. Satoh and Y. Mamiya, Jpn. J. Appl. Phys. 38(1999) 5531.

[6] J. H. Yoo, K. H. Yoon, Y.W Lee, S. J. Suh, J. S. Kim and C. S. Yoo, Jpn. J. Appl. Phys. 39(2000) 2680.

[7] M. C. Cosby and T. M. Nelms, IEEE Trans. Industry E lectronics, . 41(1994) 418.

[8] R.L .Steigerwald, IEEE Trans. On Power Electronics, 3(1988)174.

[9] M. Shoyama, K. Horikoshi, T. Ninomiya, T. Zaitsu, Y. Sasaki, IEEE PESC, (1997)715.

[10] C.A .Rosen, Proc. Electronic Comp. Symp,(1956)205

978-1-4244-1871-8/07 $25.00 © 2007 IEEE

An Internet Based Embedded Network Monitoring System

for Renewable Energy Systems

Po-Yen Chen[1,2]   Se-Kang Ho[3]   Wei-Jen Lee[1]   Chia-Chi Chu[1]   Ching-Tsa Pan[1]

[1] Dept. of Electrical Engineering National Tsing Hua University

[2] Energy and Environment Laboratories, Industrial Technology Research Institute

[3] Mechanical and Systems Laboratories, Industrial Technology Research Institute

Abstract

In this paper, an internet based embedded network monitoring system is proposed for renewable energy systems. By using a low cost network communication module (RCM 3700) as a web server, one can achieve better network security, lower power consumption, compact size, and easier to use as compared with a PC based one. Also, Java language is chosen for designing a dynamic webpage to graphically display various real time waveforms of the controlled system for multi-user at the same time. As an illustrative example, a small scale wind power generation system equipped with an EZDSP 2812 controller is adopted for demonstration. In addition, an FPGA EC10 is implemented as a bidirectional communication interface for coordinating the asynchronous data transmission modes. Experimental results from the constructed prototype verify that the proposed monitoring system can indeed achieve the desired function.

Keyword：**monitoring system, embedded system, internet.**

I. Introduction

Due to concerns regarding global warming and air pollution, there has been an international movement in the promotion of renewable energy technologies for electricity generation and development of national emissions limit. According to the national energy policy of the authors' country, it is planned to achieve 12% of the total electrical generation capacity from renewable energy in 2020. Hence, from the long term operation economics viewpoint, it is essential to have more advanced monitoring module for each, large or small, renewable energy generation unit. The conventional monitoring system based on RS485, Can Bus, etc are not only limited to local distance, expansion ability, slow speed but also can not provide interoperable capability among different platforms easily. In addition, in view of the requirement of economic dispatch among different renewable units in a micro grid system, fast and bidirectional communication capability are important considerations to achieve better distributive control. On the other hand, due to the recent rapid progress in internet technology, it is now possible to achieve much faster communication speed with much simplified architecture and low cost. Hence, in this paper, an advanced internet based embedded network monitoring system is proposed to achieve the aforementioned objectives.

II. The Proposed Monitoring System

Basically, the proposed monitoring system adopts a low cost microprocessor based core module designed for Ethernet/Internet applications, namely RCM3700, embedded with real time OS for implementing the control webpage for bidirectional communication through dual port RAM of a FPGA (Lattice EC-10) as shown in Fig. 1. It is seen from Fig.1 that the data transmission structure of the proposed system becomes much simplified compared with a conventional monitoring system equipped with network communication ability.

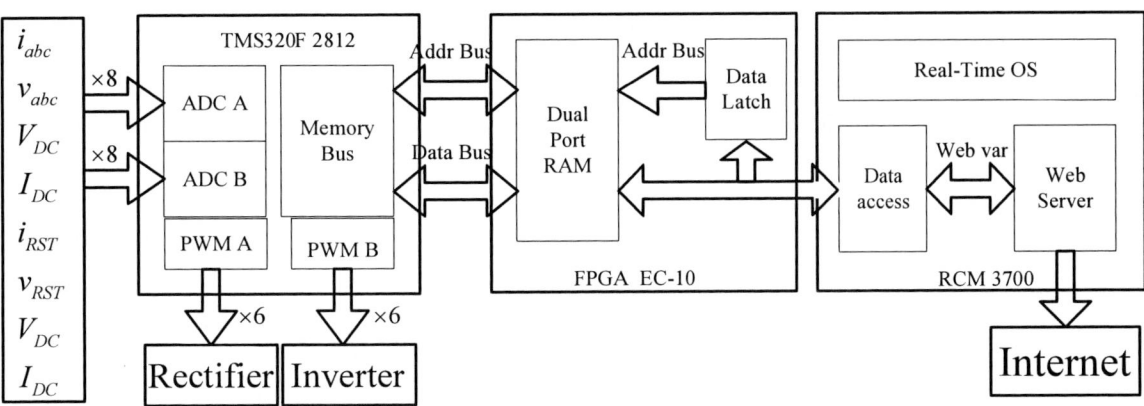

Fig.1 Data transmission structure of the proposed monitor system

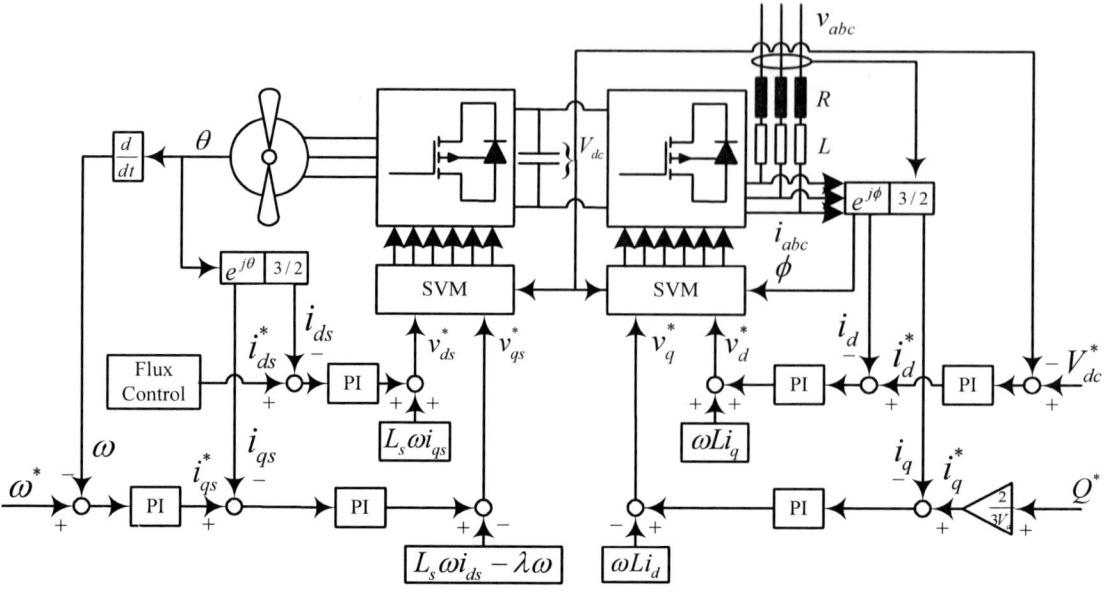

Fig.2 Surface-mounted permanent-magnet wind power generator system

In addition, Java, an object-oriented, safe and secure language that exhibits H/W independence, is adopted for the webpage design of control and monitoring to provide the interoperability among different types of equipment. As an illustrative example, a direct drive grid connected permanent magnet synchronous wind power generator system controlled through a back-to-back converter as shown in Fig.2 is chosen for experiment. Field oriented control together with the maximum output power tracking of the generator and the maximum efficiency inverter control are all implemented using a digital signal processor, namely EZDSP 2812. The dynamic model of the surface-mounted permanent-magnet generator in the synchronous reference system is as follows:

$$v_{ds} = -R_s i_{ds} - L_s \frac{di_{ds}}{dt} + L_s \omega i_{qs} \quad\quad\quad (1)$$

$$v_{qs} = -R_s i_{qs} - L_s \frac{di_{qs}}{dt} - L_s \omega i_{ds} + \omega \lambda \quad\quad (2)$$

where, v_{ds}, v_{qs}, i_{ds} and i_{qs} denote the d-axis and q-axis voltage and current respectively, R_s and L_s represent the generator resistance and inductance, and ω is the generator speed, λ is the magnet flux. The corresponding electromagnetic torque is given by $T_e = \frac{3}{2} P \lambda i_{qs}$, where P is the pole pair number. The generator torque may be controlled directly by the quadrature current component as shown in Fig.2. The dynamic model of the grid connected inverter when selecting a reference frame rotating synchronously with the grid voltage space vector is given below:

$$v_d = v_{id} - R i_d - L \frac{di_d}{dt} + \omega L i_q \quad\quad\quad (3)$$

$$v_q = v_{iq} - R i_q - L \frac{di_q}{dt} - \omega L i_d \quad\quad\quad (4)$$

where, v_{id}, v_{iq}, v_d, v_q, i_d and i_q are the inverter and power system d, q axes voltage components and current components respectively, and L, R are the inductance and ESR of three phase inductors. If the d-axis is oriented along with the supply voltage vector, then the grid voltage vector becomes $v_d + j0$. Hence, the active power P and reactive power Q may be expressed as follows:

$$P = \frac{3}{2} v_d i_d \quad\quad\quad\quad\quad (5)$$

$$Q = \frac{3}{2} v_q i_q \quad\quad\quad\quad\quad (6)$$

Hence, the corresponding active and reactive power control can be achieved by controlling the direct and quadrature current components respectively as shown in Fig.2.

III. Real-Time Network Communication

An inherent data transmission characteristic through internet is that the time sequence of transmitted packets can not be reserved when received at the client terminal. Also, it can not be guaranteed that the data to be transmitted are taken with fixed sampling time. As a result, the resulting dynamic waveform received at the client terminal will be neither real time nor correct. To solve this dilemma

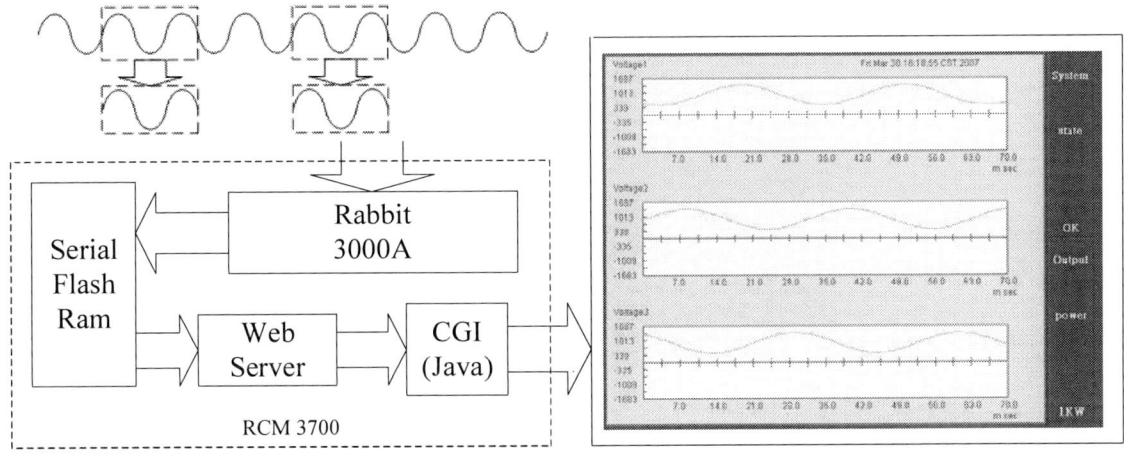

Fig.3 Illustration of data access mechanism for displaying in the dynamic webpage.

resulting from network transmission, for each channel a fixed number of data collected with fixed sampling time are stored in a serial flash RAM. Then, each time when the dynamic webpage at the client terminal is updated, these data together with the Java code are downloaded. Therefore, the time sequence of the real time data can be reserved unchanged as shown in Fig.3.

IV. Experimental Results

For demonstrating the feasibility, a small scale wind power generation system as shown in Fig.2 is constructed and the corresponding control board is shown in Fig.4. Due to limitation of space, only some experimental are given for demonstration. The speed of the network module, RCM3700, is 10Mbps. The webpage display resolution is 1280×1024 pixels. The obtained real time command signals v_a^*, v_b^* and v_c^* of the inverter from the EZDSP 2812 are shown in Fig.5(a). Also, due to fully digital processing of the proposed monitoring system, it is seen from Fig.5(b) that the corresponding three phase current waveforms, corresponding to 0.9 amp approximately, can be displayed clearly. Compared with the existing displaying technique by using a 12 bit D/A converter with output voltage range from 0 to 3 volts, the resulting waveforms of 0.9 amp amplitude, with respect to the full scale of 20 amp, can hardly be observed. As another example, in case necessary, one can also output the corresponding stator α and β current components in stationary frame as shown in Fig.5(c). For a conventional monitoring system it is usually not possible to access directly the controller signal as the proposed monitor system. Moreover, it is now very convenient to integrate the proposed monitoring system with the controller to provide fast bidirectional communication function. This will be very useful for applying to monitoring and coordinate control of micro grid systems. In addition, due to the simplified peripheral requirement, smaller volume and less power consumption, the proposed monitor system is rather cheap and attractive for other practical applications too.

Fig. 4. The control board of the proposed system

V. CONCLUSION

In this paper, a low cost microprocessor based core module designed for Ethernet/Internet application (RCM 3700) embedded with real time OS is adopted for implementing the control webpage. The control webpage can display graphically the real time waveform of various time varying quantities of the monitored system. As an illustration, a small scale wind power generation system equipped with an EZDSP 2812 controller is adopted for demonstration. In addition, an FPGA EC10 is implemented as a bidirectional communication interface for coordinating the asynchronous data transmission modes. Experimental results from the constructed prototype verify that the proposed monitoring system can indeed achieve the desired function. It is seen that the proposed internet based embedded network monitor system possesses many merits such as low cost, lower power consumption, compact size, better network security, being able to display real time dynamic waveforms graphically for multi-users. Compared with the existing PC-based monitoring systems, the proposed monitoring system seems more attractive for practical applications.

978-1-4244-1871-8/07 $25.00 © 2007 IEEE

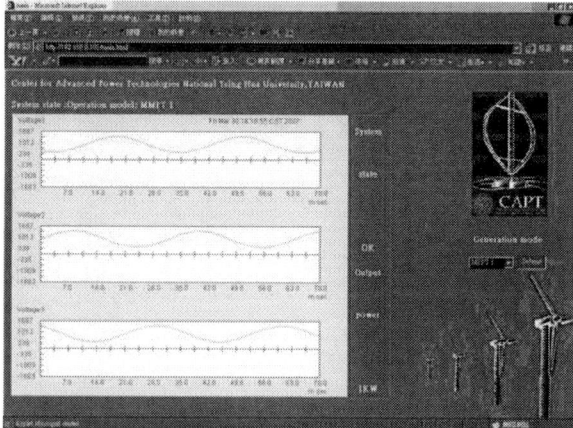

(a) Three phase stator voltage commands

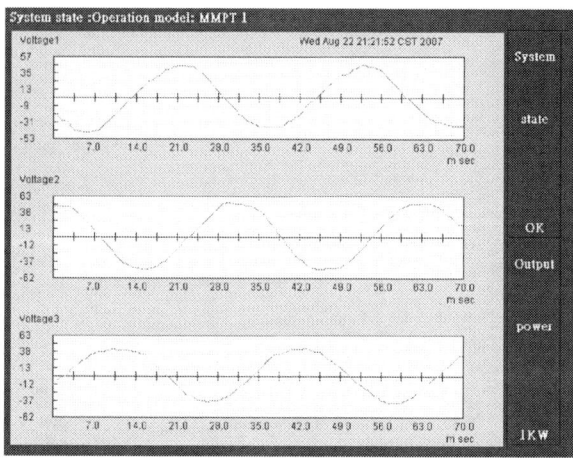

(b) Three phase stator currents in stationary frame

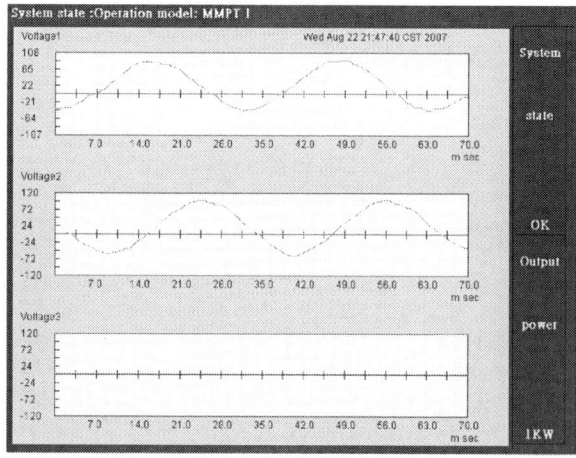

(c) α and β stator currents in stationary frame

Fig.5 Display of control webpage dynamic waveforms

ACKNOWLEDGMENT

This work was supported by the National Science Council of Republic of China under Grant NSC-95-2218-E-007-022.

References

[1] M. Davidson, "Interaction of A Wind Farm with The Distribution Network and Its Effect Voltage Quality," IEE Colloquium on the Impact of Embedded Generation on Distribution Networks (Digest No. 1996/194), pp. 9/1 - 9/5, 15 Oct. 1996.

[2] Raymond Wai-Leung Cheung; Yiu-Hon Wang; Yu-Fai Fung, "Remote Switching in Internet Network," International Conference on Advances in Power System Control, Operation and Management, Volume 2, PP. 370 - 375, 30 Oct.-1 Nov. 2000.

[3] Tsukui, R.; Beaumont, P.; Tanaka, T.; Sekiguchi, K., "Power System Protection and Control Utilising Intranet Technology," Power Engineering Journal, Volume 15, Issue 5, PP. 249 – 255, Oct. 2001.

[4] Hamamatsu, K.; Watanabe, H.; Sekiguchi, K.; Tsukui, R.; Igarashi, K.; Beaumont, P., "A New Approach to The Implementation of Intranet-Based Measurement and Monitoring," Seventh International Conference on IEE Developments in Power System Protection, PP. 102 – 105, 9-12 April 2001.

[5] Eto, H.; Matsuo, H.; Kurokawa, F.; Fukuda, M., "Network of plant remote monitoring system using Web for Windfarms," The 25th International Telecommunications Energy Conference, PP. 857 – 863, 19-23 Oct. 2003.

[6] Bordon, D.; Ribic, A.; Babnik, T., "Internet-Based Monitoring and Control of Protective Relays and Disturbance Recorders," Eighth IEE International Conference on Developments in Power System Protection, Volume 2, PP. 656 - 659, 5-8 April 2004.

[7] Brugnoni, G.; Gagliarducci, M.; Lampasi, D.A.; Podesta, L., "Modular Architecture for Remote Management of Powe Generation Plants," Proceedings of the IEEE Instrumentation and Measurement Technology Conference, Volume 2, PP. 1075 – 1078, 16-19 May 2005.

[8] Westermann , D.; John, A., "Demand Matching Wind Power Generation With Wide-Area Measurement and Demand-Side Management," IEEE Transactions on Energy Conversion, Volume 22, Issue 1, PP. 145 – 149, March 2007.

[9] Ching-Tsai Pan; Jenn-Horng Liaw, "A Robust Field-Weakening Control Strategy for Surface-Mounted Permanent-Magnet Motor Drives," IEEE Transactions on Energy Conversion, Volume 20, Issue 4, PP. 701 – 709, DECEMBER 2005.

978-1-4244-1871-8/07 $25.00 © 2007 IEEE

Fabrication of 4H-SiC Schottky Barrier Diodes with the epilayer grown by Bis-trimethylsilylmethane precursor

Myeong Sook Oh, Ho Keun Song, Jeong Hyun Moon, Jeong Hyuk Yim, Jong Ho Lee,
Han Seok Seo, Yu Jin Choi, and Hyeong Joon Kim
Department of Materials Science and Engineering, Seoul National University
San 56-1, Sillim-dong, Gwanak-gu, Seoul, 151-742
Email: vtsrc13@snu.ac.kr, hjkim@plaza.snu.ac.kr

Abstract— **The authors fabricated the 4H-SiC Schottky barrier diodes (SBDs) with the epilayers grown using the MOCVD. Bis-trimethylsilylmethane (BTMSM, [$C_7H_{20}Si_2$]) was used as a single precursor for Si and C sources.**

I. INTRODUCTION

Silicon carbide (SiC) is an excellent semiconductor for high temperature, high-frequency and high-power devices due to its excellent electrical, thermal and mechanical properties. Especially, the breakdown field in SiC, 3 MV/cm, is one order of magnitude higher than that in Si. Hence, the electrical properties of SiC Schottky barrier diode (SBD) are expected to be better than that of Si.

Although high quality α-SiC (4H-SiC, 6H-SiC) homoepitaxial films have been obtained using such methods as liquid-phase epitaxy (LPE), and vapor-phase epitaxy (VPE), there is the serious problem of high growth temperature above 1800 °C. Recently, the growth temperature was dramatically reduced to 1500 °C by using slightly off-orientated substrate. However a further reduction of growth temperature is still necessary to avoid the auto-doping problem and the redistribution of dopants.

To solve this problem, we used the alternative organo-silicon precursor, BTMSM, which contains directly bonded Si and C atoms and that decomposes at relatively low temperature. In this paper, 4H-SiC SBDs were fabricated on n-type 4H-SiC epilayer which was grown on n-type substrate using BTMSM, and its electrical characteristics will be shown for the power device application of the 4H-SiC epilayer grown by MOCVD.

II. EXPERIMENTAL DETAILS

A. Preparation of 4H-SiC epilayer

For the fabrication of the 4H-SiC SBDs, low doped 4H-SiC epilayers (target $N_d = 5 \times 10^{15}$ cm^{-3}, T_{epi}=10 µm) were grown by MOCVD using the BTMSM precursor. Because the background doping concentration is about mid 10^{14} cm^{-3}, no intentional doping was carried out. The main source of back ground dopants was the nitrogen from the ambient. The n-type SiC substrates which had different crystallographic characteristics were used for the comparison of the crystallinity effect on the electrical properties of the 4H-SiC

SBDs. Table 1 shows the crystallographic characteristics of the grown 4H-SiC epilayers and its substrates that were used for the fabrication of the SBDs, which were measured by DCD and optical microscope. Two groups of epilayers were grown, which were grown at 1370 °C and 1440 °C, respectively, and each group of epilayers were grown on three types of substrates which contain different crystallographical characteristics : the normal substrate (high quality), the substrate with micropipes, and the substrate with mosaic structure for the comparison of the crystallinity effect.

TABLE 1. The crystallographic characteristics of the 4H-SiC epilayers and substrates that was used as a fabrication of SBDs.

Sample No.	Growth temp.	Mosaic Free[1]	Micropipe Free[2]	Thickness
1	1370 °C	O	O	10 µm
2	1370 °C	O	×	10 µm
3	1370 °C	×	O	10 µm
4	1440 °C	O	O	10 µm
5	1440 °C	O	×	10 µm
6	1440°C	×	O	10 µm
7(Cree)	N/A	N/A	N/A	12 µm
8(Cree)	N/A	N/A	N/A	12 µm

[1] measured by DCD
[2] measured by optical microscope

B. Fabrication of Schottky Barrier Diodes

We fabricated SBDs with the field plate edge termination technique [1]. Fig. 1 shows the schematic of the 4H-SiC structure. Prior to the deposition, the substrates were cleaned in order to remove the surface contamination. Diluted HF cleaning was also conducted during the intervals of each step to remove the native oxide layer. Sacrificial oxide layer was grown to remove the scratch and defects in layer surface. After removing of sacrificial oxide in diluted HF, passivation and edge termination oxide (5000 Å) were deposited by STS-PECVD. The 500 Å of Ni layer for the Schottky metal was deposited using E-gun evaporator. The Schottky metal, nickel, has circular shape. For small contact resistance during the electrical measurement of SBD, gold pad metal (2500 Å) was deposited on the Schottky metal layer. Ohmic contacts were formed on the back surface of the substrates in form of $NiSi_x$ by nickel annealing at 1000 °C for 2 min. For the proper field plate edge termination, the diameter of Schottky metals were varied from 80 μm to 160 μm and the periphery of Schottky metals (2 μm ~10 μm) were overlapped on the field plate SiO_2 layer.

The electrical properties of the SBDs were measured by the C-V (HP 4140B) and the I-V (HP 4155A) measurement systems, respectively. Breakdown voltage was measured by kethly-208 system. For the precise extraction of the Schottky diode parameters, such as the ideality factor, n, and the barrier height, Φ_B, we utilized a plot of the function H(J) vs J, where $H(J)=V-n(kT/q)\ln(J/A^{**}T^2)$ (Cheung's method) [2]. From the slope of the plot of d(V)/dln(J) vs J, we can determine the values of n. Before the measurement of the breakdown voltage, the top contacts of the SBDs were observed by the optical microscope for the comparative analysis with I-V data.

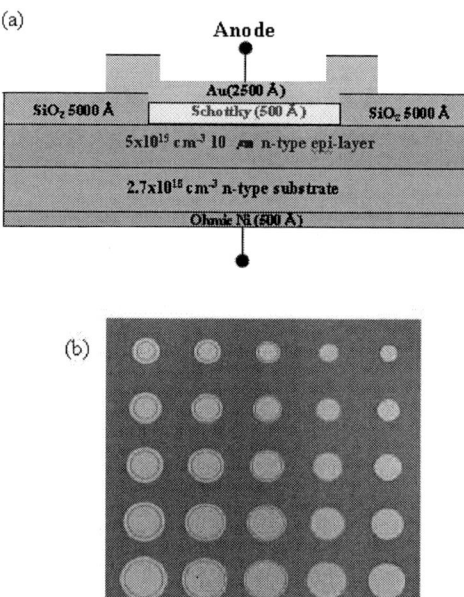

Fig. 1: (a) Vertical structure and (b) plane-view optical image of 4H-SiC SBDs having field plate edge termination.

III. RESULTS AND DISCUSSION

The ideality factor and the barrier height of the fabricated Schottky barrier diode were 1.22~1.25 and 0.85~0.96 eV, respectively. The net donor concentrations of the epilayers, which were measured by C-V technique, were about 3×10^{16} cm^{-3}. In case of the epilayer from Cree, the net donor

concentration was 7.7×10^{16} cm^{-3}, and showed higher value than the specification of the provider (5×10^{15} m^{-3}).

Fig. 2 shows the reverse I-V characteristics of the SBDs and that of the epilayers were grown at (a) 1370 °C and (b) 1440 °C, respectively. Fig. 2 (c) is the reverse I-V characteristics of the SBD which was fabricated using the Cree's epi-wafer. The breakdown voltage of the epilayers grown at 1440 °C and 1370 °C were about 450 V and 250 V, respectively. Because the crystallinity of epilayer grown at 1440 °C is better than that grown at 1370 °C, the high breakdown voltage was obtained. The origin of the high breakdown voltage was the high quality epilayer grown under the condition of high temperature and low supersaturation. In case of Cree's epi-wafer, the breakdown voltage was about 150 V and showed lower value than the epilayer grown by MOCVD using BTMSM precursor. This result may be caused by the low crystallinity and high background doping concentration.

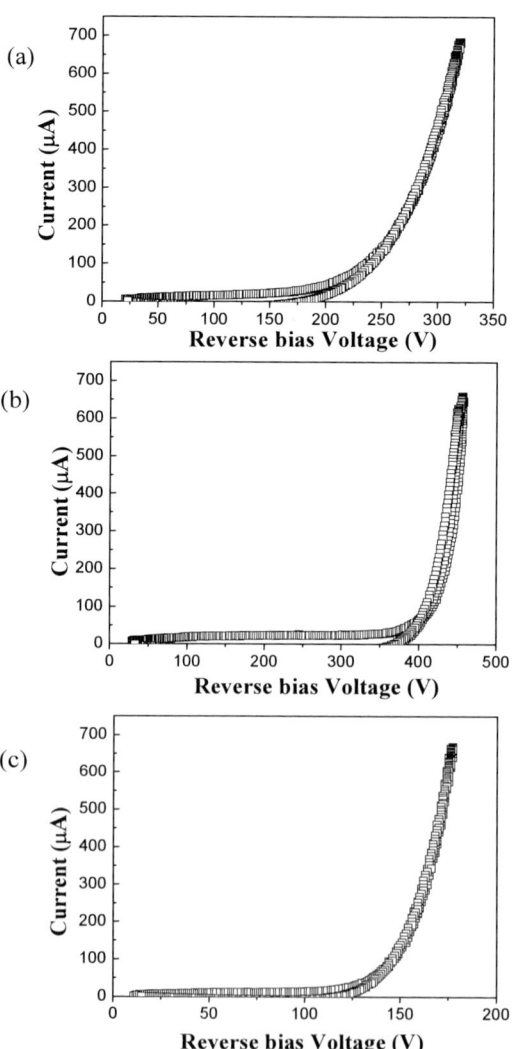

Fig. 2: Reverse I-V characteristics, epilayers were grown at (a) 1370 °C and (b) 1440 °C, respectively. (c) is that of the SBD fabricated using the Cree's epi-wafer.

The SBDs fabricated by the epilayer with mosaic structure have similar breakdown voltage to that of the high quality epilayer. However, the distribution of breakdown voltages is large due to the mosaic structure of epilayer.

Fig. 3 shows the optical images of the SBDs with micropipes, which are known as major killer defect of a SiC devices [3] [4]. The epilayers in Fig.3 were grown at ((a), (b)) 1370 °C, (c) 1440 °C, and Fig.3 (d) is the optical image of the SBD which was fabricated using a commercial epi-wafer. When the epilayer has only micropipes (not with tails), the degree of the SBD degradation is not severe (Fig.3(b) and (d)), but it has relatively small breakdown voltage (120~200 V) in comparison with the high quality epilayer with no defects (250~450 V).

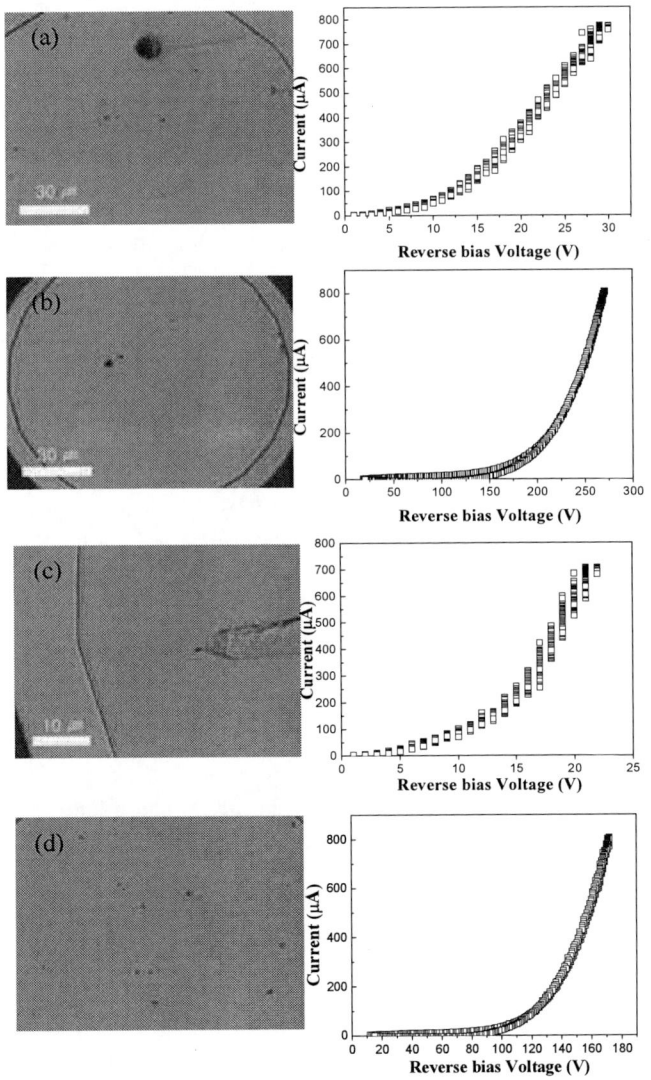

Fig. 3: The optical top-view images with micropipes and its reverse I-V plots, the epilayers were grown at 1370 °C ((a), (b)) and 1440 °C (c), respectively. (d) is that of the SBD which was fabricated using the Cree's epi-wafer.

IV. CONCLUSIONS

We fabricated SBDs with the field plate edge termination technique. The ideality factor and the barrier height of the fabricated Schottky barrier diodes were 1.22~1.25 and 0.85~0.96 eV, respectively. The breakdown voltage of the SBDs with the epilayers grown at 1440 °C and 1370 °C were about 450 V and 250 V, respectively. In case of Cree's epi-wafer, the breakdown voltage was about 150 V and showed lower value than the epilayer grown by MOCVD using BTMSM precursor. This result may be caused by the low crystallinity and high background doping concentration in comparison with the epilayer grown by MOCVD.

The SBDs fabricated by the epilayer with mosaic structure have similar breakdown voltage to that of the high quality epilayer. However, the distribution of breakdown voltages is large due to the mosaic structure of epilayer. Micropipes in Schottky contact area known as major killer defect of the SBDs also bring down the characteristics of devices. When the epilayer has only micropipes (not with tails), the degree of the SBD degradation is not severe. However it has relatively small breakdown voltage than the high quality epilayer with no defects.

ACKNOWLEDGMENT

The authors would like to acknowledge the support of "Power IT Project." in making this research and in preparation of the paper.

REFERENCES

[1] Saxena, V., Nong, J., Stackl, A.J., "High-Voltage Ni- And Pt-SiC Schottky Diodes Utilizing Metal Field Plate Termination.", IEEE Transactions on electron devices, V. 46, pp.456-160, 1999.

[2] S. K. Cheung and N. W. Cheung, *Appl. Phys. Lett.* V. 49, N. 2, pp.85-89, 1986.

[3] J. A. Powell et al, *Inst, Phys. Conf. Ser.* V. 137, pp. 161-165, 1994.

[4] W.C. Mitchel, R. Perrin, J. Goldstein, A. Saxler, M. Roth, S. R. Smith, J. S. Solomon, and A. O. Evwaraye , "Fermi level control and deep levels in semi-insulating 4H-SiC." J. Appl. Phys. V. 86, pp. 5040-5044, 1999.

High Accuracy and High Stability Magnet Power Supply by Four-Phase Buck type DC/DC Converter

S. C. Kim
Pohang Accelerator
Laboratory(PAL)/POSTECH
San 31, Hyoja-dong, Nam-gu,
Pohang, Gyeongbuk, 790-784,
Rep. of Korea
Email: schkim@postech.ac.kr

K. M. Ha
Pohang Accelerator
Laboratory(PAL)/POSTECH
San 31, Hyoja-dong, Nam-gu,
Pohang, Gyeongbuk, 790-784,
Rep. of Korea
Email: hkm@postech.ac.kr

J. Y. Huang
Pohang Accelerator
Laboratory(PAL)/POSTECH
San 31, Hyoja-dong, Nam-gu,
Pohang, Gyeongbuk, 790-784,
Rep. of Korea
Email: huang@postech.ac.kr

Abstract—Linear accelerator and Beam Transfer Line magnet power supply (MPS) in Pohang Accelerator Laboratory have been upgraded from convention scheme to switching scheme for the stable beam injection and the 4th generation light source research. New MPS consists of two-main power board (MPB), which are regulator, CPU and control power boards. The size of each board is only 100mm x 240mm. The MPB consists of full-bridge scheme switch. The functions of the one MPB are that they operate as two parallel buck (PB) DC/DC converter. Switching frequency is 42 kHz at MPB. Two-MPB is operated as synchronous four-PB DC/DC converter. Input and output of the MPS are 3p 35V_{ac} and 80A_{dc}/50V_{dc}, respectively. Its resolution and stability are 16-bit and < 20ppm, respectively. To eliminate generated switching noise from main switch, specially designed 2-stage LC output filter is adopted in output station. The 1st-stage filter is included in the MPB. The 2nd-stage filter is located in output stage of MPB as separated PCB. High accuracy CT for sensing of output current is located in the summing stage of synchronous four-PB type DC/DC converter. This paper describes the design concept and performance of the MPS with four-phase buck type DC/DC converter.

I. INTRODUCTION

Electron beam of electron accelerator is focused or corrected by DC magnet. Therefore, output stability and resolution of magnet power supply (MPS) is high closely related with quality of the electron beam. Conventional dc magnet power supplies are composed of transistor bank operating in active region, connected in series with a pre-regulated dc voltage from a phase-controlled rectifier. However, this scheme is low efficiency and large volume. In order to avoid those problems from conventional scheme, magnet power supplies by using high frequency switching scheme are widely adopted at several particle accelerators in the world, currently. Linear accelerator (Linac) and Beam Transport Line (BTL) of Pohang Accelerator Laboratory (PAL) has 22 solenoid magnets, 48 quadrupole magnets, and 28 corrector magnets. MPSs of linear regulator of SCR phase rectifier scheme have been operated for twelve years since the completion of PAL in 1994. Output current performance of these MPS has 12 bit resolution and ±1000 ppm long term stability. These MPS have been newly upgraded from convention scheme to switching scheme for the stable beam injection and the 4th generation light source research in 2006. Newly upgraded MPS adopted synchronous four-phase buck

mode operation scheme by using separated two full-bridge four-quadrant DC/DC converter. Full-bridge four-quadrant DC/DC converter is able to operate bipolar or unipolar mode by PWM switching method. Therefore, identical main power conversion PCB is commonly used at bipolar and unipolar MPS.[1][2] The high-performance MPS for particle accelerator system require a second-order low-pass filter to reduce the output current ripple within specifications. Because of the large time constant, the load magnet provides very little damping to the second-order output filter. [3] Many switching power converters have applications that require very low high-frequency noise and ripple on the output. A single-stage filter which meets the requirements can become very large and impractical, especially for boost and buck-boost converter which have pulsating output currents. Unfortunately, a second LC filter gives up to 180° additional phase delay in the control-to-output transfer function, and it can make the system unstable if improperly designed. [4] Output LC filter have to be small size at compact MPS. Output filter of the MPS is adopted two-stage LC filter. The design concept and performance of the magnet power supply with four-phase buck type DC/DC converter will be discussed.

II. BASIC CONCEPT OF FOUR-PHASE BUCK TYPE DC/DC CONVERTER

A. Basic Circuit

Fig. 1 shows basic circuit diagram of two phase buck type DC/DC converter from full-bridge four-quadrant DC/DC converter scheme. In Fig. 1, positive output current flows to load (io = i1) if Q1A is on and Q1B is off of leg A, and Q2A and Q2B are both off of leg B, and positive output current flows to load (io = i2) if Q2A is on and Q2B is off of leg B and Q1A and Q1B are both off of leg A. Twice amount output current flow to load(io = i1 + i2) more than leg A or B case if Q1A and Q2A are both on and Q1B and Q2B are both off, synchronous of leg A and B. Circuit of fig. 1 operate as buck type DC/DC converter. And output current flows to load as positive (io) or negative (-io) direction if cross two switch pair Q1A-Q2B and Q1B-Q2A is on or off. In this case, circuit of fig. 1 operates as bipolar mode. Table 1 describes relation between switch operation and output current in fig. 1.

978-1-4244-1871-8/07 $25.00 © 2007 IEEE

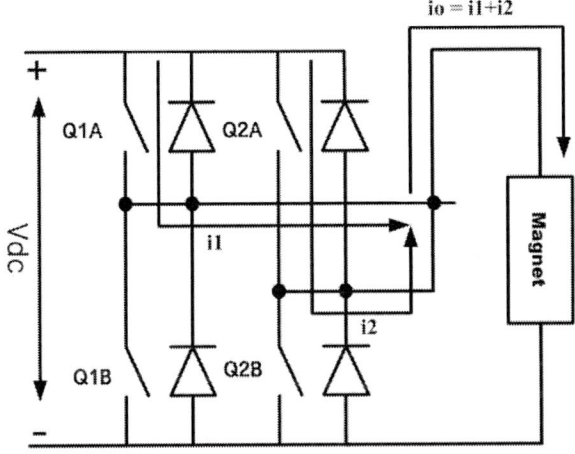

Fig. 1: Circuit diagram of two phase buck type DC/DC converter from full-bridge four-quadrant DC/DC converter scheme

TABLE 1: Relation between Switch operation and output current in fig. 1

CASE	I	II	III	IV	V
Q1A	ON	OFF	ON	ON	OFF
Q1B	OFF	OFF	OFF	OFF	ON
Q2A	OFF	ON	ON	OFF	OFF
Q2B	OFF	OFF	OFF	ON	ON
Output Current	i1	i2	i1+i2	io	-io

B. Basic circuit diagram of four phase buck type DC/DC converter

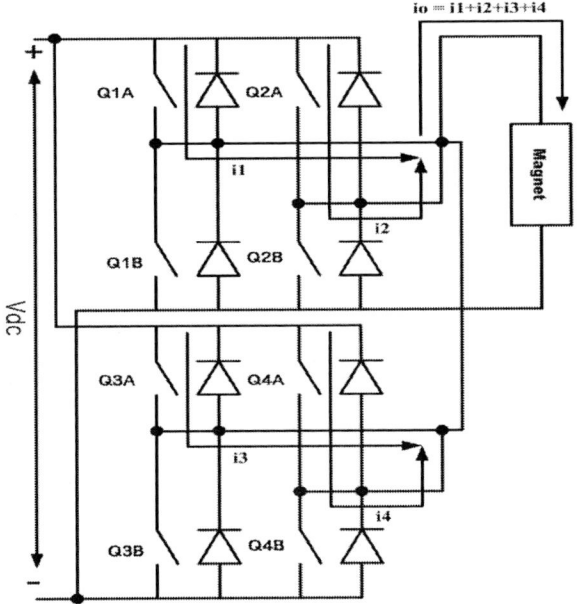

Fig. 2: Extended four phase buck type DC/DC converter from fig. 1

In case III of table 1, output current is twice amount more than case I or II. Output current is four times amount more than case I or II if two-case III circuit is parallel connected. Fig. 2 shows basic circuit diagram of extended four phase

buck type DC/DC converter from fig. 1.

III. MAGNET POWER SUPPLY BY FOUR-PHASE BUCK TYPE DC/DC CONVERTER

A. Specifications

Required maximum output current is 50A, and accumulated long stability of output current is ±50ppm at the unipolar MPS. Switching noise generated from main switch has not to be included at output current because of quality of electron beam. Table 2 shows the specifications of unipolar MPS at the PAL Linac and BTL

TABLE 2: Development specifications of MPS

Item	Spec.	unit
Size(W x H x D)	435×178×450	mm
Input	3φ 30V	V
Output	50/50	A/V
Output stability	±20ppm	< 1 hour
	±50ppm	> 10 hours
Output resolution	16	bit
Switch freq.	> 40	kHz
Output Filter Cut-off freq.	< 5	kHz

B. Configurations

MPS consists of eight parts, which are DC bus, back-plan board, control power board, CPU board, regulator board, main power module, filter board and DC Current Transducer (DCCT). Control power board, CPU Board and main power board are standardized, and PCB size is 100mm width and 250mm depth. These boards are installed in slot of back-plane board. Fig. 3 shows configuration diagram of the MPS.

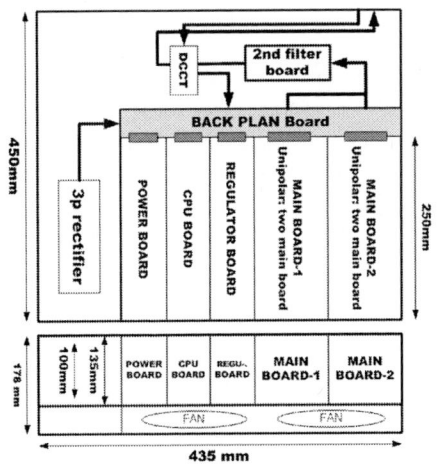

Fig. 3: configuration diagram of MPS

Fig. 4 shows circuit of the MPS by four-phase buck type DC/DC converter. Each leg of main board is operating in parallel, simultaneously. Therefore, output current of one main board is sum ($I_{1A}+I_{1B}$) of each leg current. If two main boards are parallel operating then output current is four times ($I_{1A}+I_{1B}+I_{2A}+I_{2B}$) of single-leg current. In case of parallel operation, control of each leg must be exactly same as PWM. ULTRASTAB-867 of Danfysik are used as DCCT in the MPS. VHP-4 type is selected for burden resistors. Fig. 5 shows fabricated control power board, CPU board, regulator board and main power board of the MPS.

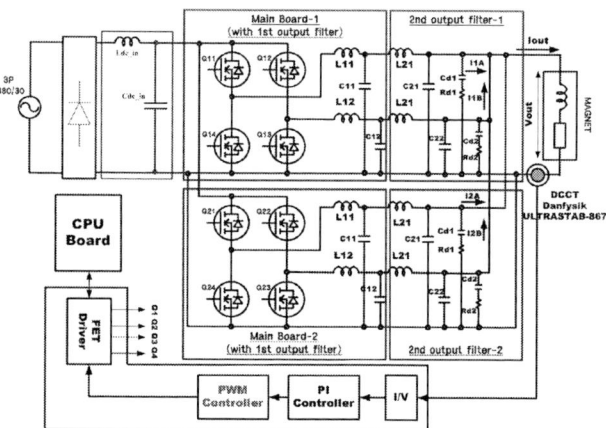

Fig. 4: Circuit of the MPS by four-phase buck type DC/DC converter

C. CPU Board

MCF5282 Cold-Fire Processor is adopted as CPU of the MPS. This process is the first micro-controller based on free-scale Semiconductor's 32-bit Cold-Fire core integrated with Ethernet, Flash, RS232 and CAN. The MCF5282 was designed to simplify embedded Ethernet networked micro-controller applications with its integrated 10/100 Mbps Ethernet MAC and network-ready applications software. Therefore, the control system of the MPS can communicate through 100 Mbps Ethernet by using Experimental Physics and Industrial Control System (EPICS).

D. Regulator Board

At regulator board, output current signal of DCCT converted voltage signal, and PWM signal. for power conversion is generated by reference value and PI controller. MPS status also is monitored at the same time.

E. Main power Board

A FET driver, a full-bridge four quadrant DC/DC converter switch and 1st L-C output filter were installed on the main power board. Switching frequency is 42 kHz. This board is doing power conversion for magnet and filtering high frequency components of the square wave output for high stability output current. Switching noise generated from main switch is almost eliminated at 1st L-C output filter.

Fig. 5: Fabricated control power board, CPU board, regulator board and main power board

F. Filter of MPS

Two-stage LC filter is adopted to eliminate switching noise generated from main switch. The configuration of the output filter has two-stage LC and RC damping filter. Fig. 6 shows circuit diagram of the output filter. 1st stage LC filter (L_{11}, L_{12}, C_{11}, C_{12}(C_1)) must be arranged within 95 mm x 70 mm at main power module PCB and 2nd stage LC filter (L_{21}, L_{22}, C_{21}, C_{22}) must be arranged within 86 mm x 110 mm as other PCB.

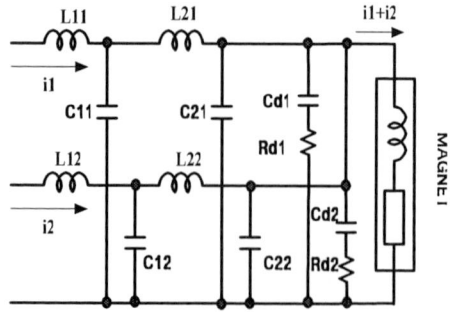

Fig. 6: Filter circuit of MPS

978-1-4244-1871-8/07 $25.00 © 2007 IEEE   234

1) Filter component selection of MPS

Key points of two-stage LC output filter design are control bandwidth, switching frequency of the power supply, and cut-off frequency of the filter. 1st filter cut-off frequency must be more about five times than that of control bandwidth, and 2nd filter cut-off frequency must be closely control bandwidth. Switching frequency must be more than ten times comparing with control bandwidth. Control bandwidth is decided in accordance with MPS application.

2) Inductance of the output LC filter

The first consideration item is inductor choice in filter design. Inductor value controls the amount of ripple current that the capacitor. Inductance is calculated by Eq. 1.

$$L = \frac{V_{out} \times (V_{in\_max} - V_{out})}{V_{in} \times f_{sw} \times I_{out} \times K} \qquad (1)$$

Here, K = 0.3 of the output current.

Desirable value of the inductor might to have a very small value of inductance greater than 30% for high current application. Recommend inductance value is several ten-μH.

In output filter, an inductor is important factors to make small size and low-cost of power supply. Normally inductor has above two-times size and above three times cost compare with capacitor.

3) Capacitance of the output LC filter

The capacitance of the output filter should be high value for cut-off at low frequency. And capacitor ESR (Equivalent Series Resistance) not only contributes to the ESR zero but also contributes significantly to the damping factor of the filter. The drawback of a high ESR capacitor is that to have low ripple and good transient response, the output filter needs to have lower ESR values.

4) Inductance and capacitance of the MPS

1st stage LC-filter is arranged at main power module PCB. The inductor of 1st and 2nd stage filter used the same core. R-material ferrite core (R-44022-EE) of Magnetics companies is selected as the core of the inductor. Switching frequency of the MPS is 42 kHz. Therefore, winding wire must consider skin effect to occur on high frequency switching. 0.2 mm thickness x 26 mm width copper foil is used as wire. Winding is 25 turns. An inductor has 72 cm^3 (4.0 x 4.0 x 4.5) volume, and DC resistance is 20 mΩ. Inductance values of 1st and 2nd filter are decided according to gap thickness. A capacitor is selected with PCB mountable type. WIMA MKS4 15 μF capacitor is adopted as a filter capacitor.

The capacitor ESR is 10 mΩ. Fig. 7 shows fabricated output filter. Output filter parameters are described in table 3.

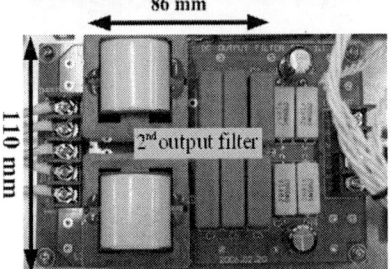

Fig. 7: Fabricated 2-stage output filter

Table 3: Output filter parameters

	inductance	capacitance	Cut-off frequency
1st stage filter	120 µH	30 µF	3.0 kHz
2nd stage filter	60 µH	45 µF	2.6 kHz

5) Frequency characteristics and effect of the output filter

A dynamic signal analyzer, Stanford Research SR780, was used for frequency characteristics measurement of filter and output current. Total cut-off frequency of the output filter is measured about 1.5 kHz. Attenuation is below than –70 dB at MPS switching frequency. Therefore, generated any noise from PS is completely eliminated and clean DC current may be supplied to the magnet. Fig. 8 shows frequency characteristics of output filter measured in MPS. Fig. 9 shows effect of output filter at output current. Any noise did not produce on the output current due to using filter circuit in MPS.

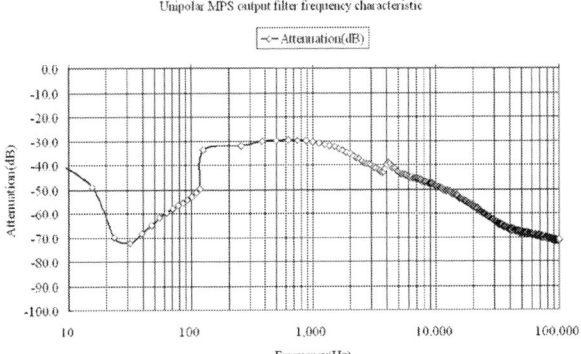

Fig. 8: Frequency characteristic of the output filter

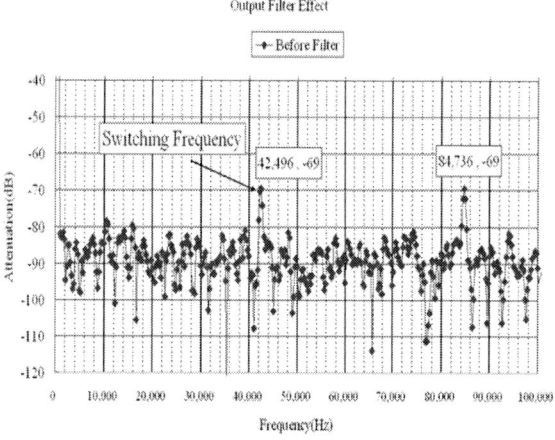

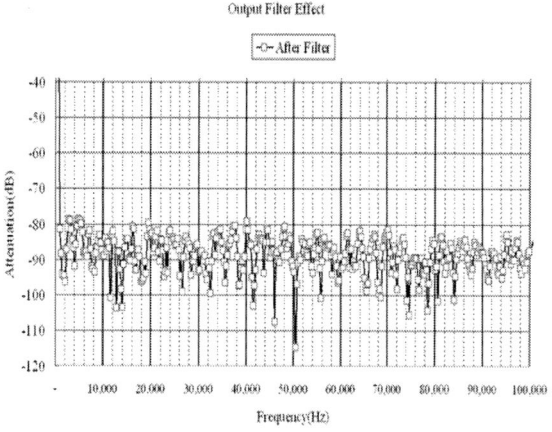

Fig. 9: Filter effect at output current

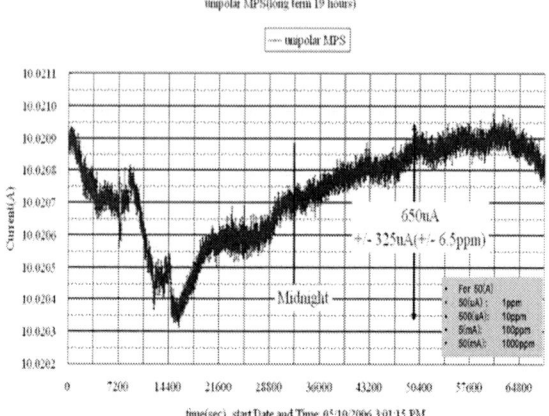

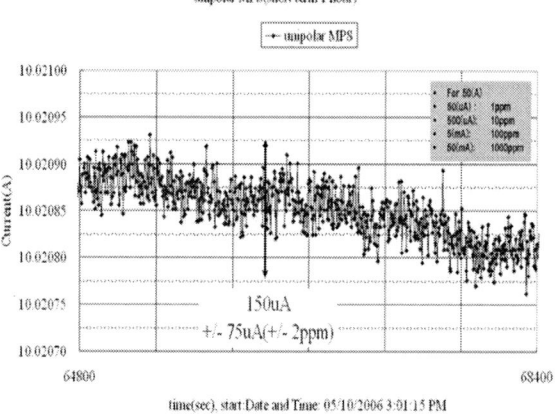

Fig. 10: Long and short-term stability of the MPS output current

G. Output current Stability of the MPS

The measured points were short and long-term stability, and frequency characteristics that are important factors for determining MPS performances. For the measurements of performance, Keithley2700 6.5digit digital multi-meter (DVM) is used. Stability for long periods of time is keeping in ±6.5 ppm during 19 hours, and stability for short periods of time is keeping in ±2 ppm during 1 hour for MPS. Fig. 10 show long and short-periods stability of the MPS output current.

H. Control system of the MPS

Eighty-nine MPSs are installed at PAL Linac and BTL, except BTL dipole MPS. For remote control and monitoring of these MPS, Ethernet 100M bps cable connected from main control room to CPU board of each MPS by the way Ethernet switching hub. Control system is developed by using Experimental Physics and Industrial Control System (EPICS). All function of the MPS function consists of EPICS Process Variable (PV). Fig. 11 shows control system configuration diagram of the PAL Linac MPS.

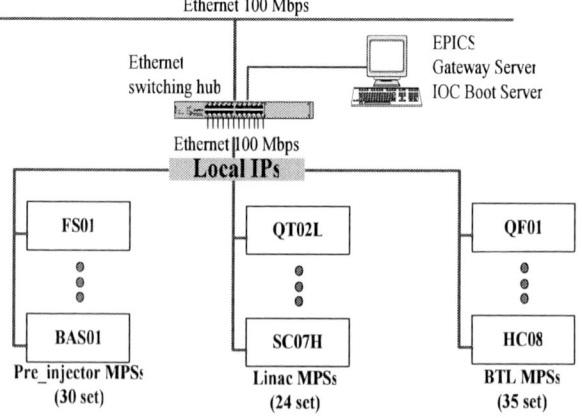

Fig. 11: Control system configuration diagram of the PAL Linac MPSs

IV. CONCLUSION

Four-phase buck type MPS from two full-bridge four-quadrant DC/DC converter scheme has been developed for PAL Linac. This MPS is compact PCB type. In order to easily maintain, mainly part of MPS is developed as standardized PCB. Switching noise generated from the power supply is perfectly eliminated to by utilize a specially designed two-stage output filter. Therefore, clean output DC current can be supplied to the magnets. Stability for long periods of time is keeping in ±6.5 ppm during 19 hours, and stability for short periods of time is keeping in ±2 ppm during 1 hour for MPS. For remote control and monitoring of MPS, EPICS control system of MPS utilizes and is also working very well through 100 Mbps Ethernet.

REFERENCES

[1] S. C. Kim, K. M. Ha, etc, "New Magnet Power Supply for PAL Linac", EPAC2006 proceeding, pp. 2685-2687

[2] S. C. Kim, K. M. Ha, etc, "Compact Output Filter for switching frequency elimination at the PLS Linac New Magnet Power Supply", PAC2007 Conference, June 25-29, NM. USA.

[3] M. R. Pavan Kumar and J. M. S. Kim, "Capacitor current feedback for output filter damping in switched-mode magnet power supplies", IEEE transactions on magnetics, vol.30, no. 4, July 1994, pp. 1778 – 1781

[4] Raymond B. Rdley, "Secondary LC Filter Analysis and Design Techniques for Current-Mode-Controlled Converters", IEEE transactions on power electronics, vol. 3, no. 4, Oct. 1988, pp. 499-507

[5] Hoyerby, M.C.W.; Andersen, M.A.E., "Envelope tracking power supply with fully controlled 4th order output filter", Applied Power Electronics Conference, APEC'06., Twenty-First Annual IEEE, 19-23 March 2006, pp.993-1000

[6] Sipex application note ANP22, Filter Design in continuous conduction mode

Automatic Interleaving Control for Paralleled Converter System and Its Ripple Estimation with Simplified Circuit Model

Teruhiko KOHAMA
Department of Electrical Engineering, Fukuoka University
8-19-1 Nanakuma, Jonan-ku, Fukuoka 814-0180, JAPAN
e-mail:kohama@fukuoka-u.ac.jp

Tamotsu NINOMIYA
Department of EESE, Kyushu University
744 Motooka, Nishi-ku, Fukuoka 819-0395, JAPAN

Abstract—Ripple estimation for paralleled converter system with automatic interleaving control is presented. Current ripple in output capacitor of multi-phase converter is estimated through a simplified circuit model which is available for any paralleled converter system under ideal out-of-phase operation. Relationships between the ripple and circuit parameters such as duty ratio, inductor and number of modules are revealed clearly which lead a simple design guideline for ripple-minimized paralleled converter system.

I. INTRODUCCTION

Paralleled DC-DC converter system is widely used to realize a low-voltage and high-current power supply. Fast dynamic response is also achieved by reducing inductance of smoothing filters of converter modules. However, small inductance causes large current ripple which results in large voltage ripple in the output. In order to reduce the voltage ripple, interleaving operation is necessary in the paralleled converter system to apply ripple cancellation technique of the current [1,2,3]. It is known that the ripple is minimized when the phases between converter modules are equal to $2\pi/N$[rad]. Where N is the number of converter modules. Normally, special design consideration or pre-adjustment is required to achieve the ideal out-of-phase operation for individual paralleled system.

We proposed an automatic interleaving control for any paralleled converter system[4,5]. It reduces voltage and current ripples in the output stage. However, the effectiveness of the ripple cancellation depends on the number of converter modules and their duty ratios. As an instance Fig.1 shows the current waveforms for two-paralleled converter system operating at different duty cycle. Even though Figs.1(a) and (b) are operating under ideal out-of-phase condition, the effectiveness of ripple cancellation is different. In this paper simple ripple estimation for any paralleled converter system under ideal out-of-phase operation is performed with simplified circuit model. Experimental results are shown to confirm the effectiveness of the model.

II. AUTOMATIC INTERLEAVING TECHNIQUE

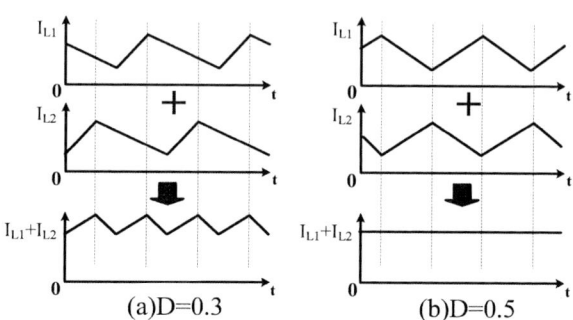

Fig.1 Current waveforms for two-paralleled converter system.

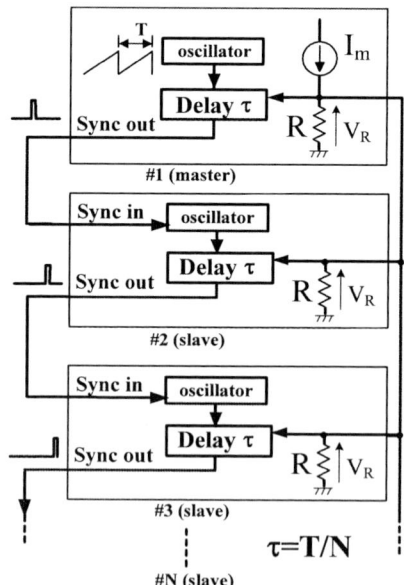

Fig.2 Principle of automatic interleaving technique.

A. Principle of automatic interleaving technique

Automatic interleaving control proposed in [5] is shown in Fig.2. Firstly, master module sends a synchronizing signal to the next slave module with time delay of τ. The slave module restarts the own switching operation by the synchronizing signal. Next, the slave module sends the synchronizing signal to the next slave module with the same delay τ. The delay τ is determined by V_R which depends on the total number of modules N including master. The master has a current source I_m in series with resistance R. Other resistances R in slave modules are connected in parallel with the master. Therefore, V_R is determined by the following equation.

$$V_R = V_m / N \qquad (1)$$

where, V_m is an amplitude of sawtooth-wave in the oscillator and it is given by

$$V_m = R I_m \qquad (2)$$

From Fig.2 there is no difference between master and slave module except for "Sync in" signal and current source I_m. Detailed circuit for slave module is shown in Fig.3(a). Each module has a sawtooth oscillator, synchronizing circuit and delay circuit. Sawtooth oscillator[6,7] consists of a capacitor C, two comparators(CP_1 and CP_2) and a charging/discharging circuit. In the normal operation, capacitor C is charged and discharged by constant current source I_{ref1} and I_{ref2}. Toggle switch SW is controlled by RS-flipflop whose state is determined by CP_1 and CP_2 with the threshold of V_{cmax} and V_{cmin} respectively. Once synchronizing signal with narrow pulse is fed into the base terminal of the transistor next to capacitor C, V_c dropped into V_{cmin} immediately and restarts the charging operation. Synchronizing signal with time delay τ is obtained by using comparator CP with the threshold of V_{TH}, which is given by

$$V_{TH} = V_R + V_{cmin}. \qquad (3)$$

In the case of N=2, for example, $V_R = V_m / 2$, and CP changes the output state when $V_c = V_m / 2 + V_{cmin}$ as shown in Fig.3(b). As a result, the next synchronizing signal is generated with time delay of τ=T/2, where T is a switching period of master module. In the case of N=3, V_R decreases to $V_m / 3$ and V_{TH} is set to $V_m / 3 + V_{cmin}$ which makes delay of T/3. Similar discussion is given for any number of N. From (1) and sawtooth-waveform in Fig.3(b), τ is determined by following equation.

$$\tau = (V_R / V_m) T = T/N \qquad (4)$$

Therefore no pre-adjustment or individual circuit consideration is necessary for any paralleled system. Delay τ is automatically determined instantly by connecting the converter modules. This control circuit is easily implemented into conventional PWM controller because of simple analog circuit compared with other interleaving technique [2].

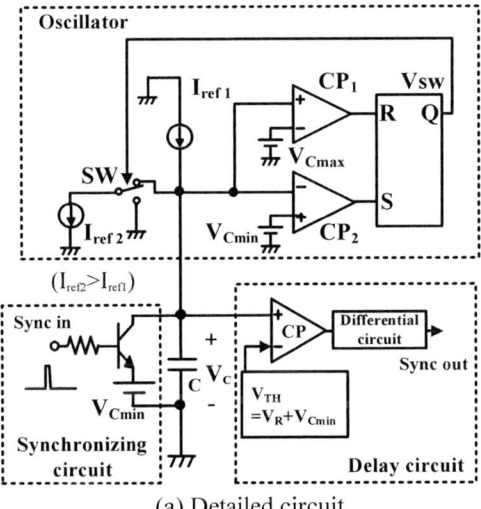

(a) Detailed circuit

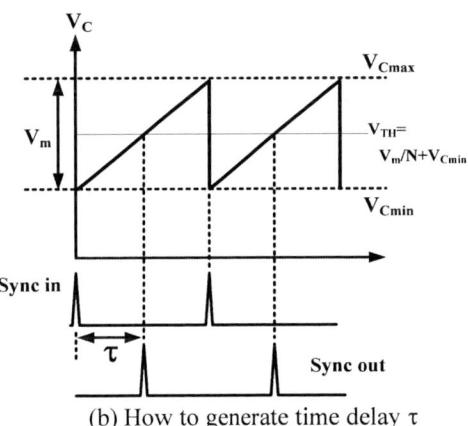

(b) How to generate time delay τ

Fig.3 Detailed circuit and its operation in slave module.

B. Experimental results

Fig.4(a) shows experimental result of sawtooth oscillators for two paralleled converter system(N=2), where τ=T/2 and ideal out-of-phase operation is achieved. Similar results are obtained for N=3, N=4 in Figs.4(b) and 4(c) respectively.

III. RIPPLE ESTIMATION

A. Simplified circuit model for ripple estimation

A module of buck type DC-DC converter in Fig.5(a) is simply represented by a pulse voltage source V_p and a smoothing LC filter as shown in Fig.5(b). The shape of V_p is determined by input voltage V_i and duty ratio D of the module. In order to simplify the circuit model all modules in paralleled converter system are assumed to be the same condition except for phase delay. In case of interleaving

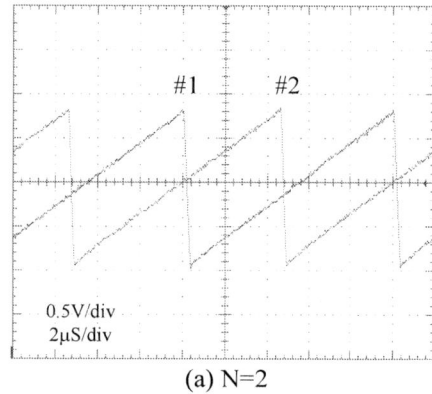

(a) N=2

0.5V/div
2μS/div

#1   #2

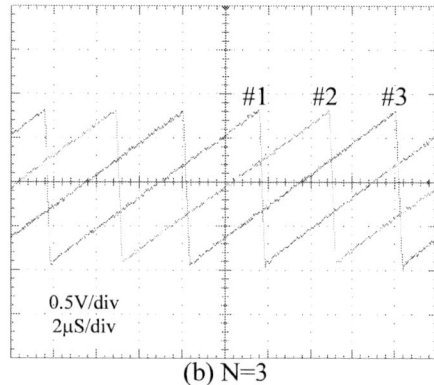

(b) N=3

0.5V/div
2μS/div

#1   #2   #3

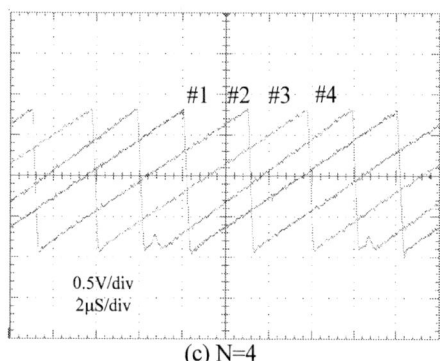

(c) N=4

0.5V/div
2μS/div

#1 #2 #3 #4

Fig.4 Waveforms of oscillator with automatic interleaving control.

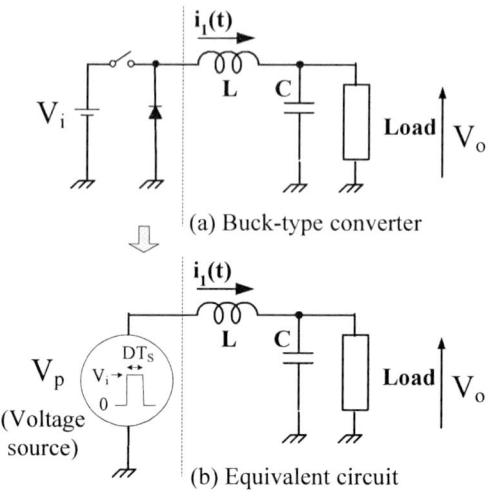

(a) Buck-type converter

(b) Equivalent circuit

Fig.5 Simplified circuit for a buck-type converter module.

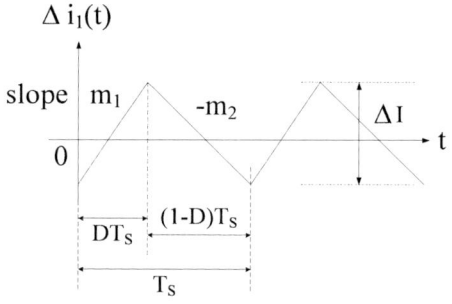

Fig.6 Current waveform in module #1

From Fig.6 amplitude ΔI of the ripple $\Delta i_1(t)$ is determined by

$$\Delta I = m_1 DT_S = m_2(1-D)T_S \qquad (7)$$

If Vo is nearly constant during a switching period of T_s, each module is represented by a triangular current source as shown in Fig.7.

The ripple $\Delta i_1(t)$ in module #1 is derived from Fig.6.

$$\Delta i_1(t) = \begin{cases} m_1 t - \Delta I/2 & (0 < t < DT_S) \\ \Delta I/2 - m_2 t & (DT_S < t < T_S) \end{cases} \qquad (8)$$

Each current ripple for module #k with ideal interleaving is also given by

$$\Delta i_k(t) = i_1\left(t - (k-1)\frac{T_S}{N}\right). \qquad (9)$$

operation, voltage ripple is small compared to DC level of V_o. Therefore the shape of module current in Fig.5 is assumed saw-tooth waveform. For example the shape of current ripple in module #1 is shown in Fig.6, where the slope m_1 and m_2 are given as follows.

$$m_1 = \frac{V_i - V_o}{L} \qquad (5)$$

$$m_2 = \frac{V_o}{L} \qquad (6)$$

978-1-4244-1871-8/07 $25.00 © 2007 IEEE   240

Where, N is the number of converter modules.

B. Ripple in output stage

Therefore the current $i_c(t)$ in the output capacitor C is given by

$$i_C(t) = \sum_{k=1}^{N} \Delta i_k(t) \qquad (10)$$

Amplitude ΔI_C of the current ripple $i_c(t)$ is given by

$$\Delta I_C = Max(i_c(t)) - Min(i_c(t)) . \qquad (11)$$

Where, Max(x) is a function to find maximum value of x(t) in one switching period.

Min(x) is a minimum value of x(t) for one switching period.

Voltage ripple $\Delta v_c(t)$ across the output capacitor is easily obtained by integrating Eq.(10) as follows.

$$\Delta v_C(t) = \frac{1}{C} \int i_C(t) dt \qquad (12)$$

Therefore the amplitude ΔVc of $\Delta v_c(t)$ is given by

$$\Delta V_C = Max(v_c(t)) - Min(v_c(t)) \qquad (13)$$

IV. DISCUSSION

A. Ripple estimation and simple design guideline

The ripples ΔIc and ΔVc are calculated with Mathcad. Figure 8 shows the relationships between the current ripples and number of modules N in the paralleled system as a parameter of duty ratio D. Current ripple ΔI for one module is assumed 5[A] as an example. Figures 8(a) and 8(b) show that the current ripple becomes zero at N=5 for D=0.2 and 0.4. In case of D=0.5, zero current ripple is achieved at N=2,4,6,8,10. On the other hand the ripple for D=0.6 becomes zero only for N=5, 10.

From Figs.8, it is clear that zero ripple operation is achieved by the condition of

$$DN = \text{int} eger . \qquad (14)$$

Equation (14) shows that zero ripple converter system can be realized by selecting D and N for design stage. Although D depends on Vi, Vo, load, and circuit topology of the module, the principle of zero ripples is essential to decrease LC filter dramatically and to achieve fast dynamic response of output voltage.The above ripple estimation is based on ideal interleaving operation. Proposed automatic interleaving control just meets the requirement.

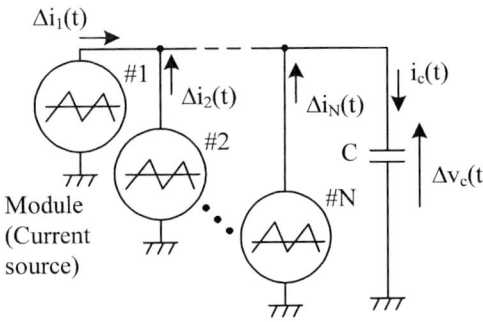

Fig.7 Simplified circuit model for N-paralleled converter system

B. Experimental results

Fig.9(a) and (b) show relationships between current ripple and duty ratio for N=2, N=3 respectively. Current ripple is normalized by the ripple in single module.

In Fig.9(a) current ripple is minimized at D=0.5 for both experimental and theoretical results.

Current ripple for N=3 is also shown in Fig.9(b). Minimum current ripple is achieved at D=0.33 and D=0.66 in the experiment that meets the estimation with simplified circuit model.

V. CONCLUSIONS

Simple ripple estimation for any paralleled DC-DC converter system with automatic interleaving control is performed through simplified circuit model. Zero ripple converter system can be achieved under specific condition of duty ratio and number of modules.

Experimental results show the effectiveness of proposed ripple estimation with simple circuit model.

REFERENCES

[1] B.A.Miwa, D.M.Otten, and M.F.Schlecht, "High efficiency power factor correction using interleaving techniques," IEEE Applied power Electronics Conference, pp.557-568, 1992.

[2] D.J.Perreault and J.G.Kassakian,"Distributed Interleaving of Paralleled Power Converters," IEEE Trans. on Circuit and Systems, Vol.44, No.8, pp.728-734, 1997.

[3] J.Wei and F.C.Lee, "A Novel Soft-Switched, High-Frequency, High-Efficiency, High-Current 12V Voltage Regulator – The Phase-Shift Buck Converter," IEEE 18th Applied Power Electronics Conference, pp.724-730, 2003.

978-1-4244-1871-8/07 $25.00 © 2007 IEEE

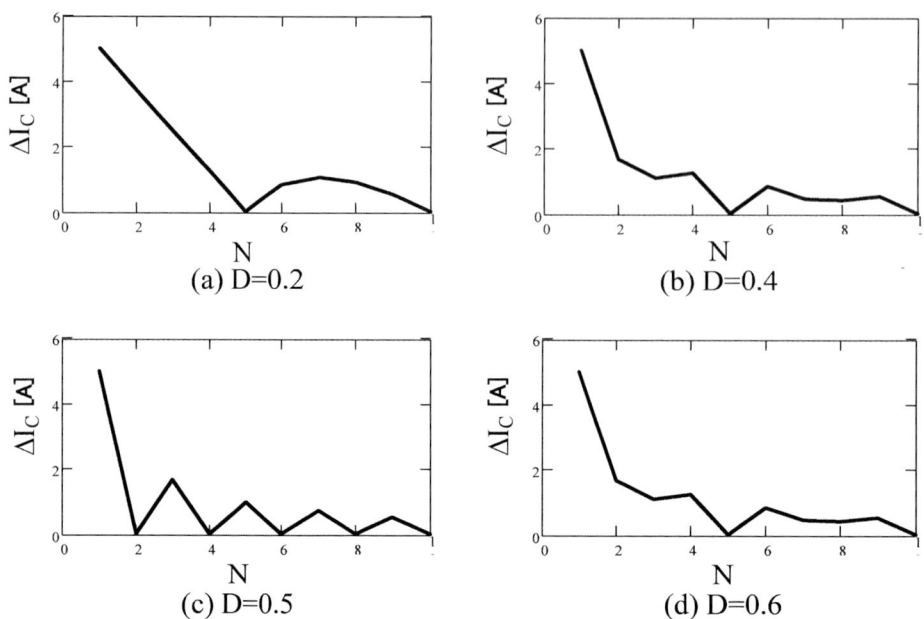

Fig.8 Current ripple estimation for N-paralleled converter system.

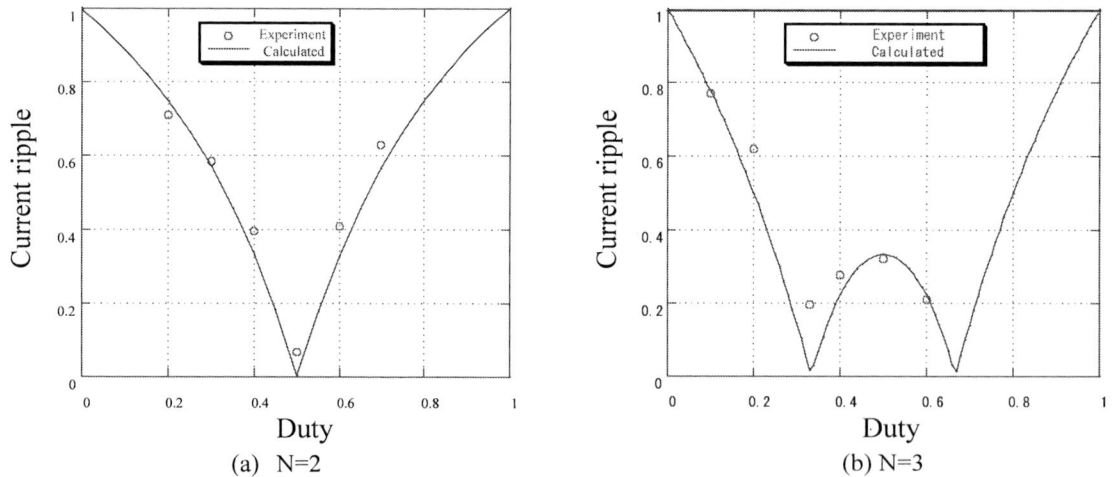

Fig.9 Current ripple vs. duty ratio of converter module.

[4] T.Kohama, G.Endo, H.Shimamori, T.Ninomiya,"New Synchronizing Circuit Suitable for Paralleled Converter System with Automatic Interleaving Operation," Proceedings of IEEE 19th Applied Power Electronics Conference and Exposition, pp.647-653, 2004.

[5] T.Kohama, G.Endo, H.Shimamori, T.Ninomiya, "Simple Multiphase Control for Paralleled Converter System", IEICE Trans. on communication, Vol.E88-B, No.12, pp.4636-4642, 2005.

[6] "High Speed PWM Controller UC3823A,B/3825A,B", Unitrode Product & Applications Handbook, pp.6-158~6-165 1995-1996.

[7] "High Speed PWM Controller UC3824", Unitrode Product & Applications Handbook, pp.6-166~6-172, 1995-1996.

978-1-4244-1871-8/07 $25.00 © 2007 IEEE

Start-up Inrush Current Reduction Technique of Asymmetrical Half-Bridge DC/DC Converter for PC Power Supply

Jae-Kuk Kim, Sung Sae Lee, Won-Sik Oh,
Jung-Eun Kim, and Gun-Woo Moon
Korea Advanced Institude of Scicence and
Technology
Department of Electrical Engineering
373-1 Guseong-Dong, Yuseong-Gu, Daejeon,
305-701, Republic of Korea
Email: jaekook@angel.kaist.ac.kr

Chang-Hyun Gil and Ja-Ryong Cho
IDKOREA
FL2 Samsung Shin hyup B/L. Samsung-Dong, Dong-Gu,
Daejeon, Korea
Email: kch1975@idkorea.co.kr

Abstract- **The asymmetrical half-bridge converter is widely used for PC power supply. However, severe inrush current produces in this converter at start-up. This paper presents start-up inrush current reduction technique of asymmetrical half-bridge DC/DC converter for PC power supply. The proposed asymmetrical half-bridge converter is composed of series connected two link capacitors, center-tapped transformer and blocking capacitor. The proposed converter can reduce the severe start-up inrush current by using the small blocking capacitor. The validity of this study is confirmed from the experimental results.**

I. INTRODUCTION

Recently electric power consumption of PC power supply is increasing due to high efficiency of CPU, mainboard, HDD, and graphic card. Especially, 350W, 400W, and 550W are needed. It is expected that consumers are upgrading their PC to high grade one, and the efficiency of the conventional power supply is very low about 70%. Therefore, the development of high power and high efficiency power supply is required without large increasing the price of power supply.

Fig. 1 is conventional asymmetrical half-bridge converter. The half-bridge converter is widely used for PC power supply because of easy conversion 110V to 220V. Moreover, this converter has only two switches and high efficiency from ZVS of two switches. Therefore asymmetrical half-bridge converter is suitable for PC power supply of which characteristic is low voltage and high current [1-4]. However, asymmetrical half-bridge converter has severe problem of start-up inrush current about 60A as shown in Fig. 2.

In this paper, start-up in-rush current reduction technique in asymmetric half-bridge converter is proposed, and to verify the proposed technique experiment and analysis is presented by producing 400W PC power supply.

II. INRUSH CURRENT REDUCTION

A. The structure of the proposed converter

The proposed converter is shown in Fig. 3. Generally only one capacitor is used for link capacitor in asymmetrical half-bridge converter, but without PFC, for converting 110V to 220V, series connected two capacitors should be used in DC/DC converter. Especially, C_B, which is called blocking capacitor, is used for eliminating the start-up inrush current.

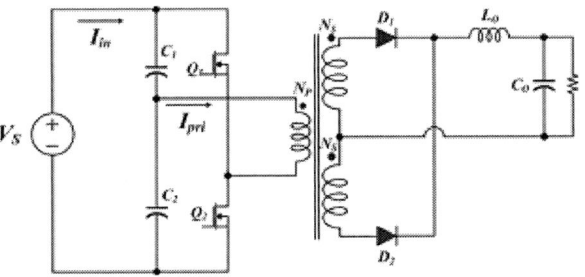

Fig. 1. Conventional converter

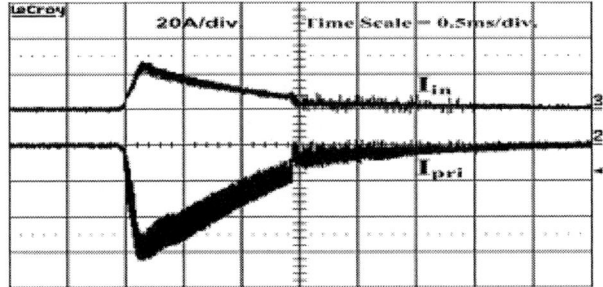

Fig. 2. Start up inrush curret

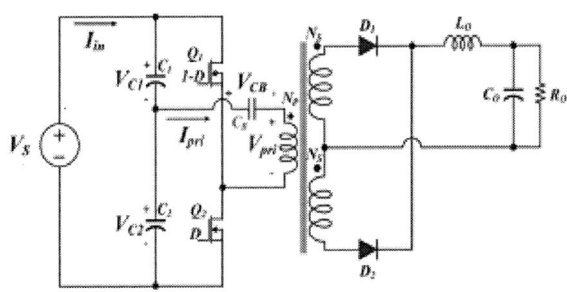

Fig. 3. Proposed Converter

B. Steady state operation of the proposed converter

Some key equations in steady state can be written as follow.

978-1-4244-1871-8/07 $25.00 © 2007 IEEE   243

$$V_{C_1} = \frac{V_S + C_T V_{C_B}}{2} \qquad (1)$$

$$V_{C_2} = \frac{V_S - C_T V_{C_B}}{2} \qquad (2)$$

$$V_{C_B} = \frac{2D-1}{C_T+2} V_S \qquad (3)$$

$$V_O = 2D(1-D)\frac{N_S}{N_P}V_S \qquad (4)$$

$$C_T = \frac{C_B}{C_{link}}, \quad C_1 = C_2 = C_{link} \qquad (5)$$

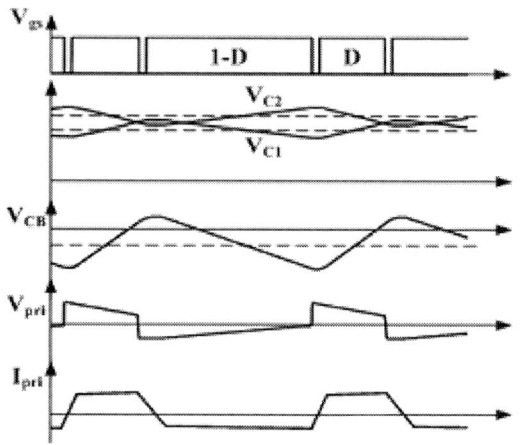

Fig. 4. Key waveforms of the proposed converter

Fig. 4 is key waveforms of the proposed converter. The steady state operation of the proposed converter is almost same as the conventional one. The difference is the variation of V_{C1} and V_{C2} because of the addition of C_B. In the conventional converter, V_{C1} is equal to DV_S and V_{C2} is equal to $(1-D)V_S$ which are dependent on duty, but V_{C1} and V_{C2} of the proposed one are changed like the above equation, which are dependent on duty and C_B. Fig. 5 and 6 show the variations of V_{C1} and V_{C2}, according to duty and C_B, compared with the conventional converter.

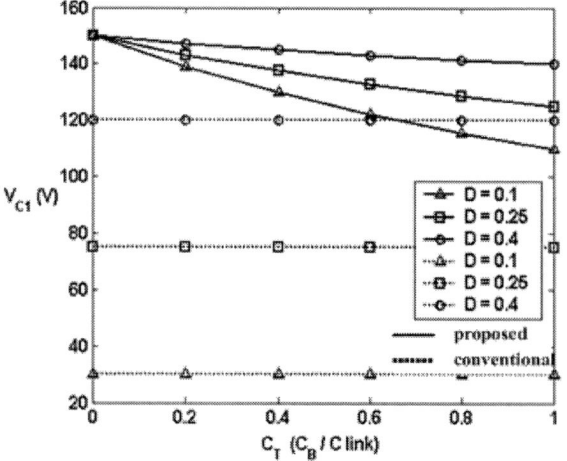

Fig. 5. V_{C1} with respect to C_T and Duty(D)

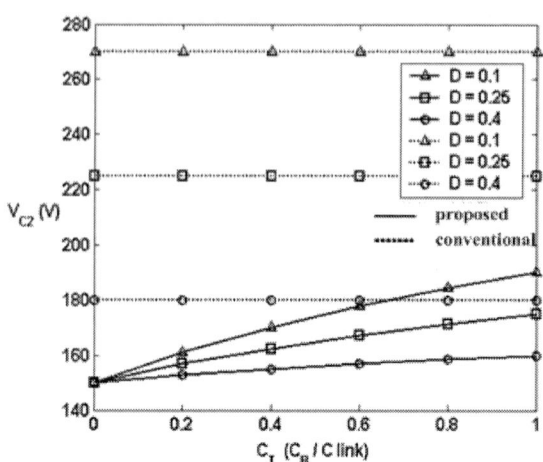

Fig. 6. V_{C2} with respect to C_T and Duty(D)

As shown in the above figures, the difference between V_{C1} and V_{C2} decreases considerably because of the addition of C_B. Therefore, we can select smaller voltage rating of C_1, C_2 compared with the conventional one.

C. Start up operation of conventional converter

Before the switch turns on, the voltage of C_1 and C_2 is $V_S/2$. However, if the duty increases slowly, which is called soft start, for delivering the power to the load gradually, undesirable inrush current occurs.

If the duty of Q_2 increases gradually from zero, for meeting voltage second balance of the primary and output inductor(L_O), the primary current increase severely. In other words, in the process to the steady state, the voltage of C_1 increases to $V_{C1}=DV_S$ and the voltage of C_2 increases to $V_{C2}=(1-D)V_S$. The big difference between the initial value of $V_{C1}=V_S/2$ and $V_{C2}=V_S/2$ and the steady state value of $V_{C1}=DV_S$ and $V_{C2}=(1-D)V_S$ causes severe inrush current as shown in Fig. 2 and 7.

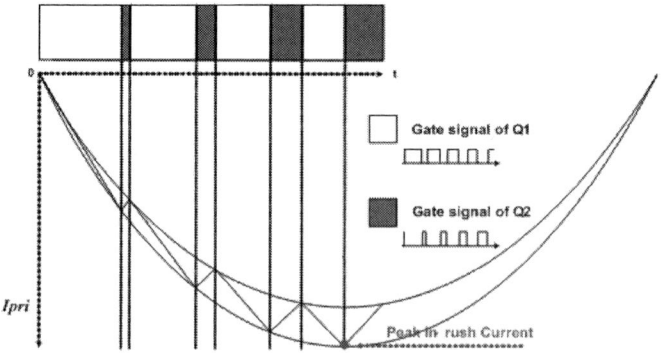

Fig. 7. The process of the inrush current build-up

D. Techniques to eliminate inrush current

First method is to use small link capacitance. Although the voltage difference is large, smaller capacitance needs smaller current to achieve the desired voltage. In experiment, when C_1 and C_2 are 2.2uF, inrush current occurs about 25A as shown in Fig. 8.

However, if the link voltage is produced from just C filters through 60Hz ac line input and full rectifier, small link capacitance can cause the link voltage to have large ripple voltage, so this imposes a heavy burden on the DC/DC converter in the aspect of line regulation. Therefore, this method can be undesirable.

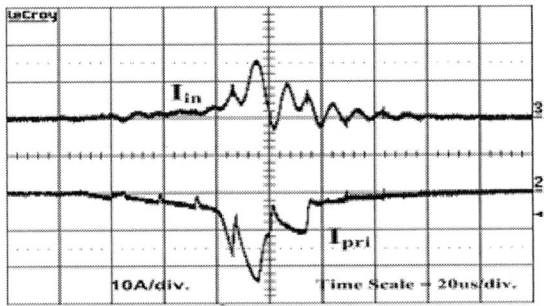

Fig.8. Start-up inrush current when C_1 and C_2 are 2.2uF

Next method is that before start-up, the voltage of C_1 falls to zero. For this, the dummy resistor paralleled with C_1 is needed, Therefore, the voltage of C_1 falls to zero with exponential slope, on the contrary, the voltage of C_2 increases to V_S. Smaller dummy resistance is better choice for faster discharge, but in an aspect of power loss or efficiency, larger dummy resistance is more desirable. In experiment, when link capacitance C_1 and C_2 are 470uF and dummy resistor is 47kΩ, it takes more than 20 seconds for V_{C1} and V_{C2} to discharge to zero. Therefore, this method can be undesirable in the aspects of the reduction of efficiency and timing problem of the voltage of link capacitors.

Finally, the addition of C_B can be a solution of the severe start-up inrush current as shown in Fig. 2. At start-up, the voltage of C_B is charged to $V_S/2$ by using small current from C_1 which had been charged to $V_S/2$ for $(1-D)T$. As a result, the primary voltage of the transformer becomes zero. Conversely when Q_1 becomes ON, the primary voltage of the transformer becomes V_S which is the sum of the voltage of $C_B(= V_S/2)$ and $V_{C2}(= V_S/2)$. Therefore, at start-up, small primary current builds up and inrush current can be eliminated. In experiment, when C_B is 220nF, as shown in Fig. 9, inrush current is eliminated.

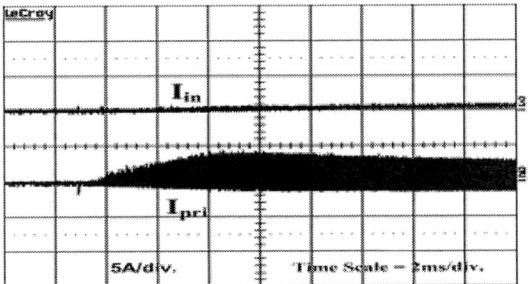

Fig.9. Start-up inrush current when C_B is 220nF

III. CONCLUSION

In this paper, start-up inrush current in asymmetric half-bridge converter is analyzed. Also, three start-up in-rush current reduction technique is proposed, and to verify the proposed technique, the experiment and analysis is presented by producing 400W PC power supply as shown in Fig. 10. Because the link voltage can oscillate with large ripple, the first method is undesirable. The second method is also undesirable because of charging or discharging timing of link capacitors and reduction of efficiency. Consequently, the addition of C_B, the third method, is verfied as most desirable solution for the severe start-up inrush current.

Fig. 10 PC Power Supply Prototype

REFERENCES

[1] Sergey Korotkov, Valery Meleshin, Rais Miftahutdinov, Simon Fraidlin, "Soft-switched Asymmetrical Half-bridge DCDC Converter: Steady-state Analysis of Switching Processes", Telecommunications Energy Special Conference, 97, 1997, pp. 177 - 184.

[2] T. Ninomiya, et al., "Static and dynamic analysis of zero-voltage switched half-bridge converter with PWM control," in Records of 1991 IEEE Power Electronics Specialists Conference (PESC'91), pp. 230-237.

[3] J. Sebastian, et al., "An overall study of the half-bridge complementary-control dc-to-dc converter," in Records of 1995 IEEE Power Electronics Specialists Conference (PESC'95), pp. 1229-1235.

[4] S. Korotkov, et al., "Small-signal modeling of soft-switched asymmetrical half-bridge dc/dc converter," in Proceedings of 1995 IEEE Applied Power Electronics Conference (APEC'95), pp. 707-711.

Comparison the Control Methods in Improvement the Performance of the DC-DC Converter

M. Namnabat
Islamic Azad University
Majlesi Branch
Department of Electrical Engineering
Email: namnabat@iaumajlesi.ac.ir

M. Bayati Poodeh
Islamic Azad University
Najafabad Branch
Department of Electrical Engineering
Email: bayati@iaun.ac.ir

S. Eshtehardiha
Islamic Azad University
Najafabad Branch
Department of Electrical Engineering
Email: eshtehardiha@gmail.com

Abstract—In this paper, two different control methods on DC-DC Converter, are compared with together. These converters, are used for the stabilization or the control of DC voltage of a battery. In addition to other applications, DC converters feed electric vehicles (trucks, electric vehicles, and subway locomotives), telephone sets and civil inverters. Lately, improvement the performance of the DC-DC converter, is one of the goals of the engineers in the industries. In this way, several control methods are used to control Buck converters. In this paper, PI and PID controllers are used to optimized the DC-DC converter performance. Also other controller, Pole placement controller, can improve the Buck converter. By the way, Genetic Algorithm (GA), is used with these control methods, to obtain the best coefficients in them. The results are shown the capability of the control methods in the improvement of the above-mentioned converter functioning.

I. INTRODUCTION

The DC converter is a device which transforms AC to DC. This device is also known as an AC to DC converter. A Chopper can be considered as a DC equivalent of an AC transformer with a convertible constant convertible in a continuous form. Like a transformer, the converter can be employed for stepwise increase or reduction of DC source voltage. The converters are wildly used for the control of motor voltage in electric cars, ceiling elevators, mine excavation etc. Their specific features are the precise control of acceleration with high efficiency and fast dynamic response.

Converters are also employed in DC motors to return the energy to its source. In this way, it results in the saving of energy in the transportation systems in prolonged stoppage. Converters are also used in DC voltage regulators along with an inductor to produce a DC current source especially for the current source inverters.

Some control methods have stated the issue of control through pole placement [1]. Another method is the use of state feedback in the control of DC-DC converters [2]. In modeling area of DC-DC converters, a variety of models are presented which comprises desirable responses by administration of control methods. Most of the articles have concentrated on controlling designs of PID and PI [3]. Of other control methods, the feedback loop is another among others, [4], [5]. The use of LQR method for the improvement of Buck converter function is the subject presented in [6].

II. CIRCUIT MODEL FOR BUCK CONVERTER

In a Buck converter the average amount of output voltage V_{out} is less than the input voltage V_{in}. The regulator circuit uses the power switch depicted in Fig. 1.

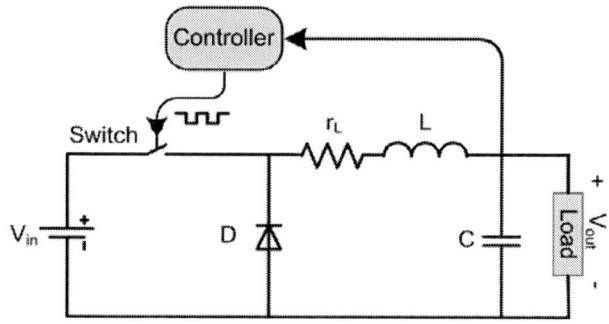

Fig.1 : Buck converter

The function of the circuit is divided into two parts. The first part starts when the switch is turned on at $t=0$. The input current which is rising passes through L filter inductor, C filter capacitor, and R charge resistance. The second part starts when the switch is turned off at $t=t_1$. Due to the presence of stored energy in the inductor, the inductor current continues passing from L, C, load and D. The inductor current declines until the second switch switching in the next cycle. On the basis of the above analysis and the state equations, the conversion function of the above converter is obtained as follows (Table. 3):

$$\frac{V_{out}(s)}{\delta(s)} = -\frac{V_{in}}{LCs^2 + s\left[\frac{L}{R} + r_L C\right] + \left[1 + \frac{r_L}{R}\right]} \quad (1)$$

III. PID CONTROL

A. PID Control of Plants

If a mathematical model of the plant can be derived, then it is possible to apply various design techniques for determining parameters of the controller that will meet the transient and steady-state specifications of the closed–loop system. However, if the plant is so complicated that is mathematical model cannot be easily obtained, than an analytical approach to the design of a PID controller is not possible. Then we must resort to experimental approaches to the tuning of PID controllers.

978-1-4244-1871-8/07 $25.00 © 2007 IEEE

The process of selecting the controller parameters to meet given performance specifications is known as controller tuning. Ziegler and Nichols suggested rules for tuning PID controllers (meaning to set values K_p, T_i, and T_d) based on experimental step responses or based on the value of K_p that results in marginal stability when only proportional control action is used. Ziegler–Nichols rules, which are briefly presented in the following, are useful when mathematical models of plants are not known. (These rules can, of course, be applied to the design of systems with known mathematical models.) Such rules suggest a set of values of K_p, T_i and T_d, that will give a stable operation of the system, as shown in Fig. 2. However, the resulting system may exhibit a large maximum overshoot in the step response, which is unacceptable. In such a case we need series of fine running until an acceptable result is obtained. In fact, the Ziegler–Nichols tuning rules give an educated guess for the parameter values and provide a starting point for fine tuning rather than giving the final settings for K_p, T_i and T_d in a single shot [7].

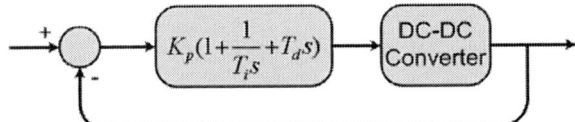

Fig. 2 : PID controls and two degrees of freedom control system

B. Ziegler–Nichols Rules for Tuning PID Controllers

Ziegler and Nichols proposed rules for determining values of the proportional gain K_p, integral time T_i, and derivative time T_d based on the transient response characteristics of a given plant. Such determination of the parameters of PID controllers or tuning of PID controllers can be made by engineers on–site by experiments on the plant. (Numerous tuning rules for PID controllers have been proposed since the Ziegler–Nichols proposal. They are available in the literature and from the manufacturers of such controllers.)

There are two methods called Ziegler–Nichols tuning rules: the first method and the second method.

1) First Method

In the first method, we obtain experimentally the response of the plant to a unit–step input, as shown in Fig. 2. If the plant involves neither integrator(s) nor dominant complex–conjugate poles, than such a unit–step response curve may look S-shaped, as shown in Fig. 3. This method applies if the response to a step input exhibits an S–shaped curve. Such step–response curves may be generated experimentally or from a dynamic simulation of the plant.

The S–shaped curve may be characterized by two constants, delay time L and time constant T. The delay time and time constant are determined by drawing a tangent line at the inflection point of the S–shaped curve and determining the intersections of the tangent line with the time axis and line $c(t) = K$, as shown in Fig. 3.

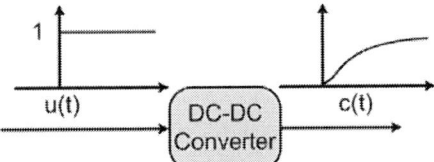

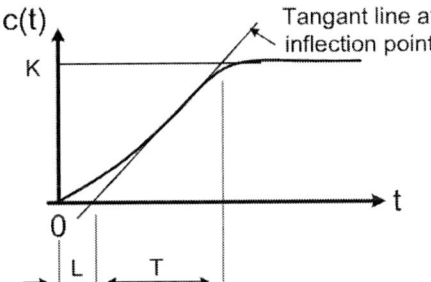

Fig. 3 : S-shaped curve of step response

Table. 1
Ziegler-Nichols tuning rules (first method)

Type of Controller	K_p	T_c	T_e
P	$\dfrac{T}{L}$	∞	0
PI	$0.9\dfrac{T}{L}$	$\dfrac{L}{0.3}$	0
PID	$1.2\dfrac{T}{L}$	$2L$	$0.5L$

The transfer function $C(s)/U(s)$ may then be approximated by a first–order system with a transfer lag as (2).

$$\frac{C(s)}{U(s)} = \frac{Ke^{-ts}}{Ts+1} \tag{2}$$

Ziegler and Nichols suggested to set the values of K_p, T_i, and T_d according to the shown in Table. 1.

Notice that the PID controller tuned by the first method of Ziegler–Nichols gives (3).

$$G_e(s) = K_p(1 + \frac{1}{T_i s} + T_d s)$$

$$= 1.2\frac{T}{L}(1 + \frac{1}{2Ls} + 0.5Ls)$$

$$= 0.6T\frac{(s+\frac{1}{L})^2}{s} \tag{3}$$

Thus, the PID controller has a pole at the origin and double zeros at $s = -1/L$.

2) Second Method

In the second method, we first set $T_k = \infty$ and $T_d = 0$. Using the proportional control action only (see Fig. 4), increase K_p from 0 to a critical set K_{er} at which the output first exhibits sustained oscillations. (If the output does not habit sustained oscillation for whatever value K_p may take, then this method does not apply,) Thus, the critical gain K_{er} and the corresponding period P_{cr} are experientially determined. Ziegler and Nichols suggested that we set the values of the parameters K_p, T_i, and T_d according to the formula shown in Table. 2.

Table. 2
Ziegler-Nichols tuning rules (second method)

Type of Controller	K_p	T_c	T_e
P	$0.5K_{cr}$	∞	0
PI	$0.45K_{cr}$	$\dfrac{1}{1.2}P_{cr}$	0
PID	$0.6K_{cr}$	$0.5P_{cr}$	$0.125P_{cr}$

Ziegler–Nichols tuning rules (and other tuning rules presented in the literature) have been widely used to tune PID controllers in process control systems where the plant dynamics are not precisely known. Over many years, such tuning rules proved to be very useful. Ziegler–Nichols tuning rules can, of course, be applied to plants whose dynamics are known. (If the plant dynamics are known, many analytical and graphical approaches to the design of PID controllers are available, in addition to Ziegler–Nichols tuning rules.)

IV. POLE PLACEMENT DESIGN METHOD

Pole placement is a method that seeks to place the poles of the closed–loop system at some predetermined locations. Although this method has some drawbacks in handling complex systems, it is still fairly sufficient for most small control systems and it gives the best introduction to the design of complex systems. The basic concept behind the method is to get K, which will satisfy the closed–loop transfer function at desired pole locations s_i, $i = 1, 2, \dots n$. Implementation of the method will be described here, through the following illustrative example in which a regulator is assumed, i.e., no reference input. (The reference input will be added after some discussion on the state estimators) [7].

Suppose the system as (4).

$$\dot{x}(t) = Ax(t) + Bu(t) \tag{4}$$

System is to be controlled by full state feedback such that (5).

$$u(t) = -Kx(t), \tag{5}$$

Where the closed–loop poles are placed at locations p_1, p_2, ... p_n. This means that the required closed–loop transfer function of the controlled system is given by (6).

$$\psi(s) = (s - p_1)(s - p_2)(s - p_3)\dots(s - p_n) = 0 \tag{6}$$

Which can be expanded as (7).

$$\psi(s) = S^n + q_n S^{n-1} + q_{n-1} S^{n-2} + \dots + q_3 S^2 + q_2 S + q_1 = 0 \tag{7}$$

Let the matrix A and input matrix B respectively, as (8).

$$A = \begin{bmatrix} a_{11} & a_{12} & \cdots & a_{1n} \\ a_{21} & a_{22} & \cdots & a_{2n} \\ \vdots & \vdots & \vdots & \vdots \\ a_{n1} & a_{n2} & \cdots & a_{nn} \end{bmatrix}, \quad B = \begin{bmatrix} b_1 \\ b_2 \\ \vdots \\ b_n \end{bmatrix} \tag{8}$$

Therefore, if the feedback matrix K is as (9).

$$K = \begin{bmatrix} k_1 & k_2 & \cdots & k_n \end{bmatrix} \tag{9}$$

Then the closed-loop system has (10), system matrix.

$$A - BK = \begin{bmatrix} a_{11} - b_1 k_1 & a_{12} - b_1 k_2 & \cdots & a_{1n} - b_1 k_n \\ a_{21} - b_2 k_1 & a_{22} - b_2 k_2 & \cdots & a_{2n} - b_2 k_n \\ \vdots & \vdots & \vdots & \vdots \\ a_{n1} - b_n k_1 & a_{n2} - b_n k_2 & \cdots & a_{nn} - b_n k_n \end{bmatrix} \tag{10}$$

Whose characteristic function is shown in (11).

$$\psi(s) = |sI - A + BK|$$
$$= \begin{bmatrix} s - a_{11} + b_1 k_1 & -a_{12} + b_1 k_2 & \cdots & -a_{1n} + b_1 k_n \\ -a_{21} + b_2 k_1 & s - a_{22} + b_2 k_2 & \cdots & -a_{2n} + b_2 k_n \\ \vdots & \vdots & \vdots & \vdots \\ -a_{n1} + b_n k_1 & -a_{n2} + b_n k_2 & \cdots & s - a_{nn} + b_n k_n \end{bmatrix} = 0 \tag{11}$$

Comparison of this characteristic equation and demanded in (7) can lead to the determination of the values of k_i and hence matrix K. However, as it can be seen, the algebra behind such a problem is very cumbersome and might in some cases be insoluble. On the other hand, however, if system (A, B) is controllable, the closed-loop system can be expressed in its controllable canonical form as (12).

$$A^* - B^* K^* = \begin{bmatrix} a^*_{11} - b^*_1 k^*_1 & a^*_{12} - b^*_1 k^*_2 & \cdots & a^*_{1n} - b^*_1 k^*_n \\ a^*_{21} - b^*_2 k^*_1 & a^*_{22} - b^*_2 k^*_2 & \cdots & a^*_{2n} - b^*_2 k^*_n \\ \vdots & \vdots & \vdots & \vdots \\ a^*_{m1} - b^*_n k^*_1 & a^*_{n2} - b^*_n k^*_2 & \cdots & a^*_{nn} - b^*_n k^*_n \end{bmatrix} \quad (12)$$

In this case, the closed-loop transfer function becomes (13).

$$\left| sI - A^* + B^* K^* \right| = \begin{bmatrix} s & -1 & \cdots & 0 \\ 0 & s & \cdots & 0 \\ \vdots & \vdots & \vdots & \vdots \\ -a_1 - k^*_1 & -a_2 - k^*_2 & \cdots & s - a_n - k^*_n \end{bmatrix} \quad (13)$$

Whose expansion can easily be determined to be (14).

$$\psi(s) =$$
$$s^n + (a_n + k^*_n)s^{n-1} + (a_{n-1} + k^*_{n-1})s^{n-2} + ... + (a_2 + k^*_2)s + (a_1 + k^*_1) \quad (14)$$

Comparison of this equation with the demanded one in (7) shows the (15).

$$a_i + k^*_i = q_i \ , \ i = 1, \ 2, \ ..., \ n \quad (15)$$

From which the elements of the feedback matrix can be computed as (16).

$$k^*_i = q_i - a_i \ , \ i = 1, \ 2, \ ..., \ n \quad (16)$$

Or in vector form as (17).

$$K^* = q - a \quad (17)$$

Where

$$q = \begin{bmatrix} q_1 & q_2 & \cdots & q_n \end{bmatrix} \quad (18)$$

$$a = \begin{bmatrix} a_1 & a_2 & \cdots & a_n \end{bmatrix} \quad (19)$$

In the emphasized again that this procedure applies only for SISO systems in controllable canonical form, and that order of the elements in vectors a, q and K^* are as shown above. Improper order of the elements will give wrong results. This matrix K^* is the feedback gain for the system in controllable canonical form, the effort is such that is express in (20).

$$u(t) = -K^* x^*(t) \quad (20)$$

Where

$$x^*(t) = P^{-1} x(t) \quad (21)$$

Therefore, for the original system (not in control canonical form) this control effort becomes as (22).

$$u(t) = -K^* P^{-1} x(t) \quad (22)$$

So that the corresponding feedback gain matrix K is;

$$K = K^* P^{-1} \quad (23)$$

V. GENETIC ALGORITHM AND ITS APPLICATION IN DETERMINING THE CONTROLLING PARAMETERS

A. Genetic algorithm

In 1960 the first serious investigation into Genetic Algorithms (GAs) was undertaken by John Holland. Genetic algorithms have become popular due to several factors. They have been successfully applied to self-adaptive control systems and to function optimization problems. They are a powerful search technique, yet are computationally simple. The search method they use is robust since it is not limited like other search methods with regard to assumptions about the search space.

The use of natural evolution method for the optimization of control system has been of interest for the researchers since a long time. The control system parameters are considered as the genes of one chromosome in this system and then with the formation of a random population of different chromosomes and calculation of the target function similar to any other chromosomes by using the promotion generation methods to reach the best response are obtained to satisfy the minimum target function [8].

B. The application of genetic algorithm for adjusting PID controlling coefficients

To optimize the system responses to the input in the least possible time, the application of genetic algorithm on K_p, T_i and T_d constants will be beneficial.

The bases for optimization are settling time t_s, rise time t_r and Mp overshoot maximum rate and also the difference between the system desired rate and its real response. The optimization target function for the above parameters is in the form of (24) [8].

$$F_{obj} = (tr + ts)e^{-1.5} + (Ess - Mp)(1 - e^{-1.5}) \quad (24)$$

The members of every individual are $K_p{}'$, T_i and T_d,
Population size $M=30$;
Crossover rate $Pc=0.8$;
Mute rate $Pm=0.01$;
The numbers of generation is 1000.

By considering Table. 3, Fig. 4, the Buck converter open loop response is obtained.

By applying PI controlling on the system and with regard to figures in Table. 4, the results of Fig. 5, are obtained which are comparable with [3].

978-1-4244-1871-8/07 $25.00 © 2007 IEEE

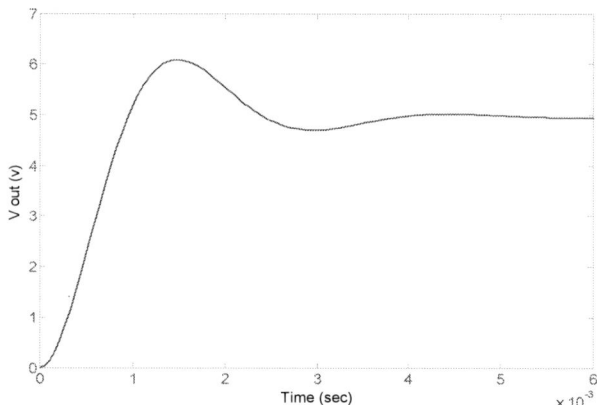

Fig.4 : Buck converter open loop response

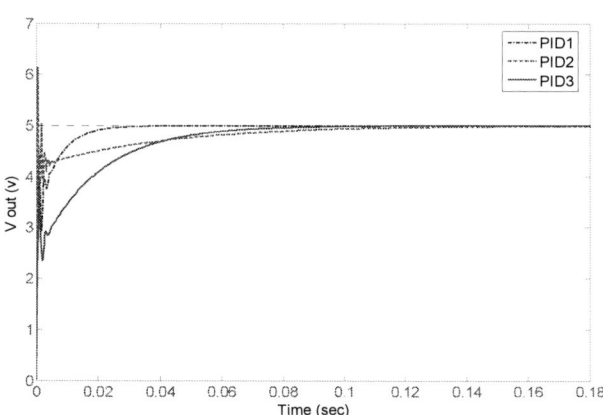

Fig. 5 : Buck converter response with PID controller

By applying genetic algorithm for the optimization of PID controller response on K_P , T_i and T_d parameters the results in Fig. 6, are obtained in which K_P =353.98 and T_i =400.63 and T_d = 493.25 . Considering the system response it is observed in Fig. 6, that the time required to reach the desired response of the converter to its possible minimum time is 2.5 nano seconds.

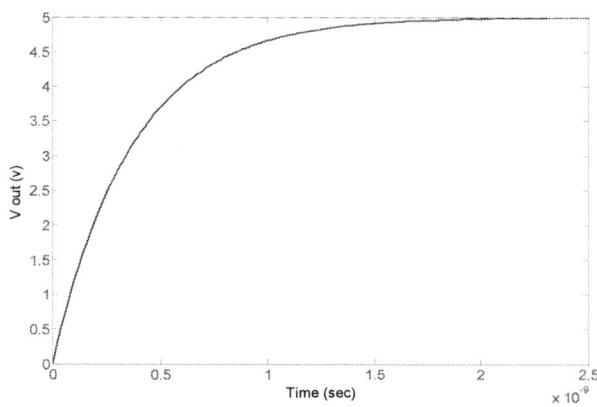

Fig. 6 : Buck converter response with PID-GA controller

C. Use of genetic algorithm for adjusting state feedback coefficients in pole placement method

To design the pole placement, it is necessary to specify the poles with regard to the system converting function at points -1000 ± 2129.2. The new location of the poles which are obtained by the system stability increase index, and reinforcement of the system responding speed in intransient conditions through trial and error. The results are presented in Table. 5, and the system response to this displacement is depicted in Fig. 7.

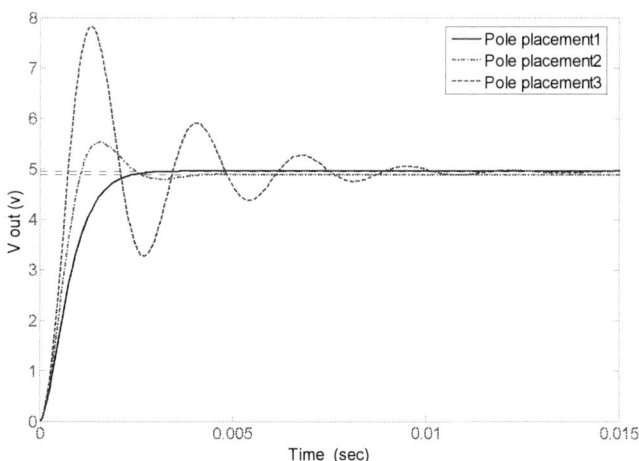

Fig. 7 : Buck converter response of pole placement

By applying genetic algorithm for placement optimization of the system poles for the improvement of converter response, we obtain the conclusions in Fig. 8.

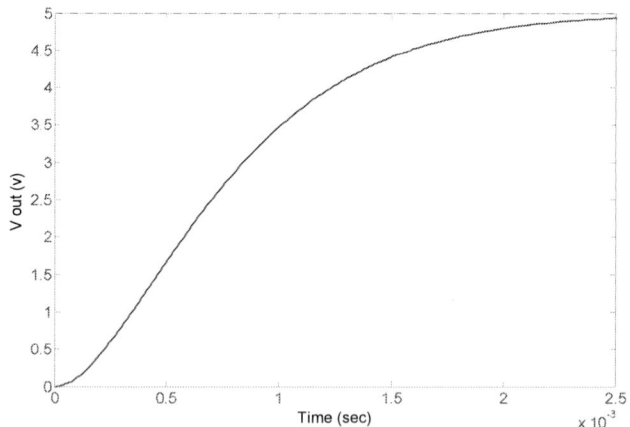

Fig. 8 : Buck converter response of pole placement-GA

With regard to the results presented in Table. 4, and Table. 5, it is observed that by applying genetic algorithm, both PID controller and pole placement controller responses will improve, but system response will be faster and more precise by employing PID-GA controller.

978-1-4244-1871-8/07 $25.00 © 2007 IEEE

VI. CONCLUSION

In this paper, two control methods are explained and used to improve the DC-DC converter performance. PID control based Ziegler and Nichols methods are useful in control the output voltage in Buck converter. Also pole placement controller has good result with converter. In this way, by using Genetic Algorithm, these controllers are optimized. The results of PID controller method are more satisfactory with genetic algorithm than the results of genetic algorithm pole placement method.

Table. 3
Buck converter parameters

Parameter	Symbol	Value
voltage source	V_{in}	12 V
output voltage	V_{out}	5 V
filter capacitance	C	100 μF
converter inductance	L	2.1mH
load resistance	R	6.8 Ω
inductor resistance	r_L	1.1 Ω
duty ratio	δ	0.48

Table. 4
PID parameters

PID 1	K_P =1.54, T_i =384.64, T_d =0
PID 2	K_P =5.11, T_i =150.89, T_d =0
PID 3	K_P =0.93, T_i =100.67, T_d =0
PID-GA	K_P =353.98, T_i =400.63, T_d =493.25

Table. 5
Poles of system with pole placement controller

Pole placement 1	$s = -54 \pm 2341.7i$
Pole placement 2	$s = -200 \pm 2342.2i$
Pole placement 3	$s = -600 \pm 2342.2i$
Pole placement GA	$s = -2256.1 \pm 633.2i$

REFERENCE

[1] Kelly, and K. Rinne, "Control of DC-DC converters by direct pole placement and adaptive feed forward gain adjustment," *IEEE 2005.*

[2] I. G. Keller, D. Lasea and J. M. A. Myrzik, "State- space Control structures For Buck Converters with/ without Input Filter," *IEEE. 2005., Power Electron, 2005 European Conf.*, Sept. 2005, 10 pp.

[3] S. Uran and M. Milanovic, "*State Controller for buck Converter,*" *EUROCON, 2003. IEEE. 2003*, Vol. 1, Sept. 2003, pp.381 – 385.

[4] S. C. Tan, Y. M. Lai, M. K. H. Cheung, and C. K. Tse, "On the practical design of a sliding mode voltage controlled buck converter," *IEEE Trans., Power Electron, 2005*, no. 20, vol. 2, pp. 425-437.

[5] Y. He, and F. L. Luo, "Study of sliding mode control for DC-DC converters," PowerCon 2004., Nov. 2004., Vol.2, pp. 1969 – 1974.

[6] F. H. F. Leung, P. K. S. Tam, and C. K Li, "An improved LQR-based controller for switching DC-DC converters," IEEE Trans., Ind Electron., Oct. 1993, pp. 521 – 528.

[7] K. Ogata, *Modern Control Engineering*, Prentice-Hall, New Jersey, 1999.

[8] R. L. Haupt and S. E. Haupt, "*Practical Genetic Algorithms,*" John wiley & Sons, Inc, New york, 2004.

BIOGRAPHIES

Majid Namnabat was born in Isfahan, Iran. He received the B.S. degree in Electronic engineering, and the M.S. degree in Control engineering from Islamic Azad University, Tehran Jonoub branch, Tehran, Iran, in 1992, and 1998, respectively. His research focuses on Modern control, specialy in industries, adaptive control, neural networks, and optimization control methods. He has working in Foolad Steel Mobarekeh Isfahan industry as a control engineer. E-mail: namnabat@iaumajlesi.ac.ir.

Mohammad Bayati Poodeh was born in Isfahan, Iran. He received the B.S. degree in Power electric engineering, and the M.S. degree in Power electric engineering from Islamic Azad University, Najafabad branch, Najafabad, Isfahan, Iran, in 2005, and 2007, respectively. His research focuses on power system control and stability, subsynchronous oscillation, FACTS devices, HVDC systems, neural networks, fuzzy modeling and control and optimization control methods. E-mail: bayati@iaun.ac.ir.

Saeid Eshtehardiha was born in Isfahan, Iran. He received the B.S. degree in Power electric engineering, and the M.S. degree in Power electric engineering from Islamic Azad University, Najafabad branch, Najafabad, Isfahan, Iran, in 2005, and 2007, respectively. His research focuses on Facts devices, specially STATCOM and reliability, adaptive control, neural networks, fuzzy controller, and optimization control methods. E-mail: eshtehardiha@hotmail.com.

Optimal Design of Bus Converter in On-Board Distributed Power Architecture

Seiya Abe
Kyushu University
Department of EESE
744, Motooka, Nishi-ku,
Fukuoka, 819-0395
Email: abe@ees.kyushu-u.ac.jp

Masahiko Hirokawa
TDK Corporation

2-15-7, Higashi-Ohwada,
Ichikawa, Chiba, 272-8558
Email: mhiroka@mb1.tdk.co.jp

Tamotsu Ninomiya
Kyushu University
Department of EESE
744, Motooka, Nishi-ku,
Fukuoka, 819-0395
Email: ninomiya@ees.kyushu-u.ac.jp

Abstract— The power supply system which requires the low-voltage / high-current output has been changing from conventional centralized power system to distributed power system. The distributed power system consists of bus converter and POL. The most important factor is the system stability in bus architecture design. The overlap between the output impedance of bus converter and input impedance of POL causes system instability, and it has been an actual problem. Increasing the bus capacitor, system stability can be reduced easily. However, due to the limited space on the system board, increasing of bus capacitors is impractical. The urgent solution of the issue is desired strongly. This paper presents the output impedance design for on-board distributed power system by means of three control schemes of bus converter. The output impedance peak of the bus converter and the input impedance of the POL are analyzed, and it is conformed by experimentally for stability criterion. Furthermore, the design process of each control schemes for system stability is proposed.

I. INTRODUCTION

Various LSI is used in the telecommunication application equipments and the driving voltage is various. On the other hand, increase of load current is also remarkable by advanced function of LSI. Since the present LSI is designed in accordance with semiconductor manufacture technology, the tolerance level of the operating voltage is very narrow. Consequently, the voltage drop by the wiring impedance of power line causes malfunction of LSI. In order to reduce the malfunction of LSI by the voltage drop, it is proposed that the converter is arranging very close to the LSI. This converter is called POL. Thus, the power supply system which requires the low-voltage / high-current output has been changing from conventional centralized power system to distributed power system. The distributed power system consists of first-stage isolated DC-DC converter as a bus converter and second-stage non-isolated DC-DC converter as a POL. However, the instability phenomenon in a distributed power system is posing a problem recently. This is instability phenomenon resulting from overlapping between the output impedance of bus converter and the input impedance of POL. Increasing the bus capacitor, the system stability can be reduced easily. However, due to the limited space on the system board, increasing of bus capacitors is impractical. The urgent solution of the issue is desired strongly, and the various discussion of system stability has been reported[1-8]. Then, we also have reported the detailed discussion of system

stability by control schemes of bus converter (Un-regulated, Semi-regulated and Full-regulated)[9-14]. However, so far, the detailed discussion of practical design of bus converter about on-board distributed power system has not been reported. This paper presents the optimal design of bus converter for on-board distributed power system by means of three control schemes of bus converter.

II. STABILITY CRITERION AND IMPEDANCE CHARACTERISTICS

The on-board distributed power system consisting of bus converter and POL. The half-bridge converter with the most popular circuit of the power–stage is used as a bus converter, and the synchronous buck converter with the most popular circuit is used as POL. Figure 1 and 2 show the circuit diagrams, respectively. Even if each converter has stable operation, the instability phenomenon may occur by connecting two converters in series. The input and output impedance is greatly concerned with the system stability.

From previous report of reference14, it is necessary to set the peak value of output impedance smaller than the low-frequency value of input impedance $|Zin(0)|$ for system stability. In this section, the stabilization design of the on-board distributed power system by means of three control schemes of bus converter is considered.

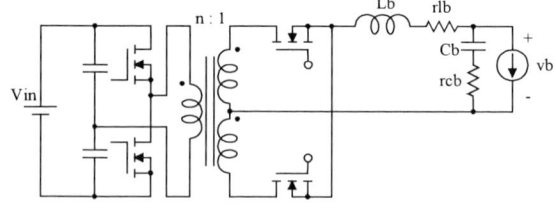

Fig. 1. Bus converter (Half-bridge converter)

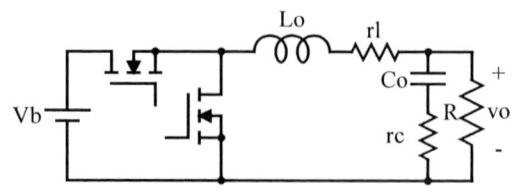

Fig. 2. POL (Buck converter)

978-1-4244-1871-8/07 $25.00 © 2007 IEEE

At first, low-frequency value |Zin(0)| of input impedance is estimated. Applying the state-space averaging method can derive the input impedance of POL as follows[15-17].

$$\frac{1}{Z_{in}(s)} = \frac{1}{Z_N(s)} \cdot \frac{T_p(s)}{1+T_p(s)} + \frac{1}{Z_D(s)} \cdot \frac{1}{1+T_p(s)} \quad (1)$$

From Eq. (1), the low-frequency value of input impedance |Zin(0)| is given by following equation.

$$\left| Z_{in}(0) \right|_{(dB\Omega)} \approx 20\log\left(\frac{R+r_L}{D^2} \right) \quad (dB\Omega) \quad (2)$$

|Zin(0)| has minimum value at rated load, so the estimation of |Zin(0)| must be at rated load. Next, the output impedance is examined. The output impedance characteristic of each control shames is different, and each bus converter has different operation. Therefore, the output impedance design suitable for the feature of each control method is required. From now, we consider the output impedance design for each control shame.

A. Unregulated

In un-regulated case, the output impedance is the same as open-loop output impedance because of this control method has no control loop. The output impedance of un-regulated bus converter can be derived as shown in following equation.

$$Z_{o\_un-regu}(s) = \frac{s^2 L_b C_b r_{c_b} + s\left(L_b + C_b r_{L_b} r_{c_b}\right) + r_{L_b}}{s^2 L_b C_b + s C_b \left(r_{L_b} + r_{c_b}\right) + 1} \quad (3)$$

From Eq. (3), the peak value of output impedance can be derived as shown in Eq. (4).

$$\left| Z_{o\_peak} \right| = \frac{L_b}{C_b \left(r_{L_b} + r_{c_b}\right)} = 20\log_{10}\frac{L_b}{C_b \left(r_{L_b} + r_{c_b}\right)}(dB\Omega) \quad (4)$$

In order to reduce the peak value of output impedance, it is effective to make inductance small or to enlarge capacitance. Generally, un-regulated bus converter is operated at maximum duty ratio. Therefore, the inductor of the bus converter can be reduced as small as possible to reduce the system instability. The peak value of output impedance is reducing with small. Figure 3 shows the experimental result of the relation between the output impedance and inductance.

Moreover, Fig. 4 shows the analytical and experimental results of the relation between the peak value of output impedance and inductance. Both results agreed well. As mentioned above, the peak of impedance is easily obtained from Eq. (4). However, this method depends on converter topology that has a double-ended circuit at secondary side such as half-bridge or full-bridge. Moreover, there are some limits such that high accurate input voltage or a POL with wide input range.

B. Semi-regulated

Semi-regulated bus converter has a control loop. However, regulation is related to variation of input voltage, therefore the output impedance is same as un-regulated case. In this case, the duty ration is changed. Therefore, the inductor of the bus converter cannot be reduced. Therefore, very large bus capacitor is needed to reduce the peak value of output impedance. Figure 5 shows the experimental result of the relation between the output impedance and capacitance. Moreover, Fig. 6 shows the analytical and experimental results of the relation between the peak value of output impedance and capacitance. Both results agreed well. In semi-regulated case, essentially it becomes very unstable and we have found that the demerit is very large capacitors are needed at the intermediate bus in order to be stable. However, it can be used at limited conditions such as wide input range (36-75V) and POL with low power (in other words, POL with very high input impedance).

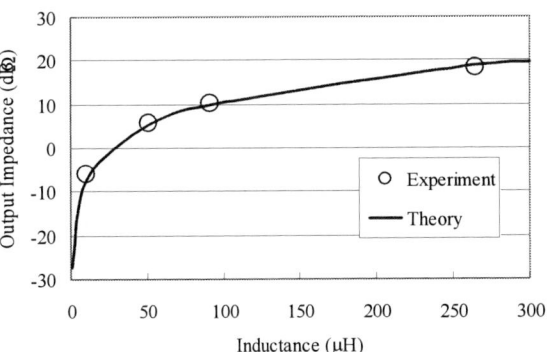

Fig. 4. Inductance and peak value of Zo.

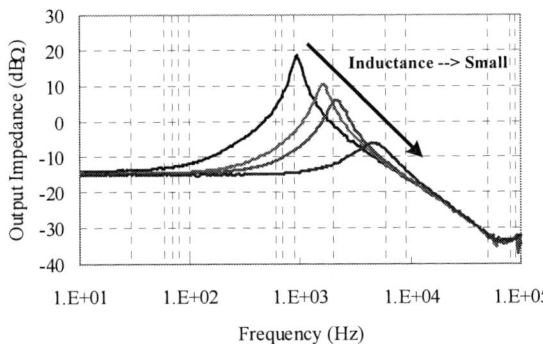

Fig. 3. Inductance and output impedance.

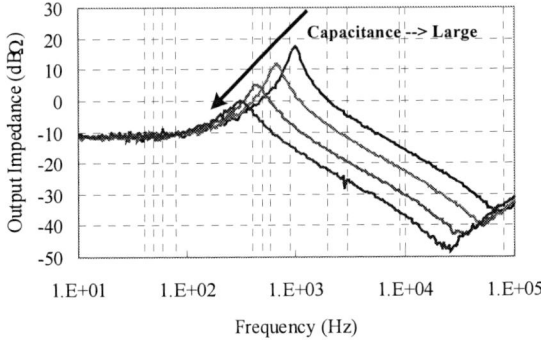

Fig. 5. Capacitance and output impedance.

C. Full-regulated

Full-regulated bus converter has a feedback loop, so the output impedance characteristic is changed as shown in Eq. (5). The output impedance peak is decreasing linearly with extending bandwidth as shown in Fig. 7. The approximation equation can be derived from these results as shown in Eq. (6).

$$Z_o(s) = \frac{Z_{op}(s)}{1+T_b(s)} \quad (5)$$

$$Z_p = Z_{p1} + 20\log\frac{f_{p1}}{f_n} \quad (dB\Omega) \quad (6)$$

where $Zp1$ is the peak value of output impedance at bandwidth $fp1$, and fpl can be discretionary. The peak value Zp can be found easily from eq. (6) by deciding bandwidth fp.

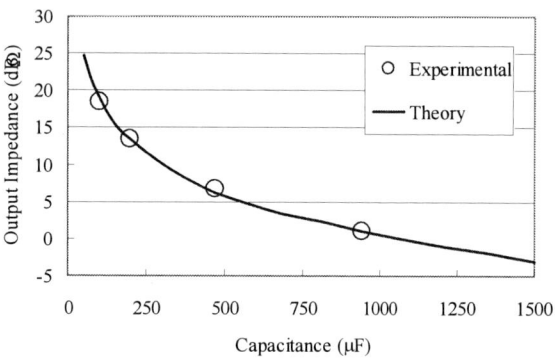

Fig. 6. Capacitance and peak value of Zo.

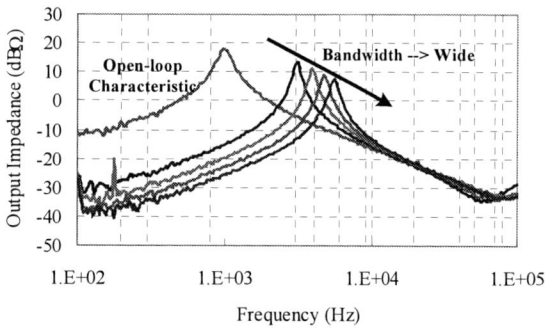

Fig. 7. Bandwidth and output impedance.

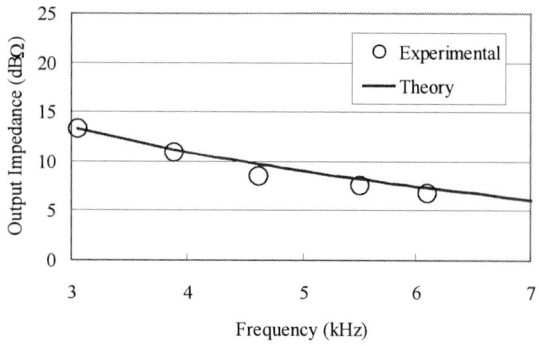

Fig. 8. Bandwidth and peak value of Zo

Figure 7 shows the experimental and analytical results of the relation between bandwidth and output impedance peak. The experimental results and analytical results are agreed well. Thereby, the design of bus converter for system stability becomes easy.

III. DESIGN CONSIDERATION

In order to evaluate the performance of this system, the experiment circuits are implemented using the specifications and parameters in Table 1. Here, the case with two POLs is discussed for actual example. The practical design process is shown below.

A. Input impedance estimation

The low-frequency value $|Zin(0)|$ of input impedance is given by equation (2). The duty ratio is D=0.275 and output resistance is R=0.66(Ω) from the relation input and output. In this case, the $|Zin(0)|$ is 20.6(dBΩ). When two POLs of same condition are connecting in parallel, $|Zin(0)|$ is 14.6(dBΩ). Figure 10 shows the experimental result of input impedance. The low-frequency value $|Zin(0)|$ is around 15(dBΩ) as shown in Fig. 10. The experimental results and analytical results are agreed well. If the stability margin is set to 6(dBΩ), then the peak value of output impedance must be set to 8.6(dBΩ).

Table 1 Circuit parameters.

	Symbol	Description	Value
Bus Converter	Vin	Input Volotage	48V
	Vb	Bus Volotage	12V
	Lb	Output Inductor of Bus Converter	270μH
	Cb	Output Capacitor of Bus Converter	100μF
	rlb	Regestance of Lb	300mΩ
	rcb	ESR of Cb	25mΩ
POL	Vo/Io	Output Condition	3.3V/5A
	Lo	Output inductor	2.8μH
	Co	Output capacitor	820μF
	rl	Regestance of Lo	25mΩ
	rc	ESR of Co	10mΩ

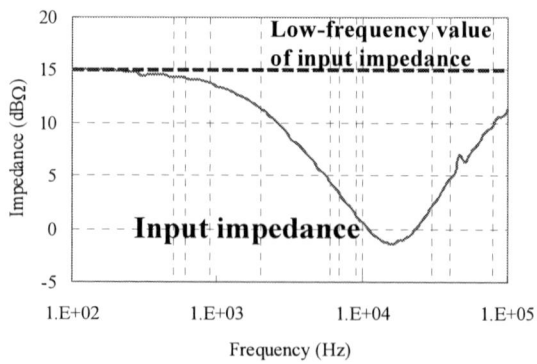

Fig. 9. Input impedance characteristic.

B. Output impedance design

Figure 10 shows the output impedance characteristic of the basic case using the parameter of Table 1. As shown in Fig. 10, the peak value of the output impedance is around 18dBW. From mentioned above calculation, the peak value of the output impedance needs to set around 8.6(dBΩ) for sufficient system stability.

In un-regulated case, the optimal inductance value is considered because the stability is improved by small inductance. From Eq. (4), the optimal inductance value can be derived as following equation.

$$L_{b\_optimal} = C_b \left(r_{L_b} + r_{c_b} \right) 10^{\frac{|Z_{o\_peak}|}{20}} \quad (7)$$

where, the unit of $|Z_{o\_peak}|$ is dBΩ.

Since the output impedance must be set to 8.6(dBΩ), the inductance value is set to around 87(μH) from Eq. (7). Figure 11 shows the experimental result of output impedance with small inductance. The inductance value is around 90 (μH), and the peak value of output impedance is around 8.5 (dBΩ). The experimental results and analytical results are agreed well.

Moreover, the output impedance does not become smaller than rL as shown in Fig. 3. Therefore, the inductance value has minimum value. From Eq. (7), the minimum value of the inductance is given by following equation.

$$L_{b\_min} = C_b \left(r_{L_b} + r_{c_b} \right) r_{L_b} \quad (8)$$

In this case, the minimum value of inductance is around 10 (μH).

In semi-regulated case, the optimal capacitance value is considered because the stability is improved by large capacitance. From Eq. (4), the optimal capacitance value can be derived as following equation.

$$C_{b\_optimal} = \frac{L_b}{\left(r_{L_b} + r_{c_b} \right) 10^{\frac{|Z_{o\_peak}|}{20}}} \quad (9)$$

where, the unit of $|Z_{o\_peak}|$ is dBΩ.

Since the output impedance must be set to 8.6(dBΩ), the capacitance value is set to around 300(μF) from Eq. (9). In this case, the influence of ESR is considered. Because the ESR becomes small when the capacitor is connected in parallel. Figure 12 shows the experimental result of output impedance with large capacitance. The capacitance value is 300 (μF), and the peak value of output impedance is around 8 (dBΩ). The experimental results and analytical results are agreed well.

Moreover, the output impedance does not become smaller than rL as shown in Fig. 4. Therefore, the capacitance value has maximum value. From Eq. (9), the maximum value of the capacitance is given by following equation.

$$C_{b\_max} = \frac{L_b}{\left(r_{L_b} + r_{c_b} \right) r_{L_b}} \quad (10)$$

In this case, the maximum value of capacitance is around 2.8 (mH).

In full-regulated case, the optimal bandwidth is considered because the stability is improved by wide bandwidth.

Figure 13 shows the experimental result of output impedance characteristic. In this case, the bandwidth is 1.9kHz, and the peak value of output impedance is 15.6(dBΩ). Since the output impedance must be set to 8.6(dBΩ), the bandwidth fn is set to 4.2kHz from Eq. (6). Figure 13 shows the experimental result of output impedance with extending bandwidth. The bandwidth is around 4.5kHz, and the peak value of output impedance is around 8 (dBΩ). The experimental results and analytical results are agreed well.

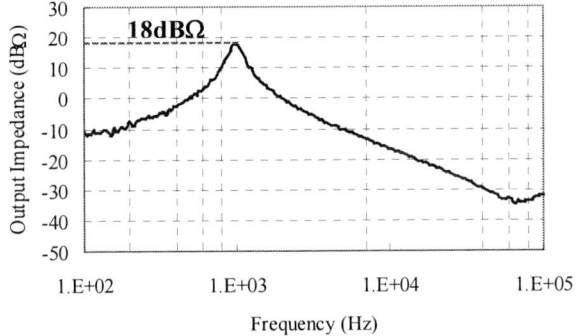

Fig. 10. Output impedance (Basic parameters).

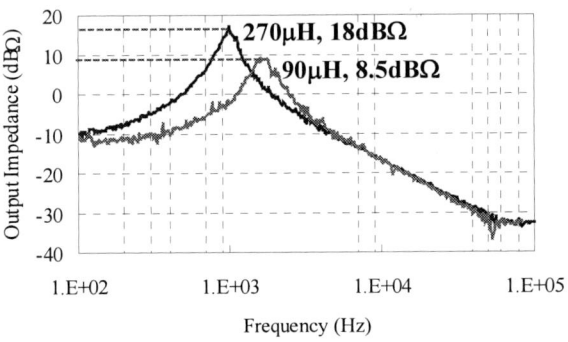

Fig. 11. Output impedance with small inductor.

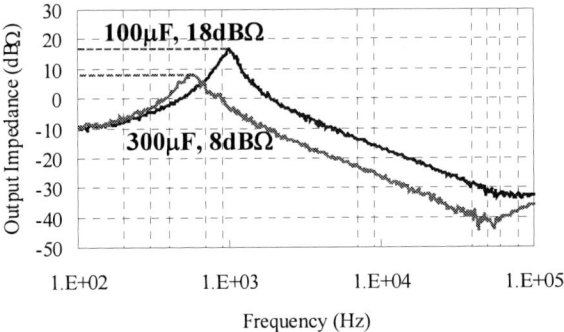

Fig. 12. Output impedance with large capacitor.

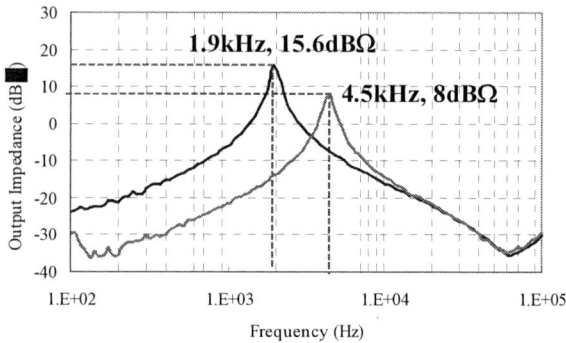

Fig. 13. Output impedance with wide bandwidth.

IV. CONCLUSIONS

This paper presented the output impedance design for on-board distributed power system by means of three control methods of bus converter. The output impedance peak of the bus converter and the input impedance of the POL were analyzed, and it was conformed by experimentally for stability criterion. As a result, the standard of the discrimination of stability on a frequency response of input and output impedance was clarified. Furthermore, the design process of each control shemes for system stability was proposed.

REFERENCES

[1] C. M. Wildrick, F. C. Lee, B. H. Cho, B. Choi, "A Method of Defining the Load Impedance Specification for A Stable Distributed Power System", IEEE Transactions on Power Electronics Vol. 10. No. 3. May 1995, pp. 280-284.

[2] X. Feng, Z. Ye, K. Xing, F. C. Lee, D. Borojevic, "Individual Load Impedance Specification for a Stable DC Distributed Power System", IEEE Applied Power Electronics Conference (APEC) 1999, pp. 923-929.

[3] X. Feng, F. C. Lee, "On-line Measurement on Stability Margin of DC Distributed Power System", IEEE Applied Power Electronics Conference (APEC) 2000, pp. 1190-1196.

[4] M. P. Sayani, J. Wanes, " Analyzing and Determining Optimum On-Board Power Architectures for 48V-input Systems", IEEE Applied Power Electronics Conference (APEC) 2003.

[5] K. Hisanaga, K. Harada, "Stability Analysis of the Distributed Power System with Intermediate Bus Converter", IEICE Technical Report, Vol.103, No.199, pp.19-24, Jul. 2003 (in Japanese).

[6] K. Hisanaga, K. Harada, "Stability Analysis of the Distributed Power System with Intermediate Bus Converter (2nd Report)", IEICE Technical Report, Vol.103, No.652, pp.7-12, Feb. 2004 (in Japanese).

[7] Y. Ren, M. Xu, K. Yao, Y. Meng, F. C. Lee, J. Guo, "Two-Stage Approach for 12V VR", IEEE Applied Power Electronics Conference (APEC) 2004.

[8] J. Wei, F. C. Lee, "An Output Impedance-Based Design of Voltage Regulator Output Capacitors for High Slew-Rate Load Current Transients", IEEE Applied Power Electronics Conference (APEC) 2004.

[9] S. Abe, M. Hirokawa, T. Zaitsu, T. Ninomiya, "Stability Design of Bus Converter Following by POLs in Distributed Power System", IASTED Circuits, Signals,and Systems (CSS), 2005, pp552-557.

[10] S. Abe, H. Nakagawa, M. Hirokawa, T. Zaitsu , T. Ninomiya, "Comparison of System Stability in Distributed Power System Based on Control Method of Bus Converter", IASTED Energy and Power Systems (EPS) 2005, pp109-114.

[11] S. Abe, H. Nakagawa, M. Hirokawa, T. Zaitsu , T. Ninomiya, "System Stability of Full-Regulated Bus Converter in Distributed Power System", International Telecommunications Energy Conference (INTELEC) 2005, pp563-568.

[12] S. Abe, H. Nakagawa, M. Hirokawa, T. Zaitsu , T. Ninomiya, " Stability Improvement of Distributed Power System by Using Full-Regulated Bus Converter ", Annual Conference of the IEEE Industrial Electronics Society (IECON) 2005, pp2549-2553.

[13] S. Abe, T. Ninomiya, M. Hirokawa, T. Zaitsu , "Stability Comparison of Three Control Schemes for Bus Converter in Distributed Power System ", International Conference on Power Electronics and Drive Systems (PEDS) 2005, pp1244-1249.

[14] S. Abe, M. Hirokawa, T. Zaitsu , T. Ninomiya, " Stability Design Consideration for On-Board Distributed Power System Consisting of Full-Regulated Bus Converter and POLs ", IEEE Power Electronics Specialists Conference (PESC) 2006, pp2669-2673.

[15] R. D. Middlebrook, "Input Filter Considerations in Design and Application of Switching Regulators", IAS'76, 1976, pp. 91-107.

[16] R.D. Middlebrook, S. Cuk, "A General Unified Approach to Modeling Switching-Converter Power Stages," IEEE Power Electronics Specialists Conference (PESC) 1976, pp. 18-34.

[17] T. Ninomiya, M. Nakahara, T. Higashi, K. Harada, "A Unified Analysis of Resonant Converters," IEEE Transactions on Power Electronics Vol. 6. No. 2. April 1991, pp. 260-270.

A New Topology for Photovoltaic 4 Series DC/DC Converter with High Efficiency Under Wide Load Range

Jong Pil Lee, Byung Duk Min, Jeong Joong Kim*, Kang Yul Ryul*, Tae Jin Kim,
Dong Wook Yoo, Eui Ho Song* and Ji Yoon Yoo**
Power Conversion & System for RES Group, Korea Electrotechnology Research Institute, Changwon, 641-600, Korea
*School of Mechatronics Control & Instrumentation Engineering, Changwon University
**School of Electrical Engineering, Korea University
e-mail : jplee@keri.re.kr, bdmin@keri.re.kr

Abstract-The paper proposes a new topology for photovoltaic dc/dc converter with high efficiency under wide input voltage range is proposed. Photovoltaic dc/dc is a very crucial part of power conditioning system(PCS). Considering that output characteristic of photovoltaic cell has wide voltage range, depending on the operating conditions of photovoltaic cell, the dc/dc converter also needs to have wide input voltage range to regulate the constant output voltage. In addition, a high rated voltage of power switch(MOSFET) for dc/dc converter is necessary in order to be compatible with the high maximum output voltage of photovoltaic cell's maximum output voltage. Also photovoltaic power conditioning system should have a high efficiency under operation condition. To satisfy PV PCS condition series connected non-isolated dc/dc converter topology is proposed. This paper examines the proposed topology employing four modules and shows that it dramatically enhances the energy efficiency from low load to high load condition by reducing the required power level for dc/dc converter by one-third of conventional dc/dc converter. A case of 25kW prototype as the PV dc/dc converter is introduced to experimentally verify the proposed topology.

I. Introduction

Since the start of the industrial age more than 150 years ago, the world economy was running on fossil fuels, which were cheap as there was no cost associated with their production, but only with their extraction and transportation. The negative effects on the environment became visible only in the last 30 years. Renewable energy resources will be an increasingly important part of power generation in the new millennium. Besides assisting in the reduction of the emission of greenhouse gases, they add the much needed flexibility to the energy resource mix by decreasing the dependence on fossil fuels. Due to their modular characteristics, ease of installation and because they can be located closer to the user, Photovoltaic (PV) Systems have great potential as distributed power source to the utilities. PV systems are installed on the roof of residential buildings and connected directly to the grid.(called Grid-Tie or Grid-Connected) A grid-tie PV system consists of three main stages : a PV module, a DC/DC converter and a DC/AC inverter as shown in Fig. 1. In these PV systems, power Conditioning System (PCS) should have high efficiency

and low cost.

This paper propose a novel topology for PV dc/dc converter with very high efficiency under wide input voltage range, and presents an approach on how to design a proposed PV dc/dc converter. also experimental results shows that the proposed topology has a good performance.

II. General Topology for non-isolated PV DC/DC Converter

In general, The standard configurations that are used to implement Photovoltaic Power Conditioning System (PV PCS) are: isolated type and non-isolated type.

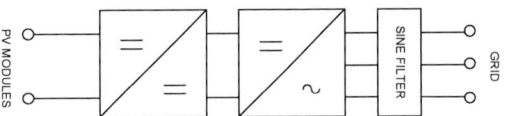

Fig. 1 PV PCS with non-isolated dc/dc converter

Especially non-isolated PCS type has a high efficiency better than isolated PCS type. because there is no need to have galvanic transformer. Fig. 1 shows a PV PCS with non-isolated dc/dc converter. This system consisted of non-isolated dc/dc converter to regulate the inverter dc link voltage and dc/ac inverter to grid tied

The famous non-isolated boost dc/dc converter for PV PCS is shown in Fig. 2. This topology is inexpensive and very simple when used in PV applications. But When the PV maximum cell voltage is over 900V, then the rated voltage of power switch need at least over 1200V. also It is very difficult to choose a suitable power switch component and design a boost inductor under high power conditions.

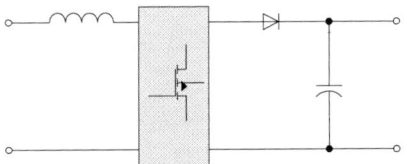

Fig. 2 non-isolated boost converter

III. New Topology for non-isolated PV DC/DC Converter

To overcome of disadvantage of former topologies for non-isolated PV dc/dc converter, new topology is proposed shown in Fig. 3 and Fig. 4.

978-1-4244-1871-8/07 $25.00 © 2007 IEEE

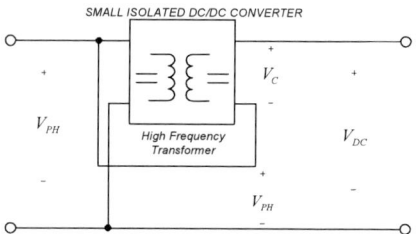

Fig. 3 a new non-isolated dc/dc converter topology

The proposed topology has isolated full bridge dc/dc converter. It is consisted of series four module isolated full bridge dc/dc converter. Especially input photovoltaic cell voltage is connected the anode of dc/dc converter rectifier diode. Then the DC link(inverter input)voltage is expressed as follows :

$$V_{DC} = V_C + V_{PH}$$

Where V_{DC} is the output voltage, V_C is the output of dc/dc converter, V_{PH} is photovoltaic cell voltage.

The advantages of the proposed topology are as follows:

- High efficiency under wide load range.(flat efficiency curve)
- Low voltage rate power switch.(series-connected)
- Need only a one-third power level dc/dc converter of conventional dc/dc converter.

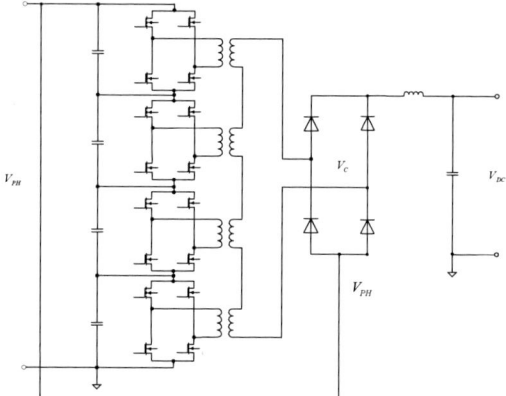

Fig. 4 The proposed four series module full bridge dc/dc converter

These converters do have the advantage of being able to increase efficiency all over the operation range. Commonly, the efficiency of full bridge ZVS dc/dc converter is not especially good, ranging from 92~93% from 45%~100% load condition. For example, when we design a 100kW PV PCS, then need same power level dc/dc converter. But If the proposed topology apply to a 100kW PV PCS, then we don't need a 100kW dc/dc converter, just need only a 20kW isolated dc/dc converter.

Efficiency of the proposed topology is expressed as follows:

$$\xi_C = \left(P_B + P_C \times \xi_{dc/dc}\right) \times 100$$

Where, ξ_C is efficiency of proposed topology converter, P_B is percentage of bypass power, P_C is percentage of converter power, $\xi_{dc/dc}$ is efficiency of isolated full bridge dc/dc converter.

Table 1. The Proposed converter efficiency

PV cell voltage	Power sharing of converter	Proposed converter Efficiency
650V	0V(0%)	100%
600V	50V(7.7%)	$(0.923+(1-0.923)\times92)\times100=99.3\%$
550V	100V(15.3%)	$(0.847+(1-0.847)\times92)\times100=98.7\%$
500V	150V(23%)	$(0.77+(1-0.77)\times92)\times100=98.16\%$
450V	200V(30.7%)	$(0.693+(1-0.693)\times92)\times100=97.5\%$
400V	250V(38.4%	$(0.616+(1-0.616)\times92)\times100=96.9\%$

Table 1. shows the proposed converter efficiency with respect to power sharing condition. For example when PV cell voltage is 650V, this is bypass mode. So converter is not working condition. When PV cell voltage is 400V, then output voltage of dc/dc converter is 250V and the percentage of the power for dc/dc converter and PV cell is 38.4 %(P_C) and 61.6%(P_B). It means that we need only 38.4% power for dc/dc converter of conventional converter system. Let see an efficiency of the isolated full bridge dc/dc converter is 92%, then we could have 96.9% efficiency of the proposed converter.

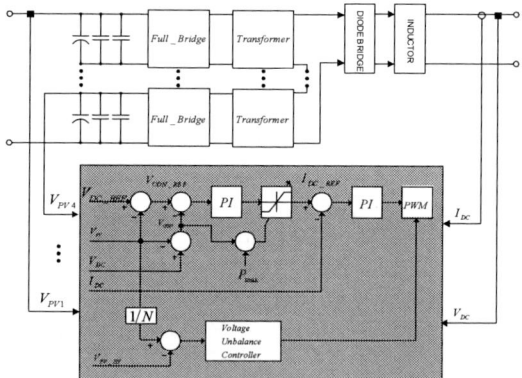

Fig. 5 Block diagram of the proposed dc/dc converter controller

Fig. 5 shows the proposed dc/dc converter controller which is consisted of voltage, current loop and power limit function. The proposed converter is series module type. It is need to control the balance of input voltage.

IV. Experimental Results

Fig. 6 shows the proposed four series connected PV dc/dc converter. The photovoltaic cell voltage range is 400~850V. DC link voltage is 650V and dc/dc output power is 25kW.

Table 2. shows the specification of the proposed converter. Input voltage is PV cell voltage.

978-1-4244-1871-8/07 $25.00 © 2007 IEEE

(a)

(b)

Fig. 6 The 25kW prototype PV dc/dc converter

Output voltage means output voltage of isolated dc/dc converter. Rated power of one module is 6.25 kW. So total rated power is 25 kW. Basic topology is Zero Voltage Switching (ZVS) method. And switching device is CoolMOS MOSFET

Table 2. The specification and main component of the proposed converter.

Classification	Descriptions
Input voltage	400~650V
Converter output voltage	0~200V
DC link voltage	650V
Topology	ZVS full bridge
Switching frequency	30 kHz
Rated power	25 kW
MOSFET	IXFN130N30

Fig. 7 shows the experimental results of balancing control. Each input module voltage keeps under 1% error tolerance.(V_{cell} :200V, V_{m1} :49.5V, V_{m2} :49.8, V_{m3} :50.2V, V_{m4} :50.5V)

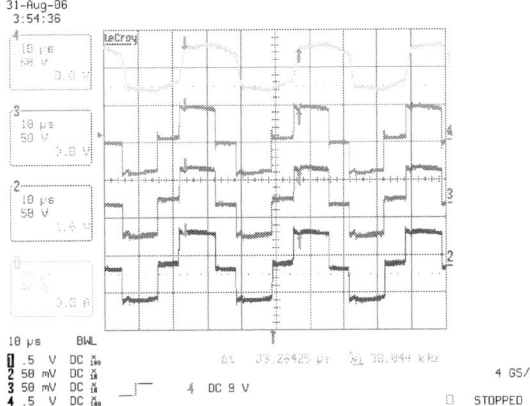

Fig. 7 Primary transformer voltage(CH2,3,4:50V/div.) and current(CH1:50A/div)

Fig. 8 shows the waveforms when PV cell voltage is 450V then DC link voltage is 650V. Output current is 120A. Output power is 21.6kW

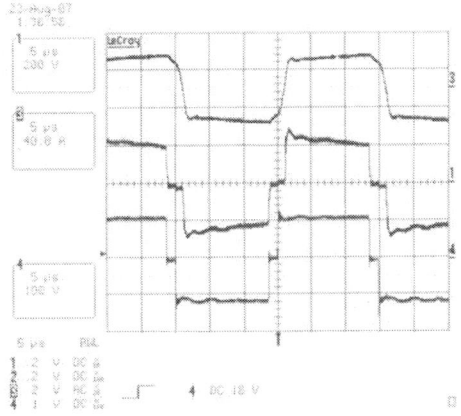

Fig. 8 Primary transformer current(CH3:40A/div.), Main switch drain-source voltage(CH1:200V/div.), Secondary transformer voltage(CH4:100V/div.)

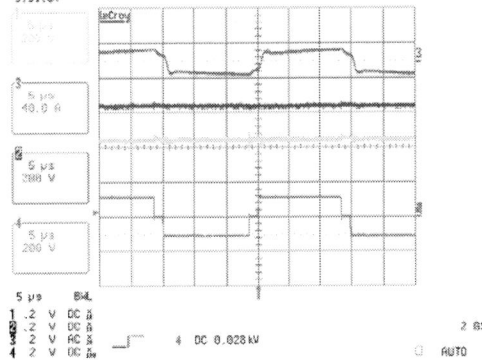

Fig. 9 Primary transformer current(CH3:40A/div.), DC Link Voltage(CH2:200V/div.), Input Cell Voltage(CH1:200V/div.), Primary transformer voltage(CH4:200V/div.)

Fig. 9 shows the waveforms when PV cell voltage is 450V then DC link voltage is 650V. It means the output voltage of dc/dc converter is 200V.

Table 3. Comparison of the efficiency(Vin:450V, Vo:650V)

% Power	Efficiency	
	DC/DC	Total
10	92.4	97.8
20	94.7	98.5
30	95.4	98.7
40	95.2	98.6
50	94.9	98.6
60	94.5	98.4
70	93.2	98.1
80	92.9	98.0
90	92.7	97.9
100	92.0	97.7

Table 4. Comparison of the efficiency(Vin:500V, Vo:650V)

% Power	Efficiency	
	DC/DC	Total
10	85.0	96.5
20	93.5	98.5
30	94.6	98.7
40	93.6	98.5
50	93.6	98.5
60	93.6	98.5
70	93.5	98.5
80	92.8	98.3
90	92.2	98.2
100	91.67	98.1

Table 5. Comparison of the efficiency(Vin:550V, Vo:650V)

% Power	Efficiency	
	DC/DC	Total
10	78.0	96.6
20	86.6	97.9
30	89.7	98.4
40	92.6	98.9
50	92.6	98.9
60	91.7	98.7
70	91.6	98.7
80	91.5	98.7
90	90.1	98.5
100	89.1	98.3

Table 6. Comparison of the efficiency(Vin:600V, Vo:650V)

% Power	Efficiency	
	DC/DC	Total
10	65.3	97.3
20	76.9	98.2
30	83.3	98.7
40	87.7	99.0
50	86.3	98.9
60	85.7	98.9
70	84.8	98.8
80	84.7	98.8
90	83.1	98.7
100	81.8	98.6

From Table 3. to 5. shows the experimental results of the efficiency between conventional DC/DC converter and the proposed converter. Fig. 10 shows the 3D curve of efficiency between conventional converter and the proposed converter with respect to load condition and cell voltage. Experimental results shows energy efficiency of ZVS dc/dc converter is around 92% at 80% load condition. In the efficiency of proposed converter is from 96% to 98% under wide PV cell voltage range. the maximum efficiency of conventional full bridge converter is around 92%. But the efficiency of the proposed dc/dc converter is almost 98% under wide load range. Nevertheless 25kW dc/dc converter, it could apply to 75kW photovoltaic power conditioning system.

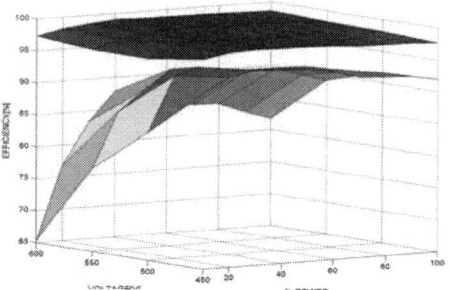

Fig. 10 Comparison of the efficiency between conventional converter and the proposed converter

Ⅴ. Conclusions

In this paper, new topology is proposed for photovoltaic dc/dc converter with very high efficiency. The proposed converter is not only cost effective because we need only one-third of conventional dc/dc converter power rated but is also highly energy-efficient, as it can run on photovoltaic converter. The experimental results show the good performances of proposed PV dc/dc converter. Also It can improve efficiency of converter for others renewable energy application. (Fuel cell, Wind generation etc.)

Ⅵ. References

[1] Qingrong Zeng, Pinggang Song, Liuchen Chang, IEEE CCECE 2002. Volume 1 (2002), 257

[2] Lopes, L.A.C. Lienhardt, A.-M., Power Electronics Specialist Conference, 2003. Volume 4 (2003), 1729

[3] Monji G. Jaboori, Mohamed M. Saied,, Adel A. R.Hanafy, IEEE Transactions on Energy Conversion, Vol. 6, No. 3 (1991), 401

[4] B.D.Min, J.P.Lee, J.H.Kim, T.J.Kim, D.W.Yoo, "A New Topology and Switching Algorithm for Photovoltaic Simulator", 21st European Photovoltaic Solar Energy Conference and Exhibition, 2006, page 2361

[5] Walker,G.R, Sernia,P.C, IEEE Transactions on Power Electronics, Vol.19, page 1130~1139

[6] "Trends In Photovoltaic Applications – survey report of selected IEA countries between 1992 and 2005" Available : http://www.iea-pvps.org/products/index.htm

[7] "Performance analysis of grid-connected PV systems, September 2006" Available: http://www.iea-pvps.org/products/index.htm

[8] Walker,G.R, Pierce,J.C, Power Electronics Specialist

978-1-4244-1871-8/07 $25.00 © 2007 IEEE

Static and Dynamic Characteristics of DC-DC Converter Using a Digital Filter

Fujio Kurokawa

Department of Electrical and Electronic Engineering,
Nagasaki University, Japan
1-14 Bunkyo-machi, Nagasaki, 852-8521 Japan
Tel: +81 95 819 2553
Fax: +81 95 819 2558
E-mail:fkurokaw@nagasaki-u.ac.jp

Masashi Okamatsu

Graduate School of Science and Technology,
Nagasaki University, Japan
1-14 Bunkyo-machi, Nagasaki, 852-8521 Japan
Tel: +81 95 819 2553
Fax: +81 95 819 2558
E-mail: d707062b@cc.nagasaki-u.ac.jp

Abstract-- This paper presents the regulation and dynamic characteristics of the dc-dc converter with the digital PID control, the minimum phase FIR filter or the IIR filter. Further, we discuss the comparison of these performance characteristics. As a result, it is clarified that the variable rate of the regulation is within 1.3%, the undershoot is less than 200mV and the transient time is less than 0.4ms in the IIR filter method.

I. INTRODUCTION

The high controllability and monitoring function are required in the switching power supply to improve the reliability of the telecommunications system, data communications system and so forth. In this case, the digital control is useful because it has the advantage of realizing both control and monitoring tasks.

The digital control circuit is composed of the A-D converter and the control part which is usually constructed by a ASIC or DSP. In this control circuit, the digital P-I-D control and the digital low-pass filter control algorithm have been programmed. So, these control methods have been reported [1]-[6]. However, the output characteristics of these dc-dc converters with the digital control have never been discussed in detail. Especially, the relationship between the output characteristics of digital filter method and the control parameters is unknown.

This paper presents the static and dynamic characteristics of the FIR and IIR filter methods compared with the digital P-I-D controlled dc-dc converter. After reviewing the fundamental configuration of digitally controlled dc-dc converter and its operation principle, we analyze the static and dynamic characteristics of three types of digitally controlled dc-dc converter. The key point of this paper is to describe the possibility to realize the digitally controlled switching power converter using DSP.

II. CIRCUIT CONFIGURATION AND OPERATION PRINCIPLE

Figure 1 shows the block diagram of the proposed digitally controlled buck type dc-dc converter using DSP. E_i is the input voltage, e_o is the output voltage and i_o is the output current. T_r is the main switch, D is the fly wheel diode, L is the energy storage reactor, C is the output smoothing capacitor and R is the load. The output voltage e_o is sent to the A-D converter through the anti-aliasing filter and is converted into digital amount N_n. The relation between the input and output values of the A-D converter is given by equation (1) when it approximately shows the linear expression by considering the width of the quantization to be small.

$$N_n = G_{AD}e_o \qquad (1)$$

where n denotes an n-th switching cycle, and the digital amount N_n is a positive integer number. G_{AD} is a gain of the A-D converter.

The digital amount N_n is sent to DSP. In DSP, the numerical value N_{Ton} that corresponds to the on-time interval T_{on} is calculated.

The relation between the on-time interval T_{on} and the numerical value N_{Ton} is shown as follows;

$$\frac{T_{on,n+1}}{T_S} = \frac{N_{Ton,n+1}}{N_{Ts}} \qquad (2)$$

where N_{Ts} is a numerical value corresponding to the switching period T_s $(=1/f_s)$. N_{Ts} is calculated in the PWM signal generation circuit which is composed of a digital comparator or a counter. According to the relation between the on-time interval T_{on} and the numerical value N_{Ton}, T_{on} is generated. This T_{on} regulates the output voltage e_o.

The numerical value N_{Ton} of each control method is represented as follows.

A. P-I-D Control Method

The on-time interval N_{Ton} of the P-I-D control circuit is represented as follows [2];

$$N_{Ton,n+1} = N_B - K_P(N_n - N_R)$$
$$- K_D(N_n - N_{n-1}) - K_I \sum (N_n - N_{INT}) \qquad (3)$$

978-1-4244-1871-8/07 $25.00 © 2007 IEEE

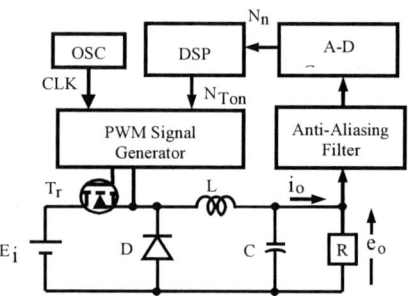

Fig.1. Block diagram of digitally controlled buck type dc-dc converter using DSP

where K_P, K_1 and K_I are the proportional, differential and integral coefficients. N_B is the numerical bias value. N_R and N_{INT} are the numerical proportional and integral desired values, these values are shown as follows;

$$N_B = N_{Ts}\left(1 + r/R\right)\frac{e_o^*}{E_i} \qquad (4)$$

$$N_R = N_{INT} = G_{AD}e_o^* \qquad (5)$$

B. FIR Filter Control Method

The on-time interval N_{Ton} of the minimum phase FIR filter control circuit is represented as follows [6];

$$N_{Ton,n+1} = N_B - \sum_{i=0}^{q} h_i\left(N_{n-i} - N_R\right) \qquad (6)$$

where h_i denotes the digital filter coefficients and q is the amount of the sampling points.

C. IIR Filter Control Method

The IIR filter is designed from an analog filter using the double primary conversion. The converted digital filter becomes the second-order IIR filters as shown in the next equation [7].

$$y_n = a_0 x_n + a_1 x_{n-1} + a_2 x_{n-2} - b_1 y_{n-1} - b_2 y_{n-2} \qquad (7)$$

where a_0, a_1, a_2, b_1 and b_2 are coefficients of IIR filter, x_n is a difference between the digital value N_n and the desired value N_R. x_n is shown in the following equation.

$$x_n = N_n - N_R \qquad (8)$$

Then, the on-time interval N_{Ton} of IIR filter control circuit is represented by the following equation.

$$N_{Ton,n+1} = N_B - y_n \qquad (9)$$

In a usual low-pass filter, a phase delay exists in the high frequency [5], [6]. The phase delay sometimes causes unstable phenomena. So, to improve the delay of the phase, the differential element is added in this IIR filter. The stabilization of the dc-dc converter is attempted by this way.

III. STATIC AND DYNAMIC CHARACTERISTICS

A. P-I-D Control Method

Figures 2(a) and (b) show the simulated and experimental static characteristics in P-I-D control. The input voltage E_i is 20V, the desired voltage E_o^* is 10V, the inductance L is 247µH, the output capacitance C is 940µF, and the number Q of bit of the A-D converter is 8bits, the dynamic range is from 0 to 20V corresponding to the output voltage of buck type dc-dc converter, respectively. The sampling frequency f_{smp} in this case is 100kHz corresponding to the switching frequency f_s of the DC-DC converter. The proportional coefficient K_P, the differential coefficient K_D and the integral coefficient K_I are 1.0, 3.0 and 0.01, respectively. Further, the other circuit parameters are f_s=100kHz, N_{Ts}=250, N_{RP} =N_{RINT} =127 and N_B =124. DSP is TMS320VC33. The calculation time of DSP is less than 4.1µs. The symbols of open and closed circles denote the observed points of stable and oscillated operations. The dotted line shows the values in open loop operation. The oscillated operation occurs in transient states between these lines. The difference of broken lines is the quantization width ΔE_{oq} of output voltage.

Figure 3 shows the experimental dynamic characteristics in P-I-D control in step change of the load R from 50Ω to 10Ω. The proportional coefficient K_P, the differential coefficient K_D and the integral coefficient K_I are 1.0, 1.0 and 0.01, respectively. The undershoot is over 200mV and the transient time T_{st} is 4.9ms. In this method, even if K_I is small and K_D is large, approximately 160mV is minimum value in the width of the vibration of the output voltage deriving from limit cycle [7]. Therefore, this method is not suitable for the digitally controlled power dc-dc converter.

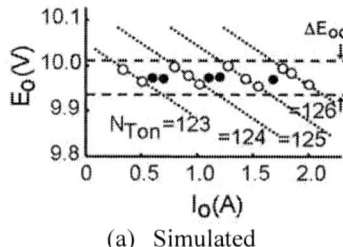

(a) Simulated

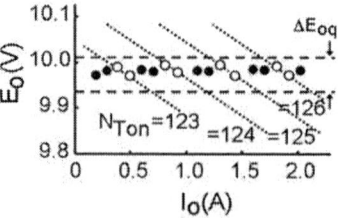

(b) Experimental

Fig. 2 Regulation characteristics in P-I-D control.

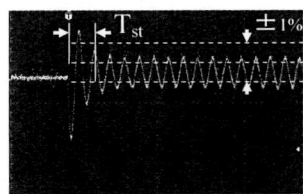

Vertical : 100mV/div.
Horizontal : 4ms/div.
Fig.3. Indicial response of digital P-I-D control.

B. FIR Filter Control Method

Figures 4(a) and (b) show the simulated and experimental regulation characteristics in the minimum phase FIR filter control. Figure 5 shows the experimental dynamic characteristics. The circuit parameters are as same as Fig. 2 except for Table 1. Figure 6 shows the frequency characteristics corresponding to Fig.5. In this method, the undershoot is over 200mV and transient time T_{st} is 2.7ms. Although the oscillation deriving from limit cycle is suppressed, the transient response is not enough.

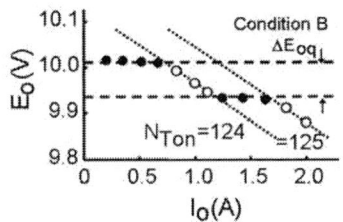

(a) Simulated

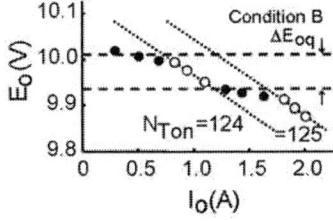

(c) Experimental
Fig.4 Regulation characteristics in FIR filter method.

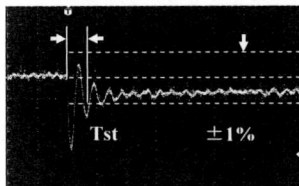

Vertical : 100mV/div.
Horizontal : 4ms/div.
Fig. 5 Indicial response of FIR filter method.

Table 1 Measured conditions corresponding to Fig. 5.

Passage Area Frequency (kHz)	5
Transition Area (kHz)	15
Passage Area Ripple (dB)	0.08
Quantity of Decrement (dB)	13

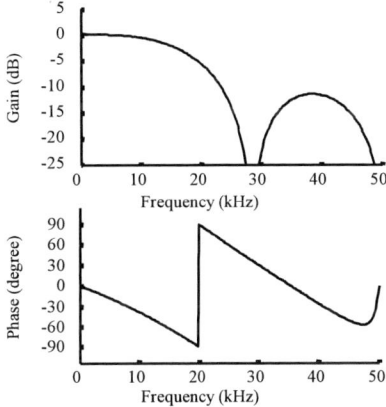

Fig. 6 Frequency characteristics corresponding to Fig.5.

C. IIR Filter Control Method

Figure 7 shows the definition of parameters in IIR filter. G_1 and G_2 are the first gains and the second gains. Further, f_a, f_b and f_c are first cut-off frequency, passing frequency and the second cut-off frequency, respectively. Table 2 shows parameters in each condition and Figs. 8 shows the frequency characteristics corresponding to each condition. The circuit parameters are as same as Fig. 2 except for Table 2.

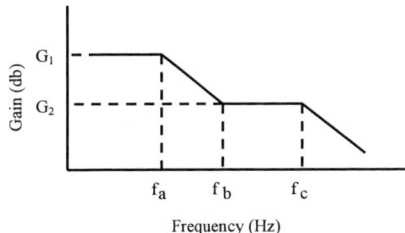

Fig.7 Definition of parameters in IIR filter.

Table 2 Parameters in each condition.

	G_1 (dB)	G_2 (dB)	f_a(Hz)	f_b(Hz)	f_c(Hz)
Condition A	0	0	—	—	1k
Condition B	10	0	1	2.5	1k
Condition C	20	0	1	9.6	1k
Condition D	40	0	1	97	1k
Condition E	40	15	1	10	40k

978-1-4244-1871-8/07 $25.00 © 2007 IEEE

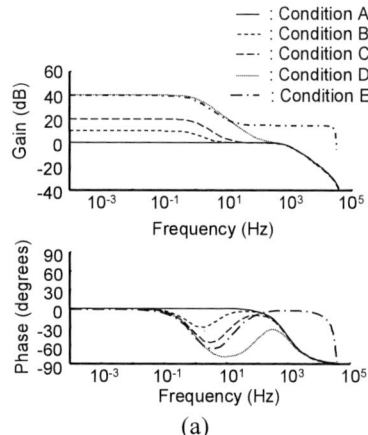

Fig.8. Frequency characteristics in IIR filter.

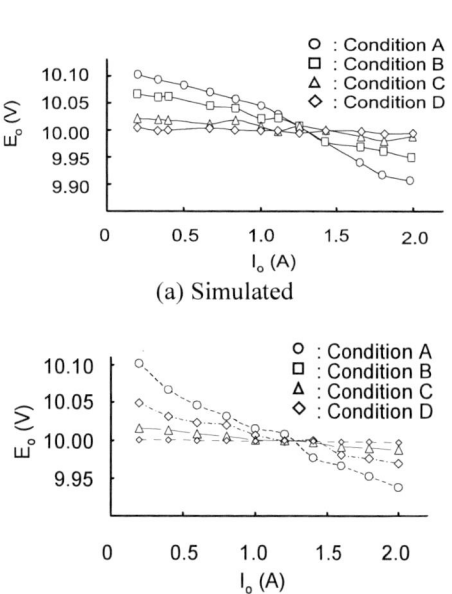

(a) Simulated

(b) Experimental

Fig. 9 Regulation characteristics in IIR filter control.

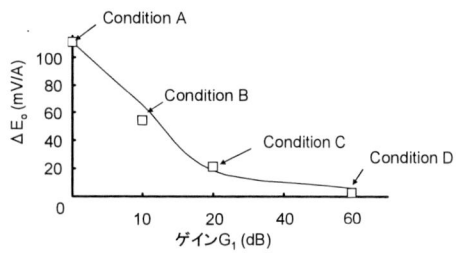

Fig. 10 Variable ratio of regulation.

Condition A is a usual first-order low-pass filter of the cut-off frequency 1kHz. In Condition B through Condition D, gain G_2 and the second cut-off frequency f_c

are assumed to be constant, and gain G_1 is enlarged. When gain G_1 is enlarged, the first cut-off frequency f_a or the passing frequency f_c moves. In this case, it designed so that f_a might reach a fixed value. It is seen in these figures that the steady-state error is suppressed because the first gain G_1 is enough high.

The simulated and experimental regulation characteristics of Condition A through Condition D are shown in Figs. 9(a) and (b). The variable rate of regulation is shown in Fig. 10. The variable rate has become small as gain G_1 is so enlarged. it is clarified that the variable rate of the regulation is within 1.3%.

Next, the experimental indicial response characteristics of Condition A and Condition C are shown in Figs. 11(a) and (b). Moreover, Fig. 11 (c) shows the indicial response in Condition E. In this condition, the gain G_1 and gain G_2 are enlarged, and then the second cut-off frequency f_c becomes high to maintain the stability. f_c is less than half of the sampling frequency 100kHz. The regulation characteristics in Condition E are same to Condition D because the first gain is equal.

From the relationship between the observed undershoot and the second gain G_2 and the relationship between the observed transient time T_{st} and the second gain G_2, it is revealed that Condition E has superior characteristics. In Fig. 11(c), the observed undershoot is less than 200mV, the observed transient time T_{st} is less than 0.4ms. Moreover, it is seen that the oscillation deriving from limit cycle is suppressed most enough in the condition E.

Therefore, it is revealed that both the undershoot and transient time are improved and the oscillation deriving from limit cycle is suppressed enough in the condition E.

This result suggests the design criterion of digital control circuit of switching power converter. It is concluded that the first gain G_1, the second gain G_2 and the second cut-off frequency f_c are relatively enlarged, and the first cut-off frequency f_a and the passing frequency f_b are relatively low in the IIR filter controlled switching dc-dc power converter. Moreover, it is important that the gain decreases extremely at the second cut-off frequency f_c as shown in Fig. 8 because f_c is always less than half of the sampling frequency based on the sampling theory.

As a result, it is revealed that the IIR filter control method has enough the regulation characteristics and that the dynamic characteristics of IIR filter control method are superior to those of the digital P-I-D control and FIR filter control ones.

978-1-4244-1871-8/07 $25.00 © 2007 IEEE   264

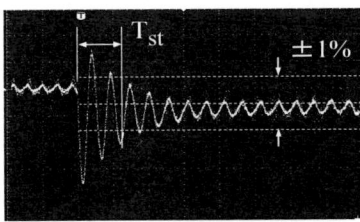

(a) Condition A

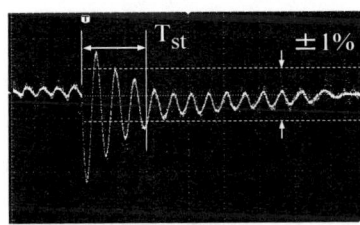

(b) Condition B

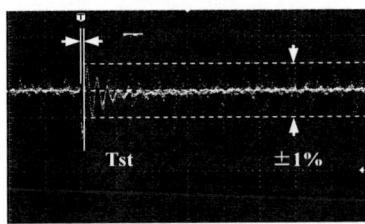

(c) Condition E

Vertical : 100mV/div., Horizontal : 4ms/div.
Fig.11 Indicial response of digital IIR filter.

IV. CONCLUSIONS

From the above discussion, it is revealed that the IIR filter control method has enough the regulation characteristics. Furthermore, it is clarify that the dynamic characteristics of IIR filter control method are superior to those of the digital P-I-D control and FIR filter control ones. Furthermore, in IIR filter, it is seen that the dynamic response is improved and the oscillation deriving from limit cycle is suppressed in the condition E. It is concluded that the first gain G_1, the second gain G_2 and the second cut-off frequency f_c are relatively enlarged, and the first cut-off frequency f_a and the passing frequency f_b are relatively low. Moreover, it is important that the gain decreases extremely at the second cut-off frequency f_c.

This work is supported in part by the Grant-in-Aid for Scientific Research (No.18310117) of JSPS (Japan Society for the Promotion of Science) and the Ministry of Education, Science, Sports and Culture.

REFERENCES

[1] L. Guo, J. Y. Hung and R. M. Nelms: "PID controller modifications to improve steady-state-- performance of digital controllers for buck and boost converters", Proceedings of Seventeenth Annual IEEE Applied Power Electronics Conference, no.9.3, pp. 381-388, March 2002.

[2] F. Kurokawa, T. Sato, H. Matsuo and H. Eto: "Output characteristics of dc-dc converter with DSP control", Proceeding of IEEE International Telecommunications and energy conference, pp. 421-426, Oct. 2002.

[3] F. Kurokawa, M. Sasaki, S. Hiura and H. Matsuo: "1MHz high-speed digitally controlled dc-dc converter", IEICE Trans. on Communications, vol. E87-B, no. 12, pp. 3437-3442, Dec. 2004.

[4] H. Matsuo, F. Kurokawa, H. Etoh, N. Suzuki and Y. Makino: "Novel digital controller with static-model reference for switching dc-dc power converters", Proceedings of IEEJ 2000 International Power Electronics Conference, vol.3, pp.1997-2001, March 2000.

[5] F. Kurokawa, H. Tanaka and H. Matsuo: "A comparison of output characteristics of dc-dc converters with digital control", Proceedings of IEE/J International Power Electronics Conference, pp.1137-1142, April 2005.

[6] F. Kurokawa, W. Okamoto and H. Matsuo: "A comparison of steady-state characteristics of buck type dc-dc converter using DSP", IEICE Trans. on Communications, vol. J89-B, no. 5, pp. 673-681, May 2006.

[7] B. Mulgrew, P. Grant and J. Thompson: "Digital signal processing", Macmillan Press Ltd., London, 1999.

978-1-4244-1871-8/07 $25.00 © 2007 IEEE

A Study on Novel Buck-Boost AC-DC Converter of High Performance by Partial Resonance Technique

[†]Dong-Kurl Kwak*, Bong-Seob Lee**, Choon-Sam Kim**, Jae-Sun Shim**

* The Graduate School of Disaster Prevention Technology, Kangwon National University, Gangwon-do, Korea
Email: dkkwak@kangwon.ac.kr
** Division of Electrical & Control Engineering, Kangwon National University, Gangwon-do, Korea

Abstract—The system efficiency of the proposed buck-boost AC-DC converter is increased by soft switching method. The converter includes to merit of power factor correction (PFC) from sinusoidal control of input current. The switching behavior of control switches operates with soft switching by partial resonance technique, and then the proposed converter has high system efficiency with decrement of switching power loss. The input current waveform in proposed converter is got to be a sinusoidal form of discontinuous quasi-pulse row in proportion to magnitude of AC input voltage under the constant duty cycle switching. Therefore, the input power factor is nearly unity. The output voltage of the converter is regulated by PWM control technique. The discontinuous mode action of current flowing into inductor makes to simplify control method and control components. The proposed novel buck-boost AC-DC converter is analyzed to compare with the conventional PFC buck-boost converter. Some computer simulative results and experimental results confirm to the validity of the analytical results.

I. INTRODUCTION

This Conventional off-line switching power supply designs usually include at their input stage a full-bridge rectifier and large input filter capacitor. The problem with this input circuit is that it produces excessive peak input currents and high level of harmonic distortion on the line. The power factor with this topology is about 0.5-0.7. By improving this factor, line noise and peak current levels can be dramatically reduced and the amount of power available to the user can be increased. The use of an on-line switching PFC topology like active power converter is the best way to get a high power factor on the line.

Recently, the PFC techniques using buck-boost ac-dc converter have received great attention and a number of new techniques have been proposed [1-4]. There are two input current control modes for the converter of this usage. One is continuous conduction mode (CCM) of dc input current and another is discontinuous conduction mode (DCM) of that. In continuous mode, ac input voltage and current are detected and the input current is formed to be sinusoidal waveform by using pulse width modulation (PWM). The control circuit for this mode is complicated one. In discontinuous mode, the ac input current is nearly sinusoidal waveform with the constant duty cycle switching. The control circuit for discontinuous mode is simple [4, 5]. In this mode, turn on of switches in converter occurs at zero current, that is zero current

switching (ZCS). Therefore switching loss caused by the switching operation is very low. However, the control devices are switched off at the maximum current and a certain level of voltage in one cycle of the carrier frequency [see Fig. 1(a)]. It causes the large current stress of the switching device and electromagnetic interference. As a result of these, the power converter brings on low efficiency. To improve these, a large number of soft switching topologies included resonant circuits have been proposed [5-8]. But these topologies increase number of switching device in circuit and complicate sequence of switching operation.

This paper describes a novel buck-boost ac-dc converter operated with new soft switching for partial resonant method and driven with DCM. The partial resonant operation makes zero current switching and zero voltage switching (ZVS) for control switches without switching losses so called "soft switching" [6,8]. Also the partial resonant circuit is driven by not continuously resonant operation but partially enforcement at only switching turn-on time and turn-off time. It reduces capacity division and stresses of resonant devices. The input current waveform in proposed converter is got to be sinusoidal waveform in proportion to magnitude of ac input voltage under the constant duty cycle switching by DCM. The input power factor is nearly unity and the control method is simple. The proposed converter achieves high efficiency and high power factor from the results mentioned above.

II. CIRCUIT CONFIGURATION AND OPERATION PRINCPLE

This Fig. 1(a) and (b) show the conventional PFC buck-boost ac-dc converter and the proposed new PFC buck-boost ac-dc converter. The proposed converter is composed of controlling devices, step up/down inductor L_r, and snubber capacitor C_r used in conventional boost converter.

The switching devices in proposed converter are operated with soft switching by partial resonance technique and with control of constant duty cycle. The partial resonant circuit makes use of a step-up inductor and a loss-less snubber capacitor. It is for the accumulated energy in snubber capacitor to regenerate into input side without the power loss of snubber circuit produced in conventional buck-boost converter. The current flowing through inductor is controlled to be discontinuous like Fig. 1(c), and then turn-on operation of the switching devices S_1 and S_2 becomes to be ZCS. And

978-1-4244-1871-8/07 $25.00 © 2007 IEEE

turn-off of switches is worked on ZVS by the partial resonant operation. Therefore the proposed converter is operated with high efficiency. Also the input current waveform is got to be sinusoidal waveform in proportion to magnitude of ac input voltage under the constant duty cycle switching by DCM, as shown Fig. 1(c). The input power factor is nearly unity and the control method is simple.

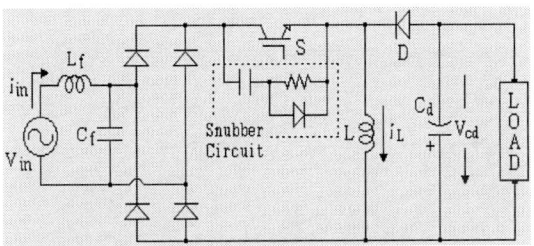

(a) Conventional PFC buck-boost AC-DC converter

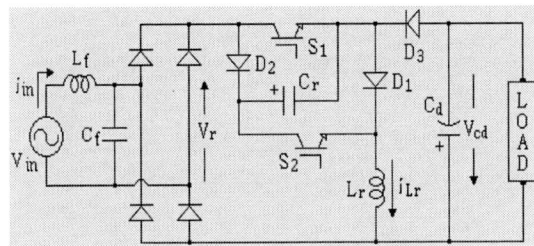

(b) Proposed PFC buck-boost AC-DC converter

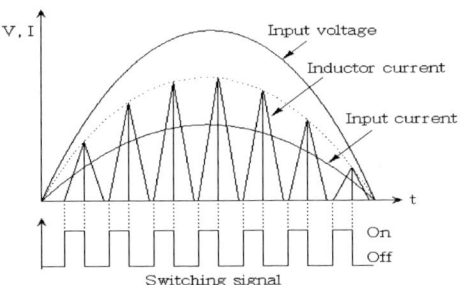

(c) Input waveforms of DCM

Fig. 1: Circuit configurations and input waveforms

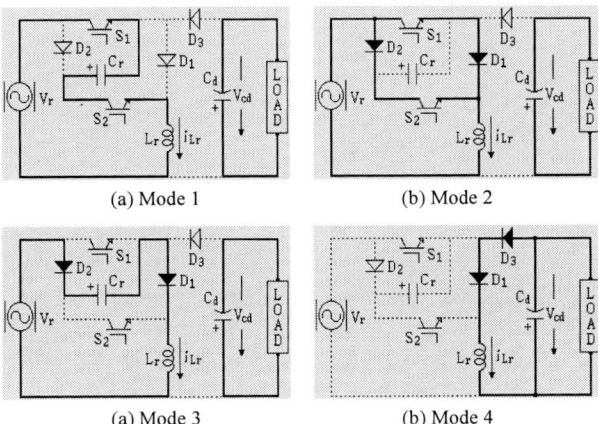

(a) Mode 1   (b) Mode 2

(a) Mode 3   (b) Mode 4

Fig. 2: Equivalent circuit in one cycle switching

Fig. 2 shows four equivalent circuits of operational modes in one cycle switching of the proposed converter. At initial condition, the current flowing through inductor L_r is zero. Switch S_1 and S_2 are off-state and capacitor C_r is charged with sum of rectifier output voltage v_r and dc output voltage V_{cd}. The ac input voltage v_{in} and the output voltage of full bridge rectifier v_r is expressed as like (1) and (2).

$$v_{in} = V_m \sin \omega_s t \qquad (1)$$

$$v_r = |v_{in}| = |V_m \sin \omega_s t| \qquad (2)$$

Mode 1 (T_1 : $t_0 \leq t < t_1$)

Mode 1 begins by turning on both S_1 and S_2 at the same time. The input voltage v_r and the capacitor voltage v_{cr} are added and applied to the inductor L_r. Then this mode takes form of a series LC resonance circuit. The capacitor C_r discharges its electric charge through the inductor L_r. Turn-on of the switching devices occurs in zero current state. Hence this is ZCS. The capacitor voltage is expressed in (3) and the inductor current increases as like (4).

$$v_{cr} = (2v_r + V_{cd}) \cos \omega_r t - v_r \qquad (3)$$

$$i_{Lr} = \frac{2v_r + V_{cd}}{X} \sin \omega_r t \qquad (4)$$

where $\omega_r = 1/\sqrt{L_r C_r}$, $X = \sqrt{L_r/C_r}$.

This mode ends when $v_{cr}=0$. And the inductor current I_1 at the end of this mode is given by

$$I_1 = \frac{1}{X} \sqrt{(2v_r + V_{cd})^2 - v_r^2} \qquad (5)$$

Mode 2 (T_2 : $t_1 \leq t < t_2$)

Mode 2 begins when the voltage across C_r achieves zero. Then the diode D_1 and D_2 start conduction. The inductor current is divided into two paths of D_1-S_1 and S_2-D_2. The inductor current is linearly increased as follows until the switches are turned off.

$$i_{Lr} = \frac{v_r}{L_r} t + I_1 \qquad (6)$$

This mode ends when both switch S_1 and S_2 are turned off, and the inductor current I_2 at the end of the mode is given by (7). Where, The period T_{on} is on-time of switch. In Mode 2, the inductor L_r stores a suitable energy.

$$I_2 = I_1 + \frac{v_r}{L_r} \left\{ T_{on} - \sqrt{L_r C_r} \cos^{-1} \left(\frac{v_r}{v_r + V_{cd}} \right) \right\} \qquad (7)$$

Mode 3 (T_3 : $t_2 \leq t < t_3$)

978-1-4244-1871-8/07 $25.00 © 2007 IEEE   267

Mode 3 begins by turning off both S_1 and S_2 at the same time. The current flowing through L_r takes a route of D_1-C_r-D_2 and charges C_r. Then this mode takes form of a series LC resonance circuit. Turn-off of S_1 and S_2 occurs in ZVS because the voltage of C_r is zero. In this mode, the voltage of C_r and the current of L_r are evaluated as follows

$$v_{cr} = v_r + \sqrt{\frac{L_r}{C_r}} I_a \sin(\omega_r t + \theta) \qquad (8)$$

$$i_{Lr} = I_a \cos(\omega_r t + \theta) \qquad (9)$$

where $I_a = \sqrt{\dfrac{C_r}{L_r} v_r^2 + I_2^2}$, $\theta = \sin^{-1}\left(-\dfrac{v_r}{\sqrt{v_r^2 + \dfrac{L_r}{C_r} I_2^2}}\right)$.

$v_{cr} = V_{cd}$ is achieved and diode D_3 begins to conduct, then this mode ends. The inductor current at the end of this mode can be assumed by the constant value I_2, because of the very short period of this mode.

Mode 4 (T_4 : $t_3 \le t < t_4$)

By the conducting of diode D_3, the inductor current flows into the load. The current is decreased linearly as following to the next equation and achieved to zero at the end of mode 4.

$$i_{Lr} = -\frac{V_{cd}}{L_r} t + I_2 \qquad (10)$$

III. COMPUTER SIMULATION RESULTS

The proposed circuit was analyzed by PSpice program. Fig. 3 shows waveforms of each part in a cycle switching in order to certify partial resonant operation and soft switching operation of control devices. Table I shows the constants of components, input voltage and the load capacity for system evaluation.

Table 1. Parameters of simulated circuit

Input voltage, v_{in} (rms)	100V, 60Hz	Resonant capacitor, C_r	50nF
Input filter inductor, L_f	2mH	Load resistor, R_L	100Ω
Input filter capacitor, C_f	3uF	Smoothing capacitor, C_d	1000uF
Resonant inductor, L_r	50uH	Switching frequency, f_c	40kHz

In Fig. 3, the controlled switches of duty factor 30% are turned on at t_0 and C_r begins to discharge. v_{cr} achieves to zero at t_1. At t_2, the controlled switches are turned off and C_r is charged with i_{Lr} and achieves to V_{cd} at t_3. At t_4, i_{Lr} achieves to zero and the controlled switches are kept off till the next cycle. T_c is one period of a cycle of switching operation. As the current flowing switches is zero at t_0, the controlled

switches are turned on with ZCS. Also, as the voltage being across switches is zero at t_2, the switches are turned off with ZVS.

The simulated results are confirmed the validity of the analytical results for each mode as previously stated.

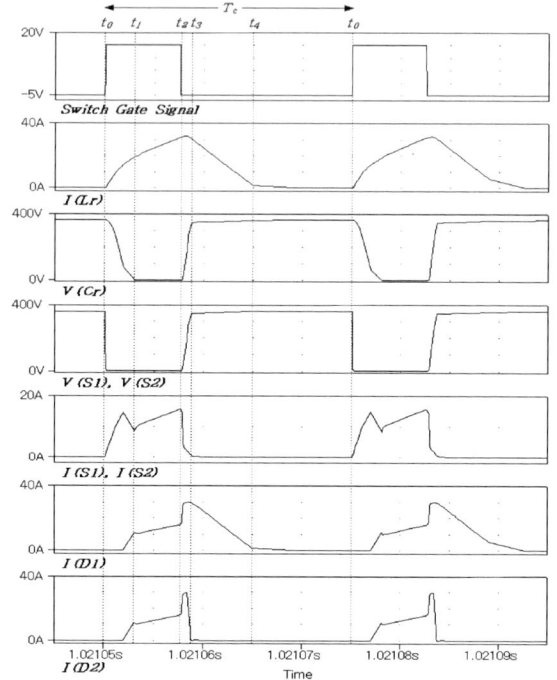

Fig. 3: Simulation waveforms of each part

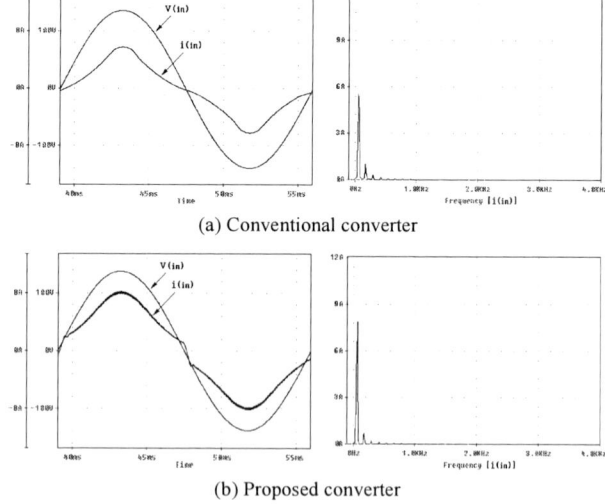

(a) Conventional converter

(b) Proposed converter

Fig. 4: Input waveforms and frequency spectra

Fig. 4 shows input waveforms and frequency spectra of input current for conventional converter (hard switching) and proposed converter (soft switching). The waveform of the current of conventional converter is smaller than it of proposed converter around the zero cross point. Hence the input current of proposed converter has quite a little of the third harmonic component. It is electrical charge of the snubber capacitor C_r to work to increment the amplitude of the current around the zero cross point. The outcome is that the current is more similar to a sinusoidal waveform.

IV. EXPERIMENTAL RESULTS

In order to confirm the feasibility, the proposed converter is experimented with power capacity 1.0kW. The experimental circuit is regulated at 300V output with AC 100V input. Fig. 5 shows waveforms of each part in one cycle switching (switching frequency 40kHz, duty factor 30%) in order to certify partial resonant operation and soft switching operation of control devices. At Fig. 5, the switches using in converter were operated with soft switching, namely turn-on at zero current and turn-off at zero voltage, according to partial resonant operation.

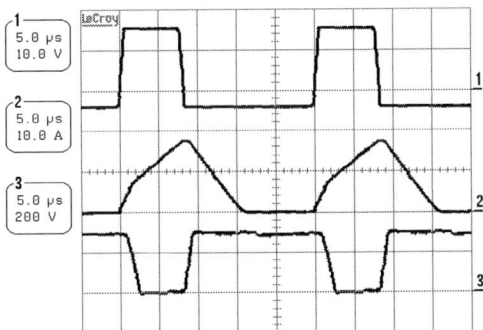

(a) Inductor current i_{Lr} and capacitor voltage v_{cr}

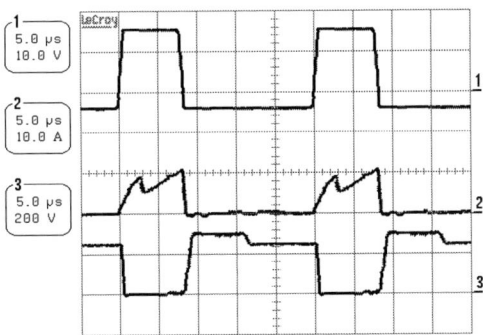

(b) Switch current i_s and switch across voltage v_s

Fig. 5: Experimental waveforms according to switching signal

To analyze the input current for convention converter and proposed converter, Fig. 6 and Fig. 7 show input waveforms and frequency spectra of input current through input low pass filter. The input current of proposed converter has quite a little of the third harmonic component.

The above experimental results are in accord with the analytical results and the computer simulation results as previously stated.

Fig. 8 shows the relation between power factor PF and duty factor D_c. The proposed soft switching converter maintains high power factor in wide operational range. The reason is that it makes to regeneration at input source of accumulated energy in snubber capacitor for a partial resonant operation of the proposed converter.

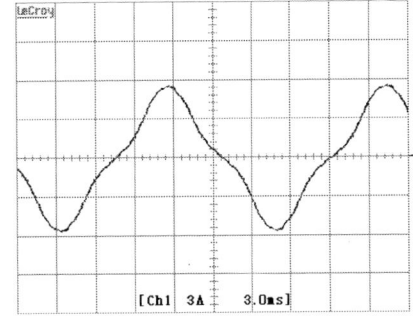

(a) Input current i_{in}

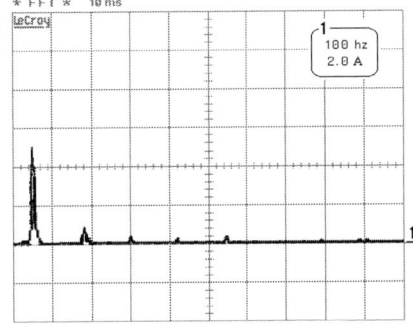

(b) Frequency spectrum of input current

Fig. 6: Waveform analyses of conventional converter

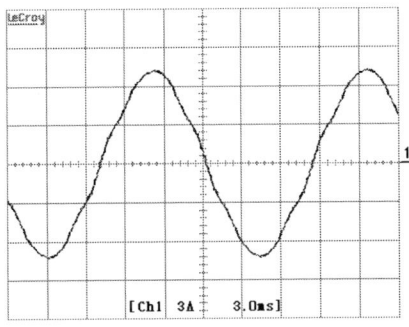

(a) Input current i_{in}

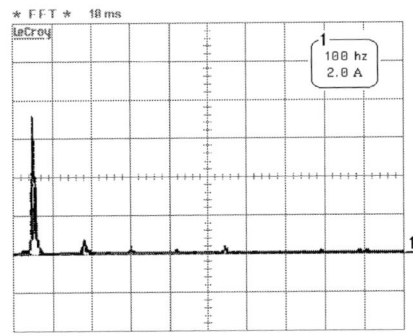

(b) Frequency spectrum of input current

Fig. 7: Waveform analyses of conventional converter

Fig. 9 shows the relation between the efficiency and output power. The efficiency of the proposed soft switching converter is increased more than that of the conventional hard switching converter.

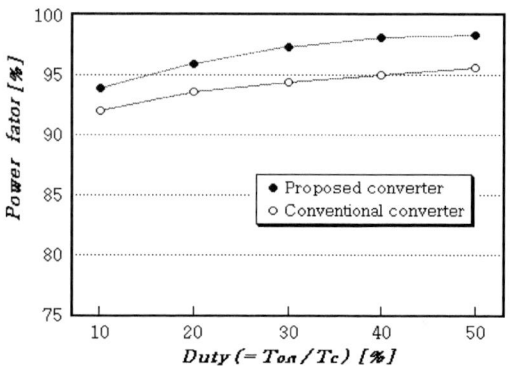

Fig. 8: Relationship between power factor and duty cycle

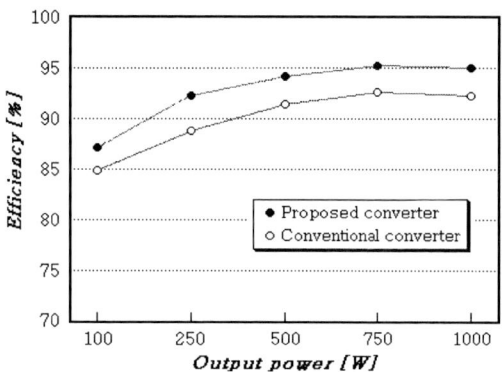

Fig. 9: Relationship between efficiency and output power

V. CONCLUSIONS

This paper described a novel buck-boost ac-dc converter of high performance by partial resonance technique. The input current waveform was got to be sinusoidal waveform in proportion to magnitude of ac input voltage under the constant duty cycle switching. The converter for DCM eliminated the complicated control circuit requirement. Therefore, the input power factor was nearly unity and the control method is simple. Also the switching devices in the converter were operated with soft switching by partial resonance technique. The partial resonant circuit made use of a step up/down inductor and a loss-less snubber capacitor. It was for the accumulated energy in the capacitor to regenerate into input side without the power loss of snubber circuit produced in conventional PFC buck-boost converter. The input power factor of proposed PFC converter for the regenerated energy was higher than that of conventional PFC converter. The proposed buck-boost converter was operated with high efficiency and high power factor from the above results.

REFERENCES

[1] W. Guo, and P. K. Jain, "A Low Frequency AC to High Frequency AC Inverter with Build-In Power Factor Correction and Soft-Switching", *IEEE Trans. on PE*, Vol.19, No. 2, pp. 430-442, March 2004.

[2] Q. Zhao, M. Xu, F. C. Lee, and J. Qian, "Single-Switch Parallel Power Factor Correction AC-DC Converters with Inherent Load Current Feedback", *IEEE Trans. on PE*, Vol. 19, No. 4, pp. 928-936, July 2004.

[3] B. Feng, D. Xu, "1-kW PFC Converter With Compound Active -Clamping", *IEEE Trans. on PE*, Vol. 20, No. 2, pp. 324-331, March 2005.

[4] O. Garcia, J. A. Cobos, et al., "Single phase power factor correction: A survey", *IEEE Trans. on PE*, Vol. 18, No. 3, pp. 749-755, May 2003.

[5] D. Sadarnac, W. Abida, and C. Karimi, "The Double Discontinuous Mode Operation of a Converter: A method for soft switching", *IEEE Trans. on PE*, Vol. 19, No. 2, pp. 453-460, March 2004.

[6] IEEJ: "Recent development on soft switching", *IEEJ Technical Report*, No. 899, pp. 4-8, Sept. 2002 (in Japanese).

[7] H. Watanabe, and H. Matsuo, "A novel high-efficiency DC-DC converter with 1V/20A DC output", *IEEE INTELEC 2002*, pp. 34-39, Sept.-Oct., 2002.

[8] Y. Tsuruta, and A. Kawamura, "A High Efficiency and High Power Chopper Circuit QRAS using Soft Switching under Test Evaluation at 8KW", *Journal of Power Electronics* in Korea, Vol. 6, No. 1, pp. 1-7, Jan. 2006.

Hybrid Ballast for Field Emission Lamp with CNT Emitter

Jong-Hyun Kim, Myung-Hyo Yoo, Dong-Wook Yoo
KERI
Changwon, Kyungnam 641-600, Rep. of Korea
Email: kimjh@keri.re.kr

Hye-Man Jung, Seok-Jin Ha
Airtecsystem
Suwon, Gyeonggi-do, 440-310, Rep. of Korea

Abstract— Hybrid ballast for field emission lamp with CNT (carbon nano tube) emitter is proposed. Hybrid ballast consists of a high voltage dc part and bi-polar pulse generation part. Field emission lamp with CNT lamp is composed of three electrodes (anode, gate, and cathode). High voltage dc part is for anode and gate and bi-polar pulse generation part is for gate and cathode in CNT triode respectively. The experimental results demonstrate that the proposed topology is good for driving CNT lamp.

I.   INTRODUCTION

A study on the new luminous source with high efficiency and high luminance is carried out lively and especially FED (Field emission display) using CNT emitter is highlighted in the new luminous source market. FED is drawing attentions as one of the most promising flat panel displays [1~2].

The technology of Spindle-type FEDs with MO emitter is being used in the field over three years. The reliability and feasibility of Spindle-type FEDs has been confirmed [3]. For the large size panel CNT FED is under development. FEDs will be used for not only the applications of small, medium size panels but also those of large-size panels. The core of FED is the principle of field emission that electrons are discharged when electric field is applied to the tip of sharp emitters in vacuum. CNT has a superior performance as emitters in FED [4~5].

In the existing driving methods for CNT FED, there are DC high voltage mode, high frequency inverter mode and high frequency pulse inverter mode and so on. In the case of the new luminous source of light which applies the field emission of CNT emitters, the only DC high voltage mode has been utilized. But, in this case we are much disadvantageous in the side of an efficiency and a life span of CNT lamp [6~8]. Therefore to improve efficiency and a life span of CNT lamp, the pulse voltage is applied to the gate and cathode in CNT triode structure [9].

Hybrid ballast for field emission lamp with CNT emitter is proposed. As shown in FIG. 1, the hybrid ballast for driving the triode CNT lamp includes a high DC power supply source which supplies a high DC voltage to an anode terminal of the triode of the triode CNT lamp in a state in which a cathode terminal is grounded and a bipolar pulse power supply source which supplies a bipolar pulse voltage to a gate terminal of the triode of the triode CNT lamp in a state in which the cathode terminal is grounded. That is, the high DC voltage is supplied to the anode terminal and the bipolar pulse is supplied to the gate terminal are simultaneously generated in one ballast.

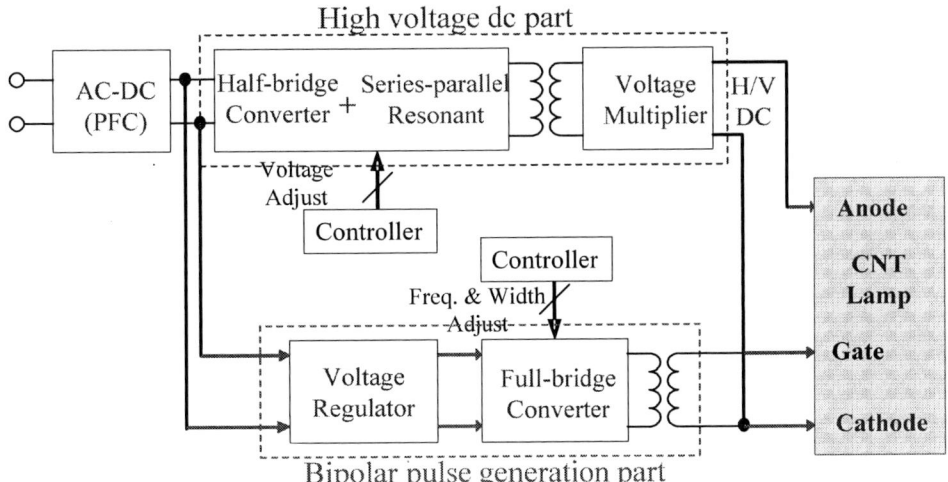

Fig. 1 Block diagrams of the hybrid ballast for field emission lamp with CNT emitter.

978-1-4244-1871-8/07 $25.00 © 2007 IEEE

II. HYBRID BALLAST FOR FIELD EMISSION LAMP WITH CNT EMITTER

Fig. 1 shows the hybrid ballast for field emission lamp with CNT emitter. Hybrid ballast consists of a high voltage dc part and bi-polar pulse generation part. These two parts produce high voltage dc and bi-polar pulse for three electrodes from line voltage through AC-DC part. AC-DC part includes PFC (power factor correction) to improve power factor and reduce harmonics.

A. High voltage dc part

Fig. 2 shows the diagram of the high voltage dc part. As can be seen in Fig. 2, V_{in} is the output voltage of the PFC part from a single phase source of 220 Vac. To transfer the energy from primary side to secondary side efficiently, parallel resonant half-bridge converter is adapted and frequency control method is applied. Output voltage of maximum 10 kV to drive the anode and the cathode can be obtained from high voltage transformer and voltage multiplier (Cockroft-Walton) circuit. To get the high voltage of 10[kV], turns ratio between primary and secondary of the transformer is about 32: 230 and it is possible to boost the voltage to 2 kV in the secondary of the transformer. Final output voltage of 10 kV applied to CNT load can be obtained from 5 times multiplier circuit.

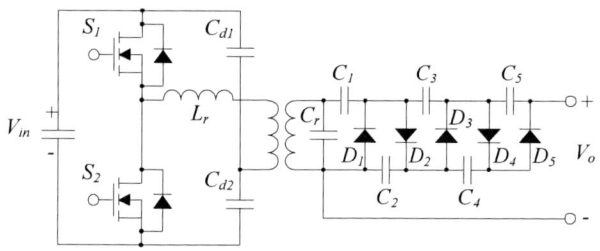

Fig. 2 High voltage dc part

B. Bipolar pulse generation part

In triode type, dc high voltage is applied to the anode and the cathode, and pulse voltage is approved between the gate and the cathode of CNT triode. Fig. 3 shows a bi-polar pulse generation part and their signals. In bi-polar pulse generator system, $+V_r$ comes to the output when switch S_1 and S_4 are turned on at the same time and $-V_r$ comes to the output when switch S_2 and S_3 are turned on at the same time as can be seen in Fig. 3(b). V_{AB} is transferred to V_{pulse} using a transformer. Due to leakage inductance of the transformer and capacitance of the gate-cathode of the CNT lamp, V_{pulse} has a semi-sinusoidal waveform from the rectangular waveform, V_{AB}.

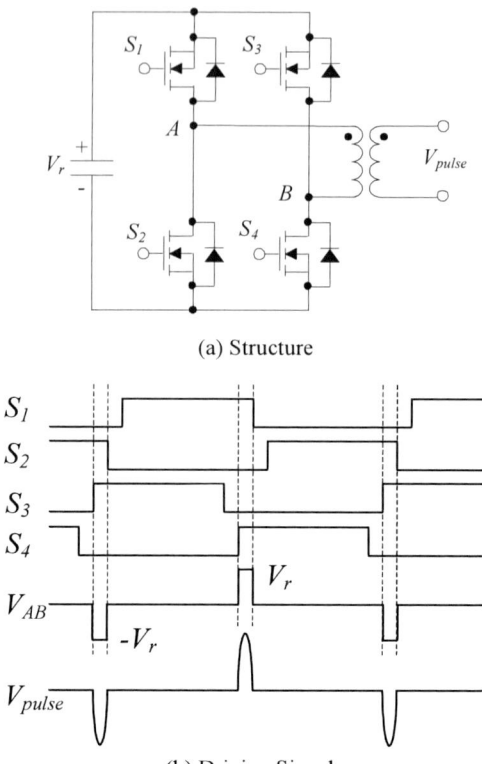

(a) Structure

(b) Driving Signals

Fig. 3 Bi-polar pulse generation part.

III. EXPERIMENTAL RESULTS

In order to verify the hybrid ballast for field emission lamp with CNT emitter, it is implemented as shown in Fig. 4 Fig. 5 shows experimental waveforms of the hybrid ballast. Fig. 5(a) shows the input voltage and current waveforms. The input current has a sinusoidal waveform and its phase agrees with the input voltage. Now the power factor is 0.98. Fig. 5(b) shows harmonic components of current. It satisfies harmonic current limitation of Class C. Fig. 6 shows the output waveforms. The high voltage DC (8.5 [kV]) is a voltage between anode and cathode. The semi-sinusoidal waveform with peak 256 [V] is applied to gate and cathode and its frequency is 15 kHz. Now the input power is 77 [W] and the output brightness is 4650 [cd/m²]. Fig. 7 is a photograph of field emission lamp with CNT emitter

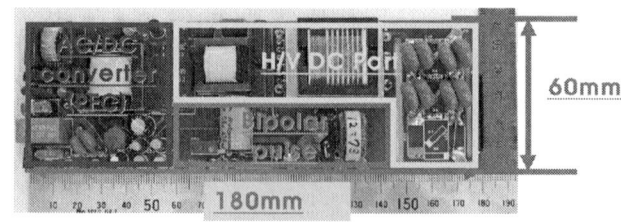

Fig. 4. The photograph of the hybrid ballast

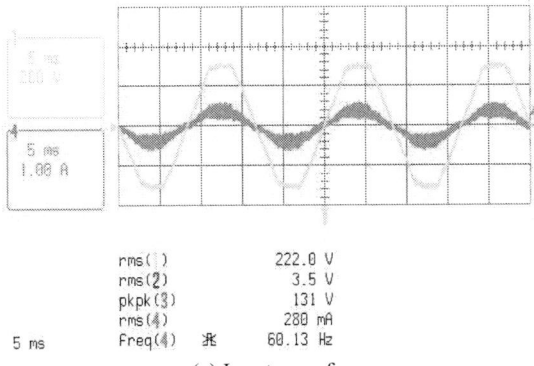

rms( )		222.0 V
rms(2)		3.5 V
pkpk(3)		131 V
rms(4)		280 mA
Freq(4)	无	60.13 Hz

5 ms

(a) Input waveforms

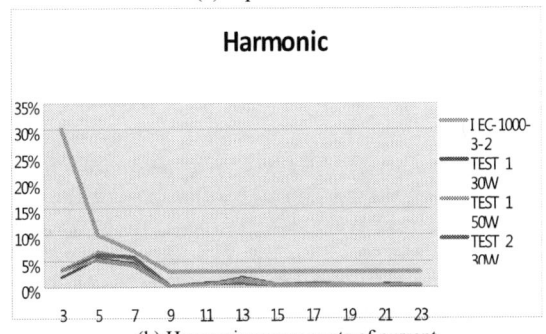

(b) Harmonic components of current

Fig. 5. Input waveforms and harmonic components of the hybrid ballast

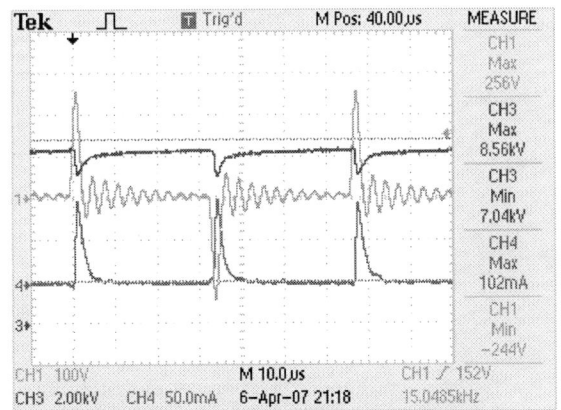

Fig. 6 Output waveforms of the hybrid ballast.

Fig. 7 The photograph of field emission lamp with CNT emitter.

IV.　CONCLUSION

Hybrid ballast for field emission lamp with CNT emitter is proposed. Hybrid ballast consists of a high voltage dc part and bi-polar pulse generation part. Field emission lamp with CNT lamp is composed of three electrodes (anode, gate, and cathode). High voltage dc part is for anode and gate and bi-polar pulse generation part is for gate and cathode in CNT triode respectively. The research is the early stage for a commercial business. And now it is the process of the research which raised system efficiency.

REFERENCES

[1] Lee Young-hee, "The Physical Property and Application of Carbon Nanotube", Sae Mulli (The Korean Physical Society), Vol.51, No. 2, pp. 84-144, August 2005

[2] Cho Jae-chul,, Kim Ji-seoung, Gu Hal-bon , "Recent Trend of FED(Field Emission Display)", The Journal of The Korean Institute of Electrical and Electronic Material Engineers, Vol. 9, No.9, pp.964-968,1996

[3] R.H. Baughman er al, "Carbon Nanotubes-the route toward application", Science 297, pp.787-792, 2002

[4] Park Chong-yun, Lee Yang-doo, Ju Byung-kwon,, Jeon Young-jin, "Development of new lighting source with high efficiency using carbon nanotube", The 18th Workshop of Energy Saving Technology, November 2003.

[5] Choi Won-bong, Lee Nae-sung, Kim Jong-min, "The Field Emission Display usnig Carbon Nanotube", The Journal of The Korean Institute of Electrical and Electronic Material Engineers, Vol. 13, No. 5, pp.44-48, 2003

[6] Chun-tao Lee, Yung-Chiang Lan, Bing-YuoTsui, and Cheng-chung Lee, "New driving method for triode CNT-FED", Technical digest of IVM 2003, Vol. 05, No. 3, pp.45-46, 2003

[7] Robert E. Neidert, Purobi M. Phillips, Sidney T. Smith, and Charles A. Spindt "Field Emission Triodes", IEEE Transaction on electron devices, Vol. 38, No. 3, March 1991

[8] Sang-su Kim, Yong-bae Kim, Sin-du Lee, and Jong-duk Lee, "Display Engineering", 2nd Edition, pp.426-486, March 2000

[9] Myung-hyo Ryu, Ju-won Baek, Jong-hyun Kim, and Dong-wook Yoo, "A study on the pulse generator for CNT lamp driving", IECON2006, pp.2683-2687.

Band-Gap Reference Voltage Control Strategy
Of Power Conditioning System for Fuel Cell Hybrid Vehicle

Young-Do Kim, Ki-Bum Park, Chong-Eun Kim and Gun-Woo Moon
Korea Advanced Institute of Science and Technology,
Department of electrical Engineering
Guseong-dong, Yuseong-gu, Daejeon, 305-701
Email: nemind@kaist.ac.kr

Abstract— Generally, the power management system of fuel cell hybrid vehicle (FCHV) requires a unidirectional DC/DC converter for the fuel cell (FC) and a bidirectional DC/DC converter for the battery. To manage the various power flows between these modules with a simpler way, a new band-gap reference voltage (BGRV) control strategy is proposed. The proposed method easily controls these variable power flows by setting the reference voltages of each converter to slightly different values, and it can be simply implemented by commercial controllers as well. The operational principle of proposed method is presented and verified experimentally by the 400W prototype

I.   INTRODUCTION

With the worldwide rising issues relevant to environmental pollution and the depletion of petroleum, an automobile industry tries to find the alternatives to the internal combustion engine, such as electric vehicle (EV), hybrid electric vehicle (HEV), and fuel cell vehicle (FCV) in order to improve fuel efficiency and to meet emission regulation.[1] Among them, FCV is the most attractive solution, since it utilizes the hydrogen as a fuel which makes no pollution and reduces the consumption of petroleum. However, high cost and a slow transient response of FC are the main technical obstacles for the commercialization of FCV. Moreover, it cannot recycle the regenerated energy from the motor since the fuel cell system does not have the storage capability. To overcome for these week points, energy storage, whether battery or ultra capacitor (UC), can be necessarily employed to reduce the cost and improve the performance of a fuel cell vehicle. FC with UC suffers from the start-up problem due to the self-discharge of UC, and utilizing both UC and battery increases the system complexity and cost. By the reason of that, the research of operating FC with battery has been mainly progressed [2, 3]. The voltage level and dynamic characteristic of the energy storage components are normally different from the primary energy sources, and dc/dc converter needs to be incorporated into the entire vehicle power conditioning system

Power conditioning system in FCHV can be used to store energy generated at fuel cell and battery and then to supplement the motor requirements at a different vehicle condition. Energy management that is handled through the dc/dc converters and battery always has power losses associated with it. However if power conditioning system is well –designed, the losses of the energy storage and discharge processes are compensated by operating the FC

and battery to an improved condition. Hence, a power conditioning system can enhance the overall fuel efficiency of a vehicle. [4]

A typical schematic diagram of the power conditioning system of a fuel cell based hybrid vehicle is shown in Fig. 1. A low voltage dc produced by the fuel cell is stepped up using a dc/dc boost converter, converter 1, to transfer the power from FC to the inverter. Battery provides an input dc voltage to the inverter during warm-up time of the fuel cell and facilitates regenerative braking through the converter 2. Once the fuel cell starts providing the required power, the battery is disconnected and the system can completely run on the fuel cell and the dual power conversion circuitry. The battery system also provides power during transient conditions.

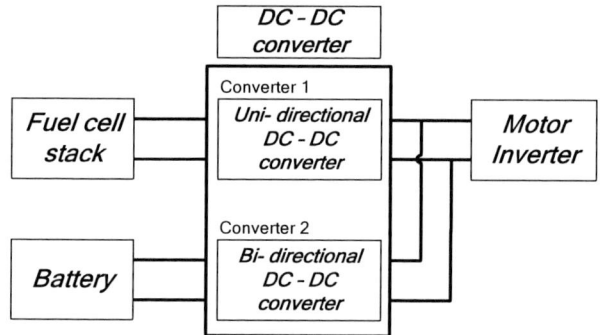

Fig. 1 schematic diagram of power conditioning system

For the efficient operation of power conditioning system, the power flow between dc/dc converters should be properly managed. Thus, many control strategies have been proposed in the literatures [5,6]. The control strategy using DSP or MICOM can improve the dynamic load response. However it features complex and high cost making the commercialization difficult. To solve these problems, a new BGRV control strategy is proposed in this paper. By simply adjusting the reference voltages of each converter to the slightly different levels, the power flow between FC, battery and inverter can be effectively controlled and it is easily implemented using widely used commercial controllers such as TL494 as well.

II.   OPERATIONAL PRINCIPLE

In our power conditioning system, the maximum power which can be supplied by FC is limited to P_{FCmax}, therefore

978-1-4244-1871-8/07 $25.00 © 2007 IEEE   274

FC supplies the power under its capability using converter 1. Since FC has a low dynamic response, no storage capability, and the limitation of supplying power, the surplus power or regenerated power from the inverter is handled by the battery using converter 2. The power flow of power management systems are as follows and flow chart is shown in Fig. 2.

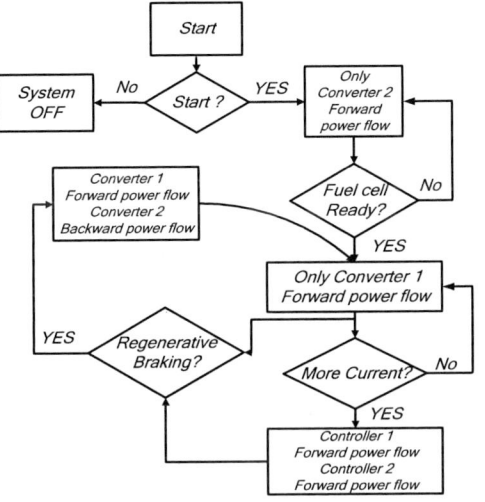

Fig. 2 Flow chart of power management system

1) Start-up mode: At the start-up period, only the battery supplies the power to the inverter for the preparing of FC.

2) Normal mode: In the normal driving, FC supplies the power to the inverter under the power limit of FC.

3) Excessive mode: When the vehicle is accelerated, the motor requires a more power than usual. If the required power is exceed over the FC's capability, FC works at its maximum power and the battery fills the surplus.

4) Regenerative mode: When the vehicle is at rest and non-loaded, the power is regenerated from the inverter to the battery.

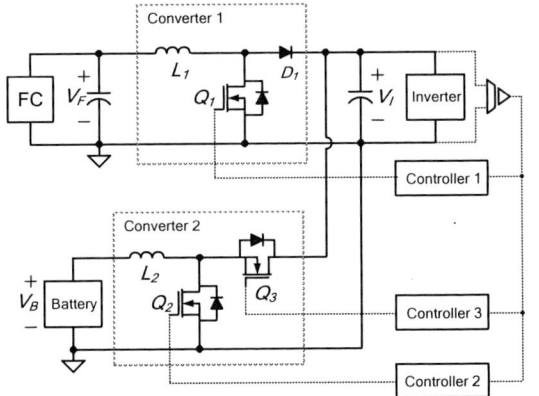

Fig. 3 Circuit diagram of power management system

Fig. 3 shows the circuit diagram of the power conditioning system. A low voltage dc produced by the fuel cell and battery is stepped up using a dc/dc boost converter to obtain the required voltage at the input of inverter and three

controllers are implemented for proper control of power conditioning system.

To regulate the inverter source voltage V_I under the mode variations, three controllers are required basically, i.e., controller 1 for the forward power flow of converter 1, controller 2 for the forward power flow of converter 2, and controller 3 for the backward power flow of converter 2 as can be seen in Fig. 3. Each controller is properly managed the dc/dc converter under mode variation.

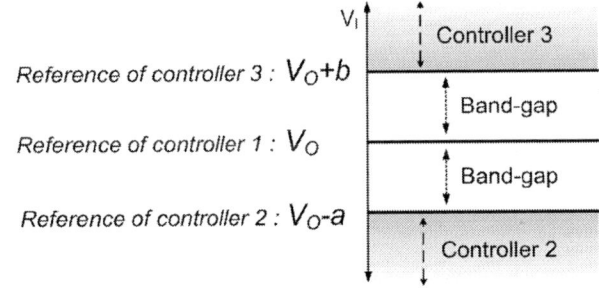

Fig. 4 Reference voltages and operating ranges of controller 2 and 3

The main concern to be focused is how effectively it can be managed so that the various power flows between the converters in a simple way using these three controllers. The proposed BGRV control strategy simply manages the power flow only by setting the slight band-gap between the reference voltages of each controller. As can be seen in Fig. 4, the reference voltage of controller 1 is set to V_O and the reference voltages of controller 2 and controller 3 are set to V_O-a and V_O+b, respectively. This band-gap between the reference voltages is set in an acceptable range for the operation of inverter

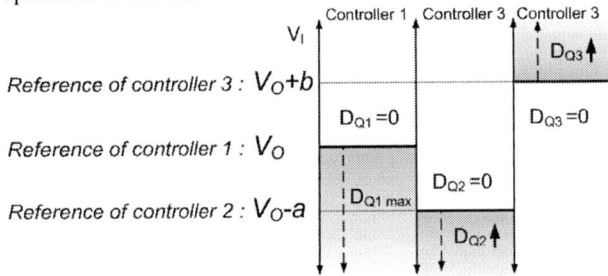

Fig. 5 Operation of controller 1, 2 and 3

The operation of each controller is shown in Fig. 5. If $V_I <$ V_O–a, which means FC can not supply the enough load current, controller 1 provides the maximum duty ratio to Q_1 (D_{Qmax}) and controller 2 start to regulate the V_I to the V_O -a. As the required load current is increased, controller 2 will raise the duty ratio of Q_2 (D_{Q2}). If V_O-a $< V_I < V_O$, V_I higher than the reference voltage of controller 2, D_{Q2} goes down and meet the zero in the end and so only the controller 1 manage the power flow of power conditioning system by controlling the duty ratio of the Q_1(D_{Q1}). If $V_I > V_O$, it means that backward current is generated from the load. So the V_I is increased and D_{Q1} is set to nearly zero. When V_I meets the V_O

+b, controller 3 starts to regulate the V_I to the V_O+b. As the backward current is increased, controller 3 will raise the duty ratio of Q_3 (D_{Q3}). After all, The proper controller according to the V_I automatically regulates V_I.

The operational principle of each mode of power conditioning system is presented as follows and key wave form of proposed control strategy is shown in Fig. 6.

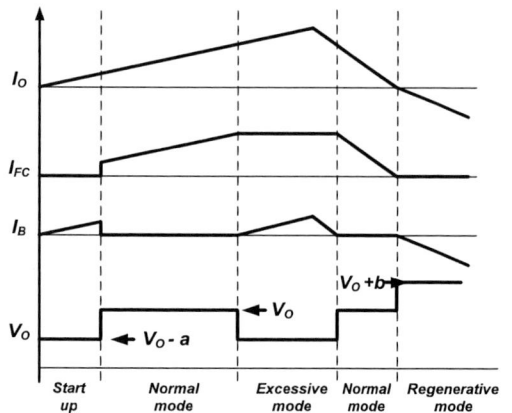

Fig. 6 Key waveform of BGRV control

1) Start-up mode: At the start-up period, only the battery supplies the power to the inverter until the preparation of FC is finished, i.e., controller 1 is off-state in some duration.

2) Normal mode: After the start-up period, controller 1 begins to work. Since the reference voltage of controller 1 is set to the higher value than that of controller 2, the power transfer by converter 1 is gradually increased to regulate V_I to V_O and the power transfer by converter 2 is decreased accordingly. After all, only converter 1 transfers the power and normal mode is operated as shown in Fig. 7(a). If the required load power does not exceed P_{FCmax} or the regenerated power does not come from the inverter, normal mode would be continued.

3) Excessive mode: If the required power from the inverter is exceed over P_{FCmax}, V_I would be decreased gradually getting out of the regulation since the supplied power by FC is less than the required power from the inverter. After V_I reaches V_O-a controller 2 starts to regulate V_I to V_O-a.

Therefore, the supplementary power is transferred from the battery to the inverter by converter 2 to regulate V_I. That is, excessive mode is operated as shown in Fig. 7(b). If the required power is decreased down to under P_{FCmax}, normal mode would be operated again regulating V_I to V_O.

4) Regenerative mode: The transition of regenerative mode is similar to that of excessive mode. In normal mode if the regenerated power comes from the inverter, V_I would be increased gradually getting out of the regulation since converter 1 can not handle the backward power flow. After V_I reaches V_O+b, controller 3 begins to regulate V_I to V_O+b. Therefore, the regenerative power is handled by converter 2 and is transferred from the inverter to the battery. That is, regenerative mode is operated as shown in Fig. 7(c). If the regenerated power is disappeared, normal mode would be operated again regulating V_I to V_O.

Besides the abovementioned operation, if the dynamic response of FC can not cope with the variation of required power, controller 2 or controller 3 would be operated automatically to regulate V_I.

III. EXPERIMENTAL RESULTS

To verify the proposed control strategy, the 400W prototype shown in Fig. 3 is built with the following
- Fuel cell voltage V_F = 40~45V (open circuit voltage=60V),
- Battery voltage V_B = 24V,
- Rated power of FC = 200W,
- Rated power of battery = 200W,
- Inverter voltage V_I = 48V,
- Band-gap 1, a= 1V,
- Band-gap 2, b = 1V,
- Controller 1,2,3 : TL494

Bi-directional converter can be shut down(SD) under excessive or regenerative mode according to the battery condition. In normal condition, battery can surplus the power in excessive mode and can absorb the power in regenerative mode. However if battery is in abnormal condition, overcharged or over discharged, it should be disconnected for protection of battery and other systems

The working state of battery can be predicted by the voltage

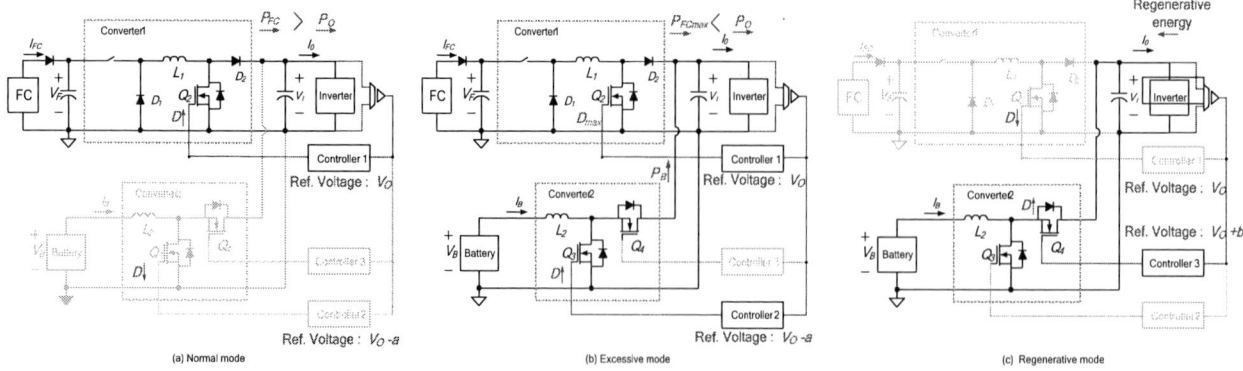

Fig. 7 Operational mode of power conditioning system

of battery, V_B. If $V_B < 20V$, it means that battery is over-discharged and should be charged. If $20V < V_B < 27V$, it shows that battery is under normal condition and can deliver and absorb power. If $V_B > 27V$, it indicates that battery is over-charged and needs to release excessive power. Table 1 shows the mode of bi-directional converter based on VB and system mode variation.

Fig. 8 presents the experimental results of proposed BGRV control strategy. After t_0, start-up mode begins and only the battery supplies the power. After t_1, controller 1 is activated and normal mode begins. Therefore, V_I is regulated to 48V by controller 1 getting out of the regulation by controller 2 and the power is supplied by FC. As the required power exceeds over P_{FCmax} at t_2, FC cannot solely handle the required power. Therefore, V_I is decreased and regulated to 47V by controller 2. After the required power is reduced under P_{FCmax} at t_3, V_I is increased and normal mode is operated again.

In brief, the inverter voltage V_I is well regulated in spite of the mode change. That is, the proper controller regulates V_I effectively according to the mode variation.

Table 1. Operational mode of Bi-directional converter

	Excessive mode	Regenerative mode
$V_B < 20V$	SD	Buck
$20V<V_B<27V$	Boost	Buck
$V_B>27V$	Boost	SD

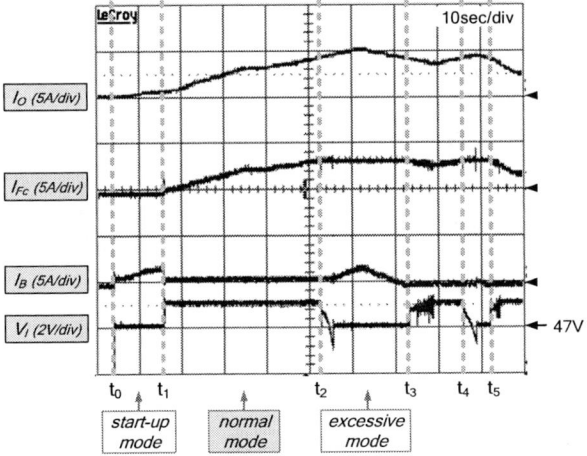

Fig. 8 Experimental waveform with mode variation

IV. CONCLUSION

A new BRGV control strategy for FCHV is proposed. By simply setting the slight band-gap between the reference voltages of each converter, the power conditioning system of FCHV effectively works. Moreover, it is easily implemented by widely used commercial controllers. Therefore, it features a simple and low cost implementation making it very promising for the commercialization of FCHV.

REFERENCES

[1] A, Emadi and S. S. Williamson, "Fuel cell vehicles: opportunities and challenges," IEEE, Power Engineering Society General meeting, 2004, pp. 1640-1645.

[2] Wenzhong Gao, "Performance comparison of a fuel cell-battery hybrid powertrain and a fuel cell-ultracapacitor hybrid powertrain," IEEE Trans. Vehicular Technology, Vol.54, No. 3, 2005, pp.1840-1845.

[3] A, Emadi ,S. S. Williamson and A. Khaligh "Power electronics intensive solution for advanced electric, hybrid electric, and fuel cell vehicular power systems," IEEE trans. Power electronics, Vol. 21, No. 3, 2006, pp. 567-577

[4] Jih- Jheng Lai and Douglas J. Nelson, "Energy management power converters in hybrid electric and fuel cell vehicles" Proc. IEEE, Vol. 95, 2006, pp. 766-777

[5] L. Berton, H.Gualous, D.Bouquain, D. Hissel, M. C. Pera, and J.M. Kauffmann, "Hybrid auxiliary power unit(APU) for automotive application," IEEE, Vehicular Technology Conference, 2002, pp. 1840-1845.

[6] Rajesh Gopinath, Sangsun Kim, Jae-Hong Hahn, Prasad N. Enjeti, Mark B. Yeary, and Jo W. Howze, "Development of a low cost fuell cell inverter system with DSP control," IEEE trans. Power Electronics, Vol. 19, No. 5, 2004, pp. 1256-1262

Development of a 72V IPMSM Drive System for Golf Carts

H.S Mok, B.C Kim
Konkuk University
Department of Electrical
Engineering
1, HwaYang-Dong, Gwangjin-Gu,
Seoul
Email: hsmok@konkuk.ac.kr

S.H Kim
Kangwon National University
Department of Electrical and
Electronics Engineering
192-1, Hyoja2-Dong, Chunchon,
Kangwon-Do
Email: kshoon@kangwon.ac.kr

S.L Ryu, Y.J Oh
HYOSUNG CORPORATION
Department of Rotating
Machinery & Application
454-2, Nae-Dong, Changwon,
GyeongSangNam-Do
Email: corido16@hoysung.com

Abstract—This paper describes the development process of a golf cart drive system using the highly efficient interior permanent magnet synchronous motor (IPMSM). The proposed drive system's properties were tested based on simulations and experiments. Vector control using the resolver is the default control mechanism of the developed IPMSM drive system, and field weakening control allows a maximum rotation speed of 4500[rpm], which is 2.5 times faster than the rated speed. The system efficiency under the rated load is about 92.3[%].

I. INTRODUCTION

Compared to the internal combustion system or the traction system using a DC motor, the interior permanent magnet synchronous motor (IPMSM) requires less consumable parts, offering the advantage of low repair and maintenance costs. Accordingly, an increasing number of applications are using the IPMSM, and the trend is expected to continue. [1]

The permanent magnet of the IPMSM is buried inside the rotor, which provides a mechanically robust rotor structure, making it adequate for high-speed operation. Furthermore, IPMSM's magnetic saliency provides a high torque generating efficiency. Such features of the IPMSM allow highly efficient operation of the drive system.

This paper describes the process of developing a golf cart drive system using the IPMSM. The system's operational characteristics and high efficiency were verified through experiments and evaluations.

II. IPMSM PROPERTIES AND MODEL

A. Characteristics of the IPMSM

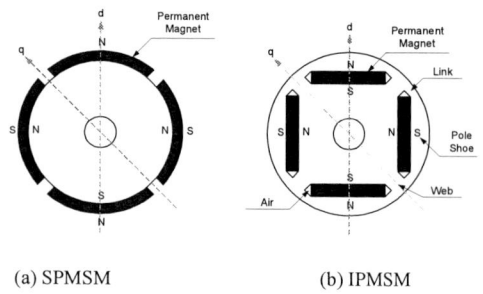

(a) SPMSM   (b) IPMSM

Fig. 1: Rotor Cross-Sections of Permanent Magnet Synchronous Motors

Unlike the SPMSM (Surface-mount Permanent Magnet Synchronous Motor), for which the permanent magnet is attached to the motor surface, the IPMSM's permanent magnet is placed inside the rotor, as shown in Fig. 1.

The structure provides the IPMSM with the following characteristics:

- Typically higher efficiency compared to induction or DC motors
- A mechanically robust rotor structure
- A high torque generating efficiency due to the reluctance torque according to magnetic saliency
- Substantial field weakening effect in the maximum output region due to relatively high rotor inductance

Based on these characteristics, the IPMSM is being widely used in applications that require a wide range of acceleration and speed variation. [2]

Fig. 2 displays the prototype IPMSM used for this study. Coils are wound in the stator as in a typical induction motor as shown in (a), and the rotor has permanent magnets secured inside without any coils as shown in (b).

(a) stator   (b) rotor

Fig. 2: Photos of Prototype IPMSM

The prototype IPMSM was designed according to the cart specifications of Appendix B, and TABLE 1 shows the IPMSM parameters.

TABLE 1
Parameters of the IPMSM

Rated Output Power	5 [kW]
Rated Voltage	72 [V]
Rated Speed	1800 [rpm]
Rated Torque	26.5 [Nm]
Rated Current	69 [A_{rms}]
Number of Poles	4 [poles]

Number of Slots	24 [slots]
Equivalent Resistance	0.008 [Ω]
L_d	0.35 [mH]
L_q	0.7 [mH]
Material of Permanent Magnet	NdFe (1.2[T])
Type of Permanent Magnet	Flat type
Maximum Torque	79.5 [Nm]
Maximum Current	207 [A$_{rms}$]

B. Model of IPMSM [3]

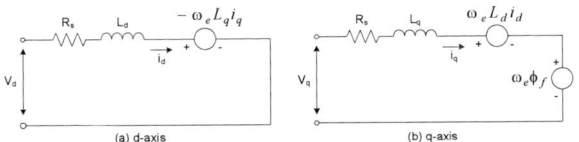

Fig. 3: Equivalent Circuit of IPMSM

The equivalent circuit of the IPMSM in the rotor-oriented synchronous coordinate system is shown in Fig. 3. The stator voltage equation and the torque equation can be expressed as (1) and (2), respectively.

$$\begin{bmatrix} v_{ds}^e \\ v_{qs}^e \end{bmatrix} = \begin{bmatrix} R_s + \dfrac{d}{dt}L_{ds} & -\omega_e L_{ds} \\ \omega_e L_{ds} & R_s + \dfrac{d}{dt}L_{qs} \end{bmatrix} \begin{bmatrix} i_{ds}^e \\ i_{qs}^e \end{bmatrix} + \begin{bmatrix} 0 \\ \omega_e \phi_f \end{bmatrix} \tag{1}$$

$$T_e = \frac{3}{2} P (\phi_f i_{qs}^e + (L_{ds} - L_{qs}) i_{qs}^e i_{qs}^e) \tag{2}$$

Here,

v_{ds}^e, v_{qs}^e : d-axis, q-aixs stator voltage,

i_{ds}^e, i_{qs}^e : d-axis, q-axis stator current,

L_{ds}, L_{qs} : d-axis, q-axis stator inductance,

R_s : stator resistance,

ϕ_f : per phase magnetic flux linkage,

P : pole pairs,

ω_e : synchronous speed of the machine.

III. DESCRIPTION OF THE CONTROL SCHEME [3]

The IPMSM has two operating regions: the constant-torque region controls the maximum torque, and the constant-power region controls field weakening.

A. Control in Constant Torque Region

Since there is magnetic saliency in the IPMSM, reluctance torque exists as indicated in (2). Accordingly, the maximum torque can be generated by controlling the d-axis current.

As shown in Fig. 4, the shortest line from the origin to an arbitrary torque T_e is the current that generates the maximum torque for usnit current.

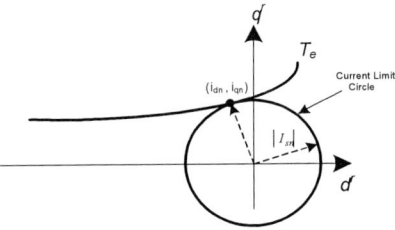

Fig. 4: Arbitrary curve and current limit circle

Controlling the maximum torque for unit current involves finding the minimum stator current according to torque variation.

The d-axis current required for generating the maximum torque for unit current can be written as (3).

$$i_{ds}^e = \frac{\phi_f - \sqrt{\phi_f^2 + 8(L_{qs} - L_{ds})^2 i_s^{e\,2}}}{4(L_{qs} - L_{ds})} \tag{3}$$

B. Control in Constant Power Region

The counter electromotive force is proportional to the rotor speed, and the maximum motor speed is limited by the amount of voltage supplied by the inverter. Therefore, a margin for the inverter control voltage can be secured by controlling the d-axis current and adequately reducing the counter electromotive force. The margin for the control voltage allows the IPMSM to operate at a higher speed.

Accordingly, field weakening control involves securing a margin for the control voltage so that the current controller does not become saturated during high-speed operation.

As shown in Fig. 5, the field weakening controller must operate along the current vector trajectory from points A to B, where the current limit circle and the voltage limit ellipse meet. The d-axis current can be expressed as (4), and the field weakening control starting point is determined by (5).

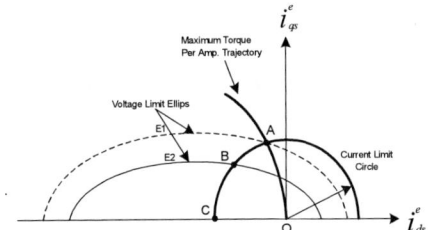

Fit. 5: Current limit circle and voltage limit ellipse

$$i_{ds}^e = \frac{L_{ds}\phi_f - \left((L_{ds}\phi_f)^2 + (L_{qs}^2 - L_{qs}^2)(\phi_f^2 + (L_{ds}i_{ds}^e)^2 - (\dfrac{V_f}{\varpi_e})^2)\right)^{\frac{1}{2}}}{L_{qs}^2 - L_{qs}^2} \tag{4}$$

$$\omega_{base} = \frac{V_f}{\sqrt{(L_{qs}i_{qs}^e)^2 + (L_{ds}i_{ds}^e)^2 + \phi_f^2 + 2L_{ds}\phi_f i_{ds}^e}} \tag{5}$$

Here, $V_f = V_{s\max} - R_s I_{s\max}$.

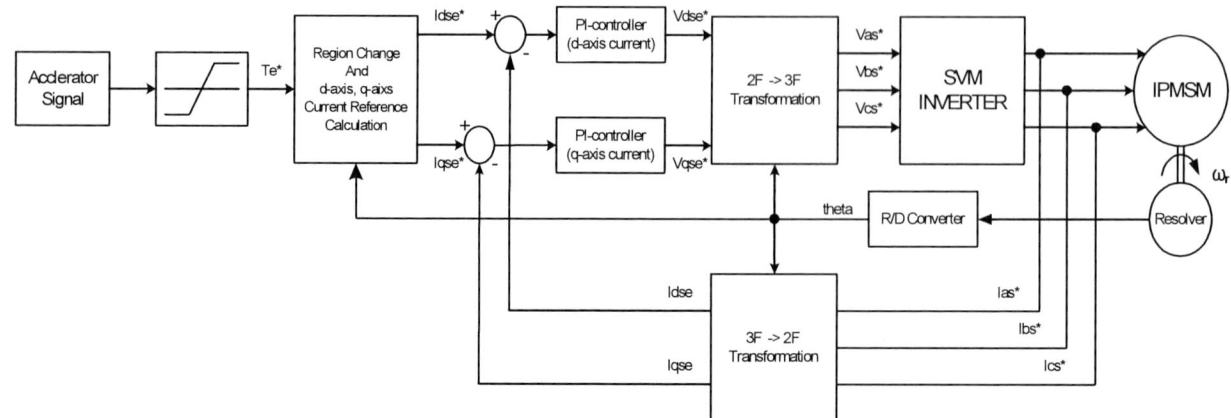

Fig. 6: Block diagram of the control scheme

IV. SIMULATION RESULT

Simulations were performed using Simplorer® from Ansoft. Fig. 6 depicts the control block diagram, and the simulation circuit diagram is shown in Appendix A.

The torque command value T_e^* is input according to the user's manipulation of the acceleration pedal. The constant torque region and the constant power region are selected based on the current operation condition, and the d-q current command values I_{dse}^* and I_{qse}^* are determined by the selected regions. I_{dse}^* and I_{qse}^* generate the d-q voltage command value through the PI current controller and the three-phase voltage command value through coordinate conversion. The voltage command value supplies three-phase voltage to the IPMSM by an SVPWM (space vector modulation) inverter. The rotor position theta used for calculating the d-q current command value, selecting the operating region and converting coordinates is obtained using the resolver and the R/D converter attached to the IPMSM.

The torque under no-load condition according to speed variation between 0 and 4500[rpm], the d-axis current and the q-axis current are shown in Fig. 7. The graph indicates that field weakening control starts at 1770[rpm]. The d-axis current under normal conditions is about -170[A].

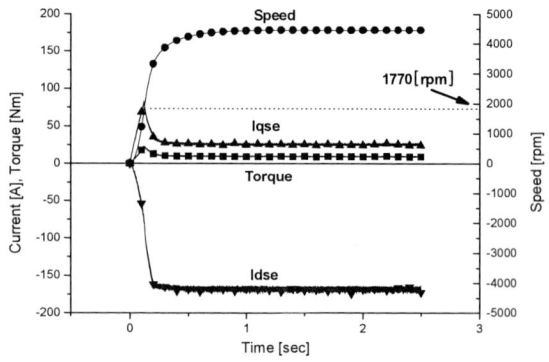

Fig. 7: Variations in d-q currents and torque at various speeds

V. EXPERIMENTAL RESULT

The chassis and body of ATT R&D's Invita were used for the experimental golf cart, the images and specifications of which are provided in Appendix B. The loading apparatus shown in Fig. 8 was used to provide variations in the load torque.

Fig. 8: The IPMSM Experimental System

Fig. 9 represents the acceleration characteristics of the IPMSM at maximum speed under a no-load condition. Controlling the d-axis current in the field weakening region allowed operation at 4500[rpm], which was 2.5 times the rated speed. Fig. 10 displays variations in the d-axis current and the q-axis current under identical conditions.

Fig. 11 depicts the normal status current waveform under the rated load (5[kW] at 1800[rpm]).

(a) shows the d-axis current, the q-axis current and the rotor position in the stationary coordinate system, and (b) indicates the d-axis current and the q-axis current in the synchronous coordinate system.

TABLE 2 summarizes the load testing results of the drive system under rated conditions.

The inverter and motor efficiencies were 98.5[%] and 93.7[%], respectively, and the efficiency of the drive system calculated with the inverter input and the motor output was about 92.3[%].

978-1-4244-1871-8/07 $25.00 © 2007 IEEE   280

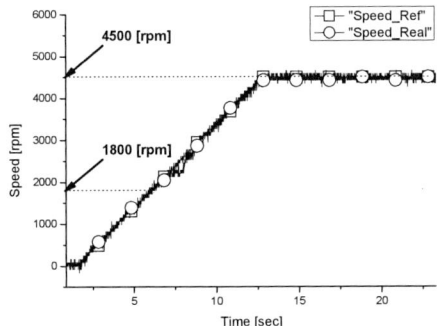

Fig. 9: No-load acceleration characteristics of the IPMSM

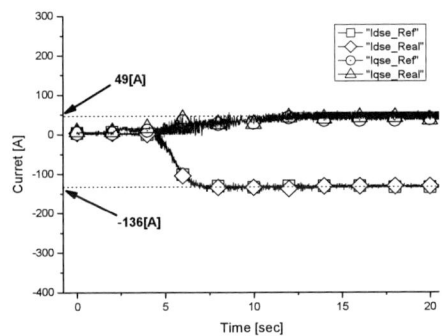

Fig. 10: Variations of d-q currents according to increase in speed

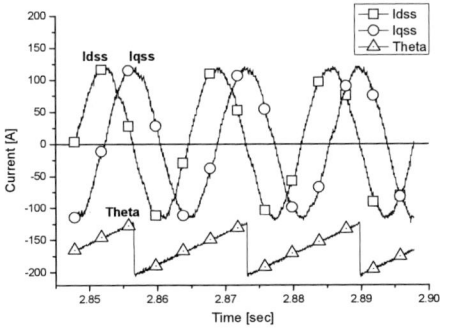

(a) d-q currents and rotor position theta (Stationary coordinate system)

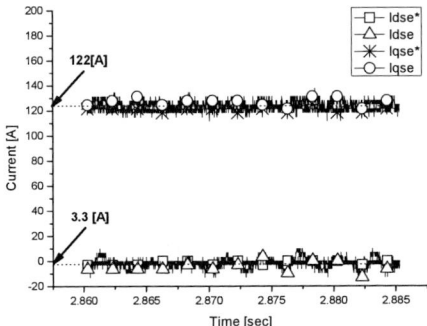

(b) d-q currents command values and actual current values(Synchronous coordinate system)

Fig. 11: d-q currents under rated load

IPMSM's operating efficiencies according to its speed and torque are indicated in Appendix C.

TABLE 2
Performance Testing Results

Drive System	Inverter Input Power [kW]	5.3
	Motor Output Power [kW]	4.9
	Efficiency [%]	92.3
Inverter	DC-Link Voltage [V]	65
	DC-Link Current [A]	81.5
	Input Power [kW]	5.3
	Output Power [kW]	5.2
	Efficiency [%]	98.5
Motor	Input Voltage [V]	27
	Input Current [V]	93.6
	Input Power [kW]	5.2
	Torque [Nm]	26
	Speed [rpm]	1800
	Output [kW]	4.9
	Efficiency [%]	93.7

VI. CONCLUSION

This paper introduced a theory for controlling a 5kW IPMSM and described the process of developing a golf cart drive system using a highly efficient IPMSM instead of a typical DC motor to achieve an efficiency rating of 92.3[%].

Moreover, the operating speed of the golf cart was improved by applying the IPMSM, which is adequate for high-speed operation.

The IPMSM can operate in a wide range of speeds from the constant-torque region to the constant-power region. Adequate control allowed the IPMSM to operate at 4500[rpm], which was 2.5 times the rated speed.

The IPMSM drive system developed for this study can be used not only in golf carts but also in other electric vehicles with similar structures.

ACKNOWLEDGMENT

This study was conducted with financial support from Hyosung Corporation to develop a golf cart drive system using the highly efficient interior permanent magnet synchronous motor (IPMSM).

REFERENCES

[1] J.F. Gieras and M. Wing, "Permanent Magnet Motor Technology", New York: Dekker, 1997.

[2] T.J. Jahns, G.B. Kliman, and T.W.Neumann, "Interior Permanent-Magnet Synchronous Motors for Adjustable-Speed Drives", *IEEE Trans. Industry Applications*, vol. IA-22,no.4,pp.738-747, July/Aug.1986.

[3] J.M. Kim and S.K. Sul, "Speed Control of Interior Permanent Magnet Synchronous Motor Drive for Flux Weakening Operation", *IEEE Trans. Industry Applications*, vol. 33, no. 1, Jan/Feb, 1997. pp. 43-48.

Appendix

A.   Simulation Scheme

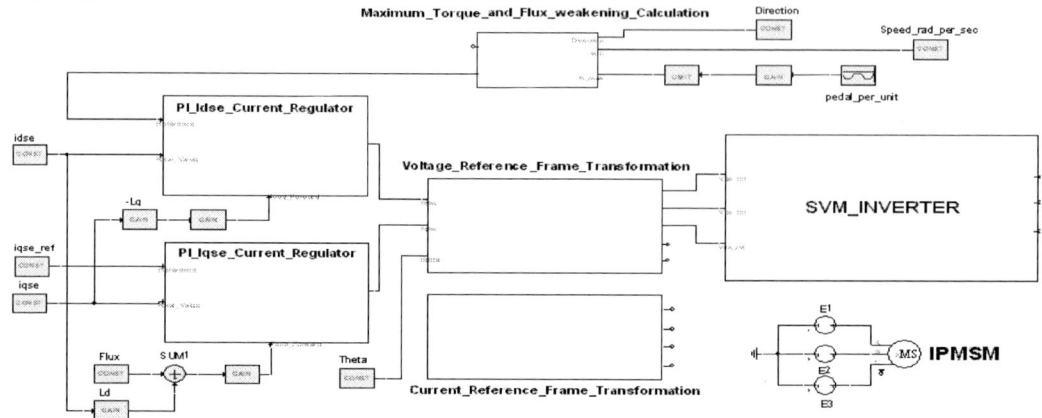

Fig.6: Simulation Circuit using Simplorer®

B.   Dimensions and Specifications of the Experimental Cart

The chassis and body used for the experimental golf cart are as follows.

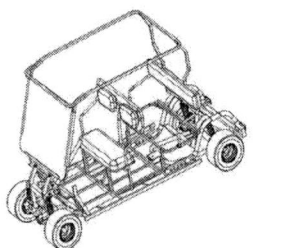

Trapezoid-shaped Chassis

4-passenger Invita LSV

The IPMSM drive system developed for this study was applied to the 4-passenger Invita golf cart with the following specifications.

Golf Cart Specifications (4 passenger capacity)	
Total Weight	915 [kg]
Speed	19 [km/h]
Tire Diameter	558 [mm]
Grade Angle	13 [deg]
Reduction Gear Ratio	10
Reduction Gear Efficiency	0.92
Rolling Resistance Coefficient	0.01

C.   Efficiency Characteristics in Operation Region

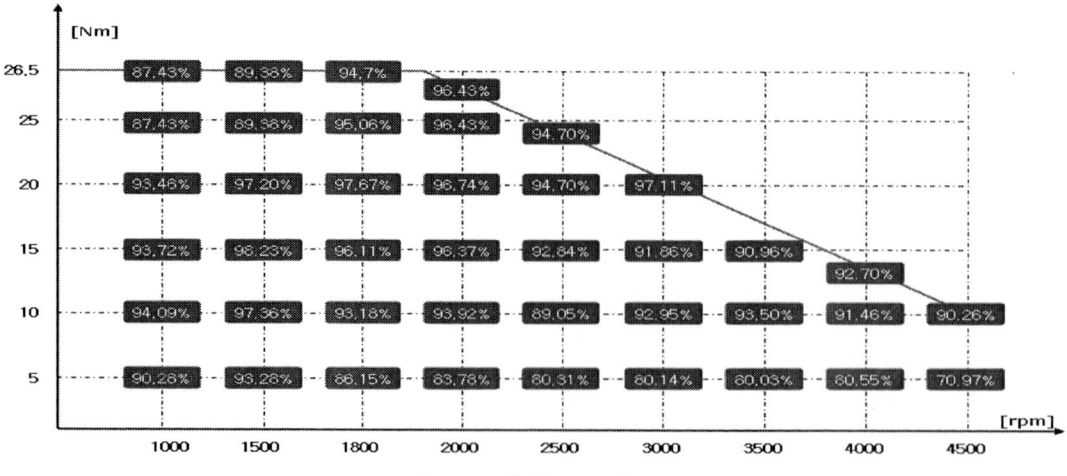

Development of Operation Algorithm for 2.5 ton Electric Forklift Using Induction Motors

N.C. Park*, J.H. Kim**, H.S. Mok**, S.H. Kim*

*Department of Electrical and Electronics Engineering
Kangwon National University
192-1, Hyoja2-Dong, Chunchon, Kangwon-Do
Email: gana007@ kangwon.ac.kr

**Department of Electrical Engineering
Konkuk University
1, Hwayang-Dong, Gwangjin-Gu, Seoul
Email: suedelover@hotmail.com

Abstract — This paper presents a drive system for a 2.5 ton electric forklift using induction motors. In this system, DC motors used in conventional electric forklifts were replaced with induction motors which are cheap, robustness, and controllable at various speeds. The high performance drive system for the induction motor was developed, and applied to a 2.5 ton electric forklift and conducted an experiment to confirm its applicability and utility.

I.   INTRODUCTION [1][2]

The use of equipment that does not emit exhaust gas in industrial sites is increasing in order to help improve work environments. Demand for such equipment is steadily increasing, particularly for forklifts which are often used indoors.

Most electric forklifts use DC motors, which are easy to control. However, DC motors are difficult to operate at high speed and periodic brush and commutator maintenance is required. To overcome these problems, AC motors are increasingly used. Induction motors are especially appropriate for electric forklifts; they are efficient, affordable and controllable at various speeds.

Regarding the speed sensor for the velocity measurement of electric forklifts, a resolver is preferred. Unlike rotary encoders, resolvers are not affected by environmental conditions such as dust and can endure temperature changes.

This study designed low voltage induction motors to replace the DC motor used in conventional electric forklifts and also developed a high performance drive system for the induction motors.

This paper applied the developed motors and drive system to a 2.5 ton electric forklift and confirmed its validity through experimentation.

II.   SYSTEM CONFIGURATION

A. Specification of Forklift

An electric forklift uses two induction motors: a traction motor to drive the electric forklift and a hydraulic motor to control oil flow. The specification of the 2.5-ton electric forklift used in this paper is shown in TABLE 1.

TABLE 1
Specifications of the forklift

Weight [kg]	5,895
Lifting Load [kg]	2,500
Tire Radius [m]	0.293
Gear Ratio	20 : 1
Rated Speed [km/h]	15

B. Traction Motor

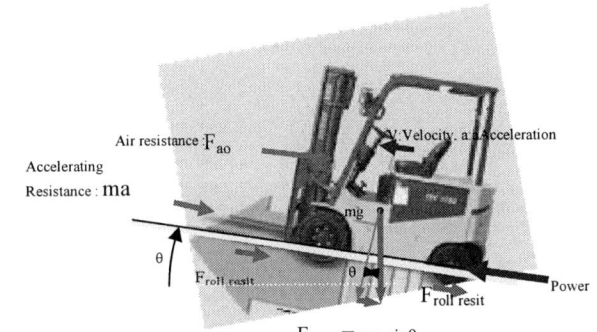

Fig. 1: Electric forklift

The running condition of the electric forklift is that the driving force of the motor must be greater than the sum of air resistance, rolling friction, gradient resistance, and acceleration resistance. However, air pressure can be ignored because it is very small. The load torque by gradient force (T_grade) is as shown in Eq. (1).

$$FL\_grade = (9.8 * M * \sin \theta)[N] \qquad (1)$$

Where, θ is maximum tilt angle, and M is mass.

To convert this load torque to motor-side torque in consideration of gear ratio and efficiency, we get Eq. (2).

$$TL\_grade = FL\_grade * R\_tire / N\_ratio / Eff \qquad (2)$$

Where, N_ratio is gear ratio, and Eff is gear efficiency.
If maximum mass were 6,945[kg], maximum climbing angle 11°, tire radius 0.293[m], gear ratio 20:1, and gear efficiency 92[%], the load torque by gradient resistance is 206.2 [Nm].

Rolling friction is mainly affected by the vehicle load, ground surface, and tire characteristics. The rolling friction was calculated to be 9.91 [Nm].

For an electric forklift to climb a maximum gradient of 11°, its overload capacity must be four times as high as the rated torque.

The specifications of the traction motor, manufactured on the basis of the above calculations, are shown in TABLE 2. And the steady state characteristics of two motors are shown in Fig.2 and 3, respectively.

TABLE 2
Specifications of the traction motor

Power [kW]	10.5
Number of Poles	4
Frequency [Hz]	60
Rated Speed [rpm]	1758
Torque [Nm]	58.7
Efficiency [%]	91.1
Voltage [V]	31
Current [A]	303.6

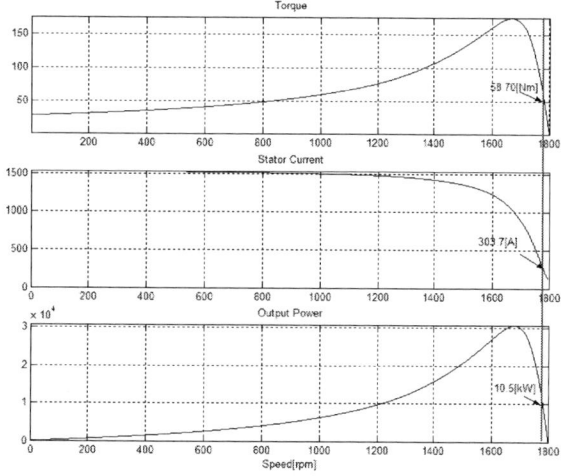

Fig. 2: Steady state characteristics for traction motor

C. Hydraulic Motor

The power (W) needed to vertically raise the lifting load (w [kg]) at the rated speed (V [m/sec]) is as follows:

$$W = 9.8 \times w \times V \qquad (3)$$

The base power for driving is 1.6 [kW]. If the lifting load is 2,500 [kg] and the full load lifting speed is 0.27 [m/sec], the full load power is 6.6 [kW]. The maximum required power is the sum of full load power and base power, which is approximately 8.2[kW]. We chose the capacity of the induction motor to be 10 [kW].

The specifications of the hydraulic motor, produced on the basis of the above calculations, are shown in TABLE 3.

TABLE 3
Specifications of the hydraulic motor

Power [kW]	10
Number of Poles	4
Frequency [Hz]	80
Rated Speed [rpm]	2347
Torque [Nm]	40.7
Efficiency [%]	91.1
Voltage [V]	31
Current [A]	249.8

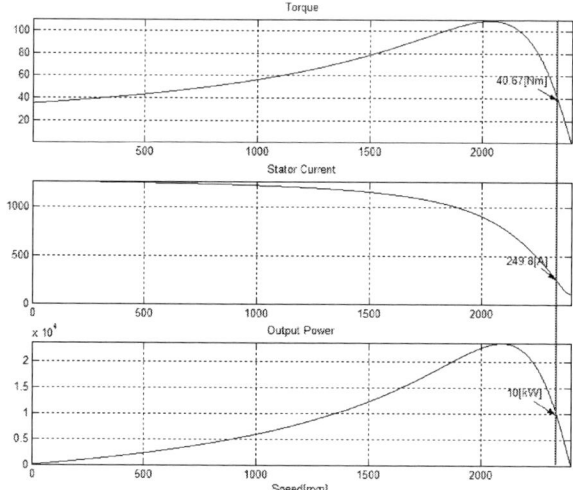

Fig. 3: Steady state characteristics for hydraulic motor

D. Inverter

An electric forklift needs a low voltage inverter and a high current to drive the system because it uses a 48 [V] battery. As shown in Fig.4, to minimize the effect of voltage drop, MOSFET (VMM1500-075P) manufactured by IXYS was used. It has small on resistance (0.55[mΩ]), with ratings of 1500 [A] and 75[V] between drain and source for the switching device. To supply stable DC link voltage, six 63 [V], 3300 [uF] capacitors in parallel was used. The efficiency of the developed inverter was 98.5[%].

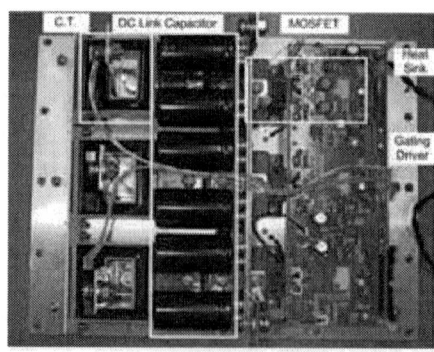

Fig. 4: Inverter configuration

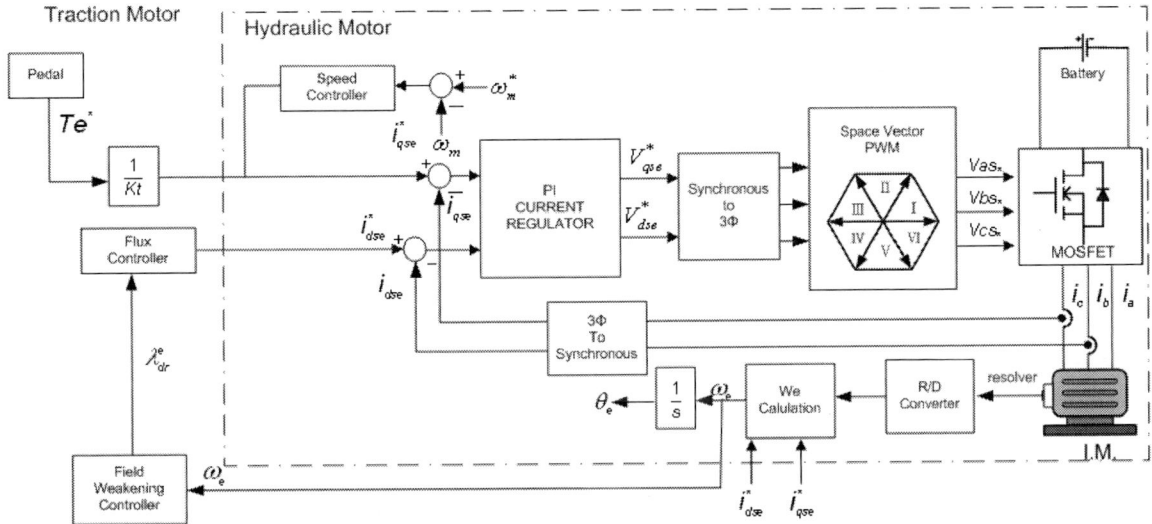

Fig. 5: Control system block diagram

III. CONTROL ALGORITHM [3][4][5][6]

As shown in Fig.6, the total system consists of two induction motors, two PWM inverters, and controllers. For the main controller 50 [MHz] DSP (TMS320C32) was used.

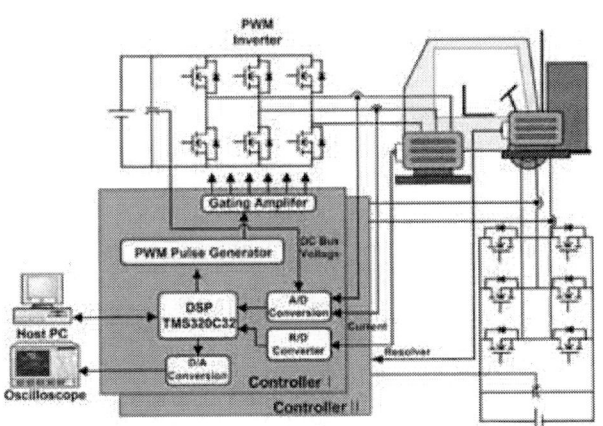

Fig. 6: System configuration

For high-performance control of induction motors, it is best to use the vector control method, which can divide stator currents into torque and flux components and independently control them. For current control, a synchronous reference frame PI current controller, and SVPWM as the voltage modulation method were used.

When the traction motor system is at its maximum speed (15[km/h]), the motor speed is around 2700[rpm]. Because the motor's rated speed is 1758[rpm], a field weakening control is needed to drive it at the maximum speed. A hydraulic motor needs the speed control of the motor to control oil flow on two levels. The whole control system for two motors are shown in Fig.5.

The hydraulic motor system only operates in a constant torque region, while the traction motor system operates up to the field weakening region I shown in Fig.7.

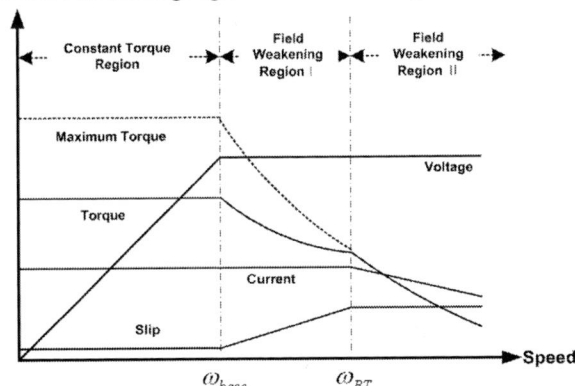

Fig. 7: Induction motor characteristic curve

In the whole drive system, while the traction motor is operated in torque control mode, the hydraulic motor is operated in speed control mode.

A traction motor is controlled by torque commands through the pedal, and forward or backward movement by the input of the direction key. If there is no input from the pedal while running, it operates in regeneration mode. The hydraulic motor drive system controls the speed of the motor on two levels by receiving input from a hydraulic lever or tilt lever. When the hydraulic lever or the tilt lever is set to level 1 (light load), the hydraulic motor operates at low speed (1800[rpm]). On the other hand, when the levers are set to level two (heavy load), it operates at high speed (2200[rpm]). In standby mode, it rotates at 350[rpm] for handle operation, etc.

B. Vector Control strategy

For high performance torque control, the indirect vector control was applied to both the traction motor and the hydraulic motor. Indirect vector control scheme controls slip angular frequency (ω_{sl}) through angular frequency (ω_e) so as to control the distribution ratio of the stator current. Thus, indirect vector control indirectly controls rotor flux so that it will exist only on the d axis ($\lambda_{qr}^e = 0$). This in turn controls the magnitudes and relative positions of flux and torque currents.

The slip angular frequency is as follows:

$$\omega_{sl} = \frac{1}{T_r} \frac{i_{qs}^e}{i_{ds}^e} \qquad (4)$$

where T_r is rotor time constant.

C. Field Weakening Operation

For high speed operation of the traction motor, a field weakening control was included. A vector controlled induction motor is convenient for the field weakening because it can easily reduce the flux current according to rotor speed.

The voltage the inverter can apply to the motor is limited by the DC link voltage and the PWM method, and the current is limited by the inverter switching device's rating and the thermal rating.

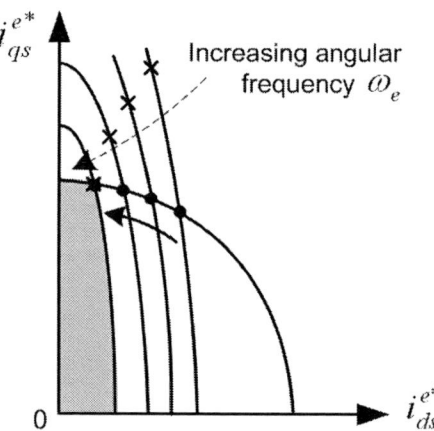

Fig. 8: Voltage-limit and current-limit boundary

As shown in Fig.8, the optimal operating point, which meets the above two limits and maximizes output torque at the same time, is given by the intersection of current and voltage limits. The optimum reference current is as shown in Eq. (5) and (6). As the speed increases, the reference current i_{ds}^{e*} decreases and i_{qs}^{e*} increases.

$$i_{ds}^{e*} = \sqrt{\frac{(V_{smax}/\omega_e)^2 - (L_2 I_{smax})^2}{L_{smax}^2 - L_s^2}} \qquad (5)$$

$$i_{qs}^{e*} = \sqrt{I_{smax}^2 - i_{ds}^{e*2}} \qquad (6)$$

D. Synchronous reference frame PI current regulator

The current, voltage and back EMF in a synchronous reference frame are all DC quantities. Through reference frame transformation, we can control the currents of an induction motor in the same way as a DC motor. To improve current control performance, back EMF compensation was added to the current regulator.

E. Flux controller

Because in transient state, the d axis rotor flux(λ_{dr}^e) and stator current(i_{ds}^e) are in linear delay where there is a delay of the rotor time constant. The rotor flux magnitude does not instantly change in proportion to i_{ds}^e in this transient state where flux changes. Therefore, when the control of rotor flux magnitude is required such as field weakening, a PI flux controller needs.

F. Speed controller

A PI speed controller was used to control the speed of the hydraulic motor. The bandwidth of the speed controller is limited by the bandwidth of the current regulator and the sampling frequency of the speed controller. The bandwidth of the current controller is 1000[rad/sec], and the period of the speed controller is 2 msec. Therefore, the bandwidth of the speed controller is selected at 12 [rad/sec].

IV. SIMULATION AND EXPERIMENT RESULT

A. Simulation Result

Fig.9. show the traction motor control model, performed using MATLAB/SIMULINK.

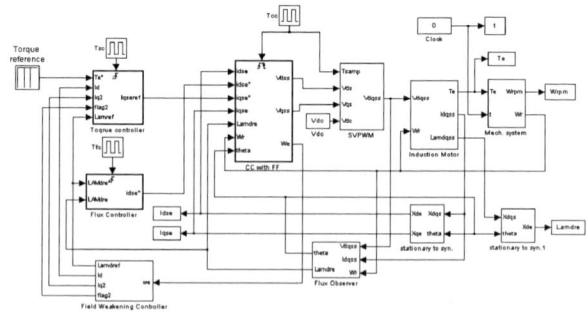

Fig. 9: Traction motor control model, simulated using MATLAB/SIMULINK

The parameters used for this simulation are shown in Appendix A.

The transient response of the rotor flux at the speed increased, under no-load condition, are shown in Fig. 10.This figure indicates that field weakening control start at 1716[rpm].

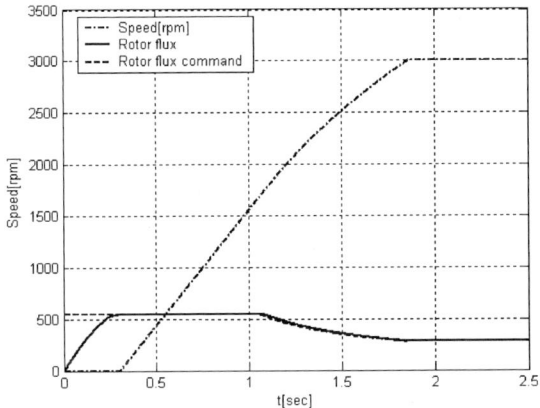

Fig. 10: Simulation result

Fig.11 represents the speed response for two levels of speed command for the hydraulic motor. (level 1: 1800[rpm], level 2: 2200[rpm]).

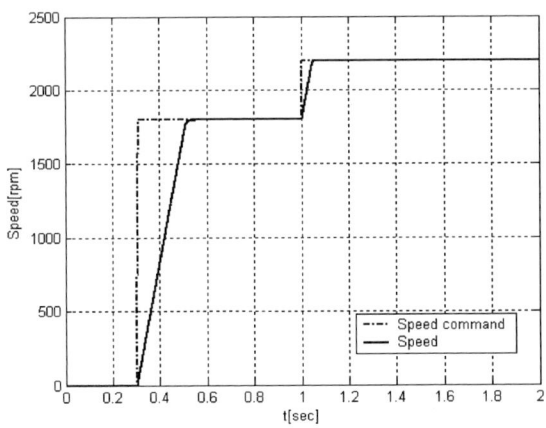

Fig. 11: Simulation result

B. Experimental Result

The designed induction motor was mounted in a 2.5 ton electric forklift and experimented by applying the developed algorithm.

Fig. 12: Actual electric forklift test

The electric forklift shown in Fig. 12 was used to test.

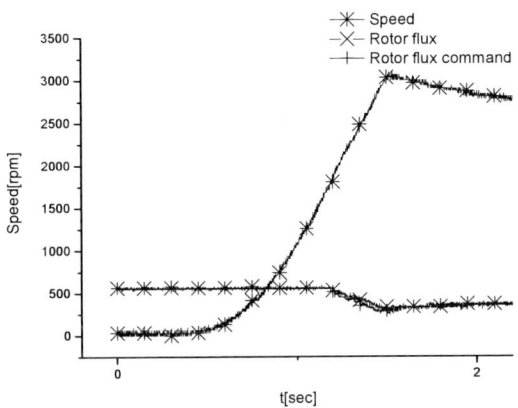

Fig. 13: Experimental result

Fig.13 shows the experimental result of the acceleration characteristics of the rotor flux at the speed increased, the rotor flux is decreased in field weakening region.

TABLE 6 summarizes the running test result for electric forklift under rated conditions.

TABLE 6
Running test result for electric forklift

Maximum Speed [km/h]	15
Maximum Climbing Angle [°]	7
Current at Maximum Climbing Angle[A]	390
Maximum Lifting Load [kg]	2,400
Lifting Speed at Maximum Lifting Load[m/sec]	0.333
Inverter Efficiency [%]	98.5

V. CONCLUSION

This paper has developed a drive system for an electric forklift that used induction motors, applied it to a 2.5 ton electric forklift and conducted an experiment. Induction motors are more robust, cheaper, and have better high-speed operation characteristics than DC motors. This paper confirmed that using an induction motor improves the performance of electric forklifts.

Accurate measurements of parameters and detuning are essential to control a low voltage motor that uses a battery such as electric forklifts because motor parameters greatly affect control characteristics.

REFERENCES

[1] J.W. Ahn, H.S. Song, D.H. Kim "Design and Drive Performance of 6/4 SRM for Pallet Truck Application" *KIPE, KIPE Transaction*, vol. 9, no.1, Feb. 2004, pp. 25-29.

[2] Song Liwei, Zheng Wei, Zhu Chunbo, Liu Weiliang, Cheng Shukang, "The Research of Induction Motor Driving System for Hybrid Electric Vehicle", Proc. *IEEE Vehicle Power and Propulsion Conference*. VPPC '06, vol. 6, no. 8, Sept. 2006 pp. 1-4.

978-1-4244-1871-8/07 $25.00 © 2007 IEEE

[3] S.H. Kim, S.K. Sul, "The Maximum Toque Control of an Induction Machine Drives in the Field Weakening Region," *IEEE Trans. Ind. Applicat.*, vol. 31, no. 4, July/Aug. 1994, pp. 787-794.

[4] D.M. Brod and D.W. Novotny, "Current control of VSI-PWM inverters." *IEEE Trans. Ind. Applicat.*, vol. IA-21, no.4, May/June, 1985, pp. 562~570.

[5] T.M. Rowan and R.J. Kerkman,"A New Synchronous Current Regulator and an Analysis of Current Regulated PWM Inverter," *IEEE Trans. Ind. Applicat.*, vol. IA-22, no.4, July/August, 1986, pp. 678~690.

[6] T.M. Rowan, R.J. Kerkman and D. Leggate, "A Simple On-line Adaptation for Indirect Field Orientation of an Induction Machine," *IEEE Trans. Ind. Applicat.*, vol. IA-27, no.4, July/August, 1991, pp. 720~727.

APPENDIX

A. Motor Parameters

	Traction	Hydraulic
Stator resistance [Ω]	0.0015	0.002
Rotor resistance [Ω]	0.0008	0.002
Stator inductance [mH]	0.365	0.310
Rotor inductance [mH]	0.365	0.310
Magnetizing inductance [mH]	0.35	0.297
Rated rotor flux [Wb]	0.055	0.05

A Monitoring System with Ubiquitous Sensors for Passenger Safety in Railway Platform

Sehchan Oh, Gilgong Kim, Hanmin Lee
Korea Railroad Research Institute
Advanced EMU Research Team
360-1, Woram-dong, Uiwang-si, Gyeongi-do, Korea.
Email: {soh, gdkim, hanmin}@krri.re.kr

Abstract— In this paper, we propose a passenger monitoring system with ubiquitous detection sensors for railway platform safety. Objectives of the system are to perceive dangerous situation in platform, such as fallen passenger on the track, and to inform central control room and local station employee and train driver about the situation. The proposed system consists of a series of sensors, such as infrared sensors, video and thermal cameras. The information acquisition unit detects and perceives dangerous factor, such as fallen passenger, disastrous fire and so on, in the monitoring area by using multiple vision and location sensors. The central data fusion unit makes more intelligent and meaningful information by using inputted the monitored results from every single camera sensor and infrared sensor for analyzing the situation. The information multicasting service unit provides different clients, such as local station employee, CCR employee and train driver with corresponding alarm message including standard operation procedure with video information about the accident situation in order to deal promptly with emergencies.

I. INTRODUCTION

Recently, advancements in information technology have enabled applying vision sensor to railway environments, such as CCTV(Closed Circuit TV). The objective of current CCTV in railway station is monitoring of passenger's movements, getting on/off train and station automation facilities. However the CCTV is a passive system that provides limited capability to maintain safety from boarding platform. Therefore when emergency situation arises, immediate recognition and response to the situation is difficult.

In this paper, we propose a passenger monitoring system with ubiquitous detection sensors for railway platform safety. Objectives of the system are to perceive dangerous situation in platform, such as fallen passenger on the track, and to inform central control room and local station employee and train driver about the situation.

As shown in Fig. 1, the proposed system consists of a series of sensors, such as infrared sensors, video and thermal cameras. Each sensor monitors its own preset monitoring area in real-time whether any dangerous factor occurs in that area or not. The central data fusion unit generates more intelligent and meaningful information by using inputted the monitored results from every single sensor.

Moreover, the system uses stereo vision technique to improve detection rate.

Fig. 1: Monitoring System with Ubiquitous Sensors

The Korea Railroad Research Institute has been developed a platform safety system as a government research and development project since 2005. The goal for this year is establishing the basic system configuration and

This paper introduces overall system overview, configuration and detection process of the system. The rest of this paper is organized as follows. The overall system configuration is described in Chapter 2. In Chapter 3, proposed detection process and algorithm are presented. We evaluate the proposed system according to the experimental results in Chapter 4. Finally, some concluding remarks and possible extension to the proposed scheme are mentioned in Chapter 5.

II. SYSTEM CONFIGURATION

As shown in Fig. 2, the proposed system can be divided into information acquisition unit, central data fusion unit and information multicasting unit.

The information acquisition unit detects and perceives dangerous factor, such as fallen passenger, disastrous fire and so on, in the monitoring area by using multiple vision and location sensors. In the system, each camera is responsible for 40 meter monitoring area and each infrared sensor covers 30 centimeter monitoring area. A detection processor of the stereo camera sensor conducts a series of process, i.e. train detection, object detection, object recognition and object tracking. A location sensor, infrared sensor catches detection and the position information of the object in its preset monitoring area.

The proposed system uses multiple vision and thermal cameras and infrared sensors for providing duality and high reliability in detection.

The fusion unit makes more intelligent and meaningful information by using inputted the monitored results from every single camera sensor and location sensor for analyzing

978-1-4244-1871-8/07 $25.00 © 2007 IEEE

the situation. According to the results from situation analysis, it generates different alarm messages for local station and CCR employees and train driver.

The information multicasting service unit provides different clients, such as local station employee, CCR employee and train driver with corresponding alarm message including SOP(standard operation procedure) for employees with video information about the accident situation in order to deal promptly with emergencies.

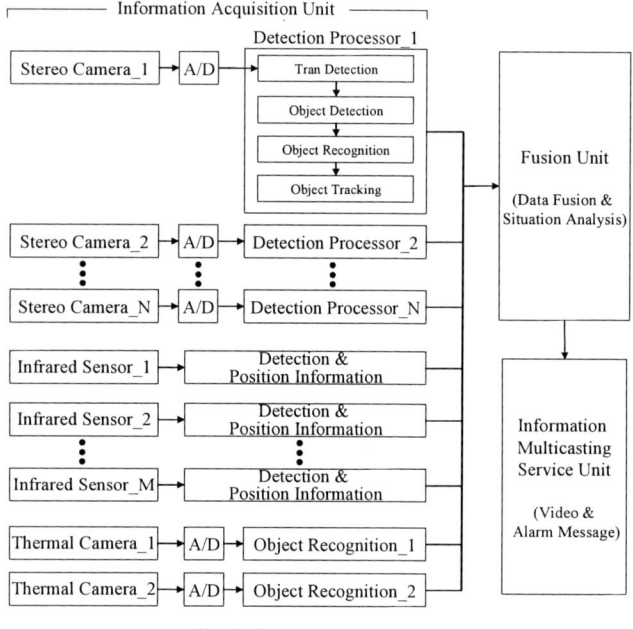

Fig. 2: System Configuration

III. DETECTION PROCESS

The detection process includes four main sub-processing steps, i.e. train detection, object detection, object recognition and object tracking. Fig. 3 shows the detection process of the proposed system.

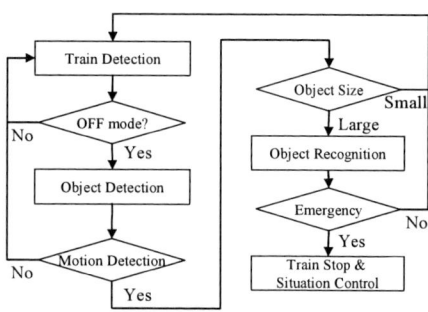

Fig. 3: Detection Process

A. Train Detection

To make a right decision of dangerous factor for fallen object in monitoring area, it is very important to find the accurate train states in the area. Finding a train state using

only location sensor is quite simple but it is difficult to distinguish between moving state and stop states of the train. Therefore, the proposed system uses camera sensor for finding train states in current monitoring area with combining the detection results of location sensors.

The train state for each stereo camera sensor can be defined as shown in table 1. The object/human detection performs in OFF modes, i.e. a state train does not exist in the monitoring area. Train movement can be defined with a series of process, i.e. frame difference, thresholding, labeling and merging.

TABLE I
Train States for Each Camera Sensor

Train States	Description
OFF	There is no train in the monitoring area
IN	Train is approaching
ON	Train is stopped
OUT	Train is pulling out of the monitoring area

The four states of train using camera sensor can be decided as shown in Fig. 4.

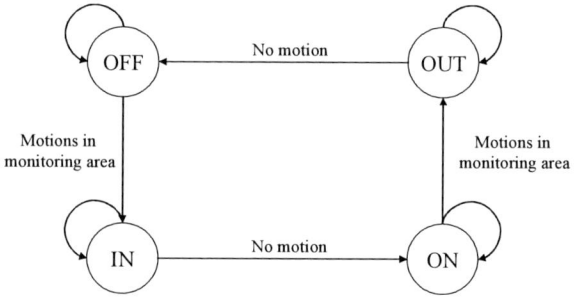

Fig. 4: System Configuration

The proposed system has four different transitions. Transition of one from another state can be defined as described in Table 2. The system uses motion of train in each monitoring area to determine the state change. The transitions OFF to IN and ON to OUT are made when more than five consecutive motion frames are occurred. To ignore the noise effect, more than five consecutive frames are analyzed.

TABLE II
Transition of Train States

Transition	Description
OFF-IN	More than five consecutive motion frames are occurred
IN-ON	More than five consecutive no-motion frames are occurred
ON-OUT	More than five consecutive motion frames are occurred
OUT-OFF	More than five consecutive no-motion frames are occurred

B. Object Detection/Recognition

The detection results of dangerous factor in platform monitoring area mainly classified two situations, i.e. a fallen object in the area and a sudden global change on the lighting conditions. To determine fallen object, the proposed system considers only movements in monitoring area in OFF state.

When an object is fallen in the monitoring area, we can observe it has fast movement while falling and stopped for a while after fallen. If a fallen object is in the preset dangerous area, the system find out whether it is a human or not by using a stereo vision technique. The system calculates height and width of the fallen object by using the stereo vision algorithm.

The process of stereo vision is described in Fig. 5. In the preprocessing step, we use DC-notch filter to compensate left and right images for different light condition effect.

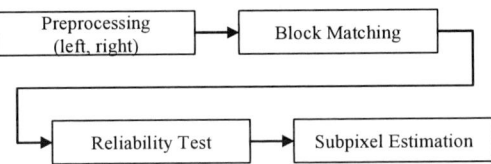

Fig. 5: Stereo Vision Process

The DC-notch filter can be described as a following expression.

$$g(x, y) = f(x, y) - \bar{f}(x, y) \qquad (1)$$

where $f(x, y)$ means an original image, $\bar{f}(x, y)$ means a partly averaged image and $g(x, y)$ is compensated image by DC-notch filter.

Basic block matching algorithm divides input image into fixed squared blocks and seeks for the best-matched block from the reference frame within search area. The most common matching criteria include sum of absolute differences (SAD) between the pixels in the windows and sum of squared differences (SSD).

$$SAD(x, y, d) = \sum_{i, j=-n}^{n} \left| L(x+j, y+i) - R(x-d+j, y+i) \right| \qquad (2)$$

where the block size is 2n+1, d means disparity, L and R represent left and right image respectively.

We can rearrange the above expression as below one to decrease calculation time.

$$SAD(x, y, d) = SAD(x, y-1, d)$$

$$+ \sum_{j=-n}^{n} \left| L(x+j, y+n) - R(x-d+j, y+n) \right|$$

$$- \sum_{j=-n}^{n} \left| L(x+j, y-n-1) - R(x-d+j, y-n-1) \right| \qquad (3)$$

We define a function U as follows:

$$U(x, y, d) = \sum_{i, j=-n}^{n} \left| L(x+j, y+i) - R(x-d+j, y+i) \right|$$

$$- \sum_{i, j=-n}^{n} \left| L(x+j, y+i) - R(x-d+j, y+i) \right| \qquad (4)$$

The process of calculation SAD is described as U function.

$$U(x, y, d) = U(x-1, y, d) \qquad (5)$$

$$\left| \begin{array}{l} L(x+n, y+n) - R(x-d+n, y+n) \\ L(x+n, y-n-1) - R(x-d+n, y-n-1) \\ L(x-n-1, y+n) - R(x-d-n-1, y+n) \\ R(x-d-n-1, y-n-1) \end{array} \right|$$

The process of reliability test uses disparity to find out a degree of reliability for each left and right images.

IV. EXPERIMENTAL RESULTS

To verify the performance the proposed system, we acquired the test sequences at aboveground Sungnae station in line number 2 and underground Heyhwa station in Seoul Metro line number 4 of Korea. A frame of the test video sequence for each station is presented in Fig. 6.

(a)

(b)

Fig. 6: Test Video Sequences; (a) a video sequence of Heyhwa station, (b) a video sequence of Sungnae station

A. Train Detection

To make decision of dangerous factor for fallen object in monitoring area, it is important to find the accurate train states in the area for every single camera. Train and dangerous areas have to be clearly defined. The preset train area and dangerous area is described in Fig. 7. The blue-lined box presents train area and red-lined box shows dangerous area.

Fig. 7: Train area and Dangerous area

978-1-4244-1871-8/07 $25.00 © 2007 IEEE

Train movement is defined as a series of process, frame difference, thresholding, labeling and merging. Fig. 8 shows the experimental result of train movement.

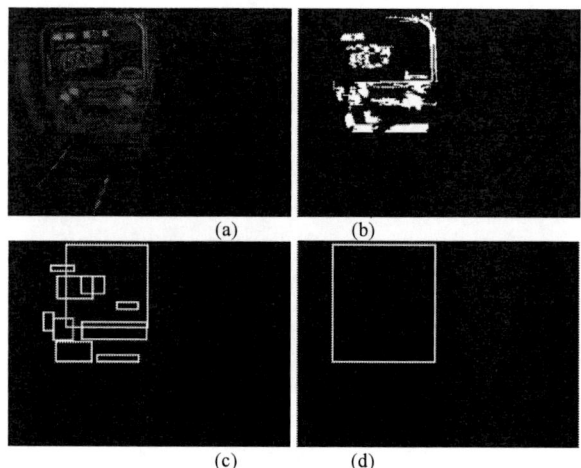

Fig. 8: Train Movement (a) result of frame difference; (b) result of thresholding; (c) result of labeling; (d) result of merging

The experimental results of train detection for the test sequences, Heyhwa and Sungnae, are presented in Fig. 9 and Fig 10. In the figure, blue-lined rectangle means train area, and redness is dangerous area. Train has relatively large potion of area in the picture. Therefore, the system regards changing relatively large potion of scene as movement of train, and ignores sudden change of light condition. To minimize noise effects, the system regards a movement of train when more than five consecutive frames have been changed.

According to the experimental results, we can see the system completely detects all of the train states with image processing technology.

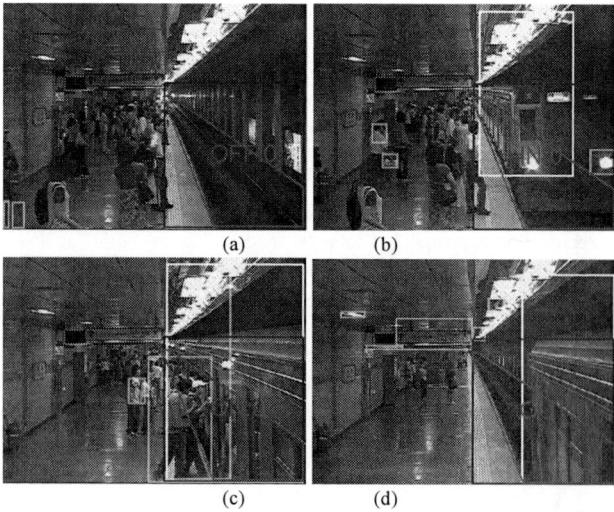

Fig. 9: Experimental results of train state transition for test sequence of Heyhwa station; (a) OFF state; (b) IN state; (c) ON state; (d) OUT state

Fig. 10: Experimental results of train state transition for test sequences of Sungnae station(a) OFF state; (b) IN state; (c) ON state; (d) OUT state

B. Object Detection/Recognition

When an object movement is detected in the monitoring area, the system can determine whether the object is fallen from platform or not and also find out it is a human or not. To determine fallen object the proposed system considers only movements in monitoring area in OFF state. In the figure, blue-lined rectangular is the area of train and red-lined polygon represents the dangerous area, and green rectangular line shows object bound. When object is in dangerous area, the system determines whether the movements come from outside dangerous area. Moreover, the system checks that the object bound is completely included in the dangerous area. If object bound is completely included in the dangerous area, system regards it as a fallen object and changes the object bound color from green to red.

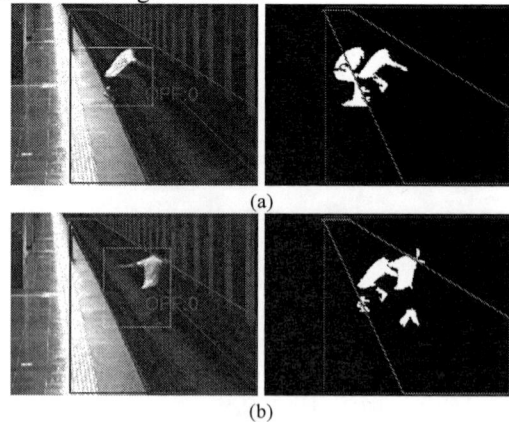

Fig. 11: Experimental results of object detection; (a) object is not completely included in the dangerous area; (b) object is completely included in the dangerous area

Fig. 12 shows a result of object recognition by using stereo matching. Fig. 13 shows the test result a series of stereo matching results for various object types. We have tested the system in various conditions using a news paper, dolls, real human and so on. According to the test result, lighting condition affects the results of stereo matching. If good illumination condition is provided we can reduce stereo matching errors.

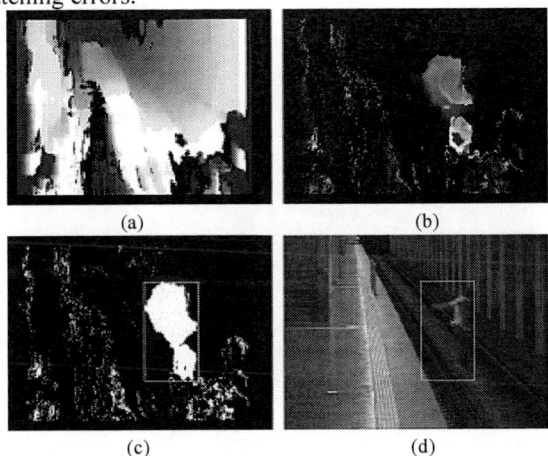

Fig. 12: Experimental results of object recognition by using stereo matching (a) disparity for background image, (b) difference image (c) object recognition by labeling (d) recognized object

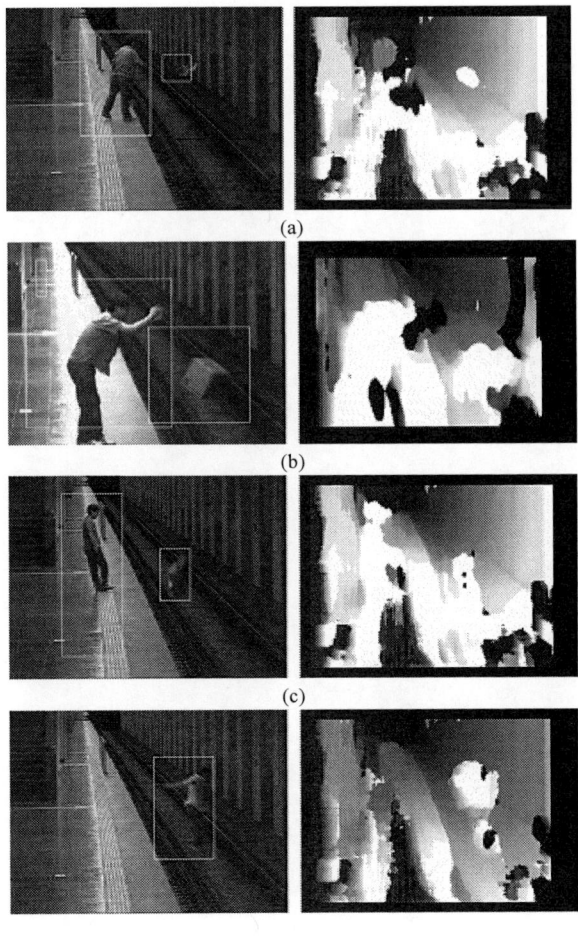

Fig. 13: Results of stereo matching for various objects; (a) case of a news paper, (b) case of a box, (c) case of a doll, (d) case of a human

Fig. 14 shows the experimental results of object detection and recognition using thermal camera. Output image of the thermal camera is less affected by illumination condition. Therefore, to use thermal camera is more suitable for object detection and recognition in insufficient lightning condition.

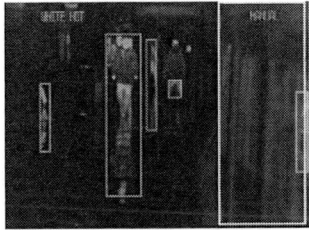

Fig. 14: Experimental result of object detection and recognition by using thermal camera

V. CONCLUSION

In this paper, we propose a passenger monitoring system with ubiquitous detection sensors for railway platform safety. The proposed system consists of a series of sensors, such as infrared sensors, video and thermal cameras. The information acquisition unit detects and perceives dangerous factor, such as fallen passenger, disastrous fire and so on, in the monitoring area by using multiple vision and location sensors. The central data fusion unit makes more intelligent and meaningful information by using inputted the monitored results from every single camera sensor and infrared sensor for analyzing the situation. The information multicasting service unit provides different clients, such as local station employee, CCR employee and train driver with corresponding alarm message including standard operation procedure with video information about the accident situation in order to deal promptly with emergencies.

We verify the system performance with experimental result in real condition. Detection of train state and object is conducted robustly by using proposed image processing algorithm.

Currently, we are pursuing an effective information transmission system for immediately dealing with the safety accidents.

REFERENCES

[1] I.Yoda, K.Sakaue. "Ubiquitous Stereo Vision for Controlling Safety on Platforms in Railroad Station," IEEJ Tr. on Electronics, Information and Systems, Vol. 124, No. 3, Mar., pp.805-811, 2004.

[2] F.Kruse, S.Milch, H.Rohling. "Multi Sensor System for ObstacleDetection in Train Applications," Proc. of *IEEE Tr.*, June, pp.42-46, 2003.

[3] Y.Sasaki, N.Hiura. "Development of Image Processing Type Fallen Passenger Detecting System, " JR-EAST Technical Review Special Edition Paper, No. 2, pp.66-72, 2003.

[4] J. Vhzquez, M. Mao, "Detection of moving objects in railway using . . vision," IEEE Intelligent Vehicles Symposium University of Parma, Parma, Italy Jun. 1447, 2004.

[5] N. Paragios and V. Ramesh. An MRF-based approach for real-time subway monitoring. In IEEE Conference on Computer Vision and Pattern Recognition, 2001.

[6] I. Yoda, "Image processing technology for advanced safety to people in railroad transportation - For railroad crossing and station platform ," IPSJ Magazine Vol.48, No.1, pp.10-16, Jan. 2007.

[7] I. Yoda, D. Hosotani, and K. Sakaue, "Multi-point Stereo Camera System for Controlling Safety at Railroad Crossings," Proc. of the IEEE International Conference on Computer Vision Systems, 2006.

[8] Shigeki Sugimoto, Hayato Tateda, Hidekazu Takahashi, Masatoshi Okutomi, "Obstacle Detection Using Millimeter-Wave Radar and Its Visualization on Image Sequence," icpr, pp. 342-345, 17th International Conference on Pattern Recognition (ICPR'04) - Volume 3, 2004

Development of Water-Cooled Heat Sink for High-Power IGBT Inverter

Minsub Han
Department of Mechanical Engineering
University of Incheon
177 Dohwa-dong, Nam-gu, Incheon, Korea 402-749
Email: mhan@incheon.ac.kr

Su-Dong Lee† , Chanook Hong, Chun-Suk Yang
and Kyung-Seo Kim
CENTRAL R&D CENTER, LS INDUSTRIAL SYSTEMS CO., LTD.
533 Hogye-dong, Dongan-gu, Anyang-si
Gyeonggi-do, Korea 431-080
† Email: sdlee1@lsis.biz

Abstract— We present the development of a water-cooled heat sink that provides reliable thermal performance for high-power IGBT inverter. The development process comprises three stages. In the concept design, the thermal performances of two design proposals are considered. The thermal system of each design is particularly analyzed using the compact model. In the detailed design stage, specific dimensions of the heat sink are determined considering the design options under given external restrictions and the results from three-dimensional heat transfer analysis. The prototype of the resultant design is made and tested on the rig for final confirmation. We emphasize the relevant use of the thermal analysis on each stage and also discuss various practical issues involved.

I. INTRODUCTION

Inverter is the device that converts the magnitude and frequency of electric voltage, and it is one of the essential elements in driving electric motors of various purposes by controlling the speed of the motor. The power module in the inverter is increasingly based on IGBT (Insulated Gate Bipolar Transistor) and FWD (Free Wheeling Diode). IGBT inverter usually generates significant amount of heat due to the operational power loss and requires immediate release of the heat to insure good performance. Then, the IGBT inverter needs an additional system that transfers the heat efficiently into environment. Heat sink is commonly found in any electrical device that needs cooling and does the efficient job of transferring the energy from the electric circuits to coolant. The high-power inverters of several Megavolts or those for electric vehicles, which also need miniaturization, require the cooling system of high efficiency and reliability and often introduce the water-cooled heat sink system.

The primary goal in the design of the cooling system of inverter is to maintain the temperature of the transistor junction in the power module below the maximum allowable temperature for normal operation. In the purpose, the heat sink and coolant system need to have enough capacity to exit the heat quickly. The heat pathways from the power module to the coolant also need to be optimized. The relevant issues are, for example, the optimum packaging of modules, minimization of the contact resistance and special arrangement for locally over-heating region.

The optimized cooling system does not only enhance the reliability of inverter. It also minimizes the capacity of heat sink and coolant and, therefore, makes the inverter design more compact. The design reduced in size and weight can significantly contribute, for example, to the performance or the packaging in engine room of electric vehicle.

We here present the development of a water-cooled heat sink that can provide a reliable thermal performance for high-power IGBT inverter. The development process is composed of three stages. In concept design, the thermal performances of the competing designs of inverter are comparatively analyzed by using the compact thermal model. We then develop a detailed design of heat sink by assessing various design parameters using the three-dimensional heat transfer models. The prototype of the final design is made and tested.

II. GENERAL DEVELOPEMNT PROCEDURE

The design process of cooling system may roughly be divided into three stages: concept design, detailed design and test. In the stage of concept design, some fundamental decision on the heat sink design is made according to the requirement of a target inverter. In this stage, it is useful to have some information on performance of the system in consideration even though the system details are not available. It may be provided by so-called compact model that focuses more on the overall performance of the system using simplified thermal models. For example, the three dimensional features like lateral distribution of thermal properties or local thermal pathways are not considered. The compact model makes it possible to analyze overall characteristics of the cooling system quickly with a reasonable accuracy before detailed design decisions are made in the concept stage.

978-1-4244-1871-8/07 $25.00 © 2007 IEEE

After major design concept on the power module inside the inverter are decided such as the module specifications, operating condition and module arrangement, the detailed dimensions of heat sink are determined. Above all, the dimensions and distribution of fins in the heat sink are determined according to the system requirement and the restrictions imposed by packaging and manufacturing issues. The three dimensional modeling process usually proceeds simultaneously. 3-D CAD (Computer Aided Design) model is produced for the uses of drawing, thermal modeling and prototype production. Also, 3-D thermal modeling is performed that includes three dimensional shape and arrangement of almost all the parts participating in the heat transfer. The 3-D model provides a more accurate picture on the thermal performance of a heat sink design that can be used in decision of the final draft for prototype production.

Finally, the prototype is produced and the test for thermal reliability is performed. Some minor modification of the design is made if necessary according to the test results. However, a major change in design would cost significantly on this stage and, therefore, relevant decisions should be made in development stage as early as possible.

III. RESULT

A. Concept Design

One of the primary design decisions to be made at the concept stage was on the power module configuration. We considered inverter systems of two different configurations. One consists of two IGBT modules of a certain type and the other consists of three modules of a different type (See Table 1). While there were other important issues regarding the decision such as inverter design and packaging, we were also interested in thermal performances of the two competing systems. Since it was an early stage of the development, the basic data for accurate analysis was not available. Also, later stages in the development, especially prototype production, usually take up a large portion of the total development period, and a comprehensive modeling of the thermal system is not practical. We instead built a compact thermal model that can be rapidly modeled and is suitable for relative comparison of the systems (Fig. 1a).

The amount of power loss is obtained that is based on the rated current of inverter using the computational tool supplied by the module manufacturer (See Table 1 & 2). The compact model consists of thermal impedances of the power module, thermal grease and heat sink. The data on the power module and thermal grease are provided by the manufacturers. The heat transfer model of the heat sink is based on the theoretical correlations of Sparrow et al [1]. The typical values for the model system are supplied in Table 2.

TABLE 1
Two Cases for Compact Model Analysis

	Power Loss per Module: steady state (kW)	Power Loss per Module: excess state (kW)	No. of Module per Inverter	Total area of modules (m^2)
Case I	0.83	1.68	2	0.040
Case II	0.35	0.71	3	0.027

TABLE 2
Comparison of Model Properties of the Two Thermal Systems

		Case I		Case II		Case II /Case I (IGBT)
		IGBT	diode	IGBT	diode	
1. Power Module Spec.	Power loss of each unit (W)	96.7	42.1	42.5	16.1	0.4
	R junciton to case (K/W)	6.0E-02	1.0E-01	1.3E-01	2.4E-01	2.2
	R case to heatsink (K/W)	4.8E-02	8.0E-02	1.0E-01	1.8E-01	2.1
2. Inverter Configuration	Number of modules	2		3		
	Unit number per module	6	6	6	6	
	Module power loss per unit area (kW/m^2)	42.1		38.4		0.9
	Total power loss	1.7		1.1		
	Total area of modules (m^2)	4.0E-02		2.7E-02		0.7
3. Thermal grease	R grease (K/W)	8.0E-03		1.7E-02		2.2
4. Heat sink	R heatsink to coolant (K/W)	4.1E-02		8.9E-02		2.2

978-1-4244-1871-8/07 $25.00 © 2007 IEEE

The temperature distribution results from the compact model analysis are compared in Fig. 1b. According to the results, the IGBT junction temperatures are given as 116.3 and 112°C for Case I and II, respectively. The temperatures are all far less than those of malfunction of the transistors in IGBT even though their accuracy are dependent on the accuracy of thermal impedances of power module and heat sink. On the other hand, the relative comparison of the two cases is more meaningful. The results indicate that Case II does a better thermal performance of the two. Or the performances of the two systems are at least comparable. The two systems are different from each other only in the configurations of power module, which is therefore the direct source of the difference. It is noted in the figures of Table 2 that the module power loss per unit area of Case II is about 10 per cent lower than that of Case I. This indicates that the system of Case II is required to transfer about 10 per cent less amount of energy into the unit area of heat sink and therefore results in a less thermal load.

The thermal performance results from the compact model analysis along with other related issues like inverter system configuration were considered in the inverter design. The results were also used in the initial phase of heat sink design, for example, when the thermal load and fin dimensions of the heat sink are to be determined tentatively.

B. Detailed Design

The detailed design of the heat sink was developed first by considering the design options available under the given restrictions. We built 3-D CAD model based on the considerations. 3-D heat transfer analysis was also performed for checking the overall thermal performance with a more realistic model of heat sink and some local 3-D effects unresolved in the compact model.

Figure 2 shows a heat-sink design. The heat sink is located on the outer face of the case of inverter (Fig. 2a). The IGBT modules are attached to the inner face of the case (Fig. 2b). The coolant flow path is determined to supply sufficient amount of coolant to the strongly heated locations considering the restrictions, which are imposed by, for example, the positions of the inlet and outlet of coolant and the bolting locations for gasket and cover that are dependent on the packaging condition. The narrow fins are densely

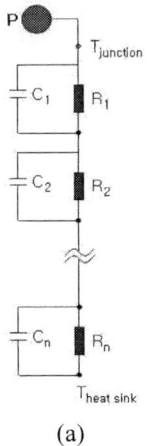

(a)

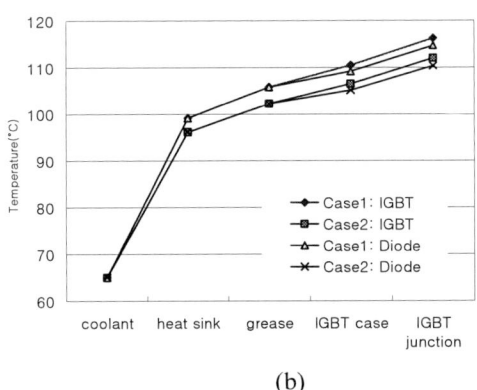

(b)

Fig. 1: (a) Compact Thermal Model and (b) Analysis Result

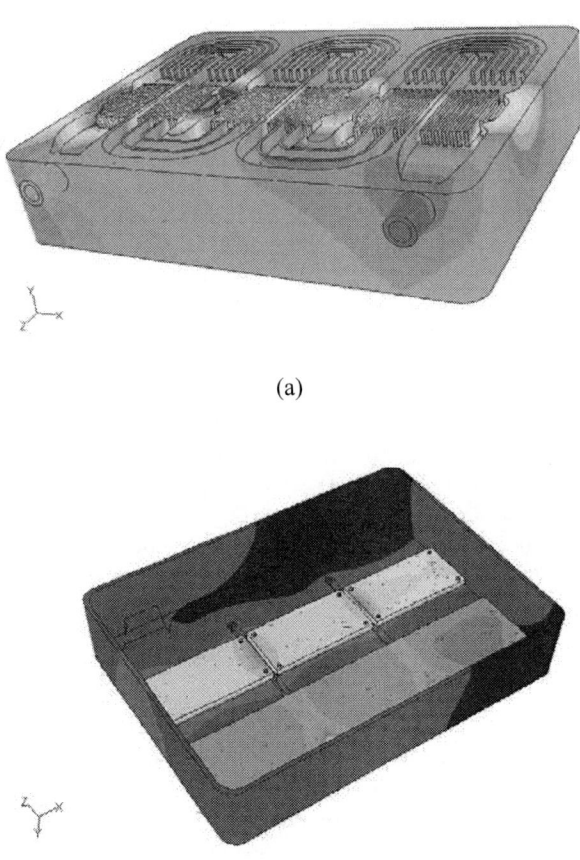

(a)

(b)

Fig. 2: 3D thermal analysis result (a) heat sink outside the inverter case
(b) IGBT locations inside the case

978-1-4244-1871-8/07 $25.00 © 2007 IEEE   297

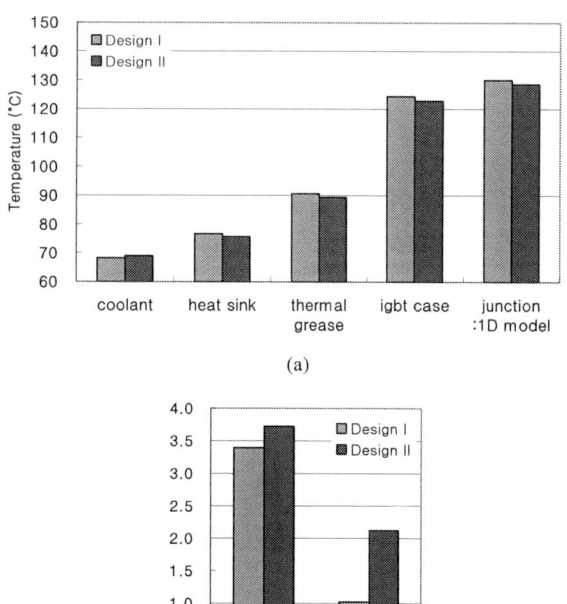

Fig. 3: 3D thermal analysis result (a) temperature distribution, (b) coolant temperature rise and pressure drop

populated in the middle where three IGBT modules are located. The rest of the flow paths do not have fins except when the flow needs to be guided. The pitch and thickness of fins are decided to maximize the heat transfer under the restriction imposed by the manufacturing issues.

The 3-D heat transfer model resolves the three dimensional heat flow from the IGBT case to heat sink by solving the energy equation and Navier-Stokes equation using a commercial package for computational fluid dynamics (CFD) [2, 3]. The convective heat transfer inside the pathways of coolant is explicitly solved while a typical value of heat transfer coefficient in natural convection is applied on the faces in contact with air. The conductive heat transfer in the heat sink and thermal grease are explicitly solved. The conduction inside IGBT module is not modeled with three-dimensional details and the compact model is used instead. The detailed thermal model of IGBT may be critical in the design of IGBT but less so in the heat sink design. It is pointed out that the thermal resistance values in the compact model provided by manufacturers are obtained by the standardized experimental methods, the results from which are usually relevant only when the cooling capacity of the heat sink is sufficient.

Figure 3a and 3b show the results of 3-D heat transfer analysis for two different designs. There is little difference between the two regarding the overall design of heat sink except for fin density. Design II has smaller fin pitch and thickness than Design I in the region of IGBT. The case in

consideration is the excess state where the maximum power loss is sustained steadily (See Table 1). The temperature distributions indicate that the coolant is sufficiently supplied to the entire region of strongly-heated area. The transistor junction temperatures are all below that of the operational failure (150℃) with large margin. Design II has a temperature 0.8℃ less than Design I at the junction. The variations between the three IGBT modules are less than 5℃.

TABLE 3
Comparison between the test result and the thermal analysis

	Coolant	Heat Sink pt #1.	Heat Sink pt #2.	IGBT Case
Test (°C)	30	35.8	35.9	41.4 (NTC Thermister)
3D model (°C)	30	34	34.6	37.2

C. Test

The prototype of the heat sink based on Design II in the section *B* was produced. It is made of aluminum using CAM machining by Machining Center.

Next, the test rig was set up for inverter performance testing. The test configuration for thermal measurement is shown in Fig 4. The inverter is run with constant voltage/frequency ratio, and a passive RL load is applied. The coolant is supplied at constant flow-rate and temperature. Temperatures were measured at more than dozen locations around the heat sink with T-type thermocouple, and the temperature of the NTC thermocouple embedded in IGBT is also monitored. A typical run takes about 30 minutes before all the monitored data are stabilized.

The measured data in the case of coolant temperature of 30℃ are provided in Table 3. While those at the heat sink and thermal grease are relatively easy to collect, the temperatures inside the IGBT module are tricky to evaluate. The thermister embedded in the module may closely represent the temperature of the module case. The measured data are also compared with the 3D simulation results in Table 3. More results on the test will be provided in future publication.

IV.   CONCLUSION

A water-cooled heat sink for high-power IGBT inverter is developed that shows a reliable thermal performance. We have presented the development process underscoring various practical issues involved. Careful considerations on the typical design conditions like main requirement, design options and engineering restrictions were made. At each stage of the development, we performed the thermal analysis resorting to the methods that reasonably satisfy the balance between the development schedule and cost and the model

978-1-4244-1871-8/07 $25.00 © 2007 IEEE   298

accuracy. More work is under way to reliably predict or estimate the temperature of the transistor junction of IGBT along with the optimization of the developed design.

ACKNOWLEDGMENT

M.H. gratefully acknowledges that this research is supported by LS Industrial Systems, Co., Ltd.

REFERENCES

[1] E.M. Sparrow, B.R. Baliga and S.V. Patankar, "Force Convection Heat Transfer from a Shrouded Fin Array with and without Tip Clearance," 1978, Journal of Heat Transfer, Vol. 100, pp. 572-579.
[2] F. P. Incropera and D.P. DeWitt, Fundamentals of Heat and Mass Transfer, 1996 John Wiley & Sons, New York.
[3] Fluent Inc., FLUENT 6.1 User's Guide, Lebanon, New Hampshire,

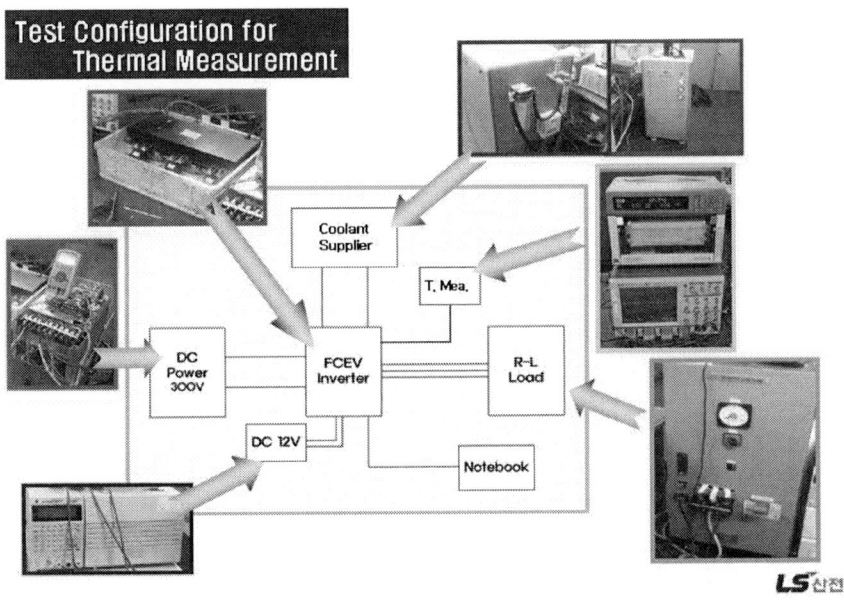

Fig. 4: Test Configuration for Thermal Measurement

2003.

The Research of Model Reference Adaptive Control of Static Var Compensator (SVC)

Zeng Guang[*,**] Zhang Lijuan[**] Wang Quanhai[**] Su Yanmin[*]

* Xi ' an Jiaotong University Xi ' an 710049 China
** Xi ' an University of Technology Xi ' an 710048 China
E-mail: g-zeng@mail.xaut.edu.cn   zljfengling@163.com

Abstract—This paper analyzed the composing and compensation principle of Static Var Compensator control system, and designed the model reference adaptive controller based on Liapunov stability theory. In the end, the dynamic and static voltage and current waveform under unbalance load condition was shown, and the feasibility of this controller was verified by experiment.

I. INTRODUCTION

The static dynamic Var compensator (SVC) is a parallel connection reactive compensating device in the end of 70s, which is applied widely in transmit electricity and distribution system at home and abroad. There are two sides of this apply in power system ,one is used to compensating reactive, and keeping the controlled node voltage on transmission line, the other is used to load compensation ,it kept voltage and correct power factor, and compensated unbalanced load, which has a capacity of unbalance compensatory[1][2].

The model reference adaptive control system is a adaptive system, which using input, state and output variable to measure one performance index, and according to comparative result of actual measurement performance targeted value and given performance targeted value, the adjustable system parameters are corrected or created a assistant input signal by adaptive system in order to keep system performance index is closed to given performance index. The model reference adaptive system is designed by the local parameter optimization, Liapunov stability theory and Popov hyperstable theory at present, because of the adaptive rule of local parameter optimization way easy lead to whole instability, but Popov hyperstable theory resolved some problems of Liapunov function just in some area, the area of application is limited, therefore, the paper applied Liapunov stability theory to design the model reference adaptive system [3][4].

II. SYSTEM COMPOSING AND COMPENSATION THEORY

A. System composing

Fig 1 is block diagram of the system structure; three-phase capacitor bank (electric power filter) and thyristor phase control reactor are linked in the transformer sublevel generatrix. The system hard core is controller, which include: testing circuit which tested required system variables and compensation variables, such as voltage and current sampling; control circuit which transmitted test signal in order to obtain required steady state and dynamic characteristics such as direct current account and model reference adaptive control; trigger circuit which generated corresponding trigger pulse by control signal of control circuit output.

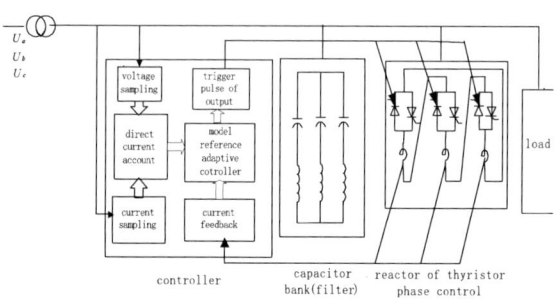

Fig. 1 The block diagram of systems structure

B. SVC compensation principle

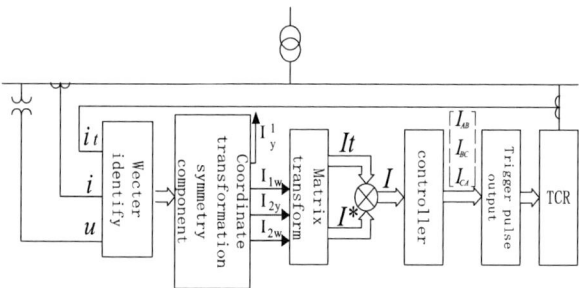

Fig. 2 Voltage and current control block diagram

We can know from fig 2, the sampled signal which is generatrix voltage, current and instantaneous value of TCR threshold current by vector identify, transformation of coordinates and symmetry component analysis, converted into positive sequence active current I_{1y}, positive sequence reactive current I_{1w}, negative sequence active current I_{2y}, negative reactive current I_{2w}. I_{1w} generated reactive power, I_{2y} and I_{2w} caused voltage fluctuation and flickering, they are all component must to eliminate. The expressions of TCR direct current I^* and feedback current I_t are gotten by matrix transform.

III. THE MODEL REFERENCE ADAPTIVE CONTROL SYSTEM IS DESIGNED BY THEORY

The transfer function of model reference adaptive control system as following:

The transfer function of controlled object is:

$$G_p(s) = \frac{Y_p(s)}{U(s)} = k_p \frac{N_p(s)}{D_p(s)} \tag{1}$$

The transfer function of reference model is:

$$G_m(s) = \frac{Y_m(s)}{R(s)} = k_m \frac{N_m(s)}{D_m(s)} \qquad (2)$$

Thereinto, $N_m(s)$, $D_m(s)$ is m steps and n steps first one coprime multinomial separately, the reference model is steady and minimum phased and $k_m>0$. The transfer function of controlled object is described by formula (1), it is set that relatively order is 2, the transfer function of reference model is described by formula (2), the relatively order of reference model is supposed to be 2, because the controlled object is 2. In this circumstance, $G_m(s)$ is not strict positive real. If a right multinomial $L(s)=s+a$ is inducted, we can make $G_m(s)L(s)$ to strict positive real as long as parameter a is right. The control system as shown in fig 3 is made up of an adjusted gain $K_n(t)$ and two assistant signal generator F_1 and F_2[5].

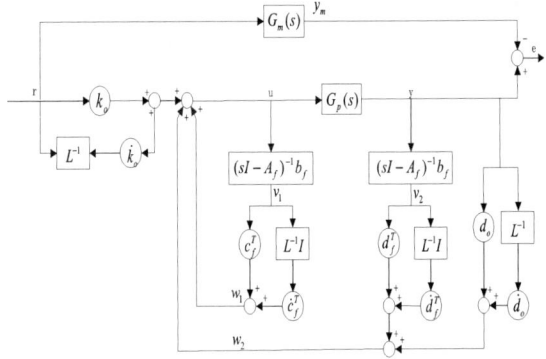

Fig.3 Block diagram of model reference

The state equation and corresponding transfer function of two signal generator in fig 3 are:

$$F_1: \begin{cases} \dot{v}_1(t) = A_f v_1(t) + b_f u(t) \\ w_1(t) = (c_f^T + \dot{c}_f^T L^{-1} I)v_1(t) \\ G_1(s) = c_f^T(sI - A_f)^{-1} b_f \\ \quad + \dot{c}_f^T L^{-1} I(sI - A_f)^{-1} b_f = \frac{N_c(s)}{D_f(s)} \end{cases} \qquad (3)$$

$$F_2: \begin{cases} \dot{v}_2(t) = A_f v_2(t) + b_f y_p(t) \\ w_2(t) = (d_f^T + \dot{d}_f^T L^{-1} I)v_2(t) + (d_0 + \dot{d}_0 L^{-1})y_p(t) \\ G_2(s) = d_0 + \dot{d}_0 L^{-1} + d_f^T(sI - A_f)^{-1} b_f \\ \quad + \dot{d}_f^T L^{-1} I(sI - A_f)^{-1} b_f = d_0 + \dot{d}_0 L^{-1} + \frac{N_d(s)}{D_f(s)} \end{cases} \qquad (4)$$

Thereinto, A_f is need to undetermined graduated stable matrix of $(n-1)\times(n-1)$, $D_f(s)$ is fist one Hurwitz multinomial, it is supposed that (A_f, b_f, c_f) and (A_f, b_f, d_f) is controlled standard type, which is to say the elements of $(n-1)$ dimension vector c_f and $(n-1)$ dimension vector d_f is the coefficients of $N_c(s)$ and $N_d(s)$, they are adjustable.

We can make the adjustable parameters vector of

adjustable system is:

$$\theta(t) = [K_0(t) \quad c_f^T(t) \quad d_f^T(t) \quad d_0]^T \qquad (5)$$

The signal vector is:

$$\phi(t) = [r(t) \quad v_1^T(t) \quad v_2^T(t) \quad y_p(t)]^T \qquad (6)$$

It is known from fig 3, control inputs u add to controlled object is:

$$u(t) = [\theta(t) + \dot{\theta}(t)L^{-1}(s)]^T \phi(t) \qquad (7)$$

If you want to make adjustable system and reference model match completely, that must have following formula:

$$G(s) = G_m(s) \qquad (8)$$

The state equation of reference is:

$$\begin{cases} \dot{x}_m(t) = Ax_m(t) + bK_{00}r(t) \\ y_m(t) = c^T x_m(t) \end{cases} \qquad (9)$$

It is ordered that:

$$\theta(t) = \theta_0 + \tilde{\theta}(t) \qquad (10)$$

The state equation of controlled object is:

$$\begin{cases} \dot{x}_p(t) = Ax(t) + b[K_{00}r(t) + (\tilde{\theta}(t) + \dot{\theta}(t)L^{-1}(s))^T \phi(t)] \\ y_p(t) = c^T x(t) \end{cases} \qquad (11)$$

It is set state error is:

$$\varepsilon = x(t) - x_m(t) \qquad (12)$$

It is set that generalized output error is:

$$e(t) = y_p(t) - y_m(t) \qquad (13)$$

The augmented state error equation is gotten by make a deduction:

$$\begin{cases} \dot{\varepsilon}(t) = A\varepsilon(t) + bL(s)\tilde{\theta}^T(t)\xi(t) \\ e(t) = c^T \varepsilon(t) \end{cases} \qquad (14)$$

Thereinto:

$$\xi(t) = L^{-1}(p)\varphi(t)$$

The transfer function corresponding with the former formula with $\theta^T(t)\xi(t)$ is input, $e(t)$ is output, and considering matching condition is:

$$G_\varepsilon(s) = c^T(sI - A)^{-1} bL(s) = \frac{k_p}{k_m} G_m(s)L(s) \qquad (15)$$

It adopt adaptive rule for the former system:

$$\dot{\theta}(t) = -\Gamma e(t)\xi(t) \qquad (16)$$

Thereinto, Γ is diagonal matrix. In this way, it can ensure the stability, state astringency or parameter astringency of designed system.

IV. THE MODEL REFERENCE ADAPTIVE CONTROLLER DESIGN OF SVC

The controlled object of model reference adaptive control system in this design is thyristor phase control rectifier TCR, we need to build controlled object that is mathematical model of TCR in order to design model reference adaptive system.

At first, it supposed that trigger angle and output current are linear in the enough short time, input is direct current I^* of trigger circuit (fig. 2), and output is effective value I_L of

978-1-4244-1871-8/07 $25.00 © 2007 IEEE

TCR phase current fundamental wave. It lead to TCR current I_L lag direct current I^*, because once the thyristor is turned on, the change of control angle is not react until this component is turned off by inverted voltage, it is set that T_d is out of control time, because it adopted separated phase control, the minimum out of control time $T_{d\min}$ is 1.67ms, the most out of control time $T_{d\max}$ is *11.67ms*, this paper takes 5ms. The TCR current must make corresponding change according to new trigger angle when trigger angle become stepped, it supposed that the delay time between those is T_Y, usually T_Y is 3~5ms, this paper takes 4ms. K_Y uses one in this paper that is coefficient of proportionality between input and output.

So the open loop transfer functions of I^* and I_L is:

$$G_p(s) = \frac{I_L}{I^*} = \frac{K_Y e^{-T_d s}}{T_Y s + 1} \tag{17}$$

The former formula is spread by Taylor series and according T_d on the small side (T_d =5ms), high order is overlooked, so the open loop transfer function of I^* and I_L is near to:

$$G_p(s) = \frac{K_Y}{(T_Y s + 1)(1 + T_d s)} = \frac{\dfrac{K_Y}{T_Y T_d}}{\left(s + \dfrac{1}{T_Y}\right)\left(s + \dfrac{1}{T_d}\right)} \tag{18}$$

It takes T_Y=4ms, T_d=5ms, K_Y=1, we can get the next formula from above:

$$G_p(s) = \frac{50000}{(s+250)(s+200)} = \frac{50000}{s^2 + 450s + 50000} \tag{19}$$

The select of reference model affected greatly stand or fall of system performance in design of model reference adaptive control system, this paper uses two orders best damping factor $\xi=0.707$, so reference model is elected as:

$$G_m(s) = \frac{\omega_n^2}{s^2 + 2\xi\omega_n + \omega_n^2} = \frac{50000}{s^2 + 316.2s + 50000} \tag{20}$$

In order to assure $G_m(s)$ strict positive real, the system is described for controlled object by (19) and reference model by (20), that multinomial is: $L_s=s+a$. The characteristic polynomial of assistant signal generator is: $D_f(s)=L(s)N_m(s)=s+a$. According to (3) and (4), the transfer function of two assistant signal generators is:

$$G_1(s) = \frac{N_c(s)}{D_f(s)} = \frac{c_1}{s+a} \tag{21}$$

$$G_2(s) = d_0 + \frac{N_d(s)}{D_f(s)} = d_0 + \dot{d}_0 L^{-1} + \frac{d_1}{s+a} \tag{22}$$

That is the next formula by (5) and (6):

$$\begin{cases} \theta(t) = [K_0(t) \quad c_1(t) \quad d_1(t) \quad d_0]^T \\ \phi(t) = [r(t) \quad v_1(t) \quad v_2(t) \quad y_p(t)]^T \end{cases} \tag{23}$$

At last, we can get the input control of controlled object by (7):

$$\begin{aligned} u(t) = &\, [K_0(t) + \dot{K}_0(t)L^{-1}(s)]r(t) + \\ &\, [c_1(t) + \dot{c}_1(t)L^{-1}(s)]v_1(t) + \\ &\, [d_1(t) + \dot{d}_1(t)L^{-1}(s)]v_2(t) + \\ &\, [d_0(t) + \dot{d}_0(t)L^{-1}(s)]y_p(t) \end{aligned} \tag{24}$$

By (16) and R is diagonal matrix, we can get the adaptive rule of parameter is:

$$\begin{cases} \dot{K}_0(t) = f_1 e(t) \dfrac{1}{s+a} r(t) \\[2mm] \dot{c}_1(t) = f_2 e(t) \dfrac{1}{s+a} v_1(t) \\[2mm] \dot{d}_1(t) = f_3 e(t) \dfrac{1}{s+a} v_2(t) \\[2mm] \dot{d}_0(t) = f_4 e(t) \dfrac{1}{s+a} y_p(t) \end{cases} \tag{25}$$

f_1, f_2, f_3, f_4, are all real numbers.

V. EXPERIMENT WAVES AND ANALYSIS

Fig 4 is main circuit schematic diagram of SVC physical simulation system. *380V* ac is converted to *69.3V* by transformer. Three-phase electric power filter and thyristor control reactor (TCR) are linked in 69.3V ac generating line; the filter contained 4th, 5th, 7th filter, which can switch in arrange in groups, each of three-phase load can switch in two different state of *20Ω, 270Ω*.

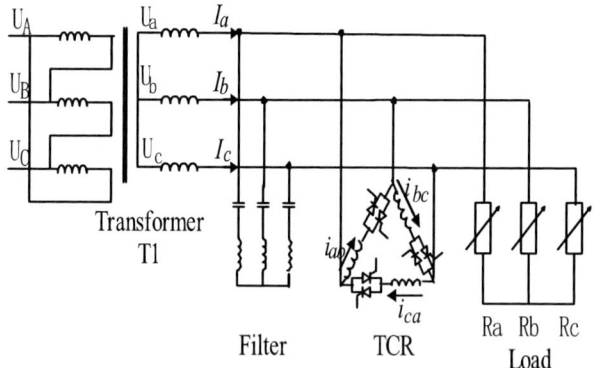

Fig. 4 The schematic diagram of main circuit

Fig 5 is phase A voltage, current waveform of three-phase load stay in one fixed unbalance state (phase A, B load are *20Ω*, phase C load is *270Ω*) before TCR is inputted, that current lead voltage because of capacitive current of filter.

Fig 6 is voltage, current waveform of phase A after TCR is inputted. Voltage and current are in phase because of the compensation of SVC. The SVC adopted model reference adaptive control has the well steady state precision by comparison between fore-and aft of compensation.

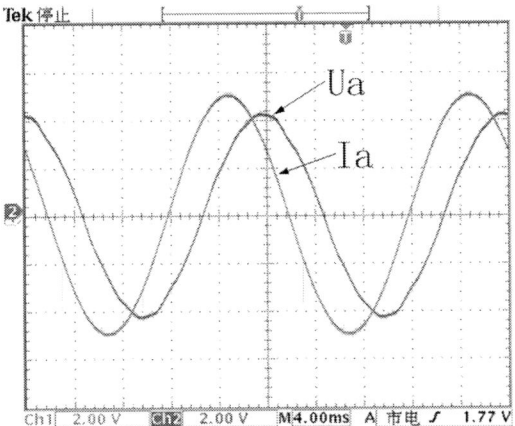

Fig. 5 The voltage and current waveform of phase A before TCR compensation

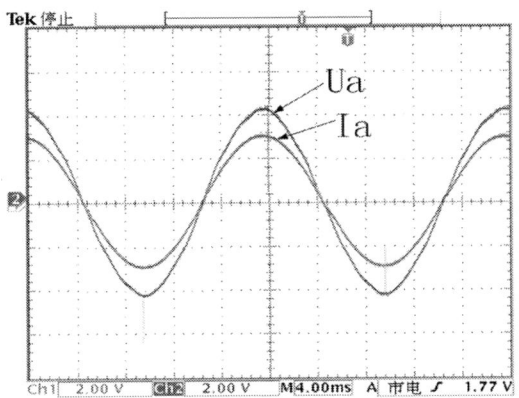

Fig. 6 The voltage and current waveform of phase A after TCR compensation

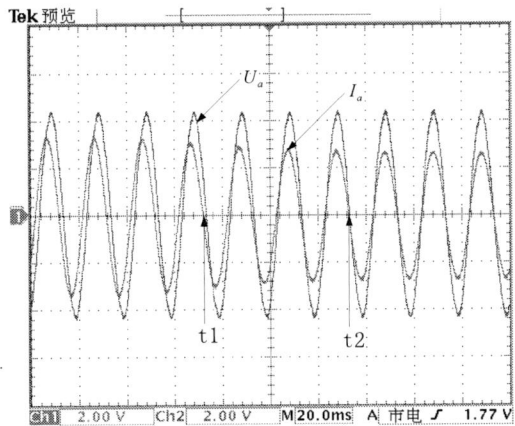

Fig. 7 The dynamic state compensative process waveform of phase when TCR is switching in

Fig 7 is phase A dynamic state compensative process waveform when TCR is switching in at the situation of 4th filter of unbalance load (phase A is 270Ω, phase B, C are all 20Ω) has run. From the figure we can know, before t1, current lead voltage because 4th filter has run; at t1, current crest value reduce and the phase discrepancy of current and voltage reduced because of TCR compensation; at t2, current and voltage are in phase and

total current crest value of phase A reached stabilization at last, the whole adjustment process in about *2.5* cycle.

Fig 8 is dynamic state compensative process wave of phase A voltage and current when 4th filter is inputted first and 5th is inputted then at the situation of unbalance (phase A is 270Ω, phase B, C are all 20Ω). From the figure we can know, before t1, the 4th filter is running only, and current and voltage are in phase because of TCR has run; at t1, the fifth filter is running, capacitive current increases, and phase of current and voltage offset and current lead voltage; at t2, the total current crest value of phase A reached stabilization at last, and current and voltage are in phase after adjustment of TCR, the whole adjustment process in about 4 cycle.

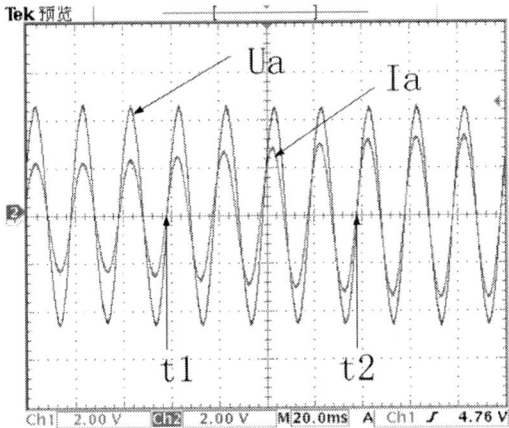

Fig.8 Dynamic waveform of phase A voltage and current when load is unbalanced

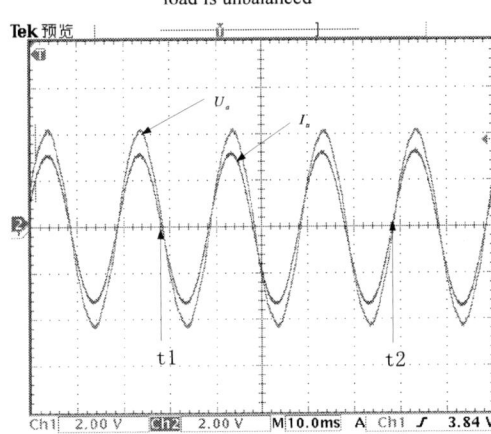

Fig.9 Dynamic waveform of A phase voltage and current when balanced load changes to unbalanced load

Fig 9 is dynamic state compensative process wave of phase A voltage and current when balance load (phase A, B, C are all 270Ω) convert into unbalance load (phase A, B are all 270Ω and phase C is 20Ω) before input the forth and fifth filter, before t1, voltage and current are in phase because of adjustment of TCR, after t1, part of phase C load is removed, at t2 we can know, phase of current and voltage offset and current lead voltage and current crest value increase; at t3, the total current crest value of phase A

978-1-4244-1871-8/07 $25.00 © 2007 IEEE   303

reached stabilization at last, and current and voltage are in phase, the whole adjustment process in about *2.5* cycle.

VI. CONCLUSIONS

The model reference adaptive controller is designed based on Liapunov stability theory in this paper, and applied in SVC experiment platform based on industrial control computer, and achieved better effect. It proved that SVC controller adopted model reference adaptive control arithmetic can carry out comprehensive compensation for reactive power and three-phase unbalance, and has better compensative precision and swift respond speed. In generally, it has definite progress and justification.

REFERENCES

[1] LEE S Y, Wu C J. Combined compensation of a static var ocompensator and an active filter for unbalanced three-phase distribution feeders with harmonic distortion. Electric Power System Research, PP. 243-250, 1998.
[2] N.G.Hingorani.FACTS-Flexible AC Transmission System. Fifth International Conference on AC and DC Power Transmission, London, pp. 1-7, 1991.
[3] Parks P C. Lyapunov redesign of model reference adaptive control systems. IEEE Trans.Autom.Control. pp. 362—367, 1966.
[4] Phillipson P H. Design medthods for model reference adaptive systems. Proc.Inst, Mech, Engrs, pp. 675—700, 1968.
[5] Li Qingquan. The adaptive system theory. design and application, The Science Publishing Company.1990.3.

The study on the Protection Circuit of the Inverter for LCD Backlight Using Digital control Method

JEONG-WOON LEE, YOUNG-CHEOL LIM
Chonnam National University
Department of Electrical Engineering
300, Yong bong-dong, Puk-gu Gwangju
Email: jungwoons@hotmail.com

SEUNG-HAK YANG
HONAM UNIVERSITY
Department of Electrical Engineering
159-1,Seo bong-dong, Gwangsan-gu, Gwangju
Email: yangsh@honam.ac.kr

Abstract— The way of operating multiple CCFLs in parallel is adopted for ensuring the evenness of brightness while reducing the price of inverters for operating backlights according to the trend of producing larger LCD televisions. However, if any of the multiple CCFLs is turned on first, parallel (circuit) resistance is shown as further resistance of a lighted lamp. In this case, resting CCFLs, which are not turned on, do not have sufficient voltage for lighting; therefore, they rest unlit, and a problem arises because of the imbalance of brightness owing to the difference of excessive electric current between each CCFL. This thesis studied the imbalance of the output voltage of the electric current that is generated from an inverter for operating LCD backlights, designed the inverter protection function for OVP (Over Voltage Protection), LCC (Limited Current Circuit), and OLP (Open Lamp Protection) as a digital type circuit based on the study, and compared the function with the existing analogue type protection circuit.

I. INTRODUCTION

The paper should be prepared in Microsoft Word Recently, the size of LCD gets bigger and thinner and the Cold Cathode Fluorescent Lamps installed as the backlight within the monitor and TV are getting thinner. As these large backlight displays have limit of lighting the appropriate brightness with one lamp, many lamps are used in parallel. As the LCD monitors and TVs get bigger, almost all the multi-lamps are being used and the current LCT TV market is getting bigger.

The discharge voltage of high initial AC is needed as the features of the discharge lamp and maintaining voltage and current are also needed. Because of these features, the high voltage generating transformer and inverter are necessary. But even with the identical manufacturing procedures, the lamps do not have identical features of lighting, frequency, voltage and current, which makes hard for the lamps to have stabilized brightness and when any one of the CCFL is lit, the resistance of the first lit lamp becomes the equivalent parallel resistance. In this case, the remaining CCFL that didn't get lit can not receive enough voltage to get lighted and allow the CCFL to stay unlighted. Also, the brightness imbalance by the over current declination of CCFL is the problem.

The inverter designed in this thesis is for the LCD TV screen and structures the protection circuit of analogue method and digital method then studied each method by comparison.

II. COLD CATHODE FLUORESCENT LAMP

A The Structure of Cold Cathode Fluorescent Lamp (CCFL)

The CCFL being used as the inverter load for LCD backlight has many advantages of high brightness, high efficiency, low energy use, long life span, low calorification, excellent durability, excellent lighting features and more, which is being used as the backlights of various displays, erasers, scanners, various lights and decorations.

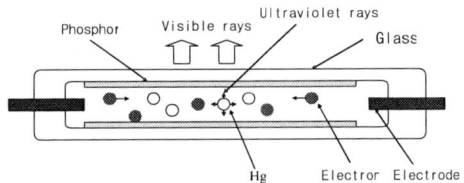

Fig.1 Construction of the CCFL

The CCFL for LCD TV was used at this thesis. Looking at the structure of CCFL, there are electrodes at the both ends of the lamp as shown on the "Fig. 1," and there are certain amounts of Hg, Ar and Ne gases mixed inside. Also, the inner surface of the lamp is painted with fluorescent substances and has identical structure with general fluorescent lamps other than thinner and different electrodes.

B The Backlight System

As presented in "Fig. 2," there are two styles of a direct method and an edge-light method in a backlight system.

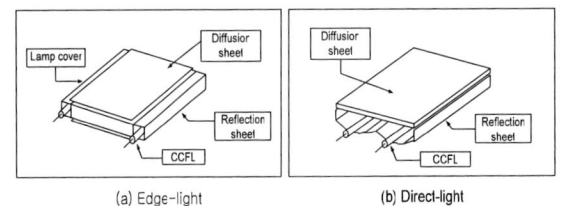

Fig. 2: Backlight system

978-1-4244-1871-8/07 $25.00 © 2007 IEEE

The edge-light cannot supply enough brightness on large LCD screens since it has the limitation on quantity of lamps physically which can be established on the edge of a panel and the size becomes bigger in accordance with the edge part. Furthermore, the edge –light requires high-priced polarized light plates to make proper brightness of a backlight unit. On the other hand the problem on brightness can be solved by designing the number of lamps according to the size of screen getting larger in a direct-light, which employs CCFL connected in an electrical series at the bottom of LCD TVs. In this paper the direct method was adopted to fabricate bright and clear screens.

III. DESIGNING THE INVERTER FOR BACKLIGHT OF LCD TV

A CCFL Operating Inverter

The inverter circuit to operate CCFL can be seen as three parts as following. It is structured with the DC/DC converter part for various input voltage and DC/AC inverter part that converts the DC to AC for discharge lamp. [1]

B Lamp Driving Method

When the 950mm lamps for 42inch TVs are used in the conventional inverter circuits, the difference on brightness arise at both ends. That is, the longer the length of lamps, the higher the operating voltage needs.

"Fig 3" shows four driving methods.

The conventional backlight inverter circuit included in LCD below 21inches usually adopts the High-Low method, which is suitable for driving individually but have the demerit that the leakage voltage among lamps are reciprocally large.

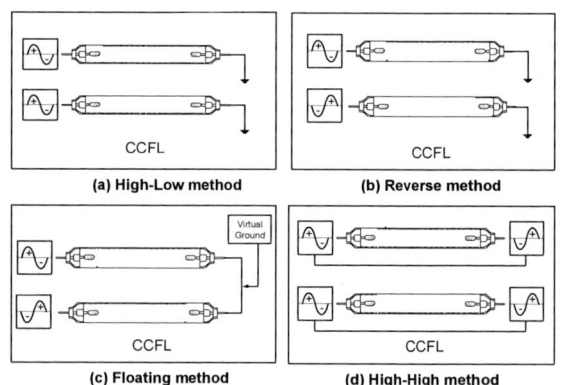

(a) High-Low method **(b) Reverse method**

(c) Floating method **(d) High-High method**

Fig. 3: Operation methods of lamps

Compared with the normal driving method, the reverse and floating driving method are excellent in the aspect of performance but the leakage voltage is large.

The High-High driving method suggested in this paper exhibits a high released voltage and it functions as equalizing

the brightness of lamps. Moreover, this method stabilizes lighting lamps for large-sized LCD and it is distinguished on emitting heat on transformers and FET. Consequently, It showed an excellent performance in improving variations generated on both sides of lamps.

C The Full-Bridge Type Operating Inverter

The topology used the inverter to operate LCD TV having 20 lamps has selected the Full-Bridge type to deliver more power as the number of lamps and length increase. The overview diagram of the inverter actually built is the "Fig 4," [2]

The circuit is structured with 2 transformers connected in series to have 4 Full-Bridge structures at the 1st point, 2 P-type MOSFET and 2 N-type MOSFET were used as the semi-conductor switch devices and has series capacitor for the resonance of boosting transformer and inductance at the 1st point of the transformer. The 2nd point of the transformer is structured with 2 high voltage series capacitor for equivalent current distribution and high voltage parallel capacitor for initial discharge voltage and power factor. The overall system has the following structures connected in parallel.

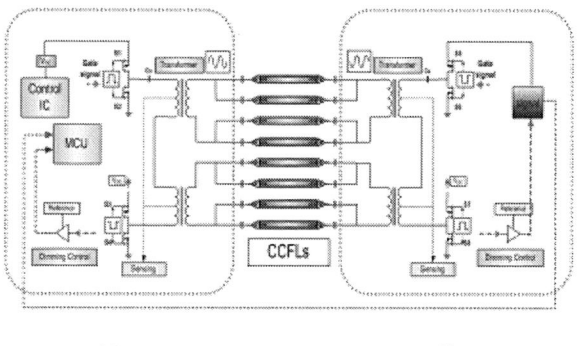

Master   Slave

Fig.4: Inverter for multi-lamp driving

The special feature is that the series capacitor is inserted between the 2nd point of the transformer and lamps with a certain value for equivalent current distribution because the lamps have different properties. This is using the properties of the discharge lamps as the voltage is important when lighting the lamps and the current is important when the lamps are lit. When comparing the inserted capacitor and the impedance, if the impedances of the lamps before lighting are some MΩ, which are very larger value than the capacitor, then the current is confirmed to all the lamps and when the impedances of the lamps are lowered to tens to hundreds kΩ after the lighting and resulting the impedance of the capacitor to have relatively larger value, it cannot have serious effect to the flowing current even when the impedances of all lamps are little different, which results the equivalent current distribution. Another special feature is the parallel capacitor insertion to the 2nd point of the transformer. [3]

978-1-4244-1871-8/07 $25.00 © 2007 IEEE

IV. PROTECTION CIRCUIT

The functions of the protection circuit used in this thesis are the OVP (Over Voltage Protection), OLP (Open Lamp Protection) and finally, the LCC (Limit Current Circuit). The OVP and OLP are to protect the lamps from any damages or bad connection. When thinking about the lamp damaged case, originally, the inverter sends the equal current and voltage to the lamps, but when the lamps are damaged, the current on other side flows to other lamps.

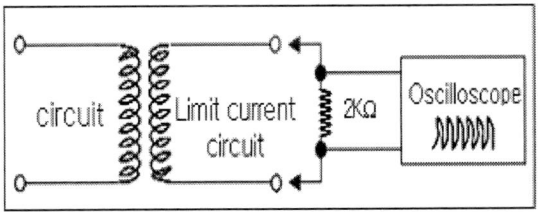

Fig.5: Measurement by Limit current circuit

When this happens, the over voltage flows to the inverter and damages the inverter. When this happens, it shuts down. The LCC has the purpose of securing the safety for the user or service agent accessible high voltage confirmed parts for normal operation status or when basic insulation is damaged or malfunction of the parts. The "Fig 5," is the overview diagram to measure the LCC. The non-inductive test was tested as like the "Fig 5," by taking the human body as 2kΩ.

The "Fig 6," shows the summery of protection circuit for analogue method and digital method.

To sense the OVP, OLP and LCC, the 2nd point of the transformer is regularly sensed and replied to IC. This replied voltage is compared with the standard voltage to shut down the inverter. The analogue method using OPAmp and digital method using MCU were applied for the comparing method.

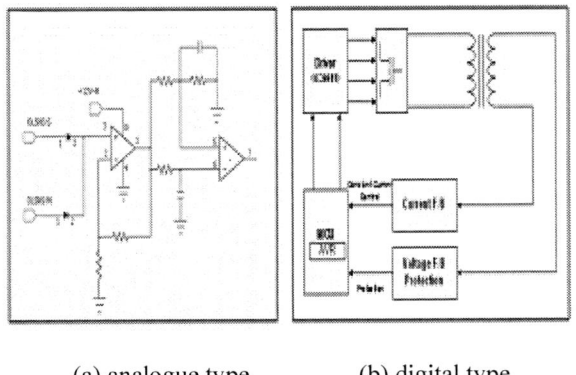

(a) analogue type   (b) digital type

Fig.6: Protection circuit

V. THE RESULT OF EXPERIMENT AND CONSIDERATION

The inverter for LCD TV backlight was used for analogue method and digital method protection circuit related tests.
The inverter was designed with the following standards.

- input voltage: 24 Vdc
- operating frequency: 65 KHz
- lamp type: 20ea (lamp voltage: 620V ±150V)
- lamp current: 6.5mA ±0.5mA)
- lamp dimensions: length: 960mm
- diameter: 4mm

"Fig 7. to 8." show a voltage and a current of a lamp. In waveforms of each lamp, channel 1 is the voltage, channel 2 is the current." Fig 8," exhibits the maximum brightness and "Fig 8," does around 50percent brightness with the burst.

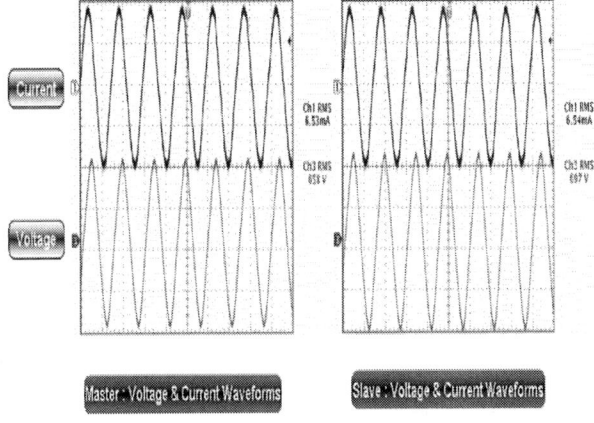

Fig. 7: output waveforms of the lamps(Max. Brightness)

"Fig 9," presented each current variation on Master and Slave in turn. There are both reasons of generating this variation.

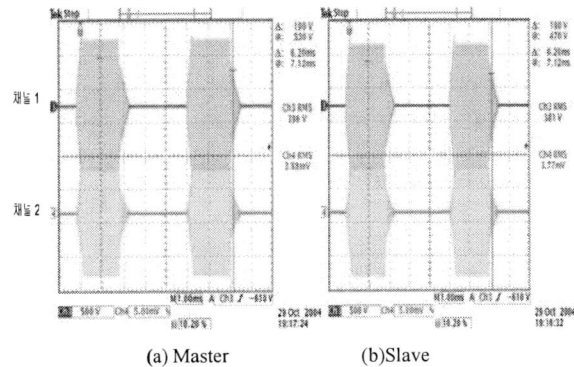

(a) Master   (b)Slave

Fig. 8: Output waveforms of the lamp (Mid. brightness)

First, due to the characteristics of discharge lamps, these discharge lamps which have different characteristics on lighting, frequency, voltage and current cause variation in spite of an identical production process.

Second, because of the impedance characteristics of transformers the variation at output signals may be generated despite same signals.

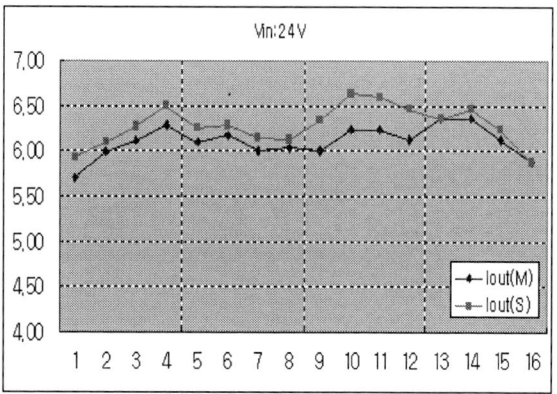

Fig. 9: Current variation

The "Fig 10," shows the OLP test waveform. It shows the measured waveform of removing 2 lamp connectors and 2nd point of the transformers of digital and analogue methods. The measuring time units are 40ms for digital method and 10ms for analogue method.

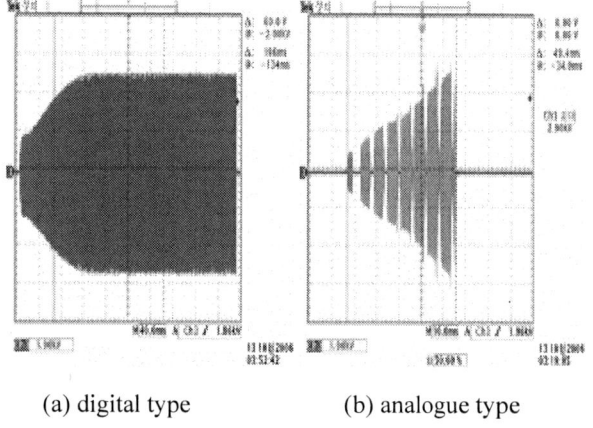

(a) digital type   (b) analogue type

Fig. 10: Waveforms of the OLP

The "Fig 11," is LCC test waveform. It shows the measured waveform of connecting the 2kΩ resistance to 2nd output point of the transformer and grounding of the inverter with inverter turned off and digital method and analogue method when turning the inverter on.

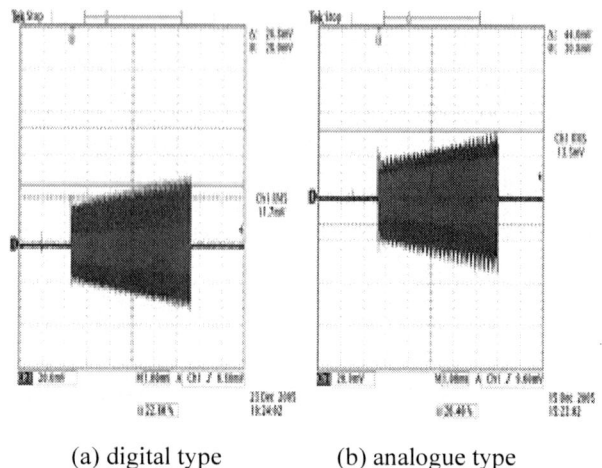

(a) digital type   (b) analogue type

Fig. 11 Waveforms of the LCC

VI.   CONCLUSION

The inverter designed in this thesis is for LCD TV and 20 lamps were operated within the TV. The methods of analogue and digital protection circuits were applied with the base of 20 lamps inserted in identical spacing as the screen base and the inverter having 2 operating IC and 4 boosting transformer. Two methods have shown identical operations.

But the analogue method compares the voltage by real-time sensing and digital method has the program executing time, which resulted slower operation than the analogue method. The analogue method was used in the actual product.

If the operation time of digital method is improved, the circuit using MCU can save the inverter's cost by simplifying the circuit with reduced number of elements.

ACKNOWLEDGEMENTS

The participants in this research received supporting funds from the second-stage BK 21.

REFERENCES

[1] Mu-Shen Lin, Wen-Jung Ho, Fu-Yuan Shih, Dan Y. Chen and Yan-Pei Wu : A Cold-Cathode Fluorescent Lamp Driver Circuit with Synchronous Primary-Side Dimming Control, IEEE TRANS. ON Industrial Electronics. Vol. 45. No. 2. pp.249-255, April 1998.

[2] Gie Hyoun Kweon, Young Cheol Lim, Seung Hak Yang, Jong Yoon Shin : An Analysis of the Characteristics of Piezoelectric Transformer and its Application, ICEE2K, 2000. 07, 489-492

[3] Abraham l. Pressman, Switching Power Supply Design. McGraw Hill. 1998, pp. 56

A Simple Partial Discharge Detector for Low-Voltage Rotating Electrical Machines

Cheng-Chi Tai
National Cheng Kung University
Department of Electrical Engineering
1 University Road, Tainan, 70101, Taiwan, ROC
Email: ctai@mail.ncku.edu.tw

Ting-Cheng Huang
National Cheng Kung University
Department of Electrical Engineering
1 University Road, Tainan, 70101, Taiwan, ROC
Email: tchelit@gmail.com

Ching-Chau Su
National Cheng Kung University
Department of Electrical Engineering
1 University Road, Tainan, 70101, Taiwan, ROC
Email: n2890141@ccmail.ncku.edu.tw

Chien-Yi Chen
National Cheng Kung University
Department of Electrical Engineering
1 University Road, Tainan, 70101, Taiwan, ROC
Email: n2893162@mail.ncku.edu.tw

Ju-Chu Hsieh
National Cheng Kung University
Department of Electrical Engineering
1 University Road, Tainan, 70101, Taiwan, ROC
Email: ea003@mail.njtc.edu.tw

Yu-Shiun Lin
Nan-Jeon Institute of Technology
Department of Electrical Engineering
178 Chau-Chin Road, Yen-Shui, Tainan, Taiwan,
ROC
Email: yuhsun.lin59@msa.hinet.net

Chung-Tzong Wang
Nan-Jeon Institute of Technology
Department of Electrical Engineering
178 Chau-Chin Road, Yen-Shui, Tainan, Taiwan, ROC
Email: wachtz@mail.njtc.edu.tw

Jeng-Hung Lai
Predict Maintenance Technology
3 F, No. 4, Alley 11, Lane 327, Sec. 2, Jhongshan
Rd. Jhonghe City, Taipei County, Taiwan, ROC
Email: lai@pmtweb.com.tw

Abstract—A new, simple partial discharge (PD) detector for low-voltage rotating electrical machines using acoustical emission (AE) technique is dseveloped in this study. Common electric components were used in the detector, which reduces the cost of inspection, comparing with the traditional PD detection methods that use expensive equipment costing from tens of thousands to several millions dollars. Experimental results by resonant type AE sensors (150 kHz) which utilize power line-cycle in a microcontroller unit (MCU) as reference to measure the PDs generated in a low-voltage motor are presented. The AE signals are then amplified by a pre-amplifier (30 kHz ~ 300 kHz, 34 dB). Since the resonant frequency of the sensor is much lower than that of the electromagnetic (EM) interferences around the motor, the effects of noise is substantially reduced by this method. In the mean time, the use of 150-kHz resonant type AE sensor also avoids the disturbance of mechanical vibration noise. According to the experiment results, the measurement system developed in this study can be used to detect the PDs' AE signals correctly. The AE measurement scheme proposed in this study provides an effective, low-cost method for PD measurements.

I. INTRODUCTION

With energy shortage befalling as a trend, many energy-conservation, electricity-saving techniques, equipment and devices have been developed in recent years. Among them, the PWM (Pulse Width Modulation) and the energy-conservation, high-efficiency rotating techniques,

including the related adjustable-frequency equipment and apparatus, such as high-speed rail's traction systems, elevators for tall buildings, hybrid cars, inverter air-conditioning systems, and so on, have been widely used.

When the low-voltage rotating electrical machines are operated under a commercial frequency, their insulation capability can arouse our concern. During the high-frequency switching period, the transient voltage greatly exceeds the endurable insulation capability [1-5] originally designed for the coils; thus, the partial discharge (PD) will occur in the long run. This will make many home appliances or commercial machines face an un-expected usage risk, such as short circuiting, electric leakage, fire, and so on.

Inside the transmission line, the impulse voltage from the PWM will generate a wave refraction and reflection, due to impedance mismatching, at the connection area of the rotating machine, when transmitted through the line, and cause a transient high voltage and high-frequency oscillation. Skibinski et al. [6, 7] indicate that the increase of cable length will cause the magnitude increase of over voltage and the decrease of oscillation frequency; when the cable reaches critical length, a near doubling of the rated voltage will arise at the connection area of the rotating machine. Although the transient high voltage can reach a magnitude two to three times the rated voltage, its over voltage will be evenly distributed on the windings of the stator under a commercial frequency of 60 Hz and, thus, will not damage the stator's insulation. However, the impulse voltage from the PWM will cause the unevenness

978-1-4244-1871-8/07 $25.00 © 2007 IEEE

of voltage distribution on the stator's windings due to its rather short rise time. Under a condition of rather short rise time (0.1–0.2 μs), the first turn of the stator's windings will bear nearly 80% voltage, resulting in a transient high-voltage impact on the nearby turns' insulation; the area with insufficient insulation is prone to become the stator windings' insulation weakness where partial discharges will thus come up.

Using acoustic emission (AE) method for PDs can effectively reduce circuitry cost and the EMI. The piezoelectric material adopted will generate a propagation wave [8, 9] when it detects a PD and convert the PD into a voltage signal. In 2005, T. Kaneko [10] published the application of AE method in detecting PDs in windings.

This study proposes using AE method to measure low-voltage rotating electrical machines' PDs. With the selection of different acoustic emission sensors and filters for various frequencies, we can avoid the interference of the vibrations from the rotating machines so as to increase the acoustic emission signal/noise (S/N) ratio for the PDs.

II. MECHANICAL VIBRATION RESPONSE

To assure that AE method can be applied to measuring low-voltage motors, it is vital to effectively reduce the interference caused by the mechanical vibrations during the operation.   Hence, we slightly shift the generator's shaft to make severe mechanical vibrations during the operation. We also place the acoustic emission sensor near the connection area between the bearing case and the machine body to acquire the maximum mechanical vibrational interference. There are four kinds of main frequency for the acoustic emission sensors, namely 30~80 kHz and 100~2000 kHz for the level type and 150 kHz and 375 kHz for the resonant type. The acoustic emission sensor is first fastened prior to the start of the motor; we measure the motor's mechanical vibrations in response to various types of acoustic emission sensors after the motor has been running for a period of time and become stable. The measured waveforms are shown in Fig. 1(a)~(d). We observe that the 150 kHz resonant type renders the best anti-vibrational interference effect.

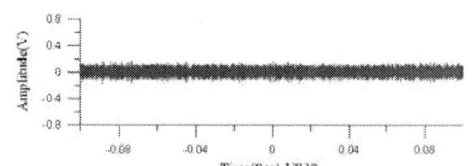

Fig. 1(a)   The level type 30~80 kHz AE response waveform under mechanical vibration

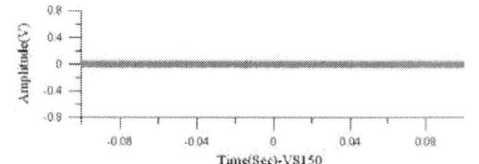

Fig. 1(b)   The resonant type 150 kHz AE response waveform under mechanical vibration

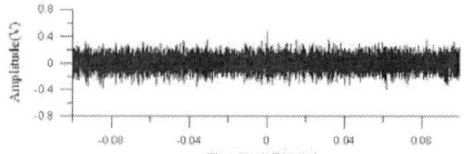

Fig. 1(c)   The level type 100~2000 kHz AE response waveform under mechanical vibration

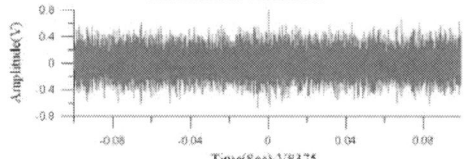

Fig. 1(d) The resonant type 375 kHz AE response waveform under mechanical vibration

III. PARTIAL DISCHARGE'S AE SIGNAL CONFIRMATION

During the preceding experiments, we adopt impulse current detection and measurement method to assure the low-voltage motor's PD signals and, with PD's features, to confirm the adequacy of the AE signals measured. A resonant type 150 kHz acoustic emission sensor is adopted and placed near the connection area between the bearing and the machine body to measure AE signals; concurrently, an HFCT sensor is used to compare and inspect the signals the waveforms of which are shown in Fig. 2(a, b).

The waveform from the resonant type 150-kHz AE sensor is shown in Fig. 2(a), while the waveform from HFCT in Fig. 2(b). In Fig. 2(a), an AE signal appears after HFCT detected a PD signal shown in Fig. 2(b). The PD signals left behind in Fig. 2(b), though existing, do not generate AE signals nevertheless. This could be caused by some PD sources occurring too deep to propagate through their conduction routes or by acoustic emission sensors with insufficient sensitivity.

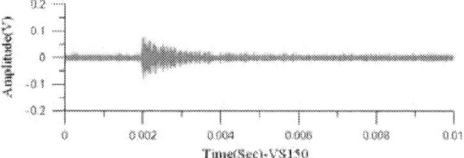

Fig. 2(a) PD acoustic-emission response waveform

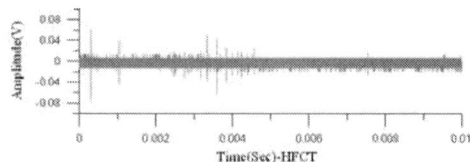

Fig. 2(b) PD HFCT response waveform

IV. SYSTEM CIRCUIT

The 60Hz-power-source frequency interference is rather severe during the low-voltage rotating electrical machines' actual operations; hence, we ought to design a high-pass filter to reduce said interference. An AE sensor is made of piezoelectric material; we thus select TI's high-frequency

978-1-4244-1871-8/07 $25.00 © 2007 IEEE   310

instrumentation amplifier INA2332 to provide a high CMRR value. We also install a first-order CR high-pass filter to reduce the interference from the 60Hz power source. The 15 kHz high-pass AE Amplifier shown in the system block diagram of Fig. 3 is for amplifying the filtered signals from the AE sensors; the circuit diagram of the amplifier is shown in Fig. 4.

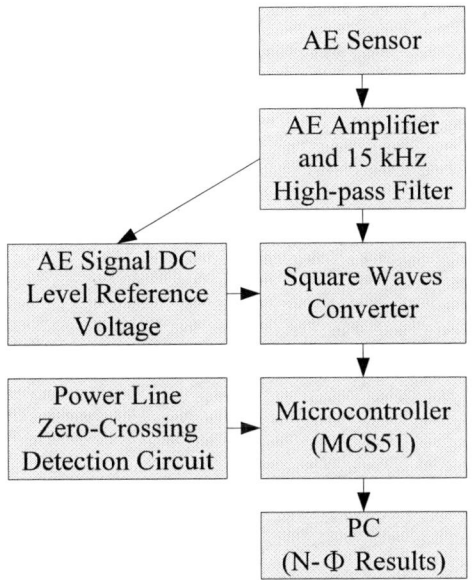

Fig. 3. The system diagram of the detector

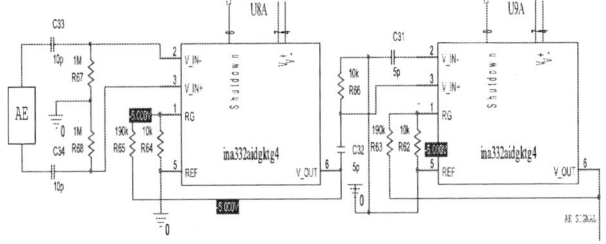

Fig. 4. A 15-kHz high-pass AE instrumentation amplifiers (INA2332) used

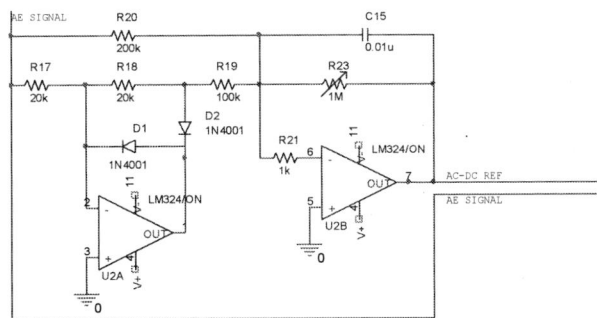

Fig. 5. An AE signal dc level voltage circuit (LM324) used for reference

To respond to various fields with different background signals and the 60 Hz interference, we cannot fix the trigger voltage to avoid the misfiring of the AE signals. We hence adopt ac/dc conversion circuit, as shown in Fig. 5, to convert the amplified AE signals depicted in Fig. 4 to dc to

provide an output dc voltage which will vary with the magnitude of the signals. The C15 in Fig. 5 will decide the speed that the output voltage varies with the signal. Here we select 0.01 uF for C15, for the frequency of the AE signal is 150 kHz and the power-source interference signal is at 60 Hz; we could only convert the 60Hz carrier, not the AE signal, into a dc voltage level. This indicates that the AE signal on the 60 Hz carrier can effectively trigger each AE signal since the reference voltage varies with the 60 Hz power source.

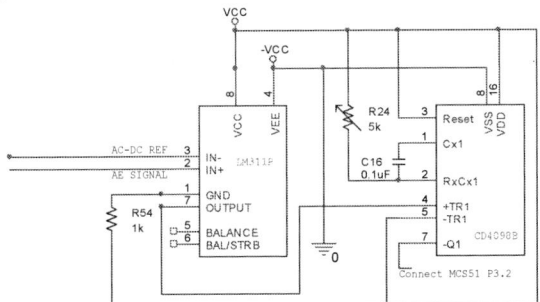

Fig. 6. The AE signals are converted into square waveforms by LM311P and CD4098B and then transmitted to an MCS51 microcontroller

Next, we connect the AE signal and the ac/dc conversion-circuit signal to a square-wave converter, as shown in Fig. 6. For the AE signal is an oscillatory waveform, it will generate a series of square waveform at the LM311 comparator's output when the reference voltage is constant. To prevent the chip from incorrectly taking an AE signal as multiple AE signals, we adopt the CD4098's pulse-delay function to convert each AE-signal duration into a square waveform to enable the MCU to clearly identify said AE signal. The C16 and R24 shown in Fig. 6 are used to decide the pulse-delay time of the CD4098. Since said signal is to be connected to an MCS51's Pin P3.2 (INTO), i.e. the negative-edge interrupt trigger, the signal, hence, is connected to Pin 7 on the CD4098.

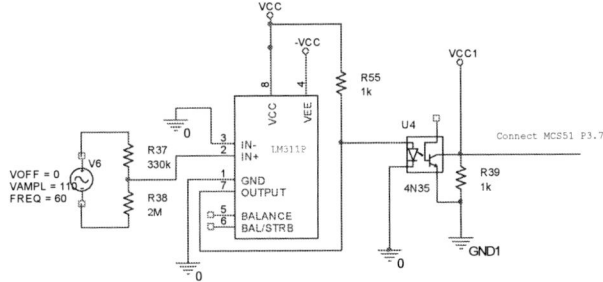

Fig. 7. Power-line zero-crossing detection circuit.

The Power-line Zero-Crossing Detection Circuit shown in Fig. 7 mainly adopts LM311 to compare the power source and the input to generate a corresponding square waveform; a 4N35 photo insulator is used to separate the MCS51 from the system power source. Hence, the right-most output of the 4N35 is connected to Pin P3.7 on the MCS51 while the 4N35's power source is provided by its own circuit rather than the input power source.

978-1-4244-1871-8/07 $25.00 © 2007 IEEE   311

With the signals from the square-wave converter and the power-line zero-crossing detection circuit, we adopt the zero-crossing signal as the trigger for the phase-angle timer. After the conversion from the AE signal into the square-waveform signal, the latter will be connected to the INT0 interrupt; said interrupt will retrieve the current phase-angle data and send the data to an RS232 to end the interrupt, awaiting the next AE signal to arise. The program residing inside the PC terminal will retrieve the phase angle outputted from the MCS51 and, after retrieving for a period of time, count the number of the AE signals occurring at each phase angle and, accordingly, plot the pattern, completing the measurement.

V. EXPERIMENT RESULTS AND DISCUSSION

We fasten an AE sensor on the surface of a motor, and have it connected to a commercial amplifier, AEP3, and to our self-made circuit. Nearby the sensor, we adopt a pencil-lead-beak testing [11] to confirm that the circuit-converted square waveform conforms to the location of the AE waveform amplified by the commercial amplifier (Fig. 8).

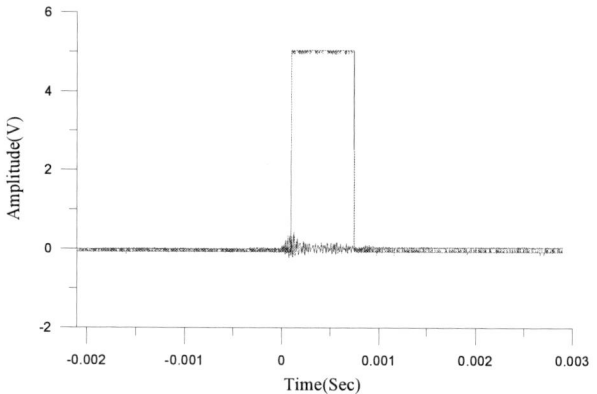

Fig. 8.   AEP3 vs AE/Square-waveform signal conversion

The AE signal measured will be converted into a dc reference voltage, for AE signal triggering, via the AC-DC reference voltage circuit, as shown in Fig. 9. The reference voltage will change with the background magnitude instead of the AE signal; such effect will prevent the low-frequency interference-waveform's AE signal, arising during the field operation, from being mis-fired due to a constant trigger voltage. Said effect also can prevent the AE signal from being unable to be triggered due to our raising up the interference trigger voltage.

The AE signals' background-signal variations could still effectively trigger the conversion and turn the variations into square waveforms, as shown in Fig. 10(a); with a small amplitude, the AE signals' wave fronts still could effectively trigger the conversion and turn the fronts into square waveforms, as shown in Fig. 10(b); with a large amplitude, AE signals' wave fronts could effectively trigger the conversion and turn said fronts into square waveforms, as shown in Fig. 10(c). The measured AE signals could,

hence, still effectively convert the signals into a square-waveform output via circuits, as shown in Fig. 10(a, b, c), regardless the background signals' variations and the difference in AE signals' magnitudes.

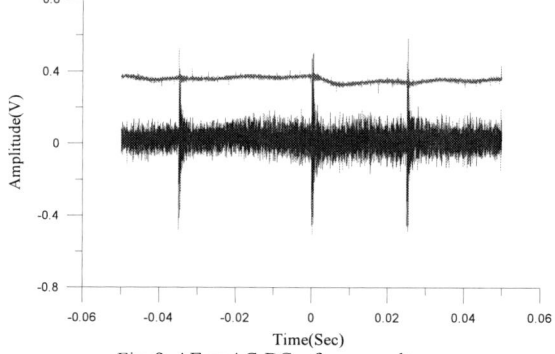

Fig. 9. AE vs AC-DC reference voltage

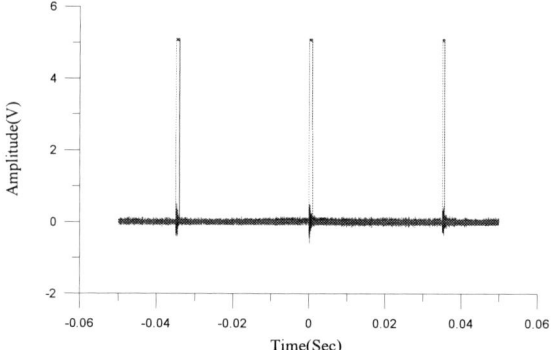

Fig. 10(a). AE vs square-waveform signal conversion

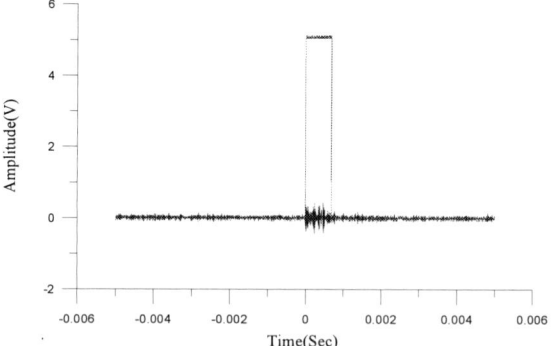

Fig. 10(b). Small AE vs square-waveform signal conversion

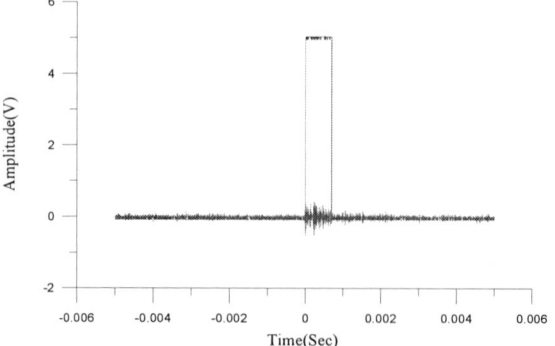

Fig. 10(c). Large AE vs square-waveform signal conversion

978-1-4244-1871-8/07 $25.00 © 2007 IEEE

The discharge pulses will become a certain degree of correlated symmetry following the variation of the power source's phase angles; hence, we could lead to a preliminary conclusion that the PD impulses and the power source's phase angles are correlated. Next, we measure the PD signals for the low-voltage rotating motor used in this study under loaded conditions. To simulate external interference signals, we randomly knock on the motor's exterior shell when it is running and record the signals. As described, we adopt, in this study, a chip for a counter, send out the outputted square waveform and the occurred phase-angle records to an RS232 interface and code a program in the computer terminal to plot the following measurement results, as shown in Fig. 11(a)~(c).

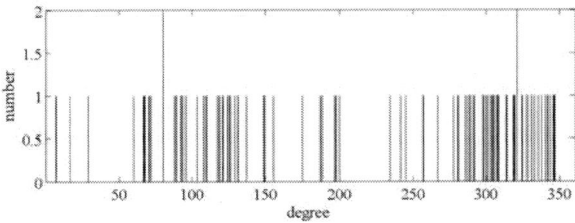

Fig. 11 (a) Noise; thus, there is no power-source-correlated distribution

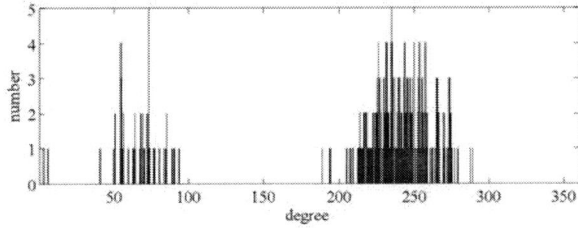

Fig. 11(b). The AE distribution with a low correlation to the power source

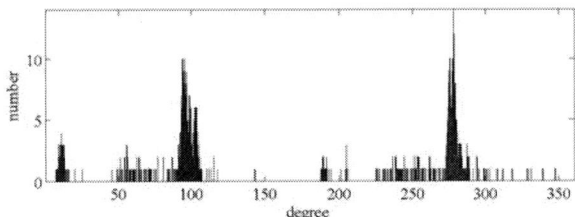

Fig. 11(c). The AE distribution correlated to the power source (the phase angles conform to the PD signal characteristics)

A.   Vibration Interferences From the Enviroment

The horizontal axis, as shown in Fig. 11(a), represents the voltage angle of the power source, while the vertical axis the number of acoustic emissions; the blue image depicts the phase angles and the number of acoustic emissions occurred. We observe from the diagram that the distribution of the AE signals' phase angles is overly scattered and represents no concentration phenomenon; the distribution should be correlated to the power-source phase angles should there be PDs. Thus, the signals are the ones produced by the external vibrations.

B. The AE Distribution Not Correlated to the Power Source

Although the distribution of the AE signals, as shown in Fig. 11(b), represents some degree of concentration, the occurring phase angles are not as concentrated as that shown in Fig. 11(c); hence, we can deduce two possible reasons: first, these are not PD-correlated AE signals, for PD-correlated AE signals shall not have an overly wide phase-angle distribution; second, there are multiple power sources for the measured object, resulting in an overly scattered PD phase angles.

C. The AE Distribution Correlated to the Power Source

The horizontal axis, as shown in Fig. 11(c), represents the voltage angle of the power source, while the vertical axis the number of acoustic emissions; the blue image depicts the phase angles and the number of acoustic emissions having occurred. The AE signals concentrate in the area between 100 and 280 degrees, as observed from the diagram. This matches the rules for arising actual PDs and, hence, could be concluded that the AE signals conform to the PDs' correlation to the power source.

VI.   CONCLUSION

A new PD detector for low-voltage rotating motors using acoustical emission technique is proposed in this study. Comparing to the existing relatively-expensive traditional PD detection methods, the proposed method uses regular components and provides an economical means in conducting PD measurements. Our experimental results, using resonant type AE sensors, are presented. A 150-kHz resonant type AE sensor is adopted to make the proposed system immune to exterenal vibrational noises. The experiment results confirm that the proposed economical measurement system can successfully detect the PDs' AE signals.

ACKNOWLEDGMENT

This research was supported by the grant from National Science Council, Taiwan, ROC and Predict Maintenance Technology (NO. NSC94-2622-E-006-047-CC3). Also, this work made use of Shared Facilities supported by the Program of Top 100 Universities Advancement, Ministry of Education, Taiwan.

REFERENCE

[1] D. Bogh, J. Coffee, G. Stone, and J. Custodio, "Partial discharge inception testing on low voltage motors," in *Petroleum and Chemical Industry Technical Conference, 2004. Fifty-First Annual Conference 2004*, 2004, pp. 241-248.

[2] M. Fenger, G. Stone, and S. Campbell, "Partial discharges on low voltage stator windings subjected to voltage surges from adjustable speed drives," in *Power Electronics, Machines and Drives, 2002. International Conference on (Conf. Publ. No. 487)*, 2002, pp. 298-302.

[3] M. Fenger, S. R. Campbell, and J. Pedersen, "Motor winding problems caused by inverter drives," *Industry Applications Magazine, IEEE*, vol. 9, pp. 22-31, 2003.

[4] D.-H. Hwang, K.-C. Lee, Y.-J. Kim, I.-W. Lee, T.-H. Lim, and D.-H. Kim, "Accessing the Insulation Characteristics for Stator Windings of Low-Voltage Induction Motors for Adjustable-Speed Drive Applications," Salt Lake City, UT, United States, 2003, pp. 432-438.

[5] N. Hayakawa and H. Okubo, "Partial discharge characteristics of

inverter-fed motor coil samples under ac and surge voltage conditions," *Electrical Insulation Magazine, IEEE,* vol. 21, pp. 5-10, 2005.

[6] G. L. Skibinski, B. M. Wood, J. J. Nichols, and L. A. Barrios, "Effect of adjustable-speed drives on the operation of low-voltage ground-fault indicators," *Industry Applications, IEEE Transactions on,* vol. 37, pp. 1423-1437, 2001.

[7] G. Skibinski and S. Breit, "Line and load friendly drive solutions for long length cable applications in electrical submersible pump applications," 2004, pp. 269-278.

[8] L. E. Lundgaard, "Partial discharge. XIII. Acoustic partial discharge detection-fundamental considerations," *Electrical Insulation Magazine, IEEE,* vol. 8, pp. 25-31, 1992.

[9] L. E. Lundgaard, "Partial discharge. XIV. Acoustic partial discharge detection-practical application," *Electrical Insulation Magazine, IEEE,* vol. 8, pp. 34-43, 1992.

[10] T. Kaneko, T. Ueda, O. Takenouchi, M. Otsubo, C. Honda, Y. Tsuruta, A. Fukura, and K. Tanaka, "Characteristics of on-line and off-line partial discharge on hydrogenerator stator windings using acoustic emission detection techniques," 2005, pp. 837-840 Vol. 3.

[11] *Standard Guide for Determining the Reproducibility of Acoustic Emission Sensor Response*, ASTM Standard E976-00.

Validation of Electrical Power System using Electrical Test Bed of GEO Satellite

Jae-Dong Choi, JA-CHUN KOO, CHEOL-HAE KOO, EUI-CHAN KIM
Korea Aerospace Research Institute
Communications Ocean and Meteolorogical Satellite Department
45, Eoeun-Dong, Yusong-Gu, Taejon 305-333, Korea
Email: jdchoi@kari.re.kr, jckoo@kari.re.kr, chkoo@kari.re.kr, eckim@kari.re.kr

Abstract—The Electrical Test Bed(ETB) integrates the test environment, which is required for acceptance tests of system level, prior to FM testing. This ETB is composed of engineering version units of spacecraft BUS, which are laid on the laboratory table. It uses for the validation of system level functions and interface between each subsystem. The ETB also supports early functional and limited performance checkout of electrical subsystems and provides the environment for the verification of the Flight Software including AOCS, EPS, and TC&R simulators.

I. INTRODUCTION

The Electrical Test Bed (ETB) is used to verify the interface in between each electrical units and its function at system level using engineering model before developing of the flight model. This ETB is configured of engineering models of satellite electrical system, which are consisted of Electrical Power System (EPS), Attitude Control System (ACS) and Telecommand/Telemetry System (TTS). And also, it is operated with connecting Electrical Ground Support Equipment (EGSE) during testing [1-4].

The ACS system on the ETB is verified the interface in between the hardware and software, then closed loop control logics are also validated at the real time. The TTS and EPS system are verified by the embedded software of On Board Computer (OBC) laid on the ETB. Actually, a newly developed electronic unit of satellite cannot be adopted as a satellite flight model within a short time because it needs to be passed for all environment and functional test. Therefore, a newly developed unit is generally demanded a lot of time to be accepted for these kinds of test.

This paper presents an integrated test for the Power Control Unit (PCU) and its management software on the ETB. The ETB and Software simulator are developed to perform verification for the power system, Telecommand/Telemetry system and attitude control system within a short time at satellite level.

A software simulator can quickly expect its results instead of a long time operation characteristics because it does not need to be perform at real time mode. And if some errors were found in the software simulator during the integration test, it can be easily corrected through the re-test.

At last, the developed software was verified and confirmed through the comparison results in between software simulation and hardware integration test.

II. ELECTRICAL TEST BED

A. System Configuration

As shown in Figure 1, the ETB is composed of Command/Telemetry Unit (CTU), Power Control Unit (PCU), On Board Computer (OBC), Vehicle Dynamic Simulator (VDS) and its support equipments.

The CTU unit is basically tested by connecting with command test equipment that receives a 50V bus voltage provided by the PCU and it has data interface via 1553B bus coupler with the OBC and VDS.

The PCU unit is connected with VDS acquisition board and CTU unit. The OBC installed the Flight Software is connected with the CTU unit, OBC monitoring unit and VDS Simulator through the 1553B bus coupler.

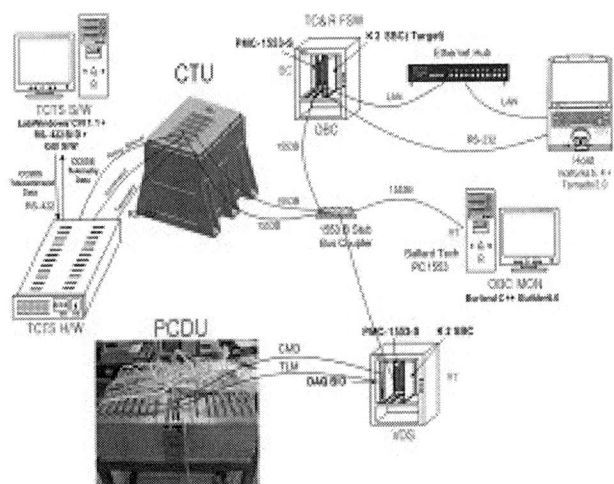

Fig. 1. Interface block diagram of developed Electrical Test Bed

B. Flight Software Configuration

The FSW loaded within the OBC unit of satellite provides telecommand and telemetry, management of satellite power, attitude control and orbit determination functions. The FSW configuration that is composed of several modules is shown in Figure 2.

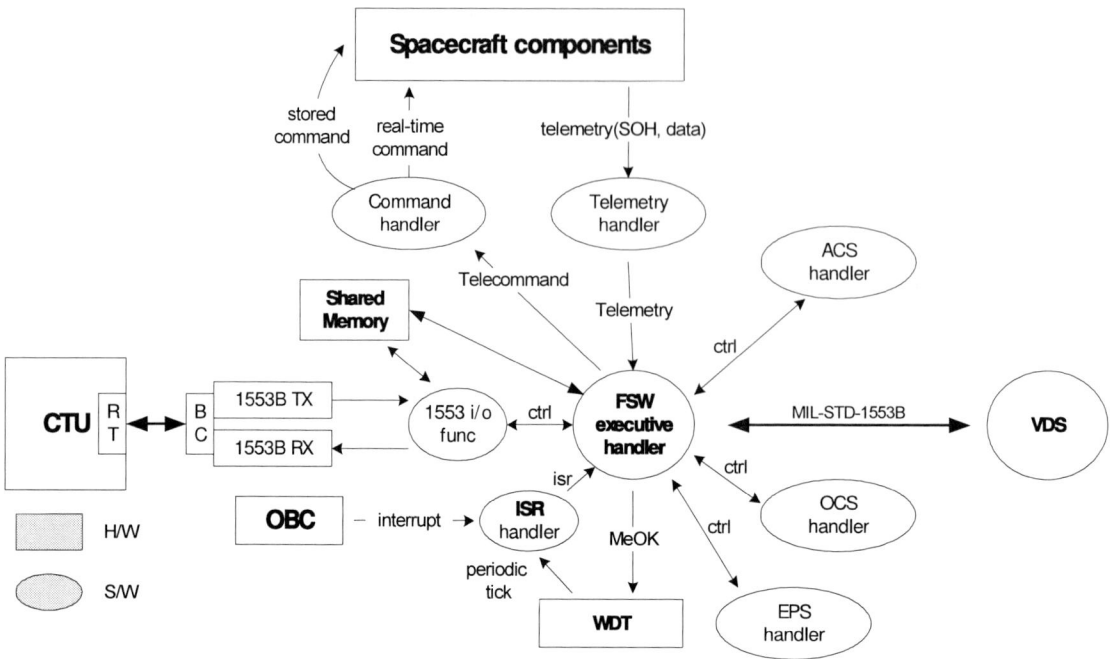

Fig. 2. Flight Software Configuration

The FSW are operated by two modes. First mode is a bus control mode which is to communicate with MIL-STD-1553B bus of OBC. Second one is a remote terminal mode having internal program which is to handle command and telemetry of the CTU unit.

The command of FSW is managed by polling method. The FSW acquires the command and telemetry through the MIL-STD-1553B data bus with 4Hz. If the command and telemetry unit receives the command data, then the collected data is handled directly. In order to transfer the data in the MIL-STD-1553B bus, the command data is firstly collected, and then uplink flag of the CTU is collected.

Command supported by the FSW is consists of a single command and a named table load command. The single command is performed one command at one time. The table upload command is used to transport an operation data of specific module of the FSW to be replaced to the other module from the ground control center. It is performed from the ground control center by the command in addition to the normal command.

Telemetry collected and formatted in the FSW is transported to the ground control center. These telemetries are generated in the internal FSW logic and processed in the FSW. And it is transported via MIL-STD-1553B bus from the satellite interface units connected with the OBC. The processed telemetries have 3 steps by the collection, formmating, and transportation to the command telemetry unit.

Regarding the attitude control software, it provides communication function over 16 Hz with the OBC and transport all data as 1/8 Hz with power control unit. And this software is interfaced with data server to be monitored and stored the parameters of Vehicle Dynamic Software (VDS) during the electric integration test on the ETB. And also, the FSW can be transferred and read the digital command and telemetry in order to drive the power control unit.

III. ETB INTEGRATION TEST

The ETB integration test is focused on internal and external function of target processor, and software command/telemetry input and output. The power system distributes power to each unit on the ETB and performs the battery charge and discharge function. And its test results are monitored through the 1553B data bus.

A. ELECTRICAL POWER SYSTEM

Figure 3 shows battery charge and discharge status of battery according to time variation. These simulation results are performed by simulator and the integration test of simulator was performed under following conditions. At first, AHCD is ENABLE (SOC=0.9) status and battery discharge is performed during 8 minutes. A charge cycle is repeatedly done by 27 minutes and 5A discharge current. Then, SOC as shown Figure 3(b) starts at full charge status and it is continuously decreased and fixed at a around 0.9.

Figure 4 shows battery charge/discharge integration test results presented discharge current, SOC, temperature, pressure and charge current.

978-1-4244-1871-8/07 $25.00 © 2007 IEEE

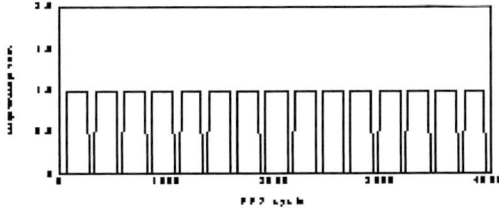

(a) Discharge Current

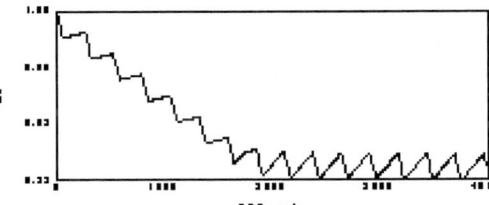

(b) SOC Variation

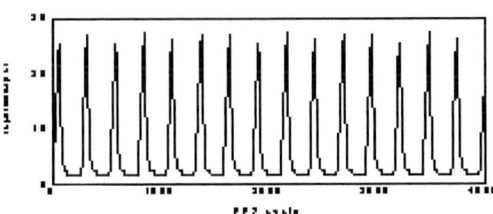

(c) Battery Temperature

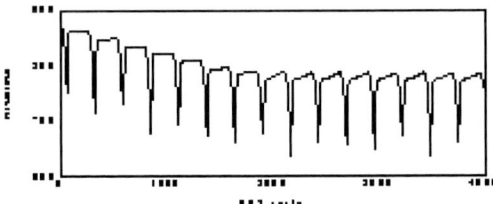

(d) Battery Pressure

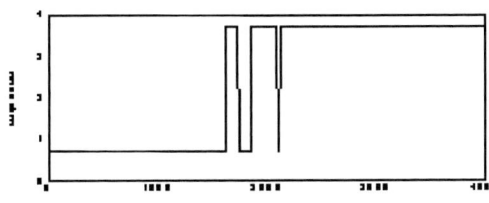

(e) Charge Current

Fig. 3. Battery Charge/Discharge Simulation Results

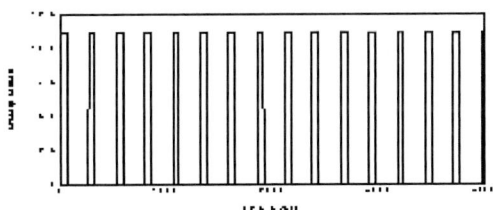

(a) Discharge Current

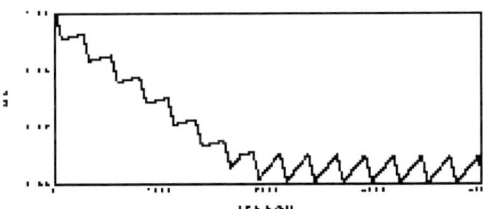

(b) SOC Variation

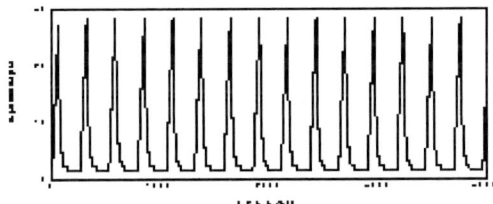

(c) Battery Temperature

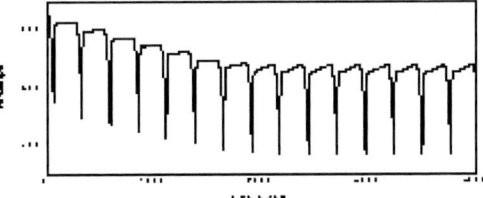

(d) Battery Pressure

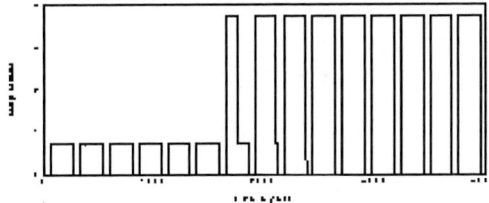

(e) Charge Current

Fig. 4. Integration Test Results for Battery Charge/Discharge status

B. Telemetry, Command Integration Test

Telemetry, Command Test Set (TCTS) was developed to verify a performance of the CTU. This TCTS provides the functions to check the input and output of the CTU.

The command generated by the ground control center which is simulated by an input generation function included in the TCTS and it is transferred to the CTU and checked for their correct function.

The monitoring function simulates the output telemetry generated from the ground control center and it validates the bi-level and relay output telemetry generated from the CTU. TCTS software performs the function to simulate the ground control center.

Figure 5 shows interface window developed to perform automatic command generation of the CTU. The user can be easily transport the command for requiring time repeatedly.

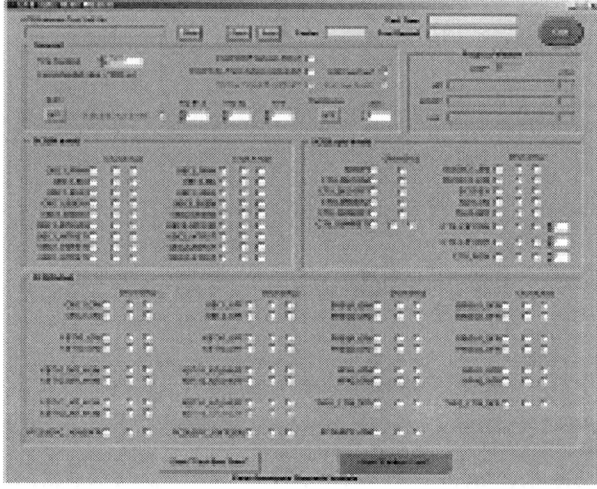

Fig. 5. Automatic command test interface of TCTS SW

IV. SUMMARY

In this paper, the ETB are composed based on the validation technology and its functions are validated through the system integration test. The TCTS, FSW, the PILS system, and the power control unit are developed for CTU verification, attitude control validation and integration test respectively. In order to verify the performance for each unit, the simulation and experimental at the unit level and system level are performed.

These simulation and experimental results are shown that the developed hardware and software are satisfied within required specification. The system level test is also verified through the ETB integration.

ACKNOWLEDGMENT

The authors wish to acknowledge their sincere thanks to Ministry of Science & Technology, Government of Korea for funding.

REFERENCES

[1] Bob Mammano and Mark Jordan, "Load Sharing with Paralleled Power Supplies," Unitrode Switching Regulated Power Supply Design Seminar Mannual, 1991, pp. 2-1 ~ 2-14.

[2] Paul M. Rostek, "Power System Design for Massive Parallel Computer Systems," IEEE APEC, pp. 808 ~ 814, 1994.

[3] Albert H Zimmerman and M. V. Quinzio, "Progress Towards Computer Simulation of NiH2 Battery Performance Over Life", The NASA Aerospace Battery Workshop, 1994, pp177-183.

[4] Jaedong Choi, and Sejin Seong, "Study on Switching Shunt Regulator for Satellite Power System", Journal of Electrical Engineering and Information Science, Vol. 3, No. 1, pp.14-20, 1998

Implement of power line communication module for an electric power energy monitoring system

Yong-Chae Jung
Namseoul University
Department of Electronic Engineering
21 Maeju-ri, Seonghwan-eup, Cheonan-city,
Choongnam, 330-707
Email: ychjung@nsu.ac.kr

Ho-Seok Jung and Jeong-Uk Kim
Nao Digital Co., Ltd.
#1106, Kolon Digital Tower Billant, Guro 3-dong,
Guro-gu, Seoul, 152-777
Email: junghs@naodigital.com, jukim@naodigital.com

Abstract— In case Power Line Communication (PLC) method is applied to an electric power energy monitoring system, PLC module should be utilized. In this paper, PLC module using ST7538Q is implemented. The detailed design procedures are presented. The operation of the implemented PLC module is verified through the experimental results.

I. INTRODUCTION

Power line communication (PLC) has received tremendous attention in recent years because of easy installation, low cost and so on [1-3]. In particular, Home Energy Monitoring System (HEMS) and Building Energy Monitoring System (BEMS) are good application fields.

In general, PLC is classified into low-speed type and high-speed type. The high-speed PLC is applied to high-speed internet modem such as 10Mbps, 100Mbps, and so forth. On the other hand, the low-speed PLC is mainly utilized as a home networking system. There are many low-speed PLC chips for implementing PLC module.

Because the PLC uses the power lines to be already existed, the installation of the PLC networks is very easy and low cost. Moreover, the enlargement of the PLC networks is also simple. Owing to these merits, the high-speed PLC has the competitive edge over the other techniques.

In this paper, the low-speed PLC module for an electric power energy monitoring system is presented using ST7538Q. This module is applied to rainwater management system to demonstrate its feasibility and effectiveness. Through the experimental results, the operation of the presented PLC module is ascertained.

II. POWER LINE COMMUNICATION MODULE

A. Rainwater Management System

In March 2007, Seoul Metropolitan Government held the Rainwater Management Equipments Exhibition to make a model city of green environment. The rainwater is impounded in some areas such as a building, a park, and so on, and it utilizes as for landscape architecture, street cleaning, and so forth. Moreover, the rainwater management plays a role of alleviating the flood damage.

In the rainwater management system, the amount of rainwater is detected using a water level transmitter. The sensing data is transferred to the main management system.

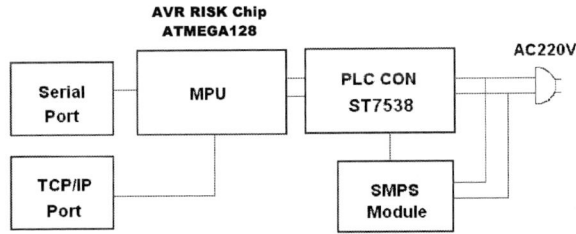

Fig. 1: Block diagram of the PLC module

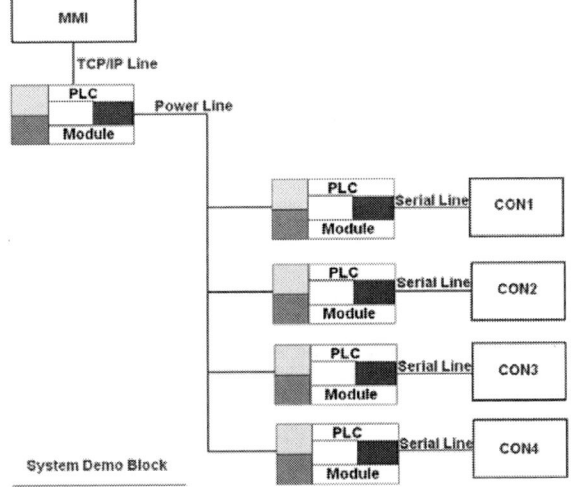

Fig. 2: Total control system structure with PLC modules

In the paper, the PLC is selected as the communication method.

B. Power Line Communication Module

As mentioned above, there are many PLC chips for a low-speed application, such as ST7538Q, ST7540, MAX2986, MAX2980, TMS320F28××, etc. Among them, ST7538Q manufactured by STMicroelectronics is selected in this paper due to easy of access. It is a half duplex frequency shift keying (FSK) transceiver. The designed carrier frequency and the baud rate are 132.5 [kHz] and 4800 [bps], respectively. Of course these two values are programmable and the largest ones.

Fig. 1 shows the PLC module block diagram. It has a serial port and a TCP/IP port. The microprocessor is ATMEGA

978-1-4244-1871-8/07 $25.00 © 2007 IEEE

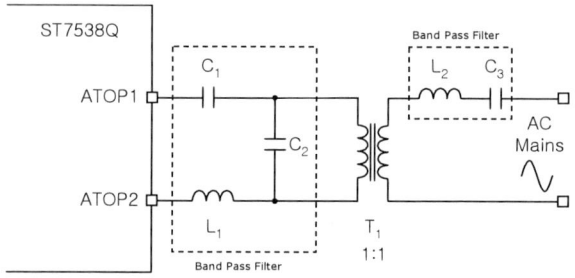

Fig. 3: Power line interface circuit

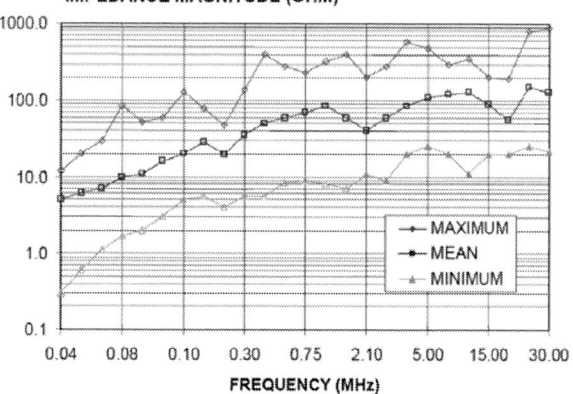

Fig. 4: European power line impedance

128 of ATMEL Corporation. The PLC module is installed in each of control unit. The several slave PLC modules are interfaced with a main computer through the power line and the master PLC module.

As can be shown in Fig. 2, the total control system applied to an electric power energy monitoring system consists of several PLC modules with the corresponding objects of control. These two parts are connected with a RS-485 serial line. A main PLC module exchanges a command with several slave modules through the power line. Also the main control center (MMI; Man-Machine Interface) communicates with the main PLC module using a TCP/IP port.

C. IC Peripheral Circuit Design

The power line interface circuit of ST7538 is shown in Fig. 3 [4]. This circuit is composed of an isolation transformer and two band-pass filters for passing the carrier frequency signal. In this figure, the left Tx band-pass filter is in charge of high-side cut-off frequency, and the right Tx band-pass filter is in charge of low-side cut-off frequency. The cut-off frequency of the Rx band-pass filter is the same as the above designed carrier frequency.

The power line impedance should be known to design the isolation transformer. Fig. 4 shows the European power line impedance according to the frequency axis [4]. It can be shown in the figure, the impedance range for the carrier frequency is roughly 1~100[Ω]. Assuming that the domestic power line impedance is similar to the above one, the transformer has been designed. The transformer has 1:1 turn

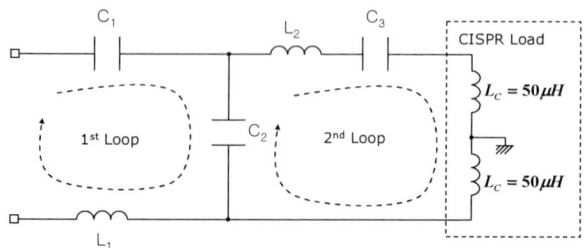

Fig. 5: Equivalent Power line interface circuit

ratio and the designed transformer impedance is about 20[Ω] at the carrier frequency.

The equivalent circuit of the power line interface circuit is depicted in Fig. 5. It is the primary side view of the transformer, and AC mains are replaced to CISPR load. In case, the first loop makes the high-side cut-off frequency of the band-pass filter and the second loop has charge of the low-side cut-off frequency. These two cut-off frequencies are calculated by (1) and (2) as follows.

$$f_h = \frac{1}{2\pi\sqrt{L_1 C_A}} \approx 160[kHz] \tag{1}$$

$$f_l = \frac{1}{2\pi\sqrt{L_B C_B}} \approx 100[kHz] \tag{2}$$

In the above equations, C_A, C_B and L_B are as follows.

$$\frac{1}{C_A} = \frac{1}{C_1} + \frac{1}{C_2} \tag{3}$$

$$\frac{1}{C_B} = \frac{1}{C_2} + \frac{1}{C_3} \tag{4}$$

$$L_B = L_2 + 2L_C \tag{5}$$

As can be known from (1) and (2), the bandwidth of the band-pass filter is 60[kHz] centering the carrier frequency, 132.5[kHz]. In practice, the two cut-off frequencies are 101.9[kHz] and 156.7[kHz], respectively.

By the way, L_C represented CISPR load in Fig. 5 is the reactive component of LISN (Line Impedance Stabilization Network). To measure the characteristics of the power line interface circuit, the power source is supplied by way of LISN. Therefore, the two reactive components have been expressed for two power lines in the figure.

III. EXPERIMENTAL RESULTS

Several PLC modules are manufactured and then applied to the rainwater monitoring system. Fig. 6 shows the exterior

978-1-4244-1871-8/07 $25.00 © 2007 IEEE

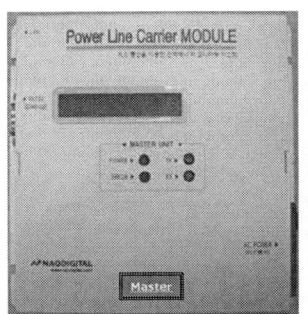 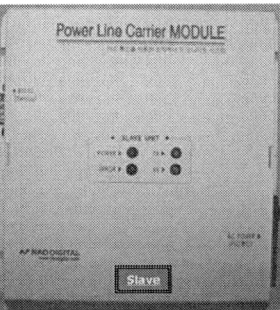

Fig. 6: Manufactured PLC modules

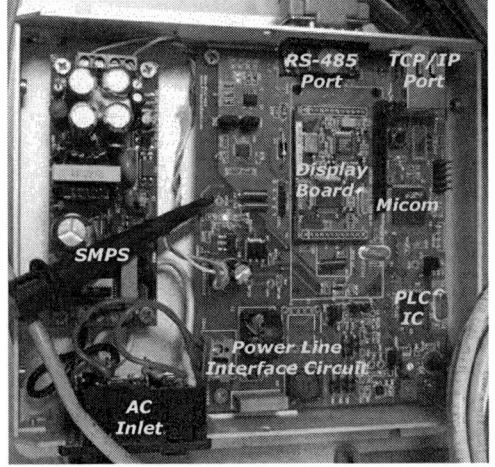

Fig. 7: The interior photograph of the main PLC module

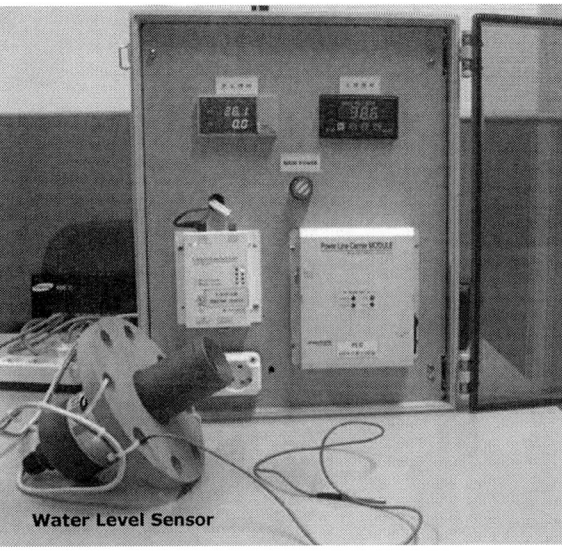

Fig. 8: The photograph of the rainwater management system

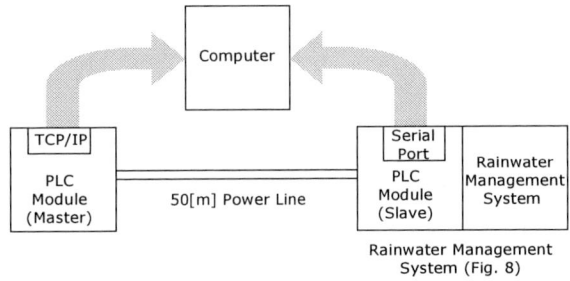

Fig. 9: Experimental configuration

views of the main PLC module and the slave one. There is distinctively a LCD window in the main PLC module. The LCD serves to display the communication status of the module. The interior photograph of the main PLC module is shown in Fig. 7. The photograph shows each block, which can be seen in Fig. 1.

The above manufactured modules are applied to rainwater management system to confirm their feasibility and effectiveness. Fig. 8 shows the photograph of the produced rainwater management system with water level sensor. This product has been displayed the Rainwater Management Equipments Exhibition held by Seoul Metropolitan Government in March 2007.

In this paper, to demonstrate the operation of the PLC modules, the experimental setting shown in Fig. 9 is built and tested. The connecting line between two modules is the electric power line utilized the communication wire in PLC. The distance from the master module to the slave one is 50[m].

At first, a data is transferred to the main PLC module from computer. Then, the main PLC module conveys the data to the slave PLC module through the long power line. The slave PLC module transmits the receiving data to the computer immediately.

Fig. 10 shows the data sending screen through the TCP/IP port toward the main PLC module. The data from a to z is transferred to the main PLC module during periods of 500[msec]. Fig. 11 shows the data receiving screen through the serial port from the slave PLC module. In this case the data is also received at intervals of 500[msec]. It can be seen in this figure that the data of 19 set has been taken without error. Therefore, the power line communication has been verified through the experimental results.

IV. CONCLUSION

In this paper, we investigate the power line communication (PLC) module applied to an electric power energy monitoring system. The constitution of the PLC module is explained in detail. And, the design procedures of the peripheral circuit of the PLC IC, ST7538Q, are presented. The PLC modules divide into two types, master and slave, respectively. So, several PLC modules are manufactured and then applied to the rainwater monitoring system to demonstrate its feasibility and effectiveness. The data transferring experiment is performed. In result, it can be confirmed that the power line communication has been operated nicely.

978-1-4244-1871-8/07 $25.00 © 2007 IEEE

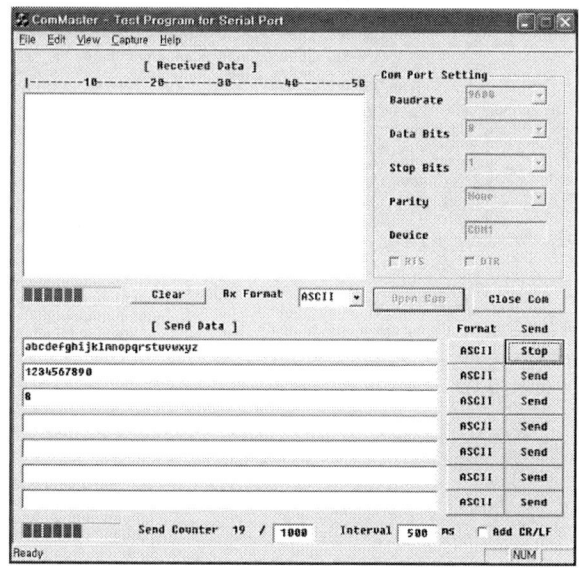

Fig. 10: Data sending screen toward the main PLC module

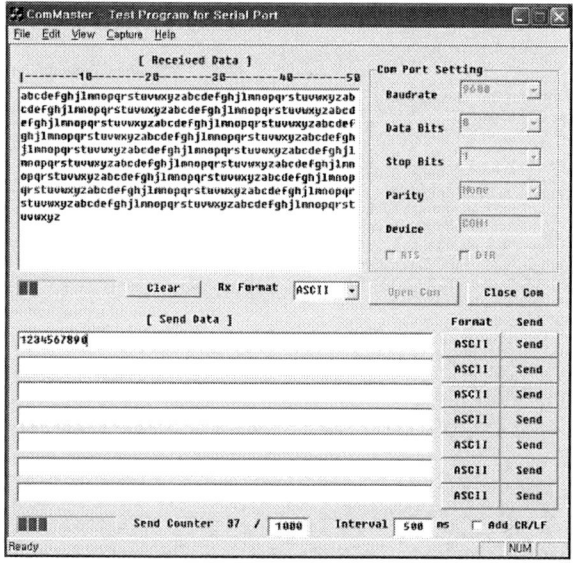

Fig. 11: Data receiving screen from the slave PLC module

From now on, we are going to apply the developed PLC modules to various kind of energy monitoring fields.

ACKNOWLEDGMENT

This work has been supported and funded by Korea Energy Management Corporation (KEMCO).
  Project Number : 2005-E-EL02-P-08-3-010-2006

REFERENCES

[1] S. Saggini, W. Stefanutti, P. Mattavelli, G. Garcea and M. Ghioni, "Power line communication in dc-dc converters using switching frequency modulation", IEEE Applied Power Electronics Conference and Exposition (APEC '06), pp.1595-1600, March 2006.

[2] M. Hagon, M. Heminger and A. Mohammed, "Power line communication for Lighting Applications using Binary Phase Shift Keying (BPSK) with a Single DSP Controller", IEEE Applied Power Electronics Conference and Exposition (APEC '06), pp.823-828, March 2006.

[3] Sam-Jin Lee, Seung-Hak Yang and Gie-Hyun Kwean, "Power Sources Control System of Home Appliances Using Power line communication", KIPE 2001 Power Electronics Annual Conference, pp.666-668, July 2001.

[4] Giuseppe Cantone, "ST7538 FSK Power-Line Transceiver Demo-kit Description", STMicroelectronics Application Note AN1714, 2003.

Electrical Interfaces Compatibility Analysis for the COMS EPS

Ja-Chun Koo
Korea Aerospace Research Institute
45 Eoeun-dong, Yuseong-gu, Daejeon, 305-333
Email: jckoo@kari.re.kr

Eui-Chan Kim
Korea Aerospace Research Institute
45 Eoeun-dong, Yuseong-gu, Daejeon, 305-333
Email: eckim@kari.re.kr

Abstract—The COMS(Communication, Ocean and Meteorological Satellite) EPS(Electrical Power Subsystem) is derived from an enhanced Eurostar 3000. The COMS EPS can offer a bus power capability of 3 kW [1].
The aim of this analysis is to verify the electrical compatibility of the interfaces which exist between COMS EPS equipments and external equipments. For each interface, this study checked the compatibility between equipments for the power links, commands, digital telemetry, analog telemetry and failure condition. In addition with this interface compatibility verification, this study outputs electrical and manufacturing constraints to be applied at harness level.

I. INTRODUCTION

The interfaces between EPS equipments and external equipments are identified in the Fig. 1 [2]. It deals with battery, solar array, SADM(Solar Array Drive Mechanism), PSR(Power Supply Regulator) and PRU(PyRotechnic Unit). For each interface, the following aspects have been studied in order to check the compatibility between equipments:

- Power links: the electrical analysis is focused on DC voltage and current compatibility. Electrical perturbations and transient are also addressed if necessary.
- Commands: the electrical analysis is mainly focused on the command voltage level and duration and the maximum load connected to the command driver.
- Digital telemetry: the electrical analysis is mainly focused on the definition of low and high levels and their correct decoding.
- Analog telemetry: the electrical analysis is mainly focused on the compatibility between the acquisition channel and the sensor operational range and on the study of AC perturbations.
- Failure conditions: the impact of failures is studied in order to check the robustness of the interface circuits and the absence of failure propagations at both ends.

In addition with this interface compatibility verification, one of the outputs of this study is electrical and manufacturing constraints to be applied at harness level:

- Maximum voltage drop for power links
- Twisted wires or shielding for analog signals

Depending on the nature of the analog signal, the AC perturbations are checked using the normalized criteria telemetry in order to identify the most sensitive analog interfaces. When immunity problems are identified, AC

perturbations are modeled more realistically then the analysis performs more accurately:

- CM(Common Mode) perturbations: the AC perturbation is modeled with a sinusoidal voltage source, its amplitude is adjusted at a level of 1 Vp and the AC analysis is performed on the frequency range 0 to 10 MHz [2]. The acceptability criterion is to have an attenuation of more than 48 dB wrt(with respect to) the full scale in order to be in line with telemetry resolution requirements which corresponds for the EPS to the 8th bit [2].
- DM(Differential Mode) perturbations: the AC perturbation is modeled with a sinusoidal voltage source, its amplitude is adjusted at a level of 1 Vp or less depending on the telemetry nature and the AC analysis is performed on the frequency range 0 to 10 MHz [2]. The acceptability criterion is identical to CM perturbation analysis [2].
- Harness coupling: the AC perturbation is modeled by a transformer in series with the perturbed wire. The transformer has a coupling ratio of 1 and is connected to a sinusoidal current source of 1 mAp amplitude in the frequency range 10 kHz to 10 MHz [2]. The acceptability criterion is identical to CM perturbation analysis. Twisting of wires ensures an attenuation of 40

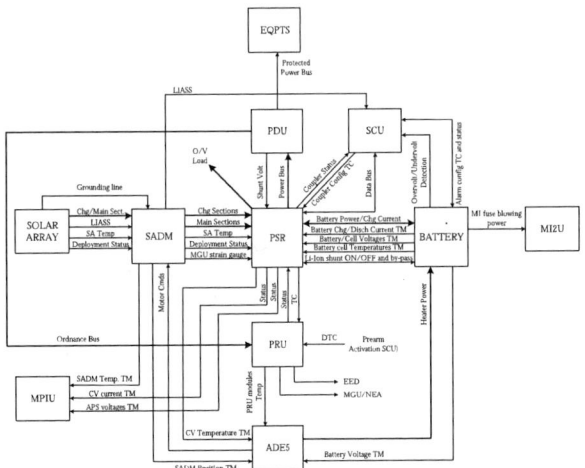

Fig. 1: COMS EPS interfaces

dB (50 turns/m) [2].

II. INTERFACES COMPATIBILITY CHECK

A. Battery Interfaces

Power Interfaces

For surge currents in the DIET(Design Interface Environment and Test for equipment), a margin ration of at least 4 shall be considered with wrt the fuse manufacturer data [3]. The required fuse blowing power transient except MI(Meteorological Imager) power interface can be calculated for a P600L 15 A fuse as Eq. 1 according to the manufacturer data sheet and gives for single 15 A fuse in worst case conditions (250 % overload and 0.3 sec fuse blow case [4]) the following result:

$$I^2 t_{max} = 4*(2.5*15)^2*0.3 = 1680 \text{ A}^2\text{s}. \tag{1}$$

COMS Li-Ion battery specifications give a requirement for a current transient capability of 3 C during more than 250 ms. COMS battery capacity is 192.5 Ah (five 38.5Ah cells in parallel) and a current transient capability of more than 80000 A^2s. This capability is largely greater than the required energy.

The weakest power elements of BDR(Battery Discharge Regulator) power path are the BDR power diodes. To access the compatibility of these diodes with transient events such as fuse blowing, the maximum temperature of a diode has been computed for the following conditions:

- worst case two BDRs are operating on the power bus
- 90 % of the fuse blowing current through a single diode of each BDR (two diodes exist in each BDR)
- same current of the fuse blowing through each BDR (i.e. half of the fuse blowing current through each BDR)
- Case temperature 95 °C by PSR thermal analysis result

The resulting diode junction temperature T_j is given in Fig. 2 and is below the maximum rating temperature 150 °C. This result demonstrates with comfortable margins for the correct sizing of the BDR modules wrt maximum COMS fuse configuration. For a fuse blowing duration lower than 100 μs, the analysis is less critical since the bus capacitor provides a significant part of the required energy wrt battery. Fuse blowing events greater than 100 ms induce a large payload shedding [2] which permits to reduce significantly permanent currents of the BDR diodes.

The battery provide power to BDR converters in eclipse conditions and to BCDR(Battery Charge Discharge Regulator) converters in charge conditions. The 50 V power bus is used to provide power to PRU boards for pyrotechnic actuators. The specified and tested input voltage ranges correspond to the 27.5 to 47 V for BDR/BCDR and 26 to 51 V for PRU [2]. The number of cells in series per battery is 10. In case of cell failure in short circuit or open circuit, no additional voltage drop needs to be added wrt the cell loss. The battery voltage range is equal to 29.1 to 42.13 V [2] and is compliant with the BDR input voltage. The 50 V power bus is compliant with the PRU input voltage.

The battery-MI2U(Meteo Imager Interface Unit)-MI interface is as Fig. 3 [2]. The MI instrument uses redundant fusing configuration. All redundant fuses are connected through a single non-redundant power input, which is connected all MI DC/DC converters or heaters. The objective of this cold redundancy fusing is to protect the nominal fuse branch, fully able by itself to handle the maximum operational current. The positive terminal of 9th cell module in battery supplies energy to the battery switch within MI2U [2]. The interface between battery and MI2U is connected directly by wires without fuse [2]. Battery can blow the MI fuses if the 42 V output drops below the positive terminal voltage of 9th cell module in battery.

When short occurs inside MI power interface, the required fuse blowing energy is 384 A^2s for two 10 A fuses in parallel (more conservative assumption than 10 A cold redundancy fusing configuration and more worse case of 200 % overload and 240 ms fuse blow) according to the LITTELFUSE characteristics curves [5]. No soft short (less than 200 % overload of 10 A fuse) to ground in the harness level and MI instrument level considered in this calculation. The capability of Li-Ion battery is largely greater than the required energy for MI fuse blowing even if the required and delivering energy are compared with different time (250 ms vs. 240 ms). This margin ratio is more than 200 which also largely compatible with DIET.

Command Interfaces

The Fig. 4 shows the forced mode and by-pass

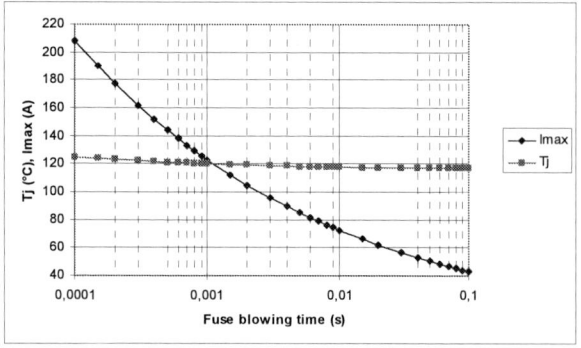

Fig. 2: Maximum BDR diode junction temperature for COMS

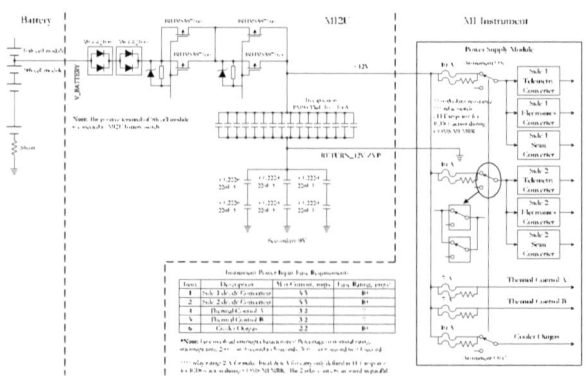

Fig. 3: Battery-MI2U-MI interface

TC(TeleCommand) interface of cell module in parallel with PRU EED(Electro Explosive Device) fire command (two optocoplers in parallel with one GP5 relay coil and one T12 non latching type coil). A maximum required current is 235 mA with minimum resistance of relay coils and maximum current of optocoplers. This value is largely compatible with PSR TC current capability of 800 mA. The minimum operate time of the GP5 relay is 5.5 ms. The PRU is full compliance with PSR TC pulse duration of 51 ms and the required derating rules of 100 % for relays. A 5 ms pulse is enough to drive optocoplers. For relay interface, the required TC interface levels are 12.1 to 15.6 V and therefore compatible with the PSR TC voltage level of 12.1 to 15.1 V. For optocoplers interface, the required TC interface levels are 9.9 to 22.9 V and therefore compatible with the PSR TC voltage level.

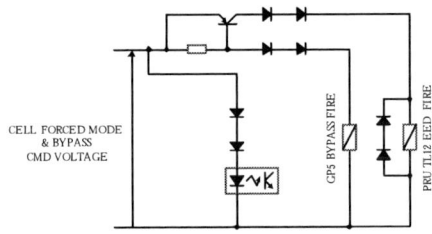

Fig. 4: Forced mode and by-pass TC interface

Telemetry Interfaces

The battery voltage is acquired by the PSR. A TM(TeleMetry) range at battery level of 0 to 47.03 V for PSR acquisition and this TM range is largely compatible with the battery voltage range computed in 29.1 to 42.13 V. The AC analysis of this acquisition channel with a CM perturbation of 1 Vp amplitude at battery return level gives a maximum perturbation of - 50 dBV with a worst case unbalance between resistors on positive and return lines, PSR resistors. The resulting curve of this AC analysis is shown in Fig. 5 and demonstrates the immunity of the battery voltage TM to CM perturbations. The results of the AC analysis of this acquisition channel with a DM perturbation of 1 Vp

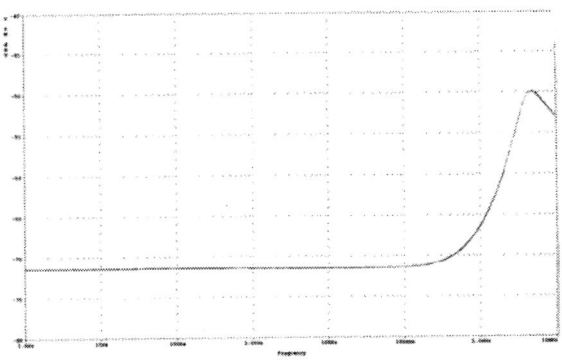

Fig. 5: AC perturbations with CM

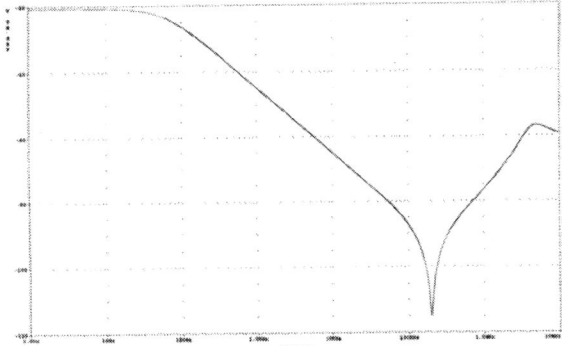

Fig. 6: AC perturbations with DM

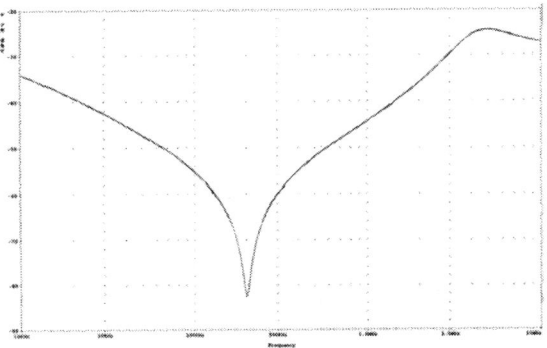

Fig. 7: AC perturbations with harness coupling

amplitude is shown in Fig. 6. For 300 Hz to 10 MHz, the perturbation level on the TM voltage is lower than -35 dBV and does not induce any problems. The domain 0 to 300 Hz corresponds to the TDMA(Time Division Multiple Access) frequency range where the TM voltage by design is more sensible to differential AC perturbations (up to -15 dBV). Same rationale as for CM is used to accept this perturbation level. The results of the AC analysis with a harness coupling current of 1 mAp amplitude is shown in Fig. 7 and gives a maximum perturbation level of -25 dBV on the TM voltage. This perturbation can be attenuated by the use of twisted wires to reduce coupling effects (attenuation of more than -40 dB for a wire of 1 m).

B. Solar Array and SADM Interfaces

Power Interfaces

The maximum current per SADM power slip ring is 11 A. The maximum current per PSR switch is 20.3 A. The PSR switch can drive up to two SADM power slip rings of which total maximum current is 20 A. The compatibility between each section and the SADM/PSR was ensured by the solar array manufacturer by analysis (beginning of life, best case, winter solstice, sunlight normal to panel).

The solar array return philosophy for the COMS platform is to have a common return for solar array section returns.

Those solar array section returns are connected on PSR power modules. Return pins dedicated to SADM return (for solar array section return) are used on PSR power modules. The PSR return studs(bus return) located on the central module of PSR are grounded to the spacecraft structure by 8 power inserts and which ensure the power return for all bus users configured in structure return. This configuration has been validated by tests. The solar array voltage is specified at 51.5 V for a bus voltage at 50 V. The maximum voltage drop is 1.5 V in end of life and worst case conditions between the solar array output and the bus output, the maximum section harness voltage drop has to be lower than 720 mV @15.2A (end of life solar array current).

Command Interfaces

The compatibility between the ADE5(Actuator Drive Electronics of 5th generation) SADE(Solar Array Drive Electronics) board and the SADM model has been demonstrated by tests. The Fig. 8 shows recorded current profiles and demonstrates compliance with the SADM specification and life test conditions. Oscillations on each step have been detected during coupling tests and have been considered as acceptable at both sides. The SADE converter is compatible with a wide voltage range (20 to 50V) and is largely compatible with the power bus voltage range even in case of degraded mode (down to 35 V in DIET requirement). Moreover a diode in series with the power line prevents from damaging of the equipment in case of an inversion of polarity. This diode prevents from rejecting power on the power bus and permits to be tolerant to bus voltage drop of greater than 1 ms.

Telemetry Interfaces

The temperatures of the solar array outboard and inboard panels are acquired by the PSR. This interface gives a temperature range of -170 to 80 °C. The wire resistance of this harness can be estimated at 2.5 Ω and represents an

important error (about 2 %) wrt the nominal value of the sensor at 40 °C. This result demonstrates the need for the inclusion of the harness resistance in the calibration curve of this TM parameter. The AC analysis of this acquisition channel with a coupling current of 1 mAp gives a maximum perturbation level of -30 dBV on the TM voltage. This result is not acceptable and has to be attenuated with the use of twisted wires. The application of a short circuit to ground on one side of both inputs do not induce any particular stress on components. The application of a voltage of ± 16 V is not compatible with the PSR circuit interface and with the temperature sensor.

C. PRU Interfaces
Power Interfaces

The PRU design is compatible with a direct connection to the power bus since the maximum input voltage of each PRU board is 51 V. The qualification tests of the PRU have shown a susceptibility of the converter for an input voltage ripple of more than 400 mVrms on the frequency range 5.7 to 10 MHz. But in that range of frequency, typical ripple measured on 50 V power bus is rather 0.4 mVrms which ensures 60 dB margin wrt PRU susceptibility. The PRU input relay is a GP250 relay; this relay is not compatible with the maximum current of 2 $I_{\text{fuse rating}}$ (i.e 20 A defined in DIET) in case of short circuit without fuse blowing. This discrepancy is nevertheless accepted as the PRU board is protected with a breaker of 8 A, and no resistive short circuit to ground has been identified at input filter which would induce a short without fuse blowing.

The compatibility of the PRU with EED or Kevlar(thermal) devices is already demonstrated with the pyrotechnic boards (see test results in Fig. 9). The Li-Ion by-pass mechanism devices are different compared to the conventional igniters used for all pyrotechnics functions. As the device is of the type "make before break", there is no power interruption during the failure passivation. For release(melting), the fuse wire is electrically heated up until is breaks. The minimum PRU specified output voltage is 19.5 V. For EED devices, a specification is taken 10 V for a current of 7 A. This specification permits to have a satisfactory margin between the PRU regulation voltage and

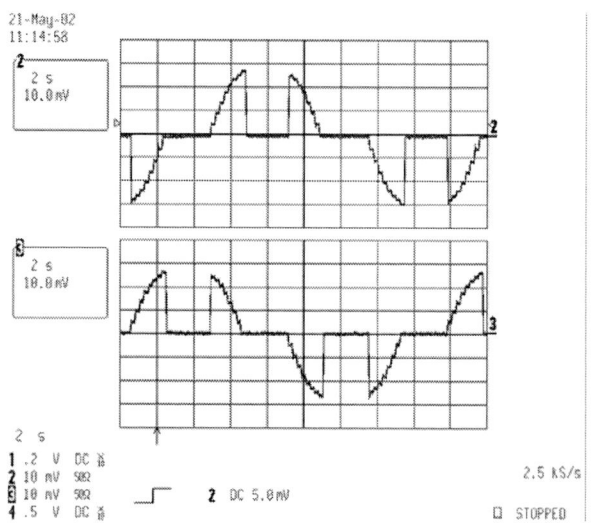

Fig. 8: SADM phase TC profiles (200 mA/div)

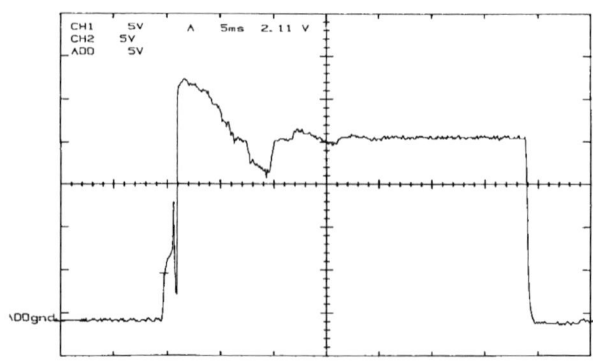

Fig. 9: EED activation of PRU power output

the initiator voltage with the PRU limitation current of 6 A ± 1 A.

Command Interfaces

The PRU TCs are mainly issued from the PSR. For PREARM activation, the SCU(Spacecraft Computer Unit) is used in order to separate functionally these TCs which connect or disconnect PRU boards from 50 V power bus. TC interface for EED mode relays activation is associated to the EED mode TCs of each pyrotechnic regulator. The TC maximum current is 38 mA and is largely compatible with the PSR TC driver characteristics of 800 mA. The required TC duration is not sized by the operate time of the relay but by the total relay activation duration which determines the PRU regulator pulse duration. This TC pulse duration of 48 ms ± 0.5ms has demonstrated the PRU full compliance wrt its performance specification. The switching voltage values are 11.9 to 17.4 V and are therefore compatible with the TC characteristics between 12.1 to 15.1V.

Telemetry Interfaces

The regulator status is associated to the regulator operating mode of each PRU board and is acquired by the PSR as Fig. 10. The state "0" corresponds to the active state of the PRU board regulator with the 2N2222 transistor in saturated conditions. The state "0" value of 0.33 V is therefore compatible with the PSR digital acquisition channels of less than 1.5 V. The state "1" corresponds to the disactive state of the PRU board regulator with the 2N2222 transistor OFF. The state "1" value is therefore compatible with the PSR digital acquisition channels of greater than 3.5 V.

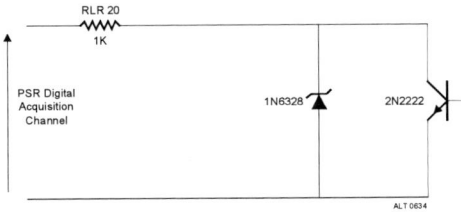

Fig. 10: PRU regulator status interface

D. PSR Interfaces

Power Interfaces

The APS(Auxiliary Power Supply) converter starts up at 20 V (first part of the transient) and operates up to 55 V which is largely compatible with the bus and battery voltages. For lower voltages, the OFF state of the PSR ensures the permanent connection of the power sources on the bus and by consequence the automatic restart in case of sunlight on the solar array. The input of each APS converter is protected with a 7.5A P600L fuse and a GP250 12V relay activated by an overtemperature protection. The converter includes an UVD(Under Voltage Detection) protection adjusted at a level lower than 19.5 V.

Each TM/TC converter is specified for a voltage range 20 to 55 V which is largely compatible with the bus and battery voltages. The input of each converter is protected with a P600L 3A fuse in series with a GP250 relay. Each relay is associated to a precharge resistor and isolates the filter from the rest of the converter when switched off. This is the reason why there is no inrush current at converter switch on. The conducted emissions of the CV TM/TC are negligible. The converter includes an UVD protection adjusted at a level lower than 19.5 V.

Each BDR is protected by fuse at the following levels:
- BDR input filter: Each capacitor is protected with a 2 A P600L fuse in series
- BDR switch: Each switch is protected with a 15 A P600L fuse in series

The BDR input relay is an EL215 relay and has been evaluated to allow BDR disconnection in case of switch short circuit with no fuse blowing(up to 30 A). The inrush current event of a BDR can damage the BDR, this is the reason why each BDR shall be switched on with the bus in sunlight regime (assembly, integration and test constraint). When the BDR filter is precharged(sunlight mode), the inrush current is less important.

Command Interfaces

Each PSR coupler TC and are connected to a SCU TC driver as Fig. 11. The TC maximum current is 315 mA and is largely compatible with the SCU TC driver characteristics of 650 mA. The maximum operate time of the GP250 and TL12 relays is 12 ms, this value is largely compatible of the TC driver characteristics (48 ms) and the required derating rules of 100 %. The switching voltage values give a voltage range from 12 to 15.6 V. The maximum SCU level of 17 V exceeds the maximum voltage of this interface. This non compliance is nevertheless acceptable since this voltage is only applied during 48 ms and can not stress relay coils in these conditions.

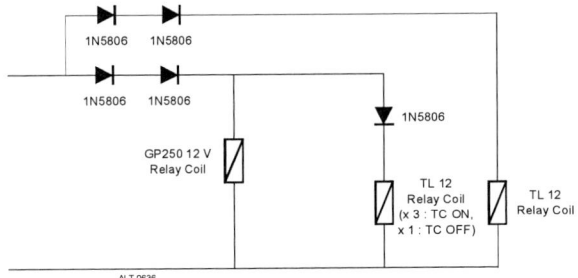

Fig. 11: PSR coupler TC interface

Telemetry Interfaces

The status of each PSR coupler is acquired by the SCU. The state "0" corresponds to close the relay with maximum resistance of 0.4 Ω, the state "1" corresponds to open the relay with minimum resistance of 10 MΩ. These values are compatible with the SCU characteristics of ON state with R < 1Ω and OFF state with R ≥ 1 MΩ.

III. SYNTHESIS AND CONCLUSIONS

The results of this electrical compatibility analysis performed on EPS interfaces have demonstrated the correct sizing and level of robustness of EPS interface circuits.

The analysis of the EPS power interfaces is satisfactory and has permitted to obtain the following results:

- Battery: the COMS electrical configuration for battery (harness, connectors, shunt) is compatible with battery capacities of up to 330 Ah. In terms of voltage, a configuration with 10 to 11 Li-Ion cells is compatible with the PSR. The PRU is connected directly to the power bus as well (400 mVrms max).
- Fuses: the use of fuses of up to two P600L 15 A fuses in parallel (single P600L 15A fuse for COMS except interface within MI) is compatible with the EPS components in terms of surge current and does not require BDR by-pass diodes.
- Power return philosophy: the COMS power return philosophy is to use structure return for power bus users, battery returns and SADM returns (solar array section return) are common with bus return by connection to each PSR power modules with dedicated wires. The PSR return studs(bus return) located on the central module of PSR is grounded to the spacecraft structure by 8 power inserts and which ensure the power return for all bus users configured in structure return.
- PRU input: the only fuses to be installed in the harness on dedicated brackets for EPS equipments are the PRU module input fuses. The derating rules lead to one 10A P600L fuse per PRU module input.

The PSR has been qualified with a TDMA level of 60 App with a duty cycle 50% (with mono PSR composed of 16 power modules) and it corresponds to a TDMA level for COMS specific PSR of 22.5 App (60 App/16*6 modules for COMS). The level of variation for COMS is around 8 App, largely covered by the PSR qualification. The worst case voltage ripple induced on bus was 109 dBμV (280 mVrms) at PSR output level, which represented an EMC margin of 11 dB wrt the conducted susceptibility qualification level of units (120 dBμV=1 Vrms in compliance with DIET). The test has been performed on a single bus, in sunlight and eclipse modes.

The analysis of the EPS TC interfaces is satisfactory. The only points to be mentioned are the maximum voltage levels of TL12 and GP250 relays which are slightly exceeded(less than 2 V) during the TC pulses (51.5 ms max), this derogation is acceptable.

The analysis of the EPS status interfaces is satisfactory. The only point to be mentioned is PRU status(bi-level or relay contact) which have a common return (matrix configuration forbidden).

The analysis of the EPS analog interfaces is satisfactory. The only points to be mentioned are the following ones:

- TM/TC converter currents: these TM are sensible to CM perturbations (up to 0 dBV). These perturbation levels are nevertheless acceptable as these TMs are only used for ground monitoring.
- Battery charge/discharge currents: these TM are sensible to DM. These perturbation levels do not have any influences on the computation of the battery state of charge. Shielded and twisted wires are used for this interface.

ACKNOWLEDGMENT

We would like to the Ministry of Science and Technology for the supply to this research.

REFERENCES

[1] J.C. Koo, *"COMS EPS Description,"* unpublished.
[2] J.C. Koo, *"COMS EPS Interfaces Compatibility Check,"* unpublished.
[3] V. Claudet, F. Chaudon, C. Ricolleau and L. Taille, *"Eurostar 3000 Design, Interface, Environment and Test Requirement Specification for Equipment (DIET),"* unpublished.
[4] *"Specification Sheet P600L Current Limiting Fuse,"* AEM INC., December 1999.
[5] *"Time Current Characteristic Curves,"* Littelfuse, February 1997.

Analysis of Effects of Inductance Component in Electrodeless Lamp on Ballast Performances

Soo-Bin Han, Sukin Park, Eugene Song, Hak-Guen Jeong, Bong-Man Jung
Korea Institute of Energy Research
71-2 JangDong YusungGu Daejeon, Korea
Email: sbhan@kier.re.kr, psi@kier.re.kr, eugenesung@kier.re.kr, Jeoung@kier.re.kr, jbm@kier.re.kr

Abstract— Ballasts for electrodeless lamp have resonant circuit as an interface means between ballast and lamp like ballasts for fluorescent lamp. Most popular resonant filter circuit is LCC and LC-C type. Their characteristics and transfer functions with electrodeless lamp are not known well although resonant filter has an important role in both start-up and steady state. In this paper, various kinds of transfer functions for electrodeless lamp including inductance component of coupling core are given and their characteristics are analyzed with the variations of coupling inductance value.

I. INTRODUCTION

Ballast design with proper reliability is not an easy task due to a lot of parameters that affect the ballast performance in real world such as circuit topology, components rating, surge voltage/current, lamp characteristics and temperature variation. Especially, ballast for electrodeless lamp is operated at much higher frequency than conventional lamp ballast up to MHz range [1-2]. In this frequency range, sensitivity of parameters to the ballast performance can be very high, which gives a great difficulty if a self-oscillation type ballast is developed by trial and error.

Main factors that determine the ballast operating point are inverter switching frequency and output resonant filter characteristics. Various types of filters can be used, but LCC type and LC-C type are most popular. For fluorescent lamp, many analysis and describing equations of filter transfer functions are researched [3-4]. In case of electrodeless lamp, coupling core is another circuit parameter which is operated as an inductor.

In this paper, various kinds of transfer functions for electrodeless lamp considering coupling core parameters are given, and their characteristics are analyzed by comparison of characteristics of conventional fluorescent lamp ballast.

II. REVIEW OF ELECTODELESS LAMP MODEL

The Endura lamp of Osram is composed of a round type lamp and two external magnetic cores [1]. The QL lamp of Philips has a coupling core in the lamp [2]. Regardless of the type of electrodeless lamp, lamp itself is not different in the discharge characteristics compared to the normal fluorescent lamp, so lamp model is considered as a resistor whose value

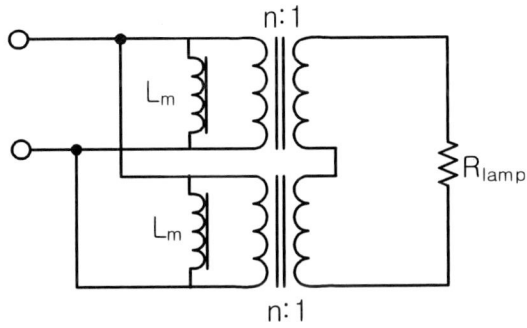

(a) Model of Osram Endura lamp

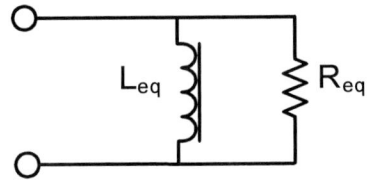

(b) Model of Philips QL lamp

is defined by lamp voltage and lamp current as in (1) [5].

$$R_{eq} = \frac{V_{Lamp}}{I_{Lamp}} \qquad (1)$$

Electrodeless lamp is shown as a transformer whose secondary is one turn of lamp plasma magnetically coupled by ferrite core [6]. Considering the electromagnetic mechanism, the model of Endura is derived as a resistance and two inductors in the ballast side as shown in Fig. 1-a [7]. The model of QL lamp is derived as a resistance and an inductor as shown in Fig. 1-b [8]. The model of Fig. 1-a can be transformed to the model of Fig. 1-b. L_m is a magnetizing inductance of a ferrite core transformer whose value can be measured as 500 μH for Endura type and as 13 μH for QL. Equivalent resistance of Endura lamp is calculated as about 220 ohm using the lamp voltage 180V and current 0.8A. Equivalent resistance of QL lamp is about 380 ohm.

III. ANALYSIS OF LCC RESONANT FILTER

One of favorite resonant filters used in ballast is LCC type as shown in Fig. 2. LCC type is used in ballast for Endura lamp. Due to the magnetizing inductance L_m of coupling core, resonant filter has another inductor component when it

978-1-4244-1871-8/07 $25.00 © 2007 IEEE   329

connects with the electrodeless lamp. Because C_p is parallel to L_m, Fig. 2 can be simplified as Fig. 3 if C_p and L_m are combined to one impedance element C_{eq}. C_{eq} is expressed as (2) if $\omega L_m > 1/\omega C_p$.

$$C_{eq} = C_p - \frac{1}{\omega^2 L_m} \qquad (2)$$

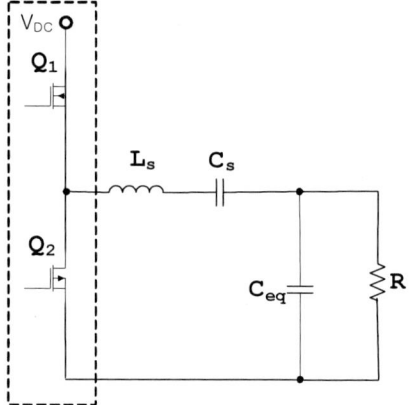

Fig. 2 LCC resonant filter with electrodeless lamp

Fig. 3 Equivalent circuit of Fig. 2

The magnetizing inductance L_m affects the equivalent parallel capacitance C_{eq}, which decreases inversely with magnetizing inductance and square of switching frequency. So, in design of the resonant filter, actual parallel capacitance value should be increased to compensate the effect by lamp magnetizing inductance for obtaining proper ignition voltage in start-up and rating power in stead state. The equation shows that if switching frequency or magnetizing inductance decreases severely, equivalent impedance can convert to an inductive component. In this case, ballast may operate in unexpected condition and go to fail.

The transfer function exactly describing the LCC filter gain characteristics is derived in (3). The equation shows that magnetizing inductance L_m affects the gain magnitude and phase. If switching frequency ω is equal to the ω_s, gain becomes to be independent on the load resistance, and gain phase becomes 0.

$$\frac{V_{Lamp}}{V_{DC}} = \frac{1}{1 + \dfrac{C_p}{C_s} + \dfrac{L_s}{L_m} - (\dfrac{\omega^2}{\omega_p^2} + \dfrac{\omega_m^2}{\omega^2}) + jQ_s(\dfrac{\omega}{\omega_s} - \dfrac{\omega_s}{\omega})} \qquad (3)$$

Where $\omega_m = \dfrac{1}{\sqrt{L_m C_s}}, \omega_s = \dfrac{1}{\sqrt{L_s C_s}}, \omega_p = \dfrac{1}{\sqrt{L_s C_p}},$

$Q_s = \dfrac{1}{\omega_s R C_s} = \dfrac{\omega_s L}{R}$.

Another useful transfer function is derived in (4), which describes the filter inductor current characteristics to DC link voltage. The function is especially useful to estimate whether the resonant inverter is operated in zero voltage switching mode.

$$\frac{i_{Ls}}{V_{DC}} = \frac{1}{R} \frac{1 + j(\dfrac{\omega}{Q_p \omega_p} - \dfrac{\omega_m}{Q_m \omega})}{1 + \dfrac{C_p}{C_s} + \dfrac{L_s}{L_m} - (\dfrac{\omega^2}{\omega_p^2} + \dfrac{\omega_m^2}{\omega^2}) + jQ_s(\dfrac{\omega}{\omega_s} - \dfrac{\omega_s}{\omega})} \qquad (4)$$

Where $Q_m = \dfrac{\omega_m L_m}{R}$

These equations allow various analysis of ballast operation with variation of important parameters. Parameters used in calculation are as follows

- Switching frequency f=250kHz
- DC link voltage V_{DC}=400V
- Series capacitor value C_s=0.22uF
- Parallel capacitor value C_p=3.3nF
- Series inductor value L_s=150uH
- Magnetizing inductance of coupled core L_m=500uH
- Lamp load R=216ohm

Fig. 4 shows ballast output voltage gain characteristics with and without magnetizing inductance L_m before ignition. Lamp has very high resistance value until ignition, so it is almost open circuit condition to ballast. In this condition, gain is designed to be very high for obtaining a sufficient high voltage to ignite the lamp. Magnetizing inductance L_m shifts the main resonant frequency to lower frequency direction if it increases and makes another resonant frequency at low side frequency. Fig. 5 shows output gain characteristics at steady state condition after ignition. Finite lamp resistance make the gain curve flatten due to lower quality factor in the resonant circuit at steady state. Main resonant frequency shifts to a little low frequency and gain decreases according to the increase of magnetizing inductance. Fig. 6 shows that the phase characteristics of voltage gain at steady state.

The inductor current of resonant circuit should be lagging to keep the ZVS condition. Fig. 7 shows that the inductor current is always lagging regardless of variation of magnetizing inductance caused by temperature if the switching frequency is selected to be higher than main resonant frequency.

978-1-4244-1871-8/07 $25.00 © 2007 IEEE   330

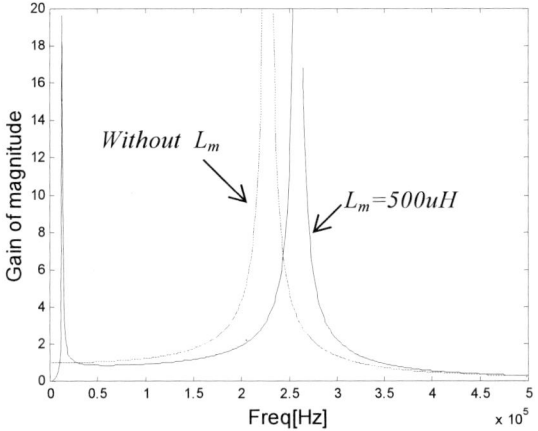

Fig. 4 Gain characteristics before ignition.

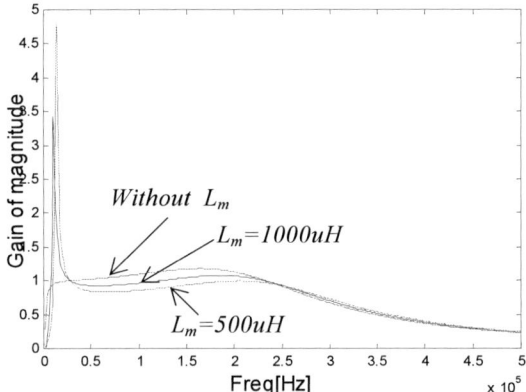

Fig. 5 Gain characteristics with variation of L_m

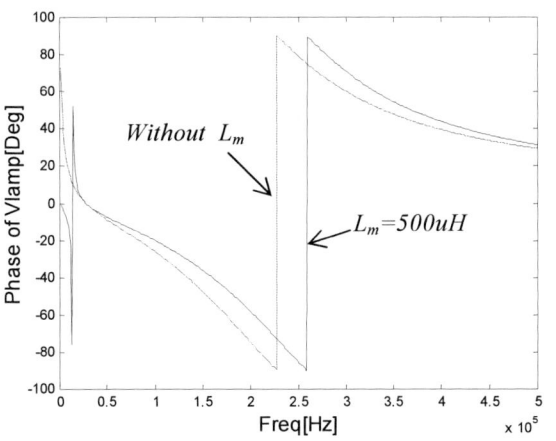

Fig. 6 Phase characteristics with variation of L_m

IV. ANALYSIS OF LC-C RESONANT FILTER

Another favorite resonant filter is LC-C type as shown in Fig. 8. LC-C type is used in ballast for QL lamp. In QL lamp, cable impedance according to its length can affect the reliable operation of lamp system because operating frequency is very high as 2.65MHz. In this case, LC-C type filter has a series

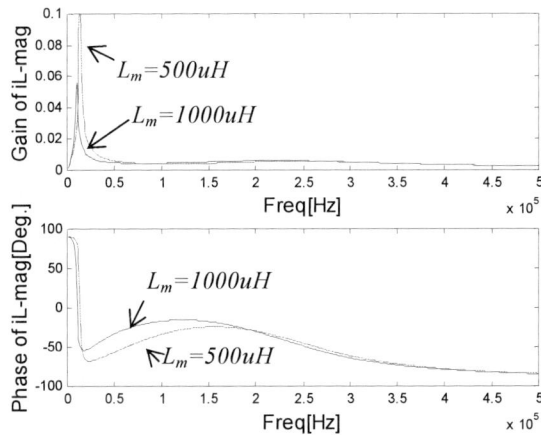

Fig. 7 characteristics of inductor current

capacitor at final output, so the capacitance can compensate the line impedance influence.

The transfer function exactly describing the LC-C filter gain characteristics is derived in (5). The equation also shows that magnetizing inductance L_m affects the gain magnitude and phase, but the equation form is more complex and hard to construct the simplified form. Current characteristics in filter inductor can also be estimated by (6).

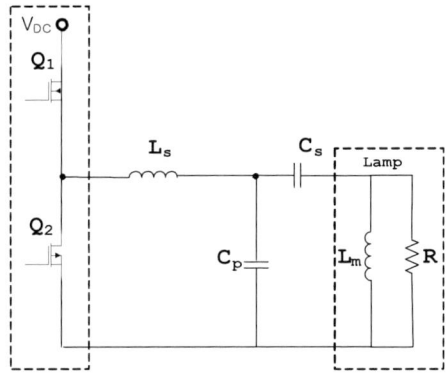

Fig. 8 LC-C resonant filter with electrodeless lamp

Parameters used in calculation are as follows

- Switching frequency f=2.65MHz
- DC link voltage V_{DC}=400V
- Series capacitor value C_s=0.01F
- Parallel capacitor value C_p=0.05nF
- Series inductor value L_s=15uH
- Magnetizing inductance of coupled core L_m=10uH
- Lamp load R=380ohm

Fig. 9 shows lamp voltage gain characteristics at steady state condition. LC-C filter gain has large variation with frequency and magnetizing inductance compared to LCC filter. Fig. 10 shows the phase variations around the main

978-1-4244-1871-8/07 $25.00 © 2007 IEEE   331

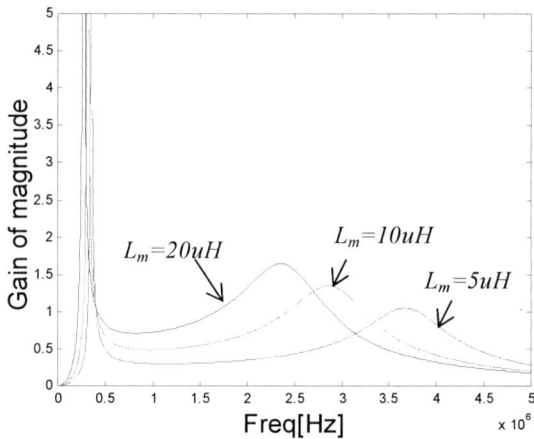

Fig. 9 Gain characteristics with variation of L_m

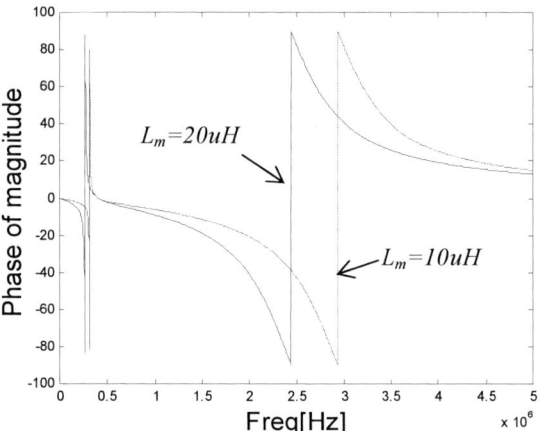

Fig. 10 Phase characteristics with variation of L_m

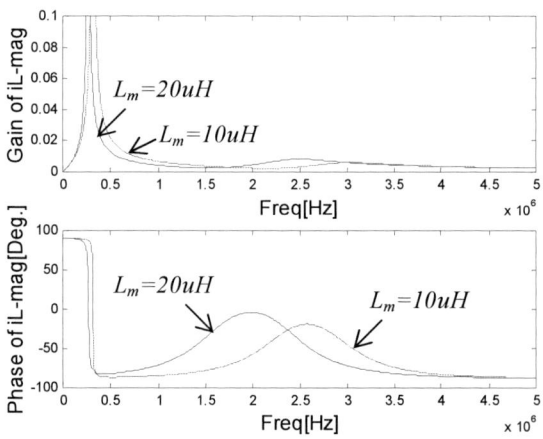

Fig. 11 Phase characteristics with variation of L_m

resonant frequency which also varies with circuit parameters. Fig. 11 shows that the inductor current is lagging, so inverter holds ZVS condition. But it is noted that the increase of magnetizing inductance of coupling core can break the ZVS condition.

V. CONCLUSION

Resonant circuit operates as an interface element between ballast and lamp, and enables the lamp to be started successfully and generated at rating power. Most popular resonant filter circuit is LCC and LC-C type, but their characteristics and transfer functions with electrodeless lamp are not known well. In this paper, lamp voltage gain characteristics and inductor current characteristics in the resonant filter are analyzed with the variations of magnetizing inductance by newly derived transfer functions for electrodeless lamp. Magnetizing inductance of the coupling core normally increases with temperature. So, ballast should be designed to be properly operated over the temperature variation range by considering the detailed ballast output characteristics.

REFERENCES

[1] Sylvania Icetron: Design Guide, Osram Sylvania, 2004
[2] Philips QL lamp systems: Product Information, Philips, 2002
[3] M. C. Cosby, et al. "A Resonant Inverter for Electronic Ballast Applications", IEEE Transactions on Industrial Electronics. Vol. 41, No. 4, pp. 418–425, 1994
[4] D. Tadesse, et. al. " A Comparison of Power Circuit Topologies and Control Techniques for a High Frequency Ballast", IEEE Conference of Industry Applications Society, pp.2341–2347, 1993
[5] M. Sun, B. L.Hesterman, "PSpice high-frequency dynamic fluorescent lamp model", IEEE Transactions on Power Electronics, Vol. 13, No. 2, pp. 261– 272, 1998
[6] L. R. Nerone, "Mathematical Modeling and Optimization of the Electrodeless, Low-Pressure, Discharge System", IEEE Conference of Industry Applications Society, pp.509–514, 1993
[7] S. Ben-Yaakov, et al. "A Behavioral SPICE Compatible Model of an Electrodeless Fluorescent Lamp", IEEE Conference of APEC, pp. 948–954, 2002
[8] L. R. Nerone, "Design of a 2.5-MHz, Soft-Switching Class-D Converter for Electrodeless Lighting", IEEE Transactions on Power Electronics, Vol. 12, No. 3, pp. 507– 516, 1997

$$\frac{V_{Lamp}}{V_{DC}} = \frac{1}{1+\dfrac{L_s}{L_m}+\dfrac{C_p L_s}{C_s L_m}-C_p L_s \omega^2 - \dfrac{1}{C_s L_m \omega^2}+j(\dfrac{L_s \omega}{R}+\dfrac{C_p L_s \omega}{C_s R}-\dfrac{1}{C_s R\omega})} \tag{5}$$

$$\frac{i_{Ls}}{V_{DC}} = \frac{L_m \omega(1-C_p L_s \omega^2 - C_s L_s \omega^2)+jR(C_p L_s \omega^2 + C_s \omega^2(L_m+L_s-C_p L_m L_s \omega^2))}{C_p R\omega + C_s R\omega - C_p C_s L_m R\omega^3 + j(C_p+C_s)L_m \omega^2} \tag{6}$$

A Study on the Resonant Inverter for Corona Generators

CHUL-YONG CHOI
PARA-ENT CO., LTD
11, Goju-Ri, Paltan-Myon,
HwaSung City, Gounggi-Do, Korea

SANGSHIN KWAK
DAEGU UNIVERSITY
School of Electronic Engineering
15 Naeri-ri, Jinryang-up, Gyungsan,
Gyungbook, 712-714
Email: sskwak@ieee.org

DAE-SIK LEE
DAEGU UNIVERSITY
School of Electronic Engineering
15 Naeri-ri, Jinryang-up, Gyungsan,
Gyungbook, 712-714
Email: dslee@daegu.ac.kr

Abstract— An algorithm for control and performance of a pulse-density-modulated (PDM) series-resonant voltage source inverter developed for corona-discharge processes is presented. The PDM inverter produces either a square-wave ac-voltage state or a zero-voltage state at its ac terminals to control the average output voltage under constant dc voltage and operating frequency. Moreover it can achieve zero-current-switching (ZCS) and zero-voltage-switching (ZVS) in all the operating condition for a reduction of switching losses. Even though the corona discharge load with a strong nonlinear characteristics, new high frequency resonant inverter is shown the wide range power control from 5% to 100%.

I. INTRODUCTION

Recently, the researches on increasing the switching time, voltage rating and current rating of power semiconductors such as GTO, power-MOSFET, IGBT, SIT and so on, have been in progress. However, in these researches a big capacity induction heater is used for fusion, surface process, or tube welding, where the speed regulation equipments and the high frequency resonant PWM inverter is wildly applying [1] [2].

The corona processes is proposed to improve the attraction of the polyethylene film, ink, glue and coating [3] [4]. Generally, for a corona surface process to maintain a stable discharge current a 10-20kV, 10-50kHz and 1-40kW AC power supply is needed. And sometimes tow or more kind of film may be used. So the AC power supply should be controllable on a wide range some 1-100%.

The corona surface process operates from an AC supply, a DC resonant inverter and increase transformer that supply as a more than 10kV voltage source on secondary circuit. Although the circuit is similar to the traditional AC power supply, because of the corona surface process is an intense nonlinear system. DC bus voltage control and frequency control is hard to achieve on the whole 1-100% range.

In this paper, we propose a 20kHz, 6kW voltage source PDM (pulse-density-modulated) series-resonant inverter. By the PDM inverter in-phase control on zero-voltage-state frequency to frequency of the series-resonant circuit can be achieved. In addition, a PLL (phase locked loop) circuit that adjusts the output signal to match the input signal constantly help to construct a ZVS (zero voltage switching) an quasi ZCS (zero current switching) PDM inverter under any load.

In the experiment a corona discharge with a power supply controllable under range 5-100% is presented.

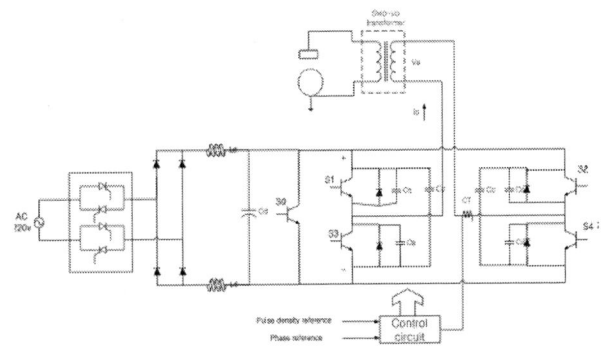

Fig. 1: Corona surface processor

II. CORONA DISCHARGE SYSTEM

A. System structure

In the corona discharge system shown in Fig. 1, a single phase H-bridge IGBT inverter is used. The rate of IGBT is 600V, 400A. Between the collector and emitter of the IGBT connects with a snub capacitor C_s (0.1uF) by witch

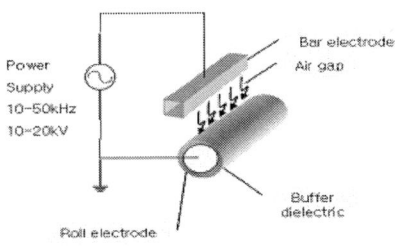

Fig. 2: Electrode structure

A 1:10 ratio transformer supplies 10-20kV AC power to secondary corona discharger. And the electrode can be change into a LCR equivalent circuit. The lossless snubbing capacitor enables the voltage-source inverter to perform ZVS when the inverter operates with a lagging load current. This means that the inverter operating frequency should be higher than the load resonant frequency. In addition, quasi zero-current switching is also achieved because a PDM inverter operates at a constant frequency in any power control conditions. Thus, zero-voltage and quasi ZCS result in a significant reduction for switching losses

B. Electrode structure

The electrode structure of corona discharge processor is shown in Fig. 2. The roll electrode that connects to ground has an inter-space of 3mm with bar electrode. When a 10-50kHz 10-20kV voltage is discharged from bar electrode to roll electrode, the film will go across the inter-space because the affinity of ink is higher at that time. Furthermore, for different ink has different thickness characteristic, an adjustable AC power supply range of 1-100% is needed. However, the supply voltage of discharge load and the current has a nonlinear relation, so when the electrode voltage is higher than the discharge interdiction voltage the current increases and vice versa.

In the equivalent circuit shown in Fig. 3, CA, Ra, LI and CD denote equivalent capacitor, gap impedance, leakage inductance and the capacitor of the buffer terminal respectively. Since Ra is a variable resistance the voltage and current has a nonlinear relation. When the input voltage is smaller than the discharge voltage Ra behaves an infinite quality and no discharge happen. Contrarily, when the input voltage is bigger, Ra will reduce and Ca will short. A resonant current can be carried out.

The DC voltage control and frequency control are wildly used in series-resonant inverter to adjust the output power of corona system. However, this kind of inverter is difficult to control the discharge power if it is small than half of full rating for the terminal voltage must higher than the discharge voltage.

C. PDM control

The PDM inverter model is shown in Fig. 4. Model 1 and model 2 shown in Fig. 4 form a square wave and the model 3 in used to force the output voltage to 0. In this model IGBT gate S3 and S4 both turn on. Although one of the IGBTs and diode in on the reverse direction, another one is on the on form, so the current io is on the free wheeling state as show in Fig. 4 (C). Fig. 5 shows the power control of PDM. PDM inverter repeats periodically on "run and stop" for adjusting resonant current of series resonant load and the in-phase average output voltage. The pulse ratio is 3/4 shown in Fig. 5. PDM inverter becomes 3 resonant cycles of square wave under model 1 and 2 with the value of E, also one cycle action as model 3 to the zero voltage state. So the average output voltage of 4 resonant cycle PDM inverter is 3/4 of the maximum voltage value.

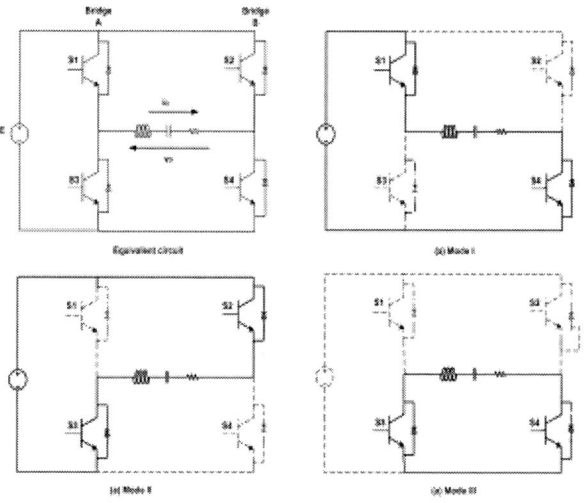

Fig. 4: PDM inverter model

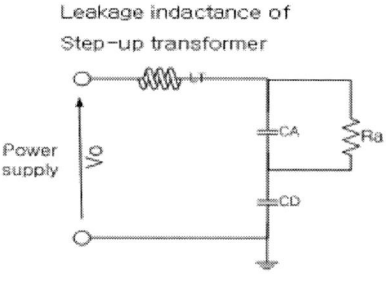

Fig. 3: Equivalent circuit of electrode

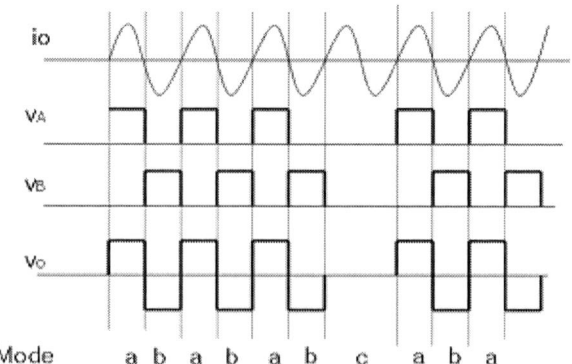

Fig. 5: PDM control technical

978-1-4244-1871-8/07 $25.00 © 2007 IEEE   334

III. CONTROL CIRCUIT

PDM inverter control circuitry shown in Fig. 6, operates from a PLL (phase locked loop) circuit witch is used to trace the resonant frequency and a PDM circuit as a zero-voltage switching.

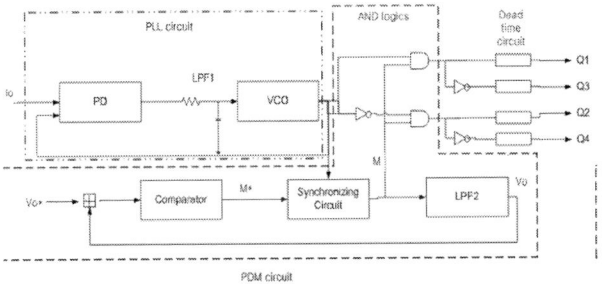

Fig. 6: Control block diagram

A. PLL circuit

PLL circuit, by what the input signal frequent can trace the output. Shown as in bellow Figure a PLL circuit consists of a phase-detector (PD), a low-pass filter (LPD1) and a voltage-controlled oscillator (VCO). PD is a correction voltage feeds the frequency of resonant current and the output of VCO in to a same phase. Loop filter LPF1 is used to eliminate the high frequency signal from PD. The VCO generates a proportional frequency of the input, and effects PDM circuit by in-phase clock signal.

B. PDM control

Fig. 7 corresponds to the PDM control circuit based on feedback control, and the control signal waveforms are shown Fig. 7 The average output voltage reference V0* adjusts the duration of the zero-voltage state. The V0 is obtained from a logic signal corresponding to the actual ac voltage state M via a LPF2. The corona discharge treatment system developed in this paper in intended to operate in a wide range of pulse density from 1/50 to 1. The inverter output voltage V0 has various low frequency components coursed by the PDM control

Comparison of V0* with V0 by means of an analog comparator generates a logic signal of the ac voltage state reference M*, choosing witch state the PDM inverter produces, either the square-wave ac voltage state or the zero-voltage state for the subsequent resonant cycle.

A synchronizing circuit consisting of a D-type flip-flop device is employed to avoid the change of the current state during the resonant cycle even when the ac voltage state reference M* changes. The synchronizing circuit takes in M* at every rising edges of the VCO output, and holds the logic signal M during the subsequent resonant cycle. A logic AND

circuit produces alternate mode I and mode II during the square-wave ac voltage state of M=1, where it produces mode III during the zero-voltage state of M=0. Signal A and B and their inverted signals are provided to four blanking time circuits which produce a time interval as short as 1μs to avoid short circuiting of the dc link input

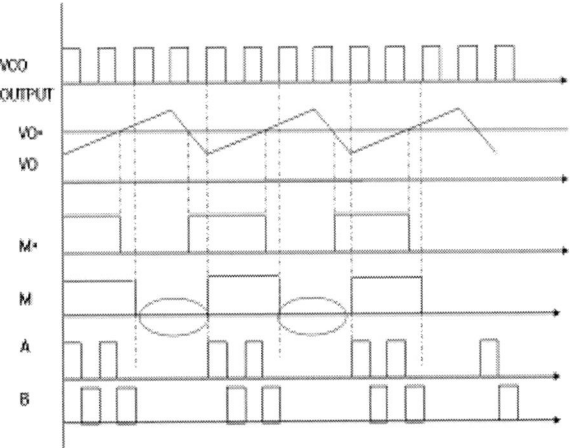

Fig. 7: PDM control signal

IV. EXPERIMENT

Fig. 8, 9 and 10 show the wave of output current (i_o) and voltage (v_o) when the DC bus is 200V

Fig. 8 shows the wave of maximum power without the zero voltage state. In evidence, the IGBT action frequency follows resonant frequency well, and quasi-ZCS is used. Wave of 1/2 pulse ratio is shown in Fig. 9, where average output voltage of PDM inverter is 1/2 of the one in Fig. 8 due to the zero voltage state appears. In addition, the DC input power is 33% of the maximum power that is 1.9kw. The quality factor of corona discharge processor LCR equivalent circuit is 10-30, but the quality factor of high-frequency heater system is much smaller as 4.5. So current converge to zero on the time of zero voltage state.

Fig. 10 shows wave of the minimum output power with a 1/50 pulse ratio. From this figure it can be seen that there is one cycle of square AC wave follows 49 cycles zero voltage state. So the 1% of power is 48W when DC input is 5.6kW.

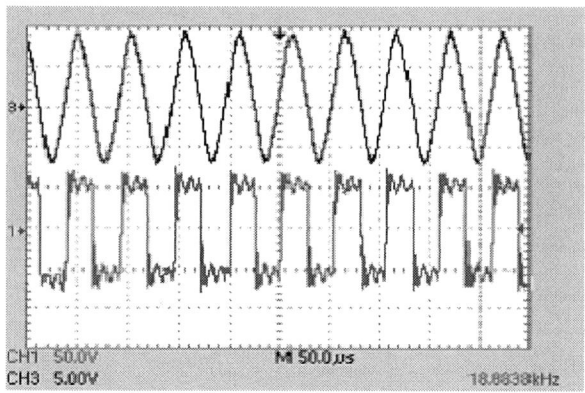

Fig. 8: Full power

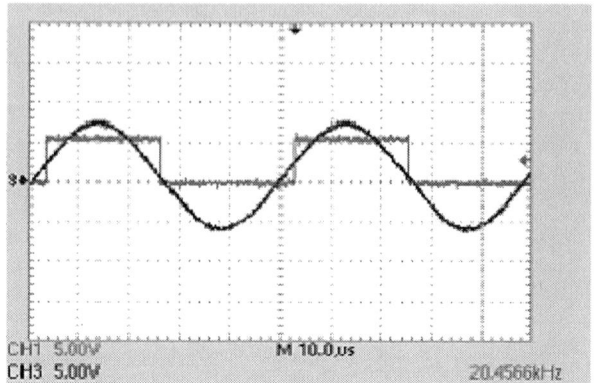

Fig. 11: Output current and gate signal

Fig. 11 shows the gate signal to IGBT and the resonant current with quasi ZCS under maximum power. It can be seen a 1μs dead time is used to avoid circuit short.

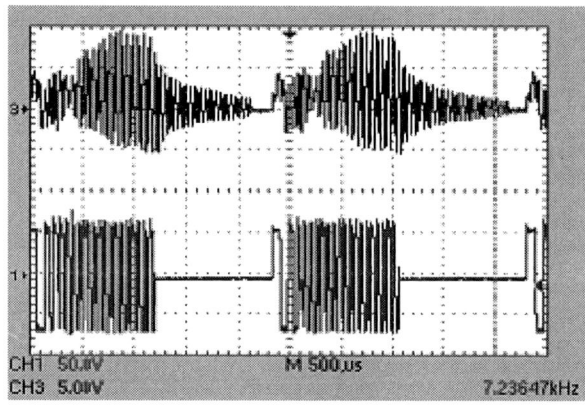

Fig. 9: Half power

V. CONCLUSION

In this paper a series-resonant PDM inverter for corona discharge system is experimented. This PDM inverter has a wide control range of discharge power of 1-100%, even the current and voltage have intense nonlinear characteristics. To reduce switching loss of PDM inverter ZVS and quasi ZCS technology have been used. The applicability of the corona discharge system can be seen in this paper.

REFERENCES

[1] H. Fujita and H. Akagi, "Pulse-density-modulated power control of a 4kW 450kHz voltage-source inverter for induction melting applications," *IEEE Trans. Industry Appli.,* vol.IA-32, no.2, pp-.279-286, 1996.

[2] S. Bottari, L. Malesani and P. Tenti, "High frequency 200kHz inverter for induction heating applications," *IEEE/PESC Rec.,* 1985, pp. 308-316 .

[3] H. Akagi and A. Nabae, "A voltage-source inverter using IGBT's for a 50kHz 10kV corona surface treater,*" IEEE/IAS Annu. Meet.,* 1989, pp. 1164-1169.

[4] L. A. Rosenthal and D. A. Davis, "Electric characterization of a corona discharge for surface treatment," *IEEE Trans. Industry Appli.,* vol.I-A-11, no.3, pp. 328-335, 1975.

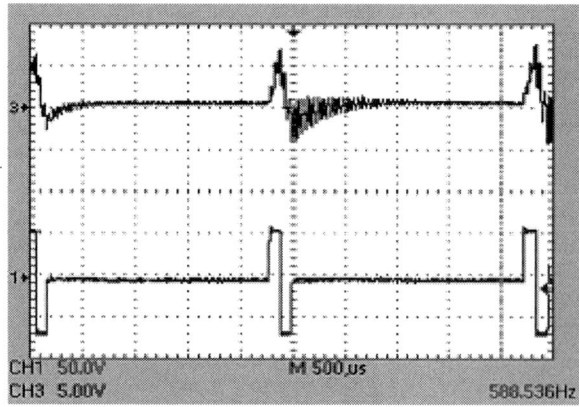

Fig. 10: Minimum power

Reduced Voltage Drop Characteristics of the Series Transformer in a Voltage Disturbance Generator

E.C. Nho, I.D. Kim, S.D. Park
Pukyong National University
Busan, Korea
Email: nhoec@pknu.ac.kr

T.W. Chun
University of Ulsan
Ulsan, Korea
Email: twchun@mail.ulsan.ac.kr

H.G. Kim
Kyungpook National Univ.
Daegu, Korea
Email: kimhg@knu.ac.kr

N.S. Choi
Chonnam National Univ.
Yeosu, Korea
Email: nschoi@chonnam.ac.kr

Abstract— **In this paper a new 3-phase voltage disturbance generator is proposed. Voltage sag, swell, outage, and unbalance can be provided by the proposed scheme. The main devices of the proposed generator are SCR thyristors, sliding type autotransformers, and transformers. Therefore, the generator has good features of high reliability, high efficiency, and low cost implementation with simple structure. Furthermore, the voltage drop on series transformer is reduced considerably compared with the conventional generator. Thus the generator can be used for the performance test of the custom power devices with improved performance. The operating principle and the characteristics of the proposed scheme are analyzed and the simulation results show the usefulness of the generator.**

I. INTRODUCTION

High quality power supplies for electrical and electronic equipments are widely used for the demand of automation, IDC (Internet Data Center), etc. Usually the equipments are sensitive to the variations of the supplied power source voltage such as voltage sag, swell, outage, and unbalance. The voltage disturbance may cause a severe problem of malfunction or shut down of the system. To provide a clean power many research works have been reported. The most popular and widely used device for the clean power is UPS. In case of high voltage and high power applications custom power devices such as DVR (Dynamic Voltage Restorer), DSTATCOM (Distribution Static Compensator), Active Power Filter, and SSTS (Solid State Transfer Switch) are considered [1-3]. For the performance test of the custom power devices a power quality disturbance generator is necessary. However, the conventional generators are too expensive to use in general. Recently a cost effective generator using thyristor controlled reactor has been suggested [4]. The demerit of the suggested generator in [4] is that it requires a huge reactive power to generate the disturbance. Another cost-effective scheme has been proposed [5-7]. The generators in [5-7] use series transformer in each phase to provide voltage sag, swell, and outage. The leakage inductance and resistance of the series transformer results in voltage drop with load current. In this paper a new voltage disturbance generator is proposed. The proposed scheme has good features of simple structure, high reliability, low cost, and reduced voltage drop of the series transformers.

II. PROPOSED CIRCUIT DIAGRAM

Fig. 1 shows the proposed circuit diagram for the voltage disturbance generation. The utility voltages v_a, v_b, and v_c are constant. In each phase series transformer T is inserted to add or subtract some voltage. The secondary voltage v_{ad} of the a-phase series transformer is determined by the sliding type autotransformer voltage v_{Ta}. T_a, T_b, and T_c means sliding type autotransformer, respectively. The magnitude of voltage sag or swell can be adjusted by moving the contact point of the autotransformer. In case of a fixed magnitude the general autotransformer with taps can be used. Switches $S_{a1} - S_{c2}$ and $S_{Ba1} - S_{Bc2}$ are constructed with anti parallel connected SCR thyristors. The SCR thyristors have low on-state loss, which makes the system efficiency high.

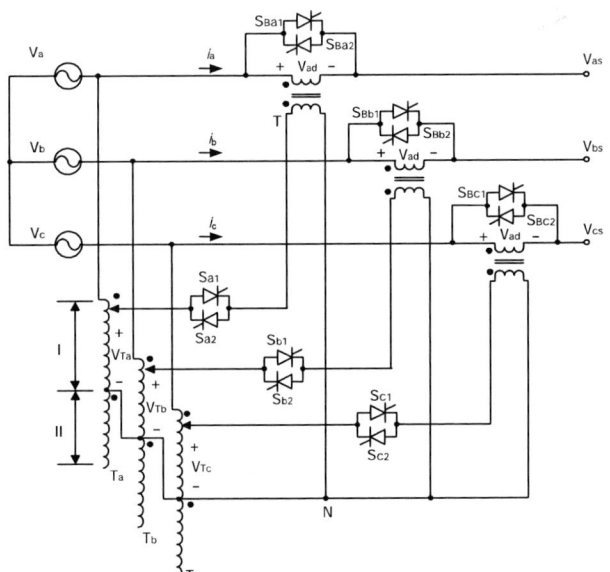

Fig. 1   Proposed voltage disturbance generator.

III. OPERATING PRINCIPLE OF THE CIRCUIT

The operating principle of each phase of the circuit in Fig. 1 is the same. Thus it is enough for the a-phase circuit operation to be described.

978-1-4244-1871-8/07 $25.00 © 2007 IEEE

Fig. 2 and 3 show phase-a equivalent circuit of the conventional and proposed scheme, respectively.

Fig. 2 Phase-a equivalent circuit of the conventional scheme.

Fig. 3 Phase-a equivalent circuit of the proposed scheme.

In Fig. 2 the output voltage Vo is described as follows.

$$V_O = V_a - \left(R_1 + j\omega L_{l1} + R_2 + j\omega L_{l2}\right)i - V_T \qquad (1)$$

- R_1, R_2 : primary and secondary winding resistance of the series transformer, respectively.
- L_{l1}, L_{l2} : primary and secondary leakage inductance of the series transformer, respectively.
- V_T : SCR thyristor voltage drop.

The leakage inductance and winding resistance of the series transformer results in voltage drop when the load current flow in the conventional scheme. However, there is no voltage drop due to the series transformer in the proposed scheme.

A. Voltage Sag Generation

To provide voltage sag the polarity of the voltage v_{ad} should be positive. The magnitude v_{ad} is determined from the primary voltage and the turn ratio n of the series transformer. The primary voltage v_{ad1} of the series transformer T is given as follows according to the switch state.

1) s_{a1} and s_{a2} are turned off, s_{Ba1} and s_{Ba2} are turned on :

$$v_{ad1} = V_T . \qquad (2)$$

2) s_{a1} and s_{a2} are turned on, s_{Ba1} and s_{Ba2} are turned off :

$$v_{ad1} = v_{Ta} . \qquad (3)$$

The case 1) means a normal operating mode, and the utility voltage v_a and the output voltage v_{as} are almost the same. To generate voltage sag the switches s_{Ba1} and

s_{Ba2} should be turned off, and the switches s_{a1} and s_{a2} should be turned on. The contact point of the autotransformer T_a is set in the region – I. Then, the primary voltage v_{ad1} is

$$v_{ad1} = n_T v_a . \qquad (4)$$

Where, n_T means the turn ratio of the transformer T_a. Therefore, the output voltage v_{as} is

$$
\begin{aligned}
v_{as} &= v_a - v_{ad} \\
&= v_a - n v_{ad1} \\
&= v_a - n n_T v_a \\
&= (1 - n n_T) v_a .
\end{aligned} \qquad (5)
$$

The generator output voltage v_{as} is determined by the turn ratio product of $n n_T$ for the given source voltage v_a. Usually n is fixed, and n_T can be adjusted by moving the contact point of the sliding type autotransformer.

B. Voltage Swell Generation

In order to generate voltage swell the output voltage v_{as} should be greater than the utility source voltage v_a, which can be achieved by addition of some voltage to the source voltage. Thus, the polarity of the series transformer primary side voltage v_{ad1} in (4) should be altered. The polarity change can be made easily by set the contact point of the autotransformer in region-II. Therefore,

$$v_{ad1} = -n_T v_a , \qquad (6)$$

and the a-phase output voltage v_{as} is

$$
\begin{aligned}
v_{as} &= v_a - v_{ad} \\
&= v_a - n v_{ad1} \\
&= v_a + n n_T v_a \\
&= (1 + n n_T) v_a .
\end{aligned} \qquad (7)
$$

The output voltage is increased by $n n_T v_a$.

C. Outage Generation

Nowadays long time outages are rarely happen, but instantaneous outages occur frequently. To generate outage the magnitude of the series transformer secondary side

978-1-4244-1871-8/07 $25.00 © 2007 IEEE   338

voltage v_{ad} should be the same with the source voltage v_a, and the phase difference between the two voltages should be exactly $180°$. In other words,

$$v_{ad} = nn_T v_a$$
$$= v_a . \tag{8}$$

The above condition can be satisfied by set the contact point in region – I and adjusting the turn ratio of the autotransformer n_T. To come back to the normal mode the switches s_{a1} and s_{a2} should be turned off, and s_{Ba1} and s_{Ba2} should be turned on.

D. Voltage Unbalance Generation

Each phase output voltage in Fig. 1 can be controlled separately. Therefore, the 3-phase output voltages can be expressed as follows.

$$v_{as} = (1 \pm nn_T)v_a \tag{9}$$
$$v_{bs} = (1 \pm nn_T)v_b \tag{10}$$
$$v_{cs} = (1 \pm nn_T)v_c \tag{11}$$

Each phase voltage can be controlled easily by adjusting n_T and setting the contact point in region – I or II.

IV. SIMULATION RESULTS

Simulations are carried out with the circuit diagram as shown in Fig. 1. The line-to-line source voltage is 220V, and the out power rating is 10kVA. The turn ratio of the series transformer n is 1. Load power factor changes from 0.8lagging to 0.8 leading.

Fig. 4 shows a-phase output voltage and current in case of 30% voltage sag for 8 cycles. Fig. 4(a) shows the a-phase output voltage. Fig. 4(b) and (c) show the output voltage and current waveforms in transient mode from normal to voltage sag and vice versa, respectively.

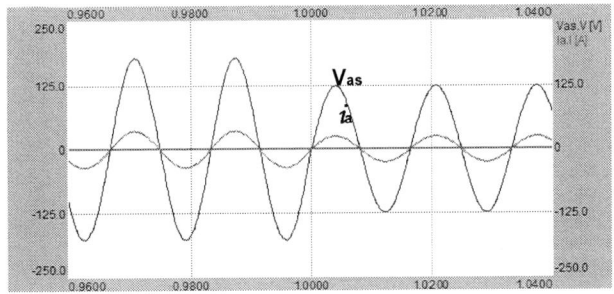

(b) From normal to sag

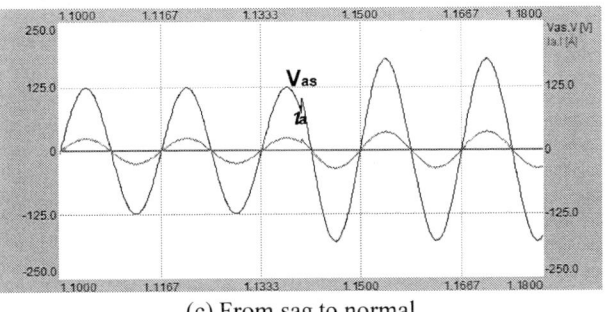

(c) From sag to normal

Fig. 4 Output a-phase voltage and current in case of 30% voltage sag. (PF=1.0)

It is found that the proposed scheme can be operated in unity load power factor without any voltage spike even in mode change.

Fig. 5 shows a-phase output voltage and current in case of 30% voltage swell for 8 cycles. Fig. 5(a) shows the a-phase output voltage. Fig. 5(b) and (c) show the output voltage and current waveforms in transient mode from normal to voltage swell and vice versa, respectively. Fig. 5(b) and (c) show the phase of the load current lag the output voltage, and the load power factor is 0.8 lagging. It is also found that there is no voltage spike during mode changes between normal and voltage sag. The magnitude of the load current depends on the output voltage magnitude and the load impedance. The magnitude of the voltage swell can be adjusted easily.

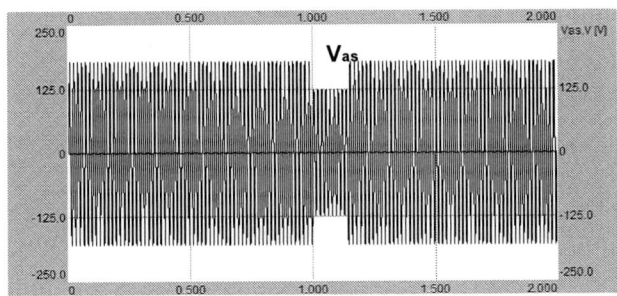

(a) a-phase output voltage

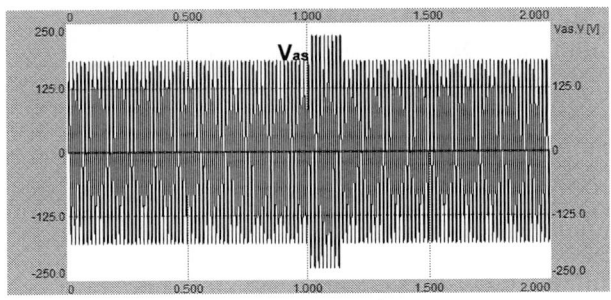

(a) a-phase output voltage

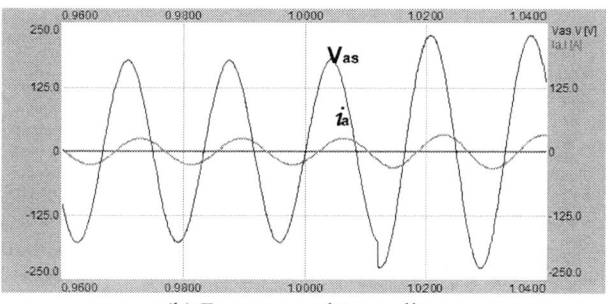

(b) From normal to swell

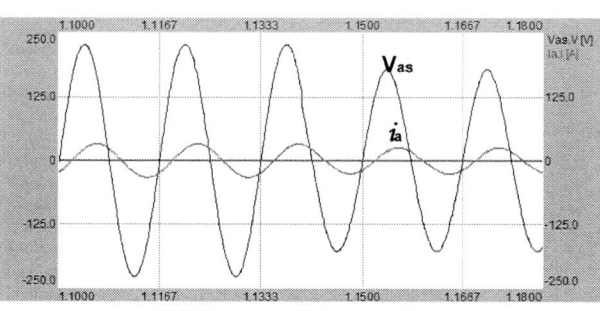

(c) From swell to normal

Fig. 5 Output a-phase voltage and current in case of 30% voltage swell. (PF=0.8 lagging)

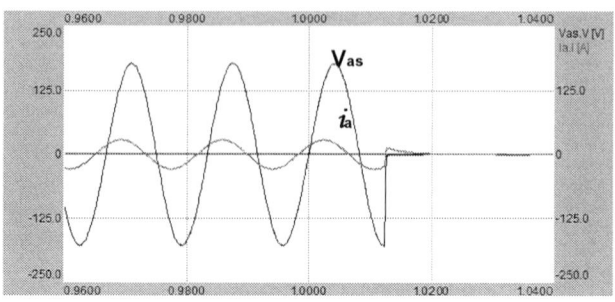

(b) From normal to outage

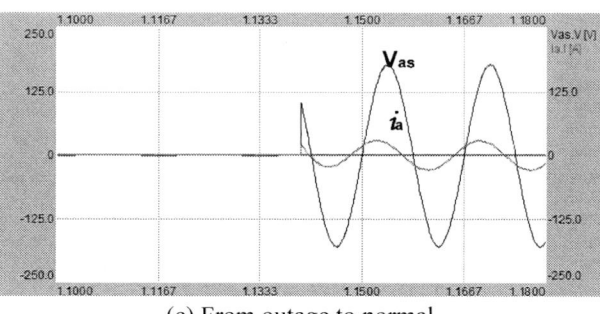

(c) From outage to normal

Fig. 6 Output a-phase voltage and current in case of outage. (PF=0.8 leading)

Fig. 6 shows the generator output voltage and current waveforms in case of the outage. Fig. 6(a) shows a-phase output voltage waveform, and Fig. 6(b) and (c) show the output voltage and current waveforms during the mode change between normal and outage. The load current leads the output voltage in phase. The outage occurs at the peak of the input utility voltage, and the normal mode comes back at half of the voltage peak without any disturbance.

Fig. 7 shows each phase output voltage in case of voltage unbalance generation. In Fig. 7 the generator output voltages v_{as}, v_{bs}, and v_{cs} are 10%, 30%, and 50% of the source voltage, respectively.

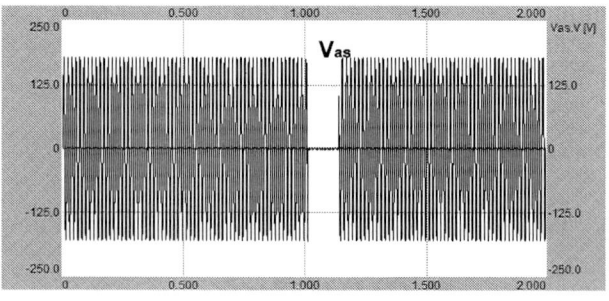

(a) a-phase output voltage

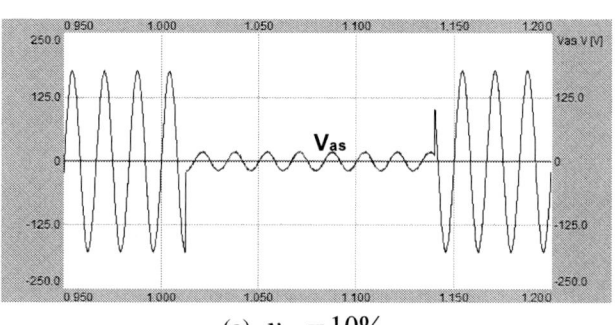

(a) $v_{as} = 10\%$

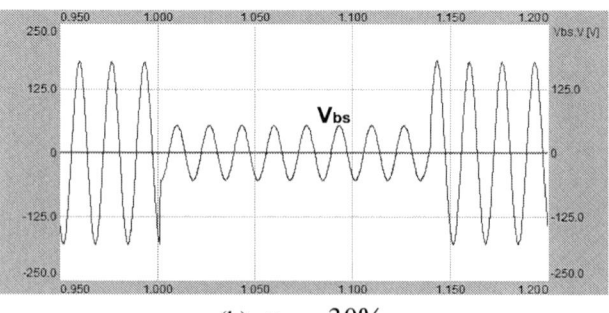

(b) $v_{bs} = 30\%$

978-1-4244-1871-8/07 $25.00 © 2007 IEEE   340

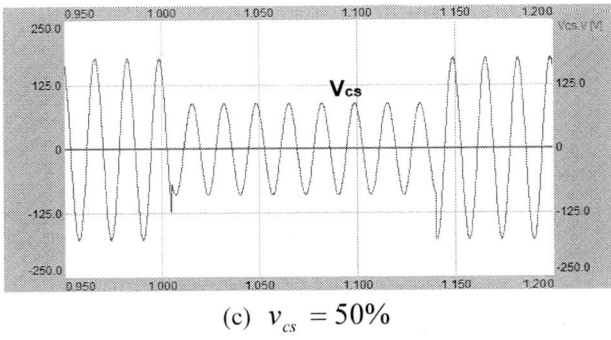

(c) $v_{cs} = 50\%$

Fig. 7 Output a-phase voltage in case of voltage unbalance.

V. CONCLUSION

A new voltage disturbance generator is proposed. The proposed generator can provide voltage sag, swell, outage, and unbalance with simple structure and easy control. The voltage disturbance generation mechanism is described in each mode, and the simulation results show the usefulness of the proposed scheme. The scheme provides disturbed output voltages regardless of the load power factor. The major characteristics of the proposed generator are summarized as follows.

- High efficiency and reliability with SCR thyristors and sliding type autotransformer.
- Considerably reduced voltage drop in the series transformer.
- Voltage disturbance generation regardless of the load power factor.
- Easy generation of the voltage sag, swell, outage, and unbalance.
- Cost effective system with simple structure.

ACKNOWLEDGMENT

This work has been supported by KESRI (R-2005-7-067), which is funded by MOCIE (Ministry of commerce, industry and energy).

REFERENCES

[1] R. S. Weissbash, G. G. Karady, P. G. Farmer. " A combined uninterruptible power supply and dynamic voltage compensator using a flywheel energy storage system," IEEE Trans. Power Delivery, vol. 16, no.6 pp. 265– 270, 2001.

[2] R. G. Lawrence, K. L. Craven, G. D. Nichols, " Flywheel UPS," IEEE IA Magazine, pp. 44–50, 2003.

[3] A. Ghosh and G. Ledwich, " Compensation of distribution system voltage using DVR," IEEE Trans. Power Delivery, vol. 17, no. 4, pp. 1030–1036, 2002.

[4] Y. H. Chung, G. H. Kwon, T. B. Park, G. Y. Lim, " Voltage sag and swell generator with thyristor controlled rectifier," IEEE Power con 2002, vol. 3, pp. 1933–1937, 2002.

[5] Eui-Cheol Nho, In-Dong Kim, Tae-Won Chun, and Heung-Geun Kim, "Cost-effective power quality disturbance generator for the performance test of custom power devices," Proceedings of IECON'04, pp. 1606-1610, 2004

[6] W.Y. Byeon, J.W. Kim, K.S. Lee, E.C. Nho, I.D. Kim, T.W. Chun, H.G. Kim, "Voltage sag-swell generator for power quality

disturbance of dynamic UPS system," Transactions of KIPE, Vol. 10. No. 1, pp. 102-107, 2005.

[7] B.C. Lee, S.H. Choi, S.H. Paeng, S.D Park, E.C. Nho, N.S. Choi, I.D. Kim, T.W. Chun, H.G. Kim, "3-phase power quality disturbance generator with phase jump function," Transactions of KIPE, Vol. 11, No. 5, pp. 463-470, 2006.

A Study on Optimal Braking Control Using Adhesion Coefficient

Hanmin Lee
Korea Railroad Research Institute
Advanced EMU Research Team
360-1, Woulam-dong, Uiwang-city, Gyeonggi-do
Email: hanmin@krri.re.kr

Gildong Kim
Korea Railroad Research Institute
Advanced EMU Research Team
360-1, Woulam-dong, Uiwang-city, Gyeonggi-do
Email: gdkim@krri.re.kr

Sunghwan Park
Pusan National University
Mechanical Engineering
Geumjeong-gu, Busasn, 609-735
Email: shpark01@pusan.ac.kr

Abstract—The brake systems of the rolling stocks are generally consisted of electrical and mechanical brake systems. Because of its inherent structure of the each brake system, the electrical brake system is mainly used at the high speed range while the mechanical brake system is used at the relatively lower speed range. It is desirable for the rolling stocks to apply the entire electrical brake system. However, since the brake force from electric brake system is not enough to stop the rolling stock within the legal stop distance. Therefore, the mechanical brake system is indispensable to rolling stocks. In general, the vast majority of the world trains are equipped with mechanical braking systems which use compressed air as the force to push block on to wheels or pads on to discs. These mechanical systems are known as air brake or pneumatic brakes. For the air brake system, basically huge scale air compressor is equipped and the long pipe line is complexively connected. Since mass of these air brake components, it is difficult to be a light weight equipment and the long pipe line raise the maintenance problem. In order to overcome these problems of air brake system, the hydraulic brake system is proposed in this research. The hydraulic brake system makes the whole weight of brake equipment be light and large braking force can be applied. Therefore, in this research, the optimal control using adhesion coefficient to apply the hydraulic brake system is reviewed.

I. INTRODUCTION

The brake systems of the rolling stocks are generally consisted of electrical and mechanical brake systems. The electrical brake system is mainly used at the high speed range while the mechanical brake system is used at the relatively lower speed range.

Two types of brake systems distribute the braking force to get it by the braking condition

Because the braking force of mechanical brake systems are made by the friction of rail and wheel, the braking responses are different by the rail conditions. The braking characteristic on the rail conditions is explained as adhesion force made by wheel and slip ratio. So the braking systems generate braking input to maintain suitable slip.

The robust slip control algorithm is designed to apply the hydraulic brake system to advanced EMU system in this paper. The response characteristics of the robust control system are evaluated by computer simulation

II. SLIP RATIO AND OPTIMAL BRAKING FORCE

When the rolling stocks run and brake, the response characteristics of adhesion force are changed by conditions of rail and slip velocity. Generally, the adhesion force of dry rail is stronger than it of wet rail. Also, the adhesion force is changed by the velocity of rolling stocks. Fig 1 shows the adhesion force by conditions of rail and rolling stocks' velocity.

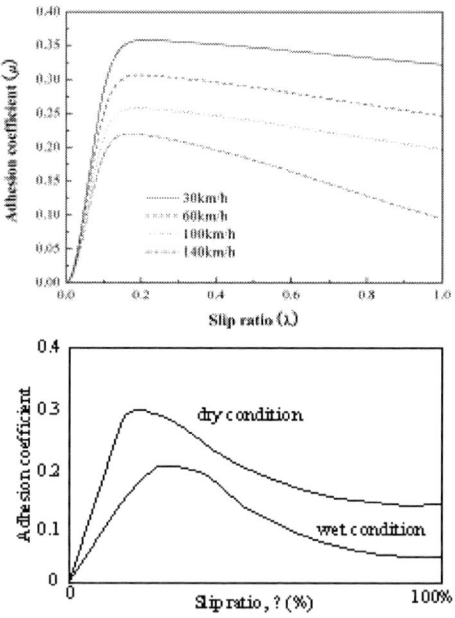

Fig 1. Relation of slip ratio and adhesion force by conditions of rail and rolling stocks' velocity

The adhesion coefficient goes up on rolling stocks' velocity. After it goes over at the specific point, it goes down as shown in Fig 1. If the braking force is larger than the adhesion force, the slip between the rail and wheel is happened. And the friction coefficient goes down. So the rolling stocks can not brake. Also the wheel might be flat. Therefore, the performance of rolling stocks is generated by using the adhesion force between rail and wheel.

978-1-4244-1871-8/07 $25.00 © 2007 IEEE

III. DESIGN CONTROLLER OF MECHANICAL BRAKE SYSTEM

Equations are used to analysis dynamic characteristics of the slip generated by braking.

$$J\dot{\omega} = rF_a - T_b - B\omega + F_u$$

$$M\upsilon = -F_a - F_r$$

$$\lambda = \frac{\upsilon - r\omega}{\upsilon}$$

Where,

J : inertial moment
r : radius of wheel
F_a : adhesion force between rail and wheel
F_u : uncertainty of wheel
B : friction coefficient
M : mass of rolling sticks
Fr : running resistance
λ : slip ratio

The adhesion force is influenced by slip velocity. Therefore, it is very important to demonstrate relation of slip velocity and adhesion force.

The dynamic characteristics of rolling stocks are very complicated.

In certain parameters are numerous. Uncertainties make numerous effects the braking system. Therefore, uncertainties are compensated by the robust control algorithm.

In this paper, we design the robust control algorithm by adaptive sliding mode control theory.

The plane of sliding mode controller is defined as following.

$$s = e + \sigma \int_0^t e\, dt$$

where,

$$e = \sigma_d - \sigma$$

$$\sigma = \lambda \upsilon$$

ρ is defined as constant. $\dot{s}$ is defined as following equation.

$$\dot{s} = \dot{\sigma}_d + \dot{\upsilon} - r\dot{\omega} + \rho e$$

From the motion equation of rolling stocks

$$\dot{s} = \dot{\sigma}_d - \frac{F_a}{M} - \frac{F_r}{M} + \frac{rF_a}{J} + \frac{T_b}{J} + \frac{B}{J}\omega + \frac{F_u}{J} + \rho e$$

$$\kappa_{min} < \left| \frac{F_r(\upsilon)}{M} + \frac{F_u}{J} \right| < \kappa_{max}$$

Generally, control input of sliding mode controller is the sum of equivalent control and robust control as following equation.

$$U = U_{eq} - U_r$$

where,

U_{eq} : equivalent input control value
U_r : robust input control value

From the equivalent about the plane of sliding mode

$$U_{eq} = -J\dot{\sigma}_d + \frac{J}{M}F_a + rF_a - B\omega - J\rho e$$

Time variable function by the equivalent control rule, $\dot{s}$ is as follow.

$$\dot{s} = \frac{F_r}{M} + \frac{F_u}{J} + U_r$$

To demonstrate robust input control value

$$\varsigma \geq \left| \frac{F_r}{M} + \frac{F_u}{J} \right| + \eta$$

$$U_r = -\varsigma\, \text{sgn}(s)$$

From U_r, switching condition is the following equation

$$s\dot{s} = s\left(\frac{F_r}{M} + \frac{F_u}{J} - \varsigma\, \text{sgn}(s) \right) \leq |s|\left(\left| \frac{F_r}{M} + \frac{F_u}{J} \right| - \varsigma \right) < -\eta s$$

Conclusive sliding mode control is as following equation.

$$T_b = U_{eq} + U_r$$

$$= -J\dot{\sigma}_d + \frac{J}{M}F_a + rF_a - B\omega - J\rho e - \varsigma\, \text{sgn}(s)$$

where,

M : variable parameter
B : parameter not to measure

$$\varphi = \left(\frac{J}{M} + r \right)F_a - B\omega = \theta^T \phi$$

Where,

$$\theta^T = \left[-B\frac{J}{M} + r \right]$$

$$\phi^T = \left[\omega \ F_a \right]$$

$$\dot{\hat{\theta}} = -\gamma\phi s$$

$$\hat{T} = -J\dot{\sigma}_d - \theta^{\hat{T}}\phi - \rho_2 e - D_2 s$$

To demonstrate stability of the proposed controller, Liapunov function is defined as following equation.

$$V = \frac{1}{2}s^2 + \frac{1}{2\gamma}\ddot{\theta}^2$$

where,

$$\tilde{\theta} = \hat{\theta} - \theta_d$$

Time variable function of Liapunov function is

$$\dot{V} = s\dot{s} + \frac{1}{\gamma}\tilde{\theta}\dot{\tilde{\theta}}$$

$$= s\left(-\frac{T_b}{J} + \dot{\sigma}_d - \theta^T\phi + \rho e \right) + \frac{1}{\gamma}\tilde{\theta}\dot{\tilde{\theta}}$$

Considering the proposed sliding mode control,

$$\dot{V} = -s\varsigma\,\mathrm{sgn}(s) + \tilde{\theta}\left(s\phi + \frac{1}{\gamma}\dot{\tilde{\theta}} \right)$$

$\dot{\hat{\theta}}$ is $-\gamma\phi s$. If $\dot{V} = -s\varsigma\,\mathrm{sgn}(s)$ is below than 0, Liapunov's stability theory is satisfied.

IV. SIMULATION RESULTS

Matlab 7.1 is used to analysis the response characteristics using Simulink diagram. The parameters for computer simulation are shown on Table 1.

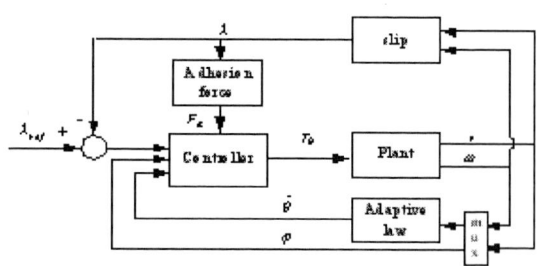

Fig 2. Simulink diagram

Table 1. System parameters

Patameter	Value
M	9.785 kg
F	0.63 υ^2
J	182.59 kgm^5
B	0.25 $kgm/\sec$

4.1 Evaluation of performance of robust controller by uncertainties

To evaluate the performance of robust controller by uncertainties, we simulate the condition that the deviation of the uncertainty is 5% regarding wheel dynamic equation at the early speed of 140km/h. The proposed controller is analyzed and compared with PI controller Zigler-Nichols proposed. Proportional gain is 250. Integral rain is 100. Simulation results of PI controller and the proposed controller are shown in Fig 3, 4.

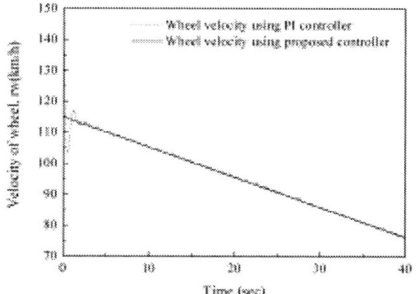

Fig 3. Wheel velocity by uncertainties

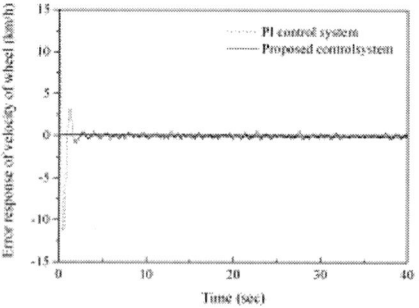

Fig 4. Error response by uncertainties

From the results, system using proposed sliding mode controller is compensated by robust control. Therefore, the proposed system is known that uncertainties don not affect the velocity of wheel.

978-1-4244-1871-8/07 $25.00 © 2007 IEEE

4.2 Evaluation of robustness by variable parameters

Mass of rolling stocks is changed in the range of 150%. The rail is wet, dry and other condition. The value of adhesion force is shown in Fig 5.

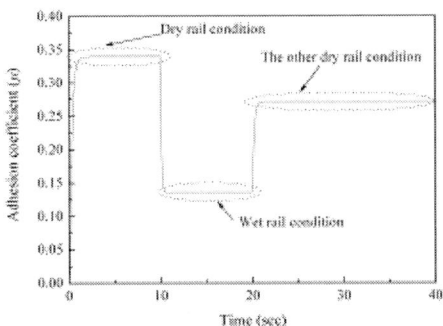

Fig 5. Adhesion coefficient by wet/dry/other rail condition

Fig 6 shows velocities of the rolling stock and the wheel using the proposed controller.

Fig 7 shows estimated parameters by variable mass of passengers. Fig 6 shows stable deceleration of the rolling stock and wheel by robust compensation of uncertainties. Also, the performance of the proposed controller shows satisfied operation, even if mass of rolling stock is changed.

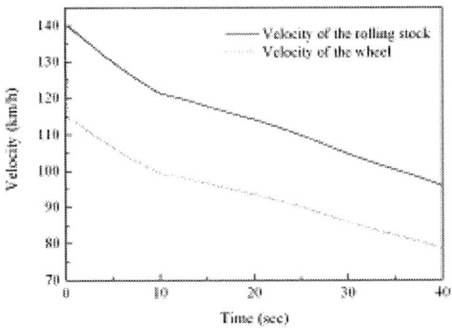

Fig 6. Velocity of rolling stock and wheel

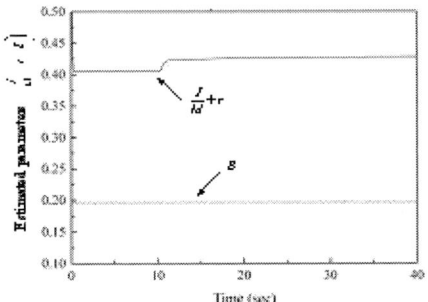

Fig 7. Estimated performance and mass of rolling stock

V. CONCLUSION

Response performance of the proposed robust control system is evaluated by computer simulations. The system using proposed sliding mode controller is compensated by robust control. Therefore, the proposed system is known that uncertainties don not affect the velocity of wheel.

Stable deceleration of the rolling stock and wheel is operated by robust compensation of uncertainties. Also, the proposed controller operated satisfied performance, even if mass of rolling stock is changed.

REFERENCE

[1] Tomoki Watanabe and Michihiro Yamashita, "Basic study of anti-slip control without speed sensor for multiple motor drive of electric railway vehicles," Proceedings of the Power Conversion Conference, Vol. 3, April, 2002, pp. 1026 ~ 1032.

[2] Satoshi, Kadowaki, Kiyoshi, Ohishi, Shinobu Yasukawa, Takashi Sano., "Anti-skid Re-adhesion control Based on Disturbance observer considering Air brake for electric commuter train," Advanced Motion Control, 2004. The 8th IEEE International Workshop Advanced Motion Control, March, 2004, pp. 607 ~ 612.

[3] Shirai, S., "Adhesion Phenomena at High-Speed Range and Performance of an Improved Slip-dectector," Quarterly Reports, Railway Technical Research Institute, Vol. 18, No. 4, 1977, pp. 189 ~ 190.

[4] Isaev, I. P. and Golubenko, A. L., "Improving Experimental Research into Adhesion of the Locomotive Wheel with the Rail," Rail International, pp. 3 ~ 10, August, 1989.

[5] Ohyama, T., "Some basic studies on the influence of surface contamination on adhesion force between wheel and rail at high speed," Quarterly Report, Railway Technical Research Institute, Vol. 30, No. 3, pp. 127 ~ 135, 1989.

Stress grading in integrated power modules

C. Duchesne, M. Mermet-Guyennet, E. Dutarde
Power Electronics Associated Research
Laboratory (PEARL), ALSTOM Transport
Rue du docteur Guinier – BP4
65601 SEMEAC, France
cyrille.duchesne@lab-pearl.com

T.Lebey, S. Dagdag
Laboratoire Plasma et Conversion d'Energie (LAPLACE),
Laboratoire de Génie Electrique de Toulouse (LGET)
Université Paul Sabatier, Bât. 3R3, 118 rte de Narbonne
31062 TOULOUSE cedex 4, France
Thierry.Lebey@laplace-univ.tlse.fr

Abstract - Power electronics packaging, like integrated power modules, constitute an advanced technology leading to power density increase, weight and volume decrease and enhancing the reliability level. However, the continuous increase of the voltage lead to questions regarding power device environment. We discuss in the following different solutions able to achieve the electric stress gradation.

I. INTRODUCTION

In integrated power modules, insulating materials often constitute the weakest points. Oversizing these materials is a basic solution to guarantee both service life and reliability of the whole system [1]. Nevertheless this oversizing is against power electronics integration. Moreover, in railways applications an increase of the voltage of semiconductor devices is observed leading to new problems at the power module level and at the packaging level. New methods and solutions have to be found to accompany the trend which consists at the same time to increase the power density and the voltage. Among other problems stress gradation must be achieved. It's within this framework that we develop the work we present in the following.

Triple points are one of the main causes of failures in high voltage power electronics modules. Such points take place where discontinuities occur in the electrical parameters (both permittivity and conductivity) leading to an enhancement of the electric stresses. In power module a classical example is the interface between the ceramic substrate, the encapsulation gel and the metallization (figure1) [3].

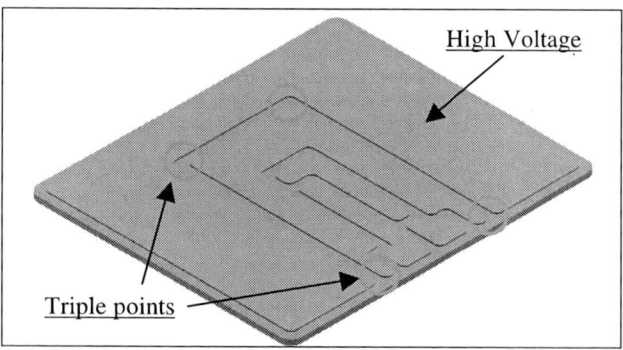

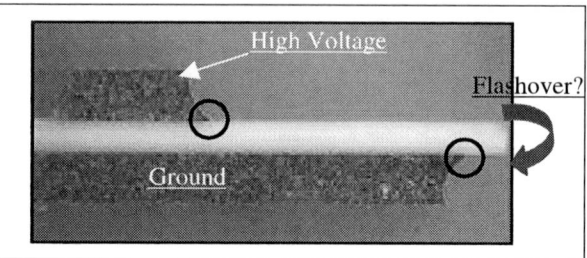

Fig.1: Triple points and flashover representation. electric stresses enhancement representation

This triple interfaces is usually named "triple point" and is a keypoint for reliability. The stresses enhancement must be controlled in order to avoid dielectric breakdown or partial discharges ignition i.e. phenomenon leading to a reduction of the service life of the module [4] [5]. Different solutions are studied. The main driving idea is to reduce the electric field magnitude to the lowest possible value: for example a value lowers than approximately 10% of the breakdown strength of the air (3kV/mm).

One of the possible solutions consists in using hybrid composites made of an organic matrix reinforced with inorganic fillers presenting a non-linear intrinsic behavior. Indeed, their characteristics generally show a non-linear behavior of the resistivity with the electrical field which may be approached by:

$$\rho = \rho_0 . \exp(-k.E) \qquad (1)$$

With ρ_0 and k, are two characteristics of the materials.

II. ELECTRIC FIELD ENHANCEMENT IN POWER ELECTRONICS MODULE

A. Case Study

Figure 2 shows a classical stacking of different materials involving a metallized substrate. Two possible problems related to creepage distances L1 and L2 are illustrated. In this study, Direct Bonded Copper substrate of Aluminum Nitride (AlN) (relative permittivity, $\varepsilon_r = 9$) are used. The thickness is respectively of 1mm for the substrate (t) and 300μm for the copper. The surface creepage distance (L1) is 2mm and the clearance distance (L2) is the sum of the creepage distance (l) and of the substrate thickness t.

978-1-4244-1871-8/07 $25.00 © 2007 IEEE   346

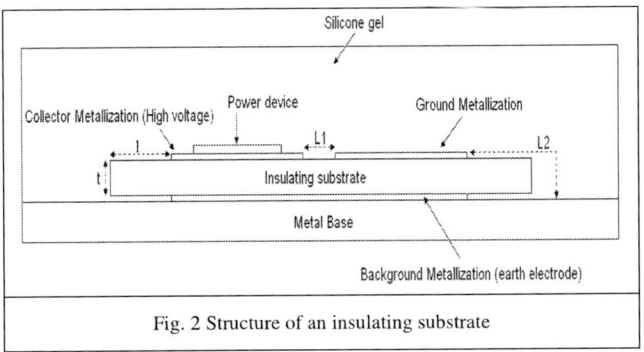

Fig. 2 Structure of an insulating substrate

B. Electric Field Calculation

A finite element method (FEM) software (Maxwell 3D [7]) is used for the simulation. In order to obtain the electrostatic field, the software solves a Poisson type partial differential equation for the electric potential unknown, Φ, with appropriate boundary conditions:

$$\nabla.(\varepsilon r.\varepsilon_0 \nabla\Phi) = -\rho v \qquad (2)$$

Where:
- $\Phi(x, y, z)$ is the electric potential scalar,
- $\varepsilon r(x, y, z)$ is the relative permittivity,
- $\varepsilon 0$ (F/m) is the permittivity of a vacuum.
- $\rho v(x, y, z)$ is the volume density of the electric charges.

Maxwell's equations and the applicable constitutive equation are then used to obtain the electric field strength, E, and the electric flux density (or the electric displacement field), D, as follows:

$$E = -\nabla\Phi \qquad (3)$$

$$D = \varepsilon r.\varepsilon_0 .E \qquad (4)$$

At the interface between two dielectrics, the normal component of the electric flux density vector presents no discontinuity if the charge density at the interface is null:

$$\varepsilon_1 .En_1 = \varepsilon_2 .En_2 \qquad (5)$$

The index "1" and "2" refer to the two dielectrics. On the contrary, when crossing a surface with non-zero charge density, the normal component of the electric flux density has a discontinuity equal to the respective local charge density.

C. Electric Field Enhancement

A general configuration of field reinforcement corresponds to an insulated electrical conductor passes through a grounded screen in an external media. This is the case, for instance, of high voltage cables, coils in motors and generators and substrate in integrated power modules. [6] In this last case, typical configuration and potential distribution is shown in Figure 3. If the stress exceeds the breakdown stress of the surrounding media, it can lead to surface discharge. The purpose of stress grading materials is to reduce this local surface stress so that nowhere does exceed the breakdown strength.

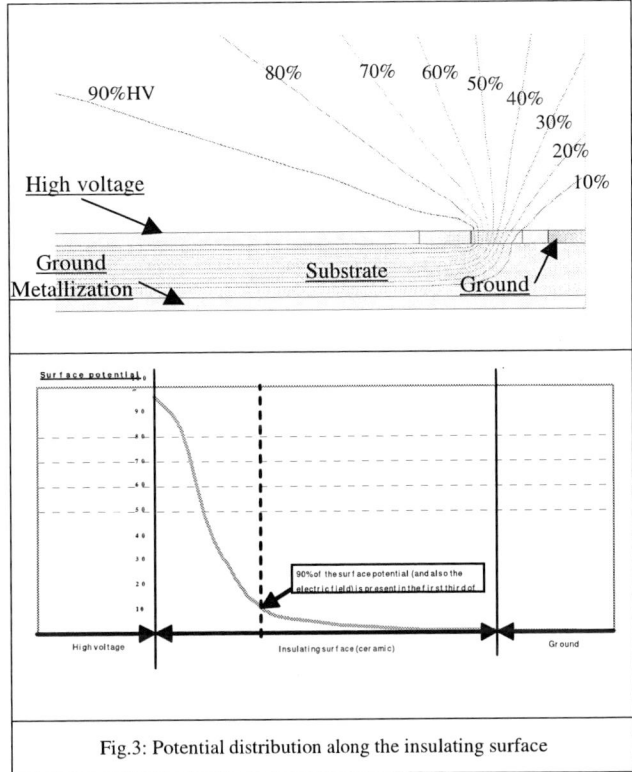

Fig.3: Potential distribution along the insulating surface

III. ELECTRICAL STRESS GRADING MATERIALS

A. Hybrid Materials Effects: Influence of a stress grading material on the voltage distribution

From a practical point of view, a geometry allowing the characterization of this type of materials and giving a visualization of the electric field distribution has to be used. Surface potential measurements under stress give a unique opportunity to obtain a mapping of the potential distribution.

Figure 4 represents the experimental set up used to measure the surface potential for the geometry under study and the figure 5 represents the results obtained on representative samples with and without stress grading materials. It clearly shows that the materials used tend to reduce the electric stresses. However, this distribution is obtained under a DC polarization which does not prejudge the AC behavior. Indeed, a homogeneous distribution can be obtained in DC simply for a material whose conductivity is high enough.

978-1-4244-1871-8/07 $25.00 © 2007 IEEE

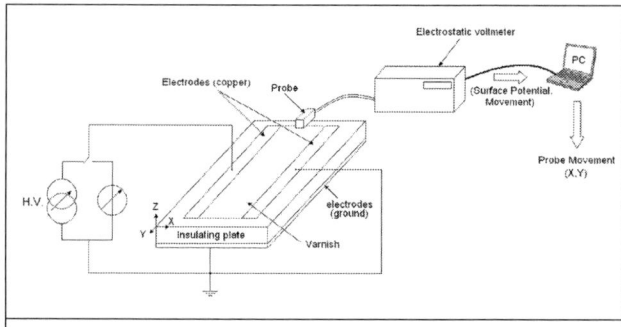

Fig.4 : Experimental device for measurements of the surface potential [8]

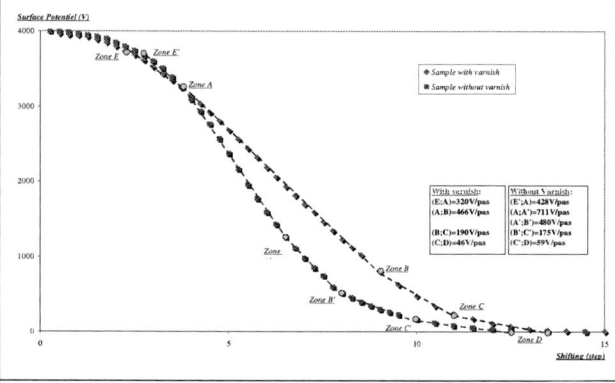

Fig.5: Voltage distribution measured in DC along axis X with (blue points) and without varnishes (red points).

B.   Measurements of grading material current/voltage characteristics.

The grading materials, generally varnishes, are classified in two different families: resistive or capacitive even if this point is still under discussion for the different authors [9] [10]. One definition rests on the comparison of the current – voltage characteristics obtained under DC and AC. Another definition assumes that the non linear electrical behavior is due to the changes in the permittivity or in the resistivity versus the voltage [8]. To our point of view, the most accurate definition relies on the measurement of the phase between current and voltage under AC conditions. If the angle is lower than 45°, the material will be considered as resistive. It will be capacitive on the contrary.

The measurement of the stress grading characteristics is relatively straightforward. Samples are most commonly realized on insulating tubes whose diameter varies between 25mm to 40mm. The grading material is applied to the surface by either painting or molding, as appropriate, and circumferential electrodes are applied, set normally 5mm apart. Either AC or DC voltage may be applied to perform measurements. The basic test configuration is shown in fig.6. A power supply of at least 10 kV is required in AC or DC. The current measurement requires a resolution of a least 10µA. Figure 7 presents the results obtained on representative samples with two different stress grading

materials. Indeed, the varnish "1" has a non linear behavior of the current versus the voltage, contrary to the varnish "2" which is linear. The distribution of the voltage measured in DC with these materials (figure 6) shows that the gradation obtained is not associated to their intrinsic nature. It is therefore difficult to conclude on the capacity of a varnish to behave correctly as a stress grading by considering its I(V) characteristics.

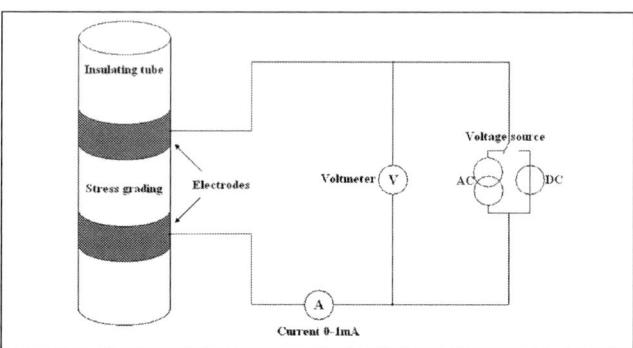

Fig.6: Experimental set up for Current/Voltage characteristics under DC and AC 50Hz

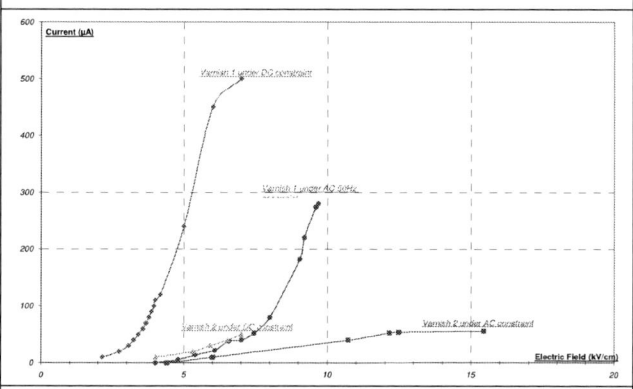

Fig.7 : Current/Voltage characteristics under DC and AC 50Hz for two different materials

C.   Partial Discharges Detection

One particular and important behavior of the dielectric materials is that their properties do not only depend on the electric field magnitude but may be dependent in some cases of the voltage value. Hence, a threshold voltage exists which defines two different areas. For a voltage value lower than this threshold, the phenomena are field dependent and are considered as intrinsic ones since they only involve the material properties. The main intrinsic phenomena are conduction mechanisms, polarization, charge injection. Beyond this threshold, extrinsic phenomena may also occur and among them Partial Discharges (PD) are the most relevant. PDs occur in insulating materials presenting voids (defects) in their bulk which are generally air filled. According to the Paschen's Law when a sufficient voltage is applied, due to the uneven distribution of the electric field, the gas breakdowns and a discharge is produced. The voids being limited in size, this discharge is said "partial" since it

978-1-4244-1871-8/07 $25.00 © 2007 IEEE

will not propagate in the material bulk and will not short the electrodes. Partial discharges produce mechanical erosion, temperature increase and lead to several gas formation (Ozone and NOx), reducing the lifetime of the dielectric material and thus of the device they are used in. The detection of the partial discharge ignition is therefore of particular importance in the targeted applications. A basic electrical circuit for their measurement is given in figure 8.

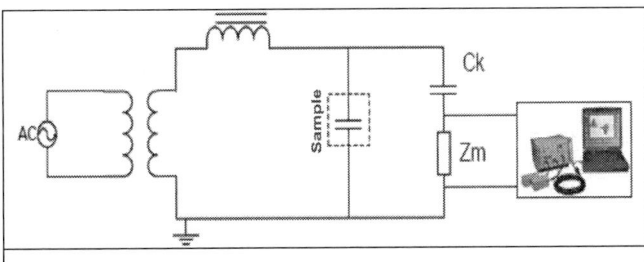

Fig.8: Electrical circuit for the partial discharges (PDs) detection [11]

D. Partial Discharge Inception

The quality of the stress grading materials under study may be estimated by their impact on the Discharge Inception Voltage (DIV). The same type of samples and of voltage configuration that the ones used for the surface potential distribution is used (with and without varnish). For measurements samples are placed in air, immersed in a dielectric liquid (FC72) or encapsulated in dielectric gel. The calibration level is 5pC. Phase Resolved Partial Discharges patterns of samples with and without varnish, measured in air and in FC72 are presented in figures 10 and 11 whereas Table 1 gives the respective DIV. For measurements on samples placed in air, an increase of the DIV of approximately 1kV is observed. The contrary is observed when the sample is immersed in the liquid since the measured DIV is higher for the sample without varnish. This difference may probably be explained by a value of the dielectric strength of the varnish which is lower than the dielectric strength of the FC.

TABLE I
PARTIAL DISCHARGES INCEPTION WITH AND WITHOUT VARNISH (RMS VALUES)

DIV/ Sample:	With grading varnish	Without grading varnish
Sample in air	2,6kV	1,6kV
Sample immersed in FC72	2,8kV	4,9kV
Sample encapsulated by dielectric gel	2,6kV	5,7kV

Last a decrease of the numbers of partial discharge with time was evidenced when using stress grading material (compare figure 10 (without varnish) to figure 11 (with varnish)).

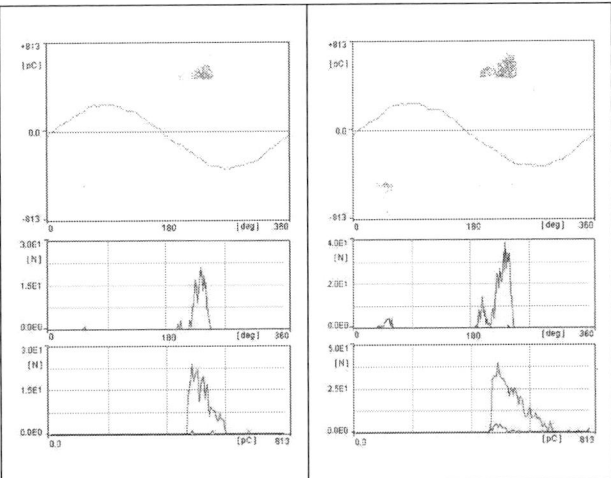

Fig.10: Changes in the partial discharges patterns vs time for sample without grading material (at to and to + 4min). Measurements performed in air and under AC 50Hz

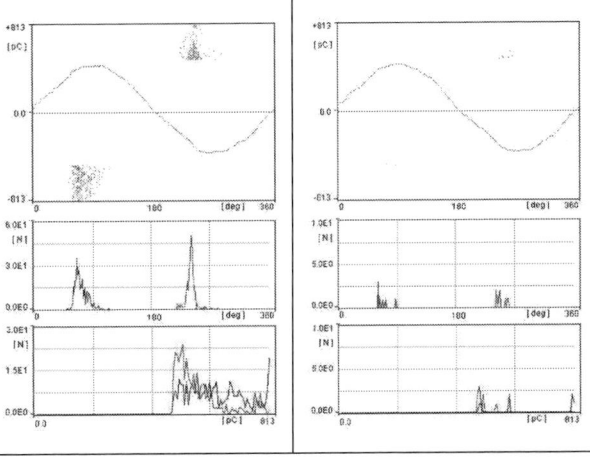

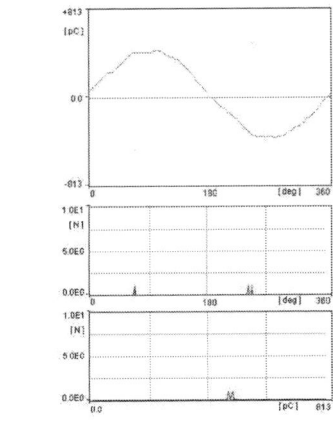

Fig.11: Changes in the partial discharges patterns vs time for sample with grading material (at to, to +2min and to + 4min). Measurements performed in air and under AC 50Hz.

IV. CONCLUSION

Studying materials able to be used for the stress grading in high voltage power electronics modules seems necessary as soon as the reliability of the systems is one of the main concerns. Indeed, triple points are the weakest points where the electrical stresses are enhanced. Therefore, new concepts able to achieve this stress grading have to be found and controlled. The validation of these systems must be made thanks to surface potential measurements under DC and AC. Moreover, it seems that the classical « current – voltage» measurements of the materials does not allow the materials choice nor to identify the phenomenon responsible for the mechanism of stress distribution, contrary to the measurement of the surface potential. We have also demonstrated that the problem had to be solved globally, ie taking into account the surrounding media (here the encapsulation gel). All these different points are still under study.

REFERENCES

[1] MAEDA Takahiko, HAGA K., MAEDA Takao, " Creepage Breakdown Characteristics of Printed Wire Board in Silicone Gel", Conference on Electrical Insulation and Dielectric Phenomena, IEEE Annual Report, 1996, pp.734-737.

[2] EBKE T., BAKIJA B., PEIER D., "Reduction of Grading Effects in Insulating Materials in Power Semiconductor Devices by Optimising the Internal Structure", Journal of Electrical Engineering, 2002, vol.2

[3] CURAMIK, "Direct Bonded Copper", Design rules of Curamik, 2005, pp.6-10

[4] MITIC G., LICHT T., LEFRANC G., "IGBT Module Technology with High Partial Discharge Resistance", IEEE Electrical Insulation Magazine, 2001, pp 1900 – 1901.

[5] ESPINO-CORTES P., CHERNEY A., JAYARAM S., "Effectiveness of Stress Grading Coatings on Form Wound Stator Coil Groundwall Insulation Under Rise Time Pulse Voltages", IEEE Electrical Insulation Magazine, December 2005, vol.20, n°4, pp 844-846.

[6] CONLEY D.J., FROST N., "Fundamentals of Semi-conductive Systems for High voltage Stress Grading", IEE Annual Report, 2005, pp.89-92

[7] ANSOFT, "Maxell Technical Notes – Electrical Field Calculation", Ansoft Co, 2007, pp.41-44

[8] RIVENC J., BIRON M., LEBEY T., "Stress grading materials: a discussion on current-voltage and surface potential technique", Charges in Solid Dielectrics, 1998, Dielectrics² Society, pp 169-181.

[9] ROBERTS A., "Stress Grading for High Voltage Motor and Generator Coils", IEEE Electrical Insulation Magazine, 1995, pp.26-31

[10] ALLISON JA., "Understanding the need for anti – corona materials in High Voltage Rotating Machines", 6th International Conference on Properties and Applications of Dielectric Materials, 2000, pp.860 – 863.

[11] BREIT F., "Contribution à l'intégration Hybride de Puissance : Etude de l'environnement diélectrique des semi-conducteurs », Thèse de doctorat, Université Paul Sabatier, Toulouse, 2003, pp.44-49.

Modelling of Iron Losses in Salient Pole
Permanent Magnet Synchronous Motors

Valentin NAVRAPESCU, Member, IEEE
POLITEHNICA University from Bucharest
Faculty of Electrical Engineering
Spl. Independentei 313, S 6, Bucuresti, ROMANIA
Email: valentin.navrapescu@plc.pub.ro

Mircea POPESCU, Member, IEEE
University of Glasgow
SPEED Laboratory
University Avenue, Glasgow G12 8QQ
UNITED KINGDOM
Email: mircea@ieee.org

Dragos Ovidiu KISCK, Member, IEEE,
Email: dragos@dsp-control.pub.ro

George ANDRONESCU
Email: g.andronescu@k.ro
POLITEHNICA University from Bucharest
Faculty of Electrical Engineering

Mariana KISCK
Cygnus Computer Bucharest
Bd. Theodor Palady, No. 4, Sector 4
Bucharest, ROMANIA
Email: cygmariana@yahoo.com

Abstract — Today, for the same rated power and speed electrical drives we have a strong competitor to the induction motors. This one it is represented by permanent magnet synchronous motors (PMSM) with salient pole rotors. In this paper are presented dedicated analytical and numerical models for the iron loss effect over the motor performance. It is presented a comprehensive comparison between no-load and loaded conditions test data for all the models presented.

I. INTRODUCTION

For permanent magnet synchronous motors analytical models we use lumped parameters equivalent circuits. Considering a fast computation, these parameters have to have constant values. In this conditions it is required an accurate prediction of the PMSM performance which assumes some aspects such as: a) an accurate estimation of the equivalent circuit parameters; b) a good and fast mathematical model for the circuits used to implement different phenomena inside the machine. In this idea, the estimation of the iron loss, which may greatly influence the motor total loss, has received various modelling approaches in the last decades. The calculation and then the measurement and implementation of the iron loss into an equivalent circuit of the PMSM represent important challenging tasks in the design of electrical motors.

II. MODELLING OF IRON LOSS

When we consider a model for permanent magnet synchronous motors we have to take note that the stator core losses occur in the teeth and in the yoke. The loss is a function of the total flux in the teeth and in the yoke. If we consider that the yoke and most of the teeth carry a flux which is proportional to the stator flux linkage, it appears to be appropriate to connect an equivalent core loss resistance (R_c) either at the terminals of the motor equivalent circuit, or more correctly inside the stator resistance R_s where the voltage across R_c would be the total induced voltage in the stator winding.

Thus, the equivalent L d-q axis circuit in Fig. 1 may be employed [8].

It should be emphasized that the equivalent iron loss resistance R_c exhibits different values with frequency variation or with current load and voltage variation through saturation effect. Also, test data allows the extraction of the sum between iron loss and stray load losses. Even tests performed at no-load operation will give only estimation for the iron loss. Thus R_{cl} is usually computed either empirically using the Steinmetz equation and coefficients or using finite element analysis.

All the following equations are written for steady-state operation when the voltages over the inductance L_d and L_q are zero.

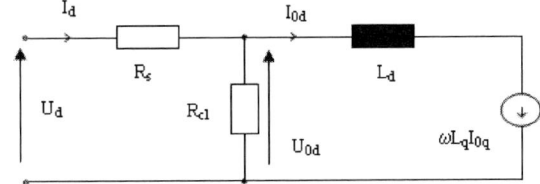

Fig.1.a: Equivalent L d axis circuit including iron loss [8]

$$\begin{bmatrix} u_d \\ u_q \end{bmatrix} = R_s \cdot \begin{bmatrix} i_{0d} \\ i_{0q} \end{bmatrix} + \left(1 + \frac{R_s}{R_{cl}}\right) \cdot \begin{bmatrix} u_{0d} \\ u_{0q} \end{bmatrix} \qquad (1)$$

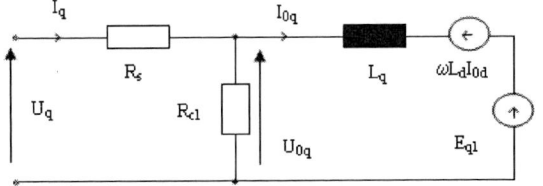

Fig.1.b: Equivalent L q axis circuit including iron loss [8]

$$\begin{bmatrix} u_{0d} \\ u_{0q} \end{bmatrix} = \begin{bmatrix} 0 & -\omega L_q \\ \omega L_d & 0 \end{bmatrix} \cdot \begin{bmatrix} i_{0d} \\ i_{0q} \end{bmatrix} + \begin{bmatrix} 0 \\ E_{ql} \end{bmatrix} \qquad (2)$$

where:

$$\begin{cases} i_{0d} = i_d - i_{cd} \\ i_{0q} = i_q - i_{cq} \end{cases} \quad (3) \quad \text{and} \quad \begin{cases} i_{cd} = -\dfrac{\omega L_q i_{0q}}{R_{cl}} \\ i_{cq} = \dfrac{E_{ql} + \omega L_d i_{0d}}{R_{cl}} \end{cases} \quad (4)$$

The terminal voltage V, the stator current I_s, and the electromagnetic torque are expressed as:

$$I_s = \sqrt{i_d^2 + i_q^2} \quad (5)$$

$$U_s = \sqrt{\left(R_s i_d - \omega L_q i_{0q}\right)^2 + \left(R_s i_q + \omega L_d i_{0d} + E_{ql}\right)^2} \quad (6)$$

$$M_e = \frac{mp}{2} \cdot \left[\frac{E_{ql}}{\omega} \cdot i_{0q} + \left(L_d - L_q\right) \cdot i_{0d} i_{0q} \right] \quad (7)$$

Thus, the iron loss is determined considering the following expression

$$P_{Fe} = R_{cl} \cdot \left(i_{cd}^2 + i_{cq}^2\right) =$$

$$= \frac{1}{R_{cl}} \cdot \left[\left(\omega L_q i_{0q}\right)^2 + \omega^2 \left(L_d i_{0d} + \frac{E_{ql}}{\omega}\right)^2 \right] \quad (8)$$

The mechanical loss is not controllable through electromagnetic design, but the copper loss and iron loss can be minimized either through an optimized design or more expensively through an optimized control strategy (e.g., vector control).

If we consider that the total iron loss may be separated into those caused by mutual flux and called core-loss and those caused by the leakage flux and called stray-load loss, the equivalent circuit (T d-q axis) in Fig. 2 has to be used [9]. The mutual flux is comprised of magnet flux and flux created by stator current (armature reaction flux).

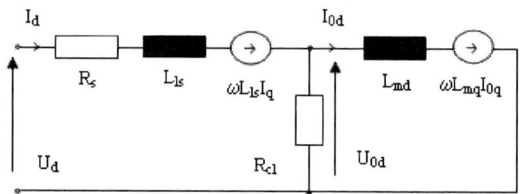

Fig.2.a: Equivalent T d axis circuit including iron loss [9]

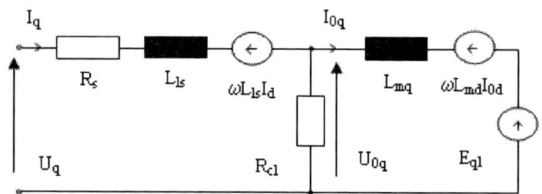

Fig.2.b: Equivalent T q axis circuit including iron loss [9]

This flux is proportional to the internal or air gap voltage. It follows that a resistor connected across the air gap voltage can represent the core-loss. For similar reasons, a resistor R_{s-l} placed across the leakage reactance ωL_{ls} will represent the

stray-load losses. Stray-load losses are comprised of several complex components, not just iron-loss due to the leakage flux.

Their physical causes are still under debate. It should be noted that the core-loss is proportional to the square of the air gap voltage and load-loss to the square of the stator current much as experiment shows them to be [2].

$$\begin{bmatrix} u_d \\ u_q \end{bmatrix} = \begin{bmatrix} R_s & -\omega L_{ls} \\ \omega L_{ls} & R_s \end{bmatrix} \cdot \begin{bmatrix} i_d \\ i_q \end{bmatrix} + \begin{bmatrix} u_{0d} \\ u_{0q} \end{bmatrix} \quad (9)$$

$$\begin{bmatrix} u_{0d} \\ u_{0q} \end{bmatrix} = \begin{bmatrix} 0 & -\omega L_{ls} \\ \omega L_{md} & 0 \end{bmatrix} \begin{bmatrix} i_{0d} \\ i_{0q} \end{bmatrix} + \begin{bmatrix} 0 \\ E_{ql} \end{bmatrix} \quad (10)$$

where:

$$\begin{cases} i_{cd} = -\dfrac{\omega L_{mq} i_{0q}}{R_{cl}} \\ i_{cq} = \dfrac{E_{ql} + \omega L_{md} i_{0d}}{R_{cl}} \end{cases} \quad (11)$$

The terminal voltage V_s and the electromagnetic torque are expressed as:

$$U_s = \sqrt{\begin{aligned} &\left(R_s i_d - \omega L_{mq} i_{0q} - \omega L_{ls} i_q\right)^2 + \\ &+ \left(R_s i_q + \omega L_{ls} i_d + \omega L_{md} i_{0d} + E_{ql}\right)^2 \end{aligned}} \quad (12)$$

$$M_e = \frac{mp}{2} \cdot \left[\frac{E_{ql}}{\omega} \cdot i_{0q} + \left(L_d - L_q\right) \cdot i_{0q} i_{0q} \right] \quad (13)$$

The iron loss is determined considering the following expression:

$$P_{Fe} = R_{cl} \cdot \left(i_{cd}^2 + i_{cq}^2\right) =$$

$$= \frac{1}{R_{cl}} \cdot \left[\left(\omega L_{mq} i_{0q}\right)^2 + \omega^2 \left(L_{md} i_{0d} + \frac{E_{ql}}{\omega}\right)^2 \right] \quad (14)$$

Another approach is to simulate the iron loss components with two equivalent resistors: R_{cv} that denotes the voltage dependent losses (similar to a conventional synchronous machine) and R_{ci} that denotes the current dependent losses.

The resistor R_{cv} is placed in parallel with the total induced voltage by the leakage and magnetization fluxes. This would be similar to the equivalent circuit in Fig. 1, but with the difference that R_{cv} takes into account only the so-called voltage dependent iron loss. Actually, the induced voltage due to the magnets determines these losses. The resultant equivalent circuit Ti d-q axis) is illustrated in Fig. 3.

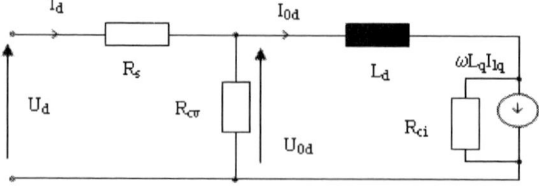

Fig.3.a: Equivalent L from Ti d axis circuit including iron loss [10]

978-1-4244-1871-8/07 $25.00 © 2007 IEEE   352

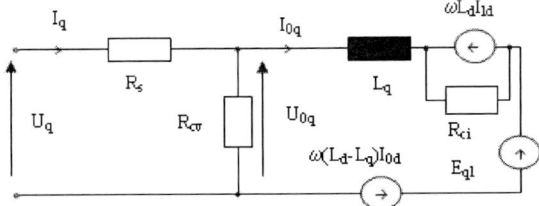

Fig.3.b: Equivalent L from Ti q axis circuit including iron loss [10]

The main drawback of this approach is that while R_{cv} can be computed or measured, R_{ci} can only be estimated through complicated measurements [10].

$$\begin{bmatrix} v_d \\ v_q \end{bmatrix} = R_s \begin{bmatrix} i_{od} \\ i_{oq} \end{bmatrix} + \left(1 + \frac{R_s}{R_{cv}}\right) \begin{bmatrix} v_{od} \\ v_{oq} \end{bmatrix} \quad (15)$$

$$\begin{bmatrix} u_{0d} \\ u_{0q} \end{bmatrix} = \begin{bmatrix} R_{echiv} & -\omega L_{echiv} \\ \omega(L_{echiv} + L_d - L_q) & R_{echiv} \end{bmatrix} \begin{bmatrix} i_{0d} \\ i_{0q} \end{bmatrix} + \begin{bmatrix} 0 \\ E_{ql} \end{bmatrix} \quad (16)$$

where:

$$\begin{cases} i_{cd} = -\dfrac{\omega L_{echiv} i_{0q}}{R_{cv}} + \dfrac{R_{echiv} i_{0d}}{R_{cv}} \\ i_{cq} = \dfrac{E_{ql} + \omega L_{echiv} i_{0d} + \omega(L_d - L_q) i_{0d}}{R_{cv}} + \dfrac{R_{echiv} i_{0q}}{R_{cv}} \end{cases} \quad (17)$$

$$\begin{cases} i_{dl} = -\dfrac{1}{\omega L_q} \cdot \left[R_{echiv} i_{0q} + \omega L_{echiv} i_{0d} \right] \\ i_{ql} = \dfrac{1}{\omega L_q} \cdot \left[-R_{echiv} i_{0d} + \omega L_{echiv} i_{0q} \right] \end{cases} \quad (18)$$

$$R_{echiv} = \frac{R_{ci}(\omega L_q)^2}{R_{ci}^2 + (\omega L_q)^2} \quad (19)$$

$$\omega L_{echiv} = \frac{R_{ci}^2(\omega L_q)}{R_{ci}^2 + (\omega L_q)^2}$$

The terminal voltage V_s and the electromagnetic torque are expressed as:

$$U_s = \sqrt{\begin{array}{l} \left(R_s i_d - \omega L_{echiv} i_{0q} + R_{echiv} i_{0d}\right)^2 + \\ + \left(R_s i_q + \omega(L_d - L_q - L_{echiv}) \cdot i_{dl} i_{ql} + E_{ql}\right)^2 \end{array}} \quad (20)$$

$$M_e = \frac{mp}{2} \cdot \left[\frac{E_{ql}}{\omega} \cdot i_{ql} + \left(L_d - L_q\right) \cdot i_{dl} i_{ql} \right] \quad (21)$$

The iron loss is determined considering the following expression:

$$P_{Fe} = P_{FeU} + P_{FeI} = \\ = R_{cv} \cdot \left(i_{cd}^2 + i_{cq}^2\right) + R_{ci} \cdot \left((i_{0d} - i_{dl})^2 + (i_{0q} - i_{ql})^2\right) \quad (22)$$

III. IRON LOSS EQUIVALENT RESISTANCE COMPUTATION

A. Empirical Method 1
Iron loss data from steel suppliers is almost always sine wave data, and may be characterized by the Steinmetz

equation with separate terms for hysteresis, eddy-current and excess loss [6–7]:

$$P_{Fe} = K_h \omega \Psi_m^{\ n} + K_e \omega^2 \Psi_m^{\ 2} + K_{exc} \omega^{3/2} \Psi_m^{\ 3/2} \quad (23)$$

It is commonly acceptable to set $n = 2.0$ and ignore the excess loss, and the iron loss power density becomes:

$$P_{Fe} = \omega^2 \Psi_m^{\ 2} \cdot \left(\frac{K_h}{\omega} + K_e\right) \quad (24)$$

B. Empirical Method 2
Iron loss equivalent resistance may be determined considering separate equivalent resistance for the core loss components: eddy current loss resistance (R_{eddy}) and hysteresis loss resistance ($R_{hysteresis}$). With a good approximation the following expression is used:

$$R_{Ecrt} = \frac{12 m \rho N^2 A}{l c^2} = \frac{12 m \rho N^2 A^2}{Vol c^2} \quad (25)$$

$$R_{Hyst} = \frac{2 \pi \omega N^2 A^2}{C_h G_{tFe}} \quad (26)$$

where A is the cross section area perpendicular to the induced eddy currents, N the number of turns per phase, c lamination width, l mean length of the flux path, ρ steel lamination resistivity, Vol iron volume, C_h hysteresis core loss coefficient and Gt_{Fe} lamination weight.

The total core loss resistance is computed as

$$R_{cl} = \left(\frac{1}{R_{Ecrt}} + \frac{1}{R_{Hyst}}\right)^{-1} \quad (27)$$

C. Finite Element method
Another option, a more accurate one is the usage of numerical analysis such as finite element method. If the tooth and yoke flux density waveforms are computed, by changing the time variable t to rotor position θ we can determine eddy-current loss as:

$$P_{Ecrt} = \frac{p\omega^2 C_e d_{Fe}}{8\pi^3 \cdot \left[V_t \displaystyle\int_0^{\frac{4\pi}{p}} \left(\frac{dB_t}{d\theta}\right)^2 d\theta + V_y \displaystyle\int_0^{\frac{4\pi}{p}} \left(\frac{dB_y}{d\theta}\right)^2 d\theta \right]} \quad (28)$$

and hysteresis-loss as:

$$P_{Hyst} = \frac{p C_h \omega d_{Fe}}{4\pi} \cdot \left(V_t B_{pk(t)}^{a + b B_{ph(t)}} + V_y B_{pk(y)}^{a + b B_{ph(y)}} \right) \quad (29)$$

where p represents the pole number, d_{Fe} material density, $V_{t,y}$ volume of the stator teeth and yoke.

The eddy current loss expression must to be corrected to consider the harmonics effect, while the hysteresis loss expression must include the minor loops hysteresis effect. An equivalent correction coefficient [9] that allows the inclusion of the minor loops effect is

$$k_{echiv} = 1 + \frac{k}{B_{pk(g)}} \cdot \frac{\displaystyle\sum_{i=1}^{N_{step}} \Delta B_i - 2 B_{pk(g)}}{2} \quad (30)$$

where k is a hysteresis loss minor loops constant in the range 0.6 to 0.7; $B_{pk(g)}$ the peak flux-density in the airgap for one electrical cycle; ΔB_i the flux-density variation between two consecutive rotor positions. The minor loops effect computed with (30) has the advantage of using the individual flux-density values, rather than the magnitude of the flux density reversals. Thus, (30) is much easier to implement than the formula proposed by Lavers et al. [9]. The total core loss resistance is computed as

$$R_{cl} = \left(\frac{E_{ql}^2}{P_{Ecrt} + P_{Hyst}} \right) \tag{31}$$

IV. RESULTS

For the practical test we considered a PMSM with 2-pole, 2-phase used in a normal refrigerator compressor at 50 Hz. The magnet material is ferrite. The cross-sectional view of the motor with the flux-lines pattern on load is presented in Fig. 4.

The stator windings exhibit a different number of turns. So, for a balanced conditions operation the applied voltages respect the ratio:

$$\frac{U_{main}}{U_{aux}} = \frac{1}{k} \tag{32}$$

where k is the effective turns ratio – auxiliary per main winding.

We performed a test at no-load for varying terminal voltage conditions [10, 12]. The iron loss was defined as the difference between the input power minus the stator copper loss and windage and friction loss. In Fig. 5 the currents variation with terminal voltage is shown. The resultant iron loss is presented in Fig. 6. The measured currents follow the usual aspect variation with a minimum value when terminal voltage equals the back emf. The back emf variation with speed is given in Fig. 7.

The iron loss at no-load exhibits a minimum at the open circuit voltage and starts to increase if the voltage is decreased beyond this point. In [12] this behaviour is explained through the additional field harmonics that are induced by the armature reaction and/or are a consequence of the PM field.

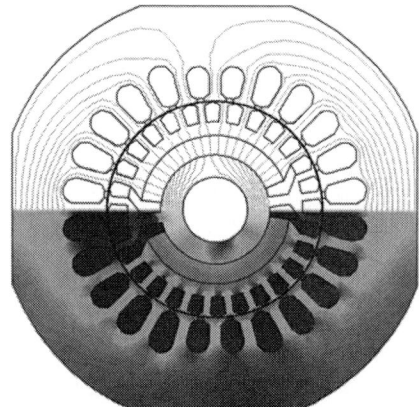

Fig. 4: Cross-sectional view of the analyzed motor with flux-lines and flux-density distribution at no-load

Thus, the same effects as saturation and magnet demagnetization must be expected due to the flux redistribution. These extra losses are modelled in the equivalent circuit from Fig. 3 by the usage of the resistor R_{ci}.

The equivalent resistances for modelling the iron loss in circuits from Figs. 1-3, are determined as

$$R_{ci} = \frac{\left(U_n^2 - E_{ql}^2 \right)}{P_{Fen}} \tag{33}$$

$$R_{cv} = R_{cl} = \frac{E_{ql}^2}{P_{Fe0}} \tag{34}$$

where W_{Fen}, W_{Fe0} represent the iron loss measured at loading conditions and no-load when the terminal voltage equals the back emf E_{ql}. Expression (34) can be used to validate the computation made with (27) or (31).

The last method analysis predicts a distribution of the iron loss in the stator and rotor lamination as in Figs. 8-13. The light coloured regions, i.e., at the stator and rotor teeth and near the airgap represent the regions where the majority of the iron loss is concentrated.

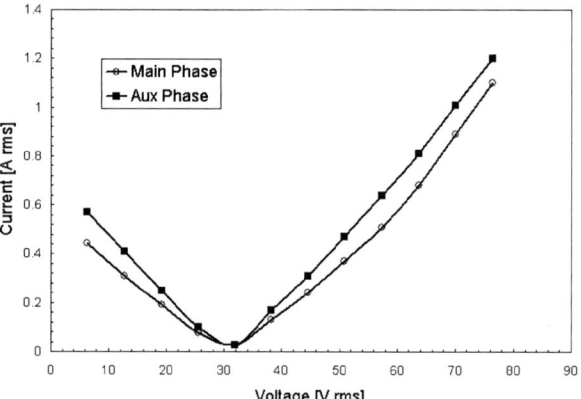

Fig. 5: Current variation with voltage – experimental data

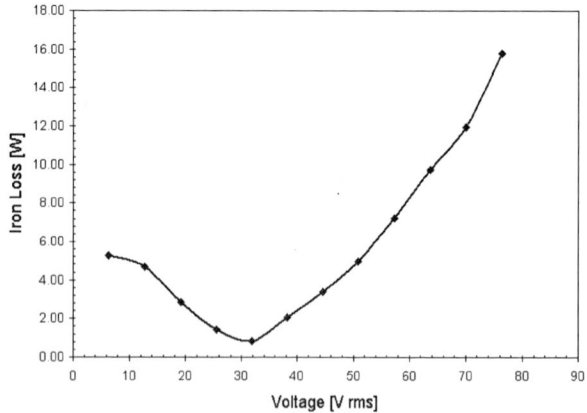

Fig. 6: Iron loss variation with voltage – experimental data

Even all the described equivalent circuits that include the iron loss effect have different approaches especially regarding the place where the core loss resistance should be connected, the predicted electromagnetic torque exhibits small differences.

978-1-4244-1871-8/07 $25.00 © 2007 IEEE

In Fig. 14 it is ilustrated a comparison of the measured torque vs. the predicted electromagnetic torque using (7).

As general patterns one may observe that: (*a*) the equivalent circuit described in Fig. 1 predicts a lower electromagnetic torque value for the whole range of load angle; (*b*) the equivalent circuit described in Fig. 2 predicts a lower electromagnetic torque value for the load angle variation between zero and the load angle corresponding to maximum torque; (*c*) the equivalent circuit described in Fig. 2 predicts a higher electromagnetic torque value for the load angle variation between the load angle corresponding to maximum torque and maximum load angle; (*d*) the equivalent circuit described in Fig. 3 predicts a higher electromagnetic torque value for the load angle variation between zero and the load angle corresponding to maximum torque; (*e*) the equivalent circuit described in Figs. 3 predicts a lower electromagnetic torque value for the load angle variation between the load angle corresponding to maximum torque and maximum load angle; (*f*) all the equivalent circuits that include core loss effect predict a lower maximum electromagnetic torque compared to the case when the core losses are neglected.

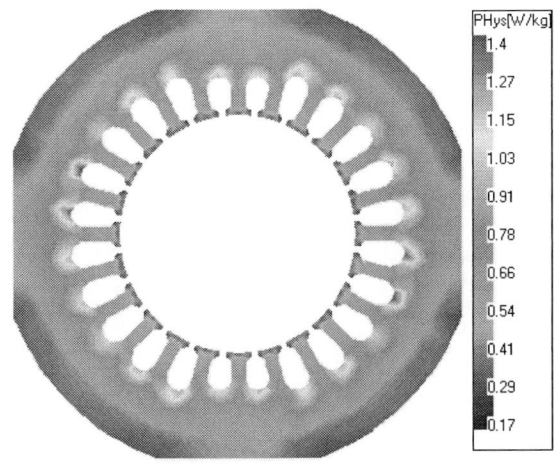

Fig. 9: Cross-sectional view of the analyzed motor stator with hysteresis iron loss distribution at no-load

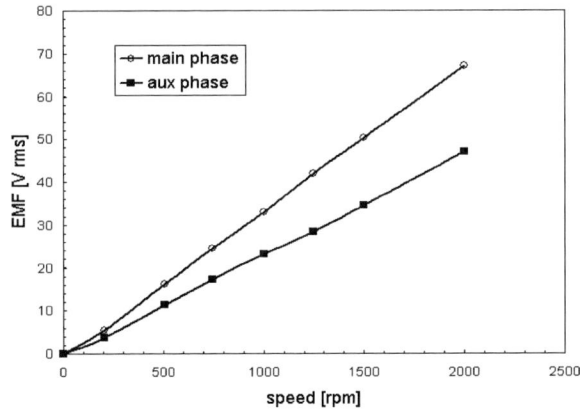

Fig. 7: Back emf variation with speed – experimental data

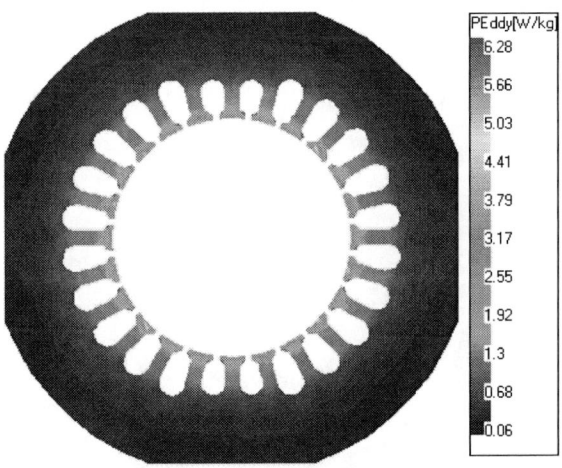

Fig. 10: Cross-sectional view of the analyzed motor stator with eddy-current iron loss distribution at no-load

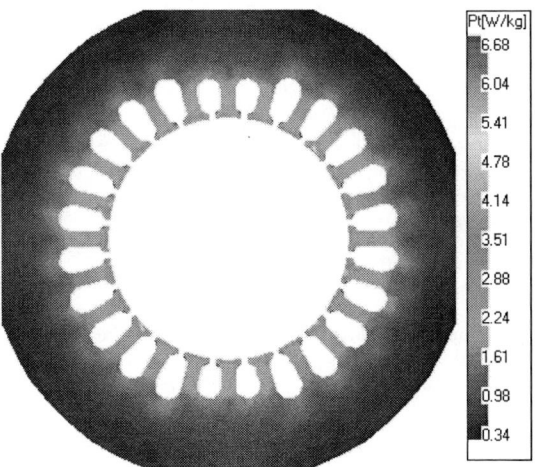

Fig. 8: Cross-sectional view of the analyzed motor stator with total iron loss distribution at no-load

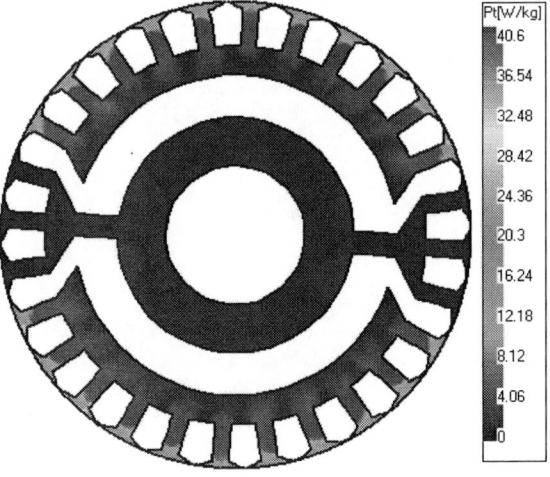

Fig. 11: Cross-sectional view of the analyzed motor rotor with total iron loss distribution at no-load

978-1-4244-1871-8/07 $25.00 © 2007 IEEE   355

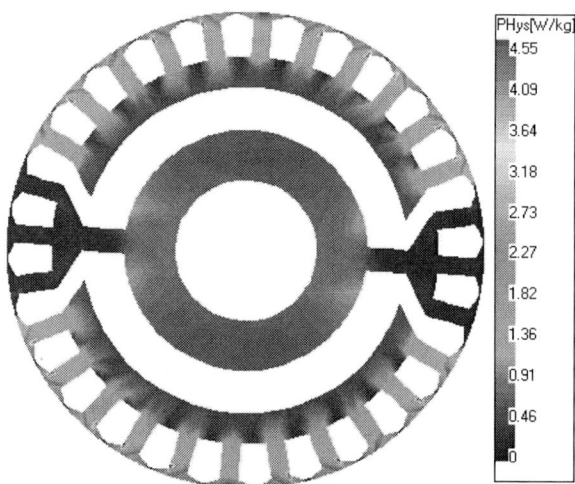

Fig. 12: Cross-sectional view of the analyzed motor rotor with hysteresis iron loss distribution at no-load

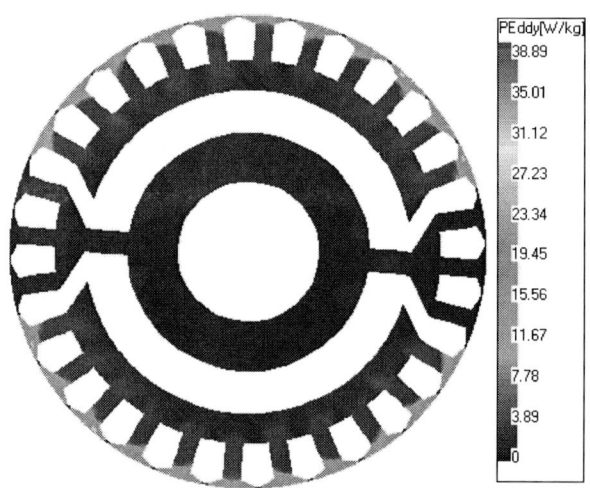

Fig. 13: Cross-sectional view of the analyzed motor rotor with eddy-current iron loss distribution at no-load

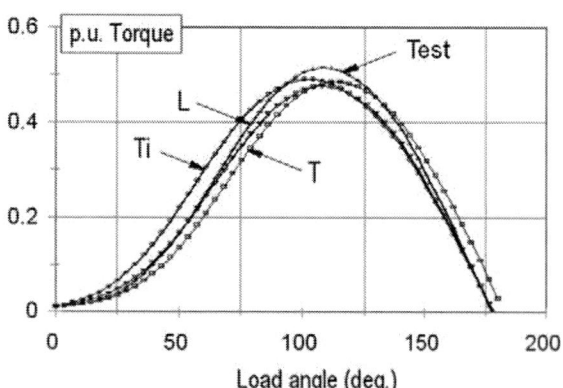

Fig.14: Torque variation with load angle when core loss effect is included

V.   CONCLUSIONS

For modelling of the iron loss in permanent magnet synchronous motors several methods can be used. Particular rotor configuration – interior permanent magnet – determines a significant level of iron loss even at no-load operation if the supply voltage is lower than the induced back emf. The flux redistribution that occurs due to the strong level of saturation may produce important time harmonics content in the air-gap flux. These harmonics determine additional losses.

As it was presented, the effect of the iron loss can be modelled accurately using fixed value for equivalent resistors (e.g. R_c). The connection of the iron loss resistor in the motor equivalent circuit may be implemented in several ways. One should note that all models have in common the idea of placing the equivalent resistor in the stator circuit. The explanation based on the fact that in salient pole PMSM, the stator iron loss is the main region where the total iron loss does occur. If the estimation of the iron loss resistor is sufficient accurate, similar motor performance is predicted with all analyzed models. The iron loss modelling becomes more important when the salient-pole PMSM is operated at high speed, low flux mode or at variable speed.

REFERENCES

[1]   Bernal-Fernandez, F., Garcia-Cerrada, A., Faure, R.: "Determination of parameters in interior permanent magnet synchronous motors with iron loss without torque measurement", *IEEE Trans. Industry Applications*, Vol. 37, No. 5, Sept./Oct. 2001. pp. 1265 – 1272

[2]   Bertotti, G., Boglietti, A.A., Chiampi, M., Chiarabaglio, D., Fiorillo, F.: "An improved estimation of iron loss in rotation electrical machines" *IEEE Trans. Magn..*, Vol. 27, pp.5007 – 5009, Nov. 1991

[3]   Consoli,A., Racitti, A. "Analysis of permanent magnet synchronous motors", *Industry Applications, IEEE Transactions on*, Volume:27, No: 2, March/April. 1991, pp. 350 – 354

[4]   Deng, F.: "An improved iron loss estimation of permanent magnet brushless machines", *IEEE Trans. on Energy Conv.*, Vol. 14, No. 4, December 1999, pp. 1391-1395

[5]   Honsinger, V.B. "Performance of polyphase permanent magnet machines", *IEEE Trans. Power Appl. Syst.*, vol. PAS-99, pp.1510-1518, July 1980

[6]   Lavers, J.D., Biringer, P.P., Hollitscher, H.: "A simple method of estimating the minor loop hysteresis loss in thin laminations" *IEEE Trans. on Magnetics*, Vol. 14, No. 5, September 1978, pp. 386 – 388

[7]   Mi, C., Slemon, G.R. Bonert, R.: "Modeling of iron loss of permanent-magnet synchronous motors", *IEEE Trans. Industry Applications*, Vol. 39, No. 3, May/June 2003. pp. 734 – 741

[8]   Miller, T.J.E., Rabinovici, R.: "Back-emf waveforms and core losses of brushless DC motors", *Proc. IEE – Elect. Power Applicat.*, Vol. 141, No. 3, pp. 144-154, 1994

[9]   Navrapescu, V., Popescu, M.: "Discrete models of the induction motors for real-time control drives", Revue Romaine des Sciences techniques – Academie Romaine, 2006, Oct-Dec, pag.489-502,

[10]   Popescu, M, Navrapescu, V.: "A method of iron loss and magnetising flux saturation modelling in stationary frame reference of single and two-phase induction machines", *8th Int. Conf. on P. Electronics and Variable Speed Drives* (IEEE Conf. Publ. No. 475), 2000, pp. 140–146

[11]   Schiferl, R., Lipo, T.A.: "Core loss in buried magnet permanent magnet synchronous motors", *IEEE Trans. on Energy Conv.* Vol. 4, No. 2. June 1989, pp. 279-284

[12]   Sebastian, T., Slemon, G.R., Rahman, M.A.: "Modeling of permanent magnet synchronous motors", *IEEE Trans. Magnetics*, Vol. MAG –22, No.:5, pp. 1069 – 1071, Sept. 1986.

[13]   Slemon, G.: "Core losses in permanent magnet motors", *IEEE Trans. on Magnetics*, Vol. 26, No. 5, Sept.1990, pp. 1653-1655

978-1-4244-1871-8/07 $25.00 © 2007 IEEE

Low cost fault detection system for inverter driven induction motor using currents signal

N.H Kim
Cheungju Univ
School of Elec. &
Info.
Cheongju, Korea.
hkim@cju.ac.kr

Oh Yabg
Cheungju Univ
School of Elec. &
Info.
Cheongju, Korea.
hkim@cju.ac.kr

M.H Kim
Yeongnam College
Dept. of EE
Daegu, Korea.
mhkim@ync.ac.kr

Hamid A. Toliyat
Texas A&M Univ.
Dept. of ECE
Texas, USA
toliyat@tamu.ece.
eud

Yongmin Oh
POSCON
Research center
Seoul, Korea
petri@poscon.co.kr

C.H. Choi
POSCON
Research center
Seoul, Korea
choi97@poscon.co.
kr

Abstract—In this paper, the induction motor rotor fault diagnosis system using current signals which are measured for axis-transformation and speed which is estimated using current information are presented. In inverter-fed motor drives the stator currents have much harmonics components and therefore fault diagnosis using stator current is difficult. The current and speed signal for rotor fault diagnosis need precise and high resolution information, which means the diagnosis system demands additional hardware such as low pass filter, high resolution ADC, encoder. Therefore, the proposed method are expected to contribute to low cost fault diagnosis system in inverter-fed motor drives without the need for encoder and any additional hardware.

In order to confirm validity of the developed algorithms, various experiments for rotor faults are tested and the line current spectrum of each faulty situation using Park transformation and speed estimation method are compared with the results obtained from fast Fourier transforms.

I. INTRODUCTION

In recent years, marked improvement based on development micro processor and power electronics has been achieved in motor drives. However, motors driven by solid state inverters have undergone serious voltage stresses because of rapid switch-on and switch-off voltage of semiconductor devices. As a result, condition monitoring and incipient fault detection technology have become an important research area in recent years to prevent systems from sudden shut-downs because of significant motor faults in the industrial manufacturing facilities

Among the fault diagnosis techniques, analyzing vibration signal with accelerometers, air-gap flux measurement with search coil and thermal analysis provide satisfactory results [1-6]. New and promising research horizons in the area of motor fault detection could be explored using the expert systems, artificial neural networks (ANNs), fuzzy systems, and genetic algorithms (GAs)[7,8] and adaptive neuron-fuzzy inference systems (ANFIS)[9]. But to extract normal and fault characteristics these techniques need offline training of diagnosis algorithm through simulations or experiments results. Also these techniques need sophisticated computational procedure.

Even though numerous successful line driven motor fault detection methods are reported in the literature, inverter fed driven motor systems still require more attention due to high speed switching noise effects in the line current data and closed loop controller bandwidths [8, 9].

Although the rotor faults are very commonly reported type with an occurrence of 8%[10][11], the diagnosis of these faults are one of the most challenging even under line driven motor case when compared to the other faults, because of the low amplitude fault signatures in the current spectrum. Fig. 1 shows the results of motor defection conducted by IEEE study.

Thus, in this work in order to detect broken rotor bar signal clarity using 12bit ADC, park transformation of measured currents and speed estimation are investigated theoretically and experimentally for inverter driven motors. To verify the proposed algorithms, a 2.2 kW induction and a TMS320F2812 DSP is used and V/F control method is adapted.

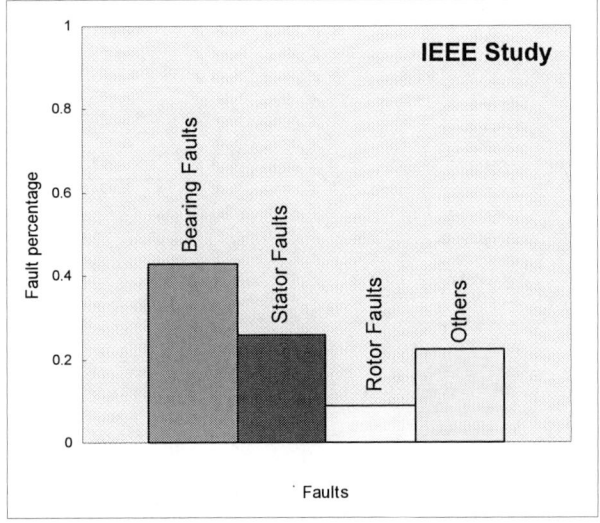

Fig. 1 Results of motor defection.

II. PROPOSED ALGORITHMS AND SYSTEM CONFIGURATION

For detection rotor bar faults [11], [12], [13], and [14] were using spectrum analysis of machine line currents, and in these papers investigate sideband components(f_b) which is around fundamental currents component for broken bar faults and equation is below.

$$f_b = (1 \pm 2s)f \qquad (1)$$

978-1-4244-1871-8/07 $25.00 © 2007 IEEE

Where, s is slip, f is supply frequency, and f_b is detectable broken rotor bar frequency.

While the lower sideband frequency presents broken rotor bar, the upper sideband frequency is due to consequent speed oscillation. [13] shows that broken bar give rise to a sequence of sideband given by (2)

$$f_b = (1 \pm 2 \cdot k \cdot s)f \qquad (2)$$

Where, $k = 1, 2, 3, \cdots$

In order to calculate specific frequency component, generally FFT(fast fourier transform) method is used, but FFT method need considerable calculation time and sophisticated algorithms and much memory. Therefore at this paper axis transformation, park transformation, and speed estimation methods using current signal and look up table, which is very simple and effective methods are presented to estimated exact sideband currents frequency components.

Axis transformation method, which is park transformation make possible a.c. to d.c. value conversion. With axis transformation against specific frequency, a.c. component and d.c. component are decoupled, i.e. if axis transformation is employed, the a.c. component(which are expressed in the stationary reference frame) are transformed into a new rotation reference frame, which rotates together with a selected frequency components expected to cause broken rotor bar and a.c. value expected to noise. Therefore axis transformation makes calculation of specific frequency component very easy and effective.

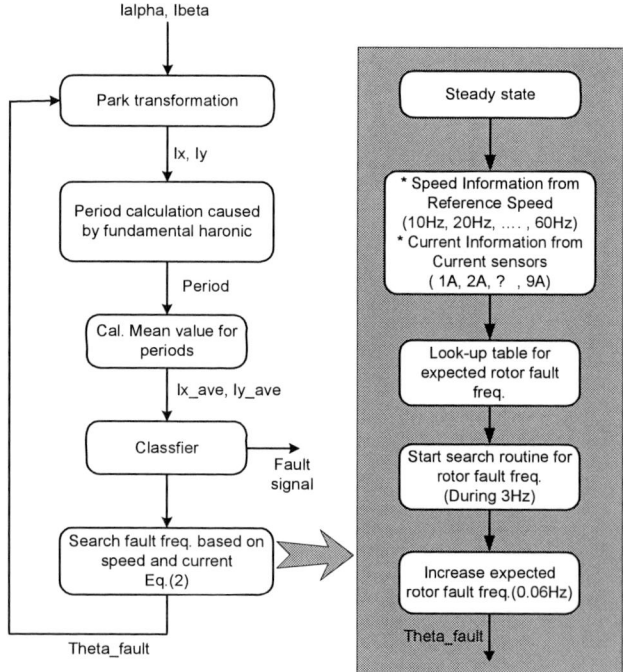

Fig. 2 Block diagram of the proposed method

Fig. 2 shows block diagram of the proposed method. Through park transformation, a.c. currents component and d.c. currents component are decoupled, and to calculate exact

d.c. component value expect to cause broken rotor fault estimate mean value for several periods. Base on previous process, ix_ave and iy_ave are calculated, and by means of classifier status of rotor is determined.

Fig. 3 shows the overall system configuration for rotor fault diagnosis. The inverter control and fault diagnosis system is implemented on the TMS320F2812 digital signal processor board from Texas Instruments. Various blocks used in the rotor bar fault diagnosis package are shown in this figure 3. Through the 12 bit on-chip ADC the current signals for motor control and fault diagnosis is collected with 4 kHz sampling frequency. Using the clark transformation and digital filter the raw current signals are transferred. To calculate the fundamental frequency and the specific frequency component which is expected fault component caused by the broken rotor bar, the Park transformation, signal conditioning, signal tracking and calculation of the average values the classifier and fault frequency estimation block are used.

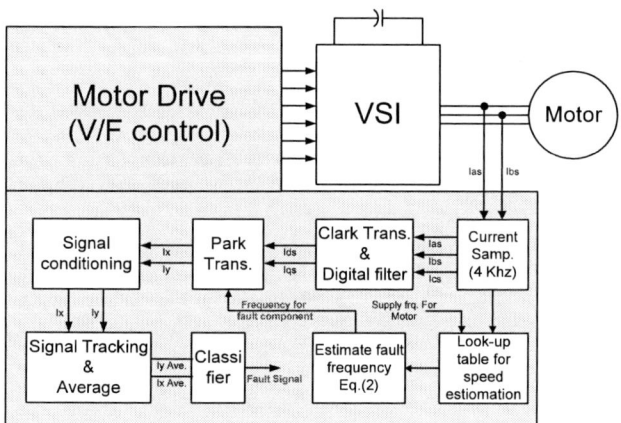

F2812 DSP board
for fault diagnosis

Fig. 3 System configuration.

A. *Clarke trans. & Digital filter*

Using measured three-phase currents of motor and clark transformation, real part current, I_α and imaginary part current , I_β, can be calculated, for calculation this current components, eq.(3) and eq.(4) are applied. Also, digital filter, which is low pass filter, are used to reduce unwanted noise.

$$I_\alpha = I_{as} \qquad (3)$$

$$I_\beta = (I_{as} + 2 \times I_{bs})/\sqrt{3} \qquad (4)$$

B. *Park trans.*

This module is used to transfer specific frequency component, which is expected to cause fault, to d.c. value, but other frequency components should be remained still a.c. value. Used equations for this transformation are (5) and (6).

$$I_x = I_\alpha \times \cos(\theta_{fault}) + I_\beta \times \sin(\theta_{fault}) \qquad (5)$$

978-1-4244-1871-8/07 $25.00 © 2007 IEEE

$$I_y = -I_\alpha \times \sin(\theta_{fault}) + I_\beta \times \cos(\theta_{fault}) \qquad (6)$$

C. Signal Conditioning

Park transformed Ix, Iy have a.c. value and d.c. value simultaneously. So we need to calculate only d.c. component. In this module, period caused by fundamental component is estimated to remove a.c. component.

D. Signal Tracking & Average. :

In this module the average values of Ix and Iy, Ix ave and Iy ave, are calculated for several period, which is coming from Signal conditioning module, and a.c. component which is contained in Iy, Ix is removed. In other word,

If $f_x = f_{ac} + f_{dc}$,

then $$\dfrac{\sum\limits_{n=1}^{k} f_{ac} + f_{dc}}{k} = f_{dc\_ave} \qquad (7)$$

Where, k = (period for a.c. component) / (sampling time)

E. Speed estimation

Fig. 4 show torque/speed curve for induction motor. For induction motor driven by V/F control, feed-forward control, motor speed depends on load torque. The certain load torque is supplied to running induction motor, motor speed vary from w_1 to w_2 and this region almost proportional to stator currents. With this simple idea, speed estimation is performed. According to each frequency speed variations depending on stator current amplitude are measured and look-up table for speed estimation is built.

Also for the proposed fault detection algorithm it is not necessary to use exact frequency which is expected to cause rotor fault, because fault detection algorithms is searching for 3 [Hz] and analyze out the fault frequency.

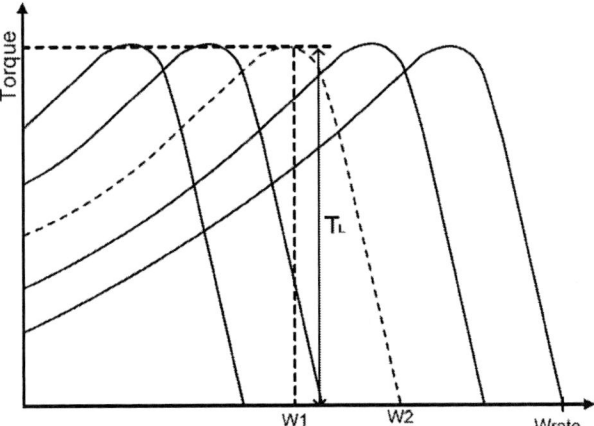

Fig. 4 Induction motor torque/speed curve

III.   EXPERIMENTAL RESULTS

In order to verify the proposed algorithms, which is axis transformation to detect specific frequency component, experiments is done. At the FFT analysis, it is very difficult to detect sideband component because FFT need much data and it take a lot of time to analyze frequency.

Table 1 Induction motor parameters

Rated Power	3Hp	Rated Speed	1760rpm
Rated Voltage	230V	Rated Current	7.6A

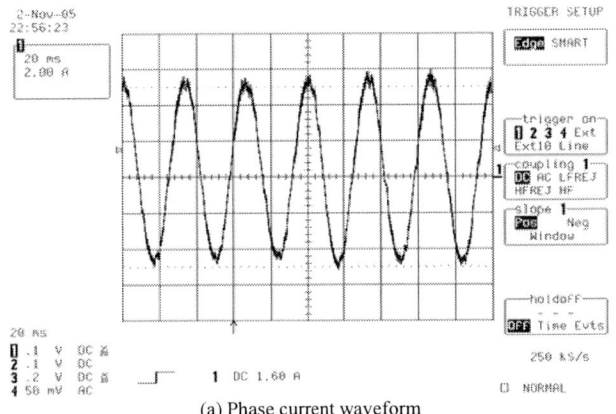

(a) Phase current waveform

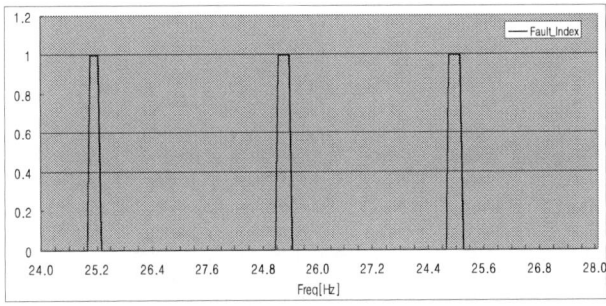

(b) Fault signal with the proposed algorithm

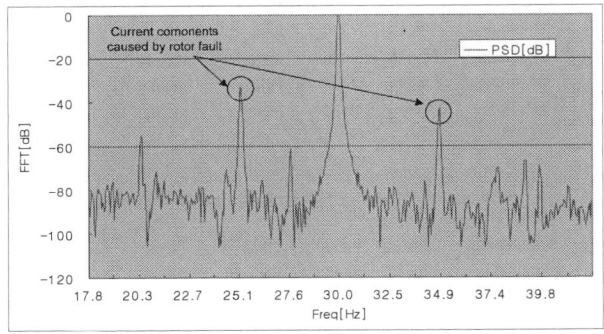

(c) PSD results of phase current

Fig. 5 Experimental results

978-1-4244-1871-8/07 $25.00 © 2007 IEEE

Figure 5(a) represents current waveform of rotor defect motor. Figure 5(b) shows PSD result of faulty motor, which has broken rotor fault current component abound 25Hz. As shown this figure, rotor fault component is -36 [dB] which means this induction motor has some problem in rotor.

Figure 5(c) shows the results of current spectrums using the proposed algorithm which is axis transformation and average method, as shown this result the proposed algorithm can detect bearing defect using 12 bit ADC. Table 1 shows motor parameters which is used at this paper.

IV. CONCLUSION

This paper has investigated the feasibility of detecting broken rotor faults using an axis transformation and speed estimation using look up table for current spectrum of an inverter driven induction machine.

An induction motor defects caused by broken rotor faults produce visible changes in the stator current spectrum at predictable frequencies. However it is very difficult to detect using FFT method because it take very long time and need pretty much data information. However, an axis transformation and speed estimation instead of FFT method is proved very effective and low cost through simulation and experimental results.

ACKNOWLEDGMENT

The authors would like to thank to the Power Application Technology Research Center of Yeungnam College of Science & Technology sponsor by the Ministry of Commerce, Industry & Energy and the Korea Electric Power Corporation.

REFERENCES

[1] G.B. Kliman, R.A. Koegl, J. Stein, R.D. Endicott, and M.W. Madden, "Noninvasive detection of broken rotor bars in operating induction motors," IEEE Trans. Energy Conv., vol. 3, pp. 873–879, Dec. 1988.

[2] R. Schoen, T. Habetler, F. Kamran, and R. Bartfield, "Motor bearing damage detection using stator current monitoring," IEEE Trans. Ind. Appl., vol. 31, no. 6, pp. 1274–1279, Nov./Dec. 1995.

[3] M.E.H. Benbouzid, "A review of induction motors signature analysis as a medium for faults detection," IEEE Trans. Ind. Electron., vol. 47, no. 5, pp. 984–993, Oct. 2000.

[4] S. Nandi, M. Bharadwaj, and H.A. Toliyat, "Performance Analysis of a Three-Phase Induction Motor Under Mixed Eccentricity Condition," IEEE Trans. Energy Conv., vol. 17, pp. 392-399, Sep. 2002.

[5] J.F. Watson, N.C. Paterson, and D.G. Dorrell, "The Use of Finite Element Methods to Improve Techniques for the Early Detection of Faults in 3-phase Induction Motors", IEEE Trans. on Energy Conversion, Vol. 14, No. 3, pp. 655-660, Sep. 1999.

[6] H.A. Toliyat, M.S. Arefeen, and A.G. Parlos, "A Method for Dynamic Simulation and Detection of Air-Gap Eccentricity in Induction Machines", IEEE Trans. on Industry Applications, Vol. 32, No. 4, pp. 910-918, Jul./Aug. 1996.

[7] F. Filippetti, G. Franceschini, C. Tassoni, and P. Vas, "Recent developments of induction motor drives fault diagnosis using AI techniques," IEEE Trans. Ind. Electron., vol. 47, no. 5, pp. 994–1004, Oct. 2000.

[8] M. A. Awadallah and M. M. Morcos, "Application of AI tools in fault diagnosis of electrical machines and drives—an overview," IEEE Trans. Energy Convers., vol. 18, no. 2, pp. 245–251, Jun. 2003.

[9] J.-S. R. Jang, "ANFIS: adaptive-network-based fuzzy inference system," IEEE Trans. Syst., Man, Cybern., vol. 23, no. 3, pp. 665–685, May/Jun. 1993.

[10] McDermid, W., "Insulation systems and monitoring for stator winding of large rotating machines," IEEE Electrical Insulation Magazine, vol. 9, no. 4, pp. 7-15, 1993.

[11] Schemp, D.E., "Predict motor failure with insulation testing," Plant Engineering, vol. 50, pp. 97-96, 1996.

[12] Stone, G.C., "Partial discharge measurements to access rotation machine insulation condition : A survey," Conference Record of the IEEE International Symposium on Electical Insulation, pp. 19-23, 1996.

[13] J.R. Cameron, W. T. Thomson and A.B. Dow, "Vibration and Current monitoring for detecting airgap eccentricity in large induction motors", Proceeding of IEE, vol. 133, Pt. B, No. 3, pp. 155-163, May 1986.

[14] J. S. hsu, "Monitoring of defects in induction motors through air-gap torque observation," IEEE Transaction on Industrial Applications, Vol. 31, No. 5, pp. 1016-1021, Sept./Oct. 1995.

Control of induction motor using IMC approach

Duong Hoai Nghia[1], Nguyen Van Nho[2]
HCM University of Technology
Faculty of Electrical & Electronics Engineering
268 Ly Thuong Kiet, Ho Chi Minh city, Vietnam
Email: dhnghia@hcmut.edu.vn, nvnho@hcmut.edu.vn

Hong-Hee Lee[3]
University Of Ulsan
Department of Electrical Engineering
680-749 San 29, Muger 2-dong, Ulsan, Korea
Email: hhlee@mail.ulsan.ac.kr

Abstract—The paper presents a method for controlling induction motor using internal model control (IMC) approach. The forward and inverse models are nonlinear in the rotor flux coordinate. Simulation results show that the control system has good performance and robust against changes in motor parameters (rotor and stator resistances, inductances, rotor inertia).

I. INTRODUCTION

The induction motor is widely used in industry because of its reliability and low cost. However, since the dynamical model of induction motor is strongly nonlinear, the control of induction motor is a challenging problem and has attracted much attention. Many methods have been proposed and can be mentioned as follows

- Field-oriented control (FOC) [4], [5], [14]
- Direct torque control (DTC) [6], [14]
- Sliding mode control [9], [10], [11], [15]
- Passivity-based control [7], [8], [15]
- Input-ouput linearization by state feedback [12], [15]
- Artificial intelligence based control [13]

etc.

In this paper, we present a new method for induction motor control. The proposed method can be considered as an extension of the linear internal model control in [1] to the nonlinear case.

The remainder of the paper is organized as follows : Section 2 introduces the model of induction motor. Section 3 describes the proposed control system. Section 4 presents simulation results. Section 5 concludes the paper.

II. MODEL OF INDUCTION MOTOR

The state space model of induction motor in the dq coordinate is given by (1)-(6) [3], [14]

$$\frac{di_{sd}}{dt} = -a_1 i_{sd} + \omega_s i_{sq} + a_2 \Psi_{rd} + a_3 \omega \Psi_{rq} + a_4 u_{sd} \quad (1)$$

$$\frac{di_{sq}}{dt} = -\omega_s i_{sd} - a_1 i_{sq} - a_3 \omega \Psi_{rd} + a_2 \Psi_{rq} + a_4 u_{sq} \quad (2)$$

$$\frac{d\Psi_{rd}}{dt} = a_5 i_{sd} - a_5 \Psi_{rd} + (\omega_s - \omega)\Psi_{rq} \quad (3)$$

$$\frac{d\Psi_{rq}}{dt} = a_5 i_{sq} - (\omega_s - \omega)\Psi_{rd} - a_5 \Psi_{rq} \quad (4)$$

$$T_e = a_6 [\Psi_{rd} i_{sq} - \Psi_{rq} i_{sd}] \quad (5)$$

$$\frac{d\omega}{dt} = a_7 [T_e - T_L] \quad (6)$$

where (i_{sd}, i_{sq}), (u_{sd}, u_{sq}), (Ψ_{rd}, Ψ_{rq}) are, respectively, the stator current (A), stator voltage (V), and rotor flux (A) normalized by the mutual inductance L_m in dq coordinate, ω is the rotor speed (rad/s), T_e is the motor torque (Nm), T_L is the load torque (Nm), ω_s is the angular frequency of the stator current (rad/s),

$$a_1 = \left(\frac{1}{\sigma T_s} + \frac{1-\sigma}{\sigma T_r}\right), \quad a_2 = \frac{1-\sigma}{\sigma T_r}, \quad a_3 = \frac{1-\sigma}{\sigma} \quad (7)$$

$$a_4 = \frac{1}{\sigma L_s}, \quad a_5 = \frac{1}{T_r}, \quad a_6 = \frac{3pL_m^2}{2L_r}, \quad a_7 = \frac{pa_6}{J} \quad (8)$$

$$\sigma = 1 - \frac{L_m^2}{L_s L_r}, \quad T_s = \frac{L_s}{R_s}, \quad T_r = \frac{L_r}{R_r} \quad (9)$$

are constants, R_s is the stator resistance, R_r is the rotor resistance, L_s is the stator inductance, L_r is the rotor inductance, L_m is the mutual inductance, p is the number of pole pairs, and J is the moment of inertia of the rotor.

Let

$$\Psi_r = \sqrt{\Psi_{rd}^2 + \Psi_{rq}^2} \quad (10)$$

The control objective is to regularize the rotor flux Ψ_r and the rotor speed ω to the set point r_Ψ and r_ω, respectively.

III. THE CONTROL SYSTEM

The control system is depicted in Fig. 1

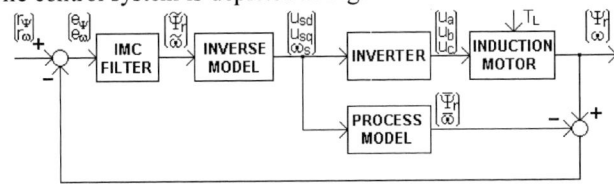

Fig. 1: The internal model control system

The process model is a model of the inverter and the induction motor. It has inputs u_{sd}, u_{sq}, ω_s and outputs $\overline{\Psi}_r$, $\overline{\omega}$.

978-1-4244-1871-8/07 $25.00 © 2007 IEEE

If the process model is exact and if there is no disturbance ($T_L \equiv 0$), then the outputs of the process model must be equal to the outputs of the induction motor

$$\overline{\Psi}_r = \Psi_r \quad \text{and} \quad \overline{\omega} = \omega \tag{11}$$

The inverse model is an inverse of the process model. It has inputs $\widetilde{\Psi}_r$, $\widetilde{\omega}$ and outputs u_{sd}, u_{sq}, ω_s. If the inverse model is exact then the inputs of the inverse model must be equal to the outputs of the process model:

$$\widetilde{\Psi}_r = \overline{\Psi}_r \quad \text{and} \quad \widetilde{\omega} = \overline{\omega} \tag{12}$$

If the process model and the inverse model are exact and if there is no disturbance then it follows from (11) and (12) that

$$\widetilde{\Psi}_r = \Psi_r \quad \text{and} \quad \widetilde{\omega} = \omega \tag{13}$$

Moreover, (11) implies that the feedback signals are null, it follows that

$$r_\Psi = e_\Psi \quad \text{and} \quad r_\omega = e_\omega \tag{14}$$

From Fig. 1, it can be seen that the transfer functions from the set point signals r_Ψ and r_ω to the output signals Ψ and ω are exactly the transfer functions of the IMC filters. We can thus assign the transfer function of the closed loop system by choosing an appropriate IMC filter.

A. The process model

The process model is derived from (1)-(6) with $T_L = 0$ (T_L is considered as disturbance)

$$\frac{d\overline{i}_{sd}}{dt} = -\hat{a}_1\overline{i}_{sd} + \omega_s\overline{i}_{sq} + \hat{a}_2\overline{\Psi}_{rd} + \hat{a}_3\omega\overline{\Psi}_{rq} + \hat{a}_4 u_{sd} \tag{15}$$

$$\frac{d\overline{i}_{sq}}{dt} = -\omega_s\overline{i}_{sd} - \hat{a}_1\overline{i}_{sq} - \hat{a}_3\omega\overline{\Psi}_{rd} + \hat{a}_2\overline{\Psi}_{rq} + \hat{a}_4 u_{sq} \tag{16}$$

$$\frac{d\overline{\Psi}_{rd}}{dt} = \hat{a}_5\overline{i}_{sd} - \hat{a}_5\overline{\Psi}_{rd} + (\omega_s - \omega)\overline{\Psi}_{rq} \tag{17}$$

$$\frac{d\overline{\Psi}_{rq}}{dt} = \hat{a}_5\overline{i}_{sq} - (\omega_s - \omega)\overline{\Psi}_{rd} - \hat{a}_5\overline{\Psi}_{rq} \tag{18}$$

$$\overline{T}_e = \hat{a}_6[\overline{\Psi}_{rd}\overline{i}_{sq} - \overline{\Psi}_{rq}\overline{i}_{sd}] \tag{19}$$

$$\frac{d\overline{\omega}}{dt} = \hat{a}_7\overline{T}_e \tag{20}$$

$$\overline{\Psi}_r = \sqrt{\overline{\Psi}_{rd}^2 + \overline{\Psi}_{rq}^2} \tag{21}$$

where $\hat{a}_1$ - $\hat{a}_7$ are defined similarly to a_1-a_7 in (7)-(9) with the estimated values of the parameters of the induction motor.

B. The inverse model

The inverse model is developed from (1)-(6) as follows

$$\frac{d\widetilde{i}_{sd}}{dt} = -\hat{a}_1\widetilde{i}_{sd} + \omega_s\widetilde{i}_{sq} + \hat{a}_2\widetilde{\Psi}_{rd} + \hat{a}_3\omega\widetilde{\Psi}_{rq} + \hat{a}_4 u_{sd} \tag{22}$$

$$\frac{d\widetilde{i}_{sq}}{dt} = -\omega_s\widetilde{i}_{sd} - \hat{a}_1\widetilde{i}_{sq} - \hat{a}_3\omega\widetilde{\Psi}_{rd} + \hat{a}_2\widetilde{\Psi}_{rq} + \hat{a}_4 u_{sq} \tag{23}$$

$$\frac{d\widetilde{\Psi}_{rd}}{dt} = \hat{a}_5\widetilde{i}_{sd} - \hat{a}_5\widetilde{\Psi}_{rd} + (\omega_s - \omega)\widetilde{\Psi}_{rq} \tag{24}$$

$$\frac{d\widetilde{\Psi}_{rq}}{dt} = \hat{a}_5\widetilde{i}_{sq} - (\omega_s - \omega)\widetilde{\Psi}_{rd} - \hat{a}_5\widetilde{\Psi}_{rq} \tag{25}$$

$$\widetilde{T}_e = \hat{a}_6[\widetilde{\Psi}_{rd}\widetilde{i}_{sq} - \widetilde{\Psi}_{rq}\widetilde{i}_{sd}] \tag{26}$$

$$\frac{d\widetilde{\omega}}{dt} = \hat{a}_7\widetilde{T}_e \tag{27}$$

$$\widetilde{\Psi}_r = \sqrt{\widetilde{\Psi}_{rd}^2 + \widetilde{\Psi}_{rq}^2} \tag{28}$$

where $\hat{a}_1$ - $\hat{a}_7$ are the same parameters as in (15)-(21). Without loss of generality we can suppose that $\widetilde{\Psi}_{rq} = 0$ then $\widetilde{\Psi}_{rd} = \widetilde{\Psi}_r$. This together with (26) and (27) yields

$$\widetilde{i}_{sq} = \frac{1}{\hat{a}_6\hat{a}_7\widetilde{\Psi}_r}\frac{s\widetilde{\omega}}{T_d s + 1} \tag{29}$$

where s is the Laplace operator, T_d is a time constant, the transfer function $s/(T_d s+1)$ is used to approximate the derivative operator d/dt. Similarly it follows from (24) that

$$\widetilde{i}_{sd} = \hat{T}_r\frac{s\widetilde{\Psi}_r}{T_d s + 1} + \widetilde{\Psi}_r \tag{30}$$

where $\hat{T}_r = \dfrac{1}{\hat{a}_5}$. Since $\widetilde{\Psi}_{rq} = 0$, (25) yields

$$\omega_s = \widetilde{\omega} + \frac{\widetilde{i}_{sq}}{T_r\widetilde{\Psi}_r} \tag{31}$$

Finally (22) and (23) become

$$u_{sd} = \frac{1}{\hat{a}_4}[\frac{s\widetilde{i}_{sd}}{T_d s+1} + \hat{a}_1\widetilde{i}_{sd} - \omega_s\widetilde{i}_{sq} - \hat{a}_2\widetilde{\Psi}_r] \tag{32}$$

$$u_{sq} = \frac{1}{\hat{a}_4}[\frac{s\widetilde{i}_{sq}}{T_d s+1} + \omega_s\widetilde{i}_{sd} + \hat{a}_1\widetilde{i}_{sq} + \hat{a}_3\widetilde{\omega}\widetilde{\Psi}_r] \tag{33}$$

The inverse model is defined by (26)-(33)

C. The IMC filter

The IMC filters are low pass filters:

$$F_\omega(s) = \frac{1}{\tau_\omega s + 1} \tag{34}$$

and

$$F_\Psi(s) = \frac{1}{\tau_\Psi s + 1} \tag{35}$$

where τ_ω and τ_ψ are time constants which determine the response time of the closed loop system. Note that the static gain of both filters is 1.

D. The rotor flux observer

In practice the rotor flux is rarely available. We often have to estimate the rotor flux to use in the control system as shown in Fig. 2. The rotor flux estimator has inputs u_{sd}, u_{sq}, ω_s, i_{sd}, i_{sq}, and output $\hat{\Psi}_r$, an estimated value for Ψ_r. The estimator can be derived from (1)-(4) as follows

$$\frac{d\hat{i}_{sd}}{dt} = -\hat{a}_1\hat{i}_{sd} + \hat{\omega}_s\hat{i}_{sq} + \hat{a}_2\hat{\Psi}_{rd} + \hat{a}_3\omega\hat{\Psi}_{rq} + \hat{a}_4u_{sd} - K_0[i_{sd} - \hat{i}_{sd}] \quad (36)$$

$$\frac{d\hat{i}_{sq}}{dt} = -\hat{\omega}_s\hat{i}_{sd} - \hat{a}_1\hat{i}_{sq} - \hat{a}_3\omega\hat{\Psi}_{rd} + \hat{a}_2\Psi_{rq} + \hat{a}_4u_{sq} - K_0[i_{sq} - \hat{i}_{sq}] \quad (37)$$

$$\frac{d\hat{\Psi}_{rd}}{dt} = \hat{a}_5i_{sd} - \hat{a}_5\hat{\Psi}_{rd} + (\hat{\omega}_s - \omega)\hat{\Psi}_{rq} - K_0[i_{sd} - \hat{i}_{sd}] \quad (38)$$

$$\frac{d\hat{\Psi}_{rq}}{dt} = \hat{a}_5i_{sq} - (\hat{\omega}_s - \omega)\hat{\Psi}_{rd} - \hat{a}_5\hat{\Psi}_{rq} - K_0[i_{sq} - \hat{i}_{sq}] \quad (39)$$

$$\hat{\Psi}_r = \sqrt{\hat{\Psi}_{rd}^2 + \hat{\Psi}_{rq}^2} \quad (40)$$

where $\hat{a}_1$ - $\hat{a}_7$ are the same parameters as in (15)-(21). Note that the terms

$$K_0\left(i_{sd} - \hat{i}_{sd}\right) \quad \text{and} \quad K_0\left(i_{sq} - \hat{i}_{sq}\right)$$

are the correction terms of the state observer [16].

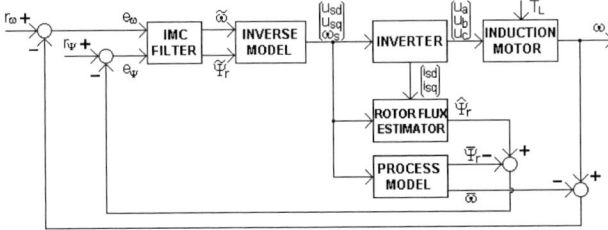

Fig. 2: The internal model control system with rotor flux estimator

IV. SIMULATION RESULTS

In the simulations, the controller is designed with the model parameters:

$$\hat{R}_s = 1.177\Omega, \quad \hat{R}_r = 1.382\Omega$$

$$\hat{L}_s = 0.119H, \quad \hat{L}_r = 0.118H, \quad \hat{L}_m = 0.113H$$

$$\hat{p} = 2, \quad \hat{J} = 0.00126 Kgm^2$$

The parameters of the controller are: the time constant of the IMC filters $\tau_\omega = 0.3s$ and $\tau_\psi = 0.05s$ (which correspond to the response times $3\tau_\omega = 0.9s$ and $3\tau_\psi = 0.15s$, respectively). Since i_{sd}, i_{sq}, ω, and Ψ_r varies slowly, we choose $T_d = 1ms$.

The constant of the rotor flux estimator is chosen $K_0 = 10$.

The motor is started time t = 0s to speed $\omega = 150rad/s$ without load. At time t = 1.5s the load torque $T_L = 3.5Nm$ is

applied. At time t = 2s the motor is decelerated to speed $\omega = 75rad/s$.

A. Nominal response

In this simulation, the parameters of the motor are equal to the ones of the model

$$R_s = \hat{R}_s, \quad R_r = \hat{R}_r$$

$$L_s = \hat{L}_s, \quad L_r = \hat{L}_r, \quad L_m = \hat{L}_m$$

$$p = \hat{p}, \quad J = \hat{J}$$

The result is given in Fig. 3. Observe that
- The response time of the rotor flux Ψ_r is about $3\tau_\psi = 0.15s$.
- The response time of the rotor speed ω is about $3\tau_\omega = 0.9s$.
- The static errors of the rotor flux and rotor speed are null.

B. Robustness against variation in R_s and R_r

In this simulation, the resistances of the motor are equal two times the corresponding value of the model

$$R_s = 2\hat{R}_s, \quad R_r = 2\hat{R}_r$$

$$L_s = \hat{L}_s, \quad L_r = \hat{L}_r, \quad L_m = \hat{L}_m$$

$$p = \hat{p}, \quad J = \hat{J}$$

The result is given in Fig. 4. Observe that there is almost no difference between Fig. 4 and Fig. 3. The controller is thus robust against variation in motor resistances.

C. Robustness against variation in L_s, L_r and L_m

In this simulation, the inductances of the motor are equal 0.8 time the corresponding value of the model

$$R_s = \hat{R}_s, \quad R_r = \hat{R}_r$$

$$L_s = 0.8\hat{L}_s, \quad L_r = 0.8\hat{L}_r, \quad L_m = 0.8\hat{L}_m$$

$$p = \hat{p}, \quad J = \hat{J}$$

The result is given in Fig. 5. Observe that there is almost no difference between Fig. 5 and Fig. 3 except the normalized rotor flux.

D. Robustness against variation in J

In this simulation, the moment of inertia of the motor is equal to 5 time the corresponding value of the model

$$p = \hat{p}, \quad J = 5\hat{J}$$

$$R_s = \hat{R}_s, \quad R_r = \hat{R}_r$$

$$L_s = \hat{L}_s, \quad L_r = \hat{L}_r, \quad L_m = \hat{L}_m$$

The result is given in Fig. 6.

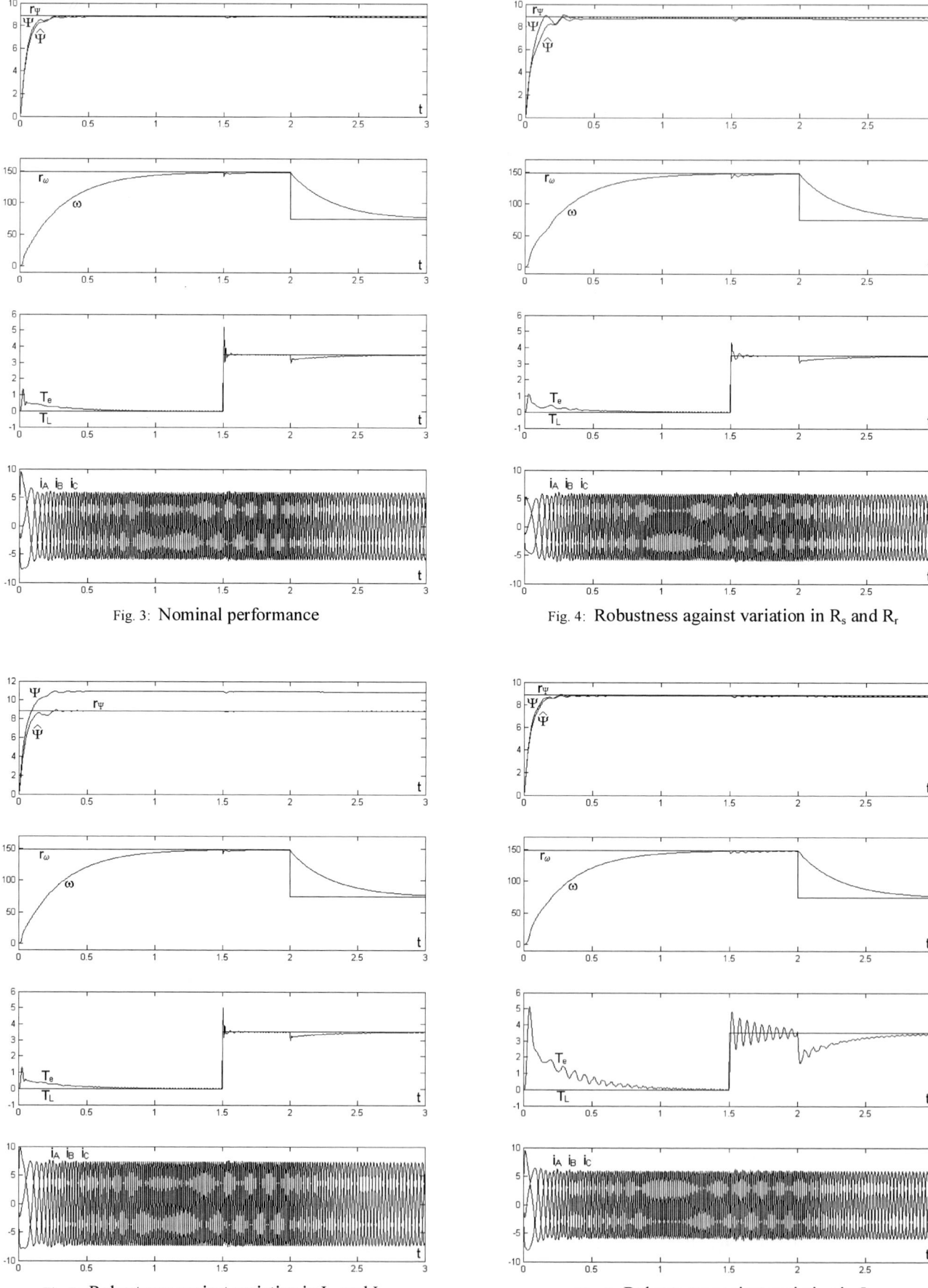

Fig. 3: Nominal performance

Fig. 4: Robustness against variation in R_s and R_r

Fig. 5: Robustness against variation in L_s and L_r

Fig. 6: Robustness against variation in J

978-1-4244-1871-8/07 $25.00 © 2007 IEEE

Observe that there is almost no difference between Fig. 6 and Fig. 3 except the torque T_e of the motor.

E.   Robustness against simultaneous variation in R_s, R_r, L_s, L_r, L_m and J

In this simulation, the resistances, the inductances, and the moment of inertia of the motor vary simultaneously

$$R_s = 2\hat{R}_s, \ R_r = 2\hat{R}_r$$

$$L_s = 0.8\hat{L}_s, \ L_r = 0.8\hat{L}_r, \ L_m = 0.8\hat{L}_m$$

$$p = \hat{p}, \ J = 5\hat{J}$$

The result is given in Fig. 7. Observe that there is almost no difference between Fig. 7 and Fig. 3 except the normalized rotor flux and the torque T_e of the motor.

F.   Influence of the time constant τ_ω

In this simulation, the parameters of the motor vary simultaneously as in previous simulation.

Fig. 8 gives the results where the time constant of the IMC filter $F_\omega(s)$ is $\tau_\omega = 0.15s$. Observe that the response time of motor speed is shorter but the control system is less robust (compare with Fig. 7).

Fig. 9 gives the results where the time constant of the IMC filter $F_\omega(s)$ is $\tau_\omega = 0.6s$. Observe that the response time of motor speed is longer but the control system is more robust (compare with Fig. 7).

V.   CONCLUSION

In this paper we describe a novel method for induction motor control based on the IMC approach. The advantage of the proposed method is that it allows the user to specify the transfer function of the closed loop system (which is exactly the IMC filter).

Simulation results shows that the control system has good performance and robust against change in motor parameters.

ACKNOWLEDGMENT

The authors would like to thank   Korea Ministry of Commerce, Industry and Energy and Ulsan Metropolitan City which partly supported this research through the Network-based Automation Research Center (NARC) at University of Ulsan.

REFERENCES

[1] M. Morari, E. Zafiriou. *Robust process control.* Prentice Hall, 1997.

[2] R. Berber. *Methods of model based process control.* Kluwer Academic Publishers, 1994.

[3] W. Leonhard. *Control of electrical drives.* Springer Verlag, 1985.

[4] J.P. Caron, J.P. Hautier. *Modelisation et commande de la machine asynchrone.* Editions Technip, 1995.

[5] A.M. Trzynadlowski. *The Orientation Principle in Control of Induction Motors.* Kluwer Academic Publishers, 1994.

[6] C. Lascu, I Boldea, F. Blaabjerg. "A modified direct torque control for induction motor sensorless drive". *IEEE Transaction on Industrial Application*, 2000.

[7] R. Ortega, A. Loria, P.J. Nicklasson and H. Sira-Ramírez, *Passivity-Based Control Of Euler-Lagrange System. Mechanical, Electrical And Electromechanical Applications.* Springer, 1998.

[8] C. Cecati, N. Rotondal. "Torque and Speed Regulation of Induction Motor Using the Passivity Theory Approach". *IEEE Transaction on Industrial Electronics*, vol. 46, no. 1, pp.119-126, April 1999.

[9] H.J. Shieh, K.K. Shyn. "Nonlinear sliding mode torque control with adaptive backstepping approach for induction motor drive", *IEEE Transaction on Industrial Electronics*, 1999.

[10] V.I. Utkin, *Sliding Modes In Control And Optimization.* Springer Verlag, 1992.

[11] V.I. Utkin, J. Guldner, J. Shi. *Sliding mode control in electromechanical systems.* Taylor & Francis, 1999.

[12] A. Benchaid, A. Rachid, E. Audrezet; "Sliding Mode Input-Output Liearization and Field Orientation for Real-Time Control of Induction Motors". *IEEE Transactions on Power Electronics*, vol 14, No.1, pp.3-13. 1999.

[13] P. Vas, *Artificial-Intelligence-Based Electrical Machines And Drives. Application of Fuzzy, Neural, Fuzzy-Neural, and Genetic-Algorithm-Based Techniques*; Oxford University Press; 1998.

[14] C. Canudas de Wit (ed.), *Modelisation, controle vectoriel et DTC.* Hermes Science. 2000.

[15] C. Canudas de Wit (ed.), *Optimisation, discretisation et observateur.* Hermes Science. 2000.

[16] H.K. Khalil. *Nonlinear systems.* Prentice-Hall, 2002.

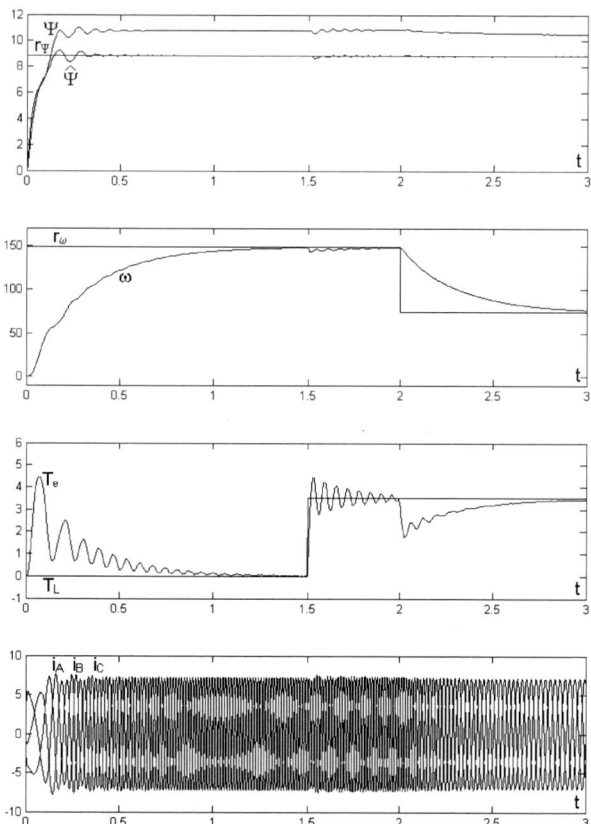

Fig. 7: Robustness against simultaneous variation in R_s, R_r, L_s, L_r, L_m and J

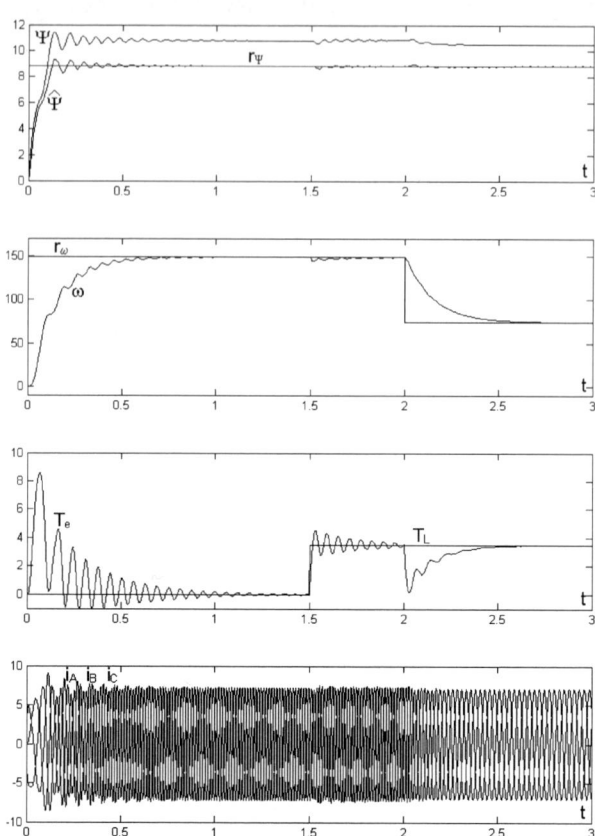

Fig. 8: Influence of the time constant of $F_\omega(s)$: $\tau_\omega = 0.15s$

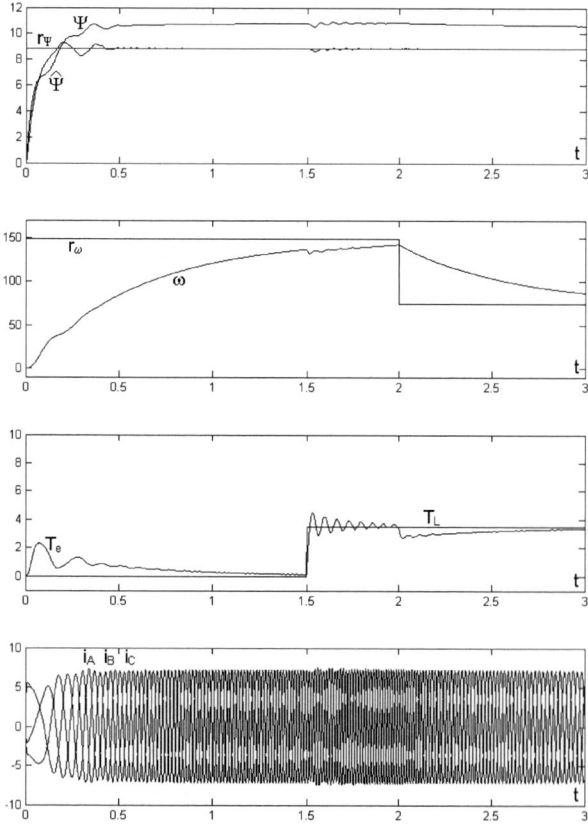

Fig. 9: Influence of the time constant of $F_\omega(s)$: $\tau_\omega = 0.6s$

978-1-4244-1871-8/07 $25.00 © 2007 IEEE

Development of HBPI Controller for High Performance Control of IPMSM Dirve

Jau Sub Ko
Sunchon National University
Department of Electrical Control Engineering
315, Maegok-dong, Suncheon, Junnam, 540-742
Email: kokos22@naver.com

Jung-Sik Choi
Sunchon National University
Department of Electrical Control Engineering
315, Maegok-dong, Suncheon, Junnam, 540-742
Email: 1108cjs@sunchon.ac.kr

Ki-Tae Park
Sunchon National University
Department of Electrical Control Engineering
315, Maegok-dong, Suncheon, Junnam, 540-742
Email: dy_tech@hanmail.net

Byung-Sang Park
Sunchon National University
Department of Electrical Control Engineering
315, Maegok-dong, Suncheon, Junnam, 540-742
Email: 23233564@hanmail.net

Dong-Hwa Chung
Sunchon National University
Department of Electrical Control Engineering
315, Maegok-dong, Suncheon, Junnam, 540-742
Email: hwa777@sunchon.ac.kr

Abstract— **This paper presents Hybrid PI(HBPI) controller of IPMSM drive using fuzzy adaptive mechanism control. In general, PI controller in computer numerically controlled machine process fixed gain. They may perform well under some operating conditions, but not all. To increase the robustness, fixed gain PI controller, HBPI controller proposes a new method based self tuning PI controller.**

HBPI controller is developed to minimize overshoot and settling time following sudden parameter changes such as speed, load torque, inertia, rotor resistance and self inductance. The results on a speed controller of IPMSM are presented to show the effectiveness of the proposed gain tuner. And this controller is better than the fixed gains one in terms of robustness, even under great variations of operating conditions and load disturbance.

I. INTRODUCTION

The Interior Permanent Magnet Synchronous Motor (IPMSM) is applied to servo drive of factory automation because torque and power density are high, also efficiency and power factor are excellent. The PI controller is generally used in an industrial application to control IPMSM[1]-[2]. Because the PI controller is simple and it has precise relation between parameter and response setting of system [1]. Also this controller is based on development of variable the newest control algorithm and technique. But, the PI controller had many problems in high performance because the IPMSM is nonlinear[3]-[4]. And the PI controller cannot realize high performance and robust control in parameter variations such as disturbance, speed and torque etc [5]. To solve these problems, adaptive fuzzy control [6], fuzzy control for parameter variation [7], new fuzzy control [8] and adaptive

(Fuzzy Neural Network) FNN [9]-[10] have been studied. But these techniques have been used to substitute for PI controller but PI controller gain cannot be tuned for good performance.

The variable adaptive control techniques have been proposed for sensorless control of drive[11] and evaluated excellent technique for good performance of speed response though a extensive movement variation. But these techniques are very complex because it takes long time to calculate and these are based on mathematical modeling. To solve these problems, this paper proposes sensorless control using (Artificial Neural Network) ANN.

In this paper, we propose the Hybrid PI (HBPI) controller using fuzzy control for speed control of IPMSM. We compare the performance of PI controller to proposed HBPI controller in this paper. Also we confirm the good performance of high performance and robust controllable HBPI controller with variable parameter variation such as speed, load torque, inertia, rotor resistance and magnetic inductance etc.

II. DRIVE SYSTEM

General method for speed control of IPMSM using the PI controller is indicated in Fig. 1. This system composition is applied to high performance system such as drive of robot, airplane and electric vehicle etc.

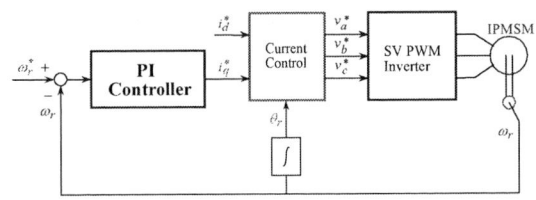

Fig. 1 Block diagram of vector controlled IPMSM for speed control

978-1-4244-1871-8/07 $25.00 © 2007 IEEE

The i_q^* and i_d^* show current of reference torque and reference flux element. v_a^*, v_b^*, v_c^* can be obtained by the current control and control the IPMSM through the (Space Vector) SV PWM inverter. The coordinate conversion utilizes position information θ_r of rotator.

The mathematical modeling of the IPMSM drive can be described by the following equation (1) to (4).

$$pi_d = \left(v_d - Ri_d + \omega_r L_q i_q\right)/ L_d \qquad (1)$$

$$pi_q = \left(v_q - Ri_q - \omega_r L_d i_d - \omega_r \phi_{af}\right)/ L_q \qquad (2)$$

$$p\omega_r = \left(T_e - T_L - B\omega_r\right)/ J \qquad (3)$$

$$T_e = \frac{3}{2}P\left[\phi_{af}i_q + (L_d - L_q)i_d i_q\right] \qquad (4)$$

III. HBPI CONTROLLER

Fig. 2 indicates the design of HBPI controller for speed control of IPMSM that is operated by indirect vector control.

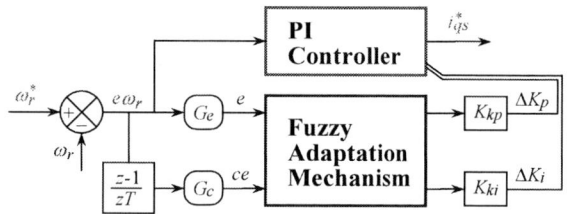

Fig. 2 The structure of HBPI controller

Two gains of PI controller are initialized by traditional method that is known. But these gains depend on estimation model of IPMSM in rating operate condition.[4]

Fuzzy algorithm that is for self-tuning of PI controller gains is used to keep of good performance in case of parameter variation. And fuzzy rule is used to generate of ΔK_p and ΔK_i.

Design of fuzzy rule is based on the fixed quantity knowledge and calculated by analysis of traditional PI controller for variable value of K_p and K_i in multiple operate condition.

The inputs are speed error e and change of speed error ce also the outputs are two gains (K_p and K_i) in fuzzy adaptation mechanism. Table 1 and table 2 show the fuzzy rule base to renew K_p and K_i. The fuzzy rule base uses the repeated triangular equation in same range.

Table 1 Rule base to update gain K_i

$\dfrac{e\omega_r}{ce\omega_r}$	NL	NM	NS	ZE	PS	PM	PL
NL	ZE	NS	NM	NL	NM	NS	ZE
NM	PS	ZE	NS	NM	NS	ZE	PS
NS	PM	PS	ZE	NS	ZE	PS	PM
ZE	PL	PM	PS	ZE	PS	PM	PL
PS	PM	PS	ZE	NS	ZE	PS	PM
PM	PS	ZE	NS	NM	NS	ZE	PS
PL	ZE	NS	NM	NL	NM	NS	ZE

Table 2 Rule base to update gain K_p

$\dfrac{e\omega_r}{ce\omega_r}$	NL	NM	NS	ZE	PS	PM	PL
NL	NL	NL	NL	NL	NM	NS	ZE
NM	NL	NL	NL	NM	NS	ZE	PS
NS	NL	NL	NM	NS	ZE	PS	PM
ZE	NL	NM	NS	ZE	PS	PM	PL
PS	NM	NS	ZE	PS	PM	PL	PL
PM	NS	ZE	PS	PM	PL	PL	PL
PL	ZE	PS	PM	PL	PL	PL	PL

Two gains are renewed by following equations.

$$K_p = K_p + K_{kp}f_1(e_n, \Delta e_n) \qquad (5)$$

$$K_i = K_i + K_{ki}f_2(e_n, \Delta e_n) \qquad (6)$$

For stability of drive, if gains reach the steady state, K_p and K_i are limited and if reference speed is changed, are initialized.

First, gain K_i is more increased in case of speed error is large in positive direct or negative direct and more decreased when change of speed error is large in positive direct or negative direct. It is improved in rising time but delayed the speed response by overshoot.

Second, gain K_p is more increased when speed error and change of speed error are large in positive direct then more decreased when are small in negative direct. It is decreased in overshoot of speed and error of steady state.

IV. SPEED ESTIMATION USING ANN

An ideal application for ANN is in the field of nonlinear system identification. Virtually any discrete-time nonlinear system may be represented by the nonlinear auto regressive moving average with exogenous(NARMAX) inputs model, which represents a system in terms of its delayed inputs and outputs.[12] The NARMAX model has the general form:

$$\mathbf{y}(k+1) = \mathbf{f}(\mathbf{y}(k), \cdots, \mathbf{y}(k-d_y), \mathbf{u}(k), \cdots, \mathbf{u}(k-d_u)) \quad (7)$$

Where d_u, d_y are the maximum delay in the input and output u, y vector respectively.

This form of model is ideal for system identification purposes because the model is expressed entirely in terms of known quantities.

An ANN may be used to identify the NARMAX model of a system by making the arguments of equation (11) the inputs to the ANN and making the output of the ANN be the one-step-ahead (predicted) output vector $\hat{y}(k+1)$ as shown in Fig. 3. TDL denotes a tapped delay line whose outputs are delayed values of its input. The ANN may then be trained to emulate the function $\mathbf{f}(\cdot)$ in equation (11) by comparing the predicted output vector with the actual output vector at time $(k+1)$, and using the error to update the ANN weight via the BPA.

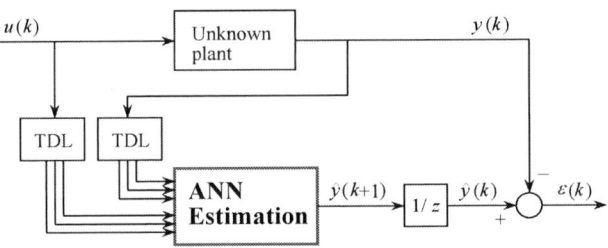

Fig. 3 System estimation using ANN

The estimated speed of IPMSM for system estimation using ANN is computed as follows

$$
\begin{aligned}
\hat{\omega}_r(k+1) &= \hat{\omega}_r(k) + \Delta\hat{\omega}_r(k) \quad (8)\\
&= \hat{\omega}_r(k) - \frac{\Delta W_2(k)}{T}\\
&= \hat{\omega}_r(k) - \frac{\eta}{T}\left\{[\phi_q(k) - \hat{\phi}_q(k)]\left[\hat{\phi}_d(k-1)\right] - [\phi_d(k) - \hat{\phi}_d(k)]\left[\hat{\phi}_q(k-1)\right]\right\}\\
&\quad - \frac{\alpha}{T}\Delta W_2(k-1)
\end{aligned}
$$

Where T is the sampling period and η is the training coefficient. The training coefficient η represents the training rate in learning procedure. More larger this coefficient, more larger the change in the weights. For practical implementation we choose a learning rate that is as large as possible without leading to oscillation. This offers the most rapid learning. The coefficient α determines the effect of the past weight changes on the current weight.

The rotor flux angle $\hat{\theta}_r$ which is necessary for derivation of the reference of stator current, is computed from the estimated speed using the proposed system.

$$\hat{\theta}_r(k+1) = \hat{\theta}_r(k) + T \cdot \hat{\omega}_r(k) \quad (9)$$

Fig.4 shows drive system of IPMSM which controlled by proposed method.

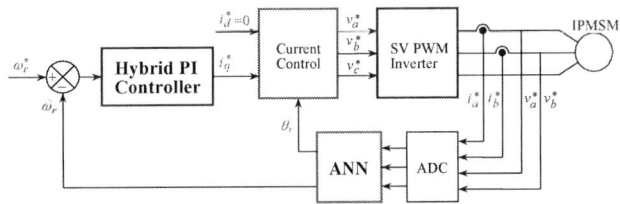

Fig. 4 Proposed Drive System of IPMSM

V. PERFORMANCE RESULT OF SYSTEM

Fig. 5 shows the response comparison with conventional PI controller and proposed HBPI controller in this paper when the reference speed is set to 1800[rpm]. K_p and K_i are changed because adaptive tuning in transient state and HBPI controller is excellent response characteristics of overshoot, stability time, steady and transient state etc more than conventional PI controller .

Fig. 6 shows the response characteristic when reference speed is set to 1800[rpm] and then -1800[rpm]. HBPI controller is a good performance than PI controller with speed change.

Fig. 7 shows the response comparison with change of load torque. When reference speed is operated as rating speed 1800[rpm], and then load torque is increased 5[N.m]) from 0.5[sec] to 0.8[sec]. Similarly, HBPI controller is a good performance than PI controller with load change.

Fig. 8 shows the response characteristic of parameter variation. Fig. 8(a) is a twice rating of inertia, Fig. 8(b) is a twice rating of rotor resistance, Fig. 8(c) is a half rating of rotor resistance and Fig. 8(d) is one and half rating of rotor inductance. HBPI controller is a good performance than PI controller with motor parameter variation.

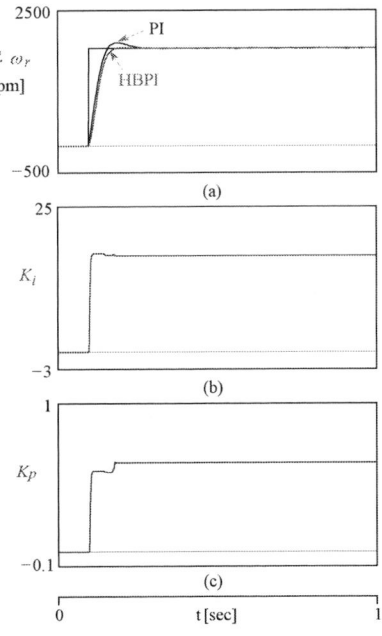

Fig. 5 The comparison of response characteristics with PI, HBPI controller

978-1-4244-1871-8/07 $25.00 © 2007 IEEE   369

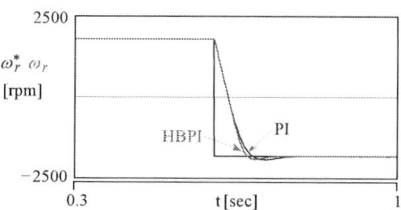

Fig. 6 Response comparison with change of step command speed

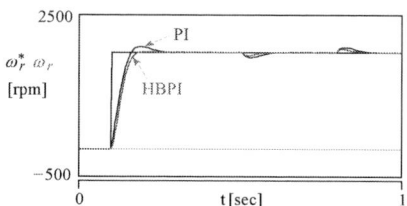

Fig. 7 Response comparison with change of load torque

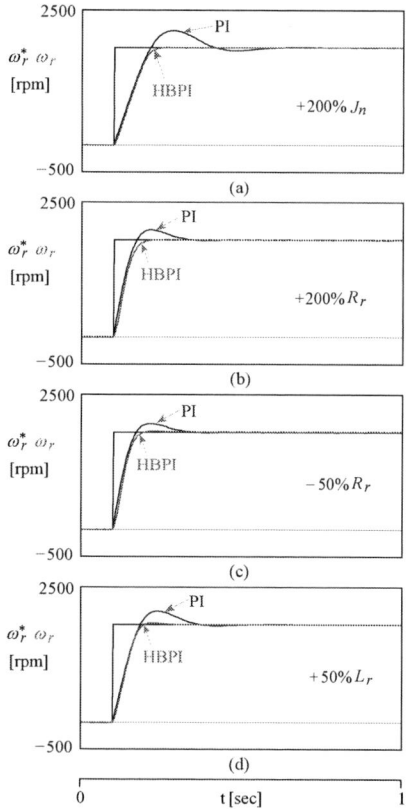

Fig. 8 Response comparison with parameter changes

Fig. 9 shows the response characteristic when operated in condition like Fig. 7. Fig. 9(a) is reference speed, real speed and estimation speed, Fig. 9(b) is error of real speed and estimation speed, Fig. 9(c) is q-axis current and Fig. 9(d) is generating torque. And estimation performance of ANN is very excellent because of speed error is within one percent.

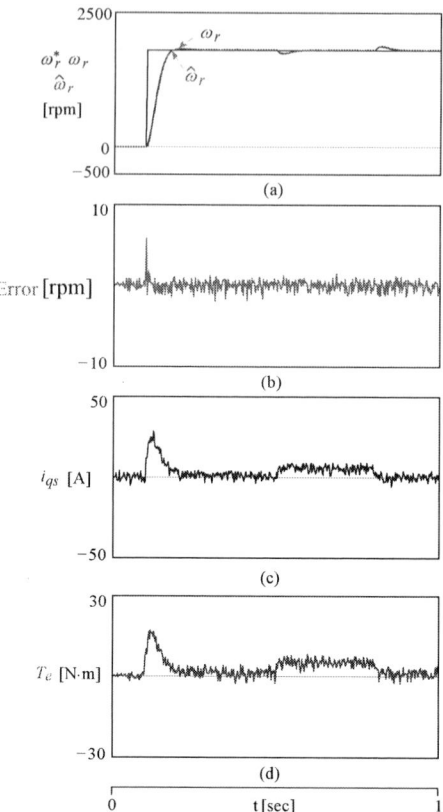

Fig. 9 speed estimation with step command speed

VI. CONCLUSION

In this paper, we proposed a HBPI controller using fuzzy control for robust control of IPMSM. HBPI controller is mixed fuzzy logic and conventional PI controller and is not used reference model but use the simple fuzzy logic adaptive mechanism. HBPI controller is automatic tuned the fixed gains of conventional PI controller using fuzzy rule base against parameter variation.

We have analyzed response characteristics in various operation conditions such as change of parameter, steady state and transient state etc. HBPI controller is indicated better response characteristics such as overshoot, rising time and stability time than PI controller. Also this controller was gotten the excellent performance and indicted robust and high performance in change of parameters.

Speed estimation using ANN was indicated a good performance with change of speed and load torque and very excellent because speed error is within one percent.

Therefore, the validity of the proposed controller, which is HBPI controller and ANN, is confirmed by performance results.

978-1-4244-1871-8/07 $25.00 © 2007 IEEE   370

REFERENCES

[1] Y. Y. Tzou, "DSP-based robust control of an AC induction servo drive for motion control," IEEE Trans. Contr. Syst. Technol., vol. 4, pp. 614-626, 1996.

[2] M. Santos and J. M. de la Cruz, "Between fuzzy PID and PID conventional controllers," NAFIPS'96, Berkley, USA, June 1996.

[3] M. Ali Unar, D. J. Murray-Smith and S. F. Ali Shah, "Design and tuning of fixed structure PID controller - A survey," Technical Report CSC-96016, Faculty of Engineering, Glasgow University, Scotland, 1996.

[4] Z. Ibrahim and E. Levi, "Comparative analysis of fuzzy logic and PI speed control in high performance AC drives using experimental approach," Proc. of IEEE IAS'2000, Rome, Italy, CD-ROM paper 46-3, 2000.

[5] C. M. Ong, "Dynamic simulation of elecctric machinery using Matlab/simulink," Upper Saddle River, NJ: Prentice-Hall, 1998. .

[6] J. C. Lee and D. H. Chung, "MRAC fuzzy control for high performance of induction motor drive," The Trans. of KIPE, vol. 7, no. 3, pp. 215-223, 2002.

[7] H. G. Lee, J. C. Lee and D. H. Chung, "Design of fuzzy controller induction drive considering parameter change," The Trans. of KIEE, vol. 51P, no. 3, pp. 111-119, 2002.

[8] H. G. Lee, J. C. Lee and D. H. Chung, "New fuzzy controller for high performance of induction motor drive," The journal of KIIS, vol. 17, no. 4, pp. 87-93, 2002.

[9] H. G. Lee, J. C. Lee and D. H. Chung, "Adaptive FNN controller for speed control of IPMSM drive," The Trans. of KIEE, vol. 41-SC, no. 3, pp. 39-46, 2004.

[10] J. C. Lee, H. G. Lee, Y. S. Lee and S. M. Nam, D. H. Chung, "Speed estimation and control of induction motor drive using hybrid intelligent control," International Conference ICPE'04, no. 3, pp. 181-185, 2004.

[11] K.J. Astron and B. Wittenmark, "Adaptive control," Addison-Wesley, 1989.

Efficiency Optimization Control of SynRM Drive by LM-FNN Controller

Jung-Sik Choi
Sunchon National University
Department of Electrical Control Engineering
315, Maegok-dong, Suncheon, Junnam, 540-742
Email: 1108cjs@sunchon.ac.kr

Jau Sub Ko
Sunchon National University
Department of Electrical Control Engineering
315, Maegok-dong, Suncheon, Junnam, 540-742
Email: kokos22@naver.com

Ki-Tae Park
Sunchon National University
Department of Electrical Control Engineering
315, Maegok-dong, Suncheon, Junnam, 540-742
Email: dy_tech@hanmail.net

Byung-Sang Park
Sunchon National University
Department of Electrical Control Engineering
315, Maegok-dong, Suncheon, Junnam, 540-742
Email: 23233564@hanmail.net

Dong-Hwa Chung
Sunchon National University
Department of Electrical Control Engineering
315, Maegok-dong, Suncheon, Junnam, 540-742
Email: hwa777@sunchon.ac.kr

Abstract— This paper is proposed an efficiency optimization control algorithm for a synchronous reluctance motor(SynRM) which minimizes the copper and iron losses. The design of the speed controller based on learning mechanism-fuzzy neural networks(LM-FNN) controller that is implemented using adaptive, fuzzy control and neural networks. The control performance of the LM-FNN controller is evaluated by analysis for various operating conditions. Analysis results are presented to show the validity of the proposed algorithm

I. INTRODUCTION

Recently, all of the worlds are increased to interest in variable speed drive because it is rose in recognition of environment problem and requested energy and nurse resource. Research of variable drive is spurred because it is improved of power semiconductor and applied to micro processor and DSP. Synchronous reluctance motor(SynRM) drive is applied to the kernel power electronic instrument from consumer electronics such as refrigerator and air conditioner to industrial instrument.

SynRM is obtained advantage big torque per ampere, small rotor loss and easy control more induction motor. Also it is excellent performance and price more induction motor. It is obtained advantage low price and available field weakening control more permanent magnet synchronous motor(PMSM). It is obtained advantage easy vector control and topology configuration of inverter more switched reluctance motor(SRM). In vector control of SynRM, d and q-axis element of stator current is mutual independence variable and characteristic torque about any speed is obtained variable sum of d, q-axis current element.[1-5] But motor efficiency is widely changed. If d axis current is big, iron loss is big. And

if d axis current is small, current and copper loss is big. Therefore, in all operation point of motor, if setting torque a concern optimum efficiency is computed, generative optimum current at this torque can be obtained.

Efficiency optimum controller is obtained loss model that get parameter value such as inductance saturation, iron loss coefficient, temperature influence, harmonic influence emerge from motor real operation. Controller at any operation point is computed efficiency optimum term, value of one or more adjust until estimate optimum value. This optimum value is used command value of drive. Availability of this research can be obtained depend on exactly loss model Research about position and speed estimation for SynRM drive variously is accomplished.[6-13] SynRM is easy realization of sensorless more induction motor because it is vary frequent inductance variation about rotor position and unique salient pole. Sensorless SynRM drive is obtained excellent advantage control performance and mechanically robust and low price. Sensorless vector control system is proposed speed estimation method such as rotor position and speed using switching ripple of inverter[7-9] at induced current and voltage [6]waveform in coil. These methods can be obtained optimum estimation at low and high speed in algorithm base on kalman filter.[10] Also variation method is proposed such as stator flux standard[11], variation ratio measurement of phase current about time[12], all speed region[13] etc. But these methods is necessary adequate process of signal about generate big error or inverter noises and necessary special process that take switching of inverter at specification condition.

In this paper, minimum of SynRM loss by LM-FNN is proposed efficiency optimum control. Estimation torque is directly computed current and parameter of motor at synchronous coordinate system and d, q-axis current that is

978-1-4244-1871-8/07 $25.00 © 2007 IEEE

output of torque controller estimate command torque. To solve nonlinear character of torque by iron loss and saturation at SynRM, in this paper is analyze compensational in iron loss considered model. This paper is obtained term of d -axis current that minimize loss and optimize efficiency using model considered iron loss. The Proposed algorithm is compensated nonlinear character of torque causal because of iron loss to addition simply iron compensation routine. Efficiency optimum control is presented good torque performance and minimized loss compare with constant control of d -axis current. In this paper is proposed result and proved propriety of efficiency optimum control.

II. MODELING OF SynRM CONSIDERED IRON LOSS

Fig. 1 shows equivalent circuit of d , q -axis considered iron loss resistance of SynRM. In equivalent circuit of SynRM, R_c considered iron loss is connected parallel with speed voltage. Current that generate torque is i_{do} and i_{qo} . And it is different a terminal current i_{ds} , i_{qs} because SynRM is generated iron loss.

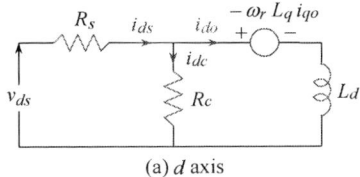

(a) d axis

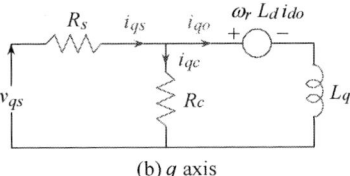

(b) q axis

Fig. 1 Equivalent circuit of SynRM

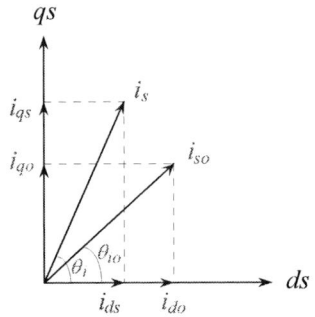

Fig. 2 Current vector diagram of SynRM

Fig. 2 shows current vector diagram of SynRM a case of considered iron loss. Voltage equation computed at equivalent circuit of Fig. 1 is as follows.

$$v_{ds} = R_s i_{ds} + L_d \frac{di_{d0}}{dt} - \omega_r L_q i_{q0} \tag{1}$$

$$v_{qs} = R_s i_{qs} + L_q \frac{di_{q0}}{dt} + \omega_r L_d i_{d0} \tag{2}$$

$$i_{ds} = i_{d0} + i_{dc} = i_{do} + \frac{1}{R_c}(L_d \frac{d}{dt} i_{do} - \omega_r L_q i_{q0}) \tag{3}$$

$$i_{qs} = i_{q0} + i_{qc} = i_{qo} + \frac{1}{R_c}(L_q \frac{d}{dt} i_{qo} + \omega_r L_d i_{d0}) \tag{4}$$

Formula (1)-(4) presented form of differential equation is as follows

$$p\begin{bmatrix} i_{d0} \\ i_{q0} \end{bmatrix} = \begin{bmatrix} \dfrac{-R_s}{L_d(1+R_s/R_c)} & \dfrac{\omega_r}{L_d}L_q \\ -\omega_r \dfrac{L_d}{L_q} & \dfrac{-R_s}{L_q(1+R_s/R_c)} \end{bmatrix} \begin{bmatrix} i_{d0} \\ i_{q0} \end{bmatrix} \tag{5}$$
$$+ \begin{bmatrix} \dfrac{1}{L_d(1+R_s/R_c)} & 0 \\ 0 & \dfrac{1}{L_q(1+R_s/R_c)} \end{bmatrix} \begin{bmatrix} v_{ds} \\ v_{qs} \end{bmatrix}$$

Generating torque is in proportion to i_{do} and i_{qo} .

$$T_e = \frac{3}{2}\frac{P}{2}(L_d - L_q)i_{d0}i_{q0} \tag{6}$$

III. LM-FNN CONTROLLER

Fig. 3 show LM-FNN and ANN controller for SynRM drive system. FNN controller is compared real performance by requested command value and observed operation of system. Learning mechanism adjust the FNN controller for agreement with requested system operation.

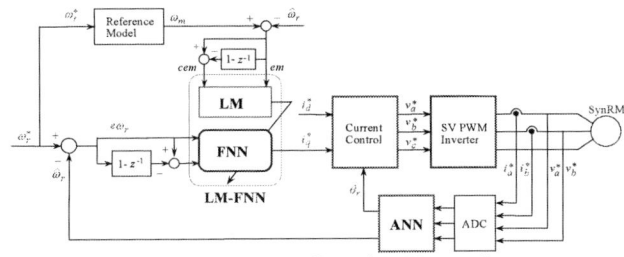

Fig. 3 LM-FNN controller and ANN controller

Fig. 4 show structure of FNN controller using learning mechanism

978-1-4244-1871-8/07 $25.00 © 2007 IEEE   373

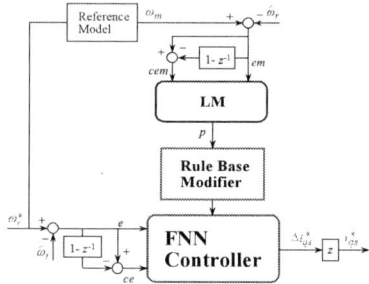

Fig. 4 FNN controller using learning mechanism

Speed performance of drive system is decided error of between reference model speed and estimation speed.

$$em(kT) = \omega_m(kT) - \hat{\omega}_r(kT) \qquad (7)$$

$$cem(kT) = em(kT) - em(kT - T) \qquad (8)$$

Rule base of FNN is changed amender of rule base according to $p(kT)$. This variation is realized adjust to the center of membership function of output at FNN controller.

$$C_i(kT) = C_i(kT - T) + p(kT) \qquad (9)$$

IV. EFFICIENCY OPTIMUM CONTROL

Because flux of SynRM is directly in proportion to current, efficiency optimum control is adjusted flux for efficiency optimum. SynRM can be maintain dynamic characteristic either in case of adjust flux. Current angle for efficiency optimum control is enough considered only electrical steady-state.

Iron loss and copper loss is obtained using formula (1)-(4). Copper loss is as follow

$$
\begin{aligned}
P_{cl} &= \frac{3}{2} R_s i_s^2 = \frac{3}{2} R_s (i_{ds}^2 + i_{qs}^2) \\
&= \frac{3}{2} R_s \{ (i_{do} - \frac{\omega_r L_q i_{qo}}{r_c})^2 + (i_{qo} + \frac{\omega_r L_d i_{do}}{r_c})^2 \}
\end{aligned}
\qquad (10)
$$

Iron loss is as follow

$$P_{ll} = \frac{3}{2} \frac{(\omega_r \lambda_m)^2}{R_c} = \frac{3}{2} \{ \frac{(\omega_r L_q i_{qo})^2}{R_c} + \frac{(\omega_r L_d i_{do})^2}{R_c} \} \qquad (11)$$

Therefore, total loss is sum of iron loss and copper loss and can be expressed frame as

$$
\begin{aligned}
P_{total} = P_{cl} + P_{ll} = \frac{3}{2} [& \{ R_s + \frac{(\omega_r L_d)^2}{R_c} + \frac{R_s}{R_c^2} (\omega_r L_d)^2 \} i_{do}^2 \\
& + \{ R_s + \frac{(\omega_r L_q)^2}{R_c} + \frac{R_s}{R_c^2} (\omega_r L_q)^2 \} i_{qo}^2 \\
& + \{ \frac{2R_s}{R_c} \omega_r (L_d - L_q) \} i_{do} i_{qo}]
\end{aligned}
\qquad (12)
$$

To express P_{total} with i_{do}, changing formula (6) is as follow

$$i_{qo} = \frac{T_e^*}{\frac{3}{2} \frac{P}{2} (L_d - L_q) i_{do}} = \frac{K}{i_{do}} \qquad (13)$$

To formula (12) substitution formula (13) is as follow

$$P_{total} = \frac{3}{2} \{ A i_{do}^2 + B i_{do}^{-2} + C \} \qquad (14)$$

Where coefficient can be expressed as follow

$$A = R_s + \frac{(\omega_r L_d)^2}{R_c} (1 + \frac{R_s}{R_c})$$

$$B = K^2 \{ R_s + \frac{(\omega_r L_q)^2}{R_c} (1 + \frac{R_s}{R_c}) \}$$

$$C = 2K \frac{R_s}{R_c} \omega_r (L_d - L_q)$$

Optimum of motor efficiency can be obtained minimum of total loss P_{total}.

$\frac{dP_{total}}{di_{do}}$ is as follow

$$
\begin{aligned}
\frac{dP_{total}}{di_{do}} &= \frac{3}{2} \{ 2A i_{do} - 2B i_{do}^{-3} \} = 3 A i_{do}^{-3} \{ i_{do}^4 - B/A \} \\
&= 3 A i_{do}^{-3} \{ i_{do}^2 + \sqrt{B/A} \} \{ i_{do} + (B/A)^{1/4} \} \{ i_{do} - (B/A)^{1/4} \} \\
&= 0
\end{aligned}
\qquad (15)
$$

If i_{do-min} that minimize total loss is computed, we can be obtained term of minimum total loss.

$$i_{do-min} = (B/A)^{1/4} \qquad (16)$$

Fig. 5 shows construction of efficiency optimum control for SynRM

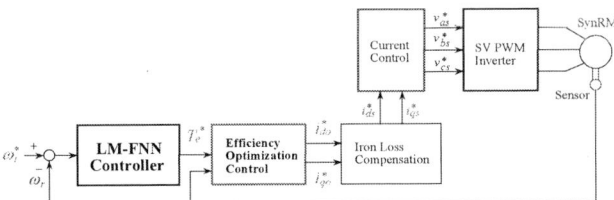

Fig. 5 An efficiency optimization control configuration for SynRM

V. PERFORMANCE RESULT OF SYSTEM

Fig. 6 is response characteristic when is operated command speed 1800[rpm] at 0.2[s] and load torque 2[N.m] at 0.5-0.7[s]. Fig. 6(a) is compare with command speed and real speed. Fig. 6(b), (c) enlarge (a) for accurate comparison between command speed and real speed.

Fig. 7 is response characteristic of quadrant operation by constant i_{ds} control and fig. 8 is response characteristic of quadrant operation by proposed efficiency optimization control.

Fig. 9 is loss result by constant i_{ds} control of traditional method and fig. 10 is efficiency optimization control. (a) is copper loss, (b) is iron loss and (c) is total loss. Fig. 11 shows total loss comparison of constant i_{ds} control and efficiency optimization control. (a) is constant i_{ds} control and (b) is efficiency optimization control. Efficiency optimization is decreased total loss more than constant i_{ds} control in steady-state. Efficiency optimization control is established decreasing loss.

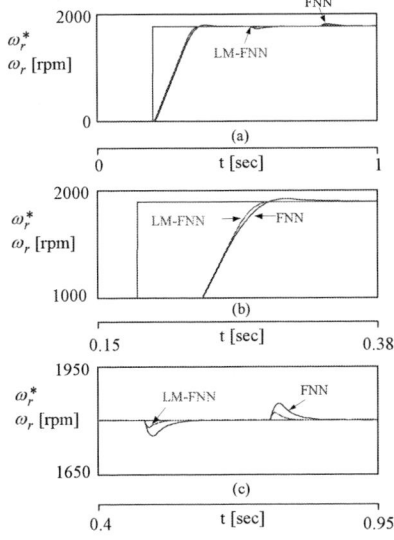

Fig. 6 Response characteristics with speed and load torque change

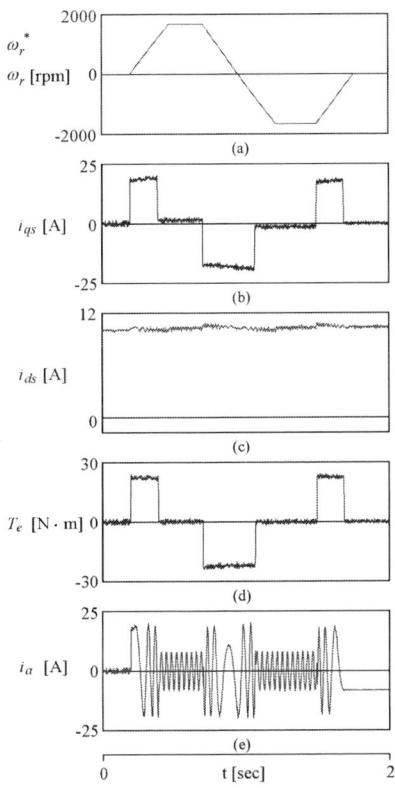

Fig. 7 The response characteristics of constant i_{ds} control

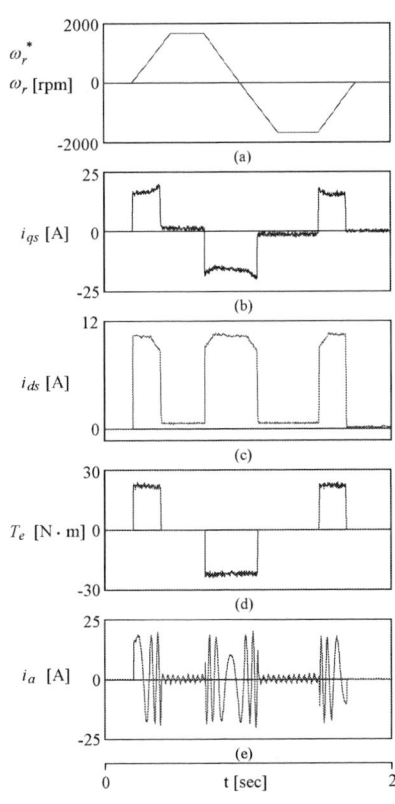

Fig. 8 The response characteristics of efficiency optimization control

978-1-4244-1871-8/07 $25.00 © 2007 IEEE   375

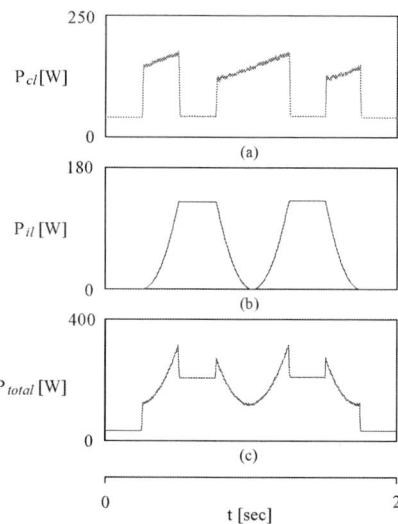

Fig. 9 Loss response of constant i_{ds} control

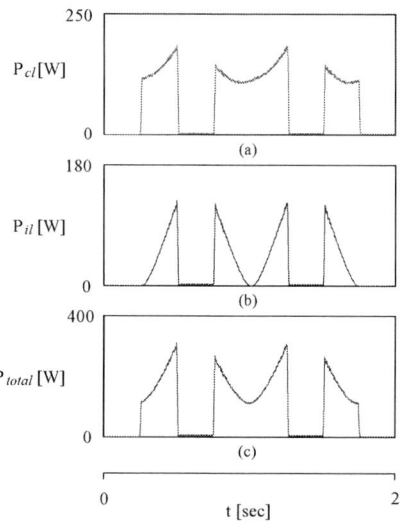

Fig. 10 Loss response of efficiency optimization control

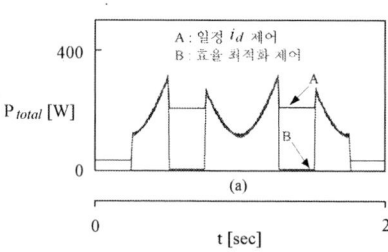

Fig. 11 Total loss comparison

VI. CONCLUSION

This paper is proposed efficiency optimization control that is minimized loss for high performance operation using LM-FNN controller. LM-FNN controller can be obtained high performance and robust control. This paper is used model of considered iron loss for solving nonlinear of torque by iron loss and saturation at SynRM.

Efficiency optimization control proposed in this paper is decreased loss more than constant i_{ds} control at various operation such as change of command speed, load torque, forward-reverse operation and quadrant operation. Therefore, efficiency very much increased and can be prove propriety of proposed efficiency optimization control.

REFERENCES

[1] T. Matsuo and T. A. Lipo, "Field oriented control of synchronous reluctance machine," in Proc. PESC, pp. 425-431, 1993.

[2] A. Fratta, A. Vagati and F. Villata, "Control of a reluctance synchronous motor for spindle application," in Proc. IPEC-Tokyo, pp. 708-715, 1990.

[3] R. E. Betz, et al., "Control of synchronous reluctance machines," IEEE Trans. Ind. Appli., vol. 29, no. 6, pp.1110-1121, 1993.

[4] L. Xu and J. Yao, "A compensated vector control scheme of a synchronous reluctance motor considering saturation and iron losses," IEEE Transaction on Industry Applications, Vol. 1A-28, no. 6, pp. 1330-1338, 1992.

[5] A. Vagati, T. A. Lipo, et al., "Synchronous reluctance motors and drives - A New Alternative," IEEE IAS Annu. Meet. Tutorial Course Note, 1994.

[6] M. S. Arefeen, M. Ehsani and T. A. Lipo, "An analysis of the accuracy of indirect shaft sensor for synchronous reluctance motor," IEEE Trans. on IA, vol. 30, no. 5, pp. 1202-1209, 1994.

[7] Y. Q. Xiang and S. A. Nasar, "Estimation of rotor position and speed of a synchronous reluctance motor for servodrives," IEE Proc EPA, vol. 142, no. 3, pp. 201-205, 1995.

[8] T. Matsuo and T. A. Lipo, "Rotor Position Detection Scheme for Synchronous reluctance motor Based on Current Measurement," IEEE Transactions on Industry Applications," vol. 31, no. 4, July/August, pp. 860-868. 1995.

[9] M. G. Jovanovic, R. E. Betz and D. Platt, "Sensorless vector controller for a synchronous reluctance motor," IEEE IAS Ann. Meet., pp. 122-129, 1996.

[10] M. Schroedl and P. Weinmeier, "Sensorless Control of Reluctance machines at Arbitrary Operating Conditions Including Standstill," IEEE Transactions on Power Electronics, vol. 9, no. 2, Narch, pp. 225-231, 1994.

[11] R. Lagerquist, I. Boldea and T. J. E. Miller, "Sensorless Control of the Synchronous reluctance motor," IEEE Transactions on Industry Applications, vol. 30, no. 3, May/June, pp. 673-682, 1994.

[12] T. Matsuo and T. A. Lipo, "Rotor Position Detection Scheme for Synchronous reluctance motor Based on Current Measurement," IEEE Transactions on Industry Applications," vol. 31, no. 4, July/August, pp. 860-868. 1995.

[13] M. Schroedl and P. Weinmeier, "Sensorless Control of Reluctance machines at Arbitrary Operating Conditions Including Standstill," IEEE Transactions on Power Electronics, vol. 9, no. 2, Narch, pp. 225-231, 1994.

A Study on Control Strategy of
Electric Power Steering Based on Fuzzy Control

Hui Zhang, Jing Ren, Yanru Zhong

Xi'an University of Technology

Department of Electrical Engineering

zhangrenhj@163.com

Jian Chen *senior member* IEEE

Huazhong University of Science and Technology

College of Electrical and Electronic Engineering

zhanghui200-20@sohu.com

Abstract-Intelligent Control, a novel stage of automatic control development, is one of the efficient strategies for the system with nonlinear or imprecise model. This paper has applied Fuzzy Control (FC) technology to AC electric power steering (EPS) system for seeking better control strategies. Proceeding with fuzzy logic, fuzzy rule and fuzzy inference, a kind of control diagram of EPS based on fuzzy control system has been presented, which can be used to solve the nonlinear or imprecise problems of electrical machine control. The paper has established the foundation for further applications of FC technology in EPS domains. The FC system is verified to be feasible by simulation and experiment.

I. INTRODUCTION

Development of power electronics and microcomputer has established the stable foundation for AC variable speed drives system with high performance. Compared with traditional and modern control strategies, intelligent control has a series of characteristics. Because of this, it is necessary that applying intelligent control into electrical drives system such as EPS in order to obtain high performance, high precision and high robust.

At the present time, traditional and modern control strategies are becoming very difficult to process controlled object with nonlinear or imprecise model. Therefore, it is possible that traditional and modern control strategies are gradually replaced by intelligent control strategy to some extent.

Fuzzy control, one of the intelligent control strategies, has many advantages. One of them is not to develop more accurate nonlinear or imprecise model, on which high-performance controller can be formed but the complex mathematical derivations will lead to very complicated control algorithms that are not suitable for practical implementation. Another is to employ heuristic reasoning based on human knowledge and experience, and such knowledge and experience is usually collected in the form of linguistic statement rules such as the expression of IF-THEN. In the latter case, no modeling is at all required, and the main work of the system design is to make fuzzy controller.

The aim of this paper is to show how fuzzy control can be used in power electronic equipment——EPS, and the first application dedicated to EPS has been presented. The derivation of a fuzzy control system will be described in detail.

II. STRUCTURE AND PRINCIPLE OF THE SYSTEM

Fuzzy control system is a kind of automatic control system. It theoretically bases on fuzzy mathematics, knowledge expression in fuzzy linguistic statement and knowledge inference in fuzzy logic, and is a kind of digital control system with closed-loop feedback. Its most notable characteristic is expresses the expert's knowledge and experience as the linguistic statement rules, and controls the system with rules. The essence of the system is fuzzy controller.

Fuzzy control system is at least composed of A/D converter, fuzzifier, inference machine (rule base and data base), defuzzifier, D/A converter, controlled object and feedback link. Among these, the fuzzy controller is composed of fuzzifier, inference machine (rule base and database), defuzzifier. This is the prime difference compared with other automatic control system. And the rest is the same as the traditional control system. The fuzzy controller will be discussed in this paper.

Fuzzy controller also can be known as fuzzy logic controller (PLC). Because the fuzzy control rules are described in fuzzy linguistic statement, the fuzzy controller is a kind of linguistic controller.

The configuration diagram of EPS based on fuzzy control, which is a dimension system discussed in this paper, is shown in Figure 1.

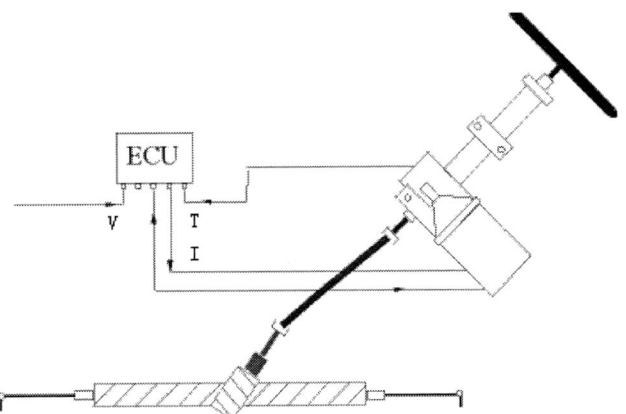

Fig1.Configuraion diagram of EPS

978-1-4244-1871-8/07 $25.00 © 2007 IEEE

Let given position of the induction motor is θ_g, the feedback signal θ_f must be compared with the input signal θ_g to form the closed-loop speed feedback. At this time, the error $e = \theta_g - \theta_f$, which will be equal to zero, can be as the input of fuzzy controller by A/D converting, then the input E will be fuzzified from crisp value to fuzzy value $\underset{\sim}{E}$; according to the compositional-rule-of-inference, the compositional operation of fuzzy set $\underset{\sim}{E}$ and fuzzy rule $\underset{\sim}{R}$ can yield a fuzzy result——fuzzy value $\underset{\sim}{W}$: $\underset{\sim}{W} = \underset{\sim}{E} \circ \underset{\sim}{R}$; at last, the output $\underset{\sim}{W}$ will be defuzzified from fuzzy value to crisp value W, which is as the output of fuzzy controller. After D/A converting, w is used to control SPWM signal generator to adjust the output frequency of the inverter so as to recover the given rotating speed of the induction motor.

On the other hand, if the given input signal θ_g has changed, then the $e = \theta_g - \theta_f$ will change with it. This can be used to adjust θ_f ——the given position of the induction motor.

Now the analysis in detail is as follows:

A. *Fuzzifier*

Let fuzzy set of the error $\underset{\sim}{E}$ is:

$$\underset{\sim}{E} = \{NL,NM,NS,ZO,PS,PM,PL\}$$

where NL——Negative Large

NM——Negative Medium

NS——Negative Small

ZO——Zero

PS——Positive Small

PM——Positive Medium

PL——Positive Large

And the error $\underset{\sim}{E}$ will be divided into thirteen grades, such as $-6,-5,-4,-3,-2,-1,0,+1,+2,+3,+4,+5,+6$, that is to say, universe of discourse of $\underset{\sim}{E}$ is:
$\underset{\sim}{E} = \{-6,-5,-4,-3,-2,-1,0,+1,+2,+3,+4,+5,+6\}$, this is shown in Figure 2.

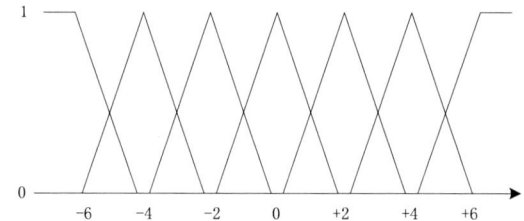

Fig2.Degree of membership

The output $\underset{\sim}{W}$ and W is the same.

$$\underset{\sim}{W} = \{NL,NM,NS,ZO,PS,PM,PL\}$$

$$W = \{-6,-5,-4,-3,-2,-1,0,+1,+2,+3,+4,+5,+6\}$$

According to expert's knowledge and experience, the degrees of membership of fuzzy set $\underset{\sim}{E}$ and $\underset{\sim}{W}$ are given in Table 1.

Table 1. Degree of membership of linguistic variable

$\underset{\sim}{E}$ and $\underset{\sim}{W}$

	NL	NM	NS	ZO	PS	PM	PL
-6	1	0.3	0	0	0	0	0
-5	0.8	0.7	0.1	0	0	0	0
-4	0.7	1	0.3	0	0	0	0
-3	0.4	0.7	0.7	0	0	0	0
-2	0.1	0.2	1	0.1	0	0	0
-1	0	0	0	0.6	0	0	0
0	0	0	0	1	0	0	0
+1	0	0	0	0.6	0	0	0
+2	0	0	0	0.1	1	0.2	0.1
+3	0	0	0	0	0.7	0.7	0.4
+4	0	0	0	0	0.3	1	0.7
+5	0	0	0	0	0.1	0.7	0.8
+6	0	0	0	0	0	0.3	1

B. Fuzzy rule

Fuzzy rule is based on the expert's knowledge and experience, and it is a kind of linguistic statement with intuition inference. Fuzzy rule is commonly expressed as the forms of IF-THEN, AND, OR and so on. According to practical control experience, the linguistic variable of fuzzy control rules can be expressed as follows:

(1) If E =NL then W =PL
(2) If E =NM then W =PM
(3) If E =NS then W =PS
(4) If E =ZO then W =ZO
(5) If E =PS then W =NS
(6) If E =PM then W =NM
(7) If E =PL then W =NL

where E and W are input and output linguistic variable.

C. Fuzzy inference

Fuzzy inference is that corresponding control rule is picked up to inference aiming at input fuzzy value. There are many methods to inference, but the most commonly used method is the compositional-rule-of-inference. The principle of the compositional-rule-of-inference is as follows:

If x is $\underset{\sim}{E}$ then y is $\underset{\sim}{W}$

now x is $\underset{\sim}{E}'$

then y is $\underset{\sim}{W}'$

Thus, the fuzzy output result $\underset{\sim}{W}'$ is:

$$\underset{\sim}{W}' = \underset{\sim}{E}' \circ \underset{\sim}{R}_{EW}$$

where $\mu_{W'}(z) = \{ \mu_{E'}(x) \circ [\mu_E(x)$

$\rightarrow \mu_W(y)]\}$

and " $\circ$ ": denotes the Sup-Star compositional-rule-of-inference. If " $\circ$ " is defined as "min-max", it denotes Mamdami method.

According to the above-mentioned Table 1 and control rules, which in fact is multiple fuzzy conditional statement, fuzzy relation $\underset{\sim}{R}_{EW}$ will get in the form of 13×13 matrix:

$$\underset{\sim}{R}_{EW} = (NL \times PL) \cup (NM \times PM) \cup (NS \times PS) \cup (ZO \times ZO) \cup (PS \times NS)$$

$$\cup (PM \times NM) \cup (PL \times NL)$$

D. Defuzzifier

Generally, the result of the fuzzy inference is fuzzy value and can't be used to control system directly. The defuzzification is just transforming the fuzzy value into crisp value. There are several methods to obtain the crisp value from the result of fuzzy inference, such as Mean-of-Maximum (MOM) method and Center-of-Area (COA) method. Because COA is the most popular, this method is adopted to defuzzify the fuzzy output variable

$\underset{\sim}{W}'$ to certain value W.

$$W = \sum \mu(W'_i) W'_i / \sum \mu(W'_i)$$

III. SIMULATION AND EXPERIMENT

MATLAB/Simulink is a kind of software in Number Crunching, and provide with an integrate environment of dynamic system. With the help of MATLAB/SIMULINK, various states such as transient-state response and steady-state error, corresponding to each control goal, can be simulated. During simulation, the performance of the fuzzy controller is found to depend, to some extent, on the size of rules. If the number of rules increases, the overall performance improves in general, but too many rules lead to longer processing time of the fuzzy controller; and vice versa. So, we must decide which one is more important so as to synthesize.

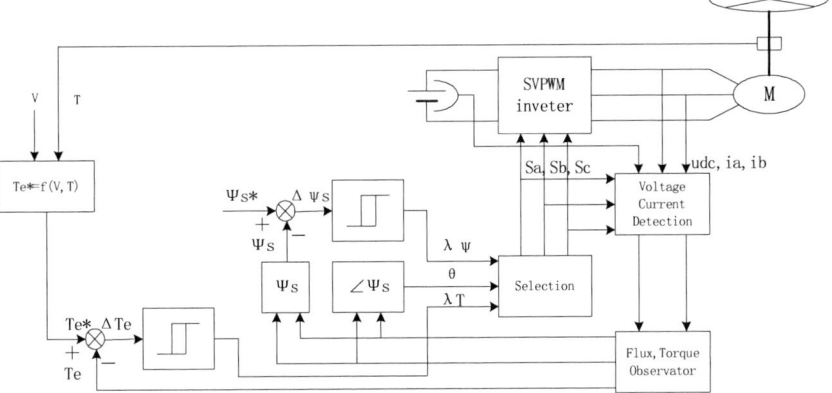

Fig.3. Control diagram for EPS

This fuzzy controller is implemented as a discrete digital system using a TI's digital signal processor (DSP) TMS320F2407A, which is mounted on an emulator. The parameters of AC variable speed drives system are the same as those used in the simulation. The experimental results coincide approximately with the simulation.

IV. CONCLUSION

According to the results of the simulation and experiment, the fuzzy control system has following advantages:

(1) The fuzzy control system is not dependent on the linear or precise model of the system, therefore, it is

978-1-4244-1871-8/07 $25.00 © 2007 IEEE   379

suitable for the nonlinear or imprecise system.

(2)The fuzzy control system, in which the knowledge expression, fuzzy rule and knowledge inference are based on the expert knowledge and experience, has intelligent and self-studied performance.

(3)The fuzzy control system, in which the essence is the fuzzy controller based on the computer, has the characteristic the same as computer control system, such as precision and flexibility of digital control.

ACKNOWLEDGMENT

The authors would like to thank the financial support of The Province Natural Science Foundation of Shaanxi (2006 E_1 28) , financial support of The Province Education Department Foundation of Shaanxi (07JK355), financial support of The University Research Foundation of XAUT(105-210704) and financial support of The University Doctor Research Foundation of XAUT (105-210622).

REFERENCES

[1] Wing-Chi So, Chi K. Tse, and Yim-Shu Lee, "Development of fuzzy logic controller for DC/DC converters: design, computer simulation, and experimental evaluation". *IEEE Trans. On PE,* vol.11, no.1, 1996, pp.24-31.

[2] Zhen ZHOU. *The theory research and application of intelligent control in AC speed-adjustable system.* Thesis of Doctor's Degree, Huazhong University of Science and Technology, 1998.

[3] Zhen-zhong DOU. *Fuzzy logic control technology and application.* Beijing, Press of Beijing University of Aeronautics and Astronautics, 1995.

[4] Jing CHU. *Fuzzy Control Principle and Application.* Beijing, Press of Mechanical Industry, 1998.

[5] Zhenyu ZHAO, Yongmao XU. *Elements and application of fuzzy theory and neural network.* Beijing, Press of Tsinghua University, 1996.

[6] Yongquan YU, Ping TANG. *Fuzzy logic software of single-chip computer.* Beijing, Press of Beijing University of Aeronautics and Astronautics, 1996.

[7] Aurobinda Routray, P.K.Dash, and Sanjeev K.Panda, "A fuzzy self-tuning PI controller for HVDC Links". *IEEE Trans. On PE,* vol.11, no.5, 1996, pp.669-679

[8] Paolo Mattavelli, *Member IEEE,* Leopoldo Rossetto, *Member IEEE,* Giorgio Spiazzi, *Member IEEE,* and Paolo Tenti, Senior *Member IEEE,* "General-purpose fuzzy controller for DC-DC converters". *IEEE Trans. On PE,* vol.12, no.1, 1997, pp.79-85.

[9] Tarun Gupta, *Member IEEE,* R.R.Boudreaux, *Student Member IEEE,* R.M.Nelms, Senior *Member IEEE,* and John Y.Hung, Senior *Member IEEE,* "Implementation of a fuzzy controller for DC-DC converters using an inexpensive 8-b microcontroller". *IEEE Trans. On IE,* vol.44, no.5, 1997, pp.661-669.

[10] Hui ZHANG, "Application of fuzzy controller in three-phase SPWM inverter system". Proceeding of the fifth conference in circuits and system, China, 1999.

[11] V.S.C. Raviraj, and P.C.Sen, *Fellow IEEE,* "Comparative study of proportional-integral, sliding mode, and fuzzy logic controllers for power converters". *IEEE Trans. On IA,* vol.33, no.2, 1997, pp.518-523.

[12] Ying-Yu Tzou, *Member IEEE,* and Shiu-Yung Lin, "Fuzzy-tuning current-vector control of a three-phase PWM inverter for high-performance AC drives". . *IEEE Trans. On IE,* vol.45, no.5, 1998, pp.782-791.

A Four-Switch and Half-Bridge Boost Converter based BLDCM Drive System For the DC-Link Voltage Unbalancing Compensation

Sang-Hoon Park, Young-Hwan Ryu, Won-cheol Lee, Jung-Hyo Lee, Jae-Sung Yu, Young-Ryul Kim, Chung-Yuen Won
Sungkyunkwan University
School of Information & Communication Engineering, Sungkyunkwan University
300 Cheoncheon-dong, Jangan-gu, Suwon,
Gyeonggi-do 440-746, Korea
E-mail: marohachi@skku.edu

Abstract - This paper proposes a controller to compensate the DC-Link voltage unbalance in the BLDC motor drive system based on the four-switch and half-bridge boost converter.

In this paper, we can see that the proposed feed?forward controller is useful for control of the neutral point voltage unbalance. The comparative experimental results show that the system became more robust by the feed-forward control of the neutral point voltage.

I. INTRODUCTION

In this paper, we operated the 3-phase BLDC motor by 4-switch inverter, which has the similar function as 6-switch inverter. However, using the 4-switch inverter to operate the 3-phase BLDC motor, we have to consider the voltage unbalance. Because one phase of the 3-phase BLDC motor is connected to the neutral point of DC-link capacitor in the half bridge boost converter, the voltage unbalance of the divide capacitor at neutral point occurs. Consequently, it causes the problems such as the capacitor destruction, increase of the current ripple of BLDC motor, and malfunction of the system.

This paper proposes the method to operate the 3-phase BLDC motor by 4-switch inverter with half bridge boost converter without voltage unbalance of divide capacitor at neutral point. We apply the control method, which is able to compensate the voltage difference by sensing the voltages of the divide capacitors in half bridge boost converter.

II. BRUSHLESS DC MOTOR

The permanent magnet motor, which is used in this paper, has a trapezoidal back electromotive force and is driven by square wave currents.

Generally, in the DC motor drive, rotating magnetic motive force (MMF) is made by the current flow through the rotor winding. However, in the BLDC motor drive, rotating MMF is made by the stator.

Moreover, BLDC motor is easy for maintenance because of no brush, and has less current peck and acoustic noise owing to no friction loss.

III. MODELING OF THE BLDC MOTOR

The modeling of BLDC motor is based on the following assumption to simplify the simulation.

1. The resistances of each phase are equal.
2. The self inductances of each phase are equal.
3. The mutual inductances of each phase are equal.
4. The iron and copper-loss are neglected.

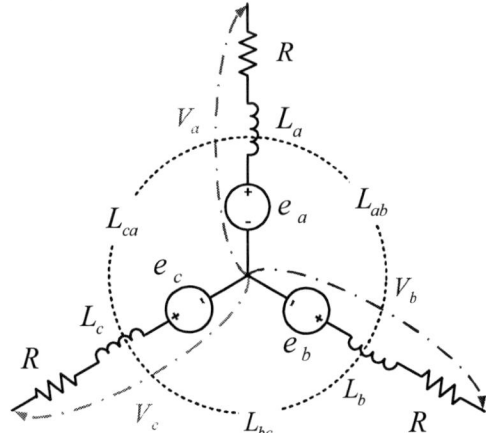

Figure1. Equivalent circuit of the BLDC motor.

Figure 1 shows the equivalent circuit of the BLDC motor. Mathematical model of the BLDC motor is described by the following equations (1).

$$
\begin{bmatrix} v_a \\ v_b \\ v_c \end{bmatrix} = \begin{bmatrix} R & 0 & 0 \\ 0 & R & 0 \\ 0 & 0 & R \end{bmatrix} \begin{bmatrix} i_a \\ i_b \\ i_c \end{bmatrix} + \begin{bmatrix} L_a & L_{ba} & L_{ca} \\ L_{ba} & L_a & L_{cb} \\ L_{ca} & L_{cb} & L_c \end{bmatrix} \frac{d}{dt} \begin{bmatrix} i_a \\ i_b \\ i_c \end{bmatrix} + \begin{bmatrix} e_a \\ e_b \\ e_c \end{bmatrix} \quad (1)
$$

Where, R, $L_{a,b,c}$, $L_{ab,bc,ca}$ are resistance, self inductance and mutual inductance. $v_{a,b,c}$, $i_{a,b,c}$ are voltage and current of each phase. $e_{a,b,c}$ is back-EMF.

Back EMF has a trapezoidal waveform. It is proportional to speed of the rotor. The θ_e is electrical position.

978-1-4244-1871-8/07 $25.00 © 2007 IEEE   381

$$e = k_e\left(\theta_e\right) \times \frac{2\omega_e}{p} \tag{2}$$

Where, k_e is proportional constant for back-EMF.

According to the relation of above equation (2), back EMF of the trapezoidal waveform is defined in 3 sections. Each section are rising, flat and falling section.

$$e = k_e(\theta_e) = \begin{cases} k_e \dfrac{6}{\pi}\theta_e & \left(0 \le \theta_e \le \dfrac{\pi}{6}\right) \\ k_e & \left(\dfrac{\pi}{6} \le \theta_e \le \dfrac{5\pi}{6}\right) \\ k_e \dfrac{6}{\pi}(\pi - \theta_e) & \left(\dfrac{5\pi}{6} \le \theta_e \le \pi\right) \end{cases} \tag{3}$$

Self inductance and mutual inductance of each phase are defined by equation (4) from equation (1) and (2). It is redefined by equation (5) using phase voltage equation.

$$\begin{aligned} L_a = L_b = L_c = L_s \\ L_{ab} = L_{ca} = L_{cb} = M \end{aligned} \tag{4}$$

$$\begin{bmatrix} v_a \\ v_b \\ v_c \end{bmatrix} = \begin{bmatrix} R & 0 & 0 \\ 0 & R & 0 \\ 0 & 0 & R \end{bmatrix}\begin{bmatrix} i_a \\ i_b \\ i_c \end{bmatrix} + \begin{bmatrix} L_s & M & M \\ M & L_s & M \\ M & M & L_s \end{bmatrix}\frac{d}{dt}\begin{bmatrix} i_a \\ i_b \\ i_c \end{bmatrix} + \begin{bmatrix} e_a \\ e_b \\ e_c \end{bmatrix} \tag{5}$$

The equation (5) is redefined as following equations using KCL.

$$\begin{bmatrix} v_a \\ v_b \\ v_c \end{bmatrix} = \begin{bmatrix} R & 0 & 0 \\ 0 & R & 0 \\ 0 & 0 & R \end{bmatrix}\begin{bmatrix} i_a \\ i_b \\ i_c \end{bmatrix} + \begin{bmatrix} L & 0 & 0 \\ 0 & L & 0 \\ 0 & 0 & L \end{bmatrix}\frac{d}{dt}\begin{bmatrix} i_a \\ i_b \\ i_c \end{bmatrix} + \begin{bmatrix} e_a \\ e_b \\ e_c \end{bmatrix} \tag{6}$$

Where L is $L_s - M$.

Then, voltages of each phase are as following equations.

$$v_a = Ri_a + L\frac{d}{dt}i_a + e_a$$

$$v_b = Ri_b + L\frac{d}{dt}i_b + e_b \tag{7}$$

$$v_c = Ri_c + L\frac{d}{dt}i_c + e_c$$

Electrical output of BLDC motor is defined by the equation of the current of each phase and the back EMF, that is, the sum of the multiplication of each back EMF by current equation (8).

$$P_e = T_e \omega_r \tag{9}$$

According to the relation of above equation (8), (9), the electrical torque is as following equation.

$$T_e = \frac{e_a i_a + e_b i_b + e_c i_c}{\omega_r} \tag{10}$$

This equation (11) is the mechanical torque equation.

$$T_m = J_m \frac{d\omega_r}{dt} + B_m \omega_r + T_L \tag{11}$$

Where, P_e, T_e, ω_r are electrical output, torque and angular speed. J_m, B_m, T_L are moment of inertia, coefficient of friction and load torque.

IV. COMPENSATION OF UNBALANCED VOLTAGE OF HALF BRIDGE BOOST CONVERTER USING FEED-FORWARD CONTROLLER.

Figure 2 shows the overall system of half-bridge boost converter and three phase BLDC motor.

In Figure 2, if the non-controlled phase of a BLDC motor uses the neutral point of half bridge boost converter, voltage unbalance between two capacitor of the DC-Link occurs.

This unbalanced voltage reduces the life time of capacitor and has influence on the ripple current of the motor.

In this paper, we apply the control method using feed forward compensation to control the neutral point voltage of the half bridge boost converter

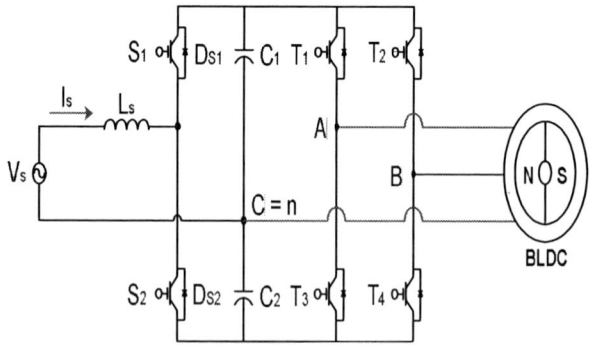

Figure2. Overall system of four-switch inverter for BLDC motor and half bride boost converter.

In the BLDC motor drive, each phase of BLDC motor need to be coupled to each pole of the inverter. However, in the BLDC motor with four-switch inverter, two phases of the BLDC motor are coupled to the inverter poles and one phase among the three is coupled to the neutral point of capacitor because four-switch inverter has only two poles. It causes the voltage unbalance problem.

Therefore, in this paper, the controller using feed-forward compensation is proposed to remove the voltage unbalance.

978-1-4244-1871-8/07 $25.00 © 2007 IEEE

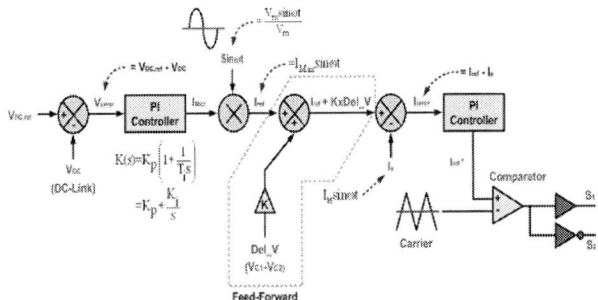

Figure3. Feed-forward control block diagram for Half-Bridge Boost Converter.

V. BLDC MOTOR USING FOUR SWITCH INVERTER

In this paper, four-switch inverter system is operated on six modes. On two among six modes, inverter is operated using switching device. However, on four modes, inverter is operated using neutral point of half bridge boost converter.

The switching pattern of four-switch inverter is expressed by the Table 1.

TABLE. 1
Switching pattern of four-switch inverter.

Operation Mode	Excited phase	Non-excited phase	Switching
Mode 1	C, B	A	T4
Mode 2	A, B	C	T1, T4
Mode 3	A, C	B	T1
Mode 4	B, C	A	T2
Mode 5	B, A	C	T2, T3
Mode 6	C, A	B	T3

As mentioned above, mode 2 and 3 use the complementary two switch of the inverter. However, the other four modes use one switch and neutral point of half bridge boost converter.

These are the analyses through the equivalent circuit of six modes.

1) Mode 1

Figure 5 shows the equivalent circuit of mode 1.

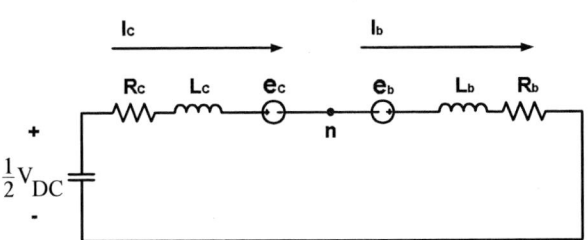

Figure5. Equivalent circuit of mode 1.

Voltage equation for the mode 1 as the figure 5 is expressed by following equation.

$$\frac{1}{2}V_{DC} = (R_c i_c + R_b i_b) + (L_c \frac{di_c}{dt} + L_b \frac{di_b}{dt}) + (e_c - e_b) \quad (12)$$

According to the equation (12), current variation of the mode-1 is expressed by

$$\frac{di_c}{dt} = -4\frac{R}{L}i_c + \frac{1}{L}(V_{DC} - 2e_{cb}) \quad (13)$$

2) Mode 2

Figure 6 shows the equivalent circuit of mode 2.

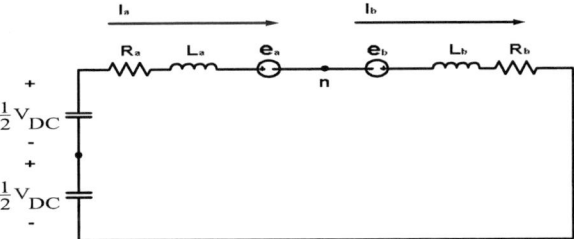

Figure6. Equivalent circuit of mode 2.

Voltage equation for the mode 2 as the figure 6 is expressed by following equation

$$V_{DC} = (R_a i_a + R_b i_b) + (L_a \frac{di_a}{dt} + L_b \frac{di_b}{dt}) + (e_a - e_b) \quad (14)$$

According to the equation (14), current variation of the mode 2 is expressed by

$$\frac{di_a}{dt} = -\frac{R}{L}i_a + \frac{1}{2L}(V_{DC} - e_{ab}) \quad (15)$$

3) Mode 3

Figure 7 shows the equivalent circuit of mode 3.

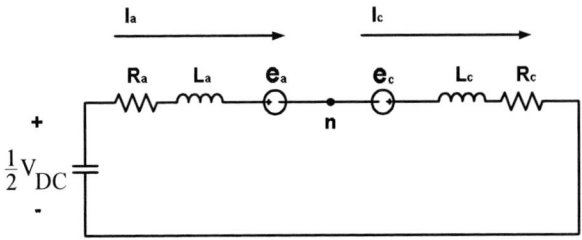

Figure7. Equivalent circuit of mode 3.

Voltage equation for the mode 3 as the figure 7 is expressed by following equation

$$\frac{1}{2}V_{DC} = (R_a i_a + R_c i_c) + (L_a \frac{di_a}{dt} + L_c \frac{di_c}{dt}) + (e_a - e_c) \quad (16)$$

According to the equation (16), current variation of the mode 3 is expressed by

$$\frac{di_a}{dt} = -4\frac{R}{L}i_a + \frac{1}{L}(V_{DC} - 2e_{ac}) \tag{17}$$

4) Mode 4

Figure 8 shows the equivalent circuit of mode 4.

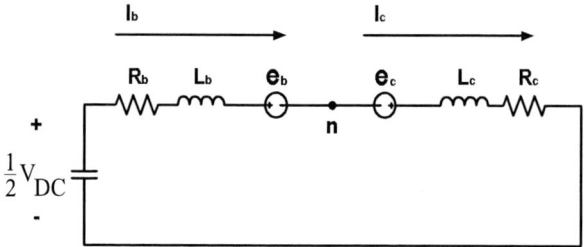

Figure8. Equivalent circuit of mode 4.

Voltage equation for the mode 4 as the figure 8 is expressed by following equation

$$\frac{1}{2}V_{DC} = (R_b i_b + R_c i_c) + (L_b \frac{di_b}{dt} + L_c \frac{di_c}{dt}) + (e_b - e_c) \tag{18}$$

According to the equation (18), current variation of the mode 4 is expressed by

$$\frac{di_b}{dt} = -4\frac{R}{L}i_b + \frac{1}{L}(V_{DC} - 2e_{bc}) \tag{19}$$

5) Mode 5

Figure 9 shows the equivalent circuit of mode 5.

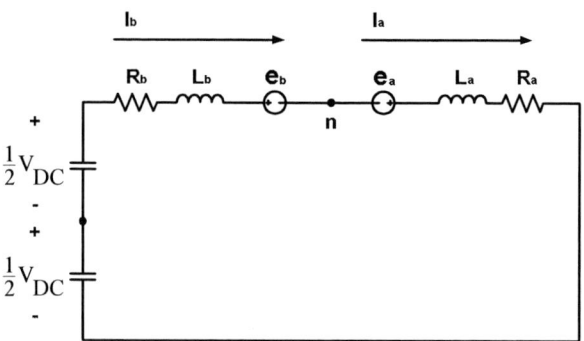

Figure9. Equivalent circuit of mode 5.

Voltage equation for the mode 5 as the figure 9 is expressed by following equation

$$V_{DC} = (R_b i_b + R_a i_a) + (L_b \frac{di_b}{dt} + L_a \frac{di_a}{dt}) + (e_b - e_a) \tag{20}$$

According to the equation (20), current variation of the mode 5 is expressed by

$$\frac{di_b}{dt} = -\frac{R}{L}i_b + \frac{1}{2L}(V_{DC} - e_{ba}) \tag{21}$$

6) Mode 6

Figure 10 shows the equivalent circuit of mode 6.

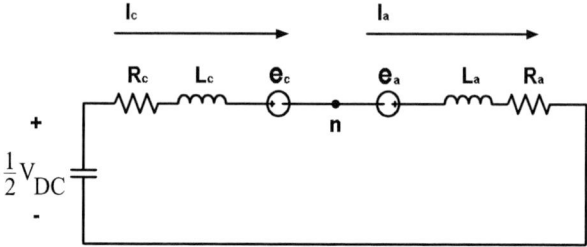

Figure10. Equivalent circuit of mode 6.

Voltage equation for the mode 6 as the figure 9 is expressed by following equation

$$\frac{1}{2}V_{DC} = (R_c i_c + R_a i_a) + (L_c \frac{di_c}{dt} + L_a \frac{di_a}{dt}) + (e_c - e_a) \tag{22}$$

According to the equation (22), current variation of the mode 6 is expressed by

$$\frac{di_c}{dt} = -4\frac{R}{L}i_c + \frac{1}{L}(V_{DC} - 2e_{ca}) \tag{23}$$

VI. SYSTEM BLOCK DIAGRAM

Figure 11 shows the control block diagram of BLDC motor using DSP TMS320F2811. In this system, we used two hysteresis current controllers.

In addition, we used voltage and current sensor for power factor compensation.

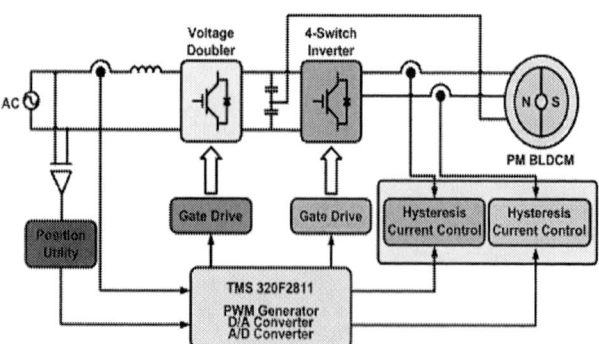

Figure11. System block diagram of 4-switch inverter for BLDCM.

978-1-4244-1871-8/07 $25.00 © 2007 IEEE   384

VII.   SIMULATION

The system is composed of half-bridge boost converter, four-switch converter, hysteresis current controller and speed controller. PSIM 6.0 software is used to simulate the system in this paper.

Figure 12 and 13 are the result of simulation. They show the unbalanced voltage waveform and compensated voltage waveform.

Figure 14 shows the voltage and current waveform of half-bridge boost converter for improved power factor.

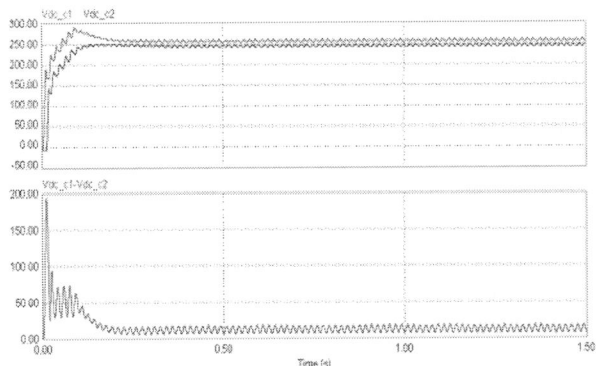

Figure12. Simulation waveform of unbalanced DC-Link voltage.

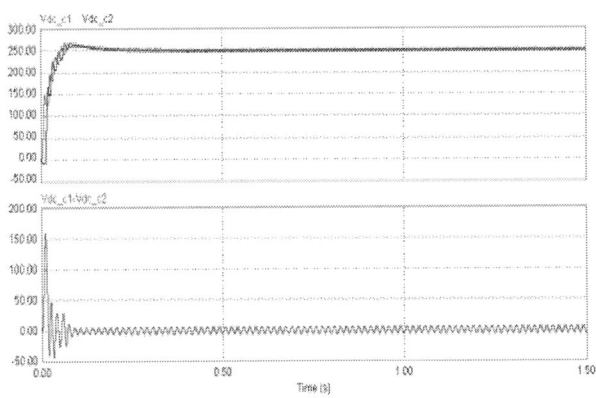

Figure13. Simulation waveform of compensated DC-Link voltage.

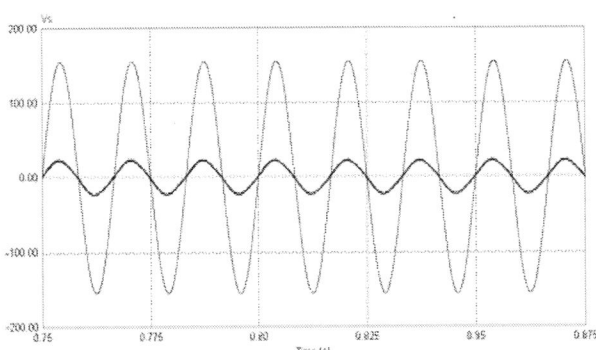

Figure14. Simulation waveform of power factor compensation.

VIII.   EXPERIMENTAL RESULTS

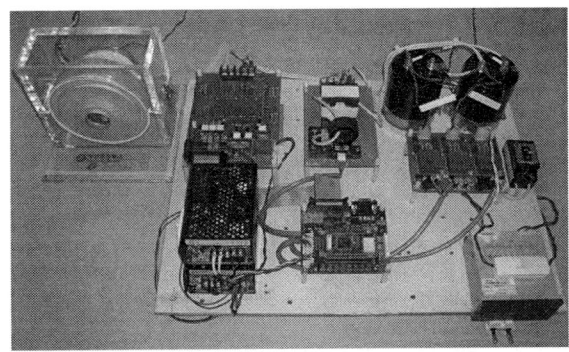

Figure15. Experimental test board.

Figure 15 is overall experimental system composed of half bride boost converter, four-switch inverter and main controller for BLDC motor. The parameters of BLDC motor are shown in table 2. The switching devices of the inverter are IPM package. The BLDC motor of 1[kW] rating for a vacuum cleaner is used in this system.

TABLE. 2
Parameters of three-phase BLDC motor

Name	Value
Stator Resistance	$0.88[\Omega]$
Rated Power	1[kW]
Rated Speed	30,000[rpm]
Rated Current	$3.6[A]/280[V]$
Number of Pole	2

Figure 16 and 17 are the experimental results. They show the unbalanced voltage waveform and compensated voltage waveform.

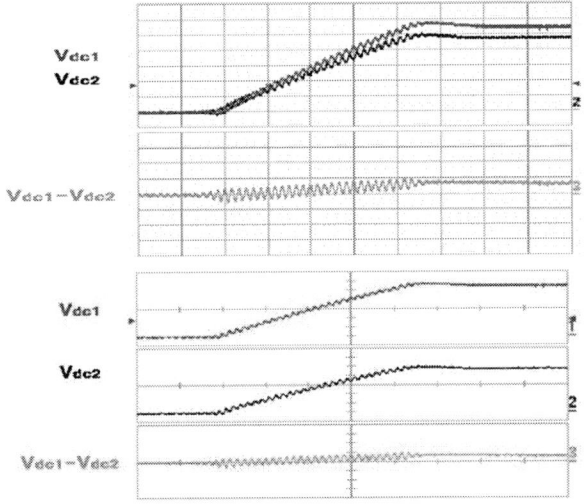

Figure16. Experimental voltage waveform before using

feed-forward controller. (50[V] / Div.)

978-1-4244-1871-8/07 $25.00 © 2007 IEEE   385

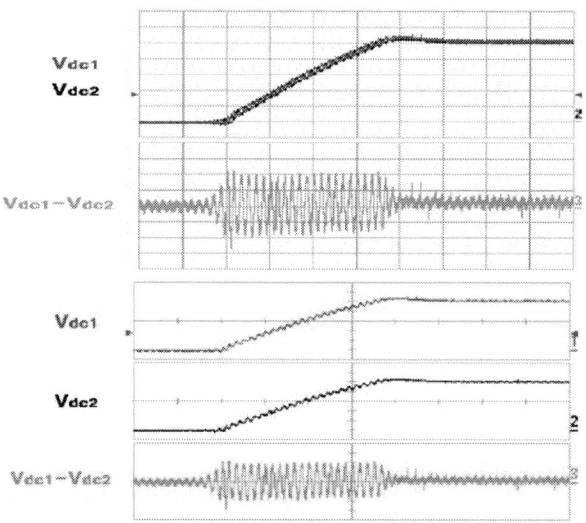

Figure17. Experimental voltage waveform after using feed-forward controller. (50[V] / Div.)

Figure18 shows the voltage and current waveform of BLDC motor for improved power.

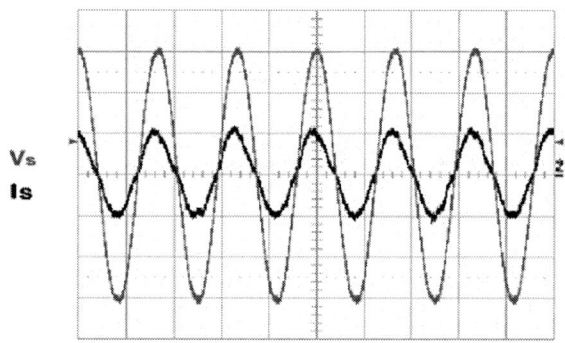

Figure18.Experimental power factor of voltage and current waveform.

(50[V] / Div., 10[A] / Div.)

IX. CONCLUSION

In this paper, we proposed the half bridge boost converter and four-switch inverter for three phase BLDC motor drive and show the result through simulation and experiment.

This four-switch inverter uses a new switching pattern, which is different from that of conventional six-switch inverter. As a result in this paper, improved power factor has acquired by the use of half bridge boost converter instead of diode rectifier.

If four-switch inverter is used for BLDC motor drive, one phase current path is needed to use the neutral point of divide capacitor in the half bridge boost converter. It generates the problem of unbalanced voltage of the capacitors. Therefore, in this paper we remove the unbalanced voltage by feed-forward controller using the voltage difference of the capacitors.

REFERENCES

[1] TJE Miller, ¡Design of Brushless Permanent Magnet Motors¡, Clarendon Press. Oxford, 1994.

[2] Takashi Kenjo, D. Eng, ¡Permanent Magnet and Brushless DC Motor¡, Sogo Electronics Publishing Company, 1984.

[3] Ramesh Srinivasan, Ramesh Oruganti, "A Unity Power Factor Converter Using Half-Bridge Boost Topology", IEEE Trans. on Power Electronics, Vol. 13, No. 3, May 1998.

[4] F. A. Huliehel, F. C. Lee, B. H. Cho, "Small-signal modeling of single-phase Boost high power factor converter with constant frequency control", IEEE Power Electronics Specialists Conf. Rec., June 1992, pp. 475-482

[5] J. T. Boys, A. W. Green, "Current-forced single phase reversible rectifier", Proc. Inst. Elect. Eng., Vol. 136, pt. B, No. 5, pp. 205-211, 1989.

[6] Byoung-Kuk Lee, Tae-Hyung Kim, Mehrdad Ehsani, "On the Feasibility of Four-Switch Three-Phase BLDC Motor Drives for Low Cost Commerical Applications: Topology and Control", IEEE Trans. on Power Electronics, Vol. 18, No. 1, January 2003.

978-1-4244-1871-8/07 $25.00 © 2007 IEEE

Sensorless Speed Control and Initial Rotor Position Estimation of an IPMSM

Woo-Taik. Joung, Jeong-Hum. Lee, Sang-Ho. Moon,
Hyo-Jin. Wang, Tai-Sik. Hwang, Chang-Ho. Choi
poscon R&D Center
101.Korea Techno Complex Building, Korea
University, Anam-Dong, Seongbuk-Gu, Seoul 136-713,
Korea
Email: jwt78@poscon.co.kr

Young-Seok Kim
Inha University
Dep. of Electrical Engineering, , 253,Yonghyun-dong,
Nam-gu, Inchon, Korea

Abstract— **This paper presents the initial rotor position estimation and the sensorless control for IPM synchronous motor. In sensorless drive, it is important to know the exact information of the initial rotor position. The initial position estimation method is based on the peak current changed according to the rotor position owing to the saliency. A speed sensorless control based on an instantaneous reactive power is proposed for the IPMSM drives. In the proposed algorithm, the current observer estimates the line currents and the estimated speed can be yielded from the voltage equations because the information of speed is included in back EMF. To minimize the estimation error, the speed estimator is compensated by using the instantaneous reactive power.**

I. INTRODUCTION

A permanent magnet synchronous motor (PMSM) is receiving the increased attention for many industrial applications because of its high torque to inertia ratio, superior power density, and high efficiency. In order to efficiently control the PMSM and generate smooth torque, it is necessary to know the absolute position of the rotor. These are ordinarily accomplished by using position encoders or resolvers which are mounted to the motor shaft. However, these position sensors increase the volume of the motor and reduce the reliability of the system. Also, they might not be allowed to be used in some cases. For the reasons mentioned above, these drawbacks may be eliminated if the rotor position and speed are estimated by a sensorless algorithm [1, 2].

Furthermore, the IPM synchronous motor is attracting attention from environmental problem, and various researches about it are advanced [3]. The sensorless control that is controlling the torque and the speed of the IPMSM without a position sensor is the most important theme of such researches. Because the algorithm for speed and position sensorless control of the IPMSM is deduced by the voltage equations in the *d-q* model, the accuracy of parameters in the IPMSM is an essential factor to determine control [4-6].

In this paper, the initial rotor position estimation and the design method of the proposed observer by using instantaneous reactive power in the rotating reference frame fixed to the rotor for the sensorless control of the IPMSM are presented. From experimental results, the proposed observer have a good estimation performance without really estimation errors.

II. SENSORLESS CONTROL USING AN INSTANTANEOUSLY REACTIVE POWER

A. Initial Rotor Position Estimation algorithm

The initial rotor position estimation method is based on the current peak measured by applying the pulse voltage and the current peak is changed according to the rotor position owing to the saliency of the rotor. The relation of between current and position can be expressed to as

$$
\begin{bmatrix} i_{\alpha u} \\ i_{\beta u} \end{bmatrix} = \sqrt{\frac{2}{3}} \frac{V_{DC}}{R} \begin{bmatrix} 1 - \frac{1}{2}(e^{-\frac{R}{L_q}T_s} + e^{-\frac{R}{L_d}T_s}) + \frac{1}{2}(e^{-\frac{R}{L_q}T_s} - e^{-\frac{R}{L_d}T_s})\cos 2\theta \\ \frac{1}{2}(e^{-\frac{R}{L_q}T_s} - e^{-\frac{R}{L_d}T_s})\sin 2\theta \end{bmatrix}
$$

$$
= \begin{bmatrix} I_s + \Delta I_s \cos 2\theta \\ \Delta I_s \sin 2\theta \end{bmatrix} = \begin{bmatrix} \sqrt{\frac{3}{2}} i_u \\ \sqrt{\frac{1}{2}}(i_v - i_w) \end{bmatrix} \qquad (1)
$$

$$
\begin{bmatrix} i_{\alpha v} \\ i_{\beta v} \end{bmatrix} = \begin{bmatrix} I_s + \Delta I_s \cos 2(\theta - \frac{2}{3}\pi) \\ \Delta I_s \sin 2(\theta - \frac{2}{3}\pi) \end{bmatrix} \qquad (2)
$$

$$
\begin{bmatrix} i_{\alpha w} \\ i_{\beta w} \end{bmatrix} = \begin{bmatrix} I_s + \Delta I_s \cos 2(\theta + \frac{2}{3}\pi) \\ \Delta I_s \sin 2(\theta + \frac{2}{3}\pi) \end{bmatrix} \qquad (3)
$$

B. Initial Rotor Position Estimation

To estimate the initial rotor position, each voltage vector $v(100)$, $v(010)$, $v(001)$ are applied. After applying the pulse voltage, each $i_{\alpha u}$, $i_{\alpha v}$, $i_{\alpha w}$ and $i_{\beta u}$, $i_{\beta v}$, $i_{\beta w}$ are detected. Fig.1 shows $i_{\alpha u}$, $i_{\alpha v}$, $i_{\alpha w}$ and $i_{\beta u}$, $i_{\beta v}$, $i_{\beta w}$ that shows Eg.(1~3) according to the position. From Eg.(1~3), Eg.(4) is derived. The sign of $\Delta i_{\alpha u}$, $\Delta i_{\alpha v}$ and $\Delta i_{\alpha w}$ is equal to the sign of $\cos 2\theta$, $\cos 2(\theta - 2\pi/3)$ and $\cos 2(\theta + 2\pi/3)$ in Eg.(4).

978-1-4244-1871-8/07 $25.00 © 2007 IEEE

$$\Delta i_{\alpha u} = \frac{1}{3}(2i_{\alpha u} - i_{\alpha v} - i_{\alpha w}) = \Delta I_s \cos 2\theta$$

$$\Delta i_{\alpha v} = \frac{1}{3}(2i_{\alpha v} - i_{\alpha w} - i_{\alpha u}) = \Delta I_s \cos 2(\theta - \frac{2\pi}{3}) \quad (4)$$

$$\Delta i_{\alpha w} = \frac{1}{3}(2i_{\alpha w} - i_{\alpha u} - i_{\alpha v}) = \Delta I_s \cos 2(\theta + \frac{2\pi}{3})$$

$$\Delta i_{\alpha vw} = \frac{1}{\sqrt{3}}(i_{\alpha v} - i_{\alpha w}) = \Delta I_s \sin 2\theta$$

$$\Delta i_{\alpha wu} = \frac{1}{\sqrt{3}}(i_{\alpha w} - i_{\alpha u}) = \Delta I_s \sin 2(\theta - \frac{2\pi}{3}) \quad (5)$$

$$\Delta i_{\alpha uv} = \frac{1}{\sqrt{3}}(i_{\alpha u} - i_{\alpha v}) = \Delta I_s \sin 2(\theta + \frac{2\pi}{3})$$

By the same method, β axis current equations are given by Eg.(6),(7). The sign of $\Delta i_{\beta vw}$, $\Delta i_{\beta wu}$, $\Delta i_{\beta uv}$ is equal with the sign of $\cos 2\theta$, $\cos 2(\theta - 2\pi/3)$ and $\cos 2(\theta + 2\pi/3)$.

$$\Delta i_{\beta u} = \frac{1}{3}(2i_{\beta u} - i_{\beta v} - i_{\beta w}) = \Delta I_s \sin 2\theta$$

$$\Delta i_{\beta v} = \frac{1}{3}(2i_{\beta v} - i_{\beta w} - i_{\beta u}) = \Delta I_s \sin 2(\theta - \frac{2\pi}{3}) \quad (6)$$

$$\Delta i_{\beta w} = \frac{1}{3}(2i_{\beta w} - i_{\beta u} - i_{\beta v}) = \Delta I_s \sin 2(\theta + \frac{2\pi}{3})$$

$$\Delta i_{\beta vw} = \frac{1}{\sqrt{3}}(i_{\beta v} - i_{\beta w}) = \Delta I_s \cos 2\theta$$

$$\Delta i_{\beta wu} = \frac{1}{\sqrt{3}}(i_{\beta w} - i_{\beta u}) = \Delta I_s \cos 2(\theta - \frac{2\pi}{3}) \quad (7)$$

$$\Delta i_{\beta uv} = \frac{1}{\sqrt{3}}(i_{\beta u} - i_{\beta v}) = \Delta I_s \cos 2(\theta + \frac{2\pi}{3})$$

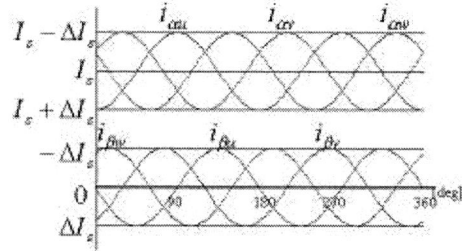

Fig. 1 Current peaks after applying of pulse voltage in three phase

Table.1 Initial position estimation equation using α axis and β axis peak current (I)

θ	$\Delta i_{\alpha u}$	$\Delta i_{\alpha v}$	$\Delta i_{\alpha w}$	Estimated angle
$-\frac{\pi}{12} \sim \frac{\pi}{12}$ $\frac{11}{12}\pi \sim \frac{13}{12}\pi$	$-$	$+$	$+$	$\dfrac{(2i_{\beta u} - i_{\beta v} - i_{\beta w})}{2(2i_{\alpha u} - i_{\alpha v} - i_{\alpha w})}$
$\frac{\pi}{12} \sim \frac{\pi}{4}$ $\frac{13}{12}\pi \sim \frac{5}{4}\pi$	$-$	$+$	$-$	$\dfrac{(2i_{\beta v} - i_{\beta w} - i_{\beta u})}{2(2i_{\alpha v} - i_{\alpha w} - i_{\alpha u})} + \frac{\pi}{6}$
$\frac{\pi}{4} \sim \frac{5}{12}\pi$ $\frac{5}{4}\pi \sim \frac{17}{12}\pi$	$+$	$+$	$-$	$\dfrac{(2i_{\beta w} - i_{\beta u} - i_{\beta v})}{2(2i_{\alpha w} - i_{\alpha u} - i_{\alpha v})} + \frac{\pi}{3}$
$\frac{5}{12}\pi \sim \frac{7}{12}\pi$ $\frac{17}{12}\pi \sim \frac{19}{12}\pi$	$+$	$-$	$-$	$\dfrac{(2i_{\beta u} - i_{\beta v} - i_{\beta w})}{2(2i_{\alpha u} - i_{\alpha v} - i_{\alpha w})} + \frac{\pi}{2}$
$\frac{7}{12}\pi \sim \frac{3}{4}\pi$ $\frac{19}{12}\pi \sim \frac{7}{4}\pi$	$+$	$-$	$+$	$\dfrac{(2i_{\beta v} - i_{\beta w} - i_{\beta u})}{2(2i_{\alpha v} - i_{\alpha w} - i_{\alpha u})} + \frac{2}{3}\pi$
$\frac{3}{4}\pi \sim \frac{11}{12}\pi$ $\frac{7}{4}\pi \sim \frac{23}{12}\pi$	$-$	$-$	$+$	$\dfrac{(2i_{\beta w} - i_{\beta u} - i_{\beta v})}{2(2i_{\alpha w} - i_{\alpha u} - i_{\alpha v})} + \frac{5}{6}\pi$

Table.2 Initial position estimation equation using α axis and β axis peak current (II)

θ	$\Delta i_{\beta vw}$	$\Delta i_{\beta wv}$	$\Delta i_{\beta uv}$	Estimated angle
$-\frac{\pi}{12} \sim \frac{\pi}{12}$ $\frac{11}{12}\pi \sim \frac{13}{12}\pi$	$-$	$+$	$+$	$\dfrac{(i_{\alpha w} - i_{\alpha v})}{2(i_{\beta v} - i_{\beta w})}$
$\frac{\pi}{12} \sim \frac{\pi}{4}$ $\frac{13}{12}\pi \sim \frac{5}{4}\pi$	$-$	$+$	$-$	$\dfrac{(i_{\alpha u} - i_{\alpha w})}{2(i_{\beta w} - i_{\beta v})} + \frac{\pi}{6}$
$\frac{\pi}{4} \sim \frac{5}{12}\pi$ $\frac{5}{4}\pi \sim \frac{17}{12}\pi$	$+$	$+$	$-$	$\dfrac{(i_{\alpha v} - i_{\alpha u})}{2(i_{\beta u} - i_{\beta v})} + \frac{\pi}{3}$

θ				
$\frac{5}{12}\pi \sim \frac{7}{12}\pi$				
$\frac{17}{12}\pi \sim \frac{19}{12}\pi$	+	−	−	$\frac{(i_{\alpha w} - i_{\alpha v})}{2(i_{\beta v} - i_{\beta w})} + \frac{\pi}{2}$
$\frac{7}{12}\pi \sim \frac{3}{4}\pi$				
$\frac{19}{12}\pi \sim \frac{7}{4}\pi$	+	−	+	$\frac{(i_{\alpha u} - i_{\alpha w})}{2(i_{\beta w} - i_{\beta u})} + \frac{2}{3}\pi$
$\frac{3}{4}\pi \sim \frac{11}{12}\pi$				
$\frac{7}{4}\pi \sim \frac{23}{12}\pi$	−	−	+	$\frac{(i_{\alpha v} - i_{\alpha u})}{2(i_{\beta u} - i_{\beta v})} + \frac{5}{6}\pi$

Table.1, 2 shows the estimation method and equation of initial rotor position.

$$\theta \cong \begin{bmatrix} \dfrac{1}{2}\tan 2\theta \\[2mm] \dfrac{1}{2}\tan 2\theta + \pi \end{bmatrix} \qquad (8)$$

Two kind of estimated position in Eq.(8) have a difference of π rad because of polarity. To identify the actual rotor position from two, the voltage vectors $v(100)$ and $v(011)$ are applied to the motor during the long time T_L at standstill under zero initial current. The voltage vector causes a magnetic saturation if the flux added to the permanent magnet flux, and does not cause a magnetic saturation, if the flux is subtracted from the permanent magnet flux. Fig.2 shows he u-phase current peak i_{uL}^{+} and i_{uL}^{-} after applying $v(100)$ and $v(011)$, respectively.

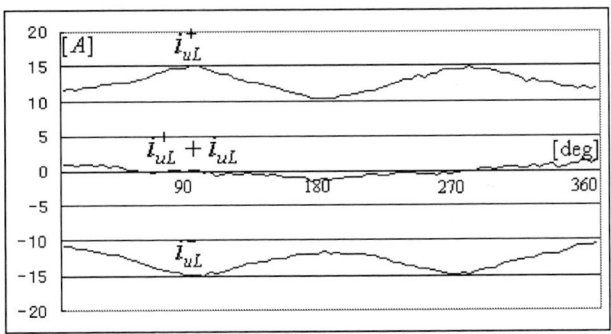

Fig. 2 u-phase Current peaks
after applying of pulse voltage vector

i_{vL}^{+} and i_{vL}^{-} are the line current peak after applying the voltage vector $v(010)$ and $v(101)$ during the long time T_L.

i_{wL}^{+} and i_{wL}^{-} are the line current peak after applying the voltage vector $v(001)$ and $v(110)$ during the long time T_L.

Table.3 shows the discrimination between two estimated positions for all domains.

Table.3 Discrimination of polarity

θ	Voltage vector	Discrimination
$-\dfrac{\pi}{6} \sim \dfrac{\pi}{6}$	$v(100)$	$i_{uL}^{+} + i_{uL}^{-} > 0$
$\dfrac{5}{6}\pi \sim \dfrac{7}{6}\pi$	$v(011)$	$i_{uL}^{+} + i_{uL}^{-} < 0$
$\dfrac{\pi}{6} \sim \dfrac{\pi}{2}$	$v(001)$	$i_{wL}^{+} + i_{wL}^{-} < 0$
$\dfrac{7}{6}\pi \sim \dfrac{3}{2}\pi$	$v(110)$	$i_{wL}^{+} + i_{wL}^{-} > 0$
$\dfrac{\pi}{2} \sim \dfrac{5}{6}\pi$	$v(010)$	$i_{vL}^{+} + i_{vL}^{-} > 0$
$\dfrac{3}{2}\pi \sim \dfrac{11}{6}\pi$	$v(101)$	$i_{vL}^{+} + i_{vL}^{-} < 0$

C. Sensorless Control

The voltage equations of the IPMSM in the stationary reference frame fixed to the stator are given by Eg.(9).

$$\begin{bmatrix} v_\alpha \\ v_\beta \end{bmatrix} = R\begin{bmatrix} i_\alpha \\ i_\beta \end{bmatrix} + p\begin{bmatrix} L_1 + L_2\cos 2\theta & L_2\sin 2\theta \\ L_2\sin 2\theta & L_1 - L_2\cos 2\theta \end{bmatrix}\begin{bmatrix} i_\alpha \\ i_\beta \end{bmatrix} + \omega K_E\begin{bmatrix} -\sin\theta \\ \cos\theta \end{bmatrix} \qquad (9)$$

where, $L_1 = \dfrac{L_d + L_q}{2}$, $L_2 = \dfrac{L_d - L_q}{2}$

In order to compose the current observer, it does coordinate conversion to eliminate the term of 2θ.

$$\frac{d\mathbf{i}_\alpha}{dt} = -\frac{R}{L_d}\mathbf{i}_\alpha - \frac{\omega(L_d - L_q)}{L_d}\mathbf{i}_\beta + \frac{K_E}{L_d}\omega\sin\theta + \frac{1}{L_d}\mathbf{V}_\alpha$$

$$\frac{d\mathbf{i}_\beta}{dt} = -\frac{R}{L_q}\mathbf{i}_\beta - \frac{\omega(L_d - L_q)}{L_q}\mathbf{i}_\alpha - \frac{K_E}{L_q}\omega\cos\theta + \frac{1}{L_q}\mathbf{V}_\beta \qquad (10)$$

where, v_α : stator D - axis voltage
v_β : stator Q - axis voltage
i_α : stator D - axis current
i_β : stator Q - axis current
ω : rotor speed
θ : rotor position
R : stator resistance
L_d : inductance of d-axis
L_q : inductance of q-axis
K_E : back EMF constant

From (10), the current observer (11) is derived.

$$\frac{d\hat{\mathbf{i}}_\alpha}{dt} = -\frac{R}{L_d}\hat{\mathbf{i}}_\alpha - \frac{\hat{\omega}(L_d - L_q)}{L_d}\hat{\mathbf{i}}_\beta + \frac{K_E}{L_d}\hat{\omega}\sin\hat{\theta} + \frac{1}{L_d}\mathbf{V}_\alpha + k_1\mathbf{e}_\alpha$$

$$\frac{d\hat{\mathbf{i}}_\beta}{dt} = -\frac{R}{L_q}\hat{\mathbf{i}}_\beta - \frac{\hat{\omega}(L_d - L_q)}{L_q}\hat{\mathbf{i}}_\alpha - \frac{K_E}{L_q}\hat{\omega}\cos\hat{\theta} + \frac{1}{L_q}\mathbf{V}_\beta + k_2\mathbf{e}_\beta \qquad (11)$$

where, $\hat{\mathbf{i}}_\alpha$: stator D - axis estimated current

$\hat{\mathbf{i}}_\beta$: stator Q - axis estimated current

k_1, k_2 : constant

$$\mathbf{e}_\alpha = \hat{\mathbf{i}}_\alpha - \mathbf{i}_\alpha$$

$$\mathbf{e}_\beta = \hat{\mathbf{i}}_\beta - \mathbf{i}_\beta \tag{12}$$

$$\frac{d\mathbf{e}_\alpha}{dt} = -\frac{R}{L_d}\mathbf{e}_\alpha - \frac{L_d - L_q}{L_d}(\hat{\omega} - \omega)\mathbf{e}_\beta + \frac{K_E}{L_d}(\hat{\omega}\sin\hat{\theta} - \omega\sin\theta) + k_1\mathbf{e}_\alpha$$

$$\frac{d\mathbf{e}_\beta}{dt} = -\frac{R}{L_q}\mathbf{e}_\beta - \frac{L_d - L_q}{L_q}(\hat{\omega} - \omega)\mathbf{e}_\alpha - \frac{K_E}{L_q}(\hat{\omega}\cos\hat{\theta} - \omega\cos\theta) + k_2\mathbf{e}_\beta$$

(13)

From (13), if the back-emf estimated errors are minimized, then the current estimation errors can be converged to zero, k_1 and k_2 are determined so that error equations stabilize and the algorithm which minimizes speed estimation errors can be developed.

D. Speed estimation based on an instantaneous reactive power

In the rotating reference frame fixed to the rotor, the voltage equations of the IPMSM can be expressed as

$$\mathbf{v}_d = (R + pL_d)\mathbf{i}_d - \omega L_q\mathbf{i}_q$$

$$\mathbf{v}_q = (R + pL_q)\mathbf{i}_q + \omega L_d\mathbf{i}_d + K_E\omega \tag{14}$$

From (14), the rotor speed (15) is derived.

$$\omega = \frac{\mathbf{V}_q - (R + pL_q)\mathbf{i}_q}{K_E + L_d\mathbf{i}_d} \tag{15}$$

If the speed of the rotor is estimated from (15), the speed estimation error between the actual speed and the estimated speed is occurred by errors due to measuring motor parameters and sensing line-currents. Therefore, (15) has to be compensated to minimize the speed estimation error. (16) is equation which compensates (15) and the value C is determined by a instantaneous reactive power.

$$\hat{\omega} = \frac{\mathbf{V}_q - (R + pL_q)\mathbf{i}_q}{K_E + L_d\mathbf{i}_d} + C \tag{16}$$

The instantaneous reactive power is defined by the cross-product of the line-current and the back-emf E_S,

$$\mathbf{q}_m = \mathbf{i}_S \times \mathbf{E}_S \tag{17}$$

where, $\mathbf{i}_S = [\mathbf{i}_d, \mathbf{i}_q]$

$\mathbf{E}_S = [\mathbf{E}_d, \mathbf{E}_q]^T$

$\mathbf{E}_d = 0, \mathbf{E}_q = K_E\hat{\omega}$

The estimated instantaneous reactive power defined by the estimated current is as follows.

$$\hat{\mathbf{q}}_m = \hat{\mathbf{i}}_S \times \mathbf{E}_S \tag{18}$$

where, $\hat{\mathbf{i}}_S = [\hat{\mathbf{i}}_d, \hat{\mathbf{i}}_q]$

In the rotating reference frame fixed to the rotor, if the estimated current has the position estimation error $\Delta\theta$ for the actual current, (17) and (18) are represented as (19) and (20)

$$\mathbf{q}_m = \mathbf{i}_d K_E\hat{\omega} \tag{19}$$

$$\hat{\mathbf{q}}_m = \hat{\mathbf{i}}_d K_E\hat{\omega} = (\mathbf{i}_d\cos\Delta\theta + \mathbf{i}_q\sin\Delta\theta)K_E\hat{\omega} \tag{20}$$

If the condition of (21) is satisfied, (20) can be approximated as (22).

$$\omega \neq 0 , \Delta\theta \cong 0 \tag{21}$$

$$\hat{\mathbf{q}}_m = (\mathbf{i}_d + \mathbf{i}_q\Delta\theta)K_E\hat{\omega} \tag{22}$$

From (19) and (22), the error of the instantaneous reactive power is derived as (23) and represented by the term which includes the information for a position estimation error.

Let us suppose that the sign of $\Delta\theta$ is positive when the estimated current leads to the actual current. The value C has to be determined by satisfying (24), (25) and adding the integral term to accomplish stable compensation.

$$\Delta\mathbf{q}_m = \hat{\mathbf{q}}_m - \mathbf{q}_m = \mathbf{i}_q\Delta\theta K_E\hat{\omega} \tag{23}$$

$$\Delta\theta > 0, K_{CP}\mathbf{i}_q\Delta\theta K_E\omega < 0 \tag{24}$$

$$\Delta\theta < 0, K_{CP}\mathbf{i}_q\Delta\theta K_E\omega > 0 \tag{25}$$

$$C = K_{CP}\Delta\mathbf{q}_m + K_{CI}\int\Delta\mathbf{q}_m d\tau \tag{26}$$

where, K_{CP}, K_{CI} : constant

If the condition of (21) is satisfied, error dynamics equations are obtained as (27).

$$\frac{d\mathbf{e}_\alpha}{dt} = -\frac{R}{L_d}\mathbf{e}_\alpha - \frac{L_d - L_q}{L_d}(\hat{\omega} - \omega)\mathbf{e}_\beta + \frac{K_E}{L_d}(\hat{\omega} - \omega)\sin\theta + k_1\mathbf{e}_\alpha$$

$$\frac{d\mathbf{e}_\beta}{dt} = -\frac{R}{L_q}\mathbf{e}_\beta - \frac{L_d - L_q}{L_d}(\hat{\omega} - \omega)\mathbf{e}_\alpha - \frac{K_E}{L_q}(\hat{\omega} - \omega)\cos\theta + k_2\mathbf{e}_\beta$$

(27)

If the speed estimation error is minimized by (26), (27) can be represented as (28) respectively. If k_1 and k_2 are determined to satisfy inequality (29), current estimation errors of (28) is converged to zero.

$$\frac{d\mathbf{e}_\alpha}{dt} = -(\frac{R}{L_d} - k_1)\mathbf{e}_\alpha$$

$$\frac{d\mathbf{e}_\beta}{dt} = -(\frac{R}{L_q} - k_2)\mathbf{e}_\beta \tag{28}$$

$$k_1 < \frac{L_d}{R}, k_2 < \frac{L_q}{R} \tag{29}$$

The line current has estimated by the observer in (11). The actual and the estimated current are transforms from the stationary reference frame to the rotating reference frame fixed to the rotor. The real reactive power and the estimated reactive power are obtained from (18) and (19), respectively. And then, (26) determines the value C from the difference of reactive power.

The speed error terms that occur due to the difference between the d-axis and the q-axis inductance in the IPMSM is eliminated when the sensorless control is used the reactive power of the rotating reference frame fixed to the rotor.

III.   EXPERIMENTAL RESULTS

The parameters of the IPMSM are listed in Table.4.

Table.4 Motor parameters

Rated power	2.5	kW
Pole number	8	Poles
Torque rating	11.9	N m

Rated speed	2000	Rpm
d-axis Inductance	1.61	mH
q-axis Inductance	1.31	mH

Fig. 3 shows u-phase current responses during the initial rotor position. The first current response are obtained for $v(100)$. Using the current peaks, two kind of estimated positions are calculated according to the Table. 1 and 2. The fourth and fifth current responses are obtained for $v(100)$ and $v(011)$. Using the sign of i_{uL}^{+} and i_{uL}^{-}, the initial rotor position is selected from two estimated position. Fig.4 and Fig. 5 shows the actual position and the estimated position from 0° to 360°. The maximum estimation error is 8.52° when using Table.1. The maximum estimation error is 9.99° when using Table.2.

Fig. 6 shows the reference and the actual speed of the sensorless control with 80% load at 500[rpm]. The reference speed is the upper side and actual speed is the lower side. The motor is started with no load and the load is changed to 80% load after 5 sec. The reference speed is well converged the actual speed. Fig. 7 (a) shows the reference and the actual speed of the sensorless control with no load at 2000[rpm] and Fig. 7 (b) shows the reference and the actual speed of the sensorless control with 80% load at 2000[rpm]. The load is changed to 80% load after 8 sec. Fig.8 shows the line current and the actual speed when the load is changed 0% to 80% at 1000[rpm]. The speed decreased when the load was applied but after 2sec, the speed reached steady state. It shows that the senseless algorithm was robust for a disturbance. Fig. 9 shows the reference and the actual speed when the speed is changed from 500[rpm] to -500[rpm]. Although the speed changes suddenly, the actual speed follows well the reference speed. Fig. 10 shows the actual speed and current of the q-axis when no load at 500[rpm]. Fig. 11 shows the reference and the actual speed of the sensorless control with no load at 100[rpm]. Fig. 12 shows the reference and the actual speed when the speed is changed from 100[rpm] to -100[rpm]. At the start in the low speed region, the actual speed has the speed error but the actual speed quickly converged to the real speed.

IV. CONCLUSION

In this paper, the estimation method of the initial rotor position and secsorless control was accomplished. The initial rotor position is estimated by using the current response because current peak is changed according to the rotor position owing to the saliency of the rotor. The accuracy of the estimation algorithm is confirmed through the experimental results of Fig.4 and Fig. 5

A speed sensorless control based on an instantaneous reactive power is proposed for the IPMSM drives. To compose current observer, voltage equations were used in the stationary reference frame fixed to the stator. Also, to eliminate the speed error terms due to the difference between the d-axis and the q-axis inductance, the reactive power of rotating reference frame fixed to the rotor was used. The accuracy of the sensorless control algorithm is confirmed through the experimental results that compare the actual speed with the reference speed.

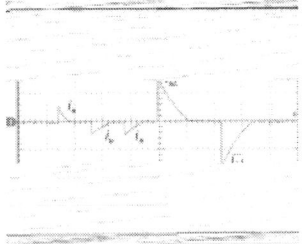

Fig. 3 shows u-phase current responses
during the initial rotor position

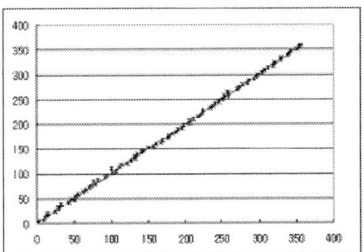

Fig. 4 actual position and the estimated position
from 0° to 360° when using Table. 1

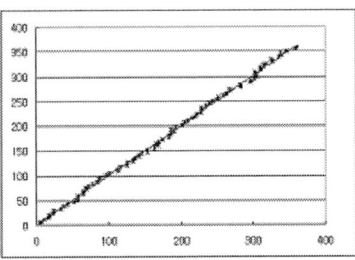

Fig. 5 actual position and the estimated position
from 0° to 360° when using Table. 2

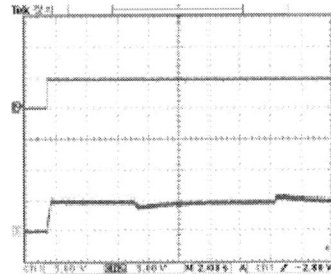

Fig. 6 The reference and actual speed when load is changed
from 0% to 80% and returns to 0% at 500[rpm]

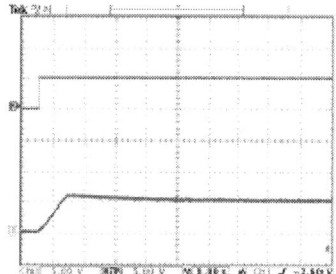

(a) The reference and the actual speed
when no load at 2000[rpm]

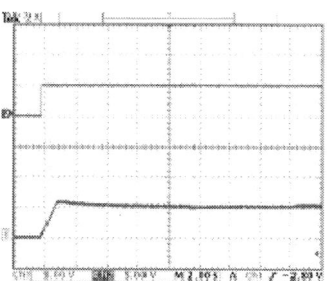

(b) The reference and the actual speed when load is changed
from 0% to 80% and returns to 0% at 2000[rpm]

Fig. 7 The reference and the actual speed when the IPMSM is
derived at 2000[rpm] with no load and 80% load

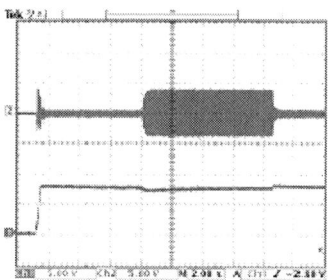

Fig. 8 The line current and the actual speed when load is
changed from 0% to 80% and returns to 0% at 1000[rpm]

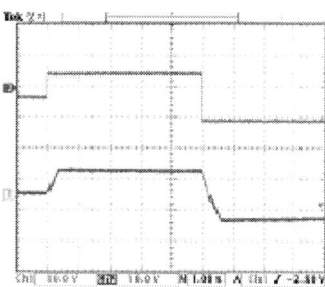

Fig. 9 The reference and the actual speed when speed is
reversed from 500[rpm] to -500[rpm]

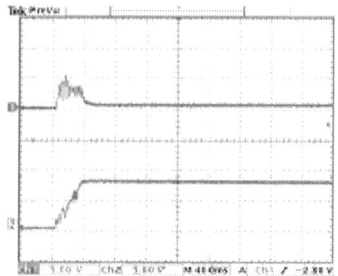

Fig. 10 The actual speed and current of the q-axis

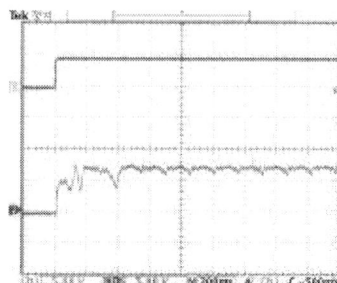

Fig. 11 The reference and the actual speed when no load at
100[rpm]

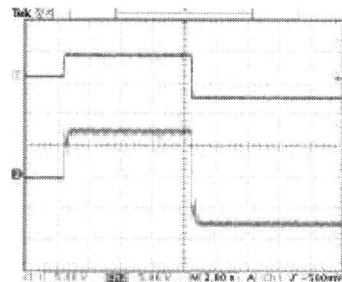

Fig. 12 The reference and the actual speed when speed is
reversed from 100[rpm] to -100[rpm]

REFERENCES

[1] R. Wu, G.R. Selmon, "A Permanent magnet motor drive without a shaft sensor", IEEE Trans. I. A., vol.27, no.5, pp 1005-1011, 1991.

[2] N. Matsui, M. Shigyo , "Brushless DC motor control without position and speed sensors", IEEE Trans, I. A., vol.28, no.1, pp.120-127, 1992.

[3] Jun Hu, Dongqi Zhu, YongDong Li, Jingde Gao, "Application of sliding observer to sensorless permanent magnet synchronous motor drive system", IEEE PESC, pp 532-536, 1994.

[4] R.B. Sepe, J.H. Lang, "Real-time observer-based(Adaptive) control of a permanent-magnet synchromous motor without mechanical sensor", IEEE Trans. I.A., vol.28, no.6, pp. 1345-1352, 1992.

[5] T. Furuhshi, S. sangwongwanich, S. Okuma "A position-and-velocity sensorless control for Brushless DC motors using an adaptive sliding mode observer", IEEE Trans. I.E, vol.39, no.2, pp.89-95, 1992.

[6] R. Dhaouadi, N. Mohan, L. Norum, "Design and Implementation of an Extended Kalman Filter for state Estimation of a Permanent Magnet Synchronous Motor", IEEE Trans. on P. E., vol.6, no.3, pp.491-497, 1991.

[7] T.S. Low, T.H. Lee, K.T. Chang, "A Nonlinear Speed Observer for Permanent-Magnet Synchronous Motor", IEEE Trans. on I.E., vol.40, no.3, pp.307-316, 1993

Rotor Position Detection Method of a Single Phase Switched Reluctance Motor Using Back EMF

Han-Geol Sun[*], Duck-Sick Shin[**], Hyong-Yeol Yang[***], Young-Cheol Lim[*]

Chonnam National University[*]
Korea Electronics Technology Institute[**]
Honam University[***]
cuteboy80@paran.com

Abstract—This paper presents the position estimation and driving method of single phase Switched Reluctance Motor(SRM) using Back EMF. To drive SRM, the information about position of a rotor must be fully needed. Generally, a rotor's position sensor is used to detect the rotor position. However, most of position sensors are complex and increase the constructing cost of the system. Moreover, these sensors tend to decrease reliability of driving system. Therefore, a detecting method of a rotor's specific position using Back EMF is supposed to solve those problems mentioned above. When a rotor pole and a stator pole are overlapped, a variation occurs in the Back EMF waveform and the instant of the overlap can be detected. Therefore the rotor position and speed can be calculated and the SRM is driven by appropriate switching. The validity of presented method is verified through simulation and experiment.

Key Words : Single phase SRM, overlap, Back EMF

I. INTRODUCTION

Switched Reluctance Motor is a motor that has the merit of high efficiency, high speed drive, simple electro-magnetic structure and easy variable speed control, and as recently the rapid technological development of power semiconductor is in full swing, its practical value gets to be elevated ever more.[7]

SRM needs the exact locating information of rotor as it works by exciting proper connecting coil upon the location of rotor. As for the sensor detector, mainly encoder, resolver, hall sensor, resolver and opto-interrupter are generally used.

However these location sensors increased the size of motor and cause to increase price, and decrease the credibility of whole system in poor surroundings of like EMI, dust, high temperature.[2]-[5] To solve these problems, the research of 3-phase SRM equipped with search coil and single-phase SRM are conducted.[6]

And the method of detecting location of rotor using search coil also is followed by inconvenience of having to install search coil unto motor and, as a result, manufacturing progress should be added to have increased producing price.

In this paper I'd like to suggest the method of driving by detecting specific location of rotor by using back EMF of steady state battery terminal voltage without additional attaching on motor. Until now, although mainly the BLDC application method has been used, there hasn't been any case

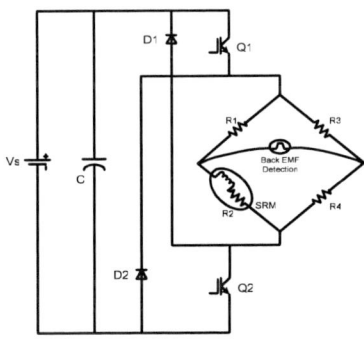

Fig. 1 Single Phase SRM Drive Circuit

of applications of method of detecting rotor location using back EMF.

Before testing a simulation was carried out using Maxwell 2D™ which is electro-magnetic analyzing program. And the possibility of detecting rotor's location by way of using unusual waveform of back EMF according to rotor's location is analyzed and verified the method suggested through testing and examination.

II. DRIVING CIRCUIT

In this paper, the asymmetric inverter is used which has excellent control capacity as single-phase 6/6 SRM motor circuit and is capable of economic drive minimizing some of switches.

Fig 1 shows a drive circuit diagram of the single-phase 6/6 SRM used in this paper. The back EMF can be obtained by composes wheatstone bridge circuit to measure the resistance value of the motor and detecting like Fig 1.

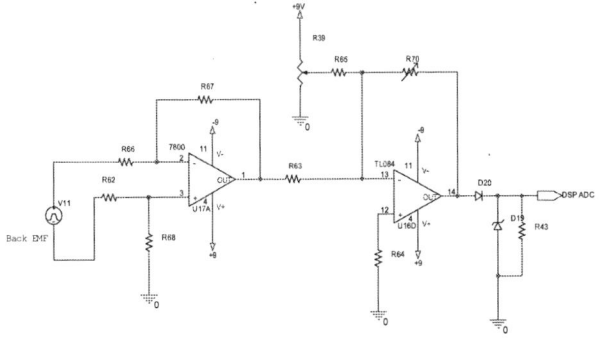

Fig. 2 Back EMF detection circuit

978-1-4244-1871-8/07 $25.00 © 2007 IEEE   393

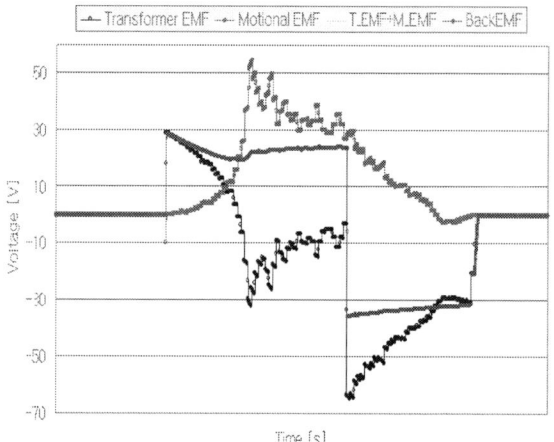

Fig. 3 Analysis waveform of the back EMF

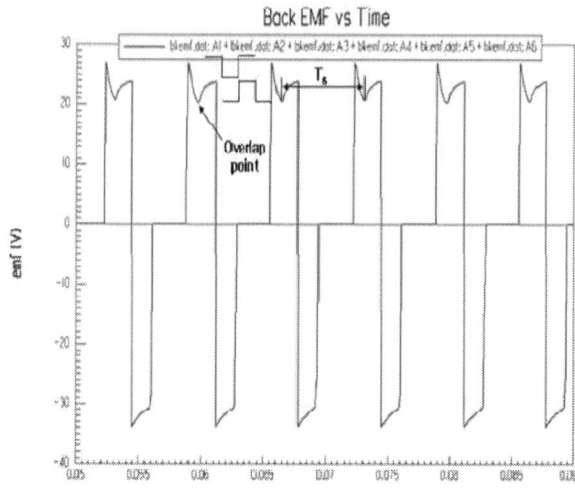

Fig. 4 Simulation waveform of back EMF

The theory of this inverter is that it can make electric current disappear and prevent following torque before current disappear and prevent following torque before inductance changed into negative stiffness inclining by restoring flowing current, that is, energy on winding toward sources of electricity while turning off both of the switches before rotor and stator aligned and by turning on Q1 and Q2 inputting DC power voltage.

III.   DETECTING METHOD OF BACK EMF

As shown in Fig 1, back EMF is obtained through detecting circuit of back EMF in Fig 2 by compose wheatstone bridge circuit and detects voltage. Back EMF can be extracted by letting only +part of back EMF into input of AD converter, regulating amplification factor set to 3V of AD converter's input voltage of DSP through OP amp in this circuit.

In this paper, the method to detect rotor's location with small amount of expense for getting back EMF through circuit composed only by resistor and OP amp without using high priced voltage sensor is suggested.

This is because, comparing with widely used hall sensor or opto-interrupter, it is less than half of price and it is not followed by additional work process of changing of apparatus part to attach sensor, so expectation of curtailment of expense is possible and the motor's size can be reduced.

IV.   THE ANALYSIS OF WAVEFORM OF BACK EMF

SRM enhances magnetic flux from the overlap point of rotor and stator's meeting, and when the rotor and stator completely aligned, it becomes maximum and from then begins to decrease.

Fig 3 shows a simulation waveform that analyzed the component of back EMF by using Maxwell 2D™. As shown in the Fig 3, back EMF is represented as the sum of Transformer EMF and Motional EMF. Therefore, we come to know that back EMF is organized by the theory of transformer and generator.

Time varying magnetic flux upon change of phase current is Transformer EMF and time varying magnetic flux happening as for change of magnetic inductance by rotating of rotor is Motional EMF.

Provided neglecting non-linear of iron core and resistance of winding while working certain angular speed of ω, it can be expressed as type(1).

$$e = \frac{d\lambda}{dt} = L\frac{di}{dt} + i\frac{dL}{dt} = L\frac{di}{dt} + i\frac{dL}{d\theta}\omega \qquad (1)$$

L : Inductance

i : Current of Phase winding

ω : rotating speed

As in the Fig 3, in the first clause of type (1), which is at first driving, the switch of Transformer EMF is on, then the electric current starts flowing rapidly resulting showing big value. But from the point of passing overlap point, the first clause gets decreased as electric current is decreasing because of increasing of inductance. On the contrary, at the second clause in type (1), because of low speed it appears with low value and it gets to be increased more and more while showing low value at first. At normal condition it shows likewise.

We get to know that the crossing point of these two EMF is overlap point and by detecting this part rotor's location can be supposed.

V.   ON-OFF METHOD BY EXTRACTING OVERLAP POINT

We come to know in Fig 4 that, when simulated, overlap point which back EMF in every pulse changing rapidly appears with back EMF waveform. So with this part detecting, the location of rotor can be presumed.

In case of 6-terminal single phase SRM, overlap point is

978-1-4244-1871-8/07 $25.00 © 2007 IEEE   394

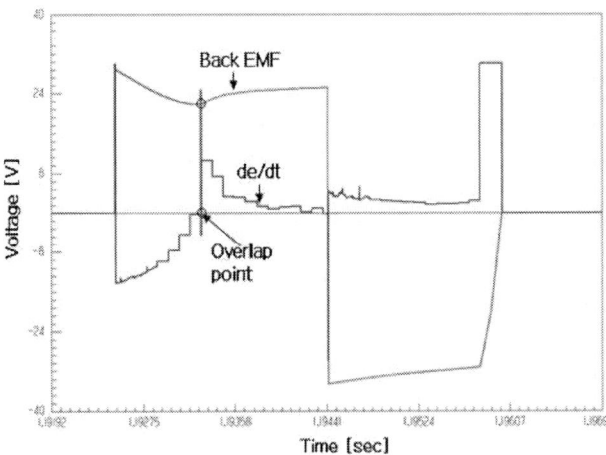

Fig. 5 Back EMF and differential waveform

be

Fig. 6 Single phase SRM for experimentation

detected in every 60°. So when 6 times of overlap point is detected, we know that it rotated one round. It shows the section ON which represent positive(+) of back EMF and only this section has information of rotor. T_s is the value that represents overlap point by time upon present and previous' ON of overlap point, and the value of division of 60° by T_s represents present speed.

Therefore, the speed is calculated by following formula.

$$\omega = \frac{d\theta}{dt} = \frac{10}{INT_s \times T_i}[rpm] \qquad (2)$$

INT_s : Interrupt number while 60 degree rotation

$T_i[sec]$: Interrupt(sampling) period

If we detect overlap point knowing present speed, switch off is possible at the angle of θ_{off} after detecting by the type (3).

$$n_{off} = \frac{INT_s}{60^o} \times \theta_{off} \qquad (3)$$

n_{off} : When becoming θ_{off} interrupt number

To get switch-on while allowing proper advance θ_{adv} in next phase current, getting switch-on will be done after interrupting to the extent of non as in type(4).

$$n_{on} = \frac{60^o - \theta_{adv}}{60^o} \times INT_s \qquad (4)$$

As for the method of detecting beginning point of torque occurring, Fig 5 shows back EMF waveform and its differential waveform. As we know from differential waveform, it's easy enough to detect the beginning point of torque occurring by way of using differential waveform. That is $\frac{de}{dt}$ value, zero crossing that transforms value from negative to positive is the overlap point. So detecting this point, we can locate starting point that rotor and stator overlap.

The back EMF inputted into controller is filtered to eliminate noise and this incline of back EMF is designed to

detected at the point that transforms from negative to positive.

VI. CONSIDERATION OF TESTING AND RESULTS

In Fig 6, electromagnet for brake is arranged to stop at location always able to stop and, as for used single-phase SRM in the test and load, ventilation fan is installed.

In Fig 7, drawing of motor system bloc, by differentiating back EMF inputted into AD converter of DSP controller, overlap point can be detected at the point that rotor and stator overlap. Detecting two times of overlap point make calculating possible and by calculating the On-Off point, we get to On-Off of inverter. I verified the precision of system by comparing presumed location of rotor using back EMF with actual location of rotor through encoder.

Fig 8 and 9 shows the drive and back EMF at normal condition, magnified back EMF, detecting signal of overlap point and waveform to phase current. Comparing the case when driving with encoder and with back EMF, we get to notice almost same waveform and as, in picture 8, because the first waveform of back EMF is in position (the situation that rotor and stator overlaps at 3°) which is able to drive when driving, it occurs in the form without overlap point. But at the second waveform we know it appears overlap point available to detect. When this back EMF amplified through OP amp, we can see overlap point appear more clearly, and at DSP drives with exact detecting this part. Therefore, with the result of test, it shows that, when driving using back EMF it is possible to drive normally without using encoder.

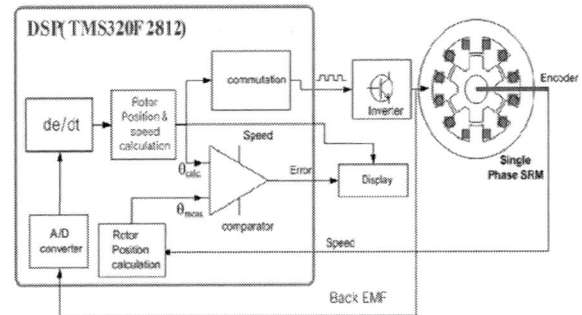

Fig. 7 System block of the experimental SRM drive

978-1-4244-1871-8/07 $25.00 © 2007 IEEE   395

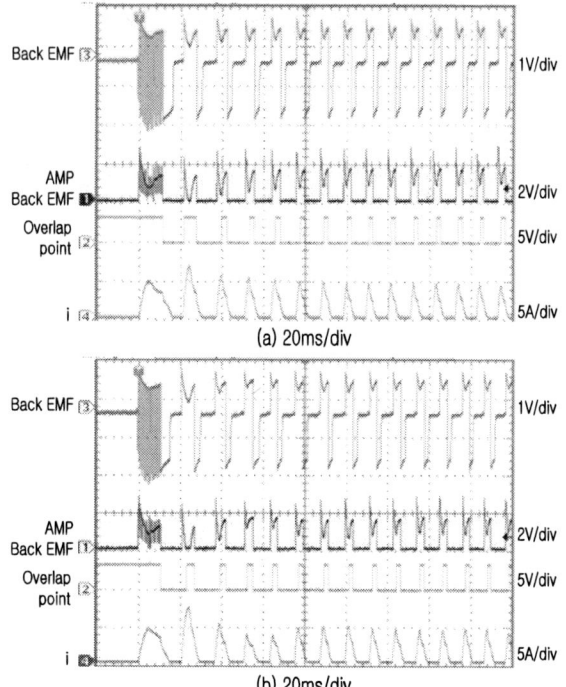

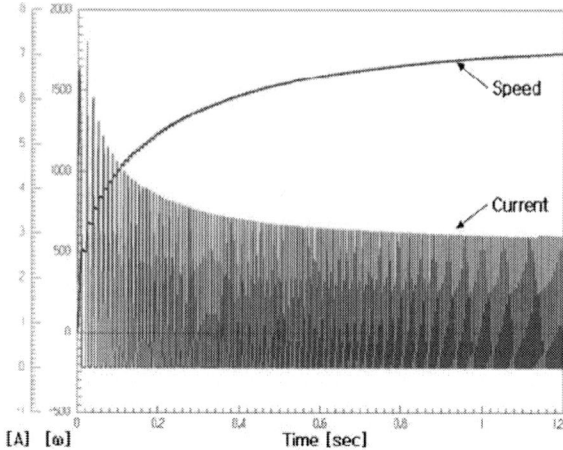

Fig. 10 Speed and current waveform of simulation

Fig 9 shows the result of driving at normal condition. Though the change of back EMF is little, we can see that it detects overlap point correctly and drives.

By using Maxwell 2D™, We simulated with the same option of motor used in the test and setting load at 0.05Nm.

Fig 10 shows that, as for simulation waveform, we can see speed curve increasing up to 1780rpm at normal condition when driving, and, at beginning, electric current flows a lot while gets stable when reached at normal condition.

Fig 11 shows estimated waveform of speed and current when driving through testing. Like the result in the simulation it can be checked that it gets increased to the normal condition through active driving at the beginning.

Fig 12 finds that, as for comparing actual location of rotor when rotating at 1780rpm with presumed rotor's location by suggested method, we know that the two results are almost agreed with each other.

Through this paper with all those testing and simulation, I showed that it is possible to drive using suggested method through testing and verified suggested method's excellence in the paper by showing the similar results with case of driving

Fig. 8 Startup : Back EMF, amplified Back EMF, Overlap point detection single and winding current
(a) Using encoder
(b) Using Back EMF

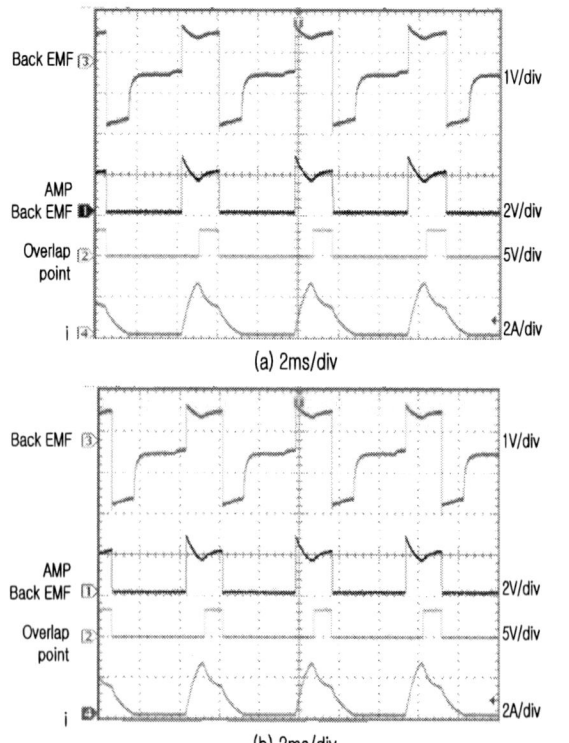

Fig. 9 Steady state : Back EMF, amplified Back EMF, Overlap point detection signal and winding current
(a) Using encoder
(b) Using Back EMF

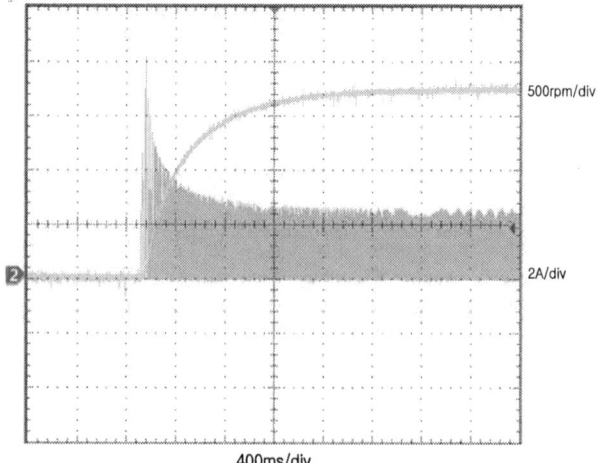

Fig. 11 Speed and current waveform of experimentation

978-1-4244-1871-8/07 $25.00 © 2007 IEEE   396

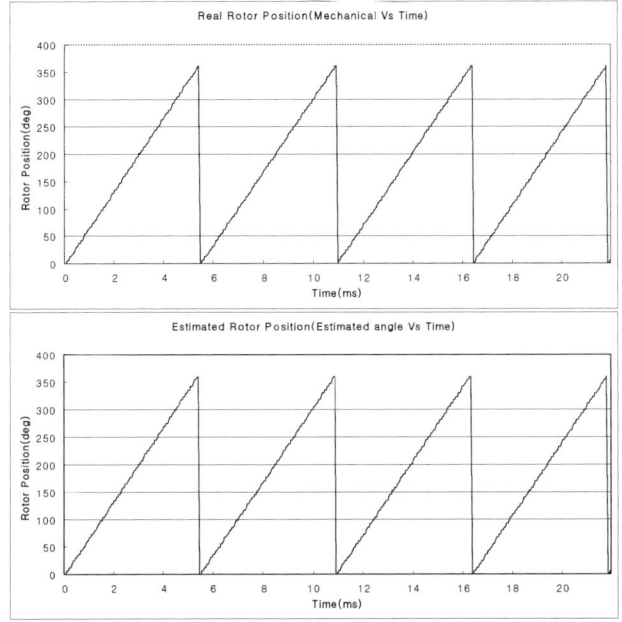

Fig. 12 Measured rotor position an encoder and estimated rotor position using encoder.

VII. CONCLUSION

This paper has suggested the method of locating rotor of simple-phase SRM by using back EMF.

Back EMF is available to detect by detecting waveform of back EMF because it has location information of rotor. Suggested method is to get location information using the change of back EMF according to location of rotor in order to assume rotor's location. The method to drive motor is suggested, that is, by detecting the part that rotor and stator begin to meet, which is overlap point, and through rapid change of Transformer EMF and Motional EMF the change of back EMF occurs, and by differentiating that part we can drive the motor. And through simulation using electromagnet translation program Maxwell 2D^{TM}, we analyzed waveform of back EMF by analyzing the change of Transformer EMF and Motional EMF we verified that it's possible to presume rotor's location.

And through testing we proved appropriateness of suggested method in this paper by comparing with method using encoder.

Because this driving method using back EMF can materialize lower price than when comparing with the case using widely used hall sensor, opt-interrupter, and doesn't accompany additional work process like changing apparatus part to attach sensor, it is considered that we can reduce the expense and size of motor and solve the problem of restricting work surrounding so as to materialize powerful character of motor driving equipment.

The method suggested in this paper is expected to draw special attention among the hard and difficult field that have trouble in using established location sensors because of vibration, high temperature, high pressure, and will be contributed to commercializing by realizing SRM's miniaturization and low price.

Acknowledgment

The participants in this research received supporting funds from the second-stage BK 21

REFERENCES

[1]   W. F. Ray and I. H. Al-Bahadly, "Sensorless method for determining the rotor position of switched reluctance motors", Fifth European Conf. on Power Electr. and Appl., Vol. 6, pp. 7-13, Sep, 1993

[2]   A. Brosse and G. Henneberger, "Different models of the SRM instate space format for sensorless control using a Kalman filter", Seventh Int. Conf. on (IEE Conf. Publ. No. 456) Power Electr. and Variable Speed Drives, pp. 269-274, Sept. 1998

[3]   A. Bellini, F. Filippetti, G. Franceschini, C. Tassoni and P. Vas, "Position sensorless control of a SRM drive using ANN-techniques", The Thirty-Third IAS Ann. Mtg. Ind. Appl. Conf., Vol. 1, pp. 709-714, Oct. 1998

[4]   H.S. Ooi and T. C. Green, "Sensorless switched reflectance motor drive with torque ripple minimization", 31st Ann. Power Electr. Specialists Conf., Vol. 3, pp. 1538-1543, June 2000

[5]   B. Fahimi, G. Suresh and M. Ehsani, "Review of sensorless control methods in switched reluctance motor drives", Conf. Rec. of the 2000 IEEE Ind. Appl. Conf., Vol. 3, pp. 1850-1857, Oct. 2000

[6]   Yang hyong-Yeol, Kim Jung-Gun, Lim young-cheol, " Rotor Position Detection of a Toroidal Switched Reluctance Motor Using Search Coils" , The Transactions of the KIPE, Vol 7, No 6, pp.537-545. 2002.

[7]   Shin Duck-Shick, Yang hyong-Yeol, Lim young-cheol, " Drive of Single-phase Switched Reluctance Motors Using Search Coils" , The Transactions of the KIPE, Vol.10, No 5, pp.488-493. 2005.

Modified Double-Modulation Signal PWM Control
for Large Motor Drive using Five-Level Converter

Noriyuki Kimura*, Tomoyuki Hamada*, Toshimitu Morizane*, Katsunori Taniguchi *
* Osaka Institute of Technology /Electrical and Electronic Systems Department, Osaka, Japan
Email: kimura@ee.oit.ac.jp

Abstract—This paper shows the newly developed modification of the double sinusoidal wave modulation used for multilevel converter. The proposed modulation makes it possible to change the modulation factor for each level of the converter to control the balance of each dc voltage and also can reduce the higher harmonics as same as sinusoidal modulation. This multilevel converter system has prospective configuration for the high voltage application such as the large motor drive without transformer.

I. INTRODUCTION

For a 3-6kV class motor drive, multi-converter system is often used. This system can reduce the converter loss and the generation of higher harmonics. However, this system requires transformers to connect converters. These transformers cause the problem of higher cost, larger space and heavier weight.

Since a multi-level converter can reduce the converter loss and the generation of higher harmonics without transformer, it is attractive for high voltage high power motor drive[1-6].

One of the most important controls for the multi-level converter is to keep the dc capacitor voltage balance. We have investigated to use the BTB converter configuration to satisfy the condition for the balance. In this paper, we propose the new PWM method for the multi-level converter suitable for the multi-level converter control.

Back-to-back configuration of 5-level converter applied for the high voltage motor drive has 4 series connected dc side capacitors. It is necessary to keep the dc capacitor voltage balance for the stable operation of the 5-level converter.

In conventional PWM method, pulse width pattern for the certain modulation factor is determined uniquely. Therefore, regulation of the current of the dc capacitor is difficult since the charging current to each capacitor cannot be controlled independently. Proposed PWM method uses two sinusoidal modulation signals. Pulse width of each voltage level is determined by comparing two shifted triangle carrier signals with these two sinusoidal modulation signals.

To have two modulation signals enables that the current of the dc capacitor can be regulated independently. However this method has one disadvantage that the total switching frequency of each device is doubled. It means

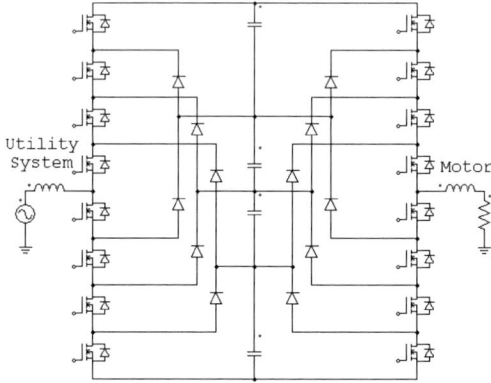

Fig.1 Single phase diagram of 5-level BTB system

that the switching loss is double of the conventional method.

To reduce the switching loss, we propose to modify the sinusoidal modulation signal. Two modulation signals have no longer sinusoidal shape, but the sum of two signals is equal to the sinusoidal shape at any moment. This modification allows having wide selection of modulation signal shape while keeping the independent capacitor charging current control and reducing the harmonics at the same time. We performed the computer simulation to verify the effectiveness by using ATP/EMTP program.

II. DOUBLE MODULATION SIGNAL PWM METHOD

Fig. 1 shows the back-to-back (BTB) configuration of 5-level converter. Voltage control method to keep each level voltage is already presented [5-7].

For 5-level converter, it is indispensable to stabilize each series connected capacitor voltage at dc side. Therefore, we have investigated the new modulation method which enables easy control of change/discharge current of each level. Waveforms in Fig.2-4 were taken by using PSIM program.

Fig. 2 shows conventional PWM method [1], which uses one sinusoidal modulation signals and four different level triangular carrier signals. Fig. 3 shows the results of our first proposed PWM method [7], which uses two sinusoidal modulation signals and two different level triangular carrier signals. Fig. 4 shows the simulation

978-1-4244-1871-8/07 $25.00 © 2007 IEEE

results of our previously proposed PWM method [7], which allows over-modulation. If over-modulation occurs in the method in Fig. 3, it causes the increase of higher harmonic components in output voltages and currents. However, in Fig. 4, two modulation signals are set to satisfy the following equation.

$$V\text{sig\_total} = V\text{sig1} + V\text{sig2} = V\text{p} \sin \omega t \quad ...(1)$$

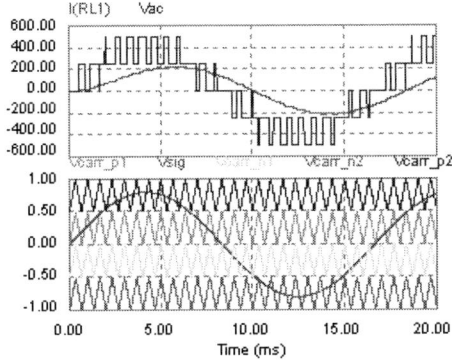

Fig.2 Results of conventional PWM method

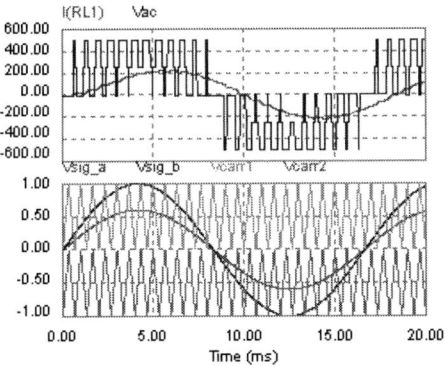

Fig.3 PWM method with double sine wave

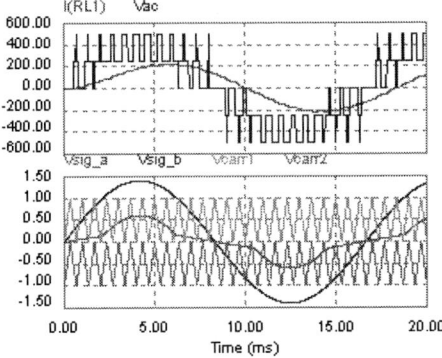

Fig.4 PWM method with modulated double sine wave

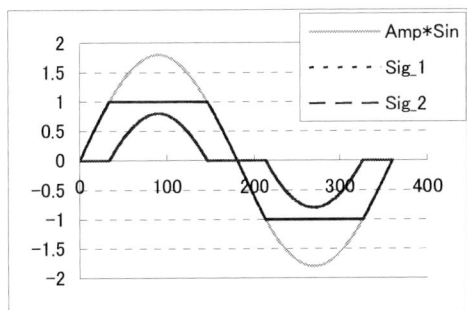

Fig.5 Modulation signal from conventional PWM method

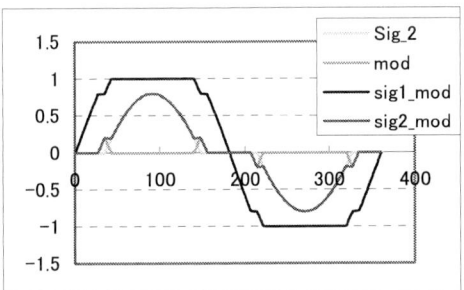

Fig.6 Modulation signal of proposed PWM method

This method can suppress the higher harmonics, since the total modulation signal is still sinusoidal.

However this method has one disadvantage that the total switching frequency of each device is almost doubled. It means that the switching loss is double of the conventional method.

III. MODIFIED DOUBLE-MODULATION SIGNAL PWM METHOD

To decrease the switching loss, the modulation signal should be kept at zero or more than 1 as long as possible. The least number of the switching is achieved by the conventional PWM method, as shown in Fig. 2. It is converted to the modulation signal of the double-modulation PWM, as shown in Fig. 5. This signal waveforms can satisfy the minimum switching condition. However, it is impossible to change the charge/discharge current of each level without changing the total modulation factor.

To make it possible to control the charge/discharge current of each level, modification signal of triangle shape (mod) is added/subtracted as shown in Fig. 6. This operation can satisfy the condition depicted by Eq. 1, and also make it possible to change the modulation factor at each level. However, this method has some disadvantages such as that the calculation of the modification is complex and ineffective when the modulation factor becomes less than 0.5.

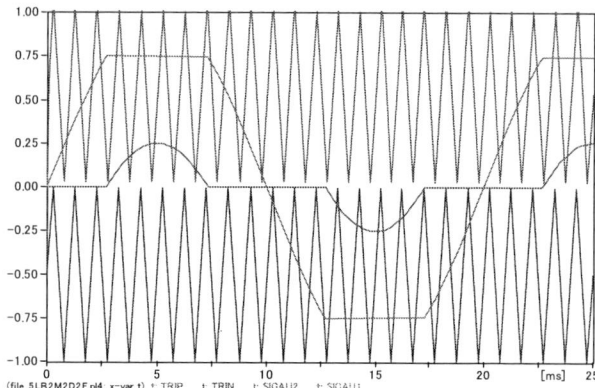

Fig.7 Modulation signal of proposed PWM
method 2

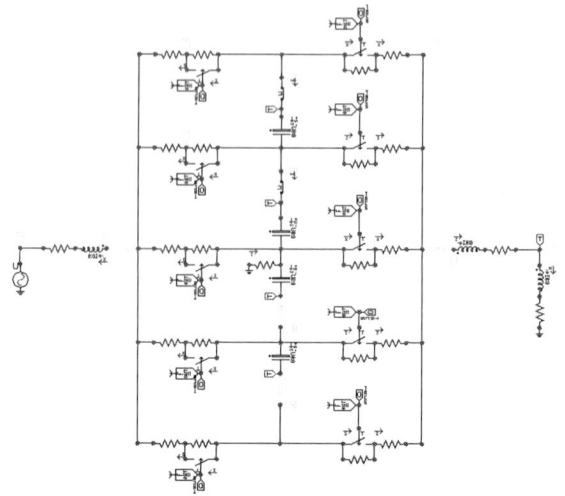

Fig.8 Equivalent circuit model of 5-level
converter in single phase

Table 1   Simulation Parameters

DC Capacitor Voltage	5	[kV]
DC Total Voltage	20	[kV]
DC Capacitance	1000 0	[uF]
AC Reactor	10	[mH]
AC Source Voltage	6.6	[kV]
Modulation Frequency	50	[Hz]
Carrier Frequency	1	[kHz]
Total Modulation Factor	1.0	
Initial Modulation Factor of centers	0.75	
Initial Modulation Factor of ends	0.25	
Load Impedance (Init.)	10	[Ω]
Load Impedance (Dec.)	5	[Ω]
Load Power Factor (end)	0.9	

In this paper, another new modification method is proposed. As shown in Fig. 7, the offset of the two modulation signals for the lower and the higher voltage level are controlled. These modulation signal waveforms are similar to the conventional ones shown in Fig. 2. The control of the offset results in the control of the charging/discharging current. It is also noticed that the switching of lower level during the higher voltage level switching occurs, and this fact cause the increase of total switching loss. The reduction of the switching loss compared with the double sinusoidal modulation signal may be half of the conventional one.

IV.   SIMULATION OF PROPOSED MODULATION

For simulation analysis, the ATP/EMTP program is used. It is noted that the switching treatment in ATP/EMTP needs some skill. It is almost impossible to treat the series connected switching devices in ATP/EMTP. Sometimes, series connected switches do not behave as theoretically expected. Hence the equivalent switching circuit shown in Fig. 8 is used in this paper and equivalence was verified from the results of 3 level converter[7].

The converter A (left side) is connected to 6.6[kV] distribution power system through the ac reactor. The converter B (right side) is connected to RL circuit, that represents the motor drive, through the ac reactor. 3 phase converter is usually used. To shorten the simulation time, single phase model is used here. The results is almost same as the three phase model except the large ripple in the dc current, which leads to the large harmonic distortion in the ac current. To reduce the influence of this ripple, larger dc capacitance than the three phase model is used. It means much smaller capacitance can be used in the real three phase converter.

DC total voltage controller and capacitor voltage balance controller are shown in Fig. 9 and 10, respectively.

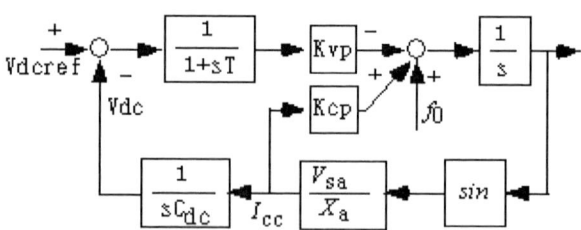

Fig.9 Total voltage controller for 5-level
converter

978-1-4244-1871-8/07 $25.00 © 2007 IEEE   400

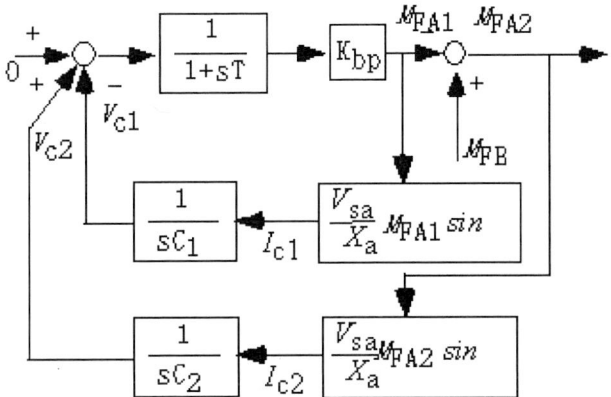

Fig.10 Capacitor voltage balance controller for 5-level converter

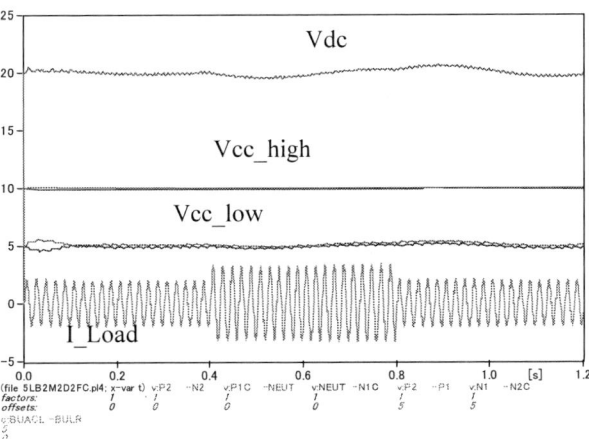

Fig.11 Total DC voltage, each capacitor voltage and AC load current

DC total voltage controller changes the PLL oscillator frequency depending on the voltage deviation. Consequently the phase difference between the converter A and the utility voltage at the connected point is changed to control the input real power, i.e. the charging current, to the converter A.

Fig. 11 shows the simulation results of the dc total voltage, the capacitor voltages and the load current. The load impedance is changed to half of the initial value at 0.4 [sec.], and recovered at 0.8 [sec.]. The dc total voltage is kept constant in stable. The capacitor voltages have a certain disturbance at the initial, there is no large disturbance at the load change.

Fig. 12 shows the simulation results of the modulation factor. The modulation factors of the converter B are kept constant to keep the output ac voltage constant. The modulation factors of the converter A are changed to keep the capacitor voltage balance. Especially, after 0.8 [sec.], the response is oscillatory.

Fig. 13 shows the simulation results of the phase difference between the converter A and the utility voltage . The phase difference looks changing largely, however the absolute value is not so large. The cause of the oscillation may be the inadequate controller gain and lack of the damping control. The oscillation of the modulation factor in Fig. 12, the cause may be the oscillation of the phase difference. Since each control block diagram shown in Fig. 9 and 10 has the other's value in the block, it is difficult to remove the interference completely. The reduction of the influence by changing the time constant or so on, is expected.

The output ac voltage waveform of the 5 level converter A is shown in Fig. 14. In Fig. 14(a), the difference of the modulation factor is minimum, and the pulse number at the higher voltage in Fig. 14(a) is more than that in Fig. 14(b), where the difference of the modulation factor is maximum.

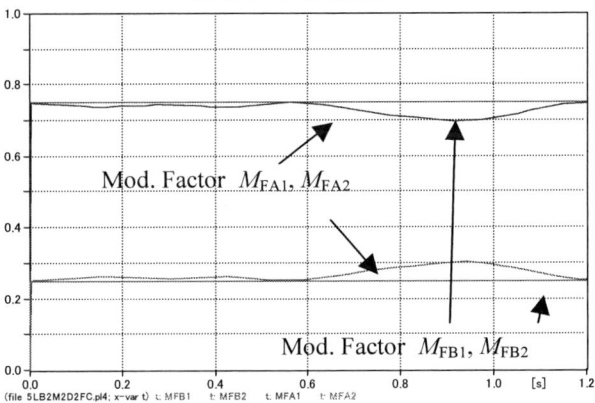

Fig.12 Simulation results of modulation factor of converter A and B

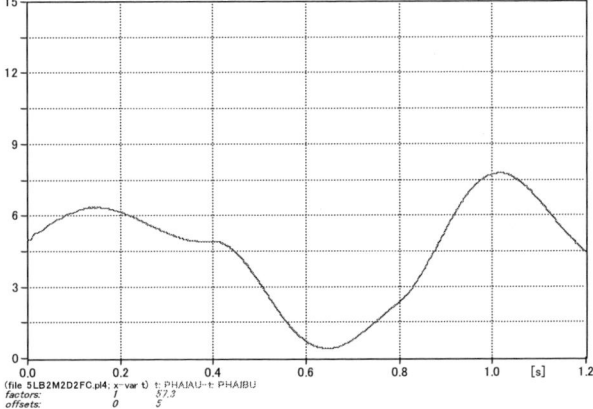

Fig.13 Phase difference between converter A and AC distribution system voltage

Therefore, it is verified that the larger difference in modulation factor cause smaller pulse number. The changes in the modulation waveforms are shown in Fig. 15.

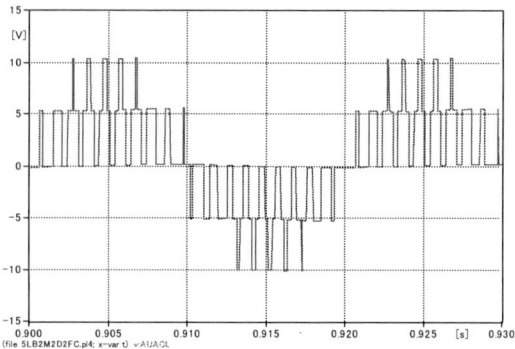

(a) Minimum difference in MF

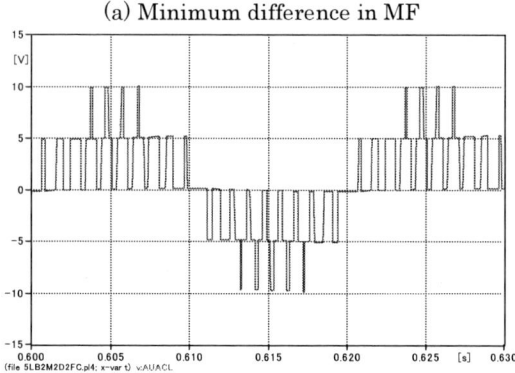

(b) Maximum difference in MF
Fig.14 5-level converter AC output voltage

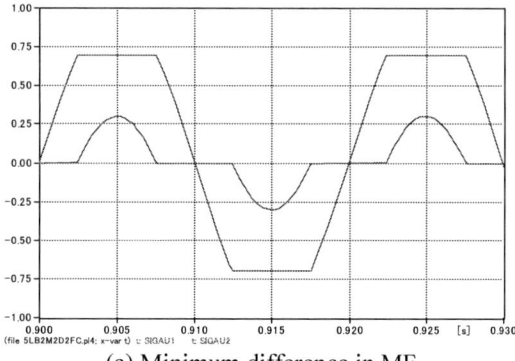

(a) Minimum difference in MF

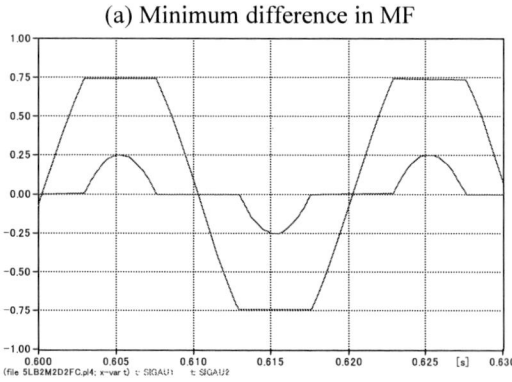

(b) Maximum difference in MF
Fig.15 Modulation signal of proposed PWM
method

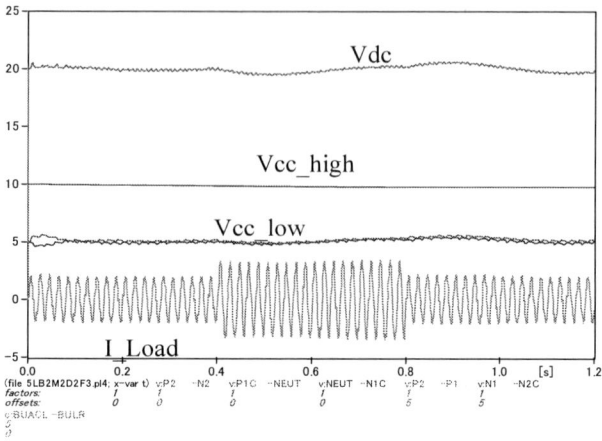

Fig.16 Total DC voltage, each capacitor voltage
and AC load current when converter B is
3-level operation

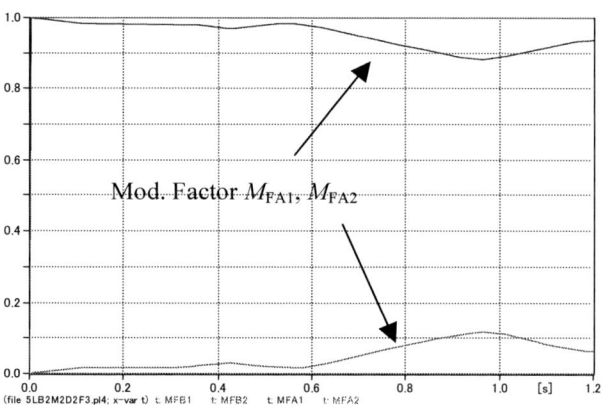

Fig.17 Simulation results of modulation factor
of converter A and B when converter B is
3-level operation

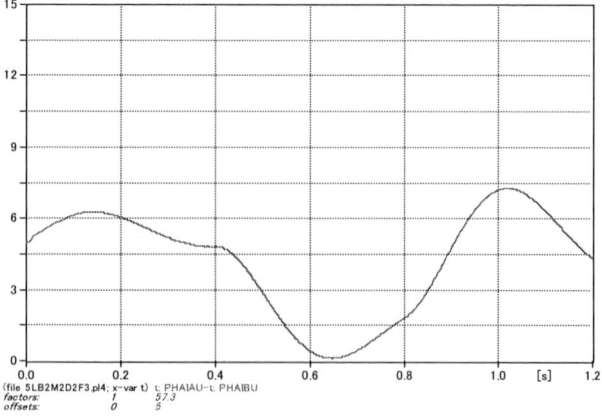

Fig.18 Phase difference between converter A
and AC distribution system voltage when
converter B is 3-level operation

978-1-4244-1871-8/07 $25.00 © 2007 IEEE   402

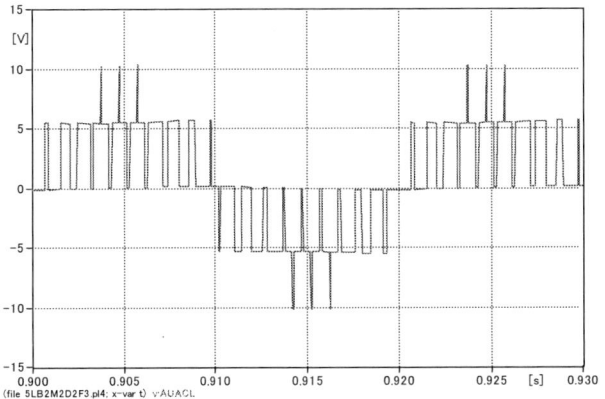

(a) Minimum difference in MF

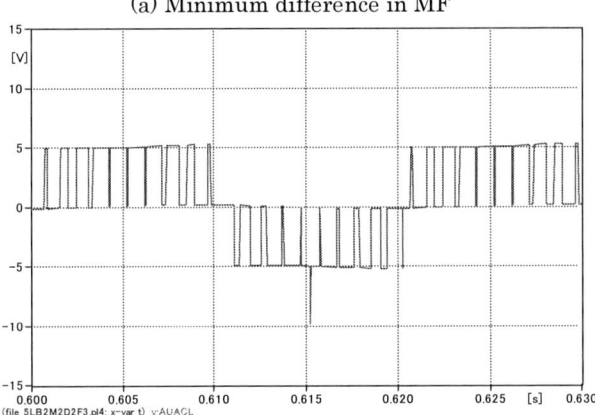

(b) Maximum difference in MF

Fig.19 5-level converter AC output voltage
when converter B is 3-level operation

If the modulation factor of lower voltage side reached to 1.0, charging to the higher voltage side capcitor is stopped. However, since the number of the switching, that is the number of the pulse, can be minimum, the operation near this condition is desirable. In this paper, to have the wider operating area, the modulation factors of the converter B are set to be 0.75 at the lower voltage side and 0.25 at the higher voltage side.

From the simulation results, it may be possible to have the modulation factors of **0.90** at the lower voltage side and **0.10** at the higher voltage side. To have wider difference between the modulation factor leads to the reduction of the number of switching, i.e. the reduction of the switching loss. It has been investigated to have the modulation factors of **1.0** at the lower voltage side and **0.0** at the higher voltage side. This operation results in the 3 level output ac voltage.

V. SIMULATION OF 3 LEVEL OUTPUT VOLTAGE AT THE INVERTER AC SIDE

The simulation of 3 level output voltage at the converter B, which works as the inverter in driving operation, has been performed.

Fig. 16 shows the dc total voltage, the each capacitor voltage and the load current. Fig. 17 shows the change in the modulation factors. Both show the stable operation of the dc and the capacitor voltages.

Fig. 18 shows the phase difference between the output ac voltage of the converter B and the distribution power system voltage. The behavior is similar to the results of 5 level output voltage at the inverter ac side. Fig. 19 the 5 level output voltage at the converter A ac side. It is seen that the number of switching is decreased comparing to the 5 level output voltage, even when the difference of the modulation factors is small.

VI. CONCLUSIONS

Proposed modulation method is simple and possible to control dc capacitor voltage stably. It is also verified that this modulation method is valid when the output voltage of the inverter side is 3 level output for lower voltage order.

In conclusion, newly proposed PWM method is advantageous in the point of the reduction of harmonic distortion. And the increase in switching loss is suppressed within the acceptable level. Another advantage of this method is wider control range of charging current. This contributes the quicker response of the converter output.

REFERENCES

[1] M.Carpita, S.Tenconi and M.Fracchia, "A NOVEL MULTILEVEL STRUCTURE FOR VOLTAGE SOURCE INVERTER", Proc. of Fourth European Conf on Power Electronics and Applications, Firenze, Italy, Vol.1, pp.90-94 Sept. 1991.

[2] A. Nabae, I.Takahashi and H.Akagi, "A New Neutral-Point-Clamped PWM Inverter", IEEE Trans. on Industry Applications, Vol.IA-17, No.5, Sep./Oct, 1981.

[3] R.W.Menzies and Yiping Zhuang, "ADVANCED STATIC COMPENSATION USING A MULTILEVEL GTO THYRISTOR INVERTER", IEEE Trans. of PD, Vol.10, No.2, April, 1995.

[4] J.S.Lai and F.Z.Peng, "Multilevel Converters-A New Breed of Power Converters", Conf. Rec. of the 1995 IEEE IAS, pp. 2348-2356, 1995.

[5] N. Kimura, A. Matsumoto, T. Morizane, K. Taniguchi, "Multilevel Converter for High Power application", International Conference on Electrical Engineering '97,1997.

[6] N. Kimura, A. Matsumoto, T. Morizane, K. Taniguchi ,"Control strategy for multilevel converter applied for electric power system", 7th European Conference on Power Electronics and Application Proceedings, 1997.

[7] N. Kimura, T. Morizane, K.Taniguchi and T. Oono, "Multi-Modulaiton Signal PWM Control for Multi-Level Converter", EPE-PEMC 2004, Paper No. A24182, (2004).

Improved Initial Pole-Position Estimation of SPMSM Sensorless Servo Drive by Absolute Integrals of Torque-Component Currents

Tae-Woong Kim
Gyeongsang National Uni.
Jinju, 660-701, KOREA
kimtw@gnu.ac.kr

P.W. Wheeler
The University of Nottingham
Nottingham, NG7 2RD, UK
pat.wheeler@nottingham.ac.uk

Chung-Il Jeong
Gyeongsang National Uni.
Jinju, 660-701 KOREA
ppukkuya@gnu.ac.kr

Jaeho Choi
Chungbuk National Uni.
Cheongju, 361-763,KOREA
choi@chungbuk.ac.kr

Abstract—In this paper, the initial pole-position estimation of a surface (non-salient) permanent magnet synchronous motor is analysed mathematically, and more accurately improved for SPMSM sensorless servo drive by the absolute accumulation (or integral) of torque-component currents. This algorithm is well carried out under the full closed loop position control without any pole sensors and is completely insensitive to any motor parameters. This estimation is based on the principle that the initial pole-position is simply calculated by the reverse trigonometric function using the two feedback currents in the full closed-loop position control. The proposed algorithm consists of the predefined reference position profile, the information of feedback currents, speed, and relative position, and the reverse trigonometric function for the initial-pole position estimation.

Compared to other existing researches, the mathematical analysis is introduced to get the more accurate initial pole-position of a surface permanent magnet motor under the closed-loop position control. It is recognized that this algorithm is simpler in implementation and is more accurate in estimation even close to standstill through the full closed control.

I. Introduction

The permanent magnet synchronous motor has been widely applied to the industrial and servo drive fields such as machine tools and semiconductor manufacturing[1][5]. With appropriate control strategies, it can result in significant energy-savings and high performance. However, it requires the precise initial pole-position information (typically obtained by a pole sensor) for a smooth start-up and a precise servo-motion control.

It may be prohibitive to install a pole sensor on the motor shaft due to compactness, low-cost requirements, and mechanical mis-installation which often yields the initial pole position (IPP) error. Depending on the type of SPMSM, some servo motor drives would require the expensive pole-sensors and often be exposed to heat, dust, electric noise, mechanical vibration, etc. such that the position sensor signals get distorted. On the other hand, if the initial pole-position cannot be accurately known, the performances of a motor itself can not be obtained. The motor may produce less torque or it may become unstable. Moreover, during start-up it may ro-

tate in the wrong direction and lose control.

Recently, several IPP estimation algorithms for SPMSM have been reported[2]-[5]; The principle of these algorithms is based on the agreement of two reference frames in the control axes and the motor axes which are carried under the current control or the speed control. They show the good estimation results, however problems such as long stroke and weakness against the mechanical disturbance and complex implementation still remain. The accuracy of the IPP estimation of SPMSM proposed by the authors[2] is dependant on the integral period of torque-component currents. The estimation accuracy should be maximized and constantly obtained by the simplified implementation for SPMSM sensorless servo drive.

In the paper, the authors introduce the absolute accumulation of q-axis current integral into the previous IPP estimation of [2] to clear the above problem. Using the absolute accumulation of torque or q-axis currents makes the accuracy of the IPP estimation not to be dependant from the integral accumulation period within the same position interval mode and the higher accuracy to keep constant. The effectiveness of the proposed IPP estimation algorithm will be verified through simulation based analysis.

II. Initial Pole-Position Estimation

A. Principle of IPP Estimation Based on Agreement of Two Coordinate Frames

The principle of the initial pole-position estimation shown in Fig. 1 was developed by the author[3]. Suppose that there are two dq reference frames; they are a control side dq reference frame (virtual reference frame) and a motor dq reference frame (actual reference frame). And the d-coordinate axis of the control side dq reference frame is fixed to 0 degree. For example, the actual IPP of a motor may be placed on a temporary control side dq reference frame and the actual IPP coincides with the d-coordinate axis of a motor side dq reference frame. The actual IPP shown in this figure is deviated from a temporary control side dq reference frame, the deviated angle of which is defined as the deviated IPP (θ_{err}). To detect the actual initial-pole position without any pole sensor, the deviated angle of the IPP should be estimated using the information of some reference or feedback signals.

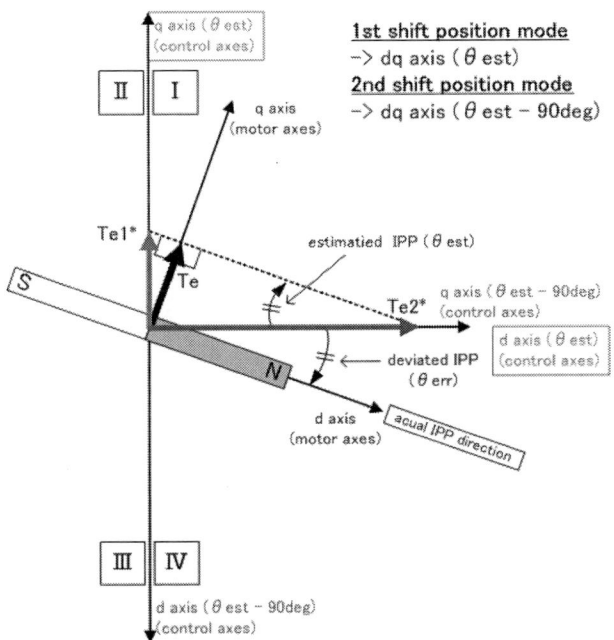

Fig. 1: Principle of initial-pole position estimation

Under the constant flux, some thrust force is necessary for a motor to be moved at the same reference speed, or to the same reference position, regardless of any deviated IPP, which is the principle of the initial pole-position estimation.

Supposing the reference thrust force T_{e1}^{*} is given at the control side q axis ($\theta_{est}+\theta_{shift1}$) shifted θ_{shift1} from the original control side q axis when the motor is controlled in agreement with its reference value; the actual thrust force T_{e1} at the motor side q axis can be expressed as (1) with the reference thrust force and the deviated position.

Supposing the reference thrust force T_{e2}^{*} is given at the control side q axis ($\theta_{est}+\theta_{shift2}$) shifted θ_{shift2} from the original control side q axis when the position of the motor is controlled in agreement with its reference value; the actual thrust force T_{e2} at the motor side q axis can be expressed as (2) with the reference thrust force and the deviated position. And the shift position functions (θ_{shift1} and θ_{shift2}) can be set to any value if the position difference between the 1st shift position θ_{shift1} and the 2nd shift position θ_{shift2} is kept at 90 degrees.

$$
\begin{aligned}
T_{e1} = T_{e1}^{shift1} &- T_{e1}^{*}\cos\left(\theta_{fb}^{shift1}-\theta_{act}\right) \\
&= T_{e1}^{*}\cos\left(\theta_{fb}-\theta_{act}+\theta_{shift1}\right) \\
&= T_{e1}^{*}\cos\left(\theta_{err}+\theta_{shift1}\right) \\
&= T_{e1}^{*}\cos\left(\theta_{err}+0\right)
\end{aligned}
\tag{1}
$$

$$
\begin{aligned}
T_{e2} = T_{e2}^{shift2} &= T_{e2}^{*}\cos\left(\theta_{fb}^{shift2}-\theta_{act}\right) \\
&= T_{e2}^{*}\cos\left(\theta_{fb}-\theta_{act}+\theta_{shift2}\right) \\
&= T_{e2}^{*}\cos\left(\theta_{err}+\theta_{shift2}\right) \\
&= T_{e2}^{*}\cos\left(\theta_{err}-90\right)
\end{aligned}
\tag{2}
$$

where $\theta_{fb}^{shift} = \theta_{fb}+\theta_{shift}$, $\theta_{shift1}=0$, $\theta_{shift2}=-90$, θ_{act} is the absolute actual position, and θ_{fb} is the incremental feedback position.

The position information in the current controller is adjusted by (3), which includes the shift function for selecting the reference thrust force direction in any control side q axis reference frame.

$$
\theta_{ref} = \theta_{fb}+\theta_{shift}+\theta_{cmp}+\theta_{default}
\tag{3}
$$

where is the compensation position that increases the estimation accuracy and is not mentioned in detail in this paper, and is the default IPP(=0).

According to the principle of the IPP estimation, each necessary thrust force of both the 1st shift position mode and the 2nd shift position mode is equal and then T_{e1}^{*} of (1) equal to T_{e2}^{*} of (2).

$$
T_{e1}^{*}\cos\left(\theta_{err}+0\right) = T_{e2}^{*}\cos\left(\theta_{err}+90\right)
\tag{4}
$$

Using the relation between torque and q axis current and solving (4), the deviated IPP can be calculated by the following equation and replaced as the estimated IPP;

$$
\theta_{est} = \tan^{-1}\left(\frac{T_{e1}^{*}}{T_{e2}^{*}}\right) = \tan^{-1}\left(\frac{I_{q1}^{*}}{I_{q2}^{*}}\right)
\tag{5}
$$

B. Improved IPP Estimation Algorithm by Absolute Integrals of q-axis Currents

Assuming that the load is constant at any position in a short moving, the load term in the motion equation can be neglected for the IPP estimation.

$$
\begin{aligned}
J\frac{d}{dt}\omega &= T_e - T_L - D\omega \\
&= T_e - D\omega
\end{aligned}
\tag{6}
$$

where D is the damping factor, T_L is the load, and J is the inertia.

Multiplying (6) by $1/s$, (7) can be produced. The shift position is set to $\pm 45°$ at each shift position mode. As (1) or (2) is put into (5) according to each shift position mode, (6) and (7) can be rewritten as (9) and (10), respectively.

$$J\omega = \int T_e\,dt - D\int \omega\,dt = \int T_e\,dt - DP \qquad (7)$$

$$\omega = \frac{1}{J}\left(\int T_e\,dt - DP\right) \qquad (8)$$

$$\omega_1 = \frac{1}{J}\left(\int T_{e1}^*\cos(\theta_{err}+45°)\,dt - DP_1\right) \qquad (9)$$

$$\omega_2 = \frac{1}{J}\left(\int T_{e2}^*\cos(\theta_{err}-45°)\,dt - DP_2\right) \qquad (10)$$

where P is the feedback relative position, and 1 and 2 are the subscripts of the 1st and 2nd shift position modes.

If $\omega_1 = \omega_2 = 0$ and $P_1 \neq P_2 \neq 0$, the following relations of (11) and (12) can be obtained from (9) and (10).

$$P_1 = \frac{1}{D}\int T_{e1}^*\,dt \cdot \cos(\theta_{err}+45°) \qquad (11)$$

$$\begin{aligned} P_2 &= \frac{1}{D}\int T_{e2}^*\,dt \cdot \cos(\theta_{err}-45°) \\ &= \frac{1}{D}\int T_{e2}^*\,dt \cdot \sin(\theta_{err}+45°) \end{aligned} \qquad (12)$$

From (11) and (12), the proportional relation can be obtained as follows;

$$P_1 : P_2 = \int T_{e1}^*\,dt \cdot \cos(\theta_{err}+45) : \int T_{e2}^*\,dt \cdot \sin(\theta_{err}+45) \qquad (13)$$

Solving (13) for the deviated position θ_{err} produces

$$\theta_{est} = \theta_{err} = -45° + \tan^{-1}\left(\frac{P_2}{P_1} \cdot \frac{\int T_{e1}^*\,dt}{\int T_{e2}^*\,dt}\right) \qquad (14)$$

As the deviated IPP is equal to the estimated IPP, the estimated IPP is calculated by (14).

Putting torque equation (15) into (14), the estimated IPP of (16) can be calculated through the q axis currents instead of torque information.

$$T_e = \frac{3}{2}p\lambda_f i_q \qquad (15)$$

$$\theta_{est} = -45° + \tan^{-1}\left(\frac{P_2}{P_1} \cdot \frac{\int i_{q1}^*\,dt}{\int i_{q2}^*\,dt}\right) \qquad (16)$$

where p is pole pair of motor, λ_f is flux, and i_q is the q-axis current in which 1 and 2 are the subscripts of the 1st and 2nd shift position modes.

In general, (16) can be used for calculating the initial pole position of SPMSM. However depending on the integral period of the q axis current within the same position interval, the accuracy of the initial pole-position estimation is not constant because the accumulated amount regarding q-axis current is different. If the integral period of q-axis current is wrongly set, the integral amount may be smaller and neglected even if the position profile is the same. To clear these problems and keep the estimation accuracy better, the accumulated amount of the q-axis should be maximized. This means that the larger the accumulated amount is, the higher the resolution for the IPP estimation accuracy is. According to the above conditions, the absolute integration of q-axis currents is introduced into (16) and the following equation of (17) is used to calculate the IPP estimation instead of (16).

$$\theta_{est} = -45° + \tan^{-1}\left(\frac{P_2}{P_1} \cdot \frac{\int |i_{q1}^*|\,dt}{\int |i_{q2}^*|\,dt}\right) \qquad (17)$$

C. Profiles of Position, Speed, and Torque

Fig. 2 shows the predefined reference position profile, which is used to estimate the IPP. The calculation of the IPP is performed after obtaining the available two feedback q-axis currents and positions, which should be obtained at the 1st and 2nd shift position modes (reference position intervals). If there is an initial pole-position error between the actual dq axes and the control dq axes, the produced thrust force will be decreased compared to the reference thrust force, as shown in Fig. 3. These relations are well expressed by (18) ~(19) and shown in Fig. 4. The speed and the position will be reduced according to the cosine function with a position error. Furthermore, if the IPP error is in the rages of 90[deg] ~270[deg], the generated thrust force is negative against the reference thrust force and the motor cannot be controlled.

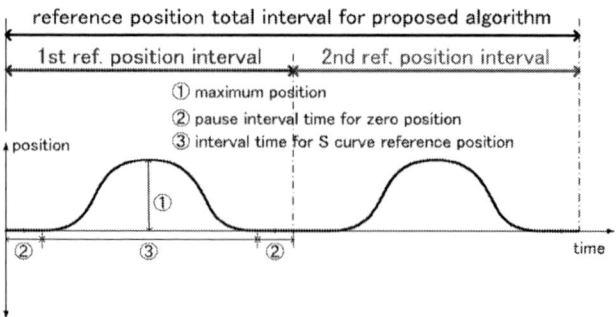

Fig. 2: Pre-defined reference position profile with S curve

$$T_{e\,loss} = T_e^* \times (1 - \cos\theta_{err}) \qquad (18)$$

$$T_e = T_e^* \times \theta_{err} \qquad (19)$$

where $T_{e\,loss}$ is a loss torque (or thrust force), T_e is a generated torque (or thrust force), T_e^* is a reference torque (or thrust force), and θ_{err} is a deviated IPP.

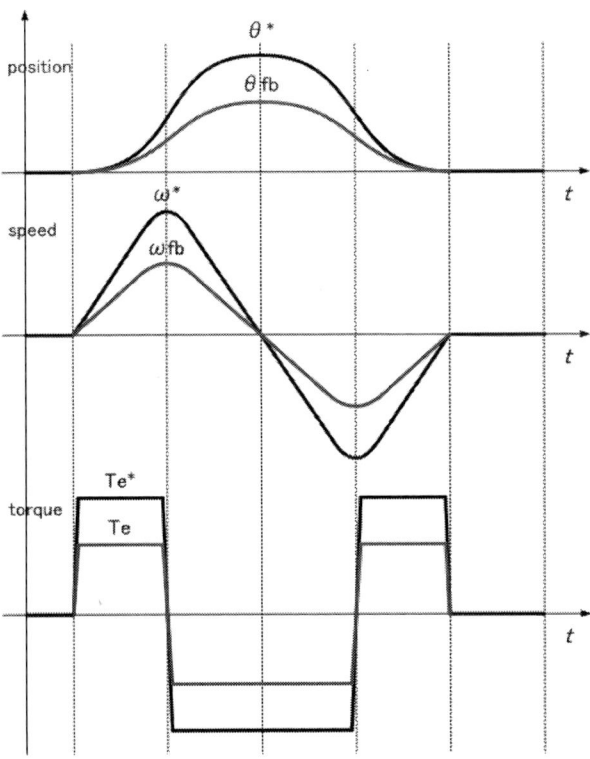

Fig. 3: Mechanical behaviours at non-zero deviated IPP; position(upper), speed(middle), and torque(bottom)

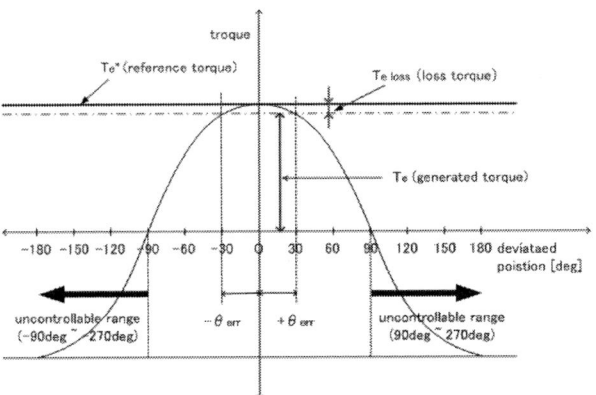

Fig. 4: Relations between generated torque and loss torque according to the deviated IPP

III. Simulation Based Analysis

A. System Configurations

The overall system configurations of simulations is shown in Fig. 5 to confirm the feasibility of the improved estimation algorithm of SPMSM. The block of proposed IPP estimation algorithm shown in this figure has the following functions; 1) to switch the position mode into the 1st shift position mode or the 2nd shift position mode, 2) to generate the reference position profile with S curve, 3) to adjust compensated position for increasing estimation accuracy, 4) finally to calculate the IPP with feedbacked incremental position and absolute currents.

B. Simulation Results and Discussions

To verify the feasibility of the proposed algorithm, the simulation based analysis is performed by PSIM simulator. Fig. 6 shows the estimation result when the IPP error is set to any given deviated IPP(±15[[deg], ±30[deg], ±45[deg]) under no load. As shown in Fig. 6, the moving distance of a motor is suppressed within the reference position peak and the estimation is performed in a very short time, and also the higher estimation accuracy is well obtained because of absolute integrals of q-axis currents accumulated during each position interval. When the non-absolute integral for (16) and the absolute integral for (17) is used individually to estimated the IPP, its results are shown in Table 1. This table shows that the estimation variance at any given deviated IPP is smaller in case of (17). Therefore, this algorithm with the absolute integrals of torque-component currents can be more effective and economic in applications such as precise servo drives without any pole sensor.

TABLE 1 Evaluation of proposed IPP estimation according to deviated IPP

evaluation item	deviated IPP						
	-45	-30	-15	0	15	30	45
[1]estimated IPP [deg]	-44.7	-29.0	-14.6	0.1	14.5	29.0	44.7
$\int \lvert i_{q1} \rvert dt$	578	597	666	816	1141	2102	5237
$\int \lvert i_{q2} \rvert dt$	5178	2097	1146	816	667	597	575
moving distance [deg]	1.87	1.89	1.89	1.88	1.87	1.88	1.87
[2]estimated IPP [deg]	-45.0	-32.7	-15.7	-1.7	17.1	31.1	45.0
$\int i_{q1} dt$	8	8	10	12	20	36	5237
$\int i_{q2} dt$	5178	30	18	13	10	9	8

note : 1) used by absolute integral, 2) is used by no-absolute integral.

Totally, it has been found that the proposed algorithm can provide excellent IPP estimation while suppressing the moving distance and in shorter estimation time.

VI. Conclusions

In this paper, the initial pole-position estimation of SPMSM is analysed mathematically, and more accurately improved for SPMSM sensorless servo drive by the absolute integrals of torque-component currents. The improved IPP estimation is verified through the simulation based analysis.

This IPP estimation is based on the principle that the initial pole-position is simply calculated by the reverse trigonometric function using the two feedback absolute currents in the full closed-loop position control. The proposed algorithm was simple in implementation and was highly accurate in estimation even close to standstill. This estimation can be widely applied to both rotating typed and liner typed surface PMSM without any limitations of their motor structures.

ACKNOWLEDGEMENT

The authors would like to thank for the project supported by R2005-B-109 of KESRI (Korea Electrical Engineering and Science Research Institute), which is funded by MOCIE (Ministry of Commerce Industry and Energy) in Korea.

REFERENCE

[1] S. Sunter and J.C. Clare, "A True Four Quadrant Matrix Converter Induction Motor Drive with Servo Performance", in IEEE Proc. on PESC96, pp. 146-151, 1996

[2] T.W. Kim, J. P.W. Wheeler, C.I Jeong, J.H. Choi and A. Kawamura, "Accurate initial pole-position estimation of Surface PM-LSM in the position control," in Proc. on EPE 2007, pp. 1-8, 2007

[3] T.-W. Kim, J. Watanabe, S. Sonoda, and J. Hirai, "New initial pole position estimation of surface PM-LSM using reference currents," in IEEE Trans. on IAS, vol. 41, no. 3, pp. 817-824, 2005

[4] J.-W. Choi, W.E. Yun, and H.-G. Kim, "Initial pole-position estimation of linear motor," in IEE Proc. on Electric Power Applications, vol. 152, no. 4, pp. 997-1002, 2005

[5] K. Ide, H.S. Song, M. Takaki, S. Morimoto, and S.-K. Sul, "Fast initial pole-position estimation for non-salient PM-LSM based on agreement of two reference frames," in IEEE Proc. on PESC06, pp. 1-7, 2006

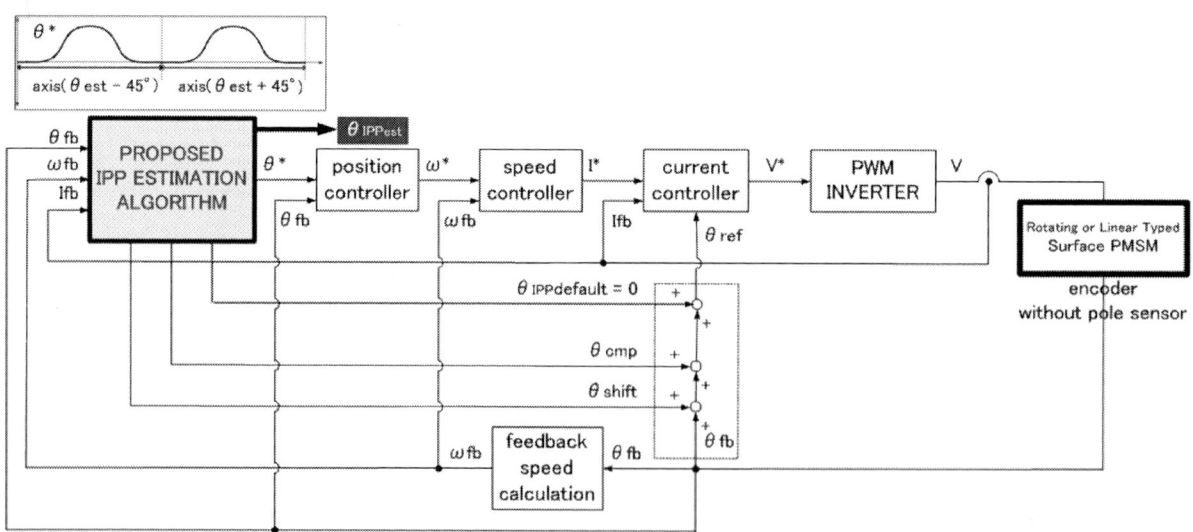

Fig. 5: Overall system configuration for calculating the IPP estimation of SPMSM

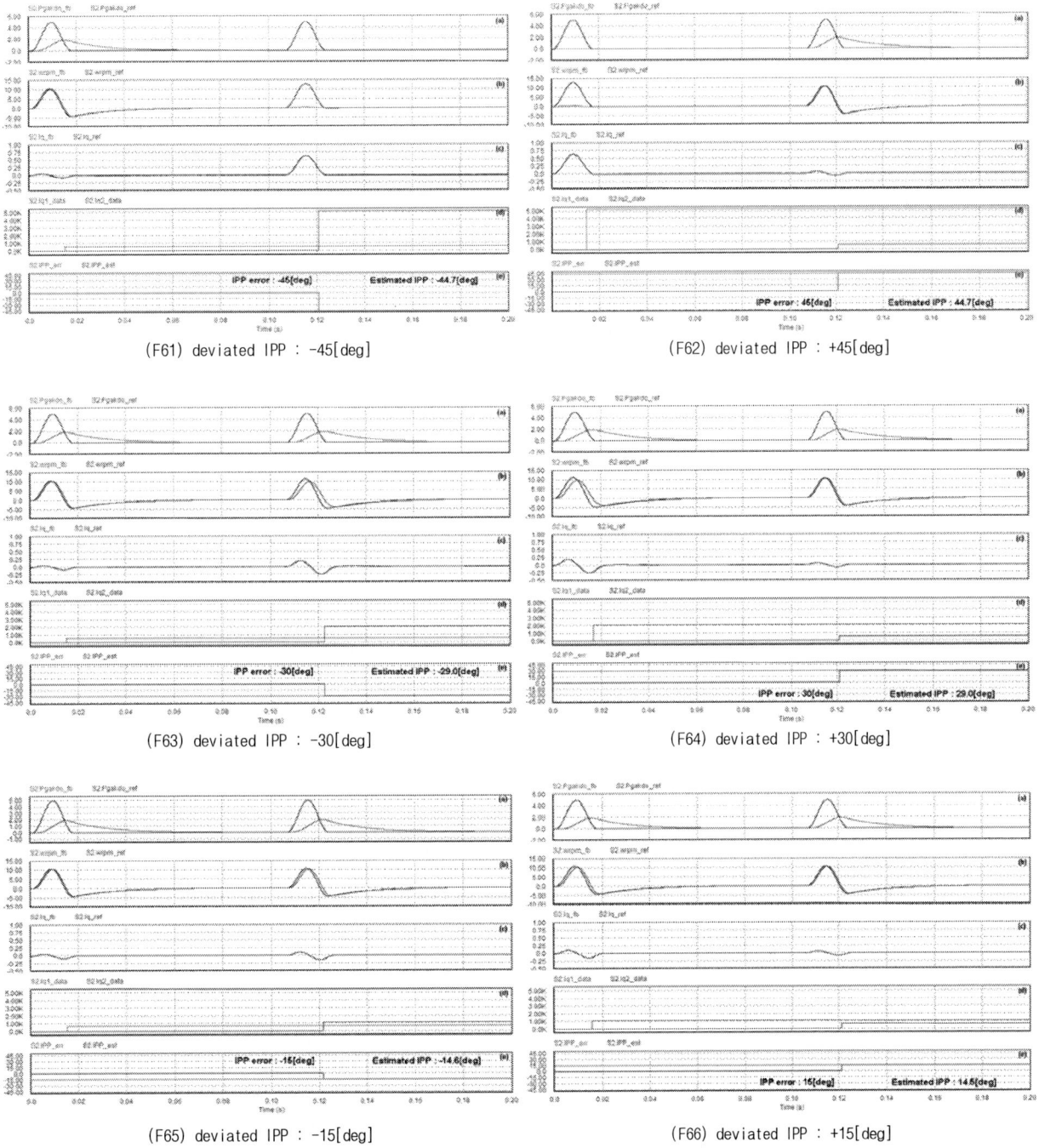

Fig. 6 Simulation results at any given deviated IPP (±45[[deg], ±30[deg], ±15[deg]) ; (a) reference position[deg] and feedback position[deg], (b) reference speed[rpm] and feedback speed[rpm], (c) q-axis reference current[A] and q-axis feedback current[A], (d) absolute integral based q-axis current[A], (e) deviated IPP[deg] and estimated IPP[deg]

978-1-4244-1871-8/07 $25.00 © 2007 IEEE

Design and development of distorted source device for circuit breakers failure analysis

Sang-Ick Lee, Jae-Geun Yoo, Hyun-Jae Jeon
Korea Electrical Safety Corporation
27, Sangcheon-ri, Cheon-ri, heongpyeong-myeon,
Gapyeong-gun, Gyeonggi-do, 477-814
Email: sangickl@kesco.or.kr

Gyu-Ha Choe
Konkuk University
Department of Electrical Engineering
1, Hwayang-dong, Gwangjin-gu, Seoul, 143-701
Email: ghchoe@konkuk.ac.kr

Abstract— Up to recently the harmonic generation has deteriorated the quality of electricity and affected the performance on the electrical installation including OA, FA, IT devices and so on. Some studies of harmonic affects in presumption according to qualitative analysis. So in order to research the harmonic affect on the electrical installation according to quantitative analysis and gather reliable data over and over again, it is necessary to develop an AC power source which is capable of generating some harmonics. In this paper, we described about realization of AC power source which can produce and compose harmonics for the analysis of accident due to harmonics.

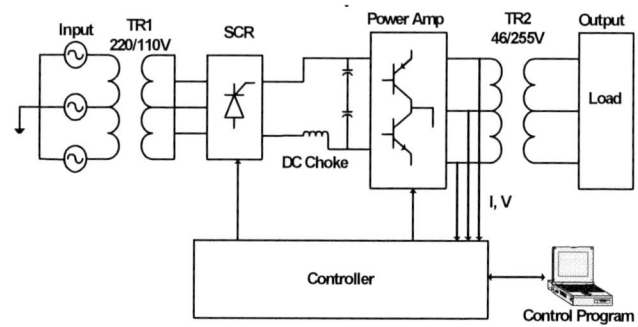

Fig. 1. Distorted source device block diagram

I. INTRODUCTION

Recently it was tested the performance and harmonic effect of an AC power source which is capable of generating some harmonics and changing the voltage and frequency. The devices which are manufactured automatically are tested about the fluctuation of voltages and frequencies before being mass produced. It is difficult to get the bottom of an accident by harmonics because the accident occurred intermittently and suddenly. So the performance of circuit breakers is not able to guarantee about the harmonic accident.

First, we questioned customers on complains of the circuit breaker performance. The 80% of respondents have experience in the circuit breaker obstacle. The performances and analysis of circuit breakers is important and necessary to examine the cause of circuit breaker defects by harmonics.

This result shows the cause of circuit breaker defects is AC source including harmonics, but the circuit breaker defects did not occur frequently. So it is difficult to define the cause of circuit breaker defects.

In this paper we researched the development of a low-priced AC source generated harmonics and had a proper capacity and performance for making a cause clear.

II. AC POWER SOURCE

We developed AC power source which consisted of the voltage input part, the converting part, the signal input part, the A/D converting part, the digital signal processing part, the D/A converting part, the controlling part, the display part, the communication part by figure 1.

A. AC Power Source Performance

The AC power source product is developed by global company P, E. But this product has to connect a series or parallel circuit for applying to several capacity and input/output performance. In this paper, we wish to develop a new harmonic AC power source that has single-phase and three phases AC voltage and electric current output, output of frequency variableness, harmonic waveform composition voltage and electric current. We decided specification that wish to develop being based on result that examine comparing outside the country 2 company's specification with table 1.

Table. 1. Spec. comparison of the foreign company

Section	Company P	Company E
Output capacity[VA]	4500	5250
Output form	1φ, 3φ	1φ, 3φ
Output max. voltage[Vrms]	135(L-N)	312(L-N)
Output current[Arms]	12 (per phase)	6.5 (per phase)
Input freq. range[Hz]	20~5000	40~5000
Harmonics	51th	-
Controller interface	RS232C, GPIB	GPIB

978-1-4244-1871-8/07 $25.00 © 2007 IEEE

Decided inputs voltage is single phase 220[V]/ three phases 220 [V], and decided output voltage is 250 [V], output current is 8 [A] by the each phase. Output frequency is able to composite and variable from 45 [Hz] to 999 [Hz]. And insert RS232C communication port to connect a PC. The developed AC source has functionally separated boards to add another function.

III. ANALYSIS AND DESIGN OF POWER APPLIER

A. Power Amplifier

Generally Class A amplifier is a simple and stable linear amplifier. But class A has most bad efficiency. When use electric cell by power supply, efficiency of amplifier and current drain become important problem at design. So various other methods are suggested Class A method need to remove bias current because 75% of supply power is exhausted by DC.

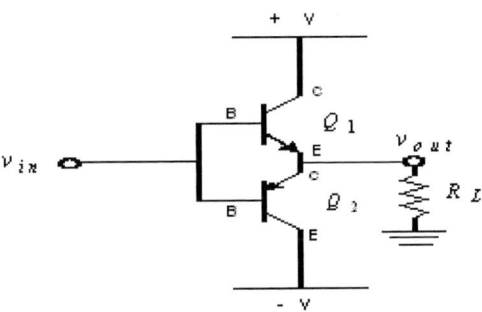

Fig. 2 Class B amplifier circuit

In class B amplifier output stage of transistor such as figure 2, Q1 is conducted when input is (+) in Q1 and Q2 form, and Q2 is conducted when input is (-). Actions of this means angle of collector electric current is 180°, and point Q is located about in cut point of DC load line and AC load line. An advantage of Class B is less power consumption of transistor and current drain is low. And when input of transistor is zero (0), Q1 and Q2 are not conducted, and power dissipation is zero (0) at stop of operation driving.

Symmetrically, electric power is same as supply from +Ve and −Ve. Therefore, all electric power is equal to equation (1). Also, load power is equation (2). Therefore, efficiency of AC amplifier is

$$P_s = P_{(+ve)} + P_{(-ve)} = 2P_{(+ve)} = \frac{2V_S A}{\pi R_L} \qquad (1)$$

$$P_L = \frac{\overline{v_{out}^2}}{R_L} = \frac{\overline{A^2 \sin^2(wt)}}{R_L} = \frac{A^2}{2R_L} \qquad (2)$$

$$\eta = \frac{P_L}{P_S} = \frac{A^2}{2R_L} \frac{\pi R_L}{2V_S A} = \frac{A\pi}{4V_S} \qquad (3)$$

In upper equations, $A \leq V_S \rightarrow \eta \leq \pi/4 = 78.5\%$. This is contrast to class A's efficiency.

B. Maximum power dissipation of power amplifier

Maximum power dissipation (P_D) is expressed to A's quadratic equation, and is equation (4)

$$P_D = \frac{4V_S^2}{\pi^2 R_L} - \frac{2V_S^2}{\pi^2 R_L} = \frac{2V_S^2}{\pi^2 R_L} \qquad (4)$$

We must consider P_D's maximum to select transistor/heat shield.

Design on basis specification of design and development objective class B amplifier design may arrive to 25[Ω], 200 [W]. A supply voltage should be bigger than A_{max}. So, we selected $V_S = 46$ [V]. Therefore, maximum power dissipation P_D (max) is as following.

$$P_{D(max)} = \frac{2V_S^2}{\pi^2 R_L} = 17.15W = 2 \times 8.6W$$

Each two output transistors must be able to diffuse safely to 8.6 [W].

We decided to use 2SA1494(PNP), 2SC3858(NPN) power TR device for AC amplifier. According to datasheet of the device, In other words, two heat shields need to be smaller than 14.375 [°C/W], or one heat shield between two should be small more than 14.375 [°C/W].

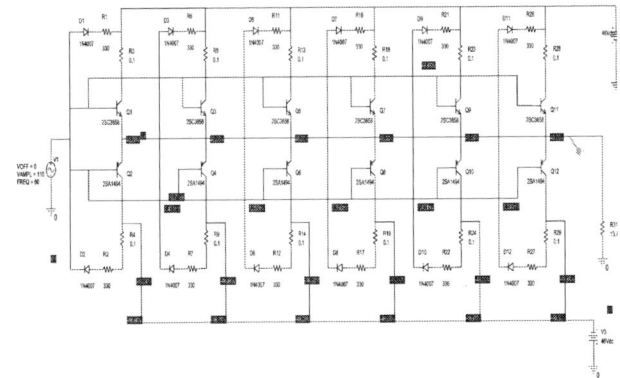

Fig.3 Series and parallel amplifier circuit

C. Simulation and result

Figure 3 displayed circuit of AC power amplifier. It consists of power transistors to amplify the current and

1N4007 diode to reflect the current. The power transistor's input is sine wave signal of maximum 110 [V] and 2SC3858(NPN), 2SA1494(PNP) of MOSPEC company is supplied voltage of maximum 46 [V].

This is the simulation to compose actually hardware. The developed AC power source is supplied the output DC ±46[V] of the capacitors part between 220/110 [V] TR1 transformer and amplifier. Figure 4 is simulation result waveform of output voltage that appears in output stage, and figure 5 did about FFT of output stage voltage waveform.

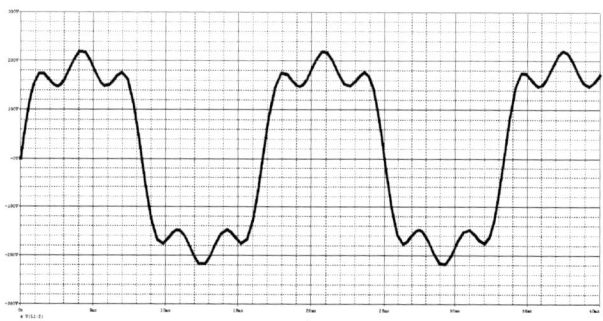

Fig.4 3rd and 5th harmonics synthesize simulation result

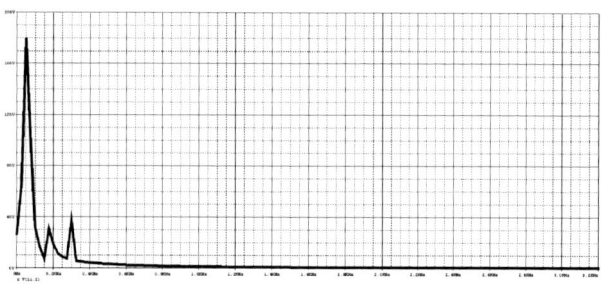

Fig.5 3rd and 5th harmonics synthesize FFT waveform

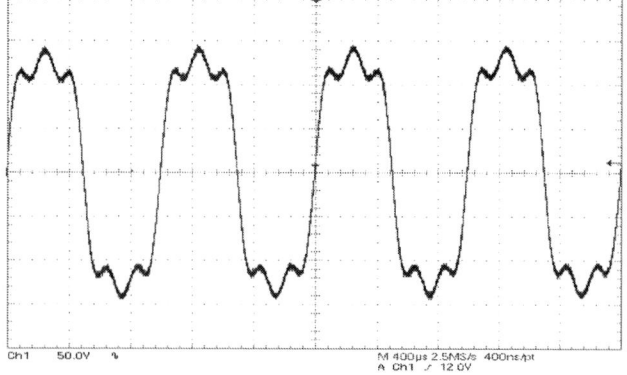

Fig.6 Measured waveform of AC power source for harmonic synthesize

D. Test result

Figure 6, 7 shows waveform measured with developed AC power source. Figure 6 is synthesized 3rd harmonic (20%) to

5th harmonic (15%). Figure 7 is synthesized 3rd harmonic (30%) to 5th harmonic (20%), 7th harmonic (20%). This results verified performances of developed AC power source.

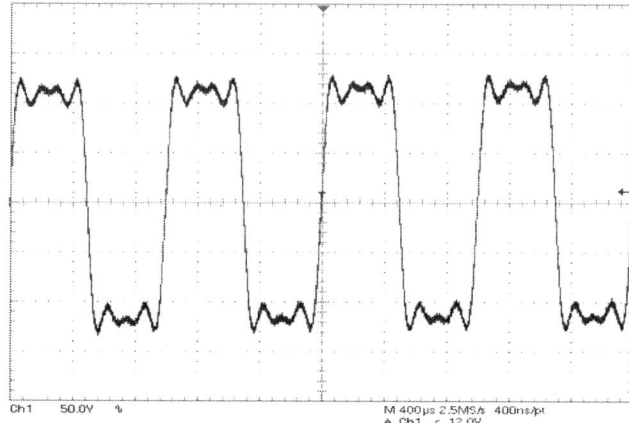

Fig.7 Measured waveform of AC power source for harmonic synthesize

IV. CONCLUSION

In this paper, we described about realization of AC power source which can produce and compose harmonics for the analysis of accident due to harmonics.

Described specification standard for development background and hardware development for device preferentially, and appeared about electric power conversion module that flows DC in AC stage and is displayed to AC stage again. The next suggested about design and development of signal generation module and control module. Finally, confirmed AC Power Source's action through measurement waveform integrated harmonics.

Hereafter, plan to heighten accuracy for harmonic waveform synthesis, and progress study of malfunction of breakers through developed AC Power Source

REFERENCES

[1] Richer, C. M., Carti, E. G., Pinheiro, H., Hey, H. L., Pinheiro, J. R. and Grundling, H. A., "A Three-Phase AC Power Source using Multi variable Repetitive Robust Model Reference Adaptive Control", IEEE, pp 2300-2305, 2003

[2] Kay-Soon Low, "A DSP-Based Variable AC Power Source", IEEE Transaction on Instrumentation and Measurement, Vol. 47, No. 4, pp 992-996, 1998

[3] Richer, C. M., Carti, E. G., Pinheiro, H., Hey, H. L., Pinheiro, J. R. and Grundling, H. A., "A Three-Phase AC Power Source using Multi variable Repetitive Robust Model Reference Adaptive Control", IEEE, pp 2300-2305, 2003

[4] C. Michael Hoff, Sarma Mulukutla, "Analysis of the Instability of PFC Power Supplies With Various AC Source", IEEE, pp 696-702, 1994

Analysis and simulation of Matrix Converter Using PSIM

Ashwin Kumar Sahoo [1], J. Meenakshi[2], S. S Dash[3], and T. Thyagarajan[4]

[1] Department of EEE, SSNCE, TN. E-mail: ashwinsahoo@ieee.org
[2] Department of EEE, SSNCE, TN. E-mail: meenajayaram@yahoo.co.in
[3] Department of EEE, SRM University, TN. E-mail: munu_dash_2k@yahoo.com
[4] Department of Instrumentation Engineering, MIT, Anna University, TN.
E-mail: thygu_vel@yahoo.co.in

Abstract—**Matrix Converter, a direct AC/AC Converter replaces the multiple conversion stages and the intermediate energy storage element of the indirect converters by a single power conversion stage, thus being a single stage converter. Matrix Converters employ an array of controlled bidirectional switches as the main power elements to create a variable output voltage system with unrestricted frequency. This paper attempts to give an overview of Matrix Converters and examines the types, features, merits and limitations of such Converters. It presents the basic operating principle of a Single Phase Matrix Converter, Two Phase to Single Phase Matrix Converter, Three Phase to Single Phase Matrix Converter and Three Phase to Three Phase Matrix Converter. The paper also introduces a topology of a Single Phase Matrix Converter , Two Phase to Single Phase Matrix Converter, Three Phase to Single Phase Matrix Converter and Three Phase to Three Phase Matrix Converter in PSIM software environment and analyses the output of the converter for different frequencies. Further, it examines the output voltage of a Single Phase Matrix Converter synthesized using the Sinusoidal Pulse Width Modulation (SPWM) technique. All simulations have been carried out in POWERSIM.**

I. INTRODUCTION

The AC/AC converters are commonly classified as direct converters and indirect converters. An Indirect converter consists of two converter stages and an energy storage element, which converts input ac to dc and then reconverts dc back to ac output with variable amplitude and frequency. The energy storage element can be either a capacitor or an inductor. However, the energy storage element is not needed in a direct converter. In general, direct converters can be classified into three topologies namely ac voltage controller, cycloconverter and matrix converter. The first and simplest topology changes the amplitude of the input ac voltage and functions by simply chopping symmetric notches out of the input waveform. The second topology is utilized if a lower output frequency is desired and approximates the desired output waveform by synthesizing it from pieces of the input waveform. The third topology namely, Matrix converter is a most versatile converter without any limit on the output frequency and amplitude. It replaces the multiple conversion stages and the intermediate energy storage element by a single power conversion stage.

A matrix converter interfaces two three phase systems with different voltage magnitudes, frequencies and phase angles. It consists of an array of bidirectional switch cells functioning as the main power circuit elements. Each switch cell is composed of two forced commutated switches in an anti parallel configuration [1]. This paper presents the different topologies of Matrix Converters and examines their basic operating principles. It also presents in detail the features of Matrix Converters. It examines the output voltage of a Single Phase Matrix Converter simulated using PSIM. Further, the simulation results of a Two Phase to Single Phase Matrix Converter, Three Phase to Single Phase Matrix Converter and Three Phase to Three Phase Matrix Converter are also obtained. The idea presented in this paper is a novel approach that uses PSIM as the simulation environment to study the behaviour of a Matrix Converter. This approach is foreign to the traditional approaches that use PSPICE and Simulink for simulation.

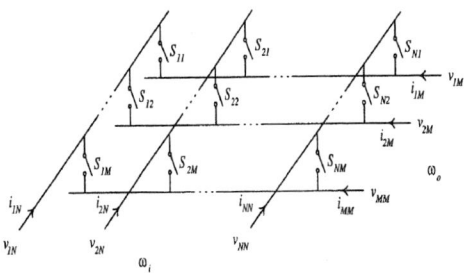

Fig. 1. An N-phase to M-phase Matrix Converter

The topology of an N-phase to M-phase Matrix Converter is shown in Fig. 1. The Converter consists of an NxM semiconductor, bidirectional switches. The frequency of the N-phase system is ω_i and that of the M-phase one is ω_o. Different topologies are derived by selecting appropriate number of bidirectional switches

II. BIDIRECTIONAL SWITCH TOPOLOGIES

The Matrix Converter requires a bidirectional switch capable of blocking voltage and conducting current in both directions. Unfortunately, there are no such devices currently available that fulfill the need, so discrete devices need to be used to construct suitable switch cells. In the bidirectional switch arrangements, devices such as IGBTs, MOSFETs, and IGCTs are used.

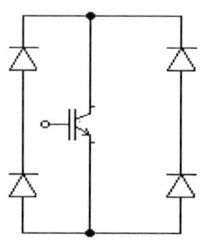

Fig. 2. Diode Bridge Arrangement

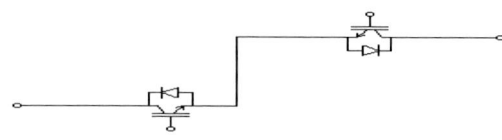

Fig. 3. Common Emitter anti parallel IGBT, Diode Pair

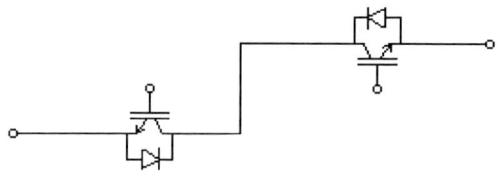

Fig. 4. Common Collector anti parallel IGBT, Diode Pair

Some of the commonly used bidirectional switch topologies [2] are listed below:

A. Diode Bridge Arrangement

This bi-directional switch arrangement consists of an IGBT at the center of a single-phase diode bridge arrangement as shown in Fig. 2. The main advantage is that both current directions are carried by the same switch device, therefore only one gate driver is required per commutation cell. Device losses are relatively high since there are three devices in each conduction path. The direction of current through the switch cannot be controlled. This is a disadvantage, as many of the higher reliability commutation methods cannot be used.

B. Common Emitter anti parallel IGBT, Diode Pair

This bidirectional switch arrangement consists of two diodes and two IGBTs connected in anti parallel as shown in Fig. 3. The diodes are included to provide the reverse blocking capability. It is possible to independently control the direction of the current. Conduction losses are also reduced since only two devices carry the current at any one time. The disadvantage is that each bi-directional switch cell requires an isolated power supply for the gate drives. Both devices however can be driven with respect to the same voltage.

C. Common Collector anti parallel IGBT, Diode Pair

This arrangement in Fig. 4 is similar to the previous one, but the IGBTs are arranged in a common collector configuration. The conduction losses are the same as the common emitter configuration. The advantage of this method is that only 6 isolated power supplies are needed to supply the gate driven signals. However this arrangement is not feasible in a practical system since inductance between commutation cells cause problems. Therefore, the common emitter configuration is generally preferred for creating the matrix bi-directional switch cells.

III. ADVANTAGES OF MATRIX CONVERTER

1) It provides direct ac-ac conversion thus eliminating the need for reactive energy storage elements.
2) The DC link components are bulky and their elimination can drastically reduce the size/footprint of the Matrix Converter system.
3) Elimination of the DC link provides the opportunity for integration of all switch cells of the Matrix Converter within one semiconductor building block (module). The semiconductor building block can be fabricated as a power integrated circuit (power IC) with significant potential for cost reduction. Furthermore, the IC can also incorporate gating/protection circuits and consequently further add to the reliability, facilitate interface to the control platform, provide modularity and more importantly reduce size and weight.
4) The Matrix Converter provides an inherent four quadrant operation
5) It provides independent control of the output voltage magnitude, frequency and phase angle and operation at lagging, unity or leading power factor.

Some of the limitations of matrix converters include-

1) Higher number of unidirectional switches can result in a higher switching loss and a lower overall efficiency.
2) Inherently, the lack of a DC link in the Matrix Converter indicates that there is a stronger coupling between the two ac sides of the Matrix Converter. Therefore, more elaborate control strategy may be required to minimize mutual interactions between the ac sides.

978-1-4244-1871-8/07 $25.00 © 2007 IEEE   414

IV. TOPOLOGIES OF MATRIX CONVERTER

Based on the number of input and output phases, a Matrix Converter can fall under the following topologies [4]-

1) Single Phase Matrix Converter
2) Three Phase to Single Phase Matrix Converter
3) Three phase to Three Phase Matrix Converter

A. SINGLE PHASE MATRIX CONVERTER

The Single-Phase Matrix Converter consists of a matrix of input and output lines with four bidirectional switches connecting the single-phase input to the single-phase output at the intersections. It comprises of four ideal switches S_1, S_2, S_3 and S_4 capable of conducting current in both directions, blocking forward and reverse voltages (symmetrical devices) and switching between states without any delays.

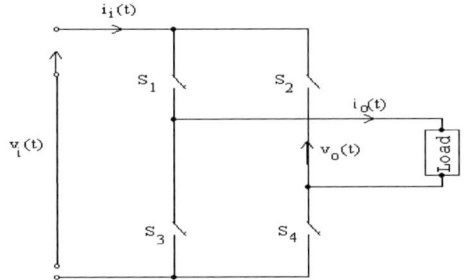

Fig. 5. Representation of a Single Phase Matrix Converter

The Single-Phase Matrix Converter is presented schematically in Fig. 5. Its instantaneous input voltage is $v_i(t)$ and its output voltage is $v_o(t)$. This topology converts the input voltage, $v_i(t)$ with constant amplitude and frequency, through the four ideal switches to the output terminals in accordance with pre calculated switching angles [5]. The input voltage of the Matrix Converter is given by

$$v_i(t) = \sqrt{2} \, V_i \cos \omega_i t, \qquad (1)$$

The matrix converter will be designed and controlled in such a manner that the fundamental of the output voltage is

$$v_{o\text{-main}}(t) = \sqrt{2} \, V_o \cos \omega_o t, \qquad (2)$$

where, V represents RMS value of the voltage and subscript i denotes input, while o denotes output. With input voltage from (1), the Matrix Converter switching angles will be calculated so that the fundamental of the output voltage will be as given in (2), keeping with the degrees of freedom specified by

$$
\begin{aligned}
v_o(t) &= v_i(t) && (S_1 \text{ and } S_4) \text{ are in on-state} && \text{Mode 1} \\
v_o(t) &= -v_i(t) && (S_2 \text{ and } S_3) \text{ are in on-state} && \text{Mode 2} \quad (3)\\
v_o(t) &= 0 && (S_1 \text{ and } S_4) \\
& && \quad \text{or} \\
& && (S_2 \text{ and } S_3)
\end{aligned}
$$

The additional states (S_1 and S_3– on) and (S_2 and S_4– on) are forbidden states (they create a short circuit of the mains).
The Single Phase Matrix Converter may operate
➢ In modes 1 and 2 or
➢ In modes 1, 2 and 3
The four power switching devices are switched at high frequency, f_s ($f_s \gg f_i$ and f_o, $f_i = \omega_i / 2\Pi$ and $f_o = \omega_o / 2\Pi$).

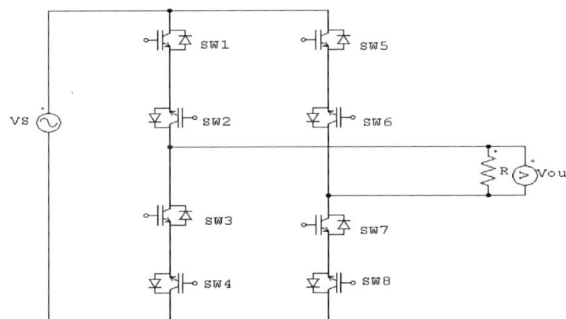

Fig. 6. Simulation Circuit of a Single Phase Matrix Converter

The simulation circuit developed in PSIM is shown in Fig. 6. Circuit operation is analyzed for a complete switching sequence assuming the load to be purely resistive. For example, to achieve a frequency of 50 Hz, S_1 and S_4 are turned on, when the supply voltage is positive and S_2 and S_3 are turned on , when the supply voltage is negative.

1) Sinusoidal Pulse Width Modulation for a Single Phase Matrix Converter

Voltage control of converters can be accomplished either external or internal to the converter. The most efficient method of controlling the output voltage is to incorporate PWM control within the converter. Though there are different PWM control techniques, the Sinusoidal Pulse Width Modulation (SPWM) is a well known shaping technique. The gating signals are generated by comparing a sinusoidal reference signal with a triangular carries wave of frequency f_c. The frequency of the reference signal f_r determines the output frequency f_o and its peak amplitude controls the modulation index and in turn the rms output voltage. To generate the SPWM patterns that are used to control the power switches, only two switches will be in the ON state (SW1 and SW7 or SW3 and SW5, say SPWM output 1) to conduct current flow during positive half of input source and at any time, only two switches will be in the ON state (SW2 and SW8 or SW4 and SW6,, say SPWM output 2) to conduct current flow during negative half of the input source.

B. TWO PHASE TO SINGLE PHASE MATRIX CONVERTER

The Fig. 7 shows a schematic of a Two Phase to Single Phase Matrix Converter. The converter is composed of two bidirectional switches S_1 and S_2.

978-1-4244-1871-8/07 $25.00 © 2007 IEEE

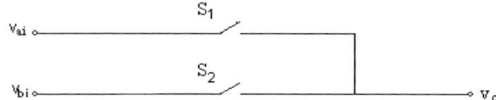

Fig. 7. Representation of a Two Phase to Single Phase Matrix Converter

Each switch connects the output line to an input phase. To avoid short-circuit in the source-side and current interruption in the load side (single-phase side), only one switch can and must be on at any time.

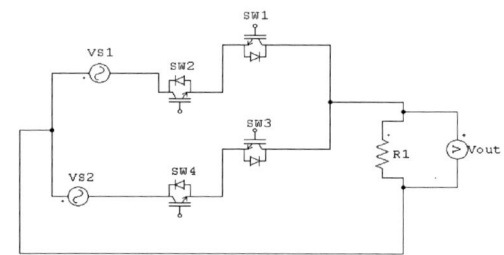

Fig. 8. Simulation Circuit of a Two Phase to Single Phase Matrix Converter

The Fig. 8 shows the simulation circuit of a Two Phase to Single Phase Matrix Converter. In steady state, both of the devices in the active bidirectional switch cell are gated to allow both directions of current flow. The upper bidirectional switch is closed when a commutation to lower bidirectional switch is required.

C. THREE PHASE TO SINGLE PHASE MATRIX CONVERTER

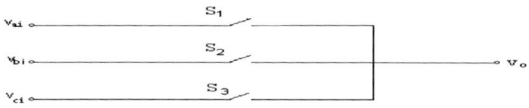

Fig. 9. Representation of a Three Phase to Single Phase Matrix Converter

A Three Phase to Single- Phase Matrix Converter is shown in Fig. 9. The converter is composed of three bidirectional switches S_1, S_2 and S_3. Each switch connects the output line to an input phase. To avoid short-circuit in the source-side (three-phase side) and current interruption in the load side (single-phase side), only one switch can and must be on at any time. The switches are turned on the off in a sequential and cyclical pattern. For the jth switching period, if t^j_1, t^j_2 and t^j_3 are the on-time intervals of S_1, S_2 and S_3, respectively, we have,

$$t^j_1 + t^j_2 + t^j_3 = T_m = 1 / f_m, \qquad (4)$$

where T_m is the switching period. The output line is connected to an input voltage for a specific period of time. Thus, the output voltage is a concatenation of segments of the three input voltages. Therefore, the output voltage waveform $v_o(t)$ is a function of the three input voltages $v_{ai}(t)$, $v_{bi}(t)$ and $v_{ci}(t)$. In general, the output voltage harmonic components depend also on the input frequency and the switching strategy.

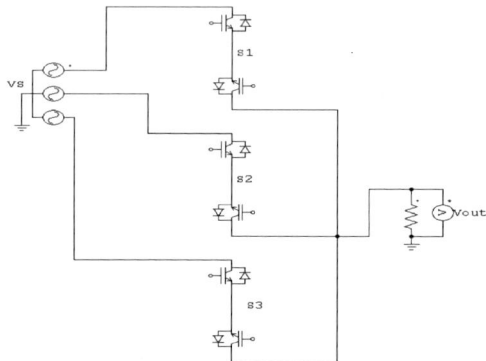

Fig. 10. Simulation Circuit of a Three Phase to Single Phase Matrix Converter

The simulation circuit of a Three Phase is Single Phase Matrix Converter is shown in Fig. 10. The analysis is started assuming that S_1 is initially on when S_2 and S_3 have been off. The Circuit operation is analyzed for a complete switching sequence assuming that S_1 is on followed by commutation to S_2, then to S_3 and finally back to S_1.

D. THREE PHASE TO THREE PHASE MATRIX CONVERTER

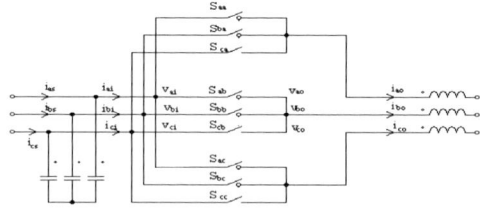

Fig. 11. Representation of a Three Phase to Three Phase Matrix Converter

A Three-Phase to Three-Phase Matrix Converter is structured based on the Three-Phase to Single-Phase Matrix Converter. If three sets of the single-output Matrix Converters are connected to the same input voltages, a three-output Matrix Converter is constructed. The structure of a three-phase to three-phase Matrix Converter is shown in Fig. 11 [3], [6].The converter consists of nine bidirectional switches (S_{aa}, S_{ba}, and S_{cc}) whose operations are coordinated by a number of switching functions.

978-1-4244-1871-8/07 $25.00 © 2007 IEEE

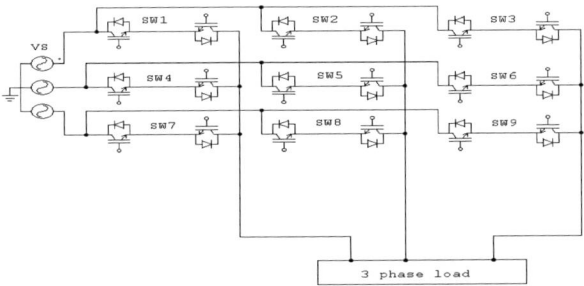

Fig. 12. Simulation Circuit of a Three Phase to Three Phase Matrix Converter

The Matrix Converter can represent a symmetric electrical system, if a proper switching strategy is used. The simulation circuit of a Three Phase to Three Phase Matrix Converter is shown in Fig. 12. S_{ab}, S_{bb} and S_{cb} must be switched with a phase delay of 120° with respect to S_{aa}, S_{ba} and S_{ca}. Also, a 240° phase delay must be considered for the switches S_{ac}, S_{bc} and S_{cc} [10] to get a three phase output.

V. SIMULATION RESULTS

Simulations were done using PSIM software package. The Single Phase Matrix Circuit was constructed using the topology shown in Fig. 6. The output voltage and current waveforms were observed for a resistive load of R = 10 Ω and an input frequency of f_i=50 Hz. Similarly, a Two Phase to Single Phase Matrix Converter, Three Phase to Single Phase Matrix Converter and Three Phase to Three Phase Matrix Converter was constructed and the output voltage and current waveforms were observed for different switching patterns. The following are sample of results:

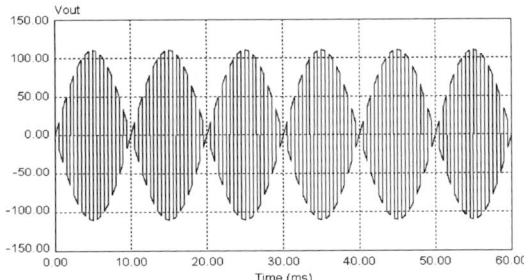

Fig. 13. Simulation result of Single Phase Matrix Converter Output Voltage for V_i=110V (Peak Value), R=10Ω and f_o=100Hz, Scale: Y:50V/div, X:10ms/div

The waveforms in Fig. 13 and Fig. 14 show the output voltage of a Single Phase Matrix Converter for output frequencies of 100 Hz and 50 Hz.

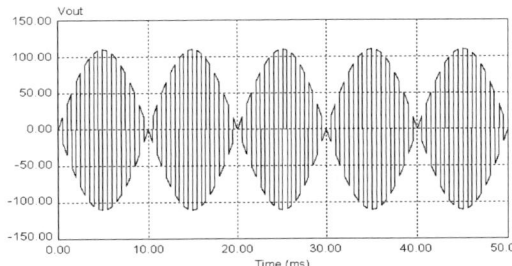

Fig. 14. Simulation result of Single Phase Matrix Converter Output Voltage for V_i=110V (Peak Value), R=10Ω and f_o=50Hz, Scale: Y:50V/div,X:10ms/div

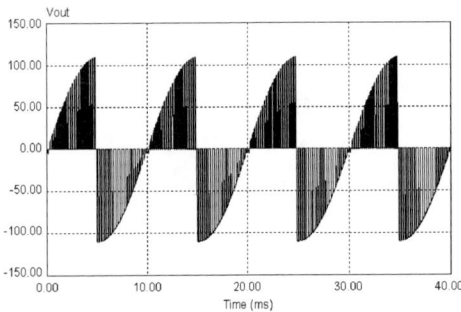

Fig. 15. Simulation result of SPWM Single Phase Matrix Converter Output Voltage for desired frequency f_r =100 Hz, V_1 =110V (Peak Value), R=10Ω and carrier frequency f_c=5 KHz with modulation index =1, Scale: Y: 50V/div, X: 10ms/div

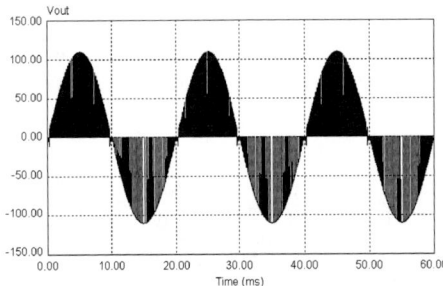

Fig. 16. Simulation result of SPWM Single Phase Matrix Converter Output Voltage for desired frequency f_r=50 Hz, V_i=110V (Peak Value), R= 10Ω and carrier frequency f_c =5 KHz with modulation index =1, Scale: Y: 50V/div, X: 10ms/div

The waveforms in Fig. 15 and Fig. 16 examine the output voltage and output current of a Single Phase Matrix Converter synthesized using the Sinusoidal Pulse Width Modulation (SPWM) technique for an output frequency of 100 Hz and 50 Hz respectively.

978-1-4244-1871-8/07 $25.00 © 2007 IEEE

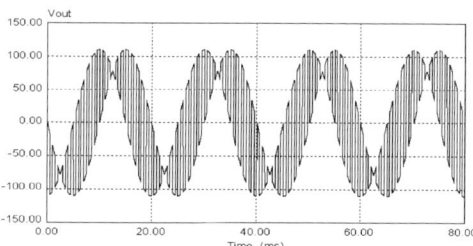

Fig. 17. Simulation result of Two Phase Single to Phase Matrix Converter Output Voltage for V_i=110V (Peak Value), R=10 Ω and f_o=50Hz, Scale: Y:50V/div,X:10ms/div

The waveform in Fig. 17 shows the output voltage of a Single Phase Matrix Converter for output frequencies of 50 Hz.

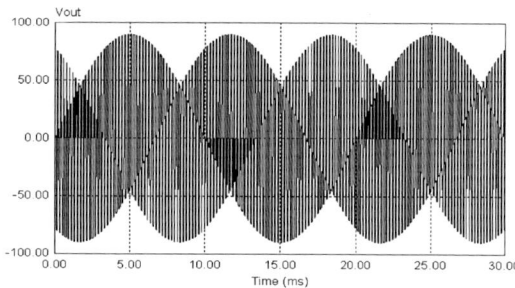

Fig. 18. Simulation result of a Three Phase to Single Phase Matrix Converter for V_i =70.7V (RMS Value), R=10 Ω and output frequency f_o=50 Hz , Scale: Y: 50 V/div, X: 10ms/div

The waveform in Fig. 18 examines the output voltage of a Three Phase to Single Phase Matrix Converter.

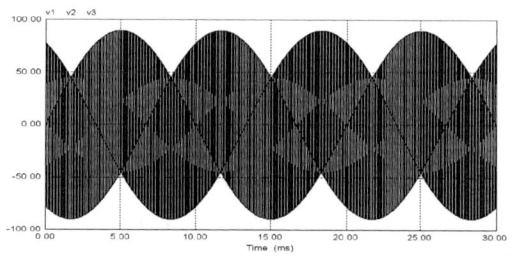

Fig. 19. Simulation result of a Three Phase to Three Phase Matrix Converter for V_i =70.7V (RMS Value), R=10Ω and output frequency f_o=50 Hz , Scale: Y: 50 V/div, X: 5ms/div

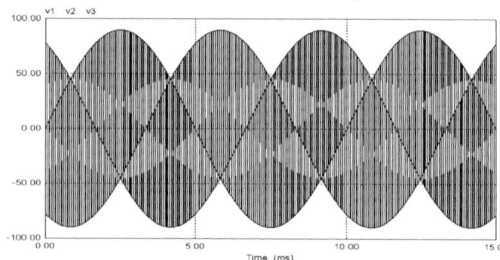

Fig. 20. Simulation result of a Three Phase to Three Phase Matrix Converter for V_i =70.7V (RMS Value), R=10 Ω and output frequency f_o=100 Hz , Scale: Y: 50 V/div, X: 5ms/div

The waveforms from Fig. 19 and Fig. 20 examines the output voltage of a Three Phase to Three Phase Matrix Converter. Sample simulations were also carried out to study the operation of a Single Phase Matrix Converter and Three Phase to Single Phase Matrix Converter that includes different switching frequencies and modulation index in case of a SPWM Single Phase Matrix Converter.

VI. CONCLUSION

Thus, in this paper a topology of a Single Phase Matrix Converter has been developed in PSIM software environment and results are obtained for different output frequencies. Further, the operation of a Single Phase Matrix Converter synthesized using the Sinusoidal Pulse Width Modulation (SPWM) technique has been examined for different reference frequencies. Simulation results for a Two Phase to Single Phase Matrix Converter, Three Phase to Single Phase Matrix Converter and a Three Phase to Three Phase Matrix Converter have also been shown.. Hence, this paper attempts to give an overview of Matrix Converters thereby providing an insight into one of the emerging trends of Power Converters. The simulation results are helpful to study the operation of a Matrix Converter, as an Unrestricted Frequency Changer. Further studies aim to analyse the operation of a Matrix Converter for different loading conditions and synthesise the output voltage using various other modulation techniques.

REFERENCES

[1] C. Klumpner, P. Nielson, I. Boldea and F. Blarabjet, "A new matrix converter motor (MCM) for industry applications," *IEEE transactions on Industrial Electronics*, vol. 49, no. 2, April 2002, pp. 325-335.

[2] P. Wheeler, J. Rodriguez, J. Clare, L. Empringham and A. Weinstein., "Matrix converters: A Technology Review," *IEEE Transactions on Industrial Electronics*, vol. 49, no. 2, April 2002, pp. 276 – 288.

[3] A. Alesina and M. Venturini, "The Generalised Transformer: A new bidirectional sinusoidal waveform frequency converter with continuously adjustable input power factor," in *Proc. of IEEE-PESC'80*, vol. 1, 1980, pp. 242-252.

[4] S. H. Hosseini and E. Babaei, " A New Generalised Direct Matrix Converter," *IEEE International Symposium of Industrial Electronics 2001*, Proc. ISIE 2001, vol. 2, pp. 1071-1076.

[5] A. Zuckerberger, D. Weinstock and A. Alexandrovitz, "Single Phase Matrix Converter," in *Proc. on Electric Power Applications*, vol. 144, no. 4, July 1997, pp. 235-240.

[6] A. Zuckerberger, D. Weinstock, A. Alexandrovitz, "Simulation of three-phase loaded matrix converter," in *Proc. on Electric Power Applications* , vol. 143, no. 4, July 1996, pp. 294-300.

[7] M. Tabone, C. S. Staines and J. Cilia, " Low Cost Three Phase to Single Phase Matrix Converter," *IEEE International Conference on Industrial Technology*, 2004, Vol. 1, Dec. 2004, pp. 474 – 479.

[8] Abdollah Koei and Subbarary Yuvrajan, " Single Phase AC-AC Converter using Power MOSTET's," *IEEE Transactions on Industrial Electronics*, vol. 35, no. 3, August 1998, pp. 442-443.

[9] Siti Zaliha Mohammad Noor, Mustafar Kamal Hamzah and Ahmad Farid Abidin, " Modelling and Simulation of a DC Chopper using Single Phase Matrix Converter Topology," *IEEE Sixth International Conference PEDS 2005*, Vol. 1, Jan 2006, pp. 827-832.

[10] M. H. Rashid, *Power Electronics: Circuits, Devices and Applications*, New Delhi : Prentice Hall of India, 2005, pp. 536.

Hybrid Three-Phase HPF Rectifier with Programmable Input Current THD, Using FPGA Device and VHDL Language

Jurandir O. Soares, Carlos A. Canesin
São Paulo State University – UNESP,
FEIS – DEE, P.O. Box 31,
Ilha Solteira(SP), Brazil
Email: jurandir@aluno.feis.unesp.br;
Email: canesin@dee.feis.unesp.br

Luiz C. de Freitas
Federal University of Uberlândia
Faculty of Elect. Engineering – FEELT
Uberlândia(MG), Brazil
Email: freitas@ufu.br

Abstract— In this paper is proposed and analyzed a digital hysteresis modulation using a FPGA (Field Programmable Gate Array) device and VHDL (Hardware Description Language), applied at a hybrid three-phase rectifier with almost unitary input power factor, composed by parallel SEPIC controlled single-phase rectifiers connected to each leg of a standard 6-pulses uncontrolled diode rectifier. The digital control allows a programmable THD (Total Harmonic Distortion) at the input currents, and it makes possible that the power rating of the switching-mode converters, connected in parallel, can be a small fraction of the total average output power, in order to obtain a compact converter, reduced input current THD and almost unitary input power factor. Finally, the proposed digital control, using a FPGA device and VHDL, offers an important flexibility for the associated control technique, in order to obtain a programmable PFC (Power Factor Correction) hybrid three-phase rectifier, in agreement with the international standards (IEC, and IEEE), which impose limits for the THD of the AC (Alternate Current) line input currents. The proposed strategy is verified by experiments.

Keywords - Digital Control, FPGA Device, Hybrid Rectifier, Active Power-Factor Correction, Hysteresis Modulation.

I. INTRODUCTION

In the last decade, due to the significant increase of the non-linear loads, including the single-phase and three-phase rectifiers, high values of harmonic distortions in the currents through the AC distribution feeders have been observed, causing significant harmonic distortions in the voltages over the point of common coupling (PCC). In this context, were established international standards such as the IEC 61000-3-2, IEC 61000-3-4, and IEEE 519, in order to impose limits on harmonic current emissions from these equipments [1-2]. Furthermore, considering the necessity to impose restrictions on the harmonic input current content, several researches have been developed passive active strategies/structures, in order to provide high input power factor, and compatible electromagnetic levels for the three-phase rectifier applications. Considering the researches developed in order to combine the advantages of passive and active rectifiers, the *"Hybrid Rectifiers"* were proposed, applied to medium and high power levels. Hybrid rectifiers are generated through special connections of one passive rectifier with PWM active rectifiers.

The passive rectifier operates in low frequency, and transfers the most part of the total active power to the load. On the other hand, the PWM active rectifier transfers only a small fraction of the total active output power, and operates in high frequency. The most important advantage of these structures is that they combine the robustness and efficiency of the passive and active rectifiers, respectively, with the imposition of an input current with reduced THD, through the appropriate control of the PWM active rectifiers [3-5]. Therefore, hybrid rectifiers, in a general analysis, behave as controlled current sources, using a very simple strategy. Thus, considering the proposed structure shown in Fig. 1 [5], imposing a special reference current to the SEPIC converters so that the driven current,s by them, when added with the driven currents by the standard 6-pulses diode rectifier, will result in a programmable input current waveform, with low harmonic distortion as desired and designed, in order to comply with the standards. So, manipulating the waveform of the reference current of each SEPIC converter, it is possible to obtain an input current with a shape of 12, 18, 24-pulses or more, and also to obtain an approximately sinusoidal input current, resulting in an almost unitary input power factor. Nevertheless, considering the proposed rectifier, shown in Fig. 1, operating with an input current of "q.6±1" pulses [6], depending on the index "q" and processed output power values, the harmonic content may not obey the limits imposed by international standards, for example the IEC61000-3-4. Furthermore, it was proposed by the authors [5] an analogical control technique that results in complex analogical circuitry, hindering the imposition of a sinusoidal input current and consequently not allowing the appropriate control of the THD at the input currents. Thus, in order to obey fully the limits imposed by international standards IEC/IEEE, it is proposed a modified digital control technique in this paper, using hysteresis modulation, offering flexibility to the implemented circuitry, that imposes the input current of the controlled converters (for example, $i_{a2}(\omega.t)$), resulting in a almost unitary power factor and small fraction of the total active power processed by controlled rectifier, allowing programmable line input currents THD. It should be noticed that, in order to implement the proposed digital control, it is necessary to make sampling and the digital

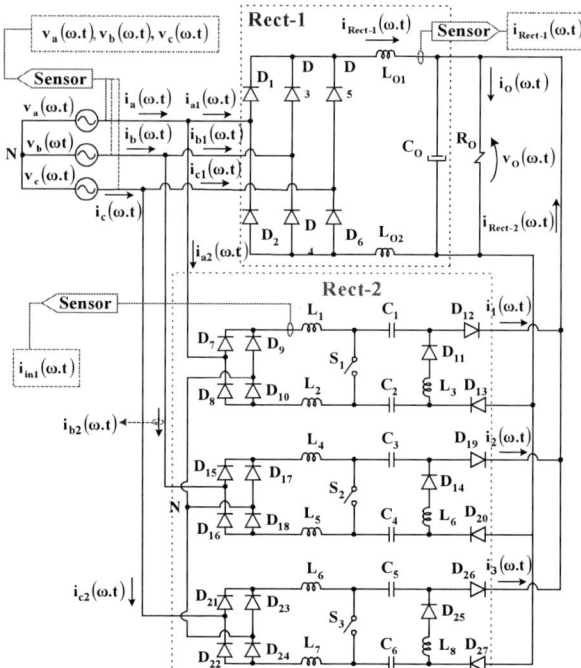

Fig. 1. Hybrid Three-Phase Rectifier with High Input Power Factor.

processing of several signals of the hybrid three-phase rectifier. In this context, the proposed digital control uses a FPGA device and VHDL language, due to its characteristics such as flexibility and concurrent processing, allowing to perform all the procedures of control in several simultaneous operations. Additionally, the use of hardware description language results in reduction of the time to design and to redesign and for implementation the control technique. Furthermore, the required logic must be optimized even using FPGA devices, in order to limit the occupation of silicon area of FPGA devices, and to improve the performance of control [7].

II. THEORETICAL ANALYSIS

In order to simplify the analysis, it will be considered a three-phase sinusoidal balanced input voltage system. Thus, it will be discussed in this paper just the control of the input current $i_a(\omega.t)$, which belongs to phase "a", as shown in Fig. 1. The main goal of the analysis presented in this section is to establish a relation between the THD (Total Harmonic Distortion) imposed to line input currents, with practically sinusoidal waveforms, as shown in Fig. 2, and the total input and output powers processed by the active and passive rectifiers.

According to Fig. 2, the input current of the active rectifier, $i_{a2}(\omega.t)$, follows a reference current generated by $i_{sin}(\omega.t)-i_{a1}(\omega.t)$. The current $i_{sin}(\omega.t)$, given by equation (1), is an imposed auxiliary sinusoidal current. It should be noticed that, the shape of $i_a(\omega.t)$ depends on a constant relation between peak values of the currents $i_{a1}(\omega.t)$ and $i_{sin}(\omega.t)$, represented through the parameter "K" according to equation (2).

$$I_{sin}(\omega.t) = K.I_{Rect-1}.sin(\omega.t) \tag{1}$$

$$K = \frac{I_m}{I_{Rect-1}} \tag{2}$$

Where:

I_m - Peak value of auxiliary sinusoidal current $i_{sin}(\omega.t)$.

I_{Rect-1} - Peak value of current $i_{a1}(\omega.t)$ (equal to average value of output current of the Rect-1).

The discontinuity of $i_{a2}(\omega.t)$, as shown in Fig. 2, occurs due to the unidirectionality of the proposed hybrid three-phase rectifier. Therefore, the time interval $\omega.\Delta t$ is given by equation (3).

$$\omega.\Delta t(K) = \begin{cases} asin\left(\dfrac{1}{K}\right) - \dfrac{\pi}{6}, & 1 \le K \le 2 \\ 0, & K > 2 \end{cases} \tag{3}$$

It is interesting to observe that, increasing the value of the parameter "K", it will result in a reduction of the THD value of the line input current $i_a(\omega.t)$, until it reaches THD=0, when K=2. However, it increases the processed power by the active rectifier (Rect-2) and reduces the overall efficiency of the hybrid rectifier.

According to analysis performed by [8] (To find a value of "K that provides the minimum power processed by the active rectifier and the maximum THD value allowed by the international standards), for a line input current THD smaller than 3%, the value of the parameter "K" increases quickly, increasing also the power processed rate by the active rectifier. Therefore, a nominal operation point with THD larger than 3% must be considered in the design, in order to reduce the power processed by the switching-mode converters.

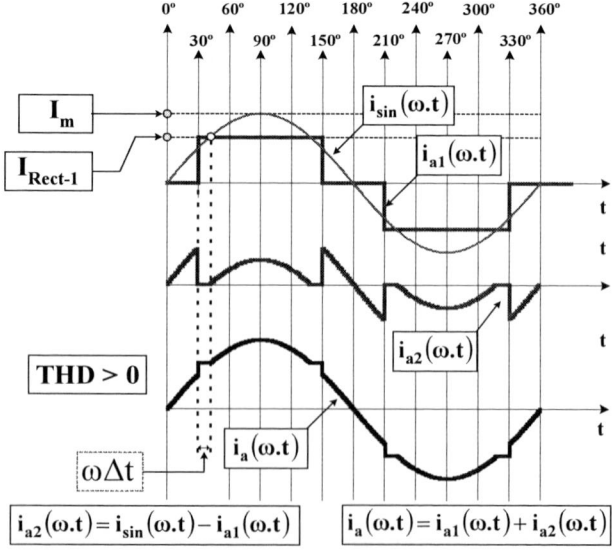

Fig. 2. Theoretical current waveforms, regarding phase "a", of the hybrid three-phase rectifier, operating with almost unitary input power factor.

978-1-4244-1871-8/07 $25.00 © 2007 IEEE   420

The definitive choice of the point of operation will impose simulation verifications, in order to analyze if the values of the line input current harmonic components are in agreement with the limits imposed by the standards (such as IEC 61000-3-2, or, IEC 61000-3-4, or, IEEE 519)..

III. PROPOSED DIGITAL CONTROL

The proposed digital control uses a Xilinx FPGA device (Spartan 2e) and VHDL language in order to obtain the control signals to the SEPIC rectifiers. In this context, in order to compose the proposed control, it is necessary to make sampling of the input current $i_{in1}(\omega.t)$ of the $SEPIC_1$ converter, and the output current $i_{Rect-1}(\omega.t)$ of the standard 6-pulses diode rectifier, as shown in Fig. 1, and additionally, to sample the input current $i_{a1}(\omega.t)$ and input voltage $v_a(\omega.t)$ in the phase "a". The basic idea of this control is that the current $i_{in1}(\omega.t)$ follows an imposed reference of current, digitally generated. Thus, the current $i_{a2}(\omega.t)$, when added to $i_{a1}(\omega.t)$, will result in a line input current $i_a(\omega.t)$ with the same theoretical waveform, as shown in Fig. 2. To operate the hybrid rectifier with almost unitary input power factor, it is digitally generated a reference of current $S_{ignal\_Ref\_sin}(n)$, as shown in Fig. 3, to impose the shape of the input current, $i_{a2}(\omega.t)$ according to Fig. 2. Therefore, in order to compose the control logic showed in Fig. 3, it will be necessary two auxiliary logical signals, $C_{ontrol\_Isin}(n)$ and $C_{ontrol\_Ia1}(n)$. These auxiliary signals are generated through an additional auxiliary logical signal $I_{a1\_sensor}(n)$ combined with logical counters (VHDL logic), as shown in Fig. 4. The signal $I_{a1\_sensor}(n)$, given by an analogical sensor, receive logical signal "1" if $i_{a1}(\omega.t)=0$, otherwise $I_{a1\_sensor}(n)$ receive logical signal "0". The width of the logical signals, $C_{ontrol\_Isin}(n)$ and $C_{ontrol\_Ia1}(n)$, depends on the time interval of discontinuity "$\omega.\Delta t(K)$", calculated by equation (3). Thus, according to Fig. 4, the signal $C_{ontrol\_Ia1}(n)$ will receive logical signal "1" if $\{30°+\omega.\Delta t(K)\}<\omega.t<\{150°-\omega.\Delta t(K)\}$ or if $\{210°+\omega.\Delta t(K)\}<\omega.t<\{330°-\omega.\Delta t(K)\}$, otherwise $C_{ontrol\_Ia1}(n)$ will receive logical signal "0". Furthermore, the signal $C_{ontrol\_Isin}(n)$ will receive logical signal "0" if $30°<\omega.t<\{30°+\omega.\Delta t(K)\}$, or if $\{150°-\omega.\Delta t(K)\}<\omega.t<150°$, or if $210°<\omega.t<\{210°+\omega.\Delta t(K)\}$, or if $\{330°-\omega.\Delta t(K)\}<\omega.t<330°$, otherwise it will receive logical signal "1". In Fig. 3, the input signals are: Control signals, $C_{ontrol\_Ia1}(n)$, $C_{ontrol\_Isin}(n)$ and $V_{a\_sensor}(n)$, and the output current $I_{Rect-1}(n)$. The purpose of the signal $C_{ontrol\_Ia1}(n)$ is to generate a digital current $I_{a1\_\omega\Delta t}(n)$ through the following logic selection. If $C_{ontrol\_Ia1}(n)=$"1", $I_{a1\_\omega\Delta t}(n)$ becomes $I_{Rect-1}(n)$, otherwise if $C_{ontrol\_Ia1}(n)=$"0", $I_{a1\_\omega\Delta t}(n)$ becomes "00000000". The signal $I_{sin\_unit}(n)$ is a rectified sinusoidal current with unitary amplitude, digitally generated, and synchronized with AC grid by the signal $V_{a\_sensor}(n)$. The signal $V_{a\_sensor}(n)$, given by an analogical sensor, receive logical signal "0" if $v_a(\omega.t)>0$, otherwise $V_{a\_sensor}(n)$ receive logical signal "1". Thus, multiplying the average value of output current $I_{Rect-1}(n)$ by the parameter "K" and $I_{sin\_unit}(n)$, it

will result in a rectified sinusoidal signal $I_{sin}(n)$ with amplitude $K.I_{Rect-1}(n)$. The signal $C_{ontrol\_Isin}(n)$ is used to obtain an interval of discontinuity "$\omega.\Delta t(K)$" in the signal $I_{sin}(n)$, resulting in the signal $I_{sin\_\omega\Delta t}(n)$, which is obtained by the following logic selection: If $C_{ontrol\_sin}(n)=$"1", $I_{sin\_\omega\Delta t}(n)$ becomes $I_{sin}(n)$, otherwise if $C_{ontrol\_sin}(n)=$"0", $I_{sin\_\omega\Delta t}(n)$ becomes "00000000". This way, the control for the line input current THD is achieved only with a desirable specification of the parameter "K", according to theoretical analysis previously discussed. Finally, the reference of current $S_{ignal\_Ref\_sin}(n)$ to the $SEPIC_1$ rectifier is generated by $I_{sin\_\omega\Delta t}(n)-I_{a1\_\omega\Delta t}(n)$, as shown in Fig. 3.

A. Digital Hysteresis Modulation

Digital hysteresis modulation has been widely applied in its standard form, where the switch turns on and turns off when the sampled current differs from its reference, considering a specification for the threshold. This solution, although very simple, has a major drawback due to the frequency jitter associated to the sampling process, causing spectrum components even in the low-frequency range [9]. In this case, the switch turn-on and the switch turn-off not provide that the current commutates exactly at the hysteresis band, resulting an additional variations in the switching frequency.

Considering that the turn-on and turn-off times are always a multiple of the sampling period, it can be needed unpredictable switching frequency in order to maintain the desired instantaneous average value of controlled current [9]. The authors in [9] investigated a digital hysteresis modulation based on switching-time prediction, which theoretically avoids frequency jitter and ensures a dynamic performance similar to that obtainable with an analog hysteresis modulation. Besides, it was used a simple frequency-stabilization algorithm, maintaining the advantages of the PWM technique [10].

Fig. 5 and 6 shows the proposed methodology using the hysteresis modulation to generate the pulses to $SEPIC_1$ rectifier, following partially the concepts applied in [9]. In the proposed hysteresis modulation, the upper reference is eliminated according to Fig. 5.

The control of the ripple and frequency range of the input current $I_{in1}(n)$ of the $SEPIC_1$ rectifier can be performed by two distinct forms. The control law of modulation in both cases is given by the states E_0, E_1, E_2, E_3 and E_4, detailed in Fig. 6.

1) Constant turn-on time (t_{on}) and variable turn-off time (t_{off}) and operation frequency: The time t_{on} occurs during the state E_0 and the P_{ulse_Gate} receive logical signal "1". This time interval is controlled by a down-up counter, which determines exactly the instant of the transition to the state E_1, avoiding errors of the comparator associated to sampling process, previously discussed. In the states E_1 and E_3 occurs the switching transitions turn-on to turn-off (P_{ulse_Gate} receive logical signal "0") and turn-off to turn-on (P_{ulse_Gate} receive logical signal "1"), respectively. In these states, also there is

not actuation of the comparator, thus, the time intervals are controlled by down-up counters, aiming for avoid an unpredictable operation of the control due the presence of switching noises.

The time t_{off} occurs during the state E_2. This time interval is controlled by the comparator within the following logic: If $I_{in1}(n) \geq S_{ignal\_Ref\_sin}(n)$, P_{ulse_Gate} will receive logical signal "0", otherwise, if $I_{in1}(n) < S_{ignal\_Ref\_sin}(n))$, P_{ulse_Gate} will receive logical signal "1" and state change to E_3.

Finally, the state E_4 has the function of verify if the signal $S_{ignal\_Ref\_sin}(n)$ changed its value during the state E_3, within the following logic: If $I_{in1}(n) \geq S_{ignal\_Ref\_sin}(n)$, the state change to E_0, otherwise, if $I_{in1}(n) < S_{ignal\_Ref\_sin}(n))$, the state E_3 is maintained. The P_{ulse_Gate} receive only logical signal "1" in the state E_4. Thus, the inherent dynamic performance of the analogic hysteresis modulation is perfectly guaranteed.

2) Constant operation frequency and variable turn-on time (t_{on}) and turn-off time (t_{off}): In this operation form is maintained the same control law applied to case 1, previously discussed. Additionally, based on switching-time prediction, the time t_{on} will be calculated to each transition of the state E_2 to E_3, according to equation (4).

$$t_{on}(m) = T_s(m-1) - t_{off}(m-1) \qquad (4)$$

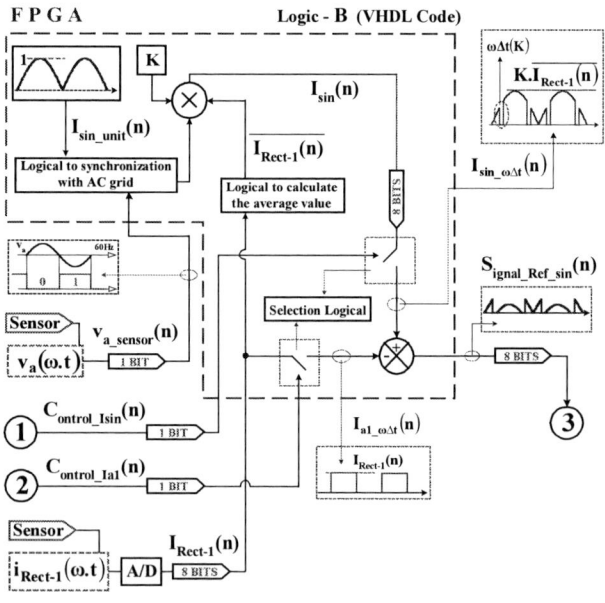

Fig. 3. Details of the logic control to generate the reference current for the SEPIC rectifiers.

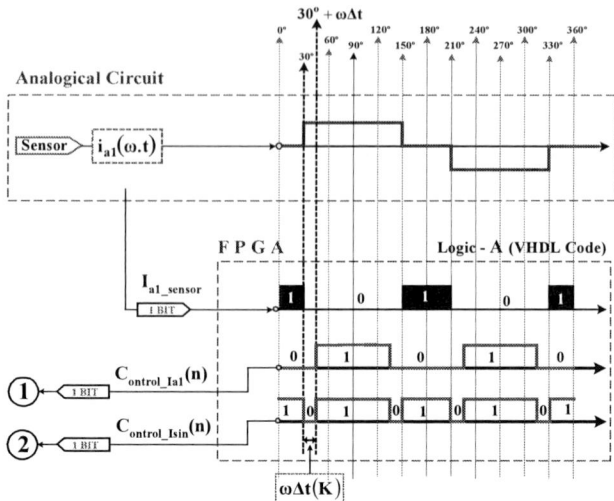

Fig. 4. Auxiliary control for the reference current, in order to generate the pulses to SEPIC rectifier

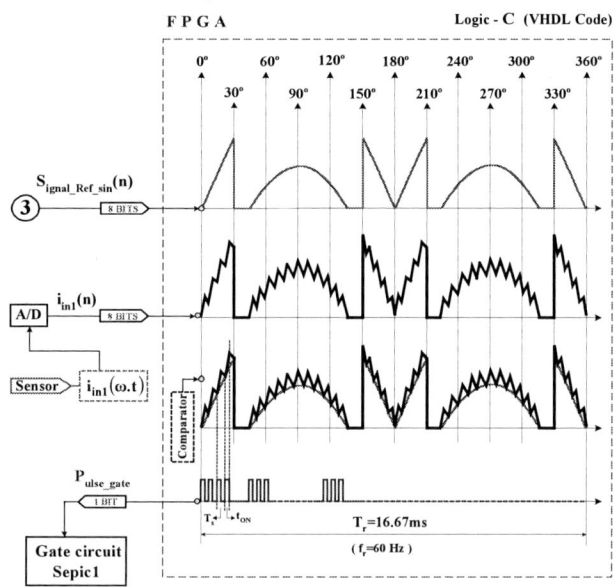

Fig. 5. Methodology to generate the pulses, in order to control the SEPIC$_1$ rectifier.

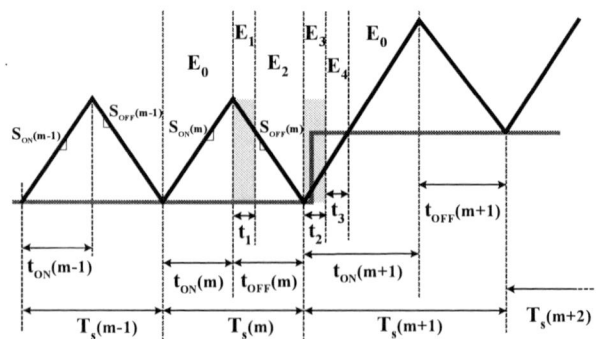

Fig. 6. Details of the methodology to generate the pulses, in order to control the SEPIC$_1$ rectifier.

IV. EXPERIMENTAL RESULTS

Considering the theoretical analysis discussed in sections II and III, applied to proposed digital control, a 3.0kW hybrid three-phase rectifier prototype was built, according to circuit depicted in Fig.1. The input and output data, including the designed parameters and components used in the prototype are shown in Table I. The waveforms shown in Figs. 7, 8 and 9, constitute preliminary experimental results of the phases "a", "b" and "c" of the hybrid three-phase rectifier operating near to 33% of load (P_O=1.0kW). Analyzing the results, showed in Fig. 7, through the Tektronix Wavestar software, one can confirm a THD=4.87% to the line input current $i_a(\omega.t)$, THD=3.42% to the line input current $i_b(\omega.t)$ and THD=3.99% to the line input current $i_c(\omega.t)$, given by harmonic spectrum, shown in Figs. 10, 11 and 12, respectively. Therefore, considering the processed line input currents values (i_{arms}=3.46A, i_{brms}=3.42A i_{crms}=3.32A) by the implemented prototype, it can be concluded that the Standard IEC 61000-3-2 is totally obeyed, being a good preliminary results. However, aiming to the application at high power (i_{arms}>16 A), and to comply with the limits imposed by Standard IEC 61000-3-4, in the future version of this paper, the previously commented results will be improved imposing an appropriate synchronism between the uncontrolled $\{i_{a1}(\omega.t)\}$ and controlled $\{i_{a2}(\omega.t)\}$ currents, in order to compose the desired waveform to line input current $i_a(\omega.t)$, in according to Fig. 2, shown in section II.

TABLE I
Parameters of the Hybrid Three-Phase Rectifiers

Input - Output requirements		Circuit's Parameters	
Parameter	Value	Parameter	Value
Vin_phase(rms)	90 V	$S_{1,2,3}$	SPW11N60S5
$V_{O\ DC}$	210 V	$D_{11,12,13,14,19}$	RHRP860
P_O	3.0 kW	$D_{20,25,26,27}$	RHRP860
fs maximum	60 kHz	$L_{1,2,4,5,6,7}$	2.5mH – EE55/21
fs minimum	28 kHz	$L_{3,6,8}$	5mH – EE55/21
		L_{O1} and L_{O2}	13mH
		Co	680uF

In Fig. 9, are depicted the phase input voltage $\{v_a(\omega.t)\}$ and line input current $\{i_a(\omega.t)\}$ waveforms, with phase-shift between its fundamental components of 0.41^{O}.

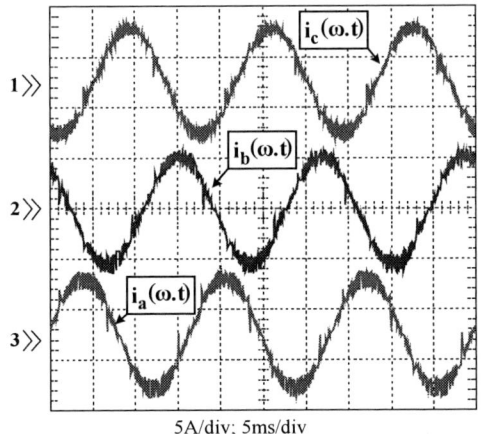

5A/div; 5ms/div

Fig. 7. Input voltage and line input current waveforms, in phases "a", "b" and "c" for the proposed hybrid three-phase rectifier.

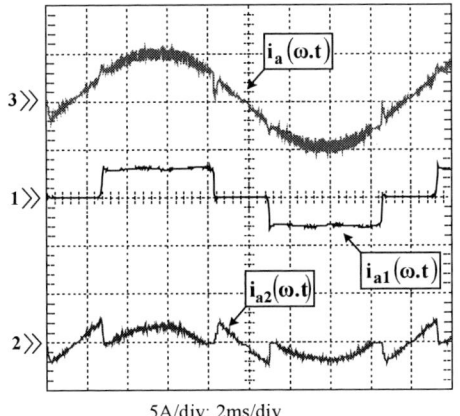

5A/div; 2ms/div

Fig. 8. Details of the line input current waveform, in the "phase a", for the proposed hybrid three-phase rectifier.

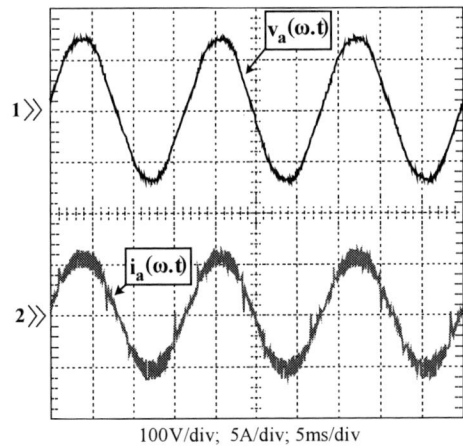

100V/div; 5A/div; 5ms/div

Fig. 9. Input voltage and line input current waveforms, in "phase a", for the proposed hybrid three-phase rectifier.

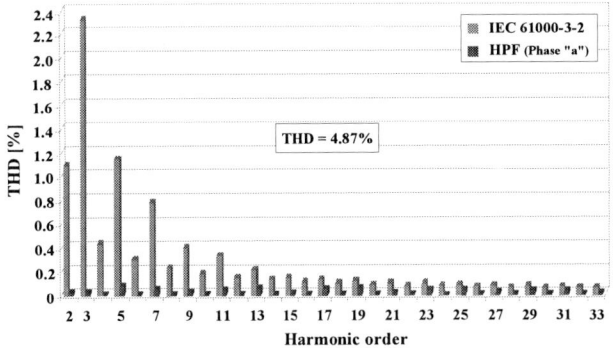

Fig. 10. Frequency spectrum of the line input current $i_a(\omega.t)$, at 33% of load. (% of the fundamental component).

978-1-4244-1871-8/07 $25.00 © 2007 IEEE

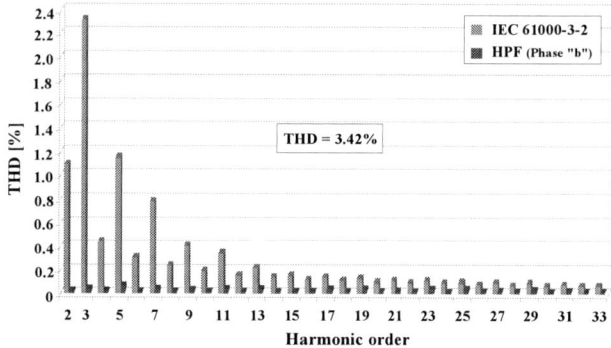

Fig. 11. Frequency spectrum of the line input current $i_b(\omega.t)$, at 33% of load. (% of the fundamental component).

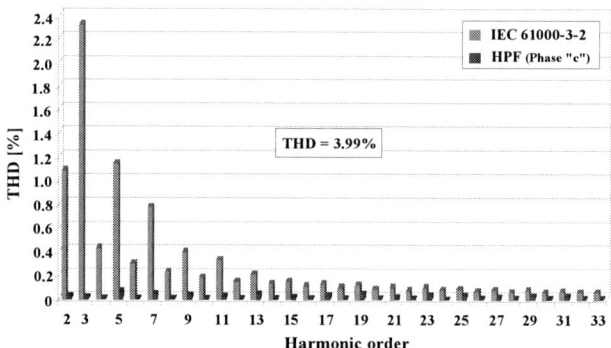

Fig. 12. Frequency spectrum of the line input current $i_c(\omega.t)$, at 33% of load. (% of the fundamental component).

It is important to emphasize that it was verified to the input voltage $v_a(\omega.t)$ a THD=2.95%, by the Tektronix Wavestar software, due to existent harmonic content in the AC grid. Nevertheless, it has caused no problem to operation of the implemented prototype, resulting an almost unitary power factor (PF=0.98).

V. CONCLUSIONS

This paper presented the analysis, and preliminary experimental results of the proposed digital control applied at a hybrid three-phase rectifier, capable to impose the line input current THD, resulting in high input power factor for the proposed converter. The digital control using hysteresis modulation was implemented through a FPGA device, using VHDL language. It was verified through analyzed experimental results that the maximum line input current THD was sustained in 4,87%, resulting an almost unitary power factor (PF=0.98), in phase "a". Through the input currents decomposition of the hybrid rectifier, in Fourier series, and using elementary mathematical relations of own circuit, it was defined in [8] a design methodology that establishes a relationship between the THD (Total Harmonic Distortion) imposed to line input currents, and the total input and output powers processed by active and passive rectifiers. Therefore, the most important aim of the discussed theoretical analysis was to obtain an maximum THD value to the input currents, whose the corresponding harmonic content is according to the limits imposed by international

standards, and consequently a minimum power value processed by active rectifier. Among the advantages of the proposed hybrid three-phase rectifier, are appointed the economic benefits of this topology, which are extremely viable to medium power installations, due to its efficiency and performance. The total active power processed by the switching-mode SEPIC rectifiers is a small fraction of the nominal output active power. The parallel rectifiers, formed by SEPIC converters, operate in continuous conduction mode and hysteresis modulation, offering reduced volume and weight to the structure. Additionally, using the proposed digital control technique, implemented in a FPGA device, allows an important flexibility and facilities to impose a desirable and specified waveform in the input currents, including sinusoidal waveforms, through VHDL code, performing a true programmable line input current TDH for the proposed hybrid three-phase rectifier.

ACKNOWLEDGMENT

The authors would like to thank to CNPq, and Xilinx for supporting this work.

REFERENCES

[1] Std 61000-3-2 and 61000-3-4, "Limits for Harmonic Current Emissions", 1998.

[2] IEEE Std 519, "IEEE Recommended Practices and Requirements for Harmonic Control in Electrical Power Systems", 1992.

[3] Singh, B. N., Chandra, A., Al-Haddad, K., Pandey, A. and Kothari, D. P., "A review of three-phase improved power quality AC-DC converters", *in IEEE Transactions on Industry Applications*, Vol. 51, n. 3, June 2004, pp. 641-660.

[4] Kolar, J. W. and Ertl, H., "Status of the Techniques of Three- Phase Rectifier Systems with Low Effects on the Mains", *in Proceedings of the 1999 IEEE International Telecommunications Energy Conference*, pp. 16-???.

[5] Freitas, L. C. G., Simões, M. G., Canesin, C. A., Freitas L. C. de, and Coelho, E. A. A., "Programmable PFC Based Hybrid Multipulse Power Rectifier for Utility Interface of Power Electronic Converters", *in Proceedings of the 2005 IEEE Power Electronics Specialists Conference*, pp. 2237-2243.

[6] R. Erickson, "Fundamentals of Power Electronics", 2nd Eddition, 861 pages.

[7] Castro, A. de, Zumel, P., Garcia, O., Riesgo, T. and Uceda, J., "Concurrent and Simple Digital Controller of an AC/DC Converter With Power Factor Correction Based on an FPGA", *in IEEE Transactions on Industry Applications*, Vol. 18 , n. 1 , January 2003, pp. 334-343.

[8] Soares, J. O., Canesin, C. A. and Freitas, L. C., "Digital Control Using FPGA Device, and VHDL Language, Applied at a Hybrid Three-Phase Multilevel HPF Rectifier", *in Proceedings of the 2006 IEEE INDUSCON*, in CD.

[9] Stefanutti, W. and Mattavelli, P., "Fully Digital Hysteresis Modulation With Switching-Time Prediction", *in IEEE Transactions on Industry Applications*, Vol. 42 , n. 3 , May/June 2006, pp. 763-769.

[10] Sonaglioni, L., "Predictive Digital Hysteresis Current Control", *in Proceedings of the 1995 IEEE IAS*, Oct 1995, pp. 1879-1886.

A Novel Soft Switched Topology for Power Factor Correction with Load Isolation Capability

M. Yazdanian
University of Tehran
Department of Electrical Engineering
University of Tehran, North Kargar St., Tehran, Iran
Email: m.yazdanian@ece.ut.ac.ir

S. Farhangi
University of Tehran
Department of Electrical Engineering
University of Tehran, North Kargar St., Tehran, Iran
Email: farhangi@ut.ac.ir

Abstract—A novel topology for single phase single stage isolated power factor correction (PFC) based on boost converter has been suggested in this paper. This PFC operates in continuous conduction mode (CCM) with constant switching frequency which injects minimum of harmonic into the power system. The leakage inductor in high frequency transformer of the insolated PFC causes voltage impulses which can damage the semiconductor switches. The proposed topology provides ZCS by control of the leakage inductor current, which suppress the aforementioned impulses, in addition of diminishing the switching losses. A simple predictive controller is used to calculate the switching times which reduces the sampling rate and complexity.

I. INTRODUCTION

The input stage of the most consumer or industrial electronic equipments is an AC/DC converter. The simplest AC/DC converter is a diode rectifier with a capacitor filter, which acts as a nonlinear load and injects harmonics into the power system. Different standards have defined limits for line harmonics, such as IEEE/ANSI 519 or IEC61000-3-2. The power factor correctors (PFC) have been introduced to improve the input power factor and satisfy the AC line harmonic standards. The PFC converters for high power consumers are commonly designed for continuous conduction mode (CCM) [1]-[4].

The isolated boost full bridge PFC is a suitable topology for high power applications [2], [3]. This topology consists of two stages as shown in Fig. 1(a). The first stage is an active PFC which allows the input current to be shaped as a sine wave. The second stage is a DC to DC converter which isolates and regulates the output voltage. In this topology the overall efficiency, which is the product of efficiency of each stage, is limited. In order to improve the overall efficiency, a single stage PFC has been introduced which is shown in Fig. 1(b) [4]. However in this topology, power switches are located between the input inductor and leakage inductance of the transformer. Hence the current discrepancy of these inductors in switching instants produces voltage impulses which can damage the circuit elements.

Different solutions have been proposed to control the generated voltage impulses as the major problem of single stage PFC. Some previous works use snubbers to absorb the voltage impulses [5]. But snubbers decrease the efficiency which is especially inconvenient in applications with high input voltages. The clamp circuit is another solution to control voltage impulses [6]. In this method a capacitor is used to store part of the input inductor energy, and then at a predetermined interval, stored energy is transferred into the transformer. In low current operating condition, this method changes the current fed topology to voltage fed converter. Therefore the total harmonics distortion (THD) increases. Some resonant topologies have presented for the PFC converters. But these resonant converters operate with variable switching frequency which also increases THD.

In this paper a soft switched PWM topology is suggested in which the switches operate under zero current switching (ZCS) condition to reduce the voltage impulse of the switches at constant switching frequency. Therefore, the switching losses are reduced and switching frequency can be increased.

II. SUGGESTED TOPOLOGY AND SWITCHING PROCESS

The suggested topology is a novel single stage isolated boost PFC which uses the soft switched PWM topology to provide ZCS and control the voltage impulse problem in conventional single stage isolated boost converters. This topology is illustrated in Fig. 2. The resonant circuit is made of the capacitor at the secondary side of the isolation transformer and the transformer leakage inductance.

The suggested topology uses two bidirectional voltage switches. These bidirectional voltage switches are a series combination of an IGBT and a diode, as shown in Fig. 2. Therefore these switches can tolerate negative voltage.

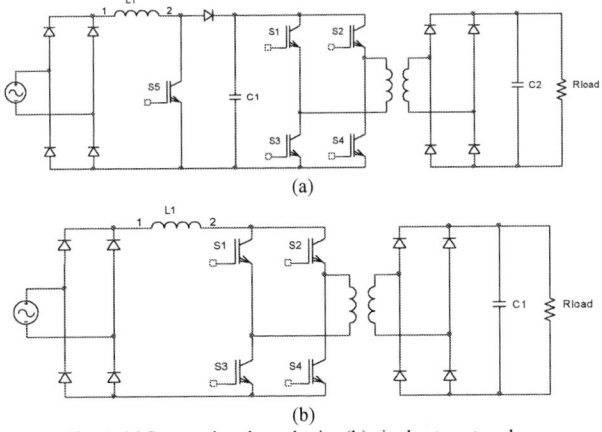

(a)

(b)

Fig. 1: (a)Conventional topologies (b) single stage topology

978-1-4244-1871-8/07 $25.00 © 2007 IEEE   425

However, their conduction losses are increased. Thanks to these bidirectional switches, it is possible to control the leakage inductor current, and provide ZCS for the IGBTs with the constant switching frequency.

The switching cycle consists of two symmetrical half cycles, which at the start of each half cycle two diagonal switches are conducting. The switching sequences over a half switching cycle are illustrated in Fig. 3. The following assumptions are made for operation:
- The PFC operates in continuous conduction mode.
- The switching frequency is much higher than the line frequency.
- The transformer leakage inductor is lumped in the primary, and magnetizing inductor effect is negligible.
- The output voltage ripple is negligible.

In the first stage S2 & S3 are turned on and a pair of output rectifier diodes is conducting (Fig. 3(a)). In this condition, the input inductor injects current into the transformer and the resonant capacitor voltage is equal to the output voltage.

In the second stage, as shown in Fig. 3(b), the switch S1 is turned on and shoots through the input inductance to store energy. However, because of the leakage inductor, the transformer current remains continuous and the reflected output voltage at the primary side decreases the leakage inductor current linearly with time to zero.

When the leakage inductor current reaches zero, the third stage begins. In this stage the diode D2 prevents reversing current in the leakage inductance, and S2 turns off under zero current switching (ZCS). As shown in Fig. 3(c) the output rectifier diodes are off and the resonant capacitor is no longer in parallel with the output capacitor.

Whenever S4 is turned on, the fourth stage starts. During this time the resonant capacitor and the leakage inductor start to oscillate at the resonant frequency. When the leakage inductor current exceeds the input current, the inverse parallel diode of S3 begins to conduct, then the S2 turns off under ZCS condition at this time. Fig. 3(d) depicts this consequent. In the oscillation process, the leakage inductor current will decrease again, and when it becomes equal to the input current value, D3 turns off. Then the leakage inductor current remains constant and first half of the switching cycle is terminated (Fig. 3(e)).

During the second half of the switching cycle, the converter operates in the same way as explained above, except that the voltage capacitor and the current inductor have opposite polarities. The main waveforms and switching commands in the ideal condition are depicted in Fig. 3(f).

III. CONTROL STRATEGY

The conventional control block diagram of a PFC

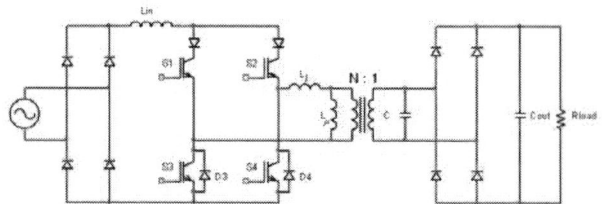

Fig. 2: Suggested topology

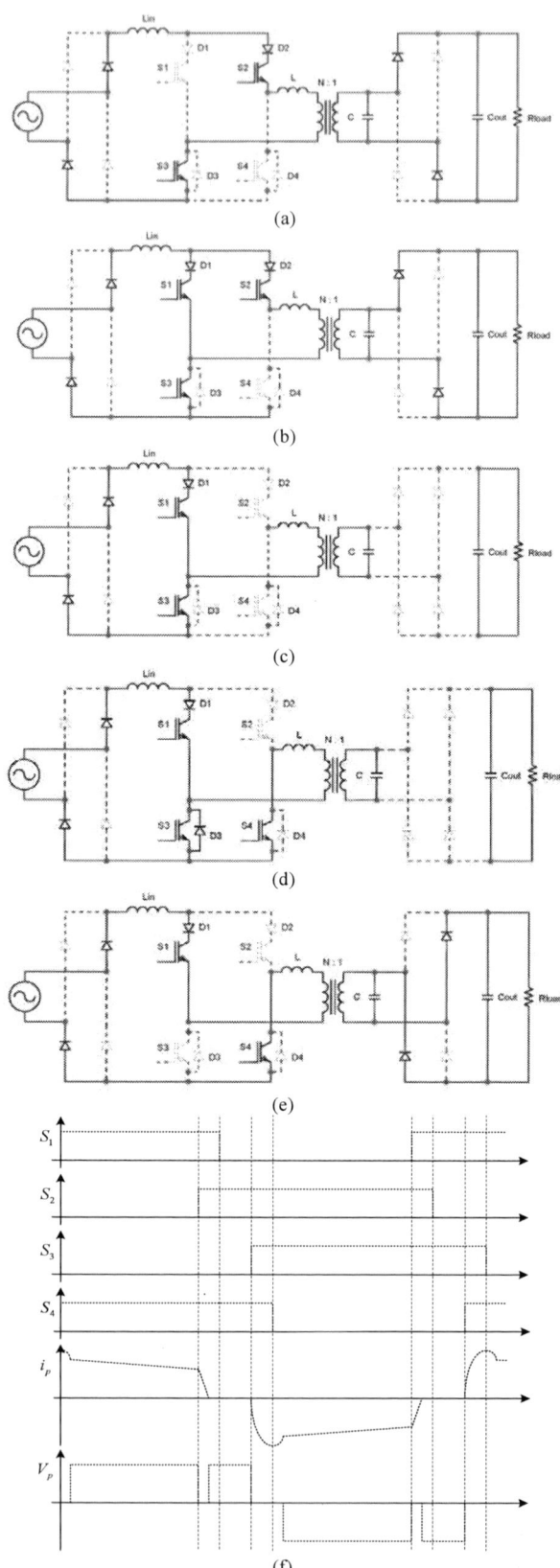

Fig. 3: (a) to (e): Five state of the switching process at half switching cycle, (f) Some of the waveforms at the ideal conditions

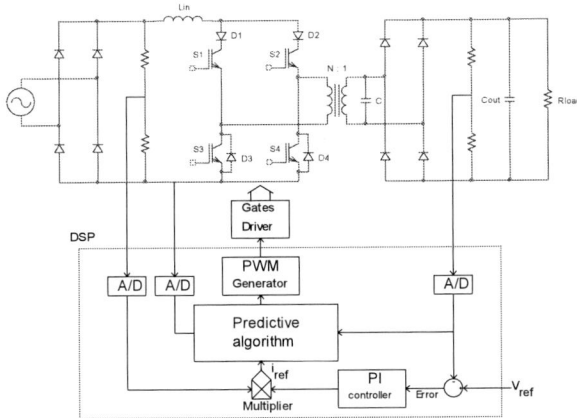

Fig. 4: Control block diagram of the proposed PFC

converter is composed of an inner line current control loop and an outer dc-link voltage control loop [2]-[4]. The inner current controller forces the input inductor current to follow the reference current, which is proportional to the rectified input voltage, in order to achieve unity power factor. Different predictive algorithms have been proposed to control the inner current control loop, because of its ability in nonlinear systems control, simple calculations and fast response. Many predictive controllers are based on valley current control strategy. Hence in this work predictive valley current control strategy has been used. The block diagram of this controller is illustrated in Fig4. In valley control strategy, the local minimum of the input current follows the reference current. The input current waveform under valley control over one line cycle is shown in Fig. 3.

Each half cycle consists of four stages as has been described in the previous section. The current of S1 over half switching cycle is depicted in Fig. 6. As can be seen in this figure, each half switching period can be expressed as the sum of four stages time intervals. The valley current controller determines the shoot through time, but other stages interval should be calculated by another algorithm. If T_n is n^{th} stage ($1 \leq n \leq 4$) the shoot through time can be written as follows:

$$d^k T_s' = T_2 + T_3 + T_4 \qquad (2)$$

T_2 and T_4 can be calculated by using equivalent resonant circuit. For the purpose of simplification all switches have been assumed ideal and an ideal transformer with a lumped leakage inductance at the primary side of transformer has been adopted as the transformer model. Then the resonant capacitor can be reflected to the primary side of ideal transformer.

In the oscillation time, the output rectifier diodes are off, hence the resonant circuit can be simplified by the leakage inductance and reflected capacitor. Therefore T_2 and T_4 can be calculated from equivalent circuit. For simplicity, the input inductor current has been assumed equal at the beginning and end of each switching cycle. Then

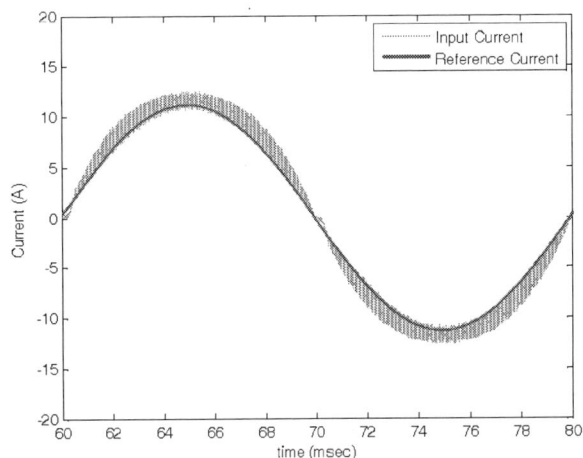

Fig. 5: Input current waveform under valley current control strategy

$$T_2 = L \frac{i_{in}^k}{n V_{out}^k} \qquad (3)$$

The oscillation equation can be written as follows:

$$i(t) = V_{out}^k \sqrt{\frac{C}{L}} \times \sin(\frac{nt}{\sqrt{LC}}) \qquad (4)$$

Hence T_4 can be obtained as follows:

$$T_4 = \frac{\sqrt{LC}}{n} \sin^{-1}(\frac{i_{in}^k}{V_{out}^k} \sqrt{\frac{L}{C}}) \qquad (5)$$

Therefore T_3 can be calculated from (2), (3) and (5). Then all the switching times can be obtained using above equations in the predictive algorithm.

IV. DESIGN CONSIDERATIONS

The resonant circuit should provide ZCS over all the line period. The minimum of duty cycle occurs at the maximum of input voltage which is worst condition. Then the resonant circuit will work properly, if it can provide ZCS in the maximum of input voltage. It can be shown, the maximum of $T_2 + T_3$ is less than half of resonant period and $T_{sw(off)}$ is turn off time for the selected switches, which is often negligible in comparison with oscillation period. Then:

$$T_{min} = T_2 + T_{sw(off)} + T_4 \simeq \pi \sqrt{LC}/n \qquad (6)$$

Where n is transformer ratio. The resonant capacitor is

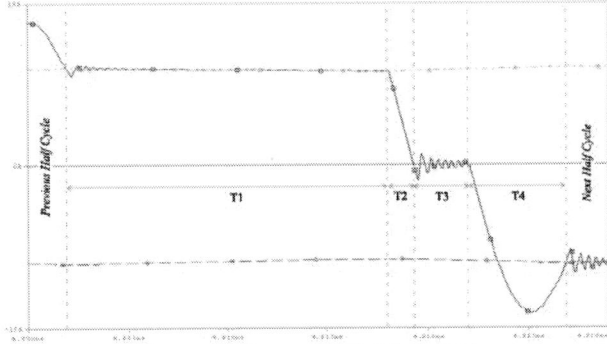

Fig. 6. Current waveform over half of the switching cycle

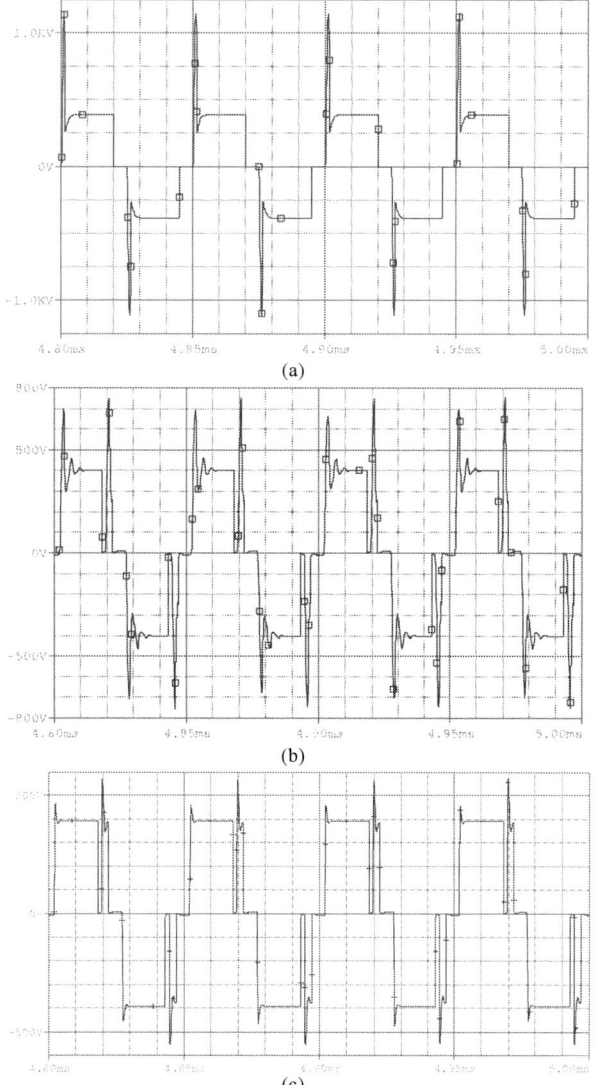

Fig. 7. Voltage waveforms in (a) hard switching (Conventional topology) with snubber (100Ω & 10nf) (b) suggested topology with out snubber (c) proposed topology with snubber (100Ω & 2nf)

chosen such that the resonance current amplitude exceeds the maximum value of input current using (4).

The shoot through time interval should be more than necessary time for inductor current reversal which is equal to calculated T_{min} from (6). This time determines the minimum of duty cycle which can be obtained from:

$$d_{min} = 1 - \frac{V_{in(peak)}}{n V_{out}} \qquad (7)$$

The chosen capacitor is acceptable if the shoot through time is longer than T_{min}, otherwise the selected capacitor value or transformer ratio should changed to provide the soft switching conditions.

V.   SIMULATION RESULTS

The simulations results for 2kw conventional topology and

TABLE 1
PFC Parameters

Parameter	Symbol	Value
Output Power	P_{out}	2 kW
Transformer ratio	n	0.6
Input Voltage	V_{in}	220 V
Line Frequency	f_{line}	50 Hz
Output Voltage	V_{out}	650 V
Half Switching Cycle	T_s	50 μH
Input Inductance	L_{in}	2 mH
Output Capacitance	C_{out}	560 μf
Resonant Capacitor	C	39 nf
Leakage Inductance	L	65 μH
Magnetizing Inductance	L_M	8.3 mH

proposed topology has been presented in this section. The specification of this converter has been presented in table 1. In order to study the operation of the control strategies, the isolated boost PFC converter, including the controller, has been simulated using ORCAD and SIMULINK.

The conventional topology without snubber, especially in high voltage cases is not applicable. Therefore snubbers have been used to limit the voltage impulses in primary side of the isolation transformer. Hence, two snubber circuits using 100Ω & 10nf are connected in parallel with S3 and S4. The transformer primary voltage of the conventional topology is illustrated in Fig. 7(a). The voltage impulse has been limited by using the snubbers, but this circuit still suffers aforementioned impulses which increase the switching losses. Also the snubber loss is considerable. Increase of the snubber capacitor value, decreases the impulse, but increases the snubber losses. Therefore using snubbers in this topology is inconvenient.

The transformer primary voltage of the proposed topology without snubber is illustrated in Fig. 7(b), which confirms remarkable decrease of voltage impulses, but still some impulses are appeared. This voltage impulse is generated by reverse recovery current of inverse parallel and series diodes. This impulse can be further decreased by using fast diodes or using snubbers. The transformer primary voltage using 100Ω

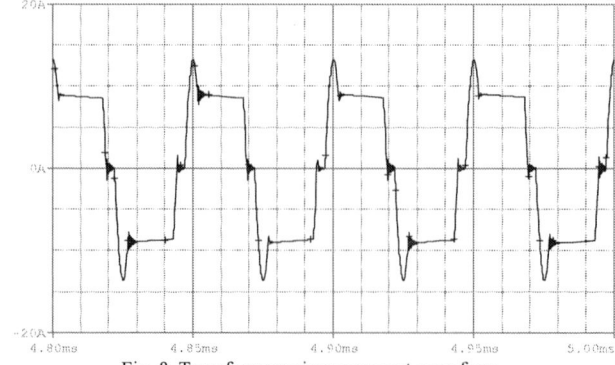

Fig. 8 Transformer primary current waveform

978-1-4244-1871-8/07 $25.00 © 2007 IEEE   428

& 2nf snubber is illustrated in Fig. 7(c). The snubber loss is proportional with the capacitor value and maximum voltage at the power of two which means considerable decrease of snubber loss. The current waveform of the transformer primary side is illustrated in Fig. 8 which shows the reverse recovery currents.

VI. CONCLUSION

Single stage isolated boost power factor correction is an attractive topology which increases the efficiency using only one stage power processing, but the voltage impulses are inconvenient which can damage the circuit elements. The snubbers can limit the voltage impulses, but decrease the overall efficiency. Therefore, using single stage topology is not reasonable. The proposed topology provides ZCS by control of the leakage inductor current, which suppress the aforementioned impulses in addition of diminishing the switching losses. The resonant capacitor design procedure for providing the ZCS in worst case condition has been presented in this paper. The simulations results show that, this topology using predictive controller satisfy the line harmonic limits.

REFERENCES

[1] Chongming Qiao, K.M. Smedley, "An isolated full bridge boost converter with active soft switching," *Power Electronics Specialists Conference*, 2001. PESC, vol. 2, pages: 896 – 903, June 2001

[2] Wanfeng Zhang, Guang Feng, Yan-Fei Liu, Bin Wu, "A digital power factor correction (PFC) control strategy optimized for DSP" *Power Electronics, IEEE Transactions* on vol. 19, issue 6, pages: 1474 – 1485, Nov. 2004

[3] V. Yakushev, V. Meleshin, S. Fraidlin, "Full-bridge isolated current fed converter with active clamp," Applied Power Electronics Conference and Exposition, APEC '99. vol. 1, pages: 560 – 566, March 1999

[4] S. Bibian, H. Jin "Digital control with improved performances for boost power factor correction circuits," in *Proc. IEEE APEC '01 Conf.*, 2001, pp.137–143.

[5] J. Sebastian, M. Jaureguizar, J. Uceda, "An overview of power factor correction in single-phase off-line power supply systems," in *proc. IEEE IECON conf.*, vol.3, pp. 1688–16 93, Sept. 1994

[6] J. Chen, A. Prodic, R. W. Erikson, D. Makimovic "Predictive digital current programmed control," *Power Electronics, IEEE Transactions* on vol. 18, pages: 411– 419, Jan. 2003

[7] Gusseme, Sype, Bossche, Melkebeek, "Digitally Controlled Boost Power-Factor-Correction Converters Operating in Both Continuous and Discontinuous Conduction Mode," *Industrial Electronics, IEEE Transactions* on vol. 52, pages: 88–97, Feb. 2005

[8] H. Benqassmi, J. C. Crebier, J. P. Ferrieux, "Comparison between Current-Driven Resonant Converters Used for Single-Stage Isolated Power-Factor Correction," *Industrial Electronics, IEEE Transactions* on vol. 47, NO. 3, pages: 518-524, Jun 2000

[9] M.K.H. Cheung, M.H.L. Chow, C.K. Tse, "Design of a 1 kW PFC Power Supply Based on Reduced Redundant Power Processing Principle," *Power Electronics Specialists Conference, 2006. PESC '06, 37th IEEE* pages: 1 - 7June 2006

A grid connected photovoltaic system with irradiation injected current control

J. B. Wang, Member, IEEE
Department of Electrical Engineering,
Ching Yun University of Technology
Email:jbw@cyu.edu.tw

Joe Chen , Ronald Li
Acbel Polytech Inc.,
Email:joe_chen@apitech.com.tw
Email:ronald_li@apitech.com.tw

Abstract-A grid connected photovoltaic (PV) system, which consists of PV modules, a boost converter with maximum power point tracking (MPPT) controller and a current controlled full bridge inverter with L-C-L filter, is presented in this paper. A simple PIC16F877 microprocessor is used to implement the InCond MPPT controller by adjusted the duty ratio of the boost converter. Then a DC bus voltage controller is designed to regulate the DC bus voltage within the specified voltage, and generated the injected current command in proportion to the sunlight irradiation. Finally, a 150W prototype grid connected PV system was built to demonstrate the performance of the proposed control schemes. In addition, some simulations and experimental results are also provided.
Keywords: PV system, MPPT, inverter

1 Introduction

A grid connected PV system involves two major functions. One is to track and regulate output power of the PV module at the MPP with respect to sunlight irradiation change. The other is to inject sinusoidal current into the utility grid. The former is often used a microprocessor to implement various MPPT algorithms and to adjust the duty ratio of the power converter to stay at maximum output power operation points [1-4]. The most popular MPPT control algorithms are perturbation and observation (P&O) and incremental conductance (InCond) methods. In addition, the P&O method has better MPPT efficiency than InCond method by optimized the control parameters with the same hardware [4].

Another important feature of the system is required an inverter to inject sinusoidal current into utility grid with low harmonic distortion. Various grid connected inverters were reviewed in the study [5]. As it pointed out that the future structure of the PV inverters is a PV module equipped with an ac-module. Using this structure, the shading effect of the PV module can be reduced significantly than a PV array with a centralize inverter [6]. Since the PV module generates DC power and injects AC power into utility grid through an inverter. Therefore, it requires inserting power-decoupling capacitors as a buffer to compensate the second harmonic power to affect the MPPT operation. Unfortunately, a simple stage power circuit is very hard to achieve aforementioned two major functions by using conventional control methods. Most of single stage schemes are adopted extra switches and separated the whole switching interval into more complex operations [5,6]. When the reliability of the grid connected PV system is the primary concerned, two stages approach may be the better choice. As a result, the control interference between

MPPT converter and inverter can be decoupled and led to design PV system more easily.

This paper designed a grid connected PV system with irradiation inject current control by using two stages approach. The system is used MPPT controlled boost converter to let PV module maintain at maximum power output, and a current control full bridge inverter to inject sinusoidal current into utility grid through a L-C-L filter and thus low THD characteristics can be obtained. Owing to the DC bus voltage of the boost converter will be varied due to irradiation change, a DC bus voltage controller is designed to generate the injected current command with irradiation change. The major purpose of this paper is to control system design and analysis of a grid connected PV system via two stages approach. In order to obtain this goal, this paper is organized as follow. Firstly, the configuration of the grid connected PV system is introduced. The designed MPPT boost converter, inverter and DC bus voltage controller are detailed and the design parameters are obtained by simulation. After simulation results have preliminarily verified the performance of the system, the hardware implementation is done accordingly. Some experimental results are provided to demonstrate the design grid connected system can track the irradiation change quickly.

2 The configuration of the proposed grid connected PV system

The proposed grid connected PV system consists of two 75W PV modules, a boost converter with MPPT control and a current control full bridge inverter with L-C-L filter as shown in Fig. 1. A PIC16F877 microprocessor is used to implement the InCond MPPT control algorithm to adjust the duty ratio of the boost converter. The synchronous AC vector is obtained from utility. When the utility line failed, an AC loss cycle detector is detected to stop boost converter operation. In words, the design grid connected PV system does not provide any power to line utility when it failed. With enough irradiation presented, the DC bus voltage of the boost converter will build up to nominal setting voltage and then the inverter starts to inject sinusoidal current into utility grid with respect to sunlight irradiation change. The inverter is used current control method with unipolar PWM switching mechanism to let the inductor's current track the sinusoidal referent current command closely and thus low THD injected current can be obtained, in addition, a DC bus voltage controller is used to regulate the decoupling capacitor voltage and reduce the second harmonic current to affect the current loop.

978-1-4244-1871-8/07 $25.00 © 2007 IEEE

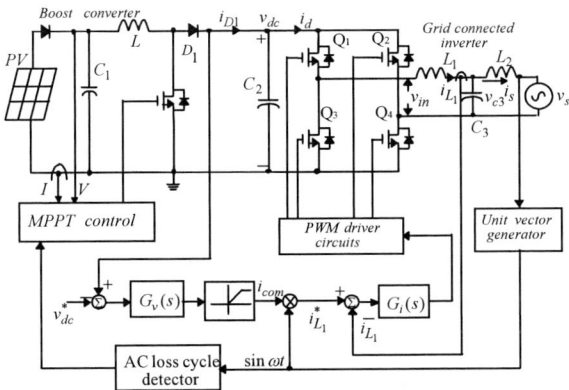

Fig. 1 The block diagram of the grid connected PV system

3 Control system design

(1)A MPPT controlled boost converter

The InCond MPPT is used the incremental conductance to discriminate the adjusting direction toward maximum output power point based on the current versus voltage curve of the PV module. These conditions are summarized as

$$\frac{dI}{dV} = -\frac{I}{V}, \quad (\frac{dP}{dV} = 0) \tag{1a}$$

$$\frac{dI}{dV} > -\frac{I}{V}, \quad (\frac{dP}{dV} > 0) \tag{1b}$$

$$\frac{dI}{dV} < -\frac{I}{V}, \quad (\frac{dP}{dV} < 0) \tag{1c}$$

where the output current, voltage and power of the PV module are I, V and P, respectively. Using (1b) and (1c) are used to determine the direction toward the MPP by added a small perturbation on duty ratio of the boost converter. A PIC16F877 microprocessor is used to implement InCond MPPT algorithm. Owing to the DC bus of the boost converter is unregulated by the MPPT controller, the DC bus voltage will vary at different sunlight irradiations. In this condition, the increased energy of the DC capacitor within a switching interval is

$$dW = \begin{cases} Li_{peak}^2 / 2 = \frac{(VD_1T_s)^2}{2L} & ,DCM \\ IVDT_s & ,CCM \end{cases} \tag{2}$$

where i_{peak} is the peak inductor current at DCM operation modes and T_s denotes the switching interval of the boost converter, in addition, the output current and voltage of the PV module are I and V, respectively. The nominal duty ratios are D_1 and D with respect to DCM and CCM operations. Then the incremental DC bus voltage Δv_{dc} without load due to previous energy (2) can be estimated approximately as

$$\Delta v_{dc} \cong \begin{cases} \frac{I_1 V T_s}{C V_{dc}} & ,DCM \ where \ I_1 = \frac{VD_1}{2L} \\ \frac{IVDT_s}{C V_{dc}} & ,CCM \end{cases} \tag{3}$$

In this case, the DC bus voltage will be fluctuated with respect to sunlight irradiation change. Therefore, a DC bus voltage controller must be designed to generate the injected current command proportion sunlight irradiation.

(2)Grid connected inverter

The grid connected current control inverter with a L-C-L filter that consists of inductor L_1, capacitor C_3 and the leakage inductor of the step up transformer L_2 is shown in Fig. 1 [7]. From Fig. 1, the designed grid connected inverter has an inner inductor current loop and an outer DC bus voltage control loop will detail as follow.

Inductor current control

Neglecting the winding resistance of the inductor and transformer, one can obtain the circuit model of the grid connected inverter with L-C-L filter as

$$\frac{di_{L_1}}{dt} = \frac{1}{L_1}(v_{in} - v_{c3}) \tag{4}$$

$$\frac{dv_{c3}}{dt} = \frac{1}{C_3}(i_{L_1} - i_s) \tag{5}$$

$$\frac{di_s}{dt} = \frac{1}{L_2}(v_{c3} - v_s) \tag{6}$$

where v_{in}, i_{L_1}, v_{c3}, i_s and v_s denote the output voltage of the full bridge converter, L_1 inductor current, C_3 capacitor voltage, injected current and grid voltage, respectively. From (4) to (6), the inductor current control block diagram is obtained and shown in Fig. 2(a). The current controller $G_i(s)$ is a proportional plus integral type controller with the transfer function as $K_p + K_i / s$. Let the PWM gain be K, one can find the transfer function from the inductor current command $i_{L_1}^*$ to the inductor current i_{L_1} is

$$\frac{i_{L_1}}{i_{L_1}^*} = \frac{K(K_p s + K_i)(s^2 L_2 C_3 + 1)}{s^4 L_1 L_2 C_3 + s^3 L_2 C_3 K K_p + s^2 (L_2 C_3 K K_i + L_1 + L_2) + sK K_p + K K_i}$$
$$\approx \frac{1}{sT_i + 1} \tag{7}$$

This transfer function can be approximate to a first order system with a pole $1/T_i$. In addition, the transfer function from the inductor current command $i_{L_1}^*$ to the injected grid current i_s is also obtained as

$$\frac{i_s}{i_{L_1}^*} = \frac{K(K_p s + K_i)}{s^4 L_1 L_2 C_3 + s^3 L_2 C_3 K K_p + s^2 (L_2 C_3 K K_i + L_1 + L_2) + s K K_p + K K_i}$$

(8)

It can be found that the steady state values of equations (7) and (8) are identical. This means that using inductor current

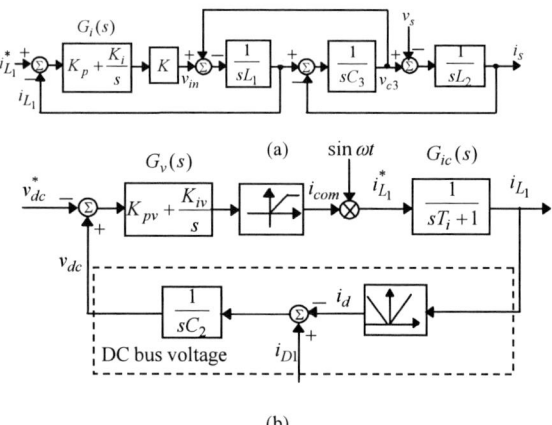

(a)

(b)

Fig. 2 The control block diagram of the inverter: (a) inductor current loop; (b)DC bus voltage loop

feedback has the same effective injected current amplitude to utility.

DC bus voltage control

The DC bus voltage controller can be treated as an irradiation injected current controller. From Fig. 1, one can find the DC bus voltage as

$$\frac{dv_{dc}}{dt} = \frac{1}{C_2}(i_{D1} - i_d)$$

(9)

where i_{D1} and i_d denote the diode current of the boost converter and DC bus current of the full bridge converter, respectively. Owing to unipolar PWM switching scheme is adopted, the DC bus current i_d of the full bridge converter can be approximate to a rectified sinusoidal pulse waveform current [8]. Hence, from (7) and (9), one can obtain the simplified DC bus voltage control block diagram as shown in Fig. 2(b), where the voltage controller $G_v(s)$ is also a proportional plus integral type controller with the transfer function as $K_{pv} + K_{iv}/s$. It is worth to note that there is an uni-direction limiter to ensure when the DC bus voltage v_{dc} is lower than voltage command v_{dc}^*, the injected current i_{com} will be diminished. Owing to the DC bus current contain 120Hz ripple current, the bandwidth of the DC bus voltage control loop must be lower than 20Hz to avoid the second harmonic distortion in current loop. If the nonlinear elements are not considered at nominal operation condition, the open loop transfer function of the DC bus voltage control loop is

$$G_{dc}(s) = \frac{sK_{pv} + K_{iv}}{s^2 C_2 (sT_i + 1)}$$

(10)

Using frequency compensation scheme and circuit simulation, the preliminary designed parameters of the inductor current and DC bus voltage controllers can be obtained.

4 Simulations and experiments

In order to verify the designed grid connected PV system, a sunlight irradiation simulator was built. Owing to the simulated sunlight irradiation is proportion to the control output voltage V_{out} so it can be used as an indication of the sunlight irradiation in following experiments. Fig. 3(a) shows the irradiation step response of the designed MPPT controlled boost converter. Fig. 3(b) shows the InCond MPPT controller can track the triangular irradiation change closely.

Having verified the performance of the proposed the MPPT controlled boost converter, the full bridge inverter with L-C-L filter was also built. The current controller was designed firstly. Fig. 4(a) shows the bode plots of the transfer functions (7), (8) and (9). It is worth to note that the inductor current transfer function can be approximate to first order system with a pole located at 3KHz. Furthermore, the desired DC bus voltage control loop gain is shown in Fig. 4(a) with bandwidth lower than 20Hz that can reduce the effect of the second harmonic

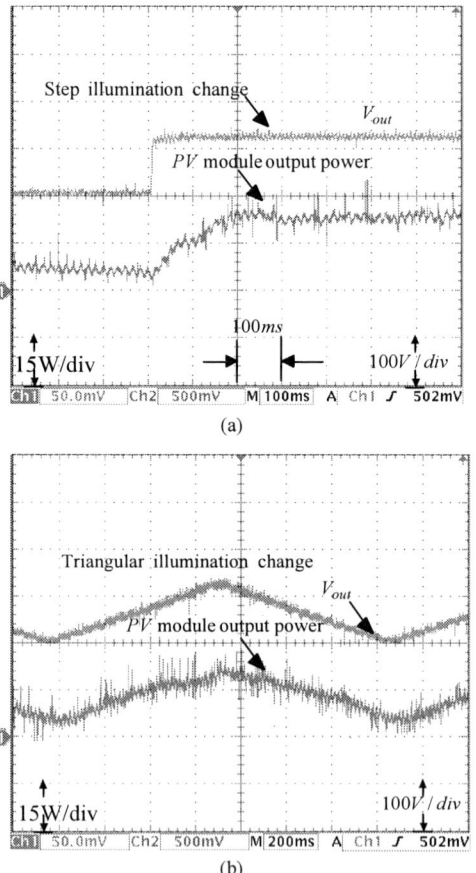

(a)

(b)

Fig. 3 The performance of the MPPT controller: (a)irradiation step change response;(b) triangular irradiation tracking response

978-1-4244-1871-8/07 $25.00 © 2007 IEEE   432

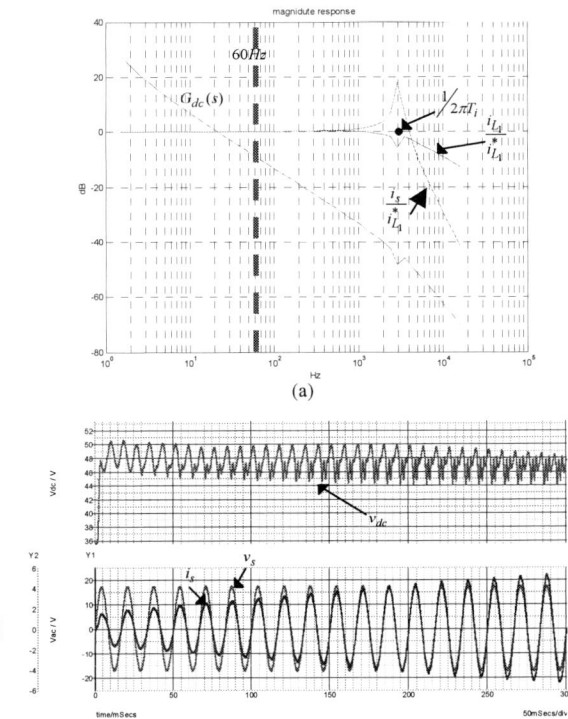

(a)

(b)

Fig. 4 (a)The frequency magnitude response of the transfer functions $i_{L_1}/i_{L_1}^*$, $i_s/i_{L_1}^*$ and $G_{dc}(s)$; (b)circuit simulation of the DC bus voltage, grid voltage and injected current

component. Using circuit simulator to further verify the control performance is depicted in Fig. 4(b). In this simulation, the boost converter is set at fixed duty ratio. The DC bus voltage is boosted up to 50 V, and then regulated to 45 V by DC bus voltage controller. The injected sinusoidal current increases gradually with DC bus voltage built up and tracks the grid voltage closely. All the key designed parameters are listed in Table. 1.

Simulation results have shown the performance of the designed grid connected PV system; the final prototype grid connected PV system shown in Fig. 1 was built accordingly Two PV modules were in series. Fig. 5(a) shows the DC bus voltage, inductor current command, inductor current and grid voltage at a fully simulated irradiation. It can find the current tracking performance is good and the DC bus voltage is regulates at 45V with a small 120Hz voltage ripples. After that, using sunlight irradiation simulator is switched to generate a slow varied triangular irradiation pattern. The injected current into the grid will increase as well as sunlight irradiation increasing as shown in Fig. 5(b). Fig. 5(c) depicts the operating points of the series PV modules at a full simulation irradiation. The result shows the designed MPPT controller can control the PV modules to operate at the MPP.

5 Conclusions

A grid connected PV system with irradiation injected current control was designed in this paper. Through proper designed

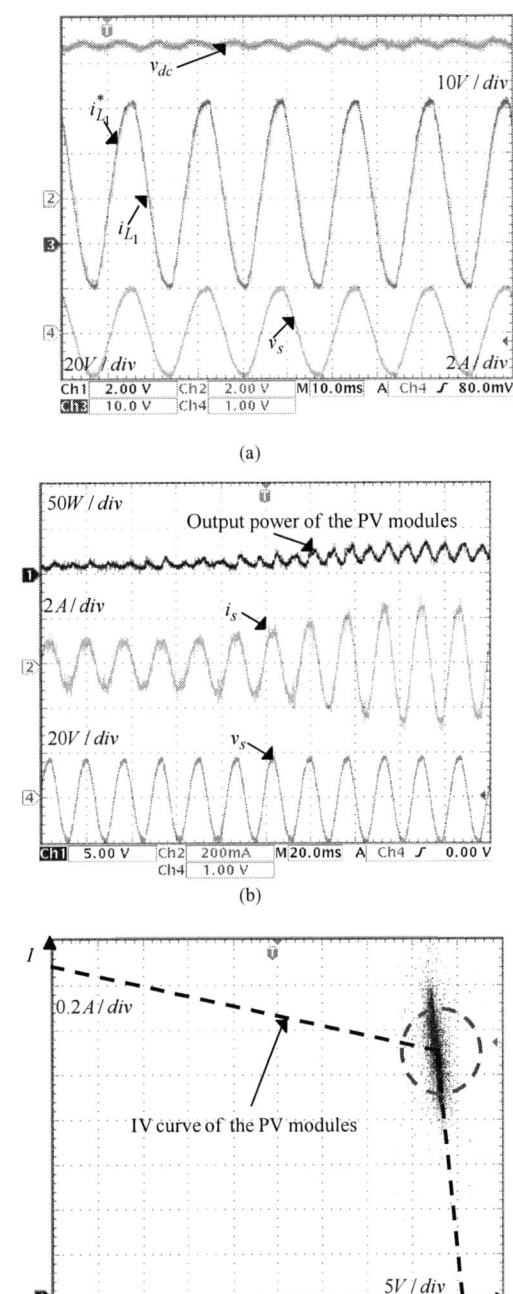

(a)

(b)

(c)

Fig. 6 (a)The waveforms of the DC bus voltage, inductor current command, inductor current and grid voltage at a steady state operation condition; (b)the waveforms of the injected current, grid voltage and output power of the PV modules at an increasing irradiation; (c)the measured MPPs of the PV modules at a fixed irradiation

MPPT, inductor current and DC bus voltage controllers, the injected sinusoidal current command can be generated in proportional to sunlight irradiation change quickly. After using simulation to demonstrate the performance of the inductor current and DC bus voltage controllers, the experimental results further show well performance of the designed system.

6 Acknowledgments

The authors wish to thank National Science Council of Taiwan (NSC 96-2221-E-231-027) for financial support.

7 Referents

[1]   Y. C. Kuo, T. J. Liang and J. F. Chen, "Novel maximum power point tracking controller for photovoltaic energy conversion system," IEEE Transactions on Industrial Electronics, Vol. 48, Issue 3, pp. 594 – 601, 2001.

[2]   J. Arias, F. F. Linera, J. Martin-Ramos, A. M. Pernia and J. Cambronero, "A modular PV regulator based on microcontroller with maximum power point tracking," Conference Record of the 2004 IEEE 39th Industry Applications Conference, vol. 2, pp. 1178 – 1184, 2004.

[3]   C. C. Hua, J. R. Lin and C. M. Shen, "Implementation of a DSP-controlled photovoltaic system with peak power tracking, "IEEE Transactions on Industrial Electronics, vol. 45, Issue 1, pp. 99-107, 1998.

[4]   D. P. Hohm, M. E. Ropp, "Comparative study of maximum power point tracking algorithms," Progress in Photovoltaics: Research and Applications, vol. 11, pp.47-62, 2003

[5]   S. B. Kjar, J. K. Pederson and F. Blaabjerg, "A review of single phase grid connected inverters for photovoltaic modules," IEEE Transactions on Industrial Applications, vol. 41, no. 5, pp. 1292~1306, 2005.

[6]   T. Shimizu, K. Wada and N. Nakamura, "Flyback-Type Single-Phase Utility Interactive Inverter With Power Pulsation Decoupling on the DC Input for an AC Photovoltaic Module System," IEEE Transactions on Power Electronics, vol. 21, no. 5, pp. 1264 – 1272, 2006

[7]   T. Abeyasekera, C. M, Hohnson, D. J. Atkinson and M. Armstrong, "Suppression of line voltage related distortion in current controlled grid connected inverters, "IEEE Transactions on Power Electronics, vol. 20, no. 6, pp. 1393~1401, 2005.

[8]   N. Mohan, T. M. Undeland and W. P. Robbins, *Power electronics: converters, applications and design*, John Wiley & Son Inc. 1989.

Table 1 Key parameters of the grid connected PV system

	Boost converter		Full bridge converter
Switching frequency	40KHz	Switching frequency	10KHz
C_1	220uF/50V	C_3	25uF/250V
C_2	220uF/50V	L_2	110uH
L	225uH	L_1	1.5mH
MOSFET	K2843	MOSFET	IRF640
Boost Diode	RURP3060	K	4.5
K_p	10	K_i	10000
K_{pv}	0.04	K_{iv}	0.4
PV modules	Siemens SP70 75W x2		

Performance Analysis of Soft-switching Inverter
for the Photovoltaic Power System

H.J.Kim[*], Y.H.Chung, K.S.Lee, Y.S.Jon, K.S.Kim

LS Industrial Systems Co., Ltd. R&D Center

533 Hogae-dong, Dongan-Gu, Anyang-Shi, Kyongki-Do 431-080, Korea

Email: hjkim3@lsis.biz

Abstract— Most PV inverters employ a PWM(Pulse Width Modulation switching) scheme to reduce harmonics. The main disadvantage of these PV inverters is the high switching loss because the power switches turn on or off the entire load current while supporting high voltage stress. Another disadvantage of the PWM schemes is EMI due to high di/dt and dv/dt caused by switching. These problems actually decrease the efficiency of the PV inverter and demand EMI filters.

To overcome these problems of the PV inverters, this paper describes the PV system with ARCP(Auxiliary Resonant Commutated Pole) inverter. Switching losses can be minimized by using ARCP inverters where the power switches turn on when the voltage across them is zero or turn off when the current through them is zero and the efficiency can be increased. The performance of the PV system with ARCP inverter is analyzed by means of the simulation. As a result, we may obtain the high performance of the PV System with ARCP inverter and expect to apply residential PV power system.

I. INTRODUCTION

In recent years, the interests to use the solar energy in applications of power distribution are growing very fast. Applications for the solar energy are in urban areas, electric drives, satellites, etc. Also, solar energy is considered as a real alternative source of energy to be used for production of electrical energy[1].

To converter solar energy into electrical energy, it is required the basic components such that solar panels, inverters to produce electric energy. The PV arrays are linked in series to achieve the sufficient dc voltage for generating an ac output voltage at the inverter. And most PV inverters

But, these PV inverters using the PWM scheme have the main disadvantages. One is the high switching loss because the power switches turn on or off the entire load current while supporting high voltage stress, another is EMI due to high di/dt and dv/dt caused by switching. These problems actually decrease the efficiency of the PV inverter and demand EMI filters.

To overcome these problems of the PV inverter, this paper describes the PV inverter with ARCP(Auxiliary Resonant Commutated Pole) circuit.

Generally, ARCP inverter has been a widely used technology in the high-frequency high power inverter application. Compared with other soft switching technologies, it has higher efficiency, lower switch stress, lower losses and hence it allows operation at higher switching frequency[4][5].

So, this paper, for increasing the efficiency of the PV system, presents PV inverter with ARCP circuit, shown in Fig. 1. This system consists of a boost chopper and a half-bridge single phase inverter with soft-switching module. And each main switch of the inverter is connected in parallel with resonant capacitor. Also, two auxiliary switches are placed in series with a resonant reactor for soft-switching capability.

II. PRINCIPLE OF OPERATION OF ARCP CIRCUIT

The main idea of the ARCP is that the resonant inductor L_R is located outside the main power flow path and the auxiliary switches are active only during the resonant commutation intervals, which represent only a fraction of the total switching period. With this topology, the main switches operate under zero voltage switching and the auxiliary ;[6].

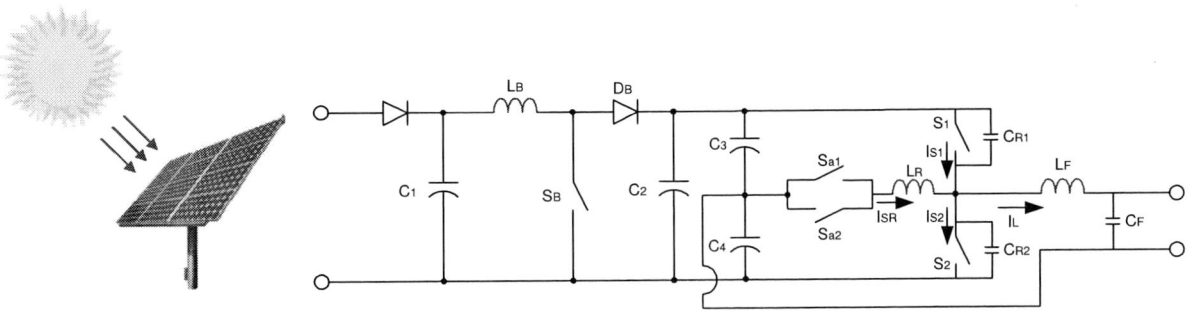

Fig. 1 : Single phase PV inverter with ARCP circuit

978-1-4244-1871-8/07 $25.00 © 2007 IEEE

Fig. 2 shows the operation waveform and Fig. 3 shows the ten different modes of ARCP inverter. To describe its operation, a positive current I_L is assumed constant during a short commutation interval.

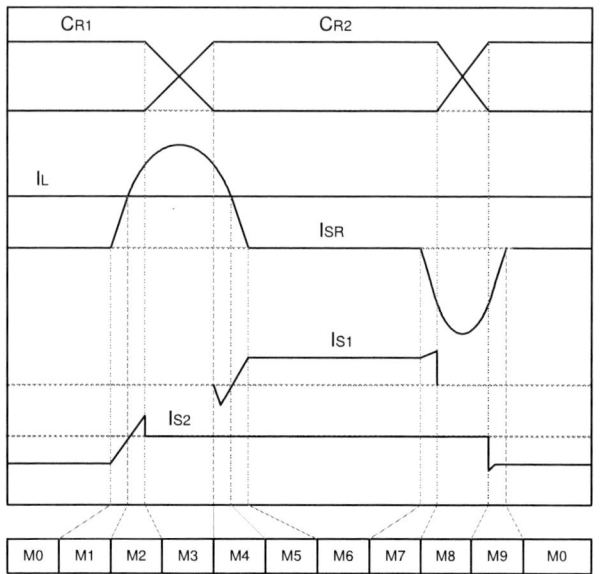

Fig. 2 : Operation waveform of ARCP circuit

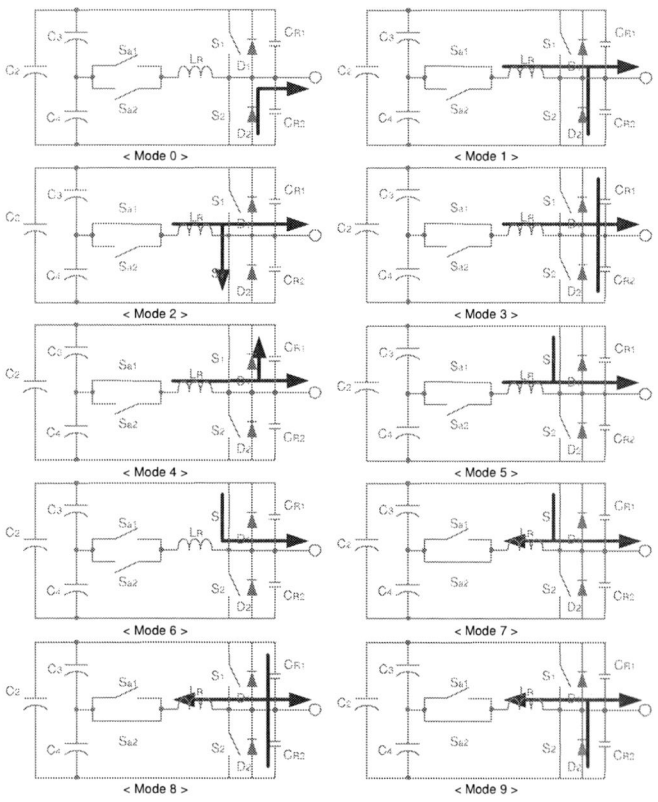

Fig. 3 : Each operation mode of ARCP circuit

Mode 0 In this mode switch S_2 is turned on and diode D_2 conducts the load current.

Mode 1 In this mode the auxiliary circuit is turned on and displaces the current in diode D_2. To initiate the commutation, switch Sa1 is turned on under ZCS condition. The inductor current I_{SR} ramps up linearly from zero with a slope $Vdc/2L_R$. Since the load current I_L is constant, the diode D_2 current decreases linearly. This period ends when current I_{SR} becomes equal to the load current I_L and diode D_2 turns off. As a result, the diode D_2 current goes to zero and the diode turns off under ZVS condition.

Mode 2 In this mode boost energy is stored in the auxiliary inductor to compensate for losses during the resonant transition. At the end of mode 1, diode D_2 turns off under ZVS condition. Switch S_2 now starts conducting. Notice that any reverse recovery current in diode D_2 contributes to the boost current. The voltage across diode D_2 is zero throughout the diode reverse recovery phase, since anti-parallel switch S_2 is on, and therefore reverse recovery losses are very small. The current in switch S_2 increases linearly with a sloe $Vdc/2L_R$. This mode ends when switch S_2 is turned off.

Mode 3 In this mode there is a resonant transition. At the end of mode 2 switch S_2 is turned off under ZVS condition. This initiates a resonant phase in which the auxiliary inductor resonates with capacitors C_{R1} and C_{R2}. During this phase the only semiconductor devices conducting are auxiliary switch S_{a1}.

Mode 4 During this mode at the end of the previous mode diode D_1 has turned on under ZVS condition. Diode D_1 current decrease linearly with a slope $Vdc/2L_R$. This mode ends when the current in diode D_1 goes to zero and the diode turns off. This happens when the resonant inductor current becomes equal to the load current I_L.

Mode 5 In this mode the load current transfers from the auxiliary inductor to the upper switch S_1. As a result, diode D_1 turns off at the end of the previous mode under ZVS condition and the reverse recovery loss is very small. The switch S_1 current increases linearly until it becomes equal to the load current I_L. This period ends when the auxiliary inductor current goes to zero.

Mode 6 In this interval switch S_1 is on and it conducts the load current I_L. This mode ends when auxiliary switch S_{a2} is turned on under ZCS condition to initiate the commutation phase.

Mode 7 In this mode it provides boost energy for the commutation. The auxiliary inductor current increases linearly and it flows through S_{a2}. This mode ends when switch S_1 is turned off under ZVS condition, initiating the resonant transition.

Mode 8 In this mode there is a resonant transition that brings voltage C_{R2} to the lower rail. This mode ends when voltage C_{R2} hits the lower rail and is clamped to it by diode D_2, which turns on under ZVS condition.

Mode 9 In this mode the voltage C_{R2} is clamped to the lower rail by diode D_2. The current in the auxiliary inductor decreases linearly to zero. This mode ends when the current in auxiliary switch S_{a2} goes to zero and the switch turns off under ZCS condition.

III. EQUIVALENT MODEL OF THE PV ARRAY

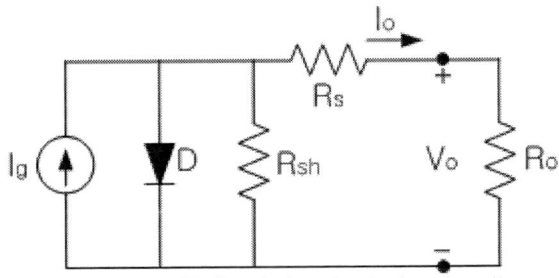

Fig. 4 : Simplified equivalent circuit of a solar cell

Solar array characteristics profoundly influences the converter and control system, therefore it will be briefly reviewed here. A solar cell is a nonlinear device and can be represented as a current source model as shown in Fig. 4.

The traditional I-V characteristic of a solar cell, when neglecting the internal shunt resistance, is given by the following equation[7].

$$I_o = I_q - I_{sat}\{\exp[\frac{q}{AKT}(V_o + I_o R_s)] - 1\} \qquad (1)$$

Where, I_q is the light generated current, I_{sat} is the reverse saturation current, q is the electronic charge, A is a dimensionless factor, K is the Boltzmann constant, T is the temperature in ${}^oK$, R_s is the series resistance of the cell.

In literature, instead of the I-V characteristic given by Eq(1) the V-I characteristic which is given below is used in many cases.

$$V_o = -I_o R_s + \frac{AKT}{q} \cdot In\left[\frac{I_q - I_o + I_{sat}}{I_{sat}}\right] \qquad (2)$$

Equation (1) was used in computer simulations with PSIM to obtain the output characteristics of a solar cell as shown in Fig. 5. And the solar array has the characteristics that the output of a solar cell is non-linear and vitally affected by the solar radiation and temperature.

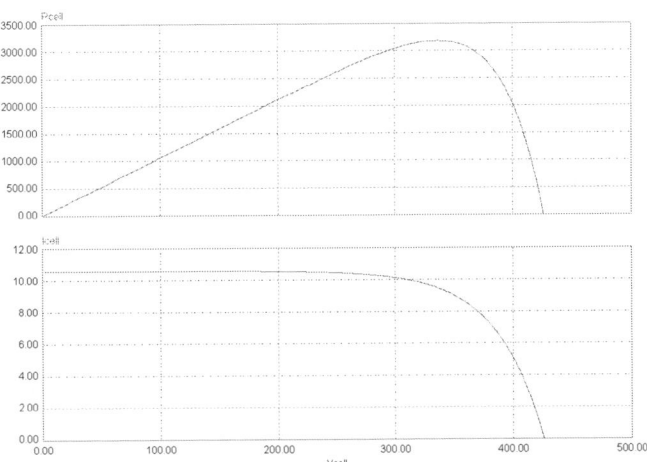

Fig. 5 : Output characteristics of PV array

IV. SIMULATION

In order to analyze the operational characteristics of PV system with ARCP circuit, computer simulation with PSIM was performed.

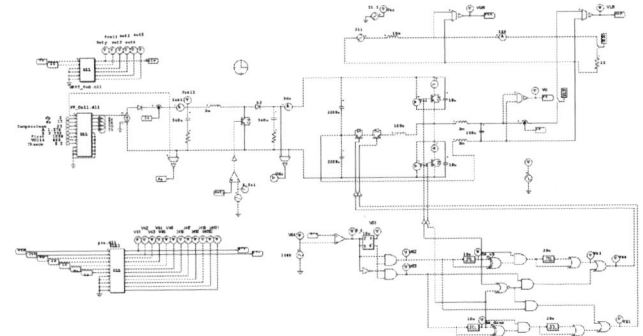

Fig. 6 : Simulation model

A simulation model with PSIM was developed for PV system, which is shown in Fig. 6. The PV system consists of a nonlinear current source as the power source, a dc/dc converter and a ARCP inverter power stage as the power processing unit, controller unit and is connected between the source and the load. The detail parameters used in simulation are shown in Table I.

TABLE 1: Simulation Parameters

Source Voltage	220V
PV Output Power	3,000W
C1	100uF
DC L	500uH
C3, C4	2200uF
AC Filter L	3mH
AC Filter C	4.7uF
Resonant Inductor	200uH
Resonant Capacitor	10uF
DC/DC Converter fs	16kHz
DC/AC Inverter fs	4kHz

Fig. 7 shows the simulation results. Fig. 7(a) shows the voltage and current of the upper and lower main switch under hard switching method without ARCP circuit and Fig 7(b) shows the voltage and current of the upper and lower main switch under soft switching method with ARCP circuit.

Fig. 7(c) shows the operation waveform of ARCP inverter. In this figure, we can see it's similar to Fig. 2. So, it's confirmed the PV inverter with ARCP circuit increases the efficiency of the PV system by reducing switching loss.

V. CONCLUSION

This paper presents PV inverter with ARCP circuit, consisted of a boost chopper and half-bridge single phase inverter with soft-switching module, for increasing the efficiency of PV system and describes the operation principles of ARCP circuit.

A simulation model with PSIM was developed to analyze performance of the proposed PV system. The simulation results confirm that the proposed PV system can reduce the switching losses of PV inverter.

REFERENCES

[1] H.Yongji, L.Deheng, "A new method for optimal output of a solar cell array" in Proc. IEEE Int. Symp. Industrial Electronics, 1992, vol. 1, pp.456-459

[2] C.Hua, J.Lin and C.Chen, "Implementation of a DSP-Controlled Photovoltaic System with Peak Power Tracking" Trans. of IEEE on Industrial Elec., vol 45, no 1, pp.99-107, 1998.

[3] Kasa.N, Iida.T, Majumdar.G, "Robust control for maximum power point tracking in photovoltaic power system" Power Conversion Conference, 2002. PCC Osaka 2002. Proceedings of the Volume 2, 2-5 April 2002. pp. 827-832 vol. 2

[4] Cavalcanti. M.C, da Silva. E.R, Jacobina. C.B, Boroyevich. D, Dong. W, "Comparative evaluation of losses in soft and hard-switched inverters", 38th IAS Annual Meeting, Conference Record of the Vol. 3, 12-16 Oct. 2003, pp1912-1917

[5] Schwarzer. U, De Doncker R.W. "Characterization of 6.5kV IGBT modules for hard-and soft-switching operation in medium voltage applications", APEC 2005. 20th Annual IEEE, Vol. 1, 6-10 March 2005, pp. 329-335

[6] Kang. X, Caiafa. A, Santi. E, Hudgins. J.L, Palmer. P.R, "Simulation of ARCP converter with physics-based circuit simulator device models", PESC 2003. Vol. 4, 15-19 June 2003, pp. 1904-1909

[7] T.Markvar, "Solar Electricity", John Willy & Sons, 1994.

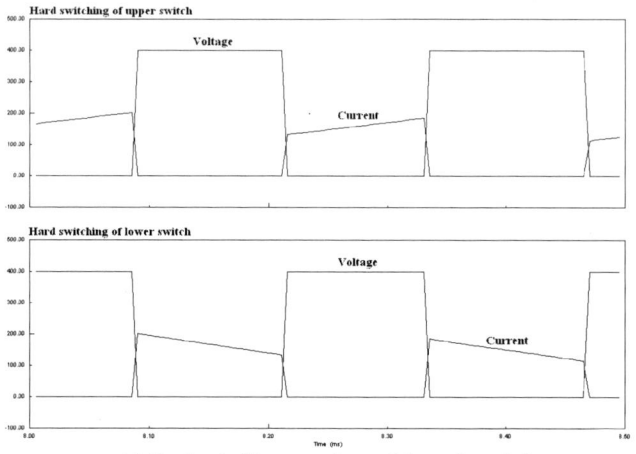

(a) Hard switching waveform of the main switch

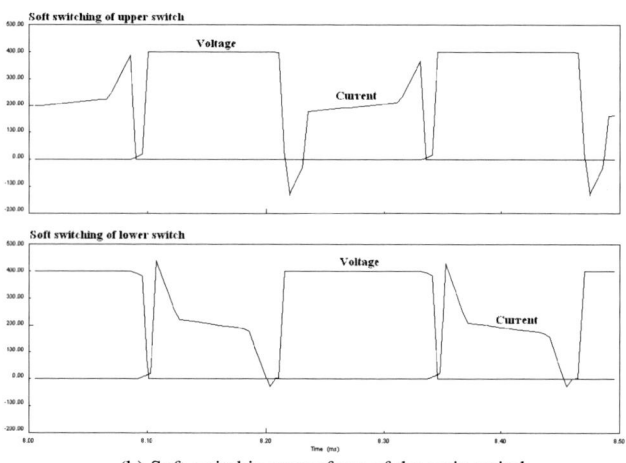

(b) Soft switching waveform of the main switch

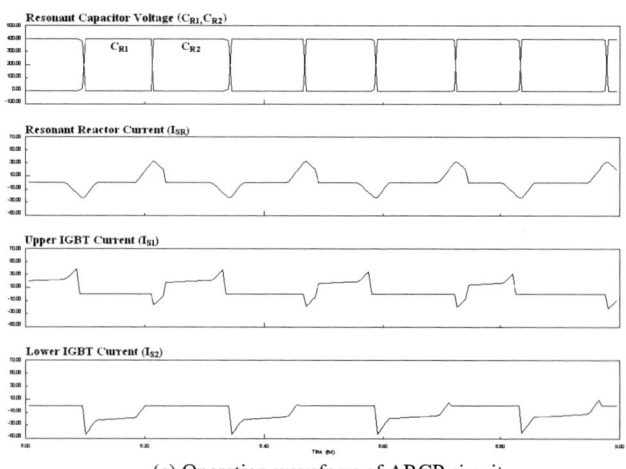

(c) Operation waveform of ARCP circuit
Fig. 7 : Simulation results

Development of the Power Conditioning System for High Power Fuel Cell System

Jin-Hee Lee, Seung-Taek Baek, Hong-Ju Jung, Ho-Hyun Kang, Joon-Mo Chung, In-Young Suh [1]

[1] The Power & Industrial Systems R&D Center, Hyosung Corporation., Korea
183-2, Hoge-Dong, Dongan-Ku, Anyang-Si, Gyeonggi-Do, Korea, 431-080
Email: jinhlee@hyosung.com

Abstract— This paper presents the design, development and performance of a power conditioning system (PCS) for application to a 250kW Molten Carbonate Fuel Cell (MCFC) generation system. A DSP controller was used to control the dc-dc and dc-ac converter operation for grid connection and power injection to the grid. The controller must also supervise the total PCS operation while communicating with the fuel cell system controller. A control method for parallel operation of dc-dc converters was proposed and verified. A 250kW prototype was successfully built and tested. Experimental performances are compared to minimum target requirements of the PCS for MCFC.

I. INTRODUCTION

Due to global energy supply and environmental pollution problems, interest in renewable and alternative energy sources have grown steadily. Especially, Fuel Cell generation systems using hydrogen energy is one of the most likely to be used among the alternative energy sources due to its high efficiency and ability accommodate a wide assortment of different fuels.

Significant market potential exists for forecasting accurate demand for energy and distributed resources, because renewable generation system such as fuel cell generation system are usually installed near the user facility.

Fuel cell power generation systems are expected to see increased usage in various applications such as stationary applications, automotive applications, and interfaces with electric utilities due to several advantages over conventional generation systems. These advantages include 1) low environmental pollution 2) highly efficient power generation 3) diversity of fuels (natural gas, LPG, methanol and naphtha) 4) usage of exhaust heat 5) modularity and 6) shorter installation time [1, 2].

A government funded research project was launched in 2005, and our scope is to develop a grid connecting power conditioning system and verified performance of 250kW PCS for Molten Carbonate Fuel Cell system.

II. DESIGN & DEVELOPMENT OF THE 250KW PCS

A. Configuration of the 250kW PCS

The PCS is composed of three modules; controller module, power converter module and distribution module. The PCS is designed considering the Fuel Cell system's output characteristic and operation mode. The single line diagram is shown in Figure 1.

The main controller of the PCS and HMI equipment are installed in the controller module. Power converter module is composed of a DC-DC converter that boosts up low voltage fuel cell DC source to high voltage source, a DC-AC inverter that converts the DC source to AC source, and a transformer that provides galvanic isolation between the PCS and the utility grid. Distribution module is used to inject energy into the utility side that is generated from the fuel cell, and is composed of many switches and sensors. The Specification of the 250kW PCS is shown in Table 1.

B. 3-Phase DC-DC Boost Converter

The DC-DC converter is composed of three parallel stacks due to high current characteristic of the fuel cell, and ability to be operated with partial load conditions during stack failure. Fig. 2 shows the DC-DC boost converter used in the 250kW PCS.

Fig. 3 shows the control block diagram of DC-DC converter. The proposed control method in this paper has the following features.

1) Regulate the dc link voltage and converter power balance.
2) Reduce input current ripple, dc-dc converter apply to interleaved-switching. [3,4]
3) The implementation achieves current comparison by modifying the measured current signal. The average value of inductor current I_{L1}, I_{L2}, and I_{L3} are measured and compared. Current reference I_{ref} uses maximum value among the measured inductor current.[3,4,5]

The proposed maximum current sharing control method is accomplished power balance of the parallel high power DC-DC converter.

978-1-4244-1871-8/07 $25.00 © 2007 IEEE

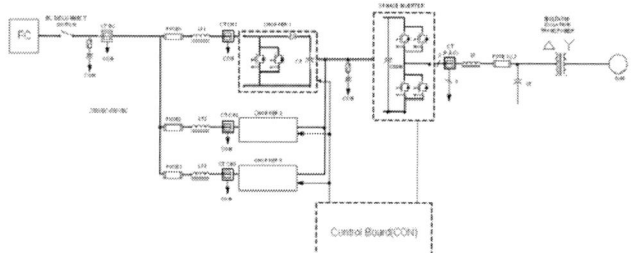

Fig. 1. Configuration of the 250 kW PCS

Table 1. Specification of the 250kW PCS

Items			Details
Rated Power			340kVA
DC-DC	Input	Voltage	$284V_{DC,nominal}$, ($250V_{DC} \sim 500V_{DC}$)
		Current	$1200A_{DC}$ (@ $284V_{DC}$)
		Ripple	Current : Less than 5% (@ < 1kHz)
			Voltage : Less than 10V (@ < 1kHz)
	Output	Voltage	$700V_{DC}$
DC-AC	Output	Voltage	3Φ 380Vac
		Current	$0\sim657A_{Max}$
		Frequency	60Hz±0.3Hz
		THD	Stand-alone : Less than 5% (Voltage) Grid-Connect : Less than 5% (Current)
		Quality	IEEE-519
		Power	250kW (340kVA)
Efficiency			Higher than 90%
Power Factor			0.9 (@ Grid-Connected point)
Protection			OV, UV, OC, OT, SC
Cooling			Air Cooling
Ambient Temp.			40℃ Max

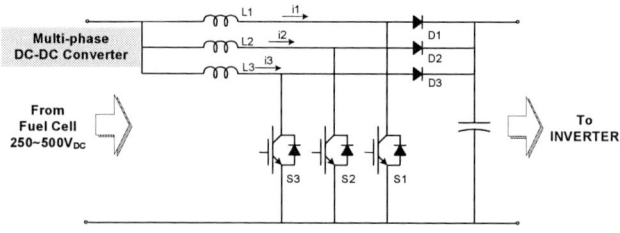

Fig. 2. 3-phase Interleaved DC-DC boost converter

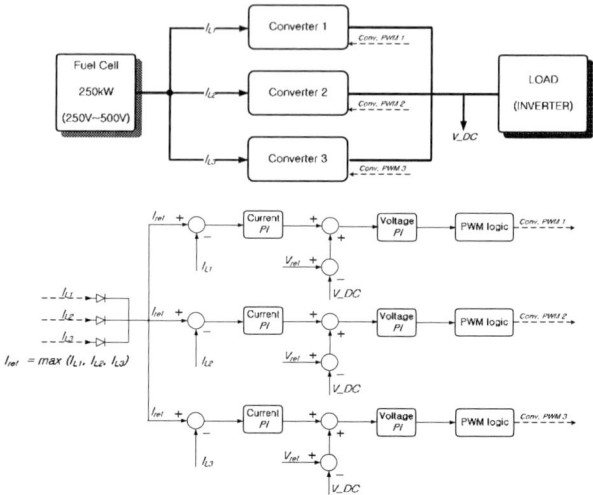

Fig. 3. Control Block Diagram of DC-DC Converter

C. DC-AC Inverter

The DC-AC inverter is composed of IGBT stacks and filter. The Filter is used to eliminate high frequency switching noise and is composed of reactors and capacitors. Switching frequency of the inverter is 3kHz. When designing the filter parameter, the following considerations need to be made.

- Define LC (f_c) value with damping ratio
- Define fundamental voltage drop at reactor (5~10%)
- Consider control stability during grid interconnection mode operation
- Consider voltage and current THD with above L, C values.

Fig. 4 shows the DC-AC inverter and isolation transformer.

DC-AC inverters consist of a stand-alone control for critical load and grid-connected control for power injection from fuel-cell generated power to utility-line.

Prior to changing the operation mode to grid connection, the magnitude and phase of the inverter output voltage in the stand alone mode needs to be synchronized with the grid voltage. For over-current suppression during grid connection, precise synchronization of the phase angle to the grid voltage is essential.

D. Controller

The controller of the 250kW PCS is composed of a DSP and other electronic components. Fig. 5. shows the main controller which performs DC-DC and DC-AC converter control.

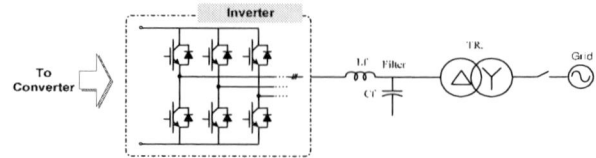

Fig. 4. DC-AC inverter & isolation transformer

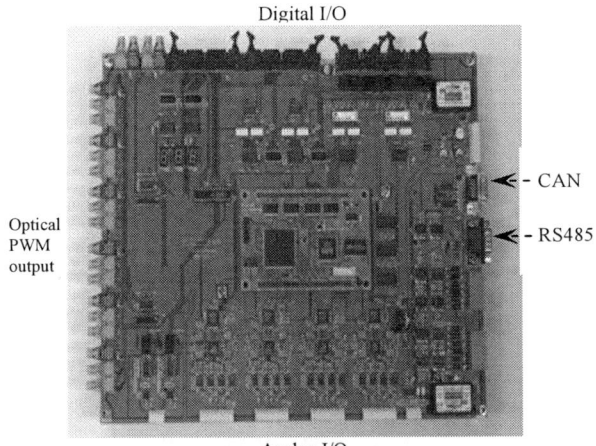

Fig. 5. **Control board of the 250kW PCS**

The DSP of main controller of the PCS is TMS320F2812 (Texas Instruments) and performs the following tasks.[6]

- Sequence Control of the PCS
- Output voltage and current sharing control and interleaved-switching of 3-phase DC-DC boost converter.
- SVPWM control of DC-AC inverter
- Constant voltage control at stand-alone mode of DC-AC inverter
- Constant power control and grid interconnection mode of DC-AC inverter
- Various protection and operation of the PCS
- Communication (To Fuel Cell System Controller)

III. EXPERIMENTAL RESULT

Fig. 6 show experimental result and performance of the 250kW PCS for MCFC.

The experimental waveforms of the 3-phase DC-DC boost converter are shown in Fig. 6(a). It is shown that the 3-phase DC-DC boost converter is operated with gate signals and inductor currents with 120° phase difference between phases by interleaved-switching of the DC-DC boost converter.

Therefore, the input current ripple of the fuel cell has an effect that is reduced by 3-phase inductor currents 120° apart.

Fig. 6(b) shows the experimental waveform for transition from stand-alone mode to grid-connected mode.

When a transfer from stand-alone mode to grid-connected mode DC-AC inverter is connected to be synchronized with the voltage magnitude and phase angle of utility-line, output current of the PCS is not generated over-current.

After grid-connection, the stand-alone voltage control mode is maintained for a fixed time and operation mode of the PCS is changed to current control mode as shown in Fig. 6(c).

Fig 6(d) shows the experimental waveforms for a PCS injecting rated power to the grid. Due to reactive power control of the DC-AC inverter of the 250kW PCS, power factor of grid-connection point is controlled to unity.

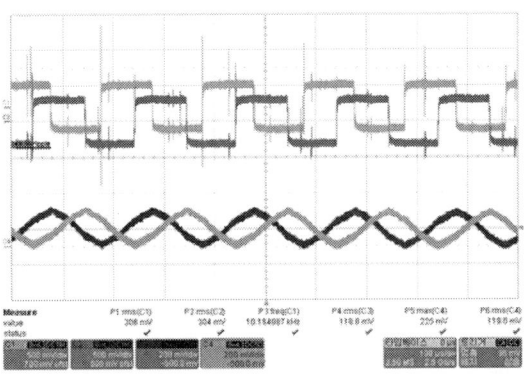

(a) Interleaved-switching and inductor current of 3-phase DC-DC converter

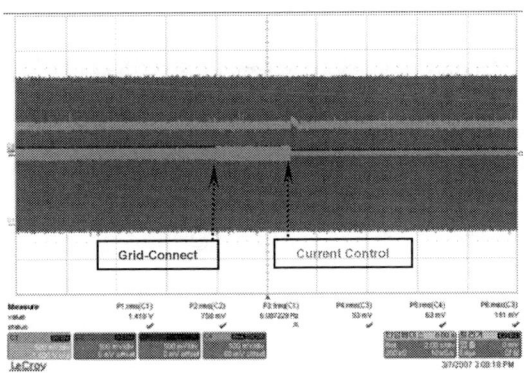

(b) Waveforms during a transition from stand-alone to grid-connected

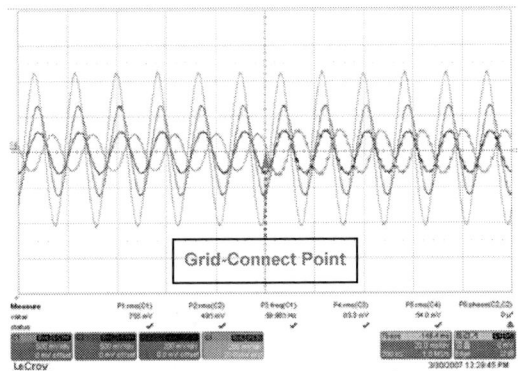

(c) Stand-alone mode to grid-connected mode

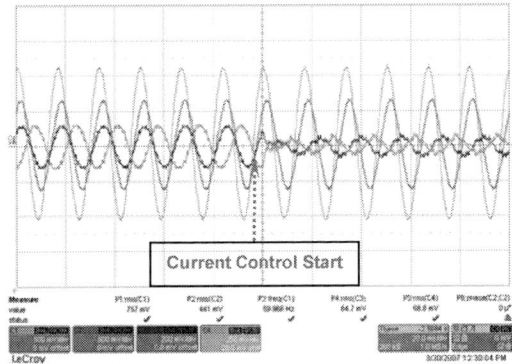

(d) Voltage control mode to current control mode

978-1-4244-1871-8/07 $25.00 © 2007 IEEE

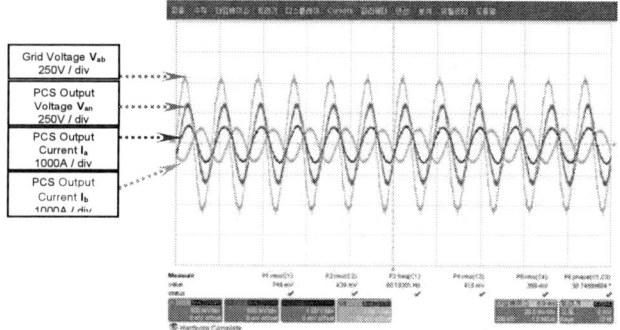

(e) Rated power operation of the 250kW PCS (@ grid-connected)

Fig. 6. Experimental results of the 250kW PCS

Fig. 7 shows the experimental waveform of an anti-islanding operation. When PCS is detected islanding condition, DC-AC inverter is operated a transfer from grid-connected mode to stand-alone mode in 1/2 period.

Fig. 8. shows efficiency of the 250kW PCS. When rated power mode is operated, efficiency of the PCS shows performance 93.68% higher than 90% of design specification.

Fig. 9. shows output current THD of the 250kW PCS. When rated power of the 250kW PCS is operated, output current THD of the PCS shows performance 3.97% less than 5% of design specification.

Fig. 10. shows the composition of the HMI control panel. It can operate real power and reactive power, distinguish the statue of the 250kW PCS.

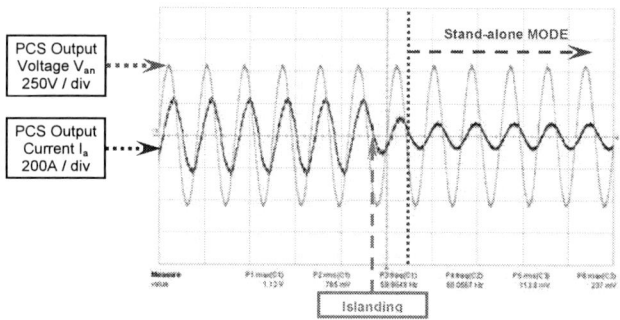

Fig. 7. Anti-islanding operation Mode

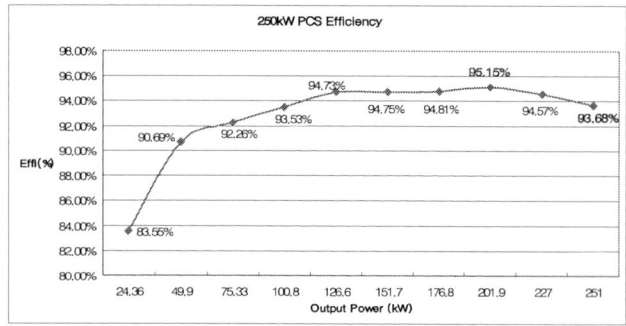

Fig. 8. Efficiency of the 250kW PCS

The Photograph of the 250kW PCS for MCFC is shown in Fig. 11.

Performance results of the 250kW PCS is better than design specification is shown in Table 2.

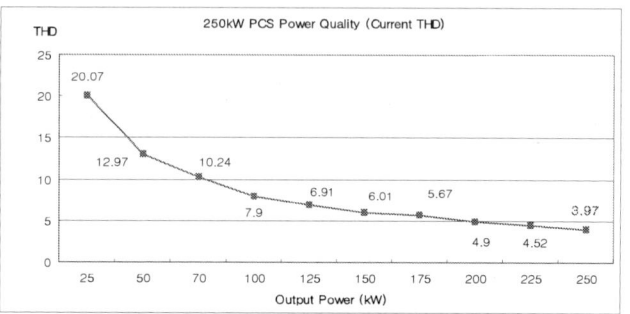

Fig. 9. Output current THD of the 250kW PCS

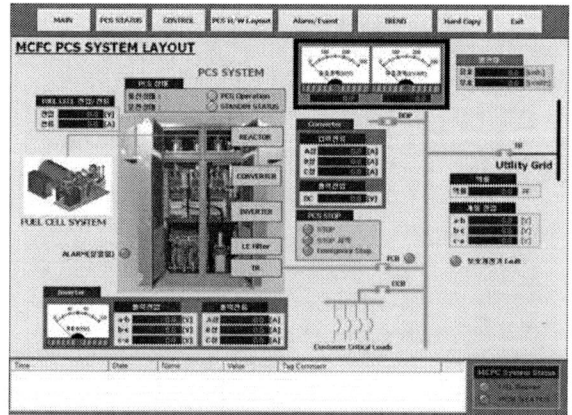

Fig. 10. The composition of HMI picture

Fig. 11. Photograph of the 250kW PCS for MCFC

Table 2. Experimental performance of the 250kW PCS (@ Rated power)

Design item	Specification performance	250kW PCS Experimental performance
Frequency	60±0.3Hz	59.8 ~60.19 Hz
THD : Current (@ Grid connect)	Less than 5%	< 3.97%
Efficiency	Higher than 90%	93.18%
Power factor	Higher than 0.9 (@ Grid-Connected)	> 0.95

978-1-4244-1871-8/07 $25.00 © 2007 IEEE

IV. CONCLUSION

This paper presents the development and performance of a power conditioning system (PCS) for application to a 250kW Molten Carbonate Fuel Cell (MCFC) generation system. A DSP controller was used to control the DC-DC converter and DC-AC inverter operation for grid connection and power injection to the grid. The controller must also supervise the total PCS operation while communicating with the fuel cell system controller. A control method for parallel operation of DC-DC converters was proposed and performance of the 250kW verified through experimental results.

V. REFERENCES

[1] R. ANAHARA, S. YOKOKAWA AND M. SAKURAI, "PRESENT STATUS AND FUTURE PROSPECTS FOR FUEL CELL POWER SYSTEMS", *PROCEEDINGS OF THE IEEE*, VOL 81, NO. 3, MARCH 1993, PP 399-407.

[2] JINHEE LEE, JINSANG JO, SEWAN CHOI, SOO-BIN HAN, "A 10-KW SOFC LOW-VOLTAGE BATTERY HYBRID POWER CONDITIONING SYSTEM FOR RESIDENTIAL USE" *ENERGY CONVERSION, IEEE TRANS.*, VOL. 21, ISSUE 2, JUNE 2006, PP 575-585

[3] GIRAL, R., MARTINEZ-SALAMERO, L., SINGER, S., "INTERLEAVED CONVERTERS OPERATION BASED ON CMC", *POWER ELECTRONICS, IEEE TRANS.*, VOL. 14, ISSUE 4, JULY 1999, PP 643-652

[4] BABU, C.S., VEERACHARY, M., "PREDICTIVE CONTROLLER FOR INTERLEAVED BOOST CONVERTER", *ISIE 2005. PROCEEDINGS OF THE IEEE*, VOL. 2, 20-23 JUNE 2005, PP 577-581

[5] V. JOSEPH THOTTUVELIL, G.G. VERGHESE, "ANALYSIS AND CONTORL DESIGN OF PARALLELED DC/DC CONVERTER WITH CURRENT SHARING", *POWER ELECTRONICS, IEEE TRANS.*, VOL. 13, NO. 4, JULY 1998, PP 635-644

[6] TEXAS Instrument, *"TMS320C28X DSP CPU and Instruction set Reference Guide"*, March 2004.

A Coordinated Control Method for Leveling Output Power Fluctuations of Multiple PV Systems

Tomonobu Senjyu[1], Manoj Datta[1], Atsushi Yona[1], and Hideomi Sekine[2]
[1]Department of Electrical and Electronics Engineering
[2]Departemt of Education Technology
University of the Ryukyus
1 senbaru, Nishihara, Okinawa 903-0213, Japan
Email: b985542@tec.u-ryukyu.ac.jp

Toshihisa Funabashi
Meidensha Corporation
Riverside Bldg 36-2, Nihonba-shi,
Hokozaki-cho, Chuo-ku, Tokyo 103-8515, Japan
Email: funabashi-t@mb.meidensha.co.jp

Abstract—A Photovoltaic (PV) system's power output fluctuates with the change of weather conditions. Frequency deviations, tie line voltage fluctuations etc. are caused by fluctuating PV power when large PV power from several PV systems is penetrated in the utility. In this paper, to overcome these problems, a simple coordinated control method for leveling the fluctuations of combined PV power from multiple PV systems is proposed. Here, output power command is generated in two steps: central and local. Fuzzy control is used to generate the central leveling output power command considering insolation, variance of insolation and absolute average of frequency deviation. In local step, a simple coordination is maintained between central power command and local power commands by producing a common tuning factor. Power converters are used to achieve the same output power as local command power employing PI control law for each of the PV generation systems. The proposed method is compared with the method where conventional Maximum Power Point Tracking (MPPT) control is used for each of the PV system's power output. Simulation results show that the proposed method is effective for leveling output power fluctuations and feasible to reduce the frequency deviations of the power utility.

I. INTRODUCTION

There are many isolated islands in the world where power are mainly provided by diesel generators. The demand for power is steadily increasing in those isolated islands. Diesel generators need heavy oil and these incur buying, transport, and storage costs. To avoid global warming, the consumption of fossil fuels like heavy oil must be reduced. Hence, the installation of renewable energy power plants has become a burning issue not only for isolated islands but also all over the world. Among various renewable energy systems, photovoltaic power generation systems (PV systems) are expected to play an important role as a clean electricity power source in meeting future electricity demands. However, the power output of PV systems fluctuates depending on weather conditions. In the future, when a significant number of PV systems will be connected to the grids of power utilities, combined power output fluctuations may cause problems like voltage fluctuation and large frequency deviation in electric power system operation [1]-[3]. Therefore, for the penetration of multiple or clustered PV system's output power in the utility without reduction of the reliability of utility power systems, suitable measures must be applied to the PV systems side. Regarding these issues, several studies have been carried out for reducing the output power

fluctuations of renewable power sources. Power characteristics of PV ensembles are presented in [4] where monitored data from 100 PV systems were used to study effects of combined power generation of these systems, compared to the characteristics of an individual system. It was claimed that a significant amount of power fluctuations disappeared, however, large amount of short term power fluctuations are remained. In addition, when the numbers of PV power systems were decreased, the power fluctuations increased. Smoothing of PV system output by tuning MPPT control is demonstrated in [5]. In this method, when the insolation increases rapidly, the operating MPPT point changes to a new point where the maximum power is not generated with the current insolation. It was reported that this method can be applied to several PV power generation systems to achieve a combined output power fluctuations smoothing. However, the condition of power utilities like frequency deviation is not considered for tuning the MPPT and for limiting the new output voltage.

To improve the contribution of distributed generation significantly in the existing electrical networks, which poses new technical and economical challenges to power system control and management, coordinated control of distributed resources is necessary. A coordinated management of a diesel power plant and a PV array is suggested in [6] in order to fully exploit the PV renewable energy. However, this coordination is not multiple PV system and it did not deal with the problems introduced in the power utility by output power fluctuations of PV power systems.

In this paper, a new and simple coordinated control method for leveling the output power fluctuations of multiple PV power generation systems considering instant insolation and frequency deviations from the power utility is proposed. In this control approach, the PV system's output power control is achieved by two levels: central and local. In central control, insolations from all PV systems are added together and an average insolation is formed. Based on this average insolatio and power system's condition, the central output power command is derived. This method uses fuzzy control [7], [8] to produce the central output power command and it has three inputs which are absolute average of frequency deviation, average insolation, and variance of insolation. Here, the central output power command is decreased in time when large frequency deviation continuously occurs. Because the frequency deviation is introduced in the power utility by the load variation and it increases when a big

978-1-4244-1871-8/07 $25.00 © 2007 IEEE

amount of PV output power with fluctuations due to rapid insolation change is given to the utility. On the other hand, central output power command increases when frequency deviation is small and this increase will not cross the maximum power available. If insolation changes rapidly in a short period, the PV system's output will be smaller than available maximum power when the new operating point of the PV array is not an optimal one. In local control, maximum power point tracking (MPPT) command is generated for each of the PV power generation systems by Perturb and Observe (P & O) method [9]. All maximum power commands are summed up together and by dividing the central power command by this sum, a common tuning factor is generated. This tuning factor is multiplied with each of the individual maximum power command and new local power commands are generated for each of the PV power generation systems. Simple PI control law is employed to control the power converter associated with each PV power generation system for extracting output power equal to the local power command from the PV array.

The proposed method is compared with the method where maximum power point tracking (MPPT) control is used for each of the PV system without any coordination and leveling. Through simulation results, it is found that the proposed method is effective in achieving the following key features: smoothing of PV output power fluctuations, reduction of frequency deviation to maintain the reliability of power utility, and supplying the possible maximum amount of PV power to the utility without reducing the reliability of the utility.

The paper is organized as follows: Section II provides the concept of a small power system, description of PV power generation system, control of power converter and a brief review of solar module characteristics. Section III describes the coordinated output power command generation system. In Section IV, effectiveness and feasibility of the proposed method is demonstrated by simulation results. Conclusions are drawn in Section V.

II. SMALL POWER SYSTEM MODEL

The concept of multiple PV systems connected to small power utility used in this paper is shown in Fig. 1. The small power utility consists of the diesel generators and ten PV systems that generate power to supply the demand.

The small power system model which consists of diesel generator in detail, PV power generation systems, and load is shown in Fig. 2 where, S_i is the insolation, N is the total number of PV power generation systems, P_{inv}^* is the central command power generated by fuzzy based central output power command generation system, ΔP_A is the combined power generated power by N PV power generation system, ΔP_d is generated power by diesel generator, R is the speed regulation, T_g is the governor time constant, T_t and T_r are the time constants, ΔP_L is the load, M is the inertia constant, D is the damping constant, and Δf is the frequency deviation of small power utility. The control of diesel generator

described in [14] is also used in this paper to fully exploit the renewable energy.

In Fig. 3, PV power generation system including solar array, inverter and PI controller is shown where V_{A1} is the first solar array voltage, I_{A1} is the first solar array current, Δf is the frequency deviation, V_{s1} is the generated supply voltage by the inverter connected to first solar array, I_{s1} is the generated supply current by the inverter connected to first solar array, and $\left| I_{inv1}^* - I_{inv1} \right|$ is the error between command current and produced current. The control algorithm for the inverter [10], [11] adopted here is very simple. The inverter output voltages and currents are sensed and transformed from 3-phase to synchronously rotating 2- phase. The command currents are generated dividing the local output power command by sensed inverter voltage. Then the error between command inverter current and actual inverter current is processed through a PI controller to generate the PWM pulses. For simple structure and less costly implementation, a Perturbed and Observed (P&O) [9] algorithm was chosen in the present structure.

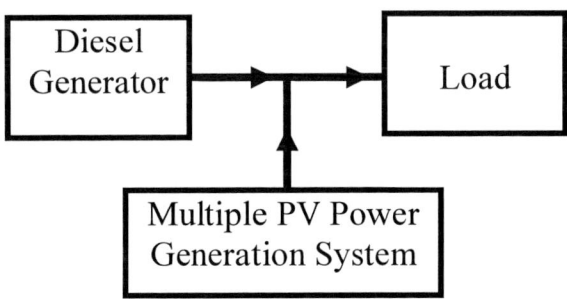

Fig. 1 Concept of small power utility with multiple PV systems.

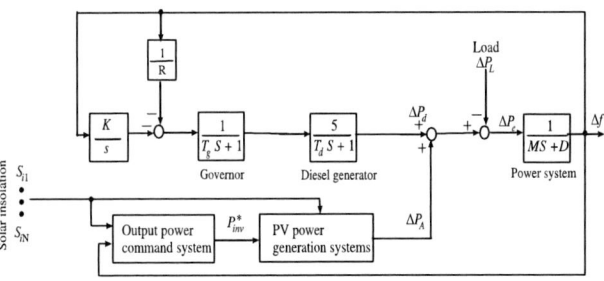

Fig. 2 Small power system model with multiple PV systems.

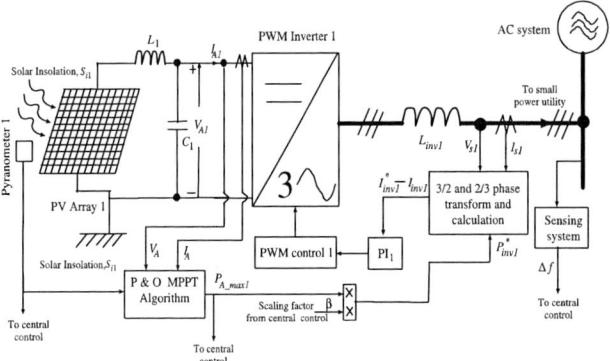

Fig. 3 Individual PV power generation system.

978-1-4244-1871-8/07 $25.00 © 2007 IEEE   445

III. COORDINATED OUTPUT POWER COMMAND GENERATION

In order to control the combined output power of PV systems considering power utility and insolation conditions, output power command is generated in two levels: central and local. Individual local output power commands are given to each of the PV power systems by maintaining a coordination between central and local levels to produce combined PV output power same as central output power command. The coordination is simple. However, it is effective for smoothing and leveling the combined output power fluctuations. Central output power command $P_{inv}^*\psi$is decided by the fuzzy based central output power command generation system shown in Fig. 6. This central output power command system consists of mainly two fuzzy controls. These fuzzy controls are described by a set of "if-then" based fuzzy rules. Fuzzy control is effective when mathematical expressions are difficult by inherent complexity or nonlinearity. Therefore, no deterministic model is required.

Firstly, Fuzzy control I is explained. There are two inputs of fuzzy control I. One is absolute average of frequency deviation Δf_s and the other is average insolation $\overline{S}_i^*\psi$. The former, which is an index to estimate power system condition, is expressed by

$$\Delta f_s = \frac{1}{T}\int_{t-T}^{t}|\Delta f|dt \qquad (1)$$

where $t\psi$is present time and $T\psi$is integral interval. Since absolute value of frequency deviation Δf is used, absolute average of frequency deviation Δf_s increases or decreases with increase or decrease in frequency deviation Δf of the power system. Therefore, (5) indicates frequency deviation quantitatively at any given time. Average insolation $\overline{S}_i^*$ is defined by

$$\overline{S}_i^* = \frac{1}{T}\int_{t-T}^{t}S_i^*dt \qquad (2)$$

where

$$S_i^* = \frac{1}{N}\sum_{N=1}^{N}S_{iN} \qquad (3)$$

where $N\psi$is the total number of PV power generation systems, $S_{iN}\psi$is instantaneous insolation of each PV power generation system, $\overline{S}_i^*$ is summation of insolation of each PV power generation system divided by N. Fuzzy rules and membership functions of Fuzzy control I are shown in TABLE I and Fig. 7, respectively. Here, central output power control of PV systems according to power system condition is accomplished by using absolute average of frequency deviation Δf_s as input of fuzzy control.

It is undesirable to increase central output power command for PV systems considerably by insolation condition because probability of solar insolation decrease at short times is high. The important thing considered in this paper is the prevention of large frequency deviations. The frequency deviation Δf of magnitude more than ± 0.2Hz is prevented rather than increase of generated power of PV systems. Thus,

membership functions are determined so that decreasing tendency is strong compared with increasing tendency for central output power command of PV systems. When frequency deviation Δf is more than ± 0.2Hz in certain times, fuzzy rules and membership functions that yield an output to decrease the output power command are defined by trial-and-error. The ith fuzzy rule is expressed as

Rule i : if Δf_s is L_x and $\overline{S}_i$ is M_y then γ_I is Z_l (4)

$$x = 1, 2, \cdots, 7, \ y = 1, 2, \cdots, 7, \ l = 1, 2, \cdots, 7$$

where L_x, M_y denote the antecedents and Z_l are consequent part. Fuzzy reasoning output γ_I is calculated by

$$\gamma_I = \sum_{i=1}^{49}w_iZ_i \Big/ \sum_{i=1}^{49}w_i \qquad (5)$$

where w_i denotes the grade for the antecedent and is obtained by

$$w_i = w_{\Delta f_s i}w_{\overline{s}_i i} \qquad (6)$$

where $w_{\Delta f_s i}$ and $w_{\overline{s}_i i}$ are the grade of antecedents for each rule.

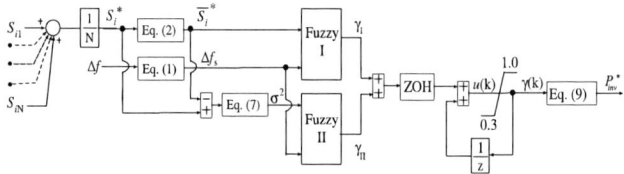

Fig. 4 Fuzzy based central output power command generation system.

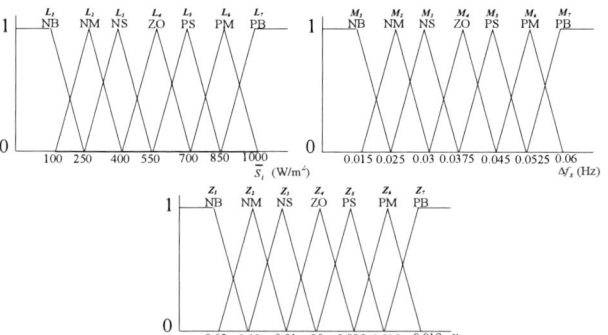

Fig. 5 Membership functions of fuzzy reasoning I

TABLE 1
FUZZY RULES OF FUZZY REASONING I

		Δf_s						
		NB	NM	NS	ZO	PS	PM	PB
$\overline{S}_i$	NB	ZO	NS	NM	NB	NB	NB	NB
	NM	PS	ZO	NS	NM	NB	NB	NB
	NS	PM	PS	ZO	NS	NM	NB	NB
	ZO	PB	PM	PS	ZO	NS	NM	NB
	PS	PB	PB	PM	PS	ZO	NS	NM
	PM	PB	PB	PB	PM	PS	ZO	NS
	PB	PB	PB	PB	PB	PM	PS	ZO

NB=Negative Big   NM=Negative Medium   NS=Negative Small
PB=Positive Big   PM= Positive Medium   PS= Positive Small
ZO= Zero

Secondly, Fuzzy control II is explained. Absolute average of frequency deviation Δf_s and variance σ^2 of average solar insolation S_i^* are used as inputs of Fuzzy control II, where variance σ^2 is expressed as

$$\sigma^2 = \frac{1}{T} \int_{t-T}^{t} \left(S_i^* - \bar{S}_i^* \right)^2 dt \tag{7}$$

Output power command that depends on power system condition rather than insolation condition is decided by using absolute average of frequency deviation Δf_s as input for both fuzzy control I and fuzzy control II. In addition, when insolation fluctuations are large, the variance of average insolation σ^2 is used as one of the inputs since the objective is to maintain decreasing tendency. Fuzzy rules and membership functions of Fuzzy control II are shown in TABLE II and Fig. 8, respectively. Setup of fuzzy rules and parameters of membership functions are determined by prioritizing to prevent boosting of frequency deviation.

As can be seen from Fig. 6, the discrete value $u(k+1)$ is obtained by the sum of outputs of Fuzzy control I, γ_I, and Fuzzy control II, γ_{II}, through zero-order-hold, and the rate of current rated output power of PV arrays. Then, new rate of central rated output power for PV arrays $\gamma(k+1)$ becomes central output power command by the following equation:

$$\gamma(k+1) = \gamma(k) + u(k+1) \tag{8}$$

Since, the rate obtained by (8) changes step, it is necessary to convert it into a smooth output power command. Linear output power command is obtained in each sampling time by using the following equation:

$$P_{inv}^* = NP_{rated} \left\{ \gamma(k) + \frac{\gamma(k+1) - \gamma(k)}{T_s} f(t) \right\} \tag{9}$$

Finally, (9) becomes central output power command P_{inv}^* to the PV power generation systems, where P_{rated} ψ is the rated output power of a single PV power generation system, N is the total number of PV power generation systems, T_s is sampling time, $f(t)$ is periodic function such that $f(t) = t(0 < t < T_s)$. After generating central output power command, local output power command for each of the PV power generation systems is produced by a simple coordinated control. The N PV power generation systems with coordinated control are shown in Fig. 9. Here the combined PV output power ΔP_A is expressed as

$$\Delta P_A = \sum_{N=1}^{N} P_{AN} \tag{10}$$

where P_{AN} is the output power of each PV power generation system. In order to achieve combined output power ΔP_A equal to the central output power command P_{inv}^* coordinated control method for each of the PV power generation systems is needed. To generate coordination between central power command and individual local power command, a common tuning factor ψ is formulated. The tuning factor ψ can be expressed as

$$\beta = \frac{P_{inv}^*}{\sum_{N=1}^{N} P_{A\max N}} \tag{11}$$

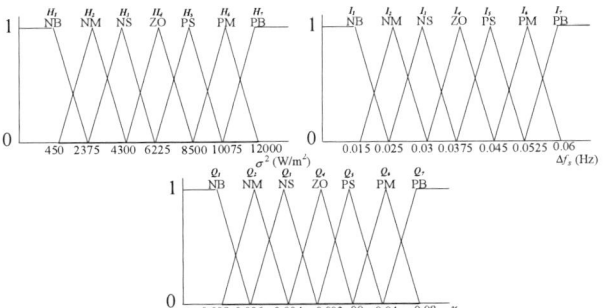

Fig. 6 Membership functions of fuzzy reasoning II

TABLE 2
FUZZY RULES OF FUZZY REASONING II

		Δf_s						
		NB	NM	NS	ZO	PS	PM	PB
σ^2	NB	PB	PB	PB	PB	PM	PS	ZO
	NM	PB	PB	PB	PM	PS	ZO	NS
	NS	PB	PB	PM	PS	ZO	NS	NM
	ZO	PB	PM	PS	ZO	NS	NM	NB
	PS	PM	PS	ZO	NS	NM	NB	NB
	PM	PS	ZO	NS	NM	NB	NB	NB
	PB	ZO	NS	NM	NB	NB	NB	NB

NB=Negative Big   NM=Negative Medium   NS=Negative Small
PB=Positive Big   PM= Positive Medium   PS= Positive Small
ZO= Zero

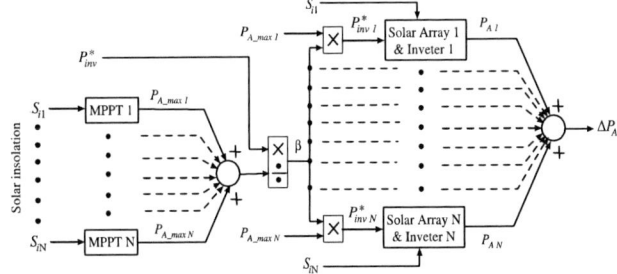

Fig. 7 Small power system model with multiple PV systems.

where $P_{A\max N}$ is maximum power point tracking (MPPT) output power command of each PV power generation system. The P& O algorithm is used to generate the MPPT output power command considering instant insolation, solar array voltage, and solar array current. The individual local output power command for each of the PV power generation systems can be obtained by the following equation

$$P_{invN}^* = P_{A\max N} \times \beta \tag{12}$$

IV.   SIMULATION RESULTS

In this paper, the effectiveness of output power fluctuations leveling of PV power generation systems and frequency deviation reduction of power utility using the proposed coordinated method is examined by simulation with system model and parameters as mentioned in [11]-[14]. In order to use parameters of real systems in [13], [14], the rated output power of each PV array is 241kW. The total number of PV power generation system used in this paper is

TABLE 3
SIMULATION PARAMETERS

Parameters of small power system	
Inertia constant, M	0.150 puMW.sec/Hz
Damping constant, D	0.008 puMW/Hz
Governor time constant, T_g	0.10 sec
Time constant, T_t	0.25 sec
Time constant, T_r	8.0 sec
Speed regulation, R	2.5 Hz/puMW

Parameters of PV array	
Rated output power	241 kW
Open circuit voltage	588.80 V
Short circuit current	520.37 A
Number of modules in series	16
Number of modules in parallel	65
Total number of cells	62,400

Parameters of PV module	
Rated output power	231.58 W
Open circuit voltage, V_{oc}	36.30 V
Short circuit current, I_{sc}	7.99 a
Shunt resistance, R_{sh}	50 Ω
Series resistance, R_s	5 Ω
Ideality factor, A	1.450
Inverse Saturation Current, I_{or}	3.047e-07 A
S.C. current temperature constant, I_t	1.73e-03 A/^{0}K
Reference temperature, T_{ref}	25 ^{0}C
Boltzman's constant, K	1.38e-23
Charge of an electron, q	1.602e-19 C
Bandgap voltage, E_g	1.11 eV
Number of cells in series, N_s	60
Standard insolation, S_i	1000 W/m^2
Dimension	0.75 m^3

Parameters of Power converter	
Inverter power rating	225 kW
Nominal ac output voltage	480 V (3-φ)
Nominal ac output frequency	50 Hz
Maximum ac line current	271 rms
Maximum dc input voltage	600 V
Maximum dc input current	781 A
Efficiency	94.5%

Parameters of PI controller	
Proportional constant, K_P	0.1
Integral constant, K_I	10

10 and the combined rated output power of ten PV power generation system is 2.41 MW. Simulation parameters for power system, PV array, power converter and PI controller are shown in TABLE III. Integral time $T\psi$ is 100s, sampling time $Ts\psi$ to obtain discrete value of output power command is 10s, and sampling time of PI controller is 1ms. Simulation time is 30 minutes.

Different solar isolations are used for PV power generation systems. Insolations for PV plants 1 and 6 are shown in Fig. 8 (a) and in Fig. 8 (b). Load is shown in Fig. 8 (c). The comparative simulation results using MPPT control and proposed coordinated method are shown in Fig. 9. Figs. 9(a) and Fig. 9 (b) show output power produced by coordinated method and available maximum power for PV plants 1, 6 respectively. From these figures, it is seen that individual smoothing of output power fluctuations is not achieved. Here, the individual output power of each PV plant is fluctuating. However, from Fig. 9(c), it can be seen that when these fluctuating powers are added together, the combined output power of ten PV plants are significantly smoothed and

leveled. Therefore, coordinated control works successfully to achieve smoothed and leveled output power from fluctuating output power of individual PV plants. Another advantage of this coordinated method is that if a PV plant's output decreases with insolation decrease, in order to compensate the shortage of power, other PV plants can produce maximum power.

Output power of diesel generator ΔP_d, are shown in Fig. 11(d). From Fig. 11(d), it is observed that the diesel generator output power produced with MPPT controlψ fluctuates in order to cancel out fluctuation of output power, ΔP_A and load, ΔP_L. On the other hand, diesel generator power produced with proposed methodψ fluctuates less. This is because the combined PV output power is controlled considering frequency deviation of the power utility and it is smoothed by the proposed method.

Fig. 10(e) shows frequency deviation Δf. From Fig 10(e), it is observed that frequency deviation produced with MPPT control deviates more than $\pm$ 0.2Hz frequently. This is a severe problem for maintaining power system reliability. Therefore, combined output power of ten PV plant produced by MPPT control has harmful effects on power system. However, frequency deviation produced with proposed method does not deviate by more than $\pm$ 0.2Hz frequently. Therefore, the proposed method is effective in maintaining power system reliability.

From t=0s to t=50s, the frequency deviation is high, and the combined output power remains steady to reduce the frequency deviation. From t=50s to t=600s, the frequency deviation is low, and the combined output power increases to supply possible maximum power. However, from t = 600 s to t =1050 s, the frequency deviation is high, therefore, the

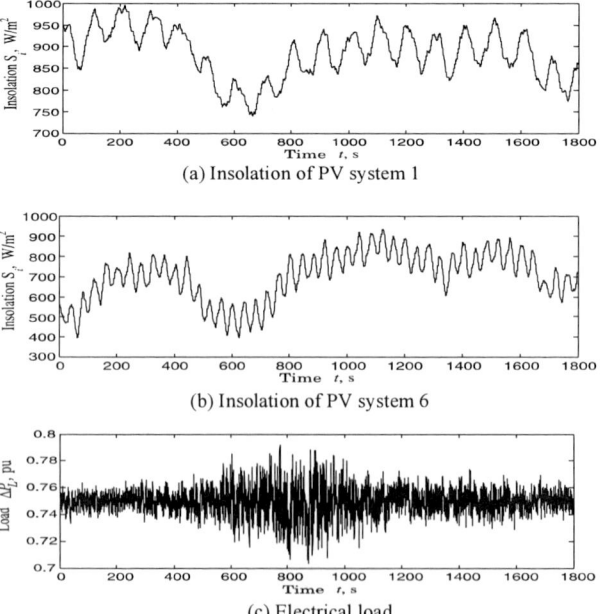

Fig. 8 Insolation and load used for the simulation. (a) Insolation used for PV system 1. (b) Insolation used for PV system 6. (c) Load.

978-1-4244-1871-8/07 $25.00 © 2007 IEEE

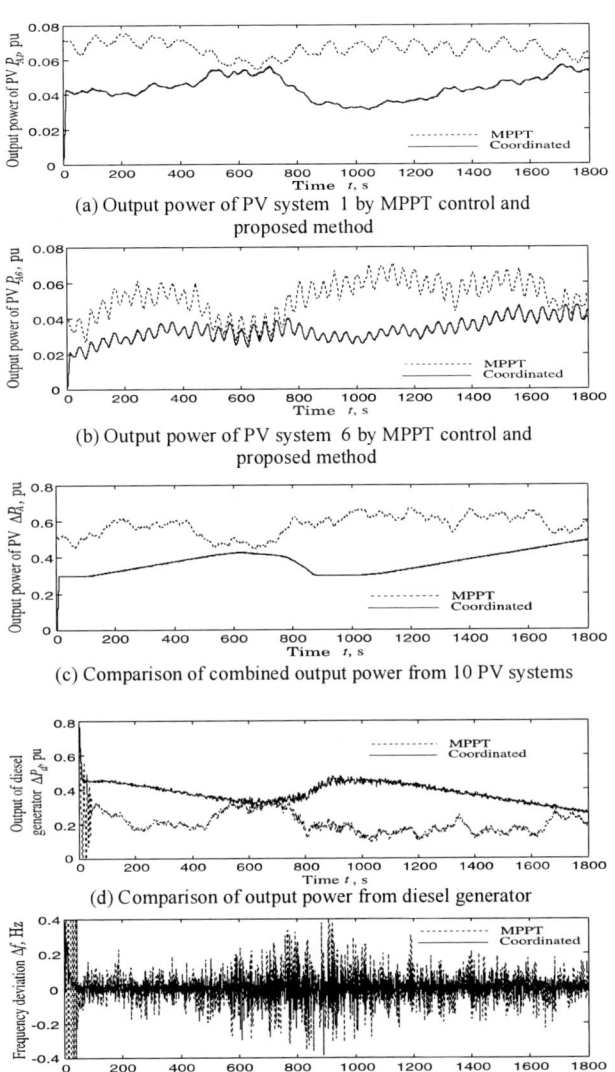

(a) Output power of PV system 1 by MPPT control and proposed method

(b) Output power of PV system 6 by MPPT control and proposed method

(c) Comparison of combined output power from 10 PV systems

(d) Comparison of output power from diesel generator

(e) Comparison of frequency deviation

Fig. 10 Comparative simulation results of MPPT control and proposed method.

combined output power first decreases, then remains steady to reduce frequency deviation. From t=1050s to t=1800s, the frequency deviation decreases, and the combined output power increases to supply possible maximum power. Therefore, it can be said that penetration of combined output power in the power utility with leveling effect is controlled by the proposed method.

V. CONCLUSION

This paper presents a coordinated control method for output power fluctuations leveling of PV plants considering power system condition and insolation condition. Central output power command is defined by fuzzy control, which has three inputs for absolute average of frequency deviation, average insolation, and variance of insolation. Setup of fuzzy rules and parameters of membership functions are determined by prioritizing to prevent increase in frequency deviation. Local output power command for each of the plants is generated by

coordination between central power command and maximum power tracking command. The PV plants are controlled with the help of PI controllers and power converters to level the output power so as not to harmfully influence the power system at times with large frequency deviation. On the other hand, at times with small frequency deviation, the proposed system can operate to increase generated power near available maximum power. From the simulation results, it can be said that the PV power generation systems with the proposed coordinated method achieve flexible output power control without introducing any harmful effects to power system, and the effectiveness of the proposed method for leveling output power fluctuations and for maintaining power system reliability is confirmed.

REFERENCES

[1] S. Yanagawa, T. Kato, K. Wu, A. Tabata, and Y. Suzuoki, "Evaluation of LFC capacity for output fluctuation of photovoltaic generation systems based on multi-point observation of insolation," in *Proc. IEEE Power Engineering Society Summer Meeting*, 2001, pp. 1652-1657.

[2] A. Woyte, V.V. Thong, R. Belmans, and J. Nijs, "Voltage fluctuations on distribution level introduced by photovoltaic systems," *IEEE Trans. Energy Convers.*, vol. 21, no. 1, pp. 202-209, March 2006.

[3] H. Asano, K. Yajima and Y. Kaya, "Influence of photovoltaic power generation of required capacity for load frequency control," *IEEE Trans. Energy Convers.*, vol. 11, no. 1, pp. 188-193, March 1996.

[4] E. Wiemken, H.G. Beyer, W. Heydenreich, and K. Kiefer, "Power characteristics of PV ensembles: experiences from the combined power production of 100 grid connected PV systems distributed over the area of Germany," *Solar Energy*, vol. 70, no. 6, pp. 513-518, 2001.

[5] N. Ina, S. yanagawa, T. Kato, and Y. Suzuoki, "Smoothing of PV system output by tuning MPPT control," *Electrical Engineering in Japan*, vol. 152, no. 2, pp. 10-17, 2005.

[6] D. Canever, G.J.W. Dudgeon, S. Massucco, J.R. McDonald, and F. Silvestro, "Model validation and coordinated operation of a photovoltaic array and a diesel power plant for distributed generation" in *Proc. IEEE Power Engineering Society Summer Meeting*, 2001, vol. 1, pp. 626-631.

[7] L. H. Tsoukalas and R. E. Uhrig, *Fuzzy and Neural Approaches in Engineering*, NewYork, USA: Jhon Wiley & Sins. Inc., 1997.

[8] C.C. Lee, "Fuzzy logic in control systems: fuzzy logic controller-part I and part II," *IEEE Trans. Syst., Man, Cybern.*, vol. 20, no. 2, pp. 404-435, Mar./April 1990.

[9] N. Femia, G. Petrone, G. Sp, "Optimization of perturb and observe maximum power point tracking method," *IEEE Trans. Power Electron.*, vol. 20, no. 4, pp. 963-973, July 2006.

[10] K. Kobayashi, I. Tanako, and Y. Sawada, "A study of a two stage maximum power point tracking control of a photovoltaic system under partially shaded insolation conditions," *Solar Energy Materials & Solar Cells*, vol. 90, pp. 2975-2988, 2006.

[11] C. Hua, J. Lin, and C. Shen, "Implementation of a DSP controlled photovoltaic system with peak power tracking," *IEEE Trans. Ind. Electron.*, vol. 45, no. 1, pp. 99-107, Feb. 1998.

[12] [T. Senjyu, T. Kaneko, A. Yona, N. Urasaki, T. Funabashi, F. Yamada, and S. Sugimoto, "Output power control for large wind power penetration in small power system," presented at the IEEE PES General Meeting, Florida, USA, 2007.

[13] [Online]. Available: http://www.powerlight.com/success/pdf/LVVWD Ronzone Reservoir.pdf.

[14] [Online].Available:http://www.xantrex.com/web/id/214/p/1840/pt/23/ product.asp.

978-1-4244-1871-8/07 $25.00 © 2007 IEEE

Performance Monitoring and Analysis of Middle Scale Grid-Connected PV System

J. H. So, B. G. Yu, H. M. Hwang, G. J. Yu
Korea Institute of Energy Research
Photovoltaic Research Center
71-2, Jang-Dong, Yuseong-Gu, Daejeon, 305-343
Email: jhso@kier.re.kr

J. Y. Choi, I. Choy
Kwangwoon University
Department of Electrical Engineering
447-1, Wolgye-Dong, Nowon-Gu, Seoul, 139-701
Email: juyeop@daisy.kw.ac.k

Abstract— **This paper presents performance results of middle scale grid-connected Photovoltaic (PV) system for monitoring periods. The performance and losses of PV system is quantitatively evaluated and analyzed using calculation model with monitored data so that various PV system technologies are development. The objective of this paper is to develop reliable and valid evaluation method of Photovoltaic (PV) system performance so that maximum output is achieved over the system lifetime with performance improvement.**

I. INTRODUCTION

As the world PV market has been increasing rapidly, the various types of PV systems have been disseminated according to design, installation and environment conditions with a wide range of users [1]. Technology development for PV system dissemination becomes the one of most important issues so that the concerns of high quality and reliable PV system have been increasing around world [1, 2]. In this paper, monitoring system is constructed for measuring and analyzing PV system performance to investigate the overall effect of meteorological conditions on operation character- istics by field test. The performance and losses of PV system is indicated in quantitative value using calculation models with monitoring data so that various PV system technologies is developed including maximum performance over lifetime. The performance of PV system has been evaluated and analyze not only for components perspective but also for global perspective and those evaluation results are reviewed.

II. SYSTEM DESCRIPTION

The installed 50kW grid-connected PV system consists of 37 parallel and 17series with 80W multi-crystalline PV module as shown in Fig. 1. The nominal capacity of PV array is 50.2 kW and PV array is curved ground mounted. Averaged tilt angle and azimuth of PV array is individually 18 degree and 5 degree. The specification of PV module at standard test conditions (STC) and power conditioning unit (PCU) were summarized in Table 1 and 2.

The PV system has been monitored continuously to evaluate and analyze the overall performance on environment conditions with meteorological sensors and electrical sensors [3]. PV systems have been monitored from September 2005 to until today. The performances of PV system are evaluated and analyzed the overall effect of environmental conditions on operation characteristics. The installed monitoring system could be remotely supervised through a wide area network [2]. The monitored data of the PV system are collected minutely so that its performance can be analyzed and evaluated exactly.

Fig. 1: 50kW grid-connected PV system

TABLE 1 PV module specifications

PV module (at STC)	SMT 80
Cell Type(c-si)	Multi
Nominal Power [W]	80
Short Circuit Current [A]	4.55
Open Circuit Voltage [V]	17.6
Maximum Power Point Current [A]	4.55
Maximum Power Point Voltage [V]	17.6

TABLE 2 PCU specifications

PCU (under rated conditions)		PV-C350S
Type		Three phase
DC Input	Input Voltage [V]	296
	Operating Voltage [V]	238-368
AC Output	Rated Power [kW]	50
	Efficiency [%]	92 or more
	Total Harmonic Distortion [%]	5 or less
	Maximum Single Harmonic [%]	3 or less

In this monitoring system, the following items are measured to evaluate and analyze the performance of PV systems.

Electrical measurement items
- DC and AC voltages

978-1-4244-1871-8/07 $25.00 © 2007 IEEE   450

- DC and AC currents
- AC active power, apparent power, and frequency
- AC power factor and total current/voltage harmonic distortion

Meteorological measurement items
- Irradiance on horizontal plane
- Irradiance on plane of PV array
- PV module surface temperature
- Ambient temperature

III. SYSTEM PERFORMANCE RESULTS

The various types of PV systems have been installed with a wide range of users. In order to accomplish general purpose results of PV systems evaluation, there is necessity for producing normalized performance indices [3-5]. Final system yield (Y_F) is PV system AC energy production and Array yield (Y_A) is DC energy production. Reference yield (Y_R) is irradiation on plane of PV array. Performance ratio (PR) is defined as ratio of the actual performance to the theoretically possible performance of PV system. PR is one of the most important performance indices to evaluate and analyze performance characteristics of PV system. Capture losses (L_C) are due to PV array losses and System losses (L_S) are due to DC to AC energy conversion [4-6]. Table 3 summarizes the performance indices used in this evaluation and analysis.

TABLE 3 Overview of derived parameter for performance evaluation

Parameter	Symbol	Equation	Unit
Final System Yield	Y_F	E_P/P_o	h/d
Array Yield	Y_A	E_A/P_o	h/d
Reference Yield	Y_R	G_A/G_S	h/d
Array Capture Losses	L_C	Y_R-Y_A	h/d
System Losses	L_S	Y_A-Y_F	h/d
Performance Ratio	P_R	Y_F/Y_R	
Array Output Energy	E_A	-	kWh/d
System Output Energy	E_P	-	kWh/d
Nominal Capacity	P_o	-	kWh/d
Irradiation on Plane	G_A	-	kWh/m².d
Irradiance at STC	G_S	1	kW/m²

The daily yields of PV system for monitoring period from September 2005 to August 2006 are shown in Fig. 2. The annual daily reference yield, array yield and final system yield is 3.7 h/d, 2.9 h/d and 2.6 h/d respectively. DC power generated by the PV array is linearly dependent on the irradiance except for lower irradiance due to the nonlinear characteristics [6]. The monthly averaged PV array conversion efficiency varies between 8.8 and 11.0% for monitoring period. The monthly averaged PCU efficiency ranges from 89 to 93%.

The daily yields of PV system performance and losses for monitoring period are shown in Fig. 3. Y_F is ranges from 1.82

to 3.34 h/d. L_C and L_S are individually ranged from 0.33 to 1.25 h/d and from 0.19 to 0.28 h/d. The annual Y_F, L_C and L_S are 2.63 h/d, 0.79 h/d and 0.23 h/d respectively. When compared with L_S, The causes of big L_C have a various troubles on performance obstruction such as shading, PV module temperature, mismatch and so on. PV array output is linearly dependent on the irradiance except for the nonlinear characteristic range of lower irradiance. Therefore, there is a necessity for carrying out a detailed temperature correction of PV module efficiency and investigated this dependence on DC output versus meteorological characteristics such as irradiance, PV array surface temperature.

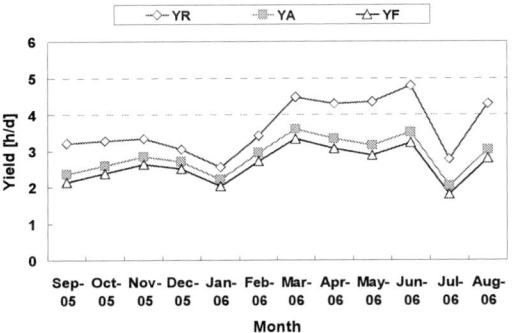

Fig. 2: Daily yields of PV system

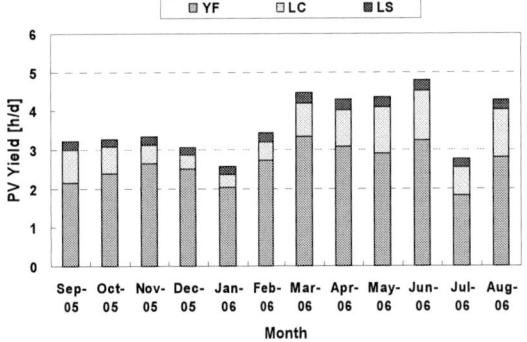

Fig. 3: Daily yield of YF, LC and LS

Fig. 4 shows the DC output power normalized characteristics of PV array according to PV module temperature variation based on measured results for a full range of irradiance. DC output power generated by PV array is linearly dependent on the irradiance except for irradiance of 200 W/m² or less due to the nonlinear characteristics. As shown in figure, the conversion efficiency of the PV array does strongly depend on module surface temperature is known from performance results. Relationship between irradiance and DC output power of the PV array on surface temperature can be approximated as the following linear regression equation [6]:

$$P_{pv,n} = k_0 \cdot Y_R + k_1 \qquad (1)$$

Where $P_{pv,n}$ is normalized PV array output power, Y_R is reference yield, k_0 and k_1 are regression coefficients respectively.

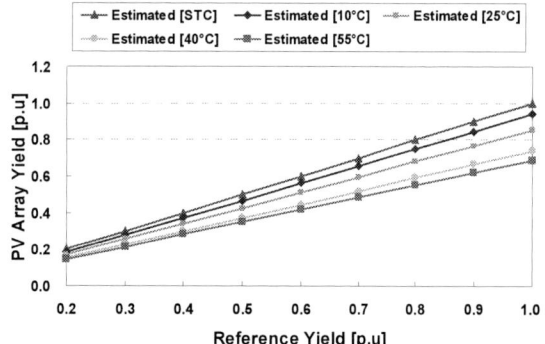

Fig. 4: Normalized characteristics of PV array

The normalized characteristics of PCU efficiency for monitoring period are shown in Fig. 5. As shown in figure, relationship between PCU input power and PCU efficiency can be approximated as the following nonlinear regression equation [6]:

$$\eta_{PCU,n} = k_0 + k_1 \cdot (1 - \exp(-k_2 \cdot P_{pv,n})) + k_3 \cdot (1 - \exp(-k_4 \cdot P_{pv,n})) \tag{2}$$

Where $\eta_{PCU,n}$ is PCU efficiency, $P_{pv,n}$ is normalized PCU input power, k_0, k_1, k_2, k_3 and k_4 are regression coefficients respectively.

The PCU starts supplying energy to the grid, when the irradiance is 30 W/m² or more and stops supplying energy when the irradiance goes down to 25 W/m². The normalized PCU efficiency for irradiance values higher than normalized PCU input power of 0.2 is approximately constant but for the lower irradiance. When normalized PCU input power is 0.2, normalized PCU efficiency is above 0.9.

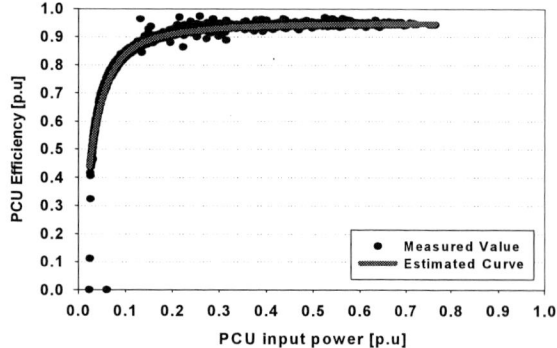

Fig. 5: Normalized characteristics of PCU

IV. SYSTEM EVALUATION AND ANALYSIS

Since PV system performance is strongly dependent on loss factors such as mismatch, PCU losses, and PV array temperature rise and so on. Therefore, there is a necessity for investigating these loss factors to disseminate high quality and reliable PV system.

Figs. 6 and 7 show daily and normalized yields of PV system performance and losses using calculation model with hourly monitored data. l_{PO} is other losses. l_{PD} is DC wiring loss of PV array. l_{PI} is loss due to irradiation variation on incident angle. l_{PM} is due to mismatch of PV sub-array and failure of maximum power point tracking (MPPT). l_{PT} is loss due to PV array temperature rise. l_P is PCU losses including conversion efficiency and stand-by. Performance ratio (PR) ranges from 0.65 to 0.82. DC wiring, incident angle and temperature rise included in L_C and PCU conversion efficiency and stand-by included in L_S. Normalized L_C and L_S are individually varied from 0.11 to 0.29 and from 0.06 to 0.08. The main aspects causing PR decrease refer to PV array temperature rise, other losses and PCU losses.

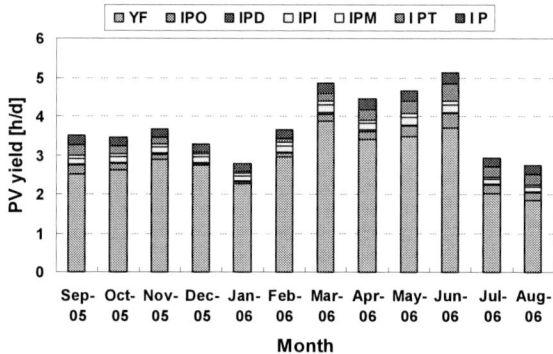

Fig. 6: Daily yield of PV system performance and losses

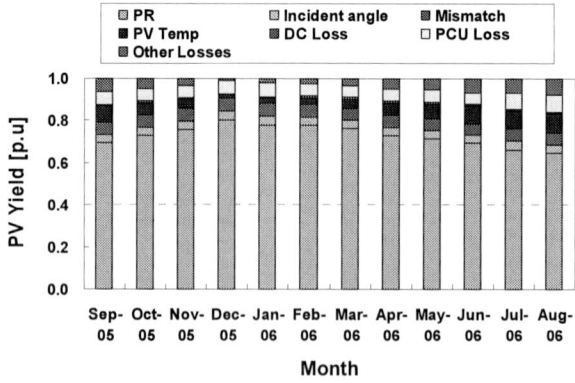

Fig. 7: Normalized yields of PV system performance and losses

Fig. 8 shows overall performance evaluation results of PV system with calculation model for monitoring period. The annual averaged PR, L_C and L_S are individually 73.2%, 20.7% and 6.1%. The detailed losses of L_C are evaluated as follows;

978-1-4244-1871-8/07 $25.00 © 2007 IEEE

incident angle is 3.9%, mismatch loss is 5.9%, DC circuit loss is 0.7%, PV temperature rise is 5.6% and other losses are 4.6%. Main reasons of PR decrease refer to other losses, mismatch, PV array temperature rise and PCU losses. From these evaluation results, it will be estimated to various loss factors on the system performance obstruction such as other losses (soil, snow, power shortage etc.), mismatch, PCU losses, etc. In general, if PR is under 75% in middle scale PV system, that means that PV system has various troubles of system performance obstructions. The decrease of PR was caused by increased L_C due to PV sub-array faults and mismatch including parallel-series unbalance, failure of MPPT on incident irradiance.

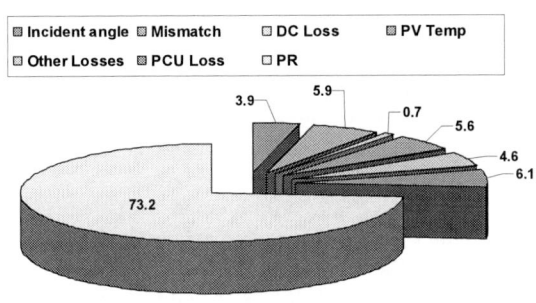

Fig. 8: Evaluation results of PV system performance

Supposing that the performance PV system has the same performance described in Table 1 and 2 and installed in fixed tilt angle of PV array, calculation models of PV components were applied to estimate the performance of PV system using simulation tool.

As a result, performance of PV system was improved with PR of 75% or more and PV system was expected to operate stably without faults. Reliable and valid evaluation method of PV system for quantitative analysis is need to be developed to establish performance improvement and optimum design of PV system.

V. CONCLUSIONS

The performance of PV system and components are analyzed and evaluated to observe the overall effect of meteorological conditions on their operation characteristics for monitoring period. To ensure high quality and reliable PV system, it is very important that reliable evaluation method of PV system performance and losses is established and indicated with quantitative analysis. Performance database of installed various PV system will be constructed continuously and improved accuracy of calculation models.

From these results, reliable evaluation method for PV system performance and losses will be planning to develop and confirm the validity.

ACKNOWLEDGMENT

This work was supported by energy technology R&D project from MOCIE (Ministry of Commerce, Industry and Energy) Republic of Korea.

REFERENCES

[1] N. M. Pearasll, H. Scholz, T. Zandowicz and C. Reise, "PV system assessment in performance-towards maximum system output." in *Conf. 21st European Photovoltaic Solar Energy Conf.*, 2006, pp. 2574-2579.

[2] J. H. So, Y. S. Jung, B. G. Yu, H. M. Hwang G. J. Yu and J. Y. Choi, "Performance results and analysis of large scale pv system." in *Proc. IEEE 4th World Conf. on Photovoltaic Energy Conv.* 2006, pp 2375-2378.

[3] S. M. Pietruszko and M. Gradzki, "Performance of a grid connected small pv system in Poland." *App. Energy,* vol. 71, no. 1-2, pp.177-183, 2003.

[4] T. Oozeki, K. Otani and K. Kurokawa, "An evaluation method for pv system to identify system losses by means of utilizing monitoring data." in *Proc. IEEE 4th World Conf. on Photovoltaic Energy Conv.* 2006, pp 2319-2322.

[5] K. Otani, K. Sakuta, T. Sugiura and K. Kurokawa, "Performance analysis and simulation on 100 japanese residential grid-connected pv system based on four year experience," in *Proc. 17th European Photovoltaic Solar Energy Conf.*, 2001, pp22-26.

[6] J. H. So, Y. S. Jung, G. J. Yu and J. Y. Choi, "Performance results and analysis of 3kW grid-connected pv systems," *Rene. Energy 32*, pp 1858-1872, 2007.

Development of On-line type Dynamic Voltage Compensation System Using Supercapacitor

Jong Hee Han
Soongsil University
Department of Electrical Engineering
1-1, Sangdo-5dong, Dongjak-gu, Seoul, 156-743
Email: wpaper@ssu.ac.kr

Jin Geun Shon
Kyungwon University
Department of Electrical Engineering
San 65, Bokjung-dong, Sujung-gu, Sungnam, 461-701
Email: shon@kyungwon.ac.kr

Il Dong Seo
Soongsil University
Department of Electrical Engineering
1-1, Sangdo-5dong, Dongjak-gu, Seoul, 156-743
Email: tjdlfehd@ssu.ac.kr

Hee Jong Jeon
Soongsil University
Department of Electrical Engineering
1-1, Sangdo-5dong, Dongjak-gu, Seoul, 156-743
Email: hjjeon@ssu.ac.kr

Abstract—In this paper, a study has been performed for optimal design and development of dynamic voltage compensation system using supercapacitor to solve voltage sag which are considered the dominant disturbances affecting power quality. Supercapacitor is employed to compensate dynamically for the voltage sag of system with sensitive load. With the prolific use of semiconductor devices in electrical equipment, modern-day loads are becoming increasingly sensitive to the sags and the disturbances prove to be costly to industries. As a technology -driven custom power installation, the Dynamic Voltage Compensation System is recognized to be most effective equipment that can be used to counter the voltage sag problem. Hence, this paper proposes the high-performance optimal design of Dynamic Voltage compensation system which can efficiently compensate the instantaneous voltage sag and instantaneous interruption of the utility voltage source. This DVR system consists of a dynamic voltage converter in series with the supply voltage and high-efficiency controller of energy storage device using Supercapacitor.

Recently, in energy storage systems, the Supercapacitor is paid attention as a new environmentally friendly energy storage element. This capacitor stores higher energy density than the conventional electrolytic capacitor. Also, this capacitor has a lot of advantages such as maintenance-free, longer life cycle and faster charge-discharge time than the conventional battery system. However, the Supercapacitor for Dynamic Voltage Compensation System must have each charge-discharge controller to control, a series-parallel switching circuit to use energy effectively with application of several compensation techniques for optimal operating.

Therefore, in this paper, optimal development of On-line type Dynamic Voltage Compensation Systems is proposed.

I. INTRODUCTION

Recently, we are growing more and more interested in power quality because of use of load equipment which is sensitive to an external effect is increased. Specially, constant control of magnitude problem of voltage is needed, because economical damage is very serious. Instantaneous voltage drop among such problem is caused by grounding fault of power distribution system, falling of a thunderbolt, operation of a large motor.[1] Constant time of instantaneous voltage drop is defined from 3seconds until in 8milli seconds. This

instantaneous voltage drop compensation system is more effective than UPS(Uninterrupted power supply) in compensation range and economical side. This system is called VSC(voltage sag compensator). But recently, name of DVR(dynamic voltage restorer) that developed in ABB is passed.[2]

As size that voltage drop happens instantaneously among voltage that this DVR system is supplied in load it series voltage injector that supply in addition in load be. And this system is used in off-line type mainly. This way measures voltage by moment and supply voltage when voltage sag event happens. This time, because of problem of reliability and correct measure about connection of two bidirectional switch elements that manage interval of power supply and supply in voltage, on-line type DVR is used. As this way is driven like continuous operating system with UPS and method that consider preferentially power and precise load operation.

Therefore, this paper is describes about development of on-line type DVR that reliability is superior. EDLC that is profitable in capacitance, life cycle, stability, environmental side is used instead of electrolytic capacitor to store energy. And this paper proposes result that experiment and develop various algorithm like high efficiency PWM and soft-starter.

II. ORGANIZATION OF DVR USING EDLC

A. Organization of off-line Type DVR

DVR system, dynamic voltage compensating equipment for instantaneous voltage drop, is used for compensating voltage of sensitive load. Therefore, DVR is system that is connected to grid through matching transformer as picture 1. Picture 2 is electrical equivalent circuit. R_f is internal resistance, L_f is inductance and C_f is capacitance. Sensitive load can be supposed R-L series load. But, as size that voltage drop happens instantaneously among voltage that is supplied in load when this DVR way produces voltage sag because of because is series voltage injector that supply in addition in load drive problem of two bidirectional switch element and problem of authoritativeness that need correct measure of

978-1-4244-1871-8/07 $25.00 © 2007 IEEE

power supply voltage and dimension of pouring in voltage and phase control, and manage interval of power supply and pouring in voltage recently upper market price in question DVR of done On-Line way use .

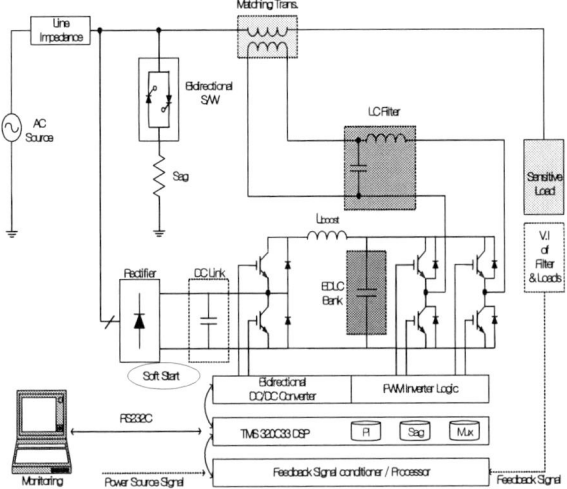

Fig 1. Block diagram of off-line type DVR system

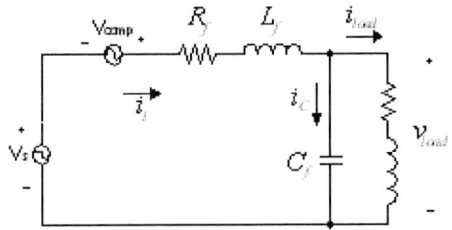

Fig 2. Equivalent circuit of off-line type DVR system

B. Organization of on-line Type DVR

It is way to operate continuous system like UPS by DVR inverter and to consider reliability of power and high sensitive load preferentially. When compose this On-line type system, it need selection of charging component, high efficiency charging-discharging control, improvement of energy utilization ratio, control of soft-starter at early charging, technique of PWM method for high efficiency operation of inverter and control of matching transformer for system isolation and voltage matching. Fig 3. is this system's organization.

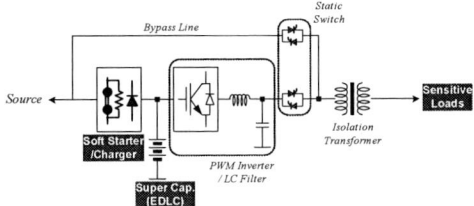

Fig 3. Organization of On-line type DVR system

Charging element appeared in Fig 3. is Electrical Double Layer Capacitor (EDLC). This is capacitor that is physical surface area of the two electrodes, the distance between them, and the type of dielectric material.

Existent DVR that used electrolytic capacitor is suitable in DVR's applications because of very fast charging and discharging speed. But, because energy save ability is very low, voltage compensation time is very short, and life-cycle is short at high temperature and because electrolyte spurts in life end, there is danger of explosion. Also, in case of use lead storage battery, charging-discharging time is very slow, rapid charging-discharging is impossible and because need periodic maintenance, economical strain is big. Specially, because can lead to very serious problem in environmental aspect, introduction of new storage system is required desperately.

C. Equivalent circuit of Supercapacitor

The equivalent circuit used for conventional capacitors can also be applied to supercapacitors. The circuit schematic in Fig. 4 represents the first-order model for an supercapacitor. It's comprised of four ideal circuit elements: a capacitance C, a series resistor Rs and a parallel resistor Rp. Rs is called the equivalent series resistance (ESR) and contributes to energy loss during capacitor charging and discharging. Rp simulates energy loss due to capacitor self-discharge, and is often referred to as the leakage current resistance.[6][7]

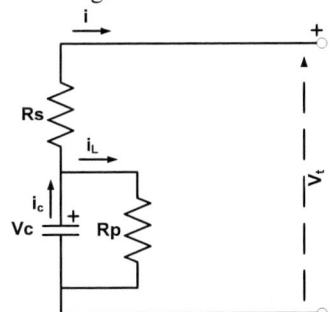

Fig 4. Equivalent Circuit of Supercapacitor

Capacitance is calculated by the following test in Fig. 5 and equation (1). Through a constant current discharge experiment, measure voltage drop, discharge time and constant current, using these data calculate capacitance of supercapacitor.

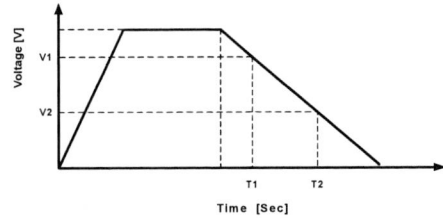

Fig. 5 constant Current Discharge Characteristics

978-1-4244-1871-8/07 $25.00 © 2007 IEEE

$$C = I \times \frac{t_2 - t_1}{V_1 - V_2} \tag{1}$$

The most demerit of supercapacitor is nature discharge by big nature discharge rate. Parallel resistor Rp that appear by these nature discharge can be calculated by nature discharge test the following in Fig. 6. Measure fluctuation of voltage by nature discharge experiment, and Rp calculates by equation (2) using linear approximation voltage.

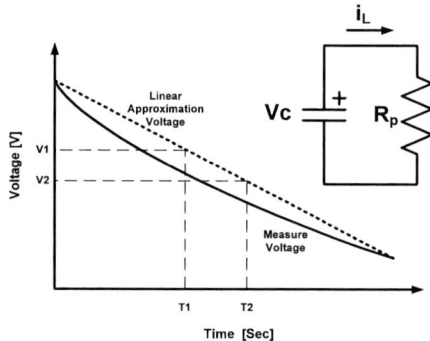

Fig. 6 Self-Discharge Characteristics

$$R_p = \frac{(V_1 + V_2)(t_2 - t_1)}{2C(V_1 - V_2)} \tag{2}$$

A series resistor Rs of supercapacitor is calculated by the following test in Fig. 7 and equation (3). Through a constant current discharge experiment, measure initial voltage drop constant current, using these data calculate series resistor Rs.

$$R_S = \frac{\Delta V}{I} \tag{3}$$

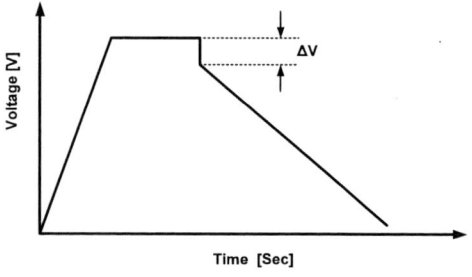

Fig. 7 Initial Voltage Drop in Constant Current Discharge

D. Charge/discharge efficiency of Supercapacitor

A equivalent circuit model can be used to describe the charging characteristics of a supercapacitor. The operation

efficiency in charge/discharge can be expressed as equation (4) and (5).

$$\eta_{discharge} = \frac{V_t I_t}{V_C I_C} = \frac{(V_C - I_t R_S) I_t}{V_C (I_t + I_p)} \tag{4}$$

$$\eta_{charge} = \frac{V_C I_C}{V_t I_t} = \frac{V_C (I_t + I_p)}{(V_C - I_t R_S) I_t} \tag{5}$$

Where V_t is the terminal voltage and It is the current input or output from the terminal. In actual operation, the leakage current I_p is usually very small and can be ignored. Thus, equation (4) and (5) can be express as equation (6) and (7).

$$\eta_{discharge} = \frac{V_C - I_t R_S}{V_C} = \frac{V_t}{V_C} \tag{6}$$

$$\eta_{charge} = \frac{V_C}{V_C - I_t R_S} = \frac{V_C}{V_t} \tag{7}$$

The above equations indicate that the loss in supercapacitor is caused by the presence of series resistance. Therefore, energy efficiency is become 50% in case achieve if charge-discharge is execute in Fig. 8 (a) constant voltage source.

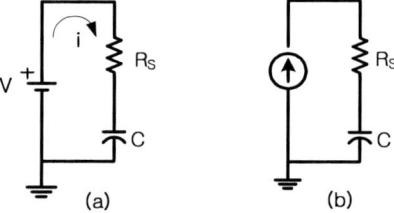

Fig. 8 Equivalent Circuit for (a) Voltage (b) current source of supercapacitor

But, in Fig. 8 (b) with current source, electric energy A that store to supercapacitor is expressed with Equation (8). Moreover, electric energy B that spent in resistance R is expressed with Equation (9).

$$A = \frac{1}{2} \cdot \frac{Q^2}{C} \tag{8}$$

$$B = I^2 R_S \cdot \Delta t = R \cdot \frac{Q^2}{\Delta t} \tag{9}$$

As a result, charge-discharge efficiency can current source control circuit by control. In addition, charge efficiency E_C and discharge efficiency E_D by constant current source are expressed by Equation (10) and (11).

978-1-4244-1871-8/07 $25.00 © 2007 IEEE

$$E_C = \frac{A}{(A+B)} = \frac{\Delta t}{(\Delta t + 2R_S C)} \qquad (10)$$

$$E_D = \frac{(A-B)}{A} = 1 - \frac{2R_S C}{\Delta t} \qquad (11)$$

E. Circuit organization of developed system

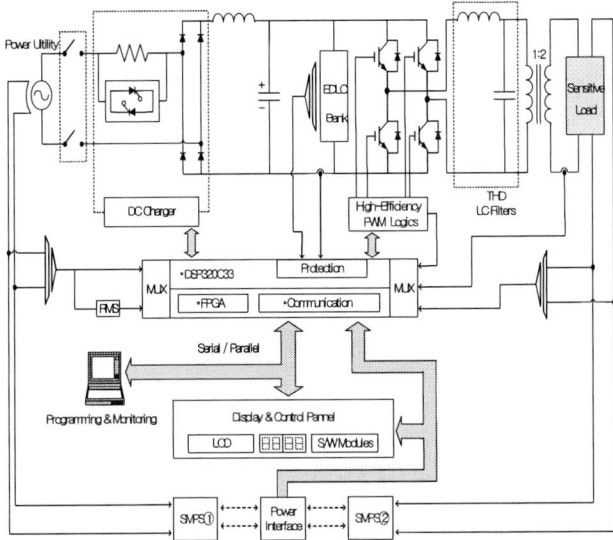

Fig 9. Overall block diagram of On-line type DVR

Fig 9. is overall block diagram about On-line type DVR system that inverter is operated continuous. Power which is supplied at early suppress charging rush current by soft-starter resistor. After EDLC is charged, SSR (Solid State Relay) which is consisted of bi-directional thyristor is on. And main power charging is started. At this time, DVR's inverter is begun to operate. In this time, control power is supplied by SMPS① and communication, FPGA and RMS operation is performed.

After this time, sensitive load is begun to supply stable power regardless of instantaneous voltage drop and instantaneous power failure. When instantaneous voltage drop and instantaneous power failure are occurred, control power is supplied by SMPS②. And DVR's compensation time is decided in proportion to established energy capacity of EDLC. DVR's inverter is operated by high efficiency PWM control scheme. And PWM is operated in interlocking discharging state of EDLC. After output voltage of DVR' inverter is passed through LC filter which reduce THD and matching transformer, main power is supplied to sensitive load. And being performed various protecting system, measuring and monitoring function, DVR's inverter is operated.

F. Energy utilization improvement and high efficiency operation of DVR system

EDLC which used in this development is energy storage equipment that has advantage of electrolytic capacitor and

lead storage battery as it is and make up for defeat (capacitance, life-cycle and stability). To use energy of EDLC effectively, it need to introduce series/parallel exchanging circuit, system protecting function. Specially, to high efficient control, charging equipment of constant current control is needed. And being discharged voltage, flexible change of PWM is needed.

Also, voltage control of inverter such as this system applies and uses PWM method. There is space vector etc. various method to this method. But, in case of simple system such as single phase system, sine wave PWM that compare deference voltage with reference voltage and power supply voltage with chopping wave is used. Because existent this method controls voltage using one error voltage and one chopping wave, all switch do on-off action and generate heat damages with switching cycle. Therefore, we applied double carrier PWM that is high effective PWM.

Embodiment of these techniques is operated by four IGBT switches as Fig. 9. When output voltage of inverter is positive, S_3 is off and S_4 is on. S_1 and S_2 perform PWM by comparison result of modulation wave and chopping wave. Also when output voltage of inverter is negative, S_1 is off and S_2 is on. S_3 and S_4 perform PWM by comparison result of modulation wave and chopping wave. In case of modulation wave is bigger than chopping wave C_1, S_1 is on. In case of modulation wave is bigger than chopping wave C_2, S_1 is off. Switching is performed on these two assumptions. Such circuit composition that use double carrier modulation method can heighten efficiency than case that four IGBT is switched continuous within one switching cycle.

Also, it need control of early soft-starter's action process which associated in EDLC's charge. To do to charge slowly EDLC, control of SSR is needed.

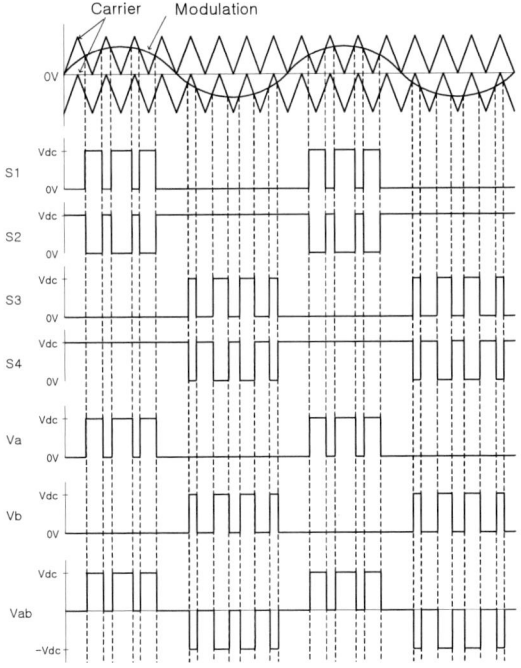

Fig 10. PWM operation of double carrier method

978-1-4244-1871-8/07 $25.00 © 2007 IEEE

III. EXPERIMENT AND RESULT

This development experimented and manufactured On-line type DVR system of 5 [kVA] capacities. Fig 11. displays external forms of EDLC and development system that explain before. EDLC which used for this development is manufactured in domestic N company and mixed by 72 series-parallel EDLC cells. Two units that have capacity of 180[V], 1.54[F] per 1 unit are mixed. Charging system is consisted to compensate 5 [kVA] loads during 1 second. Right side's reactor and capacitor are LC filter of inverter output.

Fig 11. EDLC and external form of development system

Fig 12. displays switching (+,-) signal of two arms by modulation wave of inverter.

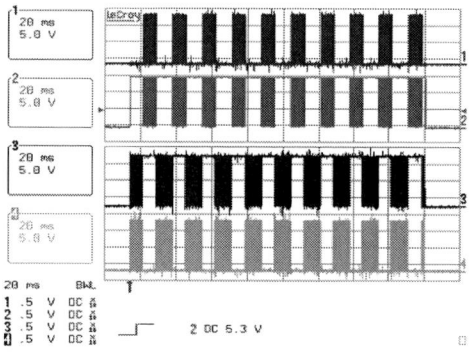

Fig 12. high efficiency switching signal of DVR's inverter

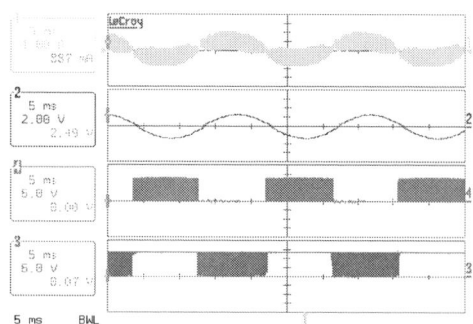

Fig 13. PWM signal and current of inverter

Fig 13. displays modulation signal of inverter(channel 2), switching signal of two arms(channel 3,4) and output current of inverter.

Fig 14. is EDLC charging state and load current wave of DVR inverter. Even if input voltage is changed in on-off state, SSR charging current of channel 2 is controlled according to state of EDLC charging voltage of channel 3. Inverter current of channel 4 is supplied properly.

When load current is changed, charging/discharging state of EDLC and protecting operation of SSR is performed well on Fig 15.

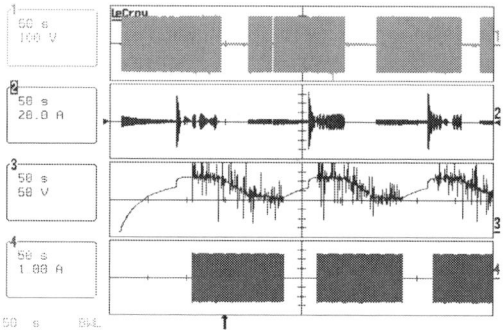

Fig 14. Charging state of EDLC and load current of inverter

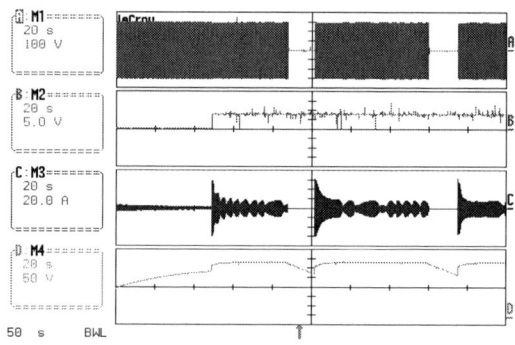

Fig 15. Charging state of EDLC and protecting operation of SSR

Fig 16. displays early start signal and protecting operation of SSR.

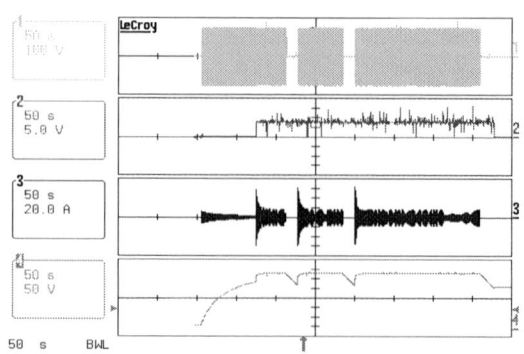

Fig 16. Early starting state and protecting operation of SSR

978-1-4244-1871-8/07 $25.00 © 2007 IEEE

Fig 17. is wave of On-line type DVR operation. When sensitive load is changed, diode input current of SSR and load current of DVR is displayed. (channel 1:input current of SSR diode, channel 2:charging voltage of EDLC, channel 3: output voltage of inverter, channel 4:load current of DVR's inverter).

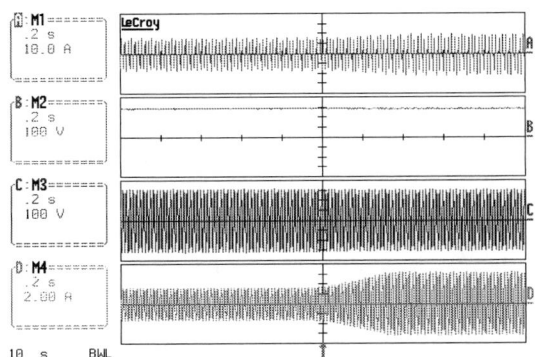

Fig 17. Operation of On-line type DVR's inverter

When load voltage is changed, operation waveform is displayed on Fig 18.

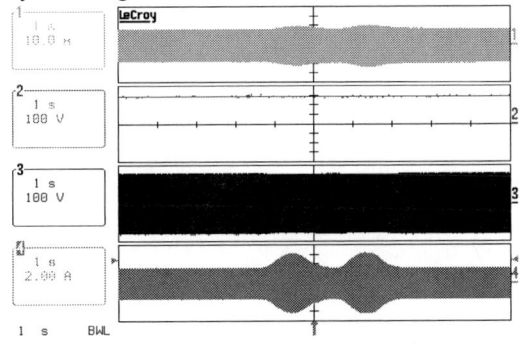

Fig 18. Operation of On-line type DVR
(Change of load voltage)

IV. CONCLUSION

Economical damage that is related with voltage sag among power quality problems is enlarging. Study about DVR which is effective alternative to solve problem of instantaneous voltage drop and instantaneous interruption is gone abuzz. Therefore, we developed On-line type DVR that is custom power installation on this study. At this time, EDLC that has advantage of electrolytic capacitor and lead storage battery is used by energy storage. And energy utilization ratio of EDLC is improved on this study. Also, we added high efficiency inverter system and soft-starter protecting algorithm to On-line type DVR. In this paper, a study has been performed for optimal design and development of DVR system using supercapacitor to solve voltage sag which are considered the dominant disturbances affecting power quality.

REFERENCES

[1] Math H. J. Pollen, Understanding Power Quality Problems, IEEE Press, 1999.

[2] W.E.Kazibwe et al, "Power quality : A Review", IEEE Computer Applications in Power, Vol.13, NO.1, pp39-42, January 1990.

[3] Florida Educational Seminars, Inc., "The 10th International Seminars on Double Layer Capacitors and Similar Energy Storage Devices", December 3-5,2001.

[4] Alexander Kara et al, "Power Supply Quality Improvement with a Dynamic Voltage Restorer(DVR)", IEEE APEC, Vol.2, pp.986-993, 1998..

[5] Mark F. McGranaghan et al, "Voltage Sags in Industrial Systems", IEEE Trans. on Industry Applications, Vol.29, No.2, pp.397-403, March/April, 1993.

[6] Halpin, S.M., "Characterization of double-layer capacitor application issues for commercial and military applications", Industrial Electronics, Control and Instrumentation, 1997. IECON 97. 23rd International Conference on Volume 3, 9-14 Nov. 1997 Page(s):1074 - 1079 vol.3

[7] Youngho Kim, "Ultracapacitor technology power electronics circuits", Power Electronics Technology, Oct. 2003, Papers from Conference Proceedings (Published):

978-1-4244-1871-8/07 $25.00 © 2007 IEEE

Development of 200kW Grid-Connected Photovoltaic Inverter

Young Chan Shin, Young Roc Kim, Byung Hun Ra, Joon Sun Moon
R&D Center of Hex Power System Co., Ltd., Korea
Email: bhraigbt@hex.co.kr

Seok Ki Kim
Korea Institute of Energy Research, Korea
Email: skim@kier.re.kr

Abstract — This paper presents development of 200kW Grid connected PV inverter. 200kW Grid connected PV inverter has the utility frequency transformer for the galvanic isolation between DC and AC , and its output is 3-phase 3-wire 60Hz, 380V. P&O algorithm is used for the Maximum Power Point Tracking (MPPT) and the active frequency drift scheme for the anti islanding protection. The inverter is controlled by 32 bit DSP. This paper explain the performance and test result of utility interface, anti-islanding, efficiency and THD of 200kW grid connected PV inverter.

I. INTRODUCTION

Recently, several MW PV power generation plant are going to be constructed triggered by the RPA (renewable energy portfolio agreement)) and feed-in tariff in South Korea. The higher efficiency and lower price grid connected Inverter is highly demanded and 200~250kW large capacity PV inverters are installed and operated in parallel for several MW PV power plant.

Over 200kW PV inverter has not been developed yet, PV system installation company depends on foreign inverter manufacturer which can produce over 200kW inverter

For the domestic market of several MW PV power plants this paper has done R&D about 200kW PV inverter. This paper presents 200kW grid connected PV inverter which converts DC power generated by photovoltaic array to 380V, 60Hz AC power and energize the three-phase three-wire utility line. Inverter doesn't have DC-DC Converter and is isolated from the utility line by the LF transformer for the galvanic isolation. The efficiency is more than 95% at rated power.

II. INVERTER SPECIFICATION AND CIRCUIT

Figure 1 is circuit of 200kW Grid connected PV inverter. Table1 is electrical specification of 200kW PV Grid connected inverter.

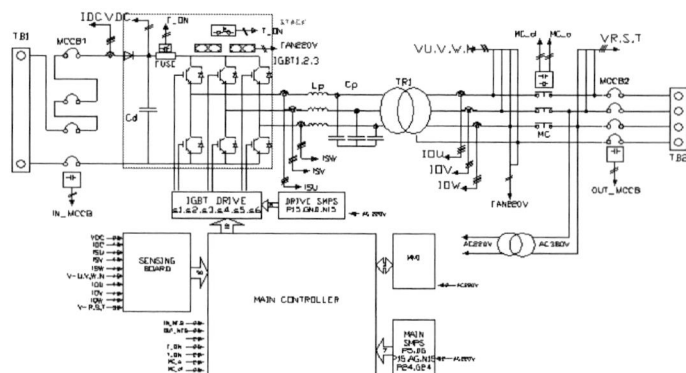

Figure 1. Electrical circuit of 200kW PV inverter

Table 1. Electrical spec. of 200kW PV inverter

Inverter Technology	PWM
Nominal AC Power Output	200kW (at 40℃)
Acceptable Utility Voltage Range	334~418 Vac (380 Vac nominal)
Acceptable Utility Frequency Range	59.3~60.5 Hz (60Hz nominal)
Current THD	< 3% (at 250kVA)
Power Factor	> 95% (at 200kVA)
AC Output Current Limit	418 Aac (380 Vac)
DC Input Voltage	450~750 [Vdc]
Peak Power Tracking Voltage	480~700 [Vdc]
PV Start Voltage	500Vdc (adjustable)
Maximum DC Current	418A
Inverter Efficiency	> 94% (at 200kVA)
Ambient Temperature (Operating)	0℃~40℃
Ambient Temperature (Storage)	-20℃ ~ 50℃
Relative Humidity	< 90%, Non-condensing
Elevation	< 2000m

III. CONTORL OF 200kW GRID-CONNECTED PV INVERTER

Figure 2 is 200kW Grid connected Photovoltaic inverter developed

Figure 2. 200kW Grid-connected Photovoltaic inverter

Figure 3 expressed CCVSI's PI control block-diagram that is three phase electric power conversion device of 200kW Grid connected Photovoltaic inverter.

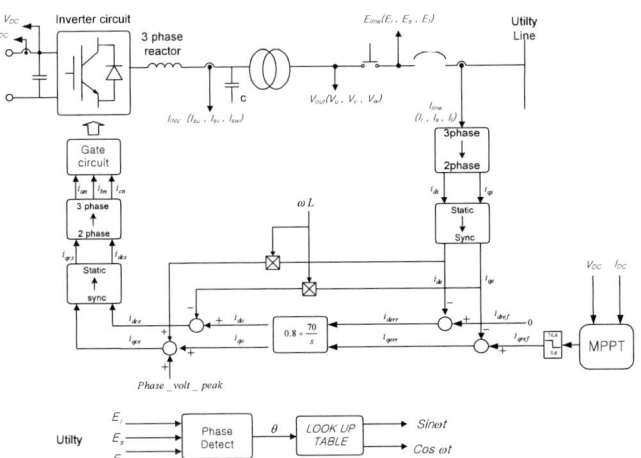

Figure 3. Control block-diagram of 3-phase PV inverter

Figure 4 displays flowchart of P&O algorithm that used in MPPT control of 200kW Grid connected Photovoltaic inverter

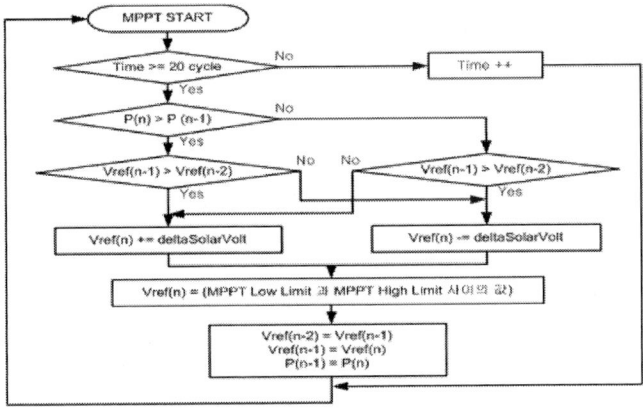

Figure 4. Algorithm of P&O method for MPPT

IV. TEST OF 200kW PV GRID-CONNECTED INVERTER

(a) and (b) of Figure 5 show output voltage and output current of 200kW Grid connected Photovoltaic inverter.

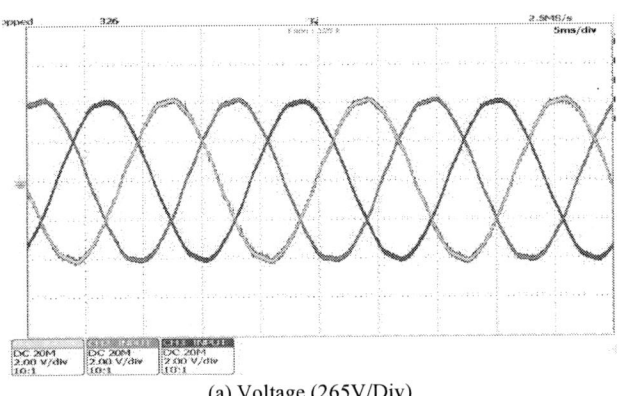

(a) Voltage (265V/Div)

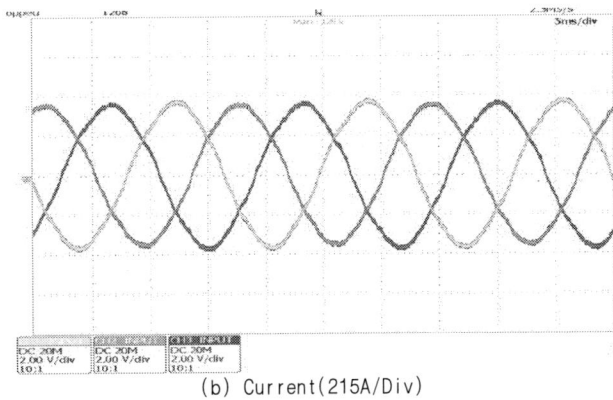

(b) Current(215A/Div)

Figure 5. Each phase output Waveform of inverter (at 100%)

Figure 6 has phase difference of 30° between L-L voltage(Vrs) and phase voltage(Vou) of Grid. And phase voltage (Vou) and phase current(Iou) can know that is controlled by more than PF 99%. Isu is the primary current of the transformer.

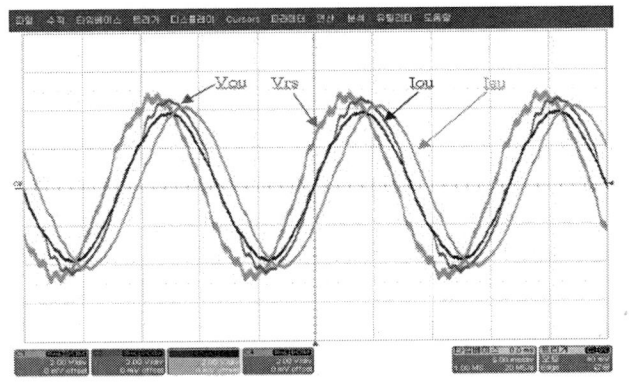

Figure 6. Output Waveform of inverter (at 100%)

The soft-start characteristic of inverter is shown in Figure 7. The output phase current is rising lineally to 100% for 2 minutes to reduce the grid voltage variations.

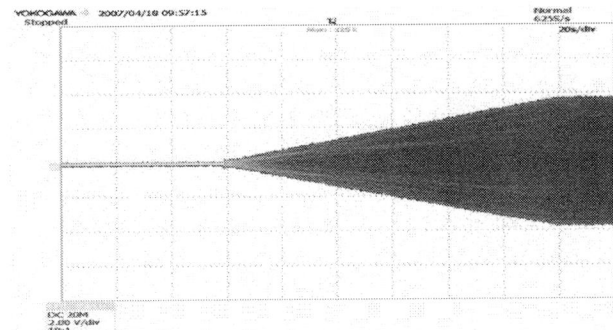

Figure7. Soft-Start characteristics of inverter

Figure 8 represents harmonic currents injected into the utility at the rated power. Each individual harmonic is in compliance with IEC61727

978-1-4244-1871-8/07 $25.00 © 2007 IEEE   461

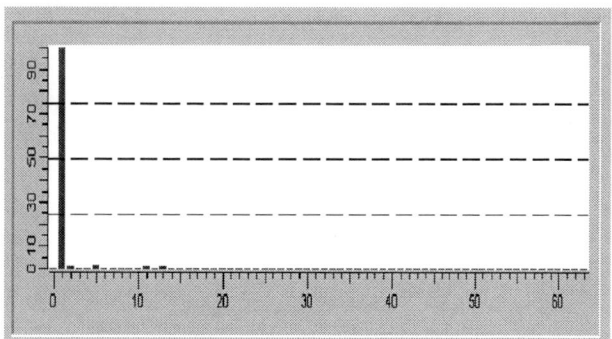

Figure 8. Harmonics at the rated output power

THD(Total Harmonic Distortion) is measured at all power test levels. See Figure 9. THD at 100kW is about 5% and below 3% at rated power

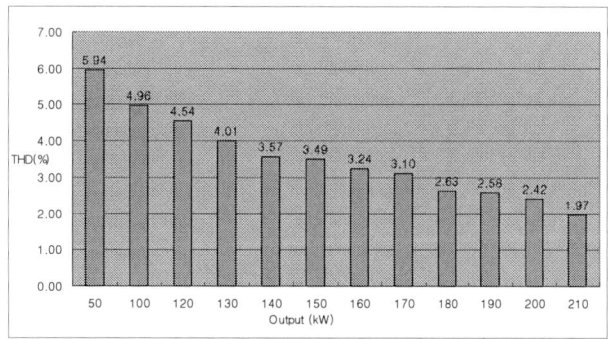

Figure 9. THD.

The static power conversion efficiency as shown in Figure 10 represents more than 96% at 50kW, and more than 95% at 200kW.

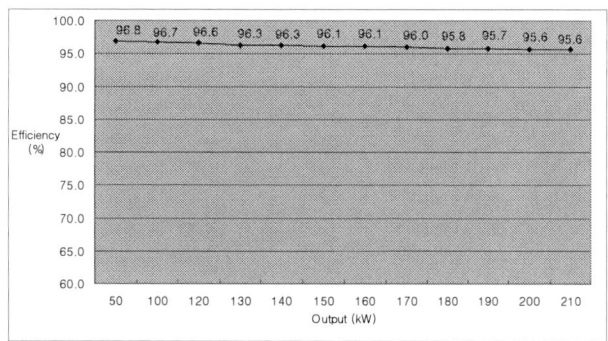

Figure 10. Efficiency at each power test level

V. CONCLUSION

The developed 200kW PV Grid connected inverter has the following performances.

1. Output Current THD is below 3% at rated power.
2. Efficiency is more than 95% at rated power.

VI. REFERENCES

[1] Y. C. Shin, Development of Unitary 120kW PCS for Grid-Connected Photo-Voltaic Power Generation, 2006 Power Electronic Annual Conference, pp188~190

[2] K.H. Hussein, I. Muta, T. hoshino and M. Osakada, "Maximum Photovoltaic Power Tracking: an Algorithm for Rapidly Changing Atmospheric Conditions", proceeding of IEE Generation, Transmission, and Distribution, Vol.142, No.1, pp. 59-46, 1995, January.

[3] Development and Testing of an Approach to Anti-Islanding in Utility-Interconnected Photovoltaic Systems SAND2000-1939

[4] IEEE Recommended Practice for Utility Interface of Photovoltaic(PV) Systems, IEEE Std 929-2000

Development of 10kW Grid-Connected Photovoltaic Inverter
with High Frequency Transformer

Young Hoon Son, Young Roc Kim, Byung Hun Ra, Sung Hwon Kim
R&D Center of Hex Power System Co., Ltd., Korea
Email: bhraigbt@hex.co.kr

Seok Ki Kim
Korea Institute of Energy Research, Korea
Email: skim@kier.re.kr

Abstract — This paper presents 10kW grid connected photovoltaic inverter with high frequency transformer. This inverter system consist of DC/DC Converters, 2 high frequency transformers in parallel, diode full bridge rectifiers and 3 phase inverter. Proto Type inverter is manufactured and experimented. High frequency link DC-DC converter increases input DC voltage and supplies power to the inverter, which can energize directly the grid without LF step-up transformer. This system is lighter, more efficient and economical than the inverter with utility frequency transformer.

I. INTRODUCTION

Three phase Grid-tied inverters for photovoltaic power generation that are produced in the South Korea, have the utility frequency transformer for the galvanic isolation and voltage step-up. There are advantages that LF inverter has simpler circuit and controller, good galvanic isolation between DC and AC. But LF inverter is bigger, heavier than transformer-less inverter and high frequency transformer inverter.

The inverter with high frequency transformer is smaller, lighter, more efficient than the LF transformer and has good galvanic isolation. But, control and System is more complex.

So, this paper proposes 10[kW] PV inverter with high frequency transformer. The efficiency of this inverter is above 94[%] at rated load 2[%] higher than the inverter with LF transformer. This inverter has the following electrical specifications and circuit diagram in Table 1 and Fig. 1

TABLE 1
Electrical specification of 10kW PV inverter

Item		High Frequency Transformer Inverter	Low Frequency Transformer Inverter
Control	CPU	TMS320F2812	TMS320C32
	Switching Frequency	5[kHz]	10[kHz]
Input	Max. Input Voltage	450[VDC]	450[VDC]
	MPPT range	200[V] ~ 420[V]	210[V] ~ 450[V]
	Input voltage range	160[V] ~ 450[V]	200[V] ~ 450[V]
Output	Phase	3	
	Nominal output power	10[kW]	
	Utility Voltage	380 [VAC] (+10[%]/-12[%])	
	Utility Frequency	60 [Hz](+0.3[%]/-0.3[%])	
	THD	< 5[%]	
	Maximum efficiency	> 93[%]	> 90[%]
	P.F(Power Factor)	> 0.95	
Mechanical	Noise level	< 50[dB]	
	Cooling	Convection cooled	Forced air cooling
	Dimensions [mm] (W x D x H)	600 x 500 x 300	550 x 600 x 1146
	Weight	25[kg]	220[kg]

II. TOPOLOGY

A. DC/DC Converter Topology

Permissible input voltage of converter is DC 160 ~ 450[V]. If input goes up to DC 160[V], converter starts operation, converts DC to 25[kHz] AC and supplies power through high frequency transformer and Diode full bride rectifier.

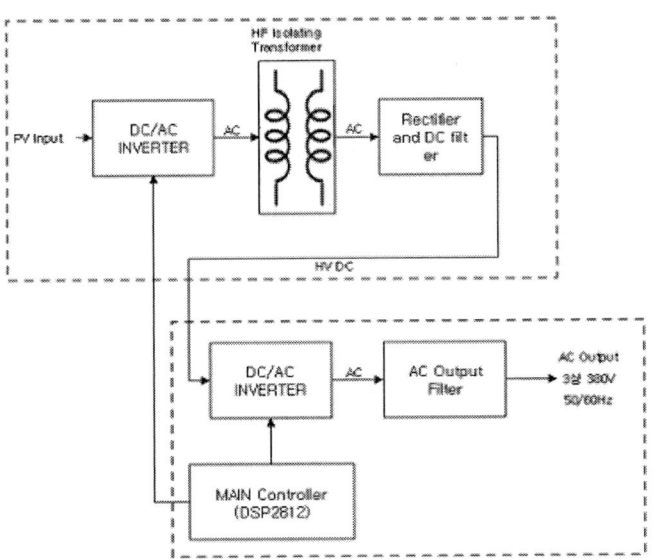

Fig. 1 Block diagram of 10kW PV inverter with High Frequency Transformer

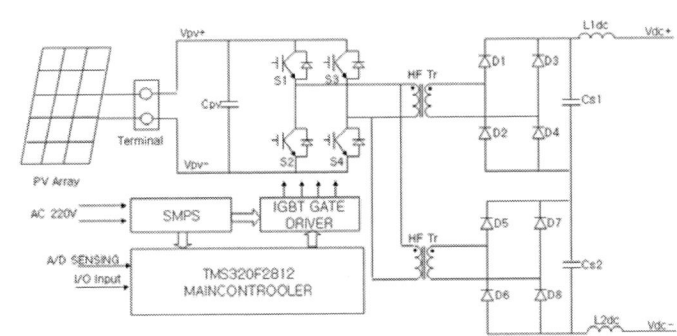

Fig. 2 DC/DC converter circuit diagram

978-1-4244-1871-8/07 $25.00 © 2007 IEEE   463

B. Inverter Topology

Inverter converts DC power supplied by DC/DC converter to AC power by PWM (Pulse Width Modulation) Method (Switching Frequency: 5 [kHz]) and energizes the grid through LC filter.

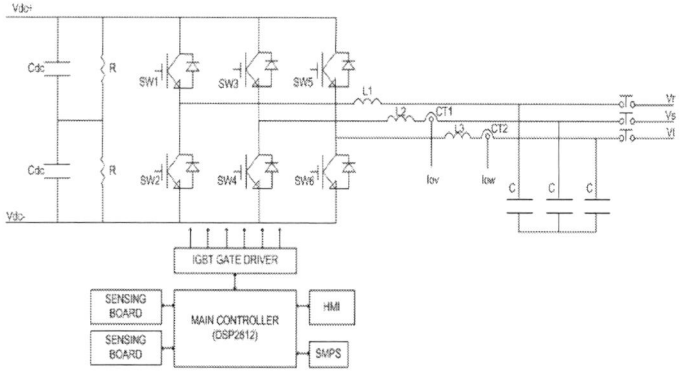

Fig. 3 Inverter circuit diagram

III. EXPERIMENTAL IMPLEMENTATION

Table 2 shows design specification of the inverter. Fig. 4 presents 10[kW] grid connected photovoltaic inverter with high frequency transformer. This inverter system consist of DC/DC Converters, 2 high frequency transformers in parallel, diode full bridge rectifiers and three phase inverter.

TABLE 2
Design specifications of the Inverter

Inverter IGBT	IKW40N120T2	1200[V], 40[A]
Converter IGBT	IKW75N60T	600[V], 75[A]
High Frequency Transformer	Core : ZP-48613-TC	A(O.D.) : 85.7[mm] B(I.D.) : 55.5[mm] C(HGT.) : 12.7[mm] AL value (nH/T2) : 2726
Rectifier Diode	DSEP30-12[A]	1200[V], 30[A]
DC filter Reactor	200[uH] (Core : CS572060)	A(O.D.) : 57.15[mm] B(I.D.) : 35.56[mm] C(HGT.) : 13.97[mm]
DC filter Capacitor	SAMHWA HE	500WV, 390uF
AC filter Reactor	8[mH] (Core : iron)	W : 95[mm] D : 90[mm] H : 170[mm]
AC filter Capacitor	C4ATJB_5100_M_	10[uF], 400[Vac]

Fig. 4 photovoltaic inverter prototype with high frequency transformer

A. DC/DC Converter

Fig. 5 High Frequency Transformer

Fig. 5 shows high frequency transformer used in the inverter. Core is Magnetics toroidal Core.

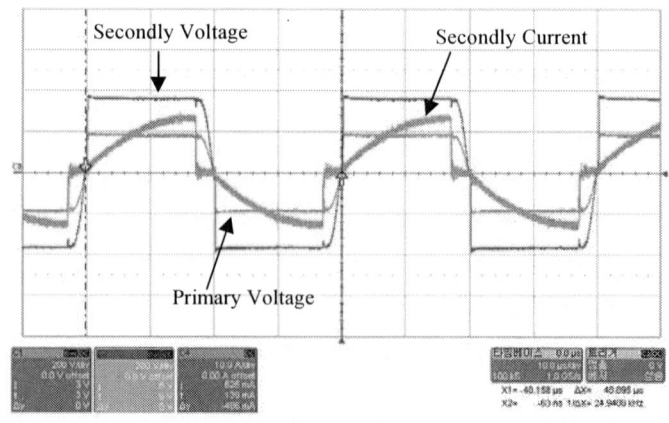

Fig. 6 Waveform of High Frequency Transformer

Converter is operated by hard switching PWM method. Switching Frequency is 25[kHz], Dead-Band Time is 3us. Fig. 6 shows waveform of high frequency transformer.

B. Inverter

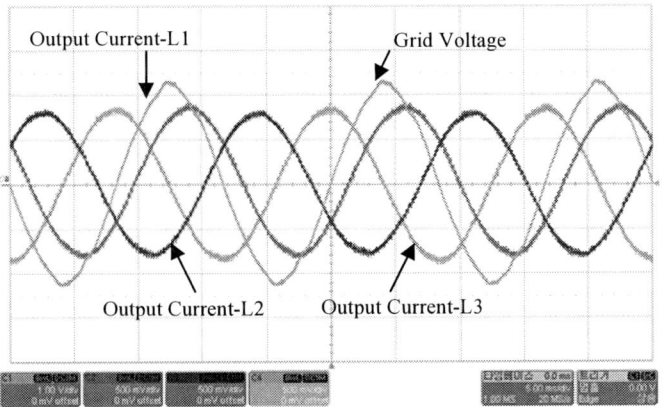

Fig. 7 Grid Voltage and Inverter output current

Because of the grid-connected inverter, output voltages of inverter and utility are in the same phase.

Fig. 7 shows line to line voltage and output phase current.

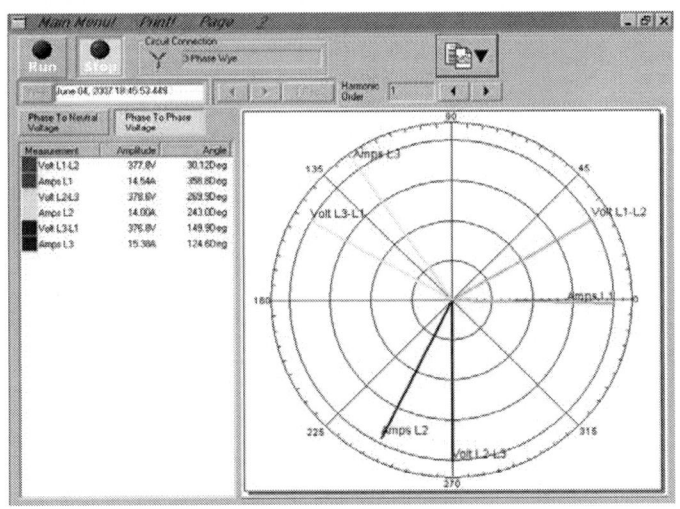

Fig. 8 vector diagram

Fig. 8 shows vector diagram of inverter output. In Fig. 9, output power of the inverter is 9.544[kW] and power factor is 0.996. Because input power is about 10.2[kW], efficiency of inverter is about 93.6[%].

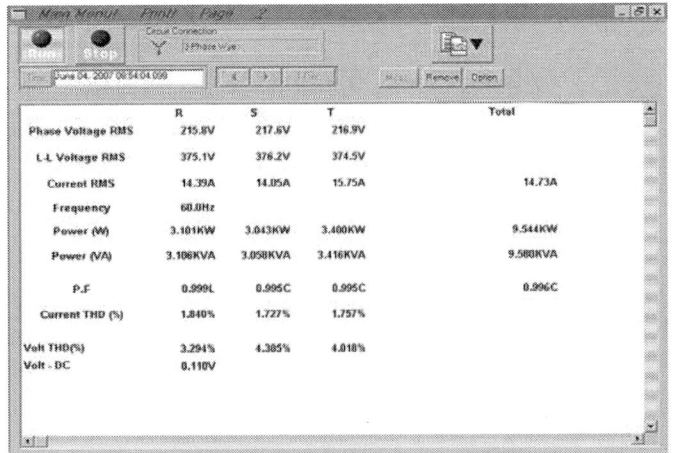

Fig. 9 Inverter output power, PF, THD

IV. CONCLUSION

Bulk and weight of this inverter is about 70[%] decreased as expected than those of low frequency transformer inverter. And this inverter can secure bigger stability than transformer-less inverter. Peak efficiency of this inverter is more than about 93[%] at rated power.

V. REFERENCES

[1] J. S. Moon, Development of 3kW Grid tied PV Inverter without Transformer, 2005 Power Electronics Annual Conference, pp. 347-349.

[2] Nasser H. Kutkut, "An Improved Full-Bridge Zero-Voltage Switching PWM Converter Using a Two-Inductor Rectifier", IEEE Trans. Industry Application, vol. 31, no. 1, pp. 119-126, Jan. 1995

[3] Yungtaek Jang, "A New PWM ZVS Full-Bridge Converter", IEEE Trans. Power Electronics, Vol. 22, no. 3, pp. 987-993, May. 2007

Z-Source Active Power Filter with a Fuel Cells Source

Jun-Hyun Oum, Young-Cheol Lim
Chonnam National University
Department of Electrical Engineering
500, Yong Bong-Dong, Kwangju, 500-757, Korea
Email: yclim@chonnam.ac.kr

Young-Gook Jung
Daebul University
Department of Electrical Engineering
72-1, Sanho-ri, Yongam-Gun, Chonnam, 526-702, Korea
Email: jyg@mail.daebul.ac.kr

Abstract—**This paper proposes a Z-source active power filter with a fuel cells source instead of conventional V-source active power filter. Z-source impedance network along with shoot through capability of the proposed system would ensure a constant dc voltage across the dc link. As the compensation and control algorithm of the Z-source active power filter, a single-phase instantaneous reactive power theory is adopted. The effectiveness of the proposed the system is verified by the PSIM simulation in the steady state and the transient state.**

I. INTRODUCTION

Recently, power factor improvement and harmonics suppression are important to obtain the high power quality because the modern industrial processes contain the sensitive devices. Active power filter[1,2] has been accepting as an efficient device for compensating the harmonics and reactive power of nonlinear loads such as thyristor controlled rectifiers or variable speed drives. As a point of view the energy storage unit connected at dc link of the inverter bridge, active power filter can be classified into two groups - V(Voltage) source and I(Current) source topologies. Especially, V-source topology is better than I-source because it easy to control and has general-purpose characteristics with the spread of a DSP(Digital Signal Processor).

Compensation performance of active power filter depends on the accuracy of compensation reference calculation. Also, stability of its energy storage unit is considerably important to supply the steady compensation energy. With the above consideration, if fuel cells or battery stack are selected as its energy storage unit, the Z-source topology[3,4] can give great promise as competitive alternatives to existing active power filter topology with many inherent advantages. It differs with conventional converters like V-source and I-source topologies due to the presence of unique X-shaped impedance network on its dc side which interfaces the source and inverter H-bridge[5]. It facilitates both voltage–buck and boost capabilities.

In this paper, a Z-source topology based active power filter is proposed, and its storage energy in storage unit can be utilized during the process of the harmonics and reactive power compensation with the use of buck-boost property of the inverter. As an energy storage unit, a fuel cells simulator[6,7] adopted and designed with the electrical characteristics of PEM((Polymer Electrolyte Membrane) fuel

cells[8,9]. It is implemented by using a simple buck converter which is controlled by simplified linear function[7]. This paper shows some verification results of the PSIM simulation, which proved the harmonics and reactive power compensation by active power filter with control, based on the Z-source topology and PEM fuel cells.

II. THE PROPOSED SYSTEM

A. System analysis

Fig. 1 shows the proposed Z-source active power filter system. Dc link stage of conventional V-source active power filter is connected with a bulky dc capacitor in stead of a Z-source network and an energy storage unit.

A bulky dc capacitor can charge or discharge by PWM switching patterns, and its dc voltage is within a certain desired limit by the PI controller with appropriate PI gains. But, in case of Z-source topology as shown in Fig. 1, a Z-source network with boost actions and an energy storage unit are used for a certain desired dc voltage of active power filter.

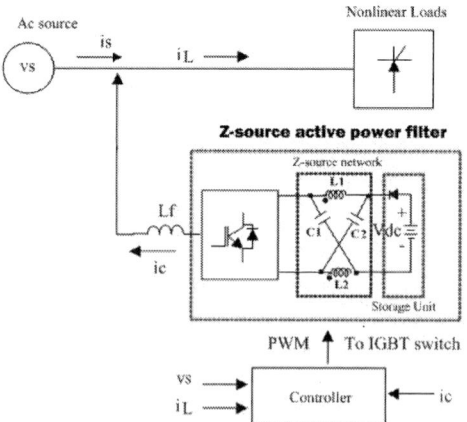

Fig. 1 Z-source active power filter system

The simplified circuit of Z-source active power filter is given in Fig. 1. A Z-source inverter has two different states – active (sw➔1) and shoot-through (sw➔2). And, Fig. 3 shows the equivalent circuit for shoot-through state. In this state the H-bridge arm is shorted. This additional state facilitates the boost actions in the impedance network. However with conventional inverters the shoot-through state has been avoided to protect the inverter switches.

978-1-4244-1871-8/07 $25.00 © 2007 IEEE

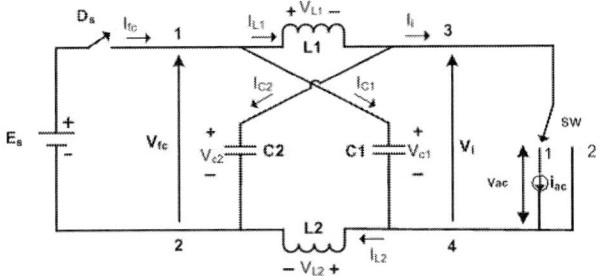

Fig. 2 Simplified circuit of Z-source active power filter

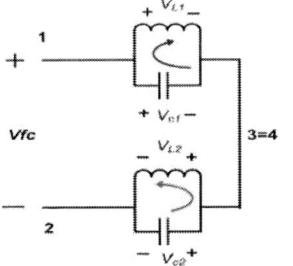

Fig. 3 Shoot-through state

The inverter's action is replaced with a current source plus a single switch. Then network becomes symmetric therefore; $V_{c1} = V_{c2} = V_c$ and $v_{L1} = v_{L2} = v_L$

For the active state, steady state equation can be obtained as in (1).

$$V_{L1} = V_{fc} - V_{c2}$$
$$V_i = V_{c1} - V_{L1} = V_{c1} - (V_{fc} - V_{c2}) = 2V_c - V_{fc} \quad (1)$$

Where V_{fc} is fuel cells dc voltage, V_i is Z-source network dc link voltage, and v_{ac} is ac output voltage of Z-source inverter.

From Fig. 3, where Z-source is shoot-through state the following equations can obtained.

$$V_L = V_c$$
$$V_{fc} = 2V_c = V_{c1} + V_{c2} \quad (2)$$
$$V_i = 0.$$

The total duration of active time and total shoot-through time are denoted by T_1 and T_0, then v_i, B and D_0 can be represented by the following equation[5].

$$v_i = \frac{T}{T_1 - T_0} V_{fc} = \frac{1}{1 - \frac{2T_0}{T}} = BV_{fc} \quad (3)$$

$$B = \frac{T}{T_1 - T_0} = \frac{1}{1 - \frac{2T_0}{T}} \geq 1 \quad (4)$$

$$D_0 = 2\frac{T_0}{T} \quad (5)$$

Where v_i is the dc link peak voltage of Z-source network, B is boost factor, and D_0 is shoot-through duty ratio.

Then, from above equations, ac output voltage of Z-source inverter v_{ac} can be written as (6).

$$v_{ac} = M\frac{v_i}{2} = MB\frac{V_{fc}}{2} \quad (6)$$

Where, M means modulation index.

Similarly if fuel cells source is a current source, ac output current i_{ac} of Z-source inverter can be written as (7).

$$i_{ac} = MB\frac{I_{fc}}{2} \quad (7)$$

B. Operation modes

Fig. 4 shows the power circuit of Z-source active power filter and Fig. 5 shows its operation modes with active state and shoot-through state. First of all, Fig. 5a represents shoot-through mode as shown in Fig. 3. In this mode, active power filter does not inject the compensating current i_c to the ac power source. In stead, the low dc voltage of storage unit is sufficiently boosted during shoot-through period. Fig. 5b shows active mode, and boosted dc voltage during shoot-through mode is inverted to ac compensating voltage. By the difference ac compensating voltage with ac source voltage, the compensating current i_c is injected to the ac power source through an inductor L_f in Fig. 5. Therefore, distorted source current i_s becomes sinusoidal waveform.

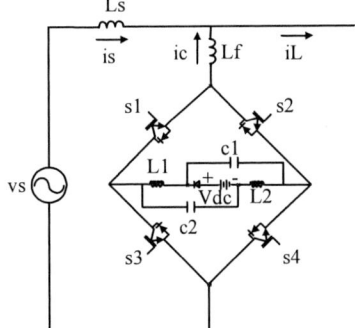

Fig. 4 Power circuit of Z-source active power filter

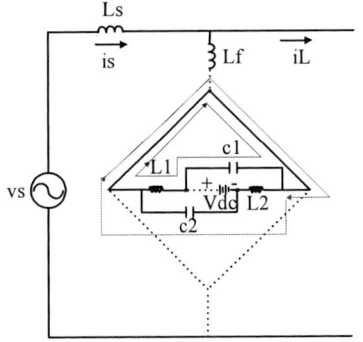

(a) Shoot-through mode

978-1-4244-1871-8/07 $25.00 © 2007 IEEE

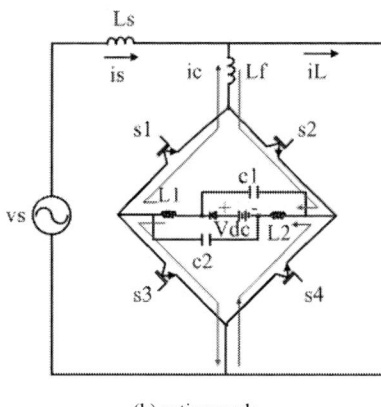

(b) active mode

Fig. 5 Operation modes

Fig. 6 shows the block diagram representation of the proposed system with feedback controller. As shown in Fig. 6a, compensation reference i_c^* is calculated by the single-phase reactive power theory[10]. And, the shoot-through circuit is shown in Fig. 6b.

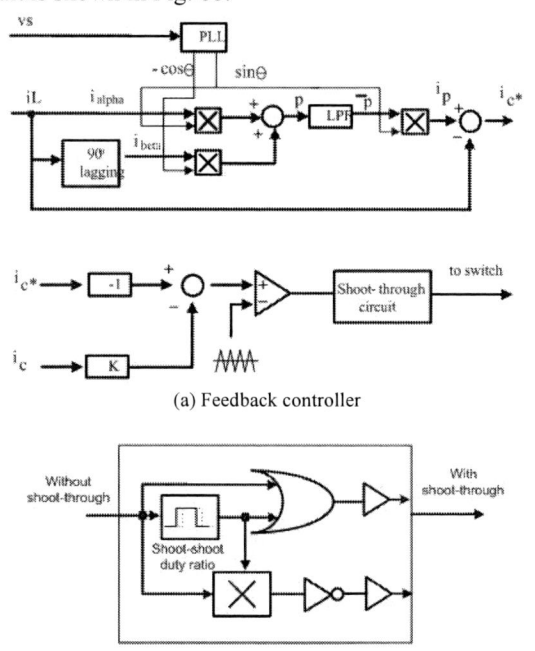

(a) Feedback controller

(b) Shoot-through circuit

Fig. 6 Block diagram of propose system with feedback controller

C. Dc energy storage unit

As dc energy storage unit in this paper, a fuel cells simulator is designed with the electrical characteristics of PEM fuel cells. It is implemented by using a simple buck converter which is controlled by simplified linear function. The linear function of the ohmic region is represented as follows[7].

$$v_{fc}(i_{fc}) = v_{fci} \ (\ 0 < i_{fc} < i_{fci}\ )$$

$$v_{fc}(i_{fc}) = (\frac{v_{fcf} - v_{fci}}{i_{fcf} - i_{fci}})i_{fc} - (\frac{v_{fcf} - v_{fci}}{i_{fcf} - i_{fci}}i_{fci} - v_{fci}) \ (8)$$

$$(\ i_{fci} < i_{fc} < i_{fcf}\ )$$

$$v_{fc}(i_{fc}) = 0 \ (\ i_{fc} > i_{fcf}\ )$$

Fig. 7 shows a typical V-I curve of the fuel cells. The output characteristics are separated into three regions : activation, ohmic, and concentration . Generally, the operation region of actual fuel cells is ohmic region.

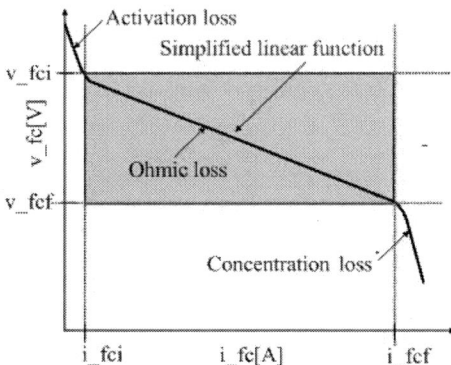

Fig. 7 V-I curve of the PEM fuel cells

III.   RESULTS AND DISCUSS

Detailed investigations of the proposed system in Fig. 1 were performed using PSIM simulation with the parameters listed in Table 1. As nonlinear load, a diode rectifier was used.

TABLE 1 System Parameters

$v_s(ac)$	155v(peak)/60Hz
i_L	40A
$V_{c1}(dc)$	280v
f_{sw}	5.4kHz, SPWM, M=0.645
D_0	0.3244
Z-source network	$L_1 = L_2 = 160uH$ $C_1 = C_2 = 1000uF$
Fuel cells stack model	PEM fuel cells, AvistaLab SR-12(0.5kW), $v_{cell} = 40.6v$, $v_{base} = 28.9v$, $i_{base} = 17.3A$

Fig. 8 shows Z-source network voltage(dc link voltage) V_{c1} proportionate to shoot-through time T_0 . In case of $T_0 = 30us$, $V_{c1} = 290v$, and during $T_0 = 21us$, $V_{c1} = 225v$. It shows that V_{c1} is controllable by T_0 .

978-1-4244-1871-8/07 $25.00 © 2007 IEEE

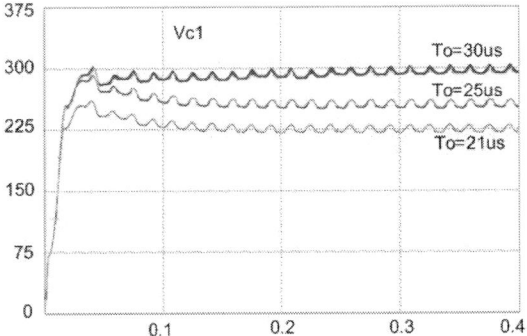

Fig. 8 Z-source network voltage(dc link voltage) V_{c1}

Fig. 9 shows comparative result of the compensating current i_c and dc link voltage of conventional V-source active power filter and the proposed Z-source active power filter. In case while the desired dc link voltage is set at 300v, the compensating current i_c and the actual dc link voltage of each system shows nearly similar. From these results, it is obvious that the proposed system can give satisfying performance as well as the conventional system.

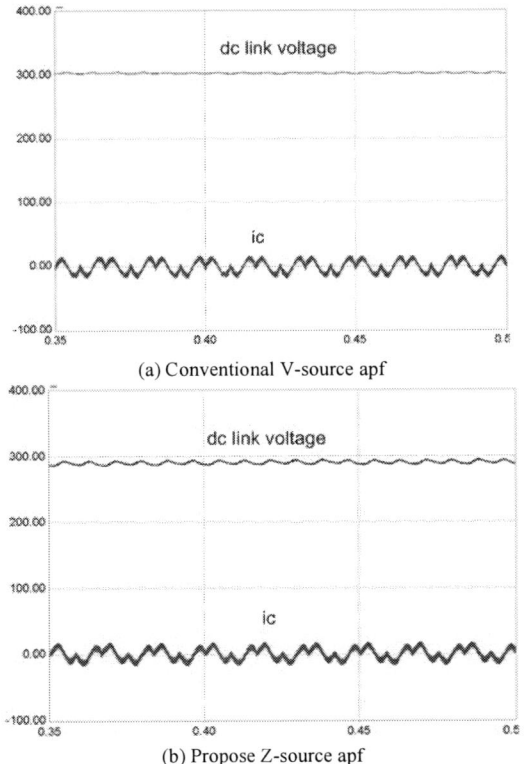

(a) Conventional V-source apf

(b) Propose Z-source apf

Fig. 9 dc link voltage and i_c

Fig. 10 shows load current i_L, compensating current i_c, and source current i_s of the proposed system. As shown in Fig. 10, a non-linear diode load generates i_L with large higher

harmonics contents, and distorted i_s is changed into a sinusoidal waveform by compensating operation of proposed system.

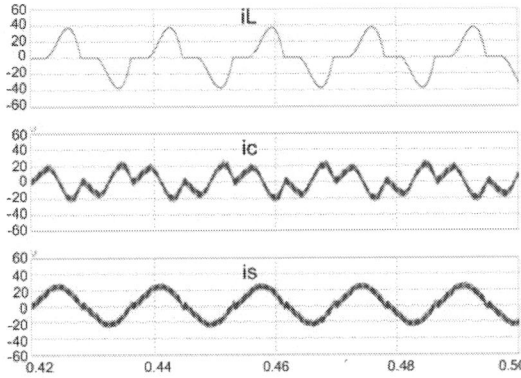

Fig. 10 i_L, i_c and i_s of proposed system

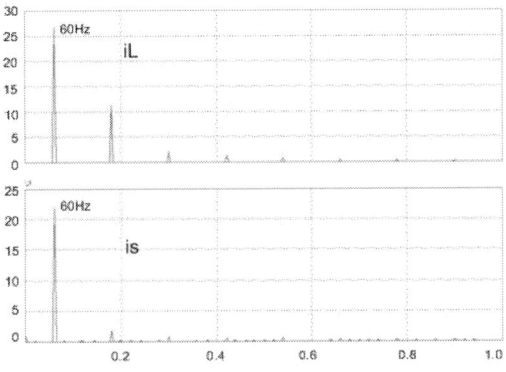

Fig. 11 Harmonic spectra of i_L and i_s
(x-axis:0.2kHz, y-axis:5A/div.)

Fig. 11 shows the FFT result of the i_L and i_s of the proposed system. As can be seen, higher harmonics content in source current i_s is nearly zero except for a fundamental component 60Hz. In result, THD(Total Harmonics Distortion) of i_s is highly improved.

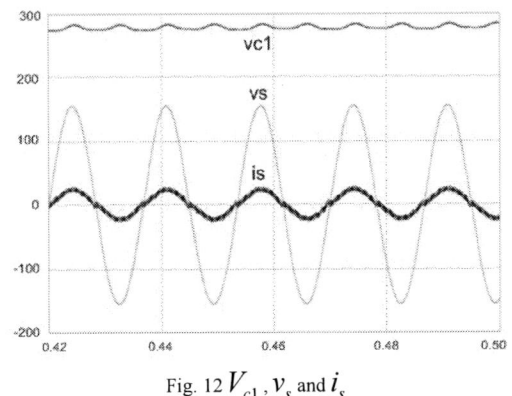

Fig. 12 V_{c1}, v_s and i_s

978-1-4244-1871-8/07 $25.00 © 2007 IEEE   469

Fig. 12 shows the source voltage v_s and source current i_s after compensated, in case of desired dc link voltage V_{c1} =280v. Also, we can know that i_s is in phase with v_s. This result causes the power factor improvement of ac source.

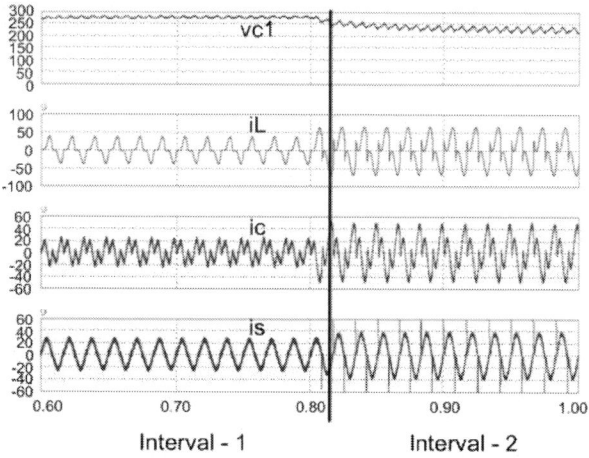

Fig. 13 V_{c1}, i_L, i_c and i_s in transient state

Fig. 13 shows the dynamic characteristics of V_{c1}, i_L, i_c, i_s in transient state, and it can be divided as two interval, that is, interval-1(0.60-0.80sec) and interval-2(0.80-1.0sec). Firstly, for a interval-1 in steady state, i_L keeps 37A(peak) and i_s is almost sinusoidal because of V_{c1} keeps nicely 280v.

But, for a interval-2 in transient state, i_L is increased suddenly from 37A to 67A. In this interval, i_s includes spikes in its waveform and is distorted because of V_{c1} is decreased to 230v. Severe dropping of V_{c1} correspond to rapid jumping of i_L is the result of a PEM fuel cells characteristics.

IV. CONCLUSION

In this study, a Z-source active power filter with a fuel cells source has been proposed. As an energy storage unit of proposed system, a PEM fuel cells is adopted and implemented by using a simple buck converter controlled by simplified linear function. Also, as compensation algorithm, the single-phase instantaneous reactive power theory is applied. Comparison of compensating current and dc link voltage of conventional system and the proposed system is done. Finally, the effectiveness of the proposed system is verified in steady state and transient state by PSIM simulation.

ACKNOWLEDGMENT

This study was supported by the Research Center for Stabilization of Power Systems with Large Distributed Generations at Chonnam National University in KOREA.

REFERENCES

[1] Y. G. Jung, Y. C. Lim and S. H. Yang, "Single-Phase Active Power Filter based on Three-Dimensional Current Coordinates", *IEE Proc. Electr. Power Appl.*, vol.147, no.6, November, pp.572-578, 2000.

[2] Y.G. Jung, W.Y. Kim, Y.C. Lim, S.H. Yang and F. Harashima, "The Algorithm of Expanded Current Synchronous Detection for Active Power Filters Considering Three-Phase Unbalanced Power System", *IEEE Trans. Ind. Electron.*,vol.50, no.5, pp.1000-1006, 2003.

[3] Fang Zheng Peng, "Z-Source Inverter", *IEEE Trans. Ind. Applicat.*, vol.39, No.2, pp.504-510, 2003.

[4] M. Shen, J. Wang, A. Joesph, F.Z. Feng, L. Tolbert, and D.J. Adams, "Maximum Constant Boost Control of the Z-Source Inverter," *in Proc., IEEE IAS'04, 2004*, pp.142-147.

[5] D. M. Vilathgamuwa, C. J. Gajanayake, P.C. Loh, and Y.W. Li, "Voltage Sag Compensation with Z-source Inverter Based Dynamic Voltage Restorer", *in Conf. Rec. of IEEE IAS'06*, 2006, pp.2242-2248.

[6] P. J. H. Wingelaar, J. L. Duarte, and M. A. M. Hendrix, "Dynamic Characteristics of PEM Fuel Cells", *in Conf. Rec. of IEEE PESC'05*, 2005, pp.1635-1641.

[7] Y. R. de Novaes, R. R. Zapelini, and I. Barbi, "Design Considerations of a Long Term Single-Phase Uninterruptible Power Supply Based on Fuel Cells", *in Conf. Rec. of IEEE PESC'05*, 2005, pp.1628-1634

[8] "Fuel Cell Seminar-Special Session on Fuel Cell Power Conditioning and International Future Energy challenge", *(Miami Florida)*, 2003

[9] Ali Emadi, " Satus Review of Power Electronic Converters for Fuel Cell Applications", *Journal of Power Electronics*, vol.1, no.2, pp.134-143, 2001

[10] J. Liu, J. Yang, and Z. Wang," A New Approach for Single-Phase Harmonic Current Detecting and Its Application in Hybrid Active Power Filter", *in proc., IEEE IECON'99, 1999*, 849-854.

A Stand Alone Type Fuel Cells Micro-Source System with a Voltage Sag Compensator

Jae-Hyun Kim, Young-Cheol Lim
Chonnam National University
Department of Electrical Engineering
500, Yong Bong-Dong, Kwangju, 500-757, Korea
Email: yclim@chonnam.ac.kr

Young-Gook Jung
Daebul University
Department of Electrical Engineering
72-1, Sanho-ri, Yongam-Gun, Chonnam, 526-702, Korea
Email: jyg@mail.daebul.ac.kr

Abstract — **This paper proposes a single-phase stand alone type fuel cells micro-source system with a voltage sag compensator for compensating the ac output voltage variations (sag or swell) of micro-source. The proposed micro-source system is consist of a PEM(Polymer Electrolyte Membrane) fuel cells Simulator, a full bridge dc converter, a 60Hz PWM VSI(Voltage Source Inverter), and a voltage sag compensator. Voltage sag compensator is similar to the configuration of hybrid series active power filter, and it is directly connected to micro-source through the injection transformer. The compensation algorithm of a sag compensator adopts a single phase p-q theory. The effectiveness of the proposed the system is verified by the PSIM simulation in the steady state and transient state which the proposed system is able to simultaneously compensate the harmonic current and source voltage sag/ swell.**

I. INTRODUCTION

Recently, lack of the fossil fuels and by increasing the pollution of atmosphere, new renewable energy sources with a green energy concept have been studying. New renewable energy generations such like photovoltaic, wind power, fuel cells and biomass are referred to as the power generation system without the environment pollution. New renewable energy sources have two functions as a low power distributed generation (DG) system and as a micro-source [1].

Generally, single-phase micro-source can be applied to domestic applications such as RPG(Residential Power Generation). Especially, in case of stand alone type micro-source, power quality such as distorted or sag/swell is completely depended on the operation performance of the micro-source unlike a grid interconnected micro-source.

In this paper, in order to improve the power quality of a stand alone type micro-source, a single-phase stand alone type micro-source with the compensation capability of the ac output voltage variation is proposed. The proposed system is consist of a fuel cells simulator, a full bridge dc converter, a PWM VSI, and a voltage sag compensator. Voltage sag compensator [2-4] consists of an injected transformer, secondary winding of which is connected in series with a micro-source, a PWM VSI connected to the primary of the injection transformer and fuel cells source connected at the dc link of the inverter bridge. For a fuel cells source, a fuel cells simulator[5] adopted and designed with the electrical characteristics of PEM fuel cells. It is implemented by using a simple buck converter which is controlled by simplified linear function[6]. The effectiveness of the proposed system

is verified by PSIM simulation in the steady and transient state.

II. THE PROPOSED SYSTEM

A. Compensation and Control Algorithm

Fig.1 shows the configuration of the proposed system. In case of voltage variations or load current rapidly increased, a voltage sag compensator connected in series with a micro-source operates to compensate the variation portion of ac output voltage. Generally, voltage sag does not go below 50 % of the nominal voltage. Also, the duration of voltage sag can be less than half line cycle to a few seconds.

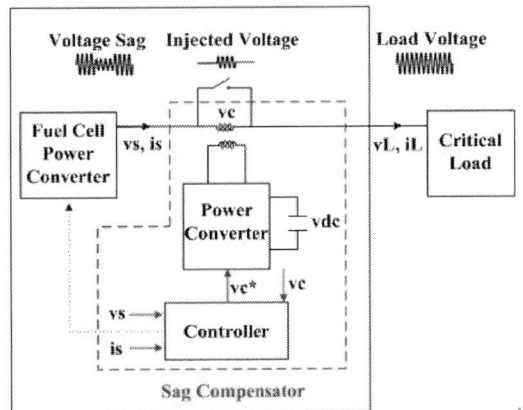

Fig. 1 Configuration of the proposed system

Output voltage v_s and current i_s of micro-source are obtained as follow.

$$v_s = \sqrt{2}E \sin \omega t \tag{1}$$

$$i_s = \sum \sqrt{2}I \sin(n\omega t - \varphi_n) \tag{2}$$

Equation (1)-(2) can be expressed by the following $\alpha\beta$ coordinates.

$$i_\alpha = \sqrt{\frac{3}{2}}\, i_s\,,\ v_\alpha = \sqrt{\frac{3}{2}}\, v_s \tag{3}$$

$$i_\alpha^{\ 1} = i_\alpha(\theta - 90^o)\,, v_\alpha^{\ 1} = v_\alpha(\theta - 90^o) \tag{4}$$

$$i_\beta = i_\alpha^{\ 1}\,,\ v_\beta = -v_\alpha^{\ 1} \tag{5}$$

978-1-4244-1871-8/07 $25.00 © 2007 IEEE

Relationship of the (3)-(5) is represented by instantaneous space vector as shown in Fig. 2. Applying (3)-(5) to the p-q theory[7,8], compensation reference current i_h can be obtained as (6).

$$i_h = \frac{v_\alpha v_\alpha^{\,1} i_\alpha^{\,1} - (v_\alpha^{\,1})^2 i_\alpha}{v_\alpha^{\,2} + (v_\alpha^{\,1})^2} \qquad (6)$$

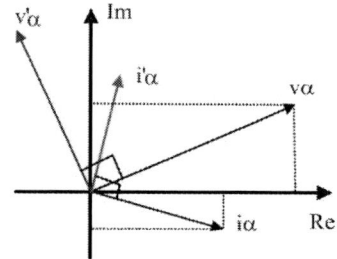

Fig. 2 Instantaneous space vector

By multiplying i_h to voltage gain k_h, compensation reference voltage v_h is can be written as (7).

$$v_h = k_h i_h \qquad (7)$$

Also, according to the voltage variation of micro source, the variation portion Δv_L of the load voltage v_L is equal to the difference between load voltage reference $v_L^{\,*}$ and load voltage v_L.

$$\Delta v_L = v_L^{\,*} - v_L \qquad (8)$$

Therefore, in order to compensate Δv_L and v_c collectively, the final compensation reference voltage $v_c^{\,*}$ can be concluded by the following equation.

$$v_c^{\,*} = k(\Delta v_L - v_h) \qquad (9)$$

The compensation and control algorithm are shown in Fig. 3.

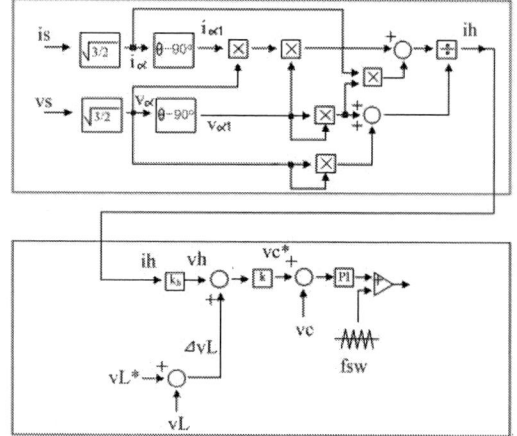

Fig. 3 Compensation and control algorithm

B. Fuel Cells Power Converter

As shown in Fig. 4-5, fuel cells power converter is consisted of a fuel cells simulator and an isolated dc-dc boost converter (dc power side) and a PWM VSI(ac power side). Because the output voltage v_{fc} of a fuel cells simulator has a low dc voltage, a full bridge dc converter is required for boost to high dc voltage. And, in order to supply 60Hz ac voltage to load, this dc boost converter should be connected to the load using a 60Hz PWM VSI.

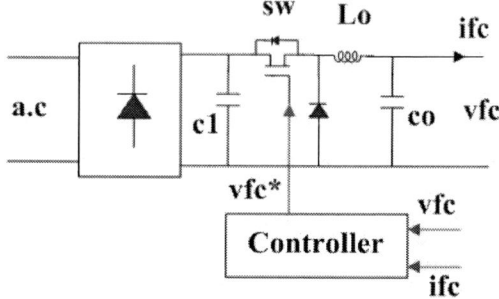

Fig. 4 Buck converter as a PEM fuel cells simulator

A fuel cells source is adopted a fuel cells simulator which is designed with the electrical characteristics of PEM fuel cells.
The characteristics voltage and current (V-I) curve for the fuel cells simulator is controlled by a simplified linear function. The linear function of the ohmic region is represented as follows[5].

$$v_{fc}(i_{fc}) = v_{fci} \quad (0 < i_{fc} < i_{fci})$$

$$v_{fc}(i_{fc}) = \left(\frac{v_{fcf} - v_{fci}}{i_{fcf} - i_{fci}}\right)i_{fc} - \left(\frac{v_{fcf} - v_{fci}}{i_{fcf} - i_{fci}} i_{fci} - v_{fci}\right) \quad (10)$$

$$(i_{fci} < i_{fc} < i_{fcf})$$

$$v_{fc}(i_{fc}) = 0 \quad (i_{fc} > i_{fcf})$$

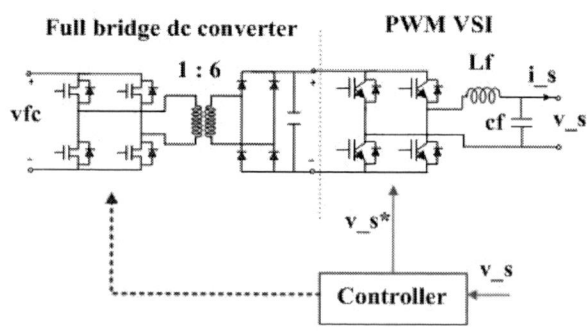

Fig. 5 Dc-dc-ac power converter

Fig. 6 shows a typical V-I curve of the fuel cells. The output characteristics are separated into three regions : activation, ohmic, and concentration . Generally, the operation region of actual fuel cells is ohmic region.

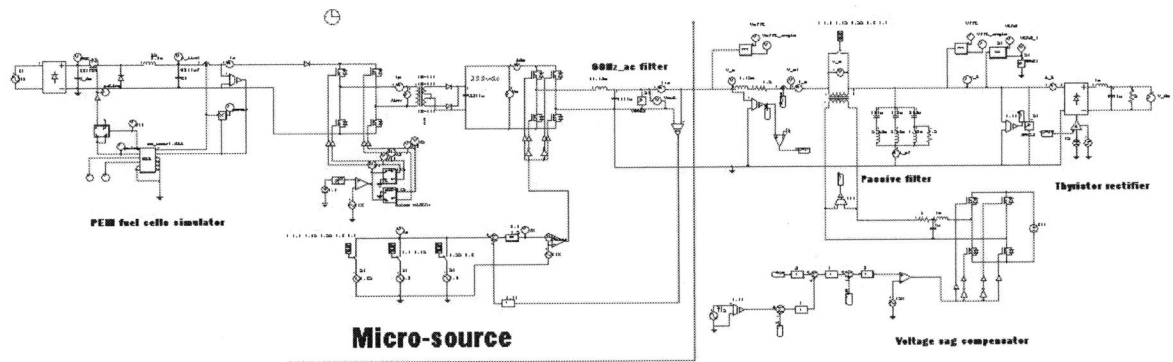

Fig. 8 PSIM model of the proposed system

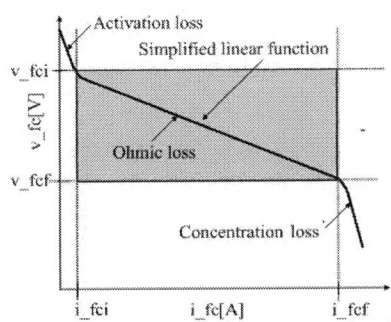

Fig. 6 V-I curve of the PEM fuel cells

C. Voltage Sag Compensator and Load

Fig. 7 shows the voltage sag compensator which is similar to the configuration of hybrid series active power filter. Voltage sag compensator consists of an injected transformer, secondary winding of which is connected in series with a micro-source, a PWM VSI connected to the primary of the injection transformer and fuel cells source connected at the dc link of the inverter bridge. For PWM modulation, 15kHz triangular wave comparison method is adopted. As a critical load, a thyristor rectifier($\alpha = 15^o$) is connected to the sag compensator.

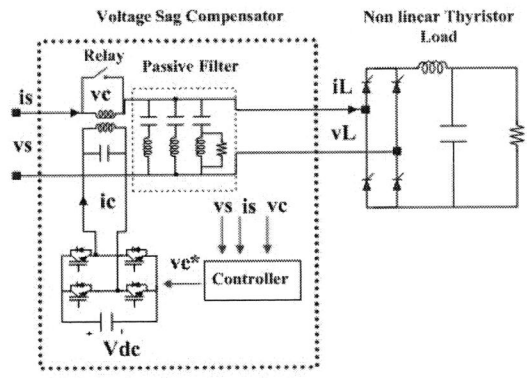

Fig. 7 Voltage sag compensator and load

III. RESULTS AND DISCUSS

Fig. 8 shows the PSIM model of the proposed system. Two cases are going to be studied in this paper ; (1) Mode 1- the voltage sag compensator always operates and (2) Mode 2- the voltage sag compensator operates in case of the output voltage v_s sag only occurs.

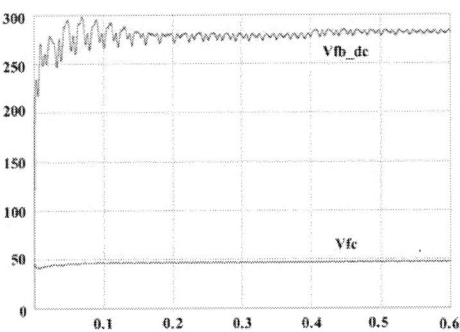

Fig. 9 Dc voltage characteristics of fuel cells power converter

Fig. 9 shows the PEM fuel cells voltage v_{fc} and the output voltage v_{fb_dc} of the dc full bridge converter. As shown in Fig. 5 and 9, we can know that v_{fb_dc} =40.6v steps up to v_{fc} =288v by using the dc full bridge converter.

A. Compensation Mode 1

Fig. 10 - Fig. 13 show compensation results in case of mode 1. In mode 1, the relay (in Fig. 1 and 7) connected in parallel with the injection transformer is continuously opened, and voltage sag compensator always operates no matter it is voltage variations occurrence. That is, the operation of sag compensator is not related to the output voltage v_s variations (sag or swell). Therefore, load voltage v_L can be compensated and restored to nominal value and the output current i_s can be sinusoidal all over the compensating

978-1-4244-1871-8/07 $25.00 © 2007 IEEE   473

periods. But, in most cases, because the battery stack can be adopted as the energy storage system of the sag compensator, it is very difficult that the compensator operates during long times.

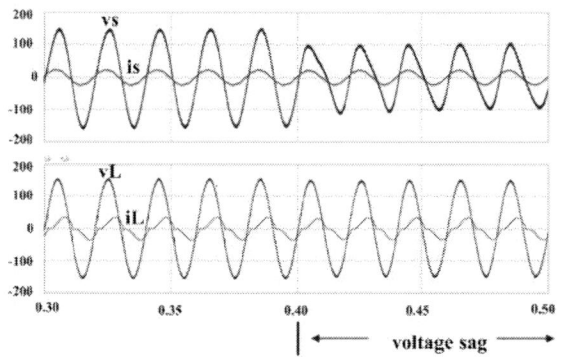

Fig. 10 Output voltage v_s, output current i_s and load voltage v_L, load current i_L in case of 35% voltage sag occurs (mode 1)

First of all, Fig. 10 shows the voltage and current waveforms while 35% voltage sag occurs 5 periods from $v_s = 155v$(peak) to $v_s = 100v$(peak).

Because the sag compensator always operates in mode 1, nominal value of v_L can be kept constantly, also v_s and i_s are displaying sinusoidal waveforms, and showing as v_s in phase with i_s. This means that power factor of micro-source can be improved by operation of the mode 1.

Fig. 11 shows the waveforms of v_L and v_s. When voltage sags occur, the peak and rms(root mean square) values of v_s is decreased. But, nominal value of v_L is kept 155v(peak) even if sag occurred.

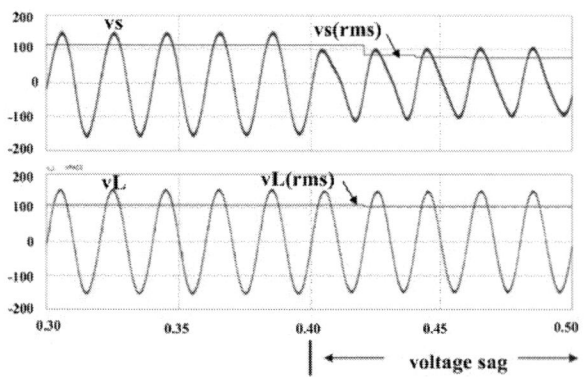

Fig. 11 Output voltage v_s and load voltage v_L in case of 35% voltage sag occurs (mode 1)

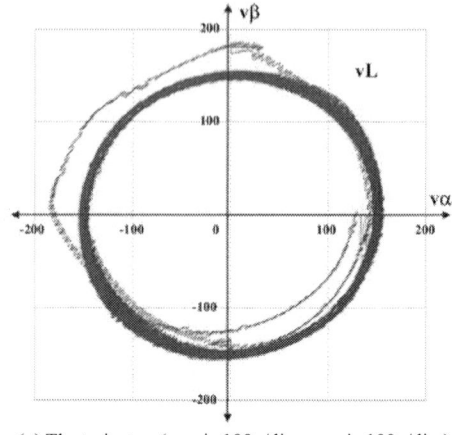

(a) The trajectory(x-axis:100v/div., y-axis:100v/div.)

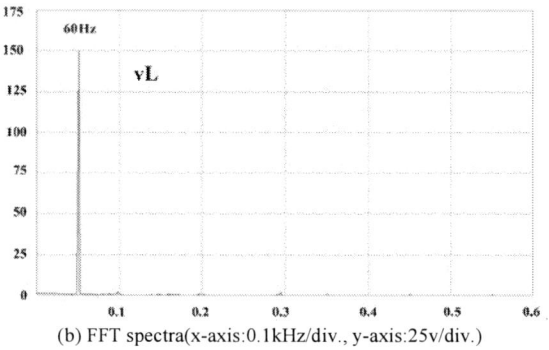

(b) FFT spectra(x-axis:0.1kHz/div., y-axis:25v/div.)

Fig. 12 Evaluation of restored load voltage v_L (mode 1)

Fig. 12 shows the $\alpha\beta$ trajectory and FFT spectra of restored v_L. The $\alpha\beta$ trajectory of restored v_L can be drew by its α and β component (v_α, v_β) and it is nearly circle shape as shown in Fig. 12a. It means the restored v_L waveform is sinusoidal with 60Hz frequency.

Also, Fig. 12b shows the FFT spectra of restored v_L. Restored v_L contains an only fundamental 60Hz component while all harmonics are reduced.

Fig. 13 shows the detail some half period waveforms of v_s and v_L. As observing the point (2), v_s step down about 55v(peak) and moreover it is distorted. But, after sags are compensated, the waveform of v_L is sinusoidal and it step up to 55v(peak).

As observing the other case – sag of v_s not occurs as shown in the point (1), the PWM switching noise remains in waveform of v_L because of the sag compensator always operating in mode 1.

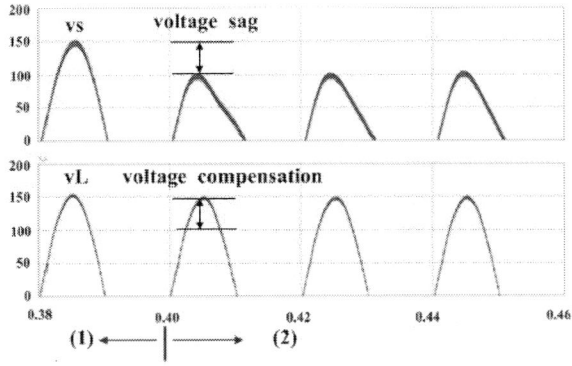

Fig. 13 Detail some waveforms of the load voltage v_L and output voltage v_s (mode 1)

B. Compensation Mode 2

Fig. 14 - Fig. 17 show the compensation results in case of mode 2. In this mode, sag compensator is only operated while sag or swell of v_s occurs. In mode 2, the relay (in Fig. 1 and 7) is closed while the variation of v_s not occurs. But, if the variation of v_s occurs, the relay must be opened. Therefore, in Mode 2, the battery stack of the sag compensator less spender than the mode 1.

Fig. 14 and Fig. 15 show the dynamic response of while 42% swell and 35% sag consecutively occurs; 155v(peak) → 220v(peak) →155v(peak) → distorted 100v(peak) →155v(peak). As shown in Fig. 14, although the peak value of v_s is varied and distorted, v_L is restored to nominal value.

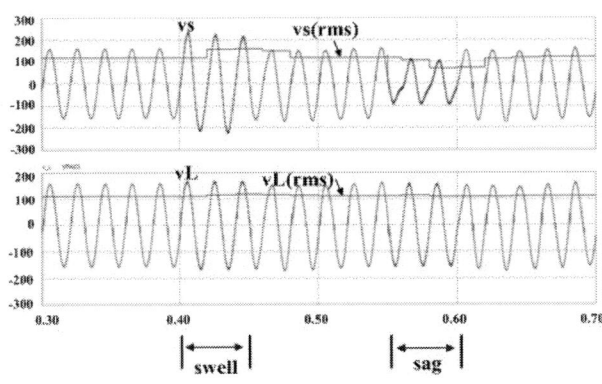

Fig. 14 Output voltage v_s and load voltage v_L in case of 42% swell and 35% sag consecutively occurs (mode 2)

Dissimilar to mode 1, i_s is not always sinusoidal all over the period as shown in Fig. 15 because the sag compensator does not operate while variation of v_s not occurs. It means that i_s turn into sinusoidal waveform (dotted line box parts in Fig. 15) in case when compensation is achieved. In particular,

we can observe that i_s is less distorted however the sag compensator does not operate. As shown in Fig. 7, this phenomenon is produced because distorted components are bypassed to R-L-C passive filter even if the sag compensator does not operate.

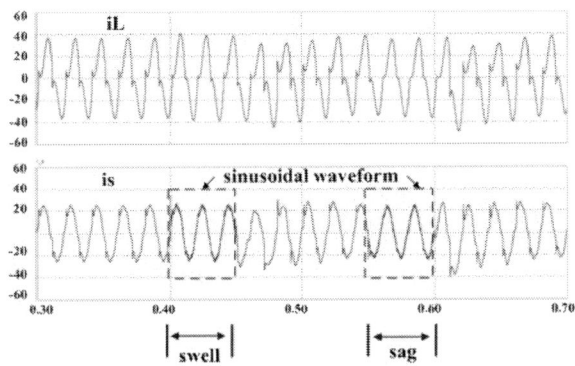

Fig. 15 Load current i_L and output current i_s in case of 42% swell and 35% sag occurs (Mode 2)

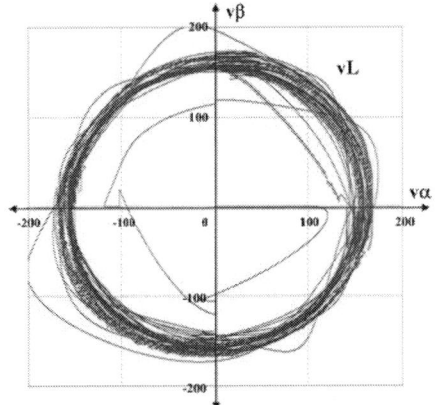

(a) The trajectory(x-axis:100v/div., y-axis:100v/div.)

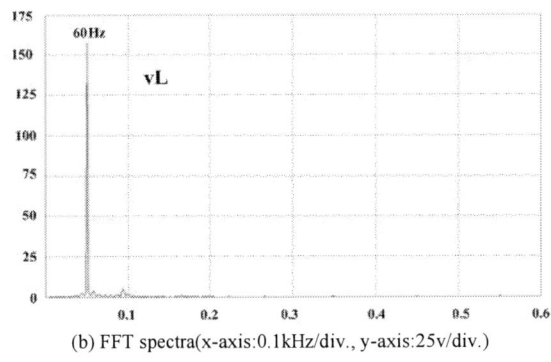

(b) FFT spectra(x-axis:0.1kHz/div., y-axis:25v/div.)

Fig. 16 Evaluation of restored load voltage v_L (mode 2)

Fig. 16 shows the $\alpha\beta$ trajectory and FFT spectra of restored v_L in case of mode 2. Similar to mode 1, $\alpha\beta$ trajectory of restored v_L generally displays circle shape,

978-1-4244-1871-8/07 $25.00 © 2007 IEEE

but it is with much oscillation. FFT spectra of restored v_L is presented in Fig. 16b. As in previous case, there is only 60Hz fundamental component of restored v_L .

Detail waveforms of the output voltage v_s and load voltage v_L are presented in Fig. 17.When voltage sag occurs at 0.40s, v_L is directly restored although there are small voltage spikes (dotted circle parts in Fig. 17). These spikes mean that the sag compensator operates at 0.40s.

From the results obtained, we can observe that sensitive equipment would then be thoroughly protected from voltage sag and harmonics situation by using the proposed system.

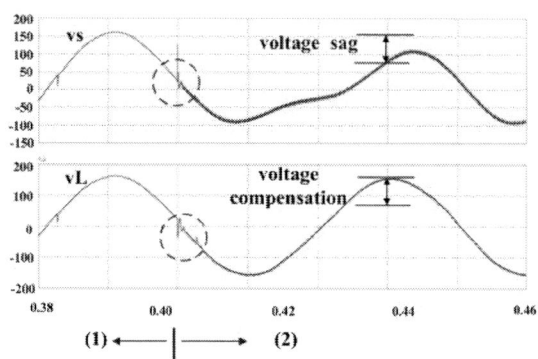

Fig. 17 Detail waveforms of output voltage v_s and load voltage v_L

(mode 2)

IV. CONCLUSION

In this paper, a stand alone type fuel cells micro-source with a voltage sag compensator was presented. Micro-source is consisted of a PEM fuel cells simulator which is controlled by simplified linear function, an isolated dc full bridge converter having a dc voltage step up capability, a 60Hz PWM voltage source inverter, and a voltage sag compensator. Two modes were studied by PSIM simulation. From the results obtained, it was observed that the proposed system effectively compensates the variation caused by sag and swell. Also, we observed that the proposed system can have a fast transient response when there was a fast dynamic change, no matter it is a voltage swell or sag.

ACKNOWLEDGMENT

This study was supported by the Research Center for Stabilization of Power Systems with Large Distributed Generations at Chonnam National University in KOREA.

REFERENCES

[1] S. Chakraborty, M. G. Simoes, "Advanced active filtering in a single phase high frequency ac microgrid", *in Conf. Rec. of IEEE PESC'05*, 2005, pp.191-197.

[2] D. M. Vilathgamuwa, C. J. Gajanayake, P.C. Loh, and Y.W. Li, "Voltage sag compensation with Z-source inverter based dynamic voltage restorer", *in Conf. Rec. of IEEE IAS'06*, 2006, pp.2242-2248.

[3] P. K. W. Chan, H. S. H. Chung, and S.Y.R. Hui, "Boundary control of dynamic voltage restorers in voltage harmonic compensation", *in Conf. Rec. of IEEE PESC'06*, 2006, pp. 7965-801.

[4] P. T. Cheng, C. C. Hung, C. C. Pan, and S. Bhattacharya, "Design and implementation of a series voltage sag compensator under practical utility conditions', *in Conf. Rec. of IEEE APEC'02*, 2002, pp.1061-1067.

[5] Y. R. de Novaes, R. R. Zapelini, and I. Barbi, "Design considerations of a long term single-phase uninterruptible power supply based on fuel cells", *in Conf. Rec. of IEEE PESC'05*, 2005, pp.1628-1634.

[6] P. J. H. Wingelaar, J. L. Duarte, and M. A. M. Hendrix, "Dynamic characteristics of PEM fuel cells", *in Conf. Rec. of IEEE PESC'05*, 2005, pp.1635-1641.

[7] M. T. Haque, and T. Ise, "Implementation of single-phase pq theory", *in Proc. PCC-Osaka'02*, 2002, pp.761-765

[8] M. T. Haque, "Single-phase pq theory", *in Conf. Rec. of IEEE PESC'02*, 2002, in CD-version

A Novel Grid-Connected PV PCS with New High Efficiency Converter

Byung Duk Min, JongPil. Lee, Jong-Hyun Kim, Tae-Jin Kim, Dong-Wook Yoo,
Korea Electrotechnology Research Institute
28-1, Sungju-dong, Changwon-si, 641-120 KOREA
Email: bdmin@keri.re.kr

Kang Ryoul Ryu, Jung Joong Kim, and Eui-Ho Song
*Changwon National University
Changwon, 641-773, KOREA
Email: ehsong@changwon.ac.kr

Abstract-: In this paper, new topology is proposed that can dramatically reduce the converter power rating and increase the efficiency of total PV system. Since the output voltage of PV module has very wide voltage range, in general, the DC/DC converter is used to get constant high DC voltage. According to analysis of PV characteristics, in proposed topology, only 20% power of total PV system power is needed for DC/DC converter. DC/DC converter used in proposed topology has flat efficiency curve at all load range and very high efficiency characteristics. The total system efficiency is the product of that of converter and that of inverter. In proposed topology, because the converter efficiency curve is flat all load range, the total system efficiency at the low power range is dramatically improved. The proposed topology is implemented for 200kW PCS system. This system has only three DC/DC converters with 20kW power rating each other. It is only one-third of total system power. The experiment results show that the proposed topology has good performance.

I. INTRODUCTION

Renewable energy resources will be an increasingly important part of power generation in the new millennium. Besides assisting in the reduction of the emission of greenhouse gases, they add the much needed flexibility to the energy resource mix by decreasing the dependence on fossil fuels. Due to their modular characteristics, ease of installation and because they can be located closer to the user, PV systems have great potential as distributed power source to the utilities.

This paper deals with new topology of grid connected PV PCS. In Chapter 2, the comparison to the several topologies is fulfilled. In chapter 3, PV characteristics are described. In chapter 4, new topology is proposed and analysis the new topology in the point of view of power rating and efficiency. In the chapter 5, the implementation and experimental data are presented to prove the usefulness of the proposed topology.

II. GENERAL TOPOLOGIES FOR PV-PCS

Fig.1 shows the general topologies of photovoltaic PCS. Figure1 (a) consists of inverter and line frequency transformer. In general, PV module has wide voltage variation range. The output voltage of inverter is restricted by the lowest PV input DC voltage. Therefore, the output voltage of inverter is much lower than grid voltage. To increase the output voltage of inverter is used the low frequency transformer. The advantage of this topology is simple. The disadvantage is very heavy and huge because of large line frequency transformer and the inverter power switches like IGBTs with large current ratings

are needed to get designed power rating because of low output voltage of inverter. The large current is also not good for the system efficiency.

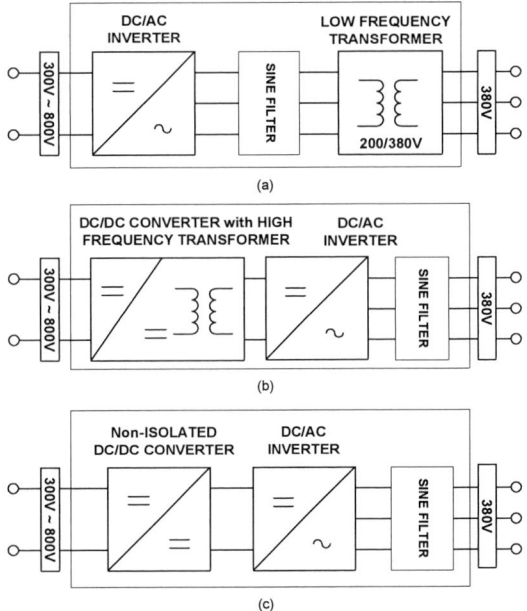

Figure 1. The topologies of photovoltaic PCS.

The topology that overcomes the disadvantages of Figure 1(a) using high frequency transformer is shown in Figure 1(b). In this case, at first, PV DC input voltage is boosted to the constant dc voltage by DC/DC converter and then the inverter generates the grid voltage using boosted high constant dc voltage. In this topology, the system size, cost and weight can be reduced dramatically because the high frequency transformer is used, and because DC input voltage of inverter is high enough to directly generate the grid voltage, the current rating of power switches of inverter is much lower than that of the power switches of Figure 1(a). But this system has also disadvantages. For high efficiency, DC/DC converter with very high efficiency is needed. Because the power rating of DC/DC converter is the same as that of total system, for high power system, DC/DC converter has also high power rating. The PV PCS using non-isolated converter is shown in Figure 1(c). This topology is very simple because the

978-1-4244-1871-8/07 $25.00 © 2007 IEEE   477

transformer for isolation is not used. In recent, this type PCS is popular because of small system size, high efficiency and low cost. The disadvantages of this topology are the same as those of Figure 1(b) topology. Figure 1(b) and Figure 1(c) system have many advantages but it is not good for high power PCS system because the converter power rating must be the same as the total system power.

III. CHARACTERISTICS OF PV MODULES

The output characteristic of PV cell is affected by irradiation and operating temperature. Figure 2 shows the PV output characteristics of PV module. At the constant temperature, if the irradiation is increased, the output power is increased. At the constant irradiation conditions, if the temperature is decreased, the output power is increased. It can be seen in figure 2 that the output voltage of PV module should be controlled in appropriate level to obtain maximum power from PV module. We call it as maximum power point tracking (MPPT). Another characteristic of PV module is that at the low output voltage, the output power is reduced significantly even if the irradiation is high. To connect the output of PV PCS to the grid, PV PCS must generate higher voltage than that of grid. In the Figure 2, it can be seen that the output voltage varies wide voltage range according to irradiation and temperature. Therefore to connect PV PCS to the grid, some technique is needed to increase the output voltage of PV PCS like boost transformer, DC/DC converter to increase the input DC voltage of inverter.

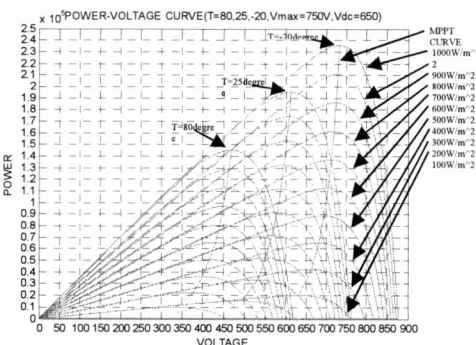

Figure 2. Characteristics of PV module

IV. NEW TOPOLOGY FOR PHOTOVOLTAIC PCS

In general, since the lowest voltage of PV module is lower than required voltage to connect to the grid directly, there is always some difference between the PV module output voltage and the required DC link voltage. In the other point of view, to obtain the required DC link voltage, it is possible if this difference voltage is added to the PV module output dc voltage. Figure 3 shows the new proposed topology to get high DC link voltage. The small isolated dc/dc converter generates only the difference voltage between the PV module voltage and the required DC link voltage of inverter. Therefore, since DC/DC converter must not generate the

whole required voltage in the proposed topology, the required power capacity of DC/DC converter is reduced dramatically. Since the output side of proposed DC/DC topology is connected in series with PV module, the current is flowing through DC/DC converter and PV module commonly.

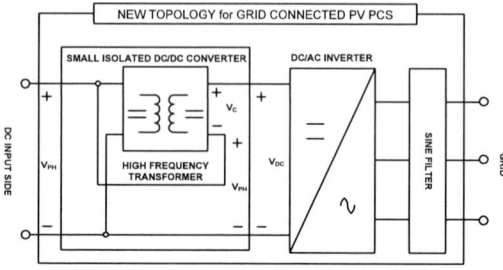

Figure 3 Proposed topology for PV PCS.

Figure 4 shows the concept of the proposed topology. The DC link power is described as follows

$$P_{DC} = V_{DC}I_{DC} = V_C I_{DC} + V_{PH}I_{DC}$$
$$= P_C + P_{PHDC}$$

Where P_c is converter power, P_{PHDC} is PV power on the DC link side. The ratio of the sharing power of converter is the ratio of the converter output voltage, V_c and DC link voltage V_{DC}. The larger the difference between PV module output voltage and the required DC link voltage is, the larger the converter power is. The closer PV module output voltage is to the required DC link Voltage, the power of converter is reduced rapidly.

In Figure 2, cyan line means the required converter power. The maximum power point of converter is at the maximum operating temperature point and maximum irradiation point. The required maximum converter power can be calculated as follows

$$I_{DCTM} = \frac{P_{PHTM}}{V_{DC}}$$
$$P_C = I_{DCTM}(V_{DC} - V_{PHTM})$$
$$= P_{PHTM}\frac{V_{DC} - V_{PHTM}}{V_{DC}}.$$

Where V_{PHTM} is the PV module output voltage at the maximum operating temperature, P_{PHTM} is PV module output power at the maximum operating temperature.

The developed PV PCS specifications are as follows

MPPT range : 450V~850V$_{DC,}$
Max. operating Temperature of PV module : 80℃,
Required DC link voltage of inverter : 650V$_{DC,}$
The output voltage of inverter : 380V$_{AC,}$
Rated PCS power = 200kW.

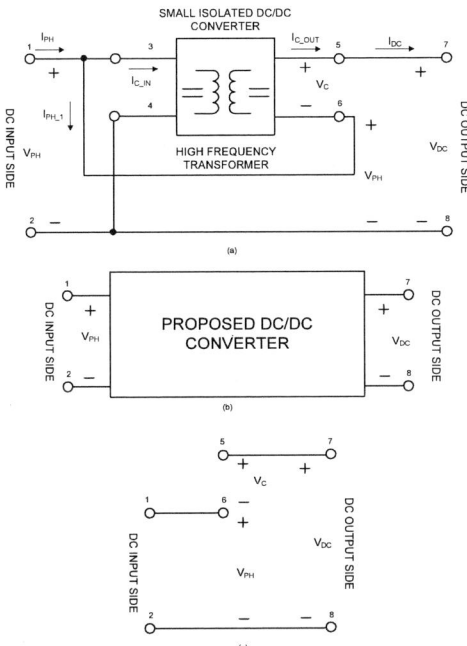

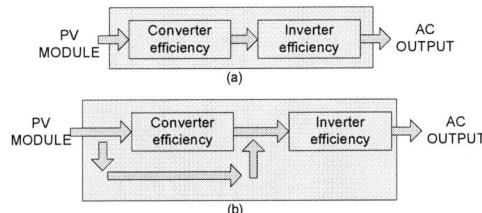

Figure 4 Explanatory diagram of the proposed topology.

Using above equations and specifications, the maxim power of converter is obtained as follows

$$V_{PHTM} = 470V(from\ Figure\ 2\ at\ Max.\ Temp),$$

$$P_{PHTM} = 145kW(from\ Figure\ 2\ at\ Max.\ Temp),$$

$$P_C = 145kW\frac{650V - 470V}{650V} = 40.15kW$$

This converter power is only 20% of the required PCS power. In the proposed topology, the required converter power is dramatically reduced. The efficiency flow diagram is shown in figure 5. Figure 5(a) shows the efficiency flow of conventional topology. The total efficiency is the product of efficiency of converter and that of inverter. Therefore, the total efficiency is affected to that of converter mainly. But the proposed topology has a different efficiency flow from conventional topology as shown in figure 5(b). The proposed converter efficiency is described as follows:

$$\eta_t = p + (1-p)\eta_C$$

$$p = \frac{P_S - P_C}{P_S}$$

Where

η_t : the total efficiency of proposed converter

p : the raio from PV module directly

P_S : Total sytem power of system

P_C : DC / DC converter power.

Figure 5 Efficiency flow diagram
(a) conventional DC/DC converter + inverter topology, (b) proposed topology

In small power sharing region at the DC/DC converter, the most power is directly provided by PV module to the load, the total efficiency of proposed converter is very high. The total PV PCS system efficiency is described as follows

$$\eta_S = \eta_t \cdot \eta_i$$

Where η_S is total PV PCS efficiency, η_i is inverter efficiency.

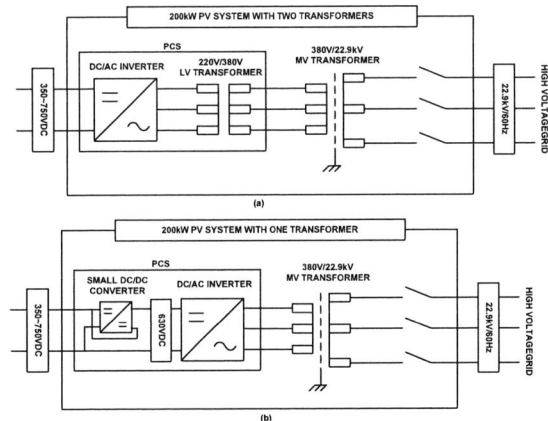

Figure 6 Comparison to the two topologies
(a) using low voltage line frequency transformer
(b) using proposed topology.

Figure 6 shows the topologies for high power PV PCS. Figure 6(a) uses the large low voltage transformer because there is no DC/DC converter. In general, high power DC/DC converter with 200kW power rating is very difficult to make. But using proposed topology, the implementation of converter with high efficiency is easy since small power capacity of converter is required.

V. EXPERIMENTAL RESULTS

The proposed system is implemented as shown in figure 7. Three DC/DC converters with 25kW power capacity each other are used to boost DC voltage. The specifications of DC/DC converter are as follows

Input voltage range : 450V~850V
Output voltage range : 0~200V

978-1-4244-1871-8/07 $25.00 © 2007 IEEE

One module power : 25kW
Max. output current of module : 130A
Total max. output current : 390A (130A x 3)
Total converter power : 75kW(25kW x 3).

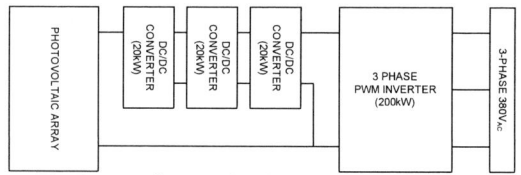

Fig. 7 system diagram of 250kW grid connected PV PCS

The specifications of PV PCS system are as follows

Rated Power : 250kW
MPPT range : 450V ~ 850V
Output Voltage : 380V$_{rms}$
Minimum DC link voltage of inverter : 650V

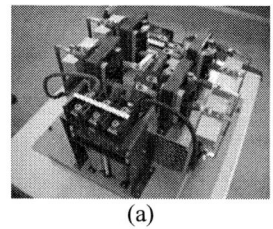

(a)

(b)

Figure 8 Implemented PV system with the proposed topology
(a) DC/DC converter, (b)inverter

Figure 8 shows the implemented system with the proposed topology. This system has three DC/DC converters. This proto type has low voltage transformer. In proposed system, the low voltage transformer is option since the DC link voltage of inverter is high enough to provide the higher ac voltage than that of grid. Some power company requires the isolation transformer. In low power PV PCS system, non-isolated PV

PCS system is used widely because that system is small and cheap.

The efficiency of proposed DC/DC boost topology is shown in Figure 9. Blue line is the efficiency curve of the small DC/DC converter. The red line is total efficiency of the proposed converter topology. We can see that the efficiency curve is flat for all load range. In figure 9, x- axis means the PV module output voltage. If PV module output voltage is high, the power sharing of DC/DC converter is low, i.e. the DC/DC converter is in the light load state. If PV module output voltage is low, the power sharing of DC/DC converter is high because the voltage difference between target DC link voltage and PV module voltage is large, i.e. the DC/DC converter is in the heavy load state. In general, at light load state, the efficiency is low. At heavy load state, that is high. But as explained before, the total efficiency is flat with no relation to the load state of converter.

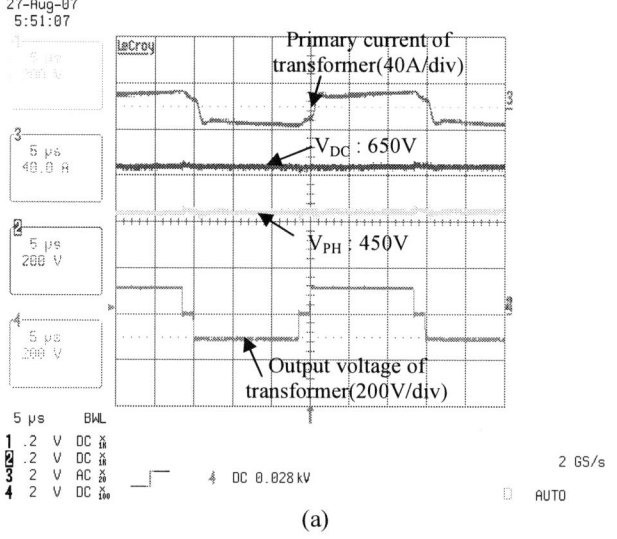

(a)

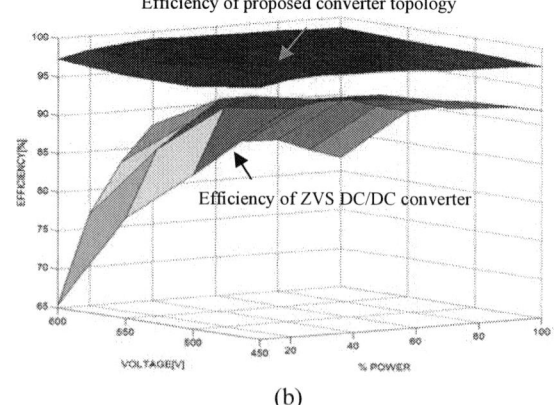

(b)

Figure 9 The experimental results of proposed converter
(a) waveforms of proposed converter, (b) Efficiency map of proposed converter

Figure 10 shows the inverter output current(red line) and grid voltage(blue line) waveforms. The output power of PCS is 100kW, THD of line currents is 3.8%.

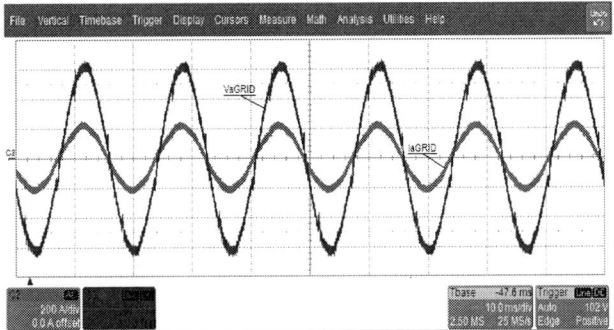

Figure 10 Experimental waveforms (blue : grid voltage,100V/div, red : inverter output current, 200A/div)

Figure 11 shows the total system efficiency curve. Using general DC/DC converter, the total efficiency is very low at the low load condition. But using the proposed converter topology, the total efficiency of inverter/converter system is very high in spite of low efficiency of general DC/DC converter.

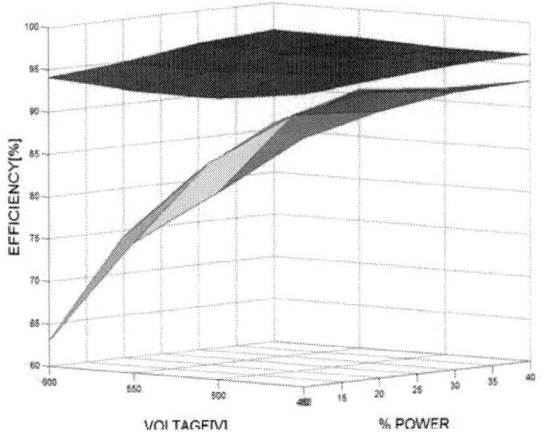

Figure 11 Experimental total system efficiency curve

VI. CONCLUSIONS

In this paper, new topology for PV PCS was proposed. The proposed topology has the flat efficiency curve throughout all load range at the converter part and the required DC/DC converter power capacity is only 20% of total rated power of PV PCS. The implementation burden of high power DC/DC converter is low. Therefore the proposed topology can be applied to all power rating PCS. Since DC link voltage can be boosted enough to provide the AC voltage higher than that of grid using the proposed topology, the current rating of IGBT is half of that of conventional PCS with low DC link voltage. It was shown that the proposed topology had good performance through experimental results.

REFERENCES

[1] Kasemsan Siri , Vahe A. Caliskan , C. Q. Lee "Maximum Power Tracking in Parallel Connected Converters" , *IEEE Transaction on Aerospace and Electronic systems* , vol. 29 , No 3 , July 1993.

[2] Mummadi Veerachary, Tomonobu Senjyu, Katsumi uezato "Voltage-Based Maximum Power Point Tracking Control of PV System" , *IEEE Transations of aerospace and electronic systems*, vol. 38, No. 1 ,Jan 2002.

[3] Guo-Kiang Hung, Chih-Chang Chang, and Chern-Lin Chen, "Automatic Phase-Shift Method for Islanding Detection of Grid-Connected Photovoltaic Inverters", *IEEE TRANSACTIONS ON ENERGY CONVERSION*, Vol. 18, No. 1, March 2003

Low-Cost Converters for Micro Wind Turbine Systems using PMSG

Hong-Geuk Park, Seok-Ho Jang, Dong-Choon Lee
Dept. of Electrical Eng., Yeungnam University
214-1, Dae-dong, Gyeongsan, Gyeongbuk, 712-749
Email: dclee@yu.ac.kr

Heung-Geun Kim
School of Elec. Eng. & Compt. Science,Kyungpook Nat Univ
1370, Sangyeok-dong, Buk-gu, Daegu, 702-701
Email:kimhg@knu.ac.kr

Abstract— This paper proposes a low-cost power converter for micro wind turbine systems using permanent magnet synchronous generators (PMSG). The proposed converter consists of a two-leg three-phase PWM inverter for the generator control and a single-phase half-bridge PWM converter which is connected to the utility grid. For the two separate DC-link voltages, a balancing control is added and the adverse effect of the DC-link voltage ripples on the inverter output voltage is compensated. The control performance of the proposed converter topology for the micro wind turbine system is shown by the simulation results using PSIM software.

I. INTRODUCTION

In recent, development of renewable energy systems is being increased rapidly since crisis of the exhaustion of fossil fuel and environmental pollution are hot issues. Among different types of renewable energy, the wind power is most promising and widely commercialized compared with other energy sources. The trend of wind power system is to employ large wind turbine systems such as multi-mega watt power capacity (3~5[MW]) and to construct the large wind farm [1]. However, micro wind turbine systems can be used for islands, mountains, and local areas which areas isolated from the utility grid. Also, the small-power wind turbine system is a good candidate for the micro-grid systems [2].

For MPPT(maximum power point tracking) control of the turbine system, some methods have been used. One is P&O (perturbation and observation) which controls the output power to the maximum by changing the rotational speed step by step [3]. The second method is to use the optimal tip-speed ratio which produces the turbine (or generator) speed reference according to the wind speed [4]. The other is to use the torque control which gives the maximum output power for the given rotational speed [5].

The conventional power converter for the small power wind turbine system is shown in Fig. 1. It consists of the diode rectifier, DC-boost converter, and full-bridge PWM converter, which is of cheap structure. Since the uncontrolled diode rectifier is connected to the PMSG, it is impossible to control the generator speed for output power control. Input current reference of the DC-boost converter is decided for the MPPT control of the turbine [6]-[8].

In this paper, a low-cost and high-performance converter for small-scale PMSG wind turbine systems is proposed, which consists of a three-phase two-leg PWM inverter for the generator control and a single-phase half-bridge PWM converter connected to the utility grid. With this topology,

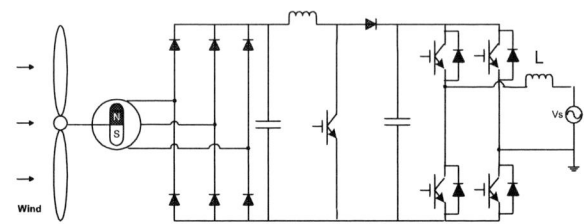

Fig. 1. Conventional small-scale wind generation system.

The vector control of the PMSG is possible, so the dynamic control performance is fast. The operation of the proposed circuit topology is shown by simulation for 3[kW] PMSG using PSIM.

II. WIND TURBINE MODEL

The power captured by the wind turbine may be written as [9][10]

$$P_{Turbine} = 0.5A\rho C_p(\lambda)\upsilon^3 [W] \qquad (1)$$

and the tip-speed ratio is defined as

$$\lambda = \frac{\omega_r R}{\upsilon} \qquad (2)$$

where,

A : blade swept area [m^2]
ρ : specific density of air [kg/m^3]
υ : wind speed [m/s]
R : radius of the turbine blade[m]
ω_r : rotating speed [rpm]
C_p : coefficient of power conversion

If the blade swept area and the air density are constant, the value of C_p is a function of λ and it is maximum at the particular λ_{opt}. Hence, to fully utilize the wind energy, λ should be maintained at λ_{opt}, which is determined from the blade design. Then, from (1),

$$P_{max} = 0.5A\rho C_{pmax}\upsilon^3 [W] \qquad (3)$$

The reference speed of the generator is determined from (2) as

$$\omega_r^* = \frac{\lambda_{opt}}{R}\upsilon \qquad (4)$$

978-1-4244-1871-8/07 $25.00 © 2007 IEEE

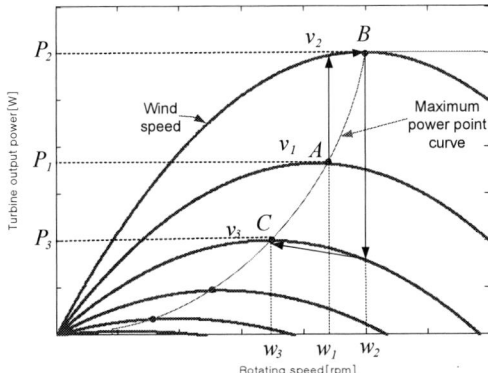

Fig. 2. Wind turbine output characteristics.

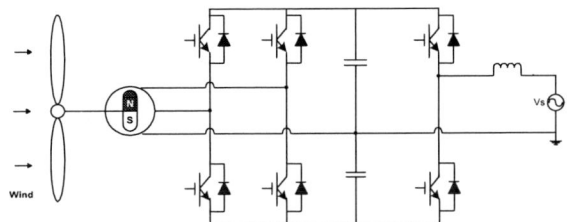

Fig. 3. Proposed converter system.

Table. I. Comparison of converter topologies

Components	Converter in Fig. 1	Converter in Fig. 3
Switch(IGBT)	5	6
Diode	11	6
DC capacitor	2	2
Inductor	2	1
Voltage sensor	3	3
Current sensor	2	3

Once the wind velocity is measured, the speed reference for the MPPT can be obtained from (4).

Fig. 2 shows the turbine power versus and the rotational speed, being the wind speed as a parameter.

III. LOW-COST CONVERTERS AND CONTROL

The proposed converter consists of the two-leg PWM inverter and the single-phase half-bridge converter as shown in Fig. 3. This type of the converter has been proposed for the three-phase induction motor drives which are supplied from single-phase utility [11]. It can be used for the three-phase PMSG to be connected to the single-phase utility grid. The comparison of the conventional converter and the proposed converter topology is given in Table I. The proposed converter requires almost the same number of components as that of the conventional type, however, it has a lot of potential for control of PMSG.

A. Control of three-phase two-leg inverters

For the PMSG control, the three-phase voltage references

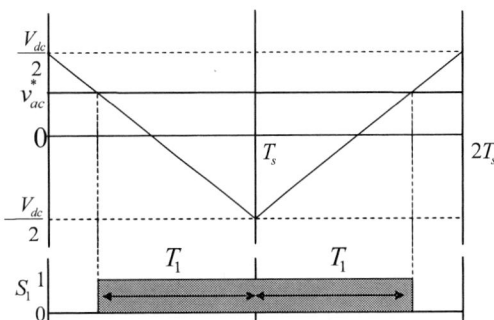

Fig. 4. Voltage modulation.

from the current controller are given in a balanced set as [11]-[13],

$$v_{as}^* = V_m cos\omega t \tag{5}$$

$$v_{bs}^* = V_m cos(\omega t - \frac{2\pi}{3}) \tag{6}$$

$$v_{cs}^* = V_m cos(\omega t + \frac{2\pi}{3}) \tag{7}$$

The two phases of the PMSG are connected to the two legs of the inverter and the third phase is connected to the neutral point of the DC-link. The line-to-line voltage instead of the phase voltage can be used for the PWM. From (5)-(7), the two line-to-line voltage references are given by

$$v_{ac}^* = v_{as}^* - v_{cs}^* = \sqrt{3}V_m cos(\omega t - \frac{\pi}{6}) \tag{8}$$

$$v_{bc}^* = v_{bs}^* - v_{cs}^* = \sqrt{3}V_m cos(\omega t - \frac{\pi}{2}) \tag{9}$$

Using the proportionality of the triangle as shown in Fig. 4, the switching time can be calculated as

$$T_1 = \frac{T_s}{2} + \frac{v_{ac}^*}{v_{dc}}T_s \tag{10}$$

$$T_2 = \frac{T_s}{2} + \frac{v_{bc}^*}{v_{dc}}T_s \tag{11}$$

where,

v_{ac}^*, v_{bc}^* : line-to-line voltage reference

T_s : sampling period

T_1, T_2 : switching time

If there is a difference between the two DC-link voltages of the upper and lower parts, which is usual because of the difference of the instantaneous waveforms, it makes a distortion of the inverter output voltage and current. The deterioration of waveform can be eliminated by compensating for the DC-link voltage ripples.

Using the compensation voltage, the switching times in (10) and (11) are modified as

$$T_1' = \frac{T_s}{2} + \frac{v_{ac}^* - v_{comp}}{v_{dc}}T_s \tag{12}$$

978-1-4244-1871-8/07 $25.00 © 2007 IEEE   483

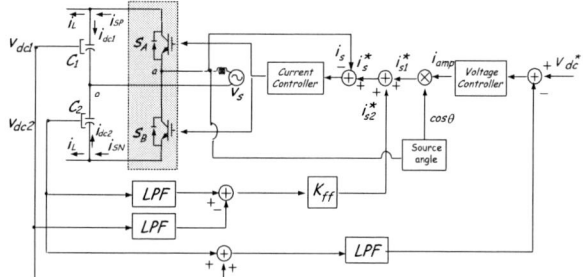

Fig. 5. Control block diagram of the grid side converter.

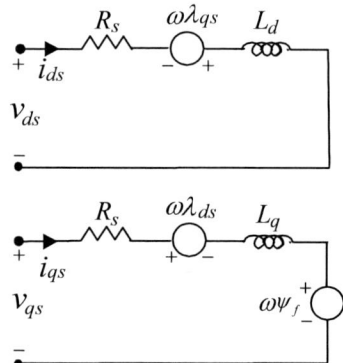

Fig. 6. D-q equivalent circuits of PMSG.

$$T_2' = \frac{T_s}{2} + \frac{v_{bc}^* - v_{comp}}{v_{dc}} T_s \qquad (13)$$

where

$$v_{comp} = \frac{v_{dc1} - v_{dc2}}{2}$$

B. Control of single-phase half-bridge rectifiers

Fig. 5 shows the control block diagram for a half-bridge PWM rectifier. For controlling the source current and the DC-link voltage, the PI regulators are employed. For a balancing control of the neutral voltage of the dc link, the difference of v_{dc1} and v_{dc2} is fed back to the current controller through the proportional gain of K_{ff}. For the DC-link voltage feedback, a cut-off frequency of the low pass filter has been chosen as 50 [Hz] and for the balancing control loop, it has as 0.01[Hz].

IV. CONTROL OF PMSG

A. PMSG modeling

Fig. 6 shows the d-q equivalent circuits of the PMSG. The voltage equation of the PMSG is expressed at synchronous reference frame [14],

$$\begin{bmatrix} v_{ds} \\ v_{qs} \end{bmatrix} = \begin{bmatrix} R_s + pL_d & -\omega_r L_q \\ \omega_r L_d & R_s + pL_q \end{bmatrix} \begin{bmatrix} i_{ds} \\ i_{qs} \end{bmatrix} + \begin{bmatrix} 0 \\ \omega\psi_f \end{bmatrix} \qquad (14)$$

where

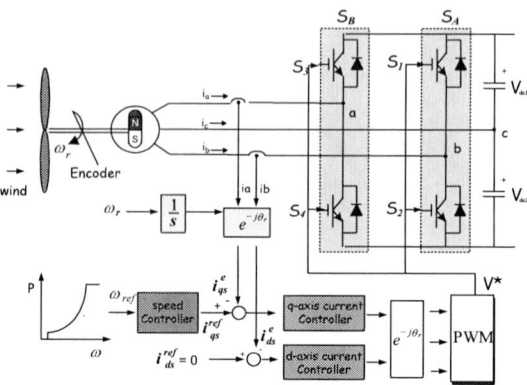

Fig. 7. Control block diagram of PMSG.

p : differential operator
v_{ds}, v_{qs} :d-q axis stator voltage
i_{ds}, i_{qs} : d-q axis stator current
L_d, L_q : d-q axis inductance
R_s : stator resistance
ω : generator speed
ψ_f : magnetic flux leakage

The electromagnetic torque is expressed as

$$T_e = \frac{3}{2} \frac{P}{2}[(L_d - L_q)i_{ds}i_{qs} + \psi_f i_{qs}] \qquad (15)$$

where
P : number of poles
If the d-axis is aligned with the magnetic flux, then

$$T_e = \frac{3}{2} \frac{P}{2}\psi_f i_{qs} \qquad (16)$$

The q-axis current component can be used for the speed control of the generator, and the d-axis current is controlled to be zero.

B. MPPT method

For MPPT control of the wind turbine system, the optimal tip-speed method is used, where the wind speed information is required.

In Fig. 2, the operating point is $A(P_1, w_1)$ at wind speed of v_1. When the wind speed is changed to v_2, the generator speed is increased to w_2, which gives the maximum power P_2. When the wind speed is changed to v_3, then the generator speed is decreased to w_3 which gives the output power P_3. In this way, the MPPT control is performed for wind speed variations

Fig. 7 shows the control block diagram of the PMSG for MPPT control.

V. SIMULATION RESULTS

Simulation has been carried out to test the performance of the proposed converter topology using PSIM. The

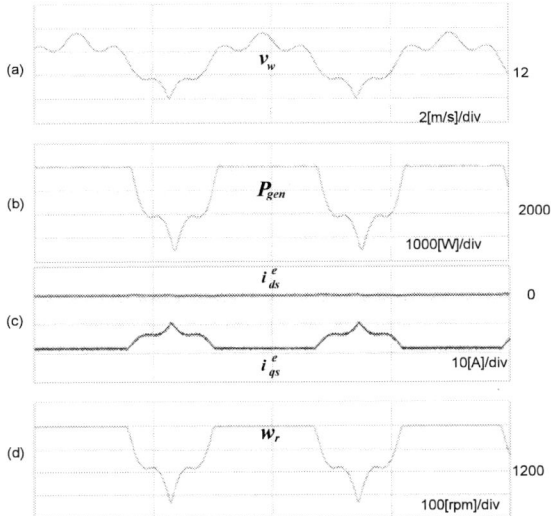

Fig. 8. Generator performance.
    (a)   wind speed[m/s]
    (b)   generator output power[W]
    (c)   d-q axis stator current[A]
    (d)   rotating speed[rpm]

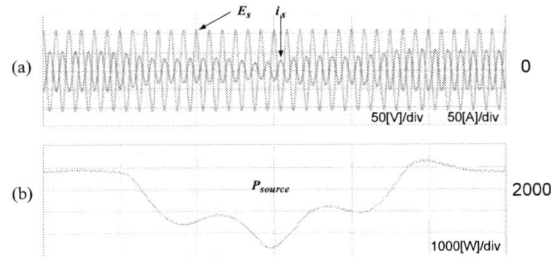

Fig. 9. Grid-side voltage and current.
    (a)   source voltage and current[V]
    (b)   source power[W]

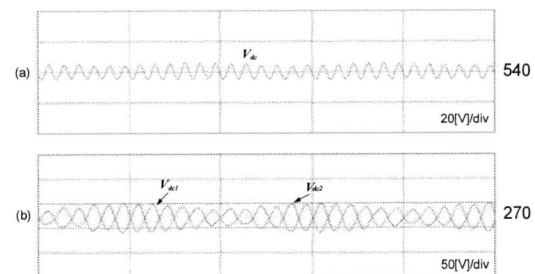

Fig. 10. DC-link voltages
    (a)   DC-link voltage[V]
    (b) v_{dc1} and v_{dc2}

parameters of the PMSG and the turbine model are listed in Table II and III, respectively. The switching frequency is 5[kHz], the input boost inductor is 3[mH], the DC-link capacitor is 3,300[μF], and the DC-link voltage is controlled to 540[V].

Fig. 8 shows the generator performance at randomly variable wind speed. The rated wind speed is 14[m/sec],

above which it is assumed that the stall control is activated so that the turbine output power is limited.

Fig. 9 shows the grid voltage and current waveforms in (a) and the power to the grid in (b), where the unity power factor operation is observed.

Fig. 10 shows the full DC-link voltage in (a) and each half-part of the DC-link voltage in (b). Due to the ripple compensation effect of the v_{dc1} and v_{dc2}, the ripple components of the full DC-link voltage have been reduced.

VI.   CONCLUSIONS

This paper has proposed a low-cost converter for small-scaled wind power generation systems, which consists of a two-leg three-phase PWM inverter and a half-bridge single-phase AC/DC converter. With this converter, the generator output power is increased a little (20W) compared with the case of the conventional converter since the vector-controlled PMSG gives a high dynamic performance according to the wind speed variation. The validity of the proposed converter system has been verified by simulation results using Psim software. Furthermore, advanced control strategies such as loss minimization or wind speed estimation can be applied to the PMSG system with the proposed converter.

APPENDIX

The parameters of the 3[kW] PMSG and turbine blade used for the simulation are listed in Table II, III.

Table II Parameters of PMSG

Parameters	Value
Rated power	3[kW]
Rated line voltage	220[V]
Stator resistance	0.49[Ω]
inductance	5.35[mH]
Number of poles	6
Rated frequency	60[Hz]

Table III Parameters of blades

Parameters	Value
Blade radius	1.26 [m]
Max. power conv. coeff.	0.45
Optimal tip-speed ratio	7
Cut-in speed	3 [m/s]
Rated wind speed	14 [m/s]
Gear ratio	1
Moment of inertia	33 [kg.m^2]

ACKNOWLEDGMENTS

This work has been supported by EIRC(I-2007-0-261-01), which is funded by MOCIE (Ministry of commerce, industry and energy) and ETEP (Electric Power Industry Technology Evaluation & Planning).

REFERENCES

[1] American public power association. Utility wind integration state of art. Annual Report, Available:http://www.uwig.org

[2] C. A. Hernandez-Aramburo, T. C. Green. and N. Mugniot, "Fuel consumption minimization of a microgrid," *IEEE Trans on. Industry Applications*, vol. 41, issue. 3, pp. 673-681, May-June. 2005.

[3] R. Datta, and V. T. Ranganathan, "A method of tracking the peak power points for a variable speed wind energy conversion system," *IEEE Trans on. Energy Conversion*, vol. 18, no. 1, pp.163-168, March. 2003.

[4] A. Abo-Khalil, D.-C. Lee, and S. -H. Lee, "Synchronization of DFIG output voltage to utility grid in wind power system," *IASTED conf proc.* in Rhodes, June. 2006, pp.372-377

[5] A. Mirecki, X. Roboam, F. Richardeau, "Architecture complexity and energy efficiency of small wind turbines," *IEEE Trans on. Industrial Electronics*, vol. 54, no. 1, pp. 660-670, Feb. 2007.

[6] S. Jiao, G. Hunter, V. Ramsden, and D. Patterson, "Control system design for a 20kW wind turbine generator with a boost converter and battery bank load," *IEEE PESC proc.*, vol. 4, 2001, pp.2203-2206.

[7] Y. Higuchi, N. Yamamura, M. Ishida, and T. Hori, "An improvement of performance for small-scaled wind power generating system with permanent magnet type synchronous generator," *IEEE IECON proc.*, vol. 2, 2000, pp.1037-1043.

[8] S.-H. Song, S.–I. Kang, and N.–K. Hahm, "Implementation and control of grid connected AC-DC-AC power converter for variable speed wind energy conversion system," *IEEE APEC proc.*, vol. 1, Feb.2003, pp.154-158

[9] A. Miller, E. Muljadi, and D. S. Zinger, "A variable speed wind turbine power control," *IEEE Trans on. Energy Conv.*, vol. 12, no. 2, pp. 181-186, 1997.

[10] A. B. Raju, K. Chatteriee, and B. G. Fernandes, "A simple maximum power point tracker for grid connected variable speed wind energy conversion system with reduced switch count power converters," *IEEE PESC proc.,* vol. 2, 2003 , pp. 748-753.

[11] D. -C. Lee, and Y. –S. Kim, "Control of single-phase-to-three-phase AC/DC/AC PWM converters for induction motor drives," *IEEE Trans. on IE*, vol. 54, no. 2, pp. 797-804, 2007.

[12] P. Enjeti, and A. Rahman, "A new single phase to three phase converter with active input current shaping for low cost ac motor drives," *IEEE Trans on. IA*, vol. 29, no. 4, pp. 806-813, 1993.

[13] M. B. de Rossiter, C. B. Jacobina, E. R. C. da Silva, and A. M. N. Lima, "A general PWM strategy for four-switch three-phase inverters," *IEEE Trans on. Power Electronics*, vol. 21, no. 6, pp.1618-1627, Nov. 2006.

[14] M. Chinchilla, S. Arnaltes, and J. C. Burgos, "Control of permanent - magnet generators applied to variable-speed wind-energy systems connected to the grid," *IEEE Trans on. Energy Conversion*, vol. 21, no. 1, pp. 130-135, March. 2006.

Study on voltage regulation method in the power distribution system

Kenichiro Yamaguchi, Kyungsoo Lee, and Kosuke Kurokawa
Tokyo University and agreculture and Technology
Department of Electrical and Electronic Engineering
2-24-16, Naka-cho, Koganei, Tokyo, 184-8588 Japan
Email: 50007645129@st.tuat.ac.jp, onnuri@cc.tuat.ac.jp, kurochan@cc.tuat.ac.jp

Abstract— This paper focuses on distribution voltage control when over-voltage or under-voltage occurs in the low-voltage (LV) distribution line. The distribution voltage control is performed by transformer and ac-ac converter. Transformer supports a part of settled voltage and ac-ac converter controls voltage disturbance. Bi-directional ac-ac converter employs digital control using Field Programmable Gate Array (FPGA). This method enables simple and fast the distribution voltage control. In this paper, voltage control concept is explained. Scaled-down experimental result is shown, when load condition is changed.

I. INTRODUCTION

In recent years, photovoltaic (PV) systems for grid connection have rapidly spread to solve energy and environmental problem. However, when the clustered PV systems connect with the power distribution system and reverse power flows, there is a possibility that the distribution line voltage increase and exceed the allowable voltage limit. This condition is called over-voltage.

In the present distribution power systems, autotransformer with line drop compensator based step voltage regulator (SVR) have been used. Also, the scheduled operation have controled distribution line voltage in the substation. However, these method are not designed for over-voltage by reverse power flow. Also, the power conditioner of PV is equipped over-voltage relay (OVR). Thus, the electric power from PV systems can not be effectively used.

Therefore, some voltage regulators used power electronics technique have been proposed [1]. Distributed-Unified Power Flow Controller (D-UPFC) is a voltage regulator and able to regulate bi-directional power flow in the distribution systems [2]. D-UPFC consists of transformer and bi-directional ac-ac converter. it can convert directly ac to ac and keep the distribution line voltage to reference voltage. This method enables simple and fast distribution line voltage control.

In this paper, voltage control concept is explained. Scaled-down experimental results are shown, when load condition is changed.

Distribution power system connected clustered PV systems is shown in Fig. 1. When the clustered PV systems connect with distribution system and reverse power flow, the distribution line voltage increase. In this condition, SVR can not control distribution line voltage level. Distribution power system controlled D-UPFC is shown in Fig. 2. D-UPFC controls the voltage to be fixed to reference voltage in allowable voltage window [3].

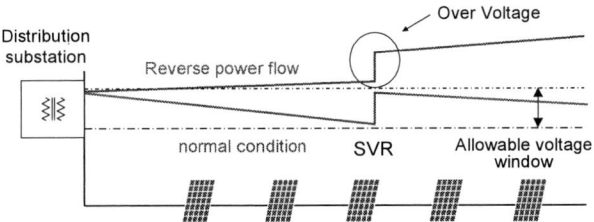

Fig. 1: Distribution power system connected clustered PV systems

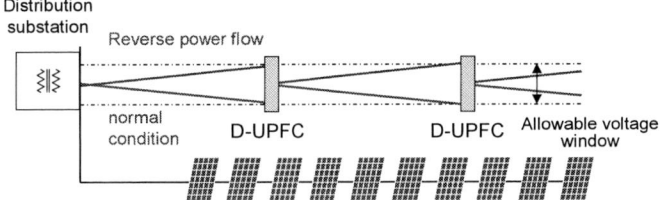

Fig. 2: Distribution power system controlled D-UPFC

II. VOLTAGE CONTROL CONCEPT

A. Voltage control circuit (D-UPFC)

Fig. 3 shows the voltage control circuit. V_s is distribution system side voltage and V_{out} is clustered PV systems side voltage. This circuit is installed between distribution the pole transformer primary side and clustered PV systems.

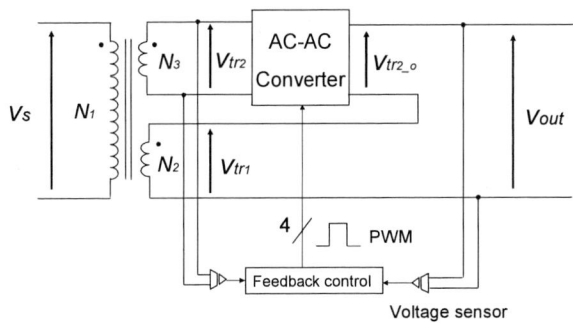

Fig. 3: Voltage control circuit of D-UPFC

Distribution voltage control is performed by transformer and bi-directional ac-ac converter. N_1, N_2, and N_3 are turns

978-1-4244-1871-8/07 $25.00 © 2007 IEEE

ratio of transformer. Normally N_2 is smaller than N_1 and N_3 is smaller than N_2. V_{tr1} and V_{tr2} are secondary voltage of transformer. Thus, V_{tr1} and V_{tr2} can be expressed,

$$V_{tr1} = \frac{N_2}{N_1} V_s \qquad (1)$$

$$V_{tr2} = \frac{N_3}{N_1} V_s \qquad (2)$$

V_{tr2_o} is output voltage of bi-directional ac-ac converter. V_{tr2_o} is controlled by pulse width modulation (PWM) signals. PWM signal is made by the feedback control block using Field Programmable Gate Array (FPGA). V_{tr2_o} can be expressed,

$$V_{tr2\_o} = D \times V_{tr2} \qquad (3)$$

Where, D is the duty cycle of PWM signal. D-UPFC output voltage V_{out} is decided by V_{tr1} and V_{tr2_o}. V_{out} can be expressed,

$$V_{out} = V_{tr1} + V_{tr2\_o} \qquad (4)$$

Therefore, V_{out} can be rewritten as follows

$$V_{out} = \frac{N_2 + (D \times N_3)}{N_1} V_s \qquad (5)$$

Thus, transformer supports a part of settled voltage and bi-directional ac-ac converter controls voltage disturbance. If voltage decrease happens in the distribution line, ac-ac converter duty ratio increases. Reversely, when voltage increase occurs in the distribution line, ac-ac converter duty ratio decreases. D-UPFC output voltage V_{out} is kept reference voltage V_{ref}. Vector diagram of D-UPFC voltage control is shown in Fig. 4.

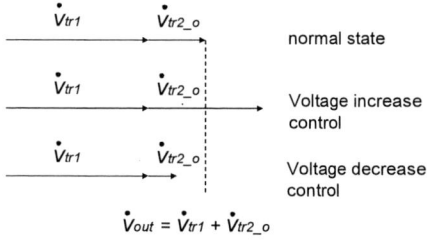

Fig. 4: Vector diagram of D-UPFC voltage control

B. Bi-directional converter

Fig. 5 shows bi-directional ac-ac converter circuit. This converter consists of four MOSFET switches (S_1, S_2, S_3, S_4), input and output filter (L_{in}, C_{int}, L_{out}, C_{out}). This converter is controlled by switching duty as a dc-dc buck converter. So, output voltage is always less than input voltage.

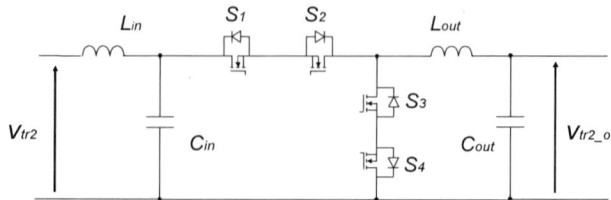

Fig. 5: Bi-directional ac-ac converter circuit

Table 1 shows the switching pattern of ac-ac converter. This switching pattern concept is proposed in referenced paper [4, 5]. Switching pattern is decided by polarity of input voltage V_{tr2}. When V_{tr2} is positive, S_1 and S_3 act PWM signal with 20[kHz]. At the same time, S_2 and S_4 turn on. Reversely, when V_{tr2} is negative, S_2 and S_4 act PWM signal with 20[kHz]. At the same time, S_1 and S_3 turn on all the while. AC-AC converter switching waveform is shown in Fig. 6.

TABLE 1 Switching pattern of the ac-ac converter

$V_{tr2} > 0$	S_1:PWM S_3:$\overline{S_1}$
	S_2:ON $\quad S_4$:ON
$V_{tr2} < 0$	S_1:ON $\quad S_3$:ON
	S_2:PWM S_4:$\overline{S_2}$

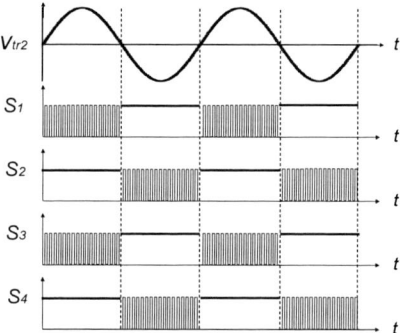

Fig. 6: Switching waveform of ac-ac converter

When the PWM signal changes, there are three operating modes in ac-ac converter [4]. These modes are called active mode, dead time mode, and freewheeling mode. Dead time mode prevents high-voltage spike in the circuit and offers safe commutation. Therefore, the protection circuit such as snubber circuit is not necessary in this converter. Moreover, these modes enable ac-ac converter to control output voltage properly when output current phase is different from output voltage phase by the four quadrant load such as inductive load or capacitive load. So, this ac-ac converter can control the bi-directional power flow in the distribution system. Fig. 7 shows switching waveform of PWM when ac-ac converter input voltage V_{tr2} is positive.

978-1-4244-1871-8/07 $25.00 © 2007 IEEE   488

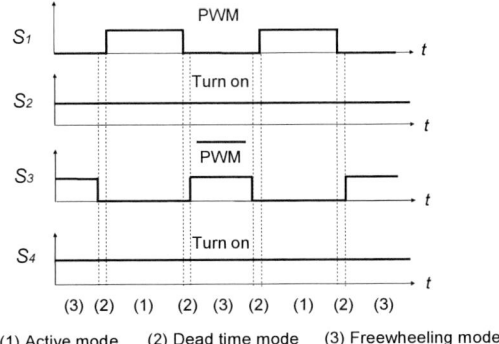

Fig. 7: Switching waveform of PWM signal when ac-ac converter input voltage V_{tr2} is positive

C. Feedback control block

D-UPFC employs digital control by field programmable gate array (FPGA). FPGA is integrated circuit (IC) which can rewire an internal circuit by very high speed integrated circuit hardware description language (VHDL). While D-UPFC is controlled, output voltage V_{out} is always kept reference voltage V_{ref}. V_{ref} is decided by distribution line voltage condition.

Fig. 8 shows feed back control block of D-UPFC. V_{out} is detected by voltage sensor. V_{out} changes DC voltage from AC voltage using rms. V_{out_dc} is transformed digital value by analog-digital (A/D) converter and inputted FPGA control block. In the FPGA, error voltage V_{error} is computed and inputted PI compensator. PI compensator gain (K_p, K_i) is depends on D-UPFC control speed and stability. V_{error} is compared triangle waveform V_{tri} with 20[kHz] to make PWM signal V_{pwm}. Finally, V_{pwm} is added dead time and inputted four switches of the ac-ac converter as Table 1.

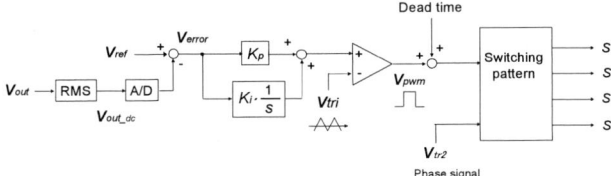

Fig. 8: Feedback control block

III. EXPERIMENTAL DESIGN

Experimental circuits of D-UPFC are shown in Fig. 9. D-UPFC input voltage V_s is supplied 10.0[V,rms]. This voltage is scale down the low-voltage (LV) distribution voltage. And, the transformer turns ratio is

$$N_1 : N_2 : N_3 = 1.0 : 0.9 : 0.2 \tag{6}$$

Thus, D-UPFC output voltage V_{out} changes from 9.0[V,rms] to 11.0[V,rms]. Also, D-UPFC load has three coditions, normal condition, under-voltage condition, and over-voltage condition. Fig. 10 shows two load conditions. In the normal condition, load is only a resistor (R_{load}). This condition tests

D-UPFC fundamental voltage control. In under-voltage condition, shunt resistance (R_{sub}) is connected to R_{load} in parallel by a switch (S_w). This condition simulates that heavy load occurred in the distribution system. R_{sub} is changed in under-voltage condition. Finally, Experimental parameters are shown in table 2.

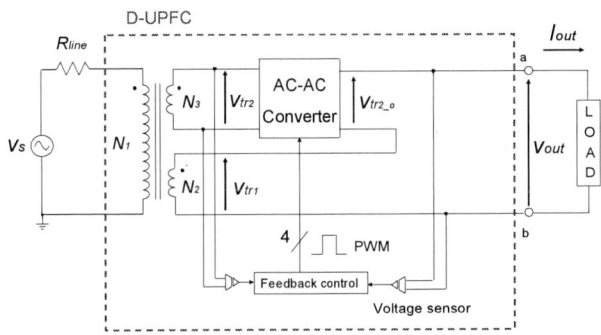

Fig. 9: Experimental circuit of D-UPFC

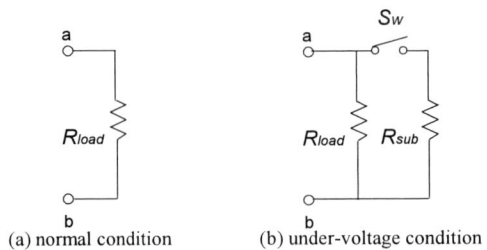

Fig. 10: Experimental conditions of load side.

TABLE 2 Experimental parameters

V_s	10.0[V,rms]
N_1:N_2:N_3	1.0:0.9:0.2
L_{in}, L_{out}	2.5[mH]
C_{in}, C_{out}	1.0[uF]
R_{line}	1[Ω]
R_{load}	1[kΩ]
K_p	4.0
K_i	0.25
Circuit frequency	50[Hz]
Switching frequency	20[kHz]

IV. EXPERIMENTAL RESULTS

A. normal condition

Fig. 11 shows waveform of D-UPFC input voltage V_s, ac-ac converter output voltage V_{tr2_o}, and D-UPFC output voltage V_{out} when reference voltage V_{ref} changed 9.5[V,rms], 10.0[V,rms], and 10.5[V,rms] in FPGA. V_{tr2_o} changed while V_{ref} was changed.

978-1-4244-1871-8/07 $25.00 © 2007 IEEE

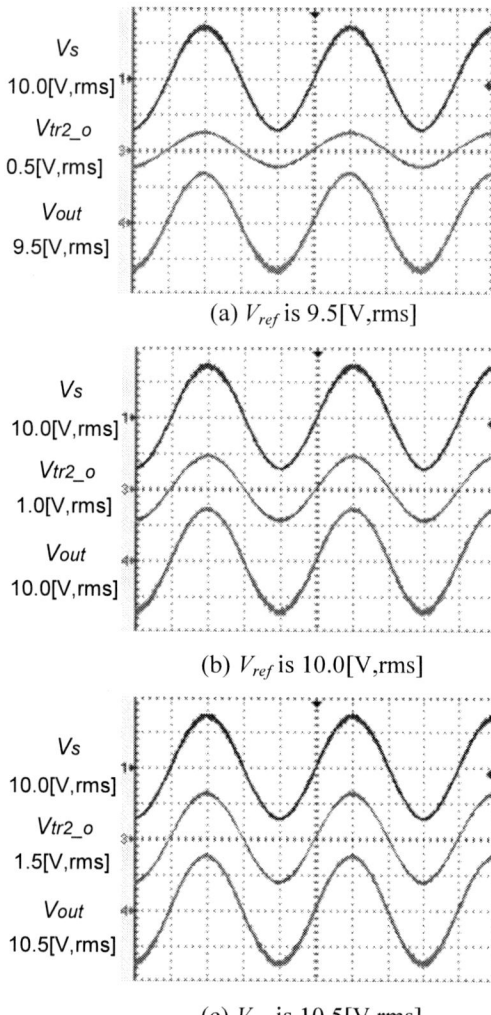

(a) V_{ref} is 9.5[V,rms]

(b) V_{ref} is 10.0[V,rms]

(c) V_{ref} is 10.5[V,rms]

Fig. 11: Waveform of D-UPFC input voltage (V_s), ac-ac converter output voltage (V_{tr2_o}), and D-UPFC output voltage (V_{out}), when the reference voltage (V_{ref}) changed in normal condition. Scales: 10V/div, 10V/div, 2V/div, 10V/div, 10ms/div.

Fig. 12 shows relationship of reference voltage V_{ref} and D-UPFC output voltage V_{out}. V_{out} increases in proportion to V_{ref}.

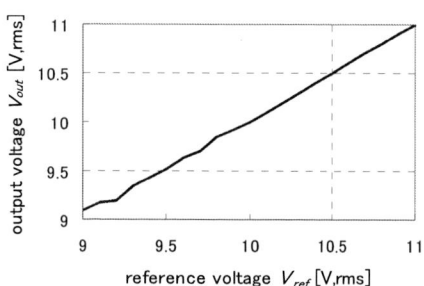

Fig. 12: Relationship of reference voltage (V_{ref}) and D-UPFC output voltage (V_{out}).

Fig. 13 show waveforms of bi-directional ac-ac converter switching pattern. This result is the same switching pattern as table 1. Fig. 14 shows waveform of PWM signal inputted S_1 and S_3, when reference voltage V_{ref} is 10.0[V]. This result shows that duty cycle 0.5 of ac-ac converter. Fig. 15 shows waveform of dead time between S_1 and S_3. In this result, dead time is 250[ns].

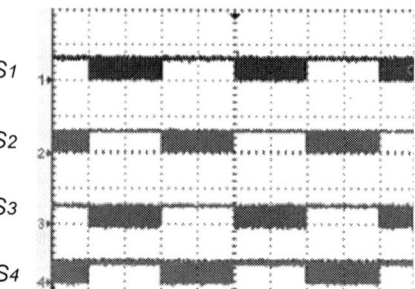

Fig. 13: Switching pattern waveform of switching pattern $S_1, S_2, S_3,$ and S_4, during converter duty cycle 0.5. Scales: 5V/div, 5V/div, 5V/div, 5V/div, 5ms/div

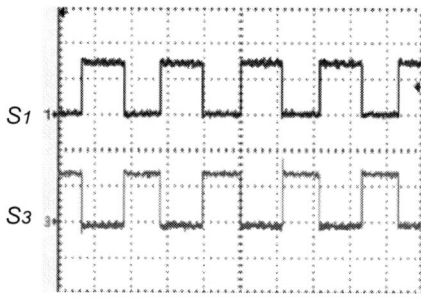

Fig. 14: Switch S_1 and S_3 waveforms during converter duty cycle 0.5. Scales: 2V/div, 2V/div, 25us/div

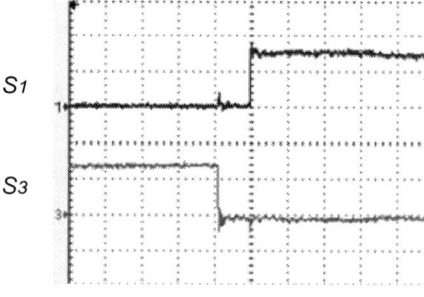

Fig. 15: Dead time waveforms of S_1 and S_3. Scales: 2V/div, 2V/div, 250ns/div.

B. under-voltage condition

Fig. 16 shows relationship of output voltage V_{out} and shunt resistance R_{sub} in under-voltage condition. Additionally, reference voltage V_{ref} was 10.0[V,rms] constant. In under voltage condition, load impedance decreases, at the same time, circuit current increases because R_{sub} is connected to

R_{load} in parallel. When circuit current increases, input voltage drop occurs in the circuit. Therefore, V_{out} decreases due to R_{sub}. When D-UPFC is not connected to the circuit, V_{out} decreases from 10.0[V,rms] to 9.75[V,rms]. However, When D-UPFC is connected to the circuit, V_{out} is constant 10.0[V,rms].

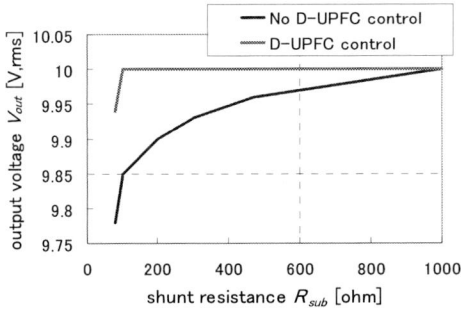

Fig. 16: relationship of reference shunt resistance (R_{sub}) and D-UPFC output voltage (V_{out})

V. CONCLUSION

This paper proposes a distribution voltage regulation method using D-UPFC when over-voltage and under-voltage occurs in the distribution line. Voltage control is performed by transformer and bi-directional ac-ac converter. Experimental results show the D-UPFC fundamental voltage control when load changes from normal condition and under-voltage condition. In the future works, D-UPFC voltage control test is experimented in over-voltage condition.

ACKNOWLEDGMENT

This research has been carried out as a part of "Autonomy-Enhanced PV cluster" project and special thanks financial support of NEDO.

REFERENCES

[1] Eddy C.Aeloiza, Prasad N.Enjeti, Luis A.Moran, Oscar C.Montero-Herandez, and Sangsun Kim, "Analysis and design of a new voltage sag compensator for critical loads in electrical power distribution systems", *IEEE Trans-Industry*, July/August 2003, pp.1143-1150.

[2] Kyungsoo Lee, Kenichiro Yamaguchi, Hirotaka Koizumi, and Kosuke Kurokawa, "D-UPFC as a Voltage Regulator in the Distribution System", Renewable energy 2006, October, 2006, pp.1756-1759.

[3] Kosuke Kuorkawa, "A CONCEPTUAL STUDY ON SOLAR PV CITIES FOR 21ST CENTURY", *WCPEC-4*, Hawaii, May, 2006, pp.2283-2288

[4] B.-H.Kwon, B.-D.Min, and J.-H.Kim, "Novel topologies of AC choppers", *IEE Proc.-Electr. Power Appl., Vol 143, No. 4*, July 1996, pp323-330.

[5] C. A. Petry, J .C. Fagundes, and I. Barbi, "New Direct Ac-Ac Converters Using Switching Modules Solving the Commutation", IEEE ISIE 2006, July 9-12, 2006, pp. 864-869.

High Efficiency Photovoltaic Pump System with Double Pumps

Fu Qing
Sun Yat-sen University
School of Physics & Engineering
Email: fuqing@mail.sysu.edu.cn

Chen Xiaogao, Xie Lei
Sun Yat-sen University
School of Physics & Engineering
Email: chenxiaogao99@163.com

Abstract—The efficiency of Photovoltaic Pump System (PVPS) varies with the solar irradiation and it is important to select the suitable pump to improve the efficiency of PVPS. Pump with higher head works longer but is in lower efficiency when solar irradiation is powerful in noon. The paper proposes a Maximum Power Trace (MPPT) controller for PVPS and a novel scheme with double pumps to improve efficiency of PVPS. The pump with higher head works when solar irradiation is weak while the pump with lower head is automatically switched to work when solar irradiation is powerful. The experiment proves the higher efficiency of PVPS with double pumps.

I. INTRODUCTION

Making use of solar energy, Photovoltaic Pump System (PVPS) take water out from underground, store it in the storage tanks, or supply it directly to people, animals and plants. Many countries attach importance to PVPS because of its great contribution to the drought area and environmental friendly [1-4].

PVPS is consisted of many components such as PV array, DC/AC converter, motor and pump. Some have also DC/DC converter especially for MPPT of solar energy. PVPS is designed according to the climate after a long-time and detailed measure of solar radialization and solar irradiation. The components are carefully designed and frequently maintained so as to keep all components efficiently. The power and voltage of all components are well designed to match with other components so as to keep a high efficiency.

The selection of pump is the key to improve the efficiency of PVPS because pump efficiency varies with solar irradiation and solar irradiation varies ceaselessly in daytime, from weak to powerful and then to weak again. It will take much time and energy to optimize the parameters of pump according to the solar irradiation curves in daytime so as to get more water. Even if the pump is designed very well, it is

hard for only one pump to work efficiently both when solar irradiation is weak and when it is powerful. So the paper proposes a PVPS with double pumps, one for weaker solar irradiation and the other for more powerful solar irradiation. The two pumps are automatically switched on and off according to solar irradiation and the experiment proves the higher efficiency of PVPS with double pumps.

II. EFFICIENCY OF PVPS

A PVPS is mainly consisted of PV array, DC/AC converter, motor and pump etc. To improve the efficiency of PVPS, the components should be well designed to improve their efficiency. Clean the PV array frequently and set a fit decline angle of PV array; Remove the objects that shadow part of PV array; Design suitable DC/AC converter (or DC/DC and DC/AC converters) with MPPT controller [5-6]; Employ suitable pump and so on.

It is much important to design a pump to improve the efficiency of the whole system. The efficiency of pump is composed of hydraulic efficiency, mechanical efficiency and cubage efficiency. Many methods can be used to improve the hydraulic efficiency and mechanical efficiency, such as optimizing impeller, increase rotate speed of pump, decrease the diameter of impeller and decrease the series of pump, etc. but the cubage efficiency can only be improved by reducing the leak of water. When the power of PV array is certain, the power of pump is certain and then it is very important to select pump.

When the power of PV array is certain and then the power of well designed pump is certain, it is important to select the head of pump. Fig.1 is the characteristic of pump (head versus flow), where the head of pump.$1^{\#}$ is higher than that of pump $2^{\#}$ and the rotate speeds of pump $n_1 \curvearrowright n_2 \curvearrowright n_3$ satisfied

$$n_1 < n_2 < n_3 \qquad (1)$$

As is well known in Fig.1, when two pump are running in

the same speed n_1, pump1#, whose head is higher, wasted the power which proportion to the area of rectangle H_0OQ_1E to get over the fact head, while pump 2# wasted the power proportion to the area of rectangle H_0OQ_3F. So more power is needed for pump 2# to get over the fact head. That is to say, when the solar irradiation is weak in morning and dusk, the pump 1# can pump water out of underground while pump 2# can't do because of leak of power. Pump 1# can work longer than pump 2# does in daytime. On the other hand, to get the same flow Q_2, the power proportion to area of rectangle H_3OQ_2C needed for pump 1# while a power proportion to area of rectangle H_1OQ_2B for pump 2#. Obviously the efficiency of PVPS employed pump1# is much lower than that employed pump 2# under this condition. Fig.2 shows the instanteous efficiency of PVPS versus the solar irradiation. When the solar irradiation is lower than a value, called water threshold, the efficiency of PVPS go suddenly to zero and no water go out of underground. There is also a peak value of the efficiency which is the goal of the optimizing design of PVPS. Pump 1# works early but is in lower efficiency when solar irradiation is powerful.

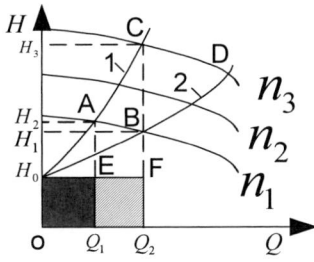

Fig.1 H-Q Characteristic of pump

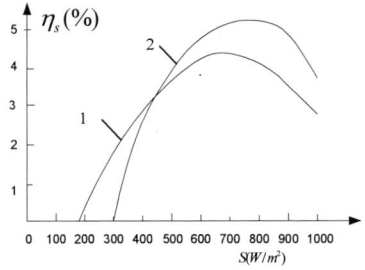

Fig.2 Instantaneous efficiency of PV pump systems

III. PVPS WITH DOUBLE PUMPS

Efficiency of PVPS varies with solar irradiation and different pump has different efficiency. The pump with higher head will work longer a day but it pump less water when the solar irradiation is very powerful at noon while the pump with lower head will work in shorter time but work more efficiently at noon. If we employ the pump with higher head to pump water in morning and dusk when the pump with lower head can not bring water out of underground, and employ the pump with lower head when solar irradiation is powerful to improve the efficiency, more water will be pumped out of underground. This is the principle of PVPS with double pumps.

In PVPS, the most expensive component is the well. It costs about several lakh yuan or more than 1 million yuan, which is more than 90 percents of the whole cost. In order to decrease the cost, PVPS with double pumps share the only one well. There are two schemes for PVPS with double pumps, the one is using two low power pumps with different heads. Pump with higher head works early when solar irradiation is much weak and then pump with lower head starts working when solar irradiation is a little stronger, and two pumps both work when solar irradiation is much strong at noon. In this scheme, the power of two pumps is either lower than the power PV array, so it is cheaper. But when the two pumps work at the same time, pump with higher head will produce a higher water pressure and influence the rest pump which only can product a lower water pressure if in the same condition. So the efficiency of PVPS is much lower when two pumps work together. The second scheme for PVPS with double pumps is employing two pumps whose power is near that of PV array, the one with high head works earlier and the other one works later when solar irradiation is much stronger. It is maybe a little more expensive than adopting the first scheme. But it is more efficient.

Fig.3 shows the configuration of PVPS with double pumps. Two pumps M1 and M2 are employed. The power of M1 is equal to that of M2 and is a little lower than the power of PV array. The head of M1 is higher than M2. A Digital Signal Processor (DSP) is employed to get the value of DC voltage and DC current from PV array through transducers. A Maximum Power Point Trace (MPPT) controller is employed

978-1-4244-1871-8/07 $25.00 © 2007 IEEE   493

to promise PV array convert maximum solar energy to electric energy.

Because the water threshold of M1 is lower then that of M2, switch K1 is normally closed so that M1 starts early in the morning and stops lately in dusk. Switch K2 is normally opened and is closed only when the solar irradiation is so powerful that DC power P_d is more than the reference of DC power P_d^*. When K2 closed, K1 would be opened, M1 stops and M2 starts. At the same time, switch signal is sent to MPPT controller to set the output of MPPT to zero and climb up slowly so as to start M2 slowly. When solar irradiation is weak during afternoon or dusk, DC power is cut down and K1 would close again and M1 starts slowly. K2 opens and M2 stops of course. In order to get maximum flux of PV pump. A hysteresis comparator is employed here to avoid switching between M1 and M2 frequently.

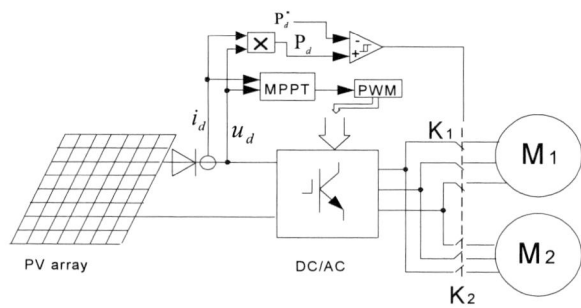

Fig.3 PVPS with double pumps

IV. MPPT CONTROLLER

As the most expensive component in PVPS, the PV array cost more than 80 percents of the cost of the whole system. It is much important to improve the efficiency of PV array so as to decrease the quantity of solar cells and then decrease the cost of whole system. Fig.4 shows the characteristics of PV array, (a) for output current vs. output voltage and (b) for output power vs. output voltage. We can see that output power of solar array vary with output voltage and we can get maximum power in a certain voltage called maximum power point (MPP)[7]. It is one of key methods to improve efficiency of the whole system.

The output voltage of solar array would vary in a range so as to get maximum power in different solar irradiation. Because the characteristics of PV array are not changeable,

the only way to change the output voltage of PV array is to change its load. As we know, a ideal DC/AC converter doesn't cost any energy. It is just like a transformer. So if trans-ratio of "transformer" is changed, the equivalent load of PV array will change. That is to say, we can control the modulate ratio of PWM DC/AC to control the equivalent load of PV array so as to control its output voltage to get maximum power.

Commonly, the MPPT controller is realized in a digital method such as DSP or single-chip computer. Suppose sample frequency is much high and current out power of PV array sampled is P(k), current modulate ratio of PWM inverter (DC/AC) if α (k), and power sampled last time is P(k-1), modulate ratio last time is α (k-1), in order to get maximum power point, the controller should follow the below rules:

① If $\Delta\alpha(k)=\alpha(k)-\alpha(k-1)>0$ and $\Delta P(k)=P(k)-P(k-1)>0$, modulate ratio α should increase to increase output power. That is

$$\Delta\alpha(k+1)=\alpha(k+1)-\alpha(k)>0;\qquad(2)$$

② If $\Delta\alpha(k)>0$ and $\Delta P(k)<0$, modulate ratio α should decrease. That is

$$\Delta\alpha(k+1)=\alpha(k+1)-\alpha(k)<0;\qquad(3)$$

③ If $\Delta\alpha(k)<0$ and $\Delta P(k)>0$, modulate ratio α should decrease. That is

$$\Delta\alpha(k+1)=\alpha(k+1)-\alpha(k)<0;\qquad(4)$$

④ If $\Delta\alpha(k)<0$ and $\Delta P(k)<0$, modulate ratio α should increase. That is

$$\Delta\alpha(k+1)=\alpha(k+1)-\alpha(k)>0;\qquad(5)$$

⑤ At the same $\Delta\alpha(k)$, if $\Delta P(k)$ is greater, that is to say the current output voltage is farther away from the maximum power point, increment of modulate ratio α should be greater. Otherwise increment of modulate ratio should be smaller.

On the other hand, a integration is added to the controller so as to improve resistant ability to the interferences. Fig.5 shows the structure of MPPT controller, where the discrete PI controller is expressed as:

$$\Delta\alpha(k+1)=K_p\left[\frac{\Delta P(k)}{\Delta\alpha(k)}-2\frac{\Delta P(k-1)}{\Delta\alpha(k-1)}+\frac{\Delta P(k-2)}{\Delta\alpha(k-2)}\right]+K_i\left[\frac{\Delta P(k)}{\Delta\alpha(k)}-\frac{\Delta P(k-1)}{\Delta\alpha(k-1)}\right]\;(6)$$

Obviously, the controller shown in Fig.5 follows all the five rules listed before. But there are two exceptional situations which have to be considered, which is $\Delta\alpha(k)=0$ or $\Delta P(k)=0$. We should restart the controller when any of them happened. At this situation, we can add a interference to modulate ratio α intentionally. The "interference" is:

$$\Delta\alpha(k+1) = sign[\Delta P(k-1)]\cdot\Delta\alpha(k-1) \qquad (7)$$

Where $sign(\bullet)$ stand for signal function, expressed as:

$$sign(x) = \begin{cases} 1 & x > 0 \\ 0 & x = 0 \\ -1 & x < 0 \end{cases} \qquad (8)$$

The proposed MPPT controller is shown in Fig.6.

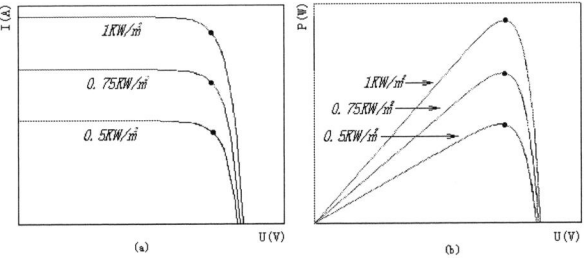

Fig.4 Characteristics of PV array

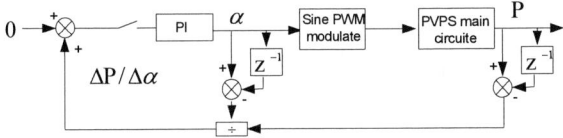

Fig.5 Prime MPPT controller

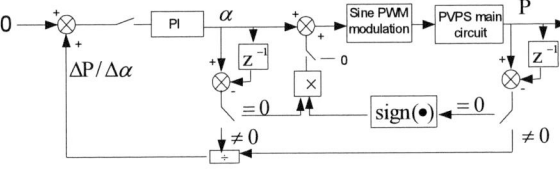

Fig.6 Proposed MPPT controller

V. EXPERIMENTAL RESULT

According to the demand of water, three PVPS are employed and each has 1 KVA PV array which is composed of 7 PV cells. The power of each PV cell is 170 Wp. Four pumps whose power are all 0.75KW are employed. One system employs the pump R95-DF-04. The second employs the pump R95-VC-09 and the third system employs pump R95-DF-04 and pump R95-VC-09 - to construct PVPS with double pumps. The characteristics of pump R95-VC-09 and pump R95-DF-04 is listed in Tab.1. All systems are working

in the similar condition except for employing different pumps. It is 17m from the water level underground to the mouth of well and the water level varies little.

The measured data of PVPS employing pump R95-VC-09 in daytime is shown in Fig.7, and the measured data of PVPS employing pump R95-DF-04is shown in Fig.8, where Idc(t) stands for the DC current out from PV array. Q(t) is the flow of PVPS and η_s^* is product of efficiency of converter, motor and pump. From the data, we know that pump R95-VC-09, with a higher head, work harder than pump R95-DF-04. It works from 8:30 to later than 15:45 while pump R95-DF-04 works in a shorter time - from 10:00 to 14:30. But pump R95-DF-04 works more powerful than pump R95-VC-09 does when the solar irradiation is powerful – from 11:00 to 13:00. So if we let a pump R95-DF-04 work from 11:00 to 13:00 and let a pump R95-VC-09 work during the other time, just like the third PVPS does, we would get much more benefit. The measured data of PVPS with double pumps is shown in Fig.9. The efficiency of PVPS with double pumps is always higher than the other two systems. The average efficiencies in one day of the three PVPS are compared in Tab.2, where Ein stands for the energy output from PV array in daytime and Eout stands for the energy output form pump. The average of efficiency of PVPS with double pumps is much higher than that of the other two systems. The experimental result proves the advantage of PVPS with double pumps.

TABLE I

H-Q CHARACTERISTIC OF TWO PUMPS

	R95-VC-09			R95-DF-04		
H(m)	31	42	53	12	17	22
Q(m³/ h)	5	4	3	10	8	6

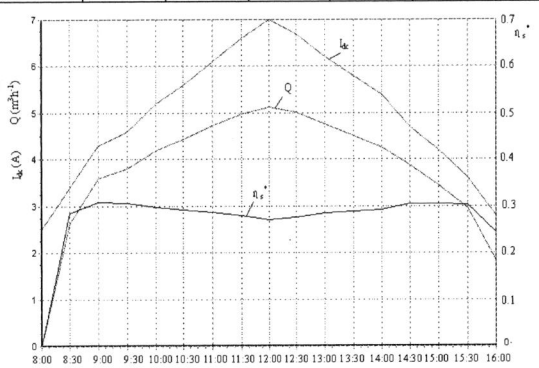

Fig.7 Measure data of PVPS employing R95-VC-09

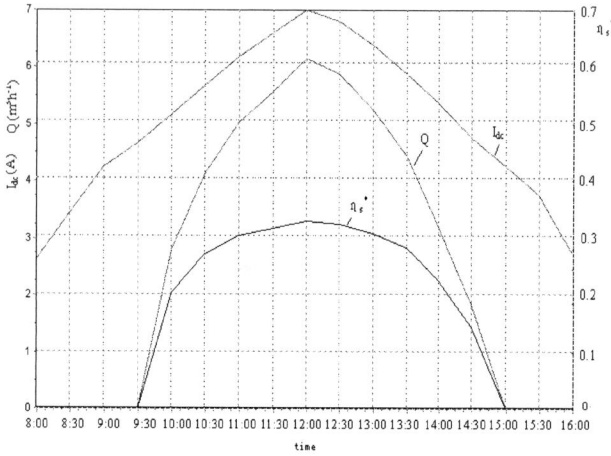

Fig.8 Measure data of PVPS employing R95-DF-04

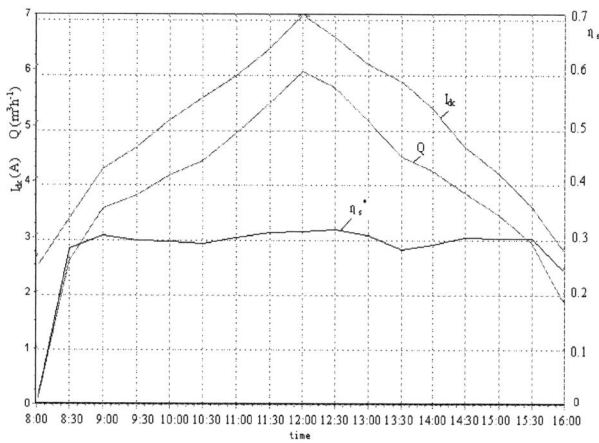

Fig.9 Measure data of PVPS with double pumps

TABLE II

EFFICIENCY OF PVPS

	Ein (Kwh)	Eout (Kwh)	η s* average	Qtotal (m³/day)
R95-VC-09	4944	1411	28.5%	29.2
R95-DF-04	3798	1012	26.6%	22.8
Double pump system	4943	1700	34.4%	31.9

VI. CONCLUSION

The efficiency of PVPS varies with the solar irradiation and it is important to select the suitable pump to improve the efficiency of PVPS. Pump with higher head works longer but

is in lower efficiency when solar irradiation is powerful in noon. The paper proposes a MPPT method and a scheme for PVPS with double pumps so as to improve the efficiency of PVPS.

REFERENCE

[1] Yu Shijie, He Huiruo, Shen Yuliang, Su Jianhui, Zhao Wei. A combine of economic and environmental benefits-photovoltaic water pumping system. JOURNAL OF HEFEI UNIVERSITY OF TECHNOLOGY(NATURAL SCIENCE, Vol.23 No.2, 2000.4: 154-159.

[2] T. S. Surendra, P. N. Prakash, J. Shrikanth. THE PV-POWERED DC FLOATING PUMP SYSTEM - A NEW REVOLUTION. the Twenty-Eighth IEEE Conference Record of Photovoltaic Specialists Conference, 15-22 Sept. 2000:1611 – 1613.

[3] Adnene, C., Moncef, J. Modelling and simulation of a PV-inverter-asychronous motor association in photovoltaic pumping systems. LESCOPE '01. Large Engineering Systems Conference on Power Engineering, 2001, 11-13 July 2001:146 – 151.

[4] Ju Hongxin; Yu Shijie; Su Jianhui; Shen Yuliang; Zhao Yan. THE PWM INVERTER BASED ON SPACE VECTOR FOR PV WATER PUMPING SYSTEM. Acta Energiae Solaris Sinica. Vol.26 No.2, 2005.4: 157-161.

[5] Chen Kunlun, Zhao Zhengming, Yuan Liqiang. Implementation of a stand-alone Photovoltaic pumping system with maximum power point tracking. Proceedings of the Fifth International Conference on Electrical Machines and Systems, 2001. ICEMS 2001. Vol.1, 18-20 Aug. 2001:612 – 615.

[6] Zheng Shicheng; Su Jianhui; Shen Yuliang; Yu Shijie. Design of digital photovoltaic pumping system with the function of true maximum power point tracking. Transactions of The Chinese Society of Agricultural Engineering, Vol.20 No.5 2004.9: 270-274

[7] Hamrouni N., Jraidi M., Cherif A., Dhouib, A.. Measurements and Simulation of a PV Pumping Systems Parameters Using MPPT and PWM Control Strategies. IEEE Mediterranean Electrotechnical Conference, 2006. MELECON 2006, 16-19 May 2006: 885 – 888.

A Variable Structure Approach to control the Active and Reactive Power for Doubly Fed Induction Generator

WON-SANG KIM, KYO-BEUM LEE*, BYOUNG-CHANG JEONG**, AND SEUNG-HO SONG***

*School of Electrical and Computer
Engineering, Ajou University, Korea
Email: kyl@ajou.ac.kr

** LS Industrial Systems Co., Ltd.,
Korea

*** Dept. of Electrical Eng.,
Kwangwoon University, Korea

Abstract—The original direct power control (DPC) is known to give a fast response under transient conditions. However, active power, reactive power and current pulsations occur in steady-state operation. A family of variable structure controllers for the doubly fed induction generator (DFIG)-based wind turbine system is presented, using the principles of an active and reactive power controller known as modified DPC, and where variable structure control (VSC) and space-vector modulation (SVM) are combined to ensure high-performance operation. VSC scheme is designed following the modified DPC philosophy, which provides robust, fast, accurate active and reactive power controls without the problem of high chattering. Simulation results demonstrate that the proposed method preserves the effectiveness and robustness during variations of active and reactive power, rotor speed and converter DC-link voltage.

I. INTRODUCTION

A DFIG system has been adopted in a large number of modern wind turbines because the topology has many advantages in terms of variable-speed operation, relatively high efficiency, four-quadrant active and reactive power capabilities and small converter size. One of the generator systems currently commercially available in the wind energy market is the DFIG with the stator winding connected directly to the grid and the sinusoidal rotor currents supplied from two back-to-back converters via slip rings. Converters are composed of a rotor-side converter and a grid side converter (Fig. 1).

The control system acting on power converters as compared with conventional constant speed systems makes possible an improved dynamic behavior of the wind generator, resulting in the reduction of the mechanical stress and electrical power fluctuations, and the increase of energy capture [1]. Control strategies and performance evaluation of DFIG have been discussed for operation in accordance with generator slip. For instance, the rotor active power proportional to generator slip and the stator active power flow from the stator to the grid for generator slip below zero and in the opposite direction for generator slip above zero. Decoupled control of active and reactive powers on wind energy applications is used in a stator flux-oriented control [2]. In this configuration, two PWM converts with a DC link capacitor are used. The grid-side converter is to provide the DC link voltage for a set of capacitor values and is responsible for controlling reactive power flow into the grid.

In these references, the grid-side converter is controlled using field-oriented control (FOC) of rotor current vector using a synchronous reference frame aligned with the supply voltage vector. The rotor-side converter is also controlled using the FOC of the rotor current on the stator flux reference frame that ensures decoupling control of stator-side active and reactive power drawn from the grid [3]. One main drawback is that its performance depends highly on accurate machine parameters such as stator resistance, rotor resistance, and inductances being known. Thus, performance degrades when actual machine parameters depart from values used in control system.

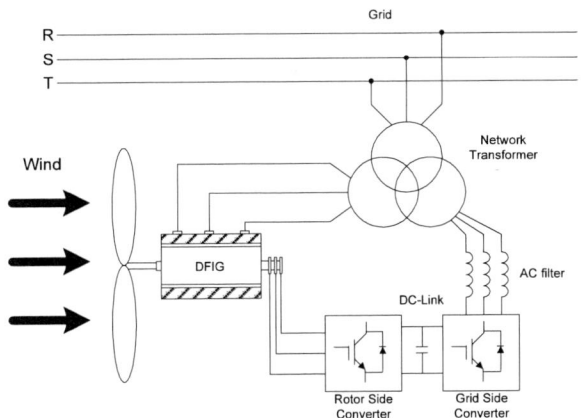

Fig. 1: Wind turbine system with doubly-fed induction generator

The DPC method directly controls stator-side active and reactive powers by selecting voltage vectors from a look-up-table using the stator active and reactive power information [4]. Direct power control (DPC) of DFIG is able to produce fast active and reactive power control with the hysteresis band and is robust with respect to the change of machine parameters and to perturbations. However, it has some drawbacks: it exhibits high active and reactive power ripples, and current ripple, produces annoying acoustical noise, operates with nonzero steady-state error, and the switching frequency is variable and lower than the sampling frequency.

Using the SVM, instead of the DPC switching logic, provides higher control resolution and helps improving the drive's behavior. Drives with SVM have excellent

978-1-4244-1871-8/07 $25.00 © 2007 IEEE

performance in terms of low active and reactive power ripples.

Linear proportional-integral (PI) active and reactive power controls using SVM is investigated in [5], in which smooth operation was obtained in the steady state, but the robustness is low due to the linear control. The variable-structure control (VSC), known also as sliding-mode control (SMC) is a discontinuous control approach, well suited for nonlinear dynamic systems with uncertainties. The control is very robust and fast, but the controlled quantities exhibit undesirable chattering [6]. Recognizing the VSC merits of using the SVM, this paper presents a VSC-DPC solution for the drives of DFIG based on wind turbine system. The modified DPC is achieved by means of VSC.

II. Principle of DPC for DFIG

A. Modeling of DFIG

The DFIG is a wound rotor induction generator, in which the rotor circuit is connected to the grid through two back-to-back converters and a DC link capacitor. The ability to subtract and supply power from the rotor makes it possible to operate the DFIG at subsynchronous or supersynchronous speed while keeping constant voltage and frequency on the stator [7]. The d-q model in the arbitrary reference frame is expressed as follows.

$$V_s = R_s I_s + \frac{d\lambda_s}{dt} + j\omega_r \lambda_s \tag{1}$$

$$V_r = R_r I_r + \frac{d\lambda_r}{dt} + j(\omega - \omega_r)\lambda_r \tag{2}$$

$$\begin{aligned}
&\lambda_s = \lambda_{ds} + j\lambda_{qs}, \quad \lambda_r = \lambda_{dr} + j\lambda_{qr} \\
&\lambda_{ds} = L_s I_{ds} + L_m I_{dr}, \quad \lambda_{qs} = L_s I_{qs} + L_m I_{qr} \\
&\lambda_{dr} = L_r I_{dr} + L_m I_{ds}, \quad \lambda_{qr} = L_r I_{qr} + L_m I_{qs} \\
&L_s = L_{ls} + L_m \\
&L_r = L_{lr} + L_m
\end{aligned} \tag{3}$$

where $I_{ds}, I_{qs}, I_{dr}, I_{qr}$ and $\lambda_{ds}, \lambda_{qs}, \lambda_{dr}, \lambda_{qr}$ are currents and flux linkages of the stator and rotor in d- and q-axis, R_s and R_r are the resistances of the stator and rotor windings, ω_r is the rotor speed.

An equivalent circuit is set up by means of the voltage and flux equations of the arbitrary reference frame, as shown in Fig. 2.

B. DPC Strategy for DFIG

In general four types of reference frames are widely used: stator fixed, rotor fixed, flux vector or synchronous rotating reference frame. In the proposed control strategy, the d-axis of the synchronous frame is fixed to the stator flux, as shown in Fig. 3. Since the stator is directly connected to the grid, and the influence of the stator resistance can be neglected, the stator flux can be considered constant. From (1), for a synchronous frame ($\omega_e = \omega_s$ - the stator flux speed, $\lambda_s = \lambda_{ds}$), the stator voltage vector is given as

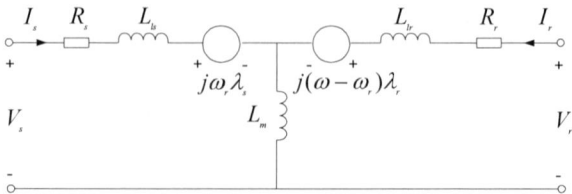

Fig. 2: An equivalent circuit in the arbitrary reference frame

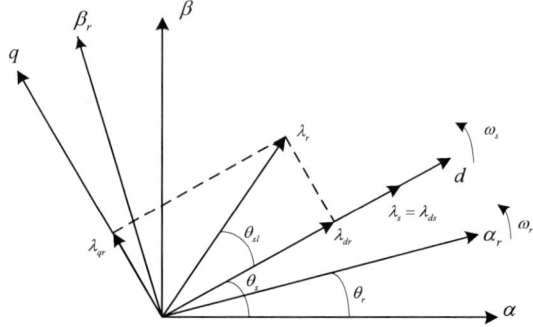

Fig. 3: Stator and rotor flux vectors in the synchronous d-q frame

$$V_s^\omega = j\omega_s \lambda_s = j\omega_s \lambda_{ds}. \tag{4}$$

Based on (3), the stator current is given as

$$I_s^\omega = \frac{L_r \lambda_s^\omega - L_m \lambda_r^\omega}{L_s L_r - L_m} = \frac{\lambda_s^\omega}{\sigma L_s} - \frac{L_m \lambda_r^\omega}{\sigma L_s L_r} \tag{5}$$

where $\sigma = (L_s L_r - L_m^2)/L_s L_r$ is the leakage factor.

The stator active and reactive power inputs from the network can be calculated as

$$\begin{aligned}
&S = P_s + jQ_s = \frac{3}{2}V_s^\omega \times \hat{I}_s^\omega \\
&P_s = -k_\sigma \omega_s \lambda_{ds} \lambda_{qr} \\
&Q_s = k_\sigma \omega_s \lambda_{ds} (\frac{L_r}{L_m}\lambda_{ds} - \lambda_{dr})
\end{aligned} \tag{6}$$

where $k_\sigma = 1.5 L_m /(\sigma L_s L_r)$. As the stator flux stays constant, according to (6), the active and reactive power changes over a constant period of T_s are given by

$$\begin{aligned}
&\Delta P_s = -k_\sigma \omega_s \lambda_{ds} \Delta\lambda_{qr} \\
&\Delta Q_s = -k_\sigma \omega_s \lambda_{ds} \Delta\lambda_{dr}.
\end{aligned} \tag{7}$$

As shown in Fig. 3, in the synchronous d-q reference frame, the rotor voltage is given by

$$V_r^\omega = R_r I_r^\omega + \frac{d\lambda_r^\omega}{dt} + j\omega_{sl}\lambda_r^\omega \tag{8}$$

where $\omega_{sl} = \omega_s - \omega_r$ is the slip frequency.

978-1-4244-1871-8/07 $25.00 © 2007 IEEE   498

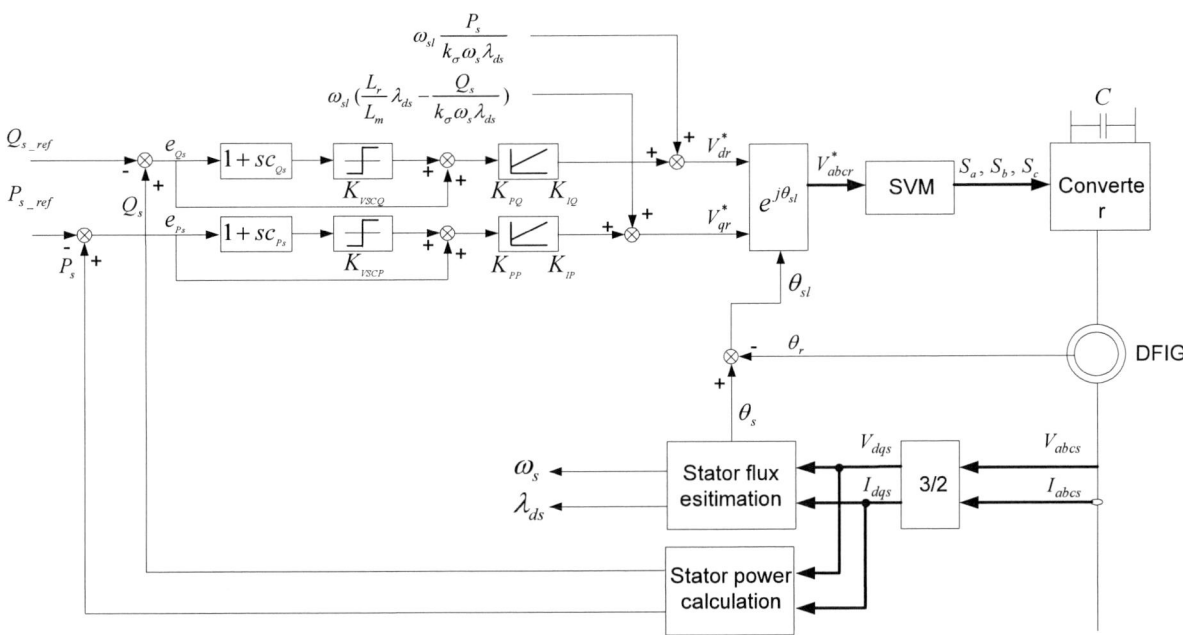

Fig. 4: Proposed VSC-DPC for a DFIG system

Combining (6), (7) and (8), neglecting the rotor resistance, within the T_s period, the rotor voltage required to eliminate the power errors in the d-q reference frame is calculated as

$$V_{dr} = -\frac{1}{T_s}\frac{\Delta Q_s}{k_\sigma \omega_s \lambda_{ds}} + \omega_{sl}\frac{P_s}{k_\sigma \omega_s \lambda_{ds}}$$

$$V_{qr} = -\frac{1}{T_s}\frac{\Delta P_s}{k_\sigma \omega_s \lambda_{ds}} + \omega_{sl}(\frac{L_r}{L_m}\lambda_{ds} - \frac{Q_s}{k_\sigma \omega_s \lambda_{ds}}).$$

(9)

The first terms on the right-hand side reduce power errors while the second terms compensate the rotor slip that causes the different rotating speeds of the stator and rotor flux. As can be see, calculations require only simple multiplications and no complicated mathematics.

III. PROPOSED VSC-DPC

The block diagram of the VSC-DPC for DFIG is shown in Fig. 4. It is named "linear and variable structure control" (LVSC) and consists of a sliding-mode controller that operates in parallel with a linear one [8]. This controller is a generalized and linear control, smooth operation, and of VSC, robustness to perturbations and modeling uncertainties. The sliding surface $S = S_{Qs} + jS_{Ps}$ is selected so as to impose sliding-mode operation with first-order dynamics

$$S = e_{Qs} + c_{Qs} \cdot s e_{Qs} + j(e_{Ps} + c_{Ps} \cdot s e_{Ps})$$ (10)

where $s = d/dt$, $e_{Qs} = Q_S^* - Q_S$ and $e_{Ps} = P_S^* - P_S$ are active and reactive errors. VSC design constants c_{Qs} and c_{Ps} are selected so as to impose the desired linear first-order behavior of the system during sliding modes, i.e., when $S = 0$.

The controller (11) produces the reference voltage vector $V_r^* = V_{dr}^* + jV_{qr}^*$ in the stator reference frame.

$$V_{dr}^* = (K_{PQ} + K_{IQ}/s)(e_{Qs} + K_{VSCQ}\,\mathrm{sgn}(S_{Qs})) + \omega_{sl}\frac{P_s}{k_\sigma \omega_s \lambda_{ds}}$$

$$V_{qr}^* = (K_{PP} + K_{IP}/s)(e_{Ps} + K_{VSCP}\,\mathrm{sgn}(S_{Ps})) + \omega_{sl}(\frac{L_r}{L_m}\lambda_{ds} - \frac{Q_s}{k_\sigma \omega_s \lambda_{ds}})$$

(11)

where K_{PQ}, K_{IQ}, K_{PP}, K_{IP} are the PI controller gains and K_{VSCQ}, K_{VSCP} are the VSC gains. An SVM block that generates the converter switching signals S_a, S_b, S_c is the output stage. The LVSC employs a switching component and a linear one. During a transient, $e_x > K_{VSCx}\,\mathrm{sgn}(S_x)$, where x is P_s or Q_s, and the linear (PI specific) behaviour is dominant. In the steady state, errors are very small and the switching (VSC specific) behavior prevails. Adequate balance between linear and switching behaviour is easily achieved by proper gain selection. PI gains are selected so that the linear control provides the desired dynamic response, while the VSC gains determine the robustness in steady-state operation.

The VSC design procedure [9] requires the gain to be large enough in order to compensate for modeling uncertainties and perturbations.

IV. SIMULATION RESULTS

Simulation of the proposed control strategy for a DFIG-based generation is carried out using Simulink/Matlab.

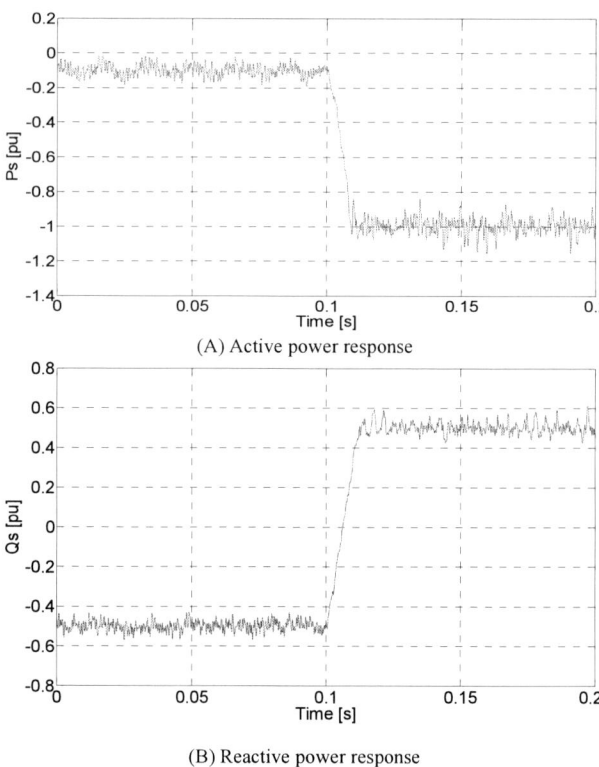

(A) Active power response

(B) Reactive power response

Fig. 5: Simulation results of active power responses (A) and reactive power response (B)

The VSC-DPC strategy is simulated for a 9MW 4poles DFIG drive at 100 μs sampling time.

The VSC-DPC controller parameters and gains are as follows:

PI gains: $K_{PQ} = 5$, $K_{IQ} = 5000$, $K_{PP} = 5$, $K_{IP} = 5000$;
VSC gains: $K_{VSCQ} = 1$, $K_{VSCP} = 1$, $c_{Qs} = 0.0001$, $c_{Ps} = 0.0001$;

Fig. 5 (A) and (B) show power responses of the VSC-DPC drive when the active and reactive powers are step changed from -0.1p.u to -1p.u and from -0.5p.u to 0.5p.u at 0.1s, respectively. The transient responses of both active and reactive power during step change of their references are within a 10ms. In the steady state, tracking errors are kept constant. In order to obtain fast power responses and to maintain low power ripples, the sliding surface gains c_{Qs} and c_{Ps} are set to very small values.

In Fig. 6 (A), simulation result shows the step power changes for the constant rotor speed 1p.u, where the synchronous speed is defined as 1p.u. Fig. 6 (B) and (C) show the stator currents and DC-link voltage while the active and reactive power references are step changed from -1 to -0.5p.u at 0.6s and -0.3 to 0.3p.u at 0.8s, respectively. These results demonstrate that the stator currents vary according to the increase and the decrease of an active power while keeping the stator voltage and DC-link voltage constant.

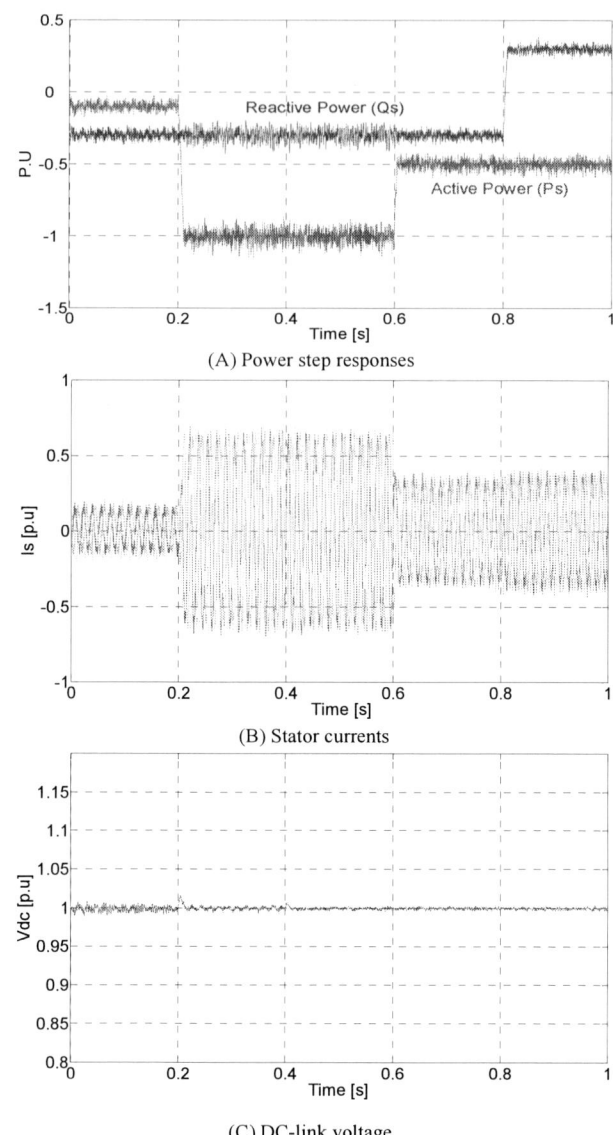

(A) Power step responses

(B) Stator currents

(C) DC-link voltage

Fig. 6: Simulation results of power step responses (A), stator currents (B) and DC-link voltage (C)

V. CONCLUSION

A variable-structure solution for DFIG drives was proposed to control the active and reactive powers directly without current controller. Simulated results show the validity of the VSC-DPC algorithm. The control strategy combines VSC and DPC principles within a simple structure and robust high performance drive. The new drive operates with low active and reactive power ripples in the steady state. During transient operation it displays good dynamic response.

REFERENCES

[1] Thomas Ackermann, 'Wind Power in Power Systems' (John Wiley and Sons, 2005)

[2] I. Boldea, 'Electric Drives' (Taylor & Francis, 2006)

978-1-4244-1871-8/07 $25.00 © 2007 IEEE

[3] S. Gomez and J. Rodriguez, "Grid Synchronisation of Doubly Fed Induction Generators using Direct Torque Control," 2002 iecon, pp. 3338-3343.

[4] Lie Xu and Phillip Cartwright, "Direct Active and Reactive Power Control of DFIG for Wind Energy Generation," IEEE Trans. Energy Conversion, vol. 21, no. 3, pp. 750-758, Sep. 2006.

[5] Dawei Zhi and Lie Xu, "Direct Power Control of DFIG With Constant Switching Frequency and Improved Transient Performance," IEEE Trans. Energy Conversion, vol. 22, no. 1, pp. 110-118, Mar. 2007.

[6] V. Utkin, J. Guldner and J. Shi, 'Sliding Mode Control in Electromechanical Systems' (Taylor & Francis, 1999)

[7] C. Lascu, I. Boldea and F. Blaabjerg, "Variable-Structure Direct Torque Control – A Class of Fast and Robust Controllers for Induction Machine Drives," IEEE Trans. Industrial Electronics, vol. 51, no. 4, pp. 785-792, Aug. 2004.

[8] A. Proca, A. Keyhani, J. Miller, "Sensorless Sliding-Mode Control of Induction Motors Using Operating Condition Dependent Models," IEEE Trans. Energy Conversion, vol. 18, no. 2, pp. 205-212, Jun. 2003.

Improved RPV (reactive-power-variation) anti-islanding method for grid-connected three-phase PVPCS

K.O. Lee, J.Y. choi, I. Cho, S.H. Song
Kwangwoon University
Department of Electrical Engineering
447-1, Wolgye-dong ,Nowon-Gu, Seoul, 139-701
Email: skalrl@nate.com

G.J. Yu, J.H. So, Y.S. Jung, B.G. Yu
Korea Institute of Energy Research,
71-2 Jang-Dong, Yuseong-Gu, Daejon, 305-343
Email: skalrl@nate.com

Abstract— As the grid-connected photovoltaic power conditioning systems (PVPCS) are installed in many residential areas, this has raised potential problems of network protection on electrical power system. One of the numerous problems is an Island phenomenon. There has been an argument that because the probability of islanding is extremely low it may be a non-issue in practice. However, there are three counter-arguments: First, the low probability of islanding is based on the assumption of 100% power matching between the PVPCS and the islanded local loads. In fact, an island can be easily formed even without 100% power matching (the power mismatch could be up to 30% if only traditional protections are used, e.g. under/over voltage/frequency). The 30% power-mismatch condition will drastically increase the islanding probability. Second, even with a larger power mismatch, the time for voltage or frequency to deviate sufficient to cause a trip, plus the time required to execute the trip (particularly if conventional switchgear is required to operate), can easily be greater than the typical re-close time on the distribution circuit. And, third, the low-probability argument is based on the study of PVPCS. Especially, if the output power of PVPCS equals to power consumption of local loads, it is very difficult for the PVPCS to sustain the voltage and frequency in an island. Unintentional islanding of PVPCS may result in power-quality issues, interference to grid-protection devices, equipment damage, and even personnel safety hazards. So the verification of anti-islanding performance is strongly needed.

In this paper, the authors propose the improved RPV method through considering power quality and anti-islanding capacity of grid-connected three-phase PVPCS in IEEE Std 1547 ("Standard for Interconnecting Distributed Resources to Electric Power Systems"). And the simulation and experimental results are verified.

I. INTRODUCTION

As the grid-connected photovoltaic power conditioning systems (PVPCS) are installed in many residential areas, this has raised potential problems of network protection on electrical power system. One of the numerous problems is an Island phenomenon. There has been an argument that because the probability of islanding is extremely low it may be a non-issue in practice. However, there are three counter-arguments: First, the low probability of islanding is based on the assumption of 100% power matching between the PVPCS and the islanded local loads. In fact, an island can be easily formed even without 100% power matching (the power mismatch could be up to 30% if only traditional

protections are used, e.g. under/over voltage/frequency). The 30% power-mismatch condition will drastically increase the islanding probability. Second, even with a larger power mismatch, the time for voltage or frequency to deviate sufficient to cause a trip, plus the time required to execute the trip (particularly if conventional switchgear is required to operate), can easily be greater than the typical re-close time on the distribution circuit. And, third, the low-probability argument is based on the study of PVPCS. Especially, if the output power of PVPCS equals to power consumption of local loads, it is very difficult for the PVPCS to sustain the voltage and frequency in an island. Unintentional islanding of PVPCS may result in power-quality issues, interference to grid-protection devices, equipment damage, and even personnel safety hazards. So the verification of anti-islanding performance is strongly needed.

They are classified as passive islanding detection techniques. Their ability of islanding detection is not guaranteed for all load conditions, especially for source-load balanced conditions. As a result, active islanding detection techniques are invented for enhancement. One simple active technique is the active frequency drift positive feedback (AFDPF) method. When the utility is disconnected, phase difference between the terminal voltage and the output current of a PV PCS depends on the load. It is detected by the internal phase-locked-loop (PLL) circuitry. To eliminate the phase difference, the frequency of PCS output current is forced to drift up or down. The intension is to make the frequency of the terminal voltage deviate from its nominal value until OFR or UFR is triggered [1]. Another is output power variation technique. There are two classified techniques at variation of active power and reactive power. The PVPCS deliberately introduces periodic variations in its power output and monitors the response of the parameters voltage, current and frequency. If the grid is stable the parameters will hardly change at all while in an islanding situation the effects of power variation can be clearly detected. This method works very well for a single inverter as the number of independent power producers in the grid increases, the reliability of islanding detection decreases. This is because the independent generators usually do not communicate. While one inverter is in its power variation phase all the others might stabilize voltage and frequency so that no change can be detected. A coordinated timing (i.e. all inverters vary their power output at the same time in the same

978-1-4244-1871-8/07 $25.00 © 2007 IEEE

manner) might eliminate this problem [5]. So, In this paper, the authors propose the improved RPV method through considering power quality and anti-islanding capacity of grid-connected three-phase PVPCS in IEEE Std 1547 ("Standard for Interconnecting Distributed Resources to Electric Power Systems"). And the simulation results are verified.

II. PVPCS SWITCHING MODEL WITH ANTI-ISLANDING CONTROL

The specifications of the three-phase PVPCS being modeled are listed in Table 1. The inverter is based on a grid-connected Inverter. There are two reasons for using a three-phase inverter to demonstrate the concept proposed by this work. First, it is close to a commercial product offering, so it is easier for technology transfer. Second, although previously the majority of grid-connected inverters were single-phase, mainly for PV applications, more and more new PV system tend to use three-phase PVPCS as grid interface. Therefore, the technology for three-phase inverter is gaining more and more practical value.

Generally, the overall power-conditioning system includes front-end conversion and regulation, for example, DC/DC conversion for prime movers with DC output (e.g., fuel cell, PV), or AC/DC conversion for prime movers with AC output (e.g. micro-turbines, sterling engines). They may have an energy-management system, such as a battery charger, at the DC bus. In either case, the input to the inverter is a regulated DC source. To simplify the discussion, the front end DC/DC or AC/DC converters are not modeled in the simulation. In the models, the input to the inverter is simplified as a DC voltage source.

TABLE 1
Data for the PVPCS, RLC load, Grid.

PVPCS			
fs	10,000	Hz	Switching frequency
Vdc	450	V	Input DC bus voltage
Lf	3E-03	H	Output inductance
Cf	20E-06	F	Output capacitance
Vac	220	V_{rms}	Line-to-line voltage
P	10000	W	Rated power
RLC load			
R	14.1376	Ω	Resistance
L	37.49E-3	H	Inductance
C	187.62E-6	F	Capacitance
			IEEE Std. 1547
Qf	1		Load quality factor
f_load	60	Hz	Load resonant frequency
Grid			

f	60	Hz	Grid frequency
Vac	220	V	Line-to-line voltage

Fig. 1 shows the PVPCS, RLC load, and grid system diagram with the inverter being modeled as a switching model. And fig. 2 shows PSIM Scheme of PVPCS, RLC load, and grid. The switching devices used for the inverter are insulated gate bipolar transistors (IGBTs). In PSIM, IGBTs can be modeled as ideal on/off switches that represent the inverter discrete switching behaviors. The controller used in the PSIM model (DLL: dynamic link library) is based on actual production code in C language. The code includes all the sensing functions, scaling and quantization. This way, once the new algorithms are coded and simulated, the same code can be readily compiled and loaded to the hardware for testing.

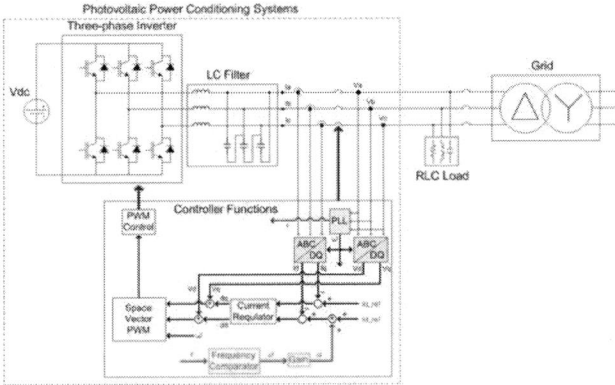

Fig. 1. PVPCS switching model with RLC load and Grid.

In this paper, the authors research the improved anti-islanding method for RPV (reactive power variation). Fig. 1 shows one of the frequency feedback schemes, from Δf to Idref (output current value of the reactive power). The scheme is implemented with the highlighted (red) path — f (frequency of grid voltage by 20 kHz sampling frequency) is passed by a FC (frequency comparator), a gain and becomes a current variation Δi adding to the Idref. Idref value measured 3% of Iqref (output current value of the active power) for anti-islanding.

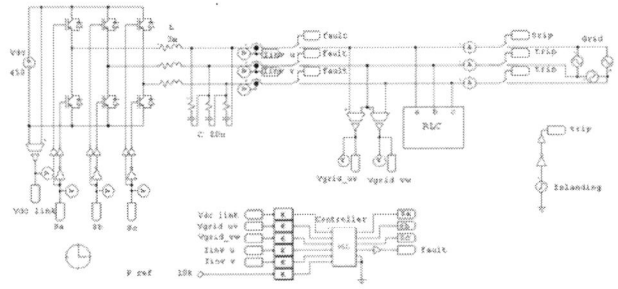

Fig. 2. PSIM Scheme of PVPCS, RLC load, Grid.

978-1-4244-1871-8/07 $25.00 © 2007 IEEE

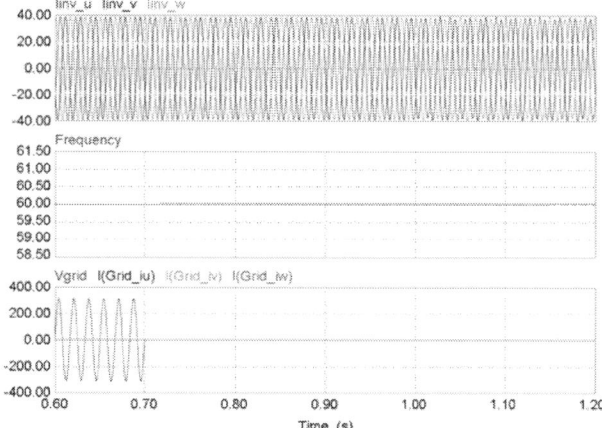

Fig 3. Without anti-islanding control

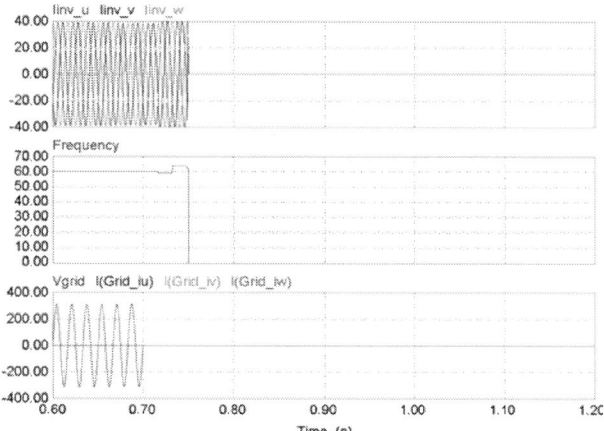

Fig. 4. With anti-islanding control

III. SIMULATION RESULT

The inverter switching model is used for the time-domain simulations. The switching model not only demonstrates the system behaviors, it also provides waveform-quality information, e.g. total harmonic distortion (THD). Besides, the code used in the switching model is based on the code for the actual hardware. Therefore, the simulations can help optimize the parameter settings for experiments and guide the experimental testing. For all simulation, the inverter AI function is enabled at 1.2 s and the grid is disconnected at 0.7 s. Before testing the effectiveness of the proposed anti-islanding schemes, a case without AI is simulated, as shown in Fig. 3. The PVPCS is operating at 10 kW. The load quality factor is 1. With close generation/load matching, the inverter can easily run on with voltage and frequency at nominal ranges. Fig. 4 shows the AI case the same power level and load quality-factor conditions. It can be seen that the frequency (Fig. 4, middle trace) becomes unstable after islanding. The feedback output, Δi (Fig. 1) is dynamically driving the change in inverter output current that leads to

frequency change. This detection of the islanding can be over frequency.

IV. CONCLUSION

This paper showed digital PLL applied to RPV feedback method, which is one of the active methods to detect islanding phenomenon, and implemented simulation under IEEE Std 929-2000 islanding prevention test. As a result, Improved RPV AI method has effectiveness in PVPCS for Output power quality such as THD and power factor.

For further work will be necessary. As islanding is still a very controversial topic there are many open questions. To reach an international consensus the following aspects have to be clarified: Definition of voltage and frequency limits for the operation of PV PCS, Definition of the allowable duration of islanding, Definition of a standard test method for islanding-prevention devices.

A valuable basis for further work and guidelines would be a probability analysis. Starting from real load patterns at low voltage distribution lines an investigation could determine the theoretical probability of islanding and the possible duration of such a condition. Such a study could act as a sound basis from which anti-islanding guidelines reflecting the real dangers could be derived. As the technical problem is absolutely identical everywhere in the world there should be no obstacle to a common approach. However, some pragmatism will be necessary to overcome the different safety philosophy which dictates widely different solutions at this moment.

REFERENCES

[1] G.J. Yu, K.O. Lee, J.Y. Choi, "Market Trend for PV System", Journal of the KARSE, vol. 23, Issue 1, pp.83~90, January, 2006.

[2] IEEE Standard for Interconnecting Distributed Resources with Electric Power Systems, IEEE Std 1547, 28 July 2003.

[3] M. E. Ropp, "DESIGN ISSUES FOR GRID-CONNECTED PHOTOVOLTAIC SYSTEMS," Doctoral Thesis, Georgia Institute of Technology, Atlanta, GA, October 12, 1998.

[4] Zhihong Ye, Amol Kolwlker, Yu Zhang, Pengwei Du, Reigh Walling, "Evaluation of Anti-Islanding Schemes Based on Nondetection Zone Concept," IEEE Trans. on Power Electronics, Vol. 19, No. 5, Sep. 2004.

[5] IEA Task V, Report IEA PVPS T5-01:1998, December, 1998.

Power Control of a Grid-Connected Hybrid Generation System with Photovoltaic/Wind Turbine/Battery Sources

Jinhong JEON
Korea Electrotechnical Research Institute
Electric Power Research Laboratory
70 Boolmosangil, Changwon, Gyoungnam, 641-120
Email: jhjeon@keri.re.kr

Seulki KIM
Korea Electrotechnical Research Institute
Electric Power Research Laboratory
70 Boolmosangil, Changwon, Gyoungnam, 641-120
Email: blksheep@keri.re.kr

Changhee CHO
Korea Electrotechnical Research Institute
Electric Power Research Laboratory
70 Boolmosangil, Changwon, Gyoungnam, 641-120
Email: chcho@keri.re.kr

Jonbo AHN
Korea Electrotechnical Research Institute
Electric Power Research Laboratory
70 Boolmosangil, Changwon, Gyoungnam, 641-120
Email: jbahn@keri.re.kr

Jangmok KIM
Pusan National University
Department of Electrical Engineering
Geumjeong-gu, Busan, 609-753
Email: jmok@pusan.ac.kr

Abstract— A grid-connected hybrid distributed generation system, composed of PV(photovoltaic) array, wind turbine and BESS(battery energy storage system), is proposed for various power transfer functions to the distribution network. The proposed system has several operation modes which are normal operation, power dispatching, and power averaging, according to coordinate control of the BESS and grid inverter. PV array and wind turbine are individually controlled to operate at the maximum power point for the most use of renewable energy. The BESS operates as an energy buffer to flexibly shift the generation from the renewable energy sources without excessively frequent shifts between battery charging and discharging. A grid interface inverter regulates the generated power injection into the grid. Developed prototype system design and experimental results are presented to demonstrate the performance and to validate the proposed system during its operation modes.

I. INTRODUCTION

Most renewable energy source such as photovoltaic and wind turbine heavily rely on the ambient environmental conditions therefore produce unpredictable output characteristics. The other renewable energy source such as a micro turbine and fuel cell system doesn't have ambient dependency but their out response characteristics are so slow that they can't hardly meet the dynamic load conditions[1]. Therefore renewable energy sourced generation systems are weak in stable and sustainable power supply since the sources are mostly dependent on weather conditions[1]. However, some of them, like solar irradiance and wind speed, have complementary profiles[2]. It has been reported that in weak grids, the wind/solar hybrid system is more reliable than single wind or PV generation since it suppresses rapid change in the output power of the single source such as the wind turbine system[3]. Grid interface of the hybrid system with battery storage or fuel cell generation may improve system stability and reliability[4][5].

This paper describes prototype system design and principle of a wind/PV/BESS hybrid system in grid interface operation and its test results by several operation modes which are normal operation, power dispatching and power

averaging. The wind and solar systems are regulated to obtain the maximum energy from given wind and solar conditions for efficient operation. The battery storage system is utilized to smooth the output fluctuation of the entire hybrid system. Energy control of the entire system and energy flow between systems and grid are based on power electronic interface. Test results present the dynamic control performance of the grid-connected hybrid generation system and also demonstrate the feasibility of output leveling with a battery storage system at the hybrid system.

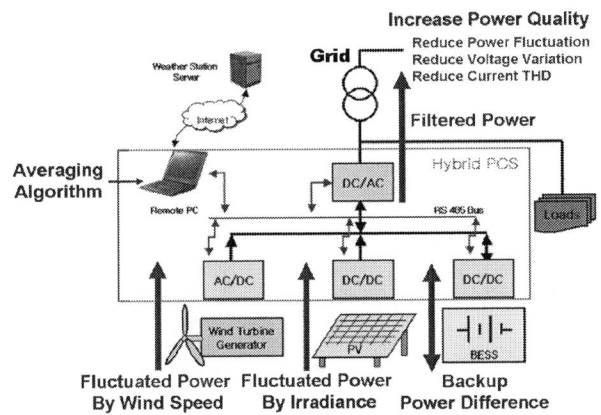

Fig. 1: A Configuration of the Proposed Hybrid Generation System and its Operation Objectives.

Figure 1 presents a configuration of the proposed hybrid generation system and its remote control system, system's operation objectives. The proposed hybrid system has a wind turbine and a photovoltaic as energy sources, battery storage as an energy buffer, power electronic converters for conditioning the power associated with the hybrid components, and a grid-connected inverter. These all components are linked with common dc bus. The supervisory control system is made up of a personal computer for remote control, system operation software, and communication network. The supervisory system monitors the entire hybrid system and coordinates power generation of the individual

978-1-4244-1871-8/07 $25.00 © 2007 IEEE

energy sources.

This hybrid system's operation objects are increasing power quality by averaging algorithm. As fluctuated input power by wind turbine and photovoltaic system, nevertheless, its output power, injected power to grid, is less fluctuated or smoother than input by averaging algorithm at remote supervising control system. As it reduce power fluctuation, and so reduce voltage variation and current harmonics. Indeed this proposed hybrid system can increase system's power qualities.

II.   PROPOSED HYBRID GENERATION SYSTEM

A.   System Configuration

Proposed grid-connected hybrid generation system has a PV array and a wind turbine as power sources, and battery as energy storage. Figure 2 is a block diagram of proposed grid-connected hybrid generation system.

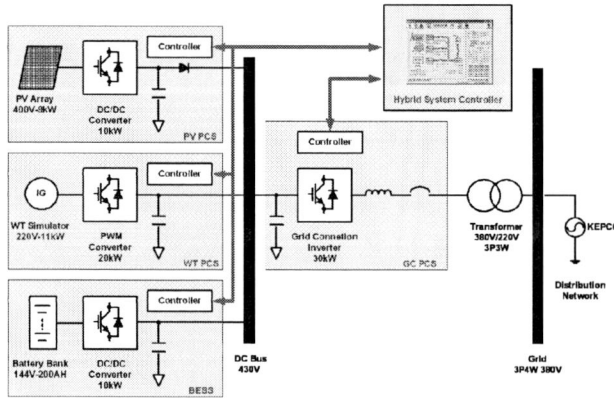

Fig. 2: Block Diagram of Proposed Hybrid Generation System

This hybrid system includes a wind turbine and a photovoltaic as energy sources, battery storage as an energy buffer, a common dc link, power electronic converters for conditioning the power associated with the hybrid components, and a grid-connected inverter.

Figure 3 shows a developed prototype system of proposed hybrid generation system. It consists of 10kW BESS, 10kW PV PCS, 20kVA Wind Turbine PCS and 30kVA Grid-Connected Inverter System from upper layer. Each PCS is controlled by its own TMS320C32 based control platform and communicate with hybrid system controller software in PC by RS485 multi-drop serial communication. The photovoltaic system consists of a 9kW photovoltaic array and a 10kW step-up dc-dc converter that raises the array voltage to a higher common dc voltage level. The PV system operates under maximum power point tracking (MPPT) control to create the maximum energy from solar irradiance variation. The wind system is composed of 11kVA induction generator based wind turbine simulator and a 20kVA variable voltage variable frequency PWM converter whose operating scheme

is to capture the maximum energy from varying wind speed by regulating the wind blade speed.

Fig. 3: Prototype of Proposed Hybrid Generation System

Figure 4 is an interface window of hybrid system controller software. System operation and control parameters setting can be performed by this software interface.

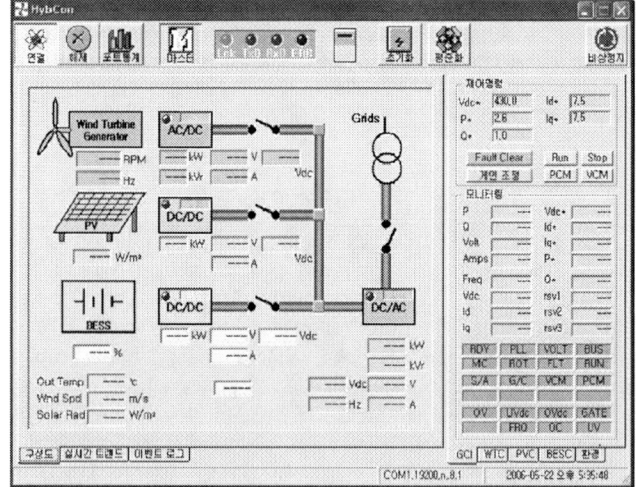

Fig. 4: Interface Window of Hybrid System Control Software

B.   System Specifications

Followings are specification of prototype hybrid generation system components.

PV Array (Total 3 PV arrays are used)
- Module Name : GMG 10530
- Maximum Open Voltage : 434V
- Maximum Output Power : 3184W

PV PCS
- Converter Type : Boost DC-DC Converter
- Switching Device and Frequency : IGBT, 10kHz
- Power Rating : 10kW
- Input PV Voltage Range : 200V ~ 400V (DC)
- Output Voltage Range : 350V ~ 600V (DC)
- Control : MPPT (Incremental Conductance Method)

Battery Bank
- Nominal Voltage : 144V(12 series connection)
- Rated Capacity : 200AH
- Battery Type : VRLA Type

BESS PCS
- Converter Type : Step Up/Down DC-DC Converter
- Switching Device and Frequency : IGBT, 10kHz
- Power Rating : 10kW
- Battery Side Voltage Range : 130V ~ 200V (DC)
- DC Link Side Voltage Range : 350V ~ 600V (DC)
- Charging Control : Bulk, Absorption, Float

Wind Turbine Simulator
- System : RTDS and MG set(Induction Machine)
- Turbine Model : RTDS
- Generator : MG Set (11kVA 220V)

Wind Turbine Model
- Type : No Pitch Control, Stall Control
- System Rating : 10kVA
- Blade Radius : 3.9m
- Air Density : 0.55 kg/m3
- Rated Speed : 74 r.p.m
- Rated Wind Speed : 12 m/sec

WT PCS
- Converter Type : 3 Phase SVPWM Converter
- Switching Device and Frequency : IGBT, 10kHz
- Power Rating : 20kVA
- Input Voltage Range : ~ 250V (AC)
- Input Frequency Range : ~ 120Hz
- Output Voltage Range : ~ 600V (DC)
- Control : MPPT, VVVF, Scalar

Grid Inverter
- Converter Type : 3 Phase SVPWM Inverter
- Switching Device and Frequency : IGBT, 10kHz
- Power Rating : 30kW
- Rated Output Voltage : 220V (AC)
- Input Voltage Range : 350V ~ 600V (DC)
- Control : DQ based Current Control

C. System Control

The proposed hybrid system has various operation modes such as normal operation, dispatching operation and power smoothing. Table 1 summarizes power control schemes for various modes.

TABLE 1. POWER CONTROL SCHEMES FOR OPERATION MODES

	Wind	PV	BESS	Grid Inverter
Normal	Variable Speed Control	Maximum Power Point Tracking	Common DC Voltage Control Within Specified Range	Constant DC Link Voltage
Dispatching				Dispatched Power Generation
Averaging				Averaged Power Generation

Normal operation

The hybrid system penetrates as much power into the grid as the PV array and wind turbine generates. In normal operation, the grid inverter is directed to maintain the common bus voltage constant so that power generation of the wind and solar sources can be delivered to the grid. The wind system and solar system have cost-free fuel, wind speed and solar irradiance. For the most efficient operation, the wind turbine and PV array are controlled at all times to produce the maximum power under the given weather conditions. Battery remains off in this mode or as battery module's SOC(Status Of Charging), it can be charging or discharging for maintaining a proper SOC.

Power dispatching

The hybrid system generates the power dispatched by a utility operator or commanded by a user for purpose of utility demand management such as peak load shaving and active load control. In this operation mode, in the same manner of normal operation mode, the PV and wind turbine modules are operating at the maximum power tracking mode but BESS module are control to maintain a specified dc link voltage so that grid connection inverter can be safe running. So mismatched power between a generation of PV, wind turbine and a grid injection of grid connection inverter are generated or absorbed by BESS.

Power averaging

The hybrid system mitigates fluctuating power generated from the PV array and wind turbine, and injects more stable (less fluctuating) power output into the grid. The purpose of this mode is to smooth power fluctuation of the renewable energy sources and so transfer more stable power into the grid. This improves the quality of power delivered to the grid. This mode of operation reduces the voltage and harmonic variation at the point of common coupling with the grid. The supervisory system should include a control block for power averaging and instruct the resulting averaged value for power command of the grid inverter. Averaging unpredictable power in real time may be a difficult problem, which requires further and comprehensive study on weather forecasting and power estimation as well as a control technique itself. In this paper, a simple technique using a low pass filter was used. This technique is simple but effective in decreasing power fluctuation. Use of battery is necessary for balancing the power production and injection, and its control is the same as

978-1-4244-1871-8/07 $25.00 © 2007 IEEE

in dispatch mode operation. The wind and PV systems are still under the maximum power control.

III. EXPERIMENTAL RESULTS

The hybrid system under different operation modes was tested to validate the performance of the proposed system. The total harmonic distortions (THD) of the current and the power variation at the grid connection point are provided for comparison. The experimental cases are classified in Table 1 and summarised experimental results are listed in Table 2.

TABLE 2. EXPERIMENT RESULTS

	Power Variation	THD Average	THD Variation	Discharging Power	Charging Power
Normal	6.7kW	3.61%	5.7%	0.0Wh	0.0Wh
Dispatching	0.6kW	3.51%	1.3%	1.7Wh	6.7Wh
Averaging	4.2kW	3.32%	2.1%	0.7Wh	5.6Wh

Figure 5 presents the experimental result of normal operation mode. The common dc voltage was maintained at 430V and the grid inverter passed all the power from the wind turbine and solar array into the distribution grid. Difference between the summed power of the wind and PV system and the grid inverter power was power loss of the hybrid PCS. THD of inverter current was reversely proportional to the power injection, since the harmonic component ratio to the fundamental component (60Hz) decreased with the greater output current.

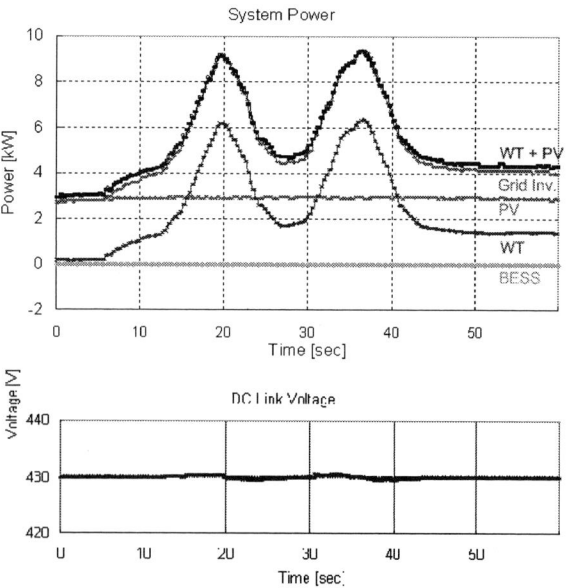

Fig. 5. Experimental Results for Normal Operation Mode

Figure 6 presents the experimental result of power dispatching operation mode where the hybrid system was dispatched to generate 5kW for the duration of test. The grid inverter injected constant power irrespective of power fluctuations of the wind and PV systems. When the summed power of the wind and solar sources was greater than the injected power into grid the battery operated in charging mode, and with the wind and solar generation less than the grid injection the battery turned in discharging mode. The dc link voltage was maintained at the upper limit in battery charging mode, and at the lower limit in discharging mode. This BESS operation scheme did not require frequent shifts of the battery operation mode that might give adverse effects on the battery lifecycle. When constant power was injected, THD of output current was constant and there were no voltage fluctuations.

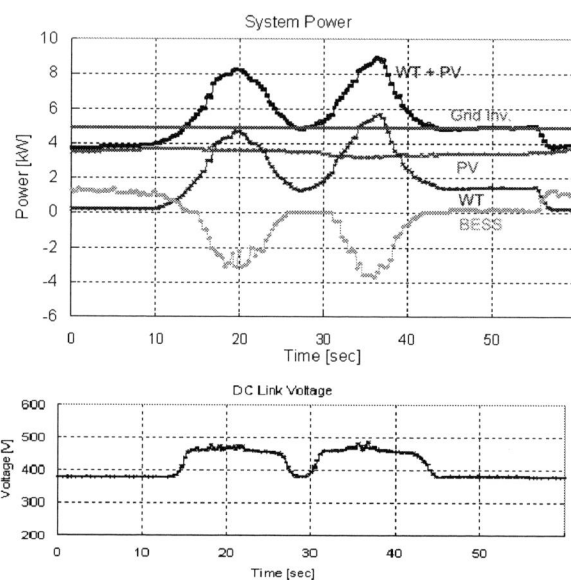

Fig. 6. Experimental Results for Power Dispatching Mode

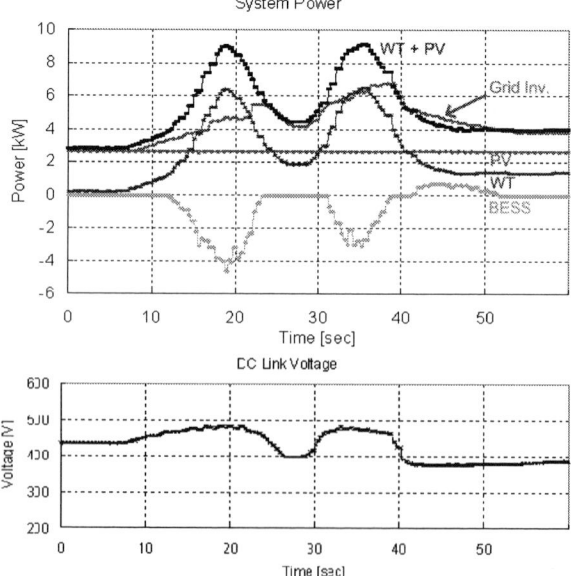

Fig. 7. Experimental Results for Power Averaging Mode

Figure 7 shows the result of power averaging operation mode where the ordered power of the grid inverter was set as low-pass filtering value of the sum of the wind and solar power. Smoothing effect in power injection into grid was clearly seen. Also, fluctuations in THD of the current and the voltage of the PCC became smoother. The battery operation strategy was the same as that of power dispatch operation mode and the dc link voltage was controlled between the pre-set limits, which did not require too many battery mode changes.

IV. CONCLUSIONS

A multifunctional grid-connected wind/PV/BESS hybrid system has been proposed. The principle of the proposed system and power control scheme for multi-operation modes were described. System performance was verified by experimental tests. A 30kW prototype was implemented and tested. Experimental results demonstrated that the proposed system was capable of supplying flexible and stable power into grid with coordination control of BESS and a grid interface inverter, and simultaneously maximizing use of the individual renewable energy sources. The proposed system has environmentally-friendly energy sources, and user- or utility-friendly operation strategy that power injection into network is dispatchable or less fluctuating. System economy, however, is another since renewable energy systems have inherently low cost-effectiveness. For optimal operation system engineering including sizing of components, particularly BESS, should be considered. In the future research, more effective algorithms for averaging of power fluctuation and operation strategies for consisting microgrids will be studied.

REFERENCES

[1] Fernando Valencaga, Pablo F. Puleston and Pedro E. Battaiotto, "Power Control of a Solar/Wind Generation System Without Wind Measurement: A Passivity/Sliding Mode Approach", IEEE Trans. Energy Conversion, Vol. 18, No. 4, pp. 501-507, December 2003.

[2] Kurozumi, Kazuhiro et al, "Hybrid system composed of a wind power and a photovoltaic system at NTT Kume-jima radio relay station", INTELEC, International Telecommunications Energy Conference 1998, pp. 785-789.

[3] Riad Chedid and Saifur Rahman, "Unit Sizing and Control of Hybrid Wind-Solar Power Systems", IEEE Trans. Energy Conversion, Vol. 12, No. 1, pp. 79-85, March 1997.

[4] Francois Giraud and Zyiad M. Salameh, "Steady-State Performance of a Grid-Connected Rooftop Hybrid Wind-Photovoltaic Power System with Battery Storage", IEEE Trans. Energy Conversion, Vol. 16, No. 1, pp. 1-7, March 2001.

[5] C. Schauder and H. Mehta, "Vector Analysis and Control of Advanced Static VAR compensators", Proc. Inst. Elect. Eng., pt. C, Vol. 140, pp. 299-306, July 1993.

[6] SungKi Sul, Electric Machinery Control, Hongneung Science Press, Korea, 2002.

[7] Z. Lubsony, Wind Turbine Operation in Electric Power Systems, Springer-Verlag, Germany, 2003.

[8] Mukund R. Patel, Wind and Solar Power Systems. CRC Press, USA, 1999.

[9] Scul-Ki Kim, Jin-Hong Jeon and Eung-Sang Kim, "Modeling of a Variable Speed Wind Turbine in Dynamic Analysis", KIEE International Trans. Power Engineering, Vol. 4, No. 2, June 2004.

[10] Minwon Park and In-Keun Yu, "A Novel Real-Time Simulation Technique of Photovoltaic Generation Systems Using RTDS", IEEE Trans. on Energy Conversion, Vol. 19, No. 1, pp. 164-169, March 2004.

[11] L Zhang, A Al-Amoudi, Yunfei Bai, "Real-time Maximum Power Point Tracking for Grid-Connected Photovoltaic Systems", Power Electronics and Variable Speed Drives, 18-19 September 2000, Conference Publication No. 475.

[12] W. A. Lynch, "Ni-Cd Battery Modeling, Evaluation, and Applications in Electric Car", Doctoral Dissertation, University of Massachusetts Lowell, Lowell, MA, June 1996.

Analysis and proposition of a PV module integrated converter with high voltage gain capability in a non-isolated topology

Samuel Vasconcelos Araújo, Peter Zacharias,
Benjamin Sahan
ISET e.V. Universität Kassel
Power Electronics Division
Königstör 59, 34119 Kassel, Germany
Email: sva@ieee.org

René P. Torrico Bascopé, Fernando L. M. Antunes
Federal University of Ceará
Department of Electrical Engineering
Energy Processing and Control Group
Fortaleza-Ceará-Brazil
rene@dee.ufc.br, fantunes@dee.ufc.br

Abstract— **Usually considered as one of the future solutions for grid connection of photovoltaic systems, module integrated converters were already the focus of several researches and projects. Most of the proposed approaches relied so far on the use of high frequency step-up transformers either in isolated operation or integrated in isolated dc-dc topologies. This paper analyses the possibility of using non-isolated topologies to achieve the necessary high-voltage gain for grid connection. Several circuits were analyzed and the best suited one for the current application was evaluated and optimized. Experimental results are presented in the final section.**

I.   INTRODUCTION

The first generation of grid connected photovoltaic systems was composed of several strings of panels associated in parallel and connected to a single inverter. Such centralized approach had as disadvantages [1] the necessity of string diodes (with their inherent power losses) and high voltage DC cabling. Furthermore, since operation was limited to only one maximum power point (MPP) for the whole array, mismatch losses reduced the system efficiency. Finally, due to the high power level of the inverter, there was little flexibility on system expansion.

At present, most of the systems are composed of a single or multiple strings of modules connected to an inverter, the so-called string and multi-string inverters. This way, in contrast with their predecessors, losses due to maximum power point tracking (MPPT) mismatch were reduced, but not totally eliminated, and string diodes are not necessary anymore.

The next expected evolution on grid connected photovoltaic systems is considered as the integration of the converter in the module and is usually named AC Module, since the output of the panel can now be directly connected to the mains. A major highlight of such an approach is the elimination of MPPT mismatches, allowing optimal coupling between panel and inverter and therefore increasing generated power per module. In addition, the small level of power and modularity allows flexibility in system expansion and low purchase investment, being considered as the best option for end-user applications [2]. Since the output of the panels can be directly connected to the grid, DC cabling and installation expertise are not necessary, allowing considerable reduction in installation expenses. Though a higher production cost per produced Watt is expected for this approach, the mass production of such small units may in the end increase competitiveness due to economy of scale. A disadvantage of

the module integrated solution is the strict requirement of a design with long lifetime under harsh ambient conditions that needs to be tackled by a highly robust power electronic design, since maintenance is much more complex than the ones of traditional string inverters.

II.   OVERVIEW AND REQUIREMENTS OF MODULE INTEGRATED CONVERTERS (MIC)

Prior to the comparison of the circuits, it is necessary to discuss some important issues regarding the proposed application.

The trinity efficiency, cost and lifetime gives in general the orientation on the topology choice and on most of the project development. It is mainly affected by the amount and rating of components in such a way that simple topologies are preferable with the condition that components are not under severe current or voltage stress.

Cost itself has been mainly one of the critical obstacles for the further expansion of module integrated solutions. Aside from the specification of the components, it is also strongly influenced by the fact that the lower the power rating is, the higher is the cost per produced kWh [3]. In order to reduce such disparity, mass production is a mandatory condition and may only be achieved by flexible solutions capable of operating with most of the available panels in the market, what leads to the necessity of high voltage gain capability as PV panels usually have output voltages around 30 and 50V.

Still regarding costs, another issue often forgotten is that, as previously explained in the introduction, module integrated converters are capable of promoting significant reduction on the installation costs. When one analyses the cost components of photovoltaic power (as depicted in Fig. 1) [4], it becomes clear that the higher expected price paid for the converter will

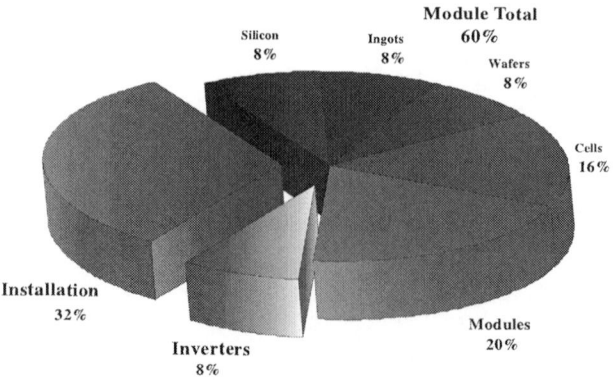

Fig. 1: Composition of the photovoltaic power costs [4].

978-1-4244-1871-8/07 $25.00 © 2007 IEEE   510

be probably justified by the superior potential reduction on the installation costs.

Lifetime is closely related to reliability and expected degree of failure of components and the converter shall be projected to endure at least the expected useful lifetime of the photovoltaic module (generally considered as 20 years). As an aggravating fact, harsh ambient conditions are expected along with a wide range of temperature variation during a single day; fact that impress mechanical stress on components and insulation materials. Among all passive or active elements, electrolytic capacitors are the ones with the shortest lifetime [5] and with the further drawback that aging increases the values of the ESR (equivalent series resistance) and consequently also the losses. As a conclusion, the chosen topology shall not require high capacitance values in order to allow the use of the film technology, especially from the polypropylene type, which has by far a longer lifetime, good thermal and electrical stability and finally low ESR.

Power density is also considered as an important factor in order to reduce installation complexity and costs and is often achieved by increasing the switching frequency. Nevertheless a trade-off shall be analyzed in order to avoid reduction of the efficiency by increasing switching losses.

A. MIC Topologies

In the last seven years, the photovoltaic modules industry introduced several new cell technologies alongside with further improvements on already existing ones. As expected, the parameters of the panels in general also changed, as can be observed in the comparison between the years 2001 and 2007.

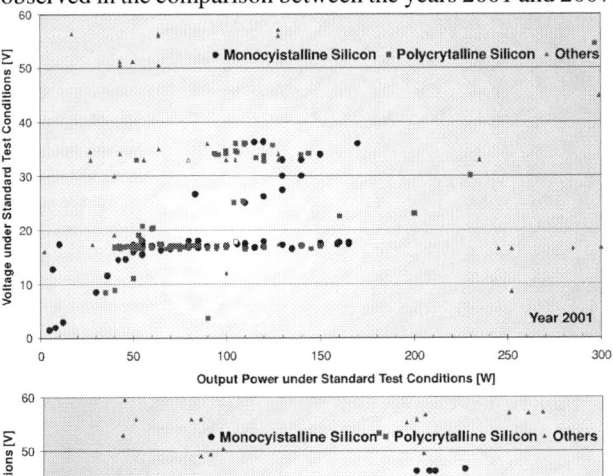

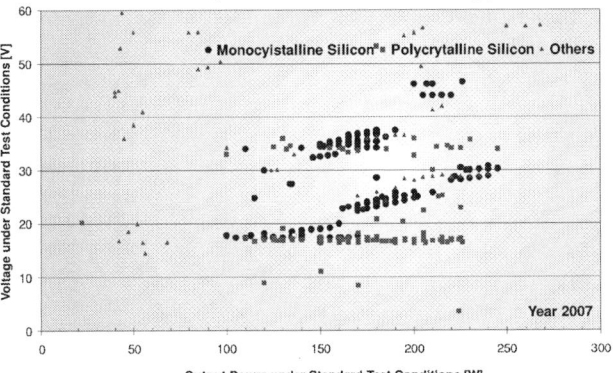

Fig. 2: Evolution of photovoltaic modules parameters [6,7].

From Fig. 2, it can be noticed an average increase in the generated power and voltage under STC (Standard Test Conditions).

Nevertheless, the output voltage level of most panels is still not high enough to directly feed an inverter for connection to the grid. Because of this fact, two approaches have been observed concerning most module integrated converters.

A. Single-stage converters coupled with photovoltaic modules with high voltage output

The first approach is rather new and consists of using specially designed modules with a high output voltage in the range of 200 to 400V, depending on the local grid voltage [8], like depicted in Fig. 3 a). This way, only a single stage is required, reducing costs and control complexity and increasing overall efficiency. While this approach looks promising it still remains that such solutions are restricted in the application spectrum to only few panels with compatible output parameters; what may not allow a large scale of production and necessary cost reduction in order to increase competitiveness.

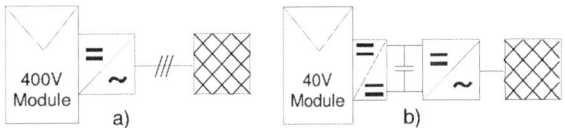

Fig. 3: Principle configurations of MIC with a) 3-phase, single-stage, b) Single-phase, dc–dc converter, dc-link capacitor and inverter.

B. Isolated Multi-stage converters coupled with modules with low voltage output

The second and most common approach is a single-phase approach using standard modules with lower voltage levels by employing multi-stage topologies (Fig. 3 b)). A high variety of different topologies were used in the past and described in the literature [1]. In most cases the topologies were characterized by high frequency step-up transformers to perform the necessary high voltage gain.

Regarding the dc-dc converter, in order to reduce the number of stages and increase efficiency and reliability, several approaches used transformers belonging to topologies derived from isolated dc-dc converters like push-pull [9, 10] and flyback [11-14]. A common drawback of such approaches in comparison to others employing a more complex switching scheme is the higher voltage stress across semiconductors. Therefore, the smaller amount of switching devices will not necessarily lead to much efficiency enhancement, since parameters like the drain-source on-resistance of MOSFETs are much affected by the voltage rating (increasing factor of 70% or more for best-in-class devices rated at 150 and 200V, for example). A further disadvantage inherent to push-pull derived topologies is the necessity of balanced magnetizing mechanisms in order to avoid transformer saturation.

In general, a clear advantage of all isolated approaches is the obtained galvanic insulation and power decoupling. Drawbacks are the lower levels of efficiency due to high frequency switching to reduce transformer size, losses on the transformer itself and increasing complexity.

C. New approach: Non-isolated dc-dc converter with high voltage gain capability.

An alternative to the previous categories that was not yet analyzed in the literature is the use of non-isolated dc-dc converters that are capable of reaching the required voltage gain with high levels of efficiency. Since galvanic insulation in photovoltaic applications is not a requirement in USA (only for this AC modules [15]) and in most of the European countries (except Spain [16]), no further limitations for the use of non-isolated converters on most photovoltaic panels are expected. However, [17] revealed that there might be lifetime issues with some thin-film technologies (e.g. CdTe) when PV panels are not grounded.

Several high voltage gain topologies were already proposed and will be evaluated on the following item.

At last but not least, the operation of a non-isolated photovoltaic system prompts attendance to further safety and regulatory requirements like limitation of the dc component on the output current, islanding detection and handling of leakage currents; though such requirements are out of the scope of this paper.

III. NON-ISOLATED DC-DC CONVERTERS WITH HIGH-VOLTAGE GAIN CAPABILITY

A. *Comparison of available topologies*

Taking into consideration that most of the mono- and polycrystalline modules available nowadays in the market have a MPP voltage level for STC under 40V, the minimum required voltage amplification to reach 400V for feeding the input of a grid-connected inverter is at least 10 and values of 20 or more may be necessary under lower irradiation levels.

Such high voltage gain can be theoretically obtained with the classical boost converter (see Fig. 4); though a serious limitation for practical operation lies on the fact that very high values of duty cycle are necessary. The operation at such conditions needs very precise control and drives for the switches in order to avoid instability, since small changes in the value of the duty will lead to great variations in the output voltage, what makes such implementation problematic and expensive.

Furthermore, the theoretical gain is normally not achievable for very high duty cycle values due to operational limitations, namely, losses at inductances and capacitors, reverse recovery and turn-on-off time of diodes, switching

transients and core susceptibility for fast changing flux intensity (what may increase losses) [18]. Therefore, for obtaining a high voltage gain it is advisable to employ special topologies that do not require very high duty cycles.

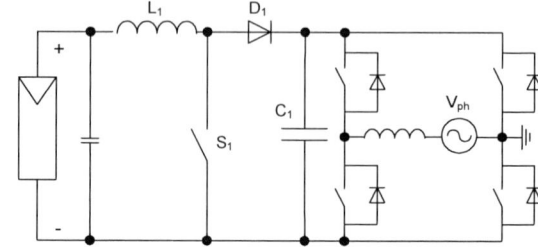

Fig. 4: Boost converter and full-bridge inverter topology.

The first analyzed circuit is depicted in Fig. 5 a) and was proposed in [19, 20] by the modification of a flyback converter with the objective of improving the coupling coefficient. Since the leakage of the coupled inductors lead to high voltage spikes across the active switch, degrading efficiency, it was necessary to employ a clamping technique. Drawbacks of this topology are the input pulsating current (what may increase the MPPT mismatch), inverted polarity of the load voltage (likewise the buck-boost converter) and finally necessity of high turns ratio in order to achieve the desired voltage gain.

Similarly to the interleaved boost, the circuit introduced in [21] and depicted in Fig. 5 b) is composed of two boost converters coupled via an autotransformer with 1:1 turns-ratio and inverted polarity in order to allow equal current share between the active switches. The output was configured as a voltage doubler rectifier. An advantage of this approach in comparison to the previous is that the input current is non-pulsating and with low ripple. As negative aspects, the voltage stress across active switches can be higher than half of the output voltage due to layout inductances.

Another topology also similar to the interleaved boost converter was proposed in [22], Fig. 5 c) with the possibility of using several stages with multiplier capacitors in order to increase the voltage gain. The voltage stress is limited to half of the output voltage for the configuration with just one multiplier stage. A snubber circuit may be required due to the sum of the reverse recovery currents of the output and multiplier diodes that increase turn-on losses on the active switches.

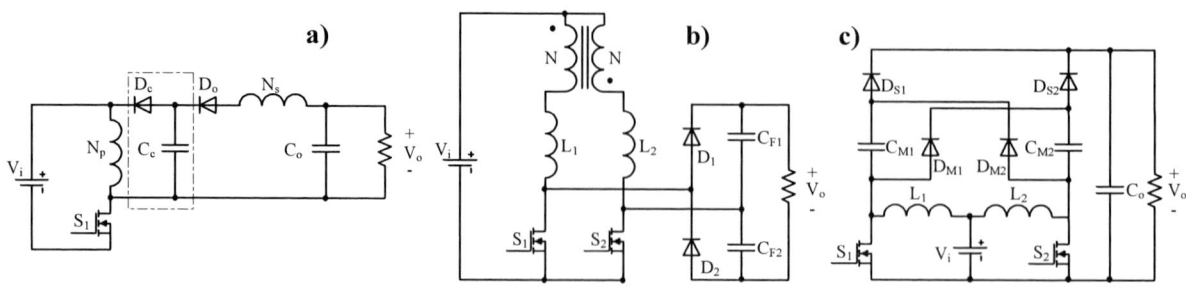

Fig. 5: Some of the analyzed topologies with high voltage gain capability.

978-1-4244-1871-8/07 $25.00 © 2007 IEEE

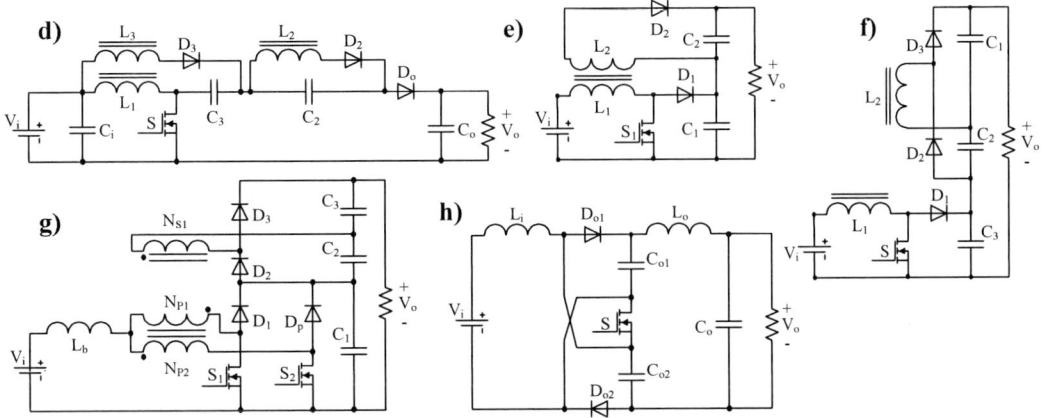

Fig. 6: Other analyzed topologies with high voltage gain capability.

A circuit composed of three coupled inductors was introduced in [23] as illustrated in Fig. 6 d), with high utilization grade of the magnetic core. Depending on the turns-ratio between the inductors, the voltage stress across the active switch can be as low as 15% of the output voltage. Like other topologies requiring coupled inductors, the high turns-ratio between windings increases system complexity and losses. A simplified variation was also proposed in [24] with just two coupled inductors, though their leakage inductance may lead to a ringing phenomenon in the voltage across diode D_2.

In Fig. 6 e), [25], a hybrid boost-flyback converter was proposed with the advantage of low voltage stress across the active switch, which is equal or lower than half of the output voltage; depending on the turns-ratio between the windings. In addition, such voltage is naturally clamped by output capacitor that additionally recycles the leakage energy to the output. As a disadvantage, the input current is pulsating and requires an additional input filter.

Another topology similar to the previous one and consisting of a boost converter with coupled inductors was introduced in [26] and is illustrated in Fig. 6 f). The difference now is that a voltage doubler rectifier is applied to the secondary winding. Drawbacks of this approach are the necessity of a filter to treat the input pulsating current since the inductor is totally discharged during normal operation and the critical charge of capacitor C_2 in order to achieve the required voltage gain.

The topology proposed in [27] and depicted in Fig. 6 g) also used coupled magnetic structures, though in a different way than previous approaches. Here, a three-state switching cell composed of two active and two passive switches employs an autotransformer with equal number of turns and inversed polarity in order to provide balanced current share proposed in [28] was used. Coupled with such transformer is a secondary winding that, with variation of turns-ratio, allows simultaneously high voltage gain and reduction of voltage stress across active switches. The use of two active switches in parallel with reduced voltage rating allows the minimization of conduction losses. A further advantage is that frequency across the inductor is the double of the switching

frequency; allowing reduction of the size without the drawback of higher switching losses. Finally, the input current is non-pulsated and with little ripple, what leads to low MPPT mismatches. The voltage level across the active switches is naturally clamped by the capacitor C_1.

Rather than employing coupled magnetic components like most of the previous topologies, [29, 30] used the switching capacitor technique, where capacitors are connected in series and in parallel in each switching cycle transition. Since the magnetic switches are not available, a higher number of active switches are necessary in [29], increasing losses and control complexity. In addition, the converter presented in [30], as illustrated in Fig. 6 h) does not have a high enough voltage gain for the proposed application.

B. Chosen topology analysis

Taking in consideration the factors presented in the item C, the circuit proposed in [27] was chosen as the best suited for the proposed application.

The operation of the converter can be divided in four stages

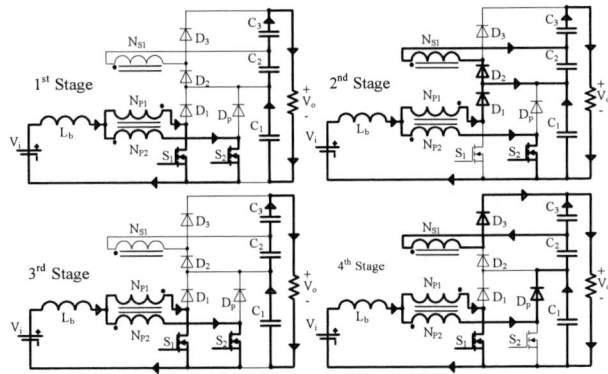

Fig. 7: Operation stages of the chosen converter.

as depicted in Fig. 7.

During the first stage, the input inductance is charged with energy provided by the source, while both switches S_1 and S_2 are turned-on and share equal amounts of current. The second stage begins as the switch S_1 is turned-off. Diode D1 and D_2 are forward biased and the capacitors C_1 and C_2 are charged (the last one by the current flowing through the secondary

978-1-4244-1871-8/07 $25.00 © 2007 IEEE   513

winding). During this stage, the inductance is being discharged. The third stage is equal to the first one, with both switches turned-on. Finally, the fourth stage starts when the switch S_2 is turned-off, forward biasing diodes D_p and D_3. The current flowing through the first diode charges once again the capacitor C_1 (for this reason its size is the half of the others). Due to the relation of the polarities, the current in the secondary winding has the opposite direction of the one during the second stage and charges diode C_3.

By observing the simplified theoretical waveforms in Fig. 8, it is possible to conclude that the converter needs to operate at the overlapping mode; what means that the duty cycle of each active switch must be higher than 50%. Such requirement comes from the fact that the input inductance is only charged during the period when both switches are simultaneously turned-on. In addition, values too near to 50% may increase the ripple at the input current, requiring larger input inductances.

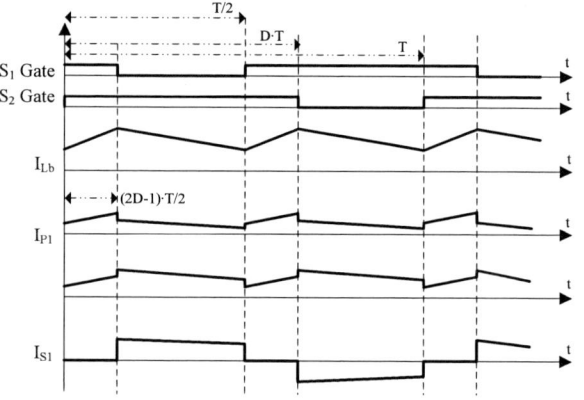

Fig. 8: Theoretical waveforms.

As a remark, the current in both primary windings is equally divided only during the first and third stage. During the second stage, in order to keep the resultant magnetic flux at zero in the autotransformer core, the current in the winding P1 is reduced by the half of relative ampere-turns of the current in the secondary, while the current on the winding P2 is increased by the same quantity. During the fourth stage, the inverse situation occurs.

Keeping in mind that such increments/decrements at the current in both primary windings are directly proportional to the relative number of turns of the secondary winding; large values of turns-ratio shall be avoided in order to improve the stability of the topology and reduce the possibility of reaching discontinuous conduction mode under low values of input power.

IV. DESIGN AND OPTIMIZATION OF THE CHOSEN TOPOLOGY

With the objective of allowing compatibility with most of the mono- and polycrystalline modules available in the market, the converter output power was chosen as 250W while maximum and minimum input voltages under nominal power are respectively 45V and 35V. As an important remark,

the influence of ambient temperature on module voltage was taken into consideration and hence, only the panels that have an output voltage lower than 40 V under STC are compatible with the projected converter.

The output voltage was chosen as 400V, in order to feed a full-bridge inverter connected to a grid with a RMS voltage level of 230V. From the selected parameters, the minimum and maximum voltage gains under nominal power are respectively 8.8 and 11.5.

For the sake of simplifying the circuit and reducing the amount of components, only one multiplier stage will be used. The voltage gain is given by (1) and depicted in Fig. 9 for some values of turns-ratio, which is represented by the letter "a" and defined as the ratio between N_1 and N_2.

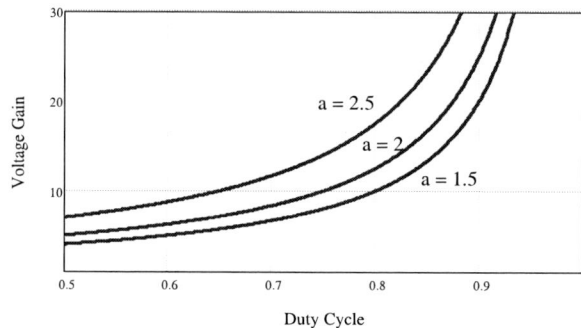

Fig. 9: Voltage gain as a function of the duty cycle and turns-ratio (a).

$$G_V = \frac{V_o}{V_i} = \frac{(a+1)}{(1-D)} \quad (1)$$

Taking into consideration that the minimum required gain will be 8.8 and that the converter shall operate on overlapping mode with a certain margin of security (a minimum duty cycle of 0.6 was chosen), the maximum allowable turns-ratio under such requirements is approximately 2.5.

The maximum possible turns-ratio value was chosen for the project since the higher it is, the lower is the voltage stress across the active switches and diodes Dp and D1, as calculated by (2). However, the voltage stress across diodes D2 and D3 increases for higher values of a, since it follows the same relation, but multiplied by "a".

$$V_{DS_{S1}} = \frac{V_o}{1+a} \quad (2)$$

Using a 20% margin of safety, it is possible to employ 150V rated MOSFETs. In order to choose the best-of-class model, the concept of Figure of Merit (FOM) [31], was used. It takes into consideration that conduction and switching losses in a MOSFET are respectively affected by the value of the drain-source on-state resistance (Rdson) and the gate charge (Qc). As a remark, Rdson and Qc are inversely proportional, so that models with very low resistances will not have an optimal switching behavior and vice versa. MOSFETs of several manufacturers were therefore analyzed, and the results are presented in the Fig. 10. Due to availability during the prototype assembly, MOSFETs FDP2535 from Fairchild were the chosen one.

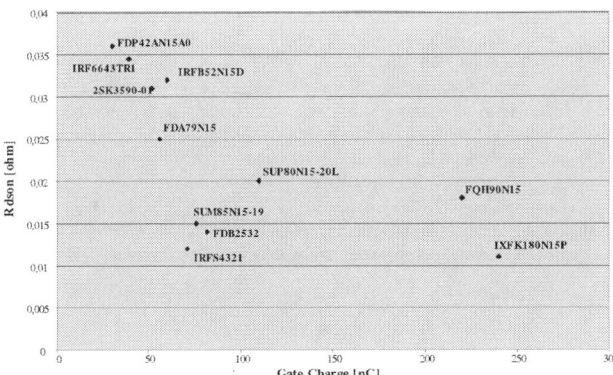

Fig. 10: 150V Rated MOSFETs characteristics.

For D1 and Dp Schottky diodes were specified rated at 150V with low forward voltage drop. The level of voltage across D2 and D3 required the use of ultra-fast diodes rated at 400 V, though 600V SiC diodes were also tested.

The calculated inductance for a maximum ripple of 20% of the input current was 274 μH. A ferrite core with the ETD geometry was chosen instead of an EE since it provides a reduction of more than 11% on the mean length per turn for the same cross-sectional area, decreasing cooper losses. In order to increase the cooper fill factor, litz wire with a rectangular profile was used.

The autotransformer was also built using an ETD core. Interleaved winding was employed, as depicted in Fig. 11, in order to reduce leakage inductance and proximity effect between the layers.

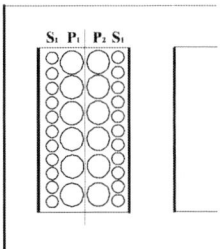

Fig. 11: Cross-section of the magnetic core depicting the interleaved winding used in the autotransformer.

The output capacitors were specified to limit the high frequency ripple to 2.5% at the output voltage. The low value of required capacitances allowed the use of polypropylene film capacitors. The low voltage rating (150V for C1 and 200V for C2 and C3) permitted further volume reduction of the specified capacitors.

The low frequency ripple due to single-phase connection will require a larger dc-link capacitor [1], but this specification is out of the scope of this paper.

V.   EXPERIMENTAL RESULTS

In order to evaluate the performance of the topology for the proposed application, a 250W prototype was built. The specified switching frequency was 20 kHz.

In Fig. 12 is depicted the measured drain-to-source voltages across the active switches S_1 and S_2 and the currents through the secondary and primary windings of the autotransformer. The obtained waveforms were a bit-different from the theoretical ones presented before because of the leakage inductance and coupling factor between the windings. Since no ringing phenomenon during turn-on of the active switches was observed, being within the expected limits, no snubber was necessary and in addition the gate resistance value was reduced in order to minimize switching losses.

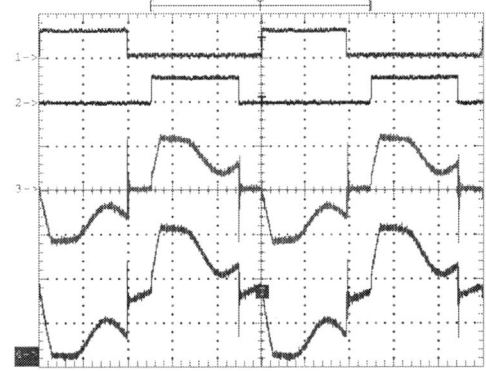

Fig. 12: Drain-source voltage across switch S1 and S2; current across secondary and primary autotransformer windings (200V/div; 2A/div; 10μs/div).

The input and output voltages and currents are presented in Fig. 13, with the remark that the desired voltage gain was reached using the specified duty cycle. The input current has low ripple content, what increases the efficiency of the MPPT and reduce size of the input capacitor.

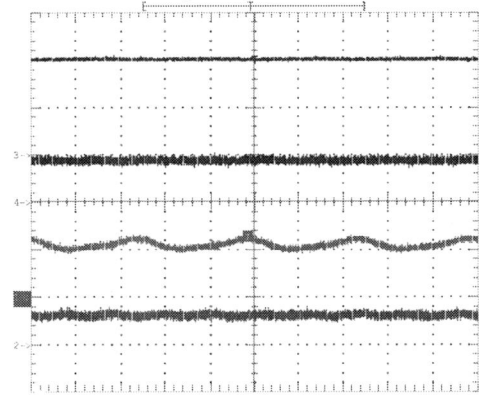

Fig. 13: Output voltage, input voltage, input current, output current, (200V/div; 50V/div; 5A/div; 1A/div;10μs/div).

In Fig. 14 are presented the efficiency curves for different levels of output power and two input voltage levels. The use of smaller gate resistances allowed the reduction of the switching losses and further enhancement of the efficiency.

SiC diodes were also tested, though for the proposed configuration their use actually reduced the efficiency a bit; possibly because of the higher forward-voltage drop value in comparison to the ultra-fast diode. In Fig. 15 is depicted the experimental set-up.

978-1-4244-1871-8/07 $25.00 © 2007 IEEE   515

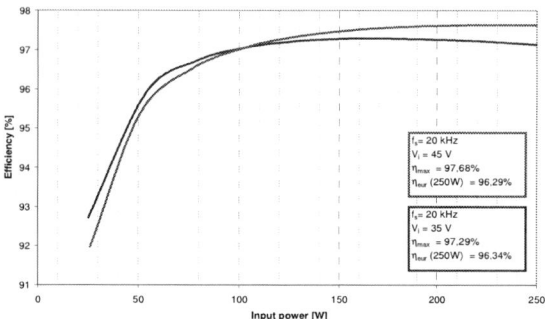

Fig. 14: Efficiency as a function of the output power.

Fig. 15: Experimental set-up.

VI.   CONCLUSIONS

This paper proposed the use of non-isolated dc-dc converters with high voltage gain integrated to a photovoltaic module. Several topologies with such capability were compared and the best suited one was projected following optimizing procedures. A 250W laboratory prototype was built and experimental results were evaluated. The converter reached a higher degree of efficiency in comparison to the original application, mainly due to the use of optimized semiconductors and other project parameters. The efficiency can be further enhanced by optimizing the circuit layout and reducing the current density in the magnetic components.

As next steps, the inverter can be designed and the interaction of the system with the photovoltaic module and grid can be examined.

REFERENCES

[1] S. Baekhoej, J. K. Pedersen, F. Blaabjerg, "A review of Single-Phase Grid-Connected Inverters for a Photovoltaic Modules", UEEE Transactions on Ind. App. Vol. 41. No. 5, September/October 2005, pp. 1292 – 1306.

[2] S. Islam, et al., "Cost effective second generation AC-modules: Development and testing aspects", Energy 31, Elsevier, 2006, pp. 1897 – 1920.

[3] P. Zacharias, B. Burger, "Overview of recent Inverter developments for grid-connected PV systems", Proceedings of the 21st European Photovoltaic Solar Energy Conference and Exhibition, Dresden, Germany, Sept. 2006, pp. 2328 – 2331.

[4] J. Conkling, M. Rogol, "The true cost of solar power: 10 ct/kWh by 2010", Photon Consulting, Solar Verlag, April 2007.

[5] N. Mohan, T. M. Undeland, W. P. Robbins, Power Electronics: Converters, Applications and Design, 2nd Edition, (John Wiley & Sons, Inc., New York, USA, 1995), pp. 727.

[6] "Marktübersicht Solarmodule", Photon: das Solarstrom-Magazin, Solar Verlag, February 2001, pp. 46 - 53.

[7] "Marktübersicht Solarmodule", Photon: das Solarstrom-Magazin, Solar Verlag, February 2007, pp. 56 – 79.

[8] B. Sahan, A. Notholt-Vergara, A. Engler, P. Zacharias, "Development of a single-stage three-phase PV module integrated converter", 12th European Conference on Power Electronics and Applications EPE, Aalborg, Denmark, September 2007, CD-ROM.

[9] U. Herrmann, H. G. Langer, H. van der Broeck, "Low cost dc to ac converter for photovoltaic power conversion in residential applications", IEEE proc. of 24th PESC, USA, June 1993, pp. 588-594

[10] D. C. Martins, R. Demonti, "Interconnection of a photovoltaic panels array to a single-phase utility line from a static conversion system", IEEE proc. of 31st PESC, vol. 3, Ireland, June 2000, pp. 1207-1211.

[11] N. P. Papanikolaou, E. C. Tatakis, A. Critsis, and D. Klimis, "Simplified high frequency converter in decentralized grid-connected PV systems: a novel low-cost solution," in Proc. EPE'03, 2003, CD-ROM

[12] T. Shimizu, K. Wada, and N. Nakamura, "A flyback-type single phase utility interactive inverter with low-frequency ripple current reduction on the DC input for an AC photovoltaic module system," in Proc. IEEE PESC'02, vol. 3, 2002, pp. 1483–1488

[13] D. Cruz Martins, R. Demonti, "Photovoltaic energy processing for utility connected system," in Proc. IEEE IECON'01, vol. 2, 2001, pp. 1292–1296.

[14] A. Lohner, T. Meyer, and A. Nagel, "A new panel-integratable inverter concept for grid-connected photovoltaic systems," in Proc. IEEE ISIE'96, vol. 2, 1996, pp. 827–831.

[15] Earley M.W. et al., National Electrical Code Handbook, 10th Edition, (National Fire Protection Association, USA, 2005), pp. 1022.

[16] "Real Decreto 1663/2000, de 29 de septiembre, sobre conexión de instalaciones fotovoltaicas a la red de baja tensión" (http://noticias.juridicas.com/base_datos/Admin/rd1663-2000.html)

[17] H. Schmidt, B. Burger, K. Kieferm, "Welcher Wechselrichter für welche Modultechnologie?" in 21. Symposium Photovoltaische Solarenergie, Bad Staffelstein, 08.-10. März 2006.

[18] J. Riatsch, "Modulintegriertes Umrichtersystem für die Netzanbindung einer einzelnen großflächigen Niederspannung-Solarzelle", P.H.D. Thesis, ETH Zürich, 2001.

[19] Q. Zhao, F. C. Lee, "High-efficiency, high-step up dc-dc converters", IEEE Transactions on Power Electronics, Vol.18, No.1, January 2003.

[20] Q. Zhao, F. Tao, F. Yongxuan, F. C., "Active-Clamp dc/dc converters using magnetic switches", APEC 2001. Sixteenth Annual IEEE Volume 2, 4-8 March 2000, pp. 946 - 952.

[21] J. Yungtaek , M. M. Jovanovic, "New two-inductor boost converter with auxiliary transformer", IEEE Trans. on Power Electronics, Vol.19, No.1, January 2004.

[22] R. Gules, L. L. Pfitscher, L. C. Franco, "An interleaved boost dc-dc converter with large conversion ratio", ISIE '03, Volume 1, June 2003 pp. 411 - 416.

[23] R.J. Wai, et al., "High-efficiency dc-dc converter with high voltage gain and reduced switch stress", IEEE Trans. on Power Electronics, Vol.54, No.1, February 2007.

[24] R.J. Wai, R.Y. Duan, "High-efficiency dc/dc converter with high voltage gain", IEEE Proc. on Power Applications, Vol. 152, No.4, July 2005.

[25] K.C. Tseng, T. J. Liang, "Novel high-efficiency step-up converter", IEEE Proc. on Power Applications, Vol. 151, No.2, March 2004.

[26] J. W. Baek, et al., "High boost converter using voltage multiplier", IECON 2005, Nov. 2005.

[27] R.P.T Bascopé, et al., "A generalized high voltage gain boost converter based on three-state switching cell", IECON´2006, Sept. 2006.

[28] G.V.T. Bascopé, and I. Barbi. "Generation of a family of non-isolated dc-dc pwm converters using new three-state switching cells", PESC'00, Vol.2, June 2000, pp. 858-863.

[29] Y. Berkovich, A. Ioinovici, "Boost converter with high voltage gain using a switched capacitor circuit", ISCAS '03. Proceedings of the ISIE'2003, vol.3, May 2003, pp. 296 - 299.

[30] Y. Berkovich, A. Ioinovici, "Transformerless dc-dc converters with a very high dc line-to-load voltage ratio", Circuits and Systems, 2003. ISCAS '03. Proc. of the ISIE'2003, Vol. 3, May 2003 pp. 435 - 438.

[31] Brown J., Moxey G., Power MOSFET basics: understanding MOSFET characteristics associated with the figure of merit, Application note AN602, Vishay Siliconix, December 2005.

978-1-4244-1871-8/07 $25.00 © 2007 IEEE

Modeling and Analysis of Power System Operation Connected with Wind Turbine Generation System and HVDC in Jeju island using PSCAD/EMTDC

Eel-Hwan Kim,Se-Ho Kim,Seong-Bo Oh
Cheju National University
Department of Electrical Engineering
Ara 1,Jejushi,690-756,Korea
Email: ehkim@cheju.ac.kr

Keong-Bin Song
Soongsil University
Department of Electrical Engineering
Sangdo-dong, Seoul, 156-743,Korea
Email: kbsong@ssu.ac.kr

Seung-Ho Song
Kwangwoon University
Department of Electrical Engineering
Nowongu, Seoul, 139-701, Korea
Email: ssh@kw.ac.kr

Abstract— This paper presents the modeling and analysis of Jeju power system operation connected with wind farm and HVDC system using PSACD/EMTDC program. It is for indicating the influence of wind farm operation in Jeju power system. For the computer simulation, there are three main models, which are 100(MW) wind farm, 300(MW) HVDC system and Jeju power load. To verify the effectiveness of the proposed method, two kinds of simulation are carried out . One is the 50 (MW) cut off and the other is 100(MW) cut off from the Jeju power grid instantaneously over the rated wind velocity . With the comparison of these results, this method shows the excellent performance. So it is very useful for understanding and applications of linking the wind farm to the power grid.

I.   INTRODUCTION

In recent years, wind power generation is very attractive to the renewable energy field because it is effectively possible to improve the capturing wind energy capability with the development of power electronics technology. Specially, with the effectuation of the Kyoto Protocol, many countries are being concerned about wind power generation. Now, trend of wind power generation is that it is composed of the large scaled wind turbine generation system(WTGS) and grid connected wind farm [1][2]. Depend on the wind velocity, the area which can be constructed the wind farm is specified. So wind farm is concentrated in the special area. Jeju island in Korea is most strong windy and have the best condition for wind power generation. And annual wind velocity in the east and west area of Jeju island is about 6-7 (m/s). This is why so many company hope to invest a lot of money to construct the wind farm in those area[3].

But there are some constrain conditions for penetration of wind power generation to the power system in Jeju island because of wind distribution and weak power load. First of all, total power generation capacity include HVDC in Jeju is almost 600(MW). This means that it is very difficult to guarantee the power system stability in Jeju if a lot of wind farm will be constructed. How much capacity of wind farm in Jeju is guaranteed the stability of power system ? Now a day in Jeju, this is a hot issue.

Consequently, using the PSCAD program which is widely favored on the analysis of power system include power electronics devises, two kinds of simulation results will be carried out for the analyzing of the stability[4][5]. One is the 50 (MW) cut off and the other is 100 (MW) cut off from the Jeju power grid instantaneously over the rated wind velocity under

the 150(MW) HVDC operation with 150(MW) Jeju load. Analyzing the variation of frequency and voltage in the transient state, it is very useful and understanding the wind power capacity in Jeju island. Finally, extensive results will be presented.

II.   CHARACTER OF POWER SYSTEM AND WIND VELOCITY IN JEJU ISLAND

A Power system in Jeju island

Although small, power system in jeju island in Fig.1 has all power system components, including generation, transmission, distribution and retail services. It shows annual load increase of $8 \sim 9$ (%), which is faster than that for the main land Korea. As the island drives to become a free-trade international city, the need for reliable supply of electric energy sources is becoming urgent.

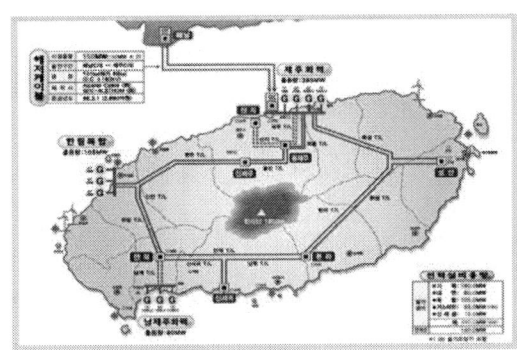

Fig. 1 Power grid network of Jeju island

`Not only the supply energy, but also the quality of power is becoming an important issue to meet the expectation of international standard in building the infrastructures. Therefore, expansion of generation facilities is an essential requirements for securing power system stability and reliability in Jeju island. Furthermore, since the island is located in a hurricane path and is known to have frequent lighting, there has been frequent contingencies because of the system heavily depending on overhead transmission and HVDC tie lines. The total installed capacity of Jeju island is

978-1-4244-1871-8/07 $25.00 © 2007 IEEE   517

587(MW), which includes 150(MW) transfer through the HVDC from the mainland. The power transfer through the HVDC is about 50(%) of the total demand in the island, and more than 40(%) in the average annually. Recently, relatively lager units are being planned and thus, the impact of failure of a unit is expected to be greater than before.

B. Wind characteristics of Jeju island

Fig.2 shows the wind distribution map of Jeju island. Considering the map, wind mean velocity in the east and west area is almost 6-7 (m/s), but north and south is 4-6 (m/s). This means that the east and west area has the very good conditions for wind power generation site than north and south, respectively. Now a days, there are many wind farms under constructions in the east and west area and will be constructed.

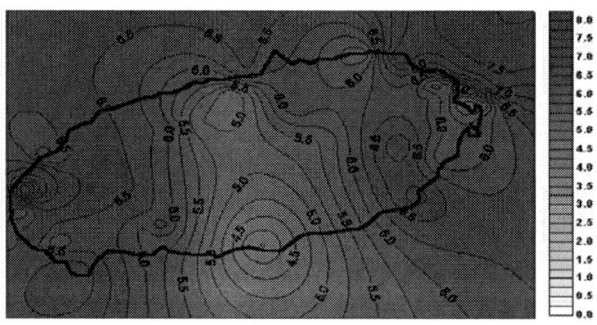

Fig. 2 Wind map of Jeju island

III. MODELING AND SIMULATION

A. Wind turbine modeling

To verify the feasibilty of the proposed control scheme, computer simulations has been carried out. 1.5(MW) NEGMICON WTGS named NM72C, 300(MW) Jeju-HaeNam HVDC system and 150(MW) Jeju power grid load are modeled and used in the simulation. Fig.3 shows the model scheme of WTGS in the PSCAD program. On the modeling of WTGS, it is very difficult to know all of the real parameter because WTGS is very complex system composed of a lot of mechanical, electrical, and aero dynamic system. This means that the results of computer simulation will be incorrect. So using the real output data of model machine located in HanGeung wind farm in the west area of Jeju island from Jan. 07 to Feb. 07, power curve versus wind velocity from cut in to cut off is obtained and made a look-up-table. While making this, the interval of data in the power curve is 0.1 (m/s) from 4.0 to 19.0 (m/s) and output data is mean value using the 15 over data according to the wind velocity.

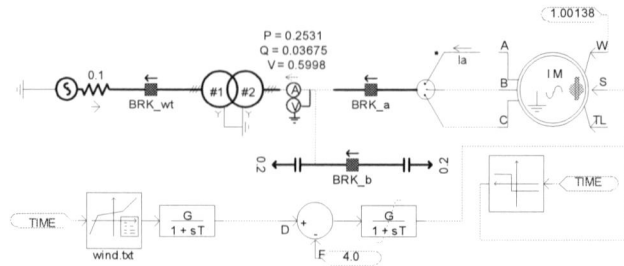

Fig. 3 WTGS model

Fig.4 shows the simulation results of model system and real mean output data measured from Jan. 2007 to Feb.2007 in HanGeung wind farm and the data is derived from the mean data with over 15 data per 0.1 (m/s) bin velocity from 4 to 19 (m/s). In Fig.4, Pwt and Qwt represent the real active and reactive power, respectively and also Pwt_s and Qwt_s are simulation results using PSCAD. Two results are almost same, so the simulation model of wind turbine generation system is good and useful for power system analysis.

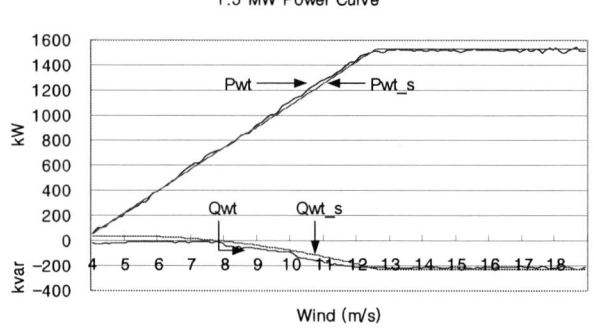

Fig.4 Power curve of model system

B. HVDC and Jeju power load modeling

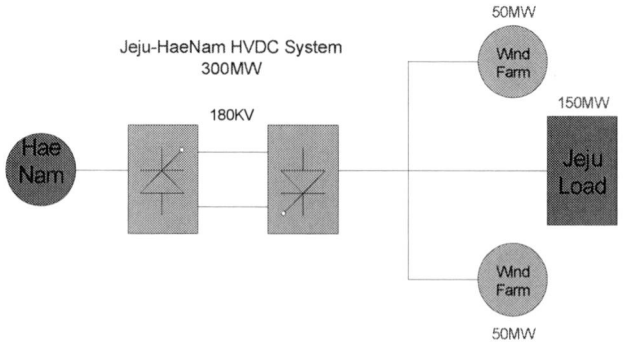

Fig.5 Schematic diagram of HVDC model

Jeju-Haenam HVDC system established in 1998 by Alstom has 300 (MW) power ratings. And this system has two poles and thyristor-controlled dual converter. To ensure the stability of Jeju power system, this system generate the half rated power in the normal times by the frequency control

978-1-4244-1871-8/07 $25.00 © 2007 IEEE   518

mode. Fig. 5 shows the HVDC schematic diagram, wind farm and Jeju load for simulation using PSCAD. Using the WTGS model in Fig.3, two 50 (MW) wind farms and 150 (MW) Jeju power load are connected with the modeled HVDC. The HVDC system which has the rated DC link voltage 180 (kV) and 1666 (A) generate the AC 154 kV with 79.2/154 (kV) power transformer. In the real Jeju power system, HVDC system cover the almost 40 (%) capacity of total consuming power. This means that output power of HVDC system is almost 150 (MW). And the response time is faster than the other generating system like a gas or steam turbine according to the variable load. So those systems are not necessary for analyzing the stability of Jeju power grid in the simulation. This is why there are no gas and steam turbine generation system.

C. Simulation results and discussion

Let us assume that one of two wind farms which has the 50 (MW) power ratings in Fig.5 generate the power according to the variable wind velocity, and mean wind value is 10 (m/s). To verify the effectiveness of proposed modeling system, Fig. 6 and Fig.7 show the characteristics of voltage and frequency response of model system when this wind farm shut down instantly over the rated wind velocity at the 72(s) time. In the Fig.6, active output power of wind generation(Pwt) is changed from 50 (MW) to 0. In that time, HVDC output power is also varied from 105 to 155 (MW).

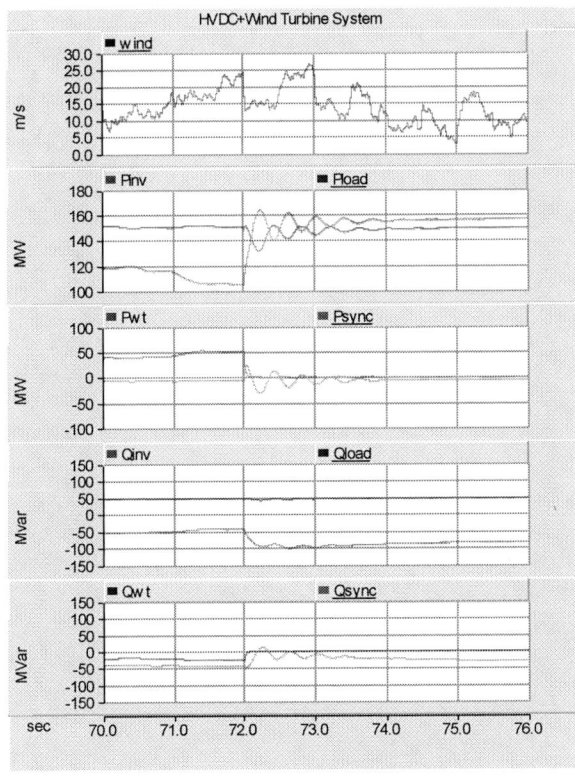

Fig.6 Results of 50 MW cut-off (Wind,P,Q)

The transient state is appeared for 1.5 (s). In the Fig.7, the variation of maximum voltage dip in the load is about 10 (%). In that time, the minimum value of frequency is almost 59.5 (Hz) during one cycle and recover it very fast. This means that HVDC system connected with renewable energy system is very useful and good for the stability of power system because of fast response time, specially in the weak power grid.

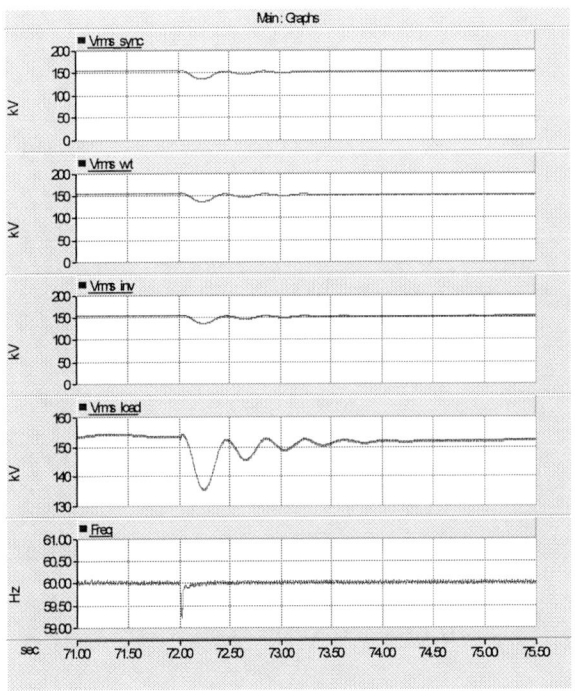

Fig.7 Results of 50 MW cut-off (V,freq)

For the simulation, supposing that wind has blow 10 and 12 (m/s) mean wind velocity in the east and west area side of Jeju island, respectively. Fig. 8 and Fig.9 show the characteristics of voltage and frequency response of model system when two 50 (MW) wind farms shut down instantaneously over the rated wind velocity at 70(s). In the Fig.8, there is a transient state for 3 (s) and two active power(Pwt,Pwt1) of wind generation are changed from 50 (MW) to 0, respectively. In that time, output power of HVDC system is varied from 50 to 155 (MW). Also, voltage dip in the HVDC and load have about 20 (%) and frequency has 2 (%) for 5 cycles. Finally, we can find that there is no problem over the power quality under the variable wind velocity in Jeju power system from 50 to 100(MW) with the HVDC system in the simulation results.

So, HVDC system will be very valued in the Jeju island that has the best condition for wind power generation in Korea.

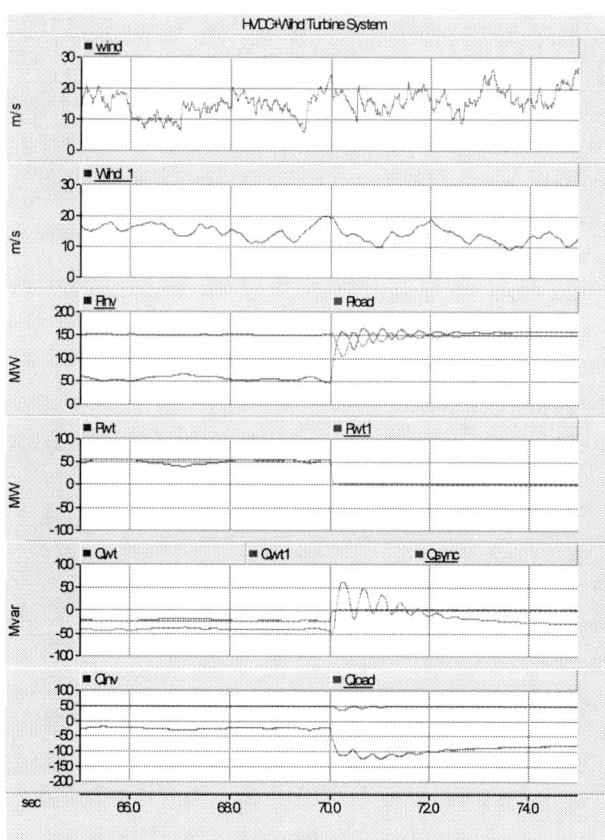

Fig.8 Results of 100 MW cut-off (Wind,P,Q)

IV. Conclusion

Modeling and analysis of wind farm operation with the steady and transient situation under the variable wind velocity in Jeju island using PSCAD/EMTDC program has been proposed in this paper. To show the validity of the proposed computer simulation, the comparative simulations were carried out based on the two wind farms. The results of simulation using the model turbine under the turn on and off situation of 50 and 100 (MW) are presented, respectively. Analyzing the simulation results, we can be expected that the proposed scheme is very useful for the establishment of how much wind power generating capacity in Jeju island for the steady state stability of Jeju power system.

REFERENCES

[1] Thomas Ackermann, "Wind Power in Power Systems", John Wiley&Sons 2004.
[2] M.Godoy Simoes and Felix A.Farret "Renewable Energy Systems," CRC Press. 2004.
[3] Kwang Y.Lee,Se-Ho Kim,Eel-Hwan Kim,Ho-Chan Kim, "Fuel Cell as an Alternative Distributed Generation Source under Deregurated Power System," in KIEE Trans. B ,Vol.55A,No12, pp.561-568. 2006.
[4] Oplimpo Anaya-Lara, F.Michael Hughes, "Influence of Windfarms on Power System Dynamic and Transient Stability", Wind Engineering Vol.30, No.2, pp.107-123, 2006.
[5] Akmatov,V. "Variable Velocity wind turbines with doubly fed induction generators" Wind Engineering ,pp.85-108.,2002.
[6] PSCAD/EMTDC Mannual,2006.
[7] HVDC Light Topology, ABB Handbook,1999.
[8] Thomas Ackermann,Wiley,"Wind Power in Power System",2004.

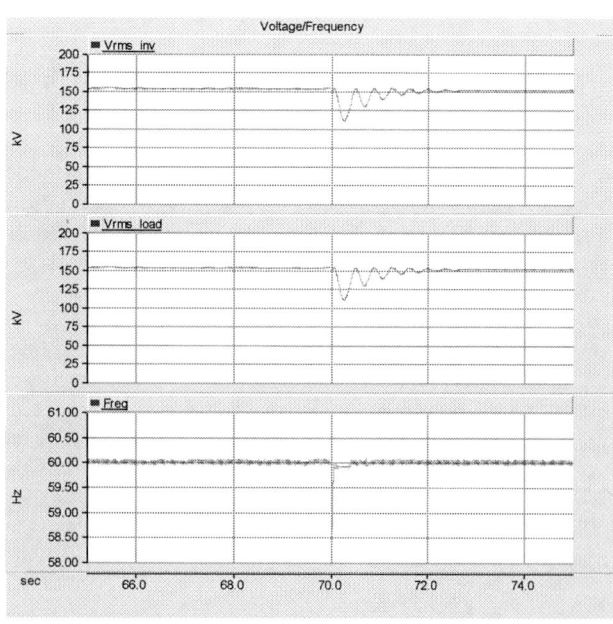

Fig.9 Results of 100 MW cut-off (V,freq)

978-1-4244-1871-8/07 $25.00 © 2007 IEEE

An Analysis of PEMFC & Photovoltaic 500W Hybrid system

[1] Hong-Jun Choi , Se-Joon Park Jeong-Sik
Choi, In-Su Cha, Jeong-Phil Yoon
Dongshin University Dept Hydrogen energy
Dept. of
Hydrogen & Fuel Cell Dongshin University,
252 Daeho,
Naju, Jeon-nam 520-714, South Korea
E-mail:chlghdwns1@naver.com

[2] Jang-Soo Suh
Dept. of Digital Electric
Songwon Junior colleage
365 Songha-dong, Nam-gu,
Gwangju, Korea 503-742
suhjs@songwon.ac.kr

[3] Sung-do Gun
Dept. of Electrical Dong-A
Junior college
Dokcheon-ri Dong-a College
Haksan-mveon Yeongarn-gun
Jeollanam-do(Seoul 526-705
Korea)
sang1212@hanmail.net

Abstract--**Renewable energy sources (solar, wind, etc) are attracting more attention as alternative energy sources to conventional fossil fuel energy sources. Hydrogen energy and Photovoltaic is friendly to the environment. In this paper a 500W PV-fuel cell hybrid generation system is designed. An electrolyzed coupled to the PV array is employed for hydrogen production. Maximum power tracking for PV array is achieved using fuzzy regression. A controller is designed to ensure continuous constant power generation through the day and after sunset via the PV and fuel cell stack. The system is simulated using LabVIEW software. The system characteristics obtained from simulation are analyzed.**

Index Terms— Hybrid system, Photovoltaic system, PEMFC., LabVIEW.

I. INTRODUCTION

The development of the solar energy is necessary since the future alternative energies that have no pollution and no limitation are restricted. Since there is such an advantage in these energies, they are being studied and developed consistently. The new-regeneration system has an advantage of unlimited and unpolluted amount of energy resource. The New-regeneration system is regarded as the main electricity supply and is connected in parallel to the supply from the local grade. Sustainable energy has relation with the creative electricity.

In this paper, we designed system hybrid photovoltaic and PEMFC system. Hybrid system is compensation a thing to supplement shortcoming of the photovoltaic and Fuel cell system.

This system being 500W class, constructed monitoring system for watch of system with LabVIEW.

II. COMPONENTS AND SPECIFICATION OF HYBRID SYSTEM

The 500W hybrid PEMFC & Photovoltaic system consists of 140W*4EA solar module, bipolar type 500W PEMFC (Polymer Electrolyte Membrane Fuel Cell), and monitoring system.

A. 500W class photovoltaic system

In Dongshin University, the 500W class array, built by serially connecting 4 pieces of 140-W class solar modules. Table 1 shows the specification of the solar cell module.

TABLE I
SPECIFICATION OF THE SOLAR MODULE

Division		Spec.	
Weight	14.5 kg	P max	140 W
Dimension	1580/795/ 45 mm	I mpp	4.39 A
Max. System Voltage	600 V	V mpp	31.9 V
Max. Series Fuse Rating	8 A	V oc	42.0 V
Silicon Type	Mono-Si	I sc	4.8 A

Figure 1 show the 500W class photovoltaic system in Dongshin University.

Fig. 1. 500W class photovoltaic system

B. 500W class PEMFC system Figure 2 show the 500W class PEMFC system in Dongshin University. Figures 3 and 4 show the PEMFC Stack.

The unit area of the membrane electrolyte assembly (MEA), which is used in making stack to assess the feature and performance of fuel cell by making PEMFC, was the same

978-1-4244-1871-8/07 $25.00 © 2007 IEEE

with 39 fuel cells with the dimension of 70 mm × 70 mm. For the electrical circuit connection to block the gas mixture between unit cells, a bipolar board is used. The anode makes the length of pipe longer for pressure difference between the inlet and outlet. The pipe for oxygen is shorter to make the density of oxygen and hydrogen the same. Likewise, this study tries to let each cell perform equally.

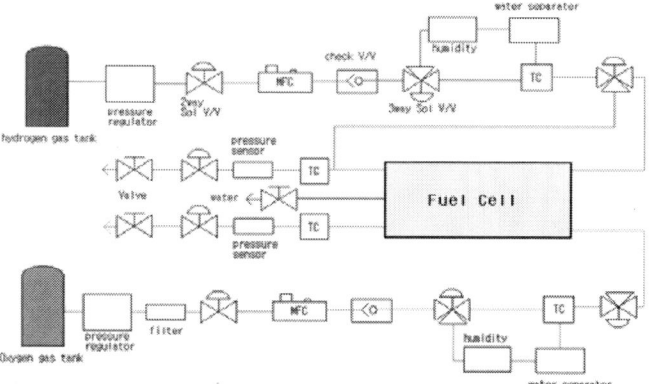

PEMFC Gas Supply Diagram System

Fig. 2. Block diagram of the 500W class PEMFC system

Fig. 3. front view of the PEMFC Stack

Fig. 4. Right side view of the PEMFC Stack

The following are the specifications of stack.

- Fuel Cell Stack Power: 500 W (max)
- Fuel Cell Stack Current: 19.5 A (max)
- Fuel Cell Stack Voltage: 27.3 V (max)
- Hydrogen Flow Rate: 8 ℓ /min
- Oxygen Flow Rate: 8 ℓ /min
- Coolant Flow Rate: 0.15 ℓ /min
- Anode Pressure: 1.1 atm
- Cathode Pressure: 1.1 atm

- Temperature of Environment: 25 ℃
- MEA Size: 70 mm * 70 mm
- Graphite Size: 100 mm * 100 mm * 3 mm
- Inner Stream: Humidity (100%)

C. Monitoring system

The monitoring system is largely divided into three components: the measurement and control part that collects and controls data from various sensors or measurement equipment the field server part that collects and processes the data from the measurement and control modules in the file; and the central integration system that manages the system with integration by collecting the data of the field server. To build the monitoring system, Compact Field Point (CFP-2010) of NI and voltage and current transducer were used. A Pentium 4 PC is used to check the result of the remote measurement.
Figure 5 shows CFP-2010, voltage and current transducer applied in monitoring in the field server.

Fig. 5. Block diagram of the 500W class PEMFC system

III. SIMULATION AND EXPERIMENTS

Figures 6 and7 show the result of simulation with the above PEMFC stack conditions. The figures indicate the change in voltage and output. The X axis is Voltage [V] and output [W], whereas the Y axis is time [min]. Figure 8 show the hybrid system (PEMFC/Photovoltaic)

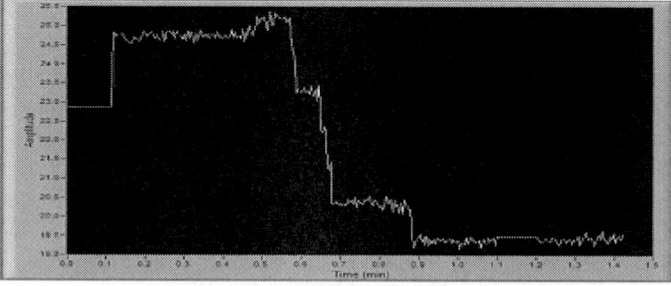

Fig. 6. simulation result the PEMFC using LabVIEW.

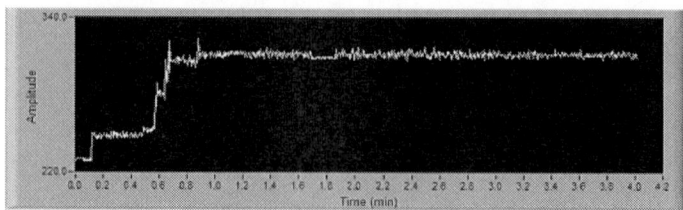

Fig. 7. Simulation result the PEMFC using LabVIEW.

978-1-4244-1871-8/07 $25.00 © 2007 IEEE

Fig. 8. PEMFC system/photovoltaic hybrid system

The following are the conditions of experiment stack.
1. Fuel Cell Stack Current: 10 A
2. Fuel Cell Stack Hydrogen Flow Rate: 5ℓ /min
3. Fuel Cell Stack Oxygen Flow Rate: 5ℓ /min
4. Hydrogen Humidification Temp: 25 ℃
5. Oxygen Humidification Temp: 25 ℃
6. No Coolant Circulation Environment Temp: 21 ℃

This study measured the temperature of the fuel cell during the operation of PEMFC by installing a temperature sensor in the inlet and outlet of gas.

With the above condition maintained, the temperature and voltage of second side of the fuel cell was measured after running 10 A by connecting cable resistance to the anode and cathode of the fuel cell.

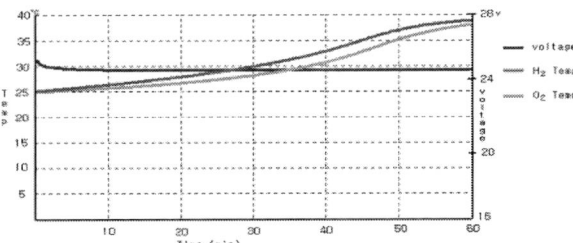

Fig. 9. Characteristics transient of the PEMFC.

After setting the temperature of hydrogen and oxygen and flow rate current value, the voltage and temperature were measured ten times. The measurement shows that the temperature of the stack rose, and the voltage fell by 1.6V. Figure 10 is the result of monitoring from sunrise to normal output. It can be checked that output is slowly increasing and reaches the normal output. The change in output due to the influence of cloud cover can be checked through the change in the monitoring result.

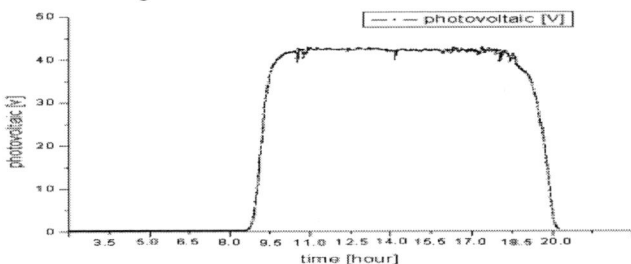

Fig. 10. The result of monitoring from photovoltaic system

Fig.11 PV PEMFC System

IV. CONCLUSIONS

In this paper, we designed system hybrid photovoltaic and PEMFC system. The 500W hybrid PEMFC & Photovoltaic system consists of 140W*4EA solar module, bipolar type 500W PEMFC and monitoring system.

We wanted to know through hybrid system if is no problem happened at same time driving of two generation system.

Fortunately, According to result by monitoring system, photovoltaic/PEMFC system displayed normal driving characteristics. But, this hybrid system has much unprepared controversial points.

We are trying to know it and supplement problem. We are expecting perfect system completion at end of this year.

ACKNOWLEDGMENT

This work was supported by grant KEPRI and KESRI R-2005-B-117

REFERENCES

[1] M.Benghanem, and A. Maafi(1997), "Data acquisition system ofr photovoltaic systems performance monitoring ", IEEE Instrumentation and Measurement Technology Conference, Ottawa, Canada, pp. 19-21

[2] Phuong T. Huynh(1999), "Design and Analysis of a Regulated Peak-Power Tracking System", IEEE Trans. On Aerospace & Electronic Systems, Vol. 35, No. 1, pp 84-91

[3] T. Takebayashi, H. Nakata, M. Eguchi, and H. Kodama(1997), "New current Feed Back Control Method for Solar Energy Inverter using Digital Signal Processor", PCC-Nagaoka

[4] Victorio Arcidiacono, Sando Corsi, Luciano Lambri, "Maximum Power Point Tracker for Photovoltaic Power Plants", IEEE. pp. 507-512, 1982.

[5] Z. Salameh, D. Taylor, "Step-Up Maximum Power Point Tracker for Photovoltaic Arrays", Solar Energy, Vol. 44, No. 1, pp. 57-61, 1990.

[6] The fuel cell's bumpy ride. The Economist, pp. 39–43. 24 March 2001.

[7] Gingrich N. An opportunities-based science budget. Science ;1303, 2000.

978-1-4244-1871-8/07 $25.00 © 2007 IEEE

A Study of Operating Characteristics for 1kW Households PEMFC System

[1] Hong-Jun Choi , Se-Joon Park Jeong-Sik Choi, In-Su Cha
Dongshin University Dept Hydrogen energy Dept. of Hydrogen & Fuel Cell Dongshin University, 252 Daeho, Naju, Jeon-nam 520-714, South Korea
E-mail:chlghdwns1@naver.com

[2] Jeong-phil Yoon
Fusion Information Technology CO.Ltd,
Usan-ri Boseong-eup Boseong-gun Jeollanamdo(
Seoul 270-19Korea)
Email:yoonip@dsu.ac.kr

[3] Jang-Soo Suh, Jeong-ill Lee
Dept. of Digital Electric Songwon Junior colleage
365 Songha-dong, Nam-gu, Gwangju, Korea 503-742
suhjs@songwon.ac.kr

Abstract — this paper discusses an actual operation of a monitoring system for fuel cells which will probably be substituted for the next generation environmental energy sources. Through this demonstration from monitoring system of 1kW fuel cell for a house, the electrical and heat efficiency of fuel cell is able to be improved, and several faults of fuel cell, such as high expense and weak durability, would also be remedied dramatically.

I. Introduction

As energy demand increases and a crude oil price suddenly raises through the world, international energy resources security competition is extremely intensifying. Also, more stable settlement toward energy structure and eco-friendly energy sources are required. As new renewable energy sources need, fuel cell technology is significantly considered more and more. Among fuel cells, macromolecule electrolyte (PEFC, PEMFC) consists of several parts – a reformer using LPG LNG to provide hydrogen, a stack supplying heat and electricity, and an inverter converting into AC, and so on. PEFC or PEMFC, which provides heat and electricity at the same time, has higher and reliable efficiency than the internal-combustion engine that is using fossil energy. The heat from PEFC or PEMFC, besides, can be converted into a heating system at home.

1kW and bigger fuel cell systems for home would be applied to flexible commercial purposes such as public offices and stores in the future because it does not make any noise and pollution. Dongshin University participated in a research with HYGAS and KOGAS, and build 1kw fuel cell for home by way of showing an example in the library. This research was divided through 3years, and 40 systems are spread over Korea, and one of those was only built among universities. This paper is highlighted at electric efficiency and disadvantages of a thermal efficiency fuel cell through actual proof driving evaluation of a fuel cell for 1kW home. This will also precede with the aim of efficiency analysis of a fuel

cell and durability test for commercial through monitoring by effort to overcome an expensive price, verification of durability, an aesthetic commercialization etc.

II. PEMFC THEORY

[1]~[2]Fuel cells (fuel cell) are energy transduction devices changing the chemical energy that fuel had by a direct electric energy by chemical reactions. While leave an electrolyte in an interval, and leave, and an electrode is located with a form of a sandwich, and leave, and hydrogen ion and oxygen ion pass through electrodes, generate electric current, and structure of a fuel cell produces lesson heat and water as by-products. Use nature gas, methanol, gasoline, fuel reformer for various fuel to be able to be used of a back, and be cleared up to hydrogen to a fuel cell do, and use. Be ordinary 0.7V, and multiplied these unit batteries like numbers to have been connected if we connect a lot of dogs in a series, and voltage is got voltage of a unit battery

A fuel cell becomes distribution according to kinds of an electrolyte so as to be various. These papers it is possible for a macromolecule fuel cell to produce electric column at the same time while reducing a discharge material development efficiency higher the existing generator liquor, and pollution as technology about the development device which converts directly a chemical energy by an energy electric by electrochemistry responses of hydrogen manufactured to things about the PEMFC which is a solid macromolecule electrolyte fuel cell.

Be cogeneration development systems a fuel cell composed to stack, fuel supply wealth, power process wealth, a peripheral device etc. convert directly a chemical energy by an electric energy unlike the existing energy transformation methods, and having high efficiency, high output, low pollutions and characteristics of the low noise, and can do eco-friendly at a discharge of pollution matter having been low, and the importance is growing gradually on a large scale larger.

III. Actual operation investigation.

III.I 1kW PEMFC System in Dongshin University

Fig.1 is an installation picture of a 1kW class PEMFC system installed in Dongshin University center libraries.

Be composed to a warm water tank, system output and monitoring systems to play stack reformer, system wealth to have been composed to BOP and cooling water role and the circulation tank which lowered temperature inner a fuel cell for a measurement environmental developments facilities me such as temperature/the outside.

It has been installed, and details refusal of a system is same as a table.1

Fig.1 A Dongshin University 1lW class fuel cell actual proof installation picture.

item	details
power generation efficiency (LHV)	30%over(efficiency 70over)
power generation time	3,000kWh over/ year
output	AC 1kWclass, 60Hz,220V
Limit efficiency	24%
fuel	natural gas
Fuel flux	4.4L - 4.8.L/min
output	1kW(grid copulative 220VAC)
Start up time	1hr soon(stop 2hrsoon)
Cogeneration	•Heat recovery :55~68℃ • hot water flux : 0~1.8LPM

Table1 I System Spec.

III.II Configuration of a monitoring system and measurement

3) The monitoring system established monitoring works through measurement systems of 2parts and evaluation / measurement, a parameter stored is same as a table 2 in these papers.

item	Unit
Electrical efficiency	%
thermal efficiency	%
overall efficiency	%
accumulation power generation	MWH
fuel amount	LPM
moving /stop test(WSS))	moving /stop possibility (WSS)
Continuous moving test	Continuous moving possibility
Electrical output test	V,I,W,Hz,THD
Hot water flux	LPM
Hot water temperature test	oC
emergency stop	emergency stop record
exhaust (gas) analysis	NOx, SOx,CO,CO2
noise analysis	dB

Table II Operating Parameter

A monitoring parameter a measurement/storage/vote working

Be made so that Fig. 2 is showing a screen main monitoring of an actual proof system, and measurement can store the amount of flowing water of each gas, total output, cumulative generation quantity., temperature, driving state, system driving efficiency etc.

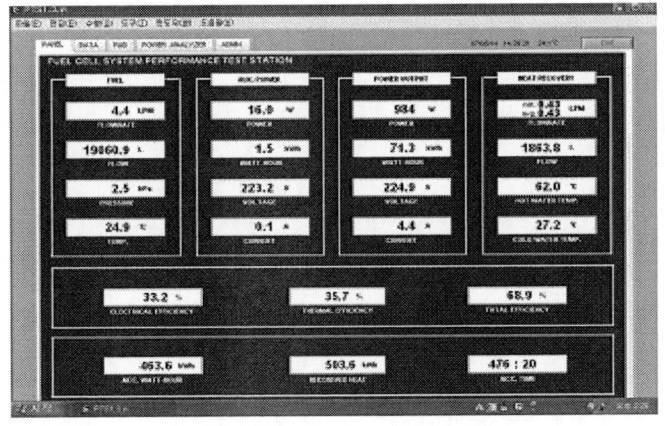

Fig.2 Screen . main monitoring.

4) Showed the files which data measured are stored in PC, and be stored with text file formats by fixed time zones, and glance, and weekly, data analysis between years by months is possible, and the

Fig.4 characteristic analysis of a development system is possible through this

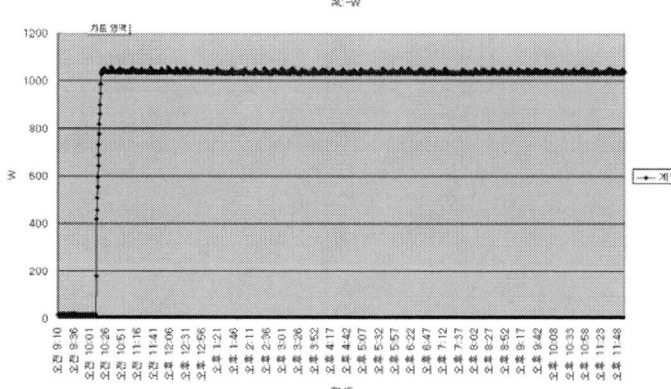

Fig.4 A storage of data measured.

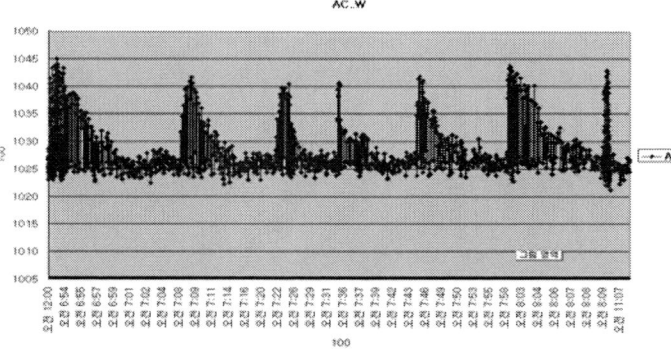

Fig .5 It takes approximately 1 hour until normal output from early operations.

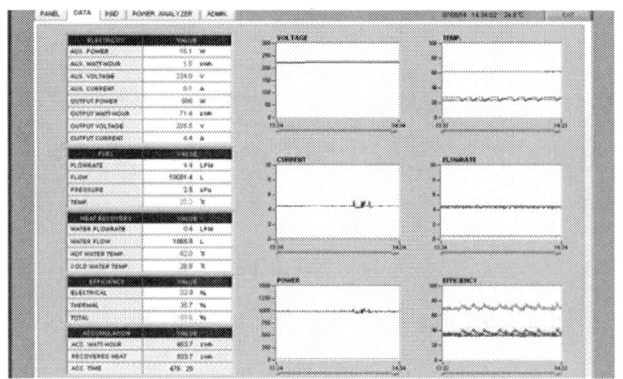

Fig.6 Output range is a more than 1kW desire in case of a normal operation.

Fig. 7 Date measurement system

⇨ 월간 자료   해양도시가스 ▾ GS-18호기 ▾   2007 ▾ 년 4 ▾ 월 검색

2007년 4월   해양도시가스(GS-18호기)

설치	전기효율 (%)	AC출력 (W)	LNG소모량 (ℓ/min)	내부온도 (℃)	에 러 상 태				
					스택	인버터	BOP	Cogen	제어DCI
2007-04-10	30.3	943.5	4.9	23.1	0	0	0	0	0
2007-04-16	35.7	1,020.6	4.4	29.1	0	0	0	0	0
2007-04-17	36.5	1,033.2	4.3	31.2	0	0	0	0	0
2007-04-18	36.8	1,037.1	4.3	33.1	0	0	0	0	0
2007-04-19	36.9	1,033.6	4.3	34.1	0	0	0	0	0
2007-04-20	36.8	1,030.3	4.3	35.1	0	0	0	0	0
2007-04-21	36.6	1,025.0	4.3	35.9	0	0	0	0	0
2007-04-23	37.1	1,095.9	4.3	35.5	0	0	0	0	0
2007-04-24	37.0	1,031.2	4.3	35.6	0	0	0	0	0
2007-04-25	36.8	1,026.7	4.3	36.1	0	0	0	0	0
2007-04-26	36.7	1,025.2	4.3	36.4	0	0	0	0	0
2007-04-27	36.8	1,023.6	4.2	36.8	0	0	0	0	0
2007-04-28	36.8	1,023.9	4.2	36.1	0	0	0	0	0
2007-04-29	36.7	1,022.9	4.2	35.9	0	0	0	0	0
2007-04-30	36.7	1,021.4	4.2	36.0	0	0	0	0	0

Fig.7 An Internet monitoring screen.

Fig.7 is a monitoring screen of the actual proof system that can confirm through the Internet. Be made so as to be able to confirm monthly production, efficiency, the real-time driving present situation etc.

IV.   Conclusion.

This monitoring system of 1kW fuel cell supplements a disadvantage for the commercial perspective as initial steps, and deduces technical efficiency; cost down, reliable data measurements, and so on. To make the appropriate the infra-structure of PEMFC, it is recommended that are collecting and analyzing data from monitoring system. Besides, the intellectual education and commercialization of fuel cell are systematically needed. From divert operational data based on monitoring system, furthermore, many synergies to spread out of fuel cell are derived from this system.

References

[1] C. S. Kim , "Recent R & D Status of Polymer ElectrolyteMembrane Fuel Cell", The Polymer Society of Korea, 15, Startpage 550,Endpage 561, Totalpage 12, 2004 f
[2] C. S. Kim, "Development of the 5kW Class Polymer Electrolyte Fuel Cell System for Residential Power Generatio", KOREA INSTITUTE OF ENERGY RESEAFCH, 1999f
[3] hygas CO.Ltd, "house hold fuel cell monitoring business", 2007 f
[4] H. J. Choi , I. S. Cha, G. Y. Jeon, "Monitoring about verification system, PEMFC on the 1kW " The Korean insritute of power electronics Jeolla district 2007f

Performance Evaluation and Analysis of 50kW Grid-Connected PV System

Jeong-Min Park, Zeong-Guo Piao, Youn-Ok Geum-Bae Cho, Hyung-Lae Baek

Dept. Electrical Engineering, Chosun University, Seo-seok Dong , Dong Gu Gwang-Ju, Korea

netpius@korea.com

Abstract—**This paper summarizes the results of these efforts by offering a photovoltaic system structure in 50kW large scale applications installed in Chosun University dormitory roof. The status of PV system components, are inter-connection and safety equipment monitoring system will be summarized as this article. This describes configuration of utility interactive photovoltaic system which generated power supply for dormitory. In this paper represent 50kW utility PV system examination result.**

I. INTRODUCTION

Photovoltaic system is using the clean, infinite energy source, no air pollution, noise and vibration. PV system operated without fuel convoy, rotation surface, high temperature and high pressure, So it is easy to maintenance, long life cycle, simple installation. PV market grows over 30% and 40% every year.

Photovoltaic power systems convert sunlight directly into electricity. A residential PV power system enables a homeowner to generate some or all of their daily electrical energy demand on their own roof, exchanging daytime excess power for future energy needs

While compare preexisted generation of electric power method with PV system, PV system is more expensive, lower solar energy density, convert efficiency and depend on the weather and natural environment.

Furthermore, fossil fuels and uranium are non-renewable resources. But the only resource needed to power a solar cell is sunlight. Solar energy is clean, quiet, abundant and a renewable energy source, which produces no pollution to the environment. With the growing concern over green-house gas emissions and other environmental issues, renewable energy sources are being increasingly adapted in many applications. Grid-Connected photovoltaic system is one of the most promising applications of photovoltaic systems. The solar panels can be effectively placed on roof tops, exterior of a building, outer walls etc. Wherever there is a good face lift towards the direct sun exposure. According to the requirement PV system needs to maintain highest possible exposure to the sunlight is established by Grid-Connected and confirmed system reliance.

And interest to make maximum efficiency into low expense is converged.[1] This paper described about composition and control of 50kW Grid-Connected photovoltaic system through load of dormitory.

In this paper presents performance evaluation and analysis of PV system from 2003 to 2006.

II. PHOTOVOLTAIC GENERATION SYSTEM

A. Photovoltaic System Composition

Fig. 1 is overall schematic diagram of dormitory supply 25kW system Grid-Connected type solar photovoltaic power generation system. Two of 25kW of PV array are located rooftop of dormitory roof. PV system supply to electric power through inverter by system Grid-Connected type middle scale 50kW system. Also, stored real-time data and operated monitoring system was installed completely. Therefore, we can analyze running condition with load. Also, using real-time monitoring system can monitor and observe stable supply of electric power. In this system, It is generate electric power of maximum 50kW because is formed 1000 array in two by array 500 in each array. Solar cell had been manufactured by single crystal silicon. Open voltage of module is 21V, short current is 3.35A, rated capacity is 53Wp, power conversion efficiency is 14%. Array can generated maximum 50kW power being consisted of module 1000.

Monitoring system consist of main 3 parts. Part I is voltage and current in each array measurement through connection group. Part II is measuring is available in PC to load power, utility power from generated power. Part III is present array's voltage and current exchanges DATA real time by PC through RTU and RS-422 communication via transduser. The generated current, voltage value and power value of load exchange through digital bidirectional integrating wattmeter.

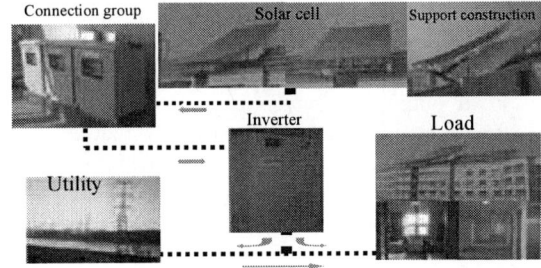

Fig. 1. 50kW Grid-Connected PV system

III. EXPERIMENT RESULT

A. PV Array Characteristic

Fig. 2 is showing conversion efficiency of PV array characteristic. Array conversion efficiency is greatly irrespective in change of quantity of solar radiation. If quantity of solar radiation reaches fixed degree, we can verify that efficiency is rising continuously.

Individually fig. 3 shows monthly solar cell Array generated electric power quantity from 2003 to 2006 that is measuring instrument.

During the measurement period that generated all electricity are 62,146kWh in 2003, 70,455 kWh in 2004 69,241kWh in 2005 and 57,894 kWh in 2006.

PV Array conversion Efficiency displayed 9.19% in 2003, 10.02% in 2004, 9.97% in 2005 and 9.70% in 2006.

According to module surface temperature, PV Array conversion efficiency changed.

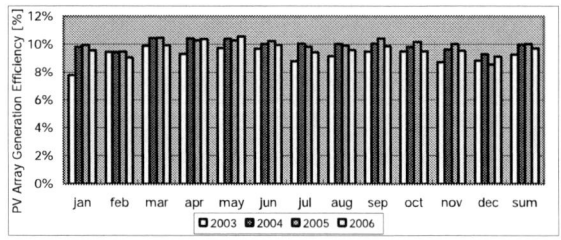

Fig. 2. PV array conversion efficiency from 2003 to 2006.

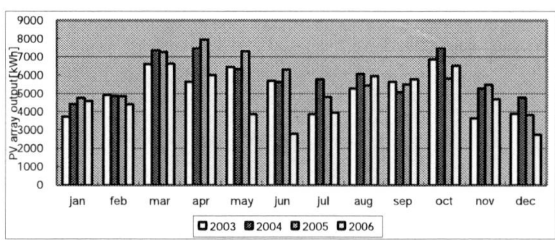

Fig. 3. PV Array Output Power Quantity.

B. PCS Characteristic

Fig. 4 is showing PCS efficiency characteristic. Because PCS efficiency is high changelessly quantity of solar radiation, if output is almost closed to rated power, PCS efficiency is kept more than 90%. This PCS that rated power does correctly MPPT control. However, we know that do not correctly MPPT control if generating power falling down. The performance ratio shows special quality such as generation efficiency and array conversion efficiency

Fig. 6 shows PCS efficiency by time. In the case of over 200W/m² sunlight conversion efficiency of Power conditioning system keeps regular characteristic. But in the case of low sunlight changes conversion efficiency.

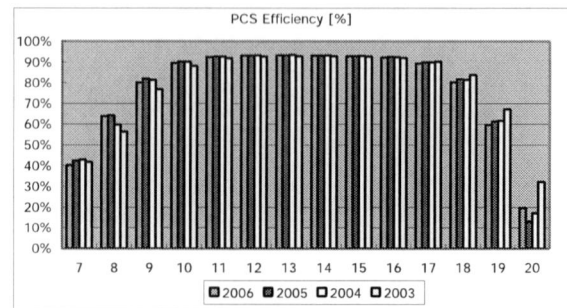

Fig. 4. PCS efficiency by month.

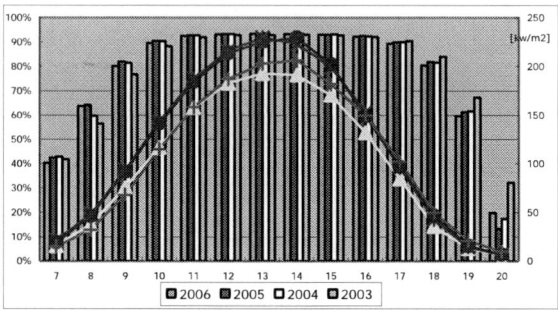

Fig. 6. PCS efficiency by time.

C. System Performance

Fig. 7 and fig. 8 show monthly output generation quantity of system and generation efficiency and characteristics of utilization ratio.

Fig. 9 shows Performance ratio.

Performance ratio has 75.09% in 2003, 81.70% in 2004, 82.56% in 2005 and 79.16% in 2006.

Damage factor is construed by shadow influence, inverter damage, MPPT miss matching, Array temperature rise etc.

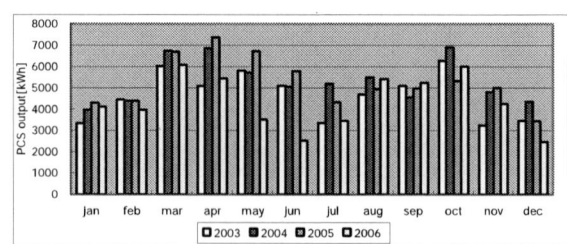

Fig. 7. Inverter Output Power quantity.

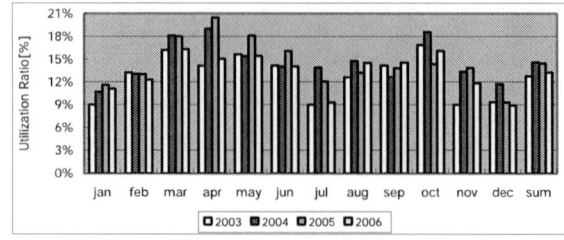

Fig. 8. Generation Efficiency and Utilization Ratio.

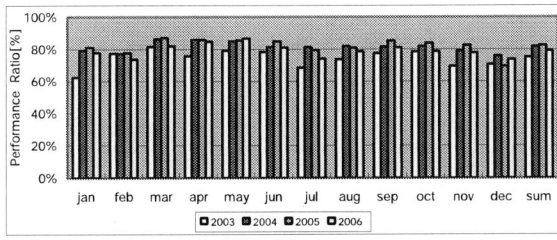

Fig. 9. Performance Ratio.

IV. CONCLUSION

In this paper, we discussed analysis driving characteristic of 50 kW Grid-Connected PV system, evaluated and achieved examination about reliability of system.

This effort is expected to lead to even higher performance research 50kW PV System in the future.

To be used by commodity establishing photovoltaic generation system, supporting research of spot application examination of full scale, it is performance existence and driving technique etc. is achieved

With this background, judged that brought photovoltaics technology elevation but is limit in that good story, authoritativeness and electric power conversion.

REFERENCES

[1] A. Bcd. And K. I. Jkl, *Introduction to PESC*, 1st ed. Aachen, Germany: Test, 2004W.-K. Chen, *Linear Networks and Systems* (Book style). Belmont, CA: Wadsworth, 1993, pp. 123–135.

[2] Ruther, R., de Silva, A.J.G., Montenegro, A.A., Salamon, I.T., Araujo, R.G., □The petrobras 45.5kwp, grid-connected pv system: a comparative study of six thin-film module types operating in brazi□ □Proceedings of the 29th IEEE Photovoltaic Specialists Conference□, pp.1440-1443 , 2002.5.

[3] Kotsopoulos, A., Duarte, J.L., Hendrix, M.A.M., Heskes, P.J.M., □Islanding behaviour of grid-connected PV inverters operating under different control schemes□,□Proceedings of the 2002 IEEE 33rd Annual IEEE Poser electronics Specialists Conference□ - Volume 3 , pp.1506-1511 , 2002.6.

[4] Kazuyoshi Tsukamoto, □Photovoltaic Power System Interconnected with Utility□, Proceedings of the American Power Conference, pp.276~ 281.

A Study on the Performance Improvement of DVR System using EDLC

Wang-Rae Kim, Chang-Jin Lim, Byung-Ki Kwon,
Chang-Ho Choi
POSCON Corporation R&D Center
Techno-Complex, Korea University, Anam-Dong
5Ga, Seongbuk-Gu, Seoul, 136-701
Email: wrkim@poscon.co.kr

Hee-Jong Jeon, Jin-Geun Shon[*]
Soongsil University
Department of Electrical Engineering
Sangdo 5-Dong, Dongjak-Gu, Seoul, 156-743
*Kyoung-won University
Email: hjjeon@ssu.ac.kr

Abstract—In this paper a novel detection technique for voltage sag and phase of an input voltage is proposed to improve the performance of DVR system. With the proposed voltage sag detection method the detection time for the voltage sag is reduced and the performance is improved of DVR system. In addition to the conventional phase detection algorithm of an input voltage, simple logic circuit is adopted to detect the input voltage direction. The erroneous information around the zero-crossing points can be successfully removed by the proposed technique and the effectiveness of the proposed technique is verified by the experiments. EDLC, environment-friendly, and the high efficiency charging/discharging technique are adopted as an energy storage device. Experimental results are presented to verify the effectiveness.

I. INTRODUCTION

In the past several decades, consumer's primary concerns were the reliable supply of the electric power. But, nowadays, power quality problem is emerging as a hot issue reflecting consumer's interests due to the proliferation of the equipments sensitive to the electrical distortion in the modern life. So, the economic loss by the power quality problem has been being increased rapidly. [1][4]

UPS has been used widely as a most realistic method for momentary interruption and voltage sag/swell. Therefore, it has been used mainly for the critical load. However, one of the drawbacks of the UPS is that its compensation method is full voltage compensation taking whole load capacity. This cause the lower efficiency and higher system cost. [3][5][6]

Recently, the DVR is researched plentifully in order to overcome the weak point of the UPS. During the normal operation, the DVR is maintained in the standby state. However, when the power quality problems occur such as instantaneous voltage sag or momentary interruption, the DVR compensates them. Since it operates only the time during the period of power quality problem, the efficiency is high. However, there is problem that it requires an electrolytic capacitor bank allowing high rates of charging/discharging currents. But, it is disadvantageous that the compensation time is not continuous than UPS with battery. Also, DVR need fast voltage measurement and phase locking for correct compensation.

In this paper, novel voltage measurement technique and phase detection algorithm for the DVR is proposed to improve the performance of the DVR. The proposed system

adopts the EDLC that is environmentally clean and novel control circuits for high efficient operation for charging/discharging EDLC.

II. COMPOSITION OF DVR SYSTEM

A. Modeling of DVR system

DVR system is used to protect sensitive load from voltage sag. Therefore, system circuit of DVR is consisted with Fig. 1. And compensation voltage (V_{comp}) is added in series to the sensitive load through matching transformer. Fig. 2 shows the system configuration of the system. R_f, L_f, C_f are internal resistance, inductance, and capacitance of LC filter. Assuming that the sensitive load is R-L as shown in the figure 1, the state equation of the system can be expressed as (1).

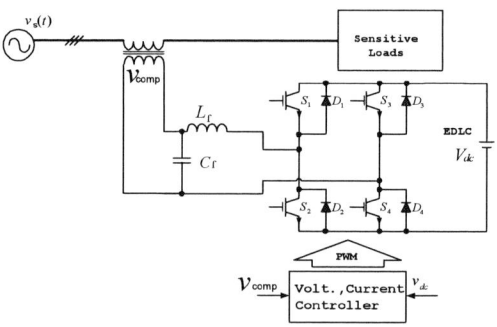

Fig. 1: Configuration of the DVR system

$$\frac{d}{dt}\begin{pmatrix} i_i \\ v_{load} \end{pmatrix} = \begin{pmatrix} -\dfrac{R_f}{L_f} & -\dfrac{1}{L_f} \\ \dfrac{1}{C_f} & 0 \end{pmatrix}\begin{pmatrix} i_{load} \\ v_{load} \end{pmatrix} + \begin{pmatrix} \dfrac{1}{L_f} & 0 \\ 0 & -\dfrac{1}{C_f} \end{pmatrix}\begin{pmatrix} v_s \\ i_{load} \end{pmatrix} + \begin{pmatrix} \dfrac{1}{L_f} \\ 0 \end{pmatrix} \cdot v_{inj} \quad (1)$$

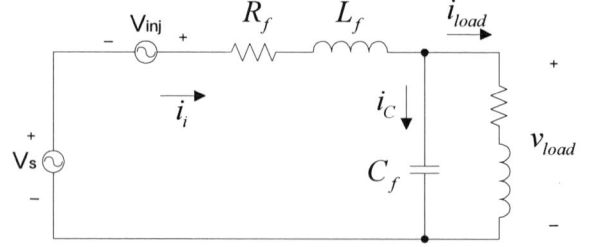

Fig. 2: Equivalent circuit of DVR system

B. Feedback control

In the DVR system voltage feedback is essential to the control the voltage to be injected to the load. Also, controller can compose double loop controller using inductor current or capacitor current to improve dynamic characteristic. Fig. 3 expresses the block diagram of the DVR controller in which transformer is combined.

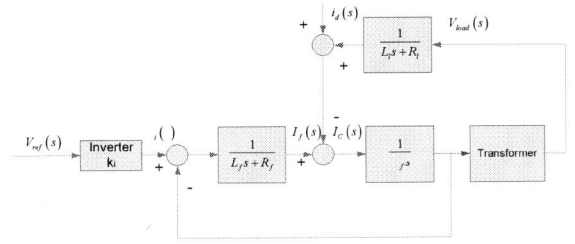

Fig. 3: Block Diagram of the DVR System Control

Inverter controller is operated considering disturbance of load current (i_d) and load voltage (V_{load}) according to the command of compensation voltage (V_{ref}). Fig. 4 is the controller model with the inductor current feedback. Fig. 5 is bode plot of inductor current feedback control. This controller shows that the low order harmonic is amplified.

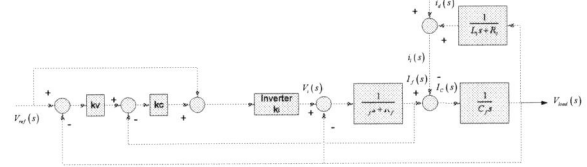

Fig. 4: Inductor current feedback control system

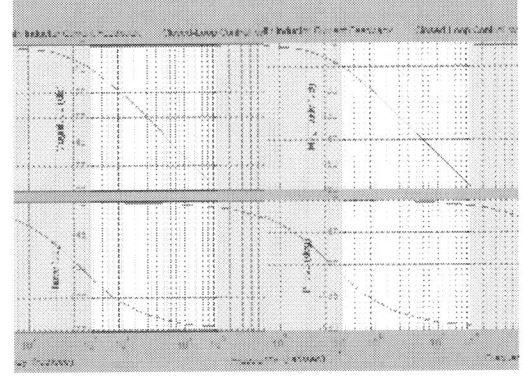

Fig. 5: Output voltage of inductor current feedback control

Capacitor current feedback controller (Fig. 6) is suitable for the DVR controller because it reduces the disturbance of load. Fig. 7 is the bode plot for the capacitor current feedback controller.

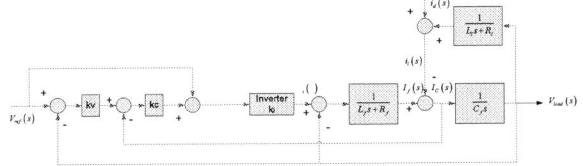

Fig. 6: Capacitor current feedback control system

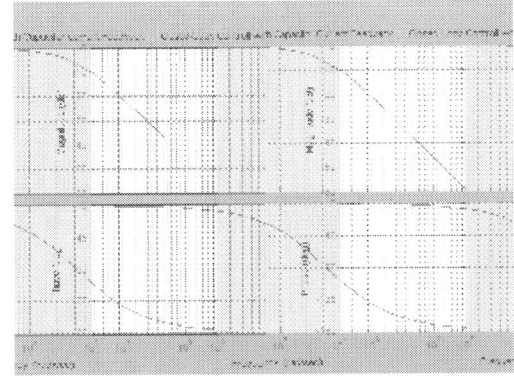

Fig. 7: Output voltage of capacitor current feedback control

III. VOLTAGE MEASUREMENT ALGORITHM

A. Measurement algorithm of instantaneous voltage sag

Usually, conventional method of the voltage detection is advantageous in that the techniques utilize the average value of the voltage. The benefit of the average measurement technique is that it is possible to correct the voltage measurement and thus it is robust in the noisy environment. However, this technique has a disadvantage that it requires at least a half period in measuring the voltage. If the periodic average of the sine wave is calculated, we need to correct zero crossing detection for time integral calculation. In order to calculate the AC in DC value with that method, equation (2) should be used.

$$V = \sqrt{\frac{V_1^2 + V_2^2 + V_3^2 \cdots + V_N^2}{N}} \qquad (2)$$

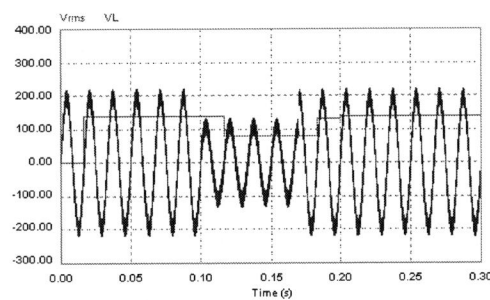

Fig. 8: Waveform results of average measurement method

Fig. 8 shows the simulation results of the conventional measuring technique of average concept. This is displayed AC to DC conversion waveform of voltage sag. It is occupying very important factor in DVR. Voltage sag detection by average concept needs time of smallest period like Fig. 8. Specially, RMS cannot be calculated within one measurement cycle. Therefore, DVR needs high speed measurement technique for the correct compensation.

In this paper, moving average techniques is proposed. Fig. 9 displays concept of instantaneous moving average techniques. Proposed moving average techniques need access

978-1-4244-1871-8/07 $25.00 © 2007 IEEE

of RMS concept to express AC amount by DC amount. Therefore, voltage measurement can detect repeatedly using (2) so that measurement value may have instantaneous concept. In the simulation, 83 samples were taken for a half period of AC waveform. Then the sampled data are averaged in every sampling period. Basic idea is the ring buffer. First sampling value of ring buffer is erased for the first time and next time sampling value is stored. Therefore, this can calculate average in each sampling time. This method can not receive correct measurement value until all 83 sample values are acquired. Also, zero crossing should be considered for correct calculation.

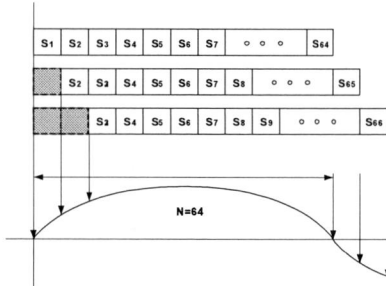

Fig. 9: Block diagram of instantaneous moving average

Fig. 10 shows the simulation result of the instantaneous moving average technique. Detection time is faster than Fig. 8.

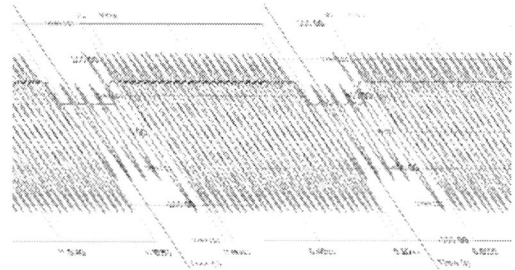

Fig. 10: Waveform results of moving average method

B. Improved phase angle detection algorithm

Correct phase angle information of power conditioning device for electric power quality compensation to utility voltage is essential. Because current or voltage command must become synchronism in phase angle of utility voltage power factor control or active power control, reactive power control and harmonic current compensation. Therefore, DVR systems are very important voltage sag detection algorithm and correct phase angle detection. From example, first sampling value of instantaneous moving average technique is deleted for the first time and calculation of RMS value moves. Correct zero crossing has to be considered for calculation that this method is correct in sampling number during first half period too. In addition, zero crossing is must considered for correct phase lock of utility voltage and DVR output voltage.

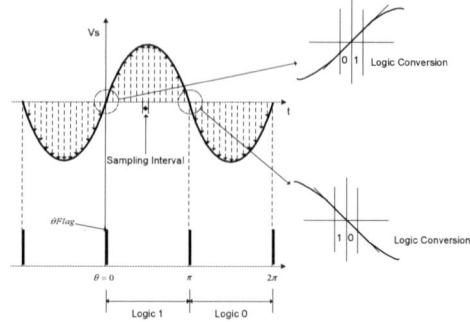

Fig. 11: Principle diagram of advanced theta angle detection

Therefore, this treatise used improved algorithm of Fig. 11. Phase angle θ at 0-degree, 180-degree and 360-degree voltage magnitude comparison for zero crossing detection is escaped. This method is reduced measurement error by noise remarkably, so that phase angle θ may compare adding logic for rising and falling at the edge moment with right of Fig 11.

IV. EDLC CHARGE-DISCHARGE CONTORL CIRCUIT

If generator is charge capacitor that is discharged perfectly likes Fig. 12(a), charging current I can pass as V/R. However, the load is short-circuited if seen from inside of the generator. Therefore, if there is no protection circuit, generator would be damaged. In addition, capacitor terminal voltage V is raised with the ratio of $V = Q/C$ by electric charge Q that is passed by charge. However, if the capacitance C is small, then the charging is finished in a extremely short time. In this case, voltage source charge is not appropriate and current source control circuit is needed to charge the EDLC. Thus, it is not proper to charge the EDLC as a voltage source.

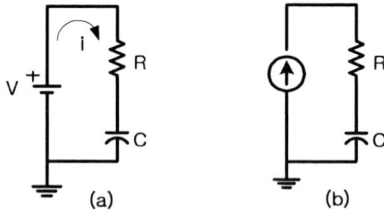

Fig. 12: Equivalent circuit for (a) voltage (b) current source of EDLC

Current i can appear by (3) in voltage source circuit such as Fig. 12(a).

$$i = \frac{V}{R}\exp\left(-\frac{t}{CR}\right) \qquad (3)$$

In addition, electric power that consumes in resistance R and express with (4). This is the same as the equation to display energy that can be accumulated to capacitor. Therefore, energy efficiency reaches 50% if

978-1-4244-1871-8/07 $25.00 © 2007 IEEE   532

charging/discharging is executed as a constant voltage source.

$$\int_0^\infty i^2 R\, dt = \frac{1}{2}CV^2 \qquad (4)$$

However, charging/discharging efficiency by current source is as Fig. 12(b). if charge or discharge for t time by constant current, electric charge Q is the same as (5).

$$Q = I \cdot t \qquad (5)$$

Therefore, electric energy A that store to EDLC is expressed with (6). Moreover, electric energy B that spent in resistance R is expressed with (7).

$$A = \frac{1}{2} \cdot \frac{Q^2}{C} \qquad (6)$$

$$B = I^2 R \cdot t = R \cdot \frac{Q^2}{t} \qquad (7)$$

If ratio k of damage and accumulative of electric energy using (6) and (7), become with (8).

$$k = \frac{B}{A} = \frac{2RC}{t} \qquad (8)$$

Ratio k of damage about storing accumulation amount and charge-discharge time t is inverse proportion in (8). As a result, charge-discharge efficiency can current source control circuit by control. In addition, charge efficiency E_C and discharge efficiency E_D by constant current source are expressed by (9), (10).

$$E_C = \frac{A}{(A+B)} = \frac{t}{(t+2RC)} \qquad (9)$$

$$E_D = \frac{(A-B)}{A} = 1 - \frac{2RC}{t} \qquad (10)$$

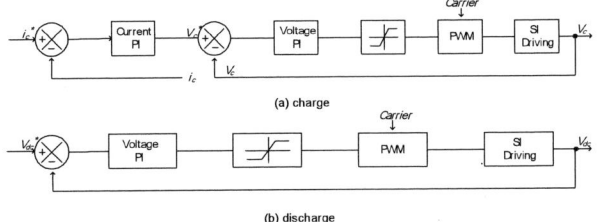

Fig. 13: Charge-discharge control system of EDLC

Therefore, EDLC charge-discharge circuit with Fig. 13 switching of charge and discharge composition of possible controller requires. Current source circuit does composition possibility by typical PI controller and voltage source circuit can get by duty ratio control of DC/DC converter.

V. AN EXPERIMENT AND RESULT CONSIDERATION

A. Measurement of voltage and phase detection

In this paper, voltage detection algorithm applied instantaneous moving average technique is proposed to Fig. 9. Fig. 14 is voltage detection result that uses such technique. Show that RMS value is following well according to variableness of input voltage. Fig. 15 is displays measurement waveform in detail magnifying time. In addition, Fig. 16 is result of improved phase detection algorithm that proposes to Fig. 11.

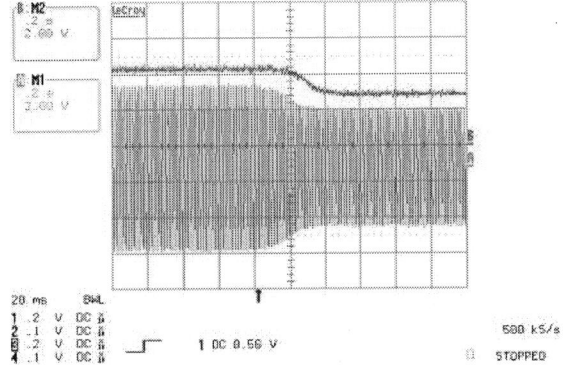

Fig. 14: RMS value with the input voltage variation

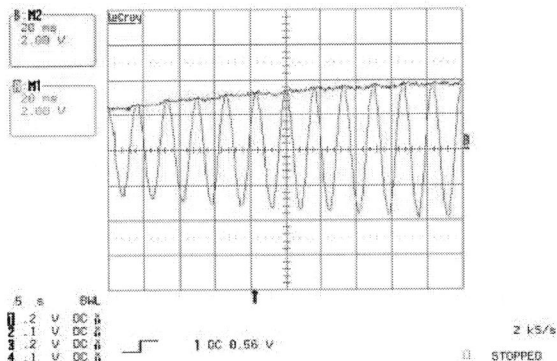

Fig. 15: RMS value and input voltage(Time expansion)

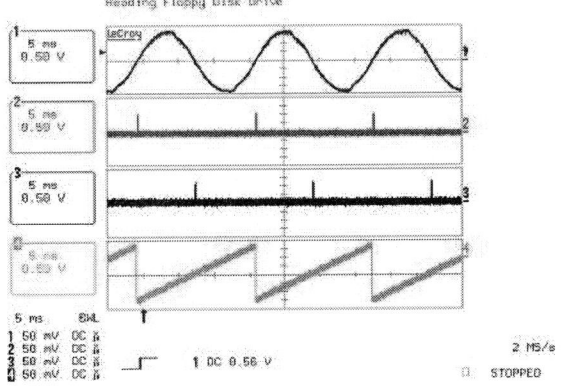

Fig. 16: Phase and edge signal of input voltage

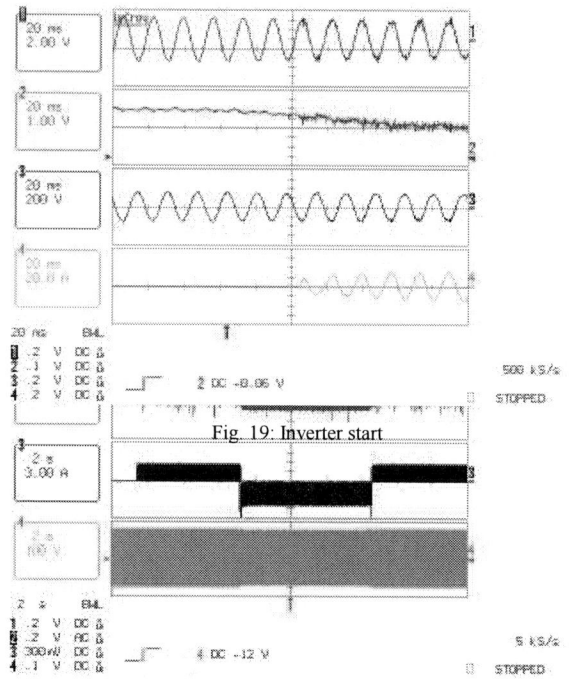

Fig. 19: Inverter start

Fig. 17: Charge-discharge of EDLC

B. Charge-discharge control circuit

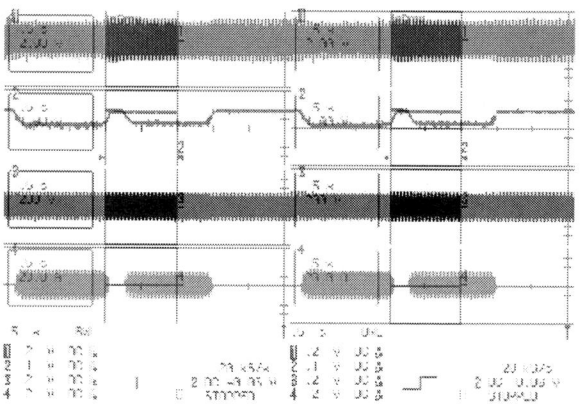

Fig. 18: Operating of DVR by voltage sag repeat

Fig. 17 is experiment result that is charge to current source and discharge to voltage source for EDLC's high efficiency operating. Normal times EDLC (channel 3) is charged by constant current. In addition, we can see that current direction changes and discharges when voltage sag is happened. Fig. 18 is DVR operating waveform when voltage sag is repeated. Fig. 19 and Fig. 20 are waveform of start and stop of DVR inverter operating.

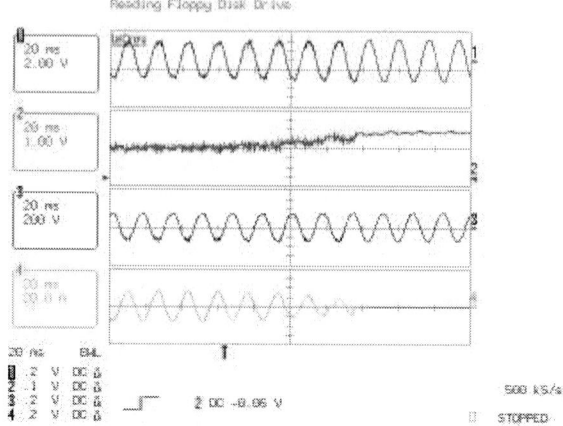

Fig. 20: Inverter stops

VI.   CONCLUSION

Voltage sag is an important problem in the electric power quality. Research about DVR has been performed vigorously to solve this quality problem. In this paper, we proposed an instantaneous moving average technique in the voltage measurement of DVR system and minimized the detection time and error. The proposed algorithm has been proven to be effective by the experiments and also accuracy of the phase detection has been improved. In addition, various charging/discharging techniques for EDLC have been proposed and the proposed technique has proven valid.

REFERENCES

[1] D. Mahinda Vilathgamuwa *et al*, "Voltage Sag Compensation With Energy Optimized Dynamic Voltage Restorer," *IEEE Trans. Power Delivery*, vol. 18, NO3, pp.928-936, July 2003

[2] Ribeiro P.F, Johnson B.K, Crow M.L, Arsoy A, Liu Y, "Energy storage systems for advanced power applications," *IEEE*, Vol. 89, Issue 12, pp.1744-1756, Dec. 2001B. Smith, "An approach to graphs of linear forms (Unpublished work style)," unpublished.

[3] W.E.Kazibwe *et al*, "Power quality : A Review", *IEEE Computer Applications in Power*, Vol.13, NO.1, pp39-42, January 1990

[4] Tounsi *et al*, "Series Compensator for Voltage Dips: Control Strategy", EPE, pp.4929-4934, 1997.

[5] M. F. Granaghan, "Dynamic Sag Corrector : Cost Effective Industrial Power·Line Conditioning", IEEE IAS, pp.1339-1344, 1999

[6] Y.Sekine *et al.*, "Present state of momentary voltage dip inferences and the countermeasure in Japan," CIGRE 36-206, September.1992.

[7] Alexander Kara *et al*, "Power Supply Quality Improvement with a Dynamic Voltage Restorer (DVR)," IEEE APEC, Vol. 2, pp.986-993, 1998.

Three-Level VSI Based Low Switching Frequency 10 MVA STATCOM in Reactive Power and Harmonics Compensation

Panu Lauttamus
Tampere University of Technology
Institute of Power Electronics
P.O. Box 692, FI-33101 Tampere, Finland
Email: panu.lauttamus@tut.fi

Heikki Tuusa
Tampere University of Technology
Institute of Power Electronics
P.O. Box 692, FI-33101 Tampere, Finland
Email: heikki.tuusa@tut.fi

Abstract—This paper examines the reactive power and harmonics compensation performance of a low switching frequency three-level voltage-source inverter based 10 MVA STATCOM. The intention is to mitigate supply current harmonics up to the 13th order using a switching frequency between 600-1200 Hz. However, due to the low switching frequency the STATCOM response time is not adequate for harmonics compensation because PWM modulation adds a notable delay to the current control loop. Four different harmonic detection methods equipped with control delay compensation feature are examined to improve the STATCOM performance. The STATCOM operation is studied by means of Simulink/Simplorer simulations in steady-state load conditions, during step load changes and in electric arc furnace compensation. The results show that the STATCOM performance can be extended to harmonics mitigation under steady-state load conditions. Under non-periodic load conditions the accuracy of harmonic detection and delay compensation decreases and reduces the STATCOM performance.

I. Introduction

A significant amount of current harmonics in power systems is generated by power electronic based loads [1]. The current harmonics impair the power quality and result in detrimental phenomena such as voltage flicker, increased thermal dissipation and malfunction of sensitive equipment. Voltage-source inverter based active compensators have proven to be a successful solution for local harmonics compensation and power quality improvement [2]. They offer good dynamics within the limits set by control system and switching frequency and can be used independently or combined with tuned passive filters [1].

An indispensable part of active compensator implementation is a proper method for harmonic detection and current reference generation [3]. The polluting currents can be detected either from the sensed supply or load currents and a variety of both time-domain and frequency-domain based methods have been studied [2-10]. Two main principles for harmonic detection are total and selective harmonic detection. The idea of total harmonic detection is to merely separate the sensed load/supply currents into fundamental and non-fundamental components [5, 6]. In contrast, the idea of selective detection is to detect only individual harmonics from the sensed currents [2, 7-10]. The advantage of total harmonics detection methods is simple implementation and fast detection speed. Selective methods can be used to decrease the required compensator rating but

the implementations are rather complex and the detection speed is often inferior to the total detection methods.

In several MVA high power applications the active compensator switching frequency is limited to some hundred Hz to prevent thermal overloading of the switching devices. Low switching frequency and PWM modulation results in significant control delay which adds to the current control loop and decreases the harmonic compensation capability. However, harmonic detection methods with control delay compensation feature can be used to overcome the control delay resulting from the low switching frequency [2, 9-11].

The delay compensation can be implemented by manipulating the phase-angle of the detected harmonics. For example, phase-lead compensators can be applied if selective harmonic detection is used and each harmonic component can be processed independently [8, 9]. In the case of total harmonic detection, e.g., instantaneous reactive power theory [4] or synchronous (d, q) reference frame based detection [5, 6], the harmonics can not be processed individually and control delay compensation requires different implementation. Possible solutions include current reference extrapolation [12] or, if the currents are strictly periodic, current waveform forward prediction using stored current data [2, 11].

This study focuses on improving the harmonics compensation performance of the low switching frequency three-level 10 MVA STATCOM shown in Fig. 1. The specified switching frequency is 600 – 1200 Hz. The aim of the study is to use the STATCOM to mitigate supply current harmonics up to the 13th order. However, the low switching frequency leads to a notable current control delay due to PWM modulation and the desired harmonics compensation performance can not be achieved. Four harmonic detection methods equipped with control delay compensation feature

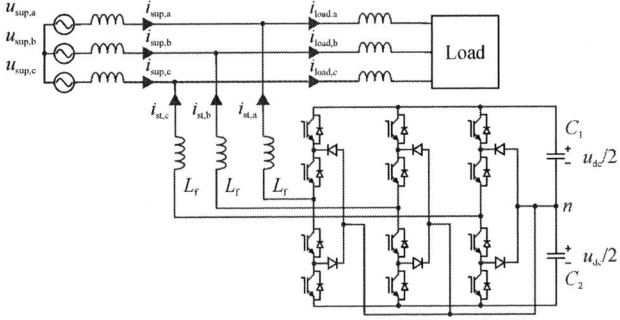

Fig. 1. Three-level VSI based STATCOM.

978-1-4244-1871-8/07 $25.00 © 2007 IEEE

are examined to overcome the current control delay; 1) Band-pass filter based selective harmonic detection [9], 2) Multiple rotating (d, q) frame based selective harmonic detection [2, 8, 9], 3) Recursive discrete Fourier transformation based selective harmonic detection [7], 4) Synchronous (d, q) frame based harmonic detection [5, 6]. The research is carried out by means of Simulink/Simplorer simulations. The supplying grid, STATCOM main circuit and load are modeled in Simplorer and the STATCOM control system in Simulink.

II. Harmonic Detection Methods

A number of different methods have been developed to determine the harmonic content of current waveform [3]. The detection method characteristics play a major role in the selection of an applicable method and, depending on the compensator application and load behavior, features such as detection speed, accuracy and complexity of implementation are emphasized [8]. The methods examined in this study are selected on the grounds of the option to compensate the current control delay.

A. Band-Pass Filter Based Selective Harmonic Detection

The first detection method studied is band-pass filter (BPF) based selective harmonic detection [9]. The principle of the method is to detect the harmonics using band-pass filters and to compensate the control delay using phase-lead compensators. A block diagram of d-axis load current selective harmonic detection is shown in Fig. 2 and includes three band-pass filters (BPF) tuned to filter 100 Hz, 300 Hz and 600 Hz signals respectively. The band-pass filter outputs are fed back to the input to improve the selectivity of individual filters [9]. The detection is carried out in the rotating (d, q) frame, where one band-pass filter with center frequency at $f_0 = nf_{sup}$ detects two three-phase system harmonics, i.e., the positive sequence harmonic $(n+1)f_{sup}$ and the negative sequence harmonic $-(n-1)f_{sup}$. For example, the 600 Hz filter detects 50 Hz three-phase system harmonics $13f_{sup}$ and $-11f_{sup}$. The continuous-time transfer function of the filters is

$$G_{BPF}(s) = \frac{H_0 \beta s}{s^2 + \beta s + (2\pi f_0)^2} \qquad (1)$$

where H_0 is filter gain, β is pass-band width (rad/sec) and f_0 is center frequency. The transfer function (1) is digitized using Tustin's approximation with prewarping, given as [13]

$$s = \frac{\omega_1}{\tan(\omega_1 h/2)} \frac{z-1}{z+1}, \qquad (2)$$

where h is sampling period and ω_1 is prewarped frequency. Bode plots of the filters in Fig. 2 are shown in Fig. 3 ($H_0 = 1$, $\beta = 2\pi 50$ rad/sec). The filters do not cause additional phase-shift to the extracted signals because the filter phase angle is zero at the center frequency f_0.

The band-pass filter outputs are led to phase-lead compensators (PLC) which compensate the control delay and the sum of the PLC outputs is the d-axis current reference. The discrete-time transfer function of first-order PLC is [14]

$$G_{PLC}(z) = K \frac{1-b}{1-a} \frac{z-a}{z-b}. \qquad (3)$$

The coefficients a ($a > b$) and b are

$$a = \frac{\cos(\phi_{max}) - \sin(\Omega_{max})}{\cos(\phi_{max} + \Omega_{max})}, \qquad (4)$$

$$b = \frac{\cos(\phi_{max}) - \sin(\Omega_{max})}{\cos(\phi_{max} - \Omega_{max})}. \qquad (5)$$

where $\phi_{max} \in (0, \pi/2)$ is the maximum phase-lead occurring at frequency $\omega_{max} = 2\pi f_{max}$ and gain K is used to set the filter gain at unity at ω_{max}. Coefficient Ω_{max} is normalized frequency at which the ϕ_{max} occurs, i.e.,

$$\Omega_{max} = \pi \omega_{max} / \omega_{nyq}, \qquad (6)$$

where ω_{nyq} is the Nyquist frequency of the discrete-time system. With a first-order PLC the maximum achievable phase-lead is 90° but several compensators can be used in series if more phase-lead is required. The q-axis harmonic detection is similar, but fundamental reactive current is also detected using a low-pass filter (LPF) with corner frequency at $f_c = 100$ Hz and summed to the current reference.

B. Multiple Reference Frame Based Detection

The second method studied is multiple reference frame (MRF) based detection where the selected harmonics are detected in individual rotating (d, q) frames [2, 8]. The principle of the method is shown in Fig. 4. The current components flowing at the frequency of the reference frames are seen as dc-quantities and can be extracted using e.g. filters or moving average calculation. The (d, q) current references $i^*_{st,dh}$ and $i^*_{st,q}$ are obtained by summing the harmonics detected. In Fig. 4, the dc-quantity of q-axis load current in the 50 Hz frame is also extracted and included in $i^*_{st,q}$ to compensate the fundamental frequency reactive current. The control delay compensation in MRF detection is included in the (d, q) coordinate transformations. The

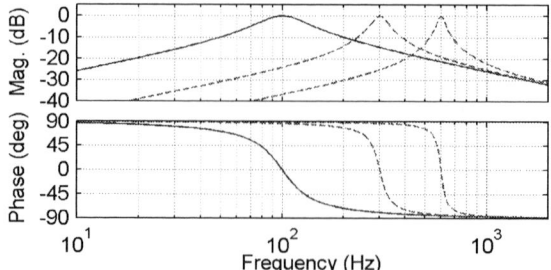

Fig. 3. Frequency responses of the band-pass filters with f_0 at 100 Hz, 300 Hz and 600 Hz.

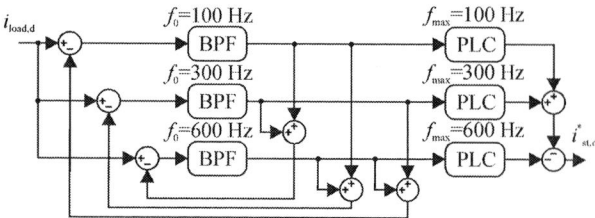

Fig. 2. Band-pass filter based selective harmonic detection.

978-1-4244-1871-8/07 $25.00 © 2007 IEEE

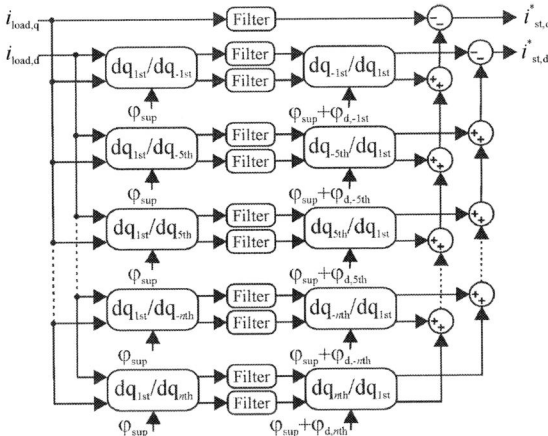

Fig. 4. MRF selective harmonic detection.

phase-delay of the detected n^{th} harmonic is compensated by adding a compensating angle $\varphi_{d,n}$ to the synchronous reference frame angle φ_{sup} when the transformation back to the 50 Hz (d, q) frame is carried out [2].

The accuracy and detection speed of the MRF method depends on how the dc-quantity extraction is implemented in 'Filter' blocks. In the case of low-pass filters good accuracy is achieved at the cost of slow response time. In contrast, moving average calculation gives good accuracy and detection speed but the implementation is computationally complex. The Bode plot in Fig. 5 shows that the moving average calculation can be thought to consist of several band-reject filters (BRF), which can be implemented as $G_{BRF} = (1 - G_{BPF})$. Thus, an operation similar to moving average calculation is achieved if a low-pass filter with corner frequency at 100 Hz and several band-reject filters are combined (Fig. 6). Such a filter offers better detection speed than low-pass filtering and requires less memory capacity than the moving average method.

C. RDFT Based Detection

The third detection method studied is based on recursive discrete Fourier transformation (RDFT). The method is similar to the MRF method shown in Fig. 4 with the exception that the dc-quantities in each reference frame are extracted using moving average calculation which corresponds to the operation of the RDFT [7]. Depending on the harmonic content of the load current, the moving average can be calculated during 1/6, one half or one full fundamental frequency period. A drawback of RDFT is the computational complexity and large memory capacity required [3]. However, the accuracy, detection speed and selectivity of the RDFT method are better compared to MRF method.

D. Synchronous (d, q) Frame Based Detection

The fourth method studied is synchronous (d, q) frame based harmonic detection [5, 6]. The method does not detect individual harmonics but simply separates all non-fundamental current components from the fundamental one. The d-axis load current harmonics are extracted by removing the dc-quantity of $i_{load,d}$ using moving average calculation (Fig. 7). The resulting fluctuating component is

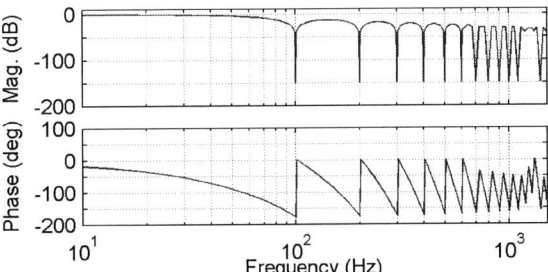

Fig. 5. Frequency response of the moving average calculation.

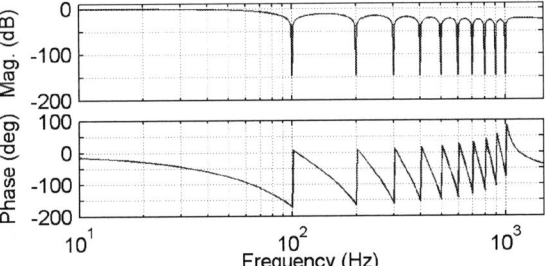

Fig. 6. Frequency response of the LPF-BRF combination.

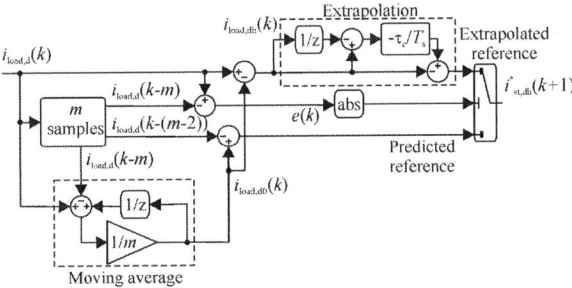

Fig. 7. Synchronous (d, q) frame based harmonic detection method equipped with delay compensation feature.

inverted to obtain the d-axis current reference $i^*_{st,dh}$. The q-axis current reference $i^*_{st,q}$ is obtained by inverting $i_{load,q}$ as a whole.

The detection method includes two control delay compensation methods (Fig. 7). The first one is used when the load currents are non-periodic and is based on linear extrapolation of current reference [11, 12]. In discrete form the extrapolated current reference is given as

$$i^*_{st,dh}(k+1) = -\frac{\tau_c}{T_{sp}}\left(i_{load,dh}(k) - i_{load,dh}(k-1)\right) - i_{load,dh}(k) \quad (7)$$

where τ_c is approximated control delay time constant and T_{sp} is sampling time [12]. The q-axis current reference extrapolation is performed similarly.

The second delay compensation method is used when the load currents are strictly periodic and is based on the forward prediction of load current behavior using sampled and stored current data [2, 11]. The d-axis load current behavior is predicted by storing m current measurement samples during a 10 ms time period, where $m = 10$ ms/T_{sp}. In this study the delay is approximated equal to two sampling intervals, i.e. $2T_{sp}$. Thus, at the kth sampling instant the current sample $i_{load,d}(k-(m-2))$ is used in harmonic detection instead of the latest sample $i_{load,d}(k)$ to compensate the control delay [2, 11].

978-1-4244-1871-8/07 $25.00 © 2007 IEEE

The load current periodicity is checked by subtracting the current sample $i_{\text{load,d}}(k-m)$ from the latest sample $i_{\text{load,d}}(k)$ and the absolute value of error $e(k)$ is used for selecting between computational and predictive delay compensation. In q-axis non-active current detection the output of the moving average calculation block is set at zero but otherwise operation is similar.

III. STATCOM CONTROL SYSTEM

The block diagram of the STATCOM control system examined is shown in Fig. 8. The control system is implemented in synchronous (d, q) reference frame (SRF) where the d-axis is synced to the supply voltage space-vector angle φ_{sup} [5]. Consequently, the fundamental frequency voltage and current components are seen as dc-quantities. The control systems consist of two control loops, i.e., the STATCOM current control loop and the dc-voltage control loop. The current control method applied is based on synchronous PID controllers and space-vector PWM (SVPWM).

The STATCOM current references $i^*_{\text{sth,d}}$ and $i^*_{\text{st,q}}$ are generated from the measured load currents $i_{\text{load,d}}$ and $i_{\text{load,q}}$ in block 'Harmonic detection' using the detection methods studied. The measured STATCOM currents, i.e., $i_{\text{st,d}}$ and $i_{\text{st,q}}$ in the SRF, are subtracted from the current references and the current errors are led to PID controllers. The controller outputs are the reference values for the supply filter voltage components $u^*_{\text{Lf,d}}$ and $u^*_{\text{Lf,q}}$. The supply filter voltage is given in SRF as

$$u_{\text{Lf,d}} = u_{\text{sup,d}} - u_{\text{st,d}} = L_f \, di_{\text{st,d}}/dt + R_{\text{Lf}} i_{\text{st,d}} - \omega_{\text{sup}} L_f i_{\text{st,q}}, \quad (8)$$

$$u_{\text{Lf,q}} = u_{\text{sup,q}} - u_{\text{st,q}} = L_f \, di_{\text{st,q}}/dt + R_{\text{Lf}} i_{\text{st,q}} + \omega_{\text{sup}} L_f i_{\text{st,d}}, \quad (9)$$

where u_{sup} is supply voltage, u_{st} is STATCOM voltage, L_f and R_{Lf} are supply filter inductance and resistance, i_{st} is STATCOM current, and ω_{sup} is supply voltage angular frequency. The cross-coupling between the supply filter voltage components through the ac inductance $\omega_{\text{sup}} L_f$ is shown in (8)-(9) and can be avoided by including decoupling terms $-\omega_{\text{sup}} L_f i^*_{\text{st,q}}$ and $\omega_{\text{sup}} L_f i^*_{\text{st,d}}$ in the final STATCOM voltage references:

$$u^*_{\text{st,d}} = u_{\text{sup,d}} - \left(u^*_{\text{Lf,d}} - \omega_{\text{sup}} L_f i^*_{\text{st,q}}\right), \quad (10)$$

$$u^*_{\text{st,q}} = u_{\text{sup,q}} - \left(u^*_{\text{Lf,q}} + \omega_{\text{sup}} L_f i^*_{\text{st,d}}\right). \quad (11)$$

Finally, the (d, q) voltage references are transformed into stationary (α, β) coordinates and led to the space-vector modulator. The capacitor voltage balancing is implemented in space-vector modulator using redundant switching states.

IV. EFFECT OF THE CONTROL DELAY

The aim of the study is to improve the STATCOM harmonic compensation performance for harmonics mitigation up to the 13th order. To fulfill the switching frequency specification 600–1200 Hz, the SVPWM carrier frequency is selected $f_{\text{mod}} = 1.5$ kHz which leads to 700–1000 Hz average switching frequency. The modulator voltage references are updated twice during a modulation

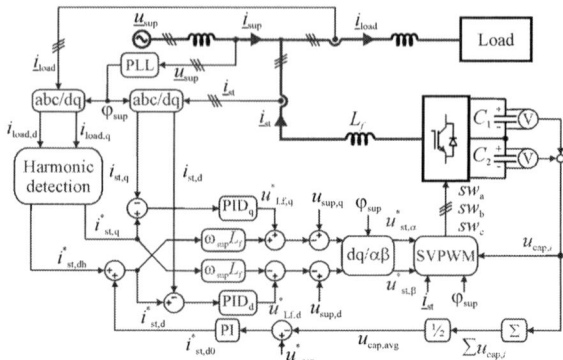

Fig. 8. STATCOM control system.

period, i.e., at the beginning and in the middle of the period, to improve the voltage reference tracking. Thus, the voltage reference sampling interval is $T_{\text{sp}} = 1/(2f_{\text{mod}}) = 333$ µs.

The current control delay in digitally controlled active compensators consists of current measurement delay, delay caused by PWM modulation, and the device response time and causes phase difference between the current reference and the actual compensator current. In this study the approximated control delay is $T_d = 2T_{\text{sp}}$. The effect of the control delay on the compensation performance can be analyzed by approximating the phase difference between the current reference and the generated compensation current. For individual harmonics the phase difference is

$$\varphi_d = 2\pi f_n T_d \left(360°/2\pi\right), \quad (12)$$

where φ_d is phase difference (deg), f_n is n^{th} harmonic frequency and T_d is control delay. It is clear that harmonic mitigation is not possible if $\varphi_d > 90°$. Phase differences of dominant three-phase harmonics in SRF are calculated in Table 1 when control delay is equal to $T_d = 2T_{\text{sp}}$. It can be seen that the selected 1.5 kHz carrier frequency leads to such a notable control delay that the compensation performance is limited in practice to fundamental frequency reactive current compensation if control delay is not compensated.

TABLE 1. PHASE DIFFERENCE BETWEEN THE CURRENT REFERENCE AND THE ACTUAL COMPENSATOR CURRENT

f_n (Hz)	f_{mod}=5 kHz T_{sp}=100 µs φ_d (deg)	f_{mod}=3 kHz T_{sp}=166 µs φ_d (deg)	f_{mod}=2.5 kHz T_{sp}=200 µs φ_d (deg)	f_{mod}=1.5 kHz T_{sp}=333 µs φ_d (deg)
100	7.2°	12°	18°	24°
300	21.6°	36°	54°	72°
600	43.2°	72°	108°	144°

The delay compensation parameters in the harmonic detection methods are selected such that the approximated $2T_{\text{sp}}$ long control delay is compensated. In selective detection methods the phase-lag of each harmonic detected is approximated using (12) and taken into account in the PLC design and MRF and RDFT coordinate transformations. In synchronous (d, q) frame based detection two sample load current forward prediction is used and in computational delay compensation the coefficient τ_c/T_{sp} is set equal to 2.

V. SIMULATION RESULTS

First, the operation of the harmonic detection methods is studied under steady-state load conditions. The compensated load is a three-phase diode bridge supplying an inductive load ($R = 77$ Ω, $L = 20$ mH) and is connected to 21 kV supply via 3 mH ac-inductors. The STATCOM dc-link voltage is $u_{dc} = 20$ kV, supply filter inductance is $L_f = 32.8$ mH and dc-capacitances are $C_1 = C_2 = 1$ mF. Load phase-a current and compensated supply currents are shown in Fig. 9. For comparison, supply phase-a current compensated using sync. (d, q) frame harmonic detection without delay compensation is shown in Fig. 9b. Ratios of the supply phase-a current harmonics after the compensation to the corresponding load current harmonics are shown in Fig. 9g. The results show that harmonic mitigation is not possible without control delay compensation. In contrast, the supply current harmonics up to the 13th order can be mitigated when detection methods with delay compensation are used.

Second, compensation of load with step changes is investigated by replacing the diode bridge with a thyristor bridge. The other load parameters remain unchanged. Load step changes occur at 20 ms (the load is connected to the supply and the thyristor gating angles are set to $\alpha = 0°$), 80 ms (load phase-a is disconnected) and 140 ms ($\alpha = 60°$ and the 3 mH inductance in load phase-b changes to the series connection of $L = 12$ mH, $R = 10$ Ω). Load currents and compensated supply currents are shown in Fig. 10. The results show that satisfactory compensation performance is achieved during steady-state load conditions. However, after each load step change the detection methods require ca. 10-15 ms settling time. Best performance during load step changes is achieved with the sync. (d, q) frame based detection method which applies the extrapolation based delay compensation method (Fig. 10e).

Finally, electric arc furnace (EAF) compensation is simulated. The load currents generated using EAF current measurement data, compensated supply currents and rectified mean values of supplied non-active power are shown in Fig. 11. The results show that the detection speed of selective methods is not adequate to solve the harmonic content of stochastic load currents and the compensation performance is impaired (Figs. 11b-d). However, the supply current waveforms are improved with each of the detection methods. The supply current waveforms and calculated rectified mean values of supplied non-active power (Fig. 11f) indicate that the best performance is achieved with sync. (d, q) frame based detection method (Fig. 11e) which uses current reference extrapolation for control delay compensation because the load currents are non-periodic.

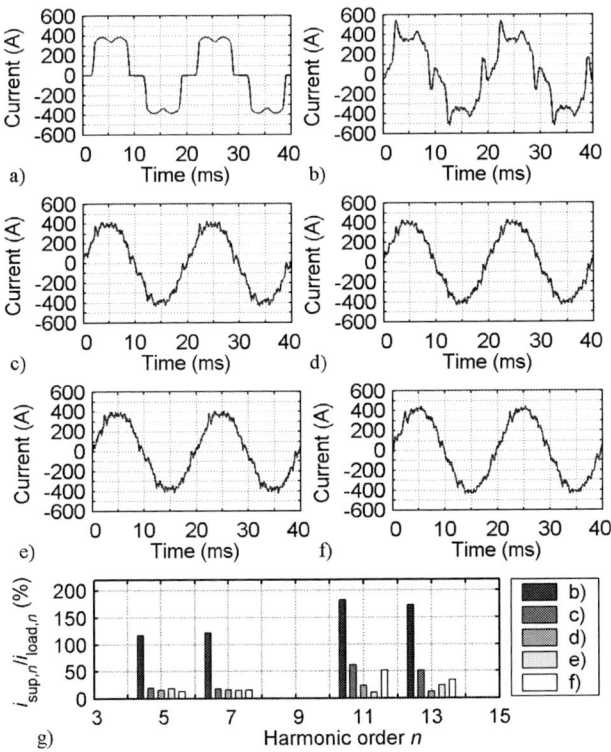

Fig. 9. Steady-state load compensation. a) Load currents, b) sync. (d, q) frame detection without delay comp., c) BPF detection, d) MRF detection, e) RDFT detection, f) sync. (d, q) frame detection with delay comp., g) supply current harmonics up to the 13th order after the compensation (%).

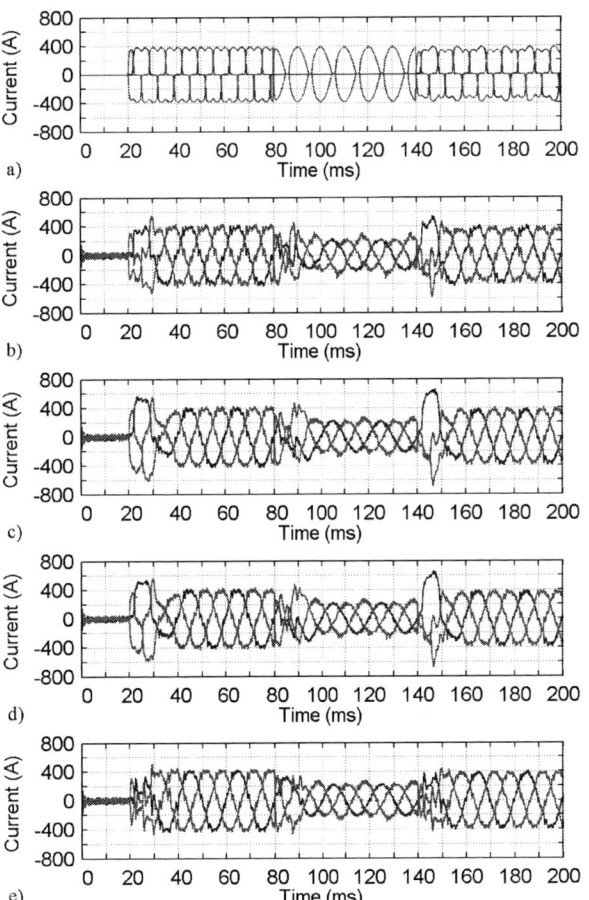

Fig. 10. Compensation of load with step changes. a) Load currents, b) BPF detection, c) MRF detection, d) RDFT detection, e) sync. (d, q) frame detection with delay compensation.

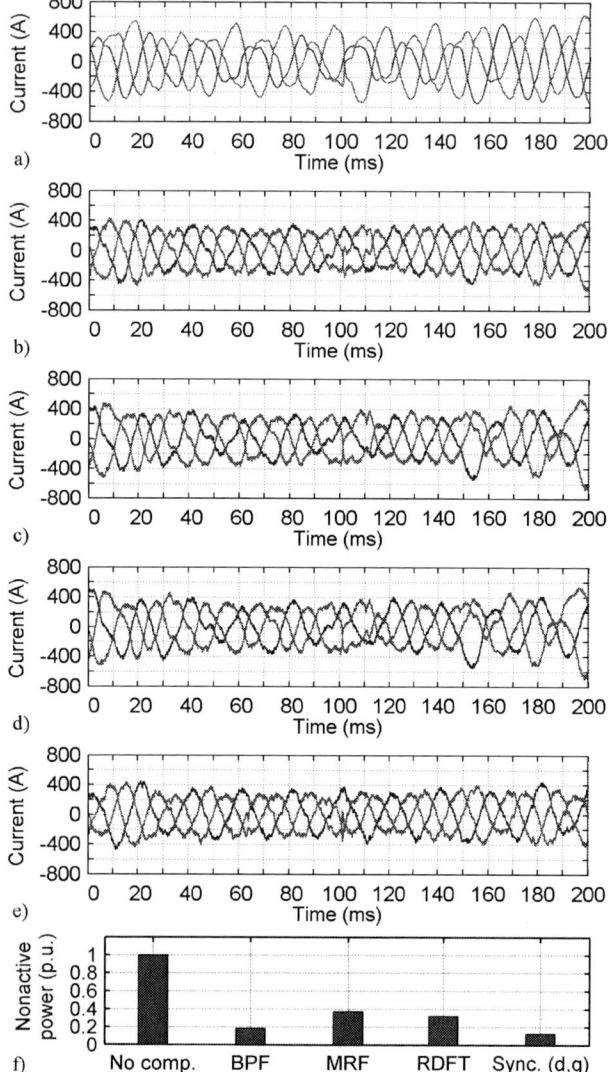

Fig. 11. EAF compensation. a) EAF currents, b) BPF detection, c) MRF detection, d) RDFT detection, e) sync. (d, q) frame detection with delay comp, f) rectified mean values (p.u.) of supplied non-active power.

was obtained using the sync. (d, q) frame based detection method. Also in EAF compensation the best performance was achieved using the sync. (d, q) frame based detection method which decreased the rectified mean value of the supplied non-active power by 87 % while the selective detection methods decreased the non-active power by 81 % (BPF), 63 % (MRF) and 68 % (RDFT). In conclusion, the harmonic detection methods studied improved the harmonic mitigation performance of a low switching frequency STATCOM under periodic load conditions. However, among the detection methods studied, only the synchronous (d, q) frame based method provided sufficient detection speed during non-periodic load conditions, but the method's performance was reliant on the accuracy of control delay compensation.

ACKNOWLEDGMENT

This study has been supported by Nokian Capacitors Ltd. The contribution of Jyri Öörni, Nokian Capacitors Ltd., and Ralf Jessler, Nokian Capacitors GmbH., is gratefully acknowledged.

REFERENCES

[1] H. Akagi, "Active harmonic filters", Proc. of the IEEE, Vol. 93, Iss. 12, pp. 2128-2141, Dec. 2005.

[2] M. Bojrup, P. Karlsson, M. Alaküla, and L. Gertmar, "A Multiple Rotating Integrator Controller for Active Filters", EPE'99, 9 p., 1999.

[3] L. Asiminoaei, F. Blaabjerg, and S. Hansen, "Detection is key - Harmonic Detection Methods for Active Power Filter Applications", IEEE Ind. Applicat. Magazine, Vol. 13, Iss. 4, pp. 22-33, July/Aug. 2007.

[4] H. Akagi, Y. Kanazawa, and A. Nabae, "Instantaneous Reactive Power Compensators Comprising Switching Devices without Energy Storage Components", IEEE Trans. Ind. Applicat., Vol. 20, Iss. 3, pp. 625-630, May/June 1984.

[5] S. Bhattacharya, T. M. Frank, D. M. Divan, and B. Banerjee, "Active Filter System Implementation", IEEE Ind. Applicat. Magazine, Vol. 4, Iss. 5, pp. 47-63, 1998.

[6] S.-G Jeong and W. Myung-Ho, "DSP-Based Active Power Filter with Predictive Current Control", IEEE Trans. on Ind. Electron., Vol. 44, Iss. 3, pp. 329-336, June 1997.

[7] S. Srianthumrong and S. Sangwongwanich, "An Active Power Filter with Harmonic Detection Method Based on Recursive DFT", Proc. of 8th International Conf. on Harmonics And Quality of Power, Vol. 1, pp. 127–132, Oct. 1998.

[8] P. Mattavelli, "A Closed Loop Selective Harmonic Compensation for Active Filters", IEEE Trans. on Ind. Applicat., Vol. 37, Iss. 1, pp. 81–89, Jan.–Feb. 2001.

[9] J. H. Allmeling, "A Control Structure for Fast Harmonic Compensation in Active Filters", PESC'02, Vol. 1, pp. 376–381, June 2002.

[10] S. Mariethoz and A. C. Rufer, "Open Loop and Closed Loop Spectral Frequency Active Filtering", IEEE Trans. Power Electron., Vol. 17, Iss. 4, July 2002.

[11] M. Routimo, M. Salo, and H. Tuusa, "A Novel Simple Prediction Based Current Reference Generation Method for An Active Power Filter ", PESC'04, Vol. 4, pp. 3215–3220.

[12] M. Salo and H. Tuusa, "A Novel Open-Loop Control Method for A Current-Source Active Power Filter", IEEE Trans. on Ind. Electron., Vol. 50, No, 2., pp. 313-321, Apr. 2003.

[13] K. J. Åström and B. Wittenmark, "Computer Controlled Systems: Theory and Design", 3rd ed., Upper Saddle River, NJ: Prentice Hall Inc., 1997.

[14] W. Messner, "Simple Formulas for Direct Design of Discrete-Time Lead Compensators", Journal of Dynamic Systems, Measurement, and Control, Vol. 122, Iss. 2, pp. 358-360, June 2000.

VI. CONCLUSIONS

This study aimed to improve the harmonic mitigation capability of a low switching frequency 10 MVA three-level STATCOM. The idea was to improve the STATCOM performance using harmonic detection methods equipped with control delay compensation feature. It was shown that the used 700–1000 Hz average switching frequency leads to a notable current control delay which forbids the harmonic mitigation. When the detection methods examined were applied to steady-state harmonic compensation the 5th and the 7th supply current harmonic were decreased by ca. 80 % and, depending on the detection method used, the 11th and the 13th harmonic by ca. 40 % or more. However, the accuracy of the selective harmonic detection methods, i.e., BPF, MRF and RTDF, impaired during non-periodic load conditions. The best performance in non-periodic load current compensation

10MVA STATCOM Installation and Commissioning

Youngseong Han, Chungchoo Chung
Hanyang University
Department of Electrical Engineering
17, Haengdang-dong, Seongdong-gu,
Seoul, 133-791
Email: yshan@hyosung.com
cchung@hanyang.ac.kr

Jongyun Choi, Daehee Kim
Hyosung Corporation
183, Hoge-dong, Donan-gu,
Anyang-si, Gyeonggi-do, 431-080
Email: choris007@hyosung.com
dhmono@hyosung.com

Jongsu Yoon
Korean Electrical Power Research Institute
103-16, Munji-dong, Yusung-gu-
Daejon-si, 305-380
Email: yoonjs@kepri.re.kr

Abstract — **10MVA STATCOM has been developed by Hyosung Corporation under a joint research project with KEPRI(Korean Electrical Power Research Institute). It has been installed and commissioned at Hyosung factory in Changwon Korea on March, 2007. This Paper describes configuration of 10MVA STATCOM system and the result of installation and commissioning.**

Keywords: FACTS (Flexible AC Transmission System), STATCOM (Static Synchronous Compensator)

I. INTRODUCTION

Power generation complexes in Korea are concentrated in coastal areas and loads are concentrated in inland metropolitan areas far away from the generated power. Increase in short circuit capacity due to the loop configuration of the system, voltage drop due to long transmission lines, and system instabilities due to the concentration of large generators have contributed to the need for power flow control to the metropolitan areas. Solutions to these problems have been addressed in several engineering projects. Prior to applying FACTS technology to the 345KV system, which is the backbone of the Korean power system, the need for a pilot plant project was acknowledged to verify reliability and operational performance through actual installation and operation of inverter type FACTS equipment. For this Hyosung Corporation and KEPRI, the research institute of Korean Electric Power Corporation (KEPCO) have developed 22.9kV, 10MVA STATCOM system, have installed and commissioned the system at Hyosung 's Factory in Changwon, Korea on March, 2007. This Paper describes configuration of 10MVA STATCOM system and the result of installation and commissioning.

II. STATCOM CONFIGURATION

A. System Configuration

Basic system configuration of 10MVA STATCOM system is shown in Fig. 1. STATCOM consists of two 3-level inverters using IGCT device, and each IGCT device used in the inverter stack is rated for 4500V, 4000A. As shown in Fig. 1 two identical twelve pulse, three level inverters are connected to a common DC capacitor bank, the combining the output through an auxiliary transformer, a twenty four pulse harmonic neutralized inverter can be configured.

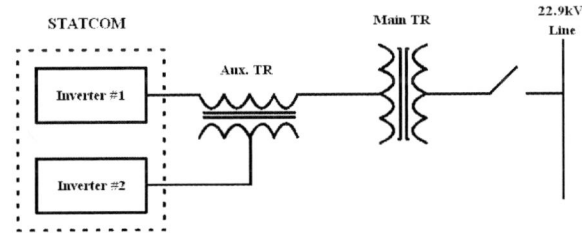

Fig 1: STATCOM Configuration

The output voltage which is combined by the auxiliary transformer is connected to 22.9kV transmission line through a main step-up transformer and CGIS system.

For convenience of movement, all equipments except transformers are installed in two containers over trailer vehicles as shown in Fig 2. All inverter poles and control system are loaded in first container, and cooling system, protective relay system and CGIS system are installed in the other container.

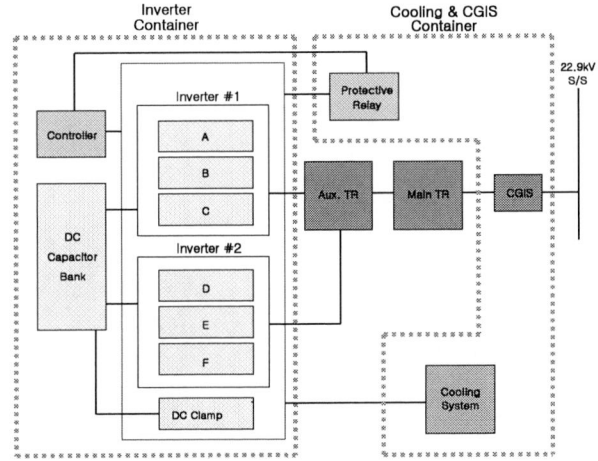

Fig 2: STATCOM Configuration

B. Components of 10MVA STATCOM

1) Inverter Pole

Fig 3 shows drawing of inverter pole for 10MVA STATCOM and it shows the actual object which is produced. The positive and negative DC input are positioned at bottom and top side of inverter pole respectively, and neutral point is

978-1-4244-1871-8/07 $25.00 © 2007 IEEE

connected to ground through bus bar in the middle of inverter pole. AC output of each inverter pole is positioned under neutral bus bar.

Each inverter pole consists of four IGCT devices, four anti-parallel diodes, clamp circuits and di/dt reactors, and all IGCT and diode devices are stacked in single clamp for compact size. In the back side, two pipes for flowing coolant to each power electronic devices are installed, and they are connected to heat-sink of each device through PVC tubes for insulation and convenience of maintenance.

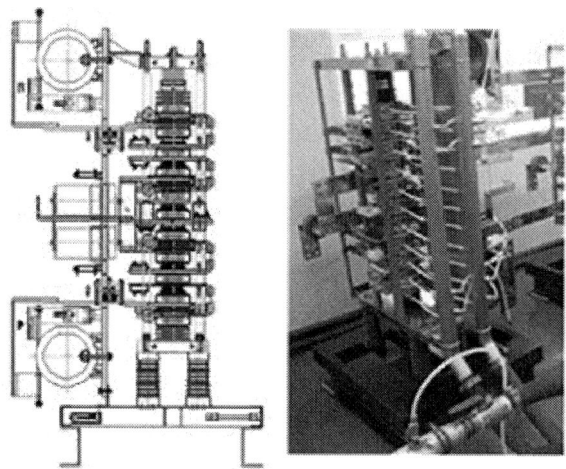

Fig 3: Inverter Pole for 10MVA STATCOM

2) DC Clamp

Fig 4 shows drawing of DC clamp for DC over-voltage protection and it shows the actual object which is produced. There are one IGCT switch and one discharging resistor in each positive and negative DC side respectively. If DC voltage rise over voltage limit, IGCT switch is turned on for discharging DC voltage through the resistor.

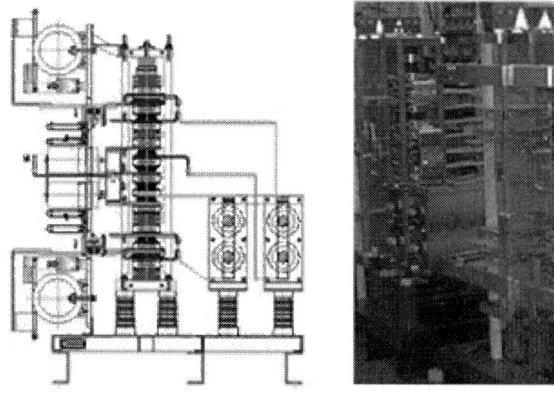

Fig 4: DC Clamp Pole for 10MVA STATCOM

3) Power Transformers

Output of inverter is connected to the grid through a transformer. There are two transformers used in the grid connection. The main transformer is 10MVA and auxiliary transformer is rated 5MVA. As the transformer is connected to inverter, DC current may be injected. When DC current of the inverter output are mixed into each transformer they can be the cause of saturation. The transformer was designed to minimize effect of DC current. As shown in Fig 1 two identical twelve pulse, three level inverters are connected to a common DC capacitor bank, the combining the output through an auxiliary transformer, a twenty four pulse harmonic neutralized inverter can be configured. By combining the output waveforms of the two inverters with this configuration, harmonic components can be reduced. The combined output is connected to the grid through the main transformer. Fig 5 shows produced picture of transformers for 10MVA STATCOM.

Fig 5: FACTS Transformers

4) Cooling System

Cooling system of 10MVA STATCOM uses water-cooling method, and two pumps operate alternatively. Thermometers, flow-meters, pressure-meters, and a resistivity sensor are installed for monitoring normal operation of cooling system. And using serial communication between main controller and control system, each sensor's value and status of pump and heat exchanger can be monitored in operator interface. Fig 6 shows pump skid and heat exchanger of cooling system

Fig 6: Pump skid and Heat exchanger

5) Controller

Master controller of 10MVA STATCOM executes the STATCOM current and voltage control algorithms in time unit <100μ sec. Master controller also carries out management of all systems such as pole protection and system monitoring through serial communication with sub-systems. Master controller consists of commercial microprocessor boards using VME bus system for convenience of system upgrade. GCU(Gate Control Unit) makes gating signal of IGCTs of each inverter pole in time unit <1μ sec , and executes hi-speed pole protection function.

C. Installation

Fig 7 shows the picture of 10MVA STATCOM installed in Hyosung's Chanwon factory. The front one in Fig1 is container loading inverter poles and controllers, and the back one is container for cooling system and CGIS.

Fig 7: Picture of installed STATCOM system

Fig 8 shows inverter poles installed in container. Inside the container, six inverter poles, one DC clamp, grounding TR, grounding resistor and DC capacitors are installed.

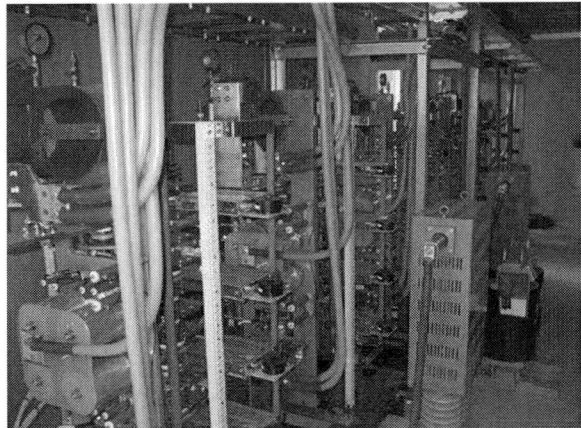

Fig 8: Inverter poles installed in container

D. Operator Interface

Operator interface of 10MVA STATCOM is designed for monitoring status of system operation and all parts of the system. It is PC based program which is connected through TCP/IP communication with master controller, so local and remote control is capable. Fig 9 and 10 show operator interface screens, operating values such as power system voltage, STATCOM current, and DC voltage are displayed in single line diagram screen. Also system start/stop command, change of operation mode and reference value of STATCOM current or system voltage by operator are done in single line diagram screen. Status of all compositions of the system can be monitored in corresponding screens as shown in Fig 10..

Fig 9: Operator interface – Single line diagram and operating screen

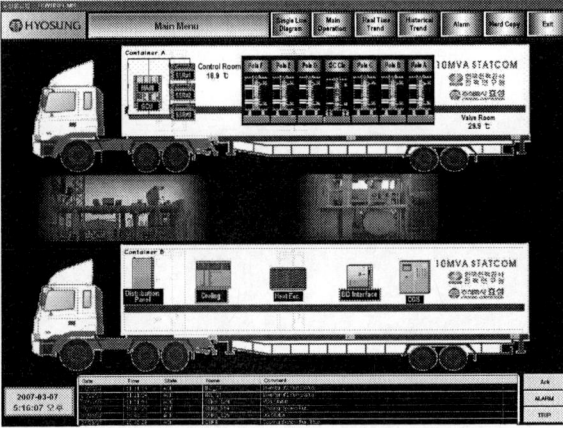

Fig 10: Operator interface – System monitoring

III. COMMISSIONING

A. Offline Commissioning

These tests are intended to verify the proper operation of the power electronics, comprising the inverter poles, the dc capacitor banks, the main coupling transformers, and controllers, before the equipment is connected to the power transmission line. Offline energization is achieved by the use of an auxiliary dc power supply permanently installed at the site.

1) DC Bus Energization and DC Clamp circuit test

These tests verify the DC capacitor banks ability to charge to rated voltage and hold the charge without loss of capacitance, and proper operation of the dc clamping circuits. Test circuit for offline energization tests is determined in Fig 11.

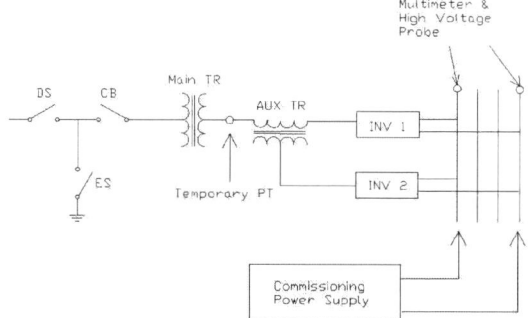

Fig 11: Test circuit for offline energizaton tests

For the DC clamp circuit test, DC voltage is applied to the DC bus by energizing the auxiliary supply taking DC voltage higher the dc clamp activation threshold 2800V. At this point the clamp valve should fire and the DC bus voltage will be discharged rapidly to a lower threshold 2600V, where the valves are again turned off. The bus should charge again through the supply impedance and clamp valves should fire again, producing an oscillation. Since the dc clamp resistors have a limited energy capacity, it is important to watch the temperature rise on these resistor during this test. Fig 12 shows DC clamp circuit operate properly.

Fig 12: DC clamp circuit test result

2) Waveform Construction test

This test is intended to important to verify inverter output voltage waveform construction and proper phasing of the inverter output voltage relative to the transmission line. During this test, inverter switching will be verified up to full voltage under no-load conditions. First af all in the low dc voltage condition, ±300Vdc, three-phase inverter output

voltage on the temporary PTs is observed, and 24-pulse voltage waveform is constructed very well as shown in Fig 13. Also it is verified that the voltage measured on the temporary PTs is in phase with the voltage measured on the main ac bus PTs.

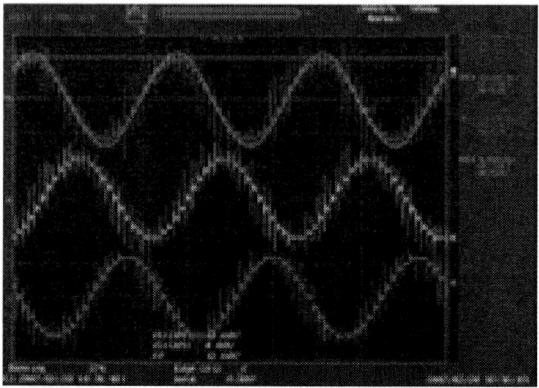

Fig 13: Waveform construction test in low dc condition

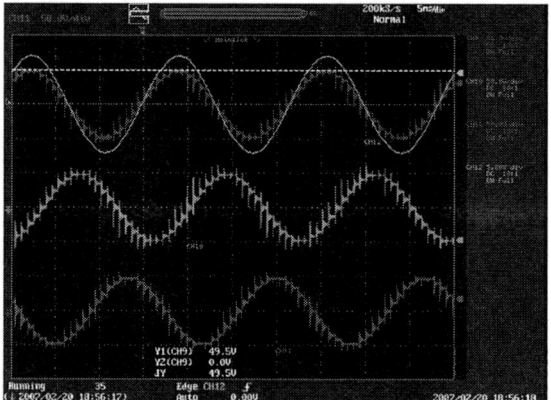

Fig 14: Waveform construction test in rated dc condition(±2300Vdc)

Once proper operation has been verified, the procedure is repeated to rated dc voltage condition, ±2300Vdc, progressively. Fig 14 shows result of test in rated dc voltage condition, and it is verified the inverter operate properly in the rated voltage condition.

3) Short-Circuit test

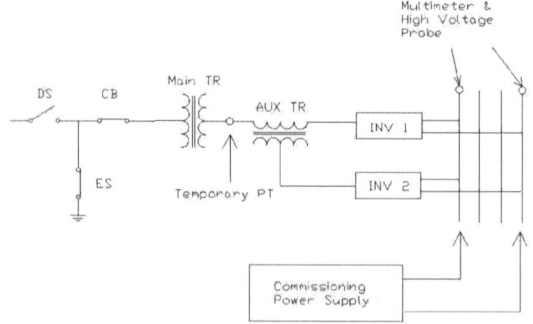

Fig 15: Test circuit for short-circuit test

This test is intended to verify proper polarity of current feedback signals and that proper operation of inverters driving high current through the ac circuit. By closing CB and ES as shown in Fig 15, high ac current can flow through inverters and transformers in low dc voltage condition. Fig 16 shows that the three phase output current is well balanced and leads 90 degrees to main ac voltage ,that means reactive circuit

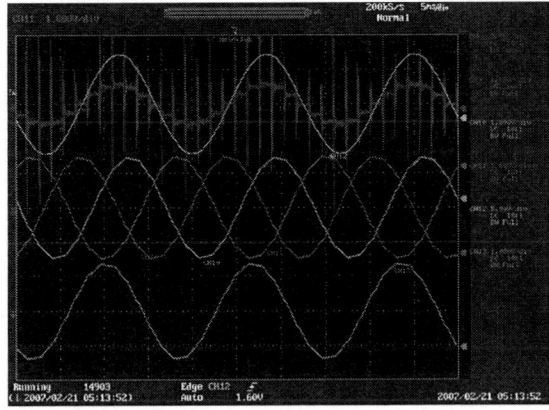

Fig 16: Short-circuit test result

B. Online Commissioning

After offline commissioning tests are completed, online commissioning tests are performed. These tests involve connection of main transformer to the transmission line and are aimed at establishing the ability of the equipment to operate on line as intended while delivering reactive power. These tests verify the connection to the ac transmission bus using transformer pre-energization procedure, and proper operation of closed-loop current controller in VAR control mode and voltage controller in voltage controller. After all online commissioning tests are done successfully, heat-run test is performed to verify stability and thermal behavior of all parts of the STATCOM under full load conditions for 8 hours.

1) Transformer pre-energization and line connection

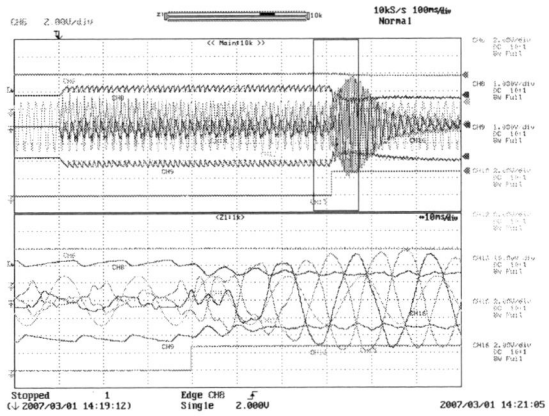

Fig 17: Result of Startup sequence

This test is intended to verify the startup sequence of the STATCOM. Simultaneously with closing main CB, DC voltage is charging and DC clamp circuit operates repeatedly, and inrush current to transformers flows in short time. After transient state passes, gating of inverter is started. Fig 17 shows change of current and DC voltage in startup sequence.

2) Current control loop test

The purpose of the test is to verify step response for change in var reference and adjust current controller gains if necessary. Fig 18 and 19 show the result of step response test. Fig 18 is the result of change of var reference from reactive 3Mvar to capacitive 3Mvar, and shows that actual current follow reference value within one cycle. Similarly Fig 19, result of change of var reference from capacitive 3Mvar to reactive 3Mvar, shows same performance.

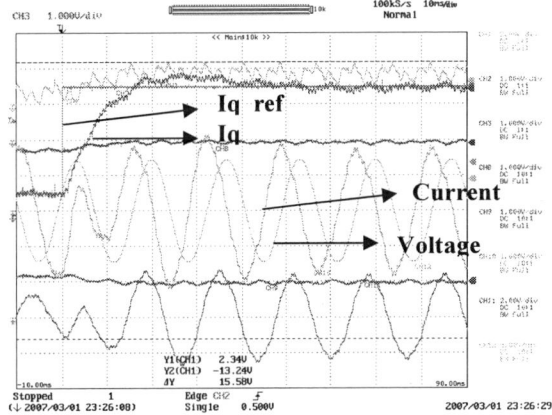

Fig 18: Step response in Var control (reactive to capacitive)

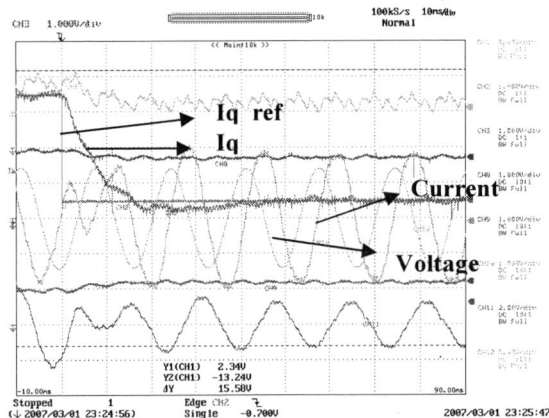

Fig 19: Step response in Var control (capacitive to reactive)

3) Heat-run test

After all offline and online tests for verifying function of the equipment of STATCOM are completed, heat-run test is performed. In this test, changes of temperature in transformer's winding and oil, temperature of coolant and inverter room, number of operating fan, etc. are checked during 8 hours. Table 1 describes record of heat-run test, it is

978-1-4244-1871-8/07 $25.00 © 2007 IEEE   545

shown that the STATCOM operate reliably, and temperature of all parts is saturated under limit value.

Table 1: Record of heat-run test

HEAT RUN TEST DATA

Time \ Item	AC Voltage (PU)	STATCOM P/Q	GPS Current (A)	outside temp. ℃	clamp resistor temp. ℃	Inv. Room temp. ℃	COOLING SYSTEM				Main Tr.				AUX Tr.			
							PUMP (gpm)	temp IN	temp OUT	#FANS	oil temp	winding temp.#1	winding temp.#	#FANS	oil temp	winding temp.#1	winding temp.#2	#FANS
Before Start(10:18)				2.4	10	13	98	4	4	x	8	5	8	x	10	5	6	x
11:40	1.03	0.06/6.2	10	5	8.8	19.4	98.9	23	22	x	8	9	13	x	13	10	11	x
12:10	1.04	0.07/7.2	10	5.5	13.9	23.1	101.3	35	33	2	10	14	18	x	14	16	15	x
12:40	1.02	0.14/10.3	10	6	14.1	24.4	101.8	34	32	2	14	28	30	x	18	29	26	x
13:10	1.02	0.15/10.14	10	6.3	16	24.7	101.7	35	33	2	16	32	34	x	22	34	31	x
13:40	1.02	0.15/10.16	10	6.8	17.5	25.5	101.7	36	35	2	26	40	42	x	30	44	41	x
14:10	1.02	0.15/10.23	10	7.5	19	27.2	102	38	36	3	30	44	44	x	32	46	43	x
14:40	1.03	0.15/10.26	10	7.6	18.4	28	101.4	30	28	3	32	48	48	x	36	50	47	x
15:10	1.04	0.15/10.47	10	7.6	18.8	28.3	101.4	29	27	3	35	50	50	x	38	52	50	x
15:40	1.04	0.15/10.33	10	7.4	19.17	28.7	101.4	31	29	3	36	52	53	x	40	54	52	x
16:10	1.05	0.15/10.47	10	7.2	20.2	29	101.7	30	28	3	38	54	54	x	42	57	54	x
16:40	1.03	0.15/10.34	10	7.2	20.2	29.5	101.7	30	28	3	39	55	55	x	44	59	55	x
17:10	1.05	0.16/10.4	10	6.8	21.36	29.8	101.6	30	28	3	40	57	57	x	46	60	57	x
17:40	1.04	0.16/10.4	10	5.3	20.81	29.9	101.8	29	27	3	40	57	57	x	46	60	57	x
18:10	1.03	0.16/10.31	10	5	19.8	29.5	101.1	28	26	3	39	57	57	x	46	60	57	x
18:40	1.02	0.16/10.2	10	3.4	19.2	28.7	101.4	27	25	3	39	57	57	x	46	60	57	x
19:10	1.03	0.16/10.31	10	2.8	18.25	28	100.9	27	25	2	39	57	57	x	46	60	57	x
19:40	1.04	0.17/10.41	10	2.3	20.08	28	101.1	32	30	2	39	57	57	x	45	60	57	x
20:20	1.04	0.16/10.38	10	1.9	20.37	28.4	102.2	34	32	2	39	57	57	x	45	60	57	x

IV. CONCLUSION

10 MVA STATCOM was successfully commissioned March 2007. The 10 MVA STATCOM shows the response time less than 1 cycle of 60 Hertz. All the modification will be applied to 100 MVA STATCOM which will be commissioned in 2008 at Migeum substation, while 10 MVA STATCOM is kept running.

ACKNOWLEDGMENT

The authors would like to acknowledge individuals in Center for Power-IT National Program and MOCIE (Ministry of Commerce, Industry and Energy) for their encouragement and continued support for this project.

REFERENCES

[1] Schauder, C., et al, "*AEP UPFC Project : Installation, Commissioning and Operation of the ±160MVA STATCOM(Phase I)*", IEEE Transactions on Power Delivery, Vol.13, No.4, Oct. 1998

[2] Gyugyi, L., Hingorani, N. "*Understanding FACTS : Concepts and Technology of Flexible AC Transmission Systems*", IEEE press, 2000

[3] Jin-Boo, Choo, et al, "*Development of FACTS Operation Technology(Phase 1 : Decision of a Type and a Spec. Draft for the Installation of a Pilot Plant)*", KEPRI Technical Report, April, 1999.

A test facility for large scale inverter valve and pole using resonant circuit

Youngseong Han, Chungchoo Chung
Hanyang University
Department of Electrical Engineering
17, Haengdang-dong, Seongdong-gu,
Seoul, 133-791
Email: yshan@hyosung.com
cchung@hanyang.ac.kr

Hyunho Yoo, Junmo, Kim
Hyosung Corporation
183, Hoge-dong, Donan-gu,
Anyang-si, Gyeonggi-do, 431-080
Email: klassica@hyosung.com
Junmokim@hyosung.com

Sooyeol Kim
Korean Electrical Power Research
Institute
103-16, Munji-dong, Yusung-gu-
Daejon-si, 305-380
Email: sykim@kepri.re.kr

Abstract—This paper proposes a test method for single phase inverter pole which is the basic building block for large scale three phase inverters. Inverter poles for utility application need to be tested at full voltage and current ratings which reach thousands of volts and amperes. This paper proposes a novel method for testing large scale inverter valves and poles using a resonant circuit which allows testing of inverter valves and poles at rated current and voltage with relatively low AC power levels. The configuration and layout of the test facility is introduced and actual test results are presented.

I. INTRODUCTION

Large scale inverters (e.g. : FACTS devices) are based on two synchronously controlled voltage sources. These are implemented as power electronic voltage source inverters based on IGCT. These inverters follow a modular design, with each inverter comprised of six three-level poles connected to a dc bus. Each three-level pole uses four valves that are switched to create a three-level ac output. Each valve composed for switching an adequate voltage rating including a significant safety margin requires connecting IGCT modules in series. The characteristics of three-level pole could be estimated with simulation, but it has to be validated test under real operation condition. Transient voltage deviation and hot spot are verified by real operation condition test. In this paper, we have designed 2 phase test facility for large scale inverter pole using resonant circuit for these tests

Phase 1 : A valve test for series connected power semiconductor using resonant circuit

Phase 2 : test facility for large scale inverter pole using resonant circuit

II. A VAVE TEST FOR SERIES CONNECTED POWER SEMICONDUCTOR USING RESONANT CIRCUIT

To perform a switching operation under very high voltage conditions, several power electronic semiconductors should be connected in series to compose a switching valve.

A. Design of valve test facility

Turn on/off performances of series connected switching valve should be verified by tests which have similar operating conditions with real system operations even though it can be expected by simulation results.

Dynamic voltage sharing and the overshoot characteristics of voltage can be reviewed by turn-off test, the energy dissipation of snubber circuit and the hard recovery characteristics of anti parallel diodes can be reviewed by turn-on test. The switching valve test facility is designed to perform these tests. It called H-bridge test facility.

the schematic diagram of the valve test equipment was shown in figure.1

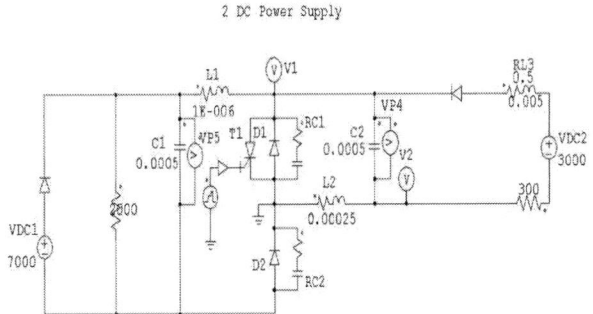

Figure.1 A valve test system circuit schematic

The proposed test facility consists of switching valve part which have switching valve, anti parallel diode and current wheeling diode, two DC power supplies, energy storage capacitor bank and load reactor. Two high-voltage low-power DC power supplies (VDC1 and VDC2) are used to precharge capacitors C_1 and C_2, while the T_1 is in the off-state. The precharged voltage on C_1 (i.e. V_{p5}) will be greater than or equal to that on C_2 (i.e. V_{p4}). L_2 is chosen so that ($V_{p4}/L_2 \leq 5*10^6$), and so that the characteristic impedance of L_2 and C_2 gives a desired peak resonant discharge current equal to the desired turn-OFF current (I_{toff}) in the T_1.

When the T_1 is turned on, C_2 discharges resonantly through T_1 and L_2. After a quarter cycle (i.e. at t = t_{pk}), C_2 is completely discharged and the current through L_2 is at a peak equal to I_{toff}. The following expressions describe the resonant build-up of current in the T_1.

978-1-4244-1871-8/07 $25.00 © 2007 IEEE

$$I_{toff} = \frac{V_{C2}}{\sqrt{\dfrac{L_2}{C_2}}} \qquad \cdots\cdots\cdots \quad (1)$$

i.e.
$$L_2 = C_2 \frac{V_{C2}^2}{I_{toff}^2} \qquad \cdots\cdots\cdots\cdots (2)$$

The resonant frequency of L_2 and C_2 is

$$f = \frac{1}{2\pi}\sqrt{\frac{1}{L_2 C_2}} \qquad \cdots\cdots\cdots\cdots\cdots (3)$$

so the time to the first current peak is

$$t_{pk} = \frac{\pi}{2}\sqrt{L_2 C_2} \qquad \cdots\cdots\cdots\cdots (4)$$

At $t=t_{pk}$, T_1 is turned off. At this time C_1 is still fully charged. Because L_2 is much larger than the stray loop inductance of the C_1-D_2 loop, the current commutates from the VUT to the C_2-D_2 loop while V_{c1} is presented to the T_1 during the turn-OFF event. The current continues to flow through C_1-D_2-L_2-C_2 for some time as it falls to zero. If the T_1 is turned on again during this interval (after minimum off-time has expired) the T_1 will experience a hard turn-ON event and D_2 will experience a hard-recovery. Figure 2 presents simulated waveforms to illustrate the operation of the proposed circuit.

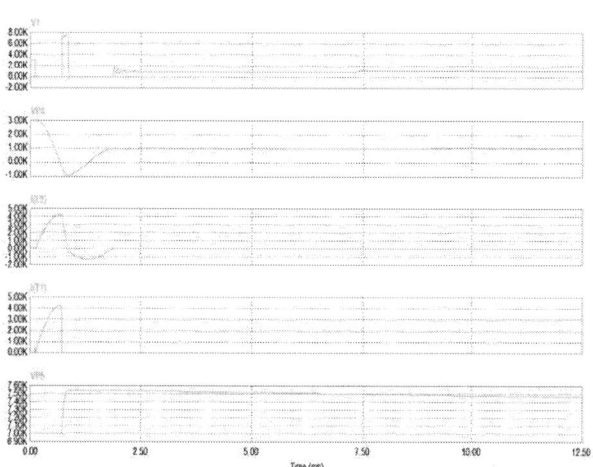

Figure.2 A valve test system circuit schematic
(L1=250uH, C1=500uF, C2=500uF, VDC1=7000V, Vc2=3000V)

B. Configuration of valve test equipment

To perform a test described in the previous chapter, the test facility is made as figure 3 to figure 5. the building block for the valve assembly is composed with IGCT rated 4500V 400A which is 5 series connected, anti-parallel diode which is 5 series connected, current wheeling diode which is 5 series connected., snubber circuit, voltage balancing resistor, and a water cooled heat sink

Figure.3 Valve assembly

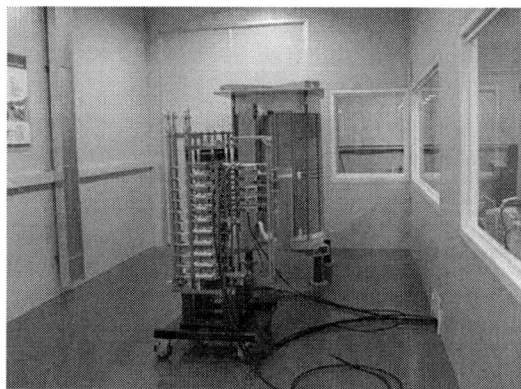

Figure.4 Valve assembly and load reactor

Figure.5 DC power supplies

978-1-4244-1871-8/07 $25.00 © 2007 IEEE

Each gate drive receives isolated power through a current transformer connection to a high frequency current loop. The current loop, which feeds all of the gate drives and control electronics on the pole, is powered by a high frequency inverter. This high frequency inverter is fed by the gating power supply . the HFI was shown in fig 6.

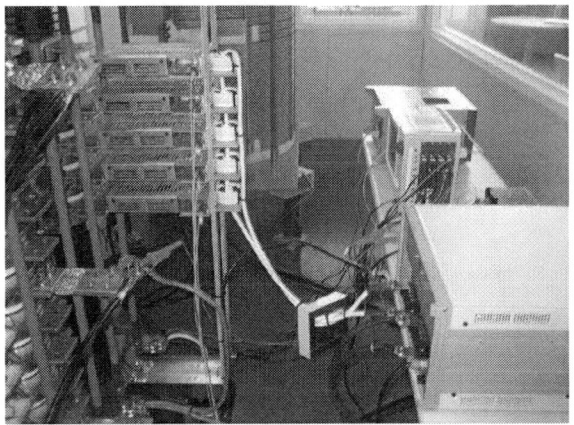

Figure.6 Gate drive power supply(HFI)

C. Test process and results

the test was accomplished in compliance with the next process.

1) Final inspection and visual check
2) check controller operation
3) HFI on
4) check IGCT gate drive LED
5) check IGCT gate drive operation
6) DC power supply 1, 2 on
7) turn on IGCT during 350usec
8) check waveform about voltage distribution in each IGCT
9) adjust DC voltage accordance to the table 1.
10) verify switching waveform

TABLE 1. DC Voltage

	DC1 Voltage	DC2 Voltage
1	1000	500
2	2500	1000
3	3500	1500
4	5000	2000
5	5000	3500
6	6000	3500
7	7000	3500
8	8000	3500
9	9000	3500
10	10000	3500

Figure 7.shows the turn on/off characteristics of 5-series connected IGCT switching valve. Load current waveform is

coincidence with the simulation results and the voltage sharing profiles shows the expected results.

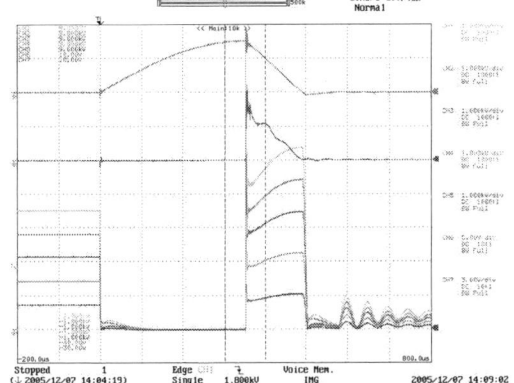

Figure.7 Turn on/off characteristic

Figure 8 the turn off waveform which is tested in DC1 10000V, DC2 3500V. The test waveform show that voltage sharing of across each IGCT was well balanced

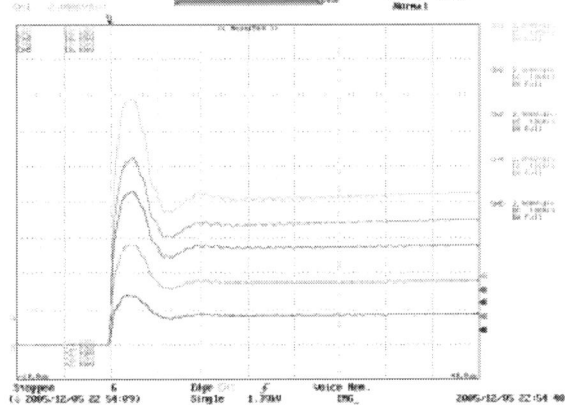

Figure.8 turn off characteristic waveform (before stray inductance modification)

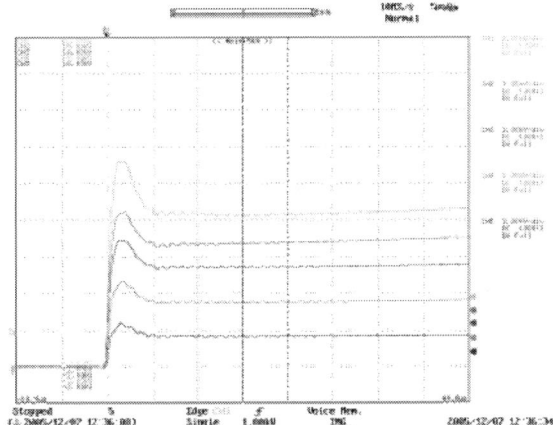

Figure.9 turn off characteristic waveform (after stray inductance modification)

978-1-4244-1871-8/07 $25.00 © 2007 IEEE

The L1 represented stray inductance which is varied by extern elements (e.g. cable length). The charged voltage peak in C1 was varied by L1.

in order to minimize L1 value consequently the load cable length minimization. We can see the minimize turn off peak voltage in figure 9.

III. A TEST FACILITY FOR LARGE SCALE INVERTER POLE USING RESONANT CIRCUIT

This test is intended to verify proper design of the 1 phase inverter pole. This test is achieved by the use of H-bridge test facility. The H-bridge test will verify the following points in particular :

Through this test, we can validate transient voltage deviation and hot spot points. We designed 2 of H-bridge and validated test facility through simultion

A. H-Bridge test facility using AC and DC power supply

This H-bridge test facility was designed using AC power supply and DC power supply. Figure 10. shows the H-bridge schematic.

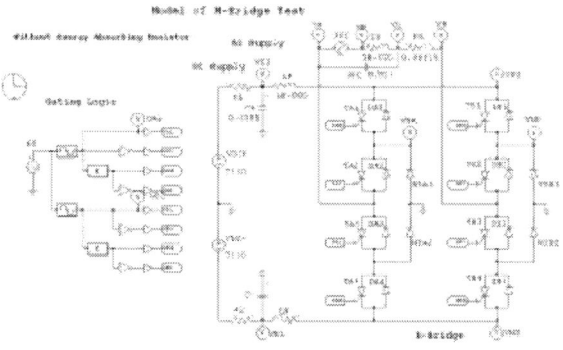

Figure.10 H-Bridge test facility using AC and DC power supply

Basic switching pattern is set up as, making time interval between two poles. AC power supply when positive switching in a moment and current flows through positive valve mostly. As result of simulation(Figure 11), we set the time for switching start with designated conducted RL current. In this simulation, switching is start when current is peak.

To save the cost of system, DC power supply and AC power supply designed as on-way. If switching pattern is set as one-way energy supply, energy will be accumulated to one capacitor, and this energy has to be consumed. To consume this energy, When voltage flowing RL is positive, C+ is charged and when voltage is negative, C+ is discharged with broadening TB as third waveform(figure 11). This energy stored DC link of AC power supply and using for switching of negative switching. Same logic is applied to switching of Negative.

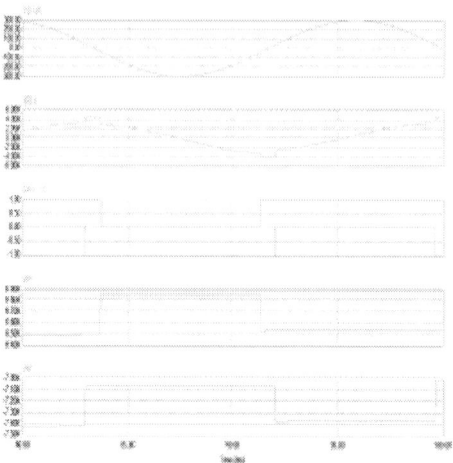

Figure.11 simulation results

B. H-Bridge test facility using resonance circuit

In case of using two power supply, very high voltage is induced to AC power supply in a moment when DC voltage is switching. Consequently, AC power supply insulation become very huge. Therefore, we have designed H-Bridge test facility using resonance circuit without AC power supply.(figure 12.)

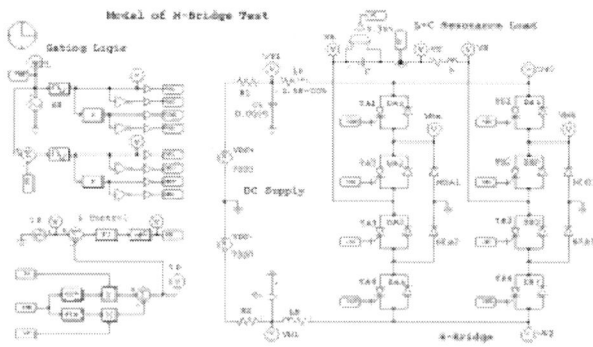

Figure.12 H-bridge test facility using resonance circuit.

AC output of H-bridge flows to a series LC resonance circuit that has 60Hz resonance frequency. Switching phase difference between two poles, current flowing this circuit is controlled. If pole switching frequency reach to 60Hz, power factor of pole voltage and current get zero. To change power factor, regulate switching frequency. Figure 4-6 show simulation waveform according to regulation of switching frequency. Using this method, a rated current switching is possible under low DC voltage

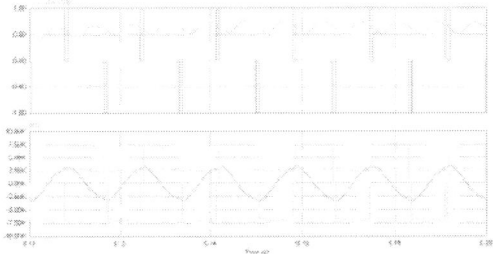

Figure.13 resonance frequency waveform : 60Hz

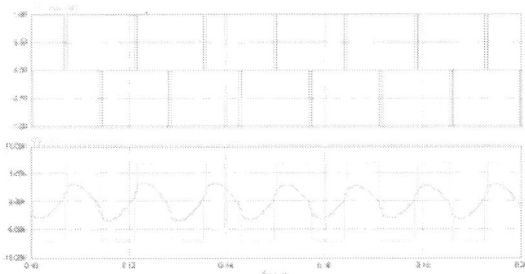

Figure.14 capacitive mode waveform : 70Hz

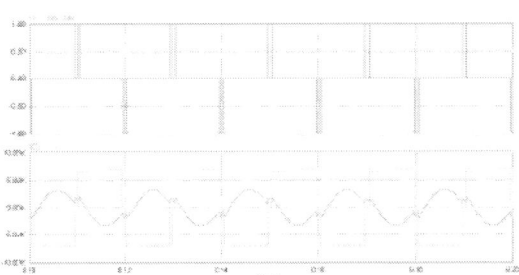

Figure.15 reactive mode waveform : 50Hz

Figure.16 3-Leve inverter pole

transformer connection to a high frequency current loop. The current loop, which feeds all of the gate drives and control electronics on the pole, is powered by a high frequency inverter, mounted in a square enclosure located on one side of the front (ac end) of the pole. These high frequency inverters, located on each pole, are fed by the gating power supply . Also mounted on the front of the pole, near the high frequency inverter is a similar enclosure that houses the pole electronics that contains the local control electronics for the pole. The pole control electronics receives inputs via *glass fiber optics* from the *central contro(GCU)l* and generates the critical timing for switching the individual IGCTs in a valve to achieve suitable voltage sharing. These switching commands are distributed via *plastic fiber optic cables* to the gate drives, which send feedback that is used to determine the health of each individual module and can be used to adjust switching times. The pole also monitors its output current via an *ac/dc current sensor* mounted on the ac output of the pole and takes protective action when an overcurrent is sensed. The pole electronics gathers this and other diagnostic information where it is sent back to the central control.

H-bridge is comprised of 2 three-level poles connected to a dc power supplies.

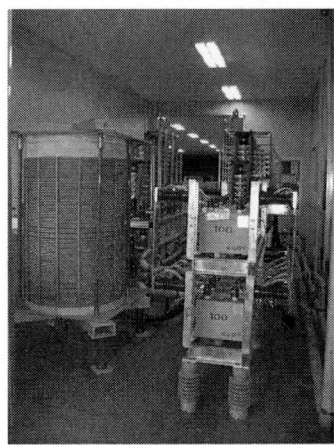

Figure.17 LC load

C. Configuration of H-Bridge test facility

We designed two H-bridge test facility. One is H-bridge test facility using AC and DC power supply the other is H-bridge test facility using resonance circuit. In side of cost, system stability, system simplicity, the later design was manufactured. Figure 16 shows the manufactured inverter pole and H-bridge facility.

The three-level pole structure includes several passive components required for the protection of the valves in the pole. Two water-cooled *di/dt limiting reactors* are required for limiting the rate of rise of current through the valves when they turn on. A *clamp* is used to limit the voltage overshoot on the valves when they turn off. The clamp includes two *clamp diode* stacks (4 devices each), two *clamp capacitors*, and four watercooled *damping resistor* assemblies.

Switching of the individual IGCTs in the valves requires both power and control electronics. Each IGCT has an associated gate drive, which turns the IGCT device on and off Each gate drive receives isolated power through a current

978-1-4244-1871-8/07 $25.00 © 2007 IEEE

Figure 17 shows LC loads. The loads used 440V 1 phase 1000kV, 14000uF capacitor and 440V 1phase 1000kVA 0.5mH inductor, and 10kV 12A DC power supply, for LC resonance. Cooling system and heat-exchanger were installed for cooling of heat that produced form switching device during switching under rated operation..

H-bridge test has to include validation of inverter stack as well safety and reliability of STATCOM controller about switching noise under rated operation.

STATCOM controller is used for fixed switching pattern considering selected switching device(IGCT) and inverter Topology. Switching pattern is decided and stored on Lookup table. For accurate voltage output, the adequately high resolution of table has to be set and to output high resolution, table access with high frequency is necessary. Considering these points, we have designed controller as figure 18.

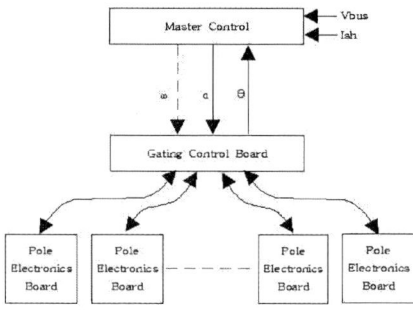

Figure.18 controller schematic

First, Master control takes charge of total system control and regulate in 65.1 usec unit. Also that performs measurement of bus voltage and current for current control.

Gating control board on next stage handles signal in usec and makes gating pattern.

At last, pole electronics boards are installed in each pole for firing and protection in pole unit.

In each PEB, minimum ON/OFF time and dead time are monitored and valve condition is determined receiving condition of each gate drive

IV.   TEST RESULTS

H-Bridge test performed under DC voltage to max Maximum 5000V of(500V interval) and electric current to Max. 1500A(50A interval)

In HMI screen, contoller performs control through fixing alpha angle of one pole and changing alpha angle of other pole accooding to change of current order. Through monitoring alpha angel, we have confirmed that current output is controlled normally according to alpha angle.

Figure 19 shows current 1500A switching waveform under DC 5000V condition. That shows similar wave form with simulation result.

    Ch1: 1/600scale current
    Ch2 : 1/110sacle voltage.

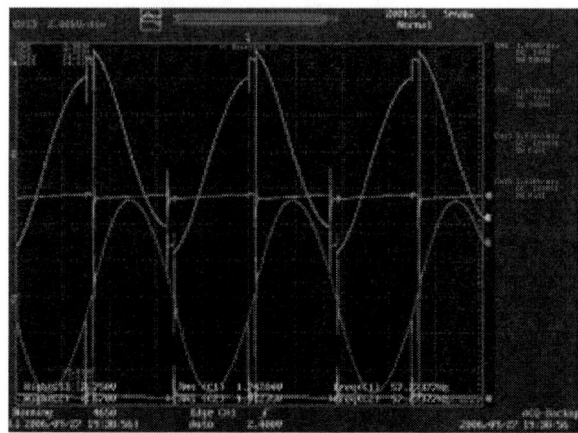

Figure.19 test waveform

After all H-bridge tests for verifying function of the equipment of inverter pole are completed, heat-run test is performed. In this test, changes of temperature in IGCT, IGCT gate drive, Diodes, snubber circuit and inverter room, etc. are checked during 8 hours. The heat-run test results shown that the inverter pole operate reliably, and temperature of all parts is saturated under limit value.

V.   CONCLUSIONS

A novel method for testing large scale inverter valves and poles using a resonant circuit is proposed in this paper. It make it possible to test inverter valves and poles at rated current and voltage with simple configuration. It is confirmed that the voltage is uniformly shared in a valve and hot spot is not detected in a pole with this facility. The performance of both pole and test facility is shown in this paper.

ACKNOWLEDGMENT

The authors would like to acknowledge individuals in Center for Power-IT National Program and MOCIE (Ministry of Commerce, Industry and Energy) for their encouragement and continued support for this project.

REFERENCES

[1] L. Gyugy, N. G. Hingorani, and P.R. Nannery et al., Advanced static var compensator using gate turn-off thyristors for utility application, CIGRE, 1990 session, 1990, pp. 237

[2] L. Gyugyu, Dynamic compensation of AC transmission lines by solid state synchronous voltage sources, IEEE Trans. P.D., vol.9, no.2, pp-904

[3] C. Schauder and H. Mehta, vector analysis and control of advanced static var compensator, IEE Proc. Vol. 140, no. 4, pp. 299

[4] Gyugyi, L., Hingorani, N. *"Understanding FACTS : Concepts and Technology of Flexible AC Transmission Systems"*, IEEE press, 2000

978-1-4244-1871-8/07 $25.00 © 2007 IEEE

An Improved Synchronization Control Scheme of A Low Cost 400Hz Power Supply for No-Break Power Transfer

Seok-Eon Joung
Hanyang University
Department of Electrical Engineering
17 Haengdang-Dong, Seongdong-Gu,
Seoul, Korea
edward@hanyang.ac.kr

Byung-Gun Park, Dong-Seok Hyun
Hanyang University
Department of Electrical Engineering
17 Haengdang-Dong,
Seongdong-Gu, Seoul, Korea
redalarm@hanyang.ac.kr, dshyun@hanyang.ac.kr

In-Ho, Cha
Ehwa Technologies Information
R&D Center
107 Munjong-Dong,
Songpa-Gu, Seoul KOREA
incha2@eti21.com

Abstract—this paper proposes an improved synchronization control scheme of a low cost 400Hz power supply for No-Break Power Transfer. In the case of Aircraft application, the 400Hz power supply called as ground power units (GPU) has been accepted as external electrical power system during stopovers in ground. When transferring from one power source to the other, there is a momentary break in supply. To transfer without a break, the two power sources are momentarily connected in parallel. The proposed synchronization control is achieved by connecting an existing synchronization bus to a voltage zero-crossing signal of a generator power with discrete logic ICs and analog circuits. Therefore, this proposed control scheme is rather simple and the cost may be decreased, compared with expensive controller such as DSP and CAN. The practical feasibility of the proposed control scheme is proved by simulation.

I.   INTRODUCTION

The 400Hz power supply used in military and ground power units for airplanes requires high reliability. These critical loads have been generally used rotating motor-generator system and, recently, supplied by external electrical power system during stopovers in ground. The 400Hz power supply called as ground power units (GPU) has been accepted as a standard due to low maintenance, low cost, high reliability, and high efficiency. The 400Hz power supply is mostly operated in parallel to improve reliability. Parallel operation is much more complex due to synchronization and load sharing. Many researches to operate in parallel have been published in [1-4]. Recently, a high performance control schemes using expensive controller such as DSP and CAN have proposed in [5]. By contrast, the low cost and simple control scheme have proposed such as concentrated controller [6-7] .

This paper proposes a low cost and robust synchronization control scheme for no-break power transfer. Interconnecting an existing synchronization bus to a voltage zero-crossing signal of the generator power, the proposed control scheme minimizes fluctuation of a voltage and frequency when connecting the other power source. Since the proposed control scheme is consisted of discrete logic ICs and analog circuits, the synchronization control is rather simple and the cost may be decreased, compared with DSP based UPS system

II.   PRINCIPLE OF NO-BREAK TRANSFER

Fig. 1 shows the block diagram of the 400Hz Power Transfer System. Generally, GPU is not simultaneously connected with generators to the same bus bar. Therefore, the synchronization circuit of GPU is simply configured because the synchronization only needs at instant of transfer.

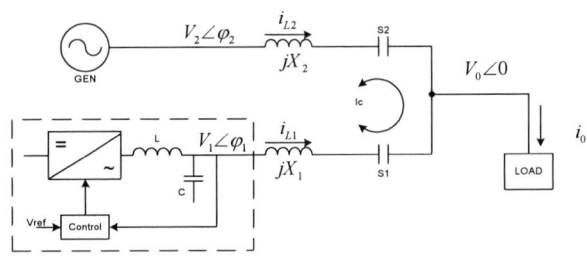

Fig. 1: Block diagram of the 400Hz Power Transfer System

Generally, the active power (kW) and reactive power (kVAR) is occurred when load changes Inverter to Generator or vice-versa with over-lapped transfer for no interrupt.

The active and reactive powers can be calculated as

$$P_1 = \frac{V_1 V_0}{X_1} \sin \varphi_1 \tag{1}$$

$$Q_1 = \frac{V_1 V_0 \cos \varphi_1 - V_0^2}{X_1} \tag{2}$$

From these equations, it can be seen that the active power flow is dominated by the angles, while the voltage amplitudes primarily influence the reactive power flow.

III.   PROPOSED CONTROL SCHEME

A. System Configuration

Fig. 2 shows the block diagram of the proposed NBPT system.

978-1-4244-1871-8/07 $25.00 © 2007 IEEE

The inverter synchronized to Generator with SYNC logic and matched output voltage to Generator with voltage sensing.

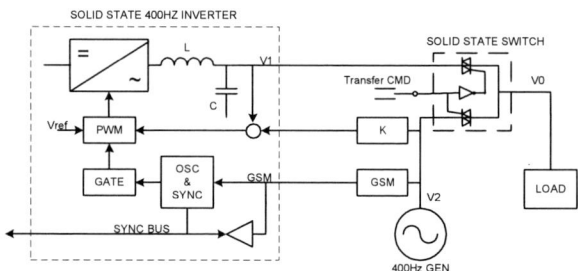

Fig. 2: Proposed 400Hz UPS system operated in parallel

B. Static Transfer Switch

The three-phase static switch is shown in Fig. 3. It consists of two thyristor blocks at the GPU and Generator, which connect the load. Thyristor block is composed of three thyritor modules for each side. In each thyristor module, two of thyristor connected in opposite direction to allow the load current flow in both directions.

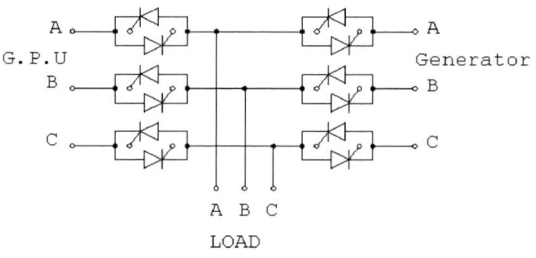

Fig. 3: Static transfer switch

Gate diver block is shown as below. It consists of three main parts including current zero-crossing, gate blocking and gate driver.

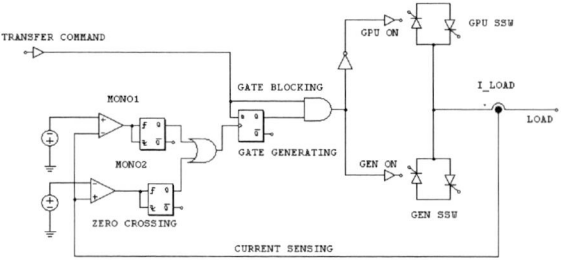

Fig. 4: Gate driver logic structure

Gate blocking function can avoid large amount of cross current when transferring even in phase voltage waveforms.

C. Inverter

The inverters are consists of two inverters and two inverter transformers as shown in Fig. 5.

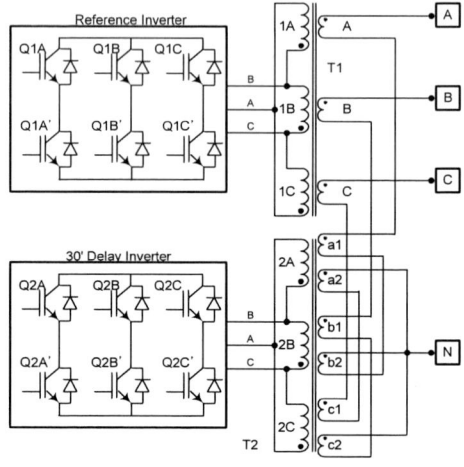

Fig. 5: Diagram of 6-level inverter

The main disadvantage of this type of topology is the large number of supplies and semiconductors required to obtain these multi-step voltage waveforms.

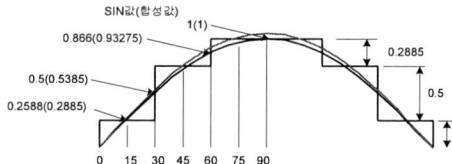

Fig. 6: Voltage waveform from 6-level inverter

But 6-level inverter can generate high quality voltage waveforms, good enough to be considered as suitable voltage template generators

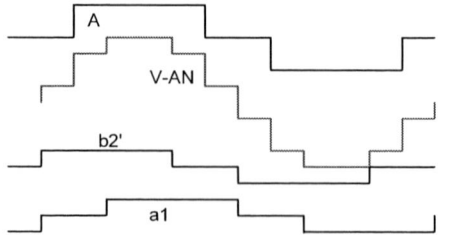

Fig. 7: Voltage waveform of basic inverter and delayed inverter

978-1-4244-1871-8/07 $25.00 © 2007 IEEE

D. *Phase Synchronization Control*

The synchronization control is achieved using 14bit binary ripple counter and stage counter in Fig. 9. The signal divided by ripple counter generates six-step signals using stage counter.

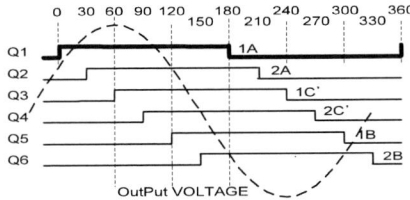

Fig. 8: Inverter Gate Signals

SYNC1 and SYNC2 is generated using gate clocks of each module. The fastest synchronization signal is automatically tracked by other inverters as the SYNC_BUS. The phase angle of each inverter is controlled by comparison between SYNC_BUS and Self-Clock.

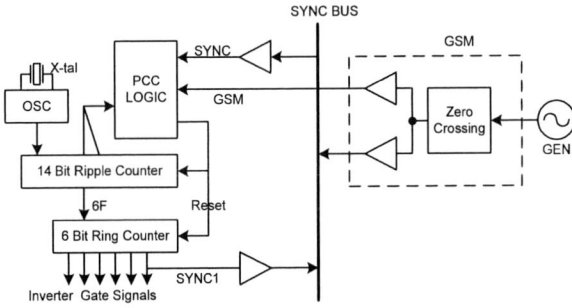

Fig. 9: Block diagram of synchronization control

As shown in Fig. 10 and Fig. 11, PCC Logic compares with the previous PCC signal and SYNC_BUS.

When the system is lagging, Reset signal is generated at Down-edge.

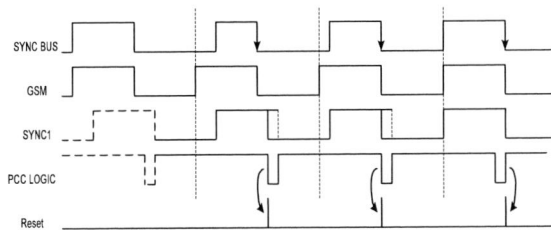

Fig. 10: Synchronization timing signals at leading of generator

But, when the system is leading, Reset signal is generated at Up-edge.

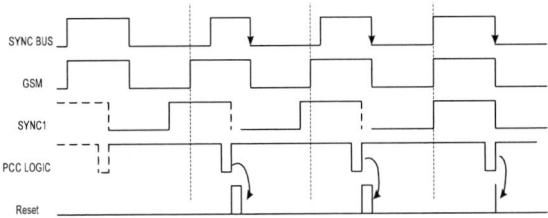

Fig. 11: Synchronization timing signals at lagging of generator

When a generator frequency is changed, the proposed synchronization control can synchronize in range of 399 ~ 401Hz as if Fig. 12.

At 399 Hz, total fundamental period is 2.506e-03sec. Power converter generates sinusoidal waveform like a negative half cycle of 1.25e-03sec and a positive half cycle of 1.256e-03. A symmetrical law of sinusoidal waveform also is 50.06 percentages.

In case of 401Hz, total fundamental period is 2.494e-03sec. Power converter generates sinusoidal waveform like a negative half cycle of 1.25e-03sec and a positive half cycle of 1.244e-03. A symmetrical law of sinusoidal waveform also is 49.94% percentages almost like 50:50.

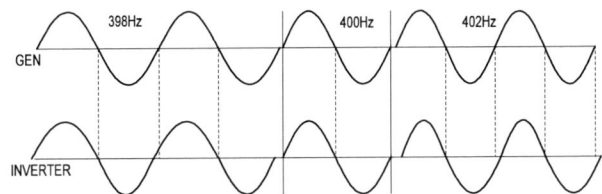

Fig. 12: Synchronized waveforms by a frequency fluctuation

IV.   SIMULATION RESULT

Simulations are performed to verify the feasibility of the proposed control scheme. The parameters of the unit listed as Table I. the current waveforms of GPU, generator, and load are depicted in Fig. 14. and Fig. 16.

Power rating	120 [kVA]
Fundamental frequency	400 [Hz]
Nominal Line-to-neutral voltage	290 [V]
Rating Current	415[A]
Internal Impedance	3.5[mΩ]

978-1-4244-1871-8/07 $25.00 © 2007 IEEE

After the current waveform of generator is synchronized with GPU, the static switch transfers output current from GPU to Generator.

The load current can stably flow without break

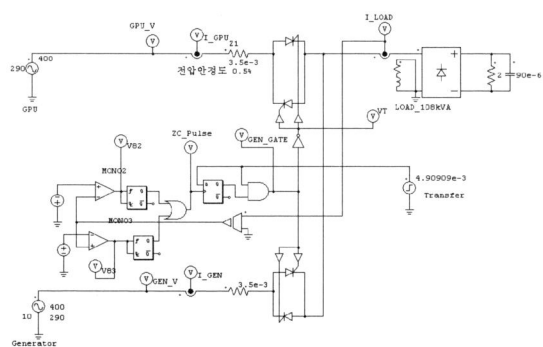

Fig. 13: Circuit for non-linear load transfer Simulation

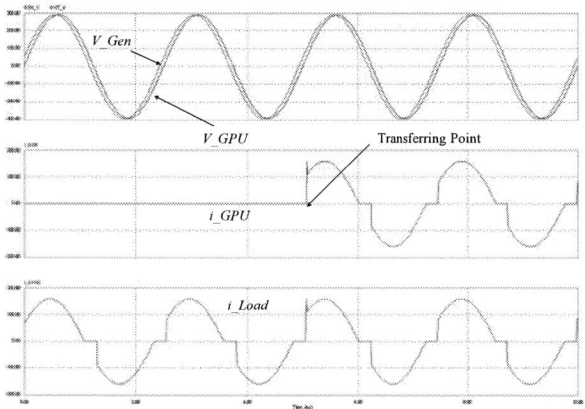

Fig. 14: Waveforms of current transferring on linear load

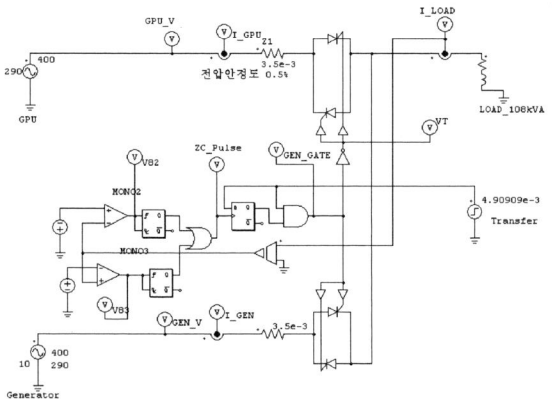

Fig. 15: Circuit for linear load transfer Simulation

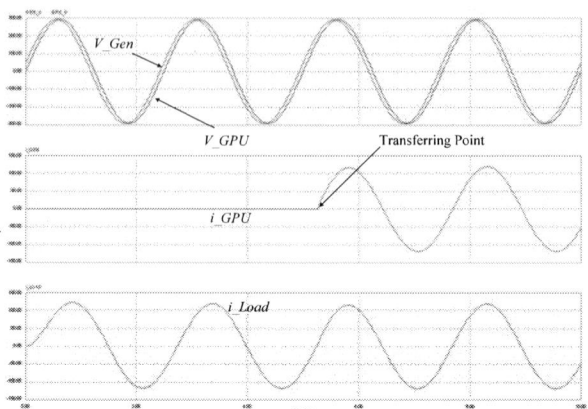

Fig. 16: Waveforms of current transferring on non-linear load

V. CONCLUSION

This paper investigates a low cost and robust synchronization control scheme for no-break power transfer. The proposed synchronization control can achieve a competitive system for cost, compared with expensive controller using DSP or CAN. Asymmetrical waveforms of a inverter is no considered by reason of stability of generator frequency. In the extreme frequency, this method can also use. The practical feasible of the proposed control scheme was verified by means of computer simulations.

REFERENCES

[1] Breit J.S., Doty J.H., "Aircraft no-break power transfer revisited," Proceedings of the IEEE NAECON 1990, vol. 1, pp. 21-25, May 1990.

[2] Jensen U.B., Blaabjerg F., Pedersen J.K., "A new control method for 400-Hz ground power units for airplanes," IEEE Transactions on Industry Applications, vol. 36, no. 1, pp. 180-187, Jan./Feb. 2000.

[3] Jiann-Fuh Chen and Ching-Lung Chu, "Combination Voltage Controlled and Current-Controlled PWM Inverters for UPS Parallel Operation," IEEE Transactions onP. E., vol. 10, no. 5, pp. 547-558, Sep. 1995.

[4] Prodanovic M., Green T.C., Mansir H., "A survey of control methods for three-phase inverters in parallel connection," IEE Conf. Publ., no. 475, pp. 472-477, Sept. 2000.

[5] Kyung-Hwan Kim, Dong-Seok Hyun, "A High Performance DSP Voltage Controller with PWM Synchronization for Parallel Operation of UPS Systems," PESC`06, vol. 3, pp. 1-7, June 2006.

[6] Zhao Qinglin, Chen Zhongying, Wu Weiyang, "Improved Control for Parallel Inverter with Current-Sharing Control Scheme," IPEMC '06, vol. 3, pp.1-5, Aug. 2006.

[7] Chen Liangliang, Xiao Lan, and Yan Yangguang, "A novel parallel inverter system based on coupled inductors," INTELEC`03, vol. 3, pp. 46-50, Oct. 2003.

978-1-4244-1871-8/07 $25.00 © 2007 IEEE

Economic Dispatch for Power Generation Using Artificial Neural Network
ICPE'07 Conference in Daegu, Korea

Sakorn Panta
Rajamangala University of Technology Lanna
Department of Electrical Engineering
128 M 1.T.Changpheuk O.Maung thailand 50300
Email: sakorn_pt@hotmail.com

Suttichai Premrudeepreechacharn
Chiangmai University
Department of Electrical Engineering
Chiangmai Thailand 50300
Email: suttic@eng.cmu.ac.th

Abstract— This paper presents an optimal economic dispatch of electrical power plants by using back-propagation neural networks. The method of economic dispatch for generating units at different loads must have total fuel cost at the minimum point. There are many conventional methods that can use to solve economic dispatch problem such as Lagrange multiplier method, Lamda iteration method and Newton-Raphson method. However, an obstacle in optimal economic dispatch of conventional methods is the changed load. They are necessary to find the optimal economic dispatch from time to time. Moreover, they need a lot of time to repeat calculation for a new solution again. This paper presents back-propagation neural networks model to carry out instead the conventional Lamda iteration method. It is compared with the experimental results of electrical power system of 3 and 10 generating units respectively.

The testing results of the back-propagation neural networks are compared with the Lamda iteration method by testing the teaching data and non-teaching data. It shows clearly that the back-propagation neural networks can find out the solutions accurately and use time to calculate less than other systems that are tested. Error of prediction will be increased slightly by the number of generating units in electrical power plants because it needs to learn a lot of input and output data in the neural network dramatically.

Index Terms-- Economic dispatch, neural network, back-propagation.

I. INTRODUCTION

ECONOMIC dispatch is a computational process where the total required generation is distributed among the generation units in operation, by minimizing the selected cost criterion, subject to load and operational constraints. Economic dispatch is used in real-time energy management power system control by most programs to allocate the total generation among the available units, unit commitment and in some other operation function.

S. Panta is with Department of Electrical Engineering, Faculty of Engineering, Rajamangala University of Technology Lanna, Chiang Mai, 50300 Thailand (e-mail: g4726097@doe1.eng.cmu.ac.th).

S. Premrudeepreechacharn is with Department of Electrical Engineering, Faculty of Engineering, Chiang Mai University, Chiang Mai, 50200 Thailand (e-mail: suttic@eng.cmu.ac.th).

A.Keawsing is with Department of Electrical Engineering, Faculty of Engineering, Rajamangala University of Technology Lanna, Chiang Mai, 50300 Thailand (e-mail: auswinmmc@hotmail.com).

In the area of economic dispatch, several methods have been proposed [1]-[12]. Recent advances were achieved by using gradient method, the recursive method, the Newton-Raphson method. These methods are required more time to converge to the correct results. Other researches in field of expert system, the unit-based genetic algorithm method, neural networks, fuzzy set theory and approximate reasoning have been investigated. However, in the conventional methods, it is difficult to solve the optimal economic if the load is changed. It needs to compute the economic dispatch each time which uses a long time in each of computation loops.

In this paper, an application of neural networks to economic dispatch is proposed. The method used a feed-forward back-propagation type of neural networks to learn different condition in operation of each unit. By changing total load condition in one day, the minimum cost operation of each unit should be selected. So, the minimum operation condition is selected with less iteration and time.

II. ECONOMIC DISPATCH PROBLEM

Figure 1 shows the configuration that will be studied in the section. This system consists of m generating units connected to a single bus-bar serving a received electrical load P_{load}. The input to each unit, show as C_i, represents the cost rate of the unit. The output of each unit, PG_i is the electrical power generated by that particular unit. The total cost rate of this system is, of course, the sum of the costs of each of the individual units. The essential constraint on the operation of this system is that the sum of the output powers must equal load demand.

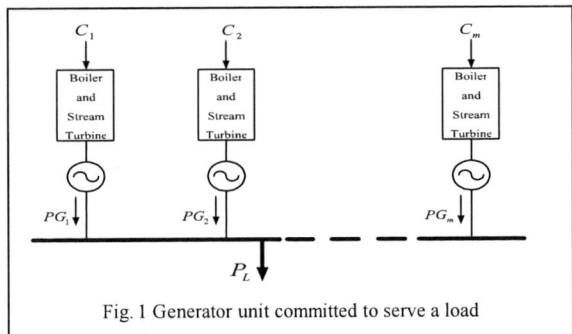

Fig. 1 Generator unit committed to serve a load

Mathematically speaking, the problem may be stated very concisely. That is, an objective function, C_T, is equal to the total cost for supplying the indicated load. The problem is to minimize C_T subject to the constraint that the sum of the power generated must equal the received load. Note that any transmission losses are neglected and any operating limits are not explicitly stated when formulating this problem. [13-14]. The objective of economic dispatch is to minimize the total generation cost throughout the target time interval by evaluation the following objection function. The economic dispatch problem can be mathematically described as follows [13]

$$C_T = C_1 + C_2 + \ldots + C_m = \sum_{i=1}^{m} C_i(PG_i) \tag{1}$$

$$PG_T = PG_1 + PG_2 + \ldots + PG_m = \sum_{i=1}^{m} PG_i \tag{2}$$

$$\phi = 0 = P_D - \sum_{i=1}^{m} PG_i \tag{3}$$

$$\underset{PG_i}{Min} \sum_{i=1}^{m} C_i(PG_i) = \underset{PG_i}{Min} \sum_{i=1}^{m} (a_i + b_i PG_i + c_i PG_i^2) \tag{4}$$

where
m : The number of generator units
i : Index of dispatchable units
C_i : The fuel generation of unit i [$/h]
C_T : The total fuel cost generation in system   [$/h]
PG_i : The power generation of unit i [MW]
PG_T : The total power generation in system [MW]
a_i, b_i, c_i : Coefficients for power generation cost of unit i

(i) Power balance

$$\sum_{i=1}^{m} PG_i = P_D + P_L \tag{5}$$

Where P_D : Total load demand   [MW]
$\quad\quad\quad P_L$: Power losses   [MW]

(ii) Generation limit of each unit
Each unit has its maximum and minimum generation limits

$$PG_{i\min} \leq PG_i \leq PG_{i\max} \tag{6}$$

Where
$\quad\quad PG_{i\min}$: the minimum generation limit of unit i [MW]
$\quad\quad PG_{i\max}$: the maximum generation limit of unit i [MW]

III. ECONOMIC DISPATCH BY LAMDA-ITERATION

The program is designed to find out the economic dispatch by Lamda iteration method witch has the process of the designed program as a flow chart that shown in Fig.2. The program is written by using Matlab program to carry out the economic dispatch by Lamda iteration method with

initial condition of λ that are constant. They are used to determine the generating power of generators in each unit and looked out the differences between loads and total generating power from all generators. They are compared with the agreeable tolerance (ε) witch has been set. If it has the difference that is higher than the agreeable tolerance. It will adjust a new λ and return to starting point to calculate the new generating power of the generators. The concept of tuning λ is considered from the summing of generating power. If it is higher value than load , it will be decreased λ. If it is lower than load value, it will be increased λ by setting $\pm 10\%$ from the old value.

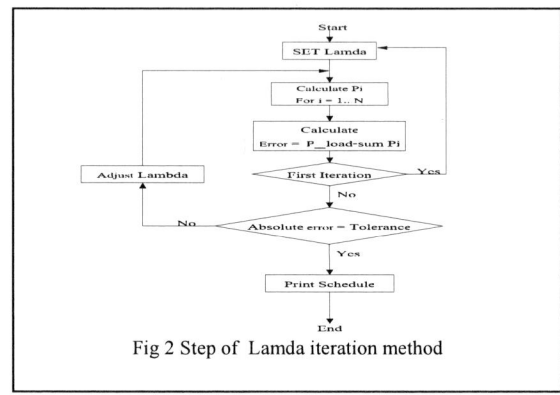

Fig 2 Step of Lamda iteration method

IV. ARTIFICIAL NEURAL NETWORK

Neural networks have self adapting capabilities which makes them well suited to handle non-linearities, uncertainness and parameter variations which may occur in economic dispatch [14]. Back propagation network is an example of nonlinear layered feedforward networks. Back propagation neural networks construct global approximations to nonlinear input-output mapping. There are capable of generalization in regions of the input space where little or no training data are available. The structure of the proposed neural network used for calculate economic dispatch is shown in Fig. 3.

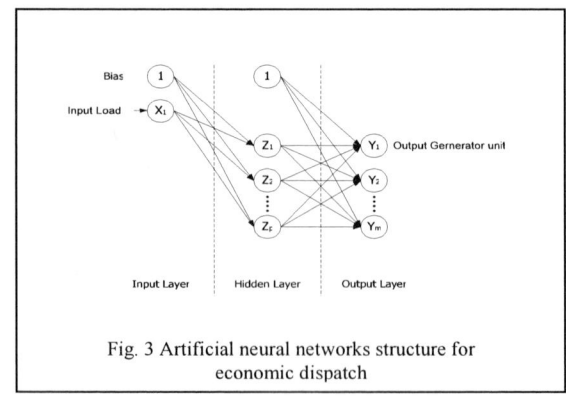

Fig. 3 Artificial neural networks structure for economic dispatch

The proposed neural networks have three layers, i.e. input layer, hidden layer and the output layer. The input layer has only one neuron which is the number of total load. The output layer has m neurons represented the number of generated power of each generators. If the system has only 3 generators, the output layer will have only 3 neurons. The hidden layer has 30 neurons. The network is fully connected, i.e. the output of each neuron is connected to all neurons in the hidden layer through a weight which is not shown in the figure. Also a bias signal is coupled to all the neurons through a weight. All the layers of neural network have a hyper tangent sigmoid transfer function. The algorithm used for training is back-propagation. The back-propagation training algorithm needs only inputs and the desired outputs to adapt the weights. A gradient descent minimization can be performed on the error function. Back-propagation training is referred to as supervised training. The input-output data for training neural networks is the variation of load as input and the optimal generated power of each generator as output which is calculated by using Lamda-iteration. The neural network was trained using MATLAB program.

V. SIMULATION RESULTS

This section discusses the simulation of neural networks for economic dispatch. In order to demonstrate the effectiveness of the proposed method, the test system with three and ten and generating units are considered. Generally, the fuel-cost of each generator is represented as quadratic function as eq. (7).

$$C_i(P_i) = a_i + b_iP + c_iP_i^2 \qquad (7)$$

where a_i, b_i, and c_i are the coefficient of quadratic function.

In this paper, the fuel-cost of each generator is not considered the loss in transmission line and fault in generation unit. This paper used computer Celeron(R) M, CPU420, 1.6GHz, and 1.24 GB Ram.

Symbol

PG : Output Power Generation of Lamda-Iteration Method (MW) ,PGn : Output Power Generation of Neural Networks Method (MW), APE : Absolute Percentage Error (%) ,MAPE : Mean Absolute Percentage Error (%)

A. 3 Generator Units.

The fuel-cost coefficients and generation limits for each unit for 3 units system is shown in table 1. In order to verify the performance of neural networks for economic dispatch, the results computed from neural networks are compared with Lamda-iteration method as shown in Table 2 and Fig. 4-5. From the example results in Table 2, this result only shows the comparison of both methods at load 105, 150, 255 350, 400, and 500 MW. When the load is changing, the optimal generated power of each generator is also changing as shown in Fig 4. The difference error of total loads

between 100 MW and 500 MW in both methods is insignificance as seen in Fig. 5. The percentage errors are shown in table 3. However, the CPU time for computing of neural network is much less than Lamda-iteration method. In three-generator system, the input matrix layers will have [401 × 1] neurons and the output matrix layers will have [401 × 3] neurons. The hidden layer has 30 neurons, and it trains with trainlm function witch has goal performance of MSE at 1e-8. The CPU time for computing to training is 252.27 sec.

TABLE 1 COST FUNCTION GENERATOR DATA OF 3 UNITS SYSTEM

Unit(i)	Coefficients for power generation cost			Limit of generation	
	a_i	b_i	c_i	Min (MW)	Max (MW)
1	328	8.66	0.01	50	250
2	60	9.76	0.012	20	100
3	137	10.05	0.014	30	150

TABLE 2.COMPUTATIONAL RESULTS OF 3 GENERATOR UNIT

Load (MW)	Power output of Lamda-Iteration Method (MW)			CPU Time (sec)	Power output of Neural Networks Method (MW)			CPU Time (sec)
	PG1	PG2	PG3		PGn1	PGn2	PGn3	
105	55.0	20.0	30.0	0.24	55.0	20.0	30.0	0.006
150	100.0	20.0	30.0	0.1	100.0	20.0	30.0	0.002
255	172.7	52.3	30.0	0.27	172.7	52.3	30.0	0.014
350	212.3	85.3	52.4	0.012	212.3	85.3	52.4	0.007
400	232.9	100.0	67.1	0.23	232.9	100.0	67.1	0.02
500	250.0	100.0	150.0	0.01	250.0	100.0	150.0	0.005

TABLE 3. APE AND MAPE OF BETWEEN LAMDA-ITERATION METHOD AND BACK-PROPAGATION NEURAL NETWORKS METHOD 3 GENERATOR UNITS

Error of Power Output		
APE Max	APE Min	MAPE
0.25	0	0.0032

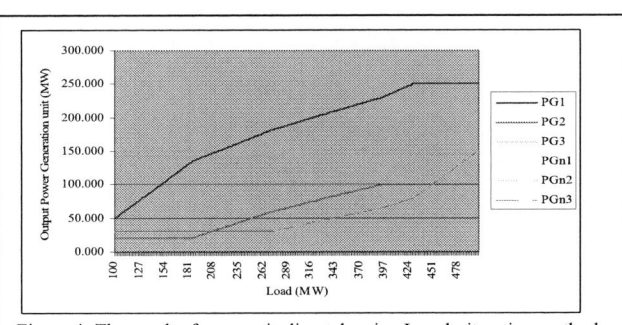

Figure 4: The result of economic dispatch using Lamda-iteration method and Back-propagation neural networks method for 3 generator units.

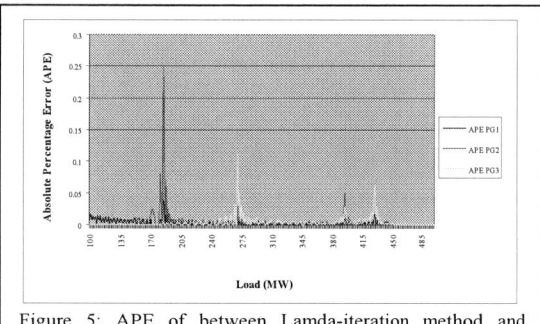

Figure 5: APE of between Lamda-iteration method and Back-propagation neural networks method 3 generator units.

B. 10 Generator Units

The fuel-cost coefficients and generation limits for each unit for 10 units system is shown in table 4. From the example results in Table 5. The result only shows the comparison of both methods at load 650, 300, and 1900 MW. When the load is changing, the optimal generated power of each generator is also changing as shown in Fig 6. The difference error of total loads between 525 MW and 2085 MW in both methods is in significance as seen in Fig. 7. The percentage errors are shown in table 6. However, the CPU time for computing of neural network is much less than Lamda-iteration method. In ten-generator system, the input matrix layers will have [1561 × 1] neurons and the output matrix layers will have [1561 × 10] neurons. The hidden layer has 60 neurons, and it trains with `trainlm` function witch has goal performance of MSE at 1e-8. The CPU time for computing to training is 407.17 sec.

TABLE 4: COST FUNCTION GENERATOR DATA OF 10 UNITS SYSTEM

Unit (i)	Coefficients for power generation cost			Limit of generation	
	a_i	b_i	c_i	Min (MW)	Max (MW)
1	1000	18.19	0.0068	150	600
2	970	19.26	0.0071	50	200
3	600	19.8	0.0065	50	200
4	700	19.1	0.005	50	200
5	420	18.1	0.00738	50	160
6	360	19.26	0.00612	20	100
7	490	17.14	0.0079	25	125
8	660	18.92	0.00813	50	150
9	765	18.27	0.00522	50	200
10	770	18.92	0.00573	30	150

TABLE 5 : COMPUTATIONAL RESULTS FOR EXAMPLE 1 GENERATOR 10 UNIT

Power output of Lamda-Iteration Method (MW)				Power output of Neural Networks Method (MW)			
Unit Gen.	Load (MW)			Unit Gen.	Load (MW)		
	650	1300	1900		650	1300	1900
PG1	150.0	230.9	421.6	PG1	150.0	230.9	421.5
PG2	50.0	70.4	200.0	PG2	50.0	70.4	200.0
PG3	50.0	50.0	193.4	PG3	50.0	50.0	193.5
PG4	50.0	132.0	200.0	PG4	50.0	132.0	200.0
PG5	65.3	160.0	160.0	PG5	65.5	160.0	160.0
PG6	20.0	81.7	100.0	PG6	20.0	81.7	100.0
PG7	125.0	125.0	125.0	PG7	125.0	125.0	125.0
PG8	50.0	103.3	150.0	PG8	50.0	103.3	150.0
PG9	59.7	200.0	200.0	PG9	59.8	200.0	200.0
PG10	30.0	146.6	150.0	PG10	30.0	146.7	150.0
CPU Time (sec)	0.3	2.05	0.1	CPU Time (sec)	0.053	0.054	0.06

TABLE 6. APE AND MAPE OF BETWEEN LAMDA-ITERATION METHOD AND BACK-PROPAGATION NEURAL NETWORKS METHOD 10 GENERATOR UNITS

Error of Power Output		
APE Max	APE Min	MAPE
0.9000	0.0000	0.018

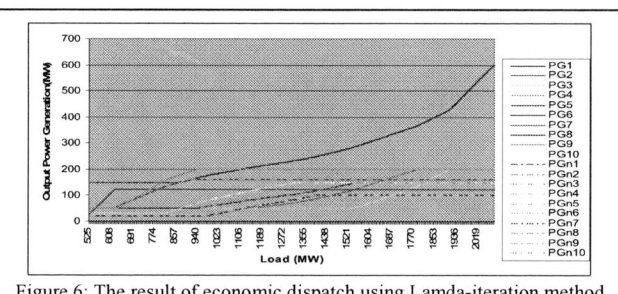

Figure 6: The result of economic dispatch using Lamda-iteration method and Back-propagation neural networks method for 10 generator units.

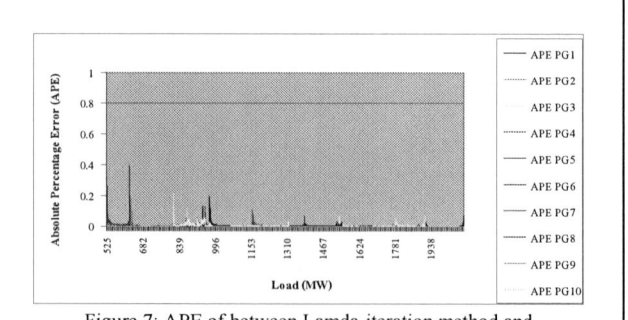

Figure 7: APE of between Lamda-iteration method and Back-propagation neural networks method 10 generator units.

978-1-4244-1871-8/07 $25.00 © 2007 IEEE

VI. CONCLUSION

This paper has presented the application of neural networks for economic dispatch for power generation units. The proposed neural networks have simulated to verify the performance. The simulation results have shown that the results from neural networks are very similar to Lamda-iteration method. However, the neural networks use less time than Lamda-iteration method. Error of prediction will be increased slightly by the number of generating units in electrical power plants because it needs to learn a lot of input and output data in the neural network dramatically.

VII. REFERENCES

[1] N. H. Danchai, M. J. Rawlins, O. Alsac, M. Paris and B. Stott, "OPF for Reactive Pricing Studies on the NGC System", *IEEE Transactions on Power System*, vol. 11, pp. 226-232, Feb. 1996.

[2] S. N. Sing and A. K. David, "Dynamic Security Constrained Congestion Management in Competitive Electricity Markets", *Power Engineering Society Winter Meeting*, vol.3, pp.1808-1813, Jan.2002

[3] X. Wang,Y. H. Song and Q.Lu, " Lagrangian Decomposition Approach to Active Power Congestion Management Across Interconnected Regions", *IEE Proceeding of Generation, Transactions and Distribution*, Vol.148 , pp.399-404, Feb. 2001.

[4] W John. Lamont and Fu Jian, " Cost Analysis of Reactive Power Support", *IEEE Transactions on Power System*, Vol.14, No.3., pp. 890-897, August 1999.

[5] Xing Wang; Yong-Hua Song and Qiang Lu, " A Coordinated Real-Time Optimal Dispatch Method for Unbundled Electricity Markets", *IEEE Transactions on Power Systems* Vol.17, pp.482 – 490, May 2002.

[6] T.T.Nguyen, " Neural Network Load-Flow", IEE Proceeding of the 4th International Conference on Advances in Power System Control, Operation and Management. APSCOM-97, Vol. 142, pp.51 - 58., Jan. 1995.97, Vol. 142, pp.51 - 58., Jan. 1995.

[7] T.T. Nguyen, " Neural Network Optimal Power Flow",*Proceeding of the 4th International Conference on Advances in Power System Control, Operation and Management, APSCOM-97.*, Vol.1, pp.266 – 271, Nov. 1997.

[8] R.S. Hartati, M.E. El-Hawary, "Optimal Active Power Flow Solutions using a Modified Hopfield Neural Network", *Canadian Conference on Electrical and Computer Engineering*, Vol.1,pp.189 – 194.,13-16 May. 2001.

[9] J. Nanda, A.Sachan, L.Pradhan, M.L.Kothari, A.K Rao, L.L.Lai and M. Prasad, "Application of Artificial Neural Network to Economic Load Dispatch", *IEEE Proceeding of the 4th International Conference on Advances in Power System Control, Operation and Management. APSCOM-97*, Vol. 2, pp.707 – 711, Nov. 1997.

[10] S. Ching-Tzong and C. Gwo-Jen, "A Fast-Computation Hopfield Method to Economic Dispatch," *IEEE Transactions on Power Systems* ,Vol. 12, pp.1759 – 1764, Nov 1997.

[11] S. Ching-Tzong and L. Chien-Tung, "New Approach with a Hopfield Modeling Framework to Economic Dispatch," *IEEE Transactions on Power Systems* ,Vol. 12, pp.541-545, May 2000.

[12] M.R.Farooqi, P.Jain and K.R.Niazi, "Using Hopfield Neural Network for Economic Dispatch of Power System," *National Power and Energy Conference*, pp.5-10, Dec 2003.

[13] A . J. Wood and B. F. Wollenberg, *Power Generation, Operation and Control 2nd ed* :John Wiley and Sons INC., 1996.

[14] F. Laurene, *Fundamentals of Neural Network Architectures* :Algorithms and Application. Prentice Hall, 1994.

1.Sakorn Panta was born in Lampang, Thailand, in 1973. He received the B.Eng in electrical engineering from Rajamangala University of Technology Thanyaburi, Thailand, in 2000, and received the M.Eng. degree in electrical engineering from Chiang Mai University, Thailand, in 2007. He is currently a lecturer at Rajamangala University of Technology Lanna, Chiang Mai Campus.

2.Suttichai Premrudeepreechacharn received the B.Eng. in electrical engineering from Chiang Mai University, Thailand and M.S. and Ph.D. in electric power engineering from Rensselaer Polytechnic Institute, Troy, NY. He is an associate professor at Department of Electrical Engineering, Chiang Mai University, Thailand. His research interests include power quality, high quality utility interface, power electronics and artificial intelligence applied power system.

3.auswin keawsing was born in Chiang Mai, Thailand, in 1977. He received the B.S. degree in electrical engineering from Rajamangala Institute of Technology, Thailand, in 2004, He is currently a lecturer at Rajamangala University of Technology Lanna, Chiang Mai Campus.

ISC-STS development for power supply to improve reliability in communication system

Hyun-Chul Jung and Dong-Suk Hyun

HANGYANG UNIVERSITY, HEANG DANG-DONG, JOUNG-GU, SEOUL, 133-791

Email: kt5100@paran.com, love001@kt.co.kr

STS's SCR installed for reliability improvement of important Load's power supply in industry field where an average of 2% fault occurs. In order to improve these problems Attached mechanical switch is parallel with SCR's both end points so that SCR can function during the 60~200[mS] for the mechanical switch action is possible. Also, an ISC (Ideal State Conditioning Interlock Device) system is developed so that the two power supply paths may not be thrown in at the same time. Therefore it removes the overlap section perfectly, raising reliability. These methods have reduced the fault rate remarkably at SCR's fatigue degree that is in operating conversion action less than 0.89 times a year. Even in the case where the SCR's fault occurred, power supply was not discontinued. It contributes to the promotion of reliability in the uninterruptible power supply device in communication system.

1. Introduction

The recent industry field has been using uninterruptible power supply (UPS) device to guarantee a continuous operation of important Load's power supply. UPS can prepare for a power failure or fault of UPS's interior system because it has a reserve power supply and a Bypass line. For a little more reliability promotion in the service in the important super high-speed communication service and IDC center server power supply situation, that is installs two independent UPSs and connects STS(Static Transfer Switch) in each output. Therefore, it supplies power to Load from STS's output.[1]-[3] This paper represents SCR's fault possibility in these used situations and introduces ISC-STS which develops for the reliability of the critical Load's power supply.

2. Actual condition analysis of STS operation in the industrial field

The power supply of CAMA(centralized automatic message account), internet Server Hosting, KONET equipment and optical communication transmission equipment becomes gradually more important in the communication service market environment than in a general industrial field. IDC center is a place that manages corporation's internet server computer and constructs a large-scale internet data center as server occupancy space like a hotel and manages all services that SOHO 's(small office home office) businessman or corporation server needs. As business area of IT field is magnified, IDC center has more and more Server Hosting on a large scale and

essential UPS still needs more bulky capacity to supply power for server. Therefore, the importance of reliability increases. In contrast, technology development speed of UPS is slow and there are problems of maintenance, administration, etc. General fault of average 12 ~ 20% happens in communication power supply equipment during any given year, and interrupted fatal damage of Load power supply of 2% happens. Like these, while KEPCO's Power Quality improves gradually, UPS's reliability falls relatively. In order to solve these problems, STS has been installed in the last output among the more important equipments within the Communication Systems since 2004 as a trial, and has been trying to install UPS in two Sources each and minimize effects of power supply discontinuance by hard failure. Even in the installed STS system, SCR that is Static power converting device of STS's interior has produced faults twice causing STS's SCR device making continuous power supply impossible even though the two UPS's are in normal operation. There is also a possibility of part controversy, but while existing UPS system is normal, fault example of uninterruptible power supply system in place that STS is installed happens much more. IGBT fault of UPS-INVERTER and DC Condenser fault of Converter output has small occurrence rate in case of UPS individual operation happens, and ripple voltage occurs in converter output and storage battery with abnormal phenomenon. harmonic value between UPS and STS increases, also.

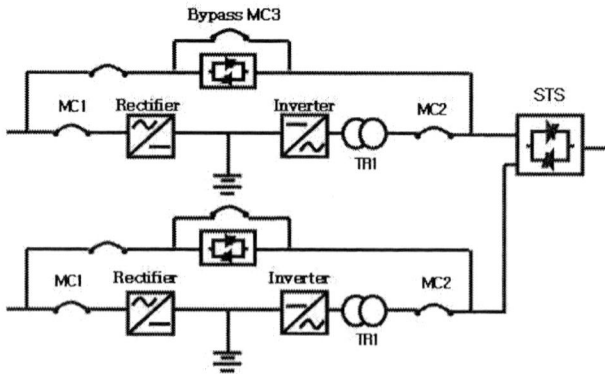

Figure 2-1 STS system that operates with two UPS output

Usually, SCR's reliability via static switching device is highly evaluated in the academic field, but it is not so in the industry field. Actually, 24% of 30 K VA ~ 500 K VA bulky UPS makes fault by power electronic device and

978-1-4244-1871-8/07 $25.00 © 2007 IEEE   562

although fault investigation of STS has been installed since 2006 has been nonsense due to short installed years, but examples of those that two STSs of 12 instruments has caused each obstacles by SCR and other effects have happened.

Year	Number of Equipment,	Total Fault	electronic device parts Fault	SCR,s Fault	IGBT's Fault
2003	992	158	18	6	12
2004	930	113	13	8	5
2005	909	90	24	9	15
Total	2831	361	55	23	32

Table 2.1 Power eletronice device fault analysis in more than 30KVA bulky UPS equipment

Since Communication System recognizes the seriousness of not tolerating 1% accident possibility it is finding a new reliable device multilaterally. Therefore, industry field presently recognizes SCR used within STS's interior as an increase of fault able device.[4]

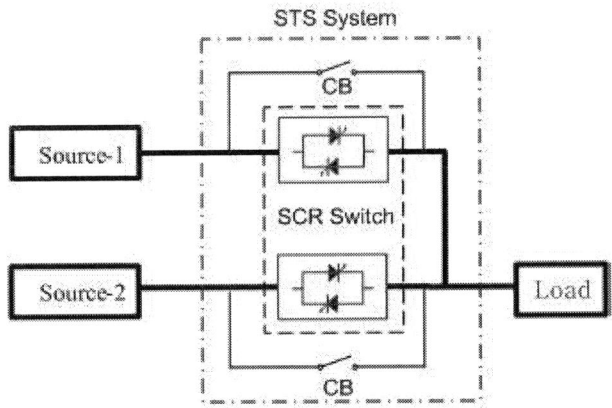

Figure 2.2 STS's structure that is applying in the industry field

 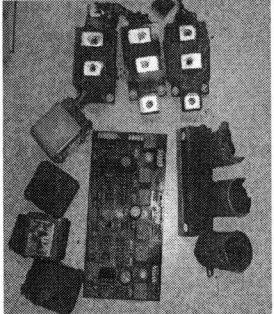

a) UPS Fire accident by SCR   b) SCR Fault

Figure 2.3 SCR Fire accident by SCR dielectric fault

3. ISC (Ideal State Conditioning Interlock Device) - STS development

a) Basic design

SCR that is used in STS has basically a zero potential switching (ZVS Zero Voltage Switching) except during a state of conversion point, the Turn-on time is 3 [uS] usually, and the Turn-off time is informed to be 200 [uS]. Therefore, a pair of SCR with zero voltage switching is known to be a point of contact switch.

However, although SCR consist of one pair, each SCR repeats Turn-on and Turn-off continuously, and one SCR fault among them supplies half-wave rectified voltage in output. In order To improve this problem A pair of machinery contact switch is attached in parallel with both end of SCR and if SCR turned on, a machinery contact switch operates with fixed interval delayed. Therefore, although SCR's Gate signal is supplied continuously, actual Turn-on and Turn-off is not achieved.[5] If SCR operates in these ways, because of SCR function at conversion point, fatigue degree decreases remarkably so that the fault rate decreases and it does not influence to Load in case of fault occurrence. SCR installed in UPS's Bypass line using in emergent conversion acts similar role to STS, it is known that SCR in Bypass line has had no example of fault from 2003 to 2005

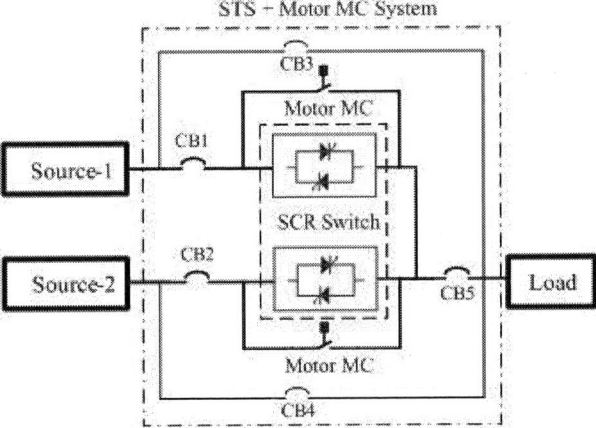

Figure 2 – 4 Machine point of contact parallel connection on SCR both end

b) Operating time difference of MC point of contact switch and SCR

When circuit is composed like figure 2 – 4, in case the first power supply which receives STS's power supply is stopped in normal operation and the second power supply, the reserve power supply, is engaged, there is no problem. But, because the first power supply has the possibility to be in a transient state when STS is engaged, there is enough possibility to have a period that two power supplies overlap. Due to SCR's Turn-on time is 3 [uS] and Turn-on time of magnetic switch is 12 ~ 20mS, there is a significant

978-1-4244-1871-8/07 $25.00 © 2007 IEEE   563

difference in operating time. Therefore, in case of using two different switch units in parallel, even if two UPS power supply is synchronized correctly, many problems can occur. These problems are shown in figure 2 - 5 and 2 - 6. Figure 2 - 5 shows waveforms that is converted to Power Source-2 when it consists of connection MC and SCR in parallel with Power Source-1. in case if short circuit accident generates in source-1 during the Transient state in MC1, source-2 is instantaneously connected to source-1 with closed circuit. Figure 2 - 6 shows waveforms in overlap period when fast magnetic point of contact style MC and SCR are used in Power Source-2. As shown in figure 2 - 5 and 2 - 6, even if reaction time is fast, mechanical point of contact has overlapping time with SCR in 24 ~ 609 [mS]. Therefore, it is impossible to remove overlapping period unless equal SCR is used.[6]

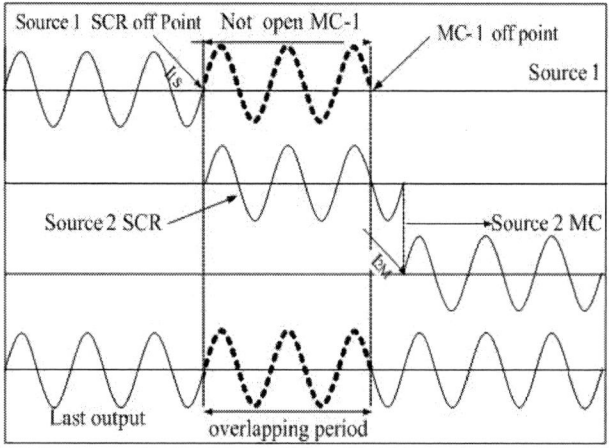

Figure 2-5 Section waveform that is overlaped each other due to SCR injection before cutting magnetic MC

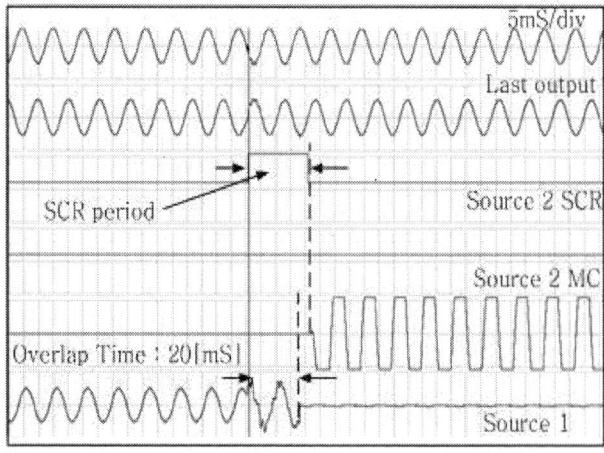

Figure 2 – 6 Waveform of period that is overlaped each other due to SCR injection before cutting magnetic point of contact style MC

c) ISC (Ideal State Conditioning Interlock Device) - STS system

To solve these problems, ISC (Ideal State Conditioning Interlock Device) STS is applied. Machine point of contact is connected on SCR's both end like basic circuit, and circuit is composed like figure 2 - 7.

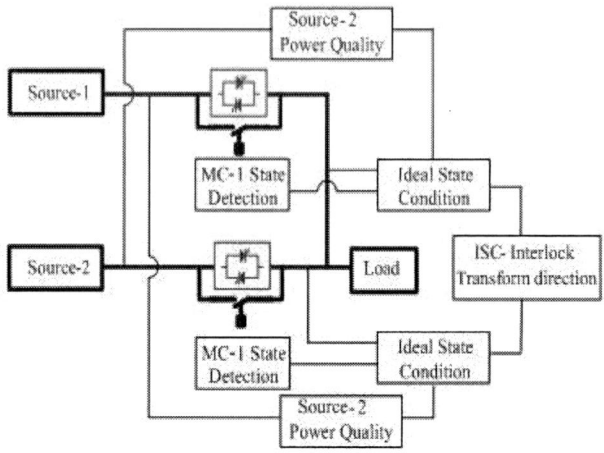

Figure 2 – 7   ISC-STS block diagram

There is MC State Detection that sense mechanical point of contact state of MC-1 in Power Source-1 and MC-2 in Power Source-2. There is MC State Detection that detect mechanical ON/Off state of MC-1 and MC-2 according to output Power Quality of MC-1/MC-2 and detection result of MC State. There is an Ideal State condition part to judge Ideal State condition through comparing and judging MC-1/MC-2 electrical ON/Off and Power Quality of Power Source-1/Power Source-2 and State Detection. It constitutes ISC-Interlock Transformer direction generating Transformer direction by comparing both-side Ideal State condition. In this composition, power is supplied to Load through SCR-1 and MC-1 of Power Source-1. At this time, if power supply of Power Source-1 is disconnected, it judge finally electrical cut-off state, and SCR-2 turns on and MC-2 turns on subsequently after 60mS. Therefore power is supplied to Load through MC-2. In this way, Power Source-1 and Power Source-2 are interlocked electrically so that it does not give strain to UPS in each Power Source. Also it has possibility that SCR-2 turns on before MC1 of Power Source-1 turns off, but there is no problem because it has not Source in Power Source-1 path. That is, it guarantees original continuous power supply that has no overlapping period. This can be expressed as flowchart such as figure 2 - 8. As shown in output and operating waveforms through ISC-STS system, STS's SCR in figure 2 - 9 dose not generate overlapping period at this point changing from Power Source-1 to Power Source-2 and it operates only before MC-2 turn-on. Afterward, power is supplied to Load through MC-2.

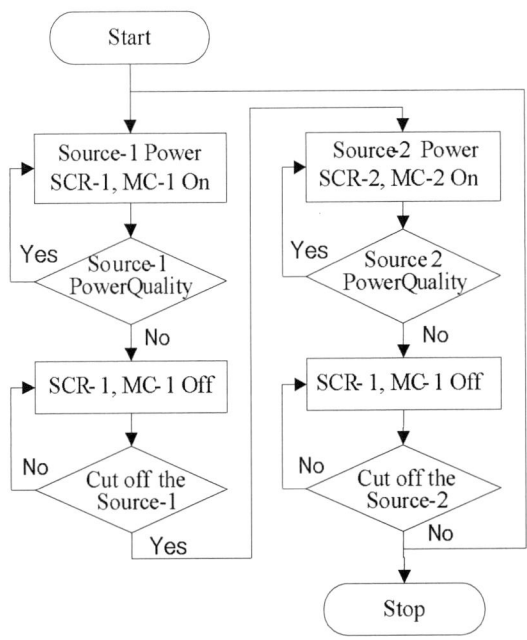

Figure 2-8 ISC-STS flowchart

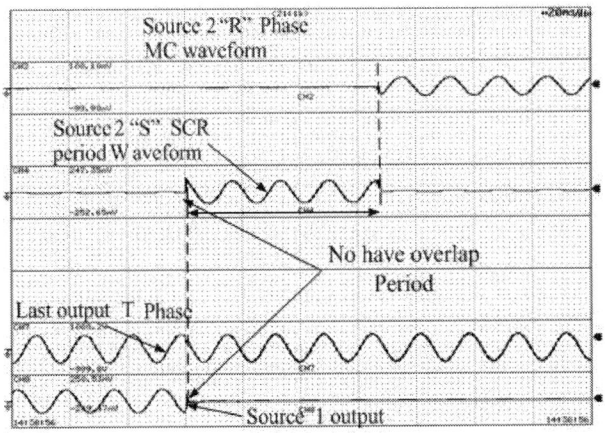

Figure 2 - 9 Conversion from STS Power Source-1 to Power Source-2

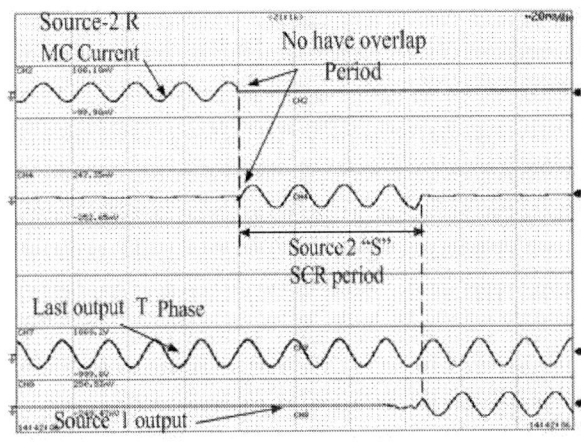

Figure 2 - 10 Conversion from STS Power Source-2 to Power Source-1

a) Parallel Motor Drive MC

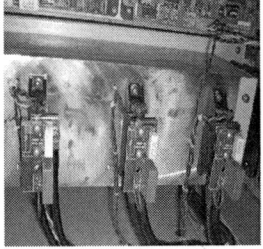
b) STS's SCR

c) Simulation System of ISC-STS

Figure 2 – 11 System that abnormal state conditional Interlock circuit (ISC-STS) is applied.

On the contrary, STS's SCR in figure 2 - 10 does not generate overlapping period at this point changing from Power Source-2 to Power Source-1 and it operates only before MC-1 turn-on. Afterward, power is supplied to Load through MC-1. In this STS system, SCR repeats turn-on and turn-off, but it actually operates during conversion time, 60 ~ 70 [mS]. Especially, with accumulation data 2005 ~ 2006 years, STS acts less than 0.89th during a year so that it can reduce SCR's fatigue degree remarkably. Figure 2 - 11 is an example of system that abnormal state conditional Interlock circuit (ISC-STS) is applied

978-1-4244-1871-8/07 $25.00 © 2007 IEEE   565

4. Conclusion

This paper focuses on reliability of uninterruptible power supply rather than cost in Communication System. If power electronic device that takes charge power converting in UPS broke down because of impedance is increased by use of long term, reserve UPS supplies power through STS system, and there is no problem. But, if STS's SCR that is last output becomes fault, there is no countermeasure. Of course, one can insist that if fault happens on one side of SCR, the other has no effect but if SCR that consist of one pair breaking down, it is general example to have an influence on the surrounding SCR and Gate driver. When see as this side Especially, in Communication System which reliability of power supply's continuity is very important, these researches are important so that the domestic representative mail-order firm selects this by technology standard in present time since 2006. Together, it requires additional research about over 20% UPS fault in UPS using STS compared to UPS supplying power independently.

REFERENCES

[1] Hossein Mokhtari, Masoud Karimi-Ghartemani, and M. Reza Iravani, "Experimental Performance Evaluation of a Wavelet-Based On-Line Voltage Detection Method for Power Quality Applications" IEEE Transactions on Power Delivery Vol.17, No.1, January 2002 Page 161

[2] H. Mokhtari, S. B. Dewan Fellow, "Performance Evaluation of Thyristor Based Static Transfer Switch" IEEE Transactions on Power Delivery Vol.15, No3, July2000 Page 960

[3] Hossein Mokhtari,, and M. Reza Iravani, Fellow, IEEE " Impact of Difference of Feeder Impedances on the Performance of a Static Transfer Switch" IEEE Transactions on Power Delivery Vol.19, No.2, April 2004 Page 679

[4] Hossein Mokhtari, Shashi B. Dewan, Fellow, IEEE, and M. Reza Iravani, Senior Member, IEEE "Analysis of a Static Transfer Switch With Respect to Transfer Time"IEEE Transactions on Power Delivery Vol. 17, No. 1, January 2002 Page 190

[5] M. N. Moschakis and N. D. Hatziargyriou, Senior Member, IEEE "A Detailed Model for a Thyristor-Based Static Transfer Switch" IEEE Transactions on Power Delivery Vol. 18, No. 4, October 2003 Page 1442

[6] A. Fulgenzl, N.Massimiani, F.Muzi, N.Polidoro, F.Torelli "UPS Interconnected with STS to Improve Power-Supply Continuity in the Gran Sasso National Laboratory" IEEE Transactions on Power Delivery No. 3, April 2002 Page 260

Optimal Current Controller in a Three-Phase Grid Connected Inverter with an LCL filter

Kui-Jun Lee, Nam-Ju Park and Dong-Seok Hyun, Fellow, IEEE
Department of Electrical Engineering
Hanyang University
17, Haengdang-dong, Seongdong-gu, Seoul, 133-791
Email:exkjule@hanyang.ac.kr

Abstract—This paper presents a robust current control in a three phase grid connected inverter using an LCL filter. This controller is based on the optimal control theory and typically designed in a complex-valued state-space. The proposed current controller has a robust tracking performance for the grid voltage harmonics and robustness to parameter error. Finally, the simulation results are presented to verify the validity of the proposed method.

I. INTRODUCTION

Over the years, power converters are used in various grid-connected applications transferring power to the grid from the renewable energy sources such as photovoltaic arrays, wind turbines and fuel cell. Also, these power sources require inverters to optimize power factor so that the current control is the important factor of the overall control system. Three major current controller include hysteresis regulator [1], proportional-integral (PI) regulator [2] and predictive regulator [3].

Besides these controllers, low-pass filters between the voltage source inverter (VSI) and the grid are used to reduce the current harmonics around the switching frequency. Among them, a L filter is a widely used form for a simple implementation, but this filter increases the switching frequency to satisfy a total harmonic distortion (THD) below 5%. So, an LCL filter having the higher harmonic attenuation can be used [4], [5]. However, an LCL form requires complex current controller because of resonance hazards disturbing the system stability.

This paper presents a robust current control in a three phase grid connected inverter using an LCL filter. This controller is based on the optimal control theory and typically designed in a complex-valued state-space. The proposed current controller has a robust tracking performance for the grid voltage harmonics and robustness to parameter error. Finally, the simulation results are presented to verify the validity of the proposed method.

II. SYSTEM MODELING

The circuit diagram of the three-phase grid connected inverter is shown in Fig. 1. An LCL filter is employed to ensure that the line current and voltage waveforms are within the harmonic specifications. A pulse-width modulation would guarantee that the inverter voltage v_o is not related to low-order harmonic distortion. So harmonic distortion in the output current originating from the grid voltage can be rejected by an appropriate control strategy and this will be explained subsequently.

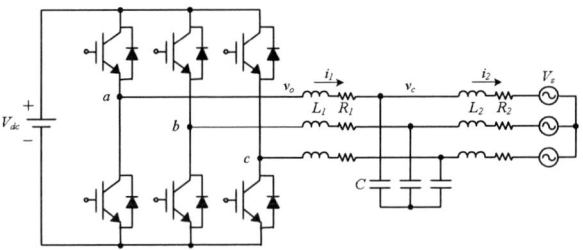

Fig. 1. Three-phase grid connected inverter.

The voltage equations in abc frame are as follows:

$$L_1 \frac{di_{1abc}}{dt} = -R_1 i_{1abc} - v_{cabc} + v_{oabc}$$

$$C \frac{dv_{cabc}}{dt} = i_{1abc} - i_{2abc}$$

$$L_2 \frac{di_{2abc}}{dt} = v_{cabc} - R_2 i_{2abc} - v_{sabc}. \qquad (1)$$

Assuming that the three-phase system are balanced, the governing equation in synchronous d-q reference frame can be written in state-space form as

$$\dot{x} = Fx + G_1 v_o + G_2 v_s \qquad (2)$$

where

$$x = \begin{bmatrix} i_{1dq} & v_{cdq} & i_{2dq} \end{bmatrix}^T$$

$$F = \begin{bmatrix} j\omega - \dfrac{R_1}{L_1} & -\dfrac{1}{L_1} & 0 \\[2mm] \dfrac{1}{C} & j\omega & -\dfrac{1}{C} \\[2mm] 0 & \dfrac{1}{L_2} & j\omega - \dfrac{R_2}{L_2} \end{bmatrix} , \quad G_1 = \begin{bmatrix} \dfrac{1}{L_1} \\[2mm] 0 \\[2mm] 0 \end{bmatrix} ,$$

978-1-4244-1871-8/07 $25.00 © 2007 IEEE

$$G_2 = \begin{bmatrix} 0 \\ 0 \\ -\dfrac{1}{L_2} \end{bmatrix} \qquad (3)$$

and

$$y = i_{2dq} = Hx, \quad H = \begin{bmatrix} 0 & 0 & 1 \end{bmatrix}. \qquad (4)$$

III. INTEGRAL CONTROL

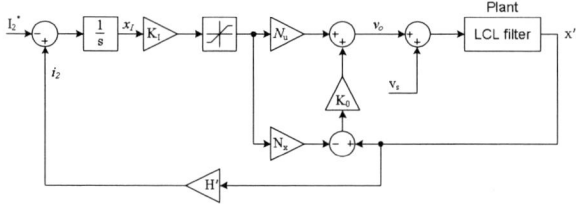

Fig. 2. Block diagram of the proposed current controller.

The proposed current controller is shown in Fig. 2. To obtain robust tracking and disturbance-rejection, integral control is designed [6]. Integral of the error satisfying following differential equation is added as additional state.

$$\dot{x}_I = Hx - r \qquad (5)$$

So, the final state equation is

$$\begin{bmatrix} \dot{x} \\ \dot{x}_I \end{bmatrix} = \begin{bmatrix} F & 0 \\ H & 0 \end{bmatrix} \begin{bmatrix} x \\ x_I \end{bmatrix} + \begin{bmatrix} G_1 \\ 0 \end{bmatrix} v_o + \begin{bmatrix} G_2 \\ 0 \end{bmatrix} v_s - \begin{bmatrix} 0 \\ 1 \end{bmatrix} r$$

$$y = H'x', \quad \left(x' = \begin{bmatrix} x \\ x_I \end{bmatrix} \right)$$

$$H' = \begin{bmatrix} 0 & 0 & 1 & 0 \end{bmatrix}. \qquad (6)$$

To determine control gains

$$v_o = -\begin{bmatrix} K_0 & K_I \end{bmatrix} \begin{bmatrix} x \\ x_I \end{bmatrix}, \qquad (7)$$

the optimal control is used because it achieves a balance between good system response and control effort. The solution is to find the control gain minimizing the performance index

$$J = \int_0^\infty \left(x^T Q x + u^T R u \right) dt. \qquad (8)$$

The Q and R are weighting matrices on the state values and inputs, respectively.

$$Q = \begin{bmatrix} \rho & 0 & 0 & 0 \\ 0 & \rho/1000 & 0 & 0 \\ 0 & 0 & \rho & 0 \\ 0 & 0 & 0 & \rho \end{bmatrix}, \quad R=1 \qquad (9)$$

With the different values of ρ, it can provide design tradeoff at particular pole locations. Gains (N_x, N_u) for reference input is determined through following equation.

$$\begin{bmatrix} N_x \\ N_u \end{bmatrix} = \begin{bmatrix} F' & G_1' \\ H' & J' \end{bmatrix}^{-1} \begin{bmatrix} 0 \\ 1 \end{bmatrix} \qquad (10)$$

Fig. 3(a) shows the closed-loop eigenvalues of overall GCI system when ρ varies from 25 to 500. Since a complex-valued state-space model was used, the pole locations are asymmetrical about the real axis. It is known that the entire system is well-damped.

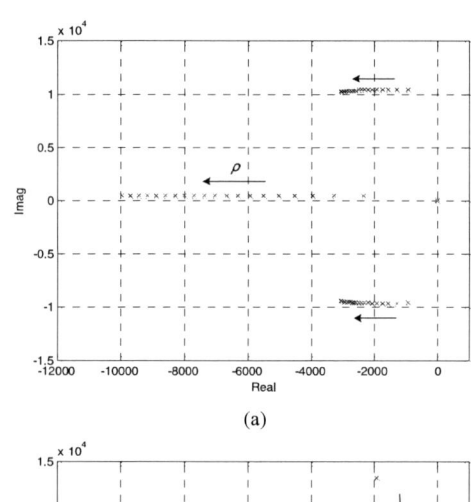

(a)

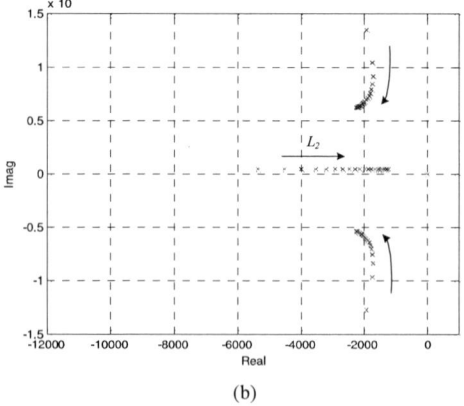

(b)

Fig. 3. Locus of closed-loop eigenvalues (a) $\rho = 25$ to 500 by 25 increment (b) $L_2 = 0.5$mH to 10mH by 0.5mH increment.

Also, Fig. 3(b) shows robustness of the controller to parameter error by varying L_2 from 0.5mH to 10mH. On the basis of previous results, $\rho = 200$ and $L_2 = 1\text{mH}$ are selected. The resulting gains are given by

$$K_0 = [\ 21.8025\quad 1.0342\quad -1.8283 - j0.0001\ ]$$
$$K_I = 22.0376 - j1.6791$$
$$N_x = [\ 0.9979 - j0.0001\quad 0.01 - j0.377\quad 1.0\quad -10.0\ ]^T$$
$$N_u = 0.0199 - j1.1294. \tag{11}$$

To simulate the GCI system, the complex-valued model must be converted to real-valued state-space model. It is achieved by using separation matrix equation.

$$W_{real} = \begin{bmatrix} Re\{M\} & Im\{M\} \\ -Im\{M\} & Re\{M\} \end{bmatrix} W_{complex} \tag{12}$$

IV. SIMULATION RESULTS

In order to verify the proposed current controller's robustness, a simulation has been performed along with an LCL filter. The circuit parameters are $L_1 = 2.0\text{mH}$, $L_2 = 1.0\text{mH}$ and $C = 15\text{uF}$.

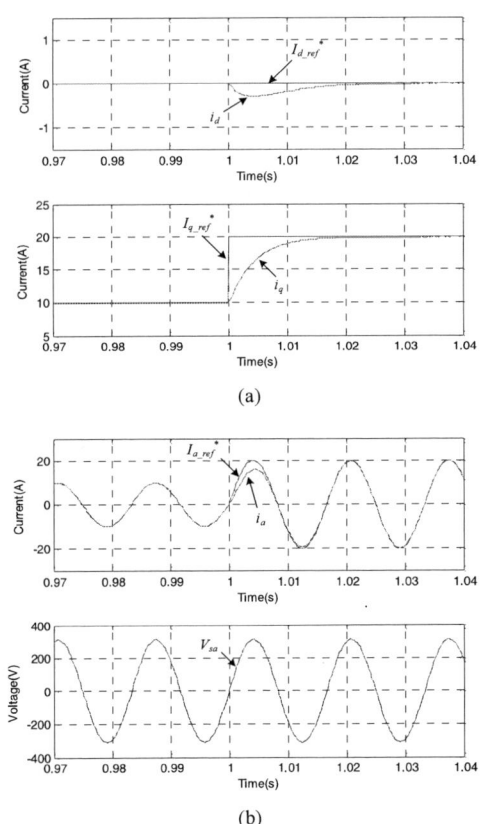

(a)

(b)

Fig. 4. Grid currents and voltage waveform for $\rho = 200$.

As shown in Fig. 4, when the load varies from 10A to 20A at 1s, the grid current tracks well the reference value within about one period. The tracking time can be reduced by

changing the ρ value.

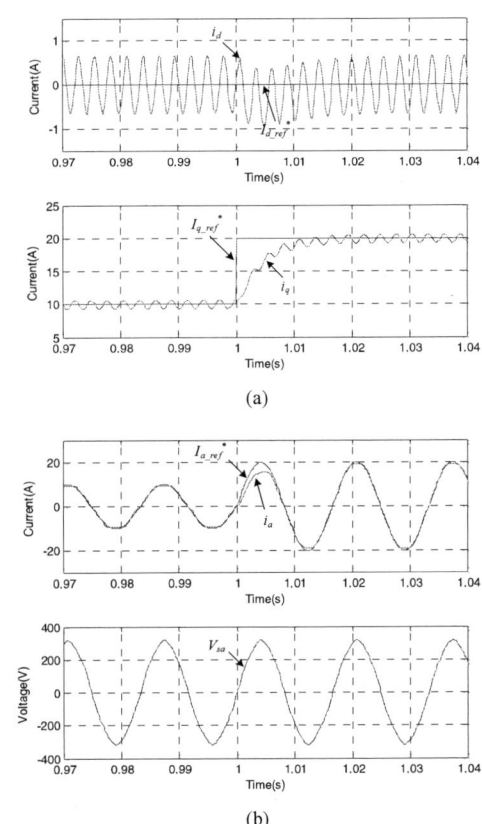

(a)

(b)

Fig. 5. Grid currents and voltage waveform for $\rho = 200$ with 2% fifth grid voltage harmonic.

Fig. 5 shows the grid current waveform when 2% fifth grid voltage harmonic exists. Although it has some oscillation, the tracking performance can be improved by selecting proper gains.

V. CONCLUSION

In this paper, a robust current controller based on optimal control theory is proposed in a three phase grid connected inverter. This current controller is designed in complex-valued state-space and it has a robust tracking performance for the grid voltage harmonics and the parameter error. The simulation results show the effectiveness of the proposed method and satisfactory performance.

REFERENCES

[1] D. M. Brod and D. W. Novotny, "Current control of VSI_PWM inverters," *IEEE Trans. Ind. Appl.*, Vol. IA-21, no. 4, pp. 562-570, Nov. 1985.

[2] T. M. Rowan and R. J. Kerkman, "A new synchronous current regulator and an analysis of current-regulated PWM inverters," *IEEE Trans. Ind. Appl.*, Vol. IA-22, no. 4, pp. 678-690, 1986.

[3] D. G. Holmes and D. A. Martin, "Implementation of a direct digital predictive current controller for single and three phase voltage source inverter," in *Proc. Annu. Meeting IEEE Trans. Ind. Appl.*, pp. 906-913, 1996.

[4] E. Twining and D. Holmes, "Grid current regulation of a three-phase voltage source inverter with an LCL input filter," *IEEE Trans. Power Electron.*, Vol. 18, no. 3, pp. 888-895, May 2003.

[5] E. Wu and P. W. Lehn, "Digital current control of a voltage source converter with active damping of LCL resonance," *IEEE Power Electron.*, Vol. 21, no. 5, pp. 1364-1373, Sept. 2006.

[6] G. F. Franklin, J. D. Powell, and A. Emami-Naeini, "*Feedback control of dynamic systems,*" 4th ed. New Jersey : Prentice Hall, 2002.

A Solar Battery Charging Module by Means of Limit-Cycle MPPT Control

Mikihiko Matsui
Tokyo Polytechnic University
Department of System Electronics and IT
1583 Iiyama, Atsugi-shi Kanagawa 243-0297 Japan
Email: matsui@seit.t-kougei.ac.jp

Kanghoon Koh
Kyungnum University
Department of Electronics & Electrical Engineering
449 Wolyoung-dong, Masan, 631-701 Korea
vwgo@kyungnam.ac.kr

Byunggyu Yu
Korea Institute of Energy Research
Photovoltaic Research Group
71-2 Jang-Dong, Yuseong-Gu, Daejeon 305-343, Korea
bgyuyu@kier.re.kr

Tatsuya Kitano
Oyama National College of Technology
Department of Electrical and Computer Engineering
771 Nakakuki, Oyama-shi, Tochigi 323-0806, Japan
kitano@oyama-ct.ac.jp

Abstract — **The objective of this paper is to develop a power converter module that charges a battery with solar power, and the design technique at the product level is introduced to realize a compact size and light weight. The main circuit is a buck-boost chopper so that it can cover wide range of charging voltage. The MPPT control is realized by a very simple algorithm of Limit-Cycle MPPT (LCMPPT) proposed by authors previously. All control functions are implemented in digital algorithm and a HITACHI H8 microprocessor is used for the control. In addition to the battery, a load can be directly connected also, which is automatically disconnected in the case of detecting a low voltage condition of the battery. RS-232C port is also available for observing the internal variables such as input/output voltages and currents through a PC. An example of obtained test results showing the relations of output power versus input/output voltages and efficiency are given. The efficiency is about 70-72% in a wide range of output power.**

Index Terms — **Solar battery charger, module, limit-cycle, LCMPPT, ultra low power microprocessor.**

I. INTRODUCTION

"Development of the circuit module technology that enables the price cutting and gains the additional system value" is becoming more important because of accelerated spread and the infiltration of the renewable energy use. From this stand point of view, this paper aims at a prototype of the power generation module that combines the power generation element such as a photovoltaic panel with the power converter and integrates them in it. The objective of this paper is to develop a power converter module that charges a battery with solar power, where the design technique at the product level is applied to realize its compact size and light weight. A MPPT (maximum power point tracking) control including the battery charge control is achieved in this prototype based on an idea of LCMPPT (Limit-Cycle MPPT) algorithm that the authors previously proposed [1-6].

II. OPERATING PRINCIPLE

Fig.1 shows the configuration of the battery charger by means of LCMPPT control scheme [1-4]. The main circuit is a buck-boost chopper so that it can cover wide range of charging voltage. The MPPT control is realized by very simple algorithm of LCMPPT. It's operating principle is shown in Fig.2. In this scheme, only the battery current I_{bat} is sensed and controlled to follow its reference value I_{bat_ref}. Suppose the input terminal of the integrator in Fig.1 is connected to the upper side positive value $1/T_{iu}$ at the beginning. Then the output of the integrator, i.e. I_{bat_ref}, gradually increases and the current I_{bat} intends to follow it until it reaches its maximum value, thus the maximum power operation is attained. Once the maximum power of the

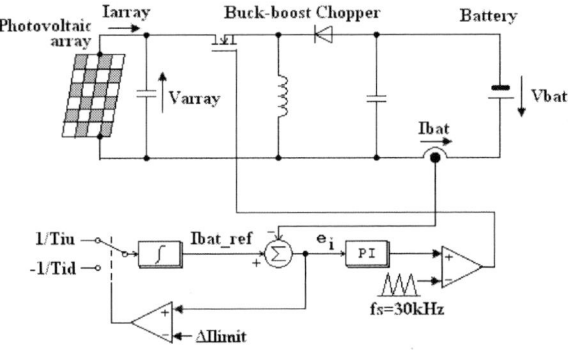

Fig.1 Battery charger by means of LCMPPT control scheme.

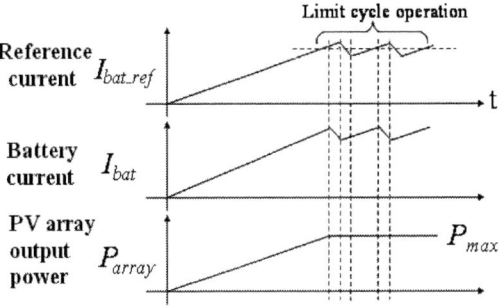

Fig.2 Operation principle of LCMPPT control scheme.

solar array is reached, the current control error ei becomes smaller than the preset value ΔI_{limit}, and then the input terminal of the integrator is switched to the lower side negative value $1/T_{id}$, which decreases I_{bat_ref}. Shortly the current control error ei gradually increases and reaches to the threshold value ΔI_{limit}, which again turns the input signal of the integrator to be positive value. In this way, a limit cycle condition occurs as shown in the figure in steady state. This is the simple mechanism of the LCMPPT scheme [1-4].

Fig.3 depicts the block diagram of the prototype module. All control are implemented in digital algorithm and a super low power HITACHI H8/38076R microprocessor is used as a controller whose stand-by power is only 0.38W. In addition to the battery charging function, a load can be directly connected also and which is automatically disconnected in the case of detecting a low voltage condition found at the battery terminal. The RS-232C port enables to observe the internal variables such as input/output voltages and currents through a PC.

The photos of the prototype module are shown in Fig.4 and Fig.5. Its dimensions are 100mm(H), 160mm(W), 38mm(D) and the weight is 280g, respectively.

III. EXPERIMENTAL RESULTS

Fig.6 shows the testing site using a solar array simulator in which the v-p patterns shown in Fig.7 are previously programmed for the convenience of repetitive testing. And Fig.8 shows the typical operating waveforms of the voltage, current and power of the PV module, respectively. At the starting up in (a), the maximum power point (MPP) is tracked within 3.0s after the start. During the steady state operation in (b), limit cycle oscillation with 2.15Hz is observed in voltage

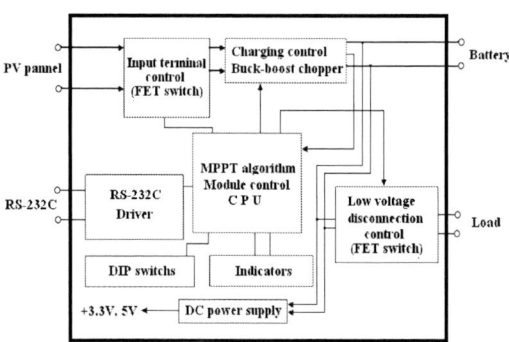

Fig.3 Block diagram of the prototypeBattery charger controlled by H8 micro-Processor.

TABLE 1
Specifications of the prototype module.

CPU	Super low power CPU H8/38076R Standby power 0.38W
Power source For CPU, RS-232C	3.3V, 5V supplied by battery
Module dimensions	100mm(H), 160mm(W), 38mm(D)
Module weight	280g

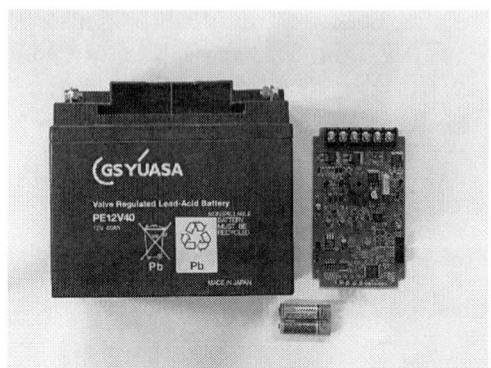

Fig.4 Prototype converter board and Batteries.

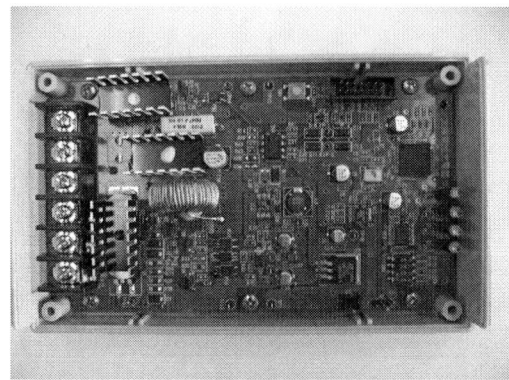

Fig.5 Closed-up view of the prototype battery charger.

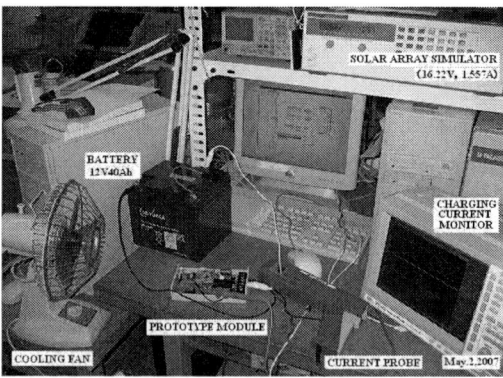

Fig.6 Testing site using a solar array simulator (SAS).

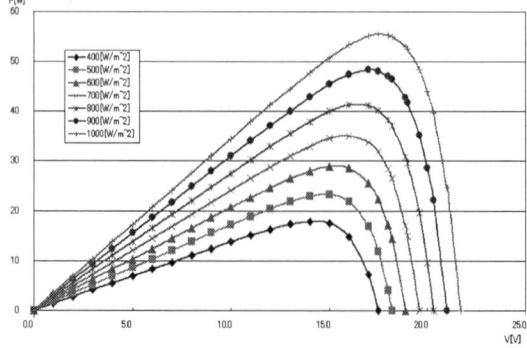

Fig.7 v-p characteristic curves of a PV array simulated in a SAS in Fig.6.

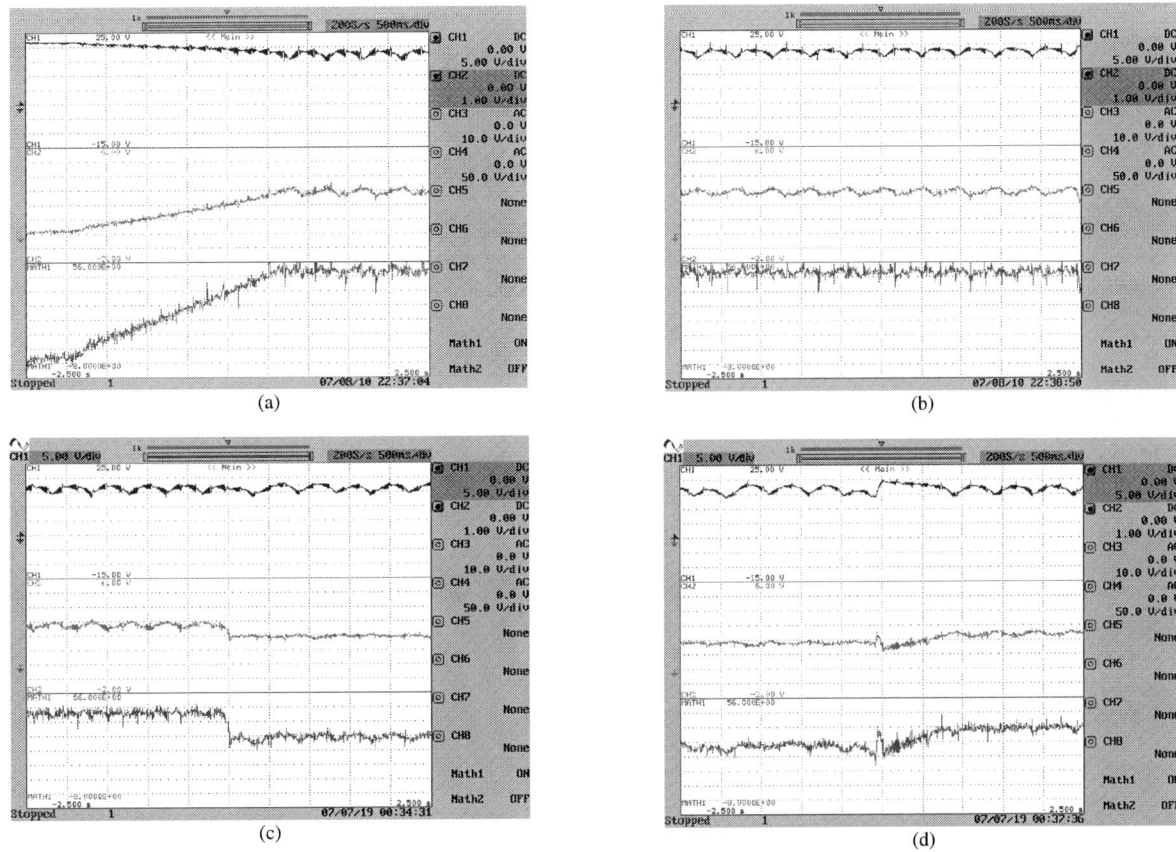

(a)

(b)

(c)

(d)

Fig.8 Operating waveforms of voltage(5V/div), current(1A/div) and power(8W/div) of photovoltaic array. (a)Starting the operation and reaching to the MPP. (b)Steady state operation upon the MPP. (c)Response to the sudden illuminant condition decrease. (d) Response to the sudden illuminant condition increase.

and current. Against the sudden illuminant condition decrease in (c), the MPP is tracked within only 0.1s. However, against the sudden increase shown in (d), the MPP is tracked about 1.0s later due to the larger time constant of the integration in Fig.1.

Fig.9 shows the output power versus input and output voltages and efficiency. The efficiency is about 70-72% in a wide range of output power from 6 to 36W. The power loss is evaluated in Fig.10 , where the switching loss of the MOSFETs and Diodes and the iron loss of the inductor core are not separated. About 60 to 70% of the loss is due to the above component. Authors are thinking that most of the loss is due to the iron loss of the inductor core because of the intense heat generation from the core. Further investigation on separating the switching loss and the iron loss is required.

IV. CONCLUSIONS

This paper has shown a prototyped solar battery charger characterized by its LCMPPT control. All digital controller using super low power microprocessor HITACHI H8/38076R

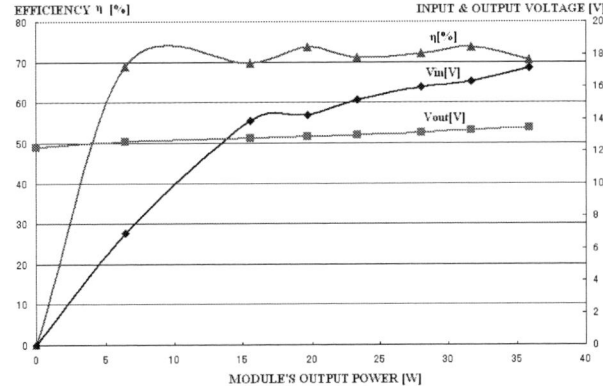

Fig.9 Output power versus input and output voltages and efficiency.

has been adopted. Expected operation of LCMPPT has been confirmed. The power conversion efficiency is about 70-72% for the most of operating output power range from 6 to 36W. The most part of the loss is considered to originate from the inductor core iron loss. Based on the experience, the authors are planning to design an ac module, or a module integrated converter (MIC), as the next step.

978-1-4244-1871-8/07 $25.00 © 2007 IEEE

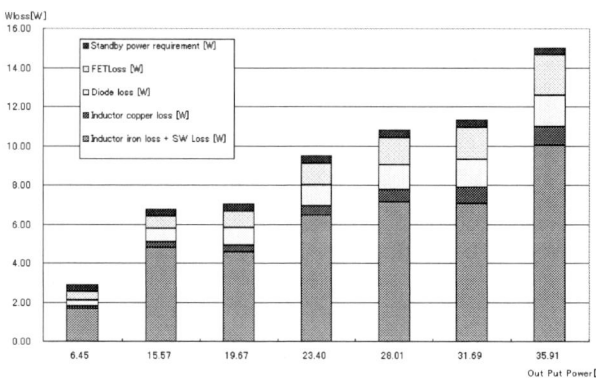

Fig.10 Power loss evaluation results.

ACKNOWLEDGMENT

This work has been partially supported by KAST (Kanagawa Academy of Science and Technology), MEXT,Japan and MOCIE,Korea. The authors would like to express their sincere thanks to whom concerned. And the special thanks goes to Mr. Yanagimura of TPU and Mr. Fukuda of Fukushima Electronic Co.,Ltd. for their help in making the prototype and doing experiment.

REFERENCES

[1] Mikihiko Matsui et al., "Proposal of simple MPPT control technique using current minor-loop error information," KIPE ICPE'04 Conf. Rec., pp.121-125,2004.

[2] Kanghoon Koh et al., "A current-control-loop error based simple MPPT controller using single current sensor, " IEEJ IPEC Niigata 2005, S.41-4, pp.1432-1437, 2005.

[3] Mikihiko Matsui, Japanese PAT.2005-069789

[4] Kanghoon Koh et al., "Enhanced LCMPPT controlled strategy for single stage power conversion," IEEJ Trans. Vol.D-127, No.5, pp.542-543, 2007.(in Japanese)

[5] Kouji Yanagimura et al., "Battery charge by means of LCMPPT control," IEEJ 2007 Annual Meeting, Conf. Rec., No.4-073, pp.110-111, 2007.(in Japanese)

[6] Kouji Yanagimura et al. , "Prototyping and evaluation of a buck-boost type solar battery charging module, " IEEJ JIASC 2007, Conf. Rec., No.1-22, pp.1-245-246, 2007. (in Japanese)

Frequency Regulation for Parallel-Connected UPS system under Independent Control

Atsuo Kawamura* , Ryota Oikawa* , Yasunari Yoshioka** and Yasushi Matsumoto**
*Yokohama National University
**Fuji Electric Advanced Technology Co.Ltd.

Abstract— This paper proposes a frequency regulation technique for the stable operation of parallel connected UPS system under the independent control proposed by authors group. The features of independent control of parallel UPS are (1) stable operation is guaranteed by sensing the output current of each UPS, and no communication is required between the UPS's, (2) The phase and amplitude of each UPS voltage are generated by a simple PI controller, (3) The lateral current between the UPS's are compensated by the lateral current compensator, (4) the load sharing is autonomously achieved, (5) the system frequency changes depending on the control gains and load. The last disadvantage is now solved by adding a simple control loop in the independent control. The paper presents the theory, simulation and experiments.

I. INTRODUCTION

A strong demand for reliable power supplies for critical loads seems inevitable to ensure a lossless operation even in a worst case scenario such as a power outage. Uninterruptible power supplies (UPS) are a key element as a backup emergency power source to assure continuous power availability. For more reliable system, the parallelism of multiple UPS is to meet these aforementioned needs, in which an emergency backup power system becomes expandable as well as more reliable by the N+1 redundancy principle. Among the control approaches for this purpose, the master-slave control provides stability and excellent load sharing [2] in spite of its lack of true redundancy due to the dependency of the master controller. To achieve redundancy, some autonomous methods for parallelism of multiple UPSs have been proposed in recent years. One of the autonomous control is a technique based on droop coefficients and its variants that relates active and reactive power flow with output voltage frequency and magnitude, respectively [3]. Another method is an independent control[5-7,9,10], where it is possible to determine the voltage variables (phase angle and amplitude) using the output current information. This independent control consists of the calculation of the active and reactive current and the determination of the output voltage phase angle and amplitude by means of a very simple scheme based on proportional and integral controllers. Theoretical studies and experimental results have demonstrated stability and promising performance under several different conditions[10]. However, in the previous studies using this control technique, the frequency is not paid so much attention, and it changes when the load varies.

This paper investigates the frequency regulation technique suitable of the independent control. The theory for the frequency regulation, and simulation and experiments will be shown in the paper.

II. CONTROL SCHEME

The scheme of the independent control is shown in Fig. 1. It shows that, with the measurement of the output current i, the controllers determine the output voltage variables (amplitude variation dv and phase angle hs).

At an instant t, the output current i is measured and the controller determines its active and reactive components, J_p and J_q, respectively, in which

$$J_p = \frac{2}{T} \int_{t-T}^{t} i(t) \cos(\omega t + hs) dt \qquad (1)$$

$$J_q = \frac{2}{T} \int_{t-T}^{t} i(t) \sin(\omega t + hs) dt \qquad (2)$$

where T is the inverse of the frequency, ω is the angular frequency; and hs is the value of the phase angle determined in the previous step calculation.

Using (1) and (2), there are some ripples during transients. Thus, a lowpass filter for each current component is employed and it results in

$$I_p = a \int_{0}^{t} (J_p - I_p) dt \qquad (3)$$

$$I_q = a \int_{0}^{t} (J_q - I_q) dt \qquad (4)$$

where I_p and I_q are the filtered current components of J_p and J_q, respectively; and a is a constant, the cutoff frequency of the lowpass filter in [rad/s].

Using I_p and I_q, controllers determine the output voltage phase angle hs and amplitude variation dv' as follows

$$hs = k_{p1}(I_{vrp} - I_p) + k_{p2} \int_{0}^{t} (I_{vrp} - I_p) dt \qquad (5)$$

$$dv' = k_q(I_{vrq} - I_q) + k_c I_q \qquad (6)$$

where I_{vrp} and I_{vrq} are the virtual rating of the active and reactive current, respectively.

A limiter maintains the voltage levels in an acceptable operation range, where

$$dv = \begin{cases} dv_{min} & \text{if } dv' \leq dv_{min} \\ dv' & \text{if } dv_{min} \leq dv' \leq dv_{max} \\ dv_{max} & \text{if } dv' \geq dv_{max}. \end{cases} \qquad (7)$$

Finally, the output voltage reference v is given by

$$v = (V_M + dv)\sqrt{2}\cos(\omega t + hs) \qquad (8)$$

where V_M is the rms value of the voltage waveform. The selection of the control parameters was already discussed in previous opportunities [5,6].

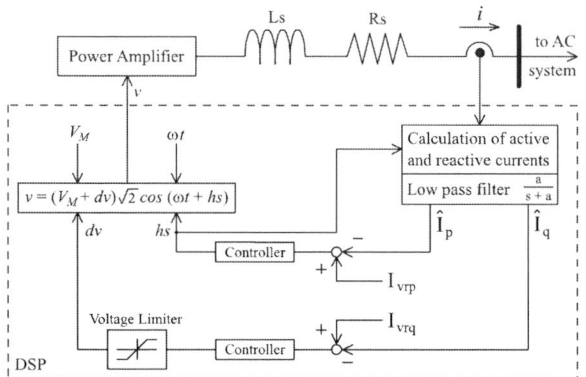

Fig. 1. Control scheme of the proposed independent control for one UPS

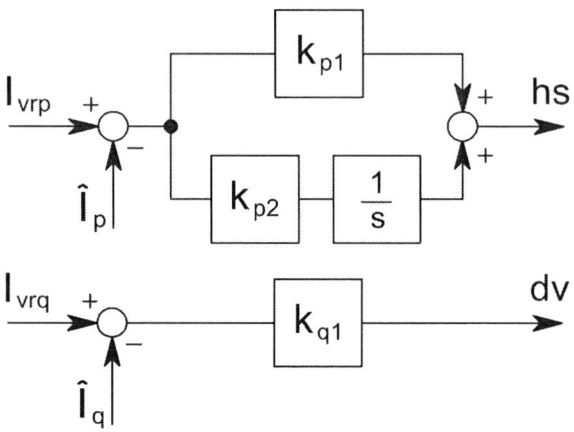

Fig. 2. Block diagram of the controllers of the proposed independent control

III. AUTO-FREQUENCY REGULATION

One of the advantages of the independent control is the robustness of the stable operation for abrupt load change, however, it is known that the frequency is not maintained at the precisely constant value [9]. Due to the decentralized control of the independent control, the frequency was a little shifted to higher or lower so that the over all stability is guaranteed. In this section the new frequency regulation control suitable for the independent control is proposed.

A. Frequency Variations

When two UPS's are operating under the independent control as shown in Fig.3, the load sharing is automatically satisfied, however, the system frequency changes as illustrated in Fig. 4-a,b and c, if the load variation occurs.

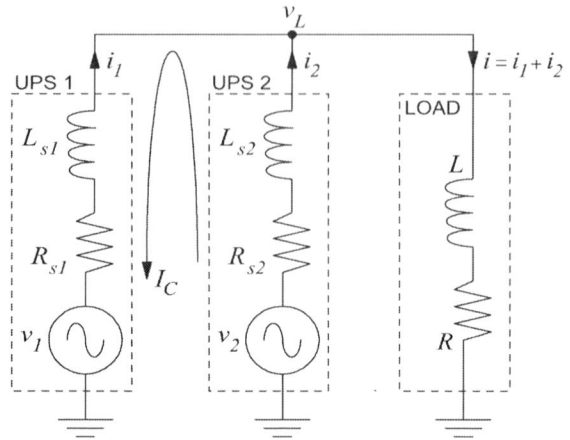

Fig. 3. An elementary parallel-connected UPS system with a resistive-inductive load

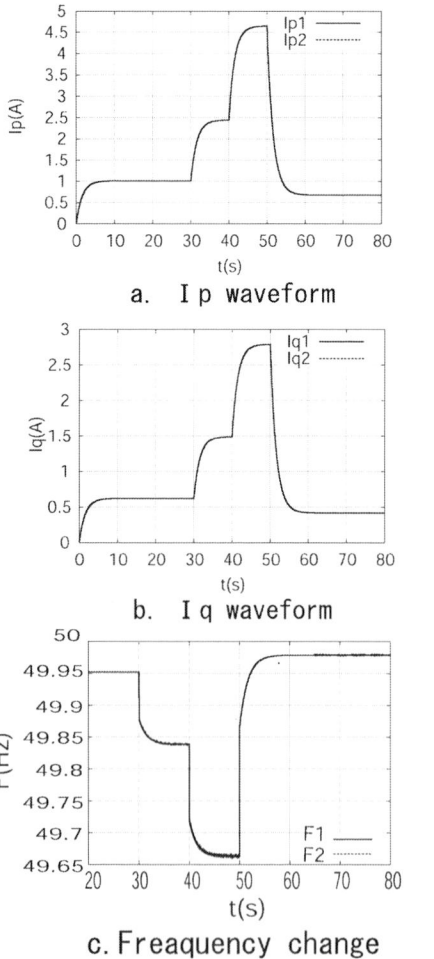

a. I p waveform

b. I q waveform

c. Freaquency change

Fig. 4. Simulation results(conventional control)

The load was abruptly changed following Table 1.

TABLE I

A LOAD CHANGE

t (s)	R (Ω)	L (mH)
0 ∼ 30	50	100
30 ∼ 40	20	40
40 ∼ 50	10	20
50 ∼	75	150

Fig.4-a and b shows the Ip and Iq, and Fig.4-c depicts the transient waveform of the frequency. From those simulation results it is apparent that the frequency changes when the load changes.

B. Frequency Regulation

The reason for the frequency variation is as follows. The phase angle hs of the UPS1 in Eq(5) continuously increases or decreases if the Ivrp is Ip. This often occurs if the load changes. The derivative of (5) becomes,

$$\frac{dhs}{dt} = K_{p2}(I_{vrp} - I_p) \qquad (9)$$

This means the frequency shift is the result of the mismatching of the Ivrp and Ip. Thus, to make the Ivrp to be equal to Ip, a new feedback loop is added, and Fig. 5 is proposed.

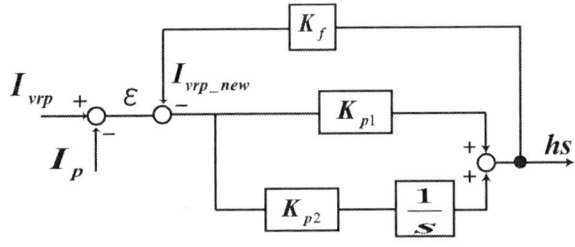

Fig. 5. The proposed feedback loop

The hs is gained by Kf and added to the error ε. The transfer function from ε to hs becomes,

$$\frac{hs}{\varepsilon}(s) = \frac{K_{p1} + \frac{K_{p2}}{s}}{1 + (K_{p1} + \frac{K_{p2}}{s})K_f}$$

$$= \frac{K_{p1}s + K_{p2}}{(1 + K_{p1})s + K_{p2}K_f} \qquad (10)$$

$$(11)$$

The derivative becomes,

$$\frac{shs}{\varepsilon}(s) = \frac{s(K_{p1}s + K_{p2})}{(1 + K_{p1})s + K_{p2}K_f} \qquad (12)$$

The final theorem says,

$$\frac{d}{dt}hs\Big|_{t=\infty} = \lim_{s \to 0} \frac{s(K_{p1}s + K_{p2})}{(1 + K_{p1})s + K_{p2}K_f} \cdot \frac{\varepsilon_{(0)}}{s} \cdot s$$

$$= 0 \qquad (13)$$

This implies that the derivative of the phase shift hs becomes zero by adding this simple control loop.

The simulation results are illustrated in Fig.6-a,b and c. The Fig.6-c indicates that the frequency was shifted when a abrupt load change occurred at the time t=30, but after that the frequency gradually recoverd to the original value 50 Hz.

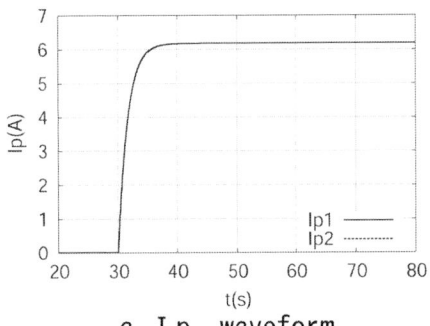

a. Ip waveform

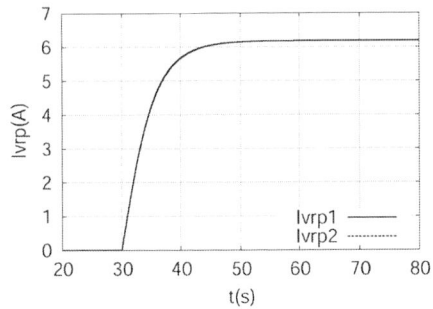

b. Iq waveform

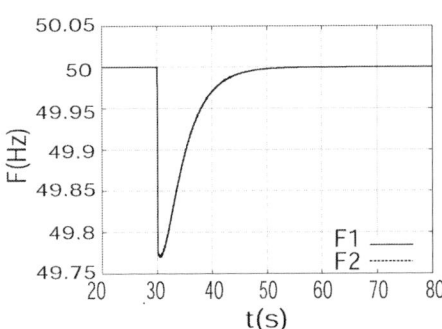

c. Freaquency change

Fig. 6. The simulation results (proposed frequency regulation)

C. Experiments

The low power experiments were carried out using the hardware as shown in Fig.7. The DSP controls the input signal to two power amplifiers following (8), which are modeled as a voltage source instead of UPS. The outputs of these power amplifiers are connected to the load through the line impedance as shown in Fig.3. The load was changes

978-1-4244-1871-8/07 $25.00 © 2007 IEEE

twice and the frequency also changed, however, the steady state frequency converges to 50 Hz.

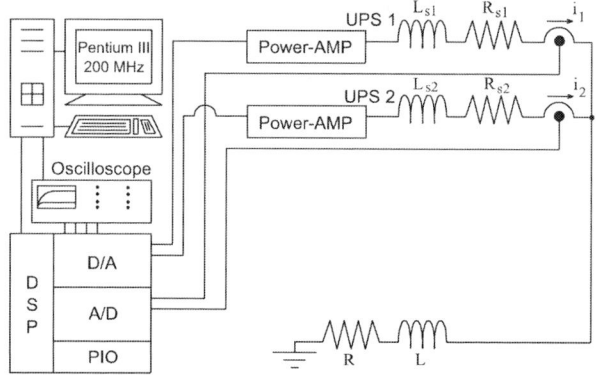

Fig. 7.   Scheme of the experimental setup for two parallel-connected UPS

Fig. 8 (a) is the Ip and (b) is the frequency transient waveform. In both graph, the measured data are off-line low pass filtered. It is observed that the frequency came back to 50 Hz after a short period.

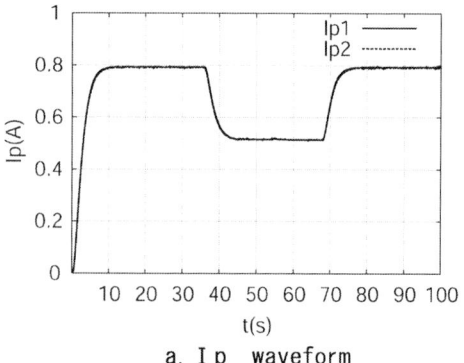

a. I p   waveform

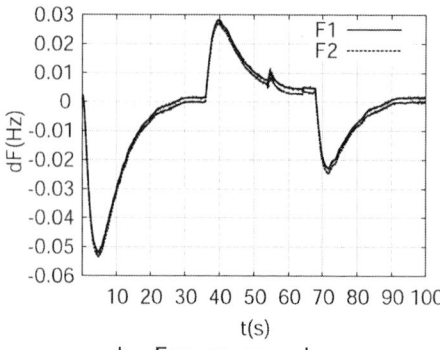

b. Freaquency change

Fig. 8.   Experimental results(single phase)

D. Discussions

Fig. 9 is the output waveforms of the case that three phase UPS's were employed at low power experiments. The Ip and frequency are almost similar to those in Fig. 8. The difference between a single phase and three phase systems are the conversion from three phases to two phases and the reverse operation in the control scheme. It is again observed that the frequency converged to 50 Hz.

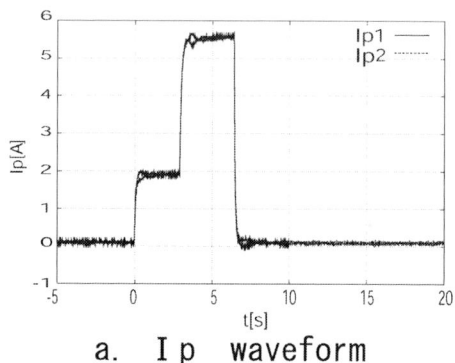

a.   I p   waveform

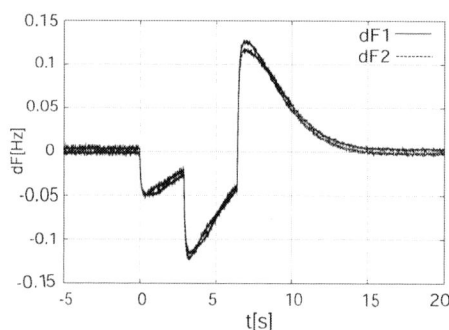

b. Freaquency change

Fig. 9.   Experimental results(three phases)

IV. CONCLUSIONS

This paper proposed the frequency regulation technique by adding a very simple control loop suitable for the independent control. The simulation and experimental results justified the proposed control. Integrating this frequency regulation, the independent control seems practical approach not only for UPS's but also even for the distributed power supply systems.

REFERENCES

[1] T. Kawabata, and S. Higashino, "Parallel operation of voltage source inverters", *IEEE Trans. Ind. Applicat.*, vol. 24, pp. 281–287, Mar./Apr. 1988

[2] J.-F.Chen, and C.-L.Chu, "Combination voltage-controlled and current-controlled PWM inverters for UPS parallel operation", IEEE Trans. Power Electron., vol. 10, pp. 547–558, Sep. 1995

[3] M. C. Chandorkar, D. M. Divan, and B. Banerjee, "Control of distributed ups systems", *Proc. of IEEE Power Electronics Specialist Conference (PESC'94)*, pp. 197–204, 1994

[4] E. A. A. Coelho, P. C. Cortizo, and P. F. D. Garcia, "Small-signal stability for parallel-connected inverters in stand-alone ac supply systems", *IEEE Trans. Ind. Applicat.,* vol. 38, pp. 533-542, Mar./Apr. 2002

[5] E. K. Sato, and A. Kawamura, "Theoretical and experimental verification of independent control for parallel-connected multi UPS", *IEICE Trans. Commun.,* Vol. E87-B, No. 12, pp. 3490-3499, Dec. 2004

[6] E. K. Sato, and A. Kawamura, "High reliability and high performance parallel-connected UPS system with independent control", *Proc. of INTELEC'05,* pp. 389-394, Sep. 2005

[7] E. K. Sato, and A. Kawamura, "Design of an autonomous control for parallel-connected uninterruptible power supplies", *IEEJ Trans. IA,* Vol. 126, No. 4, pp. 444-452, Apr. 2006

[8] A. Kusko, D. Galler, and N. Medora, "Output impedance of PWM UPS inverter - feedback vs. filters", *Proc. of IEEE-IAS Annu. Meeting,* pp. 1044-1048, Oct. 1990

[9] A. Kawamura and R. Watanabe, "Autonomously decentralized control of multiple UPS's for higher reliability", *ICPE98,* pp.836-840, 1998

[10] Eduardo K. Sato and Atsuo Kawamura, "Gain Tuning for Stability in the Low Line Impedance Region for Autonomously Controlled Parallel-Connected UPS System", *PESC06,* pp.1984-1990, 2006

[11] R.Oikawa and A. Kawamura, Y. Yoshioka, Y. Matsumoto, "Frequency regulation for paralle-connected UPS system under Independent Control ", *JIASC, IEEJ,* Y-39(YPC), 2007

Carrier Based Two-state PWM Method For Optimising Voltage Error in Multilevel Inverters

Nguyen Van Nho[1], Ta Truong Cong[2]
HochiminhCity University Of Technology
Department of Electrical Engineering
268 LyThuongKiet, District 10, Hochiminh City
Email: nvnho@hcmut.edu.vn

Hong-Hee Lee[3]
University Of Ulsan
Department of Electrical Engineering
680-749 San 29, Muger 2-dong, Ulsan, Korea
Email: hhlee@mail.ulsan.ac.kr

Abstract—In the paper, a carrier based Two-state PWM technique will be proposed. This method is advantageous for reducing switching losses compared with conventional PWM techniques. Compared with single-state PWM method, it can achieve a better linear control characteristic and harmonic distortion performance. The carrier based PWM approach of the proposed method justifies a flexiblity of carrier techniques for achieving various PWM performances in practice. Even for three-level inverter, the harmonic content is low for medium and high modulation index and more improved if the number of levels is high. The proposed method is mathematically formulated and demonstrated by simulation and experimental results.

I. INTRODUCTION

Nowadays, multilevel inverters have achieved an increasing contribution in high performance applications. Two common multilevel inverters are NPC and cascaded types (Fig.1,2). Three PWM techniques, which have been favorably used in practice are step pulse PWM, carrier based PWM and space vector PWM [1]. For less number of switchings in the fundamental period, the step pulse PWM are advantageous for high power applications, especially in power systems. Two remaining PWM techniques are commonly used in various fields because of good PWM qualities. The carrier based PWM for multilevel inverters were introduced firstly by Carrara . Then there have been followed by many modified PWM schemes, in which proper offset is added to fundamental for producing reference modulating signal. The offset function plays a key role to make carrier PWM methods become flexible for gaining various qualities such as extending voltage range, reducing the switching losses and harmonic distortion and balancing the neutral point variation in NPC inverters. Recently, a new carrier PWM modulator has been properly proposed. The main part of PWM modulator consists of active voltage and offset generators [2]-[3]. A comprehensive study of relationship between carrier PWM and space vector PWM methods for multilevel inverters in whole modulation range, including overmodulation has been also presented. The proposed circuit model of multilevel inverter enables simply to realise conventional and modified PWM techniques (SVPWM,DPWM) on conditions of variable load power factor and variable dc voltage sources.

In this paper, the previous works can be furtherly applied to approximate PWM techniques, which use less than 3 switching states in a switching state sequence, i.e. with a reduced number of switchings in sampling period [4]. The principle of proposed PWM method is based on minimised voltage error. Two voltage states in a sampling sequence will be considered.

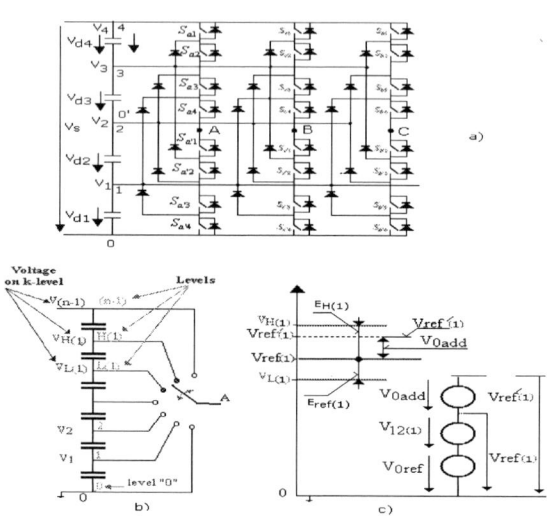

Figure 1: NPC multilevel inverter. Circuit analysis

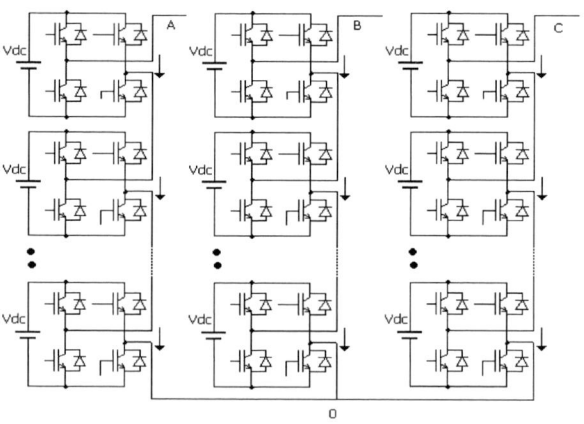

Figure 2: Cascaded multilevel inverter

II. VOLTAGE MODEL FOR MULTILEVEL INVERTER

Suppose voltages to be expressed in unit. Reference vector $\vec{v}_{ref}$ can be expressed as a sum of active $\vec{v}_{12}$ and offset v_{0ref} components as:

$$\vec{v}_{ref} = \vec{v}_{12} + v_{0ref}\vec{I} \tag{1}$$

where $\vec{v}_{ref} = \left[v_{aref}, v_{bref}, v_{cref}\right]^T$ and

$$\vec{v}_{12} = \left[v_{a12}, v_{b12}, v_{c12}\right]^T . \tag{2}$$

978-1-4244-1871-8/07 $25.00 © 2007 IEEE

Define $L_{(x)}$ and $H_{(x)}$ two integer values closest to the reference phase-leg voltage v_{xref} , $x = a,b,c$, and satisfy the conditions as:

$$L_{(x)} = \begin{cases} n_{(x)} & if \quad 0 \le v_{xref} < (n-1) \\ n_{(x)} - 1 & if \quad v_{xref} = (n-1) \end{cases} . \quad (3)$$

$$H_{(x)} = L_{(x)} + 1$$

where $n_{(x)} = Int(v_{xref}); x = a,b,c$. $\qquad (4)$

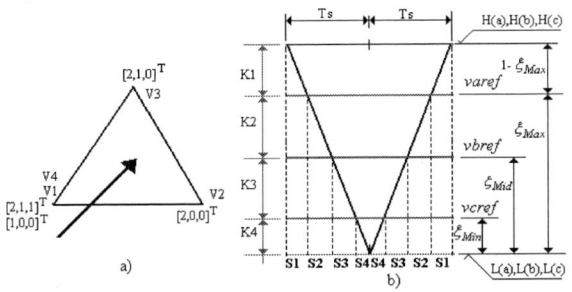

Figure 3: a) reference vector in vector triangle, b) Switching time diagrams in multicarrier PWM.

Nominal modulating signals ξ_x , $0 \le \xi_x \le 1$ *: are* defined as a subtraction of:

$$\xi_x = v_{xref} - L_{(x)}; x = a,b,c \qquad (5).$$

Define nominal switching states $\vec{s}_j; j = 1,2,3,4$, its component can be defined as below. For j=2 and 3:

$$s_{2x} = \begin{cases} 1 & if \quad \xi_x \ge \xi_{Max} \\ 0 & else \end{cases} \qquad (6)$$

$$s_{3x} = \begin{cases} 1 & if \quad \xi_x \ge \xi_{Mid} \\ 0 & else \end{cases} \qquad (7)$$

$$\vec{s}_1 = [0,0,0]^T \text{ and } \vec{s}_4 = [1,1,1]^T . \qquad (8)$$

Intersections between the nominal carrier waveform and three phase nominal modulating signals ξ_x determine a nominal switching diagram, in which nominal switching states as $\vec{s}_1, \vec{s}_2, \vec{s}_3, \vec{s}_4$ have the same switching time duties K_1, K_2, K_3, K_4 of voltage vectors in a switching state sequence.

Relationship between switching states $\vec{S}_1, \vec{S}_2, \vec{S}_3, \vec{S}_4$ of multilevel inverter and the nominal switching states $\vec{s}_j, j = 1,2,3,4$ is as:

$$\vec{S}_j = \vec{L} + \vec{s}_j \qquad (9)$$

Relevant switching time duties K_j are deduced as:

$$K_1 = 1 - \xi_{Max}; K_2 = \xi_{Max} - \xi_{Mid};$$
$$K_3 = \xi_{Mid} - \xi_{Min}; K_4 = \xi_{Min} \qquad (10)$$

where:

$$\xi_{Max} = Max(\xi_a, \xi_b, \xi_c); \xi_{Mid} = Mid(\xi_a, \xi_b, \xi_c)$$
$$\xi_{Min} = Min(\xi_a, \xi_b, \xi_c)$$

III. PROPOSED CARRIER BASED TWO-STATE PWM METHODS FOR OPTIMIZING VOLTAGE ERRORS

Carrier based two-state PWM method: For investigating the active errors, each triangle area can be divided into three sub-areas as shown in Fig.4a. Define function K_{Min} as

$$K_{Min} = Min(K_{14}, K_2, K_3). \qquad (11)$$

Function K_{Min} can be derived and drawn in three sub-areas. Reference two-state vector can be determined by a vertical projecting of reference vector on the nearest triangular side . The obtained vector $\vec{e}_{12}$ has obviously a minimum distance to it. Modified reference modulating signals can be derived from parameters of corresponding required vector and filled in Table 1. For instance, if $K_{Min} = K_3$, the modified nominal modulating signals will be set as $\xi'_{Mid} = \xi'_{Min} = 0$. It means, for conditions as $\xi_x = \xi_{Mid}$ and $\xi_x = \xi_{Min}$, modified reference modulating signals will be $v'_{xref} = L_{(x)}$.

For explanation, modulating signals in four- state PWM and relevant three-state PWM are drawn in corresponding Fig.5a and b, respectively. An additional offset for gaining three-state PWM technique will not give any change of active voltages. The final diagram of two-state PWM is shown in Fig.5c. Signal, which appears at the top (or bottom) of nominal carrier band, obviously does not need a change. Two remaining signals will be modified: one closer to boundary to be shifted to boundary level by a step of d_1 and the remaining one to be shifted by a step of d_2 . The problem is to determine optimizing values of parameters d_1 and d_2 .

Without losing of generality, suppose that $e_a = d_1; e_b = d_2$ and $e_c = 0$. The active error can be determined as :

$$\vec{e}_{12} = 2(d_1 + \vec{a}d_2)/3. \qquad (12)$$

Minimum error in active component occurs if (Fig.6d) :
1) $d_1 = K_{Min}$ and 2) $d_2 = 0.5d_1$. $\qquad (13)$

Its amplitude is determined as :

$$e_{12} = \sqrt{3}d_1/3 = \sqrt{3}M_{Min}/3 \qquad (14)$$

978-1-4244-1871-8/07 $25.00 © 2007 IEEE

If reference vector is located on the triangle side, i.e. $K_{Min} = 0$, active error attains zero value $e_{12} = 0$. If reference vector is located in the center of the vector triangle (Fig.5e), active error will get a maximum of

$$e_{12Max} = 1/(3\sqrt{3}). \qquad (15)$$

Offset reference: Two cases of offset reference have been considered: one with minimum common mode and second with medium common mode voltage. Their difference in number of switchings in relation to modulation index have been calculated for 5-level inverter and compared together in Fig.7. There is not different for overmodulation and for low modulation index (m< 0.4). For the remaining range, a better distribution of switching losses among switching pairs can be seen in PWM method with minimum common mode voltages.

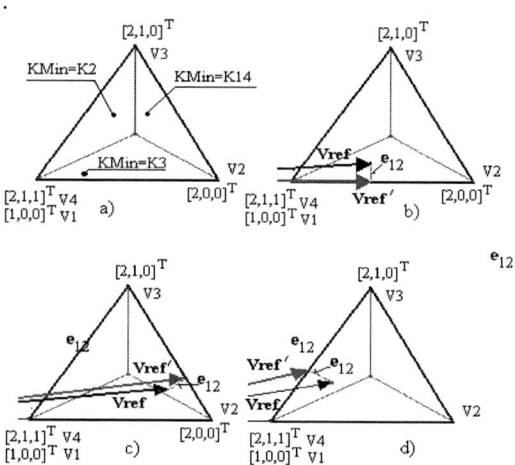

Figure 4: a) Function K_{min} in a vector triangle and B),c),d) Principle of two-state DPWM method for various locations of $\vec{v}_{ref}$

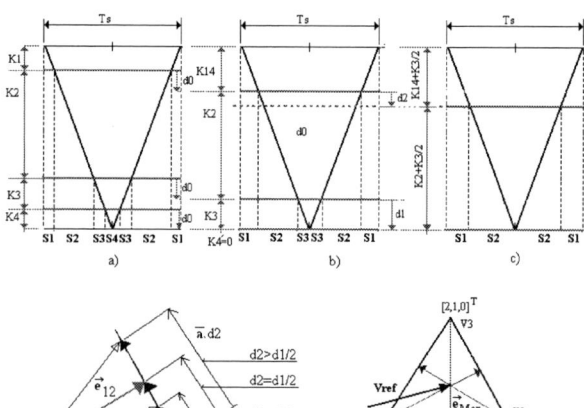

Figure 5: Explanation of two-state DPWM method

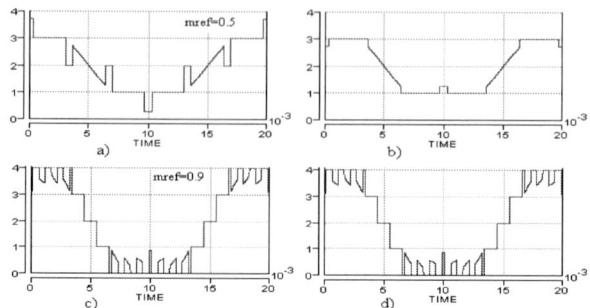

Figure 6: 5-level inverter: Diagrams of modulating signals in two-state PWM method for $m_{ref} = 0.5$: a) minimum CM offset , b) medium CM offset and $m_{ref} = 0.9$ for c) minimum CM offset and d) medium CM offset.

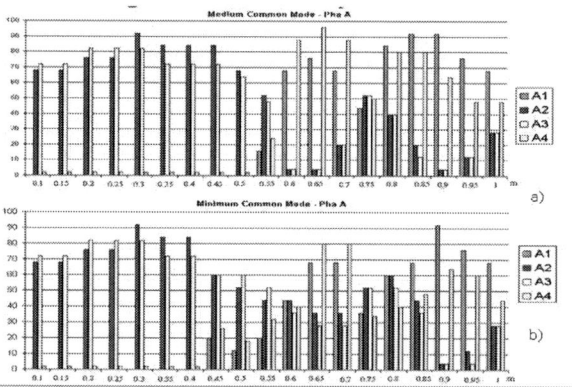

Figure 7: 5-level inverter . Comparison of number r of switching on 4 switching pairs (A1...A4) in a) medium common mode PWM control and b) minimum PWM common mode control.

IV. SIMULATIONS AND EXPERIMENTAL RESULTS

Control characteristics have been calculated for 3-,5- and 7-level inverters (Fig.8 and Fig.9). Using the principle control between two-limit trajectories to deduce the reference active voltage, the output voltage can be implemented up to a maximum amplitude of six-step mode, i.e. m_{ref}=1.1. Compared to single-state PWM method, the linearity of the control characteristic in the 3-level inverter is more improved and its range is wider. This feature can be more improved for higher level as shown for 5- and 7-level inverters. The linearity in the 7-level inverter is extended to nearly a zero modulation index.

Harmonic problems happens for low modulation index. For three-level inverter, the harmonic amplitudes in the voltage ouput can be reduced around 5% for nearly a large range of modulation index (0.45<m_{ref}<1.05) (Fig.10). An improved harmonic performance in a wider range at medium and high voltages (0.25<m_{ref}<1.05) obviously appears for 5-level inverter (Fig.11). For 7-level inverter, harmonic content is further reduced mainly for low voltages (Fig.12). The harmonic performance at higher voltages is not much changed. From this point of view can be seen that

978-1-4244-1871-8/07 $25.00 © 2007 IEEE   582

a combination between two-state PWM and conventional PWM method can be a an effective solution for controlling multilevel inverter. The conventional PWM at low modulation index can get no trouble since for instance in V/f control of induction motor drives, normally low fundamental frequency is applied for low modulation index.

Table 1

Condition	Modified reference modulating signals
$K_{Min} = K_{14}$	$v'_{xref} = \begin{cases} L_{(x)} & for \quad \xi_x = \xi_{Min} \\ v_{xref} + d_{corr} & for \quad \xi_x = \xi_{Mid} \\ L_{(x)} + 1 & for \quad \xi_x = \xi_{Max} \end{cases}$
$K_{Min} = K_2$	$v'_{xref} = \begin{cases} L_{(x)} + 1 & for \quad \xi_x = \xi_{Max} \\ L_{(x)} + 1 & for \quad \xi_x = \xi_{Mid} \\ v_{xref} + d_{corr} & for \quad \xi_x = \xi_{Min} \end{cases}$
$K_{Min} = K_3$	$v'_{xref} = \begin{cases} v_{xref} + d_{corr} & for \quad \xi_x = \xi_{Max} \\ L_{(x)} & for \quad \xi_x = \xi_{Mid} \\ L_{(x)} & for \quad \xi_x = \xi_{Min} \end{cases}$
$d_{corr} = \begin{cases} (K_1 - K_4)/2 & for \quad K_{Min} = K_{14} \\ (K_1 + 0.5 K_2) & for \quad K_{Min} = K_2 \\ -(K_4 + 0.5 K_3) & for \quad K_{Min} = K_3 \end{cases}$	

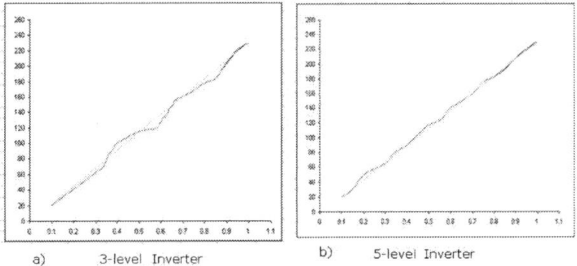

a)   3-level Inverter   b)   5-level Inverter

Figure 8: Control characteristics for a) 3-level inverter and b) 5-level inverter

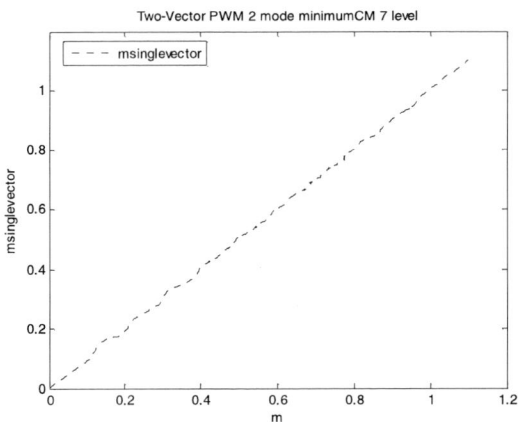

Figure 9: 7-level inverter. Control characteristic

Simulation for the proposed PWM method was implemented for 3- and 5-level inverters. Diagrams of phase and line-line load voltages were calculated and drawn in Fig.13-14.

An experimental set was built to verify the proposed method. A three-level NPC inverter was set using IGBT FMG2G100US60, gate drivers TLP250. Deadtime was set through RC circuit in a control board using IC 74HC14. The inverter was controlled by the dSPACE kit DS1104, whose program is built in MATLAB/SIMULINK tools. The sampling time period was set equal to 5kHz. Two cases of RL three-phase load connected in series with a) $\cos\varphi = 0.15$ and b) $\cos\varphi = 0.7$ were used. The advantage of two-state PWM method for high modulation index can be well followed while comparing the load voltages and currents through Fig.15-22 for corresponding modulation indexes of m_{ref}=0.6 and 0.9.

CONCLUSIONS

The paper has presented a novel carrier based PWM method for reducing switching losses with minimum voltage error. The method can be advantageously applied for high level inverters.

REFERENCES

[1]   J.Rodríguez,J.S.Lai, and F. Z. Peng,"Multilevel Inverters: A Survey of Topologies, Controls, and Applications", IEEE Transactions on Industrial Electronics, Vol. 49, No. 4, August 2002,pp.724-739

[2]   N.V.Nho, M.J.Youn," Comprehensive study on Space vector PWM and carrier based PWM correlation in multilevel invertors", IEE Proceedings Electric Power Applications, Vol.153, No.1, pp.149-158, Jan. 2006

[3]   N.V.Nho, H.H. Lee, Optimized Discontinuous PWM Algorithm With Variable Load Power Factor For Multilevel Inverters, Proceeding of the 37th IEEE Power Electronics Specialist Conference PESC 18-22nd June 2006, Jeju , Korea

[4]   Jose Rodríguez, Luis Morán, Pablo Correa and Cesar Silva,"A Vector Control Technique for Medium-Voltage Multilevel Inverters', IEEE TRANSACTIONS ON INDUSTRIAL ELECTRONICS, VOL. 49, NO. 4, AUGUST 2002

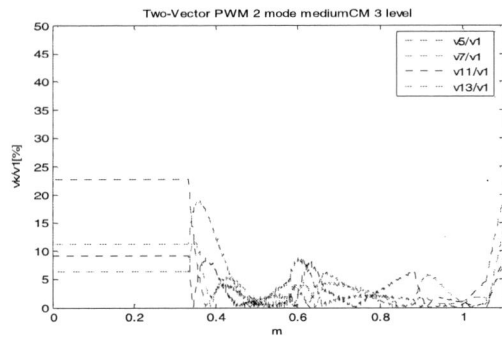

Figure 10: Harmonic content in 3-level inverter

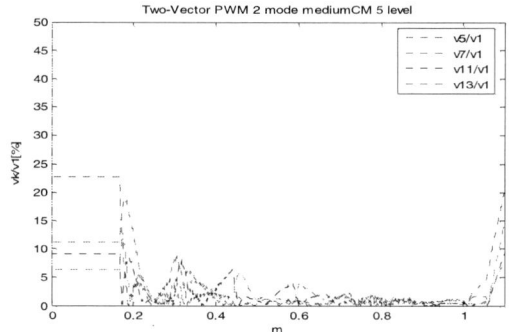

Figure 11: 5-level Inverter. Diagrams of Fourier analysis of harmonic voltage components $V_{(k)m}/V_{(1)m}$

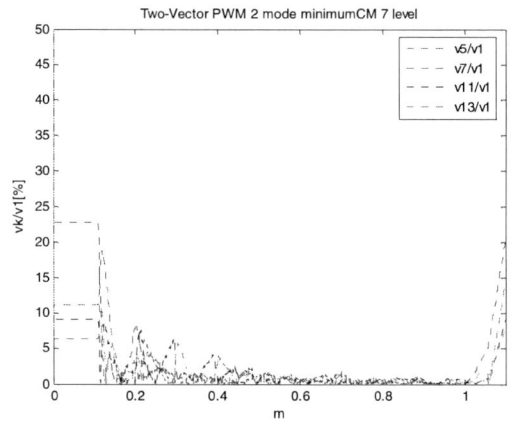

Figure 12: 7-level inverter. Fourier analysis of harmonic voltage components $V_{(k)m}/V_{(1)m}$

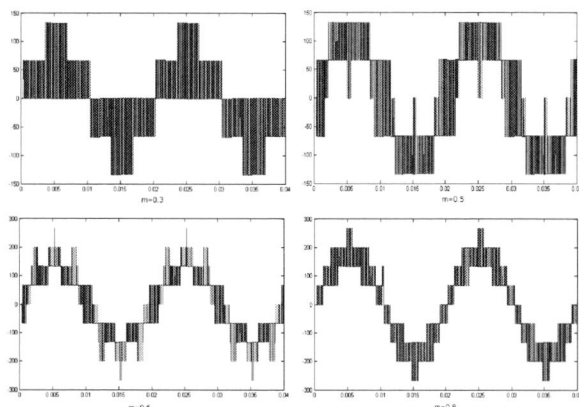

Figure 13: Simulation results. 3-level Inverter. Diagrams of phase load voltage for several modulation indexes.

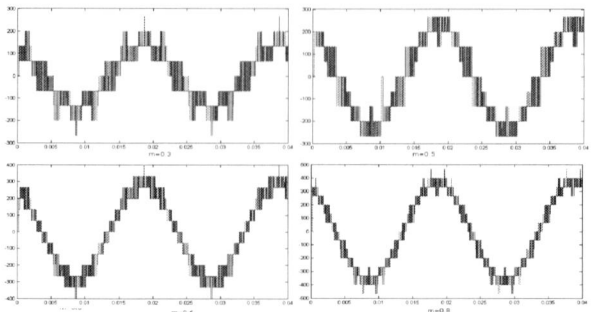

Figure 14: Simulation results. 5-level inverter. Diagrams of line-line voltages and phase voltages

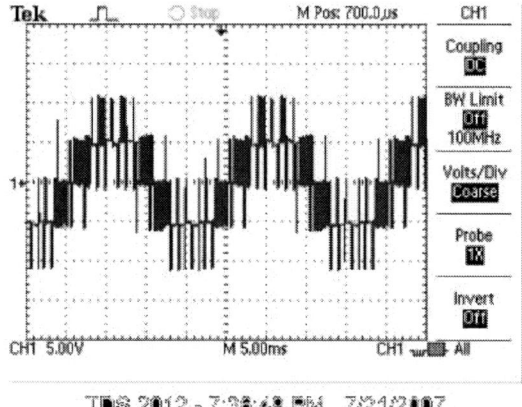

Figure 15: Experimental results. Line-line voltage, 3-level NPC inverter. m_{ref}=0.6, $\cos\varphi = 0.7$

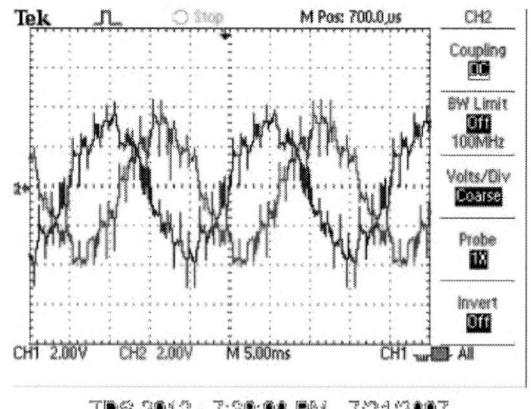

Figure 16: Experimental results. 3-level NPC inverter. M_{ref}=0.6, $\cos\varphi = 0.7$. Diagrams of load currents.

978-1-4244-1871-8/07 $25.00 © 2007 IEEE

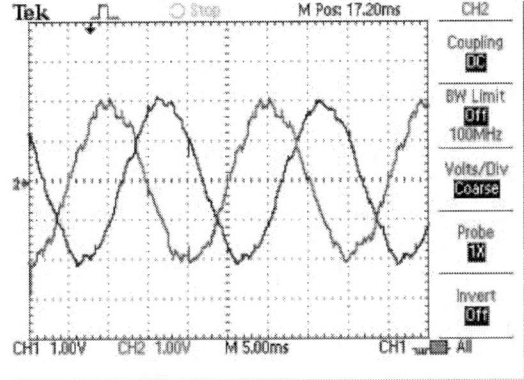

Figure 17 : Experimental results. 3-level inverter, m_{ref}=0.6, load PF $\cos\varphi = 0.15$. Diagrams of load currents.

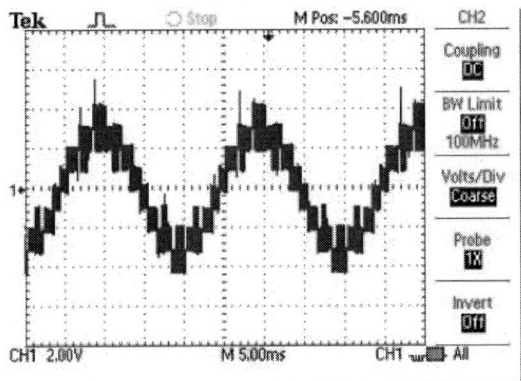

Figure 18: Experimental results. 3-level NPC inverter. m_{ref}=0.9, $\cos\varphi = 0.7$

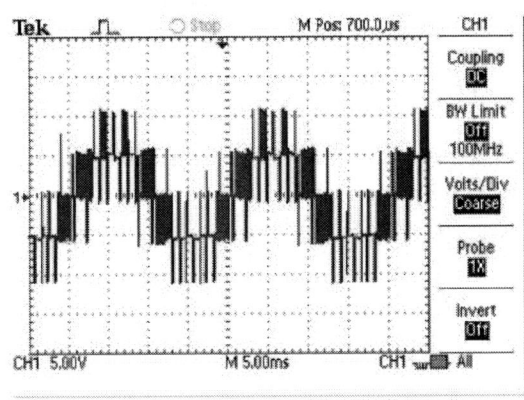

Figure 19. Experimental results. 3-level inverter, m_{ref}=0.9, load PF $\cos\varphi = 0.7$. Phase load voltage

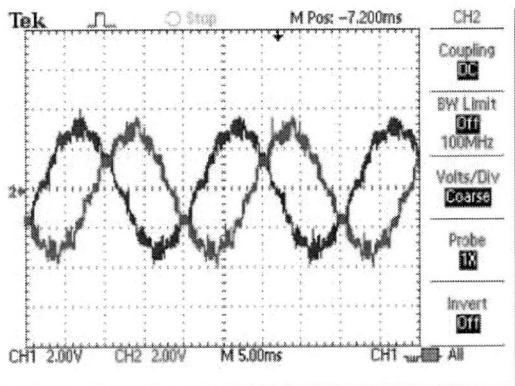

Figure 20: Experimental results. 3-level inverter, m_{ref}=0.9, load PF $\cos\varphi = 0.7$. Diagrams of load currents.

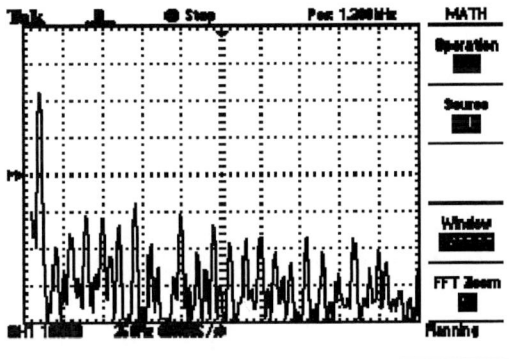

Figure 21: Experimental results . FFT analysis of load currents. m_{ref}=0.9, $\cos\varphi = 0.7$.

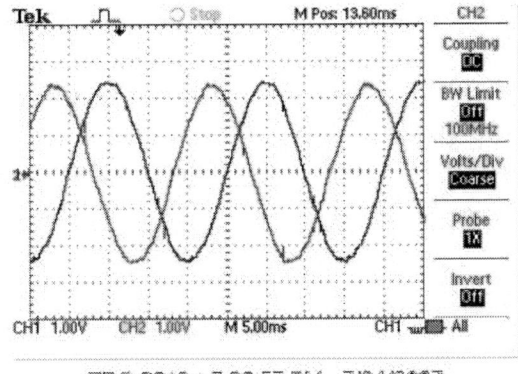

Figure 22 : Experimental results . 3-level NPC inverter, m_{ref}=0.9, , $\cos\varphi = 0.15$. Diagrams of load currents.

978-1-4244-1871-8/07 $25.00 © 2007 IEEE

Comparison of OMTHD and OHSW Harmonic Optimization Techniques in Multi-Level Voltage-Source Inverter with Non-Equal DC Sources

M. G. HOSSEINI AGHDAM, S. H. FATHI, G. B. GHAREHPETIAN
Amirkabir University of Technology
Department of Electrical Engineering
Hafez Avenue, No. 424, Tehran 15914, Iran
Email: h.aghdam@aut.ac.ir, fathi@aut.ac.ir, grptian@aut.ac.ir

Abstract— One of the major problems in electric power quality is the harmonic contents. There are several methods of indicating the quantity of harmonic contents. The most widely used measure is the total harmonic distortion (THD). This paper compares the two harmonic optimization techniques, optimal minimization of the total harmonic distortion (OMTHD) technique and optimized harmonic stepped-waveform (OHSW) technique used in multi-level voltage-source inverters. Both techniques are very effective and efficient for improving the quality of the inverter output voltage. First, we describe briefly the cascaded H-bridge multi-level inverter structure. Then, we present the switching algorithm for the inverter based on OHSW and OMTHD techniques. Finally, the results obtained by simulation for the two techniques are analyzed and compared. Simulation results verify the effectiveness of the both techniques in multi-level voltage-source inverter with non-equal DC sources, clarifying the advantages of each technique.

I. INTRODUCTION

Multi-level voltage source inverter is recently used in many industrial applications such as ac power supplies, static VAR compensators, drive systems, etc [1-8]. One of the significant advantages of multi-level configuration is the harmonic reduction in the output waveform without increasing switching frequency or decreasing the inverter output power. The output voltage waveform of a multi-level inverter is composed of a number of levels of voltages, typically obtained from capacitor voltage sources. The so-called multi-level starts from three levels. As the number of levels increases, the output THD approaches zero. The number of achievable voltage levels, however, is limited by voltage unbalance problems, voltage clamping requirement, circuit layout, and packaging constraints. Therefore, an important key in designing an effective and efficient multi-level inverter is to ensure that the total harmonic distortion (THD) in the output voltage waveform is small enough [5-6].

The well-known multi-level topologies are: 1. cascaded H-bridge multi-level inverter, 2. diode-clamped multi-level inverter, and 3. flying capacitor multi-level inverter [6].

The Multi-level inverter using cascaded-inverters with separate DC sources (SDCSs), hereafter called a cascaded multi-level inverter, appears to be superior to the other multi-level structures in terms of its structure that is not only simple and modular but also requires the least number of components. This modular structure makes it easily extensible for higher number of output voltage levels without undue increase in power circuit complexity. In addition, extra clamping diodes or voltage balancing capacitors are not necessary [3-4].

It is generally accepted that the performance of an inverter, with any switching strategies, can be related to the harmonic contents of its output voltage. Power electronics researchers have always studied many novel control techniques to reduce harmonics in such waveforms [6]. Up-to-date, there are many techniques, which are applied to inverter topologies. In multi-level technology, there are several well-known modulation techniques as follows [5-6]:

- Harmonic Optimization (HO) technique;
- Space Vector PWM (SVPWM) technique;
- Carrier-Based PWM (CBPWM) technique.

There are some advantages and disadvantages in these techniques. The harmonic optimization technique can be categorized into two methods: 1. OMTHD technique and 2. OHSW technique. OMTHD technique is based on optimal minimization of THD, where as OHSW is based on elimination o some specific harmonics. This paper focuses on OMTHD and OHSW techniques with unequal DC sources. Usually, it is assumed that the DC sources are all equal, which will probably not be the case in practice even if the DC sources are nominally equal [1-2]. Therefore, the objective here is to use harmonic optimization (HO) technique in multi-level voltage-source inverter with unequal DC sources to reduce, as much as possible, the harmonic distortion in the load voltage, working with a reduced switching frequency.

In this paper, first, we describe briefly the cascaded H-bridge multi-level inverter. Then, we present the switching algorithm for the inverter based on OHSW and OMTHD techniques. Finally, the results obtained by simulation for the two techniques are analyzed and compared. Also, several informative simulation results the effectiveness of the both techniques in multi-level voltage-source inverter with non-equal DC sources, clarifying the advantages of each technique.

II. CASCADED H-BRIDGE MULTI-LEVEL INVERTER STRUCTURE

A cascaded multilevel inverter consists of a series of H-bridge (single-phase full-bridge) inverter units. The general function of this multilevel inverter is to synthesize a desired voltage from several separate dc sources (SDCSs), which may be obtained from solar cells, fuel cells, batteries, ultra-capacitors, etc. Fig. 1 shows a single-phase structure of a cascade inverter with SDCSs [1]. Each SDCS is connected to a single-phase full-bridge inverter. Each inverter can generate three different output voltages, $+V_i$, 0 and $-V_i$ and by connecting the dc source to the ac output side according to the states of the four switching devices. The AC output of full-bridge inverters are connected in series such that the synthesized voltage waveform is the sum of all individual inverter outputs. The number of output phase voltage levels in a cascaded multi-level inverter is then $2S+1$, where S is the number of DC sources. An example phase voltage waveform for an 11-level cascaded multilevel inverter with five SDCSs ($S=5$) and five full bridges is shown in Fig. 2. The output phase voltage is given by $v_{an}=v_1+v_2+v_3+v_4+v_5$. With enough number of levels and an *appropriate* switching algorithm, the multilevel inverter results in an output voltage that is almost sinusoidal.

From the voltage waveform in Fig. 2, the voltage of the first step equals to v_1; that of the second step equals to v_2 and so on. These voltage levels are supplied by SDCSs, whose amplitudes may be different. By considering the waveform in Fig. 2, there are three possible optimization techniques to reduce the voltage THD: 1) step heights are optimized with equally spaced steps; 2) step spaces are optimized with the steps of equal height; and 3) optimizing both heights and spaces. This paper will focus on the method 3, which uses non-equal voltage amplitude and optimizes the switching angles. To achieve these optimized angles, the numerical calculation will be applied and will be presented later.

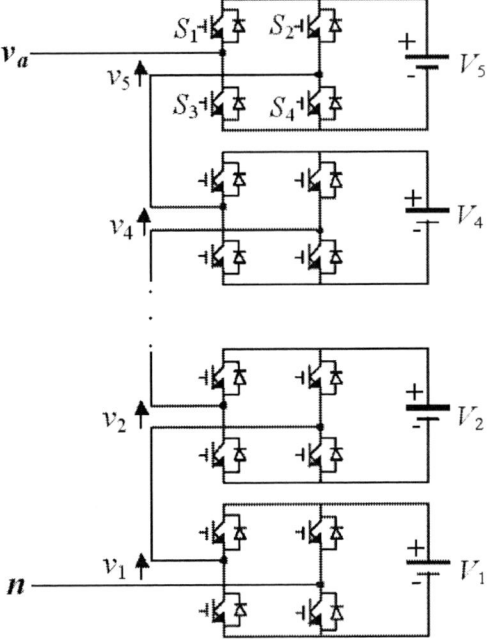

Fig. 1. Single-phase structure of a cascaded H-Bridge multi-level inverter with *unequal* DC sources.

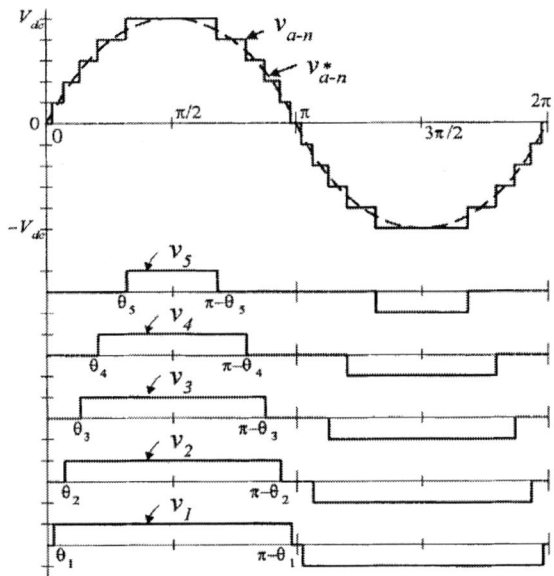

Fig. 2. Output voltage waveform of a cascaded 11-level inverter.

III. DIFFERENT TECHNIQUES FOR IMPROVING THE OUTPUT OF MULTI-LEVEL VOLTAGE-SOURCE INVERTER

Once the optimization technique of the generalized waveform has been chosen, one can derive some control techniques to improve its quality. The main techniques are:

1. The optimized harmonic stepped-waveform (OHSW) technique, for which the switching angles are chosen to eliminate a specific number of selected (generally the lowest orders) harmonics [1-6];

2. The optimal minimization of the total harmonic distortion (OMTHD) technique, for which the switching angles are determined to minimize, the most effectively possible, the waveform THD. This reduces, generally, most of the harmonic, without eliminating them completely [3-4].

These two techniques which will be briefly defined in the following sections will be applied in this paper.

IV. OPTIMIZED HARMONIC STEPPED-WAVEFORM (OHSW) TECHNIQUE

The optimized harmonic stepped-waveform (OHSW) technique is very suitable for a multi-level inverter circuit. By employing this technique along with the multi-level topology, the low THD output waveform without any filter circuit is possible. Switching devices, in addition, turn on and off only one time per cycle. This can overcome the switching loss problem, as well as EMI [5-6].

As indicated in Fig. 2, the output voltage is a stepped waveform in which $\theta_1, \theta_2, ..., \theta_s$ are switching angles and should be determined in order that some specific harmonics are eliminated. The *Fourier* series expansion of the (stepped) output voltage waveform of the multi-level inverter with *unequal* DC sources is

$$V_{out}(\omega t) = \sum_{n=1,3,5,\ldots}^{\infty} \left[\frac{4}{n\pi} \sum_{k=1}^{S} \left[V_k * \cos(n\theta_k) \right] \right] \sin(n\omega t) \qquad (1)$$

where S is the number of H-bridge cells, n is odd harmonic order, V_k is the value of the kth DC source (if all the DC sources have the same value V, then $V_1=V_2=\ldots=V_S=V$). The final aim here is to choose the switching angles $0° < \theta_1 < \theta_2 < \ldots < \theta_s < 90°$ so as to eliminate $(S-1)$ certain lower frequency harmonics from the output voltage waveform and to make the fundamental component equal to the desired value, V_f. This necessitates, mathematically, solving S equations derived from equation (1). Note that for three-phase systems, the triplen harmonics in each phase need not be eliminated, as they are automatically cancelled in the line-to-line voltage. The mathematical statement of these conditions is as follows:

$$\frac{4}{\pi} \left[V_1 \cos(\theta_1) + V_2 \cos(\theta_2) + \ldots + V_S \cos(\theta_S) \right] = V_f$$

$$\left[V_1 \cos(3\theta_1) + V_2 \cos(3\theta_2) + \ldots + V_S \cos(3\theta_S) \right] = 0$$

$$\left[V_1 \cos(5\theta_1) + V_2 \cos(5\theta_2) + \ldots + V_S \cos(5\theta_S) \right] = 0 \qquad (2)$$

$$\vdots$$

$$\left[V_1 \cos(n\theta_1) + V_2 \cos(n\theta_2) + \ldots + V_S \cos(n\theta_S) \right] = 0$$

Assume $V_{dc}=V_1+V_2+\ldots+V_S$, $m_a = V_f / (4.V_{dc}/\pi)$ and $V_{1dc}=V_1/V_{dc}$, $V_{2dc}=V_2/V_{dc},\ldots$, $V_{Sdc}=V_S/V_{dc}$. Therefore, equation can be rewritten as:

$$\left[V_{1dc} \cos(\theta_1) + V_{2dc} \cos(\theta_2) + \ldots + V_{Sdc} \cos(\theta_S) \right] = m_a$$

$$\left[V_{1dc} \cos(3\theta_1) + V_{2dc} \cos(3\theta_2) + \ldots + V_{Sdc} \cos(3\theta_S) \right] = 0$$

$$\left[V_{1dc} \cos(5\theta_1) + V_{2dc} \cos(5\theta_2) + \ldots + V_{Sdc} \cos(5\theta_S) \right] = 0 \qquad (3)$$

$$\vdots$$

$$\left[V_{1dc} \cos(n\theta_1) + V_{2dc} \cos(n\theta_2) + \ldots + V_{Sdc} \cos(n\theta_S) \right] = 0$$

S equations have been set up, from which, the switching angles $\theta_1,\theta_2,\ldots,\theta_s$ can be calculated. Unfortunately, these equations are nonlinear as well as *transcendental* in nature, which suggests a possibility of multiple solutions. There are no general methods that can be applied to solve such equations. The practical method of solving these equations is a trial and error process. A numerical technique is the best approach in solving the equations. Usually, the *Newton-Raphson* (NR) method is used to solve such nonlinear equation systems. The *Newton-Raphson* (NR) method for computing the root of an equation is a successive-approximation procedure, which is suitable for implementation in a computer program.

V. OPTIMAL MINIMIZATION OF TOTAL HARMONIC DISTORTION (OMTHD) TECHNIQUE

The basic idea for this method, developed in [4] and confirmed by recent work of [9], is to adjust switching angles in order to minimize the output voltage THD. In [4], it is assumed that the DC sources were all equal. Also, in [4], there is no control on the fundamental component of the output voltage. This means that OMTHD technique is

applied only for THD minimization with no constraint on the value of fundamental component; where as, the primary objective in any method of inverter control is adjustment of the fundamental component to the desired value. In this paper, OMTHD technique is applied to the cascaded multi-level inverter with *unequal* DC sources to minimize THD while producing the desired fundamental component at the output.

To minimize the THD, it is necessary for its *partial derivative* (PD) to be zero with respect to each switching angle. This implies that the *partial derivative* (PD) of its square is also zero because the value of THD is always positive.

After development and some mathematical simplifications, the square THD of the chosen multi-level generalized waveform (periodic with odd quarter-wave symmetric characteristic) depicted in Fig. 2 is given by [4]:

$$\text{THD}^2 = \left[\frac{\pi^2}{8} \cdot \frac{\left(\sum_{k=1}^{S} V_k \right)^2 - \frac{2}{\pi} \left[\theta_1 V_1^2 + \sum_{j=2}^{S} \theta_j \left(V_j^2 + 2V_j \sum_{i=1}^{j-1} V_i \right) \right]}{\left(\sum_{k=1}^{S} V_k \cos\theta_k \right)^2} \right] - 1$$

$$(4)$$

Differentiating THD^2 and setting equal to zero gives the value of switching angles for minimum THD.

$$\frac{\partial(THD^2)}{\partial(\theta_C)} = 0 \Rightarrow$$

$$\left[\left(V_C^2 + 2V_C \sum_{i=1}^{C-1} V_i \right) \cdot \left(\sum_{k=1}^{S} V_k \cos\theta_k \right) \right] +$$

$$\left[V_C \sin\theta_C \left[2 \left(\theta_1 V_1^2 + \sum_{j=2}^{S} \theta_j \left(V_j^2 + 2V_j \sum_{i=1}^{j-1} V_i \right) \right) \right] - \pi \left(\sum_{k=1}^{S} V_k \right)^2 \right] = 0$$

$$(5)$$

where $C=1, 2, \ldots, S$.

On the other hand, the fundamental component must have the desired value V_f. This implies that switching angles must also satisfy the following equation:

$$\left[V_{1dc} \cos(\theta_1) + V_{2dc} \cos(\theta_2) + \ldots + V_{Sdc} \cos(\theta_S) \right] = m_a \qquad (6)$$

To solve the set of nonlinear *transcendental* equations (5) and (6) an interactive method such as *Newton-Raphson* (NR) method should be used.

VI. SIMULATION RESULTS AND DISCUSSION

To analyze the harmonic performance of the two techniques for purpose of comparison, several harmonic measures are possible. The total harmonic distortion (THD) is one of these measures, which evaluates the quantity of harmonic contents in the output waveform and is a popular performance index for power converters.

For simulation, a 7-level inverter with three separate DC sources (three switching angles per quarter-cycle) is used as example. The principle parameters of this inverter which are considered for simulation are as follows:

978-1-4244-1871-8/07 $25.00 © 2007 IEEE

- DC source voltages are set to be V_1=12.56V, V_2=10.19V and V_3=12.01V.

- The operation frequency is 50Hz.

Fig. 3 shows the switching angles $\{\theta_1, \theta_2, \theta_3\}$ versus m_a for OHSW technique. Comparing Fig. 3 with the simulation and experimental results of [2] verifies the simulation results. The phase voltage THD of 7-level inverter with OHSW technique as a function of m_a, is shown in Fig. 4. As it can be seen in this figure, the minimum value of THD is at m_a=0.80, with a value of 11.61%. Fig. 5 shows the switching angles $\{\theta_1, \theta_2, \theta_3\}$ versus m_a for OMTHD technique. The phase voltage THD of 7-level inverter with OMTHD technique as a function of m_a, is shown in Fig. 6. As it can be seen in this figure, the minimum value of THD at m_a=0.80 is 0.36% which occurs. Fig. 7 shows the THD of phase voltage versus m_a for both techniques.

Simulation results have been summarized in Table I for two values of m_a. In this table, the fundamental, 3rd, 5th, 7th, 9th, 11th, and 13th components and THD of output voltage waveform for each technique has been shown. As it can be seen in table I, the value of THD with OMTHD technique is little less than that of OHSW technique for both values of m_a.

From these simulation results, it can be noticed that:

1. The THD of output voltage waveform with OMTHD technique is expectedly less than that with OHSW technique. But the difference is considerably small, especially at low values of m_a.

2. A specified number of low order harmonics are completely cancelled in OHSW technique whereas in OMTHD technique there are some residual values of all harmonic including the low orders. This results in smaller size of output filter in OHSW technique, if it required.

3. The amplitude of the first uncancelled harmonic remaining in the frequency spectrum of the output voltage waveform of the inverter with the OHSW technique is higher than of the same harmonic obtained with the OMTHD technique with same m_a.

4. The lowest harmonics eliminated by the OHSW technique have increasingly reduced amplitudes with the increase in the number of switching angles per quarter-cycle if the inverter is controlled by OMTHD method.

VII. CONCLUSION

Optimal minimization of the total harmonic distortion (OMTHD) technique can be used to minimize the THD of the output voltage waveform in a multi-level cascaded inverter. Also, the optimized harmonic stepped waveform (OHSW) technique can be used to eliminate a number lower order harmonics in a multi-level cascaded inverter.

In this paper, OMTHD and OHSW techniques have been applied to the cascaded multi-level inverter with *unequal* DC sources with control on the fundamental component of the output voltage. It has been shown that both techniques are suitable for harmonic optimization of the output voltage. When THD is the measure of comparison, OMTHD technique is better and results in a little smaller value of THD. But OHSW technique is superior in that it eliminates a number of trouble some low order harmonics, while producing a THD not very larger than that of OMTHD technique.

REFERENCES

[1] L. M. Tolbert, J. N. Chiasson, Z. Du, and K. J. McKenzie, "Elimination of Harmonics in a Multilevel Converter with Nonequal DC Sources", *IEEE Transactions on Industry Applications*, Vol. 41, No. 1, pp. 75-81, January/February 2005.

[2] L. M. Tolbert, J. N. Chiasson, K. J. McKenzie, and Z. Du, "Elimination of Harmonics in a Multilevel Converter with Non Equal DC Sources", *IEEE Applied Power Electronics (IEEE APEC 2003)*, Miami, Florida, USA, pp. 589-595, February 9-13, 2003.

[3] Y. Sahali, and M. K. Fellah, "Comparison between Optimal Minimization of Total Harmonic Distortion and Harmonic Elimination with Voltage Control Candidates for Multilevel Inverters", *Journal of Electrical Systems (JES)*, Vol. 1, Issue 3, pp. 32-46, September 2005.

[4] Y. Sahali, and M. K. Fellah, "Optimal Minimization of the Total Harmonic Distortion (OMTHD) Technique for the Symmetrical Multilevel Inverters Control", *First National Conference on Electrical Engineering and its Applications (CNEA'04)*, Sidi-bel-Abbes, May 24-25, 2004.

[5] M. G. Hosseini Aghdam, S. H. Fathi and A. Ghasemi, "Modeling and Simulation of Three-Phase OHSW Multi-Level Inverter by Means of Switching Functions", *The 6th International Conference on Power Electronics and Drive Systems (IEEE PEDS 2005)*, Kuala Lumpur, Malaysia, 28 November- 1 December 2005.

[6] S. Sirisukprasert, "Optimized Harmonic Stepped-Waveform for Multilevel Inverter", *M.Sc. Thesis*, Department of Electrical and Computer Engineering, Virginia Polytechnic Institute and State University (Virginia Tech), September 1999.

[7] A. Nabae I. Takahashi, and H. Akagi, "A New Neutral-Point-Clamped PWM Inverter", *IEEE Transactions On Industry Application*, Vol. 17, No. 5, pp. 518-523, 1981.

[8] J. S. Lai and F. Z. Peng, "Multilevel Converters- a new breed of Power Converters", *IEEE Transactions on Industry Applications*, Vol. 32, No. 3, pp. 509–517, May/January 1996.

[9] L. M. Tolbert, F.Z.Peng, T. Cunningham and J. N. Chiasson, "Charge Balance Control Schemes for Cascade Multilevel Converter in Hybrid Electric Vehicles", *IEEE Transactions on Industrial Electronics*, Vol. 49, No. 5, October 2002.

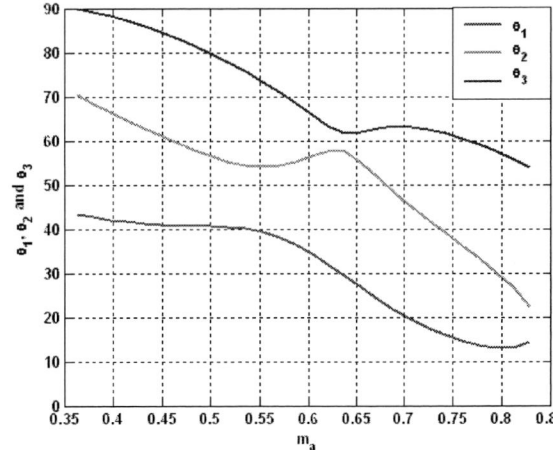

Fig. 3. Switching angles versus m_a for 7-level inverter (OHSW technique).

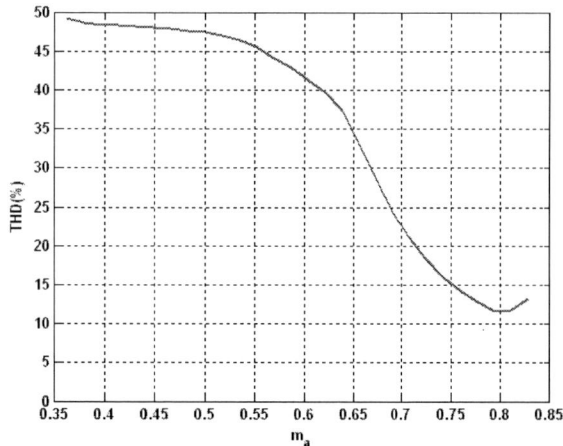

Fig. 4. Phase voltage THD as a function of m_a for 7-level inverter (OHSW technique).

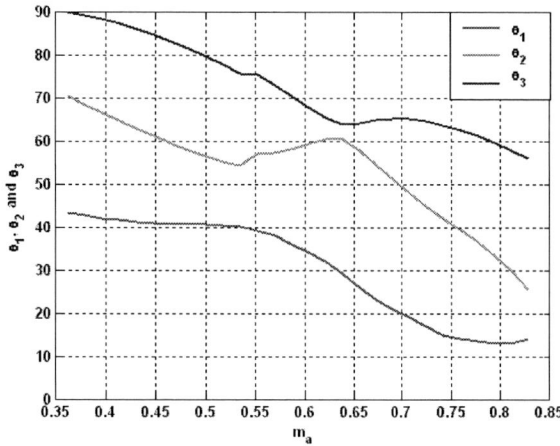

Fig. 5. Switching angles versus m_a for 7-level inverter (OMTHD technique).

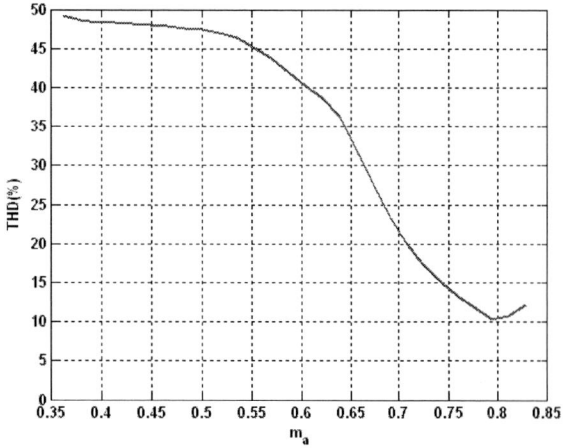

Fig. 6. Phase voltage THD as a function of m_a for 7-level inverter (OMTHD technique).

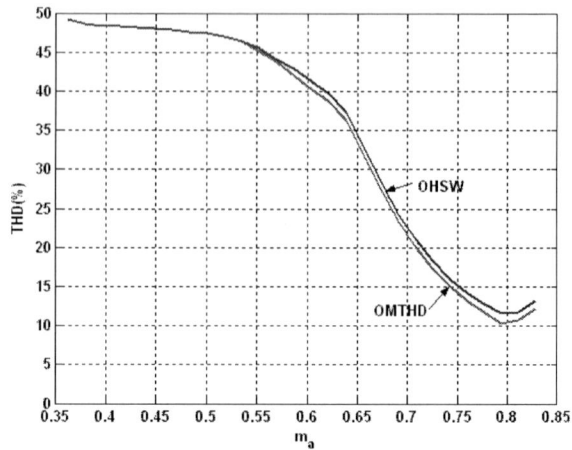

Fig. 7. Phase voltage THD as a function of m_a for 7-level inverter (OMTHD & OHSW technique).

Table I.
Low frequency harmonic components of output voltage waveform of 7-level inverter by OMTHD and OHSW techniques.

(a). m_a=0.65

Techniques ⟋ Components	OMTHD Technique	OHSW Technique
Fundamental (V)	28.78	28.78
3rd (% Fundamental)	28.89	28.88
5th (% Fundamental)	1.99	0.00
7th (% Fundamental)	2.49	0.00
9th (% Fundamental)	5.35	12.15
11th (% Fundamental)	6.17	4.07
13th (% Fundamental)	7.17	7.81
THD %	31.97	32.78

(b). m_a=0.80

Techniques ⟋ Components	OMTHD Technique	OHSW Technique
Fundamental (V)	35.00	35.00
3rd (% Fundamental)	4.77	2.79
5th (% Fundamental)	0.38	0.00
7th (% Fundamental)	0.29	0.00
9th (% Fundamental)	5.35	6.65
11th (% Fundamental)	1.36	0.17
13th (% Fundamental)	0.17	1.91
THD %	10.36	11.61

978-1-4244-1871-8/07 $25.00 © 2007 IEEE

A Modified Space Vector PWM for Common Mode Voltage Mitigation in Diode Rectifier / Z-Source Inverter System

Q.V. Tran, J.R. Ahn, T.W. Chun
University of Ulsan
Department of Electrical Engineering
Email: tqvsr@yahoo.com, passion-17
@hanmail.net, twchun@mail.ulsan.ac.kr

H.G. Kim
Kyungpook National University
Department of Electrical Engineering
Email: kimhg@knu.ac.kr

E.C. Nho
Pukyung National University
Department of Electrical Engineering
Email: nhoec@pknu.ac.kr

Abstract – In this paper, the common-mode voltage of the diode rectifier/Z-source inverter system is analyzed in details under both the shoot-through state and non-shoot-through state. Then, a modified space-vector modulation strategy is proposed to effectively mitigate the amount of negative common-mode voltage. The negative common-mode voltage can be attenuated, by eliminating the zero voltage vector V_0 in modified space-vector modulation. The proposed method is easily implemented in software. Finally, some simulation studies by PSIM and experiments with 32-bit DSP are carried out to demonstrate the benefits of the proposed algorithm.

Index terms – Common mode voltage, modified space-vector modulation, Z-source inverter (ZSI), zero-voltage vector (ZVV)

I. Introduction

As well known, the high frequency common-mode voltage (CMV) produced by the pulse width modulated inverter source causes many problems in application of ac machine drives such as the leakage current in motor winding, high shaft voltage, and radiation of the EMI noise [1,2].

Many studies have been investigated to reduce the CMV in the conventional dc boost converter/inverter system. Some of them concentrate to design several output filters to reduce both differential and common mode *dv/dt* at the motor terminal [3,4]. The other ones suggested variety of PWM schemes for reducing CMV in the diode rectifier/inverter as well as PWM converter/inverter system. These works have been focused on eliminating the zero voltage vector (ZVV) of the PWM scheme by means of using two additional active vectors instead of ZVV in that of the conventional PWM scheme[5,6]. The active-voltage vectors of inverter are shifted in order to align to those of the boost converter for disappearing one CMV pulse in every control period [7]. Even though CMV can be mitigated by the above methods, the complex extra hardware is required.

Recently, the Z-source inverter was proposed and applied to ac machine drive applications [8,9]. As the CMV in the ZSI is greater than that in the conventional PWM inverter, the suppression of CMV in the diode rectifier/ZSI system is very important. A dual Z-source inverter with three-level has been reported to diminish CMV more recently [10]. However, the effects of shoot-through state on CMV development did not analyze, and also the system needs an extra circuit, which

may increase the system cost and require a complex control algorithm.

In this paper, the CMV of diode rectifier/Z-source inverter system is analyzed in details under both the shoot-through state and non-shoot-through state. The zero voltage vector V_0 is disappeared in order to reduce the negative CMV part. Thus, a modified space-vector modulation strategy without V_0 is proposed to effectively mitigate the negative CMV portion herein. The suggested methods are verified through some simulation studies and experimental results with 32-bit DSP TMS320F2812.

II. Common-Mode Voltage of the Diode Rectifier/Z-Source Inverter System

The diode rectifier/Z-source inverter system is shown in Fig.1, where the capacitor C_1 (or C_2) of Z-source impedance network is splited to obtain the dc-bus midpoint, "m" (or "n"). The capacitor C_s plays a role of dc-voltage source to feed the ZSI. In order to satisfy the prorected requirements, neutral points of both the three-phase power source and three-phase ac machine are connected to ground "g".

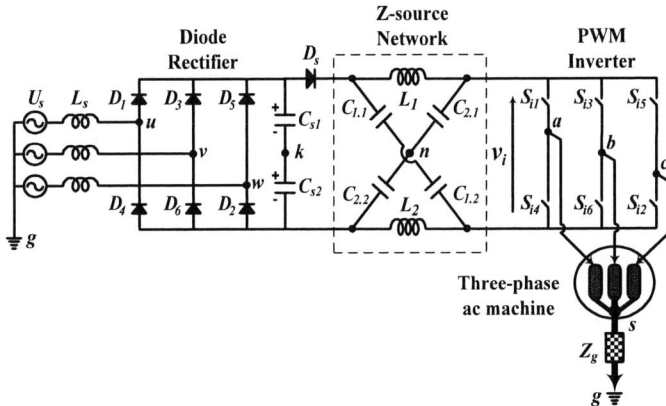

Fig. 1. Configuration of the diode rectifier/Z-source inverter system.

In principle, the CMV analysis for the diode rectifier/ZSI system fed induction motor with three-phase Y-connection stator winding is similar to that of the conventional diode rectifier/inverter system. The CMV of whole system is also constituted from the CMV of ZSI and the diode rectifier CMV.

978-1-4244-1871-8/07 $25.00 © 2007 IEEE

$$v_{sg} = v_{s*} + v_{*g} \qquad (1)$$

where the asterisk (*) represents the dc-bus midpoint "*m*" or "*n*".

The CMV terms caused by ZSI v_{s*} as well as the diode rectifier v_{*g} are the voltage differences between the neutral point of the three-phase inverter output voltage and the neutral point of the three-phase input source voltage with repect to the dc-bus midpoint "*m*" or "*n*", respectively [7].

$$v_{s*} = \begin{cases} v_{sm} = \dfrac{v_{am} + v_{bm} + v_{cm}}{3}, \text{ midpoint "m"} \\[2mm] v_{sn} = \dfrac{v_{an} + v_{bn} + v_{cn}}{3}, \text{ midpoint "n"} \end{cases} \qquad (2)$$

$$v_{*g} = \begin{cases} v_{mg} = -\dfrac{v_{um} + v_{vm} + v_{wm}}{3}, \text{ midpoint "m"} \\[2mm] v_{ng} = -\dfrac{v_{un} + v_{vn} + v_{wn}}{3}, \text{ midpoint "n"} \end{cases} \qquad (3)$$

The difference between the ZSI and conventional voltage-source inverter is that the ZSI operates under two operation states such as shoot-through state and non-shoot-through state. During shoot-through state, two switches of any switching leg conduct at the same instant to boost the capacitor voltage to desirable value regardless of dc-input voltage.

A. Common-mode voltage in shoot-through state

Assume that the shoot-though state occurs on the a-phase leg. It means that two switches S_{i1} and S_{i4} simultaneously conduct. Thus, the equivalent circuits viewed from Z-source network under shoot-through state for midpoint "*m*" and "*n*" can be illustrated in Fig. 2(a) and 2(b), respectively.

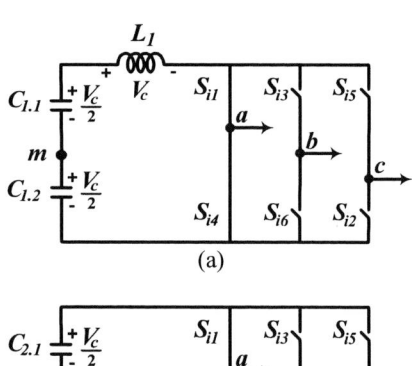

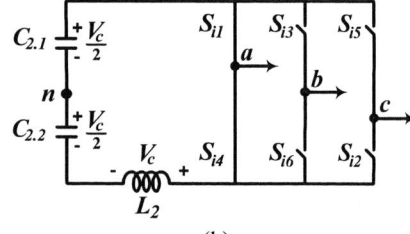

(a)

(b)

Fig. 2 The ZSI equivalent circuit in shoot-through state : (a) at midpoint "*m*", (b) at midpoint "*n*".

It should notice that the voltage across on the inductor in Z-source network equals to the capacitor voltage during shoot-through state. Thus, the voltage difference between inverter output point (*a*, *b*, *c*) and respective midpoint can be easily calculated from the equivalent circuit. Depending on the midpoint "*m*" or "*n*", these voltages become to a value of $-V_c/2$ or $+V_c/2$. For instance, if the midpoint "*n*" is chosen, the voltage v_{an} approaches to a value of $+V_c/2$. The value of voltage v_{bn} and v_{cn} are also $+V_c/2$ when upper switch is on-state and lower switch is off-state, or reversely.

From (2) the CMV v_{sn} at the midpoint "*n*" and the v_{sm} at the midpoint "*m*" can be expressed as $v_{sn} = +V_c/2$ and $v_{sm} = -V_c/2$, respectively.

Fig. 3 shows the equivalent circuit viewed from Z-source network of diode rectifier under shoot-through state. Herein, the capacitor C_s is split to get midpoint "*k*" and its voltage is V_{dc}.

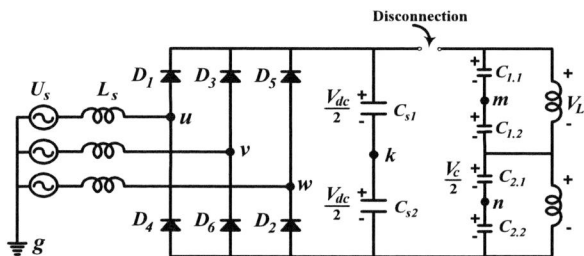

Fig. 3. The diode rectifier equivalent circuit in shoot-through state.

Due to arm-short state on a-phase leg, the diode D_s cannot conduct because of reverse bias. The input voltage completely charges on capacitor C_s at the moment. The voltage v_{*g} can be given by

$$v_{*g} = \begin{cases} v_{mg} = -(V_{rec} + v_{km}), \text{ midpoint "m"} \\[1mm] v_{ng} = -(V_{rec} + v_{kn}), \text{ midpoint "n"} \end{cases} \qquad (4)$$

where V_{rec}, v_{kn}, v_{km} are determined as follows

$$V_{rec} = \frac{v_{uk} + v_{vk} + v_{wk}}{3} \qquad (5)$$

$$v_{kn} = \frac{V_{dc}}{2} - \frac{V_c}{2} \text{ and } v_{km} = \frac{V_{dc}}{2} - \frac{3}{2}V_c. \qquad (6)$$

If three-phase source voltage is balance and large enough to neglect the voltage drop on input inductors L_s, the instantaneous waveform of V_{rec} will consist of six segments per cycle of line frequency. Each segment belongs to one of the six line-to-line voltage combinations according to conduction of each diode pair. Depending on the phase, the voltage V_{rec}

978-1-4244-1871-8/07 $25.00 © 2007 IEEE

varies from $(-V_{dc}/6)$ to $(V_{sp}/2 - V_{dc}/6)$ as shown in table I, where V_{sp} is the peak value of phase voltage.

The range of V_{rec} is well known so far, thus the CMV of the diode rectifier can be determined from (4), (5), (6) as

$$v_{*g} = \begin{cases} v_{mg} = \left(\dfrac{3V_c}{2} - \dfrac{V_{dc}}{3} - \dfrac{V_{sp}}{2} \right) \sim \left(\dfrac{3V_c}{2} - \dfrac{V_{dc}}{3} \right) \\ v_{ng} = \left(\dfrac{V_c}{2} - \dfrac{V_{dc}}{3} - \dfrac{V_{sp}}{2} \right) \sim \left(\dfrac{V_c}{2} - \dfrac{V_{dc}}{3} \right) \end{cases} \quad (7)$$

TABLE I
THE RANGE OF V_{rec}

Phase	Active diodes	V_{rec}
$-30^0 \sim 30^0$	D_5, D_6	$-V_{dc}/6 \sim (V_{sp}/2 - V_{dc}/6)$
$30^0 \sim 90^0$	D_1, D_6	$(V_{sp}/2 - V_{dc}/6) \sim -V_{dc}/6$
$90^0 \sim 150^0$	D_1, D_2	$-V_{dc}/6 \sim (V_{sp}/2 - V_{dc}/6)$
$150^0 \sim 210^0$	D_3, D_2	$(V_{sp}/2 - V_{dc}/6) \sim -V_{dc}/6$
$210^0 \sim 270^0$	D_3, D_4	$-V_{dc}/6 \sim (V_{sp}/2 - V_{dc}/6)$
$270^0 \sim 330^0$	D_4, D_5	$(V_{sp}/2 - V_{dc}/6) \sim -V_{dc}/6$

Apparently, under shoot-though condition, the CVM of ZSI can achieve a value of $V_c/2$ or $-V_c/2$ which depends on the choice of dc-bus midpoint "n" or "m", while the rectifier CMV sequentially varies according as the line frequency. The CMV of whole system v_{sg} can be derived as follows.

$$v_{sg} = \begin{cases} v_{sm} + v_{mg} = \left(V_c - \dfrac{V_{dc}}{3} - \dfrac{V_{sp}}{2} \right) \sim \left(V_c - \dfrac{V_{dc}}{3} \right) \\ v_{sn} + v_{ng} = \left(V_c - \dfrac{V_{dc}}{3} - \dfrac{V_{sp}}{2} \right) \sim \left(V_c - \dfrac{V_{dc}}{3} \right) \end{cases} \quad (8)$$

It can be seen that the CMV at dc-bus midpoint "m" is the same as that at midpoint "n".

B. Common-mode voltage in non-shoot-through state

As earlier mention, ZSI operates as a conventional voltage-source inverter during non-shoot-through state. The diode D_s conducts to couple diode rectifier to ZSI. It should noted that the voltage of each capacitor in Z-source network V_c is greater than that of source capacitor V_{dc} if shoot-though time is applied to the system [8]. The equivalent circuit under non-shoot-through state is exemplified in Fig. 4, where the inductor voltage of Z-source network V_L can be easily determined as

$$V_L = V_{dc} - V_c . \quad (9)$$

TABLE II
THE CMV OF ZSI IN NON-SHOOT-THROUGH STATE

Vector	v_{sn}	v_{sm}
V_1, V_3, V_5	$2V_{dc}/3 - 5V_c/6$	$- V_{dc}/3 + V_c/6$
V_2, V_4, V_6	$V_{dc}/3 - V_c/6$	$- 2V_{dc}/3 + 5V_c/6$
V_0	$V_{dc} - 3V_c/2$	$- V_c/2$
V_7	$V_c/2$	$- V_{dc} + 3V_c/2$

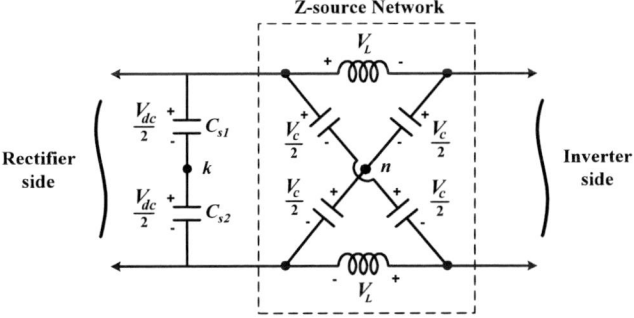

Fig. 4. Equivalent circuit of ZSI at non-shoot-through state.

Under eight space vectors $V_0 \sim V_7$ of space vector PWM scheme, the possible CMV values of ZSI corresponding to midpoint "m" or "n" are concisely synthesized in table II.

If the shoot-though time is equal to zero, the capacitor voltage V_c is identical with V_{dc} and the ZSI operates as a traditional VSI. Due to conduction of diode D_s, the voltage v_{km} is different from that at the shoot-though state, whereas the voltage v_{kn} is identical as following equations.

$$v_{kn} = \frac{V_{dc}}{2} - \frac{V_c}{2} \quad \text{and} \quad v_{km} = \frac{V_c}{2} - \frac{V_{dc}}{2} \quad (10)$$

In this case, the diode rectifier operates normally to provide a dc-voltage V_{dc} for feeding the ZSI. The voltage V_{rec} can be expressed as table I. The rectifier CMV in non-shoot-through state can be calculated from these voltage elements below

$$v_{*g} = \begin{cases} v_{mg} = \left(\dfrac{2V_{dc}}{3} - \dfrac{V_c}{2} - \dfrac{V_{sp}}{2} \right) \sim \left(\dfrac{2V_{dc}}{3} - \dfrac{V_c}{2} \right), \text{midpopint "m"} \\ v_{ng} = \left(\dfrac{V_c}{2} - \dfrac{V_{dc}}{3} - \dfrac{V_{sp}}{2} \right) \sim \left(\dfrac{V_c}{2} - \dfrac{V_{dc}}{3} \right), \text{midpopint "n"} \end{cases} \quad (11)$$

Up to now, the CMV caused by ZSI is dependent on the voltage vectors, and CMV by diode rectifier sequentially varies with the phase of input source voltage. Table III and IV show the CMV of overall system at midpoints "m" and "n", respectively.

978-1-4244-1871-8/07 $25.00 © 2007 IEEE

TABLE III
THE CMV OF DIODE RECTIFIER/ZSI SYSTEM CORRESPONDING TO MIDPOINT "m"

Vector	V_{sm}	$v_{mg}(min)$	$v_{mg}(max)$	$v_{sg}(min)$	$v_{sg}(max)$
V_1,V_3,V_5	$-V_{dc}/3 + V_c/6$	$2V_{dc}/3 - V_c/2 - V_{sp}/2$	$2V_{dc}/3 - V_c/2$	$V_{dc}/3 - V_c/3 - V_{sp}/2$	$V_{dc}/3 - V_c/3$
V_2,V_4,V_6	$-2V_{dc}/3 + 5V_c/6$	"	"	$V_c/3 - V_{sp}/2$	$V_c/3$
V_0	$-V_c/2$	"	"	$2V_{dc}/3 - V_c - V_{sp}/2$	$2V_{dc}/3 - V_c$
V_7	$-V_{dc} + 3V_c/2$	"	"	$V_c - V_{dc}/3 - V_{sp}/2$	$V_c - V_{dc}/3$

TABLE IV
THE CMV OF DIODE RECTIFIER/ZSI SYSTEM CORRESPONDING TO MIDPOINT "n"

Vector	V_{sn}	$v_{ng}(min)$	$v_{ng}(max)$	$v_{sg}(min)$	$v_{sg}(max)$
V_1,V_3,V_5	$2V_{dc}/3 - 5V_c/6$	$V_c/2 - V_{dc}/3 - V_{sp}/2$	$V_c/2 - V_{dc}/3$	$V_{dc}/3 - V_c/3 - V_{sp}/2$	$V_{dc}/3 - V_c/3$
V_2,V_4,V_6	$V_{dc}/3 - V_c/6$	"	"	$V_c/3 - V_{sp}/2$	$V_c/3$
V_0	$V_{dc} - 3V_c/2$	"	"	$2V_{dc}/3 - V_c - V_{sp}/2$	$2V_{dc}/3 - V_c$
V_7	$V_c/2$	"	"	$V_c - V_{dc}/3 - V_{sp}/2$	$V_c - V_{dc}/3$

Specifically, the system CMV varies from $(2V_{dc}/3 - V_c - V_{sp}/2)$ to $(V_c - V_{dc}/3)$, where the capacitor voltage V_c changes with the shoot-through time. It is noted that the CMV value at shoot-through state is the same as that of ZVV, V_7, which is a significant drawback for suppressing the CMV.

III. MODIFIED SPACE VECTOR MODULATION SCHEME FOR REDUCING COMMON-MODE VOLATGE

The space vector PWM techniques have widely used at current regulated PWM inverter due to lower current harmonics and a higher modulation index. The eight space vectors $V_0 \sim V_7$ are used in the SVM, where $V_1 \sim V_6$ are active vectors, and V_0, V_7 are zero vectors. The reference voltage vector V_{ref} is constituted from two adjacent vectors and two zero vectors V_0, V_7 at any sector.

unchanged. The shoot through time is evenly assigned to each phase with $T_{sh}/6$ to prevent the distortion of the inverter output voltage.

As analyzed previous part, the CMV of diode rectifier/ZSI is within from $(2V_{dc}/3 - V_c - V_{sp}/2)$ to $(V_c - V_{dc}/3)$ if conventional SVM is used at ZSI. Existence of zero voltage vectors V_0, V_7 along with voltage vector at shoot-through state V_{sh} results in a high CMV. Two voltage vectors V_7 and V_{sh} produce a large amount of positive CMV, while V_0 generates negative one.

In order to mitigate the CMV in the ZSI, the effects of these zero voltage vectors should be constrained as much as possible. The zero voltage vector, V_7 and shoot-though state V_{sh} cause a positive CMV value of $(V_c - V_{dc}/3)$. Unfortunately, the shoot-through state is a inherent state of ZSI. This positive CMV value always exists regardless of eliminating V_7.

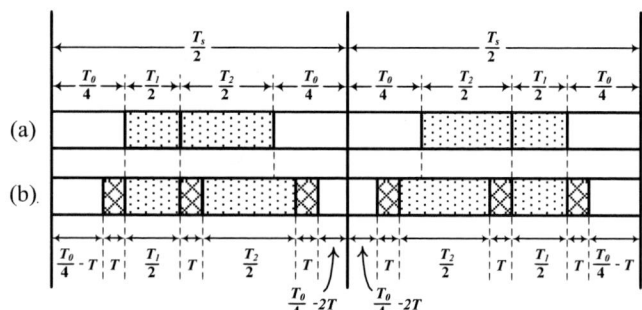

Fig. 5. The switching pattern for conventional SVM.

Fig. 5(a) shows the symmetrical pulse pattern of active vectors V_1, V_2, and zero vectors for one sampling period at traditional SVM, when the reference voltage vector is located at sector I. Fig. 5(b) shows the switching pattern for ZSI including the shoot through time ($T=T_{sh}/6$). The applied time of zero voltage vector, T_0 is diminished for generating a shoot through time T_{sh}, while the active state times T_1, T_2 are

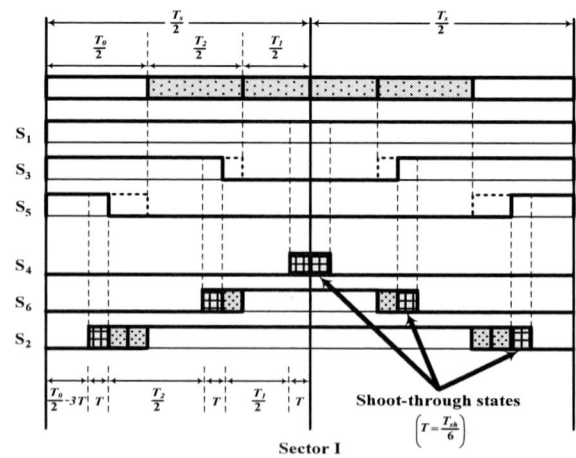

Fig. 6. The switching pattern for MSVM.

Fig. 6 illustrates a switching pattern of modified space voltage modulation (MSVM) at sector I. The vectors are

978-1-4244-1871-8/07 $25.00 © 2007 IEEE   594

distributed as V_7-V_{sh}-V_2-V_{sh}-V_1-V_{sh}, while the conventional SVM suggests another one as V_0-V_{sh}-V_1-V_{sh}-V_2-V_{sh}-V_7. It can be realized that if V_0 is disappeared, the negative CMV part will be reduced from ($2V_{dc}/3$ - V_c - $V_{sp}/2$) to ($V_{dc}/3$ - $V_c/3$ - $V_{sp}/2$). Base on above reasons, a MSVM without V_0 for mitigating the negative CMV is proposed in this literature.

The comparative switching pattern is described in the table VI.

TABLE VI
COMPARATIVE SWITCHING PATTERNS

Sector	Conventional SVM	Proposed MSVM
I	V_0-V_{sh}-V_1-V_{sh}-V_2-V_{sh}-V_7	V_7-V_{sh}-V_2-V_{sh}-V_1-V_{sh}
II	V_0-V_{sh}-V_3-V_{sh}-V_2-V_{sh}-V_7	V_7-V_{sh}-V_2-V_{sh}-V_3-V_{sh}
III	V_0-V_{sh}-V_3-V_{sh}-V_4-V_{sh}-V_7	V_7-V_{sh}-V_4-V_{sh}-V_3-V_{sh}
IV	V_0-V_{sh}-V_5-V_{sh}-V_4-V_{sh}-V_7	V_7-V_{sh}-V_4-V_{sh}-V_5-V_{sh}
V	V_0-V_{sh}-V_5-V_{sh}-V_6-V_{sh}-V_7	V_7-V_{sh}-V_6-V_{sh}-V_5-V_{sh}
VI	V_0-V_{sh}-V_1-V_{sh}-V_6-V_{sh}-V_7	V_7-V_{sh}-V_6-V_{sh}-V_1-V_{sh}

Fig. 7 depicts the negative CMV value verse shoot-though modulation index M_{sh}, when MSVM and conventional SVM are used, respectively, where M_{sh} is a ratio between shoot-through time and sampling period. Obviously, the CMV is significantly diminished in terms of the increment of shoot-through time.

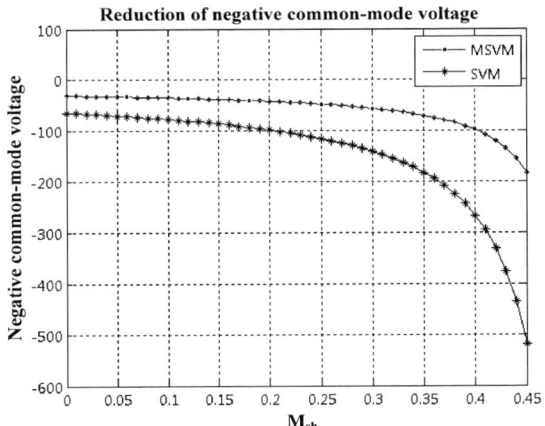

Fig. 7. The plot of the negative CMV value verse M_{sh}.

Obviously, V_0 is quite eliminated in proposed MSVM. So, the negative CMV part can be reduced if applying MSVM. Up to now, the influence of V_0 on CMV development is completely constrained.

IV. SIMUALTION AND EXPERIMENTAL RELUSTS

Some simulations and experiments are carried out to demonstrate the proposed methods. The circuit parameters of ZSI are $L_1 = L_2 = 3$mH, $C_1 = C_2 = 1$mF, $C_s = 2.2$mF and three-phase RL circuit is used as the load of the ZSI. The sampling period of MSVM is 200μs. A three-phase line voltage of 75Vrms is utilized to produce a dc voltage of 105V. The capacitor voltage is boosted to 140V by using a shoot-through time of 40μs. The peak value of inverter output voltage is controlled to 100V.

Fig. 8 shows simulation results for capacitor and dc input voltages, CMV and ac output voltage when MSVM is applied to the system at 0.4sec. the negative part of CMV can be reduced from -103V to -44V.

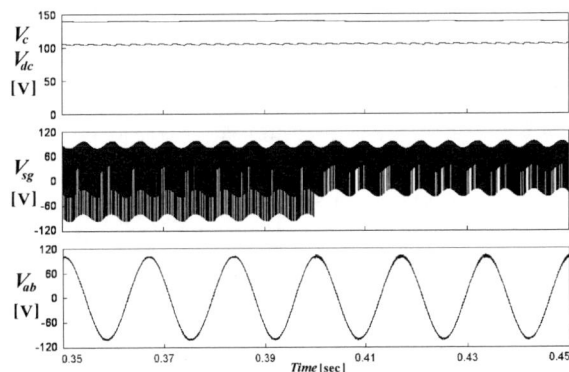

Fig. 8 Simulation result at V_{dc}=105V, T_{sh} = 40μs, V_o = 100V.

Fig. 9 shows the experiment results, which strongly verifies above theoretical analysis and proposed MSVM algorithm. The CMV values of experimental results matches with that of theoretical calculation and simulation result.

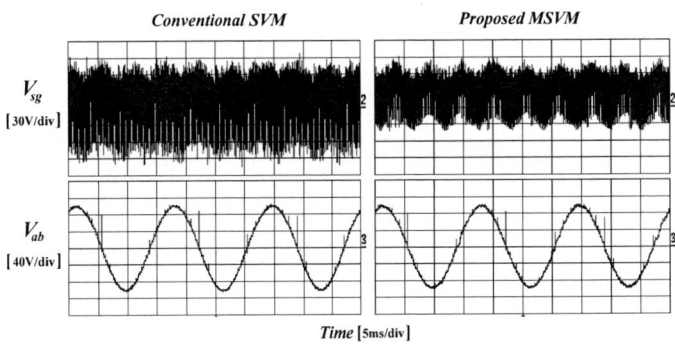

Fig. 9 Experimental result at V_{dc}=105V, T_{sh} = 40μs, V_o = 100V.

V. CONLUSION

The CMV of diode rectifier/Z-source inverter system is analyzed in details considering both the shoot-through state and non-shoot-through state. In order to mitigating CMV, a MSVM strategy without V_0 is proposed. By eliminating zero-voltage vector V_0, the negative CMV value can be decreased from ($2V_{dc}/3$ - V_c - $V_{sp}/2$) to ($V_{dc}/3$ - $V_c/3$ - $V_{sp}/2$). Finally, some simulations by PSIM and experiments with 32-bit DSP

TMS320F2812 are performed to demonstrate the benefits of the proposed algorithms.

ACKNOWLEDENT

This work has been supported by KERSI(R-2005-7-067), which is funded by MOCIE (Ministry of Commerce, Industry and Energy).

REFERENCES

[1] J. M. Erdman, R. J. Kerkman, D.W. Schlegel, and G. L. Skibinski, "Effect of PWM inverters on AC motor bearing currents and shaft voltages," IEEE Trans. on Ind. Applications., vol. 32, 1996, pp. 250–259.

[2] S. Chen, T. A. Lipo, and D. Fitzegrald, "Modeling of motor bearing currents in PWM inverter drives," in Proc. Of IEEE- IAS, 1995, pp. 388–393.

[3] Dudi A. Rend and Prasad N. Enjeti, "An improved inverter output filter configuration reduces common and differential mode dv/dt at the motor terminals in PWM drive systems," IEEE Trans.on Power Electronics, vol. 13, no.6, 1998, pp.1135-1143.

[4] M. Hongfei, X. Dianguo, and M. Lijie, "Suppress techniques of common-mode voltage generated by voltage source PWM inverter," Conf. Rec. of IPEMC, 2004, pp.1533-1538.

[5] Oriti, G., Julian, A.L., Lipo, T.A., "A new space vector modulation strategy for common mode voltage reduction in PWM invertors," in Proc. of IEEE-PESC, 1997, pp.1541–1546.

[6] Yen-Shin Lai, "Investigations into the effects of PWM techniques on common mode voltage for inverter-controlled induction motor drives," in Proc. IEEE.1999, vol. 1, 1999, pp. 35–40.

[7] Hyeoun-Dong Lee; Seung-Ki Sul, "A common mode voltage reduction in boost rectifier/inverter system by shifting active voltage vector in a control period," IEEE Trans. on Power Electronics., vol.15, no.6, 2000, pp. 1094–1101.

[8] F.Z.Peng, "Z-Source Inverter", IEEE Trans. on Industry Applications, vol. 39, no.2, 2003, pp.504-510.

[9] F.Z.Peng, X.Yuan, X.Fang and Z.Quian, "Z-Source inverter for adjustable speed drives," IEEE Tran. Power Electronics. vol.1, 2003, pp. 33-35.

[10] A. Von Jouanne, and H. Zhang, "A dual-bridge inverter approach to eliminating common mode voltages and bearing and leakage currents," IEEE Trans. on Power Electronics., vol. 14, 1999, pp. 43–48.

Development of a Current-Sensorless Multi-Loop Control for Standalone PWM Inverters

Yuan Yisheng, Liuchen Chang and Song Pinggang
East China Jiaotong University, Department of Electrical Engineering
Email: ysyuan@4y.com.cn, lchang@unb.ca, pgsong@4y.com.cn

Abstract — **A current-sensorless multi-loop control method that estimates the capacitor current is proposed for standalone inverter applications. The theory of the control loop is given, and a mathematical model is analyzed. The effect of capacitor parameter mismatch on system characteristics was emphasized and discussed through modeling and simulation. Finally, experimental results comparing the characteristics of the output voltage under parameter mismatch conditions verified the operation and merits of the proposed controller.**

I. INTRODUCTION

Digital control techniques have been widely applied to sinusoidal Pulse Width Modulation (PWM) inverters. Dead-Beat controllers, predictive controllers, traditional Proportional-Integral-Derivative (PID) controllers, and repetitive controllers have been researched and developed in recent years [1][2]. Dead-Beat control has good dynamic response, but is not robust, and is sensitive to load and filter parameters [1]. PID control is simple, but has high voltage distortion under nonlinear loads [2]. Repetitive control can improve voltage distortion under nonlinear loads.

To reduce the cost and size of inverters, development of robust current-sensorless digital control is necessary. A current observer has been proposed to eliminate current sensors [3]. But the algorithms are complex, and the effect of parameter mismatch has not been well-studied.

This paper presents a current-sensorless multi-loop controller appropriate for use in Uninterruptible Power Supplies (UPS) and standalone inverters. These can be used for electricity generation systems using renewable and non-renewable resources. The proposed controller is a current-regulated voltage controller consisting of an outer voltage loop and an inner predictive capacitor current loop. Its mathematical model is derived. The key effects caused by capacitor mismatch between the practical and ideal capacitors is discussed. Simulations and experiments prove that the controller has good performance within a reasonable parameter mismatch range.

II. CURRENT SENSORLESS MULTI-LOOP CONTROLLER

Fig.1 depicts the traditional full-bridge voltage inverter, in which the inductor L and the capacitor C construct the output filter.

Fig.2 depicts the newly-proposed current-sensorless multi-loop controller block, in which the dashed block shows the internal inverter model.

The proposed model has three control loops: a outer voltage control loop, a inner current control loop, and a feed-forward loop. These are described in the following sections.

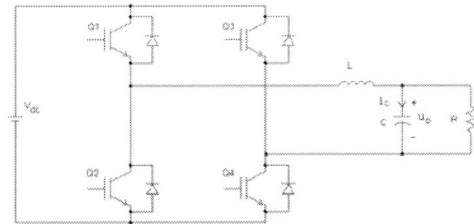

Figure 1 Traditional standalone PWM inverter

A. Voltage loop

The outer voltage loop is shown in the leftmost block of Fig. 2. This controller compares the real-time feedback output voltage $u_o(n)$ and the reference sinusoidal voltage u_{ref}. The resultant error voltage is amplified by a proportional amplifier, of gain K_{pv}, which acts as a reference signal (u_{ev}) for the inner current loop.

Because the controlled output voltage can vary, to avoid system oscillation, the voltage loop should use a proportional amplifier, not an integrator.

B. Current loop and current estimator

In an inverter controller, using capacitor current feedback is more suitable than using inductor current feedback, since the capacitor current directly reflects the charge on the capacitor and thus the output voltage across the load. Therefore, capacitor current (i_c) estimation is adopted for use in the proposed controller. Because the capacitor current is proportional to the derivative of the output voltage, using a high-speed DSP system, the capacitor current can be approximated by using the following difference equation:

$$i_C(n) = C_m \cdot \frac{u_o(n) - u_o(n-1)}{T_s} \quad (1)$$

In equation (1), C_m is the ideal capacitance used by the DSP controller, which may contain a mismatch error from the actual capacitance. T_s is the switching frequency.

As in the voltage loop case, only one proportional amplifier of gain K_{pi} is used in the current loop. To maintain the required dynamic speed, the current error amplifier gain (K_{pi}) is usually larger than the voltage error amplifier gain (K_{pv}).

978-1-4244-1871-8/07 $25.00 © 2007 IEEE

C. Disturbance feed-forward loop

The innermost loop of the proposed controller is a feed-forward controller that uses predictive output voltage control. It consists of an output voltage predictor and a proportional amplifier.

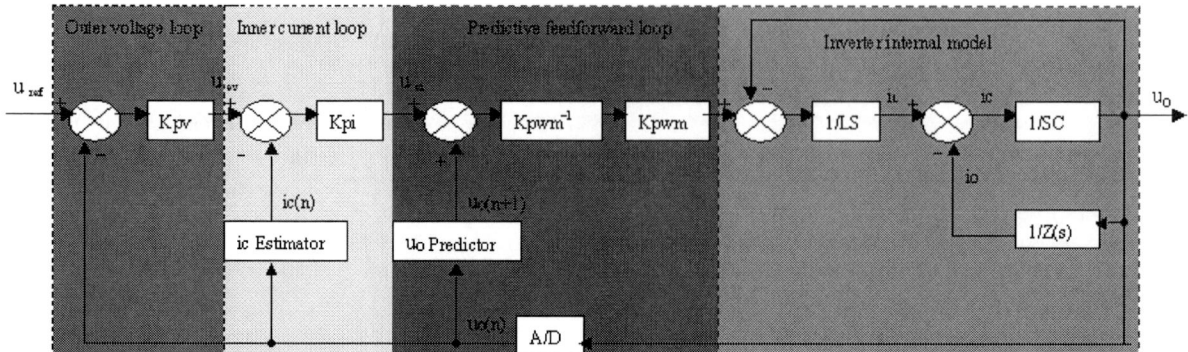

Figure 2 The proposed current-sensorless multi-loop control block

The output voltage predictor partially compensates for control delay. This predictor uses the linear estimator method [4] whose governing equation is given by:

$$u_o(n+1) = u_o(n) + [u_o(n) - u_o(n-1)] \qquad (2)$$

The predictor voltage $u_o(n+1)$ feeds forward to the output voltage u_{ev} of the current loop. By using the proportional amplifiers of gain K_{pwm}^{-1} and K_{pwm}, the feed-forward predictor voltage cancels the disturbance feedback output voltage u_o. Therefore, using the feed-forward loop simplifies the system model.

D. Transfer function of the system

Ignoring the effect of equation (2), the simplified open-loop transfer function in the S-domain is given by:

$$G_o(s) = \frac{K_{pv} K_{pi}}{LCs^2 + (\frac{L}{Z(s)} + K_{pi} C_m)s} \qquad (3)$$

where $Z(s)$ represents the impedance of the load.

Accordingly, the closed-loop transfer function in the S-domain is expressed by:

$$G_c(s) = \frac{K_{pv} K_{pi}}{LCs^2 + (\frac{L}{Z(s)} + K_{pi} C_m)s + K_{pv} K_{pi}} \qquad (4)$$

When a resistive load is considered, equation (4) can be written as the following standard equation:

$$G_c(s) = \frac{\varpi_n^2}{s^2 + 2\xi \omega_n s + \omega_n^2} \qquad (5)$$

where the natural frequency ω_n is given by:

$$\varpi_n = \sqrt{\frac{K_{pv} K_{pi}}{LC}} \qquad (6)$$

and the damping factor ξ is expressed as:

$$\xi = \frac{L/R + K_{pi} C_m}{2\sqrt{K_{pv} K_{pi} LC}} \qquad (7)$$

The L/R factor in equation (7) represents the load contribution to the damping factor ξ. The $K_{pi} C_m$ factor represents the no load contribution to the damping factor ξ. To obtain good performance at both no load and full load, the load contribution should not be more than the no load contribution.

It is important to analyze the effect of capacitance on the damping factor; and to do so, equation (7) can be used. C_m represents the ideal value used by the DSP, and C represents the actual capacitance. The mismatch between C_m and C is usually within $\pm$ 20%. Because the damping factor is inversely proportional to the square root of the actual capacitance, the damping factor varies from 0.89 to 1.09. Therefore, for a control system, the mismatch between the ideal and real capacitance is acceptable, and provided the control system is designed robust.

E. Model analysis

Because the output voltage of inverters varies over time, performance measurements should be based on the whole output frequency range. The following equation can be used to define the stability error percentage e_s of the RMS value of output voltage:

$$e_s = \frac{u_o - u_{ref}}{u_{ref}} \qquad (8)$$

978-1-4244-1871-8/07 $25.00 © 2007 IEEE   598

This index can be used to show static and dynamic performance measurements of inverter output voltage.

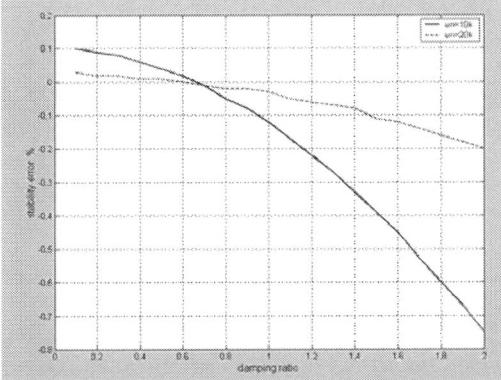

Figure 3 Stability error percentage e_s versus damping factor ξ

The analytical model used for analyzing stability error is very complex. In practical systems, the natural frequency and the damping factor must be designed in a suitable range. The simulated relationship between the stability error and the damping factor is shown in Fig.3. In Fig.3, the stability error varies from 0.2% to –0.8% as the damping factor changes from 0.1 to 2. This error is small enough for practical inverter applications. Of course, control delay and dead time both affect the stability error. However, for standalone non-parallel inverter applications, the stability error is not the most important consideration. Furthermore, as the switching frequency increases, the natural frequency ω_n also increases, which decreases the error stability and increases system stability. As a result, the proposed current-sensorless controller can be implemented by reasonable design under the mismatch between C_m and C.

III. SIMULATION RESULTS

Using an 800 W inverter as the test inverter, simulation parameters are listed in Table 1.

TABLE 1
Inverter Parameters

V_{dc} (V)	U_o (V)	P_o (W)	f_o (kHz)	K_{pwm}	R load (Ω)
360	220	800	50	1	60.5
L (mH)	C (uF)	f_s (kHz)	K_{pv}	K_{pi}	ω_n
2.5	5	20	0.04	30	10k
ξ (C_m=2.5uF)		ξ (C_m=5uF)		ξ (C_m=7.5uF)	
No load	Full load	No load	Full load	No load	Full load
0.61	0.95	1.22	1.56	1.84	2.17

For convenient comparison, three sets of controller parameters were selected for simulation studies: C_m =C=5uF,

without mismatch; C_m =2.5uF, with a mismatch of -50%; and C_m =7.5uF, indicating a mismatch of 50%.

Fig.4 depicts the Bode diagrams of the system open-loop transfer function for varying capacitance mismatch.

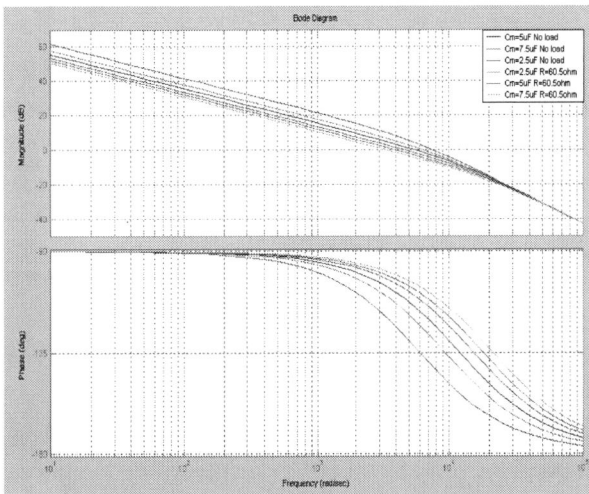

Figure 4 Bode diagram of the open-loop transfer function

Although the capacitance mismatch reaches $\pm 50\%$, the system remains stable under both no load and full load conditions. At 50% capacitance mismatch (C_m=7.5uF) and full load (R=60.5 Ω), the phase margin is maximum in Fig.4. At –50% capacitance mismatch (C_m=2.5uF) and no load, the phase margin is only $40°$, which may be regarded low for a stable system.

Based on the parameters shown in Table 1, the simulation output voltage waveforms of the complete PWM inverter are shown in Fig.5. Fig.5 shows that the oscillations in output voltage that occur when the load is removed (load off) are always larger than those that occur when the load is added (load on). This occurs because the damping factor at no load is larger than that at full load, which leads to higher under-damped response.

From Fig.5, it is obvious that the output voltage oscillations at –50% capacitance mismatch are maximum at both load on and load off,. This result agrees well with the Bode diagram, which shows that the phase margin is minimum at –50% capacitance mismatch and no load. As a result, the simulation shows that the output voltage maintains good static and dynamic performance within the $\pm 50\%$ capacitance mismatch range.

IV. EXPERIMENTAL RESULTS

A laboratory prototype inverter was designed and tested whose parameters were the same as those given in Table 1. Fig.6 depicts the waveforms of output voltage and output current under load on conditions.

Fig.6 (a) to (c) shows that the output voltage has little distortion at both no load and full load when the proposed controller is adopted. Fig.6 (d) depicts the output voltage and current waveforms when a conventional controller (with current sensor) is applied. The dynamic response under different mismatch conditions shows no obvious difference.

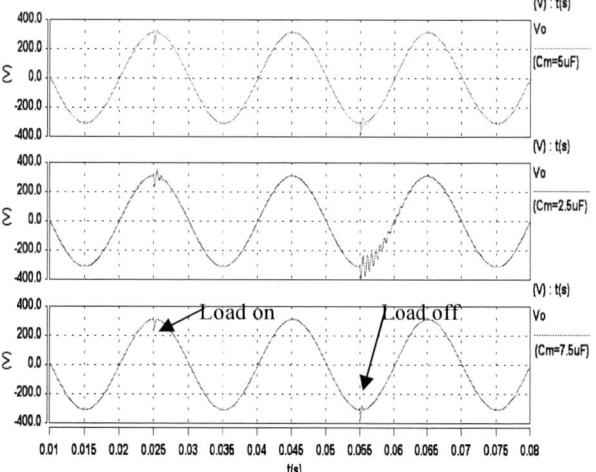

Figure 5 Transient voltage (load variation 0→100% load→0)

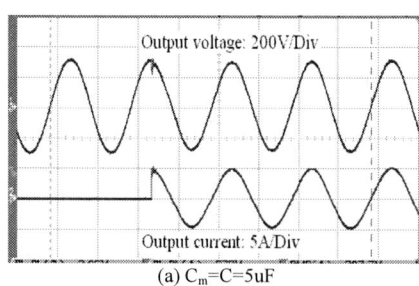

(a) $C_m = C = 5uF$

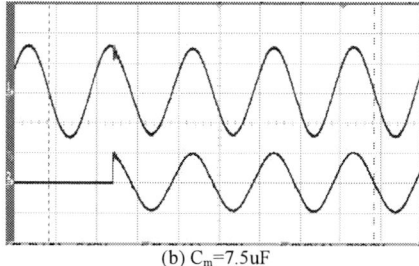

(b) $C_m = 7.5uF$

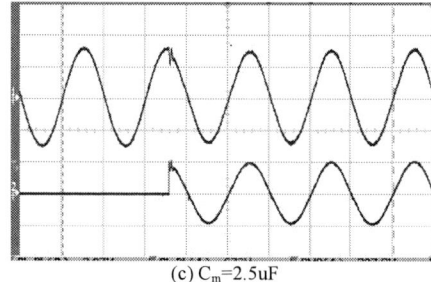

(c) $C_m = 2.5uF$

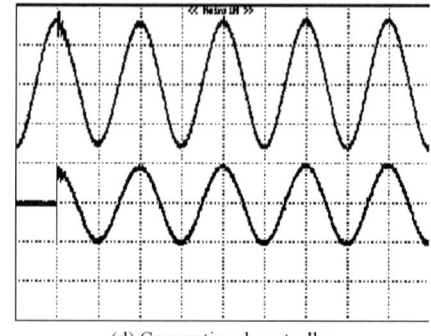

(d) Conventional controller

Figure 6 Test waveforms (10ms/ horizontal division)

At load off, the current decreases slowly because of the electrical arc effect of the load switch. This makes the output voltage change slowly at load off, so these waveforms were not included in this paper.

V. CONCLUSIONS

Using a digital controller eliminates the requirement for a current sensor, and thus reduces the cost and improves the reliability of inverters. An inverter system has been analyzed based on control theory. The impact of capacitance mismatch on the system performance has been investigated. Both the simulation and experimental results have proved that even with a mismatch of capacitance within a practical range, the proposed inverter system has very good performance in both steady and transient states.

ACKNOWLEDGMENTS

The researchers gratefully acknowledge the support of the National Natural Science Foundation of China (50577025).

REFERENCES

[1] K. P. Gokhale, A. Kawammura and R. G. Hoft, "Deadbeat microprocessor control of PWM inverter for sinusoidal output waveform synthesis" in *1985 Proc. PESC Conf.*, pp. 28–36.

[2] N. M. Abdel-Rahim and J. E. Ouaicoe, "Analysis and design of a multiple feedback loop control strategy for single-phase voltage-source UPS inverter", *IEEE Trans. Power Electronics*, vol. 11, , pp. 532–541, Jul. 1996.

[3] M. J. Ryan, W. E. Brumsickle and R. D. Lorenz, "Control topology options for single-phase UPS inverter", in *1996 Proc. PEDS Conf.*, pp. 553–558.

[4] J. H. Aylor, R. L. Ramey and G. Cook, "Design and application of a microprocessor PID predictor controller", *IEEE Trans. Industry Electronics and Control Instrumentation*, vol. 31,, pp. 133–137, Aug. 1980.

978-1-4244-1871-8/07 $25.00 © 2007 IEEE

A Novel Soft-Switching Two-Stage Step-Up DC/DC Converter

Shaowu Du[1]   Zhong Chen[1]   Liuchen Chang[1,2]

1. School of Electrical Engineering and Automation, Hefei University of Technology,
Hefei 230009, Anhui Province, China.

2. Dept. of Electrical and Computer Engineering, University of New Brunswick, Canada

Abstract—In an electric vehicle power system, a step-up converter is required to increase the voltage from 24 V to 270 V. To achieve such a high voltage boost ratio, there are several challenges that must be overcome: tackling the leakage inductance of the isolation transformer; reducing the voltage stress on the power switches; and guaranteeing fast dynamic response and high efficiency. Thus, innovative approaches are needed. A novel two-stage step-up DC/DC converter has been developed. This device has the advantages of high power density, easy control, high efficiency, high transformer utilization, and fast dynamic response.

The two-stage step-up DC/DC converter is composed of a boost converter cascaded with a push-pull converter. The boost diode and filter capacitor, as found in conventional boost converters, are eliminated. All switches achieve soft-switching by using a simple auxiliary circuit. Therefore, this novel converter is well-suited for applications where the input voltage level is significantly lower than the output voltage, for example, in the areas of electric vehicle power supplies, communication systems, aeronautic power, and power systems.

Analysis, prototype development, and experimental tests have been conducted, and the results are presented in this paper. The results confirm the performance and features of the novel converter.

Index Terms — Two-stage Converter, Soft-Switching, Step-Up, Boost Converter, Push-Pull Converter

1. INTRODUCTION

Rapid development in electric vehicle, communication system, aeronautic, and power system technologies has imposed stringent requirements on power supplies. In order to meet these new requirements, many novel methods have been proposed, of which the two-stage converter is one [1]-[9]. In some step-up conversion applications, where the input voltage is very low and the output voltage is very high (for example, ten times or more), conventional step-up converters do not meet requirements.

Generally, when the required output voltage is no more than five times the input voltage, a boost converter is used. However, such a boost converter for larger output voltages causes the following problems:

(1) Because the duty cycle is large, control is hard to realize. The required turn-off time of the switch is very short when the duty cycle is large and the frequency is high. For example, if the switching frequency is 300 kHz, and the voltage rises from 24 V to 120 V, the turn-off time is only 667 ns.

(2) If the boost ratio is larger than 5, the duty cycle, D, is larger than 0.8, so power losses and high temperatures occur because the switch turn-on time is much larger than its turn-off time. Moreover, soft-switching is hard to achieve when the turn-off time is too short.

(3) If the output voltage and current is constant, when the input voltage decreases, the average current of the inductor and the switch increases as the duty cycle increases. In this case, conduction loss occurs and the efficiency

978-1-4244-1871-8/07 $25.00 © 2007 IEEE   601

of the boost converter is reduced.

For these reasons, an isolated step-up converter must be used when the output voltage is more than five times higher than the input voltage. However, problems occur when conventional single-stage step-up converters with isolated transformer are used. The coupling of the transformer windings is not efficient because of the high turn ratio between the primary and the secondary windings, which causes high leakage inductance. The leakage inductance causes circuit oscillation, high voltage stress on the switch, decreases the electromagnetic compatibility, and decreases efficiency. Additionally, leakage inductance can increase the loss of the duty cycle and reduce the utilization of the transformer's magnetic core, which leads to sub-optimal design [10].

In summary, conventional boost converters cannot satisfy requirements when the required output voltage is much higher than the input voltage. In order to solve these problems, a novel two-stage step-up DC/DC converter is proposed. The novel converter has the advantages of easy control, high efficiency, fast dynamic response, and so on.

2. TOPOLOGY OF THE PROPOSED STEP-UP DC/DC CONVERTER

A step-up converter is researched and designed to meet requirements for providing a boost of 24 V to 270 V for an electric vehicle power system. Considering the voltage drops of the devices, the boost ratio is about 12, which is difficult to implement by using a conventional boost converter.

two-stage step-up DC/DC converter consists of a boost converter cascaded with a push-pull converter. By improving the control strategy, the boost diode and the filter capacitor that are usually present in the conventional boost converter are removed. By doing so, the two independent power modules become one integrated structure, thus reducing component count and improving dynamic response. The novel converter utilizes an energy feedback auxiliary circuit, which is composed of a coupled inductor L_r/L_s; an auxiliary switch S_2; and a feedback diode D_a. These are used to achieve the zero-voltage transition (ZVT) of the boost switch S_1, the zero-voltage switching (ZVS) turn-on, and the zero-current switching (ZCS) turn-off of the switches S_3 and S_4. The auxiliary switch S_2 also turns on with ZCS, and turns off nearly ZVS. The blocking capacitor C_2, in series with the secondary circuit of the transformer, is used to eliminate the magnetic bias problem.

The proposed two-stage step-up DC/DC converter has the following advantages:

(1) All MOSFETs are Pulse Width Modulation (PWM)-controlled. All the switches have a common source potential, thus eliminating the need for isolated gate drivers. The control circuit is simple and reliable.

(2) All MOSFETs operate in soft-switching mode, and the converter has high efficiency and produces little electromagnetic interference (EMI).

(3) The converter comprises fewer components than a conventional converter, and so requires a smaller volume. Moreover, the system dynamic response is improved by eliminating the DC filter capacitor between the two stages.

(4) Conduction losses are small because, at any one time in boost mode, only one of the switches S_1, S_3 and S_4 is on. (S_2 is ignored due to its short operation time).

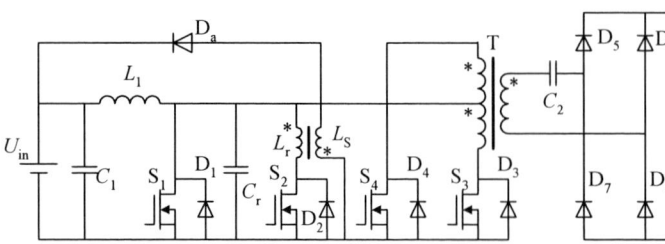

Fig.1   The topology of novel two-stage step-up DC/DC converter

As shown in Fig.1, the proposed novel

3. OPERATIONAL PRINCIPLES OF THE PROPOSED CONVERTER

To simplify analysis, the following assumptions are made. All the components are ideal, and the leakage inductance of the transformer is ignored. The input inductor L_1 is so large that i_{L1} is considered to be constant during the time that current flows through the auxiliary circuit. The output capacitor is large enough so that the output voltage U_o is considered invariant during the switching cycle.

The driving signals and key circuit waveforms are shown in Fig.2. There are eight operating intervals (or modes) during the first half of the switching cycle. The equivalent circuit for each interval is shown in Fig.3.

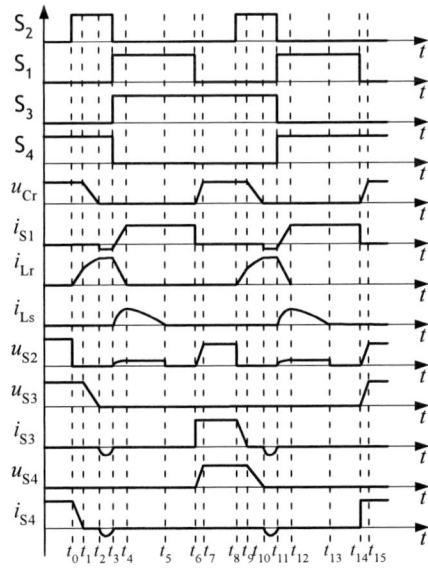

Fig.2 Driving signals and key waveforms of the proposed converter

Descriptions of the operating intervals are as follows.

1) *Interval 1 [t_0~t_1]*: Before t_0, S_1, S_2, and S_3 are off while S_4 is on. At this time, the current of the energy-storage inductance of the boost converter flows through S_4, transmitting energy to the output of the transformer. The operation is shown in Fig.3 (a). At t_0, the auxiliary switch S_2 turns on with ZCS because of the presence of the resonant inductor L_r. After this time, the current i_{S4} transfers to L_r, and i_{Lr} rises linearly

with U_o/n, where n is the turn ratio and U_o/n is the voltage reflected by the output of the transformer. At t_1, i_{Lr} rises to equal i_{L1}, and i_{S4} drops to zero. S_4 achieves ZCS turn-off. The operation of this mode is shown in Fig.3 (b).

2) *Interval 2 [t_1~t_2]*: At t_1, resonance between C_r and L_r occurs. The capacitor C_r discharges across L_r, so i_{Lr} continues to rise. At t_2, i_{Lr} rises to its maximum and u_{cr} drops to zero. The operation of this mode is shown in Fig.3 (c).

3) *Interval 3 [t_2~t_3]*: At t_2, the resonance stops. At this time, i_{Lr} begins to freewheel across the antiparallel diodes of S_1, S_3, and S_4 because i_{Lr} is greater than i_{L1}. The conduction of D_1 clamps u_{S1} to zero, and during this interval S_1 turns on with ZVS. This occurs until S_2 turns off. In this mode, i_{Lr} remains constant. The operation of this mode is shown in Fig.3 (d).

4) *Interval 4 [t_3~t_4]*: At t_3, S_2 turns off nearly ZVS. At this time, i_{Lr} drops gradually to zero. When i_{Lr} decreases, the primary voltage of the flyback transformer is negative at the top and positive at the bottom. Therefore, at the secondary side, D_a experiences positive voltage and conducts, and i_{Ls} begins to rise. The secondary voltage of the flyback transformer is clamped to U_{in} because of the conduction of D_a. The secondary voltage is reflected to the primary. The primary voltage equals U_{in}/n_1, where n_1 is the turn ratio of the flyback transformer. When S_2 turns off, u_{S2} equals U_{in}/n_1, which is quite small; in practice S_2 turns off near ZVS. At t_4, the energy of L_r entirely transfers to L_S and i_{Ls} rises to its maximum. The operation of this mode is shown in Fig.3 (e).

5) *Interval 5 [t_4~t_5]*: During this period, L_S continues to feed energy back to the input. The operation of the other parts of the circuit is the same as for the conventional boost converter when the boost switch is on. The operation of this mode is shown in Fig.3 (f).

6) *Interval 6 [t_5~t_6]*: In this mode, only S_1 is on, so the circuit operation is entirely the same as that of the conventional boost converter when

978-1-4244-1871-8/07 $25.00 © 2007 IEEE   603

the boost switch is on. The duration of this interval is decided by the duty cycle. The operation of this mode is shown in Fig.3 (g).

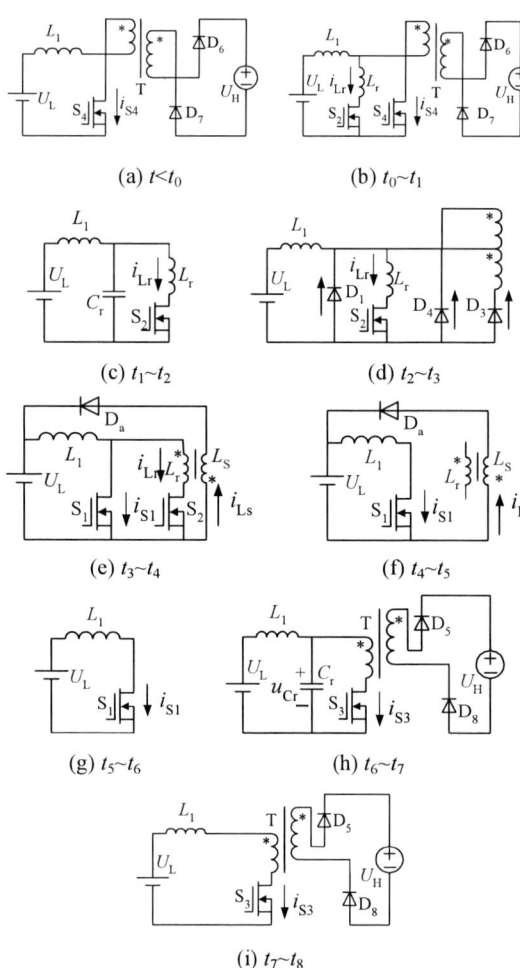

Fig.3　The equivalent circuits of each mode of the proposed converter during a half switching cycle

7) Interval 7 [t_6~t_7]: At t_6, S_1 turns off with ZVS because of the presence of capacitor C_r. C_r charges by i_{L1} and u_{cr} rises linearly. While S_1 is on, the voltage across S_3 is clamped to zero. The driving signal S_3 is applied at t_3, so S_3 turns on with ZVS when S_1 turns off. At the end of this interval, i_{L1} flows through S_3. The operation of this mode is shown in Fig.3 (h).

8) Interval 8 [t_7~t_8]: During this interval, the boost circuit transfers energy through PWM operation, and the duration of energy transfer is decided by the duty cycle. The operation of this mode is shown in Fig.3 (i).

Because the circuit is symmetrical, the operation of S_1, S_2, and S_4 in the second half-cycle is similar to that of the first half-cycle, and so is omitted.

4. DESIGN OF MAIN PARAMETERS

1) C_r Design

The resonant capacitor C_r is used to achieve the ZVS turn-off of S_1. C_r should be chosen such that u_{S1} rises slowly; otherwise it is hard to achieve ZVS turn-off. To obtain a value for C_r, usually the following formula is used [12]:

$$C_r = \frac{I_{L1max}}{U_o/n}(2\sim3)t_f \qquad (1)$$

where t_f is the turn-off time of the switch.

2) L_r Design

To mitigate its impact on normal PWM operation of the converter, the operational duration of the energy feedback auxiliary circuit should not exceed one tenth of the switching cycle T. Thus L_r is chosen for ZVS based on the following equation [12]:

$$\frac{L_r}{U_o/n}I_{L1max} + \frac{\pi}{2}\sqrt{L_rC_r} \leq 0.1T \qquad (2)$$

3) n_1 Design

So long as the energy in the auxiliary circuit can be fed entirely back to the input, the turn ratio of the flyback transformer, n_1, should be selected so that it is as large as possible. When n_1 is large, i_{Ls}, which is reflected by i_{Lr} of the flyback transformer, is low. Thus the current through the diode D_a (i_{Da}) is also quite small. Consequently, the conduction loss of the auxiliary circuit is reduced, and EMI caused by the reverse recovery current of D_a is alleviated. Moreover, when n_1 is large, u_{S2} is quite low, and thus S_2 can be considered to be in "almost ZVS turn-off".

4) C_2 Design

Placing a non-polarized capacitor C_2 in series with the secondary side of the high-frequency transformer (the blocking capacitor) can avoid DC-bias. The value of C_2 is chosen according to the following equation [11]:

$$C_2 = \frac{1}{(2\pi f_R)^2 L + 2\pi f_R R} \qquad (3)$$

where L is the leakage inductor of the secondary

978-1-4244-1871-8/07 $25.00 © 2007 IEEE

side of the transformer, R is the load resistance, and f_R is the resonant frequency of the leakage inductor and the blocking capacitor. Usually, f_R is chosen to be equal to one-fifth of the switching frequency.

5. EXPERIMENTAL RESULTS

Using the previous design methodology, a prototype converter was designed and developed.

The main experimental parameters are as follows:

Input voltage	Output voltage	Rated power	Switch Fre. of S1, S2
24V	270 V	120W	100 kHz
Switch Fre. of S3, S4	Type of S1~S4:	Type of Da:	Mag. core:
50 kHz	IRFP250	MUR1520	EE33
Number of turns:	C_r and C_2	L_r and L_s	L_1
8/8/48	25 nF、 0.47 uF	1 uH、4 uH	120 uH

The main experimental waveforms resulting from testing are shown in Fig.4.

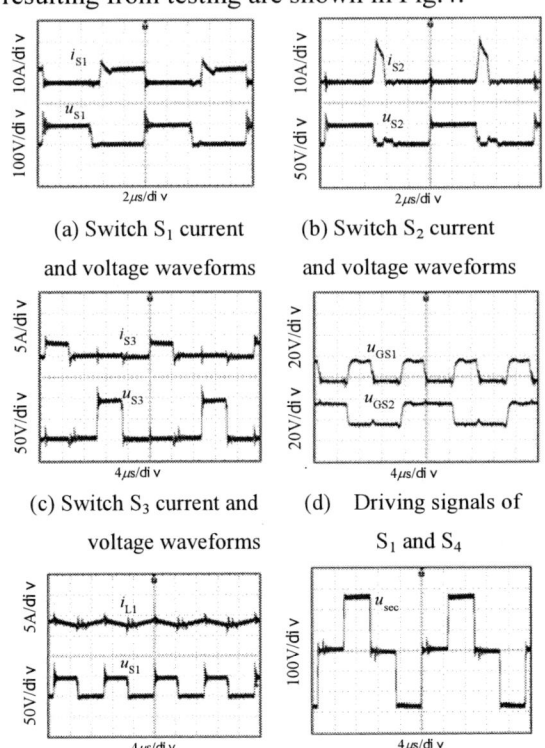

(a) Switch S_1 current and voltage waveforms

(b) Switch S_2 current and voltage waveforms

(c) Switch S_3 current and voltage waveforms

(d) Driving signals of S_1 and S_4

(e) The current waveform of L_1 and the voltage waveform of S_1

(f) The voltage waveform of the Trans. Secnd. side

Fig.4   Experimental waveforms

Fig.4 (a) shows the voltage across S_1 and the current through S_1. It is apparent that S_1 achieves ZVT. Fig.4 (b) shows the voltage across the auxiliary switch S_2 and the current through S_2. It can be seen that S_2 turns on with ZCS, and turns off near ZVS. Fig.4 (c) shows that the push-pull switch S_3 entirely achieves ZVS turn-on and ZCS turn-off. Fig.4 (d) shows the driving signals for S_1 and S_4, and shows that the switching frequency of S_1 is twice that of S_4. Fig.4 (e) shows the current of the input inductor L_1. Fig.4 (f) shows the voltage of the secondary side of the transformer.

The efficiency of the proposed converter at various loads is shown in Fig.5. This figure shows that an impressive peak efficiency of 93.8% is achieved. This is an impressive figure for such a low-power converter.

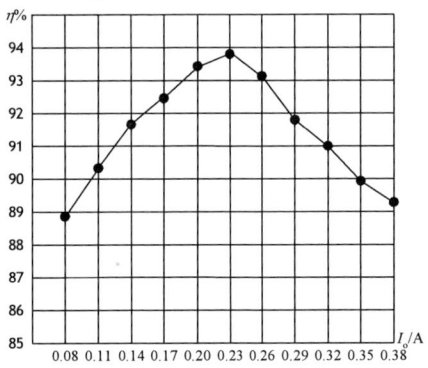

Fig.5   Efficiency of the converter

5. CONCLUSIONS

This paper proposes a novel two-stage step-up DC/DC converter that is composed of a boost converter cascaded with a push-pull converter. All its switches achieve soft-switching by introducing a simple auxiliary circuit. The proposed converter has the advantages of high power density, high efficiency, high utilization of the transformer, easy control. Therefore, this novel converter is well-suited for applications where the input voltage is significantly lower than the output voltage, for example, in the areas of electric vehicle power supplies, communication systems, aeronautic power, and power systems.

978-1-4244-1871-8/07 $25.00 © 2007 IEEE

6. REFERENCES

[1] J. Y. Zhu and B. Lehman, "Control loop design for two-stage DC-DC converters with low voltage/high current output [J]," *IEEE Trans. Power Electron.*, vol. 20, no. 1, pp.44-55, 2005.

[2] Y. C. Ren, M. Xu, K. W. Yao, Y. Meng, and F.C. Lee, "Two-stage approach for 12-V VR [J]," *IEEE Trans. Power Electron.*, vol. 19, no. 6, pp.1498-1506, 2004.

[3] J. Wei and F. C. Lee, "Two-stage voltage regulator for laptop computer CPUs and the corresponding advanced control schemes to improve light-load performance [C]," in *IEEE APEC'04*, 2004, pp. 1294-1300.

[4] Y. C. Ren, M. Xu, F. C. Lee, and P. Xu, "The optimal bus voltage study for 12V two-Stage VR based on an accurate analytical loss model [C]," in *Proc. IEEE, 35th Annual Power Electronics Specialists Conference*, 2004, pp. 4319-4324.

[5] S. Abe, J. Yamamoto, T. Zaitsu, and T. Ninomiya, "Extension of bandwidth of two-stage dc/dc converter with low-voltage/high-current output [C]," in *Proc. IEEE PESC'03 Conf.*, 2003, vol. 4, pp. 1593–1598.

[6] Y. Ren, M. Xu, K. Yao, and F. C. Lee, "Two-stage 48V power pod exploration for 64-bit microprocessor [C]," in *Proc. IEEE Applied Power Electronics Conf. Expo.*, 2003, pp. 426–431.

[7] X. C. Wang, S. Y. Xiao, and I. Batarseh, "A Novel control for two-stage DC/DC converter with fast dynamic response [C]," in *Proc. IEEE, 35th Annual Power Electronics Specialists Conference 2004,*, pp.43-48.

[8] O. Garcia, L. A. Flores, J. A. Oliver, J.A. Cobos, and J. D. L. Pena, "Bi-directional DC-DC converter for hybrid vehicles [C]," in *IEEE 36th Conference on Power Electronics Specialists*, 2005, pp. 1881-1886.

[9] R. Zhao and B. Zhang, "A double-voltage-loop controller for two-stage converter with low voltage and high current [J]," *Power Electronics*, vol. 37, no. 6, pp. 78-79, 2003.

[10] X. Y. Ren and X. B. Ruan, "Two-stage converter applicable to high voltage input low voltage output case [J]," in *Proceedings of the CSEE*, 2005, vol. 25, no. 23, pp. 153-157 .(Chinese)

[11] S. W. Du and R. J. Fang, "Research on a novel bidirectional DC/DC converter," *Power Electronics*, vol. 40, no.3, pp45-47, 2006. (Chinese)

[12] Y. Zh. Xu, "A improvement boost ZVT converter based on sliding mode control [D]," Wuhan, Huazhong university of science and technology. 2002. (Chinese)

978-1-4244-1871-8/07 $25.00 © 2007 IEEE

A new technique to achieve Zero Voltage Switching in

Resonant Reset Single Switch Forward Converter

NAGESH VANGALA
CHIRRA ELECTRONICS POWERLABS LTD
BANGALORE 560062 INDIA
Email: nagesh@chirra.in

RAYUDU MANNAM
CHIRRA ELECTRONICS POWERLABS LTD
BANGALORE 560062 INDIA
Email: rayudu@chirra.in

Abstract: Zero voltage switching (ZVS) or soft switching is gaining prominence for a variety of reasons. The benefits and the limitations of ZVS are well documented in the literature. Though the true ZVS is possible in full bridge converters, special design considerations need to be addressed in forward converters to achieve true ZVS.

In this paper a new technique is proposed to achieve true ZVS without the loss of duty cycle or increased primary circulating currents in a single switch forward converter.

Resonant reset technique is adapted for the transformer reset and the same is extended to achieve ZVS .This principle is demonstrated by developing a practical 240 watts DC to DC converter operating at 250 KHz and from a high voltage 400V DC bus. The measured data depict an improvement in the converter efficiency.

1.0 -- Introduction.

Buck derived forward converters are the most popular switching regulators in the practical DC-DC power conversion. Even now, many articles are published in the literature citing various improvements [1, 3]. With increased demand for higher power densities and larger power consumption, soft switching or ZVS/ZCS has become mandatory.

Various techniques are reported on implementing soft switching. Active clamp is one such technique in forward converters [5, 6]. In order to achieve a true ZVS in forward converter, a few limitations exist in such methods, concerning increased primary currents, erosion of effective duty cycle, etc.

In this paper, a new scheme, that results in the complete zero voltage switching of the primary switch, without sacrificing the duty cycle, is presented. In the proposed scheme, the forward converter is operated in the resonant reset mode (for power transformer reset) and the same is extended to achieve the ZVS. An auxiliary switch is incorporated in the secondary side to achieve ZVS. The auxiliary switch turns off in ZCS mode while the turn on is hard switched. With an appropriate transformer design, the main switch can also be turned off in the ZVS mode, thus virtually eliminating the switching losses. Mainly, this scheme does not warrant an extra resonant inductor and hence the duty cycle is not eroded.

2.0 -- Forward Converters--- An overview
Conventional forward converter is depicted in Fig 1.

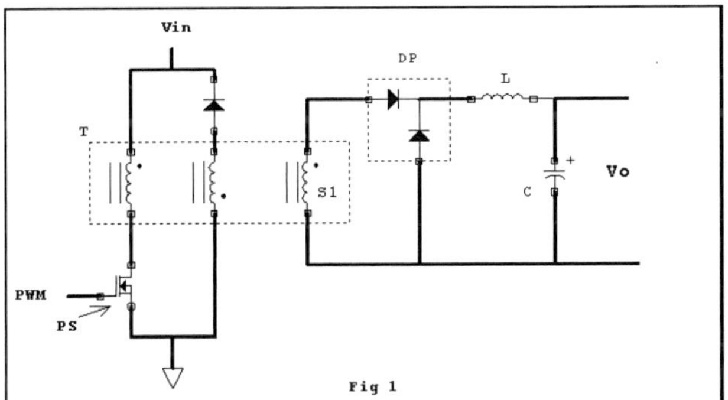

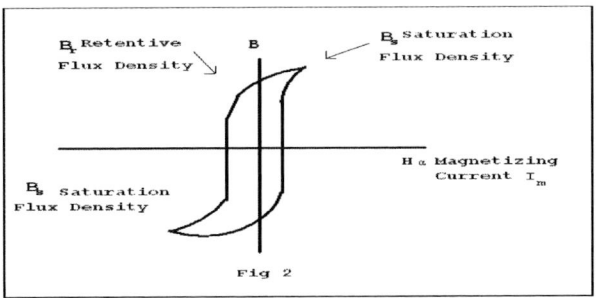

Fig 1

Forward converters are very commonly employed in the industry up to the output power levels of about 500 watts. Good understanding of the dynamics of this mode coupled with simplicity and low output ripple factors and relatively lower stresses on the semiconductor devices are a few key factors for the popularity of this topology. Also, the easy and variety of control ICs availability has further propelled the usage. We being in the business of manufacturing Switch mode power supplies and DC to DC converters mostly use only forward topology for all the commercial applications.

The forward converter has a single primary power switch PS referred to low side and an isolation power transformer to step down the input voltage. Secondary has a diode pack DP, inductor L and filter capacitor C to derive the required dc voltage.

The power transformers are helpful in transferring the power from one port to another with galvanic isolation and at varied potential levels. They are generally two or multi port devices with one of them being the input port. Input port is normally excited with alternating current voltage so that the transformer generates time varying alternating flux in the core material and generates the voltages in the secondary ports. Typical magnetic Characteristic of core materials is termed as B-H curve and the same is shown in Fig 2.

Fig 2

978-1-4244-1871-8/07 $25.00 © 2007 IEEE

The magnetizing current I_m in the primary takes the core in to 1st and 3rd quadrants of the B-H curve alternately and thus induces the voltages in the secondary of the transformer. However it is interesting to note that in the majority of the DC-DC converters, the transformer is excited with a DC voltage as is the case even with Forward converters.

Therefore, the magnetizing current I_m is only in one direction and the core remains in 1st quadrant only. Hence, it is essential to bring back the operating point to Br value at the beginning of each switching cycle. If not, the operating flux in the transformer progressively builds up in one direction only and ultimately ends up at Bs which is a saturation point and the transformer ceases to function.

The mechanism of bringing back the operating point to Br value each time before the start of a new cycle is called the Reset mechanism in forward converters. Generally there are three well known schemes adapted to achieve the transformer Reset in forward converters. They are a) Using a separate Reset winding on the core b) R-C-D clamp reset and c) Resonant Reset technique.

3.0 -- Reset Winding

A forward converter with a separate reset winding P2 is shown in Fig 3. P1 is the primary of the transformer, P2 is the reset winding and S1 is a secondary winding. Whenever the MOSFET switch Sw is turned ON, the transformer primary sees a constant input voltage and the transformer magnetizing current I_m raises linearly. The power is simultaneously transferred to secondary also. The diode D1 is reverse biased owing to the polarity of the reset winding.

When the primary MOSFET switch Sw is turned OFF, the polarities of all the windings reverse, due to fly back action and the diode D1 conducts thus clamping the reset winding to input voltage. The magnetizing current I_m gets transferred to P2 and gradually starts decaying. When the complete magnetizing energy is put back in to input, the I_m becomes zero and the core is said to be reset at this point of time. Transformer is ready to take another switching cycle.

The current and voltage waveforms are shown in Fig 4. Normally the turns-ratio between primary and the reset winding is set at 1:1 so that the maximum permissible duty cycle is 50%. It can be clearly seen that the core operates only in the 1st quadrant of B-H loop.

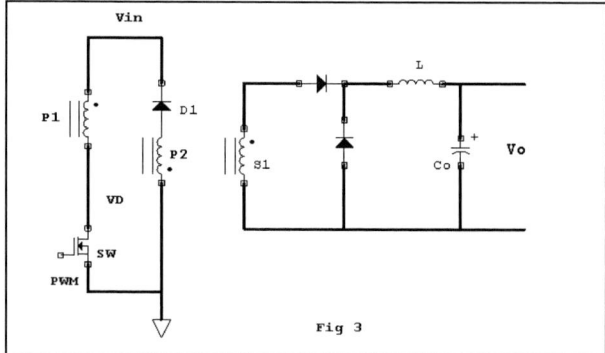

Fig 3

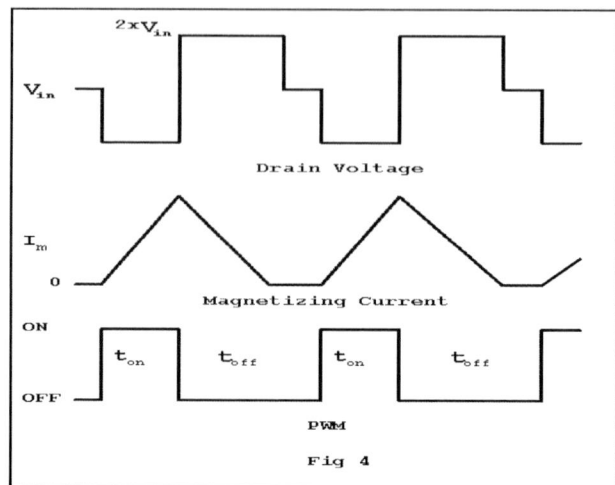

Fig 4

4.0 -- R-C-D clamp or Resonant Reset Scheme

Basically, the operating principle for both the R-C-D Clamp as well as Resonant Reset scheme is the same. Hence, the resonant reset scheme is explained since the same is extended for achieving zero voltage switching in the present work.

Resonant reset scheme is a very simple solution in terms of practical implementation since it has no extra components for reset and the transformer does not have another winding. Further, it exploits the parasitic of the circuit to achieve the reset. The basic schematic of a resonant reset forward converter is shown in Fig 5.

When the MOSFET switch, Sw is turned ON, a constant input voltage V_{in} is impressed across the primary of the power transformer. The magnetizing current I_m increases linearly and is governed by $I_m = (V_{in}/L_m) * t_{on}$, where ton is the ON time of the switch and Lm is the primary magnetizing inductance.

In Fig 5, C_{eff} is the effective capacitance connected to the drain of the switch Sw which includes the output capacitance of MOSFET, C_{oss}, transformer inter winding capacitance and any other capacitances reflected to primary side. When the power switch is turned OFF, the magnetizing current together with reflected load current flowing in primary starts charging the Ceff. The drain voltage begins to increase linearly till it reaches the level of input voltage. The raise is linear because of the load current component, reflected to primary will be fairly large compared to magnetizing current. But, once the drain voltage attains the level of input voltage, the transformer sees zero volts across the primary and hence the secondary voltage also becomes zero.

At this instant, the load current portion reflected to primary side of the transformer vanishes and only the magnetizing current remains in the primary side. The drain voltage still continues to raise, fuelled by the energy in the magnetizing inductance. But, the raise is not linear and instead is resonant. The magnetizing energy stored in the Lm is transferred in to Ceff and the drain voltage continues to increase. This will reach a peak value of V_m at which point of time the magnetizing current Im becomes zero. This is the point where in the complete magnetizing energy is transferred to Ceff. From here on, Ceff, which is charged to much higher voltage than the V_{in}, starts discharging through Lm. This also is a resonant transition, with Lm and Ceff being the key resonant elements.

The energy in the Ceff is being transferred to the Lm. It is important to note that the direction of I_m is reversed now in the Lm and the core is being taken into 3rd quadrant of the B-H curve. Ceff continues to discharge and the drain voltage starts falling in a resonant fashion. When the drain voltage reaches the value of V_{in}, resonant transition stops because the transformer effectively sees a short circuit (zero voltage across it) and the magnetizing inductance Lm is short circuited. The I_m which was a negative entity in the primary side now gets transferred to secondary side.

Assuming a continuous conduction mode for the output filter choke, both the diodes conduct in the secondary which effectively carry the difference of the load current and the reflected magnetizing current. This poses a virtual short circuit across the secondary and is in tune with the primary condition. The drain voltage at this point is equal to input voltage. The relevant wave shapes are shown in Fig 6. The I_m is negative and is constant till the start of next cycle. Though the transition is resonant, while Lm is transferring energy in to Ceff, the waveform shown is approximated to linear current decay for convenience.

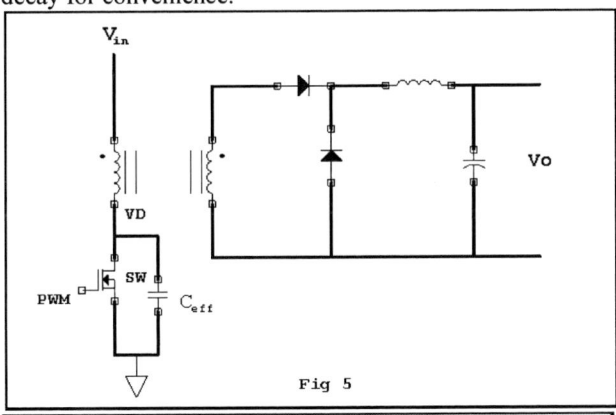

Fig 5

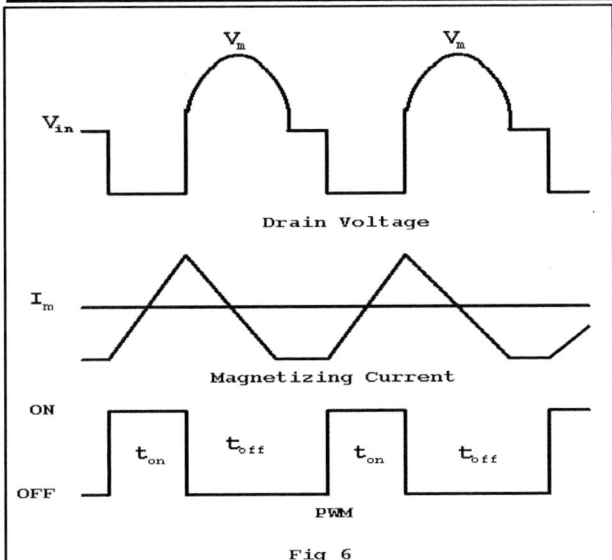

Fig 6

The maximum value of drain voltage V_m should be limited to less than the absolute maximum value of $V_{dss\ max}$ of MOSFET SW. The simplified resonant equation is obtained by equating the energy stored in the magnetizing inductance L_m and the energy transferred in to Ceff capacitance over one switching cycle. A very detailed analysis and comprehensive mathematical equations are well presented in [2, 8]. However, the following equations are presented to grasp the resonant reset component selection. These are depicted keeping in mind the practical industry personnel, rather than the academicians.

At the end of ON period of the primary MOSFET switch, t_{on}, the magnetizing current I_m is given by,

$$I_m = \frac{1}{2} \frac{Vin}{L_m} t_{on} \qquad (1)$$

The energy stored in the magnetizing inductance is given by,

$$\frac{1}{2} L_m I_m^2 \qquad (2)$$

The effective capacitance Ceff is charged in a resonant fashion, during the MOSFET switch OFF time, and its voltage raises from V_{in} to V_m.
Therefore energy transferred to the C_{eff} is given by,

$$\frac{1}{2} C_{eff} (V_m - V_{in})^2 \qquad (3)$$

Equating the energy in Lm to the energy delivered to C_{eff},

$$\frac{1}{2} L_m I_m^2 = \frac{1}{2} C_{eff} (V_m - V_{in})^2 \qquad (4)$$

$$\sqrt{L_m . C_{eff}} = \frac{t_{on}}{2 (V_m/V_{in} - 1)} \qquad (5)$$

5.0 -- True ZVS in Forward Converters

As it was discussed in earlier sections, the drain voltage of primary power switch remains at the level of input voltage, after the transformer core is reset. During the OFF period of the primary switch, when the drain voltage reaches the input voltage level, the transformer primary is effectively short circuited and thus the magnetizing inductance Lm sees zero volts. The primary magnetizing current is transferred to secondary and it remains at a constant value because of the zero volts applied across secondary. This condition abruptly stalls the resonant transitions, and hence the primary power switch voltage remains at the input level till the next switching cycle starts. This is the main reason for not achieving ZVS at turn ON in the forward converters.

To overcome this problem, generally an extra resonant inductor is added in the secondary side or in the primary side. This erodes the effective duty cycle and also demands higher primary peak currents thus causing higher conduction losses.

6.0 -- The Proposed Scheme

If the transfer of magnetizing current to secondary is prevented during the entire OFF period of the primary switch, then the energy available in the Lm can be effectively utilized to discharge the Ceff to zero volts. This creates an ideal condition for ZVS. Therefore, at the secondary an auxiliary MOSFET switch Q2 is connected in series to the forward diode D1, in the proposed scheme, as shown in Fig 7.

The auxiliary switch is controlled by the same PWM input as that of primary MOSFET switch Q1. The gate of the MOSFET is provided with an external PWM input signal.

The operation of the auxiliary MOSFET switch is also controlled by means of PWM input.

The timing diagram of PWM input signal of the primary MOSFET switch as well as PWM input signal of the auxiliary MOSFET Switch are shown in Fig 8b. The ON and OFF time phase of the PWM input of the auxiliary switch is same as the ON and OFF time of the PWM input of the primary MOSFET switch , the only difference being that the auxiliary switch is triggered OFF after a certain delay time as compared to that of the primary MOSFET switch . The amount of delay time is equivalent to the reverse recovery time of the diode D1 which is in series with the auxiliary switch.

During the time when the primary MOSFET switch is OFF, then the secondary Q2is also OFF, thereby ensuring that the transformer secondary is virtually open circuited. Therefore, during OFF time, when the drain voltage of the primary MOSFET reaches input voltage, there will be no passage available for the magnetizing current to travel from the primary winding to the secondary winding of the transformer, since the secondary winding is effectively open circuited. Therefore, the energy transferred to Lm by Ceff discharging to V_{in} from a voltage of V_m, is completely available to discharge Ceff from Vin to zero volts. In this phase, the current in Lm, (which in fact is in the third quadrant of the B-H curve) reduces the voltage on Ceff. This ensures that the drain voltage can be truly zero when the Ceff is completely discharged.

During the time when the primary MOSFET switch is ON, the secondary auxiliary switch is also ON, thereby ensuring the normal operation and not disturbing the forward power transfer to the load.

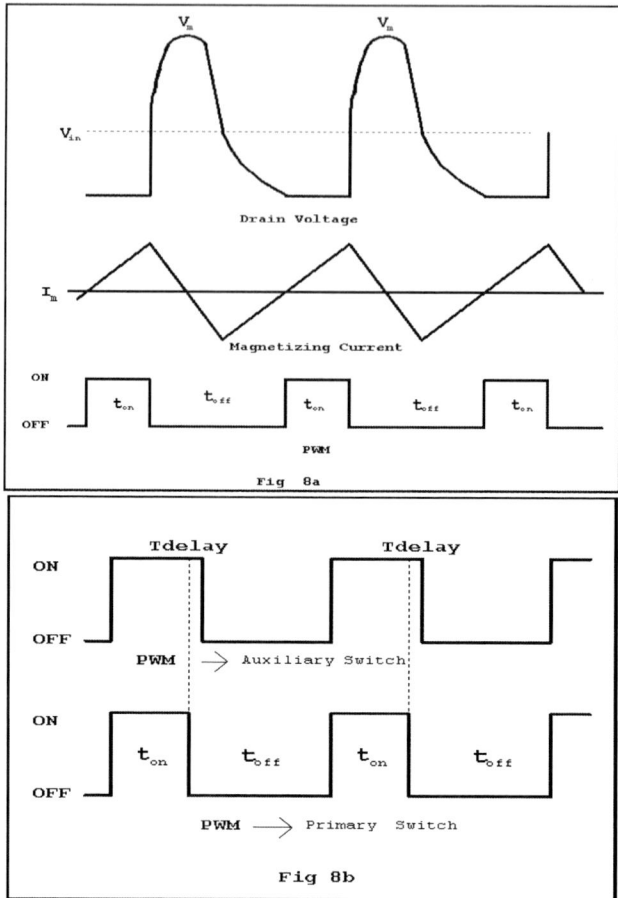

Fig 8a

Fig 8b

The turn OFF time of the auxiliary switch is delayed by certain duration of time even after the turn OFF of the primary MOSFET switch. This delay period that is provided is approximately equal to the reverse recovery time of the diode D1. During the OFF time the series diode D1 anyway is going to block the reverse current after a finite recovery time. In such conditions if the auxiliary switch is turned OFF, then it turns OFF with zero current. This avoids all the turn OFF losses in the auxiliary switch. Fig 8a shows the expected drain voltage, magnetizing current with reference to the PWM input signal. It is important to notice that the magnetizing current is starting at zero value at the instant of turn on of main MOSFET Q1.

In the practical, commercial environment, active power factor correction circuit has become mandatory in all the Offline Smps of above 100watts. This is due to strict EN norms to meet the total harmonic distortion of input current to acceptable limits.

Power factor correction circuit is a boost regulator with a typical output of 400V DC, which is fairly regulated. In such cases the down stream forward converter can be effectively optimized and the transformer can be designed to meet the requisite magnetizing inductance values to facilitate a true ZVS for the primary switch. It is also possible to add an extra capacitor across the drain and source of primary MOSFET

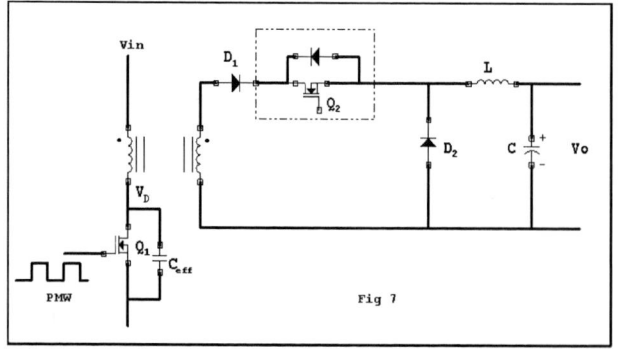

Fig 7

978-1-4244-1871-8/07 $25.00 © 2007 IEEE   610

switch and still design the power transformer to have requisite magnetizing inductance to achieve ZVS.

The addition of this capacitor would help in achieving zero voltage switching during the OFF time also of the primary switch. This virtually eliminates all the switching losses and improves the efficiency, reliability and reduces the Electromagnetic Interference (EMI).

7.0 -- Results

To evaluate the efficacy of the proposed ZVS scheme, a practical converter was built implementing the proposed scheme. Converter was a DC-DC converter with an input voltage of 400V DC. The output was an industry standard 24V DC. Output power was rated to 240 Watts. Switching frequency was selected at 250 KHZ.

Texas instruments PWM IC UCC 25706 was selected as the controller and the main power MOSFET switch was IRF PG 50. A reduction of 4.0 watts of input power is recorded with the proposed scheme. The observed data are presented below.

The waveforms thus obtained are shown in Fig 9a and 9b. An external capacitor of value 100 pf, 2kv rating was connected across drain and source of the primary MOSFET switch. Fig 9a shows the drain waveform when the ZVS is not present and Fig 9b shows the true ZVS achieved with the proposed scheme.

Fig 10 shows the photograph of the converter built with the proposed scheme.

With the proposed ZVS:

Input Voltage to converter	400V DC
Input current	0.667 A
Output Voltage	24.15 V DC
Output current	10.00 A
Efficiency	90.5 %

In the same circuit, the ZVS scheme was disabled by short circuiting the Auxiliary switch drain and source. Following data were recorded:

Input voltage to converter	400V DC
Input Current	0.676 A
Output Voltage	24.12 V DC
Output Current	10.00 A
Efficiency	89.2 %

Without ZVS, a power loss of about 9.0 watts was estimated in the primary power MOSFET IRF PG 50. With the implementation of ZVS, the power loss was estimated to be around 4.0 watts. However approximately a power loss of about 1.5 watts was estimated in the Auxiliary switch introduced in the secondary side of the converter for achieving the ZVS.

Following observations are quite relevant in the present context.

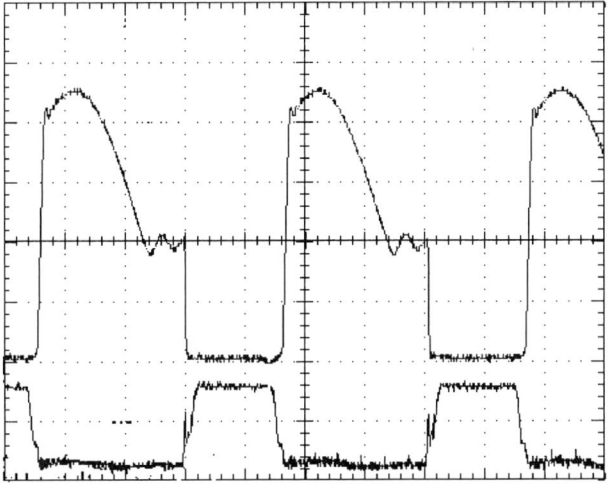

Drain Waveform without Zero Voltage Switching
Fig 9a

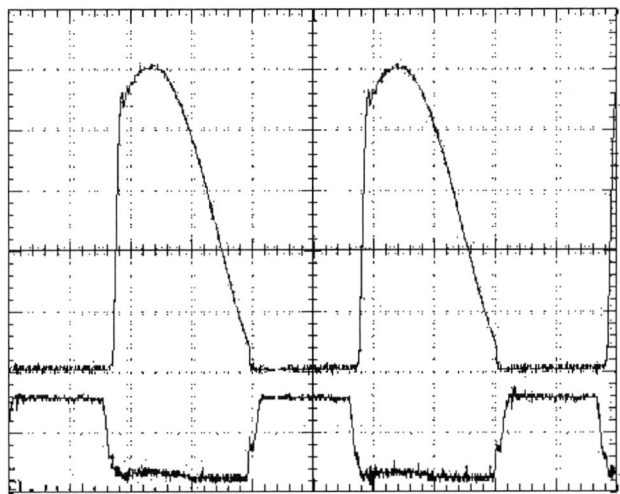

Drain Waveform with Present Zero Voltage Switching Scheme
Fig. 9b

Fig 10

1) The peak value of magnetizing current can be assumed to be half of the total swing as given in (1) in a normal resonant reset forward converter. However, with the proposed ZVS scheme, the magnetizing current starts ramping upwards from a zero value at the instance of Switch ON of the primary Switch, unlike from a negative value as in resonant reset scheme. Therefore, the magnetizing current peak value can be two times in magnitude compared to no ZVS condition. Nevertheless, the current swing is still bidirectional.

2) The maximum attained drain voltage of the primary switch is higher in the ZVS scheme compared to no ZVS condition, all other circuit and input/output parameters being the same. This is again due to higher magnitude of magnetizing current in ZVS scheme. The peak value of the drain voltage recorded practically is 1000 Volts in the ZVS condition and the same was 900 Volts without ZVS.

REFERENCES

[1]. Evandro Soares da Silva,Ernane Antonio Alves Coelho, et al, " A Soft- Single-Switched Forward Converter with Low Stresses and Two Derived Structures", IEEE Transactions On Power Electonics,Vol.19,No. 2, pp: 388 - 395 Mar 2004.

[2]. Biilent Bilgin, et al, "Resonant Demagnetization PWM Forward Converter", Turkey Journal Of Electrical Engg, Vol.11, No: 1, pp: 61-74 No 1, 2003.

[3]. Christopher Bridge, "The Implication of Synchronous Rectifiers to the design of Isolated, single-ended forward converters" Power supply design seminar Unitrode, pp: 7-1 to 7-18, 2001.

[4]. Youhao, Praveen K. Jain,et al, "A Self Core Reset and Zero Voltage Switching Forward Converter Topology", IEEE Transactions On Power Electronics,Vol.15,No 6, pp: 1192 - 1203 Nov 2000.

[5]. Bill Andreycak, " Active Clamp and Reset Technique enhances forward converter performance", Power supply design seminar Unitrode, pp: 6-1 to P6-18, 1997

[6]. Dhaval Dalal, "Design Considerations for active clamp and reset technique", Power supply design seminar Unitrode, pp: A4-1 to4 -23 1997.

[7]. Bill Andreycak, "Designing a phase shifted zero voltage transition power converter", Unitrode switching regulated power supply design seminar manual, PP3-1 to 3-15, 1993

[8]. Naoki Murakami, Mikio Yamasaki, "Analysis Of A Resonant Reset Condition for a Single- Ended Forward Converter", PESC, pp. 1018-1023 Apr 1988.

Design of Bidirectional PWM Sepic/Zeta DC-DC Converter

In-Dong Kim,
Pukyong Nat. Univ.
Dept of Elec. Eng.
Busan, Korea,
idkim@pknu.ac.kr

Young-Ho Lee, Byoung-Ho Min
Pukyong National University
Dept of Electrical Eng.
Busan, Korea,

Eui-Cheol Nho
Pukyong Nat. Univ.
Dept of Elec. Eng.
Busan, Korea,
nhoec@pknu.ac.kr

Jin-Woo Ahn
Kyungsung Univ.
Dept of Elec Eng.
Busan, Korea,
jwahn@ks.ac.k

Abstract- **Bidirectional DC-DC converters allow transfer of power between two dc sources, in either direction. Due to their ability to reverse the direction of flow of power, they are being increasingly used in many applications such as battery charger/dischargers, dc uninterruptible power supplies, electrical vehicle motor drives, aerospace power systems, telecom power supplies, etc. This paper proposes a new bidirectional Sepic/Zeta converter. It has low switching loss and low conduction loss due to auxiliary communicated circuit and synchronous rectifier operation, respectively. Because of positive and buck/boost-like DC voltage transfer function (M=D/1-D), the proposed converter is desirable for use in distributed power system. The proposed converter also has both transformer-less version and transformer one. The paper also suggest some design guide line for the proposed converter.**

I. INTRODUCTION

Bidirectional dc-dc converters allow power flow between two dc sources in either direction. They can reverse the direction of current flow, and thereby power flow while maintaining the voltage polarity of both source ends unchanged. In some necessary case, high-frequency transformer for galvanic isolation can be incorporated inside them. Due to such good features, bidirectional dc-dc converters are being increasingly used in applications such as battery charger and discharger, fuel cell applications, dc uninterruptible power supply, aerospace power systems and motor drives.

Bidirectional dc-dc converters for use in battery charger and discharger are not only to control the battery charging and discharging current, but also to regulate the output voltage of discharger to a predetermined value when again using the stored energy in the battery. In that case, if unidirectional dc-dc converter is adopted instead of bidirectional converter,

two separate unidirectional converters should be provided for either direction.

But the use of bidirectional half-bridge dc-dc converters in[1-2] for low power application and fuel cell applications, or bidirectional buck/boost converters in [4] for aerospace power systems can simplify the system structure, thus resulting in reduction of system volume, weight and cost. Especially in case of dc UPS suggested in [3], the bidirectional buck/boost converter may improve system efficiency as well.

In addition, the use of bidirectional dc-dc converter in the dc link of brushless dc motor drive dedicated to EVs allows the control of both motoring and regenerative braking operations [5]. It can contribute to significant improvement of the motor drive performance. For motoring operation, the bidirectional buck-boost converter in [5] is used to adjust the inverter input voltage and motor speed to reduce the ripple of motor current. For regenerative braking, the bidirectional converter is to control power flow reversal so that a significant amount of the vehicle kinetic energy can be recovered to battery or additional ultra-capacitor tank, thereby resulting in good system efficiency.

As examined above, bidirectional dc-dc converters make system simplification, efficiency improvement and performance upgrade a lot possible. The bidirectional dc-dc converters presented so far are as follows;

(1) Bidirectional buck/boost DC-DC converter [4],
(2) Bidirectional quasi-resonant DC-DC converter [6],
(3) Bidirectional buck-boost DC-DC converter [5],
(4) Bidirectional Cuk DC-DC converter [7, 11],
(5) Bidirectional flyback DC-DC converter [8],
(6) Bidirectional half bridge DC-DC converter [1-2],
(7) Bidirectional full-bridge DC-DC converter [9]

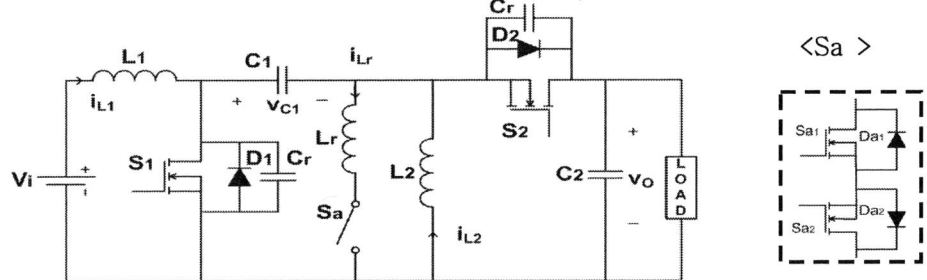

Fig. 1 Proposed bidirectional ZVS PWM Sepic/Zeta DC-DC converter

978-1-4244-1871-8/07 $25.00 © 2007 IEEE   613

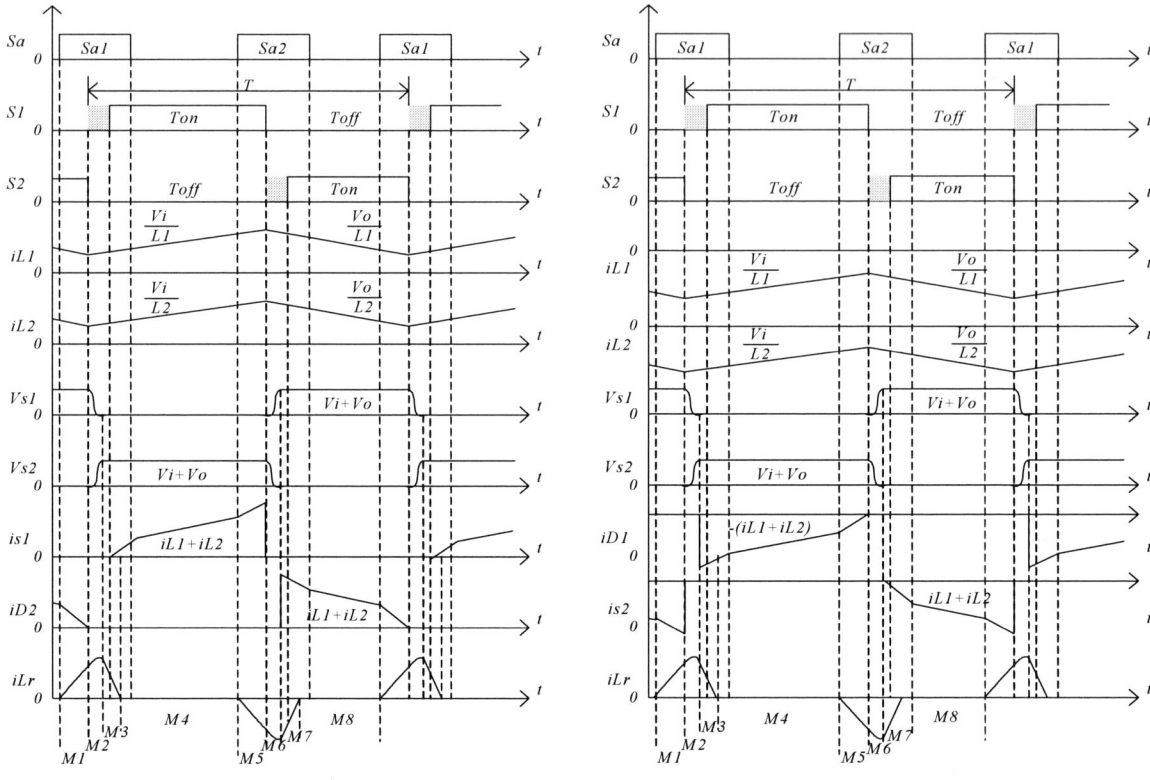

(a) Forward operation(Sepic)   (b) Backward operation(Zeta)

Fig. 2 Operating waveforms of the proposed converter

These bidirectional converters have their own features and applications. But so far, bidirectional Sepic/Zeta dc-dc converter has not fully been researched which could be regarded as a kind of combination of basic Sepic and Zeta converter. So, this paper proposes the new bidirectional Sepic/Zeta dc-dc converter. The proposed converter does not only have the capability of bilateral power flow, but also low switching loss and conduction loss due to zero voltage switching and synchronous rectifier operation. Furthermore, thanks to positive and buck/boost-like DC voltage transfer function (M=D/1-D), the proposed converter is desirable for use in distributed power system. The proposed converter also has both transformer-less version and transformer one.

II. PROPOSED BIDIRECTIONAL ZVS PWM SEPIC/ZETA DC-DC CONVERTER

The power stage circuit diagram of the proposed converter is shown in Fig. 1. The circuit can be divided into two parts, that is, hard-switched bidirectional Sepic/Zeta dc-dc converter and auxiliary resonant commutated pole circuit. The hard-switched bidirectional Sepic/Zeta dc-dc converter is also a new bidirectional dc-dc converter as a skeleton of the proposed converter. It operates like a conventional

hard-switched Sepic converter for forward power flow, and like a Zeta converter for backward power flow. The main switches S_1 and S_2 switch on and off with anti-parallel connected diodes D_1 and D_2, respectively, carrying out synchronous rectifier operation. The inductor current through L_1 and L_2 are positive for forward power flow and vice versa.

The auxiliary resonant commutated pole circuit: ARCPC) is composed of a small MOSFET St, and resonant inductor Lr. The ARCPC circuit provides the hard-switched bidirectional Sepic/Zeta dc-dc converter with ZVS operation, and is rated for a small power when compared to the output power. The ZVS operation for both the main MOSFET devices and diodes results in improved circuit efficiency. Furthermore since the ZVS switching occur during a short time when compared to the switching period, it does not have significant influence on the overall output characteristics of basic AC/DC converter such as PWM control and linear input and output conversion characteristics.

The forward operation of proposed converter is similar to the operation of Sepic converter and thus can be named Sepic mode operation here. Its operating waveforms are shown in Fig. 2(a). The backward operation of proposed converter is similar to the operation of Zeta converter. It is also named as Zeta

978-1-4244-1871-8/07 $25.00 © 2007 IEEE   614

mode operation. Its operating waveforms are shown in Fig. 2(b).

When looking into the forward(Sepic mode) and backward(Zeta) operation of the proposed converter in more detail, its operation can be divided into 8 operation modes for one switching period, respectively, as shown in Fig. 2(a) and (b).

III. DESIGN OF PROPOSED BIDIRECTIONAL ZVS PWM SEPIC/ZETA DC-DC CONVERTER

In order to verify the effectiveness of the proposed converters and suggest the design guideline in real applications, the prototype converter is properly designed with rated power 1.0[kw], controlled output voltage 100[V], input voltage range of 40~60[V], and switching frequency 40[kHz] above acoustic noise frequency.

The proposed converter has the auxiliary resonant commutated pole circuit(ARCPC) composed of a small MOSFET St, and resonant inductor Lr. But the auxiliary circuit has little effect on the overall PWM operation characteristics of hard switching PWM Sepic/Zeta converter. So, in steady-state, the average values of inductor currents i_{L1} and i_{L2} are equal to zero, resulting in the following equation (1)

$$V_i = V_{c1} \tag{1}$$

The voltage across the inductor L_2 during on-state time T_{on} and off-state time T_{off} of switch S_1 is V_i and Vo, respectively. Since the flux increment of inductor L_2 during on-state time is equal to its flux decrement during off-state time, the following equation can be obtained.

$$d\,Vi = Vo(1-d) \tag{2}$$

Thus, the voltage conversion ration V_o/V_i of proposed bidirectional ZVS Sepic/Zeta converter can be expressed as

$$\frac{Vo}{Vi} = \frac{d}{1-d} \tag{3}$$

where the duty cycle d of switch S_1 can be defined as d = Ton/T where Ton and T is the on-state time and switching period of switch S_1, respectively.

In addition, as the steady-state power balance is satisfied,

$$V_i I_{L1} = V_o I_o . \tag{4}$$

Thus,

$$\frac{V_o}{V_i} = \frac{I_{L1}}{I_o} \tag{5}$$

By combining equations (3) and (5), we can get the following equation (6).

$$\frac{I_{L1}}{I_o} = \frac{d}{1-d} \tag{6}$$

On the other hand, when S_2 is in on-state, the inductor current I_{L1} and I_{L2} flow together through S_2

to the output side, and thus the following equation can be obtained.

$$I_o = (I_{L1} + I_{L2})(1-d) \tag{7}$$

From equations (6) and (7),

$$I_{L2} = \frac{1-d}{d} I_{L1} = I_o. \tag{8}$$

Since the input voltage ranges from 40[V] to 60[V] and the output voltage is to be controlled to 100[V], the variation range of duty cycle d in practical converter control is as follows;

$$0.63 < d < 0.71. \tag{9}$$

In this condition, the average current value $I_{L1.avg}$ through the inductor L_1 has maximum value at $d = 0.63$, being expressed as

$$I_{L1.avg} = \frac{1000\,[W]}{40\,[V]} = 25\,[A]. \tag{10}$$

By using the above derived equation (8), the maximum value of the average inductor current through L_2 is $I_{L2.avg} = 15\,[A]$. When considering the ripple factor of 20% in the inductor current through L_1 and L_2, the maximum current value $I_{s.max}$ of switches S_1 and S_2 is given as

$$I_{s.max} = 1.2 \times (I_{L1.avg} + I_{L2.avg}) = 48\,[A]. \tag{11}$$

On the other hand, the switch voltage V_{s1} across S_1 is

$$V_{s1} = V_i + V_o = 160\,[V]. \tag{12}$$

Thus, MOSFETs IXFN130N30 with rated current 130[A] and rated voltage 300[V] are selected for experimental prototype by considering design margin of 100%.

The current waveform through L_1 at the duty cycle d of 2/3 is shown as Fig. 3 and the peak-to-peak value ΔI_{L1} of current ripple through L_1 is given as (13).

$$\Delta I_{L1} = \frac{V_i}{L_1} \times T_{on} \tag{13}$$

When limiting the peak-to-peak current value ΔI_{L1} within 25% of the rated current,

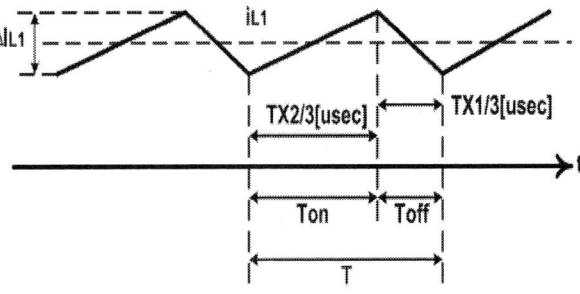

Fig. 3 Inductor current i_{L1} at d =2/3

978-1-4244-1871-8/07 $25.00 © 2007 IEEE

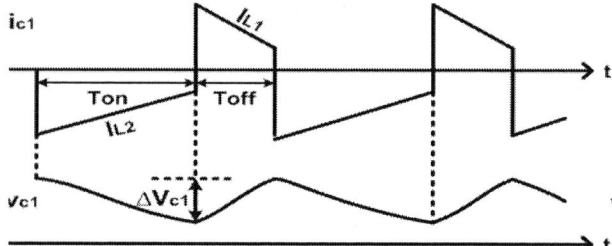

Fig. 4 Current waveform i_{C1} and voltage waveform v_{C1} of capacitor C_1

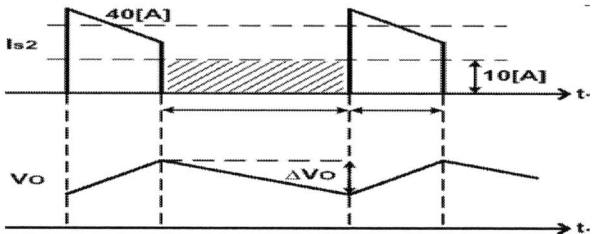

Fig. 5 Current waveform I_{s2} of MOSFET S_2 and voltage waveform V_O of capacitor C_2

$$\Delta I_{L1} = \frac{50\,[V]}{L_1} \times 25\,[\mu sec] \times \frac{2}{3} = 25\,[A] \times 0.25 \quad (14)$$

Thus, by solving (14) and considering design margin, the designed circuit parameter of inductor L_1 is selected as 133[uH] with saturation current of 25[A]. In addition, the circuit parameter of L_2 is also selected equal to that of L_1, but with the different saturation current of 15[A].

When neglecting the circuit influence of ARCPC, the voltage and current waveform of C_1 is shown as Fig. 4 .and the peak-to-peak voltage value ΔV_{c1} of capacitor C_1 can be given as (15).

$$\Delta V_{c1} = \frac{1}{C_1} I_{L2} \times T_{on} \quad (15)$$

To limit the peak-to-peak voltage value ΔV_{c1} within 25% of the input voltage, C_1 should be selected as polypropylene capacitor of 40 [uF], 12 [Arms].

Fig. 5 shows the current waveform I_{s2} through switch S_2 and the voltage waveform V_O across output capacitor C_2. The peak-to-peak voltage value ΔV_O of output capacitor C_2 is derived as (16).

$$\Delta V_O = \frac{1}{C_2} (I_{s2} - I_o)(T - T_{on}) \quad (16)$$

To limit the peak-to-peak value ΔV_O within 1% of predetermined output voltage, the output capacitor should be selected as $C_2 = 400\,[\mu F]$.
The ripple current of output capacitor C_2 can be calculated as (17)

$$I_{C_{2.ripple}} = \sqrt{\frac{1}{T} \int_0^T i_{C2}^2 dt} = 17\,[A] \quad (17)$$

Thus, the output capacitor C_2 should be designed as 400 [uF], 17 [Arms].

To ensure ZVS operation of switches S_1 and S_2, the auxiliary circuit parameters C_r and L_r of ARCPC are selected as 23.5 [nF] and 10 [uH], respectively, according to conventional design rules.

Upon turning on the auxiliary switch S_{a1} the auxiliary inductor current i_{Lr} increases linearly with the rising slope of (18).

$$\frac{di_{Lr}}{dt} = \frac{V_o}{L_r} = \frac{100\,[V]}{10\,[\mu H]} = 10\,[A/\mu sec] \quad (18)$$

When $i_{D2} = 40\,[A]$, the time interval Δt_1 of mode 1 is given as (19).

$$\Delta t_1 = \frac{40\,[A]}{20} = 2\,[\mu sec] \quad (19)$$

On the other hand, the resonant period T_r of auxiliary resonant parameters C_r and L_r is given as (20).

$$T_r = \frac{1}{f_r} = 2\pi \sqrt{L_r C_r} \quad (20)$$

Thus, the time interval Δt_2 of mode 2 can be calculated as (21).

$$\Delta t_2 = T_r/4 = 0.76\,[\mu sec] \quad . \quad (21)$$

In addition, the rising increment Δi_{Lr} of the auxiliary inductor current i_{Lr} during mode 2 is equal to (22).

$$\Delta i_{Lr} = \frac{V_i + V_o}{\sqrt{\frac{L_r}{C_r}}} = \frac{160\,[V]}{\sqrt{\frac{10\,[\mu H]}{23.5\,[nF]}}} = 7.75\,[A]. \quad (22)$$

At the beginning point t_3 of mode 3, the auxiliary inductor current i_{Lr} is equal to $i_{Lr}(t_3) = 45\,[A]$. It takes Δt_3 as (23) until the auxiliary inductor current i_{Lr} is reduced to zero.

$$\Delta t_3 = \frac{i_{Lr}(t_3)}{di_{Lr}/dt} = \frac{i_{Lr}(t_3)}{V_i/L_r}$$
$$= \frac{45\,[A]}{60\,[V]/5\,[\mu H]} = 3.75\,[\mu sec] \quad (23)$$

The switching timing diagrams of main switches S_1, S_2. and auxiliary switches S_{a1}, S_{a2} are shown as Fig. 6 to ensure ZVS operation of main switches and ZCS operation of auxiliary switches. The crossing point of reference PWM signal and triangular carrier is the criterion point to turn on and off the main and auxiliary switches. As shown Fig. 6, at the up slope of the triangular carrier, the auxiliary switch S_{a1} is turned on Δt_{be} before the crossing point, thus increasing the stored energy in the auxiliary inductor. The stored inductor energy enables ZVS turn-off operation of switch S_2. After the resonant operation, it also enables ZVS turn-on operation of switch S_1. Finally, after the stored energy is discharged to zero, the auxiliary switch S_{a1} is turned off Δt_{af} after the crossing point. At the down slope of the triangular carrier, the main and auxiliary switch S_{a2} are turned on and off in the same way as the up slope. In the prototype, Δt_{be} and Δt_{af} are properly selected as 2[usec] and 4[usec] considering the time intervals Δt_1, Δt_2, Δt_3 of mode 1, 2, and 3.

978-1-4244-1871-8/07 $25.00 © 2007 IEEE

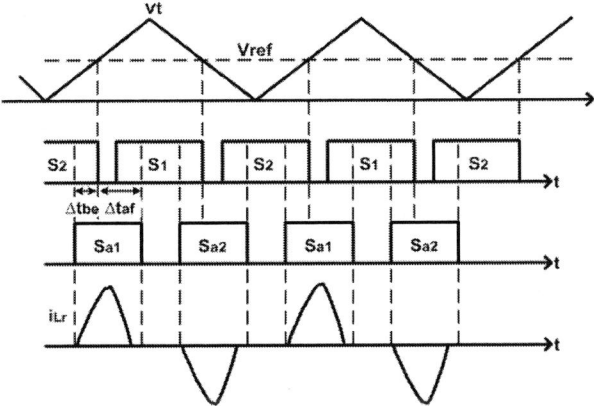

Fig. 6 Switching timing of switches S_{a1}, S_{a2} for ZVS Switching

IV. EXPERIMENTAL RESULTS

In order to verify the operation of the proposed bidirectional ZVS PWM Sepic/Zeta dc-dc converters, the prototype converter was properly implemented with rated power 1.0[kw], output voltage 100[V] and input voltage 40~60[V]. The detailed converter ratings and circuit parameters are shown in table II. The prototype converter is controlled through PWM operation with carrier frequency 40[kHz]. Fig. 7 shows the overall control block diagram of proposed converter. The control block diagram consists of two feedback loop, that is, inner current loop and outer voltage loop. The outer voltage loop is to regulate the output voltage V_o and generate the current command for the inner current loop. The inner current loop is to control the inductor current and generate the reference voltage of PWM generator which supply the PWM gate signal to each MOSFET switch.

Fig. 8(a) shows the measured voltage v_{S1} and current i_{S1} waveforms of switch S_1 in Sepic mode operation. Fig. 8(b) and (c) shows their zoomed up waveforms during turn-on interval and turn-off interval, respectively. They prove that the proposed converter achieves good ZVS switching by using ARCPC circuit. Fig. 9 shows the measured current waveforms i_{L1}, i_{L2} of inductors L_1 and L_2 in Sepic mode operation. It shows that the proposed converter is very well controlled through PWM method. Fig. 10 shows the measured current i_{Lr} of auxiliary resonant inductor Lr and gate voltage $V_{GS}(S_{a1})$ of auxiliary switch S_{a1} in Sepic mode operation. As shown in Fig. 10, the resonant inductor current increases from zero after turning on the auxiliary switch S_{a1}, and then it returns to zero before turning off it. It proves that the auxiliary switch also get good ZCS switching operation.

Fig. 11 shows the measured voltage v_{S1} of switch S_1 and current i_{D1} waveform of diode D_1 in Zeta mode operation. Fig. 12 shows the measured voltage v_{S2} and current i_{S1} waveform of switch S_2 in Zeta mode

operation. As shown in Fig. 11, 12, the proposed converter also operates very well in Zeta mode (backward power flow). Fig. 13 shows the measured efficiency of the proposed converter in both Sepic and Zeta. It shows that the proposed converter has good efficiency.

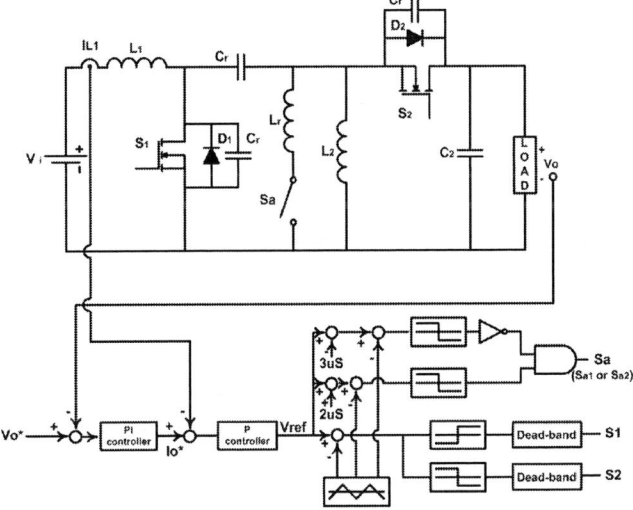

Fig. 7 Control block diagram of the proposed converter

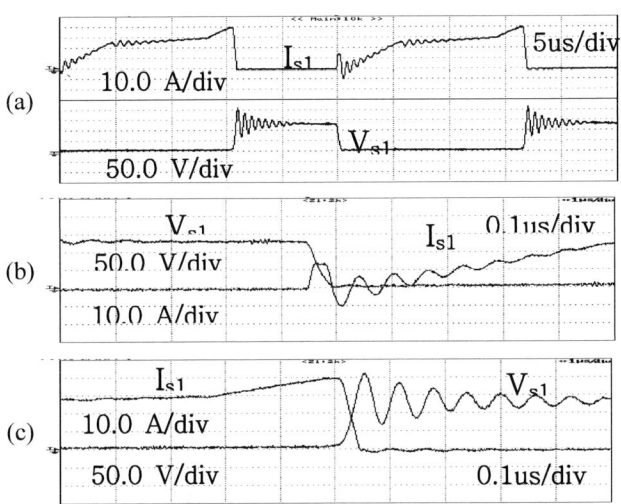

Fig. 8 (a) Voltage v_{S1} and current i_{S1} waveforms of switch S_1 in Sepic mode operation, (b) during turn-on interval, (c)during turn-off interval.

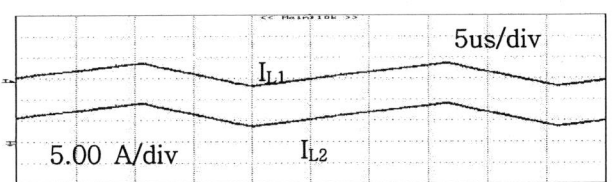

Fig. 9 Current waveforms i_{L1}, i_{L2} of inductors L_1 and L_2 in Sepic mode operation

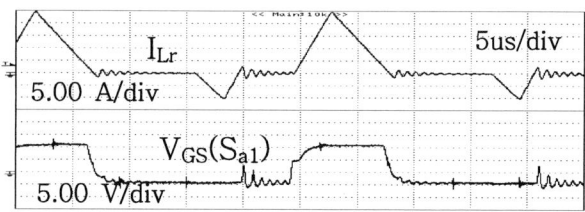

Fig. 10 Current i_{Lr} of auxiliary resonant inductor Lr and gate voltages $V_{GS}(S_{a1})$ of auxiliary switches S_{a1} in Sepic mode operation

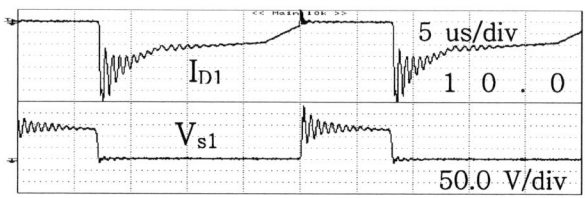

Fig. 11 Voltage V_{S1} of switch S_1 and current I_{D1} waveform of diode D_1 in Zeta mode operation

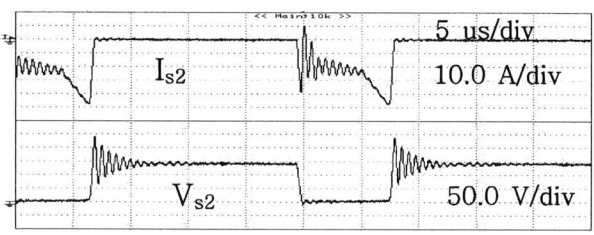

Fig. 12 Voltage v_{S2} and current i_{S1} waveform of switch S_2 in Zeta mode operation

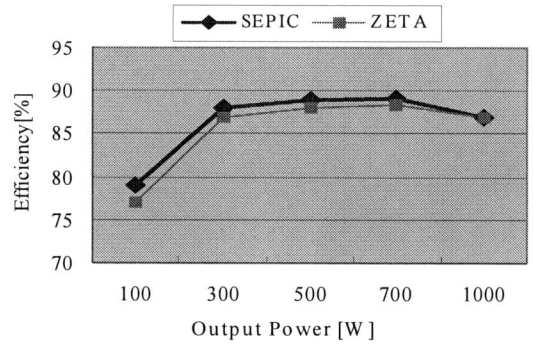

Fig. 13 The measured efficiency of the proposed converter

V. CONCLUSION

This paper proposes a new bidirectional ZVS PWM Sepic/Zeta dc-dc converter. The proposed converter does not only have the capability of bilateral power flow, but also low switching loss and conduction loss due to zero voltage switching and synchronous rectifier operation. Furthermore, thanks to positive and

TABLE 1 EXPERIMENTAL COMPONENT PARAMETERS

Power rating	1.0 [kW]	C_1	40[uF]
V_1	48[V]	C_2	470[uF]
V_2	100[V]	Cr	23.5[nF]
L_1, L_2	133[uH] 25[A}, 15{A]	Lr	10[uF]
fs	40[kHz]	MOSFET	IXYS IXFN130N30

buck/boost-like DC voltage transfer function (M=D/1-D), the proposed converter is desirable for use in distributed power system, battery charger/ dischargers, dc uninterruptible power supplies. The paper also suggest some design guide line for the proposed converter.

ACKNOWLEDGMENT

This work has been supported by KESRI (R2005-B-109), which is funded by MOCIE (Ministry of commerce, industry and energy).

REFERENCES

[1] M. Jain, M. Daniele, and P. K. Jain, "A Bidirectional DC-DC Converter Topology for Low Power Application", IEEE Trans. Power Electronics, vol. 15, no. 4, pp. 595-606, July 2000.

[2] F. Z. Peng, H. Li, G.J. Su, and J.S. Lawler, "A New ZVS Bidirectional DC-DC Converter for Fuel Cell and Battery Application", IEEE Trans. Power Electronics, vol. 19, no. 1, pp. 54-65, Jan 2004.

[3] K.-W. Ma, and Y. S. Lee, "A Novel Uninterruptible dc-dc Converter for UPS Applications", IEEE Trans. Industry Applications vol. IA-28, no. 4, pp. 808-815, July/Aug. 1992.

[4] D. M. Sable, F. C. Lee, and B. H. Cho, "A Zero-Voltage-Switching Bidirectional Battery Charger / Discharger for the NASA EOS Satellite", IEEE APEC Rec., 1992, pp. 614-621.

[5] F. Caricchi, F. Crescimbini, and A. Di Napoli, "20kW Water-Cooled Prototype of a Buck-Boost Bidirectional DC-DC Converter Topology for Electrical Vehicle Motor Drives", IEEE APEC Rec., 1995, pp. 887-892.

[6] B. Ray, "Single-Cycle Resonant Bidirectional DC/DC Power Conversion", IEEE APEC Rec., 1993, pp. 44-50.

[7] J. Majo, L. Martinez, A. Poveda, L. Vicuna, F. Guinjoan, A. Sanchez, M. Valentin, and J. Marpinard, "Large-Signal Feedback Control of a Bidirectional Coupled-Inductor Cuk Converter,"IEEE Trans. Industrial Electronics vol. 39, no. 5, pp. 429-436, Oct. 1992.

[8] G. Chen, Y.-S. Lee, S. Y. R. Hui, D. Xu and Y. Wang, "Actively Clamped Bidirectional Flyback Converter", IEEE Trans. Industrial Electronics vol. 47, no. 4, pp. 770-779, Aug. 2000.

[9] L. Zhu, "A Novel Soft-Commutating Isolated Boost Full-Bridge ZVS-PWM DC-DC Converter for Bidirectional High Power Applications," IEEE Trans. Power Electronics, vol. 21, no. 2, pp. 422-429, March 2006.

[10] G. Spiazzi and L. Rossetto, "High-Quality Rectifier Based on Coupled-Inductor Sepic Topology", IEEE PESC Rec., 1998, pp. 336-341.

[11] D. S. L. Simonetti, J. Sebastian, and J. Uceda, "The Discontinuous Conduction Mode Sepic and Cuk Power Factor Preregulators: Analysis and Design," IEEE Trans. Industrial Electronics, vol. 44, no. 5, pp. 630-637, Oct. 1997.

[12] J. J. Jozwick and M. K. Kazimierczuk, "Dual Sepic PWM Switching-Mode DC/DC Power Converter", IEEE Trans. on Industrial Electronics, vol. 36, no. 1, pp.64-70, Feb. 1989.

[13] R. W. De Doncker and J. P. Lyons, "The Auxiliary Resonant Commutated Pole Converter", IEEE IAS Rec., 1990, pp. 1228-1235.

978-1-4244-1871-8/07 $25.00 © 2007 IEEE

AUTHOR INDEX

Abass, A.	659
Abe, S.	252
Adnani, M.	816
Aguillon-Garcia, J.	1143
Ahn, H.	59
Ahn, I-R.	591
Ahn, J.	505
Ahn, J-W.	613, 649, 911
Ahn, J-Y.	90
Akagi, H.	1
Alahuhtala, J.	741
Alizadeh, A.	1101
Alizadeh, O.	748
Alkoshairy, A.	659
An, Y-J.	649
Andersen, M.	759
Andronescu, G.	351
Antunes, F.	510, 1131
Aoyama, F.	1228
Apte, N.	671
Araujo, S.	510, 1131
Bac, N.	1054
Badstuebner, U.	23
Bae, C-H.	800
Bae, H.	9
Baek, H-L.	527
Baek, J.	988
Baek, S-T.	439
Baik, W-S.	639, 723
Baljinnyam, Z.	169
Bapat, V.	671
Batista, E.	1149
Betz, R.	1217
Biela, J.	23
Borle, L.	955
Cabral Ferreira, A.	737

Caixue, C.	213
Canesin, C.	419
Castellazi, A.	1149
Cha, H.	894, 1090
Cha, I-H.	553
Cha, I-S.	521
Cha, J-S.	524
Chang, L.	597, 601
Chang, S-C.	217
Chang, S-K.	771, 800
Chang, Y-T.	624, 1085
Chen, C-Y.	309
Chen, J.	377, 430, 965
Chen, L.	1090
Chen, P-Y.	225
Chen, X.	921
Chen, Z.	601
Cho, B.	9, 988, 1171
Cho, C.	505
Cho, D.	709
Cho, G.	53
Cho, I.	502
Cho, J-H.	705
Cho, J-R.	243
Cho, K-H.	71, 203
Cho, K-M.	709, 882
Cho, K-R.	1200
Cho, M-T.	122, 217, 221
Cho, Y-O.	527
Choe, G-H.	66, 119, 169, 410
Choi, C-H.	357, 387, 723
Choi, C-L.	333
Choi, E-S.	705
Choi, G-S.	122, 217
Choi, H-J.	521, 524
Choi, I.	450

AUTHOR INDEX

Choi, J. 137, 164, 229, 404, 541

Choi, J-D. 315

Choi, J-S. 367, 372, 521, 524

Choi, J-Y. 450, 502

Choi, K-H. 221, 639

Choi, N-S. 337, 1036

Choi, S. 894, 982

Choi, W. 113

Choi, Y. 66, 71, 229

Chu, C-C. 225

Chun, T-W. 159, 337, 591

Chung, C. 541, 547

Chung, D-H. 367, 372

Chung, G-B. 164

Chung, J-M. 439

Chung, Y-H. 435

Ciappa, M. 1149

Cong, T. 580

Dagdag, S. 346

Dai, Z. 1064

Dasgupta, A. 1048

Dash, S. 413

Datta, M. 444, 828

De Freitas, L. 419

De Haan, S. 85, 192

Deng, L. 690

Dienot, J. 1149

Doi, N. 186

Du, S. 601

Duchesne, C. 346

Dugarjav, B. 169

Dutarde, E. 346

Edington, M. 921

Elgendy, O. 659

Engler, A. 1131

Eshtehardiha, S. 153, 246

Fang, Y. 921

Farhangi, S. 425, 180, 748, 1223

Fathi, S. 586

Fernandes, B. 1005

Fitchner, W. 1149

Freere, P. 899

Fukushima, K. 1064

Funabashi, T. 444, 828, 915

Funato, H. 822

Ge, L. 1154

Geetha, E. 805

Gharehpetian, G. 586

Gil, C-H. 243

Gomez, D. 935

Goto, H. 634

Griva, G. 147

Grzesiak, L. 777, 1042

Guang, Z. 300

Gun, S-D. 521

Guo, H-J. 634

Ha, K-M. 232

Ha, S-J. 271

Halasz, S. 1096

Hamada, T. 398

Hamaogi, M. 1085

Han, B-M. 1036, 1125

Han, D. 59

Han, J-H. 454

Han, M. 295

Han, M-K. 71

Han, S-B. 329

Han, S-T. 686

Han, Y. 541, 547

Haneyoshi, T. 143

Harada, Y. 1064

Hartmann, M. 1158

AUTHOR INDEX

Heising, C. 865

Heo, H-S. 66

Hidesi, K. 1228

Hiraki, E. 1183

Hirokawa, M. 252

Ho, S-K. 225

Holmes, D. 899

Hong, C. 295

Hong, S-C. 197, 713, 1125

Hosseiniaghdam, M. 586

Hou, Q. 870

Hsieh, J-C. 309

Hu, J. 850, 944

Huang, J. 53, 232

Huang, T-C. 309

Huang, X-H. 971

Hwang, H-M. 450

Hwang, L-H. 122, 217, 221

Hwang, T-S. 387

Hyun, D-S. 132, 553, 562, 567, 699, 732, 940, 950

Ichinokura, O. 634

Ion, Y-S. 435

Ise, T. 1165

Ishirobi, M. 1075

Ishizuka, T. 1228

Ito, T. 74

Jakobsen, L. 759

Jang, D-U. 800

Jang, S-H. 482

Jang, S-J. 203, 789, 976

Jang, T. 709

Jeon, H-J. 410, 454, 530

Jeon, J. 505

Jeong, B-C. 497

Jeong, C-I. 404

Jeong, H-G. 329

Jianlin, Z. 213

Jog, A. 671

Joung, S-E. 553, 940

Joung, W-T. 387

Jung, B-M. 329

Jung, H-C. 562, 950

Jung, H-J. 439

Jung, H-M. 271

Jung, H-S. 319

Jung, M. 53

Jung, S-M. 810, 1060

Jung, Y-C. 319

Jung, Y-G. 466, 471, 629

Jung, Y-S. 502

Kadar, I. 1096

Kadir, M. 1194

Kang, E-G. 62

Kang, G-I. 90

Kang, H-H. 439

Kasal, G. 1137

Kawamura, A. 575

Kennel, R. 737

Khai, N. 629

Kim, B-C. 278

Kim, B-H. 517

Kim, C-E. 274, 783, 794, 994, 1001, 1177

Kim, C-H. 783, 794

Kim, C-S. 266

Kim, D. 53, 541

Kim, D-E. 765

Kim, D-H. 90, 639

Kim, D-Y. 994

Kim, E-C. 315, 323

Kim, G. 289, 342

Kim, G-D. 203

Kim, H. 229

AUTHOR INDEX

Kim, H-G. 337, 482, 591

Kim, H-J. 435

Kim, H-S. 66, 122, 221

Kim, H-W. 810, 1060

Kim, I-D. 337, 613

Kim, J. 257, 477, 505, 547

Kim, J-B. 127

Kim, J-C. 66

Kim, J-H. 127, 271, 283, 471, 477, 1171

Kim, J-K. 243

Kim, J-S. 79

Kim, J-U. 319

Kim, J-Y. 976

Kim, K-H. 90

Kim, K-S. 295, 435

Kim, M-H. 357, 639, 723

Kim, N-H. 357, 639, 723

Kim, S. 505, 547

Kim, S-C. 232

Kim, S-H. 119, 278, 283, 463

Kim, S-J. 79

Kim, S-K. 460, 463

Kim, T. 257

Kim, T-J. 132, 477

Kim, T-S. 732, 961

Kim, T-W. 404

Kim, W-R. 530

Kim, W-S. 497

Kim, Y-C. 197, 1125

Kim, Y-D. 274

Kim, Y-G. 800

Kim, Y-R. 381, 460, 463, 929

Kim, Y-S. 169, 217, 387

Kim, Y-W. 107

Kimura, N. 398

Kisck, D. 351

Kisck, M. 351

Kitano, T. 571

Ko, J. 197, 367, 372, 713, 1036, 1125

Koczara, W. 43

Koh, K. 571

Kohama, T. 238

Kolar, J. 23, 888, 1011, 1158

Kong, T-W. 860

Kong, W. 899

Koo, C-H. 315

Koo, J-C. 315, 323

Kristoffersen, A. 1143

Kucuk, F. 634

Kumar, S. 665

Kurokawa, K. 487, 834

Kurokawa, F. 261

Kusuhara, Y. 619

Kwak, D-K. 266

Kwak, S. 333

Kwon, B-H. 79, 127

Kwon, B-K. 530

Kwon, J-M. 728

Kwon, T-S. 686, 694

Laczynski, T. 1107

Lai, J-H. 309

Lai, Y-S. 624

Lauttamus, R. 535

Lebey, T. 346

Lee, B-H. 1177

Lee, B-K. 203, 789, 976

Lee, B-S. 197, 266

Lee, C. 976, 1211

Lee, D-C. 482, 765

Lee, D-H. 649, 911

Lee, D-M. 699

Lee, D-S. 333

AUTHOR INDEX

Lee, H. 174, 289, 342

Lee, H-C. 1001

Lee, H-H. 159, 361, 580, 728

Lee, H-M. 789

Lee, H-S. 62

Lee, J. 113, 137, 164, 229, 257, 1211

Lee, J-B. 694

Lee, J-C. 132

Lee, J-D. 132, 732

Lee, J-H. 119, 381, 387, 439, 783, 794, 860, 929

Lee, J-I. 524

Lee, J-M. 1171

Lee, J-P. 169, 477

Lee, J-S. 62

Lee, J-W. 305, 1189

Lee, J-Y. 713

Lee, K. 487, 834

Lee, K-B. 497

Lee, K-J. 567

Lee, K-O. 502

Lee, K-S. 435

Lee, K-W. 882

Lee, S. 243, 955

Lee, S-D. 295

Lee, S-H. 221, 723

Lee, S-I. 410

Lee, S-J. 1171

Lee, W-C. 381, 860, 929

Lee, W-J. 225, 994

Lee, Y-H. 613

Lei, X. 492

Li, R. 430

Li, Y. 1036

Liang, J. 649

Liang, L. 690

Lijuan, Z. 300

Lim, C-J. 530

Lim, J. 71, 771

Lim, S-B. 713

Lim, S-H. 771

Lim, Y-C. 90, 208, 305, 393, 466, 471, 629

Lin, J. 1031

Lin, Y-S. 309

Liu, G. 96

Liu, H. 96

Liu, J. 33, 102, 676, 850, 870, 944

Liu, S. 1154

Liu, X. 1154

Lorenz, L. 682

Mahkhtari, H. 1101

Mahmoud, A. 659

Mannam, R. 607

Martens, A. 1107

Martinez, F. 935

Matsui, M. 571, 840, 846

Matsumoto, Y. 575

Meenakshi, J. 413

Mekhilef, S. 1194

Meng, Y. 850, 944

Mermet-Guyennet, M. 346, 1149

Miller, J. 16

Min, B. 257

Min, B-D. 477

Min, B-H. 613

Miura, Y. 1165

Mok, H-S. 119, 278, 283

Mok, J-K. 800

Mokhtari, H. 643

Moon, G-W. 274, 705, 709, 882, 961, 982, 994, 1001, 1177

Moon, J. 229, 460

Moon, S-H. 387

Moore, D. 1005

AUTHOR INDEX

Morizane, T. 398
Muesing, A. 1158
Mun, S. 1075, 1183
Mutschler, P. 1025
Na, S-K. 217, 221
Nakagawa, S. 619
Nakaoka, M. 1075, 1183
Nakayama, A. 619
Namnabat, M. 153, 246
Narikiyo, T. 107
Navrapescu, Y. 351
Nayar, C. 955
Nghia, D. 361
Nguyen, H. 159
Nguyen, N-T. 728
Nho, E-C. 337, 591, 613, 1036
Nho, N. 174, 361, 580, 1054
Ninomiya, T. 238, 252, 619, 1064, 1081
Nishida, Y. 137
Noguchi, Y. 915
Nokali, M. 59
Norigoe, I. 1064
Nussbaumer, T. 1011
Ogasawa, S. 822
Ogushi, K. 1228
Oh, H-S. 122, 217, 221
Oh, J. 794
Oh, M. 229
Oh, S. 289
Oh, S-B. 517
Oh, W-S. 243, 709, 882
Oh, Y. 278, 357
Oikawa, R. 575
Okamatsu, M. 261
Okyere, P. 1019
Oleschuk, V. 147

Omori, H. 1183
Ota, S. 1228
Oum, J-H. 466
Paek, S-H. 694
Pan, C-T. 225
Park, B-G. 553, 732
Park, B-S. 372
Park, C-H. 789
Park, D-H. 127
Park, G. 164
Park, H-G. 482
Park, H-S. 208, 367, 783, 794, 1001
Park, J. 988
Park, J-C. 169
Park, J-M. 527
Park, J-S. 810, 961, 1060
Park, K-B. 274, 1001, 1177
Park, K-T. 367, 372
Park, N-C. 283
Park, N-J. 567
Park, S. 79, 329, 342, 1211
Park, S-D. 337
Park, S-H. 381, 929, 961
Park, S-J. 521, 524
Pauta, S. 557
Peng, F. 1090
Peng, Q. 921
Peng, Y. 690
Pettersson, S. 654
Phuc, T. 1054
Piao, Z-G. 527
Ping, H. 1194
Pinggang, S. 597
Pong, C-Y. 875
Poodeh, M. 153, 246
Popescu, M. 351

AUTHOR INDEX

Pramanik, M. 1113

Premrudeepreechacham, S. 557

Profumo, F. 147

Qing, F. 492

Quanhai, W. 300

Ra, B-H. 460, 463

Raggl, K. 1011

Rahman, M. 905, 1069

Reed, J. 1119

Ren, J. 377, 965

Reza, S. 1113

Rose, J. 1119

Ryu, J-S. 732

Ryu, K. 477

Ryu, S. 59, 278

Ryu, Y-H. 381

Ryul, K. 257

Saghaleini, M. 1223

Saha, B. 1075, 1183

Sahoo, A. 413

Saito, T. 186

Salo, M. 654

Samadi, A. 180

Sasaki, N. 74

Sayeef, S. 905

Sekine, H. 444, 828, 915

Senjyu, T. 444, 828, 915

Sensama, P. 665

Sensarma, P. 1048

Seo, H. 229

Seo, I-D. 454

Seo, Y-G. 197, 1125

Seok, J-K. 1200

Shen, Y. 96

Shim, J-S. 266

Shimizu, T. 856

Shin, D-S. 393, 629

Shin, H-H. 62

Shin, Y-C. 460

Shon, J-G. 454, 530

Shoyama, M. 1081

Simanjorang, R. 1165

Singh, B. 1137

So, J-H. 450, 502

Soares, J. 419

Son, Y-H. 463

Song, B. 257

Song, E. 329, 477

Song, H. 221, 229

Song, J-H. 686, 694

Song, K-B. 517

Song, S-H. 497, 502, 517

Staudt, V. 865

Steimel, A. 865

Subrahmanyam, V. 805

Sugimura, H. 1075, 1183

Suh, I-Y. 439

Suh, J-S. 521, 524

Sul, S-K. 1206

Summers, T. 1217

Sun, H-G. 393

Sun, X-D. 840, 846

Suzuki, S. 74, 143

Tai, C-C. 309

Takahashi, F. 1085

Takeshita, T. 753

Tanabe, T. 74

Tang, Q. 1090

Taniguchi, K. 398

Tara, E. 643

Tenconi, A. 147

Thornton, R. 921

AUTHOR INDEX

Thyagarajan, T. 413, 805

Toliyar, H. 357, 723

Tomasik, J. 777, 1042

Torrical Bascope, T. 510

Tran, Q-V. 591

Tripathy, P. 1048

Tsukakoshi, K. 1064

Tu, C-C. 309

Tumbelaka, H. 955

Tuusa, H. 535, 654

Tuyen, N. 174

Uddin, M. 1113

Ueda, Y. 74

Undeland, T. 1143

Van Zwam, A. 85, 192

Vangala, N. 607

Venkataramanan, G. 1119

Wada, K. 856

Waffler, S. 888

Wang, C-T. 309

Wang, H. 387, 870, 911

Wang, J. 430

Wang, R. 102, 870

Wang, X. 33, 850, 944

Wang, Y. 85, 192

Wang, Z. 33, 676

Wheeler, P. 404

Won, C-Y. 203, 381, 789, 860, 929, 976

Wu, L. 676, 875

Wu, W. 1031

Xiao, H. 122, 217

Xiao, L-Y. 971

Xiaogao, C. 492

Yabg, O. 357

Yamaguchi, K. 487, 834

Yamamoto, Y. 822

Yanagimura, K. 846

Yang, B. 709

Yang, C-S. 295

Yang, H-Y. 393

Yang, S-H. 208, 305

Yanmin, S. 300

Yazdanian, M. 425

Yeh, C-A. 1085

Yi, K-H. 982

Yim, J. 229

Yisheng, Y. 597

Yokoyama, R. 74

Yokoyama, T. 186

Yona, A. 444, 828, 915

Yong, S-I. 686, 694

Yoo, D. 257, 271, 477

Yoo, H. 547

Yoo, J. 257

Yoo, J-G. 410

Yoo, J-H. 122, 221

Yoo, M-H. 271

Yoon, H-K. 705

Yoon, J. 521, 524, 541

Yoon, M-J. 961

Yoon, Y-D. 1206

Yoshino, T. 1211

Yoshioka, Y. 575

You, K. 1069

Yougui, G. 213

Youn, M-J. 705, 810, 1060

Yousfi, D. 816

Yrasaki, N. 915

Yu, B. 450, 502, 571, 840

Yu, G-J. 450, 502

Yu, J-S. 381, 860

Yu, Y. 690

AUTHOR INDEX

Yuan, C. 33, 944

Yun, D-J. 119

Zacharias, P. 510

Zhang, H. 377, 965

Zhi, Y-M. 122, 217

Zhong, Y. 377, 965

Zhou, G-X. 911

Zhou, Q. 1031

Zhuo, F. 676

Zoubek, H. 717

CURRAN ASSOCIATES INC.
proceedings
.com

9781424418718

2007 7th International Conference on Power Electronics

Daegu, South Korea
22-26 October 2007

IEEE Catalog Number: CFP07CPB-POD
ISBN: 978-1-42441-871-8

2007 7th International Conference on Power Electronics

Daegu, South Korea
22 – 26 October 2007

Pages 619-1232

IEEE Catalog Number: CFP07CPB-PRT
ISBN: 978-1-4244-1871-8

Copyright © 2007 by the Institute of Electrical and Electronic Engineers, Inc
All Rights Reserved

Copyright and Reprint Permissions: Abstracting is permitted with credit to the source. Libraries are permitted to photocopy beyond the limit of U.S. copyright law for private use of patrons those articles in this volume that carry a code at the bottom of the first page, provided the per-copy fee indicated in the code is paid through Copyright Clearance Center, 222 Rosewood Drive, Danvers, MA 01923.

For other copying, reprint or republication permission, write to IEEE Copyrights Manager, IEEE Service Center, 445 Hoes Lane, Piscataway, NJ 08854. All rights reserved.

***This publication is a representation of what appears in the IEEE Digital Libraries. Some format issues inherent in the e-media version may also appear in this print version.**

IEEE Catalog Number: CFP07CPB-PRT
ISBN 13:   978-1-4244-1871-8

Additional Copies of This Publication Are Available From:

Curran Associates, Inc
57 Morehouse Lane
Red Hook, NY 12571 USA
Phone: (845) 758-0400
Fax: (845) 758-2633
E-mail: curran@proceedings.com

TABLE OF CONTENTS

A Transformerless Hybrid Active Filter for Integration Into a Medium-Voltage Motor Drive with a Passive Front End..1
H. Akagi

Digital Current Mode Control Approach for the Parallel Module DC-DC Converters.............................9
B. Cho, H. Bae

Energy Storage Technology Markets and Applications: Ultracapacitors in Combination with Lithium-ion..16
J. Miller

Impact of Power Density Maximization on Efficiency of DC-DC Converter Systems.............................23
J. Kolar, J. Biela, U. Badstuebner

On the Control of Active Power Filters...33
J. Liu, X. Wang, C. Yuan, Z. Wang

Adjustable Speed Generator Systems - An Emerging Technology for Efficient Electrical Energy Generation...43
W. Koczara

Intelligent High-voltage High-power Ballast IC...53
J. Huang, D. Kim, M. Jung, G. Cho

Thermal Analysis of PT IGBT by Using ANSYS..59
S. Ryu, H. Ahn, D. Han, M. Nokali

The Effect of a Shielding Layer on Breakdown Voltage in a Trench Gate..62
J-S. Lee, H-H. Shin, H-S. Lee, E-G. Kang

Utility Interactive PV System with Improved Peak-cut Characteristics..66
H-S. Heo, G-H. Choe, Y-H. Choi, H-S. Kim, J-C. Kim

High Voltage AlGaN/GaN Schottky Barrier Diode Employing the Inductively Coupled Plasma-chemical Vapor Deposition SiO2...71
Y-H. Choi, J. Lim, K-H. Cho, M-K. Han

Optimized Operation and Stabilization of Microgrids with Multiple Energy Resources.........................74
T. Tanabe, Y. Ueda, S. Suzuki, T. Ito, N. Sasaki, R. Yokoyama

A New Converter-Inverter-Brake (CIB) Module with Shoot-through Immunity......................................79
S. Park, J-S. Kim, S-J. Kim, B-H. Kwon

Analysis of Sensitivity of the Performance of Interleaved Flyback Converter to the Principal Design Parameters...85
Y. Wang, S. De Haan, A. van Zwam

Formulation, Measurement and Analysis for the Thrust Force of HB-Type Linear Pulse Motor...90
D-H. Kim, J-Y. Ahn, G-I. Kang, K-H. Kim, Y-C. Lim

Application of Fast Lifting Wavelet Transform to the Estimation of Harmonic for Shunt Active Power Filters..96
H. Liu, G. Liu, Y. Shen

Small-Signal Analysis of the Lower-frequency Power Transfer Model...102
R. Wang, J. Liu

Generation of Differentiable Homomorphism Function Based on Self-Organizing Polynomial Data-mining Algorithm 107
Y-W. Kim, T. Narikiyo

Development of the Low-Cost Impedance Spectroscopy System for Modeling the Electrochemical Power Sources 113
W. Choi, J. Lee

3-d Education System of DCC Motor Drive Using RecurDyn and CoLink 119
H-S. Mok, J-H. Lee, G-H. Choe, S-H. Kim, D-J. Yun

A Study on the Modeling of Piezoelectric Transformer for CCFL Using a PSPICE 122
L-H. Hwang, J-H. Yoo, H. Xiao, Y-M. Zhi, H-S. Kim, H-S. Oh, M-T. Cho, G-S. Choi

An Active Gate Drive Circuit for High Power Inverter System to Reduce Turn-off Spike Voltage of Igbt 127
J-H. Kim, D-H. Park, J-B. Kim, B-H. Kwon

An Improved Strategy to Detect the Switching Device Fault in NPC Inverter System 132
J-D. Lee, T-J. Kim, J-C. Lee, D-S. Hyun

Overmodulation Strategy of NPC Type 3-level Inverter for Traction Drives 137
J. Lee, J. Choi, Y. Nishida

Discrete-time Model Following Control for PWM Inverter with Electric Double Layer Capacitor 143
S. Suzuki, T. Haneyoshi

Synchronized Pwm Control of Symmetrical Six-phase Drives 147
V. Oleschuk, G. Griva, F. Profumo, A. Tenconi

Optimized State Controller on DC-DC Converter 153
M. Poodeh, S. Esdtehardiha, M. Namnabat

A Study on Rotor Foc Method Using Matrix Converter Fed Induction Motor with Common-Mode Voltage Reduction 159
H-H. Lee, H. Nguyen, T-W. Chun

Robust Digital Controller Design for Three-phase Inverter Using CRA Method 164
J. Lee, G. Park, J. Choi, G-B. Chung

Analysis and Design of Pwm Inverter System for Flywheel Energy Storage System 169
J-C. Park, G-H. Choe, Y-S. Kim, B. Dugarjav, Z. Baljinnyam, J-P. Lee

Novel Single-state Pwm Technique in Multilevel Inverter for Unbalanced DC Sources 174
N. Nho, H. Lee, N. Tuyen

A Novel Optimization Method for Solving Harmonic Elimination Equations 180
A. Samadi, S. Farhangi

Verification of Autonomous Decentralized Control UPS Systemusing Fpga Based Hardware Controller 186
T. Saito, N. Doi, T. Yokoyama

Novel Dual-band Controller for Regulating DC-Bus Voltage That Contains a Large Voltage Ripple 192
Y. Wang, S. De Haan, A. van Zwam

A Study of the Train Running Simulation for Train Propulsion System Performance Analysis 197
Y-C. Kim, Y-G. Seo, S-C. Hong, J-S. Ko, B-S. Lee

Development of DC Line Voltage Simulator for Control of Regenerative 203
K-H. Cho, S-J. Jang, B-K. Lee, C-Y. Won, G-D. Kim

The Design and Analysis of the Piezoelectric Inverter Too the EEFL Driving of LCD Backlight 208
H-S. Park, S-H. Yang, Y-C. Lim

Direct Torque Control for Induction Machine Based on AC-AC Matrix Converter 213
G. Yougui, Z. Jianlin, C. Caixue

A Development of the Solar Position Tracker on the Program Method for the Small Typed Stand-alone PV System 217
L-H. Hwang, S-K. Na, H-S. Oh, Y-S. Kim, H. Xiao, Y-M. Zhi, M-T. Cho, S-C. Chang, G-S. Choi

A Study on Driving of 35W(T5) Fluorescent Lamp by the Electronic Ballast Using Piezoelectric Transformer 221
L-H. Hwang, J-H. Yoo, H-S. Song, S-K. Na, H-S. Kim, H-S. Oh, S-H. Lee, K-H. Choi, M-T. Cho

An Internet Based Embedded Network Monitoring System for Renewable Energy Systems 225
P-Y. Chen, S-K. Ho, W-J. Lee, C-C. Chu, C-T. Pan

Fabrication of 4H-SiC Schottky Barrier Diodes with the Epilayer Grown by Bis-trimethylsilylmethane Precursor 229
M. Oh, H. Song, J. Moon, J. Yim, J. Lee, H. Seo, Y. Choi, J. Choi, H. Kim

High Accuracy and High Stability Magnet Power Supply by Four-phase Buck Type DC/DC Converter 232
S-C. Kim, K-M. Ha, J-Y. Huang

Automatic Interleaving Control for Paralleled Converter System and Its Ripple Estimation with Simplified Circuit Model 238
T. Kohama, T. Ninomiya

Start-up Inrush Current Reduction Technique of Asymmetrical Half-bridge DC/DC Converter for PC Power Supply 243
J-K. Kim, S. Lee, W-S. Oh, C-H. Gil, J-R. Cho

Comparison the Control Methods in Improvement the Performance of the DC-DC 246
M. Namnabat, M. Poodeh, S. Eshtehardiha

Optimal Design of Bus Converter in On-board Distributed Power Architecture 252
S. Abe, M. Hirokawa, T. Ninomiya

A New Topology for Photovoltaic 4 Series DC/DC Converter with High Efficiency Under Wide Load Range 257
J. Lee, B. Min, J. Kim, K. Ryul, T. Kim, D. Yoo, B. Song, J. Yoo

Static and Dynamic Characteristics of DC-DC Converter Using Digital Filter 261
F. Kurokawa, M. Okamatsu

A Study on Novel Buck-boost AC-DC Converter of High Performance by Partial Resonance Technique 266
D-K. Kwak, B-S. Lee, C-S. Kim, J-S. Shim

Hybrid Ballast for Field Emission Lamp with CNT Emitter 271
J-H. Kim, M-H. Yoo, D-W. Yoo, H-M. Jung, S-J. Ha

Band-gap Reference Voltage Control Strategy of Power Conditioning System for Fuel Cell Hybrid Vehicle 274
Y-D. Kim, K-B. Park, C-E. Kim, G-W. Moon

Development of a 72V IPMSM Drive System for Golf Carts 278
H-S. Mok, B-C. Kim, S-H. Kim, S-L. Ryu, Y-J. Oh

Development Operation Algorithm for a 2.5-Ton Electric Forklift Using an Induction Motor 283
N-C. Park, J-H. Kim, H-S. Mok, S-H. Kim

A Monitoring System with Ubiquitous Sensors for Passenger Safety in Railway Platform 289
S. Oh, G. Kim, H. Lee

Development of Water-cooled Heat Sink for High-power IGBT Inverter 295
M. Han, S-D. Lee, C. Hong, C-S. Yang, K-S. Kim

The Research of Model Reference Adaptive Control of Static Var Compensator (SVC) 300
Z. Guang, Z. Lijuan, W. Quanhai, S. Yanmin

**The Study on the Protection Circuit of the Inverter for LCD Backlight Using Digital
Control Method** 305
J-W. Lee, Y-C. Lim, S-H. Yang

A Simple Partial Discharge Detector for Low-Voltage Rotating Electrical Machines 309
C-C. Tai, C-C. Tu, J-C. Hsieh, C-T. Wang, T-C. Huang, C-Y. Chen, Y-S. Lin, J-H. Lai

Validation of Electrical Power System Using Electrical Test Bed of GEO Satellite 315
J-D. Choi, J-C. Koo, C-H. Koo, E-C. Kim

**Implement of Power Line Communication Module for an Electric Power Energy
Monitoring System** 319
Y-C. Jung, H-S. Jung, J-U. Kim

Electrical Interfaces Compatibility Analysis for the COMS EPS 323
J-C. Koo, E-C. Kim

**Analysis of Effects of Inductance Component in Electrodeless Lamp on Ballast
Performances** 329
S-B. Han, S. Park, E. Song, H-G. Jeong, B-M. Jung

A Study on the Resonant Inverter for Corona Generators 333
C-L. Choi, S. Kwak, D-S. Lee

**Reduced Voltage Drop Characteristics of the Series Transformer in a Voltage Disturbance
Generator** 337
E-C. Nho, I-D. Kim, S-D. Park, T-W. Chun, H-G. Kim, N-S. Choi

A Study on Optimal Braking Control Using Adhesion Coefficient 342
H. Lee, G. Kim, S. Park

Stress Grading in Intergrated Power Modules 346
C. Duchesne, M. Mermet-Guyennet, E. Dutarde, T. Lebey, S. Dagdag

Modelling of Iron Losses in Salient Pole Permanent Magnet Synchronous Motors 351
Y. Navrapescu, M. Popescu, D. Kisck, G. Andronescu, M. Kisck

**Low Cost Fault Detection System for Inverter Driven Induction Motor Using Currents
Signal** 357
N-H. Kim, O. Yabg, M-H. Kim, H-A. Toliyar, Y. Oh, C-H. Choi

Control of Induction Motor Using IMC Approach 361
D. Nghia; N. Nho; H-H. Lee

Development of HBPI Controller for High Performance Control of IPMSM Drive. 367
J-S. Ko, K-T. Park, D-H. Chung, J-S. Choi, H-S. Park

Efficiency Optimization Control of SynRM Drive by LM-FNN Controller 372
J-S. Choi, K-T. Park, D-H. Chung, J. Ko, B-S. Park

A Study on Control Strategy of Electric Power Steering Based on Fuzzy Control 377
H. Zhang, J. Ren, Y. Zhong, J. Chen

**A Four-Switch and Half-Bridge Boost Converter Based BLDC Drive System for DC-Link
Voltage Unbalancing Compensation** 381
S-H. Park, Y-H. Ryu, W-C. Lee, J-H. Lee, J-S. Yu, Y-R. Kim, C-Y. Won

Sensorless Speed Control and Initial Rotor Position Estimation of an IPMSM 387
  W-T. Joung, J-H. Lee, S-H. Moon, H-J. Wang, T-S. Hwang, C-H. Choi, Y-S. Kim

Rotor Position Detection Method of a Single Phase Switched Reluctance Motor Using Back EMF 393
  H-G. Sun, D-S. Shin, H-Y. Yang, Y-C. Lim

Modified Double-Modulation Signal PWMm Control for Large Motor Drive Using Five-Level Converter 398
  N. Kimura, T. Hamada, T. Morizane, K. Taniguchi

Improved Initial Pole-Position Estimation of SPMSM Sensorless Servo Drive by Absolute Integrals of Torque-Component Currents 404
  T-W. Kim, P. Wheeler, C-I. Jeong, J. Choi

Design and Development of Distorted Source Device for Circuit Breakers Failure Analysis 410
  S-I. Lee, J-G. Yoo, H-J. Jeon, G-H. Choe

Analysis and Simulation of Matrix Converter Using PSIM 413
  A. Sahoo, J. Meenakshi, S. Dash, T. Thyagarajan

Hybrid Three-Phase HPF Rectifier with Programmable Input Current THD, Using FPGA Device and VHDL Language 419
  J. Soares, C. Canesin, L. de Freitas

A Novel Soft Switched Topology for Power Factor Correction with Load Isolation Capability 425
  M. Yazdanian, S. Farbangi

A Grid Connected Photovoltaic System with Irradiation Injected Current Control 430
  J. Wang, J. Chen, R. Li

Performance Analysis of Soft-switching Inverter for the Photovoltaic Power System 435
  H-J. Kim, Y-H. Chung, K-S. Lee, Y-S. Ion, K-S. Kim

Development of the Power Conditioning System for High Power Fuel Cell System 439
  J-H. Lee, S-T. Baek, H-J. Jung, H-H. Kang, J-M. Chung, I-Y. Suh

A Coordinated Control Method for Leveling Output Power Fluctuations 444
  T. Senjyu, M. Datta, A. Yona, H. Sekine, T. Funabashi

Performance Monitoring and Aanlysis of Middle Scale Grid-Connected PV System 450
  J-H. So. B-G. Yu, H-M. Hwang, G-J. Yu, J-Y. Choi, I. Choi

Development of On-Line Type Dynamic Voltage Compensation System 454
  J-H. Han, J-G. Shon, I-D. Seo, H-J. Jeon

Development of Grid-connected 200kW Photovoltaic Inverter 460
  Y-C. Shin, Y-R. Kim, B-H. Ra, J-S. Moon, S-K. Kim

Development of 10kw Grid-Connected Photovoltaic Inverter with High Frequency Transformer 463
  Y-H. Son, Y-R. Kim, B-H. Ra, S-H. Kim, S-K. Kim

Z-Source Active Power Filter with a Fuel Cells Source 466
  J-H. Oum, Y-C. Lim, Y-G. Jung

A Stand Alone Type Fuel Cells Micro-Source System with a Voltage Sag 471
  J-H. Kim, Y-C. Lim, Y-G. Jung

A Novel Grid-connected PV PCS with New High Efficiency Converter 477
  B-D. Min, J-P. Lee, J-H. Kim, T-J. Kim, D-W. Yoo, K. Ryu, J. Kim, E-H. Song

Low-Cost Converters for Micro Wind Turbine Systems Using PMSG 482
  H-G. Park, S-H. Jang, D-C. Lee, H-G. Kim

Study on Voltage Regulation Method in the Power Distribution System 487
K. Yamaguchi, K. Lee, K. Kurokawa

High Efficiency Photovoltaic Pump System with Double Pumps 492
F. Qing, C. Xiaogao, X. Lei

A Variable Structure Approach to Control the Active and Reactive Power for Doubly Fed Induction Generator 497
W-S. Kim, K-B. Lee, B-C. Jeong, S-H. Song

Improved RPV (Reactive-Power-Variation) Anti-islanding Method for Grid-connected Three-Phase PVPCS 502
K-O. Lee, J-Y. Choi, I. Cho, S-H. Song, G-J. Yu, J-H. So, Y-S. Jung, B-G. Yu

Power Control of a Grid-Connected Hybrid Generation System with Photovoltaic/Wind Turbine/Battery Sources 505
J. Jeon, S. Kim, C. Cho, J. Ahn, J. Kim

Analysis and Proposition of a PV Module Integrated Converter with High Voltage Gain Capability in a Non-Isolated Topology 510
S. Araujo, P. Zacharias, T. Torrical Bascope, F. Antunes

Modeling and Analysis of Jeju Power System Operation Connected with Wind Turbine System and HVDC in Jeju Island Using PSCAD/EMTDC 517
B-H. Kim, S-B. Oh, K-B. Song, S-H. Song

An Analysis of Pemfc & Photovoltaic 500W Hybrid System 521
H-J. Choi, S-J. Park, J-S. Choi, I-S. Cha, J-P. Yoon, J-S. Suh, S-D. Gun

The Study of Operating Characteristics 1KW Households PEMFC System 524
H-J. Choi, S-J. Park, J-S. Choi, J-S. Cha, J-P. Yoon, J-S. Suh, J-I. Lee

Performance Evaluation and Analysis of 50KW Grid-connected PV System 527
J-M. Park, Z-G. Piao, Y-O. Cho, H-L. Baek

A Study on Performance Improvement of DVR System Using EDLC 530
W-R. Kim, C-J. Lim, B-K. Kwon, H-J. Jeon, J-G. Shon

Three-level VSI Based Low Switching Frequency 10 MVA STATCOM in Reactive Power and Harmonics Compensation 535
R. Lauttamus, H. Tuusa

10MVA STATCOM Installation and Commissioning 541
Y. Han, C. Chung, J. Choi, D. Kim, J. Yoon

A Test Facility for Large Scale Inverter Valve and Pole Using Resonant Circuit 547
Y-S. Han, C. Chung, H. Yoo, J. Kim, S. Kim

An Improved Synchronization Control Scheme of a Low Cost 400Hz Power Supply for No-Break Power Transfer 553
S-E. Joung, B-G. Park, D-S. Hyun, I-H. Cha

Economic Dispatch for Power Generation Using Artificial Neural Network 557
S. Pauta, S. Premrudeepreechacham

ISC-STS Development for Power Supply to Improve Reliability in Communication System 562
H-C. Jung, D-S. Hyun

Optimal Current Controller in a Three-phase Grid Connected Inverter with an LCL Filter 567
K-J. Lee, N-J. Park, D-S. Hyun

A Solar Battery Charging Module by Means of Limit-Cycle MPPT Control 571
M. Matsui, B. Yu, K. Koh, T. Kitano

Frequency Regulation for Parallel-connected UPS System Under Independent Control 575
A. Kawamura, R. Oikawa, Y. Yoshioka, Y. Matsumoto

Carrier Based Two-state Pwm Method for Optimising Voltage Error in Multilevel Inverters 580
N. Nho, T. Cong, H-H. Lee

Comparison of OMTHD and OHSW Harmonic Optimization Techniques in Multi-level Voltage-Source Inverter with Non-Equal DC Sources 586
M. Hosseiniaghdam S. Fathi, G. Gharehpetian

A Modified Space Vector PWM for Common Mode Voltage Mitigation 591
Q-V. Tran, I-R. Ahn, T-W. Chun, H-G. Kim, E-C. Nho

Development of a Current-Sensorless Multi-Loop Control for Standalone PWM Inverters 597
Y. Yisheng, L. Chang, S. Pinggang

A Novel Soft-switching Two-Stage Step-Up DC/DC Converter 601
S. Du, Z. Chen, L. Chang

A New Technique to Achieve Zero Voltage Switching in a Resonant Reset Single Switch Forward Converter 607
N. Vangala, R. Mannam

Design of Bidirectional Pwm Sepic/Zeta DC-DC Converter 613
I-D. Kim, Y-H. Lee, B-H. Min, E-C. Nho, J-W. Ahn

Complete Analysis of Steady-state and Efficiency Considerations in a Forward-Flyback Mixed Converter 619
Y. Kusuhara, T. Ninomiya, A. Nakayama, S. Nakagawa

Effect of Sampling Frequency of A/D Converter on Controller Stability and Bandwidth of Digital-Controlled Power Converter 624
Y-T. Chang, Y-S. Lai

Power Spectra of the Single Phase 6/6 SRM Drives by the Separately Randomized Pulse Position (SRP) PWM Method 629
N. Khai, D-S. Shin, Y-G. Jung, Y-C. Lim

Inductance Vector Angle Based Sensorless Speed Estimation in Switched Reluctance Motor Drive 634
F. Kucuk, H. Goto, H-J. Guo, O. Ichinokura

A Position Sensorless Control System of SRM Over Wide Speed Range 639
M-H. Kim, W-S. Baik, D-H. Kim, N-H. Kim, K-H. Choi

Efficiency Map of a Switched Reluctance Motor Using Finite Element Method in Vehicular Applications 643
H. Mokhtari, E. Tara

A Novel Direct Instantaneous Pressure Control of Hydraulic Pump System with SRM Drive 649
J. Liang, D-H. Lee, J-W. Ahn, Y-J. An

Multiloop Control Strategy for Four-wire Current Source Active Power 654
S. Pettersson, M. Salo, H. Tuusa

A Novel Online Controller for Tuning Shunt Active Power Filters Based Upon Switched-Capacitors 659
O. Elgendy, A. Abass, A. Mahmoud, A. Alkoshairy

A Multiband Shunt Hybrid Active Filter with Sensorless Control 665
S. Kumar, P. Sensama

A Shunt Active Filter for Reactive Power Compensation and Harmonic Mitigation 671
N. Apte, V. Bapat, A. Jog

A Novel Control Method of Papf for Resonance Damping and Harmonics Compensation in Power System 676
L. Wu, F. Zhuo, J. Liu, Z. Wang

Power Semiconductors State-of-the-Art and Future Development Trends 682
L. Lorenz

Optimization of Power MOSFETs for Low Power Motor Drive Applications 686
J-H. Song, S-T. Han, T-S. Kwon, S-I. Yong

High-Power and High-Speed Semiconductor Switch RSD Applied in Pulsed Power System 690
L. Liang, Y. Yu, L. Deng, Y. Peng

New Smart Power Module for Low Power Motor Drives 694
T-S. Kwon, J-H. Song, J-B. Lee, S-H. Paek, S-I. Yong

Cost-effective Driving System for Plasma Displays by Replacing and Combining Functions of Power Switches 699
D-M. Lee, D-S. Hyun

A Cold Cathode Fluorescent Lamp Driving Circuit Without a Transformer for Liquid Crystal Display Backlight Unit 705
E-S. Choi, J-H. Cho, H-K. Yoon, G-W. Moon, M-J. Youn

A Novel Adaptive Dimming Technique with X-Y Channels for LED Backlight System of LCD TVs 709
W-S. Oh, K-M. Cho, D. Cho, G-W. Moon, B. Yang, T. Jang

Cim Based Soft Switching of Energy Recovery Sustain Driver for AC PDPs 713
S-B. Lim, J-Y. Lee, J-S. Ko, S-C. Hong

Automated Diagnosis of Rolling Bearing Faults in Electrical Drives 717
H. Zoubek

Rotor Fault Detection System for Inverter Driven Induction Motor Using 723
N-H. Kim, M-H. Kim, H. Toliyar, S-H. Lee, C-H. Choi, W-S. Baik

Fault Diagnosis of Induction Motor Using Decision Tree with an Optimal 728
N-T. Nguyen, J-M. Kwon, H-H. Lee

A Simple Fault Detection of the Open-Switch Damage in BLDC Motor Drive Systems 732
J-D. Lee, B-G. Park, T-S. Kim, J-S. Ryu, D-S. Hyun

Interleaved Or Sequential Switching - for Increasing the Switching 737
A. Cabral Ferreira, R. Kennel

Space Vector Modulation Method for Unidirectional Four-Wire Three-Phase/Level/Switch (Vienna) Rectifier 741
J. Alahuhtala

Voltage Balancing Technique with Low Switching Frequency for Cascade Multilevel Active Front-End 748
O. Alizadeh, S. Farhangi

Diode Rectifier Circuits with Commutation Capacitors for Three-phase 753
T. Takeshita

Two-phase Interleaved Buck Converter with a New Digital Self-Oscillating Modulator 759
L. Jakobsen, M. Andersen

Feedback Linearization Control of Three-phase AC/DC PWM Converters 765
D-E. Kim, D-C. Lee

Digital Control of Phase-shifted Full Bridge PWM Converter 771
J-G. Lim, S-H. Lim, S-K. Chang

Autonomous Variable Speed Power Generating System with Five-level Cascaded Converter 777
L. Grzesiak, J. Tomasik

Modularized Charge Equalization Converter with High Power Density and Low Voltage Stress for HEV Lithium-Ion Battery String 783
H-S. Park, C-E. Kim, C-H. Kim, J-H. Lee

Design and Control Algorithm Research of Active Regenerative Bidirectional DC/DC Converter Used in Electric Railway 789
C-H. Park, S-J. Jang, B-K. Lee, C-Y. Won, H-M. Lee

Charge Equalization Converter with Parallel Primary Winding for Series Connected Lithium-Ion Battery Strings in HEV 794
C-H. Kim, H-S. Park, C-E. Kim, J-H. Lee, J. Oh

Calculation of Regenerative Energy in DC 1500V Electric Railway Substations 800
C-H. Bae, D-U. Jang, Y-G. Kim, S-K. Chang, J-K. Mok

Robust Speed Sensorless Induction Motor Drives 805
E. Geetha, T. Thyagarajan, V. Subrahmanyam

Improved Rotor Position Estimation Employing Voltage Distortion Compensation for Sensorless PMSM Drives at Low Speed 810
S-M. Jung, J-S. Park, H-W. Kim, M-J. Youn

Indirect Position and Speed Sensing for PMSM Sensorless Control 816
D. Yousfi, M. Adnani

Hybrid Sensor-less Control of Permanent Magnet Synchronous Motor in Low-Speed Region 822
Y. Yamamoto, S. Ogasawa, H. Funato

A New Method for Smoothing Output Power Fluctuations of PV System Connected to Small Power Utility 828
T. Senjyu, M. Datta, A. Yona, H. Sekine, T. Funabashi

Case Study of Distrubution-unified Power Flow Controller (D-UPFC) in the Clustered PV System 834
K. Lee, K. Yamaguchi, K. Kurokawa

A Novel Islanding Detection Method Based on Minute Asymmetrical Current Injection for Three-Phase Grid-connected PV Inverters 840
X-D. Sun, M. Matsui, B-G. Yu

A Novel Single-voltage-sensor-based Maximum Power Point Tracking Method 846
X-D. Sun, M. Matsui, K. Yanagimura

Frequency-Domain Transformation Approaches to Develop the D-Q Synchronous-Frame Models of Three-phase Symmetrical Networks and Applications in Modeling of PWM Power Converters 850
X. Wang, J. Liu, Y. Meng, J. Hu

A Gate Drive Circuit of Power MOSFETs and IGBTs for Low Switching Loss 856
T. Shimizu, K. Wada

Load Modeling for the Drum Washing Machine System Simulation 860
J-H. Lee, T-W. Kong, W-C. Lee, J-S. Yu, C-Y. Won

Advanced Simulation Concept for the Power Train of an AC Locomotive and Its Verification 865
    V. Staudt, C. Heising, A. Steimel

Modeling Diode Reverse Recovery and Corresponding Implementation in Fast Time-Domain Simulation 870
    H. Wang, J. Liu, R. Wang, Q. Hou

A Half Bridge Flyback Converter with ZVS and ZCS Operations 875
    L-M. Wu, C-Y. Pong

A New Half Bridge Converter Without DC Offset of Magnetizing Current 882
    K-M. Cho, W-S. Oh, K-W. Lee, G-W. Moon

A Novel Low-loss Modulation Strategy for High-Power Bi-Directional Buck+Boost Converters 888
    S. Waffler, J. Kolar

Comparison of Pwm Strategies for Three-phase Current-fed DC/DC Converter 894
    H. Cha, S. Choi

Effect of AC Inductance in a Phase Shifted DC-DC Bridge Converter 899
    P. Freere, W. Kong, D. Holmes

Improved Low Speed Performance of a PI-DTC Interior Pm Machine with Compensations of Dead-time Effects and Forward Voltage Drops 905
    S. Sayeef, M. Rahman

Study on Efficiency Optimizing of PMSM for Pump Applications 911
    G-X. Zhou, H-J. Wang, D-H. Lee, J-W. Ahn

Wide-Speed-Range Optimal PAM Control for Permanent Magnet Synchronous Motors ..915
    T. Senjyu, Y. Noguchi, N. Yrasaki, A. Yona, H. Sekine, T. Funabashi

Development Issues of an ISG PM Machines and Control System 921
    X. Chen, M. Edington, R. Thornton, Y. Fang, Q. Peng

Control of IPMSM Drive System for Drum Washing Machine 929
    W-C. Lee, S-H. Park, J-H. Lee, Y-R. Kim, C-Y. Won

Fuzzy Logic Controller for Boost Converter with Active Power Factor Correction ..935
    F. Martinez, D. Gomez

A Low Cost and High Reliability Control Scheme in Parallel Operation for 400Hz Power Supply System 940
    S-E. Joung, D-S. Hyun

Frequency Characteristics of the D-Q Synchronous-Frame Current Reference Generation Methods for Active Power Filter 944
    X. Wang, J. Liu, J. Hu, Y. Meng, C. Yuan

IHTSs Development for Uninterruptible Power Supply at UPS Fault 950
    H-C. Jung, D-S. Hyun

A Grid Current-Controlling Shunt Active Power Filter 955
    H. Tumbelaka, L. Borle, C. Nayar, S. Lee

A New Battery Equalizer Based on Buck-Boost Topology 961
    S-H. Park, T-S. Kim, J-S. Park, G-W. Moon, M-J. Yoon

Design and Test of Controller in Power Conditioning System for Superconducting Magnetic Energy Storage 965
    H. Zhang, J. Ren, Y. Zhong, J. Chen

Research of Fuzzy Logic-Controlled Smes for Power System Transient Voltage Stability................971
X-H. Huang, L-Y. Xiao

A Calculation of Predicting the Expected Life of Super Capacitor Following Current Pattern of Railway Vehicles 976
J-Y. Kim, S-J. Jang, B-K. Lee, C-Y. Won, C-M. Lee

Comparative Study on a Single Energy Recovery Circuit with Dividing Energy Recovery Path for Plasma Display Panels (PDPs) 982
K-H. Yi, S-W. Choi, G-W. Moon

MFFL Driving System with Current Feedback Maintaining Glow Discharge Mode................988
J. Baek, J. Park, B. Cho

A New PWM-controlled Quasi-resonant Converter for High Efficiency PDP Sustaining Power Module................994
W-J. Lee, D-Y. Kim, C-E. Kim, G-W. Moon

Current Stress Minimizing Control Scheme for Power Factor Correction(PFC) Boost Pre-regulator................1001
H-C. Lee, C-E. Kim, H-S. Park, K-B. Park, G-W. Moon

Novel Three Phase Flux Reversal Machine with Full Pitch Winding................1005
D. Moore, B. Fernandes

Comparison of Winding Concepts for Bearingless Pumps................1011
K. Raggl, J. Kolar, T. Nussbaumer

A Multi-Winding Transformer Model for Predictive Smps Design and Analysis................1019
P. Okyere

Comparison of Topologies for Linear Drives in Industrial Material Handling and Processing Applications.................1025
P. Mutschler

AC-SPWM-cycloconverter Based on an Extended Chopper Scheme 1031
Q. Zhou, W. Wu, J. Lin

Matrix Converter with a Novel General Commutation Strategy................1036
N-S. Choi, Y. Li, B-M. Han, J-S. Ko, E-C. Nho

Novel DC Link Balancing Scheme in Generic N-Level Back-to-Back Converter System 1042
L. Grzesiak, J. Tomasik

Matrix Converter As UPFC for Transmission Line Compensation 1048
A. Dasgupta, P. Tripathy, P. Sensarma

Novel Carrier PWM Technique with Extension Range for 4-Switch Inverter................1054
N. Nho, T. Phuc, N. Bac

Current Control of PMSM in Overmodulation Region................1060
J-S. Park, S-M. Jung, H-W. Kim, M-J. Youn

Characteristics of a Pulse-Link Inverter for Fuel Cells................1064
K. Fukushima, I. Norigoe, K. Tsukakoshi, T. Ninomiya, Y. Harada, Z. Dai

Analysis of Conduction and Switching Losses of a Matrix-Z-Source Converter 1069
K. You, M. Rahman

Advanced Development of High Frequency Transformer Parasitic Inductive Components and Lossless Inductive Snubber-assisted Series 1075
B. Saha, M. Ishirobi, H. Sugimura, S. Mun, M. Nakaoka

Common-mode Noise Reduction in the Watkins-Johnson Converter................1081
M. Shoyama, T. Ninomiya

Current Ripple Reduction Technique of DC/DC Converter 1085
Y-T. Chang, C-A. Yeh, M. Hamaogi, F. Takahashi

An Improved Energy Recovery Clamp Circuit for PWM Converters with a Wide Range of Input Voltage 1090
H. Cha, L. Chen, F. Peng, Q. Tang

Power Saving of PWM Rectifier-VSI Fed Induction Machines 1096
S. Halasz, I. Kadar

A New Multi-machine Control System Based on Direct Torque Control Algorithm 1101
H. Mahkhtari, A. Alizadeh

New Current Controller for Inverter Fed Medium Voltage Drives with LC Filter 1107
T. Laczynski, A. Martens

Low Frequency Stability Study of a Three-Phase Induction Motor 1113
M. Uddin, M. Pramanik, S. Reza

Modeling of Battery Charging Wind Turbines 1119
J. Reed, G. Venkataramanan, J. Rose

Development of Simulator for DFIG-Based Wind Turbine 1125
Y-G. Seo, Y-C. Kim, J-S. Ko, S-C. Hong, B-M. Han

Lcl Filter Design for Grid-connected NPC Inverters in Offshore Wind Turbines 1131
S. Araujo, A. Engler, F. Antunes

Decoupled Voltage and Frequency Controller for an Isolated Pico Hydro System Feeding Dynamic Loads 1137
B. Singh, G. Kasal

Power Electronic Conversion Systems for Deepwater Applications 1143
J. Aguillon-Garcia, A. Kristoffersen, T. Undeland

Accurate Mixed Electrical and Electromagnetic Model of a 6,5kV Igbt Module 1149
E. Batista, J. Dienot, M. Mermet-Guyennet, A. Castellazi, M. Ciappa, W. Fitchner

Research on a Flexible Waveform Power Amplifier Adopting Switch-Linear Hybrid(SLH) Scheme 1154
X. Liu, S. Liu, L. Ge

Switching Transient Shaping of Rf Power MOSFETs for a 2.5 MHz, Three-Phase PFC 1158
M. Hartmann, A. Muesing, J. Kolar

Controlling Voltage Profile in Loop Distribution System with Distributed Generation Using Series Type BTB Converter 1165
R. Simanjorang, Y. Miura, T. Ise

A New Direct Current Internal Resistance and State of Charge Relationship for the Li-Ion Battery Pulse Power Estimation 1171
J-H. Kim, S-J. Lee, J-M. Lee, B-H. Cho

A New Single-stage PFC AC/DC Converter with Voltage-Doubler Rectified Asymmetric Half-Bridge Converter 1177
B-H. Lee, C-E. Kim, K-B. Park, G-W. Moon

Direct High Frequency Soft Switching PWM Cyclo-Converter-Fed AC-DC Converter Without DC Link for Consumer Magnetron Drive 1183
H. Sugimura, B. Saha, S. Mun, E. Hiraki, H. Omori, M. Nakaoka

An Estimation Method on the Initial Pole Position of a Z-Axis PMLSM 1189
J-W. Lee

Direct Torque Control Permanent Magnet Synchronous Motor Drive with Asymmetrical Multilevel Inverter Supply 1194
M. Kadir, S. Mekhilef, H. Ping

Robust Measurement Disturbance Observer Design for AC Motor Drive Systems with Current Measurement Errors...1200
K-R. Cho, J-K. Seok

New Flux Weakening Control for Surface Mounted Permanent Magnet Synchronous Machine Using Gradient Descent Method 1206
Y-D. Yoon, S-K. Sul

HVDC Control Development for Isolated Small Ac System with Real Time Digital Simulator...1211
T. Yoshino, J. Lee, C. Lee, S. Park

Using a Cascaded H-Bridge STATCOM for Rebalancing Unbalanced Voltages.....................................1217
R. Betz, T. Summers

Distributed Power Supply with Power Factor Correction: a Solution to Feed All Modules of Power Electronic Transformers 1223
M. Saghaleini, S. Farhangi

Development on 31.5MVA STATCOM and Digital Evaluation Tool for Voltage Flicker Compensation 1228
K. Hidesi, T. Ishizuka, F. Aoyama, S. Ota, K. Ogushi

Author Index

Complete Analysis of Steady-State and Efficiency Considerations in a Forward-Flyback Mixed Converter

Yoshito Kusuhara
Kyushu Polytechnic College
Dept. of Product. Electro. Syst. Eng.
Shi 1665-1,Kokuraminami-ku,
Kitakyushu,Fukuoka,Japan
Email: kusuhara@kyushu-pc.ac.jp

Tamotsu Ninomiya and Asahi Nakayama
Kyusyu University
Dept. of Elec. & Electro. Syst. Eng.
Motoka 6-10-1, Nishi-ku,Fukuoka,Japan
Email:ninomiya@ees.kyushu-u.ac.jp
Email:nakayama@ees.kyushu-u.ac.jp

Shin Nakagawa
FIDELIX Co.,
Matuyama 2-15-14, Kiyose
Tokyo ,Japan
Email: MXF06217@nifty.ne.jp

Abstract— This paper presents a novel DC-DC converter where both forward and Flyback actions are mixed. Previously, we proposed this converter that has prominent features of reducing the voltage of main switch and the current ripple of output inductor, and as a result, the high efficiency and the size reduction were obtained as reported in the previous paper. And, this paper presents that this converter has four operation modes when the turns ratio of the second and the third windings of the transformer is set to be a certain value. Furthermore, these operation modes are analyzed through the extended state-space averaging method, and are confirmed by simulation. These characteristics are shown by means of a three-dimensional graphics method. This paper presents the detailed analyses of static characteristics for the four operation mode used three-dimensional expression of load characteristics with another winding turns ratio by the extended state-space averaging method.

I. INTRODUCTION

The most standard circuit topology of DC-DC converter is a forward converter. This converter needs the transformer reset winding for the normal operation, and it is usually wounded in the primary side to recover the transformer magnetizing energy into the input power source. On the other hand, another way of setting the reset winding in the secondary side has been proposed before, and to transfer the magnetizing energy into the output side is better from the viewpoint of power efficiency. Here, one more improvement can be inserted, e.g., the anode terminal of the freewheeling diode is moved to a different point. As a result, a novel topology of DC-DC converter is proposed, and this converter is experimentally confirmed to have some prominent features of withstand-voltage reduction for the main switch and ripper reduction of the output reactor current. As a result, this converter achieves a higher efficiency over the conventional forward converter. Furthermore, from consideration that this converter has both operations of forward and Flyback manners, this novel converter is named "Forward-Flyback-Mixed Converter," and is abbreviated as "FFB converter." This paper describes the basic operation of this proposed DC-DC converter, and analyzes the steady-state characteristics by means of the extended state-space averaging method. The analytical results are confirmed experimentally, and an experimental result of a high efficiency of 94% was obtained for the input voltage of 140V and the output load of 18V and 4A.

II. CIRCUIT CONFIGURATION AND ITS OPERATION

The circuit configuration of the proposed FFB converter is shown in Fig.1. And the switching state of each element is shown in Table 1. The basic composition is a mixed circuit of forward and Flyback converters. In this topology, both diodes of RD_2 and RD_3 hold ON during some interval in a switching period, and then the output reactor current is constant. This results in the ripple current reduction and power loss reduction of the output reactor. The simulated waveforms of the magnetizing i_L and the output reactor current i_{L2} are shown in Fig.2. For the interval of State1, Q_1 and RD_1 are kept on, and RD_2 and RD_3 are kept off. During this interval, the magnetizing current i_L increases, and the energy is stored in the transformer. At the same time, the primary power is supplied to the secondary side. When Q_1 and RD_1 are turned off, the state changes into State2, and RD_2 and RD_3 turn on. At this time, the current i_{L2} of output reactor L_2 maintains constant. At the time when both the currents i_L and i_{L2} become equal to each other, diode RD_3 turns off, and the state changes into State3. During this interval, the stored energy in T_1 is discharged through diode RD_2 and reactor L_2.

Table 1 Switching sequence of devices.

State	Q_1	RD_1	RD_2	RD_3
State1	ON	ON	OFF	OFF
State2	OFF	OFF	ON	ON
State3	OFF	OFF	ON	OFF

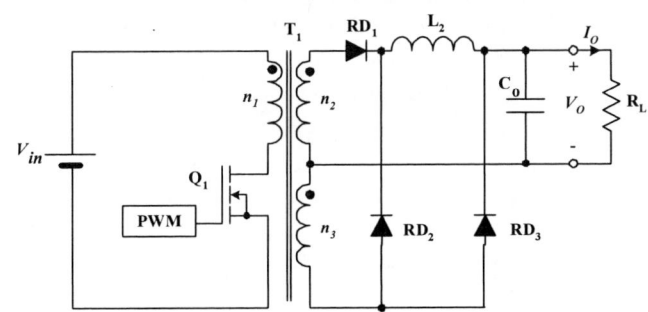

Fig. 1 Configuration of Forward-Flyback-Mixed converter.

978-1-4244-1871-8/07 $25.00 © 2007 IEEE

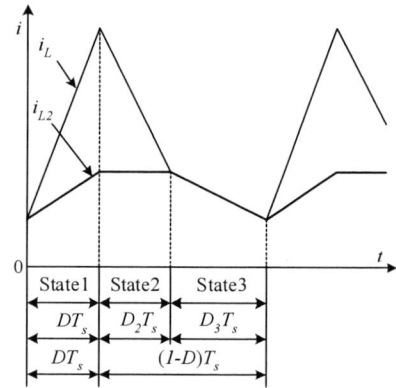

Fig.2 Waveforms of magnetizing and output reactor current.

III. STATIC CHARACTERISTIC ANALYSIS BY THE EXTENDED STATE-SPACE AVERAGING METHOD

According to the extended state-space averaging method, four operation modes are defined as FFB-CCM is assumed Mode I, and Mode II for FFB-DCM, Mode III for Flyback-DCM, Mode IV for Flyback-CCM.

A. Analysis of Mode I (FFB_CCM)

Analytical models of Mode I for three states are derived as shown in Fig.3, where the transformer magnetizing inductance L''_p referred to the secondary side, the output reactor inductance L_2, and the output capacitance C_o are chosen as the low-frequency components, and then their currents i_L, i_{L2} and the capacitor voltage v_o denote the low-frequency variables. the adjunct voltages v_L, v_{L2}. V'_{in} denotes the equivalent input voltage referred to the secondary voltage level, and n_1-n_3 denote the numbers of the transformer winding turns. From equation (1)-(4), steady state output voltage V_{OModeI} of Mode I equation (5) is obtained

B. Analysis of Mode II (FFB_DCM)

The Mode II operation is also analyzed by the extended state-space averaging method. The analytical models for the four states are derived as shown in Fig.4(a)-(d). As a result, the state-space differential equation of the low-frequency variable v_o derived as shown by equation (6).As a result ,steady state output voltage $V_{OModeII}$ of Mode II is derived as shown by equation (7).

C. Analysis of Mode III(Flyback_DCM)

It is time when this state operates only RD_3. The RD_3 operates as movement of the Flyback converter as shown in Fig. 5. This state is generated from the relation between the ratio of number n_2−n_3 of windings and duty ratios. RD_1 and RD_2 operate from the relation between the secondary voltage and the capacitor voltage and only RD_3 operates while non-operated. ModeIII is the Flyback operation in the DCM. At this time,the voltage $V_{OModeIII}$ of Flyback operation is derived as shown by equation (8).

D. Analysis of Mode IV(Flyback_CCM)

This mode remains without continuous the current of the reactor, and discharging the accumulation energy in the section of one cycle. The voltage $V_{OModeIV}$ of Flyback operation is derived as shown by equation (9).

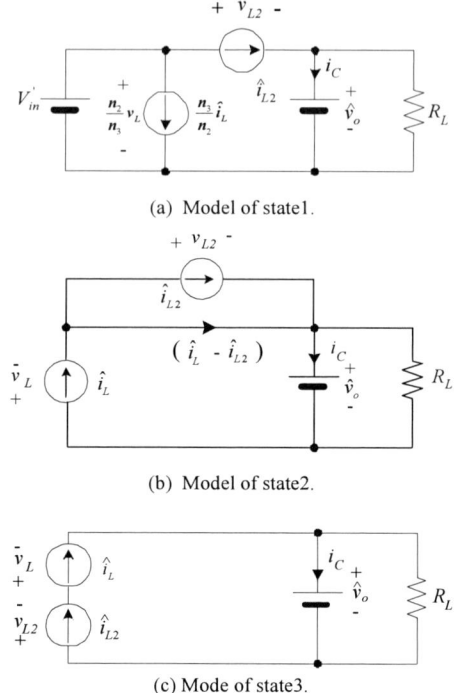

(a) Model of state1.

(b) Model of state2.

(c) Mode of state3.

Fig.3 Analytical model of each state of Mode I (FFB-CCM).

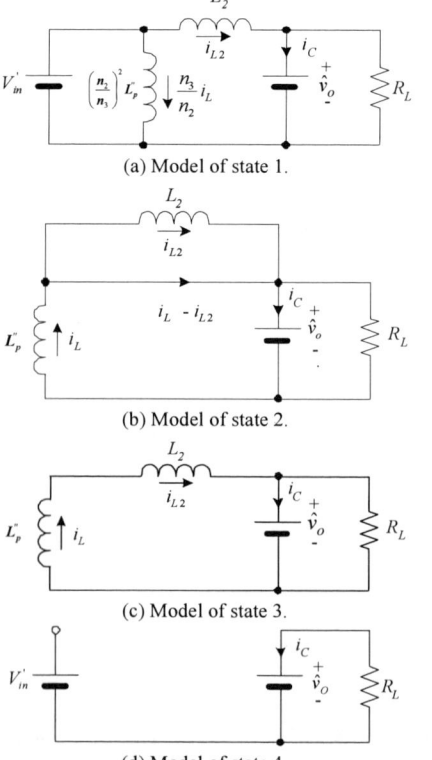

(a) Model of state 1.

(b) Model of state 2.

(c) Model of state 3.

(d) Model of state 4.

Fig.4 Analytical model of each state of Mode II(FFB-DCM).

978-1-4244-1871-8/07 $25.00 © 2007 IEEE   620

$$\frac{d\hat{i}_L}{dt} = \frac{1}{L_p''}\left\{\frac{n_3}{n_2}DV_{in}' - \left(D_2 + \frac{L_p''}{L_p'' + L_2}D_3\right)\hat{v}_o\right\} \tag{1}$$

$$\frac{d\hat{i}_{L2}}{dt} = \frac{1}{L_2}\left\{DV_{in}' - \left(D + \frac{L_2}{L_p'' + L_2}D_3\right)\hat{v}_o\right\} \tag{2}$$

$$\frac{d\hat{v}_o}{dt} = \frac{1}{C_o}\left\{D\hat{i}_{L2} + (1-D)\hat{i}_L - \frac{\hat{v}_o}{R_L}\right\} \tag{3}$$

$$D_2 = \frac{L_p''}{\hat{v}_o}\left(\frac{n_1}{n_3}\cdot\frac{V_{in}}{L_p} - \frac{\frac{n_2}{n_1}V_{in} - \hat{v}_o}{L_2}\right)D$$

$$D_3 = 1 - D - D_2 \tag{4}$$

$$V_{OModeI} = \left(\frac{n_2 + n_3}{n_1}\right)DV_{in} \tag{5}$$

$$\frac{d\hat{v}_o}{dt} = \frac{DT_S}{2L_p''L_2C_o}\left\{L_p''(V_{in}' - \hat{v}_o)(1-D_4) + \frac{n_3}{n_1}L_2V_{in}D_2\right\} \tag{6}$$

$$-\frac{\hat{v}_o}{C_oR_L}, \quad D_4 = 1 - \frac{V_{in}D}{\hat{v}_o}\left(\frac{n_2 + n_3}{n_1}\right)$$

$$V_{OModeII} = \frac{D^2T_S\left(\frac{n_3}{n_1}\right)^2 V_{in}^2\left(\left(\frac{n_2}{n_1}\right)^2 L_P + L_2\right)}{2L_p''L_2I_o + L_p''V_{in}\frac{n_2}{n_1}D^2T_S} \tag{7}$$

$$V_{OModeIII} = DV_{in}\sqrt{\frac{R_LT_S}{2L_p}} \tag{8}$$

$$V_{OModeIV} = \frac{D}{1-D}\frac{n_3}{n_1}V_{in} \tag{9}$$

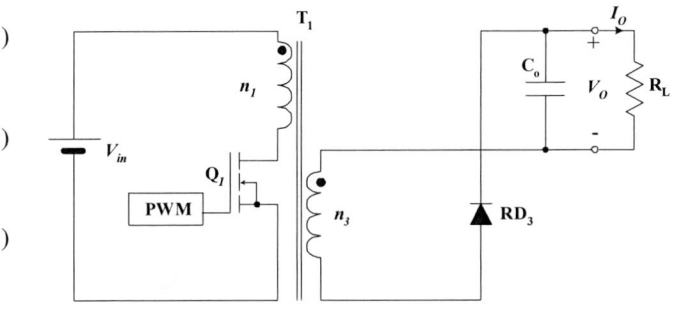

Fig.5 Circuit configuration for operating as Flyback converter(ModeIII-IV).

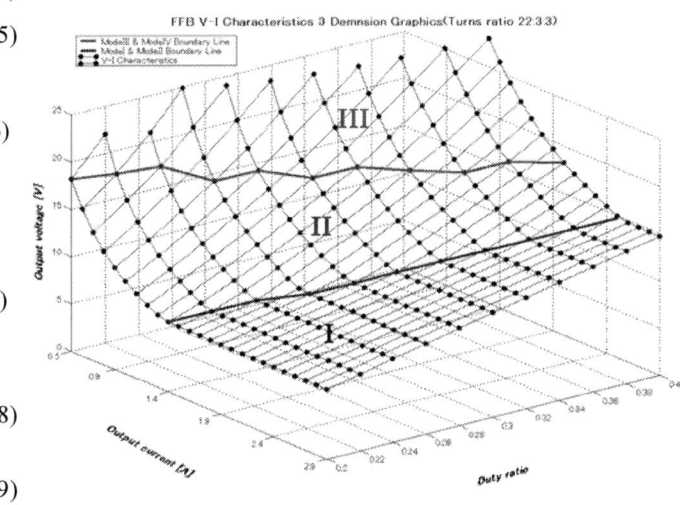

Fig.6 Three-dimensional graphics of analysis for the turns ratio of the second and the third windings $n_1 : n_2 : n_3 = 22{:}3{:}3$.

IV. THREE-DIMENSIONAL EXPRESSION OF LOAD CHARACTERISTICS WITH ANOTHER WINDING TURNS RATIO

Based on the current analysis, the duty ratio D and the output currents I_o were changed, and output voltage V_o was numerical value calculated. When a different turns ratio of transformer windings n_2 and n_3 is chosen, the FFB has four operation modes. Here, these operation modes are analyzed through the extended state-space averaging method. These characteristics are shown by means of a three-dimensional graphics method in Fig.6-Fig.8, where the number of winding turns, n_1=22, n_2=3, n_3=3,6,8. In these figures, the area I is Mode I(FFB-CCM), the area II is Mode II(FFB-DCM), the area III is Mode III (Flyback-DCM),and the area IV is Mode IV(Flyback-CCM). The dot of figure shows the output voltage corresponding to the change in the duty ratio and the output current. The borderlines between each operation modes vary correspondingly to the combination of winding turns ratios. And, it is understood that it moves up and down according to the voltage compared with the winding and the area changes.

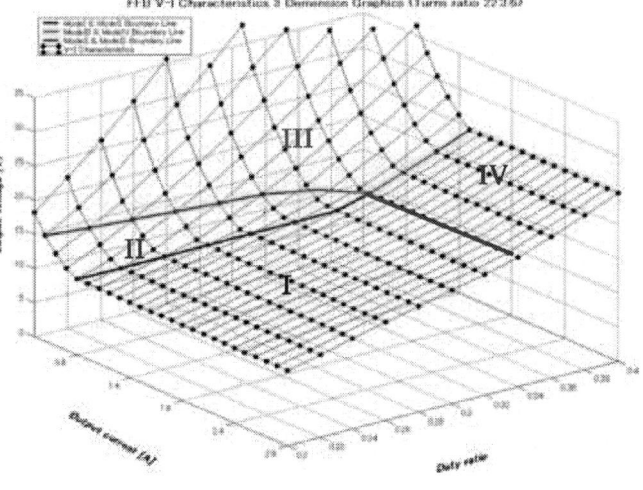

Fig.7 Three-dimensional graphics of analysis for the turns ratio of the second and the third windings $n_1 : n_2 : n_3 = 22{:}3{:}6$.

978-1-4244-1871-8/07 $25.00 © 2007 IEEE

The boundary condition is requested from the analytical equation that has been derived so far. The equation of the boundary condition of each area are shown from equation (10)-equation (13).

(i)Equation of boundary condition of ModeI and ModeII.

$$I_{oI-II} = \frac{T_S V_o^2}{2L_P L_2 (n+n')^2 V_{in}} \left\{ \frac{V_{in}}{V_o}(n^2 L_P + L_2) - nL_P \right\} \quad (10)$$

(ii)Equation of boundary condition of ModeII and ModeIII.

$$V_{oII-III} = \frac{n_2}{n_1} V_{in} \quad (11)$$

(iii)Equation of boundary condition of ModeIII and ModeIV.

$$I_{oIII-IV} = \frac{\frac{n_3}{n_1} D(1-D)V_{in} T_s}{2L_p''} \quad (12)$$

(iv)Equation of boundary condition of ModeIV and ModeI.

$$V_{oIV-I} = \frac{n_2}{n_1} V_{in} \quad (13)$$

Three-dimensional graphics of the experimental result of the turns ratio ($n_1 : n_2 : n_3 = 22:3:8$) is shown in Fig. 9. It can be understood that the experimental result shows the same characteristics as an analytical result. The value of the boundary condition obtained from the experimental result is requested, it becomes as shown in equation (14).

$$\begin{aligned} I_{oI-II} &= 0.79[A] & V_{oII-III} &= 18.4[V] \\ I_{oIII-IV} &= 2.2(D - D^2)[A] & V_{oIV-I} &= 18.4[V] \end{aligned} \quad (14)$$

V. EFFICIENCY AND LOAD CHARACTERISTICS OF EXPERIMENT FOR THE FFB CONVERTER

Furthermore, it experimented on the efficiency characteristic when changing it compared with the winding. In order to confirm analytical results of steady state characteristics of the FFB, the load characteristics above mentioned is examined by experiment. In the experimental parameter of the FFB are chosen as shown in Table 3. The efficiency of winding turns ratio 22:3:3 is shown in Fig.10, and 22:3:6 is shown in Fig.11. and 22:3:8 is shown in Fig.12. It is understood that the FFB converter maximum efficiency shifts by the number of windings of third and the load current from the result of these figure. It is shown that this is able to control maximum efficiency of the FFB according to the third winding number and duty ratio. It is higher efficiency that there are a lot of numbers of n_3 winding when the output current is small. It is thought that the reason for one with small average current of

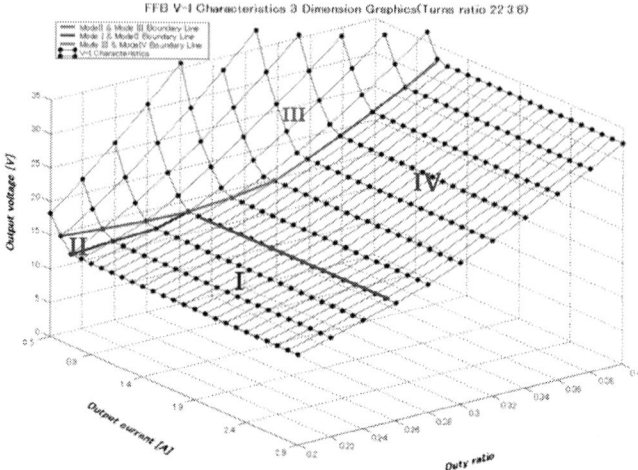

Fig.8　Three-dimensional graphics of analysis for the turns ratio of the second and the third windings $n_1 : n_2 : n_3 = 22:3:8$.

Table 2　Parameter of experiment of V-I characteristics

Parameter	symbol	constant
Input voltage	V_{in}	135[V]
Exiting inductance	L_p	400[μH]
Output reactor	L_2	40[μH]
Trun ratio	$n_1 : n_2 : n_3$	22:03:08
Frequency	f	100[kHz]
Duty ratio	D	0.17-0.35
Output current	I_o	0.5-5[A]

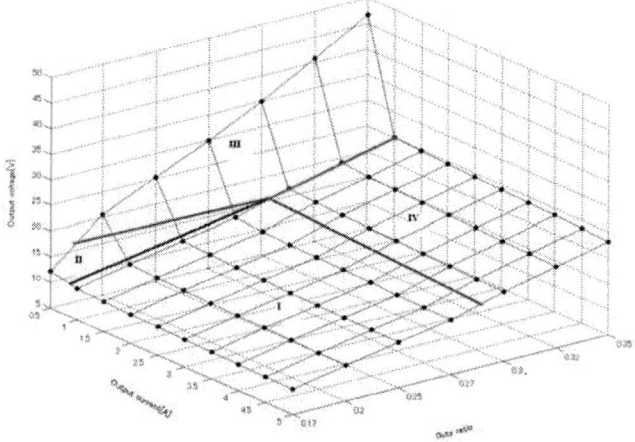

Fig.9 Three-dimensional graphics of experimental results for winding turns ratios $n_1 : n_2 : n_3 = 22:3:8$

Table 3　Parameter of experiment of efficiency characteristics

Parameter	symbol	constant
Input voltage	V_{in}	140[V]
Exiting inductance	L_p	400[μH]
Output reactor	L_2	40[μH]
Primary winding	n_1	22
Secondarily winding	n_2	3
Third winding	n_3	3,6,8
Switching frequency	f	100[kHz]

the diode RD_3 is that there is no bias at the current distribution rate in this. It is not good efficiency that there are a lot of numbers of n volume lines when the output current is large. This is because bias of the current distribution rate grow because average current I_{RD3} of diode RD_3 is small when there are a lot of numbers of n_3 winding, and It is because the peak value of the current grows to control the duty ratio small, and the power loss of the switch increases.

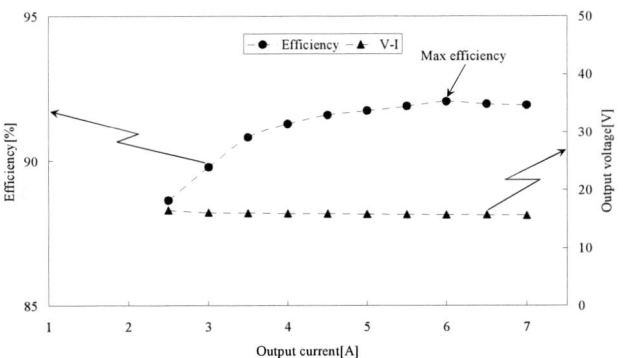

Fig.10   Experimental results of efficiency of winding turns ratios $n_1{:}n_2{:}n_3{=}22{:}3{:}3$.

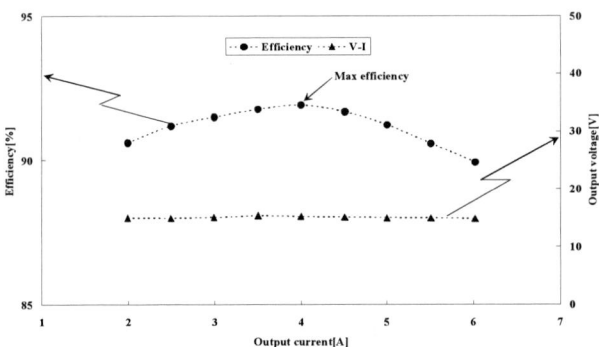

Fig.11   Experimental results of efficiency of winding turns ratios $n_1{:}n_2{:}n_3{=}22{:}3{:}6$.

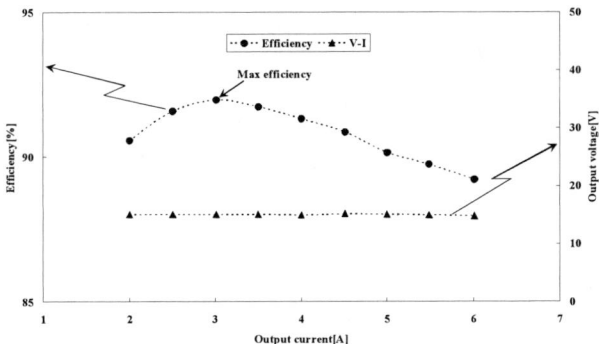

Fig.12   Experimental results of efficiency of winding turns ratios $n_1{:}n_2{:}n_3{=}22{:}3{:}8$.

VI.   CONCLUSION

The FFB converter has been analyzed by means of the extended state-space averaging method. And, the existence of four operation modes including the operation of the FFB converter and the operation of the Flyback converter was clarified by combining compared with the FFB converter winding.

Furthermore ,the efficiency characteristic when the third winding of the FFB converter was changed was compared and examined. As a result, the following issues have been clarified.

(1)The operation of the FFB converter was three-dimensionally analyzed by three dimensional graphics used the extended state-space averaging method.
(2)The FFB converter has four operation modes.
(3)The Flyback area changes when the third winding number are changed.
(4)It related to the energy distribution and the efficiency characteristic.
(5) It has been understood that the current distribution rate of the average current of output reactor L_2 and diode RD_3 influences efficiency.

Lastly, we will grope for the best range of the control where these four operation modes in the FFB can be effectively used compared with the winding.

In addition, if the circuit parameter is very setting, whether the optimum design method in an arbitrary Input/Output condition can be done is scheduled to be examined.

REFERENCES

[1]J.N.Park,T,RZaloum, "A Dual Mode Forward Flyback Converter", PESC '82 Record, pp.3-13, June 1982.
[2]S.Nakagawa,Y.Kusuhara, T.Ninomiya,"Steady state Characteristics of Mixed Forward-Flyback DC-DC Converter",IEICE Technical Report,Vol.103,No.38,E2003-5,May 2003.
[3]Y.Kusuhara,S.Nakagawa,T.Ninomiya,"Steady state Analysis of Mixed Forward-Flyback Converter",IEICE,Technical Report,Vol.103No.303,E2003-38,Septenber 2003.
[4]Y.Kusuhara,S.Nakagawa,T.Ninomiya,"Comparison of Power Loss Distribution between Forward-Flyback Type and Conventional converter",IEICE,Technical Report,Vol.104,No.406,EE2004-40,Novenber 2004.
[5]S.Nakagawa,Y.Kusuhara,T.Ninomiya,"Steady-State Characteristics of a Novel Forward-Flyback-Mixed Converter", PESC '06 Record, pp.3018-3023, June 2006.
[6]S.Nakagawa,Y.Kusuhara,T.Ninomiya,"Steady-State Characteristics of a Novel Forward-Flyback-Mixed Converter",IEICE Trans. Communications,Vol.J89-B,No.7, pp.1307-1314, July 2006.
[7]Y.Kusuhara,S.Nakagawa,T.Ninomiya,"Steady-State Analysis of Forward-Flyback Mixed Converter with Several Values of Transformer Turns Ratio",IEICE,Technical   Report,Vol.106,No.152,E2006-11,July 2006.

Effect of Sampling Frequency of A/D Converter on Controller Stability and Bandwidth of Digital-Controlled Power Converter

Ye-Then Chang and Yen-Shin Lai, Senior *Member, IEEE*

Center for Power Electronics Technology, National Taipei University of Technology

1, Sec. 3, Chung-Hsiao E. Rd., Taipei, Taiwan

Abstract—Due to the merits of flexibility, reduction of analog components and no aging issue of external components, digital-controlled power converter becomes very promising. It is required to derive the discrete model of power converter for digital-power control. The effect of sampling frequency of A/D converter on the digital-controlled power converter is also very essential to the performance. The objective of this paper is to present a discrete-time model of power converter for digital-controlled applications and to investigate the effect of sampling frequency of A/D converter stability and bandwidth of digital controller. It will be shown as the sampling frequency increases, the controller bandwidth can be increased to speed up the transient while reducing the phase margin. The results shown in this paper will provide a quantitative reference for digital controller design. Experimental results will be presented for confirmation.

I. INTRODUCTION

Fig. 1 illustrates the block diagram of digital-controlled converter using FPGA as an example. As shown in Fig. 1, voltage are sensed and fed back to the digital controller. One of the most important issues related to the digital-controlled system is the delay caused by sample and hold mechanism. The delay in any feedback system degrades the stability and damping of the system [1].

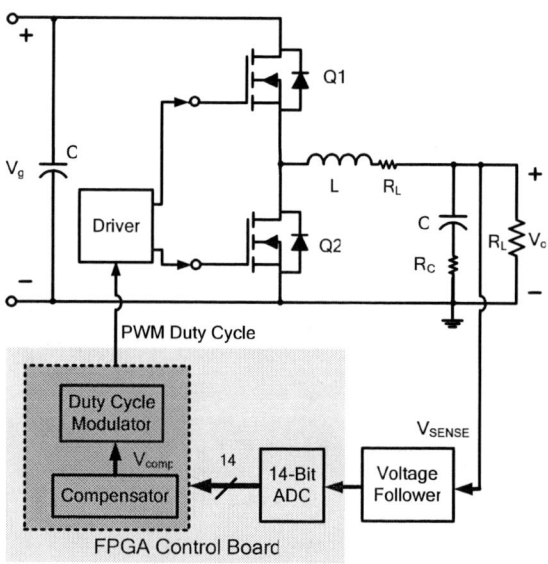

Fig. 1 Block diagram of DC/DC converter

The sampling frequency significantly affects the performance as illustrated in Fig. 2. In Fig. 2(A), the analog

signal, $u(t)$, is sampled in period T_{samp1} as $u(kT_{samp1})$. And the sampled signal, $u(kT_{samp1})$, is held by a zero order hold (ZOH) as $u^*(t)$. The average of signal $u^*(t)$ lags the original signal $u(t)$ by $T_{samp1}/2$. In Fig. 2(B), the analog signal, $u(t)$, is sampled in period T_{samp2} which is smaller than the period T_{samp1} in Fig. 2(A). As the result, the average of signal $u^*(t)$ lags $u(t)$ by only $T_{samp2}/2$ [1]. Dramatic increase of sampling frequency of A/D converter is a potential solution to cope with this issue as shown in Fig. 2 (B). However, as shown in Fig. 2 (B), there still exists a delay between original and sampled signals even though the sampling frequency, $1/T_{samp}$, is very high.

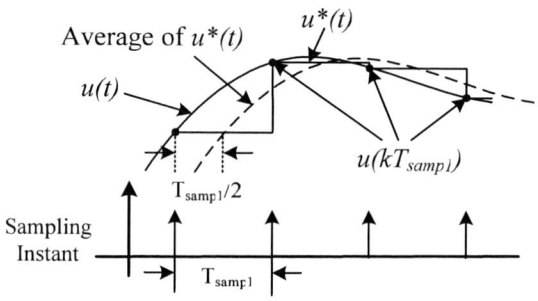

(A) Low sampling frequency

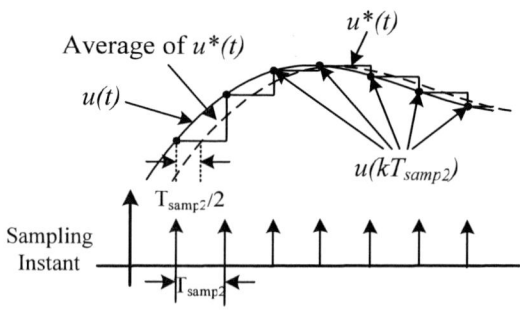

(B) High sampling frequency

Fig. 2 Effect of sampling for digital-controlled converter [1]

Fig. 3 shows the Bode plot of the transfer function for continuous-time and digital-controlled power converter. As shown in Fig. 3, the sampling frequency indeed affects the bandwidth of the transfer function. Therefore, what is the effect of A/D converter sampling frequency on controller design and how to derive the related discrete-time model for quantities analysis are two key topics to the digital controller design.

978-1-4244-1871-8/07 $25.00 © 2007 IEEE

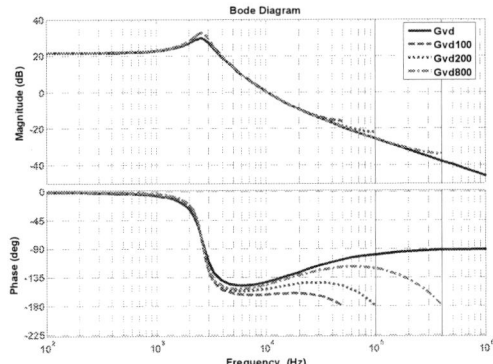

Fig. 3 Illustration of control-to-output transfer functions of continuous and sampled systems

There are many methods to model a digital-controlled system [1][2]. It has been shown in [1] that sample and zero-order hold method is the simplest way for deriving the discrete-time model. It gives the discrete-time model for the sampled data system with A/D converter which is used in the majority of digital-controlled systems. Small-signal analysis in Laplace-domain of a uniformly-sampled pulse-width modulator is presented in [3]. The delay-time caused by sample and zero-order hold and different PWM modulators is analyzed. In [4], a discrete-time domain model of buck converter with a uniformly-sampled pulse-width modulator is presented. However, in model validation stage the output capacitor is not considered. Under this circumstance, the order of buck converter model is with first order only. For practical applications, a second order model should be taken into account as considering the output capacitor.

In [5], the effect of sampling frequency of digital-controlled power converter is discussed. It was shown that low sampling frequency will cause phase delay in frequency domain. This delay can be reduced by increasing the sampling frequency up to several times of the switching frequency. However, the effect of sampling frequency on the controller stability and bandwidth is not fully explored.

This paper will discuss the discrete-time model of a buck converter including the DCR of inductor, output capacitor and its related ESR. Moreover, the effect of sampling frequency on the stability of controller and the bandwidth is discussed based upon the derived discrete-time domain model. It will be shown that the sampling frequency significantly deteriorates the bandwidth even though it is higher than the switching frequency. Moreover, the quantities analysis of its effect on the bandwidth of controller design is also included. Therefore, the results shown in this paper will provide a reference for controller design and selection of sampling frequency for digital-controlled power converter.

II. PROPOSED DISCRETE-TIME MODEL

Fig. 4 shows small signal model for a buck converter which includes DCR of output inductor and ESR of output capacitor [6] [7]. For Fig. 4, the transfer function of output voltage perturbation over duty variation can be derived as follows.

$$G_{vd} = \frac{\hat{v}_o}{\hat{d}}\bigg|_{\hat{i}_o = 0, \hat{v}_g = 0} = V_g \frac{R_L}{R_L + DCR} \frac{1 + \dfrac{s}{\omega_{esr}}}{\Delta} \quad (1)$$

where

$$\Delta = 1 + \frac{s}{Q\omega_o} + \frac{s^2}{\omega_o^2}$$

$$\omega_{esr} = \frac{1}{R_C C}$$

$$Q = \frac{1}{\omega_o} \frac{1}{\dfrac{R_L}{R_L + DCR}\left(R_C C + \dfrac{L}{R_L} + DCR\dfrac{R_L + R_C}{R_L}C\right)}$$

$$\omega_o = \sqrt{\frac{R_L + DCR}{LC(R_L + R_C)}}$$

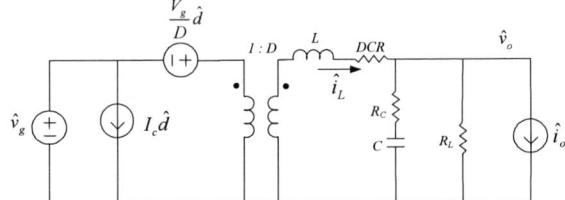

Fig. 4 Small signal model for a buck converter

The block diagram for the digital-controlled converter corresponding to Fig. 1 is shown in Fig. 5. As shown in Fig. 5, both reference and output voltages are sampled and converted to the digital counterparts via A/D converter. Fig. 6 can be further simplified by using one sampling mechanism to replace those shown in Fig. 6 to give equivalent results as mentioned in [1]. The equivalent block diagram for the digital-controlled converter corresponding to Fig. 1 is therefore derived and shown in Fig. 6. The time delay, T_d, includes conversion time of A/D converter and calculation time of digital compensator. In this paper, since the time delay, T_d, is about 67-nSec which is much lower than the sampling period 5-μSec, the time delay of T_d is negligible.

The loop gain transfer function, $T(z)$, of a sampled voltage mode controlled system can be written as (2). Different sampled frequencies will make a few differences of the z-domain transfer functions. The z-domain transfer function will be theoretical analyzed and derived.

$$T(z) = G_{ADC}C(z)F_m \frac{z-1}{z} Z\left[\frac{G_{vd}}{s}\right] \quad (2)$$

where:

G_{ADC} = A/D converter conversion ratio which is 115/1.5 in this paper

$C(z)$ = transfer function of digital controller

F_m = gain of PWM modulator which is 1/525 in this paper

$Z\left[\dfrac{G_{vd}}{s}\right]$ = the z - tranformation of $\dfrac{G_{vd}}{s}$

$\dfrac{z-1}{z}Z\left[\dfrac{G_{vd}}{s}\right]$ = the discrete - time domain model of G_{vd}

The discrete-time model of the power converter shown in Fig. 1 can be derived as shown in (3). In this model, ESR of capacitor and DCR of inductor are taken into consideration. More details about the deduction will be shown in appendix.

III. EFFECT OF SAMPLING FREQUENCY OF A/D CONVERTER ON CONTROLLER DESIGN

Based upon the derived discrete-time model shown in (2) and specifications and circuit parameters shown in Table 1, some results for the effect of sampling frequency of A/D converter on controller design are summarized in Table 2 and Table 3. As shown in Table 2, for the same bandwidth, lower sampling frequency gives less phase margin. Moreover, for the same phase margin, the reduction of sampling frequency results in smaller bandwidth and thereby deteriorating the dynamic performance shown in Table 3.

Table 1 Specifications and circuit parameters

Parameter (Unit)	Symbol	Value
Input Voltage (V)	V_g	12
Output Voltage (V)	V_o	1.5
Output Current (A)	I	10
Output Load (Ω)	R_L	0.15
Output Inductance (μH)	L	1.1
DC Resistance (mΩ)	DCR	2.24
Output Capacitance (μF)	C	3200
ESR (mΩ)	R_C	3
A/D Conversion Gain	G_{ADC}	115/1.5
PWM Modulation Gain	F_m	1/525
Switching Frequency (kHz)	f_{sw}	200

Table 2 Sampling frequency vs. PM

System	PM @ BW = 10 kHz	PM @ BW = 20 kHz
Continuous-time system	55.3°	56.1°
Sampling frequency = 100-kHz	11.9°	-30.9°
Sampling frequency = 200-kHz	35.1°	15.9°
Sampling frequency = 800-kHz	50.7°	46.7°

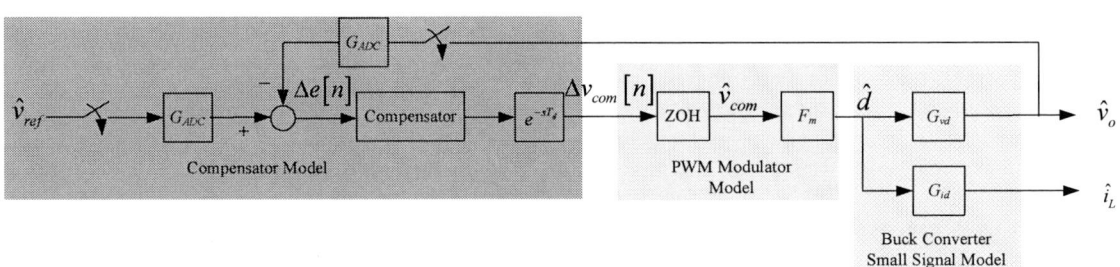

Fig. 5 Equivalent block diagram of digital-controlled converter

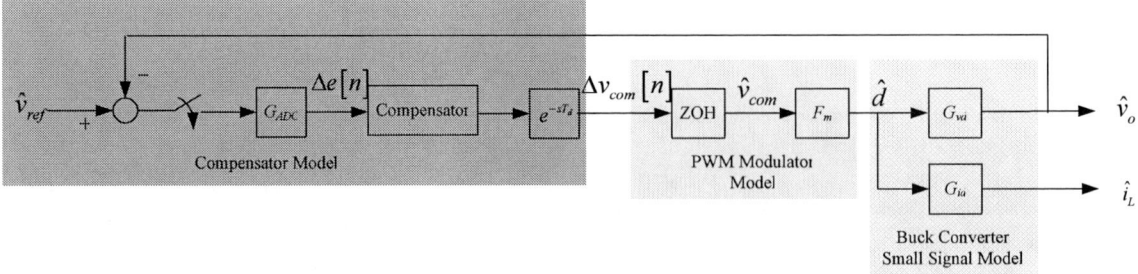

Fig. 6 Equivalent block diagram of digital-controlled converter

$$Z\left[\frac{G_{vd}}{s}\right]=V_g\frac{R_L}{R_L+DCR}\left(\frac{z}{z-1}-\frac{z^2-ze^{-\frac{\omega_o}{2Q}T}\cos\left(\left(\omega_o\sqrt{1-\frac{1}{4Q^2}}\right)T\right)}{z^2-2ze^{-\frac{\omega_o}{2Q}T}\cos\left(\left(\omega_o\sqrt{1-\frac{1}{4Q^2}}\right)T\right)+e^{-\frac{\omega_o}{Q}T}}-\frac{\frac{\omega_{esr}-2Q\omega_o}{\omega_{esr}\sqrt{4Q^2-1}}ze^{-\frac{\omega_o}{2Q}T}\sin\left(\left(\omega_o\sqrt{1-\frac{1}{4Q^2}}\right)T\right)}{z^2-2ze^{-\frac{\omega_o}{2Q}T}\cos\left(\left(\omega_o\sqrt{1-\frac{1}{4Q^2}}\right)T\right)+e^{-\frac{\omega_o}{Q}T}}\right) \quad (3)$$

Table 3 Sampling frequency vs. BW

System	BW (kHz) @ PM=45°	BW (kHz) @ PM=30°
Continuous-time system	49.3	76.2
Sampling frequency = 100-kHz	3.11	3.69
Sampling frequency = 200-kHz	3.30	13.7
Sampling frequency = 800-kHz	21.7	36.5

IV. EXPERIMENTAL RESULTS

A digital-controlled DC/DC converter is implemented which is with input voltage of 12 V, output voltage of 1.5 V, output current of 10 A, and switching frequency of 200 kHz. An FPGA, Xilinx Virtex-II Pro XC2VP30-4FF1152 [8], is used as the digital controller shown in Fig. 1. Fig. 7 and Table 4 show the measurement results of transient response for different sampling frequencies and open-loop bandwidth. As shown in Fig. 7 and Table 4, the transient response for bandwidth = 20 kHz is superior to that for 10 kHz bandwidth. However, the controller with 200-kHz sampling frequency and 20-kHz bandwidth results in undesired oscillation shown as Fig. 7 (B). Meanwhile, the phase margin will be reduced by increasing the bandwidth as shown in Fig 3. Increasing the sampling frequency can increase stability and overcome this issue as confirmed by Fig.7 (D).

V. CONCLUSION

This paper presents a novel method to model a sampled digital-controlled DC/DC converter. The presented discrete-time model takes ESR of capacitor and DCR of inductor into consideration. Moreover, the effect of sampling frequency of A/D converter on controller stability and bandwidth of digital-controlled power converter are explored. It has been shown and confirmed that the phase margin will be reduced by increasing the bandwidth and increasing the sampling frequency can increase stability. The results shown in this paper provides a quantitative reference for digital controller design.

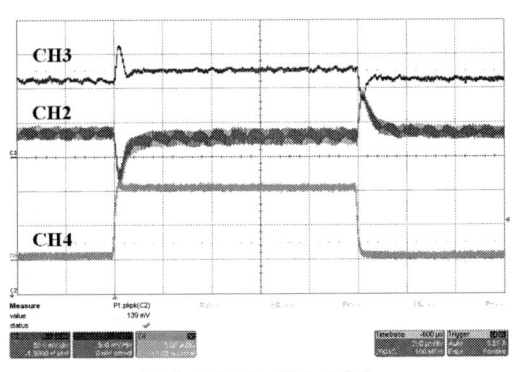

(A) f_s=200 kHz, BW=10 kHz

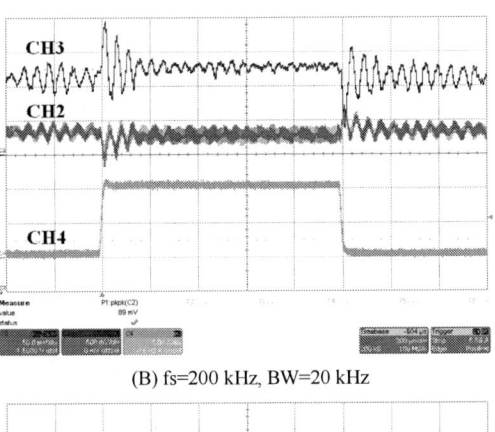

(B) fs=200 kHz, BW=20 kHz

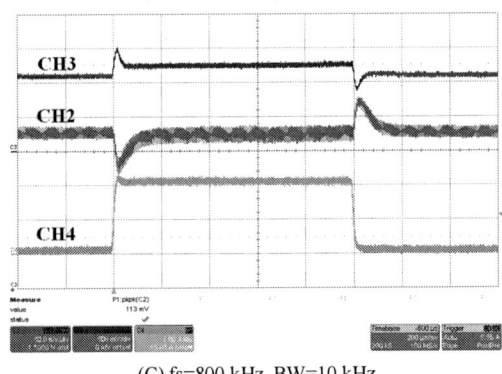

(C) fs=800 kHz, BW=10 kHz

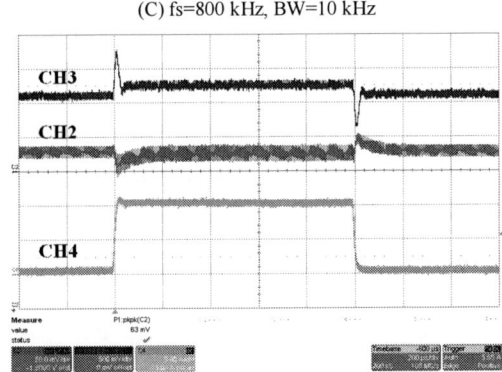

(D) fs=800 kHz, BW=20 kHz

Fig. 7 Experimental results, Ch2: Output Voltage (50 mV/Div, 1.5 V offset), Ch3: Compensator Output (500 mV/Div), Ch4: Output Current (5 A/Div)

Table 4 Experimental results, sampling frequency vs. loop gain bandwidth

Output peak to peak voltage	Loop gain bandwidth	
	10 kHz	20 kHz
Sampling frequency = 200-kHz	139 mV	89 mV
Sampling frequency = 800-kHz	113 mV	63 mV

REFERENCES

[1] G. F. Franklin, J. D. Powell, and M. L. Workman, Digital Control of Dynamic Systems, 3rd ed. Reading, MA: Addison Wesley, 1998.

[2] A Prodic, D. Maksimović, and R. W. Erickson, "Design and implementation of a digital PWM controller for a high-frequency switching DC/DC power converter," in Proc. IEEE IECON '01, vol.2, pp. 893-898, 29 Nov.-2 Dec. 2001.

[3] D. M. Van de Sype, K. De Gussemé, A. P. Van den Bossche, and J. A.

Melkebeek, "Small-signal Laplace-domain analysis of uniformly-sampled pulse-width modulators," in *Proc. IEEE PESC '04*, vol. 6, pp. 4292-4298, June 2004.

[4] D. M. Van de Sype, K. De Gussemé, F. De Belie, A. P. M. Van den Bossche, and J. A. Melkebeek, "Small-signal z-domain analysis of digitally controlled converters," *IEEE Trans. Power Electron.*, vol. 21, No. 2, pp. 470-478, Mar. 2006.

[5] L. Corradini, and P. Mattavelli, "Analysis of multiple sampling technique for digitally controlled dc-dc converters," in *Proc. IEEE PESC '06*, 18-22 June 2006.

[6] R. W. Erickson and D. Maksimović, *Fundamentals of Power Electronics*, 2nd ed., Norwell, MA: Kluwer Academic, 2001.

[7] R. B. Ridley, B. H. Cho, and F. C. Lee, "Analysis and interpretation of loop gains of multiloop-controlled switching regulators," *IEEE Trans. Power Electron.*, vol. 3, pp. 489-498, Oct. 1988.

[8] The Nallatech Limited, Eldersburg, MD. XtremeDSP Development Kit Pro User Guide, issue 1, 2004.

APPENDIX: DEDUCTION OF DISCRETE-TIME MODEL

Table A.1 Table of z-Transformation [1]

$F(s)$	$f(kT)$	$F(z)$
$\dfrac{1}{s}$	$1(kT)$	$\dfrac{z}{z-1}$
$\dfrac{s+a}{(s+a)^2+b^2}$	$e^{-akT}\cos bkT$	$\dfrac{z^2-ze^{-aT}\cos bT}{z^2-2e^{-aT}(\cos bT)z+e^{-2aT}}$
$\dfrac{b}{(s+a)^2+b^2}$	$e^{-akT}\sin bkT$	$\dfrac{ze^{-aT}\sin bT}{z^2-2e^{-aT}(\cos bT)z+e^{-2aT}}$

$$G_{vd}=V_g\frac{R_L}{R_L+DCR}\frac{1+\dfrac{s}{\omega_{esr}}}{1+\dfrac{s}{Q\omega_o}+\dfrac{s^2}{\omega_o^2}} \tag{A1}$$

Dividing (A.1) by the Laplace operator, s, and rewriting it to yield:

$$\frac{G_{vd}}{s}=\frac{V_g\omega_o^2}{\omega_{esr}}\frac{R_L}{R_L+DCR}\frac{1}{s}\frac{s+\omega_{esr}}{s^2+\dfrac{\omega_o}{Q}s+\omega_o^2} \tag{A.2}$$

Rewriting (A.2) by partial fraction to derive:

$$\frac{G_{vd}}{s}=\frac{V_g\omega_o^2}{\omega_{esr}}\frac{R_L}{R_L+DCR}\left(\frac{\dfrac{\omega_{esr}}{\omega_o^2}}{s}+\frac{-\dfrac{\omega_{esr}}{\omega_o^2}s+\left(1-\dfrac{\omega_{esr}}{Q\omega_o}\right)}{s^2+\dfrac{\omega_o}{Q}s+\omega_o^2}\right) \tag{A.3}$$

$$\frac{G_{vd}}{s}=V_g\frac{R_L}{R_L+DCR}\left(\frac{1}{s}-\frac{s+\dfrac{\omega_o^2}{\omega_{esr}}\left(\dfrac{\omega_{esr}}{Q\omega_o}-1\right)}{\left(s+\dfrac{\omega_o}{2Q}\right)^2+\left(\omega_o\sqrt{1-\dfrac{1}{4Q^2}}\right)^2}\right) \tag{A.4}$$

Rearranging (A.4) to get:

$$\frac{G_{vd}}{s}=V_g\frac{R_L}{R_L+DCR}\left(\frac{1}{s}-\frac{s+\dfrac{\omega_o}{2Q}}{\left(s+\dfrac{\omega_o}{2Q}\right)^2+\left(\omega_o\sqrt{1-\dfrac{1}{4Q^2}}\right)^2}-\frac{\dfrac{\omega_o}{2Q}-\dfrac{\omega_o^2}{\omega_{esr}}}{\left(s+\dfrac{\omega_o}{2Q}\right)^2+\left(\omega_o\sqrt{1-\dfrac{1}{4Q^2}}\right)^2}\right) \tag{A.5}$$

Rearranging (A.5) to give:

$$\frac{G_{vd}}{s}=V_g\frac{R_L}{R_L+DCR}\left(\frac{1}{s}-\frac{s+\dfrac{\omega_o}{2Q}}{\left(s+\dfrac{\omega_o}{2Q}\right)^2+\left(\omega_o\sqrt{1-\dfrac{1}{4Q^2}}\right)^2}-\frac{\omega_{esr}-2Q\omega_o}{\omega_{esr}\sqrt{4Q^2-1}}\frac{\omega_o\sqrt{1-\dfrac{1}{4Q^2}}}{\left(s+\dfrac{\omega_o}{2Q}\right)^2+\left(\omega_o\sqrt{1-\dfrac{1}{4Q^2}}\right)^2}\right) \tag{A.6}$$

By the formula of z-transformation shown in Table A.1 and (A.6), the related discrete-time transformation can be derived:

$$Z\left[\frac{G_{vd}}{s}\right]=V_g\frac{R_L}{R_L+DCR}\left(\frac{z}{z-1}-\frac{z^2-ze^{-\frac{\omega_o}{2Q}T}\cos\left(\left(\omega_o\sqrt{1-\dfrac{1}{4Q^2}}\right)T\right)}{z^2-2ze^{-\frac{\omega_o}{2Q}T}\cos\left(\left(\omega_o\sqrt{1-\dfrac{1}{4Q^2}}\right)T\right)+e^{-\frac{\omega_o}{Q}T}}-\frac{\omega_{esr}-2Q\omega_o}{\omega_{esr}\sqrt{4Q^2-1}}\frac{ze^{-\frac{\omega_o}{2Q}T}\sin\left(\left(\omega_o\sqrt{1-\dfrac{1}{4Q^2}}\right)T\right)}{z^2-2ze^{-\frac{\omega_o}{2Q}T}\cos\left(\left(\omega_o\sqrt{1-\dfrac{1}{4Q^2}}\right)T\right)+e^{-\frac{\omega_o}{Q}T}}\right) \tag{A.7}$$

978-1-4244-1871-8/07 $25.00 © 2007 IEEE

Power Spectra of the Single Phase 6/6 SRM Drives by the Separately Randomized Pulse Position (SRP) PWM Method

Nguyen Minh Khai*, Duck-Shick Shin**, Young-Gook Jung+, Young-Cheol Lim*
*Dept. of Electrical Engineering, Chonnam National University, Kwang-ju, KOREA
** Korea Electronics Technology Institute (KETI), Kwang-ju, KOREA
+Dept. of Electrical Engineering, Daebul University, Yongam-Gun, Chonnam, KOREA
jyg@mail.daebul.ac.kr

Abstract—It is known that if harmonic spectra are reduced, then acoustic noise also reduced. Hence, this paper proposes a new random switching strategy using DSP TMS320F2812 in order to reduce the harmonic spectra of a single phase 6/6 switched reluctance machine. The proposed method combines random turn-on, turn-off angle technique and random pulse width modulation technique. A Harmonic Spread Factor (HSF) is used to evaluate the random modulation technique. The proposed method is then being simulated in Matlab/ Simulink environment. Furthermore, this algorithm has also been inspected on various switching frequency. The experimental results show that the proposed strategy provides better harmonic spectra performance than conventional strategies, i.e. And a good agreement between experimental and simulation results is obtained.

Keywords – Single Phase 6/6 SRM, Power Spectra, Random Switching Technique, Harmonic Spread Factor (HSF)

I. INTRODUCTION

Switched reluctance machine (SRM) has gained a considerable attention in the areas of high-performance and adjustable speed drives. It can be a potential alternative to brushless d.c and a.c machines in various sectors of industry because of simple and robust structure, low cost and capability for high speed rotation [1,2]. However, one of the problems with the SRM is its higher acoustic noise when compared to an induction motor of the same size. The origin of the emitted acoustic noise in SRM is due to triggering mechanical resonances. It has been shown that the dominant source of the vibration and the acoustic noise in the SRM is radial vibrations of the stator [3].

In principle, acoustic noise can be reduced if the switching frequency is above 18 kHz. However, such high frequency switching results in high switching losses in power inverters. There have been some methods proposed recently to reduce the acoustic noise in SRM by modulation technique as shown in [4-6].

The first method is to vary the switching frequency randomly. The second one is to change between lagging edge and leading edge modulation. The other method is to use random pulse width modulation (RPWM) technique that is useful for induction motor [7,8]. And the last method as shown in [4] is randomly to vary the turn-on angle α_{on} and turn-off angle α_{off} within 1°-3° which is useful in the whole operating area of the SRM. However, those methods only

give a small shaping of the harmonic spectra which can reduce acoustic noise. Therefore a simple and effective method to reduce acoustic noise will be useful for practical uses.

In the paper, we propose a new method using random modulated strategy and random PWM technique for a 6/6 SRM. This technique brings acoustic noise decrease by combining the varying turn-on, turn-off angle and RPWM technique. While target of random turn-on, turn-off angle technique is to decrease amplitude of the fundamental harmonics, random pulse width modulation technique is to provide harmonic spectra intensity flatter than that obtained by conventional method. This combination will help avoiding the triggering mechanical resonances. Thus, the harmonic spectra of output voltage in SRM are reduced significantly.

II. BASIC PRINCIPLE OF OPERATION OF THE SRM

Linear analysis assumes that the inductance is unaffected by the current: that is, there is no magnetic saturation. For simplicity we also ignore the effect of fringing flux around the pole corners, and assume that all the flux crosses the airgap in the radial direction. Mutual coupling between phases is normally zero or very small, and is ignored. The voltage across a phase winding is given by [1],

$$v = Ri + \frac{d\psi}{dt} = Ri + L\frac{di}{dt} + \omega_m i \frac{dL}{d\theta} \qquad (1)$$

where v is the terminal voltage, i is the current, Ψ is the flux-linkage in volt-seconds, R is the phase resistance, L is the phase inductance, θ is the rotor position and ω_m is the rotor angular velocity in rad/s.

The instantaneous electric power vi is,

$$vi = Ri^2 + Li\frac{di}{dt} + \omega_m i^2 \frac{dL}{d\theta} \qquad (2)$$

The instantaneous electromagnetic torque is given by the following equations,

$$T_e = \frac{1}{2}i^2 \frac{dL}{d\theta} \qquad (3)$$

978-1-4244-1871-8/07 $25.00 © 2007 IEEE

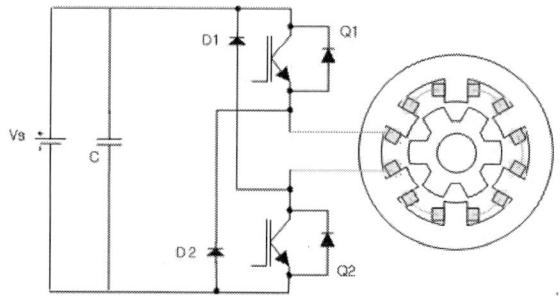

Fig. 1 Converter topology for single phase 6/6 SRM

In the single phase asymmetric bridge exciting power converter, there are two main switches and two flywheel diodes in each phase circuit. Fig. 1 show the converter topology used for drive.

The flux in the SRM is not constant but must be established from zero every stroke. In motoring operation the build-up is timed to coincide with the period when the rotor poles are approaching the stator poles of the phase that is due to be excited. The process is controlled by switching the supply voltage on at the turn-on angle and switching off at the commutation angle [2].

III. PROPOSED METHOD

In SRM the turn-on angle α_{on} and turn-off angle α_{off} can be controlled as well as the duty-cycle D [4]. The duty-cycle is normally controlled at low speed in order to reduce the current flow in the SRM.

At higher speed the current is limited by the back-emf and there is no need for using different duty-cycles. Instead the turn-on and turn-off angles are controlled. The proposed method is a combination of the random PWM technique and varying turn-on and turn-off angle that reduces the acoustic noise significantly.

The basic principle of the strategy is shown in figure 2. In figure 2 the angle α_0 is the angle where the rotor and stator pole starts overlapping physically, α_a is the aligned angle where the rotor is totally overlapped by the stator, α_{adv} is the advance angle (α_{adv} is positive if α_{on} is smaller than α_0 and conversely α_{adv} is negative) and $\Delta\alpha$ is the total conduction angle ($\alpha=0°$ is when the rotor and stator are completely unaligned).

The random strategy is then to vary the turn-on angle α_{on} and turn-off angle α_{off} randomly according to α_0 and α_a within $\Delta\alpha_r$ from 0° to 2° while $\Delta\alpha$ is kept constant. The mode 1, 2, 3, 4 as shown in Fig. 2 are practical cases for random the turn-on angle α_{on} and turn-off angle α_{off}, $\Delta\alpha_r$ is an interval for turn-on, turn-off angle control and r is the random angle value from 0° to 2°. Because the proposed method is applied not only the random PWM technique but also the random turn-on/off angle control, resultant power spectra is spread over the wide band area.

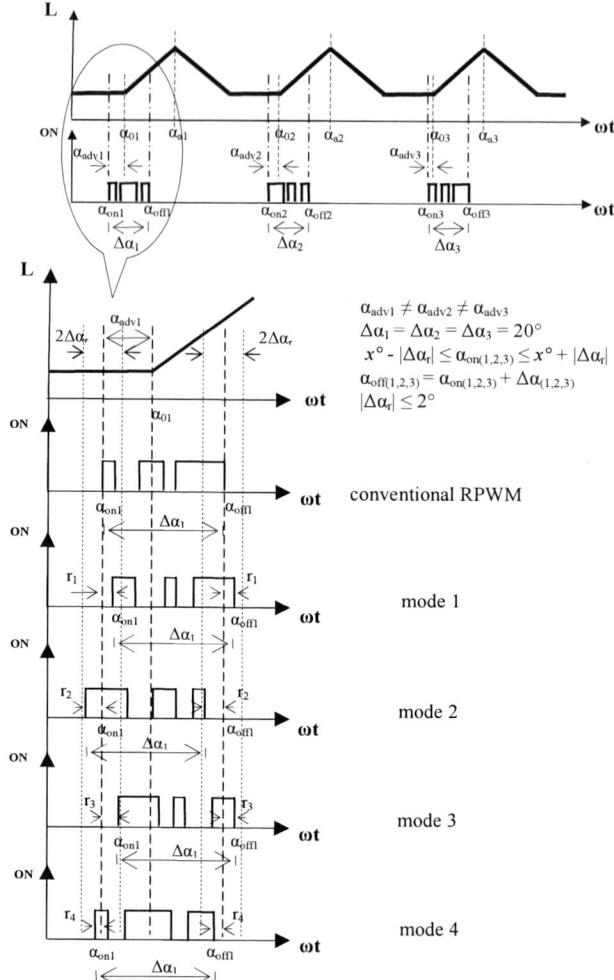

Fig. 2 The proposed method

To place the pulse position in the range of the modulation interval, there must be a determination method. A pre-defined random function, which has an even distribution, could be used for the determination method. Most microprocessor systems offer random functions generated by using a linear congruential generator (LCG) algorithm [7], which is able to generate a positive number in a predefined range. The following equation is used to generate the random number via a LCG. A positive integer random number f_{ran} within the range $[0, i_m]$ can be generated by,

$$f_{ran+1} = (f_{ran} \cdot i_a + i_c)\%i_m \qquad (4)$$

where % in the modulus function. And a floating point random number ran ranged 0 to 1 is generated by,

$$ran = (float)f_{ran}/(float)i_m \qquad (5)$$

where *"float"* produces a floating point number from an integer. An integer random number ranged j_{low} to j_{high} is generated by,

$$f = j_{low} + \frac{(j_{high} - j_{low} + 1) \cdot f_{ran}}{i_m} \qquad (6)$$

In (4)-(6), i_a, i_c and i_m are the selected coefficients and called the multiplier, increment and modulus, respectively. j_{low} and j_{high} are the minimum and maximum values of the generated random number respectively.

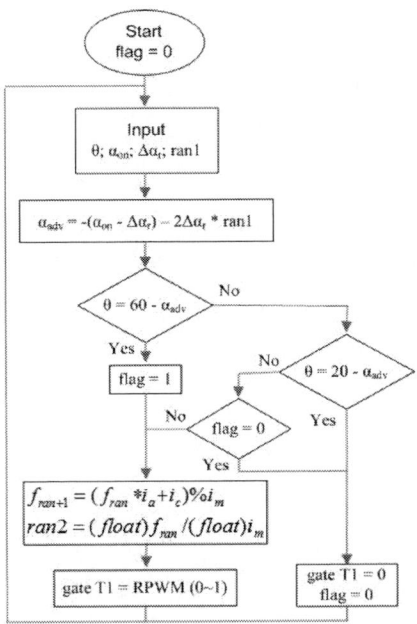

Fig. 3 The flowchart of the proposed method

Based on the flowchart of random turn-on/off angle control program, as seen in Fig. 3, the rotor position θ is input from the encoder; the reference turn-on angle α_{on} is input from speed controller; the angle $\Delta\alpha_r$ is a constant that chosen for the best performance operation system and *ran1* is generated by LCG.

The advance angle α_{adv} is calculated by these input values so as to guarantee the turn-on angle is triggered in the interval $[\alpha_{on} - \Delta\alpha_r, \ \alpha_{on} + \Delta\alpha_r]$. This value is compared with rotor position to trigger on the IGBT gate. And the turn-off angle is set based on the following formula,

$$\alpha_{off} = \alpha_{on} + 20° \qquad (7)$$

IV. EXPERIMENTAL SYSTEM

Fig. 4 shows the system block of the experimental SRM drive. It consists of the main control board which includes a micro controller DSP TMS320F2812, a single phase IGBT

inverter, an absolute position encoder, load and single phase 6/6 SRM with parameters shown in Table 1.

All of the controllers are digitally programmed with DSP, which has many useful functions for motor control such as A/D converter, Quadrature Encoder Pulse (QEP) circuit, I/O port and PWM generation. The system also contains a rheostat to control advance angle, current sensors to measure the current and a LCD for displaying speeds.

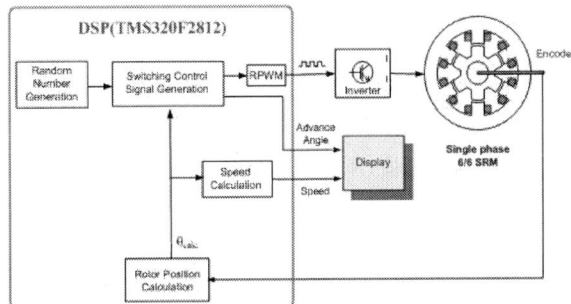

Fig. 4 System block of the experimental SRM drive.

TABLE 1
SRM PARAMETERS FOR EXPERIMENTAL CONDITION

Parameter		Value
DC bus voltage/Current		30V/1A
PWM Switching frequency (f_{sw})		6 kHz
$\Delta\alpha_r$		2 degree
Rated Power		30 W
Single phase SRM	No. of Stator Pole	6
	No. of Stator Pole	6

V. EXPERIMENTAL RESULTS

In this section, simulation and experimental results are presented and compared for three strategies respectively. The first strategy is conventional method with chopping mode which used PWM technique with fixed turn-on and turn-off angle. The second one is conventional method with RPWM technique which employed random PWM technique while turn-on and turn-off angle are fixed.

The last one is the proposed method which was analysed above. The simulation of the SRM prototype is performed based on Matlab/Simulink environment. The simulation and experimental results showed that combination of random turn-on, turn-off angle technique with random PWM technique reduce the harmonics spectra significantly.

Fig. 5 and 6 show simulation and experimental results for power spectra of output voltage at frequency span 0–10 kHz. In Fig. 5(c) and 6(c) for the proposed method, the dominant harmonic cluster at frequency around 6 kHz is reduced in comparison with conventional method as shown in Fig.5 (a)(b) and Fig.6 (a)(b).

978-1-4244-1871-8/07 $25.00 © 2007 IEEE

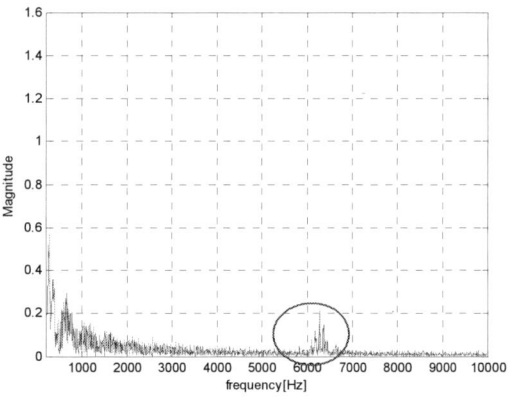

a) Conventional method with chopping mode

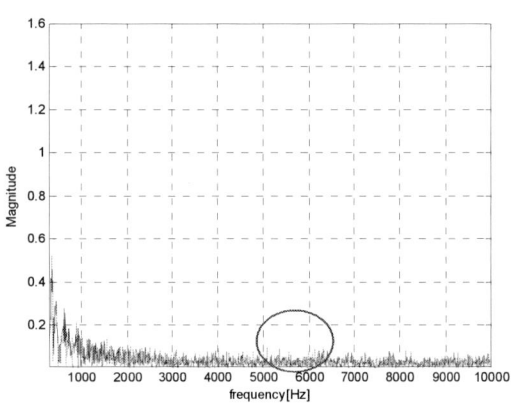

b) Conventional method with RPWM technique

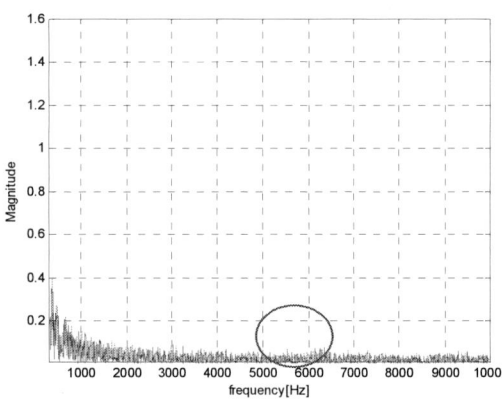

c) The proposed method ($\Delta\alpha_r = 2°$)

Fig. 5 Simulation results of the output voltage spectra at frequency span 0 – 10 kHz

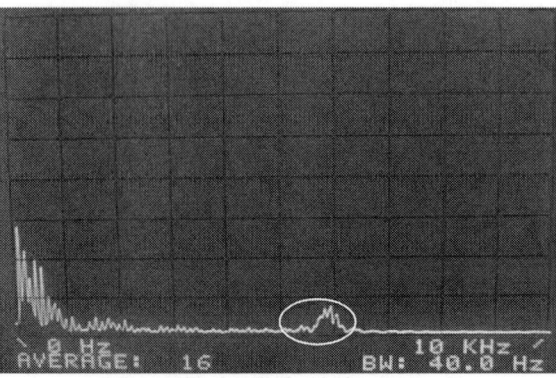

a) Conventional method with chopping mode

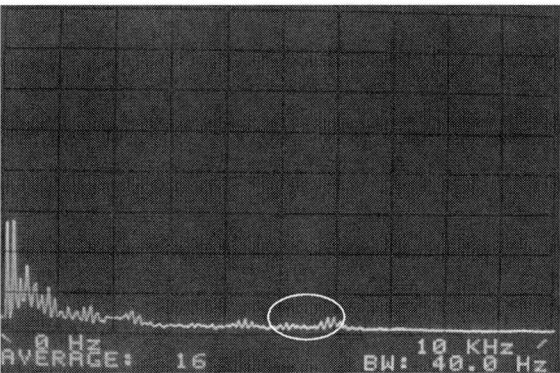

b) Conventional method with RPWM technique

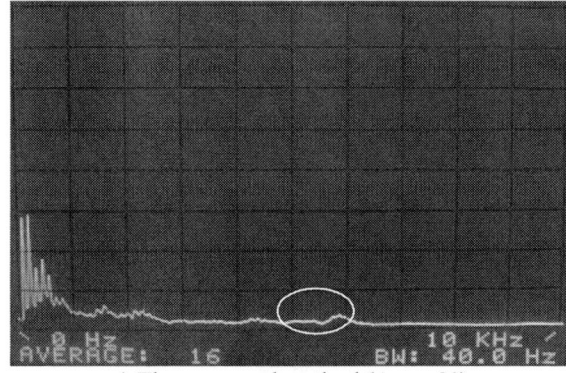

c) The proposed method ($\Delta\alpha_r = 2°$)

Fig. 6 Measured results of the output voltage spectra at frequency span 0 – 10 kHz
(x-axis: 250Hz/div.; y-axis: 50mv/div.)

Furthermore, for evaluating the random PWM technique, a simple indicator of quality of voltage spectra would be useful. For this purpose, the concept of statistical deviation can be employed and the Harmonics Spread Factor (HSF) [9] is defined as,

$$HSF = \sqrt{\frac{1}{N}\sum_{j>1}^{N}(H_j - H_0)^2} \qquad (8)$$

where N denotes the total number of frequency components considered, H_j is the amplitude of the jth component, and H_0 is the average value of all components and given by,

$$H_0 = \frac{1}{N}\sum_{j>1}^{N}H_j \qquad (9)$$

The HSF quantifies the spread spectra effect of random PWM and it should be possibly small. For ideally flat spectra of white noise, the HSF would be zero.

TABLE 2
COMPARISONS OF HSF FOR VARIOUS SWITCHING FREQUENCY

PWM types f_{sw} [kHz]	Fixed chopping mode	Conventional RPWM	Proposed method
4	38.00%	18.58%	17.05%
6	38.33%	18.31%	16.29%
8	38.15%	17.68%	17.04%
10	38.19%	17.61%	14.32%
12	38.63%	20.06%	17.32%

TABLE 3
COMPARISONS OF HSF FOR VARIOUS $\Delta\alpha_R$ FOR PROPOSED METHOD

$\Delta\alpha_r$ (degree)	1	1.5	2	2.3	3
HSF (%)	17.13	15.92	16.29	17.74	17.07

Table 2 gives the output voltage HSF for various switching frequency. The HSF of the proposed method is lower than that of the conventional methods. And the HSF of the proposed method reaches the lowest value 14.32% at switching frequency 10 kHz. For proposed method, Table 3 gives the output voltage HSF for various $\Delta\alpha_r$.

In comparing these results, we can observe that the harmonic spectra are reduced when using random modulation. It is clear that, as expected, the output voltage spectra of the proposed method both decrease the amplitude of the fundamental harmonics and a flatter provide harmonic spectra intensity than that obtained by conventional method. Therefore the proposed strategy results in an improvement in the acoustic noise reduction relative to other strategies.

VI. CONCLUSION

This paper has discussed the implementation techniques for various random PWM and random turn-on, turn-off angle strategies in single-phase 6/6 SRM. The proposed random technique has been compared with the conventional methods, to analyse their influence on harmonic spectra reduction. The HSF is used to evaluate the random PWM technique. The simulation and experimental results demonstrate that the new proposed strategies provide better harmonic spectra performance than conventional strategies in spreading the harmonic power over a wide frequency range. And the HSF of the proposed method reaches a low value

ACKNOWLEDGMENT

This study was supported by the Research Center for Stabilization of Power Systems with Large Distributed Generations at Chonnam National University in KOREA.

REFERENCES

[1] T. J. E. Miller, Ed., *Electric Control of Switched Reluctance Machines*, Newnes, Oxford, 2001.

[2] T. J. E. Miller, *Switched Reluctance Motors and Their Control*, Oxford University Press, New York, 1993.

[3] D. E. Cameron, J. H. Lang, and S. D. Umans, 'The Origin and Reduction of Acoustic Noise in Doubly Salient Variable-Reluctance Motors,' *IEEE Trans. Ind. Applicat.*, vol. 28, no. 6, pp. 1250-1255, 1992.

[4] F. Blaabjerg and J. K. Pedersen, "Digital Implemented Random Modulation Strategies for ac and Switched Reluctance Drives", *in Conf. Rec. of IEEE IECON'93*, 1993, pp.676-682.

[5] Chi-Yao Wu and Charles Pollock, "Analysis and Reduction of Vibration and Acoustic Noise in the Switched Reluctance Drive," *IEEE Trans. Ind. Applicat.*, vol. 31, no. 1, pp. 91-98, 1995.

[6] P. Pillay, R. M. Samudio, M. Ahmed and R.T. Patel, "A Chopper-Controlled SRM Drive for Reduced Acoustic Noise and Improved Ride-Through Capability Using Super-capacitors", *IEEE Trans. Ind. Applicat.*, vol.31, no.5, pp.1029-1038, 1995.

[7] S. O. Wi, Y. G. Jung, S. H. Na and Y. C. Lim, 'Separately Randomized Pulse Position PWM Technique with Fixed Switching Frequency for Power Electronic Converters', *in Conf. Rec. of ICEE'02 (KOREA)*, 2002: pp.780-785.

[8] V. Blasko, M. M. Bech, F. Blaabjerg and J. K. Pedersen, "A New Hybrid Random Pulse Width Modulator for Industrial Drives", *in Conf. Rec. of IEEE APEC'00*, 2000, pp. 932-938.

[9] R. L. Kirlin, M. M. Bech, A. M. Trzynadlowski and B. Huo, "Power and Power Spectral Density in PWM Inverters with Randomized Switching Frequency", *in Conf. Rec. of IEEE PESC'01*, 2001, pp.188-192.

Inductance Vector Angle Based Sensorless Speed Estimation in Switched Reluctance Motor Drive

Fuat Kucuk[*], Hiroki Goto[*], Hai-Jiao Guo[**] and Osamu Ichinokura[*]

* Tohoku University	** Tohoku Gakuin University
Graduate School of Engineering	Graduate School of Engineering
6-6-05, Aoba, Aramaki, Aoba-ku, Sendai, Japan	1-13-1, Chuo, Tagajo, Japan
power28@ec.ecei.tohoku.ac.jp	kaku@tjcc.tohoku-gakuin.ac.jp

Abstract—**Feedbacks signals of rotor speed and motor torque are essential in most of Switched Reluctance (SR) motor control applications. An SR motor has highly nonlinear characteristics that do not allow to be modeled by simple equations. Hence, the feedback signals can not be mathematically calculated from the model. Instead of calculation, they should be measured or estimated. In Direct Torque Control (DTC) drive, which enables easy control of torque ripple in the SR motor, position sensor is employed to obtain the feedback signals. Position sensor causes DTC drive not only less reliable but also more expensive. Estimation of feedback signals is required in order to eliminate position sensor (or encoder). This paper concerns about sensorless speed estimation under the DTC condition and presents a simple method. The proposed method is based on inductance vector angle. The inductance vector angle is obtained by applying α – β transformation to the phase inductances. A relay triggers a speed calculation circuit according to its band limits and the inductance vector angle. Inside the circuit, triggering time is kept in a memory until the next triggering. Then, rotor pole pitch is divided by the time difference between two consecutive triggerings. Finally, the estimation circuit outputs the rotor speed. Estimation method is simulated and verified experimentally to show its validity.**

I. INTRODUCTION

Switched Reluctance (SR) motors have attracted significant interest of industry due to their advantages such as simple structure, rugged behavior, and low cost manufacturing. The absence of windings and magnets on the rotor enables SR motors to run high speed and temperature. An SR motor can produce large torque in a wide speed range. However, it has some drawbacks such as torque ripple and acoustic noise. High torque ripple is inevitable unless ripple reduction strategy is applied.

Direct Torque Control (DTC) is one of the most interesting methods that allows user to control torque ripple easily. Conventional DTC has firstly and successfully applied to the ac motors. It is based on field orientation and direct self control. It can not be directly applied to an SR motor because of non-sinusoidal and independent excitation of its phases. DTC for the SR motor has firstly described and simulated within the past decade [1]. A linear equation that does not need rotor position has been used for torque estimation. It is not applicable in all operating conditions because SR motor has highly nonlinear characteristics. In addition, a new winding configuration is required. DTC with nonlinearity of the SR motor has recently described and experimented after

making some analogues with the conventional DTC [2]. Torque ripple has been controlled without any change on the winding configuration.

In order to implement the DTC, feedback signals of rotor speed and motor torque are essential. Mathematical models of some motors can be expressed with simple equations. Thus, there is a chance to calculate the feedback signals from these equations. However, the SR motor has highly nonlinear characteristics which are not possible to be expressed in a simple way. For this reason, feedback signals should be measured or estimated.

Position sensor (or encoder) is employed to obtain the feedback signals in the DTC. However, position sensor inevitably increases the cost and the motor size. On the other hand, it decreases the reliability. Estimation of the feedback signals is required in order to eliminate the position sensor.

Few studies have been reported about sensorless DTC so far. One of them includes magnetic linearization that is contrary to the SR motor nature [1]. The other is complicated for real time implementation [3]. In order to eliminate the position sensor and implement the DTC in a simple sensorless way, motor torque and rotor speed should be directly estimated from the electrical quantities.

Sensorless torque estimation has been already proposed in different studies [4], [5]. This paper concerns only sensorless speed estimation under the DTC condition and presents a novel method. Motor speed is directly estimated from inductance vector angle which is explained in the third section. Direct estimation means that position estimation is not used as an internal step.

II. DTC DRIVE FOR SR MOTOR

Neglecting mutual inductance between the motor phases, phase torque can be written as a change of co-energy with respect to the rotor position:

$$\tau_k = \frac{\partial W_k'(i_k, \theta)}{\partial \theta}\bigg|_{i_k=const}. \tag{1}$$

In equation (1), k represents number of motor phases, W_k' is the co-energy of k-th winding depending on the rotor position θ and phase current i_k.

Phase torque equation can be written in another form which was derived by White [6]:

978-1-4244-1871-8/07 $25.00 © 2007 IEEE

$$\tau_k = i_k \frac{\partial \psi_k(i_k, \theta)}{\partial \theta} - \frac{\partial W_{fk}(i_k, \theta)}{\partial \theta} \qquad (2)$$

where $\psi_k(i_k, \theta)$ and W_{fk} represent phase flux linkage and phase field energy respectively. It is firstly shown by Byrne that second term of (2) has small effect due to saturation of the SR motor and can be neglected [7]:

$$\tau_k \approx i_k \frac{\partial \psi_k}{\partial \theta}. \qquad (3)$$

The principle of DTC applied to the SR motor is based on this approximated equation [2]. Current flows in one direction and it has no effect on the direction of torque. In addition, it has first order delay with respect to $\partial \psi_k / \partial \theta$. Thus, current can be assumed constant when $\partial \psi_k / \partial \theta$ is controlled. The main purpose is to keep total torque constant. Therefore, each phase flux should be combined. The total flux linkage can be called as stator flux linkage ψ_s.

DTC scheme for the SR motor can be defined as follows:

    a) Magnitude of the stator flux linkage is kept at a constant value using hysteresis comparator.

    b) Switching command is produced according to the positive or negative change of $\partial \psi_s / \partial \theta$.

Block diagram of DTC used in this study is given in Fig. 1. Current, flux linkage and torque of each phase must be known to implement DTC. Phase currents and phase voltages are directly measured from voltage and current sensor whereas rotor position is measured from the position encoder. Phase flux linkages are calculated from the integration equation

$$\psi_k = \int (v_k - R_k i_k) dt. \qquad (4)$$

v_k and R_k represent supply voltage and resistance of the phase winding respectively.

Stator flux linkage ψ_s is obtained by applying α-β transformation to the phase flux linkages. By locating phase A on the α-axis, the transformation equations can be written for the 6/4 SR motor as

$$\begin{bmatrix} \psi_\alpha \\ \psi_\beta \end{bmatrix} = \begin{bmatrix} 1 & -\cos 60° & -\cos 60° \\ 0 & \sin 60° & -\sin 60° \end{bmatrix} \begin{bmatrix} \psi_A \\ \psi_B \\ \psi_C \end{bmatrix}. \qquad (5)$$

Stator flux linkage ψ_s and its angle δ can be found from

$$\psi_s = \sqrt{\psi_\alpha^2 + \psi_\beta^2} \qquad (6)$$

$$\delta = \tan^{-1}\left(\frac{\psi_\beta}{\psi_\alpha}\right). \qquad (7)$$

ψ_α and ψ_β in these equations are α and β components of the stator flux linkage. At the same time, a look-up table takes the rotor position and the phase current as inputs and estimates the phase torque. Total torque is the summation of each phase torque. Magnitude of the stator flux linkage and the motor torque are fed into the hysteresis comparators with their reference values. Reference torque is obtained from the output of speed controller whereas reference flux linkage is directly from the speed. Voltage vector selector determines optimal voltage vector. Finally, switching command is provided for an asymmetric half bridge converter which is commonly used for the SR motor (Fig. 2).

The equation given in (3) is an approximated equation. It is only used to determine the optimum switching command. Torque is actually provided from look-up table and then controlled by hysteresis comparator. Therefore, accuracy loss does not occur in DTC.

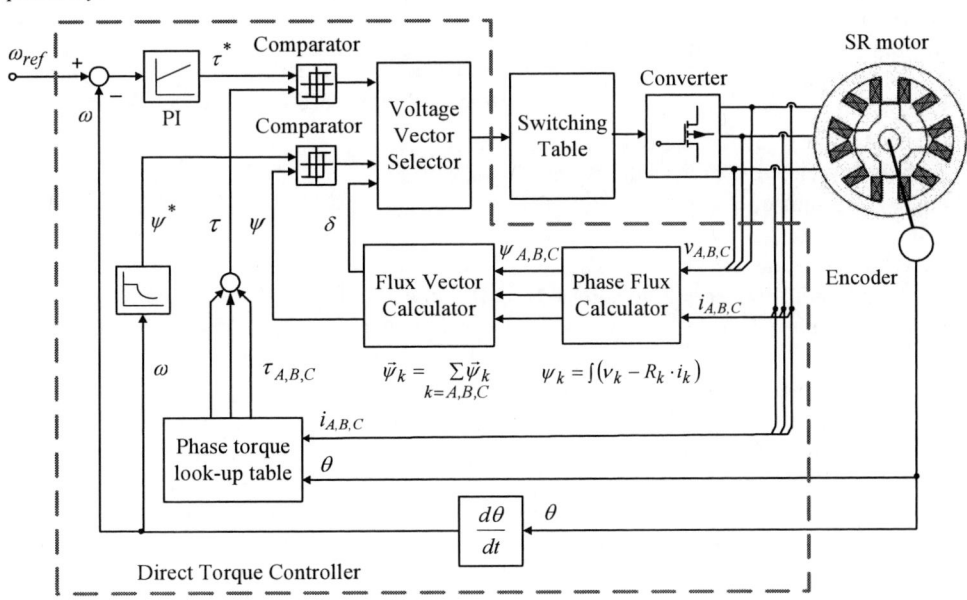

Fig. 1. Block diagram of DTC for SR motor

978-1-4244-1871-8/07 $25.00 © 2007 IEEE

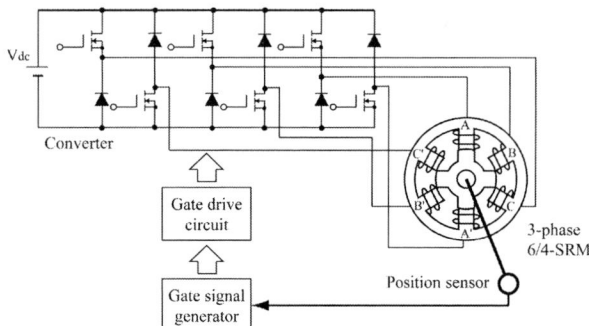

Fig. 2.   An asymmetric half bridge converter for SR motor driving.

III.   INDUCTANCE VECTOR ANGLE BASED SENSORLESS SPEED ESTIMATION

It is realized that inductance vector angle is very useful for sensorless speed estimation in the DTC. The inductance vector angle can be found by applying the α-β transformation to the phase inductances. Fig. 3 shows phase torque and inductance vector angle of the SR motor controlled by sensor – based DTC. As seen in the figure, the inductance vector angle has almost the same zero crossing time with the phase torque when it changes from negative to positive. If the time difference between two consecutive zero crossings are determined from the inductance vector angle, then rotor speed ω_m can be calculated as

$$\omega_m \cong \frac{\theta_{rp}}{\Delta t}. \qquad (8)$$

In Eq. (8), θ_{rp} is rotor pole pitch which is 90° for 6/4 SR motor and Δt is the time difference between two consecutive zero crossings. Sensorless speed estimation is based on this calculation at every zero crossing time.

Overall diagram of proposed method is shown in Fig. 4. Phase currents and voltages are already measured in sensor-based DTC. They can also be used in sensorless speed estimation. Phase flux linkages and then phase inductances can be calculated from the measured electrical quantities. Inductance vector angle is obtained by applying α – β transformation to the phase inductances. A speed calculation circuit is triggered by a relay according to its band limits and inductance vector angle. Band limits of the relay are adjusted so that only zero crossings of the inductance vector angle from negative to positive is taken into account. When the circuit is triggered, present crossing time is saved and then kept in a memory until the next triggering. Thus, speed is calculated and updated every consecutive zero crossing time.

Some oscillations may occur on the inductance vector angle and it may cause undesirable zero crossings. That is, speed estimation circuit may not work properly. This can be solved by using a simple filter. Any filter inevitably introduces time delay. However, filter delay is much smaller

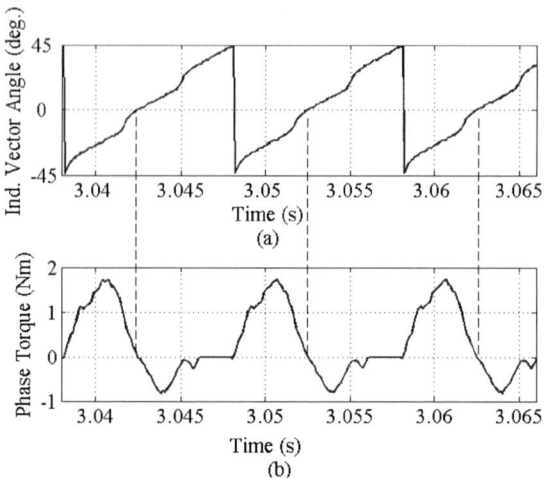

Fig. 3. Inductance vector angle and phase torque of SR motor controlled by sensor – based DTC.

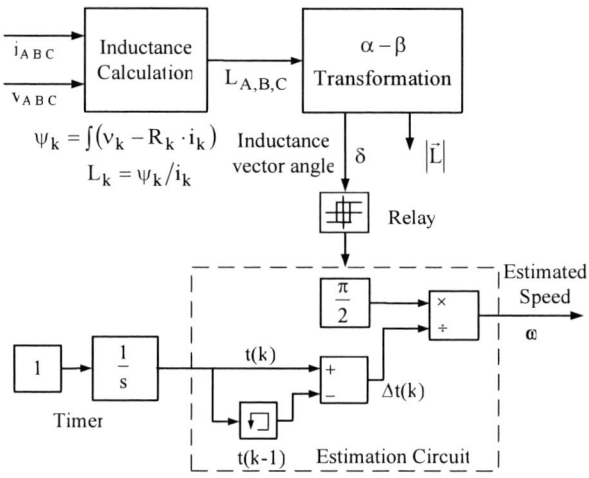

Fig. 4. Block diagram of sensorless speed estimation

than the mechanical delay. Therefore, it can be neglected when a filter is used for speed estimation.

IV.   SIMULATION RESULTS

Sensorless speed estimation is simulated and compared with the sensor-based speed measurement while a 3-phase 6/4 SR motor is being controlled by DTC. Matlab / Simulink is found to be good for performing the simulations. Simulation time and step are chosen 25 s and 50 µs respectively in all simulations. Supply voltage is kept at 50 VDC and winding resistances of all phases are assumed to be constant at 0.87Ω. Bandwidths of hysteresis comparators are adjusted to ± 0.05 Nm and ± 1.5 mWb for torque and flux linkage respectively. A proportional – integral (PI) controller is used for motor speed regulation. The SR motor is run in two different operations: A) constant speed operation, B) constant load operation. Each operation is explained in detail below.

978-1-4244-1871-8/07 $25.00 © 2007 IEEE

A. Constant Speed Operation

The SR motor is run at 1000 rpm. Load torque is firstly increased from 0.3 Nm to 0.9 Nm with 0.3 Nm step, and then decreased from 0.9 Nm to 0.3 Nm with 0.3 Nm step again (Fig. 5(a)). Reference speed is kept at 1000 rpm during the simulation. Measured speed and sensorless estimated speed in DTC are shown in Fig. 5(b) and 5(c) respectively. As seen in the figures, estimation method works well even the load change as a step function and it gives almost same result with the measurement.

B. Constant Load Operation

The SR motor is run at 500 rpm while load torque is kept constant at 0.6 Nm. Reference speed is firstly increased to 500 rpm to 1000 rpm with 250 rpm step and then decreased to 500 rpm with 250 rpm step again (Fig. 6(a)). The load torque is kept at 0.6 Nm during the simulation. Measured speed and sensorless estimated speed in DTC are given in Fig. 6(b) and 6(c) respectively. Also in this operation, the estimated method gives nearly same response with the measurement.

V. EXPERIMENTAL RESULTS

Proposed speed estimation method has been experimentally verified. It has compared with the sensor – based speed measurement while controlling the SR motor by DTC. Basic properties of the SR motor used in the experiment are given in Fig. 7. The experimental parameters are chosen the same as simulation parameters. In order to implement DTC, a DSP board named dSPACE DS1103 and Matlab / Simulink Real-time workshop are used at 50 kHz switching frequency. Fig.8 illustrates the experimental setup. The SR motor is run in two different operations as explained in the previous section. Fig. 9(a) shows the load torque change, whereas Fig 9(b) and 9(c) shows respectively the measured and estimated speed under the constant speed operation. Fig 10(a), 10(b) and 10(c) illustrate respectively the reference speed, the measured and estimated speeds under the constant load operation. As seen in the figures, sensorless speed estimation works as well as sensor – based estimation.

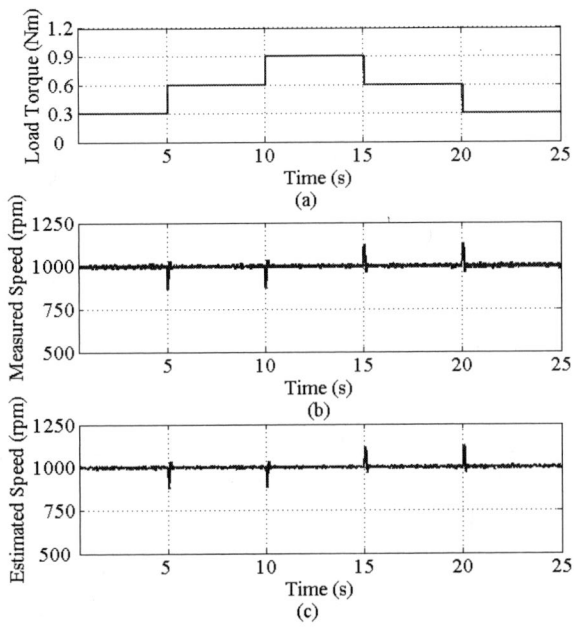

Fig. 5. Simulation results under constant speed operation (ω_{ref} = 1000 rpm).
a) Load torque b) Measured speed c) Estimated speed

Parameter	Value
Stator pole arc	30 deg.
Rotor pole arc	32 deg.
Stack length	51 mm
Rotor outer diameter	40 mm
Stator inner diameter	82.1 mm
Number of winding/pole	72 turns
Winding resistance/phase	0.87 Ω

Fig. 7. Basic properties of the 3 phase 6/4 SR motor

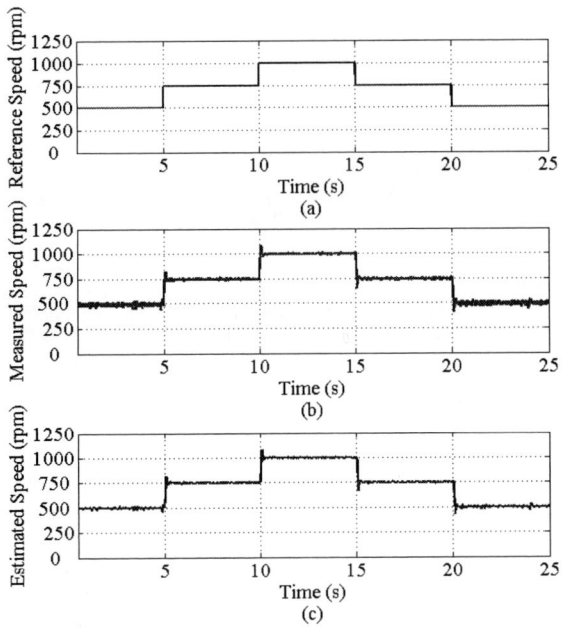

Fig. 6. Simulation results under constant load operation (T_L = 0.6 Nm).
a) Reference speed b) Measured speed c) Estimated speed

Fig. 8. Experimental setup

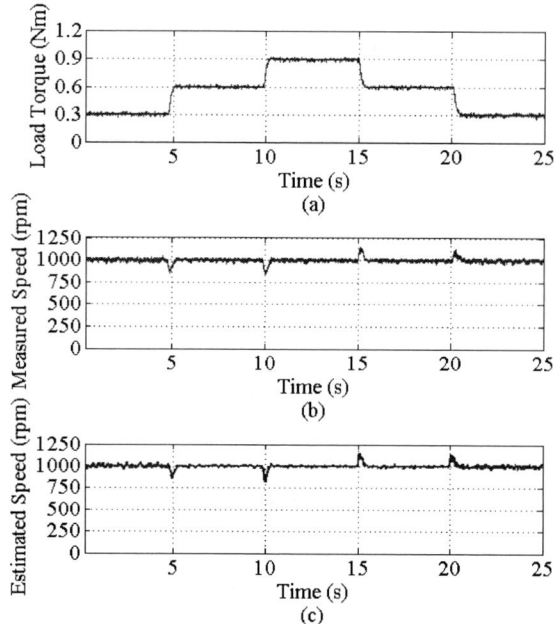

Fig. 9. Experimental results under constant speed operation (ω_{ref}= 1000 rpm).
a) Load torque b) Measured speed c) Estimated speed

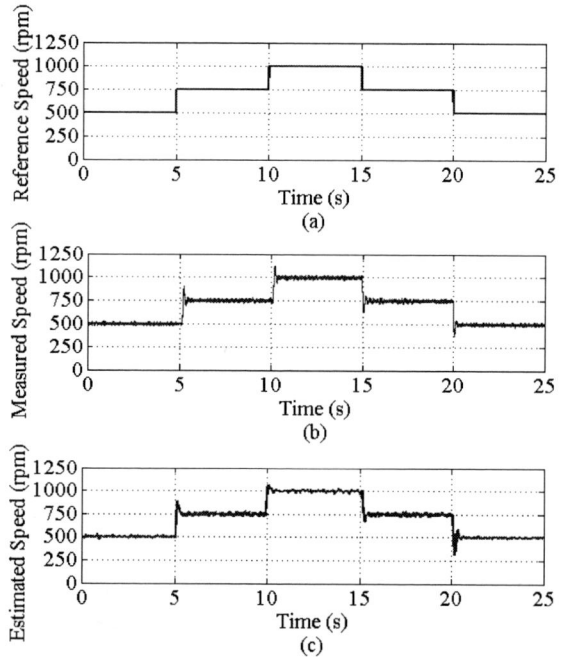

Fig. 10. Experimental results under constant load operation (T_L = 0.6 Nm).
a) Reference speed b) Measured speed c) Estimated speed

Fig. 11 gives transient speed estimation of the proposed method under constant load operation. Fig. 11(a) illustrates transient rotor speed while the reference speed is increased from 500 rpm to 750 rpm and Fig. 11(b) illustrates transient rotor speeds while the reference speed is decreased from 750 rpm to 500 rpm. Estimated speed resembles a step function because it is not continuously updated. The new value is only updated at zero crossings of inductance vector angle.

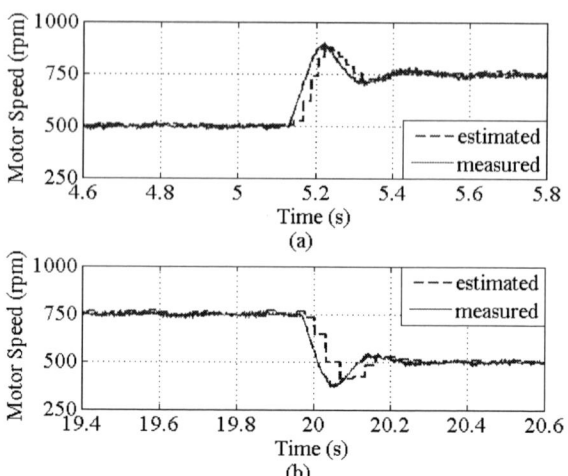

Fig. 11. Transient speed under constant load operation (T_L = 0.6 Nm).
a) Speed increasing b) Speed decreasing

VI. CONCLUSION

Sensorless speed estimation method proposed in this study has simulated and then experimented in order to investigate its performance. In addition, it has compared by sensor – based speed measurement. Both simulation and experimental results show that sensorless speed estimation works as well as the sensor - based speed measurement.

The fundamental advantages of this proposed estimation method can be summarized as below:

- it has very simple structure,
- rotor speed is estimated directly from the electrical quantities,
- position encoder and position estimation is not required,
- it may be used for other control methods if the inductance vector profile is suitable.

REFERENCES

[1] P. Jinupun and P. C.-K. Luk, "Direct torque control for sensorless switched reluctance motor drives," *7th Int. Conf. on Power Electronics and Variable Speed Drives*, pp. 329–334, 1998.

[2] A. D. Cheok and Y. Fukuda, "A new torque and flux control method for switched reluctance motor drives," *IEEE Trans. on Power Electronics*, vol. 17, pp. 543-547, 2002.

[3] M. Mitani, H. Goto, H. J. Guo and O. Ichinokura, "Position sensorless direct torque control of SR motors," *12th International Power Electronics and Motion Control Conference*, pp. 1143-1148, 2006.

[4] A. Bellini, F. Filippetti, and G. Franceschini, C. Tassoni, and P. Vas, "Position sensorless control of a SRM drive using ANN-techniques," *IEEE Industrial Application Conf., 33th IAS Annual Meeting*, vol. 1, pp. 709-714, 1998.

[5] F. Kucuk, H. Goto, H. J. Guo and O. Ichinokura, "Sensorless torque estimation in swiched reluctance drive using artificial neural networks and inductance vector angle," *IEEJ Technical Meeting on Rotating Machinery*, 2007.

[6] W. D. White, *Electromagnetic Energy Conversion*, New York: Wiley, 1959.

[7] J. V. Byrne and J. G. Lacy, "Characteristics of saturable stepper and reluctance motors," *Proc. Conf. Small Elect. Mach.*, pp. 93 – 96, 1976.

978-1-4244-1871-8/07 $25.00 © 2007 IEEE

A Position Sensorless Control System of SRM over Wide Speed Range

Min-Huei Kim
Yeungnam College of
Sci. & Tech.
Dep. of Elec. Engineering
1737, Daemyeung-dong,
Nam-gu, Daegu, 705-703
Email: mhkim@ync.ac.kr

Won-Sik Baik
Yeungnam College of
Sci. & Tech.
Dep. of Elec. Engineering
1737, Daemyeung-dong,
Nam-gu, Daegu, 705-703
Email: priwsb@hanmail.net

Dong-Hee Kim
Yeungnam University
Dep. of Electrical
Engineering
214-1, Dae-dong, GyeongSan
Gyeongbuk, 712-749
Email: dhkim@yu.ac.kr

Nam-Hun Kim
Cheongju University
School of Electronics &
Information Communication
36, Naedok-dong, Sangdang-gu,
Cheongju, Chungbuk, 360-764
Email: nhkim@cju.ac.kr

Kyeong-Ho Choi
Kyungbuk College
Dep. of Elec. Engineering
630, Hyuchon2-dong,
Yeongju, Gyeongbuk,
750-712
Email: ckh@kbc.ac.kr

Abstract—This paper presents a position sensorless control system of SRM over wide speed range. The rotor position estimation of the SRM is somewhat difficult because of its highly nonlinear characteristics. In this paper, the magnetizing curve of prototype 1-hp SRM is obtained from locked rotor test and complicated data patters are trained using Neural Network. To estimate more accurate rotor position, current-flux-rotor position lookup table over wide operating rage is generated using trained Neural Network. Experimental result for a 1-hp SRM over 16:1 speed control range is presented for the verification of the proposed sensorless control algorithm.

I. INTRODUCTION

The Switched Reluctance Motor (SRM) is considered as an attracting candidate for variable speed motor drives due to its low cost, simple structure and high efficiency over a wide speed range. In addition, the inverter of the SRM drive can prevent the shoot-through fault that exists in the induction and permanent magnet motor drives[1][2].

In order to produce motoring torque at any moment and for optimum performance, it is necessary to commutate the excitation current from phase to phase synchronously with the rotor position. This needs the use of a mechanical position sensor attached to the motor shaft. The problems associated with the mechanical position sensor are the poor reliability in a dusty environment and the high cost of the sensor.

Several sensorless control methods have been proposed. Most of these methods utilize the functional relationship between the stator flux linkage or phase inductance, phase current, and rotor position to estimate the rotor position. Due to the doubly salient structure of the SRM, the phase inductance varies along with the rotor position. Therefore, this can be used for rotor position estimation[4]-[7].

In this paper, the current-flux-rotor position lookup table based position sensorless control system over wide speed range is presented. Experimental result for a 1-hp SRM is presented for the verification of the proposed sensorless control algorithm.

II. SWITCHED RELUCTANCE MOTOR DRIVES

A. Basic principle of operation

The Switched Reluctance Motor drives are accomplished by switching the phase current on and off synchronously with the rotor position. The voltage equations of the SRM can be described as followings.

$$v = Ri + \frac{d\psi}{dt} = Ri + L\frac{di}{dt} + \omega_m i \frac{dL}{d\theta} \qquad (1)$$

where v is the phase voltage, i is the current, R is the phase resistance, L is the phase inductance, ω_m is the rotor angular velocity and θ is the rotor position.

The instantaneous electric power which is the product of phase voltage and current is as followings.

$$vi = Ri^2 + Li\frac{di}{dt} + \omega_m i^2 \frac{dL}{d\theta} \qquad (2)$$

The instantaneous electro-magnetic torque is given by the following equations.

$$T = \frac{1}{2}i^2\frac{dL}{d\theta} \qquad (3)$$

Motoring torque is produced when the phase winding is energized during the positive slope of the phase inductance variation.

B. Rotor position estimation

The rotor position of the SRM can be estimated based on the fact that the magnetic status of the SRM is a function of the angular rotor position. The variation of the magnetizing curve along with the phase current and rotor position for a 1-hp SRM is shown in Fig. 1. The highly nonlinear function of the magnetizing data can be approximated by Neural Network. Through measurement of the phase flux linkages and phase current the trained Neural Network is able to estimate the rotor position. Because of the Neural Network requires much calculation time and additional lookup tables for the transfer functions, the output data of the trained Neural Network is directly adapted for the lookup table generation.

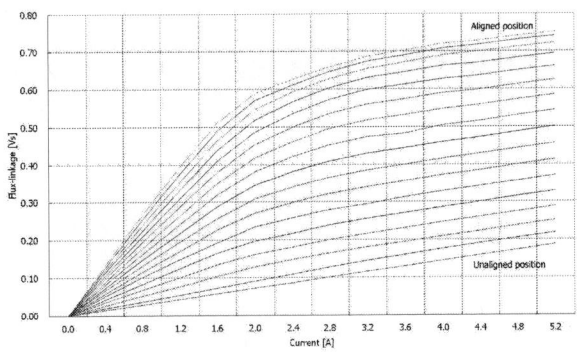

Fig. 1: Magnetizing curve of the prototype 1-hp SRM

Fig. 2 shows the flowchart for the generation of 2-d look-up table which is used to estimate the rotor position from the flux-linkage and the phase current information. Fig. 3 shows the block diagram of the feed-forward Neural Network, and Fig. 4 shows the output data patterns of the trained Neural Network.

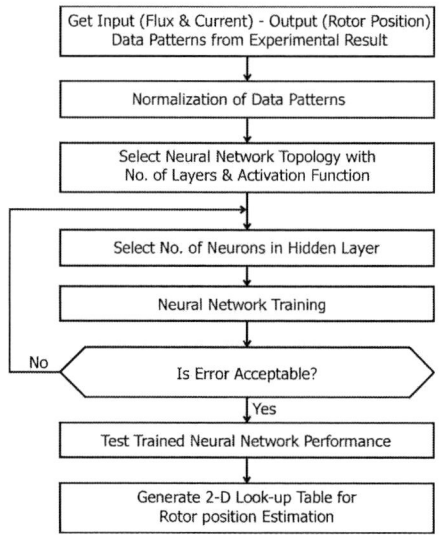

Fig. 2: Design flow of the Neural Network

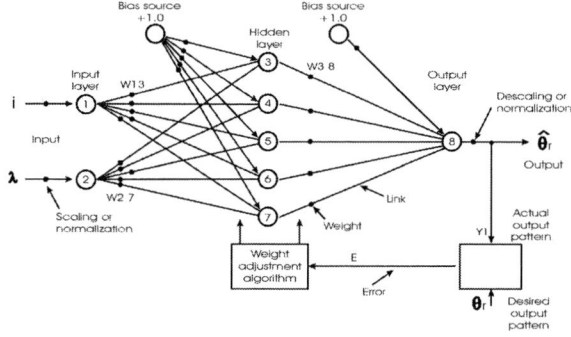

Fig. 3: Block diagram of the designed Neural Network

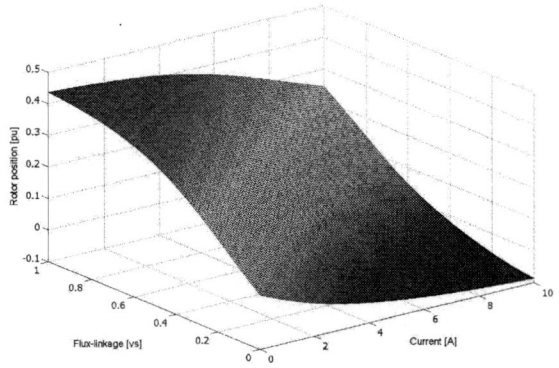

Fig. 4: Output data pattern of the trained Neural Network

III. EXPERIMENTAL RESULTS

A. System configuration

Proposed sensorless algorithm is realized using TMS 320F2812 DSP controller and 5-KVA asymmetric bridge converter. Fig. 5 shows the block diagram of the proposed position sensorless control system. In Fig. 5, the flux linkage is calculated from the DC link voltage, duty cycle and phase current data to consider the resistive voltage drop. Using calculated flux linkage and reference current data, the rotor position is estimated from 2-d look-up table. Fig. 6 shows the prototype 1-hp SRM and Table 1 shows the design parameters of the SRM. Fig. 7 shows the experimental setup.

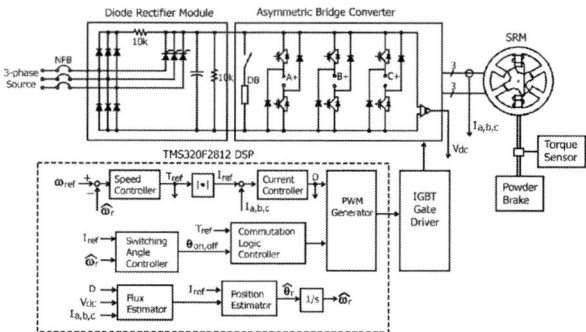

Fig. 5: Block diagram of the proposed position sensorless control system

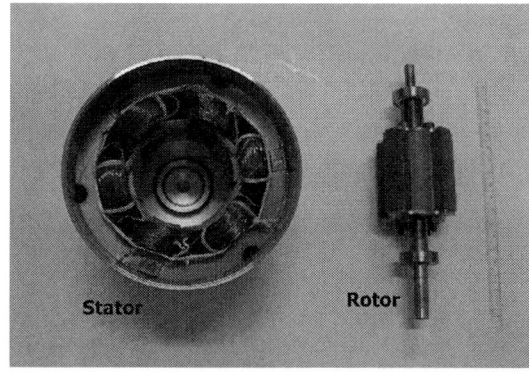

Fig. 6: Prototype 1-hp SRM

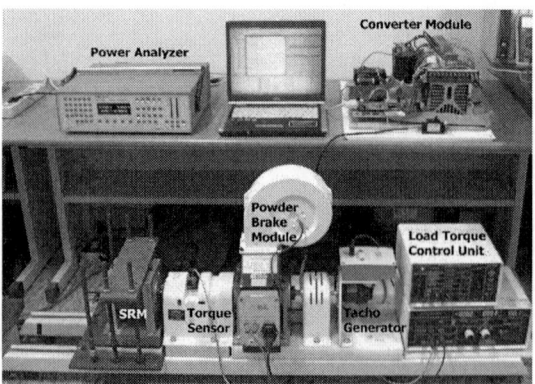

Fig. 7: Experimental system setup

TABLE 1
Design parameters of SRM

No. of phase	3	Stator pole	6
Rated output	1[hp]	Rotor pole	4
Rated output	1[hp]	Stator pole arc	30[deg]
Rated speed	3000[rpm]	Rotor pole arc	32[deg]

B. *Experimental results*

Fig. 8 to 18 illustrates the experimental results of the sensorless algorithm for wide speed range. All the waveforms are generated using software drawing program based on the experimental data which obtained from the online data communication with DSP. Fig. 8 shows the calculated flux linkage of each phase, and Fig. 9 shows the estimated rotor position at 450[rpm] and 0.5[Nm] load. The position waveforms are represented by per unit value which means one electrical period. Fig. 10~15 shows the speed, actual and estimated rotor position at different operating speed and load. Although some position errors are observed in the active phase changing interval, wide speed control range of 16:1 is accomplished. Fig. 16~18 shows the position error according to the operating condition.

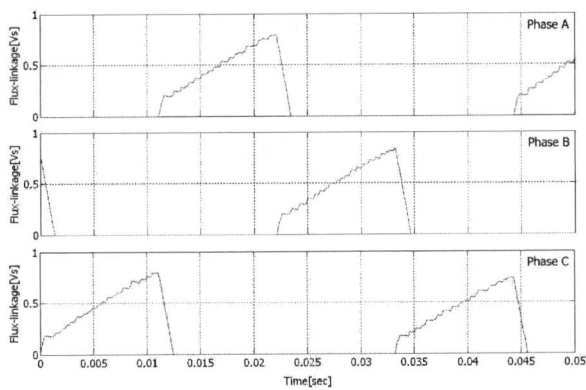

Fig. 8: Estimated flux-linkage at 450[rpm], 0.5[Nm]

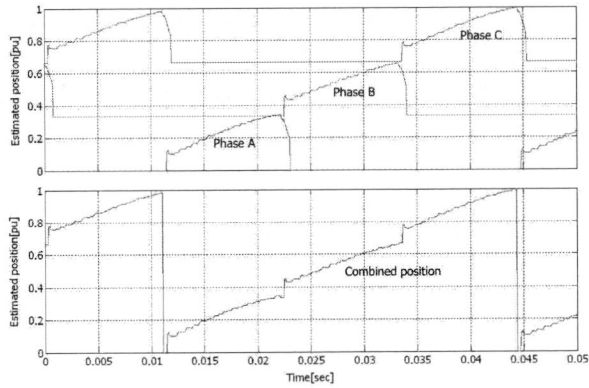

Fig. 9: Estimated rotor position at 450[rpm], 0.5[Nm]

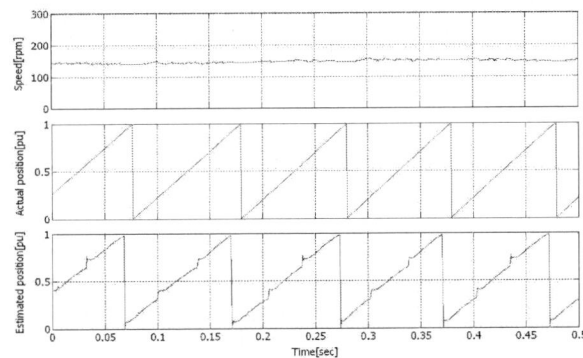

Fig. 10: Speed & rotor positions at 150[rpm], 0.5[Nm]

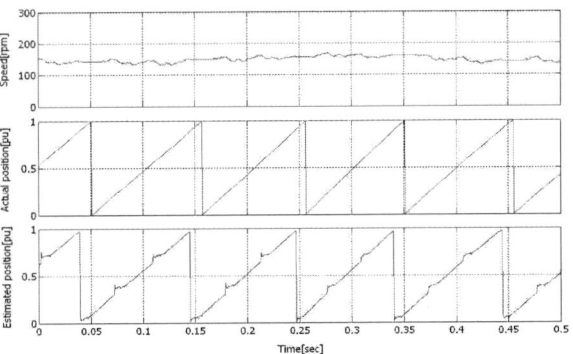

Fig. 11: Speed & rotor positions at 150[rpm], 1.5[Nm]

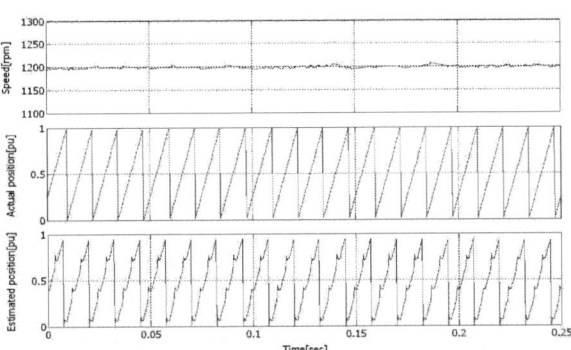

Fig. 12: Speed & rotor positions at 1200[rpm], 0.5[Nm]

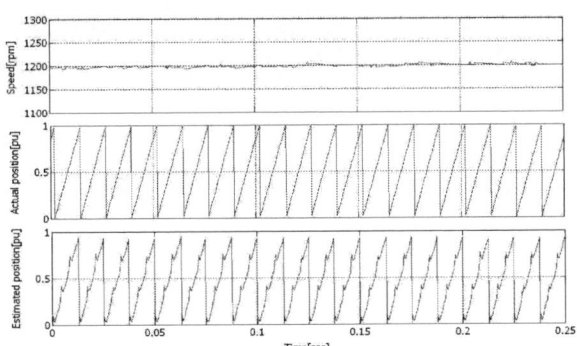

Fig. 13: Speed & rotor positions at 1200[rpm], 1.5[Nm]

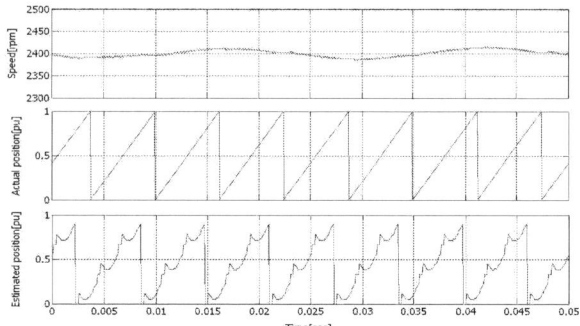

Fig. 14: Speed & rotor positions at 2400[rpm], 0.5[Nm]

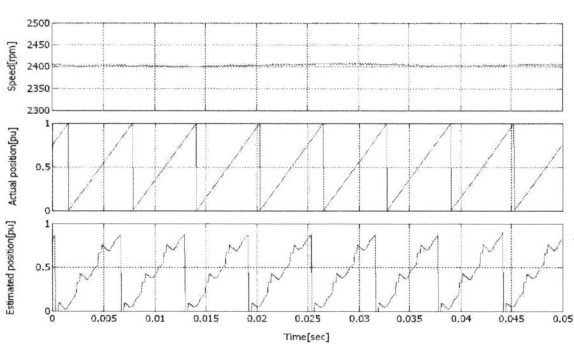

Fig. 15: Speed & rotor positions at 2400[rpm], 1.5[Nm]

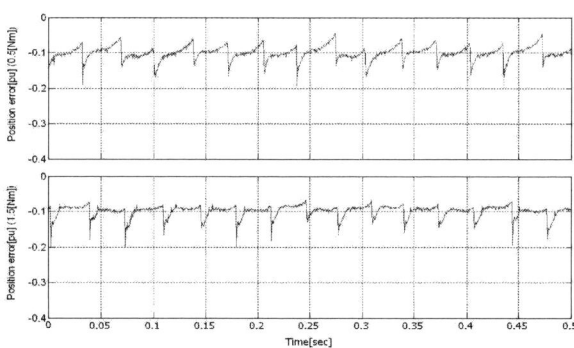

Fig. 16: Position estimation error at 150[rpm]

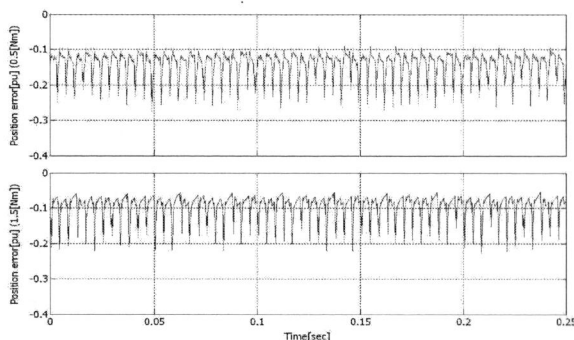

Fig. 17: Position estimation error at 1200[rpm]

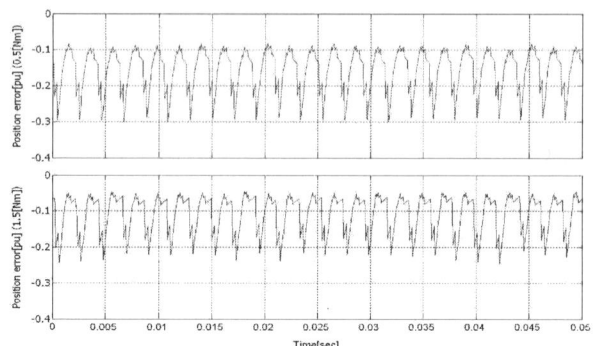

Fig. 18: Position estimation error at 2400[rpm]

IV. CONCLUSION

In this paper, a position sensorless control system of SRM over wide speed range is presented. Neural Network is adapted to approximate the highly nonlinear function of SRM. The 2-D current-flux-rotor position lookup table is generated by trained Neural Network. Using calculated flux linkage and reference current, the rotor position is estimated from 2-d look-up table. Experimental result for a 1-hp SRM is presented for the verification of the proposed sensorless algorithm. Although some position errors are observed in the active phase changing intervals, wide speed control range of 16:1 is accomplished.

ACKNOWLEDGMENT

The authors would like to thank to the Power Application Technology Research Center of Yeungnam College of Science & Technology sponsor by the Ministry of Commerce, Industry & Energy and the Korea Electric Power Corporation.

REFERENCES

[1] T.J.E. Miller, *Electronic Control of Switched Reluctance Machines*, Newnes Publication, 2003.
[2] B.K. Boss, *Modern Power Electronics and AC Drives*, Prentice Hall PRT, 2002.
[3] Peter Vas, *Artificial-Intelligence-Based Electrical Machines and Drives*, Oxford Science Pub, 1999.
[4] M.H. Kim, M.G. Simoes, B.K. Boss, "Neural Network-Based Estimation of Power Electronic Waveforms," *IEEE Trans. on PE*, Vol. 11, No. 2, pp. 383- 389, March, 1996.
[5] W.S. Baik, K.H. Choi, D.H. Kim, M.H. Kim, "A Motion Control System of SRM using Digital Hysteresis Controller," *Proceedings of ICEE*, Vol. 2, pp. 803-807, July, 2002.
[6] G.G. Lopez, P.C. Kjaer, T.J.E. Miller, "High-Grade Position Estimation for SRM Drives using Flux Linkage/Current Correction Model," *IEEE Trans. on IA*, Vol. 35, No. 4, pp. 859-869, July/Aug., 1999.
[7] E. Mese, D.A. Torry, "An Approach for Sensorless Position Estimation for Switched Reluctance Motors Using Neural Networks", *IEEE Trans. on PE*, Vol. 17, No. 1, pp. 66-75, Jan, 2002.

978-1-4244-1871-8/07 $25.00 © 2007 IEEE

Efficiency Map of a Switched Reluctance Motor Using Finite Element Method in Vehicular Applications

Hossein Mokhtari
Associated Professor
School of Electrical Engineering
Sharif University of Technology
Tehran - Iran
Email: mokhtari@sharif.edu

Ehsan Tara
M.Sc. Student
School of Electrical Engineering
Sharif University of Technology
Tehran - Iran
Email: tara@ee.sharif.edu

Abstract—**Switched Reluctance Motors (SRMs), which are simple and reliable motors, have attracted interest for electric vehicle propulsion systems. It is shown that an SRM offers a good performance at high acceleration and low speed driving conditions. Efficiency map of an electric motor is a very important characteristic specially when the motor is to be used in a wide range of speed and load variations, which is the case in electric propulsion systems. In this research, the efficiency map of an SRM is derived using a combination of the Finite Element Method (FEM) and a mathematical model. This efficiency map has been used for different driving cycles to test the SRM performance as a vehicle propulsion motor.**

I. INTRODUCTION

Simple structure, high reliability, and efficient control methods have made SRM one of the attractive candidates for electric vehicle propulsion systems. As an electric vehicle propulsion system, several features, such as simplicity, robustness, efficiency and low cost are to be considered. Among these features, efficiency is a very important issue in electric vehicles because of the vehicle limited available energy [1]. Efficiency is an interesting subject in other industries too. Industrial motors usually operate close to the rated speed and output power. Therefore, it makes sense to consider the efficiency of the motor at the nominal speed and output power as the machine efficiency. In the case of an electric vehicle propulsion system, the motor will be driven in a wide range of speed and output powers depending on the traffic, weather, road condition and driver habits. Therefore, it is not possible to use the conventional machine efficiency in such drives. The efficiency map of motor has to be used instead [2]. Efficiency map of a machine is simply a map of efficiency at different operating points.

In conventional DC and AC motors, because of the simple and linear models, efficiency map derivation is a straightforward task. However, in the case of an SRM, an extreme nonlinear behavior exists, and a linear or a definite nonlinear model is not available [3]. This is also the case for SRM motor losses. Some researches have tried to calculate SRM losses [4]–[6], but these works have not ended to an efficiency map.

In this research, a 30KW SRM is analyzed using Finite Element Method (FEM). This motor is investigated at different rotor positions and several stator currents. Flux linkage of the motor under these conditions is determined. Flux linkage profile of the motor is then used to develop a mathematical model.

The mathematical model is derived from conventional SRM equations [3]. A conventional method in iron loss calculation in an SRM is to divide the machine into different regions. Flux linkage in each region in then calculated and used for iron loss calculation [4]. Copper loss calculation is done considering skin effect of conductors. The mathematical model of the motor is solved over a wide range of speed and load variations to find flux linkage and stator currents. Flux linkage of the motor is used to estimate the iron loss, and stator currents are used to determine the copper loss. Efficiency map of the SRM is then constructed using the output power and estimated losses.

There are various driving conditions for an electric vehicle depending on vehicle type, traffic, road conditions, weather and driver habits. Theses driving conditions are characterized in various driving cycles which are widely used in vehicle studies such as emission estimation in conventional combustion engine vehicles and range estimation in battery powered electric vehicles.

In this research, efficiency map of the SRM is fitted into several driving cycles in order to estimate the SRM performance as a prime mover in a vehicle.

II. FINITE ELEMENT ANALYSIS

Different types of motors with different sizes can be used in a vehicle propulsion system [7]. In this study, a 30KW motor is chosen which is a common size in vehicle propulsion motors. Fig. 1 shows the cross section of the selected motor. Motor dimensions are given in Table I.

Flux linkage profile is the key characteristic of an SRM. This characteristic can be found by Finite Element Method (FEM) analysis. The nonlinear B-H curve of material and

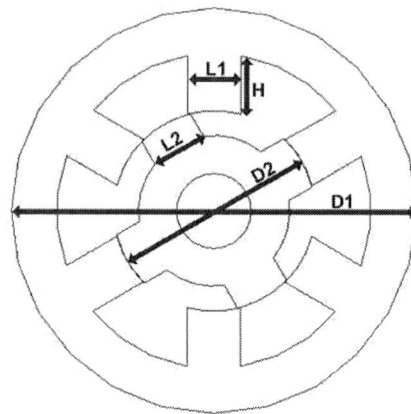

Fig. 1. SRM Cross Section

TABLE I
SRM DIMENSIONS

SRM Dimensions (in mm)	
$D1$	350
$D2$	172.93
H	50.46
$L1$	45.66
$L2$	48.35
$AirGap$	0.24

saturation phenomenon must be taken into account. The SRM under investigation is modeled and simulated using an FEM software, i.e. ANSYS. Fig. 2 shows the meshed SRM in ANSYS. Elements near the edges of the motor are meshed with finer elements while elements in farther areas and air elements are larger. There are at least two elements in the air gap. M54 steel is considered as the stator and rotor material.

Fig. 2. FEM Elements of the SRM

Flux linkage of the motor is studied under various stator currents and at different rotor positions. The derived flux linkage profile of the motor in 10A current steps is depicted in Fig. 3.

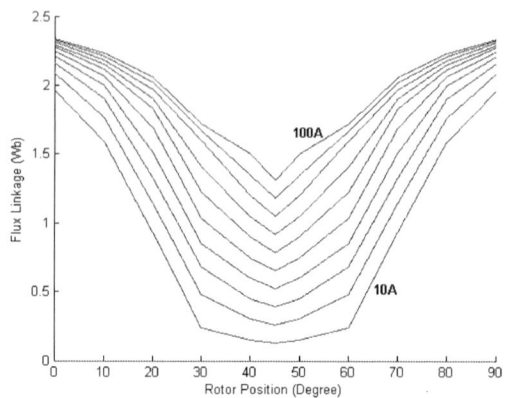

Fig. 3. SRM Flux Linkage in different rotor positions and for various stator currents

III. SRM MATHEMATICAL MODEL

Torque production in an SRM can be explained using the elementary principles of electromechanical energy conversion in a selonoid. The selonoid here is the stator winding which has N turns and excited with a current of i. Therefore, W_e, the input electrical energy can be expressed as:

$$W_e = \int eidt = \int idt \frac{dN\Phi}{dt} = \int Nid\Phi = \int \mathcal{F}d\Phi \quad (1)$$

Where e is the induced emf and $\mathcal{F}$ is the mmf.

The input electrical energy is the sum of the output mechanical energy and the stored energy in the machine magnetic field, i.e.:

$$W_e = W_f + W_m \quad (2)$$

Considering a small change in the rotor position, the incremental change of the energies can be expressed as:

$$\delta W_e = \delta W_f + \delta W_m \quad (3)$$

On the other hand:

$$\delta W_m = T_e \delta \theta_e \Longrightarrow T_e = \frac{\delta W_m}{\delta \theta} \quad (4)$$

Where T_e is the electromagnetic torque and $\delta \theta_e$ is the incremental rotor angel.

For the case of a constant excitation current one can write:

$$\delta W_m = \delta W_f' \quad (5)$$

Where W_f' is the system coenergy which is defined as:

$$W_f' = \int \Phi d\mathcal{F} = \int \Phi d(Ni) = \int \lambda(\theta, i)di = \int L(\theta, i)idi \quad (6)$$

In this equation, λ is the flux linkage and L is the phase inductance.

Hence, the electromagnetic torque of the SRM is:

$$T_e = \frac{\delta W_m}{\delta \theta} = \frac{\delta W_f'}{\delta \theta}\bigg|_{i=cte} \quad (7)$$

T_e in (7) is calculated in the model based on the flux linkage characteristics of the SRM based on FEM analysis. To take the SRM loss into account, these losses should appear into (3) which results in (8).

$$\delta W_e = \delta W_f + \delta W_m + \delta W_{loss} \qquad (8)$$

Hence (7) changes to:

$$T_e = \frac{\delta W_m}{\delta \theta} = \left. \frac{\delta W_f'}{\delta \theta} \right|_{i=cte} - \frac{\delta W_{loss}}{\delta \theta} \qquad (9)$$

Eq. (9) is the core of the SRM model. Fig. 4 shows the block diagram of the SRM model. Flux linkage is calculated using flux linkage profile of the SRM and stator currents. Current and flux linkage of the motor is used for copper and iron loss calculation. Loss calculations are described in section IV of this paper. Coenergy and losses are then put in (9) to calculate the motor output torque. This torque can be used for speed and position calculations.

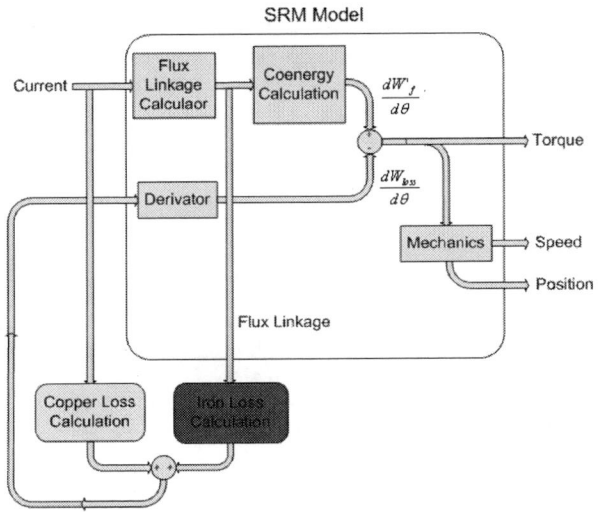

Fig. 4. Block Diagrm of SRM model

IV. MOTOR LOSSES

The losses in an SRM mainly consist of iron and copper losses [4]. Copper loss estimation is a straightforward task using the stator current, however the skin effect should be taken into account due to high frequency currents in the stator windings [8]. Iron loss in an SRM is composed of eddy current and hysteresis losses. Eddy current is a result of alternating flux in the motor. The resulting current generates heat in the motor body. Using efficient eddy current mitigation techniques like laminating the motor body, this part of loss can be minimized. In this study, the eddy current loss is neglected and the hysteresis loss is considered as the only source of core losses in the motor. However, estimation of the hysteresis loss is not straightforward due to nonlinear currents and flux of an SRM. In the proposed model, the copper loss and hysteresis loss are modeled and used for SRM efficiency calculations.

Prediction of the SRM core loss is a complicated task because unlike AC machines, the stator current and consequently the flux linkage of the SRM are nonlinear. In addition, different parts of the machine are subjected to different flux waveforms. The saturation of the material, which is common in SRM drives, makes the core loss calculation even more complicated. Accordingly, a straightforward and precise method for core loss calculation in SRMs is not available. The conventional method for SRM core loss calculation is to divide the SRM body into different zones. Each zone is subjected to a different flux waveform. Flux waveform for each zone is determined by simulation. The Fourier Transform of this waveform and the loss characteristics of the core material (W/Kg) is then used to calculate the core loss of the SRM under the study conditions. The number of zones depends on the number of rotor and stator poles of the SRM. A 12/8 SRM in [8], an 8/6 SRM in [5] and a 6/4 SRM in [4] are studied using this method. Fig. 5 shows different zones for the 6/4 SRM in this study.

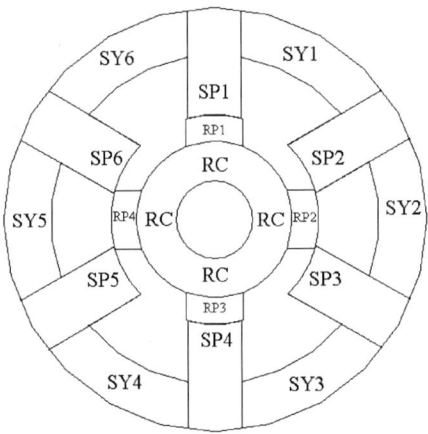

Fig. 5. 6/4 SRM Flux Zones

Considering single-phase excitation in the SRM drive, which is a common strategy, there will be two flux waveforms in the stator yoke (Φ_{sy1} and Φ_{sy2}), two in the stator poles(Φ_{sp} and 0), two in rotor poles (Φ_{rp} and 0) and one in the rotor core (Φ_{rc}). Table II shows the flux waveform in different zones assuming phase 1 of the stator is under excitation and the two other phases are left without current.

Flux waveform for each zone is calculated using the stator position and excitation current. Frequency components of each waveform is them calculated and used for core loss calculations using hysteresis loss characteristics of the core material.

Fig. 6 depicts the block diagram of the core loss estimator of the model.

V. EFFICIENCY MAP

To do the performance analysis for the SRM, an efficiency map is needed. The efficiency map shows the SRM efficiency at different operating points. In this study, motor speed is

TABLE II
FLUX WAVEFORMS IN DIFFERENT ZONES

Zones	Flux
SY_1, SY_4	Φ_{sy1}
SY_2, SY_3, SY_5, SY_6	Φ_{sy2}
RP_1, RP_3	Φ_{r1}
RP_2, RP_4	0
SP_1, SP_4	Φ_{sp}
SP_2, SP_3, SP_5, SP_6	0

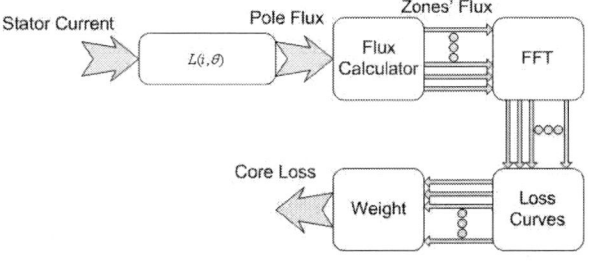

Fig. 6.   Core Loss Estimator Block Diagram

varied from zero to 5000 rpm and load is changed from 0 to 450 N.m.. A total number of 816 points are studied and the efficiency of the motor is calculated at each point. This efficiency map is shown in Fig. 7.

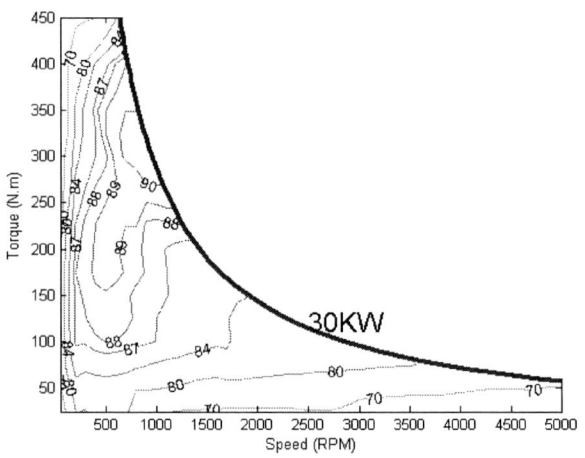

Fig. 7.   Efficiency Map

The derived efficiency map shows that the SRM operates more efficiently near the nominal speed and nominal load. Operating points with a moderate speed and moderate to relatively high loads are also efficient. However, the SRM exhibits a poor efficiency at low and extremely high speeds and also at light load conditions.

VI. SRM PERFORMANCE IN VEHICLE

There are two types of tests in electric vehicle performance estimation. The first one is the constant velocity test. This test is simple but not realistic because in reality no vehicle is driven at a constant velocity. The second type of test, more

useful and complex, is to use a profile of ever changing speed i.e. a driving cycle. Driving cycles are defined using statistic data for different driving conditions. A driving cycle shows vehicle's velocity in a test period.

FTP75 driving cycle, shown in Fig. 8, is an example of urban driving cycle. This cycle simulates an urban route of 17.77 Km with frequent stops. The maximum speed is 91.2 Km/h and the average speed is 34.1 Km/h. Highway driving cycles are also available for vehicle performance tests.

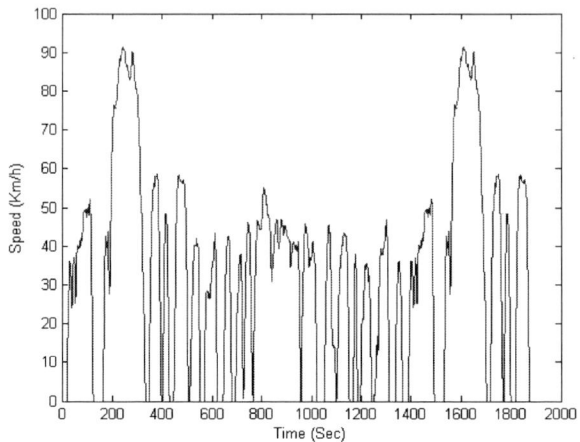

Fig. 8.   FTP75 Driving Cycle

US06 Supplemental FTP simulates aggressive highway driving. The cycle represents a 12.8 Km route with an average speed of 77.9 Km/h and maximum speed of 129.2 Km/h. Fig. 9 illustrates the speed profile of this driving cycle.

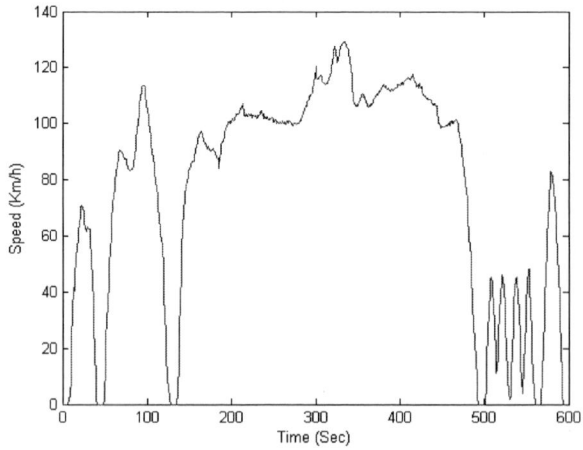

Fig. 9.   US06 Driving Cycle

FTP72, FTP75, US06, NYCC and JP 10-15 Mode driving cycles are used in performance analysis of the SRM. Characteristics of these driving cycles are summarized in Table III.

The SRM performance in a vehicle can be estimated using driving cycles and the efficiency map of the SRM. To do this estimation, SRM operating points for a driving cycle should

TABLE III
DRIVING CYCLES

Driving Cycle	Route Length (Km)	Maximum Speed (Km/h)	Average Speed (Km/h)
FTP72	12.07	91.2	31.5
FTP75	17.77	91.2	34.1
US06	12.8	129.2	77.9
NYCC	1.9	44.58	11.40
JP 10-15 Mode	4.1654	70	22.72

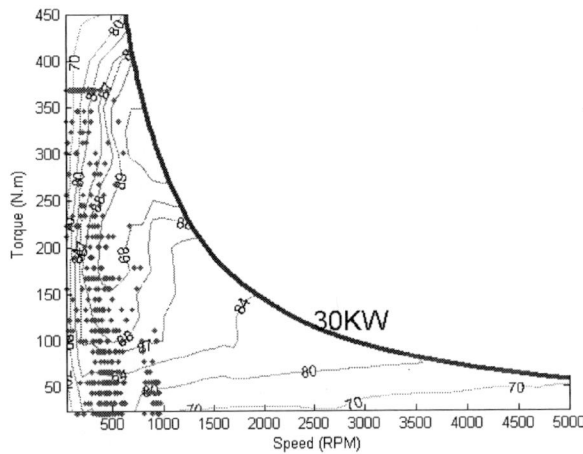

Fig. 10.   FTP75 Driving Cycle

be mapped on the SRM efficiency map. Vehicle acceleration for a driving is derived using (10).

$$a_n = \frac{V_n - V_{n-1}}{t_n - t_{n-1}} \quad (10)$$

Tractive effort of the vehicle is:

$$F = M_v.a \quad (11)$$

Where M_v is the vehicle mass.

Accordingly, the wheel torque is:

$$\tau_w = F.R_w \quad (12)$$

Where R_w is the wheel radius.

SRM operating points can be derived using the gear ratio in the transmission system.

$$\tau_m = \tau_w.i_g \quad \omega_m = \frac{\omega_w}{i_g} \quad (13)$$

According to the characteristics of electrical motors, multiple gear transmission is rarely used in electrical vehicles. A direct drive or single gear transmission is usually used instead. In direct drive designs a precise machine design is needed to meet the load demands. A single gear transmission provides more freedom by introducing an additional design parameter, i.e. gear ratio. The vehicle in this study assumed to have a direct drive transmission system.

In this study, the Performance Factor (PF) is defined as a criterion. The PF is the average of SRM performance at different operating points of driving cycle. Fig. 10 shows the scattered operating points of the SRM FTP72 driving cycle.

The SRM performance is shown in Fig. 11. This figure shows that SRM mainly operates with at least 80% efficiency. The PF for this case is 0.81.

The vehicle performance for different driving cycles is summarized in table IV.

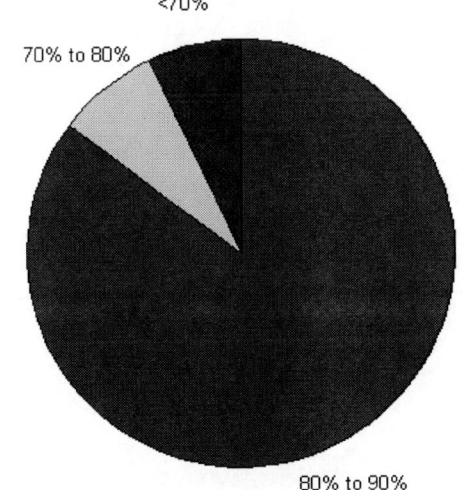

Fig. 11.   FTP75 Driving Cycle

TABLE IV
SRM PERFORMANCE IN DIFFERENT DRIVING CYCLES

Driving Cycle	Description	Performance Factor
FTP72	Urban Driving (US)	0.824739
FTP75	Urban Driving (US)	0.823957
NYCC	Urban Driving (US)	0.793228
US06	Highway Driving (US)	0.777241
JP 10-15 Mode	Urban Driving (Japan)	0.847350

Urban driving cycles include frequent stops and frequent speed changes. This speed profile demands high torques at a moderate speed. According to Fig. 7, this region of efficiency map suggests a high efficiency. Table IV also shows that SRM has a good performance in urban driving cycles. On the other hand, highway driving cycles include high speed driving with small speed changes. This speed profile needs low torque at high speeds. As it is shown in Fig. 7, this region of the efficiency map has a poor efficiency which results in a low performance for highway driving cycle in Table IV.

VII. CONCLUSION

In this paper, the performance of an SRM for vehicle traction system is investigated, and an efficiency map for this motor is derived. This efficiency map and standard driving cycles used to calculate the motor performance in a vehicle.

This study shows that SRM has a good performance in urban driving cycles which includes frequent stops and low speeds. The SRM performance is lower, but not very poor, in highway driving cycles which include high speeds.

REFERENCES

[1] J. Larminie and J. Lowry, *Electric Vehicle Technology Explained.* John Wiley and Sons Ltd, 2003.

[2] M. Ehsani, Y. Gao, S. E. Gay, and A. Emadi, *Electric, Hybrid Electric and Fuel Cell Vehicles.* CRC Press, 2005.

[3] R.Krishnan, *Switched Reluctance Motor Drives, Modeling, Simulation, Analysis, Design and Applications.* CRC Press, 2001.

[4] P. N. Materu and R. Krishnan, "Estimation of switched reluctance motor losses," *IEEE Transactions on Industry Applications*, vol. 28, no. 3, pp. 668–679, May/June 1993.

[5] T. L. Mthombeni and P. Pillay, "Lamination core losses in motors with nonsinusoidal excitation with particular reference to pwm ans srm excitation waveforms," *IEEE Transactions on Enerjy Conversion*, vol. 20, no. 4, pp. 836–843, December 2005.

[6] D. G. D. Chindurza and C. Cossar, "Assessing the core losses in switched reluctance machines," *IEEE Transactions on Magnetics*, vol. 41, no. 10, pp. 3907–3909, October 2005.

[7] K. Rajashekara, "History of electric vehicles in general motors," *IEEE Transactions on Industry Applications*, vol. 30, no. 4, pp. 897–904, July/August 1994.

[8] P. Rafajdus, V. Hrabovcova, and P. Hudak, "Investigation of losses and efficiency in switched reluctance motor," *12th International Power Electronics and Motion Control Conference*, pp. 296–301, August 2006.

A Novel Direct Instantaneous Pressure Control of Hydraulic Pump System with SRM Drive

Jianing Liang, Dong-Hee Lee, and Jin-Woo Ahn
Kyungsung University
Dept. of EE & Mechatronics
110-1, Daeyeon-Dong, Nam-Gu, Busan, 608-736
Email: jwahn@ks.ac.kr

Young-Joo An
Pukyong National University
Dept. of Electrical and Control Instrument Engineering
599-1, Daeyean 3-dong, Nam-gu, Busan, 608-737
Email: anyj@pknu.ac.kr

Abstract—A novel direct instantaneous pressure control of hydraulic pump system with SRM drive is presented in this paper. The proposed hydraulic pump system embeds pressure controller and direct instantaneous torque control (DITC) controller. DITC controller can reduce inherent torque ripple of SRM, and develop a smooth torque, which can increase stability. The proposed hydraulic pump system has also fast step response and load response. At last the proposed hydraulic pump system is verified by computer simulation and experimental results.

I. INTRODUCTION

A hydraulic pump system is very widely used in heavy-duty machine, brake system of vehicles and automatic control system of industrial applications[1]. In recent, high efficiency and high performance motor drive for hydraulic pump system is much interested in hydraulic pump system[2]. SR drive is good choice for hydraulic pump system.

Switched Reluctance Motors(SRM) is investigated for wide industrial applications due to the mechanical strength and cost advantages[3]. SRM has high power-to-weight, torque- to-weight ratios, a wide speed range and excellent starting characteristics. Therefore, it is suitable for hydraulic pump system which is frequently stopped and started with a full load condition[4].

This paper presents a novel direct instantaneous pressure control of hydraulic pump system with SRM drive. The proposed hydraulic pump system embeds pressure controller and DITC controller[5-6]. The pressure controller is made by a simple PI controller. DITC controller can reduce inherent torque ripple of SRM, and develop smooth torque to load, which can increase stability. The proposed hydraulic pump system has fast step response and load response. At last the proposed hydraulic pump system is verified by computer simulation and experimental results.

II. THE PROPOSED HYDRAULIC PUMP SYSTEM USING SR DRIVE

A. The Basic Principles of SR Drive and Hydraulic Pump

Fig. 1 shows a conventional SR drive system. The voltage equation contains resistive voltage drop, inductive voltage drop, and induced EMF.

$$v = iR + L_{(\theta_{rm})} \frac{di}{dt} + i \frac{dL_{(\theta_{rm})}}{d\theta_{rm}} \cdot \omega_{rm} \qquad (1)$$

where, θ_{rm}:rotor position, ω_{rm}: rotor speed, $L_{\theta rm}$: inductance according to rotor position.

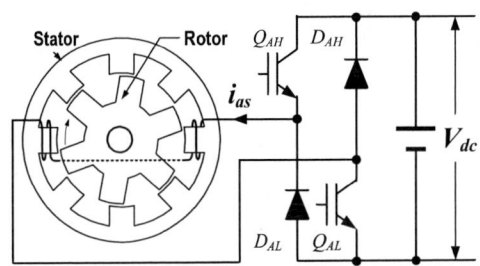

Fig. 1 SRM drive system

In the ideal case, with the constant phase current, the phase torque is produced according to square of current and inductance slope of the motor. The torque equation can be derived as follow [1-3].

$$T_m = \frac{1}{2} \cdot i^2 \cdot \frac{dL_{(\theta_{rm})}}{d\theta_{rm}} \qquad (2)$$

For ideal hydraulic pump, the flux of hydraulic pump is determined by volumetric displacement and pump speed. In fact, the internal flow leakage in pump and motor should be taken into account by the volumetric efficiency μ_v[2].

$$Q = \mu_v \cdot n_m \cdot V_p \qquad (3)$$

where, Q is output flux[m^3 /min], n_m is pump speed [rad/min], V_P is volumetric displacement [m^3/rad].

And the pressure is determined motor output torque, volumetric displacement and mechanical efficiency μ_m. For pumps pressure is given by:

$$P_p = \mu_m \cdot T_m / V_p \qquad (4)$$

where P_p is differential pressure [Pa], T_m is motor output torque [Nm].

B. Conventional Converter for SR Drive

The conventional asymmetric converter is very popular in SR drive, which can provide independent control of each

978-1-4244-1871-8/07 $25.00 © 2007 IEEE

phase and phase overlap. The asymmetric converter has three modes, which are defined as magnetization mode, freewheeling mode and demagnetization mode. They are shown in Fig. 2. The magnetization mode is defined as state 1 shown in Fig. 2(a). The freewheeling mode is defined as state 0 shown in Fig. 2(b). The demagnetization mode is defined as state -1 shown in Fig. 2(c).

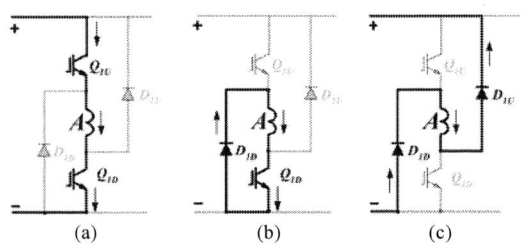

(a)   (b)   (c)

Fig. 2. Operation modes of asymmetric converter
(a) Magnetization (b) Freewheeling (c) Demagnetization

C. Proposed DITC Method for SR Drive

In order to eliminate inherent torque ripple of SR motor, DITC method is introduced. By the given hysteresis control rules, appropriate torque of each phase can be assigned and constant total torque can be obtained by the proposed DITC method. In this method, the phase inductance has been divided into 3 regions as shown in Fig. 3.

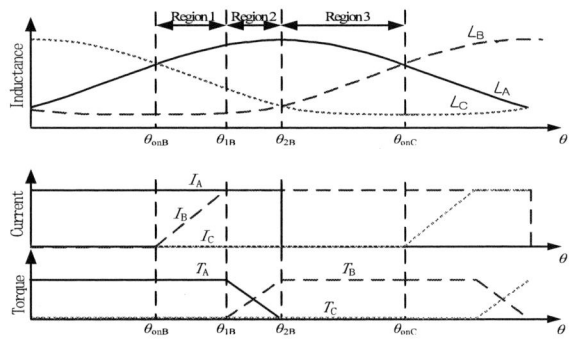

Fig. 3. Principle of DITC method with 3 regions

The regions depend on the geometrical structure and load. The boundaries of 3 regions are θ_{onB}, θ_{1B}, θ_{2B} and θ_{onC} in Fig. 3. θ_{onB} and θ_{onC} are turn-on angles in phase B and phase C respectively, which depend on load and operating speed. The θ_{1B} is a rotor position which is the inflexion of inductance in phase B. And θ_{2B} is the aligned point of inductance in phase A. Total length of those regions is 120 electrical degrees in 3 phases SRM. Let the outgoing phase to be phase A and the incoming phase to be phase B in Fig. 3. When the first 3 regions are over, the outgoing phase will be replaced by phase B in next 3 regions.

In region 1, phase A develops the outputs torque of the motor. And the excitation current of phase B will be built up easily for the torque in the next region.

In region 2, neither of phases can develop an enough output torque, so phase A and phase B produce torque together. In order to reduce negative torque in next region, the current of phase A should be cut down. Therefore, main

torque developed by phase B, and the rest of desired torque is produced by phase A.

In region 3, a large part of demagnetization energy is transferred to filter capacitor, and phase current decreased quickly. Regarding to phase B, it enters into an inductance increasing area, and can produce enough torque to overcome the negative torque and satisfy the demands of load.

In order to carry out DITC method, digital hysteresis control schemes are used. In the proposed DITC method, three control rules are assigned to the 3 regions shown in Fig. 4. The combinatorial states of phase A and phase B are shown as a pane. In order to reduce switching frequency, state variety only changes one switch state at a time. It can decrease switching loss of the power devices.

In this figure, x axis denotes state of phase A, and y axis denotes state of phase B. Each phase has 3 states, so the pane can have 9 combinatorial states. However, only the black points of pane are used in the DITC scheme of each region. Hysteresis rule of phase A and phase B are shown at the right side of pane. Real line is climbing rule, and dash line is falling rule. y axis denotes phase state and x axis denotes torque error between reference and estimated torque. Boundary of torque error is used to change the state of control scheme.

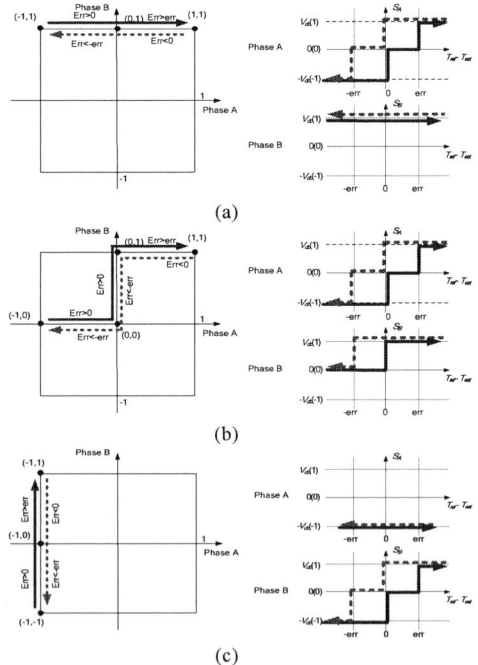

(a)

(b)

(c)

Fig.4. DITC scheme with a conventional asymmetric converter
(a) control scheme 1 (b) control scheme 2 (c) control scheme 3

D. Advance angle calculation

θ_{on} is important parameter in the proposed control method, which can be set by advance angle θ_{adv}. As shown in Fig. 5, the motor is excited at θ_{on} position advanced as θ_{adv} from the start point of positive torque region θ_1 to establish a sufficient torque current. The desired phase current is shown as dash line.

In order to ensure enough time to build-up the desire phase

current i^*, the advance angle θ_{adv} can be adjusted according to motor speed ω_{rm} and torque command T^*. The desire phase i^* can be computed by T^* using (2).

$$i^* = \sqrt{\frac{2 \cdot T^* \cdot d\theta_{rm}}{dL_{(\theta_{rm})}}} \qquad (5)$$

From the voltage equations of SRM, the proper advance angle can be calculated by the current rising time Δt as follows regardless of phase resistance at θ_1.

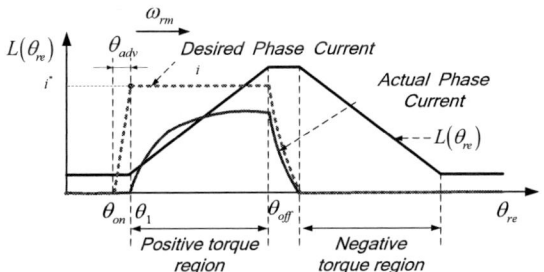

Fig. 5 The build-up of phase current of different θ_{on}

$$\Delta t = L(\theta_1) \cdot \frac{i^*}{v_{phase}} \qquad (6)$$

where, v_{phase} is the terminal voltage of each phase windings. And the advance angle is determined by motor speed and Δt as follow

$$\theta_{adv} = \omega_{rm} \cdot \Delta t \qquad (7)$$

With speed increased, the advance angle also needs increase. At the fixed turn-on position, the actual phase current denoted as solid line could not reach the desire value to produce sufficient output torque shown in Fig. 5.

E. Proposed DITC SR Drive System

Control diagram of DITC SR motor drive is shown in Fig. 6. The torque estimation block is generally implemented by 3-D lookup table according to the phase currents and rotor position. And the digital torque hysteresis controller which carries out DITC scheme generates the state signals for all activated machine phases according to torque error between the reference torque and estimated torque. The state signal is converted as switching signals by switching table block to control converter.

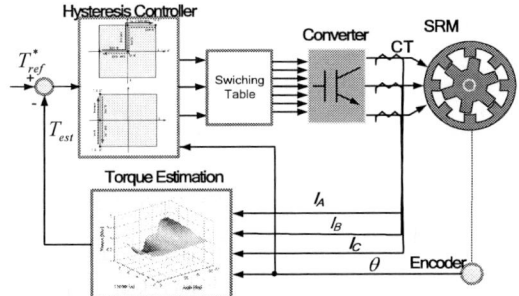

Fig. 6. Control diagram of DITC SR motor drive

III. SIMULATION RESULTS

In order to verify the proposed method, some simulations are executed. The inductance and static torque characteristic of the prototype SRM are analyzed from some experiments and Finite-Element-Method (FEM). The inductance profile of prototype SRM is shown in Fig. 7. 3-D torque table is described in Fig. 8.

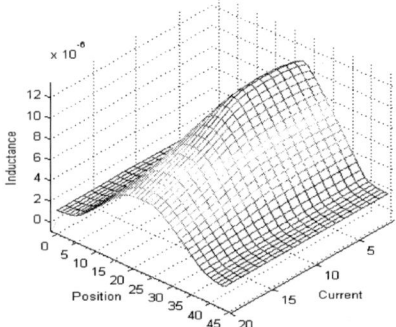

Fig. 7 Inductance profile of the prototype SRM

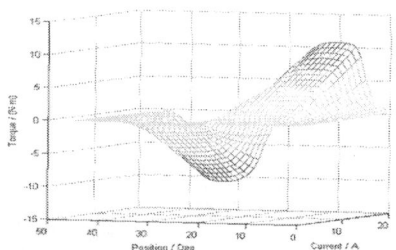

Fig.8. Torque profile of the prototype SRM

Fig. 9 and Fig. 10 show a simulation results between current chopping control(CCC) and proposed DITC control. The torque ripple of CCC method is two times higher than proposed DITC method, and the proposed DITC method obtains smoothing output pressure.

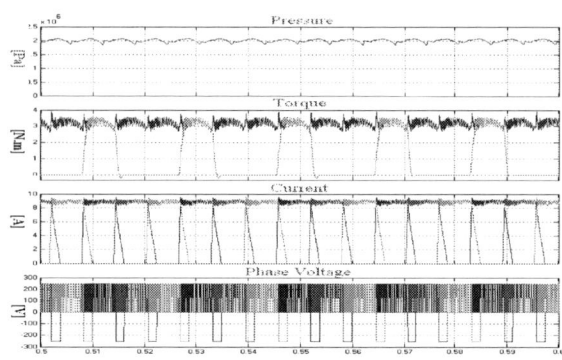

Fig. 9 Current chopping control at 2[MPa]

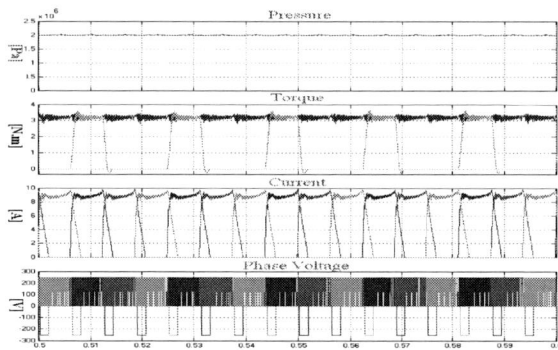

Fig. 10 Proposed DITC control at 2[MPa]

IV. EXPERIMENTAL RESULTS

Experimental setup to verity the proposed hydraulic pump system is shown in Fig. 11. The main controller for the experiments is designed by TMS320F2812 from TI(Texas Instruments) and phase current and voltage signals are feedback to 12bit ADC embedded by DSP. Sampling time of DITC controller is up to 25 [μs] and sampling time of pressure controller is 375[μs].

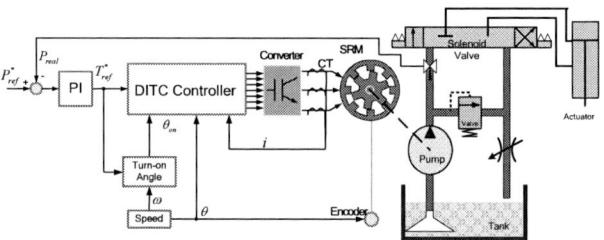

Fig. 11 Proposed hydraulic pump system

The experimental result of CCC is shown in Fig. 12. The drive motor is operate at 410[rpm], and output pressure of pump is 2[MPa]. The phase current is controlled with flat-topped. But output torque has a relatively high torque ripple. Compared with CCC, experimental result of proposed DITC is shown in Fig. 13. The smoothing torque during commutation is obtained. The torque ripple is only a half of CCC method.

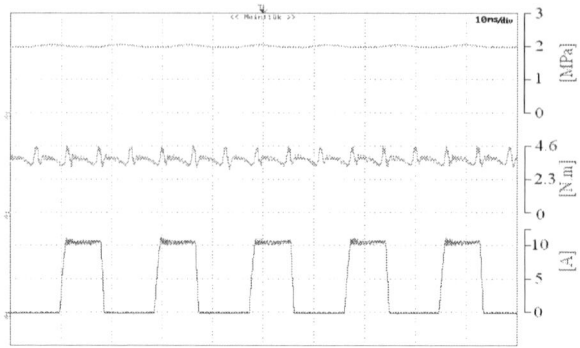

Fig. 12 Experimental result of CCC at 2[MPa]

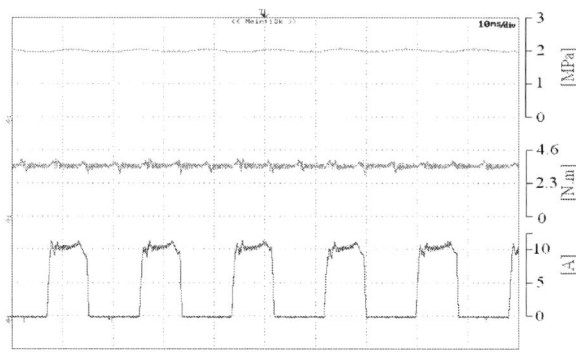

Fig. 13 Experimental result of proposed DITC at 2[MPa]

Fig. 14 shows the load test result of hydraulic pump system. The hydraulic pump system connected an actuator cylinder as load. The cylinder is double acting with single piston rod, so the extended and retracted operation needs different flux at the same pressure. From Fig. 14, when cylinder operated, the hydraulic pump system has quickly load response. The response time is less than 100[ms], so the SR motor speed has a fast step response to satisfy flux requirement.

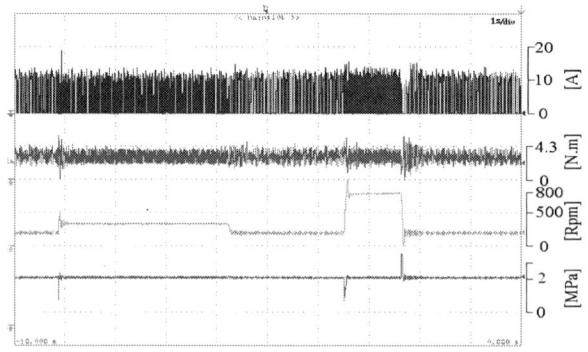

Fig. 14 Operation 2[MPa] with cylinder load

978-1-4244-1871-8/07 $25.00 © 2007 IEEE   652

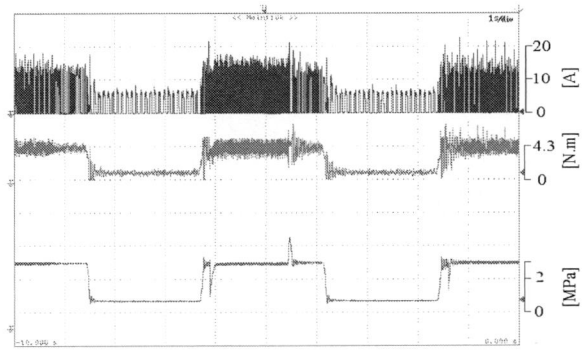

Fig. 15 Operation with cylinder in energy saving mode at 3[MPa]

Energy saving mode is very popular in home and industrial applications, which can save the power consumption at standby state. The proposed hydraulic pump system embedded energy saving mode. Fig. 15 shows cylinder operation in energy saving mode. When external standby signal input, the pump unit remains 0.5[MPa] pressure to compensate leakage of system, and saves power consumption. When the standby signal calls off, the pump unit has pressure step response to pre-set pressure value. From Fig. 15, the pressure step is less than 100[ms] from 0.5 to 3[MPa]. The appropriate setting standby signal and fast dynamic pressure can make pump system more efficient in energy mode.

V. CONCLUSION

This paper presents a novel direct instantaneous pressure control of hydraulic pump system with SRM drive. The proposed hydraulic pump system embeds pressure controller and DITC controller. The pressure controller is made by a simple PI controller. DITC controller can reduce inherent torque ripple of SRM and develop smooth torque to load, which can increase stability.

From the experiment results, the proposed hydraulic pump system operate stable, and it has fast step response and load response. And this proposed hydraulic pump system can embed energy saving mode, which can improve efficiency of whole system in operation. Therefore, the proposed hydraulic pump system has high dynamic performance and high efficiency.

ACKNOWLEDGMENT

This work has been supported by KESRI (I-2007-0-261-01), which is funded by MOCIE(Ministry of Commerce, Industry and Energy), Korea.

REFERENCES

[1] Jame L. Johnson et al, Introduction to Fluid Power, Science & Technology, 2004.
[2] Peter Chapple,"Principles of Hydraulic System design", Coxmoor Publishing Company, 2003.
[3] J. W. Ahn, "Switched Reluctance Motor", Osung Media, 2004
[4] B.C. Kim, D.H. Lee; J.W. Ahn, "Performance of SR drive for hydraulic pump" Proceedings of ICEMS 2005, Vol. 1, pp.659 - 663, 27-29 Sept. 2005.
[5] J. N. Liang, Z. G. Lee, D. H. Lee, J. W. Ahn, " DITC of SRM Drive System Using 4-Level Converter " , Proceedings of ICEMS 2006, Vol. 1, 21-23 Nov. 2006
[6] Inderka, R.B., De Doncker, R.W., "DITC-direct instantaneous torque control of switched reluctance drives", 37th IAS Annual Meeting. Vol 3, Oct. 2002, pp. 1605 – 1609
[7] R. C. Kavanagh, J. M. D. Murphy, and M. G. Egan, "Torque ripple minimization in switched reluctance drives using self-learning techniques," in Proc. IEEE-IECON Conf. Rec.'91, pp. 289-294.
[8] D. S. Schramm, B. W. Williams, and T. C. Green, "Torque ripple reduction of switched reluctance motors by phase current optimal profiling," in Proc. IEEE-PESC Conf. Rec.'92, pp. 857-860.
[9] J. C. Moreira, "Torque ripple minimization in switched reluctance motors via bi-cubic spline interpolation," in Proc. IEEE-PESC Conf. Rec.'92, pp. 851-856.
[10] Du, Hongliu, "An E/H Displacement and Power Control Design for Hydraulic Variable Displacement Pumps," in the Proceedings of the ASME International Mechanical Engineering Congress and Exposition, New York, NY,Nov. 2001.

978-1-4244-1871-8/07 $25.00 © 2007 IEEE

Multiloop Control Strategy for Four-Wire Current Source Active Power Filter

Sami Pettersson, Mika Salo, and Heikki Tuusa
Tampere University of Technology
Institute of Power Electronics
P.O. Box 692, FI-33720, Tampere, Finland
Email: sami.pettersson@tut.fi

Abstract—**Despite the fact that three-wire active power filters have been implemented using both voltage source and current source technology, the research of four-wire current source active power filters has been truly minimal. Only a few topologies have been presented over the years. This paper proposes a four-wire current source active power filter topology for improving the quality of supply in four-wire systems. The proposed topology is controlled using a multiloop control strategy suitable for a single-chip microcontroller implementation. The control strategy combines the load current feedforward connection and the cascaded closed-loop control of the supply currents and the capacitor voltages of the LC-type supply filter. By using the cascaded closed-loop control, the stability of the system can be achieved without the need to overdimension the supply filter. The performance of the proposed system is verified experimentally using a microcontroller controlled four-wire current source active power filter prototype rated for a 5 kVA load.**

I. INTRODUCTION

Four-wire active power filters provide an efficient solution to improve the quality of supply in three-phase four-wire systems. They are applicable in compensating current harmonics, reactive power, neutral current and load imbalance [1]. Four-wire active power filters have mainly been implemented using voltage source converters and the research of four-wire current source active power filters (CSAPF) has been truly minimal. However, a few CSAPF topologies for four-wire systems have been presented, e.g. in [2]-[6].

A shunt CSAPF is typically implemented so that the PWM converter bridge is connected in parallel with the load through a second-order LC-type low-pass supply filter, in which case the compensation current of the active power filter flows through the supply filter. Various control methods for this kind of implementation are presented in [7]-[16].

Reference [17] proposes a control method for a topology where the load is connected in parallel with the supply filter capacitors and the PWM converter bridge. In this case the supply filter inductor has no effect on the compensation current of the active power filter. In [17] the control method is based on the measurement of the supply currents resulting in modest filtering performance. Reference [18] aims to improve the performance of the topology presented in [17] by proposing an improved control method based on the load current feedforward connection and the closed-loop control of the supply currents. The problem with this kind of single-loop feedback control is that the resonance frequency of the LC-filter has to be set very low for stability. Because of this the LC-filter has to be unnecessarily overdimensioned.

This paper presents a four-wire CSAPF topology with a multiloop control strategy. The main circuit of the proposed

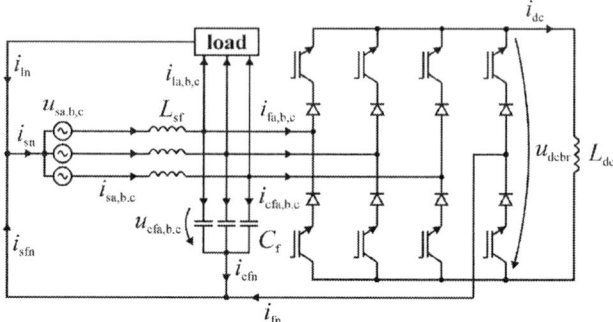

Fig. 1: Main circuit of the proposed four-wire current source active power filter topology.

topology is shown in Fig. 1. The control system is based on the load current feedforward connection and the cascaded closed-loop control of the supply currents and the filter capacitor voltages. With the filter capacitor voltage control the resonance frequency of the LC-filter can be set much higher than when only the closed-loop control of the supply currents is used. Moreover, when the closed-loop control of the supply currents is cascaded with the filter capacitor voltage control, the supply phase current can be kept sinusoidal even if the supply voltages are distorted. The performance of the proposed control system is verified with experimental tests on a microcontroller controlled four-wire CSAPF prototype rated for a 5 kVA load.

II. CONTROL SYSTEM

The block diagram of the proposed control system is presented in Fig. 2. The task of the control system is to produce active power filter currents opposite to the harmonics sensed in the load currents. The neutral current is similarly compensated by producing a neutral current opposite to the load neutral current.

The control system is implemented in the synchronous reference frame where the real axis is tied to the supply voltage vector, and therefore the fundamental quantities appear as dc components. Subscripts d and q refer respectively to the direct and quadrature components of a space vector in the synchronous reference frame, and z to a zero sequence component. In addition, subscript l refers to load, f to the PWM converter bridge, s to supply, o to fundamental components in the synchronous reference frame, ff to feedforward, cl to closed-loop and cf to supply filter capacitors. Reference values are denoted with an asterisk (*).

The control system consists of two main parts: feedforward control and closed-loop control. The feedforward control generates the compensation current

978-1-4244-1871-8/07 $25.00 © 2007 IEEE

654

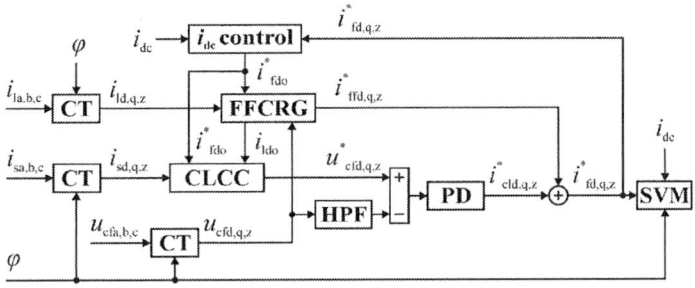

Fig. 2: Block diagram of the control system.

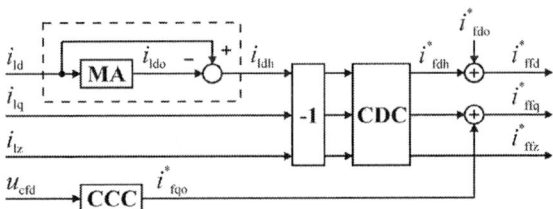

Fig. 3: Feedforward current reference generation (FFCRG).

reference on the basis of the measured load phase currents ($i_{la,b,c}$). First the measured load phase currents are transformed into the synchronous reference frame in the "CT" block. The reference frame angle φ is determined with a phase locked loop by observing zero crossings of phase-a supply voltage. The compensation current reference is generated in the "FFCRG" block, whose contents are shown in Fig. 3.

A. Extraction of Harmonics

Since the fundamental components appear as dc components in the synchronous reference frame, the extraction of harmonics can be done with a high-pass filter, which is outlined with a broken line in Fig. 3. For the active power filter also to compensate the reactive power drawn by the load, the harmonics are extracted only from the direct component of the load current. In this case the dc component of the load current direct component i_{ldo} is calculated by using a moving average (MA). The moving average algorithm can be written as follows [19]:

$$i_{ldo}(k) = \frac{1}{m}\left[i_{ld,sum}(k-1) - i_{ld}(k-m) + i_{ld}(k) \right] \quad (1)$$

where m is the number of samples stored during the supply period, k a discrete time instant and

$$i_{ld,sum}(k) = \sum_{k-(m-1)}^{k} i_{ld}(k). \quad (2)$$

The harmonics of the load current direct component are obtained by subtracting i_{ldo} from i_{ld}, as shown in Fig. 3.

B. Control Delay Compensation

Due to the sampling and calculation delay the output of the control system is late on approximately 1.5 discrete time instants, which impairs the filtering performance of the active power filter [14]. Therefore the control system is equipped with a computational control delay compensator (CDC) which is presented in [14]. The method is based on

the knowledge of the system dynamics of the active power filter: since the behavior of the active power filter current in case of a step change in the current reference is known, the reference can be corrected so that the active power filter current behaves as desired. The algorithm can be written in discrete form as follows [14]:

$$i_{out}(k+1) = \frac{\tau_c}{T_s}\left[i_{in}(k) - i_{in}(k-1) \right] + i_{in}(k) \quad (3)$$

where i_{in} is the current component entering the "CDC" block and i_{out} the current component exiting the "CDC" block, T_s the sampling interval, and τ_c the compensation time constant.

C. Capacitive Current Compensation

The filter capacitors draw capacitive current from the supply which must be taken into account in the control system if a unity power factor is pursued. The direct component of the filter capacitor current in the synchronous reference frame is

$$i_{cfd} = C_f \frac{d}{dt} u_{cfd} - \omega C_f u_{cfq} \quad (4)$$

and the quadrature component

$$i_{cfq} = C_f \frac{d}{dt} u_{cfq} + \omega C_f u_{cfd} \quad (5)$$

where C_f is the capacitance of the filter capacitors and ω the angular frequency of the reference frame. The capacitive current is mainly caused by the last term in (5) since u_{cfq} is typically rather small. Thus, the capacitive current can be compensated by adding a current reference component i^*_{fqo} to the quadrature component of the active power filter feedforward current reference, as shown in Fig. 3. On the basis of (5) the current reference i^*_{fqo} can be approximated to be

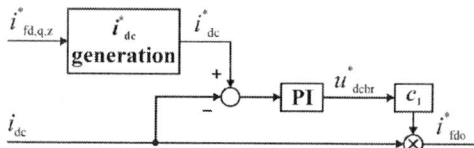

Fig. 4: Block diagram of the dc-link current control.

$$i_{fqo}^* \approx -2\pi f_s C_f u_{cfd} \qquad (6)$$

where f_s is the supply frequency. The calculation of (6) is carried out in the "CCC" block in Fig. 3.

D. Dc-link Current Control

The block diagram of the dc-link current control is presented in Fig. 4. The dc-link current i_{dc} is controlled in a closed-loop with a PI-controller, which outputs the dc-link voltage reference of the PWM converter bridge u_{dcbr}^*. If the losses of the PWM converter bridge are ignored, the ac and dc active powers are equal. Thus, the current reference i_{fdo}^*, which is added to the current reference component i_{fdh}^*, can be written as follows [14]:

$$i_{fdo}^* = \frac{2}{3u_{sd}} u_{dcbr}^* i_{dc} = c_1 u_{dcbr}^* i_{dc} \qquad (7)$$

where u_{sd} is the direct component of the supply voltage in the synchronous reference frame, which is in this case assumed to be constant. It should be noted that the control bandwidth of the dc-link current controller must be set very narrow or the supply currents will be distorted.

The dc-link current reference i_{dc}^* is calculated in the "i_{dc}^* generation" block in Fig. 4. The reference is derived as follows:

$$i_{dc}^* = c_2 \left(\sqrt{\left(i_{fd}^*\right)^2 + \left(i_{fq}^*\right)^2} + 3\left|i_{fz}^*\right| \right) \qquad (8)$$

where i_{fd}^*, i_{fq}^* and i_{fz}^* are respectively the direct, quadrature and zero sequence components of the active power filter current reference, and $c_2 \geq 1$. The expression in parentheses in (8) determines the minimum value for the dc-link current in order that the active power filter is able to realize the current reference. The value for the constant c_2 is selected based on the desired margin. In practice it is beneficial to keep the magnitude of the dc-link current as low as possible to minimize the dc-link power loss.

E. Closed-Loop Control

The task of the closed-loop control is to damp the resonance of the LC-filter and to keep the supply phase currents sinusoidal despite the shape of the supply voltages. There are two control loops: the inner loop controlling the filter capacitor voltages $u_{cfa,b,c}$ and the outer loop controlling the supply currents $i_{sa,b,c}$.

The block diagram of the closed-loop current control (CLCC) is shown in Fig. 5. The aim of the control is to produce sinusoidal and balanced supply phase currents. To accomplish this, the reference value for the direct component of the supply current is generated by summing the fundamental direct component of the load current i_{ldo} and the current reference component i_{fdo}^* from the dc-link current controller. This sum forms the current reference that

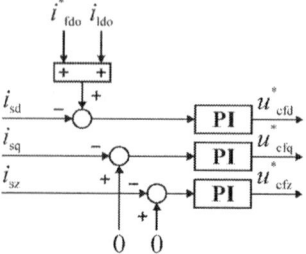

Fig. 5: Block diagram of the closed-loop current control (CLCC).

feeds the active power required by the load and the active power filter. The reference values for the quadrature and zero sequence components are zero. The supply currents are controlled with PI-controllers which output the reference values for the filter capacitor voltages $u_{cfd,q,z}^*$.

The inner voltage loop is used for damping the resonance of the LC-filter. The measured filter capacitor voltages are first transformed into the synchronous reference frame and then high-pass filtered with a first-order high-pass filter (HPF). The high-pass filtering is necessary because the voltage control loop cannot follow low-frequency references. Next, the obtained values are subtracted from the voltage references and the error values are fed into PD-controllers. Finally the output values from the PD-controllers are added to the feedforward current reference components $i_{ffd,q,z}^*$ resulting the active power filter current references components $i_{fd,q,z}^*$. The active power filter current reference is realized using space vector modulation (SVM) [6].

III. EXPERIMENTAL RESULTS

The performance of the proposed system is tested using a microcontroller controlled four-wire CSAPF prototype rated for a 5 kVA load. The control system is implemented with fixed-point arithmetic using a Freescale 32-bit MPC563 single-chip microcontroller operating at 64 MHz clock frequency. The sampling interval of the system is 50 μs and the modulation frequency 10 kHz. All currents are measured with LEM LA 55-P current transducers and the measured signals are sampled using two on-chip 10-bit A/D converter units.

The main circuit is built with Semikron SKM 50GB123D 1200-V and 50-A IGBT modules and STMicroelectronics STTA9012TV2 ultra-fast high voltage diodes in series with the IGBTs. The inductance of the dc-link inductor is selected to be 200 mH (198 % p.u.). The inductance of the supply filter is set at 2.3 mH (2.3 % p.u.) and the capacitance at 10 μF (10 % p.u.). This results in a resonance frequency of 1049 Hz. The selected capacitance value is half of the value used in [18].

The load used in the experiments is unbalanced and nonlinear. The measured load current waveforms are shown in Fig. 6. The performance of the proposed CSAPF topology is tested both in steady and transient state of the load. The measured current waveforms in the steady state are shown in Fig. 7. The total harmonic distortion (THD) and fundamental components (I_1) of the measured currents are presented in Table 1. The THD is calculated up to 2 kHz

978-1-4244-1871-8/07 $25.00 © 2007 IEEE

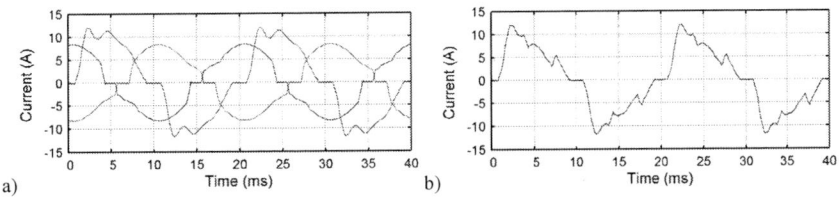

Fig. 6: Measured load current waveforms: a) phase currents and b) neutral current.

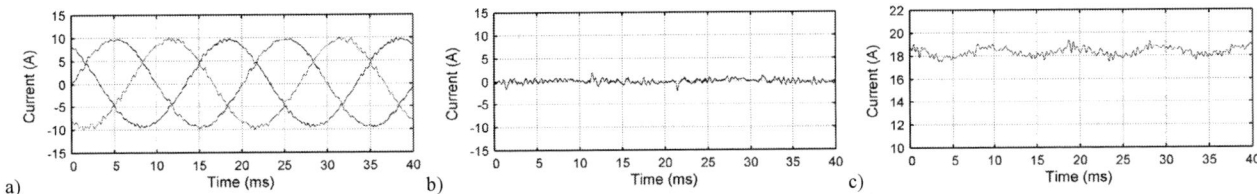

Fig. 7: Experimental results in steady state: a) supply phase currents, b) supply neutral current, and c) dc-link current.

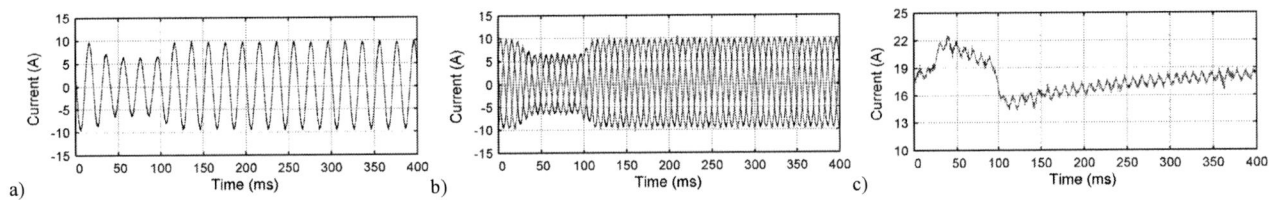

Fig. 8: Experimental results in transient state: a) supply phase-c current, b) supply phase currents, and c) dc-link current.

(THD$_{2kHz}$) and up to 20 kHz (THD$_{20kHz}$). The THD of the neutral current is calculated in relation to the average rms value of the fundamental phase currents.

Figs. 7a and 7b show that the CSAPF efficiently compensates the load phase current harmonics as well as the load neutral current, producing sinusoidal and balanced supply phase currents. In addition, the LC-filter removes the switching frequency current ripple from the supply currents. However, the magnitude of the dc-link current has to be set very high due to the neutral current compensation, as shown in Fig. 7c. For comparison, if the load neutral current is not compensated, the magnitude of the dc-link current can be decreased to about 6 A.

The performance of the system in the transient state of the load is tested so that the load in phase-b and phase-c is changed to 1/3 for a short time. As a result, the apparent power drawn by the load decreases and the load phase

TABLE 1
FUNDAMENTAL COMPONENTS AND TOTAL
HARMONIC DISTORTION OF MEASURED CURRENTS

		load i_l	supply i_s
phase-a	I_1 (A$_{rms}$)	7.53	6.81
	THD$_{2kHz}$ (%)	21.68	3.02
	THD$_{20kHz}$ (%)	21.69	3.22
phase-b	I_1 (A$_{rms}$)	5.74	6.79
	THD$_{2kHz}$ (%)	13.66	3.36
	THD$_{20kHz}$ (%)	13.69	3.60
phase-c	I_1 (A$_{rms}$)	5.80	6.70
	THD$_{2kHz}$ (%)	12.46	2.79
	THD$_{20kHz}$ (%)	12.50	3.06
neutral	THD$_{2kHz}$ (%)	30.63	4.73
	THD$_{20kHz}$ (%)	30.65	5.12

imbalance increases. The measured current waveforms during the transient state are shown in Fig. 8. Figs. 8a and 8b show that the system responds well to the changes in the load currents maintaining sinusoidal and balanced supply phase currents. Due to the increased load phase imbalance, the need for dc-link current is greater even though the apparent power drawn by the load is smaller at the time, as shown in Fig. 8c.

IV. CONCLUSIONS

In this paper a four-wire current source active power filter topology with a multiloop control strategy was presented. The control system is based on the load current feedforward connection and the cascaded closed-loop control of the supply currents and the filter capacitor voltages. With the filter capacitor voltage control the resonance frequency of the LC-filter can be set much higher than when using only the closed-loop control of the supply currents. This enables the size of the supply filter to be reduced.

The performance of the proposed system was experimentally tested both in steady and transient state of operation. The experimental results proved that the system is capable of compensating the load current harmonics and neutral current as well as removing the phase imbalance. In addition, the LC-type supply filter efficiently removes the switching frequency current ripple from the supply currents.

The drawbacks of the proposed topology are mainly the same as in [6], including a bulky dc-link inductor, high dc-link power loss, increased on-state losses due to the series diodes and a need for an overvoltage protector. In addition,

978-1-4244-1871-8/07 $25.00 © 2007 IEEE

the number of current transducers required for the control system is quite high, i.e. seven.

REFERENCES

[1] B. Singh, K. Al-Haddad, and A. Chandra, "A review of active filters for power quality improvement," *IEEE Trans. Ind. Electr.*, vol. 46, no. 5, Oct. 1999.

[2] G. Van Schoor and J. D. Van Wyk, "A study of a system of current fed converters as an active three phase filter," in *Proc. PESC '87*, 1987, pp. 482–490.

[3] A. Emadi, A. Nasiri, and S. B. Bekiarov, *Uninterruptible Power Supplies and Active Filters*, Boca Raton, FL: CRC Press, 2005, pp. 78–80.

[4] K. Hoffman and G. Ledwich, "Improved power system performance using inverter based resonance switched compensators," in *Proc. PESC '94*, vol. 1, 1994, pp. 205–210.

[5] M. Bou-Rabee, D. Sutanto, F. Barone, and G.-H. Choe, "A new technique for three-phase active power filter," in *Proc. APEC '92*, 1992, pp. 837–843.

[6] S. Pettersson, M. Salo, and H. Tuusa, "Four-wire current source active power filter with an open-loop current control," in *Proc. 4th Power Conversion Conference – PCC Nagoya 2007*, 2007, pp. 542–549.

[7] Y. Hayashi, N. Sato, and K. Takahashi, "A novel control of a current-source active filter for ac power system harmonic compensation," *IEEE Trans. Ind. Appl.*, vol. 27, no. 2, pp. 380–385, Mar./Apr. 1991.

[8] L. Malesani, L. Rossetto, and P. Tenti, "Active Power Filter with Hybrid Energy Storage," *IEEE Trans. Power Electr.*, vol. 6, no. 3, pp. 392–397, Jul. 1991.

[9] M.-X. Wang, H. Pouliquen, and M. Grandpierre, "Performance of an active filter using PWM current source inverter," in *Proc. 5th European Conference on Power Electronics and Applications — EPE'93*, 1993, pp. 218–223.

[10] L. Benchaita and S. Saadate, "Current harmonic filtering of non-conventional non-linear load by current source active filter," in *Proc. IEEE International Symposium on Ind. Electr. — ISIE '96*, 1996, pp. 636–641.

[11] Y. Sato, T. Sugita, and T. Kataoka, "A new control method for current source active power filters," in *Proc. IEEE/IAS Annual Meeting*, 1997, pp. 1463–1470.

[12] K. Nishida, Y. Konishi, and M. Nakaoka, "Current control implementation with deadbeat algorithm for three-phase current-source active power filter," *IEE Proc. – Electr. Power Appl.*, vol. 149, no. 4, pp. 275–282, Jul. 2002.

[13] K. Nishida, Y. Konishi, and M. Nakaoka, "A robust two-dimensional deadbeat current control method by using adaptive line enhancer for active power filter," in *Proc. 2002 IEEE International Symposium on Intelligent Control — ISIC '02*, 2002, pp. 384–390.

[14] M. Salo and H. Tuusa, "A novel open-loop control method for a current-source active power filter," *IEEE Trans. Ind. Electr.*, vol. 50, no. 2, pp. 313–321, Apr. 2003.

[15] H.-P. To, F. Rahman, and C. Grantham, "A current source active power filter with minimum requirement of current sensors," in *Proc. 4th International Power Electronics and Motion Control Conference — IPEMC 2004*, 2004, pp. 444–449.

[16] H.-P. To, F. Rahman, and C. Grantham, "Time delay compensation for a current-source active power filter using state-feedback controller," *IEEE/IAS Annual Meeting*, 2005, pp. 1213–1219.

[17] Y. Sato and T. Kataoka, "A current-type PWM rectifier with active damping function," *IEEE Trans. Ind. Appl.*, vol. 32, no. 3, May/Jun. 1996.

[18] M. Salo and H. Tuusa, "A new control system with a control delay compensation for a current-source active power filter," *IEEE Trans. Ind. Electr.*, vol. 52, no. 6, pp. 1616–1624, Dec. 2005.

[19] M. Routimo, M. Salo, and H. Tuusa, "A novel simple prediction based current reference generation method for an active power filter," in *Proc. PESC'04*, 2004, pp. 3215–3220.

A Novel Online Controller for Tuning Shunt Active Power Filters Based Upon Switched-Capacitors

Ossama M. Elgendy
Higher Techological Institute
Department of Electrical Engineering &
Computers
10th of Ramadan City, Egypt
Email: osaelgendy@yahoo.com

Ahmed M. A. Mahmoud
Ain Shams University, Faculty of Engineering
Electrical Power Department
Address, Cairo, Egypt
Email: aasaadm@menanet.net

A. Abass
Ain Shams University, Faculty of Engineering
Electrical Power Department
Address, Cairo, Egypt
Email: a.abbas@masrawy.com

A. D. Alkoshairy
Ain Shams University, Faculty of Engineering
Electrical Power Department
Address, Cairo, Egypt
Email: elkoshairy@gmail.com

Abstract — This paper presents a novel online controller for tuning shunt active power filters (APF). The idea of the proposed controller is based upon using a low rating switched-capacitor circuit as online estimator to extract the fundamental reactive component of the nonlinear load current. This reactive component is shifted in appropriate phase using two phase-shifters to obtain a 3φ signal. The line current is sensed and compared instantaneously with this 3φ signal using three comparators. The outputs of the comparators represent each moment the profile of the harmonic current. This harmonic current is taken as a reference signal to drive a simple closed-loop P-I controller in order to produce the necessary modulating signals for the triggering module of the APF. In the case of balanced load, which is the case under investigation in this paper, one switched-capacitor circuit is needed. In spite of being simple and straightforward, the proposed controller will be able to solve the problems arising with the p-q algorithm based controllers.

I. INTRODUCTION

Switched inductors and switched capacitors are used in power systems to control the fundamental component of the reactive current. As a result, the fundamental power factor could be considerably improved. Simply, the switched capacitor is a capacitor connected to the power line via electronic switch and a small inductor. It is clear that the switched capacitor; in this way; is not attached to the power line permanently. This helps a lot in avoiding serious unexpected resonance surges which normally occur due to fixed capacitors [1]. The innovation in this paper is that the switched-capacitor circuit will not be included as a compensator in the power system, but it will be used as a driving stage in the control loop of the shunt APF.

The multiplication of voltage and current as time functions

was the concept on which the instantaneous power theory was based.

It was possible through complex computations, axis transformations and proper filters to resolve the instantaneous power into its direct power and alternating power components [2]. The study of the optimum power flow of the system under investigation enables to derive what so called "p-q algorithm" to obtain the reference current. This reference current should be tracked by the APF to achieve the desired compensation. It is evident that the technique suffers from computational burdens and considerable delays. The first affects the cost and leads to a bulky size of the controller. The second leads to a mismatch between the APF and the power system when the load changes in a fast rate [3].

In this paper a simple and straight forward technique is presented. The proposed controller will be able to remedy those problems arising with the p-q algorithm based-controllers. This will be demonstrated in details through the following sections.

II. THEORY

According to the tetrahedron representation of the instantaneous power derived in [4], the nonlinear load current profile can be seen as if it was a "composite" time signal of three components. These components are the fundamental active current I_p, the reactive fundamental component I_q and the harmonic current component I_h. On the other hand, the frequency sets principle discussed in [5] classifies the power network into three co-related systems namely, the voltage generator system, the filtering system and the load system. As each one of those systems has its own frequency set, the reaction among them depends mainly upon the intersection set of their own frequency sets. The results out of this can be concluded as follows:

i- The filter system can not absorb the active current component I_p.

ii- The frequency spectrum of the harmonic current I_h

produced by the load system should be included into the frequency sets of both the load system and the filtering system for optimum compensation. Provided this, the harmonic current into the line will be minimal. As a consequence, the voltage drop due to the harmonic current across the voltage system impedance can be neglected.

iii- The fundamental reactive current I_q causes the supply current to move out of phase from the supply voltage, which means degrading the input power factor of the power system. In spite of being reactive, the frequency of this component is not included in the frequency set of the filter system. Moreover, it existed in an implicit way in the current reference which drives the APF controller. As that fundamental reactive component remains uncontrolled, this explains, from our point of view, why the p-q algorithm-based controllers drift into instability when the load has fast changes.

The idea of the proposed controller stems from the above conclusions. A switched-capacitor circuit is designed to estimate or extract the fundamental reactive component of the nonlinear load current. This component is subtracted via a comparator from the line current to obtain an on-line profile of the harmonic current. This harmonic profile represents the actual reference input that the APF should be tuned to track instantaneously.

III. SWITCHED-CAPACITORS CIRCUIT (SCC) MODEL AND DESIGN

There are numerous switched-capacitor configurations. The most common one is the double-switch configuration shown in Fig.1.

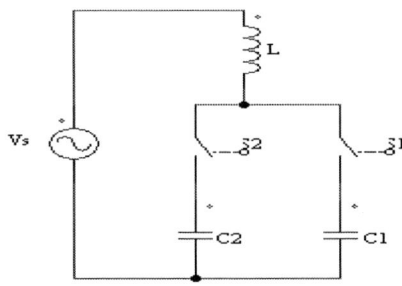

Fig. 1: Double-switch SCC.

The electronic switches S_1 and S_2 operate in an anti-phase fashion with complementary switching patterns. The switching pattern is a square pulse of a period T and a frequency of at least 20 times the mains frequency [6]. A third branch containing a resistor may be added to prevent short circuit across the capacitors if there is a possibility that the two switches overlap. The equations describing the double-switch circuit model can be written as follows:

When S_1 is closed and S_2 is open:

$$v(t) = Ri_1(t) + L\frac{di_1(t)}{dt} + \frac{1}{C_1}\int i_1(t)dt \; ; t \in [0, t_1] \quad (1)$$

When S_2 is closed and S_1 is open:

$$v(t) = Ri_2(t) + L\frac{di_2(t)}{dt} + \frac{1}{C_2}\int i_2(t)dt \; ; t \in [t_1, T] \quad (2)$$

$$i(t) = i_1(t) + i_2(t) \quad (3)$$

Where, i (t) represents the current drawn by the switched-capacitor circuit and $\lambda = t_1/T$ is the duty cycle of the switching pattern. The above set of equations was solved numerically by Runge-Kutta 4th order method to obtain appropriate values for L, C_1 and C_2 as circuit parameters so that , the circuit is able to absorb fundamental reactive current from the power system. For simplicity, C_2 is taken equal to multiple integer of C_1. Hence, $C_2 = \beta C_1$ and $\beta = 2, 3, 4,...,$ etc. With the circuit parameters $C_1 = 40 \; \mu F$, $C_2 = 80 \; \mu F$ and L = 10 mH, it was demonstrated that the inductor L saturates and the current drawn by the switched-capacitor circuit becomes pure cosine [6]. Fig. 2 shows the current drawn by the switched-capacitor circuit using the parameters mentioned above.

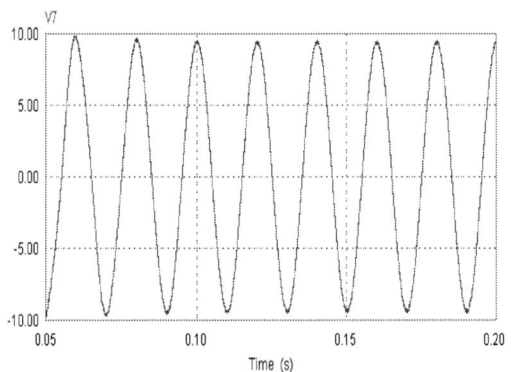

Fig. 2: The reactive current drawn by SCC ($C_1 = 40 \; \mu F$, $C_2 = 80 \; \mu F$, L = 10 mH).

The equivalent capacitance of the circuit is controlled by varying the duty cycle of the switching pattern. The equivalent capacitance of the SCC was derived in [7] and could be expressed in a compact form as follows:

$$C_{eq} = \frac{\beta C_1}{\beta \lambda^2 + (1-\lambda)^2} \quad , 0 \le \lambda \le 1 \quad (4)$$

Where β is called "the branch's ratio" defined earlier and λ is the duty cycle of the switching pattern. When the switches are allowed to operate at the optimum switching duty cycle, the fundamental reactive current approaches its maximum. The optimum duty cycle was obtained by minimizing (4) with respect to λ and found to be:

978-1-4244-1871-8/07 $25.00 © 2007 IEEE

$$\lambda_{opt} = \frac{1}{\beta + 1} \qquad (5)$$

It is obvious that at this duty cycle, the SSC possesses its maximum reactive current capacity because the equivalent path impedance is minimal.

IV. THE PROPOSED CONTROLLER

It should be noticed that the case currently investigated in this paper assumes a balanced source. The case of unbalanced source is out the scope of this paper and will be discussed in a future work. The driving stage of the proposed controller is a SSC connected across one phase of the power source. The SCC as simulated using PSIM [8] is shown in Fig. 3. Each of the switches S_1 and S_2 is a bilateral switch implemented by connecting two IGBT transistors back to back. V_{Sp} is a signal generator providing a square-wave pattern at frequency of 1 kHz. The switching pulses are coupled directly to S_1 and inverted to S_2 to realize the anti-phase switching action.

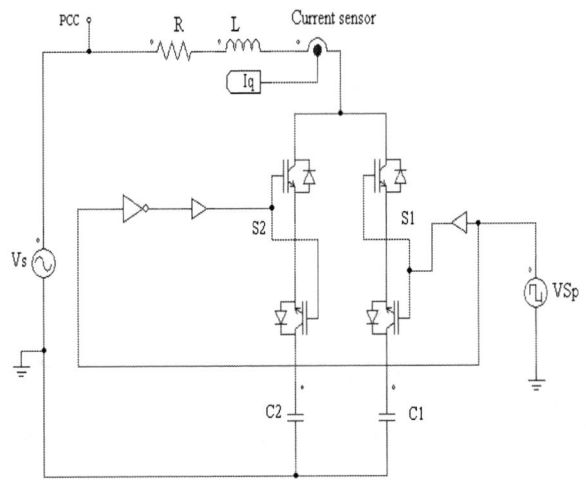

Fig. 3: The SSC as simulated using PSIM.

A current sensor is used to sense the fundamental reactive current component I_q. Figure 4 shows how the reference harmonic current can be extracted. I_q is shifted in time using two parallel time-delay blocks TD_1 and TD_2 to generate 3ϕ signal. The nonlinear load current is sensed and fed into a comparator via its non-inverting node. The 3ϕ signal representing the fundamental reactive component is coupled into the comparator via its inverting node to accomplish the subtraction process for extracting the harmonic current reference.

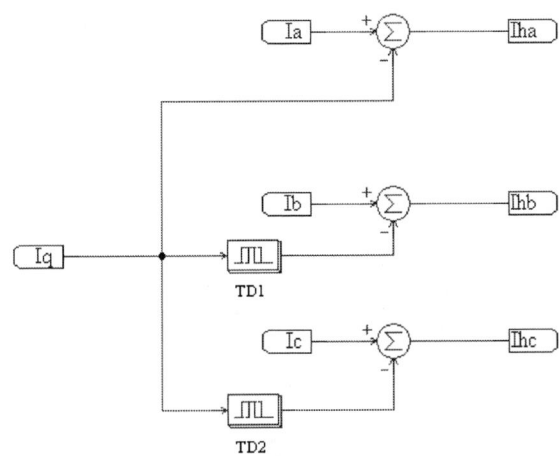

Fig. 4: Online generation of the harmonic current reference.

Figure 5 depicts the input and the output of the comparator for phase-A. The nonlinear load current is denoted V15 and the harmonic current reference denoted V16. The effect of taking the fundamental reactive component into consideration is very clear. The nonlinear load current suffers from discontinuities at the zero crossings of the mains voltage. In fact, this affects adversely the process of generating an accurate current reference for the APF controller [9].

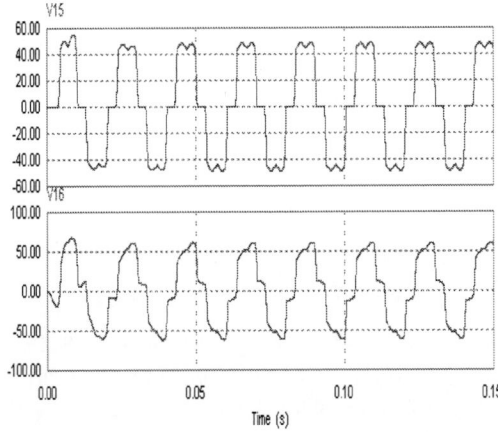

Fig. 5: The nonlinear load current and the harmonic reference current for phase-A.

It is clear that the reference harmonic current becomes continuous. This provides an inherent autonomous tendency of the controller towards a stable performance even when the load changes in a fast rate [10].

V. DESIGN OF P-I CONTROLLER

The block diagram of the P-I controller stage is shown in Fig. 6. The goal here is to modulate the inverter current of the APF to track the reference current at steady state [11].

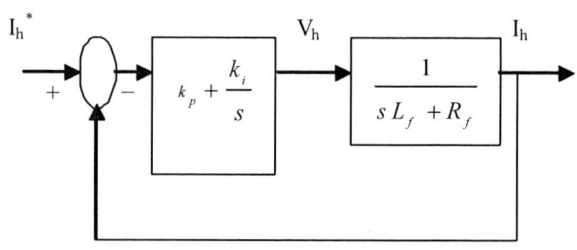

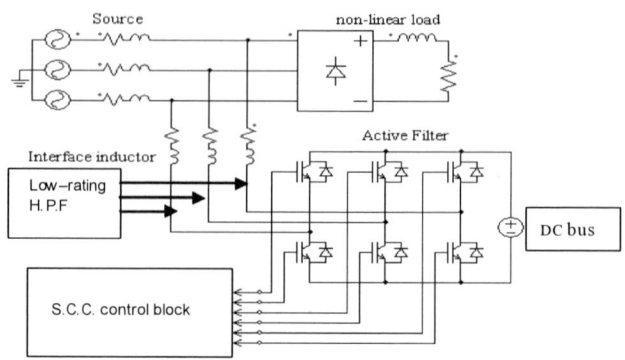

Fig. 6: The block diagram of the P-I controller.

Fig. 7: The system configuration.

Where L_f is the inductance of the interface inductor connecting the APF to the mains and R_f is its internal resistance. The transfer function of the P-I controller can be written as:

$$G(s) = \frac{k_p}{L_f} \frac{s + \dfrac{k_i}{k_p}}{s^2 + \left(\dfrac{R_f}{L_f} + \dfrac{k_p}{L_f}\right)s + \dfrac{k_i}{L_f}} \quad (6)$$

Comparing (6) to the standard transfer function of the second order system, the parameters of the P-I controllers are found to be:

$$k_p = 2\xi\omega_n L_f - R_f \quad (7)$$

$$k_i = L_f \omega_n^2 \quad (8)$$

Now, L_f and R_f for the interface inductor are assumed known. The damping ratio (ξ) is taken to be ($1/\sqrt{2}$) to minimize the percentage overshoot [12]. It is just required now to specify the natural frequency (ω_n). The dominant higher order harmonics (> 50th) in the source current spectrum are normally eliminated by the short-circuit impedance of the grid [13]. On this base, a criterion to specify (ω_n) can be defined. For the case being studied, the carrier frequency (ω_c) of the PWM modulator is selected to be ($2\pi \times$ 10K) rad/s. Accordingly, a value of ($2\pi \times 2500$) rad/s is sufficient for (ω_n) to filter out the harmonics till the 50th. Moreover, this value is used to set the cutoff-frequency of the interface filter. This filter is a low-rating high pass filter connected to the source side as shown in Fig. 7. The function of this filter is to prevent the undesired higher harmonics associated with the inverter current of being injected into the mains [14]. This way, the parameters of the P-I controller of (7) and (8) become completely defined.

VI. SIMULATION AND EXPERIMENTAL RESULTS

The system configured in Fig. 7 was simulated using PSIM. The nonlinear load represented by a full-wave rectifier driving 20 kVA inductive load. Figure 8 depicts the source voltage and current after compensation. The source current becomes sinusoidal and completely in phase with the source voltage.

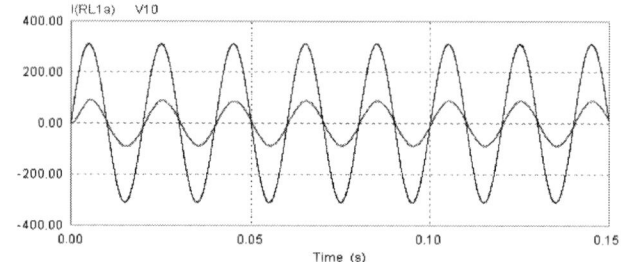

Fig. 8: The source voltage and the source current after compensation

It is obvious that the performance of the controller is excellent. The overshoot of the source current is damped and the source current starts tracking the source voltage within the first half cycle.

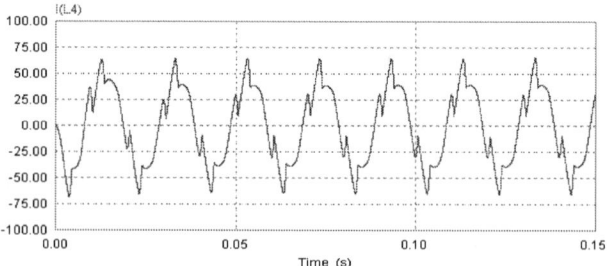

Fig. 9: The harmonic current injected by the APF into mains.

Figure 9 shows the harmonic current produced by the APF to achieve the desired compensation.

An experimental setup was implemented to verify the results of the simulation. Power IGBT transistors were used to implement the modules of the SCC and the APF. A

978-1-4244-1871-8/07 $25.00 © 2007 IEEE

microcomputer was programmed to produce the necessary switching pattern for the SCC. The pulses coupled to the IGBT switches via opto-isolators interface. A regulated DC bus of 600 V was used to supply the power inverter of the APF.

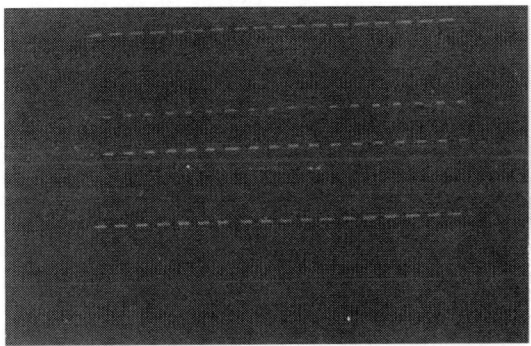

Fig. 10: The switching pattern of S_1 and S_2 of the SCC. (5 V/div)

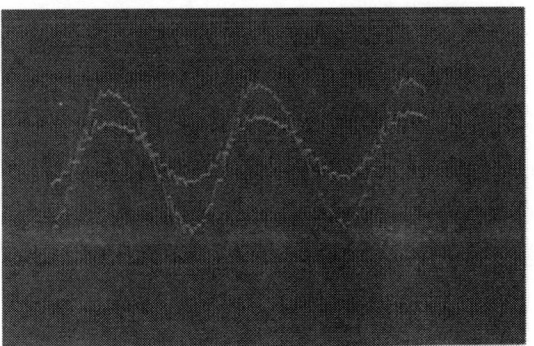

Fig. 11: The capacitor voltages V_{c1} and V_{c2} at optimum λ. (50 V/div)

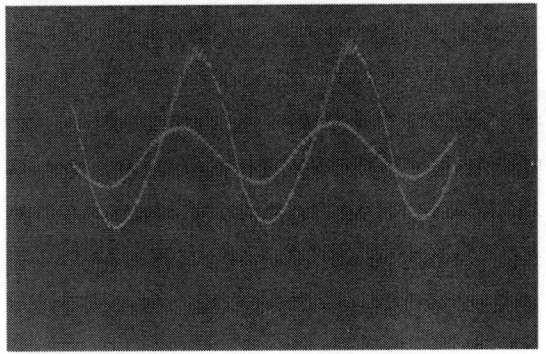

Fig. 12: The SSC reactive current (5 A/div) with the source current after compensation (20 A/div)

Figure 10 shows the switching pattern that drives the switches S_1 and S_2 of the SCC. It is clear that the switching pulses of S_2 are the logical complement of S_1. The duty cycle is chosen to be λ_{opt}. It should be noted that, according to (5), when $\beta = 2$, $\lambda_{opt} = 0.3333$. Figure 11 shows the capacitor voltages V_{c1} and V_{c2} of the SCC for the same switching duty cycle. Figure 12 shows the SCC reactive current with the source current after compensation. It is obvious that the SCC reactive current is pure cosine while the compensated source current is sinusoidal and has the same zero crossings of the source voltage.

This experimental work confirms the simulation results and verifies the versatility of the proposed controller.

VII. CONCLUSIONS

In this paper, a novel online controller for tuning shunt active power filters is presented. Assuming balanced source, the driving stage of the proposed controller is a single switched -capacitor circuit (SCC) connected across one phase of the power lines. The switching duty cycle of the double electronic switches is set to the optimum, in order to increase the fundamental reactive current capacity of the circuit. The fundamental reactive current is subtracted from the nonlinear load current instantaneously to obtain the actual harmonic reference current. A closed loop PI-controller is designed to force the APF current to track the harmonic reference current at steady state. The simulation results prove the capability of the proposed controller to respond to the fast load changes while securing a stable operation. Unlike the p-q algorithm based controllers, the proposed controller avoids axis transformations, complex computations and filtering delays. As the idea of the proposed controller is simple and straight forward, it will overcome the problems of bulky size and high cost.

REFERENCES

[1] P. Mehta, M. Darwish and T. Tomson: "Switched- capacitor filters", *IEEE Transaction on Power Electronics*, Vol. 5, No. 3, July 1990, pp.331 – 337. H. Poor, *An Introduction to Signal Detection and Estimation*. New York: Springer-Verlag, 1985, Ch. 4.

[2] H. Akagi, A. Nabae, Y. Kanazawa, "Instantaneous reactive power compensators comprising switching devices without storage components", *IEEE Trans., Ind. Appl.*, Vol. IA-20, No. 3, pp. 625–630, 1984.

[3] O. M. Elgendy, A.M.A. Abass, A. M.A. Mahmoud, A. D. Alkoshery, "An active compensation strategy for the low-voltage network in case of harmonic current-producing loads", *12th EPE/PEMC. Portorož,, Slowenia*, Aug. 30 – Sept. 1, 2006, pp. 1466–1471.

[4] E. H. Watanabe, R. M. Stephan, M. Aredes, "New concepts of instantaneous active and reactive powers in electrical systems with generic loads", *IEEE Trans. Power Delivery*, Vol. 8, No. 2, April 1993. pp. 697–703.

[5] E. Destobbeleer, L. Protin, "On the detection of load active currents for active filter control", *IEEE Trans. Power Elect.* Vol. 11, No. 6, Nov. 1996. pp. 768–775.

[6] O. M. Elgendy, A. Ahmed, A. I. Saied, A. A. Sattar, "Reactive power compensator using switched-capacitor technique", *The 3rd Regional Conference of CIGRE, Doha, Qatar*, May 25-27, 1999, Vol. 2, Paper No. 27.

[7] O. M. Elgendy, A. Ahmed, A. I. Saied, A. A. Sattar: "CAD procedures of an optimum switched-capacitor reactive power compensator" *The 8th International Middle East Power Conference "MEPCON 2000", Cairo, Egypt*, March 26-30, 2000, Vol. 2, pp. 459–466.

[8] *PSIM User's Guide, Version 6.0*, Powersim Inc., USA, June 2003.

[9] Akagi, H.: "Trends in active power line conditioners", *IEEE/IES, IECON*, pp. 19–24, 1992.

[10] M. Takeda, K. Ikeda, Y. Tominaga, "Harmonic current compensation with active filters", *IEEE/IAS, Annual meeting*, pp. 808–815, 1987.

[11] B. H. Kown, J.H. Youm, J. W. Lim, "A line-voltage – sensorless synchronous rectifier". *IEEE Trans., Power Elect.*, Vol. 14, No. 5, pp. 967–972, 1999.

[12] P. Verdelho, G. D. Marques, "An active power filter and unbalanced current compensator", *IEEE Trans., Ind. Elect.*, Vol. 44, No. 3, pp. 321–328, 1997.

[13] K. Chatterjee, B. G. Fernandes, G. K. Dubey, "An instantaneous reactive Volt-Ampere compensator and harmonic suppressor system", *IEEE Trans., Power Elect.*, Vol. 14, No. 2, pp. 381–391, 1999.

[14] E. Watanabe, M. Stephan, M. Aredes, "New concepts of instantaneous active and reactive powers in electrical systems with generic loads", *IEEE Trans., Power Del.*, Vol. 8, No. 2, pp. 697–703, 1993.

A Multiband Shunt Hybrid Active Filter With Sensorless Control

Surendra Kumar S
Samsung India Software Operations Pvt. Ltd.
No. 66/1, Bagmane Tech Park
C.V. Raman Nagar, Byrasandra
Bangalore 560 093, INDIA.
Email: ks.surendra@siso.com

Partha Sarathi Sensarma
Department of Electrical Engineering
Indian Institute of Technology
Kanpur 208016, INDIA.
Email: sensarma@iitk.ac.in

Abstract—This paper proposes a Mutiband Shunt Hybrid Active Filter(SHAF) with sensorless control. The plant is modeled in the discrete- time domain and a controller is designed using Pole shifting law in the polynomial domain. This control approach is very useful for filtering the load harmonics with reduced sensor counts where low cost solution like SHAF. Multiple Synchronous Reference Frames (MSRF) and low pass filters are used to measure 5th and 7th harmonic components separately from load as well as filter currents. Individual current controllers are designed for the 5th and 7th harmonic currents. Control is realized in the stationary, three-phase (abc) reference frame. Performance of the controller is validated through simulation, using realistic plant and controller models, as well as experimentally on a full-scale distribution system.

Index Terms Multiband Shunt Hybrid Active Filter, Harmonic extraction, Multiple Synchronous Reference Frames, Poleshift controller.

I. INTRODUCTION

Shunt Hybrid Active Filters (SHAF), consisting of an active filter and passive filters in series have been extensively reported in literature [1] [2]. This combination is placed in shunt with a non-linear load, as shown in fig. 1. This topology overcomes the disadvantages of filter overload and system resonance with a passive filter. Also, it results in significant reduction in the active filter rating. In its basic form, one SHAF

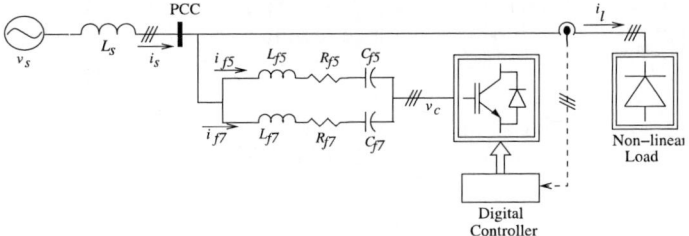

Fig. 1.   Shunt Hybrid Active Filter

branch can compensate for only one harmonic frequency. This requires multiple branches for each harmonic component in the load current. The possibility of using a single active filter in series with multiple passive filter branches have been reported [4]. The general circuit schematic of this approach is shown in fig. 1. However, this approach uses a Proportional Integral

(P-I) controller, which requires measurement of the filter capacitor voltages for feedforward compensation of disturbance and cross-coupling terms in the control law. This requires additional complexity and adds to the cost of an otherwise low-cost solution.

In this paper, a multiband SHAF is considered, in which two passive filter branches, tuned at 5th and 7th harmonic frequencies, are connected in series with the active filter. Switching harmonics of the active filter output are filtered by the passive filter impedance, which is inductive at the switching frequency. The plant is modeled in the discrete-time domain and a discrete-time control law is derived. The controller is designed using pole shifting law in the polynomial domain to radially shift the open loop system poles towards the origin i.e., more stable locations. With this Poleshift controller [5], sensing of capacitor voltages in the passive filter sections is not required. In order to damp filter resonances during transient changes in load harmonics, an active damping is introduced to greatly improve dynamic performance. Analytical control design is validated using simulation and experimental results on full-scale hardware. Multi-rate sampling is used to model the plant (fast sample-rate) and the controller (slow sample-rate). Switching of active filter devices is considered though device losses are neglected.

II. DESIGN OF PASSIVE FILTER

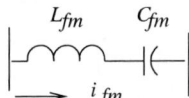

Fig. 2.   Ideal Series Resonant Circuit

A series resonant filter circuit, is shown in Fig. 2. The filter impedance is given by,

$$Z_{fm}(s) = (sL_{fm}) + \left(\frac{1}{sC_{fm}}\right) = \frac{s^2 L_{fm}C_{fm} + 1}{sC_{fm}} \quad (1)$$

Resonant frequency ω_m of the filter is given by

$$\omega_m = \sqrt{\frac{1}{L_{fm}C_{fm}}} \quad (2)$$

978-1-4244-1871-8/07 $25.00 © 2007 IEEE

where,

$$\omega_m = m.2\pi f \qquad (3)$$

In this paper, as the 5th and 7th harmonic currents are to be compensated, $m=5,7$ and $f=50Hz$ (fundamental frequency).

A. Selection of C_{fm}

The choice of C_{fm} is decided by the following factors. A high value reduces impedance at fundamental frequency resulting in passage of fundamental current, which is highly undesirable. Also a low value results in a bulky inductor in the resonant branch, which in turn increases its cost. The optimal value of the filter capacitor is found to be 10 μF. Other filter parameters are mentioned in Table I.

TABLE I
FILTER PARAMETERS

L_{f5}	40.5 mH
R_{f5}	1 Ω
C_{f5}	10.0 μF
L_{f7}	20.7 mH,
R_{f7}	0.41 Ω
C_{f7}	10.0 μF

R_{f5} and R_{f7} are the internal resistances of L_{f5} and L_{f7}.

B. Impedance Characteristic of Passive Filter Circuit

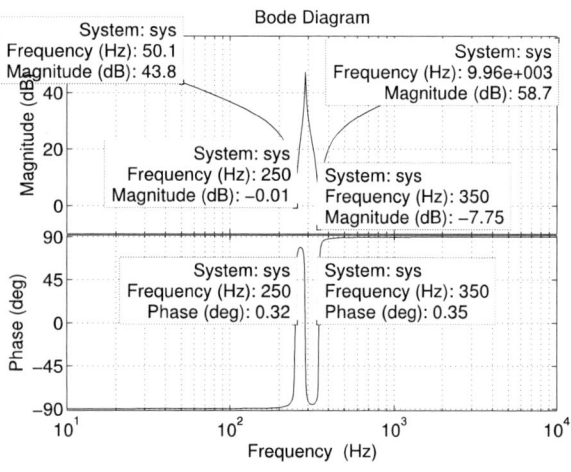

Fig. 3. Bode magnitude plot for $(Z_f(s))$

Fig. 3 shows the Bode plot (magnitude and phase) of the passive branch. It shows the fundamental component and the switching harmonics (10 kHz or more) are blocked by the passive filter, whereas at other desired 5th and 7th harmonic frequencies the filter offers low impedance. Switching harmonics of the active filter output are filtered by the passive filter impedance, which gives more than 55dB attenuation at 10kHz. Hence only the required load harmonics will pass through these branches, fundamental and switching harmonics will be blocked. This results in a great reduction of the required rating of the Active filter.

III. SYSTEM MODEL

State space model of the SHAF system for the 5th and 7th harmonic currents is given by

$$\frac{d}{dt}\begin{bmatrix} i_{fm} \\ v_{fm} \end{bmatrix} = \begin{bmatrix} -\frac{R_{fm}}{L_{fm}} & -\frac{1}{L_{fm}} \\ \frac{1}{C_{fm}} & 0 \end{bmatrix}\begin{bmatrix} i_{fm} \\ v_{fm} \end{bmatrix} + \begin{bmatrix} -\frac{1}{L_{fm}} \\ 0 \end{bmatrix}v_{cm}$$

$$y = \begin{bmatrix} 1 & 0 \end{bmatrix}\begin{bmatrix} i_{fm} & v_{fm} \end{bmatrix}^T \qquad (4)$$

where, $m = 5, 7$. From (4) transfer function between filter currents and control voltage is

$$G_{pm}(s) = \frac{I_{fm}(s)}{V_{cm}(s)} = \frac{-s/L_{fm}}{s^2 + (R_{fm}/L_{fm})s + 1/(L_{fm}C_{fm})} \qquad (5)$$

IV. HARMONIC EXTRACTION

In this paper, only 5th and 7th harmonic components are sought to be compensated. These are extracted from a measurement of the load current using Multiple Synchronous Reference Frames (MSRF) and low pass filters [3], [4]. Phase locking with the fundamental component of the PCC voltage is achieved through a PLL [6]. Denoting the fundamental phase angle as θ, transformations for each of the rotating reference frames are carried out as shown in fig. 4.

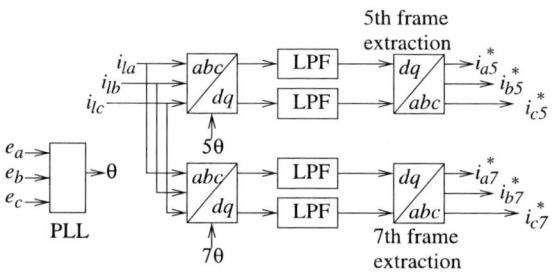

Fig. 4. Block diagram showing reference current generation using MSRF.

V. CURRENT CONTROLLER DESIGN

Individual current controllers are designed for the 5th and 7th harmonic currents. The harmonic currents, extracted above, are used as reference commands for the individual current control loops. Control is realized in the stationary, three-phase (abc) reference frame.

A. Controller for 5th harmonic current

The closed-loop model of the system is shown in fig 5. $G_{pm}(z)$ is the discrete-time (z-domain) transfer function of

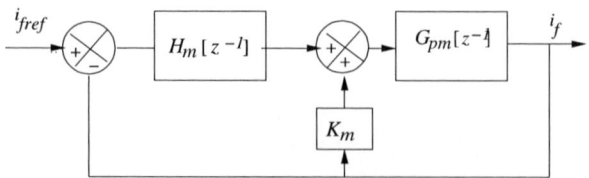

Fig. 5. Closed loop current controller schematic.

978-1-4244-1871-8/07 $25.00 © 2007 IEEE

the plant and $H(z)$ denotes the Poleshift controller. Since control is realized using a digital processor, it is appropriate to transform the plant transfer function in to its discrete-time equivalent. Zero-order-hold equivalent of the plant is obtained using the following transformation [7].

$$G_{pm}(z) = (z-1)Z\left\{\frac{G_{pm}(s)}{s}\right\}. \qquad (6)$$

Consequently, the discrete-time plant transfer function is determined as

$$G_{pm}(z^{-1}) = \frac{B_m(z^{-1})}{A_m(z^{-1})} = \frac{b_{m1}z^{-1} + b_{m2}z^{-2}}{1 + a_{m1}z^{-1} + a_{m2}z^{-2}}. \qquad (7)$$

Expressing the controller transfer function $H(z)$ as

$$H_m(z^{-1}) = \frac{S_m(z^{-1})}{R_m(z^{-1})} = \frac{s_{m0} + s_{m1}z^{-1}}{1 + r_{m1}z^{-1}} \qquad (8)$$

from fig. 5 the closed loop system transfer function is

$$i_{fm}[k] = \frac{B_m(z^{-1})S_m(z^{-1})}{A_m(z^{-1})R_m(z^{-1}) + B_m(z^{-1})S_m(z^{-1})}i_{frefm}[k]. \qquad (9)$$

The controller parameters are obtained from the solution of the following Diophantine equation.

$$A_m(z^{-1})R_m(z^{-1}) + B_m(z^{-1})S_m(z^{-1}) = T_m(z^{-1}). \qquad (10)$$

In (10), $T_m(z^{-1})$ is the closed loop system characteristic equation obtained after shifting the open loop poles. It is chosen to be of the following form.

$$T_m(z^{-1}) = A_m(\lambda z^{-1}) = 1 + \lambda a_{m1}z^{-1} + \lambda^2 a_{m2}z^{-2}, \quad 0 < \lambda < 1 \qquad (11)$$

where,
λ is the pole shift factor and determines the proximity of the closed-loop poles to the origin. A larger value of λ implies the closed loop poles are close to the open-loop pole locations. As smaller values of λ require a larger control effort, for the SHAF system, this implies a higher output voltage demand from the active filter. However, this would increase the VA rating of the active filter and would therefore undermine the cost advantage of SHAF. So, this pole shift factor is chosen here to be 0.6. Using this advanced control method, source harmonic currents are reduced below limits prescribed in IEEE 519-1992, even with distorted PCC voltages.

System parameters are mentioned in Table III. Solving

TABLE II
SYSTEM PARAMETERS

Sampling interval(T_s)	100 μs
Switching frequency (f_c)	10 kHz
Load inductance (L_d)	10.0 mH
Load resistance (R_d)	30 Ω
PCC voltage (l-l, rms)	415 V
V_{dc}	80 V

(10) and (11), the numerator and denominator polynomials of the 5th and 7th harmonic current controllers are obtained as mentioned in Tables III and IV.

TABLE III
5th HARMONIC CURRENT CONTROLLER

System num poly	b_{51} =-0.0025, b_{52} = 0.0025
System den poly	a_{51} =-1.9729, a_{52} = 0.997
Controller num poly	s_{50} =-374.2825, s_{51} =320.7645
Controller den poly	r_{51} =0.7892
Closed loop poles	0.981+0.137i, 0.981-0.137i, -0.803

TABLE IV
7th HARMONIC CURRENT CONTROLLER

System num poly	b_{71} =-0.0048, b_{72}= 0.004
System den poly	a_{71} =-1.95, a_{72} = 0.998
Controller num poly	s_{70} =-184.06, s_{71} = 162.42
Controller den poly	r_{51} =0.789
Closed loop poles	0.974+0.193i, 0.974-0.193i, -0.792

B. Active Damping

Quality factor for 5th harmonic filter is

$$Q_{f5} = \frac{\omega_5 L_{f5}}{R_{f5}} = 63.5 \qquad (12)$$

In order to damp filter resonances during transient changes in load harmonics, Q_{f5} has to be reduced. For this, series resistance of the 5th harmonic filter has to be increased. But adding physical resistance would increase losses. Instead, a "virtual resistance" [8] is added which is essentially an active damping term introduced in the control law and is given as

$$\begin{aligned} u_1 &= H_m(z^{-1})i_{frefm} - i_{fm} \qquad (13) \\ u &= u_1 + K_m i_{fm} \end{aligned}$$

where K_m is the virtual resistance and u is the control input.

C. Composite Controller

Using the above design procedure for the two controllers, analytical predictions of the closed-loop performance is reported here. The w-plane plots are analyzed to obtain the phase error (between reference command and output) for 5th and 7th harmonic currents. Fig. 6 shows the magnitude and phase plot in w-plane for the 5th harmonic controller, after transformation [7] of (9) to the w-plane, using

$$w = \frac{2}{T_s}\frac{z-1}{z+1}. \qquad (14)$$

From fig.6, the predicted phase error for 5th harmonic current is negligible ($\approx 1.7^o$). Corresponding plot for the 7th harmonic controller is also obtained in a similar fashion. Fig. 7 shows the magnitude and phase plot in w-plane for the 7th harmonic controller. From fig.7, the predicted phase error for the 7th harmonic current is also found to be negligible ($\approx 2.0^o$).

VI. SIMULATION RESULTS

Performance of closed loop system is simulated for both pure sinusoidal and distorted PCC voltage cases. The active filter is simulated using ideal switches so as to include

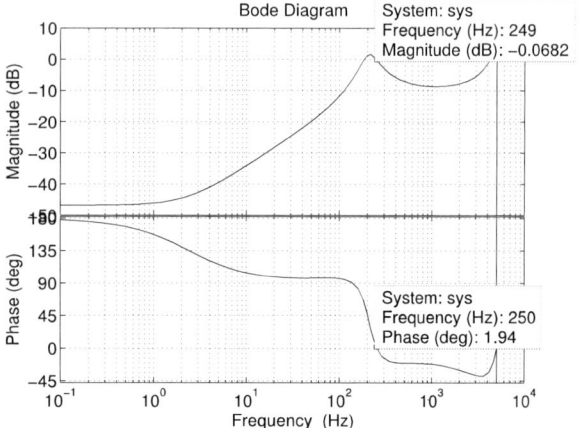

Fig. 6. Magnitude bode plot (w-plane) of closed loop system

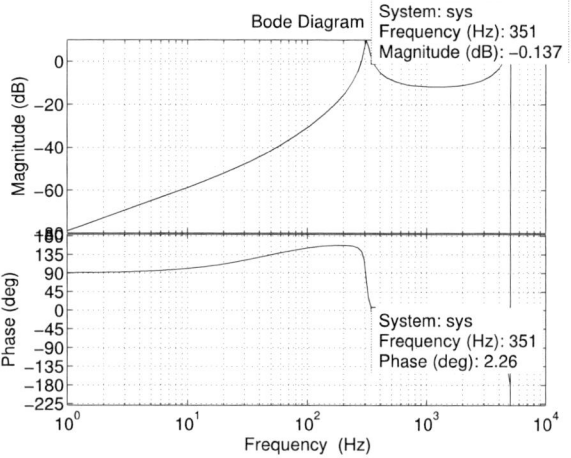

Fig. 7. Magnitude bode plot (w-plane) of closed loop system

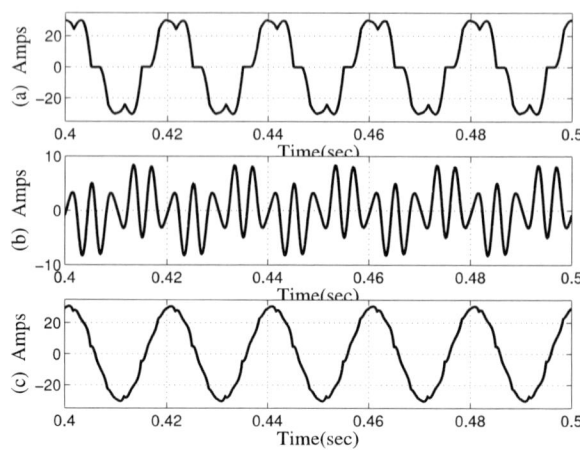

Fig. 8. Case A: (a) Load current; (b) Filter current; (c)Source current

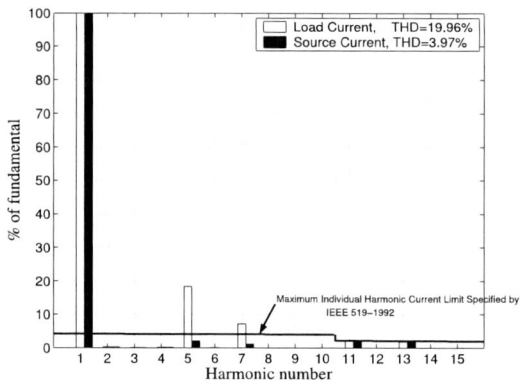

Fig. 9. Case A: FFT Comparison of Load and Source currents

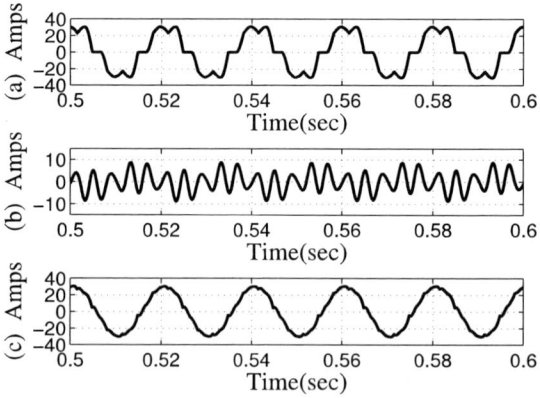

Fig. 10. Case B: (a) Load current; (b) Filter current; (c)Source current

switching action without considering switch losses. Multi-rate simulation is carried out using $2\mu s$ sampling interval for the continuous time plant and $100\mu s$ for the discrete-time controller. System parameters used are as indicated in Tables I and III. Performance of the MSHAF is tested under various operating conditions. Specifically, the performance under distorted PCC voltages and de-tuning of passive filter is investigated and reported.

A. PCC voltage without distortion

Fig. 8, shows the simulated results for pure sinusoidal PCC voltage. Fig. 9 shows the FFT analysis of source and load currents. In this 5th and 7th source harmonic currents are reduced with in IEEE 519-1992 limits and 11th and 13th source harmonic currents are not affected.

B. PCC voltage with distortion

Fig. 10, shows the simulated results for distorted PCC voltage having 5% of 5th and 4% 7th harmonics. Fig. 11 shows the FFT analysis of source and load currents. In this 5th and 7th source harmonic currents are reduced with in IEEE

519-1992 limits and 11th and 13th source harmonic currents are not affected.

C. De-tuned passive filter

Fig. 12 shows the simulated results for +5% De-tuned 5th and 7th harmonics LC components. This has shifted corner

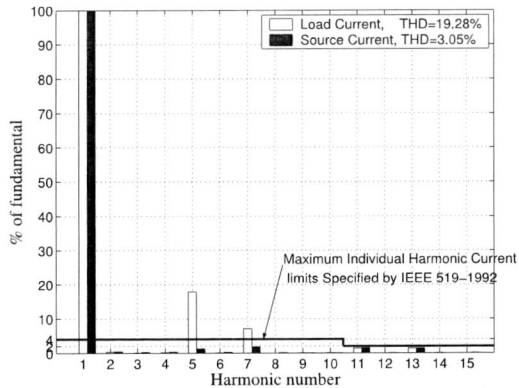

Fig. 11. Case B: FFT Comparison of Load and Source currents

frequency of the 5th harmonic filter from 250 Hz to 238 Hz. For the 7th harmonic filter the shift is from 350 Hz to 332 Hz. Fig. 13 shows the FFT analysis of source and load currents.

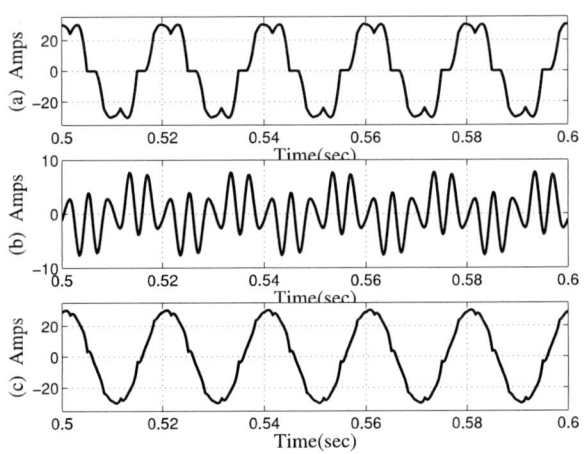

Fig. 12. Case C: (a) Filter current; (b) Load current;(c)Source current

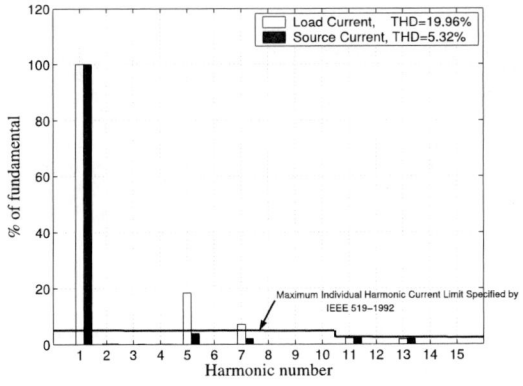

Fig. 13. Case C: FFT Comparison of Load and Source currents

In this 5th (3.92%) and 7th (1.93%) source harmonic currents are reduced with in IEEE 519-1992 limits and 11th and 13th source harmonic currents are not affected.

VII. EXPERIMENTAL RESULTS

The Poleshift control strategy has been experimentally tested on a 415 V, 3-phase, 50 Hz distribution line. The load comprised a diode rectifier feeding a resistive load through an inductive filter. A two-level, low power (400 VA) inverter (active filter) was used. Passive filter parameters used were as mentioned in Table I. The entire control algorithm including harmonic extraction and PLL was realized on a DSP platform, using TMS320F240 processor. Hall effect current sensors were used to sense the filter and load currents. As a three-wire system has been considered, only two line currents have beeen sensed. The active filter was designed along with all associated protection and gate triggering circuitry. Sensed signals was sent to on board ADC and the PWM signals having carrier frequency of 10 kHz are generated using Simple compare unit of DSP. In the experiment, inverter DC bus voltage was maintained constant with help of a diode bridge rectifier. The

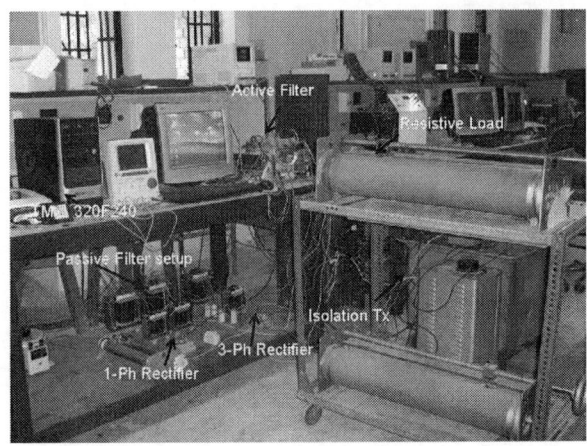

Fig. 14. MSHAF Experimental test setup

experimental test setup is shown in Fig. 14.

Using this set-up, the performance of the MSHAF was tested under different rectifier loading conditions. For practical constraints, neither arbitrary distortion in the available PCC voltage nor de-tuning of the passive filter could not be created. Fig. 15 shows a particular case where the fundamental component of the load current is 6 A (rms). The filter current, load current and source current waveforms are shown. The source current waveforms are seen to be practically sinusoidal. Fig. 16 shows the FFT analysis of source and load currents. In this 5th (2.44%) and 7th (1.57%) source harmonic currents are reduced with in IEEE 519-1992 limits and 11th and 13th source harmonic currents are reduced by a small amount. This is however purely incidental, as the control does not focus on compensation of these harmonics.

Fig. 17 shows the results for another loading condition. Here the load current (fundamental, rms) is 12 A. Fig. 18 shows the FFT analysis of source and load currents. Also in this case, the 5th (2.24%) and 7th (1.07%) source harmonic currents are reduced with in IEEE 519-1992 limits.

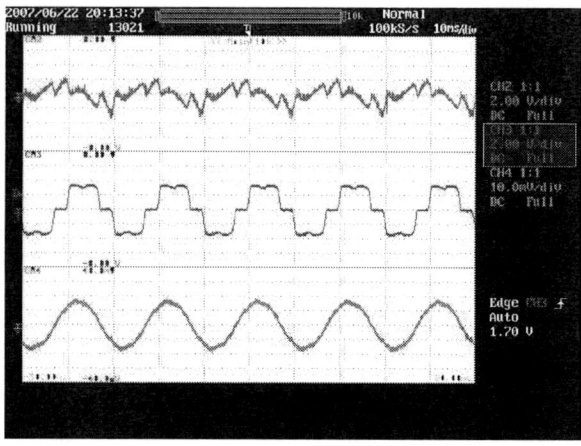

Fig. 15. Filter performance under light load conditions. (a)Filter current ; (b)Load current ; (c)Source current

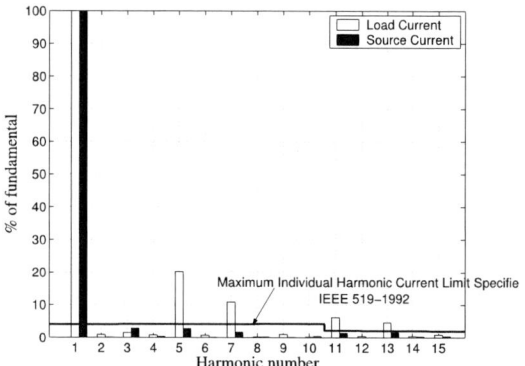

Fig. 16. Light load performance: FFT of Load and Source currents

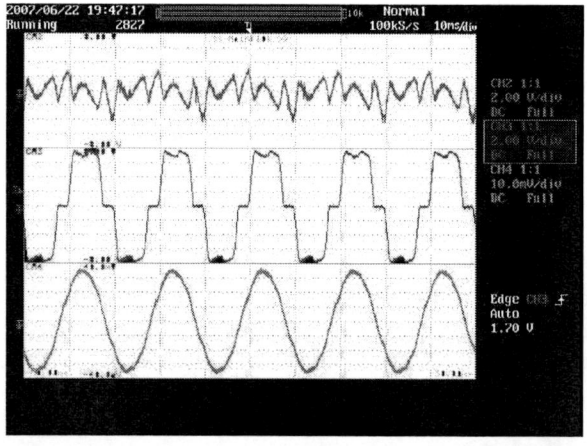

Fig. 17. Filter performance under rated load: (a) Filter current (b) Load current (c) Source current

VIII. CONCLUSION

A Multiband SHAF scheme with a Poleshift controller is shown to be an attractive realization approach for harmonic compensation. Using this advanced control method, harmonics in the source current are controlled within limits specified by

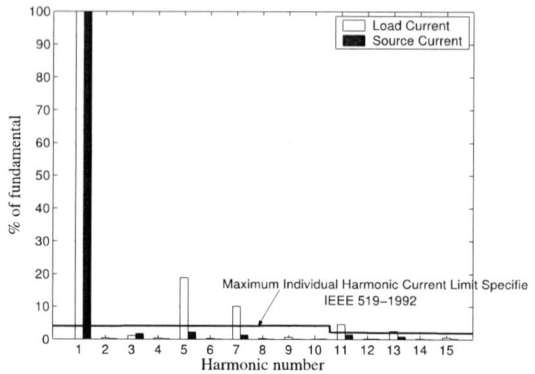

Fig. 18. Rated load performance: FFT of Load and Source currents

IEEE 519-1992. Use of reduced sensor count along with an active damping scheme ensures a low cost solution without sacrificing any performance advantage. The complete analytical design procedure of controller parameters is presented. It is verified by simulation that the proposed approach is immune to parameter variations in the passive filter and distortions in the supply voltage. Poleshift Control algorithm for 5th and 7th harmonics was verified experimentally on a full-scale diode rectifier load operating in a 3-phase, 415 V, 50 Hz system. DC bus voltage of hybrid active filter was supported by a small rating diode rectifier. Experimental results obtained completely validate the analytical and simulation results.

REFERENCES

[1] Fujita, H.; Akagi, H., "A practical approach to harmonic compensation in power systems-series connection of passive and active filters", *IEEE Transactions on Industry Applications*, Vol.27, no.6, pp. 1020-1025, Nov/Dec 1991.

[2] Po-Tai Cheng; Bhattacharya, S; Divan, D. M. , "Line Harmonics Reduction in High- Power Systems Using Square-Wave Inverters Based Dominant Harmonic Active Filter", *IEEE Transactions on Power Electronics*, Vol. 14, N0. 2, pp 265-272, Mar. 1999.

[3] Bhattacharya,S.; Divan, D. M. ; Banerjee, B. , "Synchronous Reference Frame Harmonic Isolator Using series Active Filter", *Proc. 4th EPE*, Vol.3, pp 030-035, 1991.

[4] Sukin Park; Soo-Bin Han; Bong-Man Jung; Soo-Hyun Choi; Hak-Geun Jeong. , "A Current Control Scheme Based on Multiple Synchronous Reference Frame for Parallel Hybrid Active Filter", *Power Electronics and Motion control conference 2000,Proceedings PIMEC 2000.The Thrid International*, Beijing, China, Aug. 2000.

[5] Ghosh, A.; Jindal, A.K.; Joshi, A., "Inverter control using output feedback for power compensating devices", *IEEE Region 10 Conference on Convergent Technologies for Asia-Pacific Region,TENCON 2003*, Bangalore, India, Oct. 2003.

[6] Kaura, V ;Blasko, V., "Operation of phase locked loopsystem under distorted utility conditions", *IEEE Transactions on Industry Applications* , vol.33, no.1,pp 58-63, Jan/Feb 1997.

[7] B.C. Kuo, *Digital Control Systems*, Mc-Graw Hill, 1999.

[8] Pekik Argo Dahono, "A Control Method to Damp Oscillation in the input LC filter of AC-DC PWM Converters", *IEEE Power Electronics Specialists Conference, PESC 2002*, Queensland, Australia, June 2002.

A Shunt Active Filter for Reactive Power Compensation and Harmonic Mitigation

Narayan G. Apte
Lecturer in Electrical Engineering,
Walchand College of Engg. Sangli.
India.

Dr. Vishram N. Bapat
Principal,
Annasaheb Dange College of Engineering and
Technology, Ashta, Sangli, India.

Amruta N. Jog
Engineer,
Automation & Drives, Siemens,
Kalwa, Mumbai, India.

Abstract – **This paper presents a digital simulation of a 3Φ Shunt Active Filter (SAF) for harmonic and reactive power compensation. The instantaneous harmonic and reactive compensating currents is desegregated in it's three components viz., (a) fundamental reactive current (b) oscillating harmonic reactive current (c) oscillating harmonic active current. The resultant compensating current is injected to the point of common coupling (PCC) of the ac mains by a six pulse, hysteresis current controlled, Voltage Source Inverter (VSI). A high performance hysteresis band modulation technique is proposed to maintain nearly constant device switching frequency. The hysteresis bandwidth is modulated as a function of the desired switching frequency , supply voltage, DC bus voltage and slope of the reference current. Responses of simulated model clearly reveal the desired performance of SAF.**

Index Terms: **Compensating current, Hysteresis Current Controller, Point of Common Coupling, Reactive and harmonic current, Shunt Active Filter, Voltage Source Inverter**

I. INTRODUCTION

Widespread use of nonlinear devices in modern power system is badly affecting the power supply quality. The nonlinear devices generating nonlinear currents degrade the voltage waveform on the power distribution system enough to adversely affect operation of the other electrical apparatus causing poor power factor and poor utilization of the available power [10]. Majority of nonlinear devices are based upon use of power electronics technology and the interesting fact is that the cause of the problem and its solution originate from the same technology. Active filters have been considered as an effective solution to combat the problem of harmonics. Modern Active Power Filter (APFs) not only serve as a harmonic compensator but also provide compensation for reactive power, voltage sag/swell, load imbalance etc.[2][3] APFs are current controlled, voltage source inverters (VSI) which are intended to inject a compensating current into the system to relieve the supply from harmonic and reactive currents and enhance power quality indices. APFs may be classified into pure active filters and hybrid active filters. Hybrid APFs are primarily used for harmonic mitigation. Fast switching low power loss power electronic devices and fast digital signal processing devices available at an affordable cost, it is feasible to embed a variety of functions into a pure APF to make it a power quality conditioner.

Narayan G. Apte is a faculty in Electrical Engineering in Walchand College of Engineering, Sangli, India. (E-mail: narayan_apte@yahoo.com)
Dr. Vishram Bapat is working as Principal , ADCET, Ashta, Sangli, India. (E-mail: vishram.bapat@gmail.com)
Amruta N. Jog is a postgraduate engineer working as a design engineer in Automation & Drives,Siemens, Mumbai, India.
(E-mail: vedvati.jog@siemens.com)

The proposed work aims at development of simulation model for a Three Phase Four Wire SAF based upon a 6 pulse VSI with split capacitors on DC Side. There are two sections of the SAF, a) Compensating Current Calculator used to determine instantaneous compensation current b) Current Controller which tracks the compensation current to inject the same at PCC. In the proposed work, calculation of compensation current is based upon desegregation of load current into its three components viz., a) Fundamental reactive current b) Oscillating reactive current C) Oscillating harmonic active current [5]. A high performance adaptive hysteresis band current controller is proposed which maintains nearly constant device switching frequency. Use of conventional hysteresis band current control is limited due to its inherent drawbacks such as wide variation in switching frequency which makes it unfit to implement as it is. There are techniques[6][7][8] to control the switching frequency, one of them is by modulating the hysteresis band as a function of supply voltage, slope of the command current, and other system parameters[9].

II. GENERAL DESCRIPTION OF THE SYSTEM

The 3 phase 4 wire supply system shown in fig. 1 is feeding a nonlinear load taking significant harmonic and reactive component. The compensating current calculator measures the instantaneous load currents i_a, i_b, i_c and system voltages v_a, v_b, v_c, calculates instantaneous powers p and q by performing *abc* to *αβ* transformation . The power components that are to be compensated by SAF are extracted from the instantaneous powers by using appropriate high pass filters. These powers are oscillating harmonic real power, fundamental reactive power and oscillating harmonic reactive power. Depending upon error *(V_dcref - V_dc)*, PI controller will determine the necessary active power to be absorbed by SAF from supply in order to maintain D.C. bus voltage, V_{dc} and losses in the VSI. The compensating currents i_a*, i_b*, i_c*, are then calculated by *αβ* to *abc* transformation. The compensating currents are tracked by a high performance adaptive hysteresis band current controller which dynamically adjusts the hysteresis band (*HB*) as a function of *di*(t)/dt*, DC bus voltage, V_{dc}, supply voltage *v*, and the desired switching frequency, *f_sw.*. The current controller generates the switching signals to control the ramping of current in the coupling inductor within the hysteresis band. The injected currents i_{ac}*, i_{bc}*, i_{cc}* at PCC will relieve the supply from the reactive and harmonic currents. The resulting supply current will be pure sinusoidal and in phase with the supply voltage. The compensating current generation and current control techniques are discussed in subsequent paragraphs.

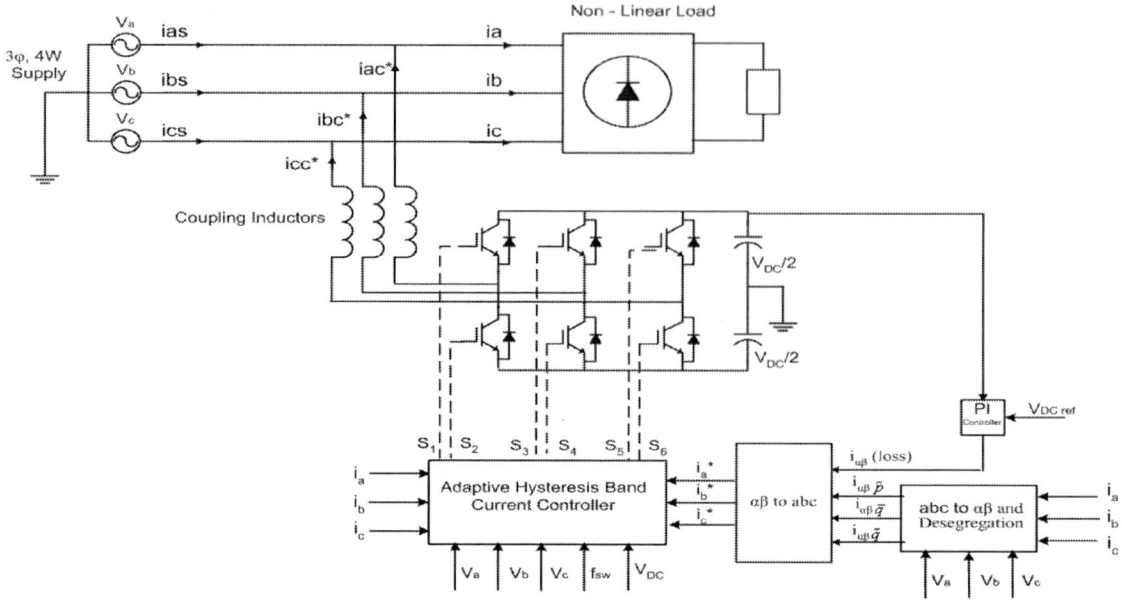

Fig. 1. General block diagram of the system

III COMPENSATING CURRENT GENERATION TECHNIQUE

Various techniques to generate SAF reference current are given in [1][2][4]. The proposed method is based upon [5] computing instantaneous load powers in $\alpha\beta$, decomposing them into four components as,

$$p_{3\phi} = p + q = \overline{p} + \tilde{p} + \overline{q} + \tilde{q} \qquad (1)$$

Where,

$p_{3\phi}$ = total 3φ instantaneous power

p = instantaneous active power component

q = instantaneous reactive power component

$\overline{p}$ = direct component of p

$\overline{q}$ = direct component of q

$\tilde{p}$ = oscillating harmonic component of p

$\tilde{q}$ = oscillating harmonic current of q

Above four components are demanded by non linear load and active filter compensator must provide $\tilde{p}, \tilde{q}, \overline{q}$. The transformation of abc into $\alpha\beta$ variables is made using the abc to $\alpha\beta$ reference frame.

$$\begin{bmatrix} v_\alpha \\ v_\beta \end{bmatrix} = \sqrt{\frac{2}{3}} \begin{bmatrix} 1 & -1/2 & -1/2 \\ 0 & \sqrt{3}/2 & \sqrt{3}/2 \end{bmatrix} \begin{bmatrix} v_a \\ v_b \\ v_c \end{bmatrix} \qquad (2)$$

$$\begin{bmatrix} i_\alpha \\ i_\beta \end{bmatrix} = \sqrt{\frac{2}{3}} \begin{bmatrix} 1 & -1/2 & -1/2 \\ 0 & \sqrt{3}/2 & \sqrt{3}/2 \end{bmatrix} \begin{bmatrix} i_a \\ i_b \\ i_c \end{bmatrix} \qquad (3)$$

The instantaneous active and reactive load powers are given by,

$$\begin{bmatrix} p \\ q \end{bmatrix} = \begin{bmatrix} v_\alpha & v_\beta \\ -v_\beta & v_\alpha \end{bmatrix} \begin{bmatrix} i_\alpha \\ i_\beta \end{bmatrix} \qquad (4)$$

Now the compensating currents in $\alpha\beta$ coordinates are given by,

$$\begin{bmatrix} i_{c\alpha} \\ i_{c\beta} \end{bmatrix} = \begin{bmatrix} v_\alpha & v_\beta \\ -v_\beta & v_\alpha \end{bmatrix}^{-1} \begin{bmatrix} -\tilde{p} + p_{loss} \\ -(\overline{q} + \tilde{q}) \end{bmatrix} \qquad (5)$$

$$\begin{bmatrix} i_{c\alpha\tilde{p}} \\ i_{c\beta\tilde{p}} \end{bmatrix} = \frac{1}{v_\alpha{}^2 + v_\beta{}^2} \begin{bmatrix} -v_\alpha \tilde{p} \\ -v_\beta \tilde{p} \end{bmatrix} \qquad (6)$$

$$\begin{bmatrix} i_{c\alpha\overline{q}} \\ i_{c\beta\overline{q}} \end{bmatrix} = \frac{1}{v_\alpha{}^2 + v_\beta{}^2} \begin{bmatrix} v_\beta \overline{q} \\ -v_\alpha \overline{q} \end{bmatrix} \qquad (7)$$

$$\begin{bmatrix} i_{c\alpha\tilde{q}} \\ i_{c\beta\tilde{q}} \end{bmatrix} = \frac{1}{v_\alpha{}^2 + v_\beta{}^2} \begin{bmatrix} v_\beta \tilde{q} \\ -v_\alpha \tilde{q} \end{bmatrix} \qquad (8)$$

The current required to replenish the DC Capacitor charge and the VSI losses is given by,

$$\begin{bmatrix} i_{c\alpha p_{loss}} \\ i_{c\beta p_{loss}} \end{bmatrix} = \frac{1}{v_\alpha{}^2 + v_\beta{}^2} \begin{bmatrix} v_\alpha \, p_{loss} \\ v_\beta \, p_{loss} \end{bmatrix} \qquad (9)$$

The respective compensating currents in the *abc* framework are obtained from (6),(7),(8),(9),

$$\begin{bmatrix} i_a * \\ i_b * \\ i_c * \end{bmatrix} = \sqrt{\frac{2}{3}} \begin{bmatrix} 1 & 0 \\ -\dfrac{1}{2} & \dfrac{\sqrt{3}}{2} \\ -\dfrac{1}{2} & \dfrac{\sqrt{3}}{2} \end{bmatrix} \begin{bmatrix} i_{c\alpha} \\ i_{c\beta} \end{bmatrix} \qquad (10)$$

The compensating currents obtained in (10) are to be tracked by the VSI. The inverter switching logic is based upon the dynamic modulation of the hysteresis band to maintain the constant switching frequency. The SIMULINK model for the reference current generation is shown in fig. 2

978-1-4244-1871-8/07 $25.00 © 2007 IEEE

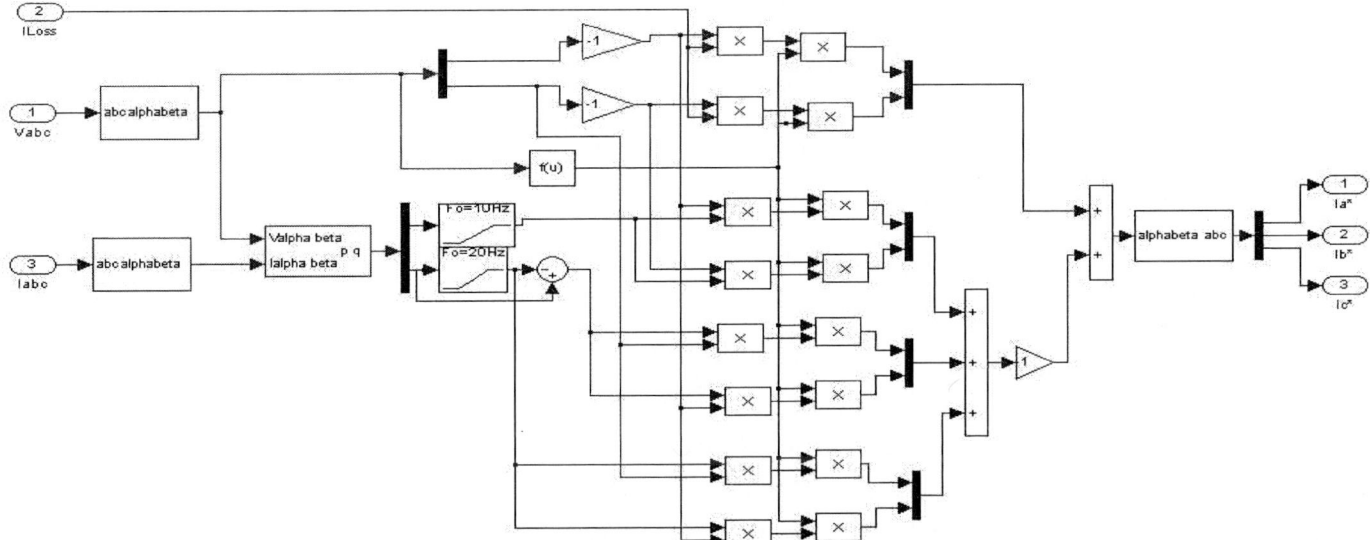

Fig. 2 Simulink diagram of the reference current generator

IV. CURRENT CONTROLLER

The proposed Adaptive Hysteresis Band Calculator varies the hysteresis band width, *HB*, according to $di_{ref}*(t)/dt$ and V_{dc}, supply voltage v. The principle was proposed by Bose [9]. The inverter current and voltage waveforms are shown in Fig. 3. When the current in the injection inductor tries to leave the hysteresis band appropriate switch is turned ON or OFF to force the ramping of the current within the hysteresis band.

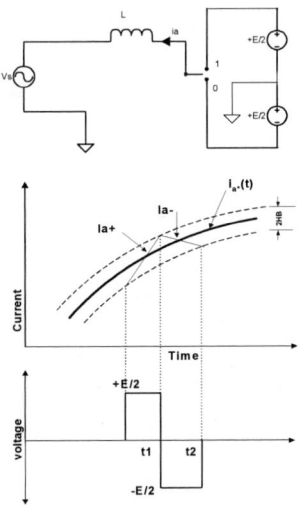

Fig. 3 Single phase VSI with HCC , current and voltage waveforms

In the interval 0 to t_1, the positive ramp of the inverter current, i_a^+ and in t_1 to t_2, the negative ramp, i_a^- can be expresses as,

$$\frac{L di_a^+}{dt} = 0.5V_{dc} + Vs \tag{11}$$

$$\frac{L di_a^-}{dt} = -(0.5V_{dc} + Vs) \tag{12}$$

Where L is the coupling inductance, and di_a^+/dt and di_a^-/dt are slopes of the respective rising and falling current segments and ia^* is the reference current.

$$\frac{di_a^+}{dt}t_1 - \frac{di_a^*}{dt}t_1 = 2HB \tag{13}$$

$$\frac{di_a^-}{dt}t_2 - \frac{di_a^*}{dt}t_2 = -2HB \tag{14}$$

$$t_1 + t_2 = T_c = \frac{1}{f_{sw}} \tag{15}$$

Where, t_1 and t_2 are the respective switching intervals, and f_{sw} is the switching frequency. Adding (13) and (14) and substituting (15), we can write,

$$t_1\frac{di_a^+}{dt} - t_2\frac{di_a^-}{dt} - \frac{1}{f_{sw}}\frac{di_a^*}{dt} = 0 \tag{16}$$

Simplifying (11), (12), (16),

$$4HB = (t_1+t_2)\frac{di_a^+}{dt} - (t_1-t_2)\frac{di_a^*}{dt} \tag{17}$$

$$HB = \frac{0.125V_{dc}}{f_{sw}L}\left[1 - \frac{4L^2}{V_{dc}^2}(\frac{Vs}{L}+m^2)^2\right] \tag{19}$$

Where m is substituted for di_a^*/dt

The equation (19) indicates that *HB* can be modulated as a function of V_{dc}, V_s, and m so that f_{sw} remains constant. The SIMULINK model for the adaptive hysteresis band calculator is given in Fig. 4

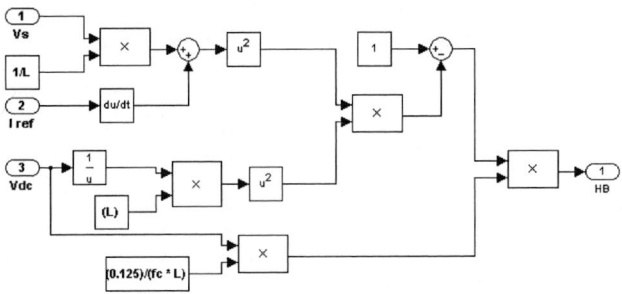

Fig. 4 SIMULINK model for adaptive hysteresis band calculator

978-1-4244-1871-8/07 $25.00 © 2007 IEEE

V. SIMULATION RESULTS

The closed loop simulation of the proposed SAF was tested for reactive and harmonic load. The test system parameters are given in Table I

TABLE I
PARAMETERS OF THE TEST SYSTEM

Supply Voltage	440 V
Supply frequency	50 Hz
VSI switching frequency f_{sw}	10Khz
DC Bus voltage, V_{dc}	1200 V
Coupling Inductor, L	2 mH
DC side capacitor	4000 µF
Non linear Load Load	6 pulse converter 31 %THD
Reactive Load-reactive power	10 kVAR
Reactive Load-active power	10 kW

The responses of the SAF indicate effectiveness of the system. Fig. 5 shows the time variation of the three supply voltages, non linear load currents and supply currents (after compensation). The supply currents are sinusoidal and in phase with the respective phase voltage. Harmonic spectrum of the load and source are shown in Fig 6 and Fig. 7. THD of the source current after compensation is 5.87 %. Performance of the adaptive hysteresis band current controller is shown in fig. 8 and Fig 9. The reference current in one phase, i_a*, the injected current, $i_{ac}*$ is shown in Fig. 8 and switching frequency, f_{sw}, in Fig. 9. The response of the DC Voltage control loop, shown in Fig. 10, responds to step change in DC bus reference within two power frequency cycles.

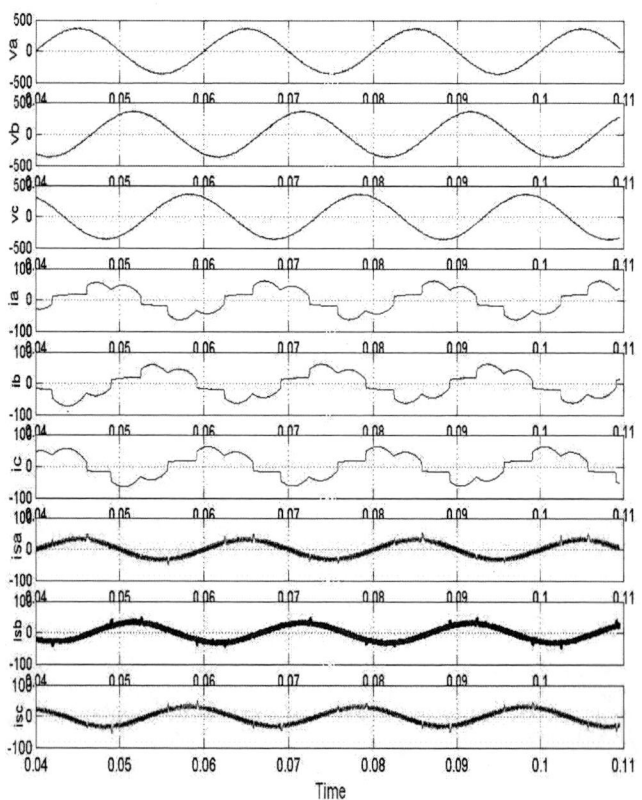

Fig.5 Performance of the system as harmonic compensator

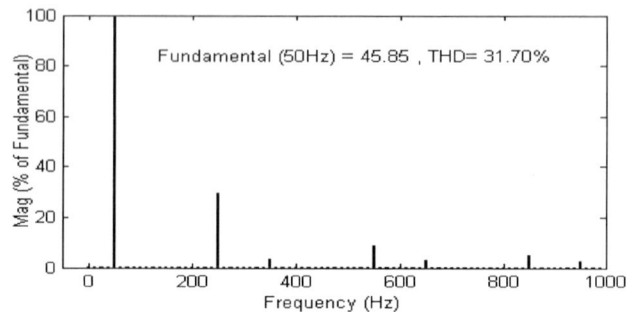

Fig. 6 Harmonic spectrum of the load current

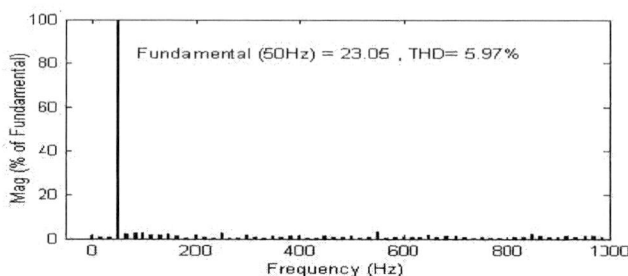

Fig. 7 Harmonic spectrum of the source current

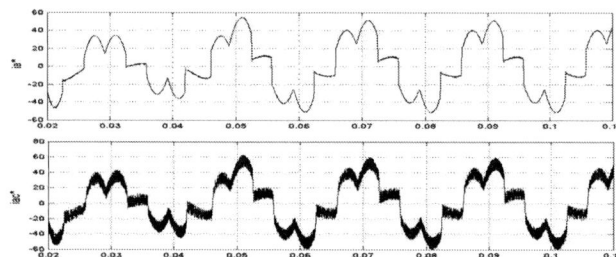

Fig. 8 Harmonic spectrum of the source current

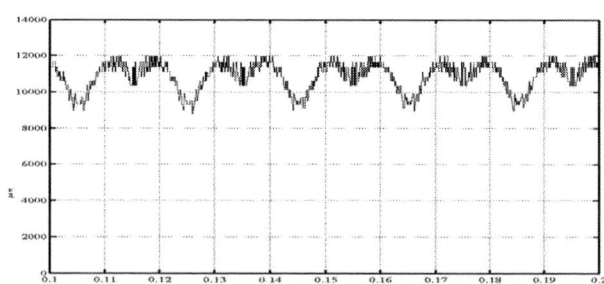

Fig. 9 Variation in switching frequency, fsw

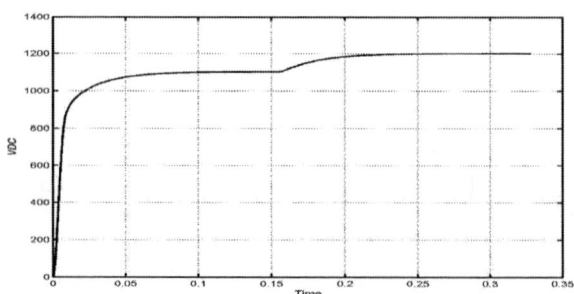

Fig. 10 Response of the DC Voltage Control loop

VI. CONCLUSIONS

The responses of the simulated model confirm the desired performance of the SAF. The proposed SAF serves as an effective solution for the harmonic compensation and reactive power mitigation. The compensated THD is 5.97 % which is slightly higher which is attributed to high frequency component in the SAF injected current that can be filtered by using a simple low cost, detuned, passive filter. The adaptive hysteresis band current controller tracks the reference current with minimum error and without any time delay. The DC bus voltage ripple is well below 5%. Practical implementation is in progress and will be published in near future.

VII. REFERENCES

[1] Akagi,H.,et.al., *"Instantaneous Reactive Power Compensation of switching Devices Without Energy Storage Components"* IEEE Trans. on Industry Applications, vol.20, no.3, May/June1984.

[2] Akagi, H., *"Active Harmonic Filters,"* Proc. Of IEEE, Vol. 93, No. 12 , pp. 2128 – 2141, December 2005.

[3] Akagi, H., *"New Trends In Active Filters For Improving Power Quality,"* Proc. Of International Conference on Power Electronics, Drives and Energy Systems for Industrial Growth Vol. 1, pp. 417 – 425, January 1996.

[4] Akagi, H.; Hyosung, Kim *"The Instantaneous Power Theory OnThe Rotating P-Q-R Reference Frames,"* Proceedings of the IEEE 1999 International Conference on Power Electronics and Drive Systems, Vol.1, pp. 422 – 427, July 1999.

[5] Nava Sagura, N.; Carmona-Hernindez, M. *"A Detailed Instantaneous Harmonic and Reactive Compensation Analysis of Three-phaseAC/DC Converters, in abc and αβ Coordinates"*, IEEE Transactions on Power Delivery, Vol. 14, No. 3,pp 1039-1045, July 1999.

[6] You , Xiaojie; Li ,Yongdong *"A Shunt Active Power FilterUsing Dead-Beat Current Control,"* proc. of IEEE 2002 28th Annual Conference of the Industrial Electronics Society, vol. 1, pp. 633 – 637, November 2002.

[7] Meo, S.; Perfetto, A *"Comparison of different control techniques for active filter applications,"* Proceedings of the Fourth IEEE International Caracas Conference on Devices, Circuits and Systems, P016-1 - P016-6, April 2002.

[8] Buso, S. et. al. *"A Dead-Beat Adaptive Hysteresis Current Control,"* IEEE Transactions on Industry Applications, Vol. 36, pp. 1174 – 1180, July-Aug. 2000.

[9] Bose, B., K., *"An Adaptive Hy steresis-Band Current Control Technique of a Voltage-Fed PWM Inverter for Machine Drivesystem,"* IEEE Transactions On Industrial Electronics, Vol. 31, No. 5,pp 402-408, October 1990.

[10] Bhattacharya, S.; Divan, D.,*"Active Filter Solutions For Utility Interface Of Industrial Loads"* Proceedings of the 1996 International Conference on Power Electronics, Drives and Energy Systems for Industrial Growth, vol.2, pp 1078-1084, 1996., November 1996.

VI. BIOGRAPHIES

Narayan G. Apte was born in Sangli, India on 2nd Nov. 1972. He received his B.E.(Electrical) in 1993 and M.E.(Electrical) in 1995 from Walchand College of Engineering, Sangli, India. Presently, he is working as a faculty in Electrical Engineering Department of the same institute since 1996. His areas of interest are Power Quality, Active Harmmonic Filters, Dynamic VAR Compensation, Embedded Microcontroller Systems and their applications.

Dr. Vishram N. Bapat was born on 10th July, 1957. He received his B.E. (Electrical) in 1980 and M.E.(Eletrical) – Control Systems in 1983 from Walchand college of Engineering, Sangli, India. He received doctorate in 1993 in Control Systems from Indian Institute of technology, Kharagpur, India. He has 26 publications in national and international journals and conferences. His research interests are Power Quality, Energy Conservation Quality Assurance in Industries.

Amruta N. Jog was born in Sangli, India on 13th January 1980. She received B.E.(Electrical) in 2003 from Walchand College of Engineering, Sangli, India and completed Post Graduate Studies in Power Systems in the same college in 2007. Presently, she is working as a design Engineer in Automation and Drives department of Siemens, Mumbai, India.

978-1-4244-1871-8/07 $25.00 © 2007 IEEE

A Novel Control Method of PAPF for Resonance Damping and Harmonics Compensation in Power System

WU Longhui, ZHUO Fang, LI Hui, LIU Jinjun, WANG Zhaoan,
School of Electrical Engineer, Xi'an Jiaotong University
Shaanxi, P.R.China

Abstract--In power system, capacitors are widely used to compensate reactive power, which generally cause resonance problems in harmonic distorted network. In this paper, parallel active power filter (PAPF) is used to damp the resonances. Firstly, the damping characteristics of PAPF with traditional load current detection methods (detecting current including or not including the capacitor current) are analyzed respectively. The analysis results show that PAPF with traditional load current detection methods is effective for parallel resonance damping, but can not damp series resonance when capacitor current is not included in the detecting current. And when capacitor current is included in the detecting current, PAPF is prone to be unstable and can not damp parallel and series resonance. Then a novel control method of PAPF is proposed. PAPF with the proposed control method has excellent performance in damping of both parallel resonance and series resonance, whether the capacitor current is included in the detecting current of PAPF or not. Also the PAPF with proposed method has strong ability in harmonic compensation. Finally, the experiment results are presented to verify the analysis.

Index Terms-- Resonance Damping, Harmonics Compensation, Capacitors, Parallel Active Power Filter.

I. INTRODUCTION

With the proliferation of non-linear loads such as diode/thyristor rectifiers, harmonic current and harmonic voltage are more and more serious in power transmission/distribution systems [1]. At the same time in many manufactories and power substations, switched capacitor banks are generally connected parallel to inductive power producing loads to improve power factor for their low cost and good compensation performance [2]. Seen from the load side, the reactive power compensation capacitor and the source impedance create a parallel resonance circuit. Thus harmonic currents in the range of the resonant frequency drawn by the load can be amplified to critical values. Looking from the source side, the reactive power compensation capacitors and the source impedance create a series resonance circuit. Hence, harmonic voltages injected by the supply side transformer can cause inadmissible high harmonic source currents. Both parallel and series resonance are very dangerous to power system [3-5].

To solve the resonance problem, inductors are connected in series with the reactive power compensation capacitors shifting the resonant frequency to an uncritical range. The rating of the reactive power compensation capacitors must be variable depending on the compensation demand. Thus, each power factor correction unit requires separate series inductors to maintain the resonant frequency out of range of possible excitation frequencies. However, the use of inductor-capacitor units noticeably escalates the costs of reactive power compensation. Furthermore, a residual risk of resonance sharpness still remains due to changing power grid configurations or unexpected excitations.

As parallel active power filters (PAPF) are powerful tools for power quality problems in power transmission/distribution systems, they are studied widely, and great developments have taken place in theory and application of PAPF [6-10]. A novel control method of PAPF is proposed in this paper for resonance damping and harmonics compensation. Firstly, the model of PAPF is built. Then, the damping characteristics of PAPF with traditional load current detection methods (detecting current including or not including the capacitor current) are analyzed respectively. Furthermore, the novel control method of PAPF is proposed and analyzed, with which PAPF has excellent performance in damping resonance and compensating harmonics. Finally, the experiment results are presented to verify the analysis.

II. MODEL OF PAPF

The control block diagram of PAPF is shown in Fig.1.

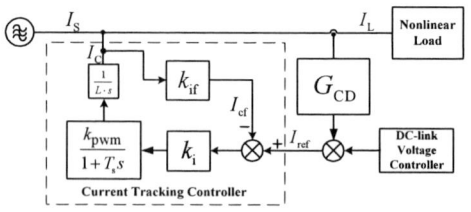

Fig.1. PAPF control system

Where G_{CD} represents the current detecting part, T_s is switching period, $k_{PWM} = U_d/U_t$ (U_d is the voltage of DC side, U_t is the amplitude of triangle carrier wave) [11], k_i is the coefficient of current tracking control part, k_{if} is feedback coefficient, L represents the connecting inductor between the PAPF and the power supply. The DC-link voltage controller is used to maintain the DC-link voltage as the reference value, which is usually a PI regulator [12].

Considering the filter and calculation delay in harmonics detection, the harmonic current detection part of PAPF can be expressed as equation (1), where k_{id} is compensation coefficient （$0 \le k_{id} \le 1$）[13].

$$G_{CD}(s) = \frac{I_{ref}(s)}{I_{lh}(s)} = \frac{k_{id}}{T_{af} \cdot s + 1} \qquad (1)$$

The current tracking control diagram of PAPF is shown in the broken line boxes in Fig.1 and its transfer function can be expressed as:

$$G_{CT}(s) = \frac{I_C(s)}{I_{ref}(s)} = \frac{k_i k_{pwm}}{T_s L s^2 + L s + k_{if} k_i k_{pwm}} \qquad (2)$$

Then transfer function of PAPF system can be written as:

$$G_{APF}(s) = G_{CT}(s) \cdot G_{CD}(s) \qquad (3)$$

III. EFFECT OF TRADITIONAL CONTROL METHODS ON RESONANCE DAMPING

Applying PAPF to the power transmission/distribution systems with the reactive power compensation capacitors, the equivalent circuit of this hybrid compensation system in harmonic frequency is shown in Fig.2. In which Z_S represents the source impedance, C represents parallel capacitors and load is treated as harmonic current source. In this hybrid compensation system, when using the load current detecting method, it may include or not include the capacitor current (corresponding detecting point in Fig.2 is point A or B respectively). The following discussion will be carried out in these two situations respectively. System parameters used in analysis are shown in Table 1. From Table 1, it can be obtained that the parallel resonance frequency in analysis is 550Hz (11th harmonics) and the series resonance frequency is 375Hz (7.5th harmonics).

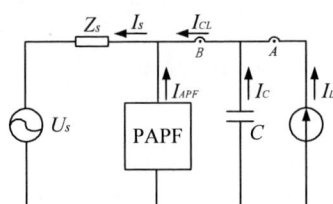

Fig. 2 Equivalent circuit of hybrid compensation system

Table 1 System Parameters

Connecting inductor	$L = 0.4\text{mH}$
Calculation delay	$T_{af} = 0.1\text{ms}$
Sampling period	$T_S = 0.1\text{ms}$
Source impedance	$Ls = 127.5\mu\text{H}$, $Rs = 45\text{m}\Omega$
Parallel capacitors	In parallel resonance analysis: $C = 656.76\mu\text{F}$, $Rc = 85\text{m}\Omega$
	In series resonance analysis: $C = 1412.76\mu\text{F}$, $Rc = 85\text{m}\Omega$
Parameters of PAPF control parts	$k_{if} = 1$, $k_i = 5$, $k_{PWM} = 128$, $0 \le k_{id} \le 1$

A. Reactive power compensation capacitor current is not included in the detected load current

When the current of reactive power compensation capacitor is not included in the detected load current, the system equivalent circuit in harmonic frequency is shown in Fig.3.

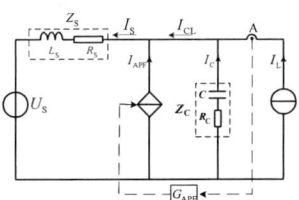

Fig. 3 Equivalent circuit when the detecting point is A

a) Damping effect on parallel resonance

The parallel resonance is mainly caused by harmonic current drawn by the nonlinear load. So in the following analysis of parallel resonance, the harmonic voltage is neglected. When the current detecting point is A and harmonic voltage is neglected, the amplitude-frequency characteristic of I_S/I_L in Fig.3 with different compensation coefficient k_{id} is shown in Fig.4. It can be seen that with the increase of k_{id}, the whole amplitude-frequency characteristic curve moves down, and the system performance for resonance damping and harmonic suppressing is improved.

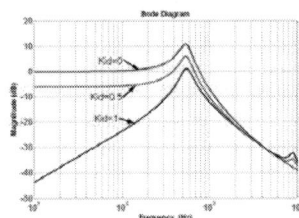

Fig. 4 Amplitude-frequency characteristic of I_S/I_L with different k_{id}

b) Damping effect on series resonance

The series resonance is mainly caused by harmonic voltage injected by the supply side transformer. So in the following analysis of series resonance, the harmonic current is neglected. When the current detecting point is A and harmonic current is neglected, output current of PAPF is zero. Then the harmonic ratio of supply current and supply voltage is:

$$G(s) = \frac{I_S(s)}{U_S(s)} = \frac{1}{Z_C(s) + Z_S(s)} \qquad (4)$$

It can be seen that the system will be resonated at frequency $f_P = f_s \sqrt{X_C / X_S}$. So in this condition PAPF has no effect on the series resonance damping.

B. Reactive power compensation capacitor current is included in the detected load current

When reactive power compensation capacitor current is included in the detected load current, the system equivalent circuit in harmonic frequency is shown in Fig.5.

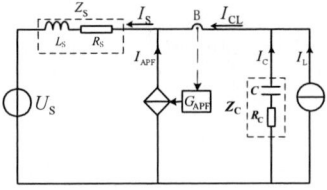

Fig. 5 Equivalent circuit when the detecting point is B

a) Damping effect on parallel resonance

978-1-4244-1871-8/07 $25.00 © 2007 IEEE

Neglecting the harmonic voltage in Fig.5, the system block diagram can be expressed as Fig.6 when the current detecting point is B.

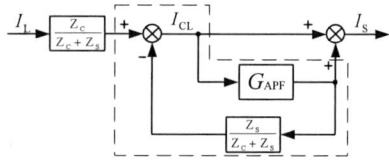

Fig.6 Block diagram in analysis of parallel resonance

In Fig.6 there is a closed loop as shown in the broken line boxes, and the open-loop Bode plots of this loop are shown in Fig.7 with different value of parallel capacitors or different compensation coefficient k_{id}.

From Fig.7 it can be concluded that, when reactive power compensation capacitor current is included in the detected load current, the system may become unstable with the change of PAPF compensation coefficient k_{id} or value of reactive power compensation capacitors.

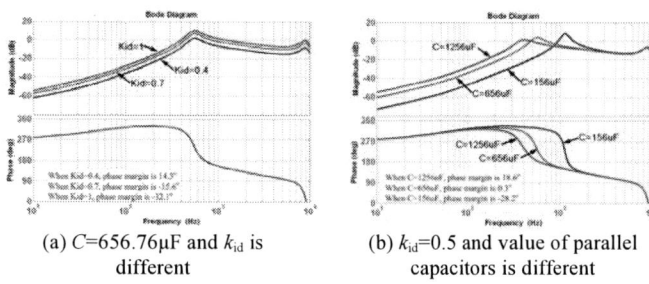

(a) C=656.76μF and k_{id} is different

(b) k_{id}=0.5 and value of parallel capacitors is different

Fig.7 Bode plots of open-loop transfer function when detecting point is B

The closed loop Bode plots of Fig.6 with different value of compensation coefficient k_{id} are shown in Fig.8. The compensation coefficient k_{id} is chosen within the stable range. In Fig.8 with the increase of k_{id}, new resonance frequency is produced, and the amplitude of harmonic current at new resonance frequency is magnified much more times than that at the former resonance frequency.

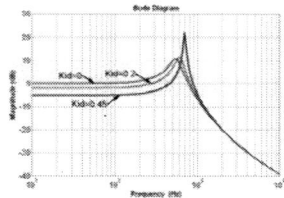

Fig.8 Amplitude-frequency characteristic of I_S/I_S with different k_{id}

b) Damping effect on series resonance

Neglecting the harmonic current of nonlinear load in Fig.5, the system block diagram can be expressed as Fig.9 when the current detecting point is B. Also there is a closed loop as shown in the broken line boxes, and the close loop is the same as the close loop shown in Fig.6. Therefore, the stability of system is necessary to be discussed.

The closed loop Bode plots of Fig.9 with different value of compensation coefficient k_{id} are shown in Fig.10. The compensation coefficient k_{id} is chosen within the stable range.

It can be seen that with the increase of k_{id}, system has new resonance frequency, and the amplitude of harmonic current at new resonance frequency is magnified much more times than that at the former resonance frequency.

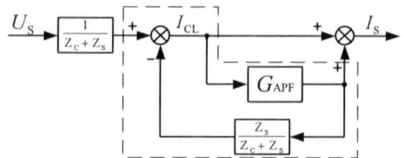

Fig.9 Block diagram in analysis of series resonance when the detecting point is B

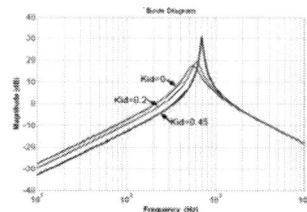

Fig. 10 Amplitude-frequency characteristic of I_S/U_S with different k_{id}

IV. EFFECT OF PROPOSED CONTROL METHODS ON RESONANCE DAMPING

From above analysis, it can be concluded that PAPF with traditional load current detection methods are ineffectual on damping resonance between the reactive power compensation capacitors and the source impedance. Moreover, system has a stability problem when reactive power compensation capacitor current is included in the detected load current.

Therefore, a novel control method of PAPF is proposed to improve system performance for resonance damping and harmonics compensation. In this control method harmonic component of both load current and supply voltage is detected to control the output current of PAPF, just as Fig.11 shows, where G_V is compensation coefficient of harmonic voltage. In the following analysis, G_V is controlled as a proportion factor $G_V=K_V$. So it is equivalent to add a "harmonic resistor" with the value of $1/K_V$ connected in parallel with supply side.

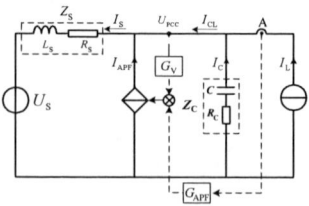

Fig. 11 Proposed control method of PAPF

A. Reactive power compensation capacitor current is not included in the detected load current

When reactive power compensation capacitor current is not included in the detected load current, the system equivalent circuit in harmonic frequency with proposed PAPF control method is shown in Fig.12. Here "$1/K_V$" is just the equivalent "harmonic resistor".

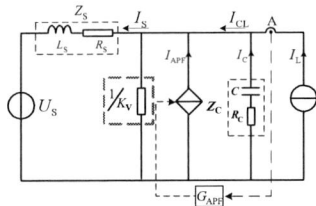

Fig. 12 Equivalent circuit when the detecting point is *A*

a) Damping effect on parallel resonance

Neglecting the harmonic voltage, the amplitude-frequency characteristic of I_S/I_L in Fig.12 with different coefficient K_V is shown in Fig.13. Here $C = 656.76\,\mu F$ and $k_{id} = 1$. It can be seen that PAPF with the proposed control method has ideal performance on parallel resonance damping as well as on harmonic compensation.

b) Damping effect on series resonance

Neglecting the harmonic current of nonlinear load, the amplitude-frequency characteristic of I_S/U_S in Fig.12 with different coefficient K_V is shown in Fig.14. Here $C = 656.76\,\mu F$ and $k_{id} = 1$. It can be seen that with the increase of K_V, the curve moves down at the resonant frequency. PAPF with the proposed control method is effective on damping series resonance when the current detecting point is A.

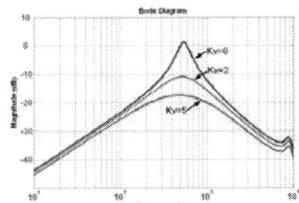 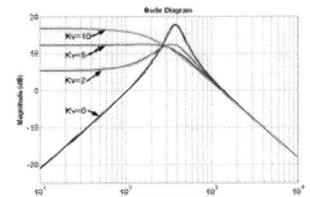

Fig. 13 Amplitude-frequency characteristic of I_S/I_L with different K_V

Fig. 14 Amplitude-frequency characteristic of I_S/U_S with different K_V

B. Reactive power compensation capacitor current is included in the detected load current

When reactive power compensation capacitor current is included in the detected load current, the system equivalent circuit in harmonic frequency with proposed control method of PAPF is shown in Fig.15.

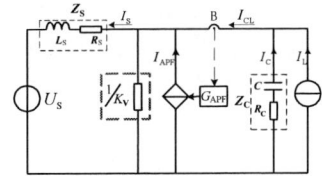

Fig. 15 Equivalent circuit when the detecting point is *B*

a) Damping effect on parallel resonance

Neglecting the harmonic voltage in Fig.15, the system block diagram can be expressed as Fig.16. Where, $Z'_S(s) = \dfrac{Z_S(s)/K_V}{Z_S(s)+1/K_V}$.

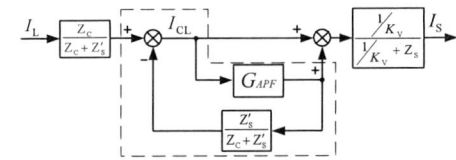

Fig.16 Block diagram in analysis of parallel resonance when the detecting point is B

In Fig.16 there is a closed loop as shown in the broken line boxes, and the open-loop Bode plots of this loop are shown in Fig.17 with different value of K_V or different value of parallel capacitors. It can be seen that the system stability can be improved by proposed PAPF control method and is insensitive to the value of capacitors.

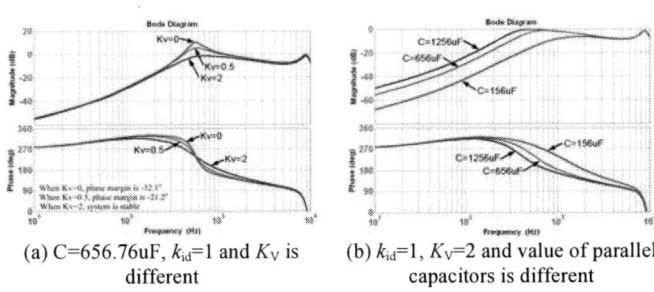

(a) C=656.76uF, k_{id}=1 and K_V is different

(b) k_{id}=1, K_V=2 and value of parallel capacitors is different

Fig.17 Bode plots of open-loop transfer function when detecting point is B

The amplitude-frequency characteristic of I_S/I_L of Fig.16 with different coefficient K_V is shown in Fig.18. Here $C = 656.76\,\mu F$ and $k_{id} = 1$. It can be seen that PAPF with the proposed control method has ideal performance on parallel resonance damping as well as on harmonic compensation even though reactive power compensation capacitor current is included in the detected load current.

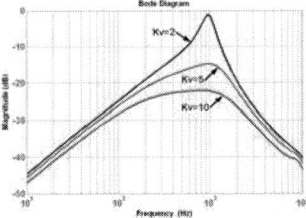

Fig.18 Amplitude-frequency characteristic of I_S/I_L with different K_V

b) Damping effect on series resonance

Neglecting the harmonic current of nonlinear load in Fig.15, the system block diagram can be expressed as Fig.19. Here, $Z'_S(s) = \dfrac{Z_S(s)/K_V}{Z_S(s)+1/K_V}$, $Z'_C(s) = \dfrac{Z_C(s)/K_V}{Z_C(s)+1/K_V}$. Also there is a closed loop as shown in the broken line boxes, and the close loop is the same as the close loop shown in Fig.16. From the analysis of Fig.16, it can be concluded that system stability is improved a lot with the proposed control method.

Fig.20 shows the amplitude-frequency characteristic of I_S/I_L with different coefficient K_V. Here $C = 656.76\,\mu F$ and $k_{id} = 1$. It can be seen that PAPF with the proposed control method has effect on damping series resonance when the current detecting point is B.

978-1-4244-1871-8/07 $25.00 © 2007 IEEE

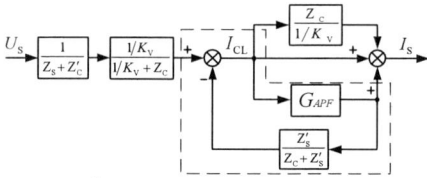

Fig.19 Block diagram in analysis of series resonance when detecting point is B

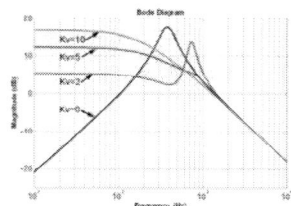

Fig. 20 Amplitude-frequency characteristic of I_S/U_S with different K_V

V. EXPERIMENTAL VERIFICATION

An experimental system was constructed to verify the above analyses and the proposed control method. Fig.21 shows the configuration of experimental system. Where, T is the transformer for voltage detection and PLL, e_{PLL} is the output of PLL, $G_C(s)$ denotes the dc-link voltage regulator (usually it is a PI regulator) [12]. Nonlinear load in experimental system is a three-phase diode bridge rectifier with a resistor in dc-side. In this experiment, harmonic current of nonlinear load is used as a stimulator for parallel resonance, and a programmable AC source is used to stimulate series resonance. The control diagram of PAPF with proposed control method is shown in Fig.22. Where PLL is phase-lock-loop, Harmonic Detector I is harmonic detector of load current, Harmonic Detector II is harmonic detector of supply voltage, LPF is 20 order digital FIR low pass filter for supply voltage detection with corner frequency of 5 kHz.

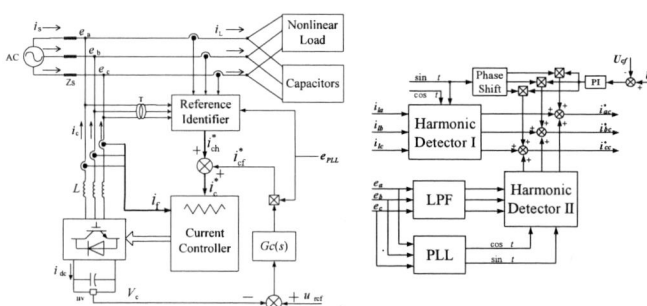

Fig. 21 Configuration of experimental system   Fig. 22 Control diagram of PAPF used in experimental system

In the following experimental results, U_S denotes supply voltage, I_S denotes supply current, I_C denotes capacitor current and I_F denotes PAPF current. The compensation coefficient k_{id} = 1.

A. Reactive power compensation capacitor current is not included in the detected load current

a) Experimental results for parallel resonance damping

When there isn't source voltage distortion, experimental results without PAPF are shown in Fig.23. It can be seen that when capacitors are connected in system, parallel resonance between source impedance and capacitors is stimulated, both I_C and I_S are magnified. Also, it causes severe distortion of U_S.

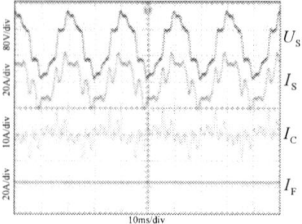

Fig.23 Experimental results of parallel resonance without PAPF

Experiment results with tradition PAPF control method and proposed PAPF control method are shown in Fig.24. It can be concluded that, when capacitor current is not included in the detected load current, PAPF with both the traditional control method and the proposed control method are effective on parallel resonance damping and harmonics compensation.

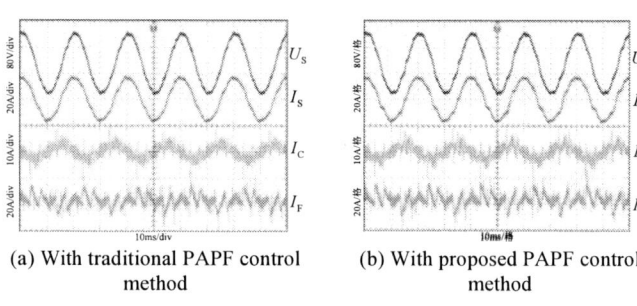

(a) With traditional PAPF control method   (b) With proposed PAPF control method

Fig.24 Experimental results with parallel resonance when detecting point is A

b) Experimental results for series resonance damping

To stimulate series resonance, 5.4% 5th, 2.7% 7th and 2.7% 11th order harmonic voltages are added in the source voltage. Experimental results without PAPF are shown in Fig.25. It can be seen that when capacitors are connected in system, series resonance between source impedance and capacitors is stimulated, both I_C and I_S are magnified.

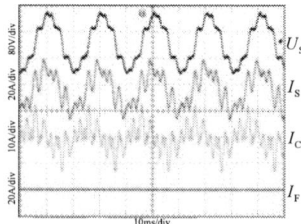

Fig.25 Experimental results of series resonance without PAPF

Experiment results with tradition PAPF control method and proposed PAPF control method are shown in Fig.26. It can be concluded that, PAPF with traditional control method has compensated harmonic current caused by nonlinear loads, however, the series resonance caused by voltage distortion can't be suppressed, I_C and I_S are still magnified a lot. However, PAPF with proposed control method has good performance for both harmonic suppression and series resonance damping.

978-1-4244-1871-8/07 $25.00 © 2007 IEEE   680

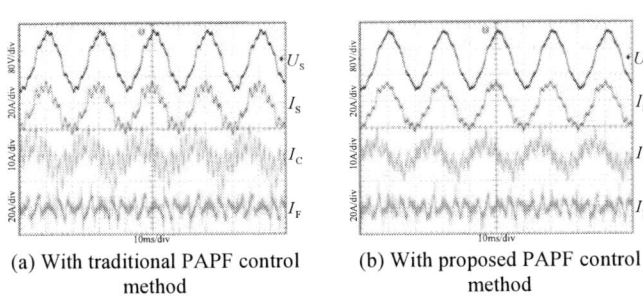

| (a) With traditional PAPF control method | (b) With proposed PAPF control method |

Fig.26 Experimental results with series resonance when detecting point is A

B. Reactive power compensation capacitor current is included in the detected load current

a) Experimental results for parallel resonance damping

Experiment results with tradition PAPF control method and proposed PAPF control method are shown in Fig.27. It can be seen that, when capacitor current is included in the detected load current, PAPF with traditional control method is unstable, whereas PAPF with proposed control method improves the system stability and has good performance for both harmonic suppression and parallel resonance damping.

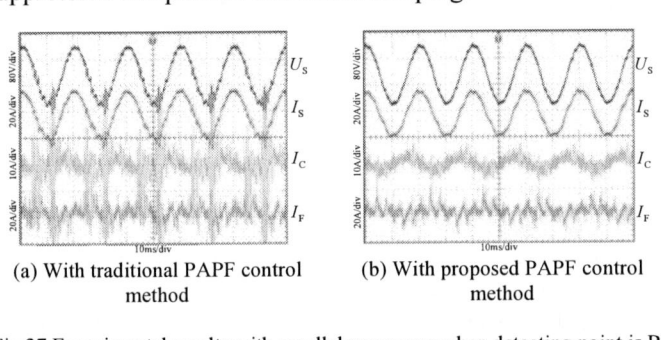

| (a) With traditional PAPF control method | (b) With proposed PAPF control method |

Fig.27 Experimental results with parallel resonance when detecting point is B

b) Experimental results for series resonance damping

When 5.4% 5th, 2.7% 7th and 2.7% 11th order harmonic voltages are added in the source voltage, experiment results with proposed PAPF control method are shown in Fig.28. It can be concluded that, when capacitor current is included in the detected load current, system is stable and PAPF with proposed control method is effective on series resonance damping as well as on harmonics compensation.

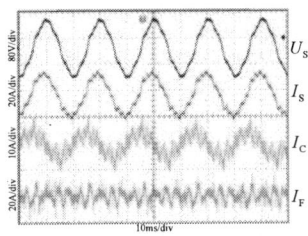

Fig.28 Experimental results with series resonance when detecting point is B

VI. CONCLUSIONS

To solve the resonance problem between reactive power compensation capacitors and source impedance, this paper firstly discusses the resonance damping effects of PAPF with traditional load current detecting methods, and then proposes a novel control method of PAPF to improve system performance on both resonance damping and harmonic

suppression. From the theoretical analysis and experiment results, the following conclusions can be obtained:

1) When the capacitor current is not included in the detected load current, PAPF with traditional control method is effective for parallel resonance damping, but can not damp series resonance caused by voltage distortion;

2) When capacitor current is included in the detected load current, PAPF with traditional control method is prone to be unstable and can not damp parallel and series resonance.

3) PAPF with the proposed control method in which both the supply voltage and load current are detected (no matter whether the capacitor current is included in the detected load current or not) can improve system stability and is effective for both series and parallel resonance damping as well as for harmonics compensation.

VII. REFERENCES

[1] Pichai Jintakosonwit, Hideaki Fujita, Hirofumi Akagi, "Control and Performance of a Fully-Digital-Controlled Shunt Active Filter for Installation on a Power Distribution System," *IEEE Trans. Power Electron.*, vol 17, pp. 132-140, Jan. 2002.

[2] Hosseini, S.H.; Taghizadegan, N. "A novel and optimal method for locating and sizing of capacitors in distribution systems based on reactive power identification", *ICECS 2003*, vol 1, pp. 364-367, Dec. 2003.

[3] Detjen, D., Jacobs, J., De Doncker, R.W., Mall, H.-G., "A new hybrid filter to dampen resonances and compensate harmonic currents in industrial power systems with power factor correction equipment" *IEEE Trans. Power Electronics*, 16(6), pp: 821-827, 2001.

[4] Rahmani, S., Al-Haddad, K., Fnaiech, F., "A Three-Phase Shunt Active Power Filter for Damping of Harmonic Propagation in Power Distribution Systems", *2006 IEEE International Symposium on Industrial Electronics*, vol 3, pp. 1760-1764, July. 2006.

[5] Fujita, H., Yamasaki, T., Akagi, H., "A hybrid active filter for damping of harmonic resonance in industrial power systems", *IEEE Trans. Power Electronics*, 15(2), pp: 215-222, 2000.

[6] Akagi H. "New Trends in Active Filters for Power Conditioning", *IEEE Trans. Ind. Appl.*, 32(6), pp: 1312-1322, 1996.

[7] Fang Zheng Peng. "Application issues of active power filters", *IEEE Trans. Ind. Appl.*, 4(5), pp: 21-30, 1998.

[8] Mattavelli, P. "A closed-loop selective harmonic compensation for active filters", *IEEE Trans. Industry Application*, 37(1), pp: 81-89, 2001.

[9] Hongyu Li, Fang Zhuo, Zhaoan Wang, et al. "A Novel Time-Domain Current-Detection Algorithm for Shunt Active Power Filters", *IEEE Trans. Power System*, 17(2), pp: 644-651, 2005.

[10] Changzheng Zhang, Qiaofu Chen, Youbin Zhao, et al. "A Novel Active Power Filter for High-Voltage Power Distribution Systems Application", *IEEE Trans. Power Delivery*, 22(2), pp: 911-918, 2007.

[11] R. Parikh, R. Krishnan. "Modeling. Simulation and Analysis of an Uninterruptible Power Supply", *IEEE-IECON'94, Bologna Italy*, vol 1, pp, 485-490, Sep 1994.

[12] Wu Longhui, Zhuo Fang, Zhang Pengbo, et al. "Analysis and Elimination Method of the Influence of Supply Voltage Fluctuation on Shunt Active Power Filter Dc-link Voltage", *37th IEEE Power Electronics Specialists Conference*, Jeju, Korea, vol 1, pp. 1577-1581, Jun. 2006.

[13] Akagi, H. "Control strategy and site selection of a shunt active filter for damping of harmonic propagation in power distribution systems", *IEEE Trans. Power. Delivery*, 12(1), pp: 354-363, 1997.

978-1-4244-1871-8/07 $25.00 © 2007 IEEE

Power Semiconductors state-of-art and development trends

Leo Lorenz
Infineon Technologies
Munich / Singapore

Abstract—System integration and high power density of monolithic and multi-chip designs are the driving force for the progress in power electronic systems. The whole system has to be considered and optimized to meet this target and to keep the overall ruggedness, sensitivity towards EMI and long term reliability, Silicon utilization system reliability and power units miniaturization are the key factors.

Power semiconductors are primarily used to control the flow of energy between the energy source and the load, and to do so with great precision, with extremely fast control times and with low dissipated power.

The drive towards rational use of energy, miniaturization of electrical systems and power management has given impetus to the revolutionary development of power semiconductors and power electronic systems over the last 20 years.

In this paper new technologies, advanced devices concepts and future system aspect for industrial segments are discussed. In these fields of applications there are huge requirements towards system dynamic characteristic, overload capability, ruggedness behaviour and reliability.

High operating temperatures is required, in the industrial high blocking voltage capabilities are needed.

I. HIGH-VOLTAGE THYRISTORS DEVELOPMENT

High-power thyristors with direct light-triggering and integrated protection functions can be utilized advantageously for applications in which several thyristors are connected in series, because this enables a significant reduction in the number of components necessary for the construction of high-power thyristor converters. This in turn leads to greater reliability and lower fabrication costs. It applies in particular to High-Voltage DC (HVDC) transmission, Static Var Compensation (SVC), converters for medium voltage drives, and also to certain pulse power applications.

B. 8-kV light-triggered thyristor with integrated protection functions

Thyristors in HVDC converter stations have to be protected against several failures that may appear under standard operation. There are three classical failure events thyristors must be protected against :

- voltage pulses with a too high amplitude (overvoltage),
- voltage pulses with a too high voltage ramp dV/dt, and

- voltage pulses appearing during the forward recovery time.

A reliable protection of the thyristor can only be achieved, if the thyristor is safely turned on in failure case. A promising concept for an integration of protection functions is therefore to utilize the Amplifying Gate (AG) structure of the thyristor in such a way that in failure case a sufficiently large internal trigger current is generated turning on the device by means of the AG [1, 2].

C. Light-triggering and overvoltage protection

Figure 1 shows the central area of the thyristor including the AG structure, a part of the main cathode area and some peculiarities incorporated for the realization of the protection functions.

The innermost AG was adjusted such that the photogenerated current provided by the integrated diode triggers the thyristor when illuminated by a 40 mW light pulse with a duration of 10 μs. The breakdown voltage of the BOD can be controlled by the curvature of the junction and by the distance to the p⁻-ring below the optical gate and its doping concentration.

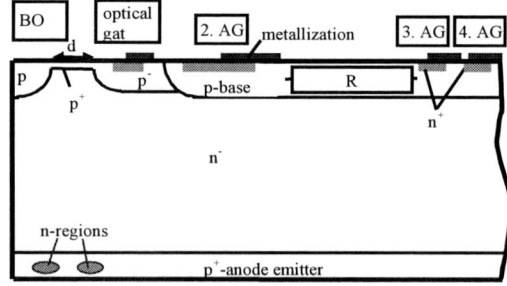

Fig. 1: Central part of the thyristor (BOD & four-stage AG structure) and part of the main cathode area (from Ref. [1,2])

D. dV/dt protection

Integration of a dV/dt protection function can be achieved by designing the AG structure and the main cathode area such, that the innermost AG has a higher dV/dt sensitivity than the other AGs and the main cathode area.

E. Forward Recovery Protection

Commutation of a thyristor leads to an extraction of the free charge carriers in the n-base and the p-base and the formation of an anode-side depletion layer in the n-base. When large regions of the thyristor have completely recovered and an

978-1-4244-1871-8/07 $25.00 © 2007 IEEE

uncontrolled forward voltage pulse appears, the remaining small region with excess carriers in the main cathode area may turn on. To avoid this, the thyristor should also be turned on in a controlled way by the AG structure if such a failure occurs. However, an increase of the carrier concentration in the AG region can be achieved by modifying the carrier's lifetime distribution so that it is lower in the main cathode area than in the AG structure.

F. 13-kV light-triggered tandem thyristor with integrated overvoltage protection

For blocking voltages of 13 kV, a new device concept consisting of a series connection of a 13-kV asymmetrical thyristor and a 13-kV diode has been developed.

Recently [3], it has been shown that the reverse recovery behaviour of 13-kV diodes can be improved significantly by applying a new field stop concept based on a deep buried field stop layer in front of the n^+ emitter. Fig. 3 shows typical time traces of the diode current and voltage during reverse recovery. The diodes were switched off under typical conditions appearing in High-Voltage DC (HVDC) converter stations with an RC-snubber circuit. The on-state current I_{on} and the current turn-off rate were 2 kA and -4 A/μs, respectively. The applied reverse voltage was -3.75 kV. Under these conditions, the diode shows a soft reverse recovery behaviour which is closely connected with the existence of the buried field stop layer:

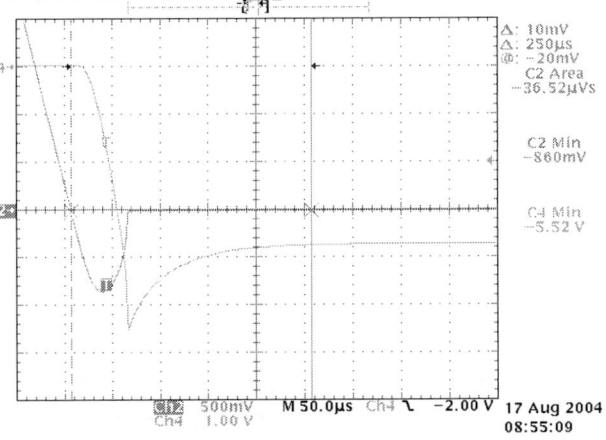

Fig. 2: Current and voltage time trace of a 13-kV diode, $V_F = 3.7$ V, $V_r = -3.75$ kV, $R = 36$ Ω, $C = 2.4$ μF.

II. IGBT AND FAST SWITCHING DIODE FOR FUTURE POWER CONVERSION

For inverters as e. g. used in automation technology or traction there is an increasing demand for a matched system consisting of the control, intelligent drivers and the power switches. Key parameters for the power switches in inverters are ruggedness, on-state and switching losses and of course cost. Emerging new developments will focus more on the indirect contributors to losses of the power switches in the inverter systems. These topics become the more important the higher the load currents are. Key factors for current developments are inherent softness of the IGBTs and free

wheeling diodes as well as enhanced controllability. Further improvement will be driven by optimized cell design and vertical structures, basically advanced field stop.

Talking about the IGBT as the key switch in these application fields in detail one must say that this technology has seen considerable innovation over the last decade regarding loss reduction as well as chip shrink resulting in more and more compact and cheaper packages as well as inverters.

Examples of this progress in Si development are shown in Fig. 3 in terms of on state voltage reduction as well as chip shrink path and in Fig. 4 regarding chip thickness reduction.

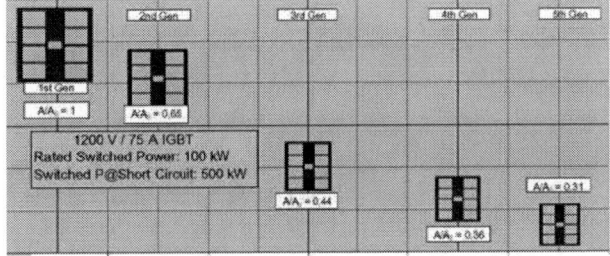

Fig. 3: Shrink and on state voltage reduction of a 75A 1200V IGBT

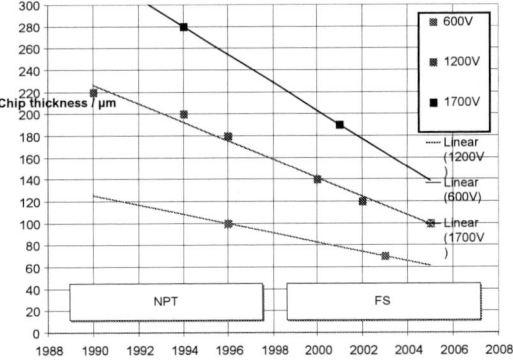

Fig. 4: Decrease of chip thickness from first 1200V NPT IGBTs up to nowadays 600-1700V FS IGBTs.

Main enablers for these improvements were the implementation of modern IC technology into the IGBT process, e.g. small planar transistor cells as well as trench transistor cells [4, 5] – and also evolutional steps of the vertical structure from PT to NPT to FS concept (see Fig. 5), [6] by challenging ultrathin wafer technology.

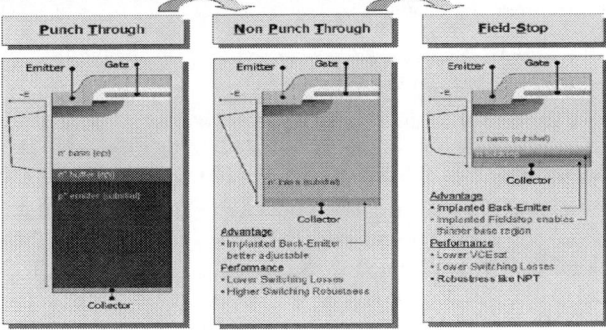

Fig. 5: Evolution of vertical IGBT concept from PT (Punch Through) to NPT (Non Punch Through) to FS (Field Stop) IGBT

Some concerns might have come up what will come next: Is the silicon switch for low to medium to high power switches for drives and traction application now at its evolutionary end? Our definitive answer is no! On the one hand the roll out of modern chip technology to highest voltage class traction IGBTs (3.3kV up to 6.5kV) is still in progress, on the other hand for the lower voltage range (600, 1200, 1700V) the next IGBT generations are already in the development phase: for low power applications there might be possible some nice advantages by a reverse conducting IGBT concept with an integrated freewheeling diode [7, 8]. Especially in the medium to high power range, significant improvements can be foreseen in the switching performance by implementing the dynamic clamping feature [9, 10, 11] as well as more advanced sub-μ trench technology and next steps in the field stop technology for lower on state losses.

Also significantly higher maximum junction temperatures may become reachable (see short circuit test of a modern 1200V IGBT at T$_{junction}$ of 200°C in Fig. 6).

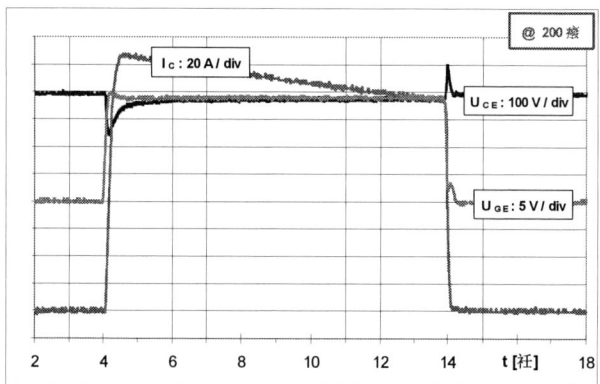

Fig. 6: Short circuit pulse of a 25A 1200V IGBT at extreme temperature conditions: Tj=200°C, UGE=19V, UCE=800V

III. ULTRA FAST SWITCHING DEVICES WITH SUPER JUNCTION TECHNOLOGIES

System miniaturization, the reduction of system size and weight, is a strong driving factor in power electronics. The power supply is often a dominating part concerning dimensions and weight of the whole system. Together with robustness the most important key driver for improvement therefore is the efficiency.

A. CoolMOS C5 – the latest generation super junction technology in volume production

The efficiency of many power supply topologies is basically determined by the device capacities of the switching MOSFETs and, especially for low line conditions, the efficiency is significantly affected by the Rds,on of the switching transistor. For standard MOSFET technology these requirements are limited by the so called silicon limit. This basic conflict could be solved with the introduction of superjunction MOSFETs [12-16]. The on-resistance is now

only a matter of technology performance and design but no longer subjected to the silicon limit. Fig. 7 shows the best commercial available standard transistor and CoolMOS C3 in comparison to its respective limits. (C5 is indicated by a blue star).

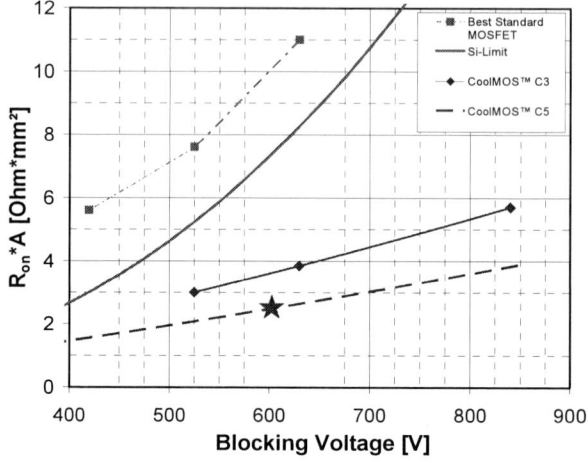

Fig. 7: Best Std. HV MOSFET (Kobayashi), Si limit line, CoolMOS™ C3 & C5

Furthermore, this new superjunction technology combines the low area specific Rds,on with half the total gate charge Qg to achieve an outstanding figure of merit, Qg * Rds,on of 5 Ohm*nC, which is less than one tenth of the standard MOSFET value (Fig 8).

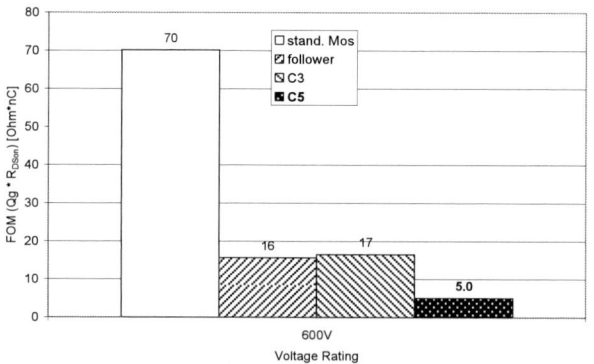

Fig. 8: Figure of merit (Rds,on * Qg) for several 600V MOSFET technologies

Fig. 9 shows the development history of the superjunction technology. On the one side there was a continuous improvement in the loss reduction – dynamic and static losses – and on the other side an unusual improvement in the overall ruggedness performance. Overload capability, short circuit ruggedness and Avalanche ratings have been increased substantially.

978-1-4244-1871-8/07 $25.00 © 2007 IEEE

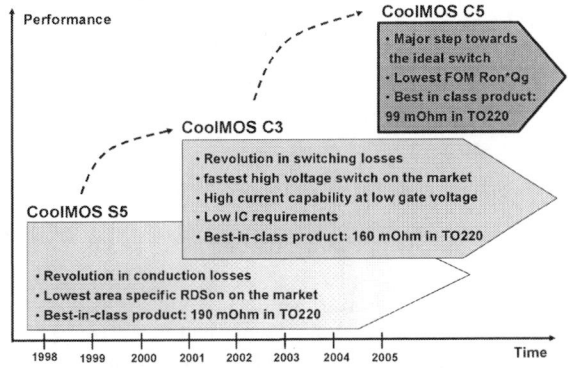

Fig. 9: System leadership - CoolMOS history

Since 2001 Infineon is providing SiC Schottky diodes in the voltage classes of 300 V and 600 V. With the availability of these virtually switching loss free rectifiers the circuit developers have gained a new degree of freedom: frequency limitations due
to rapid increasing dynamic losses are out of the way when using this new device class. What remains are only capacitive losses which are more than one order of magnitude smaller compared to even the fastest Si bipolar diodes in this voltage range. Meanwhile power supply manufacturers have released new products including PFC stages with 200 kHz operation frequency based on this new technology.

Today SiC Schottky diodes are mainly used in high end power supply applications for servers and telecom base stations (> 500 W) and in less cost sensitive solar cell inverters. Today, frequently the most critical parameter for PFC design-ins is the surge current capability of this diode. This becomes very obvious when comparing the I²t rating between SiC and standard diode (example: 6A SiC SDT06S60 → I²t=2.3 A²s; 6A Si pn IDP06E60 → I²t=4.3).

This issue is addressed in Infineon's upcoming new SiC diode generation by employing a merged pn-Schottky junction (see fig. 10).

In fact with this new approach the destructive surge current can be increased by a factor of 2 for a given current rating.

our SiC Schottky diode product family will happen during 2005 towards larger breakdown voltage (1200 V) offering the benefits of lossless switching also for UPS, converter, drive applications, etc.

REFERENCES

[1] F.-J. Niedernostheide, H.-J. Schulze, U. Kellner-Werdehausen, *Self-Protected High-Power Thyristors"*, Proc PCIM'2001, Power Conversion, Nuremberg, Germany, pp. 51-56, 2001.

[2] T. Laska, F. Pfirsch, F. Hirler, J. Niedermeyr, C. Schaeffer, T. Schmidt "1200V-Trench-IGBT Study with Square Short Circuit SOA", *Proceedings of the 10th ISPSD*, pp.433-436, 1998

[3] T.Laska, G.Miller, C.Schäffer, F.Umbach, *"Field Stop IGBTs with Dynamic Clamping Capability – A New Degree of Freedom for Future Inverter Designs?"*, to be published of EPE, 2005

[4] L. Lorenz, M. März, *"CoolMOS – A new approach towards high efficient power supplies"*, PCIM Hong Kong' 98 Proceedings

[5] L. Lorenz, *"CoolMOS – A New Approach Towards An Idealized Power Switch"*, EPE '99 Lausanne Proceedings

[6] L. Lorenz, G. Deboy, A. Knapp and M. März, *„ CoolMOS^{TM} - a new milestone in high voltage power MOS"*, Proc. PCIM 98 Nürnberg

[7] H. Mitlehner, W. Bartsch, K.-O. Dohnke, P. Friedrichs, R.Kaltschmidt, U. Weinert, B. Weis and D. Stephani,*„ Dynamic characteristics of high voltage 4H-SiC vertical JFETs"*, Proc. ISPSD 99, pp. 339-342, To-ronto 1999.

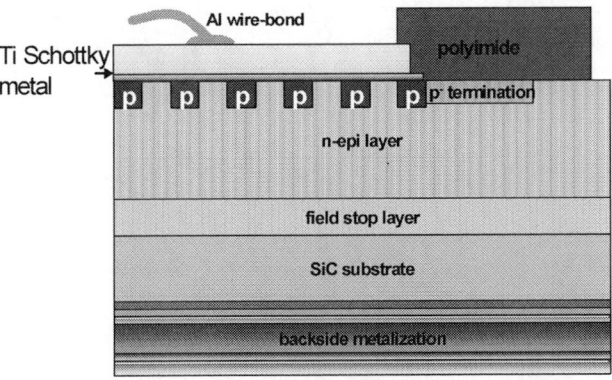

Fig. 10: Schematic construction of a merged pn-Schottky diode. The p-wells have to be contacted by a separate ohmic contact (not shown) to get the desired bipolar injection.

Furthermore, this new diode will provide stable avalanche operation for a unipolar diode in SiC for the first time. Besides these performance improvements, an enlargement of

Optimization of Power MOSFETs
for Low Power Motor Drive Applications

Jun-Ho Song
Motion Control System Team,
Functional Power Group, Fairchild Semiconductor
82-3, Dodang-Dong, Wonmi-Gu, Bucheon-Si,
Gyeonggi-Do, KOREA
E-mail : jhsong@fairchildsemi.co.kr

Tae-Sung Kwon
Motion Control System Team,
Functional Power Group, Fairchild Semiconductor
82-3, Dodang-Dong, Wonmi-Gu, Bucheon-Si,
Gyeonggi-Do, KOREA
E-mail :TaeSung.Kwon@fairchildsemi.co.kr

Sang-Tae Han
HV_Product_Engineering,
Functional Power Group, Fairchild Semiconductor
82-3, Dodang-Dong, Wonmi-Gu, Bucheon-Si,
Gyeonggi-Do, KOREA
E-mail : tangal@fairchildsemi.co.kr

Sung-Il Yong
Motion Control System Team,
Functional Power Group, Fairchild Semiconductor
82-3, Dodang-Dong, Wonmi-Gu, Bucheon-Si,
Gyeonggi-Do, KOREA
E-mail : paulyong@fairchildsemi.co.kr

Abstract –**This paper presents a method to optimize power MOSFETs for low power motor drive applications, particularly white appliances such as refrigerators and fans. For this, three essential factors are needed, which are a threshold voltage, capacitances, and body diode reverse recovery characteristics. These three factors must be controlled for reliability, stability and effective operation. Their optimization helps the application to withstand shoot-through induced by Cdv/dt and noise caused by excessively high dv/dt, and to reduce power losses. In addition, overall methods for the optimization of MOSFETs are provided in detail.**

I. INTRODUCTION

An inverter system is more energy-efficient than conventional solutions using DC motors or AC induction motors with on/off controls. In recent years, there has been a rapid increase in the demand for MOSFET driven inverters for low power motor drives such as a refrigerator and a fan since the MOSFET has strong advantages over an IGBT in low current operations.

To meet this market need, Fairchild Semiconductor has developed a 3-phase inverter module covering up to 500W in which six MOSFETs are used as a main power switch and three dedicated HVICs and one LVIC are used as driver ICs as is shown in Fig. 1. The built-in MOSFET is optimized for motor drives but a conventional MOSFET is also optimized for power supply applications. This paper describes the MOSFET optimization technology including moderate threshold voltage, capacitance ratio and reverse recovery characteristics of a body diode.

II. THE ADVANTAGES OF MOSFETS

A MOSFET is known as a better choice than an IGBT for low current operations. It is mainly chosen because the conduction loss of a MOSFET is lower in low current operations as is shown in Fig. 2. The forward characteristic of a MOSFET is ohmic and its conduction loss is proportional to the square of the drain current so that it is less than that of an IGBT under 1 ampere operating area at normal temperature. But tradeoff is not as good for the MOSFET at higher temperature. Moreover, the MOSFET's

inherent body-diode can be used as a free wheeling diode that should be employed with separate chip in IGBT inverters, resulting in reduction of components and corresponding low cost. Due to the relatively large size of the MOSFET compared to a normal diode, it has a small reverse drop voltage. And more ruggedness is obtained with the MOSFET in short circuit situations. A MOSFET can sustain short-circuit situations for several tens of microseconds under typical operating condition (V_{GS}=15V, V_{DC}=300V, T_j=125°C), while an IGBT can sustain for several microseconds. Under the worst operating condition (V_{GS}=20V, V_{DC}=400V, T_j=125°C), a MOSFET can sustain short-circuit current for 30μs. This duration time is six times longer when compared with the same current rated IGBT. Moreover, the MOSFET has no tail current at off switching instant, which reduces off switching loss. With these advantages, it is clear that a MOSFET is adequate for low power inverter applications. And also, under a surge condition, a MOSFET inverter is stronger than an IGBT inverter with the same voltage rating, which is proved by the avalanche rating of the switching device. For this reason, for 200-V utility line voltage, a 500-V MOSFET can be used, whereas a 600-V IGBT is required under the same condition [1].

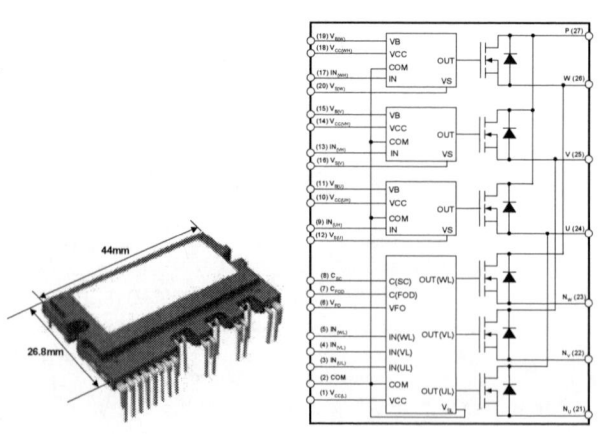

Fig.1 External view and internal block diagram of MOSFET inverter module.

978-1-4244-1871-8/07 $25.00 © 2007 IEEE

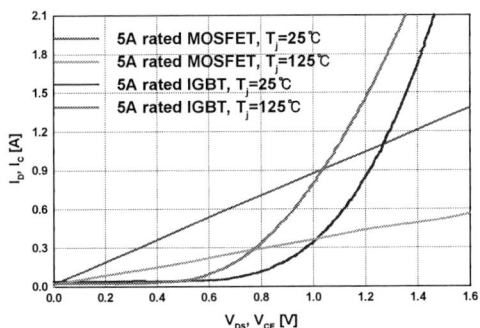

Fig. 2 On drop voltage comparison between MOSFET and IGBT with almost same die areas. (V_{GS}, V_{GE} =15V)

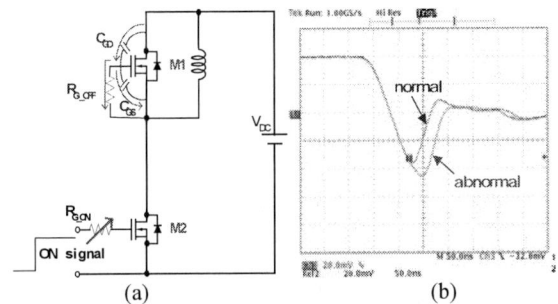

(a)   (b)

Fig. 3 Abnormal operation of MOSFET inverter under different dv/dt condition. (a) Test circuit, (b) experimental waveform showing difference between normal and abnormal recovery waveform.

III. CHALLENGES OF USING MOSFETS

A conventional MOSFET is optimized for fast switching converters such as AC/DC or DC/DC power supply applications so that it has low threshold voltage and low input capacitance for fast switching and reduced driving power. In motor drive applications, fast switching characteristics are not useful, especially considering the electromagnetic interference (EMI) caused by the high dv/dt output voltage. Additionally, low threshold voltage can cause shoot through induced by Cdv/dt in the inverter configuration. Poor reverse recovery characteristics of body diode cause more large on-switching losses, too. All of these challenges should be considered before using MOSFET for inverter applications.

IV. MOSFET OPTIMIZATION FOR INVERTER SYSTEM

A.. Capacitances

For explaining Cdv/dt effect, Figure 3(a) shows a MOSFET half-bridge circuit. The Cdv/dt induced voltage in the synchronous half bridge circuit might cause undesired turn-on of MOSFET M1 and deteriorate overall system efficiency. The Cdv/dt induced turn-on problem is caused by a fast changing voltage on the drain side of the MOSFET M1 [2]. If the turn-off gate resistance of the M1 is large and turn-on gate resistance of the MOSFET M2 is small, there might be shoot-through current due to the dv/dt induced current from the Miller capacitor C_{GD}. In this case, abnormal behavior happens as shown in Fig. 3 (b). Such abnormal operation will increase the inverter turn-on switching loss, which will eventually limit the power rating and stability of the system. A simple remedy for this could be using a large turn-on gate resistance and a small turn-off gate resistance.

However, a large turn-on gate resistance can cause longer turn-on time delay and high turn-on switching loss. Longer turn-on time delay is very critical to the current measurement system using an external shunt resistor to the negative DC bus since the ambiguity of the current measurement increases especially at low speed operation where the modulation index is small. And, small turn-off gate resistance accompanies high turn-off dv/dt switching.

Therefore the above issue cannot be solved by just adjusting the gate resistance at a time. In order to get the best performance (delay time) without instability (dv/dt induced shoot-through), another solution would be to reduce the Miller capacitance C_{GD} and to increase the gate-to-source capacitance C_{GS} in order to reduce the peak gate voltage induced by Cdv/dt. These methodologies involve design changes in the MOSFET devices that are usually not available to power system design engineers. The criterion would be based on the Q_{GD}/Q_{GS} ratio where Q_{GS} is the gate-to-source charge before the gate voltage reaches the threshold voltage. As mentioned above, lowering C_{GD} or enlarging C_{GS} will reduce Cdv/dt induced voltage. Generally, C_{GD} reduces according to the increase of voltage, V_{DS} between drain and source of MOSFET, but C_{GS} is not influenced like C_{GD} because C_{GD} is decreased according to the increase of depletion region that is varied by the magnitude of V_{DS}. C_{GD} and C_{GS} characteristics by V_{DS} are shown in Fig. 4. An internal equivalent circuit for capacitances of each area is shown in Fig. 5. It is possible to decrease C_{GD} as reducing length of poly gate that means length between Pbase and Pbase. For this reason, a poly gate pitch to get small C_{GD} has to be reduced. As shown in Fig 6, C_{GD} decreases by reducing poly gate pitch, but the C_{GS} increases according to reducing poly gate pitch because the number of gate poly regions increases according to the decrease of poly gate pitch in a same chip size.

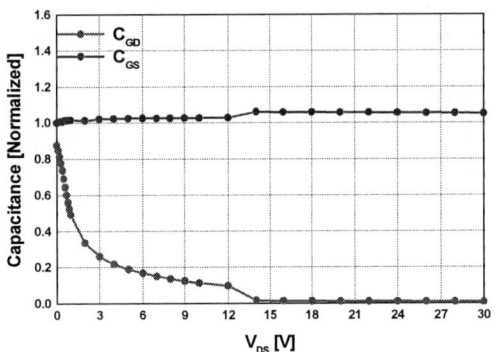

Fig. 4 MOSFET C_{GD}, C_{GS} by V_{DS}. (T_j=25℃)

978-1-4244-1871-8/07 $25.00 © 2007 IEEE   687

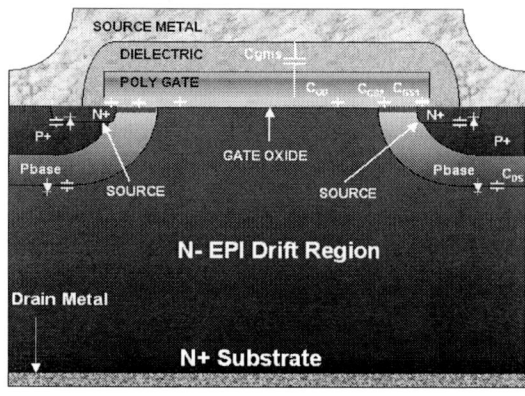

Fig. 5 Vertical structure of power MOSFET.

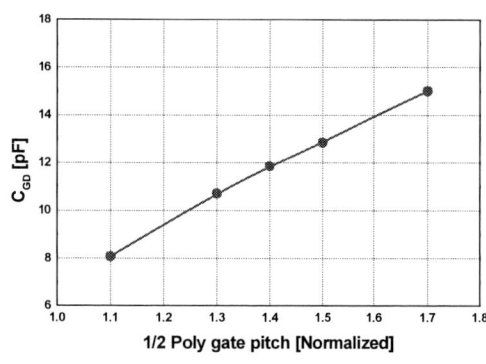

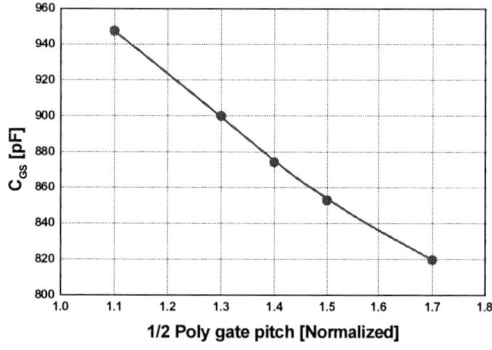

Fig. 6 C_{GD}, C_{GS} characteristics of MOSFET according to poly gate pitch (V_{DS}=40V, T_J=25℃)

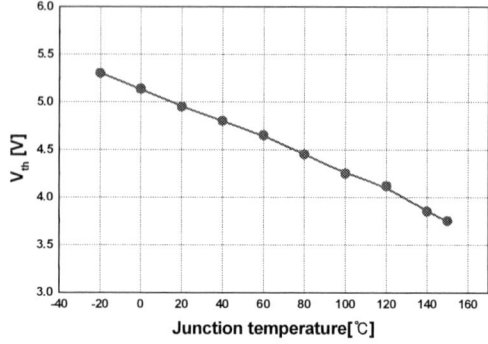

Fig. 7 Threshold voltage characteristics of MOSFET according to the variation of junction temperature.

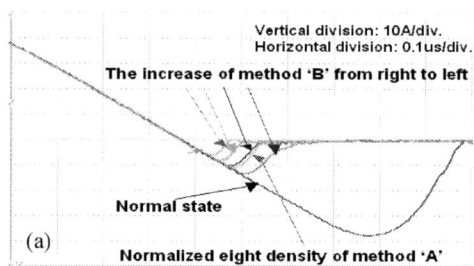

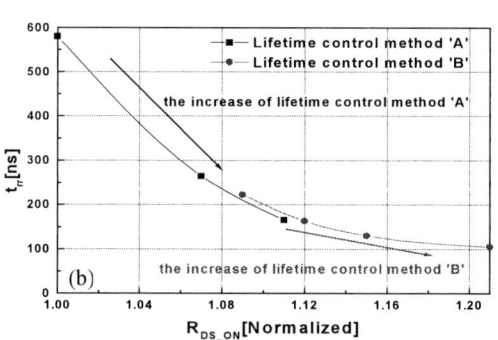

Fig. 8 t_{rr} vs. lifetime control characteristics of MOSFET.

(a) t_{rr} vs lifetime control method 'A' and 'B' waveform. (b) t_{rr} vs. R_{DS-ON}.

B. Threshold voltage (V_{th})

When MOSFETs are designed, V_{th} level at maximum guaranteed temperature should be considered because a MOSFET V_{th} level is in inverse proportion to temperature as shown in Fig. 7.

For reliable operation, it is recommended to raise the V_{th} level against possible V_{GS} surge that is caused by Cdv/dt effect. Physically, V_{th} level is determined by the Pbase doping concentration. A high V_{th} is determined according to the increase of the Pbase doping concentration. But the R_{DS-ON}, which is an equivalent resistor between drain and source of a MOSFET, is in proportion to Pbase doping concentration. Therefore, the proper V_{th} level should be determined considering conduction loss.

C. Body diode reverse recovery characteristics

The third factor for consideration is the MOSFET's body diode reverse recovery characteristics. Because of the MOSFET body diode, fewer components and reduced cost are possible. But actually it has poor reverse recovery characteristics compared with an optimized fast recovery diode (FRD). When a MOSFET is turned on, switching losses increase due to long t_{rr} characteristics of MOSFET body diode. To reduce t_{rr}, its body diode can be tuned by two kind of minority carrier lifetime control method. As is shown in Fig 8, trade-off between t_{rr} and R_{DS-ON} exists. t_{rr} is reduced according to the increase lifetime control methods, but R_{DS-ON} is slightly increased by the effect of above two methods. Additionally, the MOSFET body diode and R_{DS-ON} voltage drop increase in proportion to reduced t_{rr} as shown in Fig. 9 and 10. To reduce t_{rr}, the rate of proposed method should be determined considering switching frequency according to each application for better energy efficiency.

978-1-4244-1871-8/07 $25.00 © 2007 IEEE   688

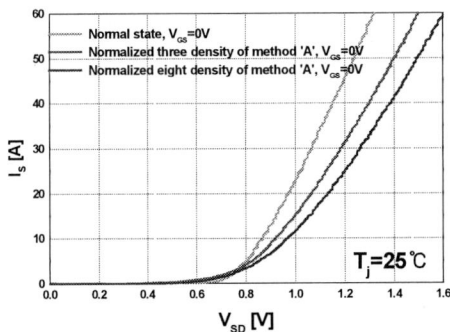

Fig. 9 MOSFET body diode I-V characteristics with differential density of lifetime control method 'A'

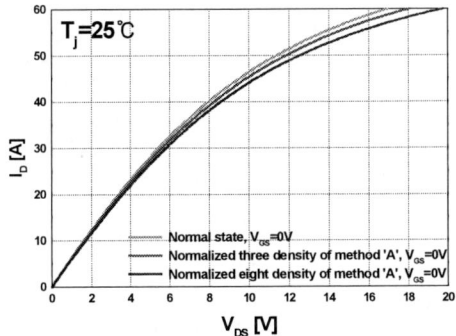

Fig. 10 MOSFET I-V characteristics with differential density of lifetime control method 'A'

V. CONCLUSION

In this paper, an overall description of methods to optimize power MOSFET is presented, which involves good performance for driving a low power inverter, the advantages of using MOSFETs and the methods for reducing challenges like t_{rr} and Cdv/dt. Improved reliability and efficiency will be gained with implementing these techniques..

REFERENCES

[1] Yo-Chan Son, Ki-Young Jang, Bum-Seok Suh, "Integrated MOSFET Inverter Module for Low-Power Drive System" APEC2005.
[2] Thomas Wu, "CDV/DT INDUCED TURN-ON IN SYNCHRONOUS BUCK REGULATORS", International Rectifier Technical Paper.

High-power and high-speed semiconductor switch RSD applied in pulsed power system

Lin Liang, Yuehui Yu, Linfeng Deng and Yabin Peng

Huazhong University of Science & Technology, Department of Electronic Science & Technology

No.1037, Luoyu Road, Wuhan 430074, China

Email: lianglin2005@163.com, yuyuehui@mail.hust.edu.cn

Abstract—Based on the controllable plasma layer conversion principle, the new-type semiconductor pulsed power switch reversely switched dynistor (RSD) turns on uniformly and simultaneously over the whole chip area. It has the advantages of high power and high speed at the same time. The blocking, turn-on and turn-off characteristics of RSD are investigated in this paper systematically. The off-state and on-state characteristics get a trade-off by an optimal design method. The equations during turn-on and turn-off process are obtained by analyzing the carrier motion on different conditions. Then two important parameters, the turn-on voltage and reverse recovery time, are calculated. In the single pulse experiment, peak currents of 21.6kA and 42.0kA (pulse widths of 30μs and 200μs) are acquired on 20mm and 40mm diameter stacks respectively, which have reached the utmost value. In the repetitive frequency experiment, output waveform of 20Hz is acquired at 4kV with a resonant triggering way.

I. INTRODUCTION

Pulsed power switch is one of the key elements in the pulsed power system. Its parameters and characteristics influence the rise time, amplitude and turn-off time of pulse. Recent years pulsed power switch has turned to semiconductor device because of its small volume, long lifetime and high reliability[1-3]. Based on the controllable plasma layer conversion, a semiconductor switch reversely switched dynistor (RSD) of new turn-on principle was invented at the Ioffe Physicotechnical Institute(IPI) in Russia[4]. Due to special device structure, RSD can turn on uniformly and simultaneously over the whole device area. Therefore, its switching characteristics are very good and it is an ideal switch applied in pulsed power system[5-7].

In this paper, the blocking and turn-on and turn-off characteristics of RSD are investigated respectively. The single pulse and repetitive frequency discharge experiments based on RSD carried out in our laboratory are also reported.

II. OPERATION PRINCIPLE

As is shown in Fig.1, RSD is a four-layer and two-electrode device. It contains thousands of alternating thyristor units and transistor units in the structure, the characteristic size of which is smaller than the width of n-base. The central collector junction J_2 is common to all units and it blocks the external main voltage u_1.

The turn-on process is as follows. When the triggering switch is closed, the reverse triggering voltage u_2 applies to

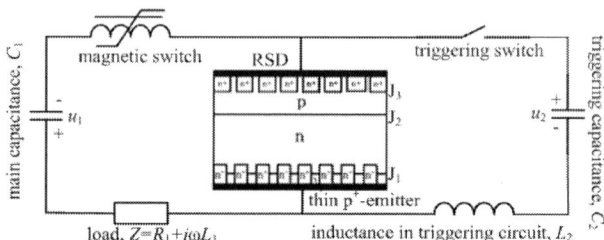

Fig.1. RSD structure and operation principle

RSD and the triggering process begins through the embedded equivalent pnn$^+$ diode units. The holes in p-base inject into n-base. The electrons in n$^+$-emitter at anode also inject into n-base and drift to the neighborhood of J_2 junction, which form a high density plasma layer together with the holes injecting from p-base. The triggering process lasts about 0.5 ~ 2.5μs. During the period the main circuit and triggering circuit are separated by the magnetic switch. After it is saturated, the main voltage u_1 applies to RSD and turn-on process begins. The electrons and holes in plasma layer near the collector junction drift to the corresponding n-base and p-base under electrical field, which leads to the minority carrier injection from p$^+$-emitter at anode and n$^+$-emitter at cathode. It brings about the base regions a strong conductivity modulation effect and makes RSD carry out the switching process.

III. CHARACTERISTICS ANALYSIS

A. Blocking characteristics

RSD doesn't have reverse blocking ability and its forward blocking voltage is mainly assumed by J_2 junction. By an optimal design method[8], an optimal factor is defined as

$$G = V_{Bo}/V_B, \tag{1}$$

then the equation can be written as

$$(1-\alpha_1)^{2/3} = 4.1 \times 10^{-3}(1/L_P)V_{Bo}{}^{7/6}\alpha_1(1+\alpha_1)^{1/2} \tag{2}$$

where α_1 is the current amplification coefficient of the equivalent transistor p$^+$np, L_p is the hole diffusion length in n-base, V_{Bo} is the forward breakover voltage, and V_B is the avalanche breakdown voltage of J_2 junction. According to V_{Bo}, α_1 can be calculated by (2), then G and V_B are obtained, and the n-base width W_n and the substrate resistivity ρ are confirmed in the end. The n-base parameters of RSD with different breakover voltage are listed in table 1.

978-1-4244-1871-8/07 $25.00 © 2007 IEEE

TABLE 1 N-BASE PARAMETERS DESIGN

Forward breakover voltage, V_{Bo} (V)	Current amplification coefficient of p^+np transistor, α_1	Optimal factor, G	N-base thickness, W_n (μm)	Substrate Resistivity, ρ ($\Omega \cdot$cm)
1000	0.9441	0.6184	147	41
1500	0.8968	0.6849	216	61
2000	0.8465	0.7317	284	82
2500	0.7970	0.7666	352	104
3000	0.7500	0.7937	419	127

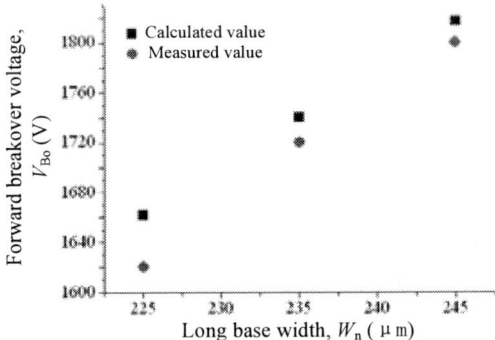

Fig.2. Forward breakover voltage of RSD with different n-base width

Fig.2 shows the comparison of forward breakover voltage between the theoretical value and measured value of RSD chips made in laboratory. The substrate resistivity is 65Ω•cm. The n-base widths are 225μm, 235μm and 245μm respectively. The measured value is the statistical average of chips from different groups. The error between theoretical and measured value is within 2.5%.

On condition that the blocking requirement is satisfied, the optimal design method guarantees the shortest n-base and thus can reduce the on-state power dissipation.

B. Turn-on characteristics

During the turn-on process, the principal mechanism responsible for conductivity modulation in n-base is bipolar drift at essentially intermediate injection levels and in field in which the bulk remains neutral. The diffusion and recombination actions of carriers can be ignored. Setting the J_2 junction as zero of x coordinate and the instant of putting on reverse triggering voltage as zero of t coordinate, the equation of plasma bipolar drift is[9]

$$\frac{\partial p}{\partial t} = -\frac{bN_d J(t)}{q[(b+1)p + bN_d]^2} \frac{\partial p}{\partial x}, \qquad (3)$$

where $p(x,t)$ is excess plasma density, N_d is n-base doping concentration, b is electron-to-hole mobility ratio in a weak field, and $J(t)$ is plasma current density.

According to (3) the expression of RSD turn-on voltage can be expressed as

$$U_F = 2W_n^2 J_F(t)/(3\mu_p\sqrt{bQ_N Q_R}) \qquad t_R < t < t_2 \quad (4)$$

$$U_F = 2W_n^2 J_F(t)/(3\mu_p\sqrt{bQ_N Q_F(t)}) \qquad t > t_2 , \quad (5)$$

where $Q_N = qN_d W_n$, $Q_R = \int_0^{t_R} J_R(t)dt$, $Q_F(t) = \int_{t_R}^t J_F(t)dt$,

$J_R(t)$ and $J_F(t)$ are the triggering loop and main loop current density respectively, t_R is the instant of choke saturation, t_2 is the instant of forward waves reaching the collector junction, and μ_p is the hole mobility.

The maximum of turn-on voltage U_{Fmax} is an important parameter to estimate the turn-on characteristics. This voltage peak is induced by a comparatively low diffusion velocity of the carriers filling the base regions during the increase of the power current. A sharp rise of U_{Fmax} means an increasing turn-on energy dissipation and a danger of current localization. Based on (4) and (5), the simulation waveforms of turn-on current and voltage on RSD can be acquired, as is shown in Fig.3. The simulation condition is C_1=600μf, L_1=1.4μH, R_1=0.113Ω, C_2=1.18μf, L_2=0.7μH, u_1=4000V and u_2=1500V.

C. Turn-off characteristics

During turn-off process, the electrical field in the device has been reduced to a low level. Therefore, the drift behavior of carriers can be ignored and the diffusion and recombination behaviors are considered mainly. Setting the J_1 junction as zero of x coordinate and the zero-crossing point of discharge current as zero of t coordinate, the equation is[10]

$$\frac{\partial p}{\partial t} = -\frac{p}{\tau} + D\frac{\partial^2 p}{\partial x^2}, \qquad (6)$$

where τ is the common lifetime of electron and hole at high-level injection, $D=2D_n \cdot D_p/(D_n + D_p)$, D_n and D_p are the diffusion coefficient of electron and hole respectively.

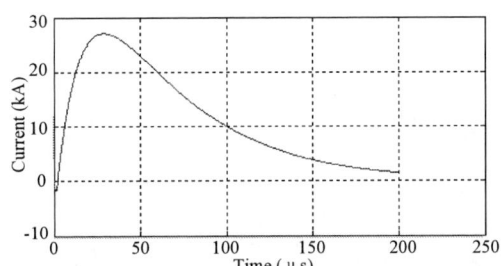

(a) Current simulation waveform

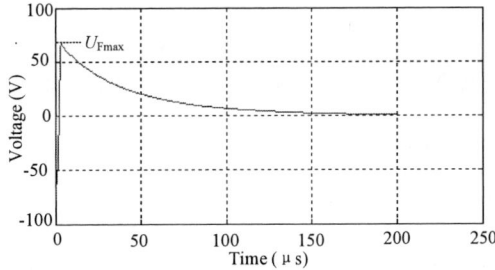

(b) Voltage simulation waveform

Fig.3. Current and voltage simulation waveforms on RSD

Assume the boundary condition is $p(x=W_n, t)=0$ and $p(x=0, t)=0$, then $p(x,t)$ is obtained by (6)

$$p(x,t) = e^{-\frac{t}{\tau}} \sum_{n=1}^{+\infty} A_n e^{-\frac{(n\pi)^2}{W_n^2}Dt} \sin\frac{n\pi x}{W_n}. \tag{7}$$

The p^+npn^+ structure can be equivalent to p^+in^+ structure at high-level injection. It is considered as the initial condition that the distribution of minority carriers in n-base at the case. That is the following equation

$$p(x,0) = n_i + \frac{J(t=0^-)\tau}{2qL}\left(\frac{\cosh[(x-x_0)/L]}{\sinh[(W_n+W_p)/2L]}\right.$$
$$\left....- \frac{D_n-D_p}{D_n+D_p} \cdot \frac{\sinh[(x-x_0)/L]}{\cosh[(W_n+W_p)/2L]}\right), \tag{8}$$

where $L = \sqrt{D\cdot\tau}$ $x_0 = (W_n+W_p)/2$, W_p is the p-base width.

Then the coefficients A_n are confirmed.

$$A_n = \frac{2}{W_n}\int_0^{W_n} p(x,0)\sin\frac{n\pi x}{W_n}dx. \tag{9}$$

The minimal duration from the instant of current zero-crossing point to that of RSD resuming blocking ability is considered as the reverse recovery time of RSD, which is denoted as t_{rr}.

Define $x=W_1$ as the coordinate of the edge of space charge region in n-base under the main loop voltage before triggering. Assume the blocking ability of RSD resumes when the concentration of minority carriers at $x=W_1$ re-reaches the thermal equilibrium level, then according to

$$p(W_1,t) = e^{-\frac{t}{\tau}} \sum_{n=1}^{+\infty} A_n e^{-\frac{(n\pi)^2}{W_n^2}Dt} \sin\frac{n\pi W_1}{W_n} = \frac{n_i^2}{N_d}, \tag{10}$$

$$W_1 = W_n - \sqrt{2\varepsilon\varepsilon_0 u_1/qN_d} \tag{11}$$

t_{rr} can be obtained. It is a natural turn-off model. Set the parameters as $n_i=1.5\times10^{10}$cm^{-3}, $D=19$cm^2/s, $N_d=10^{14}$cm^{-3}, $u_1=2000$V, $J(t=0^-)=0$. The results are listed in table 2. It is shown that the reverse recovery time is about several tens of microseconds and the n-base width affects it more than the minority carrier lifetime.

IV. EXPERIMENTS

A. Great current experiment of single pulse

In order to test the current capability of RSD, the utmost experiment is carried out at single pulse. There is an experiential formula as follow

$$I_m = KS/[f(t_p)^{1/3}] \tag{12}$$

where I_m is the peak current RSD can endure, S is the

effective chip area, K is a structural factor which is inverse proportional to the blocking voltage, f is the current curve factor and t_p is the pulse width of current.

Fig.4 shows the waveforms of utmost current experiment on RSD stacks of 20mm and 40mm diameters. The discharge currents are measured by the shunt and Rogowski coil respectively. The test conditions, experimental and theoretical value of peak current are listed in table 3. The characteristics of the former stack degenerate due to over current and the latter one keeps a good property after the experiment.

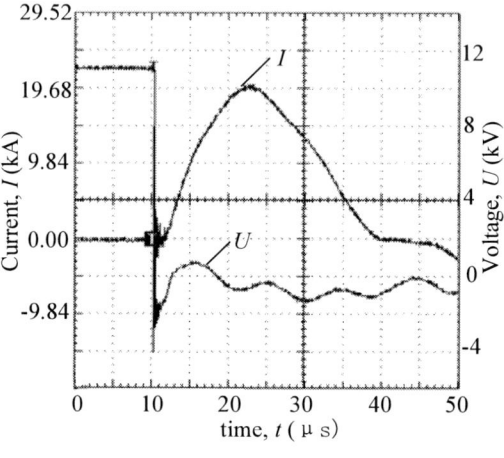

(a) Φ=20mm

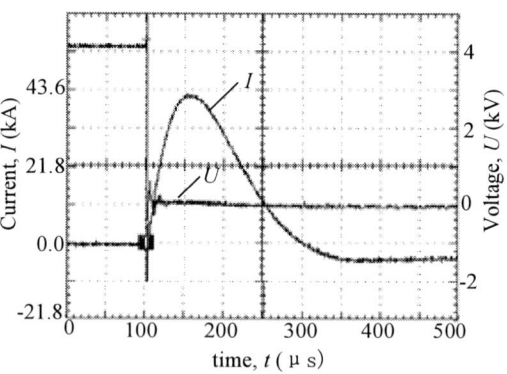

(b) Φ=40mm

Fig.4. Discharge waveforms on RSD of single pulse

TABLE 2 CALCULATION OF REVERSE RECOVERY TIME

$\tau(\mu s)$ \ $t_{rr}(\mu s)$ \ $W_n(\mu m)$	200	225	250
30	33.9	42.6	51.7
40	34.5	43.5	53.1
50	34.8	44.1	53.9

TABLE 3 PARAMETERS OF SINGLE PULSE EXPERIMENTS

Chip diameter (mm)	Test conditions	Pulse width (μs)	Experimental value of peak current (kA)	Theoretical value of peak current (kA)
20	$C_1=20\mu f$, $u_1=12.5$kV, $R_1=0.25\Omega$	30	21.6	19.8
40	$C_1=1200\mu f$, $u_1=4.2$kV, $R_1=0.063\Omega$	200	42.0	42.1

B. Repetitive frequency experiment

The repetitive frequency experiment is carried out with a resonant triggering way. Fig.5(a) shows the circuit diagram. A magnetization-reversal circuit (MRC) is used to reverse the magnetization of the saturable magnetic switch when a switching process is completed. C_2, R_2 and L_2 are the capacitance, resistance and inductance of the triggering loop, respectively. The triggering switch is a thyristor, the blocking voltage of which is 6kV and is controlled by CPLD(Complex Programmable Logic Device). When the thyristor is closed, a majority of the current produced by the resonant circuit is shunted on to RSD and forms its triggering current. Fig.5(b) is the voltage waveform on RSD when the main loop voltage is 4kV and the main loop capacitance is 250nf. The repetitive frequency is 20Hz. The result proves that RSD operates successfully.

V. CONCLUSION

The blocking, turn-on and turn-off characteristics of pulsed power switch RSD have been investigated in this paper. By an optimal method, the shortest n-base is acquired with the guarantee of blocking ability. The error between actual measured value and theoretical design value of breakover voltage is within 2.5%. Based on the plasma bipolar drift model, the expression of turn-on voltage and the simulation waveforms of turn-on current and voltage on RSD are obtained. Based on the natural turn-off model, the reverse recovery time is calculated. The results show that the n-base width influences the turn-off process a lot.

The single pulse and repetitive frequency discharge experiments have also been carried out. The ruinous utmost current experiments of single pulse are realized on RSD stacks of 20mm and 40mm diameters. The peak currents of 21.6kA and 42.0kA were acquired, respectively. (The pulse widths are 30µs and 200µs respectively.) By the way of resonant triggering, repetitive frequency output waveforms of 20Hz were acquired at 4kV.

ACKNOWLEDGMENT

This work is supported by the grant from the National Natural Science Foundation of China (No.50277016 and No. 50577028) and the Specialized Research Fund for the Doctoral Program of Higher Education of China (No. 20050487044).

REFERENCES

[1] W. H. Jiang, N. Oshima, T. Yokoo, K. Yatsui, K. Takayama, and M. Wake, "Development of repetitive pulsed power generators using power semiconductor devices," in *Proc. Int. Symp. Power Semiconductor Devices &ICs* , Santa Barbara , 2005, pp. 21-26 .

[2] H. O'Brien, W. Shaheen, Jr. R. L. Thomas, T. Crowley, S. B. Bayne, and C. J. Scozzie, "Evaluation of advanced Si and SiC switching components for army pulsed power applications," *IEEE Trans. Magnetics*, vol. 43, no. 1, pp.259-264, Jan. 2007.

[3] A. Welleman, J. Waldmeyer, and W. Fleischmann, "Semiconductor components & solid state switches for pulsed power applications," in *Proc.Int. Symp. Power Modulator*, San Francisco, 2004, pp.174-177.

[4] I. V. Grekhov. "New principles of high power switching with semiconductor devices," *Solid-State Electronics*, vol.32, no. 11, pp.923-930, Nov. 1989.

[5] S. V. Korotkov, "Switching possibilities of reverse switched-on dynistors and principles of RSD circuitry(review)," *Instruments and Experimental Techniques*, vol. 45, no. 4, pp. 437-470, Jul. 2002.

[6] Y. H. Yu, L. Liang, M. T. Li, Y. H. Liu, and L. Liu, "Study on turn-on mechanism and high-current characteristics of high-speed semiconductor switch RSD," *Transactions of China Electrotechnical Society*, vol. 20, no. 2, pp. 36-40, Feb. 2005.(in Chinese).

[7] Yu. Aristov, V. Voronkov, I. V. Grekhov, A. Zhmodikov, A. Kozlov, and S. Korotkov, et al, " Reverse switch-on dynistor switches of gigawatt-power microsecond pulses," *Instruments & Experimental Techniques*, vol. 50, no. 2, pp. 228-232, Mar. 2007.

[8] G. Liu, Y. H. Yu, and J. Q. Shi, *Semiconductor devices-power, sense, photon, microwave devices*, 1st ed. Beijing: Publishing House of Electronics Industry, 2000, pp. 12-19. (in Chinese).

[9] R. H. Dean, "Transient double injection in trap-free semiconductors," *Journal of applied physics*, vol. 40, no. 2, pp. 585-595, Feb. 1969.

[10] P. R. Palmer, E. Santi, J. L. Hudgins, X. S. Kang, J. C. Joyce, and Y. E. Poh, "Circuit simulator models for the diode and IGBT with full temperature dependent features," *IEEE Trans. Power Electronics*, Vol. 18, no. 5, pp.1220-1229, Sep. 2003.

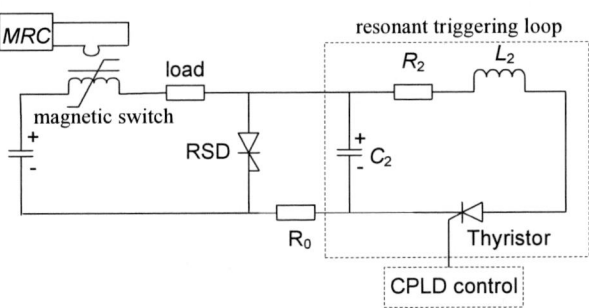

(a) Circuit diagram with the resonant triggering way

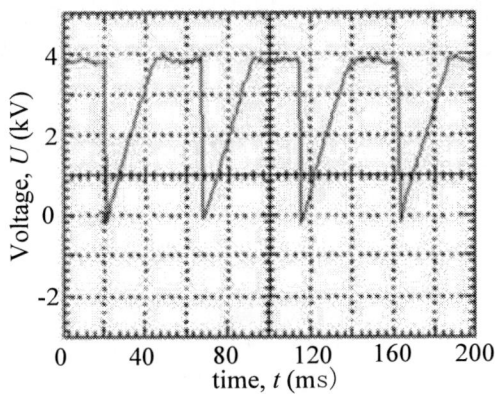

(b) Voltage waveform on RSD

Fig.5. Repetitive frequency experiment of RSD

A New Smart Power Module for Low Power Motor Drives

Tae-Sung Kwon[1], Jun-Ho Song[1], Jun-Bae Lee[2], Seung-Han Paek[1] and Sung-Il Yong[1]

[1]Motion Control System Team,
Functional Power Group, Fairchild Semiconductor
82-3, Dodang-Dong, Wonmi-Gu, Bucheon-Si,
Gyeonggi-Do, KOREA
E-mail : TaeSung.Kwon@fairchildsemi.co.kr

[2] Dept. of Electrical Engineering,
Hanyang University,
133-791, Haengdang-dong, Sungdong-Gu,
Seoul, Korea

Abstract – This paper introduces a new Motion-SPM[TM] (Smart Power Modules)module in Single In-line Package(SIP), which is a fully optimized intelligent integrated IGBT inverter module for up to 1kW low power motor drive applications. This module offers a sophisticated, integrated solution and tremendous design flexibilty. It also takes advantage of pliability for the arrangement of heat-sink due to two types of lead forms. It utilizes non-punch-through (NPT) IGBT with a fast recovery diode and highly integrated building block, which features built-in HVICs and a gate driver that offer more simplicity and compactness leading to a reduced costs and high reliablilty of the entire system. This module also provides technical advantages such as the good thermal performances through IMS (Insulated Metal Substrate), the high latch immunity. This paper provides an overall description of the Motion-SPM[TM] in SIP as well as actual application issues such as electrical characteristics, circuit configurations and power ratings.

I. INTRODUCTION

Motors are the major source of energy consumption in appliances. Since governmental and agency regulations continue to mandate reduced energy consumption, inverter technology is being increasingly accepted and used by a wide range of users in the design of their products. Power modules for inverterized motor drive applications are also part of a current trend due to the advantages that offer such as space- savings and ease of assembly. We have developed a series of SPM[®] modules for a highly efficient integrated solution. We concentrate on the development of an intelligent integrated power module using the new concept of building structure and advanced packaging technology that is the means of achieving an excellent, cost-effective solution. The first and second series of Motion-SPM[TM] in DIP packaging and Mini-DIP packaging have been successfully introduced into the market [1]-[2]. Since then, a great number of SPM[®] inverter systems have been implemented and continue to run successfully, which validates the very good results of SPM[®] reliability. Now, Fairchild Semiconductor has taken the next step with the development of low power and cost effective SPM families in SIP, which are quite specialized for IGBT inverter application range of less than 1kW. Within such an operating power, one of the most important requirements in the system is more compactness and easier mass production process with high quality and reliability resulting in more cost-effectiveness comparing to discrete inverter solutions. These modules have been fully developed as an answer to the strong demands particularly in home appliances applications such as air-conditioners, and washing machines and refrigerators, requiring higher efficiency and higher performance characteristics.

This paper describes in detail the design issues, electrical performance, and other important considerations for designing the system.

II. MAIN FEATURES OF DESIGN AND FUNCTION

A. EXTERNAL VIEW AND CIRCUIT STRUCTURE

Figure 1 and 2 show a real photograph and internal equivalent circuit of the Motion-SPM[TM] in SIP. SIP package is composed of six NPT IGBTs, six freewheeling diodes, three HVICs, one LVIC and one thermistor. The NPT IGBTs are designed with key electrical characteristics such as low conduction/switching losses over all driving conditions, short circuit withstanding capability for the motor drive, and smooth switching waveforms without EMI noises caused by rapid dV/dt. The most important characteristics of the freewheeling diodes are soft recovery behavior over the whole current, temperature range and low forward voltage drop. System reliability is further enhanced due to internal temperature detection using built-in thermistor, integrated under-voltage and short circuit lockout protection. The high-speed built-in HVIC enables the use of a single power supply without a photo coupler. This results in size and cost reduction of the inverter systems. The Motion-SPM[TM] in SIP has two types of negative DC terminals. One has three divided negative DC terminals to monitor the inverter output current by using three external shunt resistors and provide a sensor-less control solution. The other has one negative DC terminal to sense DC-link current and short-circuit protection by using one external shunt resistor.

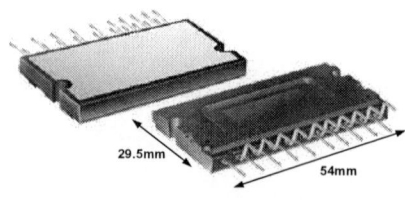

Figure 1. Photograph of Motion-SPM[TM] in SIP

978-1-4244-1871-8/07 $25.00 © 2007 IEEE

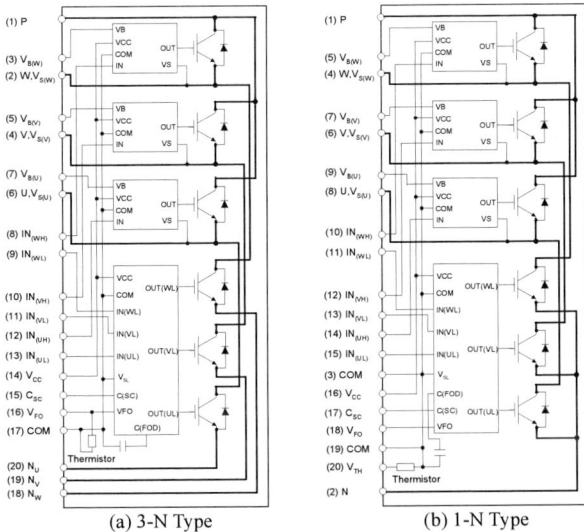

(a) 3-N Type (b) 1-N Type

Figure 2. Internal equivalent circuit of Motion-SPM™ in SIP

doing simulations and experimental tests. The total thickness of the molding is 5.5mm. Figure 3 shows the cross sectional structure of the Motion-SPM™ module in SIP packaging.

Motion-SPM module in SIP packaging provides two types of lead forms. One is Y-forms lead and the other type is L-forms lead. It takes advantage of the free arrangement of heat-sink as well as PCB design flexibility. Figure 4 shows the lead forms of the Motion-SPM™ in SIP.

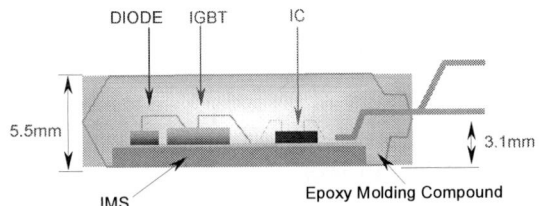

Figure 3. Cross sectional structure

B. FUNCTIONAL DESCRIPTION

The Motion-SPM™ module in SIP provides two main protective functions. One is control supply under-voltage protection and the other is short-circuit current protection.

When the control supply voltage drops under its UV detect level, the internal gating signal is blocked and a fault-out signal is generated. Once the supply voltage rises again over the UV reset level, the fault-out signal becomes high and the SPM® is operated by the command signals.

The Motion-SPM™ module in SIP packaging can monitor the inverter leg current by using an external shunt resistor. The LVIC of the SPM® in SIP monitor the low-side collector current with the C_{SC} pin. If the voltage of the C_{SC} pin exceeds 0.5V, the internal gating signal is blocked and a fault-out signal is generated. Hence, the external shunt resistor can be selected to determine the trip current level, which can be optimized depending on the field requirements.

C. PACKAGE STRUCTURE

The narrow space multi-die attach technology is used in the Motion-SPM™ in SIP. This results in reduced noise, size and less mutual interference. This package is designed to guarantee the best heat transfer from the power chips to the outer heat-sink by using an IMS package. The IMS substrate uses an aluminum plate at the base. The upper side of the substrate consists of a thermally conductive dielectric layer and a copper cladding on which the circuit etched. This takes advantage of high thermal conductivity and simple manufacturing. The lead frame structure has a 1.385mm down-set shape. This structure makes the thermal resistance low but doesn't reduce the distance between lead frame and the outer heat-sink. More down-set thickness affects the reliability and assembly process. The optimization of the bending depth has been obtained by

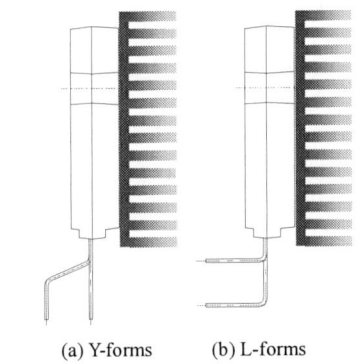

(a) Y-forms (b) L-forms

Figure 4. Lead forms of Motion-SPM™ in SIP

D. ISOLATING CONSIDERATIONS

For the design of the Motion-SPM™ module in SIP, isolation distances of pin-to-pin and pin-to-heatsink should be considered. As shown in Figure 5(a), clearance and creepage distances of pin-to-heatsink are 3.10mm and 4.08mm respectively. As shown in Figure 5(b), clearance and creepage distances of pin-to-pin are 2.35mm and 4.15mm respectively.

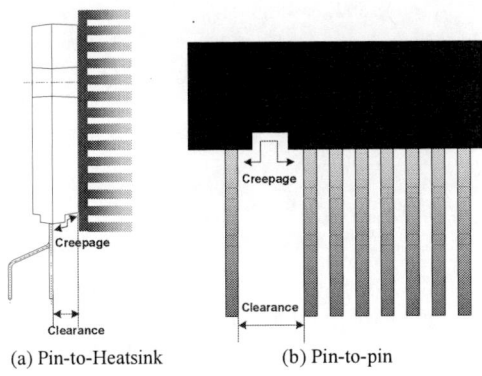

(a) Pin-to-Heatsink (b) Pin-to-pin

Fig. 5. Isolation distance.

978-1-4244-1871-8/07 $25.00 © 2007 IEEE   695

III. ELECTRICAL PERFORMANCE AND POWER RATINGS

A. ELECTRICAL CHARACTERISTICS

The NPT IGBTs and ultra fast recovery diodes are designed for fast switching without excessive ringing. The turn-on and turn-off switching waveforms at the conditions of V_{DC}=300V, I_C=10A, V_{CC}=15V and T_j=125℃ is shown in Figure 6. The Motion-SPM™ module in SIP packaging does not have current oscillation due to moderate dv/dt rate to reduce EMI problems.

Figure 7 shows the graph indicating the variation of switching dv/dt rates with switching current at the conditions of V_{DC}=300V, V_{CC}=15V and T_j=125℃ . The turn-on and turn off dv/dt have value in the range of 1.5~3kV/us and 2~7kV/us.

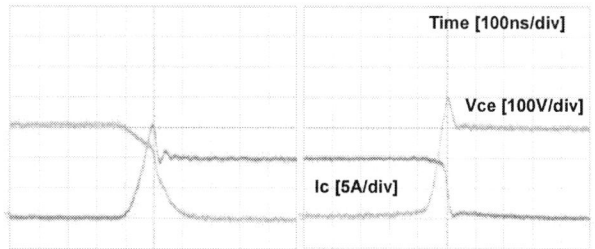

Figure 6. Switching Waveform of 600V/10A rating device at V_{DC}=300V, I_C=10A, V_{CC}=15V and T_j=125℃

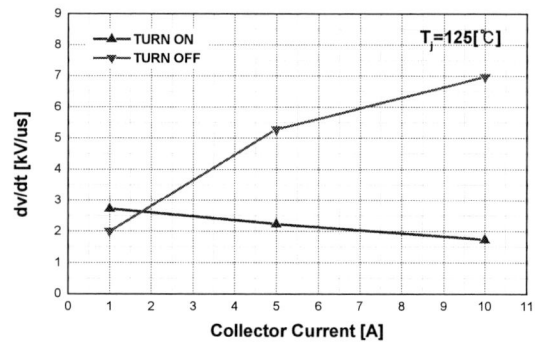

Figure 7. Switching dv/dt rates of 600V/10A rating device at V_{DC}=300V, V_{CC}=15V and T_j=125℃

Figure 8 shows that switching loss at the condition of V_{CC}=15V, V_{DC}=300V and T_j=125℃. Based on the real experimental data, the single IGBT power loss and allowable output current according to switching frequency was calculated. The procedure used for this calculation is briefly described below.

B. POWER RATINGS

The power-carrying potential of a device is dependent on the heat transfer capability of the device. The proposed module provides not only good thermal performance but also operating frequency options in accordance with the application.

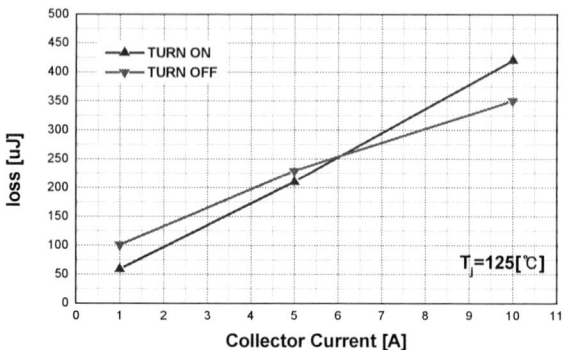

Figure 8. Switching Loss of 600V/10A rating device at V_{DC}=300V, V_{CC}=15V and T_j=125℃

The single IGBT power loss is composed of conduction and switching losses caused in the IGBTs and FRDs. The loss during the turn-off steady-state can be ignored because it is a very small amount and has little effect on increasing the temperature in the device. The conduction loss depends on the DC electrical characteristics of the device i.e., the saturation voltage. Therefore, it is a function of the conduction current and the device's junction temperature. Conversely, the switching loss is determined by the dynamic characteristics like turn-on/off time and over-voltage/current. Hence, in order to obtain the accurate switching loss, we should consider the DC-link voltage, the applied switching frequency, and the power circuit layout in addition to the current and temperature. For the detailed equations for calculating both conduction and switching losses based on a PWM-inverter system for motor control applications, refer to the references [3] and [4].

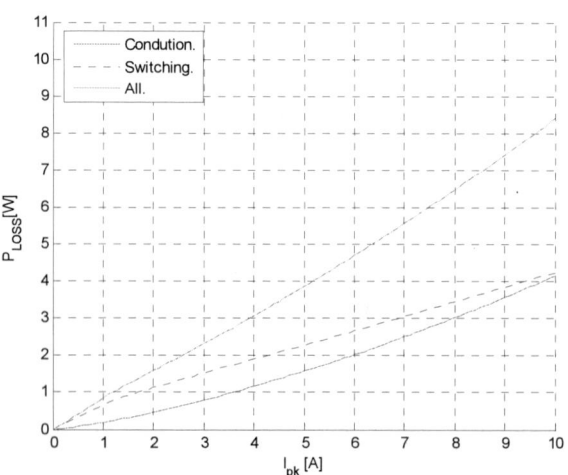

Figure 9. Power Loss of 600V/10A rating devices at V_{DC}=300V, V_{CC}=15V, f_{sw}=15kHz and T_j=125℃

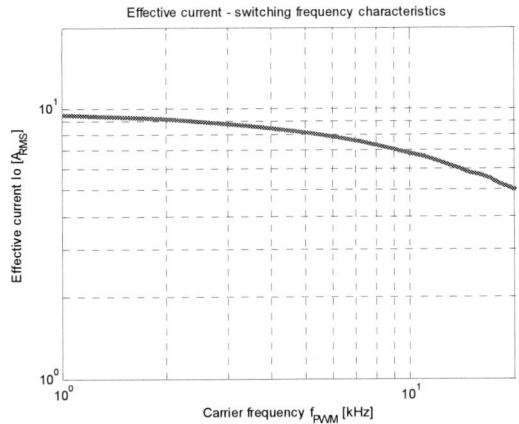

Figure 10. Allowable output current for a 600V/10A rating devices at V_{DC}=300V, V_{CC}=15V, T_C=100℃ and T_j=125℃

It should be noted that the PWM modulation index MI = 0.8, cosφ=0.9, V_{dc}=300V, V_{cc}=15V, I_{PK}=10A, T_j=125℃, f_{sw}=15kHz, sinusoidal output current are used as common simulated parameters in all the calculations. The device's junction temperature below 125℃. Figure 9 shows the single IGBT power loss characteristics of a 600V/10A rating devices. Figure 10 shows the maximum motor allowable output current versus switching frequency from 1kHz to 20kHz. These values are obtained based on typical experimental data.

C. SHORT-CIRCUIT RUGGEDNESS AND PROTECTION

The proposed module has one or three divided negative DC terminals to monitor the inverter leg current by using an external shunt resistor.

The LVIC monitors the low-side collector current with the C_{SC} pin. Figure 11 shows the actual waveforms under a short-circuit protecting situation with external shunt resistor, R_{SC}=20mΩ. Once the voltage across R_{SC} reaches 0.5V, the LVIC shuts down the gating signal after a time delay of about 1.3us. This is composed of the LVIC internal filter and the externally added low-pass filter. If there is not protection circuit, as shown in Figure 12, it is noted that the IGBT within the Motion-SPM^{TM} module in SIP package is destroyed after short-circuit withstanding time of about 8us.

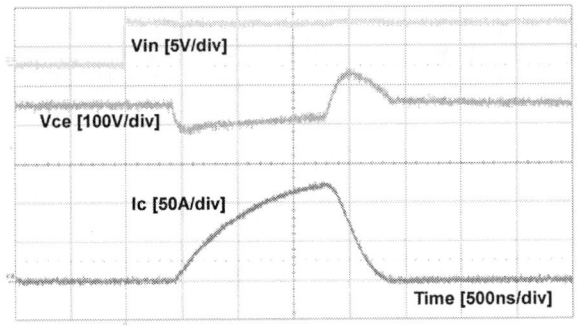

Figure 11. Protected short circuit waveform of a 600V/10A rating device at V_{dc}=450V, V_{cc}=20V, and T_j=125℃

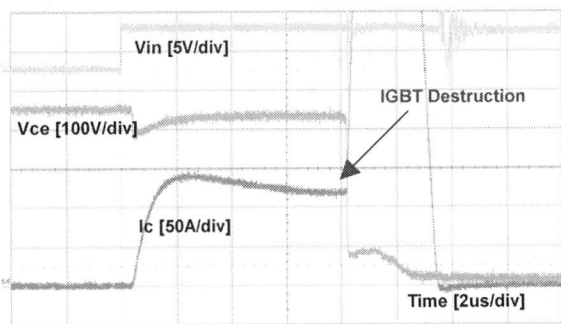

Figure 12. Destructed short circuit waveform of a 600V/10A rating device at V_{dc}=450V, V_{cc}=20V, and T_j=125℃

IV. APPLICATION CIRCUIT

A. OVERALL APPLICATION CIRCUIT

The circuit configuration for a typical application is shown in Figure 13. A single-supply 15V drives the low-side IGBTs directly and charges the bootstrap circuitry for the HVICs. The HVIC and LVIC block the command signals from the controller and generate a fault signal when the type of failure mode, i.e. the SC current failure or the supply under-voltage failure, is detected. To prevent surge destruction, the wiring between the smoothing capacitor and the P/GND pins should be as short as possible. The use of a high frequency non-inductive capacitor of around 0.1~0.22uF between the P/GND pins is recommended.

B. RECOMMENDED INPUT CONNECTION DESIGN

The integrated 5V CMOS/TTL compatible Schmitt trigger input conditioning circuit enables easy interface with a microprocessor. The input signal is High-Active Type. There is a 5kΩ resistor inside the IC to pull down each input signal line to GND. RC coupling circuits should be adopted for the prevention of input signal oscillation. The time constant of RC coupling circuits should be selected in the range 50~150ns. The recommended values of RC coupling circuits are 100Ω for the resistor and 1nF for the capacitor. V_{FO} output is open collector type. This signal line should be pulled up to the positive side of the 5V power supply with approximately 2kΩ resistor and pulled down to GND with 4.7nF capacitor for fault and temperature detection. Figure 14 shows the example of connection to microprocessor.

V. CONCLUSION

In this paper, an overall description to the Motion-SPM^{TM} module in SIP packaging is presented. This module offers tremendous advantages such as increased reliability, design flexibility, simple construction, easy assembly, and cost-effectiveness of the inverter power stage for home appliance applications. It uses six NPT 600V IGBTs, six freewheeling diodes, three HVICs, one LVIC and one thermistor.

978-1-4244-1871-8/07 $25.00 © 2007 IEEE 697

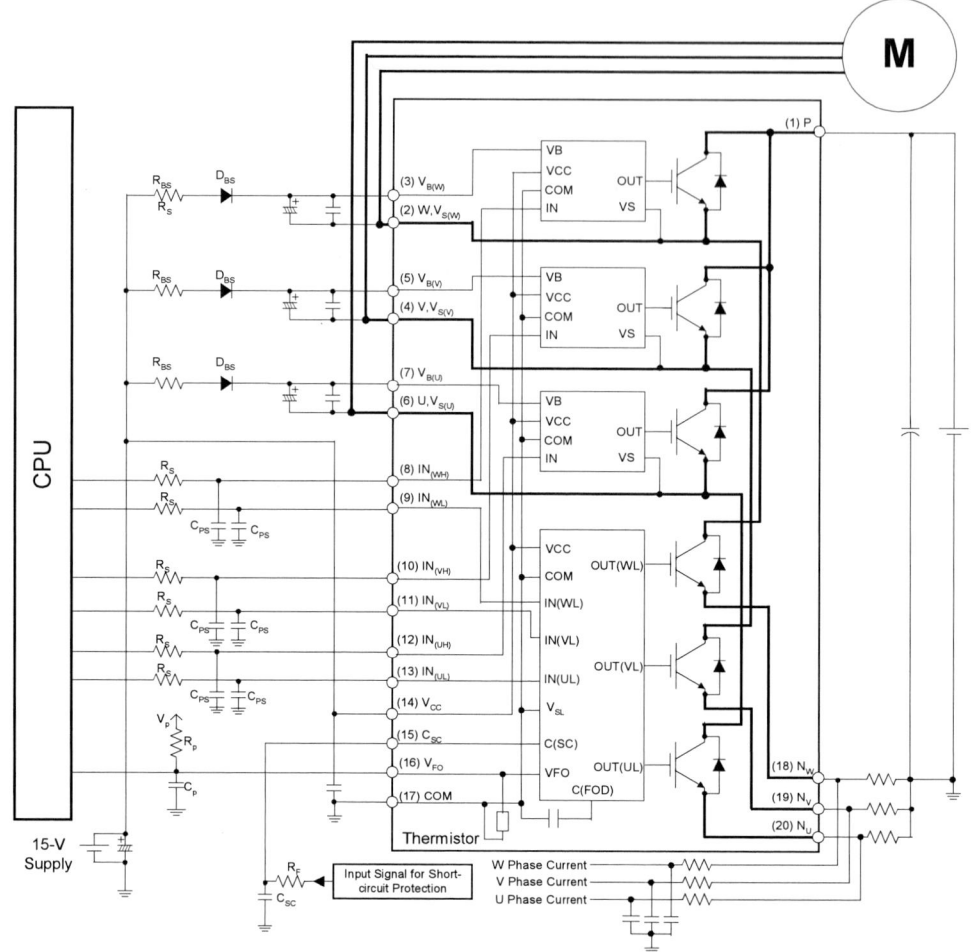

Figure 13. A typical application circuit for 3-N type of Motion-SPM™ in SIP

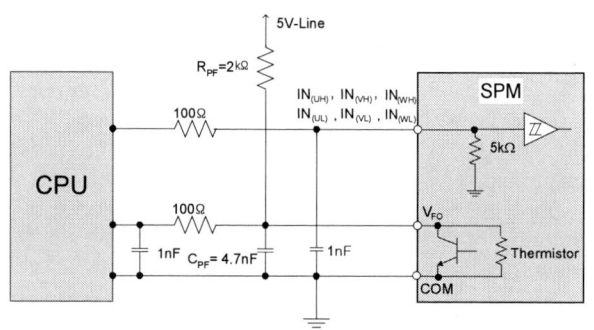

Figure 14. Example of connection to a microprocessor

For the user's design pliability, the Motion-SPM™ module in SIP packaging offers two types of negative DC terminals and lead forms. The electrical characteristics such as switching performance, single IGBT power loss, power rating and short circuit ruggedness is investigated. The Motion-SPM™ module in SIP packaging provides an excellent solution for appliance motor drives and other light industrial drive applications.

REFERENCES

[1] Smart power module motion SPM user's guide, application note AN9035, Fairchild Semiconductor

[2] JB.Lee, BC.Cho, DW.Chung, BS.Suh, "Improved Smart Power Modules for up to 7kW Motor Drive Applications", PESC 2006.

[3] F. Casanellas, "Losses in PWM inverters using IGBT's," Proc. Inst. Elect. Eng.-Elect. Power Applicant. , vol. 141, no. 5, pp. 235-239, Sept.1994.

[4] K. Berringer, J. Marvin and P. Perruchoud, "Semiconductor Power Losses in AC inverters," Conf. Rec. of IEEE IAS'95, pp. 882-888, 1995

978-1-4244-1871-8/07 $25.00 © 2007 IEEE

Cost-Effective Driving System for Plasma Displays by Replacing and Combining Functions of Power Switches

Dong-Myung Lee, and Dong-Seok Hyun
Hanyang University
Department of Electrical Engineering
17 Haengdang-dong, Seongdong-gu, Seoul, 133-791, Korea
Email: dm0503.lee@gmail.com, dshyun@hanyang.ac.kr

Abstract—**A plasma display panel (PDP) has become the most promising candidate in the market for large screen size flat panel display. PDPs have many merits such as a fast display response time and wide viewing angle. However, there are still concerns about high cost because they require complex driving circuits composed of high power switching devices to generate various voltage waveforms for three operational modes of reset, scan, and sustain. Conventional PDP driving circuits use path switches for voltage separation. In addition, there exist reset switches to initialize PDPs by regulating the wall charge conditions with ramp shaped pulses, which means the necessity of specific power devices for the reset operation. Because powers for the plasma discharge accompanied by large currents are transferred to a panel via path switches, high power rating switches are used for path switches. Therefore, this paper proposes a novel low-cost PDP driving scheme achieved by not only eliminating path switches but also merging the function of reset switches into other switches used for sustain or scan operations. The simulated voltage waveforms of the proposed topology and experimental results implemented in a 42-inch panel to demonstrate the validity of using a new gate driver that merges the functions of power switches, are presented.**

I. Introduction

Flat panel displays (FPDs) with a large screen size and high resolution, devices such as plasma display panels (PDPs) and LCDs have been widely researched and commercialized. PDPs have advantages of a fast display response time and wide viewing angle [1], [2]. However, PDPs are high costs due to complex driving circuits using high power devices. Therefore, there is high demand for developing a low cost driving circuit for PDPs. Fig. 1 shows the structure of an AC PDP having three electrodes of X, Y, and A (Address). A PDP can be modeled as a capacitor C_p and a current source representing panel discharge currents. The C_p represents the inherent capacitance due to the structure of the dielectric layer covered on the top of the sustain electrodes (X and Y). A PDP consists of many RGB cells corresponding to the horizontal and vertical resolution such as 1024 x 768. Red, green, and blue (RGB) phosphors are filled in each cell. The address electrodes are located under the RGB cells perpendicular to the X and Y electrodes. Cells in a PDP emit visible lights by the plasma discharge induced by a square-wise alternating electric field applied between X and Y electrodes. The plasma discharge ionizes the gas inside the cell, and then the excited gas generates ultraviolet lights that make the phosphors to create red, green, and blue visible lights. For charging or discharging C_p and supporting the plasma discharge, large amount of displacement and discharge currents are demanded. Hence, it is needed to use switches that can cope with high peak and RMS currents, which means high cost driving system. Many researches have been carried out to improve performance of sustain circuits and to reduce their cost [3]-[6], but this work has only focused on the sustainer. Even the sustain circuit is the main part of the driving system, the driving system needs other circuits to generate specific voltage waveforms for a gas discharge such as reset and scan. Therefore, this paper proposes a new low-cost PDP driving scheme that replaces so-called path switches with power diodes and eliminate power switches for the reset operation by merging the reset function into other switches such as sustain or scan switches.

II. Conventional Driving Method

A. Driving System

Fig. 2 shows the whole schematic diagram of a conventional driving system consisting of X, Y, scan, and address boards. Address boards are not being shown for simplification. Fig. 3 shows voltage waveforms of X, Y, and A electrodes corresponding to the voltage waveforms for the address display period separation (ADS) driving scheme that is a commonly used driving method in commercialized plasma TVs. The operational periods of PDPs in 1 TV subfield can be divided into reset, scan, and sustain period. The ramp reset is widely adopted, which initializes cells to have the same wall charge conditions by a weak gas regardless of the previous on/off state of cells.

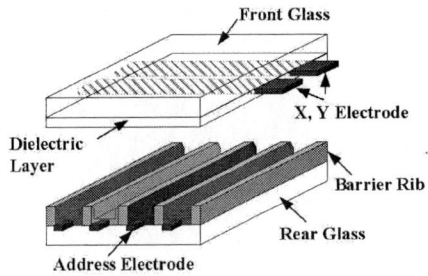

Fig. 1 The structure of three-electrode AC PDP

978-1-4244-1871-8/07 $25.00 © 2007 IEEE

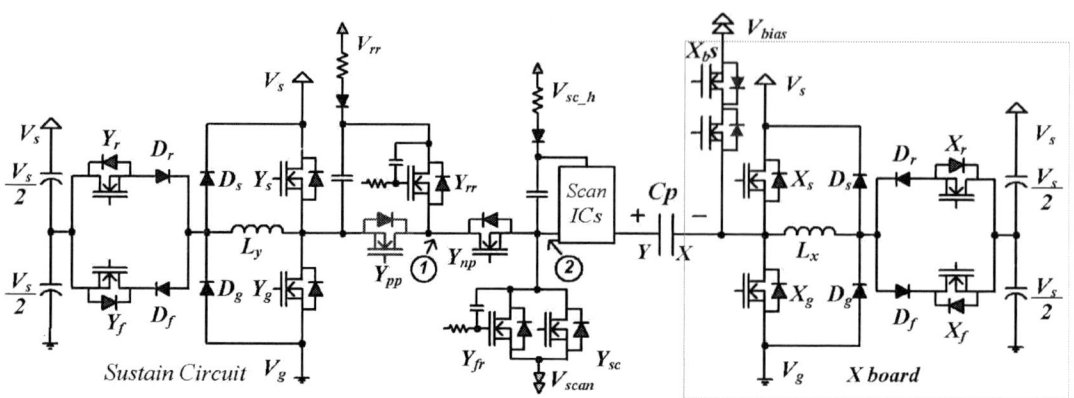

Fig. 2 Overall diagram of conventional PDP driving system

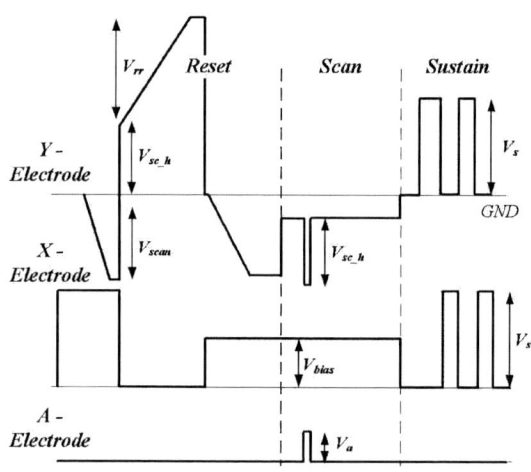

Fig. 3 Voltage waveforms of Y, X, and A electrodes for the ADS driving scheme

In the address period, scanning pulses are sequentially applied to the Y electrode, and data pulses are given to the cells to be addressed. During the display period, the addressed cells emit lights by sustain pulses imposed between X and Y electrodes in an alternating manner. The X board in Fig. 2 shown inside the square consists of sustain and bias circuits. The bias circuit applies the voltage level of V_{bias} to the X electrode for controlling wall charge conditions during the reset and scan periods. Usually, sustain circuits have a half bridge structure with energy recovery circuits (ERC) of Weber type. The Weber type ERC, which is based on an LC series resonance between an inductor and the panel capacitor, is typically used for the ERC in most commercialized PDP TVs. Under ideal condition, the resonance between the inductors and the C_p raises and decreases the panel voltage from zero to V_s level in a half resonant period, and vice versa. The series resonance is terminated after a half resonant period because reverse resonant currents are blocked by the diode D_r and D_f during sustain rising and falling, respectively.

B. Problems of Conventional Driving System

In the conventional scheme illustrated in Fig. 2, there exist Y_{np} (negative-path) and Y_{pp} (positive-path) commonly called as path switches, which are used for voltage separation during reset and scan periods. Some commercialized PDP TVs employ only one path switch, which is usually Y_{np}. By changing the reset voltage levels of the X and Y electrodes, it is possible to use only Y_{np} for voltage separation. Y_{np} isolates the circuits connected in the point 1 from that of point 2 during the ramp-down and scan periods. Unless Y_{np} is turned off during the period when the V_y goes below zero potential such as ramp down and scan periods, unexpected currents can flow through the diode inside Y_g to the Y_{fr} (ramp falling) or to the Y_{sc} (scan), respectively.

Among three operational periods of PDPs, most powers are consumed in the sustain period because of large displacement and discharge currents for generating and supporting the plasma discharge, respectively. The sustain switches as well as path switches are involved in transferring the power for plasma discharge during sustain period. Fig. 2 shows that displacement current flows via the path of $V_s/2 \rightarrow Y_r \rightarrow D_r \rightarrow L_y \rightarrow (Y_{pp} \rightarrow) Y_{np} \rightarrow C_p \rightarrow X$ board and that of the discharge current is $V_s \rightarrow Y_s \rightarrow (Y_{pp} \rightarrow) Y_{np} \rightarrow C_p \rightarrow X$ board. Since displacement and discharge currents are always flowing through the path switches, several switches are connected in parallel for path switches to cope with the large current and power dissipation occurred in the switches.

Besides path switches, conventional systems have the reset switches, Y_{rr} (ramp-rising) and Y_{fr} (falling-ramp), to initialize PDPs with the ramp shaped pulses. The reset switches operates in an active region by using the external high resistor and external capacitor connected in the switches, which make reset switches be turned on with slow switching transition. In summary, the conventional system needs specific power devices for generating reset voltage waveforms and path switches for voltage isolation. Therefore, it is apparent that eliminating path switches and reset switches can considerably lower the price of PDP driving systems.

978-1-4244-1871-8/07 $25.00 © 2007 IEEE   700

C. Proposed Scheme

Topology removing both Y_{np} and Y_{pp} has been reported [7] and it has devices replacing positive-path switches. On the other hand, this research focuses on the Y_{pp}–less scheme that aims more cost reduction by removing Y_{pp} and modifies the reset voltage waveforms to improve the addressing ability for PDPs. Fig. 4 shows the proposed driving systems that do not have path switches, which are based on Y_{pp}–less scheme. In the proposed system, the diode D_{Yg} replaces the Y_{np} that isolates the voltage of point 2 from that of point 1 to inhibit the short current from ground potential V_g to the negative voltage V_{scan} during ramp-down and scan period, respectively. The configuration of the X board of the proposed system is the same as that shown in Fig. 1. Furthermore, switches are connected in series to withstand voltages across them together, which makes it possible to use lower V-rating ones. The switches involved in reset and scan operations have to withstand the maximum voltage differences of the highest and lowest ones in the system, which is normally over 400 V for commercialized plasma TVs. Therefore, devices over 500 V are usually used for the switches in consideration of voltage safety margins. On the other hand, it can be shown from Fig. 4 of the proposed scheme that the maximum voltages across the switches of Y_r, Y_f, Y_s and Y_g are V_s level. Consequently, it is clear that the proposed scheme has advantages of using fewer power devices with lower voltage rating. For greater cost reduction, the scheme of Fig. 4 merges the function of reset switches into sustain and scan switches by the help of a new gate driver. Fig. 5(a) shows conventional gate drivers for Y_s and Y_{rr} switches in which LS/OC represents a level shifter or an opto-coupler to provide an electrical isolation between the gate signal from a control board and power side having high voltage. In order to do switching Y_s (sustain switch) with a relatively high switching speed over 200 kHz during the sustain period, the gate on/off currents for Y_s are supplied by Q_3 and Q_4 consisting the push-pull current amplifier.

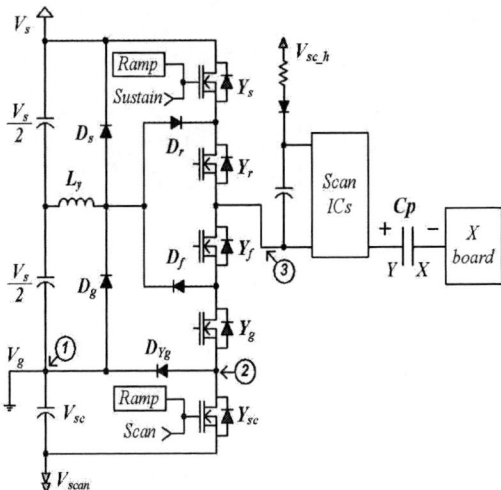

Fig. 4 Proposed low-cost systems without using Y_{rr} and Y_{fr}

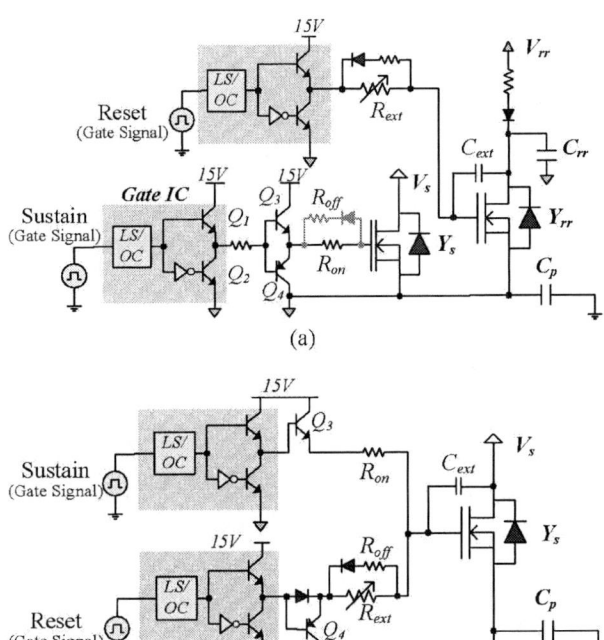

Fig. 5 Gate driver for sustain and scan in (a) conventional schemes and (b) a novel gate driver working for both reset and sustain operations

Since, the ramp-rising switch Y_{rr} does work only during the reset period and requires small gate current to operate Y_{rr} in an active region, the gate signal form the gate IC is delivered to Y_{rr} without the current amplification. Fig. 5(b) shows the novel gate driver that can merge the function of reset switches into other switches. In this case, Y_s generates both of sustain and ramp-rising pulses. It consists of two parts; a gate for sustain operation with a low gate resistor and that for reset operation with a high external resistor.

Table 1 summarizes four possible modes corresponding to on/off signals from each gate IC. In reality, there are three modes because the sustain off and reset off have the same role of turning of Y_s with the fast off process. The slow turn-on process of Y_s for ramp reset is done with the external resistor R_{ext} having high resistance. While high on/off currents for fast switching during the sustain period are supplied via the R_{on} and R_{off} having low resistance. Fig. 6 shows the timing schedule for the proposed driving scheme of Fig. 4 that has no path switch and merges the roles of power switches by using the new gate driver. The operational modes of switching cycle are described below, and current paths corresponding to the sustain period are shown in Fig. 7.

TABLE 1
GATE SIGNALS FOR FOUR POSSIBLE MODES

	Reset signal	Sustain signal
Reset ON	ON	OFF
Reset OFF	OFF	OFF
Sustain ON	ON	ON
Sustain OFF	OFF	OFF

978-1-4244-1871-8/07 $25.00 © 2007 IEEE   701

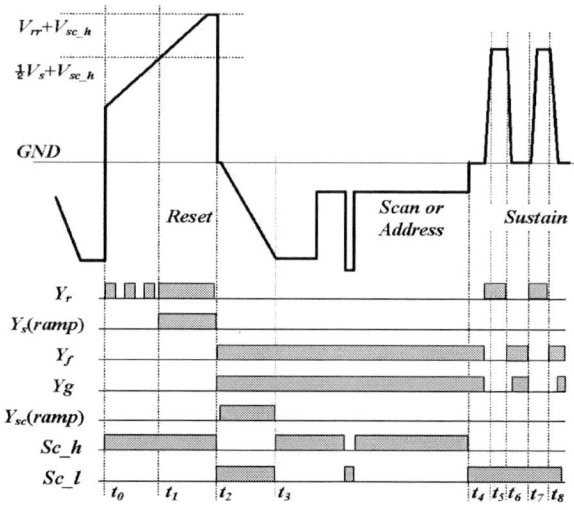

Fig. 6 Timing chart for the proposed system of Fig. 4

Reset Period (Ramp-Rising):

During $t_0 - t_1$, relatively short gate pulses are applied to Y_r until the voltage at point 3 of Fig. 4 reaches the level of $V_s/2$. Because the high side transistors of the scan ICs are turned on during this period to add the V_{sc_h} on the Y electrode voltage V_y, the level of $V_s/2$ at point 3 is equivalent to V_y of $V_s/2 + V_{sc\_h}$. Turning on the Y_r makes the C_p be charged with a current flowing through the path of $V_s/2 \rightarrow L_y \rightarrow D_r \rightarrow Y_r \rightarrow Scan\ IC \rightarrow C_p$. Neglecting parasitic circuit components and voltage drops in the circuit, the inductor current i_{Ly} during on time, t_{on}, and the voltage increment of the panel Δv_{Cp} during t_{on} can be expressed as (1)

$$
\begin{aligned}
i_{Ly}(t) &= \sqrt{C_p/L_y} \cdot \left[V_s/2 - v_{Cp}(0) \right] \cdot \sin \omega t \\
\Delta v_{Cp}(t) &= \left[V_s/2 - v_{Cp}(0) \right] \cdot (1 - \cos \omega t_{on})
\end{aligned}
\tag{1}
$$

where, $\omega = 1/\sqrt{L_y C_p}$ and $0 < t < t_{on}$. The $v_{Cp}(0)$ represents the potential V_y referred at point 3 before turning on the Y_r. Equation (1) shows that the Δv_{Cp} is varied by the t_{on} of Y_r. After V_y is increased up to $V_s/2 + V_{sc\_h}$ level, Y_s that can be operated either the sustain or reset is turned on by the reset gate signal shown as $Y_s(ramp)$ in Fig. 6 to generate the ramp-up voltage waveform.

Reset Period (Ramp-Falling):

At $t = t_2$, V_y rapidly goes to zero voltage level by turning on Y_f and Y_g. After then Y_{sc} is turned on by the reset gate signal shown as Y_{sc} (ramp).

Scan Period:

The scan ICs impose line-by-line sequential scan pulses to each Y electrode, and corresponding data are applied in data electrodes by data buffers. Turning on the switches Y_f, Y_g, and Y_{sc} is required to maintain the scan voltage V_{scan} that has negative value.

Sustain Period:

During this period, the addressed cells are lighted by sustain pulses. The Y voltage rises toward V_s with turning on the Y_r. The switch Y_s is turned on at $t = t_5$. X and Y voltages are alternating from mode 1.

Mode 1 ($t_6 - t_7$): Turning on Y_f and the switch X_r, rising switch in X board, makes the v_{cp} change from V_s to $-V_s$ through the path of $X\ board \rightarrow C_p \rightarrow Y_f \rightarrow D_f \rightarrow L_y \rightarrow V_s/2$ as shown in Fig. 7(a) with the initial condition of $v_{cp} = V_s$. Two inductors in X and Y boards are involved in the LC resonance. The panel voltage v_{cp} and the current i_{cp} during the sustain-falling period can be obtained as (2)

$$
\begin{aligned}
v_{cp}(t) &= V_s \cdot \cos \omega(t - t_6) \\
i_{cp}(t) &= -\frac{V_s}{Z} \cdot \sin \omega(t - t_6)
\end{aligned}
\tag{2}
$$

where, $Z = \sqrt{L/C_p}$, $\omega = 1/\sqrt{LC_p}$, and $L = L_x + L_y$.

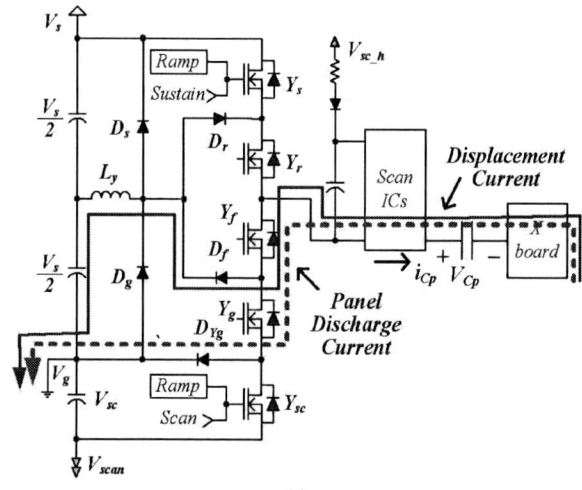

(a)

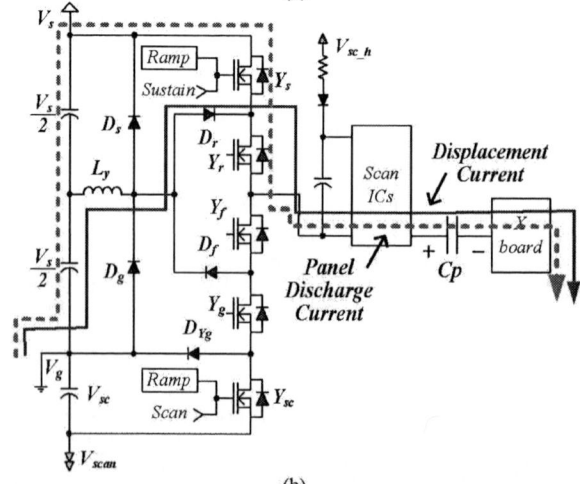

(b)

Fig. 7 The path of panel currents during period of (a) sustain –falling and (b) sustain-rising

After v_{cp} becomes $-V_s$, Y_g and X_s are turned on, and the power for plasma discharge is supplied via $X\ board \rightarrow C_p \rightarrow Y_f \rightarrow Y_g \rightarrow D_{Yg} \rightarrow GND$ as shown as the dotted line in Fig. 7(a).

Mode 2 ($t_7 - t_8$): Turning on the switches Y_r and X_f, falling switch in X board, makes the v_{cp} change from $-V_s$ to V_s through the path of $V_s/2 \rightarrow L_y \rightarrow D_r \rightarrow Y_r \rightarrow C_p \rightarrow X\ board$ with the initial condition of $v_{cp} = -V_s$. The v_{cp} and i_{cp} for the rising sustain pulse can be expressed as (3)

$$v_{cp}(t) = -V_s \cdot \cos \omega (t - t_7)$$
$$i_{cp}(t) = \frac{V_s}{Z} \cdot \sin \omega (t - t_7). \tag{3}$$

After the v_{cp} becomes V_s, the switches Y_s and X_g are turned on to support the gas discharge illustrated as the dotted line in Fig. 7(b), while Y_r remains on status. At $t = t_8$, switching operations are repeated from mode 1.

III. VERIFICATIONS

A. Simulation Results

Fig. 8 shows the Y voltage waveform for 1 TV subfield that has a main reset simulated by Pspice® with parameters L = 250 nH, C_p = 100 nF, and switching frequency = 200 kHz. It can be clearly shown that the generated Y voltage of the proposed system is identical to that of conventional driving circuit. As shown in the timing chart of Fig. 6, Y_r has been turned on/off repeatedly until $t = t_1$ when V_y rises up to $V_s/2 + V_{sc\_h}$ voltage level, and then Y_s is turned on by the ramp gate signal.

B. Experimental Results using the New Gate Driver

Experiments have been carried out using a 42-inch PDP panel to demonstrate the validity of the new gate driver. The new driver circuit for Y_s and Y_{rr} has been implemented in the PDP driving circuit having conventional topology of Fig. 3. In other words, the gate driver shown in Fig. 5(a) has been replaced by the novel driver presented in Fig. 5(b). Fig. 9 shows the ramp rising V_y and gate-source voltage waveform of Y_s using the novel gate driver.

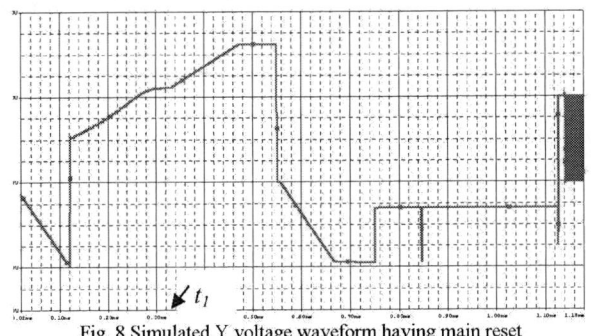

Fig. 8 Simulated Y voltage waveform having main reset
(200V/div., 100μs/div.)

Fig. 9 Experimental V_y voltage waveforms (100V/div., 100μs/div.) and gate-source voltage of Y_s (10V/div.) during the ramp-rising reset period by imposing a continuous turn-on gate signal at $t = t_1$

Two levels of gate-source voltage of Y_s appear in Fig. 9; one having the level of around threshold voltage during $t_0 - t_1$ and another having around 15 V after $t = t_1$, which is the voltage level of gate power supply. Consequently, Fig. 9 implies that Y_s has been turned on and off in two different ways by using the novel gate driver.

There is a demand that the reset pulses of V_y can have several different slopes corresponding to the ambient temperature to effectively regulate the wall charge conditions under various operating environments for the plasma TVs. It is possible to generate the ramp-rising V_y voltage waveforms with either a continuous turn-on signal or short gate pulses. Unless using additional circuits that can change circuit parameters especially the value of an external resistor and a capacitor, it is hard to generate V_y having different slopes. Therefore, to realize this function with the same circuit, the proposed scheme varies the slope of reset waveforms by changing the switching frequency or the on-time of Y_s in stead of adding additional circuits. Therefore, gate-source voltages waveforms of Y_s during $t_0 - t_1$ are shown as short gate pulses. At $t = t_1$, the gate signals from the sustain gate IC, which has around 15V gate output, immediately turns on Y_s by a continuous on signal as shown in Fig. 9. In fact, during the ramp-up period, it is not necessary to turn on Y_s with the fast on-process. Therefore, it should be noted that the experiments illustrated in Fig. 9 have been performed to verify the validity of the new gate driver by showing that the gate-source voltage of Y_s shows two voltage levels required for slow or fast switching action. It means that two operations of reset and sustain have been realized by one Y_s switch by using the new gate driver.

IV. CONCLUSION

This paper has proposed a PDP driving circuit that eliminates path switches and a new gate driver that can merge the function of reset switches to sustain or scan switches. Only five switches with a voltage rating lower than Vs have been used to generate the voltage waveforms for a gas discharge, and this scheme has been devised to realize low-cost driving circuits by considering not just sustain circuits but the entire PDP system. The simulated Y voltage waveform of the proposed scheme has been in accordance with that of conventional system. The experimental results for the new gate driver implemented in a 42-inch panel have shown that the gate drive can turn on the switch with slow or fast switching actions.

REFERENCES

[1] G. Sethi, "Plasma panels enter commercial arena," *Laser Focus World*, pp. 253–258, May 1998.

[2] G. Troussel, et al., "A new driver circuit for plasma display panels," *Proc. Soc. for Information Display*, San Diego, USA, 1996, pp. 1–5.

[3] C. W. Roh, "Novel plasma display driver with low voltage/current device stress," *IEEE Trans. on Consumer Electronics*, Vol. 49, No. 4, pp. 1360–1366, Nov. 2003.

[4] T. F. Wu, C. C. Chen; C. C. Chen, and W. F. Hsu, "Design and development of driving waveforms for ac plasma display panels," *IEEE Trans. on Plasma Science*, Vol. 31, No. 2, pp. 272–280, April 2003.

[5] H. van der Broeck, and M. Wendt, "Alternative sustain driver concepts for plasma display panels," *Power Electronics Specialist Conf.*, Aachen, Germany, 2004, pp. 2672–2677.

[6] S. S. Kwak, D. M. Lee, and J. W. Jung, "Novel driver structure based on low-cost and low-rating devices for plasma televisions," *International Conf. on Consumer Electronics*, Las Vegas, USA, 2007, pp. 1–2.

[7] D. M. Lee, S. S. Kwak, and J. W. Jung, "Cost-effective driving system for ac plasma display panels without employing voltage separation switches," *Electronics Letters*, Vol. 42, Iss. 21, pp.1211–1212, Oct. 2006.

A Cold Cathode Fluorescent Lamp Driving Circuit without a Transformer for Liquid Crystal Display Backlight Unit

Eun-Seok Choi, Je-Hyung Cho, Hyun-Ki Yoon, Gun-Woo Moon, and Myung-Joong Youn
Department of Electrical Engineering and Computer Science, KAIST,
373-1 Guseong-Dong, Yuseong-Gu, Daejeon, Korea
Email: latimee@powerlab.kaist.ac.kr

Abstract — A cold cathode fluorescent lamp driving circuit without a transformer for liquid crystal display backlight unit is proposed. Commonly, a CCFL driver has a transformer with high turn ratio to step up low input voltage to high sinusoidal lamp voltage. This high voltage and high turn ratio transformer has large size and high cost. The proposed CCFL driving circuit which is based on parallel-loaded resonant topology features simplicity, so it reduces complexity and cost of overall system. And the proposed circuit with asymmetric control has dimming capability by duty control. In this paper, the analyses of the proposed CCFL driving circuit and experimental results for BLU of 32-inch commercial LCD TV are presented.

I. INTRODUCTION

Thin film transistor liquid crystal displays (TFT-LCD) are widely used for display devices in many applications such as notebooks, laptops, cell phones, portable multimedia players, and digital televisions. Since LCD is not a self light-emissive type display, it needs backlight unit (BLU) to emit light from beneath liquid crystal layer. Commonly, LCD with cold cathode fluorescent lamp satisfies the requirements on display performance, size, and efficiency. For operation of a cold cathode fluorescent lamp, a driving circuit provides a high starting voltage of about 1,400 V_{rms} to ionize the gas discharge, and a high running voltage of about 1,000 V_{rms}, in case of 32-inch LCD TV. To maximize the lamp efficiency, a sinusoidal lamp voltage waveform operating at 60 kHz is preferable. Therefore, the resonant inverters are used to drive cold cathode fluorescent lamps and several prior circuits have been researched for LCD backlight unit. [1-5]

Since these inverters transfer low dc input voltage to high ac voltage waveform, they have high turn ratio transformers. The high voltage and high turn ratio transformer have complexity, because it is divided to a high voltage section and a low voltage section, and especially, a high voltage section is divided again to several partitions to guarantee the insulation. This constraint enforces larger size of a transformer. Furthermore, since the primary current of the transformer is high due to a high turn-ratio, core size should be larger. Larger transformers increase size, weight, and cost of the backlight unit.

For larger size screen, longer lamps are necessary in backlight unit. Longer lamps means higher starting and running lamp voltage, and also more stray capacitances. As lamps are placed side by side under liquid crystal layer, stray

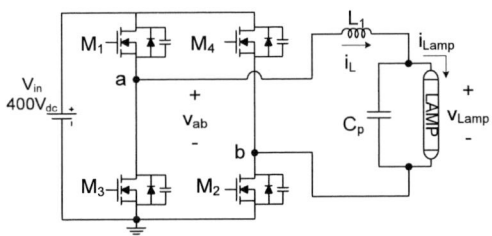

Fig. 1: Conventional CCFL Driving Circuit

capacitances exist between back plain (ground) and lamps or lamp wires, and then leakage currents flow through these stray capacitances. The drawback of transformer in prior circuits and stray capacitance problem will be more serious as the display screen size grows larger.

Dimming capability is also one of the major characteristics of lamp driving circuit. [4] For dimming control of the lamp brightness, there are two methods, one is analog dimming and other is burst dimming. Analog dimming is regulating lamp current by controlling switching frequency or duty. In burst dimming, lamp driving circuit turns on and off at 100~200 Hz.

In this paper, a cold cathode fluorescent lamp driving circuit without a transformer is proposed. The proposed circuit has input voltage of about 400Vdc which is regulated from the grid. In this proposed resonant inverter, dc voltage is transferred to a sinusoidal waveform directly through a LC parallel resonant circuit. The resonant components are composed of two resonant inductances, lamp impedance, and additional resonant capacitances. In the paper, the basic operation principle of the circuit will be given in Section II. Based on the analysis of the proposed circuit, the design considerations will be also given. In Section III, a practical design example and experimental results will also be presented.

II. PROPOSED CIRCUIT

A. Features of proposed circuit

Fig.2 shows the schematic and typical waveforms of the proposed CCFL driving circuit without a transformer. The proposed circuit provides lamp voltage using parallel LC resonance between resonant inductances and capacitances,

978-1-4244-1871-8/07 $25.00 © 2007 IEEE

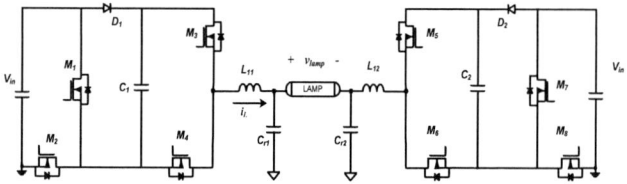

(a) Schematic of Proposed Circuit

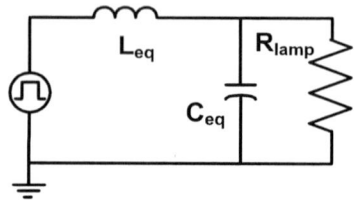

Fig. 3 : Equivalent Circuit of Proposed Circuit

B. Mode Analysis

The operation principle is discussed in this section and there are some following assumptions:

1. All the switches are ideal, but they have output capacitances.
2. The diodes are ideal and no voltage drop exists.
3. The switching frequency is higher than the resonant frequency of the resonant tank.

The mode operations are explained as follows:

Mode 1 [t_0-t_1] : Switches, M_1, M_3, M_6, and M_8 turn on and $2V_{in}$ is supplied to resonant tank. And inductor current increases, flowing through M_1, C_1, M_1, L_{r1}, lamp, L_{r2}, M_6, and M8.

Mode 2 [t_1-t_2] : M_1 and M_3 becomes turn-off, then voltage between resonant tank is zero. The inductor current flows discharging fully output capacitances of M2 and M_4 until these switches turn on. It can make zero voltage switching of M_2 and M_4.

Mode 3 [t_2-t_3] : M_6 and M_8 turn off and voltage, -$2V_{in}$ is supplied to resonant tank, then inductor current decreases. This current can discharge output capacitances of M_5 and M_7, making zero voltage switching of these switches.

Mode 4 [t_3-t_4] : M_5 and M_7 turn off and voltage supplied to resonant tank is zero. M_6 and M_8 turn on at zero voltage.

Mode 5 [t_4-t_0] : M_2 and M_4 becomes turn-off, and $2V_{in}$ is supplied to resonant tank. M_1 and M_3 turn on at zero voltage .

The proposed circuit has higher switching frequency than resonant frequency of resonant tank, so zero voltage switching of all the switches is possible. With the asymmetric control, voltage applied to resonant tank has three levels, $2V_{in}$, 0, and -$2V_{in}$.

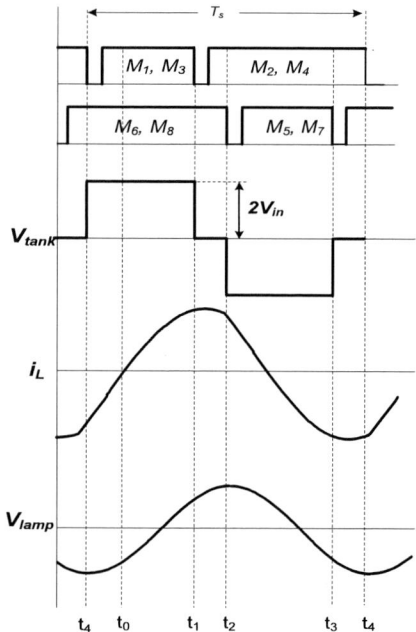

(b) Typical Waveforms of Proposed Circuit

Fig. 2 : Proposed CCFL Driving Circuit

instead of parasitic inductances of a transformer. The size of resonant inductor used in the proposed circuit is smaller than that of a transformer of prior circuits, there are following reasons. The first is no low voltage partition. And it can have smaller core, because resonant inductor current is lower than primary current in the transformer. Since resonant capacitors at both of ends of a lamp have much larger capacitances than stray capacitances, most of stray capacitances existing between wire and ground can be ignored. So the proposed circuit has no serious leakage current problem due to stray capacitances. With asymmetric control, the proposed circuit can regulate the lamp brightness by duty control.

The proposed CCFL driving circuit is based on the parallel resonant inverter, which is preferred to step up voltage. Since it has multi-level circuit, the voltage applied to the resonant LC circuit is twice as high as dc input voltage. This means the proposed circuit need not have too high quality factor for high step up input-output ratio. Too high quality factor of resonant inverter is not recommended, because it can increase the instability of the circuit.

C. Equivalent Resonant Circuit and Characteristics of Proposed Circuit

To analyze the proposed circuit, the equivalent circuit is developed. This equivalent circuit is simplified as shown in Fig.3. A square wave voltage is supplied to resonant tank. The equivalent resonant inductance and capacitances can be expressed as:

$$L_{eq} = L_{r1} + L_{r2}, \quad C_{eq} = C_{r1} + C_{r2} \qquad (1)$$

R_{lamp} is the equivalent resistance of CCFL. This equivalent circuit makes it easy to analyze and design the proposed circuit. The circuit is described by following parameters:

- The corner frequency, ω_0

$$\omega_o = 1 \Big/ \sqrt{L_{eq} C_{eq}}$$

- The characteristic impedance, Z_o

$$Z_o = \omega_o L_{eq} = \sqrt{L_{eq}/C_{eq}}$$

- The loaded quality factor, Q

$$Q = R_{lamp} \Big/ \omega L_{eq} = \omega C_{eq} R_{lamp}$$

The transfer function in frequency domain and the input-output voltage relationship is following:

$$\frac{V_o(j\omega)}{V_{in}(j\omega)} = \frac{1}{1 - \left(\dfrac{\omega}{\omega_o}\right)^2 + j\dfrac{1}{Q}\left(\dfrac{\omega}{\omega_o}\right)} \quad (2)$$

$$\frac{V_o}{V_{in}} = \frac{1}{\sqrt{\left\{1 - \left(\dfrac{\omega}{\omega_o}\right)^2\right\}^2 + \dfrac{1}{Q^2}\left(\dfrac{\omega}{\omega_o}\right)^2}} \quad (3)$$

From the above equation (3), the relationship between dc input voltage (in dc) and output voltage (lamp voltage in *rms*) can be calculated as:

$$\frac{V_{lamp(rms)}}{V_{in(dc)}} = \frac{4\sqrt{2}}{\pi}\sin(D\pi)\cdot\frac{1}{\sqrt{\left\{1 - \left(\dfrac{\omega}{\omega_o}\right)^2\right\}^2 + \dfrac{1}{Q^2}\left(\dfrac{\omega}{\omega_o}\right)^2}} \quad (4)$$

In case of symmetric control, the input-output voltage relationship is as follows:

$$\frac{V_{lamp(rms)}}{V_{in(dc)}} = \frac{4\sqrt{2}}{\pi}\cdot\frac{1}{\sqrt{\left\{1 - \left(\dfrac{\omega}{\omega_o}\right)^2\right\}^2 + \dfrac{1}{Q^2}\left(\dfrac{\omega}{\omega_o}\right)^2}} \quad (5)$$

Comparing the equation (4) with equation (5), in the circuit with symmetric control, output voltage can be controlled only by frequency. But, in case of asymmetric control, both of duty control and frequency control are enable. It means the brightness of lamps can be controlled by duty at fixed frequency in the proposed circuit. Fig. 4 (a) shows the relationship between input-output voltage ratio and quality factor, Q, and Fig. 4 (b) shows the relationship between input-output voltage ratio and duty. In Fig. 4 (a), too high quality factor makes the switching frequency range to be narrow, causing instable operations. And duty control is

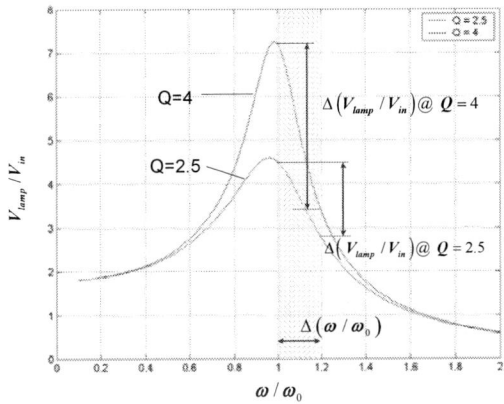

(a) Input-Output Voltage Ratio Change according to Q

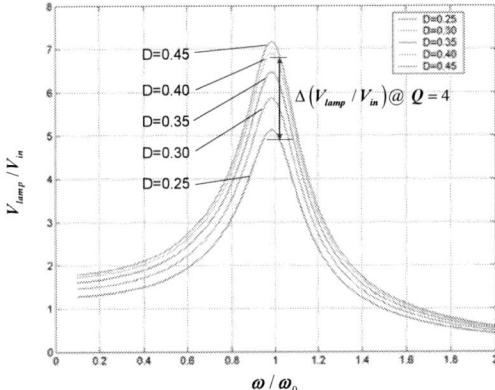

(b) Input-Output Voltage Ratio Change according to Duty

Fig. 4 : Control Factors of Proposed Circuit

more moderate than changing switching frequency, to regulate lamp brightness (current).

III. EXPERIMENTAL RESULTS

A. Design Example

Based on the above analysis, there is a design example of the proposed circuit. The lamp is a 750mm long CCFL with which is used in 32-inch LCD TV backlight unit. The detailed specifications of a lamp are follows :

- DC input voltage : 400 V_{dc}
- Switching frequency : 60 kHz
- Starting voltage : 1,400 V_{rms}
- Running voltage : 1,130 V_{rms} (at $I_{lamp} = 6.5$ A_{rms})
- Equivalent resistance of the lamp : $R_{lamp} = 160$ kΩ
- Quality factor : Q = 2.5

From the above specifications, the resonant inductances and capacitances can be calculated.

$$L_{r1} = L_{r2} = 98.5 \ mH$$
$$C_{r1} = C_{r2} = 96 \ pF$$

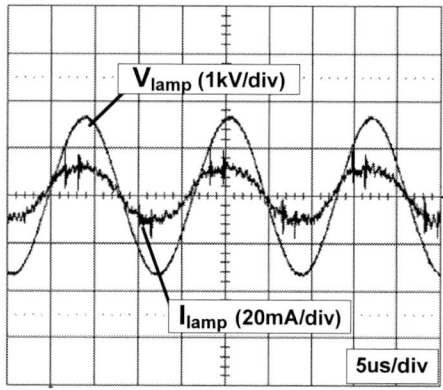

(a) Lamp Voltage and Lamp Current

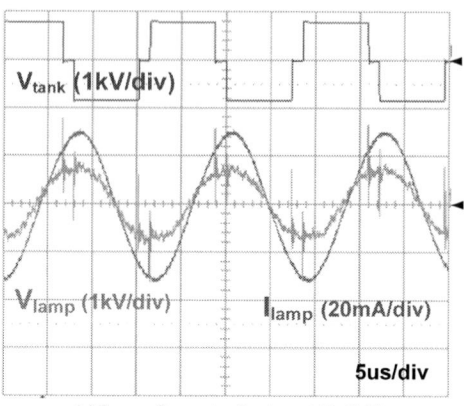

(a) Lamp Current = 9mA$_{rms}$ at D=0.42

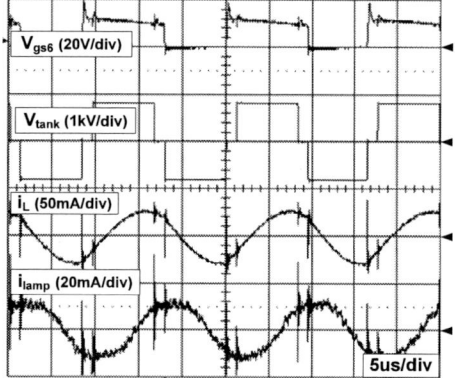

(b) Gate Signal, Resonant Voltage, Inductor Current, and Lamp Current

Fig. 5 : Experimental Key Waveforms

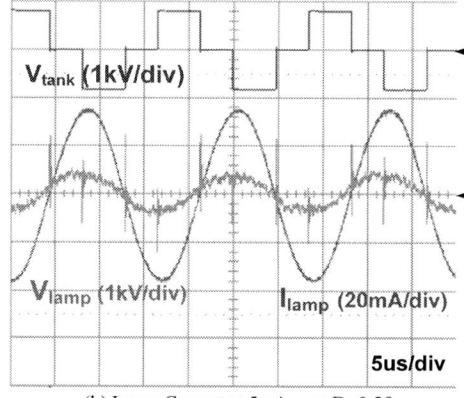

(b) Lamp Current = 5mA$_{rms}$ at D=0.28

Fig. 6 : Dimming by Duty Control

B. Experimental Results

The design example shown above was implemented and the experimental results are displayed in Fig.5. Fig.5 (a) shows lamp voltage and lamp current, and Fig.5 (b) shows resonant inductor current, resonant voltage, lamp voltage and gate signal. The brightness is 13300 cd/m^2, when 16 lamps are running for 32-inch TV, while conventional driving circuit has 13000 cd/m^2. In Fig. 6, the changes of lamp current according to duty control are shown. It means the proposed has dimming capability by duty control.

IV.   CONCLUSIONS

A cold cathode fluorescent lamp driving circuit without a transformer for LCD backlight unit is proposed in this paper. The proposed circuit has no transformer to step up the input voltage to high lamp voltage. It features simplicity and smaller size. It is based on parallel resonant inverter and multi-level circuit which are preferable for step up voltage. The resonant capacitor at ends of a lamp can reduce stray capacitance effects. And, with asymmetric control, the proposed circuit has dimming capability by duty control.

In the paper, the operation principle was analyzed and design example was presented. Experimental results were given to verify the analysis.

ACKNOWLEDGMENT

This research is supported by KAIST display research center program with Samsung Electronics.

REFERENCES

[1] D.Tadesse, et a1, "A comparison of Power Circuit Topologies and Control Techniques for a High Frequency Ballast", IEEE Transactions On Power Electronics,1993, pp.2341-2347

[2] R. L. Steigerwald, "A comparison of half-bridge resonant converter topologies," IEEE Trans. Ind. Electron., vol. 35, April. 1998, pp. 174-182.

[3] J. A. Donahue, P.E. and M. M. Jovanovic, "The LCC Inverter as a cold cathode fluorescent lamp driver", IEEE, 1994, p. 427-423

[4] G. Hsieh, C. H. Lin, C. H. Lin, H. I Hsieh, "Primary-Side Charge-Pump Dimming Controller for the Cold-Cathode Fluorescent Lamp Ballast", IEEE, 2001, p.717-723

[5] W. Chen, "Series-Parallel Resonant forward Inverter as a cold cathode fluorescent lamp driver", IEEE, 2002, p.943-947

978-1-4244-1871-8/07 $25.00 © 2007 IEEE

A Novel Adaptive Dimming Technique with X-Y Channels for LED Backlight System of LCD TVs

Won-Sik Oh[†], Kyu-Min Cho[†], Daeyoun Cho[†], Gun-Woo Moon[†*], Byungchoon Yang[‡], and Taeseok Jang[‡]

[†]Display Research Center, EE, KAIST,
373-1, Kuseong-dong, Yuseong-gu, Daejeon, 305-701, Korea
[‡]LCD Development Center, Samsung Electronics Co. LTD.,
200, Myeongam-Ri, Asan-City, 336-841, Korea
[*]Email: gwmoon@ee.kaist.ac.kr

Abstract— A novel adaptive dimming technique with X-Y channels for light emitting diode (LED) backlight system of LCD TVs is proposed. It has matrix structured LED modules with row and column switches to control the brightness of individual division block. It shows local dimming effects such as reduced power consumption and high dynamic contrast ratio with even using much fewer number of LED drivers of proposed method than that of conventional local dimming method. Therefore, low cost and compact design of LED drivers can be achieved. This paper also contains a new adaptive dimming algorithm and image compensation technique for proposed system.

I. INTRODUCTION

As modern society has been reshaped rapidly into the multimedia information society, large size of display devices is required. Thin-film-transistor liquid-crystal-displays (TFT-LCDs) are one of the most popular display devices from small to large size. TFT-LCDs need backlight source, because they are not a self luminance device. Backlight for LCDs is becoming more important with growth of the LCD size. Up to now, multiple fluorescent lamps such as cold cathode fluorescent lamp (CCFL), external electrode fluorescent lamp (EEFL) and flat fluorescent lamp (FFL) are generally used as LCD backlight source. Since, the conventional backlight of LCDs illuminates at the full luminance regardless of the images to be displayed, it wasted power and contrast ratio is row due to the light leakage in dark state. Moreover, due to the RoHS Directive's limited permission of mercury (Hg) use [1], a new LCD employed with environmentally friendly backlight system is now required.

Recently, as the luminance efficiency of light emitting diode (LED) has been improved and the cost of LED is going down, the LED is the substitutive solution for the backlight source. Moreover, since LED has many advantages such as long lifetime, wide color gamut, fast response, and so on [2], LEDs are expected to replace the conventional fluorescent lamps for backlight source of LCD in near future. Although, LED backlight driving systems have been developed and introduced to the market, further reduction of power consumption and cost reduction are still demanded to be widely used as backlight source.

To save the power consumption of backlight for LCD and enhance the image contrast, some techniques on adaptive

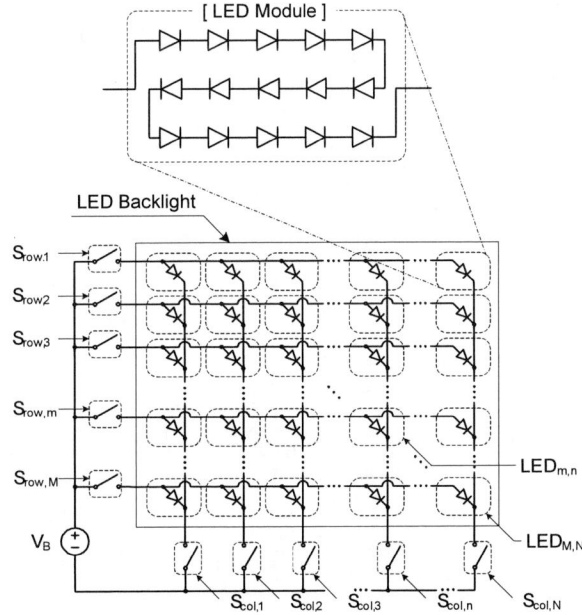

Fig. 1: Configuration of proposed LED backlight system

dimming have been introduced [3]-[9]. The adaptive dimming technique can be categorized into three classes: 0D (uniform dimming or backlight scaling), 1D (line or channel dimming), and 2D (local dimming, block dimming, or cluster dimming). The backlight scaling method [3]-[6] dims the whole backlight luminance is reduced by a dimming factor k, and, at the same time, video signal is enhanced by a factor $1/k$, according to the input video signal when the peak luminance is low. Therefore, this method can reduce the backlight illumination with image processing to retain the image fidelity. However, the power saving is limited when the peak luminance is high even with low APL.

To reduce the power consumption more effectively, channel dimming and local dimming methods are suggested. These methods also allow compensation of backlight luminance non-uniformity, enhancement of gray scale capability for low luminance images, and improvement of dark room contrast [9]. In channel dimming and local dimming methods which have usually adapted to CCFL and LED backlight source, the whole backlight is divided into several channel and block, backlight scaling is adapted to each channel and block, respectively. The more division of backlight, the more power saving can be achieved. Therefore,

978-1-4244-1871-8/07 $25.00 © 2007 IEEE

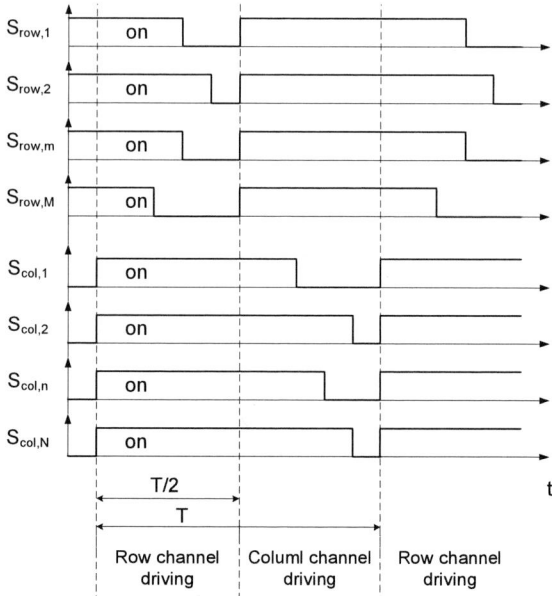

Fig. 2: Timing diagram of time division X-Y channel driving method

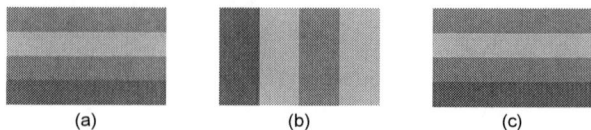

Fig. 3: Concept of proposed LED backlight system

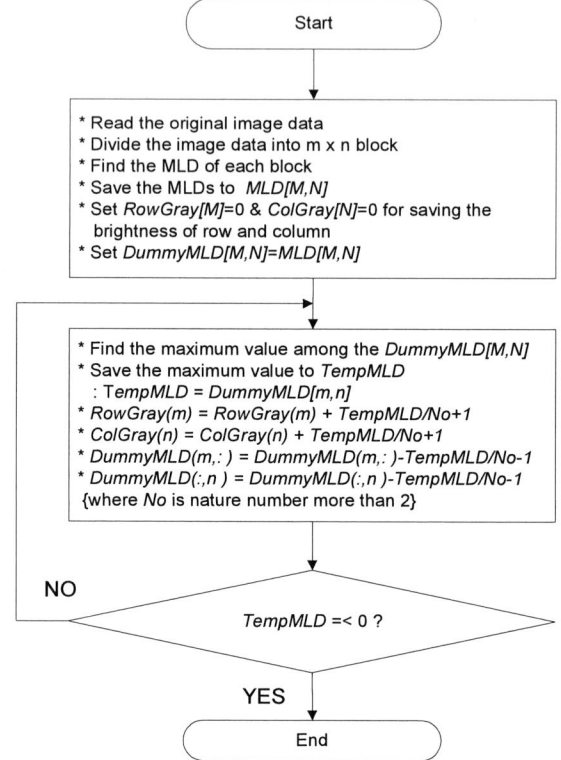

Fig. 4: Flow chart of proposed algorithm

local dimming method can save power consumption more effectively than channel dimming one. Especially, channel dimming method has limitation of power saving when the image is bright vertically. Therefore, local dimming is the best way to have local dimming effects. However, local dimming method results in huge increase in the number of drivers needed for large number of division.

To compromise between channel dimming and local dimming methods, a new LED backlight system for LCD TVs which involves the time division X-Y channel driving method that utilizes row and column switches to control the individual division screen is proposed. Throughout the proposed system, lower power consumption are successfully obtained as well as high contrast ratio even with less number of drivers than that of conventional local dimming method. This paper also contains a new adaptive dimming algorithm and image processing technique for the proposed LCD backlight system.

II. OPERATIONAL PRINCIPLES

A. Circuit Operation

The proposed LED backlight system as shown in Fig. 1 consists of LED modules which have series connected LEDs. These LED modules are connected each other as matrix structure with row channel (X channel) and column channel (Y channel). To control each channels, row and column switches are connected between each channels and an external power source. Therefore, the brightness of each division block can be controlled by a certain different combinations of row and column channels.

In this paper, time division X-Y channel method is used as shown in Fig. 2 and 3. In this method, the row switches are operating in purse width modulation (PWM) while the column switches are turned on for half cycle, and vice versa.

Fig. 3 (a) and (b) show the backlight brightness when the row switches are in PWM switching while column switches are turned on for half cycle, and when the column switches are in PWM switching while row switches are turned on for another half cycle, respectively. Therefore, total brightness is decided by temporal summation as shown in Fig. 3 (c).

B. Dimming Algorithm and Image Manipulation

In image analysis for adaptive dimming technique, the most frequent value, average value [3], [9], maximum value [6], and other key values [7] are inspected. And then backlight control level is determined based on histogram analysis. To preserve the source image, a new method using maximum value for proposed dimming method is suggested. Fig. 4 shows proposed dimming algorithm using Matlab and pseudo code.

1) Find the maximum level data (MLD) of each division block from source image.

2) Find the backlight luminance, $BL(m,n)$ which is sum of each row gray level, $RowGray(m)$, and column gray level, $ColGray(n)$, where, $BL(m,n)$ should be greater than $MLD(m,n)$ to preserve the image source.

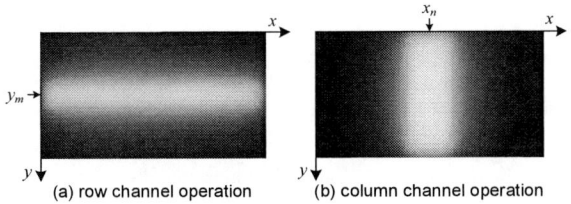

(a) row channel operation   (b) column channel operation

Fig. 5: Photograph of luminance distribution

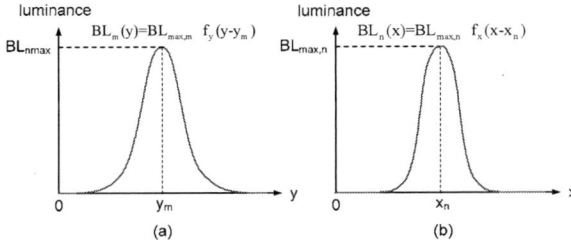

(a)   (b)

Fig. 6: Functions of luminance distribution

$$BL(m,n) = RowGray(m) + ColGray(n) \qquad (1)$$

3) Determine the duty ratios of switches using equation (2), (3).

$$D_{row,m} = RowGray(m)/255 + 0.5 \qquad (2)$$

$$D_{col,n} = ColGray(n)/255 + 0.5 \qquad (3)$$

where, $D_{row,m}$ and $D_{col,n}$ denote the duty ratio of row and column switches, respectively.

4) Manipulate the source image to preserve image information using equation (4).

$$\mathbf{cv'}(m,n) = \mathbf{cv}(m,n) \times \frac{255}{BL(m,n)} \qquad (4)$$

where, **cv** denotes vector matrix of image data.

C. Image Compensation

Fig. 5 shows the luminance distributions after diffuser plate when each a row and a column channels are turned at the full power. The vertical and horizontal positions of peak luminance are denoted by y_m and x_n, respectively. These luminance distributions are measured and plotted at Fig. 6 where $BL_{max,m}(y)$ and $BL_{max,n}(x)$ denote the peak luminance of vertical and horizontal poisons, respectively. Therefore, vertical and horizontal luminance distribution functions can be expressed as equation (5), (6), respectively.

$$\mathbf{BL_m}(y) = BL_{max,m} \times f_y(y - y_m) \qquad (5)$$

$$\mathbf{BL_n}(x) = BL_{max,n} \times f_x(x - x_n) \qquad (6)$$

And luminance functions of row and column channels can be expressed as equation (7), (8), respectively.

$$\mathbf{BL_{row}}(y) = \sum_{m=1}^{M} BL_{max,m} \times f_y(y - y_m) \qquad (7)$$

$$\mathbf{BL_{col}}(x) = \sum_{n=1}^{N} BL_{max,n} \times f_x(x - x_n) \qquad (8)$$

$BL_{max,m}$ and $BL_{max,n}$ depend on duties of row and column switches, respectively. Equeations (7) and (8) can be rewrite as

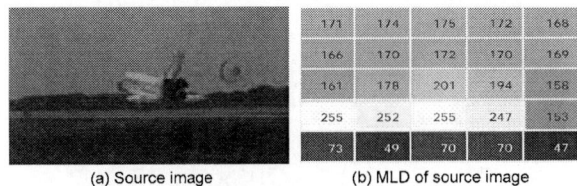

(a) Source image   (b) MLD of source image

Fig. 7: Source image and its MLD

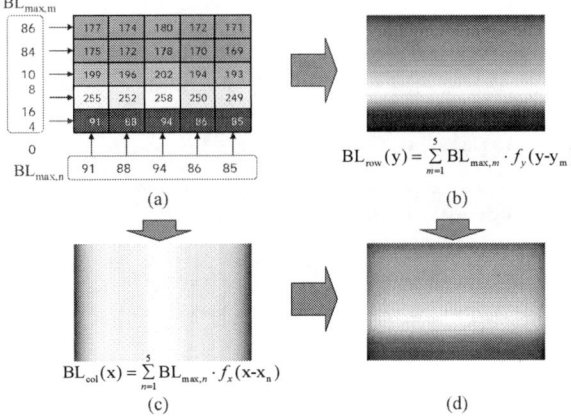

(a)   (b)

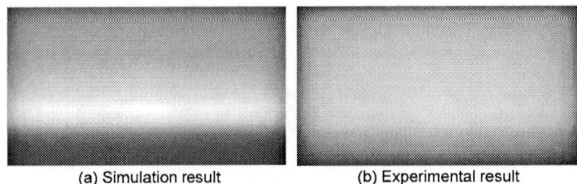

(c)   (d)

Fig. 8: Simulation process and result of backlight luminance

(a) Simulation result   (b) Experimental result

Fig. 9: Simulation process and result of backlight luminance

$$\mathbf{BL_{row}}(y) = \sum_{m=1}^{M} RowGray(m) \times f_y(y - y_m) \qquad (9)$$

and

$$\mathbf{BL_{col}}(x) = \sum_{n=1}^{N} ColGray(n) \times f_x(x - x_n) . \qquad (10)$$

Therefore, the final compensated backlight luminance can be calculated as equation (11).

$$\mathbf{BL} = \mathbf{BL_{row}}(y) + \mathbf{BL_{col}}(x) \qquad (11)$$

Now, actual backlight luminance can be expected from equation (11). Therefore, image manipulation equation (5) should be corrected to equation (12).

$$\mathbf{cv'}(m,n) = \mathbf{cv}(m,n) \times \frac{255}{\mathbf{BL}} \qquad (12)$$

III. SIMULATION AND EXPERIMENTAL RESULTS

The proposed time division X-Y channel driving backlight system is verified by simulation and experiment for 32-inch LCD TV. The RGB-LEDs are used as backlight source and the backlight is divided to 5 by 5 blocks. The three power sources for red, green and blue LED strings are set to drive the desired current: the forward current of the blocks is set to

978-1-4244-1871-8/07 $25.00 © 2007 IEEE

(a) conventional LCD backlight system

(b) Proposed LCD backlight system

Fig. 10: Pictures of luminance distribution

20mA. And 120Hz PWM signals, made from dimming algorithm, control the luminance of each block.

Fig. 7 is the source image and MLD of it for simulation and experiment. Fig. 8 shows the simulation process. The backlight luminance of simulation is well distributed. Fig. 9 shows the simulation and experimental results. Although, these results have a little color difference, the luminance distributions of simulation and experimental are well matched. Fig. 10 (a) and (b) are the experimental results of conventional LCD backlight system with fully turned RGB-LEDs on and proposed LCD backlight, respectively.

The image output of the overall system is well-displayed. In terms of power consumption, the experiment result shows that the power consumption is reduced to about 42 % as predicted.

IV. CONCLUSION

The proposed method can achieve similar power reduction effect to conventional local dimming method. However, the number of drivers or switches of proposed backlight drive method is much less than that of the conventional localized drive method. Image improvement and reduced power consumption are typical local dimming effects resulted from localized drive method. For validity of this research is confirmed from the simulation and experimental results.

ACKNOWLEDGMENT

This work was supported (in part) by Samsung LCD Business, Korea, under Display Research center program.

REFERENCES

[1] http://www.rohs.gov.uk/

[2] Ki-Chan Lee, Seung-Hwan Moon, Brian Berkeley, and Sang-Soo Kim, "LED-backlight feedback control system with integrated amorphous-silicon color sensor on an LCD panel", *journal of the SID*, vol. 14, no.2, pp. 161-168, 2006

[3] Insun Hwang, Cheol Woo Park, Sung Chul Kang, and Dong Sik Sakong, "Image Synchronized Brightness Control", *SID 01 Digest*, pp.492-493, 2001

[4] Wei-Chung Cheng, Pedram M., "Power Minimization in a backlit TFT-LCD display by concurrent brightness and contrast scaling", *IEEE Transaction on Consumer Electronics*, Vol. 50, no. 1, pp. 25-32, Feb. 2004.

[5] Naehyuck Chang, Inseok Choi, and Hojun Shim, "Dynamic Backlight Luminance Scaling of Liquid Crystal Display", *IEEE Transactions on Very Large Scale Integration System*, Vol. 12, no. 8, pp.837-846, 2004.

[6] A. Konno, Y. Yamamoto, and T. Inuzuka, " RGB Color Control System for LED Backlight in IPS-LCD TVs", *SID 05 Digest*, pp.1380, 2005.

[7] T. Shiga, S. Mikoshiba, "Reduction of LCTV Backlight Power and Enhancement of Gray Scale Capability by Using Adaptive Dimming Technique", *SID 03 Digest* , pp.1364-1367, 2003.

[8] Hanfeng Chen, Junho Sung, Taehyeun Ha, Yungjun Park, and Changwan Hong, "Backlight Local Dimming Algorithm for High Contrast LCD TV", *ASID' 06 Digest*, pp.168-171, 2006.

[9] T. Shirai, S. Shimizukawa, T. Shiga, and S. Mikoshiba, "RGB-LED Backlight for LCD-TVs with 0D, 1D, and 2D Adaptive Dimming", *SID 06 Digest*, pp.1520-1522, 2006

[10] H. J. Peng, W. Zhang, C. K. Hung, C. J. Tsai, K. W. Ng, S. L. Chen, "High Dynamic Range LCD TV Using Active Dynamic LED Backlight", *Proceeding of Asia Display 2007*, pp.429-432, 2007.

CIM Based Soft Switching of Energy Recovery Sustain Driver for AC PDPs with Reduced Sustain Voltage

Seung-Bum Lim, Jun-Yong Lee[*], Jong-Sun Ko, and Soon-Chan Hong
Department of Electrical Engineering, [*] Department of Electronics Engineering
Dankook University
San-44-1, Juckjeon-dong, Suji-gu, Yongin-si, Gyeonggi-do, 448-701, Korea
Email : schong@dku.edu

Abstract – AC-PDP(Plasma Display Panel) uses a high sustain voltage and its power consumption is also high so that circuit cost is expensive. An energy recovery circuit for AC-PDPs with reduced sustain voltage was proposed to solve these disadvantages. However, the circuit has disadvantage that the switching elements are operated with hard-switching at the start point of discharge. The reason is that the panel voltage is lower than sustain voltage at that point. In this paper, we propose the improved driving method that switching devices are operated with soft-switching by using CIM(Current Injection Method). Finally, the validity of the proposed driving method is verified by simulations and experimentation.

I. INTRODUCTION

AC-PDP(Plasma Display Panel) has high cost circuit structure due to its high voltage driving and large power consumption[1]. Recently, to solve these disadvantages, an energy recovery circuit for AC-PDPs with reduced sustain voltage was proposed[2]. However, the circuit has disadvantage that the switching elements are operated with hard-switching at the start point of discharge and sustaining region. It results from the reason that the voltage of panel is lower than the sustain voltage at that point because of the loss of switching devices and forward voltage drop of diodes at energy recovery and supply region.

In this paper, we propose an improved driving method that switching elements are operated in all region with soft-switching by using CIM(Current Injection Method). Using the method, the panel voltage can be increased to a desired sustain voltage by building up the resonant inductor current before sustain switches are turned on.

II. THE PRIOR DRIVING METHOD

If device voltage rating is lowered, we can design driving circuit of PDP economically because low-cost switching devices can be used. Moreover, the performance of driving circuit can be improved because switching devices with good frequency characteristics can be used[3].

The prior circuit has the structure that sustain drivers are installed on the left and right sides and the energy recovery circuit on the top and bottom sides, like Fig. 1[2]. Fig. 2 shows the gating signals and the waveforms of the voltage across panel and inductor currents.

The circuit was operated with 4 modes per cycle and the first half cycle is symmetrized with the next half cycle. The

operation of first half cycle is as follows.

1) Mode 1(energy recovery and supply region)

Before mode analysis, the voltage of panel is $-V_s$ at the end of mode 4. Mode 1 begins at t_b when (S_{x1}, S_{x3}) and (S_{y2}, S_{y4}) conducting in mode 4 are turn off and S_b of energy recovery circuit is turn on. At this time, (S_{x1}, S_{x3}) and (S_{y2}, S_{y4}) show zero-voltage-switching and S_b show zero-current-switching. When LC resonance starts between L_2 and C_p, the energy of panel is recovered through L_2 and the recovered energy is supplied to C_p in opposite direction. Due to this, the potentials of X and Y are exchanged. If i_{L2} is reduced to 0 at the half cycle of resonance, S_b is turned off and mode 1 is

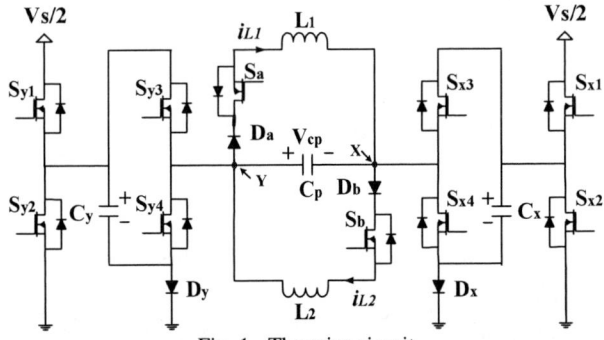

Fig. 1. The prior circuit

Fig. 2. The prior driving method

978-1-4244-1871-8/07 $25.00 © 2007 IEEE   713

finished. Therefore, S_b conducts with zero-current-switching both at turn on and turn off. The voltage of panel is ideally changed to V_s after mode 1 is finished, but the voltage of panel is lower than V_s because of the loss of switching devices.

2) Mode 2(discharge and sustaining region)

Mode 2 begins at t_c when (S_{y1}, S_{y3}) and (S_{x2}, S_{x4}) are turned on. Switching elements are operated with hard-switching because the voltage of panel is lower than V_s in previous mode. In mode 2, the potential of Y electrode is $+V_s/2$ and charges C_y as $+V_s/2$ at the same time. The potential of X electrode becomes to $-V_s/2$ because (+) terminal of C_x is grounded through S_{x2} and (-) terminal is connected to X electrode through S_{x4}. It is considered that C_x was charged as $+V_s/2$ in previous mode. Therefore $+V_s$ is applied to panel capacitor. When source voltage is applied, PDP discharge begins and the voltage of the panel capacitor, C_p, is sustained as $v_{cp} = V_s$.

This circuit has advantage that although the supply voltage reduced by half, the sustaining voltage of panel is $\pm V_s$, which satisfies gas discharge condition. However, the switching elements show hard-switching operation at the start point of mode 2 and 4. This is because that the voltage of panel is lower than the sustain voltage at that point due to the loss of switching devices and forward voltage drop of diodes at mode 1 and 3.

III.   THE PROPOSED DRIVING METHOD

To solve the hard-switching stress of switching devices, CIM(Current Injection Method) that uses build-up current before energy recovery operation has been adopted[4]. Because the inductor energy is stored in advance, the initial current value of inductor is not zero at the start point of the energy recovery and supply region and the voltage of panel is higher than the case when the initial current value of inductor is zero. Hence, the voltage of panel can be the desired value exactly by adjusting the initial current value properly.

The proposed driving method in this paper operates as 6 modes per cycle and the first half cycle and the next half cycle operate symmetrically. Fig. 3 shows the gating signals and the waveforms of the voltage across panel capacitor and inductor currents of the proposed driving method. The operation at each mode is as follows.

1) Mode 1(t_1~t_2, CIM region)

During (S_{x1}, S_{x3}) and (S_{y2}, S_{y4}) are conducting, S_b is turned on and mode 1 begins at $t=t_1$. Current path is shown in Fig. 4(a). When S_b is turned on, the current of inductor, i_{L2}, linearly increases with the slope of V_s/L_2 and it causes the energy is stored in the inductor. In mode 1, S_b performs zero-current-switching when it is turned on.

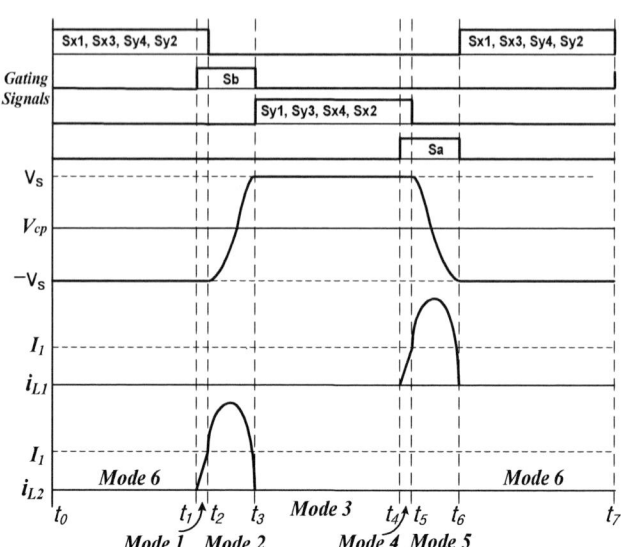

Fig. 3. Gating signals and waveforms in proposed driving method

2) Mode 2(t_2~t_3, energy recovery and supply region)

Mode 2 begins when (S_{x1}, S_{x3}) and (S_{y2}, S_{y4}) are turned off and current path is shown in Fig. 4(b). At this time, because the voltage of panel, v_{cp}, is $-V_s$, the switching elements turned off with zero-voltage-switching. When switching elements are turned off, LC resonance starts between L_2 and C_p. Mode 2 is finished because S_b is turned off when i_{L2} is zero again. Therefore, S_b performs zero-current-switching when it is turned off. The voltage of panel can be changed $+V_s$ exactly by adjusting the region of mode 1 properly.

3) Mode 3(t_3~t_4, discharge and sustaining region)

When S_{y1}, S_{y3}, S_{x4}, and S_{x2} are turned on, mode 3 begins and the current path is shown in Fig. 4(c). In mode 3, the potential of Y electrode is $+V_s/2$ and C_y is charged as $+V_s/2$ at the same time. The potential of X electrode becomes to $-V_s/2$ because (+) terminal of C_x charged as $+V_s/2$ in previous mode is grounded through S_{x2} and (-) terminal is connected to X electrode through S_{x4}. Therefore $+V_s$ is applied to panel capacitor. When source voltage is applied, discharge of PDP begins. D_x maintains off-state by the voltage polarity of C_x because S_{x2} is conduction-state. At the start point of mode 3, switching elements turned on with zero-voltage-switching because the voltage of panel is $+V_s$ at the end point of mode 2 by using CIM in mode 1.

4) Mode 4(t_4~t_5, CIM region)

During (S_{y1}, S_{y3}) and (S_{x4}, S_{x2}) are maintaining as conducting state, S_a is turned on and mode 4 begins at $t=t_4$. The current path is shown in Fig. 4(d). At this time, S_a turned on with zero-current-switching and the energy is stored in the inductor L_1. At the end point of mode 5, the voltage of panel can be $-V_s$ by adjusting the time of mode 4 properly.

978-1-4244-1871-8/07 $25.00 © 2007 IEEE   714

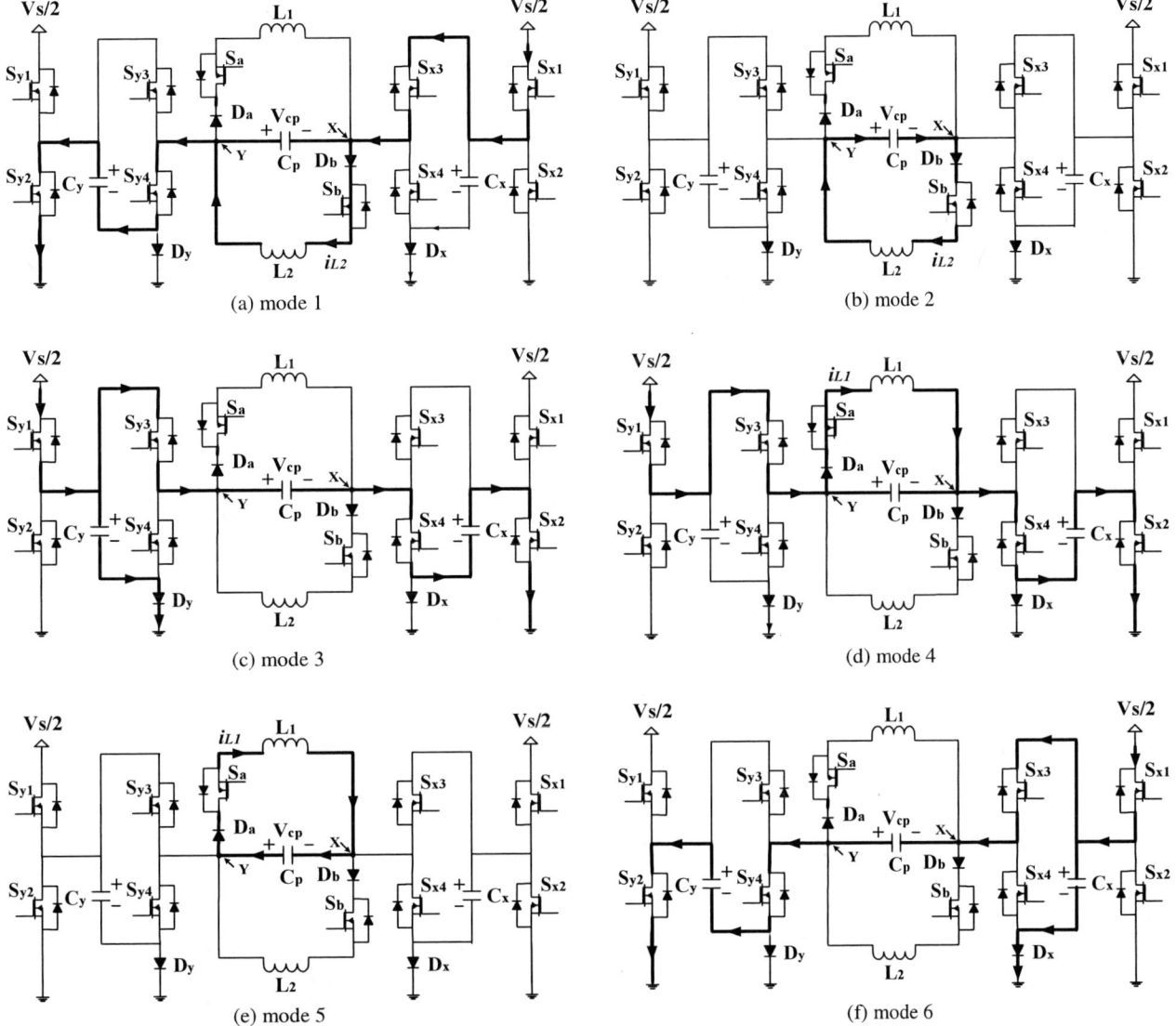

(a) mode 1   (b) mode 2

(c) mode 3   (d) mode 4

(e) mode 5   (f) mode 6

Fig. 4. Operation in proposed driving method for each mode

5) Mode 5(t_5~t_6, energy recovery and supply region)

Mode 5 begins when (S_{y1}, S_{y3}) and (S_{x2}, S_{x4}) are turned off and current path is shown in Fig. 4(e). At this time, because the voltage of panel, v_{cp}, is V_s, the switching devices turned off with zero-voltage-switching. Mode 5 is finished when i_{L1} is zero again and S_a is turned off. Therefore, S_a shows zero-current-switching when it is turned off.

6) Mode 6(t_6~t_7, discharge and sustaining region)

When S_{y2}, S_{y4}, S_{x1}, and S_{x3} are turned on, mode 6 begins and the current path is shown in Fig. 4(f). At the start point of mode 6, the switching elements turned on with zero-voltage-switching because the voltage of panel was $-V_s$ at the end point of mode 5. When source voltage is applied, discharge of PDP begins.

The proposed driving method can change between $+V_s$ and $-V_s$ exactly at the end point of energy recovery and supply region because the initial current value of inductor is not zero. Therefore, all switching elements perform soft-switching in all regions.

IV. SIMULATION RESULTS

To verify the validity of the proposed driving method, simulations are performed by using PSpice. Fig. 5 and 6 shows the simulation results of the proposed driving method. From the waveforms of i_{L1} and i_{L2} in Fig. 5, it can be seen that the current increases linearly in CIM region and the resonance current flows in energy recovery and supply region. From the waveform of v_{cp} in Fig. 5, the voltage of panel changes between $+V_s$ and $-V_s$ exactly when LC resonance is finished by storing the energy in the inductor in CIM region. From the waveforms of v_{sy3} and v_{sy4} in Fig. 6, it shows that switching elements turned on with zero-voltage-switching because v_{cp} changes to the desired sustain

978-1-4244-1871-8/07 $25.00 © 2007 IEEE   715

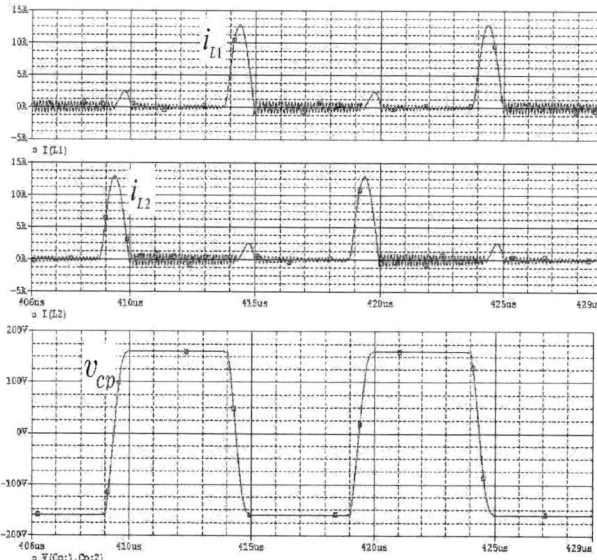

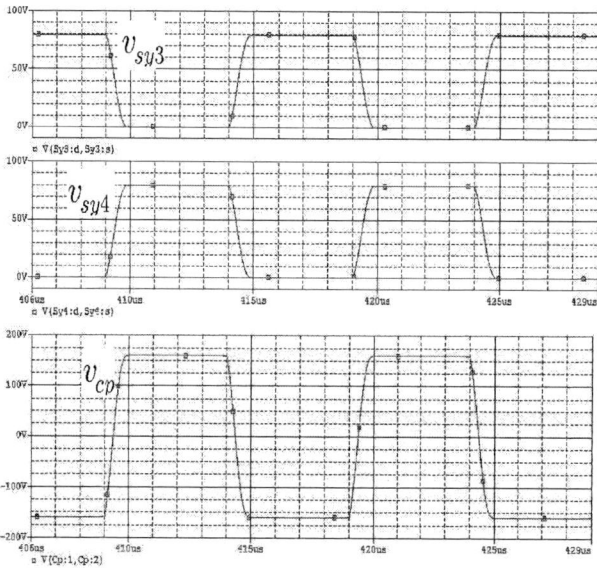

Fig. 5. Simulation results in proposed driving method

Fig. 6. The drain-source voltage of switching elements and the voltage of panel capacitor in proposed driving method

voltages at the end point of energy recovery and supply region.

V. EXPERIMENTAL RESULTS

Fig. 7 and 8 are the experimental results of proposed driving method. In Fig. 7, the i_{L2} waveform shows that the inductor current flows before energy recovery and supply region. The v_{cp} waveform shows that it is charged to $\pm V_s$ completely after LC resonance finishes. In Fig. 8, v_{sy3} shows that S_{y3} is operated with ZVS because the voltage of drain-source of S_{y3} is zero at the start point of turn-on.

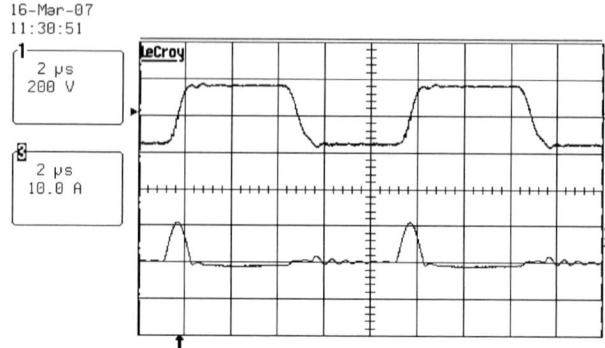

Fig. 7. v_{cp}(top) and i_{L2}(bottom) in proposed driving method

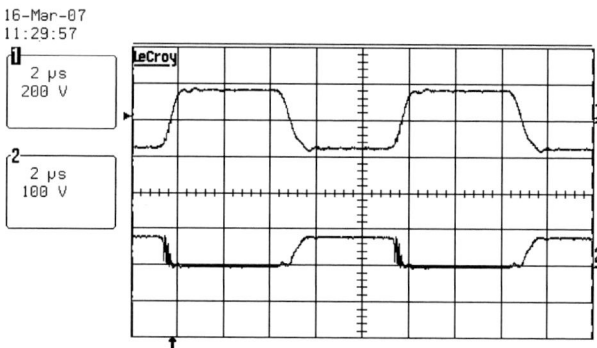

Fig. 8. v_{cp}(top) and v_{sy3}(bottom) in proposed driving method

VI. CONCLUSION

This paper proposed the new driving method that the shortcoming of the prior method was improved. Switching elements of the proposed driving method shows ZVS when they are turned-on also at the start point of the discharge and sustaining region by using CIM. The operations of the proposed circuit are analyzed at each mode and its validity is verified by the simulations and experimentation. Therefore, this method may be useful to improve switching stresses of PDP sustain driver.

REFERENCES

[1] Jong-Gul Yoo and Soon-Chan Hong, "A Study on the NPC Type Multi-Level Energy Recovery Sustaining Driver for AC Plasma Display Panel", The Transactions of the Korean Institute of Power Electronics, Vol. 10, No. 2, pp. 194-202, April 2005.

[2] Seung-Bum Lim and Soon-Chan Hong, "A Novel Energy Recovery Circuit for AC PDPs with Reduced Sustain Voltage", The Transactions of the Korean Institute of Power Electronics, Vol. 11, No. 6, pp. 494-501, December 2006.

[3] Heiju Uchiike, "液晶は動畫表示を磨く PDPは低消費電力で對抗", Nikkei Electronics, pp. 110-118, Nov. 2002.

[4] Jun-Young Lee, Jin-Sung Kim, Nam-Sung Jung, and Bo-Hyung Cho, "The Current Injection Method for AC Plasma Display Panel Sustainer", IEEE Transactions on Industrial Electronics, Vol. 51, No. 3, pp. 615-624, June 2004.

Automated Diagnosis of Rolling Bearing Faults in Electrical Drives

Henning Zoubek, Sebastian Villwock, Mario Pacas

University of Siegen

Institute of Power Electronics and Electrical Drives

Hölderlinstraße 3

57068 Siegen

Email: henning.zoubek@web.de, sebastian.villwock@uni-siegen.de, jmpacas@ieee.org

Abstract – This paper deals with a new automated diagnosis method for detecting bearing faults. The diagnosis is carried out by frequency response analysis of the mechanical system of the drive. The detection of bearing damages of a ball bearing on motor and load side of the mechanical drive system is addressed. The mechanics is assumed to be a two-inertia-system with one dominant resonant frequency The actual frequency response of the monitored system can be compared to the one obtained during the commissioning of the plant which can therefore serve as reference. The present paper addresses the investigation of two types of bearing faults: Generalized roughness and single point defects. As concerned the latter ones, inner and outer race bearing faults are topic of the research. Furthermore, based on the calculation of two tolerance bands around the calculated frequency responses, an algorithm for the automated diagnosis of both fault categories by pattern recognition is presented.

I. INTRODUCTION

Rolling bearing failures are a main factor for unplanned downtimes of machines and plants [1]. Hence, they give rise to costs for repair and maintenance. An early detection of incipient bearing failures allows therefore the reduction of these costs by initiating repair and maintenance at planned times before catastrophic failures occur.

In terms of bearing fault detection and diagnosis exist multiple techniques. Many diagnosis methods are based on vibration analysis [1], [2], [3]. A faulty bearing leads to an increase of radial movement of the bearing elements causing vibrations of the mechanics [2], [3], [4]. Spectral analysis of measured vibration data can yield information about a faulty bearing already in an early state of the bearing failure [5]. Nevertheless, vibration measurement comes along with the need for additional sensors on the bearings which are highly cost intensive [5], [6]. To avoid the use of additional sensors recent studies focus on the so-called motor current signature analysis (MCSA) for bearing fault diagnosis [3], [5], [7]. The line currents of a drive are measured for the purpose of control or for safety reasons. Radial motion of the bearing elements cause the air-gap in the electrical machine to vary. The air gap flux density is modulated and this modulation is reflected into the stator current. Thus, the spectrum of the line currents can be utilized for the diagnosis of bearing faults. However, the additional spectral components in the motor current caused by a faulty bearing are very small in contrast to other spectral current amplitudes [5], [6], [7]. Therefore, a lot of knowledge is necessary to use MCSA for bearing condition diagnosis.

This work deals with a new automated method for rolling bearing fault diagnosis without any additional sensors except the ones needed for the control of the drive.

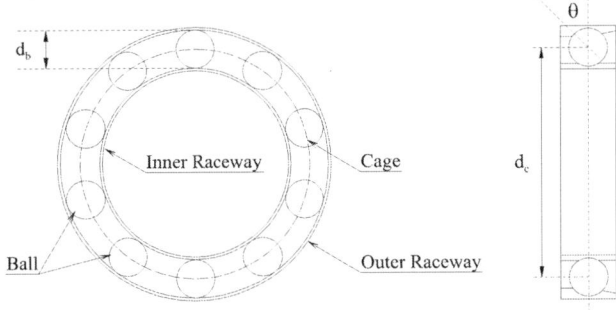

Fig. 1: Rolling bearing construction and geometrical variables

II. TYPES OF FAILURES

A ball bearing consists of the four elements inner and outer raceway, balls and the cage, shown in Fig. 1 where d_b and d_c mean the ball and cage diameter respectively and θ is the ball contact angle, which is zero for radial loaded ball bearings.

Bearing faults can be categorized in two types, which are both common in industrial applications [2]: Generalized roughness and single point defects. To maintain safety and reliability, a condition monitoring technique must be able to detect both of these faults [2], [6]. Generalized roughness faults are characterized by a degradation of the whole raceway surfaces and lead to broadband changes in the spectrum of the measured variables for diagnosis [2]. Single point defects are local faults on otherwise undamaged bearing surfaces e.g. on a raceway [1], [2]. For bearing damage detection, the most commonly reported results concern outer race faults [1].

Every time a bearing ball passes over a damaged area, a shock impulse is generated which leads to torque disturbances and speed fluctuations [5], [7], [8], [9], [10]. These periodic shocks cause the appearance of deterministic characteristic fault frequencies- f_{ORF} for an outer race fault and f_{IRF} for an inner race fault - depending on the mechanical data of a bearing with z balls [2], [3], [4]:

$$f_{ORF} = \frac{z}{2} \cdot f_n \cdot \left(1 - \frac{d_b}{d_c} \cdot \cos(\theta) \right) \qquad (1)$$

978-1-4244-1871-8/07 $25.00 © 2007 IEEE

$$f_{IRF} = \frac{z}{2} \cdot f_n \cdot \left(1 + \frac{d_b}{d_c} \cdot \cos(\theta)\right) \tag{2}$$

For a fixed outer raceway and an inner raceway connected to the rotating shaft of the drive, f_n is the rotational mechanical frequency. Besides the characteristic fault frequency, also their multiples can appear in the measured spectrum depending on the severity of the defect [1], [4]. Hence, for the purpose of single point defect diagnosis, the appearance of following spectral components is often utilized for outer or inner race defects:

$$f_{ORF,\lambda} = \lambda \cdot f_{ORF}, \quad \lambda = 1,2,3,.. \tag{3}$$

$$f_{IRF,\lambda} = \lambda \cdot f_{IRF}, \quad \lambda = 1,2,3,.. \tag{4}$$

With regard to inner race faults which rotate with the shaft of the drive, a further aspect has to be considered. If a bearing sustains a downward radial load, the damaged area rotates in and out the so called "load zone". This is a part of the bearing's circumference on which the radial load is distributed as it can be seen in Fig. 2. Inside the load zone the contact between the ball and the damage is more intensive, resulting in greater shocks impulses than outside the load zone. As the defect rotates one time per revolution of the shaft into the load zone, the intensity of the generated shock impulses becomes amplitude modulated with the rotational frequency f_n [1]. Thus, additional to the characteristic inner race fault frequency sidebands around this frequency and its multiples appear in the spectrum [1]. The width of the sidebands from the characteristic fault frequency is f_n. This effect is typical for inner race faults and will later on be used as a part of the pattern recognition for the presented diagnosis method.

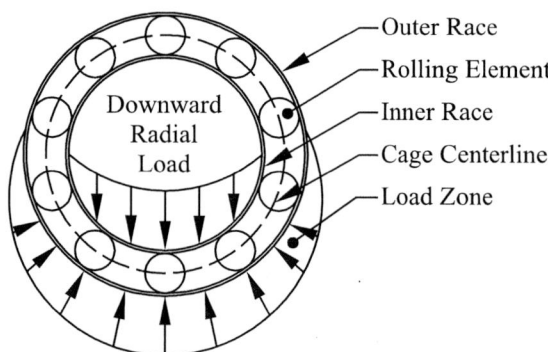

Fig. 2: Load zone of a bearing under downward radial load

Concerning generalized roughness faults, a broadband rise of the spectral amplitudes can be used to detect such type of damage [2].

III. BEARING FAULT DIAGNOSIS BY FREQUENCY RESPONSE

In this research, the bearing fault diagnosis is carried out by analysis of the frequency response $G(j\omega)$ of the mechanical system of the drive. The mechanics consist of the motor and a load inertia connected by a non-rigid shaft and can be modelled as a two-inertia system. Such a system is able to perform torsional oscillations and has a characteristic resonance frequency f_{res}. The frequency response $G(j\omega)$ of the mechanical two-inertia system is known from the literature and is given by [11]

$$G(j\omega) = \frac{1}{(T_M + T_L) \cdot j\omega} \cdot \frac{T_L T_C (j\omega)^2 + dT_C \cdot j\omega + 1}{\frac{T_M T_L T_C}{T_M + T_L} (j\omega)^2 + dT_C \cdot j\omega + 1}, \tag{5}$$

with:
- T_M: run-up time of the motor,
- T_L: run-up time of the load,
- T_C: normalized spring constant,
- d: related damping of the spring.

The identification of the system by frequency response calculation during the commissioning of the plant is necessary for reasons of safety and control to avoid excitation at its mechanical resonance frequency.

The system is excited by pseudo random binary signal (PRBS). This is a periodic and multifrequent signal which is injected at the output of the PI-speed controller in the closed speed control configuration..

The calculation of the frequency response of the mechanics $G(j\omega)$ can be achieved by applying the Welch-Method which is known from the field of communications engineering on two measured variables. Concerning the calculation of the frequency response of the mechanical system, the measured signals are the velocity of the motor n(t) and the torque generating component of the stator current $i_q(t)$. As bearing faults influence both of this variables, the presented diagnosis scheme is regarded superior to other bearing fault diagnosis methods mentioned above, which only take one affected signal into account. Furthermore, both of these variables have to be measured for the control of the drive, so that no additional sensor is needed for the presented diagnosis method.

From the commissioning of the plant, the frequency response of the healthy system is known and can serve as reference. Bearing faults will have influence on the frequency response as explained above. Single point defects affect the frequency response at the characteristic fault frequencies while generalized roughness faults will lead to a broadband change. The theoretical analysis of the failures shows that the amplitude response contains the most significant patterns for the investigated faults. Therefore, in the following the investigations will concentrate on the amplitude response $|G(j\omega)|$ alone and not on the phase diagram $\angle G(j\omega)$. The diagnosis is carried out in two steps:

Generalized roughness fault: One tolerance band around the reference amplitude response $|G_{ref}(j\omega)|$ is calculated with the width of $|G_{ref}(j\omega_i)|$ for every frequency ω_i. Thus, the limits of this tolerance band $|G_{TB1}(j\omega)|$ are given by:

$$|G_{TB1}(j\omega_i)| = \left(1 \pm \frac{1}{2}\right) \cdot |G_{ref}(j\omega_i)| \tag{6}$$

978-1-4244-1871-8/07 $25.00 © 2007 IEEE

Each point of the measured curve under influence of the bearing fault outside this tolerance band differs with more than 50% from the reference. All these points from 0 Hz up to the resonance frequency f_{res} are used as indicator for generalized roughness faults. If more than 50% of the measured data points exceed this first tolerance band, a generalized roughness fault is diagnosed. As single point damages only affect the amplitude response at the characteristic fault frequency and its multiples, they will not lead to a fault detection with the first tolerance band.

Single point damages: If no generalized roughness fault is detected for all points of the measured amplitude response $\left|G_{mes}(j\omega_i)\right|$ which are not outside the first tolerance band, the related standard deviation s is calculated by

$$s = \sqrt{\frac{1}{N_{res}-1} \cdot \sum_{i=1}^{N_{res}} \frac{\left(\left|G_{mes}(j\omega_i)\right| - \left|G_{ref}(j\omega_i)\right|\right)^2}{\left|G_{ref}(j\omega_i)\right|^2}} . \quad (7)$$

N_{res} is the number of data points from 0 Hz up to f_{res} inside the first tolerance band. A second tolerance band $\left|G_{TB2}(j\omega)\right|$ is now calculated around $\left|G_{ref}(j\omega)\right|$ with the width $2 \cdot s$:

$$\left|G_{TB2}(j\omega_i)\right| = (1 \pm s) \cdot \left|G_{ref}(j\omega_i)\right| \quad (8)$$

Every point of the measured curve which exceeds the limit of this second tolerance band is regarded as a potential indicator for a single point defect. Assuming a good and fast speed control, the rotational frequency f_n of the drive can be calculated from the motor speed set point so that the possible characteristic frequencies for an outer and an inner race fault can be computed. By comparing the points of the measured amplitude response outside the second tolerance band with the possible characteristic fault frequencies, an automated fault diagnosis for single point defects is achieved. For the diagnosis of inner race faults also the sidebands caused by the load zone can be utilized for the pattern recognition of the single point defects.

IV. EXPERIMENTAL RESULTS

All following measured results were carried out on the test rig shown in Fig. 3. The location of the investigated bearings could be either on the motor side or on the load side of the mechanics, denoted "bearing 1" and "bearing 2" respectively. A permanent magnet synchronous machine with a rated torque of 30 Nm drives the mechanics and the speed measurement is achieved with a sin-cos-encoder.

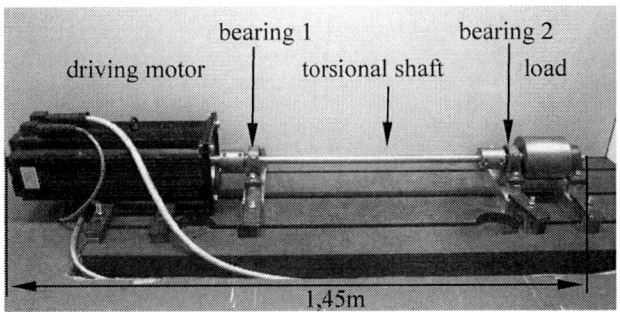

Fig. 3: Mechanical Setup

The investigated bearings are radial ball bearings with the following technical parameters: $d_b = 8,7$ mm, $d_c = 46$ mm, $z = 9$.

A. Outer race faults

To produce a single point defect on the outer bearing raceway, a hole has been eroded into the bearing as can be seen on the left side in Fig. 4. On the right hand, the installed faulty bearing on the load side inertia is shown. Under the fault's location at the 6 o'clock position a screw has been installed in order to adjust the severity of the damage.

Fig. 4: Outer race bearing fault

Fig. 5 depicts the amplitude response of the healthy bearing in green as reference and the measured curve under influence of the outer race fault in red. The faulty bearing was installed on the load side of the mechanics (bearing 2) with a dominant resonance frequency of $f_{res} = 64$ Hz. For the adjusted desired value of the motor speed of $n^* = 200$ min^{-1}, the characteristic fault frequency is $f_{ORF,1} = 12$ Hz. At this frequency, a significant deviation between the reference and the measured curve indicates the outer race fault clearly. According to the theory, differences between the two graphs also appear at multiples of f_{ORF}. All these differences therefore can serve as fault indicators for outer race bearing faults.

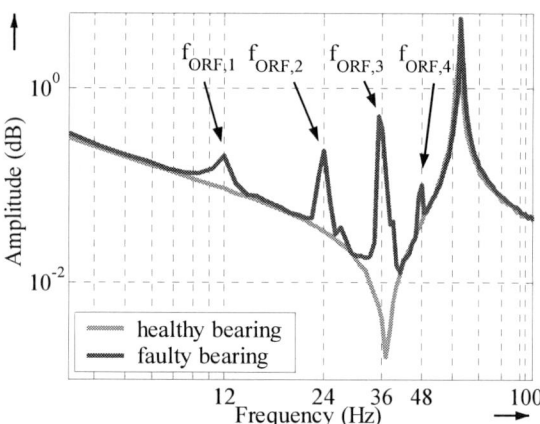

Fig. 5: Amplitude response, bearing 2, Outer race fault, f_{res}=64 Hz, $n^* = 200$ min^{-1}

A parallel measurement with a commercial acceleration sensor on the bearing housing classified the adjusted bearing fault as a progressive damage. Hence, an early damage detection is obtained.

To point out the flexibility and the reliability of the presented diagnosis method, Fig. 6 depicts the measurement results for an outer race defect on bearing 1 (motor side). The mechanical parameters of the two mass system were influenced by installing a different torsional shaft, so that the dominant mechanical resonance frequency was changed to $f_{res} = 100$ Hz. Also the set point of the motor speed has been increased to $n^* = 300$ min^{-1}.

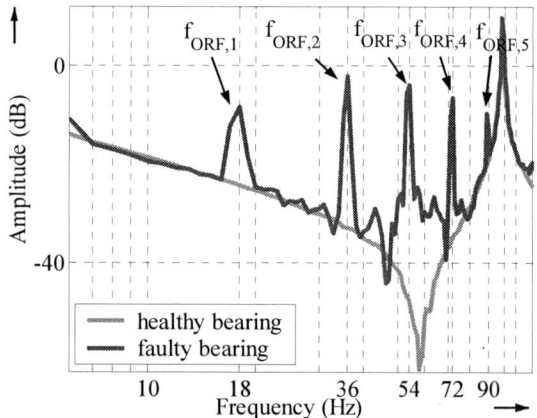

Fig. 6: Amplitude response, bearing 1, Outer race fault, f_{res}=100 Hz, $n^* = 300$ min^{-1}

For this motor speed value, the characteristic fault frequency for an outer race fault is $f_{ORF,1} = 18$ Hz. Fig. 6 points out that the influence of the outer race fault on the amplitude response becomes evident at $f_{ORF,1}$ and its further four multiples.

B. Inner race faults

For the investigations of this class of single point defects, a gap of 6.2 mm has been eroded into the inner ring of a bearing. The damaged bearing as well as the installation on the load side inertia is shown in Fig. 7. Similarly to the outer race defects, the severity of the damage can be adjusted. As the inner bearing ring rotates with the shaft, the screw for this adjustment needs to be mounted on the shaft. On the right side of Fig. 7, the installed damage is marked.

Fig. 7: Inner race bearing fault

This bearing installed on the load side of the mechanics (bearing 2) leads to the measured red curve shown in Fig. 8. In comparison to the green amplitude response as reference curve for the healthy state of the mechanical setup with $f_{res} = 64$ Hz, the influence of the defective bearing is clearly visible. Differences between the two graphs can be easily made out at the characteristic inner race fault frequency of $f_{IRF,1} = 18$ Hz for the motor speed set point of $n^* = 200$ min^{-1}. As Fig. 8 points out, the inner race fault also affects the measurement at the multiples of f_{IRF} at $f_{IRF,2} = 36$ Hz, $f_{IRF,3} = 54$ Hz and $f_{IRF,4} = 72$ Hz. According to the theory, sidebands around the characteristic fault frequency and its multiples appear in the measured curve at the sideband distance of $f_n = 3,33$ Hz. They are caused by the influence of the load zone on the rotating defect and can be utilized as a further diagnosis indicator for inner race faults.

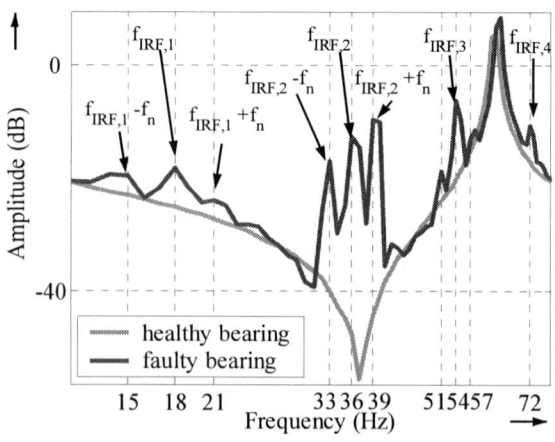

Fig. 8: Amplitude response, bearing 2, Inner race fault, f_{res}=64 Hz, $n^* = 200$ min^{-1}, $f_n = 3,33$ Hz

A measurement with the acceleration sensor again classified the adjusted inner race fault from Fig. 8 as a progressive one. An early damage detection can therefore be achieved.

C. Generalized roughness faults

To obtain an overall surface roughness, the investigated bearings were contaminated with some corundum in the laboratory. The broadband influence of such type of damage installed on bearing 2 is exemplarily depicted in Fig. 9. Over a large frequency range, the measured curve of the faulty system differs significantly from the reference amplitude response so that this kind of fault can easily be detected. The desired value of the motor speed was set to $n^* = 200$ min^{-1} during the measurement and the mechanical resonance frequency of the system was $f_{res} = 64$ Hz.

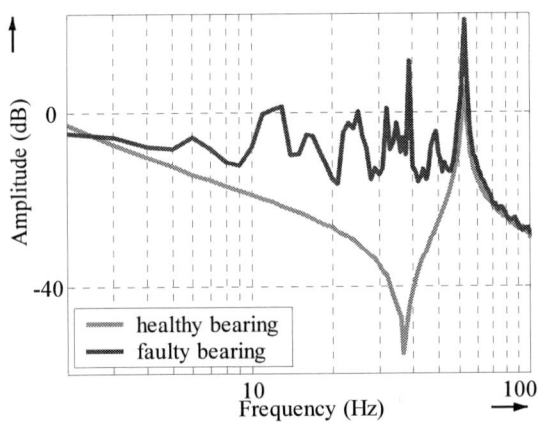

Fig. 9: Amplitude response, bearing 2, Generalized roughness fault, f_{res}=64 Hz, n˙ = 200 min^{-1}

As Fig. 10 depicts, the motor speed does not differ significantly from the motor speed set point during the 30s of data recording for the results shown in Fig. 9.

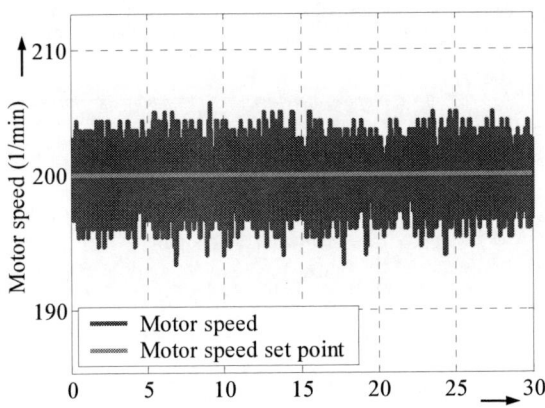

Fig. 10: Motor speed and motor speed set point during measurement

Hence, the operating behaviour of the system is not considerably affected by the faulty bearing and also for the generalized roughness fault an early damage detection is obtained.

D. Automated diagnosis

The calculated first tolerance band for the generalized roughness fault from Fig. 9 can be seen in Fig. 11. Regarding the broadband change of the measured curve in contrast to the amplitude response of the healthy system, the effect of the first tolerance band for the diagnosis becomes obvious. Definitely more than 50% of the measured data points up to the mechanical resonance frequency f_{res} are located out of the tolerance band's limit. The generalized roughness fault can easily be diagnosed. A further calculation of the second tolerance band is not required for this type of fault.

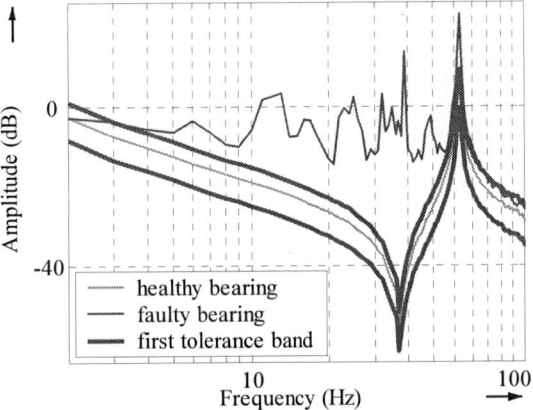

Fig. 11: Amplitude response with calculated tolerance band, bearing 2, Generalized roughness faultt, f_{res}=64 Hz, n˙ = 200 min^{-1}

For the measured outer race fault in Fig. 5 Fig. 12 shows the two calculated tolerance bands. The deviations between the reference amplitude response and the measured curve of the faulty system clearly exceed the limits of the second tolerance band at $f_{ORF,1}$ and its multiples. An algorithm can easily detect these frequencies as possible indicators for a fault. By comparing them to calculated characteristic fault frequencies with the known motor speed set point, an automated fault diagnosis is achieved.

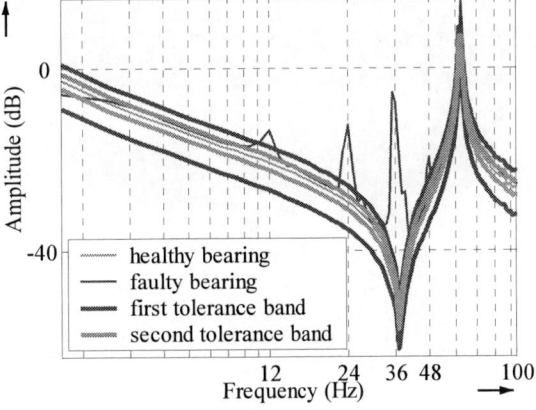

Fig. 12: Amplitude response with calculated tolerance bands, bearing 2, Outer race fault, f_{res}=64 Hz, n˙ = 200 min^{-1}

Concerning inner race faults Fig. 13 depicts the calculated tolerance bands for the inner race fault in Fig. 8. Additional to the measured deviations at the characteristic fault frequency and its multiples, the sidebands around these frequencies serve as indicators for the diagnosis of this fault. Here the importance of the second tolerance band can be seen clearly as the sidebands around $f_{IRF,1} = 18$ Hz exceed the limits of the second tolerance band but would not have been detected if only the first tolerance band had set the limits for the diagnosis.

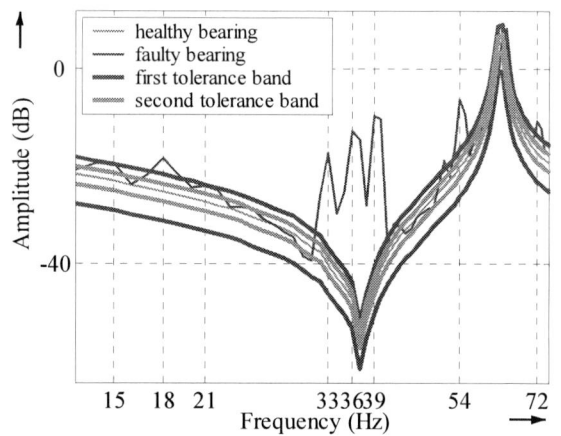

Fig. 13: Amplitude response with calculated tolerance bands, bearing 2,

Inner race fault, f_{res}=64 Hz, $n^* = 200$ min^{-1}

Fig. 14 shows the results obtained by the automated diagnosis method for 25 investigated faulty bearings. This study included all three bearing faults under different conditions as various motor speed and mechanical resonance frequency as well as both locations for the installed bearings were tested.

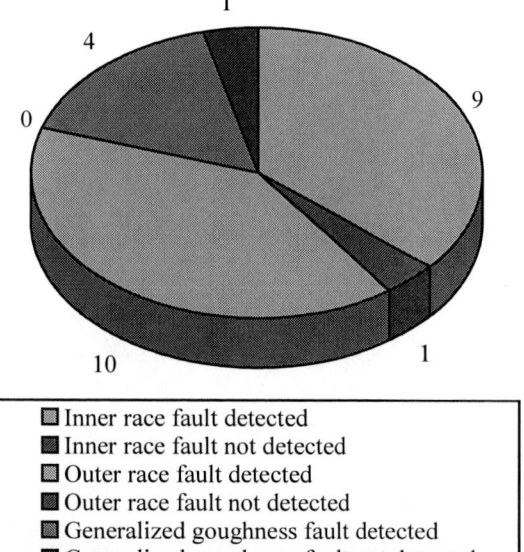

- ▨ Inner race fault detected
- ■ Inner race fault not detected
- ▨ Outer race fault detected
- ■ Outer race fault not detected
- ▨ Generalized goughness fault detected
- ■ Generalized goughness fault not detected

Fig. 14: Automated diagnosis results

It can be seen from Fig. 14 that 23 of the 25 tested bearing faults were correctly diagnosed by the automated diagnosis scheme. That value corresponds to 92% correct fault diagnosis and proofs the distinguished reliability of the presented automated bearing fault diagnosis method.

V. CONCLUSION

The present paper introduced a new automated diagnosis method for different types of rolling bearing faults. The diagnosis was carried out by the calculation of two tolerance bands around the amplitude response of the mechanical system and additional pattern recognition. Measurement results have shown that the proposed method yields excellent results for the diagnosis of the common bearing fault categories of single point defects and generalized roughness faults through their influence on the measured amplitude response.

REFERENCES

[1] J. R. Stack, T. G. Habetler, R. G. Harley: *Fault-Signature Modeling and Detection of Inner-Race Bearing Faults*, IEEE Transactions on Industry Applications, Vol. 42, No. 1, Jan./Feb. 2006

[2] J. R. Stack, Thomas G. Habetler, Ronald G. Harley: *Fault Classification and Fault Signature Production for Rolling Element Bearings in Electric Machines*, IEEE Transactions on Industry Applications, Vol. 40, No. 3, May/June 2004

[3] A. Harris: *Rolling Bearing Analysis*, Wiley-Interscience, 4th Edition, 2001

[4] B. Li, Mo-Yuen Chow, Yodyium Tipsuwan, J. C. Hung: *Neural-Network-Based Motor Rolling Bearing Fault Diagnosis*, IEEE Transactions on Industrial Electronics, Vol. 47, No. 5, Oct. 2000

[5] M. El Hachemi Benbouzid: *A Review of Induction Motors Signature Analysis as a Medium for Faults Detection*, IEEE Transactions on Industrial Electronics, Vol. 47, No. 5, Oct. 2000

[6] J.R. Stack, T.G. Habetler, R.G. Harley: *Bearing Fault Detection via Autoregressive Stator Current Modelling*, IEEE Transactions on Industry Applications, Vol. 40, No. 3, May/June 2004

[7] D. Filbert: *Diagnose von Lagerfehlern in Elektromotoren durch nichtlineare Modellierung des Motorstroms*, Forum Akustische Qualitätskontrolle, Heidelberg, Germany, 1995

[8] M. Arkan, H. Calis, M. E. Tagluk: *Bearing and Misalignment Fault Detection in Induction Motors by Using the Space Vector Angular Fluctuation Signal*, Electrical Engineering 87, pp. 197-206, 2005

[9] A. Wolfram: *Komponentenbasierte Fehlerdiagnose industrieller Anlagen am Beispiel frequenzumrichter-gespeister Asynchronmaschinen und Kreiselpumpen*, Diss. TU Darmstadt, Germany, 2002

[10] O. V. Thorson, M. Dalva: *Failure Identification and Analysis for High-Voltage Induction Motors in the Petrochemical Industry*, IEEE Transactions on Industry Applications, Vol. 35, No.4, Jul./Aug. 1999

[11] S. Villwock, J.M. Pacas, T. Eutebach: *Application of the Welch-Method for the Automatic Parameter Identification of Electrical Drives*, IEEE IECON-conference 2005, Raleigh, USA

Rotor Fault Detection System for Inverter Driven Induction Motor Using Currents Signal

N.H Kim	M.H Kim	Hamid A. Toliyat	S.H. Lee	C.H. Choi	W.S. Baik
Cheungju Univ	Yeongnam College	Texas A&M Univ.	POSCON	POSCON	Yeongnam Univ.
School of Elec. & Info.	Dept. of EE	Dept. of ECE	Research center	Research center	Dept. of EE.
Cheongju, Korea.	Daegu, Korea.	Texas, USA	Seoul, Korea	Seoul, Korea	Daegu, Korea.
hkim@cju.ac.kr	mhkim@ync.ac.kr	toliyat@tamu.ece.eud	shlee@poscon.co.kr	choi07@poscon.co.kr	priwsb@hanmail.net

Abstract—In this paper, the induction motor rotor fault diagnosis system using current signals which are measured using axis-transformation method is presented. In inverter-fed motor drives unlike line-driven motor drives the stator currents are rich in harmonics and therefore fault diagnosis using stator current is not trivial. The current signals for rotor fault diagnosis need precise and high resolution information, which means the diagnosis system demands additional hardware such as low pass filter, high resolution ADC, encoder and etc. Therefore, the proposed axis-transformation method is expected to contribute to low cost fault diagnosis system in inverter-fed motor drives without the need for encoder and any additional hardware.

In order to confirm validity of the developed algorithms, various experiments for rotor faults are tested and the line current spectrum of each faulty situation using Park transformation is compared with the results obtained from fast Fourier transforms.

I. INTRODUCTION

In recent years, marked improvement based on development micro processor and power electronics has been achieved in motor drives. However, motors driven by solid state inverters have undergone serious voltage stresses because of rapid switch-on and switch-off voltage of semiconductor devices. As a result, condition monitoring and incipient fault detection technology have become an important research area in recent years to prevent systems from sudden shut-downs because of significant motor faults in the industrial manufacturing facilities. In some factories, in order to prevent unexpected motor failures a very expensive and regular maintenance is performed using high-priced instruments. Therefore, there is a considerable demand to reduce maintenance costs and prevent unscheduled downtime. Over the past several decades substantial researches have been performed for new condition monitoring techniques for line-driven and inverter-driven motor drive.

Among these fault diagnosis techniques, analyzing vibration signal with accelerometers, air-gap flux measurement with search coil and thermal analysis provide satisfactory results [1-6]. However these methods usually need extra sensors and hardware and wirings for transmitting the signals, therefore these demands made fault diagnosis

techniques costly system. New and promising research horizons in the area of motor fault detection could be explored using the expert systems, artificial neural networks (ANNs), fuzzy systems, and genetic algorithms (GAs)[7,8] and adaptive neuron-fuzzy inference systems (ANFIS)[9]. But to extract normal and fault characteristics these techniques need offline training of diagnosis algorithm through simulations or experiments results. Also these techniques need sophisticated computational procedure.

Therefore simple and low cost protection without use of extra sensors is always the most attractive method for industrial applications. There is a tendency towards line current signature analysis due to the disadvantage of previous mentioned methods. Motor line current information of inverter fed motors is already available for control and protection purposes. Thus, by using the current sensor feedback and axis-transformation method instead of FFT(fast Fourier transform), the new trend for low-cost protection applications is achieved without using any external hardware. Even though numerous successful line driven motor fault detection methods are reported in the literature, inverter fed driven motor systems still require more attention due to high speed switching noise effects in the line current data and closed loop controller bandwidths [8, 9]. Contrary to the motor line current directly from the utility, the inverted-fed motor line current includes remarkable EMI noise. These EMI noise adversely affect the fault diagnosis due to inherent floor noise which reduces the possibility of true fault pattern recognition using line current spectrum. Therefore, one should take into consideration as much fault signatures as possible to enhance the reliability of fault diagnosis.

It is well known that adjustable speed motor drives generate sharp-edged waveforms at the output line to line voltage, which cause serious harmonics components. The harmonic contents, which are normally known as a major side effect of inverter, give rise to extra high frequency signatures and turn out to be amiss in distinguishing faulty current spectrum patterns compare with line driven drives.

Although the rotor faults are very commonly reported type with an occurrence of 5-10%[10][11], the diagnosis of these faults are one of the most challenging even under line driven motor case when compared to the other faults, because of the low amplitude fault signatures in the current spectrum. However, rotor fault detection of induction motor fed by

978-1-4244-1871-8/07 $25.00 © 2007 IEEE

inverter has not been investigated in the literature adequately and there are limited resources on the diagnosis and side effects of current spectrum floor noise that mask small fault related signals.

Thus, in this work in order to detect broken rotor bar signal clarity using 12bit ADC, park transformation of measured currents are investigated theoretically and experimentally for inverter driven motors. To verify the proposed algorithms, a 2.2 kW induction motor and a TMS320C2812 DSP is used.

II. ROTOR FAULT

Fig. 1 shows the results of motor defection conducted by EPRI and IEEE study. The study is carried out on the basis of opinion as reported by the motor manufacturer. In Fig. 1, the reasons for motor faults are classified by bearing faults, stator fault, rotor faults and others and it shows that rotor defection percentage is around 9%.

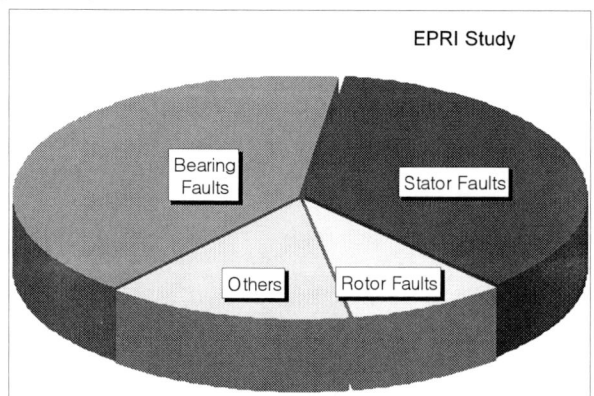

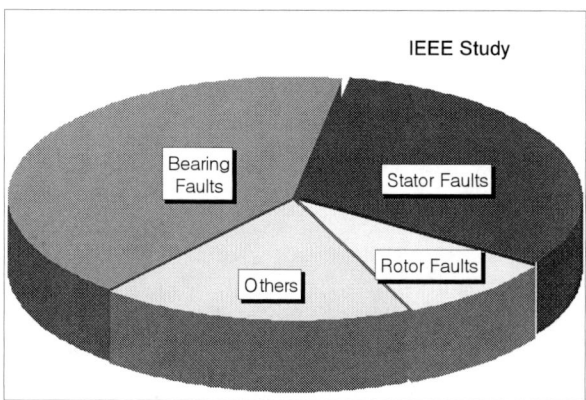

Fig. 1 Results of motor defection.

There are several reasons which cause rotor defections and these cause show below.

✓ Thermal stress, hot spot or excessive losses, sparking caused by unbalance and overload
✓ Magnetic stress due to electromagnetic forces, unbalanced magnetic pull(UMP), electromagnetic noise and vibration.
✓ Residual stress produced by manufacturing problems.
✓ Dynamic stress due to shaft torque, centrifugal forces and

cyclic stress.

✓ Environmental stress caused by abrasion of rotor material or contamination.
✓ Mechanical stresses produced by loose lamination, fatigued parts and bearing failure.

For detection rotor bar faults [11], [12], [13], and [14] were using spectrum analysis of machine line currents, and in these papers investigate sideband components(f_b) which is around fundamental currents component for broken bar faults and equation is below.

$$f_b = (1 \pm 2s)f \qquad (1)$$

Where, s is slip, f is supply frequency, and f_b is detectable broken rotor bar frequency.

While the lower sideband frequency presents broken rotor bar, the upper sideband frequency is due to consequent speed oscillation. [13] shows that broken bar give rise to a sequence of sideband given by (2)

$$f_b = (1 \pm 2 \cdot k \cdot s)f \qquad (2)$$

Where, $k = 1, 2, 3, \cdots$

III. PROPOSED ALGORITHMS AND SYSTEM CONFIGURATION

In order to calculate specific frequency component, generally FFT(Fast Fourier Transform) method is used, but FFT method need considerable calculation time and sophisticated algorithms and much memory. Therefore at this paper axis transformation, park transformation, which is very simple and effective method is presented to estimated exact sideband currents frequency components.

Axis transformation method, which is park transformation make possible conversion which is ac value to dc value. With axis transformation against specific frequency, a.c. component d.c. component are decoupled, i.e. if axis transformation is employed, the a.c. component(which are expressed in the stationary reference frame) are transformed into a new rotation reference frame, which rotates together with a selected frequency components expected to cause broken rotor bar and a.c. value expected to noise.

Fig. 2 shows block diagram of the proposed method. Through park transformation, a.c. currents component and d.c. currents component are decoupled, and in order to calculate exact d.c. component value expect to cause broken rotor fault estimate mean value for several periods. Base on previous process, ix_ave and iy_ave are calculated, and by means of classifier status of rotor is determined.

Fig. 3 shows the overall system configuration for rotor fault diagnosis. The inverter control and fault diagnosis system is implemented on the TMS320F2812 digital signal processor board from Texas Instruments. Various blocks used in the rotor bar fault diagnosis package are shown in this figure 3. Through the 12 bit on-chip ADC the current signals for motor control and fault diagnosis is collected with 4 kHz sampling frequency. Using the clark transformation and digital filter the raw current signals are transferred. To calculate the fundamental frequency and the specific frequency component which is expected fault component caused by the broken rotor bar, the Park transformation, signal conditioning, signal tracking and calculation of the average values the classifier and fault frequency estimation block are used.

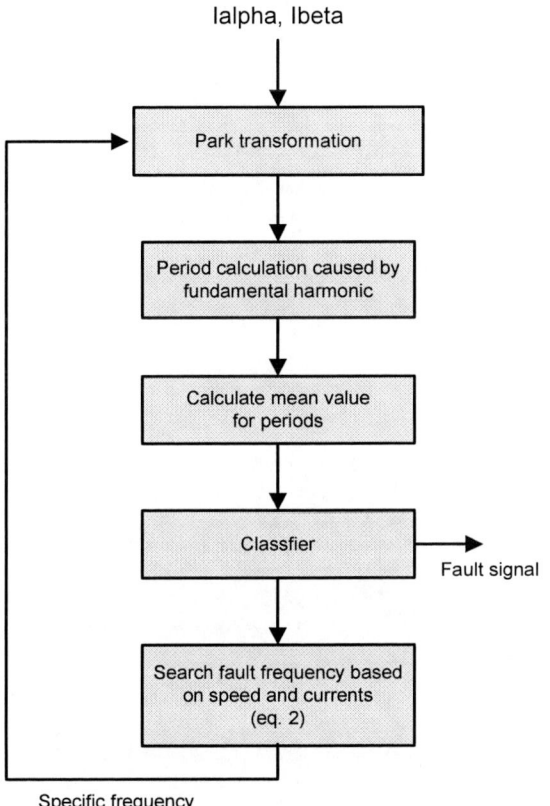

Fig. 2 Block diagram of the proposed method

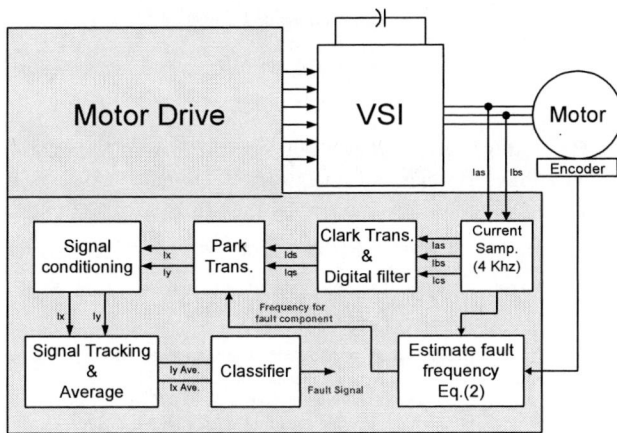

Fig. 3 System configuration.

IV. SIMULATION AND EXPERIMENTAL RESULTS

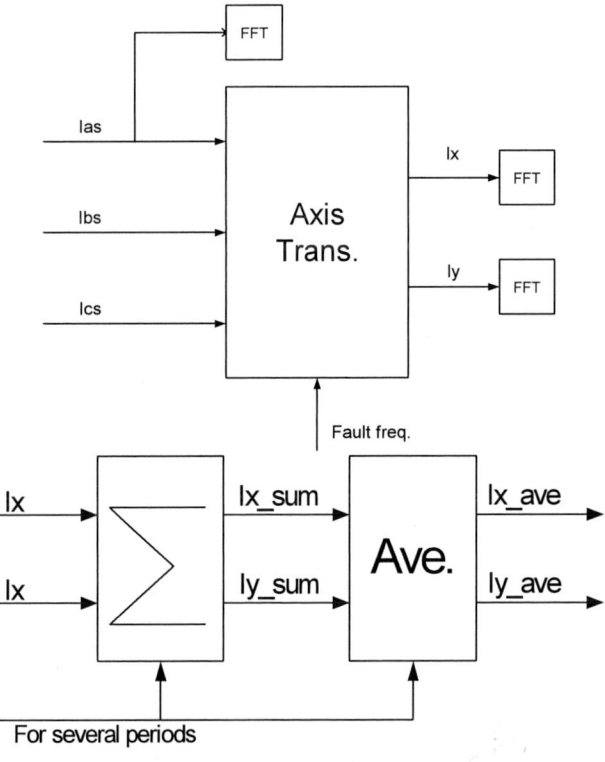

Fig. 4 System configuration of simulation

In order to verify the proposed algorithms, which is axis transformation to detect specific frequency component, simulation is done using Matlab program, and system for simulation is configured like figure 4.

To imitate current signal with side band component, below equation is used.

$$\text{Ias} = \sin(2\pi \cdot f) + 0.01 \cdot \sin(2\pi \cdot f_1) + 0.01 \cdot \sin(2\pi \cdot f_2)$$

$$\begin{aligned}\text{Ibs} = \ &\sin(2\pi \cdot f - \pi/3) + 0.01 \cdot \sin(2\pi \cdot f_1 - \pi/3) \\ &+ 0.01 \cdot \sin(2\pi \cdot f_2 - \pi/3)\end{aligned} \quad (9)$$

$$\begin{aligned}\text{Ics} = \ &\sin(2\pi \cdot f + \pi/3) + 0.01 \cdot \sin(2\pi \cdot f_1 + \pi/3) \\ &+ 0.01 \cdot \sin(2\pi \cdot f_2 + \pi/3)\end{aligned}$$

Where, $f = 60hz$, $f_1 = 56hz$, and $f_2 = 64hz$

At the FFT analysis, it is very difficult to detect sideband component because of insufficient data. However, axis transformation and average method give exact d.c. value for sideband component after 0.5 second. It is shown that the proposed algorithm can detect broken rotor defect using a 12 bit ADC.

Fig 5(a) shows current waveform that contain fundamental signal and noise signals expressed eq.(9). Fig. 5(b) represents FFT result of current signal. Fig. 5(c) present park transformation results rotating specific frequency,56hz,

978-1-4244-1871-8/07 $25.00 © 2007 IEEE   725

which is expected to cause broken rotor fault and as shown this figure, these signals are composed of d.c. value for 56hz and a.c. value for the other frequency components, 60hz and 64hz. Fig. 5(d) indicate FFT results of fig. 5(c). Fig 5(e) shows the result of the proposed algorithms using park transformation configured fig. 4 and after 0.5 sec can detect specific frequency which is 56hz component having 0.01[p.u.].

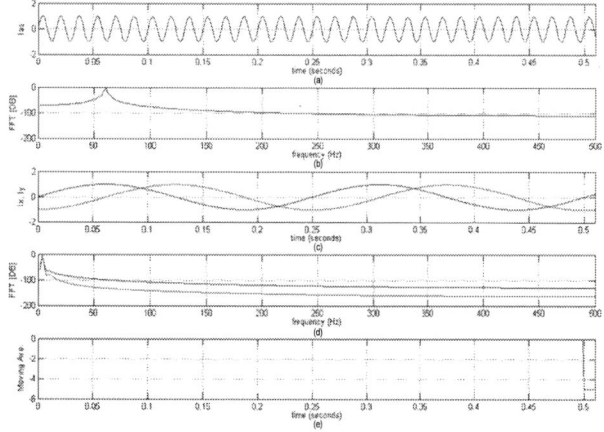

Fig. 5 Simulation results

Figure 6(a) represents current waveform of rotor defect motor. Figure 6(b) shows PSD result of faulty motor, which has broken rotor fault current component abound 24 Hz. As shown this figure, rotor fault component is -36 [dB] which means this induction motor has some problem in rotor.

Figure 6(c) shows the results of current spectrums using the proposed algorithm which is axis transformation and average method, as shown this result the proposed algorithm can detect rotor defect using 12 bit ADC. Table 1 shows motor parameters which is used at this paper.

Table 1 Induction motor parameters

Rated Power	3Hp	Rated Speed	1760rpm
Rated Voltage	230V	Rated Current	7.6A

V. CONCLUSION

This paper has investigated the feasibility of detecting broken rotor faults using an axis transformation and average method of current spectrum of an inverter driven induction machine.

An induction motor defects caused by broken rotor faults produce visible changes in the stator current spectrum at predictable frequencies. However it is very difficult to detect using FFT method because it take very long time and need pretty much data information. However, an axis transformation and average instead of FFT method is proved very effective and low cost through simulation and experimental results.

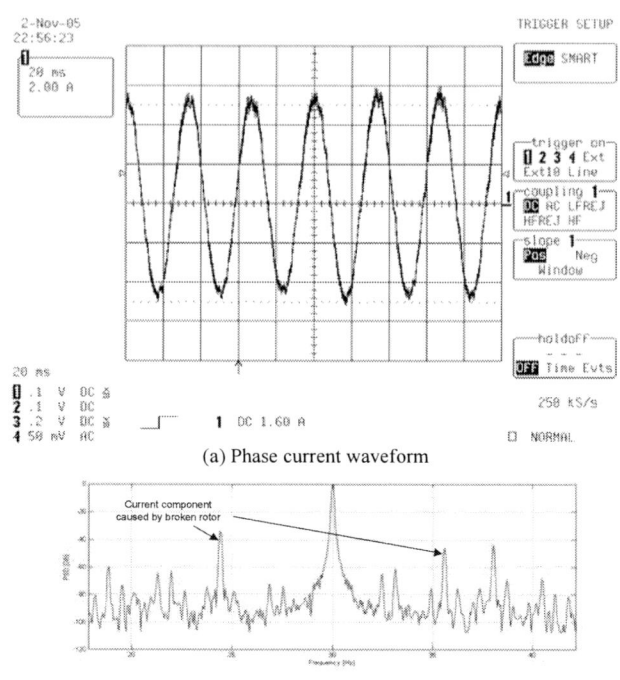

(a) Phase current waveform

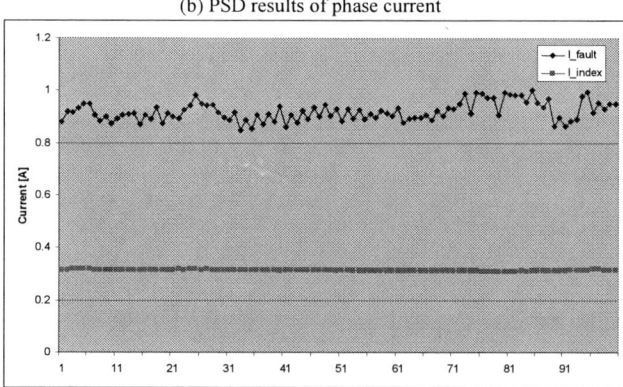

(b) PSD results of phase current

(c) Fault signal with the proposed algorithm

Fig. 6 Experimental results

ACKNOWLEDGMENT

The authors would like to thank to the Power Application Technology Research Center of Yeungnam College of Science & Technology sponsor by the Ministry of Commerce, Industry & Energy and the Korea Electric Power Corporation.

REFERENCES

[1] G.B. Kliman, R.A. Koegl, J. Stein, R.D. Endicott, and M.W. Madden, "Noninvasive detection of broken rotor bars in operating induction motors," IEEE Trans. Energy Conv., vol. 3, pp. 873–879, Dec. 1988.

[2] R. Schoen, T. Habetler, F. Kamran, and R. Bartfield, "Motor bearing damage detection using stator current monitoring," IEEE Trans. Ind. Appl., vol. 31, no. 6, pp. 1274–1279, Nov./Dec. 1995.

[3] M.E.H. Benbouzid, "A review of induction motors signature analysis as a medium for faults detection," IEEE Trans. Ind. Electron., vol. 47, no. 5, pp. 984–993, Oct. 2000.

[4] S. Nandi, M. Bharadwaj, and H.A. Toliyat, "Performance Analysis of a Three-Phase Induction Motor Under Mixed Eccentricity Condition," IEEE Trans. Energy Conv., vol. 17, pp. 392-399, Sep. 2002.

[5] J.F. Watson, N.C. Paterson, and D.G. Dorrell, "The Use of Finite Element Methods to Improve Techniques for the Early Detection of Faults in 3-phase Induction Motors", IEEE Trans. on Energy Conversion, Vol. 14, No. 3, pp. 655-660, Sep. 1999.

[6] H.A. Toliyat, M.S. Arefeen, and A.G. Parlos, "A Method for Dynamic Simulation and Detection of Air-Gap Eccentricity in Induction Machines", IEEE Trans. on Industry Applications, Vol. 32, No. 4, pp. 910-918, Jul./Aug. 1996.

[7] F. Filippetti, G. Franceschini, C. Tassoni, and P. Vas, "Recent developments of induction motor drives fault diagnosis using AI techniques," IEEE Trans. Ind. Electron., vol. 47, no. 5, pp. 994–1004, Oct. 2000.

[8] M. A. Awadallah and M. M. Morcos, "Application of AI tools in fault diagnosis of electrical machines and drives—an overview," IEEE Trans. Energy Convers., vol. 18, no. 2, pp. 245–251, Jun. 2003.

[9] J.-S. R. Jang, "ANFIS: adaptive-network-based fuzzy inference system," IEEE Trans. Syst., Man, Cybern., vol. 23, no. 3, pp. 665–685, May/Jun. 1993.

[10] McDermid, W., "Insulation systems and monitoring for stator winding of large rotating machines," IEEE Electrical Insulation Magazine, vol. 9, no. 4, pp. 7-15, 1993.

[11] Schemp, D.E., "Predict motor failure with insulation testing," Plant Engineering, vol. 50, pp. 97-96, 1996.

[12] Stone, G.C., "Partial discharge measurements to access rotation machine insulation condition : A survey," Conference Record of the IEEE International Symposium on Electical Insulation, pp. 19-23, 1996.

[13] J.R. Cameron, W. T. Thomson and A.B. Dow, "Vibration and Current monitoring for detecting airgap eccentricity in large induction motors", Proceeding of IEE, vol. 133, Pt. B, No. 3, pp. 155-163, May 1986.

[14] J. S. hsu, "Monitoring of defects in induction motors through air-gap torque observation," IEEE Transaction on Industrial Applications, Vol. 31, No. 5, pp. 1016-1021, Sept./Oct. 1995.

978-1-4244-1871-8/07 $25.00 © 2007 IEEE

Fault Diagnosis of Induction Motor using Decision Tree with An Optimal Feature Selection

Ngoc-Tu Nguyen
University of Ulsan
School of Electrical Engineering
Mugeo 2-dong, Namgu, Ulsan, 680-749
Email: nguyentu@mail.ulsan.ac.kr

Jeong-Min Kwon
University of Ulsan
School of Electrical Engineering
Mugeo 2-dong, Namgu, Ulsan, 680-749
Email: ruee006@mail.ulsan.ac.kr

Hong-Hee Lee
University of Ulsan
School of Electrical Engineering
Mugeo 2-dong, Namgu, Ulsan, 680-749
Email: hhlee@mail.ulsan.ac.kr

Abstract— Time vibration signals are measured to extract a feature set for fault diagnostics of induction motor. Feature selection by decision tree and genetic algorithm (GA) is presented in this paper to remove irrelevant information in the feature set. New data with the selected features is used to train a decision tree, which is an expert system for classification. Testing results show that systems with selected features can reliably diagnose different conditions of induction motor, which has better performance compared to original one without feature selection.

I. INTRODUCTION

Induction motors are important parts in industrial processes because of their reliability and low cost. Along with their widespread, there are many kinds of fault to be occurred. Once faults happen, they can cause the motor to be broken down or simply decrease its performance. These risks can be prevented or predicted by condition monitoring or fault diagnosis techniques. Because of that reason, machine fault diagnosis is more and more gaining importance and many researches have been developed recently. Main objective of fault diagnosis of induction motors is continuous or periodical evaluation of the healthiness of the motors.

Many systems and methods for fault diagnostics have been proposed. Expert systems have been widely used for this purpose, such as decision tree in [1-4]. Other systems are support vector machine (SVM) in [4-8], artificial neural network [9-10], k nearest neighbors (kNN) in [12], hidden markov model [13], etc. In these methods, decision tree is used because of its simplicity in structure and high accuracy. The decision tree algorithm C4.5 [11] is proposed in this paper for induction fault diagnosis.

The way of process data is an important part to improve the classification performance. Feature extraction and feature selection are two most popular way to process the data. Feature extraction is a technique that extracts the needed information from the original data to form a new smaller data. Feature selection that selects the most representative features as inputs of diagnostic system may improve its classification accuracy. This not only reduces the data dimension but also discards low quality features without changing the data. In this paper, decision tree is proposed for induction motor fault diagnosis with selecting optimal features. First, features are extracted from time-domain signals which are measured by accelerometer sensors. Then, these features are selected based on gain information criterion. After that, they are

selected again by a distance based genetic algorithm. The final selected features are used to diagnose motor conditions. The classification results are compared between the systems with feature selection and without feature selection by decision tree and kNN classifiers in experiment section.

II. TIME-DOMAIN FEATURES

Induction motor fault diagnostics is presented in this work. Five conditions of induction motor can be considered as follows: normal condition (healthy), bearing looseness, bearing damage, rotor unbalance, and stator winding fault. Time vibration signals are measured and extracted to form a feature set for diagnosing. The signals are obtained in three directions (horizontal, axial, vertical), 6 features are extracted from each direction as follows: rms (root mean square), variance, skewness, kurtosis, crest factor, and maximum value. There are total 18 extracted features.

Table 1 lists the needed equations for extracting feature set from vibration signals, and these features represent energy, amplitude, and time series distribution of the time-domain signals.

TABLE 1
Time-domain feature equations

Feature	Equation
Root mean square	$rms = \sqrt{\dfrac{\sum_{n=1}^{N}\left(x(n)\right)^2}{N}}$
Variance	$var = \sigma^2 = \dfrac{\sum_{n=1}^{N}\left(x(n)-mean(x)\right)^2}{(N-1)}$
Skewness	$skewness = \dfrac{\sum_{n=1}^{N}\left(x(n)-mean(x)\right)^3}{(N-1)\sigma^3}$
Kurtosis	$kurtosis = \dfrac{\sum_{n=1}^{N}\left(x(n)-mean(x)\right)^4}{(N-1)\sigma^4}$
Crest factor	$crest = \dfrac{\max|x(n)|}{rms}$
Maximum value	$max = \max|x(n)|$

978-1-4244-1871-8/07 $25.00 © 2007 IEEE

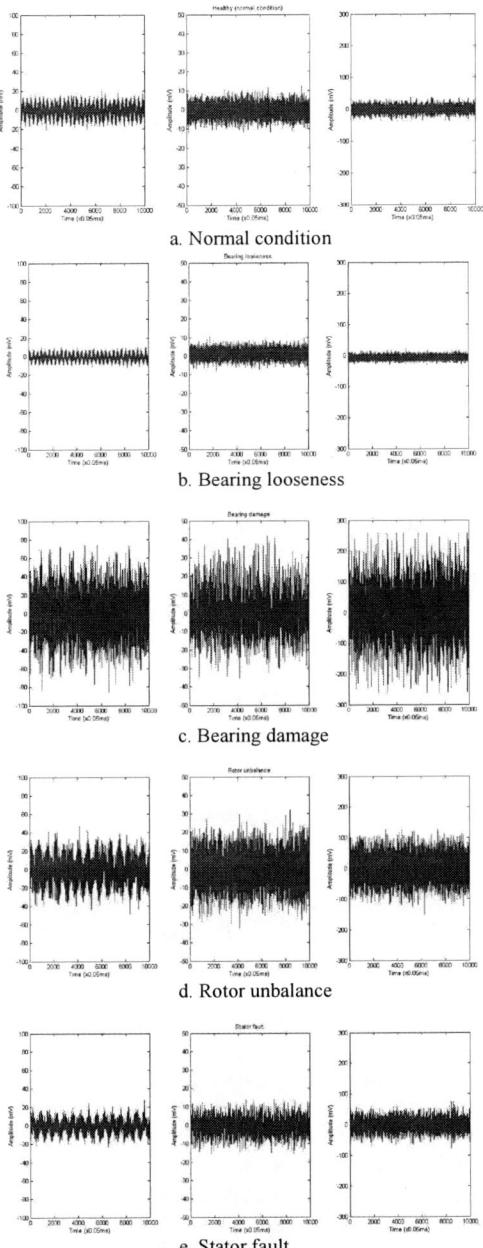

Fig. 1: Time signals measured in 3 directions (from left to right: horizontal, axial, and vertical) for 5 motor conditions: a. Normal, b. Looseness, c. Bearing damage, d. Unbalance, e. Stator fault

III. FEATURE SELECTION

A. Decision Tree

A decision tree is a diagnostic tool that builds a knowledge-based system by inductive inference from case histories. A decision tree contains leaves which represent classifications, and branches which represent conjunctions of features that lead to the classifications. The structure of decision tree highly depends on how a test is selected as the root of the tree. The criterion for selecting the root of the tree is Quinlan's information theory (information gain) [11].

According to this criterion, the information conveyed by a message depends on its probability and can be measured in bits as minus the logarithm to base 2 of that probability. The construction of decision tree is based on a training set T, which is a set of cases. Each case specifies the values for a collection of attributes and for a class. Let the classes be denoted by $\{C_1, C_2, ..., C_k\}$. Suppose there is a possible test with n outcomes that partitions the training set T into subsets denoted by $\{T_1, T_2, ..., T_n\}$. Assume S is any set of cases, freq(C_i , S) is the number of cases in S that belong to class C_i , and |S| is the number of cases in set S. If one case is selected at random from set S and it belongs to class C_j, the message has the probability

$$\frac{freq(C_j, S)}{|S|} \tag{1}$$

and the information it conveys is

$$-\log_2 \frac{freq(C_j, S)}{|S|} \text{ bits} \tag{2}$$

The expected information needed to identify the class of case in S is

$$\text{info(S)} = -\sum_{j=1}^{k} \frac{freq(C_j, S)}{|S|} \times \log_2 \left(\frac{freq(C_j, S)}{|S|} \right) \text{ bits} \tag{3}$$

When it is applied to the set of training cases, info(T) measures the average amount of information needed to identify the class of a case in T.

A similar measurement after T has been partitioned in accordance with n outcomes of a test X

$$\text{info}_X(T) = \sum_{i=1}^{n} \frac{|T_i|}{|T|} \times \text{info}(T_i) \text{ bits} \tag{4}$$

The quantity

$$\text{gain(X)} = \text{info(T)} - \text{info}_X(T) \tag{5}$$

measures the information that is gained by partitioning T in accordance with the test X. The gain criterion selects a test in order to maximize this information gain.

For induction motor fault diagnosis, each time-domain feature plays a role as a test node of the tree. At the node, only feature that has maximum information gain can be selected, and then, this feature has the largest needed information to identify a class. Therefore, the features that occur in decision tree can be selected as appreciated features for classification. The features which do not show up can be removed. So the number of features can be reduced to 6 by using c4.5 algorithm as shown in Fig. 2.

The decision tree in Fig. 2 is trained with all features that extracted from time-domain vibration signals. There are 18 features at the beginning and only 6 features which are used in the tree. They are F2, F4, F6, F8, F13, and F14. These selected features will be chosen one more times by a genetic algorithm that is described in next section.

978-1-4244-1871-8/07 $25.00 © 2007 IEEE

$$J_b = \sum_{i=1}^{c} p_i \left(m_i - m \right)^T \left(m_i - m \right). \qquad (9)$$

The fitness value of a chromosome is a measure of how good that chromosome is. Members which have good fitness value are chosen and modified to reproduce next population. Modification process includes random crossover and mutation with the bits from the chosen chromosome at random points. The solution is obtained when the terminated condition is satisfied. The selected chromosome is decoded to receive the best feature subset based on fitness function. Now, this feature subset can be used for classification. The result is showed in experiment section.

IV. EXPERIMENT RESULTS

Training set includes 775 samples, and test set has 413 samples. GA parameters are as follows: mutation rate is 0.7, 100 generations, each generation has 96 members, and 1 crossing point crossover with 0.7 crossover rate. The final feature selection by GA is 001011 which mean the features F6, F13, and F14 are selected. The performances with and without feature selection are listed in Table 2 and Table 3 by kNN and decision tree, respectively.

Performance of kNN (3 features) with k = 12 nearest neighbors is a bit better than 3 features decision tree in Table 3 but it takes a lot of calculation. For obtaining diagnostic result of a sample, kNN has to compute the distances between the testing sample and all training samples. Decision tree is much simpler than kNN when it is evaluated.

The accuracy of all states in both decision tree and kNN classifiers is increased significantly after feature selection. But for normal state in case of using decision tree, the diagnostic accuracy is decreased after feature selecting. This may be caused because some information is lost after discarding a large number of features. However, the overall diagnostic accuracy is improved for both of the classifiers.

The performance of decision tree is improved in general when decision tree is evaluated with test set as shown in Table 3 from 90.1% up to 95.9%. The efficiency of the feature selection also is illustrated by performance of kNN classifier in Table 2. The kNN accuracy increases significantly when it is evaluated with test set from 78.5% up to 97.1%.

V. CONCLUSION

Decision tree applied in this paper has a simple construction and high accuracy. The feature selection has shown to be able to improve the performance of fault classification. The classification with selected features has higher accuracy and better performance comparing to the tree without feature selection. This selection not only reduces the data dimension but also removes the redundant and irrelevant information in the original data. The results show that decision tree and feature selection are a potential way to diagnose induction motor conditions.

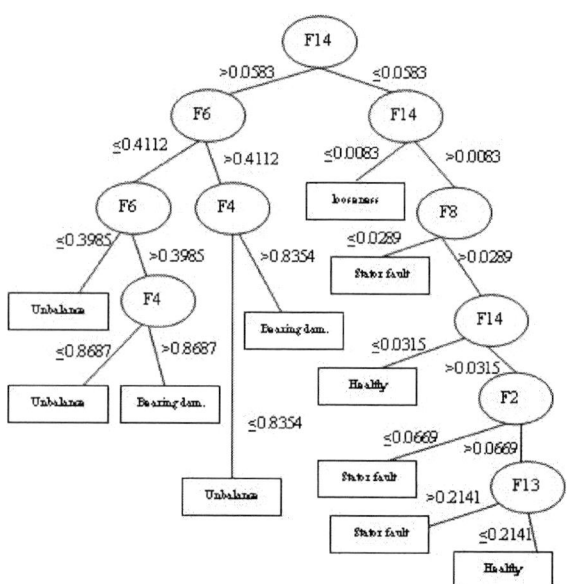

Fig. 2: Decision tree trained with all features

B. Distance based Genetic Algorithm

The features selected by decision tree can be selected again by a genetic algorithm. GA based feature selection uses distance evaluation criteria for choosing features. GA is a search algorithm which simulates natural system by following the principle of survival of the fittest. It solves problem by using the combination of selection, crossover and mutation to regenerate next generations for finding a solution.

There is a 6-bit chromosome which represents 6 features. In the chromosome, "1" represents a selected feature, and "0" represents an abandoned feature. The smallest fitness function of the GA is selected. A feature subset is obtained based on the criterion that within-class distance is smaller and between-class distance is larger. Fitness function is

$$\text{Fitness function} = \frac{\sum\limits_{\text{classes}} \text{within-class distance}}{\sum\limits_{\text{classes}} \text{between-class distance}} = \frac{J_c}{J_b} \qquad (6)$$

The within-class distance is given by

$$J_c = \sum_{i=1}^{c} p_i J_i. \qquad (7)$$

where

$$J_i = \left(1 / n_i\right) \sum_{k=1}^{n} \left(x_k^i - m_i \right)^T \left(x_k^i - m_i \right). \qquad (8)$$

Class i = 1,…, c; m_i is the mean vector of class i; n_i is the number of samples in class i; p_i is the number of samples in class i.

The between-class distance is

TABLE 2

Comparison of the performances before and after feature selection using k-NN classifier

No. of features	k	Test accuracy	Normal	Looseness	Bearing damage	Rotor unbalance	Stator fault
18 (F1 ÷ F18)	7	78.5%	80.4%	68.2%	100%	91.5%	53.8%
3 (F6, F13, F14)	12	97.1%	89.3%	100%	100%	98.9%	95.3%

TABLE 3

Comparison of decision tree performances before and after feature selection

No. of features	Size	Training accuracy	Test accuracy	Normal	Looseness	Bearing damage	Rotor unbalance	Stator fault
18 (F1 ÷ F18)	21	99.9%	90.1%	100%	100%	90.1%	93.6%	75.5%
3 (F6, F13, F14)	23	94.1%	95.9%	89.3%	100%	100%	98.9%	90.6%

ACKNOWLEDGMENTS

The authors would like to thank University of Ulsan, the Ministry of Commerce, Industry and Energy (MOCIE) and Ulsan Metropolitan City that partly supported this research through the Network-based Automation Research Center (NARC).

REFERENCES

[1] Weixiang Sun, Jin Chen, Jiaquing Li, "Decision tree and PCA-based fault diagnosis of rotating machinery", *Mechanical Systems and Signal Processing*, 2006.

[2] Bo Suk Yang, Chul Hyun Park, Ho Jong Kim, "An Efficient Method of Vibration Diagnostics for Rotating Machinery using a Decision Tree", *International Journal of Rotating Machinery*, Vol. 6, 2000, No.1, pp. 19-27.

[3] Dong Soo Lim, Bo Suk Yang, Dong Jo Kim, "An Expert System for Vibration Diagnosis of Rotating Machinery using Decision Tree", *International Journal of COMADEM*, 2000, pp. 31- 36.

[4] V. Sugumaran, V. Muralidharan, K. I. Ramachandran, "Feature selection using Decision Tree and classification through Proximal Support Vector Machine for fault diagnostics of roller bearing", *Mechanical Systems and Signal Processing* 21, 2007, pp. 930-942.

[5] Achmad Widodo, Bo-Suk Yang, Tian Han, "Combination of independent component analysis and support vector machines for intelligent faults diagnosis of induction motors", *Expert Systems with Applications* 32, 2007, pp. 299-312.

[6] J. Yang, et al., "Intelligent fault diagnosis of rolling element bearing based on SVMs and fractal dimension", *Mechanical Systems and Signal Processing*, 2006, doi: 10.1016/j.ymssp.2006.10.005.

[7] Achmad Widodo, Bo-Suk Yang, "Application of nonlinear feature extraction and support vector machines for fault diagnosis of induction motors", *Expert Systems with Applications*, 2006.

[8] Alfonso Rojas, Asoke K. Nandi, "Practical scheme for fast detection and classification of rolling-element bearing faults using support vector machines", *Mechanical Systems and Signal Processing* 20, 2006, pp. 1523-1536.

[9] B. Samanta, K. R. Al-Balushi, "Artificial Neural Network based fault diagnostics of rolling element bearings using time-domain features", *Mechanical Systems and Signal Processing* 17, 2003, pp. 317-328.

[10] B. Samanta, K. R. Al-Balushi, S. A. Al-Araimi, "Artificial neural networks and genetic algorithm for bearing fault detection", *Soft Coput.*, 2006, pp. 264-271.

[11] J. R. Quinlan, *C4.5: Programs for Machine Learning*, Morgan Kaufmann Publisher, Inc. (1993).

[12] R. Casimir, E. Boutleux, G. Clerc, A. Yahoui, "The use of features selection and nearest neighbors rule for faults diagnostic in induction motors", *Engineering Applications of Artificial Intelligence*, 2006, pp. 169-177.

[13] V. Purushotham, S. Narayanan, Suryanarayana A.N. Prasad, "Multi-fault diagnosis of rolling bearing elements using wavelet analysis and hidden Markov model based fault recognition", *NDT&E International* 38, 2005, pp. 654-664.

[14] H. Gu, Z. Gao, F. Wu, "Selection of Optimal Features for Iris Recognition", *International Symposium on Neural Networks*, 2005, pp. 81-86.

[15] Cheng-Lung Huang, Chieh-Jen Wang, "A GA-based feature selection and parameters optimization for support vector machines", *Expert Systems with Applications* 31, 2006, pp. 231-240.

[16] J. S. Rao, *Vibratory Condition Monitoring of Machines*, Alpha Science International Ltd. (2000), pp. 364-373.

A Simple Fault Detection of the Open-Switch Damage in BLDC Motor Drive Systems

Jung-Dae Lee, Byoung-Gun Park, Tae-Sung Kim, Ji-Su Ryu, and Dong-Seok Hyun

Hanyang University
Dept. of Electrical Engineering
17 Haengdang-dong Seongdong-gu, Seoul, 133-791
Korea
E-mail: skyljd78@hanyang.ac.kr

Abstract—This paper proposes a novel fault detection algorithm for a brushless DC (BLDC) motor drive system. This proposed method is configured without the additional sensor for fault detection and identification. The fault detection and identification are achieved by a simple algorithm using the operating characteristic of the BLDC motor. The drive system after the fault identification is reconfigured by four-switch topology connecting a faulty leg to the middle point of DC-link using bidirectional switches. This proposed method can also be embedded into existing BLDC motor drive systems as a subroutine without excessive computational effort. The feasibility of a novel fault detection algorithm is validated in simulation.

I. INTRODUCTION

The brushless DC (BLDC) motor is attracting much interest due to its high efficiency, high power factor, high torque, simple control, and lower maintenance requirement [1], [2]. Moreover, the use of an automated and large drive system of BLDC motor in industrial field is increasing with the development of industrial technology. The reliability of the drive system in many applications is becoming extremely important. When the faults in the drive system occur in industrial field such as aerospace, military, electrical vehicle, and robot arms, it can lead to serious consequences. For example, when one of the faults does occur in the practical applications above, the drive operation has to be stopped for a nonprogrammed maintenance schedule. The cost of this schedule can be high and this justifies the development of fault tolerant control strategies for improving reliability.

In recent years, several papers for faults in motor drive systems have been published using the following schemes: fault detection and identification methods [3]-[6], reconfiguration schemes for isolating faulty power devices [7], [8], and fault tolerant schemes [9]-[11]. However, most existing fault detection and identification methods have problems because the fault detection time takes at least one fundamental period, the process for detecting faults is complex and/or schemes to identify faults are inadequate. In addition, these methods use additional sensor for fault detection. Among them, a fault tolerant strategy for BLDC motor drive was suggested in [11]. This method introduced a fault detection algorithm. However, in this method, the fault detecting time for the fault identification must be longer than

the interval of a mode because the error times of two excited phases increase identically within a mode. And this method cannot avoid decreases of the control performance until the fault identification. To solve these problems, in novel proposed fault detection method, it is possible to detect fault within the maximum one mode and avoid decreases in the control performance.

This paper proposes a fault tolerant system that can quickly recover the control performance by fast fault detection time and reconfiguration of system topology [12]. This method is also achieved without using additional sensor and a simple algorithm, as it uses the operating characteristic of BLDC motor drive. This method can be embedded into existing BLDC motor drive system as a subroutine without excessive computational effort. The proposed fault tolerant system performs the follow tasks:

1) fault detection;
2) fault identification;
3) fault isolation;
4) reconfiguration of hardware and software.

The feasibility of the proposed novel fault detection algorithm is validated in simulation.

II. BLDC MOTOR DRIVE SYSTEM

Generally, the BLDC motor drive system can be modeled as an electrical equivalent circuit that consists of a resistance, an inductance, and back-EMF per phase. The electrical equivalent circuit and drive performance of this system are shown in Fig. 1.

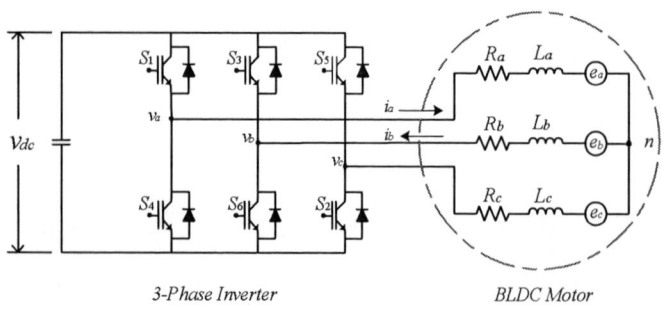

3-Phase Inverter   *BLDC Motor*

Fig. 1. The electrical equivalent circuit of BLDC motor drive system

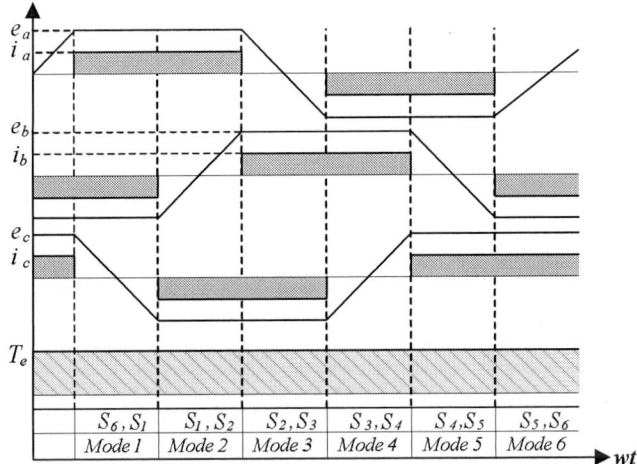

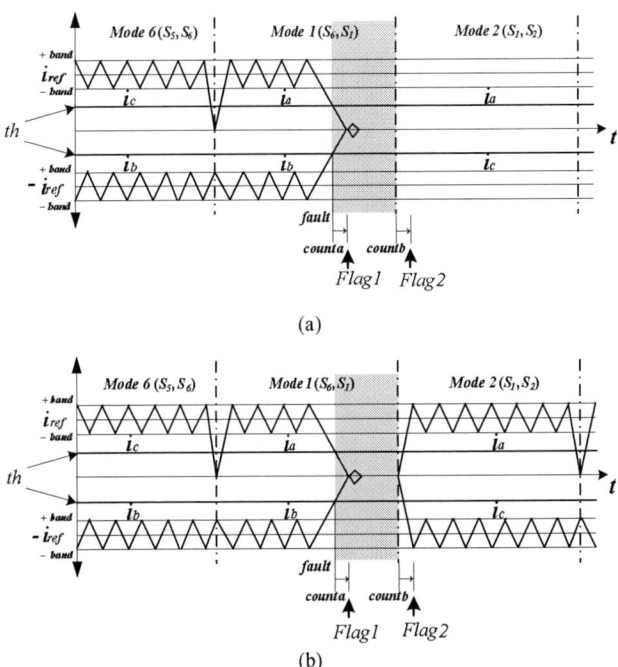

Fig. 2. Waveform of back-EMFs, phase currents, and the developing torque.

As shown in Fig. 2, BLDC motor has a trapezoid-shaped back-EMF. Because back-EMF must synchronize with square wave currents, square wave currents are injected from the same period. The BLDC motor drive system is operated by exciting two phases among the three phases. Two switches are only operated in one mode. The proposed fault tolerant system in this paper is achieved by characteristic of phase current in BLDC motor drive. The voltage equation of three phases is given by

$$
\begin{bmatrix} v_{an} \\ v_{bn} \\ v_{cn} \end{bmatrix} = \begin{bmatrix} R_a & 0 & 0 \\ 0 & R_b & 0 \\ 0 & 0 & R_c \end{bmatrix} \begin{bmatrix} i_a \\ i_b \\ i_c \end{bmatrix} + \begin{bmatrix} L_a & 0 & 0 \\ 0 & L_b & 0 \\ 0 & 0 & L_c \end{bmatrix} \frac{d}{dt} \begin{bmatrix} i_a \\ i_b \\ i_c \end{bmatrix} + \begin{bmatrix} e_a \\ e_b \\ e_c \end{bmatrix}. \quad (1)
$$

The torque equation is given by

$$
T_e = \frac{e_a \cdot i_a + e_b \cdot i_b + e_c \cdot i_c}{\omega_m}. \quad (2)
$$

where v_{an}, v_{bn}, and v_{cn} are phase voltages. R_a, R_b, and R_c are phase resistances. i_a, i_b, and i_c are phase currents. L_a, L_b, and L_c are phase inductances. e_a, e_b, and e_c are phase back–EMFs. ω_m is rotor speed.

III. CHARACTERISTIC ANALYSIS OF OPEN FAULT IN BLDC MOTOR DRIVE SYSTEM

The BLDC motor drive system is operated by exciting two phases among the three phases. When the open fault of a switch happens, the current waveform appears differently according to the position of a faulty switch. In this work, only the open circuit damage of a switch (S_1) and (S_6) will be considered. Also, the effect of this fault type in any other device is assumed to be symmetrical. For comprehension of the open fault of the switch, Fig. 3 shows the waveform of two phase currents in case of the open fault of the upper switch (S_1) or lower switch (S_6) in Mode 1.

Fig. 3. Current waveforms under open faults in *Mode 1*. (a) The upper switch fault. (b) The lower switch fault.

A. The Open Fault of the Upper Switch (S₁) in Mode 1

As shown in Fig. 3(a), when the open fault of the upper switch (S_1) in Mode 1 happens, even if the lower switch (S_6) operates normally, currents of phase A and B do not appear. Moreover, when the lower switch (S_2) operates normally by switching pattern in Mode 2, currents of phase A and C also do not appear due to the open fault of the upper switch (S_1). As a result, if the open fault of the upper switch (S_1) occurs in Mode 1, phase currents do not appear in Mode 1 and 2.

B. The Open Fault of the Lower Switch (S₆) in Mode 1

As shown in Fig. 3(b), when the open fault of the lower switch (S_6) happens in Mode 1, even if the upper switch (S_1) operates normally, phase currents of phase A and B do not appear. However, when the lower switch (S_2) operates normally by switching pattern in Mode 2, currents of phase A and C can flow despite the open fault of the lower switch (S_6). As a result, switch (S_1) and (S_2) operates normally in Mode 2, even if the open fault of the lower switch (S_6) occurs in Mode 1.

IV. PROPOSED FAULT DETECTION ALGORITHM FOR OPEN FAULT

As shown in Fig. 2, the operation of the BLDC motor drive system is classified into the six-mode. And the current waveform appears differently according to the state of each mode. This characteristic of the current waveform offers a simple algorithm to detect fault. The fault detection method employed each *Fault_flag* signal that represents the fault detection and identification. The fault detection and identification method that applies the *Fault_flag* signals is shown by the flowcharts in Fig. 4.

978-1-4244-1871-8/07 $25.00 © 2007 IEEE

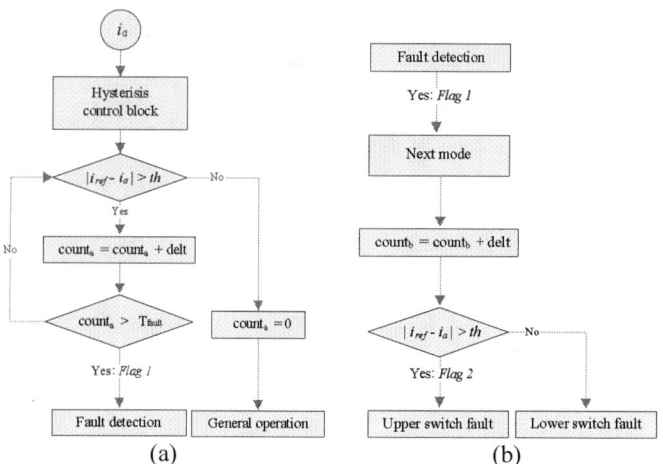

Fig. 4. (a) Flowcharts for the fault detection method. (b) Flowcharts for the fault identification.

These flowcharts are classified into two parts: Fault detection and identification. After the open fault of switch, the difference between reference and actual current increases because the actual current cannot follow the reference current. If current value of $|i_{ref} - i_{act}|$ is larger than the fixed threshold value (th), an error is detected. The others judge by normal operation. The error detection can be achieved using this difference as follows:

> **IF** $|i_{ref} - i_{act}| > th$ **THEN** *error*
>
> **ELSE** *normal*

The fault detecting time is given by

$$T_{fault} = K \times \frac{2}{p \times \omega_{ref}} \times \frac{1}{Mode} . \qquad (3)$$

where T_{fault} is the fault detecting time. K is the sensitivity for faults. p is the number of pole. ω_{ref} is the reference speed.

And then if the error time detecting the error continuously is longer than the fault detecting time, the proposed system generates *Fault_flag 1* signal. The generated *Fault_flag 1* signal represents fault condition. The algorithm for the fault diagnosis given by

> **IF** (*count a* > T_{fault}) **THEN** *Fault_flag 1*
>
> **ELSE** *normal*

where *count* is the error time detecting the error continuously.

After the *Fault_flag 1* signal is generated, the identification of the faulty switch is obtained from *Fault_flag 2* signal. *Fault_flag 1* siganl represents only fault condition after *count a*. And when the *mode* is converted to next *mode*, faulty switch is identified through *Fault_flag 2* signal after *count b*. This signal is decided, in accordance with value of $|i_{ref} - i_{act}|$. The algorithm for this fault identification method after fault detection given by

> **Next Mode,** and after *count b*
>
> **IF** $|i_{ref} - i_{act}| > th$ **THEN** *upper switch fault*
>
> **ELSE** *lower switch fault*

Table I shows how fault identification under open fault is represented in the six-mode conversion. As a result, the proposed fault detection algorithm for open faults is achieved by *Fault_flag* signals without additional sensor.

TABLE I
THE FAULT STATE OF SWITCHES IN THE SIX-MODE CONVERSION

Mode Conversion	Flag 1 (Previous mode)	Flag 2 (Next mode)	Fault Switch
Mode 1 → Mode 2	High	Low	S_5
	High	High	S_6
Mode 2 → Mode 3	High	Low	S_6
	High	High	S_1
Mode 3 → Mode 4	High	Low	S_1
	High	High	S_2
Mode 4 → Mode 5	High	Low	S_2
	High	High	S_3
Mode 5 → Mode 6	High	Low	S_3
	High	High	S_4
Mode 6 → Mode 1	High	Low	S_4
	High	High	S_5

V. ISOLATION AND RECONFIGURATION

This paper proposes a novel fault detection and identification method. To validate the reliability of this algorithm, existing isolation and reconfiguration method is introduced into [10]. The reconfigured system by the isolation of the faulty phase is shown in Fig. 5.

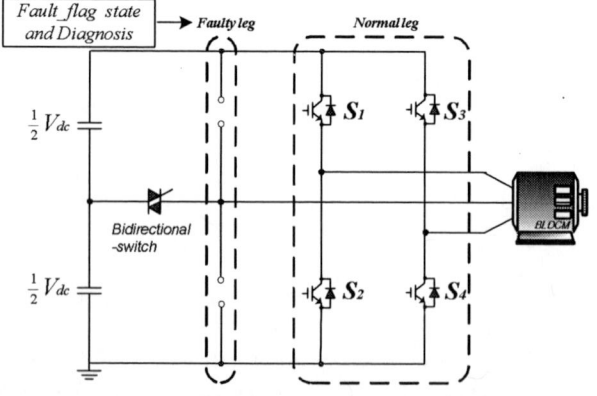

Fig. 5. The reconfigured system by the bidirectional switch and removing the switching signal of the faulty phase.

After identification of the faulty phase is achieved by *Fault_flag 2* signal, the reconfiguration scheme is achieved by four-switch topology using bidirectional switches. When the fault of a switch is identified, the bidirectional switch of the faulty phase is fired in order to connect the faulty phase to the midpoint of the DC-link and the switching signals of the faulty phase are removed at the same time. The system after isolating the faulty phase is operated by four-switch three-phase BLDC motor drives has been proposed in [12]. Because this control scheme has irregular voltage utilization, during the half DC-link voltage period, the rate of current incensement is less than the full DC-link voltage period. This irregular current shape can cause torque ripple, but it can be controllable by adjusting hysterisis band size. Although the fault tolerant system has reduced performances and reduced working time under fault, this aspect is considered to be a compromise.

VI. OVERALL FAULT TOLERANT SYSTEM

The overall structure of the proposed fault tolerant drive system is shown in Fig. 6. A novel proposed algorithm is applied to block of fault detection and diagnosis. Therefore open fault diagnosis is achieved by the proposed system in this block, as stated above.

In the case of a short-circuit fault in one or two power switches of the same leg, the control strategy interrupts the switches command after the fast active fuses blow out and isolates the respective inverter leg.

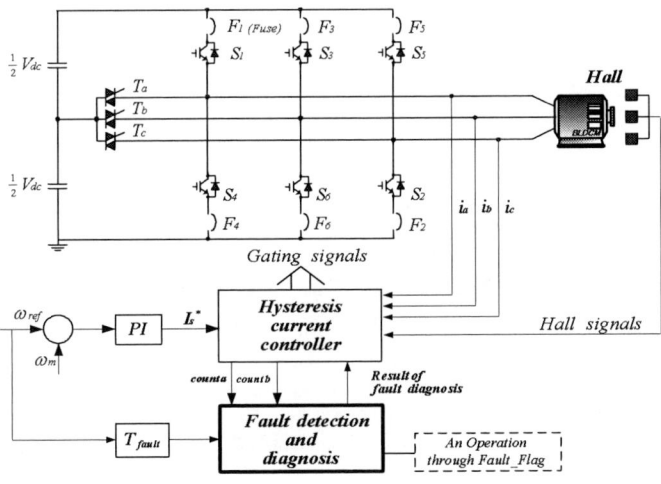

Fig. 6. The overall structure of the proposed fault tolerant drive system.

VII. SIMULATION RESULTS

In order to verify the developed fault tolerant system, a computer simulation was performed along with the existing fault tolerant system [11]. Simulations are performed for the BLDC motor with ratings of parameters in Table II. In this work, only the open fault of switch (S_1) will be considered. Also, the effect of this type of fault in any other devices is assumed to be symmetrical. And the same simulation condition of existing method applies to novel proposed method. First of all, the existing fault tolerant system was simulated.

TABLE II
RATINGS AND PARAMETERS OF BLDC MOTOR

Rated Voltage	V	300 [V]
Rated Torque	T_e	0.662 [Nm]
Rated Speed	N_r	3000 [rpm]
Resistance	R_s	11.9 [Ω]
Inductance	L_s	2.07 [mH]
Back-EMF Constant	K_e	22.9 [V/rpm]
Poles	P	4

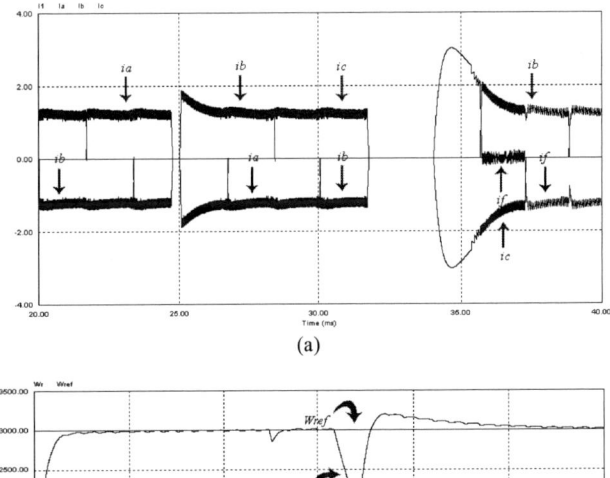

(a)

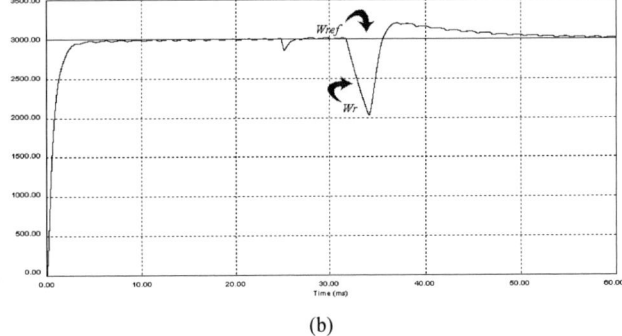

(b)

Fig. 7. Simulation results of the existing method by difference of reference and actual current. (a) Faulty current waveforms. (d) Rotor speed.

Fig. 7 shows simulation results in the existing method when the open fault occurs at 24.8ms. As shown in Fig. 7(a), although the open fault occurs at 24.8ms, existing method cannot immediately detect it. The open fault condition is detected after about one fundamental period. Therefore the method of existing system becomes unstable and requires a long time in order to detection. However the system of novel proposed method is stable and reduces the detection time without sensors and complex detection algorithm.

Fig. 8 shows the simulation results of the novel proposed method by *Fault_flag* signals when the open fault of the upper switch (S_1) happens in *mode 1*. The open fault occurs at 24.8ms. As shown in Fig. 8(a), after the open fault, *Fault_flag 1* signal represents fault condition and the fault identification is achieved by *Fault_flag 2* signal. After the fault identification, the faulty phase is promptly disconnected by removing switching signals.

978-1-4244-1871-8/07 $25.00 © 2007 IEEE   735

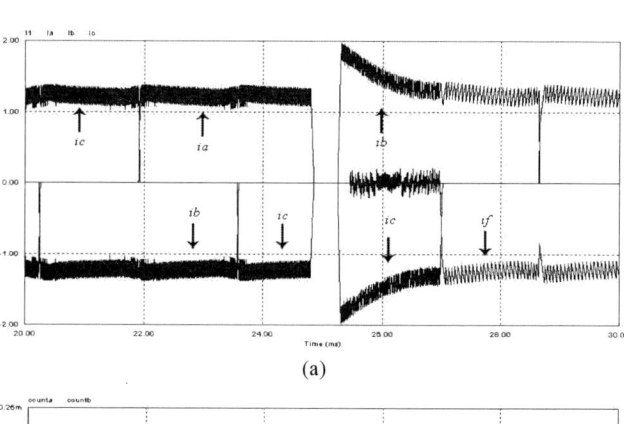

(a)

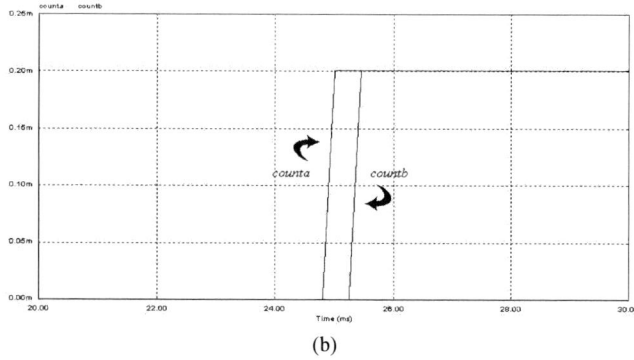

(b)

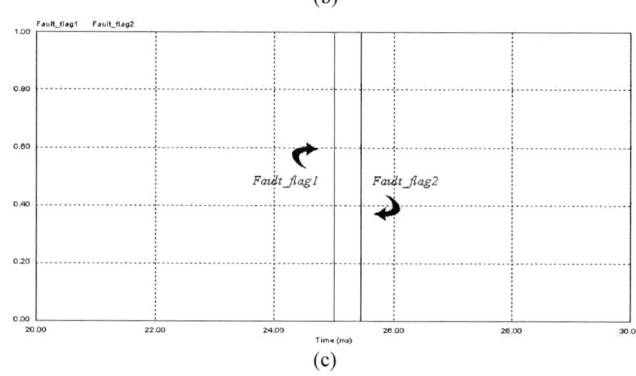

(c)

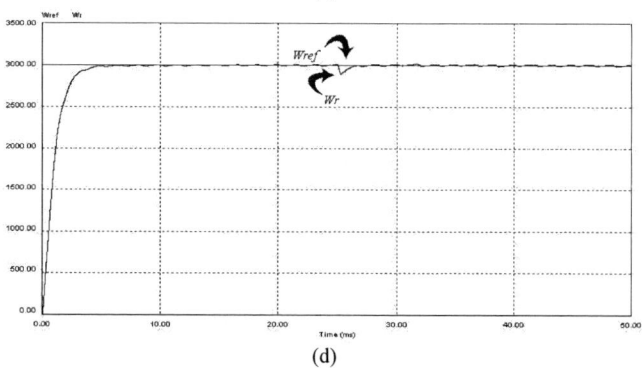

(d)

Fig. 8. Simulation results of the proposed method by *Fault_ flags*. (a) Current waveforms. (b) Counting time. (c) The fault detecting and identification time. (d) Rotor speed.

Consequently, current i_a suddenly drops to zero, while a neutral current begins to flow by bidirectional switch, with i_b and i_c. In this method, as shown in Fig. 8(d), the actual rotor speed follows the reference speed after the fault tolerant. From the detailed investigation of the simulation results, the feasibility is fully verified.

VIII. CONCLUSION

In this paper, a novel fault detection algorithm is investigated to improve the reliability of the BLDC motor drive system. The proposed method is based on *Fault_flag* signals. And the proposed fault tolerant system quickly recovers the control performance by short detecting time and reconfiguration of system topology, and then continuous free operation of the drive after the fault is available.

In comparison to the existing fault tolerant methods, the proposed method can identify fast fault condition without sensors for fault detection and identification.

Simulation results have demonstrated the feasibility of the proposed drive system continuous operation under the fault. When the proposed fault tolerant system is applied to related applications of the BLDC motor drives, the reliability and robustness of the BLDC motor drive system is improved with low cost.

REFERENCE

[1] P.Pillay and P. Freere, "Literature survey of permanent magnet ac motors and drives," *in Proc. IEEE IAS Rec.*, pp. 74–84, 1989

[2] J. R. Hendershot and T. J. E. Miller, Design of Brushless Permanent-Magnet Motors. *Oxford, UK*: Oxford Science, 1994.

[3] R. Spee and T. Lipo, "Remedial strategies for brushless dc drive failures," *IEEE Trans. Ind. Applicat.*, vol. 26, no. 2, pp. 259–266, Mar./Apr. 1990.

[4] Kenneth Scot Smith, Li Ran and Jim Penman, "Real -Time Detection of Intermittent Misfiring in a Voltage-Fed PWM Inverter Induction-Motor Drive," *IEEE Trans. Ind. Eletron.*, vol. 44, no. 4, pp. 468–476, Aug. 1997.

[5] R. Peuget, S. Courtine, and J. P. Rognon, "Fault Detection and Isolation on a PWM Inverter by Knowledge-Based Model," *IEEE Trans. Ind. Applicat.*, vol. 34, no. 6, pp. 1318–1326, Nov./Dec. 1998.

[6] R. L. A. Ribeiro, C. B. Jacobina, E. R. C. da Silva and A. M. N. Lima, "Fault Detection of Open-Switch Damage in Voltage-Fed PWM Motor Drive Systems," *IEEE Trans. Power Electron.*, vol. 18, no. 2, pp. 587–593, Mar. 2003.

[7] J. R. Fu and T. Lipo, "A Strategy to Isolate the Switching Device Fault of a Current Regulated Motor Drive," *in Conf. Rec. IEEE-IAS Annu. Meeting*, vol. 1, pp. 1015–1020, 1993.

[8] S. Bolognani, M. Zigliotto and M. Zordan, "Innovative Remedial Strategies for Inverter Faults in IPM Synchronous Motor Drives," *IEEE Trans. Energy Conversion*, vol. 18, no. 2, pp. 306–312, June. 2003.

[9] Silverio Bolognani, Marco Zordan and Mauro Zigliotto, "Experimental Fault-Tolerant Control of a PMSM Drive," *IEEE Trans. Ind. Eletron.*, vol. 47, no. 5, pp. 1134–1141, Oct. 2000.

[10] R. L. A. Ribeiro, C. B. Jacobina and E. R. C. da Silva, "Fault-Tolerant Voltage-Fed PWM Inverter AC Motor Drive Systems," *IEEE Trans. Ind. Applicat.*, vol. 51, no. 2, pp. 439–446, Apr. 2004.

[11] Byoung-Gun Park, Tae-Sung Kim, Ji-Su Ryu; Byoung-Kuk Lee, Dong-Seok Hyun," Fault Tolerant System under Open Phase Fault for BLDC Motor Drives," IEEE PESC., pp.18–22, June. 2006.

[12] Byoung-Kuk Lee, Tae-Hyung Kim and M. Ehasani, "On the Feasibility of Four-Switch Three-Phase BLDC Motor Drives for Low Cost Commercial Applications: Topology and Control," *IEEE Trans. Power Electron.*, vol. 18, no. 1, pp. 164–172, Jan. 2003.

Interleaved or Sequential Switching
– for Increasing the Switching Frequency

ANDRÉ CABRAL FERREIRA, *STUDENT MEMBER IEEE*
Wuppertal University
Electrical Machines and Drives
Rainer-Gruenter-Strasse 21 (Campus Freudenberg)
D – 42119 Wuppertal, Germany
Email: ocabral@emad.uni-wuppertal.de

RALPH M. KENNEL, *SENIOR MEMBER IEEE*
Wuppertal University
Electrical Machines and Drives
Rainer-Gruenter-Strasse 21 (Campus Freudenberg)
D – 42119 Wuppertal, Germany
Email: kennel@ieee.org

Abstract— Using several IGBTs in parallel it is possible to operate standard IGBTs with higher switching frequencies than usual. Based on this concept the current load and thermal losses are split between the paralleled devices. The free wheeling diodes, however, cannot be controlled actively and therefore do not share in sequential switching. In 3-phase VSI applications there is the danger of overloading one of the paralleled free wheeling diodes. This paper presents a concept of magnetic freewheeling control resulting in the free wheeling current following the same sequential concept as in the IGBTs. Magnetic freewheeling control makes sequential switching applicable to 3-phase voltage source inverters (VSI).

I.   INTRODUCTION

For limiting the losses of IGBTs to reasonable amounts the maximum current load of power semiconductors must be reduced when operated at high switching frequencies. When using several IGBTs in parallel (fig. 1) the switching losses can be distributed between several elements.

Interleaved switching is a well known concept for increasing the switching frequency ([1] – [9]) – especially in (single phase) boost converters. The basic idea is to use parallel power devices and to switch them sequentially – the

first pulse will switch the first device, the second pulse will switch the second device, etc. (see fig. 2). This concept distributes the current load as well as the switching losses to all devices in parallel.

II.   INTERLEAVED OR SEQUENTIAL SWITCHING ?

It has to be admitted that the expression "interleaved switching" is used in literature ([1] – [9]) in two different ways. In some papers "interleaved switching" is used to describe a concept where a power device is switched on while the corresponding device still is in on-state. The gate signal are really "interleaving". However, in most cases – especially with respect to boost converters – "interleaved switching" is used to describe the situation mentioned in the last paragraph; parallel power devices are never operated in parallel. Any power device is switched on only as long as all corresponding devices are turned off. In fact "interleaved switching" does not describe this concept properly – a more correct name is "sequential switching".

Applying the concept of sequential switching to a structure of 3 parallel IGBTs (Fig. 1) the switching frequency of each power device is only a third of the total switching frequency. As shown in fig. 2 the PWM pulses are not transferred to gate

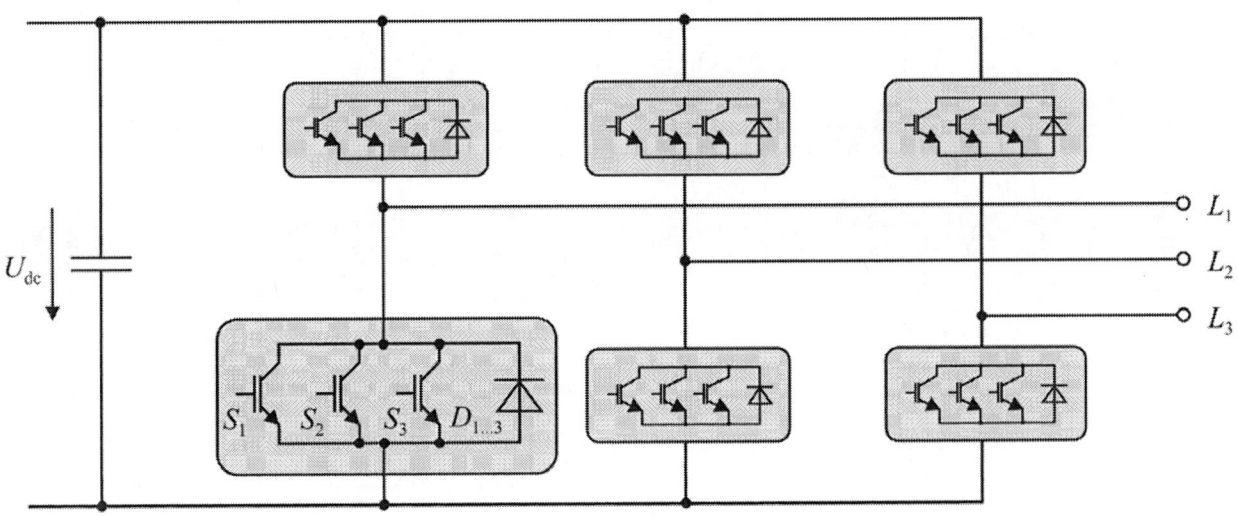

Fig. 1   basic scheme of parallel power devices in a voltage source inverter (VSI)

978-1-4244-1871-8/07 $25.00 © 2007 IEEE

voltages simultaneously, but sequentially. The pulse pattern has a switching frequency f_{ges} ; Pulse no. 1 is transferred to the gate of IGBT S_1, no. 2 to the gate of IGBT S_2 and no. 3 to the gate of IGBT S_3. In this example sequential switching reduces the switching frequency f_{igbt} to a third of the original switching frequency f_{ges}.

In general the following equations are applicable:

$$f_{igbt} = \frac{f_{ges}}{n} \qquad (1)$$

n = number of IGBTs in parallel

$$T_{igbt} = n \cdot T_{ges} = \frac{n}{f_{ges}} \qquad (2)$$

$$T_{ein} = T_{ges} = \frac{1}{f_{ges}} \qquad (3)$$

Furthermore sequential switching results in a limitation of the maximum switching-on time T_{ein} of an IGBT to a third of the switching period T_{igbt}. In fig 2 the maximum switching-on period T_{ein} of each semiconductor is limited to 33,33% of the switching period T_{igbt}

Both effects in addition result in a reduction of losses in each device enabling the operation of the system with much higher frequencies than usual (factor 5 to 7 with 3 parallel devices).

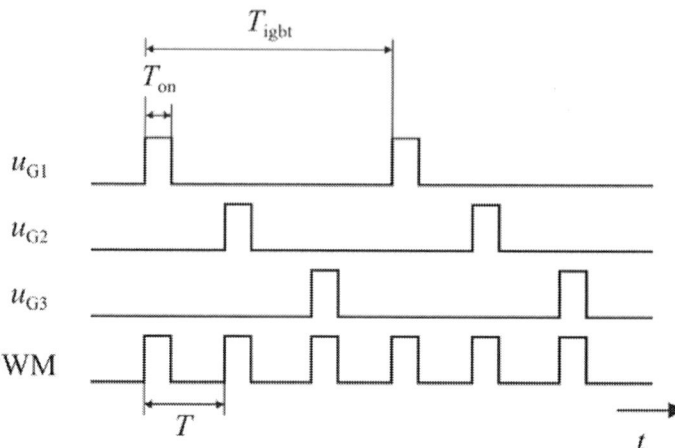

Fig. 2   Example of pulse sequence
for sequential switching of paralleled power devices

III.   SEQUENTIAL SWITCHING WITH RESPECT TO 3-PHASE APPLICATIONS

The free wheeling diodes (marked by $D_{1...3}$ in fig. 1), which are integrated in the IGBT modules and therefore connected in parallel to each other as well, cause some problems for the concept of sequential switching – problems which do not exist in single phase boost converter applications. In difference to the IGBTs the internal free wheeling diodes cannot be controlled actively. As diodes need a positive voltage between anode and cathode only for switching on, a simultaneous switching on of the parallel free wheeling diodes has to be expected.

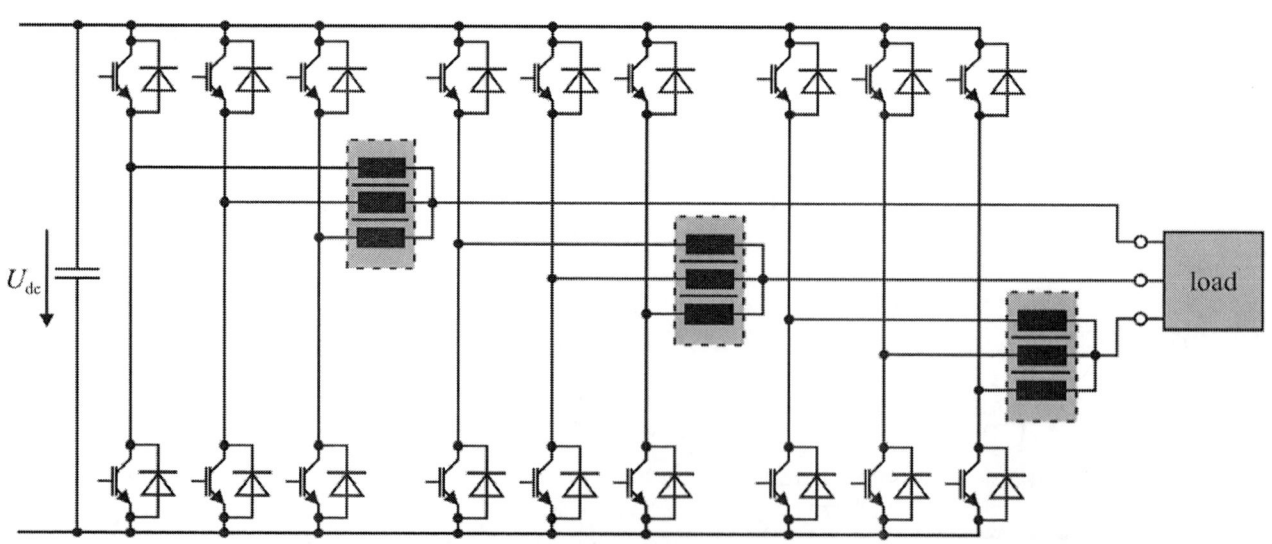

Fig. 3   sequential switching and principle of magnetic freewheeling control in a 3-phase VSI (Voltage Source Inverter)

Ideally the diodes should provide equal characteristics and share the load current balanced; with respect to statistical tolerances the diodes, however, do not have identical characteristics. Consequently the diode with lowest internal resistance will accept a greater part of the load current and heat up more than the others. With respect to the negative temperature coefficient electrical conductivity of this specific device will improve even more resulting in an even greater part of the load current. This effect will increase more and more until overload and possible destruction of the specific diode.

Of course, it would be possible to use active devices (e. g. thyristors) to guarantee sequential switching in the free wheeling circuits as well – in this case the standard IGBT modules usually available in the market could not be used any more. To make standard IGBT modules with integrated free wheeling diodes usable for sequential switching there must be an apparatus providing the free wheeling current to flow through the parallel diodes in the same sequential scheme as in the paralleled IGBTs. This can be gained by a rather simple magnetic concept. It is just necessary to adapt the inductance, which is necessary and used anyway in most applications of power electronics (fig. 3). This concept is called magnetic free wheeling control as it forces the load current to flow during the free wheeling periods exactly through the diode corresponding to that IGBT, which was switched on before.

IV. MAGNETIC FREEWHEELING CONTROL

The concept of magnetic free wheeling control is a possibility to provide sequential switching in parallel diodes. In many power electronic applications – especially in electrical drives – there is an inductive load or an inductance between inverter and load included anyway. To achieve the behaviour of magnetic free wheeling the windings of the inductances are split to several parallel coils mounted on the same magnetic core. Of course, a well designed balance between mutual inductance and stray inductance is necessary to obtain proper results.

Each IGBT/diode branch is connected separately to one of the coils; all coils of a half bridge use the same magnetic core

Fig. 4 sequential switching with magnetic free wheeling control in a buck converter

(see fig. 3). As the magnetic design of the inductance does not change, these windings can be installed on the same magnetic core as before. Consequently the size of the inductance does not increase significantly in comparison to the usual choke.

Fig. 4 shows a test set-up for simplified understanding of the magnetic freewheeling control scheme; the test circuit is a buck converter with three parallel IGBTs and diodes in sequential switching; this circuit needs only one inductance. The test set-up was operated with a switching frequency of about 18 kHz. This frequency can easily be identified in the voltage signal in fig. 5, which can be measured at all parallel free wheeling diodes.

The IGBT S_1 located most left obtains the first gate pulse of a PWM pattern. The load current starts to flow through load resistance R, the left coil L (marked by 1) and through IGBT S_1. After switching off IGBT S_1 the load current commutates to diode D_1. With respect to stray inductances between coils 1 and 2 the free wheeling current remains in coil 1. As soon as the second PWM pulse switches on the IGBT S_2, the current should commutate from diode D_1 to IGBT S_2. The commutation to IGBT S_2 is supported by the main (mutual) inductance of the common core. All three electrical coils are connected to the same magnetic circuit enabling the free wheeling current to commutate from the branch $R - L_1 - D_1$ to the branch $R - L_2 - S_2$. After commutation of the current to IGBT S_2, the current circuit includes load resistance R, the middle coil L_2 and IGBT S_2. As the current is flowing through the middle IGBT, a positive voltage occurs between anode and cathode of diode D_2 during switching-off IGBT S_2 resulting in a commutation of the current to the branch $R - L_2 - D_2$. With the next pulse occurring in the PWM pattern the whole procedure repeats with respect to IGBT S_3 and diode D_3.

In inductance design a balance between main (mutual) inductance and stray inductance is necessary to control magnetic free wheeling as described above. Fig. 5 shows measurements of current I_D (through one of the paralleld diodes) as well as the voltage U_D measured at the same diode during operation as a buck converter according to fig. 3. The respective DC-link voltage U_{dc} was 300 V and the load current I was 20 A The diagram shows clearly that the free wheeling current I_D is flowing through each diode during each third pulse only. There is only a very small current in the diode when one of the parallel diodes is in duty. As a result the switching frequency of each diode is only a third of the full frequency indeed – in the same way as obtained by active gate control in the paralleled IGBTs.

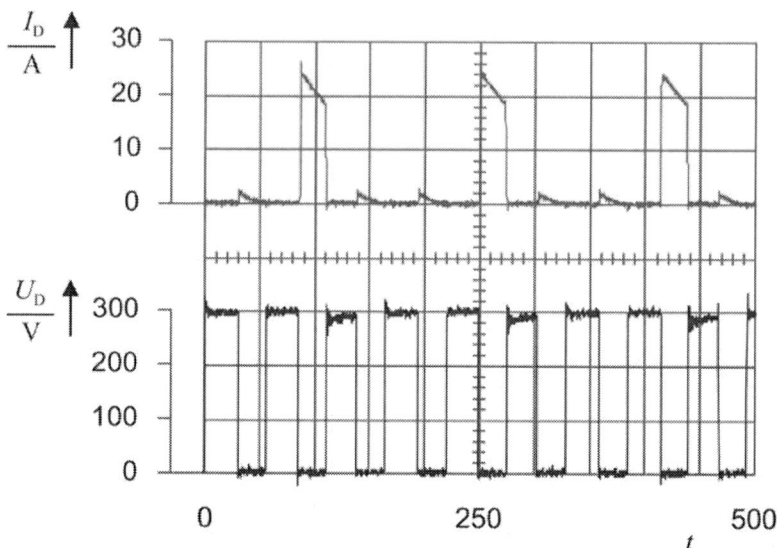

Fig. 5 measured diode current during magnetic free wheeling control

V. CONCLUSIONS

This contribution presents a concept for operating converters consisting of standard IGBT modules with much higher switching frequency than usual. As shown by the experimental results the concept of magnetic freewheeling control distributes the current – together with the switching losses – equally to the paralleled free wheeling diodes. Hereby the maximum current load is reduced in all paralleled diodes in the same way as with the IGBTs controlled by sequential switching. The concept of magnetic free wheeling control enables power electronics manufacturers to increase the power range of their products or to achieve much higher switching frequencies than usual without using specially designed power devices. Especially small and medium sized companies can use sequential switching as a tool to enhance the range of their products based on the same power electronic components (e. g. standard IGBT modules) as they use anyway without additional use of special components or devices.

ACKNOWLEDGMENT

We appreciate the financial support of FVA (Forschungsvereinigung Antriebstechnik e. V., Frankfurt, Germany) and the valuable technical inputs and supervision from the AK "Geregelte elektrische Antriebe" (working group "controlled electrical drives").

REFERENCES

[1] F. Kleveland, T. Undeland J. ,K. Langelid: "Increase of Output from IGBTs in High Power High Frequency Resonant Load Inverters", *IEEE Annual Meeting of Industry Applications Society,* Rome, 2000, CD-ROM

[2] H. Abo Zied, A-R.A.M. Makky, P. Mutschler, S. Stier: "A Modular IGBT Converter System for High Frequency Induction Heating Applications", *23rd International Conference and Exhibition on Power Electronics, Intelligent Motion, Power Quality PCIM 2002 Europe,* Nürnberg, 2002, CD-ROM

[3] S. H. Stier, P. Mutschler: "A Modular IGBT Converter System for High frequency Induction Heating Applications", *German-Korean Symposium 2004 on Power Electronics and Electrical Drives KOSEF 2004,* S.~164-171, Aachen, 2004

[4] F. J.\,Pérez-Pinal, I. Cervantes: "Simple Almost Zero Switching Losses for Interleaved Boost Converter", *3rd International Conference on Power Electronics Machines and Drives PEMD 2006,* Dublin, 2006

[5] H.-H. Chang, S.-Y. Tseng, J.G. Huang: "Interleaving boost convertsers with a single-capacitor Turn-off Snubber", *37th Annual IEEE Power Electronics Specialists Conference PESC,* Jeju, 2006, CD-ROM

[6] Yanshen Hu, Yunxiang Xie, Hao Tian, Baoming Mei: "Characterisitics Analysis of Two-Channel Interleaved Boost Converter with Intergrated Coupling Inductors", *37th Annual IEEE Power Electronics Specialists Conference PESC,* Jeju, 2006, CD-ROM

[7] Ravinder Pal Singh, A.M. Khambadkone, Ganesh S. Samudra, Yung C. Liang: "An FPGA based Digital Control Design for high-frequency DC-DC Converters", *37th Annual IEEE Power Electronics Specialists Conference PESC,* Jeju, 2006, CD-ROM

[8] Xin Zhang, Alex Huang: "MVRC and Its Tolerance Analysis For Microprocessor Power Management", *37th Annual IEEE Power Electronics Specialists Conference PESC,* Jeju, 2006, CD-ROM

[9] L. Asiminoaei, E. Aeloiza, J.H. Kim, P. Enjeti, F. Blaabjerg, L.T. Moran, S.K. Sul: "Parallel Interleaved Inverters for Reactive Power and Harmonic Compensation", *37th Annual IEEE Power Electronics Specialists Conference PESC,* Jeju, 2006, CD-ROM

Space Vector Modulation Method for Unidirectional Four-Wire Three-Phase/Level/Switch (Vienna) Rectifier

Jarno Alahuhtala
Tampere University of Technology
Institute of Power Electronics
P.O. Box 692, FI-33101 Tampere, Finland
Email: jarno.alahuhtala@tut.fi

Heikki Tuusa
Tampere University of Technology
Institute of Power Electronics
P.O. Box 692, FI-33101 Tampere, Finland
Email: heikki.tuusa@tut.fi

Abstract— **This paper presents a space vector modulation (SVM) method for four-wire Vienna rectifier (VIENNA4). The modulator uses complex plane for modulation and treats the zero-sequence voltage reference as an offset which is realized by proper weighting of redundant vectors. The main application for VIENNA4 is a transformerless ac-dc-ac converter used in uninterruptible power supplies (UPS). The applied modulation method and the control system are analyzed in detail. The analysis is verified with measurements from 10 kVA 230 V/400 V 50 Hz laboratory prototype in steady state and dynamical load conditions. The digital implementation of the modulator and the control system are realized with Freescale MPC563 microcontroller.**

I. INTRODUCTION

A three-phase/level/switch unidirectional Vienna I rectifier has been found very suitable for active rectification when only unidirectional power flow is needed [1]–[3]. It has less switches, better efficiency and smaller ac-filter size compared to the conventional six-switch full bridge PWM rectifier. If the rectifier is used in a converter system feeding independent single-phase loads, such as three-phase double-conversion ac-dc-ac UPS, the supply neutral needs to be connected to the system. In order to prevent the rectifier dc link from floating with respect to ground, the supply neutral is connected to the dc link midpoint [4]. The advantages of the Vienna I rectifier in three-phase rectification make it interesting to apply the topology in four-wire-connected systems. The main circuit of four-wire Vienna rectifier (VIENNA4) is presented in Fig. 1. The purpose of the rectifier is to draw balanced and sinusoidal currents from the utility with unity power factor.

A carrier-based modulation method for VIENNA4 was presented in [5]. Carrier-based modulation method is more suitable for applications where switching frequency is relatively high since it can be implemented with simple analogue circuitry [6]. For applications where switching frequency is lower and control is carried out digitally by means of microcontroller, the space vector modulation (SVM) can be applied.

In three-phase three-wire converters the SVM is traditionally performed in two-dimensional α-β reference frame. In four-wire connected converters the traditional two-dimensional SVM is extended to also cover the zero-sequence voltage (ZSV) component thus forming a three-dimensional α-β-0 reference frame for the modulator [7]. In [8] a simplified approach on three-dimensional SVM has been presented for a three-phase/level four-wire converter where the reference voltage is decomposed into two level converter α-β frame reference and ZSV is treated as an offset.

SVM methods for the Vienna I rectifier have been presented in [9] and [10]. This paper extends the method presented in [9] to also cover the ZSV region and studies how SVM can be applied to VIENNA4. The modulator uses the α-β reference frame for SVM and the ZSV reference is treated as an offset that is realized by proper weighting of redundant vectors in the sequences. The limits of the modulation method due to the four-wire configuration are also analyzed and the analysis is verified with measurements from 10 kVA rated laboratory prototype.

II. MAIN CIRCUIT AND PRINCIPLE OF OPERATION

The main circuit of VIENNA4 is presented in Fig. 1. It consists of three active power switches S_k, where k = a,b,c,

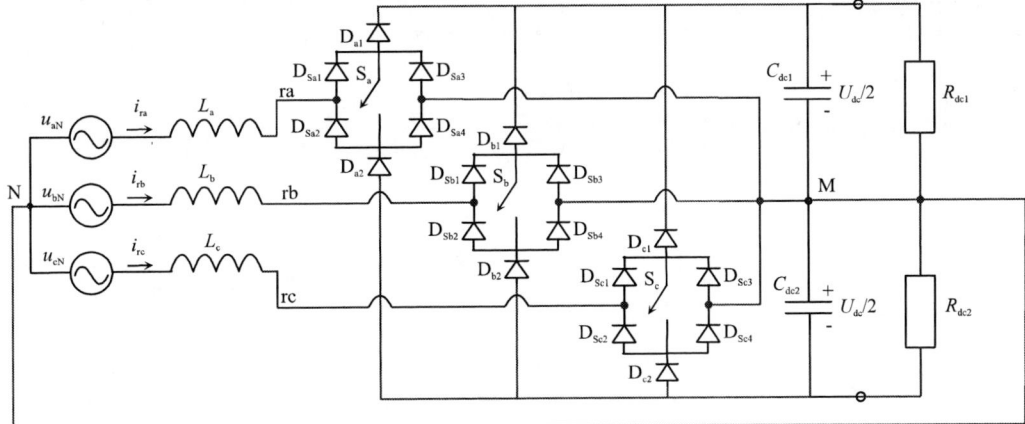

Fig. 1: Main circuit of VIENNA4.

978-1-4244-1871-8/07 $25.00 © 2007 IEEE

741

and 18 diodes each having a voltage stress of $U_{dc}/2$. The rectifier has a three-phase L-type filter L_k on the ac side and two capacitors $C_{dc(1,2)}$ in series on the dc side. Neutral wire is connected directly to the midpoint M that is formed with the split dc link capacitors.

The rectifier phase voltages u_{rkN} depend on the states of the active switches and the polarity of the phase currents i_{rk}. When an active switch S_k of a phase is turned on it forms a bidirectional current path to the midpoint M via diodes $D_{Sk(1,2,3,4)}$ and connects the rectifier phase to neutral N. If the active switch is turned off, the phase voltage is defined by the polarity of the rectifier phase current i_{rk}. Positive phase current $i_{rk} > 0$ connects the rectifier phase to positive dc link potential via diodes D_{Sk1} and D_{k1}, and negative phase current $i_{rk} < 0$ respectively to the negative dc link potential via diodes D_{Sk2} and D_{k2}. The current slope of the rectifier is defined by the voltage lying across the filter inductor i.e. the difference of supply phase voltage u_{skN} and rectifier phase voltage u_{rkN}. The resulting current slope is

$$\frac{di_{rk}}{dt} = \frac{u_{skN} - u_{rkN}}{L_k}. \tag{1}$$

The rectifier phase voltages can be presented with the help of the phase current sign and the switching functions sw_k. Switching functions determine whether the active switch S_k is conducting ($sw_k = 1$) or not ($sw_k = 0$). Assuming equal dc link capacitor voltages ($U_{Cdc1} = U_{Cdc2} = U_{dc}/2$), the rectifier phase voltages are

$$u_{rkN} = (1 - sw_k)U_{dc}/2, \text{ when } i_{rk} > 0 \tag{2}$$

$$u_{rkN} = (sw_k - 1)U_{dc}/2, \text{ when } i_{rk} < 0. \tag{3}$$

III. PROPOSED SPACE VECTOR MODULATION METHOD

A. Space Vector Modulation

A voltage space vector for arbitrary three-phase voltages $u_{(a,b,c)}$ in stationary reference frame is defined

$$\underline{u}^{stat} = u_\alpha + ju_\beta = \frac{2}{3}(u_a + \underline{a}u_b + \underline{a}^2 u_c), \text{ where } \underline{a} = e^{j2\pi/3}. \tag{4}$$

The definition of a space vector does not contain the definition of the zero-sequence component (ZSC); therefore, it has to be calculated separately. The ZSC for arbitrary three-phase voltages is defined

$$u_z = \frac{1}{3}(u_a + u_b + u_c). \tag{5}$$

Possible voltage space vectors of the VIENNA4 are presented in Fig. 2. There are a total of 19 different voltage vectors available in the Re-Im plane. The lengths of the vectors are; large vectors ($\underline{u}_1 - \underline{u}_6$): $2U_{dc}/3$, medium vectors ($\underline{u}_{12} - \underline{u}_{61}$): $U_{dc}/\sqrt{3}$, small vectors ($\underline{u}_{01} - \underline{u}_{06}$): $U_{dc}/3$.

Each small vector can be produced with two alternative switching combinations. These two combinations produce the same voltage vector in the complex plane but their ZSC

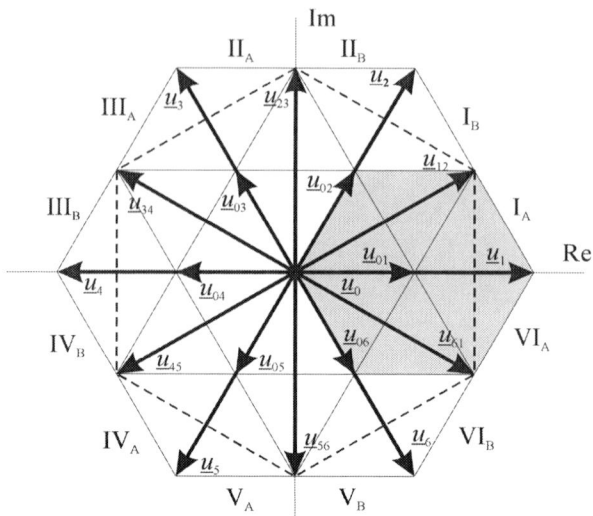

Fig. 2: Voltage space vectors of VIENNA4.

TABLE I
Voltage Space Vectors of VIENNA4

u	SIGN($\underline{swu}$)	$\underline{sw}_k$	u_z [U_{dc}]
$\underline{u}_0$	0 0 0	[1 1 1]	0
$\underline{u}_{01+}$	+ 0 0	[0 1 1]	1/6
$\underline{u}_{01-}$	0 – –	[1 0 0]	-1/3
$\underline{u}_1$	+ – –	[0 0 0]	-1/6
$\underline{u}_{12}$	+ 0 –	[0 1 0]	0
$\underline{u}_{02+}$	+ + 0	[0 0 1]	1/3
$\underline{u}_{02-}$	0 0 –	[1 1 0]	-1/6
$\underline{u}_2$	+ + –	[0 0 0]	1/6
$\underline{u}_{23}$	0 + –	[1 0 0]	0
$\underline{u}_{03+}$	0 + 0	[1 0 1]	1/6
$\underline{u}_{03-}$	– 0 –	[0 1 0]	-1/3
$\underline{u}_3$	– + –	[0 0 0]	-1/6
$\underline{u}_{34}$	– + 0	[0 0 1]	0
$\underline{u}_{04+}$	0 + +	[1 0 0]	1/3
$\underline{u}_{04-}$	– 0 0	[0 1 1]	-1/6
$\underline{u}_4$	– + +	[0 0 0]	1/6
$\underline{u}_{45}$	– 0 +	[0 1 0]	0
$\underline{u}_{05+}$	0 0 +	[1 1 0]	1/6
$\underline{u}_{05-}$	– – 0	[0 0 1]	-1/3
$\underline{u}_5$	– – +	[0 0 0]	-1/6
$\underline{u}_{56}$	0 – +	[1 0 0]	0
$\underline{u}_{06+}$	+ 0 +	[0 1 0]	1/3
$\underline{u}_{06-}$	0 – 0	[1 0 1]	-1/6
$\underline{u}_6$	+ – +	[0 0 0]	1/6
$\underline{u}_{61}$	+ – 0	[0 0 1]	0

differ from their sign and magnitude. These alternative and redundant small vectors are denoted $\underline{u}_{0n(+,-)}$, where $\underline{u}_{0n+}$ and $\underline{u}_{0n-}$ refer to the sign of the ZSC of the redundant vector. Table I summarizes all possible voltage vectors, their switching functions, and the resulting ZSC u_z of the possible VIENNA4 voltage vectors.

Each main sector I-VI presented in Fig. 2 is divided into four subsectors 1–4 (Fig. 3). The darkened area in Fig. 2 shows the realizable region of vectors when the rectifier current vector angle φ_{ir} lies between $\pi/6 > \varphi_{ir} > -\pi/6$, i.e when $i_{ra} > 0$, $i_{rb} < 0$ and $i_{rc} < 0$.

The vector modulation is based on the time averaging of the reference vector $\underline{u}_{ref}$ within a sampling period T_s. The time averaging is done by weighting the nearest three vectors (NTV) lying in the vicinity of the reference i.e. the three vectors that align the subsector where the reference is situated. For an arbitrary reference $\underline{u}_{ref}$ we can write

$$\underline{u}_{ref} = d_1 \underline{u}_1 + d_2 \underline{u}_2 + d_3 \underline{u}_3, \tag{6}$$

where $\underline{u}_{(1,2,3)}$ are the NTV of the reference and the $d_{(1,2,3)}$ are the duty cycles of the NTV. For the duty cycles we can write in the linear modulation region

$$d_1 + d_2 + d_3 = 1. \tag{7}$$

Fig. 3 illustrates the time averaging of the reference voltage vector in main sector I. To determine the duty cycles $d_{(1,2,3)}$, glancing projections u_κ and u_λ of the reference are calculated to the main vectors that outline the main sector in question. The projections are

$$u_\kappa = \left| \underline{u}_{ref} \right| \left((\cos(\varphi_{ref,sec})) - \sin(\varphi_{ref,sec}) / \sqrt{3} \right) \tag{8}$$

$$u_\lambda = \left| \underline{u}_{ref} \right| 2 \sin(\varphi_{ref,sec}) / \sqrt{3} \tag{9}$$

where $\varphi_{ref,sec}$ is the angle of the reference vector $\underline{u}_{ref}$ inside the main sector in question. The projections are scaled with the length of the small vectors ($U_{dc}/3$) to get the normalized duty cycles d_κ and d_λ which are used to identify the subsector of the reference voltage $\underline{u}_{ref}$ and the duty cycles of the NTV.

$$d_\kappa = 3 \frac{u_\kappa}{U_{dc}} \tag{10}$$

$$d_\lambda = 3 \frac{u_\lambda}{U_{dc}} \tag{11}$$

The subsectors can then be determined logically using the magnitudes of the normalized duty cycles [11] i.e.

$$\text{IF}(d_\kappa \geq 1) \rightarrow sub = 2$$
$$\text{IF}(d_\lambda \geq 1) \rightarrow sub = 4$$
$$\text{ELSE IF} \tag{12}$$
$$(d_\kappa + d_\lambda \geq 1) \rightarrow sub = 3$$
$$(d_\kappa + d_\lambda \leq 1) \rightarrow sub = 1,$$

where sub denotes the subsector.

Duty cycles $d_{(1,2,3)}$ of the NTV in the vicinity of the reference can be calculated using (6), geometrical relations of the reference vector's $\underline{u}_{ref}$ real and imaginary parts, and the normalized duty cycles d_κ and d_λ. The duty cycles for each subsector in the main sector I are presented in Table II. If the voltage reference vector is situated in other main sectors (II – VI) the vector must be virtually transformed to lie in the main sector I.

The first half of the sequence of the vectors in each subsector is presented for the 1st (odd) and 2nd (even) main sectors in Table III. The modulation period T_{mod} consists of two sample periods T_s during which the sequences of Table III are produced twice. During the first sample period the sequences are produced as in Table III, and during the second the sequence is inversed. For other main sectors the sequences must be replaced with similar vectors of the main sector in question. All the vectors are distributed in the sequences so that only one active switch state needs to be changed when changing between adjacent vectors in the sequence. This minimizes the switching losses when a continuous modulation method is used.

B. Zero-sequence Voltage Control

The resulting ZSV inside the modulation period can

TABLE II
Variable Values for (6) in the First Main Sector

i	1. sub		2. sub		3. sub		4. sub	
	d_i	$\underline{u}_i$	d_i	$\underline{u}_i$	d_i	$\underline{u}_i$	d_i	$\underline{u}_i$
1	d_κ	$\underline{u}_{01}$	$d_\kappa - 1$	$\underline{u}_1$	$1 - d_\kappa$	$\underline{u}_{02}$	$d_\lambda - 1$	$\underline{u}_2$
2	d_λ	$\underline{u}_{02}$	d_λ	$\underline{u}_{12}$	$1 - d_\lambda$	$\underline{u}_{01}$	d_κ	$\underline{u}_{12}$
3	$1 - d_1 - d_2$	$\underline{u}_0$	$1 - d_1 - d_2$	$\underline{u}_{01}$	$1 - d_1 - d_2$	$\underline{u}_{12}$	$1 - d_1 - d_2$	$\underline{u}_{02}$

TABLE III
Sequence of the Voltage Vectors in the First (Odd) And Second (Even) Main Sectors

	Subsector	sequence
Main sector I	1. sub ($\varphi_{ref,sec} < \pi/6$)	$\underline{u}_{01-} - \underline{u}_{02-} - \underline{u}_0 - \underline{u}_{01+}$
	1. sub ($\varphi_{ref,sec} > \pi/6$)	$\underline{u}_{02-} - \underline{u}_0 - \underline{u}_{01-} - \underline{u}_{02+}$
	2. sub	$\underline{u}_{01-} - \underline{u}_1 - \underline{u}_{12-} - \underline{u}_{01+}$
	3. sub ($\varphi_{ref,sec} < \pi/6$)	$\underline{u}_{01-} - \underline{u}_{02-} - \underline{u}_{12-} - \underline{u}_{01+}$
	3. sub ($\varphi_{ref,sec} > \pi/6$)	$\underline{u}_{02-} - \underline{u}_{12-} - \underline{u}_{01+} - \underline{u}_{02+}$
	4. sub	$\underline{u}_{02-} - \underline{u}_{12-} - \underline{u}_2 - \underline{u}_{02+}$
Main sector II	1. sub ($\varphi_{ref,sec} < \pi/6$)	$\underline{u}_{02-} - \underline{u}_0 - \underline{u}_{03+} - \underline{u}_{02+}$
	1. sub ($\varphi_{ref,sec} > \pi/6$)	$\underline{u}_{03-} - \underline{u}_{02-} - \underline{u}_0 - \underline{u}_{03+}$
	2. sub	$\underline{u}_{02-} - \underline{u}_{23-} - \underline{u}_2 - \underline{u}_{02+}$
	3. sub ($\varphi_{ref,sec} < \pi/6$)	$\underline{u}_{02-} - \underline{u}_{23-} - \underline{u}_{02+} - \underline{u}_{02+}$
	3. sub ($\varphi_{ref,sec} > \pi/6$)	$\underline{u}_{03-} - \underline{u}_{02-} - \underline{u}_{23-} - \underline{u}_{03+}$
	4. sub	$\underline{u}_{03-} - \underline{u}_3 - \underline{u}_{23-} - \underline{u}_{03+}$

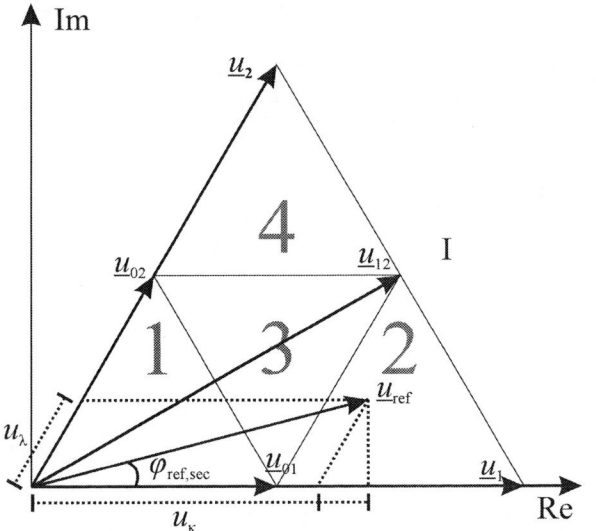

Fig. 3: Subsectors of the first main sector and the glancing projections (u_κ and u_λ) of the reference voltage vector $\underline{u}_{ref}$.

cause neutral current flow since the dc link midpoint M is directly connected to supply neutral N. In order to draw sinusoidal and balanced currents from the utility in unity power factor the ZSV needs to be controlled. This can be done by appropriate weighting of the redundant vectors present in each sequence.

To further classify the vectors in the sequence halves (Table III) they are denoted according to their ZSC and redundancy. Each sequence halve begins with negative redundant vector $\underline{u}_{0n-}$ and ends to its positive redundant combination $\underline{u}_{0n+}$. The duty cycles of these vectors are now denoted as d_{r-} and d_{r+} respectively. Between the redundant vectors there are two voltage vectors of which the one has zero ZSC $\underline{u}_z$ (duty cycle d_z) and the other has nonzero ZSC $\underline{u}_e$ (duty cycle d_e). The $\underline{u}_z$ is either null vector $\underline{u}_0$ in the 1st subsector or one of the medium vectors ($\underline{u}_{12} - \underline{u}_{61}$) in other subsectors. The $\underline{u}_z$ does not affect to resulting ZSV since its individual ZSC is zero. The $\underline{u}_e$ is one of the main vectors ($\underline{u}_1 - \underline{u}_6$) in even subsectors or one of the small vectors ($\underline{u}_{01(+,-)} - \underline{u}_{06(+,-)}$) in odd subsectors.

Redundant combinations of the small vectors are of special interest because they realize the same voltage vector in complex plane, but the ZSC changes to the opposite sign and magnitude between the two combinations. Each sequence has always two alternative small vectors $\underline{u}_{0n(+,-)}$ which can be used for the ZSV control. This is done by weighting the duty cycles of these alternative combinations.

The duty cycle d_r of the redundant vector depends on the angle and magnitude of the reference voltage vector $\underline{u}_{ref}$. As can be seen from Table III and Fig. 3, the small vector used as a redundant is always the one that has greater glancing projection of the reference to it. This enables maximal ZSV controllability and therefore in subsectors 1 and 3 two different voltage vector sequences are used depending on the angle $\varphi_{ref,sec}$ of the reference vector.

The redundant vector duty cycle d_r can be divided among the two combinations $u_{0n(+,-)}$ with a weighting coefficient r. Thus for the individual duty cycles of the two combinations we can write

$$d_{r+} = (1+r)\frac{d_r}{2} \tag{13}$$

$$d_{r-} = (1-r)\frac{d_r}{2}, \tag{14}$$

where $-1 < r < 1$. We can now rewrite (6) in terms of voltage vector's ZSC (denoted with subscript z) and weighted redundant vector combinations

$$u_{z,ref} = (1+r)\frac{d_r}{2}u_{z,0n+} + d_z u_{z,z} + d_e u_{z,e} + (1-r)\frac{d_r}{2}u_{z,0n-} \tag{15}$$

The value of r is fixed for a given voltage vector $\underline{u}_{ref}$ and ZSV reference $u_{z,ref}$. The value of r that realizes the reference $u_{z,ref}$ can be solved by taking into account the individual ZSC of each voltage vector present in the sequence. In Fig. 2 the vector diagram is divided in two types of regions aligned by the medium vectors. These regions are denoted A and B. They can be found in the subscript of the main sector indices. The value of r in terms

of duty cycles and ZSV reference in these two regions are

$$r_A = \frac{2}{3}\left(\frac{d_e}{d_r}\right) + \frac{1}{3} + \frac{4u_{z,ref}}{U_{dc} \cdot d_r} \tag{16}$$

$$r_B = \frac{4u_{z,ref}}{U_{dc} \cdot d_r} - \frac{2}{3}\left(\frac{d_e}{d_r}\right) - \frac{1}{3}. \tag{17}$$

In Fig. 2 the dashed line shows the limit inside which the ZSV u_z can be controlled in both polarities. The modulation index is defined

$$m = \frac{|\underline{u}_{ref}|}{U_{dc}/2}. \tag{18}$$

From Fig. 2 we see that the maximum length of the reference vector in the linear modulation region is $|\underline{u}_{ref,max}| = U_{dc}/2$. Thus the maximum modulation index of the reference vector in the linear modulation region is $m_{max} = 1$. This is in accordance with the results obtained in [5].

C. Current Limitation Method during Rectifier Start-up

When the partial dc link voltages are less than the peak value of supply phase voltage $U_{Cdc(1,2)} < 325$ V, the current cannot be controlled during phase voltage peaks (1). To avoid high peak currents during start-up the switches on individual phase legs are kept on the OFF state for a 60° period around phase voltage peak. During this period the rectifier's individual phase leg operates as diode bridge. The 60° turnoff period is produced by simply bypassing the negative redundant vector $\underline{u}_{0n-}$ in the sequences in region A and the positive redundant vector $\underline{u}_{0n+}$ in region B. In the modulator this is done by setting r to 1 and -1 respectively.

IV. CONTROL SYSTEM

The proposed vector modulation method is applied to a vector controlled four-wire Vienna rectifier. The space vector voltage equation of the rectifier with L-type filter in the synchronous mains-oriented reference frame rotating at ω_s is.

$$\underline{u}_s^s = L\frac{d\underline{i}_r^s}{dt} + j\omega_s L\underline{i}_r^s + \underline{u}_r^s. \tag{19}$$

Eq. 18 can be expressed in terms of direct (d), quadrature (q) and zero-sequence (z) components, i.e.

$$u_{sd} = u_{rd} + L\frac{di_d}{dt} - \omega_s L i_q \tag{20}$$

$$u_{sq} = u_{rq} + L\frac{di_q}{dt} + \omega_s L i_d \tag{21}$$

$$u_{sz} = u_{rz} + \frac{1}{3}L\frac{di_z}{dt}. \tag{22}$$

The control system is derived on the basis of (20), (21) and (22). The simplified block diagram of the control system is presented in Fig. 4.

978-1-4244-1871-8/07 $25.00 © 2007 IEEE

The phase locked loop PLL is used to track the angle φ_s of the synchronous reference frame. Measured line voltages $u_{s(a,b,c)}$ and rectifier currents $i_{r(a,b,c)}$ are transformed into the synchronous reference frame quantities $u_{s(d,q,z)}$ and $i_{r(d,q,z)}$. The total dc link voltage difference is brought to a PI controller which gives the rectifier d-component reference $i_{(d),ref}$. The balancing of the partial dc link voltages is carried out by feeding the partial dc link voltage error $U_{dc,err}$ through a PI controller that gives a reference for zero-sequence current $i_{(z),ref}$. In order to draw currents from the utility in unity power factor, the rectifier q-component current reference $i_{(q),ref}$ is set to zero. The measured rectifier currents $i_{r(d,q,z)}$ are subtracted component-wise from the references $i_{(d,q,z),ref}$, and the resulting current error is brought to PI controller that approximates the filter voltage reference $u_{L(d,q,z),ref}$. The filter voltage reference is subtracted from line voltages $u_{s(d,q,z)}$ and is brought to block "CCC" that does the cross-coupling compensation for the d- and q-component voltage equations (20) and (21). The output of the CCC block gives the rectifier voltage reference $u_{r(d,q,z)}$ in synchronous reference frame which is then transformed to stationary frame $u_{r(\alpha,\beta,z)}$ and brought to the modulator. The modulator calculates the duty cycles and the redundant vector weighting factor r and controls the three active switches.

V. EXPERIMENTAL RESULTS

A. Prototype

The 10 kVA rated laboratory prototype of VIENNA4 is presented in Fig. 5. The main circuit consists of three IXYS VUM 25-05E modules, 2200 µH (4.3% p.u.) three phase ac inductor and two series connected 3.3 mF (16.5 p.u.) aluminium electrolytic capacitors. The rectifier is fed from 230 V/400 V 50 Hz utility mains and the dc link voltage reference $U_{dc,ref}$ is set to 750 V. All main circuit components are connected via copper wires and bus bars. The prototype includes also voltage and current measurement boards that scale the measured quantities to A/D converter voltage

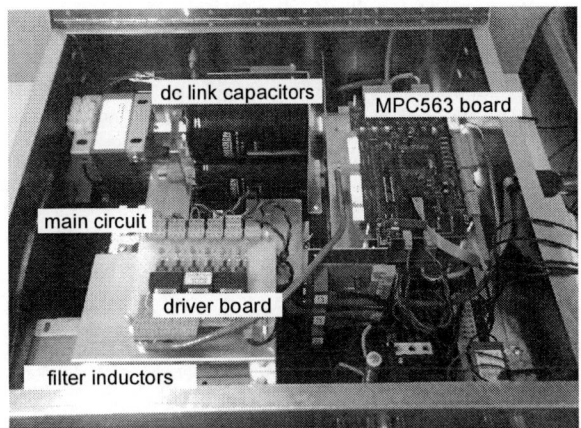

Fig. 5: Laboratory prototype of 10 kVA rated VIENNA4.

level.

The control system and the modulator are both implemented digitally with Freescale MPC563 32-bit single chip microcontroller board. The microcontroller's time processor unit TPU schedules interrupts every $T_s = 50$ µs time periods in which the modulator is updated. This results in 10 kHz switching frequency. In the microcontroller implementation, the control system and the controllers are put into discrete form.

B. Measurements

The functionality of the proposed modulation method is verified with measurements from the laboratory prototype. Rectifier feeds two resistors $R_{dc(1,2)}$ (Fig. 2) connected over the dc link. The resistor values are equal in all investigated load conditions. The prototype was tested with 5.3 kW and 10.6 kW load power levels. Dynamical behaviour was investigated with load transients between 5.3 kW and 10.6 kW.

The measured waveforms of the rectifier feeding 10.6 kW load are presented in Fig. 6. The rectifier draws sinusoidal and balanced currents $i_{r(a,b,c)}$ from the utility (Fig. 6a) and operates in unity power factor

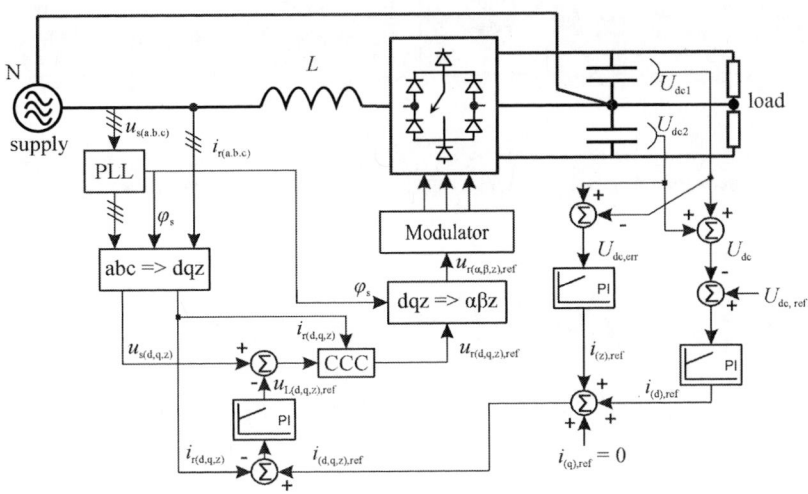

Fig. 4: The simplified block diagram of the control system for VIENNA4.

(PF = 0.996 cap). The total harmonic distortion of the phase a-current i_{ra} (Fig. 6d) calculated up to 2 kHz is

THD$_{2kHz}$ = 2.1 %. The neutral wire current i_z (Fig. 6b) mainly consists of the sum of switching frequency ripple

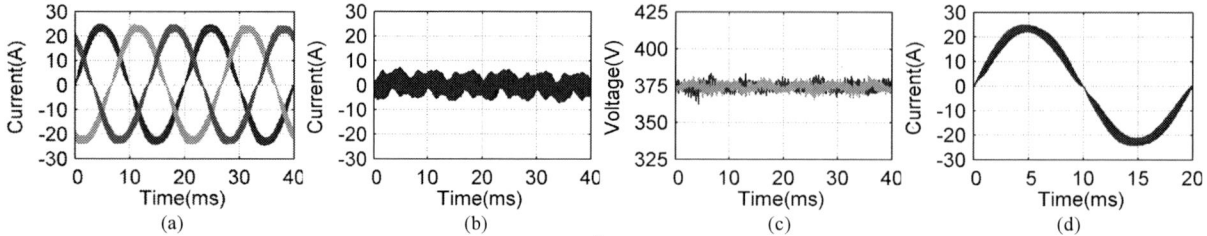

Fig. 6: Measured waveforms of the rectifier with 10.6 kW load. (a) Rectifier currents $i_{r(a,b,c)}$, (b) neutral current i_z, (c) partial dc link capacitor voltages $U_{Cdc(1,2)}$, (d) phase-a current i_{ra} THD$_{2kHz}$ = 2.1 %.

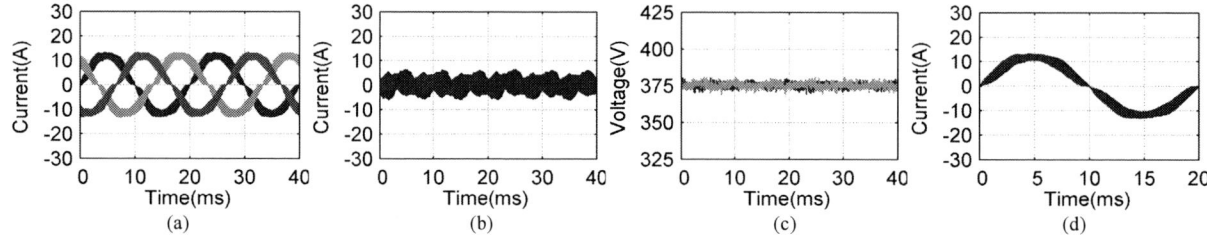

Fig. 7: Measured waveforms of the rectifier with 5.3 kW load. (a) Rectifier currents $i_{r(a,b,c)}$, (b) neutral current i_z, (c)partial dc link capacitor voltages $U_{Cdc(1,2)}$, (d) phase-a current i_{ra} THD$_{2kHz}$ = 3.2 %.

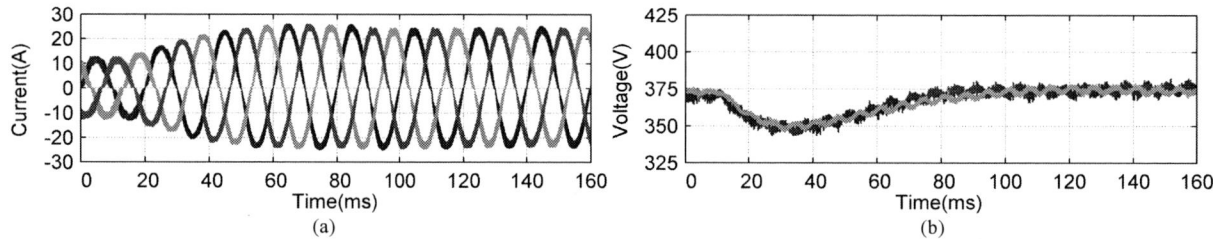

Fig. 8: Measured waveforms of the rectifier with load transition from 5.3 kW to 10.6 kW. (a) Rectifier currents $i_{r(a,b,c)}$, (b) partial dc link capacitor voltages $U_{Cdc(1,2)}$.

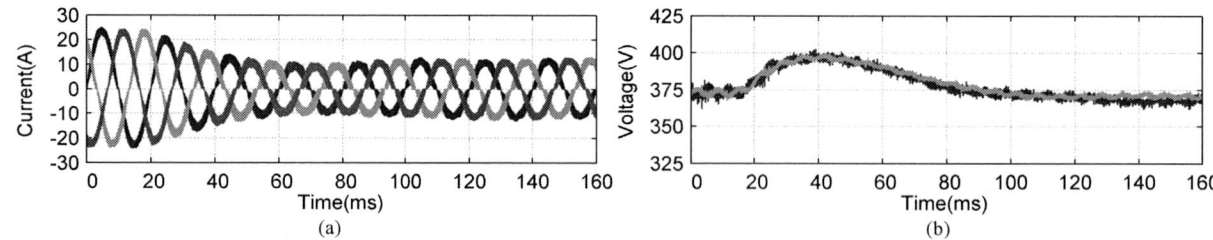

Fig. 9: Measured waveforms of the rectifier with load transition from 10.6 kW to 5.3 kW. (a) Rectifier currents $i_{r(a,b,c)}$, (b) partial dc link capacitor voltages $U_{Cdc(1,2)}$.

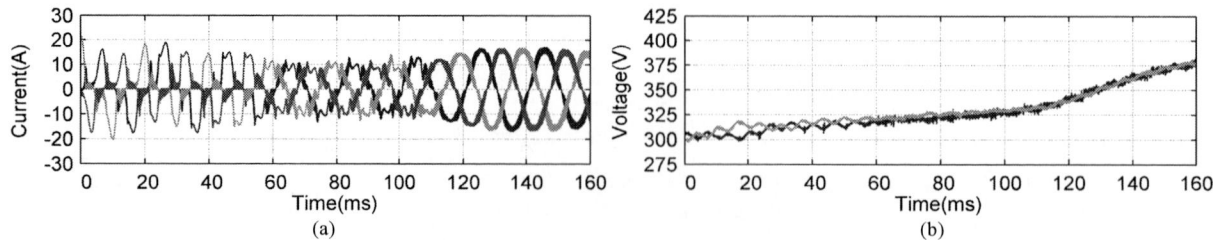

Fig. 10: Measured waveforms of the rectifier in the modulator start-up to 5.3 kW load. (a) Rectifier currents $i_{r(a,b,c)}$, (b) partial dc link capacitor voltages $U_{Cdc(1,2)}$.

978-1-4244-1871-8/07 $25.00 © 2007 IEEE

currents. Partial dc link voltages $U_{Cdc(1,2)}$ (Fig. 6c) are balanced and their sum is the desired reference voltage $U_{dc,ref} = 750$ V. The efficiency of the prototype in 10,6 kW load is $\eta = 0.960$.

The measured waveforms in 5.3 kW load condition are presented in Fig. 7. The power factor is PF = 0.989 and the THD of the phase-a current $THD_{2kHz} = 3.2$ %. The partial dc link voltages are balanced and they follow the set reference. The efficiency in 5.3 kW load is $\eta = 0.955$.

The dynamical behaviour of the rectifier was investigated with load step changes between lower (5.3 kW) and higher (10.6 kW) power levels. The measured line currents $i_{r(a,b,c)}$ and partial dc link voltages $U_{Cdc(1,2)}$ in load changes are presented in Figs. 8–9. In Fig. 8 at $t = 10$ ms the load power level is changed from lower to higher. In Fig. 9 at $t = 18$ ms the load power level is changed from higher to lower. In both load transients, the partial dc link voltages remain balanced, and the total dc link voltage recovers back to reference value in 80 ms after the transient. However, it must be noted that the load is resistive and the load level during transient varies with the total dc link voltages.

Fig. 10 shows the measured waveforms in the rectifier during bridge start-up. The modulator is enabled at $t = 0$ ms. At $t = 0$ ms $- 110$ ms the rectifier phase legs are turned off for 60° periods during supply phase voltage peaks as described in section III. After $t = 110$ ms the dc link voltage is sufficient for continuous modulation ($m < 1$) and the bridge starts operating normally.

VI. CONCLUSION

A space vector modulation method for four-wire Vienna rectifier was studied in the paper. The method treats the zero-sequence voltage reference of the rectifier as an offset which is realized by proper weighting of the redundant vectors that are present in each sequence. The applied modulation method was discussed in detail. The functioning of the method was verified with 10 kVA rated laboratory prototype.

Measurements were performed in steady state and dynamical load conditions. The rectifier is able to draw

balanced and sinusoidal currents from the utility with unity power factor (PF = 0.996 cap) and high efficiency ($\eta = 0.960$). The total harmonic distortion of the phase-a current in nominal (10.6 kW) load is $THD_{2kHz} = 2.1$ %.

REFERENCES

[1] D. Carlton, and W. G.. Dunford, "Multi-level, uni-directional AC-DC converters, a cost effective alternative to bi-directional converters," in *32nd Annual IEEE Power Electronics Specialists Conference*, Jun 2001, vol. 4. pp. 1911 – 1916.

[2] J.W. Kolar, and F.C. Zach, "A novel three-phase utility interface minimizing line current harmonics of high-power telecommunications rectifier modules," in *IEEE Trans. Industrial Electronics*, vol. 44, no. 4, Aug. 1997, pp. 456–467.

[3] R. Teichmann, M. Malinowski, and S. Bernet, "Evaluation of three-level rectifiers for low-voltage utility applications," in *IEEE Trans. Industrial Electronics*, vol. 52, no. 2, Apr. 2005, pp. 471–481.

[4] M.E. Fraser, C.D. Manning and B.M. Wells, "Transformerless four-wire PWM rectifier and its application in AC-DC-AC converters," in *IEE Proc. Electric Power Applications*, vol. 142, issue 6, Nov. 1995, pp. 410-416.

[5] J. Alahuhtala, and H. Tuusa, "Four-wire unidirectional three-phase/level/switch (VIENNA) rectifier," in *33rd Annual Conference of the IEEE Industrial Electronics Society 2006*, pp. 2420 – 2425.

[6] P. Ide, N. Froehleke and H. Grotstollen, "Investigation of low cost control schemes for a selected 3-level switched mode rectifier," in *Telecommunications Energy Conference, 1997*, pp. 413-418.

[7] R. Zhang, H. Psarad, D. Boroyevich, and F.C. Lee, "Three-dimensional space vector modulation for four-leg voltage-source converters", in *IEEE Trans. Power Electronics*, vol. 17, no. 3, May. 2002, pp. 314–326.

[8] N.Y Dai, M.C. Wong, and Y.D. Han, "Application of a three-level NPC inverter as a three-phase four-wire power quality compensator by generalized 3DSVM," in *IEEE Trans. Power Electronics*, vol. 21, no. 2, Mar. 2006, pp. 440–449.

[9] T. Viitanen, and H. Tuusa, "Experimental results of vector controlled and vector modulated Vienna I rectifier," in *35th Annual IEEE Power Electronics Specialists Conference*, Jun. 2004, vol. 6. pp. 4637 - 4643.

[10] R. Burgos, R. Lai, Y. Pei, F. Wang, D. Boroyevich, and J. Pou, "Space vector modulation for Vienna-type rectifiers based on the equivalence between two- and three-level converters: a carrier based implementation," in *38th Annual IEEE Power Electronics Specialists Conference*, Jun. 2007, pp. 2861 - 2867.

[11] S. Fukuda, Y. Matsumoto, and A. Sagawa. "Optimal-regulator-based control of NPC boost rectifiers for unity power factor and reduced neutral-point potential variations," in *IEEE Transactions On Industrial Electronics*, Vol 46, no. 3, June 1999, pp. 527–534.

978-1-4244-1871-8/07 $25.00 © 2007 IEEE

Voltage balancing technique with low switching frequency for cascade multilevel active front-end

Omid Alizadeh
University of Tehran
School of Electrical and Computer Engineering
Email: o.alizadeh@ece.ut.ac.ir

Shahrokh Farhangi
University of Tehran
School of Electrical and Computer Engineering
Email: farhangi@ut.ac.ir

Abstract— Recently cascade multilevel converters have attracted the attention of researchers for medium voltage applications such as FACTs devices and motor drives. This topology consists of several H-bridge cells in series, which has the advantages of modular structure and reduced number of components. The main challenges in these applications are limitation of switching frequency of semiconductor devices, low harmonic distortion requirement of grid current and balancing problems of the isolated dc-links. In this paper a new modulation method is proposed which produces low current distortion with low switching frequency. The balancing of the capacitors is well established, and the system has fast dynamic response. The method is simple and effective. Simulation results show the feasibility and superiority of the proposed method.

I. INTRODUCTION

Several topologies have been proposed for medium voltage converters. The main topologies are multi-pulse converter which uses coupling transformer, and transformerless multilevel topologies, diode-clamped multilevel converter (DCMLC) [1], flying capacitor multilevel converter (FCMLC) [2] and cascade multilevel converter (CCMLC) [3]. CCMLC consists of several H-bridge cells in series for each phase. It has the following advantages:

1) The structure is based on simple two level H-bridge cells, so the modularized circuit layout and packaging is possible.

2) It does not need coupling transformer, clamping diodes and flying capacitors.

3) For the same power capability, the cascade converter requires less number of power components such as switches and diodes.

The modulation methods presented for the cascade multilevel converter can be classified into staircase modulation [3]-[5], phase-shifted sinusoidal pulsewidth modulation [6]-[8] and multi-pulse optimal pulsewidth modulation [9],[10]. With the staircase modulation the converter switches operate at lower frequency. But, the current harmonics are high, and high number of H-bridge cells must be utilized in each phase to achieve low current distortion, which increases the cost of the converter. Phase-shifted sinusoidal PWM can result in lower current harmonic distortion, but a switching frequency higher than 500 Hz is needed which leads to higher switching losses. Multi-pulse optimal PWM method has a relatively low current distortion and low switching frequency. But, since the required switching angles are pre-determined by lookup table and applied in each cycle, this method can not meet the fast response for rapidly changing loads. These modulation techniques can be used in multilevel rectifiers, but they have the above mentioned problems, besides the unbalancing of the capacitor voltages in converters with high number of isolated dc-links. Some papers based on time average method are presented for rectifier with low number of series cells [11],[12].

In this paper a new modulation method is presented for the cascade multilevel rectifier, which results in low current distortion with a low switching frequency of 150 Hz. The proposed method has a fast dynamic performance, and the input current can be changed rapidly anywhere at the grid cycle. The capacitors voltage balancing is well established by this method both in transient and steady state. The predictive current control is used in current control loop [13],[14] to achieve a fast response. The method is applied to a single phase cascade rectifier with eight cells in series, and can be extended to higher number of cells and three phase rectifiers.

II. CONTROL SCHEME

A. Control system of the rectifier

Fig. 1 shows the circuit configuration of the multilevel rectifier cascading eight H-bridge cells. Each cell consists of a separated dc capacitor and four switches. The ratings of all switches are the same. Each H-bridge cell can generate three voltage levels, v_c, 0, $-v_c$ on the ac side, where v_c is the voltage of the dc capacitor. The control block diagram for the proposed converter is shown in fig.2. The reference current, i^*, is a unity power factor sinusoidal waveform, whose amplitude is the output of a PI controller that regulates the total dc-link voltage. So the dc voltage regulator determines the input active power needed to keep the total dc-link at its reference value, and the proposed modulation method is expected to balance the capacitor voltages.

B. Predictive current control

The predictive current control method is used for the current control loop. This block determines the required reference voltage for the voltage source converter, according to the actual and reference currents, and the grid voltage. The following equation can be written for the converter:

$$L \frac{di}{dt} = e - v_{conv} \qquad (1)$$

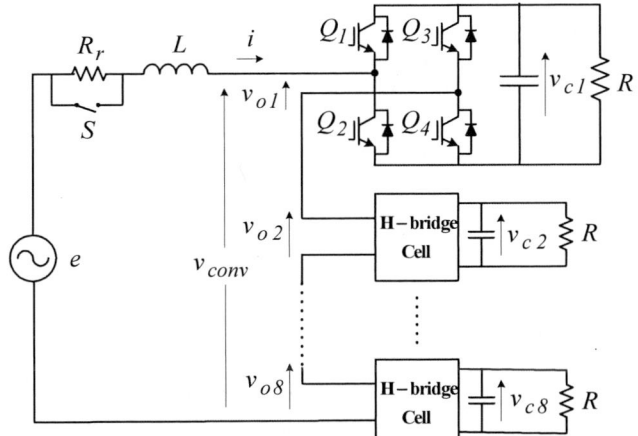

Fig.1. circuit configuration of the multilevel rectifier cascading eight cells

where, e and v_{conv} are grid and converter voltages, respectively. Equation (1) can be written as follows:

$$\Delta i = \frac{1}{L} \int_{T_s} \left(e - v_{conv} \right) dt \qquad (2)$$

where, T_s is the switching period. Equation (2) can be written as follows:

$$\bar{v}_{conv} = -\frac{L}{T_s} \Delta i + \frac{1}{T_s} \int_{T_s} e \, dt \qquad (3)$$

where $\bar{v}_{conv}$ is the average value of converter voltage over period T_s.

Discrete form of the above equation is as follows; by assuming a sinusoidal grid voltage, $e = V_m \, sin(\omega t)$:

$$\bar{v}_{conv} = \frac{L}{T_s} \left(i_k - i_{k+1} \right) + \frac{V_m}{\omega T_s} \left[cos\left(\omega t_{k+1} \right) - cos\left(\omega t_k \right) \right] \quad (4)$$

where, i_k and i_{k+1} are the converter currents at times t_k and t_{k+1} ($t_{k+1} - t_k = T_s$).

If we assume, at the end of each switching period of T_s, the actual current, i_{k+1}, can reach the reference current, i_{k+1}^*, then the desired average converter voltage, $\bar{v}_{conv}^*$, that can achieve this goal is derived as follows:

$$\bar{v}_{conv}^* = \frac{L}{T_s} \left(i_k - i_{k+1}^* \right) + \frac{V_m}{\omega T_s} \left[cos\left(\omega t_{k+1} \right) - cos\left(\omega t_k \right) \right] \quad (5)$$

This means the average reference converter voltage in (5), provides a predictive dead beat control in the first order plant of fig.2. This voltage is synthesized by the eight series H-bridge cells output voltages.

C. Proposed modulation strategy

In the proposed modulation method, the capacitor voltages are sorted from low to high according to their amplitudes, at the beginning of each converter voltage waveform half cycle. Eight voltage levels can be applied on the ac side by the converter, so that the nth level is synthesized by the first n cells based on the sorted capacitor voltages. The eight capacitor voltages are measured at the beginning of every T_s. The switching pattern in each T_s is shown in fig.3. If

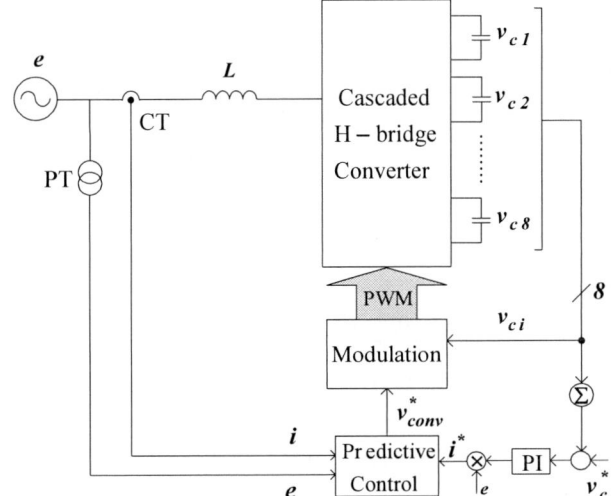

Fig.2. control block diagram of the proposed converter

$\bar{v}_{conv}^*$ is a positive voltage between the levels v_j and v_{j+1} (based on the sorting algorithm), then the cells # 1 to j are commanded to provide their dc bus, with positive polarity to the ac side, the cells after $j+1$ provide zero voltages, and the cell # $j+1$ is switched according to fig.3. When the $\bar{v}_{conv}^*$ is negative, the modulation is similar, but the commands of Q_1 and Q_3 are replaced and the cells provide negative voltages on the ac side.

Based on the presented pattern, in each half cycle of the converter voltage waveform, the cell which has the lowest capacitor voltage at the beginning of the half cycle, provide its dc bus on the ac side more than the other cells, so it is the most conducting cell. Therefore its capacitor charges more than the other capacitors. The next most conducting cell is the H-bridge which has the second lowest capacitor voltage at the beginning of the half cycle, and so on. In the next half cycle the modulation pattern is carried on by the cells with a new sorted sequence. So the balancing of the isolated dc-link voltages is kept in both transient and steady state. In regeneration mode the sorting of the capacitor voltages are reversed. In this mode, the cell which has the highest capacitor voltage must be the most conducting cell to discharge its capacitor more than the other cells. So the capacitor voltages are sorted from high to low at the beginning of each converter voltage waveform half cycle. There is a phase shift between grid and converter voltages in the rectifier that must be considered in the modulation by phase shifting of the switching patterns with respect to grid voltage.

D. Control scheme

The state-space model of the multilevel rectifier is as follows:

$$L \frac{di}{dt} = e - \sum_{i=1}^{8} q_i \, v_{ci} \qquad (6)$$

978-1-4244-1871-8/07 $25.00 © 2007 IEEE

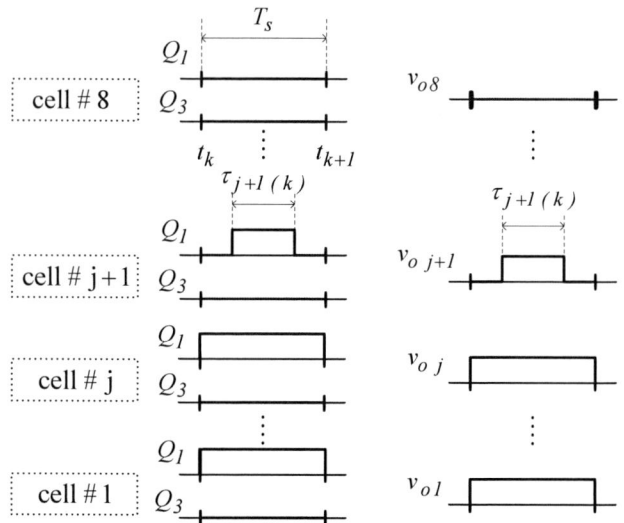

Fig.3. switching patterns in one switching period for positive converter reference voltage based on sorted capacitor voltages

$$C\frac{dv_{ci}}{dt} = q_1 i - \frac{v_{ci}}{R} \qquad i = 1,...,8 \qquad (7)$$

where q_i is +1 when cell # i provides its dc bus with positive polarity to the ac side, -1 when it supplies negative voltage, and 0 when it supplies zero voltage. If $\bar{v}_{conv}^*$ is a positive voltage between the levels v_j and v_{j+1}, then $q_1 = ... = q_j = 1$, $q_{j+2} = ... = q_8 = 0$. Solving (6) and (7) with the description of discrete-time variables, with the initial conditions i_k and $v_{ci(k)}$, $i = 1,...,8$, gives the approximate ac-side current i_{k+1} and capacitor voltages $v_{ci(k+1)}$, $i = 1,...,8$ at $t = t_{k+1}$ as (8). Where

$$X = -\frac{1}{L}T_s e^{-\frac{1}{RC}T_s} \qquad (9)$$

$$Y = \frac{1}{C}T_s e^{-\frac{1}{RC}T_s} \qquad (10)$$

$$Z = cos(\omega t_k) - cos(\omega t_{k+1}) \qquad (11)$$

and approximation $(T^2/(LC)) \simeq 0$ was used. $\tau_{j+1}(k)$ is the time duration when cell # $(j+1)$ provides its dc bus in the ac side at kth switching period, as shown in fig.3. By substituting the reference current i_{k+1}^* into i_{k+1} of (8), the time duration $\tau_{j+1}(k)$ can be obtained as (12), to be applied at kth interval.

$$\tau_{j+1}(k) = (L(i_k - i_{k+1}^*) - (1/L)T_s e^{-\frac{1}{RC}T_s}\sum_{i=1}^{j}v_{ci}(k)$$
$$+\frac{\sqrt{2}V_m}{\omega}Z)/(e^{-\frac{1}{RC}T_s}v_{cj+1}(k)) \qquad (12)$$

From (12), $\tau_{j+1}(k)$ can be determined based on i_k, i_{k+1}^*, $v_{ci(k)}$ and j. j has been obtained in section II.B according to $\bar{v}_{conv}^*$. If $\bar{v}_{conv}^*$ is a negative voltage, and $|\bar{v}_{conv}^*|$ is between the levels v_j and v_{j+1}, then the time duration $\tau_{j+1}(k)$ is obtained from (13), and the H-bridge cells provide negative voltages to the ac side.

$$\tau_{j+1}(k) = (L(i_{k+1}^* - i_k) - (1/L)T_s e^{-\frac{1}{RC}T_s}\sum_{i=1}^{j}v_{ci}(k)$$
$$-\frac{\sqrt{2}V_m}{\omega}Z)/(e^{-\frac{1}{RC}T_s}v_{cj+1}(k)) \qquad (13)$$

III. SIMULATION RESULTS

Simulation has been carried out with PSCAD/EMTDC to verify the effectiveness of the proposed method for the power circuit of fig.1. The specifications of the system are given in table I. A pre-charge circuit is used as shown in fig.1, to limit the initial charging current of the capacitors. Resistive loads are considered on the dc sides of the converter. The switching period is selected as $T_s = 1/f_s = 1/(32 \times f_{grid})$, to have the switching frequency of 150 Hz. Decreasing the switching period results in lower current distortion, with the cost of higher switching frequency and loss increment. Fig.4 shows the grid current, grid and converter voltages, and eight capacitor voltages. As shown the grid current has a unity power factor with the THD of less than 2.5% and good transient. The capacitor voltages are well balanced even during transient state. Despite a relatively high ripple which is about 10% of the reference dc voltage, the converter

$$\begin{bmatrix} i_{k+1} \\ v_{c1(k+1)} \\ \vdots \\ v_{c8(k+1)} \end{bmatrix} = \begin{bmatrix} 1 & X & \dots & X & -\frac{1}{L}\tau_{j+1}(k)e^{-\frac{1}{RC}T_s} & 0 & \dots & 0 \\ Y & e^{-\frac{1}{RC}T_s} & 0 & \dots & & & \dots & 0 \\ \vdots & 0 & \ddots & & & & & \vdots \\ Y & \vdots & & \ddots & & & & \\ \frac{1}{C}\tau_{j+1}(k)e^{-\frac{1}{RC}T_s} & \vdots & & \dots & & & & \\ 0 & \vdots & & & & \ddots & 0 & \\ \vdots & \vdots & & & & & \ddots & \\ 0 & 0 & \dots & \dots & & \dots & 0 & e^{-\frac{1}{RC}T_s} \end{bmatrix} \begin{bmatrix} i_k \\ v_{c1(k)} \\ \vdots \\ v_{c8(k)} \end{bmatrix} + \begin{bmatrix} Z \\ \frac{T_s}{C}Z e^{-\frac{1}{RC}T_s} \\ \vdots \\ \frac{T_s}{C}Z e^{-\frac{1}{RC}T_s} \\ \frac{\tau_{j+1}(k)}{C}Z e^{-\frac{1}{RC}T_s} \\ 0 \\ \vdots \\ 0 \end{bmatrix} \frac{\sqrt{2}V_m}{L\omega} \quad (8)$$

TABLE I
SPECIFICATIONS OF THE SIMULATION MODEL

grid voltage	10 kV	grid frequency	50 Hz
inductance L	10 mH	switching frequency	150 Hz
DC-link voltage v_c	2 kV	nominal active power	4 MW
DC-link capacitor	8 mF	nominal active current	400 A rms

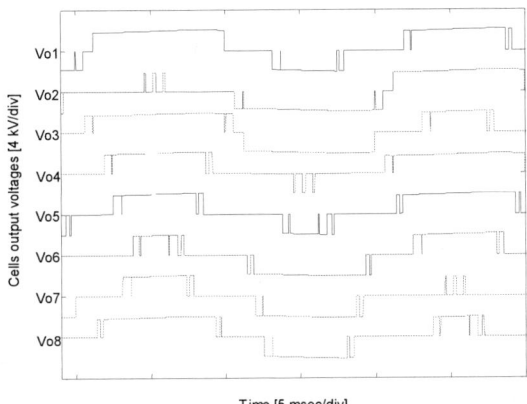

Fig.6. output voltages of eight cells.

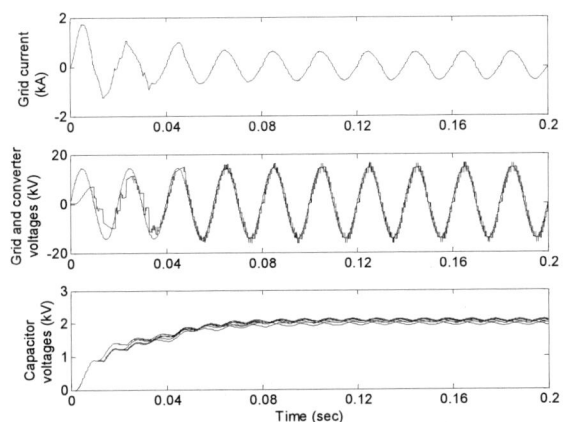

Fig.4. grid current, grid and converter voltages and eight capacitor voltages.

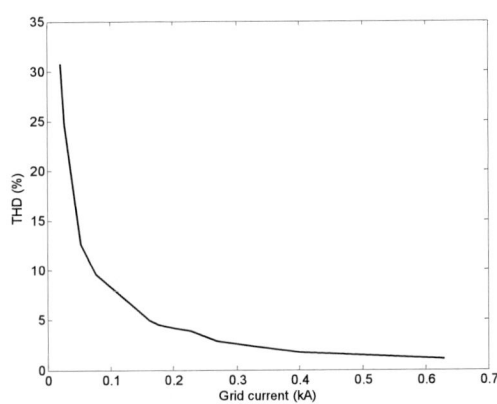

Fig.7. THD characteristic.

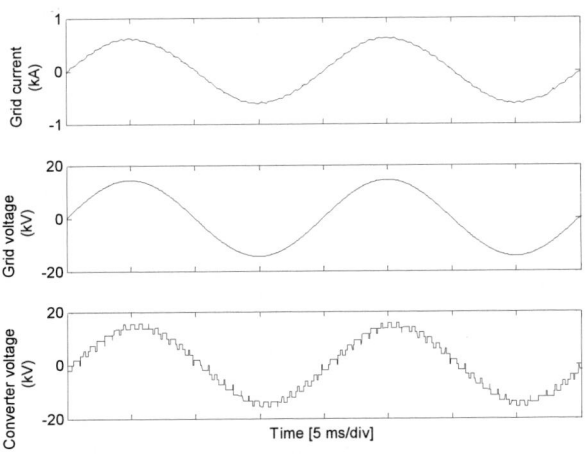

Fig.5. grid current, grid and converter voltages.

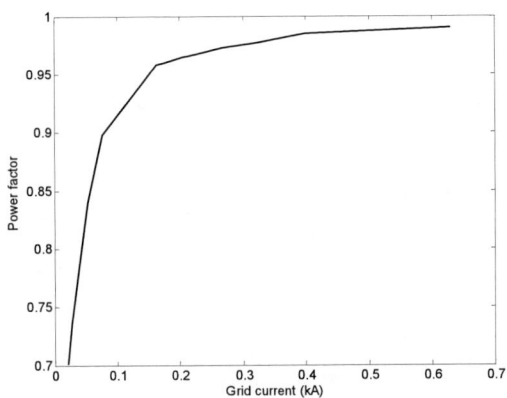

Fig.8. power factor characteristic.

current has a low current distortion and the sensitivity to this ripple is low. Fig.5 shows the grid current, grid voltage and converter voltage in steady state. The output voltages of the eight cells are shown in fig.6, which are changed in each converter half cycle based on the capacitor voltages sorting to keep the dc-link voltages balanced. Fig.7 and 8 show the characteristics of the THD and power factor versus rms grid current. As shown, THD decreases and power factor increases as the current increases. Fig.9 shows the transient

of the system from loading to regeneration, after three cycles. The system has a fast response and the grid current remains sinusoidal during transient.

IV. CONCLUSION

A new modulation method for cascade multilevel regenerative rectifier has been proposed in this paper. A capacitor voltage balancing algorithm has been presented in the modulation method, which provides safe operation for

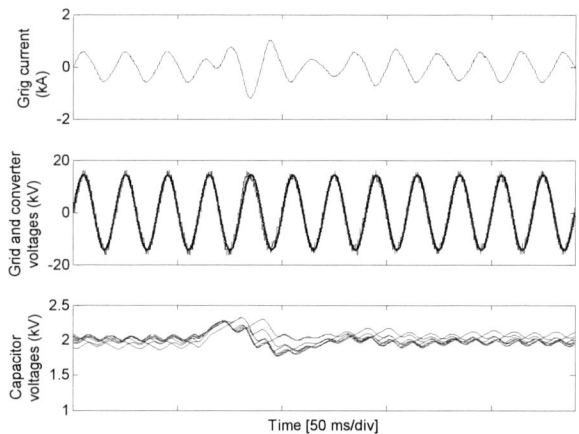

Fig.9. transient of the system from loading to regeneration.

series H-bridge cells. The simulation results show that the grid current has a sinusoidal waveform even during transient states, and the method has fast dynamic response. So it can meet the demand of rapidly changing loads. In contrast with the other presented methods, the proposed method results in lower current distortion and fast dynamic response, with low switching frequency.

REFERENCES

[1] M. Marchesoni, P. Tenca, "Diode-clamped multilevel converters: a practicable way to balance dc-link voltages", IEEE Trans. on Ind. Elec., vol. 49, no. 4, pp. 752-765, Aug. 2002.

[2] A. Meynard, H. Foch, "Multi-Level Conversion: High Voltage Choppers and Voltage-Source Inverters", in Proc. of PESC, 1992, pp. 397-403.

[3] Fang Z. Peng, J S Lai, J. W. McKeever, V. Coevering, "A multilevel voltage-source inverter with separate dc sources for static var generation", IEEE Trans on Ind. App., vol. 32, no. 5, pp. 1130-1138, Sep/Oct 1996.

[4] Fang Z. Peng, J. W. McKeever, D.J. Adams, "A power line conditioner using cascade multilevel for distribution systems", IEEE Trans on Ind. App., vol 34, no. 6, pp. 1293-1298, Nov/Dec 1998.

[5] Fang Z. Peng, J. Wang, "A universal STATCOM with delta-connected cascade multilevel inverter", in Proc. of PESC, 2005, pp. 340-346.

[6] Y. Liang, C. O. Nwankpa, "A new type of STATCOM based on cascading voltage-source inverters with phase-shifted unipolar SPWM", IEEE Trans on Ind. App., vol. 35, no. 5, pp. 1118-1123, Sep/Oct 1999.

[7] K. Fujii, R. W. De Doncker, "A novel DC-link voltage control of PWM-switched cascade cell multi-level inverter applied to STATCOM", in Proc. of Ind. App. Conf. 14th IAS Annual Meeting, vol. 2, pp. 961-967, Oct 2005.

[8] T. Yoshii, S. Inoue, H. Akagi, "Control and Performance of a Medium-Voltage Transformerless Cascade PWM STATCOM with Star-Configuration", in Proc. of IEEE Ind. App. Conf. 41st IAS Annual Meeting, vol. 4, pp. 1716-1723, Oct 2006.

[9] Q. Song, W. Liu, Z. Yuan, W. Wei, Y. Chen, "DC voltage balancing technique using multi-pulse optimal PWM for cascade H-bridge inverters based STATCOM", in Proc. of PESC, vol. 6, pp. 4768-4772, June 2004.

[10] Yu Liu, Zhong Du, A. Q. Huang, S. Bhattacharya, "An Optimal Combination Modulation Strategy for a Seven-level Cascade Multilevel Converter Based STATCOM", in Proc. of IEEE Ind. App. Conf. 41st IAS Annual Meeting, vol. 4, pp. 1732-1737, Oct 2006.

[11] C. Cecati, A. Dell'Aquila, M. Liserre, V. G. Monopoli, "A passivity-based multilevel active rectifier with adaptive compensation for traction applications", IEEE Trans on Ind. Applications, vol. 39, no. 5, pp.1404-1413, Sep/Oct 2003.

[12] C. Cecati, A. Dell'Aquila, M. Liserre, V.G. Monopoli, "Design of H-bridge Multilevel active rectifier for traction systems", IEEE Trans on Ind. Applications, vol. 39, no.5, pp.1541-1550, Sep/Oct 2003.

[13] Y. Nishida, O. Miyashita, T. Haneyoshi, H. Tomita, A. Maeda, "A predictive instantaneous-current PWM controlled rectifier with ac-side harmonic current reductuin", IEEE Trans. on Ind. Elec., vol. 44, no. 3, pp. 337-343,June 1997.

[14] V. G. Monopoli, D. Gerry, P. Zanchetta, J. C. Clare, P. W. Wheeler, "A low frequency predictive current control for multilevel active rectifiers", in Proc. of PESC, vol. 5, pp. 3553-3558, June 2004.

978-1-4244-1871-8/07 $25.00 © 2007 IEEE

Diode Rectifier Circuits with Commutation Capacitors for Three-Phase Rectangular-Waveform Distribution System

Takaharu Takeshita

Department of Electrical and Electronic Engineering,
Nagoya Institute of Technology
Gokiso, Showa, Nagoya 466-8555, JAPAN
Email: take@nitech.ac.jp

Abstract— This paper presents diode rectifier circuits with commutation capacitors in the three-phase rectangular-waveform distribution system. The commutation capacitors can improve the input total power factor and realize the zero voltage switching of the diodes. From the analysis, the authors also propose the design method of the rectifier circuit. In order to verify the effectiveness of the proposed method, the experimental characteristics of the rectifier circuit have been shown.

I. INTRODUCTION

Many industrial systems and household electric appliances have rectifier circuits for the dc voltage source of the inverter circuits. Therefore, the dc power distribution system has been proposed for the equipment with the inverter circuits. However, the isolation for the safety and the protection for the over current in the dc power distribution system are difficult compared with that in the ac power distribution system. Also, the ac distribution systems with the square-waveform [1] and the trapezoidal-waveform [2] are proposed. But the distribution system with the square-waveform cannot apply to the three-phase system. The system configuration to generate the trapezoidal waveform is complex. The authors have proposed the three-phase rectangular-waveform distribution system into the factory or the office building [3], [4]. The proposed system converts the commercial symmetrical three-phase sinusoidal voltage of 60 Hz to the three-phase rectangular voltage of 400 Hz for the distribution system.

This paper presents an analysis of steady and transient characteristics for the diode rectifier circuits with commutation capacitors in the three-phase rectangular-waveform distribution system [5]. The commutation capacitors can improve the input total power factor, and realize the zero voltage switching of the diodes. From the analysis of the rectifier circuit, the authors also propose the design method of the input inductances, the commutation capacitors and the output smoothing capacitors. In order to verify the effectiveness of the proposed method, the experimental characteristics of the rectifier circuits have been shown.

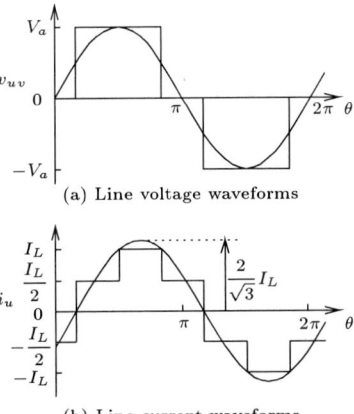

Fig. 1. Sinusoidal and rectangular waveforms

TABLE I

VOLTAGE AND CURRENT BETWEEN SINUSOIDAL AND RECTANGULAR WAVEFORMS

waveform	line voltage v_{uv} amplitude	RMS	line current i_u amplitude	RMS	power
sinusoidal	V_a	$\frac{1}{\sqrt{2}}V_a$	$\frac{2}{\sqrt{3}}I_L$	$\sqrt{\frac{2}{3}}I_L$	$V_a I_L$
rectangular	V_a	$\sqrt{\frac{2}{3}}V_a$	I_L	$\frac{1}{\sqrt{2}}I_L$	$V_a I_L$

II. RECTIFIER CIRCUIT FOR THREE-PHASE RECTANGULAR-WAVEFORM DISTRIBUTION SYSTEM

A. Comparison between sinusoidal and rectangular waveforms

Fig. 1 shows the comparison of the line voltage waveform v_{uv} and the current waveform i_u between the conventional sinusoidal- and proposed rectangular-waveform distribution systems under the unity total power factor and the same electric power of the diode rectifier circuit. In order to obtain the similar dc output voltage both diode rectifier circuits, both the amplitudes of the line voltage are selected to V_a. The amplitude of line current for the rectangular waveform is I_L. Table I shows the

978-1-4244-1871-8/07 $25.00 © 2007 IEEE

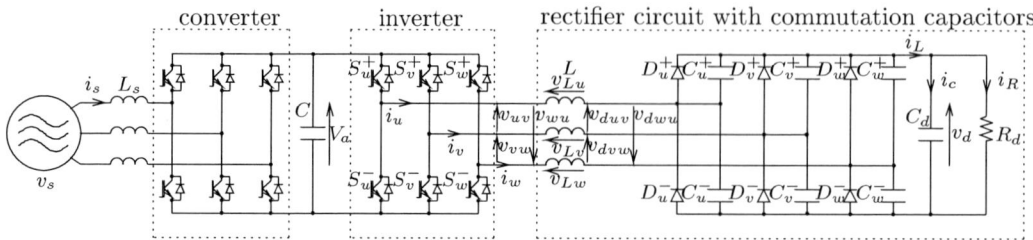

Fig. 2. System configuration of rectangular-waveform distribution system and rectifier circuit for load

amplitudes, rms values and powers for the sinusoidal and rectangular waveforms in Fig. 1. The amplitude and the rms value of the line current in the rectangular-waveform distribution system can be reduced to $\sqrt{3}/2$ of those in the sinusoidal-waveform distribution system. Therefore, the rated current of the switching devices and the capacity of the line current can be reduced in the rectangular distribution system.

B. Rectangular distribution system and diode rectifier circuits

Fig. 2 shows the system configuration of the rectangular-waveform distribution system and the rectifier circuit for load. For applying the rectangular-voltage-waveform distribution system in the consumer, the rectangular waveform voltage generator, which consists of the PWM converter and the inverter system, converts the three-phase sinusoidal-waveform voltage v_s into the three-phase rectangular-waveform voltages v_{uv}, v_{vw} and v_{wu} according to a six-step voltage inverter operation. The load has the diode rectifier circuit. Each diode in the rectifier circuit has the commutation capacitor in parallel. The commutation capacitors can improve the input total power factor, and realize the zero voltage switching of the diodes.

To present the effect of the commutation capacitors, the operations of the diode rectifier circuit without the commutation capacitors are indicated. Fig. 3 shows the line voltage and current waveforms of the rectifier circuit without the commutation capacitors. After the sign of the line current changes, the line current rises slowly up from zero because the reverse current in the diode can not flow. Therefore, the input total power factor is reduced.

Fig. 4 shows the line voltage and current waveforms of the rectifier circuit with the commutation capacitors. The input total power factor is improved because the input current waveforms i_u, i_v and i_w become the stair-step-wise waveforms. The unity input total power factor is obtained under the input current waveforms with four current levels of $\pm i_r$ and $\pm 2i_r$ [5].

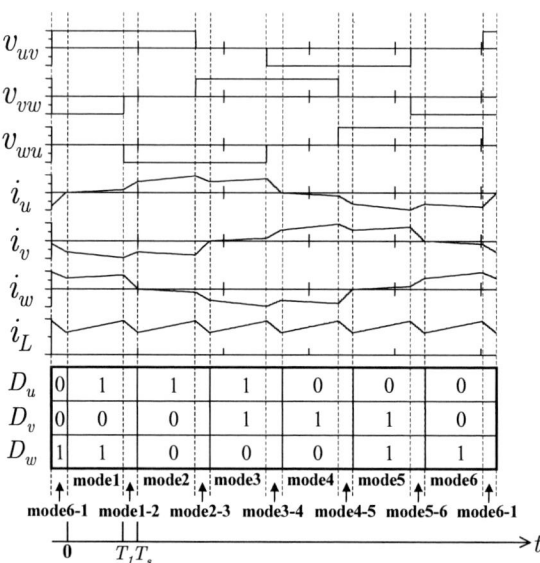

Fig. 3. Operating waveforms of diode rectifier circuit

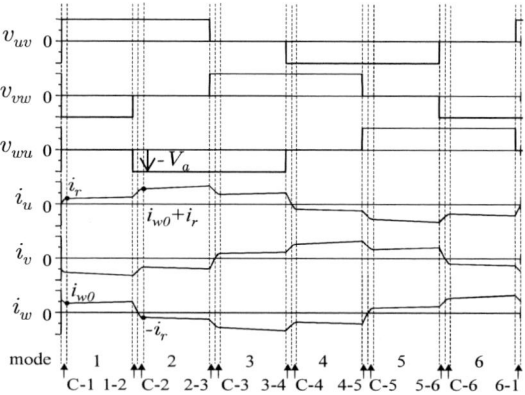

Fig. 4. Voltage and current waveforms of diode rectifier circuit with commutation capacitors

C. Steady-state analysis

By using the input inductance L and line voltages v_{duv}, v_{dvw} and v_{dwu} in Fig. 2, the voltage and current equations are given in (1) and (2).

$$
\begin{bmatrix} v_{uv} \\ v_{vw} \\ v_{wu} \end{bmatrix} = \begin{bmatrix} L & -L & 0 \\ 0 & L & -L \\ -L & 0 & L \end{bmatrix} \frac{d}{dt} \begin{bmatrix} i_u \\ i_v \\ i_w \end{bmatrix} + \begin{bmatrix} v_{duv} \\ v_{dvw} \\ v_{dwu} \end{bmatrix} \quad (1)
$$

978-1-4244-1871-8/07 $25.00 © 2007 IEEE

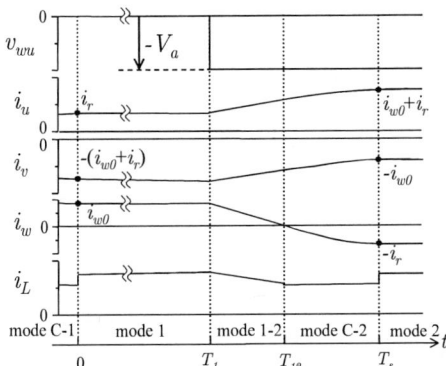

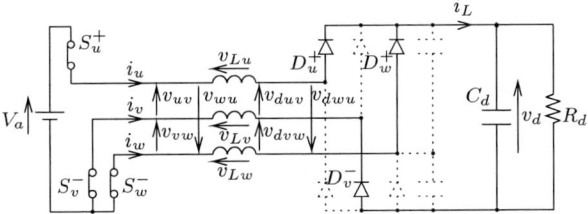

Fig. 7. Circuit in mode 1-2

Fig. 5. Extended figure of voltage and current waveforms

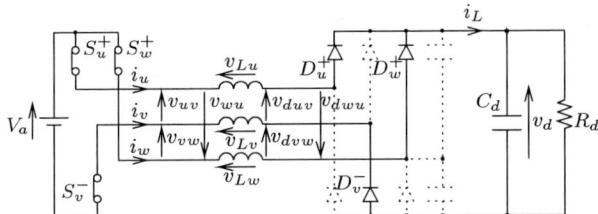

Fig. 6. Circuit in mode 1

$$i_u + i_v + i_w = 0 \qquad (2)$$

Solving for v_{uv} and v_{wu} from (1) and (2), the following voltage equation is obtained.

$$\begin{bmatrix} v_{uv} \\ v_{wu} \end{bmatrix} = \begin{bmatrix} 2L & L \\ -L & L \end{bmatrix} \frac{d}{dt} \begin{bmatrix} i_u \\ i_w \end{bmatrix} + \begin{bmatrix} v_{duv} \\ v_{dwu} \end{bmatrix} \qquad (3)$$

Fig. 5 shows the extended figure between mode 1 and C-2 during a sixth source period Ts in Fig. 4. In mode 1, the connection state as shown in Fig. 6, the inverter switches S_u^+, S_v^- and S_w^+, and the load diodes D_u^+, D_v^- and D_w^+ are continuity. Substituting the line voltage $v_{duv} = V_d$, $v_{dwu} = 0$ and initial currents $i_u(0) = i_r$, $i_w(0) = i_{w0}$ into (3), the currents i_u and i_w are obtained in (4).

$$\begin{bmatrix} i_u(t) \\ i_w(t) \end{bmatrix} = \frac{1}{3L} \begin{bmatrix} 1 & -1 \\ 1 & -1 \end{bmatrix} \begin{bmatrix} V_a \\ V_d \end{bmatrix} t + \begin{bmatrix} i_r \\ i_{w0} \end{bmatrix} \qquad (4)$$
$$(0 \le t \le T_1)$$

The currents $i_u(T_1)$ and $i_v(T_1)$ at the end of mode 1 are given as follows;

$$i_u(T_1) = \frac{V_a - V_d}{3L} T_1 + i_r \qquad (5)$$

$$i_w(T_1) = \frac{V_a - V_d}{3L} T_1 + i_{wo} \qquad (6)$$

Fig. 7 shows the connection state in mode 1-2. At $t = T_1$, the conducting switch in the inverter is changed from S_w^+ to S_w^- and the source voltage v_{wu} is changed from 0 to $-V_a$, then the mode 1-2 starts. The currents

i_u and i_w in mode 1-2 are obtained in (7).

$$\begin{bmatrix} i_u(t) \\ i_w(t) \end{bmatrix} = \frac{1}{3L} \begin{bmatrix} 2 & -1 \\ -1 & -1 \end{bmatrix} \begin{bmatrix} V_a \\ V_d \end{bmatrix} (t - T_1) + \begin{bmatrix} i_u(T_1) \\ i_w(T_1) \end{bmatrix} \qquad (7)$$
$$(T_1 \le t \le T_{12})$$

The currents $i_u(T_{12})$ and $i_w(T_{12})$ at $t = T_{12}$, when i_w becomes to zero, are given as follows;

$$i_u(T_{12}) = \frac{2V_a - V_d}{3L}(T_{12} - T_1) + i_u(T_1) \qquad (8)$$

$$i_w(T_{12}) = -\frac{V_a + V_d}{3L}(T_{12} - T_1) + i_w(T_1) = 0 \qquad (9)$$

Fig. 8 shows the connection state in mode C-2. The diodes D_w^+ and D_w^- are off state. The commutation capacitor C_w^+ is charged from zero to V_d by the positive current i_{cw}^+ and another commutation capacitor C_w^- is discharged from V_d to zero. Therefore, the zero voltage switching in diodes are realized. Using the initial values of the mode C-2,

$$\begin{bmatrix} v_{cw}^+(T_{12}) \\ v_{cw}^-(T_{12}) \end{bmatrix} = \begin{bmatrix} 0 \\ V_d \end{bmatrix}, \quad \begin{bmatrix} i_{cw}^+(T_{12}) \\ i_{cw}^-(T_{12}) \end{bmatrix} = \begin{bmatrix} 0 \\ 0 \end{bmatrix} \qquad (10)$$

the input currents i_u, i_w and the commutation capacitor voltages v_{cw}^+, v_{cw}^- are obtained as follows;

$$\begin{bmatrix} i_u(t) \\ i_w(t) \end{bmatrix} = \frac{V_a + V_d}{2} \sqrt{\frac{C_c}{3L}} \sin \frac{t - T_{12}}{\sqrt{3LC_c}} \begin{bmatrix} 1 \\ -2 \end{bmatrix}$$
$$+ \frac{V_a - V_d}{2L}(t - T_{12}) \begin{bmatrix} 1 \\ 0 \end{bmatrix} + \begin{bmatrix} i_u(T_{12}) \\ 0 \end{bmatrix} \qquad (11)$$

$$\begin{bmatrix} v_{cw}^+ \\ v_{cw}^- \end{bmatrix} = \frac{V_a + V_d}{2} \cos \frac{t - T_{12}}{\sqrt{3LC_c}} \begin{bmatrix} -1 \\ 1 \end{bmatrix} + \begin{bmatrix} \dfrac{V_a + V_d}{2} \\ -\dfrac{V_a - V_d}{2} \end{bmatrix}$$
$$(T_{12} \le t \le T_s) \qquad (12)$$

At the end of the mode C-2, $t = T_s$, the following equations are obtained from (11) and (12).

$$i_u(T_s) = \frac{V_a + V_d}{2} \sqrt{\frac{C_c}{3L}} \sin \frac{T_s - T_{12}}{\sqrt{3LC_c}}$$
$$+ \frac{V_a - V_d}{2L}(T_s - T_{12}) + i_u(T_{12}) = i_r + i_{wo} \qquad (13)$$

978-1-4244-1871-8/07 $25.00 © 2007 IEEE   755

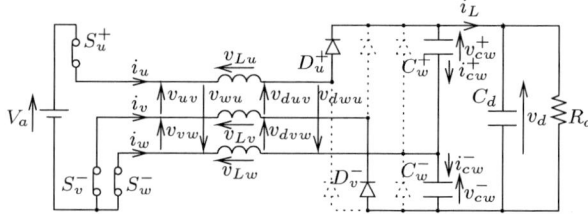

Fig. 8. Circuit in mode C-2

$$i_w(T_s) = -(V_a + V_d)\sqrt{\frac{C_c}{3L}} \sin\frac{T_s - T_{12}}{\sqrt{3LC_c}} = -i_r \quad (14)$$

$$v_{cw}^+(T_s) = -\frac{V_a + V_d}{2}\cos\frac{T_s - T_{12}}{\sqrt{3LC_c}} + \frac{V_a + V_d}{2} = V_d \quad (15)$$

From (15), the time T_{12} is given in (16).

$$T_{12} = T_s - \sqrt{3LC_c}\cos^{-1}\frac{V_a - V_d}{V_a + V_d} \quad (16)$$

Substituting (16) into (14), i_r is obtained by

$$i_r = \sqrt{4C_c V_a V_d / 3L} \quad (17)$$

where, i_r is almost constant because the variability range of the output voltage V_d is small. From (5), (6), (8), (9) and (13), the time T_1 and the initial current i_{w0} are obtained as follows;

$$T_1 = \frac{V_a + V_d}{2V_a}T_{12} - \frac{3(V_a - V_d)}{2V_a}T_s - \frac{3L}{2V_a}i_r \quad (18)$$

$$i_{w0} = \frac{V_a - V_d}{L}T_s + i_r \quad (19)$$

The time-average current I_L is obtained as follows;

$$\begin{aligned}I_L &= \frac{1}{T_s}\left\{\int_0^{T_{12}}(i_u + i_w)dt + \int_{T_{12}}^{T_s}(i_u + i_w/2)dt\right\}\\ &= \frac{1}{T_s}\left\{\frac{V_a - V_d}{12L}(3T_s^2 + 10T_sT_{12} + 3T_{12}^2)\right.\\ &\left. - \frac{V_a}{6L}(T_{12} - T_1)(2T_s + T_1 - 3T_{12}) + i_r(T_s + T_{12})|!\right\}\end{aligned}(20)$$

At the rated state, the constant current during $0 \leq t \leq T_1$ and $V_d = V_a$ are desired to realize higher total power factor. The initial current i_r in (17) and the output current I_L in (20) at rated state are expressed by I_{rn} and I_{Ln}, respectively. Then, I_{rn} and I_{Ln} are given by

$$I_{rn} = 2V_a\sqrt{\frac{C_c}{3L}} \quad (21)$$

$$I_{Ln} = 2I_{rn} - \frac{3(1 + \pi)LI_{rn}^2}{8V_aT_s} \simeq 2I_{rn} \quad (22)$$

D. Simplified analysis in steady state

The current equations above are complicated. The equations based on simplified analysis are derived in this section. For the simplified analysis, the diode with a commutation capacitor is treated as a diode with the maximum value of the reverse recovery current of $-I_{rn}$. Fig. 9 shows the input current waveforms during the commutation period. (a) shows the waveforms with commutation capacitors and (b) shows the waveforms with the revere recovery current. In (b), after the current i_w becomes to zero, the negative current i_w flows through the diode D_w^+. The mode 1-2 continues by $t = T_s$ at $i_w = -I_{rn}$.

Using (4)-(7), T_1, i_{w0} and I_L are obtained as follows;

$$T_1 = -\frac{V_a - 2V_d}{V_a}T_s - \frac{3L}{V_a}I_{rn} \quad (23)$$

$$i_{w0} = \frac{V_a - V_d}{L}T_s + I_{rn} \quad (24)$$

$$\begin{aligned}I_L &= \frac{1}{T_s}\int_0^{T_s}(i_u + i_w)dt\\ &= \frac{2(V_a^2 - V_d^2)T_s}{3LV_a} + \frac{2V_dI_{rn}}{V_a} - \frac{3LI_{rn}^2}{2V_aT_s}\end{aligned}(25)$$

The rms values of the line voltage v_{uv} and the input current i_u are given by

$$V_{uv} = \sqrt{\frac{1}{3T_s}\int_0^{T_s}\{v_{uv}^2 + v_{vw}^2 + v_{wu}^2\}\,dt} = \sqrt{\frac{2}{3}}V_a \quad (26)$$

$$I_u = \sqrt{\frac{1}{3T_s}\int_0^{T_s}\{i_u^2 + i_v^2 + i_w^2\}\,dt} \quad (27)$$

By using (25), the output power P_{Load} is obtained by

$$\begin{aligned}P_{Load} &= V_dI_L\\ &= \frac{V_d}{V_a}\left\{\frac{2(V_a^2 - V_d^2)T_s}{3L} + 2V_dI_{rn} - \frac{3LI_{rn}^2}{2T_s}\right\}\end{aligned}(28)$$

From (26) - (28), the input total power factor TPF is derived as follows;

$$\text{TPF} = P_{Load}/\sqrt{3}V_{uv}I_u$$

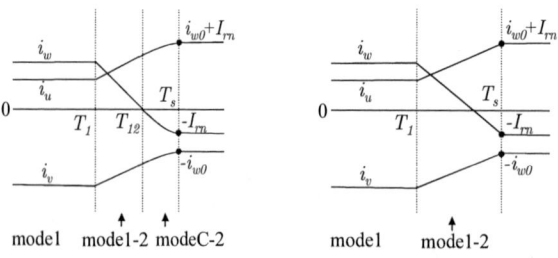

(a) Diode rectifier circuit with commutation capacitors

(b) Diode rectifier circuit with reverse recovery current

Fig. 9. Input current waveforms at commutation period

978-1-4244-1871-8/07 $25.00 © 2007 IEEE

$$= \frac{V_d}{\sqrt{2}I_u V_a^2}\left\{\frac{2(V_a^2 - V_d^2)T_s}{3L} + 2V_d I_{rn} - \frac{3L I_{rn}^2}{2T_s}\right\} \quad (29)$$

The rated input total power factor TPF_n under $V_d = V_a$ is given in (30).

$$\mathrm{TPF_n} = \left(1 - \frac{3LI_{rn}}{4V_a T_s}\right) \bigg/ \sqrt{1 - \frac{LI_{rn}}{2V_a T_s}} \quad (30)$$

E. Simplified analysis in transient state

Fig. 10 shows the simplified model of the diode rectifier circuit with commutation capacitors in transient state. Since the relation between I_L and V_d around $I_L = 0$ in (25) can treat as linear, the first-order approximate equation is obtained as follows;

$$V_a + \frac{3L}{2T_s}I_{rn} = \frac{3L}{4T_s}I_L + V_d \quad (31)$$

V_0, V_s and R_l are defined as follows;

$$V_0 = \frac{3L}{2T_s}I_{rn}, \quad V_s = V_a + V_0, \quad R_l = \frac{3L}{4T_s} \quad (32)$$

Using (32) and the inductance $L_l = 3L/2$ between V_a and V_d for transient model, that is the inductance in the mode 1 during T_s, (31) is rearranged as follows;

$$V_s = L_l \frac{di_L}{dt} + R_l i_L + v_d \quad (33)$$

where i_L and v_d are the variables added the ac components to the dc components I_L and V_d, respectively. The output voltage equation in Fig. 10 is given in (34)

$$v_d = \frac{1}{C_d}\int (i_L - i_R)dt = R_d i_R \quad (34)$$

From (33) and (34), the output current i_L and voltage v_d in rectifier circuit are obtained in (35) and (36), using the initial values $v_d(0) = v_{d0}$, $i_L(0) = i_{L0}$.

$$i_L(t) = \frac{V_s}{R_l + R_d} + e^{-Xt}\left(a\cos\omega t + \frac{b - aX}{\omega}\sin\omega t\right) \quad (35)$$

$$v_d(t) = \frac{R_d}{R_l + R_d}V_s - e^{-Xt}\bigg[\{aR_l + (b - 2aX)L_l\}\cos\omega t \\ + \left\{\frac{b - aX}{\omega}R_l - \frac{bX - aX^2 + a\omega^2}{\omega}L_l\right\}\sin\omega t\bigg] \quad (36)$$

where X, ω, a and b are given in

$$\begin{cases} X = \frac{1}{2}\left(\frac{1}{R_d C_d} + \frac{R_l}{L_l}\right) \\ \omega = \sqrt{\frac{R_l + R_d}{R_d C_d L_l} - X^2} \\ a = i_{L0} - \frac{V}{R_l + R_d} \\ b = \frac{a}{R_d C_d} + \frac{\frac{R_d}{R_d + R_l}V_s - v_{d0}}{L_l} \end{cases} \quad (37)$$

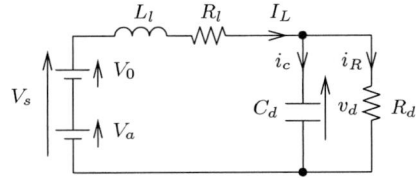

Fig. 10. Simplified model for transient analysis.

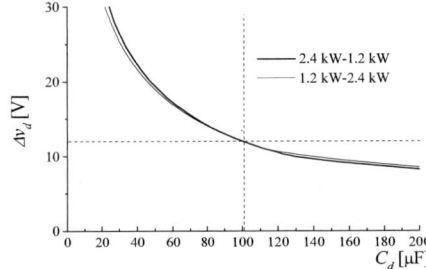

Fig. 11. Relationship between maximum dc voltage variation and dc capacitance under load step change between 50 and 100 %

III. PARAMETER DESIGN OF RECTIFIER CIRCUIT

In the rectangular-waveform distribution system of a laboratory prototype, the line to line voltage amplitude is 300 V, the frequency is 400 Hz and the rated output power is 2.4 kW. This section describes the design method for the parameters of the input inductance L, the commutation capacitors C_c and the output dc capacitor C_d under the rated output voltage $V_{dn} = V_a = 300$ V and the rated output current $I_{Ln} = 8$ A.

The relation between the rated total power factor TPF_n and the input inductance L are obtained by substituting $I_{rn} = I_{Ln}/2$ into (30). In order to realize TPF$>$0.99, the input inductance of $L = 0.5$ mH is selected.

The rated output power P_{Load} is expressed in (38), using I_{rn} and I_{Ln} in (21) and (22).

$$P_{Load} = V_a I_{Ln} = V_a^2\left\{4\sqrt{\frac{C_c}{3L}} - \frac{C_c(1 + \pi)}{2T_s}\right\} \quad (38)$$

The commutation capacitors of $C_c = 0.07\,\mu F$ in (38) is selected under $L = 0.5$ mH.

Fig. 11 shows the relationship based on (36) between the maximum dc voltage variation Δv_d and the dc capacitance C_d under the load step change between 50 and 100 %. In order to obtain $\Delta v_d < 30$ V (10 %), the dc capacitance of $C_d = 100\,\mu F$ is selected with margin due to the internal resistance without modeling of the dc capacitor.

IV. EXPERIMENTAL RESULTS

Table II shows the specifications of experimental system of a laboratory prototype in Fig. 2. The commutation capacitances C_c of $0.1\,\mu F$ whose value is larger than the designed value of $0.07\,\mu F$ are used in experi-

TABLE II
SPECIFICATIONS OF EXPERIMENTAL SYSTEM

Converter-inverter system	
source voltage v_s	200 V, 60 Hz
dc voltage V_a	300 V
switching frequency of converter f_c	20 kHz
dc capacitance C	1880 μF
frequency of output voltage f_a	400 Hz
sixth period of output voltage T_s	417 μs
input inductance L_s	4.0 mH
Diode rectifier circuit	
rated output power P_{Load}	2.4 kW
rated output voltage V_d	300 V
input inductance L	0.5 mH
commutation capacitance C_c	0.1 μF
dc capacitance C_d	100 μF

mental system because of the compensation for the line resistance.

Fig. 12 shows the experimental waveforms of the line voltage v_{uv}, the input current i_u, the output current i_L and the output dc voltage v_d in the diode rectifier circuit with commutation capacitors under the rated load. The four-level waveform of the input current similar to theoretical one is obtained. v_d and i_L are constant except during the commutation period.

Fig. 13 shows the characteristics of the output voltage V_d and the input total power factor TPF of diode rectifier circuit with commutation capacitors. The experimental output voltage V_d is reduced by 2 V compared with the theoretical one due to the forward voltage drop of the diodes. The theoretical THD is calculated under $C_c = 0.07 \mu$F in (29). The input total power factor of TPF=0.994 under rated output of $P_{Load} = 2.4$ kW is obtained.

Fig. 14 shows the theoretical and experimental waveforms of the output current i_L and the output dc voltage v_d under the load step change between 50 % and 100 %. The maximum dc voltage variation of $\Delta v_{dmax} = 25$ V are obtained. The experimental transient characteristics are coincident with the theoretical one. The errors between the theoretical and the experimental values are caused by the resistance of the output smoothing capacitor C_d.

V. CONCLUSIONS

This paper describes the diode rectifier circuits with commutation capacitors to obtain higher total power factor for the three-phase rectangular-waveform distribution system. The analysis in steady and transient states has been derived. The author has proposed the design method of the rectifier circuits. The effectiveness of the proposed method has been verified by experiments.

ACKNOWLEDGMENT

The author would like to thank for the support of "Grant-in-Aid for Exploratory Research, Japan Society for the Promotion of Science (JSPS)".

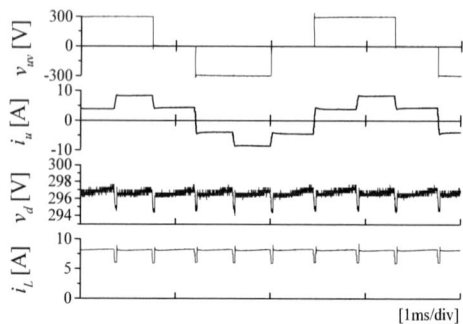

Fig. 12. Experimental waveforms at rated output power

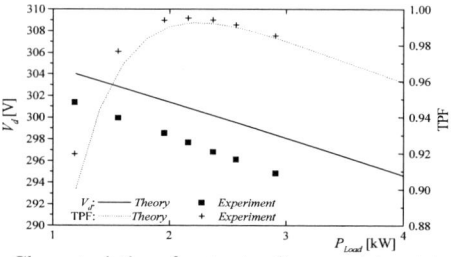

Fig. 13. Characteristics of output voltage and input total power factor of diode rectifier circuit with commutation capacitors

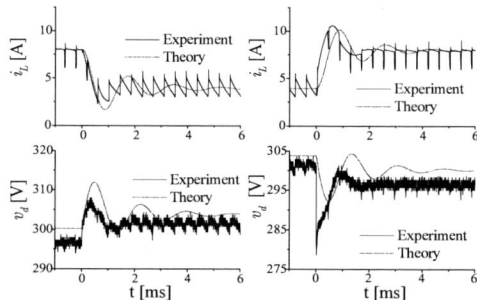

(a) From 2.4 kW to 1.2 kW   (b) From 1.2 kW to 2.4 kW

Fig. 14. Comparison between experimental waveform and theoretical waveforms

REFERENCES

[1] S. Kondo and H. Meguro : "Feasibility Study on Square-wave-voltage power Distribution System", *Trans. IEEJ*, Vol. 116-D, No. 5, pp. 556-562 (1996)

[2] Y. Sato and T. Noguchi : "Trapezoidal-Wave-Voltage Power Distribution System with UPS Function", *Proceedings of the 1999 Japan Industry Applications Society Conference*, Vol. 1, pp. 219-222 (1999)

[3] H. Matsue, T. Takeshita: "Fast Current Control of Three-Phase PWM Converter for Rectangular-Waveform Distribution System, Proceedings of 12th International Power Electronics and Motion Control Conference, EPE-PEMC 2006, pp. 212-217 (2006)

[4] T. Takeshita: "Configuration and Feature of Three-Phase Rectangular-Voltage-Waveform Distribution System", The 2006 International Conference on Electrical Machines and Systems, DS1E3-11 (2006)

[5] S. Watanabe, T. Takeshita, T. Nishida: "Characteristics of Diode Rectifier Circuits for Three-Phase Rectangular-Waveform Distribution System", *IEEJ Trans. IA*, Vol. 127, No. 3, pp.226-233, (2007-3)

Two-Phase Interleaved Buck Converter with a new Digital Self-Oscillating Modulator

Lars T. Jakobsen
Technical University of Denmark
Oersted-Automation
Email: ltj@oersted.dtu.dk

Michael A. E. Andersen
Technical University of Denmark
Oersted-Automation
Email: ma@oersted.dtu.dk

Abstract— This paper presents a new Digital Self-Oscillating Modulator (DiSOM) for DC/DC converters. The DiSOM modulator alllows the digital control algorithm to sample the output voltage at a sampling frequency higher than the converter switching frequency. This enables higher control loop bandwidth than for traditional digital PWM modulators given a certain switching frequency. A synchronised version of the DiSOM modulator is derived for interleaved converters. A prototype interleaved Buck converter for Point of Load applications has been designed and built to test the performance of DiSOM modulator. The DiSOM modulator and a digital control algorithm have been implemented in an FPGA. Experimental results show that the converter has a very fast transient response when a loadstep is applied to the output. For a loadstep of 50% of nominal output current the output voltage overshoot is less than 2.5% of the nominal output voltage and the settling time is just 8 switching periods.

I. INTRODUCTION

To design a digitally controlled DC-DC converter with fast transient response is a big challenge even though great advances have been made in the last decade. One of the main limitations in a digital controller for a DC/DC converter is the PWM modulator. Many papers have been published about digital modulators that solve either the limit cycling problems due to clock frequency quantization or the problem of how to obtain high control loop bandwidth in a digitally controlled DC/DC converter [1-3]. It can generally be said for the previously published work that the switching frequency is increased in order to increase the control loop bandwidth. This is due to the fact that the digital PWM modulators only allow the sampling frequency of the digital controller to be equal to but not higher than the switching frequency. This limitation has come into play as advanced implementations of the control law, e.g. a digital PID compensator, has reduced the computation time. Previously the main limitation was due to computational delays because microcontrollers or slow DSPs were used [4]. This paper introduces a new Digital Self-Oscillating Modulator (DiSOM) [5] that has the advantage that the digital controller can sample the output voltage at a higher sampling rate than the converter switching frequency thereby allowing the control loop bandwidth to be increased without increasing the switching frequency. A synchronised DiSOM modulator is introduced allowing the

DiSOM to be used in interleaved DC/DC converters. A prototype two-phase interleaved Buck converter has been built and tested with the synchronised DiSOM modulator.

II. DiSOM -DIGITAL SELF-OSCILATTING MODULATOR

The Digital Self-Oscillating Modulator is a new invention that covers a whole family of digital modulators, which are all self-oscillating by nature. The simplest form of the DiSOM modulator is shown in Fig. 1. It consists of a comparator with hysteresis (the red box in Fig. 1), a main forward block (MFW, the blue box in Fig. 1) configured as an integrator and a main feedback block (MFB, the green box in Fig. 1) that performs a simple multiplication. The reference signal Ref sets the duty cycle and is subtracted from the output signal of the MFB. According to the invention the MFB and MFW can both be frequency dependent and can be implemented as any kind of digital filter. By choosing other characteristics for the MFB and MFW the characteristics and performance of the DiSOM modulator can be changed. The modulator could for instance be designed to suppress quantization errors in the low-frequency band, as known from noise-shapers in class-D audio amplifiers and D/A converters. The comparator with hysteresis generates the switching output by comparing the output of the integrator to either of the hysteresis limits depending on the state of the switching output.

The output of the integrator will be a triangular waveform under steady state conditions. It can be shown that the switching frequency of the DiSOM modulator shown in Fig. 1 is given by equation (1), where n is the number of bits that is used to represent the reference input, $Window$ is the hysteresis window and T_{clock} is the period time of the clock that runs the DiSOM modulator.

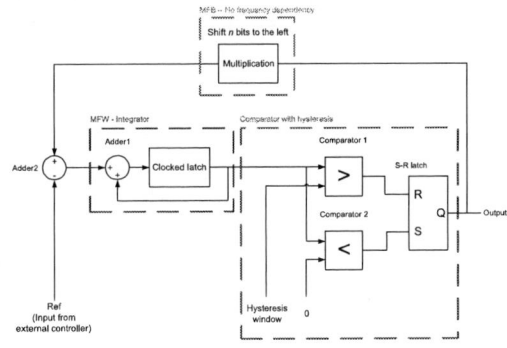

Fig. 1: A block diagram of the DiSOM modulator

978-1-4244-1871-8/07 $25.00 © 2007 IEEE

$$f_{sw}(D) = \frac{2^n}{Window \cdot T_{clock}} \cdot \left(D - D^2\right) \qquad (1)$$

Fig. 2 shows a typical plot of the switching frequency versus duty cycle, where the duty cycle (D) is normalised from 0 to 1. The big advantage of the DiSOM modulator over traditional PWM modulators is that the duty cycle command, i.e. the Ref input set by the digital compensator, doesn't directly affect the PWM output as it would in traditional digital PWM modulators based on a counter and comparator. Thereby unwanted transitions on the PWM output can be avoided when the duty cycle command is updated. If the duty cycle command for the DiSOM modulator, i.e. the Ref input, is changed during a switching cycle it will change the slope of the triangular waveform on the output of the integrator, but it will not instantaneously affect the PWM output. It is therefore possible for the digital compensator, which controls the output voltage, to sample the output voltage at a higher frequency than the switching frequency and update the duty cycle command at the same rate as the sampling frequency. By increasing the sampling frequency it will be possible to obtain a higher control loop bandwidth for the digitally controlled DC/DC converter, without increasing the switching frequency of the system.

III. Synchronisation of the DiSOM

The DiSOM modulator described in the previous section is very well suited for a typical DC/DC converter such as a Buck converter, but it isn't directly applicable to interleaved converters. The problem is that the switching frequency changes with the duty cycle, which means that the period time of the PWM signal isn't constant. The challenge then is how to ensure that the phases in an interleaved converter have the correct phase shift relative to each other, e.g. 180 degrees phase shift in a two-phase interleaved Buck converter. The solution used in the prototype converter presented in this paper is to replace the comparator with hysteresis with a clocked comparator as shown in Fig. 3. The comparator has a synchronising input, and the PWM output will transition from low to high on the positive edge of the synchronising signal. When the integrator output is equal to the hysteresis window the PWM input will transition from high to low. The principle of operation is illustrated in Fig. 4 for duty cycle values of 0.125, 0.25 and 0.375, where the "carrier" is the output of the integrator.

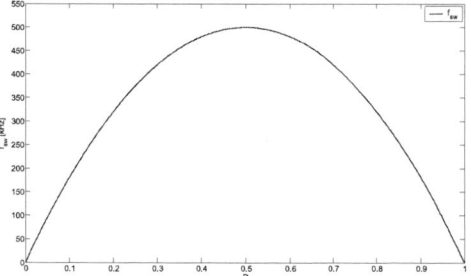

Fig. 2: Switching frequency v.duty cycle for the DiSOM

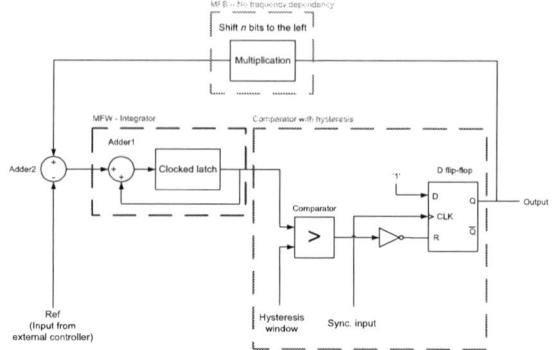

Fig. 3: DiSOM modulator with synchronisation input

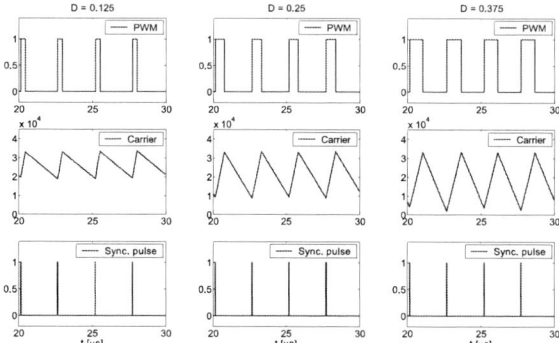

Fig. 4: Carrier and PWM signals for the synchronised DiSOM modulator

It should be noted that by adding synchronistaion to the DiSOM modulator we introduce a limitation to the PWM output. The duty cycle is now limited to the range from 0 to 0.5, which is similar to the duty cycle limitation known from analogue peak current mode control. The problem can be solved by digitally adding a negative slope to the integrator output if the converter design requires duty cycles larger than 0.5. With the new synchronised DiSOM modulator it is possible to generate phase shifted PWM signals by phase shifting the synchronisation signal for each modulator. A block diagram showing how two synchronised DiSOM modulator can be used to create interleaved control signals for two Buck stages is shown in Fig. 5. The timing is controlled by a counter and the synchronisation pulses are generated by comparing the counter value to preset phase delays. The digital PID compensator sets the same duty cycle for both synchronised DiSOMs. In most digital control schemes for interleaved Buck converters the output voltage is sampled at the same rate as the per phase switching frequency effectively limiting the control loop bandwidth to the same bandwidth as would be achievable in a single phase Buck converter switching at the same frequency. Ref [1] proposes a digital control scheme where the sampling frequency is equal to N times the per phase switching frequency, where N is the number of phases of the interleaved converter. The control loop bandwidth can in that case be significantly increased, but the simulations and measurements shown in [1] shows a lot of noise on the output voltage during transient conditions. The noise is to a large extent attributable to the very high

978-1-4244-1871-8/07 $25.00 © 2007 IEEE

sampling frequency of the output voltage control loop and the fact that the duty cycle of the separate phases in the interleaved converter differs during the transient condition.

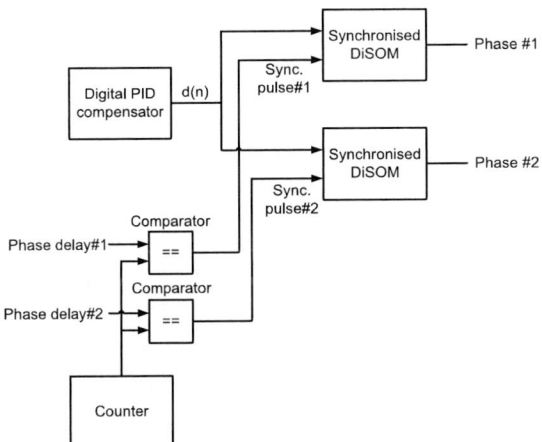

Fig. 5 Interleaving of multiple synchronised DiSOM modulators

IV. IMPLEMENTATION OF DIGITAL PID COMPENSATOR WITH SHORT COMPUTATION TIME

When designing a digital controller for a high bandwidth DC/DC converter it is important to keep the time delays to a minimum, since any delay causes negative phase shift in the control loop. The negative phase shift will have to be compensated for in the control loop design and will very often limit the performance. There are three main contributors to the delay in the digital controller. The first contributor is the PWM modulator, which for a typical uniformly sampled digital PWM modulator has the transfer function

$$G_{PWM}(s) = \frac{e^{-sDT_s}}{c_{PK}} \qquad (2)$$

where D is the duty cycle of the PWM signal, T_s is the period time of the PWM signal and c_{PK} is the amplitude of the digital carrier generated by a counter [6].

The other two contributors to the time delay in the digital controller is the ADC sampling time and the time it takes to compute the digital control law, e.g a digital PID compensator. The design of a digital PID compensator in VHDL for implementation in an FPGA can help reduce the delay time compared to implementations in a Digital Signal Processor/Controller (DSP/DSC) such as the TMS320F2801 [7].

Fig. 6 and 7 shows a block diagram and a timing diagram respectively for the digital PID compensator used in the prototype presented in this paper. The PID compensator of Fig. 6 implements the transfer function $G_c(z)$ (see equation (3)), which has to zeros and an integrator in the frequency domain.

$$G_c(z) = \frac{d(z)}{e(z)} = \frac{b_0 + b_1 \cdot z^{-1} + b_2 \cdot z^{-2}}{1 - z^{-1}} \qquad (3)$$

The digital implementation of the PID compensator calculates the difference equation (4).

$$d(n) = d(n-1) + b_0 \cdot e(n) + b_1 \cdot e(n-1) + b_2 \cdot e(n-2) \qquad (4)$$

where $d(n)$ is the duty cycle command for the two synchronised DiSOM modulators, $d(n-1)$ is the duty cycle command of the previous sample and $e(n)$, $e(n-1)$ and $e(n-2)$ is the present and two past samples of the error signal, i.e. the difference between the digital reference and the sampled output voltage.

To explain the operation of the digital PID compensator each block in the block diagram will be explained in the following description.

- Decoder – The decoder takes the input from a 10 bit ADC and reduces the resolution to 6 bits centered around the digital reference, which defines the output voltage. The ADC input is 10 bits because the FPGA development board used in the experimental work had a 10 bit pipelined ADC installed. The decoder updates the error signal $e(n)$ when the Read_ADC signal transitions from low to high.

- The sequencer is a simple state machines that generates the control signals for the remaining blocks of the PID compensator. The sequencer has a state variable that is generated by a counter, which counts from 0 to 62 and then resets to 0. The counter is clocked by a 50MHz clock and the operation of the PID compensator is synchronised to the counter. The resulting sampling frequency is equal to 794 kHz.

- The 20-bit adder and the 20-bit register works as an accumulator. The accumulator can take the previous result on the output of the accumulator and add it to the result on the output of the 6×14 bit multiplier or the sequencer can preload the duty cycle d(n-1) that is read from the limiter.

- The 6×14 bit multiplier is used to multiply an error signal with its corresponding coefficent, e.g. e(n) and b_0. The coefficients b_0, b_1 and b_2 are represented as 14 bit numbers. The 5 least significant bits represents fractions of 1 and the most significant bit is a sign bit. All calculations are performed using 2's complement binary numbers. The remaining 8 bits allows the coefficients to take on values in the range from − 256 to +256. In a typical DSP both inputs on the multiplier would have the same word length but that increases the complexity of the multiplier. The idea of limiting the word length of one input to match the error signal is presented in [8] with the purpose of reducing the

978-1-4244-1871-8/07 $25.00 © 2007 IEEE   761

complexity and cost of the digital compensator logic.

- The two multiplexers (labelled 3-1 MUX) are used to select, which error signal and coefficient is fed to the inputs of the multiplier. Both multiplexers are controlled by the 2-bit Mux_sel signal.

- The two 6-bit registers stores the error signals *e(n-1)* and *e(n-2)*. The registers reads and stores the input signal on the rising edge of the control signal, i.e. Update_e1 and Update_e2.

- The limiter reads the output of the accumulator and limits it to duty cycle values between 0 and 0.5 on the rising edge of Update_D.

The digital PID compensator has a delay of two clock cycles from the ADC input is read by the decoder until the duty cycle command is updated by the limiter. The ADC is a 10-bit pipelined ADC with a sampling delay of 6 clock cycles. Both the ADC and the digital PID compensator is clocked by a 50 MHz clock and the total delay is 8 clock cycles, which makes the delay 160 ns.

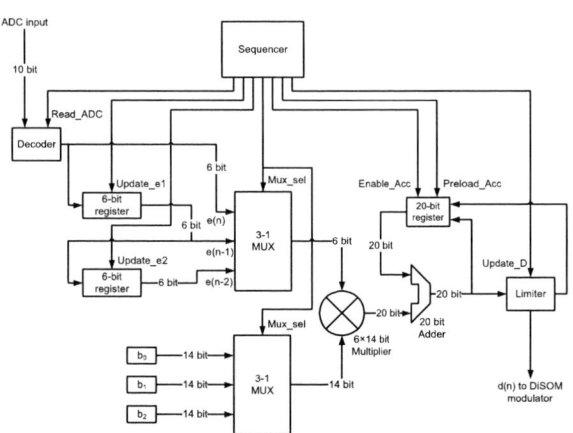

Fig. 6 Block diagram of the digital PID compensator implemented in an FPGA

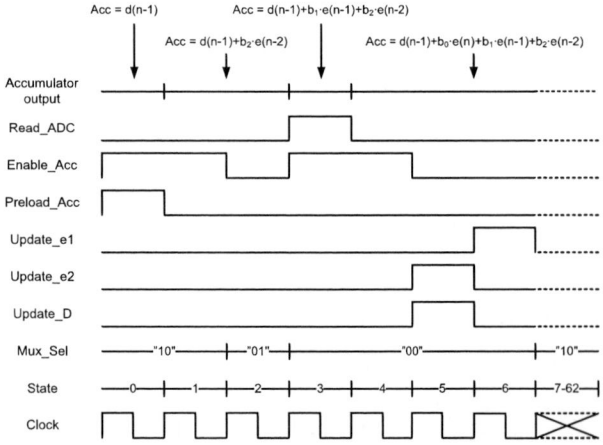

Fig. 7 Timing diagram for the digital PID compensator

The disadvantage of the digital PID compensator described above is that it can not be changed easily. If for example the

designer wants to add an extra zero or a pole to the transfer function the VHDL code will have to be modified and the design has to be simulated and tested once again. Compared to a more traditional DSP design [9], where it is a matter of rewriting a piece of assembly or C code, it is a much more challenging task to change the VHDL design.

V.   EXPERIMENTAL RESULTS

A two-phase interleaved buck converter has been designed and built to verify verify the performance of the synchronised DiSOM modulator. The specifications of the hardware prototype are given in Table 1. The output capacitor is in fact eight 100µF ceramic capacitors (X5R) for low ESR. A picture of the prototype is shown in Fig. 8.

The DiSOM modulators and the digital compensator are implemented in an LCXMO1200 FPGA from Lattice Semiconductor. The ADC used in the design is the ADC10065 from National Semiconductors.

TABLE 1

PROTOTYPE SPECIFICATIONS

Parameter	Value
Input voltage	$9 - 15V$
Output voltage	2.0V
Output current	$0 - 20A$
Inductor size	1.5 µH
Output capacitance	800 µF
Switching frequency per phase	400 kHz
Sampling frequency	800 kHz

Fig. 8: Prototype POL converter board (Right) and FPGA development board (left)

A MATLAB/SIMULINK model of two-phase interleaved Buck converter and the digital controller was used to simulate the design. The Buck converter model takes into account parameters such as capacitor ESR, inductor series resistance and the MOSFET driver propagation delay.

Fig. 9 shows a comparison of the simulated and measured open loop gain of the interleaved Buck converter. The simulated loop gain is found by injecting a sinewave into the control loop at the converter output in the same way as the loop gain is measured by a Gain/Phase analyser. The loop gain measurement shows that the open loop bandwidth of the interleaved Buck converter is 41 kHz and the phase margin is 56 degrees.

such a way that noise power is low at low frequencies and increases at high frequencies.

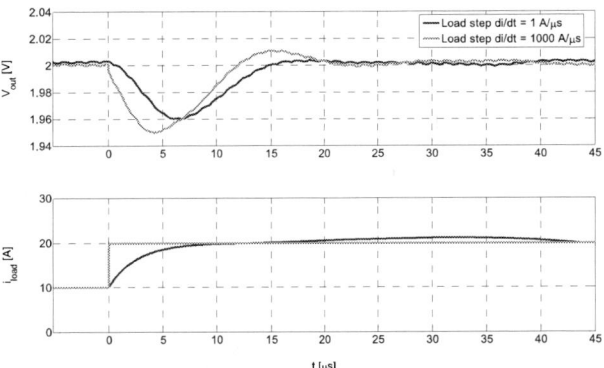

Fig. 10 Simulated load step response for $V_{in} = 12V$

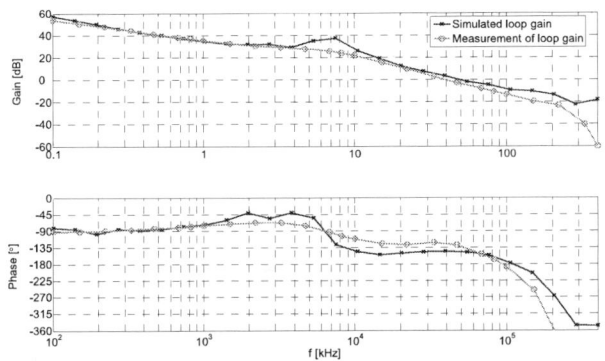

Fig. 9 Open loop gain simulation and measurement

Simulations of the output voltage response to a load step from 10 to 20A for two different current slew rates (see Fig. 10) shows that the output voltage deviation from the steady state voltage of 2.0V is 50mV for a current slew rate of 1000A/μs. If the load step has a first order lowpass response with an average slew rate of 1A/μs the output voltage deviation is approximately 40mV. The settling time is approximately 15μs for the slow load step and 20μs for the fast load step.

Fig. 11 shows the measured output voltage response to a loadstep from 10 to 20A. The output voltage deviation is less than 50mV and the settling time is less than 20μs. The electronic load used in the measurement limits the current slew to approximately 1A/μs and the measurement is comparable to the blue trace in Fig. 10. The measured output voltage deviation is slightly larger than the simulation result and the settling time is also slightly longer. The difference between simulation and measurement can be due to component tolerances and poor PCB design.

Fig. 12 shows a steady state measurement of the output voltage at nominal output current. The output voltage has some fluctuations, which isn't the ripple voltage generated by the ripple current in the inductor. Part of the explanation is the relatively low ratio between the per phase switching frequency and the clock frequency of the synchronised DiSOMs. The ratio between switching and clock frequency is 1/126, which correpsonds to a PWM duty cycle resolution of 7 bits. The resolution of the duty cycle command of the digital PID compensator is 10 bits. The reason that the fluctuations on the output voltage have an amplitude of less than 10 mV peak to peak is the nature of the synchronised DiSOM modulator. The DiSOM is a closed loop system that feeds back the PWM signal and compares it with the duty cycle command and the integrator will suppress the low frequency errors. The DiSOM is in that sense similar to sigma-delta modulators that shapes the quantisation noise in

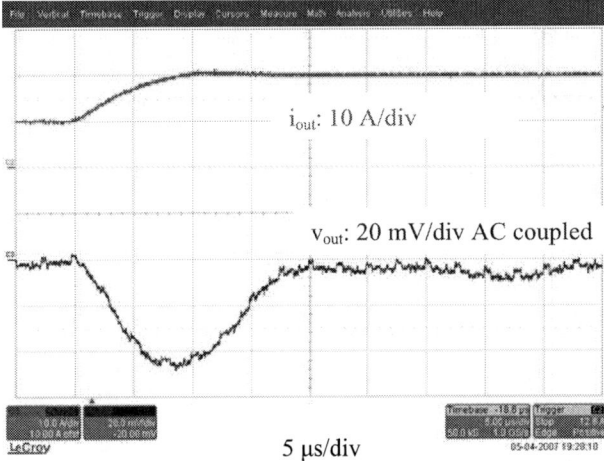

Fig. 11: Positive loadstep 10A to 20A $V_{in} = 12V$

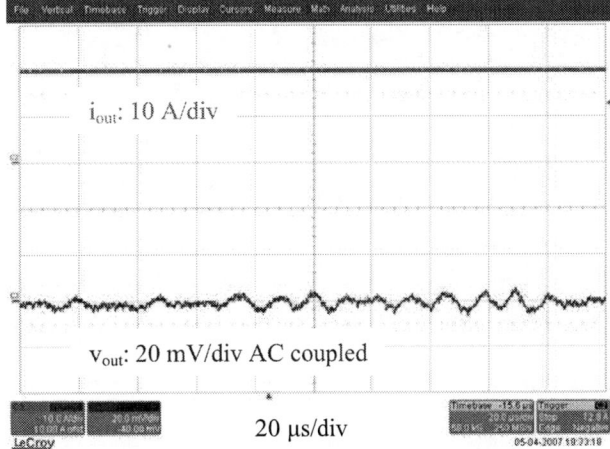

Fig. 12 Steady state measurement of the output voltage for $I_{out} = 20$ A

VI. CONCLUSION

This paper describes a two-phase interleaved buck converter with very fast transient response. A new digital modulator, called the DiSOM modulator is described and a synchronised DiSOM modulator is derived. A prototype

978-1-4244-1871-8/07 $25.00 © 2007 IEEE   763

2.0V 20A two-phase interleaved Buck converter has been built to test the proposed modulator. The output voltage response, to a loadstep of 50% of the nominal output current, is less than 2.5% of the nominal output voltage. The settling time after a loadstep is less than 20µs, which corresponds to just 8 switching cycles.

REFERENCES

[1] X. Zhang, Y, Zhang, R. Zane and D. Maksimovic, "Design and Implementation of a Wide-bandwidth Digitally Controlled 16-phase Converter", *IEEE COMPEL Workshop 2006*, pp. 106 -111, July 2006

[2] J. Pattella, A. Prodic, A. Zirger and D. Maksimovic, "High-frequency Digital PWM Controller IC for DC-DC Converters", *IEEE Transactions on Power Electronics*, vol. 18, pp. 438-446, January 2003

[3] Z. Lukić, K. Wang and A. Prodić, "High-Frequency Digital Controller for DC-DC Converters Based on Multi-Bit Σ-Δ Pulse-Width Modulation", *IEEE Applied Power Electronics Conference 2005*, vol. 1, pp. 35 - 40, March 2006

[4] L. T. Jakobsen and M. A. E. Andersen "Comparison of Two Different High Performance Mixed Signal Controllers for DC/DC Converters", *IEEE COMPEL Workshop 2006*, pp. 129-135, July 2006

[5] L. T. Jakobsen, M. A. E. Andersen, "Digital Self-Oscillating Modulators", PCT patent application no. PCT/DK2007/000104 filed March 1 2007

[6] S. Busso and P. Mattavelli, "Digital Control in Power Electronics", Morgan & Claypool Publishers 2006, ISBN: 1-59829-112-2

[7] S. Choudhury, "Designing a TMS320F280x Based Digitally Controlled DC-DC Switching Power Supply", Texas Instruments Applications Report SPRAAB3, July 2005

[8] A. Chapuis, "Digital Signal Processor Architecture for Controlling Switched Mode Power Supply", International patent application no. WO2004/073149A2, August 2004

[9] E. O'malley and K. Rinne, "A 16-bit Fixed-Point Digital Signal Processor for Digital Power Converter Control", *IEEE Applied Power Electronics Conference 2005*, vol. 1, pp. 50 - 56, March 2006

Feedback Linearization Control of Three-Phase AC/DC PWM Converters with LCL Input Filters

Dong-Eok Kim
Department of Electrical Engineering,
Yeungnam University
214-1, Daedong, Gyeongsan, Gyeongbuk, Korea
20640234@yu.ac.kr

Dong-Choon Lee
Department of Electrical Engineering,
Yeungnam University
214-1, Daedong, Gyeongsan, Gyeongbuk, Korea
dclee@yu.ac.kr

Abstract—**This paper proposes a feedback linearization control scheme of AC/DC PWM converters with LCL input filters using no damping resisters. Feedback linearization techniques use a transformation from nonlinear system models into equivalent linear models in a simpler form. The feedback linearization scheme in this paper has cascade structures unlike the usual feedback linearization, therefore it has an advantage that it is possible to limit the converter current to a certain level. To reduce the number of sensors, source voltages and currents are estimated. The performance of the proposed controller is validated with simulation and experimental results.**

I. INTRODUCTION

The three-phase AC/DC PWM converter with boost L filters has good functions such as DC-link voltage control, sinusoidal input current, unity power factor control, and bidirectional power flow. However it may cause EMI problems to other sensitive equipments connected to the PCC (point of common coupling) due to the switching-frequency related harmonic currents. Although this problem can be solved by increasing the boost inductor filter size, the cost of the inductor increases and the dynamic performance of the current control is deteriorated. To prevent these problems, LC or LCL filters can be used. With the LCL filters, the switching ripple components in grid currents are decreased with the same inductance value as that of LC-type filters while the dynamic performance is maintained [1].

For LCL filters, the resonance problem should be considered carefully. It is usually suppressed by the damping resisters connected in series with the filter capacitor. Although the resonance problem can be solved with the damping resistor, it is bulky and causes the power loss. Hence other attempts have been made such as active damping method [2] or virtual resister method [3]. However the active damping method needs the information of the filter parameters and the grid impedance. Also, the virtual resistor method needs the additional voltage sensors to measure the capacitor voltages. In [4], the active damping method applying a genetic algorithm was implemented without the parameter information and additional sensors. Also the multi-loop control method was investigated from the idea that the capacitor is a source of resonance, where the capacitor current is measured and controlled as a minor control loop [5]. In [6], another method to solve the resonance problem by using the different combination of the measured variable from the sensors was proposed. In fact, if

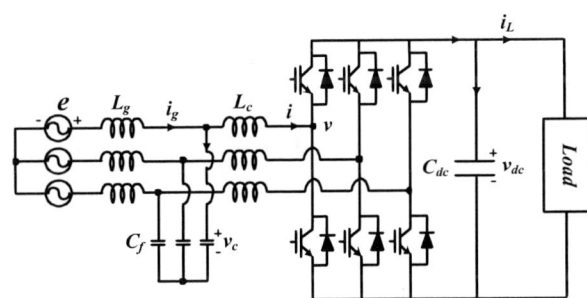

Fig. 1. Three-phase PWM converter with LCL input filters.

the converter input currents are measured and controlled, the system can be controlled stably with smaller damping resisters but it is difficult to obtain the unity power factor operation.

It is well known that feedback linearization technique is a control method to eliminate the nonlinearity of the system by using the inverse dynamics [7]. It has been applied to control the shunt active power filter [8] and the three-level NPC boost converter [9], which gives the good performance robust to disturbances as well as the fast transient responses.

In this paper, the feedback linearization is applied to control the AC/DC PWM converter with LCL input filters. The system is modeled with two state equations, of which one has the converter voltage reference as an input and the other has the converter input current as an input, and the feedback linearization technique is applied to each equation. In this control method, the converter current can be limited, and the source voltage and current are estimated to reduce the number of sensors. The performance of the proposed control scheme is validated with simulation and experimental results.

II. AC-DC PWM CONVERTERS

A. Modeling of the system

Fig. 1 shows a three-phase AC/DC PWM converter with LCL input filters. The resistance in LCL filters is assumed to be negligible. L_g is the filter inductance of the grid side, L_c is the filter inductance of the converter side, and C_f is the filter capacitance. i_g is the grid current, i is the input current of the converter side, and v_c is the filter capacitor voltage. C_{dc} is the output capacitor (DC-link capacitor) of the converter, v_{dc} is the DC output voltage, and i_L is the output current (load current).

The power variation of the DC-link capacitor is derived from the difference between the input power and the output

978-1-4244-1871-8/07 $25.00 © 2007 IEEE

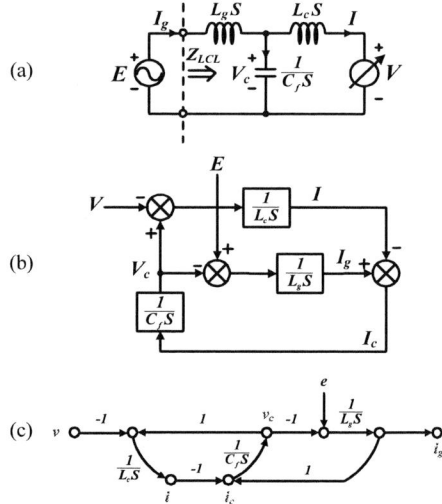

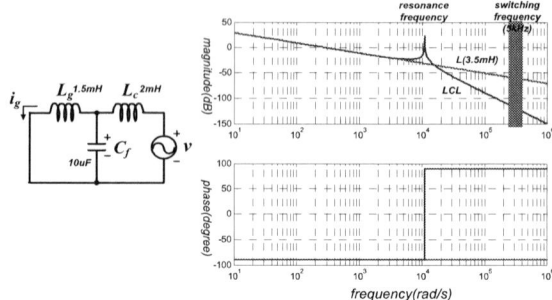

Fig. 3. Frequency characteristics of LCL and L filters.

Fig. 2. (a) a per-phase equivalent circuit of the PWM converter (b) Block diagram of transfer function of LCL filters. (c) Signal flow graph.

power. Using this, the state equation of the converter system is given by

$$
\begin{bmatrix} \dot{i}_{gd} \\ \dot{i}_{gq} \\ \dot{v}_{cd} \\ \dot{v}_{cq} \\ \dot{i}_{d} \\ \dot{i}_{q} \\ \dot{v}_{dc} \end{bmatrix} = \begin{bmatrix} 0 & \omega & -1/L_g & 0 & 0 & 0 & 0 \\ -\omega & 0 & 0 & -1/L_g & 0 & 0 & 0 \\ 1/C_f & 0 & 0 & \omega & 0 & 0 & 0 \\ 0 & 1/C_f & -\omega & 0 & 0 & 0 & 0 \\ 0 & 0 & 1/L_c & 0 & 0 & \omega & 0 \\ 0 & 0 & 0 & 1/L_c & -\omega & 0 & 0 \\ 0 & \frac{3e_q}{2C_{dc}v_{dc}} & 0 & 0 & 0 & 0 & 0 \end{bmatrix} \begin{bmatrix} i_{gd} \\ i_{gq} \\ v_{cd} \\ v_{cq} \\ i_{d} \\ i_{q} \\ v_{dc} \end{bmatrix}
$$

$$
+ \begin{bmatrix} 0 & 0 \\ 0 & 0 \\ 0 & 0 \\ 0 & 0 \\ -1/L_c & 0 \\ 0 & -1/L_c \\ 0 & 0 \end{bmatrix} \begin{bmatrix} v_d \\ v_q \end{bmatrix} + \begin{bmatrix} e_d/L_g \\ e_q/L_g \\ 0 \\ 0 \\ 0 \\ 0 \\ -i_L/C_{dc} \end{bmatrix}
$$
(1)

which is expressed in a synchronously rotating d-q reference frame. Eqn. (1), which has the converter voltage reference as an input, can be separated with two state equations. One has the converter-side current as an input and the other has the converter voltage reference as an input, as

$$
\begin{bmatrix} \dot{i}_{gd} \\ \dot{i}_{gq} \\ \dot{v}_{cd} \\ \dot{v}_{cq} \\ \dot{v}_{dc} \end{bmatrix} = \begin{bmatrix} 0 & \omega & -1/L_g & 0 & 0 \\ -\omega & 0 & 0 & -1/L_g & 0 \\ 1/C_f & 0 & 0 & \omega & 0 \\ 0 & 1/C_f & -\omega & 0 & 0 \\ 0 & \frac{3e_q}{2C_{dc}v_{dc}} & 0 & 0 & 0 \end{bmatrix} \begin{bmatrix} i_{gd} \\ i_{gq} \\ v_{cd} \\ v_{cq} \\ v_{dc} \end{bmatrix} + \begin{bmatrix} 0 & 0 \\ 0 & 0 \\ \frac{1}{C_f} & 0 \\ 0 & \frac{1}{C_f} \\ 0 & 0 \end{bmatrix} \begin{bmatrix} i_d \\ i_q \end{bmatrix} + \begin{bmatrix} e_d/L_g \\ e_q/L_g \\ 0 \\ 0 \\ -i_L/C_{dc} \end{bmatrix}
$$
(2)

$$
\begin{bmatrix} \dot{i}_d \\ \dot{i}_q \end{bmatrix} = \begin{bmatrix} 0 & \omega \\ -\omega & 0 \end{bmatrix} \begin{bmatrix} i_d \\ i_q \end{bmatrix} + \begin{bmatrix} -\frac{1}{L_c} & 0 \\ 0 & -\frac{1}{L_c} \end{bmatrix} \begin{bmatrix} v_d \\ v_q \end{bmatrix} + \begin{bmatrix} \frac{1}{L_c}v_{cd} \\ \frac{1}{L_c}v_{cq} \end{bmatrix}
$$
(3)

B. LCL filters

In order to observe the attenuation of high frequency ripple components, the frequency characteristics of the L filter and the LCL filters are investigated.

Fig. 2(a) shows the per-phase equivalent of the PWM converter in s-domain. Fig. 2(b) shows the block diagram of the transfer function of the LCL filters and (c) shows the signal flow graph of it. In Fig. 2(a), Thevenin equivalent impedance of the LCL filters is expressed as

$$
Z_{LCL}(S) = \frac{L_c L_g C_f S^3 + (L_c + L_g)S}{L_c C_f S^2 + 1}
$$
(4)

and the transfer function between the converter input voltage and the grid current is given by

$$
\frac{I_g(S)}{V(S)} = \frac{1}{L_c L_g C_f S^3 + (L_c + L_g)S}
$$
(5)

Fig. 3 shows the frequency responses of L filter and LCL filters, which have the same value of the inductance. The performance of LCL filters is better than the L filter regarding the ability of high frequency harmonic rejection. However, it has a resonance phenomenon.

III. CONTROL OF PWM CONVERTERS

A. Feedback Linearization

The feedback linearization is a method eliminating the nonlinearity of the system so that the closed-loop system dynamics are reduced to a linear form. Then the controller for the linearized system can be designed with the well-known linear control theory.

For feedback linearization process, at first, a multi-input multi-output (MIMO) system is expressed from (2) as

$$
\dot{x} = f(x) + gu + e
$$
(6)

$$
y = h(x)
$$
(7)

where the order of the output is the same as that of the input, which is two. Here, i_{gd} and v_{dc} in (2) are selected as the

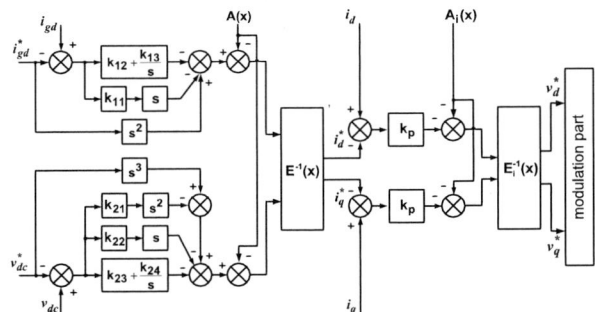

Fig. 4. Control block diagram for feedback linearization.

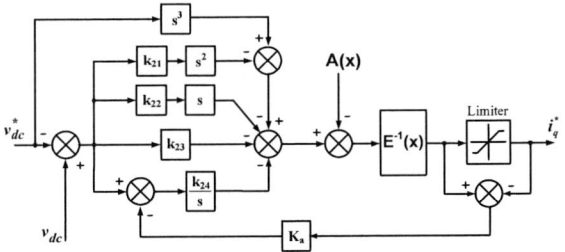

Fig. 5. Control block diagram with anti-widndup.

outputs. Next, the outputs are differentiated until the input appears in order to obtain a direct relationship between the input u and the output y. Then a new output equation is obtained as

$$y^{r_i} = A(x) + E(x)u \tag{8}$$

where, $y^{r_i} = \begin{bmatrix} \ddot{i}_{gd} & \dddot{v}_{dc} \end{bmatrix}^T$, $u = \begin{bmatrix} i_d & i_q \end{bmatrix}^T$

$$A(x) = \begin{bmatrix} \dfrac{\omega}{L_g}e_q - \dfrac{2\omega}{L_g}v_{cq} - \dfrac{1}{L_gC_f}i_{gd} - \omega^2 i_g \\[2mm] (-\dfrac{3e_q i_{gq}}{C_{dc}v_{dc}^2}\dot{v}_{dc} - \dfrac{3e_q\ddot{v}_{dc}}{2C_{dc}v_{dc}^2}i_{gq} + \dfrac{3e_q(\dot{v}_{dc})^2}{C_{dc}v_{dc}^3}i_{gq} - \dfrac{3\omega e_q}{C_{dc}L_g v_{dc}}v_{cd} \\[2mm] \quad -\dfrac{3e_q}{2C_{dc}L_gC_f v_{dc}}i_{gg} - \dfrac{3\omega^2 e_q}{2C_{dc}v_{dc}}i_{gq} - \dfrac{\ddot{i}_L}{C_{dc}}) \end{bmatrix}$$

$$E(x) = \begin{bmatrix} \dfrac{1}{L_gC_f} & 0 \\[2mm] 0 & \dfrac{3e_q}{2C_{dc}L_gC_f v_{dc}} \end{bmatrix} , \quad \dddot{v}_{dc} = \dfrac{2e_q i_{gq}}{3C_{dc}v_{dc}} - \dfrac{2e_q i_{gq}\dot{v}_{dc}}{3C_{dc}} - \dfrac{i_L}{C_{dc}}$$

By means of feedback linearization (input-output linearization), the dynamics of a nonlinear system are decomposed into an external part (input-output part) and an internal part (unobservable), and the internal dynamics should be considered for the system stability, where zero dynamics analysis is usually used. However, if the total sum of the relative degrees, which are the number for the output to be differentiated, is the same as the order of the nonlinear system, the stability of the internal dynamics may not be considered since the internal dynamic does not exist. In this case, it is said that the complete input-output linearization of the nonlinear system is achieved [10].

$E(x)$ in (8) is the decoupling matrix and it should be a non-singular matrix in the all operation range so that the input-output linearization is possible. With $E^{-1}(x)$, the control inputs v_i which makes the system be linear and decoupled can be obtained as

$$u = \begin{bmatrix} u_1 \\ u_2 \end{bmatrix} = E^{-1}(x)\begin{bmatrix} \begin{bmatrix} v_1 \\ v_2 \end{bmatrix} - A(x) \end{bmatrix} \tag{9}$$

where,

$$E^{-1}(x) = \begin{bmatrix} L_gC_f & 0 \\[2mm] 0 & \dfrac{2C_{dc}L_gC_f v_{dc}}{3e_q} \end{bmatrix}$$

Substituting (9) into (8), the linearized and decoupled model is derived as

$$y^{r_i} = \begin{bmatrix} \ddot{i}_{gd} \\ \dddot{v}_{dc} \end{bmatrix} = \begin{bmatrix} v_1 \\ v_2 \end{bmatrix} \tag{10}$$

where $v_i (i=1,2)$ is called as the decoupled input since it affects y^{r_i} only.

Then the tracking control law for the new control input is applied as

$$\begin{bmatrix} v_1 \\ v_2 \end{bmatrix} = \begin{bmatrix} \ddot{y}_{1ref} - k_{11}\dot{e}_1 - k_{12}e_1 \\ \dddot{y}_{2ref} - k_{21}\ddot{e}_2 - k_{22}\dot{e}_2 - k_{23}e_2 \end{bmatrix} \tag{11}$$

Since the feedback linearization control may not be effective for the system with parameter disturbed, the integral control needs to be added to the tracking controller as [11].

$$\begin{bmatrix} v_1 \\ v_2 \end{bmatrix} = \begin{bmatrix} \ddot{y}_{1ref} - k_{11}\dot{e}_1 - k_{12}e_1 - k_{13}\int e_1 \\ \dddot{y}_{2ref} - k_{21}\ddot{e}_2 - k_{22}\dot{e}_2 - k_{23}e_2 - k_{24}\int e_2 \end{bmatrix} \tag{11}$$

where, $e_1 = y_1 - y_{1ref}$, $e_2 = y_2 - y_{2ref}$, and y_{ref} is the tracking reference, which means the grid current reference (y_{1ref}) and the DC-link voltage reference (y_{2ref}), and k_{ij} is the gain. When the gain in (11) is determined so that the error dynamics converge to zero, the tracking control is achieved. Also, anti-windup function is added to limit properly the adverse effect of integral control. Fig. 5 shows the block diagram of the system with integral control and anti-windup [12].

On the other hand, (3) is a simple linear state equation and the relationship between the input and the output is obtained by only one-time differentiation. The procedure applying the feedback linearization control is the same as the explained before.

Fig. 4 shows the control block diagram of the proposed method, where only the proportional control is used in the inner control loop since the steady-state error does not affect the performance of the outer loop [5]. In Fig. 4, $A_i(x)$ and $E_i(x)$ are as follows:

978-1-4244-1871-8/07 $25.00 © 2007 IEEE

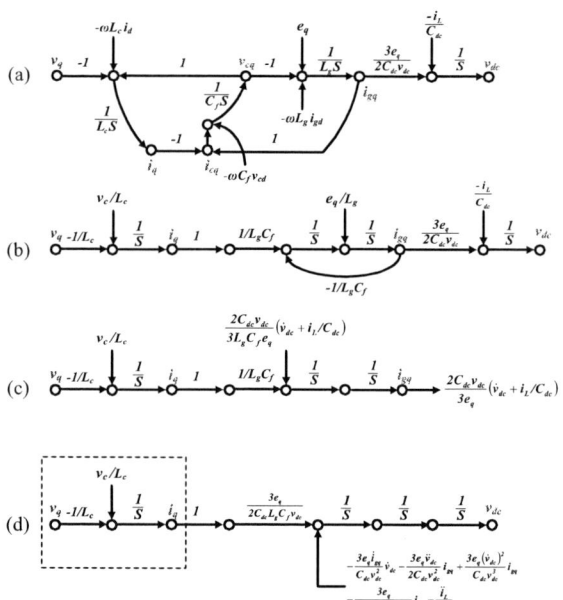

Fig. 6. Simplification process of the nonlinear system to the linear system with signal flow graph

$$A_i(x) = \begin{bmatrix} \dfrac{1}{L_c} v_{cd} + \omega i_q \\[2mm] \dfrac{1}{L_c} v_{cq} - \omega i_d \end{bmatrix}, \qquad E_i(x) = \begin{bmatrix} -\dfrac{1}{L_c} & 0 \\[2mm] 0 & -\dfrac{1}{L_c} \end{bmatrix}$$

Fig. 6 shows the simplification process of signal flow graph for the linearization of the nonlinear system. Fig. 6(a) shows the complete signal flow graph and (b) shows the graph in which the coupling terms are relocated from (a). Fig. 6(c) is the rearranged graph from (b). When the right-end term i_{gq} is moved to the left-hand side across double integrators, then the resultant form is given in (d), where the dotted block means the inner control loop part and it is assumed as the first-order delay element. In Fig. 6(d), all nonlinearities are rearranged to the front-end of the integrators, which are canceled by feedback linearization. Therefore, a simple triple-integrator relationship between the input and the output are obtained if the inner control loop part is assumed to have unity transfer function.

Fig. 7 shows the control block diagram for the linearized system in (11). It is noticed that since the derivative of the capacitor voltage is involved in feedback linearization control, it is possible to control the system stable without damping resistors.

B. Estimation of currents and voltages

In the proposed feedback linearization method, the information of the filter capacitor voltages and the converter currents are needed. Instead of measurement, the grid currents and the source voltage are estimated to reduce the number of sensors. The block diagram of the grid current estimator is shown in Fig. 8. The difference equation for the grid current

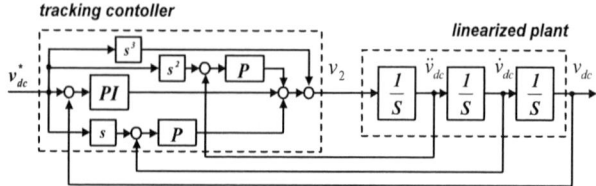

Fig. 7: Control block diagram for feedback-linearized system.

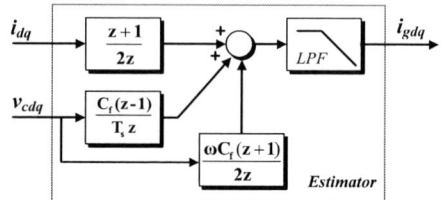

Fig. 8. Grid current estimator.

estimation is as follows [13]

$$i_{gdq} = \frac{1}{2}\left[i_{dq}(k) + i_{dq}(k-1)\right] + \frac{C_f}{T_s}\left[v_{cdq}(k) - v_{cdq}(k-1)\right]$$
$$\mp \frac{\omega C_f}{2}\left[v_{cdq}(k) + v_{cdq}(k-1)\right] \tag{12}$$

The estimation process of the source voltages is the same as in (12), which is given by

$$e_{dq} = \frac{1}{2}\left[v_{cdq}(k) + v_{cdq}(k-1)\right] + \frac{L_g}{T_s}\left[i_{gdq}(k) - i_{gdq}(k-1)\right]$$
$$\mp \frac{\omega L_g}{2}\left[i_{gdq}(k) + i_{gdq}(k-1)\right] \tag{13}$$

IV. SIMULATIONS

The simulation is performed to verify the proposed method, where the PSIM is used as a simulation tool.

First, the current waveforms are illustrated to show the elimination performance of switching ripple components of LCL filters. Fig. 9 shows the grid current waveforms, where (a) – (c) are for LCL filters with 0[Ω], 5[Ω], and 20[Ω] damping resistors, respectively, and (d) is for L filter (3.5[mH]). It is shown that the higher damping resistor gives higher switching ripple components since the LCL filters produce closer performance to the L filter as the damping

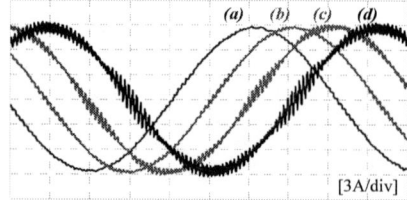

Fig. 9. Grid current waveforms. (a) 0[Ω], (b) 5[Ω], (c) 20[Ω] damping resistors in LCL filters, (d) L filters with 3.5[mH].

978-1-4244-1871-8/07 $25.00 © 2007 IEEE

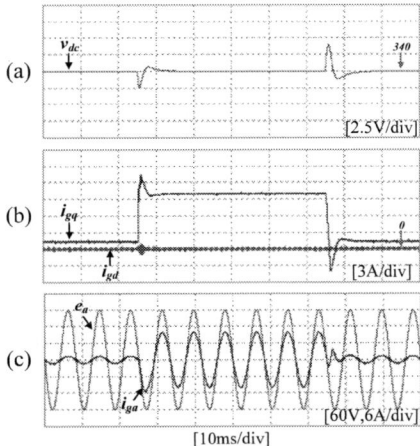

Fig. 10. PI control with load current compensation (Case 1). (a) DC-link voltage. (b) Dq-axis grid currents. (c) Source voltage and grid current.

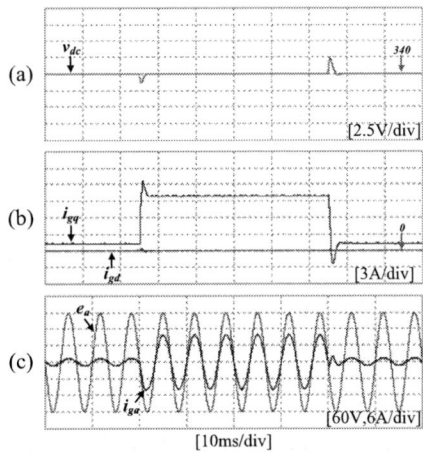

Fig. 11. Proposed control. (a) DC-link voltage. (b) Dq-axis currents. (c) Source voltage and grid current.

resistor is higher.

Next, the performance comparison between the PI controllers (Case 1) and the proposed method (Case 2) is carried out. The PI controllers are used for the grid current control and it includes the feedforward compensation of load current. The grid current and source voltage are detected and the d-axis current is controlled to zero. L_g and L_c are 1.5[mH] and 2[mH], respectively and C_f is 10[uF]. The resonance frequency of the LCL filters is 1.72[kHz] and the switching frequency is 5[kHz].

To observe the response performance in transient state, additional resister loads of 50[Ω] are applied in parallel and unloaded abruptly, while the resister loads 330[Ω] are applied during the full period of simulation from the start. Fig. 10 and 11 show the transient responses in case of PI control with load current compensation and with 5 [Ω] damping resistor, and the proposed control, respectively. For the PI control, a little resonance in the grid current is observed at the instant the load is switched on. However, with the proposed control scheme the resonance phenomenon is hardly

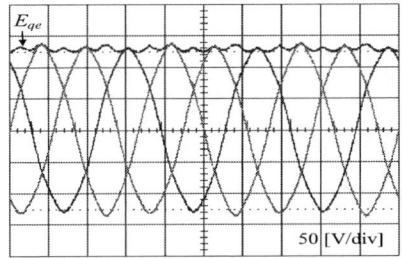

Fig. 12. Three-phase source voltage and q-axis voltage.

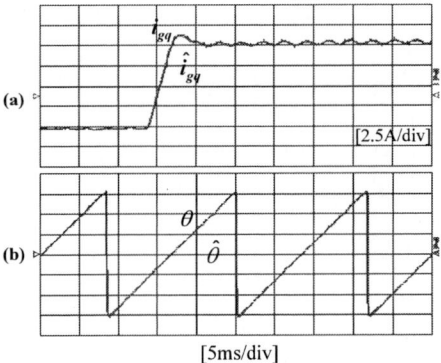

Fig. 13. (a) Q-axis real and estimated grid currents. (b) Angle generated with real and estimated source voltages.

observed in spite that no damping resistor is used. Furthermore, the transient response of the DC-link voltage is better than the PI controller as well.

For the L filter, the switching ripple component of the grid current normalized to the fundamental component is 3.94%. For the LCL filters, those of Case 1 and Case 2 are reduced to 1.26% and 0.71% respectively.

The gains used in the simulation (in Fig. 4) as follows. $K_p=6.0\times10^3$, $K_{11}=7.05\times10^3$, $K_{12}=2.0\times10^7$, $K_{13}=2.5\times10^9$, $K_{21}=1.57\times10^4$, $K_{22}=5.52\times10^7$, $K_{23}=3.24\times10^{10}$, $K_{24}=4.28\times10^{11}$

V. EXPERIMENTAL RESULTS

The experiment is performed to verify the proposed method. The switching frequency of the PWM converter is 5[kHz], and the parameters of LCL filters and damping resistor are the same value as in the simulation. The three-phase source voltage is 220Vrms, the DC-link capacitance is 1,950[uF], the reference of the DC-link voltage is 340[V], and the converter load is a resistor of 330//50[Ω].

Fig. 12 shows the source voltage waveforms in the laboratory. It is observed that a little 5th and 7th-order harmonic components and unbalance are included in the source voltage.

Fig. 13(a) shows the measured and estimated grid currents and (b) shows the phase angles which are generated from the measured and estimated source voltages, respectively, which are very close each other.

978-1-4244-1871-8/07 $25.00 © 2007 IEEE

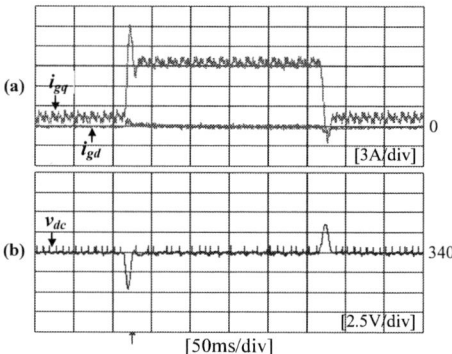

Fig. 14. PI control with load current compensation. (a) Dq-axis grid current. (b) DC-link voltage.

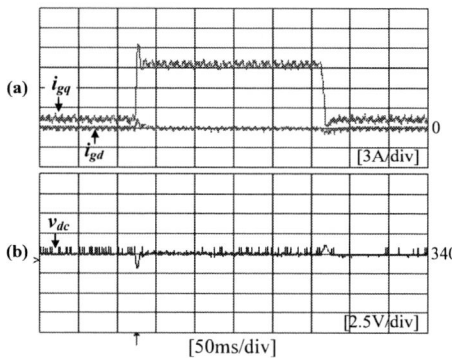

Fig. 15. Proposed controller. (a) Dq-axis grid current. (b) DC-link voltage.

Fig. 14 and 15 show the transient performance of the PI controller which controls the grid current and the proposed method, respectively. The gains of PI current controller used in the experiment are $K_p=\omega_c L_t$, $K_i=\omega_c R$ (ω_c=4000, L_t=0.035, R=0.1) and the gains of the PI voltage controller are $K_p=2\zeta\omega_v C_{dc}$, $K_i=\omega_v^2 C_{dc}$ (ζ=1, ω_v=250, C_{dc} =0.00195). The 50[Ω] resistors are loaded in parallel and unloaded abruptly while the PWM converter is operating with resistive loads of 330[Ω]. For the PI controller, the q-axis grid current has higher overshoot and larger DC-link voltage variation. However, the proposed method gives lower overshoot of the grid current response and smaller DC-link voltage variation. Also, the current ripple is lower in the proposed method.

Fig. 16 shows the source voltage and grid current waveforms measured, where the switching ripples of the grid current have been almost eliminated. Also, the unity power factor operation on the source side is performed well.

VI. CONCLUSIONS

In this paper, the feedback linearization for the control of three-phase PWM converter with LCL input filters has been proposed. Better performance in suppression of switching ripples compared to the L filter was obtained and the resonance problem of the LCL filters was solved without using damping resisters. Also, it is confirmed that the transient responses of the proposed method were better than

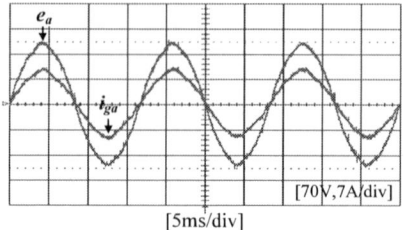

Fig. 16. Source voltage and grid current.

the PI controller. The size of filter inductors in case of the LCL filters can be made smaller than the L filter for the same filtering effect of the switching frequency-related harmonic components.

REFERENCES

[1] Jae-Seok. Noh, Jaeho. Choi, " Design of Voltage Source PWM Converter with AC Input LCL Filter", KIPE Trans, Vol. 3, pp. 490 ~ 498, 2002.

[2] P.A. Dahono, Y.R. Bahar, Y. Sato, and T. Kataoka, "Damping of transient oscillations on the output LC filter of PWM inverters by using a virtual resistor", IEEE Conf, Vol. 1, pp.403 - 407, Oct. 2001

[3] M. Liserre, A. Dell'Aquila, and F. Blaabjerg, "Stability improvements of an LCL-filter based three-phase active rectifier", IEEE PESC. proc., Vol. 3, pp. 1195-1201, June. 2002.

[4] M. Liserre, A. Dell'Aquila, and F. Blaabjerg, "Genetic Algorithm-Based Design of the Active Damping for an LCL-Filter Three-Phase Active Rectifier", IEEE Trans Power Electron, Vol. 19, pp.76-86, Jan. 2004.

[5] Twining. Erika, and D.G. Holmes, "Grid current regulation of a three-phase voltage source inverter with an LCL input filter", IEEE Trans. Power Electron., Vol. 18. No. 3, pp. 888-895, May. 2003.

[6] R. Teodorescu, F. Blaabjerg, M. Liserre, and A. Dell'Aquila, "A stable three-phase LCL-filter based active rectifier without damping", IEEE IAS Conf. proc., Vol. 3, pp. 1552-1557, Oct. 2003.

[7] Tzann-Shin. Lee, "Input–Output Linearization and Zero-Dynamics Control of Three-Phase AC/DC Voltage-Source Converters", IEEE Trans Power Electrons, Vol. 18, Jan. 2003.

[8] N. Mendalek, K. Al-Haddad, F. Fnaiech, and L.A. Dessaint, "Nonlinear control technique to enhance dynamic performance of a shunt active power filter," IEE proc.- Electr. Power Applicat., vol. 150, pp. 373 – 379, July. 2003.

[9] L. Yacoubi, K. Al-Haddad, L.-A. Dessaint, and F. Fnaiech, ""Linear and nonlinear control techniques for a three-phase three-level NPC boost rectifier,"" IEEE Trans. Ind. Electr., vol. 53, pp.1908–1918, Dec. 2006.

[10] J. E. Slotine and W. Li, "Applied Nonlinear Control", Prentice Hall, pp.207-271, 1991.

[11] D.-C. Lee, G-myoung Lee, and Ki-Do. Lee, "Dc-bus voltage control of three-phase AC/DC PWM converters using feedback linearization," IEE Trans. Ind. Applicat., vol. 36, no. 3, pp. 826-833, 2000.

[12] M. M. Seron, S. F. Graebe, and G. C. Goodwin, ""All stabilizing controllers, feedback linearization and anti-windup: a unified review,"" in American Control Conference, pp.1685-1689, 1989.

[13] S. Derksen, and D. Reefman, "Load current estimation for control algorithms in buck converter", Power Electron and Appl, 2005 European Conference, 10 pp., Sept. 2005

Digital Control of Phase-Shifted Full Bridge PWM Converter

Jeong-Gyu Lim, Soo-Hyun Lim, Se-Kyo Chung[†]

Department of Control and Instrumentation Engineering
Gyeongsang National University
900 Gazwa-Dong, Jinju, Gyeongnam 660-701, Korea
[†]Email: skchung@gnu.ac.kr

Abstract—This paper presents the modeling and design of the digital controller for a phase-shifted full-bridge converter (PSFBC) in the discrete-time domain. The discretized PSFBC model is first derived and then analyzed considering the sampling effect and the system parameters. Based on this model, the digital controller was directly designed in discrete-time domain. The simulation and experimental results are provided to verify the proposed modeling and controller design.

I. INTRODUCTION

Recently, numerous soft switching techniques for the switching power converters have been proposed. These techniques reduce the switching losses enabling high frequency operation and consequently reducing the overall system size. Among those, the phase-shifted full-bridge converter (PSFBC) has been used as one of the most popular topologies for many applications.

Digital control offers a number of advantages over analog control. It is easier to implement the computational functions and more flexible in designing and modifying the controller. Also, it is less sensitive to the noise and environmental variations. As the price/performance ratio of digital processors continues to decrease, digital controllers are becoming a viable and competitive option. The studies on the digital control of the PSFBC were recently presented [1],[2]. However, they considered only the implementation of the digital controller.

This paper presents the analysis and design of the digital controller for the PSFBC in the discrete-time domain using the small signal model presented in [3],[4]. The transfer function in discrete-time domain is first derived. Then, the effects on the sampling time and parameter variations are investigated. Based on this model, the digital controller is directly designed in discrete- time domain. The simulation is carried out to verify the validity of theoretical analysis and design. The experimental verifications are performed for the prototype PSFBC with a rating 100W/100kHz. The control algorithms designed by the proposed approach are implemented on a Texas Instruments TMS320F2812 digital signal processor (DSP). The simulated and experimental results show the validity of the proposed model and design approach.

II. PHASE-SHIFTED FULL-BRIDGE CONVERTER

A. Phase-Shifted Full-Bridge Converter

Fig.1 shows the digital control system of the PSFBC, which consists of the power stage, the DSP controller and signal conditioning circuit.

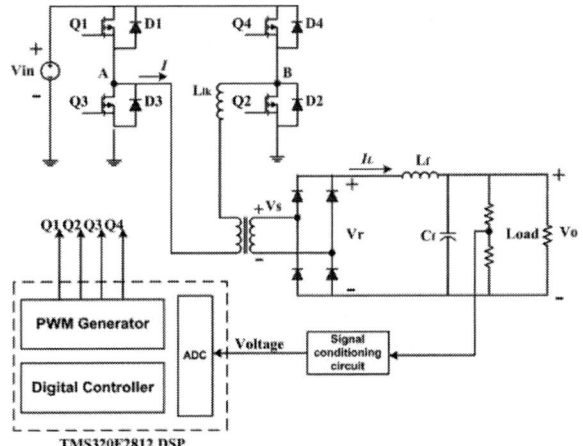

Fig. 1 Phase-shifted full bridge PWM converter

To achieve zero voltage switching (ZVS), this converter utilizes the transformer leakage inductance L_{lk} and MOSFET's output capacitance. Also, it has the switching pattern for ZVS, which is different from conventional one. The operation of the PSFBC is described in [3].

Fig. 2 shows the waveforms of the current I and voltage V_{AB} in primary and the voltage across the secondary of the power transformer. The load range at which the converter operates with ZVS increases with the leakage inductance. However, the large leakage inductance reduces the effective duty cycle of the secondary voltage D_{eff}, which can be express as [5]

$$D_{eff} = D - \Delta D \tag{1}$$

where D and ΔD are the duty cycle of the primary voltage set by the control and the loss of the duty cycle, respectively. The loss of the duty cycle is also given as

$$\Delta D = D - D_{eff} = \frac{2nL_{lk}}{V_o T_s}\left(2I_L - \frac{(1-D)V_o T_s}{2L_f} \right) \tag{2}$$

978-1-4244-1871-8/07 $25.00 © 2007 IEEE

where n, T_s and I_L are the transformer turns ratio, the switching period and the output filter inductor current, respectively.

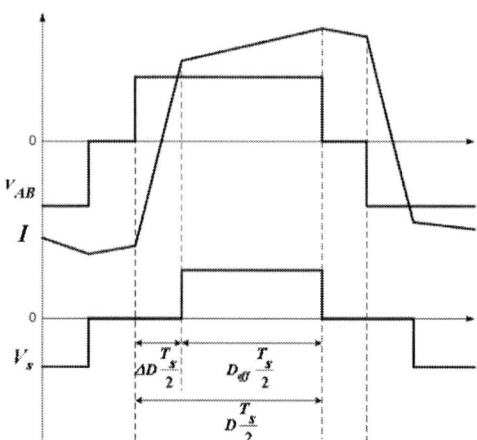

Fig. 2 Waveforms of voltage and current

B. Small signal model

Fig. 3 shows the small-signal circuit model of PSFBC, which can be derived from the averaged small-signal model of the PWM buck converter considering the change of the duty cycle caused by the change of filter inductor current I_L and input voltage V_{in} [5]. The effective duty cycle of the secondary voltage can be expressed as

$$d_{eff} = D_{eff} + \hat{d}_{eff} . \tag{3}$$

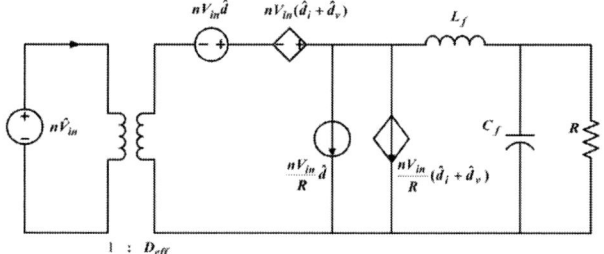

Fig. 3 Small-signal equivalent circuit

The effective duty cycle depends not only on the duty cycle d of the primary voltage but also on the current I_L of the output filter inductor, the leakage inductance L_{lk}, the input voltage V_{in}, and the switching frequency f_s. Assuming that the leakage inductance and the switching frequency are constant, the total change of the effective duty cycle is given as

$$\hat{d}_{eff} = \hat{d} + \hat{d}_i + \hat{d}_v \tag{4}$$

where $\hat{d}_i$ and $\hat{d}_v$ are the duty cycle variations due to the change of the filter inductor current and the input voltage, respectively. And they are represented as [5],

$$\hat{d}_i = -\frac{4nL_{lk}f_s}{V_{in}}\hat{i}_L \tag{5}$$

$$\hat{d}_v = \frac{4nL_{lk}f_s}{V_{in}^2}\hat{v}_{in} . \tag{6}$$

From the small-signal model of Fig. 3, the control-to-output transfer function is given as

$$G(s) = \frac{\hat{v}_o}{\hat{d}} = \frac{nV_{in}}{LC} \frac{1}{s^2 + s\left(\frac{L}{R} + R_dC\right) + \left(\frac{R_d}{RLC} + \frac{1}{LC}\right)} \tag{7}$$

where $R_d = 4n^2L_{lk}f_s$.

III. ANALYSIS AND DESIGN OF THE DIGITAL CONTROLLED SYSTEM

There are, in general, two approaches in designing a digital controller: the digital redesign and direct-digital approaches [6]. In this paper, the controller is designed using the direct-digital approach and then it is compared with that using the digital redesign approach. The parameters used in this section are summarized in Table 1 and also used in Section IV.

TABLE 1 System parameter

Parameter	Symbol	Value	Unit
Input voltage	V_{in}	48	V
Output voltage	V_O	12	V
Power	-	100	W
Turn ratio (Ns/Np)	n	0.5	-
Switching frequency	f_S	100	kHz
Output filter inductance	L_f	30	uH
Output filter capacitance	C_f	150	uF
Transformer leakage inductance	L_{lk}	5	uH

A. PSFBC model in the discrete-time domain

The PSFBC model in the continuous-time domain is first transformed into that in the discrete-time domain. Then, the controller is directly designed in the discrete-time domain. Fig. 4 shows the block diagram of the digital control loop, where $G_c(z)$ is the transfer function of the digital controller, ZOH denotes the zero order holder, $G(s)$ is the transfer function of the PSFBC and , K_d and T_s denote the voltage sensing gain and sampling period. $H(s)$ is the transfer function of the computation delay and it is given as

$$H(s) = e^{-sT_d} \tag{9}$$

where T_d denote the computation delay.

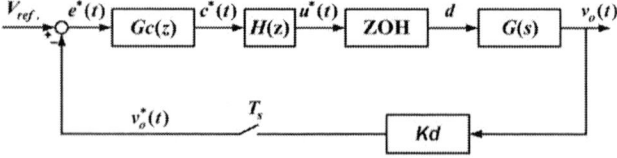

Fig. 4 Block diagram of digital control loop

The discrete-time transfer function $G(z)$ of the PSFBC considering the ZOH and K_d is derived as

$$G(z) = Z\left[\frac{1-e^{-sT_s}}{s} \cdot G(s) \cdot K_d\right] = \frac{\gamma}{\beta} \cdot \frac{b_1 z + b_0}{z^2 + a_1 z^1 + a_0} \quad (8)$$

where $Z[\cdot]$ denotes the z-transform of $[\cdot]$, and

$$a_1 = -2e^{-\frac{\alpha}{2}T_s} \cos\sqrt{\beta-(\alpha/2)^2}\, T_s\,,$$

$$a_0 = e^{-\alpha T_s}\,,$$

$$b_1 = 1 - e^{-\frac{\alpha}{2}T_s}\cos\sqrt{\beta-(\alpha/2)^2}\,T_s - \frac{\alpha/2}{\sqrt{\beta-(\alpha/2)^2}}e^{-\frac{\alpha}{2}T_s}\sin\sqrt{\beta-(\alpha/2)^2}\,T_s\,,3$$

$$b_0 = e^{-\alpha T_s} - e^{-\frac{\alpha}{2}T_s}\cos\sqrt{\beta-(\alpha/2)^2}\,T_s + \frac{\alpha/2}{\sqrt{\beta-(\alpha/2)^2}}e^{-\frac{\alpha}{2}T_s}\sin\sqrt{\beta-(\alpha/2)^2}\,T_s\,,$$

$$\alpha = \frac{1}{RC} + \frac{R_d}{L}\,,\quad \beta = \frac{R_d}{RLC} + \frac{1}{LC}\,,\quad \gamma = \frac{nV_{in}}{LC}\,.$$

$H(s)$ is not included in (8) and it is considered in Section III-B.

B. Effects of leakage inductance

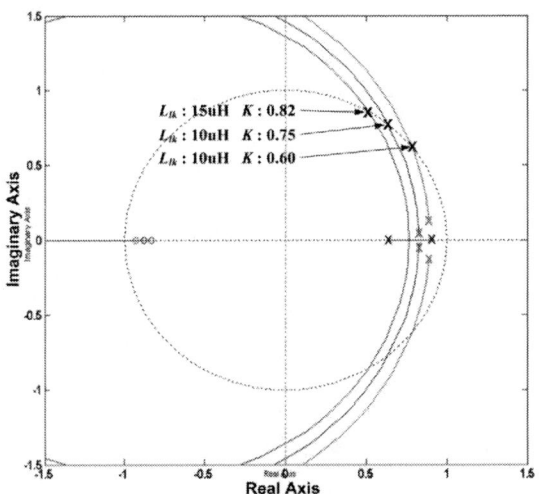

Fig. 5 Root loci of the closed-loop system
(Characteristic equation : $1+K\cdot G(z)=0$)

Fig. 5 shows the root loci of the closed-loop system using (8). Also, to examine the effect of the leakage inductance L_{lk}, the root loci for various L_{lk} are added. In this section, the PI controller is not included in the control loop and the loop gain K is just included. In Fig. 5, as the leakage inductance L_{lk} increases, the critical value of the loop gain K also increases. It is noted that the increase of the leakage inductance brings about the increase of the relative stability.

On the other hand, the leakage inductance also causes the duty loss. As shown Fig. 2, the finite slope in the rising and falling edges of the primary current reduces the duty cycle available in the secondary. As the increase of the leakage inductance enlarges the finite slope, consequently, it induces more duty loss.

C. Effects of time delay

The computation delay is the time delay between the sampling instant of the A/D converter and the updated instant of the subsequent PWM duty ratio [7]. It must be inevitably considered for the design of controller. Therefore, the transfer function $H(s)$ of the computation delay is included into the discrete-time transfer function $G(z)$ of (8). If the delay is an integral multiple of the sampling period T_s, the transfer function of the computation delay in the discrete-time domain can be simply expressed as

$$H(z) = z^{-k} \quad (9)$$

where $k = 1, 2, 3, \cdots$. Thus, the discrete-time transfer function $G'(z)$ with the delay of one sampling period is given as

$$G'(z) = \frac{\gamma}{\beta} \cdot \frac{b_1 z + b_0}{z(z^2 + a_1 z^1 + a_0)}\,. \quad (10)$$

The root loci of the closed-loop system with the computation delay of one sampling period are shown in Fig. 6. Like Section III-B, in the control loop, the loop gain K is only included. The root loci are affected by the addition of the open-loop pole at $z=0$. The root of the closed-loop system asymptotically approaches the 90° and 270° asymptotes that intersect at $z=1.36$ and the critical value of the loop gain K is 0.284. Thus, the computation delay causes the reduction of the stability margin of the system.

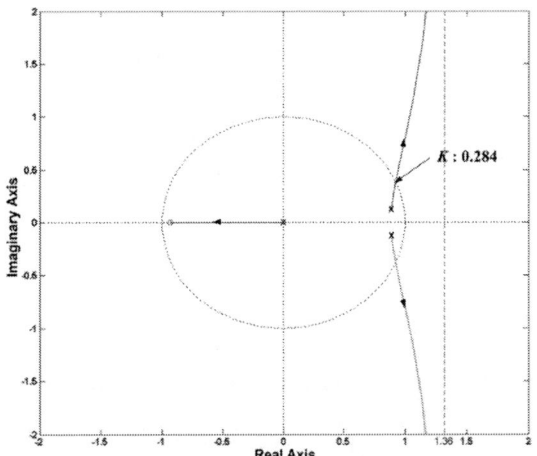

Fig. 6 Root loci of the closed-loop system considering computation delay
(Characteristic equation : $1+K\cdot G'(z)=0$)

D. Effects of sampling time

In this section, using the characteristic equation that the controller is included in, the effects of sampling time are investigated. The digital PI controller is used to regulate the output voltage and can be expressed as

$$G_c(z) = \frac{K_P z - (K_P - K_I T_s)}{z-1}\,. \quad (11)$$

The characteristic equation of the closed-loop system including controller is given as

$$d_3 z^3 + d_2 z^2 + d_1 z + d_0 = 0 \quad (12)$$

where

$$d_3 = 1,$$

978-1-4244-1871-8/07 $25.00 © 2007 IEEE

$$d_2 = \frac{\gamma}{\beta} K_P \left(1 - e^{-\frac{\alpha}{2}T_s} \cos\sqrt{\beta-(\alpha/2)^2} T_s - \frac{\alpha/2}{\sqrt{\beta-(\alpha/2)^2}} e^{-\frac{\alpha}{2}T_s} \sin\sqrt{\beta-(\alpha/2)^2} T_s \right)$$

$$-2e^{-\frac{\alpha}{2}T_s} \cos\sqrt{\beta-(\alpha/2)^2} \; T_s - 1,$$

$$d_1 = \frac{\gamma}{\beta} K_P \left(e^{-\alpha T_s} + \frac{\alpha}{\sqrt{\beta-(\alpha/2)^2}} e^{-\frac{\alpha}{2}T_s} \sin\sqrt{\beta-(\alpha/2)^2} \; T_s - 1 \right)$$

$$+ K_I T \left(1 - e^{-\frac{\alpha}{2}T_s} \cos\sqrt{\beta-(\alpha/2)^2} T_s - \frac{\alpha/2}{\sqrt{\beta-(\alpha/2)^2}} e^{-\frac{\alpha}{2}T_s} \sin\sqrt{\beta-(\alpha/2)^2} T_s \right)$$

$$+ e^{-\alpha T_s} + 2e^{-\frac{\alpha}{2}T_s} \cos\sqrt{\beta-(\alpha/2)^2} \; T_s,$$

$$d_0 = -e^{-\alpha T} - \frac{\gamma}{\beta}(K_P - K_I T)\left(e^{-\alpha T} - e^{-\frac{\alpha}{2}T_s} \cos\sqrt{\beta-(\alpha/2)^2} \; T_s \right.$$

$$\left. + \frac{\alpha/2}{\sqrt{\beta-(\alpha/2)^2}} e^{-\frac{\alpha}{2}T_s} \sin\sqrt{\beta-(\alpha/2)^2} T_s \right).$$

Fig. 7 shows the stable region of the closed loop system for various proportional gain K_P and sampling time T_s, which is obtained by using (10). Also, in this figure, K_I is 0. When the sampling time is 0.01ms, the boundary value of K_P for the stable operation of the digital control system is 0.6.

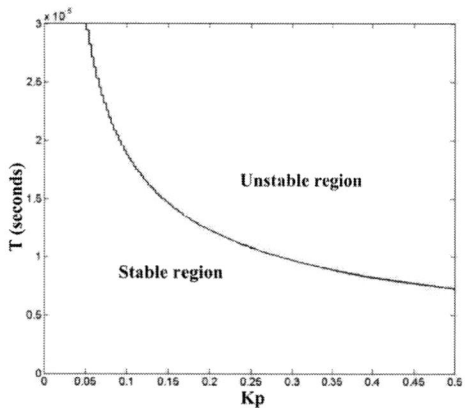

Fig. 7 Stable region for various K_p and T_s values

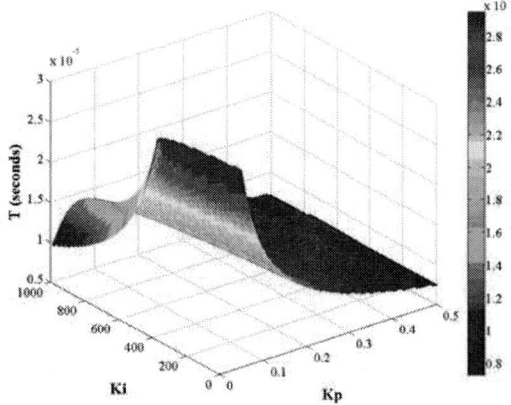

Fig. 8 Stable region for various K_I, K_P and T_s values

Fig. 8 presents the stable region of the closed loop system for various controller gain K_P, K_I and sampling time T_s. It is noted in this figure that the sufficiently fast sampling time below 0.01ms is required to guarantee the stability in a wide range of K_P, K_I.

E. Design of digital controller

The control algorithms of the experimental system implemented in Section IV have the computation delay of one sampling period. Therefore, the digital controller must be designed using (10). The PI controller of (11) is used to improve the steady-state performance and, at the same time, improve the relative stability. The digital controller is designed by frequency response method in the w-plane.

$$z = \frac{2 + wT}{2 - wT}. \tag{13}$$

Applying the w-transformation of (13) to the obtained discrete-time transfer function, the plant $G'(z)$ is transformed to

$$G'(w) = \frac{0.017w^3 + 8.904 \times 10^4 \; w^2 - 3.769 \times 10^{10} \; w + 3.833 \times 10^{14}}{0.027 \; w^3 + 7.945 \times 10^5 w^2 + 1.643 \times 10^{10} w + 2.176 \times 10^{14}}. \tag{14}$$

Fig. 9 shows the bode plots of the uncompensated and compensated systems, where the closed loop bandwidth is set at 2.16 kHz with a phase margin of 54°. The gain margin is 10.1dB at 4.45 kHz. The transfer function of the PI controller obtained directly in discrete-time domain is given as

$$G_{c1}(z) = 0.0432 \frac{z - 0.859}{z - 1}. \tag{15}$$

where K_P, and K_I are 0.0432 and 574, respectively. In this design, the sampling time and maximum overshoot of the control system can be predicted as 0.6ms and 10.9%, respectively.

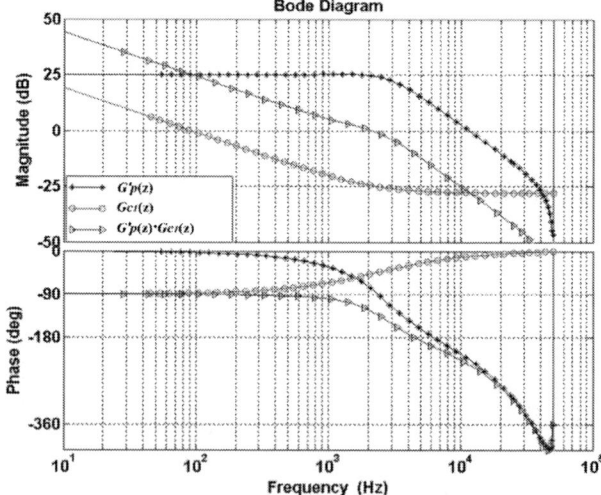

Fig. 9 Bode plots of discrete-time PSFBC system with PI controller

In order to compare with the direct digital design, the PI controller of the similar specification was designed using the digital redesign approach. The frequency response of this control system is shown in Fig. 10 (plot 1), where the system bandwidth is set at 2.79[kHz] with a phase margin of 52°. The gain margin is 7.9dB at 4.47 kHz. Therefore, the controller in continuous-time domain for the same system is given as

$$G_{c2}(s) = \frac{K_P(s - K_I/K_P)}{s} = 0.056 \frac{s - 13210}{s}. \tag{16}$$

where K_P and K_I is 0.056 and 739, respectively. The settling time and maximum overshoot can be predicted 0.39ms and 10.4%, respectively. For the implementation of the digital control system, the analog controller is transformed into a digital form and there are the several discretization methods [6]. The controller of (16) is discretized using the Step Invariant method among the discretization methods and it is shown as

$$G_{c2}(z) = 0.056 \frac{z - 0.867}{z - 1}. \qquad (17)$$

Plot 2 of Fig. 10 is the frequency response for $G'(z)$ with the redesigned controller $G_{c2}(z)$. The system bandwidth and the phase margin are 2.71[kHz] and 37.3°, respectively. In figure, the difference of the phase margin is 15.9°. It is because of the fact that the sample and hold and the computation delay is not considered in the redesigned system. Consequently, the performance of the designed and actual system is different and the results will be presented in latter section IV.

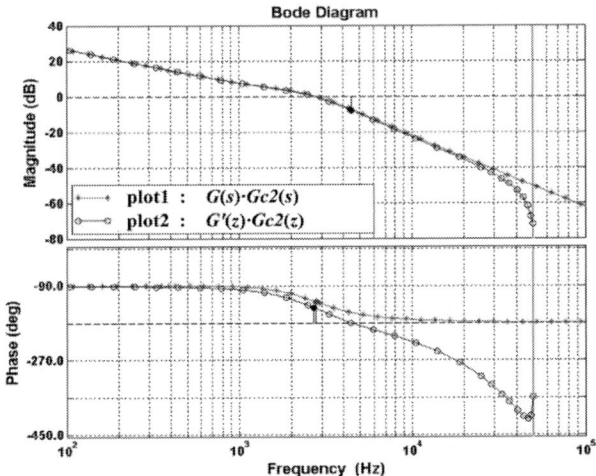

Fig. 10 Frequency response of the digital redesigned system

IV. SIMULATION AND EXPERIMENT

A. Simulated results

Simulations were performed in PSIM7 simulator. Fig. 11 shows the output voltage waveforms of the control system using the digital redesigned(V_{O1}) and direct redesigned(V_{O2}) controllers under the step load changes. The parameters of controllers are shown in (15) and (17), respectively. The load current was abruptly changed from 4[A] to 8[A]. As shown in Fig. 11, the settling time (2% of steady-state voltage) and the maximum overshoot of the output voltage V_{O1} is 10ms and 11%. In (15), the settling time and the maximum overshoot of the digital redesigned system were predicted as 0.3ms and 10.4%. These differences in the designed and actual control performance are caused by the fact that the effect of sample and hold and the computation delay is ignored. The settling time (2% of steady-state voltage) of the output voltage V_{O2} is 0.7ms and the maximum overshoot was 12% of the output voltage. It is noted that the control system has a satisfactory time response.

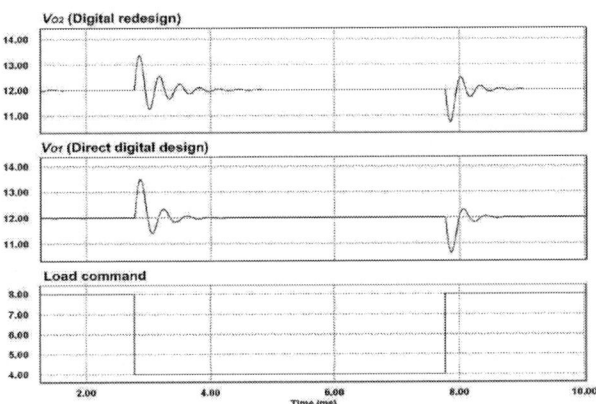

Fig. 11 Simulated step-load response

B. Experimental results

Fig. 12 shows the experimental system of the PSFBC prototype with a rating of 100W/100kHz, where the design parameters are presented in Table. 1. The digital controller is implemented on a TMS320F2812 DSP development board, which is shown the left side of Fig. 12.

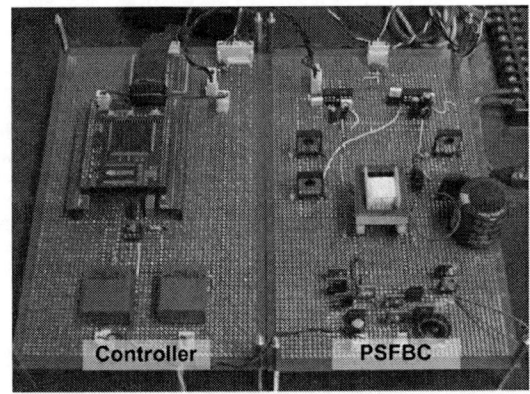

Fig. 12 Photograph of the PSFBC system

Fig. 13 shows the measured experimental results, where a trace on Ch2 is the voltage V_r across the secondary rectifier in Fig. 1 and traces on Ch3 and Ch4 are the primary voltage and current of the transformer.

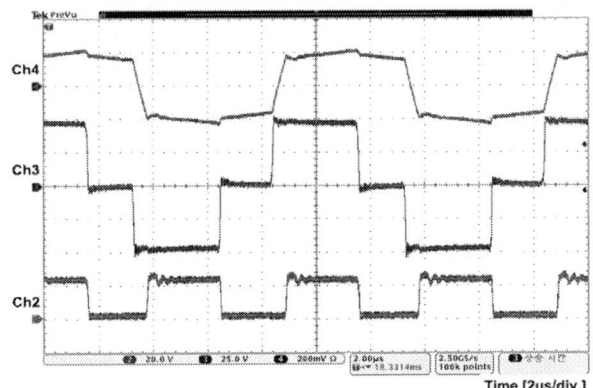

Fig. 13 Measured primary voltage and current of the transformer and voltage of the secondary rectifier (Ch4 : current [2A/div.], Ch3 : voltage [25V/div.] Ch2 : voltage V_r of the rectifier [20V/div.])

978-1-4244-1871-8/07 $25.00 © 2007 IEEE

Fig. 14 shows the drain and gate voltages of switch Q2, respectively. The ZVS for the switches Q1 and Q3 of the leading leg in the PSFBC is easily achieved by the energy of the large filter inductor L_{lf}. But, the switches Q4 and Q2 of the lagging leg use only the energy restored in the leakage inductance for operating the ZVS [3]. Therefore, the ZVS in the lagging leg is lost for a low load current. In Fig. 14, traces on Ch1 and Ch2 are the drain and gate voltages of switch Q2, respectively. The minimum range of the ZVS in this PSFBC system is calculated at 3.5 [A] [3]. As shown in Fig. 14, after the drain voltage of Q2 becomes low, the gate voltage becomes high. It is noted that the ZVS is achieved for the lagging switch Q2.

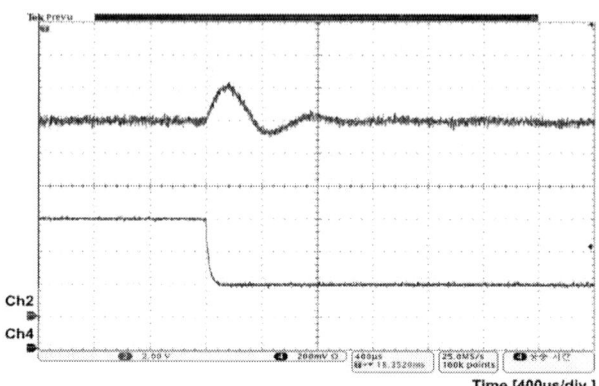

Time [400us/div.]

Fig. 16 Measured step-load response of the direct designed system (Ch1 : output voltage[2V/div.], Ch4 : output current[2A/div.])

V. CONCLUSIONS

In this paper, the digital control of the PSFBC is presented. The transfer function in the discrete-time domain is first derived using the small-signal model, and the effects on the sampling time and converter parameter are then analyzed. It is noted from this analysis that the relative stability is improved as the leakage inductance increases. However, since the large leakage inductance may cause the duty loss and narrow ZVS range, it is carefully selected. The proposed controller was directly designed in the discrete-time domain using the discretized PSFBC model and implemented using the DSP-based digital controller. The performance of the proposed controller is compared with the conventional controller using the digital redesign approach. It was shown from the simulation and experimental results that the proposed controller provides better performance than the conventional controller using the digital redesign approach, because the time delay and sampling effects were considered in the controller design.

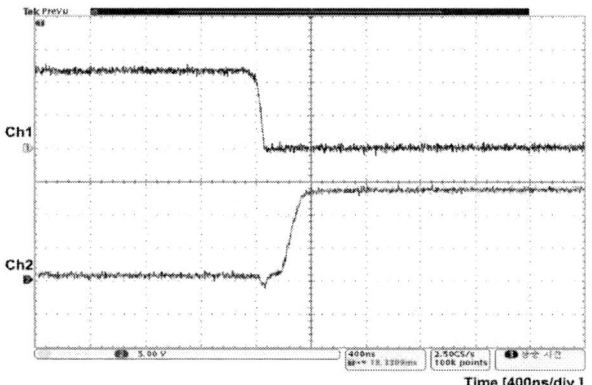

Time [400ns/div.]

Fig. 14 Measured drain and gate voltage of switch Q2 (Ch1 : drain voltage [20V/div.], Ch2 : gate voltage [5V/div.])

Fig. 15 and 16 show the output voltage responses for the step load change in the both digital redesign and proposed approaches, respectively, where the trace on Ch2 is the output voltage and Ch4 is the output current. In Fig. 15, after the load current is changed from 8[A] to 4[A], the settling time (2% of steady-state voltage) of the output voltage is stabilized in about 1.6ms and the maximum overshoot is 16%

ACKNOWLEDGMENT

This work was financially supported by MOCIE through IERC program.

REFERENCES

[1] P. F. Kocybik, K. N. Bateson, "Digital Control of a ZVS Full-Bridge DC-DC Converter", *IEEE APEC '95 Proc.*, pp. 687 – 693, 1995.

[2] P. H. Tang, G. J. Yang, T. C. Li, Y. J. Liu, "A Novel FB ZVS PWM DC-DC Converter Based on Digital Control", *IEEE WCICA 2006 The Sixth World Congress on* , vol. 2, pp. 8457-8461, 2006.

[3] J. A. Sabate, V. Vlatkovic, R. B. Ridley, F. C. Lee, and B. H. Cho, "Design considerations for high-voltage high-power full-bridge zero-voltage-switched PWM converter", *IEEE APEC'90 Proc.*, pp. 275-284, 1990.

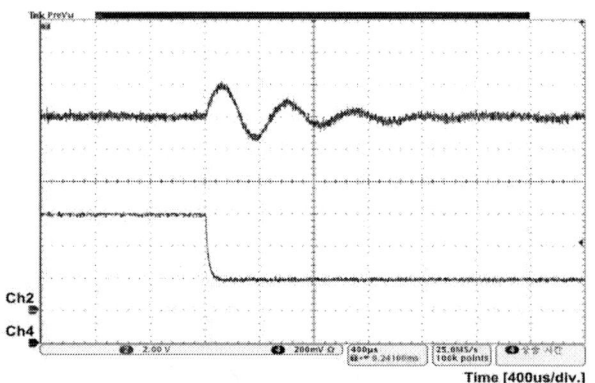

Time [400us/div.]

Fig. 15 Measured step-load response of the digital redesigned system (Ch1 : output voltage[2V/div.], Ch4 : output current[2A/div.])

In Fig. 16, the settling time (2% of steady-state voltage) of the output voltage is 1ms and the overshoot is 18%. The settling time presents about 60% of that using the digital redesign approach. The difference of the maximum overshoot can be ignored considering the results predicted in Section III-E.

[4] M. J. Schutten, D. A. Torrey, "Improved Small-Signal Analysis for the Phase-Shifted PWM Power Converter", *IEEE Transaction on power electronics*, vol. 18, No.2, pp. 659-669, March, 2003.

[5] J. A. Sabate, V. Vlatkovic, R. B. Ridley, F. C. Lee, and B. H. Cho, "Small-signal analysis of zero-voltage switched full-bridge PWM converter", *High Frequency Power Conversion Conf.'90 Rec.*, 1990, pp. 262-272.

[6] Y. Duan, H. Jin, "Digital Controller Design for Switchmode Power Converter", *IEEE APEC '99 Proc.*, vol. 2, pp. 967-973, March, 1999.

[7] Shamim Choudhury, "Designing a TMS320F280x Based Digitally Controlled DC-DC Switching Power Supply", Texas Instruments Application Report, SPRAAB3, July, 2005.

978-1-4244-1871-8/07 $25.00 © 2007 IEEE

Autonomous variable speed power generating system with five level cascaded converter

Lech M. Grzesiak
Warsaw University of Technology
Institute of Control and Industrial Electronics
Warsaw 00-662, Poland
Email: L.Grzesiak@isep.pw.edu.pl

Jacek G. Tomasik
Warsaw University of Technology
Institute of Control and Industrial Electronics
Warsaw 00-662, Poland
Email: tomasikj@webmail.co.za

Abstract—The paper presents a novel autonomous power generating system with a three-phase five-level cascaded inverter (5LCI) topology fed from a diesel engine with multiple-winding permanent magnet synchronous generator set (DE/PMSG) for UPS application. The power electronics system consists of six isolated windings which supply six full-bridge diode rectifiers with boost converters stabilizing the DC link voltages for a particular cell of the 5LCI. The five-level current mode control algorithm provides the high quality PWM waves which mitigate the high harmonic content and satisfy a robust output voltage regulation even under the unbalanced load conditions. The interfaced battery energy storage (BES), which constitutes an integral part of the entire "gen-set" system, it supplies the DC link capacitor voltages to particular cells of the 5LCI in a standby emergency mode, when the DE/PMSG are off-line. The topology under discussion can be applied to medium or high power UPS systems.

I. INTRODUCTION

This n a modern electrical energy conversion the back-up system plays a very important role. It not only facilitates a transfer of an electrical energy from an input to the output but it also conditions the characteristics of the source in such a manner that the output characteristics are equal to those required by the standards of a utility and the appliances connected into it. Thus it can also serve as an interface between the utility and compensate for undesirable effects such as: unbalanced loads, nonlinear loads, impact loads which can cause the dips or a harmonic distortion in the utility. Many applications were a result of such a research in this field, predominantly with the standard two-level converters incorporated in such systems [1], [2], [3], [4]. With an advent of multi-level converter technologies the UPS system can further be explored with a potential of increasing power levels, especially in a medium and a high voltage applications. Moreover, the advantages of a multi-level modulation over a conventional two-level one [5], make such a system an attractive alternative. Therefore one of the goals of this paper was to introduce a new structure of a back-up system with an extension to the multi-level converters based on the five-level cascaded topology.

The autonomous power generating system with multi-level converters has previously been introduced in back-to-back structure, which was required to handle the voltage imbalance in the DC link network. In such operation the back-end active-rectifier was interfaced to constant frequency/voltage input of the utility [6] or variable frequency/voltage input of the generator [7], [8], [9]. The front-end constituted a voltage source inverter that was able to operate with symmetrical as well with asymmetrical loads in all three-phases.

In one of the alternative topologies serving power generation and energy conversion by means of the DE/PMSG and the AC/DC/AC converters, the multiple isolated windings generator can supply the multi-level inverter.

Generally comparing the n-level inverter topologies i.e. diode-clamped with cascaded, they both will consist of exactly the same number of main switching devices but the latter does not require the clamping diodes. The main disadvantage in devising the cascaded inverter is its inherent requirement for isolated DC power supplies. However the diode clamped inverter must have a mechanism for balancing the input DC link voltages, which often requires the isolated winding power supply and rectifiers. Yet the modular structure of the cascaded topology constitutes an important advantage over the diode clamped topology, because the former is constructed from the standard H-bridges, which it makes easier maintenance of the entire system.

II. SYSTEM TOPOLOGY

The novel power generating system topology consists of the diesel-engine as a prime-mover that drives on the common shaft the permanent magnet synchronous generator, which constitutes the "gen-set" (DE/PMSG) {see the Figure 1}.The PMSG consists of the multiple single-phase windings ($W_1 \div W_6$) which feed the six consecutive full-bridge diode rectifiers ($BD_1 \div BD_6$) to form the DC voltage supplying inputs to the six boost converters. The role of six step-up choppers ($SU_1 \div SU_6$) is to regulate the DC link voltage on predetermined level for the particular full-bridge cell ($FBC_1 \div FBC_6$) of the cascaded inverter topology. Each phase-leg of the three-phase five-level cascaded inverter (5LCI) consists of such two symmetrical full-bridge cells. The system synthesizes the five-level output PWM waveforms in each phase so that the high harmonic content is significantly mitigated. The low-pass LC filter added on the output of the 5LCI provides the sinusoidal and rigid output voltage regulation. The neutral leaders (N) of each phase-leg are connected together to form the path for a neutral current flow. Therefore, in such configuration the unbalanced load operation is feasible. The most important feature of the topology under discussion is its modularity and symmetry.

978-1-4244-1871-8/07 $25.00 © 2007 IEEE

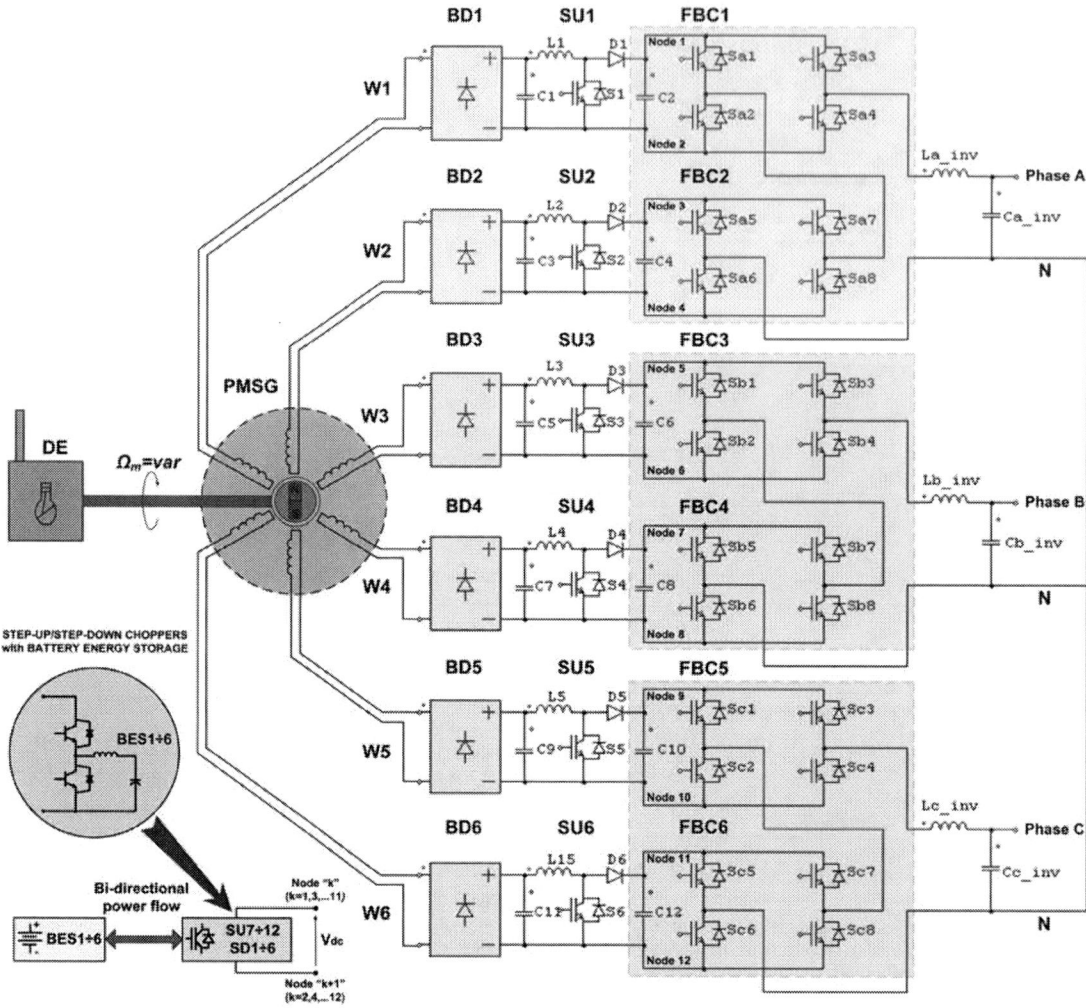

Fig. 1: Block diagram of autonomous power generating system with five-level cascaded inverter.

III. CONTROL STRATEGY

The control law for each of the DC-DC converters (operating in boost and buck modes) i.e. the $SU1 \div SU6$ fed from full-wave bridges $BD1 \div BD6$ as well as the $SU7 \div SU12$ fed from the battery energy storages $BES1 \div BES6$ and the $SD1 \div SD6$ fed from the DC link voltages, all can generally be depicted by:

$$\begin{cases} i_{DC}^*(t) = \begin{cases} 0 & \text{for } v_{DC}^* > v_{DC} \\ K_{1p}\left(v_{DC}^*(t) - v_{DC}(t)\right) + K_{1i}\int\left(v_{DC}^*(t) - v_{DC}(t)\right)dt & \text{for } v_{DC}^* < v_{DC} \end{cases} \\ e_{DC}(t) = K_{2p}\left(i_{DC\_MAX} + i_{DC}^*(t) - i_{DC}(t)\right) + K_{2i}\int\left(i_{DC\_MAX} + i_{DC}^*(t) - i_{DC}(t)\right)dt \end{cases} \quad (1)$$

where: e_{DC} – instantaneous value of error signal in DC-DC controller, v^*_{DC}, v_{DC} – instantaneous values of reference and actual output voltages from choppers, i^*_{DC}, i_{DC}, i_{DC_MAX} – instantaneous value of reference inductor current, actual current and limit current in choppers, K_{1p}, K_{1i}, K_{2p}, K_{2i} – proportional and integral gains in DC voltage and current loops.

The switching conditions for DC-DC converters are given by:

$$if \begin{cases} e_{DC}(t) > v_{TA}(t) \Rightarrow S_x(t) = 1 \\ e_{DC}(t) < v_{TA}(t) \Rightarrow S_x(t) = 0 \end{cases} \quad (2)$$

Analogically to what has been presented in [7] and [8], the continuous control law for a anyone phase ($x=a,b,c$) of the 5LCI is given by:

$$\begin{cases} i_{x\_inv}^*(t) = K_{3p}\left(v_{x\_out}^*(t) - v_{x\_out}(t)\right) + K_{3i}\int\left(v_{x\_out}^*(t) - v_{x\_out}(t)\right)dt \\ e_{x\_inv}(t) = K_{4p}\left(i_{x\_inv}^*(t) - i_{x\_inv}(t)\right) + K_{4i}\int\left(i_{x\_inv}^*(t) - i_{x\_inv}(t)\right)dt \end{cases} \quad (3)$$

where: e_{x_inv} – instantaneous value of error signal in 5LCI controller, $v^*_{x\_out}$, v_{x_out} – instantaneous values of reference and actual output voltage, $i^*_{x\_inv}$, i_{x_inv} – instantaneous value of reference inductor current and actual phase current in 5LCI, K_{3p}, K_{3i}, K_{4p}, K_{4i} – proportional and integral gains in voltage and current loops.

978-1-4244-1871-8/07 $25.00 © 2007 IEEE

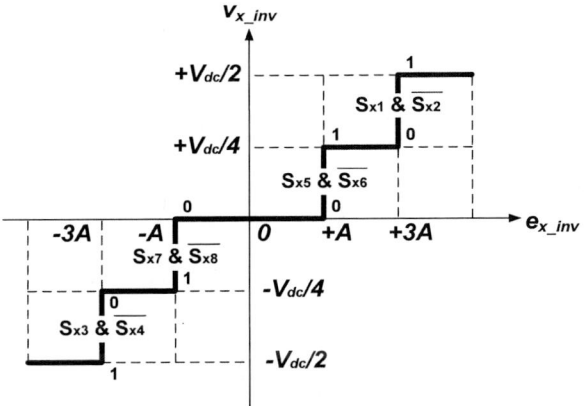

a) Five-level PWM modulator (x=a,b,c)

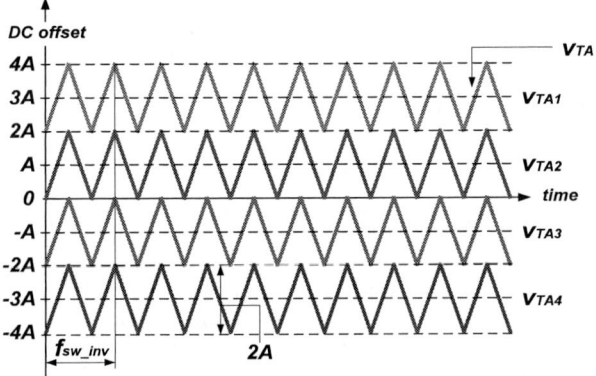

b) Five-level PWM modulation in cascaded inverter

Fig. 2: Principle of modulation in cascaded converter:
a) Five-level modulator, b) Multi-carrier PWM.

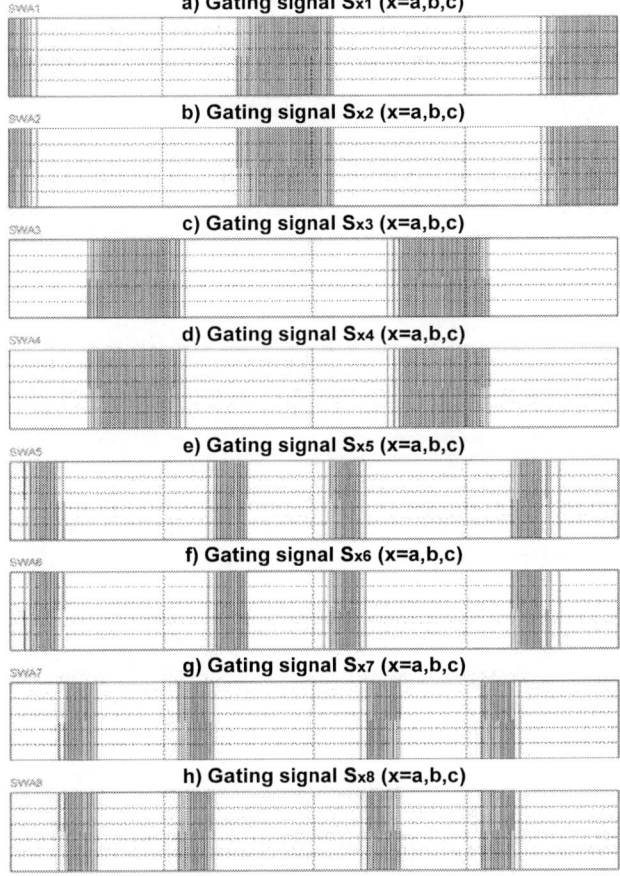

Fig. 3: Gating signals in single phase (x=a,b,c) of 5LCI :
a)÷ d) Top H-bridge, e) ÷ h) Bottom H-bridge.

Definition of the five-segment switched waveform in the 5LCI, which is determined by the control error in (3), is derived from:

$$(4)$$

a) Positive output $\left(\dfrac{V_{dc}}{2} \leq v_{x\_inv}(t) \leq \dfrac{V_{dc}}{4} \right)$

$$if \begin{cases} e_{x\_inv}(t) > v_{TA1}(t) \Rightarrow \left(S_{x1}(t)=1 \right) \& \left(S_{x2}(t)=0 \right) \\ e_{x\_inv}(t) < v_{TA1}(t) \Rightarrow \left(S_{x1}(t)=0 \right) \& \left(S_{x2}(t)=1 \right) \end{cases}$$

b) Positive output $\left(\dfrac{V_{dc}}{4} \leq v_{x\_inv}(t) \leq 0 \right)$

$$if \begin{cases} e_{x\_inv}(t) > v_{TA2}(t) \Rightarrow \left(S_{x5}(t)=1 \right) \& \left(S_{x6}(t)=0 \right) \\ e_{x\_inv}(t) < v_{TA2}(t) \Rightarrow \left(S_{x5}(t)=0 \right) \& \left(S_{x6}(t)=1 \right) \end{cases}$$

c) Negative output $\left(0 \leq v_{x\_inv}(t) \leq -\dfrac{V_{dc}}{4} \right)$

$$if \begin{cases} e_{x\_inv}(t) > v_{TA3}(t) \Rightarrow \left(S_{x7}(t)=1 \right) \& \left(S_{x8}(t)=0 \right) \\ e_{x\_inv}(t) < v_{TA3}(t) \Rightarrow \left(S_{x7}(t)=0 \right) \& \left(S_{x8}(t)=1 \right) \end{cases}$$

d) Negative output $\left(-\dfrac{V_{dc}}{4} \leq v_{x\_inv}(t) \leq -\dfrac{V_{dc}}{2} \right)$

$$if \begin{cases} e_{x\_inv}(t) > v_{TA4}(t) \Rightarrow \left(S_{x3}(t)=1 \right) \& \left(S_{x4}(t)=0 \right) \\ e_{x\_inv}(t) < v_{TA4}(t) \Rightarrow \left(S_{x3}(t)=0 \right) \& \left(S_{x4}(t)=1 \right) \end{cases}$$

The DC offset displacement variables in the control scheme on the 5LCI side PWM modulator are defined by:

$$\begin{cases} v_{TA1}(t) = 3A(t) + v_{TA}(t) \\ v_{TA2}(t) = A(t) + v_{TA}(t) \\ v_{TA3}(t) = -A(t) + v_{TA}(t) \\ v_{TA4}(t) = -3A(t) + v_{TA}(t) \\ V_{TAp-p\,min} = 2A_{min} \\ V_{TAp-p\,max} = 2A_{max} \\ A_{min} \leq A \leq A_{max} \end{cases} \qquad (5)$$

where: A, v_{TA} – instantaneous values of DC offsets and triangular carriers on 5LCI side, A_{min}, A_{max} – constants determined from conditions on minimum and maximum value of triangular carrier slope on 5LCI side, $V_{TAp-p\,min}$, $V_{TAp-p\,max}$ – minimum and maximum triangular carrier peak-to-peak slope required for linear five-level PWM modulation on 5LCI side, v_{TA1}, v_{TA2}, v_{TA3}, v_{TA4} – instantaneous values of adjacent triangular carriers in 5LCI PWM.

The five-level comparator and four carrier waves are shown in the Figures 2a and 2b, respectively. In such configuration the top H-bridge synthesizes the outermost positive and negative sections of the control error signal whereas the

978-1-4244-1871-8/07 $25.00 © 2007 IEEE   779

five-level PWM modulator utilizes outer carriers v_{TA1} and v_{TA4}.

In result, the complementary pairs of gating signals S_{1x} & $NOT(S_{2x})$ {see the Figures 3a÷b}, contribute to the outer positive section of the voltage switched waveform v_{x_inv} and the complementary pairs of gating signals S_{3x} & $NOT(S_{4x})$ {see the Figures 3c÷d}, contribute to the outer negative section of the voltage switched waveform v_{x_inv}.

Analogically, the bottom H-bridge synthesizes the innermost positive and negative sections of the control error signal whereas the five-level PWM modulator utilizes inner carriers v_{TA2} and v_{TA3}.

In result, the complementary pairs of gating signals S_{5x} & $NOT(S_{6x})$ {see the Figures 3e÷f}, contribute to the inner positive section of the voltage switched waveform v_{x_inv} and the complementary pairs of gating signals S_{7x} & $NOT(S_{8x})$ {see the Figures 3g÷h}, contribute to the inner negative section of the voltage switched waveform v_{x_inv}.

Hence, the symmetrical five-segment voltage PWM waveform v_{x_inv} is generated on the output of the 5LCI {see the Figures 4c÷e and 5c÷e}.

On line-to-line basis, the side-band harmonic content eminent in five-segment waveform is cancelled, so that the nine-segment voltage switched waveforms are obtained {see the Figures 4f÷h and 5f÷h}.

IV. Simulation Model Parameters

Model of the proposed autonomous power generating system was implemented in PSIM 6.0 package with the following parameters shown below in the Table 1.

TABLE I
Simulation model parameters

Inverter section	
Nominal power P_{max}	15[kW]
Nominal phase current I_{out}	22[A]
Nominal phase voltage V_{out}	230[V]
Nominal frequency f_{out}	50[Hz]
Nominal DC link voltage V_{DC}	400[V]
Output filter inductance L_{inv}	1[mH]
Output filter capacitance C_{inv}	60[μF]
Switching frequency f_{sw}	5[kHz]
Step-up/down converter section	
Input rectified DC voltage per cell V_{IN}	150[V]
Battery DC voltage per cell V_{BAT}	48[V]
Regulated DC link voltage per cell V_{CELL}	200[V]
Current limit per DC-DC cell I_{DC}	100[A]
Boost/buck converter inductance L_{DC}	1[mH]
Boost/buck converter capacitance C_{DC}	100[mF]
Switching frequency f_{sw}	5[kHz]

Through the Figures 4 and 5, there are shown simulation results of the system operating under asymmetrical and symmetrical loads respectively.

In the Figure 4a, there are shown six DC link voltages supplying each H-bridge cell and six DC currents in the SU1÷6. Two such H-bridge cells constitute a single phase of the three-phase 5LCI. It can be noticed that due to imbalanced phases on the output of the three-phase 5LCI, different magnitudes of the currents from the DC-DC converters are drawn, which are proportional to the power required by the output load. The discernible high frequency ripple in inductor current is related to 5[kHz] switching frequency and the low frequency ripple is related to the fundamental frequency 50[Hz] of the inverter.

In the Figure 5a, there is shown case of the 5LCI operating with symmetrical loads in three-phases. Similarly, to above, the high frequency ripple in inductor current is related to 5[kHz] switching frequency and the low frequency ripple is related to the fundamental frequency 50[Hz] of the inverter. Moreover, due to unequal duty cycles of the switches in both consecutive cells per each phase of the 5LCI, it is evident that the currents drawn by the top DC-DC converter will always differ from the current drawn by the bottom DC-DC converter irrespective whether the load is symmetrical or asymmetrical in three-phases of the 5LCI.

In both cases, the DC link voltages across each H-bridge cell remain regulated even under asymmetrical load conditions in the 5LCI, which proves effectiveness of the control in the DC-DC converters.

In the Figure 4b, there are shown output phase-to-neutral voltages in three phases of the 5LCI after the LC filter and the inductor currents in three phases under unbalanced load conditions. The high frequency ripple is related to 5[kHz] carrier waves in modulators. It shows that even under unbalanced conditions the output voltages in the 5LCI remain stable. Similarly to above, in the Figure 5b, there is shown case of symmetrical loads operation.

In the Figures 4c÷e and Figures 5c÷e, there are shown phase-to-neutral switched voltage waveforms in three-phases of the 5LCI. Five steps in the PWM waveforms prove the five-level modulation process in the 5LCI. It shows that in both cases of loads, asymmetrical as well as the symmetrical, the inverter voltages remain stable.

In the Figures 4f÷h and Figures 5f÷h, there are shown line-to-line switched voltage waveforms in three-phases of the 5LCI. Similarly, nine steps in the PWM waveforms prove the five-level modulation process in the 5LCI. It shows that in both cases of loads, asymmetrical as well as the symmetrical, the inverter voltages remain stable.

V. Summary

In this paper a new topology of autonomous power generating system with the multi-level converters has been presented. The proposed system can be alternative solution to the problem with capacitor imbalance in the five-level diode clamped topology as reported in [7], [8] "in-press", [9] "unpublished". The novel system has been simulated with five-level cascaded inverter operating at variable voltage and frequency on the input and constant voltage and frequency on the output. The results prove effectiveness of the control methods under symmetrical as well as asymmetrical load conditions. In conclusion, the proposed autonomous power generating system with the multi-level converters can be used in medium or high voltage applications.

978-1-4244-1871-8/07 $25.00 © 2007 IEEE

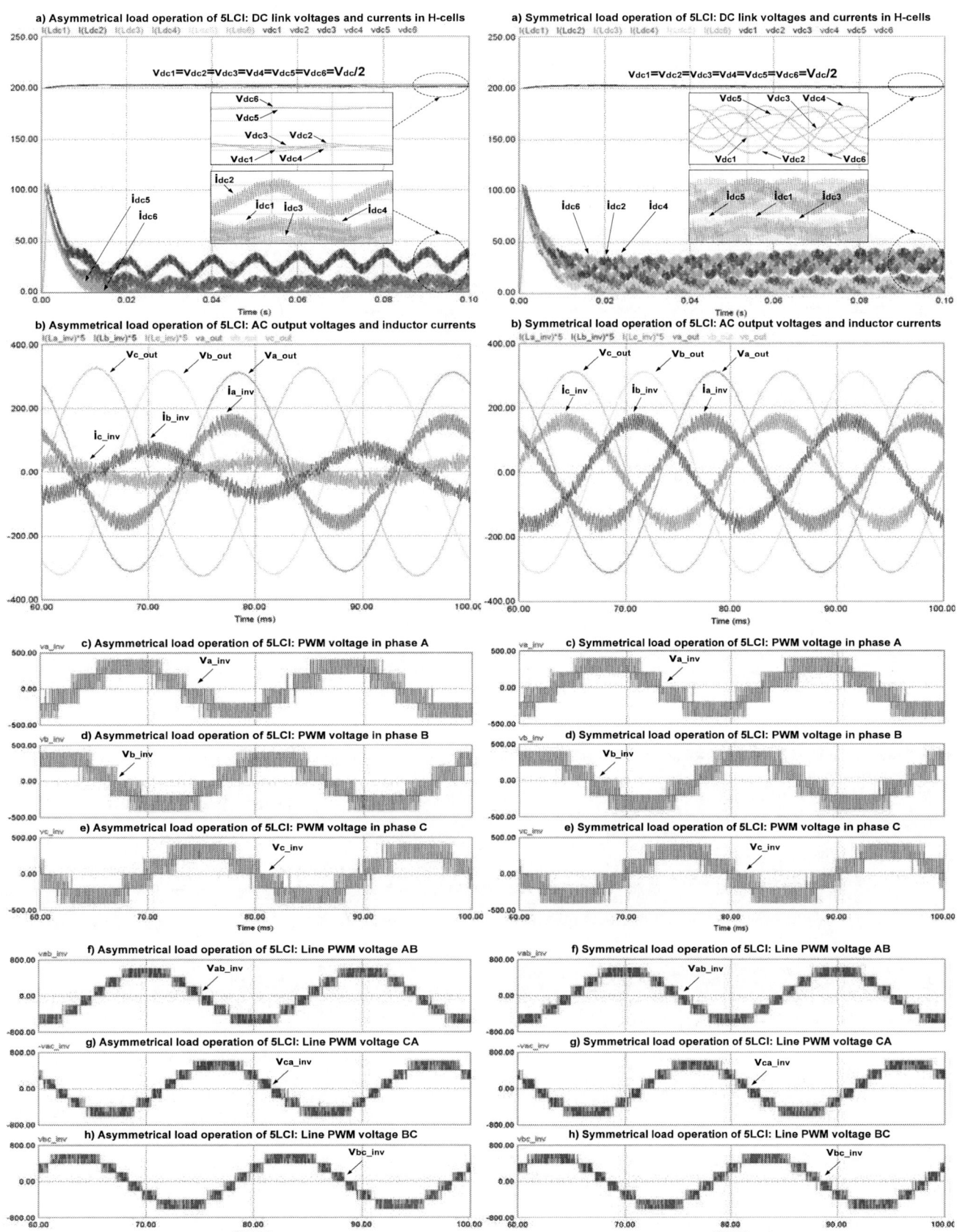

Fig. 4: a) ÷ h) Unbalanced load operation of 5LCI.

Fig. 5: a) ÷ h) Unbalanced load operation of 5LCI.

ACKNOWLEDGMENT

The authors would like to thank the Volt-Ampere (Pretoria, RSA) for financial support of this work.

REFERENCES

[1] Grzesiak L., Koczara W., da Ponte M.: "Novel hybrid load-adaptive variable speed generating system" in *Proceedings IEEE International Symposium on Industrial Electronics ISIE'98*, Pretoria, South Africa 7-10 July 1998, pp. 271-276.

[2] Paciura, K., Hancke, G.P., Grzesiak, L., Koczara, W., De Ponte, M.: "A battery energy storage system for the Hygen electricity generator and power supply system" in *AFRICON, 1999 IEEE,Volume 2, 28 Sept.-1 Oct. 1999*, pp. 949-954

[3] Tomasik J., Grzesiak L., Hancke, G.: "Digital signal processor control of a three-phase inverter for a high quality power supply system with permanent magnet alternator" in *Proceedings SAUPEC 2001 Conference, Cape Town, South Africa*, pp. 226-230

[4] Ghali F.M.A., Arafah S.H.: "Dynamic analysis of hybrid wind/diesel system with three-level inverter" Power Conversion Conference, 2002. PCC Osaka 2002. *Proceedings of the; Volume: 2, 2-5 April 2002.* pp. 727-732

[5] Loh P., Bode G.H., Holmes D.G., Lipo T.A.: "A time-based double-band hysteresis current regulation strategy for single-phase multi-level inverters" in *IEEE Transaction on Industry Applications, Vol. 39, No. 3, May/June 2003*, pp. 1994-2001

[6] Grzesiak L.M., Tomasik J.: "DC link balancing method in back-to-back UPS system with multi-level converters" in *Industrial Electronics, 2006 IEEE International Symposium on, Volume 2, July 9-12, 2006, Montreal, Quebec, Canada*, pp. 908-913

[7] Grzesiak L.M; Tomasik J.G.: "Autonomous power generating system with multi-level converters" in *IEEE Industrial Electronics, IECON 2006 - 32nd Annual Conference on, November 7-10, 2006, Paris, France*, pp. 2815-2820

[8] Grzesiak L.M; Tomasik J.G.: "Novel DC link balancing scheme in 5-level back-to-back converter system" the International Conference on Computer as a tool, the IEEE Region 8 EUROCON 2007, September 9-12, 2007, Warsaw, Poland (final paper accepted)

[9] Grzesiak L.M; Tomasik J.G.: "Autonomous power generating system with generic n-level back-to-back converters" the 8th International Power Engineering Conference - IPEC2007, 3 - 6 Dec 2007, Meritus Mandarin Singapore (paper provisionally accepted).

Modularized Charge Equalization Converter with High Power Density and Low Voltage Stress for HEV Lithium-ion Battery String

Hong-Sun Park, Chong-Eun Kim, Chol-Ho Kim,
Bong-Chul Kim, and Gun-Woo Moon
Div. of Electrical Engineering, Korea Advanced
Institute of Science and Technology
373-1, Guseong-dong, Yuseong-gu, Daejeon, Korea
Email: gwmoon@ee.kaist.ac.kr

Joong-Hui Lee
HEVB System Development TF, SK Institute of
Technology
140-1, Wonchon-dong, Yuseong-gu, Daejeon, Korea
Email: hui@SKenergy.com

Abstract—This paper proposes the modularized charge equalization converter for hybrid electric vehicle (HEV) lithium-ion battery cells, in which the intra-module and the inter-module equalizer are implemented. Considering the high voltage HEV battery pack, approximately above 300V, the proposed equalization circuit modularizes the entire *M*N* cells; in other words, *M* modules in the string and *N* cells in each module. With this modularization, low voltage stresses of all the electronic devices, roughly below 64V, can be obtained. In the intra-module equalization, the current-fed type DC/DC converter with cell selection switches is employed for concentrated charging of the specific under charged cells. On the other hand, the inter-module equalizer makes use of the voltage-fed type DC/DC converter for bi-directional equalization. In the proposed circuit, high power density can be achieved by employing the optimal power rating design rule. In addition, small size and low cost can be accomplished by sharing the MOSFET switch by the intra-module and inter-module equalizer and employing no additional reset circuitry at the inter-module equalizer. Experimental results of an implemented prototype show that the proposed equalization scheme has the promising cell balancing performance for the 7Ah HEV lithium-ion battery string while maintaining low voltage stress, low cost, small size, and high power density.

I. INTRODUCTION

The Hybrid electric vehicle (HEV) has become one of the most promising cars in the automobile industry due to its ability of energy saving and low emission of harmful pollutants [1]-[5]. This is because HEV can recover energy from the wheels, which have been wasted in the past, and reuse it to propel the vehicle at low speeds or boost extra power required at high acceleration. Another reason is that a battery powered HEV does not emit air pollutants.

Most of HEVs on the streets of these days use nickel-metal hybrid (Ni-MH) batteries. Recent development of lithium-ion battery and its test results show that lithium-ion battery has higher power and energy density, lower self-discharge rate, and higher single cell voltage than the Ni-MH battery. Therefore, it has great potential of taking the place of Ni-MH battery in the near future [4], [5]. However, realizing this possibility needs reliability and safety of a lithium-ion battery; in other words, the lithium-ion battery should be maintained within the allowed ranges of voltage and current limits to prevent permanent deterioration of characteristics and, in the worst case, explosion or fire in the vehicle.

Conventionally, continual charge and discharge of the series connected battery cells can cause charge imbalance. The problem is that when they are left in use without any control such as cell equalization, the unwanted circumstance more frequently occurs. For example, in the regenerative braking mode the highly charged cells cannot capture an optimal amount of renewable energy. And in the battery powered driving mode the deeply discharged cells cannot sufficiently provide the stored energy. Therefore, charge equalization for the series connected battery cells is essential to prevent these undesirable situations, accomplish the maximum utilization of battery, and prolong the lifetime [4]-[17]. A protection circuitry is, of course, important in battery management, but it is beyond the scope of this paper.

To achieve charge equalization for the battery string, plenty of schemes in algorithm and topologies in circuitry have been developed [5]-[16] and are well summarized [6]. Among them, a resistive current shunt, which is connected across each cell and controlled by a micro-processor, has been presented, Simple implementation, low production cost, and stable operations are great advantages, but energy dissipation is a major drawback of this circuit [7].

To improve the energy consumption, non-dissipative charge equalization methods have been applied, in which charge-type, discharge-type, and charge- and discharge-type equalization are involved [8]-[16]. In the charge-type equalization, any under charged cells can receive energy from the overall battery stack automatically [8] or selectively [9]. This scheme is suitable for the case where a few of batteries are frequently under charged. On the other hand, in the discharge-type equalization, extra energy from any over charged cells is regenerated into the whole battery string [10] or removed to the adjacent cells [11]. This is beneficial to the case that only a few of cells are over charged. In charge- and discharge-type equalization, in other words, bi-directional equalization, energy from any over charged cells flow into the other under charged cells through the current-fed type DC/DC converters [12]-[14] or the voltage-fed type DC/DC converters [15], [16]. It is more profitable to the phenomenon, in which some batteries are under charged while others are over charged.

All of these non-dissipative cell balancing schemes show good equalization performance by employing high efficiency DC/DC converters. However, designing a cell balancing circuit using the above schemes for a HEV battery stack,

978-1-4244-1871-8/07 $25.00 © 2007 IEEE

approximately above eighty cells with higher than 320V, will cause the severe problems. For example, high voltage stresses of electronic devices and high circulating energy can be experienced [10], [12]. In addition, establishing multiple windings within a single common core is a hard task [8], [9], [16]. Even if not so, cell balancing performance will degrade due to lots of energy conversions since over charged cells and under charged cells are sometimes far away from each other [11] and [13]-[15].

To overcome these problems, the modularized charge equalization converter is proposed in this paper. In the proposed circuit, the intra-module and the inter-module equalizer are constructed for low voltage stresses of all the electronic devices. Both the intra-module and the inter-module equalizer commonly use the MOSFET switch so that small size and low cost can be obtained. Besides, to achieve the further size reduction, the current-fed type DC/DC converter with cell selection switches is employed for intra-module cell balance, and the voltage-fed type DC/DC converter with no additional reset circuitry for inter-module equalization. The optimal power rating selection guide is also applied to achieve high power density.

This paper is organized as follows. The modularized charge equalization converter for HEV lithium-ion battery string is proposed in Section II, where the intra-module and the inter-module equalization scheme are carefully described. Then, the optimal power rating design rule is considered for intra-module cell balancing. Section III shows the experimental results of a prototype of the modularized equalization scheme by employing the optimal power rating design rule. Finally, concluding remarks are summarized in Section IV.

II. MODULARIZED CHARGE EQUALIZATION CONVERTER

A. Circuit descriptions

Fig. 1 shows the proposed modularized charge equalization converter for a HEV battery pack, in which the inter-module and the intra-module equalizer are employed. For easy explanation, the proposed equalization scheme will be applied to the lithium-ion battery string of 4 cells. As shown in Fig. 1, the battery string is modularized into two modules and each module consists of two cells. In the intra-module equalizer, each cell has the flyback DC/DC converter with a cell selection switch at the primary side. These two flyback converters are in parallel coupled while commonly utilizing one MOSFET switch. On the other hand, the inter-module equalizer employs the voltage-fed type DC/DC converter, where bi-directional equalizing current can be observed. Especially in this equalizer, there is no additional reset circuitry for the magnetizing current. That is because it can be reset by the under charged cells of which the corresponding cell selection switches are turned on.

The desirable features of the proposed charge equalization

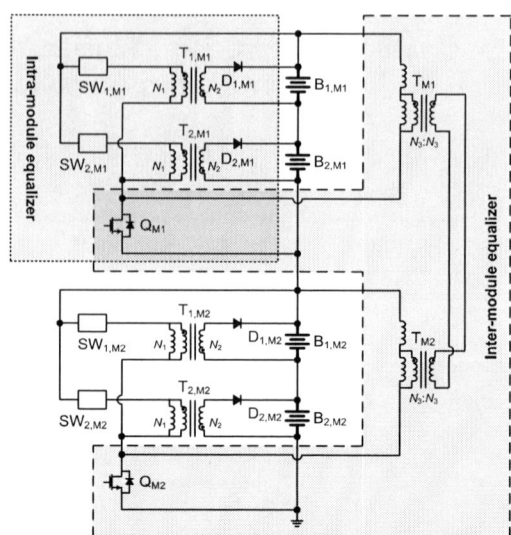

Fig. 1. Proposed modularized charge equalization converter.

converter are as follows. First of all, by modularizing the battery stack, low voltage stresses of all the electronic components can be obtained. For example, the larger number of modules and the smaller number of cells in the module, the lower voltage stress. Second, the intra-module and the inter-module equalizer share the MOSFET switch to achieve low cost and size reduction. In addition, the intra-module equalizer employs the cell selection switch in place of MOSFET switch and drive circuit. And inter-module equalizer has no additional reset circuit. With these effects, the proposed circuit can be implemented in the more compact size. Third, high power density equalizer can be constructed by using the optimal power rating design rule which is proposed in [17]. Lastly, efficient equalization strategy and short balancing time can be accomplished owing to the cell selection switches; in other words, the intra-module equalizer can charge the several under charged cells at the same time by conducting the corresponding cell selection switches.

B. Operational principles

The operational principles of the modularized cell balancing circuit are composed of the superposition of the intra-module and inter-module equalizer. The operational modes of the proposed equalization circuit can be separated into four parts. Before describing mode 1, it is assumed that the upper module, M_1, is the more under charged than the lower module, M_2, and the first cell in M_1 is the most under charged in the string. The cell selection switch, $SW_{1,M1}$, is turned on before the MOSFET switches, Q_{M1} and Q_{M2}, is turned on.

• **Mode 1**(t_0-t_1): In this mode, Q_{M1} and Q_{M2} are turned on at the same time. As shown in Fig. 2 and Fig. 3, the inter-module equalizing current transfers from M_2 to M_1 in the voltage-fed type. The magnetizing current builds up at the magnetizing inductor of the first flyback DC/DC converter. The primary currents of the inter-module equalizer, $I_{kg,M1}$ and

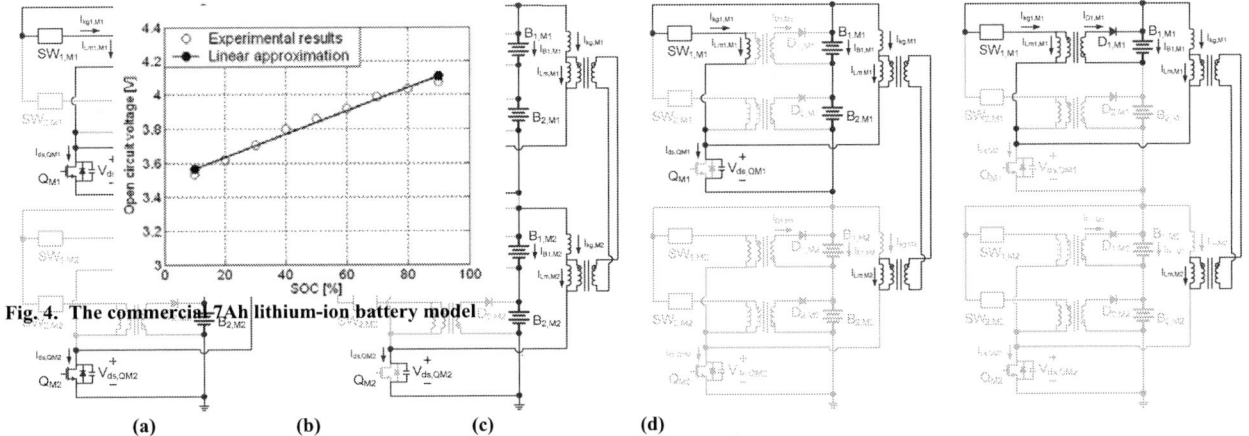

Fig. 4. The commercial 7Ah lithium-ion battery model

Fig. 2. Operational modes of the proposed equalization converter. (a) Mode 1. (b) Mode 2. (c) Mode 3. (4) Mode 4.

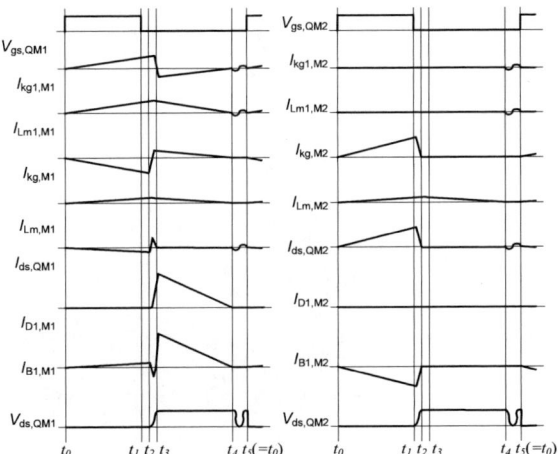

Fig. 3. Key waveforms of the proposed equalization converter.

$I_{kg,M2}$, and the primary current of the intra-module equalizer, $I_{kg1,M1}$, can be given by:

$$I_{kg,M1}(t) = -\left(\frac{V_{M2}-V_{M1}}{L_{kg,M2}+L_{kg,M1}}\right)t \qquad (1)$$

$$I_{kg,M2}(t) = \left(\frac{V_{M2}-V_{M1}}{L_{kg,M2}+L_{kg,M1}}\right)t + \frac{L_{kg,M2}V_{M1}+L_{kg,M1}V_{M1}}{L_{kg,M2}+L_{kg,M1}}\cdot\left(\frac{1}{L_{m,M2}}+\frac{1}{L_{m,M1}}\right)t \qquad (2)$$

$$I_{kg1,M1}(t) = \left(\frac{V_{M1}}{L_{m1,M1}}\right)t, \qquad (3)$$

where two leakage inductors, $L_{kg,M1}$ and $L_{kg,M2}$, represent all the parasitic components within the transformers, T_{M1} and T_{M2}, respectively. The values of $L_{kg,M1}$ and $L_{kg,M2}$ are much smaller than those of the magnetizing inductors, $L_{m,M1}$ and $L_{m,M2}$. V_{M1} and V_{M2} are the voltages of M_1 and M_2.

• **Mode 2**(t_1-t_2): In this mode, Q_{M1} and Q_{M2} are synchronously turned off. The output capacitor of Q_{M2} is fully charged.

• **Mode 3**(t_2-t_3): Mode 3 begins when the output capacitor of Q_{M1} begins to be charged. As shown in Fig. 2 and Fig. 3,

Q_{M1} is completely turned off later than Q_{M1}. This is because only the leakage inductor of the first flyback DC/DC converter contributes to the turn-off process of Q_{M1}. On the other hand, in case of Q_{M2}, there exists much energy which is come from two leakage inductors and two magnetizing inductors of the inter-module transformers.

• **Mode 4**(t_3-t_4): In this mode, the rectifier diode, $D_{1,M1}$, is turned on. The magnetizing current, $I_{Lm1,M1}$, flows into the first cell which is previously assumed to be the most under charged. This is the intra-module equalization process. In addition, the magnetizing currents of the inter-module transformer are reset through the first cell voltage, which are the additional equalizing currents.

C. Design of the optimal power rating

In this subsection, to obtain high power density of the proposed circuit, the optimal power rating design rule will be applied. As presented in [17], the optimal power rating selection guide can provide the minimal size of cell balancing circuit while achieving equalization within cell balancing time. In this paper, this power rating design rule will be applied only to the intra-module equalizer, not the inter-module equalizer since the inter-module equalization can be achieved automatically by using the voltage-fed type DC/DC converter. Before applying the optimal power rating selection guide, it is noted that the proposed cell balancing circuit will be designed for 40 cells which are modularized into 5 modules. In addition, we assume that although nearly all the cells in the string match well with each other, only a few cells are occasionally under charged.

Fig. 4 shows the battery model used for designing the optimal power rating of the proposed equalization circuit. The open circuit voltage of the commercial 7Ah lithium-ion battery is described according to state of charge (SOC). The dot symbol shows the experimental results from 10% to 90% in SOC and the solid line is the linear approximation of the experimental observations in a least square sense. Among them, the recommended operation regions, approximately from SOC=30% to SOC=70%, are mainly considered.

978-1-4244-1871-8/07 $25.00 © 2007 IEEE

Fig. 5 shows the simulation results of the optimal power rating design for the proposed modularized charge equalization circuit, where equalization time is plotted according to the input current into the intra-module equalizer, and net regenerated current which flows into the first cell. For the more information on this procedure, refer to the paper [17]. The various efficiencies of the intra-module equalization circuit are considered as simulation parameters. Initial SOC difference between the under charged cell and the other cells is assumed to be 10% or 20%.

The simulation results show that the larger input current into the intra-module equalizer takes the shorter equalization time with the higher equalizing current. In addition, the larger SOC difference, the longer equalization time, and the higher efficiency of a balancing circuit, the shorter equalization time. For example, when applying the balancing circuit of η=80%, equalization time of about 2 hours is required to achieve charge equalization provided that input current of 0.12A flows into the intra-module equalizer for the SOC difference of 20% as shown in Fig. 5(c). In that case, the net equalizing current into the first cell is approximately 600mA.

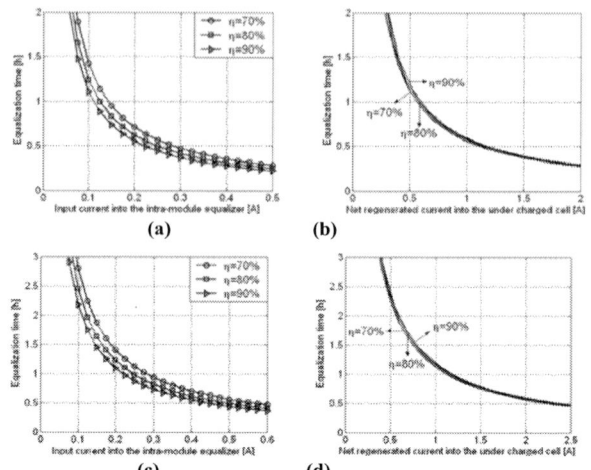

Fig. 5. Simulation results of the optimal power rating design. (a) Equalization time vs. input current for SOC difference of 10%. (b) Equalization time vs. regenerated current for SOC difference of 10%. (c) Equalization time vs. input current for SOC difference of 20%. (d) Equalization time vs. regenerated current for SOC difference of 20%.

III. EXPERIMENAL RERULTS

To verify the operational principles of the proposed cell balancing circuit and the usefulness of the optimal power rating design rule, a prototype is implemented. Its photograph is shown in Fig. 6. The intra-module equalizer is employed for charge balance among 8 cells, and the inter-module equalizer for module balance among 5 modules. The intra-module equalizer is designed using the optimal power raring selection guide. In this prototype, the commercial 7Ah lithium-ion batteries of 40 cells are used for the HEV battery system. The parameters used for constructing the prototype are summarized in Table I.

Fig. 7 shows experimental key waveforms of the implemented prototype, where only two modules are driven to carefully verify the operational principles of the intra-module and inter-module equalizer. The battery environment for this experiment is as follows. The voltage of the first module is 29.6V with SOC=28.9% and the second module is 30.9V with SOC=49.4%. The first cell within the first module is the most under charged so that the cell selection switch, $SW_{1,M1}$, is turned on in advance. As shown in Fig. 7(a), the magnetizing current of the intra-module equalizer flows into the first cell through the rectifier diode, $D_{1,M1}$, during the turn-off period of a MOSFET switch. In addition, the magnetizing currents of the inter-module equalizer are reset by transferring into the first cell. The inter-module charge equalization can be also observed as shown in Fig. 7(b). The equalizing current flows from the second module into the first during the turn-on period of the MOSFET switches. The maximum voltage stress of the proposed equalization circuit will reside at the MOSFET switches and its value does not exceed 80V including the

Fig. 6. Photograph of the implemented prototype.

voltage spikes. From these results, the proposed balancing circuit has the advantageous features such as the low voltage stress due to modularization and efficient equalization process during the entire equalization time.

To show the cell balancing performance of the modularized charge equalization circuit, equalization test is conducted for the commercial 7Ah lithium-ion battery cells. In this test, 5 modules are realized for 40 cells and each module has 8 cells. There are the most under charged two cells; the SOC of one cell is 24.8% and the SOC of the other is 33.5%. In this equalization test, an intelligent battery management system (BMS) is employed and its control strategy is as follows. First, the BMS detects the most under charged cell, and then charges it until its SOC increase to the average SOC of the other cells. Second, after finishing the first step, the BMS will charge the next most under charged cell in the same way. Lastly, these two steps are repeated until all the voltages of the cells are within 50mV.

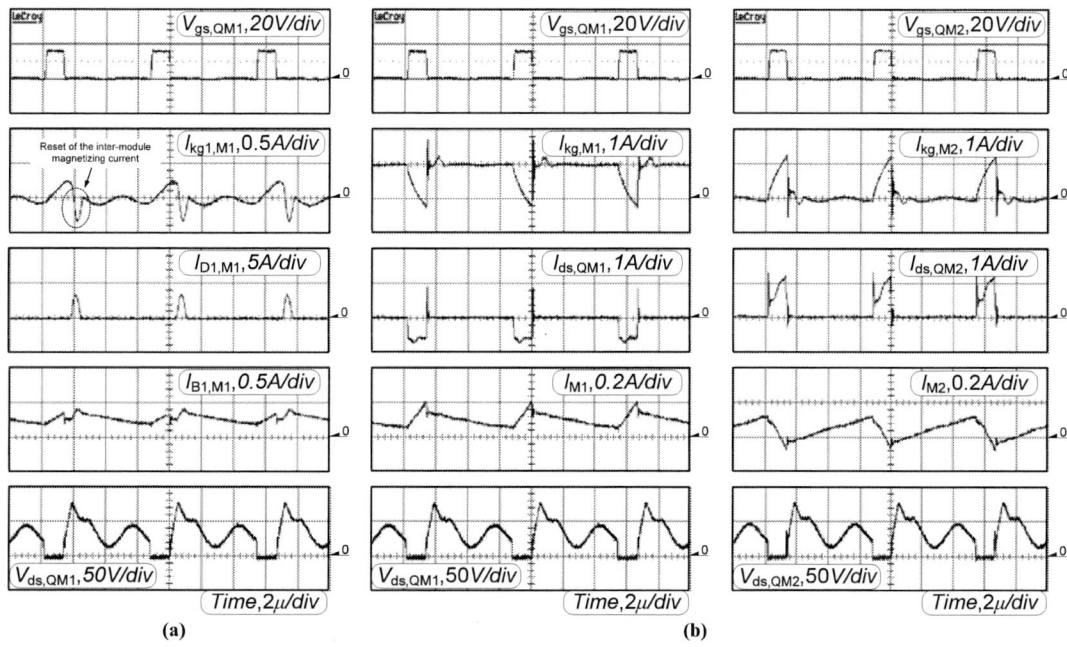

Fig. 7. Experimental key waveforms of the prototype for the first two modules.

Fig. 8 shows the equalization performance of the proposed modularized balancing circuit; the equalization time of the first 2 hours is used for charging the 39th cell, the next 47 minutes for the 10th cell, and then the last 18 minutes for the other most under charged cell. As shown in Fig. 8, the initial

SOC difference of 21.9% decreases to 5.4% at the end of equalization. It is remarkable that in the case of the 39th cell, the average equalizing current is measured to be 600mA. This value is much similar to the simulation result shown in Fig. 5(d). From this result, the proposed charge equalization circuit, implemented by using the optimal power rating design rule, shows the outstanding charge balancing performance within a high power density.

Table I. Parameters used for the implemented prototype.

Parameters			Value
Intra-module equalizer	Selection switch, $SW_{1,M1}$- $SW_{8,M5}$		PS710EL-A
	Diode, $D_{1,M1}$- $D_{8,M5}$		STPS2L30A
	MOSFET, Q_{M1}-Q_{M5}		IRF7495
	Floating Gate Driver		HCPL-3140
	Transformer, $T_{1,M1}$- $T_{8,M5}$	Core	CM102173
		$N_1:N_2$	39:5
		L_m, L_{kg}	127µH, 2.8µH
Inter-module equalizer	Transformer, T_{M1}- T_{M5}	Core	RN41812
		$N_3:N_3$	14:14
		L_m, L_{kg}	500µH, 200nH
Lithium-ion battery string	Capacity		7Ah
	SOC of B_{39} at $t=0$		24.8%
	SOC of B_{10} at $t=0$		33.5%
	SOC of the other cells at $t=0$	Max.	46.7%
		Min.	24.8%
		Mean	40.0%

IV. CONCLUSIONS

When designing a cell balancing circuit for the high voltage HEV lithium-ion battery system, voltage stress of electronic components, circuit size, production cost, and implementation problem are the major design parameters. To satisfy these requirements, the modularized charge equalization converter is proposed in this paper. In the proposed circuit, low voltage stresses of all the electronic devices are achieve by modularizing the entire cells. And small size and low production cost can be obtained by commonly using the MOSFET switches of the intra-module and the inter-module equalizer. For the more size reduction, cell selection switches at the intra-module equalizer and no additional reset circuit at the inter-module equalizer are employed. By applying the optimal power rating design rule, the proposed balancing circuit can be constructed in high power density. Finally, all the magnetic components can be easily implemented, and simple equalization strategy intended for only the intra-module equalization can be used. From these good points, the modularized charge equalization

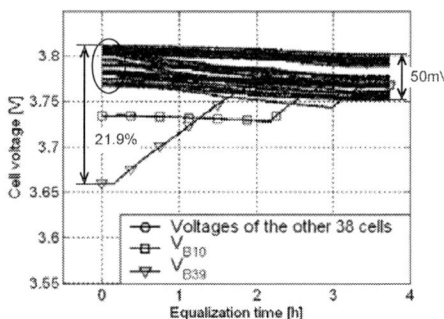

Fig. 8. Equalization performance of the proposed cell balancing circuit.

converter proposed in this paper can be widely applied to the high voltage series connected lithium-ion battery stack such as HEV and EV.

ACKNOWLEDGMENT

This work was supported through research fund by SK Corporation, Republic of Korea.

REFERENCES

[1] A. Emadi, S. Williamson, and A. Khaligh, "Power electronics intensive solutions for advanced electric, hybrid electric, and fuel cell vehicular power systems," IEEE Trans. Power Electron., vol. 21. pp. 567-577, May 2006.

[2] S. Williamson, A. Emadi, and K. Rajashekara, "Comprehensive efficiency modeling of electric traction motor drives for hybrid electric vehicle propulsion applications," IEEE Trans. Veh. Tech., vol. 56. pp. 1561-1572, July 2007.

[3] J. M. Miller, "Hybrid electric vehicle propulsion system architecture of the e-CVT type," IEEE Trans. Power Electron., vol. 21. pp. 756-767, May 2006.

[4] A. Affanni, A. Bellini, G. Franceschini, P. Guglielmi, and C. Tassoni, "Battery choice and management for new-generation electric vehicles," IEEE Trans. Ind. Electron., vol. 52, pp. 1343-1349, Oct. 2005.

[5] B. T. Kuhn, G. E. Pitel, and P. T. Krein, "Electrical properties and equalization of lithium-ion cells in automotive applications," in Proc. 2005 IEEE Vehicle Power and Propulsion Conf., Chicago, USA, Sep. 2005, pp. 55-59.

[6] N. H. Kutkut and D. M. Divan, "Dynamic equalization techniques for series battery stacks," in Proc. 18th Annu. Int. Telecommunications Energy Conf., Boston, USA, Oct. 1996, pp. 514-521.

[7] B. Lindemark, "Individual cell voltage equalizers (ICE) for reliable battery performance," in Proc. 13th Annu. Int. Telecommunications Energy Conf., Kyoto, Japan, Nov. 1991, pp. 196-201.

[8] N. H. Kutkut, H. L. N. Wiegman, D. M. Divan and D. W. Novotny, "Design considerations for charge equalization of an electric vehicle battery system," IEEE Trans. Ind. Appl., vol. 35, pp. 28-35, Feb. 1999.

[9] M. Tang and T. Stuart, "Selective buck-boost equalizer for series battery packs," IEEE Trans. Aerosp. Electron. Syst., vol. 36, pp. 201-211, Jan. 2000.

[10] D. C. Hopkins, C. R. Mosling, and S. T. Hung, "Dynamic equalization during charging of serial energy storage elements," IEEE Trans. Ind. Appl., vol. 29, pp. 363-368, Mar.-Apr. 1993.

[11] N. H. Kutkut, H. Wiegman, and R. Marion, "Modular battery charge equalizers and method of control," U.S. Patent 6 150 795, Nov. 21, 2000.

[12] H. Schmidt and C. Siedle, "The charge equalizer-a new system to extend battery lifetime in photovoltaic system, U.P.S. and electric vehicle," in Proc. 15th Annu. Int. Telecommunications Energy Conf., Paris, France, Sep. 1993, pp. 144-151.

[13] Y. -S. Lee and G. -T. Cheng, "Quasi-resonant zero-current-switching bidirectional converter for battery equalization applications," IEEE Trans. Power Electron., vol. 21, pp. 1213-1224, Sep. 2006.

[14] N. H. Kutkut, "A modular non dissipative current diverter for EV battery charge equalization," in Proc. 13th Annu. Appl. Power Electron. Conf. and Exp., Anaheim, USA, Feb. 1998, pp. 686-690.

[15] P. T. Krein, S. West, and C. Papenfuss, "Equalization requirements for series VRLA battery," in Proc. 16th Annu. Battery Conf. on Applications and Advances, Long Beach, USA, Jan. 2001, pp. 125-130.

[16] N. H. Kutkut, "Non-dissipative current diverter using a centralized multi-winding transformer," in Proc. 28th Power Electron. Specialists Conf., St. Louis, USA, June1997, pp. 648-654.

[17] H. -S. Park, C. -E. Kim, G. -W. Moon, J. -H. Lee, and J. K. Oh, "Two-stage cell balancing scheme for hybrid electric vehicle Lithium-ion battery strings," in Proc. 38th Power Electron. Specialists Conf., Orlando, USA, June 2007.

Design and Control Algorithm Research of Active Regenerative Bidirectional DC/DC Converter used in Electric Railway

Chan-Heung Park[1], Su-Jin Jang[1], Byoung-Kuk Lee[1], Chung-Yuen Won[1] and Han-Min Lee[2]

[1] Department of Information and Communication Engineering
Sungkyunkwan University, Su-won, Korea
[2] Korea Railroad Research Institute, Uiwang, Korea
Email: colorbook@skku.edu

Abstract—The regeneration power which is generated braking period of railway vehicles is increasing voltage of DC-line, and excessive voltage of DC-line is generating several problems. So, it's very important issue to DC traction system about dealing with regenerate power. Regeneration energy storage system is one of the methods to use the regeneration energy efficiently and to improve the problems. In this paper, we proposed the design and efficient control algorithm of the bidirectional DC/DC converter used in electric railway for utilization of regeneration energy. And as realization of prototype model, we verified the efficiency of bidirectional DC/DC converter.

I. INTRODUCTION

The regeneration energy caused increasing the DC-line voltage because of flow the regeneration energy into the DC-line. And excessive voltage in DC-line is caused many problems to overall system. So, DC-line voltage must be controlled as maintaining regulated value. It is possible to use bidirectional DC/DC converter. The current which flow into the DC-line is the high value current and changing frequently. So, super capacitor is used in system to charged regeneration power. Super-capacitor can store the current which any amount of that. But super-capacitor has the weak point like high cost and rated voltage is so low. So, we proposed the converter algorithm which could operate two modes. First mode is buck converter mode, and second mode is boost converter mode. By the buck-boost operation, bidirectional converter could control the charge and discharge operation. Regeneration energy storage system is able to increase the reliability as controlling the bidirectional DC/DC converter. In this paper, we are proposing a new control algorithm of converter and proved the efficiency of our algorithm through PSIM simulation, and experiment with prototype model.[1-3]

II. THE OPERATION OF BIDIRECTIONAL DC/DC CONVERTER OF REGENERATIVE ENERGY STORAGE SYSTEM

Regenerative energy which is increasing the DC-line voltage controlled by bidirectional DC/DC converter through charge/discharge operation of super-capacitor. Energy flow by generating regenerative energy can be seen in Fig. 1.

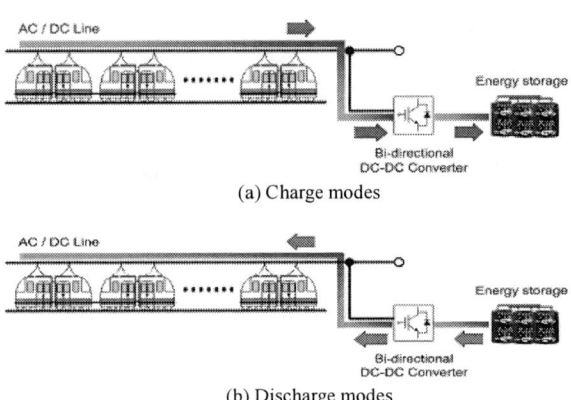

(a) Charge modes

(b) Discharge modes

Fig. 1 Schematic power flow diagram of regenerative energy.

Bidirectional DC/DC converter controlled power flow of regenerative energy by variation of DC line voltage. This paper used non-isolation half-bridge type converter to realize the bidirectional DC/DC converter because half-bridge converter has a low inductance value and low capacitance of passive elements and low rated voltage of active elements appropriated in energy storage system in DC traction application. Comparison of converter topologies shows in appendix of last paper. Bidirectional DC/DC converter proposed in regenerative energy storage system using super-capacitor can be seen in Fig. 2.

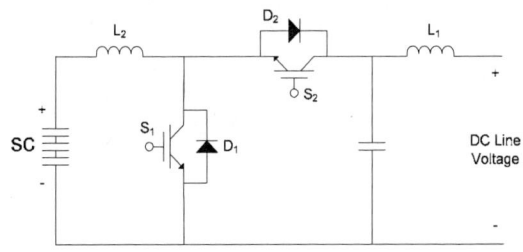

Fig. 2 Non-isolation half-bridge bidirectional DC/DC converter.

Half-bridge DC/DC converter operated as buck converter when charge mode and operated as boost converter when discharge mode.[4-8] Half-bridge DC/DC converter prototype design parameters can be seen in Table 1.

978-1-4244-1871-8/07 $25.00 © 2007 IEEE

TABLE 1
DESIGN PARAMETER OF BIDIRECTIONAL DC-DC CONVERTER.

Means	Value
Maximum Discharge Power	4 [kW]
Voltage range of Super-capacitor	50~100[V]
Maximum current of Super-capacitor	50 [A]
Input voltage of Converter (DC Line voltage)	311 [V]
Input current of Converter	10 [A]
Switching Frequency	10 [kHz]

A. Charge mode

Due to the increasing of DC-line voltage bidirectional DC/DC converter operating as buck converter and charging a increasing value in super-capacitor. Switch 2 is operating as active switch element and Switch 1 is operating diode. During the charging mode Switch 2 is operating on/off switching operation frequently as duty ratio DT_S in a T_S cycle. Charge mode operation of bidirectional DC/DC converter can be seen in Fig. 3[4-8]. And output of bidirectional DC/DC converter in charge mode can be seen in Fig. 4 and according to the Fig. 4, balance condition of $Vol \cdot \sec$ shows in equation (1) and duty ratio of bidirectional DC/DC converter when charge mode can be expressed in equation (2).

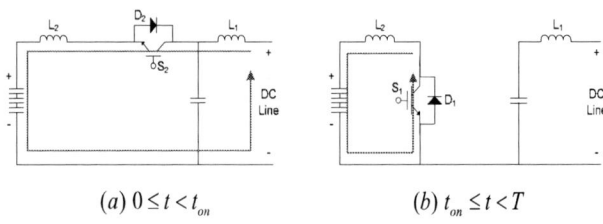

(a) $0 \leq t < t_{on}$ (b) $t_{on} \leq t < T$

Fig. 3 Charge mode operation of bidirectional DC-DC converter.

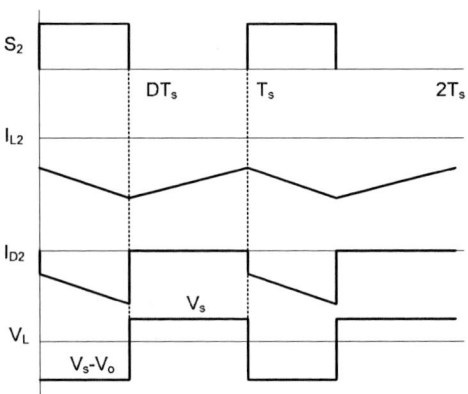

Fig. 4 Output waveform of charge mode.

$$(V_O - V_S)DT_S = V_S(1 - D)T_S \qquad (1)$$

$$D = \frac{V_S}{V_O} \qquad (2)$$

B. Discharge mode

When the DC-line voltage decreased or increasing a retrograde vehicles operation, regenerative energy charged in super-capacitor is supplied into the DC-line through a bidirectional DC/DC converter. This mode was designated as Discharge mode. In this mode, bidirectional DC/DC converter operating as boost converter and Switch 1 is operated as active switch and Switch 2 is operated as diode contrary to the charge mode. During the discharging mode Switch 1 is operating on/off switching operation frequently as duty ratio DT_S in a T_S cycle. Fig. 5 and Fig. 6 can be seen such as discharge mode operation of bidirectional DC/DC converter and output of discharge mode operation.[4-8] Similar to the charge mode discharge mode equation can express according to the Fig. 6, balance condition of $Vol \cdot \sec$ as follows. And duty ratio of bidirectional DC/DC converter when discharge mode can be expressed in equation (3) and (4).

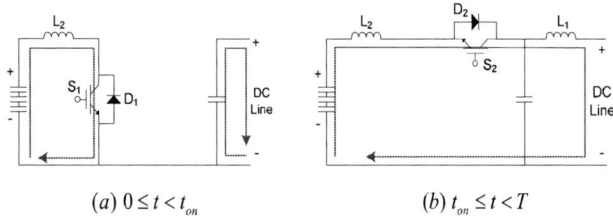

(a) $0 \leq t < t_{on}$ (b) $t_{on} \leq t < T$

Fig. 5 Discharge mode operation of bidirectional DC-DC converter.

$$V_S \cdot DT_S = (V_O - V_S) \cdot (1 - D)T_S \qquad (3)$$

$$D = \frac{V_O - V_S}{V_O} \qquad (4)$$

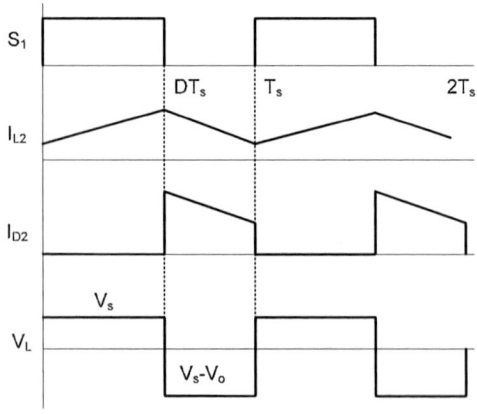

Fig. 6 Output waveform of discharge mode.

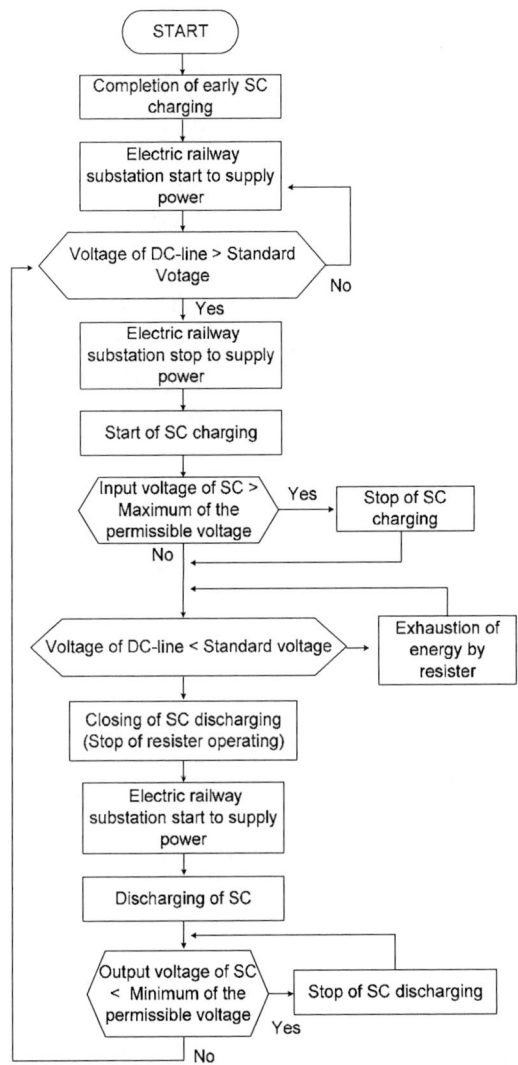

Fig. 7 Control algorithm of regenerative energy storage system.

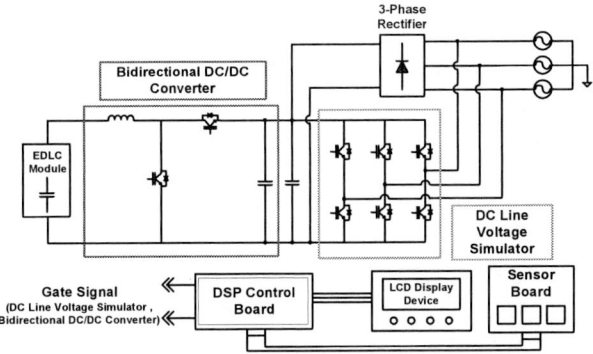

Fig. 8 The prototype model of regenerative energy storage system.

III. SIMULATION RESULTS

The simulation composition of the prototype model and operational standard value Fig. 9 and table 2 as follows.

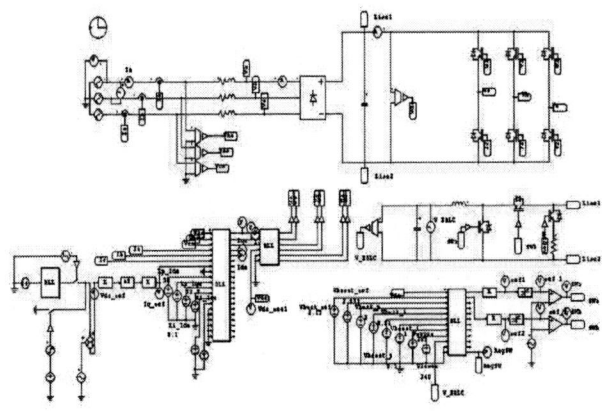

Fig. 9 Scheme of regenerative energy storage system by PSIM.

TABLE 2
CONDITION OF SIMULATION

Condition	Value
Input Voltage	220 [V]
Charge	Over the 370[V]
Discharge	Down the 350[V]
Resistor Operation	DC Line : Over the 370[V]
	EDLC : Over the 100[V]

This paper proposes a control algorithm of bidirectional DC/DC converter which according to the characteristics of super-capacitor. Fig. 7 shows the diagram of control block to control the regenerative energy storage system. In this algorithm, super-capacitor needs initial charging time because the system demanding a large voltage in early stage. And charging operation to the super-capacitor is starting when all vehicles were generating the regenerative energy and when the retrograde vehicles were nonexistence at the moment because that cases caused DC-line voltage rising. If the regenerative energy is generating continuously over the capacity of super-capacitor, regenerative energy which over the capacity of super-capacitor sent into the resistor and dissipated as heat energy.

The prototype model of regenerative energy storage system is shown in Fig.8.

Buck mode operations of the bidirectional DC/DC converter that simulated DC-line voltage will rise above 370V it charges in EDLC. And when the DC-line voltage falls below 350V, EDLC energy discharging, it is risen DC-line voltage 350V by bidirectional DC/DC converter boost mode operation. When the voltage of EDLC is charged to 100V completely and the DC-line voltage is over the 370V, charging operation into the EDLC is impossible anymore. Therefore regenerative energy is consumed by the resistor in this case.

978-1-4244-1871-8/07 $25.00 © 2007 IEEE

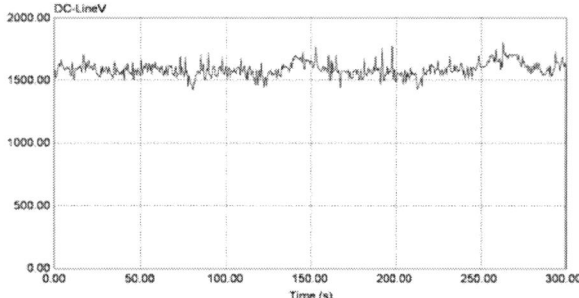

Fig. 10 Variation of DC-line voltage according to the station of Sillim.

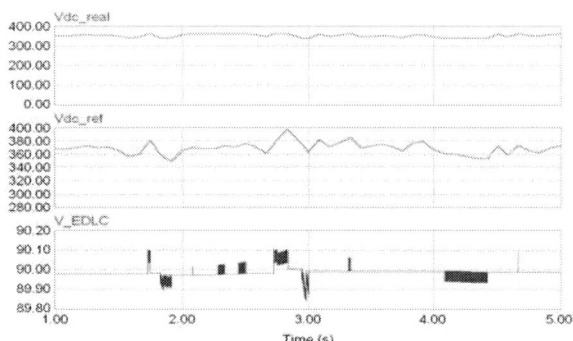

Fig. 11 The graph of simulation result.

Fig. 10 is DC-line voltage data of the Sillim station that was measured. Simulation is simulated by using measured DC-line voltage. Fig. 11 is the wave of simulation result. The first graph is waveform of the DC-line voltage which has become stable. The second waveform is the DC-line voltage that bidirectional DC/DC converter is not applied. The third graph is voltage of EDLC which is charged and discharged.

IV. EXPERIMENT SETUP AND RESULTS

Fig. 12 is the actual composition set of the prototype model. Fig. 13 through 16 shows charging and discharging current and voltage waveform by bidirectional DC/DC converter. The DC-line voltage charged into the EDLC dropped at 100V by the bidirectional DC/DC converter, and it can be seen in Fig. 13. Fig. 14 showed the charging current of EDLC and input current of bidirectional DC/DC converter.

In discharging mode, EDLC voltage is operating with 311V which boosted from 100V by switching operation of boost mode, and it can be seen in Fig. 15.

Fig. 17 is shown the output voltage of the DC-line voltage simulator and Fig. 18 shows the result waveform that stabilized by bidirectional DC/DC converter. As you can see in Fig. 17-18, when the DC-line voltage increasing over 370V, bidirectional DC/DC converter operating as buck converter to charge the voltage into the EDLC, and when the DC-line voltage decreasing down 350V, bidirectional DC/DC converter operating as boost converter to discharge the voltage of EDLC into the DC-line. And through those operations, DC-line voltage keeps up the voltage of 350V.

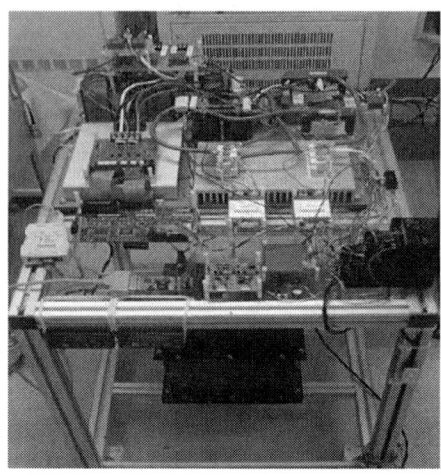

Fig. 12 Proposed prototype model.

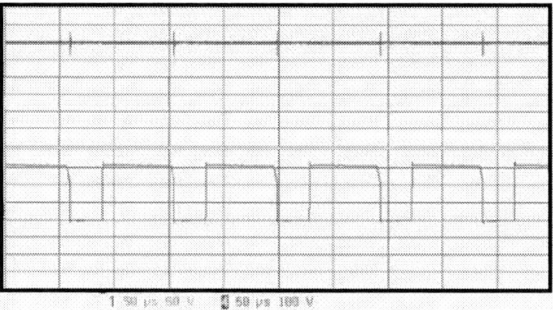

Fig. 13. Charging mode (Buck mode).
Top : Supercapacitor charging voltage, Bottom : Switch voltage.
(50V/div, 100V/div, 50us/div)

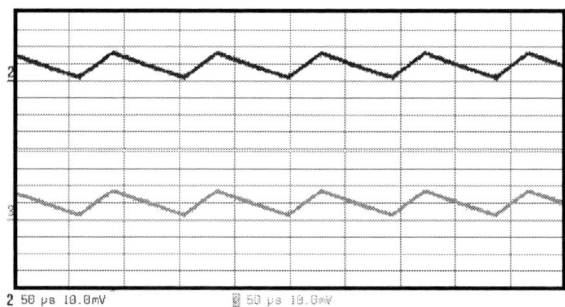

Fig. 14 Charging mode (Buck mode).
Top : Supercapacitor charging current, Bottom : Input current.
(4A/div, 50us/div)

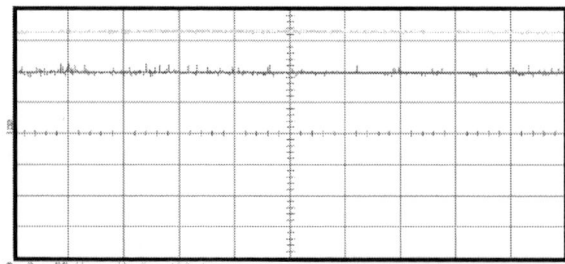

Fig. 15 Discharging mode (Boost mode).
Top : Output voltage , Bottom : Supercapacitor discharging voltage.
(Top: 100V/div, Bottom: 50V/div, 2s/div)

978-1-4244-1871-8/07 $25.00 © 2007 IEEE

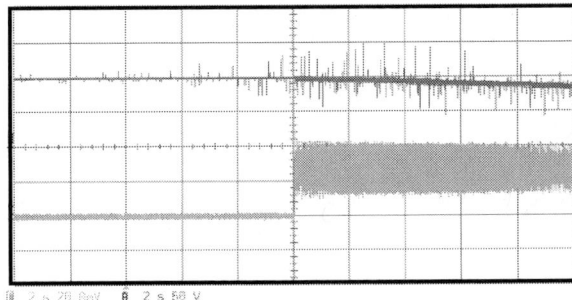

Fig. 16 Discharging mode (Boost mode).
Top : Supercapacitor discharging voltage, Bottom : discharging current.
(4A/div, 50us/div)

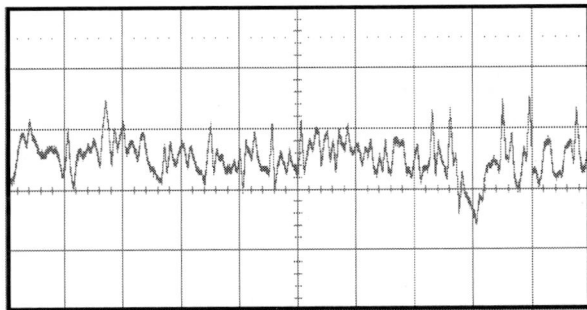

Fig. 17 Output voltage of the DC-line voltage simulator.

(30V/div, 10s/div)

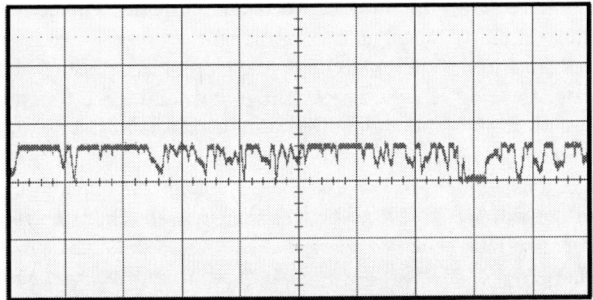

Fig. 18 The DC-line voltage that the bidirectional DC/DC converter is applied. (30V/div, 10s/div)

V. CONCLUSION

There is the method which using storage system for efficiently usage of regenerative energy that generated by DC traction.

In this paper, we propose the half bridge type bidirectional DC/DC converter. And efficient control algorithm and design method of bidirectional DC/DC converter which applied for regenerative energy storage system used in DC traction system are proposed also, for improvement of the efficiency of whole system.

Through the simulation and experimental results, it can be confirmed that regenerative energy use safely, and DC-line is stable. Consequently, it is expected that proposed converter topology and control algorithm will improve the efficiently

usage of regenerative energy in real regenerative energy storage system.

APPENDIX

TABLE A1
VOLTAGE AND CURRENT COMPARISON AMONG THE BIDIRECTIONAL DC-DC CONVERTER TYPE

	Half-bridge	Cuk	Sepic
Switch V_{peak}	V_o	$V_{in}+V_o$	$V_{in}+V_o$
Diode V_{peak}	V_o	$V_{in}+V_o$	$V_{in}+V_o$
Capacit V_C or	$\cdot$	$V_{in}+V_o$	V_{in}
Switch I_{peak}	$H.B < Cuk = Sepic$	$H.B < Cuk = Sepic$	$H.B < Cuk = Sepic$

TABLE A2
ACTIVE ELEMENTS COPACITY COMPARISON AMONG THE BIDIRECTIONAL DC-DC CONVERTER TYPE

	Half-bridge	Cuk	Sepic
L_1	Small	Large	Large
L_2	$\cdot$	$Cuk = Sepic$	$Cuk = Sepic$
C_1	Large	Small	Large
C_2	$\cdot$	$Cuk = Sepic$	$Cuk = Sepic$

REFERENCES

[1] MinChul Kim, "A study of utilizing regenerative energy in DC traction system by using energy storage equipment" A master's thesis for a degree in Sungkyunkwan University, Suwon, Korea, 2006.

[2] JinSang Jo, SangMin Jung,, JinHee Lee, SeWan Choi, and SooBin Han,"A Control Method of Bidirectional DC-DC Converter for Fuel Utilization and Durability Improvement in Fuel Cell Vehicles" The transactions of the Koream institute of power electronics,pp. 428-435, Korea, October, 2005.

[3] OhJung Kwon, ChangKwon Park, ByeongSoo Oh, "A Study of Increasing Regeneration Energy and Braking Using Super Capacitor(EDLC)" The transactions of the Korean Society of Automotive Engineers, Vol. 14, No.6, pp 24-33 2006

[4] R. M. Schupbach and J. C. Balda, "35kW Ultracapacitor Unit for Power Management of Hybrid Electric Vehicles:Bi-directional DC-DC Converter Design", IEEE PESC Conf. Rec., pp. 2157-2163, 2004.

[5] M. Cacciato, F. Caricchi, F. G. Capponi and E. Santini, "A Critical Evaluation and Design of Bidirectional DC/DC Converters for Super-Capacitors Interfacing in Fuel Cell Application", IEEE IA Conf. Rec., Vol. 2, pp. 1127-1133, 2004.

[6] R. L. Steigerwald, "A Comparison of Half-Bridge Resonant Converter Topologies", IEEE Trans. on Power Electronics, Vol. 3, No. 2, pp. 174-182, 1988.

[7] F. A. Himmelstoss, "Analysis and Comparison of half-bridge Bidirectional DC-DC Converters", IEEE PESC Conf. Rec., Vol. 2, pp. 922-928, 1994.

[8] J. Zhang, R. Y. Kim and J. S. Lai, "High-Power Density Design of a Soft-Switching High-Power Bidirectional DC-DC Converter", IEEE PESC Conf. Rec., pp. 1-7, 2006.

Charge Equalization Converter with Parallel Primary Winding for Series Connected Lithium-Ion Battery strings in HEV

Chol-Ho Kim[†], Hong-Sun Park[†], Chong-Eun Kim[†]
and Gun-Woo Moon[†]
Dept. of EECS, Korea Advanced Institute of Science
and Technology
[†]Email: gwmoon@ee.kaist.ac.kr

Joong-Hui Lee[‡] and Jeon Keun Oh[‡]
Corporate R&D Center, SK Corporation
[‡]Email: hui@skcorp.com

Abstract—In this paper, a charge equalization converter with parallel-connected primary windings of transformers is proposed. The proposed work effectively balances the voltage among Lithium-Ion battery cells despite each battery cell has low voltage gap compared with its SOC. The principle of the proposed work is that the equalizing energy from all battery strings moves to the lowest voltage battery through the isolated dc/dc converter controlled by the corresponding solid state relay switch. In this paper, a prototype of four Lithium-Ion battery cells is optimally designed and implemented, and experimental results show that the proposed method has excellent cell balancing performance.

Index Terms—Hybrid electric vehicle (HEV), charge equalization, lithium-ion battery

I. INTRODUCTION

Recently, HEV has become a proper vehicle for energy saving and an eco-friendly vehicle, since HEV has employed with the battery as an alternative power source [1]. In HEV, the energy from wheel, which have been wasted in the past, converts the electrical energy to charging the battery and then it is reused from battery to propel the vehicle at low speeds or boost extra power required in high acceleration. In addition, there is no emission of CO_2 during utilizing the battery as a power source. Therefore, in HEV, employing the battery power source is one of the major issues and the battery having long life time is economically beneficial in automotive market.

These days, various batteries, which are agreeable to the above requirements, have been discussed to employ the battery as alternate power source in HEV. Among them, nickel-metal hybrid (Ni-MH) batteries have the greater part of HEVs on the streets [1]. Recent development of lithium-ion battery and its test results show that the lithium-ion battery has higher power, higher energy density, lower self-discharge rate, and higher single cell voltage than the Ni-MH battery such that it has the potential of taking the place of Ni-MH battery in the HEVs of the future [2]. However, to realize this possibility, reliability and safety is first of all ensured; in other words, the lithium-ion battery should be maintained within the ranges of allowed voltage and current limits to prevent permanent deterioration of characteristics and, in the worst case, explosion or fire in vehicle [1], [2].

Furthermore, in HEV employed with series-connected battery strings to achieve a high voltage for driving electric motors, to improve battery lifetime in battery strings, an effective charge equalization method to avoid the voltage imbalance among battery strings is essential. This is because the voltage imbalance will decrease the total storage capacity and total life cycle of the battery [3].

Therefore, to avoid the possible risks and enhance the battery lifetime, the charge equalization method for Lithium-Ion battery strings is significantly necessary in HEV.

Charge equalization methods for series-connected battery strings have been presented in [3]-[28]. One of them is an energy dissipative method such as a resistive current shunt [4], [5]. The resistive current shunt, which consists of a resistor and an active switch, is in parallel connected across the each battery cell as shown in Fig. 1. This dissipative method is attractive to balance charge among battery cells because of simple and easy implementation. However, energy dissipation and long equalization time are critical drawbacks of this circuit compared with the other cell balancing circuits. Furthermore, this method is not suitable to balance the under-charged battery cells due to the energy dissipative type. For this reason, such a cell balancing scheme is not strongly recommended for charge equalization of the high power system such as a HEV.

To obtain more effective charge balancing, non-dissipative methods, which is lossless charge equalization method, are presented in [6]-[25]. In general, non-dissipative charge equalization schemes are divided into three categories: charge-type, discharge-type, and charge- and discharge-type. Charge-type equalization scheme is intended for the environment of a few of under-charged cells among the whole batteries [6]-[9] and discharge-type is for a few of over charged cells [10]-[12]. Bi-directional charge equalization scheme has both characteristics; charge- and discharge-type [13]-[25]. Among these equalization schemes, in case under-charged cells are occurred in battery strings frequently, the charge-type scheme is attractive in HEV.

Charge-type cell equalization converter can be classified into two styles as shown in Fig. 2, where the control scheme for the cell balancing circuit is mainly taken into account. In the centralized control based equalization shown in Fig. 4(a) [6], [7], every cell in the string is in parallel coupled with the corresponding secondary winding and all the secondary windings are realized at the single common core as well as the primary winding. In this centralized control based equalization scheme, only the single power switch at the primary side is controlled in pulse width modulation (PWM) mode to achieve charge equalization automatically. This centralized control

978-1-4244-1871-8/07 $25.00 © 2007 IEEE

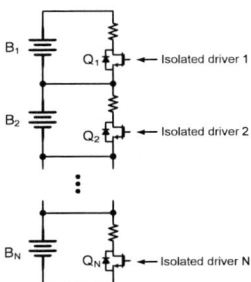

Fig. 1. Resistive current shunt.

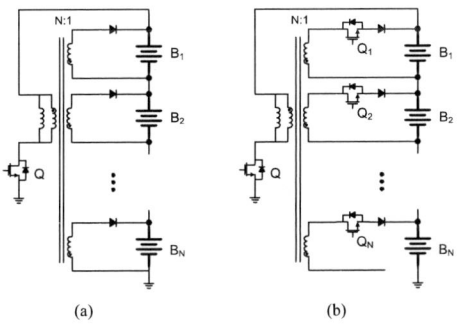

Fig. 2. Charge-type cell equalization converter. (a) Centralized control based. (b) Cell control based.

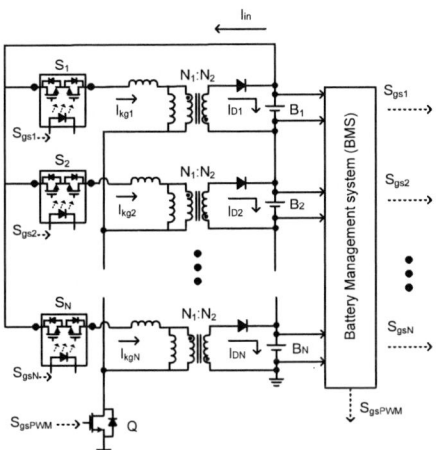

Fig. 3. Proposed charge equalization converter with parallel primary winding

based cell balancing circuit has a merit of the simple structure of a controller, but it can not show good equalization performance for a large number of cells since the unbalancing parasitic components appear at the secondary sides of the single common transformer. In detail, equalization performance is limited due to the parasitic leakage inductances among the mismatched transformers. Moreover, from the implementation point of view, it is not easy to establish a large number of secondary windings, equal to the number of cells, in a single common core.

To improve cell balancing performance of the centralized control based equalization scheme, cell control based equalization scheme is discussed as shown in Fig. 2(b) [8], [9]. Compared with the centralized control based scheme, the cell control based scheme employs the external switches in all the secondary sides of the transformer to efficiently control the equalizing current; that is, these external switches can govern cell balancing current to selectively charge the specific under-charged cells. The prime advantage of this control scheme is to obtain of high quality cell balancing performance especially for the lithium-ion battery. However, like the centralized control based scheme, it has still implementation problem with a large number of cells due to a single common core. In addition, since the switch current rating in secondary side of transformer is higher than the switch in primary side, the solid state relay, which can be controlled easily, is limited to employ. Hence, cell control based scheme has complexity to control these isolated switches, which are placed at all the secondary sides.

To directly apply to Lithium-Ion battery equalization of

HEV, unlike the previous balancing scheme, the charge equalization method has requirements as excellent cell balancing, simple controlling and easy implementation. To meet these requirements, this paper proposes a charge equalization converter with parallel-connected primary windings of transformers. The principle of the proposed work is that the equalizing energy from all battery strings moves to the lowest voltage battery through the isolated dc/dc converter controlled by the corresponding bi-directional switch.

In proposed circuit, by using the solid state relay switch, which can control easily, the efficiency of charge equalization is enhanced even at the low voltage gap between the over-charged and the under-charged cells in Lithium-Ion battery strings and total equalization time is reduced. Furthermore, the proposed circuit can be easily implemented with own magnetic core individually.

In this paper, a prototype of four Lithium-Ion battery cells employing the proposed method is optimally designed and implemented, and experimental results show that the proposed method has excellent cell balancing performance.

II. PROPOSED CHARGE EQUALIZATION CONVERTER

A. *Circuit descriptions*

Fig. 3. shows the proposed charge equalization circuit with parallel primay winding for the HEV lithium-ion battery cells, in which solid state relays can select the individual battery cell for balancing by Battery Management System (BMS).

The proposed circuit is effective against the case that some battery cells are frequently under-charged. In the series connected battery pack, a under-charged battery cell could be occurred during charging the battery pack. This is because the charging current from battery charger could be stoped desipite of remaining under-chrged battery cells due to some battery cell reached over-charged condition.

This scheme is to balance the battery stack that the equalizing energy from overall battery stack moves to the lowest voltage battery. To realize this cell balancing scheme, each cell has its

978-1-4244-1871-8/07 $25.00 © 2007 IEEE   795

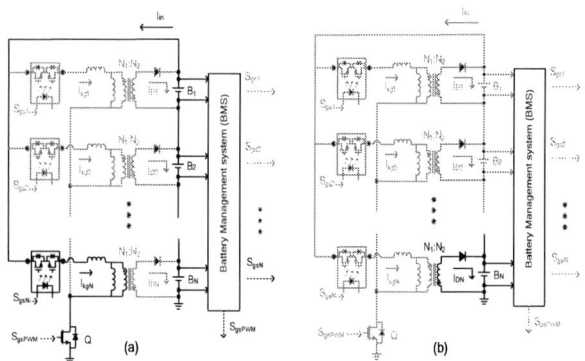

Fig. 4. Operational modes of the proposed circuit

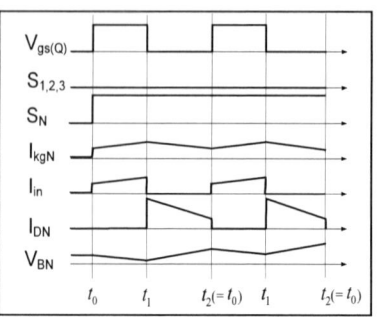

Fig. 5. Key waveforms of the proposed circuit.

own flyback DC/DC converter. The flyback DC/DC converter is parallel-connected to each battery in output stage and input stage of them are coupled together to be connected across the voltage of strings, where the solid state relays, $S_1 \sim S_{N-1}$, are placed in series with the primary sides as shown in Fig.3. To show the operation of the proposed circuit, it is assumed that a battery management system (BMS) always monitors SOC of every battery in the strings. A MOSFET switch, Q, is controlled by the BMS with fixed duty, and $S_1 \sim S_N$ are operated by ON or OFF signal with respect to the state of charge (SOC).

B. *Operational principles*

The operational principles of the proposed charge equalization converter with parallel primary winding circuit are very simple and easy. Basically, the circuit operations are similar to flyback DC/DC converters, where flyback DC/DC converters operate individually at the selected battery cell by the solid state relay switch, Sn. In detail, the flyback DC/DC converter is driven with a fixed duty ratio by the centralized control of a BMS, and also the relay switch control signal is taken by BMS, which checks the lowest voltage cell and selects the corresponding relay switch.

C. *Modal Analysis*

The proposed equalization circuit operates in two modes according to the switching states of the primary single MOSFET and the secondary diodes. The operational modes and the key waveforms are presented in Fig. 4 and Fig. 5. Before starting mode 1, it is assumed that only the last cell, B_N, is under-charged.

• Mode 1(t_0-t_1): mode 1 begins when Q is turned on, where $S_1 \sim S_{N-1}$ are turned off and S_N is turned on. In this mode, the primary current of the Nth cell, I_{kgN}, builds up, as shown in Fig. 4 (a).

• Mode 2(t_1-t_2): mode 2 starts when Q is turned off, as shown in Fig. 4 (b). In mode 2, the corresponding rectifier diode, D_N, is turned on so that the magnetizing current flows into the B_N.

It is similar to individual flyback converter operation. With this process of a conventional flyback converter, equalizing currents from the overall cells can be transferred to the lowest voltage cell effectively.

III. THE OPTIMAL POWER RATING DESIGN SCHEME

This section presents the optimal power rating design rule for the charge equalization converter considering equalization time for a given SOC distribution of imbalance [26].

Before presenting the power rating design scheme, the estimation of SOC in the lithium-ion battery willing to be taken into account with a relationship of battery voltage. Through the estimation of SOC of battery, BMS is considered as the charging and discharging level of battery cell. However, the estimation of SOC can be obtained using a lot of variables in a very complex algorithm [29]. In this paper, SOC is represented with only open circuit voltage of battery cell for simplicity.

In Fig. 6, the dot symbol represents the experimental results and the solid line is the linear approximation of those. In this figure single cell voltage is plotted according to SOC of a commercial 7Ah battery. From 30% to 70% of SOC is the recommended operational range of safety and this paper shows the experimental result in the safety range.

• $Q_n(t)$: charge quantity of the nth cell at time t,
• $V_n(t)$: voltage of the nth cell at time t,
• I_{in}: constant input current from the overall cells to the converter,
• I_{out}: constant output current from the converter to the overall cells,
• $P_{in}(t)$, $P_{out}(t)$: input power, output power of an equalization circuit at time t, respectively,
• $P_{in,avg}$, $P_{out,avg}$: average input power, average output power of an equalization circuit, respectively, and
• η: overall efficiency of an equalization circuit.

To obtain the optimal power rating of the cell balancing circuit while achieving cell balancing within equalization time under the given SOC distribution of one under charged cell, the following simultaneous equations should be satisfied:

$$\frac{1}{N-1}\sum_{n=2}^{n=N} Q_n(t) = Q_N(t), \qquad (1)$$

$$P_{out,avg} = \eta P_{in,avg}, \qquad (2)$$

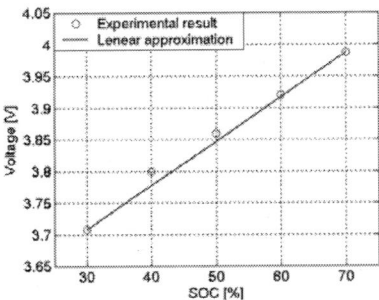

Fig. 6. Cell voltage according to SOC of a commercial lithium-ion 7Ah battery over the ranges from 30% to 70% of SOC.

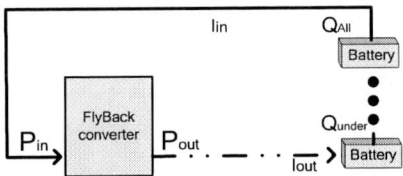

Fig. 7. Simple block diagram of proposed scheme.

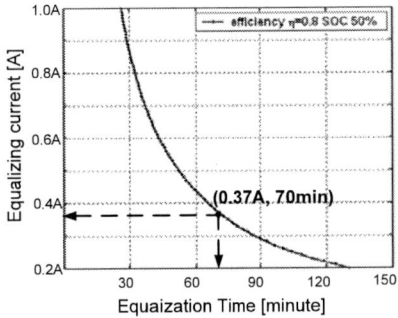

Fig. 8. The equalizing current according to the equalization time of the DC/DC converter.

Where it is noted that by (1) and (2), the left side of equation (1) is the average charge quantity of overall battery at equalization time t and the right side is the charge quantity of selected battery cell due to the lowest battery voltage or SOC in the battery string. In equation (2), it shows that the average power flow to the selected battery cell is equal to the average power of overall battery cells converted power by cell balancing circuit; efficiency of the cell balancing circuit is η. Through the block diagram of proposed scheme as shown in the Fig. 7, this equation can explain more easily. This block diagram indicates the charge transferring from all battery cells to selected battery cell. This charge transferring is occurred by flyback DC/DC converter and the input power of flyback DC/DC converter come from all battery cells. The output power of flyback DC/DC converter flows into the selected cell for charging the battery cell and that is the equalizing current for balancing the lowest voltage or SOC cell.

Furthermore, amount of average charge remained in the under-charged cell, $Q_N(t)$ and average output power of DC/DC converter that also indicate the input power from output power of DC/DC converter to under-charged cell, $P_{out,avg}$ is given by, respectively

$$Q_N(t) = Q_N(0) + (I_{out} - I_{in}) \cdot t, \qquad (3)$$

$$\begin{aligned}P_{out,avg} &= \frac{1}{t}\int_0^t P_{out}(\tau)d\tau = \frac{1}{t}\int_0^t V_N(\tau)(I_{out} - I_{in})d\tau \\ &= \frac{1}{t}\int_0^t\left(V_N(0) + \frac{(I_{out} - I_{in})\cdot\tau}{C}\right)(I_{out} - I_{in})d\tau \\ &= \left(V_N(0) + \frac{1}{2C}(I_{out} - I_{in})\cdot t\right)(I_{out} - I_{in}), \quad (4)\end{aligned}$$

where C is capacitance of the 7Ah lithium-ion battery used in this paper and the negative sign means that the current flow from cell to DC/DC converter. Therefore, the output current of DC/DC converter, I_{out}, is the charging current for under-charged cell and the difference between input and output current of DC/DC converter, $I_{out} - I_{in}$, is the equalization current that actually utilize for balancing the cell voltage at under-charged cell. Especially in the design of power rating of a cell balancing circuit, average power is taken into account since the cell voltages, even a little bit, can change during the equalization process.

Fig. 8 shows simulation result of the equalizing current versus equalization time. This result comes from solving above equations. As expected, the shorter equalization time will be taken for the higher equalizing current and also higher power rating of balancing circuit.

To obtain the optimal power rating of the cell balancing circuit, first of all, the equalization time should be decided. Then by the result in Fig. 8, the corresponding equalizing current can be obtained. Hence, the power rating of DC/DC converter is calculated by equalizing current and voltage of battery cell. For instance, if the equalization time is 70min, at that time the equalization current is about 0.37A. When the input current from overall battery to DC/DC converter, I_{in} is 0.13A, the output current, I_{out}, is about 0.5A. since the voltage of battery cell is about 4V and the efficiency is assumed about 80% as shown in Fig. 7, the power rating of DC/DC converter can be obtained about 2.5W.

IV. EXPERIMENTAL RESULTS

To verify the operational principles of the proposed cell balancing circuit and the usefulness of the proposed power rating design rule of optimum, a prototype is implemented and its circuit diagram is shown in Fig. 9. In this prototype, the commercial 7Ah lithium-ion batteries of four cells are used for the HEV battery system and the flyback DC/DC converters, constructed by utilizing the proposed power rating design rule.

In the detail, for this prototype system 4W flyback converter is utilized at unbalanced SOC gap 20%, where the voltage of over-charged cells is 3.92V at SOC 60% and the voltage of

978-1-4244-1871-8/07 $25.00 © 2007 IEEE

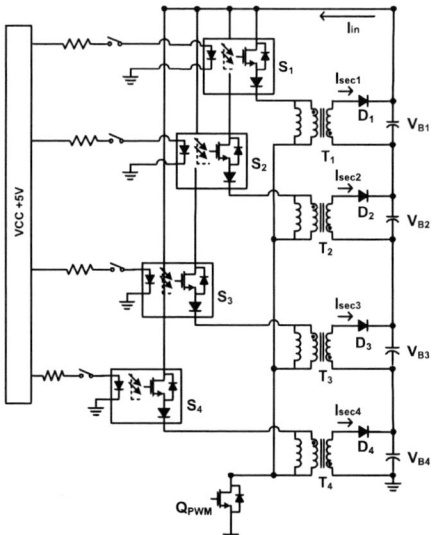

Fig. 9. Proposed charge equalization converter with parallel primary winding

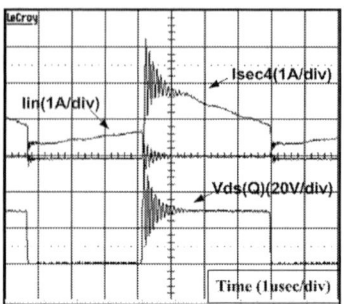

Fig. 10. Experimental waveforms of the prototype at the fourth cell.

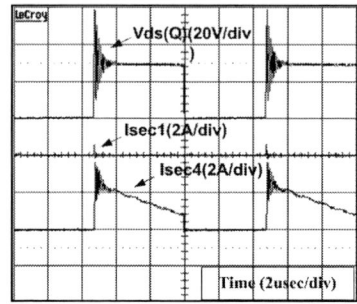

Fig. 11. Comparative experimental waveforms of the prototype between the first cell and fourth cell.

single under-charged cell is 3.80V at SOC 40%.

The parameters used for realization of the prototype are described in TABLE 1.

In this experiment, the initial SOC difference between the under charged cell and the others is approximately 20%, where the fourth cell has the lowest SOC among them. In this circuit, instead of using BMS signal for select the lowest voltage battery cell, VCC 5V of power supply is used as selection signal of the mechanical relay switch; this signal can operate the corresponding solid state relay switch at the lowest voltage battery cell. If Solid state relay is connected by the mechanical relay switch at the lowest voltage battery cell, the corresponding flyback DC/DC converter for release of the equalizing energy is driven in PWM mode with fixed duty ratio. This equalizing energy comes from overall battery cell.

Fig. 10 and Fig. 11 show the experimental waveforms of the implemented prototype. In Fig. 9, it is assumed that the fourth cell of the battery string is under-charged and the only selection switch, S4, is turned on. At this time, when the MOSFET switch is operated with on time duty, the input current, I_{in},

TABLE 1
PARAMETERS FOR THE PROPOSED CHARGE EQUALIZATION CIRCUIT

Parameters			Value
Balancing circuit	Solid state relay, S_1-S_4		PS710B-1A
	MOSFET switch		IRF7452
	Diode, D_1-D_4		SS24
	Transformer	Core	EP41313
		N_1:N_2	22:7
		L_m	130μH,
		L_{kg}	750nH
Lithium-ion battery	Capacity		7Ah
	SOC of B_1 at t=0		60%
	SOC of B_2 at t=0		60%
	SOC of B_3 at t=0		60%
	SOC of B_4 at t=0		40%

builds up at fourth battery cell. After the MOSFET switch is turned off, the input current is transferred to the secondary side of transformer as the output current, I_{out} or the secondary current, $I_{sec\,4}$, at the fourth battery cell in Fig. 10. Fig.11 show the difference results of two current waveform at the secondary diode, $I_{sec\,1}$ and $I_{sec\,4}$. At the first battery cell, there is no current in the secondary diode. This is because the selection switch, S1, is not operated at the first battery cell.

To show the cell balancing performance of the proposed charge equalization circuit, the equalization test cooperated with the lithium-ion battery cells is conducted and its result is plotted in Fig. 9. In this test, to achieve charge equalization within 70 minutes under the SOC difference of 20% between the under- charged cell and the other cells, the average input current, I_{in}, of 0.15A is designed to flow into the DC/DC converter of the fourth cell. With this charged current, I_{out} of 0.5A, equalizing current of 0.35A which is the measured value in this experimental environment is recovered into the overall battery pack. In addition, the measured efficiency of the DC/DC converter is 81%. As shown in Fig. 12, the prototype built by the proposed optimal power rating design rule shows the equalization performance to be similar that of the simulator as an expected result in Fig. 7. Furthermore, charging voltage of under-charged cell linearly increases despite of voltage gap between an over charged cell and an under-charged cell. This is because the charging current has uniform value and this characteristic can achieve the effective balancing of the battery string at a low voltage gap between the over-charged cells and the under-charged cells.

978-1-4244-1871-8/07 $25.00 © 2007 IEEE

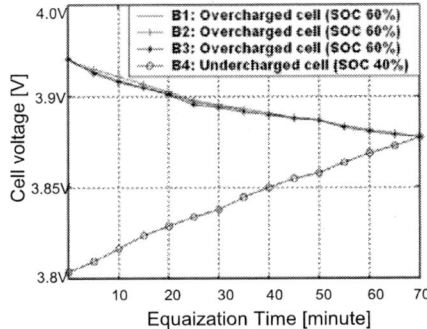

Fig. 12. Charge equalization performance of the proposed cell balancing circuit.

From the above experimental results, the proposed charge equalization circuit and the optimal power rating design scheme proposed in this paper shows the outstanding cell balancing performance with simple operation and control.

V. CONCLUSIONS

In this paper, charge equalization converter with parallel primary winding for HEV lithium-ion battery cells is proposed with the optimal power rating design and implementation of the prototype.

This propose work can achieve the excellent charge balancing for an under-charged cell, in which much more current can be flow to the under-charged batteries by controlling the solid state relay switch. Moreover, it can be directly applied to Lithium-Ion battery equalization of HEV, since it is simple to control the equalizing current at the low voltage gap between the over-charged and the under-charged cells and the implementation of one common core with the multiple secondary winding is easy for the series-connected battery cells about fifty and over.

Furthermore, in this paper, a prototype employing the optimal power rating design method is implemented for verifying the usefulness of the proposed power rating design rule of optimum and excellent performance of proposed cell balancing circuit.

The proposed cell balancing circuit is expected to be widely applicable to the series connected the lithium-ion battery string of HEV or EV.

ACKNOWLEDGMENT

This work was supported through research fund by SK Corporation, Republic of Korea.

REFERENCES

[1] D. S.- Repila and J. E. W. Poxon, "Hybrid electric vehicles: current concepts and future market trends," *J. Barcelona IEEE Student Branch*, vo. 13, pp. 5-30, Mar. 2006.

[2] N. Takeda, S. Imai, Y. Horii, and H. Yoshida, "Development of high-performance lithium-ion Batteries for hybrid electric vehicles," *Mitsubishi Motors Technical Review*, pp. 68-72, Jan. 2003.

[3] S. West and P. T. Krein, "Equalization of value-regulated lead-acid batteries: issues and life test results," in *Proc. IEEE INTELEC '00*, Phoenix, USA, pp. 439-449, Sep. 2000.

[4] J. Rouillard, C. Comte, R. A. Hagen, O. B. Knudson, A. Morin, and G. Ross, "Equalizer system and method for series connected energy storing devices," U.S. Patent 5 952 815, Sep. 14, 1999

[5] Y. Fukuyama, "Series connection circuit for secondary battery," U.S. Patent 5 602 481, Feb. 11, 1997

[6] N. H. Kutkut, H. L. N. Wiegman, D. M. Divan and D. W. Novotny, "Design considerations for charge equalization of an electric vehicle battery system," *IEEE Trans. Ind. Appl.*, vol. 35, pp. 28-35, Feb. 1999.

[7] D. M. Divan, N. H. Kutkut, D. W. Novotny, H. L. Wiegman, "Battery charging using a transformer with a single primary winding and plural secondary windings," U.S. Patent 5 659 237, Aug. 19, 1997

[8] M. Tang and T. Stuart, "Selective buck-boost equalizer for series battery packs," *IEEE Trans. Aerosp. Electron. Syst.*, vol. 36, pp. 201-211, Jan. 2000.

[9] Y. Xi and P. K. Jain, "A forward converter topology with independently and precisely regulated multiple outputs," *IEEE Trans. Power Electron.*, vol. 18, pp. 648-658, March. 2003.

[10] C. S. Moo, Y. C. Hsieh, and I. S. Tsai, "Charge equalization for series-connected batteries," *IEEE Trans. Aerosp. Electron. Syst.*, vol. 39, pp. 704-710, Apr. 2003.

[11] T. -C. Chen and Z. -J. Guey, "Charge equalizer for series of connected battery strings," U.S. Patent 6 008 623, Dec. 28, 1999.

[12] N. H. Kutkut, H. Wiegman, and R. Marion, "Modular battery charge equalizers and method of control," U.S. Patent 6 150 795, Nov. 21, 2000.

[13] S. Anzawa and K. Kobayashi, "Voltage equalizer for battery elements," U.S. Patent 2005/0140336 A1, June 30, 2005.

[14] Y. -S. Lee and M. -W. Cheng, "Intelligent control battery equalization for series connected lithium-ion battery strings," *IEEE Trans. Ind. Electron.*, vol 52, pp. 1297-1307, Oct. 2005.

[15] K. Nishijima, H. Sakamoto and K. Harada, "A PWM controlled simple and high performance battery balancing system," in *Proc. IEEE PESC '00*, Galway, Ireland, June 2000, pp. 517-520.

[16] N. H. Kutkut, "A modular non dissipative current diverter for EV battery charge equalization," in *Proc. IEEE APEC '98*, Anaheim, USA, Feb. 1998, pp. 686-690.

[17] C. Pascual and P. T. Krein, "Switched capacitor system for automatic series battery equalization," in *Proc. IEEE APEC '97*, Atlanta, USA, Feb. 1997, pp.848-854.

[18] J. Kimball and B. Kuhn, "Battery equalization to extend useful operating life of rechargeable battery packs", Battery power products & technology, vol. 8, pp. sep. 2004.

[19] J. P. Keller, "Systems and methods for battery charging and equalization," U.S. Patent 6 771 045 B1, Aug. 3, 2004.

[20] D. L. Miller, "Battery charge balancing system having parallel switched energy storage elements," U.S. Patent 6 064 178, May 16, 2000.

[21] N. H. Kutkut, "Non-dissipative current diverter using a centralized multi-winding transformer," in *Proc. PESC '97*, Missouri, USA, June 1997, pp. 648-654

[22] Y. C. Hsieh, C. S. Moo, and W. Y. Ou-Yang, "A bi-directional charge equalization circuit for series-connected batteries," in *Proc. PEDS '05*, Kuala Lumpur, Malaysia, Nov.-Dec. 2005, pp. 1578-1583.

[23] C. -H. Lee and P. -C. Lin, "Equalizer for series of connected battery strings," U.S. Patent 2005/0140335 A1, June. 30, 2005.

[24] D. R. Racholok, D. T. Fouchard, and W. B. Ebner, "Charge equalization of series connected cells or batteries," U.S. Patent 5 594 320, Jan. 14, 1997.

[25] S. Patel, C. D. Shelton, and L. S. Tyson, "Cell equalizing circuit," EP 1 388 921 A2, Feb. 11, 2004.

[26] H. –S. Park, C. –E. Kim, G. –W. Moon, J. –H. Lee, and J. K. Oh, "Two-stage cell balancing scheme for hybrid electric vehicle Lithium-ion battery strings," in *Proc. IEEE PESC '07*, Orlando, USA, June 2007.

[27] N. H. Kutkut and D. M. Divan, "Dynamic equalization techniques for series battery stacks," in *Proc. IEEE INTELEC '96*, Boston, USA, Oct. 1996, pp. 514-521.

[28] S. W. Moore and P. J. Schneider, "A review of cell equalization methods for lithium ion and lithium polymer battery systems," in *Proc. SAE World Congress*, Detroit, USA, Mar. 2001, Doc. No 2001-01-0959.

[29] J. Chiasson and B. Vairamohan, "Estimating the state of charge of a battery," *IEEE Trans. Contr. Syst. Technol.*, pp. 465-470, May. 2005.

Calculation of regenerative energy in DC 1500V electric railway substations

Chang-han Bae   Dong-uk Jang   Yong-gi Kim   Se-ky Chang   Jai-kyun Mok

Korea Railroad Research Institute
360-1, Woram-dong, Uiwang, 437-757, Korea
Tel : +82-31-460-5417, Fax: +82-31-460-5649
Email: chbae@krri.re.kr

Abstract - DC electric railway system with diode rectifier operates in the first quadrant of voltage-current plane and thus requires regenerative inverters which transfer the surplus regenerative power caused by regenerative braking of electric train sets into the grid. In order to install a regenerative inverter with optimum capacity, it is necessary to investigate the consumed and regenerative energy of the electric traction substations in advance.

In this paper, analysis results of regenerative energy in two substations operating in Seoul line 2 and Kwangju line 1 are presented. DC line voltage and feeder currents are measured for a day to calculate consumed and regenerative power for four feeders. We calculated an amount of regenerative energy consumed in other feeders and estimated the cost reduction in energy consumption due to the reuse of regenerative energy.

I. INTRODUCTION

Most of the DC 1500V electric railway substations have adopted diode rectifier method to supply stable DC power. However, to transmit excess regenerated power following the regenerative braking of railcars to the system, an additional regenerative inverter facility is needed[1~4]. In DC 1500V electric railway substations do not have a regenerative inverter or storage facility as yet. When a railcar performs regenerative braking, the regenerated braking energy is used in a powering on the same or another feeder or consumed as heat by the on-board resistor. If a regenerative inverter, which is a regenerated power absorbing device, is installed in inverse parallel with the diode rectifier, it can absorb surplus regenerated energy that was otherwise absorbed and consumed by the on-board resistor and transmit it to the high-voltage distribution end for reuse. Hence an energy saving effect can be expected. In addition, it can improve the electric braking of the railcar and ATO(Automatic Train Operation) performance by increasing the absorption rate of regenerated power by the catenary line, extend the replacement cycle of brake shoes by reducing the use of brake shoes, reduces fine dust caused by the abrasion of brake shoes in the underground space, and suppress rise in temperature[5,6]. In subways and light railway transits to be constructed recently, consideration on the installation of regenerative inverters has begun, and the measurement and analysis of the amount of regenerated energy that can be actually absorbed by railway substations.

This paper analyzes the catenary line voltage and the current for each feeder of Seolleung substation in Seoul and Yangdong substation in Kwangju measured for a day. The operational characteristics of diode rectifiers were checked by obtaining the voltage-current characteristic curves of these substations from measurement results, and reduction in the cost of electric energy was calculated by calculating the average value of consumed and regenerated power on each feeder and the amount of regenerated power transmitted to another feeder from each feeder. Besides, under the assumption that the distance from which regenerated power can be interchanged among substations is a half of the distance to the next substation, the total regenerated power from a substation was calculated with an approximate formula based on the regenerated power of a trainset of railcars. Using the values obtained as such and the measured regenerated power interchanged among feeders, regenerative energy consumed in the resistor bank on the railcars was calculated, and regenerated energy that can be recycled after the installation of a regenerative inverter was obtained approximately.

II. REGENERATIVE ENERGY

In a DC electric railway system, when a railcar brakes down, regenerated power is created while kinetic energy is converted to electric energy. Regenerated power increases voltage at the pantograph of the railcar, making it possible to interchange power with other power rail transits following the voltage differential. That is, regenerated power, created by a regenerative braking railcar on each feeder divided for up and down lines, may be consumed by exchanging among feeders, creating an effect of reducing operational power supplied by substations. If there exists no powering rail transit that can consume such regenerated power, it raises catenary line voltage, making the system unstable, and when it reach a certain limit, it is consumed as heat in the resistor of the railcar.

In addition to the above, it is possible to install a regenerative inverter in each substation so that the inverter can change DC power to AC to be used at the high-voltage distribution system of the substation, thereby stabilizing the catenary line voltage and conserving energy. Responses to excess regenerated power on the train and on the ground are as shown in Table 1.

978-1-4244-1871-8/07 $25.00 © 2007 IEEE

TABLE I
Responses to excess regenerated power

On-board	consumption	braking chopper resistor
Ground	consumption	braking chopper resistor
	conversion	regenerative inverter
	storage	battery, supercapacitor, flywheel

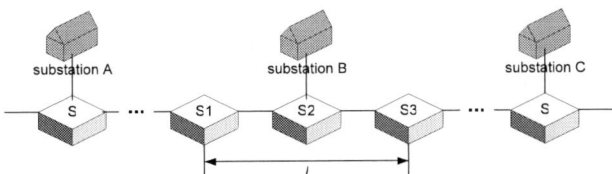

Fig. 1 Stations and substations of a DC electric railway

When electrical motors of trains perform regenerative braking between substation A and B, it is supplied to both the substations as reverse load. Assuming that the exchange of regenerative power among feeders, the number of trainsets operated at commercial speed v_s with headway of h is obtained as in (1) with regard to a section l.

$$N = \left\lceil \frac{l}{v_s \times h} \right\rceil \tag{1}$$

where $\lceil a \rceil$ means the integer larger than a, while distance l, headway h and commercial speed are expressed in the units of meters, minutes and km/h, respectively.

The total regenerated energy that takes place in a day in the section l can be approximated as in the following equation.

$$E_g = 2T_0 N(k-1)E_1 / h \tag{2}$$

where T_0 is the operation time for a day in the unit of minutes. k is the number of stations in the section l, and E_1 is the amount of regenerated energy created by a single trainset in a regenerative braking. By use of the daily inter-feeder exchange of regenerated energy measured at the DC panel of the substation and the daily total regenerated energy approximately obtained from (2), it is possible to estimate surplus regenerative energy lost in the substation. The regenerative energy created in the section l covered by a substation can be expressed as (3). It means that the regenerative energy in a substation is divided into the energy shared and utilized among feeders and the energy that raises catenary line voltage or is consumed as heat in the vehicle as shown in (3).

$$E_g = E_t + E_r \tag{3}$$

where E_g is the total regenerative energy that occurred in the section l, and E_t is the measurement of regenerative energy interchanged among feeders. E_r is the energy that raises the catenary line voltage or gets lost in the braking resistor, including energy loss caused by the internal resistance of the feeder lines. From (3), we can approximately calculate the amount of energy restorable by installing regenerative inverters. However, there is a problem that the calculated amount of restorable regenerative energy includes loss caused by the internal resistance of the feeder lines. Such loss in power caused by internal resistance not only becomes larger with the increase of power consumption but also increases in proportion to the regenerated power interchanged among feeders.

III. MEASUREMENT OF SUBSTATIONS

In Seolleung and Yangdong substations, voltage and current at the DC panel were measured and recorded for a day. As shown in Fig. 2, for measurement in each substation, the DCCT outputs of the protection relays installed on 4 feeders and the DCPT outputs of the power recorders of the DC panel were recorded on a notebook computer connected via DAQ cards. We used Labview program which is capable of dividing and recording data for every 2 hours .

A. Voltage-current characteristic curves

Fig. 6 and 7 show the voltage-current characteristic curves at the DC side of Yangdong and Seolleung substations. As load current increase, voltage drop by the internal impedance of the rectifier increases and DC voltage falls down from no-load voltage 1,650V. In the case of Seolleung substation, since the voltage-current characteristics of diode rectifiers operated in parallel do not coincide among themselves, the DC side voltage does not decrease with a constant slope but has a bending point. That is, the initial sharp decrease down to about 500A represents a non-uniform voltage drop in the rectifiers operated in parallel, and this problem disappears if

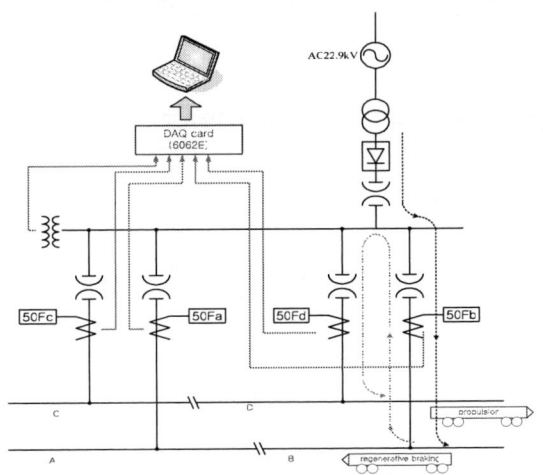

Fig. 2 Substation measurement schematic diagram.

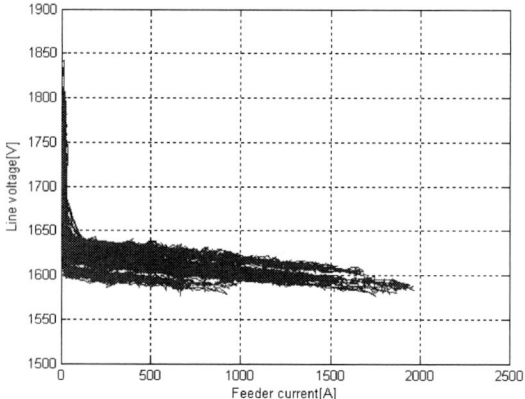

Fig. 6 Total current and catenary voltage characteristic curve in Yangdong Substation (Measured time : 06:47 ∽ 08:47)

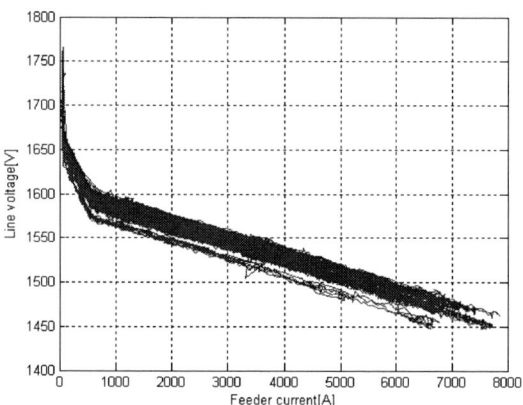

Fig. 7 Total current and catenary voltage characteristic curve in Seolleung Substation (Measured time : 06:50 ∽ 08:50)

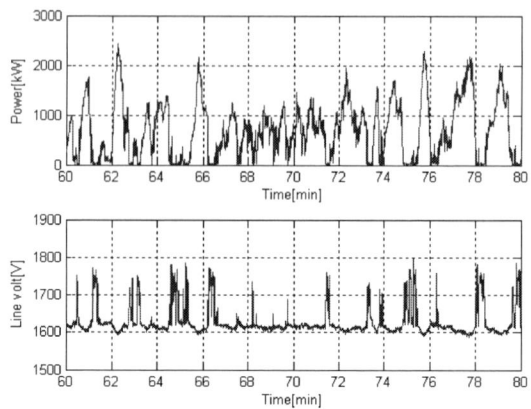

(a) Total consumed power and catenary voltage waveform

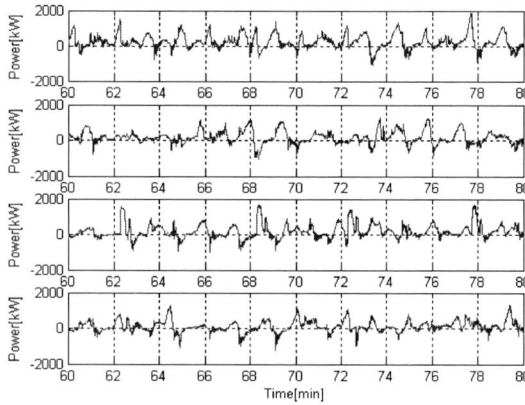

(b) Consumed power waveform of feeder 1 through 4
Fig. 8 Consumed power and catenary voltage waveform in Yangdong substation

the voltage-current characteristics of the rectifiers operated in parallel coincide among themselves.

On other hand, since Yangdong and Seolleung substation do not have a regenerative inverter installed, we confirmed that the total current of DC side can never have a negative value, and instead, catenary line voltage may increase over 1,800V. If a regenerative inverter is installed and the characteristics of the diode rectifiers operated in parallel are adjusted similarly, we would be able to obtain an ideal voltage-current curve of a DC substation.

B. Consumed and regenerated power

Fig. 8 and 9 show the total consumed power, catenary voltage, and consumed and regenerated power waveform of each feeder in Yangdong and Seolleung substation. Fig. 8 and 9(a) show total consumed power and catenary voltage waveform, in which, when no power is consumed, catenary voltage rises owing to the regenerated power by railcars in most cases. Also in the case of Yangdong substation, because railcars of the same type are operated with 8-minute

or 10-minute headway, we can clearly confirm the timing of rise in catenary voltage when regenerative braking. On the other hand, in the case of Seolleung Substation, because railcars of various types are operated with shorter headway, change in catenary voltage is irregular. Fig. 8 and 9(b) represent consumed power waveform of each feeder and shows that regenerated power from one feeder is recycled in another feeder as consumed power. If there is no interchange of regenerated power among feeders, in other words if no load that can consume regenerated power exists on another feeder, then catenary voltage rises.

IV.   ANALYSIS OF REGENERATIVE ENERGY

The ratio of average regenerated power to average consumed power for various time periods are as shown in Fig. 10 and 11. In the case of Seolleung substation, we can confirm regenerative energy interchange depending on the length of railway line and gradient characteristics and fluctuations are similar among feeders. In Seolleung substation whose headway is shorter than Yangdong substation, powering

978-1-4244-1871-8/07 $25.00 © 2007 IEEE   802

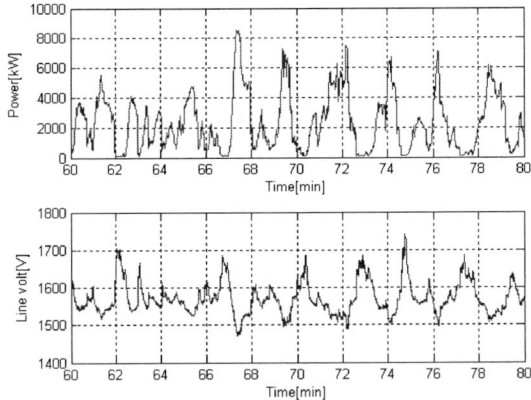

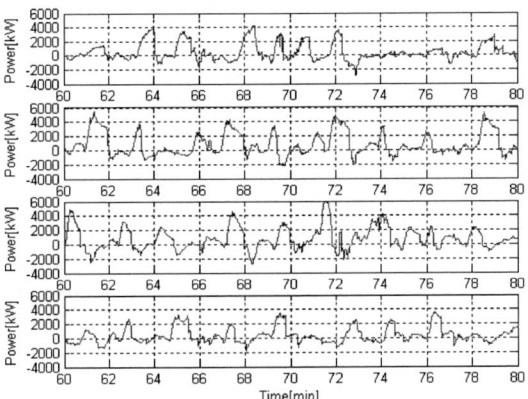

(a) Total consumed power and catenary voltage waveform

(b) Consumed power waveform of feeder 1 through 4
Fig. 9 Consumed power and catenary voltage waveform
in Seolleung substation

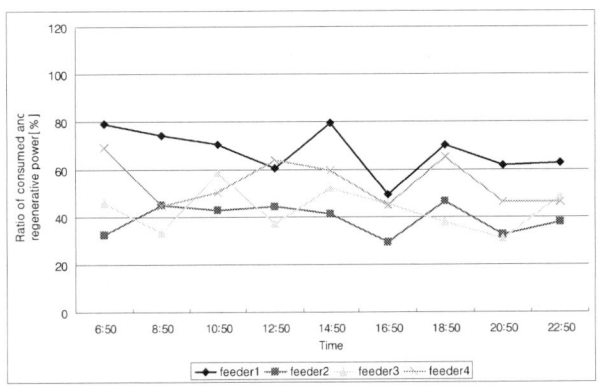

Fig. 10 Ratio of consumed and regenerated power of each feeder in Seolleung substation

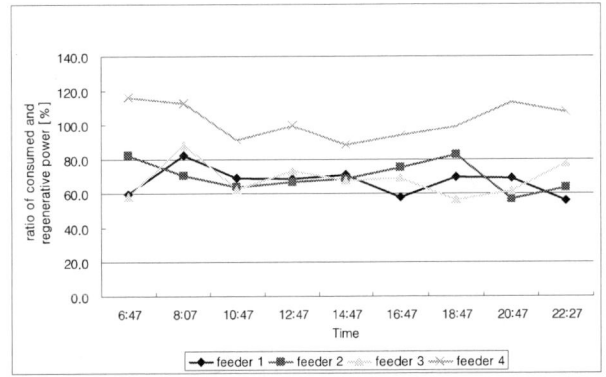

Fig. 11 Ratio of consumed and regenerated power of each feeder in Yangdong substation

trams on the same feeder consume more regenerated power. In the case of Yangdong substation, average regenerated power on Feeder 4 is sometimes larger than average reverse-direction power. This means that at such a time Feeder 4 creates regenerated power which is larger than consumed power and it is transmitted to powering trains on another feeder. Because Feeder 3 and 4 of Yangdong substation are connected to Ssangchon substation which is the terminal substation of the phase 1 of Kwangju Line No. 1, the only substation for Yangdong substation to interchange regenerated power is Ssangchon substation, and its Feeder 3 and 4 has less trainsets of railcars to interchange regenerated power than Feeder 1 and 2. Besides, the railway from Ssangchon substation to Honam University station is a continuously negative gradient section and has no other substation to interchange regenerated power, thereby creating high regenerated power.

The calculated results of reverse-direction and regenerated power for each feeder calculated for a day are shown in Table 2 and 3. The consumed energy for train traction calculated from the data measured in Yangdong substation is 11,093kWh/day and the regenerated energy recycled by energy interchange among feeders is 4,416kWh/day. This

value corresponds to 39% of the consumed energy, the cost of which is about ₩309,000 per day based on the unit price of ₩70/kWh. Table 4 shows the same calculation for Seolleung substation where the consumed energy for train traction is 53,180kWh/day and the ratio of regenerated energy to consumed energy for train traction is 21%. As a result, about ₩1,015,000 can be saved per day from energy interchange among feeders based on the unit price of ₩70/kWh.

TABLE II

Energy of each feeder in Yangdong Substation(06:47 – 24:08))

	traction energy [kWh/day]	regenerative energy [kWh/day]	consumed energy [kWh/day]	
feeder-1	3,277.7	712.9	2,564.8	average consumed power measured at seolleung substation (September)
feeder-2	3,584.0	700.4	2,883.6	
feeder-3	2,360.2	1,426.1	934.1	
feeder-4	1,871.9	1,576.9	295	
total	11,093.8	4,416.3	6,677.5	6,650

TABLE III

Energy of each feeder in Seolleung Substation(06:50 – 25:50)

	traction energy [kWh/day]	regenerative energy [kWh/day]	consumed energy [kWh/day]	
feeder-1	13,846	4,470	9,376	average consumed power measured at seolleung substation (December)
feeder-2	20,481	3,376	17,105	
feeder-3	18,305	3,175	15,130	
feeder-4	15,060	3,491	11,569	
total	67,692	14,512	53,180	69,580

In order to estimate excess regenerated power that could be absorbed if a regenerative inverter is installed, we approximately calculated total regenerated energy based on the coverage distance of substations, headway, regenerated energy for stopping once, commercial speed, etc., and calculated absorbable regenerated energy by subtracting regenerated energy interchanged which was actually measured from total regenerated energy.

Table 4 shows various conditions applied to the calculation of total regenerated energy. In the case of Yangdong Substation, we considered the section between Geunnamno 4-ga and Ssangchon stations. The reason we did not select the station at the midway between the two substations is because Ssangchon substation is the terminal substation of Kwangju Line No. 1, and Feeder 3 and 4 of Yangdong substation exhibit higher regenerated energy than other substations in Kwangju Line No. 1. Railcars are operated with 8-minute headway, and for the amount of recovered energy by regenerative braking a trainset, we referred to the railcar design data. In the case of Seolleung substation, we performed calculation based on the condition that railcars are operated with 5-minute headway in the section between Samseong and Gangnam stations.

Table 5 shows total regenerated energy obtained under the condition of Table 4, regenerated energy interchanged among feeders as actually measured, and lost regenerated energy. The Table 5 shows that Yangdong substation loses 5,043kWh/day while Seolleung loses 15,310kWh/day. However, E_r contains not only lost regenerated energy but

TABLE IV

Conditions for approximate calculation of total regenerative energy.

conditions	Seolleung	Yangdong
distance	3.3km (Samsung ~ Gangnam)	4.6km (Geunnamno 4-ga ~ Ssangchon)
headway	5min	8min
regenerative energy of a trainset	22.8kWh	9.13kWh
commercial speed	35km/h	35km/h
time	17:21	18:10

Table V

Analysis results of regenerative energy.

substation	regenerative energy[kWh]			regenerative brake energy of a trainset
	E_g	E_t	E_r	
Yangdong	9,459	4,416	5,043	9.13
Seolleung	29,822	14,512	15,310	22.8

also power loss while current flows in feeder lines, and in general, raises catenary line voltage more than loss by the resistance of feeder lines and is smaller than excess regenerated power lost in railcars.

V. CONCLUSIONS

We measured the catenary voltage and current of each feeder in DC electric railway substations and presented the voltage-current characteristic curves and the analysis results on consumed and regenerated energy. In the cases of seolleung Substation in Seoul and Yangdong substation in Kwangju, interchanged regenerated energy among 4 feeders was 39% and 21% of consumed energy for train traction. In addition, we approximately formulated total regenerated energy from each substation owing to regenerative braking of railcars and approximately calculated lost regenerated energy using regenerated energy interchanged among feeders.

REFERENCES

[1] Y.S. Tzeng, R.N. Wu and N. Chen, "Electric network solutions of DC transit systems with inverting substations," *Vehicular Technology IEEE Trans.*, vol. 47, no. 4, Nov. 1998. pp. 1405 ~ 1412

[2] H.D. Fuchs, P.H. Henning and H. du T Mouton, "Development and implementation of a 1.5MW Inverter and Active Power Filter for the Injection of Regenerated Energy In a Spoornet Substation", *IEEE AFRICON*, pp.1119 ~ 1124, 2004.

[3] E.J. Goh, K.N. Chu and N.K. Ng, "1500V DC Traction system for the North East Line", *International Conference on Power System Technology-POWERCON*, pp. 1904~1909, Nov. 2004.

[4] P.J. Rdndewijk and J.H. R. Enslin, "Inverting DC traction substation with active power filtering incorporated," *IEEE-PESC Conf. Rec.*, vol. 1, pp.360-366, 1995.

[5] Electric Railway DC Power Supply System Investigation Committee, "*Phenomena of Power Supply System Including Regenerative Cars and Future Directions*", Technological Report No. 296 of Japanese Institute of Electrical Engineers, 1989.

[6] Korea Railroad Research Institute, "*Development of power supply system for light railway transit*," 2004.

ROBUST SPEED SENSORLESS INDUCTION MOTOR DRIVES

Geetha E.K
RMKEC Chennai
gita_ramadas@yahoo.com

T.Thyagarajan, Professor MIT
Chromepet, Chennai
Member, IEEE

Vedam Subrahmanyam
Professor, RMKEC, Chennai
Member, IEEE

Abstract- Induction motors are widely used in industry due to the fact that they are relatively cheap, rugged and maintenance free. As a consequence, much attention has been given to the motor torque and speed control. The control schemes available today require information regarding speed of the motor, which can either be obtained by using speed sensors or without speed sensors. Speed sensors have several disadvantages from the standpoint of drive cost, reliability, inertia and noise immunity. Advantages of eliminating speed sensors thus have been a strong motivation to develop speed sensor less induction motor drives for industrial drives. Several control strategies of sensor less control are available in literature. This paper is an attempt to explore the possibility of estimation of rotor speed with the help of extended Kalman filter trained recurrent Neural Network. The speed estimation is made robust by simultaneously adapting the rotor resistance and rotor flux which are also done by the same Neural Network. The training is very fast as it requires only one iteration. The proposed scheme is studied on an induction motor and it gives better performance as compared to the already existing algorithms in the literature.

Key words: Kalman Filter, Neural Networks, Robust Speed sensor less operation, Vector control

1. INTRODUCTION

Field Oriented control of an Induction motor is best suited for dynamic performance improvement. Information regarding the speed of induction motor is essential for field oriented control. Conventional field oriented control schemes use electromechanical speed sensors (Fig.1). Speed sensor has several disadvantages from the standpoint of drive cost, reliability, inertia, and noise immunity .Also it may not be possible to maintain perfect decoupling due to rotor time constant for some reason or other. . This has led to the concept of speed sensor less field oriented control (Fig.2) in which speed is estimated from the terminal voltages and currents [1].

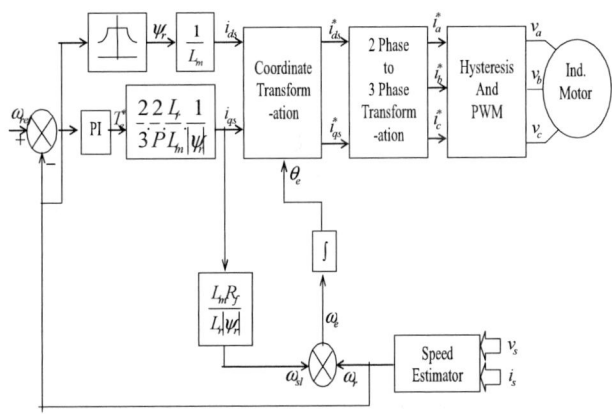

Fig-2. Sensorless Vector Control

One of the most well known methods used for speed sensor less field oriented control of induction motor is by using the Extended Kalman Filter (EKF) [2, 3]. Also, the same EKF is used for estimation of rotor resistance and ,rotor flux along with the estimation of rotor speed there by making the performance robust[4].Neural networks, having the ability to learn and process information in the same way as brain does, can be advantageously used for speed estimation[5,6]. Extended Kalman filtering techniques (EKF) can be used to train the neural network which can be used for accurate estimation of speed and rotor resistance, if necessary, simultaneously. This method takes least time to train neural network and hence, can be used for real time application with robust performance. Kalman filter is an optimum speed estimator. A block diagram of speed estimation based on a Kalman filter is shown in Figure 3. By adapting this algorithm, rotor resistance and speed can be accurately estimated without additional computational effort (Fig 4)

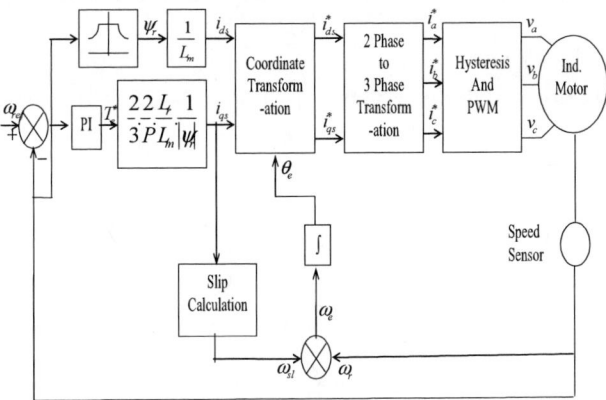

Fig-1. Conventional Indirect Vector Control

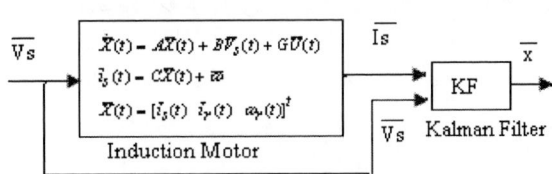

Fig.3 Block Diagram Representation of the proposed Scheme

978-1-4244-1871-8/07 $25.00 © 2007 IEEE   805

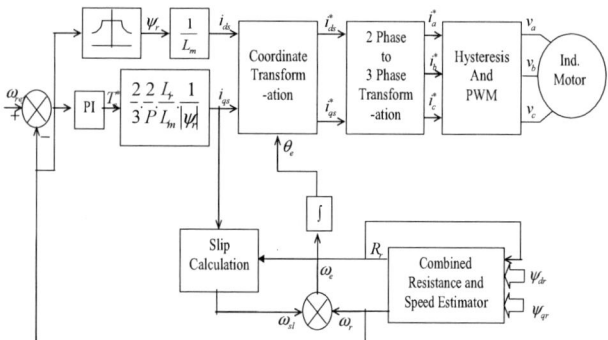

Fig-4. Sensorless Control with Rotor Resistance identification

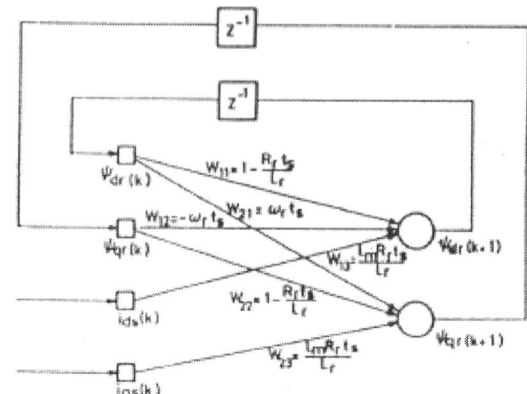

Fig 5 Real Time recurrent Neural Network

The aims of this paper are

1. To develop a speed identification using Kalman Filter(KF) trained recurrent neural network.
2. To make the speed estimation robust by simultaneously estimating the rotor resistance and applying the same to the speed estimation. This will enable perfect decoupling.

II SPEED IDENTIFICATION PRINCIPLE

The speed estimation using Kalman filtering techniques is based on the following two expressions.

$$\frac{d}{dt}\begin{bmatrix}\psi_{dr}\\\Psi_{qr}\end{bmatrix}=\frac{L_r}{L_m}\begin{bmatrix}V_{ds}\\V_{qs}\end{bmatrix}-R_s\frac{L_r}{L_m}\begin{bmatrix}i_{ds}\\i_{qs}\end{bmatrix}-\sigma\frac{L_sL_r}{L_m}\frac{d}{dt}\begin{bmatrix}i_{ds}\\i_{qs}\end{bmatrix} \quad (1)$$

$$\frac{d}{dt}\begin{bmatrix}\Psi_{dr}\\\Psi_{qr}\end{bmatrix}=\begin{bmatrix}\dfrac{-R_r}{L_r} & -\omega_r\\\omega_r & \dfrac{-R_r}{L_r}\end{bmatrix}\begin{bmatrix}\Psi_{dr}\\\Psi_{qr}\end{bmatrix}+L_m\frac{R_r}{L_r}\begin{bmatrix}i_{ds}\\i_{qs}\end{bmatrix} \quad (2)$$

The former is solved for flux variables from voltage and currents. The latter is solved for flux variables from speed and currents. Flux calculated from both the equations are compared and speed in second equation is adjusted till both the flux variables match. Eqn.(2) is descretised and represented as real time neural network as given in Fig. 5. The descretised equation is

$$\begin{bmatrix}\Psi_{dr}(k)\\\Psi_{qr}(k)\end{bmatrix}=\begin{bmatrix}1-\dfrac{R_rt_s}{L_r} & -\omega_rt_s\\\omega_rt_s & 1-\dfrac{R_rt_s}{L_r}\end{bmatrix}\begin{bmatrix}\Psi_{dr}(k-1)\\\Psi_{qr}(k-1)\end{bmatrix}+$$

$$\begin{bmatrix}R_rL_mt_s & 0\\0 & R_rL_mt_s\end{bmatrix}\begin{bmatrix}i_{ds}(k-1)\\i_{qs}(k-1)\end{bmatrix} \quad (3)$$

The outputs of the network are the fluxes calculated using (2) and the desired outputs are those calculated using (1). The errors between the two are used for updating

the weights of the network. The outputs of the network can be expressed as

$$\psi_{dr}=f_1(net_1)$$
$$\psi_{qr}=f_2(net_2) \quad (4)$$

where

$$(net_1)=W_{11}\psi_{dr}(k-1)+W_{12}\psi_{qr}(k-1)+W_{13}i_{ds}(k-1)$$

$$(net_2)=W_{21}\psi_{dr}(k-1)+W_{22}\psi_{qr}(k-1)+W_{31}i_{qs}(k-1)$$

And f_1 and f_2 are activation functions of the output neurons, which are linear with unity gain .Because only two out of the six weights contain speed term, it is enough to train only these two weights. However, to estimate rotor resistance, along with speed, the other terms containing rotor resistance also should be trained.

III PARAMATER BASED EKF ALGORITHM FOR TRAINING NEURAL NETWORK

Parameter estimation problem can considered as a neural network training problem. The estimation scheme is formulated based on the concept of minimizing the error vector obtained as the difference between functions of the network output nodes and desired values of these functions. Let the desired value at k_{th} instant be

$$d[k]=[d_1(k) \quad d_2(k) \quad K K \quad D_N(k)]^T \quad (5)$$

and the vector

$$f(k)=[f_1(net_1) \quad f_2(net_2) \quad K K \quad f_N(net_N)]^T \quad (6)$$

Denote a vector of functions of the network outputs. The training cost function is given by

$$E(k)=0.5e(k)^T \quad s(k) \quad e(k) \quad (7)$$

Where s(k) is user specified non negative weighting matrix and e (k) = d (k)- f (k) is the error vector.

The trainable weights of the network are arranged into M ×N dimensional vector W (k) .An error covariance matrix p(k) is stored and updated at every step. This is used to model the correlation between each pair of weights in the network. The matrix p (k) is stored and updated at every step. This is used to model the correlation between each pair of weights in the network. The matrix p (0) is initialized as a diagonal matrix.

The training procedure is summarized as given below.

1. The input signals and the recurrent node outputs are propagated through NN and the functions f(k) are calculated.

2. The error vector e(k) is calculated and the dynamic derivatives of each component of f(k) are formed with respect to the current weight estimates W (k). These derivates are arranged in to M ×N matrix H (k).

3. then the W(k) and p(k) are updated using the following global EKF recursion formulae

$$A(k) = [\eta(k) \quad s(k)^{-1} + H(k)^T \quad p(k) \quad H(k)]^{-1}$$

$$K(k) = p(k) \quad H(k) \quad A(k)$$

$$W(k+1) = w(k) + K(k) \quad e(k) \tag{8}$$

$$p(k+1) = p(k) - K(k) \quad H(k)^T \quad p(k) + Q(k)$$

Where η (k) is a scalar learning rate parameter, which along with s (k) establishes the learning rate. The matrix A (k) is established as the inverse of M ×N matrix. It is used for calculating the Kalman gain matrix. The weights are updated by adding the previous weight matrix, the product of K (k) and e (k). K(k) is used for updating the error co variance matrix.Q (k) is a diagonal covariance matrix that provides a mechanism by which the effects of artificial process noise are included in the Kalman recursion.

In the present work, it can be identified that

$$d(k) = \begin{bmatrix} \psi_{dr}(k) & \psi_{qr}(k) \end{bmatrix}^T \tag{9}$$

which is evaluated using Eqn.(1) at k^{th} instant.

$$f(k) = \begin{bmatrix} f_1(net_1) & f_2(net_2) \end{bmatrix}^T \tag{10}$$

Error vector is

$$e(k) = d(k) - f(k) \tag{11}$$

The trainable weight matrix is

$$W(k) = \begin{bmatrix} w_{11}(k) & w_{12}(k) & w_{13}(k) & w_{21}(k) & w_{22}(k) & w_{13}(k) \end{bmatrix}^T$$

The dynamic derivative matrix is obtained as

$$H(k) = \begin{bmatrix} \dfrac{\delta f_1}{\delta w_{11}} & \dfrac{\delta f_1}{\delta w_{12}} & \dfrac{\delta f_1}{\delta w_{13}} & 0 & 0 & 0 \\[2mm] 0 & 0 & 0 & \dfrac{\delta f_2}{\delta w_{21}} & \dfrac{\delta f_2}{\delta w_{22}} & \dfrac{\delta f_2}{\delta w_{23}} \end{bmatrix} \tag{12}$$

Where

$$\frac{\delta f_1}{\delta w_{11}} = \frac{\delta f_1}{\delta w_{(net_1)}} \cdot \frac{\delta f_{(net_1)}}{\delta w_{11}} = \psi_{dr}(k-1)$$

$$\frac{\delta f_1}{\delta w_{12}} = \frac{\delta f_1}{\delta w_{(net_2)}} \cdot \frac{\delta f_{(net_2)}}{\delta w_{11}} = \psi_{qr}(k-1)$$

$$\frac{\delta f_1}{\delta w_{13}} = \frac{\delta f_1}{\delta w_{(net_1)}} \cdot \frac{\delta f_{(net_1)}}{\delta w_{13}} = i_{ds}(k-1)$$

$$\frac{\delta f_2}{\delta w_{21}} = \frac{\delta f_2}{\delta w_{(net_2)}} \cdot \frac{\delta f_{(net_2)}}{\delta w_{21}} = \psi_{dr}(k-1)$$

$$\frac{\delta f_2}{\delta w_{22}} = \frac{\delta f_2}{\delta w_{(net_2)}} \cdot \frac{\delta f_{(net_2)}}{\delta w_{22}} = \psi_{qr}(k-1)$$

$$\frac{\delta f_2}{\delta w_{23}} = \frac{\delta f_2}{\delta w_{(net_2)}} \cdot \frac{\delta f_{(net_2)}}{\delta w_{23}} = i_{qr}(k-1)$$

Rotor resistance and rotor speed can be calculated as

$$R_r(k) = (1 - w_{11}(k)) \frac{L_r}{t_s}$$

$$w_r(k) = \frac{w_{21}(k)}{t_s} \tag{13}$$

IV RESULTS

Simulation and experimental tests have been carried out using the indirect field oriented control scheme of Fig.4. Actual speed is measured from the experimental set up Terminal voltages and currents are evaluated at every sampling instant (simulation). The same terminal voltages and currents are given as inputs to the neural network speed estimator. The effectiveness of speed estimation scheme has been tested for different operating conditions. It is found that in all cases, the estimated speed agrees with the actual speed. The applied load has characteristics as shown in Fig.6 (a).At the moment of application of load, there is a dip in speed. Similarly, at the moment of removal of load, there is a small overshoot in speed. But, during both the above transient conditions, recovery to original speed is very quick. There is no error between the between the measured speed and the estimated speed (Fig.6(b) & Fig 6 (c)).

In order to test the rotor resistance estimation, it is assumed that the value of rotor resistance is not known exactly. Simulation is started with a resistance value in estimation and control parts different from the actual value used in machine modeling. It is found that the combined speed and rotor resistance estimation process resulted in fast convergence to its actual value. This confirms to parameter insensitivity of rotor resistance variation with respect to speed estimation. The characteristics obtained from simulation and experiment are exactly the same and hence error is zero (Fig.7(a) & Fig.7 (b)).

The procedure is repeated for reverse speed operation also. Reference speed is changed from 300rad/s to zero -300rad/sec and then back to 300rad/s. The estimated along with actual speed is shown in Fig 8(a)

978-1-4244-1871-8/07 $25.00 © 2007 IEEE   807

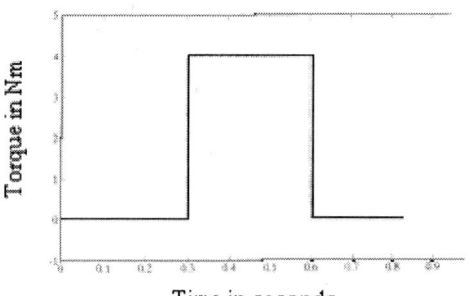

Fig.6(a) Applied Torque Characteristics.

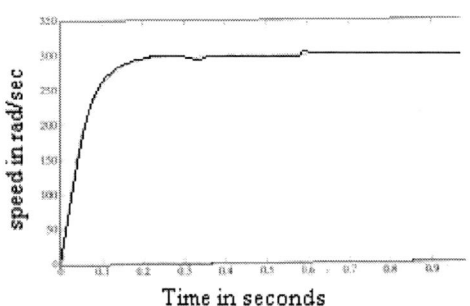

Fig.6 (b) Actual speed

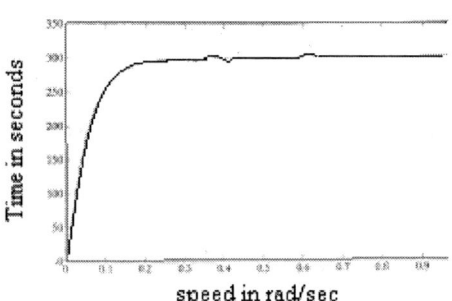

Fig.6 (c) Estimated Speed.

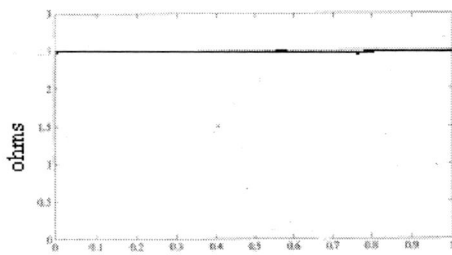

Fig.7(a) Actual & estimated rotor resistance

actual value
estimated value

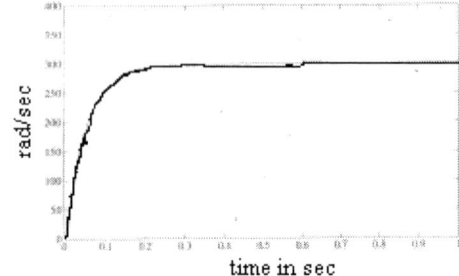

Fig 7(b) Actual and estimated speed

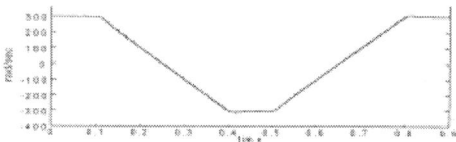

Fig.8 Actual and Estimated Speed

V.CONCLUSIONS

The Kalman Filter trained real time recurrent NN is found to provide accurate estimation of speed as well as rotor resistance. This makes the speed estimation robust and provides perfect decoupling in field oriented control schemes. The main features of the proposed method over the conventional method are

1. Does not depend on time consuming off line training.
2. Estimation process is complete in one cycle of training algorithm. Hence, method is faster in real time applications.
3. Separate rotor resistance identifier is not required since speed estimation and rotor resistance identification are carried out simultaneously.

EXPERIMENTAL SET UP

Motor Details: 1.1kW, 3 phase , 60 Hz, 4 pole,200V, R_0 = 3.3534 ohms, R_r = 1.991 ohms , L_0 = 170.67mH, L_m =163.73mH, L_r = 163.73mH , J = 0.005n-m^2/s

ACKNOWLEDGMENT

The authors acknowledge the financial support rendered by the Management of RMK Engineering College, Chennai, India. Also, the authors express their sincere gratitude to Mr. T Ashok Kumar and Mr. E Kaliappan , faculty, Electrical Department, RMK Engineering College, for helping them in preparing the final manuscript.

978-1-4244-1871-8/07 $25.00 © 2007 IEEE

REFERENCES

[1] Toshiyuki Kanmachi and Isao-Takshashi, "Sensor less speed control of an Induction Motor" IEEE Industry Applications Magazine Jan/Feb 1995.

[2] A. Mendoza, R. Arnanz, M.A. Pacheco, and J.R. Perán (Spain) "Diagnosis and Control in Induction Motor with Extended Kalman Filter" Proceeding (377) Modeling, Identification, and Control - 2003

[3] Salomon Chavez Velázquez, Ruben Alejos Palomares,Alfredo Nava Segura "Speed estimation of an Induction Motor using the Extended Kalman Filter" Proceedings of the 14th International Conference on Electronics, Communications and Computers (CONIELECOMP) 2004 IEEE p.63

[4] G. Garcia Soto,[1] E. Mendes,[1] and A. Razek "Reduced-order observers for rotor flux, rotor resistance and speed estimation for vector controlled induction motor drives using the extended Kalman filter technique",IEE Proceedings - Electric Power Applications -- May 1999 -- Volume 146, Issue 3, p. 282-288

[5] Razik, H.; Rezzoug, A. "Neural networks applied to the control of an induction motor" Power Electronics Congress, 2000. CIEP2000. VII IEEE International Volume, Issue, 2000 Page(s):60 – 64

[6] H.A.F. Mohamed and S. Yaacob "Identification and Control of an Induction Motor using Artificial Neural Networks From Proceeding (529) Control and Applications – 2006

Improved Rotor Position Estimation employing Voltage Distortion Compensation for Sensorless PMSM Drives at Low Speed

Shin-Myung Jung[1], Jin-Sik Park, Hag-Wone Kim[2], and Myung-Joong Youn

[1]Department of Electrical Engineering and Computer Science
Korea Advanced Institude Science and Technology
373-1 Guseong-Dong, Yuseong-Gu, Daejeon, Korea
caesarju@powerlab.kaist.ac.kr

[2]Digital Appliance Lab., LG Electronics Co., Ltd., Seoul 153-023, Korea
khw@lge.com

Abstract— The PMSM is inherently electronically controlled and requires rotor position information for proper drives. However, the problems of the space, cost, and reliability of rotor position sensors have motivated research in the area of position sensorless PMSM drives. Numerous rotor position sensorless control techniques have been proposed for PMSM drives, but most rely directly or indirectly on the estimation of the rotor flux linkage vector.

The performance of this method is poor due to the voltage distortions caused by the nonlinear behavior of a pulsewidth-modulation (PWM) voltage source inverter (VSI). In this paper, the nonlinearity in a PWM VSI is analyzed and a new online estimation method is proposed to compensate the voltage distortion. To prove effectiveness of the voltage distortion compensation method to the improvement of the rotor position estimation, the proposed method is applied to PMSM drive system and implemented in a digital manner using a digital signal processor.

I. INTRODUCTION

Different strategies for position sensorless operation of the PM synchronous motor are discussed in many papers. Most sensorless control schemes, however, rely directly or indirectly on the estimation of the flux linkage vector. In this method, the flux linkage is calculated using measured voltages and currents. The fundamental idea is to take the voltage equation of the machine and by integrating the applied voltage and current, flux linkage can be estimated. From the initial position, machine parameters and the flux linkages' relationship to rotor position, the rotor position can be estimated. This method, however, has significant estimation error at low speed region. The reasons are dc offset and drift components in the acquired feedback signals, the increased sensitivity against model parameter mismatch, and voltage distortion caused by the nonlinear behavior of the PWM VSI. Among those reasons, only the imperfections of the PWM VSI are considered in this paper.

II. PMSM MODEL

A PM synchronous motor considered in this paper consists of a 3-phase stator winding and the permanent magnet on the

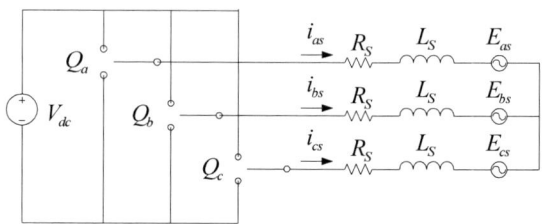

Voltage Source Inverter

Fig. 1: Simplified voltage source inverter and PMSM model

rotor. The stator voltage equation of a surfaced mounted PM synchronous motor can be expressed as follows [1]:

$$
\begin{bmatrix} v_{as} \\ v_{bs} \\ v_{cs} \end{bmatrix} = \begin{bmatrix} R_S & 0 & 0 \\ 0 & R_S & 0 \\ 0 & 0 & R_S \end{bmatrix} \begin{bmatrix} i_{as} \\ i_{bs} \\ i_{cs} \end{bmatrix} + \frac{d}{dt} \begin{bmatrix} \lambda_{as} \\ \lambda_{bs} \\ \lambda_{cs} \end{bmatrix}
\tag{1}
$$

where $\begin{bmatrix} i_{as} & i_{bs} & i_{cs} \end{bmatrix}^T$ represents the phase currents, $\begin{bmatrix} \lambda_{as} & \lambda_{bs} & \lambda_{cs} \end{bmatrix}^T$ represents the flux linkages of respective phases, and R_S represents the equivalent resistance of a phase winding. If the flux linkages from the permanent magnet are assumed to be pure sinusoids with constant amplitude, the flux linkage in equation (1) can be expressed as follows:

$$
\begin{bmatrix} \lambda_{as} \\ \lambda_{bs} \\ \lambda_{cs} \end{bmatrix} = \begin{bmatrix} L_S & 0 & 0 \\ 0 & L_S & 0 \\ 0 & 0 & L_S \end{bmatrix} \begin{bmatrix} i_{as} \\ i_{bs} \\ i_{cs} \end{bmatrix} + \lambda_m \begin{bmatrix} \sin(\theta_e) \\ \sin(\theta_e - 2\pi/3) \\ \sin(\theta_e + 2\pi/3) \end{bmatrix}
\tag{2}
$$

where L_S is the equivalent inductance of a phase winding, θ_e is the electrical angular position of the rotor, and λ_m is the amplitude of the flux linkage which is established by the permanent magnet.

Fig. 1 shows the simplified VSI and PMSM model. In this figure, E_{as}, E_{bs}, and E_{cs} represent the voltages caused by the flux linkage of the permanent magnet of rotor in each phase.

978-1-4244-1871-8/07 $25.00 © 2007 IEEE

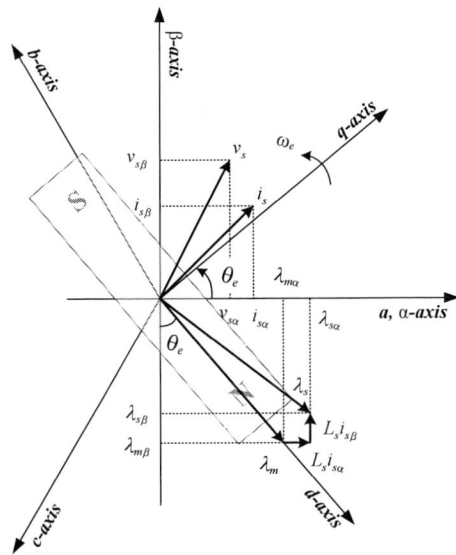

Fig. 2: Phasor diagram of PM synchronous motor

$$\begin{bmatrix} v_{qs} \\ v_{ds} \end{bmatrix} = \frac{2}{3} \begin{bmatrix} \cos\theta_e & \cos\left(\theta_e - \frac{2\pi}{3}\right) & \cos\left(\theta_e + \frac{2\pi}{3}\right) \\ \sin\theta_e & \sin\left(\theta_e + \frac{2\pi}{3}\right) & \sin\left(\theta_e + \frac{2\pi}{3}\right) \end{bmatrix} \begin{bmatrix} v_{as} \\ v_{bs} \\ v_{cs} \end{bmatrix} \quad (3)$$

and the result becomes

$$\begin{bmatrix} v_{qs} \\ v_{ds} \end{bmatrix} = \begin{bmatrix} R_S + L_S \dfrac{d}{dt} & L_S \omega_e \\ -L_S \omega_e & R_S + L_S \dfrac{d}{dt} \end{bmatrix} \begin{bmatrix} i_{qs} \\ i_{ds} \end{bmatrix} + \begin{bmatrix} \lambda_m \omega_e \\ 0 \end{bmatrix} \quad (4)$$

where v_{qs} and v_{ds} are the q- and d-axis voltages, i_{qs} and i_{ds} are the q- and d-axis currents, and ω_e is the electrical angular velocity of the rotor.

In order to analyze the characteristics of three phase quantities such as the voltages, currents, and fluxes, a change of variables is often utilized to reduce the complexity of the three phase variables. Using the Park's transformation or synchronous dq transformation, the variables of the PM synchronous motor can be regarded as DC quantities and it is, thus, easy to manipulate the voltages, currents, and fluxes. The voltage equation in the synchronously rotating reference frame can be obtained from the voltage equation in (1) using the relationship as follows [2]:

III. FLUX OBSERVER AND POSITION ESTIMATION

A. Estimation of the Flux Linkage Vector

The phasor diagram of a PM synchronous motor is shown in Fig. 2, the current vector i_s and the voltage vector v_s being derived from the measured phase current and voltage, and the transformation matrix from the abc reference frame to the stationary $\alpha\beta$ reference frame. The vector λ_s represents the resultant stator flux linkage, while the vector λ_m represents the excitation flux linkage due to the permanent magnets, which is in phase with the rotor d-axis. Hence, the rotor position can be obtained from the estimated phase angle $\hat{\theta}_e$.

From the equation (1), λ_s is observed as follows:

$$\lambda_s = \int_0^t (v_s - R_S i_s) \cdot d\tau + \lambda_s(0). \quad (5)$$

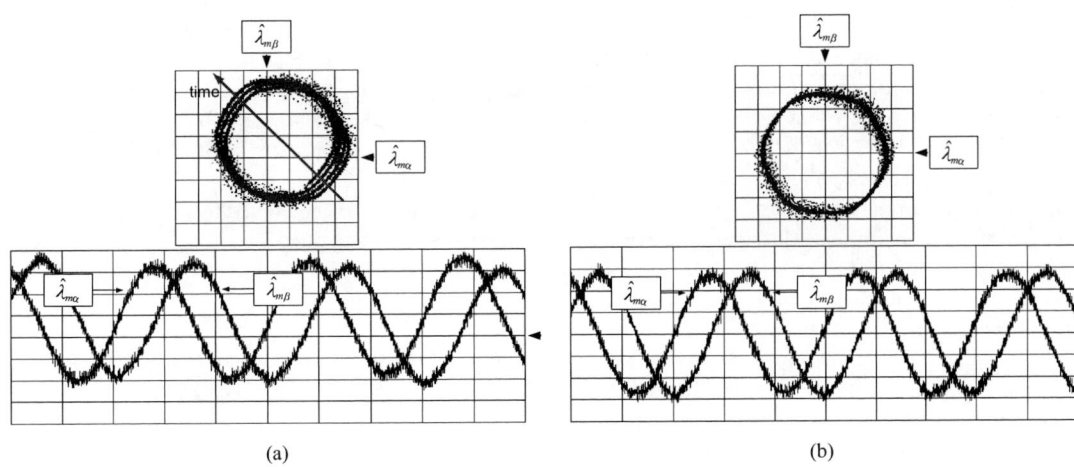

(a)　　　　　　　　　(b)

Fig. 3: Offset Problem in Flux Observer
(a) Using a pure integrator
(b) Using a high pass filter

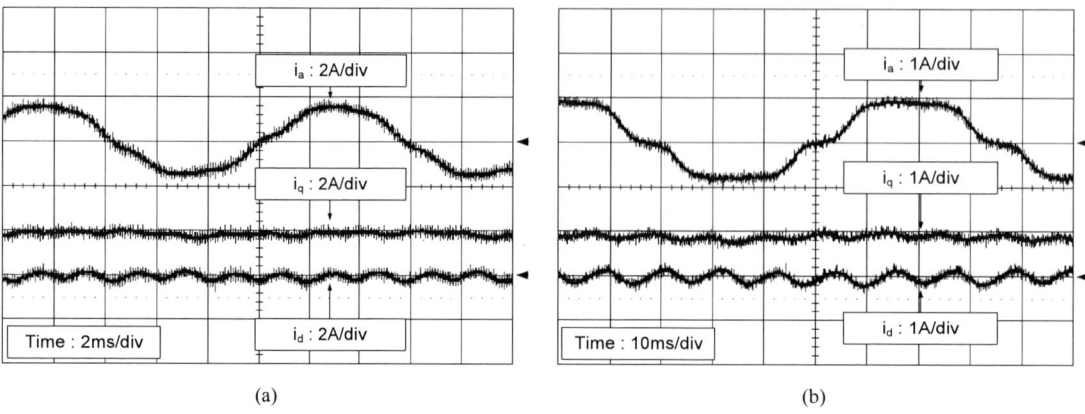

(a)   (b)

Fig. 4: Voltage distortion in Flux Observer
(a) 1200 rpm   (b) 200 rpm

with zero current in the stator windings, the stator flux linkage vector is simply the excitation flux linkage vector, i.e., $\lambda_s(0) = \lambda_m$, which is obtained by initially aligning the rotor before the flux observer is applied.

From Fig. 2, it is seen that λ_m can be calculated from the stator flux linkage vector λ_s, the current vector i_s, and winding inductance. Further, if surface-mounted type PM synchronous motor is used, saliency can be neglected, $L_d \cong L_q \cong L_S$, and λ_m is simply calculated as follows:

$$\lambda_m = \lambda_s - L_S i_s. \tag{6}$$

In the $\alpha\beta$ reference frame, λ_m is expressed by the projections of $\lambda_{m\alpha}$ and $\lambda_{m\beta}$ on the α- and β- axes as shown in Fig. 2. Therefore, the rotor position is obtained as follows:

$$\theta_e = \arctan\frac{\lambda_{m\alpha}}{\lambda_{m\beta}}. \tag{7}$$

B. Offset Error Problem

The estimation of the stator flux linkage vector λ_s according to equation (5) requires performing the integration in real time. The use of a pure integrator has not been reported in the literature. The reason is that an integrator has an infinite gain at zero frequency. The unavoidable offsets contained in the integrator input then make its output gradually drift away beyond limits as shown in Fig. 3(a). The integrator amplifies any dc offset or error in the voltages, currents, and resistances until saturation is reached. Clearly, it is not possible to deduce the rotor position from such a flux vector. Thus, a high pass filter, having the transfer function $s/(s+\omega_0)$, is applied to the variables to be integrated [3]. Since the transfer function of a pure integrator is $1/s$, the resultant transfer function is $1/(s+\omega_0)$, which is equivalent to replacing the integrator by a low pass filter. A low pass filter has finite dc gain which eases the drift problem. Hence, equation (5) becomes

$$\lambda_s = \int_0^t \left[-\omega_0 \lambda_s + \left(v_s - R_s i_s\right)\right] \cdot d\tau + \lambda_s(0). \tag{8}$$

With the cutoff frequency ω_0, the observed stator flux vector λ_s remains stable, so as to λ_m, as shown in Fig. 3(b). However, the filter introduces phase delay and amplitude error around its cutoff frequency.

The induced voltage, which is the signal to be integrated for flux vector estimation, is obtained as the difference between the stator voltage and the resistive voltage drop across the machine windings. When a PWM VSI is used to feed the machine, the stator voltages changes very rapidly, yielding a rich, high-frequency spectrum content. These are digitally acquired at a high, though limited sampling rate [4]. The limited bandwidth of such sampling process may fail to establish the exact volt-second equivalent between the actual and the acquired signals and, hence, produce an error. The error is more significant in low speed region. To avoid this problem in a PWM VSI, it is preferred to replace the actual stator voltage by the reference voltage vector. This very simple method yields good results, except when operating in the low speed region since there is voltage distortion between actual stator voltage and reference voltage as shown in Fig. 4. As can be seen in Fig. 4, the voltage distortion is more severe at 200 rpm operation than 1200 rpm operation. The voltage distortion is caused by the imperfection of PWM VSI. If voltage distortion is corrected by a compensation method, the offsets could be reduced so that the phase delay problem introduced by the filter could also be reduced.

IV.   OBSERVATION AND COMPENSATION OF VOLTAGE DISTORTION

In a PWM VSI, the voltage distortion exists between the reference voltage and the output voltage. This distortion is partly caused by the blanking time which is inevitable to prevent the shoot-through phenomenon. The other reasons are the inherent characteristics of the switching devices such as the voltage drops, turn on/off time, and output voltage

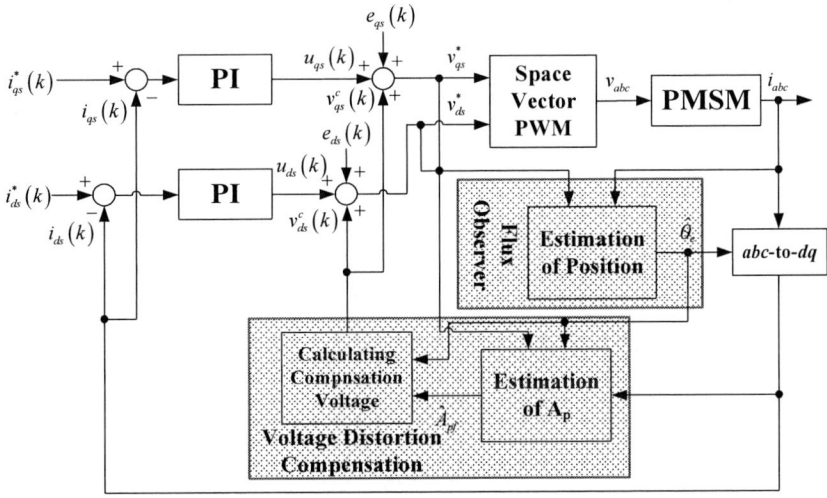

Fig. 5: Overall block diagram of flux observer with voltage distortion compensation

transition time. These non-ideal characteristics of the devices vary with the operating conditions such as the temperature, DC bus voltage, and phase current level. To compensate this distortion, the non-ideal characteristics of the power devices have been considered in some papers [5]-[8]. Some approaches require off-line experiments and the compensation is based on these experiments [5], [6]. However, it is difficult to compensate perfectly the time varying voltage distortion using the results from these experiments. The other approach requires the terminal voltage sensing hardware [7]. It is, however, also difficult to sense the on-drop of power devices, switching delay, and transition time. In [8], the on-line observing method for the voltage distortion due to the blanking time and non-ideal characteristics has been proposed. In this method, the voltage distortion is directly estimated by a time delay control observer. Since this voltage distortion is abruptly changed at some instant, however, the phase delay exists between the real distortion voltages and observed one. This delay originated from a low-pass filter of the observer and causes a degradation of compensation characteristics.

To solve this sensitivity problem related to the nonlinearities of the PWM VSI, the mechanism of the nonlinearities in analyzed in detail and a new robust online nonlinearity estimation algorithm is proposed [9]. From the analysis results, the nonlinearities of the VSI can be calculated using the newly defined intermediate value (A_p) which is a function of the operating condition of VSI. The output voltage errors considered as disturbance voltages are estimated by a disturbance observer and fed to voltage references in order to compensate these nonlinearities. The disturbance voltage observer is based on time delay control technique and this observer estimate the newly defined intermediate value related to the inverter nonlinearities. This method is applied to the flux-observer-based rotor position estimation method of PM synchronous motor drive system

and implemented in a digital manner using a digital signal processor.

TABLE 1
Specifications of Test Motor

Rated power	750[W]	Number of poles	8
Rated torque	2.4[Nm]	Rated speed	3000[rpm]
Stator resistance	0.49[Ω]	Rated current	6.0[A]
Linkage flux	0.0667[Wb]	Stator inductance	6.9[mH]

TABLE 2
Specifications of Drive System

DC link voltage	310[V]	Switching period	120[μsec]
Dead time	3[μsec]	Switching device	IGBT
Turn-on time[*]	0.8-2.0[μsec]	Turn-off time[*]	2.0-2.9[μsec]
Saturation voltage[*]	1.8-2.7[V]	Forward voltage[*]	2.2-3.3[V]

[*]:Mitsubishi data sheet(PM50RSA060)

V. EXPERIMENTAL RESULTS

In this section, experimental results are shown to prove the effectiveness of the voltage distortion observer and compensation method for sensorless control of PM synchronous motor at low speed region. The parameters related to the PMSM and PWM VSI are represented in Table 1 and Table 2, respectively. The overall block diagram of flux observer with voltage distortion compensation is shown in Fig. 5. The experimental current waveforms without voltage distortion compensation, and with compensation are shown in Fig. 6. The current waveforms of a-, q-, and d-axis without

978-1-4244-1871-8/07 $25.00 © 2007 IEEE   813

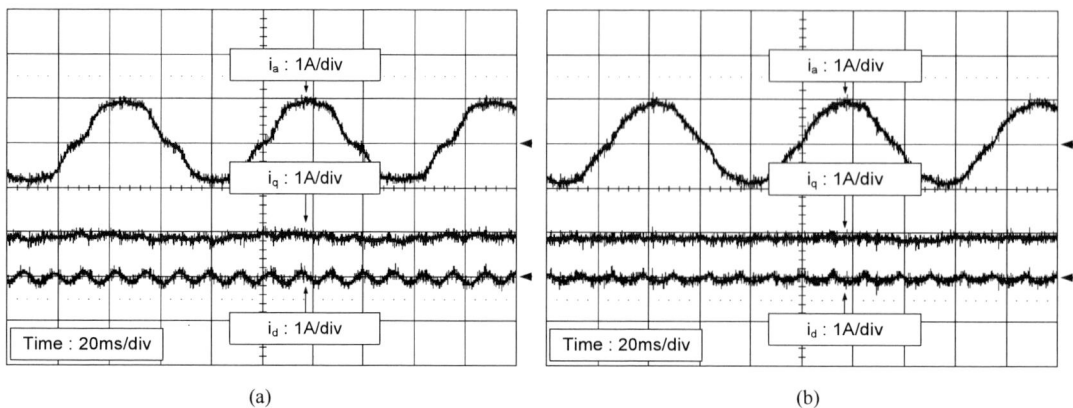

Fig. 6: Current waveform comparison (200 rpm)
(a) Without voltage distortion compensation
(b) With voltage distortion compensation

the voltage distortion compensation under 200 rpm operation are shown in Fig. 6(a). These current waveforms are distorted by the voltage distortion effects. The current waveforms with the compensation are shown in Fig. 6(b). As can be seen in this figure, phase current waveforms are almost sinusoids.

Experimental results for rotor position estimation under 600 rpm and 200 rpm are shown in Fig. 7, and Fig. 8, respectively. As can be seen in these figures, the rotor position error is existed in both cases when operating without voltage distortion compensation. However, the voltage distortion compensation improves the estimation of the rotor flux linkage vector, hence the estimated rotor position angle is almost coincided with the real rotor position angle as can be seen in Fig. 7(b) and Fig. 8(b).

VI. CONCLUSION

In this paper, the flux observer and position estimation method are presented. The rotor position information is estimated by using machine variables such as voltages, currents, resistances, and inductances. However, the performance of this method is poor due to the offsets caused by a pure integrator. The integrator amplifies any dc offset or error in the voltages, currents, and resistances until saturation is reached. Therefore, a high pass filter is applied to the results of the integration to eliminate offsets. However, the filter introduces severe phase delay and amplitude errors at frequencies around its corner frequency.

The offsets problem is more severe at low speed region, because there is more error in the voltages to be integrated. The error of voltages is caused by the imperfection of PWM VSI, namely voltage distortion. If voltage distortion is corrected by a compensation method, the offsets could be reduced so that the phase delay problem introduced by the filter could also be reduced.

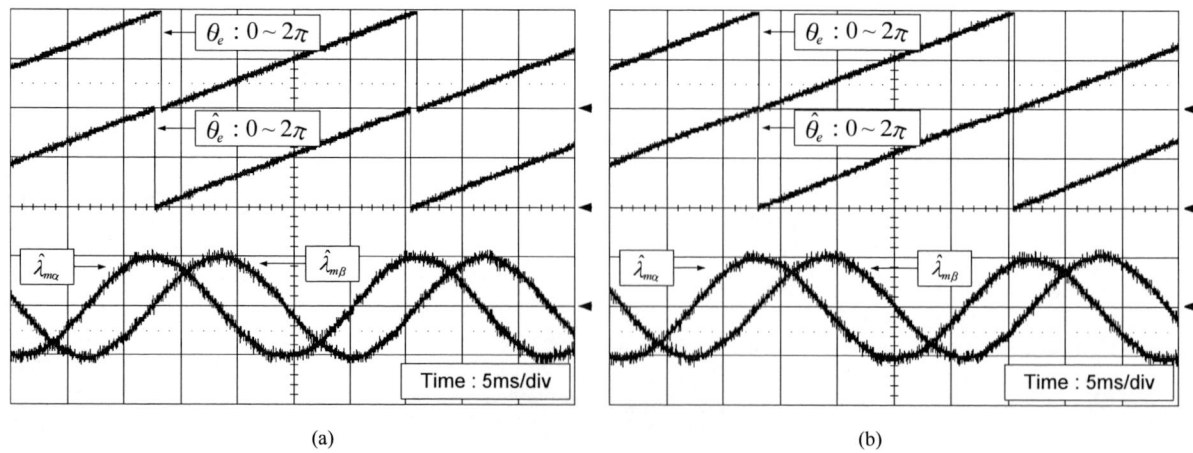

Fig. 7: Real and estimated rotor position angle (600 rpm)
(a) Without voltage distortion compensation
(b) With voltage distortion compensation

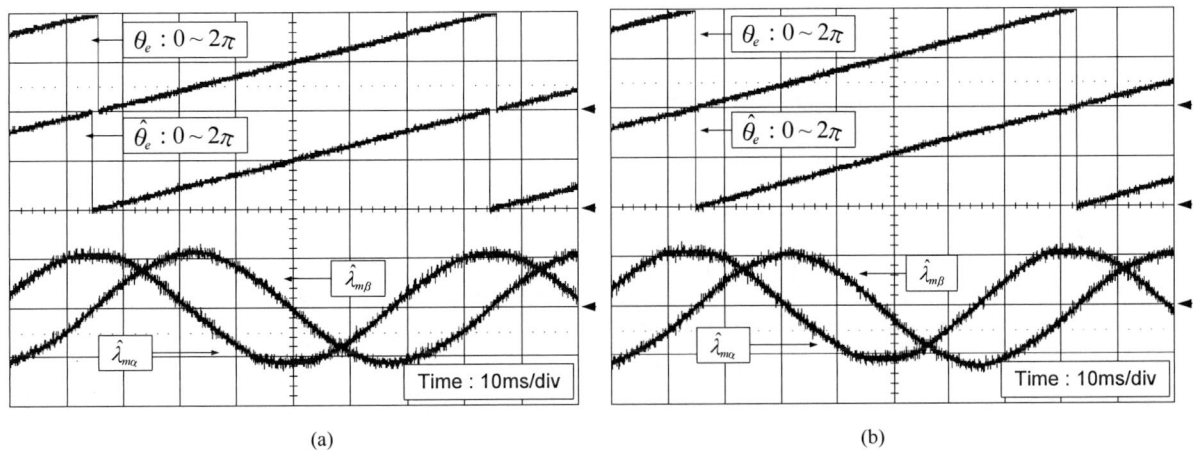

Fig. 8: Real and estimated rotor position angle (200 rpm)
(a) Without voltage distortion compensation
(b) With voltage distortion compensation

The voltage distortion of PWM VSI is caused by the intended blanking time and the inherent characteristics of the switching devices. Based on the analysis results, the voltage distortion observing method is presented using time delay control observer with the intermediate value, A_p, instead of a direct observing and compensation of voltage distortion. The experimental results show that this compensation method improves the estimation of rotor position angle and performance.

REFERENCES

[1] A. E. Fitzgerald, C. Kingsley Jr., and S. D. Umans, *Electric Machinery*. New York: McGraw-Hill, 1992

[2] P. C. Krause, *Analysis of Electric Machinery*. New York: McGraw-Hill, 1986

[3] H. Tajima and Y. Hori, "Speed sensorless field-orientation control of the induction machine", *IEEE Trans. Industry Applications*, Vol.29, pp.175-180, 1993

[4] X. Xu and D. W. Novotny, "Implementation of direct stator flux oriented control on a versatile DSP based system", *IEEE Trans. Industry Applications*, Vol.29, pp.694-700, 1991

[5] J. W. Choi and S. K. Sul, "Inverter output voltage synthesis using novel dead time compensation," *IEEE Trans. On Power Electronics*, Vol. 11, No.2, pp.221-227, 1996.

[6] A. R. Munoz and T. A. Lipo, "On-line dead-time compensation technique for open-loop PWM-VSI drives," *IEEE Trans. On Power Electronics*, Vol. 14, No. 4, pp.683-689, 1999.

[7] F. Blaabjerg, J. K. Pederson, and P. Thoegersen, "Improved modulation techniques for PWM-VSI drives,", *IEEE Trans. on Industrial Electronics*, Vol. 44, No. 1, pp.87-95. 1997

[8] H. S. Kim, K. H. Kim, and M. J. Youn, "On-line dead-time compensation method based on time delay control," *IEEE Trans. on Control System Technology*, Vol. 11, No.2, pp.279-285, 2003

[9] H. W. Kim, M. J. Youn, and K. Y. Cho, "New voltage distortion observer of PWM VSI for PMSM," *IEEE Trans. on Industrial Electronics*, Vol.52, No. 4, pp.1188-1192, 2005

Indirect Position and Speed Sensing for PMSM Sensorless Control

Driss Yousfi, *Member, IEEE*, and Mustapha El Adnani

Ecole Nationale des Sciences Appliquées - Marrakech
MESESYP Laboratory, Control Systems, Power Electronics and Electric Dives Group
Avenue Abdelkrim Elkhattabi, B.P. 575 Guéliz, Marrakech
Tel: +212 24 43 47 45 / 46, Fax: +212 24 43 47 40.
Email: dr_yousfi@yahoo.com

Abstract—This paper present a position and speed estimation algorithm based on measurement of phase currents and voltages and a pre-knowledge of the machine model. The aim is to elaborate an accurate estimator in accordance with some specifications, especially the ability to cancel any initial position error and robustness toward a dc-disturbance measured within the electric measurements. In earlier work, an old version of this method was investigated for open-loop position and speed estimation of permanent magnet motor. The present work provides complete description and Real-Time implementation of the method incorporated into sensorless vector control.

I. INTRODUCTION

There has recently been considerable interest in using the Permanent Magnet Synchronous Motor (PMSM) drives for industry applications. For successful operation of these drives, rotor position and speed are of critical importance. Due to the drawbacks of conventional mechanical sensors, indirect position and speed sensing draws wide attention [3,7-15]. This paper presents an estimation algorithm based on measurement of phase currents and voltages and the machine model. In earlier works [7,8], an old version of this method was investigated for an open-loop estimation of the mechanical quantities. The present work provides complete description and Real-Time implementation of the method incorporated into sensorless drive.

Furthermore, the problem due to a possible dc-disturbance measured within the motor back-emf is also taken into account. The proposed technique is based on flux linkage estimation and PI-controller. The accurate flux linkages are derived by integrating the back-emf. However, the flux linkages obtained by the integral operation of the stator quantities usually hold integral error. An improved integrator based on Low-Pass Filter is introduced to overcome this problem. Thanks to the algorithm conception no initial knowledge of either the position or the flux linkage is necessary. Thus the position and speed estimation can be correctly achieved.

Experimental results are included in order to demonstrate the robustness and the effectiveness of the drive under different operating conditions.

II. PMSM MODEL

According to Fig. 1, the PMSM models in the stationary frame (α-β) and the rotating frame (d-q) are respectively [1,2]:

$$v_\alpha = Ri_\alpha + \frac{d\psi_\alpha}{dt}$$
$$v_\beta = Ri_\beta + \frac{d\psi_\beta}{dt} \tag{1}$$

$$\psi_\alpha = L_\alpha i_\alpha + \psi_m \cos(\theta)$$
$$\psi_\beta = L_\beta i_\beta + \psi_m \sin(\theta) \tag{2}$$

and

$$v_d = Ri_d + \frac{d\psi_d}{dt} - w\psi_q$$
$$v_q = Ri_q + \frac{d\psi_q}{dt} + w\psi_d \tag{3}$$

$$\psi_d = L_d\, i_d + \psi_m$$
$$\psi_q = L_q\, i_q \tag{4}$$

where ψ_m is the maximum phase flux linkage of the permanent magnet, $L_{\alpha\beta}$ and L_{dq} the inductances in $\alpha\beta$ and dq frames, R the winding resistance and θ the actual rotor angle.

These models form the basis for the presented rotor position and speed estimator. As the considered PMSM is a non-salient machine, all the inductances in those equations are equal .i.e. $L_\alpha = L_\beta = L_d = L_q = L$.

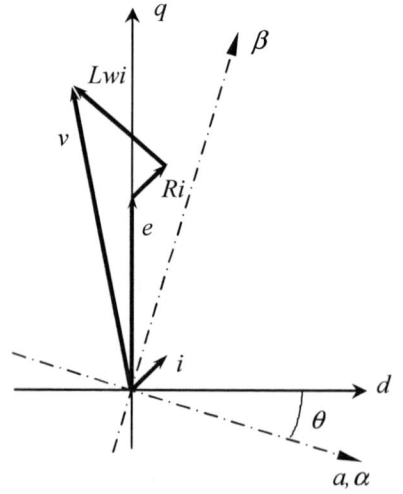

Fig. 1: Different frames and phasor diagram of PMSM.

III. POSITION AND SPEED ESTIMATION TECHNIQUE

The key equation in this estimation technique arises from the flux dependency on both the position and the current i.e.:

$$\Delta\psi = \frac{\partial\psi}{\partial\theta}\Delta\theta + \frac{\partial\psi}{\partial i}\Delta i . \qquad (5)$$

$\Delta\psi$, $\Delta\theta$ and Δi are flux, position and current errors.

In the aimed estimation technique, only the q-current error will be used, since the current error in the d-direction does not contain substantial information about an angular deviation.

Then, two coordinate systems are defined as shown in Fig.2: an actual frame $(d\text{-}q)$ identified by the angle θ, and a hypothetical one $(\hat{d}-\hat{q})$ identified by the angle $\hat{\theta}$ and the deviation $\Delta\theta = \hat{\theta} - \theta$ from the rotor position.

The projection of the PMSM flux linkage onto these frames leads to an expression similar to the partial derivatives (5):

$$\Delta\psi_{\hat{q}} = \psi_{\hat{q}} - \hat{\psi}_{\hat{q}} = \psi_m\Delta\theta + L_q\Delta i_q \qquad (6)$$

where

$$\Delta i_q = i_q - i_{\hat{q}} \qquad (7)$$

Position is estimated by assuming that the algorithm reduces the flux error to zero. Thus $\Delta\theta$ is derived from:

$$\Delta\theta = -\frac{L}{\psi_m}\Delta i_q \qquad (8)$$

This equation indicates that if the error Δi_q is controlled to be zero, the position error vanishes automatically.

A. Position and Speed Estimation Algorithm

The estimation algorithm illustrated schematically in Fig. 3. involves the following steps [8]:

Step 1:

Estimation of the flux linkage in $\alpha\beta$-frame by measuring the phase voltages and currents:

$$\hat{\psi}_{\alpha\beta} = \int (v_{\alpha\beta} - Ri_{\alpha\beta})dt \qquad (9)$$

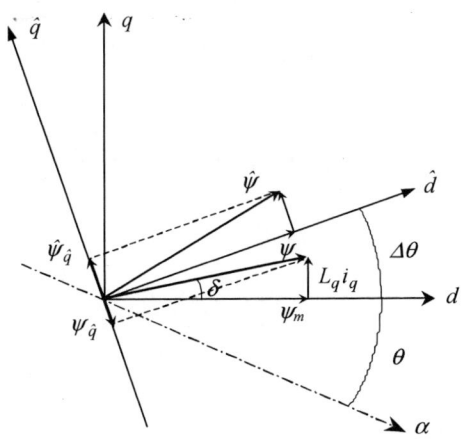

Fig. 2: Flux diagram under angular error.

Step 2:

Estimation of the motor currents $\hat{i}_{\alpha\beta}$:

$$\begin{aligned}
\hat{i}_\alpha &= \frac{1}{L}[\hat{\psi}_\alpha - \psi_m\cos(\hat{\theta})] \\
\hat{i}_\beta &= \frac{1}{L}[\hat{\psi}_\beta - \psi_m\sin(\hat{\theta})]
\end{aligned} \qquad (10)$$

Step 3:

Calculation of the difference between the actual currents and the estimated ones in $\alpha\text{-}\beta$ and $d\text{-}q$ frames using the Park transformation:

$$\Delta i_{\alpha\beta} = i_{\alpha\beta} - \hat{i}_{\alpha\beta} \qquad (11)$$

$$\Delta i_q = -\Delta i_\alpha\sin(\hat{\theta}) + \Delta i_\beta\cos(\hat{\theta}) \qquad (12)$$

This last current error is passed through a Low Pass filter to reduce the effect of the noise due to the differentiation.

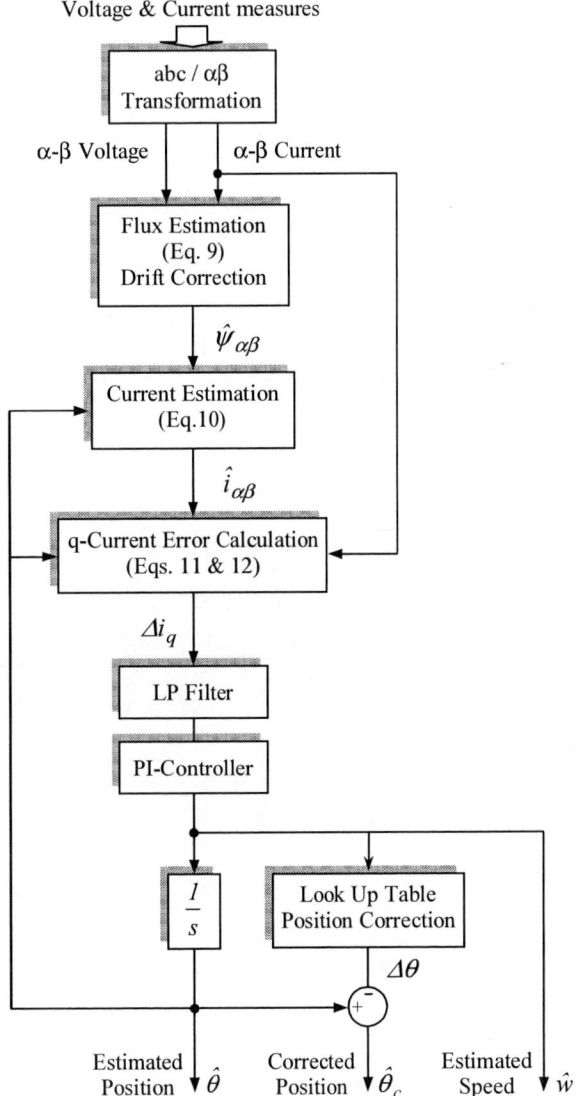

Fig. 3: Flowchart of the position and speed estimation algorithm.

Step 4:
Finally, a PI-Controller is used to cancel the obtained error by accelerating or decelerating the $\hat{d}-\hat{q}$ frame. Thus, the PI-Controller output is the estimated speed $\hat{w}$. The position estimate $\hat{\theta}$ can be deduced through an integration and corrected using Look Up Table procedure that gives a correction term $\Delta\theta$; so:

$$\hat{\theta}_c = \hat{\theta} + \Delta\theta \qquad (13)$$

The use of the LUT procedure in the estimation algorithm will be clarified in section IV.

The major problems witch must be resolved in this method results from the implementation of the flux integrator (Step 1) and the PI-controller (Step 4).

B. Flux Linkage integration

In practice, the implementation of an integrator for flux estimation is problematic because of the ramp-drift and the dc-offset produced on the output signal. A dc-component in measured motor back-emf is inevitable in experiment. This dc-component, no matter how small it is, can finally drive the pure integrator down to saturation. Furthermore, the initial value problem associated with the pure integrator occurs from the integral value at the initial instant [4,6].

To solve these problems and avoid complexity at the same time, a Low Pass Filter based Block (LPF-Block) is used. The block is illustrated in the Fig. 4-(a) and has obviously a very low cut-off frequency.

When the dc-component measured within the back-emf is weak, only one LPF-Block serves well for estimating the flux motor. In the opposite case, we are confronted with the problem of an offset persisting in the output $\psi_{\alpha\beta}$. At this stage, a second LPF-Block is used. The complete necessary arrangement is illustrated in Fig. 4-(b).

C. PI-controller with anti windup

In the 4th step, the aspect of the position correction system is understood from a linear model described by the expression (8). There are, however, some nonlinear phenomena that must be taken into account. There is typically limitation in the actuator: the motor has limited speed. For control system with a wide range of operating speed and a controller containing an integral action, it may happen that the control variable $\hat{w}$ reaches the actuator limits. When this happens the feedback loop is broken and the system runs in open-loop because the actuator will remain at its limit independently of the process output as long as the actuator remains saturated. The position estimation, and consequently, the motor flux synchronism deteriorates. This phenomenon is colloquially referred to as integrator windup. There are many ways to protect against windup. Tracking is a simple method which is illustrated in the block diagram of Fig. 5. The system has an extra feedback path around the integrator. The signal e_{sat} is the difference between the nominal controller output and the saturated controller output. This difference is fed to the input of the integrator through logical NOT. The signal e_{sat} is zero when there is no saturation. Under these circumstances it will not have any effect on the integration action. When the actuator saturates e_{sat} becomes different from zero and it disables the integrator. This implies that controller output is kept less than or equal to the saturation limit and integral windup is avoided.

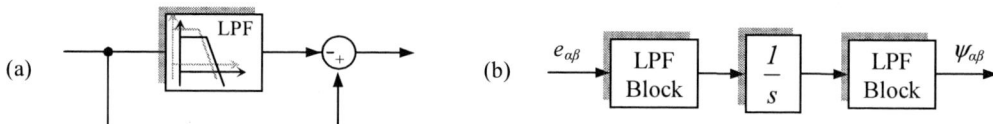

Fig. 4: (a) LPF-Block and (b) Layout made to correct the flux integration.

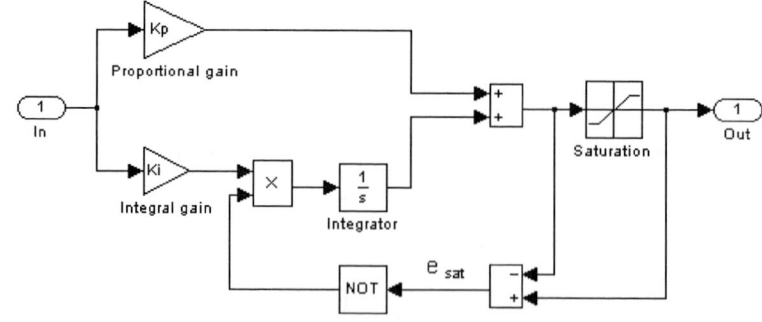

Fig. 5: PI-controller with anti windup.

978-1-4244-1871-8/07 $25.00 © 2007 IEEE   818

IV. EXPERIMENTAL RESULTS

The test stand consists of 6 poles PMSM, with parameters given in TABLE 1, assembled with the following elements:

- DS1104 controller DSP board (made by dSpace).
- Three-phase Voltage-Source-Inverter designed with six Insulated Gate Bipolar Transistors (IGBTs).
- LEM Hall-effect current sensors. Currents are measured in two phases only since the motor is Y connected.
- Voltage transducers built around isolation operational amplifiers.

Rotor position is measured by means of incremental encoder providing two-channel quadrature incremental signals with index pulse. The resolution is $360°/4096 \approx 0.088°$. Motor speed is calculated from position using a backward difference approximation.

Although position and speed are measured, they are used only for comparison with the estimates.

The sensorless vector control exemplified in Fig. 6 is implemented on the DSP board using measured currents and position and speed estimates feedback. The equations of the motion controller and the estimation algorithm are discretized using a simple first-order Euler approximation.

The cut-off frequency of the LPF-Block added to correct the flux estimation is taken very low (*5 rad/s*) to achieve good detection and elimination of the dc-components.

Concerning the startup, sensorless schemes are not self-starting. Many methods [16,17] are elaborated to detect the initial rotor potion. However, we opt for an open-loop starting to keep the algorithm simple since the initial position is not critical in the considered application. So, in order to sense the currents and voltages, the motor must first be started and brought up to a certain speed. In practice, open-loop starting is accomplished by generating a rotating stator field with a certain increasing frequency profile. After the speed reaches the required level, the back-emf can be detected, providing the position and speed information and the system switches to synchronous commutation mode.

Several experimental tests were performed to verify the transient and the steady state behaviour of position and speed estimation.

With rotor placed arbitrarily inside *360°electrical* (at *120°electrical* in this experiment, that is equivalent to *40°mechanical*), the PMSM drive system is run up to *25rpm* speed as stepper motor with sinus-wave excitation (Fig. 7). Then the estimation program initiated (at $T_{est}=2.5s$). Control is transferred to the sensorless signals afterwards (at $T_{SVC}=5s$). The speed estimate follows closely the actual one. The position estimation, however, presents a relatively large error ($\Delta\theta \approx 20°$) especially at such low speed. The machine continues to operate, but the control will not be efficient.

A simple analysis proves that the position error depends principally on the speed, slightly of the load (Fig. 8-without correction). Therefore, a Look Up Table with speed as input is performed and the deficiency is corrected by subtracting the error (Fig. 8-with correction). Henceforth, the correction routine will be inserted.

The results given in Fig. 9 show successful completely sensorless starting up from standstill to *800rpm* ($T_{est}=0.5s$ and $T_{SVC}=0.6s$).

The dynamical behaviour of the sensorless closed-loop system in response to positive and negative speed steps (*600-1100rpm*) is given in Fig. 10. The speed estimate always provides excellent tracking and the persisting position error is very small ($\Delta\theta \leq 0.8°$).

Finally, the close match between the actual and the estimated speed and position ($\Delta\theta \leq 1.5°$) in Fig. 11, justifies the efficiency of the method in considerable load case: *50%* of the rated torque at *400rpm* and *100%* at *950rpm*.

Unities: Speed in rpm, current in A and position (position error also) in degree-mechanical unless specified.

TABLE 1: PMSM PARAMETERS.

Rated Power	$P=500\ W$
Rated Current	$I=3.4\ Arms$
Rated Voltage	$V=230\ Vrms$
Winding resistance	$R=1.59\ \Omega$
Winding inductance	$L_d=L_q=3.3\ mH$
Maximum magnet flux	$\psi_m=0.052\ Wb$
Inertia (motor and load)	$J=35.73\cdot10^{-4}\ kgm^2$
Friction coefficient	$f_r=0.47\cdot10^{-3}\ Nm/rad.s^{-1}$

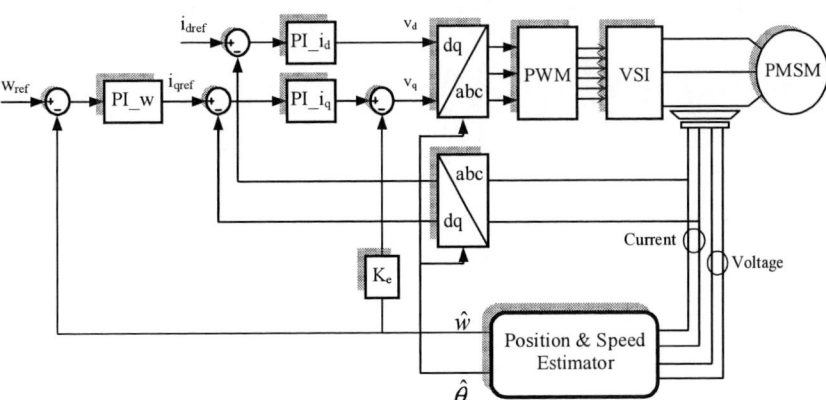

Fig. 6: Sensorless vector control using the Indirect Position and Speed Sensing.

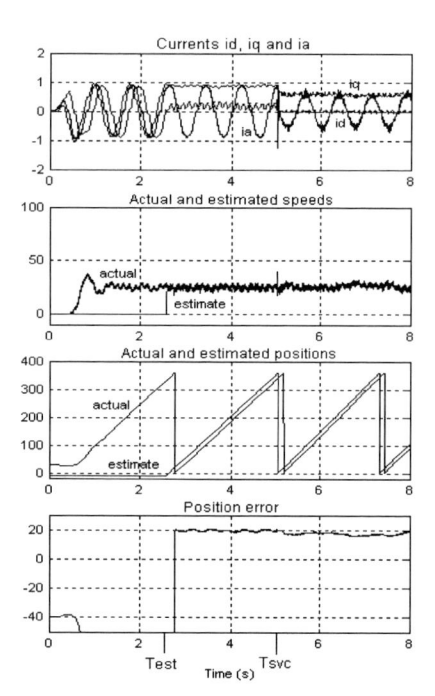

Fig. 7: Events chronology at the motor stating up.

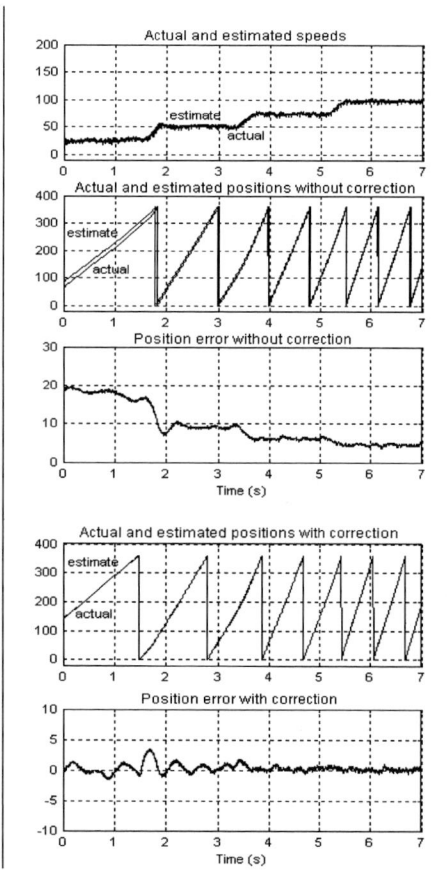

Fig. 8: Position error correction using LUT procedure.

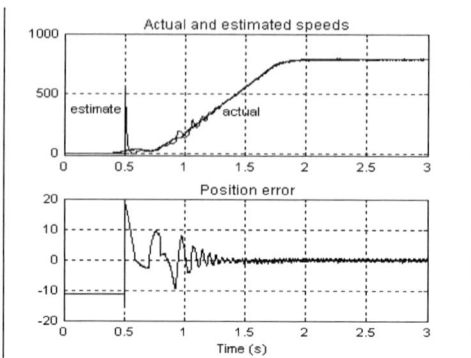

Fig. 9: Starting up to 800rpm at 20% of the rated torque.

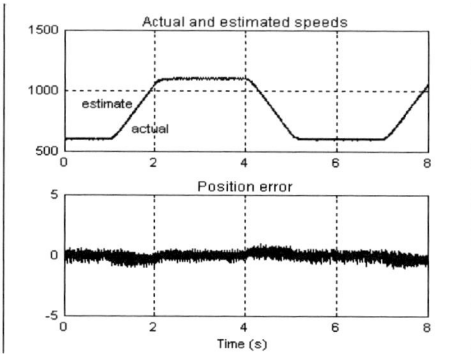

Fig. 10: Negative and positive speed steps (600→1100rpm).

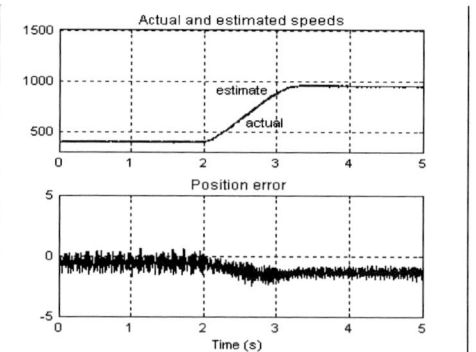

Fig. 11: Motor operation at 50% (400rpm) and 100% (950rpm)
of the rated torque.

V. CONCLUSION

The sensorless PMSM drive is demonstrated to operate in satisfactory manner with the presented method in wide speed range. The speed response is very good and the positional error is measured to be insignificant i.e. within *+/-1.5°* in all conditions.

The main advantages of the method arise from its self-adjusting feature gained thanks to the improved integrator and the PI-controller. They are:

− Robustness towards dc-component measured with the back-emf.

− No need for initial conditions, especially the initial value of the motor flux linkage.

− Ability to cancel any initial rotor position error.

− Efficiency in a low speed operation (*25 rpm*).

− Efficiency in both steady state and dynamic operation.

REFERENCES

[1] F. Piriou, A. Razek, "Modeling of synchronous motor drives using PWM controlled current for robotic applications", *Proc. of European Conference on Power Electronics and Application (EPE'87)*, pp. 1131-1134, 1987.

[2] J. P.Caron, J. P. Hautier, "Modélisation et commande de la machine synchrone," *Proc. of Journées 3.El'95*, France, 1995.

[3] L. Xu, M. Fu, "A novel sensorless control approach for PM motors using DSP," *Proc. of NAECON'97 Conference*, Dayton, July 1997.

[4] B. Terzic, M. Jadric, "Sensorless Brushless DC Motor Drive with Improved Speed Estimation Accuracy Using Stator Resistance Estimation," *Proc. of European Conference on Power Electronics and Application (EPE'99)*, Lausanne, Switzeland, Sep. 1999.

[5] M. Niemelä, J. Pyrhönen, "Drift Correction Method of the stator Flux Linkage in DTC Synchronous Motor Drives," *Proc. of European Conference on Power Electronics and Application (EPE'99)*, Lausanne, Switzerland, Sept. 1999.

[6] M. H. Shin, D. S. Hyun, S. B. Cho, S. Y. Choe, "An Improved Stator Flux Estimation for Speed Sensorless Stator Flux Orientation Control of Induction Motors," *IEEE Trans. Power Electron.*, vol. 15, no. 2, pp. 312-318, 2000.

[7] D. Yousfi, M. Azizi, A. Saâd, "Robust Position and Speed Estimation Algorithm for Permanent Magnet Synchronous Motor Drives," *Proc. of Annual Industry Applications Conf.*, vol. 3, pp.1541-1546, Rome (Italy), Oct. 2000.

[8] D. Yousfi, M. Azizi, A. Saâd, "New Position and Speed Estimation Technique for PMSM with Drift Correction of the Flux Linkage," *Electric Power Components and Systems Journal, Taylor & Francis Ed.*, vol. 29, no. 7, pp. 597-613, 2001.

[9] T. Lixin, M.F. Rahman, "A direct torque controlled interior permanent magnet synchronous machine drive with compensation of the forward voltage drops of the power switches", *Proc. of Industry Applications Conf.* , vol. 1, pp. 625- 631, Oct. 2003.

[10] M. Boussak, "Implementation and experimental investigation of sensorless speed control with initial rotor position estimation for interior permanent magnet synchronous motor drive," *IEEE Trans. Power Electron.*, vol. 20, no. 6, pp. 1413-1422, 2005.

[11] T.D. Batzel, K.Y. Lee, "Electric Propulsion With Sensorless Permanent Magnet Synchronous Motor: Implementation and Performance", *IEEE Trans. Energy Conversion*, vol. 20, no. 3, pp. 575-583, 2005.

[12] O. Wallmark, L. Harnefors, "Sensorless Control of Salient PMSM Drives in the Transition Region," *IEEE Trans. Industrial Electron.*, vol. 53, no. 4, pp. 1179-1187, 2006.

[13] N. Bianchi, S. Bolognani, F. Luise, "High speed drive using a slotless PM motor," *IEEE Trans. Power Electron.*, vol. 21, no. 4, pp. 1083-1090, 2006.

[14] T. D. Batzel and K. Y. Lee, "Electric propulsion with the sensorless permanent magnet synchronous motor: model and approach," *IEEE Trans. Energy Convers.*, vol. 20, no. 4, pp. 818–825, 2005.

[15] K.Y. Cho, S. B. Yang and C. H. Hong, "Sensorless control of a PM synchronous motor for direct drive washer without rotor position sensors," *IEE Proc., Electric Power Applications*, vol. 151, no. 1, pp. 61-69, 2004.

[16] D. W. Chung, J. K. Kong and S. K. SUI, "Initial rotor position detection of PMSM at standstill without rotational transducer," *Proc. of International Electric Machines and Drives Conference, (IEMDC'99)*, pp.785-787, Seattle, USA, May. 1999.

[17] T. W. Kim; J. Watanabe, S. Sonoda and J. Hirai, "New initial pole-position estimation of surface PM-LSM using reference currents," *IEEE Trans., Industry Appl.*, vol. 41, no. 3, pp. 817-824, 2005.

Hybrid Sensor-less Control of Permanent Magnet Synchronous Motor in Low-speed Region

Yasuhiro Yamamoto
Core Technology Department Division
Meidensha Corporation
Higashimakado 515, Numazu, Shizuoka, Japan,
410-8588
yamamoto-ya@mb.meidensha.co.jp

Satoshi Ogasawara
Division of Systems Science and Informatics
Hokkaido University
Kita 14, Nishi 9, kita-ku, Sapporo, Japan, 060-0814
oga@ist.hokudai.ac.jp

Hirohito Funato
Department of Electrical and Electronics
Engineering, Faculty of Engineering
Utsunomiya University
Yoto 7-1-2,Utsunomiya, Tochigi, Japan, 321-8585
funato@cc.utsunomiya-u.ac.jp

Abstract— **This paper proposes a method of improving the stability in sensor-less control of permanent magnet synchronous motors. The control method for low-speed region is divided into two: One is a high frequency method, which involves a problem of reverse rotation once misdetection of the permanent magnet polarity should occur, and another one is a current drive method, which has a problem that phase and speed oscillations are caused by quick speed changes.**

Hence, authors propose adoption of the current drive method for the basic control system with added compensation of stabilization by means of the high frequency method. This combination secures stable control with no risk of reversal and less vibration.

In addition, authors have also considered a frequency separation filter of a shorter delay time so that current control performance will not lower even when high frequencies are introduced. This filter has achieved simplified compensation using repetitive characteristic through the utilization of the periodicity of high frequency current.

Simulation and experiment have been conducted to verify that the stable performance of this system is improved.

I. INTRODUCTION

For sensor-less control of Interior Permanent Magnet Synchronous Motors (IPMSM), the control methods in a low-speed region are divided into two categories. One method is to supply constant amplitude current to the motor to revolve the rotor while forcibly attracting the magnetic poles of the rotor, which we call the current drive method here. Another method is to inject a high frequency voltage or current into the motor in order to estimate the phase of magnetic poles from the information of the detected high frequency components. Speed detection and vector control have been realized by using such estimated phase [1]-[4]. We call this technique a high frequency method.

A control system in which the high frequency method is applied can directly estimate the phase of magnetic poles. Use of such estimated phases helps attain stable control with minor phase fluctuations. However, this method has a critical problem: Once the function of polarity estimation should fail to correctly identify the N- and S-poles of the field magnet, or should it determine the polarity opposite, acceleration in the direction of reverse rotation would arise. The major cause of this phase error is the influence of the dead time. Since current amplitude becomes smaller to near zero, a large distortion occurs in the zero crossing of the multiple-phase current waveforms. This distortion causes disturbance in the phase estimation, likely to fail in the polarity detection. At a start-up stage, another means of measurement is needed to discriminate the N- and S-poles, and normally the characteristic of motor magnetic saturation is used for this purpose. Therefore, the high frequency method cannot be applied to the motors that are designed to have low magnetic saturation [5].

On the other hand, the current drive method is free of reversing problem caused by polarity misdetection because it does not use magnetic pole discrimination. However, it has a problem of stepping out when the motor is overloaded because the allowable load torque is low. In addition, IPMSM is subject to vibration of rotational phase when the speed or the load changes because they do not have any damper winding. In addition, IPMSM without damper windings are subject to vibration of rotational phase when the speed or the load changes. IPMSM is characterized by this weak point of being prone to vibration although they have the advantage of low moment of inertia.

Authors discussed the vibration problem in the current drive method, and as a result, this paper offers a method of additional feedback control to restrain vibration. This vibration restraining control uses phase estimation by the previously mentioned high frequency technique, but it does not directly use the estimated phase for control. The Phase is differentiated, and only the component that is equivalent to the speed vibration is extracted, which is then used only for vibration damping. More specifically, this control method has a strong point that it does not involve the risk of reverse rotation runaway, and it also provides a strong damping effect on vibration.

The high frequency technique requires a filter that splits detected current into high frequency components and other components. However, this filter contains a time delay, and therefore, the response characteristic of the current control system is lowered. As a measure against this weak point, we propose a separation filter that has a shorter delay time, in which we applied simple learning prediction that uses the high frequency periodicity. At the end, the effect of vibration restraint was evaluated by means of simulation and experiment.

II. The Principle of Our Proposed Control Method

A. Basic current control system

A control block diagram of the entire system is given in Fig. 1. This diagram contains the basic current control system and the stability compensation function by the added high frequency technique. In the current control, two current command components have been entered, including the amplitude $|I_1^*|$ and the frequency ω_1^*. This frequency ω_1^* is integrated to give an angle reference θ_γ, and the current regulator is implemented in a rotating reference frame (γ and δ axes) locked to this angle θ_γ. The reverse rotating conversion and two-phase/three-phase conversion convert the output voltage commands of the current regulator to three-phase voltage in the fixed frame for the PWM. In addition, a coordinate conversion that has a characteristic reversal to the voltage command was applied to the three-phase detected current i_u, i_v, and i_w, which were converted into orthogonal biaxial components i_γ and i_δ.

This current control system should set the sampling period to 250 µs or so to ensure that it is realized by a low-cost CPU. Consequently, the performance of current response will be 500 to 1000 rad/s or so. We adopt a new filter system that can be achieved even at this performance limit in the high frequency technique. The frequency separation filter of the high frequency technique has a feature to deteriorate in the separation characteristic if the input quickly changes. Accordingly, a current command first passes through the rate-of-change limiter, and then it is supplied to the current control.

B. Pole estimation using high frequency method

Normal IPMSM has the negative saliency ($L_d < L_q$) in the inductance of d-axis and q-axis. The phase estimation principle of high frequency technique utilizes that the inductance of the magnetic pole axis (d-axis) is the smallest.

In the case of low current control performance like in this situation, the method of detecting a high frequency current by supplying a high frequency voltage has an advantage in the accuracy. For high frequencies, a simple harmonic motion wave shape is adopted in the input voltage in order to reduce the torque pulsation components due to the high frequency current [1]. Fig. 2 shows the loci of the steady voltage vector v_{1_0}, the steady current vector i_{1_0}, and the high frequency trajectories that superimpose on the former vectors.

This section describes the principle of the magnetic pole phase estimation, which uses high frequency. Fig. 3 shows a diagram in which components are re-plotted so that the centers of high frequency loci of voltage and current will fall on the same point for clearly indicating the phase relation of high frequencies.

High frequency voltage, which is used in this system, supplies eight samples of voltage commands per cycle, with

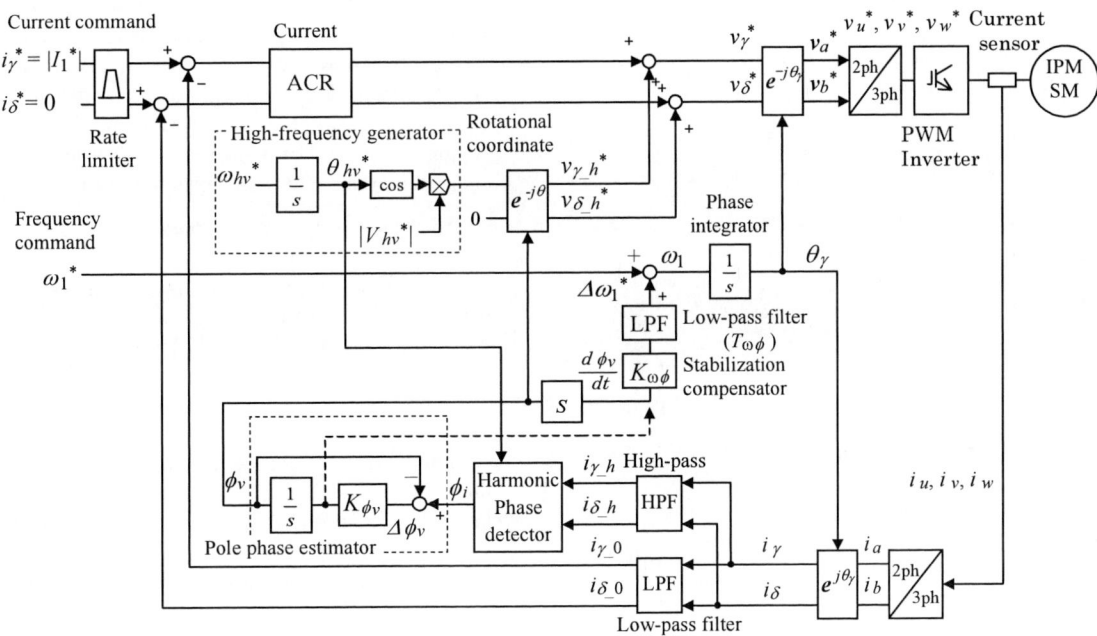

Fig. 1: Control block diagram of proposed system.

978-1-4244-1871-8/07 $25.00 © 2007 IEEE

the frequency $\omega_{hv}{}^*$ set to 500 Hz. As shown in Fig. 1, the high frequency command $\omega_{hv}{}^*$ is integrated to calculate the phase $\theta_{hv}{}^*$, and then a simple harmonic motion voltage command $v_{d\_h}{}^*$ is generated by the cosine function and the amplitude command $|V_{hv}{}^*|$. The reference axis θ_γ of current control is offset by ϕ_d from the magnetic pole axis, and here we assume that this phase offset can be estimated as the offset phase of high frequency voltage axis ϕ_v. The voltage v_{d_h} of d-axis component undergoes rotating coordinates by the degree of this ϕ_v, and it is converted into the components of $v_{\gamma\_h}{}^*$ and $v_{\delta\_h}{}^*$ in the γ-δ frame, then they are added to the steady voltage command.

Since the above operation generates a high frequency current, a frequency separation filter is used, which will be described in the next section, to extract the high frequency components from the current detection. Because IPMSM has negative saliency, the phase axis ϕ_i of the high frequency current appears in d-axial direction with smaller inductance from the voltage axis ϕ_v. Use of this feature enables us to find the magnetic pole phase that has the minimum inductance. Specifically, we apply the integral feedback, which can correct the previously mentioned high frequency voltage phase ϕ_v into the direction of the current generation axis ϕ_i. This feedback system leads to a convergence of phase in

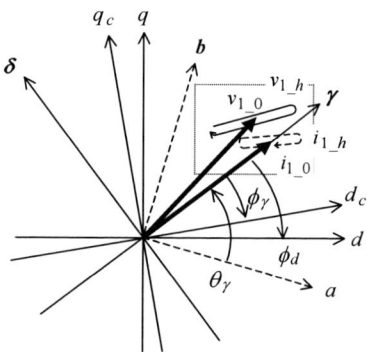

Fig. 2: Definition of the current control coordinates and the magnetic pole coordinates, and loci of the high frequency components.

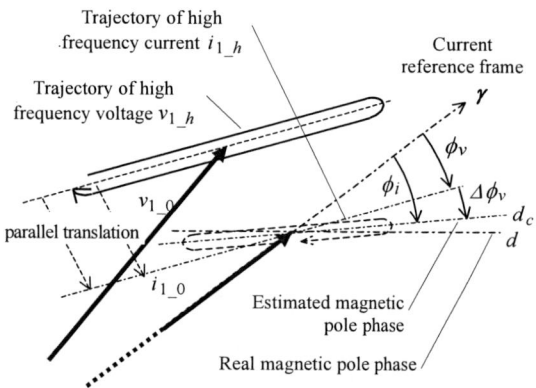

Fig. 3: Injected high-frequency voltage, and pole phase estimated from locus of the detection current.

which the axes of high frequency voltage and high frequency current meet each other, and this is the estimated phase of magnetic pole.

The separation filter of high frequency components i_{γ_h} and i_{δ_h} from the current detection i_γ and i_δ after the coordinate conversion is shown in Fig. 1. Furthermore, in the harmonic phase detection block, the amplitude component of high frequency current i_{γ_f} and i_{δ_f} is calculated according to the Fourier integral using the phase $\theta_{hv}{}^*$ that is used in generation of the voltage command. Then, the amplitude of these two axes causes the high frequency current axis phase ϕ_i by means of the arctangent function. Since the phase of high frequency current generation has been detected in the above manner, the phase ϕ_v can be corrected by means of integration of the phase difference $\Delta\phi_v$ between the voltage phase ϕ_v and the current phase ϕ_i. Consequently, when the phase difference $\Delta\phi_v$ is converged on zero as previously mentioned, ϕ_v means the pole phase ϕ_d of either the N-pole or the S-pole.

C.   Stabilization control

Because the pole phase ϕ_d can be estimated, it is possible to apply the compensation that corrects the reference phase θ_γ of the current control in proportion to ϕ_v, in order to restrain the speed vibration. This is constituted by the components of a pole phase estimator, a differentiation, a stabilization compensator, and a low-pass filter in Fig. 1.

Since distinction of the N-pole and the S-pole is not performed, an error of 180 degrees may be involved in the estimated pole phase. In order to eliminate the effects of this error, only a changed component has to be taken out from the estimated phase. Therefore, the compensator needs to differentiate the estimated phase once, and to integrate it in order to return to the phase component.

In practice, however, signals can be extracted at the point before the integrator ϕ_v to obtain the differential components, just as the route given by broken lines in Fig. 1. The integration can use the section of conversion from speed to phase, by adding the compensation speed $\Delta\omega_1{}^*$ to the speed command $\omega_1{}^*$. Use of this configuration does not require any other special differentiation or integration.

D.   High frequency separation filter

The current drive method requires a moderate level of current control performance for prevention of the current from fluctuation affected by speed changes. However, a delay time will take place if a low pass filter is used to eliminate high frequency components. The current control gain has to be reduced by such a delay time. Therefore, the simplified repetitive compensation that utilizes the periodicity of high-frequency components is applied to the filter. The structure of this frequency separation filter is shown in Fig. 4, and the operation time chart in Fig. 5. The structure in Fig. 4 is expressed in a discrete system that uses a delay operator z^{-1}, consisting of the following three elements.

First, the current detector samples the eight point data $i_\gamma[n]$, which correspond to one cycle, and calculate the moving average thereof. This moving average eliminates the high-pass components, and outputs the average value *ave* (i_γ[n]: i_γ[n-7]), as given in Fig. 5. However, the outputs contain the detection delay, therefore it cannot be used for current control. Next, the i_{γ_h}[n] is calculated based on the difference between the present value i$_\gamma$[n] and the moving average *ave*(i_γ[n]: i_γ[n-7]). It is supplied to the stabilization control as the high-pass components of filter, and at the same time, the value is saved internally for the coming compensation. Eventually, the component without the delay used in current control can be obtained by the following approach. Since the high-frequency components have periodicity, we assume that i_{γ_h}[n-8], in which eight samples of the present detected values i_{γ_h}[n] are stored, can be an approximation of the present high-frequency component. The present detected value i_γ[n], from which i_{γ_h}[n-8] is subtracted, outputs i_{γ_0}[n] from which high-frequency components have been eliminated, as shown in Fig. 5. This low-pass component is used in the current control.

As shown by the route given by broken lines in Fig. 4, both the high frequency components and the high frequency-eliminated components do not pass through the delay operator, which reveals that the separation filter has no time delay.

III. EVALUATION OF TRANSIENT CHARACTERISTICS

A. Evaluation by simulation

The effect of using this system was evaluated by means of simulation and experiment. The parameters of the motor and the control gains are listed in TABLE 1. In the simulation and experiment, the moment of inertia of the motor alone is used so that the vibration frequency will be the highest.

At first, the characteristic of frequency separation filter of current was examined, and the results are shown in Fig. 6. The amplitude of high frequency current was set to approximately 5%, and then, the separation characteristics were checked when the γ-axis current command i_γ^* was changed from 0 to 50%.

Since the separation characteristic of this filter deteriorates if there is any discontinuity, a rate-of-change limit of 100% per 10 ms or so is applied to the current command. The current control response was set to approximately 700 rad/s. The three-phase currents i_u, i_v, and i_w containing high

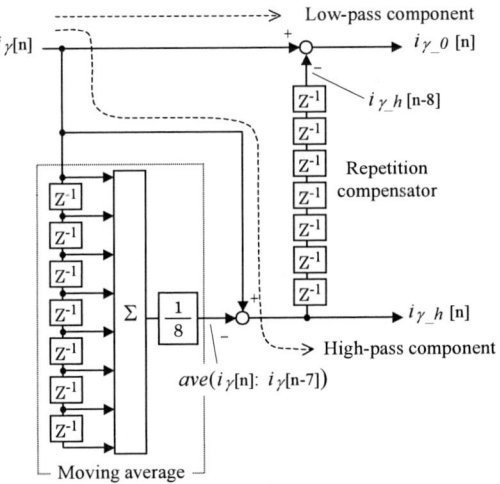

Fig. 5: Separation filter of high and low frequency component with repetitive compensation.

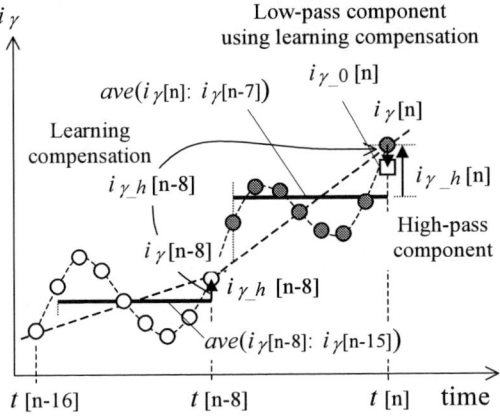

Fig. 6: Timing chart for principle description of high frequency separation using repetitive characteristics.

TABLE 1
Motor Parameters and Control Gain for Experiments

Parameter	Value
Power	37 kW
Pole	6 pole
Rated speed	3400 min^{-1}
Rated voltage	340 V
Rated Current	78 A
Resistance	28 mΩ
Inductance Ld	0.8 mH
Lq	2.4 mH
Moment of Inertia	0.029 kg-m^2
Control Gain	$K\phi_v$ = 0.0083 p.u. $K\omega\phi$ = 5.7 p.u./rad $T\omega\phi$ = 0.01 s

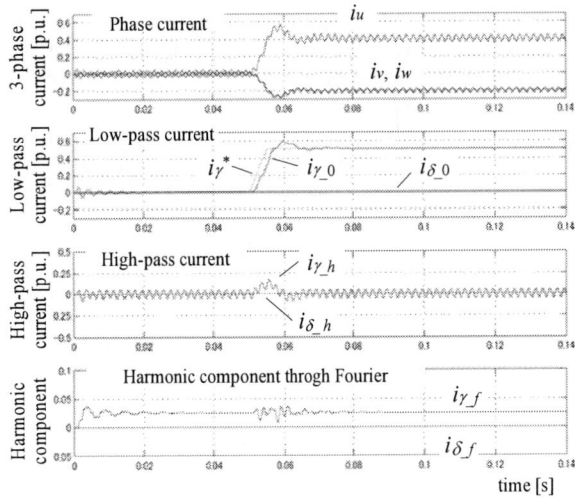

Fig. 4: Separation characteristics of detection current to high frequency and fundamental component.

978-1-4244-1871-8/07 $25.00 © 2007 IEEE

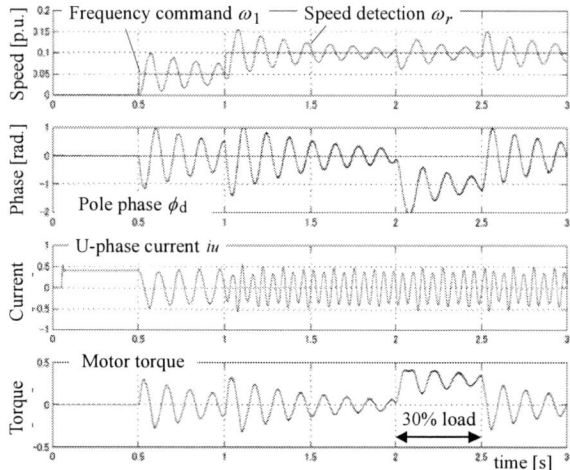

Fig. 7: Unstable characteristics without Stabilization compensator by Simulation.

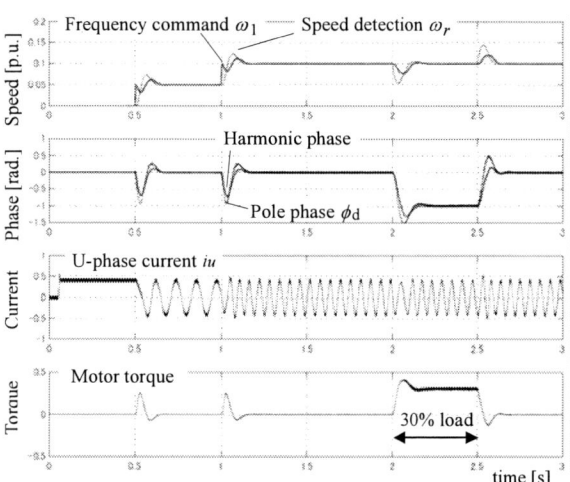

Fig. 8: Transient stability characteristics with proposed stabilization compensator by Simulation.

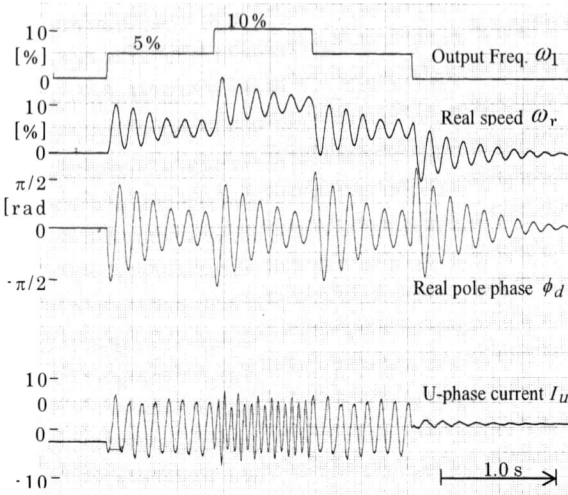

Fig. 10: Experimental result of transient characteristics without stability compensation.

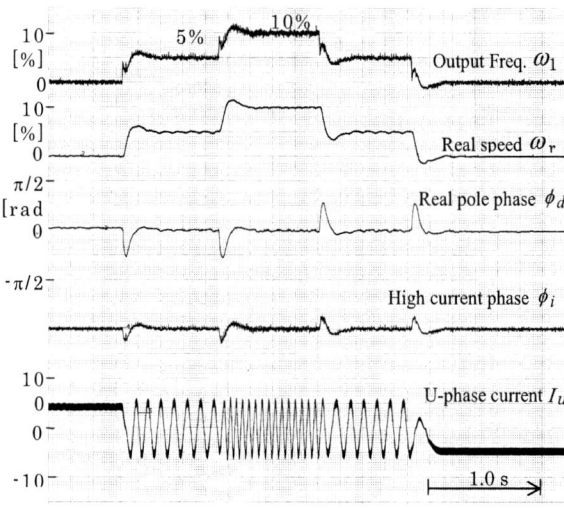

Fig. 9: Experimental result of transient characteristics with stability compensation.

frequencies were separated into each component of i_{γ_0} and i_{γ_h} by means of coordinates conversion and the separation filter. Here, the δ components (i_{δ_0}, i_{δ_h}) is zero because of the no load condition ($\phi_v = 0$). From this result, we confirmed that the separation characteristic of the filter is good and the delay time is short. Although DC components are generated in the high-pass current i_{γ_h} when the current changes, but the fundamental harmonic component $i_{\delta f}$ converted using Fourier integral did not find any influence. In other words, it is considered that little effects are seen on the high frequency current phase ϕ_i.

Comparison of vibration restraint effects is shown in Fig. 7 and Fig. 8. Control gains were set as described below. If we combine integrate and feedback, an equivalent low-pass filter results. Since the vibration frequency is about 7 Hz or so in our experiment, $K_{\phi v}$ was set so that a time constant will be 30 ms or so, which is a little longer than this time constant. The control gain $K_{\omega\phi}$ is a parameter of importance in the stability,

and its value was adjusted so that vibration will be smaller in our experiment. $T_{\omega\phi}$ was set to 10 ms so that it will restrain the residual ripple components of high frequency filter output.

Regarding the running conditions, we assume that the moment of inertia of load is small. The amplitude of current command i_γ^* is set to 50%. At two time points of 0.5 s and 1.0 s, the speed command was increased by 5%. Here, the speed command is subjected to the rate-of-change limit of 100% per 0.1 s. During the period from 2.0 s to 2.5 s, a step load torque of 30% is imposed. Fig. 7 shows that the actual speed and the actual phase are vibrating, and the current amplitude is also distorted affected by the electromotive force.

In contrast with the above, the application of vibration restraint triggers a correction so that the output frequency ω_1 will follow the actual speed ω_r, when the speed command fluctuates, as shown in Fig. 8. This correction can restrain the amount of overshoot of actual speed, offering a much more high-speed damping characteristics. When load fluctuation occurs, the degree of speed lowering becomes larger contrariwise to secure stability, but the characteristic of quickly damping the vibration is still offered.

In contrast with the above, the application of vibration restraint triggers a correction so that the output frequency ω_1 will follow the actual speed ω_r, when the speed command fluctuates, as shown in Fig. 8. This correction can restrain the amount of overshoot of actual speed, offering a much more high-speed damping characteristics. When load fluctuation occurs, the degree of speed lowering becomes larger contrariwise to secure stability, but the characteristic of quickly damping the vibration is still offered.

B. Evaluation by experiment

To verify the effects of simulation, acceleration and deceleration test was performed using an actual system. To detect the real phase and speed, position sensor was mounted on the motor. With the motor kept running alone, the speed and the phase fluctuation were measured while quickly changing the speed command because the condition for motor alone was the highest vibration frequency. The speed command was changed by three levels of 0%, 5% and 10%, and the rate-of-change limit of speed command were set to the same as in the simulation.

Fig. 9 shows the characteristics of the conventional current drive method, which does not apply any vibration restraint. This diagram indicates that vibration is occurring in the real speed and the real phase, which is similar to the simulation. Fig. 10 shows the characteristics of a case with added vibration restraint control by the high frequency technique. It is shown that the stability is improved, too, similar to the simulation.

The above results of experiment have confirmed that our proposed method is effective.

IV. CONCLUSION

Our study used the hybrid control, which combined the high frequency technique with the current drive method, and improved the stability. The key is to restrain speed vibration by utilizing the changing portion of the *d-axis* phase, without performing estimation of the N-pole and the S-pole. For this reason, our system can be applied to the motors that do not have any magnetic saturation characteristic.

Although this method has some shortcomings such as the small startup torque and the tendency of stepping out when overloaded, the method can be applied to a wide variety of motors. The position sensor-less control of IPMSM is limited in its possible applications because the contents of motor design affect the control performance. We expect that our suggestion may help users apply the sensor-less control to much more motors.

REFERENCES

[1] T. Aihara, A. Toba, and T. Yanase, "Sensor-less Torque Control of Salient-Pole Synchronous Motor at Zero Speed Operation," the 1996 Japan Industry Applications Society Conference, No.170, pp.1-2 (1996) (in Japanese).

[2] P. L. Jansen and R. D. Lorenz, "Transducerless position and velocity estimation in induction and salient AC machines," *IEEE Trans. Ind. Applicat.*, vol. 31, no. 2, pp. 1162-1169 (1995-3/4).

[3] M. Schroedl, "Sensorless control of AC machines at low speed and standstill based on the 'INFORM' method," *Conf. Rec. of IEEE IAS. Annual Meeting*, pp. 270-277 (1996-10).

[4] S. Ogasawara, and H. Akagi, "An approach to real-time position estimation at zero and low speed for PM motor based on saliency," in *Proc. IEEE Ind. Appl. Soc. Annual Meeting*, San Diego, CA, Oct. 5-10, 1996, pp.29-35.

[5] N. Bianchi, S. Bolognani, "Influence of Rotor Geometry of an IPM Motor on Sensorless Control Feasibility," *IEEE Trans. Ind. Applicat.*, vol. 43, no. 1, pp. 87-96 (2007).

A New Method for Smoothing Output Power Fluctuations of PV System Connected to Small Power Utility

Tomonobu Senjyu[1], Manoj Datta[1], Atsushi Yona[1], Hideomi Sekine[2]
[1]Department of Electrical and Electronics Engineering
[2]Departemt of Education Technology
University of the Ryukyus
1 senbaru, Nishihara, Okinawa 903-0213, Japan
Email: b985542@tec.u-ryukyu.ac.jp

Toshihisa Funabashi
Meidensha Corporation
Riverside Bldg 36-2, Nihonba-shi,
Hokozaki-cho, Chuo-ku, Tokyo 103-8515, Japan
Email: funabashi-t@mb.meidensha.co.jp

Abstract—A Photovoltaic (PV) system's power output is not constant and fluctuates depending on weather conditions. Fluctuating power causes frequency deviations in the power utilities when PV power penetration is large. Using a battery is one feasible measure to stabilize a PV system's power output, but it requires additional costs and results in additional waste of used batteries. In this paper, to overcome these problems, we propose a new method for leveling the fluctuations in a PV system's power output. By means of the proposed method, output power control of PV system considering the conditions of power utilities becomes possible and the conflicting objective of output power leveling and maximizing energy capture are achieved. Here, fuzzy reasoning is used to generate the output leveling power command. This fuzzy reasoning has three inputs of average insolation, variance of insolation, and absolute average of frequency deviation. Power converter is used to achieve output power same as command power employing PI control law. The proposed method is compared with the method where extracted maximum power is given to the utility without leveling. The simulation results show that proposed method is effective for leveling output fluctuations and feasible to reduce the frequency deviation of the small power utility.

I. INTRODUCTION

Mitigating the possibility of global warming is of great interest worldwide. The consumption of fossil fuels must be reduced, and clean and renewable energy sources must be introduced. Among various renewable energy systems, photovoltaic power generation systems (PV systems) are expected to play an important role as a clean power electricity source in meeting future electricity demands. However, the power output of PV systems fluctuates depending on weather conditions, season, and geographic location. In the future, when a significant number of PV systems will be connected to the grids of power utilities, power output fluctuation may cause problems like voltage fluctuation and large frequency deviation in electric power system operation [1], [2]. Therefore, for the penetration of large PV system's output power in the utility without reduction of the reliability of utility power systems, suitable measures must be applied to the PV systems side.

Statistical smoothing of power delivered to utilities has been reported in [3]. However, it is not helpful to reduce the real output power fluctuations and frequency deviation. Moreover, it is a statistical approach rather than a technical approach. On the PV system side, batteries can be used as smoothing device for a PV system's output. There have been investigations aimed at improving the performance of PV

systems equipped with batteries [4]-[7]. However, the additional cost of batteries is a barrier to the large scale installation of PV systems. In addition, used batteries must be disposed of without causing environmental problems. On the other hand, these investigations focused on peak power generation problem, load management, and outage attributes of PV power generation system. They have not given any emphasis on reduction of frequency deviation and the conflicting objectives of output power leveling and acquisition power increase cannot be achieved by them.

To overcome these shortcomings, in this paper, we propose a new method for leveling the fluctuations in a PV power generation system's output. This method uses fuzzy reasoning [8] to produce output power command and it has three inputs, which are absolute average of frequency deviation, average insolation and variance of insolation. Here, the output power command is decreased in time when large frequency deviation continuously occurs. Because the frequency deviation increases when big amount of output power with fluctuations due to rapid insolation change is given to the utility. On the other hand, output power command increases when frequency deviation is small. Hence, it is possible to control the output power command corresponding to power system condition by the proposed method. It is also important to consider the insolation condition besides the power system condition for realizing the conflicting objectives of both output power leveling and acquisition power increase. If insolation changes rapidly in a short period, the PV system's output will be smaller than available maximum power when the new operating point of PV array is not an optimal one.

Simple PI control law is employed to control the power converter for extracting output power equal to command power from the PV array. The proposed method is compared with the method where maximum available power is always fed to the utility without leveling. The simulation results using the actual detail model of PV generation system show the effectiveness and feasibility of the proposed method.

The paper is organized as follows: Section II provides concept of small power system, description of PV power generation system, control of power converter and brief review of solar module characteristics. Section III describes the command power generation system. In Section IV, effectiveness and feasibility of the proposed method is demonstrated by simulation results. Conclusions are drawn in Section V.

978-1-4244-1871-8/07 $25.00 © 2007 IEEE

II. SMALL POWER SYSTEM

The concept of small power utility in this paper is shown in Fig. 1. The small power utility consists of the diesel generators and PV systems that generate power to supply the demand. In addition, it is assumed that the small power utility is not connected to large power utility and it is always operated independently.

The small power system model which consists of diesel generator in detail, PV power generation system and load is shown in Fig. 2 where S_i is the insolation, P_{inv}^* is the command power generated by output power command system, ΔP_A is the generated power by PV power generation system, ΔP_d is generated power by diesel generators, R is the speed regulation, T_g is the governor time constant, T_t and T_r are the time constants, ΔP_L is the load, M is the inertia constant, D is the damping constant, and Δf is the frequency deviation of small power utility. As a frequency control method of power system, flat frequency control method which is used in majority of small power system is adopted.

In Fig. 3, PV power generation system including solar array, inverter and PI controller is shown where V_A is the solar array voltage, I_A is the solar array current, Δf is the frequency deviation, V_s is the generated supply voltage by the inverter, I_s is the generated supply current by the inverter, and $\left|I_{inv}^* - I_{inv}\right|$ is the error between command current and produced current. The control algorithm for the inverter [9] adopted here is very simple. The inverter output voltages and currents are sensed and transformed from 3-phase to synchronously rotating 2- phase. The command currents are generated dividing the output power command by sensed inverter voltage. Then the error between command inverter current and actual inverter current is processed through a PI controller to generate the PWM pulses. For maximum power extraction, the output power command is generated by maximum power point tracking algorithm. For simple structure and less costly implementation, a Perturbed and Observed (P&O) [10] algorithm was chosen in the present structure.

As the design of power converter and the control system is significantly influenced by the solar module characteristics, these will briefly reviewed here. The solar module is a nonlinear device and can be represented as a current source model, as shown in Fig. 4. The traditional *I-V* characteristics of a solar module is given by the following equation [11]:

$$I_0 = N_p I_g - N_p I_{sat}\left\{\exp\frac{qV_0}{AKT_a}\left(V_0 + \frac{N_s R_s I_0}{N_p}\right) - 1\right\} - I_{rsh} \quad (1)$$

where I_0 and V_0 are the output current and output voltage of the solar module, respectively, I_g is the generated current under a given insolation, I_{sat} is the reverse saturation current, q is the charge of an electron, K is the Boltzmann's constant, A is the ideality factor, T_a is the temperature (K), N_p is the number of cells in parallel, N_s is the number of cells in series and I_{rsh} is the current due to intrinsic shunt resistance of the solar module.

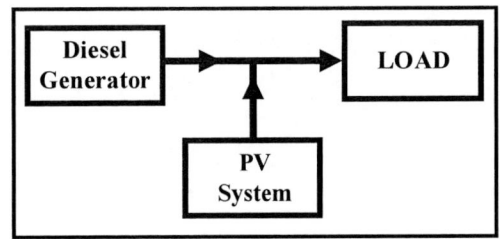

Fig. 1 Concept of small power utility.

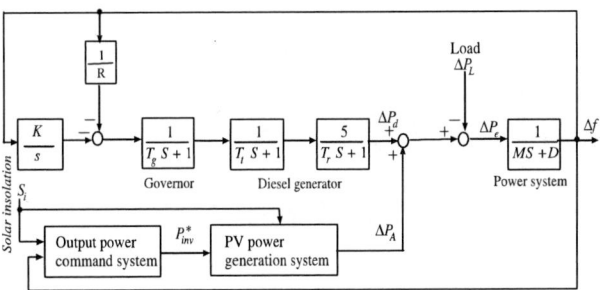

Fig. 2 Small power system model.

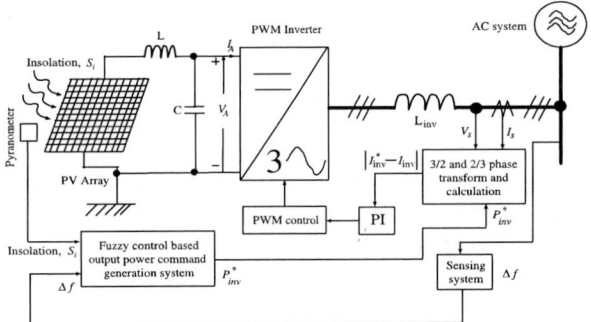

Fig. 3 Photovoltaic power generation system including inverter.

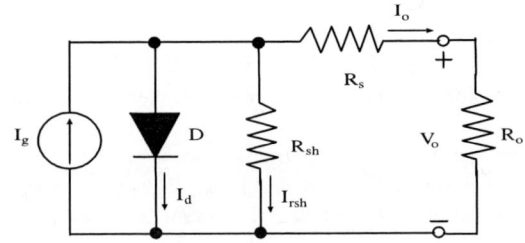

Fig. 4 Equivalent circuit of a solar module

The saturation current (I_{sat}) of the solar module varies with temperature according to the following equation [11]:

$$I_{sat} = I_{or}\left[\frac{T_a}{T_{ref}}\right]^3 \exp\left[\frac{qE_g}{KT_a}\left(\frac{1}{T_{ref}} - \frac{1}{T_a}\right)\right] \quad (2)$$

$$I_g = \left[I_{sc} + I_t(T_a - T_{ref})\right]\frac{S_i}{1000} \quad (3)$$

where I_{or} is the saturation current at T_{ref}, T_a is the temperature of the solar array (K), T_{ref} is the reference temperature, E_g is the band gap energy, I_t is the short circuit current temperature coefficient, I_{sc} is the short circuit current of solar module, and S_i is the insolation .

978-1-4244-1871-8/07 $25.00 © 2007 IEEE

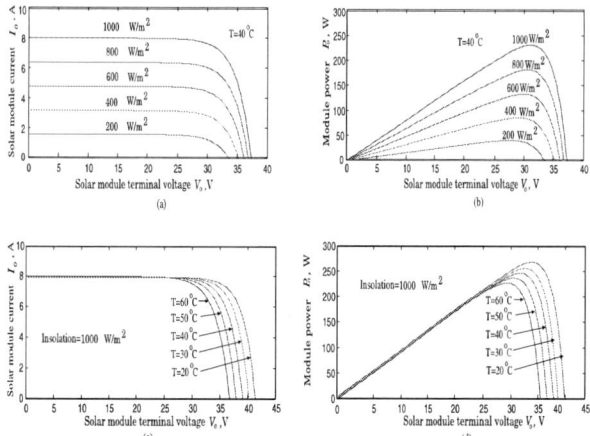

Fig. 5 Solar module characteristics curves. (a) Current voltage curve at constant temperature. (b) Power-voltage curves at constant temperature. (c) Current-voltage curve at constant insolation. (d) Power-voltage curve at constant insolation.

The current due to the shunt resistance is given by the following equation [11]:

$$I_{rsh} = \frac{V_0 + N_s I_0 R_s}{N_s R_{sh}} \tag{4}$$

where R_{sh} $\psi$is the internal shunt resistance of the solar module.

The solar module output power is given by the following equation:

$$P_0 = V_0 I_0 \tag{5}$$

Equations (1)-(5) are used in the development of computer simulations for the solar module. The MATLAB/SIMULINK is used. Figs. 5(a) and (b) show the simulated ampere-volt and power-volt curves for the solar module for different insolations at constant temperature. Figs. 5(c) and (d) show the simulated ampere-volt and power-volt curves of the solar module for different temperature at constant insolation. From these curves, it is observed that the output characteristics of the solar module are nonlinear and vitally affected by the variation of insolation. However, the variation of temperature does not affect the output curves significantly. Therefore, for the present case study, only insolation variation effects are taken in to account.

III. OUTPUT POWER COMMAND GENERATION SYSTEM

In order to control output power of PV system considering power system condition, output power command P_{inv}^* is decided by output power command system shown in Fig. 6. Output power command system consists mainly of two fuzzy reasoning and the rate of rated output power for PV array is decided by these fuzzy reasoning. Fuzzy reasoning are described by a set of "if-then" rules based on fuzzy rules. Fuzzy reasoning is effective when mathematical expressions are difficult by inherent complexity or nonlinearity.

Firstly, Fuzzy reasoning I is explained. There are two inputs of fuzzy reasoning I. One is absolute average of

frequency deviation Δf_s and the other is average insolation $\overline{S}_i$. The former, which is an index to estimate power system condition, is expressed by

$$\Delta f_s = \frac{1}{T} \int_{t-T}^{t} |\Delta f| dt \tag{6}$$

where t is present time and $T\psi$is integral interval. Since absolute value of frequency deviation Δf is used, absolute average of frequency deviation Δf_s increases or decreases with increase or decrease in frequency deviation Δf of the power system. Therefore, (6) indicates frequency deviation quantitatively at any given time. Average insolation $\overline{S}_i$ is defined by

$$\overline{S}_i = \frac{1}{T} \int_{t-T}^{t} S_i dt \tag{7}$$

where S_i $\psi$is instantaneous insolation. Fuzzy rules and membership functions of Fuzzy reasoning I are shown in TABLE I and in Fig. 7, respectively. Here, output power control of PV system according to power system condition is accomplished by using absolute average of frequency deviation Δf_s as input of fuzzy reasoning. It is undesirable to increase output power command of PV system considerably by insolation condition because probability of insolation decrease in short times is high. The important thing considered in this paper is prevention of large frequency deviations. The frequency deviation Δf of magnitude more than ± 0.2Hz is prevented rather than increase of generated power of PV system. Thus, membership functions are decided so that decreasing tendency is strong compared with increasing tendency of output power command of PV array. When frequency deviation Δf deviates by more than ± 0.2Hz in certain times, fuzzy rules and membership functions that yield an output to decrease or increase the output power command are defined by trial-and-error. The ith of fuzzy rules is expressed as

Rule i : if Δf_s is L_x and $\overline{S}_i$ is M_y then γ_I is Z_l

$$x = 1, 2, \cdots, 7, \ y = 1, 2, \cdots, 7, \ l = 1, 2, \cdots, 7 \tag{8}$$

where L_x, M_y denote the antecedents and Z_l are consequent part. Fuzzy reasoning output γ_I is calculated by

$$\gamma_I = \sum_{i=1}^{49} w_i Z_l \bigg/ \sum_{i=1}^{49} w_i \tag{9}$$

where w_i denotes the grade for the antecedent and is obtained by

$$w_i = w_{\Delta f_s i} w_{\overline{s}_i i} \tag{10}$$

where $w_{\Delta f_s i}$ and $w_{\overline{s}_i i}$ are the grade of antecedents for each rule.

Secondly, fuzzy reasoning II is explained. Absolute average of frequency deviation Δf_s and variance of insolation σ^2 are used as inputs of fuzzy reasoning II, where variance of insolation σ^2 is expressed as

$$\sigma^2 = \frac{1}{T} \int_{t-T}^{t} (S_i - \overline{S}_i)^2 dt \tag{11}$$

978-1-4244-1871-8/07 $25.00 © 2007 IEEE   830

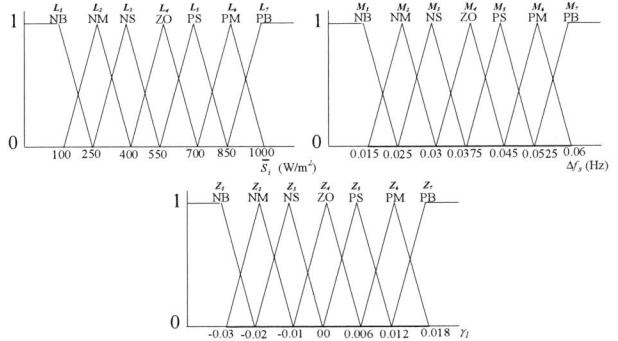

Fig. 7 Membership functions of fuzzy reasoning I

TABLE 1
FUZZY RULES OF FUZZY REASONING I

		Δf_s						
		NB	NM	NS	ZO	PS	PM	PB
	NB	ZO	NS	NM	NB	NB	NB	NB
	NM	PS	ZO	NS	NM	NB	NB	NB
	NS	PM	PS	ZO	NS	NM	NB	NB
$\overline{S}_i$	ZO	PB	PM	PS	ZO	NS	NM	NB
	PS	PB	PB	PM	PS	ZO	NS	NM
	PM	PB	PB	PB	PM	PS	ZO	NS
	PB	PB	PB	PB	PB	PM	PS	ZO

NB=Negative Big   NM=Negative Medium   NS=Negative Small
PB=Positive Big   PM= Positive Medium   PS= Positive Small
ZO= Zero

Output power command that depends on power system condition rather than insolation condition is decided by using absolute average of frequency deviation Δf_s as input for both fuzzy reasoning I and fuzzy reasoning II. In addition, when insolation fluctuations are large, the variance of insolation σ^2 is used as one of the inputs since the objective is to decrease frequency deviations. Fuzzy rules and membership functions of fuzzy reasoning II are shown in TABLE II and in Fig. 8, respectively. Setup of fuzzy rules and parameters of membership functions are determined to prevent the boosting of frequency deviation. The oth of fuzzy rules is expressed as

Rule o : if Δf_s is H_c and σ^2 is I_z then γ_{II} is Q_h (12)

$$c = 1, 2, \cdots, 7, z = 1, 2, \cdots, 7, h = 1, 2, \cdots, 7$$

where H_c, I_z denote the antecedents and Q_h are consequent part. Fuzzy reasoning output γ_I is calculated by

$$\gamma_{II} = \sum_{i=1}^{49} w_o Q_h \Big/ \sum_{i=1}^{49} w_o \quad (13)$$

where w_o denotes the grade for the antecedent and is obtained
by

$$w_o = w_{\Delta f_s o} w_{\sigma^2 o} \quad (14)$$

where $w_{\Delta f_s o}$ and $w_{\sigma^2 o}$ are the grade of antecedents for each rule.

As can be seen from Fig. 6, the discrete value $u(k+1)$ is obtained by the sums of output of fuzzy reasoning I, γ_I, and fuzzy reasoning II, γ_{II} through zero-order-hold, and the rate

of current rated output power of PV array. Then, new rate of rated output power of PV array $\gamma(k+1)$ becomes output power command by the following equation:

$$\gamma(k+1) = \gamma(k) + u(k+1) \quad (15)$$

Moreover, since the rate obtained by (14) changes step, it is necessary to convert it into a smooth output power command. Linear output power command is obtained in each sampling time by using the following equation:

$$P_{inv}^* = P_{rated} \left\{ \gamma(k) + \frac{\gamma(k+1) - \gamma(k)}{T_s} f(t) \right\}$$
$$= P_{rated} \gamma_l \quad (16)$$

Finally, (16) becomes output power command P_{inv}^* to the PV power generation system, where P_{rated} is rated output power, T_s is sampling time, $f(t)$ is periodic function such that $f(t) = t$ for $0 < t < T_s$ and γ_l is linear output power command.

IV. SIMULATION RESULTS

In this paper, the effectiveness of output power leveling of PV array and frequency deviation reduction of power system using the proposed method is examined by simulation with system model and parameters as mentioned in [11]-[13]. In order to use parameters of real PV system given in [13], [14], the rated output power of the PV array is 241kW. Simulation parameters of power system, PV array, power converter and PI controller are shown in TABLE III. Here, integral time T is 100 s, sampling time T_s to obtain discrete value of output power command is 10 s, and sampling time of PI controller is 1 ms. Simulation time is 30 minutes and the averaging sample time of insolation is 20 s.

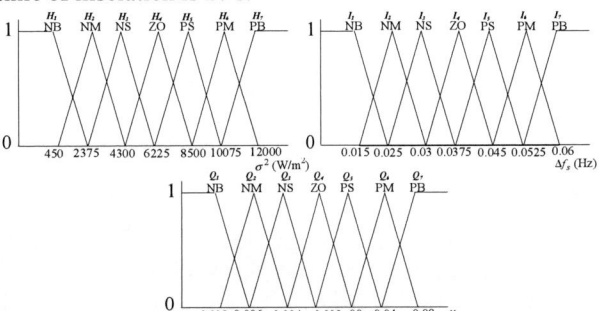

Fig. 8 Membership functions of fuzzy reasoning II

TABLE 2
FUZZY RULES OF FUZZY REASONING II

		Δf_s						
		NB	NM	NS	ZO	PS	PM	PB
	NB	PB	PB	PB	PB	PM	PS	ZO
	NM	PB	PB	PB	PM	PS	ZO	NS
	NS	PB	PB	PM	PS	ZO	NS	NM
σ^2	ZO	PB	PM	PS	ZO	NS	NM	NB
	PS	PM	PS	ZO	NS	NM	NB	NB
	PM	PS	ZO	NS	NM	NB	NB	NB
	PB	ZO	NS	NM	NB	NB	NB	NB

NB=Negative Big   NM=Negative Medium   NS=Negative Small
PB=Positive Big   PM= Positive Medium   PS= Positive Small
ZO= Zero

TABLE 3
SIMULATION PARAMETERS

Parameters of small power system	
Inertia constant, M	0.150 puMW.sec/Hz
Damping constant, D	0.008 puMW/Hz
Governor time constant, T_g	0.10 sec
Time constant, T_t	0.25 sec
Time constant, T_r	8.0 sec
Speed regulation, R	2.5 Hz/puMW

Parameters of PV array	
Rated output power	241 kW
Open circuit voltage	588.80 V
Short circuit current	520.37 A
Number of modules in series	16
Number of modules in parallel	65
Total number of cells	62,400

Parameters of PV module	
Rated output power	231.58 W
Open circuit voltage, V_{oc}	36.30 V
Short circuit current, I_{sc}	7.99 a
Shunt resistance, R_{sh}	50 Ω
Series resistance, R_s	5 Ω
Ideality factor, A	1.450
Inverse Saturation Current, I_{or}	3.047e-07 A
S.C. current temperature constant, I_t	1.73e-03 A/^{0}K
Reference temperature, T_{ref}	25 ^{0}C
Boltzman's constant, K	1.38e-23
Charge of an electron, q	1.602e-19 C
Bandgap voltage, E_g	1.11 eV
Number of cells in series, N_s	60
Standard insolation, S_i	1000 W/m^2
Dimension	0.75 m^3

Parameters of Power converter	
Inverter power rating	225 kW
Nominal ac output voltage	480 V (3-φ)
Nominal ac output frequency	50 Hz
Maximum ac line current	271 rms
Maximum dc input voltage	600 V
Maximum dc input current	781 A
Efficiency	94.5%

Parameters of PI controller	
Proportional constant, K_P	0.1
Integral constant, K_I	10

Fig. 9 shows insolation, average insolation, variance of insolation, absolute average of frequency deviation, and load used for simulation study. Fig. 9(a) shows insolation S_i by solid line and average insolation $\bar{S}_i$ by dotted line. Fig. 9(b) shows variance of insolation σ^2. The variance is very high as insolation changes so rapidly. Load ΔP_L is shown in Fig. 9(c). Fig. 9(d) shows absolute average of frequency deviation. As the variation of load is high from 600 s to 1000 s, absolute average of frequency deviation is also high at these times.

The comparative simulation results of conventional MPPT control and proposed control are shown in Fig. 10 to evaluate their performance in reducing power system frequency deviation. Fig. 10(a) shows the output power ΔP_A of MPPT control by dotted line and output power of proposed control by solid line. From Fig. 10(a), it can be seen that PV power produced by MPPT control fluctuates randomly with the insolation variation. On the other hand, PV power produced by the proposed method is smooth. However, power loss is introduced with the proposed method. As the main target of

the proposed method is to reduce frequency deviation, loss in output PV power cannot be removed by the proposed method. Fig. 10(b) shows the output power of diesel generator ΔP_d with MPPT control by dotted line and with proposed control by solid line. From Fig. 10(d), it can be said that, since output power ΔP_A produced by the proposed method is smooth, diesel generator output power with the proposed method is mainly controlled corresponding to the fluctuation component for load ΔP_L Therefore, diesel generator output power produced with the proposed method becomes smooth compared with diesel generator output power produced with MPPT control. Fig. 10(c) shows frequency deviation Δf of MPPT control by dotted line and proposed method by solid line. From Fig. 10(c), it is observed that frequency deviation of proposed method is not maintained within the range of ± 0.2 Hz. However, frequency deviation of proposed method is maintained within the range of o0.2Hz. From Fig. 10(a), it is seen that, from t=200s to t=550s, the frequency deviation is small, so PV output power produced by the proposed method has increasing tendency. From t=550s to t=1,000s, the load variation is high, therefore, PV output power$\psi$is reduced and maintained at constant value to decrease the frequency deviation. From t=1,000s to t=1,800s, the frequency deviation produced by load is decreased, so the output power command is increased maintaining a correlation with maximum power point. Therefore, output power control of PV system accoding to insolation and power utility condition is possible by the proposed method.

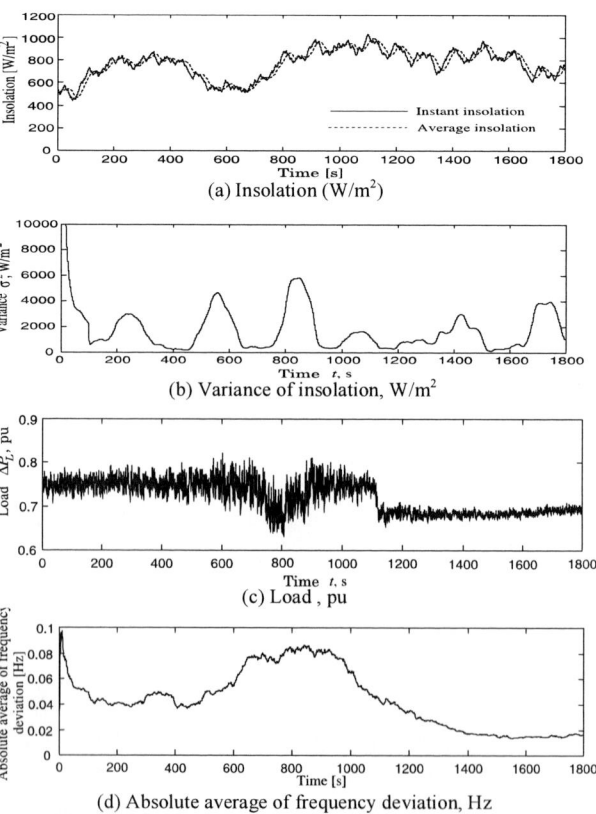

(a) Insolation (W/m^2)

(b) Variance of insolation, W/m^2

(c) Load , pu

(d) Absolute average of frequency deviation, Hz

Fig. 9 Insolation pattern, load, and frequency deviation. (a) Insolation and average insolation. (b) Variance of insolation. (c) Load. (d) Absolute average of frequency deviation.

Outputs of Fuzzy reasoning I, II and the rate of linear rated output power command of PV array γ_l are shown in Fig. 10(d), (e), (f), respectively. It is confirmed that output of Fuzzy reasoning I, II change in response to each of the inputs, which are absolute average of frequency deviation $\overline{\Delta f_s}$, average insolation $\overline{S_i}$, and variance of insolation σ^2. However, sum of fuzzy reasoning I, II is discretized, and this sum is smoothed by using (16).

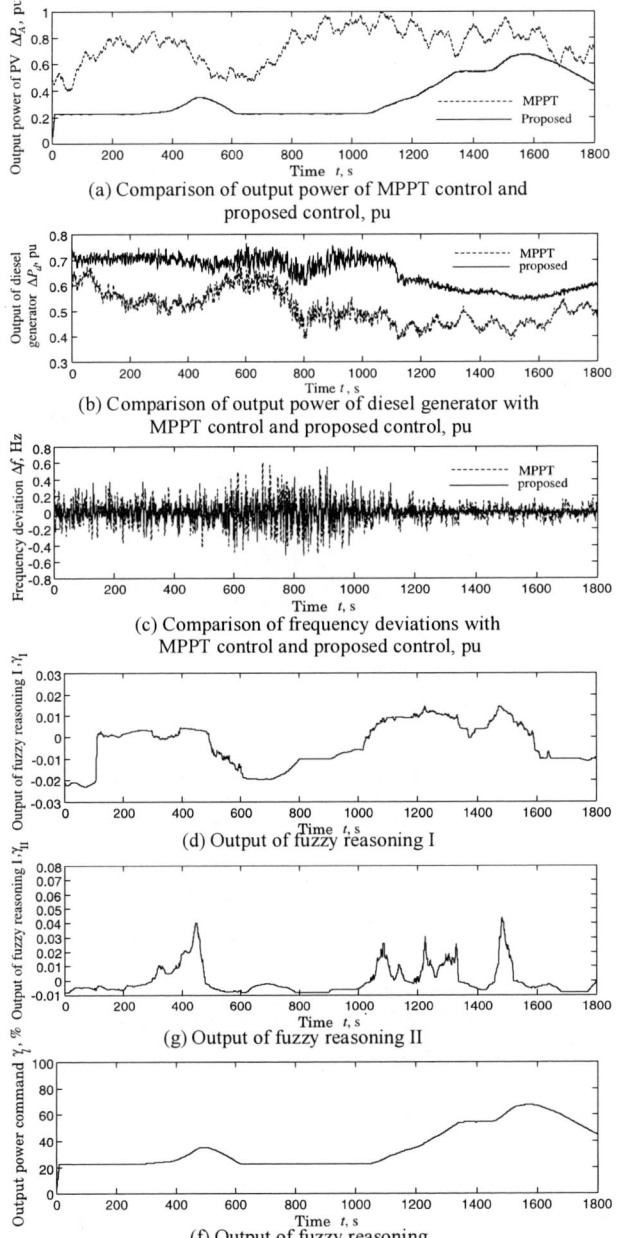

(a) Comparison of output power of MPPT control and proposed control, pu

(b) Comparison of output power of diesel generator with MPPT control and proposed control, pu

(c) Comparison of frequency deviations with MPPT control and proposed control, pu

(d) Output of fuzzy reasoning I

(g) Output of fuzzy reasoning II

(f) Output of fuzzy reasoning

Fig. 10 Comparative simulation results of MPPT control and proposed method.

V. CONCLUSIONS

This paper presents output power leveling of PV array considering power system condition and insolation condition. Output power command is defined by fuzzy reasoning which has three inputs of absolute average of frequency deviation, average insolation, and variance of solar insolation. Setup of fuzzy rules and parameters of membership functions are determined by prioritizing to prevent increase of frequency deviation. It is possible to control output power of PV array by the proposed method in all possible operating regions. The PV array is controlled with the help of a PI controller and a power converter to level the output power for reducing harmful influence on the power system at times with large frequency deviation. On the other hand, at times with small frequency deviation, the proposed system can operate to increase generated power near available maximum power. From the simulation results, it can be said that the PV power generation system with the proposed method achieves flexible output power control without introducing any harmful effects to power system, and the effectiveness of the proposed method is confirmed.

REFERENCES

[1] S. Yanagawa, T. Kato, K. Wu, A. Tabata, and Y. Suzuoki, "Evaluation of LFC capacity for output fluctuation of photovoltaic generation systems based on multi-point observation of insolation," in *Proc. IEEE Power Engineering Society Summer Meeting*, 2001, pp. 1652-1657.

[2] A. Woyte, V.V. Thong, R. Belmans, and J. Nijs, "Voltage fluctuations on distribution level introduced by photovoltaic systems," *IEEE Trans. Energy Convers.*, vol. 21, no. 1, pp. 202-209, March 2006.

[3] K. Otani, A. Murata, K. Sakuta, J. Minowa, and K. Kurokawa, "Statistical smoothing of power delivered to utilities by distributed pv systems," in *Proc. 2nd World Conference and Exhibition on Photovoltaic Solar Energy Conversion*, Vienna, Austria, 1998.

[4] T. Kinjo, T. Senjyu, N. Urasaki, and H. Fujita, "Output levelling of renewable energy by electric double layer capacitor applied for energy storage system," *IEEE Trans. Energy Convers.*, vol. 21, no.1, pp. 221- 227, March 2006.

[5] M. H. Rahman, S. Yamashiro, and K. Nakamura, "Development of an advanced grid-connected PV-ECS system considering solar energy estimation," *T. IEE Jpn.*, vol. 125-B, no.4, pp. 399-407, 2005.

[6] J. P. Barton and D. G. Infield, "A probabilistic method for calculating the usefulness of a store with finite energy capacity for smoothing electricity generation from wind and solar power," Journal of Power Sources., vol. 162, pp. 943-948, 2006.

[7] R. Wanger, "Large lead/acid batteries for frequency regulation, load levelling and solar power applications," Journal of Power Sources, vo. 67, pp. 163-172, 1997.

[8] L. H. Tsoukalas and R. E. Uhrig, Fuzzy and Neural Approaches in Engineering, New York, USA: Jhon Wiley & Sins. Inc., 1997.

[9] K. Kobayashi, I. Tanako, and Y. Sawada, "A study of a two stage maximum power point tracking control of a photovoltaic system under partially shaded insolation conditions," Solar Energy Materials & Solar Cells, vol. 90, pp. 2975-2988, 2006.

[10] N. Femia, G. Petrone, G. Spagnuolo, and M. Vitelli, "Optimization of perturb and observe maximum power point tracking method," IEEE Trans. Power Electron., vol. 20, no. 4, pp. 963-973, July 2006.

[11] C. Hua, J. Lin, and C. Shen, "Implementation of a DSP controlled photovoltaic system with peak power tracking," IEEE Trans. Ind. Electron., vol. 45, no. 1, pp. 99-107, Feb. 1998.

[12] T. Senjyu, T. Kaneko, A. Yona, N. Urasaki, T. Funabashi, F. Yamada, and S. Sugimoto, "Output power control for large wind power penetration in small power system," presented at the IEEE PES General Meeting, Florida, USA, 2007.

[13] [Online]. Available: http://www.powerlight.com/success/pdf/LVVWD Ronzone Reservoir.pdf.

[14] [Online]. Available http://www.xantrex.com/web/id/214/p/1840/pt/23/ product.asp.

Case study of distrubution-unified power flow controller (D-UPFC) in the clustered PV system

Kyungsoo Lee, Kenichiro Yamaguchi, and Kosuke Kurokawa
Tokyo University of Agriculture and Technology
Department of Electronics and Information Engineering
2-24-16, Naka-cho, Koganei, Tokyo, 184-8588 Japan
Email: onnuri@cc.tuat.ac.jp, 50007645129@st.tuat.ac.jp, kurochan@cc.tuat.ac.jp

Abstract—This paper shows the case study of D-UPFC in the clustered PV system. D-UPFC mainly controls distribution voltage and thus, it is installed in the low-voltage distribution system connects with clustered PV system. Proposed D-UPFC topology is shown and compared with existing topology. The proposed topology can decrease the transformer capacity compared with existing topology. In the case study, voltage control and dynamic characteristic of D-UPFC are analyzed. Forward power flow, reverse power flow, and load power factor change conditions are simulated using ATP-EMTP tool.

I.   INTRODUCTION

In the power flow condition, faults occuring in power distribution systems or facilities in plants generally cause the voltage sags or swells. Also, power systems supply power for a wide variety of different user applications, and sensitivity to voltage sags and swells varies widely for different applications [1].

A few voltage control methods have been developed. Static var compensator (SVC) regulates over- and under-voltage conditions by controlling its reactive power. Autotransformer with line drop compensator based step voltage regulator (SVR) selects suitable voltage using a switch during voltage change. Also, scheduled operation controls distribution line voltage in the substation [2].

These voltage control methods concerns during forward power flow condition. Also, they are performed not the low-voltage distribution system but the high-voltage distribution system.

Reverse power flow happens when clustered PV system connects with distribution system. Voltage increase phenomenon happens due to reverse power flow. When the voltage increase occurs in the low-voltage distribution system, it affects to stop generating power from clustered PV system or to trouble distribution system equipments.

This paper shows the case study of D-UPFC in the distribution system. D-UPFC is a voltage controller in the low-voltage distribution system. When the voltage decrease is happened due to load consumption power low, D-UPFC controls the voltage rapidly. Also, when voltage increase condition due to reverse power flow from clustered PV system occurs, D-UPFC regulates the distribution voltage.

Proposed D-UPFC topology is compared with existing D-UPFC topology. Proposed topology can decrease transformer size, weight, and capacity because it uses only

one secondary side winding. Using this proposed topology, the case study of the distribution model is performed. Distribution voltage control during RL load condition and reverse power flow condition is simulated. The dynamic voltage control due to load power change and reverse power flow is also verified using ATP-EMTP software.

II.   D-UPFC THEORY

A.   Proposed Topology

D-UPFC consists of transformer and bi-directional ac-ac converter in the low-voltage distribution system. The transformer of the existing D-UPFC topology is divided one primary side and two secondary sides. Bi-directional ac-ac converter is connected in the upper side of the secondary. Fig. 1 shows the existing D-UPFC topology [2,3].

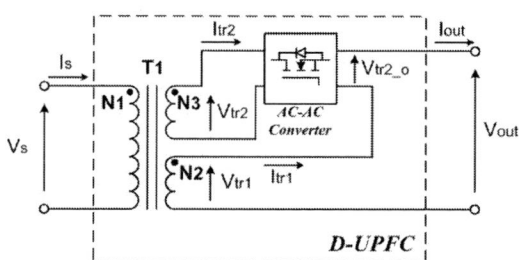

Fig. 1: Existing D-UPFC topology

D-UPFC output voltage equation can be expressed,

$$V_{out} = \frac{N_2 + (D \times N_3)}{N_1} \times V_s = V_{tr1} + (D \times V_{tr2}) = V_{tr1} + V_{tr2\_o} \quad (1)$$

Where, D is duty ratio of the bi-directional ac-ac converter.

Transformer of the proposed topology is divided one primary side and one secondary side. Tap voltage N_1' of primary side is added to secondary side voltage. It is similar to auto-transformer topology. Proposed D-UPFC topology is shown in Fig. 2.

D-UPFC output voltage equation can be expressed,

$$V_{out} = \frac{N_1' + (D \times N_2)}{N_1} \times V_s = V_{tr1} + (D \times V_{tr2}) = V_{tr1} + V_{tr2\_o} \quad (2)$$

978-1-4244-1871-8/07 $25.00 © 2007 IEEE

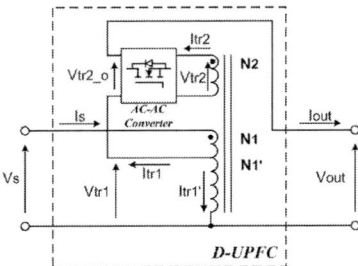

Fig. 2: Proposed D-UPFC topology

Transformer turn's ratio N_1' is the tap which is located in the primary side. Thus, Proposed D-UPFC topology can decrease the transformer size, weight and capacity compared with the existing topology.

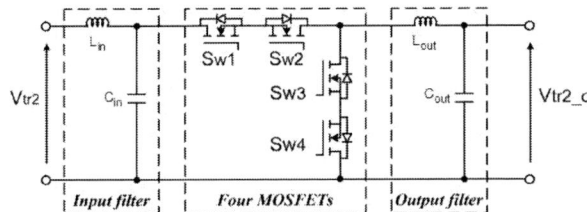

Fig. 3: Bi-directional ac-ac converter circuit

Bi-directional ac-ac converter from the Fig. 2 is shown in Fig.3. The equation of this converter is,

$$V_{tr2\_o} = D \times V_{tr2} \qquad (3)$$

B. Voltage Control Method

D-UPFC input voltage V_{in} is always controlled by reference voltage V_{ref_dc}. V_{in} is changed from ac to dc through RMS function. V_{ref_dc} is 202[V,rms], which is low-voltage distribution system voltage. Error voltage V_{error} between V_{in} and V_{ref_dc} is through PI compensator. V_{ref_duty} which is the reference duty of ac-ac converter is added to V_{error_pi}. V_{pwm} compares with V_{tri} in the PWM function. Switches S_{w1} to S_{w4} are operated by PWM function. D-UPFC voltage control block is shown in Fig. 4.

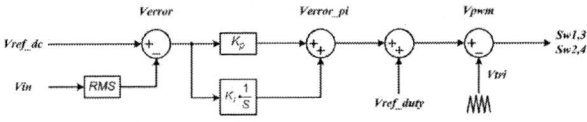

Fig. 4: D-UPFC voltage control block

Bi-directional ac-ac converter can directly transfer ac power to ac power without large energy storage devices. Also, it can control the voltage during power flow change and load change conditions. These conditions are realized using the switching patterns [3, 4].

Considering the clustered PV system connects with low-voltage distribution system, the power flow and load power factor should be changed. Forward power flow, reverse power flow and load power factor change can be expressed using phase diagram. Fig. 5 shows the phase diagram of input voltage and output current relation in the bi-directional ac-ac converter.

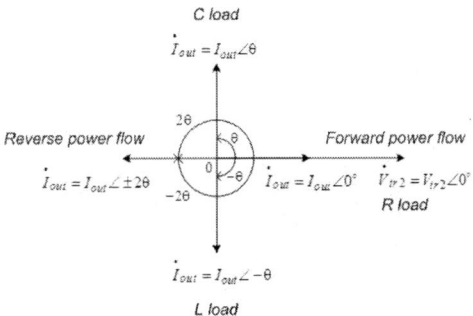

Fig. 5: AC-AC converter input voltage and output current phase diagram considering power flow and load power factor

III. CASE STUDY

D-UPFC voltage control in the low-voltage distribution system is shown. Voltage decrease and increase control are simulated considering load power factor change. D-UPFC dynamic voltage control also simulated using ATP-EMTP.

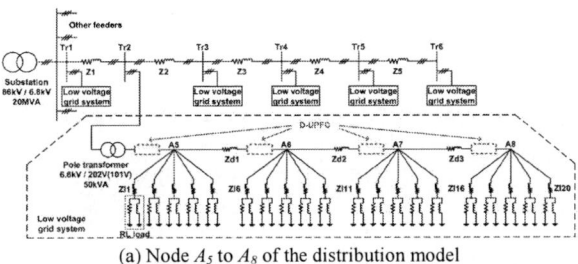

(a) Node A_5 to A_8 of the distribution model

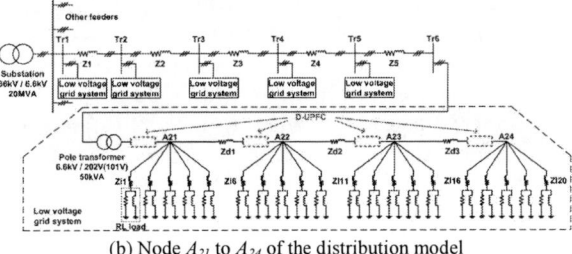

(b) Node A_{21} to A_{24} of the distribution model

Fig. 6: Distribution model for voltage decrease condition in the RL load

D-UPFC simulation models are shown in Fig. 6 and 7. They are assumed the residential area of Japan. Detailed model explanation is shown in reference [2]. Fig. 6 shows voltage decrease model due to heavy load condition.

978-1-4244-1871-8/07 $25.00 © 2007 IEEE

D-UPFC installation is shown from A_5 to A_8 of Tr_2, from A_{21} to A_{24} of Tr_6 pole transformer.

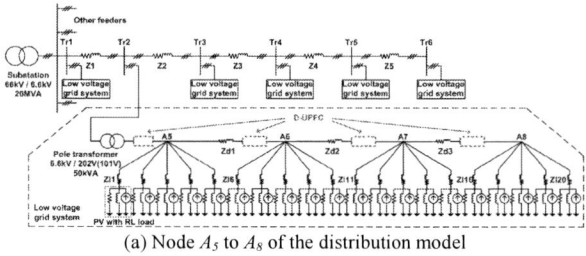

(a) Node A_5 to A_8 of the distribution model

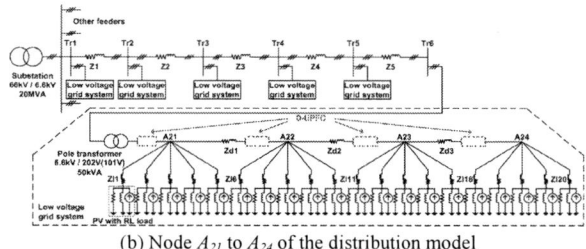

(b) Node A_{21} to A_{24} of the distribution model

Fig. 7: Distribution model for voltage increase condition in the RL load with clustered PV system

Fig. 7 shows the voltage increase model due to reverse power flow from clustered PV system. D-UPFC installation is shown from A_5 to A_8 of Tr_2, from A_{21} to A_{24} of Tr_6 pole transformer.

Table 1 shows the distribution model parameters. Power factor is assumed 0.9 in the RL load condition. Capacity of a PV system is regarded as 3[kW].

TABLE 1. Distribution model parameters

Substation		66kV/6.6kV, 20MVA
Pole transformer		6.6kV/202V(101V), 50kVA
HV line impedance(Z_1 to Z_5)		0.626+j0.754[Ω/2km]
LV line impedance(Z_{d1} to Z_{d3})		0.025+j0.020[Ω/40m]
Lead-in wire imp.(Z_{i1} to Z_{i20})		0.0552+j0.037[Ω/20m]
Total load	Light load	4.08+j2.028[Ω] (p.f. 0.9)
	Heavy load	1.02+j0.507[Ω] (p.f. 0.9)
Each PV power		3[kW]

D-UPFC parameters are shown in Table 2. The maximum D-UPFC voltage control range is ±20.2[V,rms] from transformer voltage tap N_2. Input and output LC filters reduce input current and output voltage harmonics [5]. D-UPFC output is the same as distribution system voltage 202[V,rms] during normal mode.

TABLE 2. D-UPFC parameters

V_S	202[V,rms]	C_{in} & C_{out}	50[μF]
N_1: N_1': N_2	1:0.9:0.2	$V_{ref\ dc}$	202[V,rms]
V_{tr1}	181.8[V,rms]	PI gain	K_p=0.025 K_i=0.001
V_{tr2}	40.4[V,rms]	Switching freq.	20[kHz]
L_{in} & L_{out}	50[μH]	$V_{ref\ duty}$	0.5

A. Voltage decrease control

Voltage decrease control is shown in Fig. 8. These simulation results are performed when heavy load is connected to the distribution system. Heavy load parameters are shown in Table 1.

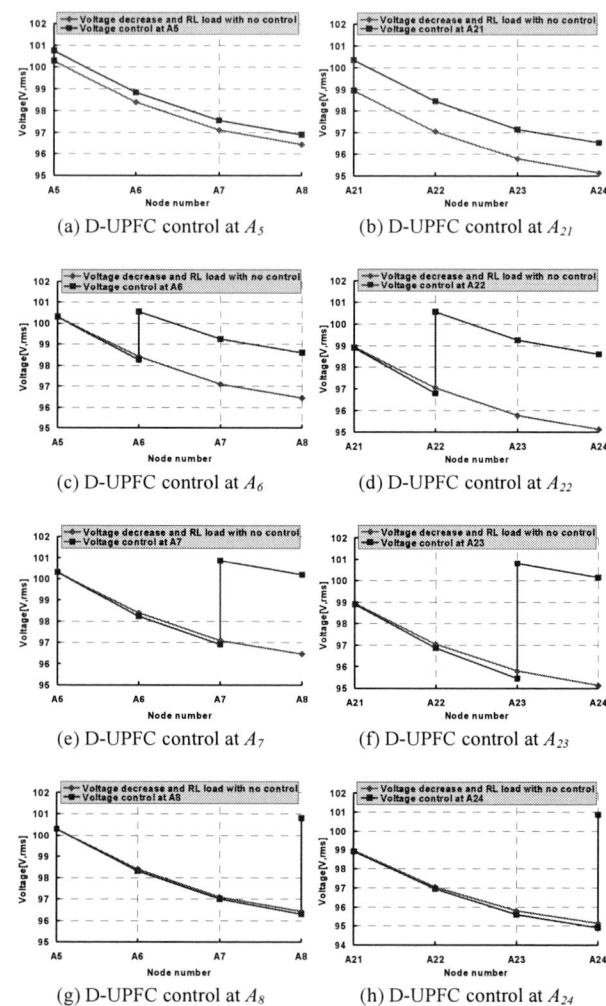

(a) D-UPFC control at A_5 (b) D-UPFC control at A_{21}

(c) D-UPFC control at A_6 (d) D-UPFC control at A_{22}

(e) D-UPFC control at A_7 (f) D-UPFC control at A_{23}

(g) D-UPFC control at A_8 (h) D-UPFC control at A_{24}

Fig. 8: Voltage decrease control in the RL load condition

As the Fig. 6, D-UPFC is installed in the secondary of the pole transformer. The low-voltage distribution voltage range is 202±20 (101±6)[V,rms]. Fig. 8 shows the 100[V,rms] line results. Before the D-UPFC control voltage decrease from node A_5 (a) to A_8 (g) is from 100.3[V,rms] to 96.5[V,rms], respectively. However, D-UPFC controls 100.7[V,rms] to 100.8[V,rms] from node A_5 to A_8, respectively. Also, voltage decrease from node A_{21} (b) to A_{24} (h) is from 99[V,rms] to 95.2[V,rms], respectively. These voltage decreases are controlled from 100.4[V,rms] to 100.9[V,rms], respectively. Thus, D-UPFC controls distribution voltage to reference voltage at the installation site.

Fig. 9 shows D-UPFC inner voltage (V_{tr1}, V_{tr2}, V_{tr2_o}), output voltage (V_{out}), output current (I_{out}), and ac-ac conveter switches (S_{w1} to S_{w4}) current waveforms at node A_{24}. When the voltage decrease occurs caused by heavy load the bi-directional ac-ac converter output voltage V_{tr2_o} increases in the (a) of Fig. 9. D-UPFC output current phase lags to output voltage in the (b) of Fig. 9. Also, switches from S_{w1} to S_{w4} show the current waveforms without any problem in the (c) and (d) of Fig. 9.

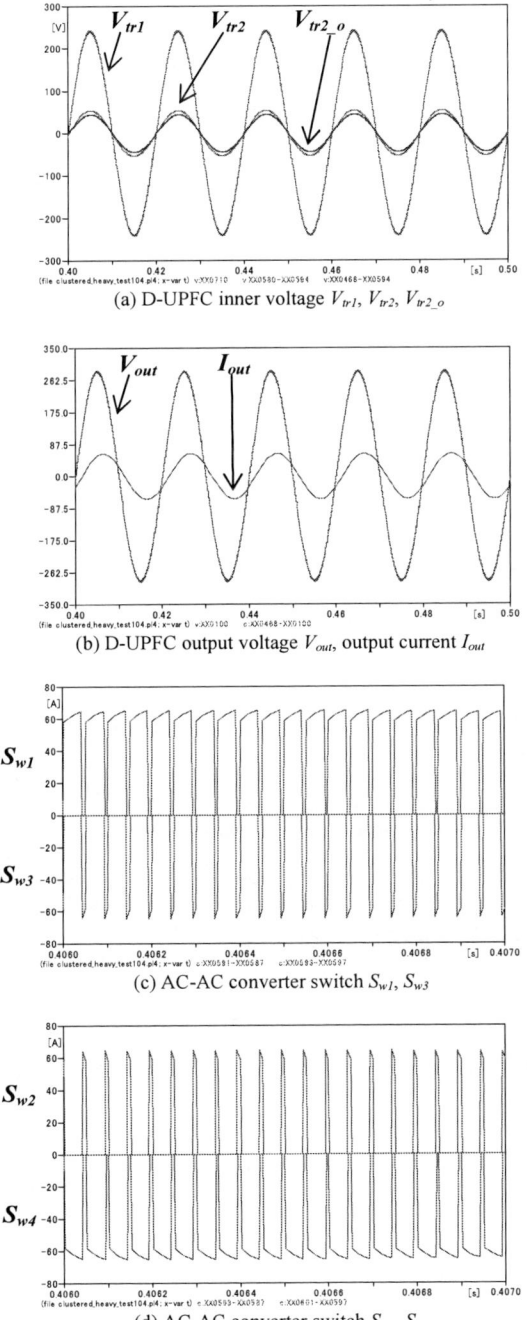

(a) D-UPFC inner voltage V_{tr1}, V_{tr2}, V_{tr2_o}

(b) D-UPFC output voltage V_{out}, output current I_{out}

(c) AC-AC converter switch S_{w1}, S_{w3}

(d) AC-AC converter switch S_{w2}, S_{w4}

Fig. 9: Voltage and current waveforms during voltage decrease at A_{24}

B. Voltage increase control

Fig. 10 shows the D-UPFC voltage control during voltage increase from PV reverse power flow. As shown in Table 1 and 2, parameters are used in the simulation. Voltage increase from node A_5 (a) to A_8 (g) is 101[V,rms] to 105.3[V,rms], respectively. These node voltages are controlled through D-UPFC from 101.3[V,rms] to 101.1[V,rms], respectively. Also, voltage increase from node A_{21} (b) to A_{24} (h) is 101.3[V,rms] to 105.6[V,rms], respectively. D-UPFC controls the voltages from 101.5[V,rms] to 101.3[V,rms], respectively. From the Fig. 10, D-UPFC controls the distribution voltage at the installation site as the voltage decrease control.

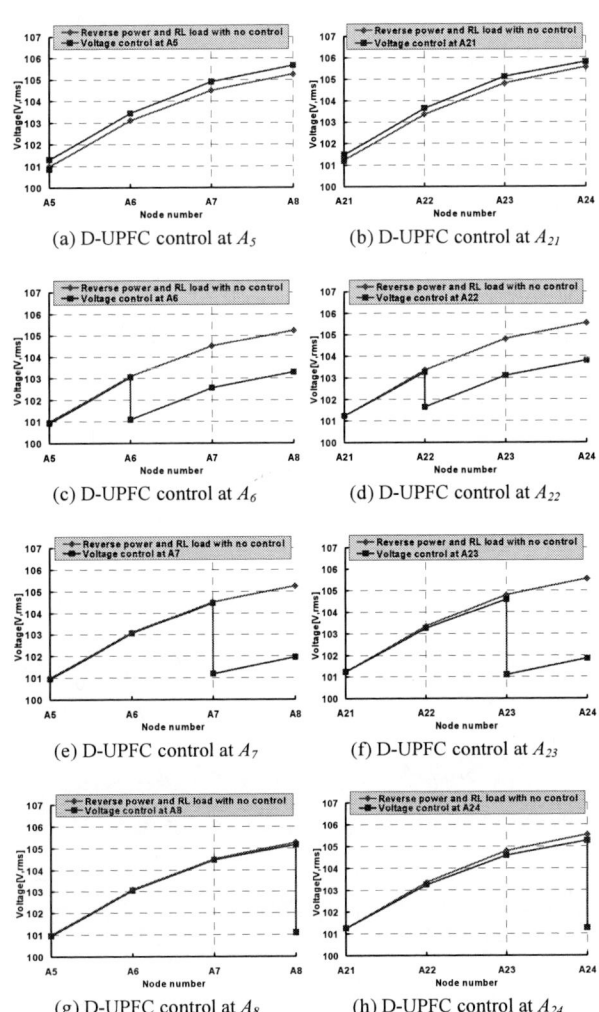

(a) D-UPFC control at A_5

(b) D-UPFC control at A_{21}

(c) D-UPFC control at A_6

(d) D-UPFC control at A_{22}

(e) D-UPFC control at A_7

(f) D-UPFC control at A_{23}

(g) D-UPFC control at A_8

(h) D-UPFC control at A_{24}

Fig. 10: Voltage increase control in the reverse power flow condition

Fig. 11 shows D-UPFC inner voltage (V_{tr1}, V_{tr2}, V_{tr2_o}), output voltage (V_{out}), output current (I_{out}), and ac-ac conveter switches (S_{w1} to S_{w4}) current waveforms at node A_{24}. In the D-UPFC control, ac-ac converter output voltage V_{tr2_o} is decreased in the (a) of Fig. 11. The second waveforms show the V_{out} and I_{out} of D-UPFC. Here, I_{out} phase is 180° different

978-1-4244-1871-8/07 $25.00 © 2007 IEEE

from V_{out} due to reverse power flow condition. Even though the load power factor is 0.9, I_{out} is not affected by load current because PV output current was large and grid-connected.

Switches which are from S_{w1} to S_{w4} current waveforms (c) and (d) of Fig. 11 perform with no problem during operation.

power and PV reverse power. Thus, D-UPFC should control this rapid voltage change in order to prevent voltage problem in the distrution line. Dynamic voltage control during the voltage decrease condition is shown in Fig. 12.

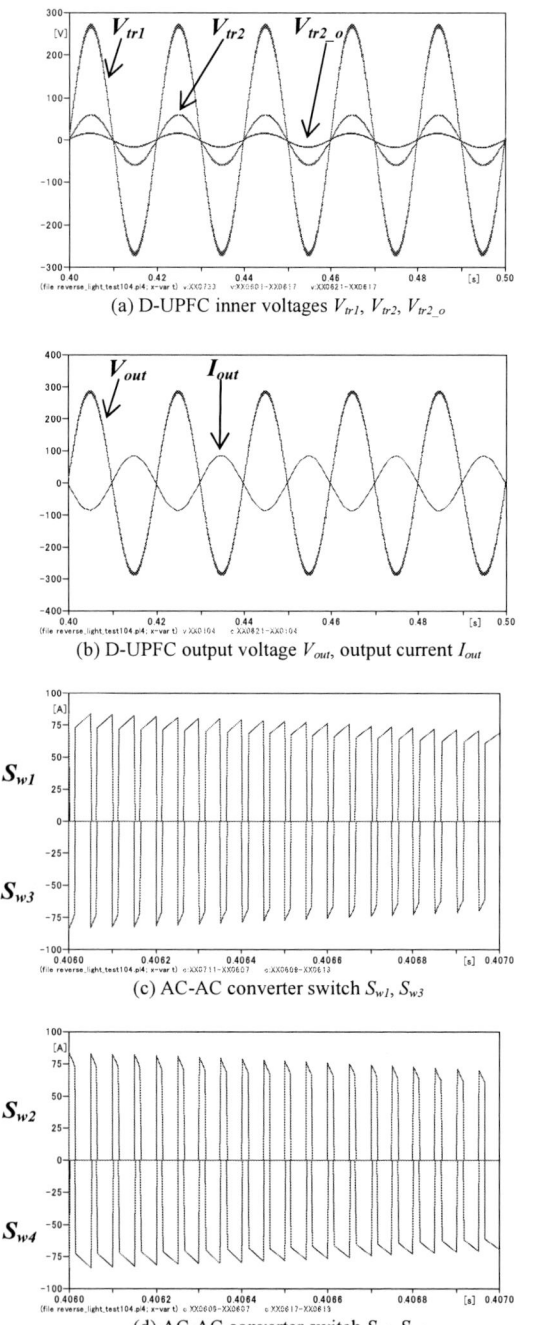

(a) D-UPFC inner voltages V_{tr1}, V_{tr2}, V_{tr2_o}

(b) D-UPFC output voltage V_{out}, output current I_{out}

(c) AC-AC converter switch S_{w1}, S_{w3}

(d) AC-AC converter switch S_{w2}, S_{w4}

Fig. 11: Voltage and current waveforms during voltage increase at A_{24}

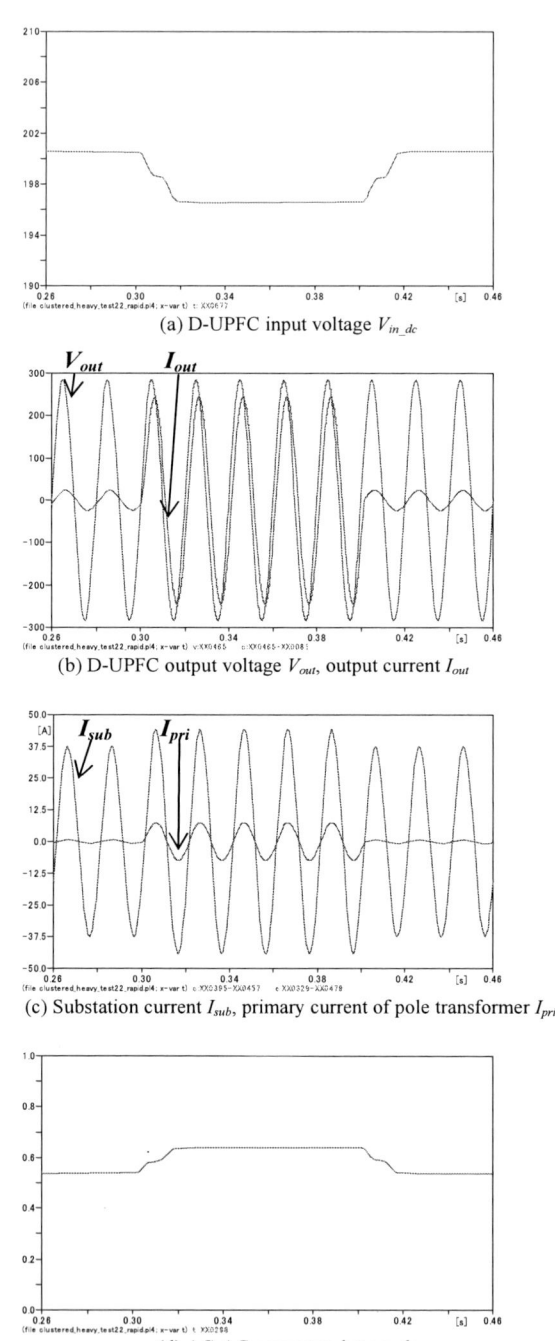

(a) D-UPFC input voltage V_{in_dc}

(b) D-UPFC output voltage V_{out}, output current I_{out}

(c) Substation current I_{sub}, primary current of pole transformer I_{pri}

(d) AC-AC converter duty cycle

Fig. 12: Load consumption power change from 0.3[s] to 0.4[s] and D-UPFC installation at node A_6

C. Dynamic voltage control

Rapid voltage change is always occurred in the low-voltage distribution system due to load consumption

From the Fig. 6 (a), load consumption power changes 4.9[kW] to 34.6[kW] between 0.3[s] and 0.4[s] at pole transformer. In the Fig. 12 (a), distribution voltage at node A_6 changes 200.5[V,rms] to 196.5[V,rms] during voltage

978-1-4244-1871-8/07 $25.00 © 2007 IEEE

decrease period. Also, it shows the voltage decrease as the rms value. Fig. 12 (b) shows the D-UPFC output voltage, current waveforms and D-UPFC controls the voltage to 201.1(100.55)[V,rms] during 0.3[s] to 0.4[s]. Fig. 12 (c) shows the substation output current and pole transformer primary current. Here, current from the substation increases due to heavy load during 0.3[s] to 0.4[s]. Fig. 12 (d) shows the ac-ac converter duty ratio and it controls the D-UPFC output voltage during 0.3[s] to 0.4[s].

Dynamic voltage control during the voltage increase condition is shown in Fig. 13. From the Fig. 7 (a), PV reverse power increases 9[kW] to 47.6[kW] at pole transformer. PV reverse power flows at random from 0.32[s] to 0.48[s]. The maximum reverse current of PV output is 236[A,rms]. Distribution voltage at node A_6 changes 200.5[V,rms] to 206.14 (103.07)[V,rms] during reverse power flow period. Fig. 13(a) shows the voltage increase as the rms value. Fig. 13 (b) shows the D-UPFC output voltage, current waveforms and D-UPFC controlling the voltage is 203.3 (101.65)[V,rms]. Fig. 13 (c) shows the substation output current and pole transformer primary current. Substation output current decreases 0.32[s] to 0.48[s], because the reverse current of PV output flows to other pole transformers. Finally, Fig. 13 (d) shows the ac-ac converter duty ratio and it controls the D-UPFC output voltage during 0.32[s] to 0.48[s].

IV. CONCLUSION

This paper shows distribution voltage control using the proposed D-UPFC topology in the distribution model. D-UPFC controls the distribution voltage during voltage decrease and increase conditions. Also, dynamic characteristic of D-UPFC voltage control is verified through the simulation results. D-UPFC protection study will be performed soon.

ACKNOWLEDGMENT

This research has been carried as a part of "Autonomy-Enhanced PV Cluster" project and special thanks for financial support of NEDO.

REFERENCES

[1] D. M. Lee, T. G. Havetler, R. G. Harley, T. L. Keister, and J. R. Ronstron, "A Voltage Sag Supporter Utilizing a PWM-Switched Autotransformer," *IEEE Trans. Power Electron.*, vol. 22, no. 2, pp. 626~635, March 2007.

[2] K. S. Lee, K. Yamaguchi, and K. Kurokawa, "Proposed Distribution Voltage Control Method due to Connecting the Clustered PV System," *Journal of Power Electronics*, in press.

[3] K. S. Lee, K. Yamaguchi, and K. Kurokawa, "D-UPFC as a Voltage Regulator in the Distribution System," *Renewable Energy 2006*, pp. 1756~1759, Oct. 2006.

[4] J. H. Youm and B. H. Kwon, "Switching Technique for Current-Controlled AC-to-AC Converters," *IEEE Trans. Ind. Electron.*, vol. 46, no. 2, Aprl 1999.

[5] E. C. Aeloíza. P. N. Enjeti, L. A. Morán, O. C. Montero-Hernandez, and S. Kim, "Analysis and Design of a New Voltage Sag Compensator for Critical Loads in Electrical Power Distribution System," *IEEE Trans. Ind. Appli.*, vol. 39, no. 4, pp.1143~1150, July/Aug. 2003.

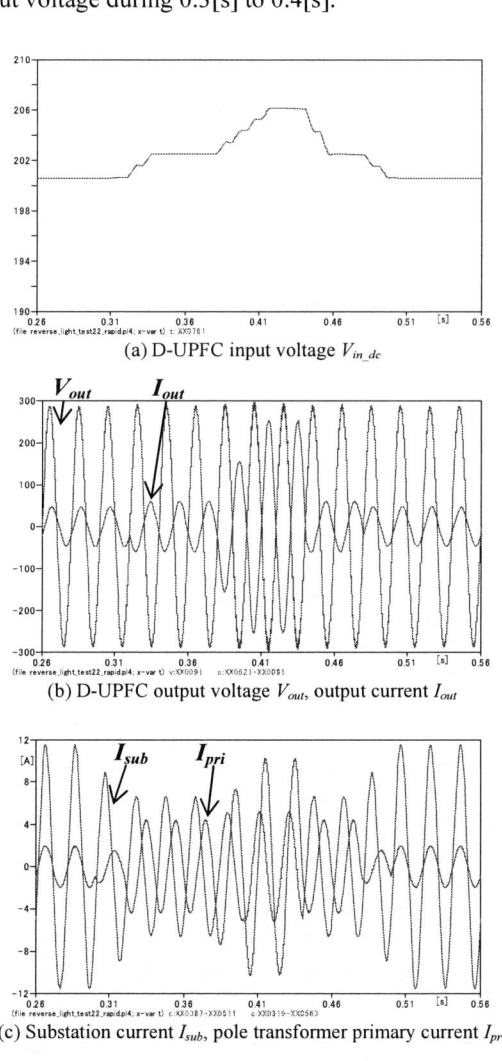

(a) D-UPFC input voltage V_{in_dc}

(b) D-UPFC output voltage V_{out}, output current I_{out}

(c) Substation current I_{sub}, pole transformer primary current I_{pri}

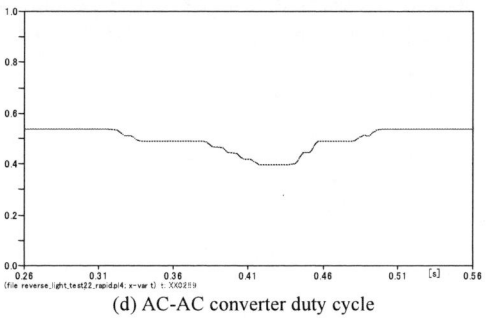

(d) AC-AC converter duty cycle

Fig. 13: PV reverse power flow during 0.32[s] to 0.48[s] and D-UPFC installation at node A_6

A Novel Islanding Detection Method Based on Minute Asymmetrical Current Injection for Three-Phase Grid-Connected PV Inverters

Xiang-Dong Sun, Mikihiko Matsui

Tokyo Polytechnic University

1583 Iiyama Atsugi-Shi Kanagawa, Japan

Email: sxd1030@163.com

Byung-Gyu Yu

Korea Institute of Energy Research

71-2, Jang-dong, Yuseong-gu, Daejeon, 305-343, Korea

Email: bgyuyu@kier.re.kr

Abstract—An active anti-islanding method based on the current control for a three-phase grid-connected PV inverter is proposed in this paper. The current of phase a is synchronous with the corresponding voltage at the common coupling point in its positive half cycle, and a zero-current zone is inserted into the end of the cycle in a negative half cycle. As for phase c, the zero-current zone is inserted in the positive half cycle of the current, and the current of phase c in a negative half cycle is synchronous with the corresponding voltage. Therefore the currents of phase a and c in one cycle become asymmetrical slightly. Before the islanding takes place, because of the operation of utility voltage the positive and negative half cycles of three-phase voltages are symmetrical. While the islanding occurs, the time differences between the positive and negative half cycles of the voltages of phase a and c will generate and the islanding is detected in accordance with the repetitive cycles fulfilling the conditions of the time differences. On the basis of the above analysis for three-phase four-wire power systems, three-phase three-wire systems are designed. The experiment on a 0.3kW PV inverter system connected to 200V/50Hz utility are carried out, and it is verified that the method can detect the islanding for the parallel RLC loads even if the voltage frequency is within the non-detection zone. It is seen from the analysis of the detecting principle that there is no influence on the proposed method even if a large load connected in parallel frequently starts up or stops.

I. INTRODUCTION

The islanding will result in safety, reliability, power quality and many other problems, therefore the anti-islanding protection is required by IEEE Std. 929-2000 and IEEE Std. 1547-2003.

Many anti-islanding detection methods have been discussed in the past decades, as a whole they include passive and active methods. Passive methods depend on measuring the certain system parameters and do not interfere with the operation of PV system. For example, the islanding is detected based on measuring the range of frequency and amplitude of the point of common coupling voltage. Although the method is simple, it fails to detect the islanding when the power of PV system matches with the demand of the loads [1]. The harmonic contents of the voltage are calculated to know the occurrence of the islanding, the problem of this method is how to set an appropriate harmonic threshold [2]. The phase jump detection method fails to detect the islanding when the load

power factor is unity, and it is easy to detect the islanding by mistake when the load such as a motor starts to run [3]. In the active methods, the output current or power of the inverter is designed to facilitate the islanding detection. For instance, both active frequency drift (AFD) and sliding mode frequency shift (SMS) change the frequency of output current of the inverter to detect the islanding, they become ineffective when the phase angle of a load matches with the phase offset or starting phase angle [4-6]. In addition, when they are applied to three-phase systems, if the large loads start up and stop frequently so that the frequency of utility voltage changes slightly, they maybe detect the islanding by mistake. Islanding detection is offered by power-factor control, however the method may be inappropriate when the mismatch between the generated power and the load demand is significant [7]. The islanding is detected by means of combining the disturbance of reactive power and real power, and the non-detection zone (NDZ) still exists in this method [8].

A novel islanding detection method is proposed based on a minute asymmetrical current injection in this paper. The positive and negative half cycles of the output currents of phase a and c in the inverter are slightly asymmetrical so that the corresponding voltages also become asymmetrical after the islanding occurs. Calculating the time differences between two half cycles of phase a and c, and if the repetitive cycles fulfilling the conditions of the time differences are the same as the presetting ones, the islanding can be identified effectively.

II. ISLANDING DETECTION APPROACH

A. Disadvantages of AFD

The system of PV inverter is modeled as a current source and the load is represented by the equivalent parallel RLC components as shown in Fig.1. When the islanding takes place, the impedance and its angle of parallel RLC are expressed in (1) and (2), respectively. The zero-current phase angle θ_z in one cycle for AFD method is expressed by (3), t_z is the total time of a zero-current zone in one cycle, ω_o is the output voltage frequency of the inverter.

$$Z_{load} = \frac{jR\omega_o L}{R - \omega_o^2 LCR + j\omega_o L} \qquad (1)$$

$$\theta_{load} = -\tan^{-1}\frac{Q \cdot (\omega_o^2 - \omega_r^2)}{\omega_o \omega_r} \qquad (2)$$

978-1-4244-1871-8/07 $25.00 © 2007 IEEE

$$\theta_z = \omega_o t_z \qquad (3)$$

Where $Q = R\sqrt{\dfrac{C}{L}}$, $\omega_r = \dfrac{1}{\sqrt{LC}}$, $\omega_o = \dfrac{2\pi}{T_o}$.

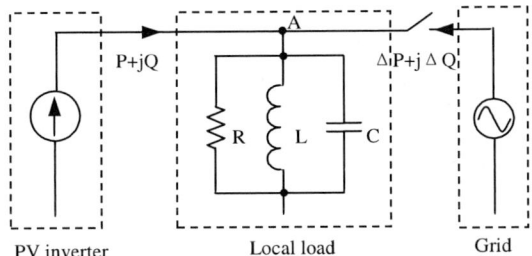

PV inverter   Local load   Grid

Fig.1. One-phase equivalent circuit of the PV system.

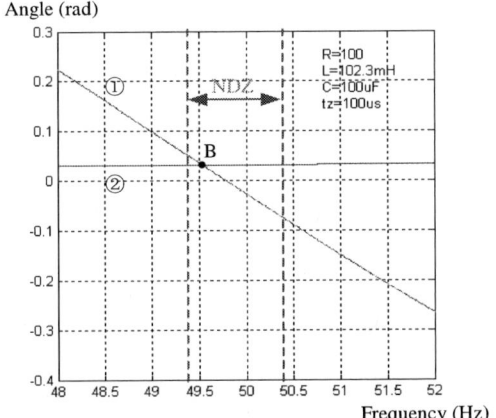

Fig.2. Relationship between θ_{load} and θ_z .

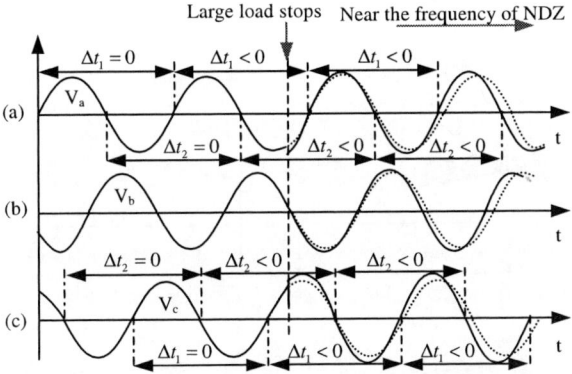

Fig.3. The changes of three-phase voltages in the condition of stopping

process of the large load. (a), (b), (c) Voltages of phase a, b and c.

Fig.2 is obtained from (2) and (3) in the condition of certain parameters. The line ① is the change curves of θ_{load} after the occurrence of the islanding. The line ② is the curve of θ_z . When the islanding occurs, if the frequency of output voltage of the inverter reaches to the frequency at the point B, at the point the system reaches a stable state, because the voltage frequency belongs to the frequency range of the permission (49.42Hz~50.42Hz for 50Hz are obtained in proportion to 59.3Hz~60.5Hz for 60Hz) at this time, the

islanding can not be identified by AFD method.

Moreover, the changes of three-phase voltages in the condition of the stopping process of the large load are shown in Fig.3, the dot curves are the ideal voltage tracks. Δt_1 is the time difference between the negative and the positive half cycles of the voltage, and Δt_2 is the time difference between the next positive half cycle and the last negative half cycle in two different cycles. When the large load stops in the terminal of a power system, because of the influence of the large load on the utility voltage, Δt_1 and Δt_2 between the positive and negative half cycles of the voltages of phase a and c are not equal to zero as shown in Fig.3, therefore AFD method changes the frequencies of output currents. If the voltage frequency is close to the frequency limit, AFD method maybe detects the islanding by mistake.

B. Principle of the Novel Islanding Detection Method for Three-Phase Four-Wire Systems

The islanding is not only detected by means of measuring the voltage frequency at the common coupling point, but also identified in accordance with the repetitive cycles fulfilling the conditions of the time differences between the positive and negative half cycles of the voltages of phase a and c for the proposed asymmetrical current injection method.

Now consider the utility phase voltage of a three-phase four-wire power system defined as:

$$\begin{cases} V_a = V_m \sin \omega_o t \\ V_b = V_m \sin(\omega_o t - \dfrac{2\pi}{3}) \\ V_c = V_m \sin(\omega_o t - \dfrac{4\pi}{3}) \end{cases} \qquad (5)$$

Assuming the power factor is unity, the reference output currents of the inverter in the positive half cycle and those in the negative half cycle can be expressed in (6) and (7), respectively, and the current of phase b is expressed by (8).

$$\begin{cases} i_{aref} = I_m \sin \omega_o t, \quad 0 \le t \le \dfrac{T_o}{2} \\ i_{cref} = \begin{cases} I_m \sin(\omega_i t - \dfrac{4\pi}{3}), \quad 0 \le t \le \dfrac{T_o}{2} - t_z \\ 0, \quad \dfrac{T_o}{2} - t_z < t \le \dfrac{T_o}{2} \end{cases} \end{cases} \qquad (6)$$

$$\begin{cases} i_{aref} = \begin{cases} I_m \sin \omega_i t, \quad \dfrac{T_o}{2} < t \le T_o - t_z \\ 0, \quad T_o - t_z < t \le T_o \end{cases} \\ i_{cref} = I_m \sin(\omega_o t - \dfrac{4\pi}{3}), \quad \dfrac{T_o}{2} < t \le T_o \end{cases} \qquad (7)$$

$$i_{bref} = -(i_{aref} + i_{cref}) \qquad (8)$$

Where, $\omega_i = \dfrac{2\pi}{T_o - t_z}$.

t_z should be chosen properly for reducing the DC component of output currents of the inverter and avoiding islanding detection by mistake, therefore $100\mu s$ is selected which is 0.5% of one cycle in this paper.

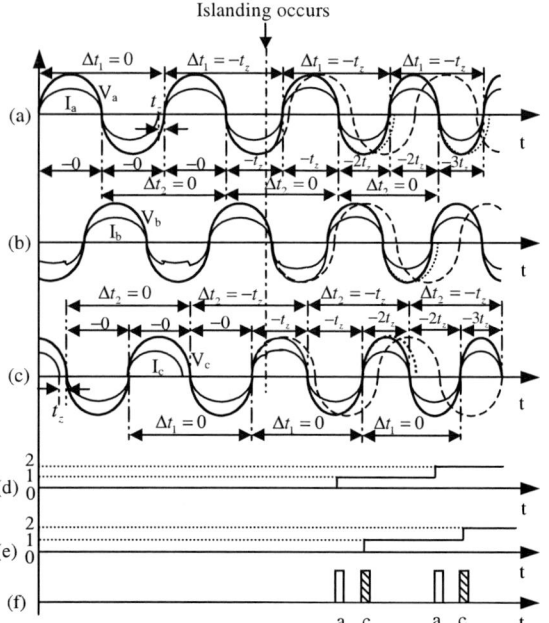

Fig.4. The principle of islanding detection. (a), (b), (c) Voltages and currents of phase a, b and c. (d), (e) Asymmetrical voltage cycles of phase a and c. (f) Detected status.

Fig.4 shows the principle of islanding detection. The time of the positive half cycle is $0.5T_o$ and that of the negative half cycle is also $0.5T_o$ for the voltage of phase a in the first cycle in Fig.4 (a), and the time difference Δt_1 is zero, thus detected status of phase a is false "0". The duration of the positive half cycle is $0.5T_o$ and that of the negative half cycle is $0.5T_o-t_z$ in the second cycle, and the time difference Δt_1 becomes $-t_z$, and the difference Δt_2 between the third positive half cycle and the second negative half cycle is zero, at this time detected status of phase a is true "1" and this is the first asymmetrical cycle. In the third cycle Δt_1 is also $-t_z$, Δt_2 between the fourth positive half cycle and the third negative half cycle is zero, thus detected status of phase a is also true "1" and this is the second asymmetrical cycle as shown in Fig.4 (d) and (f). The detected process of phase c is the same as that of phase a shown in Fig.4 (c), (e) and (f), it is different that detected status is true in the condition of $\Delta t_1 = 0$ and $\Delta t_2 = -t_z$ for phase c. When the detected statuses of both phase a and c are true "1" as shown in Fig.4 (f), the islanding maybe occurs. In order to avoid the islanding detection by mistake, the comparing thresholds of Δt_1 and Δt_2 are selected as $t_z/10$ or $-t_z/2$ shown in Tab.1. The detected statuses are judged repeatedly and as long as detected statuses of phase a and c are always true for several cycles or the asymmetrical cycles of both phases reach to the preset value, the islanding is identified. The flowchart is shown in Fig.5.

TABLE I
CONDITIONS OF ISLANDING DETECTION

Phase	Time difference	Status
a	$\Delta t_1 > -t_z/2$ or $\left\|\Delta t_2\right\| > t_z/10$	False "0"
a	$\Delta t_1 \le -t_z/2$ and $\left\|\Delta t_2\right\| \le t_z/10$	True "1"
c	$\Delta t_2 > -t_z/2$ or $\left\|\Delta t_1\right\| > t_z/10$	False "0"
c	$\Delta t_2 \le -t_z/2$ and $\left\|\Delta t_1\right\| \le t_z/10$	True "1"

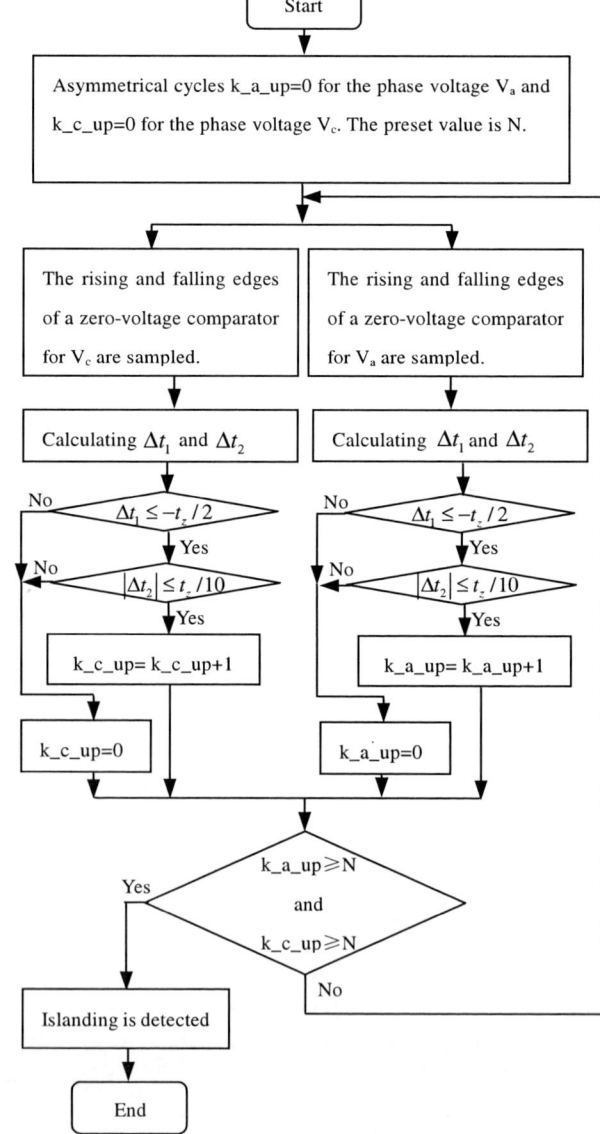

Fig.5. Flowchart of islanding detection.

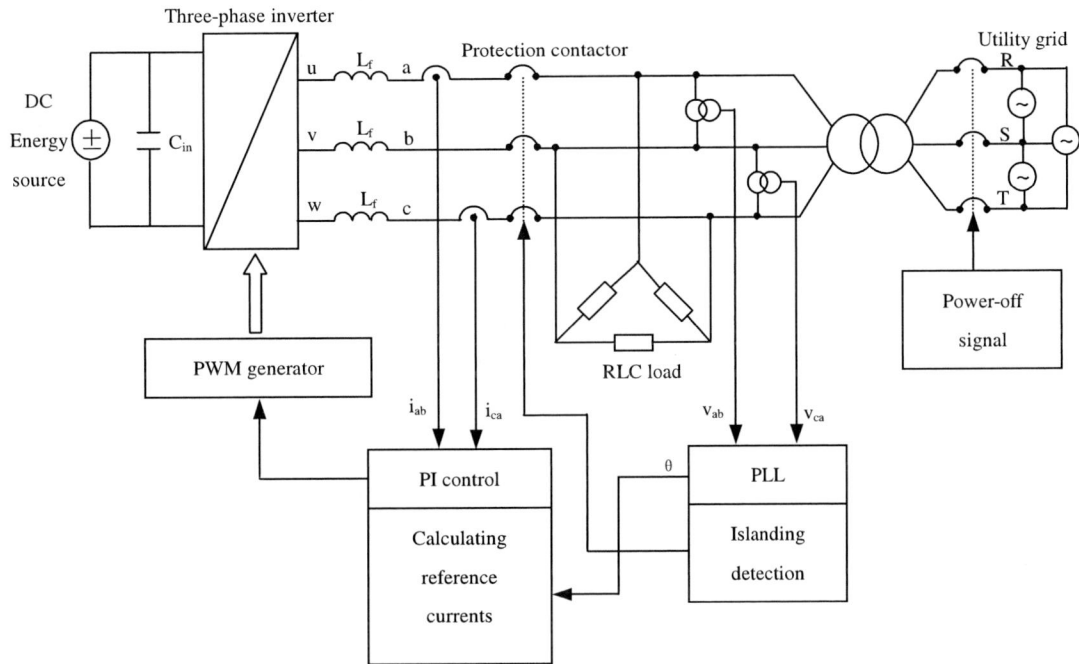

Fig.6. Block diagram of the islanding detection system.

When the large load stops in the terminal of a power system, because the time differences Δt_1 and Δt_2 between the positive and negative half cycles of the voltages of phase a and c are different from those of the proposed method, detected statuses of both phase a and c are not true. Therefore the islanding is not identified by the proposed method at this time.

C. Application to Three-Phase Three-Wire Systems

The three-phase three-wire power system is shown in Fig.6. The islanding is detected by measuring the line voltages V_{ab} and V_{ca}. The phase angle between the line voltage and the phase current is accordant, the phase voltages V_a, V_b, V_c in Fig.4 can be replaced by the line voltages V_{ab}, V_{bc} and V_{ca}, thus the same detecting method as the three-phase four-wire power system is implemented. The synchronously rotating angle θ in terms of DQ transformation is calculated [9]. The relationship between the phase angle of the voltage of phase a and θ is expressed by (9), and then the phase angle of the current of phase a is deduced by (10), according to (6) and (7) three-phase reference line currents are calculated by (11), (12) and (13). The reference currents and the feedback currents are used to calculate the reference values of PWM generator by PI regulators. When the power-off signal is effective and utility grid is cut off, the islanding is detected according to the conditions of Tab.1 and the protection contactor is cut off.

$$\theta_{V_a} = \theta + \frac{\pi}{2} \tag{9}$$

$$\theta_{i_a} = \theta_{V_{ab}} = \theta_{V_a} + \frac{\pi}{6} = \theta + \frac{\pi}{2} + \frac{\pi}{6} \tag{10}$$

$$i_{ab\_ref} = i_{aref} - i_{bref} \tag{11}$$

$$i_{ca\_ref} = i_{cref} - i_{aref} \tag{12}$$

$$i_{bc\_ref} = -(i_{ab\_ref} + i_{ca\_ref}) \tag{13}$$

III. SIMULATION RESULTS

The simulation is carried out in PSIM software. The parameters are given as follows: Line voltage 200V/50Hz; Filter inductance L_f =5mH; Load R=24 Ω, L=102.3mH, C=100μF ; t_z=300μs and100μs ; Winding ratio of the transformer 1:1; Repetitive cycles fulfilling the conditions in Tab.1 N=15.

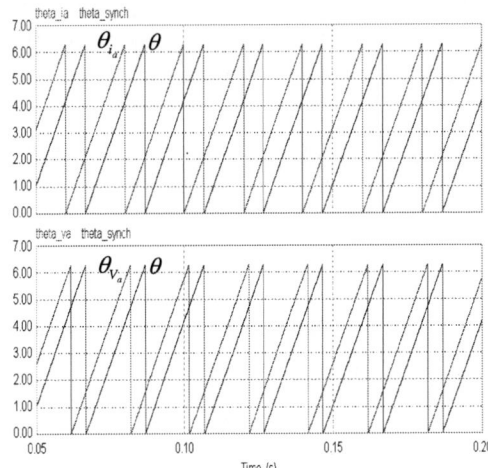

Fig.7. Relationships among θ and θ_{i_a} and θ_{V_a}.

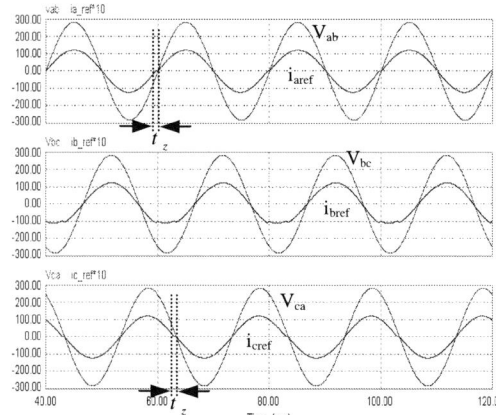

Fig.8. Line voltages and reference phase currents under the condition of $t_z = 300\mu s$.

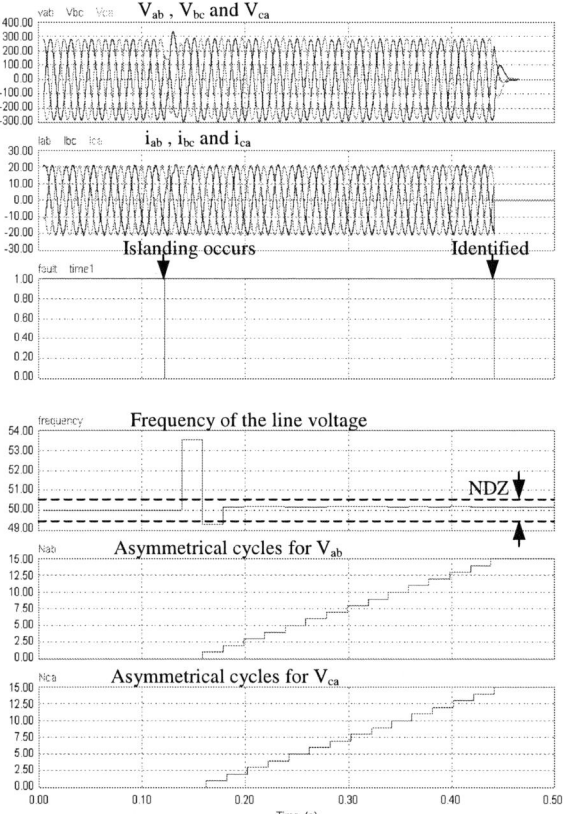

Fig.9. Simulation result of islanding detection under the condition of $t_z = 100\mu s$.

Fig.7 verifies the relationships among the synchronously rotating angle θ and the angle θ_{V_a} of the phase voltage V_a and the angle θ_{i_a} of the phase current I_a.

Fig.8 shows the waveform of line voltages and reference phase currents under the condition of t_z=300μs. It is seen that the phase angles of reference phase currents are synchronous with the corresponding line voltages and the islanding can be judged by measuring the line voltages after

the islanding happens.

Fig.9 shows the simulation result of islanding detection. After the islanding occurs, line voltages V_{ab} and V_{ca} become asymmetrical. When the repetitive cycles fulfilling the conditions in Tab.1 reach to 15, the islanding is identified. During the islanding detection, the frequency of line voltage is within the NDZ when it reaches to a stable state.

IV. EXPERIMENTAL VERIFICATION

The inverter system is built based on DSP TMS320C32. The parameters are the same as the simulation circuit basically, but the followings are different: t_z=100μs ; Winding ratio of the transformer 1:2.4; Repetitive cycles fulfilling the conditions in Tab.1 N=3.

Fig.10 shows the waveforms of the line voltage and phase current of one-phase load in the start-up process. Their phase angles become identical basically after one cycle. It is seen from Fig.10 that it is possible to detect the islanding by means of measuring the line voltages at the common coupling point when the minute zero-current zones are injected into the phase currents instead of the line currents.

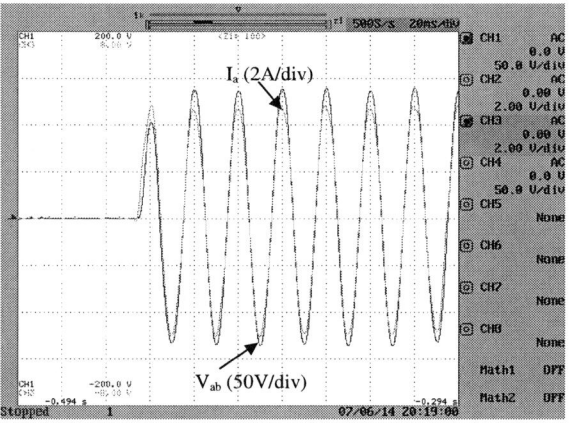

Fig.10. Line voltage and one-phase current of local load.

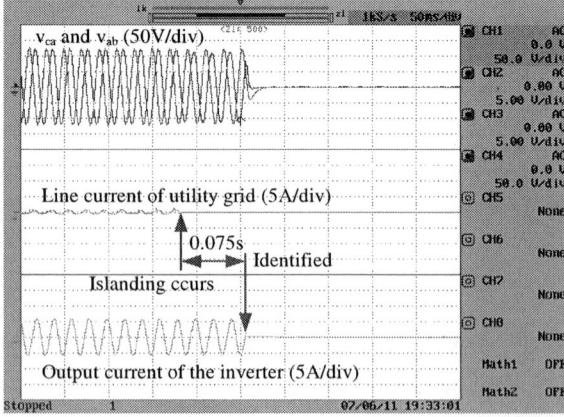

Fig.11. Experimental result of the islanding detection.

978-1-4244-1871-8/07 $25.00 © 2007 IEEE   844

The line current of utility grid is so small that it shows $\triangle P \approx 0$ and $\triangle Q \approx 0$ in Fig.11. When the islanding takes place, the frequency of line voltage at the common coupling point changes from 50Hz to 50.15Hz, which belongs to the frequency range of NDZ until the protection is implemented. After repetitive cycles fulfilling the conditions in Tab.1 reach to 3, the islanding is identified and the detecting time is about 0.075s.

V.   CONCLUSIONS

A novel anti-islanding method based on the minute asymmetrical current injection for the three-phase grid-connected PV inverter is proposed. The conclusions are drawn as follows:

(1) When output power of PV inverter matches with the demand of the loads, the islanding can be detected reliably and quickly even if the frequencies of the voltages at the common coupling point is within the range of NDZ;

(2) When the large load starts up and stops frequently, even if the frequency of utility voltage changes, there is no influence on the proposed method of islanding detection, because the changes of voltage frequencies are not accordant with the conditions of islanding detection.

REFERENCES

[1] Khanchai Tunlasakun, Krissanapong Kirtikara, Sirichai Thepa and Veerapol Monyakul, "A microcontroller based islanding detection for grid connected inverter," *47th IEEE International Midwest symposium on Circuits and Systems*, 2004, pp.267-269.

[2] A. Kitamura, M. Okamoto, F. Yamamoto, K. Nakaji, H. Matsuda and K. Hotta, "Islanding phenomenon elimination study at Rokko test center," in *Proc. 1st IEEE World Conf. Photovoltail Energy Conversion*, 1994, pp.759-762.

[3] H. Kobayashi, K. Takigawa, and E. Hashimoto, "Method for preventing phenomenon of utility grid with a number of small scale PV systems," in *Proc. 22nd IEEE photovoltaic specialists Conf.*, 1991, pp.695-700.

[4] G. A. Smith, P.A. Onions and D.G. Infield, "Predicting islanding operation of grid connected PV inverters," *IEE Proc.-Electr. Power Appl.*, Vol. 147, No.1, 2000, pp.1-6.

[5] Jaeho Choi, Youngseok Jung and Gwonjong Yu, "Novel AFD with pulsation of chopping fraction for islanding prevention of grid-connected photovoltaic inverter," *EPE 2005-dresden*, pp. p.1-p.10.

[6] G. Hung, C. Chang, and C. Chen, "Automatic phase-shift method for islanding detection of grid-connected photovoltaic inverters," *IEEE Trans. Energy Conv.*, Vol.18, 2003, pp.169-173.

[7] S. J. Huang and F. S. Pai, "Design and operation of grid-connected photovoltaic system with power-factor control and active islanding detection," *IEE Proc.-Gener, Transm. Distrib.*, Vol.148, No.2, 2001, pp. 243-249.

[8] Chuttchaval Jeraputra and Prasad N. Enjeti, "Development of a robust anti-islanding algorithm for untility interconnection of distributed fuel cell powered generation," *IEEE Trans. Power Elect.*, Vol. 19, No.5, 2004, pp. 1163-1170.

[9] A. B. Rey, S. de Pablo, J. M. Ruiz, "Decoupled current source vector control of a 3-phase grid-connected inverter using a novel vector utility observer for synchronization," *IEE Michael Faraday House*, 2004, pp.609-614.

A Novel Single-Voltage-Sensor-Based Maximum Power Point Tracking Method

Xiang-Dong Sun, Mikihiko Matsui, Kouji Yanagimura
Tokyo Polytechnic University
1583 Iiyama Atsugi-Shi Kanagawa, Japan
Email: sxd1030@163.com

Abstract—A novel maximum power point tracking method for photovoltaic systems based on a single voltage sensor is proposed in this paper. By means of the simulation studies with PSIM software, it has been found that there are similar relationship curves between the expression $(C-d)\cdot(-\Delta V_{PV})$ and the actual power P_{PV} of a photovoltaic array regardless of the chopper stage topology connected to the array. Therefore the maximum power point tracking based on P_{PV} is replaced with the expression $(C-d)\cdot(-\Delta V_{PV})$. For the purpose of diminishing the power fluctuation in the steady state, the averaged voltages of a photovoltaic array and the average duty ratios are used for the calculation. Experimental results obtained on a 50W prototype of a lead acid battery charger show high performance such as the less power fluctuation and the less ambient, i.e. irradiation and temperature, influence.

I. INTRODUCTION

In recent years the maximum power point tracking (MPPT) of the photovoltaic power systems have been studied widely. The perturbation and observation method (P&O)[1] and the increment conductance method [2] are presented on the basis of the power measurement, the disadvantage of the P&O method is the bigger power fluctuation and the drawback of both methods is that the perturbation step is difficult to choose for the better performance. Reference [3] tracks the MPP by using the sign of the differential coefficient dP_{PV}/dI_{PV} of power P_{PV} with respect to the current I_{PV} in the $P_{PV}-I_{PV}$ characteristics. And the algorithm [4] searches for the MPP by means of the sign of the differential coefficient dP_{PV}/dV_{PV} of power P_{PV} versus the voltage V_{PV} in the $P_{PV}-V_{PV}$ characteristics. Both methods are carried out based on the P&O so that they also have power fluctuation and power tracking loss. Checking the sign of the incremental conductance dI_{PV}/dV_{PV}, the MPP is tracked in the method [5], it is difficult to find out the MPP in the lower irradiation regions. The limit cycle method searches for the equilibrium power point while making the current and voltage in the range of the predetermined limit [6], but the drawback of the method is also the bigger power fluctuation. The optimal current reference needed to converge the output current on the optimal operation point of the prediction line is determined by means of linear relationships between the maximum power and between the optimal current and the short-circuit current and the optimal current. The hill-climbing method is done under various weather conditions. But it is also difficult to find out the MPP in lower irradiation regions and the power fluctuation is inevitable [7].

In this paper the relationship between the expression $(C-d)\cdot(-\Delta V_{PV})$ and the duty ratio d of a chopper is analyzed by means of the simulated results with PSIM software and a single-voltage-sensor MPPT method based

on this special expression is proposed. A boost chopper for a battery charger is implemented and the experimental results under the simulative conditions of different irradiations and temperatures verify the proposed method is feasible.

II. MPPT CONTROL SCHEME

A. Relationship Between the Voltage V_{PV} and Duty Ratio d of a Chopper

The block diagram of constructing the special expression for PV systems is shown in Fig.1 and the typical circuits such as buck, boost and buck-boost choppers are simulated respectively in PSIM software.

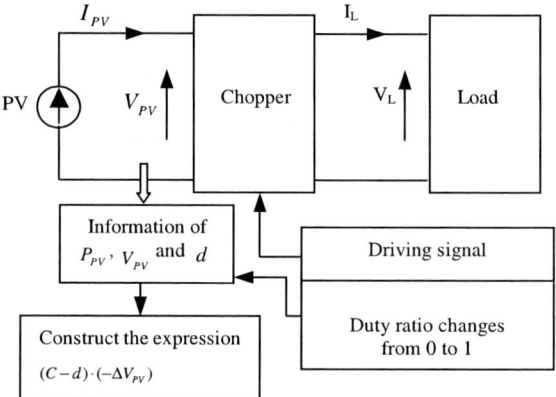

Fig.1. Block diagram of constructing the special expression.

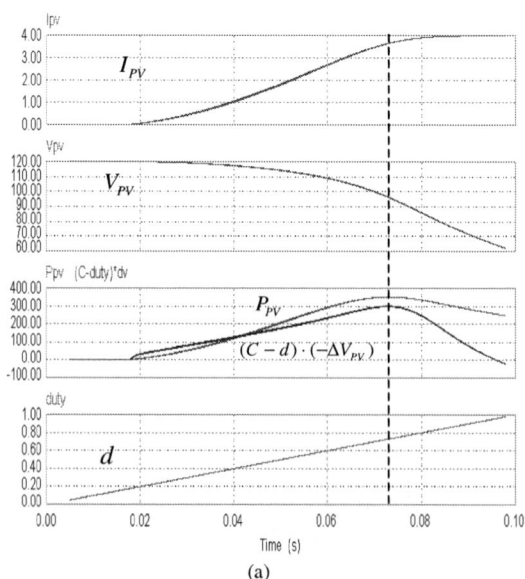

(a)

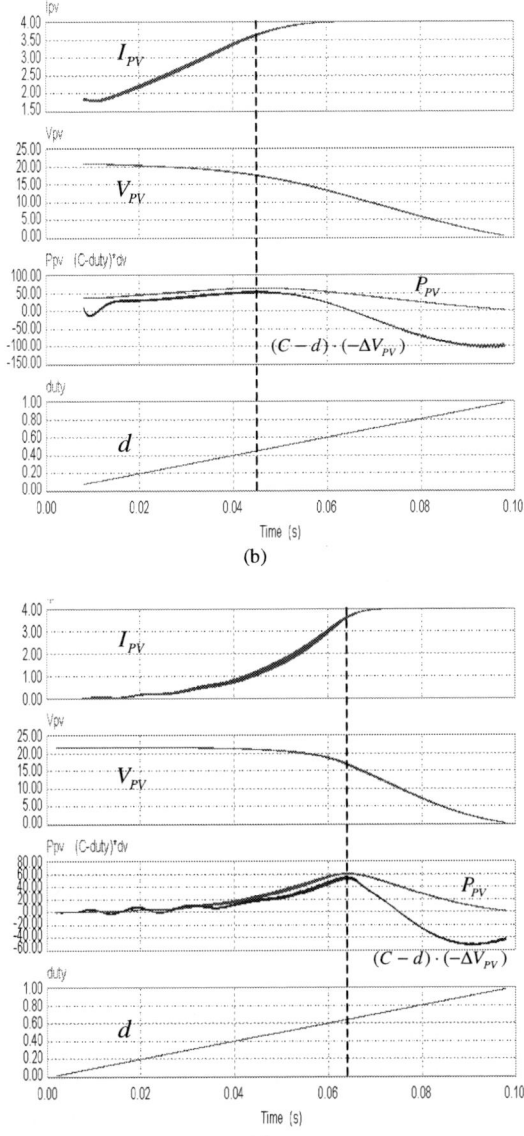

(b)

(c)

Fig.2. Simulated results in PSIM software. (a) Simulated result of a buck chopper. (b) Simulated result of a boost chopper. (c) Simulated result of a buck–boost chopper.

When the duty ratio changes from 0 to 1 slowly, the simulated waveforms are shown in Fig.2 (a), (b) and (c). The curves between the power of a PV array and the expression $(C-d)\cdot(-\Delta V_{PV})$ is very similar, and there is the same duty ratio basically at the MPP and the maximum point of $(C-d)\cdot(-\Delta V_{PV})$, therefore the MPPT is equivalent with searching for the maximum point of the curve of $(C-d)\cdot(-\Delta V_{PV})$. Where, ΔV_{PV} is a difference value of the sampled PV array voltages, and C is a constant ($C>0$). Usually C has to be decided by the same simulation as the method shown in Fig.1 in advance for a certain circuit.

B. Integral Operation Method Based on the Difference of the Expression $(C-d)\cdot(-\Delta V_{PV})$

According to the above method, Fig.3 is drawn.

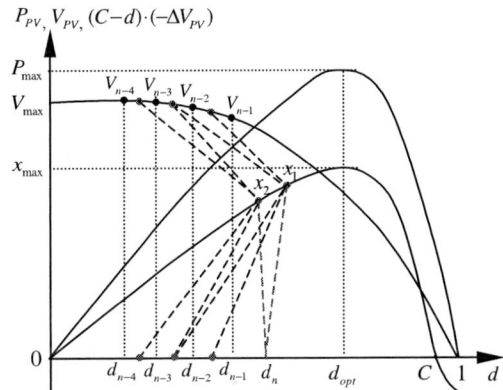

Fig. 3. Characteristics of P_{PV}, V_{PV} and $(C-d)\cdot(-\Delta V_{PV})$ versus d .

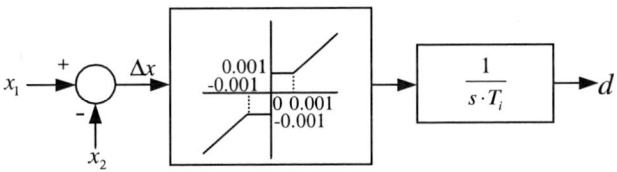

Fig.4. The control block diagram.

The expression $(C-d)\cdot(-\Delta V_{PV})$ is expressed by x . Four continuously sampled voltages V_{n-4} , V_{n-3} , V_{n-2} and V_{n-1} are used for diminishing the voltage fluctuation, and the corresponding duty ratios are shown in Fig.3. The averaged voltages and the averaged duty ratios are used to calculate x_1 and x_2 expressed by

$$x_1 = \frac{V_{n-3}-V_{n-1}}{2}\cdot(C-\frac{d_{n-3}+d_{n-1}}{2}) \tag{1}$$

$$x_2 = \frac{V_{n-4}-V_{n-2}}{2}\cdot(C-\frac{d_{n-4}+d_{n-2}}{2}) . \tag{2}$$

In the left-side area of the maximum point of the $d-x$ curve, x_1 is always larger than x_2 , on the other hand x_1 is always smaller than x_2 in the right-side area, and x_1 is approximately equal to x_2 at the maximum point. Therefore, x_1 is taken as the reference variable and x_2 is looked upon as the feedback variable. The duty ratio is obtained by the integral operation of the difference of two variables with keeping the difference Δx limited in order to prevent the invalidation of the integral operation as shown in Fig.4.

III. EXPERIMENTAL VERIFICATION

A 50W boost chopper has been built for a 20V storage battery charger with the equivalent series resistor $9.7\,\Omega$ and capacitor 360F shown in Fig.5. The solar array simulator E4351B Hewlett Packard is taken as the input power source. The relationship between the optimal voltage and the open-circuit voltage is $V_{opt}=0.85\cdot V_{oc}$, and that of the optimal current and the short-circuit current is designed as $I_{opt}=0.8\cdot I_{sc}$. The circuit parameters are as follows:

978-1-4244-1871-8/07 $25.00 © 2007 IEEE   847

switching frequency 10kHz, sampling frequency of the voltage 1kHz, inductor 1.2mH, filter capacitor 1000μ F, constant $C = 0.75$.

The simulative conditions are shown as Fig.6 (a) and (b). In Fig.6 (a) the open-circuit voltage is 21.7V and the short-circuit currents are changed from Mode A (0.6A), B (1.5A) to C (3.0A) in turn which shows the simulative irradiation change in the condition of a certain temperature. In Fig.6 (b) the short-circuit current is 1.5A and the open-circuit voltages are changed from Mode D (24V), E (21.7V) to F (18V) in turn which shows the simulative temperature change under the condition of a certain irradiation.

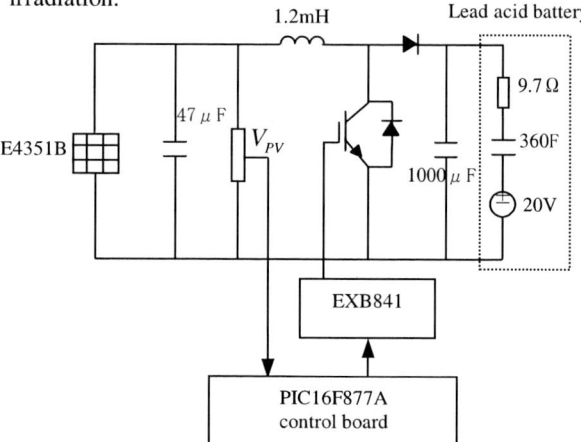

Fig.5. Sketch of the experimental circuit.

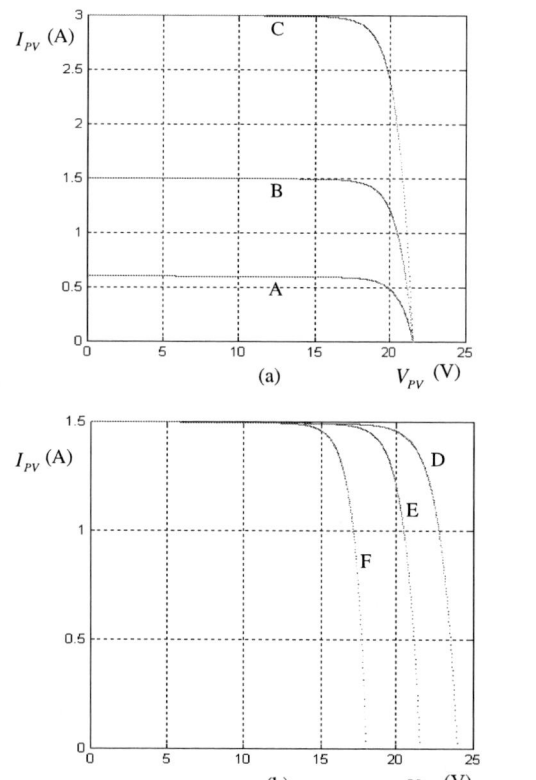

Fig.6. Simulative conditions of different irradiations and temperatures. (a) Case of different irradiations. (b) Case of different temperatures.

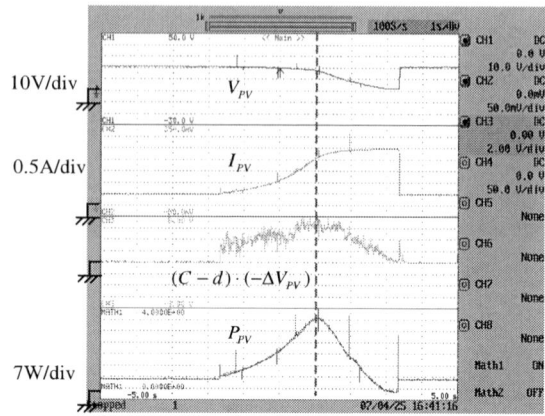

Fig.7. Waveforms of the variables V_{PV}, I_{PV}, $(C-d)\cdot(-\Delta V_{PV})$ and P_{PV}.

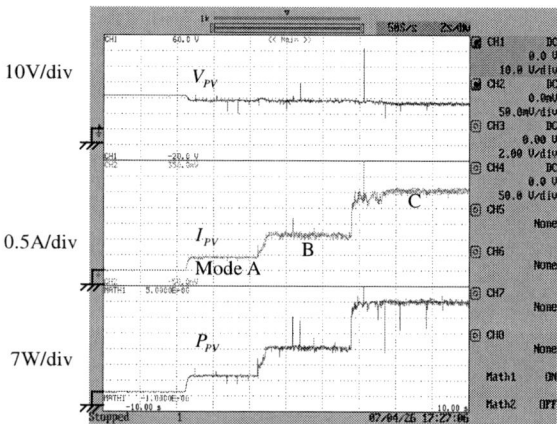

Fig.8. Waveforms under the conditions of different short-circuit currents and constant open-circuit voltage.

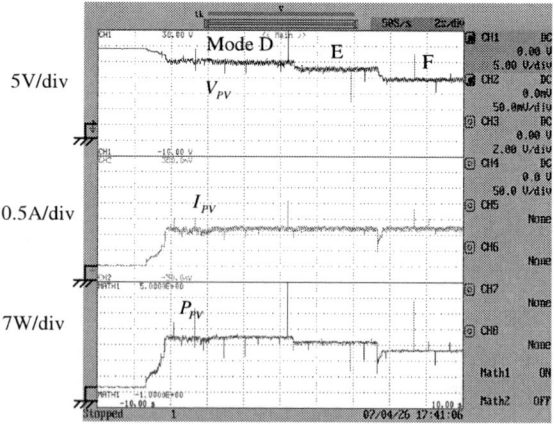

Fig.9. Waveforms under the conditions of different open-circuit voltages and constant short-circuit current.

Fig.7 shows the waveforms of the expression $(C-d)\cdot(-\Delta V_{PV})$ and P_{PV} obtained by changing the duty ratio from 0.05 to 1. It is observed that both values take their maximum points at the same instant shown by a dotted line. The result agrees well with the simulation result previously given in Fig.2 (b).

Fig.8 shows the waveforms with the simulative

irradiations in accordance with Fig.6 (a). The less influence of the irradiation change on the proposed method is verified.

Fig.9 shows the waveforms with the simulative temperatures in the same conditions as Fig.6 (b). The proposed method is basically independent of the temperatures. Moreover, Fig.8 and Fig.9 show the power fluctuation is small.

IV. CONCLUSION

A single-voltage-sensor-based MPPT method with high performance is proposed. The constant C is selected previously by simulation software so that the actual power P_{PV} and the expression $(C-d)\cdot(-\Delta V_{PV})$ correspond well against duty ratio change. The less influence of the irradiations and the temperatures on the proposed method is verified and the power fluctuation is small for a battery load. It is feasible that the maximum value of the special expression is tracked instead of the MPPT.

REFERENCES

[1] C. Hua, J. Lin, and C. shen, "Implementation of a DSP-controlled photovoltaic system with peak power tracking," *IEEE Trans. Ind. Electron.*, vol.45, no.1, pp.99-107, 1998.

[2] K. H. Hussein, I. Muta, T. Hoshino, and M. Osakada, "Maximum photovoltaic power tracking: An algorithm for rapidly changing atmospheric conditions," in *Proc. IEE Generation, Transmiss., Distrib.*, vol.142, pp.59-64, 1995.

[3] T. Ouchi, H. Fujikawa, S. Masukawa, and S. Iida, "A control scheme for three-phase current source inverter in interactive photovoltaic system," *Trans. Inst. Electr. Eng. Jpn.*, vol.120-D, no.2, pp.230-238, 2000.

[4] T. Kitano, M. Matsui, and D. Xu, "A maximum power point tracking control scheme for PV system based on power equilibrium and its system design," *Trans. Inst. Electr. Eng. Jpn.*, vol.120-D, no.12, pp.1263-1269, 2001.

[5] Y. C. Kuo, T. J. Liang, and J. F. Chen, "Novel maximum-power-point-tracking controller for photovoltaic energy conversion system," *IEEE Trans. Ind. Electron.*, vol.48, no.3, pp.594-601, 2001.

[6] T. Kitano, M. Matsui, and D. Xu, "A MPPT control scheme for PV system utilizing limit cycle operation and its system design," *Trans. Inst. Electr. Eng. Jpn.*, vol.122-D, no.4, pp.382-389,2002.

[7] N. Mutoh, M. Ohno, and T. Inoue, "A method for MPPT control while searching for parameters corresponding to weather conditions for PV generation systems," *IEEE Trans. Ind. Electron.*, vol.53, no.4, pp.1055-1065, 2006.

Frequency-Domain Transformation Approaches to Develop the *d-q* Synchronous-Frame Models of Three-Phase Symmetrical Networks and Applications in Modeling of PWM Power Converters

Xiaoyu Wang, *Student Member*, *IEEE*, Jinjun Liu, *Member*, *IEEE*, Yuji Meng, Jinku Hu, *Student Member*, *IEEE*,
Chang Yuan, *Student Member*, *IEEE*

School of Electrical Engineering
Xi'an Jiaotong University
28 West Xianning Road, Xi'an, Shaanxi 710049 China
Email: xywang@ieee.org

Abstract— Although state-space averaging small-signal modeling techniques are successfully established in DC-DC converters and three-phase sinusoidal PWM converters, these approaches are limited only to low-order systems, and the modeling process is very complex. This paper proposes two equivalent approaches in frequency domain to develop the *d-q* synchronous-frame models from stationary-frame models for arbitrary three-phase symmetrical circuits. The first modeling approach is illustrated with frequency transformation, by which the target synchronous-frame models could be easily obtained by frequency transformation of the prototype stationary-frame models. The second modeling approach is based on the reactive elements substitutions. These two approaches can be proven to be equivalent, and the first approach is suitable for input-output models, while the second approach is preferable for state-space models. As illustrative examples, these approaches are applied to develop the synchronous-frame models of large three-phase systems with PWM converters. They are proven valid with the comparison with current published models.

I. INTRODUCTION

The power electronics system is time-variant nonlinear system due to the inherent switching behavior. It is attractive to develop the linear model of the power electronics system. The state-space averaging modeling techniques have been successfully established for the DC-DC converters [1][2][3]. Later on, they were extended to three-phase pulse width modulation(PWM) power electronics converters in *d-q* synchronous frame, by which the three-phase sinusoidal PWM converters are transformed to time-invariant linear systems because the quantities are transformed to DC quantities in synchronous frame [4][5]. In these modeling process, ideal DC transformers serve as equivalent circuits to the switches [6][7]. However, state-space averaging modeling approaches are limited to low-order systems with a few reactive elements because the manipulating matrixes are too complex in high-order systems [7]. Some approaches are proposed to simplify the modeling development process [7][8][9].

This paper proposes two equivalent approaches in frequency domain that may greatly simplify the modeling process of three-phase power converters. The principle ideas of these approaches come from the research of the frequency analysis of *d-q* synchronous-frame control methods [10][11]. Amplitude modulation (AM) theory is employed to

intuitively explain the synchronous transformation with frequency conversion by the authors [12].

In many applications, it is preferable to model the three-phase system in synchronous frame if the controller executes in synchronous coordinate. In a view of system, the key issue to model a power electronics system is the modeling of arbitrary electrical networks because the power converters may be viewed as sources. This paper proposes several equivalent approaches based on the principles of frequency transformation [12], with which the model of arbitrary three-phase symmetrical electrical networks is easily transformed to *d-q* synchronous-frame model. Combination of this synchronous-frame model of electrical networks and the averaging model of power switches yields the system model.

With the proposed approaches, the modeling processes are significantly simplified. And it is possible to directly model large systems even with multi-phase power converters.

In section II, the traditional modeling process in time domain is reviewed and one of the new modeling approaches is employed to get the same result as a demonstration of the new methods. Then in section III, the two new modeling approaches are given in detail. A variation of approach II is also given as direct circuit transformation, which is much easy and practical for electrical engineers.

As illustrative examples, these approaches are applied to develop the synchronous-frame models of large three-phase systems with PWM converters. They are proven valid with the comparison with current published models.

II. REVIEW OF THE TRADITIONAL MODELING PROCESS AND DEMONSTRATION OF A NEW MODELING APPROACH

Before detailed description of the new approaches in section III, this section reviews the tradition state-space averaging modeling process and gives a demonstration of the new modeling process.

Fig. 1a) shows the popular three-phase boost PWM rectifier, which is time-variant and nonlinear system. With the averaging techniques of switches, average models of boost rectifiers can be obtained as time invariant nonlinear system. Fig. 1b) shows the corresponding average model of Fig. 1a). Stimulated with three-phase sinusoidal sources (sinusoidal voltage source and duty ratio), the averaging model is still nonlinear system. The key focus of this paper is

978-1-4244-1871-8/07 $25.00 © 2007 IEEE

to obtain the small-signal linear models of three-phase system in d-q synchronous coordinate from the averaging models like Fig. 1b).

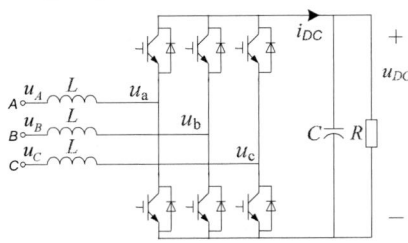

(a) Three-phase boost rectifier

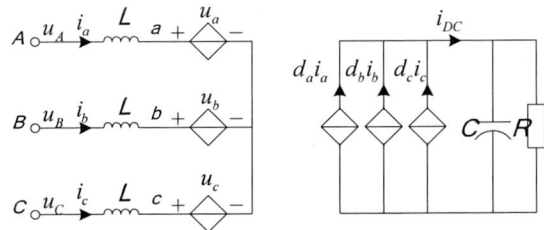

(b) Average model of three-phase boost rectifier.
Fig. 1 Average model of three-phase boost rectifier.

As the start point of analysis, the state-space equations of a three-phase boost rectifier as Fig. 1a) are given by

$$\begin{cases} L \cdot \dfrac{d}{dt} \begin{bmatrix} i_a(t) \\ i_b(t) \\ i_c(t) \end{bmatrix} = \begin{bmatrix} u_A(t) - d_a(t)u_{DC}(t) \\ u_B(t) - d_b(t)u_{DC}(t) \\ u_C(t) - d_c(t)u_{DC}(t) \end{bmatrix} \\ \dfrac{du_{DC}(t)}{dt} = \dfrac{1}{C}i_{DC}(t) - \dfrac{1}{R \cdot C}u_{DC}(t) \end{cases} \quad (1)$$

in time domain, or (2) in terms of Laplace transformation.

$$\begin{cases} \begin{bmatrix} u_A(s) - d_a(s)u_{DC} \\ u_B(s) - d_b(s)u_{DC} \\ u_C(s) - d_c(s)u_{DC} \end{bmatrix} = sL \cdot \begin{bmatrix} i_a(s) \\ i_b(s) \\ i_c(s) \end{bmatrix} \\ i_{DC}(s) = \left(s \cdot C + \dfrac{1}{R} \right) \cdot u_{DC}(s) \end{cases} \quad (2)$$

A. Traditional approach of modeling in time domain.

Applying the d-q transformation to three-phase equations in (2), the model of boost rectifier in d-q synchronous coordinate is obtained as

$$L \cdot \frac{d}{dt} \left(\mathbf{T}_{dq/abc} \begin{bmatrix} i_d(t) \\ i_q(t) \end{bmatrix} \right) = \left(\mathbf{T}_{dq/abc} \begin{bmatrix} u_{1d}(t) - d_d(t)u_{DC}(t) \\ u_{1q}(t) - d_q(t)u_{DC}(t) \end{bmatrix} \right) \quad (3)$$

with the transformation operator $\mathbf{T}_{abc/dq0}$ given by

$$\mathbf{T}_{abc/dq0} = \sqrt{\frac{2}{3}} \begin{bmatrix} \cos \omega t & \cos\left(\omega t - \frac{2\pi}{3}\right) & \cos\left(\omega t + \frac{2\pi}{3}\right) \\ -\sin \omega t & -\sin\left(\omega t - \frac{2\pi}{3}\right) & -\sin\left(\omega t + \frac{2\pi}{3}\right) \end{bmatrix}$$

and

$$\mathbf{T}_{dq/abc} = \mathbf{T}_{abc/dq}^H$$

where ωt is the synchronous angle obtained with phase-lock loop (PLL).

The left side of (3) can be simplified as

$$L \cdot \frac{d}{dt} \left(\mathbf{T}_{dq/abc} \cdot \begin{bmatrix} i_d(t) \\ i_q(t) \end{bmatrix} \right)$$

$$= L \cdot \left(\frac{d}{dt} \left(\mathbf{T}_{dq/abc} \right) \cdot \begin{bmatrix} i_d(t) \\ i_q(t) \end{bmatrix} + \mathbf{T}_{dq/abc} \cdot \frac{d}{dt} \begin{pmatrix} i_d(t) \\ i_q(t) \end{pmatrix} \right) \quad (4)$$

$$= L \cdot \left(\begin{bmatrix} 0 & -\omega_0 \\ \omega_0 & 0 \end{bmatrix} \cdot \mathbf{T}_{dq/abc} \cdot \begin{bmatrix} i_d(t) \\ i_q(t) \end{bmatrix} + \mathbf{T}_{dq/abc} \cdot \frac{d}{dt} \begin{pmatrix} i_d(t) \\ i_q(t) \end{pmatrix} \right)$$

Substitution of (4) to (3) yields the state-space averaging model as

$$\frac{d}{dt} \left(\begin{bmatrix} i_d(t) \\ i_q(t) \end{bmatrix} \right) = \frac{1}{L} \begin{bmatrix} u_{1d}(t) - d_d(t)u_{DC}(t) \\ u_{1q}(t) - d_q(t)u_{DC}(t) \end{bmatrix} - \begin{bmatrix} 0 & -\omega_0 \\ \omega_0 & 0 \end{bmatrix} \begin{bmatrix} i_d(t) \\ i_q(t) \end{bmatrix} \quad (5)$$

Perturbing i_d, i_q, u_{1d}, u_{1d} at the operating point I_d, I_q, U_{1d}, U_{1d} by $\tilde{i}_d, \tilde{i}_q, \tilde{u}_{1d}, \tilde{u}_{1d}$ yields the small-signal linear model of three-phase boost rectifier as

$$\frac{d}{dt} \left(\begin{bmatrix} \tilde{i}_d(t) \\ \tilde{i}_q(t) \end{bmatrix} \right) = - \begin{bmatrix} 0 & -\omega_0 \\ \omega_0 & 0 \end{bmatrix} \begin{bmatrix} \tilde{i}_d(t) \\ \tilde{i}_q(t) \end{bmatrix}$$

$$+ \frac{1}{L} \begin{bmatrix} \tilde{u}_{1d}(t) - \tilde{d}_d(t) \cdot U_{DC} - \tilde{u}_{DC}(t) \cdot D_d \\ \tilde{u}_{1q}(t) - \tilde{d}_q(t) \cdot U_{DC} - \tilde{u}_{DC}(t) \cdot D_q \end{bmatrix} \quad (6)$$

B. Demonstration of a new modeling approach in frequency domain.

Even for the simple boost rectifier shown as Fig. 1a), the modeling process is complicated. If the system has more reactive elements (such as three-phase LCL filters) or multi converters, the modeling process would be very complicated. The approaches proposed in this paper tremendously simplify the modeling process.

Before detailed description of the new approaches in next section, here gives a demonstration to model the previous three-phase boost rectifier with a new approach. The equivalent input-output model in frequency domain can be easily obtained.

Fig. 2 shows the three-phase connection networks of boost rectifier, which can be modeled as (7) in terms of Laplace transformation.

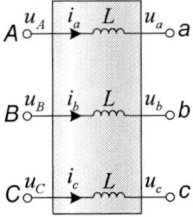

Fig. 2. Connection networks of three-phase boost rectifier.

$$\begin{bmatrix} u_A(s) - u_a(s) \\ u_B(s) - u_b(s) \\ u_C(s) - u_c(s) \end{bmatrix} = sL \cdot \begin{bmatrix} i_a(s) \\ i_b(s) \\ i_c(s) \end{bmatrix} \quad (7)$$

where $u_x(s)$ is the Laplace transformation of

$$u_x(t) = d_x(t) \cdot u_{DC}(t), \quad x = a,b,c.$$

As will be given in the next section, the synchronous model of (7) can be easily obtained with the substitution of s with

$$s \rightarrow \begin{bmatrix} s & -\omega_0 \\ \omega_0 & s \end{bmatrix} \quad (8)$$

Substitution of (8) to (7) directly yields the d-q model as

$$\begin{bmatrix} u_{1d}(s) - u_{2d}(s) \\ u_{1q}(s) - u_{2q}(s) \end{bmatrix} = L \cdot \begin{bmatrix} s & -\omega_0 \\ \omega_0 & s \end{bmatrix} \begin{bmatrix} i_d(s) \\ i_q(s) \end{bmatrix} \quad (9)$$

where u_{1d}, u_{1q} denote the voltage of terminal ABC in d-q coordinate, and u_{2d}, u_{2q} denote the voltage of terminal abc in d-q coordinate. It is easy to prove that (9) is the Laplace transformation of the model as (5) in traditional modeling approach.

With the same perturbation as traditional approach, the small-signal linear model is obtained as:

$$\begin{bmatrix} \tilde{u}_{1d}(s) - \tilde{d}_d(s) \cdot U_{DC} - \tilde{u}_{DC}(s) \cdot D_d \\ \tilde{u}_{1q}(s) - \tilde{d}_q(s) \cdot U_{DC} - \tilde{u}_{DC}(s) \cdot D_q \end{bmatrix} = L \cdot \begin{bmatrix} s & -\omega_0 \\ \omega_0 & s \end{bmatrix} \begin{bmatrix} \tilde{i}_d(s) \\ \tilde{i}_q(s) \end{bmatrix}$$

which is equivalent with the model as (6) obtained with the traditional approach.

III. SYNCHRONOUS-FRAME MODELING APPROACHES FOR THREE-PHASE SYMMETRICAL NETWORKS

In many applications, it is preferable to model the three-phase system in synchronous frame if the controller executes in synchronous coordinate. In a view of system, the PWM converters acts as one of the stimulations, so the key issue to model a PWM converter system is the modeling of arbitrary networks. This section proposes two equivalent simple approaches to develop the synchronous models of arbitrary three-phase symmetrical networks.

Fig. 3a) represents an arbitrary three-phase network with impedance function of $Z(s)$.

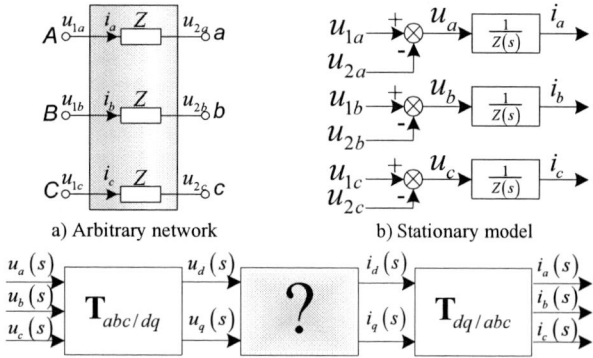

a) Arbitrary network   b) Stationary model

c) Synchronous-frame d-q model

Fig. 3. Modeling of arbitrary three-phase networks in a-b-c stationary and d-q synchronous frames.

Its stationary model in a-b-c coordinate is easy to obtain from the equations of linear circuits:

$$\begin{bmatrix} I_a(s) \\ I_b(s) \\ I_c(s) \end{bmatrix} = \frac{1}{Z(s)} \begin{bmatrix} U_a(s) \\ U_b(s) \\ U_c(s) \end{bmatrix} = \frac{1}{Z(s)} \begin{bmatrix} U_{1a}(s) - U_{2a}(s) \\ U_{1b}(s) - U_{2b}(s) \\ U_{1c}(s) - U_{2c}(s) \end{bmatrix} \quad (10)$$

To design the synchronous regulators in d-q coordinate, it is important to model the power stage in d-q coordinate shown as Fig. 3c). So the key target here is to established the transfer function of arbitrary networks in d-q coordinate from the stationary model as

$$\begin{bmatrix} I_d(s) \\ I_q(s) \end{bmatrix} = \mathbf{Z}_{dq}(s) \begin{bmatrix} U_d(s) \\ U_q(s) \end{bmatrix} \quad (11)$$

A. Approach I: Input-Output Modeling with Frequency Transformation of Stationary-Frame Model.

Applying the linear $\alpha\beta$ transformation as

$$T_{abc2\alpha\beta} = \sqrt{\frac{2}{3}} \begin{bmatrix} 1 & -\frac{1}{2} & -\frac{1}{2} \\ 0 & -\frac{\sqrt{3}}{2} & \frac{\sqrt{3}}{2} \end{bmatrix} \quad (12)$$

to the stationary model (10) of arbitrary network yields the model in $\alpha\beta$ coordinate:

$$\begin{bmatrix} I_\alpha(s) \\ I_\beta(s) \end{bmatrix} = \frac{1}{Z(s)} \begin{bmatrix} U_\alpha(s) \\ U_\alpha(s) \end{bmatrix} \quad (13)$$

The synchronous voltages and current in time domain are

$$\begin{bmatrix} u_d(t) \\ u_q(t) \end{bmatrix} = \begin{bmatrix} \cos\omega_0 t & \sin\omega_0 t \\ -\sin\omega_0 t & \cos\omega_0 t \end{bmatrix} \begin{bmatrix} u_\alpha(t) \\ u_\beta(t) \end{bmatrix} \quad (14)$$

$$\begin{bmatrix} i_d(t) \\ i_q(t) \end{bmatrix} = \begin{bmatrix} \cos\omega_0 t & \sin\omega_0 t \\ -\sin\omega_0 t & \cos\omega_0 t \end{bmatrix} \begin{bmatrix} i_\alpha(t) \\ i_\beta(t) \end{bmatrix} \quad (15)$$

Applying Laplace transformation to (14) and (15) yields

$$\begin{bmatrix} U_d(s) \\ U_q(s) \end{bmatrix} = \frac{1}{2} \begin{bmatrix} 1 & -j \\ j & 1 \end{bmatrix} \begin{bmatrix} U_\alpha(s - j\omega_0) \\ U_\beta(s - j\omega_0) \end{bmatrix} + \frac{1}{2} \begin{bmatrix} 1 & j \\ -j & 1 \end{bmatrix} \begin{bmatrix} U_\alpha(s + j\omega_0) \\ U_\beta(s + j\omega_0) \end{bmatrix}$$

(16)

$$\begin{bmatrix} I_d(s) \\ I_q(s) \end{bmatrix} = \frac{1}{2} \begin{bmatrix} 1 & -j \\ j & 1 \end{bmatrix} \begin{bmatrix} I_\alpha(s - j\omega_0) \\ I_\beta(s - j\omega_0) \end{bmatrix} + \frac{1}{2} \begin{bmatrix} 1 & j \\ -j & 1 \end{bmatrix} \begin{bmatrix} I_\alpha(s + j\omega_0) \\ I_\beta(s + j\omega_0) \end{bmatrix}$$

(17)

Combination of (16), (17) and the model in $\alpha\beta$ coordinate as (13) yields the d-q synchronous-frame model as

$$\mathbf{Z}_{dq}(s) = \begin{bmatrix} \frac{1}{2}(Z(s + j\omega_0) + Z(s - j\omega_0)) & \frac{j}{2}(Z(s + j\omega_0) - Z(s - j\omega_0)) \\ -\frac{j}{2}(Z(s + j\omega_0) - Z(s - j\omega_0)) & \frac{1}{2}(Z(s + j\omega_0) + Z(s - j\omega_0)) \end{bmatrix}$$

(18)

Detail derivation of (18) is given in the appendix. With (18), the synchronous impedance in d-q coordinate of three-phase networks can be directly obtained from the frequency transformation of stationary impedance in $\alpha\beta$ coordinate. This approach greatly simplifies the process to develop the input-output models.

B. Approach II: State-Space Modeling With Reactive Elements Substitutions in Stationary-Frame Circuits.

An alternative modeling approach is based on the

978-1-4244-1871-8/07 $25.00 © 2007 IEEE

substitutions of reactive elements. Given the impedance of arbitrary networks in terms of Laplace transforms as

$$Z(s) = \frac{M(s)}{N(s)}$$

where $M(s)$ and $N(s)$ are polynomial expressions of s, the synchronous model in d-q coordinate can be directly obtained by substitution as

$$Z(s) \xrightarrow{\ \ s \rightarrow \begin{bmatrix} s & -\omega_0 \\ \omega_0 & s \end{bmatrix} \ \ } Z\left(\begin{bmatrix} s & -\omega_0 \\ \omega_0 & s \end{bmatrix}\right) \qquad (19)$$

The derivation of (19) is similar with those in [14], and is not given in this paper. But it can be proven equivalent to the first approach with mathematical inductive method. And this approach is easy to employ for state-space modeling.

A variation of approach II is direct circuit transformation, which is much easy and practical for electrical engineers. From (19) and the comparison of the circuits in *a-b-c* stationary frame and *d-q* synchronous frame, it is found that the equivalent circuits in *d-q* frame can be directly obtained by substitutions of reactive elements in the averaging circuit in *a-b-c* frame.

Because the approach II as (19) only transforms the variable of *s* which represents the reactive elements, this approach only transforms the reactive elements in the system. So the circuit in d-q synchronous frame can be directly obtained from the average model in *a-b-c* stationary frame with this rule:

- The resistors in *d-q* frame remain the same as *a-b-c* frame.
- The sources are obtained with synchronous transformation.
- The inductors are transformed as Fig. 3a). And the capacitors are transformed as Fig. 3b).

Actually, similar idea of circuit transformation has been reported from different point of view, and more elements are discussed in [7].

With the equivalent approaches in this section, the large signal model of three-phase system can be easily obtained by direct transformation of the stationary-frame model.

Perturbation to the large signal model yields the linear small-signal model. Examples will be given in next section to illustrate these approaches.

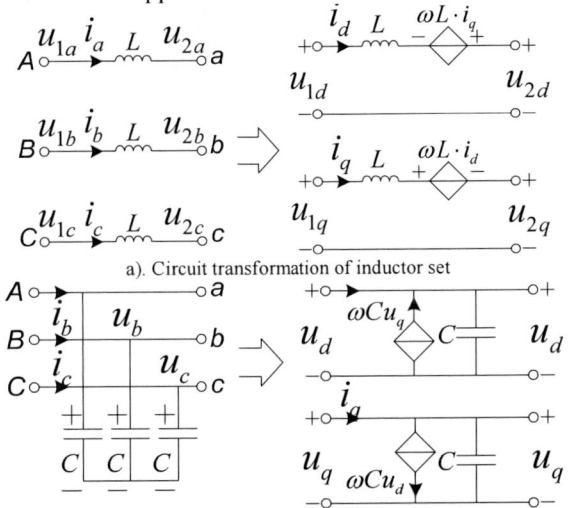

a). Circuit transformation of inductor set

b). Circuit transformation of capacitor set

Fig. 4 Practical method of approach II with circuit transformation of the reactive elements.

IV.   APPLICATION IN POWER CONVERTERS MODELING

As illustrative examples, the frequency transformation approaches are applied to develop the synchronous models of several typical three-phase power converters. The obtained models are compared with current publications to verify the validity of the modeling approaches.

A. *Modeling of three-phase voltage source inverter (VSI).*

Fig. 5a) shows the three-phase buck rectifier. Its connecting electrical network is shown as Fig. 5b) with the assumption that the load is resistor.

Because the connecting electrical network contains 6 reactive components, the modeling in synchronous frame is complicit with the traditional approaches. Here both of the proposed approaches in this paper are employed to develop the synchronous model of VSI as illustrative examples.

TABLE 1. FREQUENCY TRANSFORMATION FROM AVERAGE MODELS IN *A-B-C* FRAME TO MODELS *D-Q* FRAME

	Model in *a-b-c* frame	Model in *d-q* synchronous frame with frequency transformation
Approach I	$Z(s)$	$\mathbf{Z}_{dq}(s) = \begin{bmatrix} \frac{1}{2}\big(Z(s+j\omega_0)+Z(s-j\omega_0)\big) & \frac{j}{2}\big(Z(s+j\omega_0)-Z(s-j\omega_0)\big) \\ -\frac{j}{2}\big(Z(s+j\omega_0)-Z(s-j\omega_0)\big) & \frac{1}{2}\big(Z(s+j\omega_0)+Z(s-j\omega_0)\big) \end{bmatrix}$
Approach II	$Z(s)$	$\mathbf{Z}_{dq}(s) = Z\left(\begin{bmatrix} s & -\omega_0 \\ \omega_0 & s \end{bmatrix}\right)$
Circuit Transformation	Average circuit in *a-b-c* stationary frame.	• The reactive elements are transformed as Fig. 3. • The sources are obtained with synchronous transformation • The other elements are the same as in *a-b-c* stationary frame.

978-1-4244-1871-8/07 $25.00 © 2007 IEEE

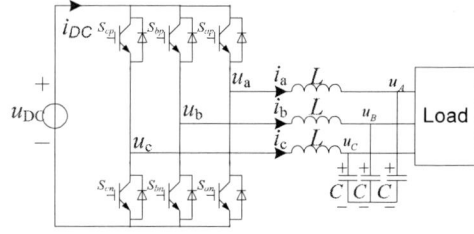

a) System configuration.

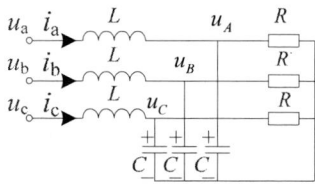

b) Connecting electrical network.
Fig. 5 Three-phase voltage source inverter (VSI).

1) Modeling with approach I based on transfer function.

The transfer function of the electrical network is given by

$$\begin{bmatrix} u_a(s) \\ u_b(s) \\ u_c(s) \end{bmatrix} = \left(s^2 LC + s\frac{L}{R} + 1 \right) \cdot \begin{bmatrix} u_A(s) \\ u_B(s) \\ u_C(s) \end{bmatrix} \quad (20)$$

where the variables are indicated in Fig. 5. The converter output is determined by the product of the duty ratio and the DC voltage as

$$\begin{bmatrix} u_a(t) \\ u_b(t) \\ u_c(t) \end{bmatrix} = \begin{bmatrix} d_a(t) \\ d_b(t) \\ d_c(t) \end{bmatrix} u_{DC}(t) \quad (21)$$

With approach I, the synchronous-frame model is obtained according to the frequency transformation in (18) as:

$$\begin{bmatrix} u_d(s) \\ u_q(s) \end{bmatrix} = \begin{bmatrix} LC(s^2 - \omega^2) + \dfrac{L}{R}s + 1 & -\left(2\omega LC \cdot s + \dfrac{L}{R}\omega + 1 \right) \\ 2\omega LC \cdot s + \dfrac{L}{R}\omega + 1 & LC(s^2 - \omega^2) + \dfrac{L}{R}s + 1 \end{bmatrix} \begin{bmatrix} u_D(s) \\ u_Q(s) \end{bmatrix} (22)$$

Perturbing the input and output variables in (22) yields small-signal linear transfer function of VSI as

$$\begin{bmatrix} \tilde{d}_d(s) \\ \tilde{d}_q(s) \end{bmatrix} U_{DC} = \begin{bmatrix} LC(s^2 - \omega^2) + \dfrac{L}{R}s + 1 & -\left(2\omega LC \cdot s + \dfrac{L}{R}\omega + 1 \right) \\ 2\omega LC \cdot s + \dfrac{L}{R}\omega + 1 & LC(s^2 - \omega^2) + \dfrac{L}{R}s + 1 \end{bmatrix} \begin{bmatrix} \tilde{u}_D(s) \\ \tilde{u}_Q(s) \end{bmatrix}$$

(23)

2) Modeling with approach II

The state-space model of the electrical networks as Fig. 5b) is given by (24) and (25):

$$\begin{bmatrix} u_A(s) \\ u_B(s) \\ u_C(s) \end{bmatrix} = \begin{bmatrix} u_a(s) \\ u_b(s) \\ u_c(s) \end{bmatrix} - sL \cdot \begin{bmatrix} i_a(s) \\ i_b(s) \\ i_c(s) \end{bmatrix} \quad (24)$$

$$\begin{bmatrix} i_a(s) \\ i_b(s) \\ i_c(s) \end{bmatrix} = \left(sC + \frac{1}{R} \right) \cdot \begin{bmatrix} u_A(s) \\ u_B(s) \\ u_C(s) \end{bmatrix} \quad (25)$$

Substitution of s to the matrix as (8) yields the synchronous-frame model as

$$\begin{bmatrix} u_D(s) \\ u_Q(s) \end{bmatrix} = \begin{bmatrix} u_d(s) \\ u_q(s) \end{bmatrix} - \begin{bmatrix} sL & -\omega_0 L \\ \omega_0 L & sL \end{bmatrix} \begin{bmatrix} i_d(s) \\ i_q(s) \end{bmatrix}$$

$$\begin{bmatrix} i_d(s) \\ i_q(s) \end{bmatrix} = \left(\begin{bmatrix} sC & -\omega_0 C \\ \omega_0 C & sC \end{bmatrix} + \frac{1}{R} \right) \begin{bmatrix} u_D(s) \\ u_Q(s) \end{bmatrix} \quad (26)$$

Perturbing to this synchronous state-space model yields the small-signal linear model of three-phase buck rectifier as

$$\begin{bmatrix} \tilde{u}_D(s) \\ \tilde{u}_Q(s) \end{bmatrix} = -\begin{bmatrix} sL & -\omega_0 L \\ \omega_0 L & sL \end{bmatrix} \begin{bmatrix} \tilde{i}_d(s) \\ \tilde{i}_q(s) \end{bmatrix} + \begin{bmatrix} \tilde{d}_d(s) \\ \tilde{d}_q(s) \end{bmatrix} U_{DC}$$

$$\begin{bmatrix} \tilde{i}_d(s) \\ \tilde{i}_q(s) \end{bmatrix} = \left(\begin{bmatrix} sC & -\omega_0 C \\ \omega_0 C & sC \end{bmatrix} + \frac{1}{R} \right) \begin{bmatrix} \tilde{u}_D(s) \\ \tilde{u}_Q(s) \end{bmatrix} \quad (27)$$

This small-signal model in term of Laplace transform is equivalent to the model given in the publication in term of derivation equations.

Fig. 6 shows this small-signal circuit model in d-q synchronous frame, which can also easily be obtained with the circuit transformation approach as Fig. 4.

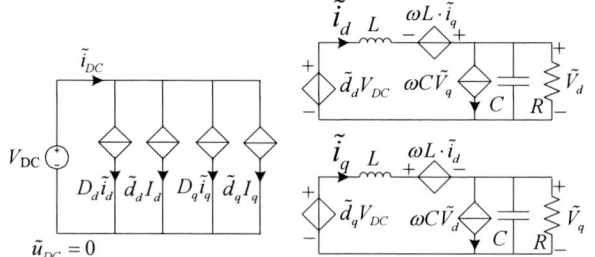

Fig. 6 Small-signal linear model of voltage source inverter (VSI).

B. Modeling of three-phase current-source inverter (CSI).

Fig. 7a) shows the three-phase current-source inverter (CSI). Its connecting electrical network is shown as Fig. 7b) with the assumption that the load is resistor.

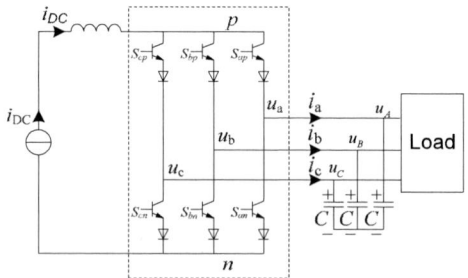

a) System configuration.

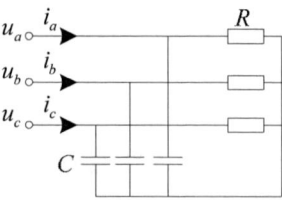

b) Connecting electrical network.
Fig. 7 Three-phase buck rectifier.

The relationship between the current and voltage of the connecting network is given by (28) in stationary a-b-c frame.

$$\begin{bmatrix} i_a(s) \\ i_b(s) \\ i_c(s) \end{bmatrix} = \left(sC + \frac{1}{R}\right) \cdot \begin{bmatrix} u_a(s) \\ u_b(s) \\ u_c(s) \end{bmatrix} \qquad (28)$$

Substitution of s to the matrix as (8) yields the synchronous-frame model as

$$\begin{bmatrix} i_d(s) \\ i_q(s) \end{bmatrix} = \left(C \begin{bmatrix} s & -\omega_0 \\ \omega_0 & s \end{bmatrix} + \frac{1}{R} \right) \begin{bmatrix} u_d(s) \\ u_q(s) \end{bmatrix} \qquad (29)$$

Perturbing to the (29)system yields the small-signal linear model of three-phase buck rectifier as

$$\begin{bmatrix} \tilde{i}_d(s) \\ \tilde{i}_q(s) \end{bmatrix} = \left(C \begin{bmatrix} s & -\omega_0 \\ \omega_0 & s \end{bmatrix} + \frac{1}{R} \right) \begin{bmatrix} \tilde{d}_d(s) \cdot V_{DC} + D_d(s) \cdot \tilde{V}_{DC} \\ \tilde{d}_q(s) \cdot V_{DC} + D_q(s) \cdot \tilde{V}_{DC} \end{bmatrix} \qquad (30)$$

V. Discussion and Comparisons of the Approaches

This paper proposed several model approaches based on frequency transformation of the stationary-frame models. With mathematical inductive method, these approaches can be proven to be equivalent. Yet in practical applications, approach I is more convenient for input-output model in frequency domain, while the approach II in favorable for the state-space modeling.

In comparison with the traditional modeling approaches in time-domain, the proposed approaches in frequency are much simply and easy to employ. For large system modeling, the traditional time-domain state-space averaging modeling approaches may be unpractical because the manipulating matrixes are too complex in high-order system. In contrast, the proposed approaches simplify the modeling of large system by directly transformation of the stationary-frame models, which is much easy to get. Besides the simplicity, the physical meanings of the frequency transformation are explained with amplitude modulation, which is much intuitive for practical engineering.

VI. Conclusion

This paper proposed two approaches based on frequency transformation and reactive elements substitution to develop the synchronous-frame models from stationary-frame models, with which the modeling processes are significantly simplified. To verify their validity, the synchronous-frame models of several power converters are obtained with frequency conversion and compared to the publications.

References

[1] G. W. Wester and R. D. Middlebrook, "Low-Frequency Characterization of Switched dc-dc Converters," Aerospace and Electronic Systems, IEEE Transactions on, vol. AES-9, no. 3, pp. 376-385, 1973.

[2] R. D. Middlebrook and S. Cuk, "A general unified approach to modeling switching converter stages," in *IEEE Power Electronics Specialists Conf. Rec.*, 1976, pp.18-34.

[3] V. Vorperian, "Simplify PWM converter analysis using a PWM switch model," *Power conversion & Intelligent Motion*, vol. 16, no. 3, pp. 8-11, 1990.

[4] S. Hiti, D. Boroyevich, and C. Cuadros, "Small-signal modeling and control of three-phase PWM converters," in *Industry Applications Society Annual Meeting, 1994., Conference Record of the 1994 IEEE 1994*, pp. 1143-1150.

[5] Y. Yang, M. Kazerani, and V. H. Quintana, "Modeling, control and implementation of three-phase PWM converters," Power Electronics, IEEE Transactions on, vol. 18, no. 3, pp. 857-864, 2003.

[6] V. Vorperian, R. Tymerski, and F. C. Y. Lee, "Equivalent circuit models for resonant and PWM switches," IEEE Transactions on Power Electronics, pp. 205-214, 1989.

[7] C. T. Rim, D. Y. Hu, and G. H. Cho, "Transformers as equivalent circuits for switches: General proofs and *D-Q* transformation-based analyses," IEEE Transactions on Industry Applications, vol. 26, no. 4, pp. 777-785, 1990.

[8] H. Bin, R. Burgos, F. Wang, and D. Boroyevich, "D-Q-0 Synchronous Frame Average Model for Three-Phase Arrays of Single-Phase PFC Converter Loads," in Computers in Power Electronics, 2006. COMPEL '06.IEEE Workshops on 2006, pp. 83-88.

[9] M. Hengchun, D. Boroyevich, and F. C. Y. Lee, "Novel reduced-order small-signal model of a three-phase PWM rectifier and its application in control design and system analysis," Power Electronics, IEEE Transactions on, vol. 13, no. 3, pp. 511-521, 1998.

[10] Zmood, D.N. Holmes, D.G. Bode, G.H. , "Frequency-domain analysis of three-phase linear current regulators," Industry Applications, IEEE Transactions on, vol. 37, no. 2, pp. 601-610, 2001.

[11] Newman, M.J. Zmood, D.N. Holmes, D.G. , "Stationary frame harmonic reference generation for active filter systems," Industry Applications, IEEE Transactions on, vol. 38, no. 6, pp. 1591-1599, 2002.

[12] Xiaoyu Wang, Jinjun Liu, Jinku Hu, Yujimeng, and Chang Yuan, "Frequency Characteristics of the *d-q* Synchronous-Frame Current Reference Generation Methods for Active Power Filter," in 7th international conference on power electronics (ICPE), Korea 2007.

[13] Fred Lee. Modeling and control of three-phase PWM converters. Course notes. Blacksburg, Virginia, United States of America: CPES, Virginia Tech, 2000.

[14] Peng FZ, Akagi H, Nabae A. Compensation characteristics of the combined system of shunt passive and series active filters. Industry Applications, IEEE Transactions on, vol. 29, no. 1, 1993: 144-152

Appendix

Here gives the derivation of (31).

Because of these equations as

$$\begin{bmatrix} 1 & -j \\ j & 1 \end{bmatrix}\begin{bmatrix} 1 & j \\ -j & 1 \end{bmatrix} \equiv \mathbf{0}, \begin{bmatrix} 1 & -j \\ j & 1 \end{bmatrix}^2 = 2\begin{bmatrix} 1 & -j \\ j & 1 \end{bmatrix} \text{ and } \begin{bmatrix} 1 & j \\ -j & 1 \end{bmatrix}^2 = 2\begin{bmatrix} 1 & j \\ -j & 1 \end{bmatrix},$$

multiplied by

$$\begin{bmatrix} 1 & -j \\ j & 1 \end{bmatrix} \text{ and } \begin{bmatrix} 1 & j \\ -j & 1 \end{bmatrix},$$

(16) becomes

$$\begin{bmatrix} 1 & -j \\ j & 1 \end{bmatrix}\begin{bmatrix} U_d(s) \\ U_q(s) \end{bmatrix} = \begin{bmatrix} 1 & -j \\ j & 1 \end{bmatrix}\begin{bmatrix} U_\alpha(s - j\omega_0) \\ U_\beta(s - j\omega_0) \end{bmatrix}$$
$$\begin{bmatrix} 1 & j \\ -j & 1 \end{bmatrix}\begin{bmatrix} U_d(s) \\ U_q(s) \end{bmatrix} = \begin{bmatrix} 1 & j \\ -j & 1 \end{bmatrix}\begin{bmatrix} U_\alpha(s + j\omega_0) \\ U_\beta(s + j\omega_0) \end{bmatrix} \qquad (32)$$

Combination of (32) yields

$$\begin{bmatrix} U_d(s) \\ U_q(s) \end{bmatrix} = \frac{1}{2}\begin{bmatrix} 1 & -j \\ j & 1 \end{bmatrix}\begin{bmatrix} U_\alpha(s - j\omega_0) \\ U_\beta(s - j\omega_0) \end{bmatrix} + \frac{1}{2}\begin{bmatrix} 1 & j \\ -j & 1 \end{bmatrix}\begin{bmatrix} U_\alpha(s + j\omega_0) \\ U_\beta(s + j\omega_0) \end{bmatrix}$$

$$= \frac{1}{2}\begin{bmatrix} 1 & -j \\ j & 1 \end{bmatrix}\begin{bmatrix} I_d(s) \\ I_q(s) \end{bmatrix} Z(s - j\omega_0) + \frac{1}{2}\begin{bmatrix} 1 & j \\ -j & 1 \end{bmatrix}\begin{bmatrix} I_d(s) \\ I_q(s) \end{bmatrix} Z(s + j\omega_0)$$

$$= \underbrace{\begin{bmatrix} \frac{1}{2}(Z(s + j\omega_0) + Z(s - j\omega_0)) & \frac{j}{2}(Z(s + j\omega_0) - Z(s - j\omega_0)) \\ -\frac{j}{2}(Z(s + j\omega_0) - Z(s - j\omega_0)) & \frac{1}{2}(Z(s + j\omega_0) + Z(s - j\omega_0)) \end{bmatrix}}_{=\mathbf{Z}_{\alpha\beta}(s)}\begin{bmatrix} I_d(s) \\ I_q(s) \end{bmatrix}$$

A Gate Drive Circuit of Power MOSFETs and IGBTs for Low Switching Losses

Toshihisa Shimizu
Tokyo Metropolitan University
Dept. of Electrical and Electronics Engineering
1-1, Minami-ohsawa, Hachioji, Tokyo, 192-0397
Email: shimizut@tmu.ac.jp

Keiji Wada
Tokyo Metropolitan University
Dept. of Electrical and Electronics Engineering
1-1, Minami-ohsawa, Hachioji, Tokyo, 192-0397
Email: kj-wada@tmu.ac.jp

Abstract—In order to increase the power density of power converters, reduction of the switching losses at a high-frequency switching condition is one of the most important issues. This paper presents a new gate drive circuit that enables to reduce the switching loss on both the Power MOSFET and the IGBT. A distinctive feature of this method is that both the turn on loss and the turn off loss can be decreased simultaneously without using the conventional ZVS circuit, such as quasi-resonant adjunctive circuit. Some experimental results of the switching loss of Power MOSFET and IGBT used on the buck-chopper circuit is shown and confirmed the effectiveness of the proposed circuit.

I. INTRODUCTION

Increase in the power density of power converters have been one of the most important issue among the power electronics authorities. Since high frequency switching in power conversion makes the volume and the size of magnetic component small, then raising the switching frequency with keeping low switching loss is the effective way for realizing high power density converter. However, there is a tradeoff between the switching frequency and the switching losses. This paper presents a new gate drive circuit for Power MOSFETs and IGBTs which provides low switching loss in both the turn on and turn off transition at high frequency switching operation. A distinctive feature of the gate drive circuit is that both the accelerated turn on action and the ZVS turn off operation can be realized simultaneously by adding only a small number of adjunct components on to the gate drive circuit. In the first section, a mechanism of increase of turn on loss caused by the mirror effect on the Power MOSFET is described and the circuit configuration of the proposed gate driver is described. In the next section, a circuit configuration and the operation principle of the proposed gate drive circuit are presented. Followed by the theoretical discussion, some experimental results on both the switching devices are shown, and discussed the effectiveness of this method. A conclusion of this paper is also given.

II. OPERATION PRINCIPLE AND CIRCUIT CONFIGURATION

Figure 1 shows a conventional gate drive circuit for MOSFET or IGBT devices. It is well known that an inductance component, L_G, and a resistance component,

R_G, the of the wiring cable, which connects the gate/source terminal with the drive circuit, harms high speed charge/discharge action of gate drive voltage and results in increase of switching loss. Especially at the turn on transition, gate charge energy supplied from the gate drive circuit leaks from gate terminal to drain terminal thorough a mirror capacitor, C_{GD}, as shown in Fig.2. This means that the supplied energy is by-passed thorough mirror capacitor. What is worse, the amount of leak charge increases because

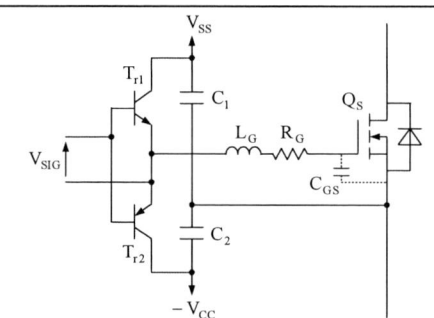

Fig. 1: Basic gate drive circuit.

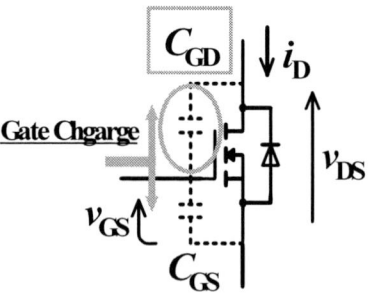

Fig. 2: Parasitic capacitor and the mirror effect.

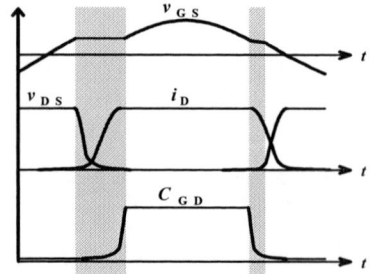

Fig. 3: The operation waveforms on the switching transition.

the mirror capacitor increases very much at low drain-source voltage condition. Then the gate-source voltage, V_{GS} does not increase that we expected. Then the on-state voltage, $V_{DS(on)}$, at the turn on transition decreases slowly as shown in Fig.3 and hence the resultant turn on loss increases very much. This phenomenon is called as a mirror effect.

If we can compensate the gate charge for input capacitor, C_{GS}, by proper method we can decrease the turn-on loss. Then we focused on the fact that the change of drain-gate voltage, V_{DG}, occurs at the same instance with the change of voltage of the drain-source voltage, V_{DS}. This means that the gate charge leak occurs during the fall time of drain-source voltage, V_{DS}. It is clear that we cannot collect the charge that is flowing through the mirror capacitor. But we can simulate the leak current if we detect the discharge current on the capacitor that is connected between the drain and the source terminal. Further more, we can charge the gate terminal more quickly if we can transfer this current into the input capacitor, C_{GS}. Based on this principle, we propose a new gate drive circuit as shown in Fig, 4. This circuit is composed of a CT current feedback circuit and a conventional gate drive circuit.

The detail operation of the gate drive circuit is explained as follows. At the turn on transition, the discharge current of C_S is flowing thorough the primary side of the current transformer, and the current energy is transferred to the secondary winding of the CT. A small part of secondary current turns the transistor, T_F, on, and most of the secondary current flowing thorough the gate-source terminal and charges the input capacitor, C_{GS}. Then, the turn on transition is shortened and the resultant turn on loss can be decreased. Since the transistor, T_F, is turned off automatically after the turn on transition, the secondary winding of the CT is disconnected from the gate drive terminal. Then the CT is prevented from a magnetic saturation, and we can induce the required voltage into the gate-source terminal. Another advantage of this circuit is that the capacitor, C_S, and the diode, D_S, make the ZVS turn-off snubber unit. Hence, the rise time of V_{DS} at turn off transition of the main switch, Q_S, can be decreased, and the resultant turn off loss can also be decreased simultaneously.

III. EXPERIMENTAL RESULTS

Figure 5 shows the circuit configuration of the buck-chopper circuit in which the proposed gate drive circuit is used, and Table 1 shows the circuit parameters of the test circuit. Figs 6(a) and (b) show the switching waveforms on the Power MOSFET at turn-on and turn-off transients, respectively. As shown in Fig. 6(a), the on state voltage, V_{DS}, at the turn on transient takes high value on the conventional gate drive condition. On the contrary, the voltage is remarkably reduced on the proposed gate drive condition. It should also be noted that the increase rate of drain current at the turn-on transient is much larger than the conventional one because reverse recovery current on the FWD increases. This fact means that the turn on loss on the MOSFET cannot be reduced that we expected. However, it will be possible to reduce the loss by using a fast recovery diode as FWD. At the turn off transition, as shown in Fig. 6(b), the rate of increase of the drain voltage on the proposed gate drive condition is reduced rather than the one on the conventional condition. It is clear that the turn-off loss on the proposed system can be simply reduced. Hence, the switching on the MOSFET can totally be reduced by the proposed drive circuit.

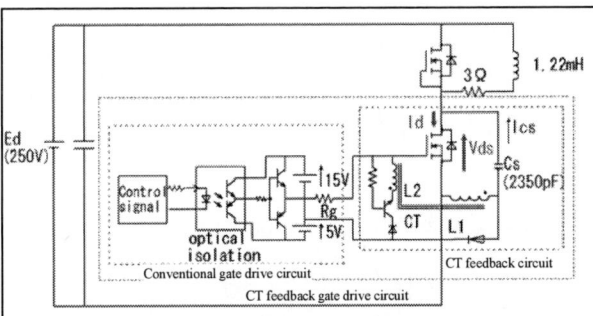

Fig. 5: A buck-chopper circuit for the switching loss measurement.

Table 1. Circuit parameters.

Device	Label	Specifications
MOSFET	Q_S	2SK2382 (TOSHIBA)
CT	L_1, L_2	1.1 μH (Feroomolybdenum)
Inductor	L_G	1.3 μH (Ni-Zn Ferrite)
Transistor	T_F, T_{r1}	2SC3596 (SANYO) 60 V/0.3 A
	T_{r2}	2SA1402 (SANYO) 60 V/0.3 A
Resistor	R_B	62 Ω, 1/4 W
Diode	D_F	11DQ04 (IR-Japan) 40 V/1.57 A
	D_S	SDP06S60 (INFINEON) 600 V/6.0 A
Capacitor	C_1, C_2	100 μF
	C_S	180 pF
DC Power Source	E_d	120 V

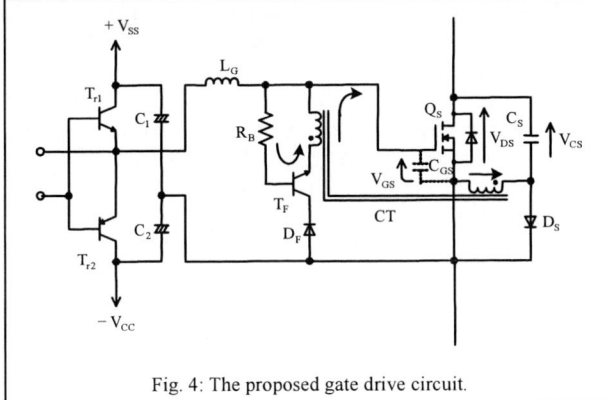

Fig. 4: The proposed gate drive circuit.

978-1-4244-1871-8/07 $25.00 © 2007 IEEE

drain current is appeared, the overall turn-off loss on the IGBT can be reduced compared to the conventional condition.

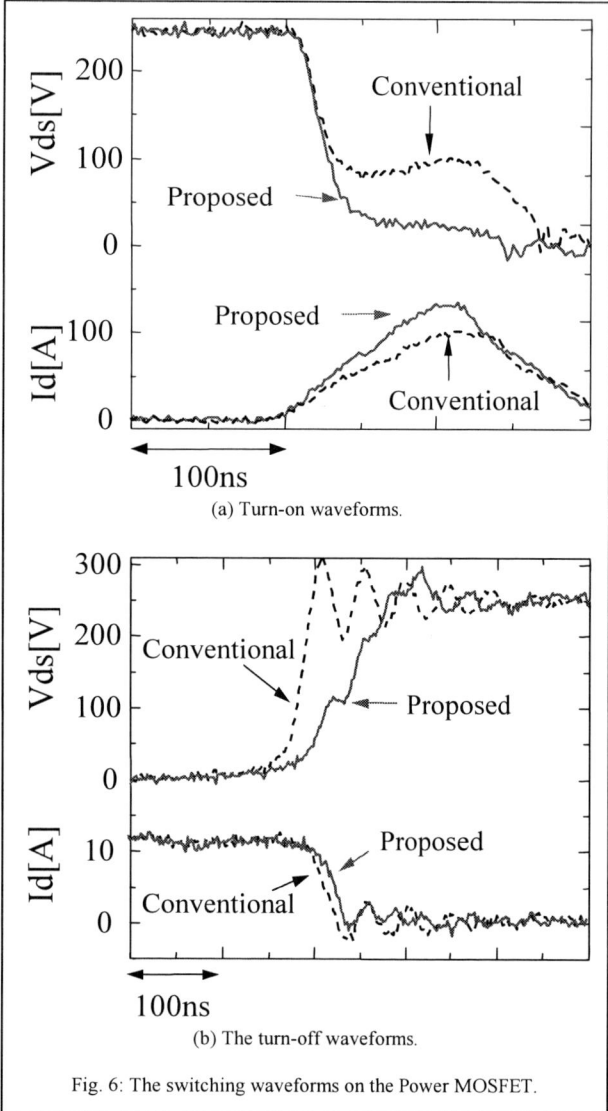

(a) Turn-on waveforms.

(b) The turn-off waveforms.

Fig. 6: The switching waveforms on the Power MOSFET.

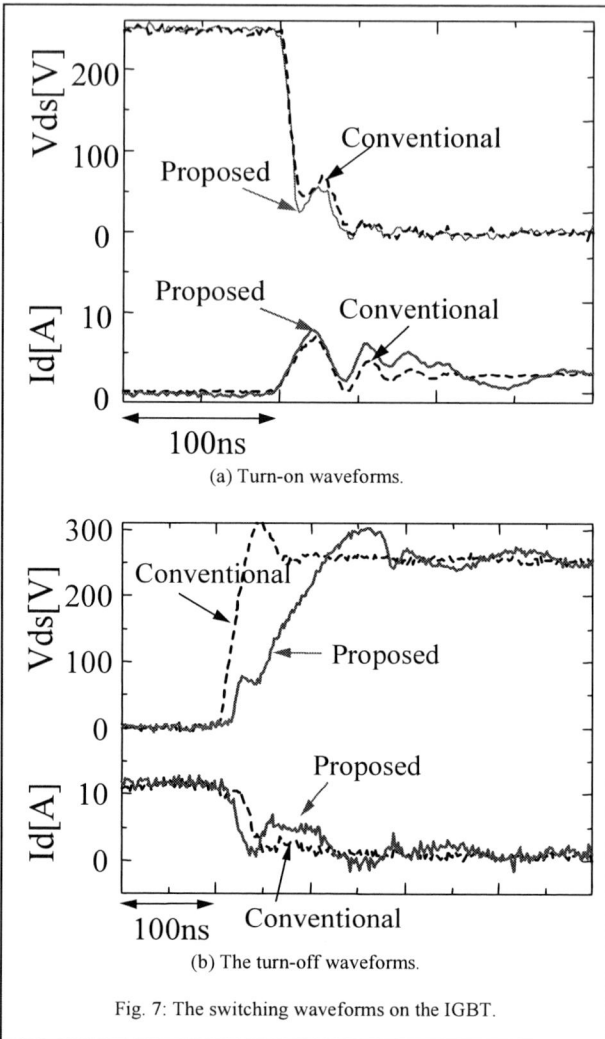

(a) Turn-on waveforms.

(b) The turn-off waveforms.

Fig. 7: The switching waveforms on the IGBT.

Figs. 7(a) and (b) show the switching waveforms on the IGBT at turn-on and turn-off transients, respectively. Fig. 7(a) shows the drain voltages at the turn-on transient on both the conventional and the proposed gate drive conditions. It is clear that those two waveforms coincide well, and this is completely different situation on the MOSFET. This is because that the turn-on transient on the IGBT is mainly influenced by the carrier injection phenomenon in the IGBT structure and hence it does not influenced by the mirror effect. Hence, the turn-on loss on the IGBT can not be decreased by the proposed gate drive circuit. Fig. 7(b) shows the drain voltages at the turn-on transient on both the conventional and the proposed gate drive conditions. We can see that the rate of increase of the drain voltage on the proposed gate drive condition is reduced. However, there exist small amount of current rise in drain current at the turn-off transient on the proposed gate drive condition. This current rise is caused by the minor carrier ejection in the IGBT junction which depend on the IGBT structure. Even though the current rise in the

Figure 8(a) and (b) show the measured loss at turn on and turn off transients on the Power MOSFET and the IGBT in many current conditions, respectively. As shown in Fig.8(a), the turn on loss on the Power MOSFET is constantly decreased of about 500[uJ] compared to the loss on the conventional gate drive condition. On the contrary, the value of turn-on loss on the IGBT cannot be decreased by the proposed gate drive circuit. But the turn-on loss on the IGBT is much smaller that the one the Power MOSFET. The turn-off loss on both the Power MOSFET and the IGBT are shown in Fig. 8(b). The turn off loss is also decreased compared to the loss on the conventional one. What we are surprising is that the reduced turn-off loss on the IGBT is much higher than the one on the Power MOSFET. Differ from the turn-on loss, decrease of turn off loss on the IGBT is almost in proportion to the turn off current, and about 30-50% of the switching loss can be decreased. We can say

978-1-4244-1871-8/07 $25.00 © 2007 IEEE

that the proposed gate drive circuit is effective on both the Power MOSFET and the IGBT.

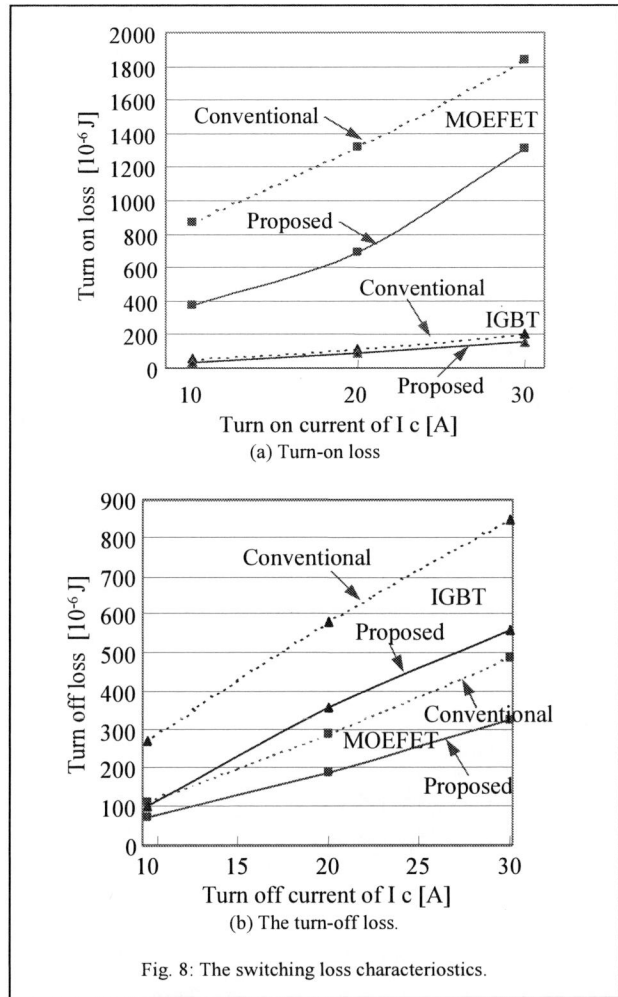

(a) Turn-on loss

(b) The turn-off loss.

Fig. 8: The switching loss characteriostics.

The turn off loss might have been decreased much more if we use larger capacitance in Cs. But in this case, the energy regenerated from the capacitor, C_S, increases too much and cause the over charging of gate voltage, V_{GS}. Then, in this case, we need to transfer the surplus of regenerated energy into some appropriate power consumption.

We can conclude that the proposed gate drive circuit is one of the effective methods for reduction of switching loss on both the Power MOSFET and the IGBT.

IV. CONCLUSIONS

A novel gate drive circuit of MOSFETs and IBGTs useful for decreasing the switching loss at high frequency operation is proposed. The experimental result in the buck-chopper circuit is shown and discussed the effectiveness of the proposed circuit.

ACKNOWLEDGMENT

The author whishes to thank Mr. Hirokazu Matshuno for his contribution in the experimental work.

REFERENCES

[1] K.Takao, et.al, "Novel exact power loss design method for high output power density converter," Conference Record of IEEE PESC'06, pp.2651-2655, 2006.

[2] T.Shimizu, H.Kinjyo, K.Wada, "A Novel High-frequency Current Output Inverter based on an Immittance Conversion Element and a Hybrid MOSFET-SiC Diode Switch," Conference Record of IEEE PESC'03, pp.2003-2008, 2003.

[3] S.Finney, B.Williams, T.Green, "RCD Snubber Revised," IEEE Trans. on Industry Applications, vol.32. No.1, pp.155-160, 1996.

[4] R.Chokhawala, J.Catt, B.Pelly, "Gate Drive Circuit for IGBT Module," IEEE Trans. on Power Electronics, Vol.31, No.3, pp.603-611, 1995.

Load modeling for the drum washing machine system simulation

Jung-Hyo Lee, Tae-Woong Kong, Won-Chul Lee, Jae-Sung Yu, Choong-Yuen Won
Power Electronics laboratory
Dept. of Information & communication, Sungkyunkwan Univ.
24206 Cheoncheon-dong, Jangan-gu, Suwon, 440-746
Email: daumin@skku.edu

Abstract—In this paper, we explained the load modeling of drum washing machine for determining the rating of motor drive device. The motor drive should supply enough current to motor in spite of load variation. However, drum washing machine load torque is not only affected by the weight. Besides, drum washing machine load torque has the frequency which is influenced by the drum velocity.

We analyze the forces that affect the laundry. And the load torque is designed by the mathematical equation. Finally, we compare the conventional constant torque and the proposed load torque simulation.

I. INTRODUCTION

Washing machine is classified by the washing method, the pulsator washing machine, the agitator washing machine, and the drum washing machine. The pulsator washing machine bring vortex to pass by forward and backward rotating pulsator in the bottom of the washing machine. Pulsator washing machine has demerits that damaging the laundry, and washing the laundry inequality. The agitator washing machine has an agitator in the middle of the washer. The agitator rotates right and left to roll over the laundry for washing. The agitator type has merits in large amount of laundry. However, this machine is expensive and it easily breaks down. The drum washing machine has demerits that long washing time and weak washing power. Nevertheless, it hardly damages the laundry and above all, it uses small amount of the water than other washing machine. Environmental restriction gets stricter all over the world, therefore, the drum washing machine demand increases every year. [2]

Drum washing machine washes the laundry by the mechanical force that the laundry is raised by the drum and impacted by the gravity on the drum plane. Therefore, drum washing machine load torque is different to the other machine load and more complex. [3]

Although the motor drive should consider the load torque because of the rating of motor drive devices, the load torque modeling has some problems in drum washing machine. [1] First, it does not simply determine the weight. Second, it has transient state when the rotating direction is changing. Consequently, this paper proposes the load model by the load characteristics as follow.

- Load variation has periodicity which is determined by the drum velocity.
- Falling motion of the laundry makes instantaneous velocity variation.

- When the drum velocity direction is changed, load torque is increased by the load inertia.

The above enumeration is affected by the force of gravity, friction and centrifugal. [4]

The laundry needs three processes, wash the dirt of the laundry (wash mode), removing the cleaner from the laundry (rinse mode), removing the water from the laundry (spin mode). As the purpose of each mode is different, the load torque variation is also different.

Among the three modes, wash mode load torque is the most violent fluctuations, the reason is that the amount of water is small and the drum velocity is slow for washing. [2] Therefore, this paper simulates the load torque modeling of wash mode.

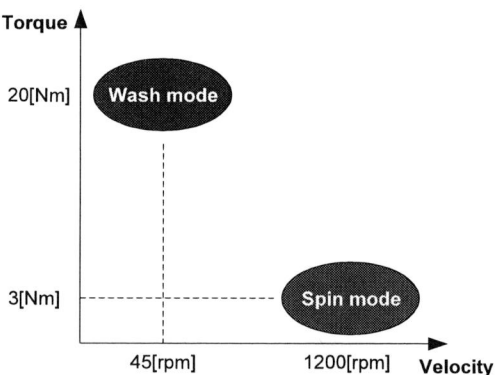

Fig. 1 Comparison of the wash mode and the spin mode

II. ANALYSIS OF THE WASH MODE

The wash mode is the mode of removing dirt of the laundry. And the wash mode drum velocity is nearly 45[rpm]. General method of the wash mode is as follow. First of all, the water is supplied minimum height in the drum. The motor rotates the fixed velocity and measure the period until the motor reaches the fixed velocity. From the acceleration time, measure the load amount. And the washing type is determined by according to the load amount.

The laundry is raised by the baffle and the friction force. If it raises enough height, the laundry is falling down on the drum. Finally, the load torque is rapidly increased and kept up until the load is falling again. Therefore, the load variation associates with the point of the laundry which is fallen.

978-1-4244-1871-8/07 $25.00 © 2007 IEEE

Because the amount of water supplied is minimized in wash mode, only the minor effect of water affects the load torque. Also, due to the large viscosity between the drum and the laundry which is caused by wet, the friction force tends to increase.

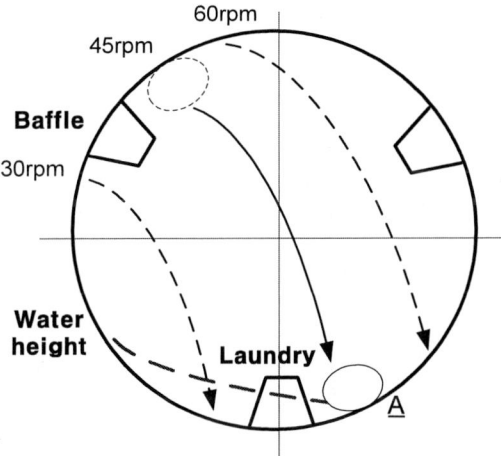

Fig. 2 Falling motion traces of the laundry in wash mode

Fig. 2 shows that falling traces of the laundry is determined by the velocity. The laundry drop point (A) is determined by drum velocity. The laundry motion trace is longer and longer, the period that only small water load is granted on the drum washing machine is longer.

To compare the rinse mode, falling motion traces of the laundry in rinse mode is Fig. 3.

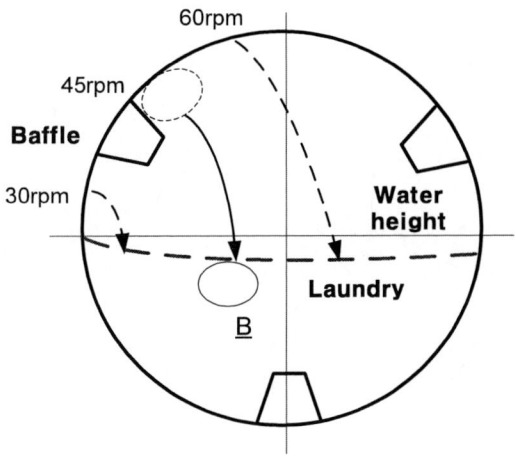

Fig. 3 Falling motion traces of the laundry in rinse mode

The rinse mode drum velocity is almost same as the wash mode. However, the water height is maximized in rinse mode. Therefore, the laundry is falling on water plane. (B) And the motion traces are shorter than wash mode.

As a result, the load variation of drum washing machine is also smaller. And the rinse mode load torque is easily estimated than wash mode.

III. MECHANICAL CHARACTERISTICS OF DRUM WASHING MACHINE LOAD

A. Load variation by the falling motion

Drum washing machine load is raised by the baffle and velocity.

The raised load is affected by the force of gravity, friction and centrifugal. Gravity force is able to divide the centrifugal direction component and tangential component of the velocity direction.

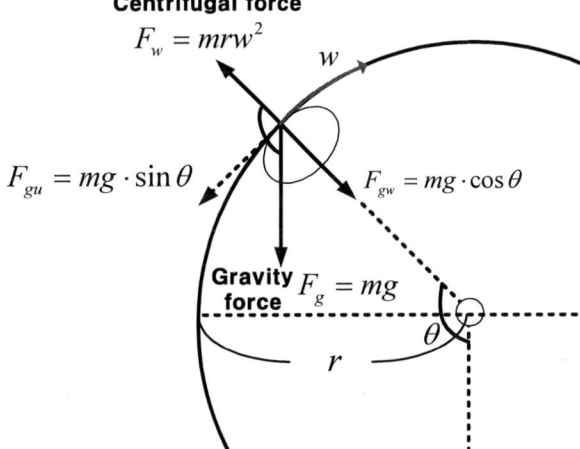

Fig. 4 Forces of the laundry at fixed velocity

Fig.4 shows the force direction of centrifugal and gravity. Analyze the relationship of the force, time of the falling motion is determined by equation (1).

$$F_w + \Delta f = F_{gw} \qquad (1)$$

Where F_w is the centrifugal force and Δf is the friction force and F_{gw} is the reverse centrifugal direction component of the gravity force.

As the equation (1) explained, the start time of the falling motion is when the sum of friction force and centrifugal force is equal to the reverse centrifugal direction component of gravity force.

If the friction force is neglected and the radius of drum is considered, equation (1) is approximately substituted by equation (2).

$$mr\omega^2 = mg\cos\theta \qquad (2)$$

From the equation (2), determine the start position of the falling motion at fixed velocity θ is able to represent the equation (3).

978-1-4244-1871-8/07 $25.00 © 2007 IEEE

$$\theta = \cos^{-1}\left(\frac{r\omega^2}{g}\right) \qquad (3)$$

Position θ is the times of velocity and time. From equation (3), a period of load variation at fixed velocity is able to be obtainable.

$$T_0 = \frac{\cos^{-1}(\frac{r\omega^2}{g})}{\omega} \qquad (4)$$

The load pattern is able to describe the function of constant period and this load variation is shown that equation (5).

$$T(t) = M \sum_{k=0}^{\infty}\left\{u(t-2kT_0) - u(t-T_0-2kT_0)\right\} \qquad (5)$$

B. Transient state by the inertia

The transient state of drum washing machine load pattern is associated with the inertia of water flow. Fig. 5 shows the transient state by reverse direction load of water and laundry.

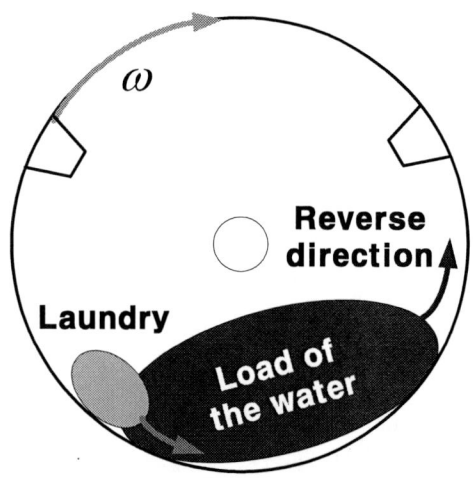

Fig. 5 The load of transient state

Transient state is occurred when the direction of the drum velocity is changed radically. The load of water and laundry has the inertia from previous velocity. And this load of reverse direction causes radical load variation to drum washing machine. The strict friction force causes. that the magnitude of peak load is two times of the washing load in steady state.

If the transient state has a characteristic that reduction of natural response, it can be described by equation (6).

$$T_{tr1}(t) = (M + M_{water})\bullet(e^{-\frac{t}{\tau}}) \qquad (6)$$

Where M is the total torque of the load variation and τ is the reduction coefficient.

IV. SIMULATION

This paper presents the drum washing machine load pattern simulation of wash mode that uses the longest time in all of the other modes and this load pattern is easily able to extend the rinse mode.

Table 1 shows the washing machine parameters that uses in load pattern. And Fig.6 shows the load simulation block diagram using the Matlab/Simulink design tool.

TABLE 1
THE PARAMETERS FOR
DRUM WASHING MACHINE LOAD MODELING

Load torque by the laundry	4[N·m]
Load torque by the water	2.6[N·m]
Velocity of the washing machine	45[rpm]
Radius of the drum	33[cm]
Friction coefficient	0.64

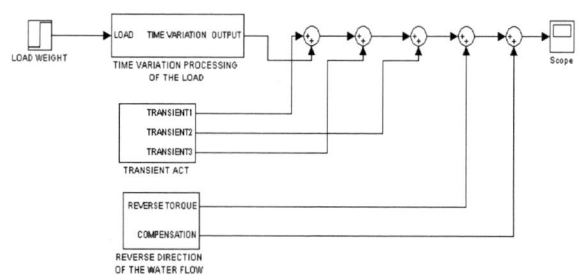

Fig. 6 Block diagram of the drum washing load modeling simulation

Because the wash of drum washing machine is radical falling motion of the load, constant laundry load can approximate the pulse waveform that constant period at fixed velocity (Fig.7).

In wash mode, drum washing machine is driven by forward and backward. Consequently, state of initial load torque should have two times of the steady state load. And this is described by transient state (Fig.8).

Besides, the reason that initial state of water movement is unstable makes reverse load torque by the viscosity between laundry and water (Fig.9).

Following figures are simulation waveforms that illustrate these load characteristics.

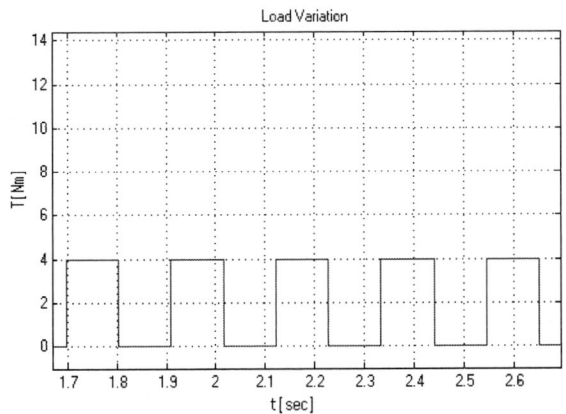

Fig. 7 Load variation of the drum washing machine

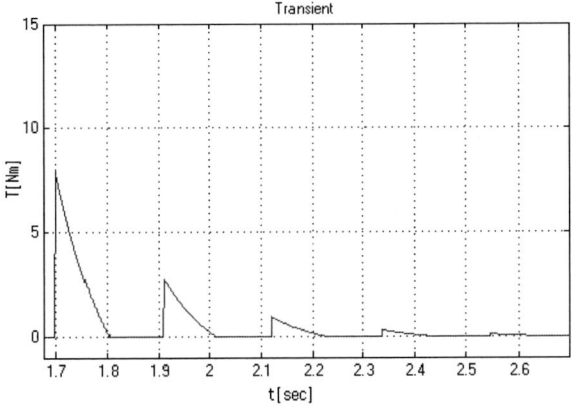

Fig. 8 Load of the transient state

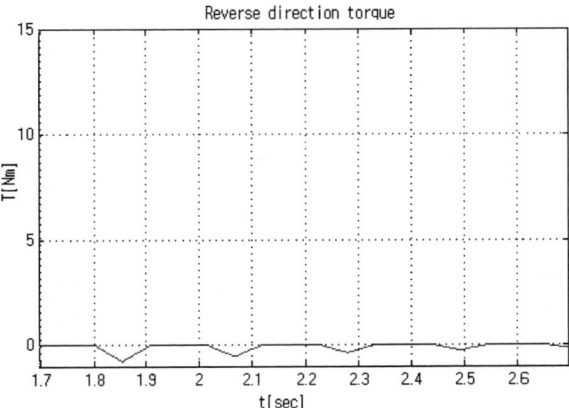

Fig. 9 Reverse direction torque by the water

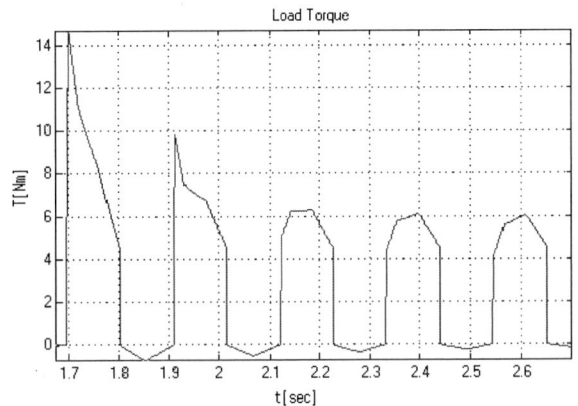

Fig. 10 Load torque of the drum washing machine

To apply the simulated load to the motor, we simulated the brushless dc (BLDC) motor drive system by following name plate.

TABLE 2
NAME PLATE OF THE BLDC MOTOR

Phase resistance	0.75[Ω]
Phase inductance	3.05[mH]
Coefficient of back EMF	0.215[V/(rad/sec)]
Coefficient of torque	0.215[N·m/A]
Moment of inertia	8.26×10^{-5}[kg·m^2]
Rated voltage	310[V]
Rated Power	1.2[kW]
Viscous factor	0[N·m/(rad/sec)]
Pole number	24[pole]

Fig.11 shows the simulated control block diagram of the BLDC motor drive system. Simulation of BLDC motor control uses hysteresis current controller and permission range is 5% of the reference current.

In block diagram, difference between reference velocity ω_r^* and real velocity ω_r goes through the PI controller. The output of the PI controller is the reference current I_s^*. And difference between reference current and real current has application to the switch voltage V_{abc}^* which pulse width is modulated. To determine the phase current that should be compared by the reference, rotor position data θ should be known by the hall sensor signals H_{abc} from the motor.

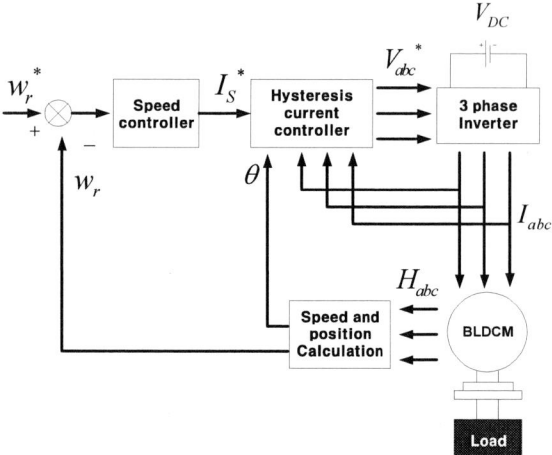

Fig. 11 Control block of the BLDC motor simulation

Fig.12 shows the phase current from the simulation that constant period load is connected by the motor. The assumed maximum load is 4[N·m] that is equal to the laundry load.

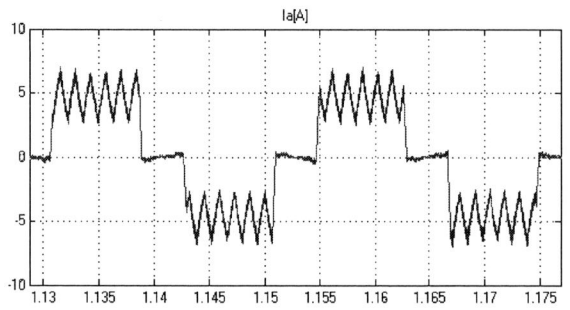

Fig. 12 Phase current when the motor is connected by the load of constant period

Fig.13 is the phase current from the proposed simulation that the load of drum washing machine connects the motor. The laundry load is the same as previous simulation. The figure shows that the more phase current in transient state should be supplied to the motor than the current of constant load simulation.

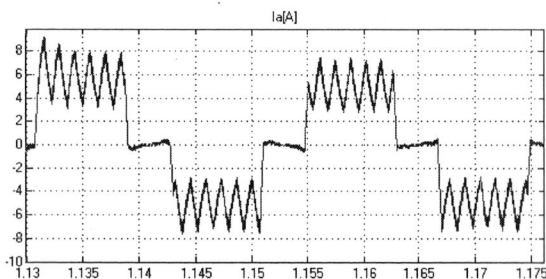

Fig. 13 Phase current when the motor is connected by the load of proposed modeling

V.   CONCLUSION

In motor drive, one of the most important considerations is the load characteristic and variation. Generally, the motor drive should be made enough current for the load variation and it should be able to control the motor overcome the load weight. However, the drum washing machine's load variation is irregular and large because of the water flow and reverse load torque. Therefore, this paper describes the drum washing machine's load pattern modeling by the mathematical theory.

Also, this paper presents the simulation of BLDC motor system. The load of conventional method and proposed method is applied to this BLDC motor simulation. As a result, the simulation using proposed load needs more current when the transient state.

ACKNOWLEDGMENT

The authors wish to acknowledge their sincere thanks to Korea Electronics Technology Institute (KETI) for funding.

REFERENCE

[1]   T. J. E. Miller, "Brushless Permanen-Magnet and Reluctance Motor Drives", Clarendon Press, Oxford, 1989.

[2]   T.Shigeya, N.Kazuo, "Motors and inverter for home appliances" Toshiba Review, vol.55, no.4, pp.25-27, 2000.

[3]   Wang C., He Z., Guo J., "The application of a novel motor in washing machines", Electrical Machines and Systems, ICEMS 2001. Proceedings of the Fifth International Conference, vol 2, pp. 1030 ~ 1033, Aug. 2001.

[4]   Harmer K. , Mellor P.H. , Howe D. , "An energy efficient brushless drive system for a domestic washing machine ", Power Electronics and Variable-Speed Drives, Fifth International Conference, pp. 514 ~ 519 Oct 1994.

[5]   Bianchi A. , Buti L. , "Three-phase A.C. motor drive and controller for clothes washer", Appliance Magazine, June 2003.

[6]   Krause P.K., Wasynczuk O., Sudhoff S.D., "Analysis of electric machinery and drive systems", 2nd Edition, IEEE Press 2002.

Advanced Simulation Concept for the Power Train of an AC Locomotive and its Verification

Dr.-Ing. Volker Staudt
Ruhr University Bochum
Institute for Electrical Power
Engineering and Power Electronics
D-44780 Bochum, Germany
Email: staudt@eele.rub.de

Dipl.-Ing. Carsten Heising
Ruhr University Bochum
Institute for Electrical Power
Engineering and Power Electronics
D-44780 Bochum, Germany
Email: heising@eele.rub.de

Prof. Dr.-Ing. A. Steimel
Ruhr University Bochum
Institute for Electrical Power
Engineering and Power Electronics
D-44780 Bochum, Germany
Email: steimel@eele.rub.de

Abstract— Recent AC railway traction vehicles contain two main power electronic components in the power train, the four-quadrant converter and the three-phase machine-side inverter. They are joined by a common DC-link capacitor and – often – a resonant circuit tuned to twice the supply frequency. Converter and inverter are mainly controlled separately. The overall performance of the traction vehicle can be improved considerably by an integrated control of converter and inverter, explicitly taking into account the dynamics of the DC-link filter.

This requires, however, enhanced control and simulation effort. The simulation has to deal with a hybrid system, described on the one hand by continuous differential equations and on the other hand by converter switching functions. The continuous part of the system is described by state-space equations solved by the Bulirsch-Stoer implementation of the Richardson Extrapolation, the switchings are modeled using Petri Nets. This combination allows high-speed simulations of the complete system at suitable precision.

For verification the simulation results are compared to measurements made at a 50-kW laboratory test set-up of the complete power train. At this early stage of the project, simple control schemes are used for first tests of simulation and test set-up. Measured and simulated quantities are in very good agreement, allowing to proceed to more sophisticated control schemes.

I. INTRODUCTION

Presently, AC railway traction vehicles contain two main power electronic components in the power train, the four-quadrant line-side converter and the three-phase machine-side inverter (Fig. 1) [1]. The four-quadrant converter is linked to the power supply via contact line and pantograph by a transformer reducing the high line voltage (15...25 kV) to suitable values; the converter allows freely both directions of energy flow. Due to two-wire operation, oscillations of energy flow with twice the railway power-supply system frequency (in Central Europe 16.7 Hz) are unavoidable. Stationarily, they are cancelled in all 16.7-Hz traction vehicles by a tuned series-resonant circuit, in parallel to the DC-link capacitor. The induction traction machines (IM) are fed by a PWM inverter connected to this DC link.

It is state of the art to control four-quadrant converter and PWM inverter in principle separately, with the four-quadrant converter taking care of the DC-link voltage, the line-current harmonics and the dynamics on the power-supply-side, including grid stability concerns. The PWM inverter feeds the induction machines with only very limited regard to DC-link voltage, power-supply-side harmonics and grid stability.

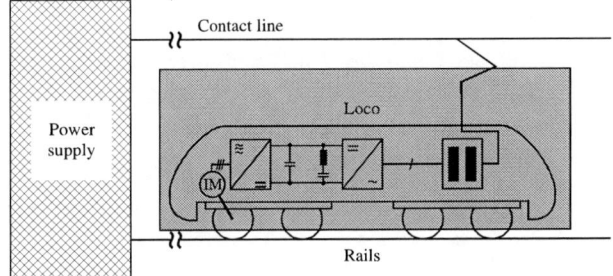

Fig. 1: Schematic of the traction system under analysis

In railway traction, power-supply side harmonics are an important safety issue due to interference with railway signaling systems. Also, the reaction of the locomotive to power-supply side conditions caused, e.g., by other traction vehicles or converter substations is an important issue [2]. Power-supply interaction is not only influenced by the four-quadrant converter, but also by the PWM inverter and its interaction with the DC link. High torque dynamics on the IM side (e.g. at load shedding) can excite the resonant circuit in the DC link. They make a high-power DC-link chopper necessary, to damp the occurring DC-link over-voltage.

Obviously, the overall performance of a traction vehicle can be improved considerably by an integrated control of converters and inverters, explicitly taking into account the dynamics of the complex DC-link filter and the requirements concerning the power-supply side. This demands, however, enhanced control and simulation effort and is therefore a new topic of research.

The simulation needed for such a configuration and its control has to deal with a hybrid system. This is described on the one hand by continuous differential equations modeling the supply, the transformer, the DC-link filter and the induction machine. On the other hand discontinuous elements, like the switching of the converters, have to be modeled, taking, e.g., into account the different sets of parameters describing IGBT and diode devices (voltage drop, switching delay). Both components, continuous and discontinuous, have to be modeled with suitable precision in the simulation. Due to the chosen concept, nonlinearities of the equations and the parameters linking the state variables, e.g., saturation effects, can also be covered.

The simulation is completely realized using C code, directly including the code used for the control of the test set-up, leading to very short simulation times. In this way the

978-1-4244-1871-8/07 $25.00 © 2007 IEEE

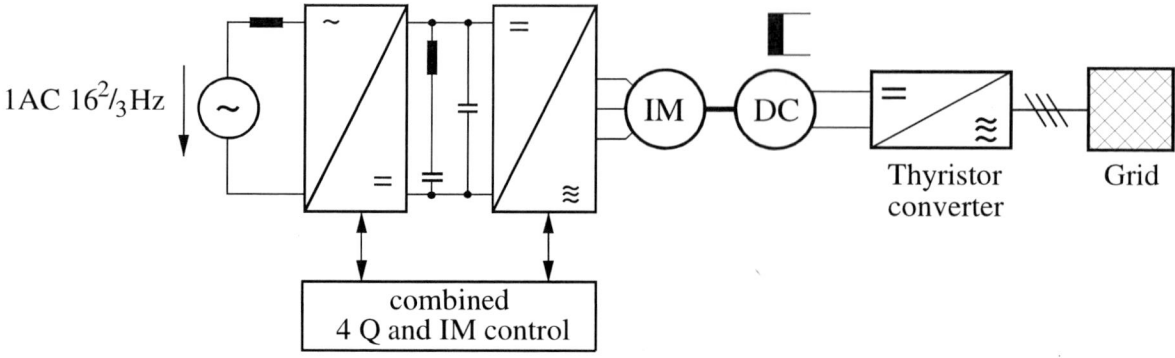

Fig. 2: Simplified equivalent system of AC traction system

control software can be developed, optimized and tested using the simulation and then directly verified at the test set-up (and finally transferred to a vehicle control in industrial application).

In the following chapters the structure of the test set-up and the associated simulation system is described in more detail, concluded by a verification of the simulation by measurements.

II.   THE SYSTEM UNDER ANALYSIS

Based on Fig. 1, a simplified equivalent system can be deduced (Fig. 2). It shows the main components to be simulated and also the main components making up the 50-kW test set-up.

The 16.7-Hz voltage source representing the grid (H-bridge inverter with filter in lab, ideal source in the first step of simulation) is applied to a four-quadrant converter which feeds the DC-link capacitor and the resonant circuit, such representing the input stage of the locomotive. A 50-kVA four-quadrant IGBT converter with a resulting switching frequency of 2 kHz (1 kHz per valve) was built. A suitable working point with 250 V transformer output rms voltage and 200 A transformer rms current has been selected. A DC-link chopper damping overvoltages in the DC-link (not shown in the figure) is provided to guarantee safe operation during tests.

The three-phase IGBT inverter feeds the IM with a state-of-the-art speed-sensorless control [3, 4, 5]. It is rated at 95 kVA. With a switching frequency of up to 2 kHz, a (star) rms voltage of 220 V and a rms current of 150 A in each leg the 50-kW IM (380 V, 95.5 A) can be fed.

Stationary and dynamic behaviour of locomotive and train on track are represented by a DC load machine, fed conventionally by a line-commutated antiparallel thyristor converter.

In the test set-up the four-quadrant converter and the PWM inverter are controlled commonly by a standard Personal Computer running a Real-Time Operating System, with a second PC featuring interface functionality. Details on this Target-Host concept can be found in [6].

III.   SIMULATION METHOD FOR CONTINOUS COMPONENTS: BULIRSCH-STOER

In a first step a set of ordinary differential equations is transformed into a set of N coupled first-order differential equations. To solve these resulting differential equations numerically, certain boundary conditions have to be defined. In general there are two types of boundary conditions: two-point boundary value problems and initial-value problems. Because only initial value problems are important for this time-based simulator of power electronic systems, the set of coupled first-order differential equations is solved taking into account initial-value boundary conditions.

The basic idea of every numerical solving method is the substitution of differential formulations by discrete, piece-wise linear differential quotients. As a result, the former differential equation is transformed to an algebraical problem which is easy to solve with modern computer systems. The way of solving numerical problems could, in the most simple case, be formulated mathematically using the Euler Method. However, then the derivative information is only used at the beginning of each integration time interval. In consequence the error grows quickly with increasing simulation time interval, allowing only short time steps. Therefore the Euler method is not used in professional applications.

Three major numerical methods are commonly used for solving initial-value problems for ordinary differential equation systems: the Runge-Kutta methods, the Predictor-Corrector methods and the Bulirsch-Stoer implementation of the Richardson Extrapolation [7]. The Runge-Kutta methods combine several Euler-Method steps and work like a Taylor-Series expansion. These methods are used very often and are even suitable for extremely nonstiff differential equations containing eigenvalues which vary quickly and over a wide range. The Predictor-Corrector methods calculate the solution of the ordinary differential equation one step ahead. So the derivatives to both sides, one step back and one step ahead, can be used for the solution, because symmetrical information around the actual integration point is available for the calculating algorithm.

The Richardson Extrapolation uses principles of extrapolation. As a result, the error of the simulation is as small as it would have been if a smaller step-size was used.

978-1-4244-1871-8/07 $25.00 © 2007 IEEE   866

For the described simulator of power-electronic problems the Bulirsch-Stoer implementation of the Richardson Extrapolation has been chosen, with an adaptive step-size algorithm to achieve short simulation times. The most important information for the step-size control is the estimated error generated by the last solving step. If this is too big the integration algorithm will be started once again and the last step of integration will be repeated with a smaller step-size; typically the integration interval is halved. For further performance enhancement, the step-size will be cut into subsections depending upon the ratio of the estimated error and a set reference value.

The Bulirsch-Stoer algorithm is based on the consideration of the result of a numerical calculation as a function of an adjustable parameter, namely the step-size (Fig. 3).

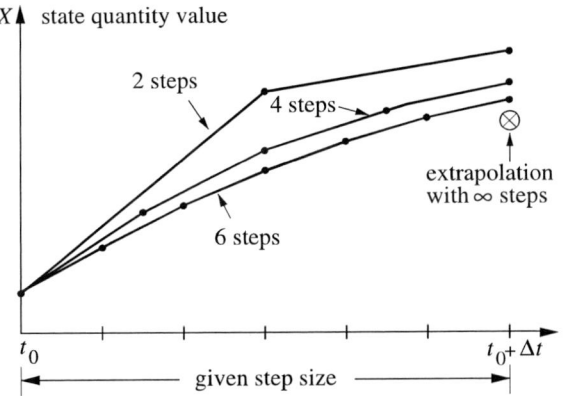

Fig. 3: Step-size used as an adjustable parameter

The algorithm primarily combines the already described idea of the step solution as an analytic adjustable function, the strength of the good convergence of rational functions or polynomial extrapolations and the modified-midpoint method, which advances a vector of dependent variables one step ahead by a sequence of N substeps.

Typical differential equations found in power-electronic systems do not have singularities within the interval of integration. So the Bulirsch-Stoer implementation of the Richardson Extrapolation suits best for simulation of power-electronic problems and leads – in combination with an adaptive step-size algorithm – to short simulation times [7].

IV.   SIMULATION METHOD FOR SWITCHED SYSTEMS: PETRI NETS

While the continuous part of the system is solved by the Bulirsch-Stoer algorithm, the switchings of the converters are modeled using Petri Nets which are one possible mathematical representation of discrete systems.

Though finite state machines are also a viable mathematical representation, Petri Nets have been chosen because of two main reasons:
1. Based on the net-array description, they are easier to compute with modern personal computers.
2. Self-commutated power-electronic converter systems are dominantly composed from several two-quadrant conver-

ters (phase legs) as "building blocks", which are connected in parallel at the same DC link. While the modeling effort of finite state machines increases exponentially by the number of parallel branches, the modeling effort of Petri Nets accumulates only in an additive way.

A Petri Net consists of places, transitions, directed arcs and tokens. Tokens marking the actual state (dot) are located in places p (circles) and are only able to move along the directed arcs to the next place, when the transition t on their way is enabled by commands or enabling events f (Fig. 4).

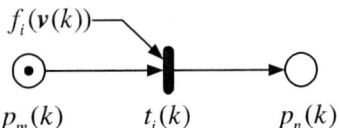

Fig. 4: Basic elements of Petri Nets

Because the events $v(k)$ and the number of enabled transitions are nondeterministic, the built-up system is time-variant and therefore nonlinear [8].

The basic Petri-Net structure of power-electronic components is given by the polymorphic current paths which depend on the directions of the current and the selected voltage. In case of the basic two-quadrant converter there are five possible states or places, respectively: positive current combined with positive and negative voltage, negative current combined with positive and negative voltage and no current. The enabling events of the transitions depend on the commanded switchings, e.g., of the IGBT, and the commutations of the antiparallel diodes (Fig. 5).

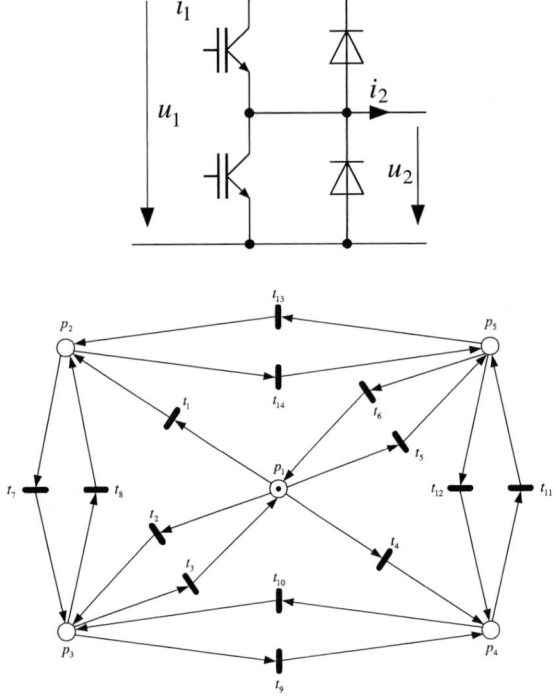

Fig. 5: Converter phase leg and its Petri Net representation

978-1-4244-1871-8/07 $25.00 © 2007 IEEE   867

In short, Petri Nets are a powerful and efficient mathematical representation of discrete systems like power-electronic converter branches, which allow high-speed simulation of complex power-electronic topologies. In combination with the described Bulirsch-Stoer implementation of the Richardson Extrapolation, Petri Nets enable the simulation of extended hybrid systems, especially power-electronic systems like the power train of locomotives with inverter-fed three-phase drives.

The main advantage of the combination of Bulirsch-Stoer and Petri Nets realized in C-Code is the very efficient simulation leading to short simulation time even in case of very complex systems with low-frequency frequency components (here, e.g., the 16.7-Hz supply) in combination with high-frequency nondisregardable switchings. Simulation extends over several seconds and is completed within few minutes.

V. MODEL OF THE IM AND THE INVERTER

In this early stage of the project not all components of the test set-up are in full function, sophisticated control schemes have yet to be implemented or developed and tested, first in simulation, then in reality. For verification of the concept described in this paper a model of an induction machine fed by the PWM inverter is used. The machine (Fig. 6), as a continuous system, is modeled by differential equations [9].

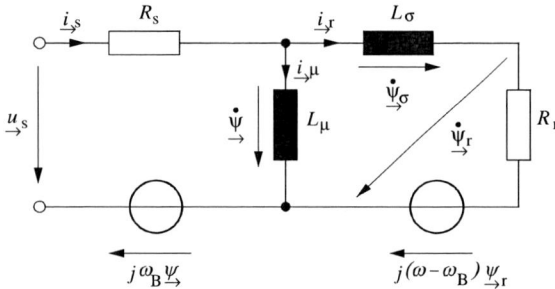

Fig. 6: Γ-ECD of IM in space-vector representation

The PWM inverter (Fig. 7) is described by three Petri Nets of two-quadrant converters (Fig. 5) connected in parallel at the DC-link side.

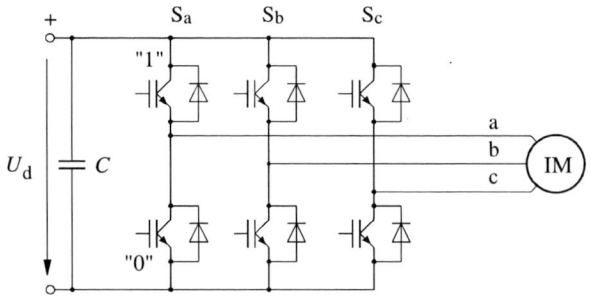

Fig. 7: Equivalent scheme of PWM inverter and machine

VI. SIMULATION RESULTS IN COMPARISON TO MEASUREMENTS

For verification, the simulation results are to be compared with measurements made at the described 50-kW laboratory test set-up. At this early stage of the project, a comparatively simple V-f-control scheme is used for first inverter tests of simulation and test set-up.

Fig. 8 shows simulated and measured steady-state phase current of the PWM machine-side inverter, with the parameters given below, and the current error. The error shows a dominant six-pulse component. This component still exists because the inverter voltage errors (due to voltage drop and delay time, e.g. [10]) are not yet compensated.

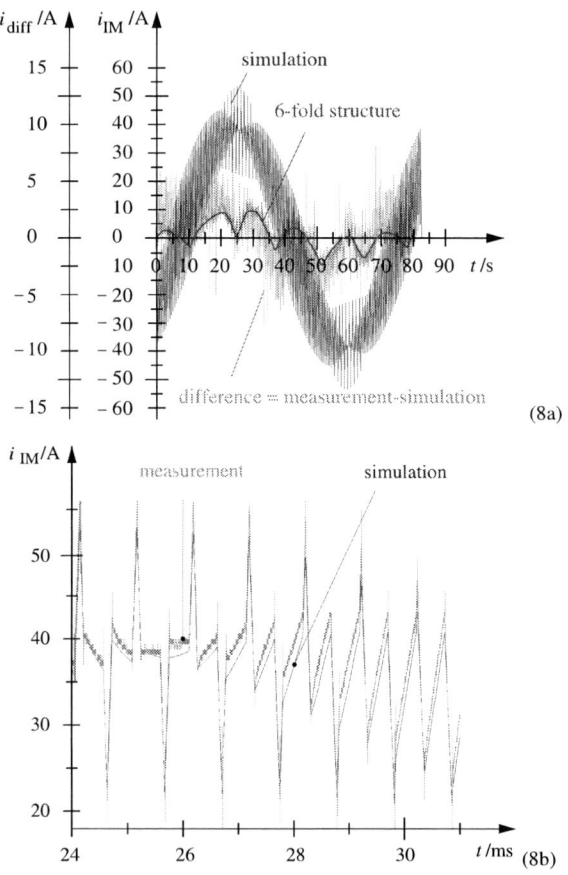

Figs. 8a/b: Comparison of measured and simulated phase current of V/f-controlled IM

It can clearly be seen that the magnetizing current amplitude and the harmonics are correctly modeled.

The conditions for the results shown in Fig. 8 are: Normalized voltage 0.25, stator frequency 12.5 Hz, leakage factor $L_\sigma/(L_\sigma+L_\mu)$ = 6%, no load, I_μ = 29 A; switching frequency 1 kHz, Symmetrized Sinusoidal Modulation [3, 5].

978-1-4244-1871-8/07 $25.00 © 2007 IEEE

VII. CONCLUSION

The paper describes a C-code-based simulation concept for hybrid systems, specially tailored to simulate the complex interactions between four-quadrant converter, DC-link with resonant circuit, PWM inverter and induction machine found in the power train of AC railway traction vehicles. Suitable concepts for fast and efficient simulation of nonlinear continuous and discontinuous components of the system are presented, allowing also for simple modifications based on a modular concept. The C-code used for the control in simulation is identical to that of the PC-based control of the test set-up, allowing efficient tests.

For first verification of the simulation, a model of an induction machine fed by a PWM inverter is used. As shown in detail in section VI, measured and simulated quantities are in very good agreement, allowing to proceed to more sophisticated control schemes.

Next steps will be: add the other system components to the simulation, integrate the speed-sensorless Indirect Stator-quantity Control system (ISC) of the IM, implement a control scheme taking into account all components of the power train and test these modifications on the test set-up.

Finally, the expected improvement of the control performance of the complete power train, taking into account the resonant circuit, too, is to be verified, giving improved grid stability, a reduction of harmonics on the supply side and improving the interoperability of traction vehicles in general.

REFERENCES

[1]   Steimel, A.: Line Interference by A.C. Traction Material Controlled by Power Electronic Systems. KAIST-Workshop "New Transportation Systems", 1993, Taejon/Korea, Proc. pp. 30-37

[2]   Meyer, M.; Schöning, J.: Netzstabilität in großen Bahnnetzen.
Eisenbahn-Revue 7–8/1999, pp. 312–317

[3]   Hoffmann, F., Jänecke, M.: Fast Torque Control of an IGBT-Inverter Fed Three-Phase A.C. Drive in the Whole Speed Range – Experimental Results. 6th European Power Electronic Conference (EPE), Sevilla 1995, pp. 3.399-404

[4]   Depenbrock, M.; Foerth, Ch.; Koch, St.: Speed Sensorless Control Of Induction Motors At Very Low Stator Frequencies. 8th European Power Electronic Conference (EPE), Lausanne 1999

[5]   Depenbrock, M.; Steimel, A.: Speed-sensorless stator-flux-oriented control of induction motor drives in traction. German-Korean Symposium 2004 on Power Electronics and Electrical Drives, Aachen 2004

[6]   Kail, M.; Staudt, V.; Steimel, A.: PC-based control hardware for 200-kW double three-level converter. 10th International Power Electronics and Motion Control Conference, Cavtat/ Croatia 2002

[7]   Press, W. H., Teukolsky, S. A., Vetterling, W. T., Flannery, B. P.: Numerical Recipes in C. Cambridge University Press, 1988

[8]   Lunze, J.: Automatisierungstechnik. Oldenburg Wissenschaftsverlag München, 2003

[9]   Staudt, V.: AC Machine Dynamics described by Space Vectors using State Space Notation. ETEP Vol. 9 (1999), No. 1, pp. 17–25

[10]   Weschta, A.; Weberskirch, W.: Nonlinear Behaviour Of Voltage Source Inverters With Power Transistors. 3rd European Power Electronic Conference (EPE), Aachen 1999.

Modeling Diode Reverse Recovery and Corresponding Implementation in Fast Time-Domain Simulation

Hao Wang, Jinjun Liu, Runxin Wang and Qinsan Hou

School of Electrical Engineering, Xi'an Jiaotong University, Xi'an, CHINA

E-mail: haowang@mail.xjtu.edu.cn

Abstract-- Existing power diode models in most time-domain circuit simulators cannot accurately present the reverse recovery characteristics of actual diode devices with fast simulation speed. Ma-Lauritzen model is a relatively simple solution for modeling the reverse recovery of diode with reasonable accuracy but so far little result has been reported on the methodology and process for applications in circuit simulators. Based on Ma-Lauritzen model, this paper proposes a configuration and implementation methodology for building diode models with reverse recovery characteristics in an ordinary time-domain circuit simulator like SIMPLIS. A silicon power diode 8ETX06 is selected for simulation speed and accuracy evaluation regarding the proposed modeling methodology and implementation approach. Furthermore, simulation and hardware test results on a switching power converter are shown to verify that the diode model following the proposed building method is reasonably accurate and fast in dynamic waveform emulation and component stress estimation for the reverse recovery process.

I. INTRODUCTION

Power diode is one of the widely used components in the power electronic area. Nowadays while the power rating goes larger and switching speed goes faster, power loss caused by the diode reverse recovery is significant and cannot be neglected. For most time-domain circuit simulation tools, diode models in them normally cannot accurately present the reverse recovery characteristic of actual diode devices with fast enough simulation speed. Then when the researchers want to do the simulation in device behavior level, they will meet the problems with the not accurate reverse recovery characteristic diode model. How to accurately modeling the diode with reverse recovery characteristics with fast simulation speed comes to an important issue for the circuit simulation in device behavior level.

Most of the time-domain circuit simulators include the power diode model [1]. Those diode models are based on the original charge control model [2-5], which includes the effects of the charge storage during reverse recovery but does not provide reverse recovery of the diode [6-7]. Then it will cause large error in power loss comparison between hardware test and simulation results of the circuit with power diode.

Lauritzen and Ma have proposed a simple diode model with reverse recovery, which is relatively simple than other power diode models with reverse recovery [8-10]. Following the Lauritzen and Ma theory, in this paper, method of how to model the diode with reverse recovery is presented. With this method, particular diode models can be built with the standard components and all the model parameters can be derived based on the data in the diode datasheet or measurement data. Silicon diode 8etx06 is selected for an example. Simulation in SIMPLIS and hardware tests of a switching power converter with 8etx06 are done for the simulation speed and the model accuracy evaluations.

II. DIODE MODELING APPROACH AND MODEL PAPRAMETERS ABSTRACTION

The model Lauritzen and Ma proposed has the foundations on the semiconductor theory. It also has the assumption of symmetric P-i-N structure and equal electrons/holes mobility. Some equations are used for the model parameters derivation, which are shown below.

$$i(t) = (q_E - q_M) / T_M \qquad (1)$$

$$0 = \frac{dq_M}{dt} + \frac{q_M}{\tau} + \frac{(q_E - q_M)}{T_M} \qquad (2)$$

$$q_E = I_s \cdot \tau \cdot [\exp(v/(n \cdot V_T)) - 1] \qquad (3)$$

$$I = I_s /(1 + T_M / \tau) \cdot [\exp(v/(n \cdot V_T)) - 1] \qquad (4)$$

$$\tau_{rr} = 0.1 \cdot (t_{rr} - I_{RM} / a) \qquad (5)$$

$$T_I = (I_F - I_{RM}) / a \qquad (6)$$

$$I_{RM} = a \cdot (\tau - \tau_{rr}) \cdot [1 - \exp(-T_I / \tau)] \qquad (7)$$

$$1 / T_M = 1 / \tau_{rr} - 1 / \tau \qquad (8)$$

Variables definition is shown in table 1.

A. Diode Model Structure and Working Principle

Based on the Lauritzen and Ma model, diode model can be built and its structure is shown in Fig. 1.

Diode model structure includes two voltage-controlled-voltage-sources (E1 and E2), one current-controlled-voltage-source (H1), one voltage-controlled-current-source (G1), one piecewise- linear resistor (R2), two fixed resistors (R1, Rs) and one fixed capacitor (C1).

Those components build up an equivalent circuit of the Ma-Lauritzen diode model, which bases on (1) - (4). Just as the description in (1), instant diode current is controlled by the difference of two charges q_E and q_M. In the diode model, q_E and q_M are presented with two voltage signals and the diode instant current are got with a voltage controlled current source G1.

978-1-4244-1871-8/07 $25.00 © 2007 IEEE

TABLE 1.   VARIABLES DEFINITION

Instant current of diode	$i(t)$	Time begins from the diode turning off	t
Temporary charge variable	q_E	Temporary charge variable	q_M
Diffusion time of diode	T_m	Life time of diode	τ
Reverse saturation current of diode	I_S	Voltage drop across diode at static state	V
Thermal voltage, equal to KT/q	V_T	Diode current at static state	I
Reverse recovery time constant of diode	τ_{rr}	Reverse recovery time of diode	t_{rr}
Peak value of diode reverse current	I_{RM}	Current ramp down slope of diode current	a
Time when diode meets peak reverse current	T_1	Forward current of diode	I_F
Diode model parameters		$G_1, H_1, E_1, E_2, C_1, R_1, R_2, R_S$	

As the description in (2), q_M can be derived with E1, R1 and C1 working together. As the description in (3), q_E can be derived with E2, H1 and R2 working together. In the model structure, R_s presents the contact resistance of the diode.

B. Diode Model Parameters Abstraction

Parameters of the components in diode model structure shown in Fig. 1 should be derived for the diode model building. Parameters abstraction process of the proposed diode model is shown below.

From the datasheet of diode, reverse recovery time t_{rr}, static V-I characteristic and the peak reverse current value in different forward current and different di/dt conditions can be abstracted. In the τ_{rr} abstraction procedure, an assumption is drawn that using 1/10 of a part of t_{rr} (the reverse recovery time) to be τ_{rr} (the reverse recovery time constant), which is shown in (5).

With the data collected from the datasheet of diode, T_1 can be calculated using (6). τ (lifetime of diode) can be derived using (7). T_m (diffusion time of the minority carries) can be derived using (8). τ_{rr}, τ and T_m derived in different forward current and different turning off current slope should be averaged respectively to get the averaged τ_{rr}, τ and T_m value for the final model parameters derivation.

R_s (contact resistance of diode) can be derived from the V-I characteristic. R_s is equal to the $\Delta V/\Delta I$ on the very high forward current case.

R_2 is a piecewise-linear resistor, which represents the diode static V-I characteristics after subtracting the voltage contribution due to R_s.

Because the gain of E_2 is set to 1, the output voltage of E_2 has the same value as the voltage drop across the diode except that due to R_s, which is supplied on the R_2.

Because the gain of H_1 is set to 1, the output voltage of H_1 has the same value as the static current flowing through the diode.

q_E can be derived using (3) and (4).

$$q_E = I \cdot (\tau + T_M) \qquad (9)$$

Using (1), (2) and (9), the gain of E_1 can be derived as $\tau/(\tau+T_M)$.

With the particular components connection way in the diode model structure, a derivation calculus like that in (2) is emulated and q_M can be derived with E_1, R_1 and C_2 working together.

There, product of R_1 and C_2 values is used to generate the time constant of the integration, which should be equal to the averaged value of reverse recovery time constant τ_{rr} in different I_F and di/dt conditions.

Using (1), (9) and the deduction result of q_M, the gain of G_1 can be derived as $(T_M+\tau)/T_M$.

Thus, all the necessary component parameters in the proposed diode model structure are derived. A list of the model parameters derivation results is shown in Table 2. The equations in Table 2 will be used throughout the whole diode model building procedure.

TABLE 2. DIODE MODEL PARAMETERS ABSTRACTION

G_1	H_1	E_1	E_2
$\dfrac{T_M + \tau}{T_M}$	1	$\dfrac{\tau}{T_M + \tau}$	1
C_1	R_1	R_2	R_S
1nF	$\dfrac{\tau_{rr}}{C_1}$	Abstracted from static V-I curve of diode	$\Delta V/\Delta I$ on the very high I_F case

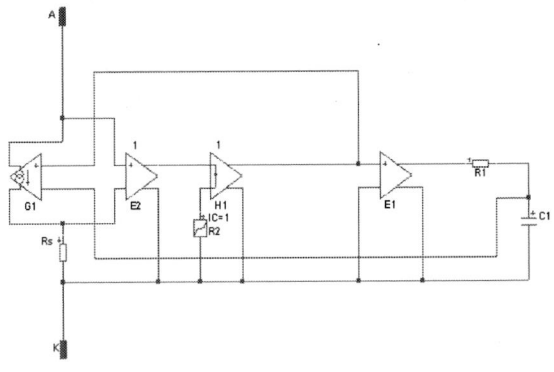

Fig. 1:   Diode model with reverse recovery

With the parameters derivation process, a diode modeling procedure can be proposed, which includes 6 steps as following.

Step1: Collecting t_{rr} and I_{RM} values in different I_F and di/dt conditions from the datasheet of diode;

Step2: Using (5) and (6) to calculate T_1 and τ_{rr} in different I_F and di/dt conditions;

Step3: Using (7) and (8) to calculate τ and T_m in different I_F and di/dt conditions;

Step4: Averaging τ_{rr}, τ and T_m in different I_F and di/dt conditions. Using the averaged τ_{rr}, τ, T_m values and the equations in Table. 2 to abstract diode model parameters G_1, H_1, E_1, E_2, C_1 and R_1 values;

Step5: Abstracting the static V-I characteristic chart of diode. Abstracting R_s through calculating $\triangle V/\triangle I$ with the data in very high I_F condition on the static V-I characteristic chart;

Step6: Abstracting R_2 by eliminating the voltage drop contribution due to R_s in the static V-I characteristic of the diode.

An obvious diagram of the proposed diode modeling procedure is shown in Fig.2.

III. SIMULATION TIME AND ACCURACY VERIFICATION

A. Simulation Time

A diode model of 8etx06 in SIMPLIS is built following the process proposed in the last section. A demo board circuit of PFC (power factor correction) controller UC3854 is selected for the simulation time estimation in SIMPLIS, which is shown in Fig. 3.

Simulations are done respectively for the cases that the initial diode model or the 8ETX06 diode model built with the proposed diode model building procedure is used as the rectifier diode in the simulation circuit.

Simulation time in these two cases are respectively 1.4 minutes and 1.7 minutes, which means the diode model built with the proposed process is fast enough and will not affect the simulation time a lot.

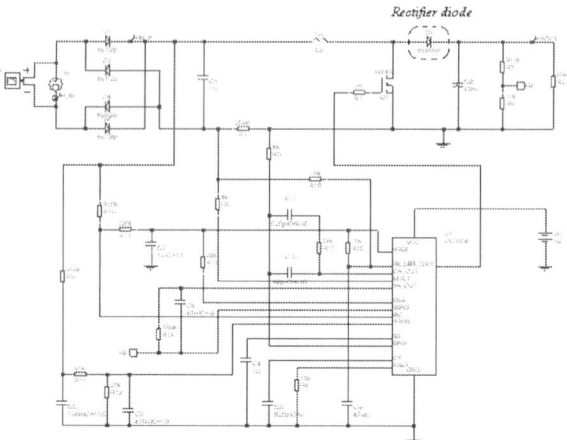

Fig. 3: UC3854 demo board circuit in SIMPLIS

B. Peak Reverse Recovery Current Estimation

A Boost circuit in SIMPLIS is used for the diode model peak reverse recovery current estimation. The simulation circuit diagram is shown in Fig. 4.

In this simulation circuit, value of load resistor R_1 can be changed to set I_F of the diode. Value of the current source I_1 should be changed accordingly when R_1 is changed. Through changing the value of resistor R_G, di/dt of the diode current during turning off can also be controlled.

With the simulation circuit, simulations are done for verifying the peak reverse current value of the diode model in different forward current and di/dt conditions.

Peak reverse recovery current value comparison between simulation results and data collected from datasheet of 8etx06 was done and the comparison results are shown in Table 3.

In Table 3, comparison results show that the biggest error of the peak reverse recovery current value is about -11.45%. Taking into account that data of the peak reverse recovery current values of 8etx06 in the datasheet are all typical values, which are marked with "90% confidences," the biggest -11.45% error are reasonable.

So the comparison results prove that the diode model built with the proposed diode model building procedure has reasonable accuracy in the peak reverse recovery current estimation.

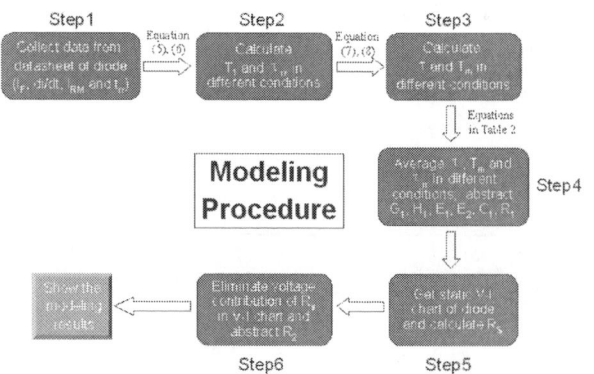

Fig. 2: Diode model building process diagram

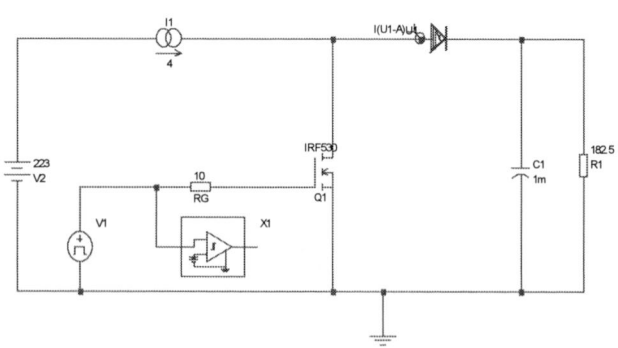

Fig. 4: Peak reverse recovery current estimation circuit

978-1-4244-1871-8/07 $25.00 © 2007 IEEE

TABLE 3. PEAK REVERSE CURRENT ESTIMATION

I_F (A)	di/dt (A/us)	Irm (A) (Datasheet)	I_{RM} (A) (Simulation)	Error (%)
4	200	5.0	4.808	-3.84
4	400	8.0	7.547	-5.66
4	600	10.5	9.626	-8.32
4	800	12.8	11.335	-11.45
8	200	5.5	5.766	4.84
8	400	8.9	9.608	7.96
8	600	11.9	12.584	5.75
8	800	14.5	15.078	3.99
16	200	6.5	6.426	-1.14
16	400	10.8	11.532	6.78
16	600	14.5	15.67	8.07
16	800	17.8	19.203	7.88

C. Comparison with Hardware Test Results on a Power Converter

For evaluating the practice use of the proposed diode model building procedure, hardware tests have been done with an evaluation test board. For the compassion purpose, corresponding simulation of the tests done on the evaluation board have also been done.

Photo of the evaluation test board is shown in Fig.5.

The evaluation test board is a demo board of PFC controller IR1150, which has a universal input voltage range, 400Vdc output voltage and 300W output power at full loading condition. Topology of this evaluation test board is boost converter and the control strategy is "one cycle control". Silicon diode 8etx06 is the UUT (unit under test), which is used as rectifier diode of the boost converter in the evaluation test board.

For the diode model accuracy verification purpose, additional bridge rectifier and electrolytic capacitors bank are used to rectify the ac input voltage. Thus, rectified 270V voltage in dc was supplied to the evaluation test board and this PFC converter worked as DC-DC converter.

With this rectified dc voltage input, the evaluation test circuit worked in a steady state at full loading condition. Thus, the waveform of diode current, which is a part of the input current of the converter, is regular and periodic. Then it will ease the difficulty of the diode current measurement during the hardware tests.

Estimation tests of the diode 8etx06 had been done with the evaluation test board. During those tests, the rectified 270V dc voltage from the electrolytic capacitors bank is used as the input voltage. Resistors with particular values and power ratings are used to be the load of the converter and then the converter can work at full loading condition.

In the last two sections, based on the proposed diode model building procedure, diode model of 8etx06 in SIMPLIS has been built. The peak reverse recovery current estimation simulation results prove the reasonable accuracy of the diode model comparing with that in the datasheet. Then this diode model of 8etx06 can be used in the comparison simulations.

Corresponding simulation circuit of the evaluation test board in SIMPLIS is built with the 8etx06 diode model. The simulation circuit diagram is shown in Fig. 6.

With this simulation circuit, corresponding simulations of the diode model estimation tests on the evaluation test board have been done. The simulation results are used for the diode mode accuracy estimation purpose comparison with the hard ware test results.

Test waveform and simulation waveform in one test condition, which is with 0.7A forward diode current (I_F) and 280A/us diode turning off slope (di/dt), are respectively shown in Fig. 7 and Fig. 8.

In Fig.7, test result shows that the peak reverse current of the diode in this test condition is -3.4A. In Fig.8, simulation result shows that the peak reverse current of the diode model in this test condition is –3.8A. Through the calculating comparison, there is an error of 11.7% between the simulation result and the hardware test result in this test condition.

The comparison result between the test waveform and the simulation result proves that the diode model built with the proposed diode model building procedure has reasonable accuracy to meet with the practice diode device behaviors in the experimental tests.

Fig. 5: Evaluation test board

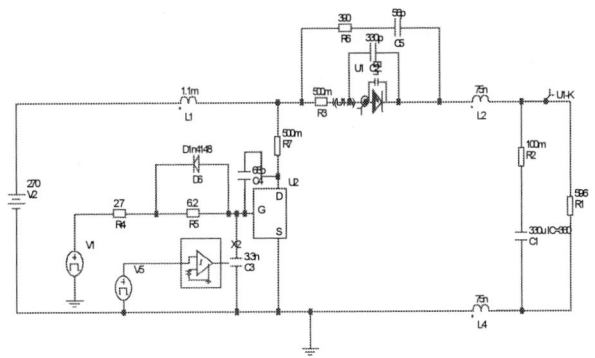

Fig. 6: Simulation circuit of test board

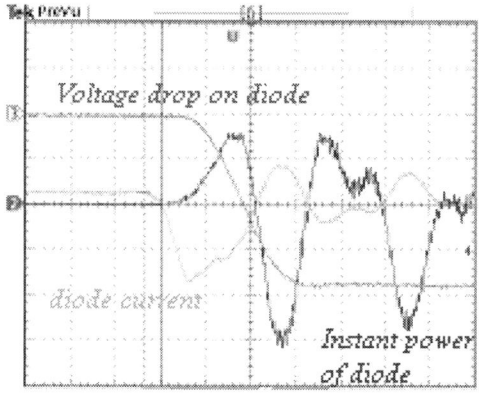

Fig. 7:   Test results when diode turns off

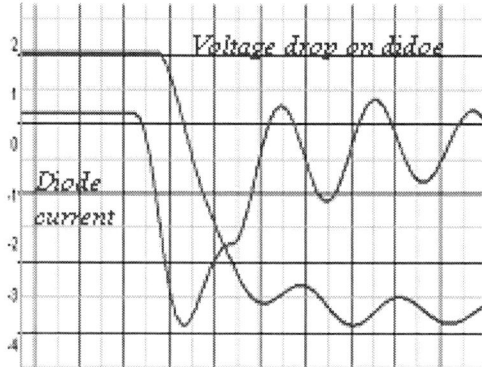

Fig. 8:   Simulation results when diode turns off

IV. CONCLUSION

Based on the Ma-Lauritzen model theory, a diode model is presented. The diode model working principle and the diode model parameters abstraction method are discussed in this paper. With the deduced parameter abstraction method, a corresponding diode model building procedure and the diode model implementation in time-domain circuit simulators are also proposed. With the proposed diode model building procedure, a diode model of 8etx06 is set up in SIMPLIS for the simulation time and model accuracy estimation. Comparison between the simulation results and the hardware test results proves that the correspondingly established diode model in SIMPIS is very fast and reasonably accurate in dynamic waveform emulation and component stress estimation including reverse recovery process. Similarly the proposed diode model building method can be implemented in other time-domain circuit simulators.

ACKNOWLEDGEMENT

The authors thank Astec Custom Power Ltd. for the devices and the evaluation test board supports.

The authors thank Transim Technology Corporation for SIMPLIS software supports.

REFERENCES

[1]   R. Krauss and H.J. Mattausch, "Status and Trends of Power Semiconductor Device Models for Circuit Simulation", IEEE Transaction on Power Electronics, vol.13, Issue3, May 1998.

[2]   P. Antogneti and G. Massobrio, "Semiconductor Devices Modeling with SPICE", New Yok: McGraw-Hill, 1988.

[3]   Y-C. Liang and V.J. Gosbell, "Diode forward and reverse recovery model for power electronic SPICE simulations" IEEE Transactions on Power Electronics, vol. 5, July 1990.

[4]   A.G.M. Strollo, " A new SPICE subcircuit model of power p-i-n diode ", IEEE Transaction on Power Electronics, vol.9, Nov 1994.

[5]   Buiatti, G.M.; Cappelluti, F.; Ghione, G., "Power PiN diode model for PSPICE simulations", Applied Power Electronics Conference and Exposition, Volume 3, March 2005.

[6]   Peter O. Lauritzen and Cliff L. Ma, "A Simple Diode Model with Reverse Recovery", IEEE Transactions on Power Electronics, Vol. 6, No. 2, April, 1991.

[7]   H. Goebel and K. Hofhan, "Power diode HYBRID model with forward and reverse recovery for use in circuit simulators ", Applied Power Electronics Conference and Exposition, Feb 1992.

[8]   C. L. Ma, P. O. Lauritzen, P. Y. Lin, "A Physically-based Lumped-charge P-v-N Diode Model", in The European Power Electronics Association, 1993.

[9]   Ma, C.L.; Lauritzen, P.O.; Sigg, J, "Modeling of power diodes with the lumped-charge modeling technique", IEEE Transactions on Power Electronics, Volume 12, Issue 3, May 1997.

[10]   Ma, C.L.; Lauritzen, P.O, "A simple power diode model with forward and reverse recovery" , IEEE Transactions on Power Electronics, Volume 8, Issue 4, Oct. 1993.

978-1-4244-1871-8/07 $25.00 © 2007 IEEE

A Half Bridge Flyback Converter with ZVS and ZCS Operations

Li-Ming Wu
Ching Yun University
Department of Electrical Engineering
229, Chien-Hsin Rd., Jung-Li, Taiwan 320, R.O.C.
Email: lmwu@cyu.edu.tw

Chen-Yin Pong
Ching Yun University
Department of Electrical Engineering
229, Chien-Hsin Rd., Jung-Li, Taiwan 320, R.O.C.
Email: m9512014@cyu.edu.tw

Abstract—An asymmetrical half bridge flyback converter without output inductor is presented in this paper, in which the power switches can be turned on with ZVS and the output diode can be turned on and off with ZCS. While the ZVS is obtained from the magnetizing current, the ZCS is achieved from the resonant current. The ZVS and ZCS conditions can be maintained for full load range. The characteristics of the proposed converter rely on the utilization of resonant inductor and blocking capacitor as resonant energy delivery devices rather than linearly charging and discharging components. The operational principle and circuit analysis of the proposed converter is described in this paper. The experimental test of a 90W prototype converter is performed and the results are presented.

I. INTRODUCTION

For reasons of simple structure and robust characteristics, the flyback converter has been one of the most widely used dc-dc power supplies in the small power areas. The multiple-output and cost effective design makes the flyback converter attractive for applications such as personal computers, assorted home appliances, and various office equipments, etc. Major disadvantage of flyback converter attributes to the hard switching operation. In general, the hard switching operation of the power switch results in a high switching loss, high EMI noise, and high switch voltage stress. To overcome these drawbacks, various kinds of soft switching techniques have been proposed [1~9]. Among them, the active-clamp converters [3~5] make use of simple active clamp network to achieve ZVS operation of the power switches. However, the power switch still suffers from high voltage stresses since the sum of input voltage and the transformer primary voltage is imposed on them. Therefore, the asymmetrical half bridge flyback converters [6~9], which utilizes the half bridge structure and a blocking capacitor to achieve ZVS operation of the power switches as well as low switch voltage stresses, are gaining popularity. From previous works [7,8], it had been shown that, by reducing the magnetizing inductor and allowing duty cycle constraint beyond 50%, the half bridge flyback converter can achieve ZVS operation for power switches from no load to full load conditions. However, the rectifier suffers from high di/dt at turn off, which results in significant energy loss from high reverse recovery current.

Another research [9] used the same power circuit as in [8] but added an inductor as output filter to obtain the same secondary diode current waveforms as the conventional forward converter does. The output rectifier is complex and has no soft switching operations as well.

In this research, a half bridge flyback converter without using output inductor is studied, in which both the power switches and output power diode can achieve soft switching operations. The voltage stress of power switches is not greater than the input voltage. The power switches can be turned on with ZVS and the output diode can be turned on and off with ZCS. The ZVS and ZCS conditions can be maintained for the full spectrum of load range. The characteristics of the proposed converter rely on the utilization of resonant inductor and blocking capacitor as resonant energy delivery devices rather than a pair of linearly charging and discharging components.

The operational principles, circuit analysis, and design equations of the proposed converter are described in this paper. The experimental test of a 90W prototype converter is performed and the test results are presented to confirm the validity of the proposed converter.

II. OPERATIONAL PRINCIPLES

Figure 1 shows the simplified circuit diagram of the proposed half bridge flyback converter. The resonant inductor L_k consists of total leakage inductance of transformer T_r and external stray inductances. The ESRs in passive components and internal resistance of semiconductors are not included in the diagram for simplification. Fig. 2 shows the eight operating modes of the converter during a switching cycle, and Fig. 3 illustrates the key waveforms in time intervals corresponding to each of the eight modes. Prior to describing the basic operational principles, several assumptions are made first:

* The converter is operating in steady state.

* The resonant inductor L_k is much less than the magnetizing inductor L_m.

* The resonant period of C_b and L_k is comparable with the off-time of switch Q_l.

* In mode 2 and 6, the circuit is assumed to behave linearly.

* In mode 3, the resonant current is assumed to be the same as the magnetizing current.

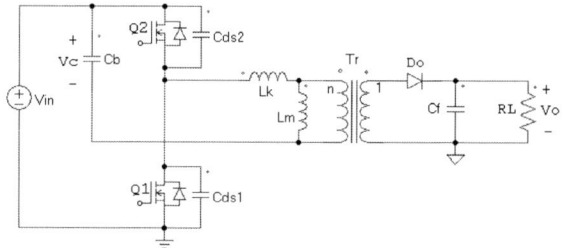

Fig. 1 Simplified schematic of the half bridge flyback converter

To proceed, the operational analysis of the asymmetrical half bridge flyback converter can be explained as follows.

Mode 1 ($t_o \sim t_1$): At t_o, Q_1 is on and Q_2 is off. The output diode D_o is reversely biased. Components C_b, L_m and L_k form a series resonant tank with voltage source V_{in} providing the input energy. Since the length of time interval (t_1-t_o) is short compared with the time constant of the resonant tank, the power source V_{in} charges C_b, L_m and L_k in a linear fashion. The leakage and magnetizing currents start to decrease from zero. The circuit governing equations are:

$$(L_m + L_k)\frac{di_{L_k}}{dt} = -V_{in} + v_C \qquad (1)$$

$$C_b \frac{dv_C}{dt} = i_{L_k} = i_{L_m} \qquad (2)$$

This mode ends when Q_1 turns off.

Mode 2 ($t_1 \sim t_2$): At t_1, Q_1 is turned off and Q_2 remains off. The components C_b, L_m, L_k and C_{ds1} form a new series resonant network, and C_b, L_m, L_k and C_{ds2} form another series resonant network. The power source V_{in} charges C_{ds1} and C_b with half of the magnetizing current. The other half of the magnetizing current charges C_b through discharging C_{ds2}. The circuit governing equations are:

$$(L_m + L_k)\frac{di_{L_k}}{dt} = -V_{in} + v_C + v_{ds1} \qquad (3)$$

$$C_b \frac{dv_C}{dt} = i_{L_k} = i_{L_m} \qquad (4)$$

$$C_{ds1}\frac{dv_{ds1}}{dt} = \frac{1}{2}i_{L_k} = \frac{1}{2}i_{L_m} \qquad (5)$$

$$C_{ds2}\frac{dv_{ds2}}{dt} = -\frac{1}{2}i_{L_k} = -\frac{1}{2}i_{L_m} \qquad (6)$$

By assuming a linear magnetizing current, the voltage v_{ds1} starts to increases from zero in a quadratic fashion. This mode ends when v_{ds1} reaches to the value of ($V_{in} + V_{fd}$) and the body diode of Q_2 starts to conduct. The notation V_{fd} stands for diode forward voltage drop.

Mode 3 ($t_2 \sim t_3$): In this period, both Q_1 and Q_2 are remaining off. The body diode of Q_2 starts to conduct and the voltage across the magnetizing inductor changes polarity. By assuming the resonant current the same as the magnetizing current, the governing equations can be written as:

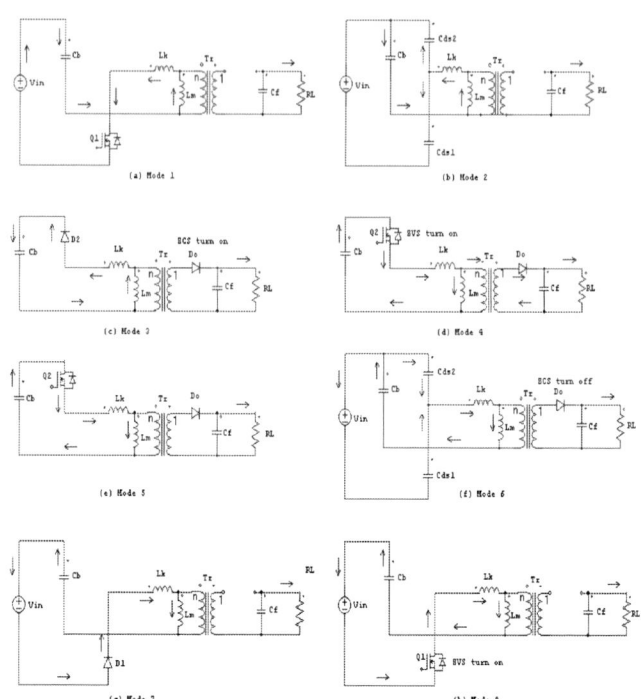

Fig. 2 Half bridge flyback converter topological states

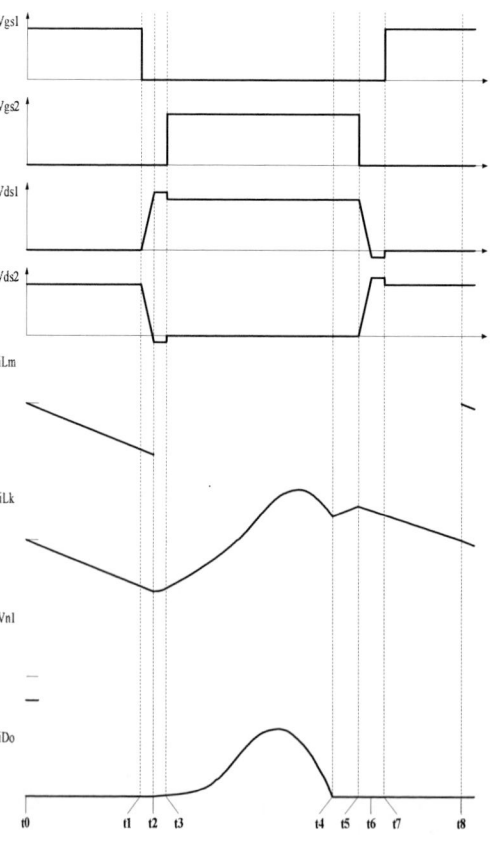

Fig. 3 Steady state waveforms of the half bridge flyback converter with ZVS and ZCS operations

$$(L_m + L_k)\frac{di_{L_k}}{dt} = v_C + V_{fd} \tag{7}$$

$$C_b \frac{dv_C}{dt} = i_{L_k} = i_{L_m} \tag{8}$$

Since the difference between the resonant current and the magnetizing current is zero at this stage, the output diode is not conducting and no voltage is reflected from the secnodary to the primary side of transformer. This mode ends when Q_2 is ZVS turned on and the dead time is over.

Mode 4 ($t_3 \sim t_4$): In this period, Q_2 is on and Q_1 is off. The output diode is conducting and the converter output voltage is reflected to the transformer primary side. The resonant tank formed by C_b, L_m and L_k starts to develop a resonant current which is different from the magnetizing current. It is the difference between the resonant current and the magnetizing current that determines the output diode current. This mode ends when the resonant current meets the magnetizing current and the output diode turns off with zero current. The governing equations in this mode include:

$$L_m \frac{di_{L_m}}{dt} = nV_0 \tag{9}$$

$$L_k \frac{di_{L_k}}{dt} = v_C - nV_0 \tag{10}$$

$$C_b \frac{dv_C}{dt} = i_{L_k} \tag{11}$$

$$i_{D_0} = n(i_{L_k} - i_{L_m}) \tag{12}$$

Mode 5 ($t_4 \sim t_5$): In this mode, states of Q_2 and Q_1 remains the same as in mode 4. The resonant network stops resonance because the resonant current can not be smaller than the magnetizing current. The output diode is forwardly biased with zero current conducted. The governing equations become:

$$(L_m + L_k)\frac{di_{L_k}}{dt} = v_C \tag{13}$$

$$C_b \frac{dv_C}{dt} = i_{L_k} = i_{L_m} \tag{14}$$

This mode ends when Q_2 is turned off.

Mode 6 ($t_5 \sim t_6$): In this period, Q_1 and Q_2 are both off. At t_5, half of the magnetizing current starts to charge C_{ds2} from zero voltage, the other half discharges C_{ds1} from voltage V_{in}. The governing equations are:

$$(L_m + L_k)\frac{di_{L_k}}{dt} = v_C - v_{ds2} \tag{15}$$

$$C_b \frac{dv_C}{dt} = -i_{L_k} = -i_{L_m} \tag{16}$$

$$C_{ds2}\frac{dv_{ds2}}{dt} = \frac{1}{2}i_{L_k} = \frac{1}{2}i_{L_m} \tag{17}$$

$$C_{ds1}\frac{dv_{ds1}}{dt} = -\frac{1}{2}i_{L_k} = -\frac{1}{2}i_{L_m} \tag{18}$$

This mode ends when v_{ds1} drops to $-V_{fd}$ and the body diode of Q_1 starts to conduct.

Mode 7 ($t_6 \sim t_7$): In this mode, both Q_1 and Q_2 are still off. At t_6, the voltage across the magnetizing inductor changes polarity. The body diode of Q_1 starts to conduct and the magnetizing current starts to decrease. The governing equations are:

$$(L_m + L_k)\frac{di_{L_k}}{dt} = -V_{in} + v_C - V_{fd} \tag{19}$$

$$C_b \frac{dv_C}{dt} = -i_{L_k} = -i_{L_m} \tag{20}$$

This mode ends when the dead time is over and Q_1 is ZVS turned on.

Mode 8 ($t_7 \sim t_0$): At t_7, Q_1 turns on under ZVS. The magnetizing current keeps decreasing in a linear fashion and C_b discharges energy back to the voltage source V_{in}. This mode ends when the magnetizing current decreases to zero. Then the input power source begins to charge C_b, L_m and L_k again, and another switching cycle repeats.

III. CIRCUIT ANALYSIS

In this section, key characteristics of the studied converter are analyzed based on the above operational modes. However, to simplify the analysis, the dead time between Q_1 and Q_2 and the switch output capacitances have been neglected. The simplification should not affect the truth of the analysis. Fig. 4 shows the simplified waveforms of the resonant inductor current, the magnetizing inductor current, the voltage of blocking capacitor, and the output diode current.

(1) Voltage transfer ratio

In steady state, the voltage transfer ratio depends on duty ratio D and turns ratio n of transformer T_r. It can be obtained using the flux balance principle in L_m and L_k. The equations describing these relationships are:

$$k_1(V_{in} - V_C)D - nV_o(1 - D) = 0 \tag{21}$$

$$k_2(V_{in} - V_C) - (V_C - nV_o)(1 - D) = 0 \tag{22}$$

where $k_1 = L_m/(L_m + L_k)$, and $k_2 = L_k/(L_m + L_k)$.
Solving (21) and (22) simultaneously, the voltage transfer ratio is obtained as follows:

$$\frac{V_o}{V_{in}} = \frac{k_1 D}{n} \approx \frac{D}{n} \tag{23}$$

(2) Average voltage across blocking capacitor

The average voltage V_C across the blocking capacitor is proportional to the duty ratio D. The relationship can be obtained from solving (21) and (22) simultaneously:

$$V_C = DV_{in} \tag{24}$$

978-1-4244-1871-8/07 $25.00 © 2007 IEEE   877

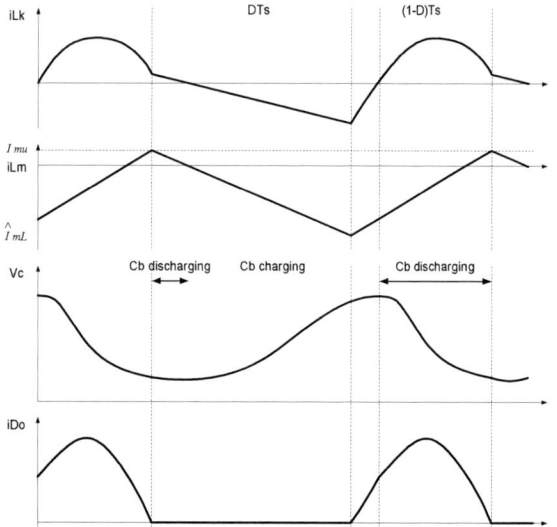

Fig. 4 Simplified waveforms of inductor currents, output diode current, and voltage across blocking capacitor.

Equation (23) and (24) reveal that, depends on V_{in} and V_o, the value of duty ratio D can be larger or less than 0.5. For fixed V_{in}, the smaller the V_o, the shorter the duty ratio.

(3) Average magnetizing current

Referring to Fig. 4, the average magnetizing current I_m can be derived as follows:

$$I_m = \frac{1}{T_s}\int_0^{T_s} i_{L_m}(t)dt = \hat{I}_{mu} - \frac{V_{in}D(1-D)T_s}{2(L_m+L_k)} \quad (25)$$

where $\hat{I}_{mu}$ is the upper peak current of magnetizing inductor.

(4) Average load current

The output diode current $i_{Do}(t)$ is determined by the turns ratio n of T_r and the difference between the resonant current i_{Lk} and the magnetizing current i_{Lm}. The average load current I_o can be obtained as follows:

$$I_o = \frac{1}{T_s}\int_0^{(1-D)T_s} n(i_{L_k}(t)-i_{L_m}(t))dt$$
$$= n\left[\frac{d}{\omega_r T_s}\sin(\omega_r(1-D)T_s-\phi)+\sin\phi - \hat{I}_{mL}(1-D)-\frac{nV_o}{2L_m}(1-D)^2T\right]$$

$$(26)$$

where

$$d^2 = \hat{I}_{mL}{}^2 + \left[\frac{(nV_o - v_C(DT_s))}{Z}\right]^2, \quad z = \sqrt{\frac{L_k}{C_b}}, \quad (27)$$

$$\omega_r = 1/\sqrt{L_k C_b}, \quad (28)$$

$$\phi = \tan^{-1}\left(\frac{nV_o - v_C(DT_s)}{Z\hat{I}_{mL}}\right) + \pi. \quad (29)$$

and $\hat{I}_{mL}$ is the lower peak current of magnetizing inductor.

Utilizing the charge balance of C_b over one switching cycle yields:

$$\int_0^{T_s} i_{L_k}(t)dt = 0 \quad \Rightarrow$$

$$\frac{d}{\omega T_s}\left[\sin(\omega(1-D)T_s-\phi)+\sin\phi\right] = \frac{V_{in}(1-D)D^2T_s}{2(L_m+L_k)} - \hat{I}_{mu}D$$

$$(30)$$

Bringing (30) into (26), one obtains

$$\frac{I_o}{n} = -\hat{I}_{mu} + \frac{V_{in}D(1-D)T_s}{2(L_m+L_k)} \quad (31)$$

Substituting (31) into (25) gives the simple relationship between I_m and I_o:

$$I_m = -\frac{I_o}{n} \quad (32)$$

(5) Resonant frequency

During the time interval $DT_s \le t \le T_s$, the resonant current $i_{Lk}(t)$ is a sinusoidal function and the magnetizing current $i_{Lm}(t)$ increases linearly. The difference between the resonant current and the magnetizing current generates the output diode current. Under full load condition, the output diode turns off with ZCS only if the following criterion is satisfied:

$$i_{L_k}(T_s) \le i_{L_m}(T_s) \quad (33)$$

where

$$i_{L_k}(t) = d\cos(\omega_r(t-DT_s)-\phi) \quad (34)$$

$$i_{L_m}(t) = \hat{I}_{mL} + \frac{nV_o}{L_m}(t-DT_s) \quad (35)$$

That is,

$$d\cos(\omega_r(1-D)T_s-\phi) \le \hat{I}_{mL} + \frac{nV_o}{L_m}(1-D)T_s \quad (36)$$

Since parameter d is a function of ω_r, (36) is highly nonlinear in ω_r. Analytically, it is difficult to solve (36) for ω_r. To find the resonant frequency, the following empirical formula can be used:

$$\omega_r > \frac{\pi}{(1-D)T_s} \quad (37)$$

Satisfaction of (37) means that the resonant current will continue to develop for at least half of its resonant period during the time interval $0\sim(1-D)T_s$ in order to have ZCS turn off for the output diode. It is also noted that once the output diode turns off with ZCS under full load condition, the ZCS will continue to effect for light load because satisfaction of criterion (33) is guaranteed.

(6) ZVS conditions

The ZVS operation of power switches in the studied converter depends on the magnitude of magnetizing current. The ZVS conditions of Q_2 can be easily reached because the magnetizing current at time t_1 is at its lower peak value $\hat{I}_{mL}$, which is large enough in magnitude to fully discharge the energy stored in the output capacitor C_{ds2}. According to (31), $\hat{I}_{mL}$ can be obtained as follows:

$$\hat{I}_{mL} = -\frac{V_{in}(1-D)DT_s}{2(L_m+L_k)} - \frac{I_o}{n}, \quad (38)$$

and the ZVS condition for Q_2 is

$$\frac{1}{2}L_m \hat{I}_{mL}^2 \geq \frac{1}{2}C_{ds2}V_{in}^2 . \tag{39}$$

The discharged energy from C_{ds2} is charged back to the blocking capacitor C_b. The dead time between Q_1 and Q_2 must be long enough to allow C_{ds2} fully discharged.

The ZVS conditions of Q_1 depends on the upper peak current $\hat{I}_{mu}$ of the magnetizing inductor. This peak current must be positive and large enough in magnitude to fully discharge the energy stored in the output capacitor C_{ds1}. Repeating (31), one is requiring

$$\hat{I}_{mu} = \frac{V_{in}D(1-D)T_s}{2(L_m+L_k)} - \frac{I_o}{n} \geq 0 , \tag{40}$$

which leads to

$$(L_m+L_k) \leq n\frac{V_{in}D(1-D)T_s}{2I_o}. \tag{41}$$

Inequality constraint (41) must be satisfied for Q_1 to have ZVS. Another necessary condition for Q_1 to have ZVS is

$$\frac{1}{2}L_m \hat{I}_{mu}^2 \geq \frac{1}{2}C_{ds1}V_{in}^2 \tag{42}$$

It should be noted that the time used to discharge C_{ds1} (mode 6) is longer than that of C_{ds2} (mode 2). This is because the upper peak current $\hat{I}_{mu}$ of the magnetizing inductor is smaller in magnitude compared with that of the lower peak current $\hat{I}_{mL}$. The minimum dead time between Q_1 and Q_2, therefore, should be determined by the time required to fully discharge C_{ds1}.

(IV) DESIGN CONSIDERATION

To validate the characteristics of the studied converter, a prototype converter has been designed based on the following specifications:

- Input voltage V_{in}: 156V DC
- Output voltage V_o: 12V
- Maximum output power P_{max}: 90W
- Switching frequency f_s: 50kHz

(1) Duty ratio

The duty ratio D can be determined according to (23). Depending on the transformer turns ratio n, the value of D can exceed 0.5 when voltage mode control is applied. The larger the D value, the more energy stored in L_m, L_k and C_b, which implies a bigger input capacitor with longer hold-up time is required. On the other hand, the voltage stress on output diode is smaller when the duty ratio is larger.

(2) Transformer turns ratio

The transformer turns ratio n is related to duty ratio D and magnetizing inductance L_m through (23) and (41). A larger value of n reduces the current on the primary side of transformer and increases the current on the secondary side. This is beneficial to the reduction of conduction losses on the primary side. The drawback is a higher current stress on the rectifier. A compromise between selection of n, D and L_m are recommended.

(3) Transformer magnetizing inductor

The design formula used to determine the magnetizing inductor for conventional flyback converter operating in continuous conduction mode can be referred here. It is described as

$$L_m \leq \left[B_{sat} \times 10^{-8} - \frac{V_{in}D(1-DT_s)}{2N_pA_e} \right] \times N_pA_e \frac{n}{I_0} \tag{43}$$

where B_{sat} is the saturated flux density of the transformer core, N_p is the number of turns of primary winding, and A_e is the effective cross section area of the transformer core. However, Eq. (43) does not guarantee the satisfaction of ZVS conditions. It is recommended that, after using (43), Eq. (40) and (42) should be checked to ensure the existence of ZVS operation for switch Q_1. A preliminary checking formula for ZVS is (41).

(4) Resonant inductor

The resonant inductor L_k is mainly used to develop a resonant current during the period energy being transferred from the primary to the secondary side of the transformer. To ensure the ZCS operation of the output diode, the resonant frequency $\omega_r = 1/\sqrt{L_kC_b}$ can be estimated from satisfying Eq. (37). Since the transformer leakage inductor is the major part of the resonant inductor, and the solution of (37) usually leads to a higher frequency compared with switching frequency, the leakage inductor should be determined as small as possible. It has been noted that the resonant inductor L_k is closely related to the ZCS operation of the rectifier and has nothing to do with the ZVS operations of the power switches.

(5) Blocking capacitor

Depending on L_k, the block capacitor C_b can be determined by

$$C_b \leq \frac{1}{L_k\left(\dfrac{\pi}{(1-D)T_s}\right)^2} \tag{44}$$

Limiting the voltage ripple of the blocking capacitor smaller than the average voltage V_c generates another constraint for the determination of C_b:

$$C_b \geq \left(\frac{d}{DV_{in}}\right)^2 L_k, \tag{45}$$

where parameter d comes from (27).

IV. EXPERIMENTAL RESULTS

Based on the design consideration in the preceding section, a prototype converter operated at frequency of 50kHz has been constructed. Table 1 lists the components that are used in the prototype of the studied converter. Fig. 5 shows the key waveforms of a simulated converter at full load condition. Fig. 6 shows the key waveforms measured from the prototype converter under full load condition. The resonant inductor current $i_{Lk}(t)$ and the output diode current $i_{do}(t)$ in Fig. 6 are in good agreement with the theoretical and simulation analysis in Fig. 3 and Fig. 5, respectively. The ZVS operations of Q_1 and Q_2 can be observed from zero to full load, as indicated in the waveforms shown in Fig. 7.

Table 1. Components used in the prototype converter

Magnetizing inductor L_m	0.3mH
Resonant inductor L_k	2.58μH
Turns ratio	7:1
Blocking capacitor C_b	0.68μF/250V
Filter capacitor C_f	4700μF/35V
Power switch Q_1 & Q_2	K2843×2
Power diode D_o	MOSPEC S30C40C
Switching frequency f_s	50kHz
Load R_{load}	1.6 Ω
Driver ICs	UC3843+IR2111
Opto-coupler	4N25

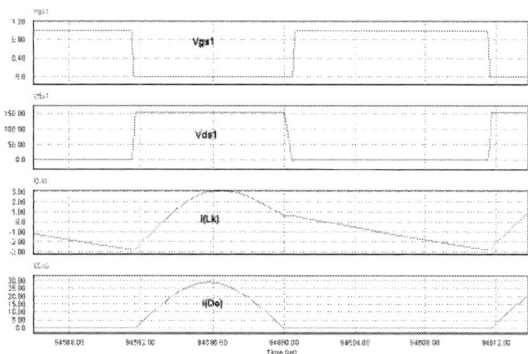

Fig. 5 Key waveforms from simulation

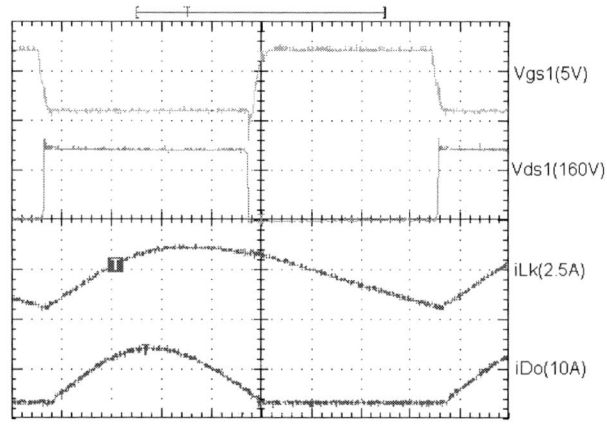

Fig. 6 Key waveforms from experiments

V. CONCLUSION

This report has presented the analysis and experimental results of an asymmetrical half bridge flyback converter without output inductor. The power switches can be turned on with ZVS and the output diode can be turned on and off with ZCS. It has been found that the ZVS operation of the power switches is obtained from the magnetizing current, the ZCS operation of the output power diode is achieved from the

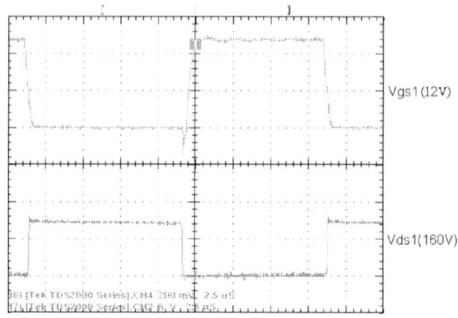

(a) Q_1 at zero load

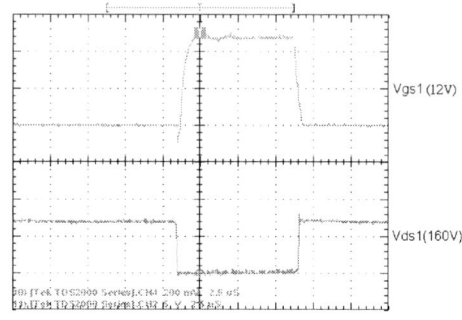

(b) Q_1 at full load

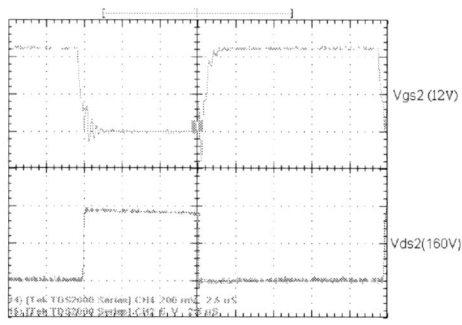

(c) Q_2 at zero load

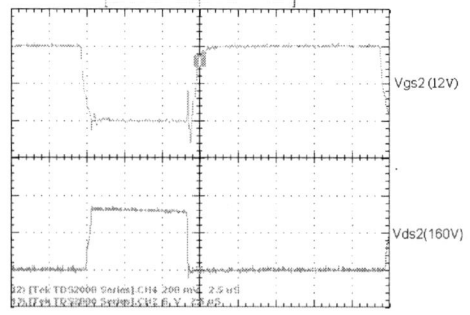

(d) Q_2 at full load

Fig. 7 Experimental waveforms of ZVS operations

resonant current. The ZVS and ZCS conditions can be maintained for the full load range. The operational principles have been presented in the mode analysis, and the design equations have been derived in steady state analysis. The necessary conditions for ZVS and ZCS operations have been obtained as well. Based on the design equations, a 90W prototype converter is constructed and tested by experiment. The experimental results of the prototype converter have shown the characteristics that are in good agreement with simulation and theoretical analysis.

ACKNOWLEDGMENT

This work was supported under the contract number NSC 95-2221-E-231-043 by the National Science Council, Taiwan, R.O.C.

REFERENCES

[1] A. Aqik and I. Cadirci, "Active clamped ZVS forward converter with soft-switched synchronous rectifier for high efficiency, low output voltage applications," IEE Proc. Electr. Power Appl., Vol. 150, No. 2, March 2003.

[2] Yilei Gu, Zhengyu Lu, etc., "A Novel ZVS Resonant Reset Dual Switch Forward DC–DC Converter," IEEE Transactions on Power Electronics, Vol. 22, No. 1, January 2007.

[3] Robert Watson. Fred C. Lee, Guichao C. Hua, "Utilization of an Active-Clamp Circuit to Achieve Soft Switching in Flyback Converters," IEEE Trans. on Power Electronics, Vol. 11, No. I, 1996, pp.162-169.

[4] JI, H.K., and KIM, H.J., "Active clamp forward converter with MOSFET synchronous rectification," IEEE Power Electron. Spec. Conf. Rec., 1994, pp. 895-901.

[5] WATSON, R., LEE, F.C., and HUA, G.C.: 'Utilization of an active clamp circuit to achieve soft switching in flyback converters," IEEE Tran. On Power Electron., 1996, **11, (I),** pp. 162-169.

[6] YOSHIDA, K., ISHII, T., and NAGAGATA, N.: "Zero voltage switching approach for flyback converter," Proceedings of IEEE INTELEC, 1992, pp. 324~329.

[7] SEO, D.H., LEE, O.J., LIM, S.H., and PARK, J.S.: 'Asymmetrical PWM flyback converter," Proceedings of IEEE PESC, 2000, pp. 848-852.

[8] T. M. Chen, C. L. Chen, "Analysis and Design of Asymmetrical Half Bridge Flyback Converter," Electric Power Applications, IEE Proc., Vol. 149, 2002, pp. 433-420.

[9] Sung-Sae Lee, Sang-Kyoo Han, and Gun-Woo Moon, "Analysis and Design of Asymmetrical ZVS PWM Half Bridge Forward Converter with Flyback Type Transformer," Proceeding of the 35th Annual IEEE Power Electronics Specialists Conference, 2004, p. 1525~1530.

A New-Half Bridge Converter
without DC offset of magnetizing current

Kyu-Min Cho, Won-Sik Oh, Keun-Wook Lee and Gun-Woo Moon

Department of Electrical Engineering and Computer Science, KAIST
373-1 Guseong-Dong, Yuseong-Gu Daejeon 305-701,Korea
Email: negative@angel.kaist.ac.kr

Abstract— A new half bridge converter without DC offset of magnetizing current is proposed. The ZVS operation of proposed converter is maintained by only reflected load current because it doesn't have DC offset of magnetizing current as the duty changes. Therefore it has high efficiency and high power density especially when it has a wide input range. The operational principle, large signal modeling and ZVS analysis are presented. Experimental results demonstrate that the proposed converter can achieve a large ZVS range and significant improvement in the efficiency for a 280W (12V, 23.3A) prototype converter.

I. INTRODUCTION

PC power supply unit (PSU) is the device that converts the AC voltage to DC voltages needed by the personal computer. Since the introduction of IBM PC/XT, there have been about a dozen of different PC standards (such as AT, ATX, SFX, LFX, and so on) that differ by their form factors, connector and voltage/current ratings. Output power of a typical PSU for PC is 180W to 450W. PSUs over 450W are usually used in servers or industrial PCs.

Fig. 1 shows the entire configuration of ATX12V PC power supply system which is widely used. It consists of input filter & rectifier stage, main power stage, auxiliary power stage, control stage and supervisor & protection stage. Generally, the selectable switch is used because it doesn't have the power factor correction (PFC) circuit. Moreover, the PC power supply should maintain output regulation for minimum 17 ms despite a loss of input power. Therefore, it has a very wide input range. It makes two big problems. One is a very small duty ratio in a nominal operation. It makes it difficult to use the phase shift full bridge (PSFB) converter because the PSFB has a large conduction loss in the freewheeling period when duty ratio is small. The other is a wide variation of duty ratio. It makes it difficult to use the asymmetric half bridge (AHB) converter because the AHB has a large DC offset of magnetizing current when duty ratio is small.

As a result, conventional PC PSUs generally use the symmetric half bridge converter or the active clamp forward converter. In the conventional symmetric half bridge converter, it has no DC offset current of magnetizing current. But it can not guarantee the ZVS operation of all switches. Therefore it has very low efficiency. In the conventional active clamp forward converter, it can guarantee the ZVS operation of all switches. But it has a very high voltage stress

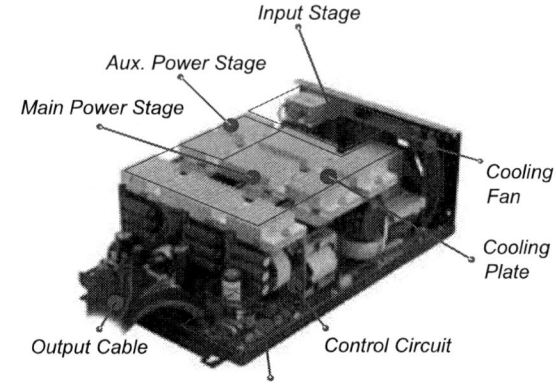

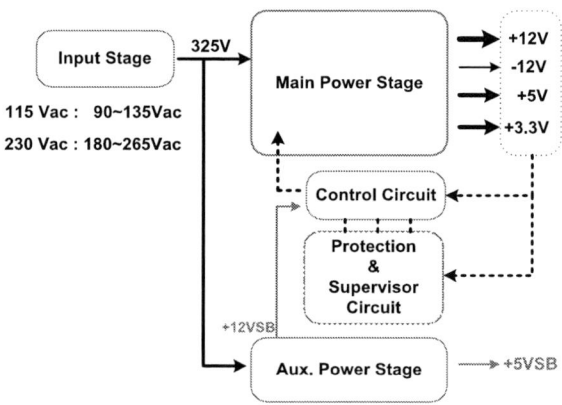

Fig. 1 Entire Configuration of PC Power Supply

of switches which is over 800V. Consequently, they usually use two switches in series, which has low reliability because sometimes they have the different voltage stress.

In this paper, a new half bridge converter is proposed. The proposed converter has no DC offset of magnetizing current and guarantees the ZVS operation of switches. Therefore it has high efficiency and high power density although it has a very wide input range. The operational principle, analysis and experimental results are presented to confirm the validity of the proposed converter.

II. OPERATIONAL PRINCIPLE

Fig. 2 shows the circuit diagram of the proposed converter. The circuit configuration of the proposed converter is similar

978-1-4244-1871-8/07 $25.00 © 2007 IEEE

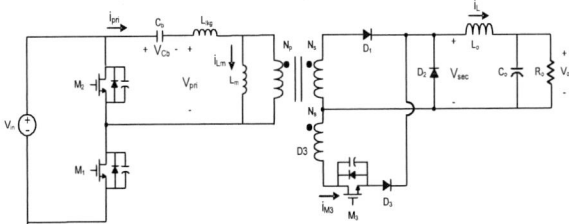

Fig. 2 Schematic of the Proposed Converter

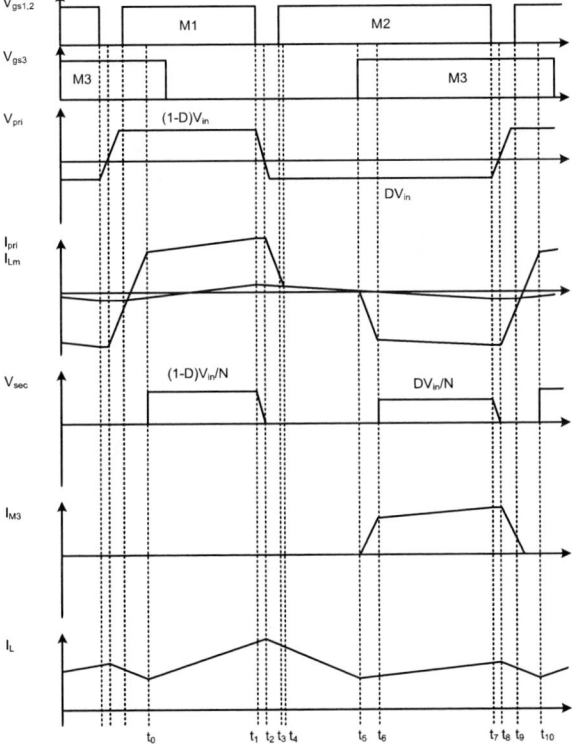

Fig. 3 Key Waveforms for Mode Analysis

to that of the asymmetric half bridge converter except the auxiliary switch M_3. Fig. 3 shows key waveforms of the proposed converter in steady state. Each switching period is subdivided into ten modes and their topological stages are shown in Fig. 4. For the convenience of the analysis of the steady state operation, several assumptions are made as follows:

(a) All parasitic components except for the leakage inductor are neglected,

(b) M_1, M_2 and M_3 are ideal expect for output capacitor $C_{oss1}=C_{oss2}=C_{oss3}=C_{oss}$ and internal diode,

(c) The blocking capacitor C_b and the output capacitor C_o are large enough to be considered as a constant voltage source DV_{in} and V_o respectively,

(d) The transformer magnetizing current $i_{Lm}(t)$ is assumed to be constant during the time intervals $t_1 \sim t_2$ and $t_7 \sim t_8$.

(e) Turn ratio of transformer is $n=N_p/N_s$.

Mode 1($t_0 \sim t_1$) : Mode 1 begins when the commutation of secondary diode current is completed. Then D_3 is turned off and D_1 is turned on, respectively. Since M_1 is on state and M2 is off state, $V_{in}-V_{Cb}$ is applied to L_m+L_{lkg}. The current flowing through L_m and L_{lkg} can be expressed as follows:

$$i_{pri}(t) = nI_o + i_{Lm}(t) \tag{1}$$

$$i_{Lm}(t) = \frac{(1-D)V_{in}}{L_m}t + i_{Lm}(t_0) \tag{2}$$

where

$$i_{Lm}(t_0) = -\frac{(1-D)DV_{in}}{2L_m}T_s$$

During this mode, M3 is turned off with ZCS condition.

Mode 2($t_1 \sim t_2$) : When M_1 is turned off, this mode begins. Since $V_{ds2}(t)$ is higher than V_{Cb}, diode D_2, D_3 are still reverse biased. Therefore C_{oss1} and C_{oss2} are linearly charged and discharged, respectively. From the assumption (d), $i_{pri}(t)$ and $V_{ds1}(t)$ can be expressed as follows:

$$i_{pri}(t) = \frac{I_o}{n} + i_{Lm}(t_1) \tag{3}$$

$$V_{ds1}(t) = \frac{i_{pri}(t_1)}{2C_{oss}}t \tag{4}$$

where

$$i_{Lm}(t_1) = \frac{(1-D)DV_{in}}{2L_m}T_s$$

Mode 2 ends at t_2 when $V_{ds1}(t)$ becomes equal to $V_{in}-V_{Cb}$

Mode 3($t_2 \sim t_3$) : After $V_{ds1}(t)$ increase to $V_{in}-V_{Cb}$, the output inductor current, $i_L(t)$ begins to freewheel through D_2. Since the primary voltage across the transformer is 0V, $V_{ds1}(t)$ and $V_{ds2}(t)$ are charged and discharged, respectively, in matter of the resonance between L_{lkg} and $C_{oss1}+C_{oss2}=2C_{oss}$. $i_{pri}(t)$ and $V_{ds1}(t)$ can be expressed as follows:

$$i_{pri}(t) = i_{pri}(t_2)\cos(\sqrt{\frac{1}{2L_{lkg}C_s}}t) \tag{5}$$

$$V_{ds1}(t) = i_{pri}(t_2)\sqrt{\frac{L_{lkg}}{2C_s}}\sin(\sqrt{\frac{1}{2L_{lkg}C_s}}t) + (1-D)V_{in} \tag{6}$$

After $V_{ds1}(t)$ and $V_{ds2}(t)$ reach V_{in} and 0V, respectively, $i_{pri}(t)$ flows through the output diode of switch M_2 and zero voltage across M_2 is maintained. Therefore M_2 can be turned on with ZVS condition at the next mode.

Mode 4($t_3 \sim t_4$) : When M_2 is turned on, this mode begins. Since the commutation of secondary diodes does not complete, the voltage across the transformer is still 0V and V_{Cb} is all applied to the leakage inductor. Therefore, the leakage inductor current rapidly decreases as follows:

$$i_{pri}(t) = -\frac{DV_{in}}{L_{lkg}}t + i_{pri}(t_3) \tag{7}$$

where

978-1-4244-1871-8/07 $25.00 © 2007 IEEE 883

$$i_{pri}(t) = -\frac{I_o}{n} + i_{Lm}(t) \tag{10}$$

Mode 8($t_7 \sim t_8$) : When M_2 is turned off, this mode begins. The operation of this mode is similar to that of mode 2. Since $V_{ds2}(t)$ is smaller than V_{Cb}, diode D_1, D_2 are still reverse biased. Therefore C_{oss1} and C_{oss2} are linearly charged and discharged, respectively. From the assumption (d), $i_{pri}(t)$ and $V_{ds2}(t)$ can be expressed as follows:

$$i_{pri}(t) = -\frac{I_o}{n} + i_{Lm}(t_7) \tag{11}$$

$$V_{ds2}(t) = \frac{i_{pri}(t_7)}{2C_s}t \tag{12}$$

where

$$i_{Lm}(t_7) = -\frac{(1-D)DV_{in}}{2L_m}T_s$$

Mode 8 ends at t_8 when $V_{ds2}(t)$ reaches to V_{Cb}.

Mode 9($t_8 \sim t_9$) : The operation of this mode is similar to that of mode 3. After $V_{ds2}(t)$ increases over V_{Cb}, D_1 and D_2 are turned on. Since the voltage across the transformer is 0V, $V_{ds1}(t)$ and $V_{ds2}(t)$ are discharged and charged respectively, in manner of the resonance between L_{lkg} and $C_{oss1}+C_{oss2}=2C_{oss}$. Therefore $i_{pri}(t)$ and $V_{ds2}(t)$ can be expressed as follows:

$$i_{pri}(t) = i_{pri}(t_8)\cos\left(\sqrt{\frac{1}{2L_{lkg}C_S}}t\right) \tag{13}$$

$$V_{ds2}(t) = i_{pri}(t_8)\sqrt{\frac{L_{lkg}}{2C_S}}\sin\left(\sqrt{\frac{1}{2L_{lkg}C_S}}t\right) + DV_{in} \tag{14}$$

After $V_{ds1}(t)$ and $V_{ds2}(t)$ reach 0V and V_{in}, respectively, $i_{pri}(t)$ flows through the output diode of the switch M_1 and zero voltage across M_1 is maintained. Therefore M_1 can be turned on with ZVS condition at the next mode.

Mode 10($t_9 \sim t_{10}$) : When M_1 is turned on, this mode begins. The operation of this mode is similar to that of mode 4. Since commutation of secondary diodes does not complete, the voltage across the transformer is still 0V and V_{in} applied to the leakage inductor. Therefore, the leakage inductor current rapidly increases as follows:

$$i_{pri}(t) = \frac{V_{in}}{L_{lkg}}t + i_{pri}(t_9) \tag{15}$$

where

$$i_{pri}(t_9) = \sqrt{i^2_{Lm}(t_8) - \frac{2C_s(DV_{in})^2}{L_{lkg}}} \tag{16}$$

Mode 10 ends at t_{10} when $i_{pri}(t)-i_{Lm}$ reaches I_L/n.

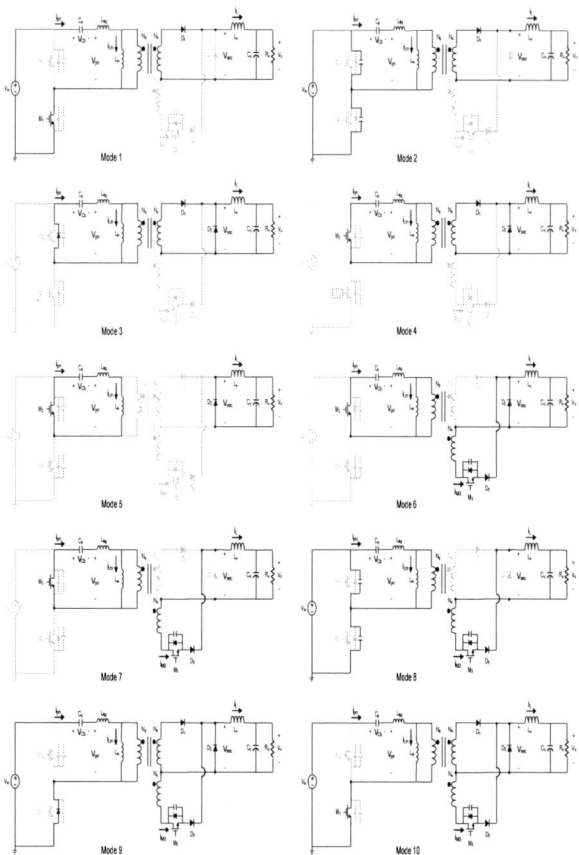

Fig. 4 Equivalent Circuit of the Proposed Converter

$$i_{pri}(t_3) = \sqrt{i^2_{Lm}(t_2) - \frac{2C_s\{(1-D)V_{in}\}^2}{L_{lkg}}}$$

Mode 5($t_4 \sim t_5$) : When $i_{pri}(t)$ reaches $i_{Lm}(t)$, the output diode current completes its freewheeling. D_1 is turned off and output current freewheels through D_2. Since V_{Cb} is applied to $L_{lkg}+L_m$, $i_{Lm}(t)$ can be expressed as follows:

$$i_{pri}(t) = -\frac{DV_{in}}{L_m}t + i_{pri}(t_4) \tag{8}$$

Mode 6($t_5 \sim t_6$) : When M_3 is turned on, this mode begins. Since the secondary voltage of the transformer is zero, $-V_{Cb}$ is applied to the leakage inductor. Therefore, $i_{pri}(t)$ rapidly decreases to the reflected load current. $i_{pri}(t)$ can be expressed as follows:

$$i_{pri}(t) = -\frac{DV_{in}}{L_{lkg}}t + i_{pri}(t_5) \tag{9}$$

Mode 7($t_6 \sim t_7$) : When $-[i_{pri}(t)-i_{Lm}(t)]$ becomes equal to I_L/n, this mode begins. The operation of this mode is similar to that of mode 1. $i_{pri}(t)$ can be expressed as follows:

III. ANALYSIS OF THE PROPOSED CONVERTER

In this section, the large signal modeling, DC offset of magnetizing current and the comparison of ZVS conditions between the proposed converter and the asymmetric half bridge converter are presented.

A. Large signal modeling

The large signal modeling can be obtained by averaging currents and voltage over one switching cycle. The stage variables are the magnetizing current $i_{Lm}(t)$ and output inductor current $i_L(t)$. The slope of state variables and the freewheeling times D_1T_s and D_2T_s of the output inductor current are depicted in Fig. 5. Therefore, the resulting large signal equations can be obtained as follows:

$$\frac{di_{Lm}(t)}{dt} = \frac{V_{in}-V_{Cb}}{L_m+L_{lkg}}(D_1+D_{eff}) - \frac{V_{Cb}}{L_m+L_{lkg}}(1-D_{eff}-D_1)$$

(17)

$$\frac{di_{Lm}(t)}{dt} = \frac{(V_{in}-V_{Cb})/n-(V_o+V_{fd})}{L_o}D_{eff} + \frac{V_{Cb}/n-(V_o+V_{fd})}{L_o}D_{eff}$$
$$- \frac{V_o+V_{fd}}{L_o}(1-2D_{eff}-D_1) - \frac{V_o+V_{fd}}{L_o}D_{eff} \quad (18)$$

where V_{fd} is the forward voltage drop of the secondary diode. From (17) and (18), the voltage conversion ratio of the proposed converter can be derived as follows:

$$V_o = \frac{D_{eff}}{n}V_{in} - V_{fd} \quad (19)$$

B. DC offset of magnetizing current

The DC offset of magnetizing current can be calculated by applying the current-second balance law to the blocking capacitor C_b. The current thorough the blocking capacitor is depicted in Fig. 6. Therefore, DC offset of magnetizing current can be expressed as follows:

$$D_{eff}(I_{Lm,dc}+\frac{I_o}{n}) + (1-2D_{eff})I_{Lm,dc} + D_{eff}(I_{Lm,dc}-\frac{I_o}{n}) = 0$$

$$\therefore I_{Lm,dc} = 0 \quad (20)$$

C. Comparison of ZVS conditions

(a) ZVS condition of the asymmetric half bridge converter

The conventional asymmetric half bridge converter has DC offset of magnetizing current according to duty ratio. It makes it more difficult to design transformer and to obstruct ZVS operation. The DC offset of magnetizing current $i_{Lm,DC}$ can be derived using the current-sec balance law of the blocking capacitor C_b and can be expressed as follows:

$$I_{Lm,dc} = \frac{(1-2D)}{n}I_o \quad (21)$$

Fig. 7(a) shows the primary current of the asymmetric half bridge converter according to duty ratio. The mechanism of ZVS operation is as same as that of the proposed converter. Therefore the ZVS condition can be expressed respectively as follows:

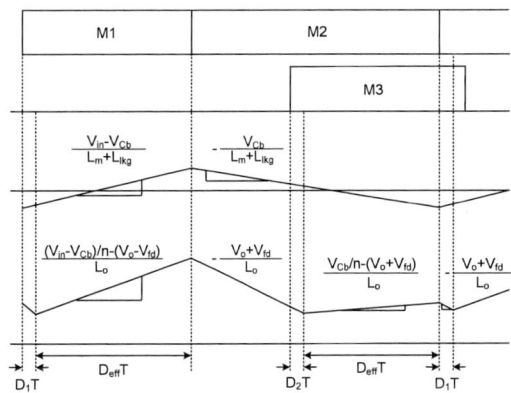

Fig. 5 The Waveforms of the Stated Variables for the Large Signal Analysis

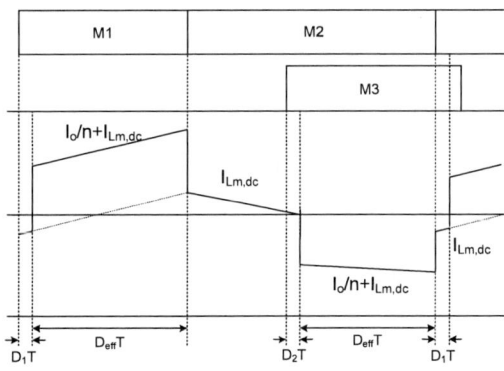

Fig. 6 The Waveforms for Calculation of DC offset of Magnetizing Current

$$\text{M1}: \quad \frac{1}{2}L_{lkg}\left(\frac{2(1-D)I_o}{n}+\frac{D(1-D)V_{in}}{2L_m}\right)^2 > \frac{1}{2}(2C_{oss})(DV_{in})^2 \quad (22)$$

$$\text{M2}: \quad \frac{1}{2}L_{lkg}\left(\frac{2DI_o}{n}+\frac{D(1-D)V_{in}}{2L_m}\right)^2 > \frac{1}{2}(2C_{oss})(1-DV_{in})^2 \quad (23)$$

From (21) and (22), we can see that DC offset of magnetizing current induces the ZVS condition to be poor when duty ratio is small.

(b) ZVS condition of the proposed half bridge converter

Fig. 7(b) shows the primary current according to duty ratio. Since the proposed half bridge converter has no DC offset of magnetizing current, all reflected load current can be used the ZVS operation of main switches. Therefore the proposed half bridge converter has better the ZVS condition than the conventional half bridge converter. The ZVS condition of the proposed half bridge converter can be expressed respectively as follows:

$$\text{M1}: \quad \frac{1}{2}L_{lkg}\left(\frac{I_o}{n}+\frac{D(1-D)V_{in}}{2L_m}\right)^2 > \frac{1}{2}(2C_{oss})(DV_{in})^2 \quad (24)$$

$$\text{M2}: \quad \frac{1}{2}L_{lkg}\left(\frac{I_o}{n}+\frac{D(1-D)V_{in}}{2L_m}\right)^2 > \frac{1}{2}(2C_{oss})(1-DV_{in})^2 \quad (25)$$

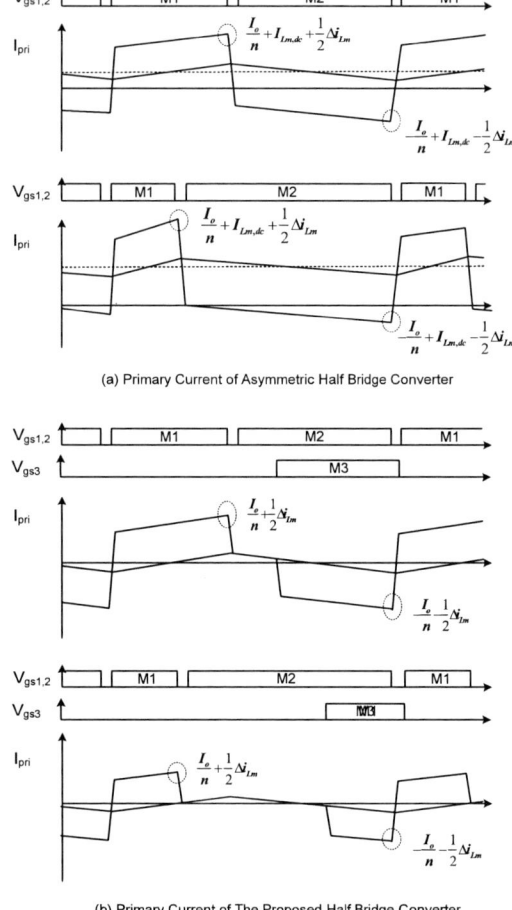

(a) Primary Current of Asymmetric Half Bridge Converter

(b) Primary Current of The Proposed Half Bridge Converter

Fig. 7 ZVS Conditions of Asymmetric Half Bridge Converter and The Proposed Half Bridge Converter

IV. EXPERIMENTAL RESULTS

Based on the analysis of the proposed converter, a prototype circuit has been designed with the ATX12V specification as follows:

- Input voltage V_{in}: $180V_{ac} \sim 265V_{ac}$
- Output voltage V_o: 12V
- Maximum output power $P_{o,max}$: 280W
- Switching frequency f_s : 90kHz
- Hold up time : 17ms
- Maximum duty ratio : D_{max}=0.25 at V_{in}=180V_{ac}, full load

Table 1 shows components that are used for the prototype of the 12V, 280W converter operated at 90kHz has been built. Fig. 8 shows the key waveforms of the proposed converter at V_{in}=180V_{ac}, V_{in}=265V_{ac} respectively. The primary current $i_{pri}(t)$ and the output inductor current $i_L(t)$ are well agreed with the theoretical analysis. We can see that there are no DC offset of magnetizing current in all input range. It is noted that the ZVS operation of main switches is achieved by the

reflected load current in all input range. Fig. 9 shows the efficiency of the proposed converter and that of the symmetric half bridge converter. We can see that the efficiency of the proposed converter always is about 4% higher than that of the symmetric half bridge converter in all input range.

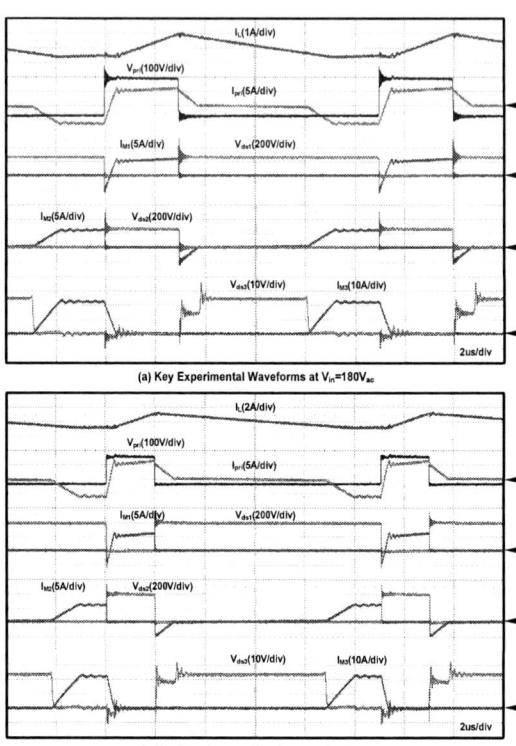

(a) Key Experimental Waveforms at V_{in}=180V_{ac}

(b) Key Experimental Waveforms at V_{in}=265V_{ac}

Fig. 8 Key Experimental Waveforms

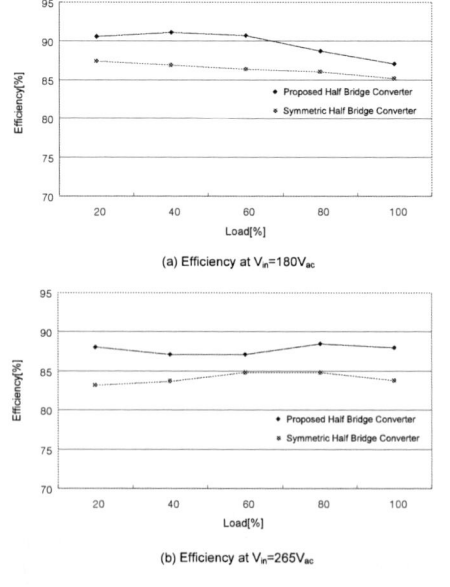

(a) Efficiency at V_{in}=180V_{ac}

(b) Efficiency at V_{in}=265V_{ac}

Fig. 9 Measured Efficiency with input variation

978-1-4244-1871-8/07 $25.00 © 2007 IEEE

Table 1. Components List for the Prototype Converter

Switching frequency (fs)	90kHz
Main switches (M_1,M_2)	FQP13N50
Auxiliary switch (M_3)	IRFP064
Diode (D_1,D_2)	63CTQ100
Diode (D_3)	MUR2020CT*2
Turn ratio (N_p:N_s)	20 : 5
Leakage inductance (L_{lkg})	5.7uH
Magnetizing inductance (L_m)	1.7mH
Blocking capacitor (C_b)	10uF
Output inductance (L_o)	60uH

V.   CONCLUSION

This paper has presented the analysis and experimental results of the new half bridge converter without DC offset of magnetizing current. Since it has no DC offset of magnetizing current according to duty ratio, the ZVS operation can be achieved by reflected load current. Therefore, it has a good ZVS condition, especially when nominal duty ratio is small. Moreover, it has the low conduction loss because it has no freewheeling current in the primary side. This characteristic is more remarkable when duty ratio is small. Therefore, it can achieve high efficiency when duty ratio is small although auxiliary switch M3 makes conduction loss in powering mode. The experimental results of a 280W (12V, 23.3A) prototype converter has proved the key characteristics of the proposed converter in the wide input range. The efficiency of the proposed converter is obtained about 87% at the full load condition of $180V_{ac}$ input and 88% at the full load of $265V_{ac}$ input.

REFERENCES

[1] Hong Mao, "A new duty-cycle-shifted PWM control scheme for half-bridge DC-DC converters to achieve zero-voltage-switching", Eighteenth Annual IEEE Applied Power Electronics Conference and Exposition, APEC 2003, Vol. 2, pp. 629-634.

[2] Hong Mao, "Zero-voltage-switching half-bridge DC-DC converter with modified PWM control method", IEEE Transactions on Power Electronics, Vol. Issue 4, July 2004, pp.947 - 958.

[3] Jiangtao Feng, "ZVS analysis of asymmetric half-bridge converter", PESC, 2001, pp.243-247.

[4] Sebastian, J. "An overall study of the half bridge complementary control DC to DC converter", PESC, 1995, pp.1229-1235.

978-1-4244-1871-8/07 $25.00 © 2007 IEEE

A Novel Low-Loss Modulation Strategy for High-Power Bi-directional Buck+Boost Converters

S. Waffler and J. W. Kolar
Power Electronic Systems Laboratory
ETH Zurich
8092 Zurich, Switzerland
Email: waffler@lem.ee.ethz.ch

Abstract — A novel low-loss, constant-frequency, zero-voltage-switching (ZVS) modulation strategy for bi-directional, cascaded, buck-boost DC/DC converters, used in a hybrid electrical vehicle (HEV), is presented and its benefits over state-of-the-art converters and soft-switching solutions are discussed in a comparative evaluation. To obtain ZVS with the purposed modulation strategy, the buck+boost inductance is selected and the switches are gated in a way that the inductor current has a negative offset current at the beginning and the end of each pulse period. This allows the MOSFET switches to turn on when the anti-parallel body diode is conducting. As the novel modulation strategy is a software-only solution, there are no additional expenses for active or passive components compared to conventional modulation implementations. Furthermore, an analytical and simulation investigation predicts an excellent efficiency over the complete operating range and a higher power density for a multi-phase converter equipped with the low-loss modulation. Experimental measurements performed with a converter prototype verify the mode of operation and the ZVS principle.

I. INTRODUCTION

Drive systems for Hybrid Electrical Vehicles (HEVs) include energy storage elements such as batteries, and these are typically connected by DC/DC converters to a common DC link that supplies the main motor inverter. For feeding back the braking energy, a bi-directional energy flow has to be implemented. Depending on the cell number and characteristics of the battery, the battery voltage range may overlap with the nominal DC link voltage range. In this case, the DC/DC converter has to be able to function in both the buck and boost operating modes.

Furthermore, a converter for this application has to meet the prevalent automotive requirements, such as being a low cost design, and minimizing the component sizes and count. Fixed frequency operation is desired due to EMI restrictions and a high efficiency over a wide output power range, as well as a highly compact design and a low overall weight are required.

One commonly used converter topology for this application is the hard-switched, cascaded, buck+boost converter [1][2]. In order to improve efficiency, silicon carbide (SiC) technology could be applied [3]. Furthermore, there is a wide variety of soft-switching auxiliary circuits to extend this converter topology, including the Auxiliary Resonant Commutated Pole (ARCP) [4][5], Zero Current Transition (ZCT) [6][7] or Snubber Assisted Zero Voltage and Zero Current Transition (SAZZ) [8] methods. An alternative approach to achieve ZVS of the cascaded buck+boost converter (Fig. 1) is to implement a novel low-

loss modulation strategy. This is a software-only solution that does not need any additional active or passive components and therefore offers an increased efficiency while maintaining a low component count and simplicity of the power electronics circuit.

In this paper, the state-of-the-art soft-switching additions for the cascaded buck+boost converter, and the further soft-switching concepts and benefits of the proposed alternative modulation method are discussed (**Section II**). In **Section III** a detailed description of the operating principle of the novel low-loss modulation method, the requirements for ZVS operation and an optimized switching strategy are presented. Analytical results, including an efficiency comparison of the conventional hard-switched and the proposed soft-switched buck+boost converter and results of a converter volume optimization are given in **Section IV**. Simulation and experimental results, which verify the method of operation, are presented in **Section V**.

II. TOPOLOGIES

A problem of the cascaded buck+boost converter, shown in Fig. 1, operated in continuous conduction mode (CCM) and with conventional pulse width modulation (PWM) is the significant switch turn-on loss caused by the reverse recovery of the anti-parallel diode of the complementary switch in the half-bridge. These higher losses typically result in an overall efficiency of approximately 92% [2].

In the literature, several ways to improve the loss behavior can be found: First of all, the internal switch anti-parallel diodes (body diodes) may be substituted by low recovery charge silicon carbide types to avoid switching losses caused by reverse recovery. In combination with multi-phase technology, greater efficiency and power density could be achieved [3]. Besides the fact that SiC

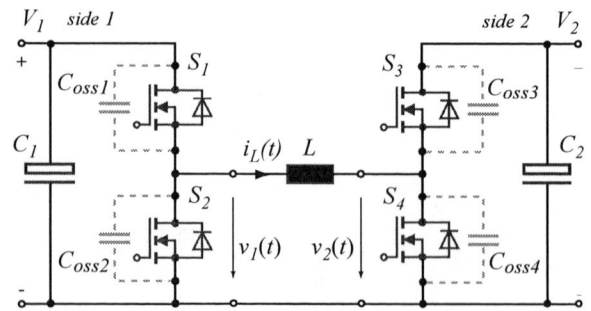

Fig 1: Cascaded buck+boost converter for bi-directional power-flow, shown with parasitic MOSFET output capacitances $C_{oss,I}$.

978-1-4244-1871-8/07 $25.00 © 2007 IEEE

technology is expensive, a hard-switched converter is often not a preferred solution because of increased EMI emissions caused by high rate of voltage change (dV/dt), which involves the filter design. Therefore, secondly, the transistors should be operated such that

  a) Zero Current Switching (ZCS)
  b) Zero Voltage Switching (ZVS)
  c) and/or low switching losses

are achieved. When lowering switching losses by these soft-switching techniques, the rate of voltage change is reduced at the same time. Thus, lower EMI emissions are also expected. On the other hand the switching frequency f_s may be increased without severe impact on switching losses, resulting in a smaller volume of the passive components.

A. Auxiliary Circuitry

A variety of soft-switching solutions applicable to the half bridges used in the cascaded buck+boost converter have been presented in the literature [4]-[9]. These are based on the following fundamental ZVS or ZCS concepts:

- Synchronous Resonant DC Link (SRDCL)
- Auxiliary Resonant Commutated Pole (ARCP)
- Zero Current Transition (ZCT)
- Snubber Assisted Zero Voltage and Zero Current Transition (SAZZ).

In these concepts, the basic operating principle is to conduct the main inductor current in an additional resonant circuit such that the main switch is operated with ZVS and/or ZCS, where the resonant process is initiated by auxiliary switches.

The SRDCL [9] suffers from additional components in the main current path, high peak currents in the resonant circuit and a resonant overshoot of the DC link voltage of approximately 1.5 times the DC link voltage. ARCP [4][5], ZCT [6][7] and SAZZ [8] do not have the drawback of a resonant overshoot, but the auxiliary switch peak currents are greater than the main inductor current.

All of the mentioned solutions require additional active and passive components as well as gate drive circuitry. For instance, four auxiliary switches are employed in the case of ARCP, ZCT or SAZZ plus the resonant capacitors and inductors and four additional diodes in the case of SAZZ, when applied to the cascaded buck+boost converter. Another drawback is the PWM duty cycle limitation caused by the time consumed by the resonant transition.

B. Resonant Converters

Besides soft-switching extensions to well known hard-switched converters, there are a number of independent resonant, soft-switching, bi-directional buck-boost converter concepts like the constant-frequency zero-voltage-switching quasi-square-wave (CF-ZVS-QSC) converter [10], SEPIC converter [1][11] or the zero-voltage zero-current switching (ZVZCS) converter presented in [11].

Compared to the proposed converter with low-loss modulation method, the main drawbacks of these converters are a doubled switch blocking voltage stress and diode recovery losses in case of the SEPIC [11], a larger number of passive components as well as a larger inductance value

for the main inductor in the SEPIC and ZVZCS topology [10]. Furthermore, SEPIC and ZVZCS use a capacitive energy transfer that performs badly in high-power applications. Other drawbacks of resonant converters are variable switching frequency, which complicate EMI filter design, or limitations in operating range for soft-switching.

III. LOW-LOSS MODULATION STRATEGY

To overcome the above-mentioned drawbacks like additional component effort, capacitive energy transfer or variable switching frequency, a new constant-frequency zero-voltage switching modulation strategy is proposed.

A. Operating Principle

In conventional PWM for the cascaded buck+boost converter only one of the two half bridges is switched and static control signals are applied to the switches of the other bridge leg [2].

In contrast, with the proposed new modulation method, each of the four switches S_1 to S_4 is turned on and off exactly once per pulse period $T_p = 1 / f_s$. Turn-on is accomplished under ZVS when the anti-parallel body diode is conducting. A negative inductor offset current $-I_0$ is needed to fulfill this condition (Fig. 2). The pulse period T_p can be divided into four modes according to the four switching states of S_1 to S_4, as shown in Fig. 2 and Fig. 3.

At the beginning of time period $t_0 \le t < t_1$ the switch S_2 is conducting and therefore is turned off under ZVS. The negative inductor current $i_L(t)$ charges the parasitic drain-source capacitance C_{oss2} of S_2 and discharges C_{oss1} respectively. The body diode of S_1 takes over $i_L(t)$ and S_1 can now switched on under ZVS and conducts the current when $i_L(t)$ becomes positive (Fig. 3 (a)). Due to the applied inductor voltage $v_L(t) = V_1$ the inductor current rises in this time period.

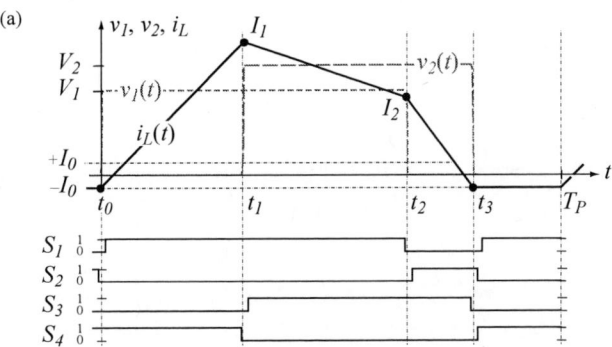

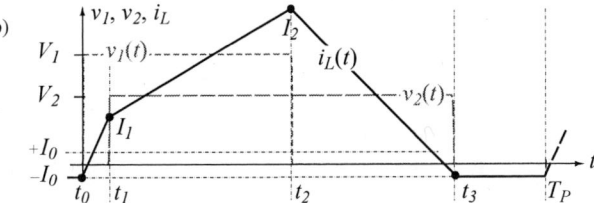

Fig. 2: Basic timing diagram for switch S_1 to S_4 control signals and inductor current i_L for boost operation ($V_2 > V_1$) (a) and basic timing diagram for buck operation ($V_2 < V_1$) (b).

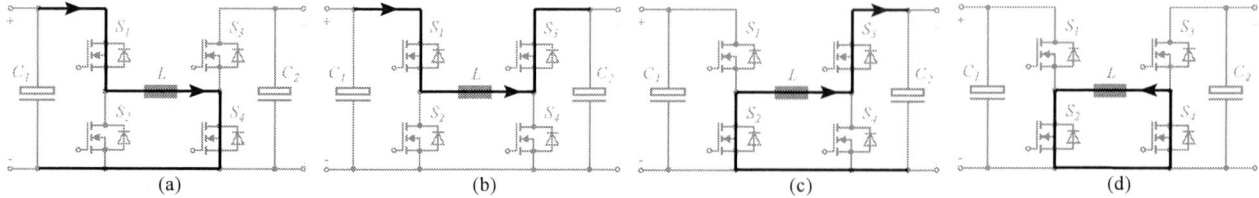

(a)	(b)	(c)	(d)

Fig. 3: Inductor current flow for time intervals $t_0 < t < t_1$ (a), $t_1 < t < t_2$ (b), $t_2 < t < t_3$ (c) and $t_3 < t < T_p$ (d). Current direction reverses in (a) and (c) as can be seen in Fig 2, which is not shown for the reason of simplicity.

At $t = t_1$, S_4 is conducting, therefore turned off under ZVS and $i_L(t)$ charges C_{oss4} and discharges C_{oss3} respectively. Again, S_3 can be turned on under ZVS. The applied inductor voltage $v_L(t) = V_1 - V_2$ either causes i_L to rise (buck operation) or fall (boost operation) during the time period $t_1 \leq t < t_2$ depending on V_1 and V_2.

The ZVS principle also applies to the remaining two switching actions at $t = t_2$ (S_1 is turned off and S_2 is turned on) and $t = t_3$ (S_3 is turned off and S_4 is turned on).

In the last time period $t_3 < t < T_p$ the switches S_2 and S_4 are turned on and the switches S_1 and S_3 are turned off, so that the current $i_L(t)$ circulates in the power circuit as shown in Fig. 3 (d). This is to keep the switching frequency constant and to provide the negative offset current needed for ZVS at the beginning of the subsequent pulse period.

B. Calculation of the switching times

Starting with the basic inductor current waveforms as depicted in Fig. 2 for buck and boost operation, assuming constant voltages V_1 and V_2 and neglecting resistances of the inductor and the switches S_i, the differential equation

$$v_L(t) = v_1(t) - v_2(t) = L \cdot \frac{d}{dt} i_L(t) \tag{1}$$

is solved in the four time intervals and yields:

$$t_1 - t_0 = L \cdot \frac{I_0 + I_1}{V_1} \tag{2}$$

$$t_2 - t_1 = L \cdot \frac{I_2 - I_1}{V_1 - V_2} \tag{3}$$

$$t_3 - t_2 = L \cdot \frac{I_0 + I_2}{V_2} . \tag{4}$$

The average power P_{tr} transferred from converter *side 1* to *side 2* is calculated from

$$P_{tr} = \frac{1}{T_p} \cdot \int_{t_0}^{T_p} v_1(t) \cdot i_1(t) dt . \tag{5}$$

The integral (5) can be solved for the current waveform depicted in Fig. 2 and $t_0 = 0$:

$$P_{tr} = \frac{V_1}{T_p} \cdot \int_0^{t_2} i_L(t) dt = \frac{V_1}{2 \cdot T_p} \left((I_1 + I_2) \cdot t_2 - (I_0 + I_2) \cdot t_1 \right) . \tag{6}$$

The set of equations (2)-(4), (6) with unknowns t_1, t_2, I_1, I_2 may be solved to calculate the switching times t_1, t_2 for the desired operating point (V_1, V_2, P_{tr}) and a given time t_3. Furthermore, the maximum of P_{tr} is calculated as

$$P_{tr,max}(t_3) = \frac{V_1 V_2 \cdot \left(I_0^2 L^2 - 2 I_0 L \cdot (V_1 + V_2) \cdot t_3 + V_1 V_2 \cdot t_3^2 \right)}{2 \cdot L \cdot T_p \cdot \left(V_1^2 + V_1 V_2 + V_2^2 \right)} . \tag{7}$$

The inductance of the buck+boost inductor L has to be derived from (7) for a given operating voltage and power range and $t_3 = T_p$ in such a manner that $P_{tr,max}$ will never be exceeded, otherwise the constraints for soft switching will be violated. The maximum inductance L depends on the minima of V_1 and V_2 and the rated converter peak power $P_{max} = P_{tr,max}(T_p)$ as shown, exemplarily, in Fig. 4.

C. Optimized Switching Pattern

As stated before, there is the degree of freedom to choose the switching time t_3 for a given converter operating point. An example is depicted in Fig. 5: Assuming a exemplary current waveform (marked by $-I_0$, I_1, I_2), there are two possibilities to increase P_{tr}. Firstly, the switching time t_3 could be kept constant and the times t_1 and t_2 adjusted (waveform B). Secondly, the switching time t_3 could be shifted towards the end of the switching period T_p (waveform A). This is preferred because of a lower peak and RMS inductor current and thus lower inductor losses.

Besides the considerations for a low RMS current, a continuous variation of $t_1 \dots t_3$ without any steps is desired over the operating voltage and power range for the reason of steady control behavior of the modulator and a low-error

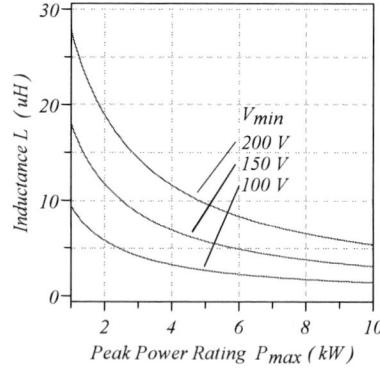

Fig. 4: Maximum inductance L, shown for $V_{1,min} = V_{2,min} = V_{min}$, $I_0 = 10A$, and $f_s = 100kHz$ as function of the converter peak power rating P_{max}.

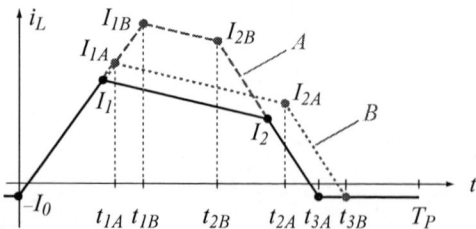

Fig. 5: Possibilities to increase transferred power P_{tr}: Keeping t_3 constant (waveform A) and shifting t_3 towards the end of the switching period T_p (waveform B).

implementation of a three-dimensional, lookup-table-based interpolation of the switching times. To address both requirements, a time calculation scheme is derived for a given power P_{tr}:

- The minimum of t_3 is calculated by differentiation of V_1 and V_2 at the limit of ZVS operation ($I_1 = +I_0$ for buck operation and $I_2 = +I_0$ for boost operation respectively) and is used as a initial value $t_{3,min}$ for t_3. The initial value is normally applicable at the upper ends of the operating voltage range.
- If the energy $E_{tr} = P_{tr} \cdot T_p$ could not be converted within $t_0 < t < t_{3,min}$, t_3 is shifted towards the end of the pulse period at the limit of ZVS operation until the transfer of E_{tr} for any combination of V_1 and V_2 is possible (cf. Fig. 5, waveform A).
- When t_3 is shifted that far to match T_p, t_1 and t_2 are adjusted as depicted in Fig. 5, waveform B.

The calculated, normalized values for $t_1 \ldots t_3$ for a voltage transfer ratio of $V_2 / V_1 = 2$ and $V_{2,max} = 450V$ are shown in Fig. 6.

D. Ensuring ZVS condition

The minimum absolute value needed for I_0 depends on the resonant circuit formed by L and the MOSFET output capacitances C_{oss} of the affected half-bridge as well as the converter input and output voltage levels V_1 and V_2, which excite the LC-circuit:

$$I_0 \geq \max(V_{1\max}, V_{2\max}) \cdot \sqrt{\frac{C_{oss}}{L}} \qquad (8)$$

If the condition (8) is true, the energy stored in the inductor L at turn-off of a switch is large enough to completely transfer the charge between the both parasitic MOSFET output capacitances of the half-bridge, so that the complementary switch of the half-bridge can be turned on at zero voltage. The same applies to the absolute value of $i_L(t)$ at $t = t_1$ and $t = t_2$ inducing the conditions:

$$I_1 \geq I_0 \quad \text{and} \quad I_2 \geq I_0 \qquad (9)$$

It should be noted that (8) is an approximation, since C_{oss} shows a nonlinear, voltage-dependent characteristic. Furthermore, the number of paralleled MOSFETs has to be taken into consideration.

A drift of i_L below $-I_0$ in the time period $t_3 < t < T_p$ due to component nonlinearity or inexact timing must be avoided

to keep conduction losses low. An accurate high-speed zero-crossing detection could be used to exactly determine the time t_3.

Another option, which is implemented in the converter prototype, is to monitor the half-bridge voltages $v_1(t)$ and $v_2(t)$ with high-speed analog comparators to detect a zero voltage level. For instance, i_L is falling in time period $t_2 < t < t_3$. When i_L becomes negative the body diode of S_4 starts conducting and $v_2(t)$ steps to zero, causing the comparator output to change. This information is then used to turn on S_4 under ZVS.

IV. ANALYTICAL RESULTS

A. Efficiency Comparison

For comparison of the novel modulation strategy to the hard-switched cascaded buck+boost topology operated with conventional PWM, the semiconductor conduction and switching losses are analytically calculated based on measured switching losses of two MOSFETs in half-bridge configuration and $V_{max} = 450V$. Figure 7 shows an improvement in efficiency of 3% and an excellent efficiency over the full input and output voltage range at nominal power as effect of the soft-switching. Furthermore, the efficiency for part load conditions (better than 93% at 10% nominal power) is a great advantage for automotive applications and could even be improved with a multi-phase converter concept and by changing the number of operating phases in dependency on the required output power.

(a)
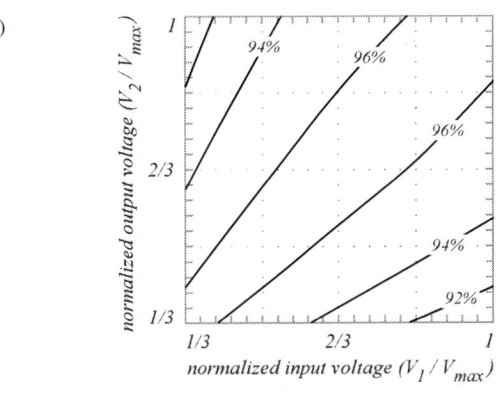

(b)
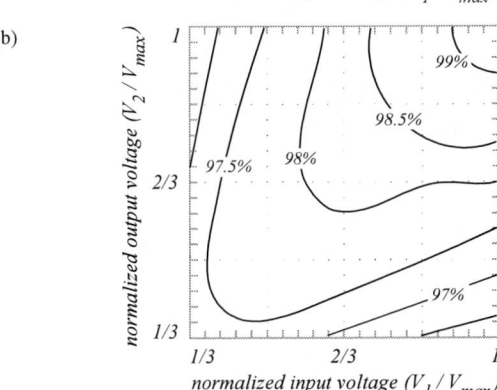

Fig. 7: Calculated semiconductor efficiency for cascaded buck+boost converter in hard-switched operation (a) and operated with the novel low-loss modulation strategy (b).

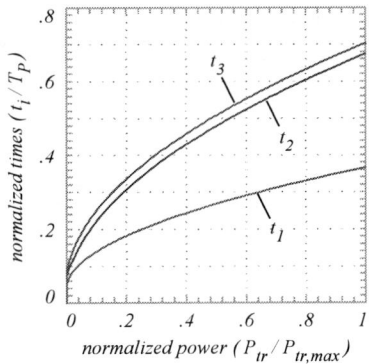

Fig. 6: Normalized output power $P/P_{tr,max}$ and times t_i/T_p for $V_2=V_{2,max}$ and a voltage transfer ratio of $V_2/V_1=2$.

B. Expected Power Density

A multi-phase converter concept is a means to minimize the overall converter volume. To optimize power density, the individual converter components such as inductor, filter or number of semiconductors are investigated in terms of phase count and switching frequency.

The worst case inductor peak current I_{max} (cf. I_1 in Fig. 2 (a) for boost operation) is found at a voltage transfer ratio of $V_2 / V_1 = 2$ and for the rated peak power P_{max}. With the assumption that the energy $\frac{1}{2} \cdot L \cdot I_0^2$ is much smaller than the energy $P_{max} \cdot T_p$ transferred within a switching period, the negative offset current $-I_0$ could be neglected and I_{max} is approximated to:

$$I_{max} \approx \cdot \sqrt{\frac{P_{max}}{f_s \cdot L}} \qquad (10)$$

The volume of the inductor in a first step is assumed proportional to the stored energy

$$E_L = \frac{1}{2} \cdot L I_{max}^2 = \frac{P_{max}}{f_s}, \qquad (11)$$

which is calculated from (10). That is why the volume of L is constant for any number of phases N at a certain switching frequency f_s and decreases with increasing f_s.

On both sides of the converter the filter structure shown in Fig. 8 is used. For the filter design, the phases are modeled as current sources that connect to a Π-filter with a damped common capacitor C_0 and the load R_L. A split filter design is necessary because of the requirement of a low-impedance connection of the phase output capacitors $C_{2,i}$ to the half-bridge. When N is increased, the ripple current calculated from the superposition of the N current sources decreases. In this case and in the case of a higher f_s, smaller reactive filter component sizes are needed and therefore a lower volume is achieved.

Furthermore, the volume occupied by the MOSFET switches including a liquid cooler is calculated with an iterative optimization algorithm. Starting with one MOSFET for each of the switches S_1 to S_4, the number of paralleled MOSFETs for the worst performing switch is incremented consecutively until the calculated efficiency is sufficient. It turns out and can also be seen from the modes shown in Fig. 3 that the RMS currents are higher for the switches S_1 and S_3. Thus, a ratio for the switch counts of approximately $S_1 / S_2 \approx S_3 / S_4 \approx 4 / 3$ is determined.

The results of the phase count optimization are shown in Figs. 9 and 10 for a typical peak power rating of P_{max}=70kW and V_{max}=450V. The volume is calculated for IXFB82N60P MOSFETs in a TO-264 package, film capacitors and the inductor L built from planar EILP ferrite cores.

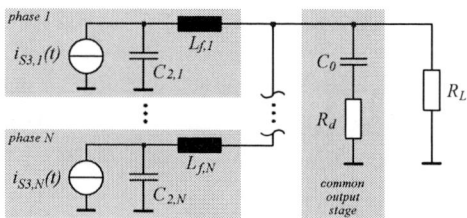

Fig. 8: Phase interconnection and input/output filter structure shown for converter side 2.

The volume share of the converter sections is shown in Fig. 9 for $f_s = 100$ kHz. The total converter volume decreases with higher f_s whereas a minimum occurs at a phase count $4 \leq N \leq 6$ depending on f_s (Fig. 10). A value of $N = 6$ is chosen for the reason of the greater flexibility for partial load operation. In this case, for a liquid-cooled converter with ideally packing of the converter sections and a volume ratio of $V_{cooler}/V_{semi} = 1.4$ for the semiconductors, including mounting and the cooler, the calculated power density for N = 6 is 27 kW/dm³ for $f_s = 100$ kHz and 30 kW/dm³ for $f_s = 150$ kHz respectively. Integration of converter and motor housing allows a saving in the cooler volume. By applying a thermal optimization strategy as presented in [12], it is possible to even reduce the inductor size in comparison to a standard design. It is estimated that with new inductor cooling techniques the power density could be increased to 40 kW/dm³ for a switching frequency of $f_s = 100$ kHz and 42 kW/dm³ for $f_s = 150$ kHz.

The calculated overall efficiency of a single converter module and the related loss components for the worst case operating point (boost operation 225V→450V) are listed in Tab. 1. The high efficiency more than compensates for the drawback of a higher filtering effort caused by the increased inductor RMS and peak current.

TABLE 1
Worst case loss distribution and efficiency

Total Losses	171 W	
Conduction	112 W	66 %
Switching	26 W	15 %
Winding	18 W	11 %
Core	9 W	5 %
Filter	6 W	3 %
Efficiency ≈ 96%		

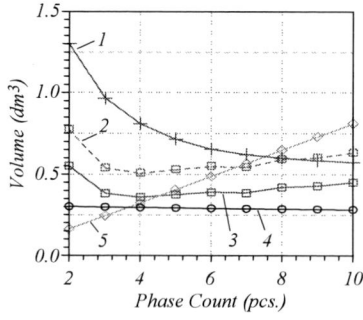

Fig. 9: Volume share of the converter sections for f_s = 100kHz, input and output filter including capacitors C_1 and C_2, (1), liquid cooler (2), semiconductors (3), inductor L (4), gate drive and control circuit (5)

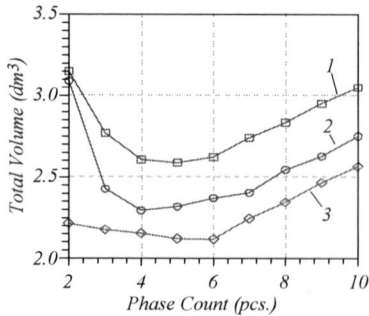

Fig. 10: Total converter volume (volume of liquid cooler not included) for switching frequencies f_s of 50 kHz (1), 100 kHz (2) and 150 kHz (3).

978-1-4244-1871-8/07 $25.00 © 2007 IEEE

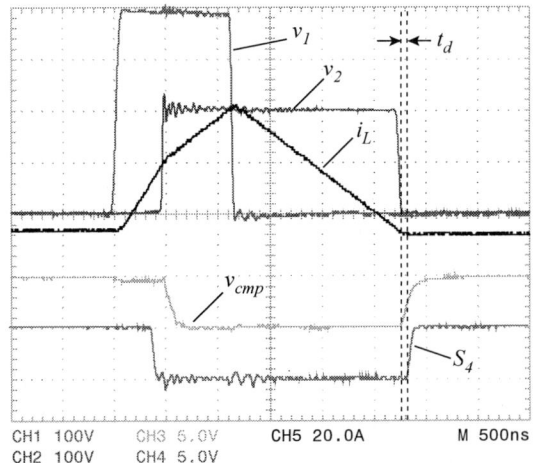

Fig 11: Ensuring ZVS by monitoring of S_4 drain-source voltage $v_2(t)$ with a comparator signal $v_{cmp}(t)$; switch S_4 control signal.

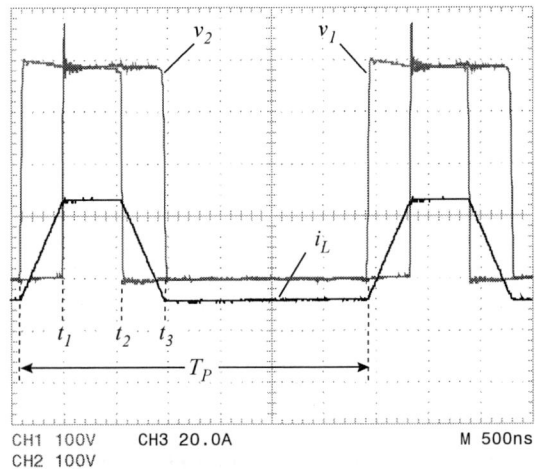

Fig. 12: Waveforms for equal input and output voltage $V_1 = V_2 = 200V$ supplying a resistive load at 1.5 kW.

V. EXPERIMENTAL VERIFICATION

To demonstrate the new modulation strategy, a prototype dimensioned for $P_{max} = 10$ kW and $V_{max} = 450$ V (that requires an inductance of $L = 6.5$ µH) has been built. The switching times t_1 to t_3 are pre-calculated offline for different operating points and given values of L and I_0. These are interpolated in three-dimensions for V_1, V_2 and P_{tr} by the controller software and provided to a CPLD-implemented state-machine, which generates the switching patterns.

Figure 11 shows an experimentally measured inductor current $i_L(t)$ and the half-bridge voltages $v_1(t)$ and $v_2(t)$ for the case of buck operation and a voltage transfer ratio $v = V_2 / V_1 = ½$. As described in section III, comparators are used to monitor the half-bridge voltages $v_1(t)$ and $v_2(t)$ to ensure ZVS conditions. For instance, turn-on of the switch S_4 is derived from the digital comparator output signal $v_{cmp}(t)$ and a certain dead-time t_d as depicted in Fig. 11.

The case of operation with identical input and output voltages $V_1 = V_2 = 200V$ is shown in Fig. 12 for a power of $P_{tr} = 1.5$ kW. Due to the fact of the control-optimized switching pattern, the times t_2 and t_3 are not placed towards the end of the pulse period.

VI. CONLUSION

In this paper a comparison between different converter topologies and soft-switching circuitry for a bi-directional buck+boost DC/DC converter, which could be used in an automotive application, are discussed. Major disadvantages of state-of-the-art converters are low efficiency, expensive technology or complex power circuitry.

To deal with those drawbacks, a novel constant-frequency, soft-switching modulation strategy for a cascaded buck+boost topology is presented, along with a detailed description of the operating principle.

The proposed new modulation strategy not only provides an excellent overall efficiency of at least 96% at nominal power but also features a higher efficiency for partial load operation. In addition, it results in a simple power circuitry

and a high power density, which is proved by analytical calculations considering a multi-phase converter design.

Furthermore, the novel modulation strategy is successfully implemented in a prototype and measurements verify the control concept.

REFERENCES

[1] R. M. Schupbach, J. C. Balda, "Comparing DC-DC Converters for Power Management in Hybrid Electric Vehicles", in Proc. IEMDC, vol. 3, pp. 1369-1374, June 2003.

[2] F. Caricchi, F. Crescimbini, A. Di Napoli, "20 kW Water-cooled Prototype of a Buck-Boost Bidirectional DC-DC Converter Topology for Electrical Vehicle Motor Drives", in Proc. APEC, vol. 2, pp. 887-892, March 1995.

[3] B. Eckardt, M. März: "A 100kW Automotive Powertrain DC/DC Converter with 25kW/dm3 by using SiC", in Proc. PCIM, May 2006.

[4] W. McMurray, "Resonant Snubbers with Auxiliary Switches", IEEE Trans. on Industry Applications, vol. 29, no. 2, pp. 355-362, March 1993.

[5] F. R. Dijkhuizen, J. L. Duarte, "Pulse commutation in nested-cell converters through auxiliary resonant pole concepts", in Proc. IAS, vol. 3, pp. 1731-1738, September 2001.

[6] L. Solero, D. Boroyevich, Y. P. Li, F. C. Lee, "Design of Resonant Circuit for Zero-Current-Transition Techniques in 100-kW PEBB Applications", IEEE Trans. on Industry Applications, vol. 39, no. 6, pp. 1783-1794, November 2003.

[7] Yong Li, F. C. Lee, "A Comparative Study of a Family of Zero-Current-Transition Schemes for Three-phase Inverter Applications", in Proc. APEC, vol. 2, pp. 1158-1164, March 2001.

[8] Y. Tsuruta, A. Kawamura, "QRAS and SAZZ Chopper for HEV Drive Application", in Proc. PCC, Nagoya, Japan, pp. 1260-1267, April 2007.

[9] D. M. Divan, L. Malesani, p. Tenti, V. Toigo, "A synchronized resonant DC link converter for soft-switched PWM", IEEE Trans. on Industry Applications, vol. 29, no. 2, pp. 940-948, September 1993.

[10] B. Ray, A. Romney-Diaz, "Constant frequency resonant topologies for bidirectional DC/DC power conversion", in Proc. PESC, Seattle, USA, pp. 1031-1037, June 1993.

[11] K. Chong-Eun, H. Sang-Kyoo, Y. Kang-Hyun, L. Woo-Jin, M. Gun-Woo, "A New High Efficiency ZVZCS Bi-directional DC/DC Converter for 42V Power System of HEVs", in Proc. PESC, Recife, Brazil, pp. 792-797, September 2005.

[12] J. W. Kolar, U. Drofenik, J. Biela, M. L. Heldwein, H. Ertl, T. Friedli, S. D. Round, "PWM Converter Power Density Barriers", in Proc. PCC, Nagoya, Japan, pp. P-9-P29, April 2007.

978-1-4244-1871-8/07 $25.00 © 2007 IEEE

Comparison of PWM Strategies for Three-Phase Current-fed DC/DC Converter

Hanju Cha   Soonho Choi

Chungnam National University
Department of Electrical Engineering
220, Gung-dong, Yuseong-gu, Daejeon 305-764, Korea

Abstract- **In this paper, three kinds of PWM strategies for a three-phase current-fed dc/dc converter are proposed and compared on the basis of reduced switching losses and improved efficiency. Each PWM strategy is described graphically and their switching losses are analyzed. Digital signal processor (DSP: TI320LF2407) and a field-programmable gate array (FPGA: EPM7128) board are used to generate PWM patterns for three-phase bridge and clamp MOSFETs. 500W prototype converter is built and its experimental results verify the validity of the proposed PWM strategies.**

I. Introduction

As an interest in clean energy sources increases significantly in recent years, more effort is being put into fuel-cells, photovoltaic, and wind generation. The rated voltage of a fuel cell is usually lower than 60V [1]. In order to generate 220V ac voltage, at least 400V dc is required for the inverter's input voltage. Therefore, dc/dc converter is essential for boosting the fuel-cells output voltage to 400V dc voltage [2-4]. At present, single-phase isolated boost dc/dc converter is widely used to interface low dc voltage with high dc voltage. To enlarge the power transfer capability, single phase dc/dc converter is extended to three-phase dc/dc converter [5-8]. Resulted advantages by the three phase configuration are : higher power density caused by three-phase power transfer; a smaller input current and output voltage ripple due to an increase of effective frequencies by a factor of three; lower RMS current through the inverter switches; reduction in size of the reactive (filter) components; better transformer copper and coil utilization. Therefore, this converter is suitable for an interface between a low dc voltage from fuel cells and a high dc voltage for a cascading inverter stage. Fig.1 shows the configuration of the converter and it consists of fuel cells, boost inductor, active clamp, three-phase dc/dc converter, three-phase transformer, three-phase diode rectifier [9].

II. Comparison of PWM Strategies

Since the system discussed here has a three-phase construction, possible strategies for the three-phase current-fed dc/dc converter are increased. Even though it is the same converter, the voltage transfer ratio and the efficiency could be different according to the PWM strategy applied to the converter. Therefore, the conventional PWM strategy for the three-phase current-fed dc/dc converter is analyzed first. Then, two additional PWM strategies are proposed and compared to find the new PWM strategy that improves the efficiency and increases the voltage transfer ratio.

A. PWM A

Fig.2 shows the conventional PWM strategy which exhibits gate signals for $S_1 \sim S_6$, S_c and its waveforms of input current I_d, bridge switch current I_{s1}, transformer line current I_A, and clamp capacitor current I_{Sc} [9].

Before t_0, all six switches $S_1 \sim S_6$ are turned on and the boost inductor L_{dc} charges energy from fuel cells V_d.

Mode 1 [$t_0 \sim t_1$]: At t_0, four switches S_2, S_3, S_4, and S_5 are turned off except S_1 and S_6. The bridge voltage V_{PN} reaches the clamp capacitor voltage V_c and the S_c's body diode conducts the boost inductor current I_d. The current through the leakage inductance L_{lk} increases as a slope determined

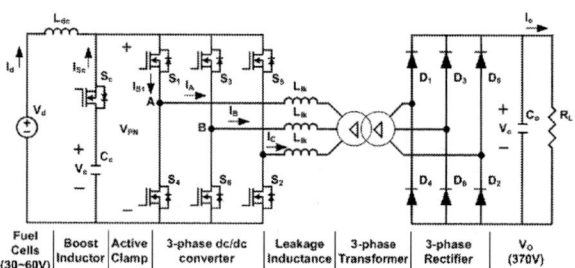

Fig. 1 Three-phase current-fed dc/dc converter with active clamp

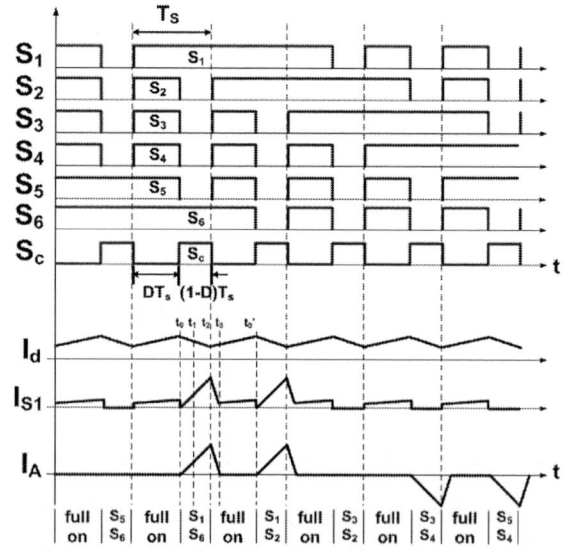

Fig. 2 Waveforms of input current I_d, bridge switch current I_{s1}, and transformer line current I_A with *PWM A*

978-1-4244-1871-8/07 $25.00 © 2007 IEEE   894

by a voltage difference between the clamp voltage V_c and the reflected output voltage V_o'. Phase A current I_A starts flowing to the output. To facilitate ZVS for S_c, the switch S_c is turned on before the clamp current I_{Sc} reverses at t_1.

Mode 2 $[t_1 \sim t_2]$: The clamp current I_{Sc} reverses and flows through MOSFET S_c. I_{Sc} provides the difference between the increasing I_A and constant boost inductor current I_{dc}.

Mode 3 $[t_2 \sim t_3]$: At t_2, the active clamp switch S_c is turned off and the energy stored in L_{lk} discharges output capacitances of S_3 and S_4, and then the body diodes of S_3 and S_4 begin to conduct. Therefore S_3 and S_4 can be turned on with zero voltage. I_A now decreases at a linear rate determined by the reflected output voltage and the value of L_{lk}. When I_A decreases to the value of the boost inductor current I_d, S_3 and S_4 begin to conducting a half of difference between the two currents.

Mode 4 $[t_3 \sim t_0']$: At t_3, I_A decreases to zero. All switches $S_1 \sim S_6$ are turned on and the boost inductor L_{dc} charges energy. At t_0', four switches S_3, S_4, S_5, and S_6 are turned off and the same operation repeats again.

When switches are turning off, voltage across switch falls on the current, resulting turn-off losses. Switching losses of the bridge can be obtained by following.

$$P_{Q,S} = \frac{1}{12} V_c I_d t_{sw} f_s \quad (1)$$

Where V_c is a voltage across switch, I_d is an input current, t_{sw} is a switch turn-off transition time, and f_s is a switching frequency. Since the three-phase bridge has six switches, the total switching losses of the bridge switches are,

$$P_{Q,Stotal} = 6 \times P_{Q,S} \quad (2)$$

While current flows through the bridge switches, there are conduction losses because of the on-resistance of the MOSFET switch. Therefore, the conduction losses are obtained by following.

$$P_{Q,C} = I_{Q,RMS}^2 \times R_{DS} \quad (3)$$

Where, $I_{Q,RMS}$ is a RMS current through a bridge switch, and R_{DS} is an on-resistance of the MOSFET switch. The total conduction losses of the bridge switches are,

$$P_{Q,Ctotal} = 6 \times P_{Q,C} \quad (4)$$

Since the current waveform of the clamp switch is different from those of the bridge switches, the conduction losses of the clamp switch are,

$$P_{Q,Cclamp} = I_{C,RMS}^2 \times R_{DS} \quad (5)$$

Where, $I_{C,RMS}$ is a RMS current through a clamp switch. The switching losses of the clamp branch switch are,

$$P_{Q,Sclamp} = \frac{1}{2} V_c I_d t_{sw} f_s \quad (6)$$

Therefore all the losses at the converter switches are,

$$P_{Loss} = P_{Q,Stotal} + P_{Q,Ctotal} + P_{Q,Cclamp} + P_{Q,Sclamp} \quad (7)$$

B. PWM B

When switches are turning off, there are turn-off losses. So it is predictable that the efficiency of the converter would be increased by reducing the number of turn-offs. Accordingly, new PWM strategy is proposed as shown in Fig.3. It is similar to *PWM A* but it has difference that one switching operation is removed to reduce switching losses during switch turn-off transition time. The switch losses could be obtained by the same procedure in the section above.

The main voltage and current waveforms are almost the same as the *PWM A* but the magnitude of the current through bridge switches during the full turn-on interval is different. With *PWM B*, the current through the bridge switches is 1/2 of the input current I_d, while *PWM A* flows a third of I_d. Consequently the total switching turn-off losses are the same as with the *PWM A*. Furthermore, the conduction losses increase because the RMS current of each bridge switch is larger than those with the *PWM A*.

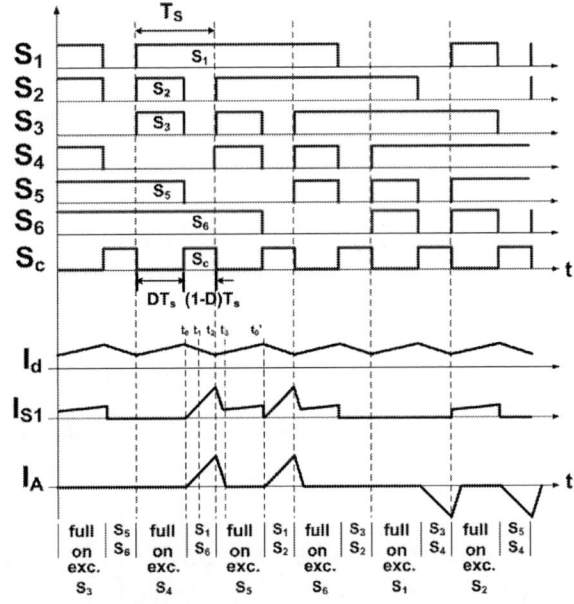

Fig. 3 Waveforms of input current I_d, bridge switch current I_{s1}, and transformer line current I_A with *PWM B*

C. PWM C

In *PWM C*, the basic concept to minimize the switch losses is the same as in *PWM B* but the way to approach is different from *PWM B*. Instead of eliminating one switching operation, just one turn-off operation is canceled as shown in Fig.4. *PWM A* and *PWM B* use two switches at a time while transferring energy to the secondary of the transformer. However, three switches are used at a time with the *PWM C*. As a result, the current waveform of each switch is different because only three bridge switches are turned off after full turn-on. Since the bridge current flows through two ways, each switch connected in parallel takes charge of half of the transformer input current. Therefore, the RMS current of the bridge switch decreases and conduction losses are reduced. In addition, the turn-off losses are reduced to 3/4 of those with *PWM A* by eliminating one turn-off operation at each bridge switch under the same input current.

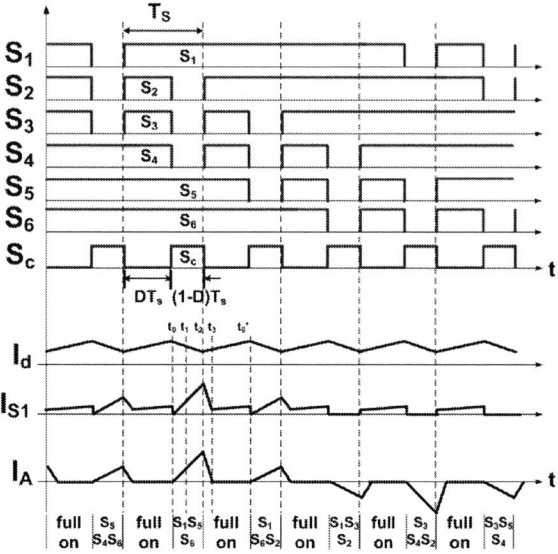

Fig. 4 Waveforms of input current I_d, bridge switch current I_{s1}, and transformer line current I_A with *PWM C*

III. CALCULATION OF SWITCHING LOSSES

Three-phase current-fed dc/dc converter is simulated with the parameters in Table I. Fig.5 shows the input current I_d, bridge switch current I_{s1}, and transformer line current I_A with each PWM strategy. The input current I_d increases during DT_S and decreases during $(1-D)T_S$. While input current I_d is decreasing, the energy stored in input inductor L_{dc} is being transferred to the output. The bridge switch current I_{s1} is 1/3 of input current I_d with the *PWM A* and *PWM C*, while 1/2 with the *PWM B*, which is important because the turn-off losses are determined by the current.

TABLE I
Parameters used in simulation

Input voltage V_d	30V
Input inductance L_{dc}	330µH
Leakage inductance L_{lk}	7µH
Clamp capacitance C_c	120µF
Output capacitance C_o	470µF
Turn ratio $n(=N_2/N_1)$	4.15
Duty D	0.75
Switching frequency f_s	25kHz
Magnetizing inductance L_m	6mH
Load R_L	270Ω

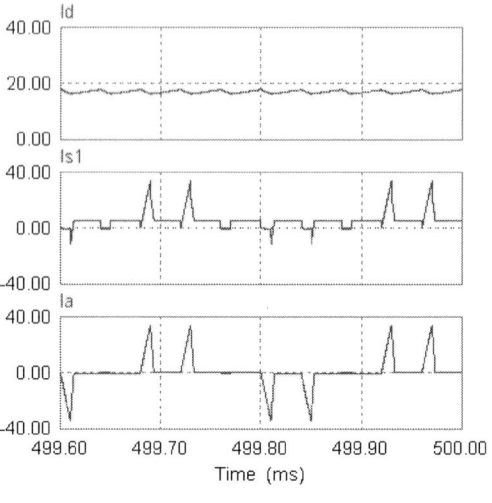

(a)

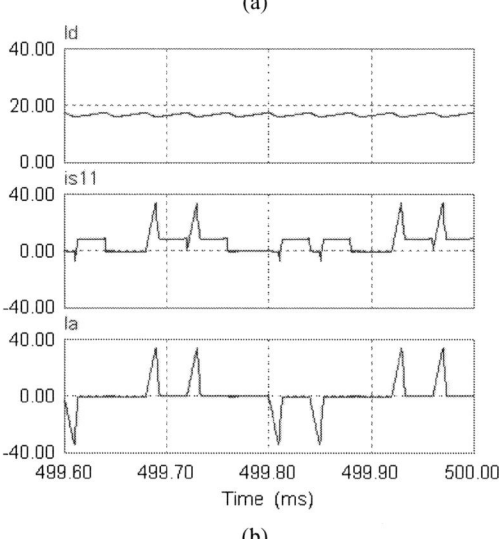

(b)

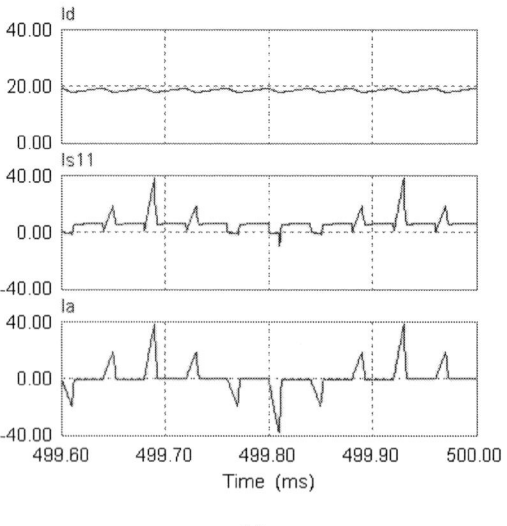

(c)

Fig. 5 Simulation results - input current I_d, bridge switch current I_{s1}, transformer line current I_A: (a) *PWM A*, (b) *PWM B*, (c) *PWM C*

Table II shows the switching losses obtained at 500W three-phase current-fed dc/dc converter. It appeared that the conduction losses are about 80% of the total switch losses in each case. So it is important to reduce the conduction losses of the switch. With *PWM B*, the switching losses are almost the same as with *PWM A* but the conduction losses are increased compared with the *PWM A*. Consequently, the total losses are increased. However, with the *PWM C*, both conduction and switching losses are decreased. Therefore, it is expected that the efficiency of the converter would be improved just by applying the *PWM C* strategy to the conventional three-phase current-fed dc/dc converter.

TABLE II
Comparison of switch losses at each PWM strategy

	Conduction losses(W)		Switching losses(W)		Total losses (W)
	bridge	clamp	bridge	clamp	
PWM A	8.02	0.51	1.28	1.33	11.1
PWM B	9.36	0.51	1.27	1.33	12.4
PWM C	6.53	0.56	1.15	1.20	9.4

IV. IMPLEMENTATION AND EXPERIMENTAL RESULTS

Fig.6 shows the schematic diagram of PWM switching realization for the three-phase current-fed dc/dc converter. First, the digital signal processor (DSP: TI320LF2407) generates the one full-on signal and six gate PWM signals. Next, the field-programmable gate array (FPGA: EPM7128) modifies the signals from DSP and creates dead-time to facilitate the zero-voltage switching (ZVS) of the bridge switches and clamp switch of the converter. In addition, the gate driver board is added to protect the DSP and FPGA from the surge of the converter.

500W prototype three-phase current-fed dc/dc converter is built and tested as shown in Fig.7. It consists of a digital signal processor, a field-programmable gate array board, gate driver board, three-phase bridge and clamp MOSFET, delta-delta wound three-phase transformer, three-phase

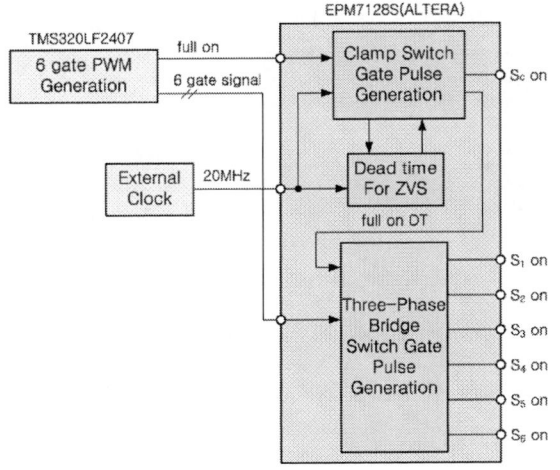

Fig. 6 Generation of gate signal for the bridge and active clamp MOSFET switches

rectifier and load. The following waveforms are measured under the parameters in Table I.

Fig. 7 500W prototype three-phase current-fed dc/dc converter

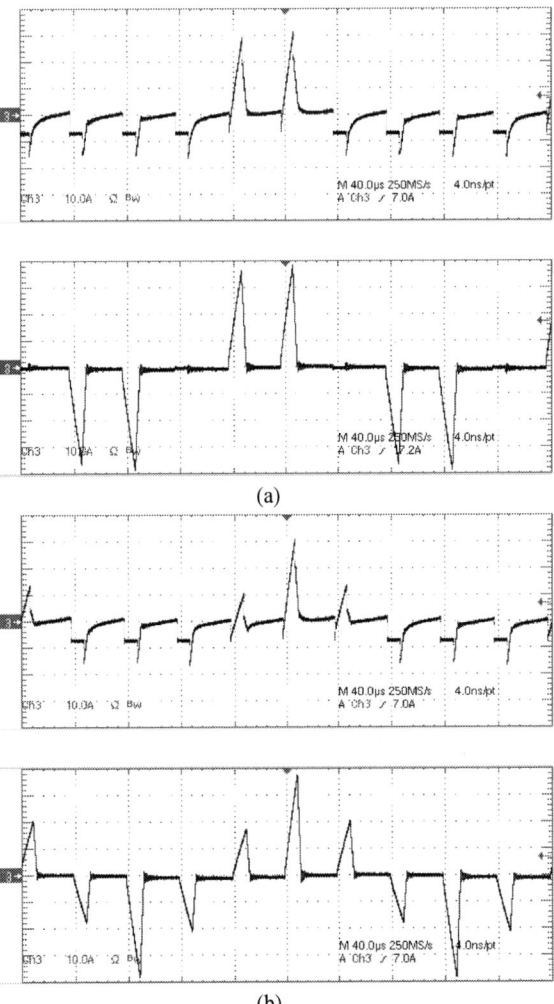

Fig. 8. Bridge switch current I_{s1} and transformer primary line current I_A (10A/div, 40usec/div): (a) *PWM A*; (b) *PWM C*

Fig.8 shows the bridge switch current I_{s1} and transformer primary current I_A with the *PWM A* and *PWM C* strategies respectively. The experimental results are in a good agreement with the simulation results shown in Fig.5. Two of the bridge switches are turned on during $(1-D)T_s$ with the *PWM A*. However, with the *PWM C*, three of the bridge switches are turned on and the input current I_d divided into half flows through two bridge switches in parallel. Therefore, the RMS current of each bridge switch is lowered, which results in the reduction of conduction losses.

V. CONCLUSION

In this paper, three PWM strategies have been compared and analyzed to improve the total efficiency. *PWM A* generates biggest switching losses because of more number of switching during one switching period. To minimize switching losses, *PWM B* and *C* strategies are proposed and tested. *PWM B* has the advantage of a reduced switching number for one period but because of the increased switch current, total switching losses are almost same. *PWM C* removes just one turn-off of each switch instead of eliminating one switching operation. It makes both conduction losses and switching losses decreased and results in improving the efficiency of the converter. Simulation and experimental results are addressed to verify the proposed PWM strategy.

REFERENCES

[1] M.W. Ellis, M.R. Von Spakovsky, D.J. Nelson, "Fuel cell systems: efficient, flexible energy conversion for the 21st century", Proceedings of the IEEE Volume 89, Issue 12, Dec. 2001 Page(s):1808 - 1818

[2] V. Yakushev, V. Meleshin, S. Fraidlin, "Full-bridge isolated current fed converter with active clamp", Applied Power Electronics Conference and Exposition, Vol. 1, pp. 560-566, 1999

[3] W.C.P. De Aragao Filho, I. Barbi, "A comparison between two current-fed push-pull DC-DC converters-analysis, design and experimentation", 18th International Telecommunications Energy Conference, 1996, 6-10 Oct. 1996 Page(s):313 – 320

[4] Kunrong Wang, Lizhi Zhu, Dayu Qu, H.Odendaal, J. Lai, F.C. Lee, "Design, implementation, and experimental results of bi-directional full-bridge DC/DC converter with unified soft-switching scheme and soft-starting capability", IEEE 31st Annual Power Electronics Specialists Conference, 2000. Volume 2, 18-23 June 2000 Page(s):1058 - 1063 vol.2

[5] Jr.de Souza Oliveira, I. Barbi, "A three-phase ZVS PWM DC/DC converter with asymmetrical duty cycle for high power applications", IEEE Transactions on Power Electronics, Volume 20, Issue 2, Mar 2005 Page(s):370 – 377

[6] J.Jacobs, A. Averberg, R. De Doncker, "A novel three-phase dc/dc converter for high-power application", IEEE 35th Annual Power Electronics Specialists Conference, Volume 3, 20-25 June 2004 Page(s):1861 - 1867 Vol.3

[7] A.R. Prasad, P.D. Ziogas, S. Manias, "Analysis and design of a three-phase offline DC-DC converter with high-frequency isolation", IEEE Transactions on Industry Applications, Volume 28, Issue 4, July-Aug. 1992 Page(s):824 – 832

[8] Changrong Liu, A. Johnson, Jih-Sheng Lai, "A novel three-phase high-power soft-switched DC/DC converter for low-voltage fuel cell applications", IEEE Transactions on Industry Applications, Volume 41, Issue 6, Nov.-Dec. 2005 Page(s):1691 – 1697

[9] Hanju Cha, Prasad Enjeti, "A novel three-phase high power current-fed DC/DC converter with active clamp for fuel cells", IEEE 38th Annual Power Electronics Specialists Conference, 2007. Volume 3, 18-22 June 2007 Page(s):2485-2489

978-1-4244-1871-8/07 $25.00 © 2007 IEEE

Effect of AC Inductance on a Phase Shifted DC-DC Bridge Converter

Peter Freere
Monash University
Dept. Electrical Engineering
Vic. 3800, Australia
email: peter.freere@eng.monash.edu.au

Wang Kong
Monash University
Dept. Electrical Engineering
Vic. 3800, Australia
email: wang.kong@eng.monash.edu.au

Donald Grahame Holmes
Monash University
Dept. Electrical Engineering
Vic. 3800, Australia
email:
grahame.holmes@eng.monash.edu.au

Abstract—Phase shifted DC-DC converters are a well established topology for producing an isolated DC output supply. However, the conventional topology with a DC output LC filter has some practical difficulties when issues such as diode reverse recovery effects, diode voltage ratings for wide ranging input voltages, and common mode EMI caused by the switched transformer output are considered. In many applications, these difficulties can be mitigated by moving the output filter inductor to the AC side before the bridge rectifier. However, this leads to alternative constraints of increased filter series voltage drop and higher filter capacitor current ripple.

This paper presents design equations and a design process to determine which of these alternative filter arrangements better suits a particular DC-DC converter application.

I. INTRODUCTION

A well established topology to create an isolated DC-DC supply is a full bridge converter feeding through a high frequency isolating transformer into a diode rectifier and output LC filter. The converter is typically controlled by modulating the two phase legs with square waves that are phase shifted by a delay angle α, to produce variable duty cycle three level output pulses that are filtered and averaged by the output LC filter to create the final DC output voltage. However, the conventional topology as shown in Fig. 1 has a number of practical difficulties, as follows:

Firstly, the reverse recovery characteristics of the rectifying diodes can cause a significant voltage ring across the reverse blocking diodes at turn off, because of resonance between the diode junction capacitance and the transformer/filter leakage inductance [1]. This usually requires snubbing to resolve, leading to efficiency loss and heat dissipation issues.

Secondly, the peak diode blocking voltage for this topology is the transformer switched output voltage, which for lower duty cycles will be significantly higher than the average output voltage. This has particular relevance for applications with a wide ranging input supply voltage, which require a matching wide range in duty cycle to maintain a constant output voltage. Consequently, higher voltage rating diodes may be required, with a corresponding increase in reverse recovery stored charge that leads in turn to a higher magnitude of turn off voltage ring.

Lastly, since the negative rail of the DC output toggles between the positive and negative peaks of the transformer switched output voltage, the output has a significant common mode voltage referenced to ground, which can lead to significant EMI problems. This effect is often exacerbated when there is significant coupling capacitance between the primary and secondary windings of the high frequency transformer, which is typically the case for higher power converter systems.

The issues of the high diode voltage rating [2] and effects of diode reverse recovery characteristics [3] can be significantly mitigated by moving the dc inductance to the ac side of the diode bridge, as shown in Fig 2(a). However, this introduces other issues, such as increased inductor voltage drop, which requires a higher transformer secondary voltage, and increased current ripple into the output filter capacitor, which must be considered in any particular application.

The common mode voltage problem can be addressed by separating the filter inductance into two sections, located on

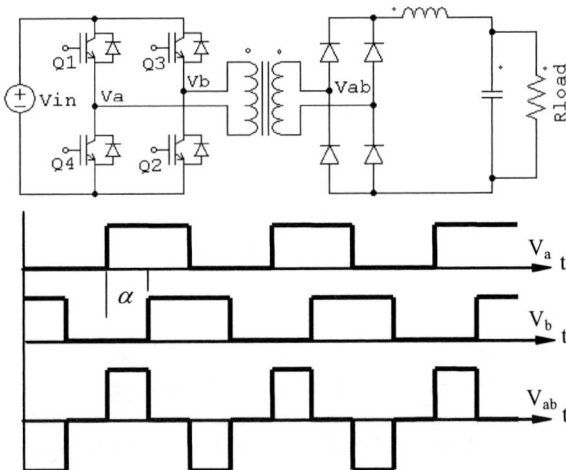

Fig. 1: Conventional Full Bridge DC-DC Converter controlled by Phase Shifted Square Wave Modulation on each Phase Leg.

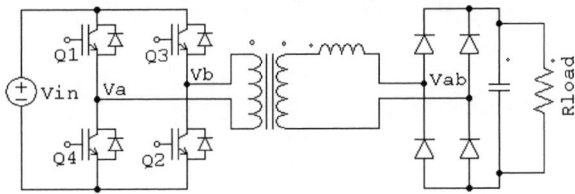

Fig.2(a): Full Bridge DC-DC Converter with filter inductor on the AC side of the rectifier

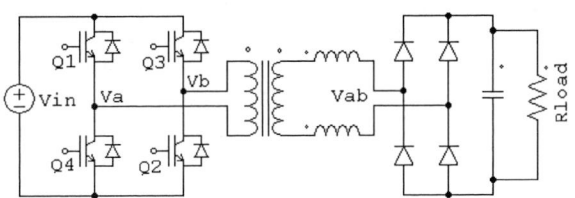

Fig.2(b): Full Bridge DC-DC Converter with 2 filter inductors on the AC side of the rectifier

978-1-4244-1871-8/07 $25.00 © 2007 IEEE

the forward and return paths respectively as shown in Fig. 2(b). This better separates the switched common mode voltage from the output filter capacitor, and makes it much easier to shield from the outside world [4][5].

Surprisingly, there appears to have been little discussion in the literature relating to possible benefits of locating the filter inductor on the AC side, except in the context of soft switching topologies. However, there are obviously tradeoffs to be considered between these configurations for particular applications. The objective of this paper is to investigate these tradeoffs and benefits.

II.   INITIAL EXPERIMENTAL INVESTIGATIONS

Initial investigations were conducted with a simplified system, consisting of an existing H-bridge converter connected to a diode rectifier and output filter through approximately 200mm of loose wire (to represent unavoidable transformer leakage inductance). Fig. 3(a) shows the experimental switching performance of this system, where a substantial reverse recovery induced overvoltage ring can be seen across the diode rectifier. Conventional design processes would require significant snubber circuits to adequately damp this overvoltage.

Fig. 3(b) shows the effect of moving the main filter inductor to the AC side, where it can been seen that the overvoltage ring across the diode has been significantly reduced even without any snubber circuits. However, not unexpectedly, the DC output voltage has also reduced from 7.89V to 5.67V, because of the additional voltage drop across the AC filter inductance.

Fig. 3(c) shows the matching performance of the PSIM model that was developed to allow the tradeoffs between the two filter alternatives to be further investigated. Considerable effort was taken to precisely model all important second order parasitic effects, so that various parameters could be varied in simulation with confidence that their influence would reflect back into experimental reality without difficulty. The close match between the results of Fig. 3(a) and Fig. 3(c) confirms the detail accuracy of the simulation representation of the experimental system

III. VOLTAGE DROOP CAUSED BY DC AND AC INDUCTANCE

The next stage of the investigation was to explore the relationship between the input and output voltages of the simplified system with either a DC or an AC side inductance, as the duty cycle of the switched waveform was varied from 0 to 1 (i.e. the delay angle α varies from 0 to π). Each relationship was developed theoretically, and then confirmed experimentally.

A.   DC Inductance

For steady state operation, the integral of the filter inductor voltage across is zero over each periodic cycle, ie.

$$0 = \int_0^T V_{inductor} \, dt \qquad (1)$$

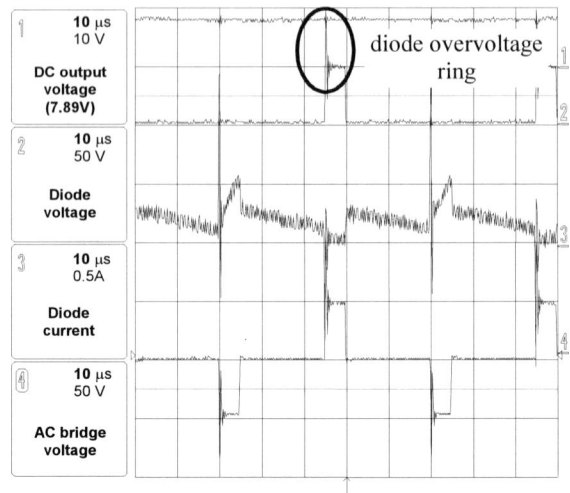

Fig. 3(a): Initial experimental results with 270μH DC filter inductance. 23.7Ω load, 50° phase shift delay, continuous conduction.

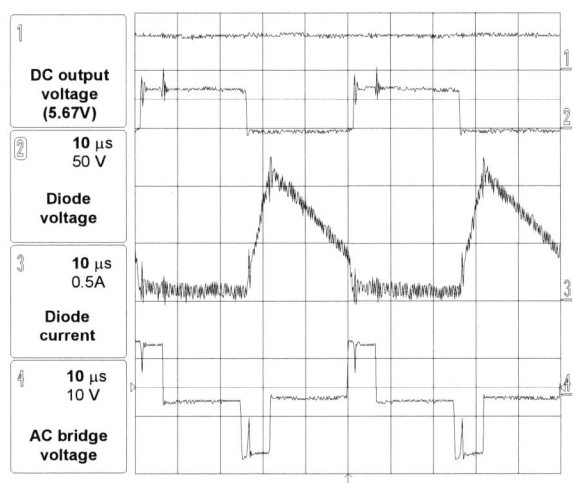

Fig. 3(b): Initial experimental results with 270μH AC filter inductance. 23.7Ω load, 50° phase shift delay, continuous conduction

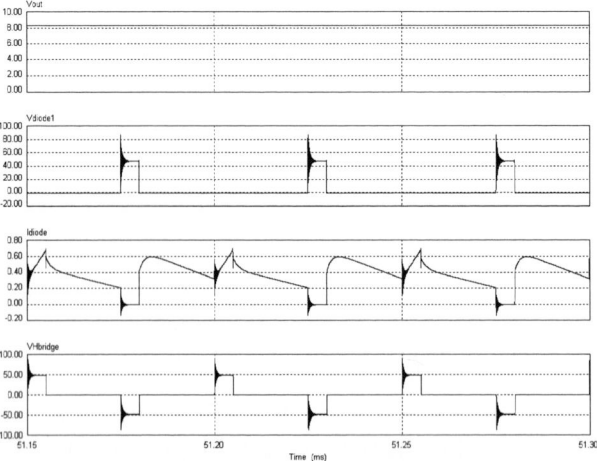

Fig. 3(c): Matching simulation results compared to Fig. 3(a).
DC Filter inductance and resistance: 270μH, 1Ω
Diode junction capacitance and resistance: 80pF, 1Ω
Stray earth capacitance and resistance: 200pF, 10Ω
Inverter to inductor/rectifier connection wire: 1.4μH, 1Ω

978-1-4244-1871-8/07 $25.00 © 2007 IEEE   900

During the active output pulse period, the inductor voltage is the difference between the dc supply voltage (V_{dc}) and the load voltage (V_{load}), $V_{dc} - V_{load}$ [6], while during the freewheeling period the inductor voltage is $-V_{load}$. Hence,

$$\int_0^\alpha (V_{dc} - V_{load})dt - \int_\alpha^\pi V_{load}dt = 0 \rightarrow V_{load} = \frac{\alpha}{\pi}V_{dc} \quad (2)$$

This simple (and not unexpected) relationship is illustrated in Fig. 4(a).

B. AC Inductance

When the filter inductor is on the AC side of the rectifier, it now presents a series impedance to the circuit whose magnitude varies with frequency. However, its effect on the output voltage of the converter is adequately predicted by considering only the fundamental component of the inverter switched output voltage, as follows:

The square wave output of each phase leg can be readily expressed as a Fourier series, i.e.

$$V_a = \frac{V_{dc}}{2} + \frac{2}{\pi}V_{dc}\sum_{n=1,3,5..}^\infty \frac{1}{n}\sin(n\omega t) \quad (3)$$

$$V_b = \frac{V_{dc}}{2} + \frac{2}{\pi}V_{dc}\sum_{n=1,3,5..}^\infty \frac{1}{n}\sin(n(\omega t - \alpha)) \quad (4)$$

where f_s = switching frequency, $\omega = 2\pi f_s$ and V_{dc} is the inverter DC supply voltage.

The inverter output voltage is $V_a - V_b$, which becomes

$$V_a - V_b = \frac{2}{\pi}V_{dc}\left[\sum_{n=1,3,5..}^\infty \frac{1}{n}[\sin(n\omega t) - \sin(n(\omega t - \alpha))]\right]$$

$$= \frac{2}{\pi}V_{dc}\left[\sum_{n=1,3,5..}^\infty \frac{1}{n}\left[2\sin\left(\frac{\alpha}{2}\right)\cos\left(\frac{2\omega t - \alpha}{2}\right)\right]\right] \quad (5)$$

Hence the peak amplitude of the fundamental component is

$$V_{ab(1)} = \frac{4}{\pi}V_{dc}\sin\frac{\alpha}{2} \quad (6)$$

Since the filter capacitor remains on the DC side of the output rectifier, it has no effect on the AC side of the circuit, and the load resistance can be reflected directly back through the bridge rectifier as R_L. Hence the fundamental peak current through the AC part of the circuit becomes

$$I_{load(peak)} = \frac{V_{ab(1)}}{\sqrt{R_L^2 + (\omega L)^2}} \quad (7)$$

The average voltage across the load resistor is then given by

$$V_{load(av)} = \frac{2}{\pi}I_{load(peak)}R_L \quad (8)$$

which gives

$$V_o = V_{dc}\frac{8}{\pi^2}R_L\frac{\sin\frac{\alpha}{2}}{\sqrt{R_L^2 + (\omega L)^2}} \quad (9)$$

Fig. 4(b) illustrates this relationship, and shows excellent confirmation between theory and experiment. Furthermore,

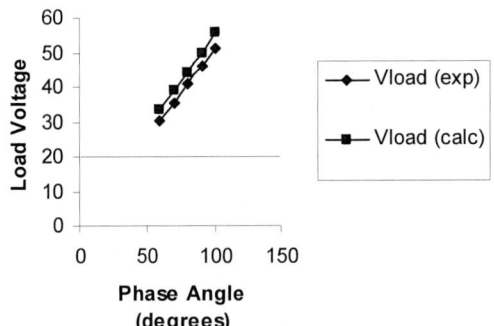

Fig. 4(a) Calculated and experimental relationship between supply load voltage as delay angle varies – **DC filter inductance**, $V_{dc} = 100V, L=540\mu H, C_{dc}=2200\mu F, R_L=23.7\Omega, f=20kHz$.

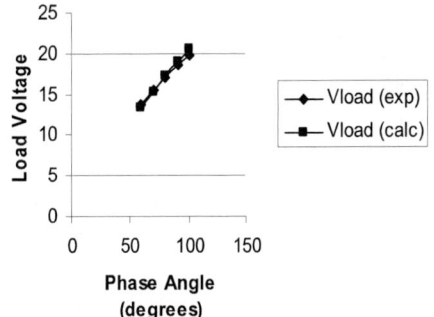

Fig. 4(b) Calculated and experimental relationship between supply load voltage as delay angle varies – **AC filter inductance**, $V_{dc} = 100V, L=540\mu H, C_{dc}=2200\mu F, R_L=23.7\Omega, f=20kHz$

this relationship identifies that (as expected) the output voltage decreases for the same phase delay angle as the ac inductance is increased. Hence for any particular converter design, the use of an AC filter inductor requires either an increased phase delay angle (which may limit the output voltage range), or an increased transformer turns ratio, to achieve the same output voltage as an equivalent system with a DC filter inductor.

IV. EFFECT OF VARYING AC AND DC INDUCTANCE

A further issue that was explored was the effect of combinations of AC and DC inductance. An investigation was conducted for a 5 kW system operating from a variable DC input voltage of 75 – 150 V, switching through a high frequency transformer at 20 kHz to produce a 400V DC output voltage. The parameter that was considered was the required diode reverse blocking voltage, as AC inductance was substituted for DC inductance. Fig. 5 shows the results of this study, where the sum of the AC and DC inductance has been kept constant at 500 μH as the AC inductance varies from nil to 500 μH (note that the transformer turns ratio has been adjusted for each set of vertical points, so that the system achieves an output of 400V for that particular value of AC inductance over the full range of input voltage variation). The results show that invariably, the combination of an AC and a DC inductance result in a higher diode reverse voltage than either an AC or a DC inductance alone. Hence a combination of the two inductances (such as is

978-1-4244-1871-8/07 $25.00 © 2007 IEEE

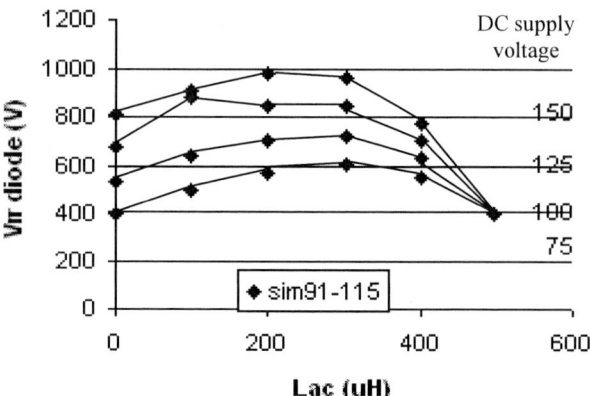

Fig. 5. Rectifier diode reverse voltage as AC inductance varies (sum of AC and DC inductance maintained at 500µH, transformer turns ratio adjusted for each value of AC inductance)

conventionally used for a soft switching converter) is sub-optimal, and should be always avoided.

However, as the AC inductance value increases, the transformer turns ratio must increase (more secondary turns) to overcome the increasing AC voltage drop across the filter inductor. Hence the transformer primary current and inverter phase leg current must increase proportionally. Fig. 6 shows this effect, where it can be seen how the phase leg current increases with increasing AC inductance. Hence it is desirable to minimise the value of AC inductance as much as possible in a practical design.

Of course, the benefit of using an AC inductor is that the current transitions are slower during phase leg switching, and hence switching resonances across the diode rectifiers are essentially absorbed by the load without requiring additional snubber components. Consequently lower voltage diodes rated closer to the load voltage can be used, with potential cost benefits and reduced reverse recovery issues. For example, the lack of overvoltage ringing achieved with an AC filter inductance, such as was observed in Fig 3(b), could well allow 600V diodes to be used in place of 1200V devices for the 400V output system.

V. SYSTEM DESIGN ISSUES

From the studies that have been undertaken it was found that the following issues need to be considered for any particular system to use an AC filter inductance.

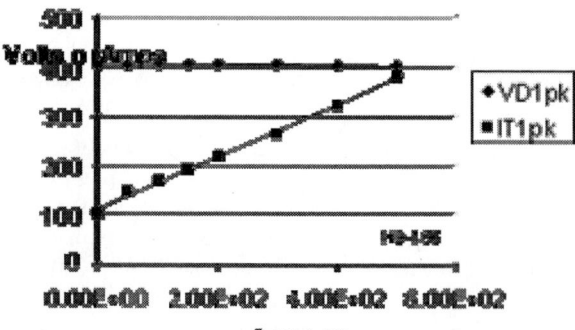

Fig. 6. Peak inverter phase leg current and diode reverse voltage as a function of AC inductance.

- whether to use an AC or a DC inductance
- transformer turns ratio
- inductor size
- filter capacitance size

A. AC or DC Inductance

The consideration whether to use AC or DC inductors is primarily determined by the magnitude of the secondary voltage that is consequently required from the transformer if an AC inductor is used. If it is too high, insulation issues become more difficult. The secondary consideration is how much the primary side current increases, in terms of phase leg device current rating and losses.

B. Transformer Turns Ratio

Once the transformer maximum secondary voltage has been decided, the turns ratio is readily determined from the maximum primary voltage available to the system.

C. Inductor Size

Irrespective of the filter inductor configuration, maximum peak current for any DC-DC converter occurs when the input voltage suddenly rises from minimum to maximum, before the control circuit can adjust the phase delay angle to compensate for this change. Under these conditions, the output voltage is essentially constant (for a short while) and the current peak is only limited by the filter inductance. Two conditions need to be considered.

1) Discontinuous Conduction

In discontinuous conduction, the current starts from zero in each switching period. If the worst case is assumed ($\alpha = 180°$ before the input voltage change), the maximum peak inductor current is

$$I_{peak(max)} = \frac{V_{dc(max)} \times N_{trans} - V_{out}}{L_{min}} \bullet \frac{T_s}{2} \quad (10)$$

giving a total minimum inductance of

$$L_{min} = \frac{V_{dc(max)} \times N_{trans} - V_{out}}{I_{peak(max)}} \bullet \frac{T_s}{2} \quad (11)$$

2) Continuous Conduction

For a DC filter inductance, the step change in current will added to the current already flowing at the start of the switching cycle,

$$I_{peak(max)} = I_{start} + \frac{V_{dc(max)} \times N_{trans} - V_{out}}{L_{min}} \bullet \frac{T_s}{2} \quad (12)$$

where: I_{start} = current at start of the switching period.

The upper bound for I_{start} is I_{av} = average load current.

Hence $L_{min} = \dfrac{V_{dc(max)} \times N_{trans} - V_{out}}{\left(I_{peak(max)} - I_{av}\right)} \bullet \dfrac{T_s}{2} \quad (13)$

Note that the minimum inductance required is larger for the continuous conduction case when using DC inductors.

For the AC inductance case, the inductance is selected so that with the minimum input voltage, the voltage drop across the inductance is small enough to maintain the required output voltage. Under these conditions, the phase

978-1-4244-1871-8/07 $25.00 © 2007 IEEE

delay angle α is 180°, and (9) can be used to calculate the load voltage for the selected AC inductance value. If spare voltage headroom is available, consideration can be given to reducing the transformer turns ratio and repeating the inductance calculation

D. Capacitor Size

Once the filter inductance is determined, the DC capacitor size is selected to limit the DC output ripple voltage.

For the case of a DC inductor, standard step down converter relationships hold [6], and the filter capacitor is determined by

$$C = \frac{\Delta I}{8 f_s \Delta V_c} \qquad (14)$$

where ΔV_c is the allowable ripple voltage.

For the case of an AC inductor, from (9) the phase delay angle for any particular operating condition is given by

$$\alpha = 2\sin^{-1}\left[\frac{V_o}{N_{trans}V_{dc}}\frac{\pi^2}{8}\frac{\sqrt{R_L^2 + (\omega L)^2}}{R_L}\right] \qquad (15)$$

and hence the on-period is given by $t_{on} = \dfrac{\alpha}{180}T_s$ (16)

If the variation in inductor current during the on period is assumed to be entirely absorbed by the filter capacitor, the change in capacitor voltage is given by:

$$\Delta V_c = \frac{1}{C}\int i_C \, dt = \frac{1}{C}\int_0^{t_{on}} \frac{(V_{dc} \times N_{trans}) - V_o}{L} t \, dt \qquad (17)$$

and hence

$$C = \frac{1}{\Delta V_c}\left[\frac{[(V_{dc} \times N_{trans}) - V_o]t_{on}^2}{2L}\right] \qquad (18)$$

VI. WORKED EXAMPLE – 200W SYSTEM

The concepts described in Section V will now be applied to the experimental system of Section VII to show how they can be used. This system has operating parameters of an input voltage range of 75V-150VDC, a switching frequency of 20kHz, an output voltage of 400V (for subsequent inversion to 230V, 50Hz) and a power rating of 2000W.

For the maximum input voltage of 150V, the maximum transformer secondary voltage is arbitrarily set to 1000V. This gives a turns ratio of $N_s/N_p = 6.67$. For the minimum input voltage of 75V, this gives a transformer secondary voltage of 500V, which is above the required output voltage of 400V and hence suggests this turns ratio is adequate.

For the maximum inverter input current of 30A, the peak inductor current will be 4.5A. From (11) and (13), this gives filter inductances of $L_{min\,dc}=3.7$mH and $L_{min\,ac} = 3.3$mH. However, for an AC inductance of 3.7mH, the load voltage using (9) will only be 373V, which means that this AC inductance is too large to achieve the required 400V.

The output voltage ripple is arbitrarily set to 1% (4V p-p), under continuous conduction. For the DC filter inductance, ΔI is calculated using standard step down converter

relationships [6] and (14), to give $C_{dc}=1.1\mu F$. For the AC filter inductance (9) gives a phase delay angle of $\alpha=59.13°$, $t_{on} = 8.21\mu s$ and $C_{dc}=1.4\mu F$ using 18). (Note however that this assumes a 400V output, which has already been shown not to be possible with this size AC inductance.)

Fig. 7 shows PSIM results for this converter with these firs pass design filter values, where $V_{ripple} = 2.9V$ with a 400V DC output voltage. It can be seen also that for the AC inductance with a 75V input, the output voltage is 366V, below the required 400V as expected. Hence either a smaller AC inductance or an increased transformer turns ratio will be required, with a corresponding increased peak inverter and transformer primary current. It should be noted also that while the secondary voltage of the transformer resonates above the 1kV nominal design target to nearly 1.5kV, the diode voltages remain well behaved.

VII. DETAILED SYSTEM SIMULATION

Fig. 8 shows the final system that was developed to investigate the tradeoffs between a DC and an AC filter inductance. This simulation included a complete transformer model with interwinding capacitance, which was required to achieve sufficient accuracy to match the experimental system. In addition, the experimental inverter had Hall effect LEM current sensors in each phase leg output, which were found to have a small but significant inductance of 2.1μH. This parameter also needed to be included into the simulation for it to generate adequate results.

Figs. 9 and 10 show matching simulation and experimental results for this system at the minimum input voltage of 75V. The close match obtained between these two sets of results confirms the validity of the design and simulation process, and supports the use of this procedure for investigations of alternative filter arrangements.

VIII. CONCLUSION

Conventional phase shifted DC-DC converters have a simple LC filter located on the output DC side of the secondary side rectifier bridge. For some applications, this topology has some practical difficulties in relation to diode reverse voltage rating, reverse recovery transients and output EMI caused by common mode switched voltages.

For some applications, these issues can be mitigated by

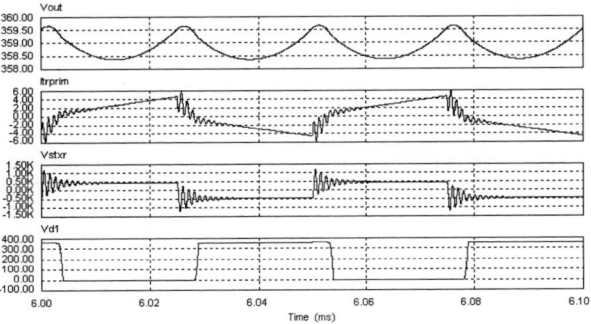

Fig. 7 Simulation with the analytical parameter values calculated above. Traces top to bottom: V_{load}, transformer primary current, transformer secondary voltage, diode voltage.

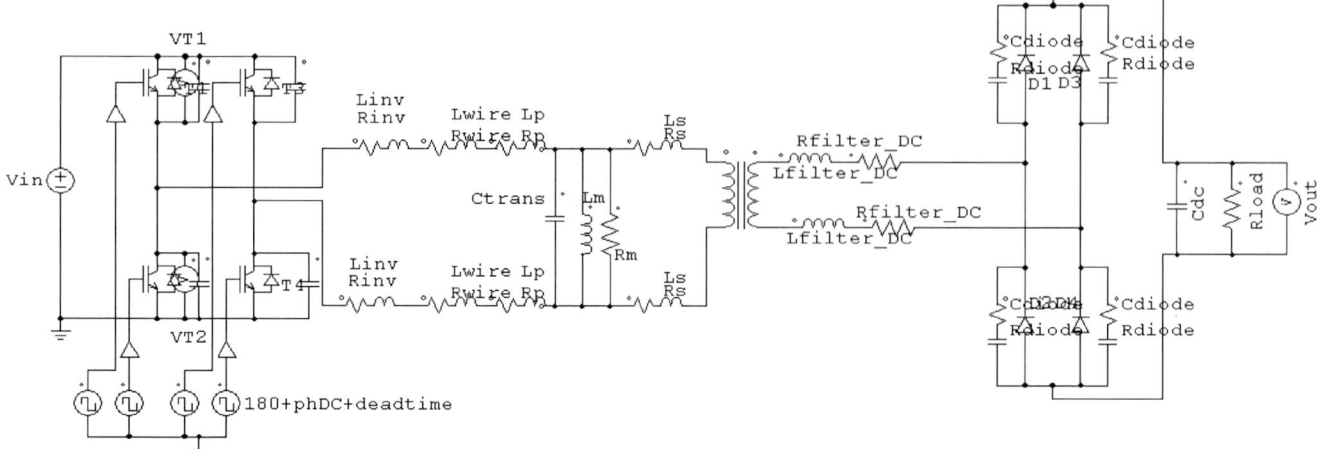

Fig. 8. Detailed circuit of complete simulated inverter system.

Deadtime = 2μs, fsw=20kHz, Vdiode (IGBT)=1.0V, Vsat=2.0V (IGBT), Cigbt=73pF, Linv=2.1μH, Rinv=0.05Ω, Rwire=0.1Ω, Lwire=0.07μH, Np=24, Ns=149, Rp=0.025 Ω, Rs = 3.06 Ω, Lm=5.04mH, Ctrans=4nF, Rm=500Ω, Rdiode=10 Ω, Cdiode=80pF, Lfilter_DC=3.67mH, Cdc=1.7μF, Rload = 974 Ω.

moving the filter inductor to the ac side of the rectifier, so that the voltage rating of the rectifier diodes can be reduced to the filtered output voltage, and overvoltage transients caused by reverse recovery effects can be significantly reduced. However, moving the inductor in this way has the disadvantages of needing to increase the output transformer turns ratio to overcome the voltage drop across the AC filter inductor, and an increased magnitude of voltage ring across the transformer output and filter inductor. Furthermore, for the same system DC output current, the phase shifted inverter current will be increased, since a significant reactive current component now flows through the system.

This paper has presented some of the design issues that must be considered for a converter system with an AC filter inductor, and has illustrated the level of simulation detail that is required to achieve a reasonable match with experimental results. Equations have also been presented that give some guidance as to how to go about the design process for such a filter system.

IX REFERENCES

[1] Karadzinov, L, Hamill, D., "DH phenomenon in DC/DC converters", IEEE Conference Proceedings, Power Electronics Specialists Conference, 2001, pp 296-301.

[2] Salmon, J., "Circuit topologies for pwm boost rectifiers operated from 1 phase and 3 phase ac supplies and using either single or split dc rail voltage outputs", IEEE APEC Conference Procs., 1995, vol 1, pp 473-479.

[3] Lee, Woo Jin, Kim, Chong Eun et al, "A new high efficiency phase shifted full bridge converter for sustaining power module of plasma display panel", IEEE Power Electronics Specialist Conference, 2005, pp 2630-2634.

[4] Rockhill, A., Lipo, T., Julian, A., "High voltage buck converter topology for common mode voltage reduction", IEEE APEC Conference, vol 2, 1998, pp940-943.

[5] [Cochrane, D., Chen, D., Boroyevic, D., "Passive cancellation of common mode noise in power electronic circuits", IEEE transactions on Power Electronics, vol./ 18, No. 3, May 2003, pp756-663.

[6] Rashid, "Power Electronics", Pearson Education, ISBN 81-7808-730-8, 2003.

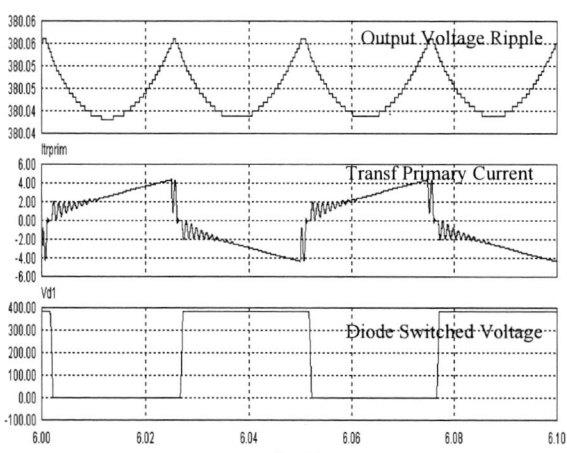

Fig. 9. Detailed Simulation Results, Vout = 380.5V at Vin = 75V: phase delay angle = 180°, 2μs deadtime, 974Ω load. Rdiode=10Ω Lac=0.773mH Transformer turns ratio=24:149.

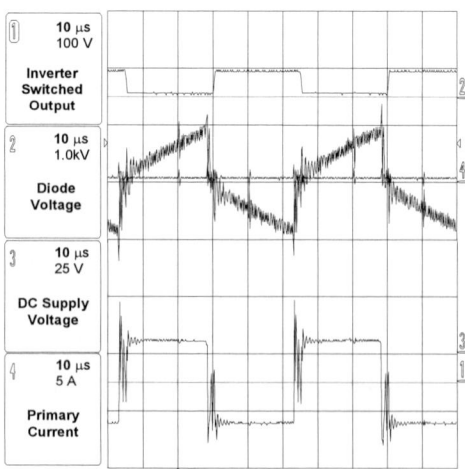

Fig. 10. Experimental Results, C_{dc}=4700μF, V_{in}= 75V, Vout = 389V

978-1-4244-1871-8/07 $25.00 © 2007 IEEE   904

Improved Low Speed Performance of a PI-DTC Interior PM Machine with Compensations of Dead-Time Effects and Forward Voltage Drops

Saad Sayeef
University of New South Wales
School of Electrical Engineering and
Telecommunications
Sydney, NSW 2052, Australia
Email: s.sayeef@student.unsw.edu.au

M. F. Rahman
University of New South Wales
School of Electrical Engineering and Telecommunications
Sydney, NSW 2052, Australia
Email: f.rahman@unsw.edu.au

Abstract— The performance of direct torque control (DTC) interior permanent magnet (IPM) machines at low speed is poor due to a few reasons, namely limited accuracy of stator voltage acquisition and the presence of offset and drift components in the acquired signals. Due to factors such as forward voltage drop across devices in the three phase inverter and dead-time of switching devices, the voltage across the machine terminals differ from the reference voltage vector used to estimate stator flux and electromagnetic torque. This can lead to instability of the IPM drive during low speed operation. Compensation schemes for forward voltage drops and dead-time are implemented, resulting in better performance of the PI DTC IPM drive.

I. INTRODUCTION

Direct torque control (DTC) of Interior Permanent Magnet Synchronous Machine (IPMSM) was proposed and developed in the late 1990's [1, 2]. Some of the advantages of DTC over the conventional Field Oriented Control (FOC) include fast torque and stator flux dynamics, less machine parameter dependence, the elimination of the dq-axis current controllers, associated coordinate transformation, the rotor position sensor requirement and separate voltage pulse width modulator.

However, a major drawback of DTC is its poor performance at low speed which researchers have not been able to solve. The poor performance at low speed is due to a few reasons, namely limited accuracy of stator voltage acquisition and the presence of offset and drift components in the acquired signals.

In this paper, the effect of forward voltage drop and dead-time of power devices on a space vector modulation (SVM) integrated direct torque controlled IPMSM drive system is investigated. It is found that the voltage drop across power switches and power diodes is not negligible for flux and torque estimation at low speeds. The effect of forward voltage drop and dead-time becomes more significant for flux estimation when the machine operates at low speeds as it becomes comparable to the low applied voltage on the machine terminals. Holtz and Quan have examined and compensated the effect of forward voltage drop on a vector controlled induction motor drive [3]. In this paper, a compensation method using additional voltage vectors is used and experiments were carried out to test the compensation scheme. Experimental results show that a better low-speed flux control has been achieved.

II. DIRECT TORQUE CONTROL OF AN IPMSM

The basic principle of DTC is to directly select stator voltage vectors according to the differences between the reference and actual torque and stator flux linkage. The current controller followed by a pulse width modulation (PWM) comparator used in vector control is not required in DTC systems, and the parameters of the motor are also not used, except the stator resistance. Therefore, DTC has advantages of lesser parameter dependence and fast torque response when compared with the torque control via PWM current control [4].

Fig. 1 shows the system diagram of a DTC IPM drive integrated with space vector modulation (SVM), which would be referred to as PI-DTC drive in this paper. It includes flux and torque estimators, a Proportional + Integral (PI) torque controller, reference flux vector calculator and a space vector modulation block. A DC bus voltage sensor and two output current sensors are needed for the flux and torque estimation. Speed sensor is not necessary for the torque and flux control.

The PI torque controller outputs the reference angular frequency of the stator flux linkage from the difference between the actual and reference torque commands at its input. The reference flux vector is then calculated using the reference angular frequency, position of the estimated stator flux linkage and amplitude of the reference flux linkage, as shown in Fig. 2 [4]. The voltage space vectors and their duration are then selected and calculated according to the error flux linkage vector to reduce the error to zero.

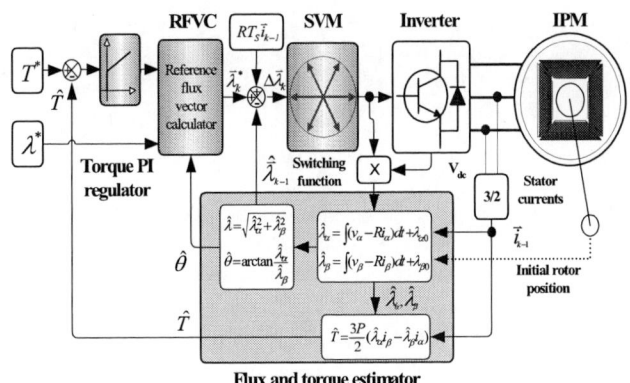

Fig. 1 System diagram of the PI DTC drive

978-1-4244-1871-8/07 $25.00 © 2007 IEEE   905

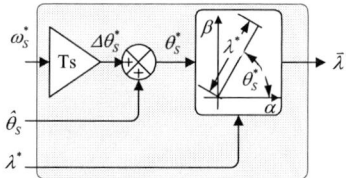

Reference flux linkage vector calculator

Fig. 2 Signal flowchart of the RFVC block

In order to estimate the stator flux and torque, an integrator is used in (1), of which the scalar form is (2). The voltage vector v can be obtained by multiplying the DC bus voltage with the selected voltage space vector and the current vector can be obtained by measuring any 2 of the 3 phase currents. The torque is estimated by (3).

$$\lambda = \int (v - Ri)dt + \lambda_0 \qquad (1)$$

$$\lambda_\alpha = \int (v_\alpha - Ri_\alpha)dt + \lambda_{\alpha 0}$$
$$\lambda_\beta = \int (v_\beta - Ri_\beta)dt + \lambda_{\beta 0} \qquad (2)$$

$$T = \frac{3}{2} p(\lambda \times i) = \frac{3}{2} p(\lambda_\alpha i_\beta - \lambda_\beta i_\alpha) \qquad (3)$$

It is seen from (1), (2) and (3) that the key for a good performing DTC is flux estimation. If the estimated flux drifts from the actual value of the flux linkage, the estimated torque will be wrong. In order to estimate the flux linkage accurately, accurate values for voltage vector, stator resistance, and initial rotor flux linkage and current vector must be known.

In the space vector modulated (SVM) DTC drive, the reference stator voltage vector V_{ref} is used for the estimation of stator flux, instead of the actual stator voltage vector seen at the machine terminal, to eliminate errors and offsets caused by sensing hardware and electric isolation. Sensing voltage vectors from the machine terminals has the problem

of high harmonics which would require large filters and hence would introduce unacceptable delay. The V_{ref} signal is readily available in the Proportional + Integral (PI) direct torque controller and is free from harmonic components. However, it does not exactly represent the voltage vector at the machine terminals due to the distortions introduced by the dead-time and forward voltage drops inside the inverter [3]. The compensation must be made to the reference voltage vector which is used by the observers or estimators as an input for stator flux estimation.

At low speeds, the back EMFs are too small to be used to obtain accurate stator flux estimation. The stator flux estimation using pure integrator or a programmable low pass filter is still needed for start-up and very low speed operation close to zero. DC offset compensation is an important factor in a practical implementation of the integration since drift can cause large errors of the amplitude and position of the stator flux linkage space vector, which may make the drive system unstable in the very low speed region.

III. NON-LINEAR INVERTER MODEL

A three-phase two level IGBT inverter similar to that shown in Fig. 3 would normally be used for a three phase IPM motor. Voltage drops across the power switches in the inverter reduce the terminal voltage seen by the machine. At very low speeds, the voltage drop can be higher than the induced voltage and hence introduce severe disturbance.

According to the forward characteristic of the power devices illustrated in Fig. 4, the forward voltage drop across the IGBT and diode (v_{CE}) can be modelled by an average threshold voltage v_{CE0} and a current-dependent $r_{CE}i$, where r_{CE} is the differential resistance and i is the forward current of the device. This characteristic is expressed in (4) and represented by the dotted line in Fig. 4.

$$v_{CE} = v_{CE0} + r_{CE} \bullet i \qquad (4)$$

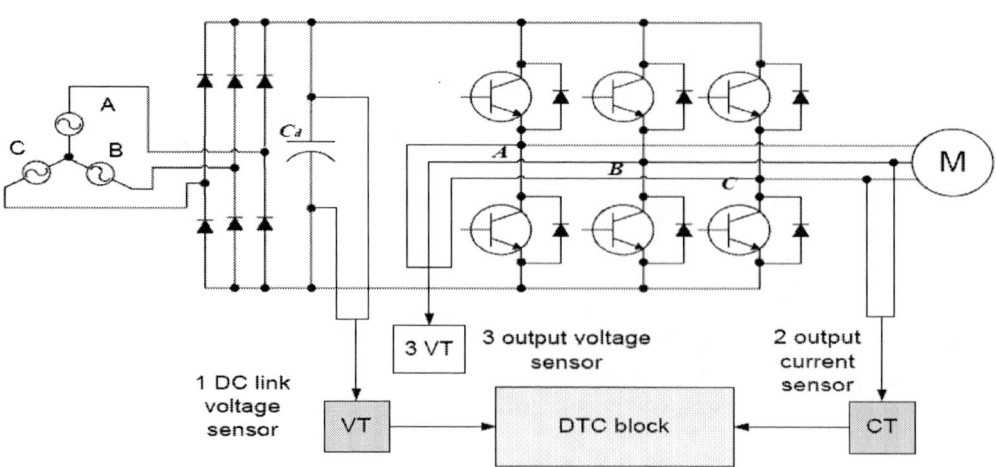

Fig. 3 Circuit diagram of the three phase two level inverter with a DTC Controller

978-1-4244-1871-8/07 $25.00 © 2007 IEEE

It is well known that forward characteristics of the IGBT and its freewheeling diode are non-linear, as illustrated in Fig. 4. The focus in this paper is to compensate the error caused by v_{CE0}.

In order to examine the error in the voltage vector caused by neglecting the forward voltage drop, the switching modes of the inverter with different voltage vector and polarity is investigated. The operation mode of one inverter leg is analysed in Fig. 5. The letter s in Fig. 5 indicates the switching function of the inverter leg. Output voltage of the inverter leg with positive current (into the load) and negative current (from the load) are compared under the conditions of $s=1$ and $s=0$. It is seen that when the current is positive, the actual voltage is shifted down by v_{CE} and when the current is negative, the waveform of the actual output voltage is shifted up by v_{CE}. As a result, the error caused by the forward voltage drop of the switches of the inverter leg A can be expressed as in (5). The error voltage vector of the three inverter legs can be represented by (6).

$$\Delta v_A = s * v_{DC} - v_A = v_{CE0} * sgn(i_A) \tag{5}$$

$$V_{fvd} = \frac{2}{3} v_{CE}[sgn(i_A) + \alpha * sgn(i_B) + \alpha^2 * sgn(i_C)] \tag{6}$$

where $\alpha = e^{j2\pi/3}$

From (6), it is seen that the voltage error vector only depends on the polarity of the three-phase currents and the switching status does not affect the error vector. Hence, for the purpose of compensating the error vector, only the polarities of the three phase currents are needed.

There is a time in each switching cycle where both the high and low side switches in the same leg of the inverter are off and the current flow is through diodes. This is known as the dead-time which is inserted into the gate signal of the switch that is to be turned on to avoid direct short circuit across the DC bus voltage source. This dead-time delay causes a change in and distortion of the stator voltage vector applied to the machine [5].

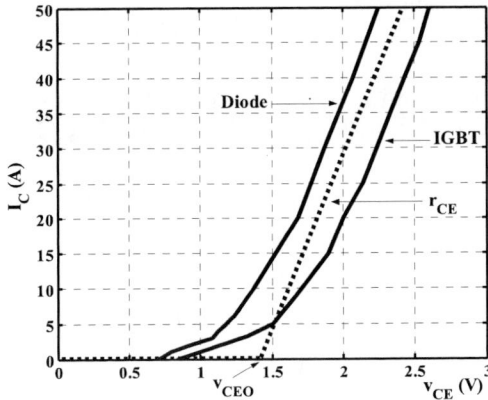

Fig. 4 Forward characteristics of the IGBT and diode in the three phase inverter

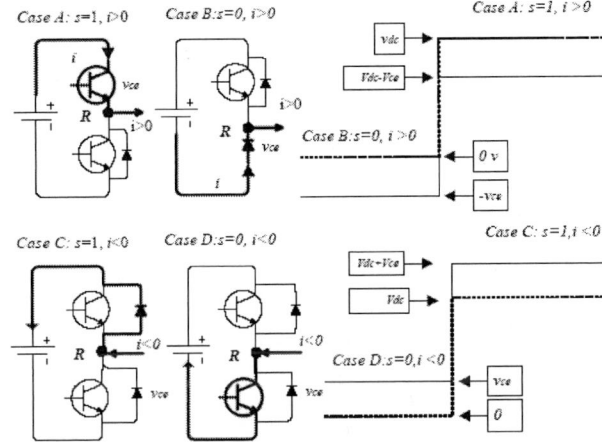

Fig. 5 Analysis of forward voltage drop of an inverter leg

Fig. 6 shows the PWM control signal as an example and the switching signals of a single inverter leg. The control signals for the upper and lower switches are shown together with the dead-time. The turn-on and turn-off time of the power switches are also considered in the figure. In the last two subplots of Fig. 6, the expected and the actual output voltages are compared. It is seen that due to the effect of dead-time t_d, the turn-on delay t_{ON} and the turn-off delay t_{OFF}, there is distortion between the expected and the actual output voltages.

The dead-time error vector between the expected output and actual voltage vector is given by (7) for different directions of the current vector. This error can be compensated by detecting the polarity of the output current.

$$V_{dt} = \frac{2}{3} v_{dead}[sgn(i_A) + \alpha * sgn(i_B) + \alpha^2 * sgn(i_C)] \tag{7}$$

$$V_{dead} = \left(\frac{t_{dead} + t_{on} - t_{off}}{T_s} \right)(V_{dc} + V_{diode} - V_{sat}) \tag{8}$$

where $t_{dead}, t_{on}, t_{off}, V_{sat}, V_{dc}$ and T_s are dead time, turn-on time, turn-off time, average on-state voltage, dc-link voltage and switching period respectively.

The compensation in (9) must be made to the reference voltage vector which is used by the estimators as the input $V_{\alpha,\beta}$ for more accurate observation. In order for the machine to run at very low speed region smoothly, the fundamental reference voltage must be compensated for the inverter device forward voltage drops and the dead-time loss using the estimated value of ΔV.

$$V_{\alpha,\beta} = V_{ref} - (V_{fvd} + V_{dt}) = V_{ref} - \Delta V \tag{9}$$

978-1-4244-1871-8/07 $25.00 © 2007 IEEE

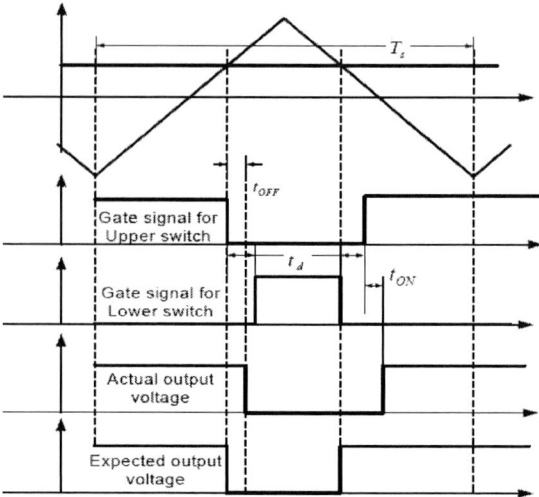

Fig. 6 PWM switching signals of the inverter with positive current considering the effect of dead-time and switching delays in the power switches

From (6) and (7), it is noticed that the line to neutral voltage reduction in the stationary reference frame is a six-step waveform which is in phase with the corresponding phase current and has step values of $\frac{4}{3}\Delta V$, $\frac{2}{3}\Delta V$, $-(\frac{2}{3}\Delta V)$ and $-(\frac{4}{3}\Delta V)$. Fig. 7 illustrates the compensation voltage vectors and six sectors. Table 1 shows the compensation voltages as a function of phase current directions. For example, if the current vector is in region II as shown in Fig. 7, the error vector will be

$$\Delta \vec{v} = -\frac{4}{3} v_{CE0} \alpha^2 \tag{10}$$

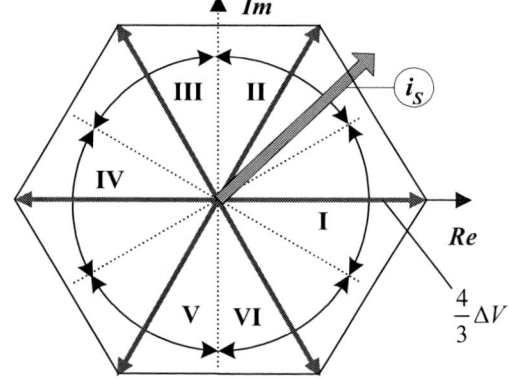

Fig. 7 Compensation voltage vectors and six sectors.

TABLE 1
Deadtime and forward voltage drop compensation voltages in sectors.

$sgn(i_A)$	$sgn(i_B)$	$sgn(i_C)$	Sector	$V_{\alpha,com}$	$V_{\beta,com}$
+	-	-	I	$\frac{4}{3}\Delta V$	0
+	+	-	II	$\frac{1}{2}\left(\frac{4}{3}\Delta V\right)$	$\frac{\sqrt{3}}{2}\left(\frac{4}{3}\Delta V\right)$
-	+	-	III	$-\frac{1}{2}\left(\frac{4}{3}\Delta V\right)$	$\frac{\sqrt{3}}{2}\left(\frac{4}{3}\Delta V\right)$
-	+	+	IV	$-\left(\frac{4}{3}\Delta V\right)$	0
-	-	+	V	$-\frac{1}{2}\left(\frac{4}{3}\Delta V\right)$	$-\frac{\sqrt{3}}{2}\left(\frac{4}{3}\Delta V\right)$
+	-	+	VI	$\frac{1}{2}\left(\frac{4}{3}\Delta V\right)$	$-\frac{\sqrt{3}}{2}\left(\frac{4}{3}\Delta V\right)$

IV. EXPERIMENTAL RESULTS

The effectiveness of the compensation schemes for the forward voltage drop (FVD) and dead-time was tested experimentally. The compensation schemes are software oriented and no additional hardware is required. A DS1104 DSP card is used to carry out the real-time algorithm. A three-phase insulated gate bipolar transistor (IGBT) intelligent power-module is used for an inverter. Coding of real time control software was done using C language. The experimental setup is shown in Fig. 8. Space vector PWM signals were generated on the DS1104 board. An incremental encoder is used to detect the rotor speed and a DC machine is used to load the PM machine. Table 2 shows the parameters of the IPM motor used.

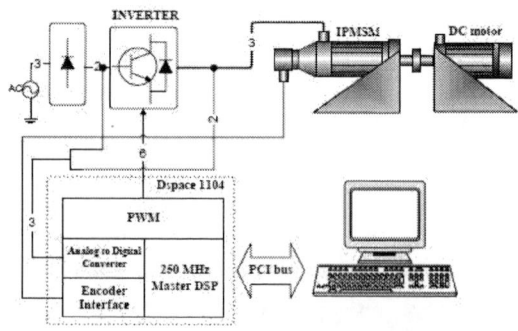

Fig. 8 Experimental setup used for IPM drive control

The performances of the PI-DTC IPM drive at 500rpm with and without the proposed compensation are compared in Figs. 9 and 10 respectively. It is seen that the estimated values of torque and flux are close to the actual values without compensation.

The performance at 186rpm is shown in Figs. 11-14. It is seen in Fig. 11 that there is a significant deviation of the estimated torque and stator flux from the actual ones in the absence of FVD and dead-time compensation. This deviation is eliminated with the implementation of the compensation schemes, as shown in Fig. 12.

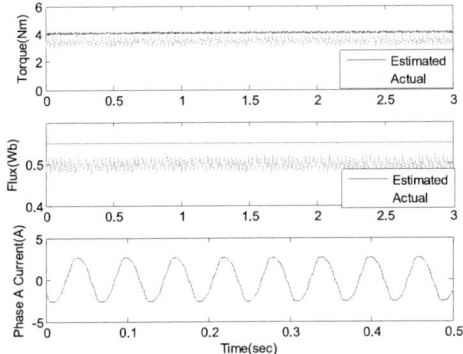

Fig. 9 Flux linkage, torque and current at 500rpm without FVD and dead-time compensation

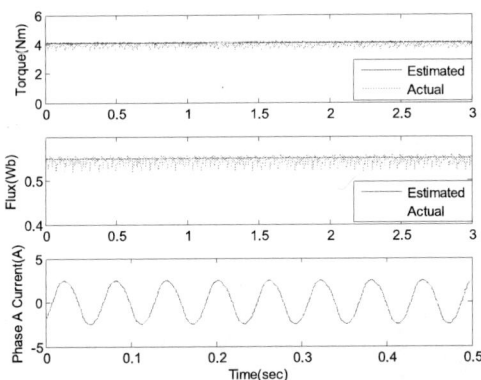

Fig. 10 Flux linkage, torque and current at 500rpm with FVD and dead-time compensation

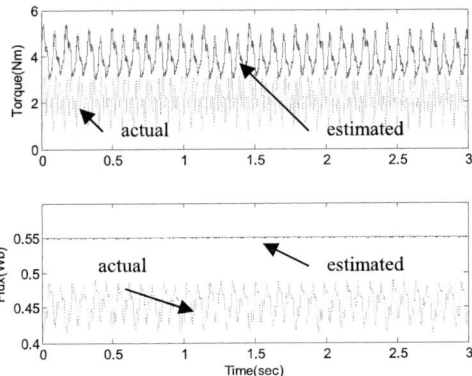

Fig. 11 Responses of torque and stator flux without FVD and dead-time compensation at 186rpm with 4Nm load

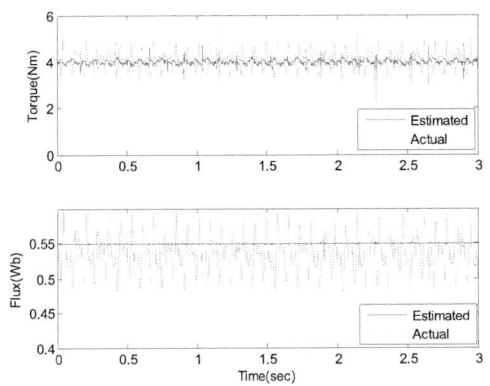

Fig. 12 Responses of torque and stator flux with FVD and dead-time compensation at 186rpm with 4Nm load

Figs. 13 and 14 show the corresponding phase current, reference voltage (V_α) and speed responses. The lowest speed at which the PI-DTC IPM drive can run smoothly without FVD and dead-time compensation is 186rpm (19.5rad/s). The compensation voltage for accurate flux and torque estimation at 186rpm load is also shown in Fig. 14.

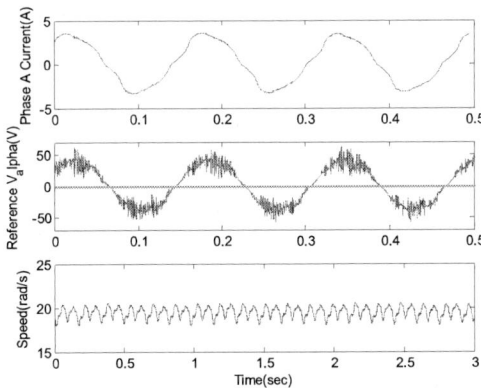

Fig. 13 Responses of phase current, reference voltage and rotor speed without FVD and dead-time compensation

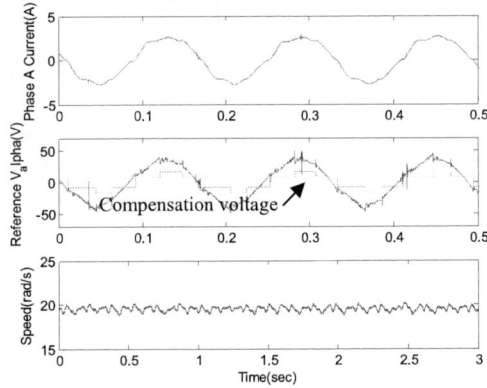

Fig. 14 Responses of phase current, reference voltage and rotor speed with FVD and dead-time compensation

It is seen from Figs. 13 and 14 that the phase current is a smoother sinusoidal waveform when the effects of forward voltage drop and dead-time are compensated for.

The torque, flux and speed response of the compensated PI-DTC IPM drive at 100rpm with 4Nm load is illustrated in Fig. 15. The lowest speed at which the loaded PI-DTC IPM drive can run with the integration of the compensation schemes is found to be 52.5 rpm (5.5 rad/s). Fig. 16 shows the torque, flux and speed responses of the PI-DTC IPM drive during operation at 52.5rpm with 2Nm load. Therefore, a better low speed performance of the PI-DTC IPM drive is achieved with the implementation of compensations for effects of forward voltage drops and dead-time.

V. CONCLUSION

In this paper, the effects of the forward voltage drop of power switches and dead-time on the performance of space vector modulation integrated direct torque controlled IPMSM drive system are analysed and investigated. It is seen that the forward voltage drop and dead-time of power devices cause errors in the estimation of flux and torque, especially during very low speed operation.

The proposed compensation scheme reduces the mentioned errors significantly without the need for any extra hardware. Experimental results show that a better low speed performance of the PI-DTC IPM drive system is achieved with the integration of compensation for the effects of forward voltage drop and dead-time.

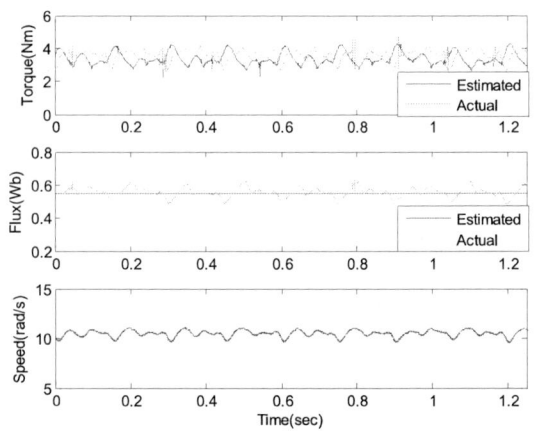

Fig. 15 Responses of torque, flux and speed with FVD and dead-time compensation at 100pm

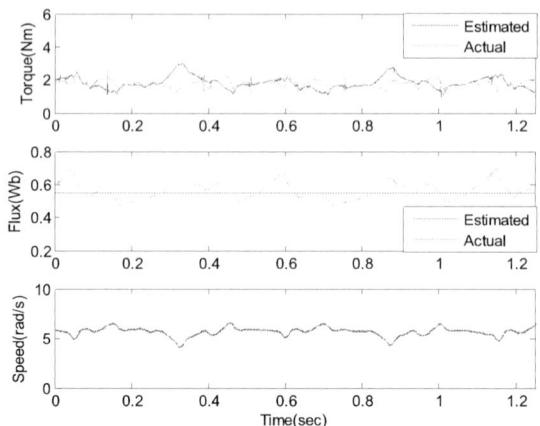

Fig. 16 Responses of torque, flux and speed with FVD and dead-time compensation at 52.5rpm

TABLE 2
Parameters of the IPM motor used

Number of pole pairs	P	2
Rated power	P_r	1 kW
Phase Voltage	V	132V
Stator resistance	R	5.8 Ω
Magnet flux linkage	λ_f	0.533 Wb
d-axis inductance	L_d	0.0448 H
q-axis inductance	L_q	0.1024 H
Phase current	I	3 A
Base speed	w_b	1260 rpm
Rated torque	T_b	6 Nm

REFERENCES

[1] Zhong, L., Rahman, M.F., Hu. Y, *Analysis of direct torque control in permanent magnet synchronous motor drives.* Power Electronics, IEEE Transactions on, 1997. **12**(3): p. 528-536.

[2] French, C. and P. Acarnley, *Direct torque control of permanent magnet drives.* Industry Applications, IEEE Transactions on, 1996. **32**(5): p. 1080-1088.

[3] Holtz, J. and Q. Juntao, *Drift- and parameter-compensated flux estimator for persistent zero-stator-frequency operation of sensorless-controlled induction motors.* Industry Applications, IEEE Transactions on, 2003. **39**(4): p. 1052-1060.

[4] Lixin, T., Zhong L., Rahman, M.F., Hu, Y., *A novel direct torque controlled interior permanent magnet synchronous machine drive with low ripple in flux and torque and fixed switching frequency.* Power Electronics, IEEE Transactions on, 2004. **19**(2): p. 346-354.

[5] Xu, Z., "*Robust and Improved Flux Estimator and Controller Designs for Direct Torque Controlled Interior Permanent Magnet Synchronous Motors,*" PhD Thesis, Electrical Engineering and Telecommunication, The University of New South Wales, 2005.

978-1-4244-1871-8/07 $25.00 © 2007 IEEE

Study on Efficiency Optimizing of PMSM for Pump Applications

Guang-xu Zhou
Kyungsung University
Department of Electrical
and Mechatronics
Engineering, Busan,
Korea
E-mail:zhouguangxu1124
@163.com

Hui-jun Wang
Kyungsung University
Department of Electrical
and Mechatronics
Engineering, Busan,
Korea
E-mail:huijun024@gmail.
com

Dong-Hee Lee
Kyungsung University
Department of Electrical
and Mechatronics
Engineering, Busan,
Korea
E-mail:leedh@ks.ac.kr

Jin-Woo Ahn
Kyungsung University
Department of Electrical
and Mechatronics
Engineering, Busan,
Korea
E-mail:jwahn@ks.ac.kr

Abstract—In view of the operation characteristics of pump application, pump system should have high efficiency and energy saving during the whole operating process. A novel efficiency-optimization control strategy is presented to meet the demand of high efficiency of variable speed high efficiency PMSM. The core of this strategy is the excellent integration of mended maximum torque to current control algorithm, which is based on the losses model during the dynamic and the grade search method with changed step by fuzzy logic during the steady. A TMS320LF2407A DSP-based digital control inverter for variable speed high efficiency PMSM is developed. The performance experiments for control system of variable speed high efficiency PMSM have been completed. The tested result verified that the system can reliably operate with the different control strategy during the dynamic and steady operation, and the system has more perfect performances using the efficiency-optimization control.

I. INTRODUCTION

Loss minimization control strategies can fundamentally be summarized into two main categories: "loss model control" strategies and "search control" ones [1-6]. The "loss model control" technique is based on the development of a mathematical model, which allows for estimating the energy losses of the motor [7]. While, the "search control" algorithm is based on a searching procedure that can successfully find a maximum efficiency operating point. The latter one's advantage is that it doesn't need to know the mathematical model of the motor and its parameters [8-9]. However major drawback of this technique is that additional devices (such as voltage and current transducers) are required on the dc-link in order to obtain a measurement of the power absorption of the electrical drive, which results in increasing the system complexity and costs.

According to the operation characteristics of pump application it should have high efficiency and energy saving during the whole operating, a novel efficiency optimization control strategy is presented to meet the demand of high efficiency, which is required in variable speed high efficiency PMSM. The core of this strategy is the excellent integration of mended maximum torque to current control method based on the losses model during the dynamic period and the gradient search method with variable step by fuzzy logic during the steady operation. A TMS320LF2407A

DSP-based digital control inverter for variable speed high efficiency PMSM is developed. The performance experiments for control system of variable speed high efficiency PMSM have been completed.

II. THE EFFICIENCY OPTIMIZING STRATEGY DURING THE DYNAMIC STATE

During the dynamic process, the parameters equations can be achieved from the dynamic mathematical model including the losses. Then the classical maximum torque to current is improved in the light of the load characteristics of pump applications.

A. PMSM dynamic mathematical model include the losses

As shown in [4], assuming that the spatial distribution of the magnetic flux in the air gap is sinusoidal and the magnetic circuit is linear, the PMSM dynamic d and q axis equivalent circuits, which accounts for the iron losses, can be drawn as shown in Fig. 1.

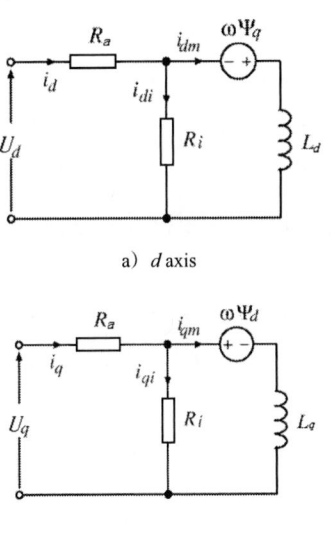

a) d axis

b) q axis

Fig. 1: d-q axes equivalent circuit for IPMSM

With reference to Fig.1, the state space equations of the dynamic model of a PMSM, which takes the losses into account, is expressed by the following relations:

$$P_{in} = (u_d i_d + u_q i_q) = \left(R_a(i_d^2 + i_q^2) + \frac{\omega^2(\psi_d^2 + \psi_q^2)}{R_i} + \omega\left[\psi_f i_{qm} + (L_d - L_q)i_{dm}i_{qm}\right]\right) \quad (1)$$

Where, P_{in} is the input power, i_d and i_q is the d and q axis components armature current. The d-q axes currents are divided into iron loss currents (i_{di}, i_{qi}) and magnetizing currents (i_{dm}, i_{qm}). R_a is the armature winding resistance. ω is electrical angular velocity.

With the increasing of load, the current is becoming more and larger. The stray loss in (2) is approximately proportional enhancement with the square of i_s, where, i_{sN} is the rated current and P_{sN} is the stray loss at rated load. Getting the linear expression of mechanical loss by experimental method (K_{ms} is a constant).

$$\begin{cases} P_s = \dfrac{i_d^2 + i_q^2}{i_{sN}^2} P_{sN} \\ P_{ms} = K_{ms}\Omega \end{cases} \quad (2)$$

According to the above equations, the torque expression is arranged as (3).

$$T_m = p\left[\psi_f i_{qm} + (L_d - L_q)i_{dm}i_{qm}\right] - K_s(i_d^2 + i_q^2) - K_{ms} \quad (3)$$

where, T_m is the electromagnetic torque.

B. The mended maximum torque to current method

Integrating (2) and (3), the salvations of i_d and i_q can't be obtained by the traditional maximum torque to current expression. Then the assistant function (4) is established.

$$F = i_d^2 + i_q^2 + \lambda\left\{p\left[\psi_f i_{qm} + (L_d - L_q)i_{dm}i_{qm}\right] - K_s(i_d^2 + i_q^2) - K_{ms} - T_m\right\} \quad (4)$$

where, λ is the Lagrange operator. Then the non-linear equations is obtained by the partial derivatives of i_d and i_q and λ from (4). The solutions of this equation are gotten by Newton method. In the actually application, according to the load characteristic of pump application, the i_d and i_q current at different torque and speed are calculated by the proposed method. Then the maximum torque to current control strategy is realized by looking up table.

III. THE EFFICIENCY OPTIMIZED STRATEGY DURING THE STEADY STATE

During the steady operation, the gradient searching method is used to get the loss minimization point. The initial point is one dynamic value. For enhancing the convergence speed, the step is altered by fuzzy logical method.

A. The gradient searching method with fuzzy logic

The implement steps of gradient searching method are following:

First, during the search process, the slope is determined as i_d decreased until the sign of the two slopes is reversed as shown in Fig.2.

$$S_{21} = (P_2 - P_1)/(i_{d2} - i_{d1}) \quad (5)$$

$$S_{32} = (P_3 - P_2)/(i_{d3} - i_{d2}) \quad (6)$$

Second, a new i_{d4} can be desired from (7),

$$i_{d4} = i_{d3} + 0.5(i_{d2} - i_{d3}) \quad (7)$$

Third, the slop S_{42} can be calculated using (8). Then compare the sign of S_{32} and S_{42}. If the sign is different then execute the forth step, else execute the fifth step.

$$S_{42} = (P_4 - P_2)/(i_{d4} - i_{d2}) \quad (8)$$

Forth, substitute the marker $\{2,4,3\}$ by $\{1,2,3\}$ in Fig.2. Then execute the second step.

Fifth, substitute the marker $\{1,2,4\}$ by $\{1,2,3\}$, and then execute second step.

Repeat the above steps, until the convergence point is found.

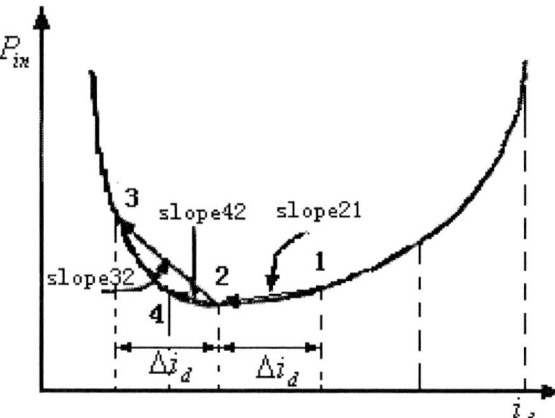

Fig. 2: Different slops of the gradient method

B. The step size changing rule by fuzzy logic

In order to increase the convergence speed, the fuzzy logical method is used to real change the steps of gradient search method on time. According to the load condition and the last slop variation ratio, the new d axis current increment is obtained automatically. Fig.3 is the membership functions for variables fuzzy sets.

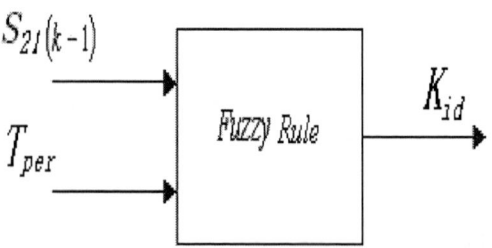

a) Control strategy of fuzzy logical controller

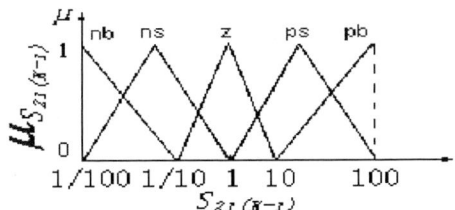

b) Membership function for slop

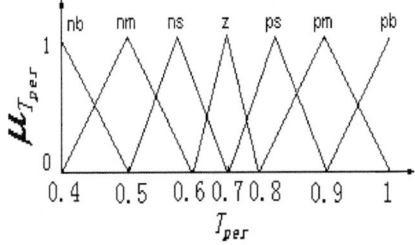

c) Membership function for torque

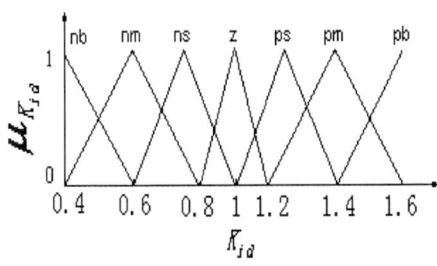

d) Membership function for step

Fig. 3: Membership functions for variables fuzzy sets

TABLE 1 is the fuzzy logical rules of different variables fuzzy sets.

TABLE 1
Fuzzy rules based matrix

K_{id}		T_{per}						
		PB	PM	PS	Z	NS	NM	NB
$S_{21}(K-1)$	PB	PB	PB	PB	PM	PS	Z	NS
	PS	PB	PB	PM	PS	Z	NS	NM
	Z	PB	PM	PS	Z	NS	NM	NB
	NS	PM	PS	Z	NS	NM	NB	NB
	PB	PS	Z	NS	NM	NB	NB	NB

C. The estimation of running states

In the proposed efficiency optimized control strategy different methods are adopted during different states, dynamic and steady states. For exactly getting the operation conditions of the whole system, first the speed error is checked in the main program. Second, for the stability during the condition switching, the error hysteresis comparison is used. Through distinguishing the above two terms, the operation state is captured for self-adaptive adjust i_d.

IV. EXPERIMENTAL TEST BENCH AND RESULTS

A test bench for the validation of the loss minimization algorithm (LMA) has been set up for implementing the proposed control system. Fig. 4 is the tested motor and load motor. Fig. 5 is the frequency spectrum of line voltage. Fig. 6 is the frequency spectrum of current. A simplified block diagram of the test bench is shown in Fig. 7. Extensive laboratory tests, obtained by using the above mentioned work bench, were carried out aiming to verify the optimized efficiency in a PMSM drive. Fig.8 is the power factor curve of PMSM experimented. Fig.9 is the efficiency of whole system.

Fig. 4: The experimental set up

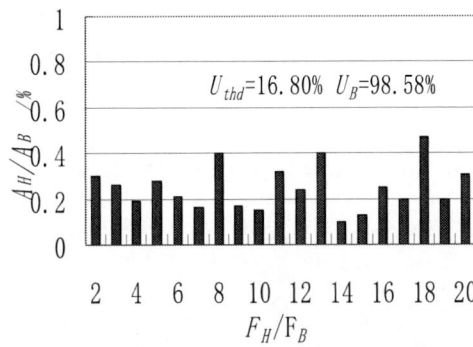

Fig. 5: Frequency spectrum of the line voltage

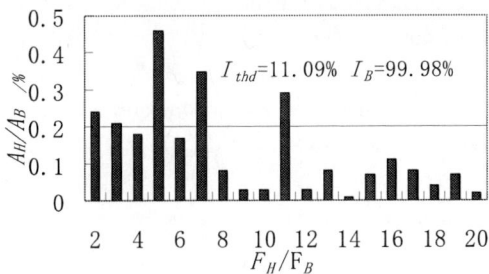

Fig. 6: Frequency spectrum of the current

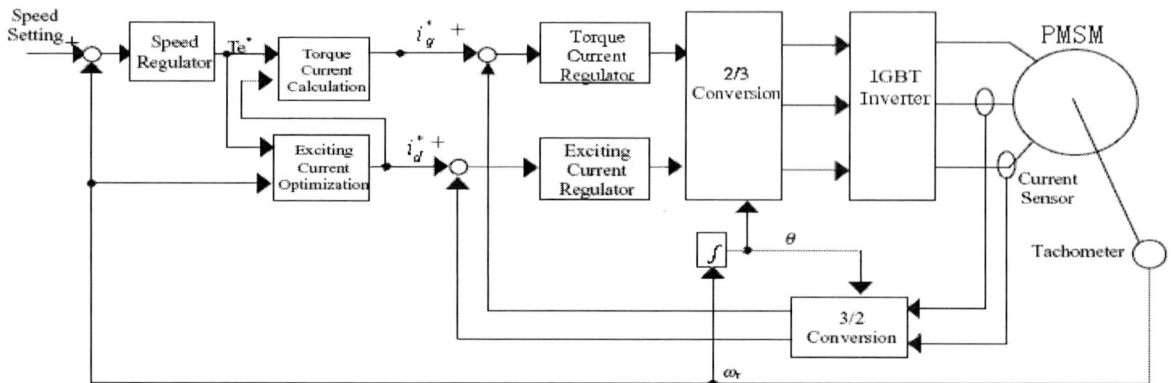

Fig. 7: Simplified block diagram of the PMSM efficiency optimization drive system

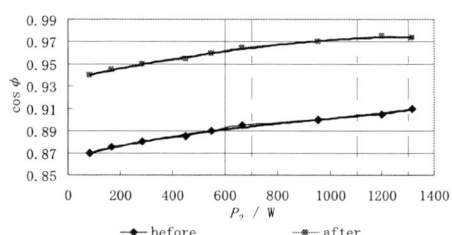

Fig. 8: Power factor curve of PMSM

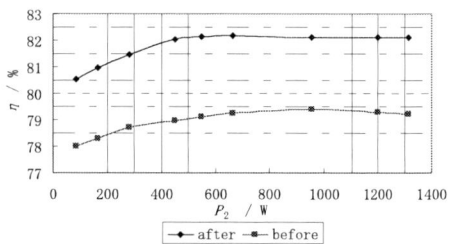

Fig. 9: Efficiencies of whole system

Form Fig.8 and Fig.9, we can see that after the efficiency optimization the characteristics of efficiency and power factor are improved. The efficiency increase is about 5.5% in the whole system, about 4.3% in average. The improvement of power factor is 0.056 maximum and about 0.038 in average. It is obvious that the efficiency is increased with the proposed optimizing strategy.

Form the Fig. 9, we can see the efficiency of system is over 70% during the whole operation scope and the efficiency of the motor is up to 82.2%, while that of the induction motor is about 76% at the same power level.

V. CONCLUSION

In this paper, a novel efficiency optimization control strategy is proposed to meet the demand of high running efficiency of variable speed high efficiency PMSM. The key of this strategy is the excellent integration of mended maximum torque to current control algorithm based on the losses model during the dynamic operation and grade search method with varied step by fuzzy logic during steady operation. The performance experiments for control system of variable speed high efficiency PMSM have been completed. The tested result verified that the system can reliably run with the different control strategy during the dynamic and steady operation, and the system has better performances with the efficiency optimization control.

REFERENCES

[1] R. S. Colby and D. W. Novotny, Efficient Operation of Surface Mounted PM Synchronous Motors. *IEEE Trans.Ind.Appl.*,1987,23(6):1048-1054.

[2] R.F. Schiferl, R. S. Colby and D. W. Novotny. Efficiency Consideration in Permanent magnet Synchronous Motor Drives. *Proc. Electric Energy Conference (eecon'87)*. Adelaide, Australia.1987:286-291.

[3] D. S. Kirschen, D. W. Novotny, and T. A. Lipo. On-line Efficiency Optimization of a Variable Frequency Induction Motor Drive. *Conf. Rec. 1984 IEEE-IAS Ann. Meeting*. 1984：488-493.

[4] G.D. Sousa，B.K.Bose, G.C.John. Fuzzy Logic Based On-line Efficiency Optimization Control of an Indirect Vector-controlled Induction Motor Drives. *IEEE Trans. on Industry Electronics*,1995, 42(2)：192-195.

[5] Fernandez-Bernal F，Garcia-Cerrada A，Faure R．Model-based Loss Minimization for DC and AC Vector-controlled Motors Including Core Saturation[J]. *IEEE Trans Ind Appl*, 2000, 36 (3)：755-763.

[6] Yi Tong, Shigeo Morimoto, Yoji Takeda and Takao Hirasa. Maximum Efficiency Control for Permanent Magnet Synchronous Motors. *IECON'91*,1991:283-288.

[7] Chunting Mi, Gordon R. Slemon, and Richard Bonert.Modeling of Iron Losses of Permanent-Magnet Synchronous Motors. *IEEE Transactions on Industry Application*,2003,39(3)：734-742.

[8] Naomitsu Urasaki, Tomonobu Senjyu, and Katsumi Uezato. A Novel Calculation Method for Iron Loss Resistance Suitable in Modeling Permanent-Magnet Synchronous Motors. *IEEE Transactions on Energy Conversion*,2003,18(1)：41-47.

[9] Naomitsu Urasaki, Tomonobu Senjyu and Katsumi Uezato. Investigation of Influences of Various Losses on Electromagnetic Torque for Surface Mounted Permanent Magnet Synchronous Motors. *IEEE Transactions on power elcetronics*,2003,18(1)：131-139.

978-1-4244-1871-8/07 $25.00 © 2007 IEEE   914

Wide-Speed-Range Optimal PAM Control for Permanent Magnet Synchronous Motors

Tomonobu Senjyu, Yohei Noguchi,
Naomitsu Urasaki, and Atsushi Yona
University of the Ryukyus
Department of Electrical and
Electronics Engineering
Faculty of Engineering
1. Senbaru, Nishihara-cho,
Nakagami 903-0213
Email: b985542@tec.u-ryukyu.ac.jp

Hideomi Sekine
University of the Ryukyus
Department of Technology Educations
Faculty of Education
1. Senbaru, Nishihara-cho,
Nakagami 903-0213
Email: h_sekine@hotmail.com

Toshihisa Funabashi
Meidensha Corporation
36-2, Nihonbashi, Hakozaki-cho,
Chuo-ku, Tokyo 130-8515
Email: funabashi-t@mb.meidensha.co.jp

Abstract— Generally, PWM voltage source inverter (VSI) has been used for the variable speed permanent magnet synchronous motor (PMSM) drives. However, the PWM-VSI fed PMSMs has two disadvantages. Firstly, PMSMs cannot operate at the speed at which the motor terminal voltage exceeds voltage constraint imposed by the DC-link voltage. Secondly, the distortion of the voltage commands increase at the low speed, because the DC-link voltage is significantly higher than the motor line voltage. Because of this, the armature currents are distorted and then the torque ripple increases. Consequently, higher DC-link voltage influences the speed control performance at the low speed range.

This paper proposes an optimal PAM control which can adjust the inverter DC-link voltage by using the buck-boost DC-DC converter. Further, considering the electrical vehicle application, the converter enables bi-directional power flow. At low speed range, the proposed system can reduce the total harmonic distortion (THD) of the voltage commands, and then the armature currents waveforms would be improved. Also, the allowable speed range would be extended. In order to verify the proposed method, simulation results are presented.

I. INTRODUCTION

Recent years, the variable speed controlled permanent magnet synchronous motors (PMSMs) are developed by evolving the power electronics technology and utilizing the rare-earth magnets. PMSMs are widely applied in industry due to the high energy density, high torque to ampere ratio, robustness and high efficiency.

Generally, PWM voltage source inverter (PWM-VSI) has been used for variable speed PMSM drives. However, the PWM-VSI fed PMSMs has mainly two disadvantages. Firstly, the PWM-VSI DC-link voltage limit the magnitude of the PMSM terminal voltage. Consequently, the motor speed is also limited because it is necessary that the terminal voltage is increased proportional to the motor speed. In order to overcome this problem, flux-weakening control[1,2] or boosting the PWM-VSI DC-link voltage according to motor driving conditions have been proposed. Secondly, in low speed range, the PWM-VSI modulation index decreases due to the high terminal voltage to DC-link voltage ratio. This problem causes the distortion in the armature currents and produced torque, as a result the speed control performance deteriorates. So far, for the

control performance at low speed range, it seems that the approach which focuses on the PWM-VSI DC-link voltage is few. Improving these disadvantages is important in the application which requires high performance at wide speed range, such as the electrical vehicles (EVs).

This paper proposes an optimal PAM control system for PMSM, in order to overcome the above disadvantages. Further, the proposed system allows bi-directional energy flow. The system has a bi-directional buck-boost DC-DC converter[5] in primary side of PWM-VSI. This paper describes the determination of an optimal DC-link voltage command and the duty factors of the converter. The proposed method can achieve the follow things. Firstly, the THDs of the commanded voltages are improved at low speed range due to the bucked DC-link voltage. Consequently, the THDs of armature currents and produced torque are improved and then, desired speed control performance are achieved under no load condition which is influenced significantly. Secondly, the allowable driving range is extended due to the boosted DC-link voltage. Finally, braking performance is improved due to boost operation at regenerative braking.

This paper is organized as follows. In Section II, an optimal DC-link voltage are considered. Section III roughly explains the operation principle of the bi-directional DC-DC converter. The proposed system is shown in Section IV, and this section explains about switching control of the used DC-DC converter. In order to verify the proposed method, simulation results are presented in Section V.

II. DETERMINATION OF AN OPTIMAL DC-LINK VOLTAGE COMMAND

The PMSM terminal voltage V_l is expressed as

$$V_l = \sqrt{\frac{3}{2}} V_{amp} \tag{1}$$

where V_{amp} is the amplitude of the phase voltage. Since $V_{amp} \leq V_{DC}/2$ in the triangular wave comparison method, V_l is imposed following constraint by the DC-link voltage V_{DC}:

978-1-4244-1871-8/07 $25.00 © 2007 IEEE

$$V_l \le \sqrt{\frac{3}{8}} V_{DC}. \qquad (2)$$

In order to make PMSMs drive at constant torque region, V_{DC} should be determined so that V_l does not exceed the constraint. However, too large V_{DC} reduces the PWM-VSI modulation index and then, it is difficult that the PWM-VSI realizes the required phase voltages. Further, the a-phase disturbance voltage due to dead time of switching devices increases proportional to V_{DC} as follows:

$$v_a{}^{dis} = \frac{T_d}{T_s} V_{DC} \, \mathrm{sgn}(i_a) \qquad (3)$$

where T_d, T_s and i_a are respectively dead time, switching period of the PWM-VSI and a-phase current. Similar disturbance voltages also occurs at b and c-phases. In order to maintain the maximum modulation index and minimum disturbance voltages at any speed range, it is necessary that V_{DC} maintains an optimal value. Therefore, the commanded DC-link voltage V_{DC}^* is calculated as follows after calculation of the commanded dq axes voltages:

$$V_{DC}{}^* = \sqrt{\frac{8}{3}} V_l^*. \qquad (4)$$

By the above calculation, V_{DC}^* is determined on-line according to V_l^*, and can correspond to load variation. It is important that V_l is greater than the nominal value because the PWM-VSI output voltage includes the harmonic components due to several nonlinearity of the PWM-VSI. Therefore, in order to give a margin, V_{DC}^* is finally determined as follows:

$$V_{DC}{}^* = \sqrt{3} V_l^*. \qquad (5)$$

III. BI-DIRECTIONAL BUCK BOOST DC-DC CONVERTER

In order to buck and boost the DC-link voltage, the proposed system uses a bi-directional buck-boost DC-DC converter[5] shown in Fig. 1. In this converter, the terminal voltages v_1 and v_2 are expressed as

$$\left.\begin{aligned} v_2 &= \frac{D_1}{1 - D_2} v_1 \\ v_1 &= \frac{D_3}{1 - D_4} v_2 \end{aligned}\right\} \qquad (6)$$

where $D_i\{i \in 1, 2, 3, 4\}$ is respectively the duty factors for switches $S_i\{i \in 1, 2, 3, 4\}$. From (6), the bi-directional buck-boost operation can be achieved by using the switches S_1 and S_2 when the power flows from v_1 to v_2, while using the switches S_3 and S_4 when the power flows from v_2 to v_1. The variables in Fig. 1, i_L, v_1 and v_2, are then expressed as the following differential equations:

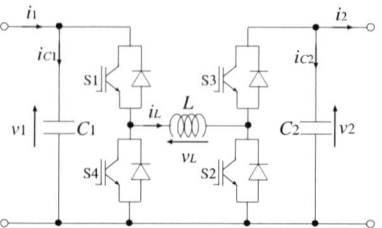

Fig. 1 Bi-directional buck-boost DC-DC converter.

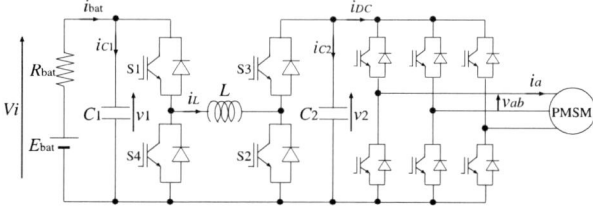

Fig. 2 Variable DC-link voltage circuit.

$$\left.\begin{aligned} \dot{i}_L &= \frac{v_L}{L} \\ \dot{v}_1 &= \frac{i_{C1}}{C_1} \\ \dot{v}_2 &= \frac{i_{C2}}{C_2} \end{aligned}\right\}. \qquad (7)$$

Note that the variables, v_L, i_{C1} and i_{C2}, are determined according to the switches states and the initial value of i_L.

IV. OPTIMAL PAM CONTROL

In this section, control method for the bi-directional buck-boost DC-DC converter will be explained. A variable DC-link voltage circuit are established as shown in Fig. 2 by using the circuit shown in Fig. 1.

For motoring mode, if $V_{DC}^* < V_i$ then the duty factors for each switch is determined as

$$\left.\begin{aligned} D_1 &= K_p(V_{DC}{}^* - V_{DC}) \\ D_2 &= 0 \\ D_3 &= D_4 = 0 \end{aligned}\right\} \qquad (8)$$

where K_p is the proportional gain. The converter outputs the optimal V_{DC} by the step-down operation according to motor speed. If $V_{DC}^* > V_i$ for motoring mode, then the duty factors for each switch are determined as

$$\left.\begin{aligned} D_1 &= 1 \\ D_2 &= 1 - \frac{V_i}{V_{DC}{}^*} \\ D_3 &= D_4 = 0 \end{aligned}\right\}. \qquad (9)$$

In this case, the converter performs the step-up operation according to motor speed. For regenerative mode, the battery charge current are controlled by boosting up V_{DC} to over V_i. The duty factors for each switch are determined as

978-1-4244-1871-8/07 $25.00 © 2007 IEEE

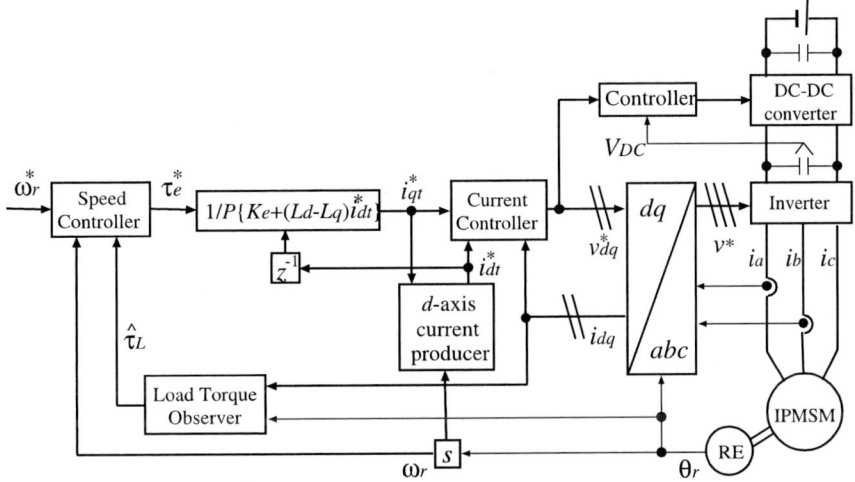

Fig. 3 Control system for IPMSM drive.

$$\left.\begin{aligned} D_1 &= D_2 = 0 \\ D_3 &= 1 \\ D_4 &= 1 - \frac{V_{DC}}{v_1^*} \end{aligned}\right\}. \qquad (10)$$

V. SIMULATION

A. Simulation conditions

In order to verify the proposed system, an optimal PAM control simulations are carried out for an interior PMSM (IPMSM) whose data is given in Table 1. The IPMSM mathematical model with iron loss is used in this simulation. Further, this system assumes a PWM-VSI with triangular wave carrier in the programming, and simulated PWM-VSI considers the dead time of switching devices. Also, a bi-directional buck-boost DC-DC converter is modeled by the differential equations so that the instantaneous analysis is allowed. The specifications of the IPMSM, PWM-VSI and DC-DC converter are shown in Table I.

The IPMSM control system is composed of the speed and current control loops, and assumes that the actual rotor position and three phase currents can be obtained by the rotary encoder and current sensors. Also, a dead time compensation method[3] is included with the designed control system to realize the commanded voltages as actual motor voltages. Then, using the calculated v_{dq}^*, the DC-DC converter controller determines the duty factors based on Section IV.

Secondly, the proposed control system are shown in Fig. 3. This system is composed of the PMSM model with iron loss[7], and commanded dq axes currents are determined by the maximum efficiency control strategy[2]. Further, this system has the load torque compensation in order to minimize the drastic drop in the motor speed due to load variation. The system then includes a disturbance torque estimation observer[9] as follows:

TABLE I
Specifications of machines

IPMSM	rated power	0.35 kW
	rated speed	800 rpm (V_{DC} = 42 V)
	rated torque	1.60 N·m
	armature resistance	0.57 Ω
	iron loss resistance	$67.61 + 0.41|\omega_e|$ Ω
	d-axis inductance	7.73 mH
	q-axis inductance	$28.3 - 0.657|i_q|$ mH
	emf constant	1.08×10^{-1} V·s/rad
	pole pair	2
	inertia moment	7.25×10^{-4} Kg·m^2
	viscous coefficient	2.81×10^{-4} N·m·s
PWM-VSI	career frequency	5 kHz
	dead time	3 μs
Bi-directional DC-DC converter	switching frequency	25 kHz
	supply voltage	42 V
	coil	300 μH
	input/output capacitor	5,500 μF

$$\left.\begin{aligned} G_1 &= -(\gamma_1 + \gamma_2 + \gamma_3 + D/J) \\ G_2 &= \gamma_1\gamma_2 + \gamma_2\gamma_3 + \gamma_3\gamma_1 - G_1\frac{D}{J} \\ G_3 &= \gamma_1\gamma_2\gamma_3 J \end{aligned}\right\} \qquad (11)$$

$$\begin{bmatrix} \dot{\hat{\theta}}_r \\ \dot{\hat{\omega}}_r \\ \dot{\hat{\tau}}_L \end{bmatrix} = \begin{bmatrix} -G_1 & 1 & 0 \\ -G_2 & -\frac{D}{J} & -\frac{1}{J} \\ -G_3 & 0 & 0 \end{bmatrix} \begin{bmatrix} \hat{\theta}_r \\ \hat{\omega}_r \\ \hat{\tau}_L \end{bmatrix}$$

$$+ \begin{bmatrix} G_1 \\ G_2 \\ G_3 \end{bmatrix} \theta_r + \begin{bmatrix} 0 \\ \frac{1}{J} \\ 0 \end{bmatrix} \tau_{ecul} \qquad (12)$$

where G_1, G_2, and G_3 are observer gains, γ_1, γ_2, and γ_3 are observer poles, J and D are respectively inertia moment and viscous coefficient, and θ_r and ω_r are respectively rotor position and rotor speed. Then, τ_{ecul} is the torque calculated by the measured dq axes currents.

978-1-4244-1871-8/07 $25.00 © 2007 IEEE

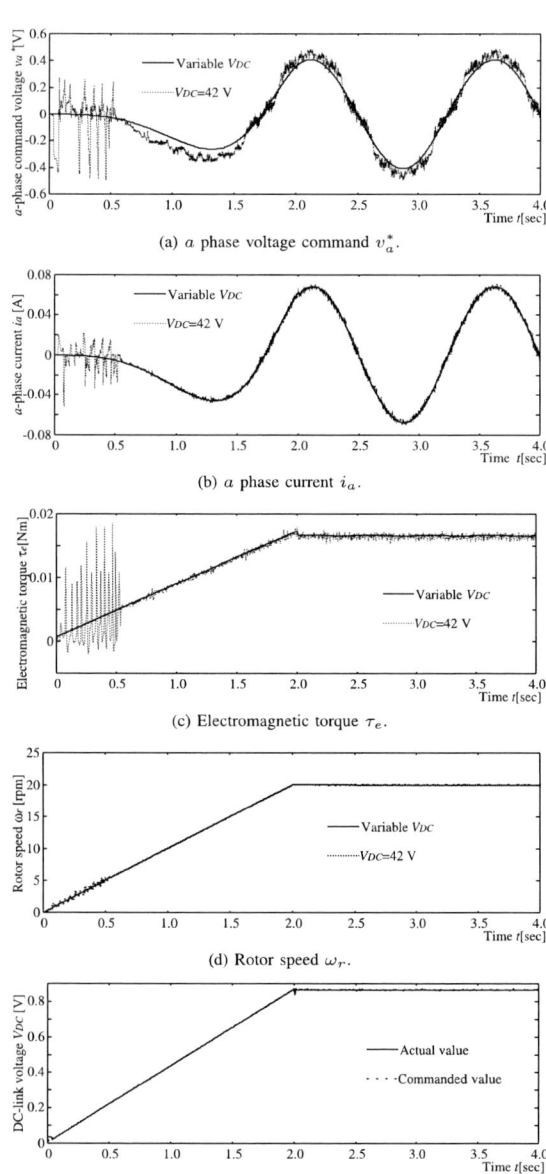

(a) a phase voltage command v_a^*.

(b) a phase current i_a.

(c) Electromagnetic torque τ_e.

(d) Rotor speed ω_r.

(e) DC-link voltage V_{DC}.

Fig. 4 Comparison between variable and fixed DC-link voltage for low speed operation (20rpm).

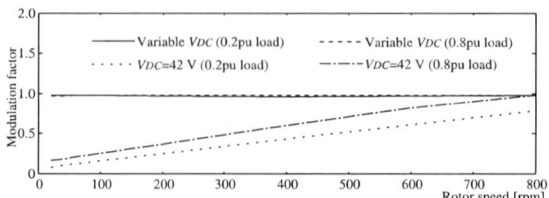

Fig. 5. Modulation factor for rotor speed.

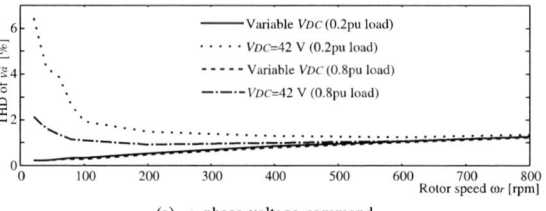

(a) a phase voltage command.

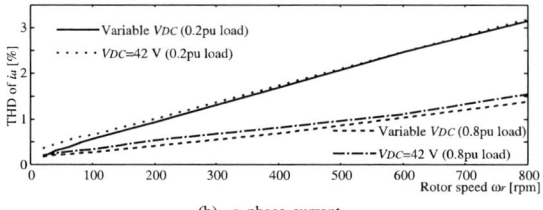

(b) a phase current.

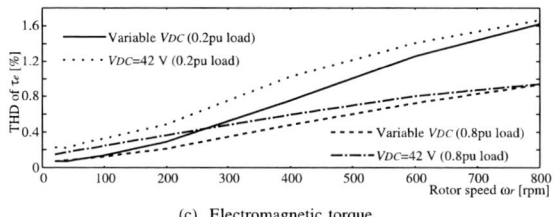

(c) Electromagnetic torque.

Fig. 6. Total harmonic distortion for rotor speed.

B. Simulation results

Fig. 4 shows the simulation results for low speed operation. In this simulation, IPMSM is accelerated from 0 to 20rpm. Then, load torque increases in proportional to the rotor speed so that it becomes rated load at 2,000rpm. For comparison, the simulation result for the fixed DC-link voltage operation ($V_{DC} = 42$V) is also shown in this figure. It can be confirmed from Fig. 4(a)~(c) that the desired commanded voltage are produced by the proposed method, and the distortion of armature currents and pulsation of torque are improved. Consequently, the proposed method achieves the desired low speed control performance even

under light load. Further, it can be confirmed from Fig. 4(e) that DC-link voltage is controlled according to driving condition.

Fig. 5 shows the PWM-VSI modulation indexes for the proposed and $V_{DC} = 42$V systems under 0.2pu(light load) and 0.8pu(heavy load) load. Although the modulation index for $V_{DC} = 42$V system decreases at low speed, the proposed system has approximately maximum modulation index regardless of any driving conditions.

Fig. 6 shows THDs of a phase commanded voltage; a phase current and produced torque for each rotor speed. THDs are respectively evaluated by

$$
\left.
\begin{aligned}
THD_{v_a^*} &= \frac{\sqrt{V_a^{*2} - V_{a1}^{*2}}}{V_{a1}^*} \\
THD_{i_a} &= \frac{\sqrt{I_a^2 - I_{a1}^2}}{I_{a1}} \\
THD_{\tau_e} &= \frac{\sqrt{T_e^2 - T_{ed}^2}}{T_{ed}}
\end{aligned}
\right\} \quad (13)
$$

where V_a^* and V_{a1}^* are root mean square (RMS) values of a phase commanded voltage and fundamental component

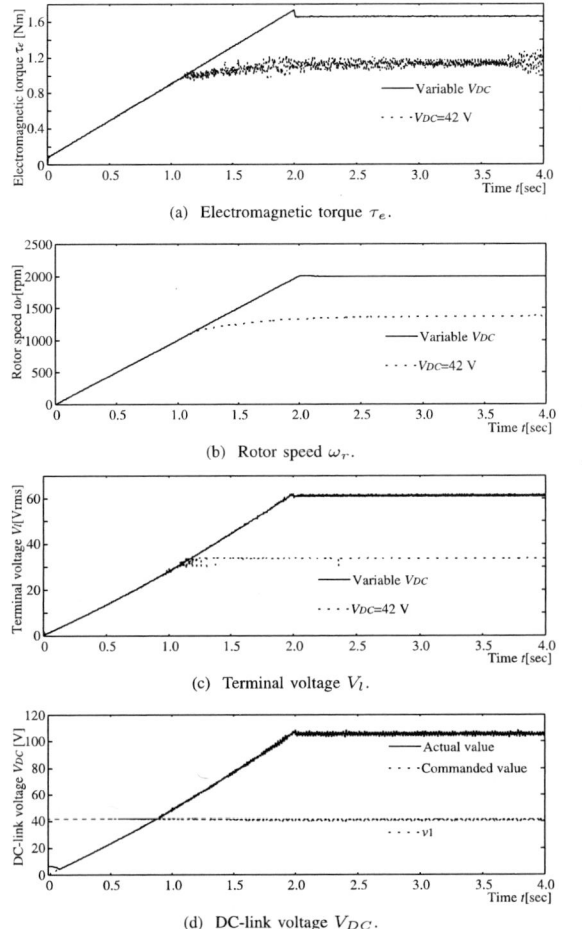

(a) Electromagnetic torque τ_e.

(b) Rotor speed ω_r.

(c) Terminal voltage V_l.

(d) DC-link voltage V_{DC}.

Fig. 7. Comparison between variable and fixed DC-link voltage for high speed operation (2,000rpm).

(a) Electromagnetic torque τ_e.

(b) Rotor speed ω_r.

(c) DC-link voltage V_{DC}.

(d) Load torque τ_L.

Fig. 8. Transient performance for instant injection of lated load at low speed (20rpm).

of that, I_a and I_{a1} are RMS values of a phase current and fundamental component of that and T_e and T_{ed} are RMS values of produced torque and DC component of that, respectively. Comparing with the $V_{DC} = 42V$ system, THDs are improved at low speed with light load. Particularly the THD of the commanded voltage is significantly improved.

Fig. 7 shows the simulation results for high speed operation. In this simulation, IPMSM is accelerated from 0 to 2,000rpm. Then, load torque increases in proportional to the rotor speed so that it becomes rated load at 2,000rpm. For comparison, the proposed method and DC-link voltage constant ($V_{DC} = 42V$) simulation results are shown. For fixed DC-link voltage, IPMSM cannot accelerate over the speed of 800rpm because of deficiency of output torque. While for proposed system, the allowable drive range is extended due to boosted DC-link voltage.

Fig. 8 shows the control performance for instant injection of rated load at low speed. When IPMSM is controlled at 20rpm under 0.01pu load, the load torque suddenly changes to 1.0pu at 4.0sec. It can be confirmed that the proposed system accepts instant injection of rated load

in low speed drive. The DC-link voltage is controlled according to load variation in spite of rapid increase of the commanded value. Further, the disturbance torque observer also demonstrates desired performance.

Fig. 9 shows the simulation results for regenerative braking operation. IPMSM is accelerated to 500rpm, and then, applied -0.3pu torque command at 6.0sec. After that, in order to prevent reverse, 0.0pu torque command is applied below 10rpm. For comparison, the simulation result for the fixed DC-link voltage operation ($V_{DC} = 42V$) is also shown in this figure. From Fig. 9(a), the proposed system achieves faster braking than $V_{DC} = 42V$ system due to boosted DC-link voltage. Fig. 9(b) and (c) show the battery currents for the $V_{DC} = 42V$ and proposed systems. It can be confirmed that the one of the proposed system is larger than that of the $V_{DC} = 42V$ system at regeneration term.

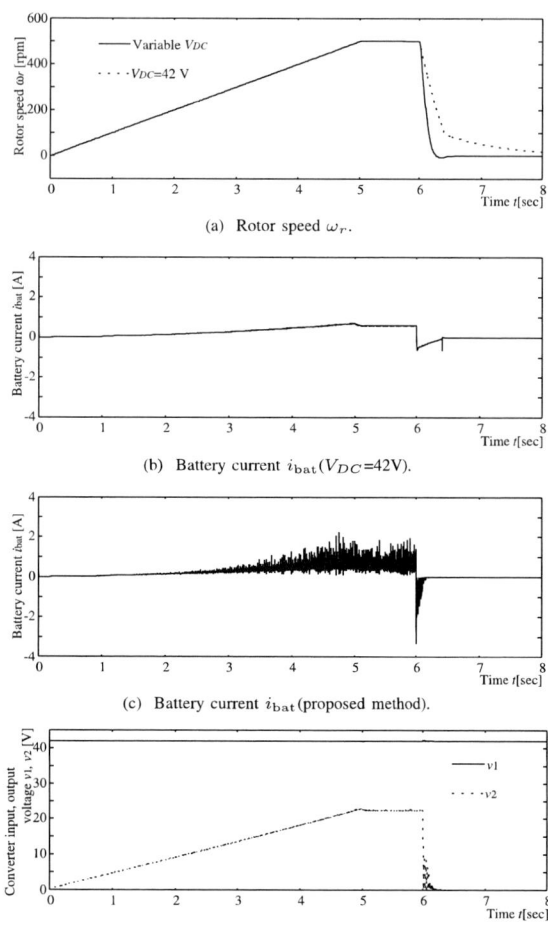

(a) Rotor speed ω_r.

(b) Battery current $i_{\text{bat}}(V_{DC}=42\text{V})$.

(c) Battery current i_{bat}(proposed method).

(d) Converter input, output voltage v_1, v_2.

Fig. 9. Regenerative breaking performance at 500rpm.

VI. CONCLUSION

This paper proposes an optimal PAM control system for PMSM drives which can adjust the PWM-VSI DC-link voltage according to the driving condition at the range from low to high speed. This system uses a bi-directional buck-boost DC-DC converter and this paper describes the determining methods of an optimal DC-link voltage and the converter duty factors.

It can be confirmed that the THDs of the commanded voltages, armature currents and produced torque are improved at low speed range, the allowable driving range is extended and the regenerative breaking performance is improved by using the proposed method. Further, it can be confirmed that the proposed system allows instant injection of rated load for low speed control.

REFERENCES

[1] G. Gallegos-López, F. S. Gunawan and J. E. Walters, "Optimum Torque Control of Permanent-Magnet AC Machines in the Field-Weakened Region," *IEEE Trans. Ind. Applicat.*, vol. 41, pp. 1020-1028, July/August, 2004.

[2] S. Shinnaka, "New Practical Optimal Current Control Methods for Energy-Efficient Wide-Speed-Range Operation of Salient-Pole Permanent Magnet Synchronous Motor with Core-Losses," *IEEJ, Trans. IA,* vol. 123, no. 11, pp. 1359-1370 (2003).

[3] N. Urasaki, T. Senjyu, K. Uezato, and T. Funabashi, "A dead-time compensation strategy for permanent magnet synchronous motor drive suppressing current distortion," in *Proc. IEEE IECON'03*, Roanoke, Virginia USA, pp. 1255-1260, November 2003.

[4] N. Urasaki, T. Senjyu, T. Kinjo, T. Funabashi, and H. Sekine, "Adaptive Dead-Time Compensation of Voltage Source Inverter for Variable Speed Drives," *IEEJ, Trans. IA,* vol. 126, no. 7, pp. 872-880 (2006).

[5] F. Caricchi, F. Crescimbini, F. G. Capponi, and L. Solero, "Study of Bi-Directional Buck-Boost Converter Topologies for Application in Electrical Vehicle Motor Drives," in *Proc. APEC1998*, Anaheim, CA, USA, 1998.

[6] Frederick D. Kieferndorf, Matthias Förster and Thomas A. Lipo, "Reduction of DC-Bus Capacitor Ripple Current With PAM/PWM Converter," *IEEE Trans. Ind. Applicat.*, vol. 40, pp. 607-614, Mar. /Apr. 2004.

[7] N. Urasaki, T. Senjyu, and K. Uezato, "Relationship of Parallel Model and Series Model for Permanent Magnet Synchronous Motors Taking Iron Loss Into Account," *IEEE Trans. on Energy Convers.*, vol. 19, no. 2, pp. 265-270, June. 2004.

[8] Y. Tong, J. Morimoto, S. Morimoto, Y. Takeda, and T. Hirasa, "High Efficiency Control of Brushless DC Motors for Energy Saving," *IEEJ, Trans. IA,* vol. 112, no. 3, pp. 285-291 (1992).

[9] T. Senjyu, T. Shingaki, and K. Uezato, "Sensorless Vector Control of Reluctance Motor using Speed and Disturbance Torque Observer," *IEEJ Proc. of the 1999 Japan Industry Appl. Society Conf.*, vol. 1, pp. 5-8 (1999).

Development Issues of An ISG PM Machine and Control System

Xiaojiang Chen, Murray Edington, Roger Thornton
Ricardo UK Ltd, Westbrook Centre
Milton Road, Cambridge CB4 1YG, UK
Email: Xiaojiang.chen@ricardo.com
Murray.Edington@ricardo.com

Yunzhou Fang, Qingfeng Peng
Chery Automotive Co. Ltd
No. 8 Changchun Rd, Wuhu, China, 241009

Abstract—This paper focuses on addressing some practical considerations in developing an ISG PM system for hybrid electric-vehicle (HEV) applications. Extensive practical test results and theoretical analyses are presented and analyzed. Firstly, the ISG PM machine design is discussed in term of the machine size, pole-pair number, permanent magnet flux, inductance, etc. A cost-effective machine design example, which was successfully demonstrated in a mild ISG HEV application, is analyzed. Its performance test results are evaluated. Subsequently, the impacts of the position sensors on the ISG performance are investigated exclusively. Three prominent position sensor candidates are studied in detail to examine their limitation and feasibility for ISG applications. Furthermore, the control related issues are discussed. The torque control of the ISG machine is implemented through vector control algorithm. Flux-weakening control is employed to achieve a wide speed range of constant-power operation. The impact of the inverter DC-link voltage variation is explored. In addition the digital control delay influence and its effective compensation strategy are addressed.

I. INTRODUCTION

Today almost all the leading automotive manufactures are involving in developing hybrid electric vehicles (HEV). A traction electric machine is employed in the HEV powertrain to integrate with an engine in order to improve the overall fuel efficiency and reduce engine exhaust emission. The electric machine in HEV basically acts as an integrated starter and generator (ISG) to provide assistant torque to the powertrain, recover mechanical energy during braking by charging a battery, provide fast engine start, etc. The continuous growth of the HEV developments unavoidably creates new application opportunities for the electric machines and their drive systems in the automotive industry.

This paper focuses on addressing some practical considerations in developing a permanent-magnet synchronous machine (PMSM) system for ISG applications. First of all, the practical design criteria of an ISG PM machine are discussed. Several important factors including machine technology selection, pole-pair number, permanent magnet flux, inductance, torque capability should be carefully considered in order to design an appropriate PM machine with high power density and high efficiency for ISG HEV applications. A practical design example of a high-performance and low-cost ISG interior PM machine (IPM) is discussed. The practical test result of the machine performance is revealed.

An absolute position sensor is one of most important components in the ISG machine system. Its accuracy has significant impact on the control performance of the ISG PM machine. Three low-cost viable position sensors including resolver, linear Hall sensor based position sensor and eddy-current type sensor are studied in detail. Their practical test results are shown respectively to explore the sensors' limitation and feasibility for ISG applications.

The major function of ISG for HEV applications is to deliver required torque [4-5]. The feedback current control based on the vector theory is applied to implementation torque control of the ISG PM machine. The required wide constant-power operation for the ISG PM machine can be achieved by the flux-weakening technique. The impact of the DC-link voltage variation due to the battery is analyzed. The impact of the digital control delay on the ISG control performance is discussed.

The objective of this paper is mainly to analyze some significant issues facing for designing the ISG system and improving the system performance. Extensive practical test results and theoretical analyses are presented.

II. ISG MACHINE TECHNOLOGY

An ISG machine is required to be designed with high efficiency, high power/weight ratio, and high reliability with high-speed operation. The power rating of an ISG machine can range from a few kilo-watts for a mild HEV application to more than 35kW for a full HEV application demanded on the EV mode operation. For the applications with this range of the machine power rating, permanent magnet synchronous AC machines (PMSM) have been demonstrated to be one of most prominent candidates for HEV applications because of their competitive efficiency and power density [1-3].

There are two types of PMSM machines that have been practically used on HEV so far: one is a surfaced mounted PM machine (SPM), which is used on the Honda's early HEV cars such as Insight and Civic [1]. The other is an interior PM machine (IPM), which is used in the Honda's latest Civic and Accord, and Toyota Prius HEV [2-3].

Fig. 1 shows a SPM machine installed on the engine shaft for a mild ISG HEV application, developed by Ricardo and Chery. The machine has a 15-kW peak power and 10kW continuous power rating. The machine was designed in thin pancake shape with the overall length of less than 65mm in order to meet the tight mounting restriction and be integrated

978-1-4244-1871-8/07 $25.00 © 2007 IEEE

on the engine shaft.

The design factors for an ISG PM machine are mainly confined by the machine pole number, permanent magnet flux linkage, dq-axis inductance, peak torque, torque ripple, operational speed. In addition, the machine design is significantly influenced by the mounting space restriction and stringent cost target as well. Realistically it is difficult to design a PM ISG machine that can satisfy all the design requirement factors, which are usually contradictory in each other. The design compromises are usually made to balance the machine performance, operation safety and control convenience.

The peak torque definition of an ISG machine mainly relies on the requirement of engine cranking control. The instant engine start is necessary to achieve the start-and-go function in HEV. Fig. 2 reveals the test result of a fast engine cranking control demonstrated in a mild ISG HEV car with a 1.3L engine. The closed-loop speed control of the ISG SPM machine was employed during engine cranking start. As we can see that the engine was started instantly. Empirically the engine starting time within 250ms~400ms is unnoticeable to drivers. The peak torque rating of the ISG machine should be large enough to provide reliable cold start within its current rating range.

Fig. 1: A SPM ISG installed on engine shaft

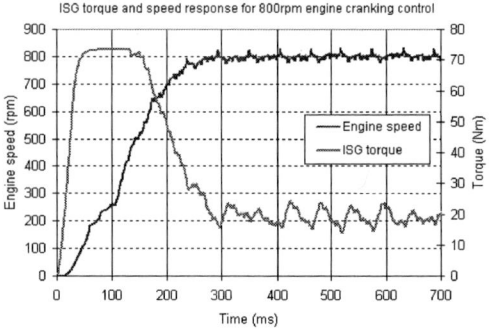

Fig. 2: A rapid engine cranking test driven by an SPM ISG machine

The back EMF constant of the ISG PM machine is one of major criteria for the machine design. The back EMF is governed by the machine pole number and permanent magnet flux, i.e. $K_E = \psi_m p$, where, ψ_m is permanent magnet flux, p is the pole pair number. The increase in K_E may result in the reduction on copper loss and efficiency improvement at low speed. However, flux-weakening control capability and uncontrolled generation due to flux-weakening failure at high speed should be taken into account for selecting K_E. For the safety consideration at the uncontrolled condition, the back EMF at the maximum operational speed is preferably chosen to be less than the inverter peak operational voltage.

Due to the requirement of the high-speed operation (over 6000rpm) for an ISG, the pole number of the PM ISG machine should be optimized to justify the loss and control capability. High pole number has the benefit for smooth torque, more compact machine design and less end winding (less copper loss). However it would causes inductance and flux-weakening control capability decrease. Furthermore, the high pole number causes the increase of the fundamental frequency of the controlled current. This will cause significant difficulty in controlling ISG stably at high speed.

Torque ripple is one of important concerns for many PM machine designs. But for ISG applications, it is not a major issue because the HEV drivetrain is basically a large inertia system. Apparently an ISG PM machine with less torque ripple is favorable. However, the remedy measures for compressing torque ripple including stator/rotor skewing, magnet shaping, dummy slotting in stator, fractional slot winding, will unavoidably increase the machine manufacturing cost.

IPM machines have received increasing attention worldwide. Their abilities to deliver the very high power density and wide constant power speed range with very high efficiency make them become more attractive than SPM for HEV/EV applications. We have been studying and developing the high-performance low-cost traction machine technologies for many years. Fig. 3 shows an IPM ISG design example with a 15kW peak-power rating for the application in a mild ISG hybrid car. In this application, the IPM ISG machine has to be designed to meet with a tight size restriction due to very limited mounting space available.

The cost limit can bring significant impact on how to design the ISG machine. Using the low-grade lamination steel, rectangular-shape magnet, compact machine structure, concentrated winding layout, etc, may result in a low-cost design of a ISG PM machine. However the machine performance is usually forced to be compromised in order to justify the cost target. Low-cost lamination steel in the stator/rotor cores is adopted in designing the IPM ISG shown in Fig. 3. Even with the severe constraints imposed, we have successfully designed this 15-kW IPM ISG machine with peak efficiency close to 94%. Fig. 4 shows the measured torque-speed-efficiency map of this IPM ISG machine.

Fig. 3: An IPM for mild ISG HEV applications

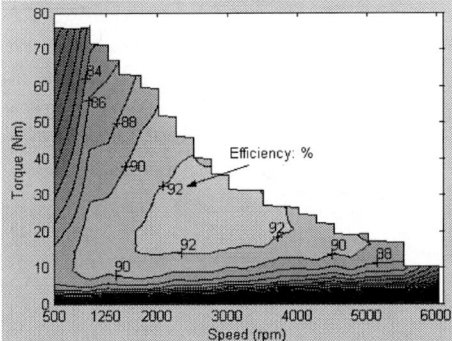

Fig. 4: Measured torque-speed-efficiency map of an IPM ISG

This cost-effective IPM ISG machine has 6 pole-pairs. Its back EMF constant was specially tailored not to exceed the battery voltage limit at its maximum operational speed of 6500rpm in order to provide safe operation of the ISG machine at high speed and avoid excessive uncontrolled generation during unexpected flux-weakening control failure. Its reluctance torque can account for more than 40% of its overall output torque. This has the benefit to reduce the back EMF constant significantly without sacrificing efficiency. Comparatively a SPM ISG will be difficult to be designed with such a low back EMF constant and high efficiency for the similar torque and power rating to this IPM ISG.

III. Position Sensor Issues

The well-known vector control is normally employed to achieve high-performance control of ISG PM machines. The absolute position knowledge of the rotor PM magnetic field is required for successfully implementing advanced control algorithms. Accurate position measurements can result in improved torque control dynamics, control torque ripple reduction, wider speed control range at flux-weakening control region, higher control efficiency, control stability improvement at high speed, etc. However, accurate absolute position sensors are normally subject to high cost. For ISG applications, a trade-off has to be made by considering sensor cost, reliability and resolution.

The position-sensor pole has to match with the ISG machine pole correctly. For an example, a 20-pole PM machine requires an absolute position encoder be able to deliver 10, 5, 2, or 1 cycle position information per mechanical rotation otherwise the ambiguity in absolute position decoding is introduced.

The zero-position of the absolute position sensor should be fixed with respect to the back EMF signals of the ISG PM machines. This prerequisite must be satisfied in order for vector control algorithm to be implemented without phase error. Fig. 5 illustrates the required relationship of the absolute position and ideal 3-phase back EMF signals of the ISG PM machine. If the position sensor is mounted randomly relative to the ISG machine, a misalignment angle of the position sensor must be detected and compensated in order to achieve the synchronous control of vector algorithm.

The accuracy of the positional measurements has significant impact on the control performance of ISG machines. Fig. 6 illustrates the performance test results of a SPM ISG machine mounted with an absolute position sensor of the ±1-mechanical-degree resolution. At the motoring mode, the ISG SPM machine could only safely run up to 4500rpm. The machine output power decreased dramatically over 4000rpm. The position error caused the distortion of control current and eventually leads to unstable control at the high speed over 4500rpm.

Fig. 7 shows the comparative test result for the same ISG SPM machine equipped with a more accurate absolution position sensor (about ±0.5 mechanical degree error). The SPM ISG machine could be controlled stably and reliably up to its maximum operation speed 6500rpm with the 10kW continuous power output. The efficiency of the overall ISG system including machine and inverter has been improved.

The proper selection of the absolute position sensor is crucial for achieving reliable and efficient control of the ISG PM machine. Hereinafter three types of low-cost and reliable absolute position sensors are studied and their limitations and feasibilities for ISG applications are investigated.

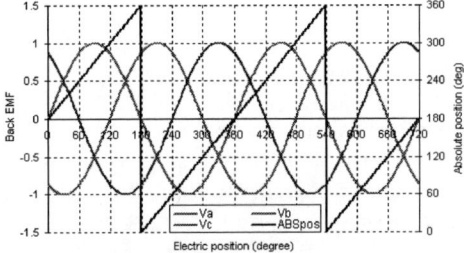

Fig. 5: Required relationship of position and back EMF signals

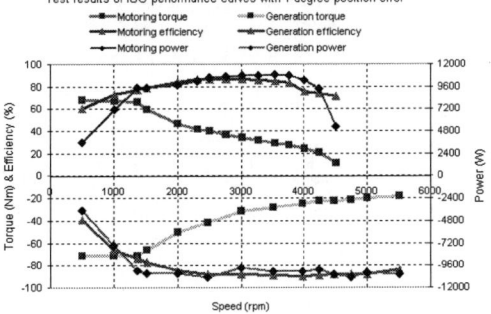

Fig. 6: ISG SPM torque-speed curve on a ±1-degree position sensor

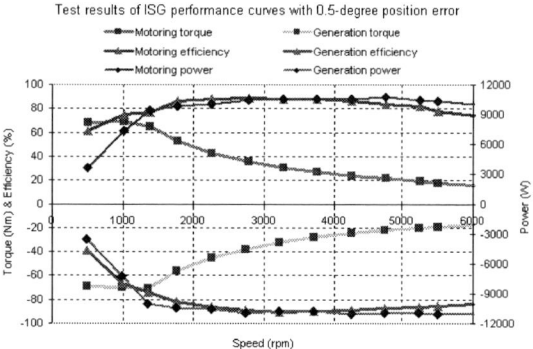

Fig. 7: ISG torque-speed curve on a ±0.5-degree position sensor

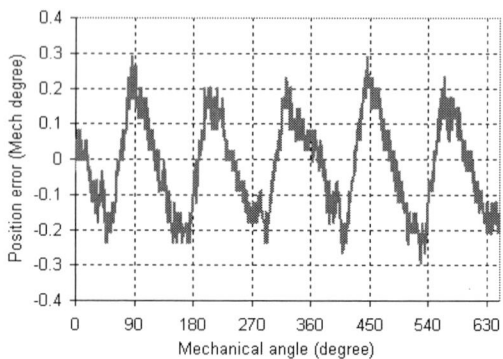

Fig. 8: Measured resolver position error

A. Resolver

Resolvers are widely used as absolute position sensors in industrial servo drive systems because of their rugged construction and very high reliability. They are basically rotating transformers. Their sinusoidal input references are amplitude-modulated with the sine and cosine of the mechanical angle. The modulated positional sine/cosine output signals from a resolver are demodulated to obtain the angular position through a so-called resolver-to-digital converter (RDC). The resolvers for ISG applications are required to be designed with very low cost and large bore diameter. The successful application examples of the resolvers employed in HEV are the Tamagawa's framelss VR type thin resolvers, which have been widely adopted in Toyota Prius and Honda's latest ISG HEVs [2].

The resolver has a very simple structure that consists of a stator and a rotor. The stator windings are displaced mechanically at 90° to each other. Its rotor consists of only shaped electro-magnetic steel sheets in order to provide the electromagnetic interaction when the stator winding are injected with modulated signals. The frequency of its reference modulation signal can be set with 10~20kHz by a RDC ASIC chip. The RDC can give a 10-bit to 12-bit position conversion resolution. Because of the lack of the fixed supporting frame for the resolver stator and rotor parts, its very thin flat shape tends to be prone to the position resolution degradation if no tight mounting is applied. Fig. 8 indicates a test results on a VR resolver with the 12-bit RDC resolution setup. However, the actually achieved position measured accuracy for the resolver is only limited within ±0.3 mechanical degree. The more accurate position measurement from the resolver might be achievable if the strict mounting can be ensured.

Fig. 9 shows the measurements of the modulated sine/cosine signals from the resolver and resulted 3-phase control current by applying vector control. The sine/cosine modulation signals have symmetrical and uniform envelope. This implies that the rersolver was installed reasonably and acceptable position detection is achieved. Consequently the control current has less harmonics and distortion.

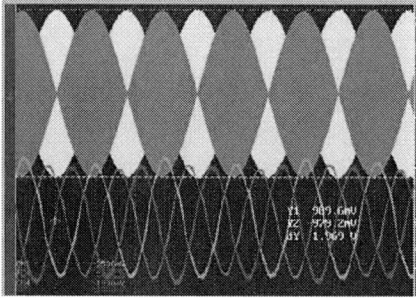

Fig. 9: Sine/cosine modulation signals from good resolver mounting and resulted control current

The measurements in Fig. 10 represent a poor mounting of the resolver. The sine/cosine modulation signals have unbalanced envelope, which is mainly induced by eccentric mounting of the resolver. As a result, the 3-phase control currents have variations in amplitude. This would unavoidably introduce extra torque oscillation and unstable control to the ISG PM machines at high speed.

The dominant limitation of the resolvers for HEV applications is due to the extra cost on its RDC chip and extra excitation circuits. A low-cost solution for RD conversion can be achieved by fully exploring the potential of the DSP/micro-controllers, which are used in implementing high-performance digital control of ISG machines. The start-of-art high-speed motor controller such as DSP or microcontroller can be directly employed to achieve the RDC task [6].

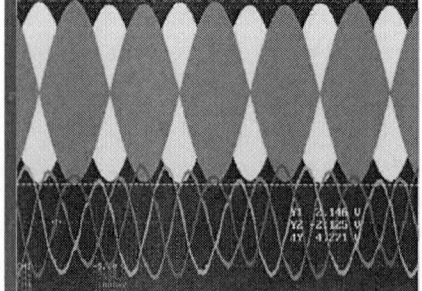

Fig. 10: Unsatisfactory resolver assembly and resulted control current

B. Hall Sensor Based Position Sensor

This type of absolute position sensors consists of two linear Hall sensors mounted on the stator and an excitation ring magnet with sinusoidal flux distribution assembled on the rotor shaft. The linear Hall sensors are used to sense the sinusoidal flux pattern of the ring magnet. They are assembled on the ISG stator with a 90-electric-degree displacement with respect to the ring magnet flux pattern. The pole number of the ring magnet should match with that of the PM machine.

If the flux distribution of the ring magnet is sinusoidal, the two linear Hall sensors will deliver the quadrature sine/cosine positional signals required for absolute position decoding. The position determination accuracy will highly depend on the flux pattern of the ring magnet and sensor mounting discrepancy. Fig. 11 illustrates a test result of the measured sine/cosine signals, decoded mechanical position and computed position measurement error based on a Hall-type position sensor. This position sensor was used to control a 20-pole SPM ISG machine for a mild ISG HEV demonstration. Its peak-to-peak position error is more than 2 degree. Such a big position error deteriorated the control performance of the machine at high speed, as shown in Fig. 6. As we can see from Fig. 11, the position error signal contained a distinctive harmonics of one mechanical cycle, which would be mainly due to the eccentric mounting of two Hall sensors. With the careful treatment on the sensor mounting, the linear-Hall-sensor based position sensor can still be feasible to deliver a position measurement resolution within ±0.5-mechanical degree, which was proved to be viable for HEV applications. One major drawback of the linear Hall sensors is that their DC offsets drift with ambient temperature. The uncompensated extra DC offset will cause extra position error. The DC-offset drift can be compensated through an appropriate algorithm.

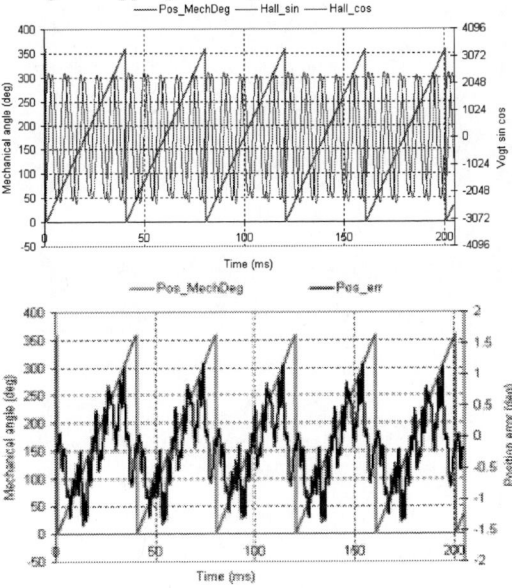

Fig. 11: Position error from a linear Hall position sensor

C. Eddy-Current Position Sensor

This type of absolute position sensor is based on the eddy-current principle. It can offer contactless and reliable position determination. It is less sensitive to electro-magnetic fields, and immune to temperature influences. It can be used as a substitute to a resolver. The principle of this absolute encoder is based on the losses of "eddy current" in the target (tooth wheel).

The positional sine/cosine signals from an eddy-current-type position sensor are decoded through an ASIC chip. The output signals of the ASIC depend on the covered area of the planar coils by the target wheel, as illustrated in Fig. 12. That means the form of the target determines the output signals. If the target wheel is designed with the sinusoidal structural shape, the required quadrature sine/cosine positional signals can be generated.

Fig. 13 shows our validation test results on an eddy-current type positions sensor. Similar to the resolver, the eddy-current position sensor can feasibly deliver the position resolution within around ±0.5 mechanical degree. Based on the position measurement from an eddy-current position senor, an ISG HEV was fully demonstrated. The adopted 12-pole IPM ISG machine was capable of running reliably up to the peak operational speed 6500rpm under control.

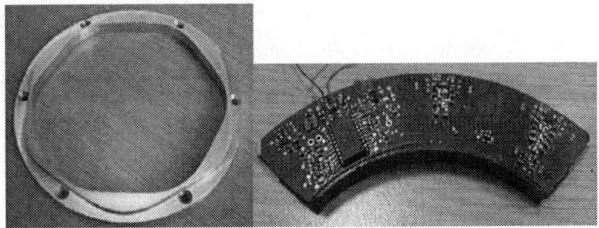

Fig. 12: Target wheel and ASIC circuit board of an eddy-current position sensor

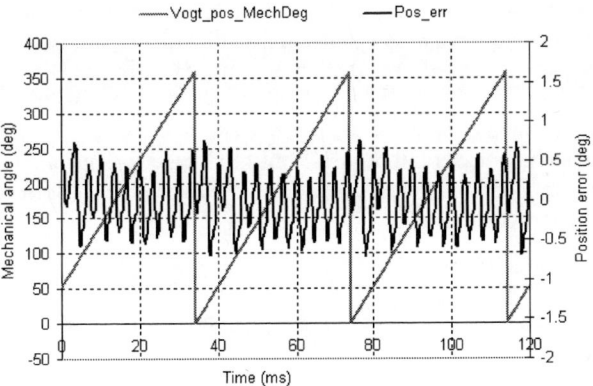

Fig. 13: Measured position error from an eddy-current position sensor

978-1-4244-1871-8/07 $25.00 © 2007 IEEE

IV. ISG CONTROL CONSIDERATIONS

A. ISG Control Strategy

The major function of ISG for HEV applications is to deliver required torque. Therefore to achieve torque control is the major task for implementing effective HEV powertrain. The instantaneous electro-magneto torque generated by a PM machine is governed by the equation of:

$$T_e = 1.5p[\psi_m i_q + (L_d - L_q) i_d i_q] \qquad (1)$$

where ψ_m is the permanent magnet flux linkage; L_d and L_q are dq-axis inductances; i_d and i_q are dq-axis currents. The torque in (1) contains two components: The first term $1.5p\psi_m i_q$ is due to permanent magnet, which is linear to i_q. The second term $1.5p(L_d - L_q) i_d i_q$ is called as the reluctance torque, which is due to the saliency of the PM machine.

For a SPM ISG machine, the dq-axis inductances are almost constant against control current. Magnetic saturation in SPM is not severe. Therefore for an SPM, $L_d = L_q$ can be assumed. In such a case, the SPM torque can be simply controlled by adjusting i_q.

An IPM ISG has dramatic difference in torque control. The reluctance torque of IPM can account for more than 50% of its overall torque. Therefore the reluctance torque has to be controlled correctly to meet the control demand. In addition, IPM machines are usually designed with magnetic saturation in favor of reluctance torque and sizing benefit. The magnetic saturation has significant impacts on L_q but very little influence on L_d (assuming $i_d < 0$). The knowledge of L_d and L_q is important for torque determination and torque control of an IPM. Reference 4 introduces an effective method for dq-axis inductance detection.

Fig. 14 illustrates the block diagram of a torque control strategy for an ISG PM machine. The dq-axis current control is achieved by the well-known vector control algorithm. The dq-axis current references for the current-loop controllers are computed based on the required torque command. The optimal current vector is determined based on the pre-stored Maximum-Torque-Per-Ampere (MTPA) lookup profile table. In addition, the inverter saturation level is checked to estimate the extra flux-weakening current component i_{dFW} based on the inputs of the inverter DC-link voltage V_{dc}, ISG machine speed, and dq-axis control voltage u_d and u_q. The q-axis current reference i_{qref} is verified accordingly to meet the torque demand T_{ref}.

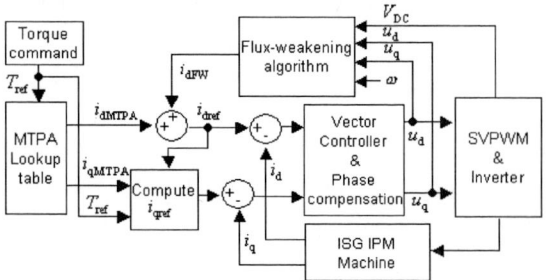

Fig. 14: Schematics of torque control for IPM ISG

B. DC-Link Voltage Impact and Flux-Weakening Control

If the DC-link terminals of ISG inverter are directly connected to the HEV high-voltage battery, the DC-link voltage will vary with the battery state-of-charge condition. Due to the voltage restriction in DC-link, the ISG motor may not be able to provide the required control power if no flux-weakening control is introduced above the machine base speed. In addition, the variation of DC-link voltage V_{dc} will have significant influence on the ISG performance.

Fig. 15 shows the estimated performance for a 20-pole SPM ISG machine with 15kW power rating, 300A peak current limit and no flux-weakening control, i.e. $i_d = 0$. The machine control power and torque decrease quickly with the speed increase and V_{dc} drop. Its maximum operational speed varies significantly with the battery-voltage level.

Flux-weakening control is commonly required for extending the ISG operational speed and improving ISG efficiency at high speed. The principle of flux-weakening for PM machines is to employ to negative i_d to reduce the d-axis flux-linkage, and essentially counteract the impact of the back EMF build-up due to permanent magnet.

The performance of flux-weakening control of PM machines highly depends on the machine design. A PM machine with small L_d will require large negative flux-weakening current i_d, which result in high loss in the machine. Both SPM and IPM machines can be specifically designed to provide a wide-speed constant-power operation if their characteristics current is close to the machine's rated current, i.e..

$$I_{d\text{-}FW} = -\psi_m / L_d \approx I_{rate} \qquad (2)$$

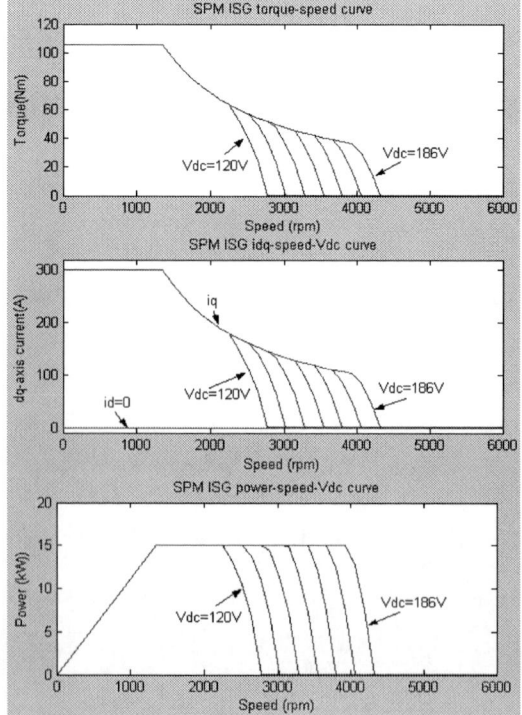

Fig. 15: Performance of a SPM ISG without flux-weakening control

978-1-4244-1871-8/07 $25.00 © 2007 IEEE

Fig. 16 shows the improved performance for the same SPM ISG machine under flux-weakening control. Above 1350rpm, the machine power output is clamped to 15kW. The required flux-weakening negative d-axis current for maintaining 15kW power changes largely with the available DC-link voltage. The lower the battery voltage is, the higher the flux weakening current is required. Consequently, the lower DC-link voltage will lead to more copper loss caused by increased flux-weakening current.

Flux-weakening control can be implemented by either an open-loop strategy or a closed-loop control. The open-loop strategy employs a pre-stored i_d look-up table based on the knowledge of the PM machine parameters and inverter DC-link voltage. The machine parameter uncertainties may cause under-controlled or over-controlled flux-weakening. The excessive flux-weakening i_d current will introduce unnecessary extra copper loss. The under-controlled flux-weakening may reduce the control power output and even cause the flux-weakening failure, which may trigger harmful uncontrolled generation. Furthermore, the inverter DC-link voltage variation and position sensor errors may bring significant difficulty in achieving reliable open-loop flux-weakening control. Therefore, a more robust feedback flux-weakening strategy against the system uncertainties and V_{dc} variation can be utilized [4]. However, dynamics of the feedback flux-weakening loop is limited [5]. Sudden system variation may trigger the instability of current control. Therefore the combination of the feed-forward open-loop i_d, with a feedback method should achieve an optimal trade-off between control response and control loss.

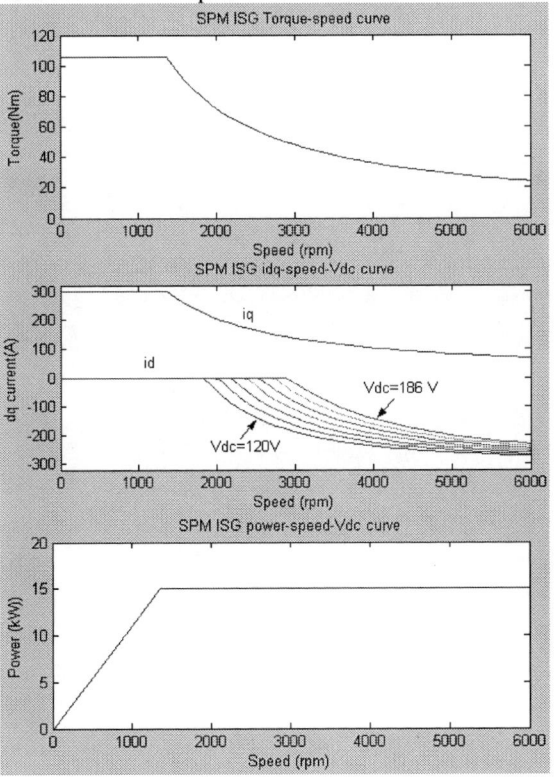

Fig. 16:   Performance of the SPM ISG with flux-weakening control

C. Digital Control Delay

The time delay is unavoidably induced due to the digital implementation of the current control and Pulse-Width-Modulation (PWM). The control delay will cause the phase deviation and voltage control error. At the low speed, such a delay is not so serious to influence the control performance. However when ISG machines runs to the high speed, the digital delay may cause the instable control of ISG PM machines. For a 20-pole PM machine, the fundamental frequency of phase currents is 1kHz at 6000rpm. If an 1.5-PWM cycle-period delay (which is reasonable due to the PWM update delay in the DSP or microcontrollers and ADC sampling delay) is assumed, the overall phase control error due to digital delay can reach up to 54° (assuming 10kHz PWM frequency) at 6000rpm. Such a large phase error would seriously deteriorate the ISG control performance and even drive the ISG control system to an instable region. Therefore the digital delay compensation has to be applied to maintain the robustness of the ISG control system.

A simple phase-delay compensation method can be introduced to improve the current control. The phase delay of the 1.5-PWM cycle-period delay can be computed by:

$$\theta_{\text{delay}} = 1.5\, p\omega\, T_{\text{PWM}} \qquad (3)$$

where p is the pole-pair number of the ISG PM machine; ω is the ISG rotor speed; T_{PWM} is the PWM cycle period. The delay compensation can be applied into Inverse Park transformation of the vector control algorithm.

V.   CONCLUSIONS

To develop an ISG PM system for HEV applications, many design criteria are required to be evaluated and synthesised. The system is usually developed with an optimal compromise solution, which would balance the system cost, performance and reliability as well. This paper summarises some of the important design experiences gained through our long HEV development history.

The ISG PM machine design is extensively involved in the machine parameter optimization in term of the machine size, pole-pair number, permanent magnet flux, inductance, etc. A cost-effective solution is more favourable for HEV applications. The 15kW low-cost IPM discussed in this paper has been successfully demonstrated in a mild ISG HEV application. It is a good example that has successfully achieved the trade-off between the performance and cost.

The position sensor problem is usually ignored in the academic papers dealing with the ISG machine control. However, this paper presents the evidence that the position measurement accuracy has significant impacts on the performance, stability and even safety of the ISG system operation. Three prominent position sensors are evaluated and analysed. Reliable and accurate position determination can bring significant benefits for improving torque control dynamics, reducing control torque ripple, extending operational speed, increasing ISG control efficiency and stability, etc.

The torque control of ISG PM machines is achieved through vector control algorithm. Flux-weakening control is usually necessary for achieving constant-power operation at high speed. The ISG machine should be stably controlled to cope with the voltage variation at the inverter DC-link. The maximum ISG power may have to be de-rated accordingly to avoid the inverter saturation, which may cause feedback control failure. The digital control delay is another important factor having significant influence on the ISG control efficiency and stability. A simple phase delay compensation method is introduced.

ACKNOWLEDGMENT

The authors wish to express sincere gratitude to the Ricardo Hybrid Vehicle Development team and Chery Automotive Co. Ltd for strong support of preparing and reviewing the paper.

REFERENCES

[1] K Yamamoto, "Development of Motors for EV FCV and HEV", 2005 International Power Electronics Conference, pp. 1489-1494, 2005.

[2] S. Abe, M. Murata, "Development of IMA Motor for 2006 Civic Hybrid", 2006 SAE World Congress, SAE num 2006-01-1505.

[3] M. Kamiya, "Development of Traction Drive Motors for the Toyota Hybrid System", 2005 International Power Electronics Conference, , pp. 1474-1481, 2005.

[4] X. Chen, R. Thornton, E. Murray, "Accurate Torque Control of IPM machines for ISG Hybrid Vehicle Applications", 14th Asia Pacific Automotive Engineering Conference (APAC-14), Aug 2007.

[5] N. Bianchi, S. Bolognani, M. Zigliotto, "High Performance PM Synchronous Motor Drive for an Electrical Scooter", 2000 IEEE Industry Application Conference, pp. 1901-1908, 2000.

[6] M. Staebler, "TMS320F240 DSP Solution for Obtaining Resolver Angular Position and Speed", Texas Instrument application node, SPRA605, Feb 2000.

Control of IPMSM drive system for drum washing machine

Won-Cheol Lee[a], Sang-Hoon Park[a], Jung-Hyo Lee[a], Young-Ryul Kim[b], Chung-Yuen Won[a*]

[a]School of Information & Communication Eng., Sunkyunkwan Univ., Korea
[b]Department of Electronic & Electric Engineering Anyang Univ., Korea

Abstract - The drum washing machine needs high torque than the other types of washing machine because of the large load torque variation. Therefore, in this paper, we applied a interior permanent magnet synchronous motor(IPMSM) to the drum washing machine system. However, IPMSM operate at lower output torque than the other permanent magnet synchronous motor(PMSM) when the motor control is performed by the conventional control method. This paper suggests adaptive control method for IPMSM to obtain better torque performance and apply the control method in actual experiments of the drum washing machine system.

I. INTRODUCTION

This paper describes the control system for drum washing machine applications. The system target is to control the IPMSM (Interior permanent magnet synchronous motors), that is adopted to the drum washing system by home appliance company, with better torque performance.

Drum washing machine needs to operate at washing speed of 40~50 rpm and at spin speed of 800~1200 rpm, and its speed ratio is about 25:1.

At the washing mode, motor demands the high starting torques and high operating torques at low speed. On the other hand, at the spinning mode, the demanding torque is lower than at the washing mode, but the output power of the motor must be maximum value.

IPMSM has wider operating ranges and has more abilities to produce torque compared with the other types of PMSM with the same frame size. Therefore, IPMSM is more attractive for use in domestic appliance including drum washing machine.[1]

However, if the conventional advanced angle control or FOC(field oriented control) are applied to IPMSM control,

IPMSM has lower torque than the conventional SPMSM which adopted those control.

Therefore, in this paper, we present new control algorithm applicable to IPMSM drive for drum washing machine.

The proposed control algorithm is verified through the experimental results.

II. THE DEVELOPED TORQUE OF IPMSM

Field oriented IPMSM drive systems are increasingly used in appliance industry. Because it has many benefits of smaller size, reduced noise and energy saving.[4]

Fig. 1 shows inductance distributions of PMAC motors as a function of the saliency. There are three motor types, that is, SynRM(Synchronous Reluctance Motor) which does not include any permanent magnets in rotating frame, SPMSM(Surface Permanent Magnet Synchronous Motor) which has surface magnets on surface of rotor, IPMSM which has interior permanent magnets in the rotor.

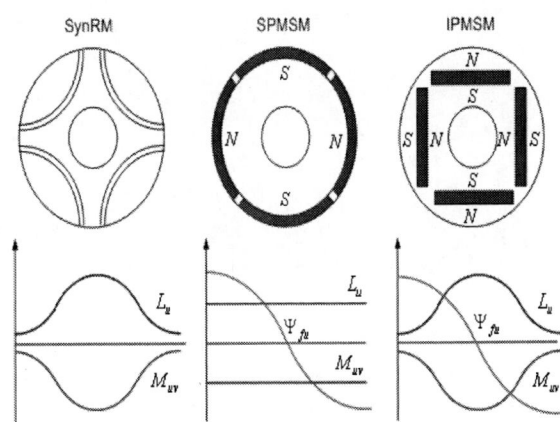

Fig. 1. Inductance variation of the PMAC motors

978-1-4244-1871-8/07 $25.00 © 2007 IEEE

Because of saliency of SynRM, its L(Self inductance of stator windings) and M(Mutual inductance) are changed by rotor position. Therefore, the magnetic energies stored in the air gap are transformed to the mechanical energies with its changes. Generally, the torques generated by those energy transform of the torque mechanism are Reluctance Torques.

When the torque generating mechanisms of SynRM is expressed mathematically, it is as following equation.

$$T_r = \frac{1}{2} i_u^2 \frac{\partial L_u}{\partial t} + \frac{1}{2} i_v^2 \frac{\partial L_v}{\partial t} + \frac{1}{2} i_w^2 \frac{\partial L_w}{\partial t}$$
$$+ i_u i_v \frac{\partial M_{uv}}{\partial t} + i_v i_w \frac{\partial M_{vw}}{\partial t} + i_w i_u \frac{\partial M_{uu}}{\partial t} \qquad (1)$$

Because the relative permeability of permanent magnet and air gap is almost same and surface magnets are arranged at same pitch, there are constant reluctances at any rotor positions. Therefore, L and M are constant values. However, stator flux generated by permanent magnets have a sinusoidal pattern with rotating angle of rotor, there are mechanical torques using those stator flux.

When the torque generating mechanisms of SMPMSM is expressed mathematically, it is as following equation.

$$T_m = i_u \frac{\partial \Psi_u}{\partial \Theta} + i_v \frac{\partial \Psi_v}{\partial \Theta} + i_w \frac{\partial \Psi_w}{\partial \Theta} \qquad (2)$$

$$\begin{bmatrix} V_u \\ V_v \\ V_w \end{bmatrix} = \begin{bmatrix} R_a + pL_u & pM_{uv} & pM_{uu} \\ pM_{uv} & R_a + pL_v & pM_{vw} \\ pM_{uu} & pM_{vw} & R_a + pL_w \end{bmatrix} \begin{bmatrix} i_u \\ i_v \\ i_w \end{bmatrix} \qquad (3)$$
$$- \begin{bmatrix} \omega \Psi_f \sin\Theta \\ \omega \Psi_f \sin(\Theta - \frac{2\pi}{3}) \\ \omega \Psi_f \sin(\Theta + \frac{2\pi}{3}) \end{bmatrix}$$

$$\begin{bmatrix} v_d \\ v_q \end{bmatrix} = \begin{bmatrix} R_a + pL_d & -\omega L_q \\ \omega L_d & pL_q \end{bmatrix} \begin{bmatrix} i_d \\ i_q \end{bmatrix} + \begin{bmatrix} 0 \\ \omega \Psi_a \end{bmatrix} \qquad (4)$$

In the IPMSM, stator flux generated by buried permanent magnets have sinusoidal variations as in the SPMSM. Because the permanent magnets are buried in the rotor of IPMSM, even if rotor has symmetric mechanical structures, it's electrical distributions are not symmetric. Therefore, self inductance and mutual inductance of IPMSM rotate sinusoidally at twice the

speed than the rotating speed of stator flux.

Self inductance and mutual inductance of stator and the stator flux generated by magnets are changed by the revolution of rotor, and those changes affect torque of IPMSM.

Voltage equations to express torque of IPMSM at synchronous d-q reference frame are as following equation (3). Torque of IPMSM is expressed by equation (4) at d-q reference frame.

Torque of IPMSM is expressed by equation (4) at d-q reference frame. Voltage equations at d-q reference frame in equation (4) are changed to torque equation as equation (5).

$$T = P_n \Psi_a i_q + P_n (L_d - L_q) i_d i_q \qquad (5)$$

In equation (5) there are two torque components, one is generated by flux of permanent magnets and the other is generated by the differences of reluctance. The first component is due to the rotor magnet flux, while the second component is reluctance torque.

Therefore, IPMSM can produce 15% more torque than SPMSM of the same size. The additional torque is due to the saliency of rotor that produces reluctance torques. In Fig. 2, torque curves vs current phase are shown.

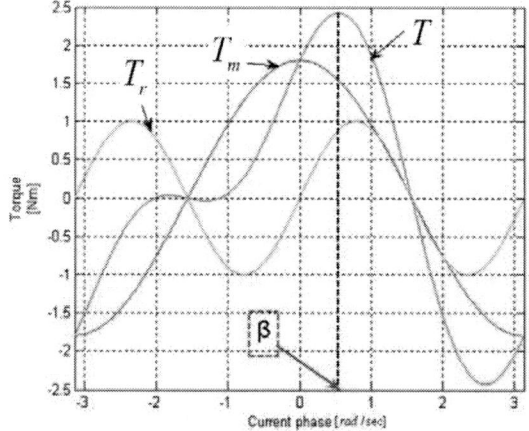

Fig. 2. Torque curve of the IPMSM

III. CONTROL OF IPMSM

3.1 Maximum torque control of IPMSM

In fig. 2, because the torque angle(β) varies with the current phase, the phase angle to maximize the output torque at any given current phase is required.

Therefore, if the conventional field oriented control is applied to IPMSM drive system, the IPMSM motor produces same output torque as SPMSM. Moreover, if the torque angle(β) is leaded at certain miss-matched time, output torque of the IPMSM is decreased.[2][3]

Therefore, in IPMSM drive system, to produce maximum output torque, the torque angle(β) must be lagged properly, and to find optimal torque angle(β), equivalent output torque is expressed at d-q synchronous reference frame.

In fig. 3, the red line is the maximum current line in IPMSM and I_a is the vector sum of d-axis current and q-axis current. It is possible to maximize the output torque using a certain d-axis current at a given total current.

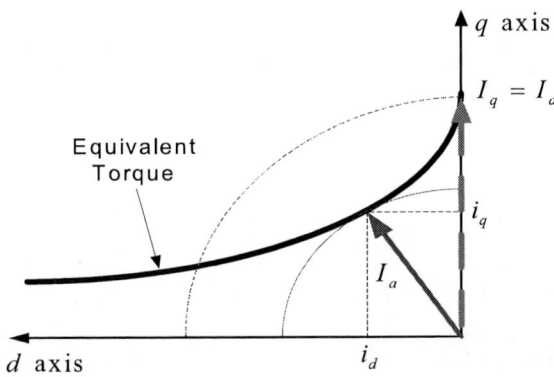

Fig. 3. Compare of the field oriented control and efficiency control

d-axis current and q-axis current as a function of torque angle β are defined in the equation below.

$$i_d = -I_a \sin\beta \qquad (6)$$
$$i_q = I_a \cos\beta$$

If equation (6) is applied into equation (5), the output torque is a function of d-axis current, q-axis current and torque angle β.

$$T = P_n \left\{ \Psi_a I_a \cos\beta + \frac{1}{2}(L_q - L_d)I_a^2 \sin2\beta \right\} \qquad (7)$$

Therefore, to find the maximum output point, partial differential of torque angle β is performed as in equation (7), and maximum torque angle point when the slope is zero are shown in the equation below.

$$\frac{\partial T}{\partial \beta} = P_n \{ -\Psi_a I_a \sin\beta + (L_q - L_d)I_a^2 \cos2\beta \} \qquad (8)$$
$$= P_n \{ -\Psi_a I_a \sin\beta + (L_q - L_d)I_a^2(1 - 2\sin^2\beta) \}$$

$$\beta = \sin^{-1}\left(\frac{-\Psi_a + \sqrt{\Psi_a^2 + 8(L_q - L_d)^2 I_a^2}}{4(L_q - L_d)I_a} \right) \qquad (9)$$

If operating angle β is applied into equation (6), there are optimal compounds of d-axis currents and q-axis currents.

$$i_d = \frac{\Psi_a - \sqrt{\Psi_a^2 + 8(L_q - L_d)^2(i_d^2 + i_q^2)}}{4(L_q - L_d)} \qquad (10)$$

$$i_d = \frac{\Psi_a}{2(L_q - L_d)} - \sqrt{\frac{\Psi_a^2}{4(L_q - L_d)^2} + i_q^2} \qquad (11)$$

3.2 Field weakening control of IPMSM

D-axis current of IPMSM as a function of the speed and load conditions must be controlled by the field weakening algorithms to avoid that the motor voltage is larger than the inverter maximum.

In the field weakening algorithms shown in fig. 4, the voltage magnitude of the target motor is measured and the reference d-axis current is calculated, then the d-axis current is regulated.

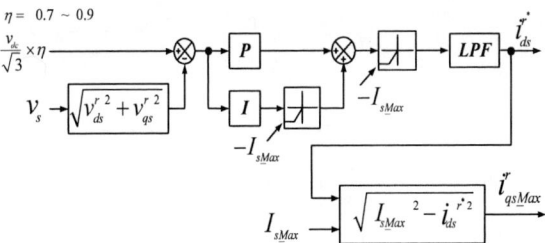

Fig. 4 Field weakening control by voltage feedback loop

IV. EXPERIMENTAL RESULTS

4.1 System implementation of field oriented IPMSM drive system for washing machine

The specifications and machine parameter of the prototype IPMSM are listed in Table 1. The block diagram of the overall control system of IPMSM for washing machine is shown in Fig.5.

It consists of the power conversion unit, the micro-controller unit for control algorithm implementation, and the hardware interface unit.

The power conversion unit is composed of power converter, 6-switch inverter, and voltage regulating circuits. The micro-controller unit chose the DSP chip TMS320F2811 for high dynamic performance.

The current controller, the speed controller, and the field weakening algorithm were implemented in software. The hardware interface unit includes SVPWM, current sensors, and speed sensors

An overall system block for the complete drum washing machine system is shown in fig. 6.

The overall system is composed of power conversion unit, μ-controller unit for control algorithm and hardware interface unit.

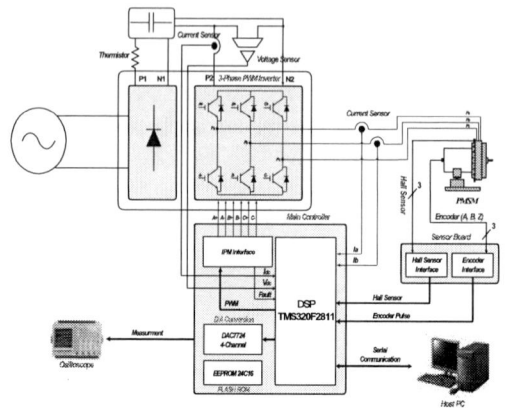

Fig. 6. Block Diagram of complete System

It include measuring circuits, an intelligence power module, integrates power rectifier, gating driver and inverter switches. There are 2 hall ICs for speed control and 2 current sensors for current control, respectively.

Specification of prototype IPMSM	
Rated speed	400 r/min
Rated current	5.0 A_{rms}
Back EMF constant	450 V_{pk}/krpm
Number of Poles	24
Phase resistance	3.1 Ω
Ld	42.5 mH
Lq	48.5 mH

Table 1 Specification of prototype IPMSM

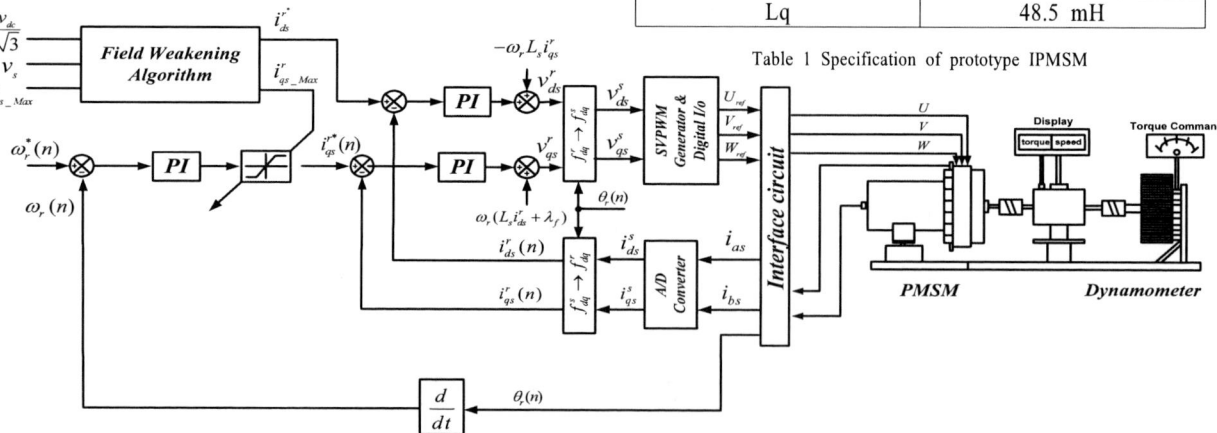

Fig. 5 IPMSM drive system block

4.2. Experimental results

Fig. 7 shows the acceleration characteristics. The motor speed reaches to +50% of the rated speed from -50% of the rated speed only in 0.1 second at no load condition. By using FOC, the proposed drum washing machine has good acceleration characteristics.

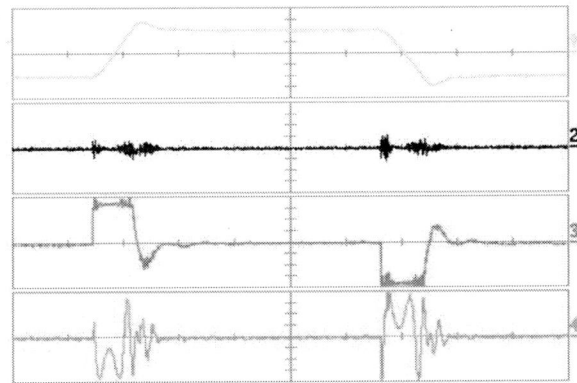

Fig. 7. Reference Speed(-200[rpm]~200[rpm]~-200[rpm]), d-axis Current, q-axis Current, Phase-a Current

Also fig. 8 shows the step response to 100% of the rated torque(23 Nm) at washing mode speed(45rpm).

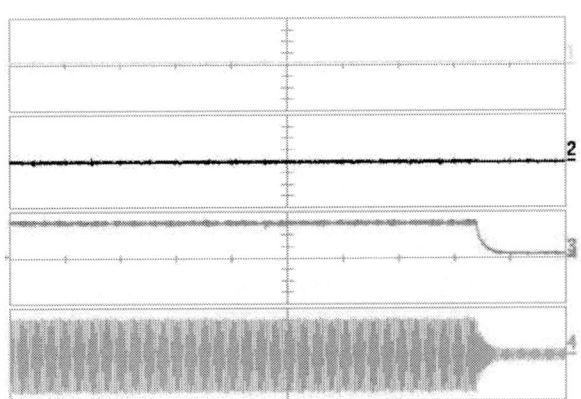

Fig. 8. Load torque test(45[rpm], 23[N.m])
(a)motor Speed, (b)d axis Current, (c)q axis Current, (d)Phase Current

Fig. 9 shows the speed and torque characteristics in the IPMSM drive system of the proposed method. The speed reference is 300 r/min. Initially, the load torque of 23 [Nm] is applied, and then, after about 30 seconds, the load is removed from the system. Fig. 9(a) shows that there is no variations of the real speed when the load torque is applied to the system.

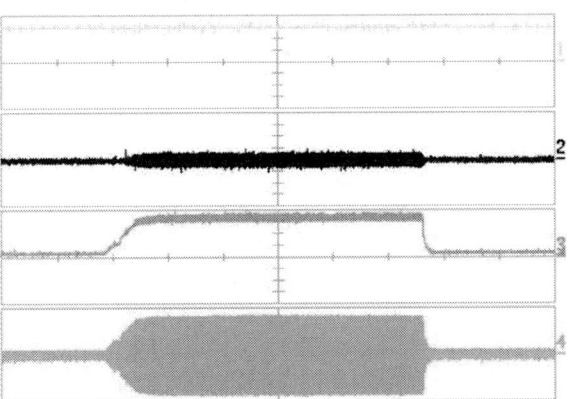

Fig. 9. Load Response Characteristic (300[rpm], 23[N.m])
(a)Rotor Speed, (b)d-axis Current, (c)q-axis Current, (d)Phase Current

Fig. 9(b) shows that the d-axis current is stable only with the small variations when the load torque is applied and removed.

4.3. Field weakening control

Fig. 10 show the acceleration characteristics in field weakening region. The motor speed reaches to +200% of the rated speed from -200% of the rated speed in 12 second at no load condition.

By using field weakening algorithm with voltage feedback loop, the proposed washing machine has good acceleration characteristics.

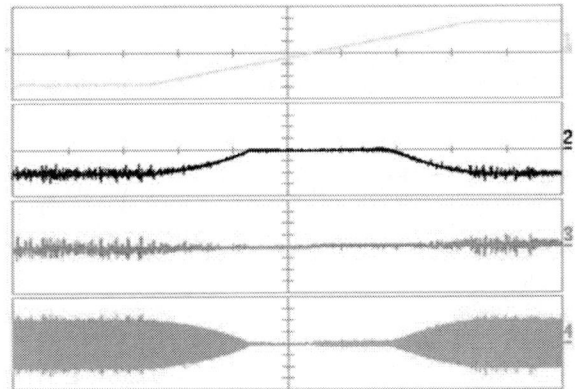

Fig. 10. Field weakening test
(a)Motor Speed, (b)d-axis Current, (c)q-axis Current, (d)Phase Current

V. CONCLUSIONS

The complete implementation of a field oriented IPMSM for drum washing machine has been presented in this paper. The application of field oriented control to IPMSM for washing machines provides many application benefits including high power density and good dynamic performance both at low speed in washing mode and at high speed in spin mode.

The dynamic performances of the system has been investigated at different dynamic operating conditions both theoretically and experimentally. Therefore, it can be concluded that the proposed IPMSM drive system with FOC has shown superior performance over the SPMSM drive system for drum washing machine.

ACKNOWLEDGEMENTS

This work was supported by Energy resources R&D Projects, under the contract number 2006-11-01070000-0-000, sponsored by Korea Energy Management Corporation

REFERENCE

[1] T. M. Jahns, G. B. Kliman, T. W. Newmann , "Interior permanent-magnet Synchronous motors for Adjustable-speed Drives", IEEE trans. on Ind. App. IA-22, No.4, pp 738-747, (1986)

[2] S. Morimoto, Y. Tong, Y. Takeda, T. Hirasa, "Loss minimization control of permanent magnet synchronous motor drives", IEEE trans. on Ind. App. IE-41, No.5, pp.511-517, (1994)

[3] B. K. Bose "A high performance inverter-fed drive system of an interior permanent magnet synchronous machine", IEEE Trans. on Ind. App, IA-24, No.6, pp.987-997, (1998)

[4] S.A. NASAR, I. BOLDEA, L.E. UNNEWEHR,　"Permanet magnet, reluctance, and self synchronous motors", CRC Press,　Inc. 1993

Fuzzy Logic Controller for Boost Converter
with Active Power Factor Correction

Ing. Esp. Fredy H. Martínez S.
Distrital University Francisco José de Caldas
Technological Faculty – Technology in Electricity
Carrera 7 No 40 - 53, Bogotá D.C., Colombia
Email: fhmartinezs@udistrital.edu.co

Ing. Teg. Diego F. Gómez M.
Distrital University Francisco José de Caldas
Technological Faculty – Technology in Electricity
Carrera 7 No 40 - 53, Bogotá D.C., Colombia
Email: dgomezm@codensa.com.co

Abstract—**This paper presents the successful use of fuzzy logic to derive a practical control scheme for boost converter with active power factor correction. Aspects of sudden changes in the load and the input voltage are discussed and presented. The fundamentals governing the design, control and performance are also illustrated. Properties of the proposed controller are: (1) converter with active power factor correction, (2) good performance of transient responses under varying loading conditions and/or input voltage, and (3) robustness around the operating point.**

I. INTRODUCTION

In the eagerness to reduce the complexity of the circuits when implementing traditional solutions for the control of converters with active power factor correction, different control techniques are presented [1], [2], [4]. In classical approach, fuzzy controllers are designed by using the output voltage error and the change of voltage error, or a mixture of these variables. However, to reconstruct the current (power factor correction) and to improve the dynamic answer of the converter; it is necessary for the control to have information about the absorbed energy of the network; in order to get that, the error of the input current is included like a variable [5].

In this scheme, the proposed fuzzy controller uses only two input variables: (1) Output voltage error (*Eu*) that is used for the regulation of the output voltage, and (2) Inductor current error (*Ei*) that is used to improve the dynamic answer of the converter, to correct the power factor and to limit the current as a protection. It is important to point out that while the

output voltage reference is usually considered as an external signal, the inductor current reference depends on the operating point; for this reason, it is computed by means of a low-pass filter in the assumption that the DC value of the current is automatically adjusted by the converter according to the power balance condition [3]. This DC reference is used by the control unit to construct a sine waveform signal reference in phase with the input voltage (a detector of crossing by zero is used). A block diagram of the fuzzy controller structure is shown in Fig. 1.

The proposed control scheme has been verified on a Boost rectifier of 200 W, output voltage DC of 400 V and nominal input voltage of 120 V at 60 Hz. The power factor reached was 0.924 with a total harmonic distortion of 21.1%.

The basic theory of the developed design, the collected data, conclusions as well as suggestions are presented in order to solve the identified problems in the implementation of the scheme.

II. THEORY

A. Fuzzy Logic Controller

In this work, a fuzzy control is applied in the feedback loop in order to use all the possibilities that this structure of control can offer to the active power factor correction. As an advantage, it is not necessary to use a mathematical model of the converter, which supposes a quite complex work.

The control action in a fuzzy logic controller is determined by a set of linguistic rules. It means, a certain knowledge of the converter behavior is required (knowledge of the expert),

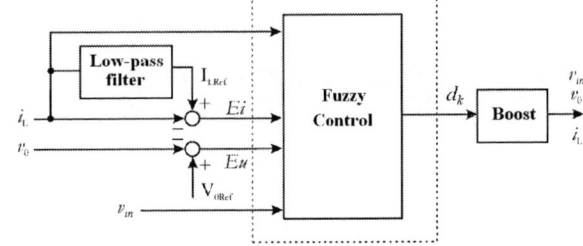

Fig. 2: Detail of the proposed controller

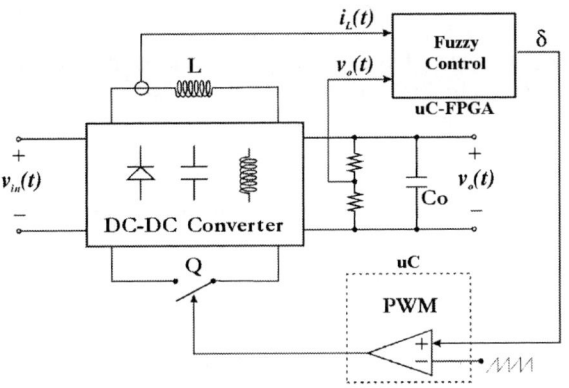

Fig. 1: Converter with proposed controller

sufficient to create the linguistic control rules, but it is not necessary to create the small signal model. Nevertheless, the fuzzy control cannot provide a better dynamic answer than the obtained by a very well tuned traditional system. The great advantage of the scheme is the possibility of considering simultaneously in a simple control scheme, regulation, dynamic answer and power factor correction aspects.

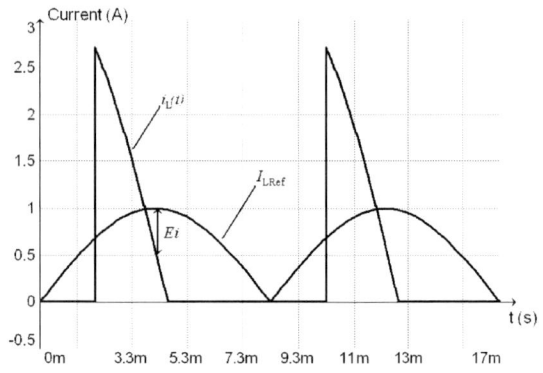

Fig. 3: Current error

In the classical approach, fuzzy controllers are designed by using the output voltage error and the change of voltage error, or a mixture of these variables. However, to reconstruct the current (power factor correction) and to improve the dynamic answer of the converter; it is necessary for the control to have information about the absorbed energy of the network. In order to obtain that, the error of the input current is included like a variable. This way, the basic information of the current loop of the traditional schemes of control is included, but it allows the inclusion of more information in the control, thanks to the fuzzy control scheme.

Fig. 2 shows the configuration of the fuzzy logic control. The output voltage of the boost converter is scaled and compared with the reference value, which is inside a microcontroller. The current reference value is created by sensing the inductor's current, after being filtered through an analog low-pas filter, since this value must change based on the converter's load (operating point). This is supposed, if the DC value of the current is automatically adjusted by the converter according to the power balance condition.

This DC value of the current is used to construct the sine signal reference, which finally will be the current reference to

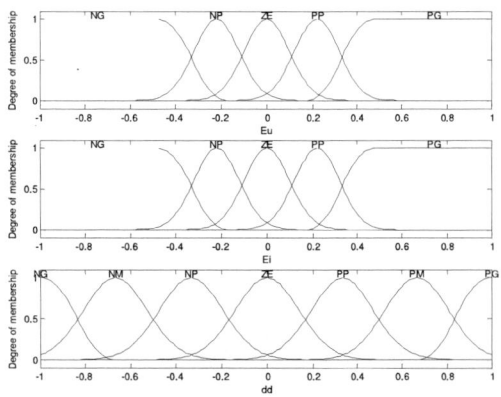

Fig. 4: Membership functions

compare directly by the microcontroller with the instantaneous current value sensed through the inductor. This sine reference is the signal that allows the fuzzy control to make the power factor correction (Fig 3).

These input variables to the fuzzy control block are defined as:

$$Eu[n] = \frac{V_{0Ref} - v_0[n]}{\alpha} \quad \text{and} \quad Ei[n] = \frac{I_{LRef}[n] - i_L[n]}{\beta} \quad (1)$$

Where α y β are constants used to normalize the voltage and current error. The output variable of the fuzzy control is change of duty cycle in (2) [i.e., δd]. The actual duty cycle is determined by adding $d[n-1]$ to the calculated change,

$$d[n] = \eta \times \delta d[n] + d[n-1] \quad (2)$$

Where η represents the effective gain of fuzzy control. Other signals that are introduced to the control block from the microcontroller are: the value of the current sensed in the inductor, in order to protect the system in case of over currents; and a crossing by zero pulse, used as a reference to create the sine signal reference (sine reference, I_{LRef}).

B. Fuzzification

Fuzzificattion is to map *Eu* and *Ei* in (1) into suitable linguistic values. Five fuzzy levels are defined for *Eu* and *Ei*, including negative big (NG), negative small (NP), zero (ZE), positive small (PP), and positive big (PG). Each input variable is assigned a membership value μ to each fuzzy set, based on a corresponding membership function. Fig. 4 shows the membership functions, which are Gaussian fuzzy-set values.

Besides, as a way of protecting, an exclusive fuzzy set is used for the input current; it is called LIMIT and is really a restricted point set from 0.7. Whenever, the current is below 0.7, normal conditions are assumed and the developed theory of control is applied. Otherwise, if the current is over 0.7 (for example, starting point) a special operation in the control is required to avoid overshoots. This is described in detail when the fuzzy rules are defined.

In order to facilitate the analysis, the input fuzzy variables have been described by normalized singleton values. This way the measured values for these variables are used in the inference process and it is not necessary to use the fuzzification stage.

C. Rules of fuzzy control

Construction of fuzzy control rules is a heuristic process, and in this particular case it is based on the following conditions:

- In normal conditions ($i_L(t)$ is not LIMIT):

When the output voltage is far from the set point ($Eu(t)$ is NG or PG), the corrective action must be strong to minimize the response time (a huge value in the duty cycle increase and reach the set point quickly).

978-1-4244-1871-8/07 $25.00 © 2007 IEEE   936

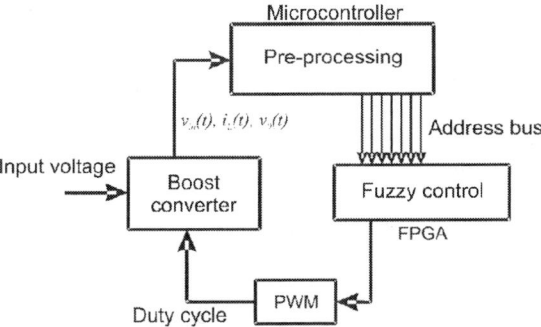

Fig. 5: Basic control structure

When the output voltage is close to the set point ($Eu(t)$ is NP, ZE or PP), the corrective action must be small or zero to minimize the overshoot possibility.

When the output voltage is on the set point ($Eu(t)$ is ZE) and the output is still changing, the duty cycle must change just a little, in order to avoid the strong change in the output.

When the output voltage achieves the set point ($Eu(t)$ is ZE), and the output is stationary, the duty cycle must kept constant.

When the output is over the set point, the change in the duty cycle must be negative and vice versa.

When the inductor's current is below the reference, duty cycle must be increased in order to reconstruct the current waveform; and vice versa, if the inductor's current is over the reference, duty cycle must be reduced.

- In limit current conditions, ($i_L(t)$ is LIMIT):

If $Eu(t)$ is PG, duty cycle increase must be zero, so that the value of the current is limited.

If $Eu(t)$ is ZE, duty cycle increase must be negative to avoid non desired overshoots (for example, low load starting).

In agreement with these conditions, control tables have been constructed and simulated in MatLab® for each and every possibility, in order to get all the possible fuzzy control outputs. The simulation results were codified in binary and stored in a FPGA.

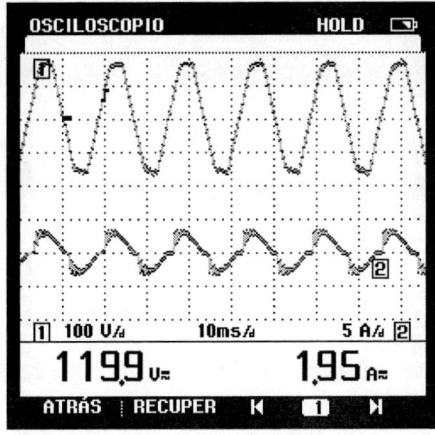

Fig. 6: Steady-state input voltage and current waveforms (Ch1: input voltage 100V/div and Ch2: input current 5A/div, time base 10ms/div)

D. Decision-Making

Decision-Making infers a fuzzy control action from knowledge of the fuzzy rules and the linguistic variable definition. The control rules that determined the output of the fuzzy logic control are based on the general knowledge of the system behavior or intuition of the process being controlled.

Fuzzy control inference works depending on the conditions defined by the expert, in the rules of fuzzy control. Table 1 shows the fuzzy model based on fuzzy rules for the boost rectifier. The entries of the table are the normalized singleton values of the change of $d[n]$, stored in a digital device. The inferred output of each rule is obtained by Mandani's min fuzzy implication.

TABLE 1
Control rule of the fuzzy logic control

Ei \ Eu	NG	NP	ZE	PP	PG
PG	NG	PP	PM	PG	PG
PP	NG	ZE	PP	PG	PM
ZE	NG	NP	ZE	PP	PM
NP	NG	NG	NP	ZE	PP
NG	NG	NG	NM	NP	ZE

E. Defuzzification

Defuzzification is to convert the inferred fuzzy control action to a nonfuzzy control action.

For the change of $d[n]$ output, seven fuzzy sets are used (PG, PM, PP, ZE, NP, NM and NG) in order to get a uniform control action. (the larger the numbers of fuzzy sets, the higher resolution in the fuzzy control is). Fig 4 shows the Gaussians membership functions used in the fuzzy control. Each input-output variable is normalized between [-1, 1] by using suitable scales factors.

During this operation, a crisp value for $\delta d[n]$ is calculated by using the center of gravity method.

III. EXPERIMENTAL DESIGN

The fuzzy controller has been implemented by a Xilinx's FPGA SPARTAN 3. Based circuit and fuzzy computations have been pre-calculated in MatLab® and stored in the FPGA. The A/D conversion and the supervision of all operation are made by Motorola's M68HC908GP32 microcontroller.

Fig 5 shows the implemented control design schematically. It has three basic sections: (1) A pre-processing section, which evaluates the control's inputs. (2) A fuzzy control based on a table with the stored duty cycle set points as a function of the input variables. (3) A PWM that applies the final duty cycle to the boost converter. Basically, the control was designed as easily as possible, using commonly used equipment and devices.

Input signals are digitalized in the fuzzy control block, and injected to the configured ports in the FPGA, because these outputs represent the concatenated digital address of the set value of the duty cycle. The signals were codified in 8 bits, for this reason, it was necessary to store the information in a FPGA.

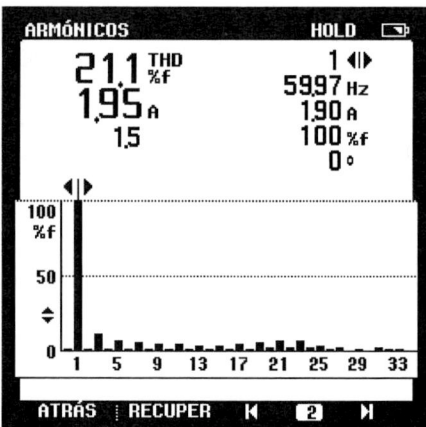

Fig. 7: Harmonic distortion in steady-state

This control is tested using a boost converter at 50 Khz with a 2mH inductor, 330uF capacitor and resistive load R_L of 800Ω. The microcontroller takes 18us to digitalize each analog signal.

IV. DATA ANALYSIS AND CONCLUSION

Fig. 6 shows the experimental verifications in steady-state on a 200 W boost rectifier (desired DC output voltage: 400 V, nominal input voltage: 120 V at 60 Hz), which reaches a 0.924 power factor and THD of 21.1%. Fig 7 shows the signal harmonic content. It is possible to observe how the signal of the current is in phase with the voltage signal (a power factor close to the unit is obtained).

In order to investigate the behavior with load changes, when the system is in steady state, a step change in the load is applied with a constant input voltage (R_L = 1500Ω to 800Ω). The output voltage waveforms in the transient are showed in Fig 8.

The output shows an overshoot near 12% and a settling time of 48ms, within the 5% band. Then, the resistance comes back to the 1500Ω value. The new transient is showed in Fig 9. It presents an overshoot of about 18% and a settling time of 70ms.

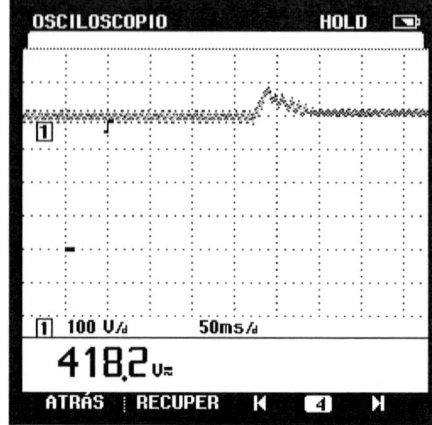

Fig. 9: Output voltage (50V/div). R_L is changed from 750Ω to 1500Ω

Concerning a large-signal supply voltage change in v_{in}, a test was made to change the input voltage from 120V to 80V, keeping the R_L value constant. The transitory in the output voltage is showed in Fig 10. The input current is shown in Fig 11.

The output voltage, in this case, presents a drop of voltage of 16%, and it returns to its nominal value after 80ms. It is important to point out that the input current is a sine signal in this period of time.

In conclusion, this design presents a fuzzy control scheme, in order to develop a pre-regulator of voltage with active power factor correction, and satisfy some international standards like IEC 61000-3-2 with the purpose of implementing it in low and median power electronic devices. The results obtained experimentally show the viability of the scheme.

The fuzzy control scheme allows to work on three different problems at the same time: voltage regulation, dynamic response and active power factor correction; this is possible because the fuzzy rules could be assigned for different operation areas.

Future developments must be dedicated to optimize the fuzzy rules to improve the converter performance. Also, it is

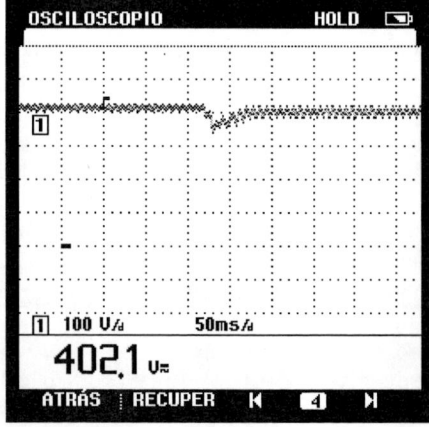

Fig. 8: Output voltage (50V/div). R_L is changed from 1500Ω to 750Ω

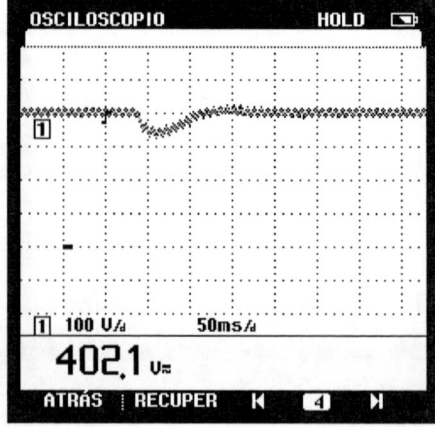

Fig. 10: Output voltage (50V/div). v_{in} is changed from 120V to 80V

suggested to shift the control structure located inside the microcontroller to the FPGA in order to use just one control device and increase the speed in the processing time, which is the main reason for the high distortion in the current presented.

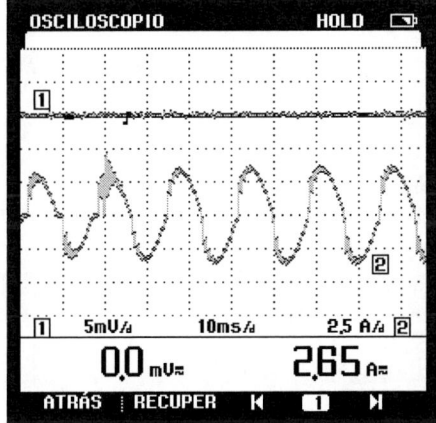

Fig. 11: Input current (2.5A/div). v_{in} is changed from 120V to 80V

Power electronics future is not restricted to the power density, to reduce EMI and to obtain high efficiency, it also involve the development of new strategies of intelligent control that could be adapted to the conditions of the new applications and market demands.

ACKNOWLEDGMENT

This work was supported by the group of investigation ARMOS of the Distrital University of Colombia, and its Scientific Search and Development Center (CIDC),

REFERENCES

[1] H. S. H. Chung, E. P. W. Tam, and S. Y. R. Hui, "Development of a fuzzy Logic Controller for Boost Rectifier with Active Power Factor Correction," in *30th Annual IEEE Power Electronics Specialists Conference*, 1999, PESC 99, pp. 149-154.

[2] A. Khoshooei, and J. S. Moghani, "Implementation of a Single Input Fuzzy Controller for a High Power Factor Boost Converter," in *IEEE AFRICON, 7th Africon conference in Africa, Technology innovation*, 2004, Gaborone, Botswana, pp. 69-72.

[3] A. Rubaai, M. F. Chouikha, "Design and Analysis of Fuzzy Controllers for DC-DC Converters," in *First International Symposium on Control, Communications and Signal Processing*, 2004, pp. 479- 482.

[4] E. Vidal-Idiarte, L. Martinez-Salamero, F. Guinjoan, J. Calvente, and S. Gomariz, "Sliding and Fuzzy Control of a Boost Converter using an 8-bit microcontroller," in *IEE Proceedings Electric Power Applications*, 2004, pp. 5-11.

[5] F. Martínez, "Evaluación de estructuras digitales para el diseño de sistemas de control difuso (Unpublished work style)," unpublished, Distrital University of Colombia.

A Low Cost and High Reliability Control Scheme
In Parallel Operation for 400Hz Power Supply

Seok-Eon Joung
Hanyang University
Department of Electrical Engineering
17 Haengdang-Dong, Seongdong-Gu, Seoul, Korea
edward@hanyang.ac.kr

Dong-Seok Hyun
Hanyang University
Department of Electrical Engineering
17 Haengdang-Dong, Seongdong-Gu, Seoul, Korea
dshyun@hanyang.ac.kr

Abstract—In this paper, a low cost and high reliability control scheme is proposed for 400Hz UPS system operated in parallel.

The proposed control scheme is consisted of two parts which are synchronization and load sharing control. The synchronization control is achieved by discrete logic ICs and analog circuit. The load sharing control is realized by current transformers (CTs) without any controller.

Therefore, this proposed control scheme is rather simple and the cost may be decreased, compared with control scheme using expensive controller such as DSP and CAN.

The practical feasibility of the proposed control scheme is proved by simulation..

I. INTRODUCTION

The 400Hz Uninterruptible Power Supply (UPS) systems are a new family unit that is specifically aimed at Military and airport installations.

These systems have advantages that eliminated or reduced the amount of harmonic distortion, thus reducing the size of the output filter.

Generally, parallel operation of UPS system is known as an appropriate solution to improve the power capacity, reliability, and stability in power electronic equipment. However, parallel operation is much more complex due to synchronization and load sharing.

Many researches to control for UPS system operated in parallel have been published in [1-3]. Recently, a high performance control scheme using an expensive controller such as DSP and CAN has proposed to reduce the circulating current flowing each UPS[4].

By contrast, the low cost and simple control schemes have proposed such as concentrated controller[5], load sharing control using magnetic circuit without current control loop[6,7].

Also, control technique about nonlinear and changing load was presented as well as synchronization and load sharing[8].

In this paper, a robust and low cost control scheme is proposed to improve synchronization and load-sharing for 400Hz UPS system operated in parallel.

Since the proposed control scheme is consisted with discrete logic ICs and analog circuit, the synchronization control is rather simple and the cost may be decreased, compared with DSP based UPS system. The load sharing control is achieved by current transformers (CTs).

In the following section, the fundamental theory of connecting AC power units in parallel will be described.

The principle of sharing current and synchronization are adapted using sub-signal and CTs.

Finally, simulations show the feasibility and performance of the proposed 400Hz UPS system operated in parallel

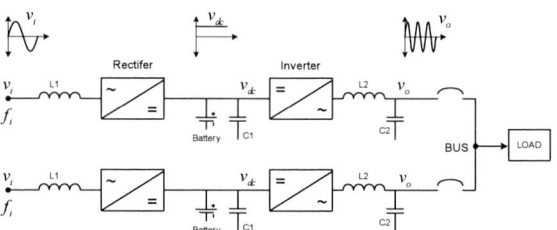

Fig. 1: Parallel operation of two UPS system

II. PRINCIPLE OF PARALLEL OPERATION

Fig. 1 shows a single-phase generalized equivalent circuit for parallel operation of two UPS system.

Since the inverter is shown as a voltage source with an output impedance Xi (i = 1,2) connected in series.

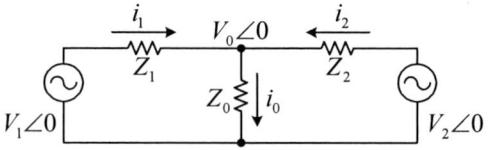

Fig. 2: Equivalent circuit for parallel operation of two UPS system

Generally, the active power (kW) and reactive power (kVAR) is unbalanced when output voltages of parallel inverters are different.

The active and reactive powers can be calculated as

$$P_1 = \frac{V_1 V_0}{X_1} \sin \varphi_1 \qquad (1)$$

$$Q_1 = \frac{V_1 V_0 \cos \varphi_1 - V_0^2}{X_1} \qquad (2)$$

978-1-4244-1871-8/07 $25.00 © 2007 IEEE   940

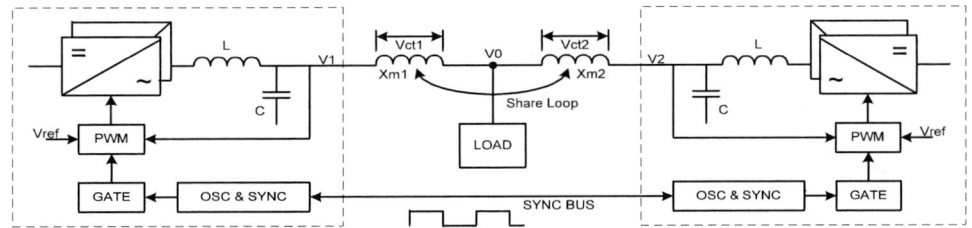

Fig. 3: Proposed 400Hz UPS system operated in parallel

From these equations, it can be seen that the active power flow is dominated by the angles, while the voltage amplitudes primarily influence the reactive power flow.

III. PROPOSED CONTROL SCHEME

A. System Configuration

As shown in Fig. 2, the proposed system is consisted with multi-step inverter which can be operated by low switching frequency, load sharing circuit controlling output impedance, and voltage controller embedded synchronization circuit

B. Inverter

The inverters are consisted with two inverters and two inverter transformers as shown in Fig. 6.

The main disadvantage of this type of topology is the large number of supplies and semiconductors required to obtain these multi-step voltage waveforms but 6-level inverter which can generate high quality voltage waveforms, good enough to be considered as suitable voltage template generators.

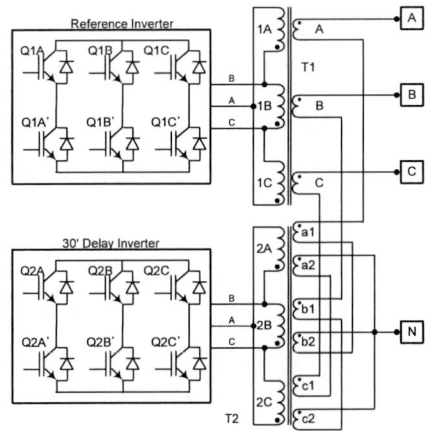

Fig. 6: Diagram of 6-level inverter

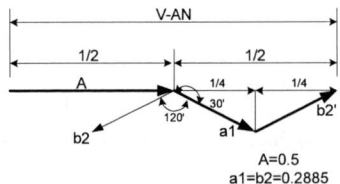

A=0.5
a1=b2=0.2885

Fig. 7: Vector diagram of 6-level inverter output voltage

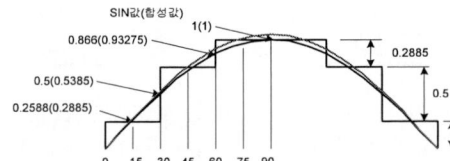

Fig. 4: Voltage waveform from 6-level inverter

C. Phase Synchronization Control

The synchronization control is achieved using 14bit binary ripple counter and stage counter in Fig. 3. The signal divided by ripple counter generates six-step signals using stage counter.

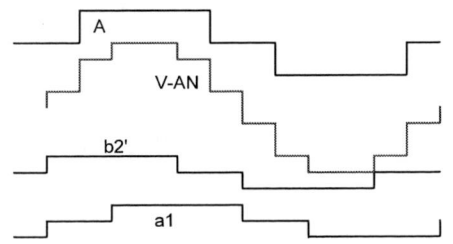

Fig. 5: Voltage waveform of basic inverter and delayed inverter

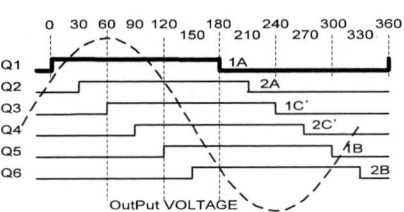

Fig. 8: Inverter Gate Signals

978-1-4244-1871-8/07 $25.00 © 2007 IEEE   941

SYNC1 and SYNC2 is generated using gate clocks of each module. The fastest synchronization signal is automatically tracked by other inverters as the SYNC_BUS. The phase angle of each inverter is controlled by comparison between SYNC_BUS and Self-Clock.

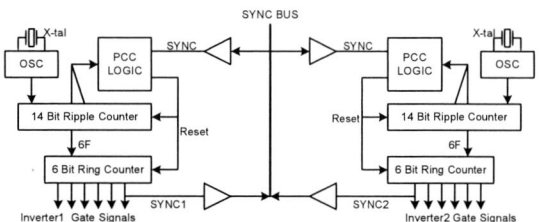

Fig. 9: Block diagram of synchronization control

As shown in Fig. 4, PCC Logic compares with the previous PCC signal and SYNC_BUS.

When the system is lagging, Reset signal is generated at Down-edge. But, when the system is leading, Reset signal is generated at Up-edge.

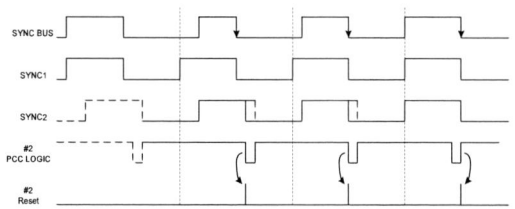

Fig. 10: Synchronization signal to control phase

D. Load-Sharing Current Control

The load current-sharing control of the parallel UPS is based on current transformers (CTs), which are provided in each UPS to sense the output currents.

As shown in Fig 5. All CTs secondary windings are interconnected to form a current sharing.

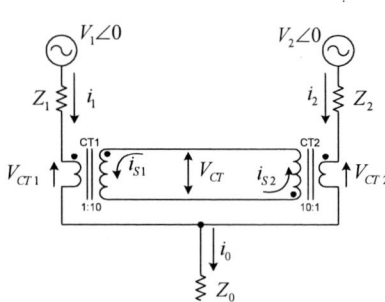

Fig. 11: Equivalent circuit for Load-sharing control

The relation equation of Each CTs given as

$$V_{CT1} \cdot i_1 = V_{CT} \cdot i_{S1} \qquad (3)$$

$$V_{CT2} \cdot i_2 = V_{CT} \cdot i_{S2} \qquad (4)$$

Because secondary windings are interconnected, Eq. (5) is derived by Eq. (3), (4) as follows.

$$V_{CT1} \cdot i_1 = V_{CT2} \cdot i_2 \qquad (5)$$

The voltage drop of primary winding is not occurred by the current flowing through secondary windings, because CT have an ignorable leakage inductance.

On the contrary, when the current flowing through secondary winding is zero, becomes infinite value by Eq. (3) and then the voltage drop of primary winding is occurred by self inductance of CT. When changes to different value, is changed to satisfy Eq. (5).

As a result, the currents following through primary windings have identical value. The load current sharing control can be achieved without any controller.

IV. SIMULATION RESULT

Simulations are performed to verify the feasibility of the proposed control scheme. The parameters of the unit listed as Table I.

Table I.

Power rating	5 [kVA]
Fundamental frequency	400 [Hz]
Nominal Line-to-line voltage	115 [V]
DC-link voltage	100
Internal Impedance	10%

Fig. 12 and Fig.15 shows the simulation circuits for linear and non-linear load condition. The sharing start at 20mSec and load changes at 30mSec.

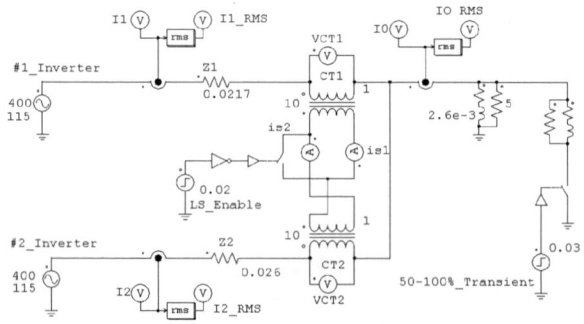

Fig. 12: Circuit for linear load Simulation

Fig. 13 shows the currents and voltage of current transformers before and after current share circuit activate.

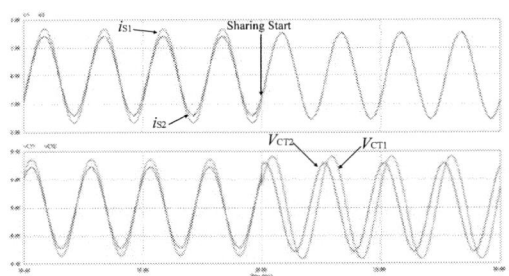

Fig. 13: Waveforms of Current Transformer

Fig. 14 shows the output currents controlled by the proposed load-sharing current control when load is changed from 50% to 100%. The load-sharing current control is well achieved despite of changed the linear load.

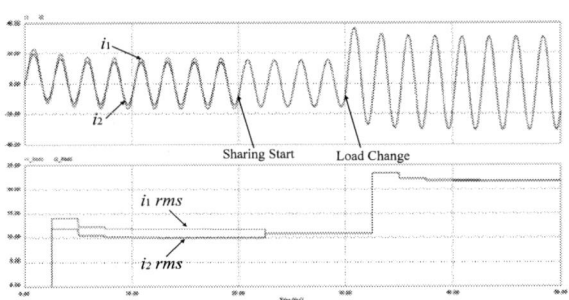

Fig. 14: Waveforms and RMS value of Load Currents on linear load

Fig. 16 also shows the output currents controlled by the proposed load-sharing current control when load is changed from 50% to 100%. The load-sharing current control is well achieved despite of changed the non-linear load.

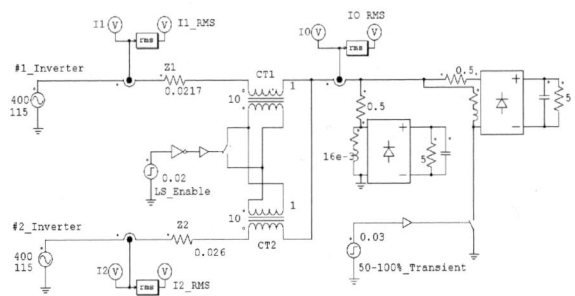

Fig. 15: Circuit for non-linear load Simulation

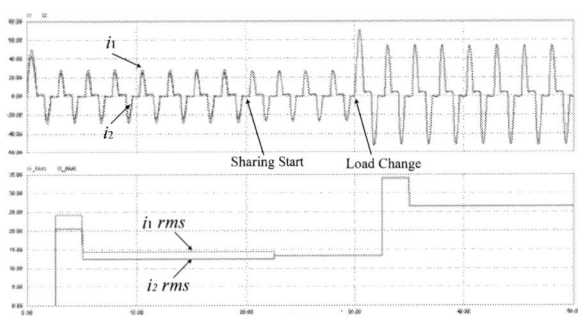

Fig. 16: Waveforms and RMS value of Load Currents on non-linear load

V. CONCLUSION

This paper investigates a low cost and high reliability control scheme for 400Hz UPS system operated in parallel.

Synchronization control can achieve a competitive system for cost, compared with expensive controller using DSP or CAN. And load sharing control scheme separated with system controller gives a high performance both stationary and dynamically at linear and nonlinear load. The performance of the proposed control scheme was verified by means of computer simulation

REFERENCES

[1] Jiann-Fuh Chen and Ching-Lung Chu, " Combination Voltage Controlled and Current-Controlled PWM Inverters for UPS Parallel Operation", IEEE Transactions on PE, vol. 10, no. 5, September 1995, pp.547-558.

[2] Prodanovic, M.; Green, T.C.; Mansir, H "A survey of control methods for three-phase inverters in parallel connection", Power Electronics and Variable Speed Drives, 2000. Eighth International Conference on (IEE Conf. Publ. No. 475)18-19 Sept. 2000, Page(s):472 - 477

[3] Jensen, U.B.; Blaabjerg, F.; Pedersen, J.K , "A new control method for 400-Hz ground power units for airplanes", Industry Applications, IEEE Transactions on Volume 36,Issue 1,Jan.-Feb. 2000 Page(s):180 - 187

[4] Kyung-Hwan Kim, Dong-Seok Hyun, "A High Performance DSP Voltage Controller with PWM Synchronization for Parallel Operation of UPS Systems", Power Electronics Specialists Conference, 2006. PESC '06. 37th IEEE Publication Date:18-22 June 2006 On page(s): 1-7

[5] Zhao Qinglin, Chen Zhongying, Wu Weiyang, "Improved Control for Parallel Inverter with Current-Sharing Control Scheme," in IPEMC '06. CES/IEEE 5th International Volume 3,Aug. 2006 Page(s):1 - 5

[6] Chen Liangliang; Xiao Lan; Yan Yangguang, "A novel parallel inverter system based on coupled inductors": International Telecommunications Energy Conference, 2003. INTELEC apos;03. The 25th InternationalVolume , Issue , 19-23 Oct. 2003 Page(s): 46 – 50

[7] Mak, C.; Bolster, L. "Bus-tie synchronization and load share technique in a ring bus system with multiple power inverters", Mak, C.; Bolster, L. Applied Power Electronics Conference and Exposition, 2005. APEC 2005. Twentieth Annual IEEEVolume 2, Issue , 6-10 March 2005 Page(s): 871 - 874 Vol. 2

[8] Uffe Borup, Frede Blaabjerg and Prasad N. Enjeti, "Sharing of Nonlinear Load in Parallel-Connected Three-Phase Converters" Industry Applications, IEEE Transactions on Volume 37, Issue 6, Nov/Dec 2001 Page(s):1817 - 1823

978-1-4244-1871-8/07 $25.00 © 2007 IEEE

Frequency Characteristics of the *d-q* Synchronous-Frame Current Reference Generation Methods for Active Power Filter

Xiaoyu Wang, *Student Member, IEEE*, Jinjun Liu, *Member, IEEE*, Jinku Hu, *Student Member, IEEE*, Yuji Meng, and Chang Yuan, *Student Member, IEEE*

School of Electrical Engineering
Xi'an Jiaotong University
28 West Xianning Road, Xi'an, Shaanxi 710049 China
Email: xywang@ieee.org

Abstract—The *d-q* harmonics detection algorithms are dominant methods to generate current reference for active power filter (APF). They are often processed in synchronous frame and time domain. This paper researches the frequency characteristics of d-q synchronous transformation which are quite related to the design of close loop regulator and analysis of the performance. Intuitively, the synchronous transformation is explained with amplitude modulation (AM) in this paper. Then the synchronous filter is proven as time-invariant linear transformation and modeled with transfer functions in stationary frame and frequency domain. The frequency-domain analytical models imply that the synchronous transformation has an equivalent effect of frequency transformation, with which the synchronous filter achieves a band pass filter from the low pass filter in synchronous frame. To simplify the analytical model, an instantaneous positive/negative sequence frame is proposed as an expansion of symmetrical components theory. Based on these frequency analytical models, the synchronous filter is compared with the traditional filters in stationary frame. Further more, the d-q harmonics detection methods are improved to get rid of the inherent coupling of synchronous transformation. Typical examples are given to verify previous analysis and comparison. Simulation and experimental results are also given.

I.   INTRODUCTION

Various solutions have been proposed to suppress power system harmonics and improve power quality, in which active power filter (APF) has been one of the most competitive solutions[1][2]. For the control of APF, current reference generation scheme plays a crucial role [2]. Existing approaches to compute the reference current for APF can be divided into time-domain and frequency-domain methods.

After the proposal of instantaneous power theory [3][4], the dominate algorithms are mostly based on synchronous frame in time domain [5], such as "*p-q* method" based on *p-q* theory [4], "*d-q* method" in *d-q* synchronous frame [6][7], and flux-based control method etc [8]. Although these time-domain methods are popular and easy to implement especially in digital controller, there are still several design-related theoretical questions that has not been clearly answered:

- Since the essential aim of *d-q* synchronous-frame current reference generation methods for APF is separating the fundamental current and the harmonic current, what are their filtering characteristics in frequency domain?

- What are the differences and relationship between *d-q* synchronous-frame methods and the traditional filter which are generally designed in frequency domain?
- How to design the low pass filter (LPF) in synchronous frame to satisfy the harmonics detection requirements in stationary frame?
- The loop gain is very important for design and analysis of the close-loop performance. Then how to evaluate the close-loop gains of the control systems based on *d-q* synchronous frame?

All these questions are related to the frequency characteristics of synchronous transformation. In 1993, a frequency analytical model of *p-q* method was derived and employed to the stability analysis of series active power filter (SAPF) [9]. But it is too complicated and too academic for electrical engineers in practical application. In the research of current close-loop regulators, the synchronous proportional plus integral (PI) regulator was analyzed in frequency domain, and a stationary PI (or resonant-PI) regulator was subsequently proposed [10][11]. But this discussion focused mainly on current regulators and there is no intuitive explanation with physical meaning.

In this paper, the synchronous transformation is intuitively explained with amplitude modulation (AM) in communication theory. Then the synchronous filter is proven as time-invariant linear transformation and modeled with transfer functions in stationary frame and frequency domain. The frequency-domain analytical models imply that the synchronous transformation has an equivalent effect of frequency transformation, by which several bandwidth pass filters (BPF) are achieved with the spectral transformation of two low pass filters (LPF) in synchronous frame.

With this stationary frequency analytical model, the frequency characteristics of *d-q* methods are investigated in detail. And to simplify the analytical model, an instantaneous positive/negative sequence frame is proposed as an expansion of symmetrical components theory. Then the simple frequency analytical model is obtained in instantaneous positive/negative sequence frame.

Based on these frequency analytical models, the *d-q* synchronous-frame current reference generation methods are compared with the traditional filters, and improved to get rid of the inherent coupling of *d-q* methods.

Typical examples are given to verify previous analysis and comparison. Simulation and experimental results are also given.

II. REVIEW AND NOVEL INTUITIVE EXPLANATIONS OF SYNCHRONOUS FRAME CURRENT GENERATION METHOD

Fig. 1 shows the widely-used d-q Synchronous-Frame Current Reference Generation Methods for Active Power Filter, where i_L is the load current, i_c^* is the current reference to APF, and the transformation operators are

$$C_{32} = \sqrt{\frac{2}{3}} \begin{bmatrix} 1 & -\frac{1}{2} & -\frac{1}{2} \\ 0 & -\frac{\sqrt{3}}{2} & \frac{\sqrt{3}}{2} \end{bmatrix}, \quad C = \begin{bmatrix} \cos\omega_0 t & \sin\omega_0 t \\ -\sin\omega_0 t & \cos\omega_0 t \end{bmatrix},$$

$$C^{-1} = \begin{bmatrix} \cos\omega_0 t & -\sin\omega_0 t \\ \sin\omega_0 t & \cos\omega_0 t \end{bmatrix}, \quad C_{32} = C_{32}^T. \quad (1)$$

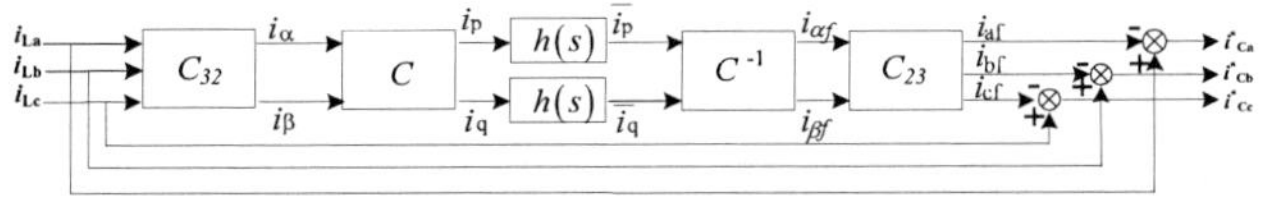

Fig. 1. The d-q Synchronous-Frame Current Reference Generation Methods.

In the synchronous $d–q$ reference frame and flux-based controllers, voltage and current signals are transformed to a synchronously rotating frame, in which fundamental quantities become dc quantities, and then the harmonic compensating commands are extracted. The key operation of synchronous transformation is the multiple of the signals with trigonometric function, which is much like the process of amplitude modulation. Before amplitude modulation theory is employed to explain the synchronous frame current generation scheme, basic amplitude modulation is reviewed.

Let $F(\omega)$ represents the spectrum of the original signal $f(t)$, then the spectrum of the amplitude-modulated signal becomes

$$\begin{aligned} S(\omega) = \mathscr{F}(s(t)) &= \mathscr{F}(f(t) \cdot \cos\omega_0 t) \\ &= \mathscr{F}\left(f(t)\frac{e^{j\omega_0 t} + e^{-j\omega_0 t}}{2}\right) = \frac{1}{2} \cdot \mathscr{F}\left(f(t)e^{j\omega_0 t} + f(t)e^{-j\omega_0 t}\right) \\ &= \frac{1}{2}\left(F(\omega - \omega_0) + F(\omega + \omega_0)\right) \end{aligned} \quad (2)$$

Fig. 2 compares the original signal and its corresponding modulated signal with amplitude in both time domain and frequency domain. With amplitude modulation, the spectrum of the original signal $F(\omega)$ is up transformed to 50Hz higher as shown in Fig. 2.

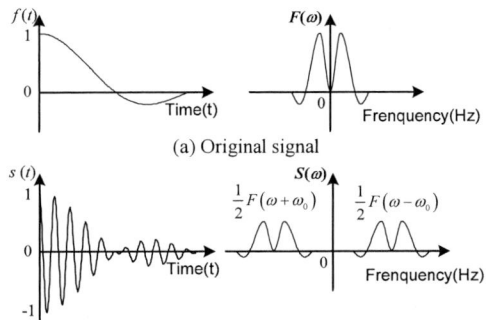

(a) Original signal

(b) Modulated signal with amplitude modulation (AM).

Fig. 2. Illustration of amplitude modulation (AM) with waveforms and spectrum plots.

Similar to the amplitude modulation, the d-q harmonics detection algorithm achieved stationary band pass filters with two low pass filters in synchronous frame. Fig. 3 intuitively illustrates the d-q synchronous frame current generation scheme with amplitude modulation. The first synchronous transformation down transforms the current spectrum to 50Hz lower, which is much like amplitude demodulation. After the process of lower pass filtering in the synchronous frame, the current spectrum is up transformed to 50Hz higher, which is much like amplitude modulation. In this way, the low pass filter in synchronous frame actually filters out the signal at the fundamental frequency (50Hz), so the equivalent effect of synchronous-frame low pass filter is band width filter in stationary frame. Detail analysis and precise derivation will be given in the next section.

III. FREQUENCY-DOMAIN ANALYTICAL MODEL IN STATIONARY FRAMES

With the intuitive explanation with amplitude modulation, a frequency analytical model is derived in this section. The kernel of the d-q synchronous-frame current reference generation is the synchronous filter to detect the fundamental component of currents, shown as Fig. 4.

In the view of system shown as Fig. 4, the synchronous filter could be modeled with transfer function in frequency domain if it could be proven as a time-invariant linear system. Because the $\alpha\beta$ transformation is linear transformation, the key of the analysis is the analytical model of synchronous transformation.

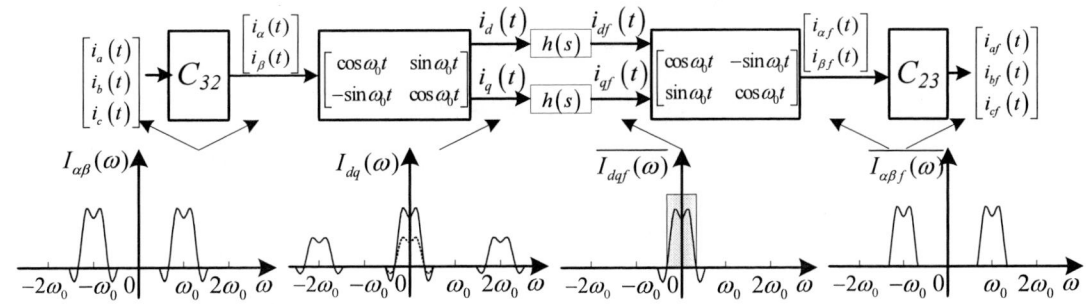

Fig. 3 Novel and intuitive explanation of d-q synchronous frame current generation scheme with amplitude modulation (AM).

978-1-4244-1871-8/07 $25.00 © 2007 IEEE

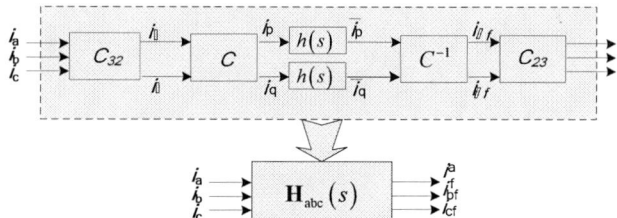

Fig. 4 Frequency model of synchronous filter.

Consider the transfer function $\mathbf{H}_{\alpha\beta}(s)$ in $\alpha\beta$ stationary frame from $\mathbf{I}_{\alpha\beta f}(s) = \begin{bmatrix} i_{\alpha f}(s) & i_{\beta f}(s) \end{bmatrix}^{\mathrm{T}}$ as input vector to $\mathbf{I}_{\alpha\beta}(s) = \begin{bmatrix} i_{\alpha}(s) & i_{\beta}(s) \end{bmatrix}^{\mathrm{T}}$ as output vector shown as Fig. 3:

$$\begin{bmatrix} i_{\alpha f}(s) \\ i_{\beta f}(s) \end{bmatrix} = \mathbf{H}_{\alpha\beta}(s) \begin{bmatrix} i_{\alpha}(s) \\ i_{\beta}(s) \end{bmatrix} \quad (3)$$

In most applications, $\mathbf{I}_{\alpha\beta f}(s)$ is the fundamental component of $\mathbf{I}_{\alpha\beta}(s)$. In appendix I, the synchronous filter is proven to be time-invariant linear transformation, and the transfer function $\mathbf{H}_{\alpha\beta}(s)$ in $\alpha\beta$ stationary frame is derived as (4).

$$\mathbf{H}_{\alpha\beta}(s) = \begin{bmatrix} \frac{1}{2}\left[h(s+j\omega_0)+h(s-j\omega_0)\right] & -j\frac{1}{2}\left[h(s+j\omega_0)-h(s-j\omega_0)\right] \\ j\frac{1}{2}\left[h(s+j\omega_0)-h(s-j\omega_0)\right] & \frac{1}{2}\left(h(s+j\omega_0)+h(s-j\omega_0)\right) \end{bmatrix}$$
$$= \begin{bmatrix} a & -jb \\ jb & a \end{bmatrix} \quad (4)$$

with a and b given by

$$a(s) = \frac{1}{2}\left(h(s+j\omega_0)+h(s-j\omega_0)\right) \quad (5)$$

$$b(s) = \frac{1}{2}\left(h(s+j\omega_0)-h(s-j\omega_0)\right) \quad (6)$$

where $h(s)$ is low pass filter in d-q synchronous frame. This means the LPF of $h(s)$ is transformed to a bandwidth pass filter with frequency conversion to 50Hz higher.

Fig. 5 illustrates the stationary frequency analytical model of the synchronous transformation. From Fig. 5, unexpected coupling is found in synchronous frame methods, which makes the design of close-loop regulator complicated. Some modifications have been proposed to decouple in [12].

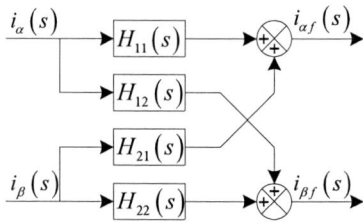

Fig. 5. Stationary frequency model of the synchronous transformation.

It is more direct to investigate the transfer functions of $\mathbf{H}_{abc}(s)$ of d-q synchronous-frame current reference generation methods in a-b-c stationary frame:

$$\begin{bmatrix} i_{af}(s) \\ i_{bf}(s) \\ i_{cf}(s) \end{bmatrix} = \mathbf{H}_{abc}(s) \begin{bmatrix} i_a(s) \\ i_b(s) \\ i_c(s) \end{bmatrix} \quad (7)$$

In a-b-c stationary frame, the frequency-domain analytical model $\mathbf{H}_{abc}(s)$ can be obtained from $\mathbf{H}_{\alpha\beta}(s)$ as

$$\mathbf{H}_{abc}(s) = C_{23}\mathbf{H}_{\alpha\beta}(s)C_{32}$$
$$= C_{23} \begin{bmatrix} a & -jb \\ jb & a \end{bmatrix} C_{32} \quad (8)$$
$$= \begin{bmatrix} \frac{2}{3}a & -\frac{1}{3}a-j\frac{\sqrt{3}}{3}b & -\frac{1}{3}a+j\frac{\sqrt{3}}{3}b \\ -\frac{1}{3}a+j\frac{\sqrt{3}}{3}b & \frac{2}{3}a & -\frac{1}{3}a-j\frac{\sqrt{3}}{3}b \\ -\frac{1}{3}a-j\frac{\sqrt{3}}{3}b & -\frac{1}{3}a+j\frac{\sqrt{3}}{3}b & \frac{2}{3}a \end{bmatrix}$$

with a and b given by (5) and (6).

It is clear in (8) that the synchronous filter has coupling effects among three phases, which means the synchronous filter is only suitable for symmetrical systems.

IV. FREQUENCY-DOMAIN ANALYTICAL MODEL IN POSITIVE/NEGATIVE SEQUENCE FRAME

To simply the previous frequency analytical models, define an instantaneous positive/negative sequence frame as

$$\begin{bmatrix} i_p(s) \\ i_n(s) \\ i_0(s) \end{bmatrix} = \frac{1}{\sqrt{3}} \begin{bmatrix} 1 & -\frac{1}{2}+\frac{\sqrt{3}}{2}j & -\frac{1}{2}-\frac{\sqrt{3}}{2}j \\ 1 & -\frac{1}{2}-\frac{\sqrt{3}}{2}j & -\frac{1}{2}+\frac{\sqrt{3}}{2}j \\ 1 & 1 & 1 \end{bmatrix} \begin{bmatrix} i_a(s) \\ i_b(s) \\ i_c(s) \end{bmatrix}, \quad (9)$$

$$\begin{bmatrix} i_a(s) \\ i_b(s) \\ i_c(s) \end{bmatrix} = \frac{1}{\sqrt{3}} \begin{bmatrix} 1 & 1 & 1 \\ -\frac{1}{2}-\frac{\sqrt{3}}{2}j & -\frac{1}{2}+\frac{\sqrt{3}}{2}j & 1 \\ -\frac{1}{2}+\frac{\sqrt{3}}{2}j & -\frac{1}{2}-\frac{\sqrt{3}}{2}j & 1 \end{bmatrix} \begin{bmatrix} i_p(s) \\ i_n(s) \\ i_0(s) \end{bmatrix} \quad (10)$$

where $i_a(s)$, $i_b(s)$ and $i_c(s)$ are three-phase currents in terms of Laplace transformation, and $i_p(s)$, $i_n(s)$ and $i_0(s)$ are the positive, negative and zero sequence separately.

This instantaneous transformation is an expansion of symmetrical components theory. In this positive/negative sequence frame, the frequency analytical model becomes very simple as (11) without coupling.

$$\begin{cases} i_{f+}(s) = h(s-j\omega_0) \times i_+(s) \\ i_{f-}(s) = h(s+j\omega_0) \times i_-(s) \end{cases} \quad (11)$$

V. COMPARISONS OF REFERENCE GENERATION METHODS IN FREQUENCY STATIONARY FRAME

Based on the frequency-domain analytical models, the d-q synchronous-frame harmonics detection algorithms can be compared with the traditional bandwidth pass filters which are generally designed in frequency-domain.

A. Comparison with traditional band-pass filter

The tradition frequency-domain band-pass filters includes many different types of either analog filters or digital filters. One of the most practical design methods of notch filter or

978-1-4244-1871-8/07 $25.00 © 2007 IEEE

band-pass filter is spectral transformation from a low pass filter, with the mapping function given by

$$s \rightarrow Q\left(\frac{s}{\omega_0} + \frac{\omega_0}{s}\right). \tag{12}$$

where ω_0 is the center frequency, and Q is the quality factor of the target band-pass filter.

In comparison of the spectral transformation in the design of traditional band-pass filter, the d-q synchronous-frame harmonics detection algorithm also achieves band-pass filters with lower pass filters. Although both of these two methods transform the prototype low-pass filter to band-pass filter, the transformation happens in different stages: the band-pass filter is transformed from lower pass filter in the design stage, while the *d-q* method is in the run-time.

Fig. 6 illustrates the bode plots of these two band-pass filters. The prototype low pass filter to design the traditional notch filter is the same as the low pass filter of synchronous filter. Fig. 6 shows the spectrum of synchronous filter is similar with notch filter. The spectrum of notch filter is geometrical symmetrical in log scale frequency axis because the notch filter transforms the low pass filter with the mapping function given by (12), while the spectrum of *d-q* synchronous scheme is mathematically symmetrical in linear scale frequency axis.

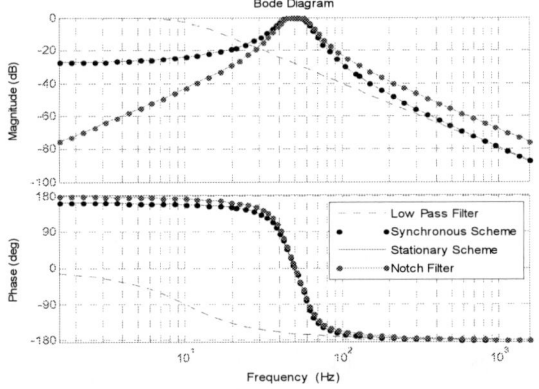

Fig. 6. Comparison with the bode plots of different band-pass filters.

B. Stationary frame reference generation scheme

Based on the frequency analytical model of *d-q* method, a stationary frame reference generation scheme is proposed with the same frequency characteristics of synchronous frame.

Because of the advancement of cheap digital signal processor (DSP), some popular interpretations of synchronous schemes should be updated.

VI. ANALYTICAL EXAMPLES AND SIMULATION RESULTS

For the demonstration of previous analysis, design examples and simulation results are given in this section.

A. Analytical example of varies bandwidth pass filter.

Fig. 7a) shows the spectrum of the low pass filter in synchronous frame with transfer functions given by

$$H_{dq}(s) = \frac{1}{\left(\dfrac{s}{\omega_C}\right)^2 + \sqrt{2}\,\dfrac{s}{\omega_C} + 1} \tag{13}$$

where $\omega_C = 2\pi \cdot 50$ rad/s. Fig. 7b) shows the corresponding Nyquist plots of the synchronous lower pass filter.

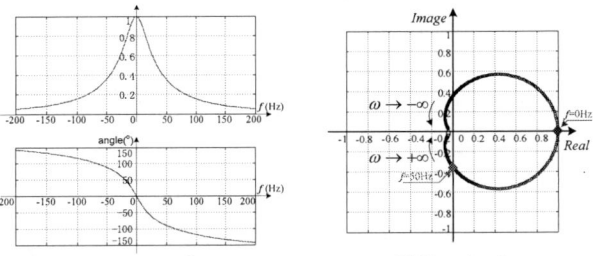

(a) Spectrum plots   (b) Nyquist plots

Fig. 7. Spectrum and Nyquist plots of synchronous low pass filter.

Fig. 8 shows the spectrum and Nyquist plots of

$$\mathbf{H}_{11}(s) = \frac{i_{\alpha f}(s)}{i_\alpha(s)}, \text{ and } \mathbf{H}_{22}(s) = \frac{i_{\beta f}(s)}{i_\beta(s)}$$

With the analytical model of (4) in $\alpha\beta$ stationary frame, the spectrum of $\mathbf{H}_{11}(s)$ can be obtained by the frequency transformation of the spectrum in Fig. 8.

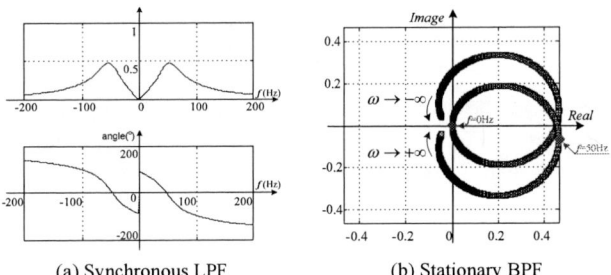

(a) Synchronous LPF   (b) Stationary BPF

Fig. 8. Spectrum and Nyquist plots of $\mathbf{H}_{11}(s)$ in $\alpha\beta$ stationary frame.

Similarly, Fig. 9 shows the spectrum and Nyquist plots of the coupling item:

$$\mathbf{H}_{12}(s) = \frac{i_{\alpha f}(s)}{i_\beta(s)}, \text{ and } \mathbf{H}_{21}(s) = \frac{i_{\beta f}(s)}{i_\alpha(s)}.$$

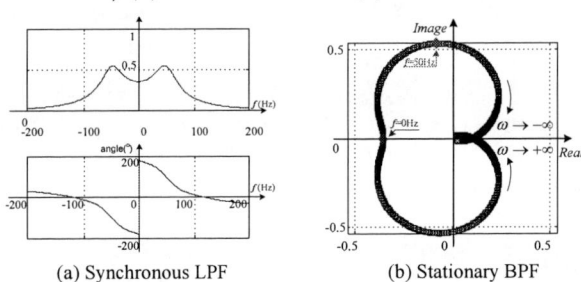

(a) Synchronous LPF   (b) Stationary BPF

Fig. 9. Spectrum and Nyquist plots of $\mathbf{H}_{12}(s)$ in $\alpha\beta$ stationary frame.

B. Simulation results

Fig. 10 comparatively shows the simulation result of each filter.

978-1-4244-1871-8/07 $25.00 © 2007 IEEE   947

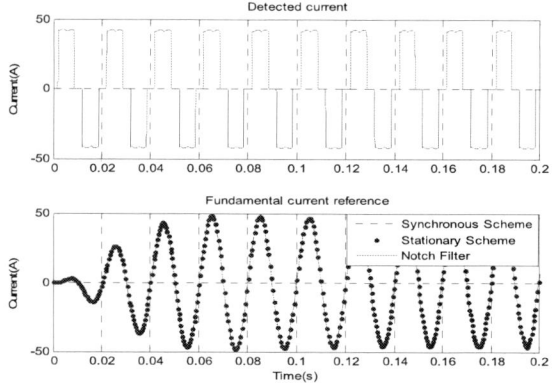

Fig. 10. Comparison with time domain simulation.

VII. EXPERIMENTAL RESULTS

To verify the previous analysis, the synchronous filter, the traditional notch filter and the proposed stationary filter are realized in DSP (TMS320F2812) and their performances are compared within the experiments. Fig. 11 shows the experiments configuration.

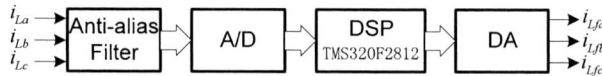

Fig. 11. Configuration of the experiment equipments.

The input currents are three-phase rectifier with inductor, which is the typical harmonic source of power quality problems. Fig. 12 shows the experimental result in symmetrical system. All the filters have the similar performance in both transient and steady states. Fig. 13 shows the result when the input current are unsymmetrical. In this case, the synchronous filter can only generate

symmetrical output because of the unexpected coupling effects, while the traditional notch filter and the proposed stationary filter works well in unsymmetrical system.

VIII. CONCLUSION

With amplitude modulation, the synchronous transformation was intuitively explained and the transfer functions of synchronous filter were derived in frequency domain. These frequency-domain analytical models imply that the *d-q* synchronous-frame current reference generation methods obtains the current harmonics with spectral transformation of low pass filter at runtime. To simplify the analytical model, an instantaneous positive/negative sequence frame was proposed as an expansion of symmetrical components theory.

Based on these frequency analytical models, the synchronous filter was compared with the traditional filters in stationary frame. Further more, the stationary filters were proposed to get rid of the inherent coupling of synchronous transformation. Typical examples are given to verify previous analysis and comparison. Simulation and experimental results are also given.

REFERENCES

[1] B. Singh, K. Al-Haddad, and A. Chandra, "A review of active filters for power quality improvement," Industrial Electronics, IEEE Transactions on, vol. 46, no. 5, pp. 960-971, 1999.

[2] Ortega, J.M.M. Esteve, M.P. Payan, M.B. Exposito, A.G. Franquelo, L.G. , "Reference current computation methods for active power filters: accuracy assessment in the frequency domain," Power Electronics, IEEE Transactions on, vol. 20, no. 2, pp. 446-456, 2005.

[3] H. Akagi, et. al, "Instantaneous Reactive Power Compensators Comprising Switching Devices without Engergy Storage Components," IEEE Trans on IA, vol. 20, no.3, pp. 625-630, 1984.

[4] H. Akagi, et al, "Control Strategy of Active Power Filters Using multiple Voltage-Source PWM converters," IEEE Trans. on IA,

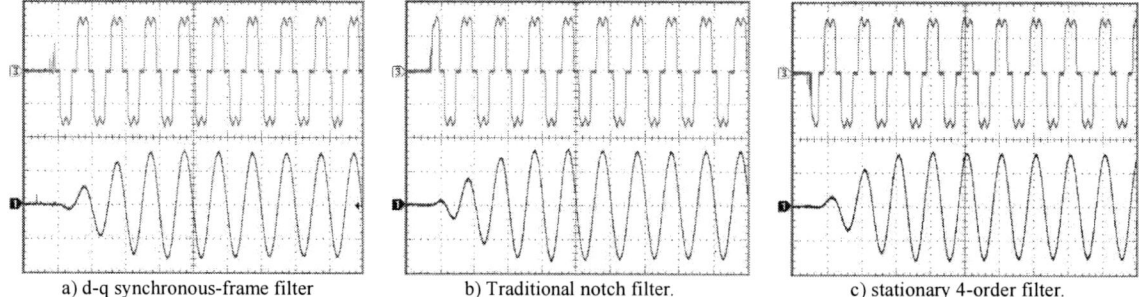

a) d-q synchronous-frame filter b) Traditional notch filter. c) stationary 4-order filter.

Up curve: load current of phase a (2A/div); Down curve: detected fundamental current of phase a (2A/div). Horizon axis: time: (20ms/div).

Fig. 12. Experimental results in symmetrical system.

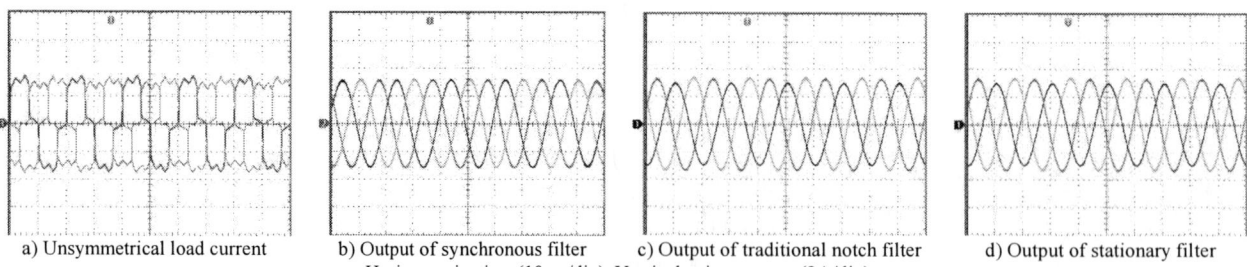

a) Unsymmetrical load current b) Output of synchronous filter c) Output of traditional notch filter d) Output of stationary filter

Horizon axis: time (10ms/div). Vertical axis: currents (2A/div).

Fig. 13. Experimental results of three-phase current in unsymmetrical system.

978-1-4244-1871-8/07 $25.00 © 2007 IEEE   948

vol.22, no.3, pp.460-465, 1986.

[5] C. Donghua and X. Shaojun, "Review of the control strategies applied to active power filters," in Electric Utility Deregulation, Restructuring and Power Technologies, 2004.(DRPT 2004).Proceedings of the 2004 IEEE International Conference on, 2 ed 2004, pp. 666-670.

[6] V. Soares, P. Verdelho, and G. Marques, "A control method for active power filters under unbalanced nonsinusoidal conditions," in Power Electronics and Variable Speed Drives, 1996.Sixth International Conference on (Conf.Publ.No.429) 1996, pp. 120-124.

[7] V. Soares, P. Verdelho, and G. D. Marques, "An instantaneous active and reactive current component method for active filters," Power Electronics, IEEE Transactions on, vol. 15, no. 4, pp. 660-669, 2000.

[8] J. S. Tepper, J. W. Dixon, G. Venegas, and L. Moran, "A simple frequency-independent method for calculating the reactive and harmonic current in a nonlinear load," Industrial Electronics, IEEE Transactions on, vol. 43, no. 6, pp. 647-654, 1996.

[9] F. Z. Peng, H. Akagi, and A. Nabae, "Compensation characteristics of the combined system of shunt passive and series active filters," Industry Applications, IEEE Transactions on, vol. 29, no. 1, pp. 144-152, 1993.

[10] Zmood, D.N. Holmes, D.G. Bode, G.H. , "Frequency-domain analysis of three-phase linear current regulators," Industry Applications, IEEE Transactions on, vol. 37, no. 2, pp. 601-610, 2001.

[11] Newman, M.J. Zmood, D.N. Holmes, D.G. , "Stationary frame harmonic reference generation for active filter systems," Industry Applications, IEEE Transactions on, vol. 38, no. 6, pp. 1591-1599, 2002.

[12] R. Pedro, P. Josep, B. Joan, J. I. Candela, P. B. Rolando, and B. Dushan, "Decoupled Double Synchronous Reference Frame PLL for Power Converters Control," Power Electronics, IEEE Transactions on, vol. 22, no. 2, pp. 584-592, 2007.

APPENDIX

The transfer function $\mathbf{H}_{\alpha\beta}(s)$ in $\alpha\beta$ stationary frame is derived in detail in the appendix. Define the stationary frame transfer function $\mathbf{H}_{\alpha\beta}(s)$ as

$$\begin{bmatrix} i_{\alpha f}(s) \\ i_{\beta f}(s) \end{bmatrix} = \mathbf{H}_{\alpha\beta}(s) \begin{bmatrix} i_{\alpha}(s) \\ i_{\beta}(s) \end{bmatrix}$$

where the input vector is $\mathbf{I}_{\alpha\beta}(s) = \begin{bmatrix} i_{\alpha}(s) & i_{\beta}(s) \end{bmatrix}^{\mathrm{T}}$ and the output vector is $\mathbf{I}_{\alpha\beta f}(s) = \begin{bmatrix} i_{\alpha f}(s) & i_{\beta f}(s) \end{bmatrix}^{\mathrm{T}}$ which is the fundamental component of input vector.

First, the time domain $\mathbf{I}_{dq}(s) = \begin{bmatrix} i_d(s) & i_q(s) \end{bmatrix}^{\mathrm{T}}$ is transformed to (14).

$$\begin{aligned}
\begin{bmatrix} i_d(t) \\ i_q(t) \end{bmatrix} &= \begin{bmatrix} \cos\omega_0 t & \sin\omega_0 t \\ -\sin\omega_0 t & \cos\omega_0 t \end{bmatrix} \begin{bmatrix} i_{\alpha}(t) \\ i_{\beta}(t) \end{bmatrix} \\
&= \begin{bmatrix} \dfrac{e^{j\omega_0 t}+e^{-j\omega_0 t}}{2} & \dfrac{j}{2}\left(e^{-j\omega_0 t}-e^{j\omega_0 t}\right) \\ -\dfrac{j}{2}\left(e^{-j\omega_0 t}-e^{j\omega_0 t}\right) & \dfrac{e^{j\omega_0 t}+e^{-j\omega_0 t}}{2} \end{bmatrix} \begin{bmatrix} i_{\alpha}(t) \\ i_{\beta}(t) \end{bmatrix} \\
&= \frac{1}{2}e^{j\omega_0 t}\begin{bmatrix} 1 & -j \\ j & 1 \end{bmatrix}\begin{bmatrix} i_{\alpha}(t) \\ i_{\beta}(t) \end{bmatrix} + \frac{1}{2}e^{-j\omega_0 t}\begin{bmatrix} 1 & j \\ -j & 1 \end{bmatrix}\begin{bmatrix} i_{\alpha}(t) \\ i_{\beta}(t) \end{bmatrix}
\end{aligned} \quad (14)$$

In terms of Laplace transformation, (14) becomes

$$\begin{bmatrix} i_d(s) \\ i_q(s) \end{bmatrix} = \frac{1}{2}\begin{bmatrix} 1 & -j \\ j & 1 \end{bmatrix}\begin{bmatrix} i_{\alpha}(s-j\omega_0) \\ i_{\beta}(s-j\omega_0) \end{bmatrix} + \frac{1}{2}\begin{bmatrix} 1 & j \\ -j & 1 \end{bmatrix}\begin{bmatrix} i_{\alpha}(s+j\omega_0) \\ i_{\beta}(s+j\omega_0) \end{bmatrix} \quad (15)$$

Applying the low pass filter in synchronous frame to (15) yields

$$\begin{aligned}
\begin{bmatrix} i_{df}(s) \\ i_{qf}(s) \end{bmatrix} &= \begin{bmatrix} h(s) & 0 \\ 0 & g(s) \end{bmatrix}\begin{bmatrix} i_d(s) \\ i_q(s) \end{bmatrix} \\
&= \frac{1}{2}\begin{bmatrix} 1 & -j \\ j & 1 \end{bmatrix}\begin{bmatrix} h(s)\cdot i_{\alpha}(s-j\omega_0) \\ g(s)\cdot i_{\beta}(s-j\omega_0) \end{bmatrix} + \frac{1}{2}\begin{bmatrix} 1 & j \\ -j & 1 \end{bmatrix}\begin{bmatrix} h(s)\cdot i_{\alpha}(s+j\omega_0) \\ g(s)\cdot i_{\beta}(s+j\omega_0) \end{bmatrix}
\end{aligned} \quad (16)$$

Similar to the first step, the output vector $\mathbf{I}_{\alpha\beta f}(s)$ is obtained in terms of Laplace transformation

$$\begin{bmatrix} i_{\alpha f}(s) \\ i_{\beta f}(s) \end{bmatrix} = \frac{1}{2}\begin{bmatrix} 1 & -j \\ j & 1 \end{bmatrix}\begin{bmatrix} i_{df}(s+j\omega_0) \\ i_{qf}(s+j\omega_0) \end{bmatrix} + \frac{1}{2}\begin{bmatrix} 1 & j \\ -j & 1 \end{bmatrix}\begin{bmatrix} i_{df}(s-j\omega_0) \\ i_{qf}(s-j\omega_0) \end{bmatrix} \quad (17)$$

Substitution of (15) and (16) to (17) yields (18).

$$\begin{bmatrix} i_{\alpha f}(s) \\ i_{\beta f}(s) \end{bmatrix} = \frac{1}{2}\begin{bmatrix} 1 & -j \\ j & 1 \end{bmatrix}\begin{bmatrix} i_{df}(s+j\omega_0) \\ i_{qf}(s+j\omega_0) \end{bmatrix} + \frac{1}{2}\begin{bmatrix} 1 & j \\ -j & 1 \end{bmatrix}\begin{bmatrix} i_{df}(s-j\omega_0) \\ i_{qf}(s-j\omega_0) \end{bmatrix} \quad (18)$$

Because of the equation

$$\begin{bmatrix} 1 & -j \\ j & 1 \end{bmatrix}\begin{bmatrix} 1 & j \\ -j & 1 \end{bmatrix} \equiv \mathbf{0} \quad (19)$$

(18) is simplified to (21), and varies forms of the transfer function in $\alpha\beta$ stationary frame are obtained as (20). Each of the expression can be explained with specific physical meanings, but here in this paper the expression of (4) is selected for simplify.

$$\begin{cases}
\mathbf{H}_{\alpha\beta}(s) = \dfrac{1}{2}\begin{bmatrix} 1 & -j \\ j & 1 \end{bmatrix}\begin{bmatrix} h(s+j\omega_0) \\ g(s+j\omega_0) \end{bmatrix} + \dfrac{1}{2}\begin{bmatrix} 1 & j \\ -j & 1 \end{bmatrix}\begin{bmatrix} h(s+j\omega_0) \\ g(s-j\omega_0) \end{bmatrix} \\[12pt]
\mathbf{H}_{\alpha\beta}(s) = \begin{bmatrix} \dfrac{1}{2}\left[h(s+j\omega_0)+h(s-j\omega_0)\right] & -j\dfrac{1}{2}\left[g(s+j\omega_0)-g(s-j\omega_0)\right] \\ j\dfrac{1}{2}\left[h(s+j\omega_0)-h(s-j\omega_0)\right] & \dfrac{1}{2}\left(g(s+j\omega_0)+g(s-j\omega_0)\right) \end{bmatrix} \\[12pt]
\mathbf{H}_{\alpha\beta}(s) = \mathcal{L}\left\{\begin{bmatrix} \cos\omega_0 t & -\sin\omega_0 t \\ \sin\omega_0 t & \cos\omega_0 t \end{bmatrix}\begin{bmatrix} h(t) \\ g(t) \end{bmatrix}\right\}
\end{cases} \quad (20)$$

$$\begin{aligned}
\begin{bmatrix} i_{\alpha f}(s) \\ i_{\alpha f}(s) \end{bmatrix} &= \frac{1}{4}\begin{bmatrix} 1 & -j \\ j & 1 \end{bmatrix}\begin{bmatrix} 1 & -j \\ j & 1 \end{bmatrix}\begin{bmatrix} h(s+j\omega_0)\cdot i_{\alpha}(s) \\ g(s+j\omega_0)\cdot i_{\beta}(s) \end{bmatrix} + \frac{1}{4}\begin{bmatrix} 1 & -j \\ j & 1 \end{bmatrix}\underbrace{\begin{bmatrix} 1 & j \\ -j & 1 \end{bmatrix}}_{=0}\begin{bmatrix} h(s+j\omega_0)\cdot i_{\alpha}(s+j2\omega_0) \\ g(s+j\omega_0)\cdot i_{\beta}(s+j2\omega_0) \end{bmatrix} \\
&\quad + \frac{1}{4}\begin{bmatrix} 1 & j \\ -j & 1 \end{bmatrix}\underbrace{\begin{bmatrix} 1 & -j \\ j & 1 \end{bmatrix}}_{=0}\begin{bmatrix} h(s-j\omega_0)\cdot i_{\alpha}(s-j2\omega_0) \\ g(s-j\omega_0)\cdot i_{\beta}(s-j2\omega_0) \end{bmatrix} + \frac{1}{4}\begin{bmatrix} 1 & j \\ -j & 1 \end{bmatrix}\begin{bmatrix} 1 & \overset{=0}{j} \\ -j & 1 \end{bmatrix}\begin{bmatrix} h(s-j\omega_0)\cdot i_{\alpha}(s) \\ g(s-j\omega_0)\cdot i_{\beta}(s) \end{bmatrix} \\
&= \underbrace{\begin{bmatrix} \dfrac{1}{2}\left[h(s+j\omega_0)+h(s-j\omega_0)\right] & -j\dfrac{1}{2}\left[g(s+j\omega_0)-g(s-j\omega_0)\right] \\ j\dfrac{1}{2}\left[h(s+j\omega_0)-h(s-j\omega_0)\right] & \dfrac{1}{2}\left(g(s+j\omega_0)+g(s-j\omega_0)\right) \end{bmatrix}}_{\mathbf{H}_{\alpha\beta}(s)}\begin{bmatrix} i_{\alpha}(s) \\ i_{\beta}(s) \end{bmatrix}
\end{aligned} \quad (21)$$

IHTS development for Uninterruptible power supply at UPS fault

Hyun-Chul Jung and Dong-Suk Hyun

HANGYANG UNIVERSITY, HEANG DANG-DONG, JOUNG-GU, SEOUL, 133-791

Email: kt5100@paran.com, love001@kt.co.kr

This paper is about high speed conversion device in UPS(Uninterruptible Power Supply System). When interrupted, the situation is finally sensed in UPS output: this device detects such interruption preventing any fault in load and converts to Bypass line to prevent interruption of electric power. This system is composed of additional independent system that is operational in an uncontrollable situation of UPS system, compared to conventional UPS system that detects output power and converts when UPS system is out of the reference voltage. Also, UPS interior short circuit accident appears most greatly as an effect of fault in industry field. So this paper proposes UPS-IHTS (uninterruptible power supply-Isolation control high speed transfer switch) to consider a point of contact state at conversion to minimize the effect of fault in load and grid.

1. Introduction

The recent industry field has been using uninterruptible power supply (UPS) device to guarantee a continuous operation of important Load's power supply. UPS can prepare for a power failure or fault of UPS's interior system because it has a reserve power supply and a Bypass line. For a little more reliability promotion in the service in the important super high-speed communication service and IDC center server power supply situation, two independent UPS's and STS(Static Transfer Switch) is installed in each output suppling power to Load from STS's output.[1]-[3] Hence, it proves that existing UPS's reliability drops so much. In this paper, the problem of control function for Bypass conversion in existing UPS system is presented. UPS independent control high speed conversion system explains that it can prevent phenomenon influencing to load and system at UPS interior short circuit accident which is becoming an issue in general UPS's conversion system.[4] Also, it has an advantage that possible wrong may not exist in power supply of load even if a critical situation including control function fault of UPS's interior happens.[5]

2. UPS's problem in industrial

a)   Accident of the industrial field

Actual condition analysis of STS operation in an industrial field. The power supply of CAMA(centralized automatic message account), internet Server Hosting, KONET equipment and optical communication transmission equipment becomes gradually more important in the communication service market environment than in a general industrial field. IDC center is a place that manages corporation's internet server computer and constructs a large-scale internet data center as server occupancy space like a hotel and manages all services that SOHO 's(small office, home office) businessman or corporation server needs. As business area of IT field is magnified, IDC center has more and more Server Hosting on a large scale and essential UPS still needs more bulky capacity to supply power for server. Therefore, the importance of reliability increases. In contrast, technology development speed of UPS is slow and there are problems of maintenance, administration, etc. General fault of average 12 ~ 20% happens in communication power supply equipment during any given year, and interrupts fatal damage of Load power supply of 2% happens. Like these, while KEPCO's Power Quality improves gradually, UPS's reliability falls relatively. To solve these problems, STS was installed in the last output of a little more important equipment among several Communication Systems since 2004 as a trial and tried to install each UPS's to two Sources and to minimize effect of power supply interruption by hard failure. While the phenomenon of UPS supplies the power supply, commercial power ought to be supplied directly by bypassing any internal obstacle or an instant overload in order for the power supply to operate normally. However, it is a reality that is not so in an actual field. 24% fault occurrence in 30KVA ~ 500KVA bulky UPS were fault of UPS internal power electronic device. Remaining fault lies on in and outside disability, and within the whole system the power supply not having bypassed as normally caused 1.3% parts

Year	Number of Equipment,	Number of Fault	Number of parts Fault	electronic device Fault	Number of Load accident
2003	992	158	95	18	15
2004	930	113	78	13	10
2005	909	90	56	24	11
Total	2831	361	229	55	36

Table 2-1 More than 30KVA bulky UPS equipment fault analysis

978-1-4244-1871-8/07 $25.00 © 2007 IEEE   950

and interruptions. That is 13 out of 1000 equipments exposed to the problem which does not supply power continuously in load during one year. Communication System does not tolerate 1% accident probability and considers it to be a serious problem that two parallel UPS system is constructed, installation of STS in an output, a type of an OFF line, a type of a module, and a dynamic type is being carefully examined at an added much expense for a least defective power shortage alternative.

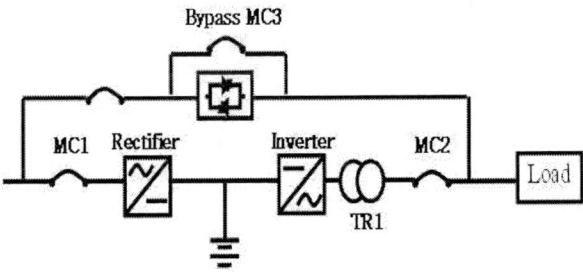

Figure 2-1 General Double Conversion UPS composition

b) Problem of the UPS

If UPS internal obstacle happens or external overload happens, UPS detects an inverter output voltage or an output waveform during the last output and intercepts the MC2 breaker and conducts SCR in Bypass, and turns on MC3 breaker to supply power continuously in load. However, UPS system dose not guarantee continuous output power when it is interrupted due to mechanical defect of MC in the last output, TR fault in last output and Main Controller does not operate normally[6]. To solve these problems, first, Bypass method is adopted when inverter output voltage are below reference voltage. However, in following cases:

i) Open due to mechanical defect of MC in the last output with normal operation of Inverter

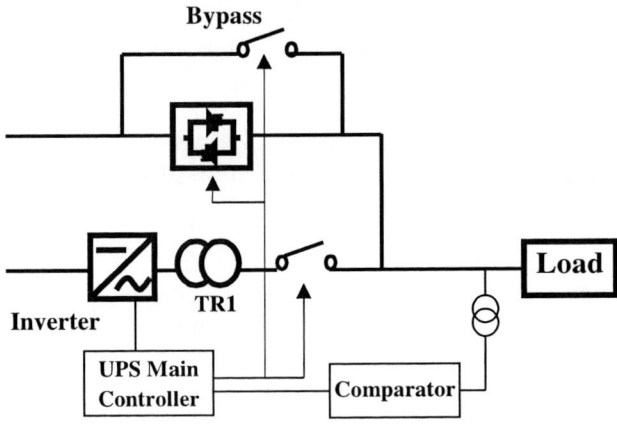

Figure 2-2 block diagram of general UPS's output power supply detection and control

ii) Having TR1, reactor as harmonic filter, Condenser in inverter output. as shown in fig 2-3 "a", conversion instruction becomes late because detection is delayed until inverter output voltage are below constant value.Second, in several UPS, it is composed with Figure 2-2 to improve preceding problemes and it is operated with detecting power quality in the last output and giving a conversion instruction to Bypass after comparing with reference voltage through the comparator.

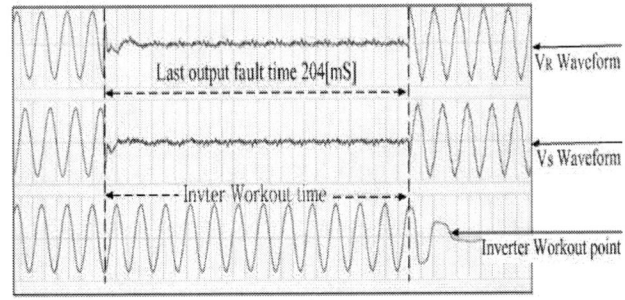

a) In case output MC open by mechanical fault

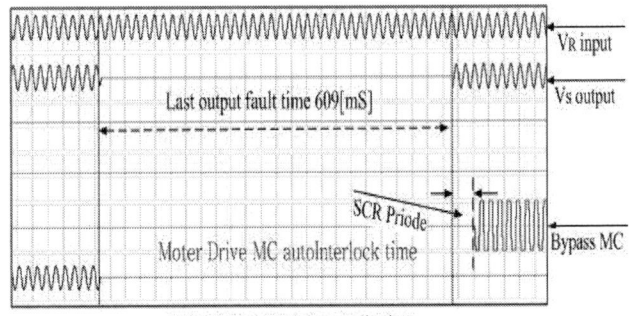

b) In case Inverter Main controller fault

Figure 2-3 Output waveform with obstacle of Main CPU in UPS and forced interception in last output

By the way, in case of obstacle of Main Board/Controller and interruption of control power supply due to UPS's fault, it frequently happens that UPS does not operate entirely. Those problem as shown in fig 2-3 "b"

Third example gave the biggest damage on communication power system. figure 2 – 1 represents uncontrollable state due to MC2 fault in the last output. This generation of short circuit current due to Inverter IGBT burn-out so that MC2 point of contact is melted and stuck, and the Main controller order STS/MC3 injection of Bypass line and MC2 intercepts at the same time. As a result, accident is generated not only in all systems of load but also in all equipments that is associated to grid. These phenomenons are general cases in UPS that is applied in the industry field. We can see the overlapped waveform in figure 2 - 4 (a) and figure 2 - 4 (b) shows accident occurred MC. These particular accident cases have had no alternative preventions until present time that interruptions to the supplying power to load occurred at every accident.

978-1-4244-1871-8/07 $25.00 © 2007 IEEE

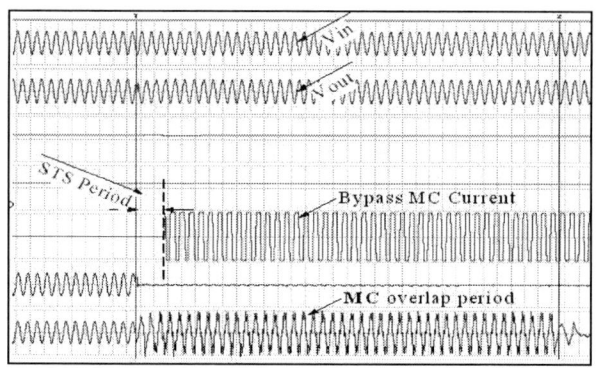

Figure 2-4 waveforms in overlapped period and short circuit accident in MC2 point of contact

3. UPS-IHTS (uninterruptible power supply-Isolation control high speed transfer switch) system development

a) Establishment in reference voltage range

UPS in general observes the power quality in the last output and converts to Bypass when output value is out of regular range. Therefore, it observes voltage waveform in output power continuously and judges whether the waveform reaches the reference voltage or not through the comparator and finally decide on Bypass conversion. Reference voltage is usually established with V1 = Vref $\pm \triangle$V. UPS-IHTS also has a similar dynamic characteristic. As shown in figure 4-1, IHTS chooses comparator value, V2= Vref $\pm(\triangle$V$\pm\alpha)$, a little more wide and set to 4.16~8.32[mS](1/4~2/4[Cycle]) longer than 4mS, as shown in fig 3-2 . which is a standard conversion time in communication power system. With this composition, when it can not convert even if it exceeds range(V= Vref $\pm\triangle$V) due to failure of existent control system, IHTS intercepts MC2 in the range of V = Vref $\pm$ ($\triangle$ V $\pm$ α).

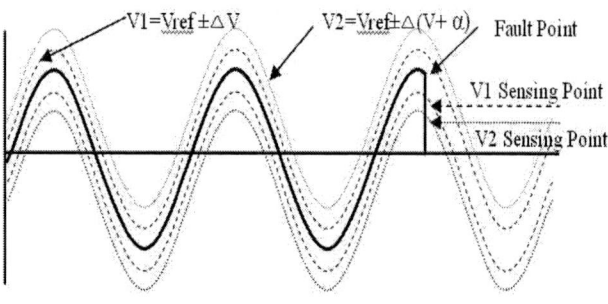

Figure 3 - 1 Example of general control functions in UPS and UPS-IHTS reference voltage settlement

As a result of testing Server equipment, FDSU(Fiber Digital Service Unit, optical transmission equipment), exchange

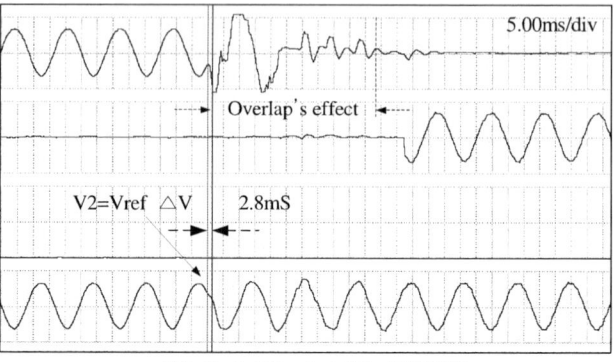

a) V2=Vref $\triangle$V, unInstall the UPS-IHTS

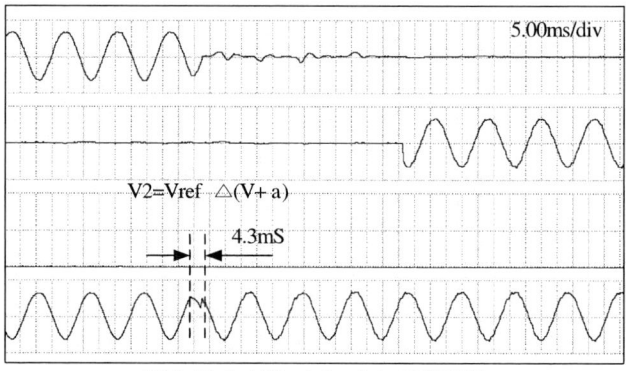

b) V2=Vref $\triangle$(V+ a), Install the UPS-IHTS

Figure 3 - 2 Compare with Conversion waveforms in UPS-IHTS reference voltage

equipment of TDX interrelated, reference value of conversion time in communication power supply does not generate specific problems in system operation until 12[mS] of STS is consisted of magnetic Relay in point of contact. It is regarded as effect of residual charge by reactance and conductance on grid.

b) Set of the Ideal State Conditioning
IHTS supplies power to comparative part that judges problem in load when it converts to Bypass and three power supplies independently to prepare UPS-Main Board obstacle. IHTS's operating characteristic is that if MC2 is intercepted exactly with advance recognition of power quality state, on the basis of the result, it turns on SCR of STS in Bypass and MC3 in order. Operating procedure of IHTS in detail is as follows: As shown in figure 3 - 3, If it proves safe for MC2 to open up precisely for it to convert to bypass, STS would then turn on as the late responding MC3 follows. If it is supposed to permissible to load side when it converts to Bypass after MC2 becomes open exactly, STS turns on and MC3 that has late response turns on in order. Lastly, it supplies power to load through MC-3. Then, UPS-Inverter and Bypass are interlocked perfectly without overlapped

978-1-4244-1871-8/07 $25.00 © 2007 IEEE   952

period such as figure 3 - 4 "b" from overlapped period generated in figure 3-4 "a". Also, when UPS's internal short circuit accidents generate, it can minimize effect to electric power distribution and reduce restoration expenses and time.

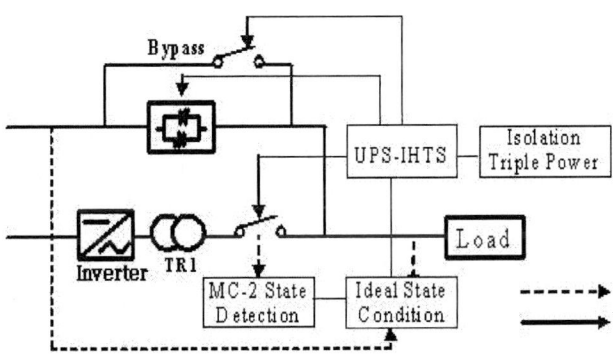

Figure 3 - 3 UPS-IHTS block diagram

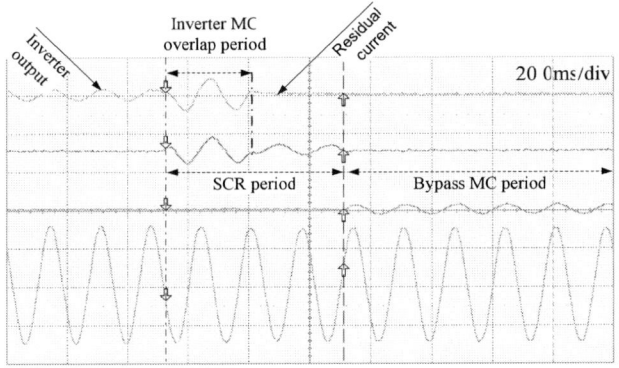

a) Did not install the IHTS, have the overlap period

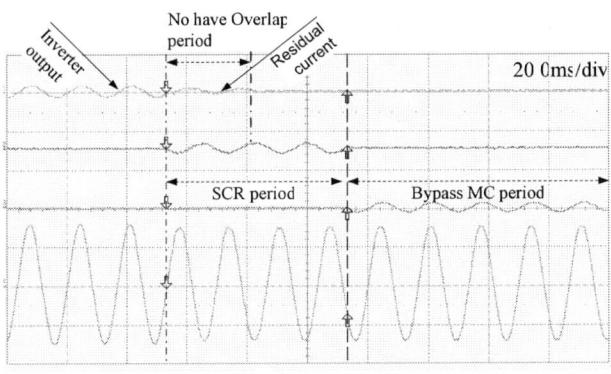

b) the IHTS install and using Ideal State Condition

Figure 3 - 4 Functional test waveforms in Ideal State Conditioning

c) System configuration

UPS-IHTS is developed with synthesizing two systems that are expressed on preceding section. Figure 3 - 5 is an example of UPS system that IHTS is applied

a)   IHTS Test Model UPS

Figure 3 - 5 IHTS Simulation Model UPS System

b) Isolation Triple Power Board and IHTS Main Board

4. Conclusion

In this paper, uninterruptible power supply focused on reliability improvement is achieved in communication system. It is developed to supply power to load continuously when UPS's electric power device fault when an impedance increase by use of long term or interior obstacle happens unexpectedly, or an external short circuit accident is generated instantly. Since July 2005, the proposed system has been applied and operated in communication power supply system. Due to a ground fault in utility input in April 2007, a strong surge which flowed to UPS occurred. As a result of the accident, all control systems including UPS input filter, Input transistor for Main Control Board power supply and Power Board, etc. failed and all control functions of UPS were interrupted but the supplying power to load achieved normally. Besides, several effects are attained. Beside the stated facts, several other effects were attained. The ultimate development purpose of UPS is in using it effectively and safely in the industry field. Therefore, fault type occurring in real system with various forms should be examined, and researches

978-1-4244-1871-8/07 $25.00 © 2007 IEEE   953

about reliability improvement in UPS system should be continued.

REFERENCES

[1] Hossein Mokhtari, Masoud Karimi-Ghartemani, and M. Reza Iravani, "Experimental Performance Evaluation of a Wavelet-Based On-Line Voltage Detection
Method for Power Quality Applications" IEEE Transactions on Power Delivery Vol.17, No.1, January 2002 Page 161

[2] H. Mokhtari, S. B. Dewan Fellow, "Performance Evaluation of Thyristor Based Static Transfer Switch" IEEE Transactions on Power Delivery Vol.15, No3, July2000 Page 960

[3] Hossein Mokhtari,, and M. Reza Iravani, Fellow, IEEE " Impact of Difference of Feeder Impedances on the Performance of a Static Transfer Switch" IEEE Transactions on Power Delivery Vol.19, No.2, April 2004 Page 679

[4] Hossein Mokhtari, Shashi B. Dewan, Fellow, IEEE, and M. Reza Iravani, Senior Member, IEEE "Analysis of a Static Transfer Switch With Respect to Transfer Time" IEEE Transactions on Power Delivery Vol. 17, No. 1, January 2002 Page 190

[5] M. N. Moschakis and N. D. Hatziargyriou, Senior Member, IEEE "A Detailed Model for a Thyristor-Based Static Transfer Switch" IEEE Transactions on Power Delivery Vol. 18, No. 4, October 2003 Page 1442

[6] A. Fulgenzl, N.Massimiani, F.Muzi, N.Polidoro, F.Torelli "UPS Interconnected with STS to Improve Power-Supply Continuity in the Gran Sasso National Laboratory" IEEE Transactions on Power Delivery No. 3, April 2002 Page 260

A Grid Current-controlling Shunt Active Power Filter

Hanny H. Tumbelaka
Electrical Engineering
Department
Petra Christian University
Surabaya, Indonesia
Email: tumbeh@petra.ac.id

Lawrence J. Borle
School of Electrical and
Electronics Engineering
University of Western Australia, Western Australia
Email: lborle@ee.uwa.edu.au

Chem V. Nayar
Electrical and Computer
Engineering Department
Curtin University of Technology, Western Australia
Email: C.V.Nayar@
exchange.curtin.edu.au

Seong Ryong Lee
School of Electronic and
Information Engineering
Kunsan National University
Kunsan, Korea
Email: srlee@kunsan.ac.kr

Abstract— In this paper, the implementation of a three-phase shunt active power filter is presented. The filter is essentially three independent single-phase current-controlled voltage source inverters (CC-VSI) with a common DC bus. The CC- VSI is operated to directly control the AC grid current to be sinusoidal and in phase with the grid voltage. The APF consists of a current control loop, which uses polarized ramptime current control and a voltage control loop, which employs a simple Proportional Integral control. The experimental results indicate that the active filter is able to handle predominantly the harmonics, as well as the unbalance and reactive power, so that the grid currents are sinusoidal, in phase with the grid voltages and symmetrical.

I. INTRODUCTION

Non linear loads, especially power electronic loads, generate harmonic currents and voltages in power systems. They cause a low power factor, increase the losses and reduce the efficiency of the power system, and lead to voltage distortion. Passive LC filters can be used to eliminate harmonic currents. However, bulk passive components, series and parallel resonance, a fixed compensation characteristic are the main disadvantages of passive filters. To overcome passive filter problems, for many years, various active power filters (APF) have been developed [1][2][3].

Conventionally, the power inverter as a shunt APF is controlled in such a way as to inject equal-but-opposite harmonic and reactive compensation currents based on calculated reference currents. Hence, the current sensors are installed on the load side. Then, their output signals will be processed to construct the reference or desired currents, which consist of harmonic and reactive components as well as negative- and zero-sequence components for unbalance compensation. Once the desired reference currents have been established, the currents must be injected into the grid accurately using the power inverter with a current control mechanism. The actual inverter currents must attempt to follow the harmonic-rich reference currents.

However, the construction of a reference current waveform will introduce distortion or inaccuracies due to filter and extensive calculations with inherent delays and errors. Furthermore, load or power system changes take time to be included by the reference current waveforms. Hence, the reference current created for the inverter current will have not only significant steady-state error, but also transient error. The distortion and inaccuracies can be significantly reduced if these computational, filtering and control problems could be avoided. Therefore, in this paper, the idea of directly controlling the grid/source current to track a three-phase balanced sinusoidal reference current rather than the inverter current to follow the harmonic-rich reference current using a current controller will be presented. In addition, for selection of a current control technique, capability of minimizing ripple current using a fixed switching frequency technique is greatly desirable so that there is no additional error. Hence, polarized ramptime current control (PRCC) based on zero average current error (ZACE) will be chosen for this application.

II. ACTIVE FILTER CONFIGURATION

The three-phase shunt active power filter is a three-phase four-wire current-controlled voltage-source inverter (CC-VSI) with a mid-point earthed split capacitor (C_1 and C_2) in the DC bus and inductors (L_{inv}) in the AC output. The APF consists of two control loops, namely a current control loop and a voltage control loop. The current control loop is PRCC that shapes the grid currents to be sinusoidal by generating a certain pattern of PWM for continuous switching of the inverter switches. The voltage control loop is a simple Proportional Integral (PI) control to keep the DC-bus voltage constant and to provide the magnitude of reference current signals. Fig. 1 describes the shunt active power filter configuration.

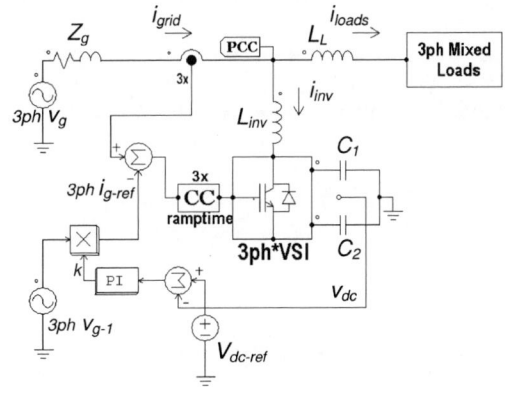

Fig. 1: Shunt active power filter configuration

A. Series Inductance

Another component of this system is a series inductance L_L, which value of the reactance X_L is comparable to the effective

---

978-1-4244-1871-8/07 $25.00 © 2007 IEEE

grid impedance, Z_g [4][5]. Basically, the inductor provides the required voltage decoupling between load harmonic voltage sources and the grid. It also enhances the controllability of the current controller by reducing the current slope of the loads.

Without a series inductor, the point of common coupling (PCC) is tied to the load harmonic voltage source. The load harmonic voltage considerably characterizes the harmonic voltage at the PCC. As a result, there would still be harmonic voltages across the grid impedance, which would continue to produce harmonic currents and could not be compensated for by a shunt APF. Currents from the shunt APF do not significantly change the harmonic voltage at the loads.

With a series inductor, for a sinusoidal grid current, the CC-VSI output current is equal-but-opposite to the unwanted load current. These two currents generate identical harmonic voltages across the series inductance and the inverter inductance (in relative proportion to the inductances). In doing so, the CC-VSI generates the output harmonic voltage, which is equivalent to the load harmonic voltage. For perfect filtering, when no harmonic current flows through the grid impedance, a combination of the inverter and the load will be seen as infinite impedance for the grid harmonic voltages. The circuit equivalent from the harmonic point of view is shown in Fig. 2.

If a fundamental grid current component flows through the grid impedance, the voltage across the grid impedance will also be a fundamental component. The voltage across the grid impedance is represented by the $jI_{g-1}X_g$ phasor (assuming the resistance is negligible). With the V_{g-1} (the fundamental component of the grid voltage) fixed, the grid harmonic voltage V_{g-h} is represented in vector form by a circular region added to V_{g-1}. The voltage at the PCC (V_{pcc}) can be illustrated in Fig. 3. Hence, any harmonic voltage at the PCC actually mirrors the grid harmonic voltage. For a harmonic-free grid voltage, the voltage at the PCC only contains a fundamental component.

B.   Direct Control of the Grid Current

As seen from Fig. 1, a node, which is a point of common coupling (PCC), is created with three connections, one each to the load, the grid and the inverter. Accordingly, all three currents – i_L, i_g, and i_{inv} – (for three or four wires) are potentially accessible to be directly controlled by the CC-VSI, following the basic current summation rule:

$$i_g = i_{inv} + i_L \qquad (1)$$

Thus, for the CC-VSI operated to directly control the grid current, the current sensors are located on the grid side. The grid current is sensed and directly controlled to follow a symmetrical sinusoidal reference signal, which is in phase with the grid voltage. For perfect tracking, the shunt APF automatically provides the harmonic, reactive, negative- and zero-sequence currents for the load according to (1) without measuring and determining the unwanted load current components. Hence, the shunt APF has also the ability to balance the asymmetrical currents.

Moreover, the controllability of the grid current can be achieved using bipolar PWM switching. The upper and lower power switches of each half-bridge are switched on a complementary basis. As a result, the inverter output current, as well as the grid current, can always be controlled to ramp up and down continuously. Therefore, the direct control of the grid current is feasible because the switching action will have a direct, immediate and predictable effect on the AC grid current, and hence provide the controllability.

By directly controlling the grid currents, the shunt APF can provide complete compensation for many loads at the PCC instead of compensating for each load individually. The system is simple and efficient because only one current sensor per phase is required, located on the grid side.

In addition, controlling the grid current rather than the inverter current allows us to create a sinusoidal current reference (for the grid current), rather than having to create a harmonic- and transient-rich current reference (for the inverter current). The idea to obtain the desired grid current waveform instantaneously without calculation is easily fulfilled by using an active power balanced technique. The active power is maintained balanced among the grid, the load and the DC bus of the power inverter.

III.   THE CURRENT CONTROL LOOP

In the current control loop, the current sensors on the grid side detect the grid currents. The outputs of the sensors are compared to the three-phase symmetrical sinusoidal reference signals, which are in phase with the grid voltages. The current error signals, which are the differences between the actual currents (grid currents – i_g) and the reference signals – i_{g-ref}, are processed using polarized ramptime current control (PRCC) to generate PWM signals. The pulse signals drive the switches so that the VSI produces currents for compensation.

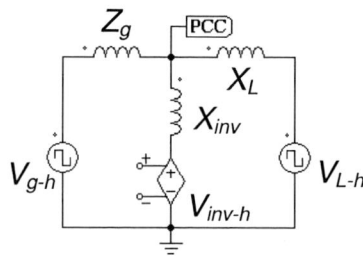

Fig. 2: Circuit equivalent for harmonics

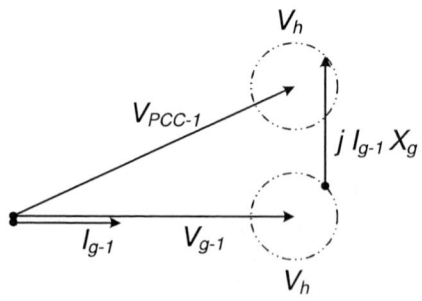

Fig. 3: Phasor diagram of voltages at the PCC for successful compensation

The PRCC technique has an important role in enhancing the performance and effectiveness of the filter to control the current loop of the CC-VSI. The operation principle of PRCC is based on ZACE (zero average current error) [6][7]. The current error signal is forced to have an average value equal to zero with a constant switching frequency. The PRCC maintains the area of positive current error signal excursions equal to the area of negative current error signal excursions, resulting in the average value of the current error signal being zero over a switching period (Fig. 4). The switching period (or frequency) is also kept constant based on the choice of switching instants relative to the zero crossing times of the current error signal. The PRCC has a high bandwidth with a fast transient response that can quickly follow the rapid changes in non linear loads.

In order to observe the current-control operation, the single-phase equivalent circuit as shown in Fig. 5 is examined.

A. Operating condition requirement

Neglecting the losses in the inverter, the output current for each phase of the inverter through the inductance L_{inv} can be expressed in a switching function (s) as:

$$\frac{di_{inv}}{dt} = \frac{1}{L_{inv}}(v_{pcc} - sv_{C1} - (s-1)v_{C2}) \tag{2}$$

$s = 1$ if the upper switch is closed, and $s = 0$ if the upper switch is open. It must be assumed that v_{pcc}, v_{C1} and v_{C2} are constant over the switching period. The switches are operated on a complementary basis. For $\dfrac{di_{inv}}{dt} > 0$, then

$$v_{pcc} - sv_{C1} - (s-1)v_{C2} > 0 \tag{3}$$

To satisfy this condition, the relationship between DC-capacitor voltages and the voltage at the PCC in terms of the switching function is given by:

$$\begin{matrix} \text{If } s = 1, \text{ then } v_{C1} < v_{pcc} \\ \text{If } s = 0, \text{ then } -v_{C2} < v_{pcc} \end{matrix} \tag{4}$$

Values of v_{C1} and v_{C2} are always positive. For $s = 1$, the system does not work during the negative half-cycle of v_{pcc}. For $s = 0$, any value of v_{C2} will satisfy (4) during the positive half-cycle of v_{pcc}. For the negative half-cycle of v_{pcc}, the inverter functions well, as long as $|v_{C2}| > |v_{pcc}|$.

For $\dfrac{di_{inv}}{dt} < 0$, then

$$v_{PCC} - sv_{C1} - (s-1)v_{C2} < 0 \tag{5}$$

To satisfy (5), the relationship between the DC-capacitor voltages and the voltage at the PCC in terms of the switching function is given by:

$$\begin{matrix} \text{If } s = 1, \text{ then } v_{C1} > v_{pcc} \\ \text{If } s = 0, \text{ then } -v_{C2} > v_{pcc} \end{matrix} \tag{6}$$

In this case, the system is unlikely to be operated at $s = 0$ during the positive half-cycle of v_{pcc}. For $s = 1$ during the negative half-cycle of v_{pcc}, any value of v_{C1} will satisfy (6). During the positive half-cycle of v_{pcc}, the inverter functions well as long as $|v_{C1}| > |v_{pcc}|$.

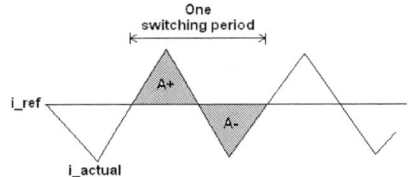

Fig. 4: Zero average current error (ZACE)

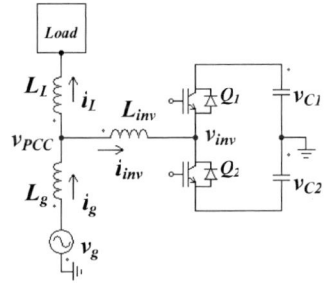

Fig. 5: Single-phase equivalent circuit

Therefore, for both cases, the inverter always generates currents as long as the magnitude of both DC-capacitor voltages (v_C) is greater than the peak value of the PCC voltage ($v_{pcc-peak}$). If this condition is not achieved, then the required operating condition for the system is not provided, and the compensation fails completely.

B. Controllability condition

The PRCC has characteristics similar to a sliding mode control [6][8]. Therefore, the current error signal ε, which is a controlled parameter, can be defined as a sliding surface.

$$\varepsilon = i_g - i_{g\text{-}ref} \tag{7}$$

To ensure that the system can remain on the sliding surface and maintain perfect tracking, the following condition must be satisfied:

$$\varepsilon \,\dot{\varepsilon} \leq 0 \tag{8}$$

where $\dot{\varepsilon}$ is derived from (7):

$$\frac{d\varepsilon}{dt} = \frac{di_g}{dt} - \frac{di_{g-ref}}{dt} \tag{9}$$

From the switching operation implementation, the $\dot{\varepsilon}$ is able to be controlled such that a positive value of the error signal produces a negative derivative of the error signal ($\dot{\varepsilon} < 0$), and a negative value of the error signal produces a positive derivative of the error signal ($\dot{\varepsilon} > 0$). Controlling the sign of $\dot{\varepsilon}$ is associated with controlling the sign of di_g/dt to be positive or negative.

Referring to the filter operation by controlling the grid currents, and combining (1) and (9), $d\varepsilon/dt$ can be expressed as:

$$\frac{d\varepsilon}{dt} = \frac{di_L}{dt} + \frac{di_{inv}}{dt} - \frac{di_{g-ref}}{dt} \tag{10}$$

Since the sign of di_{inv}/dt and di_g/dt are matching, perfect tracking can be achieved when:

978-1-4244-1871-8/07 $25.00 © 2007 IEEE   957

$$\left|\frac{di_{inv}}{dt}\right| > \left|\frac{di_L}{dt} + \frac{di_{g-ref}}{dt}\right| \qquad (11)$$

As long as equation (11) is not satisfied, then the system is moving away from the sliding surface, and the CC-VSI loses its controllability. This means that zero crossing of the current error signal will not occur at the end of a half-switching period, since by definition, the system is controllable if there exists a (piecewise continuous) control signal that will take the state of the system from any initial state to any desired final state in a finite time interval [9]. Refer to (2), di_{inv}/dt is normally determined by L_{inv} dan capacitor voltages.

IV.   THE VOLTAGE CONTROL LOOP

In order to satisfy the operating condition requirement of the current control loop, the DC-capacitor voltages have to be forced greater than $v_{pcc-peak}$. In addition, the DC-bus voltage has to be kept constant by regulating the active power balance of the system, and then deciding the amplitude of the grid currents. The voltage control loop employs a Proportional Integral (PI) controller to maintain the desired DC-bus voltage level.

A.   Power flow

Power flowing in the grid, the load and the inverter is expressed in terms of real power p, imaginary power q, and zero sequence power p_0 [10][11][12], which consist of an average value and an oscillating value.

Considering a non-sinusoidal and unbalanced system, the load power can be decomposed into:

$$p_L = \bar{p}_L + \tilde{p}_L$$
$$q_L = \bar{q}_L + \tilde{q}_L \qquad (12)$$
$$p_{L-0} = \bar{p}_{L-0} + \tilde{p}_{L-0}$$

Due to successful compensation for the unwanted currents of the load, the grid currents will be sinusoidal, balanced and in-phase with the grid voltages. As a result, if the grid voltage is non-sinusoidal and/or unbalanced, the powers p_g, q_g and p_{g-0} generated from the grid become:

$$p_g = \bar{p}_g + \tilde{p}_g$$
$$q_g = \tilde{q}_g \qquad (13)$$
$$p_{g-0} = 0$$

The power developed by the inverter is calculated by subtracting the power supplied by the grid and the power consumed by the load, given by:

$$\begin{bmatrix} p_{inv} \\ q_{inv} \\ p_{inv-0} \end{bmatrix} = \begin{bmatrix} \bar{p}_g - \bar{p}_L \\ -\bar{q}_L \\ -\bar{p}_{L-0} \end{bmatrix} + \begin{bmatrix} \tilde{p}_g - \tilde{p}_L \\ \tilde{q}_g - \tilde{q}_L \\ -\tilde{p}_{L-0} \end{bmatrix} \qquad (14)$$

From (14), the inverter controls the whole imaginary power associated with the load $\bar{q}_L + \tilde{q}_L$ and the grid $\tilde{q}_g$ (in a small value). However, the DC bus does not contain imaginary power. The CC-VSI generates q_{inv} but it does not flow out of or into the DC-bus capacitors. According to Watanabe [11] and Peng [12], the imaginary power circulates among the phases due to the switching of the inverter. In other words, instantaneously, the imaginary power required by one phase can be supplied by the other phase.

For the average value $\bar{p}$ of the real power, the inverter supplies the zero sequence average (active) power $\bar{p}_{L-0}$ needed by the load. To supply the load zero sequence active power, the inverter has to take an active power from the grid because the inverter has no DC source. Neglecting the losses in the power converter, in steady state, the active power consumed by the load is equal to the active power supplied by the grid, and total active power flowing to the inverter is zero. The active power balance according to (14) can be stated as: $\bar{p}_{inv} + \bar{p}_{inv-0} = \bar{p}_g - \bar{p}_L - \bar{p}_{L-0} = 0$. Thus, the active power taken from the grid by the inverter, which is used to support the zero sequence power delivered to the load, is: $\bar{p}_{inv} = \bar{p}_g - \bar{p}_L = \bar{p}_{L-0}$. Additional active power consumption is required to compensate for the losses, so that $\bar{p}_{inv} = \bar{p}_{L-0} + \bar{p}_{loss}$. The active power balance is regulated by the voltage control loop to retain the DC-bus voltage around its reference level.

The inverter supplies $\tilde{p}_L$ as well as $\tilde{p}_{L-0}$ and consumes $\tilde{p}_g$ (in a small value) using DC capacitors as an energy storage element. The DC capacitors absorb (release) energy when $\tilde{p}$ is positive (negative). This power does not affect the DC-bus voltage level since its mean value equals to zero. However, it will appear in the DC-bus voltage as a ripple. The ripple becomes small if the value of DC capacitors is high.

B.   DC-bus voltage control system

For successful compensation, the grid current is the same as its reference. To obtain the reference signal, only one phase of the three-phase grid voltage is detected as the reference phase. Afterwards, a three-phase symmetrical sinusoidal waveform is generated using a phase lock loop (PLL) circuit. Only the magnitude of the grid current needs to be determined.

As mentioned above, in steady state and ignoring the losses, the active power consumed by the load is equal to the active power supplied by the grid, and $\bar{p}_{inv}$ will be zero. With no power flow into the inverter, the average DC-bus voltage thus can be maintained at the reference voltage level.

When a load variation occurs, the active power balance between the load and the grid will cease to be maintained. The inverter immediately supplies (absorbs) the active power mismatch between the grid and the load, since the voltage control loop cannot respond instantaneously to provide the appropriate grid reference current magnitude. This yields a DC-bus voltage deviation (ΔV_{dc}).

Due to active power balance, the amplitude of grid active currents must be adjusted appropriately to compensate for the active power charged/discharged from DC capacitors of the inverter. The required change in grid currents will come as soon as the voltage control loop responds to change (increase or decrease) in the magnitude of the grid currents. The output of the PI controller, which is a gain k, can determine the amount of ΔV_{dc} that corresponds to the grid current amplitude (Fig. 6). The total active power flowing to the inverter will go

978-1-4244-1871-8/07 $25.00 © 2007 IEEE

to zero when the grid current amplitude approaches its final value. The average DC bus voltage is then recovered and stays at the reference voltage. Finally, the active power supplied from the grid is matched to that consumed by the load. A new steady state has been achieved with a new grid current amplitude. Hence, the sinusoidal grid current reference signal is given by:

$$i_{g\text{-}ref} = k \, v_{grid\text{-}1} \tag{15}$$

where $v_{grid\text{-}1}$ is the fundamental component of the grid voltage, and obtained from a phase-lock-loop (PLL) circuit detecting the grid voltage. The value of k is obtained from the output of a simple PI controller in an voltage control loop regulating the CC-VSI DC-bus voltage.

The voltage control loop block diagram is shown in Fig. 6. Considering a perfect tracking current control loop, the grid current is the same as its reference. The inverter DC-bus voltage is detected and reduced by a gain K_f to the level of a signal. Since the DC-bus voltage contains ripples, a first-order low-pass filter (LPF) is added to the feedback loop to obtain a smooth gain k. The output of the PI controller, which is a gain k, is multiplied by $v_{g\text{-}1}$, which is the fundamental component of the grid voltage obtained from a PLL circuit and then used as a reference waveform. The inverter currents will flow through the switches to the DC capacitors to develop the DC-bus voltage. K_C is a power conversion factor between the AC side and the DC side of the power converter.

V. THE THREE-PHASE SHUNT ACTIVE POWER FILTER FOR MIXED LOADS

The system in Fig. 1 is tested using laboratory experiment to verify the shunt APF concepts. The three-phase grid voltages contain harmonics, and the mixed loads consist of single- and three-phase linear and non-linear loads. The linear loads are resistive and inductive loads, while the non-linear loads are a rectifier type of loads. The loads represent the distributed linear and non-linear loads, which exist in a typical electrical distribution system such as in commercial buildings. The three-phase current waveforms along with their harmonic spectrums of the mixed loads, as well as the neutral current from the laboratory experiment, are shown in Fig. 7. It shows clearly that the currents are not sinusoidal. The load phase-currents are also unbalanced and contain reactive components. The significant third-harmonic current flows in the neutral wire.

Fig. 8 demonstrates the steady-state performance of compensation results. It can be seen that the shunt APF is successfully able to compensate for the total mixed loads. Although the grid voltage contains harmonics, it does not distort the grid currents. The PRCC can force the grid currents to follow accurately the sinusoidal reference waveforms without additional low order harmonics. The grid currents become both sinusoidal and in phase with the grid voltages (with insignificant phase leading by approximately 5° due to AC filter capacitors (C_{ac}) – in Fig. 9, only phase A of the grid voltage is shown). The amplitude is determined by the active power required by the system. The PRCC VSI is capable of controlling the low order harmonics due to ZACE with a fixed switching frequency. However, it produces a high frequency switching current ripple. To avoid the current ripple flowing to the grid, small AC filter capacitors (C_{ac}) are installed on the grid side.

After compensation, the grid currents are symmetrical both in magnitude and phase. As a result, the neutral current at the grid is also reduced to zero. The grid currents are balanced because the CC-VSI is able to force the grid currents to follow a three-phase balanced sinusoidal reference signal. Then, the inverter creates the inverse of the negative- and zero sequence currents automatically to balance the unbalanced loads, without measuring and determining the negative- and zero sequence components. From Fig. 10, it is obvious that the CC-VSI is able to generate three different currents for each phase as well as the neutral current. Hence, the inverter not only generates harmonics to eliminate the load harmonics but also provides balancing to create the symmetrical grid currents.

VI. CONCLUSION

This paper explains the implementation of a three-phase four-wire shunt active power filter (APF) operated to directly control the AC grid current to be sinusoidal and in phase with the grid voltage. By doing this, the three-phase shunt APF automatically provides compensation for harmonics, reactive power and unbalance without measuring/sensing the load currents. The computational, filtering and control problems can be avoided so that the distortion and inaccuracies problems can be significantly minimized. The experimental results prove the validity of the concept.

The polarized ramptime current control (PRCC) is very effective to shape the grid to be sinusoidal without additional low order harmonics due to the concept of zero average current error (ZACE) with fixed switching frequency. Thus, it is suitable for the grid current-controlling shunt APF.

There are many advantages to directly control the grid current. Firstly, it is easy to create a simple sinusoidal reference for the grid current using the active power balance method. The reference current is an appropriate reference to minimize the grid harmonic currents. Secondly, the grid currents produced will be sinusoidal, balanced and in phase with the grid voltage regardless of grid voltage conditions. Thus, it prevents (more) pollution of the electrical system from non-linear loads. Moreover, the control mechanism becomes very simple.

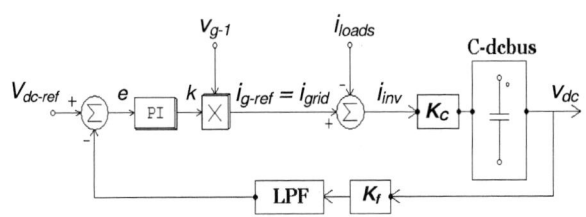

Fig. 6: Voltage control loop

REFERENCES

[1] M. El-Habrouk, M. K. Darwish, and P. Mehta, "Active Power Filter: A Review," *IEE Proc. Electr.Power.Appl*, pp. 403-413, Sept 2000

[2] B. Singh, K. Al-Haddad, and A. Chandra, "A Review of Active Filter for Power Quality Improvements," *IEEE Trans. on Industrial Electronics*, pp. 960-971, Feb 1999

[3] H. Akagi, "New Trends in Active Filters for Power Conditioning," *IEEE Transactions on Industry Applications*, **32**(6), p. 1312-1322, 1996.

[4] F. Z. Peng, "Application issues of active power filter," *IEEE Industry Applications Magazine*, **4**(5): p. 21-30, 1998

[5] H. H. Tumbelaka, L. J. Borle, and C. V. Nayar, "Analysis of a Series Inductance Implementation on a Three-phase Shunt Active Power Filter for Various Types of Non-linear Loads," *Australian Journal of Electrical and Electronics Engineering, Engineers Australia*, **2**(3): p. 223-232, 2005

[6] L. J. Borle, "Zero average current error control methods for bidirectional AC-DC converters," PhD Thesis, Dept. Elect. Eng., Curtin University of Technology, Western Australia and the Australian Digital Theses Program: http://adt.caul.edu.au/, 1999

[7] L. J. Borle, and C. V. Nayar, "Ramptime Current Control", *in Conf. Proc. 1996 IEEE Applied Power Electronics Conference (APEC'96)*, p. 828-834.

[8] J. Y. Hung, W. Gao, and J. C. Hung, "Variable Structure Control: A Survey," *IEEE Trans. on Industrial Electronics*, **40**(1), p. 2-22, 1993

[9] Franklin, G.F., J.D. Powel, and A. Emami-Naeni, *Feedback Control of Dynamic Systems*. 3rd ed., Addison-Wesley, 1994

[10] H. Akagi, Y. Kanazawa, and A. Nabae, "Instantaneous Reactive Power Compensators Comprising Switching Devices without Energy Storage Components," *IEEE Transactions on Industry Applications*, **20**(3), p. 625-630, 1984

[11] E. H. Watanabe, R.M. Stephan, and M. Aredes, "New concepts of instantaneous active and reactive powers in electrical systems with generic loads," *IEEE Transactions on Power Delivery*, **8**(2), p. 697-703, 1993

[12] F. Z. Peng, G. W. Ott, and D.J. Adams, "Harmonic and Reactive Power Compensation based on the Generalized Instantaneous Reactive Power Theory for Three-phase Four-wire Systems," *IEEE Transactions on Power Electronics*, **13**(6), p. 1174-1181, 1998

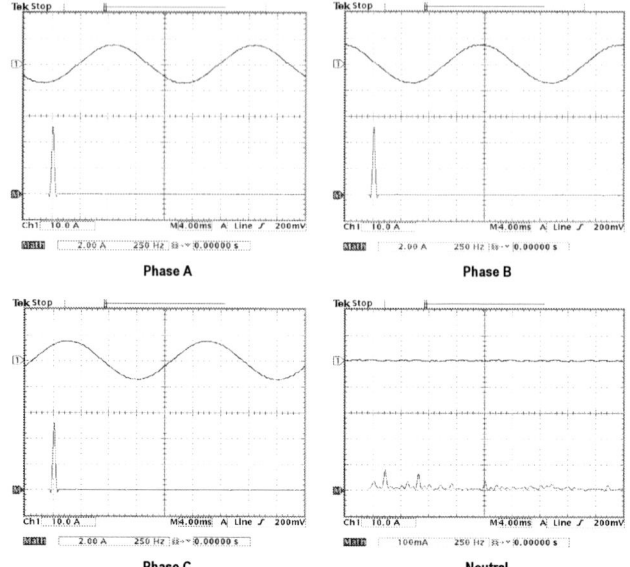

Fig. 8: phase and neutral currents of the grid after compensation

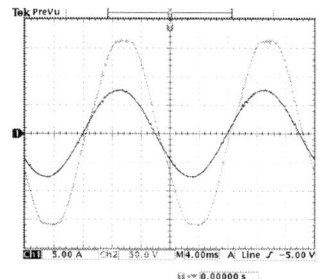

Fig. 9: phase-A grid voltage and current after compensation

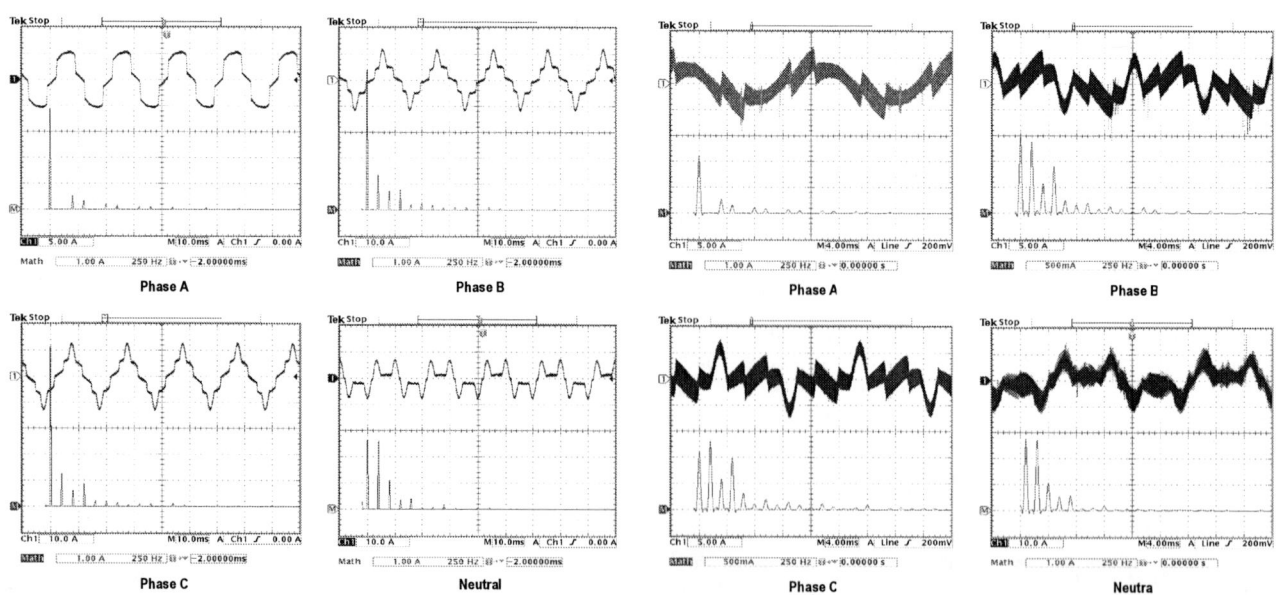

Fig. 7: phase and neutral currents for mixed loads

Fig. 10: phase and neutral currents of the CC-VSI

A New Battery Equalizer Based on Buck-boost Topology

Sang-Hyun Park, Tae-Sung Kim, Jin-Sik Park, Gun-Woo Moon and Myung-Joong Yoon
Korea Advanced Institute of Science and Technology
Department of Electrical Engineering and Computer Science
373-1 Guseong-Dong, Yuseong-Gu, Daejeon, 305-701, Republic of Korea
Email: gbs@powerlab.kaist.ac.kr

Abstract— A new technique based on buck-boost topology is proposed to equalize a series-connected battery stack in this paper. The proposed scheme transfers the energy from fully charged battery cell to the weakest charged battery using buck-boost operation. This operation maintains batteries at the same charge and voltage level. Unlike previous battery equalizing schemes, the new battery equalizer uses only one magnetic component, resulting low cost and small size. Experiment results are provided to verify the operation of the proposed battery equalizer.

I. INTRODUCTION

Series-connected battery stacks are being utilized to supply high voltage in many applications such as artificial satellite, hybrid electric vehicles (HEVs), electric vehicles (EVs), uninterruptible power supplies (UPS) and photovoltaic (PV) systems. However, series-connected battery stacks are prone to reduction in life and potential damage due to chemical differences among cells in stack and differences of charging and discharging circumstance. These differences lead to large non-uniformities in cell charge level after several cycles of charge and discharge. Charging in a series-connected battery stack, some cells can be fully charged before others. Such a process leads to overcharge in a subset of battery cells. If these cells are charged into the gassing phase after severe overcharging, there can be significant degradation in the battery life. Maintaining same charge level of cells is needed to enhance battery life. Therefore, the battery equalizing schemes are required.

A simple way to equalize the battery cells in a stack is to use shunt circuitry. The bypass resistive shunt is connected across each cell. The amount of current drawn by the shunt is proportional to the cell voltage which results in more current being diverted to the shunt as the cell voltage increases. But this method can't regulate the shunt current and the additional energy is being converted into additional loss in the shunt elements.

To regulate the shunt current, a new scheme using the individual circuit equalizer (ICE) has been proposed. In this case, each cell has its own equalizer circuit which will bypass the excess floating current. The ICE is a voltage controlled current shunt which diverts the current away from the cell during trickle charging. This scheme prevents fully charged cells from getting overcharged. Similar to the first approach,

this scheme also has additional losses in the resistor of ICE [2].

In order to prevent the additional losses, many new battery equalizers have been proposed such as battery equalizer using centralized multi-winding transformer, battery equalizer using a modular non-dissipative current diverter, and battery equalizer using DC-DC converter. But these battery equalizers need more magnetic components, as the number of battery cells increase in battery stack [3]-[5].

To solve this problem, a new battery equalizer is proposed in this paper. The proposed battery equalizer uses only one magnetic component, and is proposed in Section II. The analysis of the proposed scheme is given in Section III. In Section IV, a practical design example and experimental results are presented.

II. PROPOSED BATTERY EQUALIZER

The proposed battery equalizer is shown in Fig. 1. Each battery terminal has two unidirectional passes. One pass connects to node A of inductor L and the other connect to node B. These unidirectional passes have unidirectional switches. The first and last passes have normal switches or diodes instead of unidirectional switches. By controlling these unidirectional switches or normal switches, current passes from the battery cells to the inductor can be selected.

For example, if battery B_1 is fully charged and battery B_2 is the weakest charged, switch S_{1d} and S_{1d2c} is turned on and energy is stored into the inductor L. When switch S_{1d} is turned off and switch S_{2c3d} is turned on, the stored energy in the inductor is delivered to battery B_2. The average current drawn by the equalizer circuit is the same as the charging current I_S. Consequently, the average current into battery B_1 is zero and battery B_1 maintains its state of charge. The details of analysis are presented in next Section.

978-1-4244-1871-8/07 $25.00 © 2007 IEEE

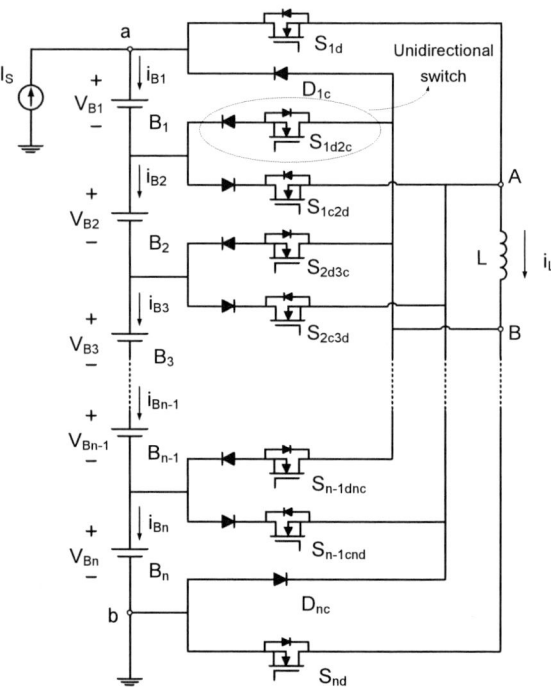

Fig. 1: Proposed Battery Equalizer

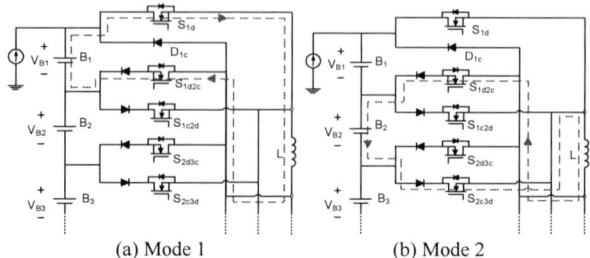

(a) Mode 1   (b) Mode 2

Fig. 2: Current Passes from B_1 to B_2

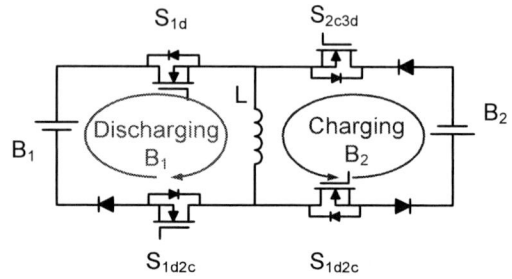

Fig. 3: Equivalent Circuit from B_1 to B_2

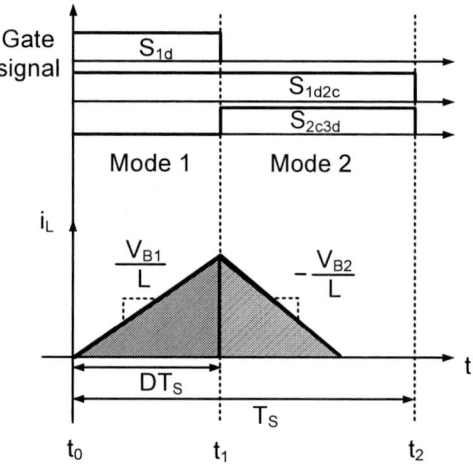

Fig. 4: Gate Signal and Inductor Current

III. OPERATIONAL PRINCIPLE

When battery B_1 is fully charged and battery B_2 is the weakest charged during charging operation, the current passes to equalize the battery stack are shown in Fig. 2. In this case, the equivalent circuit and inductor current wave form are shown in Fig. 3 and Fig. 4. The equivalent circuit is similar to the buck-boost DC-DC converter. The operation is also similar to the buck-boost DCM operation.

In the operation principle, the following assumptions are made:

1. All the switches are ideal.
2. The diodes are ideal and no voltage drop exists.
3. The battery cells have constant terminal voltages for a cycle T_S.

The mode operations are explained as follows:

Mode 1 [t_0-t_1] : Switches S_{1d} and S_{1d2c} are turned on and the current of inductor L increases with constant slope.

$$i_L(t) = \frac{V_{B1}}{L}t \qquad (1)$$

$$i_{peak} = \frac{V_{B1}}{L}DT_S \qquad (2)$$

During mode 1, the energy which is transferred from battery B_1 to inductor L is

$$E_{B1,discharge} = \frac{1}{2}i_{peak} \times DT_s = \frac{1}{2}\frac{V_{B1}}{L}\left(DT_s\right)^2 \quad (3)$$

Mode 2 [t_1-t_2] : Switch S_{1d} is turned off and switch S_{2c3d} is turned on. The current of inductor decreases with constant slope.

$$i_L(t) = \frac{V_{B1}}{L}DT_S - \frac{V_{B2}}{L}t \qquad (4)$$

Since the current passes have unidirectional switches, inductor current I_L will not be negative. During mode 2, the energy which is transferred from inductor to battery B_2 is

$$E_{B2,charge} = E_{B1,discharge} = \frac{1}{2}\frac{V_{B1}}{L}(DT_s)^2 \qquad (5)$$

In the same way, the current passes of charging battery and discharging battery can be selected by controlling the gate signal of switches. Each current pass is shown in Fig. 5.

When two or more battery cells are fully charged, those battery cells have to be discharged turn and turn about to avoid overcharging. Consider a battery stack with N battery cells. If (N-1) battery cells are fully charged and only one battery cell doesn't reach full state of charge, the average current from each (N-1) battery cell is $(N-1)I_S$.

$$\frac{i_{peak} \times DT_S}{2} \geq (n-1)T_S \times I_S \qquad (6)$$

$$i_{peak} = \frac{V_{B1}}{L}DT_S \qquad (7)$$

For the worst case, inductance of L is

$$L \leq \frac{V_B D^2 T_S}{2(n-1)I_S} \qquad (8)$$

IV. EXPERIMENTAL RESULTS

A. Experimental specifications

To verify the operation of the proposed battery equalizer, experiment is performed for a three cell battery stack. The detailed specifications of the experiment are as follows:

- Battery :
  Sealed Lead-acid battery (12V, 4.0Ah)
  Max average current : 1.0A
  Max voltage under charging : 17V
- Target voltage : 13.3V (80% state of charge)
- Switching frequency : 20kHz
- Switching duty D : 0.4
- Charging current I_S : 0.2A
- Inductance L : 83.33uH
- Initial terminal voltage of battery cells :
  B_1 : 12.90V
  B_2 : 13.03V
  B_3 : 13.20V

Considering the reaction time of the lead-acid battery, the equalizing process is kept for 4 minutes and the dead-time is kept for 2 minutes.

B. Experimental results

Fig. 6(a) shows the current of the inductor, switches, and diodes, when only one battery B_3 reaches the target stage of charge. When battery B_2 and B_3 reach the target state of charge at once, the current of the inductor, switches, and diodes are shown in Fig. 6(b). The terminal voltages of each battery cell are shown in Fig. 7. All the terminal voltage of each battery cell is the same in about 20 minutes.

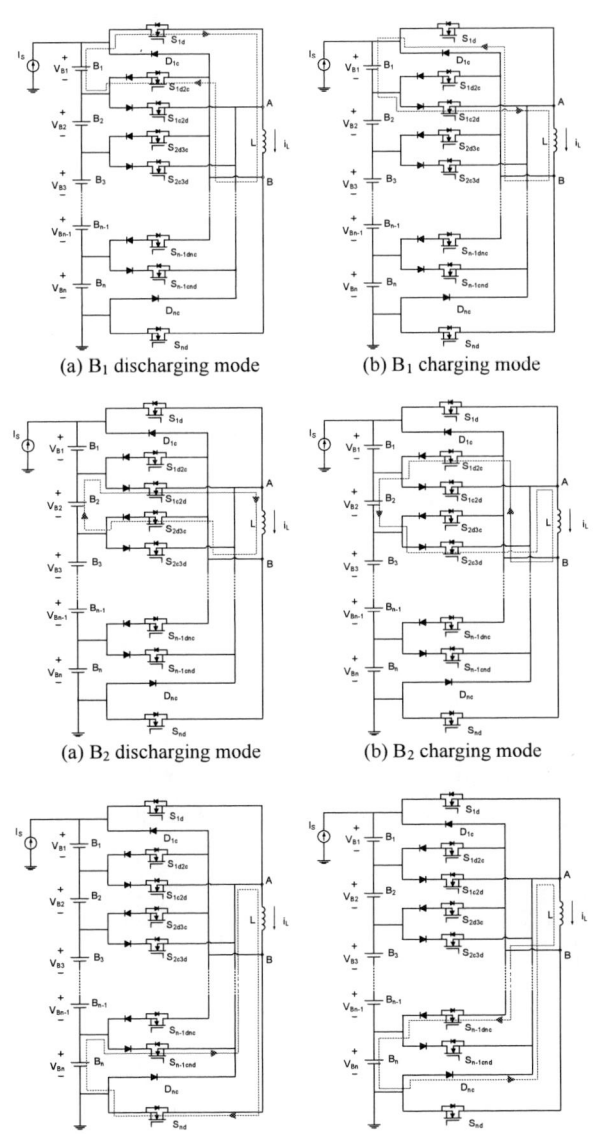

(a) B_1 discharging mode   (b) B_1 charging mode

(a) B_2 discharging mode   (b) B_2 charging mode

(a) B_n discharging mode   (b) B_n charging mode

Fig. 5: Current passes of Discharging and charging operation

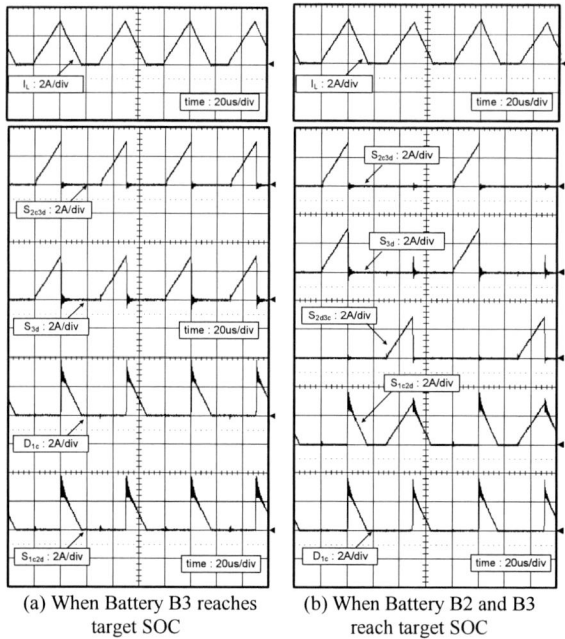

(a) When Battery B3 reaches target SOC

(b) When Battery B2 and B3 reach target SOC

Fig. 6: Current of Inductor, Switches and Diodes

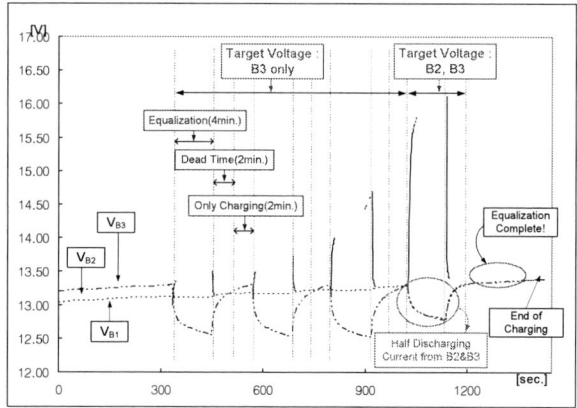

Fig. 7: Terminal Voltages of Battery Cells with Charging and Equalization

V.   CONCLUSIONS

A new battery equalizer is proposed in this paper. This battery equalizer prevents battery cells from overcharging by buck-boost operation. The proposed battery equalizer uses only one magnetic component. Therefore this proposed scheme leads to advantages in low cost and small size.

In this paper, the operation principle is analyzed and experimental results are given to verify the analysis.

REFERENCES

[1] D. Bjork, "Maintenance of Batteries - new trends in batteries and automatic battery charging," INTELEC Conf. Proceedings, pp. 355-360, 1986.

[2] B. Lindemark, "Individual Cell Voltage Equalizers (ICE) for Reliable Battery Performance," INTELEC Conf. Rec., pp. 196-201, Kyoto, Japan, 1991.

[3] Lee, Y.S., and Cheng, G.T.: 'ZCS bi-directional dc-to-dc converter application in battery equalization for electric vehicle'. Proc. IEEE PESC, 35th Annual Meeting, Aachen, Germany, 2004, pp. 2766–2772

[4] N.H. Kutkut, "Non-Dissipative Current Diverter Using A Centralized Multi-Winding Transformer," Power Electronics Specialists Conference, PESC '97.28th Annual IEEE, Vol. I, 1997, pp. 648-226

[5] N.H. Kutkut, "A Modular Non-dissipative Current Diverter for EV Battery Charge Equalization," Applied Power Electric Conference and Exposition. APEC '98, Thirteenth Annual, vol.2,1998, pp. 686 -690.

[6] M. Tang, T. Stuart, "Selective buck-boost equalizer for series battery packs," IEEE Trans. Aero. Electronic Systems, vol.36, no. I, pp. 201-211, January 2000.

[7] R. Perez, "Lead-Acid Battery State of Charge vs. Voltage," Home Power, vol.36, pp. 66-69, August/September, 1993

Design and Test of Controller in Power Conditioning System for Superconducting Magnetic Energy Storage

Hui Zhang, Jing Ren, Yanru Zhong
Xi'an University of Technology
Department of Electrical Engineering
zhangrenhj@163.com

Jian Chen *senior member* IEEE
Huazhong University of Science and Technology
College of Electrical and Electronic Engineering
zhanghui200-20@sohu.com

Abstract-**A prototype, double-DSPs controlled Power Conditioning System (PCS) for Super-conducting Magnetic Energy Storage (SMES), is used for experimental research in dynamic simulation test of power system. The function of SMES is to improve the dynamic performance of a single-machine infinite bus for power system. The experimental results show that the energy transfer of SMES is done almost immediately depending on the system requirement, and SMES is able to restore system stability for short circuit. The control algorithms for the SMES are simple and implementation has required very a little hardware.**

I. INTRODUCTION

In recent years constraints imposed by environment, right of way and energy costs have resulted in power system operating with considerably reduced stability margins. As a result, modern power systems rely heavily on stabilizing devices to maintain reliable and stable operation. These devices, such as SMES, can provide appropriate damping in the system during the transient period following a system disturbance.

A SMES can be used for leveling the load variation, damping the low-frequency oscillation and improving the transient stability in a power system. It has to regulate the active and reactive power rapidly and independently.

The power system operates with constant voltage and variable current, while the SMES with variable voltage and current. The interconnection between these two systems

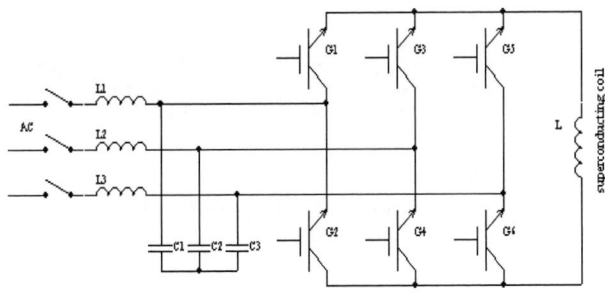

Fig.1 Current Source Converter of PCS

requires a device that matches the characteristics of both systems. The power conditioning system (PCS), which is one of the most important parts in the SMES, is used as this interconnection. It generally may have two kinds of types: current source converter (CSC) PCS and voltage source converter (VSC) PCS. The configuration of CSC-PCS is a current source converter and the configuration of VSC-PCS

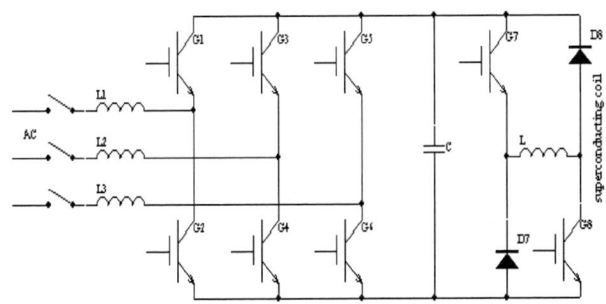

Fig.2 Voltage Source Converter of PCS

is voltage source converter in series with a DC/DC chopper, shown in Fig.1 and Fig.2 as follows.

In this paper, a kind of configuration of the PCS including a VSC and a DC/DC chopper is utilized in the prototype SMES in Xi'an University of Technology, in which the VSC is used as converting of AC/DC while the chopper as converting of DC/DC. The objective of research in this paper is to prove that the SMES can effectively enhance the capability in improving stability and damping oscillation of the power system. Through the operation of the entire SMES system, both in MATLAB based simulation and DSP based experiment, the results from simulation and experiment study in the laboratory clearly demonstrate the feasibility of the various operating modes of PCS.

II. VOLTAGE SOURCE CONVERTER

The VSC-PCS is with two parts—4-quadrant voltage source converter and 2-quadrant chopper. The 4-quadrant voltage source converter is utilized to accomplish the power transformation between three-phase AC power utility and the DC bus.

The basic converter scheme is shown in Fig.3. It consists of three-phase IGBT (G1～G6) full-bridge with three input inductors (L) on the AC side and a filter capacitor (C) on the DC side. In order to control the power flow of the system, the input inductors' currents must be accurately controlled following the power required P and Q.

A. The mathematical model of VSC

Before analyzing the voltage source converter, several assumptions are proposed firstly:

1) utility is symmetrical Power sinusoidal voltage source.

$$u_{sa} = U_m \sin \omega t$$
$$u_{sb} = U_m \sin(\omega t - 2\pi / 3) \qquad (1)$$
$$u_{sc} = U_m \sin(\omega t + 2\pi / 3)$$

978-1-4244-1871-8/07 $25.00 © 2007 IEEE

where u_{sa}, u_{sb}, u_{sc} are threephase voltage of the power utility.

U_m is the peak value of the three phase voltage.

ω is the angular frequency.

2) The all switches operate at the constant frequency, and the switching frequency is much higher than the power utility frequency.

3) The input three inductors L are linear and symmetrical, and no difference occurs among them.

4) The overall losses are represented as three symmetrical resistors R.

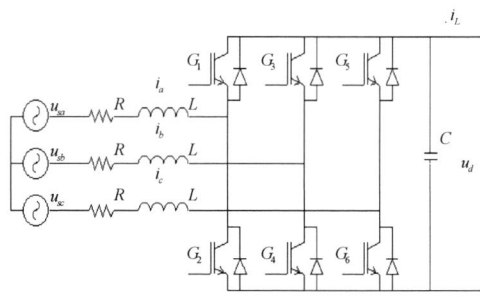

Fig.3 Topological configuration of VSC

Based on the above assumptions, system mathematical model of VSC for three-phase stationary frame is given as follows:

$$
\begin{cases}
L\dfrac{di_a}{dt} = -i_a R + u_{sa} - (S_a^* - \dfrac{S_a^* + S_b^* + S_c^*}{3})u_d \\
L\dfrac{di_b}{dt} = -i_b R + u_{sb} - (S_b^* - \dfrac{S_a^* + S_b^* + S_c^*}{3})u_d \\
L\dfrac{di_c}{dt} = -i_c R + u_{sc} - (S_c^* - \dfrac{S_a^* + S_b^* + S_c^*}{3})u_d \\
C\dfrac{du_d}{dt} = S_a^* i_a + S_b^* i_b + S_c^* i_c - i_L
\end{cases}
\tag{2}
$$

$S_K^* (K = a, b, c)$ is switching function of phase K.

where $S_K^* = 1$ expresses the up switch of phase K is on.

$S_K^* = 0$ expresses the down switch of phase K is on.

From (2), the current of every phase is not only determined by state of local switching devices, but also by state of other switching devices. This is difficult for analysis of control system under three-phase stationary frame.

Because of $\sum\limits_{K=a,b,c} i_K = 0$, the system has only two independent variables. The system mathematical model can be simplified by PARK transformation from three-phase to two-phase.

According to definition of PARK transformation, the transforming matrix $T_{ABC \to dq0}$ is:

$$
T_{ABC \to dq0} = \sqrt{\dfrac{2}{3}}
\begin{pmatrix}
\sin\varpi t & \sin(\varpi t - \dfrac{2}{3}\pi) & \sin(\varpi t - \dfrac{2}{3}\pi) \\
\cos\varpi t & con(\varpi t - \dfrac{2}{3}\pi) & \cos(\varpi t + \dfrac{2}{3}\pi) \\
\dfrac{1}{\sqrt{2}} & \dfrac{1}{\sqrt{2}} & \dfrac{1}{\sqrt{2}}
\end{pmatrix}
\tag{3}
$$

then the system mathematical model can be obtained as follows:

$$
\begin{pmatrix}
\dfrac{di_d}{dt} \\
\dfrac{di_q}{dt} \\
\dfrac{du_d}{dt}
\end{pmatrix}
=
\begin{pmatrix}
-\dfrac{R}{L} & \varpi & -\dfrac{S_d^*}{L} \\
-\varpi & -\dfrac{R}{L} & -\dfrac{S_q^*}{L} \\
\dfrac{S_d^*}{C} & \dfrac{S_q^*}{C} & 0
\end{pmatrix}
\begin{pmatrix}
i_d \\
i_q \\
u_d
\end{pmatrix}
+
\begin{pmatrix}
\dfrac{1}{L} & 0 & 0 \\
0 & \dfrac{1}{L} & 0 \\
0 & 0 & -\dfrac{1}{C}
\end{pmatrix}
\begin{pmatrix}
u_{sd} \\
u_{sq} \\
i_L
\end{pmatrix}
\tag{4}
$$

The investigation for VSC is based on synchronous frame—the above mentioned equality (4).

B. *The equivalent circuit of VSC*

The concept of complex-space-vectors is applied to analyze the model of VSC. It will give possibility representing the three-phase quantities with one space vector.

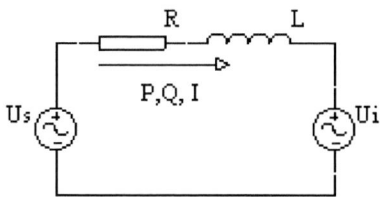

Fig.4 Equivalent circuit of VSC

According to the above assumptions, the equivalent circuit can be expressed in Fig.4. In the circuit, R represents the overall losses (including conductive loss on the inductor and the switching loss on the IGBTs). L is the inductor on the AC side. $\dot{U}_s$ is the complex-space-vector representing the power utility voltage. $\dot{U}_i$ is the complex-space-vector representing the fundamental component of the input voltage for the bridge.

According to the Fig.4, equivalent circuit of the VSC, the current can be obtained as:

$$
\dot{I} = (P - jQ)/\dot{U}_s \tag{5}
$$

$$
\dot{U}_i = \dot{U}_s - \dot{I}(R + jX) \tag{6}
$$

This shows that the current can be controlled through varying the amplitude and phase of U_i and that the amplitude and phase of U_i can be accurately adjusted by the switching function $S_K^* (K = a, b, c)$.

By the assumption that the power utility is constant, according to equality (5), the active current I_q and reactive current I_d are proportional to the active power P and reactive power Q.

$$
\begin{cases}
i_q = \dfrac{P}{\left|\dot{U}_s\right|} \\
i_d = -\dfrac{Q}{\left|\dot{U}_s\right|}
\end{cases}
\tag{7}
$$

If I_q and I_d are accurately controlled, then P and Q can be independently controlled.

C. The control strategy of VSC

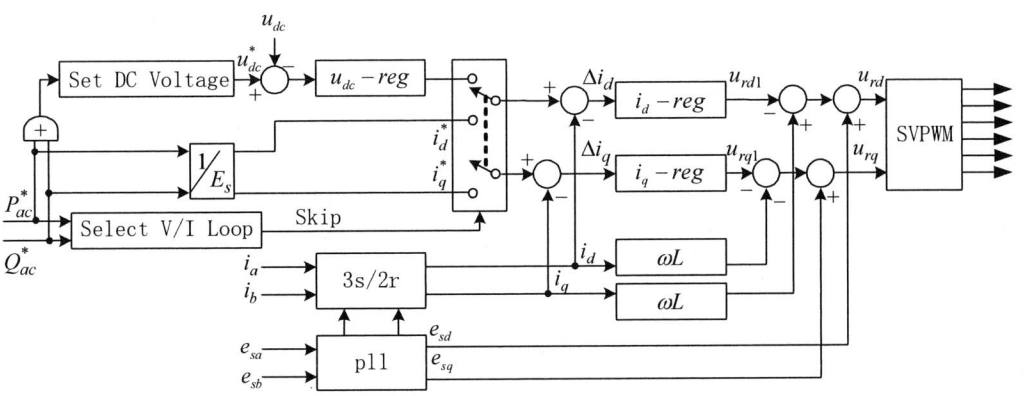

Fig.5 Control diagram for VSC

The diagram of functional control strategy of the voltage source converter is presented in Fig.5. Complete control algorithm with the AC input current and the DC output voltage feedback loops are accomplished in a DSP. The senses of the phase voltages u_a and u_b generate the phase and the amplitude of the power utility voltage. The Park (3S/2R) transformation block transforms the three-phase currents (i_a, i_b, i_c) from the stationary frame based quantities to the synchronous frame based quantities. According to the definition of the transformation, i_q and i_d represent the active power component and the reactive power component respectively. Thus, synchronous reference frame based proportional-integral current regulators regulate the active power component and reactive power component of the current independently. The command of the active power component is the sum of the output of the voltage loop and the active power demand, and the command of the reactive power component is the reactive power demand. S_d, S_q can be obtained from the outputs of the current regulators and the feedforward components. With the SV-SPWM algorithm, the PWM signals of the full bridge can be obtained.

III. TWO- QUADRANT CHOPPER

In order to match the power flow demand of the system, the magnitude and polarity of the voltage across the coil must be controlled.

The DC/DC chopper is operated to control the DC bus voltage on the capacitor and provide the required DC current flowing in the DC super-conducting coil. It has two legs in parallel in order to distribute the current in the IGBTs, and corresponding coupled inductors are used to share the current in two legs. To estimate the number of parallel paths required, it is necessary to perform the same current level calculation, and a smooth transition from charge mode to discharge mode and vice versa is insured by the two-quadrant configuration.Fig.6 shows the circuit diagram of the chopper.

A. Principle of chopper

For the two-quadrant chopper, there are three kinds of work states in the operation of PCS: charging state, waiting state and discharging state. In order to match the power flow of active power P and reactive power Q of the system, the magnitude and the polarity of the voltage across the coil must be adjusted. The two-quadrant chopper in this paper is used to control the DC voltage on the capacitor and the DC current flowing in the super-conducting coil.

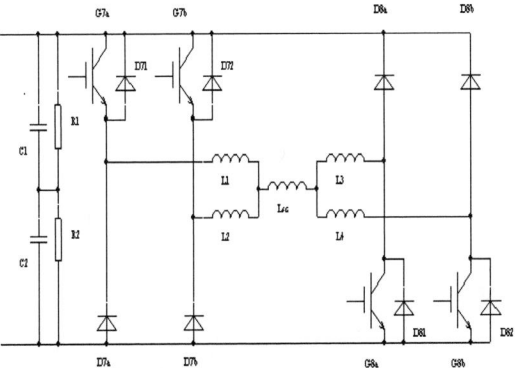

Fig.6 Topological configuration of the chopper

As shown in Fig.6, in the experimental study, the two-quadrant chopper is mostly made up of four IGBTs (CM600HA-24H, 600/1200V, in parallel per two IGBTs) and four DIODEs (MEO450-12DA, 450/1200V, connect in the same with IGBTs). Every phase of the two-quadrant chopper is composed of two legs in parallel with each other. At the same time, the couple inductor at the output of each two legs is used to share the current in two IGBTs.

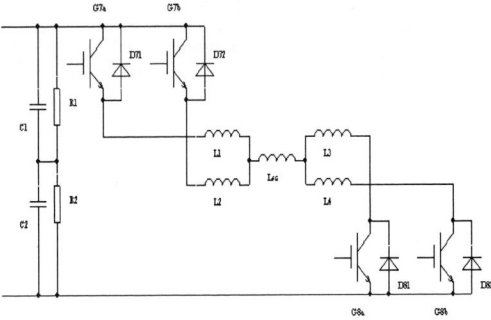

Fig.7 Charging mode

1) Charging

There are two kinds of operation modes in the charging state. They are charging mode and current-continued mode.

Among these, G7a, G7b are always "on", and G8a, G8b alternate between "on" and "off". G8a, G8b are controlled by monitoring I_{sc}.

Mode 1: Charging (Fig.7)

In this mode, the three-phase bidirectional converter operates as a rectifier. For the two-quadrant chopper, G7a, G7b are always "on", and G8a, G8b are also "on" to charge to the coil. In this process, the capacitor C charges to coil L so that U_c drops (C charging to L) and I_{sc} rises (charging magnetism).

Let voltage drop $\triangle U_c$ =13V , then charging magnetism time T1=0.26ms, so current rise $\triangle I_{sc}$ =0.42 A during this time.

In order to match the capacity of converter and limit the fluctuation of the DC voltage, the duty cycle of the two-quadrant chopper must be limited.

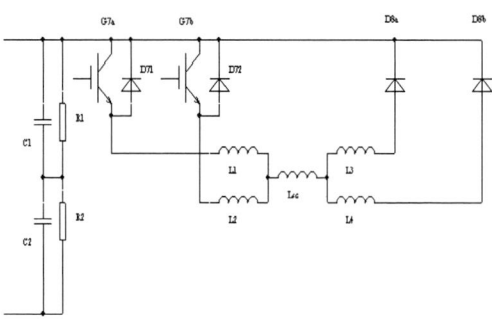

Fig.8 Current-continued mode I

Mode 2: Current-continued (Fig.8)

This mode is an assistant mode of mode 1. In this mode, the coil is bypassed through G7a, G7b and D8a, D8b, so the I_{sc} gradually reduces. At this time, the power charges the capacitor C, so the U_c rises by degrees.

Because of current drop $\triangle I_{sc}$ =0.42 A , then waiting time T2=33.44ms, so the charging current of power must be I =3.9 A

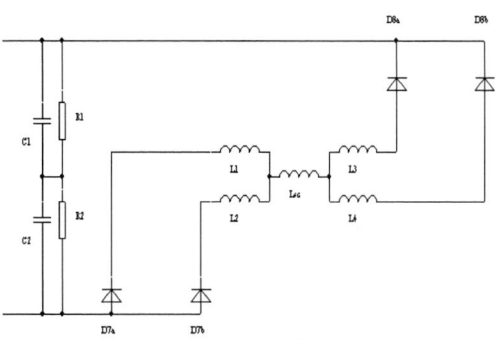

Fig.9 Discharging mode

2) Waiting

This state is a special state of charging state. In this state, the super-conducting coil current is approximate the rating value. The PCS is just used to convert energy of utility to compensate the loss of super-conducting coil. The two kinds of modes of the two-quadrant chopper are the same as that in charging state.

3) Discharging

There are also two kinds of operation modes in the discharging states. They are discharging mode and current-continued mode. Among these, G7a, G7b are always "off", and G8a, G8b alternate between "on" and "off".

Mode 3: Discharging (Fig.9)

In this mode, the three-phase bidirectional converter operates as an inverter. For the two-quadrant chopper, G7a, G7b are always "off", and G8a, G8b are also "off" to discharge from the coil. In this process, the coil L discharges to capacitor C so that U_c rises (L discharging to C) and I_{sc} drops (discharging magnetism).

Let voltage rise $\triangle U_c$ =13 V , then discharging magnetism time T3=0.26ms, so current drop $\triangle I_{sc}$ =0.42 A during this time.

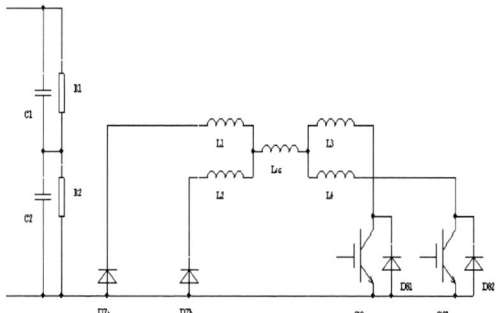

Fig. 10 Current-continued mode II

Mode 4: Current-continued (Fig.10)

This mode is an assistant mode of mode 3. In this mode, the coil is bypassed through D7a, D7b and G8a, G8b, so the I_{sc} gradually reduces. At this time, the capacitor C charges the power, so the U_c reduces by degrees.

Because of current drop $\triangle I_{sc}$ =0.42 A , then waiting time T4=33.44ms, so the discharging current of power is also I =3.9 A .

B. The control strategy of chopper

The control strategy of the two-quadrant chopper, which will decide the total performance of the system, is composed of two aspects according to the operating state. If the chopper operates at charging or waiting state, the coil current I_{sc} will be controlled to keep constant. If the chopper operates at discharging state, capacitor voltage U_c will be controlled to keep constant.

Fig.11 is the block diagram of the control strategy in realization.

1) $\triangle$Pg $\leqslant$0

Port 1 is switched on. G7a, G7b are constantly open while G8a, G8b are chopped. By monitoring the I_{sc}, stagnant loop of current is used to control "on or off" of G8a, G8b. When $\triangle I_{sc}$ $\geqslant$+10 A , G8a, G8b are "off". When $\triangle I_{sc}$ < −10 A , G8a, G8b are "on". In this way, the chopper's frequency (f)=1.25Hz.

2) $\triangle$Pg>0

Port 2 is switched on. G7a, G7b are constantly close while G8a, G8b are chopped. By monitoring the U_c, stagnant loop of voltage is used to control "on or off" of G8a, G8b. When $\triangle U_c$ $\geqslant$+13 V , G8a, G8b are "on".

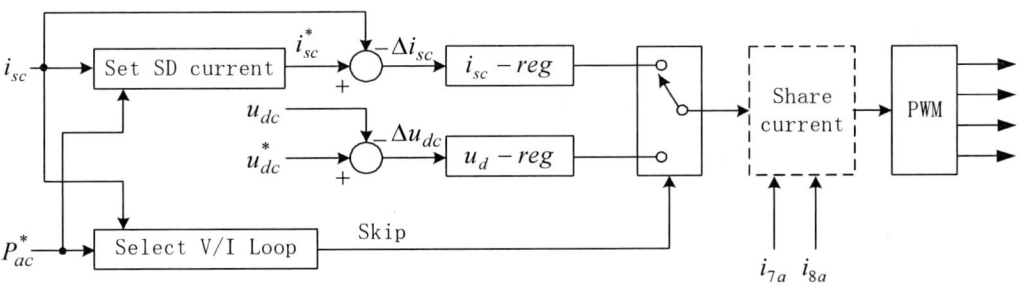

Fig.11 Control diagram for DC-DC

When $\triangle U_c < -13V$, G8a, G8b are "off". In this way, the chopper's frequency (f)=100Hz.

IV. EXPERIMENTAL RESULTS

The experiment of the system with super-conducting energy storage coil is quite expensive, so the experiment is carried out in three stages.

The first stage is adopting the storage battery to replace super-conducting energy storage coil. In this case, the power flow is bidirectional and the cost is lower, but the disadvantage is the time duration of test is limited since the energy in the battery is limited.

control element is the TI's newest DSP TMS320F240. The rating energy storage of the super-conducting coil is $50\,kJ$.

The current flowing in the super-conducting coil is $500\,A$. The inductors sharing current are $16\,mH$.

Fig.12～Fig.15 are the given power instruction curve and power response curve without or with SMES.

ACKNOWLEDGMENT

The authors would like to thank the financial support of The Province Natural Science Foundation of Shaanxi ($2006\,E_1 28$), financial support of The Province Education Department Foundation of Shaanxi (07JK355) and financial support of The University Research Foundation of XAUT (105-210704).

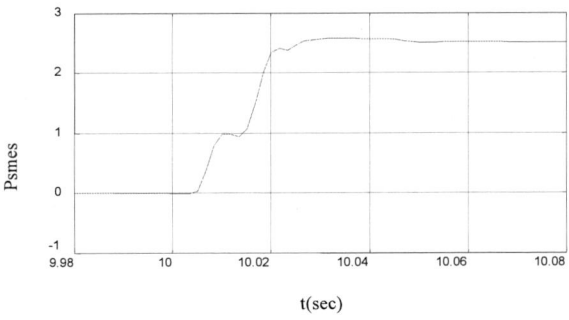

Fig.12 response of power for positive step

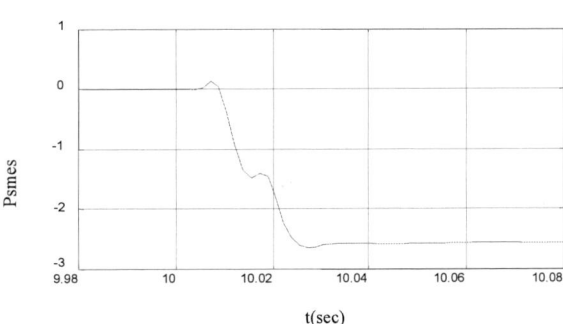

Fig.13 response of power for negative step

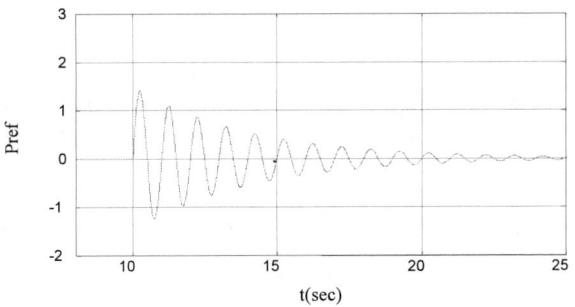

Fig.14 power instruction with sine waveform

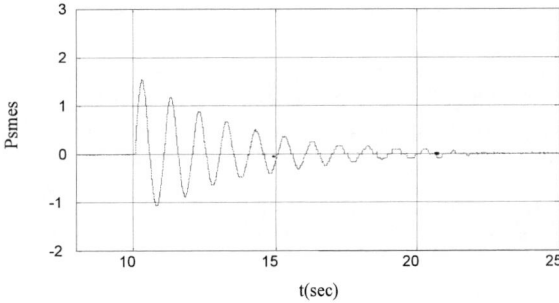

Fig.15 power response with sine waveform

The second stage is adopting general energy storage coil to replace super-conducting energy storage coil.

The third stage is adopting super-conducting energy storage coil.

The rating power of the VSC in prototype SMES is $15\,kVA$. The value of capacitor on the DC bus is $1650\,\mu F$. The value of the inductors on the AC side is $5\,mH$. The switching frequency of the VSC is $3.6\,kHz$. The main

REFERENCES

[1] Warren Buckles and William V. Hassenzahl, "Superconducting magnetic energy storage", *IEEE Power Engineering Review*, p16-20, May 2000.

[2] William V Hassenzahl, "Superconductivity, an enabling technology for 21st century power systems?". *IEEE Trans. on Applied Superconductivity*, Vol.11, No.1, p1447-1453, March 2001.

[3] R.W.Boom and H.A.Peterson, "Superconductive energy storage for power system". *IEEE Trans. on Magnetics, Vol.*MAG-8, p701-703, 1972.

[4] Chung-Shih Hsu, Wei-Jen Lee, "Superconducting magnetic energy storage for power system applications". *IEEE Trans. on Industry Applications*, Vol.29, No.5, p990-996, September/October 1992.

[5] Ibrahim D.Hassan, Richard M.Bucci and Khin T.Swe, "400MW SMES power conditioning system development and simulation". *IEEE Trans. on Power Electronics*, Vol.8, No.3, p237-249, July 1993.

[6] N.Mohan, "*Supercinductive energy storage inductors for power systems*". Ph.D. dissertation, Electrical and Computer Engineering Dept., Univ. of wisconsin, Madsin, May 1973.

[7] Satoshi Suzuki, Jumpei Baba, Katsuhiko Shutoh, and Eisuke Masada, "Effective application of super-conducting magnetic energy storage (SMES) to load leveling for high speed transportation system". *IEEE Trans. on Applied Superconductivity,* Vol.14, No.2, p713-716, June 2004.

[8] Zhang Hui, iZhu Pengcheng, Kang Yong, Chen Jian, "Improvement of generator damping through super- conducting magnetic energy storage (SMES) in dynamic simulation test of power system". *Conference Record - IAS Annual Meeting (IEEE Industry Applications Society)*, vol.2, p890-895, 2002.

[9] Zhang Hui, Kang Yong, Zhu Pengcheng, Kong Xuejuan, Liu Ping, Chen Jian, "Enhancement of generator transient stability using Superconducting Magnetic Energy Storage (SMES) in dynamic simulation test of power system". Proceedings of the International Conference on Power Electronics and Drive Systems, vol.1, p208-213, 2001.

[10] Zhang Hui, Liu Ping, Zhu Pengcheng, Kong Xuejuan, Kang Yong, Chen Jian, "Study of DSP controlled super-conducting magnetic energy storage in dynamic simulation test of power system". *PESC Record -IEEE Annual Power Electronics Specialists Conference*, vol.4, p2185-2190, 2001.

[11] Jun Li, Yunpeng Shi, Dehong Xu, "Control design of modular current source power conditioning system for high temperature SMES". 2006

[12] R.H.Lasseter and S.G.Jalali, "Power conditioning system for superconductive magnetic energy storage". *IEEE Trans. on Energy Conversion, IECON Proceedings (Industrial Electronics Conference)*, vol.1, p531-537, 2003.

[13] A.M.Hava, "*Carrier based PWM-VSI drives in the overmodulation region*". Ph.D. dissertation, University of Wisconsin, 1999.

[14] Li Hui, Baldwin Thomas L., Luongo Cesar A., Zhang Da, "A multilevel power conditioning system for superconductive magnetic energy storage". *IEEE Transactions on Applied Superconductivity*, vol.15, no.2 PART II, p1943-1946, June 2005.

[15] Peng Dengming, Lee Dong-Ho, Lee Fred C., Borojevic Dusan, "Modulation and control strategies of ZCT three-level choppers for SMES application". *PESC Record -IEEE Annual Power Electronics Specialists Conference*, vol.1, p 121-126, 2000.

[16] Ahmed El-Sayed M., "DSP based robust fuzzy logic controller for firing scheme for SMES power conditioning system". *Proceedings-2004 International Conference on Electrical, Electronic and Computer Engineering, ICEEC'04, ICEEC'04*, p367-372, 2004.

[17] Mohd. Hasan Ali, Toshiaki Murata, and Junji Tamura, "A fuzzy logic-controlled superconducting magnetic energy storage for transient stability augmentation". *IEEE Trans. on Control System Technology*, Vol.15, No.1, p144-150, January 2007.

[18] S.Nagaya, N.Hirano, H.Moriguchi, K.Shikimachi, H.Nakabayashi, S.Hanai, J.Inagaki, S.Ioka, and S.Kawashima, "Field test results of the 5 MVA SMES system for bridging instantaneous voltage dips". *IEEE Trans. on Applied Superconductivity,* Vol.16, No.2, p632-635,June 2006.

[19] IEEE Task Force on Benchmark Models for Digital Simulation of FACTS and Custom-Power Controllers, T&D Committee, "Detailed modeling of superconducting magnetic energy storage (SMES) system". *IEEE Trans. on PowerDelivery* Vol.21, No.2, p699-710, April 2006.

[20] Toshifumi Ise, Masanori Kita, and Akira Taguchi, "A hybrid energy storage with a SMES and secondary battery". *IEEE Trans. on Applied Superconductivity,* Vol.15, No.2, p1915-1918, June 2005.

[21] Li Hui, Steurer Michael, Cartes Dave, "Investigations on a 5-Level VSI-Chopper for a superconductive magnetic energy storage (SMES) power conditioning system".

[22] Li Jun, K.W.E.Cheng, D.Sutanto, and Dehong Xu, "A multimodule hybrid converter for high-temperature superconducting magnetic energy storage systems (HT-SMES)". *IEEE Trans. on Power Delivery*, Vol.20, No.1, p475-480, January 2005.

[23] S.Nagaya, N.Hirano, M.Kondo, T.Tanaka, H.Nakabayashi, K.Shikimachi, S.Hanai, J.Inagaki, S.Ioka, and S.Kawashima, "Development and performance results of 5 MVA SMES for bridging instantaneous voltage dips". *IEEE Trans. on Applied Superconductivity,* Vol.14, No.2, p699-704, June 2004.

[24] Klaus-Peter Juengst, Rainer Gehrin, Andrej Kudymow, Grigory Kuperman, and Edmund Suess, "25MW SMES-based power modulator". *IEEE Trans. on Applied Superconductivity*, Vol.12, No.1, p758-761, March 2002.

[25] S.Kolluri, "Application of distributed superconducting magnetic energy storage system (D-DMES) in the energy system to improve voltage stability". 2002

[26] Jun Li, Yunpeng Shi, Dehong Xu, Yuejin Tang, Xiaotao Peng, Shijie Cheng, "Control design of a modular current source power conditioning system for high temperature SMES". 2006

978-1-4244-1871-8/07 $25.00 © 2007 IEEE

Research of Fuzzy Logic-Controlled SMES for Power System Transient Voltage Stability

Huang Xiao-hua

Graduate School of the Chinese Academy of Science

Chinese Academy of Science

Institute of Electrical Engineering

Key Laboratory of Applied Superconductivity

P.O. Box 2703, Beijing China, 100080

Email:hxh80@mail.iee.ac.cn

Xiao Li-ye

Chinese Academy of Science

Institute of Electrical Engineering

Key Laboratory of Applied Superconductivity

P.O. Box 2703, Beijing China, 100080

Email: xiao@mail.iee.ac.cn

Abstract—Superconducting Magnetic Energy Storage (SMES) is a power electronic based device with a superconducting coil that has the potential to bring essential functional characteristics to the utility transmission and distribution systems. This paper presents a new approach and corresponding experiments for fuzzy logic-controlled SMES to improve the transient voltage stability of electric power system. The SMES coil built for power utility applications has the following unique design characteristics: 0.2H, 20kW/100kJ, 1kA dc current. Based on the result of SMES prototype experience, a mathematic model of power system and SMES are adopted. A fuzzy controller is assigned. Bus voltage is taken as the input signal to fuzzy controller. The fuzzy rule-bases are designed and explained. The validity of the suggested control strategies are confirmed by simulation and experimental tests. From the results they are demonstrated that transient voltage stabilizing effect by mean of SMES was very significant.

I. INTRODUCTION

Voltage stability refers to the ability of a power system to maintain steady voltages at all buses in the system after being subject to a disturbance from a given initial operating conditions[1]. Voltage instability have become one of the important concerns in the power industry, as this has been the major reason for several major blackouts that have occurred throughout the world including the recent Northeast Power outage in North America in August 2003. It is the premise of mid-term and long-term voltage stability to keep transient voltage stability, whose study period of interest is in the order of several seconds. The way to save the system from voltage instability is to reduce the reactive power load or add additional reactive power. Canizares and

Faur studied the effects of SVC and TCSC on voltage collapse[2]. Natesan and Radman introduced the effects of STATCOM, SSSC and UPFC on voltage stability[3]. Superconducting magnetic energy storage (SMES) with a suitable control strategy has the potential to significantly increase the system transient voltage stability.

Superconducting magnetic energy storage (SMES) systems have been the subject of active investigation for the past 30 years. Since the early 1970's active programs of research are begun by the University of Wisconsin and by the department of energy at Los Alamos National Laboratory. SMES has a long life time benefit on power system because the superconducting magnet does not have degradation like the battery. It has good properties such as high efficiency, quick response and lagging and leading phase control by means of IGBT. Meanwhile, electric energy can be stored in SMES in the form of magnetic energy. Thus, the superconducting coil has a high efficiency because its electrical resistance is almost zero and little energy is dissipated in it. Therefore SMES systems have received much attention in power system applications, such as diurnal load demand leveling, frequency control, automatic generation control, uninterruptible power supplies etc [4].

The effective use of SMES unit greatly depends on its control strategy. Many kinds of controllers for SMES unit have been proposed in literature[5-6]. As an alternative to these controls, the concept of fuzzy control was first introduced by Zadeh[7]. Fuzzy logic is a powerful problem-solving methodology with a myriad of applications in embedded control and information processing. This work focuses on the application of a fuzzy logic-controlled SMES unit for providing auxiliary service to enhance the transient voltage stability of power system. In this paper, mathematic models for power systems and superconducting magnetic energy storage (SMES) are introduced. The fuzzy rule-bases are designed and explained through practical experiences and heuristic decision rules. The simulation and experimental results are presented and discussed. The results show that the proposed fuzzy logic-controlled SMES is a very effective device for improving the transient voltage

978-1-4244-1871-8/07 $25.00 © 2007 IEEE

stability of power system.

II. Model System

The model system as shown in Fig.1-2 has been used in the present work. Fig.1 shows that the model system consists of a synchronous generator, feeding an infinite bus through a transformer and double circuit transmission line. Three-phase-to-ground fault occurs at point F3 at line#L₁. Fig.2 shows that the model system consists of four synchronous generators two areas which are interconnected by double circuit 500kV, 280km long transmission line. With the loads in area 1 being kept constant, the load in area 2 will be sudden increased. The various parameters of the generator used for the simulation are shown in Table I. The location of the SMES controllers is shown in Fig.1-2.

III. Modeling of SMES

The energy stored in the SMES enables an exchange of real power and reactive power with the system for a short period of time. SMES unit consists of isolated transformers, LC filters, a dc-dc chopper, dc-ac bridge converters with turn-off capable IGBT and a superconducting coil or inductor of 0.2H as shown in Fig.3. Fig.4 shows a photograph of SMES in the field.

The dc-ac converters generate a controllable ac voltage source, and the variable voltage generates the exchange of real and reactive power flow between the SMES and the network. The chopper is used to maintain a constant dc voltage across the capacitance. In this way, real power is exchanged between the network and the superconducting coil. At steady state, SMES should not consume any real or reactive power.

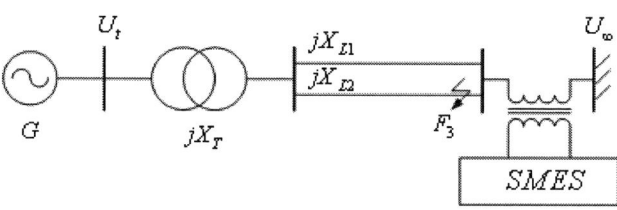

Fig. 1 Single machine power system model.

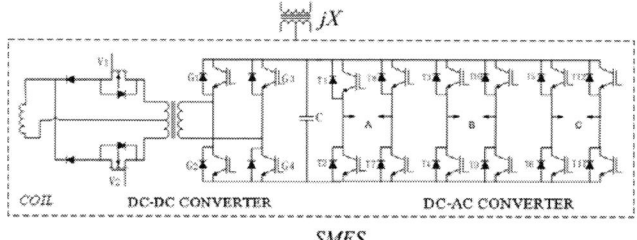

Fig. 3 The configuration of SMES

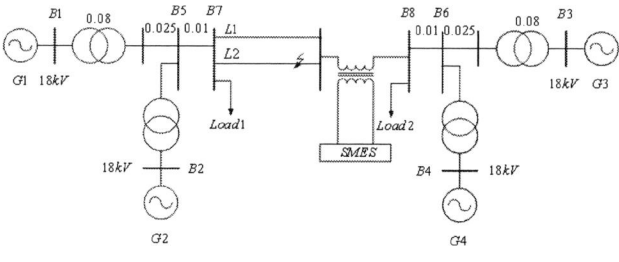

Fig. 2 Four machine power system model.

TABLE I

Parameters Of Power System

Sₙ MVA	1000	x$_d'$ p.u	0.32
Vₙ kV	15.7	x$_q'$ p.u	0.32
x$_d$ p.u	1.896	x$_d''$ p.u	0.213
x$_q$ p.u	1.896	x$_q''$ p.u	0.213
x$_2$ p.u	0.26	t$_d'$ s	1.083
x$_0$ p.u	0.0914	t$_q'$ s	1.1
r$_a$ p.u	0.00242	t$_d''$ s	0.135
J kg-m^2	10^5	t$_q''$ s	0.135

Fig. 4 A photograph of SMES in the field

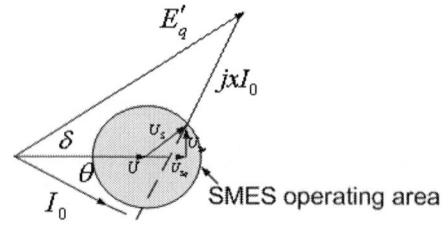

Fig. 5 Vector diagram of a SMES connected to power system.

978-1-4244-1871-8/07 $25.00 © 2007 IEEE   972

The vector diagram of a SMES connected to power system (Fig.1) is presented in Fig.5.

The energy stored in the superconducting inductor is

$$W = W_0 + \int_0^t P d\tau \qquad (1)$$

where W_0 is the initial energy in the inductor. P is the real power absorbed or delivered by the SMES can be given by

$$P = \frac{E_q' U_S}{X} \sin(\delta - \alpha) \qquad (2)$$

where E_q' is given value of generators. δ is the rotor angle. Reactance X represents the sum of generator transient reactance, transformer leakage reactance and the equivalent reactance of the parallel lines. $U_S \angle \alpha$ is the SMES output voltage. When U_S lags the current by $90°$ ($\alpha = \theta - 90°$), P becomes

$$P = \frac{E_q' U_S}{X} \cos(\delta - \theta) \qquad (3)$$

Thus the electrical output power P_e of the generators can be written as

$$P_e = P_{e0} + P \qquad (4)$$

Here P_{e0} is the generators output power without SMES ($U_S=0$) and is given by:

$$P_{e0} = \frac{E_q' U}{X} \sin \delta \qquad (5)$$

where U is the infinite bus voltage.

From Fig.5 the general equation of the current can be written as

$$I = \frac{E_q' - U_S - U}{jX} = \frac{E_q' - U}{jX} + \frac{-U_S}{jX} = I_0 + \Delta I \qquad (6)$$

Here I_0 is the current of the power system without SMES ($U_S=0$). ΔI is an additional current term caused by the SMES voltage $U_S \angle \alpha$.

The angle θ of the current can be written as

$$\theta = \tan^{-1}(\frac{U - E_q' \cos \delta}{E_q' \sin \delta}) \qquad (7)$$

IV. DESIGN OF FUZZY LOGIC CONTROLLER

Fuzzy logic controller is one of the most practically successful approaches for utilizing the qualitative knowledge of a system to design a controller. The operation of fuzzy logic controller does not rely on how accurate the model, parameters and operation conditions are, but rather, on how effective the linguistic rules of the fuzzy controller are. Fig.6 shows the proposed control strategy in SMES unit. The design of fuzzy logic controller is described as three steps that involves fuzzification, rule base and defuzzification in the following.

A. Fuzzification

The fuzzification procedure is a process that the input variable is mapped onto fuzzy linquistic variables. For the

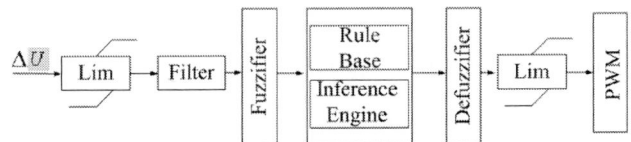

Fig. 6 The proposed fuzzy logic controller

design of proposed fuzzy logic controller, change in bus voltage (ΔU) at the output of generator and magnitude of SMES series voltage are selected as the input and output respectively. Triangular membership functions for input and output are shown in Fig.7 to 8, in which the linguistic variables NB, NS, Z, PS, and PB stand for Negative Big, Negative Small, Zero, Positive Small and Positive Big respectively.

B. Fuzzy Rule Base

The inference engine is a program which uses the rule base and the input data of the controller to draw the conclusion. The rule base is the heart of a fuzzy controller, which is determined from the viewpoint of practical system operation and by trial and error and is shown in Table 3. The min-max inference is applied to determine the degree memberships for output variables [4, 5].

C. Defuzzification

A defuzzification method is also required to transform the fuzzy output of the controller into a crisp output value. For the incremental fuzzy logic controller, using the center of mass defuzzification method the output $\Delta u_f(k)$ is :

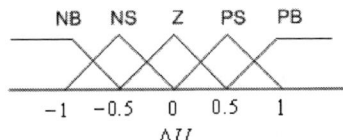

Fig. 7 Membership of input for SMES

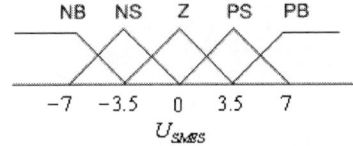

Fig. 8 Membership of output for SMES

TABLE II

PARAMETERS OF SMES

ΔU(pu)	U_{SMES}
NB	NB
NS	NS
Z	Z
PS	PS
PB	PB

978-1-4244-1871-8/07 $25.00 © 2007 IEEE   973

$$\Delta u_{f(K)} = \frac{\sum_{j=1}^{9} u_j c_j(k)}{\sum_{j=1}^{9} u_j} \qquad (8)$$

where $c_j(k)$ is the value of control output corresponding to the membership value of input equal to unity.

V. SIMULATION RESULTS AND DISCUSSIONS

The impact of the proposed fuzzy-controlled strategy of SMES on transient voltage stability was studied through simulation.

In the following, simulations were carried out considering two cases. Time step and simulation time had been chosen as 0.00005s and 2s, respectively. The switching losses in the SMES system were neglected.

A. Case 1

In this case it is considered that a balanced (3LG: Three-phase-to-ground) fault occurs at 0.1 sec at point F1 near the asynchronous generator at Line #L_1 as shown in Fig.1.

Fig.9 shows the voltage response at the load bus U. It is clear that the proposed SMES can maintain the terminal voltage at rated level. It effectively enhances the transient voltage stability. If the simulation adds real power control, it can own much better voltage response result.

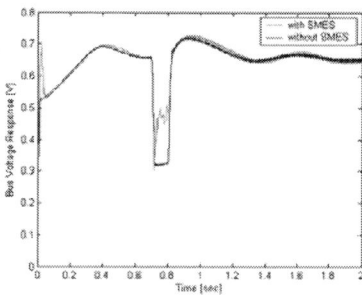

Fig. 9 3LG load angle response

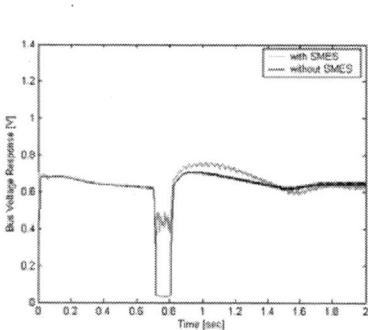

Fig. 10 1LG rotor speed response

B. Case 2

In this case it is to study the effect of sudden increase of dynamic loads on the voltage stability. The static model (ZIP-30% constant impedance, 40% constant current, 30% constant power) adopted in this case. Three-order induction motor model is adopted to model the dynamic loads. It is also considered that the fault occurs from 0.7s to 0.8s.

Fig.10 shows the simulation voltage response result of bus 8 for this case. It is clearly seen that from the response SMES effectively enhance the transient voltage stability. The α of SMES output voltage is lagging the line current by $112.5°$.

From the above observations, the simulation results of case (1) and case (2) show that with the bus voltage control, the proposed fuzzy logic-controlled SMES unit is effective in enhancing the transient voltage stability of power system. It can be concluded that with the real power and bus voltage coordinating control, the effect of SMES on voltage stability will be significantly improved.

VI. EXPERIMENTAL RESULTS

This part shows the experimental results of improving the power system voltage stability using the SMES system which consists of a 100kJ NbTi magnet and 20kW IGBT converter. Experiments are performed in three-phase-to-ground (3LG) fault at load bus as shown in Fig.1. SMES system series to the transmission line near protected bus with three set of isolated transformers. The voltage of capacitance is maintained to a constant value through the dc-dc chopper connected to a superconducting coil.

Fig.11 shows the load bus voltage response for this 3LG fault experiment. When a fault occurred, the bus voltage dropped to a half value compared to the steady voltage (green). With the proposed control of SMES, the fault voltage can be compensated to the steady value without phase shift and time delay (blue). It is clear from the voltage response that a controlled SMES effectively enhanced the voltage stability.

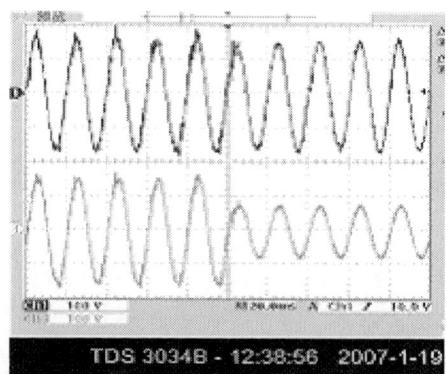

Fig. 11 Voltage response(with SMES(blue),without SMES(green))

VII. CONCLUSION

Three-phase-to-ground fault and sudden increase of dynamic loads influence the transient voltage stability profoundly. Especially in the condition of heavy loads and low voltage, disturbances can cause voltage instability easily, leading to insecure domain. This paper dealt with the transient voltage stability enhancement of electric power system by fuzzy logic-controlled SMES unit. The conclusions and suggestions are as follows:

1) Time-domain simulations verified that a SMES with proposed fuzzy logic controller can improve system voltage performance both for short circuit fault and sudden increase of dynamic.

2) Control is the most effective in increasing the dropping voltage. The experimental result shows that the fuzzy-logic -control compensate the fault voltage so as to improve the voltage stability.

3) It is short of reliable rules of the characteristics, proportion, capacity and increase style of sudden loads. Model validation and database should be built basing on the actual data of system, available for further research and determination.

ACKNOWLEDGMENT

The authors would like to thank all co-leagues of superconducting group in IEE/CAS and all members of our laboratory for their great help in the experiments.

REFERENCES

[1] IEEE/CIGRE Joint Task Force on Stability Terms and Definitions, "Definition and classification of Power System Stability", *IEEE Transactions on Power System,* Vol. 19, No.2, May 2004

[2] Canizares, C.A. and Z.T. Faur, Analysis of SVC and TCSC controllers in voltage collapse. *IEEE Transactions on Power Systems,* 1999. **14**(1): p. 158-165.

[3] Natesan, R. and G. Radman. *Effects of STATCOM, SSSC and UPFC on voltage stability.* 2004.

[4] A. Demiroren and H. L. Zeynelgil, "The Transient Stability Enhancement of Synchronous Machine with SMES by Using Adaptive Control," *Electric Power Components and Systems,* Vol. 30, pp. 233-249, March 2002

[5] S. Banerjee, J. K. Chatterjee and S. C. Tripathy, "Application of magnetic energy storage unit as load-frequency stabilizer", *IEEE Trans. on Energy Conversion,* Vol. 5, No. 1, pp. 46-51, March 1990.

[6] C. J.Wu arid Y. S. Lee, "Application of superconducting magnetic energy storage unit to improve the damping of synchronous generator", *IEEE Trans. on Energy Conversion,* Vol. 6, No. 4, pp.573-578, December 1991.

[7] L. A. Zadeh, "Fuzzy Sets", Inform. contr. Vol. 8 pp. 338-353, 1965

[8] Eskandar Gholipour and Shahrokh Saadate,"Improving of Transient Stability of Power Systems Using UPFC," *IEEE Transactions on Power delivery,* 2005. **20**(2): p. 1677-1682.

[9] V.K.Chandrakar and A.G.Kothari, "Fuzzy Logic based Static Synchronous Series Compensator (SSSC) for Transient Stability improvement," *2004 IEEE International Conference on Electric Utility Deregulation, Restructuring and Power Technologies.* p.240-245.

[10] Ali, M.H., T. Murata, and J. Tamura, "A Fuzzy Logic-Controlled Superconducting Magnetic Energy Storage for Transient Stability Augmentation," *IEEE Transactions on Control Systems Technology,* 2007. **15**(1): p. 144-150.

A Calculation of Predicting the Expected Life of Super-capacitor Following Current Pattern of Railway Vehicles

Jong-Yoon Kim[1], Su-Jin Jang[1], Byoung-Kuk Lee[1], Chung-Yuen Won[1] and Chang-Mu Lee[2]

[1] Department of Information and Communication Engineering
Sungkyunkwan University, Su-won, Korea
[2] Korea Railroad Research Institute, Uiwang, Korea
Email: tigerjhmc@skku.edu

Abstract— Recently it is widely conducted about the usage of regenerative energy produced during brake periods of railway vehicles. The storage system for regenerative energy using super-capacitor (EDLC:Electric Double Layer Capacitor) is one of those research areas. In the storage systems of the regenerative energy, super-capacitor (EDLC) is used for storage device of modular energy.

It is critical to estimate the proper replacement moment for the existed super-capacitor (EDLC) due to the safety of the system. However, there is no standard method to predict the expected life of super-capacitor (EDLC). In this paper, therefore, we propose the method to estimate expected life of super-capacitor by evaluating of internal heating which caused by regenerative current pattern of railway vehicles. With that, we improve the reliability of whole system. Furthermore, it is also addressed to prove that our research is reasonable.

I. INTRODUCTION

Regenerative energy is produced based on the drive pattern during the brake periods of railway vehicles, and this energy causes the increase of DC line voltage. Since it is possible that this surplus voltage of DC line voltage might cause malfunction or destruction of rectifier or any other power conversion devices, DC line voltage needs to be stable in all the time. In current railway systems, DC line voltage is controlled either by exhausting the energy using resistors attached to the vehicles, or by transferring regenerative energy to the trains heading to the opposite direction. Those methods, however, have limitations in a way that the resistors increase the energy waste by making the vehicle heavier, and it is required to have a counter directing vehicle at the same time. Therefore, it is worth to study how to use this extra energy more efficiently, and many researches have conducted recently. The regenerative energy storage systems using super-capacitor (EDLC) is one of those methods. It is a method to charge regenerative energy with super-capacitor (EDLC) and to discharge to the DC line when the energy is needed, consequently improving the efficiently of regenerative energy. Especially, various virtues of the super-capacitor (EDLC) allow it a good candidate of the storage system. General characteristics of super-capacitor (EDLC) as an energy storage system are as follows [1-2].

- It has rapid response by fast charging and fast discharging capability.

- It has high charge ability and it is easy to maintain.
- It has a long cycle life because of no chemical reaction by using ion adsorption reaction.

Super-capacitor (EDLC) in regenerative energy storage system is the core device charging/discharging the regenerative energy. Run-out super-capacitor (EDLC) bank is easily replaceable so its usability is reasonably high. Therefore expecting lifetime of the super-capacitor (EDLC) bank is the critical process to increase the entire reliability in the storage system.

In this paper, we propose a method to predict more accurate temperature change by examining internal heating of the super-capacitor (EDLC). Using the heat transfer equation and Arrhenius equation related to the temperature variation, we propose to evaluation more precise expected life time. Adding that, using the experiment environment with prototype model of regenerative energy storage system, we predict the internal heating values caused by charging/discharging operation. Finally, we prove our theory that it is important to expect internal heating values and improve the reliability of regenerative energy storage system.

In this paper, EDLC, an example of the super-capacitor, is used for realization of the prototype regenerative energy storage system. EDLC can be organized through either series or parallel connection module depending on its purpose. At first, we structure the EDLC module to fit the capacity of the regenerative energy storage system prototype. Then, we propose a method to evaluate the expected life time of the EDLC module. And finally, we suggest a method capable of applying to the actual system. This paper is organized as follows: First of all, presented the method which can predict the lifetime of EDLC bank. And constituted the regenerative energy storage system using super-capacitor (EDLC) and through the experiment, showed the reasonability of this method. Then conclusion and future works are followed.

II. CALCULATING THE LIFE TIME OF EDLC BANK

A. Predicting the lifetime of EDLC bank

It is required to install the regenerative energy storage system at the selected substations where regenerative energy is frequently occurred. In case of the domestic subway systems in Korea, it is efficient to install at the substations since one substation supplies the electricity to the nearby 3~4 subway stations. However it is vary that regenerative energy produce pattern and the consequent charging current of EDLC module. For this reasons, we need a general approach to evaluate the life time of the EDLC bank installed at each substation. In general, major factors effecting to the EDLC life time are temperature factor and applied voltage factor. It is as follows that the method to expect the life time of EDLC bank presented by the manufacturers [3].

Expected Life Time

=Specified Life Time×Temperature Factor×Voltage Factor

$$Temperature\ Factor = 2^{\left(\frac{T_1 - T_2}{10}\right)}$$

(T_1 : Guaranteed Temperature, T_2 : Operating Temperature)

$$Voltage\ Factor = V_1 / V_2$$

(V_1 : Specified Voltage, V_2 : Applied Voltage)

However, the over rated voltage is controlled uniformly since the EDLC modules assembled by the manufacturers are equipped with protection circuit in each cell. So, it naturally addresses that the temperature factor is the bigger factor evaluating the life time.

The manufacturers are generally calculating the life time under the standard circumstance with 25 degrees and 2.7V in applied voltage. Then, they explain that the temperature variation in installed environment can change the life time. However, if it is used for regenerative energy storage system, not only the nearby temperature but also the internal heat temperature is needed to be considered. Internal heating values are varying depending on the installed area since the regenerative currency patterns are different in diverse circumstances. The resulting charging current also shows different characteristics. To predict more precise life time, therefore, it is needed to predict the entire operating temperature variation cumulating the environmental temperature and internal heat temperature [4-7].

B. Equivalent circuit of EDLC module

To predict the lifetime of EDLC bank, we have to analyze the equivalent circuit of EDLC module. Fig. 1 shows the first-order equivalent circuit of EDLC [2][8][9].

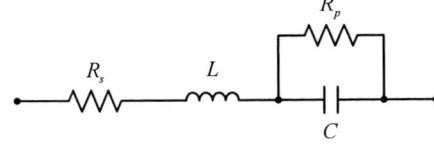

Fig .1 First-order equivalent circuit of EDLC.

As you can see in the Fig. 1, ideal equivalent circuit of EDLC is composed with four elements. R_s is called the equivalent series resistance (ESR). It's including the resistances of electrode, double-layer, electrolyte and contact resistance of lead, and collector. R_s is contributed to power loss during the charging and discharging operations. R_p is called the self-charge resistance, and often referred as the leakage current resistance. And value of R_p is always much higher than R_s in practical capacitors. So, it can be neglected in high-power applications. L results primarily from the physical construction of the EDLC. And value of L is usually small. Thus, L can be neglected too. And C is the capacitance of EDLC. Consequently, Power dissipation of EDLC is associated to impedance (Z_{sc}) which is shown in (1)

$$Z_{sc} = R_s + j\frac{1}{\omega C} \qquad (1)$$

In the ripple of charge current by DC traction patterns, capacitance must be considered. Because the charge current is including low frequency ingredients, power dissipation will be considered as impedance. It can be proved through the experiment in the end of this paper. In this case, power dissipation of EDLC can be presented by (2).

$$Q = I^2 \cdot Z_{sc} \qquad (2)$$

Using (2), internal power dissipation can be calculated and transformed to inner heating values using the heat transfer equation.

C. Heating transfer equation

Self heating is generated when the ripple current passes the internal impedance of EDLC module. EDLC module can be cooled by a combination of convection and radiation shown in (3).

$$Q = Q_{conv} + Q_{rad}\ [W] \qquad (3)$$

The models of convection heat transfer can be seen in (4). And (5) shows the convection heat transfer coefficient h. G term is adjusted to achieve a better fit for EDLC module and height L equals H for banks shows in (6).

$$Q_{conv} = h \cdot A \cdot \Delta T \qquad (4)$$

$$h = G \cdot \left(\frac{\Delta T}{L}\right)^{0.25} \qquad (5)$$

$$G = 1.2\ (For\ EDLC\ bank)$$
$$L = Height\ H\ (For\ EDLC\ bank) \qquad (6)$$

The surface area A is defined by (7). For EDLC bank, A equals the area of the bounding box dimensions. It can be seen in Fig. 2.

$$A = 2WH + 2ZH + 2ZW [m^2]$$
$$(For\ EDLC\ bank)$$
(7)

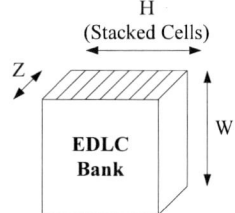

Fig. 2 Surface area of EDLC bank.

The models of radiation heat transfer can be seen in (8). e is presented as emissive of EDLC module. It can be shown in (9). σ, which equals Stefan-Boltzmann constant, can be seen in (10). Area A is assumed to be equal to area (7).

$$Q_{rad} = e \cdot \sigma \cdot A \cdot (T^4_{surf} - T^4_{amb})$$
(8)

$$e = Emissivity = 0.85\,(For\ EDLC\ bank)$$
(9)

$$\sigma = Strfan - Boltzmann\ Constant$$
$$= 5.669 \times 10^{-8} (W/m^2\,{}^\circ K^4)$$
(10)

Using (3) ~ (10), surface rising temperature ΔT can be estimated. According to the energy conservation law, internal heating values of EDLC module are represented to the increase of surface temperature. This surface temperature increment is expressed after added to the nearby temperature.

$$T_{element} = T_{amb} + \alpha \cdot \Delta T [K]$$
(11)

($\alpha = 1.5$: Constant of rising temperature)

Using the above equation, internal heating temperature is predicted. Then the result is used as an input of the Arrhenius equation related to the temperature and the life time, to finally calculate the changes in the expected life [4-7].

D. Calculating the lifetime of EDLC modules

As well as the conventional capacitor life time evaluation, the Arrhenius equation is also used in evaluating the EDLC life time (12). L represents reaction life time, U is a constant of equation, σ equals to Stefan-Boltzmann constant of (10), T represents absolute temperature, and the last E represents reacting activation energy.

$$L = U \cdot e^{\frac{E}{\sigma T}}$$
(12)

Most of the manufacturers are using this equation to predict the life time not only conventional capacitors but also EDLC.

And (13) is describing the relation between nominal life time and temperature variation to predict the expected life of the regular capacity.

$$\frac{L_2}{L_1} = 2^{\left(\frac{T_1 - T_2}{10}\right)}$$
(13)

(L_1 : Rated lifetime of capacitor, L_2 : Expected lifetime of capacitor)

Using (13), the expected life of EDLC bank can be calculated [4-7].

III. REGENERATIVE EHERGY STORAGE SYSTEM USING SUPER-CAPACITOR(EDLC)

Fig. 3 shows the regenerative energy storage system using super-capacitor for the experiment. This system is to be classified into three parts and each part is explained as follows.

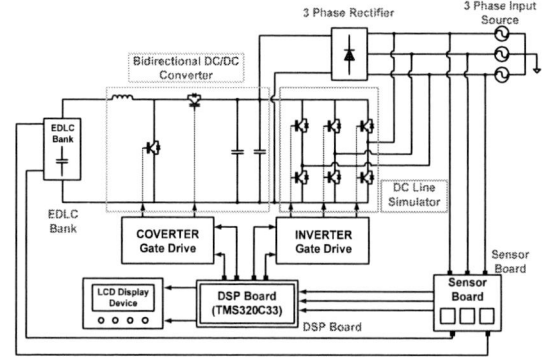

Fig. 3 Regenerative energy storage system using super-capacitor.

A. DC Line simulator

Depending on the actual DC line voltage changes, it is determined the charge/discharge mode of super-capacitor. However it is impossible to simulate without an extra system. So, DC line simulator is proposed to control the energy storage system and the bidirectional DC/DC converter efficiently.

Fig. 4 shows the block diagrams of DC line simulator.

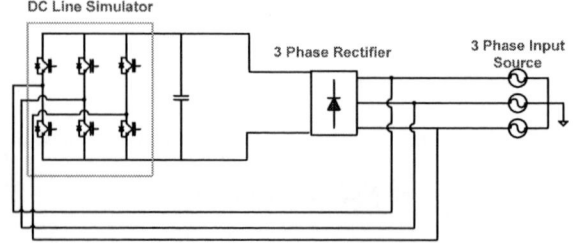

Fig. 4 Block diagram of DC Line simulator.

B. Bidirectional DC/DC converter

Surplus regenerative energy which increases the DC line voltage is controlled by bidirectional DC/DC converter which charges/discharges the EDLC module at needs. In this experiment, we used the non-isolation half-bridge converter to simulate bidirectional DC/DC converter.

The non-isolation half-bridge converter has low inductance value, low capacitance of passive element, and low late voltage of active elements. Fig. 5 shows half-bridge type bidirectional DC/DC converter used in regenerative energy storage system. The design specifications are indicated in Table 1.

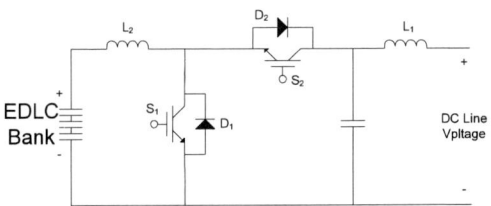

Fig. 5 Non-isolation bidirectional DC/DC converter.

TABLE 1
Design specifications of bidirectional DC/DC converter

Means	Value
Maximum Discharge Power	4[kW]
Voltage range of EDLC Module	50 ~ 100[V]
Maximum Current of EDLC Module	50[A]
Input Voltage of Converter (DC Line Voltage)	311[V]
Input Current of Converter	10[A]
Switching Frequency	10[kHz]

C. EDLC bank

Super-capacitors (EDLC) represent a compromise between batteries and conventional capacitors. Although power density of super-capacitors is lower than capacitors and energy density is lower than capacitors, they are still interesting because of their long lifetime and rapid respond by fast charging/discharging capability in DC traction system. The capacity design of EDLC bank is shown (15). Prototype model of regenerative storage system is designed by the EDLC bank voltage as 100[V], one cell voltage in EDLC module is 2.7[V](Using voltage is 2.5[V]). So, 40 cells are needed to make module and cells must be connected in series.

$$\frac{100[V]}{2.5[V]} = 40[ea] \quad (14)$$

The maximum regenerative time is arranged 30 seconds, and the capacity of EDLC module can be calculated in (16) using Table 1 parameters.

$$C = \frac{dt}{\left(\dfrac{dv}{i} - R\right)} = \frac{15[A] \times 30\,sec}{100[V] - 50[V]} = 9[F] \quad (15)$$

Using 40 EDLC cells which is capable of 600[F] in series, the storage system is organized only with one series low since it is 15[F].

Fig. 6 shows the figure of EDLC bank composed for experiment made by NESSCAP CO. while Table 2 shows the parameters of EDLC module.

Fig. 6 EDLC bank.
[ESHLP-0600C0-002R7(2.7V_600F)*40]

TABLE 2
Parameters of EDLC module

Means	Value
Capacitance	16.4[F]
DC-ESR	19.2[$m\Omega$]
Self-discharge Voltage	77.31[V]

IV. EXPERIMENTAL RESULTS

The stable ranges of DC line voltage in prototype regenerative energy storage system can be determined voluntarily. This paper determines the stable ranges as 355~365[V] for experiment, and presents charging /discharging conditions including the input voltage in Table 3.

TABLE 3
Conditions of experiment

Means	Value
Input Voltage	220[V]
Charging Condition	Over the 365[V]
Discharging Condition	Down the 355[V]

Fig. 7 shows the figures of prototype regenerative energy storage system. TMS320VC33 DSP board is used as control board to manage DC line simulator and bidirectional DC/DC converter. Fig. 8 shows the realization of DC line voltage variation of voluntary three stations which can be established the regenerative storage system by DC line simulator.

978-1-4244-1871-8/07 $25.00 © 2007 IEEE

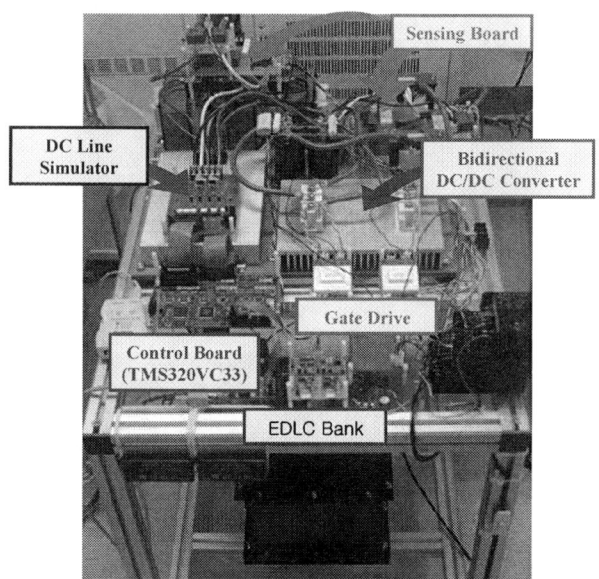

Fig. 7 Experimental set of regenerative energy storage system.

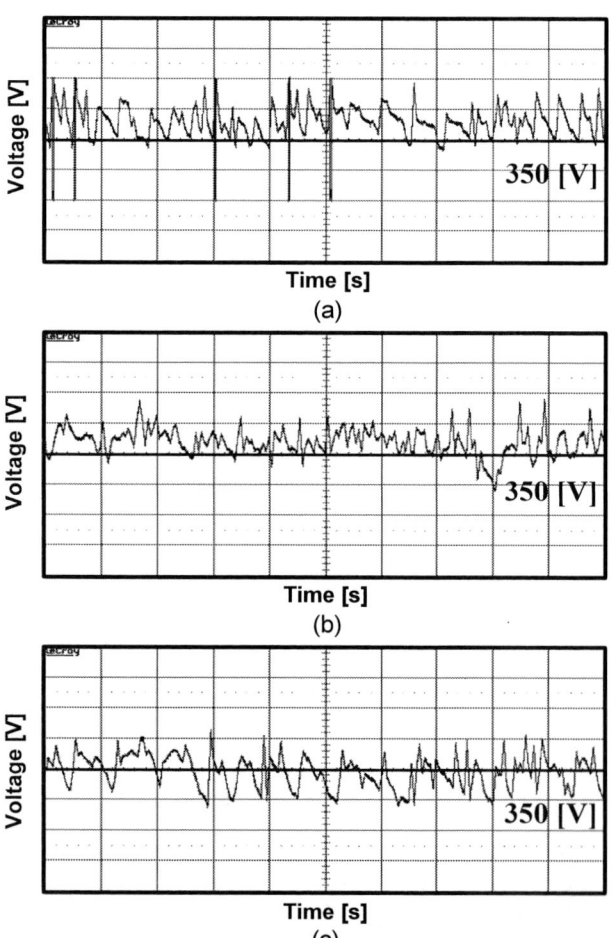

Fig. 8 Variation of DC line voltage according to the stations. (a) Station of Nakseongdae in line No.2 (b) Station of Sillim in line No.2 (c) Station of Seocho in line No.2.(5V/div, 10s/ div)

As you can see in Fig. 8, regenerative energy flow is different depending on the stations, and charging current of EDLC bank is different as well. Therefore, consequent internal heating of EDLC bank is different depending on the stations.

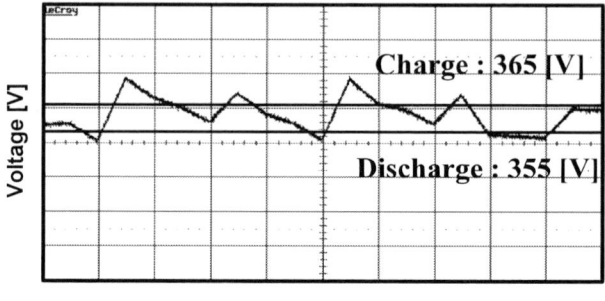

Fig. 9 Enragement waveforms of DC line voltage in Nakseongdae station in line No.2. (5V/div, 1s/ div)

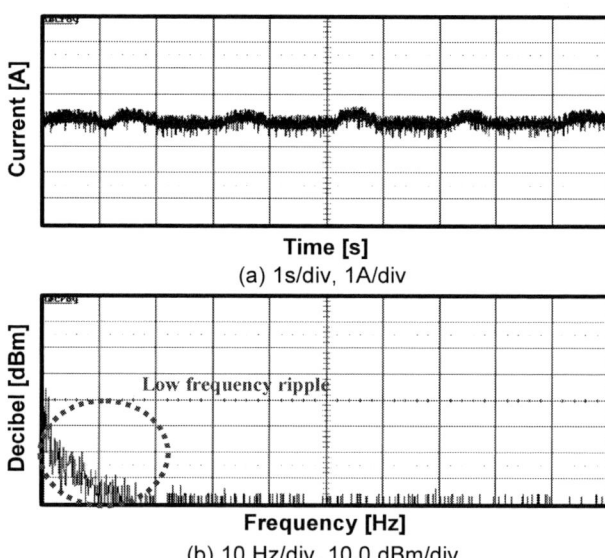

Fig. 10 FFT Analysis of charging current of EDLC module in Seocho station line NO.2.

Fig. 9 shows the enlarged waveform of Nakseongdae station presented in Fig. 8. And Fig. 10 shows the charging current of EDLC bank and its FFT analysis waveform of Nakseongdae station presented in Fig. 8. As you can see through the waveforms in Fig. 9 and Fig. 10, the variation values of DC line voltage have different patterns depending on the station, and charging current of EDLC bank has included the low frequency ripple components. It can be prove the adequacy of method proposed in this paper.

According to the manufacturer's datasheet, internal parameters are little related to the temperature variation, and impedance values are also uniform regardless of the internal temperature increase if it is a room temperature [10].

Therefore, we are able to find that the major factor effecting to the increase of internal heating is charging current depending on to the DC line voltage pattern, or ripple

978-1-4244-1871-8/07 $25.00 © 2007 IEEE

components of current. Hereafter, if real DC line voltage patterns, which of stations that established the regenerative energy storage system are evaluated more exactly, we can evaluate the power dissipation of EDLC bank through the internal impedance value. And using that value, we can evaluate the internal heating values using heat transfer equation. Finally, using the Arrhenius method, it is possible to calculate life time of EDLC bank more accurately.

V. CONCLUSION

This paper proposes the method about prediction the life time of EDLC module more exactly, using the temperature of internal heating values. Internal heating values are different by charging current according to the flow of regenerative energy. Research about selecting the stations which will be established the regenerative energy storage system and evaluating about generating pattern of regenerative energy in selecting station are progressing now.

If the result is calculated based on the method proposed in this paper, we can estimate the lifetime of EDLC bank which will be installed in each station. With that estimation, we can predict the replacing period of EDLC bank, consequently, the entire reliability of regenerative system can be improved.

VI. FUTURE WORKS

To predict life time of EDLC bank more precisely, it is needed to analyze the ripple components of charging current which flow into the EDLC bank. It is also required to calculate the power loss of impedance by ripple components of charging current. At last, it is currently progressing to improve the reliability of the suggested method by comparing the ideal value and the actual experiment result.

REFERENCES

[1] Lai, J.S, Levy, S, Rose, M.F "High Energy Density Double-Layer Capacitors for Energy Storage Applications" Aerospace and Electronics System Magazine, IEEE Volume 7, Issue 4, April 1992 Page(s):14-19.

[2] Y.H, Kim Director of Product Development, NESSCAP Co. Ltd, Korea "Ultracapacitor Technology Powers Electronic Circuits" Oct 1 2003. Power Electronics Technology Magazine pp 34~39.

[3] NESSCAP Co. Ltd Technology Paper July 2007.

[4] D.C Lee, Research Report "Prediction of elect electrolysis capacitor used in inverter system" Yeungnam Univ. September 2000.

[5] M.L. Gasperi, "Life Prediction Model for Aluminum Electrolytic Capacitors," IEEE IAS, Annual Meeting, pp. 1347-1351, 1996, October.

[6] M.L. Gasperi, "A Method for Predicting the Expected Life of Bus Capacitors," IEEE IAS, Annual Meeting, pp. 1042-1047, 1997 October.

[7] Gasperi, M.L, Gollhardt, N "Heat Transfer Model for Capacitor Banks" Industry Application Conference, 1998. Thirty-Third IAS Annual Meeting. The 1998 IEEE Volume 2, 12-15 Oct. 1998 Page(s): 1199-1204 vol.2

[8] Spyker, R.L, Nelms, R.M "Classical Equivalent Circuit Parameter for a Double-Layer Capacitor" Aerospace and Electronics System, IEEE Transactions on Volume 36, Issue 3, Part 1, July 2000 Page(s):829-836.

[9] Dandan Zhang, man Luo, Jin Li, Junjia He "Surveying Into Some Aspects of Internal Resistance of Super Capacitor" Electrical Insulation and Dielectric Phenomena, 2005 CEIDP '05. 2005 Annual Report Conference on 16-19 Oct. 2005 Page(s):637-640

[10] NESSCAP Co. Ltd Ultracapacitor Technical Data ESHSP-1700-002R7.

Comparative Study on a Single Energy Recovery Circuit with dividing energy recovery path for Plasma Display Panels (PDPs)

Kang-Hyun Yi, Seong-Wook Choi and Gun-Woo Moon

Department of Electrical Engineering,
Korea Advanced Institute of Science and Technology,
373-1 Guseong-Dong, Yuseong-Gu, Daejeon, 305-701, Republic of Korea,
Phone: +82-42-869-3475, Fax: +82-42-869-8520

Abstract- **Comparative study on a low cost sustaining driver with single and dual path energy recovery circuits for plasma display panels (PDPs) is shown in this paper. The cost of PDPs has been still high and about half of the cost has been occupied by driving circuit. A simple sustaining driver is proposed to reduce the cost and size of driving circuit. The proposed driver has small number of devices and reactive components and there are two methods for charging and discharging PDPs such as single and dual path energy recovery circuits. A comparative research on two-types of energy recovery path is practiced to evaluate performance. As a result, the dual energy recovery path circuit has low power consumption, low surge current and high performance. To verify those results, experiment will be shown with 42-inch HD panel.**

I. INTRODUCTION

Plasma display panels (PDPs) are expected to be the best display for high definition (HD) television among flat panel display (FPD) due to wide view angel, large screen size, high contrast ratio, thinness, lightness. However, PDPs has been still expensive so there is new approach to reduce the cost such as reducing the address voltage [1], a single-scan method [2] and decreasing the number of electrical parts.

Generally, PDPs have been driven by the address display separation (ADS) method. The driving operation is divided into three periods: reset, addressing, and sustaining periods as shown Fig 1. During the reset period, a high voltage is forced between electrode X and electrode Y to initialize all the cells in order to obtain the same conditions for all the cells. In the addressing period, wall charges are accumulated in the cells which display the image. In the last period, sustaining, the discharges occur in the cells addressed previously and the desired image can be obtained on the panel. There are many power circuits to make three steps such as resetting, addressing and sustaining in driving the PDP.

During the sustaining period of Fig. 1, the panel voltage waveform is a rectangular pulse from V_S to 0 in electrode X and Y, alternatively. When the electrode X or Y becomes V_S, the light is emitted by the gas discharge. Therefore, most of the electrical power is consumed in sustaining period because a gray scale is made by the number of sustaining pulse [3]. Also, since the PDP is regarded as a capacitive load (C_p),

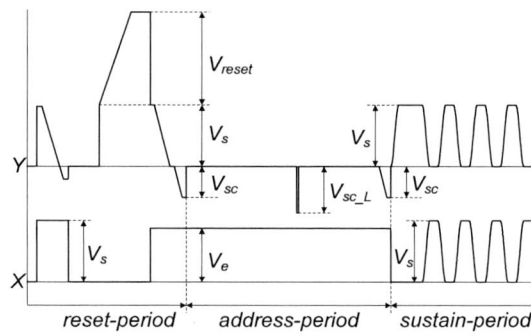

Fig. 1. Voltage waveforms applied to X and Y electrodes in ADS driving method

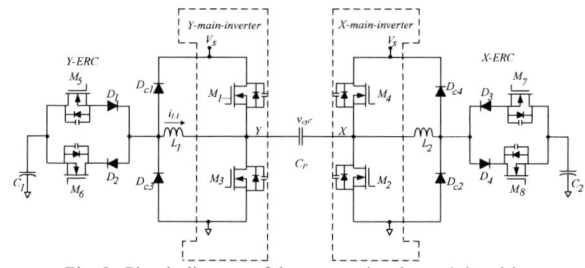

Fig. 2. Circuit diagram of the conventional sustaining driver

minimizing the considerable energy loss of $2C_pV_s^2$ and electromagnetic interference (EMI) noise for each cycle, and reducing the reactive power loss for charging or discharging the panel capacitor through an energy recovery circuit (ERC) [4-11] are much demanded. The best well-known sustaining driver is Weber & Wood circuit which has been composed H-bridge inverter to make high frequency rectangular pulse and series resonant type ERC in recent PDPs as shown in Fig. 2 [10]. This circuit has good performance, high efficiency and flexibility. However, many MOSFET switches, power diodes and reactive components such as inductors and capacitors make the structure of the sustaining driver and series resonant ERC complex and bulk. Above all, the worst demerit of the conventional driver comes from the high cost.

978-1-4244-1871-8/07 $25.00 © 2007 IEEE

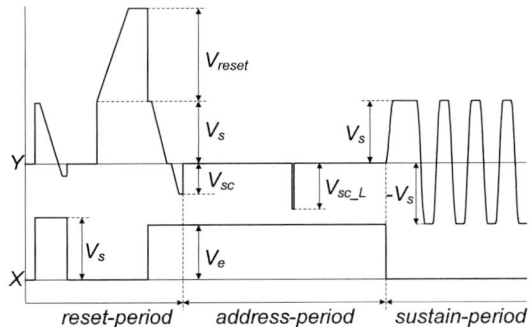

Fig. 3. Voltage waveforms applied to X and Y electrodes in ADS driving method with the low cost sustaining driver

To capture the TV consumer marker and maintain a lead over other FPD, a new sustaining driver with a half-bridge inverter and a parallel resonant type ERC [11] are favorable due to the simple structure and small number of components: half the number of components in the conventional sustaining driver is used in the proposed. Furthermore, it can be driven with single and dual path ERCs. In this paper, a comparative study on the single and dual path ERCs is shown to obtain desirable performance and feasibility to commercialize the proposed driver. The mostly focused is the performance compared between the single and dual path ERCs with respect to power consumption and heat dissipation in ERC components. The dual path ERC achieves superiority over the single path ERC in efficiency and performance. To verify this result, experiment for the comparison of two circuits is performed with 42-inch HD PDP.

II. A LOW COST SUSTAINING DRIVER WITH SINGLE AND DUAL PATH ERCs

To reduce the cost of the sustaining driver, the single sustain waveform for PDPs has been studied as shown in Fig. 3 [12]. During the sustain-period, X-electrode is fixed in to the ground and Y-electrode is enforced by V_S or - V_S, alternatively. To make the rectangular waveform on Y-electrode, the half-bridge inverter has been used. Since all switching devices can achieve a zero-current-switching (ZCS), IGBT switches can be used in the half-bridge inverter to overcome voltage stress of devices. Fig. 4 (a) and Fig. 5 (b) show the sustaining driver with half-inverter and parallel resonant type ERC, respectively. The proposed sustain driver is composed of small number of components compared with the conventional sustaining driver. As shown in Fig. 4 and Fig. 5, there are two types of circuits according to the energy recovery path. The energy recovery operation in two type circuits is almost the same. The operation principle of the two type circuits is shown in this chapter.

A. The sustaining driver with single path ERC

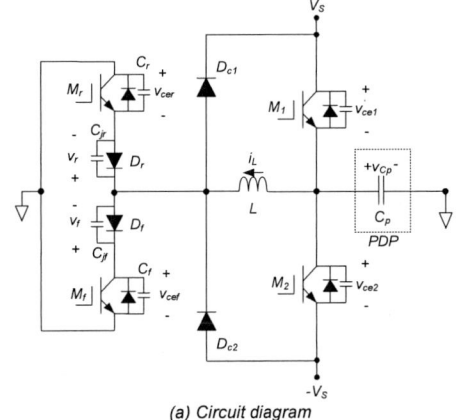

(a) Circuit diagram

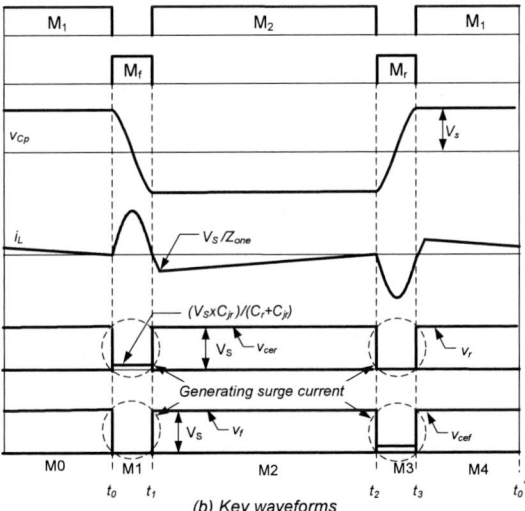

(b) Key waveforms

Fig. 4. Circuit diagram and key waveforms of the sustaining driver with single path ERC

Fig. 4 shows the circuit diagram and key waveforms of the sustaining driver with the single path ERC. One inductor is used for charging and discharging the panel capacitor.

Mode 0(before t_0): Since switch M_1 is turned on, panel voltage, v_{Cp}, is clamped to V_S.

Mode 1($t_0 \sim t_1$): When the switch M_1 is turned off and switch M_f is turned on, mode 1 begins. The panel voltage, v_{Cp}, is changed from V_S to $-V_S$ with resonance of panel capacitor C_P and ERC inductor, L. In this mode, a collector-emitter voltage of switch M_r and a reverse bias voltage of diode D_r can be expressed as following:

$$v_{cer}(t) = v_r(t) = V_S \frac{C_{jr}}{C_r + C_{jr}} \quad (1)$$

where output capacitance of switch C_r and junction capacitance of diodes C_{jr}. There are surge current to charge or discharge output capacitor of switch and junction capacitor of diodes.

978-1-4244-1871-8/07 $25.00 © 2007 IEEE   983

Mode 2($t_1 \sim t_2$): When the panel voltage is $-V_S$, switch M_2 is turned on at t_2. The panel voltage is sustained to $-V_S$ and gas discharge occurs in mode 2.

Mode 3($t_2 \sim t_3$): Mode 3 begins when the switch M_2 is turned off and switch M_r is turned on. The panel voltage, v_{Cp}, is changed from $-V_S$ to V_S with resonance of panel capacitor C_P and ERC inductor, L. In this mode, a collector-emitter voltage of switch M_f and a reverse bias voltage of diode D_f can be expressed as follow:

$$v_{cef}(t) = v_f(t) = V_S \frac{C_{jf}}{C_f + C_{jf}} \tag{2}$$

where output capacitance of switch C_f and junction capacitance of diodes C_{jf}. There are surge current to charge or discharge output capacitor of switch and junction capacitor of diodes like mode 1.

Mode 4($t_3 \sim t_0{}'$): Switch M_1 is turned on and the gas discharge occurs at t_3. The panel voltage is maintained in V_S like mode 0.

B. *The sustaining driver with dual path ERC*

Fig. 5 shows the circuit diagram and key waveforms of the sustaining driver with the dual path ERC. Two inductors are used for charging and discharging the panel capacitor, each other.

Mode 0(before t_0): Since switch M_1 is turned on, panel voltage, v_{Cp}, is clamped to V_S.

Mode 1($t_0 \sim t_1$): When the switch M_1 is turned off and switch M_f is turned on, mode 1 begins. The panel voltage, v_{Cp}, is changed from V_S to $-V_S$ with resonance of panel capacitor C_P and ERC inductor, L_f. In this mode, a collector-emitter voltage of switch M_r and a reverse bias voltage of diode D_r is resonance with other inductor L_r. There is no surge current to charge or discharge output capacitor of switch and junction capacitor of diodes due to inductor L_f.

Mode 2($t_1 \sim t_2$): When the panel voltage is $-V_S$, switch M_2 is turned on at t_2. The panel voltage is sustained to $-V_S$ and gas discharge occurs in mode 2.

Mode 3($t_2 \sim t_3$): Mode 3 begins when switch the switch M_2 is turned off and M_r is turned on. The panel voltage, v_{Cp}, is changed from $-V_S$ to V_S with resonance of panel capacitor C_P and ERC inductor, L. In this mode, a collector-emitter voltage of switch M_f and a reverse bias voltage of diode D_f is resonance with other inductor L_f. There is no surge current to charge or discharge output capacitor of switch and junction capacitor of diodes due to inductor L_r like mode 1.

Mode 4($t_3 \sim t_0{}'$): Switch M_1 is turned on and the gas discharge occurs at t_3. The panel voltage is maintained in V_S like mode 0.

The energy recovery operation is same in single path ERC and dual path ERC, but the behavior in parasitic capacitor is different. As shown in Fig. 4 (b) and Fig. 5(b), there are some difference between the collector-emitter voltage of ERC switches and the reverse bias voltage of ERC diodes in single

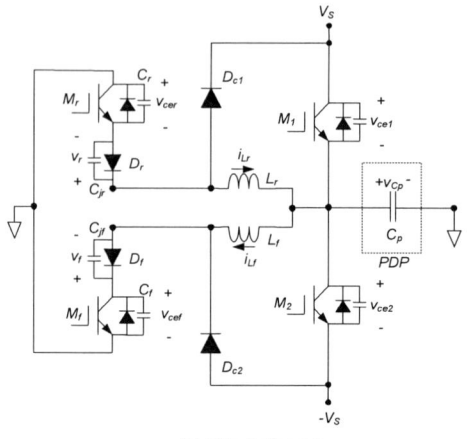

(a) Circuit diagram

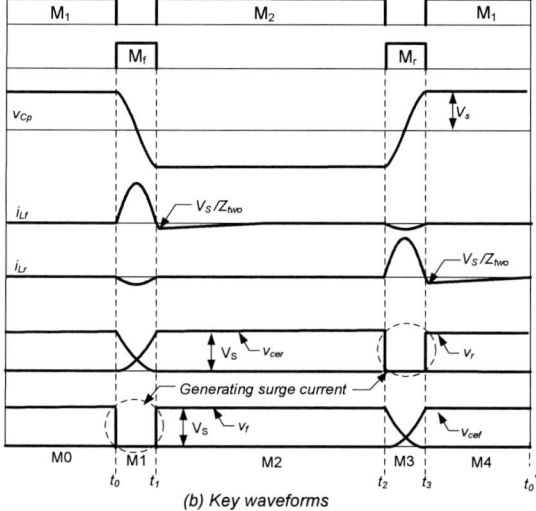

(b) Key waveforms

Fig. 5. Circuit diagram and key waveforms of the sustaining driver with dual path ERC

and dual path ERC. Also, the number of generating surge current in the single path ERC is two times greater than that in the dual path ERC. After finishing energy recovery action, quantity of free-wheeling current for charge the parasitic capacitor in ERC devices is not same in two circuits. Those results in different quality of performance, heat dissipation and power consumption in two circuits.

III. COMPARATIVE STUDY ON SINGLE AND DUAL PATH ERCs

Regarding to achieving energy recovery action, the operation of the two circuits, single and dual path ERCs, is almost the same. However, effects by parasitic components of ERC devices between single and dual path ERCs are some different after finishing energy recovery.

A. *The quantity of free-wheeling current*

978-1-4244-1871-8/07 $25.00 © 2007 IEEE

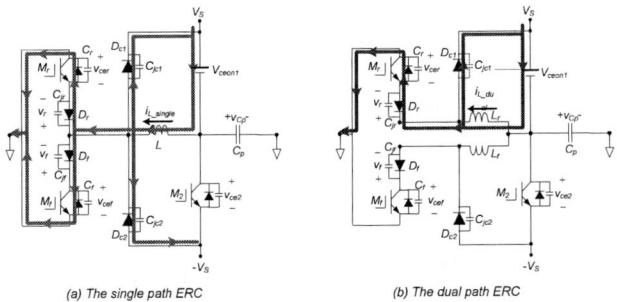

(a) The single path ERC   (b) The dual path ERC

Fig. 6. The current path for charging the parasitic capacitor in the single and dual path ERCs

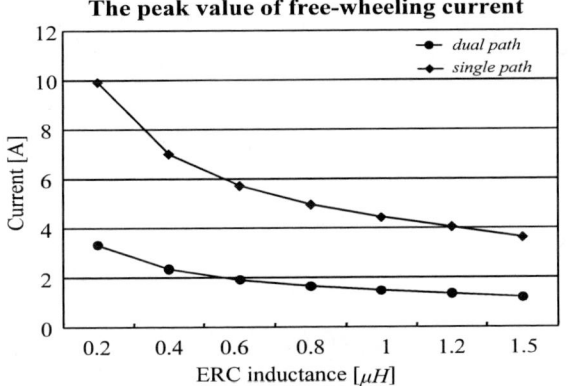

Fig. 7. Estimation of the peak value of free-wheeling current in two circuits

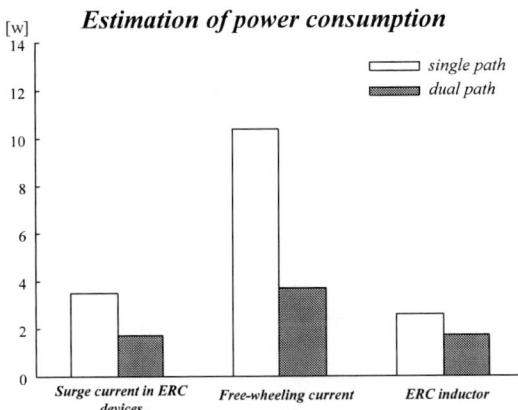

Fig. 8. Estimation of power consumption

After completing the charging the panel capacitor, additional current is needed to charge parasitic capacitors of switching components by ERC inductors. Remained current is free-wheeling through clamping diode such as D_{C1} and D_{C2} as shown in Fig 4, (b) and Fig. 5 (b). This free-wheeling current results in additional conduction loss. Fig. 6 is a visualized example of the current path for charging the parasitic capacitor of ERC device after the panel voltage becomes V_S. Compared with six parasitic capacitors for charging in the single path ERC, there are small numbers of parasitic capacitors for charging in dual path ERC. As results, there are larger current for charging additional capacitors in the single path circuit than that of dual path it. The peak value of free-wheeling current in single path circuit can be obtained as follows:

$$I_{L\text{-}free\text{-}one} = V_S / Z_{ONE} \tag{3}$$

where $Z_{ONE} = \sqrt{L/(C_{oss} + C_j + 2C_{jc})}$, output capacitance of ERC switches, $C_{oss} = C_r = C_f$, junction capacitance of diodes, $C_j = C_{jr} = C_{jf}$ and junction capacitance of clamping diodes, $C_{jc} = C_{jc1} = C_{jc2}$. In the same manner, the peak value of free-wheeling current is obtained in dual path circuit as follows:

$$I_{L\text{-}free\text{-}two} = V_S / Z_{TWO} \tag{4}$$

where $Z_{TWO} = \sqrt{L/(C_j + C_{jc})}$, $L = L_r = L_f$, junction capacitance of clamping diodes, $C_{jc} = C_{jc1} = C_{jc2}$. As it is shown in equations, (3) and (4), the peak value of current for charging the parasitic capacitor is smaller and free-wheeling current is lower in dual

path ERC as shown in Fig. 5 (b). For example, when FGA180N30D is used as the ERC switches with C_{oss}=400pF, DSEC59-03AQ is used as the ERC diodes with C_j=20pF, FFAF40UDN is used as clamping diodes with C_{jc}=35pF and the sustaining voltage, V_S, is 200V, the peak value of the free-wheeling current can be estimated according to the ERC inductance as shown in Fig. 6. This figure shows that the dual path ERC has smaller the peak free-wheeling current than the single path ERC.

B. Heat dissipation and size of ERC inductors

In the commercialized driving circuit, air-cored inductor has been used as ERC inductors. The inductance of air-cord inductor can be designed as follows

$$L = \frac{(d^2 \times n^2)}{18d + 40l} \tag{5}$$

where L is inductance in micro Henrys and d is coil diameter in inches, l is coil length in inches, and n is number of turns. In conventional ERC inductors like the single path ERC, the ERC inductor has been used in parallel to release heat dissipation. Therefore, the ERC inductance has to be designed as two times greater than the desired inductance. However, since the dual path ERC can use a single inductor, the size of the ERC inductance can be reduced compared with that of the single path ERC.

Also, power loss is observed in ERC inductor because the ERC inductor has a resistive component. The resistance of ERC inductor can be obtained in aired core inductor with copper material wire as follows:

$$R = 1/\sigma \times \left(\ell /(\pi^2 \times (D\varepsilon - \varepsilon)^2) \right) [\Omega] \tag{6}$$

where conductivity of copper, σ, length of wire, l and skin depth , $\varepsilon = 6.62/\sqrt{f}$ [cm] in f=frequency. In case of dual path ERC, since the wire length of the ERC inductance is shorter, smaller resistance can be obtained and the ERC currents are flowing with two inductors, the power consumption in the ERC inductors can be reduced and heat in inductor is distributed. In addition, since the smaller free-wheeling

978-1-4244-1871-8/07 $25.00 © 2007 IEEE

TABLE I
COMPONENT LIST

42-inch HD PDP (C_P)	About 100nF
Switching frequency (f_s)	200kHz
Inverter switches (M_1, M_2)	IXGQ240N30
ERC switches (M_r, M_f)	FGA180N30D
ERC diodes (D_r, D_f)	DSEC59-03AQ
Transition time	1 μs
ERC inductance (Single path)	2 μH(2EA)
ERC inductance (Dual path)	1 μH(2EA)

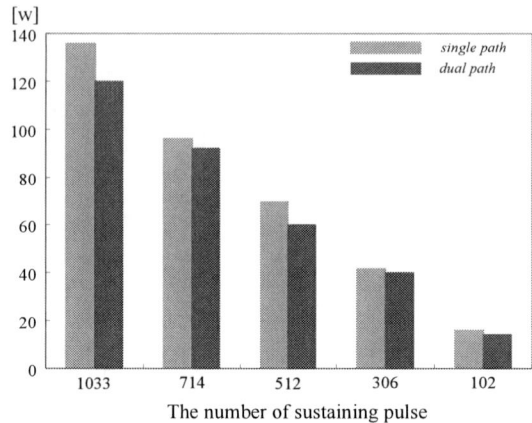

Fig.10. Power consumption

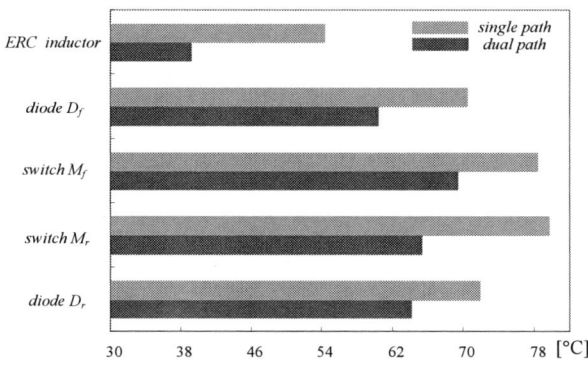

Fig.11. Temperature in ERC components

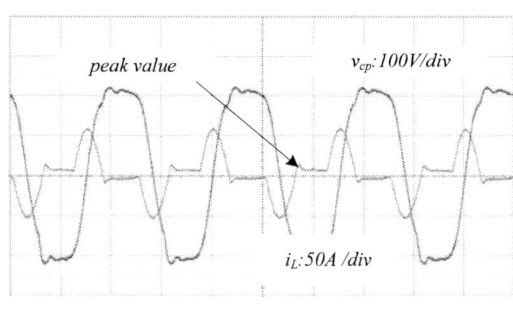

(a) The single path ERC

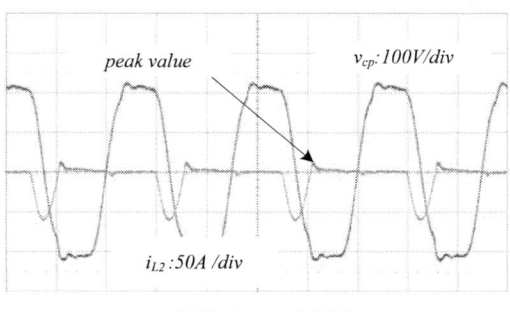

(b)The dual path ERC

Fig. 9. Waveforms of panel voltage and ERC inductor currents

current is flowing in ERC inductor, the power loss in ERC inductor can be minimized in dual path ERC.

C. *The number of generating surge current*

Additional, there are four times of generations of excessive surge current in single path ERC during one switching cycle as shown in Fig. 4 (b) because the ERC switches can not be achieved zero voltage switching. However, the excessive surge current is generated as two times in dual path ERC during one switching cycle as shown in Fig. 5 (b). The collector-emitter voltage of turned-off switch and diode voltage of junction capacitor are charged or discharged by resonance of parasitic capacitors and another ERC inductor

without surge current. Comparing with single path ERC, it results in smaller surge current and lower power consumption in dual path ERC.

Fig. 8 shows estimation of power consumption according to three factors three factors (i.e. hard switching, free-wheeling current and ERC inductor) as mentioned above. This estimation is determined by parameters of used power devices during one switching cycle. As shown in graph in Fig. 8, the dual path ERC has lower power consumption and higher performance than that of single path ERC because of lower surge current, smaller free-wheeling current and reducing the resistance of ERC inductors.

IV. EXPERIMENTAL RESULTS

The experiment shows the comparisons between the two circuits, single and dual path ERCs, 42-inch HD PDP according to the number of sustaining pulse in 1 TV-field (per 16.6ms) without gas discharge. Table 1 shows the component list in the experimental circuit. Fig. 9 shows waveforms of panel voltage in two circuits. While the energy recovery action is almost same in two circuits, the peak value of free-wheeling current is 11.2A in dual path circuit and it is 15.4A in single path circuit. The conduction loss by free-wheeling current is greatly reduced in dual path circuit. According to the number of sustaining pulse, the power consumptions are measured as shown in Fig. 10. The number of sustaining pulse can be 1033 to obtain the brightest image. As mentioned above, since the dual path ERC has low surge current and free-wheeling current, it can be obtained higher efficiency than that of single path ERC as shown Fig. 10. The temperature in components is very important in terms of reliability of PDP. Fig. 11 shows temperature of ERC components such as inductor, switches and diodes when the number of sustaining pulse is same in two circuits. Temperature of all components is lower in dual path ERC. Since ERC inductor has small resistive component in dual path circuit, the temperature of that is much lower than that of single path circuit. Also, reliability of switches and diodes can be guaranteed in two-path circuit. Through these experiment results, the detailed analyses on the comparison between the two circuits are successfully done.

V. CONCLUSION

The low cost sustaining driver is discussed in this paper to reduce the cost and to get high performance with single path and dual path energy recovery circuits in this paper. The power consumption is analyzed in respect to parasitic effects in two ERCs. The experimental tests show difference and comparison of two circuits well. Due to lower surge current, smaller free-wheeling current and lower power consumption in dual path ERC, the dual path ERC shows better performance. The proposed sustaining circuit with dual path ERC can be expected to be suitable for the low cost and high performance sustaining driver for PDPs.

REFERENCES

[1] L.F. Weber,: 'The promise of plasma displays for HDTV', in *Proc. SID Dig. 2000*, pp. 402-405

[2] Y. Takeda, M. Ishii, T. Shiga and S. Mikoshiba, 'A technique for reducing data pulse voltage in ac-PDP's using metastable-paticle priming', in *Proc. IDW Dig.*, 1999, pp. 747-750

[3] Sobel, A.: 'Plasma Display', in *IEEE Trans. Plasma Sci*, 1991, (6), pp. 1032-1047

[4] J.Y. Lee, J.S. Kim, N.S. Jung, and B. H. Cho 'The Current Injection Method for AC Plasma Display Panel (PDP) Sustainer', in *IEEE Trans. Ind. Electronics*, vol. 51, NO. 3 pp. 615-624, June, 2004.

[5] Sang-Kyoo Han, Jun-Young Lee, Gun-Woo Moon, Myung-Joong Youn, Chang-Bae Park, Nam-Sung Jung, and Jung-Pil Park, "A New High-Efficient Energy-Recovery Circuit for Plasma Display Panel" Journal of Power Electronics, Vol 7, No. 2, pp. 121-127, Apr. 2002.

[6] S. K. Han, S. W. Choi, G. W. Moon and M. J. Youn: 'Zero-Voltage and Zero-Current Switching Energy Recovery Circuit for Plasma Display Panel', in *Proc.IEEE PESC'04*, pp. 782-786, 20-25 June. 2004 Aachen, Germany

[7] Sang-Kyoo Han, Jun-Young Lee, Gun-Woo Moon, Myung-Joong Youn, Chang-Bae Park, Nam-Sung Jung, and Jung-Pil Park, 'A New High-Efficient Energy-Recovery Circuit for Plasma Display Panel' in *Journal of Power Electronics*, Vol 7, No. 2, pp. 121-127, Apr. 2002.

[8] Sang-Kyoo Han, Gun-Woo Moon, and Myung-Joong Youn 'A Novel Current-fed Energy Recovery Sustaining Driver for Plasma Display Panel' in *Journal of Power Electronics*, Vol. 4, No.1, pp39-45, Jan. 2004.

[9] Kang-Hyun Yi, Sang-Kyoo Han, Seong-Wook Choi, Chong-Eun Kim and Gun-Woo Moon, "A Simple ZVZCS Sustain Driver for Plasma Display Panel," Journal of Power Electronics, vol. 6, pp. 298-304, Oct. 2006.

[10] Webber, L.F. and Wood, M.B.: 'Energy recovery sustain circuit for the AC plasma display', in *S.I.D.*, 1987, pp 92-95

[11] M. Ohba and Y. Sano: 'Energy recovery driver for a dot matrix ac plasma display panel with a parallel resonant circuit allowing power reduction', U.S. Patent 5 670 974, Sep, 1997

[12] Byung-Gwon Cho, Heung-Sik Tae, Kazuhiro Ito, Nam-Sung Jung, and Kwang-Sik Lee.: 'Study on discharge stability of cost-effective driving method based on Vt closed-curve ananlysis in AC plasma display panel', *IEEE Trans. Elec. Devices*, vol. 53, NO. 3, pp. 1032-1047, May 2006

MFFL Driving System with Current Feedback Maintaining Glow Discharge Mode

Jong Bok Baek*, Joung Hu Park, Bo Hyung Cho
Seoul National University
Department of Electrical Engineering
San 56-1, Sillim-dong, Gwanak-gu, Seoul, 151-742
*Email: bjbsla@pesl.snu.ac.kr

Abstract—Recently, a mercury-free flat fluorescent lamp has been developed, utilizing glow discharge mode instead of discharge contraction. This paper proposes a lamp-driving system with a feedback loop which prevents the discharge contraction and stabilizes the operation of the lamp ignition and radiation. By measuring the current that flows through the lamp, the loop can adjust the current level in a normal operation level and suppress the long-term excitation causing the operation transfer to the contraction. The proposed method has been verified by hardware experiments which are compared to a conventional open-loop circuit by a discharge contraction time and the change of luminance.

I. INTRODUCTION

Recently, a mercury-free flat fluorescent lamp (MFFL) has been developed, utilizing glow discharge mode by pulse driving due to long-gap discharge instead of discharge contraction [1].

Fig. 1 shows the simple DBD(Dielectric Barrier Discharge) panel with two soda lime glass. Sustaining electrodes insulated by 80μm thick dielectric layer are separated 70mm apart. White phosphor layers are placed both on front and rear plates. The total pressure of Ne-Xe mixture charges 10 to 110 Torr with 2~7% Xe concentration. The discharge is generated and sustained by alternating short pulse voltages. [1-2]

When the driving voltage arrived at the gas breakdown voltage, the gas becomes plasma state. And then, the discharge of the flat light source starts to occur. This local ignited discharge expands to the whole panel as the increased voltage is applied. Within the normal voltage margin, the diffused glow discharge occurs and accomplishes uniform luminance. Usually, diffused glow discharge is influenced by a few factors like input power, gas conditions or discharge

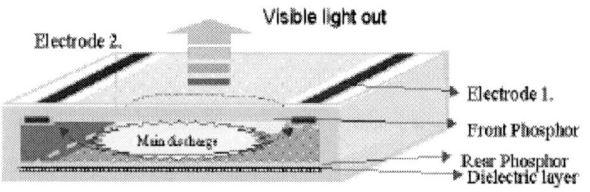

Fig. 1: Structure of the flat light source panel [1]

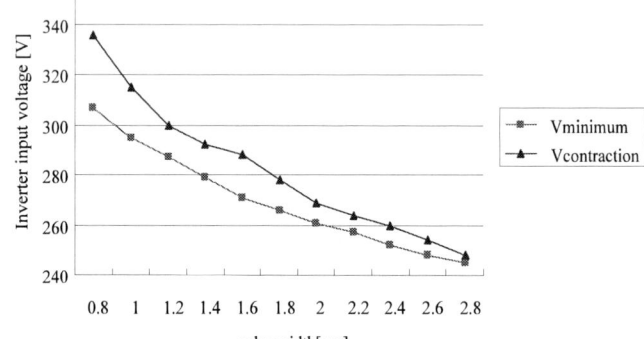

Fig. 2: Relationship of input voltage and pulse width

volume geometry etc.

However, an excessive perturbation changes the diffused glow discharge to a contracted state and restricts the uniform emissions. It means a failure state of a normal discharge.

Fig. 2 shows the relationship between the inverter input voltage and the pulse width. From Fig. 2, it can be shown that the operating region of maintaining the glow discharge state between the ignition voltage(Vminimum) and Vcontraction changes according to driving frequency and pulse width [2-6]. The relationship between the lamp driving voltage and luminance is shown in Fig. 3(a). It shows that the improvement of luminance needs a high lamp driving voltage (Vdriving). As shown in Fig. 3(b) and Fig. 3(c), the frequency and pulse width also have similar relationships as that of the driving voltage. In other words, when the transmission of the power increases by a change of the lamp characteristics, the luminance of the lamp also increases.

However, the higher the input voltage of the lamp increases, the narrower the glow discharge operation region becomes [7-8].

Furthermore, on a long-term operation, the possibility of discharge contraction increases because of the change of lamp discharge states such as temperature, lamp voltage, lamp currents, etc. From those reasons, the time of maintaining the glow discharge is generally reduced as the driving voltage or the current increases.

Fig. 4 shows a situation of the conventional driving operation. In case of using an open-loop driving circuit, when the conventional method has constant dc-link voltage, it has a tendency for the lamp power consumption increases according to radiation time. Because of this characteristic, it has the possibility of a normal glow discharge failure or a reduction in the normal operating time.

978-1-4244-1871-8/07 $25.00 © 2007 IEEE

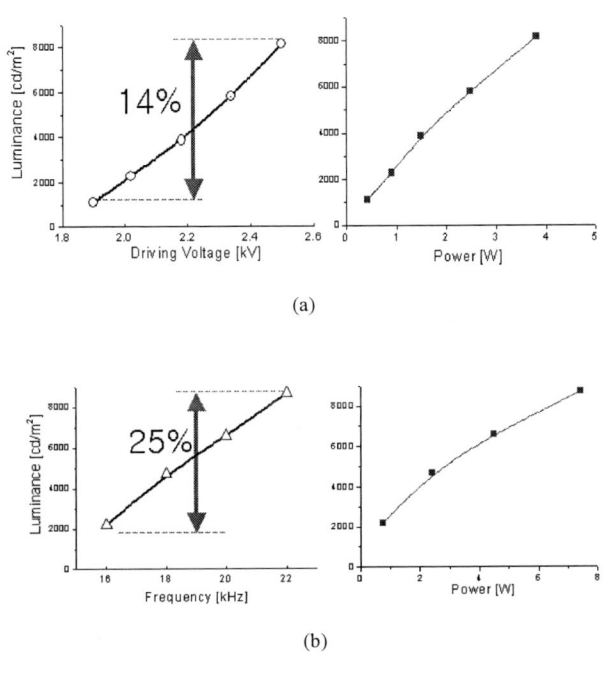

(a)

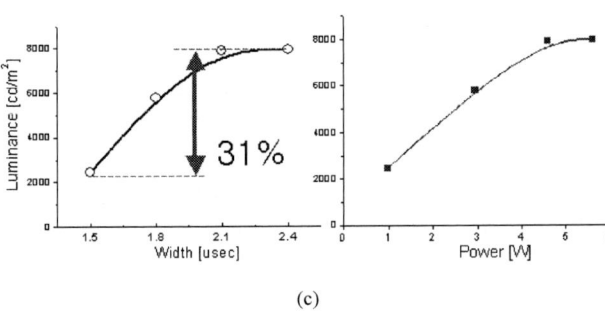

(b)

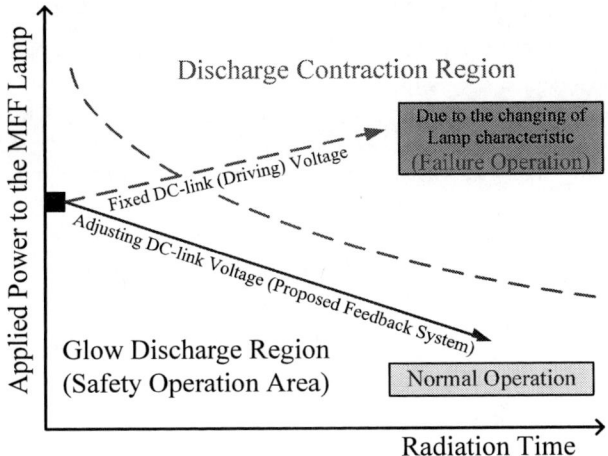

(c)

Fig. 3 MFFL Luminance and operation parameters [8]
(a): Luminance vs. drive voltage and lamp system power
(b): Luminance vs. driving frequency and lamp system power
(c): Luminance vs. pulse width and lamp system power

One of the best ways is to control the lamp driving through lamp state feedback, in order to prevent the occurrence of

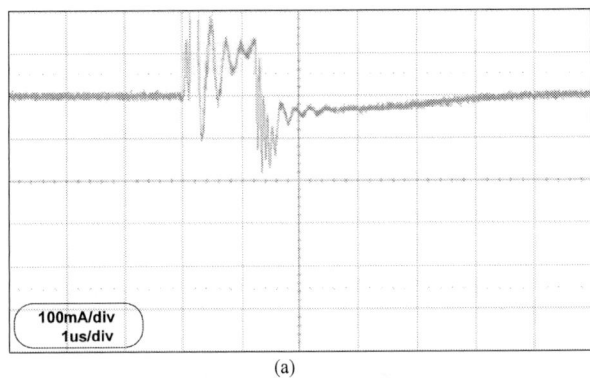

(a)

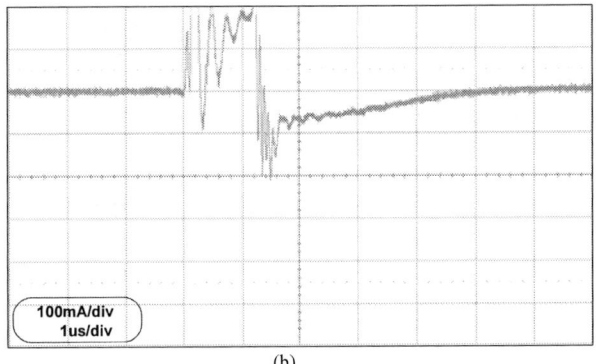

(b)

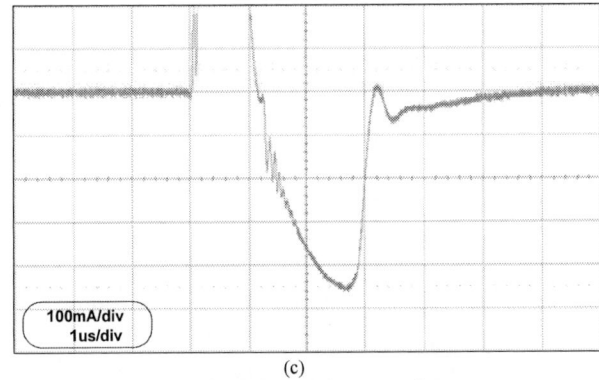

(c)

Fig. 5 MFFL current waveform
(a): Glow discharge state
(b): Long-term excitement
(c): The discharge contraction state

discharge contraction. In this paper, feedback control for stable MFFL driving is proposed. Fig. 4 shows a situation of the proposed driving operation. The proposed system consists of a sensing part for feedback information and a control part for appropriate regulation of the lamp driving circuit. To verify the proposed method, hardware experiments measuring the discharge contraction transfer time in both the closed loop and in open loop systems are performed and compared.

Fig. 4: Relationship between applied power to the MFFL and operation time

II. PRINCIPLE OF THE CURRENT FEEDBACK SYSTEM

For the feedback system, the symptom of discharge

contraction should be considered. To observe this symptom, the lamp current is measured. Fig. 5(a) shows the lamp current waveform of the initial glow discharge. Fig. 5(b) shows the long-term excitement in the glow discharge mode. Fig. 5(c) shows the discharge contraction. As shown in Fig. 5, after the lamp is ignited, the lamp current increases with the change of the discharge state due to changes of temperature, humidity, and driving condition. As the operating time flows, lamp current slowly increases as well. Finally, the lamp state becomes discharge contraction with a sudden in-rush current. As shown in Fig. 5(c), it can be determined whether or not the lamp state is in the discharge contraction state through the observation of the lamp current variation.

Therefore, sensing the electric characteristic of the lamp can be a reasonable solution for stable operation by maintaining the glow discharge through the control of the driving circuit. Fig. 6 shows the overall the MFFL system configuration with the proposed feedback method, which is based on the principle of current feedback. The outer-loop (current-loop) of the proposed system generates a reference for the inner-loop (voltage-loop, frequency-loop, pulse width-loop) by the regulation of the lamp current. For controlling the lamp power, the inner-loop can be controlled by a combination of adjustments of the voltage reference, frequency reference, pulse width reference etc.

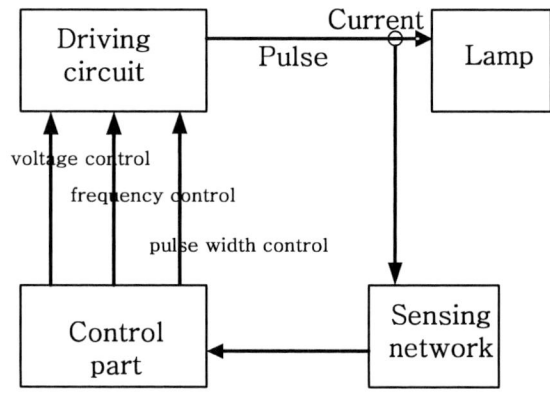

Fig. 6: Configuration of feedback system

III. FEEDBACK CIRCUIT CONFIGURATION

The current-feedback circuit can be divided by the sensing part and the control part. The sensing part has various circuit

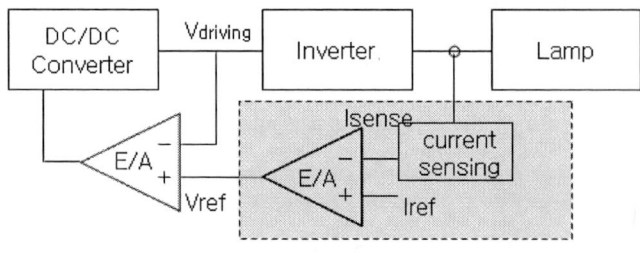

Fig. 7: Block diagram of the proposed feedback system

versions according to the sensing position, and the sensing method etc.

The control part also has various circuit versions according

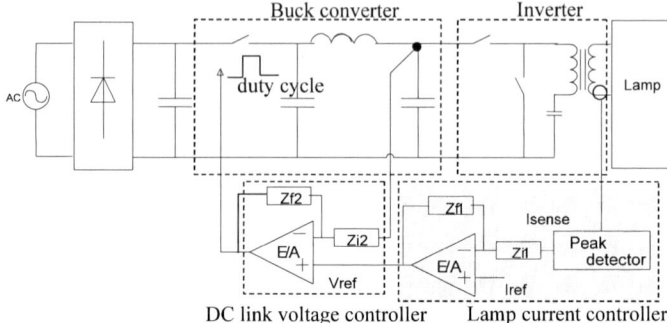

Fig. 8: The proposed prototype hardware schematic

to the driving voltage, the frequency, the pulse width, etc. That there is a DC/DC converter in front of inverter, it assumed.

Fig. 7 shows an example of the proposed feedback system controlling the lamp current by controlling the dc-link voltage. The system consists of two parts, namely the sensing part and the control part. Since the conventional system is open-loop without feedback in the dark box, shown in Fig. 7, it operates with a constant input voltage. However, the proposed system can change the input voltage through the voltage reference for the regulation of the lamp current.

Fig. 8 shows the detailed description of the power stage. First, the lamp current is sensed to obtain the discharge state information. The sensed current is compared with the current reference, thus regulating the limitation of a normal operation region. This sensed current is given to the input of the lamp current controller. The output of the lamp current controller can be used for the control of the driving voltage. By adjusting the dc-link voltage, the glow discharge state is maintained. Also, the driving frequency and pulse width can be used for maintaining a glow discharge. In the following section, the sensing and control parts will be explained in detail.

A. Sensing

The transition symptom of discharge contraction can be observed at several positions on the circuit, such as the input current, primary current, secondary current, etc. In this hardware experiment, the sensing part was implemented using a resistor sensor in the secondary transformer coil, which conducts the lamp current, because this point shows the most accurate relationship of the transition symptom. The sensed current maintains information of the lamp state through different forms. For example rms, average, or peak values maintain information of the lamp state. In this hardware experiment, information of current is maintained through a quasi-peak detector.

B. Control

As previously mentioned, the control part has various circuit versions according to the driving voltage, the

978-1-4244-1871-8/07 $25.00 © 2007 IEEE 990

frequency, the pulse width, etc. A role of the control part is to control the power which is transferred from the buck converter to the inverter, or another power level which is transferred from the inverter to the lamp using a different control variable. Through such control the glow discharge can be maintained. In this hardware experiment, the control part was implemented by a method to change the driving voltage (dc-link voltage). The control part adjusts the lamp current by regulating the output of the buck converter. The voltage reference determining the dc-link voltage is given by the current–loop amplifier. In order to stabilize and to regulate the lamp current, a compensator is designed as a proportional integrator. The output of the compensator is used as the reference of the buck converter. For example, when the lamp current increases, compensator decreases the voltage reference of the buck converter. Then lamp current is also decreased. Through this negative-feedback action, the proposed feedback system controls the lamp state for normal operation, even for long-term operation.

IV. EXPERIMENTAL RESULTS

It has been known that the lamp current can be different according to the gas mixture ratio and the fluorescence material through the experimentation of various panels. In this research, a 2 by 3 panel was used. The open-loop and the closed-loop were tested under the conditions of an operating

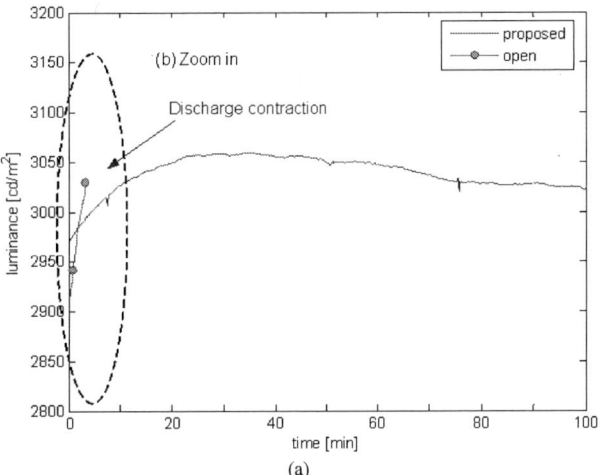

(a)

frequency of 12.5kHz, a duty of 1.38% and a turn ratio of 11:132.

First, the experimental conditions are as follows. The driving voltage is maintained at the highest level in which the discharge contraction does not occur.

When the lamp is operated by a conventional open-loop system, the value of the sensed current and the luminance are measured according to changes in time. In the proposed system, a current reference is decided by a sensed current in the open-loop system. Also, in the closed-loop experiment, the value of the sensed current and luminance are measured.

Fig. 9 shows the change of current according to time variation. In the case of the conventional system, the lamp current increases suddenly and the discharge contraction

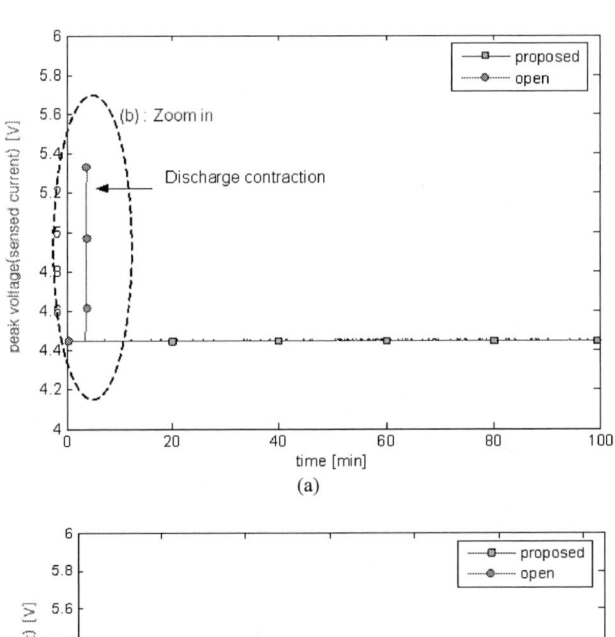

(a)

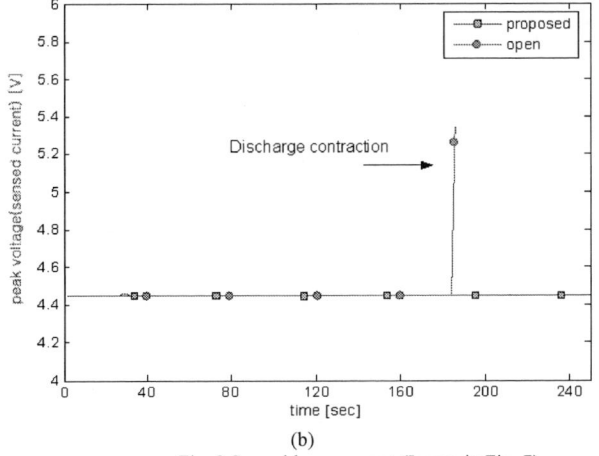

(b)

Fig. 9 Sensed lamp current (Isense in Fig. 7)
(a): Lamp current variation
(b): Local focusing (zooming) of (a)

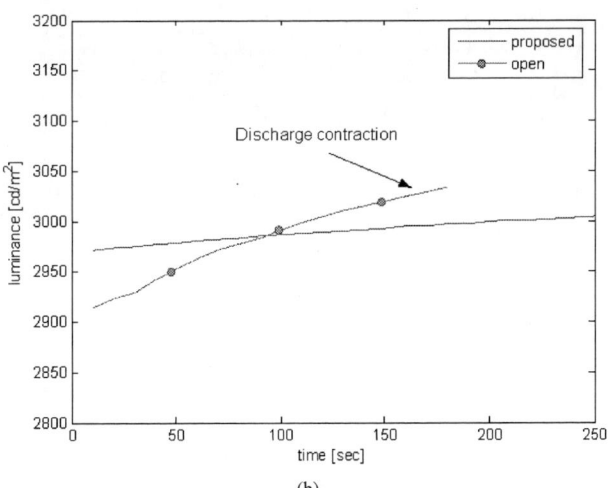

(b)

Fig. 10 Comparison of the luminance change
(a): Luminance variation
(b): Local focusing (zooming) of (a)

(b)

(a): a photo of MFFL in glow discharge
(b): a photo of MFFL in discharge contraction state
Fig. 11: a photo of 2 by 3 MFFL discharge state of lamp

occurs in 5 minute operation. This means that a normal glow discharge failed. However, in the case of the proposed system, the discharge contraction does not occur and a normal glow discharge is maintained even longer than 100 minute in radiation.

Fig. 10 shows the change of luminance according to time variation. In the case of the conventional system, from initial discharge the luminance increases continuously and then it goes into the discharge contraction mode suddenly, which indicates a failure of normal discharge. However, in the case of the proposed system, the luminance increases to a lower slope than the one of the conventional system. The luminance is then maintained at a proper value with a balanced state. Fig.

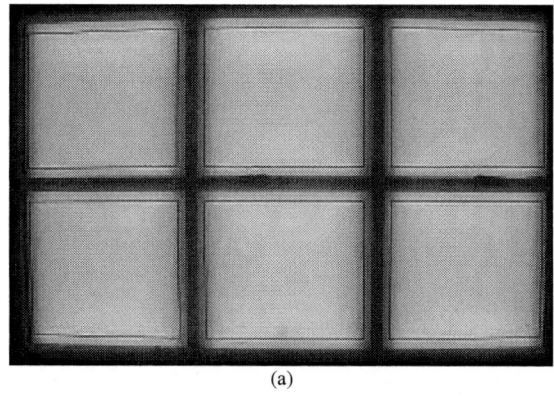

(a)

10(b) shows the difference of the luminance change as well.

Fig.11 shows the photos of MFFL. Fig. 11(a) shows a stable glow discharge. In case of the Fig. 11(b), a normal operation failed. It shows the discharge contraction state.

When the performance of luminance is considered, it can be shown that the proposed system has another merit in terms of the dimming function, as well as maintaining a glow discharge. Therefore it can be used for dimming by adjusting a current reference.

Fig. 12 shows the result of an additional experiment using an aged lamp. When the aged lamp was operated by the conventional method, the lamp state was very unstable and eventually entered the discharge contraction mode. However, in the case of the proposed system, the aged lamp was

maintained more stable, resulting in normal glow discharge without discharge contraction. From the result, the proposed system has another advantage, which is stable operation.

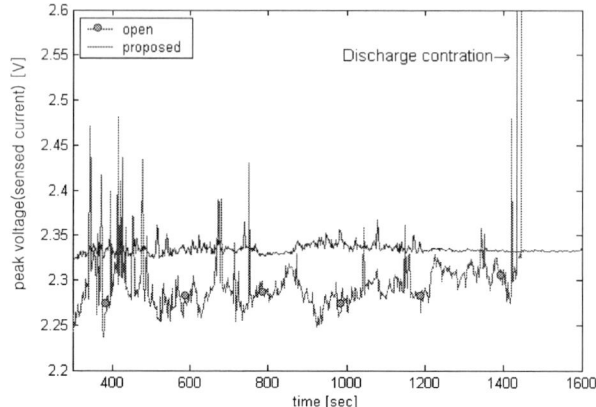

Fig. 12: Sensed lamp current using aged lamp

V. CONCLUSIONS

This paper proposes a driving system with current feedback for maintaining the glow discharge of the MFFL. The discharge contraction is prevented by feedback control and through the regulation of the lamp current. The relationship between the lamp discharge state and the lamp current has been shown and the principle of adjusting the lamp current has been presented. The proposed systems are verified by a 2×3 cell MFFL with a DC/DC converter and a pulse type inverter. The results show that the closed-loop system has very stable operation such that the glow discharge is maintained without the occurrence of the discharge contraction despite long-term radiation.

978-1-4244-1871-8/07 $25.00 © 2007 IEEE

REFERENCES

[1] J. H. Park, I.K Lee, B. H Cho, Ju Kwang Lee, Ki-Woong Whang, "High Efficiency Inverter Systems for Driving Mercury-free Flat Fluorescent Lamps," *PCC '07*, pp. 717-720, 2007.

[2] Ju kwang Lee, Tae Jun Kim, Hae youn Jeong and Ki-Woong Whang, "High efficiency Mercury-free flat light source for LCD Backlighting," *SID 05 Digest*, pp. 1309-1311, 2005.

[3] S. Mikoshiba, "Xe discharge Backlights for LCDs," *SID 01 Digest*, pp. 286-289, 2001.

[4] M. Ilmer, R. Lecheler and M. Seibold, "Efficiency Enhancement of Hg-free Fluorescent PLANON Backlights by CAF," *IDW '99*, pp. 1107-1108, 1999.

[5] T. Shiga, S. Mikoshiba and S. Shinada, "No-Mercury Flat discharge Lamp for LCD Backlighting," *IDW '99*, pp. 347-350, 1999.

[6] Y. Ikeda, T. Shiga, S. Mikoshiba, "Mercury-Free, Simple-Structured Flat discharge LCD Backlights Ranging from 0.5 to 5.2-in. Diagonals," *SID 00 Digest*, pp. 938-941, 2000.

[7] Ju kwang Lee, Byungjoo Oh, Jaechul Jung, and Ki-Woong Whang, "High luminous efficiency Mercury-free flat light source for LCD BLU," *IMID 05 Digest*, pp. 1161- 1164, 2005.

[8] Ju kwang Lee, Jaechul Jung, Byungjoo Oh, Inwoo Seo, Joongkyun Kim and Ki-Woong Whang, "The Electro-optic Characteristics of MFFL," *SID 06 Digest*, pp. 1422-1424, 2006

A New PWM-Controlled Quasi-Resonant Converter For High Efficiency PDP Sustaining Power Module

Woo-Jin Lee, Duk-You Kim, Chong-Eun Kim, and Gun-Woo Moon
Dept. of Electrical Engineering and Computer Science,
Korea Advanced Institute of Science and Technology (KAIST),
373-1 Guseong-Dong, Yuseong-Gu, Daejeon, 305-701, Republic of Korea
Phone: +82-42-869-3475, Fax: +82-42-861-3475
Email : gwmoon@ee.kaist.ac.kr

Abstract— A new PWM-controlled quasi-resonant converter for high efficiency PDP sustaining power module is proposed in this paper. The load regulation of the proposed converter can be achieved by controlling the ripple of the resonant voltage across the resonant capacitor with bi-directional auxiliary circuit, while the main switches are operating at the fixed duty ratio and fixed switching frequency. Hence, the waveforms of currents can be expected to be optimized from the view-points of the conduction loss. Furthermore, the proposed converter shows the good ZVS capability, simple control circuits, no high voltage ringing problem of rectifier diodes, no DC offset of the magnetizing current and low voltage stresses of power switches. In this paper, operational principles, features of the proposed converter, analysis and design considerations are presented. Experimental results demonstrate that the output voltage can be controlled well by the auxiliary circuit as PWM method.

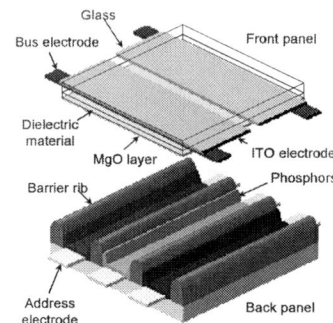

Fig. 1: Simplified structure of PDP with three electrodes

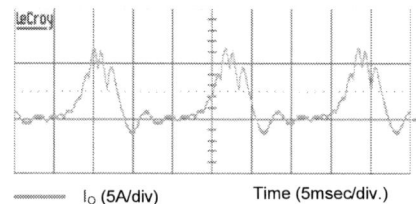

Fig. 2: Load current of sustaining power module

I. INTRODUCTION

Plasma display panels (PDPs) have been considered as the best candidate for flat panel display because of its wide view angle, high contrast ratio, and long life time. Fig. 1 shows a simplified structure of PDP with three electrodes. It consists of transparent X and Y sustain electrodes covered with a dielectric layer on the front panel, address electrodes perpendicular to the sustain electrodes on the back panel as shown in Fig. 1. Due to the existence of dielectric layer, PDP is purely capacitive load in respect to circuit operation [1]. Since, not only it features pure capacitive load characteristics but also it is driven by the address display separation (ADS) method, the load variation of the sustaining power module is very wide and abrupt in case of a full-white screen as shown in Fig 2. In ADS method, the operation of PDP can be divided into three periods such as resetting, addressing, and sustaining period. Also, the power dissipated by a PDP is the maximum in this case. However, in real PDP TVs, the load condition is strongly dependent on the average pixel level (APL) such as a concept that defines the total light output of a given TV image as a percentage of the total light output of a full-white image. Since TV signals typically have an APL of 20% or less [4], the sustaining power module is usually operating on light load conditions. Moreover, all PDPs have an automatic power control (APC) system that limits the power consumptions to some maximum level by automatically reducing the luminance of the PDP. Thus, the lower power is dissipated even the full load condition which corresponds to the full-white screen [4]. Although the power dissipated during sustaining period is lower than the maximum power, it is still the most of power driving the PDP compared with that dissipated during resetting and addressing periods. Therefore, the sustaining power module is mainly responsible for overall system efficiency [1-3]. In addition, when the PDP is operating on TV signals, the high efficiency is needed significantly at light load conditions. Until now, several DC/DC converters which can realize the high efficiency and low cost, have been proposed for sustaining power module of PDP. Among them, the resonant converters have been investigated to realize prominent features of miniaturization, high efficiency, and low noise [7-8]. However, since a large variation of switching frequency is needed to control the output voltage, those converters have some difficulties from the view-points of size reduction and noise problem [9]. To overcome above problems, recently a half bridge LLC resonant converter has been discussed because it shows many unique characteristics and improvements over previous topologies [5-6]. Fig. 3 (a) shows a circuit diagram of the half bridge LLC resonant converter with voltage doubler rectifier. As shown in this figure, it shows simple structure, and low voltage stress on primary power switches. Moreover, since there is no secondary filter inductor, the voltage across the secondary

978-1-4244-1871-8/07 $25.00 © 2007 IEEE

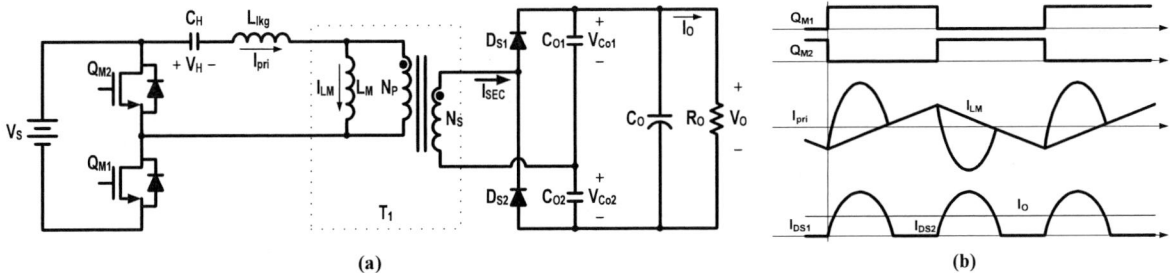

(a)   (b)

Fig. 3: Half bridge LLC resonant converter with voltage doubler rectifier, (a) Circuit diagram (b) Key waveforms

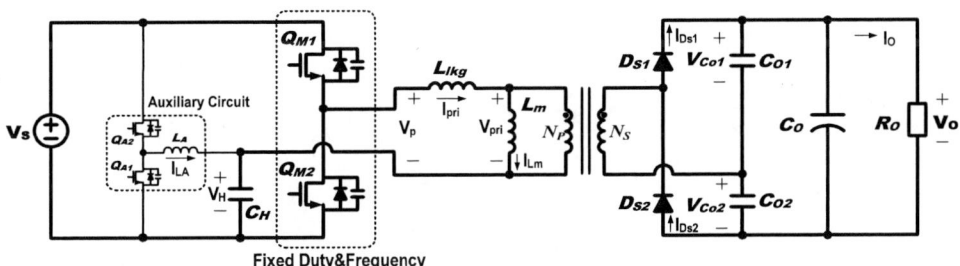

Fig. 4: Circuit diagram of proposed converter

rectifier can be effectively clamped to the output voltage. Employing the rectifier diodes of low voltage rating, the conduction loss can be greatly reduced. And also its zero-voltage-switching (ZVS) capability is excellent from zero to full load condition [6]. These features make the half bridge LLC resonant converter to be very suitable for PDP sustaining power module. However, this converter has a small magnetizing inductance in order to have a narrow variation of switching frequency. This results in not only considerably higher circulating energy of the primary side of the transformer, but also more conduction loss especially in the below resonance mode as shown in Fig. 3 (b). In case of light load conditions, high circulating energy can be one of serious problems that reduce the system efficiency. In addition, a variable frequency control method makes the control circuits much more complicated than those of the pulse width modulation (PWM) control method. To resolve these problems effectively, we propose a new PWM-controlled quasi-resonant converter which has the simpler control circuits and less conduction loss compared with those of half bridge LLC resonant converter at light load conditions. As shown in Fig. 4, the proposed converter is similar to the half bridge LLC resonant converter except for the auxiliary circuit which is needed to control the output voltage. In the proposed converter, the output voltage can be regulated by controlling the voltage across the resonant capacitor while two main switches are operating at fixed duty ratio and fixed switching frequency. Therefore, the waveforms of both primary and secondary currents can be expected to be optimized from the view-points of conduction loss and current stress. Simultaneously, since the auxiliary circuit controls some portion of the voltage ripple on the resonant capacitor, the output voltage can be regulated well at whole load condition. Thus, keeping the good characteristics of half bridge LLC resonant converter, the proposed

converter is expected to overcome sufficiently the above mentioned problems such as the higher current stress and high circulating energy, and can realize the high power density, high performance, and high efficiency.

II. FEATURES OF PROPOSED CONVERTER

Fig. 5 shows a variation of the voltage ripple across the resonant capacitor, C_H, according to load conditions in the half bridge LLC resonant converter which is frequency controlled converter. As the load is changed from full load to light load, the variation of the resonant voltage ripple of C_H is getting smaller. This is because not only the change of switching frequency causes the same effect to change the impedance of resonant tank which is composed of the resonant capacitor and inductor, but also the amount of the primary current which is reflected the load current is getting smaller. Thus, in the proposed converter, the bi-directional auxiliary circuit is operating in order to change the resonant voltage ripple with the PWM method while two power switches Q_{M1} and Q_{M2} are operating with the constant duty ratio (D=0.5) and constant switching frequency. Since the load regulation can be achieved by the auxiliary circuit, the magnetizing inductance of the proposed converter is larger

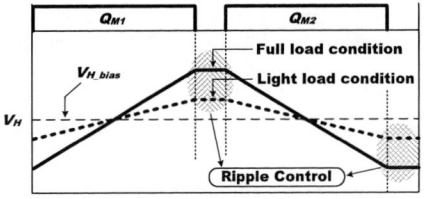

Fig. 5: Variation of voltage ripple across resonant capacitor, C_H according to load conditions

978-1-4244-1871-8/07 $25.00 © 2007 IEEE   995

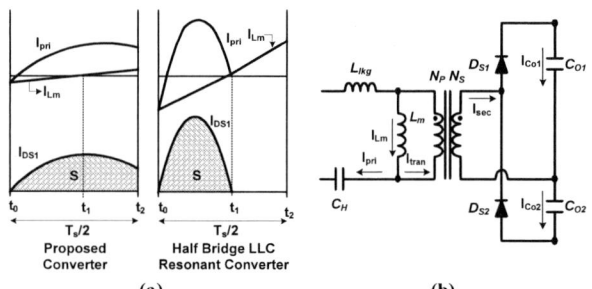

Fig. 6: Simplified waveforms and circuit (a) Comparison of current waveforms (b) Zero DC offset of magnetizing current

than that of the half bridge LLC resonant converter. Thereby, the circulating energy of the proposed converter is considerably reduced at light load conditions. Additionally, due to the operating of Q_{M1} and Q_{M2} with the fixed duty ratio and fixed frequency, the primary current can be optimized from the view-points of conduction loss. Fig. 6 (a) shows the comparison of current waveforms in case of the proposed converter and half bridge LLC resonant converter. In each case, the load current is the same as the averaging value of I_{DS1} because of the capacitive output filter. Therefore, the peak value of I_{DS1} in proposed converter must be smaller than that of I_{DS1} in conventional converter on the same load condition. This can be similarly applied to the primary side of the transformer. Thus, to reduce the conduction loss, the waveforms of current in proposed converter is very reasonable. On the other hand, the proposed converter employs the voltage doubler type rectifier which has no output inductor. Due to a lack of the output inductor, there is no high voltage ringing across the rectifier diode. Also, by choosing the proper capacitance of C_{o1} and C_{o2}, additional resonant voltage ripple of C_{o1} and C_{o2} helps the variation of the resonant voltage ripple of C_H, which is controlled by the auxiliary circuit. Moreover, no DC offsets of the magnetizing current and magnetic flux can be achieved. Considering the DC value of the current through the capacitor is 0A in steady state, the DC values of I_{pri}, I_{Co1}, and I_{Co2} ($<I_{pri}>$, $<I_{Co1}>$, and $<I_{Co2}>$, respectively) are all 0A, where $<\bullet>$ means the DC value of '$\bullet$'. As shown in Fig. 6 (b), I_{sec} is equal to I_{Co1}-I_{Co2}, $<I_{sec}>=<I_{Co1}>-<I_{Co2}>=0A$. Thus, $<I_{Lm}>=<I_{pri}>+<I_{tran}>=0A$, because $<I_{tran}>=<I_{sec}/n>=0A$. This means that DC offsets of the transformer magnetizing current and magnetic flux are completely blocked. Therefore, the transformer magnetic core is fully utilized, and its power density can be considerably increased while the heat generation of the transformer can be greatly reduced. Also, the control circuits which generate the gate signals for all power switches, can be easily implemented by using TL494.

III. OPERATIONAL PRINCIPLES

Fig. 7 shows the key waveforms of the proposed converter. The operation of the proposed converter can be divided into ten modes. One switching cycle of the proposed circuit is divided into two half cycles, $t_0 \sim t_5$ and $t_5 \sim t_{10}$. Since the operational principles of two half cycles are symmetric, only the first half cycle is explained. A half cycle can be divided into 5 modes and its equivalent circuits are shown in Fig. 8. The switches of Q_{M1} and Q_{M2} are turned on and off alternately with the constant duty ratio (D=0.5) and constant frequency. And the auxiliary switches such as Q_{A1} and Q_{A2} are turned on and off in duty cycle controlled, where $D_E T_S$ is the operational conduction time of the auxiliary switches. To illustrate the steady state operation, several assumptions are made as follow:

- The power switches such as Q_{M1}, Q_{M2}, Q_{A1}, and Q_{A2} are ideal except for their internal diodes and output capacitors, C_{oss}.
- The rectifier diodes D_{S1} and D_{S2} are ideal except for their junction capacitors, C_j.
- The output voltage V_o is constant during a switching period.

Mode 1 ($t_0 \sim t_1$): After the ZVS turned on of Q_{M1} is achieved and the commutation between D_{S1} and D_{S2} is completed, Mode 1 begins. The primary current I_{pri}, which rises with resonance between the leakage inductor and resonant capacitor, is given by

$$I_{pri}(t) = \frac{1}{nZo}\left[\frac{V_S}{n} - \frac{V_H(t_0)}{n} - V_{Co1}(t_0)\right]\sin \omega_r(t-t_1) + I_{Lm}(t). \quad (1)$$

Concurrently, the magnetizing current, I_{Lm}, also rises with the resonance between the magnetizing inductor L_m, and rectifier capacitors $C_{o1}//C_{o2}$. On the other hand, since the resonant frequency, f_m, decided by L_m and $C_{o1}//C_{o2}$, is much slower than the switching frequency, I_{Lm} can be linearly

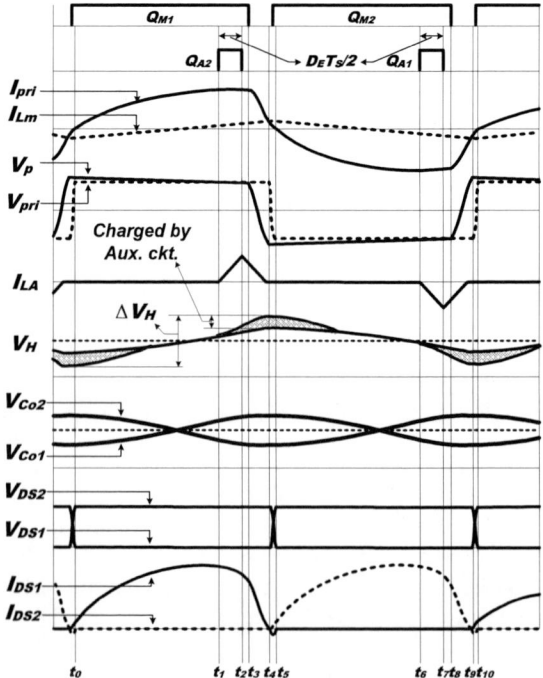

Fig. 7: Key waveforms of proposed converter

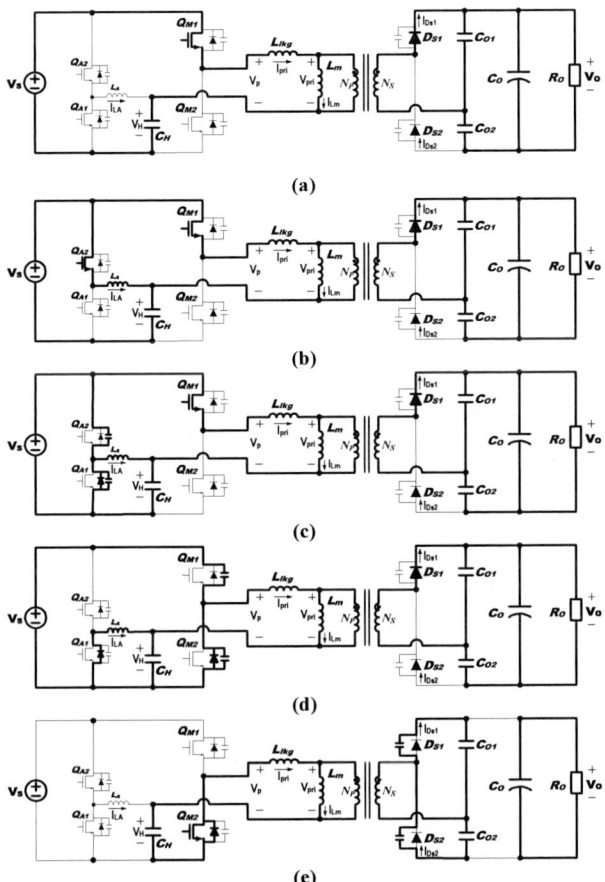

Fig. 8: Equivalent circuits of proposed converter, (a) Mode 1 (b) Mode 2 (c) Mode 3 (d) Mode 4 (e) Mode 5

approximated as follows:

$$I_{Lm}(t) = I_{Lm}(t_1) + \frac{nV_{Co1}(t)}{L_m}(t - t_1). \qquad (2)$$

where, $\omega_r = \dfrac{1}{\sqrt{L_r C_r}}$, $Z_0 = \sqrt{\dfrac{L_r}{C_r}}$, $L_r = \dfrac{L_{lkg}}{n^2}$, $C_r = \dfrac{n^2 C_H \times C_{o1} // C_{o2}}{n^2 C_H + C_{o1} // C_{o2}}$, $n = \dfrac{N_p}{N_s}$.

The current of rectifier diode D_{S1}, I_{DS1}, is flowing through C_{o1} and equivalent load resistor, while the rectifier capacitor C_{o1} and C_{o2} is charged and discharged respectively.

Mode 2 ($t_1 \sim t_2$): When Q_{A2} is turned on, mode 2 begins. During this mode, the resonant capacitor, C_H, is additionally charged from the input source, V_S, through the auxiliary inductor L_A operating in the discontinuous conduction mode (DCM). It is assumed that L_A is so large enough to approximate I_{LA} is increased and decreased linearly. The slope of I_{LA} can be obtained as follows:

$$\frac{dI_{LA}(t)}{dt} = \frac{V_S - V_H(t)}{L_A}. \qquad (3)$$

The primary and secondary side of the transformer operates similarly to Mode 1.

Mode 3 ($t_2 \sim t_3$): After Q_{A2} is turned off, I_{LA} starts to charge and discharge the output capacitors of Q_{A2} and Q_{A1}

respectively. When the voltage across Q_{A1} becomes 0V, I_{LA} begins to flow through the internal diode of Q_{A1}. Since the voltage across the C_H, V_H is applied to L_A reversely, I_{LA} is decreased. During this interval, C_H is still charged and both of transformer sides operate similarly to mode 1.

Mode 4 ($t_3 \sim t_4$): When Q_{M1} is turned off, mode 4 begins. Since I_{pri} starts to charge and discharge the output capacitors of Q_{M1} and Q_{M2} respectively, the voltage across the primary side of the transformer V_P, is decreased to $-V_S$. Since the rectifier diode D_{S1} is still conducting, the voltage across L_m, V_{pri} is maintained to be nV_{Co1}. Thus, the negative voltage which is the same as the difference between V_P and V_{pri}, is applied to the leakage inductor, L_{lkg}. Thereby I_{pri} is decreased rapidly. Also, in this interval, C_H is continuously charged until I_{LA} becomes 0A. After the voltage across Q_{M2} becomes 0V, I_{pri} starts to flow through the internal diode of Q_{M2}. Thus, the ZVS condition of Q_{M2} is satisfied. mode 4 is finished when I_{pri} is equal to I_{Lm}.

Mode 5 ($t_4 \sim t_5$): When I_{pri} is smaller than I_{Lm}, the current of secondary side of the transformer flows reversely through the junction capacitors of D_{S1} and D_{S2}. Since the voltage across each diode is increased and decreased respectively, the commutation between D_{S1} and D_{S2} is started. During this mode, the ZVS turned on of Q_{M2} can be achieved. After D_{S2} is fully conducting, Mode 5 is finished.

The circuit operation of $t_5 \sim t_{10}$ is similar to that of $t_0 \sim t_5$. Subsequently, the operation from t_0 to t_{10} is repeated.

IV. ANALYSIS & DESIGN CONSIDERATIONS

A. Variations of V_{Co1}, V_{Co2}, and V_H

Fig. 9 shows the equivalent circuit and its key waveforms for the analysis. Since both time duration of $t_2 \sim t_3$ and dead time are much smaller than switching period T_S, they can be discarded for simplicity of analysis. During the half switching period $T_S/2$, the charging current of C_{o1} is equal to the load current I_O. This is because C_{o1} must supply as much as I_O during last half switching period. In addition, since the sum of V_{Co1} and V_{Co2} is always equal to the output voltage V_O, the bias voltage of V_{Co1} and V_{Co2} is $V_O/2$ as shown in Fig. 9

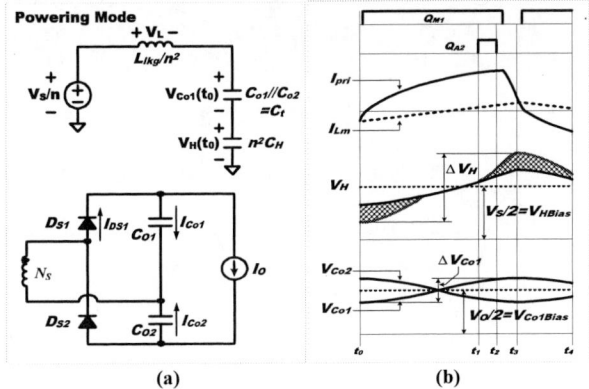

Fig. 9: Powering Mode, (a) Equivalent circuits (b) Key waveforms

978-1-4244-1871-8/07 $25.00 © 2007 IEEE

(b). Thus, by using the equation $I_C = C(dV_C/dt)$, $V_{Co1}(t_0)$ can be obtained as follows:

$$dV_{Co1}(t) = \frac{1}{C_{o1}} \int I_{Co1}(\tau)d\tau = \frac{TsIo}{2C_{o1}}, \tag{4}$$

$$where, \quad \frac{2}{Ts} \int I_{Co1}(\tau)d\tau = Io.$$

$$V_{Co1}(t_0) = \frac{Vo}{2} - \frac{TsIo}{4C_{o1}}. \tag{5}$$

Similarly $V_{Co2}(t_0)$ can be obtained as follows:

$$V_{Co2}(t_0) = \frac{Vo}{2} + \frac{TsIo}{4C_{o2}}. \tag{6}$$

As mentioned previously, since both Q_{M1} and Q_{M2} are operating with the constant duty ratio (D=0.5), the bias voltage of V_H is the same as $V_S/2$. Thus, $V_H(t_0)$ can be represented as follows:

$$V_H(t_0) = \frac{Vs}{2} - \frac{\Delta V_H}{2}. \tag{7}$$

B. Turns Ratio of Transformer and Load Current

By using the equations (5) and (7), the current through D_{S1}, I_{DS1} can be easily obtained from the equivalent circuit as shown in Fig. 9 (a) as follows:

$$I_{DS1}(t) = \frac{1}{Zo}\left[\frac{Vs}{n} - \frac{1}{n}\left(\frac{Vs}{2} - \frac{\Delta V_H}{2}\right) - \left(\frac{Vo}{2} - \frac{TsIo}{4C_{o1}}\right)\right]\sin \omega_r t. \tag{8}$$

Turn ratio of the transformer can be determined at no load condition in order to maintain V_O to be desired value without the operation of auxiliary circuit. Thus, it can be obtained by using the equation (8) with these conditions such as I_O=0A, I_{DS1}=0A, ΔV_H=0V, and $\sin \omega_r t \neq 0$.

$$n = \frac{Vs}{Vo}. \tag{9}$$

Since both the charging current of C_{o1} and the discharging current of C_{o2} flow through D_{S1}, the averaging value of I_{DS1} during $T_S/2$ is equal to the two times of I_O. From this fact, I_O can be represented as follows:

$$Io = \left[\frac{Vs}{n} - \frac{1}{n}\left(\frac{Vs}{2} - \frac{\Delta V_H}{2}\right) - \left(\frac{Vo}{2} - \frac{TsIo}{4C_{o1}}\right)\right]$$
$$\times \frac{1}{ZoTs\omega_r}\left[1 - \cos\left(\frac{Ts\omega_r}{2}\right)\right]. \tag{10}$$

From the equation (10), ΔV_H can be expressed as following equation.

$$\Delta V_H = 2n\frac{IoZoTs\omega_r}{1-\cos(Ts\omega_r/2)} - 2n\left(\frac{Vs}{2n} - \frac{Vo}{2} + \frac{TsIo}{4C_{o1}}\right). \tag{11}$$

As shown in Fig. 10 (a) and (b), $\Delta V_H/2$ can be plotted by using the equation (11) according to I_O in different conditions. Fig. 10 (a) shows the $\Delta V_H/2$ with the fixed value of C_{o1} and C_{o2} while C_H is varied. Similarly, Fig. 10 (b) shows it with fixed value of C_H while C_{o1} and C_{o2} are varied. As can be seen in these figures, to regulate the output voltage according to load conditions, $\Delta V_H/2$ is getting larger as the load goes to full load.

C. Resonant Frequency vs. Switching Frequency

As mentioned above, the waveform of I_{pri} should be similar

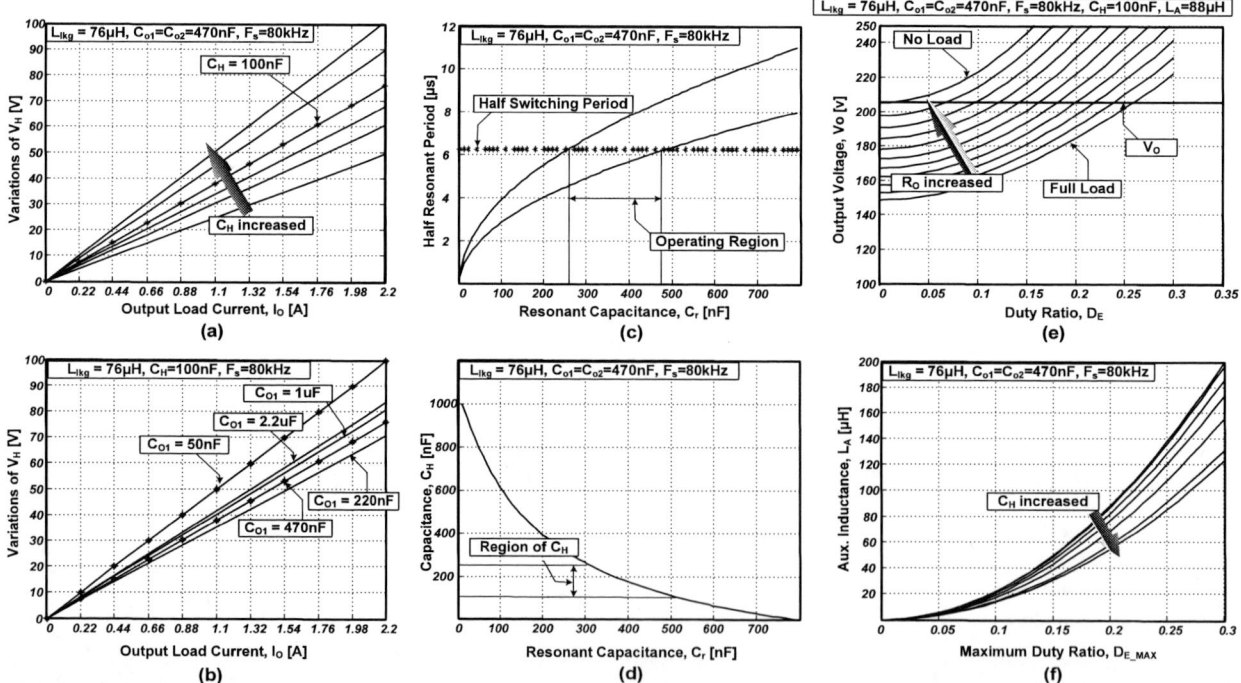

Fig. 10 Figures for analysis,
(a) V_H according to load with different C_H
(b) V_H according to load with different C_{o1} and C_{o2}
(c) Desirable region of resonant frequency
(d) Selection of C_H
(e) Output voltage according to duty ratio D_E with different load
(f) Selection of L_A and maximum duty ratio D_{E_MAX}

with that of the proposed converter as shown in Fig. 6 (a) so as to reduce the conduction loss. Therefore, the resonant frequency, which is decided by the resonant inductance and capacitance, should be selected according to the switching frequency. Fig 10 (c) shows the desirable region of resonant frequency when the switching frequency is fixed as 80kHz. To be optimized about the conduction loss, the resonant frequency should be selected within the operating region. After the resonant frequency is selected, C_H can be decided with the resonant inductance 76μH, as shown in Fig. 10 (d). Also, the values of C_{o1} and C_{o2} can be obtained properly by using the equation (11) and Fig. 10 (a) and (b).

D. DC conversion ratio and Maximum duty ratio according to auxiliary inductor, L_A

ΔV_H, which is controlled by the auxiliary circuit, can be obtained in similar way as mentioned in subsection *A* of the part IV.

$$\Delta V_H = \frac{1}{C_H L_A}\left[\frac{V_S}{2} - \frac{T_S I_o}{2 n C_H}\right](T_S D_E)^2 . \qquad (12)$$

From the equation (10) and (12), the steady state voltage conversion ratio of the overall system can be derived as follows:

$$\frac{V_o}{V_S} = \frac{\dfrac{1}{2} + \dfrac{(T_S D_E)^2}{4 C_H L_A}}{\dfrac{T_S}{R_o}\left[\dfrac{n Z_o \omega_r}{1 - \cos(T_S \omega_r / 2)} - \dfrac{n}{4 C_{o1}} + \dfrac{(T_S D_E)^2}{2 C_H L_A}\right] + \dfrac{n}{2}} . \qquad (13)$$

Using the equation (13), the output voltage can be plotted as shown in Fig. 10 (e). In this figure, the output voltage can be obtained without the operation of auxiliary circuit at no load ($D_E = 0$). As the load goes to full load, D_E is increased to get the desired output voltage. From this figure, the output voltage can be regulated well at whole load condition by the auxiliary circuit. Fig. 10 (f) shows the proper auxiliary inductance to achieve the load regulation according to the maximum duty ratio $D_{E\text{-MAX}}$. As shown in this figure, when L_A is selected as the larger value to reduce the peak value of

Table 1: Parameters of prototype circuit

Item	Symbol	Value/Part
Input Voltage	V_{in}	390~410V
Output Voltage	V_O	205V
Max. Power Rating	P_{max}	451W
Switching Frequency	F_s	80kHz
Turn Ratio	n:1	1.96:1
Leakage inductance	L_{lkg}	76uH
Magnetizing inductance	L_m	1.2mH
Capacitance of half bridge	C_H	100nF
Capacitance of doubler cell	C_{o1}, C_{o2}	470nF
Output Capacitance	C_o	680uF/250V
Main Power Switches	Q_{M1}, Q_{M2}	IRFPC60
Auxiliary Power Switches	Q_{A1}, Q_{A2}	FQP2N60
Auxiliary Inductor	L_A	88uH
Rectifier Diodes	D_{S1}, D_{S2}	15ETH03

I_{LA}, the more maximum duty ratio is needed.

V. EXPERIMENTAL RESULTS

A 450W prototype of the proposed converter has been built for the experiment. The parameters of this prototype circuit are listed in Table 1. Fig. 11 shows the experimental waveforms at 20%, half, and full load. As can be seen in Fig. 11 (a), the waveform of I_{pri} is similar with that of below resonance mode. This results in less conduction loss and lower peak values of I_{pri}, I_{DS1}, and I_{DS2}. Fig. 11 (b) shows V_H and I_{LA} at each load. As can be seen in this figure, C_H is additionally charged or discharged by I_{LA} to regulate the output voltage. The ZVS operation of Q_{M1} and Q_{M2} at 10% and full load is shown in Fig. 12. In order to have the waveforms of current in BRM, the leakage inductance is rather large compared with that of half bridge LLC resonant converter. Due to the large leakage inductance, the ZVS operation of Q_{M1} and Q_{M2} is easily achieved even 10% load as shown in Fig. 12. Fig. 13 shows the ZCS operation of D_{S1}

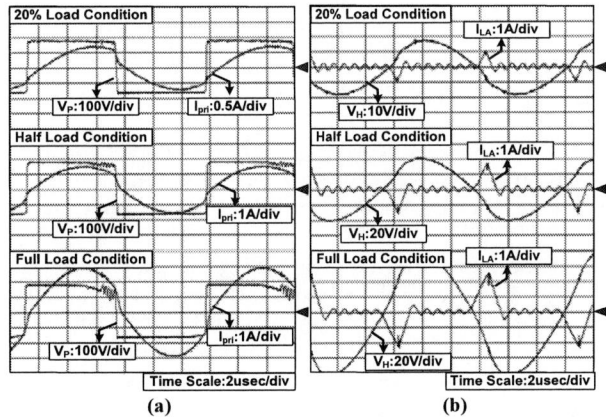

Fig. 11: Experimental waveforms at 20%, half, and full load, (a) V_P and I_{pri} (b) V_H and I_{LA}

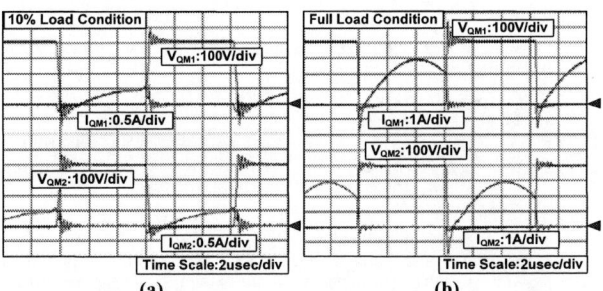

Fig. 12: ZVS of Q_{M1} and Q_{M2}, (a) At 10% load (b) At full load

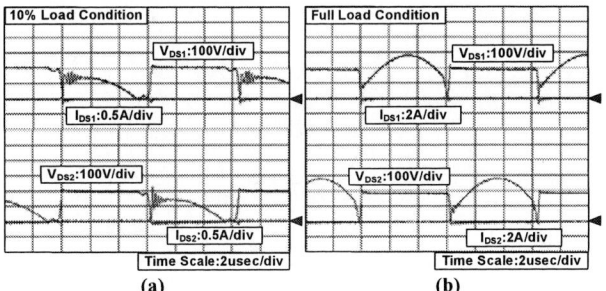

Fig. 13: ZCS of D_{S1} and D_{S2}, (a) At 10% load (b) At full load

and D_{S2}. Also, in this figure, the voltage across the D_{S1} and D_{S2} can be clamped to the output voltage without the high voltage ringing.

VI. CONCLUSIONS

A new PWM-controlled quasi-resonant converter for high efficiency PDP sustaining power module is proposed in this paper. Since the load regulation of the proposed converter can be achieved by the auxiliary circuit, the waveforms of current can be optimized from the view-points of the conduction loss especially at the light load conditions. Moreover, employing the voltage doubler type rectifier, additional resonant ripple of the voltage across the rectifier capacitors helps the operation of the auxiliary circuit. Besides, DC offsets of the magnetizing current and magnetic flux can be completely blocked. From the experimental results, good ZVS capability of the power switches Q_{M1} and Q_{M2} is also proved. A prototype was used in experiments to prove the validity of the proposed converter. Fig. 14 shows the measured efficiency. The measured efficiency with 10%~40% load range, is higher than that of the half bridge LLC resonant converter. As mentioned in introduction, when the PDP is operating on TV signals, the sustaining power module is usually operating on light load conditions. Thus, the proposed converter is expected to be suitable for the sustaining power module of PDP. On the other hand, the measured efficiency is decreased as the load goes to full load. This is because the power dissipated by the auxiliary circuit is increased. However, the measured efficiency along wide load ranges shows as high as around 94%. Therefore, the proposed converter demonstrates its suitability as a sustaining power module owing to its simple control circuits, low noise and high efficiency.

REFERENCES

[1] C. W. Roh, H. J. Kim, S. H. Lee, and M. J. Youn, "Multilevel Voltage Wave-Shaping Display Driver for AC Plasma Display Panel Application", *IEEE Journal of Solid-State Circuits*, Vol. 38, No. 6, June 2003, pp.935-947.

[2] S. K. Han, G. W. Moon, and M. J. Youn, "Current-fed Energy-Recovery Circuit for Plasma Display Panel", *Electronics Letters*, 10th July 2003, Vol. 39, No. 14, pp.1035-1036.

[3] S. K. Han, J. Y. Lee, G. W. Moon, M. J. Youn, C. B. Park, N. S. Jung, and J. P. Park, "A New Energy-Recovery Circuit for Plasma Display Panel", *Electronics Letters*, 18th July 2002, Vol. 38, No. 15, pp790-792.

[4] Larry F. Webber, "Do LCD TVs Really Last Longer than PDP TVs?" *Information Display*, Society for Information Display. Aug. 2004, Vol. 20, No. 8, pp12-17.

[5] Lazar, J. F.; Martinelli, R., "Steady-state analysis of the LLC series resonant converter" *Applied Power Electronics Conference and Exposition*, 2001. APEC 2001. Sixteenth Annual IEEE, Volume: 2, 2001 Page(s): 728-735 vol.2.

[6] Bo Yang; Lee, F. C.; Zhang, A. J.; Guisong Huang, "LLC resonant converter for front end DC/DC conversion," *Applied Power Electronics Conference and Exposition*, 2002. APEC 2002. Seventeenth Annual IEEE, Volume: 2, 2002 Page(s): 1108-1112 vol.2.

[7] F. C. Lee: "High-Frequency Quasi-Resonant Converter Technologies", *Proc. of the IEEE*, Vol. 76, No. 4, pp.377-390, April 1988.

[8] K. Liu and F. C. Lee: "Zero-Voltage Switching Technique in DC/DC Converters", *IEEE PESC'86*, Record, pp.58-70, June 1986.

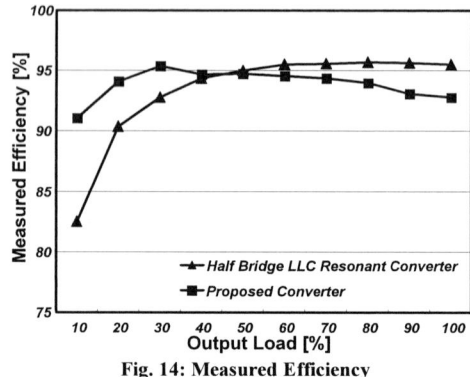

Fig. 14: Measured Efficiency

[9] Tanaka, H.; Ninomiya, T.; Shoyama, M.; Zaitsu, T.: "Novel PWM-controlled resonant converter" *Telecommunications Energy Conference*, 1996. INTELEC'96, 18th international 6-10 Oct. 1996 Page(s):823-828.

Current Stress Minimizing Control Scheme
for Power Factor Correction (PFC) Boost Pre-regulator

Hee-Chul Lee, Chong-Eun Kim, Hong-Sun Park, Ki-Bum Park and Gun-Woo Moon
Korea Advanced Institutes of Science and Technology (KAIST)
Department of Electrical Engineering
373-1 Guseong-Dong, Yuseong-Gu, Daejeon, 305-701, Republic of Korea
Phone:+82-42-869-8075, Fax:+82-42-869-8520
Email: gwmoon@ee.kaist.ac.kr

Abstract—A simple technique for PFC circuit is presented using UC3854. This technique is about current peak controlling by a reference current generator. Decreased peak currents of the boost pre-regulator reduce circuit current stress and so rated currents of circuit elements are minimized. Simulation and experimental results will verify the viability of the new scheme.

I.   INTRODUCTION

Power factor correction (PFC) circuits have been developed to use electric devices during past few years. A conventional capacitive bridge rectifier induces much harmonic currents, therefore it needs to rule world-wide current harmonic limits. The IEC Std. 61000-3-2 is representative regulation agreement for PFC [1].

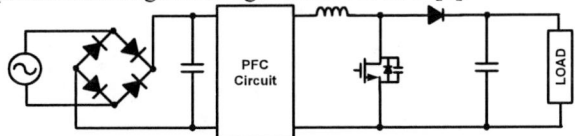

Fig. 1 Circuit Diagram of PFC Pre-regulator

There are several types of PFC circuits and here we refer two-stage active PFC which composed of both boost pre-regulator circuit and DC-DC converter. Generally the boost circuits are controlled with current mode control. And these are classified with CCM, BCM and DCM.

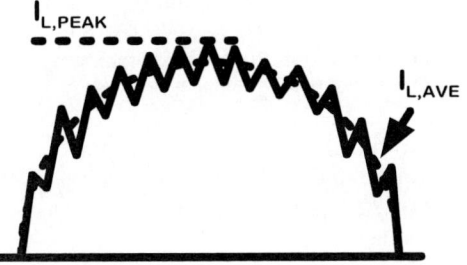

Fig. 2 Current Waveform of CCM Boost PFC

PFC boost pre-regulator in CCM shows low peak current characteristic and so conduction loss and conduction noise decrease more than BCM, DCM [2]. However, PFC circuit controlled by CCM has still high peak current therefore effects harms for current stress in overall circuit which is composed of inductor, switch, and diode [3].

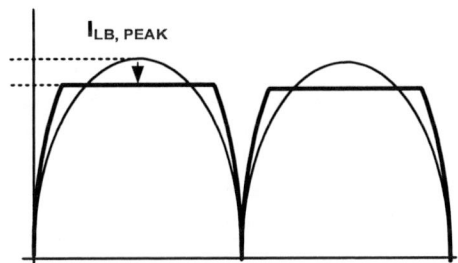

Fig. 3 Proposed Inductor Current Waveforms

To suppress the current stress more, the novel peak reducing technique for PFC is proposed and it can be achieved by only inserting a zener diode. Therefore, the current rating of semiconductor and magnetic devices can be reduced. Moreover, the output capacitor voltage ripple can be also reduced. For validity of proposed scheme, simulation and experimental results are presented.

II.   OPERATIONAL PRINCIPLE

Theoretically reference current follows line voltage waveform in conventional PFC circuit with UC3854 controller.

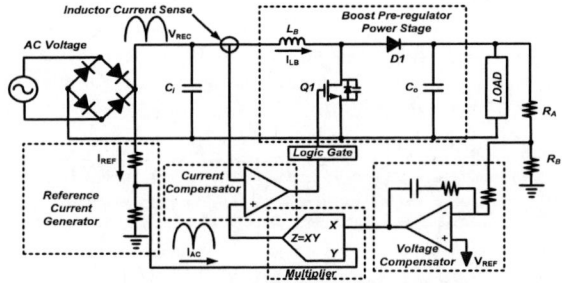

Fig. 4 Block Diagram Of UC3854 Controller

Proposed control technique is to make a reference current I_{REF} which clamped and decreased peak m× $I_{REF, PEAK}$ by using a zener diode. We define m as scaling factor of peak value of I_{REF} and also define t as scaling factor of I_{IN}. Of course, I_{IN} may follow I_{REF} waveforms very well by UC3854.

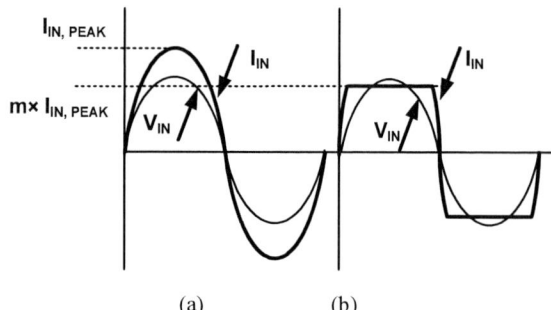

(a)　　　　(b)

Fig. 5 Line Current and Voltage Waveforms,
(a) Conventional Circuit,
(b) Proposed Circuit

Generally boost PFC circuit has multiplier, voltage compensator, current compensator and reference current generator. This multiplier creates the current programming signal I_{AC} by multiplying the rectified line voltage V_{REC} with the output of the voltage error amplifier so that the current programming signal I_{AC} has the shape of the input voltage V_{IN} and average amplitude which controls the output voltage V_{OUT}.

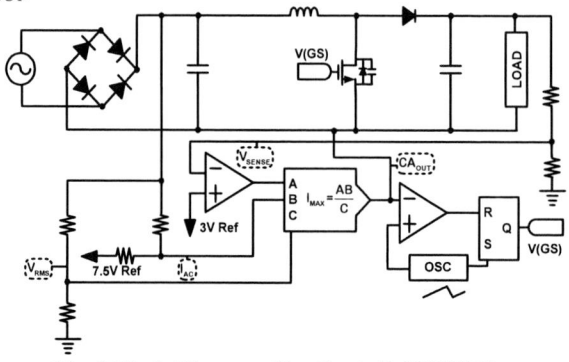

Fig. 6 Block Diagram of I_{AC} Controlled UC3854

We especially replace the reference current generator to get decreased current peak. It shows reference current I_{REF} is induced from V_{REC} and then it flows to I_{AC} of UC3854. A zener diode clamps V_{REC} and this clamped current flows to I_{AC} in proposed circuit.

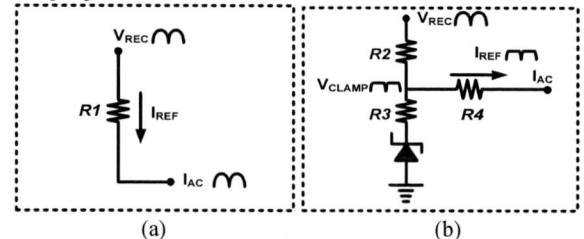

(a)　　　　(b)

Fig. 7 Reference current generator circuit,
(a) Conventional circuit,
(b) Proposed circuit

These clamped sine current waveforms are different from pure sine waveforms. Theoretically when value k is smaller, the harmonic currents become larger. Correlation of both k and m according to duration time and simulation values are compared with the IEC limiting values.

$$m = \frac{\pi \times k}{2(t - \frac{\sin 2t}{2} + 2k\cos t)}$$

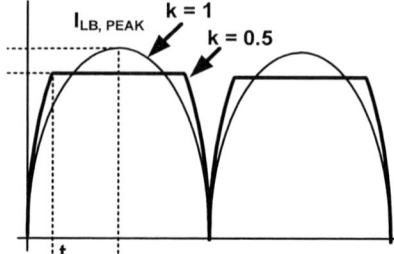

Fig. 8 Correlation of Value K and Waveform

Simulation proves that proposed controller is acceptable for harmonic current limits at k=0.55 and has reduced peak current in case of 200W, 90 V_{AC} ~ 270 V_{AC} universal line input.

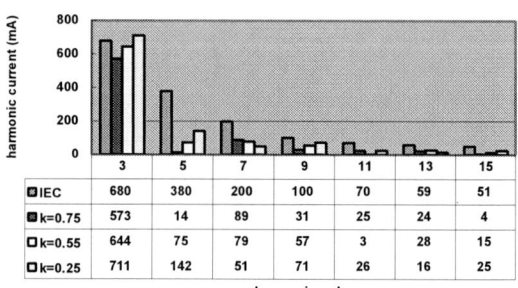

	3	5	7	9	11	13	15
IEC	680	380	200	100	70	59	51
k=0.75	573	14	89	31	25	24	4
k=0.55	644	75	79	57	3	28	15
k=0.25	711	142	51	71	26	16	25

harmonic order

Fig. 9 Harmonic currents comparison according to k,

III. EXPERIMENTAL RESULTS

Experiment of 200 W, 90 V_{AC} ~ 270 V_{AC} PFC circuit is implemented. Conventional, proposed circuits are implemented by the same system specification and circuit parameters. At first, line input voltage V_{IN} values 90 VAC ~ 270 VAC. Output Voltage V_O is 400[V] and Maximum output power $P_{O,MAX}$ is 200[W]. Switching frequency of UC3854 Fs is 100[kHz]. Next, Inductor L_b is designed according to inductor ripple as 600[u H]. Therefore this is satisfied with CCM condition at the condition of low line voltage 90[V]. And used MOSFET switch is IRFP460P with $R_{DS,ON}$ = 0.27[Ω](500[V], 20[A]).
Bridge Diode D25XB60 with V_F = 1.05[V] (600[V], 25[A]) and Boost Diode 15ETL06 with V_F=0.99[V](600[V], 5[A]) are used for experiment.
Output capacitor which valued 450[V], 20[u F] is used.
Conventional line current waveforms are experimented and follow the line voltage well as CCM boost PFC.

978-1-4244-1871-8/07 $25.00 © 2007 IEEE

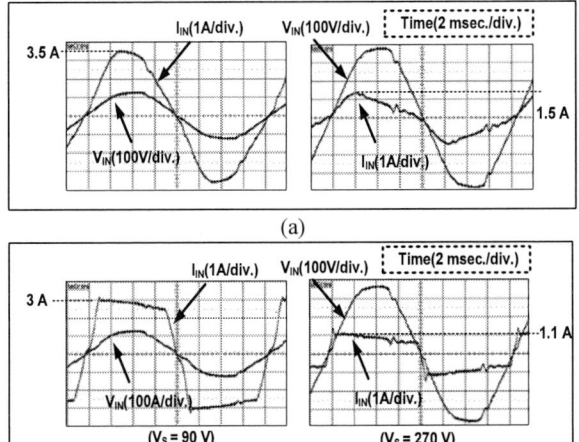

(a)

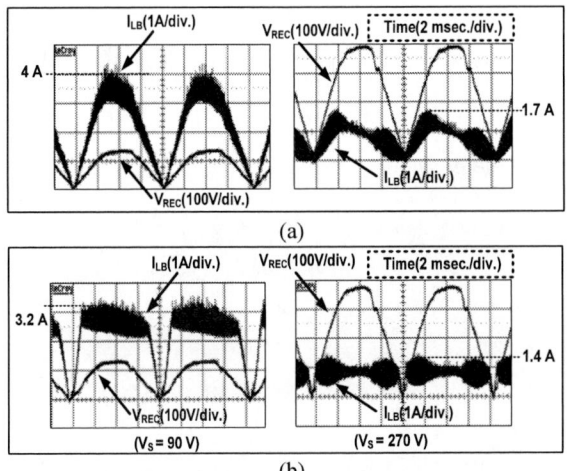

(b)

Fig. 10 Line Current and Voltage Waveforms
(a) Conventional Circuit
(b) Proposed Circuit

Experiment of 200 W, 90 V_{AC} ~ 270 V_{AC} PFC circuit is implemented. It shows conventional and proposed waveforms, and it shows line current I_{IN} peak value is improved by proposed controller. Data of Both 90 V_{AC} and 270 V_{AC} have same experimental results concerning as peak decreasing. Line input current I_{IN} decrease 14% from 3.5 A to 3 A at 90 V_{AC}.

(a)

	I_{IN}	$I_{INDUCTOR}$	V_{RIPPLE}
CONV	3.5	4	9
PROP	3	3.2	6
DIFF	0.5	0.8	3

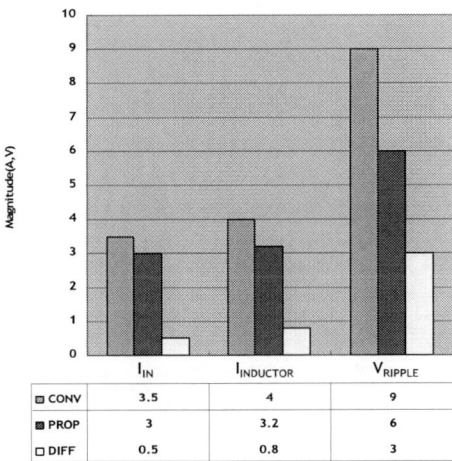

(b)

	I_{IN}	$I_{INDUCTOR}$	V_{RIPPLE}
CONV	1.5	1.7	8
PROP	1.1	1.4	6.4
DIFF	0.4	0.3	1.6

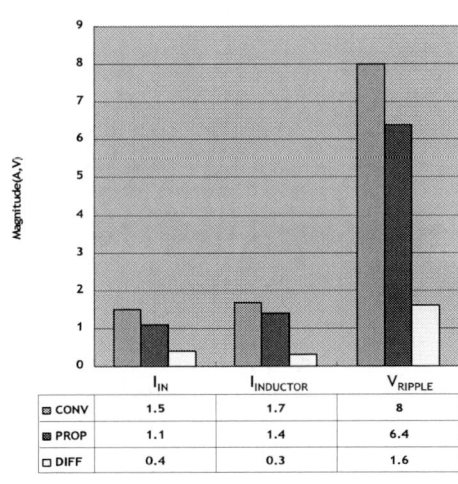

(a)

(b)

Fig. 9 Inductor Current and Rectified Voltage Waveforms
(a) Conventional Circuit
(b) Proposed Circuit

The boost inductor current I_{LB} decreases 20% from 4 A to 3.2 A at 90 V_{AC} in Fig. 6 like the line current I_{IN} does.

Moreover, output capacitor can be minimized due to voltage ripple decrease 20% from 8 V to 6.4 V at 270 V_{AC}. We can tell controller is superior to conventional controller in case of current peak reducing and output voltage ripple.

Fig. 16 Amplitude comparison of Current peak
(a) V_S = 90[V]
(b) V_S = 270[V]

Next, harmonic currents are compared both of controllers with the IEC regulation values. Even though the proposed controller increases harmonics more or less, it satisfies the IEC Std. well Both 90 V_{AC} and 270 V_{AC}.

978-1-4244-1871-8/07 $25.00 © 2007 IEEE   1003

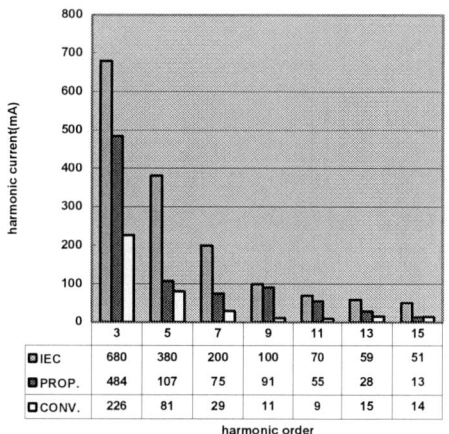

harmonic order	3	5	7	9	11	13	15
☐ IEC	680	380	200	100	70	59	51
☐ PROP.	484	107	75	91	55	28	13
☐ CONV.	226	81	29	11	9	15	14

(a)

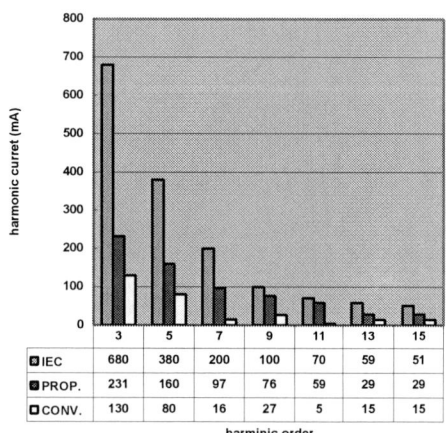

harminic order	3	5	7	9	11	13	15
☐ IEC	680	380	200	100	70	59	51
☐ PROP.	231	160	97	76	59	29	29
☐ CONV.	130	80	16	27	5	15	15

(b)

Fig. 17 Harmonic Current comparison
(a) $V_S = 90[V]$
(b) $V_S = 270[V]$

IV. CONCLUSIONS

For better PFC boost converter, this digest shows the new technique to decrease the current peak and voltage ripple. Decreasing peak currents of the boost pre-regulator influences on reducing circuit current stress and then rated

current of circuit elements can be minimized. The practical guideline is suggested for application for high power consumption such as 800[W]-1[k W].

REFERENCES

[1] Jin-Sheng Lai, " Design Consideration for Power Factor Correction Boost Converter Operating at the Boundary of Continuous Conduction Mode and Discontinuous Conduction Mode", IEEE 1993, pp 267-273.

[2] Miaoson Shen, " A unified average model for single stage and two stage power factor correction converters", IEEE 2001, pp 125-129.

[3] A.Pridenzi, " IEC Std. 61000-3-2 Harmonic Current Emission Limits in Practical System: Need of Considering Loading Level and Attenuation Effects", IEEE 2001, pp 27-282.

Novel Three Phase Flux Reversal Machine with Full Pitch Winding

D.S.More*, and B.G.Fernandes**

*Research Scholar, Electrical Engineering Department, Indian Institute of Technology Bombay.
**Department of Electrical Engineering, Indian Institute of Technology Bombay,
Powai, Mumbai - 400 076, INDIA.
Phone : +91-022-25767428, Fax : +91-022-25723707
Email : dsmore@ee.iitb.ac.in
Email : bgf@ee.iitb.ac.in

Abstract— This paper proposes a concept of fictitious 'Electrical Gear' in analyzing the flux pattern of Flux Reversal Machine (FRM). Concept of novel full pitch winding is presented for FRM. Finite Element Method (FEM) analysis of conventional concentrated stator pole winding FRM and proposed full pitch winding FRM are carried out to validate the concept. Generator performance of full pitch winding FRM is compared with conventional concentrated stator pole winding FRM. Full pitch winding arrangement increases the inductance of the machine which results in poor voltage regulation. Series capacitive compensation is provided to the machine for better voltage regulation. Output power of compensated full pitch winding FRM is approximately twice as that of the conventional concentrated stator pole winding FRM. Physical dimensions, electrical and magnetic loadings are kept same in both the machines.

Index Terms— Flux reversal machine, full pitch winding, doubly salient permanent magnet electrical machine, finite-element method (FEM) analysis.

I. INTRODUCTION

SINGLE phase FRM was first introduced in 1997 by R. P. Deodhar et al [1] for an automobile generator to replace the standard claw pole alternator. It has numerous advantages such as simple construction, low inertia, high power density and is suitable for high speed application due to permanent magnets on the stator pole. This single phase configuration was fully explored as a generator and not as a motor. Three phase FRM was introduced by C. Wang et al [2] in 1999. The design of the machine was optimized to (i) increase the peak to peak coil PM flux linkage variation, (ii) reduce the cogging torque and PM weight, (iii) restrain the self and mutual inductance. The basic machine configuration has an 8 salient pole rotor and 6 pole stator with concentrated windings. Permanent Magnets are mounted on stator pole. Fig. 1 shows the machine configuration.

FRM for low-speed servo drive application was introduced by Ion Boldea et al [3] in 2002. This low speed machine has 28 rotor poles and 12 stator poles with two permanent magnet pairs on each stator pole. This machine was designed for 128 rpm at 60 Hz. High torque density with less than 3% torque pulsation with vector control was achieved. In order to reduce the cogging torque, rotor teeth pairing method was proposed [4]. Attempts were made to reduce the leakage flux by providing flux barrier on the rotor poles at its edges

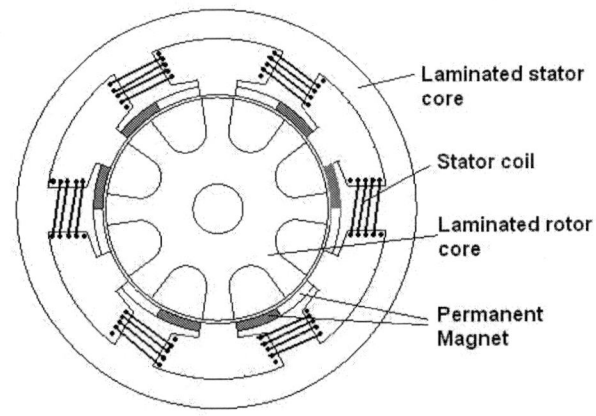

Fig. 1. Cross-section of conventional concentrated stator pole winding FRM.

[5]. Power density comparison of doubly salient permanent magnet electrical machines were made. FRM has higher power density in comparison with other machines in the same class [6].

In this paper, concept of fictitious 'Electrical Gear' and full pitch winding applicable to FRM is presented. 2-D FEM analysis [7] is carried out on the conventional concentrated stator pole winding FRM and proposed full pitch winding FRM. Optimized machine dimensions are obtained from C. Wang et al [2]. The important dimensions of FRM are given in Table I for ready reference.

Section II describes the flux pattern linking to the stator winding of the machine and from there the concept of fictitious 'Electrical Gear' and full pitch stator winding arrangement is discussed. Section III describes the 2-D FEM analysis results obtained for full pitch FRM and concentrated stator pole winding FRM. The results are compared with the results of paper [2]. Section IV describes the FEM analysis on load to find the voltage regulation of full pitch winding FRM. Effects on machine parameters due to full pitch winding and method used to improve the voltage regulation is also discussed. Section V compares the power density of full pitch winding FRM with PMSM and finally conclusions are drawn.

978-1-4244-1871-8/07 $25.00 © 2007 IEEE

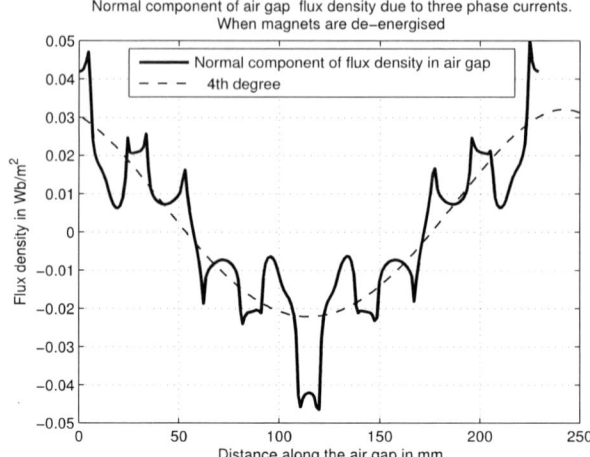

Fig. 4. Normal component of armature flux density along the air gap at full load with magnets de-energized

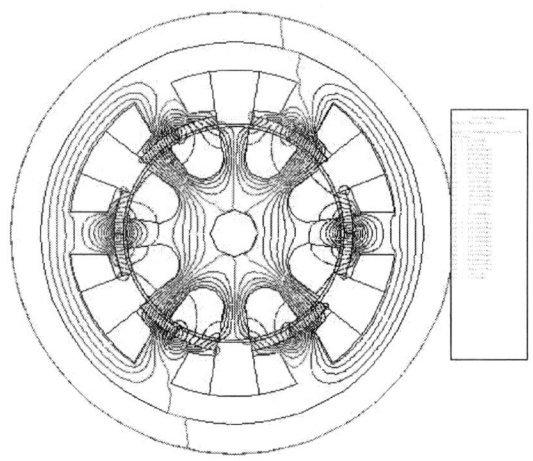

Fig. 2. Flux plot of machine at no load

II. FLUX PATTERN LINKING TO THE STATOR WINDING AND FULL PITCH STATOR WINDING

Geometry of three phase flux reversal machine (as per Table I) is shown in Fig. 2. The flux plot of the machine at no load is shown in the same Fig. FRM machine has 6 stator poles and 8 pole variable reluctance rotor. The normal component of flux density at the middle of stator pole along the periphery of the machine is shown in Fig. 3. The observation of this normal component of flux density plot reveals that the machine has two pole flux pattern. In other words machine has two effective poles.

Phase flux linkage of FRM is sinusoidal and hence induced voltage is also sinusoidal [2]. Considering a linear load the phase current is sinusoidal. Normal component of armature reaction along the air gap at one instant of time is shown in Fig. 4. This flux pattern also reveals that machine has two poles. In general, machine has same number of stator poles

and rotor poles.

The frequency and speed relationship for FRM [3] is given by

$$f = \frac{n \times n_r}{60} \quad (1)$$

where, n = rotor speed in rpm.
n_r = Number of rotor teeth (poles).
f = frequency in Hz.
FRM machine has two effective poles, and hence in one cycle, flux pattern completes one revolution. Thus when rotor speed is n rpm, the flux pattern completes $n_r \times n$ revolutions per minute. Hence flux pattern rotates n_r times the shaft speed. In conventional synchronous machine, flux pattern speed and rotor speed is same. Pictorial representation of FRM generator and permanent magnet synchronous generator is shown in Fig. 5. Both machine representation is for same speed and output frequency. Flux pattern in FRM rotating n_r times rotor speed is represented as a fictitious step-up gear and is called as 'Electrical Gear'. Conventional synchronous machine should have $2 \times n_r$ poles for same output frequency.

FRM has 6 slots and two pole flux pattern, hence electrical angle per slot is 60°. Conventional concentrated stator pole

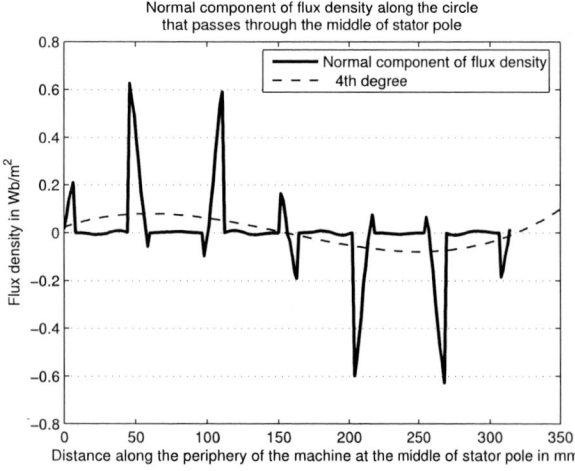

Fig. 3. Normal component of flux density

TABLE I
DIMENSIONS OF FRM

Sr. No.	Description	Symbol	Value
1	Air gap in mm	g	1
2	Magnet thickness in mm	h_{pm}	3
3	Rotor pole span angle	β_r	16.2°
4	Stator pole span angle	β_s	42.6°
5	Stator pole span in mm	τ_{ps}	27.8
6	Rotor pole span in mm	τ_{pr}	10.3
7	Stator pole height in mm	h_{ps}	15
8	Rotor pole height in mm	h_{pr}	18
9	Outer diameter of rotor in mm	D_i	72
10	Outer diameter of stator in mm	D_o	129
11	Number of turns /phase	N_{ph}	52
12	Stack length of machine in mm	l_{sk}	86

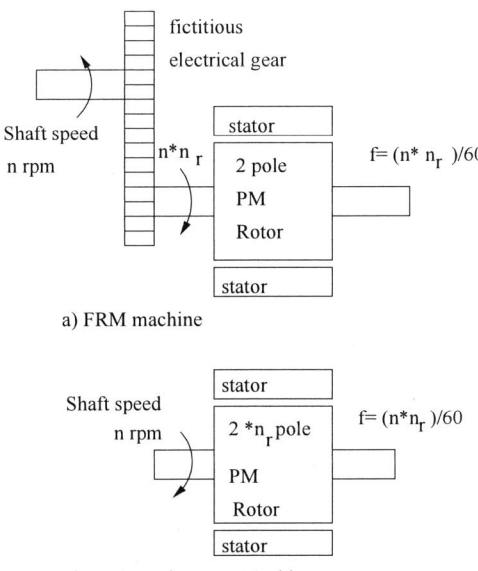

a) FRM machine

b) PM synchronous Machine

Fig. 5.   Generator representation of FRM and PMSM

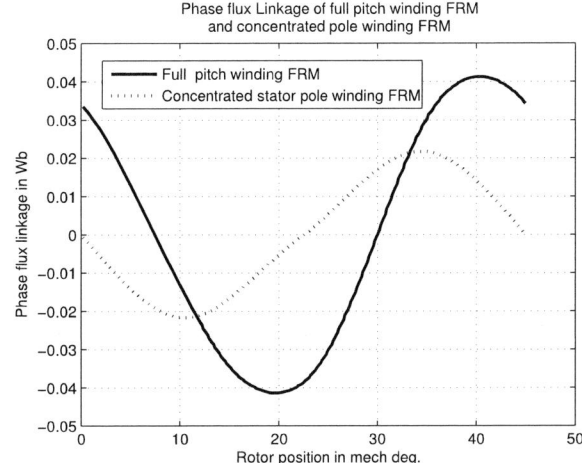

Fig. 7.   Phase flux linkage of full pitch winding FRM and concentrated stator pole winding FRM

winding has a coil span of $60°$. Fundamental pitch factor of the stator winding is 0.5. As electrical angle between the slots is $60°$, full pitch winding is possible. The arrangement of full pitch winding is shown in Fig. 6, which results in unity pitch factor.

III. FEM ANALYSIS OF FULL PITCH WINDING FRM AT NO LOAD

FRM machine design data is obtained from [2] and given in the Table I for ready reference. Physical dimensions of the machine and number of turns are kept same in full pitch winding FRM and conventional concentrated stator pole winding FRM. Only the winding arrangement is changed. FEM analysis is carried out to find the phase flux linkage variation of both the machines with rotor position. Phase flux linkage variation of both the machines without skewed rotor is shown in Fig. 7. This variation is shown for one rotor pole pitch (i.e. mechanical $45°$). Fig. 7 clearly shows that full pitch winding FRM stator flux linkage is approximately twice

as that of the flux linkage of conventional concentrated stator pole winding FRM.

Simulations are carried out at different speeds on both machines to determine the induced voltage on no-load. Induced voltage at different speed for both machines without skewed rotor is shown in Fig. 8. The simulated result shows that no load induced voltage in full pitch winding FRM is approximately twice as that of the voltage induced in the conventional concentrated stator pole winding FRM. No-load phase voltage waveforms of both configurations at 8000 rpm are shown in Fig. 9.

Stator phase inductance is calculated using FEM as per procedure given in [2]. First magnet flux linking the stator winding is obtained and then positive mmf to the winding, total flux linkage to the winding is obtained. The self inductance is obtained by subtracting the PM flux linkage from total flux linkage. The same procedure is used to determine the self inductance when negative stator mmf is applied. The

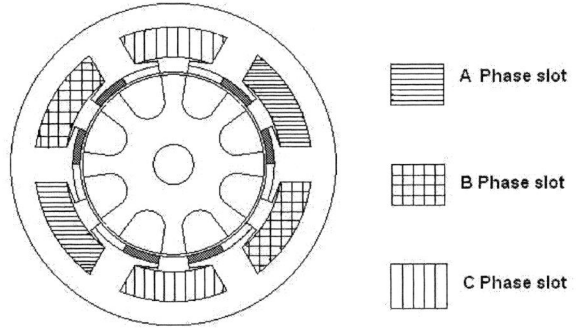

Fig. 6.   Full pitch winding arrangement

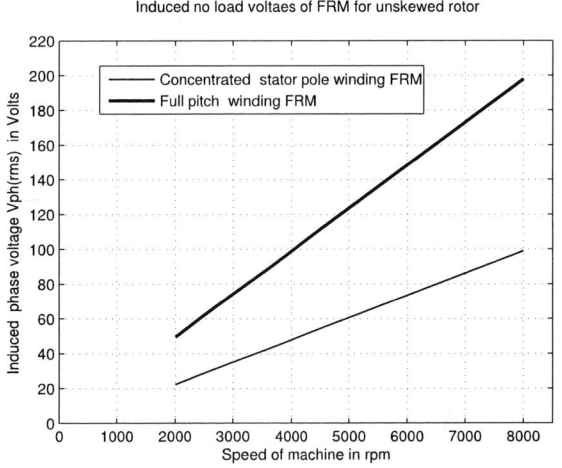

Fig. 8.   No load phase voltages of full pitch winding FRM and conventional concentrated stator pole winding FRM

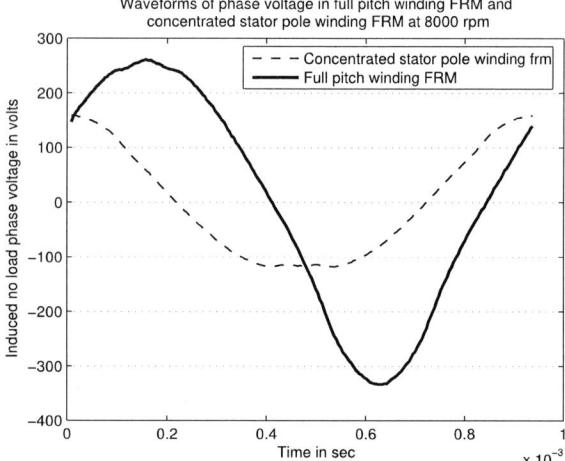

Fig. 9.   No load phase voltage waveform at 8000 rpm

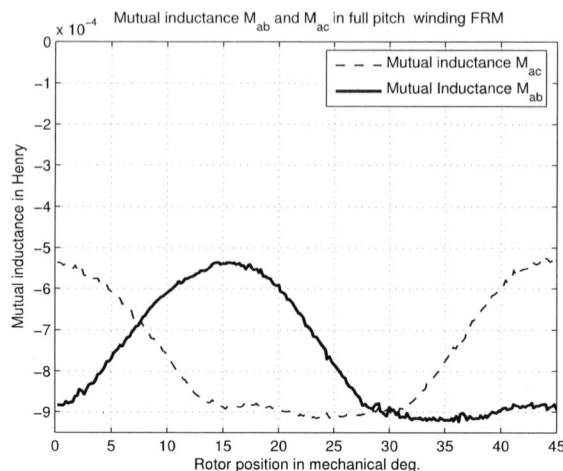

Fig. 11.   Variation of full pitch winding FRM mutual inductance with rotor position.

average of these two self inductances represents the actual value of self inductance. The stator phase inductance variation for both machines with rotor position is shown in Fig. 10. Self inductance variation of full pitch winding FRM is 2.42 mH to 2.52 mH, whereas this variation for conventional concentrated stator pole winding FRM is 0.78 mH to 1.1 mH. Full pitch winding arrangement increases the self inductance of winding by 2.5 to 3 times depending upon the saturation in the machine. If saturation is neglected, then self inductance increases by 3 times, whereas variation of self inductance with rotor position is very small in both cases.

Under normal operating conditions some parts of magnetic circuit gets saturated in the full pitch winding FRM. Mutual inductance variation is 0.54 mH to 0.91 mH. Under unsaturated condition, this variation is from 0.835 mH to 0.93 mH. Mutual inductance of full pitch winding FRM is approximately 0.33 times that of its self inductance. Mutual inductance variation with rotor position is shown in Fig. 11.

IV.   FEM ANALYSIS OF FULL PITCH WINDING FRM AT STEADY STATE ON LOAD

FEM simulation is carried at different load condition to determine the voltage regulation of machine at 2000 rpm for both types of winding configuration. Terminal voltage variation with load current is shown in Fig. 12. It is seen that regulation of full pitch winding FRM is poor as compared to concentrated stator pole winding FRM.

Regulation of full pitch FRM can be improved by using series capacitive compensation [9]. The arrangement for capacitive compensation is shown in the Fig. 14. The value of capacitor is given by [8].

$$C = \frac{1}{\omega^2 \times L_s} \qquad (2)$$

where, $\omega=$ Induced EMF frequency in rad/sec.
$L_s =$ synchronous inductance of the machine in Henry.

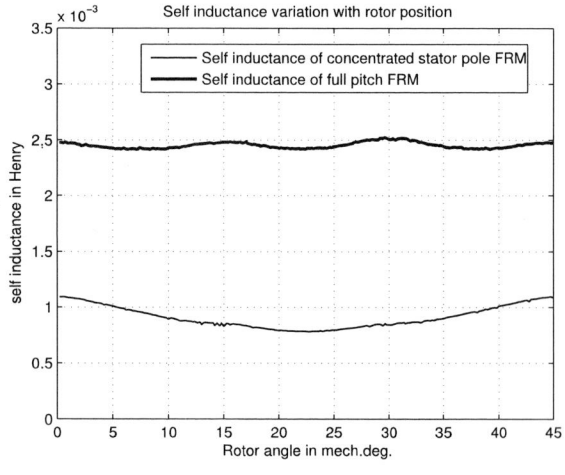

Fig. 10.   Variation of Self inductance of full pitch winding FRM and concentrated stator pole winding FRM with rotor position

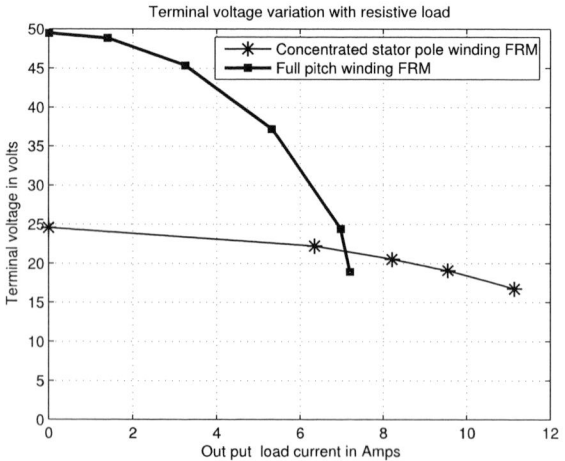

Fig. 12.   Terminal voltage variation with resistive load

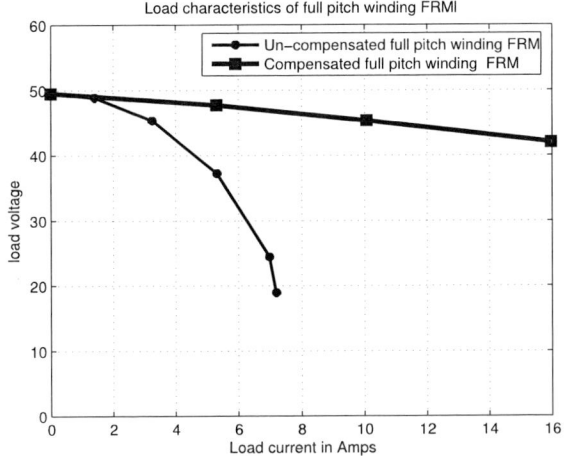

Fig. 13. Terminal voltage variation with resistive load for full pitch winding FRM

Fig. 15. Flux distribution in FRM and PMSM for same output voltage and frequency at common shaft speed.

C = capacitance for series compensation in Farad.

Synchronous inductance L_s is obtained from regulation results of uncompensated full pitch FRM shown in Fig. 13. The value of L_s is 3.76 mH. Series compensation capacitor of $94.7\mu F$ is chosen for a speed of 2000 rpm. FEM simulation of compensated full pitch FRM is carried out at different loads to access the regulation of the machine. Terminal voltage variation at different loads is obtained. Terminal voltage variation for compensated and uncompensated full pitch FRM is shown in Fig. 13. Compensated FRM gives better regulation which is less than 15%. Obseving the external (load) characteristics from Fig. 12 and Fig. 13, it can be concluded that compensated full pitch winding FRM has approximately double output power as compared to conventional concentrated stator pole winding FRM for same electrical and magnetic loading.

V. POWER DENSITY OF FULL PITCH FRM

The flux linkage in full pitch winding FRM is shown in the Fig. 7. From this Fig. it can be observed that maximum phase flux linkage is 0.041 Wb. The relationship between maximum flux linkage $\psi_1(max)$ and maximum flux density $Bm_1(max)$ is given by [11].

$$\psi_1(max) = \frac{Bm_1(max) \times l_s \times \tau \times 2 \times N_{ph}}{\pi} \quad (3)$$

where, l_s = stack length of the machine in meter.

τ = pole pitch distance in meter.

N_{ph} = number of turns per phase.

$\tau = 0.014137$ m for equivalent PMSM. (Euivalent PMSM has 16 poles.)

$\tau = 0.113097$ m for full pitch FRM.(Full pitch FRM has effective 2 poles.)

Using the above equation $Bm_1(max)$ in FPFRM is 0.127 Wb/m^2, whereas $Bm_1(max)$ in PMSM is 1.0186 Wb/m^2. Flux density distribution is assumed to be sinusoidal as flux linkage is sinusoidal. Calculated flux density distribution along the air gap for both machines is shown in the Fig. 15.

Flux density distribution pattern of FRM rotates at $n \times n_r$ rpm, whereas flux density distribution pattern of PMSM rotates at n rpm for a common shaft speed of n rpm. Both flux pattern generates same output voltage and frequency. Hence equivalent conventional PMSM should have a flux density of 1.0186 Wb/m^2 for the same output power as full pitch winding FRM. Maximum possible air gap flux

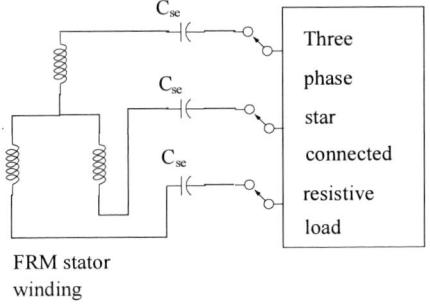

Fig. 14. Circuit diagram for series compensation

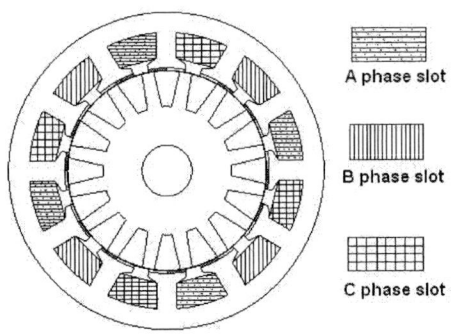

Fig. 16. Full pitch winding arrangement for low speed FRM with 12 stator poles and 16 rotor poles

density in PMSM is $0.9\ Wb/m^2$ [10]. Hence full pitch FRM (equivalent PMSM) has 13.17% higher power density than conventional PMSM in low speed, low power range.

VI. FRM CONFIGURATIONS AND FULL PITCH WINDING

FRM machine configurations for low speed are presented by Ion and et.al. [3]. The configurations are

- low speed FRM with pole PMs on stator.
- low speed FRM with inset-PMs on stator.

These configurations have 12 stator poles (slots) and flux pattern of 4 poles. Full pitch winding arrangement is also possible for both configurations. Full pitch winding arrangement for low speed FRM machine with 16 rotor poles and 12 stator poles with 2 PM pole pieces per stator pole is shown in Fig. 16. End turn length will be less in higher pole machine which decreases the end turn leakage inductance and copper loss. Hence full pitch winding concept is applicable to all FRM configurations.

VII. CONCLUSION

Full pitch winding concept for FRM is introduced. Output power of compensated full pitch winding FRM is approximately twice as that of conventional concentrated stator pole winding FRM. Full pitch winding FRM is fully analyzed and simulated with flux-2D FEM software [7]. Full pitch winding arrangement increases the self inductance of machine by 2.5 to 3 times depending upon saturation in the machine which results in poor voltage regulation. Series capacitive compensation is provided to improve the regulation of full pitch winding FRM. Fictitious 'Electrical Gear' concept is introduced which makes it possible to compare the power density of full pitch winding FRM with PMSM. It is found that power density of compensated full pitch winding FRM is 13.75% higher than conventional PMSM. Full pitch winding FRM is a competitive candidate for low speed, low power generator and motor applications.

VIII. ACKNOWLEDGMENT

D. S. More is faculty member in Electrical Engineering Department of Walchand College of Engineering Sangli, 416 415, (India). The financial assistance provided by Walchand College of Engineering Sangli for registration and travel is highly acknowledged.

REFERENCES

[1] R. P. Deodhar, Savante Anderson, Ion Boldea and T. J. E. Miller, "The flux reversal machine : A new doubly salient permanent magnet machine," *IEEE Trans. Industry Applications.*, vol. 33, No. 4, pp. 925-934, July/August 1997.

[2] C. Wang, S. A. Nasar, I. Boldea "Three phase flux reversal machine (FRM)," *IEE Trans. Electrical power application .*, vol. 146, No. 2, pp. 139-146, March 1999.

[3] Ion Boldea, Jichum Zhang, S. A. Nasar, "Theoretical characterization of flux reversal machine in low speed servo drives-The pole PM configuration." *IEEE Trans. Industry Applications.*, vol. 38, No. 6, pp. 1549-1557, November/December 2002.

[4] Tae Heoung Kim, Sung Hong Won, Ki Bong and Ju Lee, "Reduction in cogging torque in flux reversal machine by rotor teeth pairing." *IEEE Trans. on Magnetics.* vol. 41, No. 10, pp 3964-3966, october 2005.

[5] Tae Heoung Kim and Ju Lee "A study of the design for the flux reversal machine," *IEEE Trans. on Magnetics.*, vol. 40, No. 4, pp. 2053-2055, July 2004

[6] Jianzhong zhang, Ming Cheng, Wei Hua and Xiaoyong Zhu, "New approach to power equation for comparison of doubly salient electrical machines," in *in Proc.IEEE Industry Applications Annu. meeting .*, pp. 1178-1185, 2006.

[7] "Flux 2-D FEM Software, CEDRAT, France".

[8] C. Wang, I. Boldea, S. A. Nasar "Characterization of three-phase flux reversal machine as an automotive generator. ," *IEEE Trans. on Energy Conversion.*, vol. 16, No. 1, pp. 74-80, March, 2001.

[9] Nobuyuki Naoe "Voltage compensation of permanent Magnet generator with capacitors". *Proc. of IEEE electrical Machines and Drives .*, pp. wb2.14.1- wb2.14.3, 1997.

[10] Miller T.J.E. "Brushless Permanent Magnet and Reluctance Motor Drives. Clarendon Press. Oxford-1989 .

[11] Gieras J. F. and M. Wing " Permanent Magnet Motor Technology. Design and Applications. Marcel Dekker Inc. 2002 .

Comparison of Winding Concepts for Bearingless Pumps

K. Raggl, J. W. Kolar
Power Electronic Systems Laboratory
ETH Zurich
8092 Zurich, Switzerland
Email: raggl@lem.ee.ethz.ch

T. Nussbaumer
Levitronix GmbH
Technoparkstrasse 1
8005 Zurich, Switzerland
Email: nussbaumer@levitronix.com

Abstract— Bearingless slice motors as employed in semiconductor, pharmaceutical and medical industry perform magnetic levitation by radial bearing forces and rotation by tangential forces. This requires bearing and drive windings, which can be realized as separate bearing and drive coils or as identical, concentrated coils on the stator claws. In this paper, a detailed comparison between these two winding concepts is undertaken, where the coil losses, the coil volume, the power electronics requirements and the achievable rotation speed are evaluated. Furthermore, practical features such as control complexity, cabling effort and manufacturability are taken into account. Finally, the trade-off between losses, volume and realization effort will be discussed in order to give a guideline for the selection of the appropriate winding concept for a specific application of a bearingless pump.

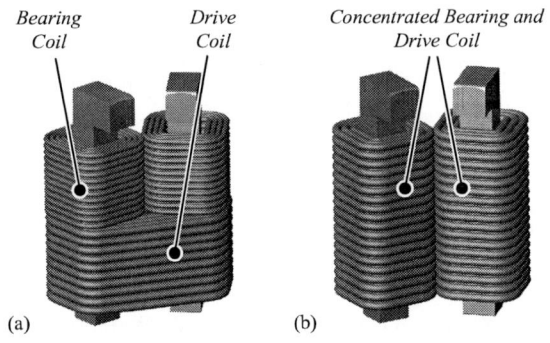

Fig. 1: Basic setup of a bearingless slice motor for a centrifugal pump with a diametrically magnetized rotor

I. INTRODUCTION

Bearingless motors have been extensively investigated during the past years and have been employed successfully in high purity environments (e.g. in semiconductor, pharmaceutical and medical industry e.g. as pumps or mixers) due to their great variety of benefits such as non–contact bearing capability and the lack of mechanical wearing, lubricants and seals [1], [2], [3].

However, new applications require a further optimization of the pump systems, i.e. the maximization of the hydraulic pump pressure while minimizing the pump volume, the power losses and the costs of the pump system [4]. Within this optimization, a crucial element that determines the performance of the pump system is the motor part. Here, the required bearing and drive forces can be generated for the same mechanical motor setup by different winding concepts of the bearing and drive coils. As will be shown in this paper, this may have an influence on the resulting losses, the coil volume and the pump performance.

The comparison is based on a pump system with a diametrically, one pole pair rotor (cf. Fig. 1). The optimum pole pair number for the drive system ($p_d = 1$) and the bearing system ($p_b = 2$) for this setup has been discussed extensively in literature [5], [6], [7], [8]. Furthermore, it has been found that for achieving full control freedom for the drive and bearing system at least eight stator claws are needed [5]. With this, the motor setup as shown in Fig. 1 is defined for the comparison in this paper.

Fig. 2 shows the two winding concepts that will be analysed in the following (two of the eight stator claws are shown). For the conventional setup (cf. Fig. 2(a)) two separate coils are used to impress the bearing and the drive forces separately [1], [2], [3] where each bearing coil is wound over a claw and each drive coil is wound over two claws.

Fig. 2: Winding concepts for the bearingless pump — separated coils (a) and concentrated coils (b) (detailed view on two of the eight stator claws)

An alternative way is to apply only one coil on each claw (cf. Fig. 2(b)) as it is done eg. in 1–phase bearingless motor configurations [9], [10]. This concepts results in easier assembly and therefore lower price but on the other hand in a bigger control and power electronics effort since each current has to be calculated and impressed in each coil separately.

The discussion of these two winding concepts is based on the following assumptions:

Same motor setup (iron circuit, rotor magnet size and magnetization)
Same required bearing force and torque for both concepts
Same max. allowable current density
Same winding factor

After a short explication of the generation of the bearing and drive forces in section II, in the subsequent sections the

978-1-4244-1871-8/07 $25.00 © 2007 IEEE

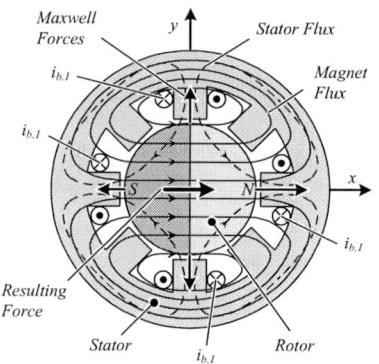

Fig. 3: Maxwell Force generation in magnet direction

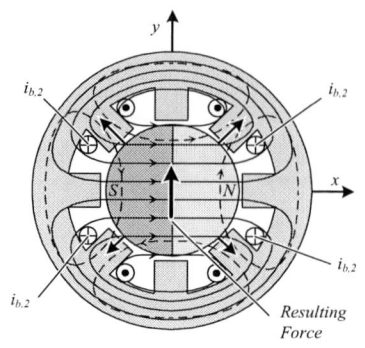

Fig. 4: Maxwell Force generation perpendicular to magnet direction

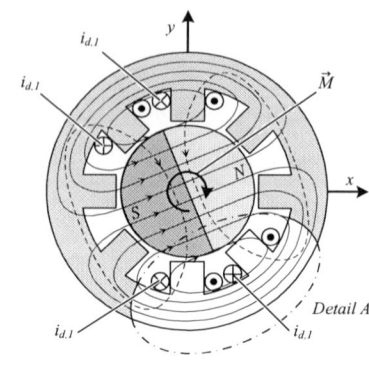

Fig. 5: Torque generation with one phase — the second phase is rotated by 90 deg

detailed comparison is performed, i.e.:

  Section III: Copper losses
  Section IV: Coil volume
  Section V: Power electronics requirements
  Section VI: Maximum achievable rotation speed

Finally, experimental measurements on an existing pump system with separated coils verify the correctness of the considerations in section VII.

II. FORCE GENERATION

In this section, the generation of the bearing and drive forces is explained for the two winding concepts based on the motor setup at hand (cf. Fig. 1). The fundamentals are essential for the understanding of the comparison in the subsequent sections.

The principle of the bearing force generation is shown in Fig. 3 for resulting forces in magnet direction and in Fig. 4 for resulting forces perpendicular to magnet direction. For sake of clearness, only Maxwell–forces are shown in these figures. The magnetized rotor is impressing a magnetic field into the stator (solid lines in Fig. 3) directed from the magnet north pole through the stator to the magnet south pole. By applying a current $i_{b\,1}$ in the shown coils a superposed magnetic field (dashed lines) is generated. This field, impressed by the coils, leads to an attenuation of the permanent magnet field in the left air gap and a reinforcement of the magnet field in the right air gap. According to the Maxwell–equations these two magnetic fields causes forces on the magnet surface directed from the medium with higher permeability to the medium with lower permeability, which is proportional to the air gap field $\vec{F} \propto \vec{B}$. With the shown current direction of i_{b1} in Fig. 3 a resulting force in positive x–direction is built up.

Additionally, forces based on the Lorentz–equations appear in this setup [5], [11]. As generally known, the force on a conductor, in which a current is flowing and which is positioned in a magnetic field, is given by $\vec{F} = i \cdot (\vec{l} \times \vec{B})$. The appearing Lorentz–forces are superposed to the Maxwell–forces and directed in the same direction.

Fig. 4 shows the bearing force generation perpendicular to the magnet direction. In this case, Maxwell– and Lorentz–forces are superposed again and with currents i_b flowing in

the direction shown in Fig. 4 a resulting force in positive y–direction is built up.

With this application, it is possible to generate forces in x– and y–direction independent of the angular rotor position.

The torque generation due to one drive phase is depicted in Fig. 5. The impressed current $i_{d\,1}$ in the shown coils leads to a magnetic field orthographic to the magnetic rotor field for this specific angular rotor position. This leads to a resulting torque on the rotor as shown in Fig. 5. By applying a sinusoidal current i_d with a phase shift of 90 with respect to $i_{d\,1}$ on the four remaining stator claws in an analogous manner a constant torque can be built up.

A. Separated Coils

Since the same currents (e.g. $i_{b\,1}$ in Fig. 3 and i_b in Fig. 4) are flowing through the shown coils, they can be directly connected in series. Therefore, only two bearing phases are needed to ensure levitation in x– and y–direction. Another speciality of that winding configuration is that the appearing induced voltage in the bearing phases is eliminated due to the symmetry. This leads to a higher available coil voltage and hence to an increased bearing dynamics.

The drive coils shown in Fig. 5 can be connected in series too, hence in total four independent full–bridges are needed to generate autonomous forces in x– and y–direction and a torque in z–direction for this coil setup. Since always two neighbored coils carry the same current, the drive windings can be wound over two claws, as it is shown in Fig. 2. As it

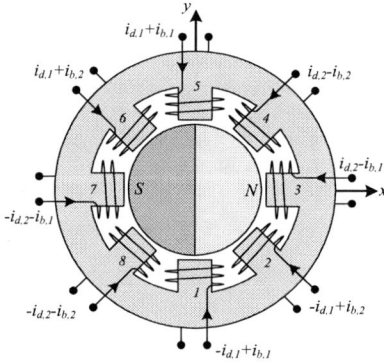

Fig. 6: Needed currents in each coil in case of concentrated coils.

978-1-4244-1871-8/07 $25.00 © 2007 IEEE

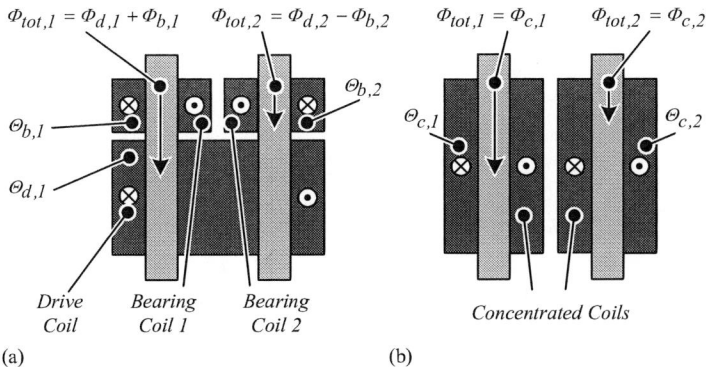
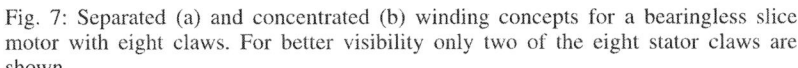

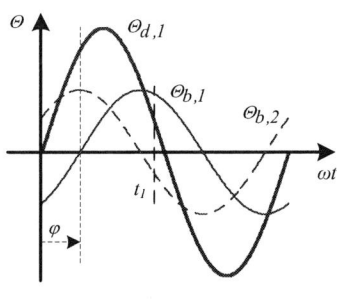

Fig. 7: Separated (a) and concentrated (b) winding concepts for a bearingless slice motor with eight claws. For better visibility only two of the eight stator claws are shown.

Fig. 8: General curves of ampere-turns in bearing and drive with a global force acting on the rotor

will be shown later (cf. section III-B) this slightly reduces the average winding length and therefore the copper losses of the separated coils.

B. Concentrated coils

Out of the Figures 3, 4 and 5 the needed currents per coil can be ascertained. The resulting situation is shown in Fig. 6. One can see easily that in case of concentrated coils a simplification by connecting two or more coils in series is not possible anymore due to a different current needed in each coil. Consequently, eight full–bridges are needed in case of concentrated coils to ensure the same performance. However, this does not necessarily mean a disadvantage. The power electronics effort will be evaluated separately in section V.

III. COPPER LOSSES

For the subsequent analysis of the copper losses in the bearing and drive coils we exemplary look at one stator claw pair (cf. *Detail A* in Fig. 5 and stator claws 1 and 2 in Fig. 6). The situation at the claws –8 can be analysed in an analogous way by symmetry considerations. In Fig. 7 the magnetic flux generation in these claws through the bearing and drive ampere–turns is illustrated for the case of separated (cf. Fig. 7(a)) and concentrated (cf. Fig. 7(b)) coils.

Due to the winding arrangement of the separated coils [1], the flux phase difference between two neighbored bearing coils (cf. Fig. 8) has to be 90 to ensure levitation [5]. As a matter of fact, the needed ampere–turns in the bearing have the same frequency as the drive ampere–turns. However, in case of a centrifugal pump, a steady asymmetric force is acting on the rotor directed to the hydraulic outlet of the pump due to a pressure loss at the outlet, which causes a load angle φ between the drive and bearing ampere–turns (cf. Fig 8). Exemplarily, we are looking now at a typical operation state at $ = _1$. Here, the impressed magnetic flux of the bearing coil is operating against the impressed flux of the drive coil in the right iron claw (Fig. 7(a)). Therefore, the resulting magnetic flux Φ_{tot} in that claw is reduced due to counteraction of the individual magnetic fluxes. Obviously, in the case of concentrated coils (Fig. 7(b)) the same resulting flux Φ_{tot} in the right claw can be built up by less impressed

current, wherefore lower copper losses occur in this setup for the same ampere–turns. This loss reduction will be quantified in the following.

A. General loss reduction

Generally, the copper losses can be written as

$$P_{cu} = R \cdot {}_{rms} \tag{1}$$

and can be transformed with $R = N \cdot \frac{\rho_{cu} l_m}{A_{cu}}$ and $\Theta_{rms} = N \cdot {}_{rms}$ to

$$P_{cu}(\Theta) = N \ \cdot \frac{\rho_{cu} l_m}{A_{tot} k_f} \cdot \frac{\Theta_{rms}}{N} = \frac{\rho_{cu} l_m}{A_{tot} k_f} \cdot \Theta_{rms} \tag{2}$$

dependent on the *rms* value of the impressed ampere–turns Θ_{rms}. In (2) k_f means the winding factor, A_{tot} the whole coil cross area, A_{cu} the wire cross area, ρ_{cu} the copper density, l_m the average winding length per claw and N the winding number. The wire cross area is given by

$$A_{cu} = \frac{A_{tot} \cdot k_f}{N}. \tag{3}$$

With this, the needed ampere–turns

$$\begin{matrix} \Theta_{b\,1} \\ \Theta_{b} \\ \Theta_{d\,1} \\ \Theta_{d} \end{matrix} = \begin{matrix} \Theta_{b} \cdot & (\omega & \varphi) \\ \Theta_{b} \cdot & (\omega & \varphi) \\ \Theta_{d} \cdot & (\omega &) \\ \Theta_{d} \cdot & (\omega &) \end{matrix} \tag{4}$$

and (2) the copper losses can be calculated for each claw and coil separately in case of separated coils. The total copper losses for all coils are given by

$$P_s = \ \cdot P_{cu}(\Theta_{b\,1}) + \ \cdot P_{cu}(\Theta_{b} \ ) + \ \cdot P_{cu}(\Theta_{d\,1}) + \ \cdot P_{cu}(\Theta_{d} \ ). \tag{5}$$

In (5) the same average winding length l_m for drive and bearing coils has been assumed, i.e. the drive windings are wound around the stator claws individually in order to simplify the calculations in the first instant. By winding the drive coils around two stator claws the average winding length can be reduced by a geometry dependent factor k_{lm}. The influence of this factor on the copper losses will be analysed separately in section III-B.

978-1-4244-1871-8/07 $25.00 © 2007 IEEE   1013

In case of concentrated coils the needed ampere–turns per claw can be written as (cf. Fig. 6)

$$
\begin{matrix}
\Theta_{c\,1} \\
\Theta_{c} \\
\Theta_{c} \\
\Theta_{c\,4} \\
\Theta_{c\,5} \\
\Theta_{c\,6} \\
\Theta_{c} \\
\Theta_{c\,8}
\end{matrix}
=
\begin{matrix}
\Theta_{b\,1} & \Theta_{d\,1} \\
\Theta_{b} & \Theta_{d\,1} \\
\Theta_{b\,1} + \Theta_{d} \\
\Theta_{b} + \Theta_{d} \\
\Theta_{b\,1} + \Theta_{d\,1} \\
\Theta_{b} + \Theta_{d\,1} \\
\Theta_{b\,1} & \Theta_{d} \\
\Theta_{b} & \Theta_{d}
\end{matrix}
. \qquad (6)
$$

The resulting copper losses are given by

$$
P_c = \sum_{i=1}^{8} P_{cu}(\Theta_{c\,i}) \qquad (7)
$$

with (2),(4) and (6). By calculating the copper losses P_s and P_c the loss reduction by implementing concentrated coils is derived by

$$
\frac{P_c}{P_s} = \frac{\Theta_b + \Theta_d}{\left(\Theta_b + \Theta_d\right)} \qquad (8)
$$

Equation (8) is valid under the assumption of the same current density for the separated bearing and drive coils

$$
J_{s\,rms} = \frac{\Theta_{b\,rms}}{A_b} = \frac{\Theta_{d\,rms}}{A_d} \qquad (9)
$$

and the assumption that the total coil volumes (i.e. cross areas) are the same for the separated and the concentrated setup:

$$
A_c = A_d + A_b. \qquad (10)
$$

With this assumption in (10) it becomes clear, that the loss reduction for the concentrated coils is achieved through a lower current density $J_{c\,rms} < J_{s\,rms}$. This fact can also be utilized for a volume reduction as will be discussed separately in section IV. However, the results in this section are based on the equality of the coil volumes (10).

An interesting result is the independence of the copper loss reduction on the load angle φ. While the individual losses of claw pairs (cf. Fig. 7) show a dependency on φ, this dependency disappears by summation of the losses of all coils.

The calculated copper loss ratio is shown in Fig. 9 (solid line with $k_{lm} = 1$) with dependency on the ampere–turns ratio Θ_d/Θ_b. One can see that there is a maximum loss reduction of about $50\,\%$ at an ampere–turn ratio of $\Theta_d/\Theta_b = 1$. The ampere–turn ratio Θ_d/Θ_b is dependent on the operating point and is typically in the range of 2–. In this range a copper loss reduction of about 0 % is possible. A realistic value of improvement will be around 0 % due to the fact that non–sinusoidal, non–repetitive forces will always be present in the system to a certain extent, e.g. due to noise in the sensor signals.

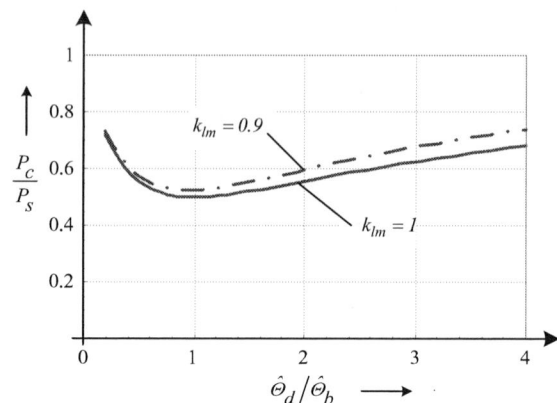

Fig. 9: Copper loss ratio P_c/P_s over ampere–turn ratio. The solid line indicates the losses for $k_{lm} = 1$ the chain dotted line for $k_{lm} = 0.9$

B. Influence of the average winding length factor

Due to the fact that two neighbored drive coils can be combined in case of separated coils (cf. Fig. 7) the average winding length and thus the resulting copper losses can be reduced. This amount of reduction will be calculated in the following. Based on the geometry shown in Fig. 10 the lengths

$$
l_1 = \frac{1}{}\left(\sqrt{2}\,b_c + \overline{\left(2 \quad \sqrt{2}\right) r_c} \right) \qquad (11)
$$

and

$$
l = \overline{\left(2 \quad \sqrt{2}\right) r_c}. \qquad (12)
$$

can be calculated. Here it is assumed that the available space between the stator claws is completely used, i.e. filled with coils as depicted in Fig. 10. The average winding length l_m of a combined drive coil is therefore given by

$$
l_m = 2\,l_1 + 2\,l = \frac{b_c}{\sqrt{2}} + \frac{1}{2}(+)\overline{\left(2 \quad \sqrt{2}\right) r_c}. \qquad (13)
$$

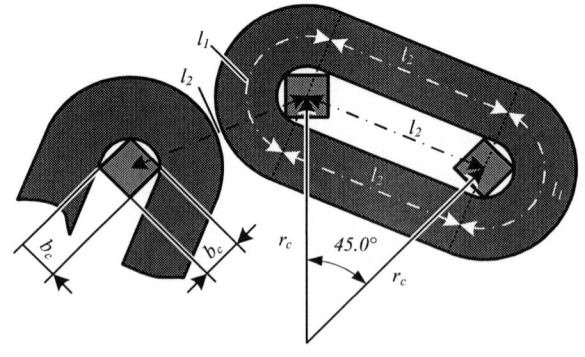

Fig. 10: Geometries for the calculation of the average winding length of a combined drive coil in case of separated coils.

If all drive coils were separately wound around the claws, the average winding length of one drive coil would be

$$
l_{m\,1} = \left(\sqrt{2}\,b_c + \overline{\left(2 \quad \sqrt{2}\right) r_c} \right). \qquad (14)
$$

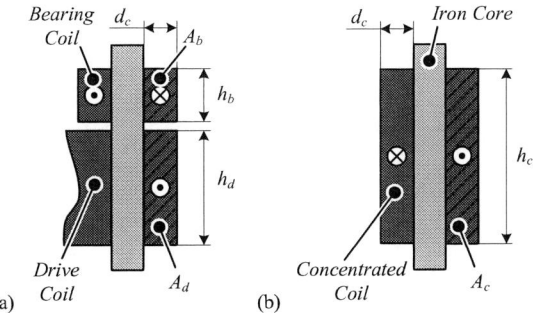

(a) (b)

Fig. 11: Coil dimensions

Hence, the average winding length reduction by combining the coils can be written to

$$\frac{l_m}{l_{m\,1}} = k_{lm} = \frac{\sqrt{2}\ b_c/r_c + (\ +\ )\sqrt{2\ \sqrt{2}}}{2\ \left(\sqrt{2}\,b_c/r_c + \sqrt{2\ \sqrt{2}}\ .\right)} \quad (15)$$

Usually, the factor b_c/r_c is in the range of $0.\quad 0.5$ resulting in a typical value of $k_{lm} = 0.9$. This leads to a slight decrease of the copper losses of the separated coils when taking the average winding length reduction k_{lm} into consideration in (8). The corrected copper loss ratio can be calculated by

$$\frac{P_c}{P_s} = \frac{\Theta_b + \Theta_d}{\left(\Theta_b + \Theta_d\right)\left(\Theta_b + k_{lm}\,\Theta_d\right)}. \quad (16)$$

Fig. 9 shows the copper loss ratios for $k_{lm} = 1$ and $k_{lm} = 0.9$. It can be seen that the factor k_{lm} for a realistic value of $k_{lm} = 0.9$ reduces the copper loss improvement of the concentrated coils by approximately 5%.

IV. COIL VOLUME

The previous section was based on the same iron circuit configuration, which leads to the same total coil volume (resp. total coil area $A_c = A_d + A_b$, cf. Fig. 11). Hence, the resulting current density for concentrated coils was lower in some coils causing the before mentioned loss reduction. Releasing the constraint of fixed iron circuit and dimensions may lead to a smaller coil volume in case of concentrated coils as will be shown in the following.

The current densities in the concentrated coils can be calculated with

$$J_{c\,i\,rms} = \frac{\Theta_{c\,i\,rms}}{h_c\,d_c}. \quad (17)$$

Using (6) and (4) for evaluating (17) leads to dependencies of the current densities on the load angle φ, i.e. for a certain load angle different current densities $J_{\Theta\,c\,i\,rms}$ will occur in the coils. Exemplarily, in Fig. 12 (lower graph) it is illustrated, how the coil volume could be reduced if only the maximum current density of coil 1 would be kept, i.e. $J_{c\,1\,rms} = J_{max}$. However, as can be seen in the upper graph, in the area of the volume decrease the current densities in the other coils increase drastically $J_{c\,i\,rms}\quad J_{max}$, which can lead to local overheating and should be avoided generally (area of local current density exceedence in Fig. 12).

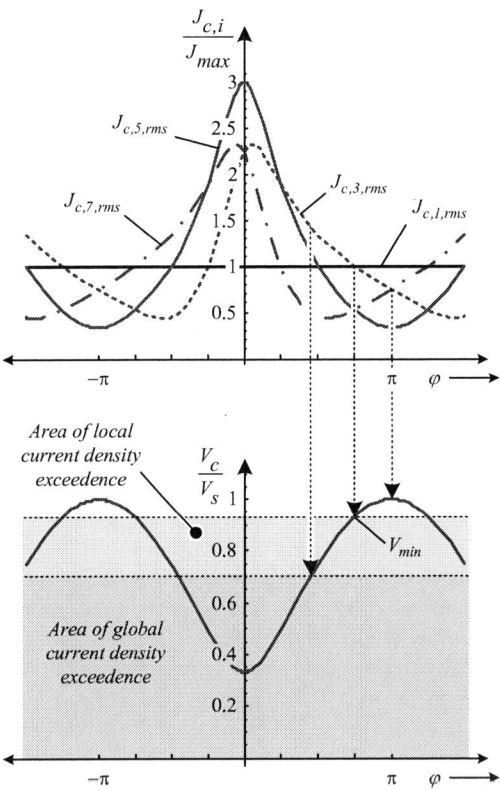

Fig. 12: Local current densities in the concentrated coils for the condition $J_{c,\ ,rms} = J_{max}$ (upper curve); resulting coil volume reduction for that condition in dependency on the load angle φ and a fixed ampere–turn ratio $\hat{\Theta}_d/\hat{\Theta}_b = 2$.

Therefore, if $J_{c\,i\,rms} \leq J_{max}$ shall be maintained in all coils, the minimum volume V_{mi} is given at $\varphi = \quad/\ $, which has to be set by orientating the outlet of the pump accordingly. For the case shown in Fig. 12 ($\Theta_d/\Theta_b = 2$), this leads to a small volume reduction by 7% as compared to the separated coils. At the same time, the copper losses are increasing slightly, as it is depicted in Fig. 13.

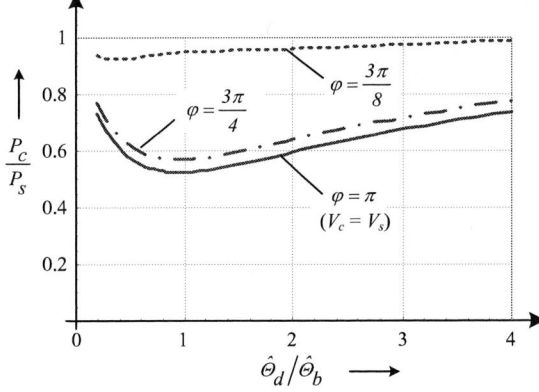

Fig. 13: Copper loss ratio for different load angles φ (and therefore different coil volumes according to Fig. 12) and $k_{lm} = 0.9$.

Hence, as long as the current density condition is kept, there is only an insignificant volume decrease possible. An interesting point is if locally higher current densities at some

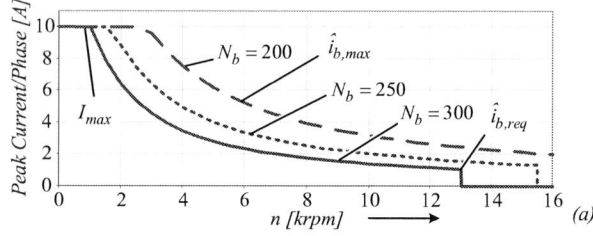

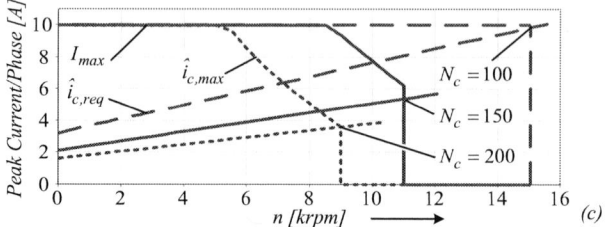

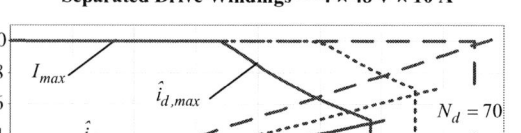

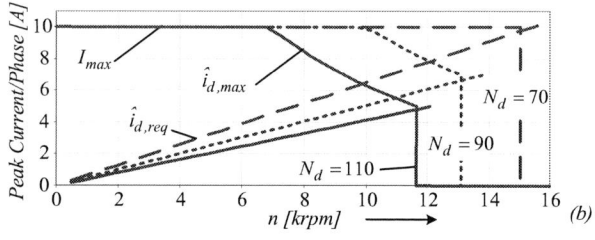

Fig. 14: Maximum achievable current and speed for separated bearing (a), drive (b) and concentrated coils.

coils are allowed (e.g. due to a good thermal coupling with coils with lower current densities) and the mean value of the current densities in the coils is kept below the maximum allowable value, i.e. $1/8 \sum_{i=1}^{8} J_{c\,i\,rms} < J_{max}$. The load angle φ that has to be set in order to achieve this point is in the area of $/2 \ldots /$ (cf. Fig. 12) depending on the ampere–turns ratio Θ_d/Θ_b. As can be seen in Fig. 13, e.g. for $\varphi = /$ the total copper losses (in all concentrated coils) are approximately equal the copper losses of the separated coils, while the volume is reduced to about 70 % (area of local current density exceedence in Fig. 12). However, this also means a smaller heat sink volume, which may complicate the thermal situation in addition to the unequal current distribution, wherefore a thermal modeling of the system is indispensable in that case.

V. POWER ELECTRONICS REQUIREMENTS

For estimating the needed power electronics volume the *components load factor (CLF)* calculation [12] can be used. This factor is an indicator for size, costs and appearing losses in a system [12]. Two systems with the same *CLF* factor can be compared directly. According to [12], the power electronics approximately scales with $CLF \propto U^* \cdot {}^*$. Of course, this factor does neither consider the effort for additional circuitry (such as gate driver circuits) nor it is based on a detailed loss analysis. However, is serves as a rough indicator for the total power electronics effort.

As mentioned before, the concentrated winding setup requires eight full–bridges while the separated coils require only four full–bridges. Therefore, in order to achieve the same *CLF*, either U_{DC} or $_{max}$ has to be doubled for the separated setup. With this, exemplarily possible combinations are given in Table I for $U_{DC} = 2 \ V/ \ 8\,V$ and $_{max} = 5\,A/10\,A$. The same component ratings are used for bearing and drive in case of separated coils even though the bearing requirements would permit lower current ratings.

Hence, regarding the power electronics requirements the two winding concepts are equivalent, as long as the same *CLF* is chosen in the design. As will be shown in the next

Low power equal setups			
Separated coils	$\times 2$	$V \times 10\,A$	
Separated coils	$\times 8$	$V \times 5\,A$	
Concentrated coils	8×2	$V \times 5\,A$	

High power equal setups			
Separated coils	$\times 8$	$V \times 10\,A$	
Concentrated coils	8×2	$V \times 10\,A$	
Concentrated coils	8×8	$V \times 5\,A$	

TABLE I: Possible power electronics setups with same *CLF*

section, the equality of the *CLF* leads to a fair comparison, since it means the same power being delivered to the motor.

VI. MAXIMUM ACHIEVABLE ROTATION SPEED

A. Separated coils

For a stable operation it has to be ensured that the minimum required bearing current given by $i_{b\,req} = \Theta_{b\,req}/N_b$ for a chosen number of windings N_b per coil can be impressed into the bearing coils at any time. The maximum applicable current in a coil is given by

$$i_{b\,max} = \frac{u_{i\,d} \cdot R \pm \overline{(R + \omega_{el} L\,)\,U_{DC} \quad \omega_{el} L\,u_{i\,d}}}{R + \omega_{el} L}$$
(18)

which is a function of the induced voltage $u_{i\,d}$, the total bearing phase inductance $L \propto N_b$, the supply voltage U_{DC} and the angular frequency $\omega_{el} = 2 \ n/60$, where n is the rotation speed in rpm. A stable operation of the bearingless motor with separated coils is therefore given for the condition $i_{b\,req} < i_{b\,max}$. For the separated coils, where no induced voltage is appearing in the bearing phase ($u_{i\,d} = 0$, cf. section II), the maximum achievable current $i_{b\,max}$ is approximately decreasing with $1/n$ and limited by $_{max}$ given by the power electronics.

Fig. 14(a) shows the curves of the available peak currents per phase for different winding numbers N_b and for $U_{DC} = 8\,V$ and $_{max} = 10\,A$. It can be seen that for lower number

978-1-4244-1871-8/07 \$25.00 © 2007 IEEE

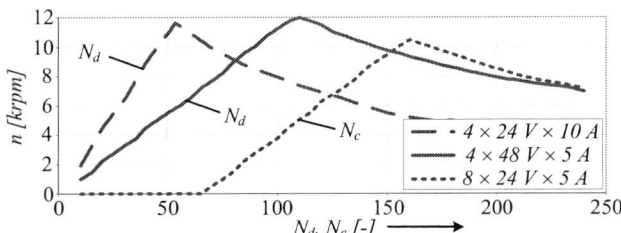

Fig. 15: Maximum achievable speed dependent on number of turns for low power (left picture) and high power (right picture).

of turns higher rotation speeds are feasible, but higher currents are needed and therefore higher losses in the power electronics occur. Additionally, it has to be ensured that the minimum needed drive current $i_{d\,req}$ can be impressed at any time. Analogously to (18), $i_{d\,max}$ can be calculated leading to the condition $i_{d\,req} < i_{d\,max}$. In case of a pump usually the drive requirements are more restrictive than the bearing ones. The pump performance is measured by the hydraulic pressure at a certain flow rate $Q = cons$. Based on the scaling properties of pumps the pressure is scaled with

$$Y \propto n . \tag{19}$$

Therefore, a high rotation speed n is desireable in order to increase the hydraulic pressure and the power $P_{hyd} = Y \cdot Q$ of the pump.

However, also the drive current requirements are increasing with high rotation speeds according to

$$i_{d\,req} \propto M = \frac{P}{n \cdot 2 \quad /60} \propto n. \tag{20}$$

Out of this requirement the minimum needed current $i_{d\,req}$ can be calculated. Curves for three different winding numbers are plotted in Fig. 14(b). The maximum achievable rotation speed is given by the crossing of the $i_{d\,req}$ and the $i_{d\,max}$ curves, which are both dependent an the number of turns N_d.

B. Concentrated Coils

Due to $\Theta_c = \Theta_b + \Theta_d$ in case of concentrated coils the condition

$$i_{c\,req} \cdot N_c = i_{b\,req} \cdot N_b + i_{d\,req} \cdot N_d < i_{c\,max} \cdot N_c \tag{21}$$

has to be satisfied. Therefore, the minimum needed current $i_{c\,req}$ is a superposition of the minimum needed bearing ampere–turns and the minimum needed drive ampere–turns with (20). The needed bearing current is almost independent of the rotation speed and is causing the offsets in the current curves in Fig. 14(c), which are plotted for a *CLF* equivalent system as compared to Fig. 14 (a)/(b). Due to the superposition of the drive and bearing ampere–turns the required current $i_{c\,req}$ is now higher than for the separated case (cf. Fig. 14(a)/(b)).

Thus, the maximum achievable speed is depending on the winding numbers, which is shown in Fig. 15 for both separated and concentrated coils for the load cases mentioned in section V. Since the bearing system is not the limiting factor in case of separated coils, only the dependency of the

maximum achievable speed on the drive winding number N_d is plotted.

One can see that especially at low power (left picture in Fig. 15) with separated coils higher rotation speeds can be achieved. The main reason for this is that there is no induced voltage appearing in the bearing phase and therefore only the needed drive current has to be built up against the induced voltage. At high power (right picture in Fig. 15) this benefit disappears and approximately the same maximum speed can be achieved with both winding concepts.

VII. CONCLUSIONS

In this paper two different winding concepts for bearingless pumps have been discussed. The comparison was based on the assumptions of the same needed force and torque in both cases, same dimensions, same iron circuit, same magnet size and magnetization, same maximum current density and finally the same winding factor in the coils. The concepts have been compared concerning the copper losses, the total coil volume, the maximum achievable speed and the requirements for the used power electronics.

A qualitative compilation of the results is given in Tab. II. The most remarkable difference between the two concepts is the aspect of the copper losses, where a realistic reduction of about 0 % can be achieved for the concentrated coils. Regarding the coil volume, there is a general trade–off between the coil volume and the copper losses, i.e. a smaller volume will always lead to higher losses. However, if the maximum current densities in all coils shall not be exceeded, there is no significant volume decrease possible for the concentrated coils. If local exceedence of current densities is allowed in some claws (and the mean value of the current densities is still below the maximum allowable current density), the volume can be reduced to about 70 %. However, in this case the benefit of the loss reduction for the concentrated coils is lost and the coil losses are equal for the both cases.

The calculations for the volume and loss comparison are based on the assumption of sinusoidal bearing and drive currents and a load angle φ, which is occurring between the bearing and the drive currents and is dependent on the orientation of the outlet of the centrifugal pump. In an experimental setup (cf. Fig. 17) these assumptions have been validated for two outlet configurations (cf. Fig. 16)

Basically, for the concentrated coils twice the number of

978-1-4244-1871-8/07 $25.00 © 2007 IEEE

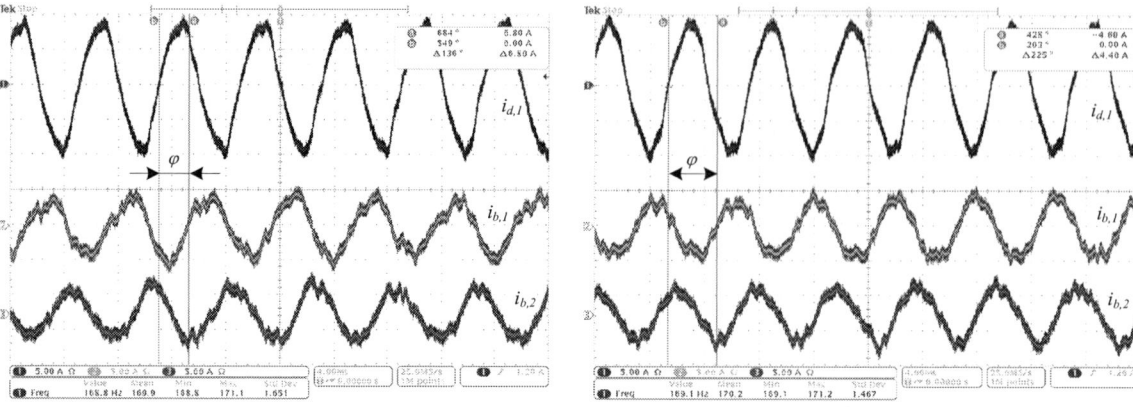

Fig. 16: Current measurements for two outlet configurations (left: $\varphi = 3\,\pi/4$, right: $\varphi = 5\,\pi/4$) at $10'000\ rpm$, $14\,l/min$ hydraulic flow rate and $1.6\,bar$ outlet pressure. (Current scale $5\,A/div$, time scale $4\,ms/div$)

Fig. 17: Experimental setup with separated coils and power electronics with four full bridge inverters.

	Sep. coils	Conc. coils
Copper losses		+
Coil volume	✓	✓
Power electronics	✓	✓
Maximum speed	+	✓
Manufacturability	✓	+
Control effort	✓	
Cabling effort	✓	

TABLE II: Comparison of separated and concentrated winding concepts

full–bridges is required. However, if the same power shall be delivered to the pump in order to perform a fair comparison, the concentrated coils can be driven with either half the current or half the voltage. This leads, in consequence, to approximately the same power electronics requirements.

Regarding the maximum achievable speed the two concepts show similar performance for high power delivered to the pump (e.g. $8\,V/10\,A$ for the separated coils and $2\ V/10\,A$ for the concentrated coils). For lower power (e.g. $8\,V/5\,A$ for the separated coils and $2\ V/5\,A$ for the concentrated coils) the maximum achievable speed is noticeably lower in the case of concentrated coils. This is due to the fact that the induced voltages are appearing in all concentrated coils and are reducing the coil voltages (in case of separated coils the induced voltages are not present in the bearing system due to the winding configuration).

As to manufacturability, there is an apparent advantage for the concentrated coils, both in terms of number and uniformity. On the other hand, the calculation and control of the eight required currents leads to a clearly higher effort for the concentrated coils. Finally, also the cabling effort has to be taken into account, which is doubled for the concentrated coils.

Therefore, the decision about the preferable winding concept has to be taken from case to case depending on the application, since both concepts have their advantages in different aspects. However, for given specifications, the calculations and considerations detailed in this paper can serve as a guideline for the selection of the appropriate winding concept.

REFERENCES

[1] M. Neff, "Bearingless pump system for the semiconductor industry," in *6th International Symposium on Magnetic Suspension Technology*, 2001, pp. 169–173.

[2] ——, "Bearingless centrifugal pump for highly pure chemicals," in *Proc. 8th ISMB*, Aug. 2002, pp. 283–287.

[3] "Levitronix website," 2007. [Online]. Available: http://www.levitronix.com/

[4] T. Nussbaumer, K. Raggl, P. Boesch, and J.W.Kolar, "Trends in integration for magnetically levitated pump systems," in *PCC Nagoya*, 2–5 April 2007, pp. 1551–1558.

[5] N. Barletta, "Der lagerlose scheibenläufer," Ph.D. dissertation, Electrical Engineering and Design Laboratory, ETH Zurich, 1998.

[6] R. Schoeb, "Principle and application of a bearingless slice motor," *JSME International Journal*, vol. 40, no. 4, pp. 593–598, 1997.

[7] G. Schweitzer, "Active magnetic bearings — chances and limitations," *6th International Conference on Rotor Dynamics*, pp. 1–14, Oct. 2002.

[8] J. Bichsel, "The bearingless electrical machine," *International Symposium on Magnetic Suspension Technology*, vol. 2, pp. 1–14, Aug. 19–23 1991.

[9] S. Silber, W. Amrhein, P. Bösch, R. Schöb, and N. Barletta, "Design aspects of bearingless slice motors," *IEEE/ASME Transactions on Mechatronics*, vol. 10, no. 6, pp. 611–617, 2005.

[10] W. Amrhein, S. Silber, and K. Nenninger, "Levitation forces in bearingless permanent magnet motors," *IEEE Transactions on Magnetics*, vol. 35, no. 5, pp. 4052–4054, 1999.

[11] M. Takemoto, M. Uyama, A. Chiba, H. Akagi, and T. Fukao, "A deeply–buried permanent magnet bearingless motor with 2–pole motor windings and 4–pole suspension windings," in *Conference Record of the 38th Industry Applications Conference (IAS)*, vol. 2, 2003, pp. 1413–1420.

[12] B. Carsten, "Converter component load factors; a performance limitation of various topologies," in *PCI*, June 1988, pp. 31–49.

978-1-4244-1871-8/07 $25.00 © 2007 IEEE

A Multi-winding Transformer Model for Predictive SMPS Design and Analysis

Dr. Philip Kwaku Fosu Okyere
Siemens Transportations Systems,
Rail Automation, TS RA PC ED 1
Ackerstr. 22, D-38126 Brunswick, Germany
E-mail: philip.okyere@siemens.com

Abstract— This paper outlines the applicability of a multi-winding transformer model for the predictive analysis and design of switched-mode power supplies (SMPSs). The transformer model is PSPICE-compatible and its parameters are obtained from the physical dimensions and material properties of the transformer. Skin and proximity effects in the conducting layers, as well as transformer parasitic capacitances and non-linear core properties, are considered. The model has proven very useful for the predictive design and analysis of SMPS transformers, especially those of forward-converters and enables the prediction of conducted EMI in SMPSs at an early design stage, even before the construction of the first prototypes.

I. INTRODUCTION

Designers of modern switched-mode power supplies (SMPSs) are faced with the challenges of the high-frequency behavior of SMPS components, especially magnetic ones such as multi-winding transformers, which are used in isolated switching power supplies. The behavior of such transformers is affected by the core material used, the skin and proximity effects within the windings, the operating temperature, coupling and the parasitic capacitances between the individual windings, which on the one hand affect the efficiency and on the other contribute to electromagnetic noise (EMC/EMI) in SMPSs. Experience has shown that SMPS designers spend most of the design/development time optimizing the transformer switching behavior, often by time-intensive "trial and error". Computer-aided tools, especially SPICE-based tools, are often used for simulations to accelerate SMPS design. However, the transformer models available in the PSPICE library are inadequate for studying high-frequency effects in multi-winding transformers. In recent years, however, several models for multi-winding transformers have been presented; see [1] to [6]. While the model parameters in [1] and [2] are determined on the basis of curve fitting of the measured short-circuit impedances of a transformer, those of [3] are calculated using time-consuming commercial FEM (finite element method) tools. The models presented in [4] to [6] are based on an analytical approach, but winding parasitic capacitance in the transformers and non-linear core properties are not considered.

This paper outlines the applicability of a multi-winding

transformer model [10] for the prediction and analysis of the high-frequency behavior of multi-winding transformers at an early SMPS design stage, taking account of skin and proximity effects in the conducting layers, winding parasitic capacitances and non-linear magnetic core properties. The model parameters are extracted from the physical geometry and material properties of the windings and magnetic core. The model can be incorporated into well-known simulators such as SPICE and SABER for both frequency and time domain analysis.

II. MULTI-WINDING TRANSFORMER MODEL

A multi-winding transformer is described as a series of individual winding layers separated by insulation layers [1][5]. In real multi-winding transformers the conducting layers are connected in series and/or in parallel. Fig. 1 is a schematic diagram of the model for a multi-winding transformer with $u_1 \cdots u_n$ and $i_1 \cdots i_n$ representing the terminal voltages and currents of the individual conducting layers [10][12].

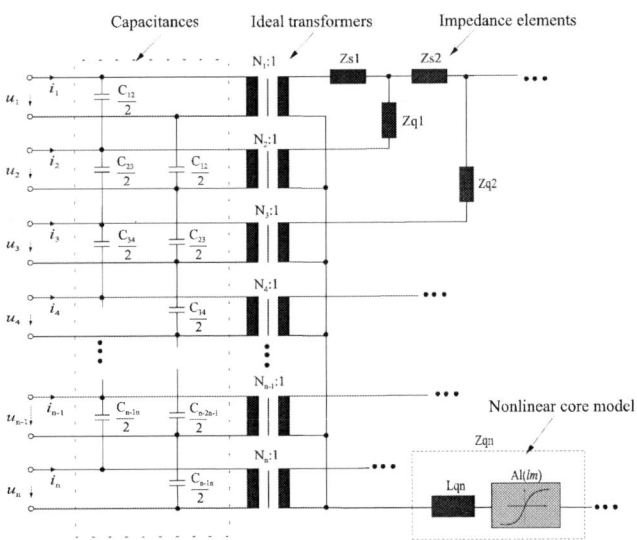

Fig. 1: Model of a multi-winding transformer

The model comprises the following:

- *Ideal transformers*, with the transformation ratios $N_1:1... N_n:1$ representing the number of winding turns of the individual conducting layers.

978-1-4244-1871-8/07 $25.00 © 2007 IEEE

- *Parasitic interlayer capacitances* between adjacent conducting layers.
- *Impedance elements* $Z_{s1} \cdots Z_{qn-1}$, which model the electrical characteristics of the conducting and insulation layers.
- *Impedance element* Z_{qn}, which models the magnetic core.

III. MODEL COMPONENT: IDEAL TRANSFORMERS

The number of winding turns of the individual conductor layers is modeled by the transformation ratios of the ideal transformers. A schematic of a PSPICE model for an ideal transformer and its corresponding subcircuit netlist are given in Fig. 2.

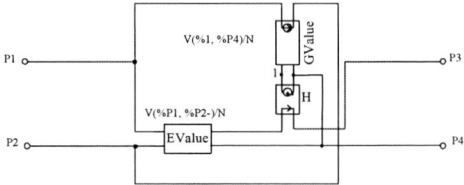

```
PSPICE subcircuit netlist of the ideal transformer
.subckt id_trafo p1 p2 p3 p4 params: N=1
  E1      1    p4    Value={V(p1,p2)*1/N}
  Vsen  1    p3    0
  Hpoly  3    p4    Poly(1) Vsen 0 1
  G1         p1 p2  Value={V(3,p4)*1/N}
ends;   End of subcircuit.
```

Fig. 2: PSPICE model of an ideal transformer

IV. MODEL PARAMETER: PARASITIC CAPACITANCES

Parasitic capacitance in multi-winding transformers occurs between:

- Adjacent conducting layers (interlayer capacitance)
- Winding turns of the same winding layer (interwinding capacitance)
- Individual conducting layers and the magnetic core.

The parasitic capacitances between the windings and core are usually small, and negligible compared to the interlayer capacitance.

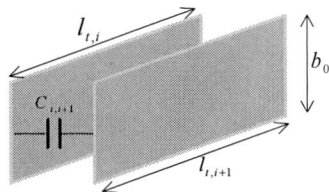

Fig. 3: Determination of the interlayer capacitance

The determination of the interlayer capacitance values is simplified by assuming that:

- The insulation material between adjacent conducting layers is homogeneous.

- The conductivity of the insulation material and dielectric losses are negligible.
- Adjacent conducting layers form a plate capacitor, as depicted in Fig. 3. The values of the interlayer capacitances are determined by (1):

$$C_{i,i+1} = \varepsilon_I \frac{b_o}{h_i} \left(\frac{l_{t,i} + l_{t,i+1}}{2} \right) \tag{1}$$

V. MODEL PARAMETER: IMPEDANCE ELEMENTS

The impedance functions $Zs1(p)$, $Zsi(p)$, $Zqi(p)$ and $Zqn(p)$ are linear complex functions in the frequency domain and are dependent on the geometrical parameters and properties of the transformer. They are written in complex form as (see [5][7][12]):

$$z_{s1}(p) = \frac{l_{tL1}\alpha_1}{\kappa_1 b_0} \coth(\alpha_1 h_{L1}) + \frac{l_{tL2}\alpha_2}{\kappa_2 b_0} \tanh\left(\frac{\alpha_2 h_{L2}}{2}\right) + \frac{\mu \, l_{tI1} h_{I1}}{b_0} p$$

$$z_{si}(p) = \frac{l_{tLi}\alpha_i}{\kappa_i b_0} \tanh\left(\frac{\alpha_i h_{Li}}{2}\right) + \frac{\mu \, l_{tIi} h_{Ii}}{b_0} p + \frac{l_{tLi+1}\alpha_{i+1}}{\kappa_{i+1} b_0} \tanh\left(\frac{\alpha_{i+1} h_{Li+1}}{2}\right)$$

$$z_{qi}(p) = \frac{l_{tLi}\alpha_{i1}}{\kappa_i b_0} \left(\coth(\alpha_i h_{Li}) - \tanh\left(\frac{\alpha_i h_{Li}}{2}\right) \right)$$

$$z_{qn}(p) = \frac{l_{tLn}\alpha_n}{\kappa_n b_0} \tanh\left(\frac{\alpha_n h_{Ln}}{2}\right) + pA_L$$

where

$\alpha_i = \sqrt{\kappa_i \mu p}$	skin constant
κ_i, μ :	conductivity, permeability
l_{tLi}, l_{tIi} :	mean turn length of the ith conductor layer (index L), and i^{th} insulation region (index I)
h_{Li}, h_{Ii} :	height of the ith conductor layer (index L), and of the ith insulation region (index I)
b_0	winding width
A_L	inductance factor

The time domain simulation with network analysis tools such as SPICE requires the replacement of the impedance functions with equivalent linear electrical network elements. However, these complex functions without approximations would result in equivalent models with an infinite number of network elements. The approximation of the functions by means of continuous fractions yields linear networks consisting of linear inductors and resistors. Hence the accuracy of the model depends on the order of approximation.

The next section outlines the equivalent network elements of the impedance function obtained by "zero-order approximation" and "first-order approximation".

A. Impedance Network: Zero-order Approximation

The zero-order approximation yields a so-called "quasi-static" model of the multi-winding transformer [6][9], representing its low-frequency behavior only, i.e. without any consideration of skin and proximity effects.

The equivalent linear network elements of the impedance

functions obtained by zero-order approximation consist of linear resistors and inductors, as shown below:

---

1. Zero-order approximation of $Zs1(p)$

$$Z_{s1}(p) \approx \underbrace{\frac{l_{tL1}}{\kappa_1 b_0 h_1}}_{R_{s10}} + p \underbrace{\left(\frac{l_{tL1}\mu h_1}{3 b_0} + \frac{l_{tL2}\mu h_2}{2 b_0} + \frac{l_{tl1}\mu h_{l1}}{b_0} \right)}_{L_{s10}}$$

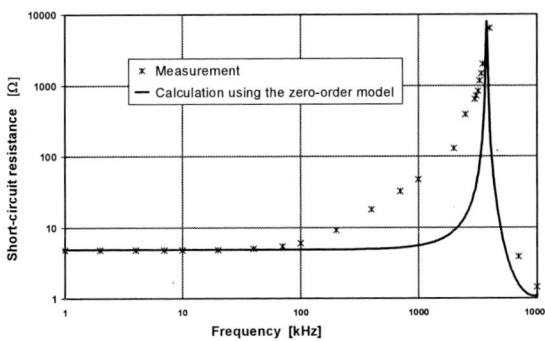

---

2. Zero-order approximation of $Zsi(p)$

$$Z_{si}(p) \approx p \underbrace{\left(\frac{l_{tLi}\mu h_i}{2 b_0} + \frac{l_{tLi+1}\mu h_{i+1}}{2 b_0} + \frac{l_{tli}\mu h_{li}}{b_0} \right)}_{Lsi}$$

---

3. Zero-order approximation of $Zqi(p)$

$$Z_{qi}(p) \approx \underbrace{\frac{l_{tL(i+1)}}{k_{i+1} b_0 h_{i+1}}}_{Rqi} - p \underbrace{\frac{l_{tL(i+1)}\mu h_{i+1}}{6 b_0}}_{Lqi}$$

---

4. Zero-order approximation of $Zqn(p)$

$$Z_{qi}(p) \approx p \frac{l_{tLn}\mu h_n}{2 b_o} + p A_L ; \qquad AL = \mu_e \cdot \frac{A_{eff}}{l_{eff}}$$

$(A_L\text{-value})$

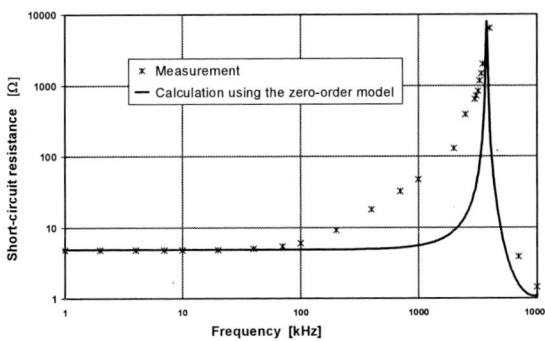

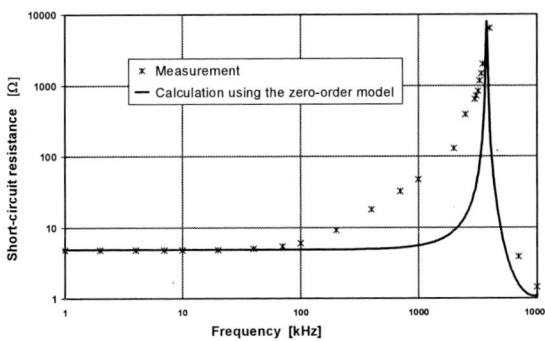

Fig. 4: Short-circuit resistance using the quasi-static model

The result of the calculation using the transformer model based on the zero-order approximation, i.e. the quasi-static model, is depicted in Fig. 4. A comparison with measurements clearly demonstrates the accuracy of the model for low-frequency analysis, but reveals deviations at higher frequencies, thereby justifying the use of the quasi-static model for an analysis not taking account of skin and proximity effects. The resonance observed is caused by the leakage inductance and parasitic capacitance of the transformer.

B. Impedance Network: First-order Approximation

The first-order approximation yields a so-called "quasi-stationary" transformer model. Unlike the zero-order model, the first-order model is applicable for analyses at higher frequencies and is used when skin and proximity effects in the windings are to be taken into account [10]. The corresponding equivalent network approximations with linear resistors and inductors for the first-order approximation of the impedance function are shown below (see [6][12]):

---

1. First-order approximation of $Zs1(p)$

$$z_{s1}(p) \approx \frac{\dfrac{l_{tL1}}{\kappa_1 b_0 h_1} + \dfrac{3}{7}\dfrac{l_{tL1}\mu h_1}{b_0} p + \dfrac{1}{105}\dfrac{l_{tL1}\kappa_1 \mu^2 h_1^{3}}{b_0} p^2}{1 + \dfrac{2}{21}\kappa_1 \mu h_1^{2} p}$$

$$+ \frac{\dfrac{1}{2}\dfrac{l_{tL2}\mu h_2}{b_0} p + \dfrac{1}{120}\dfrac{l_{tL2}\kappa_2 \mu^2 h_2^{3}}{b_0} p^2}{1 + \dfrac{1}{10}\kappa_2 \mu h_2^{2} p} + \dfrac{l_{tl1}\mu h_{l1}}{b_0} p$$

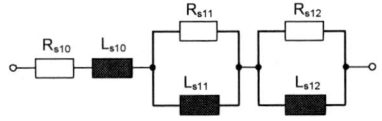

$$R_{s10} = \frac{l_{tL1}}{\kappa_1 b_0 h_1} \qquad L_{s10} = \frac{l_{tL1}\mu h_1}{10 b_0} + \frac{l_{tL2}\mu h_2}{12 b_0} + \frac{l_{tl1}\mu h_{l1}}{b_0}$$

$$R_{s11} = \frac{49 l_{tL1}}{20\kappa_1 b_0 h_1} \qquad L_{s11} = \frac{7 l_{tL1}\mu h_1}{30 b_0}$$

$$R_{s12} = \frac{25 l_{tL2}}{6\kappa_2 b_0 h_2} \qquad L_{s12} = \frac{5 l_{tL2}\mu h_2}{12 b_0}$$

---

2. First-order approximation of $Zsi(p)$

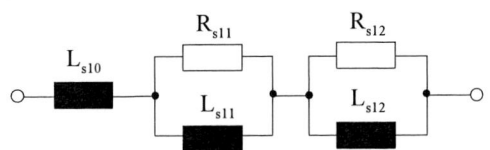

$$R_{s11} = \frac{25}{6} \cdot \frac{l_{tLi}}{\kappa_i b_0 h_i} \quad ; \quad R_{s12} = \frac{25}{6} \cdot \frac{l_{tL(i+1)}}{\kappa_{1+1} b_0 h_{i+1}}$$

$$L_{s10} = \frac{1}{12} \frac{l_{tLi}\mu h_i}{b_0} + \frac{1}{12} \frac{l_{tL(i+1)}\mu h_{i+1}}{b_0} + \frac{l_{tli}\mu h_{li}}{b_0} \; ;$$

$$L_{s11} = \frac{5}{12} \frac{l_{tLi}\mu h_i}{b_0} \; ; \; L_{s12} = \frac{5}{12} \frac{l_{tL(i+1)}\mu h_{i+1}}{b_0}$$

3. First-order approximation of $Z_{qi}(p)$

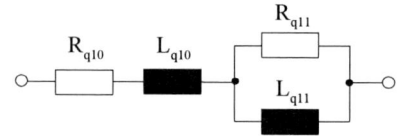

$$R_{q10} = \frac{l_{tL(i+1)}}{\kappa_{i+1} b_0 h_{i+1}} \; ; \; L_{q10} = \frac{11}{620} \cdot \frac{l_{tL(i+1)}\mu h_{i+1}}{b_0} \; ;$$

$$R_{q11} = -\frac{7}{4} \cdot \frac{l_{tL(i+1)}}{\kappa_{i+1} b_0 h_{i+1}} \; ; L_{q11} = -\frac{343}{1830} \frac{l_{tL(i+1)}\mu h_{i+1}}{b_0}$$

4. First-order approximation of $Z_{qn}(p)$

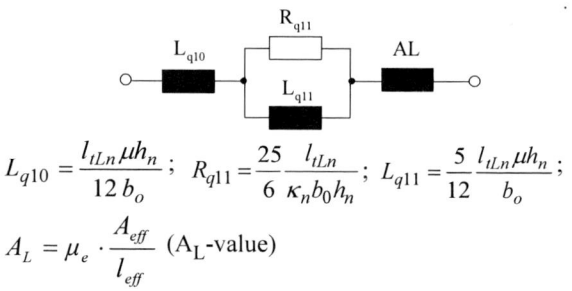

$$L_{q10} = \frac{l_{tLn}\mu h_n}{12\, b_o} \; ; \; R_{q11} = \frac{25}{6} \frac{l_{tLn}}{\kappa_n b_0 h_n} \; ; \; L_{q11} = \frac{5}{12} \frac{l_{tLn}\mu h_n}{b_o} \; ;$$

$$A_L = \mu_e \cdot \frac{A_{eff}}{l_{eff}} \quad \text{(A}_L\text{-value)}$$

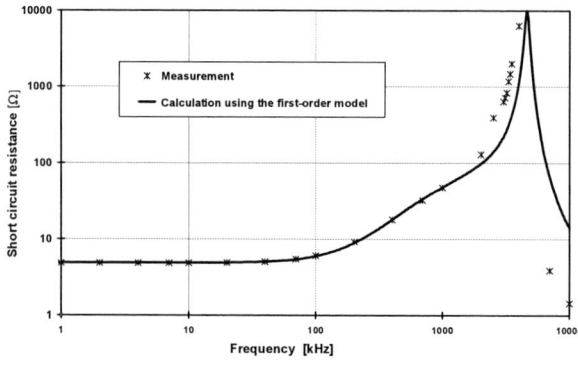

Fig. 5: Short-circuit resistance

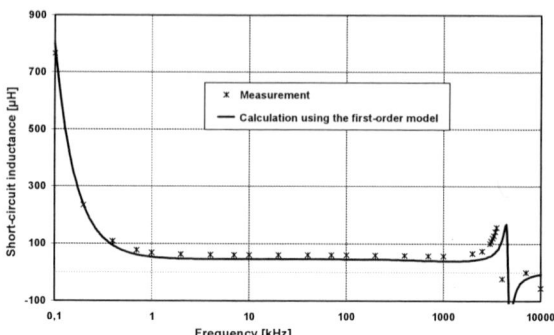

Fig. 6: Leakage inductance

The short-circuit resistance and leakage inductance depicted in Figs. 5 and 6 respectively demonstrate the applicability of the first-order approximation for analysis which take account of the skin effect in multi-winding transformers.

At lower frequencies the resistance is constant and corresponds to the short-circuit DC resistance of the transformer. As the frequency is increased, the skin and proximity effects dominate and an increase in the winding resistance is observed. This behavior is accurately predicted by the transformer model using the first-order approximation.

VI. CORE PROPERTIES

So far, the magnetic core has been assumed to have linear characteristics and its behavior to be independent of the magnetization current. The value of the inductance factor A_L was also assumed to be constant [5]. Non-linear magnetic core properties can, however, also be included in the transformer model.

The impedance element $Z_{qn}(p)$ has essentially two parts, see Fig. 7. The first is determined by the geometrical parameters of the outermost winding layer and the second by the inductance factor A_L of the core. To consider nonlinear core properties, the constant inductance factor (A_L-value) is replaced by one which is dependent on the magnetization current $A_L(Im)$.

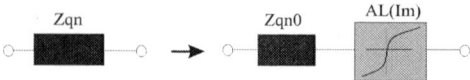

Fig. 7: Including nonlinear core properties

The current-dependent core model from Jiles-Atherton [8] for magnetic core material was incorporated into the multi-winding transformer model. Fig. 8 shows the result of the analysis at core saturation. Core loss due to hysteresis is neglected.

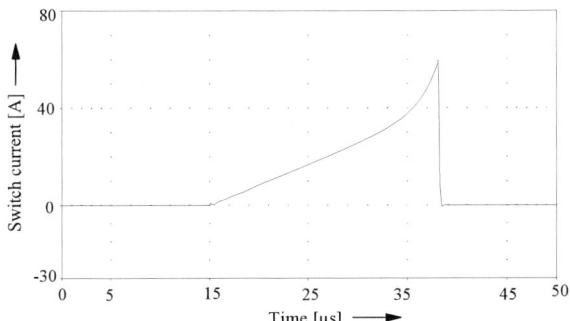

Fig. 8: Effect of core saturation on the switch current

VII. PREDICTIVE ANALYSIS

The applicability of the transformer model for predictive analysis was investigated on a transformer constructed with an ELP 22 planar core, N87 core material and a 12-layer PCB with eight primary layers and four secondary layers. Each layer is made up of one winding turn. A 70 µm copper-thickness is used for the conducting layers and a 210 µm thickness of FR4-base material for the insulation between the conducting layers of both the primary and the secondary.

The effect of winding interleaving on the skin und proximity effects and of the transformer parasitic capacitance on the switching performance of switching power supplies was analyzed.

A. Skin and Proximity Effects

As is well known, interleaving secondary and primary layers reduces copper losses resulting from the proximity effect in multi-winding transformers [13]. The interleaved winding configurations 8P-4S, 4P-4S-4P and 2P-2S-4P-2S-2P, shown in Fig. 9, were investigated using a PSPICE version of the multi-winding transformer.

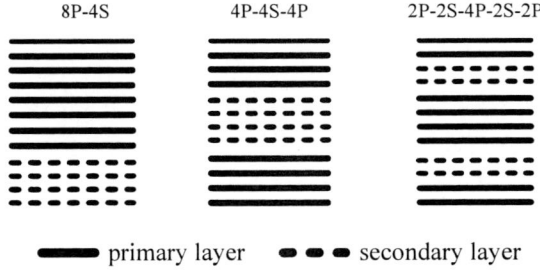

Fig. 9: Interleaved winding configurations

1.	8P-4S:	8P - eight primary layers 4S - four secondary layers
2.	4P-4S-4P:	4P - four primary layers 4S - four secondary layers 4P - four primary layers

		2P - two primary layers 2S - two secondary layers
3.	2P-2S-4P-2S-2P:	4P - four primary layers 2S - two secondary layers 2P - two primary layers

The short-circuit resistance (secondary shorted) was analyzed. Fig. 10 shows the results of the analysis. The results confirm the reduction in AC resistance by interleaving the secondary and primary layers.

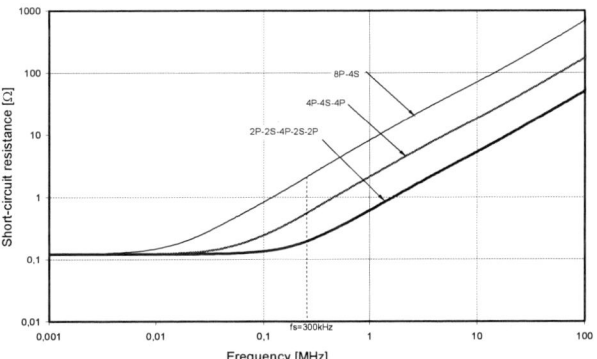

Fig. 10: AC resistance of the transformer with the different interleaved winding configurations

While the winding configuration 8P-4S exhibited the worst AC resistance rise, the configuration 2P-2S-4P-2S-2P with most interleaving, as expected, yielded the smallest increase in AC resistance at the switching frequency fs=300 kHz, which is why it was chosen for the design of the transformer.

B. Effect of Parasitic Capacitances

A time domain analysis was performed with the transformer model to examine the effect of the transformer parasitic capacitances on the switching behavior of switched-mode power supplies.

A typical DC-to-DC forward converter with an input voltage of 36-72 V and an output voltage of 12 V was selected; see Fig. 11. The converter operates at a switching frequency of 300 kHz and is designed for an output power of 60 W.

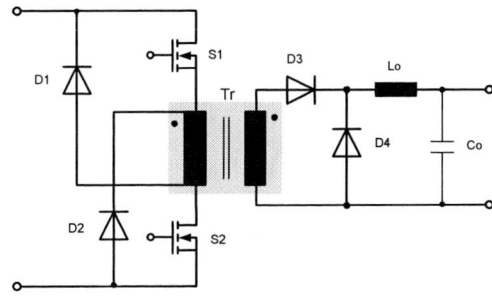

Fig. 11: Circuit diagram of a two-transistor forward converter

978-1-4244-1871-8/07 $25.00 © 2007 IEEE

The analysis was performed using the multi-winding transformer model:
- <u>without</u> capacitances and
- <u>with</u> capacitances.

Fig. 12 shows the voltage across the switch (drain-source voltage of the MOSFET) and Fig. 13 the demagnetization current of the transformer.

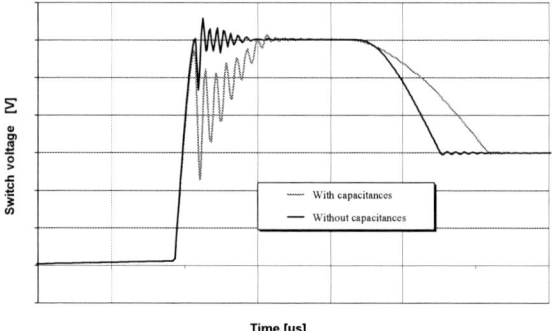

Fig. 12: Switch drain-source voltage

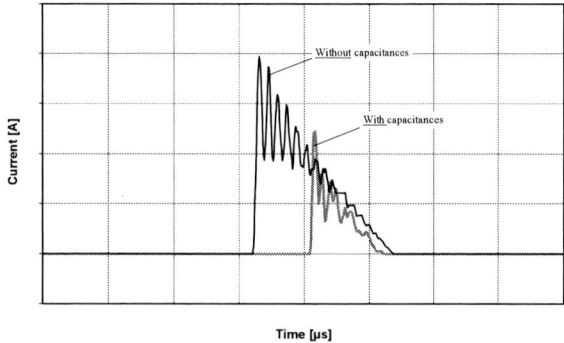

Fig. 13: Demagnetization current

The waveforms clearly show the effect of parasitic capacitances on the switch voltage and demagnetization behavior. In forward converters, normally one would expect the demagnetization current to start flowing immediately after the switch is turned off. Transformer parasitic capacitances, however, cause a delay in the demagnetization of the core. This delay is due to the discharging of the parasitic capacitances as the switch is turned off.

VIII. SUMMARY

A novel PSPICE-compatible multi-winding transformer model for frequency and time domain analysis and the design of switched-mode power supplies has been presented, with formulas for determining the model parameters from the physical dimensions and material properties of the transformer. The paper discussed the applicability of the transformer model for the prediction of the AC resistance increase

due to skin und proximity effects and the analysis of the effect of transformer parasitic capacitances on the behavior of switching power supplies. The capability of the model to include nonlinear core properties was also demonstrated. The transformer model permits the predictive analysis and design of switched-mode power supplies for optimum performance, since all its model parameters are obtained from the physical dimensions and material properties of a transformer.

REFERENCES

[1] V. A. Niemela, H.A. Oven, T. G. Wilson; Frequency-Independent Element Cross-Coupled-Secondaries Model for Multiwinding Transformers; IEEE Power Electronics Specialists Conference 1992

[2] H.A. Oven, V. A. Niemela, T. G. Wilson; Enhanced Cross-Coupled-Secondaries Model for Multiwinding Transformers; IEEE Power Electronics Specialists Conference 1992

[3] R. Asensi, J.A. Cobos, O. Garcia, R. Prieto, J. Uceda; A Full Procedure to Model High Frequency Transformer Windings; IEEE Power Electronics Specialists Conference 1994

[4] A. Schellmanns; P. Fouassier; J. P. Keradec; J. L. Schanen; 1D-Propagation Based Equivalent Circuits for Transformers: Accounting for Mul-tilayer Structure of Windings and Ferrite Losses; IEEE Industrial Application Society Annual Meeting 1997

[5] L. Heinemann, R. Ullrich, B. Becker, H. Grotstollen, State Space Modeling of High Frequency Multiwinding Transformers; IEEE Power Electronics Specialists Conference 1993

[6] L. Heinemann; A novel SPICE-compatible high frequency multi-winding transformer model; IEEE Industrial Application Society Annual Meeting 1998

[7] L. Heinemann; Simulation and Design of High Frequency Magnetics Including the Skin Effect, Nonlinear Core properties and thermal effects" IEEE Applied Power Electronics Conference 1996

[8] Jiles, D.C.; Atherton, D. L.: Theory of ferromagnetic hysteresis. Journal of Magnetism and Magnetic Materials, 61, 48, 1986

[9] P. F. Okyere, L. Heinemann: Computer-Aided Analysis and Reduction of Conducted EMI in Switched-Mode Power Supplies. IEEE Applied Power Electronics Conference 1998

[10] P. F. Okyere, L. Heinemann: An Advanced SPICE-Compatible Model for High Frequency Multiwinding Transformers, Power Electronics Specialist Conference PESC 1999

[11] P. F. Okyere, Habiger, Ernst: A novel physically-based PSPICE-compatible model for common-mode chokes, EMC Tokyo 2000

[12] P. F. Okyere, E. Habiger: Effektivierung und Qualifizierung des EMV-gerechten Entwurfs von Schaltnetzteilen. Forschungskurzbeitrag für das EMC Kompendium 2000. München.

[13] Mohan, N.; Undeland, T. M.; Robbins W. P.: Power Electronics: -converters, applications and design; John Wiley & Sons, USA 1995

[14] Billings, K. H.: Switch-mode power supply handbook. New York: McGraw-Hill Publishing Company 1989.

[15] Pressman, A.: Switching Power Supply Design. McGraw-Hill. New York 1998.

[16] Mohan, N.; Undeland, T. M.; Robbins W. P.: Power Electronics: -converters, applications and design; John Wiley & Sons, USA 1995.

[17] Grandi, G.; Kazimierczuk, Massarini, A.; Reggiani, U.: Stray capacitances of single-layer air-core inductors for high-frequency application. IAS '96, vol.3 Page(s): 1384 –1388.

[18] R. W. Erikson, D. Maksimovic: Fundamentals of power electronics. Kluwer Academic Publishers. 2nd Edition 2001.

978-1-4244-1871-8/07 $25.00 © 2007 IEEE

Comparison of Topologies for Linear Drives in Industrial Material Handling and Processing Applications.

Peter Mutschler

Technische Universität Darmstadt
Department of Power Electronics and Control of Drives
Landgraf Geotg Str. 4; 64283 Darmstadt, Germany
Email: pmu@srt.tu-darmstadt.de

Abstract— For integrated material handling and processing in industrial applications, permanent magnet (PM) linear drive systems are proposed. Solutions with Short Primary and Long Primary are compared. The Short Primary type uses a passive track and an active vehicle that is supplied by a contact less energy transmission and is operated by an on- board controller and inverter. The Long Primary type uses an active track and passive, lightweight vehicles. The active track consists of a large number of segments, each operated by an individual controller and inverter. Experimental results of the Long Primary type are presented.

I.   INTRODUCTION

Linear drives are known in many different designs and with a variety of different properties.

In *industrial* applications, typically linear drives are used for straight-line movement of a single vehicle along a limited distance. Today the application of linear drives in industry is preferred for movements with no backlash, high accuracy of positioning and high dynamics (acceleration). The main application areas of linear drives in industry today are in advanced machine tools.

Besides industrial applications, linear drives are used in *transportation* applications. There are the well known high speed maglev trains and several projects on using linear drives for urban transportation systems (people movers), e.g. [1], [2], [3] etc..

This paper deals with linear drives aimed for integrated material handling and processing in industrial applications.

II.   EXTENSION OF THE FUNCTIONAL RANGE OF LINEAR DRIVE SYSTEMS.

In industrial production plants, materials are processed e.g. in processing stations and transported between these stations. Using the **same** linear direct drive for <u>processing</u> and <u>transportation</u> without releasing and re-adjusting the work pieces,

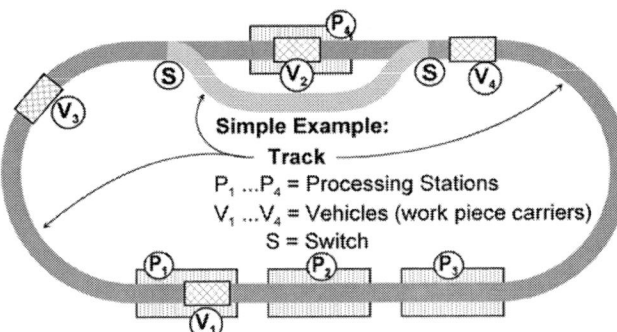

Figure 1: Simple example of combined transportation and processing of materials with a linear drive system

will result in many benefits:

  ➢ High productivity
  ➢ High dynamic and high precision (few μm)
  ➢ No mechanical transmission ⇨ reduced wear and maintenance

Figure 1 shows a simple example of combined transportation and processing of materials with a linear drive system. In such applications, the following properties are necessary for the linear drive system:

- On a carriage way (track) several vehicles must travel with a high degree of independency. Each vehicle has to be controlled very precisely (few μm) when the vehicle operates within a processing station.
- The carriage way must allow for horizontal and vertical curves and for closed paths.
- In a future step also switches to work with different processing stations in parallel shall be included.

As a high thrust- force density in combination with a rather large air- gap is mandatory for the application, we concentrate on permanent **m**agnet **s**ynchronous **l**inear **m**achines (PMSLM).

A circulating linear transport system for concatenation of machine tools was proposed already by [4]. But the transverse flux <u>induction</u> machine which was designed for this project, needed very close air gap tolerances i.e. a high precision of the carriage way. Additionally the reactive power of the inverter had to be rather large due to the large air gap and thus the system was not optimal.

In an application like in Fig. 1, the track passes through processing stations (P_1....P_4). There, typically high precision positioning and high forces are necessary. For high precision positioning, position sensors cannot be avoided. But outside of the processing stations, typically a lower precision in positioning is sufficient. In these parts of the track, motion control without using sensors for speed or position should be realized

III.   TOPOLOGIES OF LINEAR DRIVES FOR INDUSTRIAL MATERIAL HANDLING AND PROCESSING.

Concerning suspension, two types of linear drives are distinguished: Magnetically levitated vehicles and wheel based vehicles. For special applications e.g. in clean rooms, magnetically levitated vehicles are highly interesting, as no mechanical wear will pollute the environment. But in this paper, we will not go into details of this technology; instead we concentrate on wheel based vehicles. There exist a large variety of possible topologies for PMSLMs using wheel

978-1-4244-1871-8/07 $25.00 © 2007 IEEE   1025

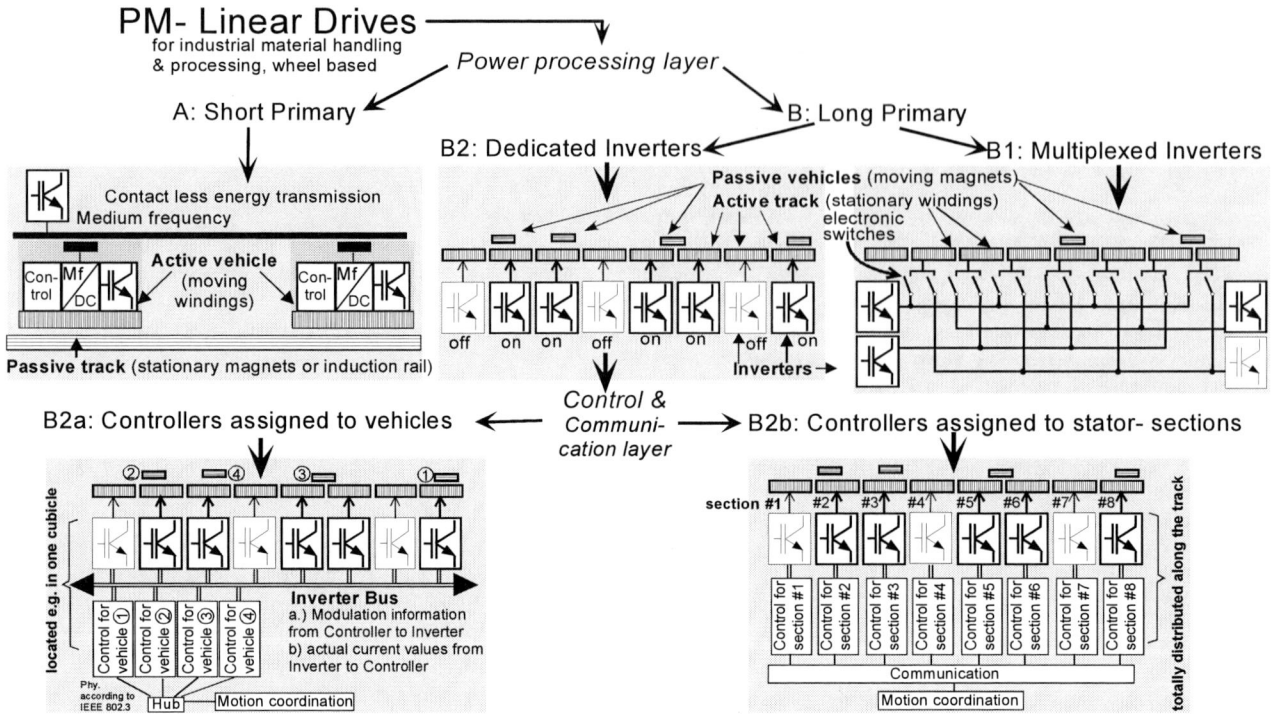

Figure 2: Comparison of topologies for PM-linear drives in industrial material handling and processing applications.

based vehicles for industrial material handling and processing. Two basic layers are distinguished in Figure 2:

> Power processing layer and
> Control and communication layer

At the power processing layer, machines with a Short-Primary (moving winding) and machines with a Long- Primary (moving magnets) are distinguished.

A. Short Primary

The **Short- Primary** type needs an active vehicle, i.e. power- and information processing equipment is located on board of the vehicle, see Fig. 2-A. To transfer the energy to the vehicle, either a sliding contact like a pantograph or a contact less, inductive energy transmission may be used. A cable run is not applicable, as the vehicle has to travel long distances and closed paths. In many industrial production environments, sliding contacts should be avoided, due to reasons of safety for workers and maintenance. Therefore, in Fig. 2-A a contact less energy transmission is proposed. In this case, the stationary part of the contact less energy transmission system is fed with constant (medium to high) frequency by some soft or resonant switching inverter and the reactive power for the energy transmission system is generated by passive elements. The second stage of power electronic energy conversion is located on board of the vehicles and supplies the moving windings with a voltage having variable frequency and amplitude. This conversion stage may use rectification and voltage or current source inverters or even single-phase to three-phase matrix converter (resonant ac-link converters). The on board converter needs an on board control unit for all closed control loops, i.e. loops for

current, speed and position. But in addition, there must be a contact less transmission of information to supply the vehicle's controller with position reference values and to send back status information.

Main benefits of this topology are:

a) The number of the converters and control units is equal to the number of vehicles plus the (big) converter feeding the contact less energy transmission.

b) Position sensing is much easier on board of an active vehicle than with a totally passive vehicle.

c) The passive track plus contact less energy transmission may be cost completive to a long active track, especially when some sections of the track need only a low thrust force. Then, at these sections, operation may change from synchronous machine to induction machine, saving the magnets along the low- force sections of the track. But this needs still more detailed investigations.

The main drawbacks are: Due to power limitation of contact less energy transmission (all vehicles may accelerate at the same time) and due to higher weight and volume than with passive vehicles, the dynamics (acceleration) of the vehicles is limited. On board energy storage with super- caps may increase peak power, but weight, volume and complexity is increased as well.

Therefore, Short- Primary may be a good solution for applications with a long track, low number of vehicles and low acceleration.

B. Long Primary

For applications, where very high acceleration is mandatory, passive, lightweight vehicles using an active track, i.e.

the Long Primary type is very interesting. In order to allow for individual motion control of several vehicles, the active track is separated into many sections, which are fed by numerous inverters.

1) Multiplexed Inverters

In order to keep the number of inverters small, multiplexing of inverters is an option. In high speed transportation like maglev trains (Transrapid) multiplexing of inverters is applied, where the multiplexer is made up by mechanical vacuum switches. In industrial material handling with much more vehicles on a much shorter track, mechanical switches are not applicable due to the very high number of switching actions necessary. Instead, electronic switches are considered. The minimum number of inverters necessary is twice the number of vehicles, as two inverters feed each vehicle during the transition between two sections and all vehicles may be in a transition state simultaneously. Fig. 2-B1 shows a simple example with 2 vehicles and 4 inverters. But as can be seen, there are a lot of restrictions concerning the independent motion of the two vehicles. It must be strictly avoided that one inverter feeds two vehicles simultaneously as in such a situation they cannot be controlled individually. A higher degree of independency is achieved, if more inverters and more switches are added, but the complexity and cost will grow very fast. Additionally, the realization of the electronic switches is a problem as well. It is highly questionable weather blocking triacs or antiparallel thyristors can withstand the high dv/dt generated by the IGBTs in the inverters. Adding snubbers will not completely solve this problem, as capacitive current charging the snubbers will additionally stress the IGBTs in the inverters. Alternatively, the electronic switches of the multiplexer may use two IGBTs per phase. But this would end up with a much higher number of IGBTs and gate-drivers than using one dedicated inverter per section of the track. Concluding, we state that in the case, where the multiplexer cannot use mechanical (vacuum) switches, multiplexing of inverters is not recommended.

2) Dedicated Inverters

Fig. 2-B2 shows the Long Primary with one dedicated inverter per stator-section. The length of one stator-section largely depends on the application. To give a rough idea, one section may be in the range of one meter up to a few meters. With a DC-link voltage of 560V, the currents of the inverters are in the range of $10...30A_{rms}$. With short stator-sections quite a large number of inverters will be necessary for an industrial application. Therefore cost effective, highly integrated inverters like IPMs are interesting. This will be discussed later on in more detail. Only those sections, where a vehicle is located are energized, all other inverters are turned off.

Concerning the control and communication layer, two alternatives are shown in Fig. 2

a) Controllers assigned to vehicles

The maximum number of vehicles is 50% of the number of inverters and stator-sections. But in this situation, the independency of motion of the individual vehicles is highly restricted. Typically, the number of vehicles should be much lower, may be in the range of 30% of the number of stator-sections. This is a motivation to assign one controller to one vehicle as in Fig. 2-B2a, in order to have the lowest number of controllers. A lot of computing power is necessary in the controllers, e.g. for sensorless positioning using hf-signal injection. Therefore, the cost for a controller can be higher than for a IPM- based inverter with a rating of 10A...30A. In Fig. 2-B2a there are four controllers for four vehicles, but each controller must be able to communicate with each inverter. An „inverter- bus" achieves this.

For the development of all the necessary control algorithms, we build a set-up using the topology of Fig. 2-B2a. In our experimental set-up, we use a circular PM linear machine with 8 double-sided stator-sections, as shown in Fig. 3. In the center of the circular track, there is a high-torque load machine, which can inject load forces into the vehicle. For our experimental set –up, we developed an "inverter bus", which allows exchanging all necessary data between 10 controllers and all the inverters within each 100µs [5]. The controller calculates the modulation data, i.e. numbers which represent the time when the next switching in the inverter has to be performed. This information is passed on the "inverter bus" and received by a CPLD in the inverter's interface. There, counters are implemented in the CPLD to generate the gate signals. The CPLD controls A/D-converters which sample the currents in the middle of the zero- vector interval and send the actual current values to the controller via the "inverter bus".

The simplified block diagram of our experimental set-up in Fig. 4 shows at the bottom just two of the eight stator-sections and a vehicle -represented by the magnets- passing the border of the sections. During such a transition, two inverters power the vehicle and the PC calculates two d-q-current controllers. There are a total of three nested control loops: Position-, speed-, force- and current-control. Due to the winding arrangement of the stator-sections, there

Figure 3: Experimental set-up with one vehicle on a circular stator with 8 stator-sections, controlled by a PC attached to an inverter-bus according Fig 2-B2a.

978-1-4244-1871-8/07 $25.00 © 2007 IEEE

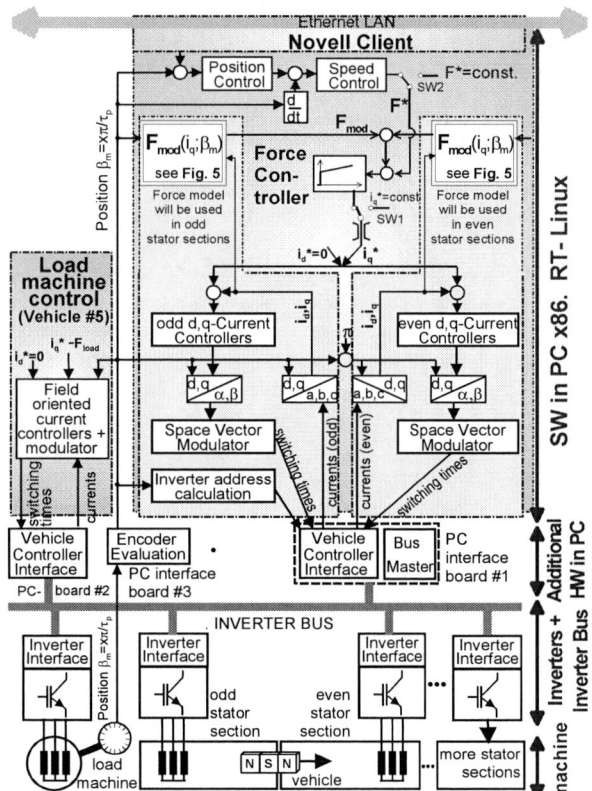

Fig. 4: Simplified block diagram of experimental set-up

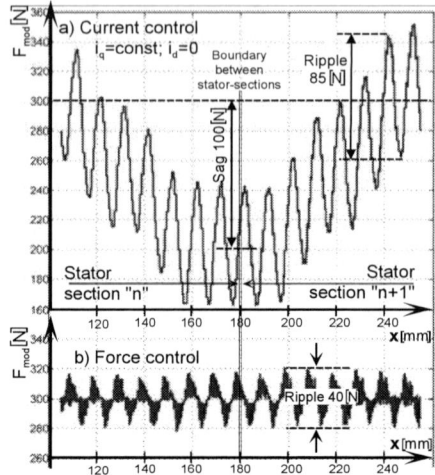

Fig. 6: Force ripple. Experimental result.

is a sag in magneto-motive force at the ends of the stator sections that will

cause a drop of force during transition of the vehicle. In addition, the machine shows a rather high cogging force ripple. To reduce these effects, a force model of the machine by means of look up tables have been provided off- line by a large number of finite element calculations [6]. These tables deliver the calculated force depending on current and position as shown in Fig. 5 and are used in the feedback path of the force controller. Fig. 5 shows the force sag between two stator-sections as well as the cogging force. The effect of using this model $F_{mod}(i_q,\beta_m)$ is shown in Fig. 6. In Fig. 6a, constant current mode is applied, whereas in Fig. 6b the force controller is used. Which eliminates the force sag at the boundary between two sections completely and reduces the force ripple greatly [7].

In our experimental set – up, we calculate all control algorithms including EMK- based sensor less control within a PC (800MHz) under RT-Linux. One PC can calculate the con-

trollers for about three vehicles within the control cycle time of 100µs, several PCs can be attached to the "inverter bus" in order to expand the system.

Using the PC during the development phase of a project provides a lot of advantages. Typically, development starts with one vehicle, but with a stator having several sections and inverters. Using the topology of Fig 2-B2a, just one controller is necessary. But the complexity of the control algorithms is unknown at the beginning. When using a PC as controller, an enormous amount of computing power is available, which is sufficient even for complex algorithms. Additionally, the large memory of the PC is used as "transient recorder", which supports debugging.

The topology of Fig. 2-B2a has many benefits for small applications e.g. with 10-(15) vehicles and 30- (45) inverters. But it cannot be scaled freely to arbitrary large systems, as the bandwidth of the inverter bus will always be a limiting fact, even if in future busses with higher bandwidth will be available.

b) Controllers assigned to stator- sections

To overcome the bottleneck of an inverter- bus that has to enable communication between each controller and each inverter within a time slot of 100µs, a totally distributed control and power-processing system is proposed in Fig. 2-B2b. But the price that has to be paid for this topology, which can be expanded to arbitrary large systems, is the large number of inverters and individual controllers. Therefore, a key question is, how to get down the cost for these components. In practice, such a system will be accepted only, if reasonable costs together with high reliability can be achieved. Fig.7 shows an overview of the topics discussed here in order to minimize expenses.

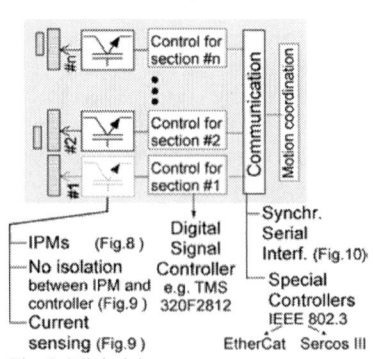

Fig. 7: Minimizing expenses

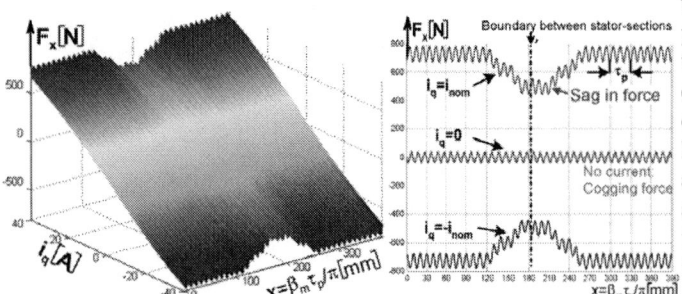

Fig. 5: Thrust force $F(i_d=0,i_q,x)$ obtained by FEM- calculations

978-1-4244-1871-8/07 $25.00 © 2007 IEEE

It is essential to note, that most of these topics are proposals only up to now. Experimental results are available for "Current sensing" [8] only. The rest is planned as future research, depending on funding.

Intelligent Power Modules (IPMs): In recent years, inverters based on low cost IPMs in the range of 0.5…3 kVA spread out in home appliances. Using the "transfer mold" technology for insulation and housing, mass production of these devices is well established for modules with 600V- IGBTs. Since a few years transfer molded IPMs in dual in- line packages with 1200V-IGBTs and currents up to 25A became available; the internal structure is shown in Fig.8. The high side drivers use level shifters and bootstrap circuits for supply. For linear drive systems with light vehicles (20 - 40 kg) and short stator sections (1-1,5m) the power rating of such IPMs is sufficient and low cost inverters seem possible using such devices

No isolation between IPM and controller: In industrial electronics, controllers are located at potential of Protection Earth (PE) typically. On the other hand, the signal ground for the driver circuits within the IPM is the negative dc-link potential. Therefore, the control signals have to be transmitted by isolating devices like optocouplers from the controller to the IPM and the status signal (fault indication) is transmitted from the IPM to the controller by isolating device too. Moreover, current measurement is done in industrial drive electronics by isolating current sensors like Hall- effect sensors. The isolating devices, especially the current sensors contribute up to about 50% to the electronic material-costs of the power section (without control and communication) of a small inverter in the range of 0.5- 4kVA. One way to reduce costs is to omit most isolating devices. This means, that the signal ground of the controller will be at negative dc-link potential. Only the serial communication link to the superimposed "Motion Coordination", which also contains the human interface, has to be isolated. Fig. 9 shows the structure without isolation between IPM and Controller.

Current sensing with shunt resistors: In order to provide full protection for the inverter and to reconstruct the phase currents even at very low modulation index, we propose current sensing by three shunt resistors. Phase current reconstruction at low modulation index (standstill) is essential for sensorless position control, where the saturation dependent current re-

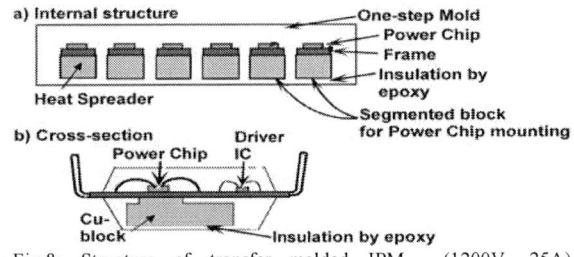

Fig.8: Structure of transfer molded IPM (1200V, 25A) in dual in line package (Mitsubishi)

sponse caused by injected Hf-voltage signals has to be evaluated. Phase currents are measured in the middle of the "Null-Vector" with the switching state "0,0,0". For high modulation index, the duration of the state "0,0,0" gets short and the following method [8] is used: For current measurement use the two legs having the longest "0"-state and apply modulation without "1,1,1" states, i.e. discontinuous zero sequence "bus clamping" modulation. [8] shows, that the additional harmonics due to this modulation are quite small and that the full linear modulation range can be used.

Communication between controllers (C/C-link): When a vehicle moves from one stator section to the next one, the control task has to be carried out by the next controller. Moreover, during transition, both controllers have to control currents in such a way, that the necessary resulting force is developed at the vehicle and velocity and position control is transferred bump-less to the next controller. During this phase, information exchange between the two controllers is necessary in each control-cycle of 100μs. As all vehicles may be in such a transition state simultaneously, a huge amount of overall communication bandwidth will be necessary in this situation.

Communication between controllers and superimposed motion coordination (C/M-link): As the motion control is located at PE-potential and the controllers are at negative DC-link potential, there must be a safety relevant isolation ($v_{iso} \geq 4$ kV) in this communication link. This communication link has to provide:

1.) Cyclic exchange of commanded position values and status information between motion control and all controllers. Cycle time is about 4ms.

2.) Monitoring, Diagnostics. Cyclic transfer of selected variables from selected controllers to human interface, im-

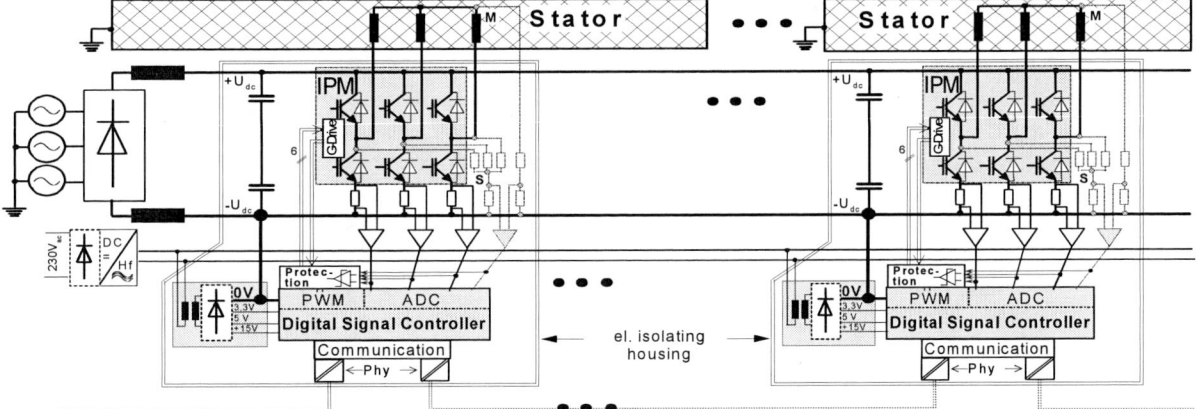

Fig.9: No isolation between IPM and controller, Current sensing by shunt resistors

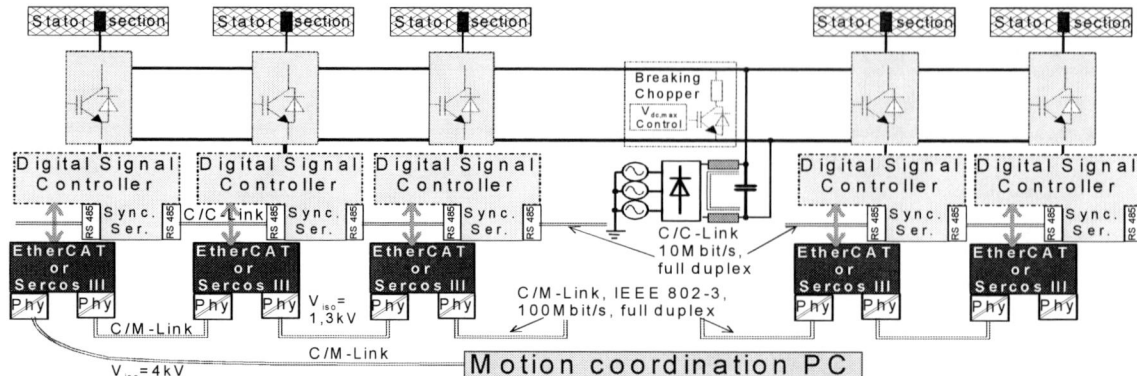

Fig. 10: One of several possibilities for the communication topology

plemented at the motion control PC.

3.) Download of control programs and parameters from motion control PC to controllers (off-line)

Communication topology: Fig. 10 shows one of several possibilities for the communication topology.

The *C/C-links* are short (1-2m) point-to-point connections between two neighboring controllers. The TMS320F2812 controllers provide two synchronous full duplex serial interfaces (SPI and McBSP) which can operate at >10Mbit/s, using internal FIFOs. At the physical layer, RS 485 standard claims to transmit >10Mbit/s for short point-to-point connections, but it needs more investigation concerning noise and common mode voltages which will appear in the proposed system, operating at negative dc-link potential.

For the *C/M-link* a modern field-bus like EtherCat or Sercos III may be used. Their physical layer is based on IEEE 802.3 and provide 100Mbit/s full duplex, isolated by signal transformers (up to V_{iso}=4 kV). With EtherCat, the frame sent by the master (motion coordination PC) passes through all slaves (controllers) while the slaves extract and insert their messages from and to the data stream on the fly (100Mbit/s). The last slave returns the bit-stream on the second twisted pair, forming a ring topology.

With the above-proposed links, the communication demand in a rather large system of linear drives can be satisfied.

IV.   CONCLUSION

For integrated material handling and processing in industrial applications, permanent magnet (PM) linear drive systems are proposed. Solutions with Short Primary and Long Primary are compared. The Short Primary system (Fig. 2-A) is best suited for applications with

➢ demand for auxiliary energy on board of the vehicles.
➢ a small number of vehicles per length of the track
➢ low accelerations and dynamics of the vehicles

For the Long Primary PM linear drive, different topologies concerning power- and information- processing were compared. Only solutions with one inverter per stator section (dedicated inverters, Fig. 2-B2) are recommended. For applications, which need an arbitrary large number of vehicles, a solution with one inverter and one controller for each stator section should be used (Fig.2-B2b). The large number of inverters and controllers in such a system will only be accepted, if reasonable costs and reliability can be achieved.

Concerning cost reduction, use of transfer molded IPMs, omitting isolation between controllers and IPMs as well as low cost current sensing is proposed. The demand for communication between the numerous controllers and a superimposed motion coordination is outlined and a proposal for a topology is discussed.

Such a system (Fig.2-B2b) is best suited for applications with

➢ very high accelerations and dynamics of the vehicles
➢ a large number of vehicles per length of the track (high traffic density)

For applications, where very high accelerations and dynamics of the vehicles are necessary too, but only a small number (10-15) of vehicles are a required, a system where one controller is assigned to one vehicle (Fig. 2-B2a) is best suited. There, an Inverter- Bus provides communication between each controller and each inverter within each control cycle (100μs). For such a system, an experimental set- up was build and experimental results are presented.

REFERENCES

[1] Andreas Pottharst, Markus Henke, Horst Grotstollen: "Power Supply Concept of the Longstator Linear Motor of the NBP-Test Track". *EPE-PEMC 2002 Dubrovnik & Cavtat*

[2] Bo Yang, Horst Grotstollen: Control Strategy for a Novel Combined Operation of Long Stator and Short Stator Linear Drive System". *EPE 2003 Toulouse*

[3] Junha Kim, Bon-Gwan Gu, Gubae Kang, and Kwanghee Nam "Attractive Force Reducing Strategy of LIM for PRT Systems*" IEMDC 2003 Madison*

[4] N.A.Duffi, R.D.Lorenz, J.L.Sanders: High performance LIM based material transfer, Proc. *NSF Design and Manufacturing Systems Conf.,* Atlanta,GA, Jan 8-10, 1992, S.1027-1030

[5] Benavides, R.; Mutschler, P.: „Controlling a System of Linear Drives". *IEEE- Power Electronics Specialists Conference PESC 2005,* 13.-17. June 2005, Recife, Brazil

[6] Benavides,R.; Mutschler, P.: „Compensation of Disturbances in Segmented Long Stator Linear Drives using Finite Element Models" . *IEEE International Symposium on Industrial Electronics ISIE ,* 7.-13. Juli 2006 Montreal Kanada

[7] Benavides, R.; Mutschler, P. "Detent force compensation in Segmented Long Stator Permanent Magnet Linear Drives using Finite Element Models." *12th European Conference on Power Electronics and Applications EPE* 2-5. Sept. 2007, Aalborg, Denmark

[8] A.Banerjee, R.Leidhold, R.Benavides and P.Mutschler: "Integration of the Inverters with the Segmented Stators in Long Stator Permanent Magnet Linear Synchronous Motor Drives" *6th International Symposium on Linear Drives for Industrial Applications LDIA*, Sep. 16-19, 2007, Lille, France

978-1-4244-1871-8/07 $25.00 © 2007 IEEE

AC-SPWM-Cycloconverter Based on An Extended Chopper Scheme

Qianzhi Zhou SM IEEE, Weihua Wu, Jianwei Lin

Key Laboratory of Power Electronics & Motion Control
Anhui University of Technology
Maanshan, P. R. China (243002)
Email: azhouqz@hotmail.com, wwh@ahut.edu.cn

Abstract— A novel cycloconverter based on chopper scheme is proposed, which is entirely different from the former cycloconverters based on phase-shift control, so leading to higher power factor than that of the latter, also in price of minimum power device numuber. Two kinds methods are derived for determining the relationships between Duty-Cycle and other parameters by means of different Area-Equivalent criteria. The single and 3-phase topologies and the ac-SPWM control strategy are introduced, which can realize variable frequency and variable voltage operation at the same time. Also some key simulations and experimental results show the validity of the conversion as well as the applicable prospect of the new cycloconverter-category.

I . INTRODUCTION

Due to 4-quadrant operation and direct AC-AC conversion without DC link-unit, cycloconverters have being widely used in low speed control since 70 years of last century. However, more active devices must be adopted. Also low power factor of them comes from phase-shift controlling SCRs. AC-AC conversion based on chopper scheme with less devices, high power factor and simple control strategy, almost being developed at the same period with traditional cycloconverters, but only operating in a fixed frequency [1]. For further reducing active devices, 3 phase-3 device ac chopper with freewheeling path of passive elements was just reported in 2006[2]. The present paper extends the function of an ac chopper for changing the output voltage in operation frequency with the output voltage together. Hence, a duel-frequency waveform-control strategy is introduced to produce an ac-SPWM-cycloconverter with minimum device, high power factor and simple control mode.

II . CHOPPER-CYCLOCONVERTER SCHEME

A. Fundamental Variable Frequency Principle

The waveforms in Fig.1 show the traditional ac chopper operating in a fixed frequency of electrical network, where the pulse-voltage with a sinusoidal envelope on the load varies with the Duty-Cycle in proportion. To change the frequency, a special waveform-control strategy is introduced as in Fig.2 (a) and (b). In (a), 3 half-period network waveforms with positive envelopes covering $0°$ to $180°$ and another 3 half-period network waveforms with negative envelopes covering $180°$ to $360°$ are chopped by

switching frequency, F_s (5kHz) correspondingly, the negative chopping procedure between $0°$ and $180°$ and the positive chopping procedure between $180°$ and $360°$ are plugged by a controlled rectangular pulse in step of the network frequency, F_i. Here, the time-sequence of the rectangular pulse series is set according to divide the network frequency, and the dividing factor is expressed by M (M is an integral number). While M is an odd number, the Output voltage frequency $F_o = M \cdot F_i / (2N-1)$, N expresses the number of wave-heads versus Fi. While M is an even number, $F_o = Fi/2N$. Obviously, Fig.2(a) is just an example of odd dividing, where $F_o = 50/(2 \times 3 - 1) = 10Hz$, ($M=5$, $N=3$). Fig.2(b) shows an even dividing situation, where $F_o = 50/(2 \times 2) = $, ($M=4$, $N=2$). Fig.3 shows a traditional topology of an ac chopper with fixed operation frequency, originally, only changing the output voltage with the successive equal pulse-width Duty-Cycle, where just one power device is adopted.

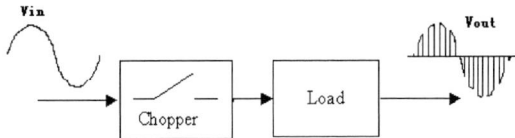

(a) Fundamental principle

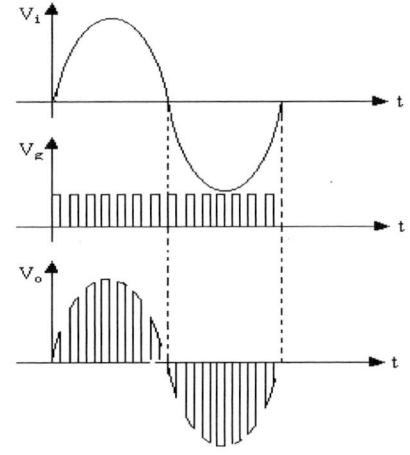

(b) Typical waveforms

Fig.1: Traditional ac chopper scheme

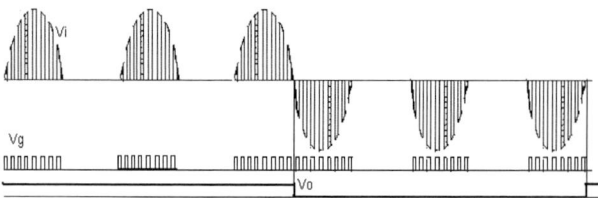

(a) Odd dividing net-frequency (F_o=10Hz)

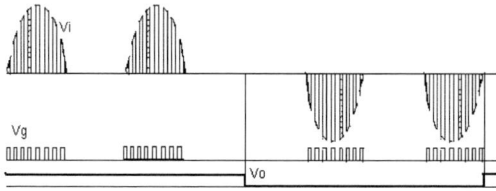

(b) Even dividing net-frequency (F_o=12.5Hz)

Fig.2: Variable frequency control strategy by chopper scheme

If the chopper switching Duty-Cycle is controlled by above plugged pulse-series and modulated rectangular waveforms with desired operation frequency F_o, then the frequency of the output voltage V_o can be varied simultaneously with its voltage variation. However, merely by previous measures, the envelope waveforms of the output voltage, V_o are shaped as flat tops. Furthermore, for better sinusoidal waveforms, while the chopper switching Duty-Cycle of each wave-head under the same envelope curve is set according sinusoidal regulation, not equally distributed, i.e. setting the Duty-Cycle of the first wave-head equal to that of Nth wave-head and smallest, setting the Duty-Cycle of the second wave-head equal to that of N^{th-1} wave-head and smaller. Set the Duty-Cycle of central wave-head maximum and proportional to the peak value of desired voltage. Then the envelope curve of the output voltage, V_o can be equivalent sine shape. It can be named after ac-SPWM modulation. In other words, while a traditional ac chopper changes its normal control pulse-series from successive and equal width into periodically plugging, and sinusoidal width-distribution, it becomes an ac-SPWM cycloconverter. Fig.4 shows its output voltage waveforms with sinusoidal envelope.

To implement above regulation, CPLD is adopted to determine the operation frequency by means of the wave-head numbers with the network frequency.

B. Relationship of Vo-D

According to certain relationship between the output voltage, V_o and Duty-Cycle, D, ac-SPWM cycloconverter can operate with variable voltage and variable frequency simultaneously, which is also realized inside CPLD. Take odd-dividing frequency mode for ex., this relationship can be derived by means of Area Equivalent Criterion (i.e. Impulse-Equivalent Criterion) as the follow.

The network voltage can be expressed as

$$v_i = V_{im} \sin\omega_i t \qquad (1)$$

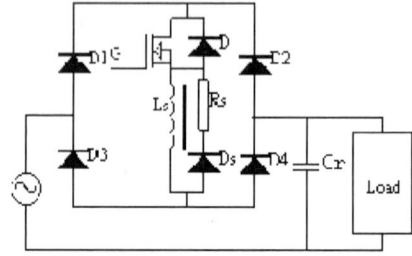

Fig.3: AC-AC conversion topology with single power device

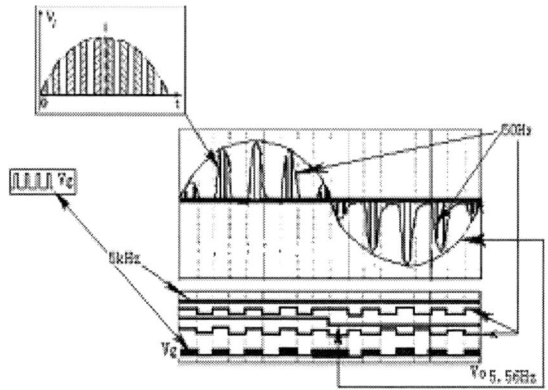

Fig.4: Typical waveforms from an ac-SPWM cyloconverter

While v_i is chopped by switching frequency, F_i with Duty-Cycle, D, a series of voltage pulses with v_i's envelope are expressed as in Fig.5.

From Fig. 5, the sum of areas, Si covered by a series of voltage pulses under the half-cycle of the sinusoidal envelop can be integrated by the following equation

$$
\begin{aligned}
S_i &= \sum_{n=0}^{(T_i/2)/T_s-1} \int_{nT_s}^{(n+D)T_s} V_{im} \sin\omega_i \, dt \\
&\approx \sum_{n=0}^{(T_i/2)/T_s-1} (DT_s) \bullet V_{im} \sin(\omega_i \bullet nT_s) \\
&= DT_s V_{im} \sum_{n=0}^{(T_i/2)/T_s-1} \sin(\omega_i \bullet nT_s)
\end{aligned}
\qquad (2)
$$

where T_s is the switch cycle, T_i is the network cycle, ω_i is the network angle frequency, n expresses the nth integrated unit.

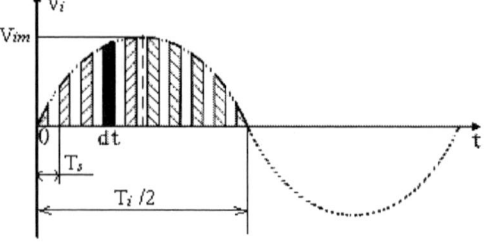

Fig.5: Equivalent area-sum of voltage pulses under half-cycle net-voltage

978-1-4244-1871-8/07 $25.00 © 2007 IEEE   1032

There are two methods to derive the required results.

a) Sub-unit Area-Equivalence

Fig.6 shows an area equivalent criterion, where for a 3 wave-head output voltage of an ac SPWM cycloconverter with D, it can be set that

$$S_{i1}=S_{o1}, \; S_{i2}=S_{o2}, \; S_{i3}=S_{o3}.$$

This could be extended to N wave-heads situation. So it follows

$$S_{ip}=S_{op}, \; (p=1, 2, 3, \; \cdots\cdots N).$$

Calculate the pth sub-area

$$S_{op} = \int_{\frac{1}{N}\frac{T_o}{2}(p-1)}^{\frac{1}{N}\frac{T_o}{2}p} V_{om}\sin(\omega_o t)dt \qquad (3)$$

$$= (V_{om}/\omega_o)[\cos(\frac{p-1}{N}\pi) - \cos(\frac{p}{N}\pi)]$$

From (2), the equivalent area in the pth wave-head should be expressed as

$$S_{ip} = D_p T_s V_{pm} \sum_{n=0}^{(T_i/2)/T_s-1}\sin(\omega_i \bullet nT_s) \qquad (p=1,2,3......N) \; (4)$$

Since $S_{ip}=S_{op}$, from (3) and (4),

$$D_p T_s V_{im} \sum_{n=0}^{(T_i/2)/T_s-1}\sin(\omega_i \bullet nT_s)$$

$$= (V_{om}/\omega_o)[\cos(\frac{p-1}{N}\pi) - \cos(\frac{p}{N}\pi)]$$

(5)

Hence, the Duty-Cycle of pth wave-head covered under the network voltage envelope is

$$D_p = \frac{V_{om}\left[\cos(\frac{p-1}{N}\pi) - \cos\frac{p}{N}\pi\right]}{\omega_o T_s V_{im} \sum_{n=0}^{(T_i/2)T_s-1}\sin(\omega_i \bullet nT_s)} \qquad (6)$$

(6) illustrates D_p versus V_{om}, ω_i, ω_o, n, T_s, and $D_p \propto V_{om}$, which can be available for CPLD-program.

b) Full Area-Equivalence

In this situation, for the expression in Fig.6, it should be set the whole area covered by the half-cycle of sine-waveform with operation frequency, S_o to equal to the area of the sum of S_{i1}, S_{i2} and S_{i3}, i.e.

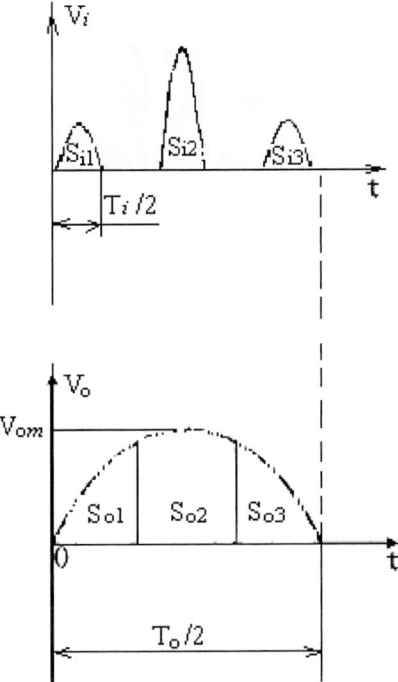

Fig. 6: Area-Equivalent to each sub-unit

$$S_o = (S_{i1}+ S_{i2} + S_{i3})$$

which can be extended as

$$S_o = (S_{i1}+ S_{i2} + S_{i3} + \cdots\cdots + S_N) = \sum_{p=1}^{N} S_{ip} \qquad (7)$$

(7) expresses the total energy delivered by chopped wave-heads of sub-unit with number N via the converter should be equivalent to that by half-cycle of the expected output waveforms with desired frequency.

Also, from (1)

$$S_o = \int_0^{T_o/2} V_{om}\sin(\omega_o t)dt = \frac{2V_{om}}{\omega_o} \qquad (8)$$

Then, the voltage peak value of the pth wave-head can be expressed as

$$V_{pm} = V_{im}\sin\left[\frac{\pi}{2(2N-1)}(4p-3)\right] \qquad (9)$$

It can be obtained from (4), (7), (8) and (9) that

$$\frac{2V_{om}}{\omega_o} = \sum_{p=1}^{N} \left\{ D_p T_s V_{im} \sin\left[\frac{\pi}{2(2N-1)}(4p-3)\right] \sum_{n=0}^{(T_i/2)/T_s-1} \sin(\omega_i \bullet nT_s) \right\}$$

$$T_s V_{im} \left[\sum_{n=0}^{(T_i/2)T_s-1} \sin(\omega_i \bullet nT_s)\right] \bullet \sum_{p=1}^{N} \left\{ D_p \sin\left[\frac{\pi}{2(2N-1)}(4p-3)\right] \right\}$$

(10)

From (10),

$$D_p = \frac{2V_{om}}{\omega_o T_s V_{im} \left[\displaystyle\sum_{n=0}^{(T_i/2)/T_s-1} \sin(\omega_i \bullet nT_s)\right] \bullet \displaystyle\sum_{p=1}^{N} \sin\left[\frac{\pi}{2(2N-1)}(4p-3)\right]}$$

(11)

(11) specifies that as long as the expected voltage amplitude is set, the equivalent Duty-Cycle of the pth wave-head can be confirmed versus ω_o, ω_i, T_s etc., which presents fundamental calculation for CPLD.

Obviously, previous two kinds of equivalent criteria can be conveniently available to implement V/F regulation in motor speed control, operating below the half-frequency of the network, also realize the variable voltage speed control at network frequency. The two control modes can be smoothly switching by software to make a continuous speed adjustment from ultra-low frequency to the net-frequency.

III. EXTENSION

Previous single-phase scheme can be extended into 3 phase circuits. Fig.7 and Fig.8 introduce two kinds of voltage regulators based on ac-chopper scheme with minimum active devices [1], [2]. For extending them into cycloconverters, just apply previous ac-SPWM control strategies to them. For high performance applications, large filter can be added to previous circuits getting continuous waveforms. Another way is introducing 3-device system for single phase and 9 device- system for 3 phases, so as to get Fig.9's continuous waveform.

The authors' present work is just developing Fig.8's topology for ac-SPWM cycloconverter applications. Besides saving devices and simplifying the control strategy, the parameters in the π-LC unit can be acted in multi-functions, which is either a filter for switch frequency, or a freewheeling path of the current to release the energy from the inductors while all the power devices are in off-state, as well as a snubber for improving the state of switching on and off, while synthetically considering the effect of the discrete inductance. Also, a natural commutation can be turned from the procedure of delivering main power into that of freewheeling without forcing transient and required dead time like in Fig.7's topology. In this condition, coupling applies to LC parameters, which requires the parameter selection to satisfy the following formula [2].

$$(L_f C_f)^{1/2} >> T_s/2\pi >> (L_s C_r)^{1/2}$$

Thus, soft switching can be realized at the same time without increasing auxiliary active devices like in other soft switching circuits. Here, non-crystallized material should be selected to manufacture the low-pass filter-inductor for reducing the switching consumption. And the practice verified that the electrical and CBB capacitors could be used in the π-LC unit.

It would be predicted that with the development of the passive components, the system size of ac-SPWM cycoloconverters could be increased gradually.

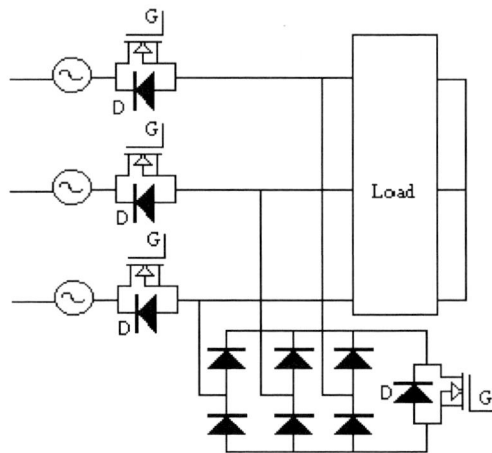

Fig. 7:   3 phase-chopper with 4 devices

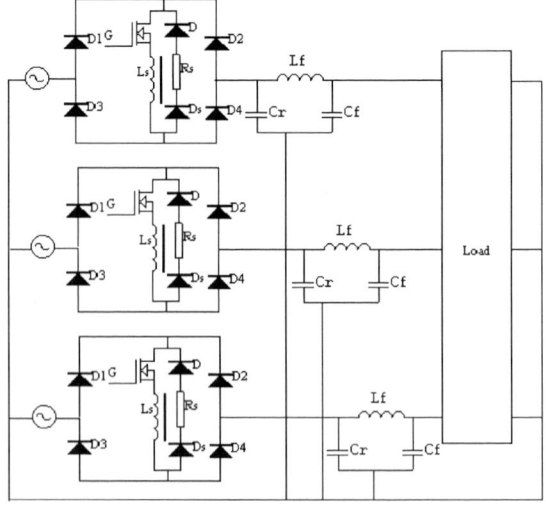

Fig.8:   3-phase chopper with 3devices

978-1-4244-1871-8/07 $25.00 © 2007 IEEE   1034

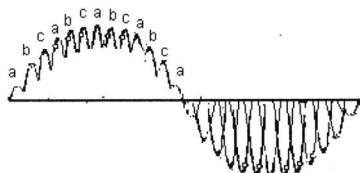

Fig.9: V_o formed by 3 phases of supplies

IV. SIMULATIONS AND EXPERIMENTS

Fig.10 shows the simulation results of the voltage waveforms on a resistive load in one phase of Fig.8's circuit, operating at 5.56Hz with 5kHz of chopper frequency (C_f=13μF, C_r=0.22μF, L_f=10mH, load R=200Ω). It can be seen that discontinuous pulse-series with equal width are periodically plugged by rectangular waveforms with net-frequency, and modulated by a rectangular waveform with desired frequency.

A fan and a light load were connected to Fig.3's circuit respectively, controlled by a CPLD chip, operating from 0.5Hz to 25 Hz in ac-SPWM mode. Fig.11 shows typical experimental waveform series, illustrating that the lower the operating frequency is, the better the THD performance of ac-SPWM cycloconverter is. The output voltage waveforms on load in (c) seem to be continuous since the low-pass filter action becomes stronger in lower operation frequency.

The wave-envelopes in (b) and (d) apparently demonstrate the harmonics contents due to the influences from the motor parameters, which is not hard to smooth by modifying the Duty-Cycle value, letting each peak value of the wave-head to be able to land on the sinusoidal envelope curve. Also, (a), (b) and (d) are got from Sub-unit Area-Equivalent Criterion, while (c) is got from Full Area-Equivalent Criterion, which makes the latter obtain better sinusoidal envelope than that of the former, however, the voltage pulses width in the first wave-head and the $(N-1)^{th}$ wave-head in the latter condition will be very narrow, so leading to the energy-delivery much difficult. Fig.12 shows π-LC unit with special parameter-match resulting in quasi-zero switching.

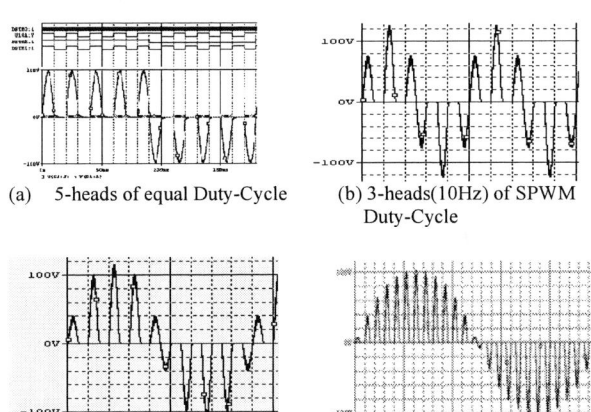

(a) 5-heads of equal Duty-Cycle

(b) 3-heads(10Hz) of SPWM Duty-Cycle

(b) 5-heads (5.56Hz) of SPWM Duty-Cycle

(c) 13-heads (2Hz) of SPWM Duty-Cycle

Fig.10: Simulation waveforms in one phase

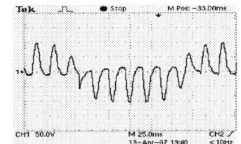 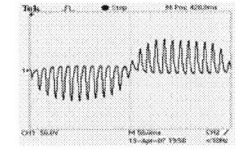

(a) 7wave-heads (38.46Hz) for fan   (b) 13 wave-heads (2Hz) for fan

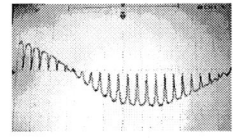 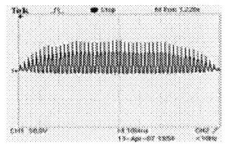

(c) 17 wave-heads (1. 5Hz) for light   (d) 51 wave-heads (0.5Hz) for fan
Fig.11: Experimental out-voltage waves with ac-SPWM regulation

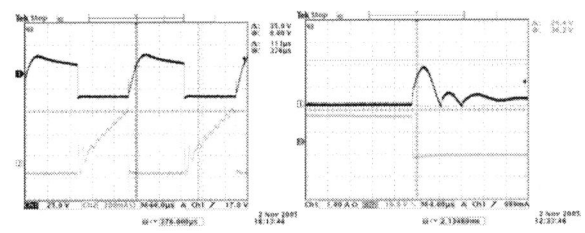

(a) Quasi-ZVS validity   (b) Quasi-ZCS validity
Fig.12: Soft-switching relying on parameter match

V. CONCLUSION

A new cycloconverter principle based on ac-SPWM chopper scheme has been verified that a traditional ac chopper can become a cyclocoverter while the waveform-control strategy is changed from successive equal pulse-width Duty-Cycle control into variable pulse-width Duty-Cycle control with net-frequency plugging and desired frequency modulation. Much more power devices can be reduced in ac-SPWM cycloconverter than that in traditional cycloconverter, as well as better power factor. Due to a good performance / cost value, ac-SPWM cycloconverters would have a wide application prospect in air conditioners, refrigerators under half-net-frequency operation hybridizing with pure voltage regulation at net-frequency, for saving electrical energies, or in 3 phase-system, like motor starters for getting high power factor and low losses, and elevators, hoists, or ac servo system for utilizing their good ultra-low frequency performance with the function of bi-direction delivering of energy.

ACKNOWLEDGMENT

The authors thank to Mr. Gang Li and Wei Duang for their diligent work in experiments.

REFERENCES

[1] Donato Vincenti, Hua Jin, Phoivos Ziogas, "Design and implementation of a 25-kVA three-phase PWM ac Line Conditioner," *IEEE Trans. Power Electronics*, vol. 9, no. 4, pp.384-389, 1994.

[2] Qianzhi Zhou, Wenhua Hu and Bin Wu, "3 Phases-3 devices ac voltage regulator with quasi-zero switching," in *Proc. International Conf. Power Electronics and Motion Control*, Shanghai, China, 2006, vol.1, pp. 397-400.

Matrix Converter with a Novel General Commutation Strategy

Nam-Sup Choi[1*], Yulong Li[2], Byung-Moon Han[3], Jong-Sun Ko[4], Eui-Cheol Nho[5]

[1] Departement of Elctrcial Enginerring, Chonnam National University, Yeosu, Korea, nschoi@chonnam.ac.kr
[2] Departement of Elctrcial Enginerring, Chonnam National University, Yeosu, Korea, gerrylee930@gmail.com
[3] Departement of Elctrcial Enginerring, Myungji University, Korea, erichan@mju.ac.kr
[4] Dankook University, Korea, jsko@dku.edu
[5] Pukyong National University, Korea, nhoec@pknu.ac.kr

Abstract—A novel general commutation strategy is proposed for *n*-phase to *m*-phase matrix converter applications. The proposed circuit with a simple RC snubber as commutation aid requires no extra sensing information allowing conventional dead time commutation. In this paper, detailed commutation procedure of the proposed circuit is analyzed; the choice of specific semiconductor switches and snubber components is also derived and given. The operation of the proposed circuit is very stable, reliable and robust. Finally, experimental results support the validity of the proposed structure and circuit.

I. INTRODUCTION

The matrix converter is an array of controlled semiconductor switches that connect directly the three-phase source to the three-phase load. This converter has several attractive features that have been investigated in recent years. Among them are:

1) simple and compact power circuit without any large energy storage elements

2) generation of output voltage with variable magnitude and frequency

3) sinusoidal input and output current

4) controlled input power factor

5) regeneration capability.

Due to these advantages, the matrix converter finds its application in variable power conversion, AC drive and renewable power generation [1].

The key element in a matrix converter is the fully controlled four-quadrant bidirectional switches, which are capable of conducting bidirectional currents and blocking bidirectional voltages. The normally utilized bidirectional switches are the common collector or common emitter structures [2]. Meanwhile, reliable commutation in matrix converters is very important and practical. All the switches have to be preciously controlled in order to guarantee continuous current flow for inductive loads. With the common collector and common emitter structures, an output current direction based four-step commutation method is proposed in [3]. Also, a method with measuring the voltage across the individual semiconductor device to perform a so-called two-step semi-soft commutation is proposed [4]. Another technique based on the knowledge of the relative magnitudes of the input voltages is given in [5]. These techniques can all solve the commutation problem. However, these methods all require extra sensing information. The above mentioned methods also have some drawbacks such as relatively long commutation time, which may cause high output voltage loss and distortion. In addition, due to noise and possible perturbation around zero-crossing point where the direction of current or voltage is changed, there may be some sensing ambiguity.

In order to overcome the above described problems, in this paper, a novel general commutation strategy is proposed for *n*-phase to *m*-phase matrix converter application. The proposed strategy utilizes a simple RC circuit as commutation aid, eliminating the need of current or voltage sensing. Resultantly, dead-time commutation can be easily applied, leading to stable and reliable operation.

In this paper, a brief review of existed bidirectional switch structures and commutation techniques is given. Then for the proposed technique, detailed circuit analysis is performed and commutation procedure is illustrated. Meanwhile, equations for determinations of rated semiconductor devices and circuit components are also investigated. Finally, experimental results show the verification and validity of the proposed circuit.

II. REVIEW OF BIDIRECTIONAL SWITCHES AND COMMUTATION TECHNIQUES FOR MATRIX CONVERTER

A. Bidirectional switches

The matrix converter requires a bidirectional switch capable of blocking voltage and conducting current in both directions. Unfortunately there are few discrete semiconductor devices currently available that meets this requirement. Discrete devices need to be used to construct suitable bidirectional switch cells. Generally, there are two arrangements commonly used to create this bidirectional switch. Fig. 1 shows these two bidirectional switch structures: common emitter in Fig. 1(a) and common collector in Fig. 1(b). The two kinds of structure shown in Fig. 1 both involve a pair of switching devices and diodes in a back-to-back arrangement. For the switching devices, insulated gate bipolar transistors (IGBTs) can be used, but other forced commutated semiconductor switches can also be used in the same way. Fig. 2 shows another bidirectional switch configuration which utilizes the commercial IGBT and diode modules. With the switch structure in Fig. 2, the internal diode with relatively slow reverse recovery characteristics can be substituted by the fast recovery diode.

B. Commutation techniques

Reliable current commutation between switches in matrix converters is more difficult than in conventional

978-1-4244-1871-8/07 $25.00 © 2007 IEEE

voltage source inverters since there are no natural free wheeling paths. The commutation has to be actively controlled at all times with respect to two basic rules. These rules can be visualized by considering just two switch cells on one output phase of a matrix converter. It is important that no two bidirectional switches are switched on at any instant, as shown pictorially in Fig. 2(a). This would result in line-to-line short circuits and the destruction of the converter due to over currents. Also, the bidirectional switches for each output phase should not all be turned off at any instant, as shown in Fig. 2(b). This would result in the absence of a path for the inductive load current, causing large over voltages. These two considerations cause a conflict since semiconductor devices cannot be switched instantaneously due to propagation delays and finite switching times.

Various solutions for commutation were proposed. In [3], a four step commutation strategy is given. By sensing the output current direction and gating the switches in a specific determined sequence, direction of current flow through the commutation cells can be controlled. However, this kind of current direction information based technique has some problems because it is difficult to reliable determine the output current direction in a switching power converter, especially at low current levels. To avoid this problem, in [4], a technique for using the voltage across the bidirectional switch to determine the current direction has been developed. This method is more reliable than the current sensing method, but a relatively expensive digital gate array such as a complex programmable logic device (CPLD) or field programmable gate array (FPGA) has to be used to implement the control logic. Another alternative approach in [5] technique relies on the knowledge of the relative magnitudes of the input voltages instead of considering the direction of the output current.

Whether the current direction based commutation methods or the voltage level based techniques all require peripheral sensing circuit. The extra circuit part may add complexity to the overall system. And possible sensing error or sometimes sensing mistake may cause damage to the system or devices.

In order to overcome the above problems, a general

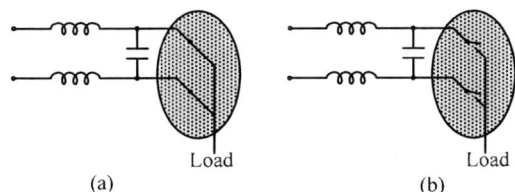

Fig. 3(a) Avoid short circuits on the matrix converter input lines; 3(b) Avoid open circuits on the matrix converter output lines

commutation technique is proposed in this paper.

III. PROPOSED COMMUTATION CIRCUIT

A. Two phase to single phase matrix converter

Fig. 4(a) shows the circuit topology of a two phase to single phase matrix converter. The converter mainly consists of two bidirectional switches, S_A and S_B namely. v_{sa} and v_{sb} are the input voltage sources and i_o is the output current. The bidirectional switches S_A and S_B can be of any structure such as the common-collector or common-emitter arrangement or the arrangement shown in Fig. 3(b). Extra elements C_{sA}, C_{sB}, R_s are added for commutation purpose. The gating signals for bidirectional switches are in the type of Pulse Width Modulation (PWM). The gating signal is shown in Fig. 4(b). It should be noted that a short dead time t_d is inserted.

B. Three phase to three phase matrix converter

Fig. 5 shows the three phase to three phase matrix converter with the proposed commutation circuit. As seen in Fig. 5, the converter mainly contains nine bidirectional switches, which are S_{11}, S_{12}, S_{13}, S_{21}, S_{22}, S_{23}, S_{31}, S_{32} and S_{33}. C_{s11}, C_{s21}, C_{s31}, C_{s12}, C_{s22}, C_{s32}, C_{s13}, C_{s23}, C_{s33}, R_{s1}, R_{s2} and R_{s3} are added as commutation circuits.

C. n-phase to m-phase matrix converter

The same technique can be easily applied to the general

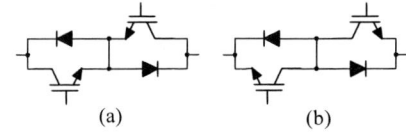

Fig. 1 Bidirectional switch structures:
(a) Common emitter; (b) Common collector

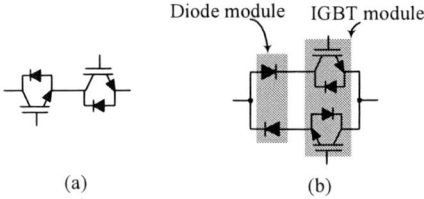

Fig. 2 Bidirectional switch structures:
(a) Commercial IGBT module; (b) Bidirectional switch with commercial IGBT module and diode module

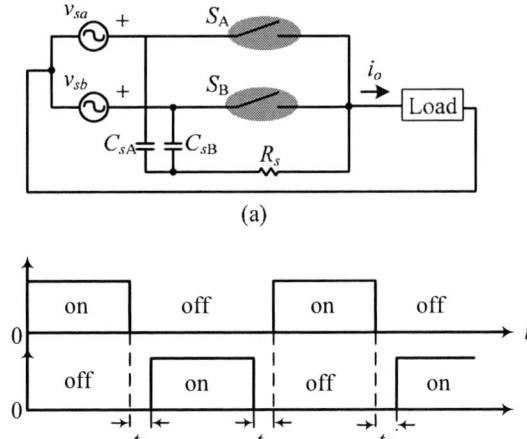

Fig. 4 Two phase to single phase matrix converter.
(a) Circuit; (b) Switches' control signal.

978-1-4244-1871-8/07 $25.00 © 2007 IEEE   1037

n-phase to *m*-phase matrix converter. Fig. 6 shows the *n*-phase to *m*-phase matrix converter with the proposed commutation circuit. It can be seen from Fig. 4 to Fig. 6 that the proposed strategy can be considered as a general approach to solve the matrix converter commutation problem.

IV. COMMUTATION PROCEDURE

In order to explain the commutation principle, it is enough

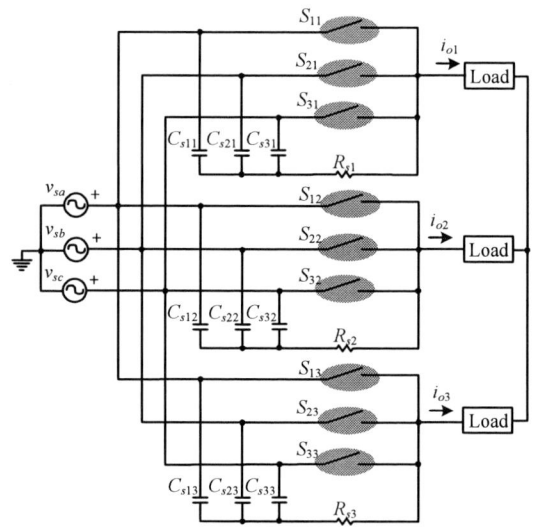

Fig. 5 Three phase to three phase matrix converter

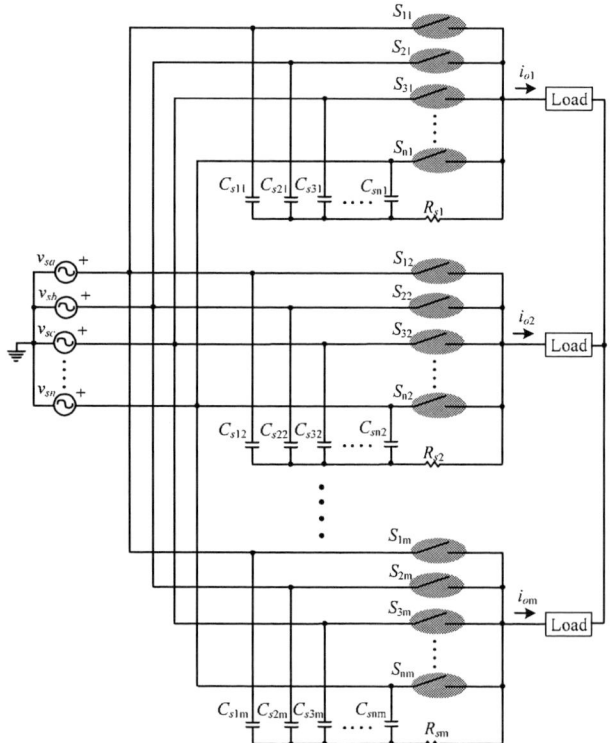

Fig. 6 n-phase to m-phase matrix converter

to deal with a two-phase to single-phase matrix converter. Fig. 7(a) is the schematic of a two-phase to single-phase matrix converter with the proposed commutation circuit. And Fig. 7(b) shows the gating signal for the IGBTs (S_1 S_2, S_3 and S_4). It should be noted that a short period of dead-time t_d is inserted in the PWM signal. IGBT switches S_1 and S_2 share the same gating signal, so do S_3 and S_4. Hence, compared with conventional commutation technique, the gating logic for completing commutation will be quite easy.

According to the input voltage polarity and output current direction, there exist 4 commutation cases. The condition when the input voltage and output current are both positive is dealt as an example. Operation under other condition will be quite similar. It is assumed that the switching frequency f_s is much higher than the source frequency f_i and output frequency. Also, for convenience, it is defined that $C_{s1}=C_{s2}=C_s$, $R_{s1}=R_{s2}=R_s$. The input voltage and output current are defined in (1) and (2) respectively.

$$v_s(t) = V_i \sin(\omega_i t) = V_i \sin(\theta) \tag{1}$$

$$i_L(t) = I_L \sin(\omega_o t + \phi) = I_L \sin(\theta + \varphi) \tag{2}$$

where $\varphi = (\omega_o - \omega_i)t + \phi$. The notations V_i and I_L will be the magnitude of input voltage and output current respectively.

A. Turing off $S_3(S_4)$ and turning on $S_1(S_2)$

Fig. 8 shows a set of typical operating waveforms. Fig. 9 shows the commutation modes, and the practical current path is shown with bold line in Fig. 8.

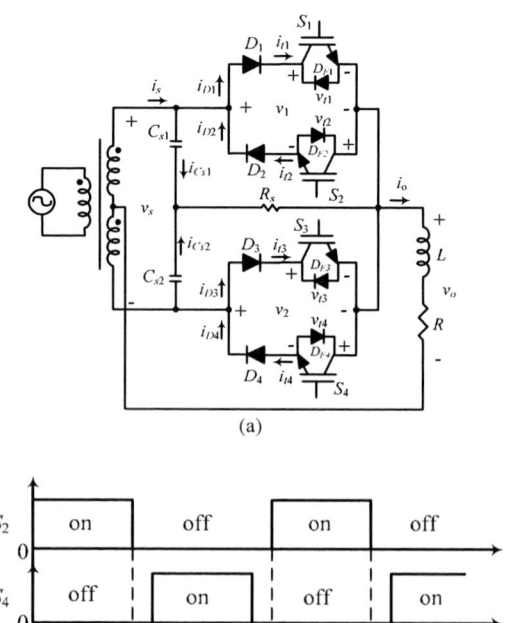

(a)

(b)

Fig. 7 Two phase to single phase matrix converter schematic.
(a) Circuit; (b) IGBTs' gating signal.

978-1-4244-1871-8/07 $25.00 © 2007 IEEE   1038

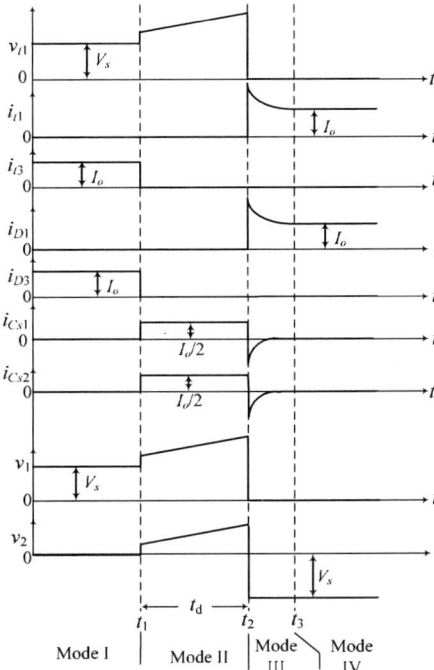

Fig. 8 Commutation waveforms (set I)

Mode-I: Mode-I is the initial circuit state. S_1 and S_2 are off, S_3 and S_4 are on. Output current I_o flows through D_3-S_3. Initially, $v_1=V_s$, $v_2=0$.

Mode-II: When dead time begins at t_1, all the four IGBTs are turned off together and mode-II begins. Since C_{s1}-R_s and C_{s2}-R_s imply same impendence, I_o will flow through these two loops equally and the two snubber capacitor (C_{s1}, C_{s2}) will be charged by this current.

At the end of dead time, S_1 and D_2 will have the maximum voltage stress,

$$v_{\max} = V_s + \frac{1}{2}R_s \cdot I_o + \frac{I_o}{2C_s} \cdot t_d \qquad (3)$$

From (3), it can be seen that a smaller dead time will be preferred to reduce the voltage stress across semiconductor devices. Actually, choice of dead time will depend on the turn-off characteristics of IGBT. It is recommended that the dead time be set just enough to safely turn off the IGBT.
In (3), it is worth noting that the notations V_s and I_o will be instant sampled values of (1) and (2) respectively.

Hence, by substituting (1) and (2) into (3) and collecting terms, (4) can be obtained.

$$V_{\max} = V_i + \frac{I_L}{2} \cdot \left(R_s + \frac{t_d}{C_s} \right) \cdot \qquad (4)$$

(4) determines the voltage rating of semiconductor devices. It can be seen from (8) that $V_{\max}$ is proportional to R_s and inverse to C_s.

Mode-III: When S_1 and S_2 are simultaneously turned on at t_2, dead time ends and mode-III starts. I_o commutates to D_1-

S_1 loop. v_1 jumps to zero and v_2 jumps to $-V_s$ immediately.

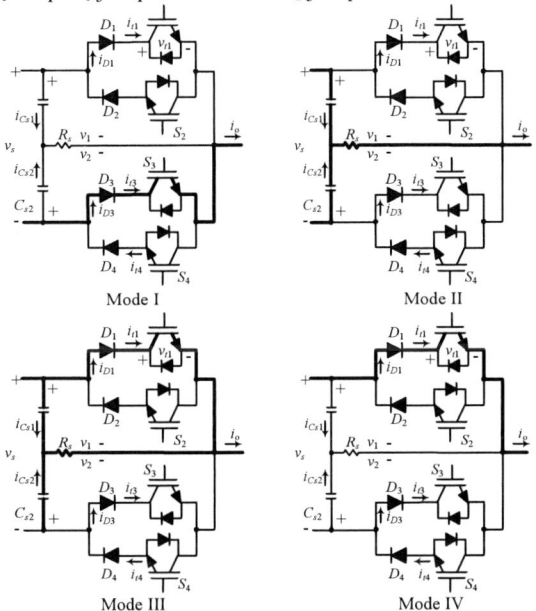

Fig. 9 Commutation modes (set I)

The previously charged C_{s1} will discharge through C_{s1}-D_1-S_1-R_s. C_{s2} will charge toward $-V_s$ by the source. In this mode, there exist three current components in the circuit; the load current, C_{s1} discharging current and C_{s2} charging current.

At the beginning of mode-III, i_{t1} and i_{D1} have the maximum value $i_{\max}$,

$$i_{\max} = \frac{1}{R_s}\left(2V_s + \frac{I_o t_d}{C_s} \right) + \frac{3}{2}I_o \cdot \qquad (5)$$

Since V_s and I_o will be instant sampled values of (1) and (2) respectively, by substituting (1) and (2) into (5) and collecting terms, (6) can be obtained,

$$I_{\max} = \frac{2V_i}{R_s} + (\frac{3}{2} + \frac{t_d}{R_s C_s}) \cdot I_L \cdot \qquad (6)$$

(6) determines the current rating of semiconductor devices. Seen from (6), $I_{\max}$ is inverse proportional to both R_s and C_s.

By (4) and (6) and choosing proper snubber component value, specific IGBT and diode can be selected.

Mode-IV: After discharging C_{s1} and charging C_{s2}, all the commutation procedure ends at t_3. The output current flows through D_1-S_1. Finally, $v_1=0$ and $v_2=-V_s$.

The commutation power loss in resistor R_s can be approximately calculated by considering the summation of capacitor C_s discharging energy.

$$E = 2 \times \frac{1}{2}C_s \sum_{n=1}^{k/4} v_{sn}^2 = 8 \times \frac{1}{2}C_s \sum_{n=1}^{k/4}\left(4f_i V_i \cdot \frac{n}{f_s} \right)^2 \qquad (7)$$

$$\approx \frac{1}{3}V_i^2 \cdot C_s \cdot k$$

978-1-4244-1871-8/07 $25.00 © 2007 IEEE   1039

where $k = \dfrac{f_s}{f_i}$.

Then, the resistor power will be

$$P = E \cdot f_i = \frac{1}{3} f_i \cdot V_i^2 \cdot C_s \cdot k \cdot \qquad (8)$$

In general, with n-phase input voltage source, (8) will be

$$P_n = \frac{n}{6} f_i \cdot V_i^2 \cdot C_s \cdot k \cdot \qquad (9)$$

B. Turing off $S_1(S_2)$ and turning on $S_3(S_4)$

Fig. 10 shows a set of typical operating waveforms. Fig. 11 shows the commutation modes, and the practical current path is shown with bold line in Fig. 10.

Mode-1: Mode-1 is the initial circuit state. S_1 and S_2 are on, S_3 and S_4 are off. Output current Io flows through D_1-S_1. Initially, $v_1=0$, $v_2=-V_s$.

Mode-2: When dead time begins at t_1, all the four IGBTs are turned off together and mode-2 begins. I_o will flow C_{s1}-R_s loop and C_{s2}-R_s loop equally, C_{s1} and C_{s2} will be charged by this current.

Mode-3: When S_3 and S_4 are simultaneously turned on at t_2, dead time ends and mode-3 begins. v_1 jumps to V_s and v_2 jumps to 0 immediately. C_{s1} will continue to charge by source to V_s. C_{s2} will discharge toward 0. At t_2, since v_2 implies a negative value, D_3 will be reverse blocked. Current will flow through S_4-D_4.

Mode-4: At t_3, D_4 is reversed blocked and D_3 conducts to flow current.

Mode-5: After charging C_{s1} and discharging C_{s2}, all the

commutation procedure ends at t_4. The output current flows through D_3-S_3. Finally, $v_1=V_s$ and $v_2=0$.

V. EXPERIMENT

An experimental setup was made to verify the operation of the proposed commutation circuit. The converter circuit is shown in Fig. 7(a) with the circuit parameters as follows:

V_i=100V, L=5mH, R=2.5Ω, f_s=5kHz,
R_s =10Ω, C_s=0.22μF, t_d=1μs.

Fig. 12 shows the overall converter operation with the same output frequency 60Hz. The upper two waveforms are the input and output voltage waveforms, while the lower two the input and output currents. As seen in Fig. 12, the converter operates well. The output current curve is quite continuous and smooth.

Fig. 13 is the experimental voltage and current waveforms of commutation aid components when the load current commutates from the lower switches to the upper switches. Fig. 14 is the commutation aid component experimental voltage and current waveforms when the load current commutates from the upper switches to the lower switches. It can be seen from Fig. 13 and Fig. 14 that the experiments results are identical to the analysis. In Fig. 13

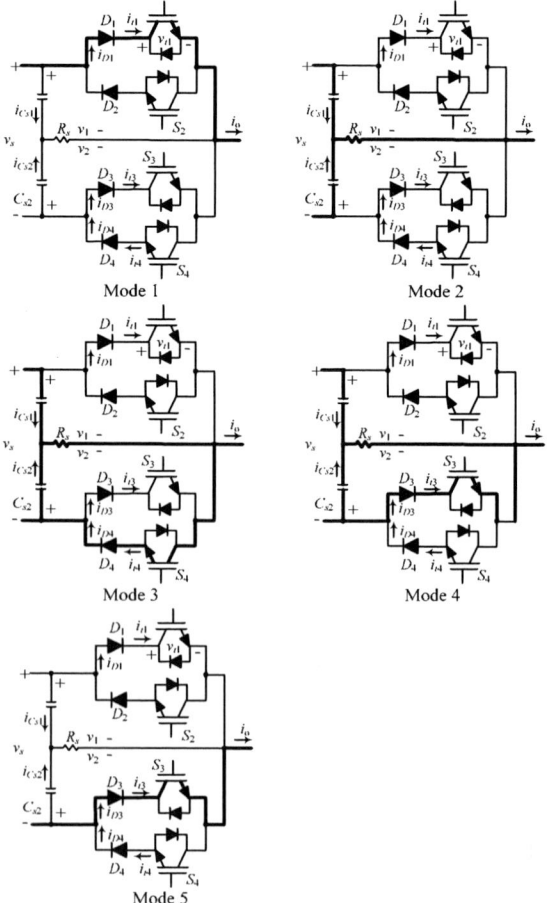

Fig. 11 Commutation modes (set II)

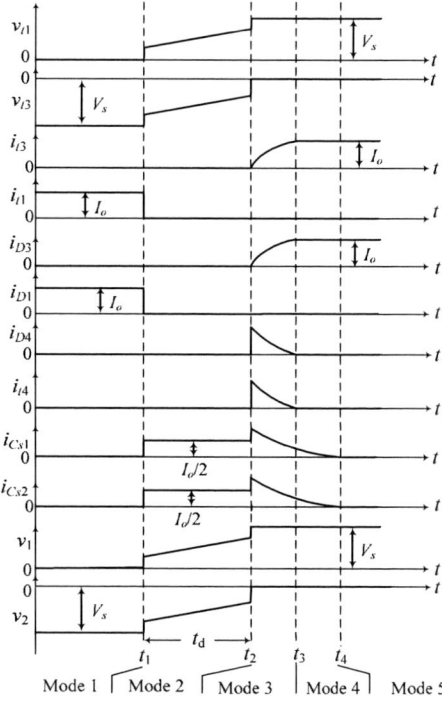

Fig. 10 Commutation waveforms (set II)

978-1-4244-1871-8/07 $25.00 © 2007 IEEE   1040

and Fig. 14, it can be seen that i_{cs1} and i_{cs2} contains some resonant waves. The lower frequency part is due to the stray inductance of the switch power circuit. The high frequency part is because of the diode forward recovery phenomena and line insertion of DC current probes.

Finally, seen from the experiment results, it can be concluded that the proposed circuit is quite available and very practical. It solves the matrix converter current commutation problem in a simple and reliable way.

VI. CONCLUSION

A novel general commutation strategy is proposed for matrix converter application. The proposed circuit enables safe commutation without extra sensed circuit information. Thus, the operation is very stable, reliable and robust. Operating principles are described and modes analysis is

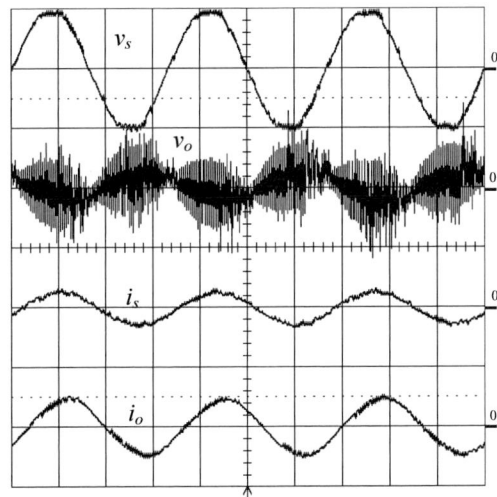

Fig. 12 Overall operation
(v_s: 100V/div, 5ms/div; v_o: 100V/div, 5ms/div;
i_s: 10A/div, 5ms/div; i_o: 10A/div, 5ms/div)

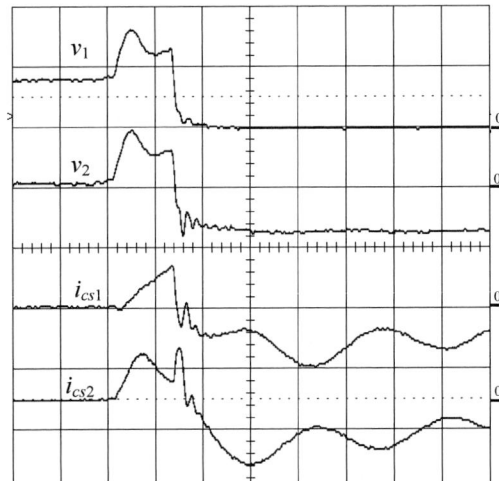

Fig. 13 Voltage and current waveforms when current commutates from the lower switches to upper switches
(v_1: 500V/div, 0.5μs/div; v_2: 50V/div, 0.5μs/div;
i_{cs1}: 5A/div, 0.5μs/div; i_{cs2}: 5A/div, 0.5μs/div)

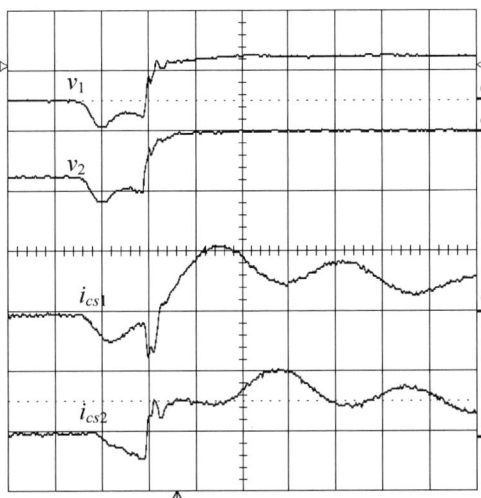

Fig. 14 Voltage and current waveforms when current commutates from the upper switches to lower switches
(v_1: 50V/div, 0.5μs/div; v_2: 50V/div, 0.5μs/div;
i_{cs1}: 5A/div, 0.5μs/div; i_{cs2}: 5A/div, 0.5μs/div)

given. Guidelines for choosing semiconductor devices and commutation components are also investigated. Finally, experiment results support the verification of the design

ACKNOWLEDGMENT

This study was financially supported by Chonnam National University(2006).

REFERENCES

[1] Gyugi, L. and Pelly, B.; "Static power frequency changers: theory, performance and applications"; *John Wiley and Sons*, 1976.
[2] Wheeler, P.W.; Rodriguez, J.; Clare, J.C.; Empringham, L.; Weinstein, A.; "Matrix converters: a technology review", *Industrial Electronics, IEEE Transactions on,* Volume 49, Issue 2, April 2002, Page(s): 276-288.
[3] Burany, N.; "Safe control of four-quadrant switches", *Industry Applications Society Annual Meeting*, 1989., vol.1, Page(s):1190 - 1194.
[4] Empringham, L., Wheeler, P. W., and Clare, J. C., "A Gate drive level intelligence and current sensing for matrix converter current commutation", *IEEE Transactions on Industrial Electronics*, Vol. 49, No. 2, Page(s): 382-389., April 2002.
[5] Ziegler, M.; Hofmann, W., "Performance of a two steps commutated matrix converter for AC-variable-speed driver", *Proceeding of the European Power Electronics Conference*, September. 1999, CD-ROM.

Novel DC link balancing scheme in generic n-level back-to-back converter system

Lech M. Grzesiak
Warsaw University of Technology
Institute of Control and Industrial Electronics
Warsaw 00-662, Poland
Email: L.Grzesiak@isep.pw.edu.pl

Jacek G. Tomasik
Warsaw University of Technology
Institute of Control and Industrial Electronics
Warsaw 00-662, Poland
Email: tomasikj@webmail.co.za

Abstract— Paper presents a novel DC link balancing scheme for a generic n-level back-to-back system with multi-level diode clamped topologies. The proposed algorithm is generalization of the control strategy formerly introduced with the five-level back-to-back system and it relays on measurement of adjacent capacitor voltages which provide information about the potential variation in consecutive nodes of the n-level DC link network. Subsequently the adequate control signals are used to modify the multi-carrier PWM modulators on the rectifier side as well as on the inverter side respectively. In result, the voltages across all the capacitors in DC link network are effectively maintained balanced. The new balancing scheme has been demonstrated in simulations with 5-level, 7-level and 9-level single phase back-to-back systems and it can be extended to any n-level back-to-back system.

I. INTRODUCTION

In an AC-DC-AC energy conversion, generalized with a multi-level diode clamped topology, the capacitor imbalance problem can be handled in two ways: firstly, via supplying a particular voltage level from an isolated winding transformer and a four full-wave bridge rectifiers, whereas a primary is interconnected with the input AC-side [1] or secondly, via a back-to-back connection of a front-end AC-DC multi-level converter with a back-end multi-level DC-AC converter.

The first solution is impractical from a size-wise and a cost point of view and additionally the system without a power factor correction generates a non-negligible amount of harmonics which are restricted by the utility regulations.

The second alternative solution many-times reported in a literature e.g. [2], [3] seems to be a suitable one since it was shown in an operation with the back-end active-rectifier interfaced to a constant frequency/voltage input of the utility [4] or variable frequency/voltage input of the generator [5], [6] "in press". The front-end constitutes a voltage source inverter that is able to operate with symmetrical as well with asymmetrical loads in all three-phases. The back-end boost active-rectifier with power factor correction (PFC) on an input AC-side, meets a criteria of the utility regulations in an aspect of the harmonic distortion.

II. GENERIC BACK-TO-BACK SYSTEM TOPOLOGY

The topology shown in the Figure 1 consists of two half-bridge legs with their neutral point N linked with a centre tap COM of a split-capacitor DC link network $C_1, C_2, \dots C_{(n-1)}$.

Each phase-leg $(x=a,b,c)$ of n-level topology contains $2(n-1)$ number of main switches with free-wheeling diodes and $2(n-2)$ number of clamping diodes. Top switches with odd numbers are denoted as $SW_{x1}, SW_{x3}, \dots SW_{x(2n-3)}$ as well as top diodes with odd numbers are denoted as $D_{x1}, D_{x3}, \dots D_{x(2n-5)}$. Similarly, bottom switches with even numbers are denoted as $SW_{x2}, SW_{x4}, \dots SW_{x(2n-2)}$ as well as bottom diodes with even numbers are denoted as $D_{x2}, D_{x4}, \dots D_{x(2n-4)}$. The pairs of switches included in the same branch with clamping diodes are complimentary i.e. $SW_{x1} \| SW_{x2}, SW_{x3} \| SW_{x4}, SW_{x5} \| SW_{x6}$ and so forth.

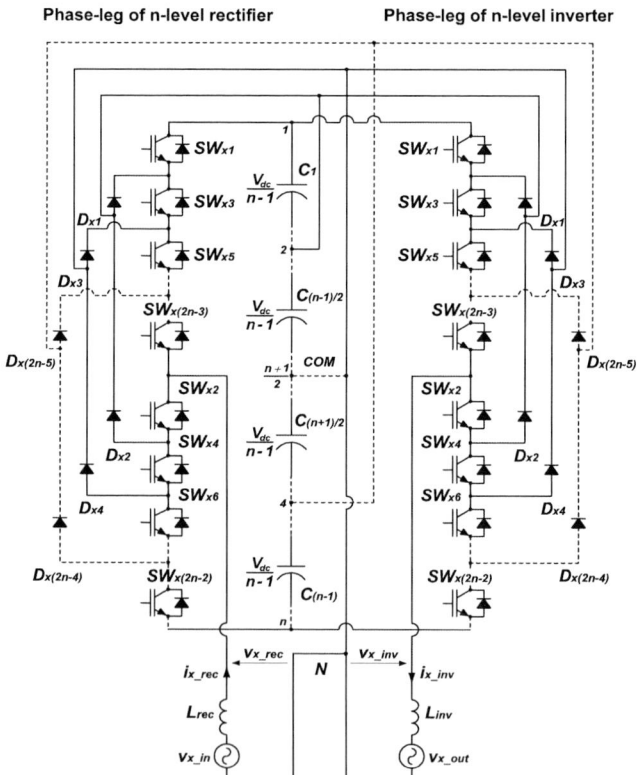

Fig. 1: Schematic diagram of generic n-level back-to-back converter system.

It should be noted that only those multi-level topologies which contain odd number of levels are considered here i.e. 3-level, 5-level, 7-level, due to the fact that only such structures are able to operate with neutral wire connected to the centre tap COM which symmetrically divides the split capacitor DC link network.

978-1-4244-1871-8/07 $25.00 © 2007 IEEE

III. CONTROL ALGORITHM FOR N-LEVEL CONVERTER

The generalized control law for anyone phase $(x=a,b,c)$ of the n-level diode clamped rectifier is defined by:

$$\begin{cases} i^*_{x\_rec}(t) = \left[K_{1p}\left(v^*_{dc}(t) - v_{dc}(t)\right) + K_{1i}\int\left(v^*_{dc}(t) - v_{dc}(t)\right)dt \right] v_{x\_in}(t) \\ e_{x\_rec}(t) = K_{2p}\left(i^*_{x\_rec}(t) - i_{x\_rec}(t)\right) + K_{2i}\int\left(i^*_{x\_rec}(t) - i_{x\_rec}(t)\right)dt \end{cases} \quad (1)$$

where:

$e_{x\ rec}$ − instantaneous value of error signal in rectifier controller, v^*_{dc}, v_{dc} − instantaneous values of reference voltage and actual DC link voltage, v_{x_in} − instantaneous value of phase-to-neutral input AC-side voltage, $i^*_{x\_rec}$, i_{x_rec} − instantaneous value of reference inductor current and actual phase current in rectifier, K_{1p}, K_{1i}, K_{2p}, K_{2i} − proportional and integral gain in voltage and current loops.

The definition of the n-segment PWM switched waveform in multi-level rectifier is given by:

$$(2)$$

$$\begin{cases} \bullet \text{ Positive output } \left(\dfrac{(n-3)V_{dc}}{2(n-1)} \le v_{x\_rec}(t) \le \dfrac{V_{dc}}{2} \right) \\ if \begin{cases} e_{x\_rec}(t) > v_{TA1}(t) \Rightarrow \left(SW_{x1}(t)=1\right) \& \left(SW_{x2}(t)=0\right) \\ e_{x\_rec}(t) < v_{TA1}(t) \Rightarrow \left(SW_{x1}(t)=0\right) \& \left(SW_{x2}(t)=1\right) \end{cases} \\ \vdots \\ \bullet \text{ Positive output } \left(\dfrac{V_{dc}}{n-1} \le v_{x\_rec}(t) \le 0 \right) \\ if \begin{cases} e_{x\_rec}(t) > v_{TA(n-1)/2}(t) \Rightarrow \left(SW_{x(n-2)}(t)=1\right) \& \left(SW_{x(n-1)}(t)=0\right) \\ e_{x\_rec}(t) < v_{TA(n-1)/2}(t) \Rightarrow \left(SW_{x(n-2)}(t)=0\right) \& \left(SW_{x(n-1)}(t)=1\right) \end{cases} \\ \bullet \text{ Negative output } \left(0 \le v_{x\_rec}(t) \le -\dfrac{V_{dc}}{n-1} \right) \\ if \begin{cases} e_{x\_rec}(t) > v_{TA(n+1)/2}(t) \Rightarrow \left(SW_{xn}(t)=1\right) \& \left(SW_{x(n+1)}(t)=0\right) \\ e_{x\_rec}(t) < v_{TA(n+1)/2}(t) \Rightarrow \left(SW_{xn}(t)=0\right) \& \left(SW_{x(n+1)}(t)=1\right) \end{cases} \\ \vdots \\ \bullet \text{ Negative output } \left(-\dfrac{(n-3)V_{dc}}{2(n-1)} \le v_{x\_rec}(t) \le -\dfrac{V_{dc}}{2} \right) \\ if \begin{cases} e_{x\_rec}(t) > v_{TA(n-1)}(t) \Rightarrow \left(SW_{x(2n-3)}(t)=1\right) \& \left(SW_{x(2n-2)}(t)=0\right) \\ e_{x\_rec}(t) < v_{TA(n-1)}(t) \Rightarrow \left(SW_{x(2n-3)}(t)=0\right) \& \left(SW_{x(2n-2)}(t)=1\right) \end{cases} \end{cases}$$

Similarly, the generalized control law for anyone phase $(x=a,b,c)$ of the n-level diode clamped inverter is defined by:

$$\begin{cases} i^*_{x\_inv}(t) = K_{3p}\left(v^*_{x\_out}(t) - v_{x\_out}(t)\right) + K_{3i}\int\left(v^*_{x\_out}(t) - v_{x\_out}(t)\right)dt \\ e_{x\_inv}(t) = K_{4p}\left(i^*_{x\_inv}(t) - i_{x\_inv}(t)\right) + K_{4i}\int\left(i^*_{x\_inv}(t) - i_{x\_inv}(t)\right)dt \end{cases} \quad (3)$$

where:

e_{x_inv} − instantaneous value of error signal in inverter controller, $v^*_{x\_out}$, v_{x_out} − instantaneous values of reference voltage and actual output AC-side voltage, $i^*_{x\_inv}$, i_{x_inv} − instantaneous value of reference inductor current and actual phase current in inverter, K_{3p}, K_{3i}, K_{4p}, K_{4i} − proportional and integral gain in voltage and current loops.

The definition of the n-segment PWM switched waveform in multi-level inverter is given by:

$$(4)$$

$$\begin{cases} \bullet \text{ Positive output } \left(\dfrac{(n-3)V_{dc}}{2(n-1)} \le v_{x\_inv}(t) \le \dfrac{V_{dc}}{2} \right) \\ if \begin{cases} e_{x\_inv}(t) > v_{TB1}(t) \Rightarrow \left(SW_{x1}(t)=1\right) \& \left(SW_{x2}(t)=0\right) \\ e_{x\_inv}(t) < v_{TB1}(t) \Rightarrow \left(SW_{x1}(t)=0\right) \& \left(SW_{x2}(t)=1\right) \end{cases} \\ \vdots \\ \bullet \text{ Positive output } \left(\dfrac{V_{dc}}{n-1} \le v_{x\_inv}(t) \le 0 \right) \\ if \begin{cases} e_{x\_inv}(t) > v_{TB(n-1)/2}(t) \Rightarrow \left(SW_{x(n-2)}(t)=1\right) \& \left(SW_{x(n-1)}(t)=0\right) \\ e_{x\_inv}(t) < v_{TB(n-1)/2}(t) \Rightarrow \left(SW_{x(n-2)}(t)=0\right) \& \left(SW_{x(n-1)}(t)=1\right) \end{cases} \\ \bullet \text{ Negative output } \left(0 \le v_{x\_inv}(t) \le -\dfrac{V_{dc}}{n-1} \right) \\ if \begin{cases} e_{x\_inv}(t) > v_{TB(n+1)/2}(t) \Rightarrow \left(SW_{xn}(t)=1\right) \& \left(SW_{x(n+1)}(t)=0\right) \\ e_{x\_inv}(t) < v_{TB(n+1)/2}(t) \Rightarrow \left(SW_{xn}(t)=0\right) \& \left(SW_{x(n+1)}(t)=1\right) \end{cases} \\ \vdots \\ \bullet \text{ Negative output } \left(-\dfrac{(n-3)V_{dc}}{2(n-1)} \le v_{x\_inv}(t) \le -\dfrac{V_{dc}}{2} \right) \\ if \begin{cases} e_{x\_inv}(t) > v_{TB(n-1)}(t) \Rightarrow \left(SW_{x(2n-3)}(t)=1\right) \& \left(SW_{x(2n-2)}(t)=0\right) \\ e_{x\_inv}(t) < v_{TB(n-1)}(t) \Rightarrow \left(SW_{x(2n-3)}(t)=0\right) \& \left(SW_{x(2n-2)}(t)=1\right) \end{cases} \end{cases}$$

IV. GENERALIZED N-LEVEL BALANCING SCHEME FOR BACK-TO-BACK CONVERTER SYSTEM

\ The abovementioned balancing scheme, formerly introduced with the back-to-back connected five-level diode clamped rectifier (5LDCR) and five-level diode clamped inverter (5LDCI), can subsequently be adapted with 7-level, 9-level and further generalized for n-level diode clamped topology. The principle of capacitor voltages balancing scheme, in the DC link network that consists of $(n-1)$ capacitors, relays on redistribution of the charge on the rectifier side and the inverter side so that the average currents flowing into the nodes $1,2,3,.....,n$ are equal to the average currents drawn from these nodes. It can be noticed that DC voltages across the capacitors $C_{(n-3)/2}$, $C_{(n-1)/2}$, $C_{(n+1)/2}$, $C_{(n+3)/2}$ are balanced in exact same way as in case of the 5LDCR/5LDCI [4], [5], however the outer capacitor voltages in higher than 5-level structure are further maintained equal using balancer structure from the Figure 2.

$$\begin{cases} \Delta v_{dc12}(t) = v_{dc2}(t) - v_{dc1}(t) \\ \vdots \\ \Delta v_{dc2(n-3)/2}(t) = v_{dc(n-3)/2}(t) - v_{dc2}(t) \\ \vdots \\ \Delta v_{dc(n-3)/2\,(n-1)/2}(t) = v_{dc(n-1)/2}(t) - v_{dc(n-3)/2}(t) \\ \Delta v_{dc(n+1)/2\,(n+3)/2}(t) = v_{dc(n+1)/2}(t) - v_{dc(n+3)/2}(t) \\ \vdots \\ \Delta v_{dc(n+3)/2\,(n-2)}(t) = v_{dc(n+3)/2}(t) - v_{dc(n-2)}(t) \\ \vdots \\ \Delta v_{dc(n-2)\,(n-1)}(t) = v_{dc(n-2)}(t) - v_{dc(n-1)}(t) \end{cases} \quad (5)$$

978-1-4244-1871-8/07 $25.00 © 2007 IEEE

Whereas the PI compensator errors can be derived from:

$$
\begin{cases}
\delta_1(t) = K_{5p}\Delta v_{dc12}(t) + K_{5i}\int \Delta v_{dc12}(t)dt \\
\vdots \\
\delta_2(t) = K_{5p}\Delta v_{dc2\,(n-3)/2}(t) + K_{5i}\int \Delta v_{dc2\,(n-3)/2}(t)dt \\
\vdots \\
\delta_{(n-3)/2}(t) = K_{5p}\Delta v_{dc\,(n-3)/2\,(n-1)/2}(t) + K_{5i}\int \Delta v_{dc\,(n-3)/2\,(n-1)/2}(t)dt \\
\delta_{(n-1)/2}(t) = K_{5p}\Delta v_{dc\,(n+1)/2\,(n+3)/2}(t) + K_{5i}\int \Delta v_{dc\,(n+1)/2\,(n+3)/2}(t)dt \\
\vdots \\
\delta_{(n-4)}(t) = K_{5p}\Delta v_{dc\,(n+3)/2\,(n-2)}(t) + K_{5i}\int \Delta v_{dc\,(n+3)/2\,(n-2)}(t)dt \\
\vdots \\
\delta_{(n-3)}(t) = K_{5p}\Delta v_{dc\,(n-2)\,(n-1)}(t) + K_{5i}\int \Delta v_{dc\,(n-2)\,(n-1)}(t)dt
\end{cases}
\tag{6}
$$

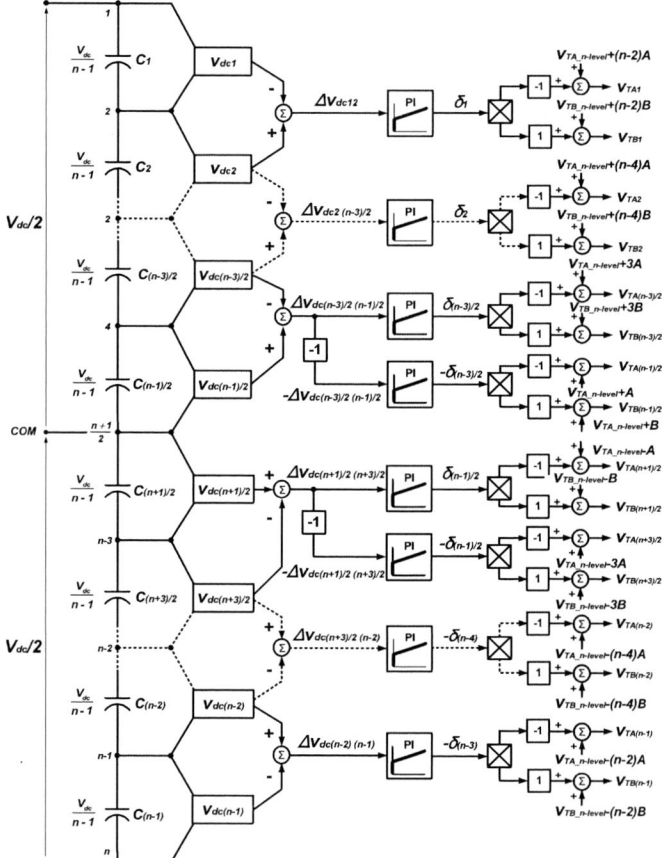

Generalized n-level balancer

Fig. 2: Structure of generalized balancer for back-to-back system.

If the triangular carrier signals in the *n*-level converter are referred to the magnitude of v_{TA} in the 5LDCR and the magnitude of v_{TB} in the 5LDCI, they can be defined by:

$$
\begin{cases}
v_{TA\_n-level}(t) = \left(\dfrac{4}{n-1}\right)v_{TA}(t) \\
v_{TB\_n-level}(t) = \left(\dfrac{4}{n-1}\right)v_{TB}(t)
\end{cases}
\tag{7}
$$

It implies that the ripple current in the multi-level converter should decrease proportionally as the number of the *n*-level converter will increase, thus the harmonic content should significantly be mitigated.

Therefore, from comparison between adjacent two DC link voltages, the following modifications are proposed to the control strategy in the *n*-level diode clamped rectifier {see the Figure 3a}, which satisfy the following conditions:

$$
\tag{8}
$$

$$
\begin{cases}
if(\Delta v_{dc12}(t)<0) \Rightarrow v_{TA1}(t) = (n-2)A(t) + \left(\dfrac{4}{n-1}\right)v_{TA} - \delta_1(t) \\
\vdots \\
if(\Delta v_{dc(n-3)/2\,(n-1)/2}(t)<0) \Rightarrow v_{TA(n-3)/2}(t) = 3A(t) + \left(\dfrac{4}{n-1}\right)v_{TA} - \delta_{(n-3)/2}(t) \\
if(\Delta v_{dc(n-3)/2\,(n-1)/2}(t)>0) \Rightarrow v_{TA(n-1)/2}(t) = A(t) + \left(\dfrac{4}{n-1}\right)v_{TA} + \delta_{(n-3)/2}(t) \\
if(\Delta v_{dc(n+1)/2\,(n+3)/2}(t)>0) \Rightarrow v_{TA(n+1)/2}(t) = -A(t) + \left(\dfrac{4}{n-1}\right)v_{TA} - \delta_{(n-1)/2}(t) \\
if(\Delta v_{dc(n+1)/2\,(n+3)/2}(t)<0) \Rightarrow v_{TA(n+3)/2}(t) = -3A(t) + \left(\dfrac{4}{n-1}\right)v_{TA} + \delta_{(n-1)/2}(t) \\
\vdots \\
if(\Delta v_{dc(n-2)\,(n-1)}(t)<0) \Rightarrow v_{TA(n-1)}(t) = -(n-2)A(t) + \left(\dfrac{4}{n-1}\right)v_{TA} + \delta_{(n-3)}(t)
\end{cases}
$$

Similarly, from comparison between two adjacent DC link voltages, the following modifications are proposed to the control strategy in the *n*-level diode clamped inverter {see the Figure 3b}, which satisfy the following conditions:

$$
\tag{9}
$$

$$
\begin{cases}
if(\Delta v_{dc12}(t)<0) \Rightarrow v_{TB1}(t) = (n-2)B(t) + \left(\dfrac{4}{n-1}\right)v_{TB} + \delta_1(t) \\
\vdots \\
if(\Delta v_{dc(n-3)/2\,(n-1)/2}(t)<0) \Rightarrow v_{TB(n-3)/2}(t) = 3B(t) + \left(\dfrac{4}{n-1}\right)v_{TB} + \delta_{(n-3)/2}(t) \\
if(\Delta v_{dc(n-3)/2\,(n-1)/2}(t)>0) \Rightarrow v_{TB(n-1)/2}(t) = B(t) + \left(\dfrac{4}{n-1}\right)v_{TB} - \delta_{(n-3)/2}(t) \\
if(\Delta v_{dc(n+1)/2\,(n+3)/2}(t)>0) \Rightarrow v_{TB(n+1)/2}(t) = -B(t) + \left(\dfrac{4}{n-1}\right)v_{TB} + \delta_{(n-1)/2}(t) \\
if(\Delta v_{dc(n+1)/2\,(n+3)/2}(t)<0) \Rightarrow v_{TB(n+3)/2}(t) = -3B(t) + \left(\dfrac{4}{n-1}\right)v_{TB} - \delta_{(n-1)/2}(t) \\
\vdots \\
if(\Delta v_{dc(n-2)\,(n-1)}(t)<0) \Rightarrow v_{TB(n-1)}(t) = -(n-2)B(t) + v_{TB}(t) - \delta_{(n-3)}(t)
\end{cases}
$$

where:

v_{dc1}, v_{dc2},… $v_{dc(n-1)}$ – instantaneous value of voltages across capacitors in *n*-level DC link network, Δv_{dc12}, Δv_{dc34},… $\Delta v_{dc(n-2)(n-1)}$ – differences of adjacent capacitor voltages in *n*-level DC link network, δ_1, δ_2,… δ_{n-3} – error signals for correction of capacitor voltages in *n*-level DC link network, K_{5p}, K_{5i} – proportional and integral gain in capacitor balancing regulators, v_{TA}, v_{TB} – triangular carriers in 5LDCR and 5LDCI modulators, v_{TA1}, v_{TA2},… $v_{TA(n-1)}$ – triangular carriers in *n*-level rectifier modulators, v_{TB1}, v_{TB2},… $v_{TB(n-1)}$ – triangular carriers in *n*-level inverter modulators.

978-1-4244-1871-8/07 $25.00 © 2007 IEEE

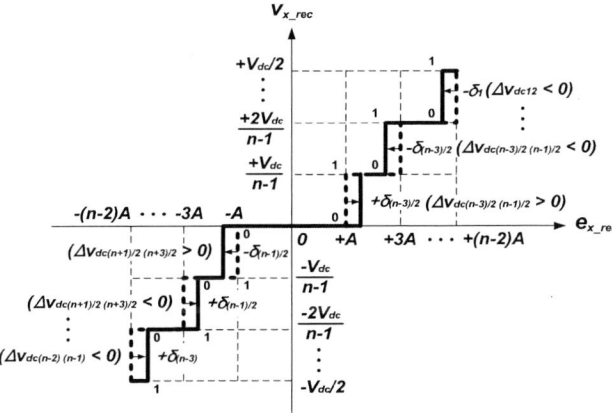

a) Modified operation of n-level diode clamped rectifier

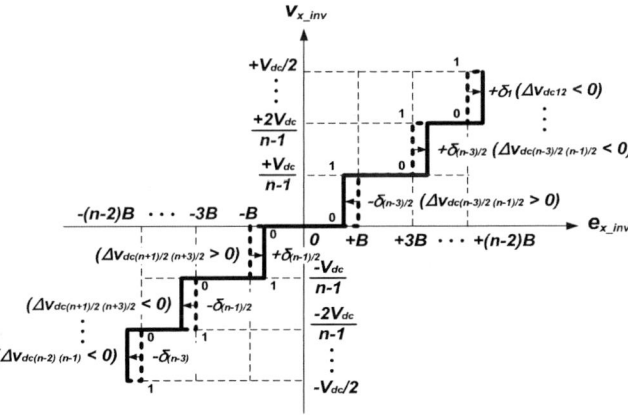

b) Modified operation of n-level diode clamped inverter

Fig. 3: Generalized multi-level PWM modulators for back-to back operation with balancing scheme: a) n-level diode-clamped rectifier side, b) n-level diode-clamped inverter side.

V. SIMULATION RESULTS

In order to verify the proposed balancing scheme the simulations were performed on single-phase back-to-back 5-, 7-, 9-level diode clamped topologies respectively.

Through the Figures 4a, 5a, 6a, there are shown the balanced sets of voltages across capacitors in 5-level, 7-level and 9-level DC link network respectively using the generalized n-level balancing scheme. The voltages across capacitors tend to converge to the point where they are equal. In case of the 5-level topology the balance is maintained at $V_{dc}/4$, for the 7-level it is equal to $V_{dc}/6$ and in case of the 9-level topology it is maintained at $V_{dc}/8$.

Subsequently, through the Figures 4b÷c, 5b÷c, 6b÷c, there are shown the modified carrier signals on the rectifier side as well as on the inverter side of the 5-level, 7-level and 9-level diode clamped topologies respectively. The triangular carriers are adjusted accordingly by the signals coming from

the error amplifiers which synthesize two adjacent voltages in the DC link network so that to the voltage balance is maintained across all the respective pairs of the capacitors.

In the Figures 4d, 5d, 6d, there are shown the rectifier switched phase-to-neutral voltages which represent the 5-segment, 7-segment and 9-segment PWM waveforms respectively.

Through the Figures 4e, 5e, 6e, there are shown harmonic spectrums of the above switched waveforms. The predominant harmonics are the 1st fundamental at 60[Hz] and the 1st carrier at 5[kHz] representing the switching frequency and their multiples. It is evident that as the converter level increases also the harmonic content decreases in the switched voltage waveform. For instance, the 1st carrier of the 7-level converter is twice smaller then the 1st carrier of the 5-level converter.

In the Figures 4f, 5f, 6f, there are shown the inverter switched phase-to-neutral voltages which represent the 5-segment, 7-segment and 9-segment PWM waveforms respectively. It's evident that the switched voltages on the inverter side tend to commutate to the contiguous levels. It is due to the balancing scheme. However, the voltage stress across the switching devices is never exceeded due to the clamping diodes used. Only the harmonic content is larger in such "non-optimum" PWM waveforms if comparison is made with pure multi-level PWM waveforms {see the Figures 4e versus 4g, and the Figures 5e versus 5g and the Figures 6e versus 6g}. Again, as the converter level increases also the harmonic content tends to decrease in the switched voltage waveform on the inverter side.

Through the Figures 4h, 5h, 6h, there are shown the input inductor current waveforms in the 5-level, 7-level and 9-level rectifier respectively. It again proves the fact, that as the converter level increases also the harmonic content in the rectifier current decreases {see the Figures 4i, 5i, 6i}. For instance, the ripple current in the 7-level rectifier is twice smaller then the ripple current with same parameters in the 5-level rectifier.

In the Figures 4j, 5j, 6j, there are shown the output inductor current waveforms in the 5-level, 7-level and 9-level inverter respectively. Similarly, the harmonic content of the inductor current in inverter will be larger {see the Figures 4k, 5k, 6k}, if compared with the inductor current in rectifier {see the Figures 4i versus 4k, and the Figures 5i versus 5k and the Figures 6i versus 6k).

It again proves the fact that as the converter level increases also the harmonic content in the inverter current decreases. For instance, the ripple current in the 7-level inverter is twice smaller then the ripple current with same parameters in the 5-level inverter.

978-1-4244-1871-8/07 $25.00 © 2007 IEEE

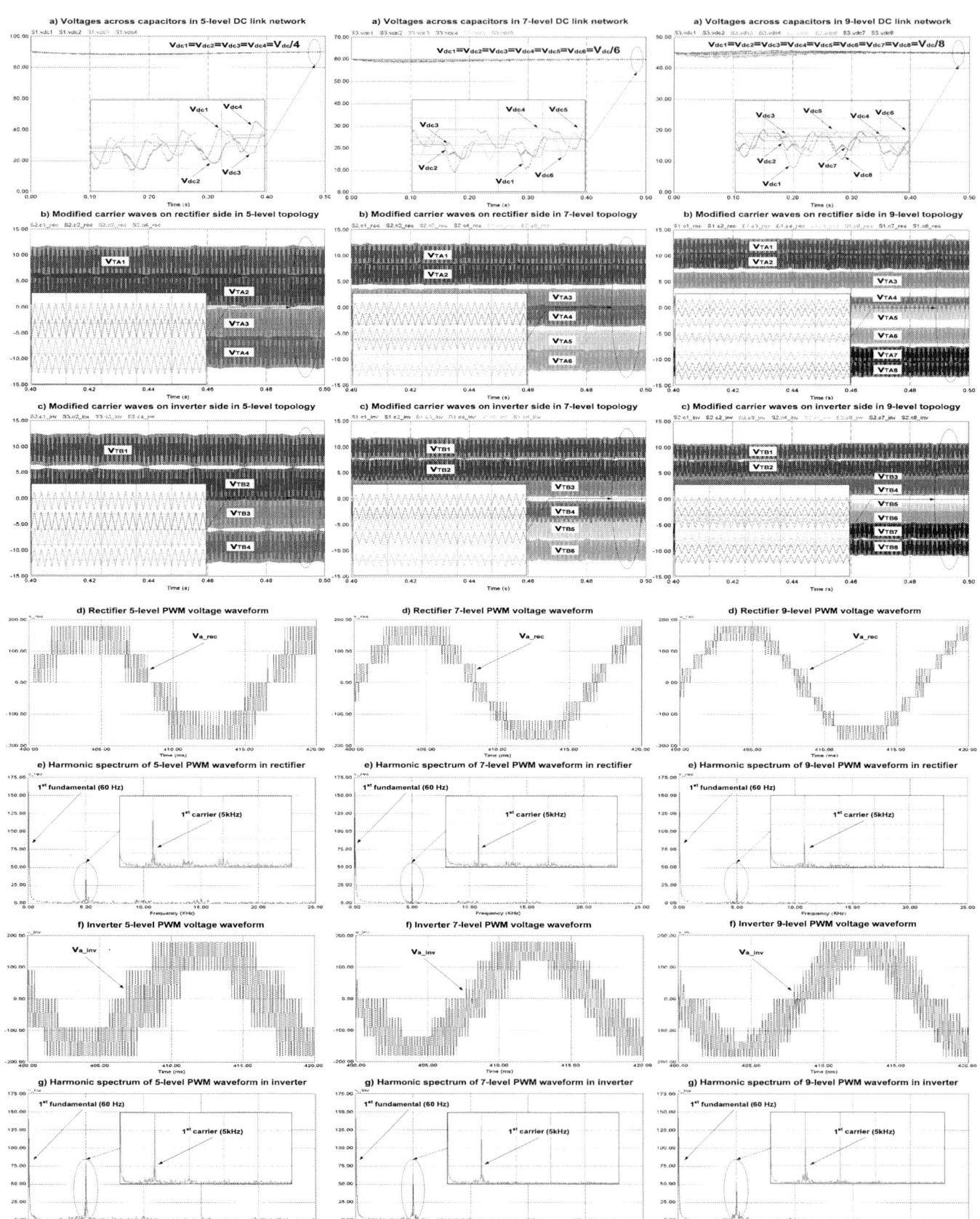

Fig. 4: (a)÷(g) Simulation results in 5-level topology. Fig. 5: (a)÷(g) Simulation results in 7-level topology. Fig. 6: (a)÷(g) Simulation results in 9-level topology.

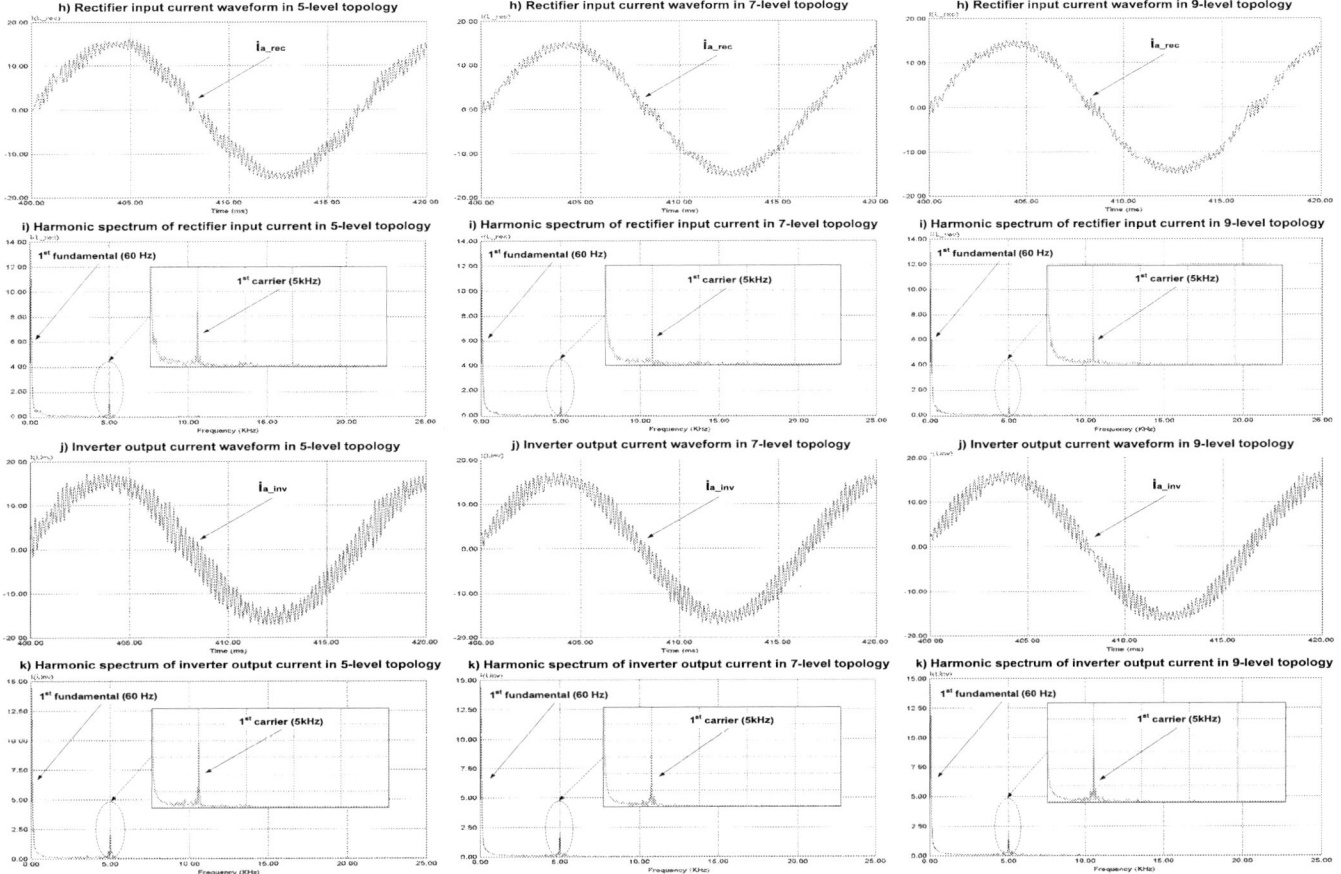

Fig. 4: (h)÷(k) Simulation results in 5-level topology.   Fig. 5: (h)÷(k) Simulation results in 7-level topology.   Fig. 6: (h)÷(k) Simulation results in 9-level topology.

VI.   SUMMARY

In this paper a novel balancing scheme for generic n-level back-to-back converter system with diode clamped topologies has been presented. The comprehensive mathematical analysis has been conducted and the proposed control strategy has been verified in the simulations with 5-level, 7-level and 9-level single phase back-to-back system. In conclusions, due to the simplicity and generality of the presented scheme, it can be extended to solve imbalance problem in any n-level DC link network with back-to-back connected diode clamped topologies. Although the analysis and results have been verified in the single phase system, the presented algorithms can be adapted with three-phase back-to-back systems with neutral leaders connected together and irrespective whether the load on the output of the inverter is symmetrical or asymmetrical. Additionally, it can be used in back-to-back systems operating with either constant or variable voltage/frequency configurations of the front-end inverter and back-end rectifier. Hence the proposed scheme is not constrained by equal modulation indexes in both converters, thus it will be effective at any input/output voltage relation in the converter at default.

ACKNOWLEDGMENT

The authors would like to thank the Volt-Ampere (Pretoria, RSA) for financial support of this work.

REFERENCES

[1] Newton S.; Sumner M.: "A novel technique for maintaining balanced internal DC link voltages in diode-clamped five-level inverters" *IEE Proc.-Electr. Power Appl., Vol.146, No.3, May 1999*, pp. 341 -349

[2] Marchesoni, M.; Tenca, P.: "Diode-clamped multilevel converters: a practicable way to balance DC-link voltages" *Industrial Electronics, IEEE Transactions on, Volume: 49 , Issue: 4 , Aug. 2002*, pp. 752 -765

[3] Ishida, T.; Matsuse, K.; Miyamoto, T.; Sasagawa, K.; Huang, L.: "Fundamental characteristics of five-level double converters with adjustable dc voltages for induction motor drives" *Industrial Electronics, IEEE Transactions on,Volume: 49 , Issue: 4 , Aug. 2002*, pp. 775-782

[4] Grzesiak L.M., Tomasik J.: "DC link balancing method in back-to-back UPS system with multi-level converters" in *Industrial Electronics, 2006 IEEE International Symposium on, Volume 2, July 9-12, 2006, Montreal, Quebec, Canada*, pp. 908-913

[5] Grzesiak L.M; Tomasik J.G.: "Autonomous power generating system with multi-level converters" in *IEEE Industrial Electronics, IECON 2006 - 32nd Annual Conference on, November 7-10, 2006, Paris, France*, pp. 2815-2820

[6] Grzesiak L.M; Tomasik J.G.: "Novel DC link balancing scheme in 5-level back-to-back converter system" the International Conference on Computer as a tool, the IEEE Region 8 EUROCON 2007, September 9-12, 2007, Warsaw, Poland (final paper accepted)

Matrix Converter as UPFC for Transmission line Compensation

Anindya Dasgupta
Department of Electrical Engineering
Indian Institute of Technology
Kanpur 208016, India
Email: anindyad@iitk.ac.in

Praveen Tripathy
Department of Electrical Engineering
Indian Institute of Technology
Kanpur 208016, India
Email: ptri@iitk.ac.in

Partha Sarathi Sensarma
Department of Electrical Engineering
Indian Institute of Technology
Kanpur 208016, India
Email: sensarma@iitk.ac.in

Abstract—**Rapid developement of high power semiconductor devices with fast control features have given birth to several Flexible AC Transmission Systems (FACTS) controllers for improving power system performances. One of the advanced FACTS Controllers is Unified Power Flow Controller (UPFC), which can control power flows, both real and reactive in Transmission line. This paper presents performance analysis of UPFC based on d-q axis theory. So far the standard practise of realizing the UPFC has been confined to connecting two inverters back to back with a DC link capacitor in between. This capacitor brings with it several disadvantages which affects the overall reliability of the UPFC. The Matrix Converter (MC) topology comes without any such bulky reactive element except for lighter components as filters. This paper examines the effectiveness of MC as UPFC and simulation results shows a comparable performance with the traditional back to back constructs.**

I. INTRODUCTION

With the developement of new and advanced Flexible AC Transmission Systems(FACTS) devices several innovative concepts has been introduced which has resulted in more flexible power flow control.In addition to power flow control, FACTS controllers can also facilitate in reducing power flows in heavily loaded lines resulting in increase loadability, low system loss, improved stability of system and reduced cost of production.In principle, the UPFC offers new horizons in terms of power system control, with great potential to independently control three power system parameters viz. line active power, reactive power flow and bus voltage. Provided no operating limits are violated, the UPFC regulates all three variables simultaneously or any combination of them. Using controllable components of UPFC, the line flows can be changed in such a way that thermal limits are not violated, losses minimized, stability margin increased, contractual requirement fulfilled etc, without violating specified power dispatch. With these features, UPFC is probably the most powerful and versatile FACTS device which combines the properties of Thyristor Controlled Phase Angle Regulator (TCPAR) and Static Compensator (STATCOM).For simplification of control analysis and to improve the dynamic performance of UPFC, various control strategies including d-q axis control have been reported in the literature.The dynamic modeling of UPFC with conventional proportional integral (PI) and proportional integral derivative (PID) based control techniques has been

reported in [1], [2], [3]. All papers have used the back to back inverter model which uses the DC Link. This DC link voltage has to be maintained constant which requires an extra control handle, furthermore this DC Link capacitor decides the life time of the converter thereby affecting its overall performance in addition to greatly increasing the weight of the converter. Keeping these issues in mind an attempt has been made in this work to investigate the competence of matrix converters which comes without the dc link as shown in fig. 1, as an alternative to the inverters. In this context only one

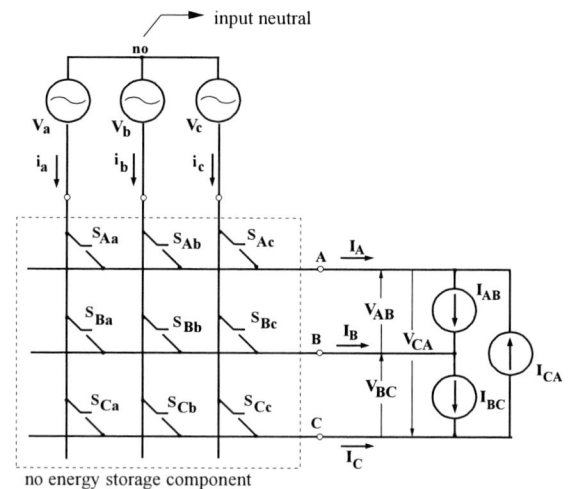

Fig. 1. 3ϕ to 3ϕ matrix converter

work has been reported [6] where the Matrix converter (MC) is placed in series with the transmission line. This results in the whole power being transmitted through MC thereby demanding higher device ratings.The author has advocated this series placement of MC for it's ability to link two regions operating at different frequency, which is however, a very rare situation in power system. Furthermore, this construct may also force the converter to carry high currents during faulty conditions.

In this work the MC is placed in shunt combination as shown in fig.2 . This construct using MC as UPFC has not been reported any where so far. The simulation of the system has

978-1-4244-1871-8/07 $25.00 © 2007 IEEE   1048

been carried out using MATLAB/Simulink.The control of bus voltage has not been attempted in this work. However the input displacement factor of the matrix converter has been maintained close to unity.

II. SYSTEM MODELLING WITH MATRIX CONVERTER

The equivalent circuit of a transmission system equipped with the Matrix Converter as UPFC is shown in fig.2.

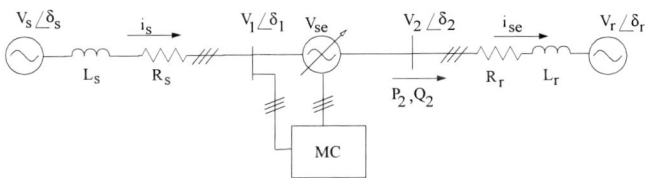

Fig. 2.   Matrix Converter placed as UPFC in the system

Here the matrix converter is injecting series voltage (v_{se}) so as to maintain the required flow of active and reactive power over the transmission line . R_s, L_s, R_r and L_r are transmission line resistance and inductance at the sending and receiving end respectively. The resistances and leakage inductances of the transformer is neglected. Performing standard axis transformation ([1], [2], [3]) the normalized currents (in p.u.) through the transmission line are described by eqn.1 and eqn.2.

$$\frac{d}{dt}i_{sed} = -\frac{R_r}{L_r}i_{sed} + \omega i_{seq} + \frac{1}{L_r}[v_{2d} - v_{rd}]. \qquad (1)$$

$$\frac{d}{dt}i_{seq} = -\frac{R_r}{L_r}i_{seq} - \omega i_{sed} + \frac{1}{L_r}[v_{2q} - v_{rq}]. \qquad (2)$$

The Matrix Converter is modelled using ideal switches in Simulink. Space vector modulation using the Indirect Transfer Function approach outlined by Huber and Borojevic [5] has been used. In this approach the MC is conceptually decoupled into a voltage source rectifier(VSR) and voltage source inverter(VSI) interconnected by a fictitous DC link. The schematic of this construct is shown in fig.3. Here space vector

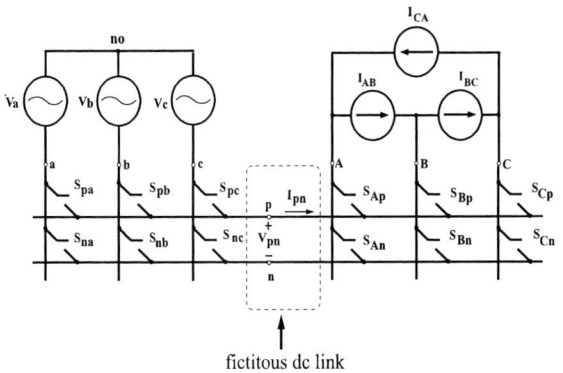

Fig. 3.   MC decoupled into a VSR and VSI

modulation is separately applied to the VSR and VSI parts. This gives the duty cycle for the six switches in each parts

which are then mapped into the duty cycle of the nine switches for the actual matrix converter. Filters are employed both at the input side and output of the matrix converter to filter out the high frequency components of input current and output voltage beacuse of switching. The input side is connected to the transmission line by a Y - Y transformer and the ouput through a $\Delta - \Delta$ transformer. This is shown in fig. 4

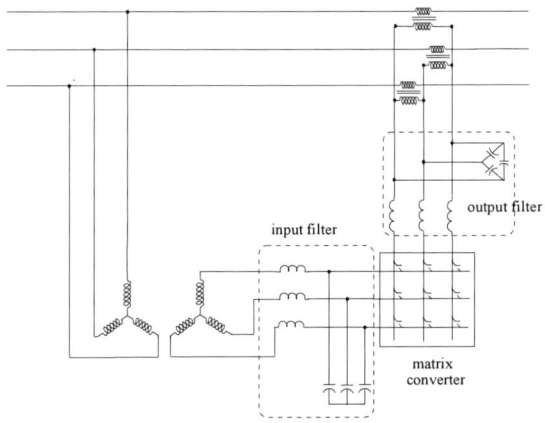

Fig. 4.   Connecting the matrix converter to the transmission line

The matrix converter can be modelled as a current stiff load when seen from the source side and a voltage stiff source when viewed from the ouput side as shown in fig. 5. Here V_{co} is the filtered output voltage of the converter which is same as V_{se} (shown in fig. 2) in the present context. The converter is fed by a voltage source V_s and has a current stiff load i_L at the output.

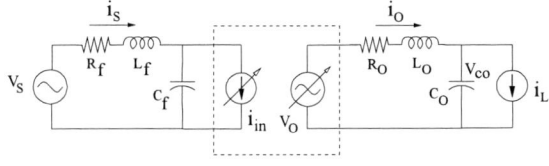

Fig. 5.   Equivalent circuit model of matrix converter

III. CONTROL SCHEME

The overall control scheme is shown in fig.6. Realization of power flow control is done by using hierarchial control wherein the outer loop is a current control block and the inner control loop controls the series injected voltages, v_{sed} and v_{seq}. The reference currents for the outer loop are determined by the values of active and reactive power demand (P_2^{ref}, Q_2^{ref}) in the transmission line at node 2 as shown in fig.2. Governing equations for the current references are

$$i_{sed}^{ref} = \frac{P_2^{ref} * v_{2d} - Q_2^{ref} * v_{2q}}{v_{2d}^2 + v_{2q}^2} \qquad (3)$$

$$i_{seq}^{ref} = \frac{P_2^{ref} * v_{2q} + Q_2^{ref} * v_{2d}}{v_{2d}^2 + v_{2q}^2}. \qquad (4)$$

978-1-4244-1871-8/07 $25.00 © 2007 IEEE

The current reference values thus obtained are used to generate the references for V_{se} (V_{co} , the filtered output voltage as mentioned in the earlier section) which in turn gives the references for the output voltage of the matrix converter , V_O in fig. 5. The entire control scheme has been shown in fig. 6

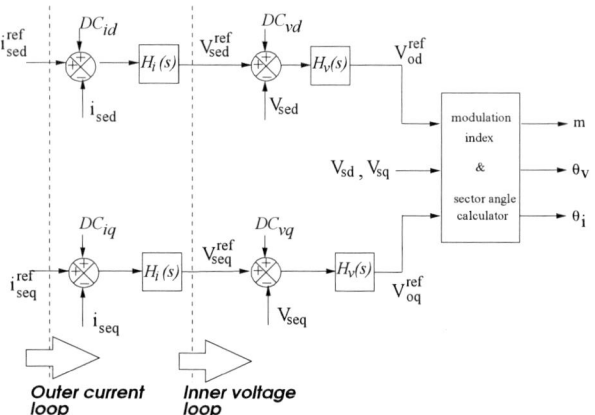

Fig. 6. Control scheme : DC_i and DC_v are decoupling and disturbance rejection terms

Here an inherent system coupling exists between the d and q axis components of current (shown in eqns 1 and 2) as well as voltages. This makes it difficult to achieve good performance in terms of settling time and overshoot. The currents i_{sed} and i_{seq} are measured and compared with the reference currents along d and q axes. These error signals are fed to the current controller to regulate the injected series voltage (v_{sed}, v_{seq}) to the transmission line. Here v_{rd} and v_{rq} are the remote signal sensed for implementing the control law [4]. These signals along with the measured variables bus voltages (v_{1d},v_{1q}) and line currents (i_{sed},i_{seq}) are subtracted with appropriate gains as feedforward terms to nullify their effect in the system and thereby achieving both disturbance rejection and decoupling of d and q axis terms.The values of DC_{id} and DC_{iq} are given by

$$[DC_i]_{d,q} = \mp \omega L_r [i_{se}]_{d,q} - [v_1 - v_r]_{d,q}.$$

However to realize perfect decoupling the reactance of the line must be known accurately. A similar treatment for decoupling and disturbance rejection is also employed for the inner voltage control block. Here DC_{vd} and DC_{vq} are

$$[DC_v]_{d,q} = -\omega^2 L_o C_o [V_{co}]_{d,q} \mp \omega C_o (2 L_o p + R_o)[V_{co}]_{q,d}$$

$$\mp \omega L_o [i_L]_{q,d} + (L_o p + R_o)[i_L]_{d,q}$$

where p is the operator $\frac{d}{dt}$. The derivative terms in the above equations are first passed through a low pass filter before being differentiated to prevent amplification of high frequency noises. In the simulation which will be dealt with in the next section , the dynamics of machines has been neglected as a result of which the system bandwidth is very high. The bandwidth of the plant is at 3720 rads/sec. The MC is operated at a switching frequency of 10 KHz(62832

rads/sec) which requires that the bandwidth of the voltage control block should be less than about 4000 rads/sec to ensure sufficient attenuation at the switching frequency. This makes the bandwidth of the voltage loop close to the the bandwidth of the outer current control block. So to implement the hierarchial control the outer current block is slowed down by using a PI controller with Kp=0.2 and Ki=62 which makes the bandwidth at 744 rads/sec. The bode diagram of the plant with the current controller (H_i) is shown in fig. 7. The voltage controller, which is a third order controller has been designed to make the bandwidth (at 2900 rads/sec) of the inner loop 4 times faster than the outer current loop. The bandwidth, gain margin(GM) and phase margin (PM) of the outer loop and inner loop are tabulated in the Table I. The bode diagram of the inner voltage loop with the controller is shown in fig. 8. The hierarchial strategy was found to be working fine with these controllers.

TABLE I
FREQUENCY RESPONSE

	Bandwidth (rads/sec)	GM (dB)	PM (deg)
Outer current loop	744	infinity	90
Inner voltage loop	2930	7.44	47.6

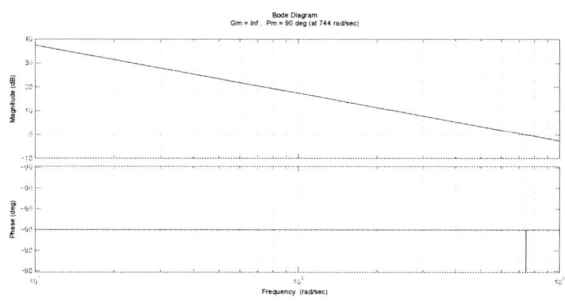

Fig. 7. Bode diagram of outer current loop with controller H_i

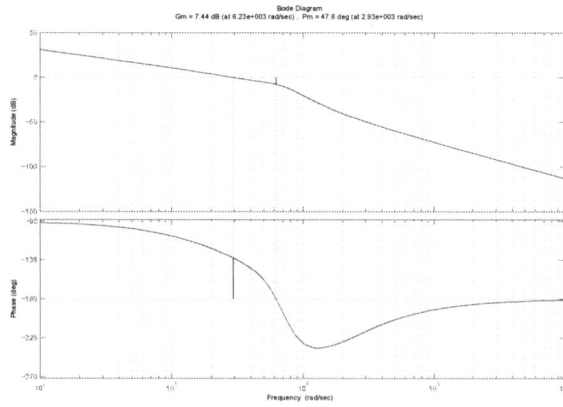

Fig. 8. Bode diagram of inner voltage loop with controller H_v

The references for the matrix converter output voltages

(V_{od}^{ref} and V_{oq}^{ref}) along with the converter input voltages (V_{sd} and V_{sq}) are used to generate the modulation index (m) and the sector angles required for voltage modulation. As has been mentioned before - the input displacement factor is maintained close to unity in this work - for this the modulation index (m) is given by

$$m = \frac{2}{3} \frac{\sqrt{(V_{od}^{ref})^2 + (V_{oq}^{ref})^2}}{\sqrt{(V_{sd})^2 + (V_{sq})^2}}$$

θ_V and θ_i are the sector angles for the VSI and VSR parts which have been briefly mentioned in the earlier section. A detailed formulation for finding the duty ratios from these can be found in [5].

IV. SIMULATION RESULTS

Here for performance analysis of UPFC controller, this paper has neglected the dynamic of synchonous machine. The results are shown for two cases.

The test system taken for simulation study of UPFC is shown in fig.2, which is a single-line circuit diagram of UPFC installed in a power system. Two 220 kV, 100 MVA lines are considered for simulation study. First one is 150 km long and the other is 100 km long. Specifications of the system taken for testing the simulation study are given in Table II.

TABLE II
SYSTEM PARAMETERS

Parameters	Value
Base frequency	50 Hz
R_s	0.01 pu
X_s	0.15 pu
R_r	0.0166 pu
X_r	0.0843 pu

A. Choice of ripple filter capacitor and transformers ratio

The indirect modulation strategy ensures that the fundamental component of the MC input current i_{in} is in phase with the input voltage. This current contains switching frequency ripples which has to be filtered out by adding a low pass filter at the input side as shown in fig. 5 and also in fig. 9. The capacitor C_f gives a phase lead to the source current i_s as can be seen in fig. 9.b.

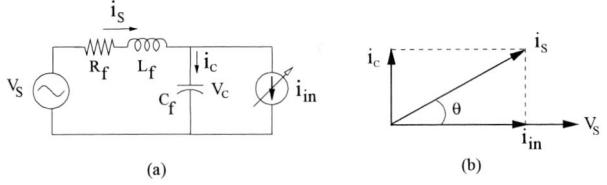

Fig. 9. Effect of filter capacitor on the input displacement factor

A limit for the capacitor size for a specified input displacement factor(IDF) has been mentioned in [7]. For a displacement angle of θ the maximum filter capacitor is given by

$$C_f = \frac{1}{Z_b} \frac{\hat{i}_{in_{max}}}{\omega . \hat{V}_{s_{max}}} tan(\theta)$$

where Z_b is the base impedance and ω the base frequency in radians. For the given system (Table II) value of Z_b is about 1450. Furthermore, normally the series voltage required to be injected is quite small with respect to the input voltage of the converter. This makes the modulation index very low and with a given value of current at the output, the input current of the converter which is a function of the modulation index and the ouput current also becomes very small. This results in a very low $\hat{i}_{in_{max}}$ to $\hat{V}_{s_{max}}$ ratio. All these brings down the maximum farad limit to a great extent(to the order of tens of nano farad). Now with a desired corner frequency of the filter this low farad rating necessitates use of large inductor resulting in very low damping ratio and high input current overshoots.

Now, the transformers (stepped down at the input and stepped up at output side) connecting the MC with the line reduces Z_b by a factor of n*n where n:1 is the turns ratio at the input. This helps to achieve a high IDF with a reasonably high C_f and higher damping The value of n is taken to be 10 and the filter capacitor value chosen as 10 μF for simulation. The bode diagram of the V_s to V_{Cf} transfer function is shown in fig. 10

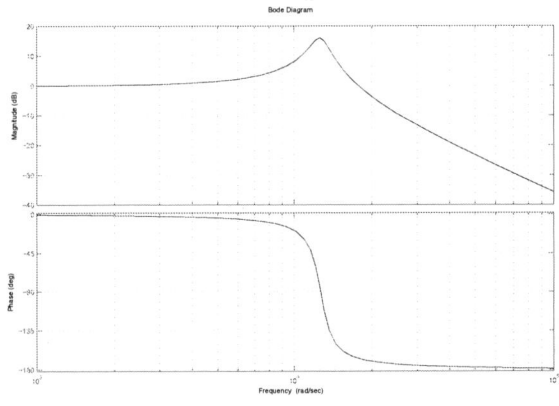

Fig. 10. Bode diagram of V_s to V_{Cf} transfer function

B. Case I :

In the first case P_{ref} was subjected to a step change from 0 to 2 pu and Q_{ref} from 0 to 0.25 pu at time 0.05 second. The active power flow in this case is shown in fig. 11 which has an overshoot of about 5% and a settling time of one cycle. The reactive power flow is shown in fig.12. where the overshoot is a bit high but the settling time is very good, around one cycle. The voltage and currents at the matrix converter terminals are shown in fig. 13. It could be seen from fig. 13(a), the voltage at the input terminal remains almost constant . The input current

978-1-4244-1871-8/07 $25.00 © 2007 IEEE

as seen in fig. 13(b), remains almost in phase with the input voltage even though it's amplitude is very low and remains so from the instant when the step change in power is applied.Here the active power drawn by MC is positive after the step input is applied (fig.14) . Fig 13(c) shows the output voltage of the matrix converter which is injected in the line . The output current of the converter is shown in fig. 13(d).

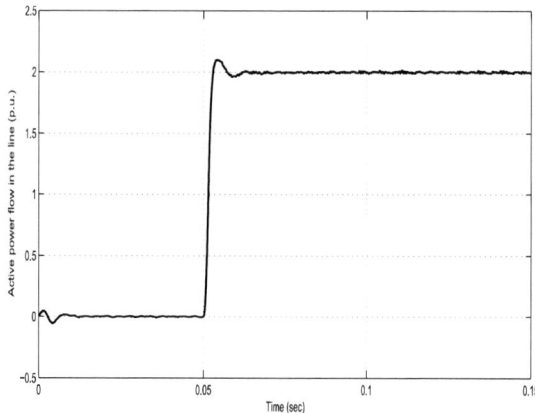

Fig. 11.   Active power flow through the line for a change in active power reference from 0 to 2.0 p.u

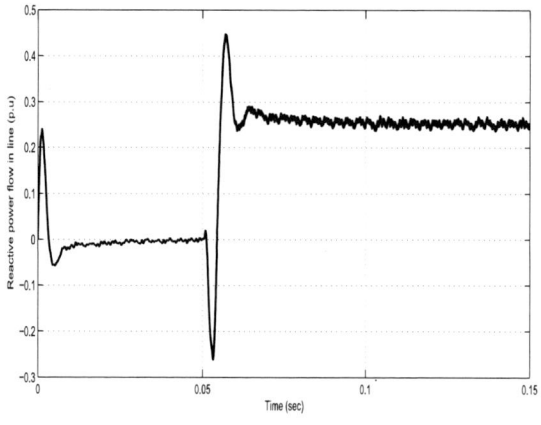

Fig. 12.   Reactive power flow through the line for a change in reactive power reference from 0 to 0.25 p.u

C. Case II :

In this case P_{ref} was subjected to a step change from 0 to 0.5 pu and Q_{ref} from 0 to 0.5 pu at time 0.05 second. It can observed from fig.15 that the overshoot in active power flow was about 20% and it took about one cycle to settle.

In case of reactive power (fig.16) the overshoot was about 10 % and the settling time was well within one cycle.The voltage and current waveforms of the matrix converter are

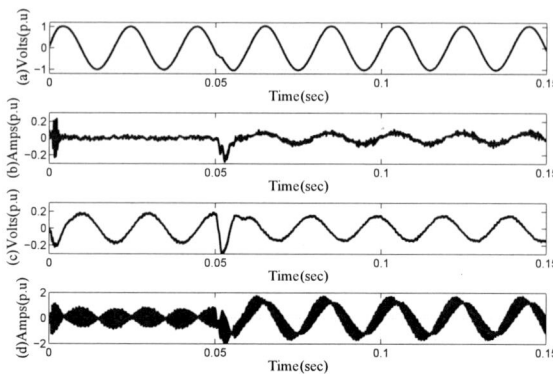

Fig. 13.   Voltage and currents of the matrix converter in per unit for phase-a : (a)Input voltage (b)Input current, i_s (c)Output voltage (v_{se}) (d)Output current, before filtering, i_o.

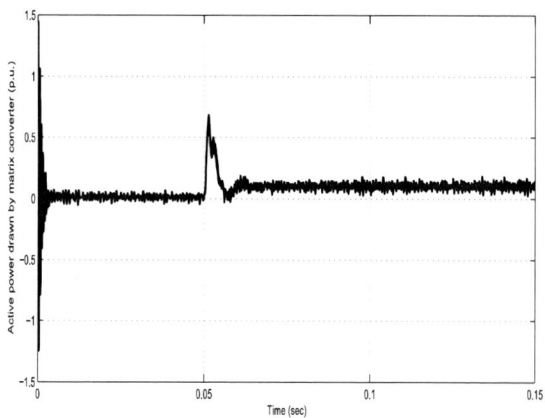

Fig. 14.   Active power drawn by matrix converter

given in fig.17 and the power drawn by the converter from the line is shown in fig.18, which, in this situation is negative . In this case too the transients in the input current and the output voltages are limited to pretty low values as can be seen in fig.17 when the step change in power reference is given at 0.05 seconds. Here too the input current (fig.17.b) is out of phase(the power drawn being negative) with the input voltage (fig.17(a)) by almost 180° resulting in an IDF close to -1.

V. CONCLUSION

A scheme using Matrix converter as UPFC has been proposed in this work. The scheme has been tested with the step change in reference power. The steady state performance was found to be comparable to the works referred to where back to back inverters have been used. Hence similar performance was achieved with the matrix converter which, comes with reduced control complexity owing to absence of DC link and increases the overall life time of the system.

978-1-4244-1871-8/07 $25.00 © 2007 IEEE

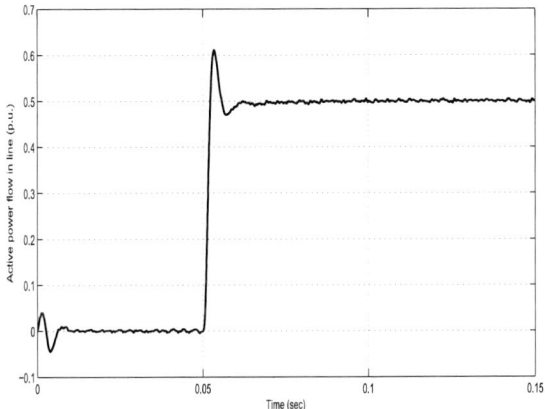

Fig. 15. Active power flow through the line for a change in active power reference from 0 to 0.5 p.u

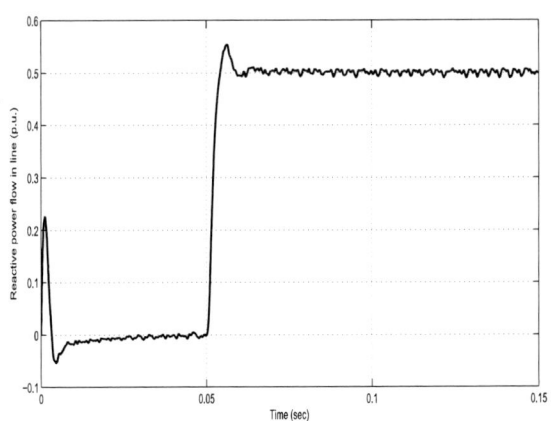

Fig. 16. Reactive power through the line for a change in reactive power reference from 0 to 0.5 p.u

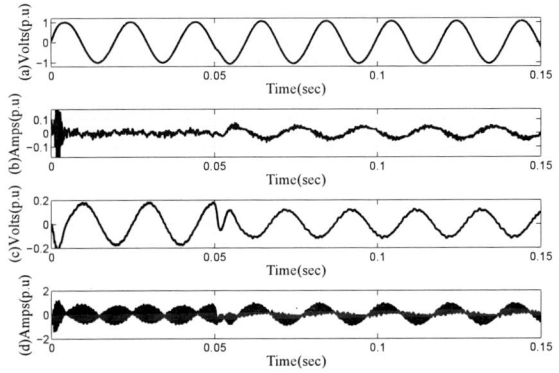

Fig. 17. Voltage and currents of the matrix converter in per unit for phase-a : (a)Input voltage (b)Input current (c)Output voltage (v_{se}) (d)Output current

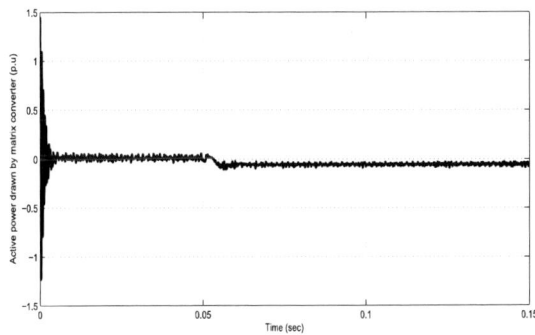

Fig. 18. Active power drawn by matrix converter

PESC '96 Record, 27th Annual IEEE, Vol. 1 , 23-27 June 1996 Page(s): 502-507

[7] V Vlatkovic, D Borojevic and F. C. Lee "Input filter design for power factor correction circuits", *IEEE Transactions on Power Electronics*, Vol. No. 11, Jan 1996, Page(s): 199-205.

REFERENCES

[1] S. D. Round, Q. Yu, L. E. Norum and T. M. Underland, "Performance of a unified power flow controller using a d-q control system", *Conf. Proc. No. 423 of the IEE Sixth International Conference on AC and DC Power Transmission*, London, April-May 1996, pp.357-362.

[2] K. R Padiyar and A. M Kulkarni, "Control Design and Simulation of Unified Power Flow Controller", *IEEE Transactions on Power Delivery*, Vol.13, No.4, Nov. 1997, pp.1346-1354.

[3] O. P Dwivedi, J. G Singh and S. N Singh, "Power Flow Control Using Multi-Converter FACTS Controller", *International Conference on Power Systems, ICPS2004*, Kathmandu, Nepal, Nov.2004, pp.711-718.

[4] Q. Yu,S. D. Round,L. E. Norum and T. M. Underland, "Dynamic control of a Unified Power Flow Controller", *Power Electronics Specialists Conference, 1996, PESC '96 Record, 27th Annual IEEE*, Vol. 1 , 23-27 June 1996 Page(s): 508-514.

[5] L Huber and D Borojevic, "Space Vector modulated Three phase to Three phase Matrix Converter with Input Power factor Correction", *IEEE Transactions on Industry Applications*, Vol. No. 6, 1995.

[6] B. T Ooi and M Kazerani, "Unified Power Flow Controller Based on Matrix Converter", *Power Electronics Specialists Conference, 1996,*

Novel Carrier PWM Technique with Extension Range For 4-Switch Inverter

Nguyen Van Nho[1], To Huu Phuc[2], Nguyen Xuan Bac[3]
[1,2,3]HCMC University Of Technology
Department of Electrical Engineering
268 LyThuongKiet, District 10, HCM City, Vietnam
Email: nvnho@hcmut.edu.vn, thphuc@ieee.org; nxbac@hcmut.edu.vn

Abstract— Three-phase 4-switch voltage source inverters are advantageous for reduced number of power switches. For these topologies, space vector Pulse Width Modulation (SVM) methods have been commonly used for controlling the circuit. This paper presents a novel carrier based PWM method for 4-switch inverters, which is simple and flexible. Linear overmodulation will be also investigated. The proposed method will be mathematically analyzed and demonstrated by simulation and experimental results.

I. INTRODUCTION

Three-phase Four switch inverter has been known as a possible hardware solution for reduced cost in some electric drive applications. One of the drawbacks is reduction of available maximum output voltage [1]. To supply motor of higher rated voltage, the inverter requires a high AC voltage source or to be deployed with a PWM rectifier, these however will give rise to voltage stress on switching devices. Another problem of this unsymetric inverter is the unbalance between two dc voltage cells, which deforms the space vector diagram and further confines the load voltage range. Voltage ripples for definite capacitors further reduces the inverter performance [2],[3]. The level of unbalance between two dc cells varies depending on output frequency and load currents. Therefore, a good control of overmodulation can contribute to improving of inverter performance.

As a result, properly utilising the dc voltages for overmodulation range can help to improve the voltage control range. In the paper, a novel carrier based PWM method for 4-switch inverter on condition of unbalanced dc sources will be proposed. The method bases on the modifying of modulating signals, depending on the dc source variation [4]. Two-mode overmodulation using the principle control between two limit trajectories will be presented [5].

II. PROPOSED PWM TECHNIQUE

A particular characteristics of PWM control in 4-switch inverter is that offset control is not available since voltage level of the first phase is fixed as shown in Fig.1.
a) Define the voltages on dc sources and their total value as

$$V_{c1}, V_{c2} \text{ and } V_d = V_{c1} + V_{c2}.$$

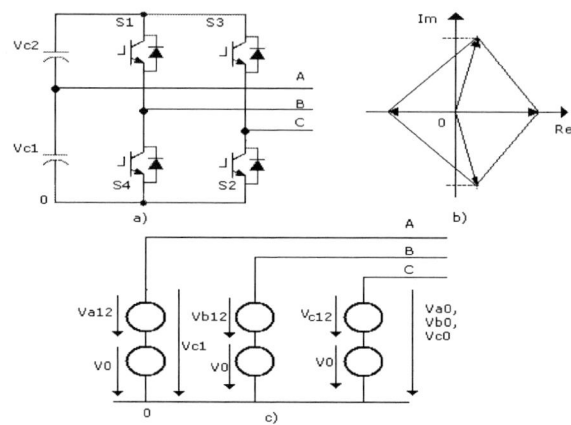

Figure 1: Four switch inverter. Circuit diagram a), voltage space vector diagram for unbalanced dc sources b) and proposed circuit model c).

b) Analysis of circuit [2]

Phase-leg voltages can be described in a general form as:

$$V_{x0} = V_{x12} + V_0 \text{ ; x=a,b,c} \tag{1}$$

From Fig.1, it can be seen that:

$$
\begin{aligned}
V_{a0} &= V_{a12} + V_0 = V_{c1} \\
V_{b0} &= V_{b12} + V_0 = V_{c1} + (V_{b12} - V_{a12}) = V_{c1} + V_{ba12} \\
V_{c0} &= V_{c12} + V_0 = V_{c1} + (V_{c12} - V_{a12}) = V_{c1} + V_{ca12}
\end{aligned}
\tag{2}
$$

The phase leg voltage is confined by the total dc voltage as:

$$0 \leq V_{x0} \leq V_d \tag{3}$$

For generating the control signals, it is enough to measure the dc source as V_{c1}, and to determine two active line-line voltage components V_{ba12} and V_{ca12}.

For undermodulation range, the reference voltages can be deduced as:

$$V_{ba12,1} = V_{L12m} \cos(\theta - 5\pi/6)$$
$$V_{ca12,1} = V_{L12m} \cos(\theta - 7\pi/6)$$
(4)

In the fundamental period, let's define the parameter d as

$$d = Min(V_{c1}, V_{c2}) = Min(V_{c1Min}, V_{c2Min}).$$
(5)

where the condition for the line-voltage amplitude is:

$$V_{L12m} \le d$$
(6)

Linear overmodulation: From (6), overmodulation happens if $V_{L12m} > d$. Because of the unbalance and oscilation of dc voltage sources, the boundaries between under- and over-modulation values will continuously change. To simplify the algorithm for determining the reference voltages on the condition of variable dc sources, some nominal voltages will be performed first, and the reference voltage on condition of unbalance will be deduced.
The principle control between two limit trajectories will be applied [5]. For two-mode overmodulation, three limit modulation indexes are considered as

$$m_{1N} = 0.5,$$
(7)

$$m_{2N} = 0.52645$$

and $m_{3N} = 0.55125$,

The standard modulating signals described below are produced for the balanced condition of dc sources :

$$V_{c1} = V_{c2} = V_{CN}/2.$$
(8)

For case of $m_{1N} = 0.5$:

$$V_{ba12,1N} = 0.5 V_{CN} \cos(\theta - 5\pi/6);$$
$$V_{ca12,1N} = 0.5 V_{CN} \cos(\theta - 7\pi/6)$$
(9)

For case of $m_{2N} = 0.52645$
(10)

$$V_{ba12,2N} = \begin{cases} 0.5 V_{CN}(\theta - \pi/3) & for \quad 0 \le \theta < 2\pi/3 \\ -0.5 V_{CN}(\theta - 4\pi/3) & for \quad \pi \le \theta < 5\pi/3 \\ 0.5 V_{CN} & for \quad 2\pi/3 \le \theta < \pi \\ -0.5 V_{CN} & for \quad else \end{cases}$$
(11)

$$V_{ca12,2N} = \begin{cases} 0.5 V_{CN}(\theta - 2\pi/3) & for \quad \pi/3 \le \theta < \pi \\ -0.5 V_{CN}(\theta - 5\pi/3) & for \quad 4\pi/3 \le \theta < 2\pi \\ 0.5 V_{CN} & for \quad \pi \le \theta < 4\pi/3 \\ -0.5 V_{CN} & for \quad else \end{cases}$$
(12)

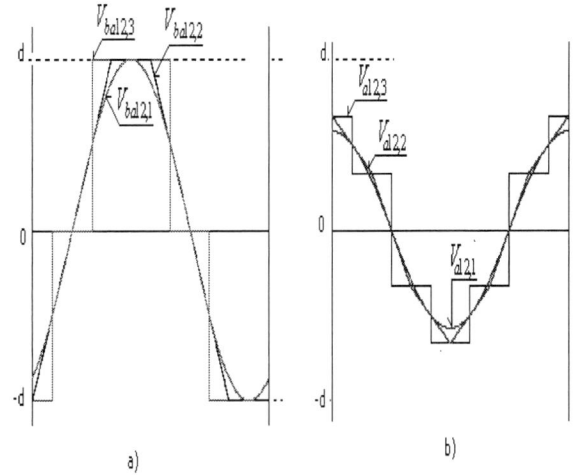

Figure 2: Diagrams of a) Line-line voltages and b) corresponding phase voltages for three limit modulation indexes

For case of $m_{3N} = 0.55125$,

$$V_{ba12,3N} = \begin{cases} -0.5 V_{CN} & for \quad (0 \le \theta < \pi/6) or (3\pi/2 \le \theta < 2\pi) \\ 0.5 V_{CN} & for \quad \pi/2 \le \theta < 7\pi/6 \\ 0 & for \quad else \end{cases}$$
(11)

$$V_{ca12,3N} = \begin{cases} -0.5 V_{CN} & for \quad (0 \le \theta < \pi/2) or (1 \, 7\pi/3 \le \theta < 2\pi) \\ 0.5 V_{CN} & for \quad 5\pi/6 \le \theta < 3\pi/2 \\ 0 & for \quad else \end{cases}$$
(12)

For the unbalanced dc voltage sources, the corresponding limit modulation indexes will be:

$$m_{d1} = \frac{2 m_{1N} d}{V_{CN}},$$

$$m_{d2} = \frac{2 m_{2N} d}{V_{CN}}$$

and $m_{d3} = \dfrac{2 m_{3N} d}{V_{CN}}$.
(13)

The active voltages can be deduced as follows:

$$V_{ba12,m} = \frac{k_d m_{HN} - m}{m_{HL}} V_{ba12,LN} + \frac{-k_d m_{LN} + m}{m_{HL}} V_{ba12,HN}$$

$$V_{ca12,m} = \frac{k_d m_{HN} - m}{m_{HL}} V_{ca12,LN} + \frac{-k_d m_{LN} + m}{m_{HL}} V_{ca12,HN}$$
(14)

where $k_d = 2d/V_{CN}$, $m_{HL} = m_H - m_L$
(15)

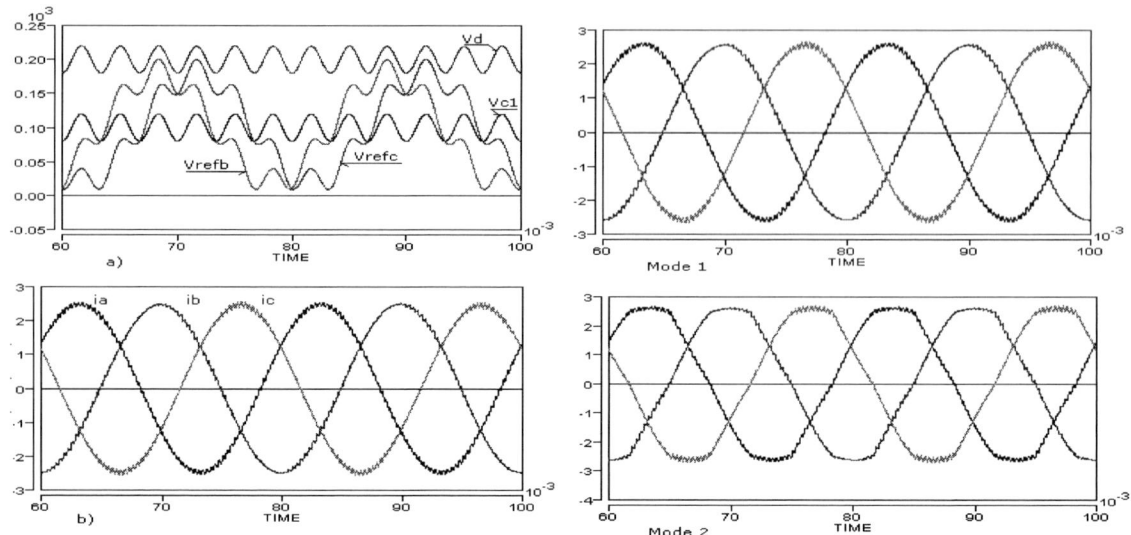

Figure 3: Simulation results. Unbalanced dc sources $V_{c2}=100V; V_{c1}=100+20\sin(12\pi ft+\pi/3)[V]$; d=80V; R=10 Ω ;L=50mH,f_{out}=50Hz; f_{sw}=3kHz; Diagrams of dc voltages and reference leg voltages and corresponding load currents for undermodulation a) and diagrams of currents in overmodulation mode 1 $\eta = 0.8$ (upper) and mode 2 - $\eta = 0.8$ (below) b).

Functions and parameters for two-mode overmodulation are described in Table 1 as:

Table 1: Parameters in two mode overmodulation

Conditions	m_{LN}	m_{HN}	m_{HL}
$m_{d1} < m \leq m_{d2}$ (Mode 1)	0.02645	0.5264	0.5
$m_{d2} < m \leq m_{d3}$ (Mode 2)	0.02480	0.55125	0.52645

Table 2: Functions in two-mode overmodulation

Conditions	$V_{ba12,LN}$	$V_{ba12,HN}$
Mode 1: $m_{d1} < m \leq m_{d2}$	$V_{ba12,1N}$	$V_{ba12,2N}$
Mode 2: $m_{d2} < m \leq m_{d3}$	$V_{ba12,2N}$	$V_{ba12,3N}$

Carrier based PWM method:

For implementing carrier based PWM method on condition of variable dc voltage V_d, the change must be compensated by transfoming the leg voltages to nominal modulating signals as:

$$v_{b0} = \frac{V_{b0}}{V_d}; \qquad v_{c0} = \frac{V_{c0}}{V_d} \qquad (16)$$

where V_{b0} and V_{c0} are determined by (2).

Two possible carrier PWM techniques as PS PWM and PD PWM can be implemented

Simulation results:

Simulation results for oscilating dc voltages were implemented to demonstrate the proposed principle. It was shown a good performance with compensated output voltages in undermodulation, overmodulation mode 1 and mode 2 (Fig.3).

Experimental results:

An experimental hardware using integrated devices IRAMS20UP60A and control scheme kit DSPACE DS1104 was built to validate the theory. Control algorithm was set up in MATLAB/SIMULINK (Fig.4).

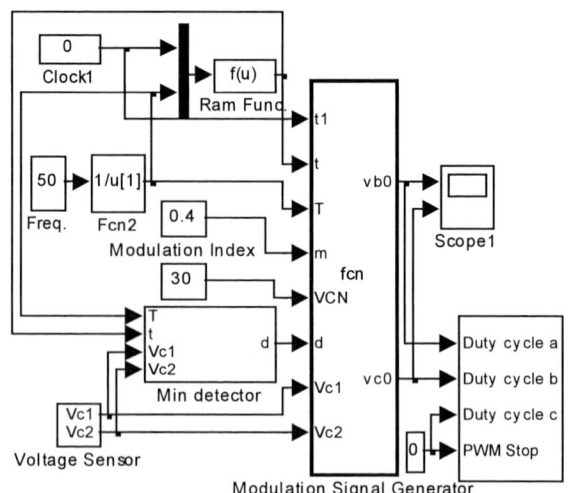

Figure 4: MATLAB/Simulink based control scheme for DS1104

The dc voltages were measured using LEM LV25 NP, their data were were sampling each triangle waveform and transfered to DS1104 through DA converters. Control

978-1-4244-1871-8/07 $25.00 © 2007 IEEE

parameters and digrams were monitored and viewed also on DS1104 Control Desk Developer (Fig.5). The frequency of triangle carrier waveform was set equal to 5kHz. For varifying the algorithm of balancing output voltages, two cases with output frequencies of 10Hz and 50Hz were considered. A set of controllable single phase AC voltage, diode rectifier and dc capacitor filter produced a voltage value of about 50Vdc on each capacitor. For setting modulating index m, the normal voltage was defined as $V_{CN}=50V_{dc}$. The osilation on the dc capacitors was evaluated and the value of V_{CMIN} was updated each fundamental period.

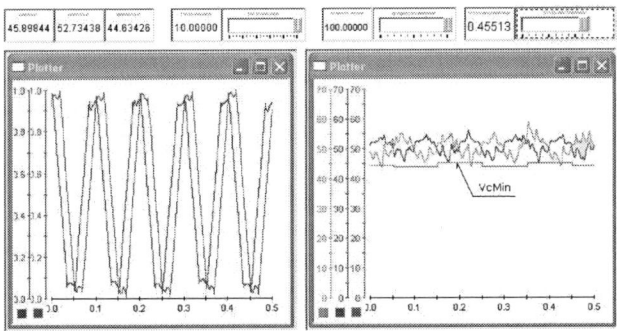

Figure 5: Diagrams of Modulating signlas (left), two dc voltages and V_{cmin} on Control Desk for low output frequency of $f_{out}=10Hz$.

For low output frequency of $f_{out}=10Hz$, a strong oscilating of capacitor voltages caused a reduction of selected value of V_{cmin} and related undermodulation range. As a result, inverter was still in undermodulation for m=0.3, but overmodulation mode 1 and mode 2 soonly appeared at corresponding modulation indexes of 0.455 and 0.5 as shown in Fig.6 and Fig.7.

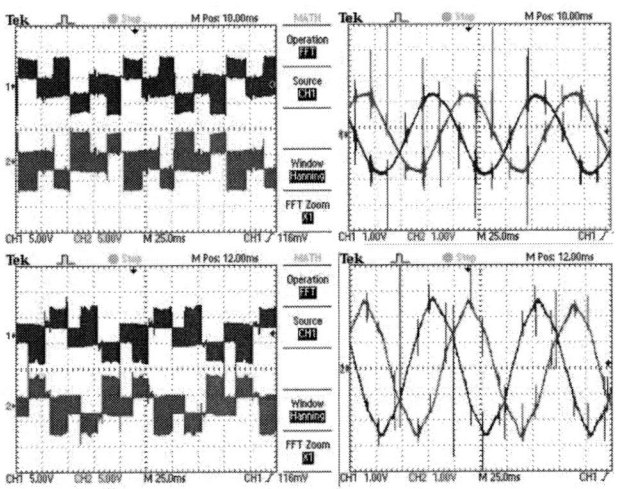

Figure 6: Influence of low output frequency of 10Hz. Diagrams of voltages and currents for a)m=0.3 and b)m=0.455.

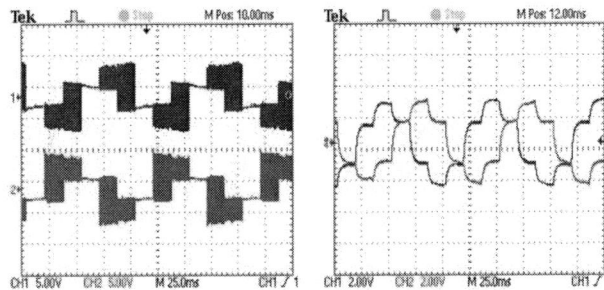

Figure 7: Diagrams of voltages and currents in overmodulation mode 2 for m=0.5 and $f_{out}=10Hz$. Average dc voltages $V_{c1}=Vc2=50V$, $f_{out}=50Hz$, $f_{sw}=5KHz$.

At high frequency of 50Hz, the oscilation of dc voltages reduced (Fig.8). Compared to the previous case for m=0.5 and on the same dc power supply, the inverter was still in undermodulation mode (Fig.9). The overmodulation mode 1 occurred for m=0.52 and changed into mode 2 for m=0.54 (Fig.10). The undermodulation range was extended for an increased amplitudes of dc sources (and V_{cMin}) as shown in Fig.11.

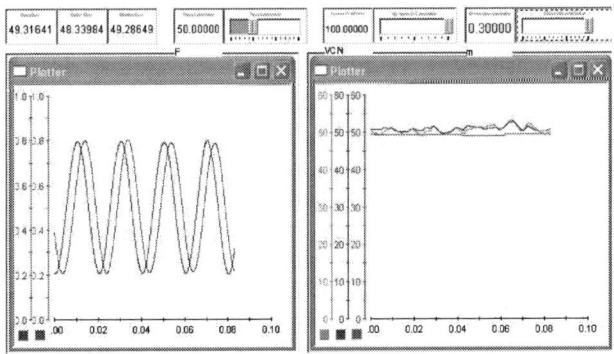

Figure 8: Diagrams of Modulating signlas (left), two dc voltages and V_{cmin} on Control Desk for high output frequency of $f_{out}=50Hz$ and m=0.3.

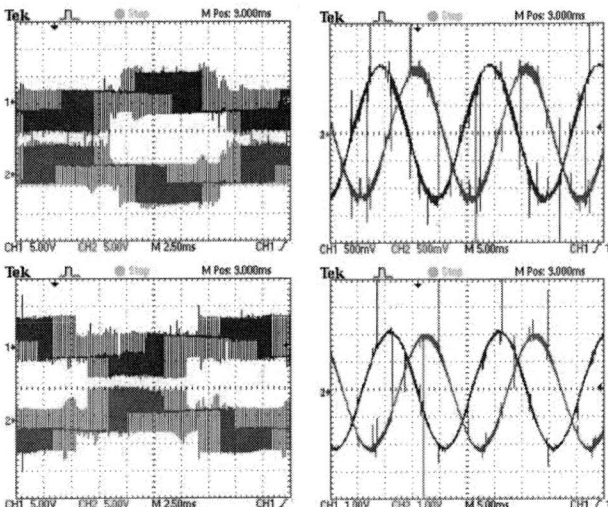

Figure 9: Diagrams of voltages and currents in undermodulation for a) m=0.3 and b) m=0.5. Average dc voltages $V_{c1}=Vc2=50V$, $f_{out}=50Hz$, $f_{sw}=5KHz$.

978-1-4244-1871-8/07 $25.00 © 2007 IEEE

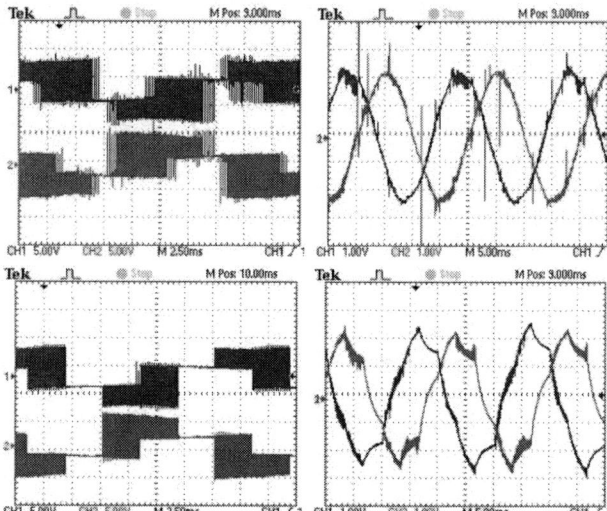

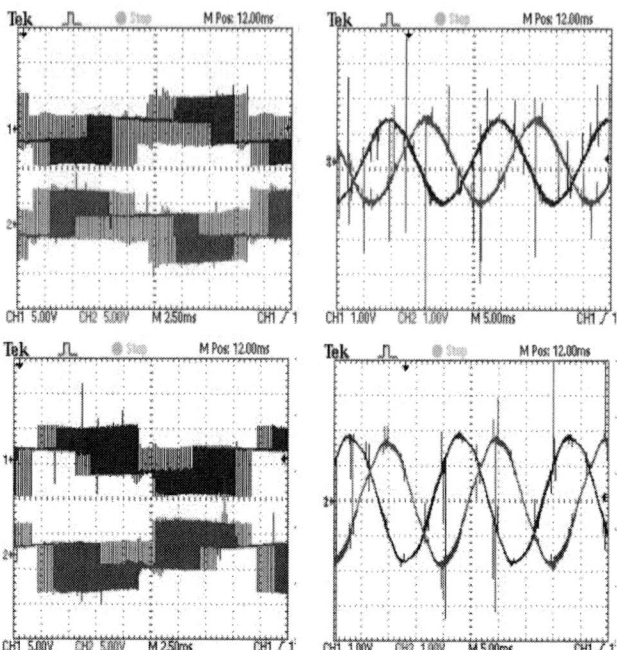

Figure 10: Diagrams of voltages and currents in overmodulation a)mode 1 for m=0.52 and b) mode 2 for m=0.54. Average dc voltages $V_{c1}=V_{c2}=50V$, $f_{out}=50Hz$, $f_{sw}=5KHz$.

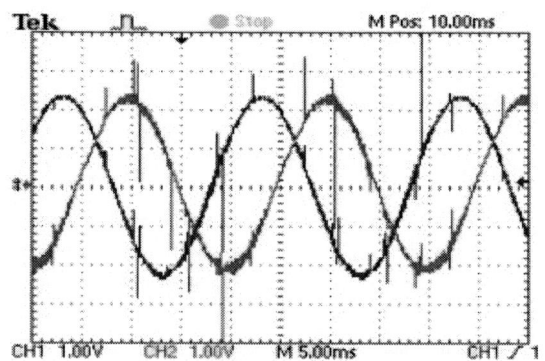

Figure 11: Diagrams of two phase currents in undermodulation for m=0.54. Average dc voltages are increased to $V_{c1}=V_{c2}=60V$, $f_{out}=50Hz$, $f_{sw}=5KHz$.

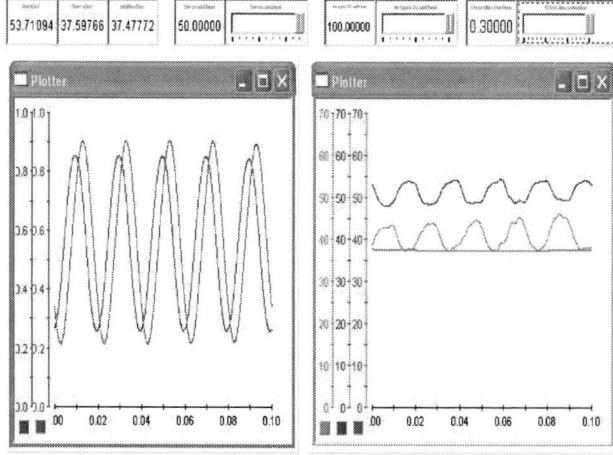

Figure 12: Diagrams of a) two modulating signals and b) voltages on dc capacitors and VcMin for unbalanced dc voltages are increased to $V_{c1}=50V$, $Vc2=40V$, $f_{out}=50Hz$, $f_{sw}=5KHz$

Figure 13: Influence of unbalanced dc sources as Vc1=50V and Vc2=40V. Diagrams of voltages and currents in a) undermodulation for m=0.3 and b) overmodulation mode 1 for m=0.4.

The validity of algorithm was also demonstrated for unbalanced dc voltages, while two dc sources were set as 40V and 50V (Fig.12). The reduced value of V_{cMin} got a lower undermodulation range as described in Fig.13 and Fig.14.

CONCLUSIONS

The paper has presented a new carrier PWM method for controlling 4-switch inverters. Its flexible ability for adaptive changing of under- and overmodulation boundary is based on the measured dc voltage sources. Linear PWM control can be obtained for the whole modulation index range. The proposed method has been proved by simulation and experimental results.

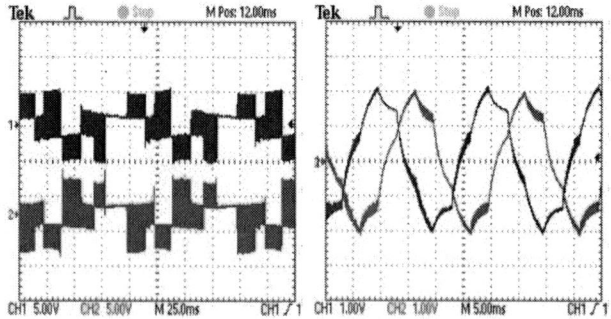

Figure 14: Influence of unbalanced dc sources as Vc1=50V and Vc2=40V. Diagrams of voltages and currents in a) overmodulation mode 2 for m=0.5.

978-1-4244-1871-8/07 $25.00 © 2007 IEEE

REFERENCES

[1] H. W. Van Der Broeck and H. CH. Skudelny, Analytical Analysis of the Harmonic Effects of a PWM AC Drive, IEEE Transactions On Power Electronics, Vol. 3, No. 2, April 1988, pp.216-223

[2] F.Blaabjerg, D.O. Neacsu and J. K. Pedersen, Adaptive SVM to Compensate DC-Link Voltage Ripple for Four-Switch Three-Phase Voltage-Source Inverters, IEEE Transactions On Power Electronics, Vol. 14, No. 4, July 1999, pp.743-752

[3] P. N. Enjeti, A. Rahman, A New Single phase to three phase converter with active input current shaping for low cost ac motor drives , IEEE Transactions On Industry Applications, Vol.29,No.4, August 1993

[4]N.V.Nho,H.H. Lee,Generalized Carrier PWM Algorithms For Multilevel Inverters With Unbalanced DC Voltages,Proceeding of the 37th IEEE Power Electronics Specialists Conference PESC 18-22nd June 2006, Jeju , Korea

[5] N.V.Nho,M.J. Youn, Comprehensive Study On SVPWM and Carrier Based PWM Correlation In Multilevel Inverters, IEE- Proceedings Electric Power Applications, Jan. 2006, Vol.153, No.1,pp.149-158

978-1-4244-1871-8/07 $25.00 © 2007 IEEE

Current Control of PMSM in Overmodulation Region

Jin-Sik Park[1], Shin-Myoung Jung, Hag-Wone Kim[2], and Myoung-Joong Youn

[1]Department of Electrical Engineering and Computer Science, KAIST,
373-1 Guseong-Dong, Yuseong-Gu, Daejeon, Korea
Email: galaxy@powerlab.kaist.ac.kr
[2]Digital Appliance Lab., LG Electronics Co., Ltd., Seoul 153-023, Korea
khw@lge.com

Abstract — This paper presents a study on a current control of a PM synchronous motor which is fed by two-level voltage source inverter in the large modulation index area. In order to increase dynamic response and extend speed limit caused by voltage limit, overmodulation method is generally used. But, the voltage ripple generate current ripple in overmodulation region. This current ripple reduce current controller bandwidth. In this paper, an improved current controller has been proposed for overmodulation region. The methods are described and analyzed and the experimental results are presented.

I. INTRODUCTION

Permanent magnet synchronous motors.(PMSM) fed by voltage source inverter (VSI) have been much attention because of the inherent advantage of high power density and maintenance free. Generally PM motors are widely used for industrial applications, home appliance and servo applications. The fast dynamics and speed operations are important to obtain better performance. To obtain fast dynamic, the overmodulation is required [1] - [3].

In linear modulation area, space vector modulation, carrier based pwm or several PWM methods are used[4][5]. There are two kinds of overmodulation method. One is based on space vector plane, the other is carrier comparison method. CPWM-overmodulation method can be extended continuously from linear region to overmodulation region. In the view of implementation, CPWM method is easier than SVPWM method. A simple solution for carrier based overmodulation has been introduced[6].

As fast current control loop is mandatory for a vector control drive, the problem of current control in the overmodulation range comes forth. How the do we deal with harmonic currents? This problem is being addressed in this paper. the problem is different from the problem [7], We deal with the compensation of the effects of harmonic voltage in the current control loop. this problem has little attention, because generally low pass filter are used. But, LPF generates time delay, and current loop has slow dynamics.

In this paper, to solve this problem, current ripple model have been proposed for reducing the effects of harmonic voltage in the overmodulation region.

This paper presents the methodology how to apply voltage in overmodulation region. It shows overmodulation process with CPWM (linear modulation region ~ six-step operation). In addition to this, current controller with ripple compensator is applied. The proposed controller achieve to get fast current control loop dynamics.

II. PMSM AND INVERTER MODEL

A. PM synchronous motor

The voltage eqaution of a SPMSM can be expressed as follows:

$$\mathbf{v}_{abcs} = -\mathbf{r}_s \mathbf{i}_{abcs} + \frac{d}{dt}\lambda_{abcs}$$

$$\lambda_{abcs} = \mathbf{L}_s \mathbf{i}_{abcs} + \lambda_m' \qquad (1)$$

$\mathbf{v}_{abcs} = \begin{bmatrix} v_{as} & v_{bs} & v_{cs} \end{bmatrix}^T$: phase voltage vector

$\mathbf{i}_{abcs} = \begin{bmatrix} i_{as} & i_{bs} & i_{cs} \end{bmatrix}^T$: phase current vector

$\mathbf{r}_s = diag\begin{bmatrix} r_s & r_s & r_s \end{bmatrix}$: phase resistance matrix

λ_{abcs}' : phase flux linkage

λ_m': the amplitude of the flux linkages by the permanent magnet

Fig. 1 shows the simplified VSI and PMSM model. In this figure, e_a, e_b, e_c represent the voltages cased by the flux linkage of permanent magnet of rotor in each phase.

Using the Park's transformation, the voltage equation in the synchronously rotating reference frame can be obtained from the voltage equation (fig. 1).

The electrical equation of PMSM can be expressed as follows:

$$v_{qs} = r_s i_{qs} + L_s \frac{d}{dt} i_{qs} + \omega_e L_s i_{ds} + \omega_e \lambda_m$$

$$v_{ds} = r_s i_{ds} + L_s \frac{d}{dt} i_{ds} - \omega_e L_s i_{qs} \qquad (2)$$

i_{qs}, i_{ds} represents the qd-axis current, ω_e represents the electrical speed, R_s represents the resistance, L_s represents the inductance of a qd-axis, λ_m represents the amplitude of the

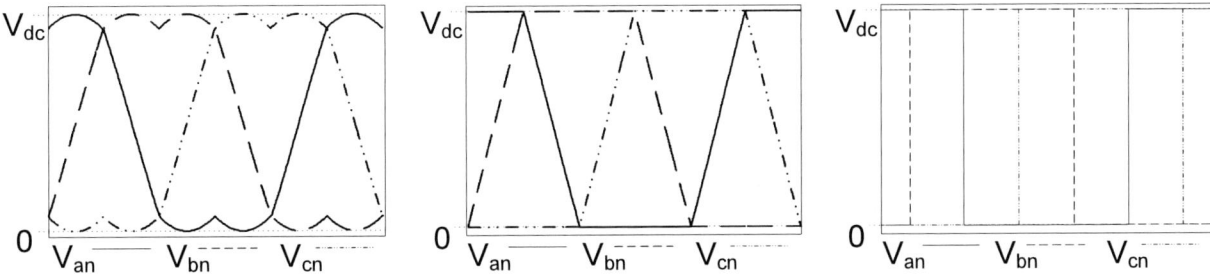

Fig 2. Limit Trajectories C1(m_i=0.907), C2(m_i=0.951), C3(mi=1.0)

flux linkage which is established by the permanent magnet.

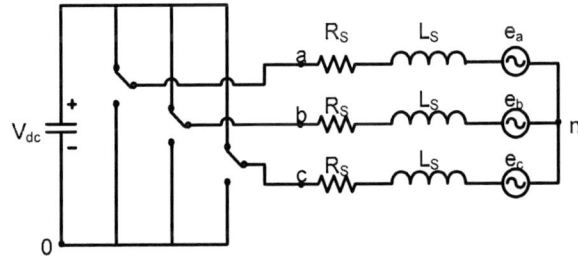

Fig 1. Voltage Source Inverter and PMSM

B. Inverter model

To control motor, variable frequency and magnitude AC voltage is necessary. The pulse width modulated voltage source inverter (PWM-VSI) is generally used to supply the variable AC voltage. The operation of VSI can be studied by considering the equivalent circuit shown in Fig. 1. In this figure, Q_a, Q_b, Q_c represents the leg of the voltage source inverter. V_{a0}, V_{b0}, V_{c0} represents the phase voltage, V_{an}, V_{bn}, V_{cn} represents the node voltage (pole voltage).

III.   OVERMODULATION USING CPWM

A. Modulation Index

The modulation index (m_i) is defined as a ratio between fundamental component of current state and fundamental component of six-step operation.

$$m_i = \frac{\left| \vec{v}_s^* \right|}{\frac{2}{\pi} V_{dc}}$$ (3)

m_i : 0 ~ 0.907   (linear region)
m_i : 0.907 ~ 0.951   (overmodulation region1)
m_i : 0.951 ~ 1.0   (overmodulation region2)

There are three limit trajectories in PWM-VSI. The limit trajectories are represented in Fig 2. C1 is the largest inner circle in SV plane., C2 is the hexagon in SV-plane, C3 is discontinuous six-step. The m_i of C1 is 0.907, the m_i of C2 is 0.951, the m_i of C3 is 1.

B. CPWM using Limit trajectories

A linear overmodulation can be obtained by a linear combination of limit trajectory. If command modulation index is between x_1 and x_2, the command voltage can be represented as follows:

$$y = (1 - \xi_{12})x_1 + \xi_{12}x_2$$

$$\xi_{12} = \frac{m_i - m_{i\_x_1}}{m_{i\_x_2} - m_{i\_x_1}}$$ (4)

y: generated voltage command
x_1:   limit trajectory 1
x_2:   limit trajectory 2
m_{i_x1}:   modulation index of x_1
m_{i_x2}:   modulation index of x_2

In general, overmodulation modes can be classified into two modes. The region between C1 and C2 exist some magnitude error, and The region between C2 and C3 exist larger magnitude and phase error. The magnitude voltage and phase voltage error generate current ripple. This current ripple increase voltage command variation. Consequently current controller bandwidth grow narrower.

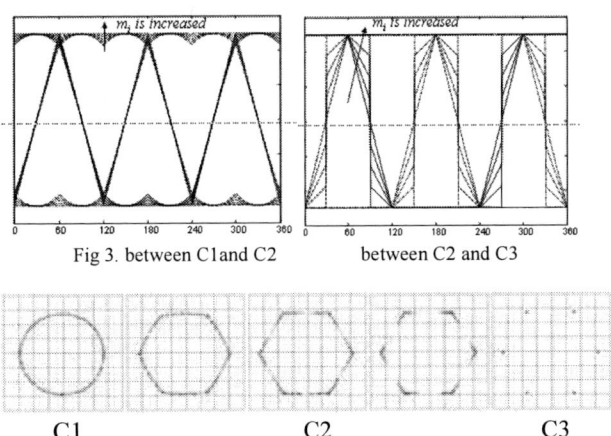

Fig 3. between C1and C2   between C2 and C3

C1   C2   C3
Fig 4. Voltage command in Space vector plane

978-1-4244-1871-8/07 $25.00 © 2007 IEEE   1061

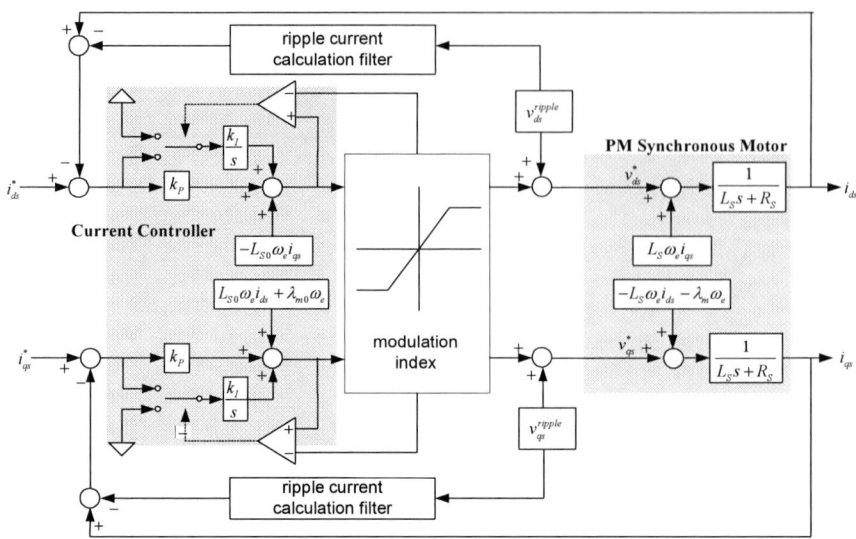

Fig 5. Proposed current controller for overmodulation region

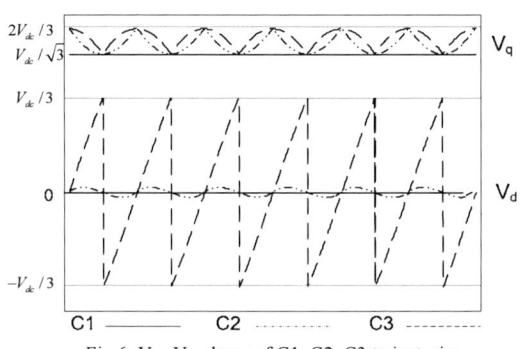

Fig 6. V_{qs}, V_{ds} shape of C1, C2, C3 trajectories

IV. CURRENT CONTROL IN OVERMODULATION REGION

Generally, to get fast dynamics, high-band current controller is necessary. Fast current response can cause severe torque ripple in overmodulation region. The current ripple model for PMSM is used to estimate current ripple. By this method, the fast response current controller can be obtained without limitation of overmodulation performance.

Current dynamics of PMSM can be represented as follows:

$$L_S s \Delta i_{qs}(s) = \Delta v_{qs}(s) - R_S \Delta i_{qs}(s) - L\omega_e \Delta i_{ds}(s)$$
$$L_S s \Delta i_{ds}(s) = \Delta v_{ds}(s) - R_S \Delta i_{ds}(s) + L\omega_e \Delta i_{qs}(s) \quad (5)$$

In practical experiment, cross coupled term neglected. Current ripple model expressed as single-input-single-output (SISO) with current output.

There is another problem for overmodulation operation. In control near to limitation, it is important to prevent integral windup phenomenon. Windup leads to a large overshoot and a large settling time of the process output. So, Conditional or tracking anti-windup is necessary[8]. As a current controller, synchronous frame decoupled PI controller is used.

Totally, the estimated current is used to compensate the harmonics present in the measured current before it is fed to the current controller. The current controller now operates under low noise condition.

V. SIMULATION AND EXPERIMENTAL RESULTS

Simulation tool is Psim6.0. Fig. 7, shows that voltage command is maximum (m_i=1.0), but d-axis current reduce motor speed. Fig. 8 shows good characteristics. Overmodulation increase the utilization of the installed capacity of the inverter. So, We obtain faster dynamics and high speed.

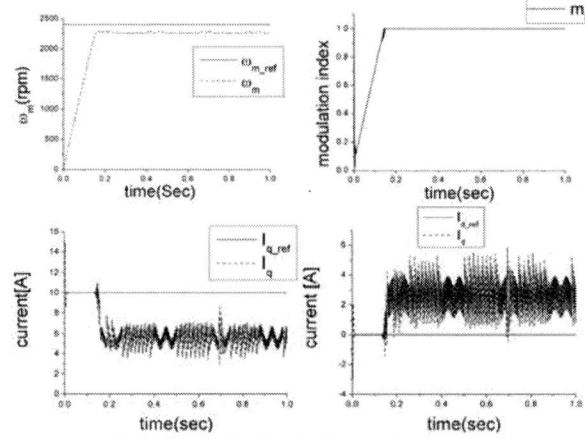

Fig 7. Simulation Results (conventional method)

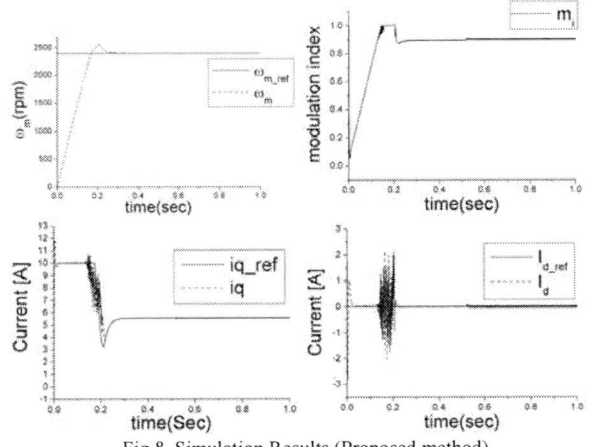

Fig 8. Simulation Results (Proposed method)

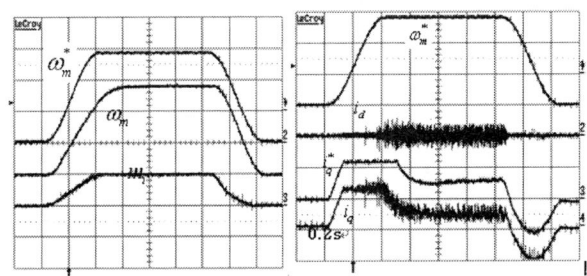

Fig 11. Six-step operation using the proposed method

ω m*: Angular velocity command(krpm/div) id : d-axis current (5A/div)
ω m*: Angular velocity (krpm/div) iq*: q-axis current com. (5A/div)
m_i: modulation index(1.0/div) iq : q-axis current (5A/div)

The scheme was implemented on a TMS 320C31. Motor type is surface mounted type. Inductance is 6.9mH. Experimental results for conventional current controller without consideration of overmodulation and for proposed method are show in fig. 10, fig. 11. In the case of conventional case, because current ripple take place by the overmodulation, output current has much variation. Furthermore, d-axis current is increased by the phase error. Consequently, motor speed is reduced. However, the proposed method shows good characteristics as fig. 10, fig. 11.

In the case of the proposed method, current ripple estimator prevent current controller to response voltage distortion in overmodulation, and anti-windup guarantee that d-axis average current remain new 0A current command. By that reason good speed operation characteristics can be obtained using the proposed current controller in overmodulation region

VI. CONCLUSION

We have shown that overmodulation provides a gain of 10% in fundamental output voltage, and hence inverter output. However, in vector-controlled drives, where current control is mandatory, the presence of harmonic current deteriorates the performance of the current controller, especially mode2 (m_i=0.951~1.0). The proposed method improve performance using current ripple estimator. The proposed scheme can also be easily implemented ion a state-of-the-art DSP. The proposed method help us to design high gain current controller.

REFERENCES

[1] Joachim holtz, Wolfgang Lotzkat, and Ashwin M,. "The Overmodulation Range Including the Six-Step Mode,: IEEE Trans. on Power Electronics, Vol. 8, No.4, Oct. 1993, pp. 546-553

[2] Hava, A.M., Kerkman, R.J., and Lipo, T.A., "Carrier-based PWM-VSI overmodulation strategies: analysis, comparison, and design," IEEE Trans. on Power Electronics, Vol.13 No.4, July, 1998, pp. 674-689

[3] Jul Ki Seok, Joohn Sheok Kim, and Seung Ki Sul, "Overmodulation Strategy for Hi-Performance Torque Control," IEEE Trans. on Power Electronics, Vol. 13, No.4, Jul, 1998, pp.786-792

[4] N.V. Nho and M.J. Youn, " A novel simple linear PWM technique in two-lelvel VSI," the fifth Int'l IEEE PEDS Conference 2003, pp. 1241-1244, 2003

[5] C.B. Jacobina, A>M.N Lima, and E.R.C da Silva," PWM space vector based on digital scalar modulation," Proc. Conf. Rec. PESC, pp.606-611, 1997

[6] N.V. Nho and M.J. Youn,"Two-mode overmodulation in tow-lelvel VSI" Proc. Conf. Rec. IEEE PEDS Conference 2003, pp. 1274-1279, 2003

[7] Ashiwin M. Khambadkone and Joachim Holtz,,"Compensated Synchronous PI Current Controller in Overmodulation Range and Six-Step Operation of Space-Vector-Modulation-Based Vector-Controlled Drives," IEEE Transactions on Industrial Electronics, Vol. 49, No. 3, June, 2002, pp. 574-580.

[8] Youbin Peng, Damir Vrancic, and Raymon Hanus,"Anti-Windup, Bumpless, and Conditioned Transfer Techniques for PID Controllers:, Control Systems magazines, IEEE, Vol. 16, No.4, Aug. 2002, pp. 47\8-57

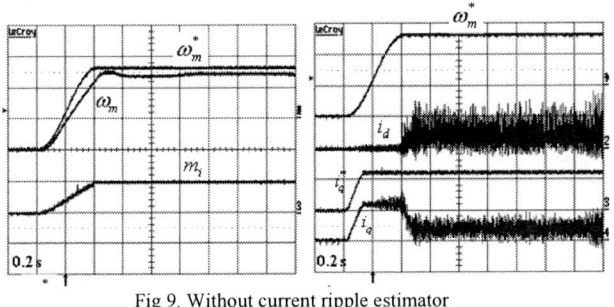

Fig 9. Without current ripple estimator

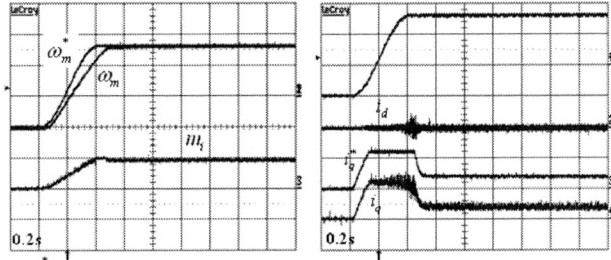

Fig 10. With proposed current ripple estimator

978-1-4244-1871-8/07 $25.00 © 2007 IEEE

Characteristics of a Pulse-link Inverter for Fuel Cells

Kentaro Fukushima
Kyushu University
Department of Electrical and Electronic Systems
Engineering
744, Moto-oka, Nishi-ku, Fukuoka, 819-0395
Email: fukushima@ckt.ees.kyushu-u.ac.jp

Isami Norigoe
EBARA DENSAN .LTD
11-1 Asahi-machi, Ota-ku, Tokyo, 144-8510
Email: norigoe@attglobal.net

Kenta Tsukakoshi
EBARA DENSAN .LTD
11-1 Asahi-machi, Ota-ku, Tokyo, 144-8510
Email: tsukakoshi.kenta@ed.ebara.com

Tamotsu Ninomiya
Kyushu University
Department of Electrical and Electronic Systems
Engineering
744, Moto-oka, Nishi-ku, Fukuoka, 819-0395
Email:ninomiya@ees.kyushu-u.ac.jp

Yosuke Harada,
EBARA DENSAN .LTD
11-1 Asahi-machi, Ota-ku, Tokyo, 144-8510
Email: harada.yosuke@ed.ebara.com

Zheng Dai
EBARA DENSAN .LTD
11-1 Asahi-machi, Ota-ku, Tokyo, 144-8510
Email: dai.zheng@ed.ebara.com

Abstract— This paper proposes a novel DC-AC converter topology for fuel cells. This topology provides pulsed voltage to PWM inverter directly. As the results, it will be possible to reduce the size of this component because of no need for smoothing circuit between boost converter and PWM inverter. Furthermore, this topology is expected to achieve ZVS of the PWM switches because pulsed voltage has certainly zero voltage period in the switching period. This paper analyzes the novel topology called Pulse-link inverter. And it is shown the advantage the converter by the experimental results.

I.   INTRODUCTION

Today, environmental issues such as global heating have become an international crisis. And clean energy system is strongly demanded. Therefore, several clean energy system have been researched and developing, recently. Among them, the generating systems used by fuel cells have come to attention. Fuel cells convert chemical energy directly to electricity and thermal energy. As the result, fuel cells are high efficiency about energy conversion. And now, the home co-generation system using both electricity and thermal energy from fuel cells are researched actively [1].

The voltage provided from fuel cells is DC, so DC-AC converter is needed in home application. The specifications of DC-AC converter for fuel cells are shown below:

1.   It boosts input voltage that is from fuel cells to a level of commercial voltage.
2.   It can be isolated between fuel cells and load stage.
3.   It does not make the input current ripple.

The voltage of fuel cells is very low, so the converter needs to boost input voltage up to commercial voltage. Second thing is needed from the safety. And third thing is particular to fuel cells. In the fuel cells, current-ripple not only affects the fuel cell capacity, but also damages the fuel consumption and life-span [2, 3 and 4].

The conventional DC-AC converter for fuel cells is shown Fig. 1. The conventional DC-AC converter is constructed

two stages. First stage is isolated DC-DC converter and second stage is PWM inverter. Between the stages, large capacitance of capacitor is inserted to reduce input current-ripple from commercial frequency. However, this capacitor causes the difficulty in small size of the unit.

To overcome the problem, this paper proposes a novel DC-AC converter topology shown Fig. 2. This topology boosts input voltage to the level of commercial voltage by pulse waveform at the first stage. And it provides the boost-pulse-voltage directly to PWM inverter. This concept has known as AC link or pulse link [5, 6]. Here, we call this topology as Pulse-link inverter. Pulse-link inverter needs neither smoothing circuit nor large capacitor between first stage and second stage. Furthermore, it is clear that this topology can achieve zero-voltage-switching (ZVS) of the PWM inverter switches. As the results, it will be possible to reduce the size of the component. This paper analyzes the steady-state characteristics of Pulse-link inverter topology. And it makes clear the advantage of this topology by the experimental results.

II.   OPERATING STATES

Fig. 2 shows the proposed circuit topology. This topology provides boosted voltage pulse that is controlled by switch Q_1 directly to PWM inverter. The switching sequence of this converter is shown as Fig. 3. This converter has 5 switches.

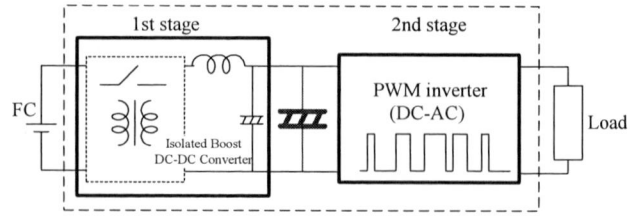

Fig. 1. The conventional DC-AC converter for fuel cells.

978-1-4244-1871-8/07 $25.00 © 2007 IEEE   1064

As mentioned before, switch Q_1 controls the boost pulse from input voltage. And, from S_1 to S_4 are PWM switches. S_1 and S_4 are controlled to make output voltage sinusoidal waveform, while S_2 and S_3 are decided the plus/minus of output voltage.

And control combination of S_1, S_3 and S_2, S_4 is a pair. Here, it is analyzed when the output voltage is in the positive semicircle period. So, there are 3 states in one switching period.

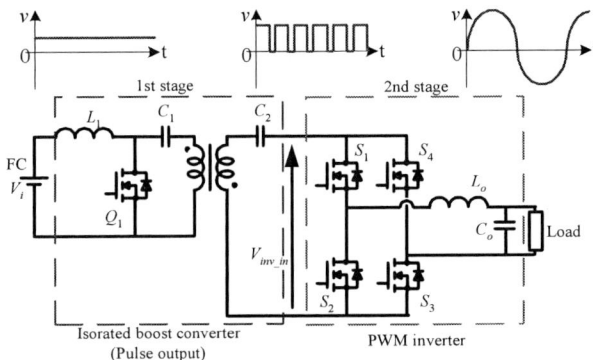

Fig. 2. The proposed circuit topology.

A. State 1 (Q_1:ON, S_1:ON, S_3:ON)

Fig.4 (a) shows the equivalent circuit of state 1. State equations on state 1 are written below:

$$\begin{cases} v_{L_1} = V_i - r_{Q_1}\hat{i}_{L_1} - nr_{Q_1}\hat{i}_{L_o} \\ v_{L_o} = \hat{v}_{C'} - \hat{v}_{C_o} - nr_{Q1}\hat{i}_{L_1} - \left(r_{s1} + r_{s3}\right)\hat{i}_{L_o} \\ i_{C'} = -\hat{i}_{L_o} \\ i_{C_o} = \hat{i}_{L_o} - \dfrac{\hat{v}_{C_o}}{R_o} \end{cases} \quad (1)$$

, where, $C' = \dfrac{C_1 C_2}{C_1 + n^2 C_2}$.

B. State 2 (Q_1:ON, S_1:OFF, S_3:ON)

Fig. 4 (b) shows the equivalent circuit of state 2. State equations on state 2 are written below:

$$\begin{cases} v_{L_1} = V_i - r_{Q_1}\hat{i}_{L_1} \\ v_{L_o} = -\hat{v}_{C_o} - \left(r_{D2} + r_{s3}\right)\hat{i}_{L_o} \\ i_{C'} = 0 \\ i_{C_o} = \hat{i}_{L_o} - \dfrac{\hat{v}_{C_o}}{R_o} \end{cases} \quad (2)$$

C. State 3 (Q_1:OFF, S_1:OFF, S_3:ON)

Fig. 4 (c) shows the equivalent circuit of state 3. State equations on state 3 are written below:

$$\begin{cases} v_{L_1} = V_i - \dfrac{1}{n}\hat{v}_{C'} - \dfrac{1}{n}\left(r_{D1} + r_{D2}\right)\hat{i}_{L1} \\ v_{L_o} = -\hat{v}_{C_o} - \dfrac{1}{n}r_{D1}\hat{i}_{L_1} - \left(r_{D2} + r_{s3}\right)\hat{i}_{L_o} \\ i_{C'} = \dfrac{1}{n}\hat{i}_{L_1} \\ i_{C_o} = \hat{i}_{L_o} - \dfrac{\hat{v}_{C_o}}{R_o} \end{cases} \quad (3)$$

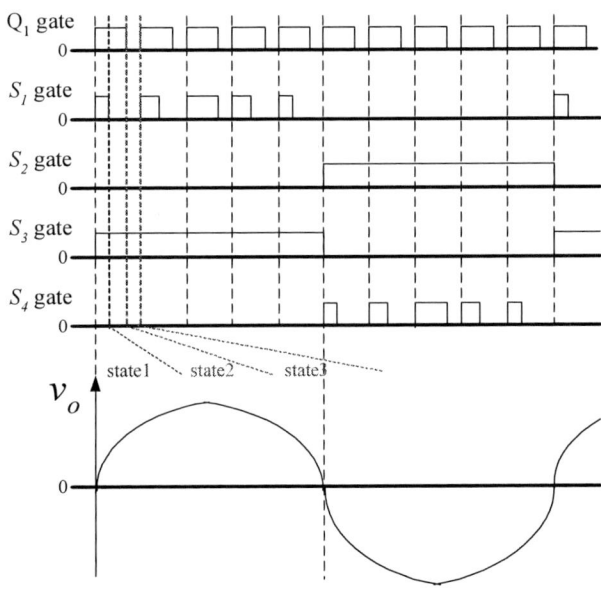

Fig. 3. Switching sequences.

D. Steady State

From the above equations, the state-averaging-vector is shown by using state space averaging method. Here, on-resistance or conduction losses are ignored.

$$\frac{d\mathbf{X}}{dt} = \begin{bmatrix} 0 & 0 & -\dfrac{1-D_{Q_1}}{nL_1} & 0 \\ 0 & 0 & \dfrac{d_s}{L_o} & -\dfrac{1}{L_o} \\ \dfrac{1-D_{Q_1}}{nC'} & -\dfrac{d_s}{C'} & 0 & 0 \\ 0 & \dfrac{1}{C_o} & 0 & -\dfrac{1}{C_o R} \end{bmatrix} \mathbf{X} + \begin{bmatrix} \dfrac{1}{L_1} \\ 0 \\ 0 \\ 0 \end{bmatrix} V_i$$

$$(4)$$

, where $X = \begin{bmatrix} \hat{i}_{L1} & \hat{i}_{Lo} & \hat{v}_{C'} & \hat{v}_o \end{bmatrix}^T$

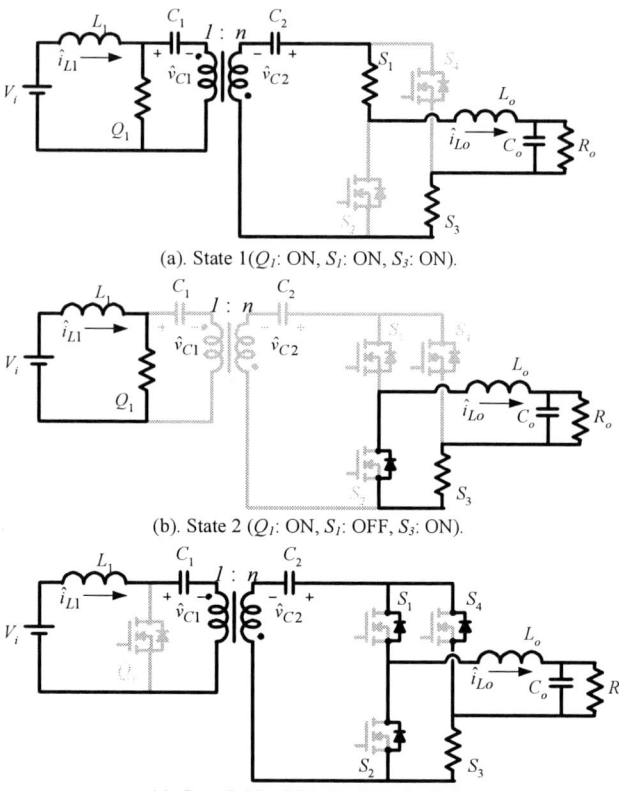

(a). State 1(Q_1: ON, S_1: ON, S_3: ON).

(b). State 2 (Q_1: ON, S_1: OFF, S_3: ON).

(c). State 3 (Q_1: OFF, S_1: OFF, S_3: ON).

Fig. 4. Equivalent circuit of each state.

And from (4), the steady-state characteristics are shown below:

$$V_{c'} = \frac{n}{1-D_{Q_1}} V_i \qquad (5)$$

$$V_o = \frac{nd_s(t)}{1-D_{Q_1}} V_i \qquad (6)$$

Furthermore, from (5), the peak voltage pulse that is input to PWM inverter (v_{inv_in}) is written below

$$v_{inv\_in} = \frac{n}{1-D_{Q_1}} V_i \qquad (7)$$

Here, D_{Q_1} is duty ratio of switch Q_1. And, $d_{s1}(t)$ is duty ratio of PWM inverter switch of S_1/S_4. $d_{s1}(t)$ is changed shown as (8) in order to make output voltage to be sinusoidal waveforms.

$$ds(t) = d_{s1\_max} \cdot \sin(2\pi \cdot 50t) \qquad (8)$$

Moreover, the relationship of D_{Q_1} and d_{s1_max} is limited (9), because PWM inverter is provided voltage only when Q_1 is ON.

$$D_{Q_1} \geq d_{s1\_max} \qquad (9)$$

TABLE I. CIRCUIT PARAMETER VALUES.

Symbol	Description	value
Vi	Input voltage	20[V]
L1	Input inductor	400[uH]
LM	Magnetizing inductor	38.8[uH]
C1	primary-side capacitor	3[mF]
C2	Secondary-side capacitor	330[uF]
n	Turn ratio	3
Lo	Output inductor	3[mH]
Co	Outout capacitor	9.4[uF]
fs	Switching frequency	30[kHz]
DQ1	Duty ratio of Q1	0.7
ds_max	Maximimum duty ratio of S1/S4	0.7
Ro	Output resistance	100[Ω]

At next chapter, it is shown the experimental results and it is considered the characteristics of the proposed converter.

III. EXPERIMENTAL RESULTS

A. Setting of parameters

In order to evaluate the performance of the proposed circuit, the experimental circuit is implemented by using the specifications and parameters in Table 1. From table 1, C_1 is 3[mF], and it is aluminum electrolytic capacitor. C_1 is decided from the allowable current. Primary-side is flown large current, so capacitance of C_1 becomes large value. If C_1 is film capacitor or conductive polymer electrolytic capacitor, the value will become less. However, the capacitor these types are expensive, so the cost of this unit becomes higher. Furthermore, primary-side is low voltage, so the size of aluminum electrolytic capacitor is not so large even if the value is large because withstand-voltage is low. Therefore, large value of aluminum electrolytic capacitor is used at C_1 in this experiment.

B. Output voltage waveform and input current waveform

Fig. 5 shows the experimental waveforms of output voltage (v_o) and input current (i_i). And table II shows the experimental measurement. From those results, it is considered that output voltage is achieved to output commercial voltage. Furthermore, it is considered that the waveform of v_o becomes distorted around zero cross point. This is caused from the characteristics of the body-diode of S_2 and S_3.

However, it is observed that input current ripple is not enough to be reduced. Input current ripple is 3[Ap_p]. So, at next chapter, it is mentioned a control method to reduce input current ripple.

Fig. 7 shows the waveforms of input voltage to PWM inverter (v_{inv_in}), drain-source voltage of switch Q_1 (v_{dsQ1}) and drain current of switch Q_1 (i_{dQ1}). From the waveforms of v_{inv_in} and v_{ds}, it is considered that boosted voltage pulse v_{inv_in} is generated when switch Q_1 is ON. Furthermore, the peak voltage of the voltage pulse v_{inv_in} is shown from (7). In this experiment, duty ratio of Q_1 is 0.7, so the peak voltage of v_{inv_in} is 200[V] by (7). And the peak voltage of output voltage is 140[V] calculated from equation (6). This peak voltage is corresponding to the peak voltage of commercial alternating voltage in 100[V].

Furthermore, the waveform of v_{inv_in} and i_{dQ1} is occurred surge. This surge is occurred by charging parasitic capacitance of MOSFET for PWM inverter. To overcome this surge, clamp capacitor (C_{clamp}) and auxiliary MOSFET (Q_2) are inserted shown fig. 7. Q_2 is turned ON just before Q_1 is turned ON, and duty ratio of Q_2 is short.

Figure 8 shows the experimental waveforms of v_{inv_in}, v_{dsQ1} and i_{dQ1} when clamp-circuit is inserted. From the figure, v_{inv_in} and i_{dQ1} are not occurred surge. As the result, improvement circuit is able to use the MOSFET which is lower withstanding voltage than first topology, so it is expected to improve the efficiency.

IV. IMPROVEMENT METHOD OF INPUT CURRENT-RIPPLE

A. Control method

In order to control input current, duty ratio of switch Q_1 (D_{Q1}) is controlled. From fig. 5, it is observed that when output voltage is around zero crossing, input current is minimum point. This means that input power is not needed, so extra power is about to go back to input stage.

To avoid the above thing, it is controlled D_{Q1} by detecting input current. Fig. 9 shows the experimental component of current detecting. In the experiment, FPGA is used and it is controlled by digital-control with A/D converter.

In the experiment, current transformer detects input current. 1[A] is converted to 0.125[V] used by current transformer. And the converted voltage is input to A/D converter. 0.01[V] is corresponding to 1 binary data at A/D converter.

By using the binary data, the duty ratio signal of switch Q_1 ($Q_{1signal}$) is calculated by below equation:

$$Q_{1signal} = Q_{1signal\_ref} + k\left(I_{in} - I_{ref}\right) \quad \cdots (10)$$

, where $Q_{1signal\_ref}$ is corresponding to the binary data that duty ratio of switch Q_1 (D_{Q1}) is 0.7, and I_{ref} is converted reference input current to binary data.

B. Experimental result

Fig. 10 shows the experimental waveforms of v_o, and i_i.. From the figure, it is considered that output voltage is output commercial voltage, and input current has fewer ripples. This means that the input current regulation is achieved.

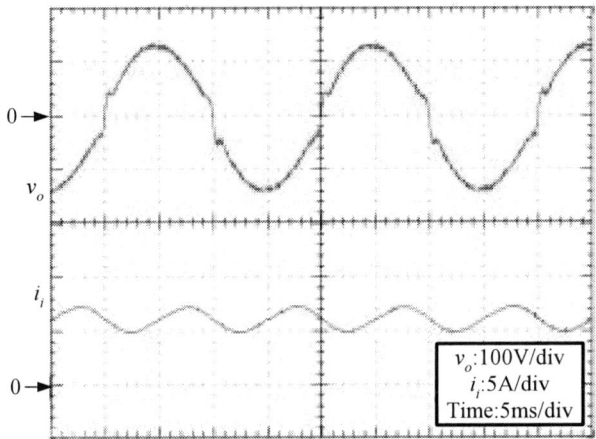

Fig. 5. Experimental waveforms of v_o and i_i.

TABLE II. EXPERIMENTAL MEASUREMENT.

Symbol	Description	value
Vi	Input voltage	20[V]
Ii	Input current	6.0[A]
Vo	Output voltage	98.2[V(rms)]
R_o	Output resistance	100[Ω]
η	Efficiency	80.4[%]

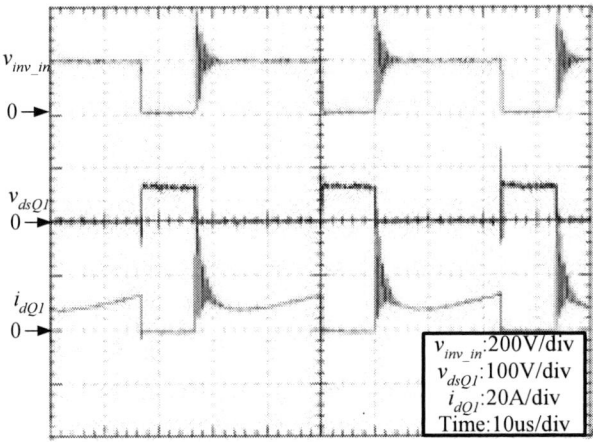

Fig. 6. Experimental waveforms of v_{inv_in}, v_{dsQ1} and $idQ1$

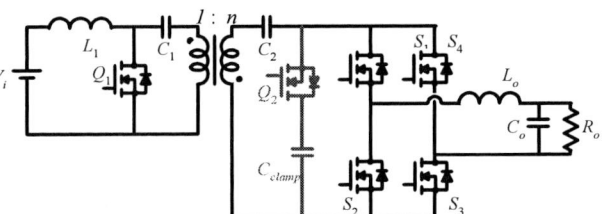

Fig. 7. Improvement of circuit topology.

978-1-4244-1871-8/07 $25.00 © 2007 IEEE

This topology is expected to be used for fuel cells. The optimum parameters of this proposed topology and improvement of the efficiency is future work.

V. CONCLUSION

This paper proposes a novel DC-AC converter for fuel cells. This converter provides the boosted voltage pulse directly to PWM inverter. We call this topology as Pulse-link inverter. And this paper is shown that this topology is achieved to convert DC to AC. Furthermore, the surge reduction circuit is inserted, improvement circuit topology is shown.

Moreover, in order to reduce input current ripple, the control method of this topology is shown. As the result, the input current is improved to be fewer ripples.

The optimum parameters of this proposed topology and improvement of the efficiency is future work.

.

REFERENCES

[1] S. Sumiyoshi, H. Omori, and Y. Nishida, "Power Conditioner Consisting of Utility Interactive Inverter and Soft-Switching DC-DC Converter for Fuel-Cell Cogeneration System," PCC-Nagoya 2007, pp.455-462, Apr. 2007.

[2] S. Moon, J. Lai, S. Park and C. Liu, "Impact of SOFC Fuel Cell Source Impedance on Low Frequency AC Ripple, " Power Electronics Specialists Conference, Proc. of IEE PESC 2006, pp.2037-2042, Jun. 2006.

[3] W. Choi, P.N. Enjeti and J.W. Howze, "Development of an Equivalent Circuit Model of a Fuel Cell to Evaluate the Effects of Inverter Ripple Current," Proc. of IEEE APEC 2004, pp. 255-361, Feb. 2004.

[4] G. Fontes, C. Turpin, R. Saiset, T. Meynard, and S. Astier, "Interactions between fuel cells and power converters Influence of current harmonics on a fuel cell stack," Proc. of PESC 2004, pp. 4729-4735, 2004.

[5] P. T. Kerin, R. S. Balog, and X. Geng, "High-Frequency Link Inverter for Fuel Cells Based on Multiple-Carrier PWM," IEEE Transaction on PE, Vol. 19, No. 5, pp. 1279-1288, Sep. 2004.

[6] D. Chen and L. Li, "Novel Static Inverters With High Frequency Pulse DC Link," IEEE Transaction on PE, Vol. 19, No. 4, pp. 971-978, Jul. 2004.

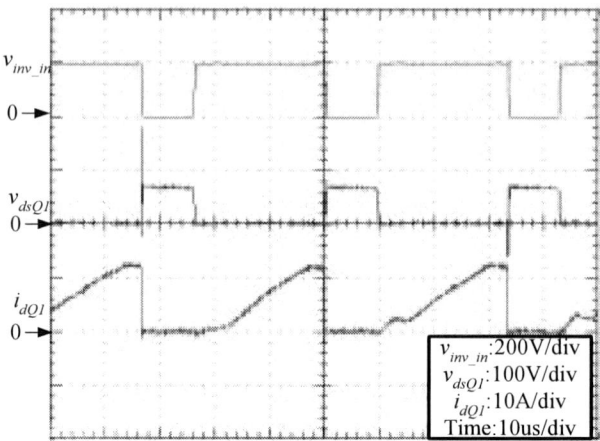

Fig. 8. Experimental waveforms of v_{inv_in}, v_{dsQ1} and $idQ1$

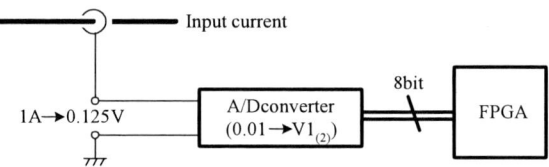

Fig. 9. Experimental component of current sensing.

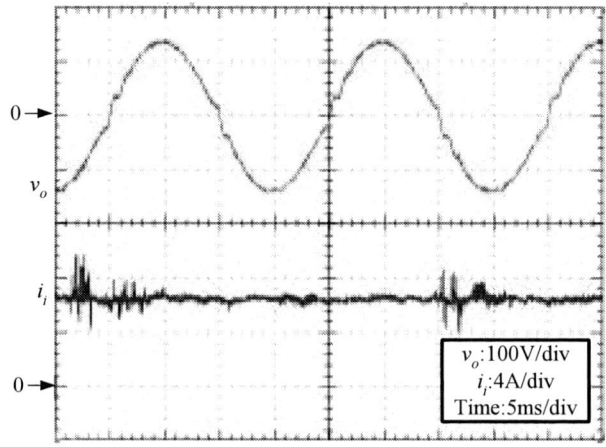

Fig. 10. Experimental waveforms of v_o and i_i.

Analysis of Conduction and Switching Losses of A Matrix-Z-Source Converter

Keping You
The University of New South Wales
The School of Electrical Engineering and
Telecommunication
Sydney Australia
Email: youkeping@ieee.org

M. F. Rahman
The University of New South Wales
The School of Electrical Engineering and
Telecommunication
Sydney Australia
Email: F.Rahman@unsw.edu.au

ABSTRACT

The analytical model of the conduction and switching losses of matrix Z-source converter (MZC) was investigated in this paper. The model provides a measurable way to approximate the total loss of MZC.

I. INTRODUCTION

The purpose of this paper is to investigate the analytical models for the approximation of conduction and switching losses of the power-switch network in the Matrix-Z-source converter (MZC), topology of which will be briefly described in section II. Investigation of the conduction and switching losses is one of the fundamental requirements in the thermal design of any power converter. Analytical models of conduction and switching losses can provide the circuit designer with a measurable way to approximate the total loss of the given power converter at different operating points so as to estimate the trend of the loss versus the change of the operating points. Furthermore, the analytical models of conduction and switching losses can be the fair measures in the comparison of power loss between different topologies (in the same category of power conversion) over a range of operating points. Hence the analytical models of conduction and switching losses for the MZC are desirable.

The models of conduction and switching losses for IGBT and free-wheeling diode are introduced in section III as the base for analysis. The conduction and switching losses of the power-switch networks for the DC-AC inversion mode of MZC are investigated in section IV, for AC-DC rectification in section V.

II. TOPOLOGY AND MODULATION STRATEGIES OF MATRIX-Z-SOURCE CONVERTER

Figure 1(a) shows the configuration of MZC. Each power-switch of the switch-network of the configuration consists of two pairs of IGBT and diode. Figure 1(b) shows the configuration when MZC is in DC-AC inversion mode. The power-switch network in this mode is equivalent to the standard one in VSI. One of the IGBT in each power-switch is set at ON-state all the time so as to provide the current path of the free-wheeling diode. Figure 1(c) shows the configuration when MZC is in AC-DC rectification mode. The configuration is equivalent to that of an AC-DC matrix converter [6].

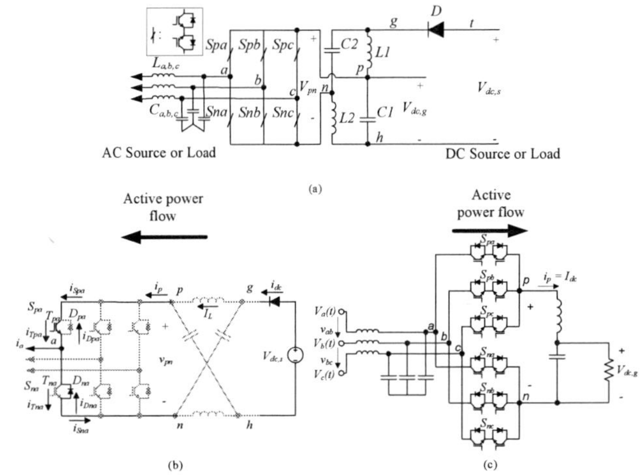

Figure 1 The proposed MZC topology and operation principles. (a) Basic configuration, (b) The structure in DC-AC inversion mode, where every topological switch always has one IGBT in On-state all the time, (c) The structure in AC-DC rectification mode.

In DC-AC inversion mode, the reduced form of Alesina - Venturini optimum PWM (AV-optimum-PWM) algorithm for VSI (voltage source inverter), which is equivalent to the third-harmonic injection PWM, is used. Maximum constant boost control method is accommodated into the modulation algorithm. In AC-DC rectification mode, the reduced form of Alesina -Venturini optimum PWM (AV-optimum-PWM) algorithm for AC-DC matrix converter is used.

A. The modulation strategy in DC-AC inversion mode of MZC

$$\begin{bmatrix} V_a(t) \\ V_b(t) \\ V_c(t) \end{bmatrix} = \begin{bmatrix} d_{pa}(t) & d_{na}(t) \\ d_{pb}(t) & d_{nb}(t) \\ d_{pc}(t) & d_{nc}(t) \end{bmatrix} B \begin{bmatrix} V_{dc}/2 \\ -V_{dc}/2 \end{bmatrix} \quad (1)$$

$$\begin{bmatrix} I_{link} \\ -I_{link} \end{bmatrix} = \begin{bmatrix} d_{pa}(t) & d_{pb}(t) & d_{pc}(t) \\ d_{na}(t) & d_{nb}(t) & d_{nc}(t) \end{bmatrix} \begin{bmatrix} I_a(t) \\ I_b(t) \\ I_c(t) \end{bmatrix}$$

$$d_{pa}(t) = M[(1/2)\cos(\omega^\bullet t) - (1/12)\cos(3\omega^\bullet t)] + 1/2$$

$$d_{pb}(t) = M[(1/2)\cos(\omega^\bullet t - 2\pi/3) - (1/12)\cos(3\omega^\bullet t)] + 1/2$$

$$d_{pc}(t) = M[(1/2)\cos(\omega^\bullet t + 2\pi/3) - (1/12)\cos(3\omega^\bullet t)] + 1/2 \quad (2)$$

$$d_{na}(t) = 1 - d_{pa}(t)$$

$$d_{nb}(t) = 1 - d_{pb}(t)$$

$$d_{nc}(t) = 1 - d_{pc}(t)$$

$$B = 1/(\sqrt{3}M - 1)$$
$$1/\sqrt{3} < M < 2/\sqrt{3} \quad (3)$$

where and hereafter B denotes the voltage boost factor.

B. The modulation strategy in AC-DC rectification mode of MZC

$$\begin{bmatrix} V_{dc}/2 \\ -V_{dc}/2 \end{bmatrix} = \begin{bmatrix} d_{pa} & d_{pb} & d_{pc} \\ d_{na} & d_{nb} & d_{nc} \end{bmatrix} \begin{bmatrix} V_a(t) \\ V_b(t) \\ V_c(t) \end{bmatrix} \quad (4)$$

$$\begin{bmatrix} I_a(t) \\ I_b(t) \\ I_c(t) \end{bmatrix} = \begin{bmatrix} d_{pa} & d_{na} \\ d_{pb} & d_{nb} \\ d_{pc} & d_{nc} \end{bmatrix} \begin{bmatrix} I_d \\ -I_d \end{bmatrix}$$

$$d_{pa}(t) = \frac{\sqrt{3}}{2} M \begin{bmatrix} (1/2)\cos(\omega t) \\ +(7/36)\cos(2\omega t) \\ -(1/36)\cos(4\omega t) \end{bmatrix} + 1/3$$

$$d_{pb}(t) = \frac{\sqrt{3}}{2} M \begin{bmatrix} (1/2)\cos(\omega t - 2\pi/3) \\ +(7/36)\cos(2\omega t + 2\pi/3) \\ -(1/36)\cos(4\omega t) \end{bmatrix} + 1/3 \quad (5)$$

$$d_{pc}(t) = 1 - d_{pa}(t) - d_{pb}(t)$$

$$d_{na}(t) = \frac{\sqrt{3}}{2} M \begin{bmatrix} (-1/2)\cos(\omega t) \\ +(7/36)\cos(2\omega t) \\ -(1/36)\cos(4\omega t) \end{bmatrix} + 1/3$$

$$d_{nb}(t) = \frac{\sqrt{3}}{2} M \begin{bmatrix} (-1/2)\cos(\omega t - 2\pi/3) \\ +(7/36)\cos(2\omega t + 2\pi/3) \\ -(1/36)\cos(4\omega t) \end{bmatrix} + 1/3$$

$$d_{nc}(t) = 1 - d_{na}(t) - d_{nb}(t)$$

$$0 < M < 2/\sqrt{3} \quad (6)$$

III. MODELS OF CONDUCTION AND SWITCHING LOSSES FOR SEMICONDUCTOR SWITCHES

The approximation models of switching loss and On-state parameters reported in [3] are used in the following analysis. The junction temperature is assumed at 120°C.

A. Switching losses approximation

The energy loss w of one single switching action has dependency on switched voltage u and current i, is approximated by using (7) (8) (9).

$$w_{Toff} = K_{Toff1}ui + K_{Toff2}ui^2 + K_{Toff3}u^2 + K_{Toff4}u^2i + K_{Toff5}u^2i^2 \quad (7)$$

$$w_{Ton} = K_{Ton1}ui + K_{Ton2}ui^2 + K_{Ton3}u^2 + K_{Ton4}u^2i + K_{Ton5}u^2i^2 \quad (8)$$

$$w_{Doff} = K_{Don1}ui + K_{Don2}ui^2 + K_{Don3}u^2 + K_{Don4}u^2i + K_{Don5}u^2i^2 \quad (9)$$

where $_T$ and $_D$ denote IGBT and diode respectively, and K_i is IGBT switching loss Parameter as in Table I. The turn-on switching loss of diode is ignored.

Table I Switching loss model parameters at 120 °C junction temperature

Tj		K1	K2	K3	K4	K5
120 °C	T_{off}	179	-1.31	0.650	-0.116	0.00348
	T_{on}	70.0	2.94	0.518	0.102	0.00155
	D_{off}	97.7	-3.73	0.488	0.140	0.00427
Units		nWs(VA)-1	nWs(VA)-1	nWs(VA)-1	nWs(VA)-1	1 nWs(V-2A2)-1

B. On-state parameters

Table II IGBT On-state Parameter

Tj	r_T	r_D	v_T	v_D
120°C	0.0787	0.038	0.0768	0.0732
Units	V/A	V/A	V	V

where r_T is on-state resistance of IGBT; r_D is on-state resistance of free-wheel diode; V_T is voltage drop of IGBT; V_D is forward voltage drop of free-wheel diode.

IV. AVERAGE CONDUCTION AND SWITCHING LOSSES MODEL OF MZC IN DC-AC INVERSION MODE

In DC-AC operation mode, the MZC operates as a Z-source inverter. The model of conduction and switching losses of Z-source is first investigated in subsections A and B. The model of conduction and switching losses for MZC in DC-AC inversion mode then is worked out in section C with extra attention on the series connected devices.

The calculation results will be converted into the form of "mWatts per Watt" as given by (10) for the convenience of visualization.

$$P'_{loss} = \frac{P_{loss}}{P_o} \times 1000 \quad (10)$$

where, P'_{loss} is the result to be visualized; P_{loss} is the calculated loss at certain operating point indicated by a pair of modulation index and power factor in Table III; P_o is defined in Table III.

Table III Conditions for the calculation of power loss using the analytical model of conduction and switching power loss of MZC

Symbol	Description	In DC-AC mode	In AC-DC mode
pf	Power factor	0.7 < pf < 1.0	0.7 < pf < 0.9
M	Modulation index	0.866<M< 1.155	0.5<M< 1.155
Vdc	DC voltage	36 V	42 V
Po	Active power	2000 W	2000 W
f_{sw}	Switching frequency	10 kHz	10 kHz

A. Average Conduction loss in Z-source inverter

1) Conduction loss Caused by IGBT T_{pa} in Z-source Inverter

When the AC load current i_a is in the positive-half cycle, $i_a > 0$, within every switching cycle, there are two components of current crossing the T_{pa}. They are shoot-through current and AC load current within shoot-through duration and active-ON duration respectively. The average current of T_{pa} over the switching cycle is given by (11). The R.M.S. current of i_{Tpa} in the switching cycle is then given by (12).

$$\langle i_{Tpa} \rangle = D_0 \frac{2}{3} I_L + d_{pa} i_a \quad (11)$$

$$i^2_{Tparms} = D_0 (\frac{2}{3} I_L + \frac{i_a}{2})^2 + (d_{pa} - \frac{D_0}{2}) i_a^2 \quad (12)$$

where, $\langle i_{Tpa} \rangle$ is the average current over switching cycle; i_{Tparms} is the R.M.S current over switching cycle; D_0 is the shoot-through duty-cycle; I_L is the average Z-source inductor current; d_{pa} is the duty-cycle of T_{pa}, i_a is the AC load current.

In negative-half cycle, i.e., $i_a < 0$, only shoot-through current happens within every switching cycle. The average current and R.M.S. current over a switching cycle in the negative-half cycle of AC current is given by (13) and (14) respectively.

$$\langle i_{Tpa} \rangle = D_0 (\frac{2}{3} I_L + \frac{i_a}{2}) \quad (13)$$

$$i_{Tparms}^2 = D_0 (\frac{2}{3} I_L + \frac{i_a}{2})^2 \quad (14)$$

Thus the average conduction loss caused by T_{pa} over line cycle is given by (15).

978-1-4244-1871-8/07 $25.00 © 2007 IEEE

$$P_{Tpa,c} = \frac{1}{2\pi} \int_{-\frac{\pi}{2}+\Phi}^{\frac{3\pi}{2}+\Phi} (r_T i^2{}_{Tparms} + V_T \langle i_{Tpa} \rangle) d\varphi$$

$$= r_T \left[\frac{1}{2\pi} \int_{-\frac{\pi}{2}+\Phi}^{\frac{3\pi}{2}+\Phi} D_0 (\frac{2}{3} I_L + \frac{i_a}{2})^2 d\varphi + \frac{1}{2\pi} \int_{-\frac{\pi}{2}+\Phi}^{\frac{\pi}{2}+\Phi} (d_{pa} - \frac{D_0}{2}) i_a{}^2 d\varphi \right] \tag{15}$$

$$+ V_T \left[\frac{1}{2\pi} \int_{-\frac{\pi}{2}+\Phi}^{\frac{\pi}{2}+\Phi} (D_0 \frac{2}{3} I_L + d_{pa} i_a) d\varphi + \frac{1}{2\pi} \int_{\frac{\pi}{2}+\Phi}^{\frac{3\pi}{2}+\Phi} D_0 (\frac{2}{3} I_L + \frac{i_a}{2}) d\varphi \right]$$

where, $\varphi = \omega t$; $i_a = I_a(t)$. The shoot-through duty-cycle, D_0, can be calculated by modulation index, M, due to the use of maximum constant control as given by (16).

$$D_0 = 1 - \frac{\sqrt{3}}{2} M \tag{16}$$

Replacing variables of d_{pa}, i_a, in (15) by those in (1) and (2), and D_0 by (16), after some trigonometric operation, the average conduction loss caused by IGBT T_{pa} can be approximated as given by (17).

$$P_{Tpa,c} = \left[\begin{array}{c} r_T \left(I_{pk}{}^2 \left(\frac{1}{8} + \frac{M \cos \Phi}{3\pi} - \frac{M \cos 3\Phi}{90\pi} \right) + I_L{}^2 \frac{2(2 - \sqrt{3}M)}{9} \right) + \\ V_T \left(I_{pk} \left(\frac{1}{2\pi} + \frac{M \cos \Phi}{8} \right) + I_L \left(\frac{2 - \sqrt{3}M}{2} \right) \right) \end{array} \right] \tag{17}$$

where, the average Z-source inductor current I_L, is given by (18), and I_{pk} by (19) from (3.60) of Chapter 3.

$$I_L = \frac{P_o}{V_{dc}} \tag{18}$$

$$I_{pk} = \frac{4}{3} \frac{\sqrt{3}M - 1}{M \cos \Phi} \frac{P_o}{V_{dc}} \tag{19}$$

where, P_o is the active power; the modulation index M is $1/\sqrt{3} < M < 2/\sqrt{3}$.

2) Conduction loss caused by free-wheel diode D_{na} in Z-source Inverter

The free-wheeling diode in the structure of Z-source inverter, shown in Figure 1, does not carry current during shoot-through intervals. The average current and the R.M.S. current of i_{Dna} over switching cycle is given by (20) and (21).

$$\langle i_{Dna} \rangle = (1 - d_{pa} - \frac{D_0}{2}) i_a = (d_{na} - \frac{D_0}{2}) i_a \tag{20}$$

$$i^2{}_{Dnarms} = (d_{na} - \frac{D_0}{2}) i_a{}^2 \tag{21}$$

where, $\langle i_{Dna} \rangle$ is the average current; i_{Dnarms} is the R.M.S current; d_{na} is the duty-cycle of D_{na}, i_a is the AC load current.

In the negative-half line cycle, the D_{na} does not carry any current. The average conduction loss caused by D_{na} over line cycle is given by (22).

$$P_{Dna,c} = \frac{1}{2\pi} \int_{-\frac{\pi}{2}+\Phi}^{\frac{\pi}{2}+\Phi} (r_D i^2{}_{Dnarms} + V_D \langle i_{Dna} \rangle) d\varphi$$

$$= r_D \left[\frac{1}{2\pi} \int_{-\frac{\pi}{2}+\Phi}^{\frac{\pi}{2}+\Phi} (d_{na} - \frac{D_0}{2}) i_a{}^2 d\varphi \right] + V_D \left[\frac{1}{2\pi} \int_{-\frac{\pi}{2}+\Phi}^{\frac{\pi}{2}+\Phi} (d_{na} - \frac{D_0}{2}) i_a d\varphi \right]$$

$$\tag{22}$$

where, $\varphi = \omega t$; Φ is AC current phase angle; $i_a = I_a(t)$. Replacing variables of d_{na}, i_a, in (22) by those in (1) and (2), after some trigonometric operation, the average conduction loss caused by free-wheel diode D_{na} can be approximated as given by (23).

$$P_{Dna,c} = \left[\begin{array}{c} r_D I_{pk}{}^2 \left(\frac{1 - D_0}{8} - \frac{M \cos \Phi}{3\pi} + \frac{M \cos 3\Phi}{90\pi} \right) + \\ V_D I_{pk} \left(\frac{1 - D_0}{2\pi} - \frac{M \cos \Phi}{8} \right) \end{array} \right] \tag{23}$$

where, I_{pk} is given by (19).

3) Total average conduction loss in Z-source Inverter

Since the loss is evenly distributed in the power-switch network of the Z-source inverter, the total conduction loss can be approximated as given by (24).

$$P_{z,c} = 6 \times (P_{Tpa,c} + P_{Dna,c})$$

$$= 6 \times \left[\begin{array}{c} r_T I_{pk}{}^2 \left(\frac{1}{8} + \frac{M \cos \Phi}{3\pi} - \frac{M \cos 3\Phi}{90\pi} \right) + r_T \left(\frac{2}{3} I_L \right)^2 D_0 + \\ V_T I_{pk} \left(\frac{1}{2\pi} + \frac{M \cos \Phi}{8} \right) + V_T I_L D_0 + \\ r_D I_{pk}{}^2 \left(\frac{1 - D_0}{8} - \frac{M \cos \Phi}{3\pi} + \frac{M \cos 3\Phi}{90\pi} \right) + \\ V_D I_{pk} \left(\frac{1 - D_0}{2\pi} - \frac{M \cos \Phi}{8} \right) \end{array} \right] \tag{24}$$

where,

$$D_0 = 1 - \frac{\sqrt{3}}{2} M; \quad I_{pk} = \frac{4}{3} \frac{\sqrt{3}M - 1}{M \cos \Phi} P_o; \quad I_L = \frac{P_o}{V_{dc}}$$

B. Switching loss model of Z-source Inverter

1) Switching loss caused by IGBT T_{pa} in Z-source Inverter

When AC load current is positive, i.e., $i_a > 0$, the T_{pa} in the Z-source inverter has four switching actions. Two of them are the same as that in an ordinary VSI, in which the switched voltage is the peak DC-link voltage and the switched current is the AC load current. They occur within non-shoot-through switching states between the active and zero switching states. The other two switching actions are imposed by the unique shoot-through actions in the Z-source inverter: shoot-through turn-on and shoot-through turn-off. The switched voltage is still the peak DC-link voltage; the switched current is $\frac{2}{3} I_L + \frac{i_a}{2}$. Thus, the energy loss, w_I, within switching cycle can be approximated using (7) and (8), as given by (25).

$$w_1 = w_{Tonoff}\left(V_{pn}, \left(\frac{2}{3}I_L + \frac{i_a}{2}\right)\right) + w_{Tonoff}(V_{pn}, i_a) \quad (25)$$

where w_1 is the energy loss of T_{pa} in a switching cycle when i_a is positive. w_{Tonoff} has the same form of $w_{off}(u,i)$ or $w_{on}(u,i)$ where the coefficients K_{Toni} or K_{Toffi} becomes $K_{Tonoffi} = K_{Toni} + K_{Toffi}$, $i = 1, 2, ..5$; V_{pn} is the peak DC-link voltage; I_L and i_a were defined

When AC load current is negative, i.e., $i_a < 0$, the T_{pa} still has four switching actions; all of them are for shoot-though. Two of them occur in a manner of zero-current turn-on/off. The logic signal for driving T_{na} and the current waveform of i_{Tna} illustrates the two zero-current turn-off of T_{na} between intervals z and v.

Using the approximation model in (7) and (8), the energy loss, w_2, per switching action of T_{pa} in the negative-half cycle is given by (26).

$$w_2 = w_{Tonoff}\left(V_{pn}, \left(\frac{2}{3}I_L + \frac{i_a}{2}\right)\right) \quad (26)$$

where, w_{Tonoff} was defined in (25). Thus the average switching loss in the line cycle is

$$P_{Tpa,sw} = f_{sw}\left[\frac{1}{2\pi}\int_{-\frac{\pi}{2}+\Phi}^{\frac{3\pi}{2}+\Phi}(w_1 + w_2)d\varphi\right]$$

$$= f_{sw}\begin{bmatrix}\frac{1}{2\pi}\int_{-\frac{\pi}{2}+\Phi}^{\frac{3\pi}{2}+\Phi} w_{Tonoff}\left(V_{pn}, \left(\frac{2}{3}I_L + \frac{i_a}{2}\right)\right)d\varphi \\ +\frac{1}{2\pi}\int_{-\frac{\pi}{2}+\Phi}^{\frac{\pi}{2}+\Phi} w_{Tonoff}(V_{pn}, i_a)d\varphi\end{bmatrix} \quad (27)$$

Replacing i_a by that in (1) and (2) and with some trigonometric operation, the average switching loss of T_{pa} can be approximated by (28).

$$P_{Tpa,sw} = f_{sw}\begin{bmatrix}\frac{3}{2}K_{T3}V_{pn}^2 + (K_{T1}V_{pn} + K_{T4}V_{pn}^2)(\frac{3}{2\pi}I_{pk} + \frac{2}{3}I_L) \\ + (K_{T2}V_{pn} + K_{T5}V_{pn}^2)(\frac{1}{\pi}I_{pk}I_L + \frac{5}{16}I_{pk}^2 + \frac{4}{9}I_L^2)\end{bmatrix} \quad (28)$$

where $K_{Ti} = K_{Ton,i} + K_{Toff,I}$, $i=1,2,..5$; f_{sw} is the switching frequency; I_{pk} is the peak AC current; I_L the Z-source network inductor current; I_L is given by (19). From maximum constant boost control, peak DC-link voltage V_{pn} is given by (29).

$$V_{pn} = BV_{dc} = \frac{1}{1 - 2D_0}V_{dc} = \frac{1}{\sqrt{3}M - 1}V_{dc} \quad (29)$$

2) Switching loss cause by Diode D_{na} in Z-source Inverter

The Diode D_{na} has four switching actions. The turn-on switching loss of a diode is usually neglected. The energy loss caused by switching action of diode D_{na} can be approximated using (9) as given by (30).

$$w_{Doff} = w_{Doff}(V_{pn}, i_a) \quad (30)$$

The average switching power loss caused by diode over line cycle is given by (31).

$$P_{Dna,sw} = f_{sw}\left[\frac{1}{2\pi}\int_{-\frac{\pi}{2}+\Phi}^{\frac{3\pi}{2}+\Phi} w_{Dn}d\varphi\right] = f_{sw}\left[\frac{1}{2\pi}\int_{-\frac{\pi}{2}+\Phi}^{\frac{\pi}{2}+\Phi} w_{Doff}d\varphi\right]$$

$$= f_{sw}\left[\frac{1}{2\pi}\int_{-\frac{\pi}{2}+\Phi}^{\frac{\pi}{2}+\Phi} w_{Doff}(V_{pn}, i_a)d\varphi\right] \quad (31)$$

Replacing i_a by the $I_a(t)$ in (1), the average switching power loss of the free-wheeling diode can be approximated by (32).

$$P_{Dna,sw} = f_{sw}\begin{bmatrix}K_{D3}V_{pn}^2 + (K_{D1}V_{pn} + K_{D4}V_{pn}^2)\frac{1}{\pi}I_{pk} \\ + (K_{D2}V_{pn} + K_{D5}V_{pn}^2)\frac{1}{4}I_{pk}^2\end{bmatrix} \quad (32)$$

where $K_{Di} = K_{Doff,I}$, $i=1,2,..5$; from maximum constant boost control, peak DC-link voltage V_{pn} is given by (29).

3) Total average switching loss in Z-source Inverter

Since the switching power loss is evenly distributed, the total average switching loss can be approximated by

$$P_{z,sw} = 6 \times (P_{Tpa,sw} + P_{Dna,sw})$$

$$= 6 \times f_{sw}\begin{bmatrix}\frac{3}{2}K_{T3}V_{pn}^2 + (K_{T1}V_{pn} + K_{T4}V_{pn}^2)(\frac{3}{2\pi}I_{pk} + \frac{2}{3}I_L) + \\ (K_{T2}V_{pn} + K_{T5}V_{pn}^2)(\frac{1}{\pi}I_{pk}I_L + \frac{5}{16}I_{pk}^2 + \frac{4}{9}I_L^2) + \\ K_{D3}V_{pn}^2 + (K_{D1}V_{pn} + K_{D4}V_{pn}^2)\frac{1}{\pi}I_{pk} + \\ (K_{D2}V_{pn} + K_{D5}V_{pn}^2)\frac{1}{4}I_{pk}^2\end{bmatrix} \quad (33)$$

where, $V_{pn} = BV_{dc} = \frac{1}{\sqrt{3}M - 1}V_{dc}$ by the

C. Average conduction and switching losses of MZC in DC-AC inversion

In DC-AC operation mode, the MZC operates as a Z-source inverter while each switch consists of two devices in series as shown in Figure 2(a). Thus the equivalent configuration of each switch is that shown in Figure 2(b). Figure 2(c) shows the equivalent model of the whole switch during the ON-state of the active IGBT. For this reason, the V_T, V_D, r_T, and r_D, used in (17) - (24) for the approximation of average conduction loss are replaced by V_{T+D}, V_{D+T}, r_{T+D}, and r_{D+T} respectively. V_{T+D}, V_{D+T}, r_{T+D}, and r_{D+T} are given by (34) and (35).

$$V_{T+D} = V_{D+T} = V_T + V_D \quad (34)$$
$$r_{T+D} = r_{D+T} = r_T + r_D \quad (35)$$

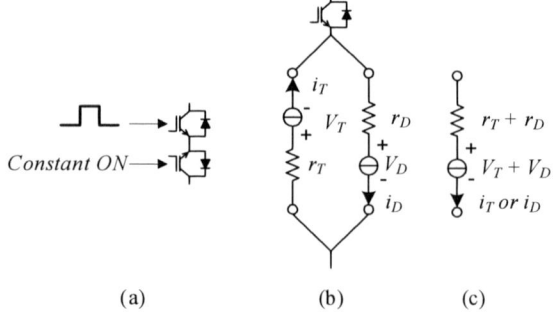

(a)   (b)   (c)

Figure 2   IGBT and free-wheeling equivalent model in MZC when MZC is in DC-AC inversion mode

1) Average conduction loss of MZC in DC-AC inversion mode

Equations (17) and (23) share the same parameters, hence the average conduction loss of Z-source inverter given by (24) can be arranged in (36)

$$P_{mzc-dcac,c} = 6 \times \left[\begin{array}{l} r_{T+D}I_{pk}^2 \left(\dfrac{2-D_0}{8} \right) + r_{T+D}\left(\dfrac{2}{3}I_L \right)^2 D_0 + \\ V_{T+D}I_{pk}\left(\dfrac{2-D_0}{2\pi} \right) + V_{T+D}I_L D_0 \end{array} \right] \quad (36)$$

where,

$$I_L = \frac{P_o}{V_{dc}}; \quad I_{pk} = \frac{4(\sqrt{3}M-1)P_o}{3MV_{dc}\cos\Phi}; \quad D_0 = 1 - \frac{\sqrt{3}M}{2}$$

The calculation result using the given values for DC-AC inversion in Table III is visualized in Figure 3.

2) Average switching loss of MZC in DC-AC inversion mode

MZC in DC-AC inversion mode operates as Z-source inverter. The average switching loss then can be approximated from (28), (32), and (33). It is given by (37).

$$P_{mzc-dcac,sw} = 6 \times (P_{mzc-dcacTpa,sw} + P_{mzc-dcacDna,sw}) = equation(33) \quad (37)$$

where,

$$K_{Ti} = K_{Ton,i} + K_{Toff,i} + K_{Doff,i} \quad (38)$$

$$I_L = \frac{P_o}{V_{dc}}; \quad I_{pk} = \frac{4}{3}\frac{\sqrt{3}M-1}{M\cos\Phi}P_o; \quad V_{pn} = \frac{1}{\sqrt{3}M-1}V_{dc}$$

The calculation result using the given values for DC-AC inversion in Table III is visualized in Figure 4.

3) Total average conduction and switching loss of MZC in DC-AC mode

By the result in above two sub-sections, the total average loss caused by power semiconductors in DC-AC mode of MZC can be approximated as given by (39).

$$P_{mzc-dcac} = P_{mzc-dcac,c} + P_{mzc-dcac,sw} \quad (39)$$

where, $P_{mzc-dcac,c}$ and $P_{mzc-dcac,sw}$ are given in (36) and (37) respectively. The calculation result using the given values for DC-AC inversion in Table III is visualized in Figure 5.

V. AVERAGE CONDUCTION AND SWITCHING LOSSES MODEL OF MZC IN AC-DC RECTIFICATION MODE

A. Average Conduction loss of AC-DC matrix converter

The average conduction loss is irrelative to the AC power factor but dominated by DC load current as in (40), where the square bracketed item is the conduction loss per output phase.

$$P_{mc,c} = 2[(V_T + V_D)I_d + (r_T + r_D)I_d^2] \quad (40)$$

The calculation result using the given values for AC-DC rectification in Table III is visualized in Figure 6.

B. Average Switching loss of MZC in AC-DC rectification mode

It is presumed that only the non zero-current switching action of a semiconductor contributes its switching loss.

The sequencer controllers, which are essential to the current continuity of any matrix converter, naturally lead to semi-zero-current switching during each switching cycle.

Thus for a specific semiconductor switch, the chance of zero-current turn-off (ZC-off) and non zero-current turn-off

(NZC-off) is equal in its turn-off, so does zero-current turn-on (ZC-on) and non zero-current turn-on (NZC-on) in its turn-on. The occurrence of ZC-off/on or NZC-off/on is determined by the polarity of the input line-line voltage viewed from the semiconductor's terminals during the current commutation.

The balanced three-phase input voltages naturally have six 60°-length segments within a line cycle. By simple investigation one can see the fact that among all semiconductors of one phase output there are always 4 instances of non zero-current switching: two NZC-on and two NZC-off in any one of the six segments of the input AC line cycle.

With the above analysis, the average switching loss energy of the four NZC switching actions of one output phase can be estimated in (41) within a segment $(-\pi/6, \pi/6)$ of the AC line cycle.

$$P_{mc,sw,per-phase} = \int_{-\pi/6}^{\pi/6} F(u_{ab}, u_{bc}, I_d, K_i)d\varphi \quad (41)$$

where $F(u_{ab}, u_{bc}, I_d, K_i)$ can be calculated by (7), (8), and (9). Then one has the average switching power loss of the MZC in AC-DC rectification mode as in (42).

$$P_{mc,sw} = F(K_i, M, V_{dc}, I_d)$$

$$= f_p \left[\begin{array}{l} \dfrac{2.12K_3V_{dc}^2}{3/4M^2} + I_d \left(\dfrac{1.91K_1V_{dc}}{\sqrt{3}/2M} + \dfrac{2.12K_4V_{dc}^2}{3/4M^2} \right) \\ + I_d^2 \left(\dfrac{1.91K_2V_{dc}}{\sqrt{3}/2M} + \dfrac{2.12K_5V_{dc}^2}{3/4M^2} \right) \end{array} \right] \quad (42)$$

where V_{dc} is the expected DC voltage; Φ is the input phase displacement of AC current; M is modulation index, $0 < M < 2/\sqrt{3}$. The calculation result using the given values for AC-DC rectification in Table III is visualized in Figure 7.

C. Total average conduction and switching loss of MZC in AC-DC rectification mode

By the result in above two sub-sections, one may estimate the total average loss caused by power semiconductors in DC-AC mode of MZC as in

$$P_{mzc-acdc} = P_{mzc-acdc,c} + P_{mzc-acdc,sw} \quad (43)$$

The calculation result using the given values for AC-DC rectification in Table III is visualized in Figure 8

VI. CONCLUSION

The conduction and switching loss models for MZC have been established for its DC-AC and AC-DC operation modes respectively.

The models can be used in comparing loss with other topologies for the same application – the single-stage three-phase bi-directional conversion.

And also, the model can be used in the estimation of the thermal requirement during the design work based on MZC, , or in the investigation of the thermal performance of certain new power semiconductor to be used in MZC.

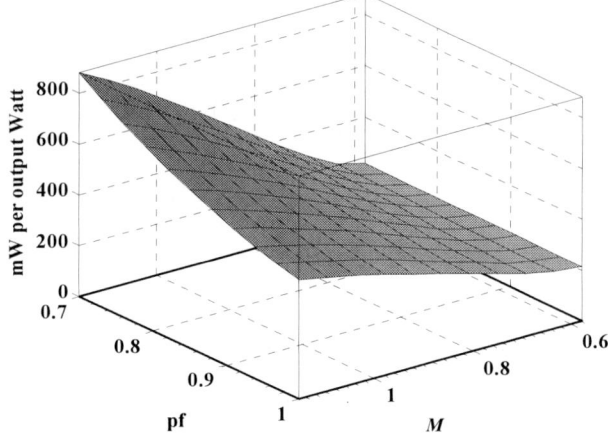

Figure 3 Conduction loss in DC-AC operation mode

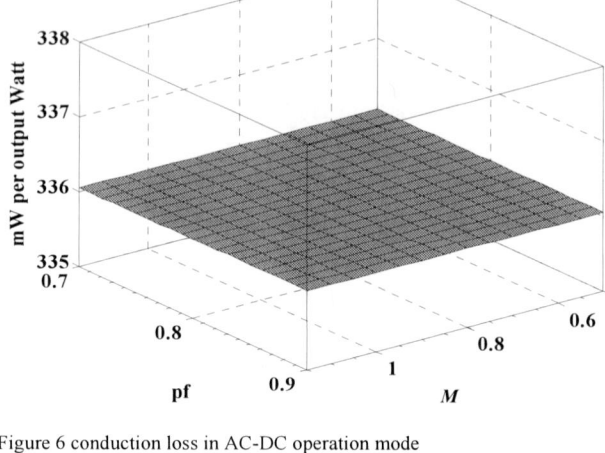

Figure 6 conduction loss in AC-DC operation mode

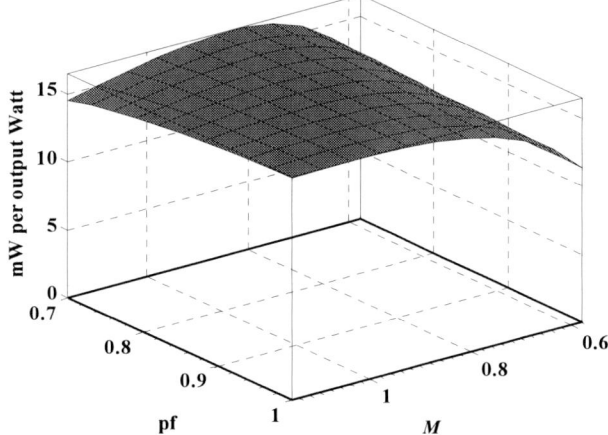

Figure 4 Switching loss in DC-AC operation mode

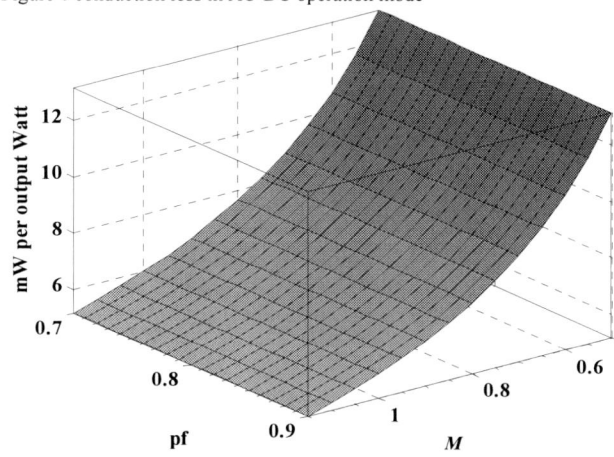

Figure 7 Switching loss in AC-DC operation mode

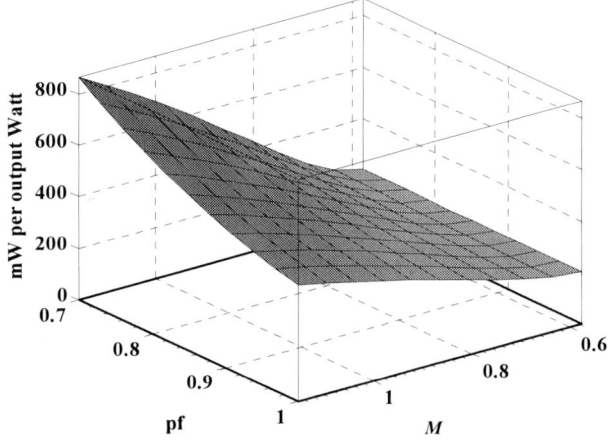

Figure 5 Total conduction and switching loss in DC-AC operation mode

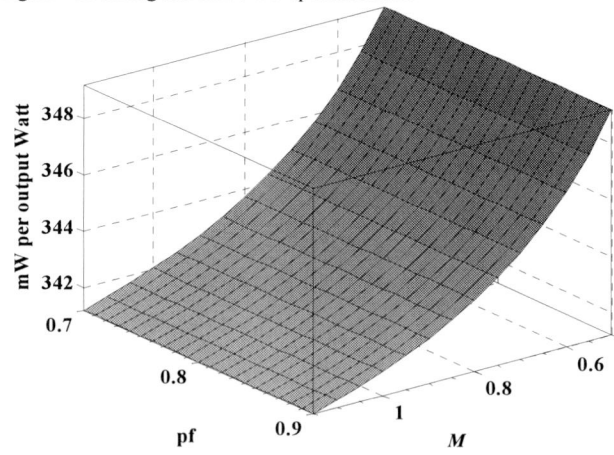

Figure 8 Total conduction and switching loss in AC-DC operation mode

Reference:

[1] Alesina, A.; Venturini, M.G.B.; "Analysis and design of optimum-amplitude nine-switch direct AC-AC converters",Power Electronics, IEEE Transactions on , Volume: 4 , Issue: 1 , Jan. 1989,Pages:101 - 112

[2] Holmes, D.G.;"A unified modulation algorithm for voltage and current source inverters based on AC-AC matrix converter theory", Industry Applications, IEEE Transactions on , Volume: 28 Issue: 1 , Jan.-Feb. 1992, Page(s): 31 -40

[3] Kolar, J.W.; Ertl, H.; Zach, F.C.,"Influence of the modulation method on the conduction and switching losses of a PWM converter system", Industry Applications, IEEE Transactions on Volume 27, Issue 6, Nov.-Dec. 1991 Page(s):1063 – 1075

[4] Kolar, J.W.; Baumann, M.; Schafmeister, F.; Ertl, H., "Novel three-phase AC-DC-AC sparse matrix converter", Applied Power Electronics Conference and Exposition, 2002. APEC 2002. Seventeenth Annual IEEE Volume 2, 10-14 March 2002 Page(s):777 - 791 vol.2

[5] Robert W. Erickson, Dragan Maksimovic', Fundamentals of Power Electronics Second Edition, Kluwer Academic Publishers, Massachusetts USA, 2001

[6] Keping You, M. F. Rahman; "A New Power Converter for ISA 42 V PowerNet System using Matrix converter theory and Z-source concepts" VPPC 2005 - IEEE Vehicle Power and Propulsion (VPPC'05) Conference, proceedings of, September 7-9, 2005, Chicago, Illinois, USA, page 461-467, ISBN: 0-7803-9281-7

Advanced Development of High Frequency Transformer Parasitic Inductive Components and Lossless Inductive Snubber-Assisted Series Resonant ZCS-PFM DC-DC Converter for High Power Microwave Generator

Bishwajit Saha[1], Manabu Ishitobi[2], Hisayuki Sugimura[1], Sang Pil Mun[1], Mutsuo Nakaoka[1, 3]

[1] Kyungnam University, Masan, KOREA
[2] Nara National College of Technology, JAPAN
[3] Industrial College of Technology-University, JAPAN

Abstract—The conventional series-resonant power frequency modulated DC-DC converter with a high-frequency transformer link designed for driving the magnetron of microwave ovens has the problem of the hard switching commutation at turn-on and turn-off the active power switching devices due to the influence of the magnetizing current of the high-frequency transformer. This paper presents a novel prototype of high-frequency transformer parasitic parameters with a lossless inductive snubber and a series resonant capacitor assisted series-resonant zero current switching pulse frequency modulated DC-DC power converter, which is designed for industrial use of high power magnetron for microwave ovens. In order to implement a complete and efficient soft switching commutation, the performance of the new converter topology is practically confirmed and evaluated in the prototype of power microwave generator for plasma application from practical point of view.

Keywords—series resonant inverter, high frequency transformer link, voltage doubler rectifier, magnetron, ZCS-PFM

I. INTRODUCTION

With great advances of the power semiconductor switching devices such as MOSFETs, IGBTs, ESBTs and B-SITs as well as high-frequency circuit components, the latest research and development of the high-frequency inverter type switching mode DC-DC power conversion circuits and systems have attracted special interests for high voltage DC power applications [1]-[5]. From the downsizing point of view, the high-frequency resonant soft switching DC-DC converter using MOS gate power semiconductor devices are actively introduced in this new particular field of power electronic applications. On the other hand, IGBTs or MOS gate bipolar transistors are more suitable for zero current switching (ZCS) commutation rather than zero voltage switching (ZVS) due to their associated inherent tail currents at turn off switching commutations [6]-[12]. Thus, the series resonant inverters and converters operating under the principle of ZCS commutation are practically preferred from the switching losses point of view.

The conventional series-resonant circuit topologies of high frequency inverters can operate under a commutation principle of ZCS based on a discontinuous current mode (DCM) control. However, in the high frequency transformer link topologies it is difficult to implement the DCM in the conventional type circuit due to the influence of the magnetizing current that flows through the high-frequency transformer primary side. Especially, the influence of the magnetizing current is remarkable for the DC-DC converters with capacitor type output-smoothing filter and a constant high-voltage load as the power magnetron of the microwave appliances.

Under above technological background, this paper presents a high-frequency transformer parasitic parameters and a lossless inductive snubber assisted series-resonant DC-DC power converter with a voltage doubler rectifier circuit, which can operate under ZCS commutation based on a pulse frequency modulation (PFM) in order to improve the problem of the transformer magnetizing-inductive components-based hard switching commutation effectively. This new circuit topology can actively utilize the transformer parasitic circuit components as leakage and magnetizing inductance to achieve soft switching operation with the aid of a lossless inductive snubber and series resonance capacitor in the high voltage transformer primary side. The performance evaluations as the switching voltage and current waveforms of the power semiconductor switches, DC power regulation and power conversion efficiency characteristics are practically confirmed and evaluated as applied to the power magnetron of the microwave generator for plasma application.

II. CHARACTERISTICS AND ELECTRICAL EQUIVALENT CIRCUIT MODEL OF MAGNETRON

The heart of any microwave generator is the high voltage system. Its purpose is to generate microwave energy. The high-voltage components accomplish this by stepping up AC line voltage to high voltage, which is then changed to an even higher DC voltage. This DC power is then converted to the RF (microwaves) energy that cooks the food. Microwaves are very short waves of electromagnetic energy that travel at the speed of light. Microwaves used in microwave ovens are in the same family of frequencies as the signals used in radio and television broadcasting. Electromagnetic forms of energy, such as microwaves, radar waves, radio and TV waves, travel millions of miles through the emptiness of space without the need of any material medium through which to travel.

Figure 1 indicates the input voltage vs. input current characteristics of the power magnetron. As shown in this

978-1-4244-1871-8/07 $25.00 © 2007 IEEE

figure, the magnetron has non-linear input characteristics, what can be represented with the piecewise linear approximation which includes high resistance area in non-oscillating region and low resistance area in oscillating region for magnetron. When the voltage between anode and cathode exceeds about 7.4 kV (cut-off voltage), the magnetron anode current begins to flow from anode to cathode. On the other hand, when the voltage between anode and cathode is lower than the cut-off voltage, the anode current does not mostly flow. The cut-off voltage of magnetron a little fluctuates with the operating temperature which depends on the characteristics of the ferrite magnet in the magnetron, which are not considered here for the approximate v-i characteristics.

The electrical equivalent circuit model of the magnetron can simply represent by using pure resistances R_0 and R_1, an ideal diode D, an ideal battery VZ (cut-off voltage) and an ideal switch as shown in Fig. 2. As illustrated in this figure, the position of the ideal switch in the equivalent circuit of the magnetron is selected whether the voltage between anode

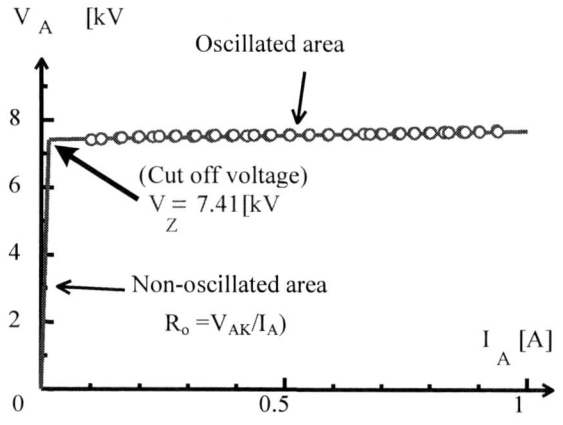

Fig. 1. Experimental characteristics of magnetron.

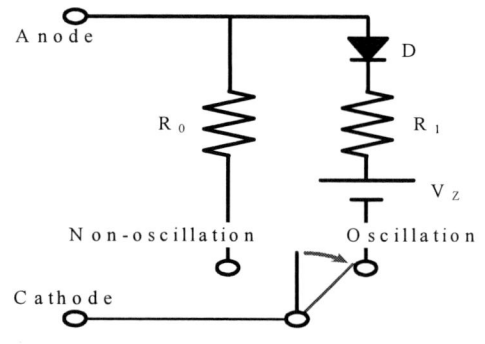

Fig. 2. Electrical equivalent circuit model of magnetron.

and cathode is higher than the cut off voltage or not. In this paper, the stable microwave power is required for the semiconductor manufacturing production for industrial applications and the magnetron is used under the condition of continuous oscillation. Therefore, the only stable oscillation state is considered for the converter operation.

III. SERIES-RESONANT DC-DC CONVERTER WITH HIGH-FREQUENCY TRANSFORMER LINK

A. Operation in Discontinuous Current Mode (DCM)

Figure 3 shows the schematic circuit configuration of the conventional type DC-DC power converter with a high-frequency transformer link and voltage doubler circuit, which is designed for a high power magnetron drive in industrial power applications (this circuit is called circuit 1 in the following). This DC-DC converter circuit is composed of high frequency high voltage transformer, resonant high-frequency full bridge inverter with a series resonant capacitor C_r in the primary side of the high frequency transformer, full-wave voltage doubler type rectifier circuit, current smoothing inductor L_o, and a magnetron to generate microwave power in the high frequency transformer secondary side. The magnetron of Fig. 3 is represented by its electrical equivalent circuit shown in Fig. 2 under the condition of continuous oscillations. The circuit parameters and the design specifications of this circuit are indicated in table I.

TABLE I
DESIGN SPECIFICATIONS AND CIRCUIT PARAMETERS OF CIRCUIT 1.

Equivalent DC Voltage of Input Source	E	283 [V]
Leakage Inductance of HF-Transformer	L_ℓ	2.3 [μH]
Primary Self-Inductance of HF-Transformer	L_1	26.4 [μH]
Turns-Ratio of HF-Transformer	N_2/N_1	20
Series Resonant Capacitor	C_r	467.3 [nF]
Output DC Filter Capacitor	C_1, C_2	11 [nF]
Output DC Filter Inductor	L_o	0.3 [H]
On Time of Gate Voltage Pulse	T_{on}	5.16 [μs]
Cut-off Voltage of Magnetron	V_Z	7.41 [kV]
Equivalent Resistance in Magnetron	R_1	266 [Ω]

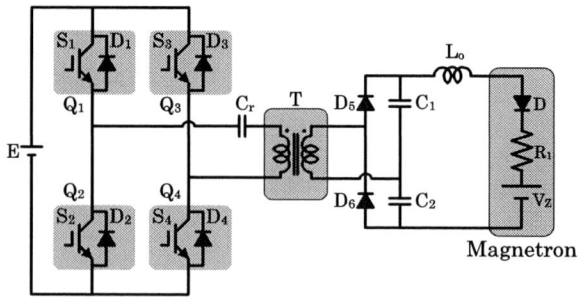

Fig. 3. High frequency link series resonant DC-DC converter (Circuit 1).

In the oscillating state of the magnetron, its input side DC voltage can be c nearly considered constant since it is represented by a linear v-i characteristic with small slope as indicated in Fig. 1. Thus, the microwave power from the magnetron is proportional to its anode current. Therefore, the output power of magnetron can be regulated by controlling the anode current with a closed feedback loop. Consequently, the runaway of the magnetron is prevented and the stable high microwave power can be effectively generated and controlled.

978-1-4244-1871-8/07 $25.00 © 2007 IEEE

The steady state voltage and current waveforms in a DCM operation in the previously-developed series resonant DC-DC converter are illustrated in Fig. 4. In this case, the current through transformer primary winding is discontinuous. As shown in Fig. 4, the gate pulse signal sequences of IGBTs are designed to regulate the pulse frequency under constant on-time condition. The IGBTs are turned on with ZCS and turned off with hybrid ZVS&ZCS in all power regulation range when the switching frequency of this DC-DC converter is less than half of the resonant frequency decided by the resonant capacitor and the leakage inductance of high-frequency transformer. If the switching frequency is increased more than half of the resonant frequency, bridge current of this converter becomes continuous waveform and this converter operates in continuous current mode (CCM). Consequently, the operation becomes hard switching.

However, in case of using a high-frequency transformer with leakage and magnetizing parasitic circuit parameters, it is actually difficult to realize the DCM operation to achieve a zero current soft commutation. In order to implement this operating mode, it is necessary that the magnetizing current through the high-frequency transformer primary winding to be nearly zero. Nevertheless, the magnetizing current remarkably flows in the DC-DC converter with a capacitor input type output-smoothing circuit and a constant high voltage load as the magnetron is equal to zero. Because the output rectifier circuit is cut off when the secondary side voltage of the high-frequency transformer is less than the output-smoothing capacitor voltage, and the series resonant inverter is isolated separately from the voltage doubler rectifier with the capacitor input filter. In this case, only magnetizing current of the transformer circulates through the circuit of transformer primary side during the period of the cut off mode of the voltage doubler type rectifier. Consequently, the discontinuous current operating mode (see Fig. 5) of this converter can not appear and realize practically.

B. Magnetizing Inductor Hard Switching Operating Mode

Figure 5 shows the measured operated waveforms of the current flowing through each IGBT switch and the voltage across it of the circuit 1 (see Fig. 3). In comparison with the simulation voltage and current waveforms in Fig. 4, the measured switch current waveforms observed in Fig. 5 are distorted. It is easily proven that this power converter operates under a hard switching mode due to the remarkably confirmed high voltage surges in the operating current and voltage waveforms.

At turn-on switching transition, the high-frequency transformer primary side current has an initial value due to the magnetizing current through the high-frequency high-voltage transformer.

Therefore, the current flowing through the switches jumps to this initial value, and IGBTs in the bridge arms of the series transformer resonant inverter has hard switching commutation at turn-on in all power regulation setting range. Extremely high voltage surges actually occur with this high

di/dt stress. At turn-off switching transition, the case which occurs while the current flowing through the IGBT switches is forcibly cut off to zero before the zero current crossing point. Consequently, the IGBT switches are turned-off at hard switching commutations.

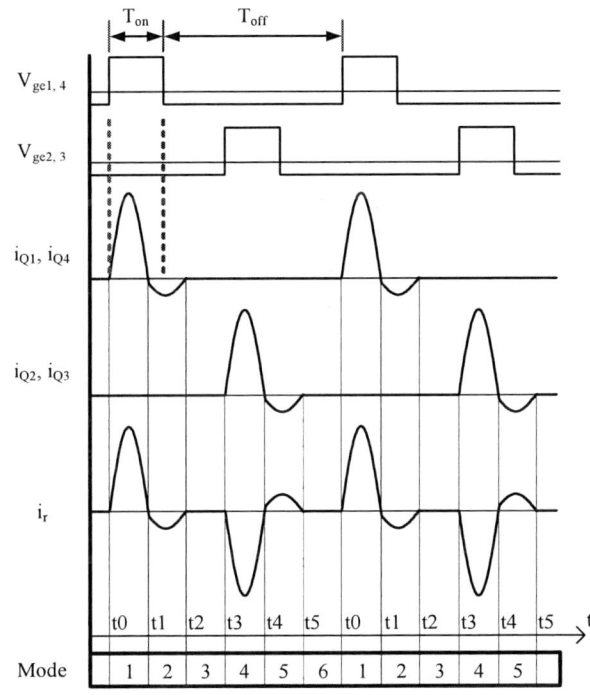

Fig. 4. Operating waveforms in discontinuous current mode.

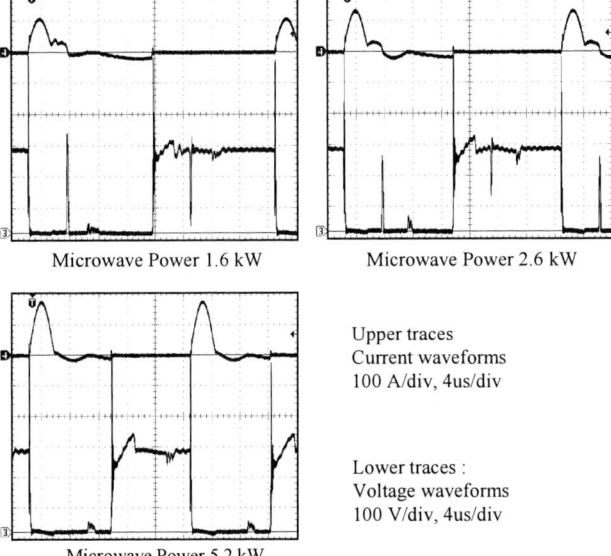

Microwave Power 1.6 kW

Microwave Power 2.6 kW

Microwave Power 5.2 kW

Upper traces
Current waveforms
100 A/div, 4us/div

Lower traces :
Voltage waveforms
100 V/div, 4us/div

Fig. 5. Experimental voltage and current waveforms of Q1 (Circuit 1).

IV. IMPROVED SERIES TRANSFORMER RESONANT DC-DC POWER CONVERTER WITH LOSSLESS INDUCTIVE SNUBBER

In order to solve the significant problems mentioned above, a single inductive snubber assisted series-resonant

978-1-4244-1871-8/07 $25.00 © 2007 IEEE

ZCS-PFM DC-DC power converter with a high-frequency transformer link is proposed in Fig. 6 (this circuit is called circuit 2 in the following). This DC-DC power converter has a single inductive lossless snubber L_s in the input DC busline of the full bridge inverter. The current flowing through IGBT switches rises gradually at turn-on with the aid of this inductive snubber, and ZCS turn-on commutation can be achieved completely.

The gate pulse timing sequences of the IGBT switches of this DC-DC converter are illustrated in Fig. 7. The duty cycle of each gate pulse signal of IGBT is designed for a constant duty cycle of 50%. The pulse width of each gate signal varies with the switching frequency. As shown in Fig. 8, the IGBT switches can be always turned off while a current continues to flow through the antiparallel diodes and the primary winding of the transformer. Thereby, IGBT switches can be turned off with ZVS&ZCS hybrid commutation. Furthermore, even though the dead time between two gate pulse signals is set to zero, the input DC busline of this newly developed DC-DC converter can not be shorted due to the effect of the single lossless inductive snubber connected in the input side of the full bridge inverter. Therefore, this DC-DC converter circuit is possible to perform zero current soft switching commutation over wide power regulation range in PFM strategy.

TABLE II
DESIGN SPECIFICATIONS AND CIRCUIT PARAMETERS OF CIRCUIT 2.

Equivalent DC Voltage of Input Source	E	283 [V]
Lossless Inductive Snubber	L_s	1 [nH]
Leakage Inductance of HF-Transformer	L_ℓ	2.3 [nH]
Primary Self-Inductance of HF-Transformer	L_1	26.4 [nH]
Turns-Ratio of HF-Transformer	N_2/N_1	20
Series Resonant Capacitor	C_r	467.3 [nF]
Output DC Filter Capacitor	C_1, C_2	11 [nF]
Output DC Filter Inductor	L_o	0.3 [H]
Cut-off Voltage of Magnetron	V_Z	7.41 [kV]
Equivalent Resistance in Magnetron	R_1	266 [Ω]

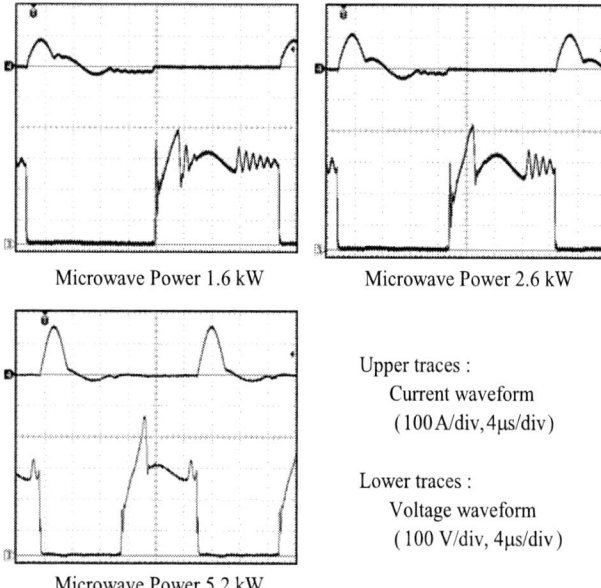

Microwave Power 1.6 kW   Microwave Power 2.6 kW

Microwave Power 5.2 kW

Upper traces :
Current waveform
(100 A/div, 4 μs/div)

Lower traces :
Voltage waveform
(100 V/div, 4 μs/div)

Fig. 8. Experimental voltage and current waveforms of Q1 (Circuit 2).

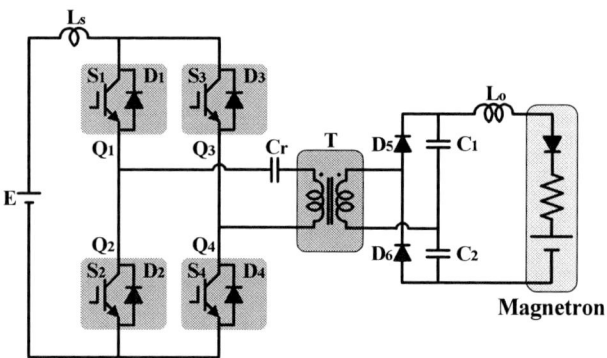

Fig. 6. Proposed soft switching DC-DC power converter (Circuit 2).

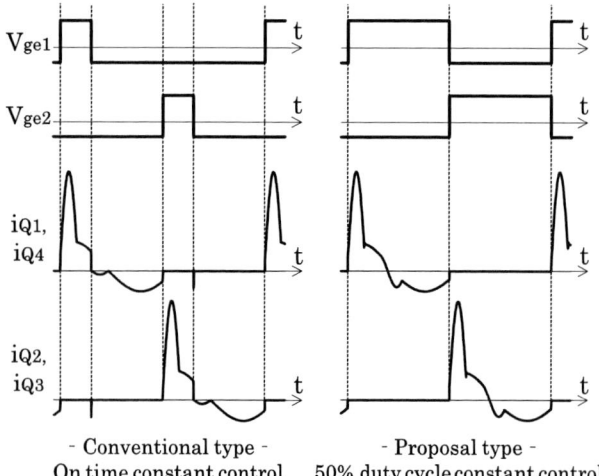

- Conventional type -   - Proposal type -
On time constant control   50% duty cycle constant control

Fig. 7. 50% Duty cycle constant PFM control scheme.

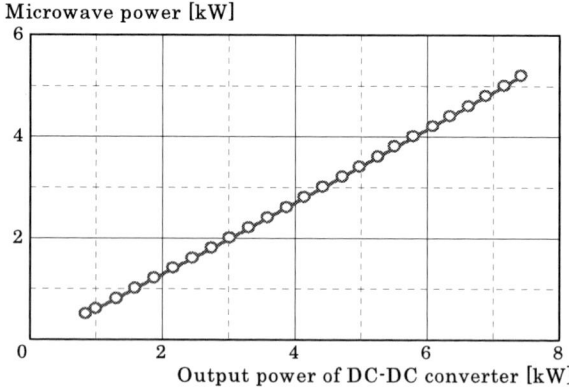

Fig. 9. Output power of DC-DC converter vs. microwave power.

The design specification of this DC-DC converter in this experiment is indicated in table II. The observed voltage and current operating waveforms of IGBT switches in the proposed converter are shown in Fig. 8. Observing Fig. 8, it is easy to prove that this proposed DC-DC converter can operate under a principle of ZCS operation at both turn-on and turn-off transitions.

978-1-4244-1871-8/07 $25.00 © 2007 IEEE   1078

The microwave power characteristic is shown in Fig. 9, while the measured efficiency characteristics for the output power of the circuit 1 and 2 are comparatively shown in Fig. 10. The power conversion efficiency of the proposed soft switching DC-DC converter is improved in comparison with that of the conventional DC-DC converter in the wide range of the high power region.

V. REDUCING PEAK CURRENT STRESSES

As stated in section III, when the switching frequency is more than half the resonant frequency, the conventional DC-DC converter operates on hard-switching transition even in the good condition. However, the proposed DC-DC converter operates on soft-switching transition even if the switching frequency is more than half the resonant frequency, because its soft-switching operation bases on continuous current mode.

With a proper design of the circuit parameters, the peak value of the currents through the IGBT switches can be significantly reduced. Fig. 11 shows the observed voltage and current waveforms of the IGBT switches on the proposed converter with the proper circuit parameters (this circuit is called circuit 3).

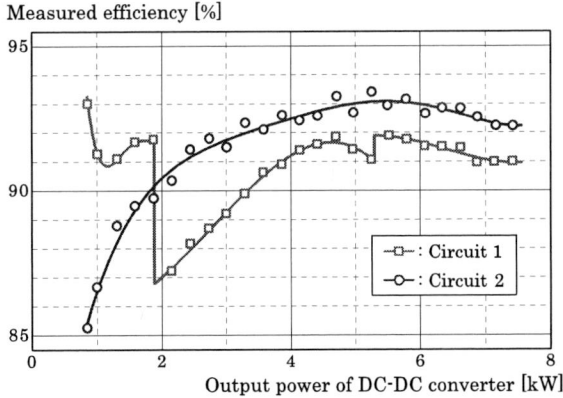

Fig. 10. Comparison of power conversion efficiency vs. output power characteristics for two converter topologies.

TABLE III
DESIGN SPECIFICATIONS AND CIRCUIT PARAMETERS OF CIRCUIT 3.

Items	Symbol	Value
Equivalent DC voltage of input source	E	283 V
Lossless inductive snubber	Ls	3.5 mH
Leakage inductance of HF transformer	Lℓ	11.3 mH
Primary self-inductance of HF transformer	L_1	35.4 mH
Turn ratio of HF transformer	N_2 / N_1	20
Series resonant capacitor	Cr	274.8 nF
Capacitance of output filter capacitor	C_1, C_2	11 nF
Inductance of output filter inductor	Lo	0.3 H
Cut-off voltage of magnetron	Vz	7.41 kV
Equivalent resistance of magnetron	R_1	266 Ω

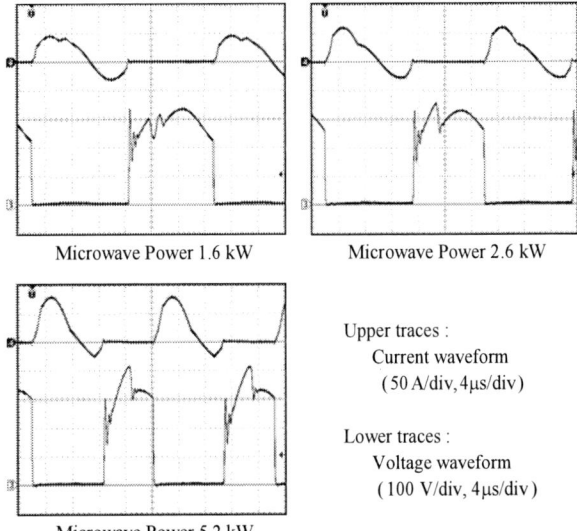

Microwave Power 1.6 kW

Microwave Power 2.6 kW

Microwave Power 5.2 kW

Upper traces :
Current waveform
(50 A/div, 4μs/div)

Lower traces :
Voltage waveform
(100 V/div, 4μs/div)

Fig. 11. Experimental voltage and current waveforms of Q1 Circuit 3).

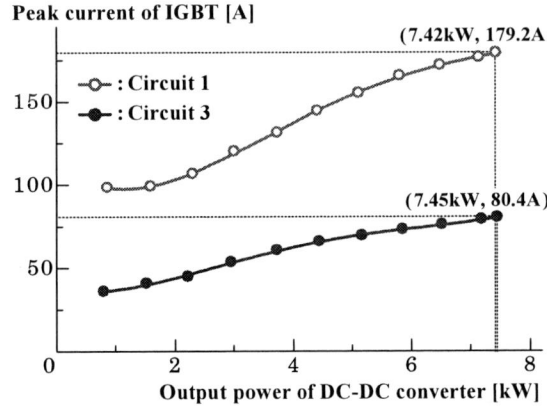

Fig. 12. Comparison of peak values of current flowing through IGBT.

The measured peak current values for a wide output power control range of the conventional and proposed DC-DC converter are indicated in Fig. 12. It is proven from this figure that the peak current value of IGBT on circuit 3 is lower than the half of that on previous type over all power regulation

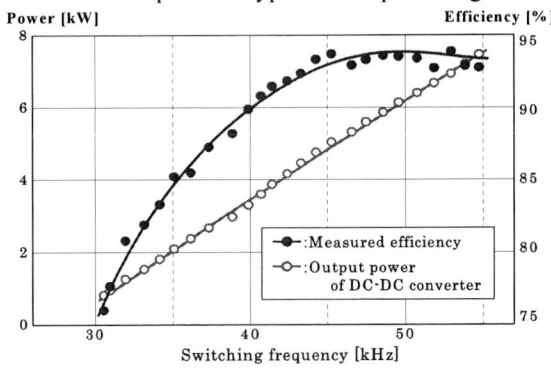

Fig. 13. Output power and measured efficiency vs. switching frequency characteristics of circuit 3.

The design specification of this circuit is indicated in table III. These circuit parameters are designed to actively utilize the continuous current mode, in other words, the maximum of switching frequency on circuit 3 is set near the resonant frequency of the resonant circuit to increase the utilization factor of electric power. Observing Fig. 11, it is confirmed that the peak values of the current flowing through the IGBT switches of circuit 3 are much lower than those on the previous converter circuits (circuit 1 and circuit 2).

The measured peak current values for a wide output power control range of the conventional and proposed DC-DC converter are indicated in Fig. 12. It is proven from this figure that the peak current value of IGBT on circuit 3 are lower than the half of that on previous type over all power regulation range. The improvement of the efficiency can be expected by improving the characteristics of IGBT switches with reduced peak current values and rated values. Fig. 13 shows the power conversion efficiency and the output power characteristics of the experimental prototype of proposed DC-DC converter designed under the specification of table III. The maximum value of the efficiency of circuit 3 shown in Fig. 13 is about 94.3[%], and this value is higher than that of circuit 2.

VI. CONCLUSIONS

In this paper, a transformer parasitic parameter and a lossless inductive snubber assisted series-resonant ZCS-PFM DC-DC converter has been proposed in order to improve the significant problems of the hard switching commutation at turn-on and turn-off the active power switching devices in series-resonant PFM controlled DC-DC power converter with a high-frequency high-voltage transformer link. Based on the experimental results of the proposed pulse frequency modulated DC-DC power converter with a high-frequency high-voltage transformer, it was actually confirmed that all the active power switches could achieve ZCS commutation operation. The unique features of the proposed DC-DC power converter applied to a high power magnetron drive could be stated in the following:

ZCS commutation could be implemented in the all DC power regulation range and high power conversion efficiency could be performed in a wide power regulation range.

The transformer parasitic circuit components as leakage and magnetizing inductive components were effective to achieve soft switching (ZCS) operation with the aid of a single lossless inductive snubber.

The peak values of the current flowing through IGBT switches could be reduced more than the half of that on conventional DC-DC converter over all power regulation ranges by actively utilizing the continuous current mode operation with a proper design selection of circuit parameters.

REFERENCES

[1] A. Harada, "How to Use C.W. Magnetrons", *Proceedings of Microwave Effect and Application Symposium*, Japan, August, 2001.

[2] K. Harada, T. Ninomiya, "Fundamentals of Switched-Mode Converters", *Corona Book Publishing Co., Ltd.*, February, 1992,

[3] T. Matsushige, E. Miyata, M. Ishitobi, and M. Nakaoka, "Voltage-Clamped Soft-Switching Inverter-Fed DC-DC Converter for Microwave Oven and Utility AC Side Harmonic Current Evaluations", *Proceedings of IEEE-IAS International Appliance Technical Conference, IATC*, pp. 185-195, Kentucky, USA, May, 2000.

[4] T. Myoi, M. Ishitobi, L. Gamage, and M. Nakaoka, "Lossless Inductive Snubber-Assisted Series Resonant DC-DC Converter with ZCS-PFM Control Scheme", *Proceedings of International Power Electronics and Motion Control Conference (EPE-PEMC)*, Dubrovnik, Croatia, September, 2002.

[5] E. Hiraki, M. Nakaoka, "Practical Power Loss Analysys Simulator Development of Switching Mode Power Converter Using Measured Characteristic Values of Power Semiconductor Devices", *The Trans.of Institute of Electrical Engineering of Japan, IEEJ*, vol. 122-D, no. 12, December, 2002

[6] Yuzurihara, A. Takayanagi, and H. Fujikawa, "Microwave Generator System for Plasma Application (vol. 2)", *Kyosan Technical Circular*, vol. 53, no. 3, 2002.

[7] S. Furuyama, "Transformer for Electronics", Book, The Nikkan Kogyo Shimbun, Ltd., February, 1966

[8] S. Tuboshima, M. Hada, "Transformer", *Tokyo Denki University Book Press*, December, 1981

[9] Shibata, "Industrial Microwave Power Engineering", Denkishoin Publishers Co., Ltd., December, 1986.

[10] K. Takahashi, "Base and Application of High Frequency", *Tokyo Denki University Book Press*, October, 1990

[11] Shibata, S. Misawa, "Energy Conversion System Engineering", *Morikita Book Publishing Co., Ltd.*, December, 1990

Common-mode Noise Reduction in the Watkins-Johnson Converter

Masahito Shoyama, Tamotsu Ninomiya*

Graduate School of Information Science and Electrical Engineering, Kyushu University,*

744 Moto-oka, Nishi-ku, Fukuoka 819-0395, Japan

shoyama@ees.kyushu-u.ac.jp

Abstract --- Common-mode conducted noise in switching converters frequently causes radiated noise emission from their power code, so it is important to reduce this noise to meet EMC. In this paper common-mode conducted noise is much reduced in the Watkins-Johnson converter by current cancellation. The mechanism of the noise reduction is investigated, and the condition for the noise reduction is obtained by experiments.

I. INTRODUCTION

Common-mode conducted noise in switching converters frequently causes radiated noise emission from their power cord, so it is important to reduce this noise to meet EMC. Because conventional switching converters are usually using unbalanced circuit topologies, parasitic capacitance between the drain of an MOS-FET and the frame ground through its heat sink may generate the common-mode conducted noise.

We proposed a concept of balanced switching converter circuit, which is an effective way to reduce the common-mode conducted noise by current cancellation. As an example, a boost converter version of the balanced switching converter

were presented and its effectiveness in the common-mode noise reduction was shown by experiments [1][2].

In this paper, at first, an example of a balanced switching converter is briefly reviewed. Then the mechanism of reducing the common-mode conducted noise is applied to the Watkins-Johnson converter.

II. BALANCED BOOST CONVERTER CIRCUIT (REVIEW)[1]

Fig. 1 (a) shows a conventional unbalanced boost converter circuit. A heat sink is usually used with the MOS-FET Q, so the parasitic capacitance Cs is formed between the drain of the MOS-FET and the frame ground FG through its heat sink. In the case where a small size and a low profile are strongly required to the converter, the metal frame ground itself may be used as a heat sink, so that Cs becomes very large. Because the drain voltage changes very rapidly in the switching time, a large current pulse flows through the parasitic capacitance Cs. Consequently this causes the large common-mode noise current to lead to serious problems.

In order to solve this problem, we proposed a balanced switching converter circuit as shown in Fig. 1 (b). Its basic circuit operation is essentially the same as that of the conventional unbalanced boost converter. The winding of the inductor is equally split into two parts, thus the total winding turns are the same as those of the conventional unbalanced boost converter. In the balanced boost converter, the drain voltage and the source voltage of the MOS-FET Q change complementarily, that is, by the same amount but in an

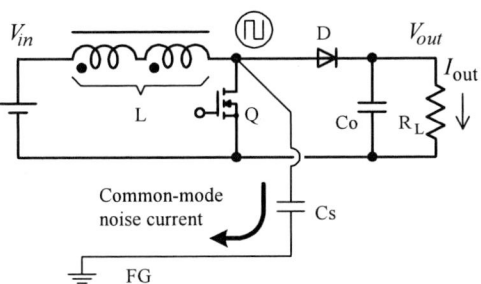

(a) Conventional unbalanced boost converter

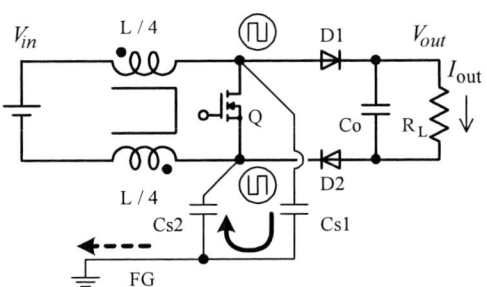

(b) Proposed balanced boost converter

Fig. 1. Common-mode noise reduction by balanced boost converter [1].

(for review)

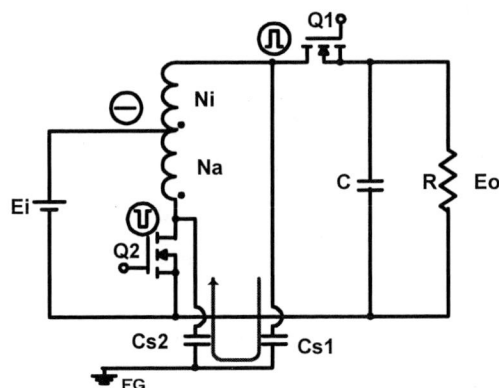

opposite polarity in the switching time. This is ideally performed by the equally split winding inductor. As a result, the common-mode conducted noise is much reduced by current cancellation in the case where Cs1=Cs2. This common-mode noise reduction mechanism is applied to the Watkins-Johnson converter in the next section.

III. THE WATKINS-JOHNSON CONVERTER

Fig. 2 shows a Watkins-Johnson converter circuit [3]. This converter is suitable for a large step-down ratio, and the output voltage is positive for the duty ratio $D_1>0.5$. Q2 is originally a diode, however, a synchronous rectifier of MOS-FET is used in Fig. 2 instead. So, the soft switching can be realized with a dead time between Q1 and Q2. Fig. 3 shows some key operation waveforms.

In the steady-state condition, the output/input voltage ratio is simply obtained by neglecting internal resistances as:

$$\frac{E_o}{E_i} = \left[1 + \left(1 - \frac{1}{D_1}\right)\frac{N_i}{N_a}\right]. \quad \text{-------------(1)}$$

If Ni=Na, this relation can be written as:

$$\frac{E_o}{E_i} = 2 - \frac{1}{D_1}. \quad \text{-------------(2)}$$

Because the heat sinks are usually used with the MOS-FET Q1 and Q2, the parasitic capacitance Cs1 and Cs2 are formed between the drains of the MOS-FETs and the frame ground FG through the heat sinks. If Ni=Na, the drain voltages of Q1 and Q2 change complementarily, that is, by the same amount but in an opposite polarity in the switching time. In this case, if Cs1=Cs2, their noise currents are well canceled, and this leads that the common-mode noise current is much reduced.

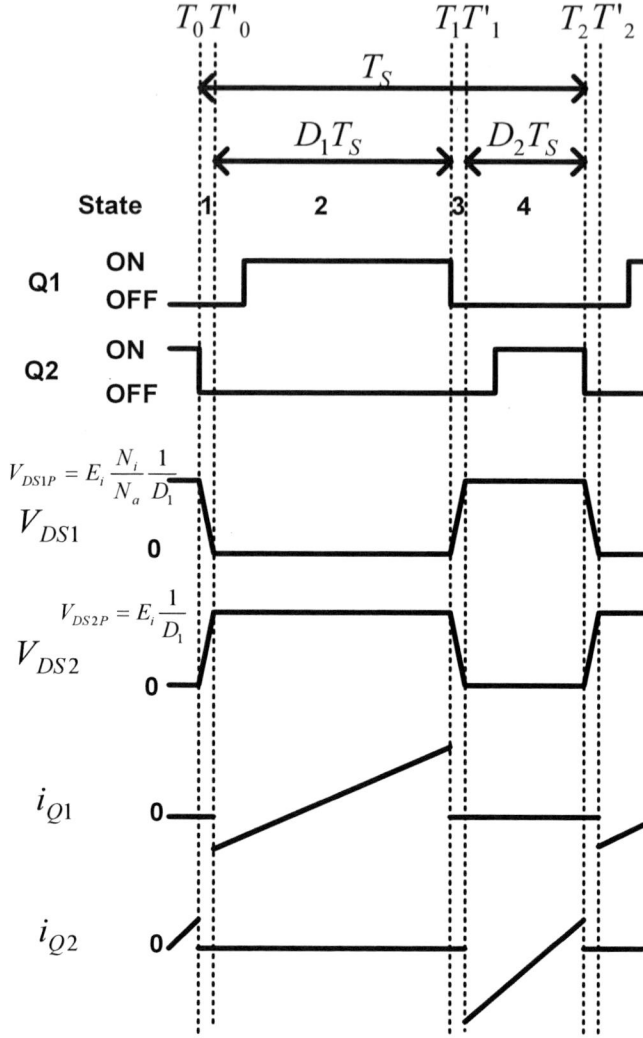

Fig. 3. Key operation waveforms of the Watkins-Johnson converter.

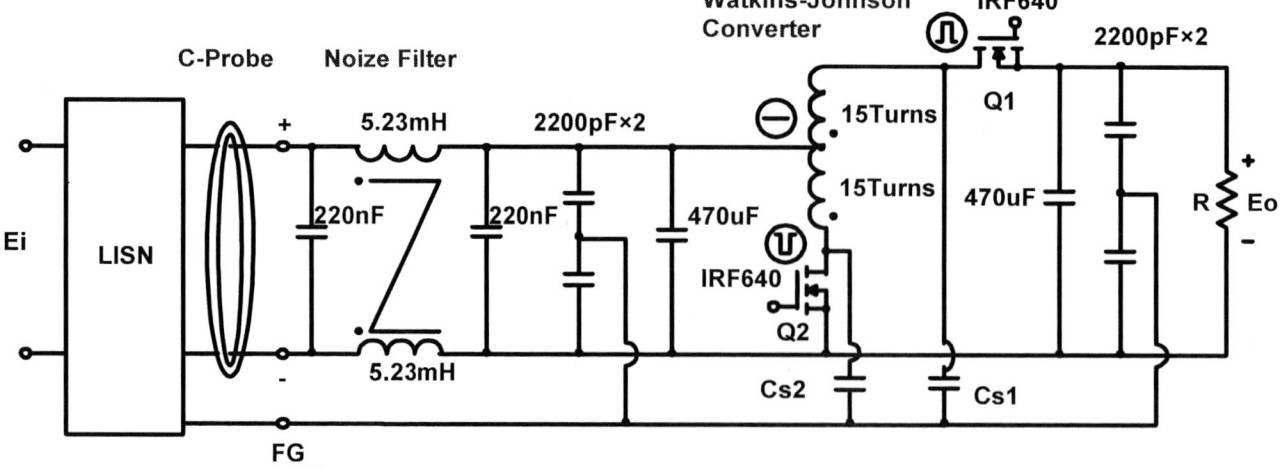

Fig. 4. Experimental circuit for common-mode noise current measurement. Conditions: fs=100kHz, Ei=42V, D=0.6, Eo=11.7V, Io=1A. A high frequency current probe (C-Probe) was used to sense the common-mode noise current.

IV. EXPERIMENTS

In order to confirm the mechanism of the noise reduction, we made an experimental circuit as shown in Fig. 4. A high frequency current probe (C-Probe) was used to directly sense the common-mode noise current, which is measured by a spectrum analyzer. External capacitors are used for the parasitic capacitances Cs1 and Cs2 in this experiment.

Fig. 5 shows experimental results of the common-mode noise current for some different values (0pF, 220pF, 470pF, 1000pF) of the parasitic capacitor Cs2, where Cs1 is

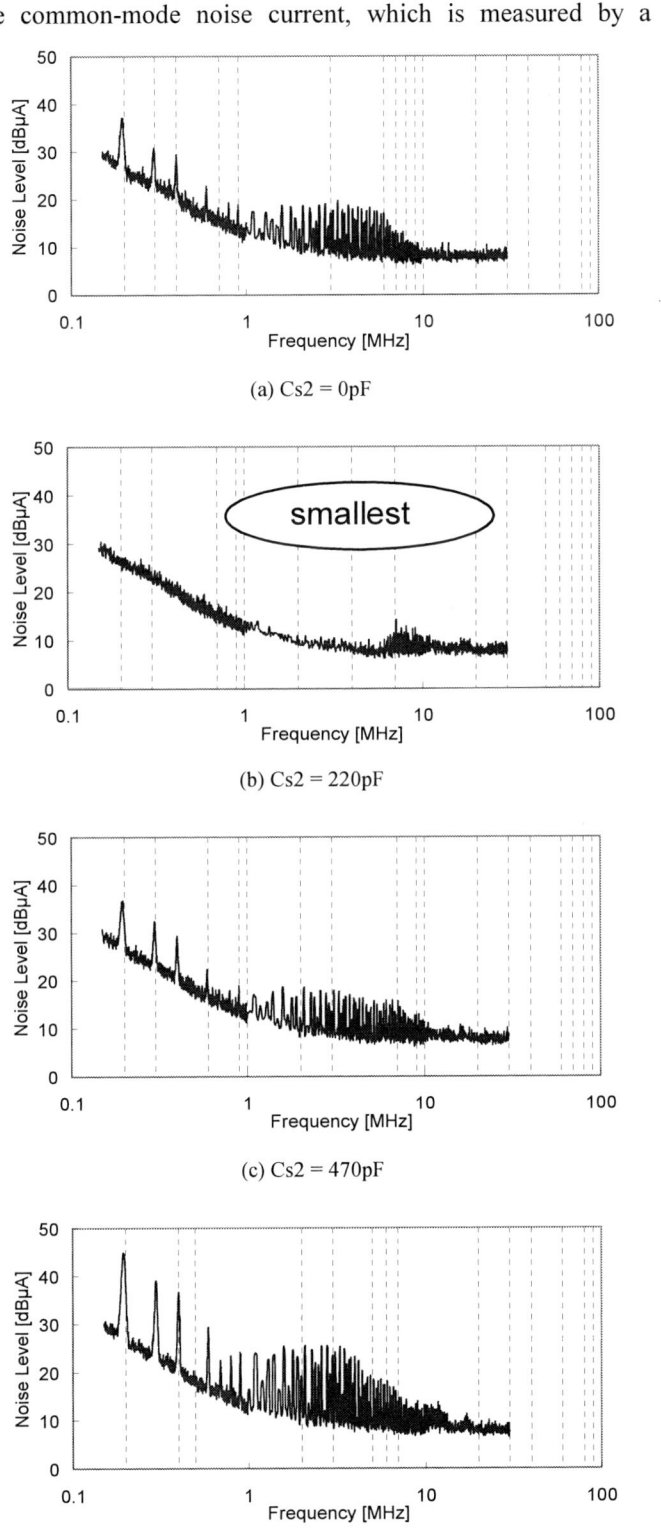

Fig. 5. Experimental results of common-mode noise current. Cs1=220pF for all cases.

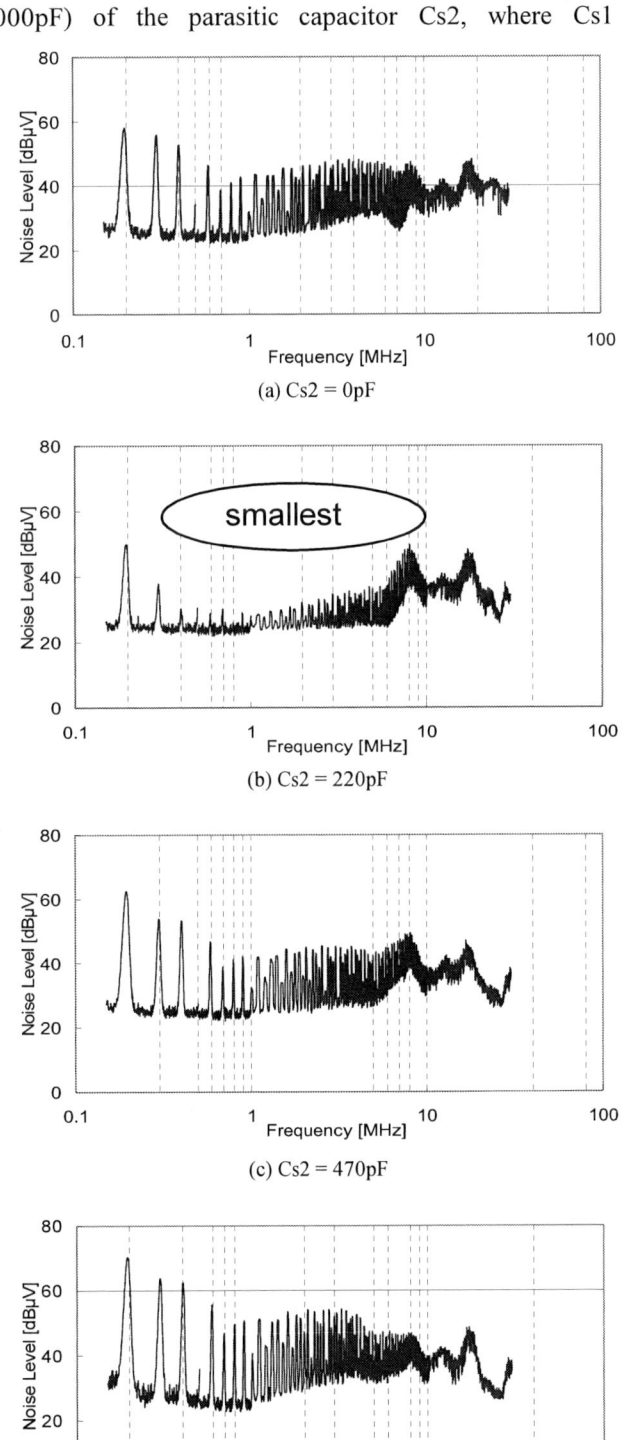

Fig. 6. Experimental results of noise current using LISN. Cs1=220pF for all cases.

commonly set at 220pF. From this figure, it is found that the common-mode conducted noise is most reduced in the case where Cs1=Cs2(=220pF) in the Watkins-Johnson converter.

Fig. 6 shows experimental results of noise current using LISN. This noise voltage includes both the common mode and the normal mode conducted noise. From this figure, it is found that the conducted noise is well suppressed in the case where Cs1=Cs2(=220pF) in the Watkins-Johnson converter.

The mechanism of the common-mode noise reduction in the Watkins-Johnson converter is shown in Fig. 7 using an equivalent circuit. The switching noise currents through Cs1 and Cs2 are well canceled by each other, so that the common-mode noise current is much reduced.

V.   CONCLUSION

In conclusion, the condition for the noise reduction of common-mode conducted noise is investigated in the Watkins-Johnson converter. As a result, it is found that, if Ni= Na, the common-mode noise current is much reduced by current cancellation when the condition Cs1=Cs2 is met.

REFERENCES

[1]   M. Shoyama, T. Okunaga, G. Li, T. Ninomiya: "Balanced Switching Converter to Reduce Common Mode Conducted Noise," IEEE PESC 2001 Record, pp. 451-456, Jun., 2001.

[2]   M. Shoyama, T. Tsumura, T. Ninomiya: "Mechanism of Common-Mode Noise Reduction in Balanced Boost Switching Converter," IEEE PESC 2004 Record, pp.1115-1120, Jun. 2004.

[3]   Y. Darroman, A. Ferre, "42V/3V Watkins-Johnson Converter for Automotive Use," IEEE Trans. on Power Electronics, Vol.21, No.3, pp.592-602, May 2006.

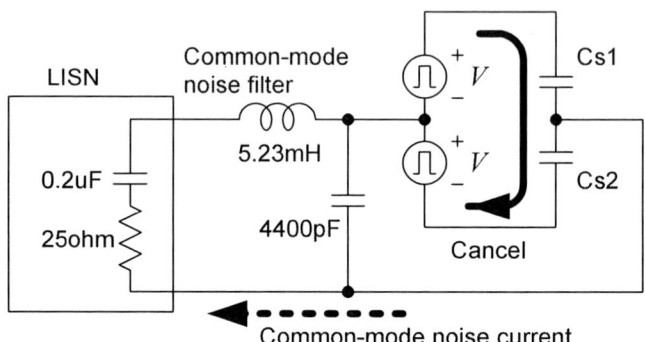

Fig. 7. Equivalent circuit to show the mechanism of common-mode noise reduction in the Watkins-Johnson converter.

Current Ripple Reduction Technique of DC/DC Converter

Ye-Then Chang, Chia-An Yeh, Masahiro Hamaogi*, Fumikazu Takahashi**,

and Yen-Shin Lai, Senior *Member, IEEE*

Center for Power Electronics Technology, National Taipei University of Technology

1, Sec. 3, Chung-Hsiao E. Rd., Taipei, Taiwan

* Hitachi Computer Peripheral Co. Ltd., Japan

**Hitachi Research Lab., Japan.

Abstract-This paper presents a novel method to reduce the current ripple caused by the variation of inductor for a DC/DC converter. The presented method is to increase the switching frequency to keep constant current ripple when the inductance changes. Therefore, no extra component is required. The problem will be described first and the related method is presented. An experimental DC/DC converter is implemented which is with input voltage of 12 V, output voltage of 1.5 V, output current of 10 A, and switching frequency of 200-250 kHz. Finally, experimental results derived from the designed digital-controlled DC/DC converter are included for confirmation.

I. INTRODUCTION

Due to the development of high performance and high density micro processors, it is required to reduce the supplied voltage while increasing the current rating. The point-of-load (POL) converter is usually designed and implemented to provide well regulated voltage to microelectronic load [1-2]. The allowed output voltage ripple becomes smaller for such applications.. However, the current slew rate is increased. For such applications, the effect of parameter variation caused by temperature and load variation etc. becomes very essential to converter design and implementation.

It is well known that the inductor, L, shown in Fig. 1 affects the current ripple and thereby causing the variation of output voltage ripple and efficiency. The lower its inductance, the larger the current ripple will be. However, the inductance reduces as the load current increases since the increases of load current forces the inductor to go toward saturation. Figure 2 shows the B-H curve of inductor [3]. As shown in Fig. 2, the magnetic intensity, H(t) can be classified as DC component and AC component, H_0 and ΔH, respectively. As the load current increases, the DC operation point increases from H_{0_L} to H_{0_H} and thereby reducing the slope of the related B-H curve. This will result in the reduction of inductance and increasing the current ripple when the inductor is applied to buck converter.

Fig. 3 illustrates the relationship between load current and inductance. As shown in Fig. 3, the inductance is reduced as the load current increases. Fig.4 illustrates the relationship between output current and inductor current ripples. As shown in Fig. 4, under all kinds of load

conditions, the inductor current ripple is constant in ideal case. Actually, the ripple increases under heavy load condition. Although the issue has been discussed [4-5], to our best knowledge, no cost-effective method has been presented to deal with it.

Digital power provides some particular advantages over analog approach [6-7]. These advantages include: no aging issue for control loop, low power consumption, easy to change topology and control algorithm, fast time to market and easy to design the associated ASIC. In this paper a novel method to reduce the current ripple is presented for digital-controlled point-of-load buck converter. The presented method is to change the switching frequency to keep constant current ripple when the inductance changes. An experimental DC/DC converter is implemented which is with input voltage of 12 V, output voltage of 1.5 V, output current of 10 A, and switching frequency of 200-250 kHz. Finally, experimental results derived from the designed digital-controlled DC/DC converter are presented for confirmation.

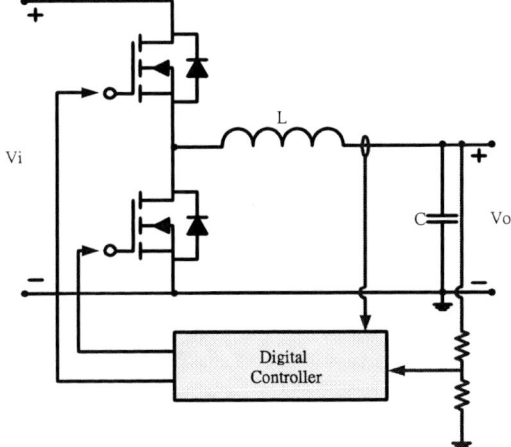

Fig.1 Block diagram of a digital-controlled DC/DC converter

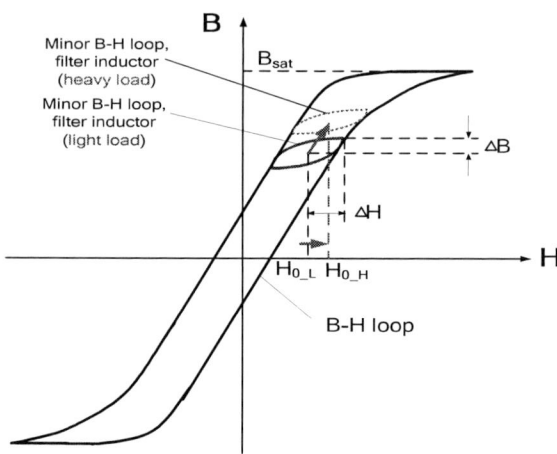

Fig. 2 Filter inductor minor B-H loop [3]

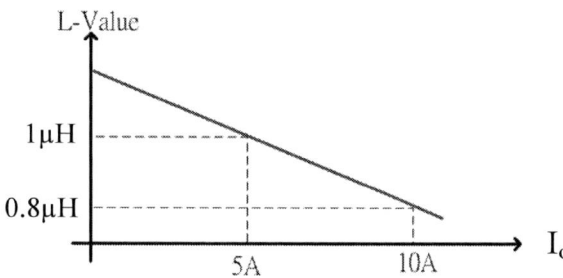

Fig. 3 Illustration of inductance vs. load current

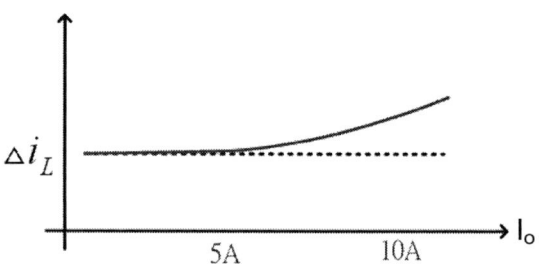

Actual current ripple variation: Solid line ———
Ideal current ripple variation: Dash line ·········

Fig. 4 Relationship between output current and inductor current ripple

II. PROPOSED CURRENT RIPPLE REDUCTION TECHNIQUE

Fig. 5 illustrates the system configuration of the proposed technique using a buck converter as an example. As shown in Fig. 5, a block entitled "adaptive frequency control" is added to control the switching frequency and sampling rate as the load current increases.

As shown in Fig. 3, the inductance of output choke is reduced as the load current increases. And the relationship between the decrease of inductance and load current is

approximately linear in order to avoid magnetic saturation. Therefore, the inductance can be represented by:

$$L^* \cong L \cdot (1 - K \cdot I_o) \qquad (1)$$

where

L^* = actual inductance

L = inductance under no load

K = coefficient of inductance reduction

I_o = output current

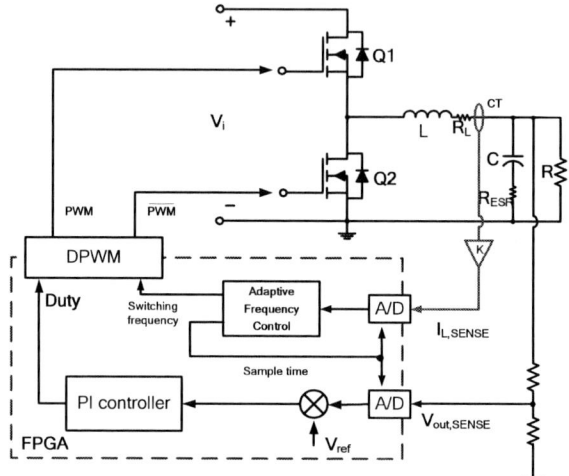

Fig. 5 Proposed digital control system block diagram

For the buck converter illustrated in Fig. 5, when the duty is on, the current ripple is determined by (2) for continuous current mode.

$$\Delta I_L = D \cdot \frac{V_{in} - V_o}{L^*} \cdot \frac{1}{f_s} \qquad (2)$$

And its counterpart when duty is off can be calculated by:

$$\Delta I_L = (1 - D) \cdot \frac{V_o}{L^*} \cdot \frac{1}{f_s} \qquad (3)$$

where

ΔI_L = inductor current ripple

D = duty

V_o = output voltage

V_{in} = inputvoltage

f_s = switching frequency

As shown in (2) and (3), for a given switching frequency, the current increases as the inductance decreases. In this paper, a novel current ripple reduction technique is proposed. The method is to calculate the required switching frequency in order to keep constant current ripple when the inductance changes. By substituting (1) into (2) and (3) gives the current ripple as shown in (4) and (5) when PWM duty is on and off, respectively.

978-1-4244-1871-8/07 $25.00 © 2007 IEEE

$$\Delta I_L = (1-D) \cdot \frac{V_o}{L \cdot (1-K \cdot I_o)} \cdot \frac{1}{f_s} \tag{4}$$

$$\Delta I_L = D \cdot \frac{V_{in} - V_o}{L \cdot (1-K \cdot I_o)} \cdot \frac{1}{f_s} \tag{5}$$

By (4) and (5), the switching frequency can be determined by (6) and (7), respectively.

PWM duty = On:

$$f_s = (1-D) \cdot \frac{V_o}{L \cdot \Delta I_L} \cdot \frac{1}{(1-K \cdot I_o)}$$

$$= K_{a1} \cdot \frac{1}{(1-K \cdot I_o)} \tag{6}$$

PWM duty = Off:

$$f_s = D \cdot \frac{V_{in} - V_o}{L \cdot \Delta I_L} \cdot \frac{1}{(1-K \cdot I_o)}$$

$$= K_{a2} \cdot \frac{1}{(1-K \cdot I_o)} \tag{7}$$

where

$$K_{a1} = (1-D) \cdot \frac{V_o}{L \cdot \Delta I_L}$$

$$K_{a2} = D \cdot \frac{V_{in} - V_o}{L \cdot \Delta I_L}$$

Under steady state $K_{a1} = K_{a2} = K_a$ and therefore (6) and (7) can be merged to give:

$$f_s = K_a \cdot \frac{1}{(1-K \cdot I_o)} \tag{8}$$

As shown in (8), the switching frequency can be determined by the load current and the coefficient of inductance reduction, K. Fig.6 illustrates the relationship between load current and frequency in three kinds of K. As shown in Fig. 6, more increase of switching frequency is required to keep constant current for larger coefficient of inductance reduction. It becomes more relevant as the load current increases.

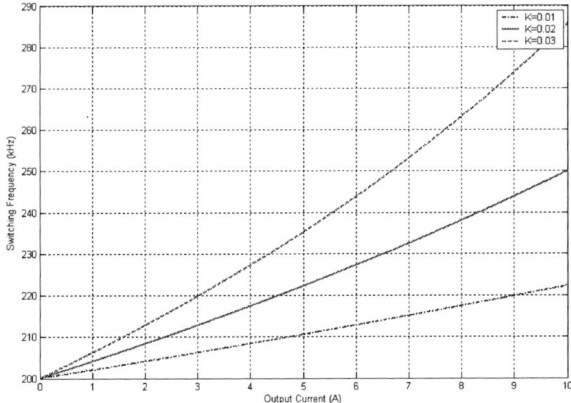

Fig. 6 Switching frequency vs. K

III. EXPERIMENTAL SYSTEM AND RESULT

A. Hardware configuration

Fig. 7 shows the experimental system which consists of a buck converter with synchronous rectification and its digital controller using FPGA. The FPGA is Virtex2 xc2v4000-ff1152 which generates the digital PWM control signal based upon the control method presented in this paper. Both output voltage and inductor current are sensed to provide the feedback signals for the digital controller.

Table 1 shows the circuit parameters and the specifications of the experimental system. As shown in Table 1, the input voltage is 12 V and its output voltage is 1.5 V. The switching frequency is 200-250 kHz and will be determined by the load condition. As shown in Fig. 8, under light load condition, the duty and switching frequency is low. And the switching frequency is increased as the load increase. Meanwhile, the duty will be adjusted to regulate the output voltage. The parameters of controllers are shown in Table 2.

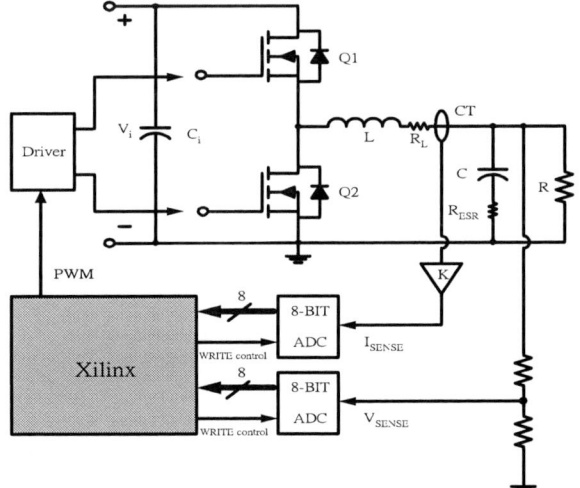

Fig. 7 Experimental system

Table 1 Circuit parameters for the experimental system

Component/Specification	Value
Clock Rate, CLK	50 MHz
Input Voltage V_i	12 V
Output Voltage V_O	1.5 V
Load Current I_O	10 A
Output Inductor, L	1.1 µH @ 10 A
Output Capacitor, C	5630 µF
Switching Frequency, f_{SW}	200-250 kHz

978-1-4244-1871-8/07 $25.00 © 2007 IEEE

Table 2 Parameters of controller

Parameter of controller	Value
Kp	13
Ki	67200
K	0.02
Ka	200000

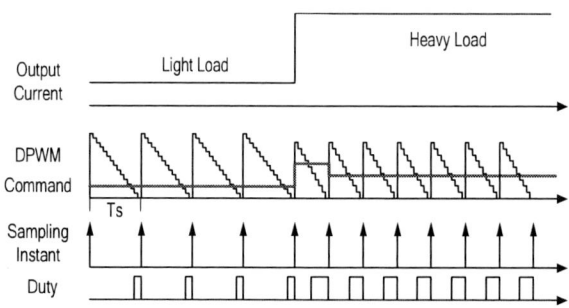

Fig. 8 Schematic diagram of control as load changes

B. Experimental results

Fig. 9 shows the measured results of inductance of output inductor under different load conditions. As shown in Fig. 9, the inductance is reduced from 1.1μH under no load to 0.8μH for full load. Fig. 10 and Table 3 show the measured method of the load current and its ripple for the case of fixed switching frequency and the presented method. As shown in Fig. 10 and Table 3, the current ripple can be maintained at almost constant. The current ripple can be reduced by increasing the switching frequency as illustrated using switching frequency = 250 kHz as an example. However, the current ripple is improved at the cost of more switching losses. Moreover, the current ripple can **not** be kept constant even the switching frequency is increased. Fig. 11 shows the experimental results for the conventional method and the presented method; both are under full load conditions. As shown in Fig. 11 (A) and Fig. 11 (B), the presented method indeed can reduce the current ripple and thereby improving the output voltage ripple.

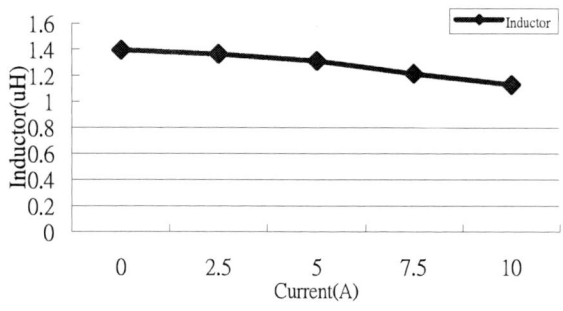

Fig. 9 Variation of inductance

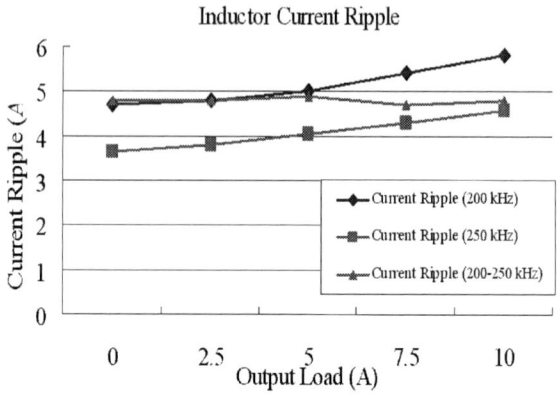

Fig. 10 Experimental results, current ripple vs. load current

Table 3 Experimental results, current ripple vs. load current

Current Ripple (A)	Output Load (A)				
	0	2.5	5	7.5	10
200 kHz	4.7	4.8	5	5.4	5.8
250 kHz	3.66	3.8	4.06	4.3	4.6
Presented method (200~250 kHz)	4.8	4.8	4.9	4.7	4.8

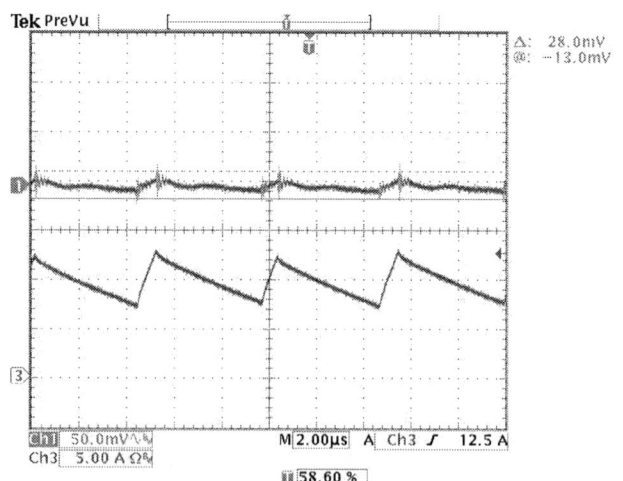

(A). Conventional method

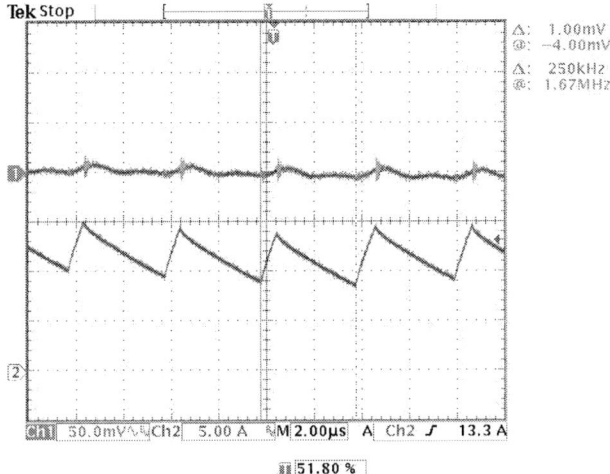

(B). Presented method

Fig. 11 Experimental results, Ch1: Voltage ripple (50 mV/Div), Ch2: Current ripple (5 A/Div)

CONCLUSION

This paper presents a novel method to keep the current ripple to be constant despite of the load current condition for DC/DC converter. The presented method doesn't require any extra circuit component. An experimental DC/DC converter is implemented which is with input voltage of 12 V, output voltage of 1.5 V, output current of 10 A, and switching frequency of 200-250 kHz. Experimental results are presented to fully support the claims.

ACKNOWLEDGEMENT

This work is sponsored by Hitachi Research Lab. and Hitachi Computer Peripheral Co. Ltd., Japan.

REFERENCE

[1] M. Merisio, and E. D. Bartolomeo, "Advanced dc-dc controller design solutions to manage new application requirements in Non-Isolated-Point-of-Load Converters (niPOLs)," *IEEE ISIE,* 2005, pp.463 – 467, June, 2005.

[2] K. Y. Lee, C. A. Yeh and Y. S. Lai, "Design and implementation of fully digital controller for non-isolated-point-of-load converter with high current slew rate," *IEEE IECON,* pp.2605 – 2610, Nov., 2006.

[3] R. W. Erickson, and D. Maksimovic, Fundamental of Power Electronics, 2nd edition, Kluwer Academic Publishers, 2001.

[4] M. Ludwig, M. Duffy, T. O'Donnell,P. McCloskey and S. C. Ó Mathùna "PCB integrated inductors for low power DC-DC converter, " *IEEE Transactions on,* Vol. 18, pp.937 – 945, Jul. 2003.

[5] I. Nishikawa, M. Uasao, Y. Ishizuka, H. Matsuo and J. Saito, "Dynamic characteristics of pulse rate control of a POL converter, " *International Telecommunications Energy Conference,* pp.1-6, 2006.

[6] M. Shirazi, R. Zane, D Maksimovic, L. Corradini and P. Mattavelli, "Autotuning techniques for digitally-controlled point-of-load converters with wide range of capacitive load," *IEEE APEC,* pp.14-20, Feb., 2007.

[7] A. R. Oliva, S. S. Ang, and G. E. Bortolotto, "Digital control of a voltage-mode synchronous buck converter," *IEEE Trans. On, Power Electronics,* Vol.21, No.1, pp.157-163, Jan., 2006.

978-1-4244-1871-8/07 $25.00 © 2007 IEEE

An Improved Energy Recovery Clamp Circuit for PWM Converters with a Wide Range of Input Voltage

Honnyong Cha, Lihua Chen, Fang Z. Peng
Michigan State University
ECE Department
East Lansing, Michigan 48824 USA
Email: chahonny@egr.msu.edu

Qingsong Tang
Beijing Jiaotong University
School of Electrical Engineering
Beijing, China
Email: tangqs@teg.cn

Abstract—A full-bridge dc-dc converter employing diode rectifier in the output experiences severe voltage overshoot and oscillation problem across the diode rectifier caused by the interaction between the junction capacitance of the rectifier diode and the leakage inductance of transformer. The pronounced reverse-recovery current of high power diodes enforces these issues significantly by increasing power loss and voltage overshoot. Conventional energy recovery clamping circuits suffer from high voltage overshoot if the input voltage of converter is wide. In this work, a novel energy recovery clamp circuit is proposed to overcome this problem. Performance of the proposed circuit is verified both theoretically and experimentally with a 70 kW dc-dc converter.

I. INTRODUCTION

Full-bridge PWM dc-dc converters have been widely used in high power and high voltage dc-dc converters since they have several advantages over resonant converters. However, a full-bridge dc-dc converter employing a diode rectifier in the output experiences severe voltage overshoot and oscillation across the diode rectifier since the rectifier diode is located between two current sources, i.e. transformer leakage inductance and output filter inductor [1]-[3]. Therefore, it increases the diode voltage rating and cost, and causes EMI problems. As the output voltage of the dc-dc converter increases, higher voltage diodes are required. However, the use of a higher voltage diodes increase power loss and voltage overshoot in the diode since higher voltage diodes has poor recovery characteristics.

In order to reduce voltage spike in rectifier diodes, several techniques have previously been proposed. The conventional method is the use of an RCD snubber circuit which is shown in Fig. 1 [1]. However, the power loss in snubber resistor is very high as the output power increases. As a result, it degrades system efficiency. The active clamp method shown in Fig. 2 can solve efficiency degradation problem and the voltage overshoot can be clamped but it increases system complexity and degrades system reliability [2]. Therefore, it is not desirable in high-power applications.

To overcome the aforementioned problems, several energy recovery clamp circuits (ERCC) have been proposed recently [4]-[8]. Fig. 3 shows one example of an ERCC. However, the voltage stress across rectifier diodes could be very high for PWM converters with wide ranges of input voltage, especially when duty cycle is less than 0.5. This paper proposes a novel ERCC that improves the previously

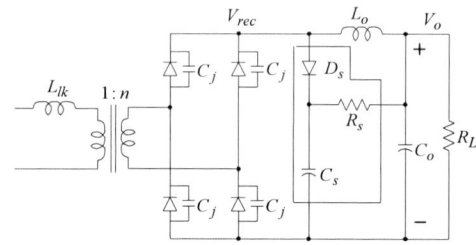

Fig. 1 RCD snubber circuit

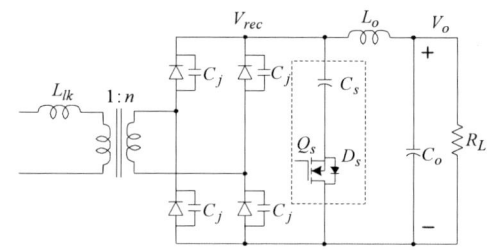

Fig. 2 Active clamp snubber circuit

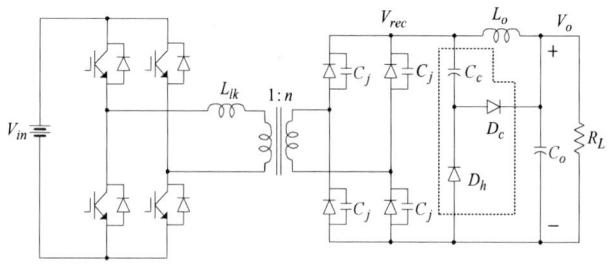

Fig. 3 Dc-dc converter using energy recovery clamp circuit.

proposed circuit. The proposed circuit employs a simple auxiliary circuit in which neither lossy components nor active switches are used.

In section II, some previous solutions are reviewed and their associate problems are pointed out. In section III, a novel ERCC is proposed and its principle operation is described in detail. The simulation and experimental results of the proposed ERCC is shown in Section IV. Performance of the proposed ERCC is verified by applying it to a 70 kW PWM dc-dc converter.

978-1-4244-1871-8/07 $25.00 © 2007 IEEE

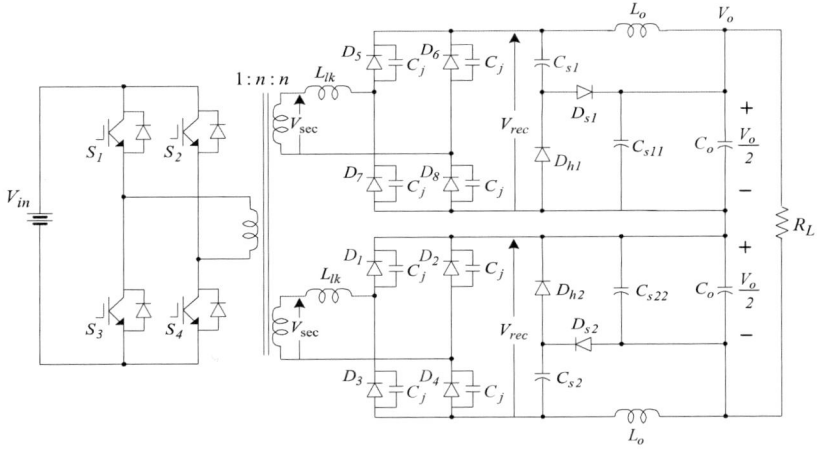

(a)

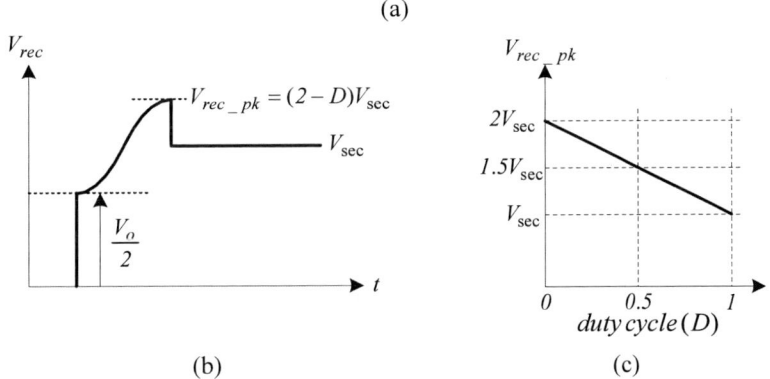

(b)   (c)

Fig. 4 ERCC proposed in [6]

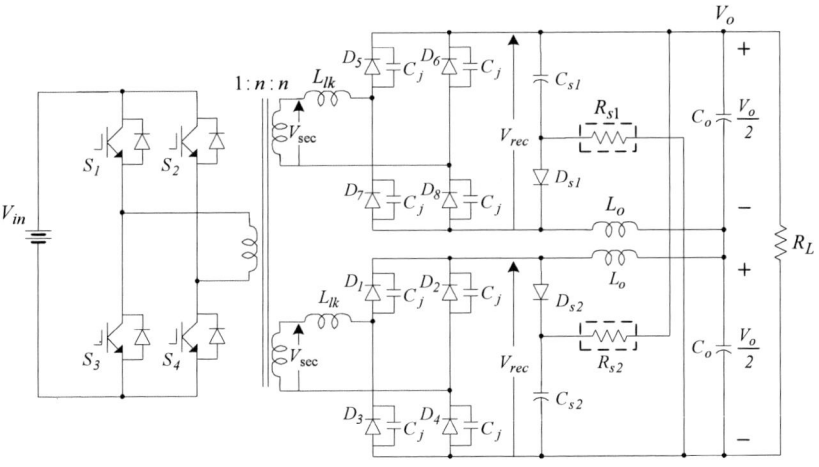

Fig. 5 ERCC proposed in [8]

II. REVIEW OF PREVIOUSLY PROPOSED ERCC

Fig. 4 (a) shows an ERCC modified from the circuit shown in Fig. 3. To achieve high output voltage, the transformer secondary winding is split into two windings.

Two rectifier bridges are used and their outputs are connected in series. With this configuration, each bridge needs to sustain only one half of the output voltage. Transformer turns ratio is set to 6:7:7 (N_p:N_{s1}:N_{s2}) by considering the duty cycle loss and others. The operation of this circuit is the same as that of Fig. 3. Fig. 4 (b) depicts the peak voltage across

978-1-4244-1871-8/07 $25.00 © 2007 IEEE   1091

rectifier diodes (V_{rec_pk}) without considering diode reverse recovery current and is expressed as (1).

$$V_{rec\_pk} = 2(V_{sec} - \frac{V_o}{2}) + \frac{V_o}{2} = 2V_{sec} - \frac{V_o}{2} = (2-D)V_{sec} \qquad (1)$$

where, V_{in} is input voltage, n is the transformer turns ratio ($n=N_{s1}/N_p$ or N_{s2}/N_p), V_{sec} is the voltage in transformer secondary winding and D is converter duty cycle. Fig. 4 (c) shows the changes of V_{rec_pk} as D changes. In this paper, the input voltage (V_{in}) range of the converter is 333 ~ 666 Vdc and output voltage (V_o) is 750 V. With this voltage range, the minimum duty cycle is determined as (2).

$$D_{min} = \frac{\dfrac{V_o}{2}}{V_{sec}} = \frac{\dfrac{750}{2}}{\dfrac{7}{6} \times 666} \approx 0.483 \qquad (2)$$

From (1) and (2), V_{rec_pk} can be calculated as (2-0.483)×7/6×666=1180V. Considering the reverse recovery current in the rectifier diodes, the voltage stress would be higher than 1200 V. Thus, we cannot use standard 1200 V diodes. The use of higher voltage diodes increases power loss and voltage overshoot across diodes because higher voltage diodes have poor recovery characteristics. Therefore, the ERCC shown in Fig. 4 (a) is not applicable to the system described in this paper, although it has the advantages of resetting circulating current in the primary side and achieves zero-voltage and zero-current switching (ZVZCS) in switching devices using the phase shift PWM control method [6].

Fig. 5 shows another example of an ERCC proposed in [8]. This circuit works only when V_{sec} is less than V_o (D > 0.5) and the V_{rec_pk} is clamped to V_o. However, when V_{sec} is higher than V_o (D < 0.5), this circuit does not work.

One possible way is to insert additional snubber resistors (R_{s1} and R_{s2}) as shown in dotted box in Fig. 5. By inserting R_{s1} and R_{s2}, some amount of energy stored in transformer leakage inductance (L_{lk}) is dissipated in R_{s1} and R_{s2}. The total power loss in the snubber resistors are calculated as (3).

$$P_{Rs} = 2\frac{(V_{sec} - V_o)^2}{R_s} = 2\frac{V_{sec}^2}{R_s}(1-2D)^2 \qquad (3)$$

where, $R_s = R_{s1} = R_{s2}$.

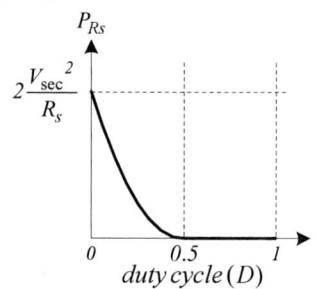

Fig. 6 Power loss in R_s

As shown in Fig. 6, power loss in snubber resistors increase significantly as duty cycle decreases (V_{in} increases), thus decreasing system efficiency.

III. PRINCIPLE OPERATION OF PROPOSED ERCC

The two snubber circuits discussed (Fig. 4 and Fig. 5) have limitations and are not applicable to the systems with wide ranges of input voltage or duty cycle, especially when D < 0.5.

Fig. 7 shows the proposed ERCC and its theoretical waveforms are sketched in Fig. 8. The output filter inductor (L_o) current is assumed as a constant value (I_o). To control the output voltage, either duty cycle control or phase shift PWM control can be used. In this paper, the duty cycle control method is used. The operational modes of this circuit are explained as follows. For the sake of simplicity, only the diode rectifier located at the bottom side is considered since top and bottom sides operate in the same manner. The operational modes are analyzed by assuming D < 0.5.

Mode 1 (~t0): S1~S4 turned off and secondary rectifier diodes are in freewheeling period. V_{sec} remains zero. D1~D4 are on and carry one half of output current (I_o).

Mode 2 (t0~t1): S1 and S4 turn on. V_{sec} changes from zero to nV_{in} and transformer current builds up linearly with the slope of nV_{in}/L_{lk} until it reaches I_o. The current in D1 and D4 are increasing, while the current in D2 and D3 are decreasing.

Mode 3 (t1~t2): reverse recovery period of D2 and D3. Current builds up with the same slope of nV_{in}/L_{lk} until D2 and D3 turn off at point t2. At t2, the current in L_{lk} is $I_o + 2I_{rr}$, where I_{rr} is the reverse recovery current of the rectifier diode. Until this mode, V_{rec} remains zero since D2 and D3 are still on (conducting).

Mode 4 (t2~t3): D2 and D3 snap off at point t2 and its junction capacitors (C_j) start resonance with L_{lk}. During this mode, V_{rec} and I_{Llk} are obtained as follows with the initial conditions $V_{rec}(0)=0$, $I_{Llk}(0)=I_o+2I_{rr}$.

$$V_{rec}(t) = nV_{in}\left[1 - \cos(w_o t)\right] + (I_R Z_c)\sin(w_o t) \qquad (4)$$

$$I_{Llk}(t) = I_o + \frac{nV_{in}}{Z_c}\sin(w_o t) + I_R\cos(w_o t) \qquad (5)$$

where, $Z_c = \sqrt{\dfrac{L_{lk}}{2C_j}}$, $w_o = \dfrac{1}{\sqrt{L_{lk}(2C_j)}}$, $I_R = 2I_{rr}$

Mode 5 (t3~t4): When V_{rec} reaches V_o at t3, D_{s1} starts conducting and there is another resonance between C_{s1} and L_{lk}. Since C_{s1} is much bigger than C_j, the current flowing through C_j can be ignored in this mode analysis. C_{s11} is added to minimize the circuit stray inductance and can be assumed large enough since it is connected in parallel with the output capacitor. During this mode, V_{rec} and I_{Llk} are expressed as (6)

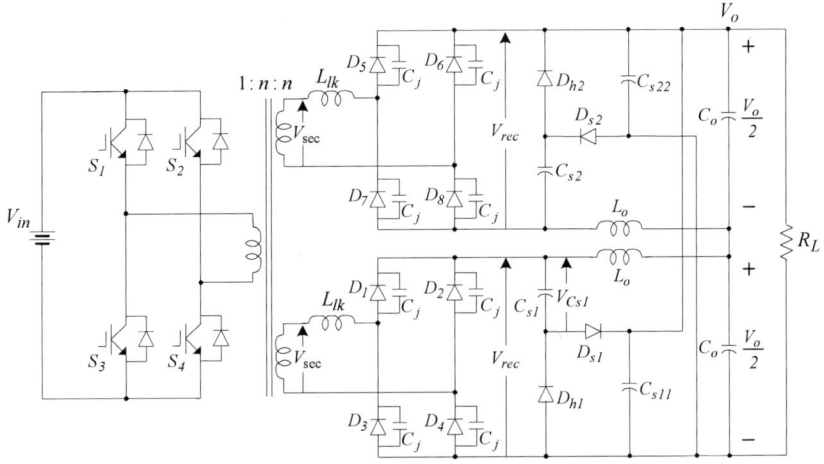

Fig. 7 Configuration of proposed ERCC

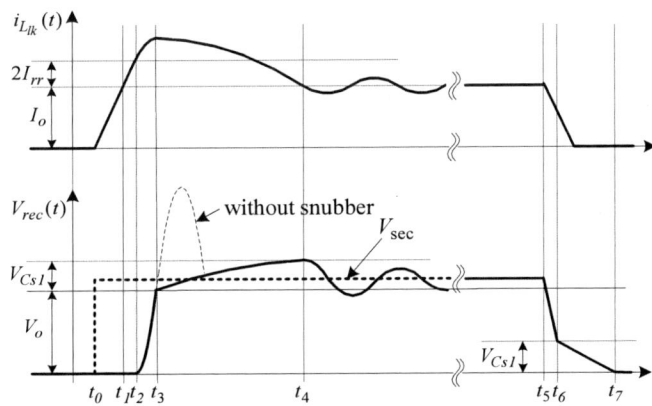

Fig. 8 Theoretical waveforms. Secondary transformer current (top), rectifier diode voltage (bottom)

and (7) with the initial conditions $V_{rec}(0)=V_o$, $I_{Llk}(0)=I_p$. I_p can be calculated from (4) and (5).

Mode 6 (t4~t5): when I_{Llk} decreases to I_o at t4, D_{s1} stops conducting and there is a resonance between L_{lk} and C_j. This resonance is similar to that of Mode 4 and V_{rec} gradually decaying and finally converges to nV_{in}. At point t4, V_{rec} reaches its peak value and the voltage in C_{s1} is kept constant as V_{Cs1}.

Mode 7 (t5~t6): S1 and S4 turned off. I_{Llk} and V_{rec} start decreasing.

Mode 8 (t6~t7): at point t6, V_{rec} is equal to V_{Cs1} and D_{h1}

starts conducting. C_{s1} is discharged through D_{h1} and supplies a portion of the load current I_o. V_{Cs1} is fully discharged at point t7 and D1~D4 turn on and start freewheeling.

The operational mode analysis shown above is applied to the condition of D < 0.5 ($V_{sec} > V_o$). For D > 0.5, V_{rec} ($=V_{Cs1}+V_o$) is almost equal to V_o because current in the transformer leakage inductance is not sufficient to charge C_{s1}. Therefore, only a small voltage is applied in primary transformer leakage inductance. This voltage is not enough to reset primary current to zero and to achieve ZVZCS in phase shift PWM control. For D < 0.5, there is a reduction in transformer primary current.

From the results discussed above, V_{rec_pk} is expressed as

$$V_{rec}(t) = V_o + (nV_{in} - V_o)[1 - \cos(w_d t)] + \frac{Z_d}{Z_c}\sqrt{(I_R Z_c)^2 + 2nV_{in}V_o - V_o^2}\sin(w_d t) \qquad (6)$$

$$I_{Llk}(t) = I_o + \frac{nV_{in} - V_o}{Z_d}\sin(w_d t) + \frac{\sqrt{(I_R Z_c)^2 + 2nV_{in}V_o - V_o^2}}{Z_c}\cos(w_d t) \qquad (7)$$

$$\text{where,} \quad Z_c = \sqrt{\frac{L_{lk}}{2C_j}}, \ Z_d = \sqrt{\frac{L_{lk}}{C_{s1}}}, \ w_d = \frac{1}{\sqrt{L_{lk}C_{s1}}}$$

(8) and (9) for D. It should be pointed out that the reverse recovery current of diode is not included in this plot for simplicity.

1) When D < 0.5,

$$V_{rec\_pk} = 2(V_{sec} - V_o) + V_o = 2(1 - D)V_{sec} \quad (8)$$

2) When D > 0.5,

$$V_{rec\_pk} \approx V_o = 2 \times \frac{V_o}{2} = 2DV_{sec} \quad (9)$$

Fig. 9 plots V_{rec_pk} as a function of D with the proposed ERCC and compared with V_{rec_pk} in Fig. 4 (c). By using proposed circuit, V_{rec_pk} was clearly reduced with duty cycle range from 0 to 2/3 where diode voltage rating is determined.

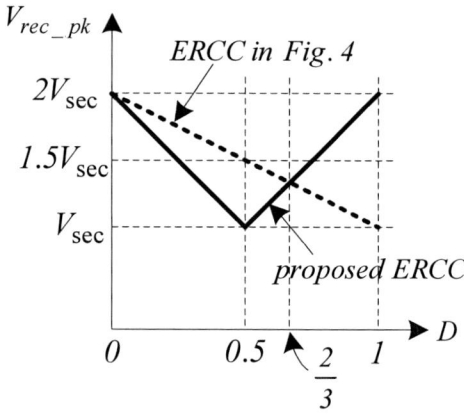

Fig. 9 Vrec_pk vs. D

TABLE 1
Operational Conditions and Circuit Parameters of dc-dc Converter

Input Voltage	333 ~ 666 Vdc
Output Voltage / Current	750 Vdc / 93 A
Switching frequency	13 kHz
IGBT	Powerex CM600HU-24F (1200V, 600 A)
Rectifier Diode	Powerex QRD1230T30 (1200 V, 150 A) trr=130 ns, Irr=30 A
Transformer turns ratio	6:7:7
Transformer leakage inductance (Llk)	1 uH
Cj	1 nF
Cs1, Cs2	100 nF
Cs11, Cs22	470 nF
Ds1, Ds2, Dh1, Dh2	IXYS DSEI 2×61-12
Lo	500 uH

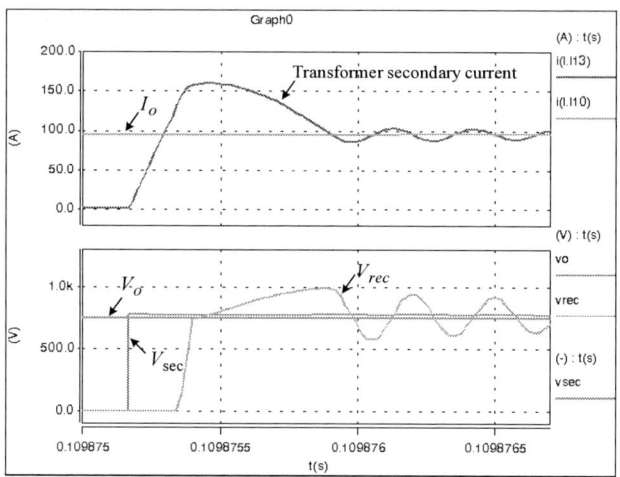

Fig. 10 Simulation results when D=0.483

IV. SIMULATION AND EXPERIMENTAL RESULTS

A 70 kW prototype dc-dc converter employing improved ERCC has been built and tested to verify the principle of operation and is compared with the simulation results. Table I shows the operational conditions and circuit parameters of the dc-dc converter tested in this paper. Fig. 10 and 11 show the simulation results with D=0.483 and 0.8, respectively. Simulation waveforms shown in Fig. 10 are consistent with theoretical ones shown in Fig. 8. When D > 0.5, V_{rec_pk} is clamped to V_o as expected (see Fig. 11).

Fig. 12 shows rectifier voltage waveforms using the circuit shown in Fig. 5. Since the duty cycle with maximum voltage is less than 0.5, 500 ohm resistors are used as R_{s1} and R_{s2} in this experiment. The rectifier voltage is effectively clamped to 1100 V. Under the test conditions of Vin=666V, fs=13 kHz, n=7/6, Vo=750V and Po=70 kW, the total power loss in the resistors is about 200 W. Fig. 13 shows the experimental waveforms of transformer primary current and rectifier diode voltage using the proposed snubber circuit under the same test conditions. The voltage and current waveforms are compatible to those of theoretical waveforms

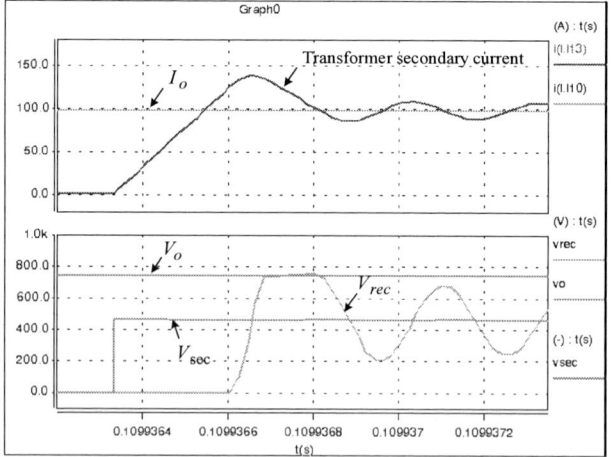

Fig. 11 Simulation results when D=0.8

shown in Fig. 8. Diode voltage is also effectively clamped to 1100 V.

978-1-4244-1871-8/07 $25.00 © 2007 IEEE

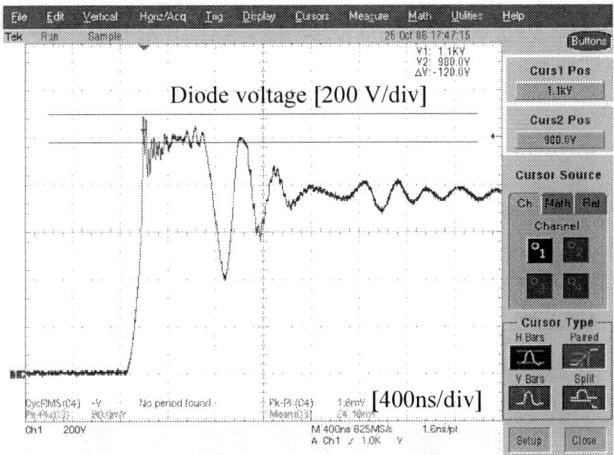

Fig. 12 Diode voltage waveform

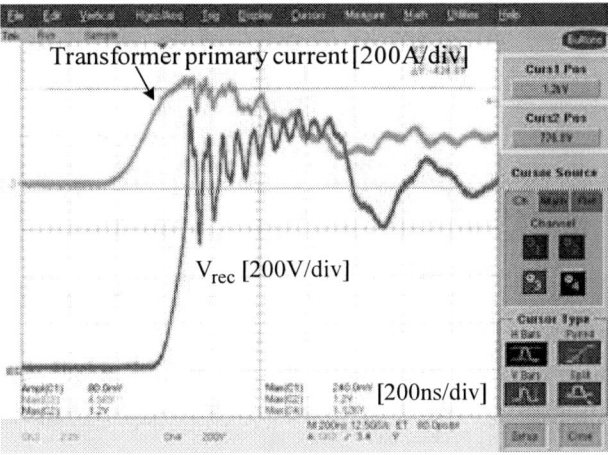

Fig. 13 Experimental waveforms using proposed ERCC
Primary transformer current (top), diode voltage (bottom)

V. CONCLUSIONS

This paper introduced an improved ERCC for PWM dc-dc converters with wide ranges of input voltage. A 70 kW prototype dc-dc converter employing the proposed ERCC has been built and tested to verify the principle of operation. The peak rectifier diode voltage of proposed circuit was compared with traditional circuits. The proposed circuit consists of two small capacitors and two diodes in each bridge. Neither lossy components nor additional active switches are used to clamp diode voltage. Therefore, the proposed ERCC is very promising for high voltage and high power systems with wide ranges of input voltage.

REFERENCES

[1] L. H. Mweene, C. A. Wright, and M. F. Schlecht, "A 1kW 500 kHz front-end converter for a distributed power supply system," *IEEE*

Transactions on *Power Electronics*, vol. 6, no. 3, pp. 398-407, July 1991

[2] J. A. Sabate, V. Vlatkovic, R. B. Ridley, and F. C. Lee, "High-voltage, high-power, ZVS, full-bridge PWM converter employing an active snubber," in Proc. IEEE APEC, 1991, pp. 158-163.

[3] S. Lin and C. Chen, "Analysis and design for RCD clamped snubber used in output rectifier of phase-shift full bridge-bridge ZVS converters," *IEEE Transactions on Industrial Electronics*, vol. 45, no. 2, pp. 358-359, 1998.

[4] E. S. Kim, K. Y. Joe, M. H. Kye, Y. H. Kim, and B. D. Yoon, "An Improved Soft Switching PWM FB DC/DC Converter for Reducing Conduction Losses," *IEEE Transactions on Power Electronics*, vol. 14, no. 2, pp. 258-264, March 1999

[5] J. G. Cho, J. W. Baek, D. W. Yoo, H. S. Lee, and G. H. Rim, "Novel zero-voltage and zero-current-switching (ZVZCS) full bridge PWM converter using transformer auxiliary winding," *IEEE Transactions on Power Electrononics*, vol. 15, pp. 250–257, March 2000.

[6] J. G. Cho, J. W. Baek, C. Y. Jeong, and G. H. Rim, "Novel Zero-Voltage and Zero-Current-Switching (ZVZCS) Full Bridge PWM Converter Using A Simple Auxiliary Circuit," *IEEE Transaction on Industry Applications*, vol. 35, Issue 1, pp. 15-20, Jan.-Feb. 1999.

[7] E. S. Kim and Y. H. Kim, "A ZVZCS PWM FB DC/DC Converter Using a Modified Energy-Recovery Snubber," *IEEE Transactions on Industrial Electronics*, vol, 49, no. 5, pp. 1120 – 1127, October 2002

[8] A. Bendre, S. Norris, D. Divan, I. Wallace, and R.W.Gascoigne, "New high power DC-DC converter with loss limited switching and lossless secondary clamp," *IEEE Transactions* on *Power Electronics*, vol. 18, no. 4, pp. 1020-1027, July 2003.

Power Saving of PWM Rectifier-VSI Fed Induction Machines

S. Halász

I. Kádár

Budapest University of Technology and Economics, Department of Electric Power Engineering
H-1521 Budapest, Egry József u. 18, Hungary

Email: shalasz@eik.bme.hu

Email: ikadar@eik.bme.hu

Abstract- The inverter fed ac drives with transistorized grid rectifier are investigated from the point of view of steady-state power saving. It is shown that the optimal power saving control of frequency controlled induction motor drive provides saving about 0.5 to 3% of rated power by flux control less then the rated flux. This valid only for motor torques less than 50% of the rated load. In case of high modulation frequency about the same energy can be saved for both the inverter and sinusoidal operations. For the rated load some power saving (0.5-1.0%) is also possible but only by the increasing of the motor flux over the rated value. Generally, the energy saving control is economical only when the technology requires the frequency control.

Keywords: Energy saving, frequency control, induction motor drive, efficiency optimization.

I. INTRODUCTION

The inverter fed ac drives become very common and the control of their efficiency can be easily achieved by varying the supplied voltage and frequency providing the possibility of optimal power loss control [1]. In the last years instead of diode rectifier a transistorized rectifier is often used therefore the investigation of PWM rectifier-VSI fed induction motor system becomes important. The system control providing the minimum of the electric energy consumption - for a given steady-state operating point **with fixed *W* speed and *M* torque** values - becomes very important too.

The theoretical investigation of the optimal power loss control of induction motors started 80 years ago. Especially these investigations have got practical importance with wide use of thyristors and transistors. Today numerous companies propose thyristor voltage control of induction motors emphasizing energy saving control. These drives are able to control the speed and the current of the motor. However, the energy saving ability of these drives is restricted. Numerous transistorized frequency controlled systems with different energy optimizers have been also developed but practical applications of these drives are questionable. The system control is very complex [2-6] and can it be economically rational only for high power drives. From this point of view

a system with PWM rectifier possibly gives more advantages.

The investigated system together with the measuring instruments is presented in Fig. 1. The PWM rectifier provides the control of the dc voltage, which must be higher than that in case of the diode rectifier. Therefore the system with PWM rectifier can control the motor flux in higher region of the motor frequencies and the optimal control of a normal series motors can be realized at least up to the rated frequency.

The total system losses consist of the motor fundamental and harmonic loss as well as the rectifier and inverter ones.

II. MOTOR FUNDAMENTAL LOSSES

The motor equivalent circuit is drawn in Fig. 2, the parameters are given in p.u. system. The X_l stator leakage, the X_{rl} rotor leakage and the X_m magnetizing reactances are given for the rated frequency, while w_1 is the actual frequency or the electric angular frequency in p.u., W is the motor speed (in p.u.) and $\beta = w_1 - W$ is the actual absolute slip. The base values of the p.u. system are the rated voltage, the rated current, the rated apparent power and the rated frequency, the torque base is calculated from the last two values.

The P_i iron stator losses are taken into account by R_m iron resistance:

$$R_m = \frac{(w_1 \psi_m)^2}{P_i} = \frac{(w_1 \psi_m)^2}{\psi_m^2 (k_h w_1 + k_e w_1^2)} = \frac{w_1}{k_h + k_e w_1}, \quad (1)$$

where ψ_m is the mutual magnetic flux, k_h is a factor representing hysteresis losses and k_e is a factor of eddy current losses.

From Fig. 2 for a given M torque the rotor current losses are calculated as:

$$I_r^2 R_r = M\beta \quad (2)$$

and neglecting the current across R_m the stator current losses are:

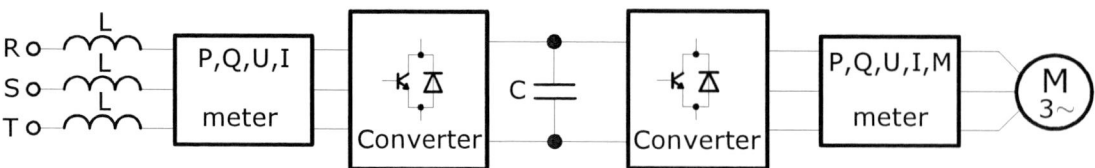

Fig. 1. The transistorized rectifier and inverter system (with electrical power, voltage, current and torque meter)

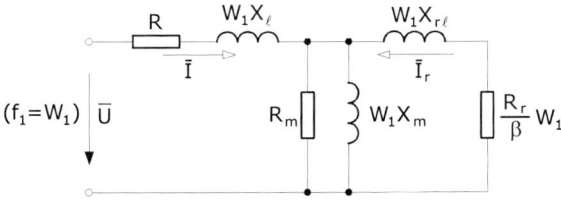

Fig. 2. Motor equivalent circuit

(R=0.042, R_r=0.041, X_l=X_{rl}=0.10, X_m=3.24,
k_h=0.011, k_e=0.012)

$$I^2 R = M \frac{R}{R_r} \left[\beta \left(\frac{X_r}{X_m} \right)^2 + \frac{R_r^2}{\beta X_m^2} \right], \qquad (3)$$

where $X_r = X_m + X_{rl}$.

The ΔP_M fundamental losses consist of the sum of the stator and rotor winding losses and of the iron losses as well. For a given speed the friction losses are constant therefore in the optimal slip computation they can be omitted. The result related to the torque is [1]:

$$\frac{\Delta P_M}{M} - \left[\begin{array}{l} \dfrac{w_1 - W}{R_r} \left[R_r + R \dfrac{X_r^2}{X_m^2} + X_{rl}^2 \left(k_e w_1^2 + k_h w_1 \right) \right] \\ + \dfrac{R_r}{w_1 - W} \left(\dfrac{R}{X_m^2} + k_e w_1^2 + k_h w_1 \right) \end{array} \right]. \quad (4)$$

These losses exhibit minimum for each fixed W speed and M torque values as function of w_1 or slip frequency $\beta = w_1 - W$. The computed optimal β_{opt}, η_{opt} efficiency and the relative value of motor losses $\Delta P_M / M$ are shown in Fig. 3 as function of motor speed. The dashed lines in Fig. 3 give the same variables for the minimal value of stator current, which arises for the slip $\beta_i = R_r / X_r$. It will be seen in latter case for the region of higher speed that the minimal current control produces considerably inferior results.

The stator and rotor currents for the optimum operation and the main magnetising flux is given in Fig. 4 - in comparison to $\sqrt{M}$ - as a function of w_1.

It should be noted that computation of optimal slip is simplified if in (4) by the very small change of w_1 is neglected [1]. The error in the optimal slip will be below 1% while the error in the losses of optimal point is lower than 0.01%.

These theoretical results are in accordance with tests only if the saturation effect of the machines is taken into account [3]. For investigated motor the magnetized reactance at the rated flux was 244%.

For constant motor parameters the optimal β_{opt} depends only on motor speed and is independent of the motor torque.

III. MOTOR HARMONIC LOSSES

The switching frequency of today's commercial inverters of rated power under 100kW is between 3-20kHz. The switching frequency in our test drive was 7kHz for the motor side inverter and 10kHz on the network side converter. The harmonic current losses at this switching frequency are usually 0.05-1% in the stator and 1-2% in the rotor (relatively to the rated losses of the stator and rotor, respectively).

The motor current harmonic losses can be characterized by the generalized loss-factor [9-10] which is expressed as follows:

$$G = \left(\Delta I \cdot L' \cdot f_c \right)^2 ,$$

where ΔI is the square of the rms. value of the current harmonics, L' is the motor transient reactance and f_c is the carrier frequency (related to the rated frequency of the motor).

In case of space vector modulation the generalized loss-factor is presented in Fig. 5. It is seen that this loss-factor has the maximal value at modulation index

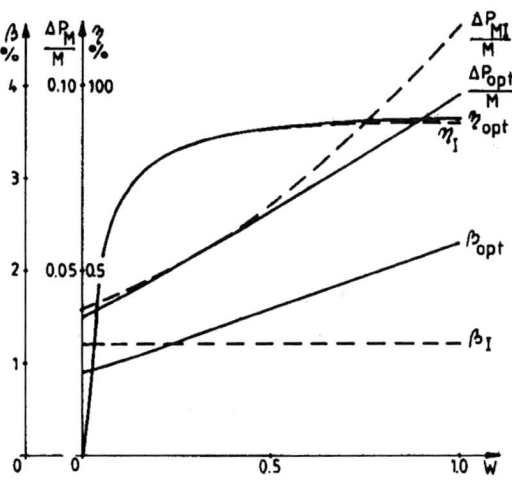

Fig 3. Optimal power losses and slip vs. motor speed

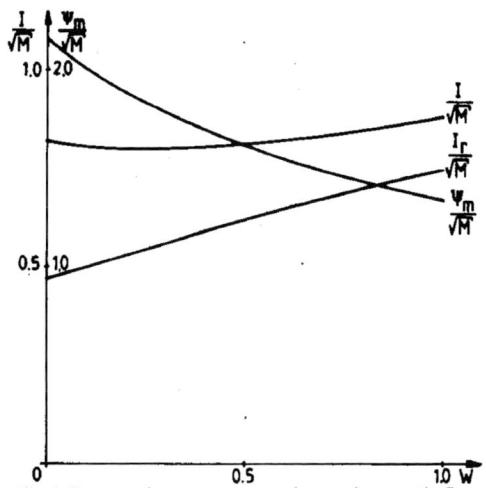

Fig. 4. Stator and rotor currents and mutual magnetic flux vs. speed in optimal point

978-1-4244-1871-8/07 $25.00 © 2007 IEEE

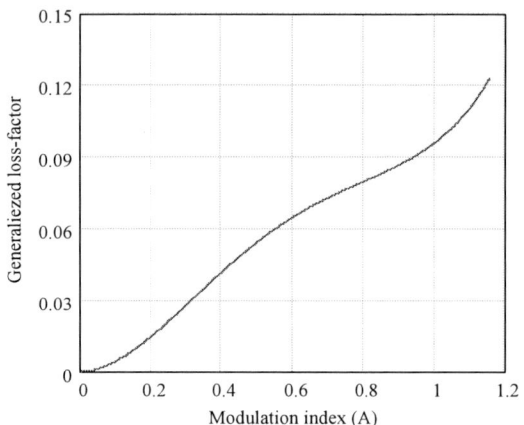

Fig. 5. Generalized loss-factor of space vector PWM.

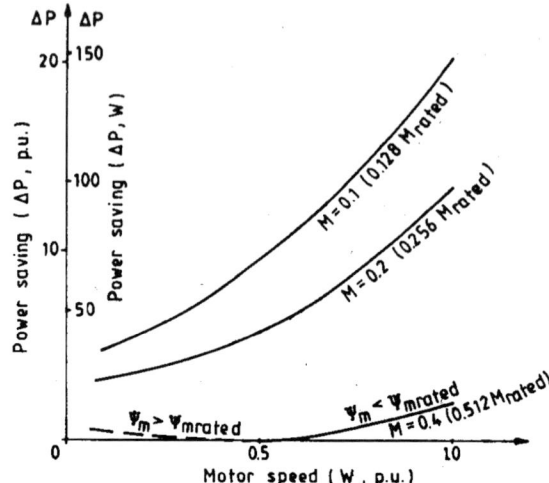

Fig. 6. Theoretical power saving of the drive for inverter and sinusoidal supply

$A = 2 / \sqrt{3} = 1.1547$ where $G = 0.123$. The square of the rms. value of the current harmonics in this point is as follows:

$$\Delta I^2 = \frac{G}{\left(L_{skin}^{'}\right)^2 f_c^2} = \frac{0.123}{0.01\left(7000/50\right)^2} = 0.06\%, \quad (5)$$

where $L_{skin}^{'} = 0.1$ is the motor transient reactance considering the skin effect and $f_c = 7000/50$ is the relative value of the carrier frequency. Hence, the stator current harmonic losses are less than 0.06 % of the rated fundamental losses while in the rotor they are approximately 0.9% of the rotor rated losses even if the skin effect is taken into account. Thus, the motor harmonic current losses are very small if the carrier frequency is higher than 5kHz.

The motor harmonic iron losses are mainly equal to the eddy current losses. Its computation methods are given in [7-8] but the experimental checking of these methods still is insufficiently proved.

Therefore a rough approach gives satisfactory result in that case when the harmonic winding and iron losses are taken proportional to the square of its harmonic flux which, in turn, proportional to $(U_1 / w_1)^2$. Neglecting the stator resistance the last term can be expressed by motor torque:

$$M \approx \left(\frac{U_1}{w_1}\right)^2 \left(\frac{X_m}{X_m + X_l}\right)^2 \frac{1}{\frac{R_r}{\beta} + \frac{\left(X_r^{'}\right)^2}{\beta}},$$

where $X_r^{'}$ is the transient inductance of the rotor.

Expressing the fundamental motor voltage by the motor torque the harmonic losses have the form as follows:

$$\Delta P_{Mv} = c_1 M \frac{X_m^2}{\left(X_m + X_l\right)^2} \left[\beta \frac{\left(X_r^{'}\right)^2}{R_r} + \frac{R_r}{\beta} \right], \quad (6)$$

where c_1 depends on the switching frequency of the inverter. According to (5) for modulation frequency over

5kHz in (6) $c_1 \rightarrow 0$. It should be noted that this statement is not entirely justified by test results.

IV. RECTIFIER AND INVERTER LOSSES

These ΔP_c losses include all the losses between ac grid and motor (transistor conduction and switching losses of two converters, losses of transistor protection elements, L inductance and as well the power losses of control circuits). Theoretical computation of these losses is difficult. Measurements show that for a given torque and speed losses are virtually independent of the slip but monotonously increase with the motor torque.

Therefore these losses for a given torque and speed are taken into account as constant which don't affect the optimum slip.

V. TOTAL POWER LOSSES

The $\Delta P_\Sigma = \Delta P_M + \Delta P_{Mv} + \Delta P_c$ total power losses in a PWM inverter fed induction drive are the sum of the motor fundamental and harmonic losses and of the inverter and rectifier losses as well. For selected different W and M the optimal β_{opt} values can be analytically computed by minimizing the $\Delta P_\Sigma(\beta)$ function. Similarly is done the computation of optimal slip in case of fundamental supply. The result of computation is summarized in Fig. 6 where the possible power saving is shown for three different loads. The motor saturation was taken into account. It should be noted that there is no sensible difference in power saving between results of sinusoidal and inverter supply.

It is seen from theoretical investigations that even for small load only approximately 2% of apparent power can be saved by flux control of induction motor. At the same time for 50% of the rated load the optimal control requires the decrease of the motor flux below the rated value for high speed and increase of the flux over rated value for low speed. The use of PWM rectifier gives advantages only on

978-1-4244-1871-8/07 $25.00 © 2007 IEEE

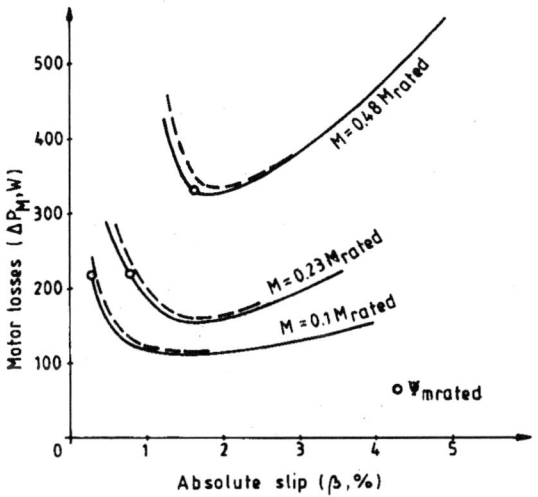

Fig. 7. Motor power losses, $W = 2/3$,
— inverter supply, - - - sinusoidal supply

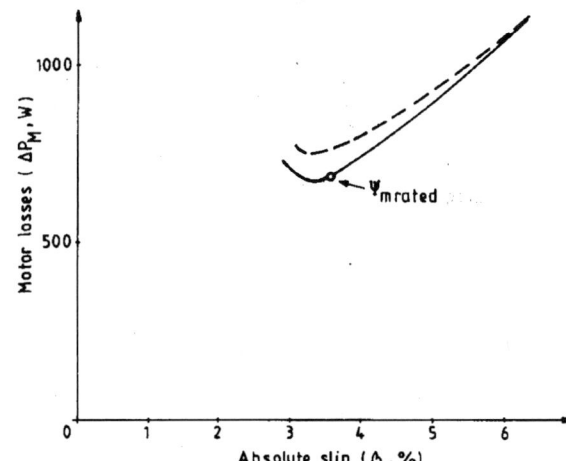

Fig. 8. Motor power losses for rated load, $W = 2/3$,
- - inverter supply, - - - sinusoidal supply

high speed since the dc voltage is higher than in case of diode bridge. This allows the increase of the motor flux over rated value in wider range. At low speed and high torque the optimal control usually leads to operation with motor flux over the rated value.

VI. TEST RESULTS

For the speed controlled test system (Fig. 1) a PWM rectifier was installed with network friend control. Thus the grid current is sinusoidal and it is in phase with the grid voltage. The inverter has rotor flux oriented control so the flux control is performed by control of the current component of the rotor flux direction. The test program contains measurement of the motor and the system losses at

different speeds. For each set speed the losses were measured at different motor torques (approximately 0.1; 0.25; 0.5 of the rated torque). The dc voltage was controlled at minimal necessary value: the peak of line voltage plus 5%. In this case the losses of two inverters were minimized.

The load was a dc pendel generator with mechanical torque measurement possibilities. At the same time the electromagnetic torque was measured by electrical torque meter.

In Fig. 7 the motor losses are presented for $W = 2/3$ ($n = 1000\,rpm$) and at three different loads: $M = 17.5Nm$ ($0.48M_{rated}$), $M = 8.6Nm$ ($0.23M_{rated}$), and $M = 3.6Nm$ ($0.1M_{rated}$). The point with rated stator flux is marked by circle. It is seen that at this speed the energy optimized control can save approximately 100W for 3.6Nm

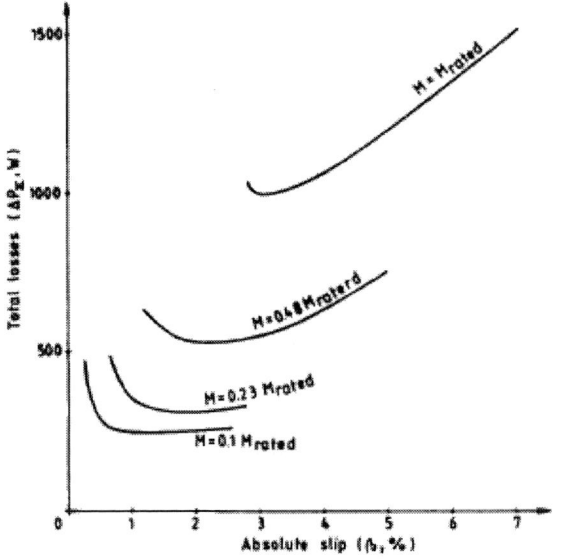

Fig. 9. The measured total power losses, $W = 2/3$

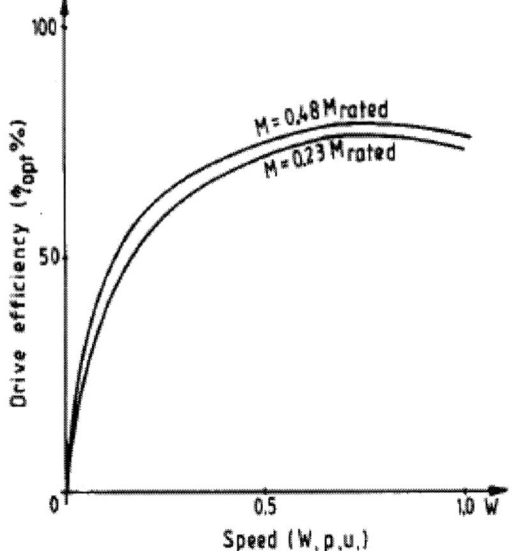

Fig. 10. Optimal drive efficiency vs. speed

978-1-4244-1871-8/07 $25.00 © 2007 IEEE

load, 70W (0.01p.u.) in case of $M = 8.6 Nm$ and virtually nil for $M = 17.5 Nm$. In case of network supply and small load the motor flux is a little higher than the rated flux. Therefore some better results are obtained if this phenomenon is taken into account, approximately 100W can be saved for $M = 8.6 Nm$ and 40-50W for $M = 17.5 Nm$.

For rated torque and $W = 2/3$ ($n = 1000 rpm$) the motor power losses are drawn in Fig. 8. In this case the power saving control requires increase of the stator flux over the rated value. However, due to the saturation of the machine and increase of magnetizing current a notable power saving is impossible.

Similar results can obtain for several different motor speeds.

In Figs. 7-8 the motor power losses are also given for inverter supply. As it was discussed above the additional motor losses owing to inverter operation are really small in case of 7 kHz modulation frequency. These losses have measurable value only in case of higher motor fluxes.

The theoretical characteristics in Figs. 3-5 are producing correct values only for unsaturated machine that is for $M < 0.4 M_{rated}$. At the same time the friction torque of the drive (which is not measured by mechanical torque meter) must be considered. In case of $n = 1000 rpm$ the friction power is $\approx 50W$ ($\approx 0.5 Nm$).

The rectifier and the inverter losses for theoretical investigation were taken as independent of the slip. According to measurements in case of 1000rpm these losses were 300W for the rated torque, 200W for $M = 0.48 M_{rated}$, 138W for $M = 0.23 M_{rated}$ and 120W for $M = 0.1 M_{rated}$. For different slip the deviation from above values in all cases was inside 10W.

The measured total power losses are presented in Fig. 9 for $W = 2/3$ ($n = 1000 rpm$). These losses contain really total system losses since the control circuit is supplied from the intermediate dc voltage according to Fig. 1. The rectifier and the inverter losses given above contain about 60W constant basic losses. The other losses are the losses of two converters with firing circuit losses and the losses of main circuits in Fig. 1 as well. The optimal drive efficiency as function of the motor speed is shown in Fig. 10. It is seen that this characteristic differs from the motor efficiency in Fig. 3 especially in the low speed region the reason of that is the rectifier and inverter losses P_c which don't decrease proportionally to the motor speed therefore the drive efficiency at low speed is considerably lower than the motor efficiency. The small decreasing of the optimal drive efficiency at high speed is explained by increasing of the friction loss.

VII. CONCLUSIONS

The optimal power saving control of frequency controlled induction motor drive permits about 0.5 to 3% of rated power saving compared with the rated flux operation for motor torques that are below 50% of the rated torque. For

presently used switching frequency over 5kHz this is valid for both the sinusoidal and the inverter supplies. For rated load the power saving (0.5-1.0%) is possible only at low speed and with the increase of motor flux over the rated flux. Generally, the energy saving control is economical when the technology requires a frequency control. The use of PWM instead of diode rectifiers allows increasing of the flux control region only in case of the high motor torque and speed, however in this case it is impossible to reach sensible energy saving.

ACKNOWLEDGMENT

The research was supported by the Hungarian National Scientific Fund (OTKA #T 042866) for which authors express their sincere gratitude.

REFERENCES

[1] S. Halász, M. Hunyár, T. Molnár, "Power Saving in the VSI Fed Induction Motor Drive," *In Proc. of the Aegean Conference on Electrical Machines and Power Electronics.* 1992. Kusadasi, Turkey.

[2] D.S. Kirchen, D.W. Novotny, T.A. Lipo: "On line efficiency optimization of a variable frequency induction motor drive," *IEEE Trans. on Ind. App.,* Vol. IA-21, No. 4, 1985.

[3] P. Thogersen, M. Tonnes, U. Jaeger, S.E. Nielsen, "New high performance vector controlled ac-drive with automatic energy optimizer," *In Proc. of EPE'95 Conference.* Sevilla, Spain, 1995, pp. 3.381-3.386.

[4] G.-S. Kim, I.-J. Ha, M.-S. Ko, „Control of induction motors for both high dynamic performance and high power efficiency," *IEEE Transactions on Ind. Electronics,* Vol. 39, No.4, August 1992, pp. 323-333.

[5] I. Kioskerdis, N. Margaris, "Loss minimization in scalar-controlled induction motor drives with search controllers," *IEEE Transactions on Ind. Electronics,* Vol. 11, No.2, March 1996, pp. 213-220.

[6] G. Dong, O. Ojo, "Efficiency optimizing control of induction motor using natural variables", *IEEE Transaction on Ind. Electronics,* vol. 53, No 6, December 2006, pp. 1791-1798.

[7] N. Hildebrand, H. Röhrdanz, "Losses due to the voltage harmonics in PWM converter-fed three-phase motors," *in Proceedings of ICEM'96,* Vigo, Spain, pp. 327-331.

[8] A. Ruderman, R. Welch, "Electrical machine PWM loss evaluation basics," *in Proceedings of Energy Efficiency in Motor Driven System Conference,* 5-8 September, 2005, Heidelberg, Germany, pp. 1-4.

[9] H. van der Broeck: Analysis of the harmonics in voltage fed inverter drives caused by PWM schemes with discontinuous switching operation, *in Proceedings of 3rd European Conference on Power Electronics and Applications,* Florence, Italy, 1991, pp.3-261-266.

[10] S.Halász, B.T.Huu: Generalized harmonic loss-factor as a novel important quality index of PWM techniques, *PCC-Nagaoka,* Japán, Aug. 3-6. 1997. pp. 787-792.

A New Multi-Machine Control System Based On Direct Torque Control Algorithm

Hossein Mohktari
Associated Professor
School of Electrical Engineering
Sharif University of Technology
Tehran, Iran
Email: mokhtari@sharif.edu

Abdollah Alizadeh
M.Sc. Student
School of Electrical Engineering
Sharif University of Technology
Tehran, Iran
Email: alizadeh.ab@gmail.com

Abstract—**In this paper, a new and simple control method based on Direct Torque Control (DTC) of induction motors is proposed for a multi-machine system. Similar to a conventional DTC, the proposed method has two separate control loops. In the torque control loop, before selection of optimum voltage from the DTC look-up table, the system overall requirement is determined based on requirements of motors torque. Also, Switchable Master-Slave control is used in the flux control loop. The method, which is simulated for a two-parallel induction machine system, can be extended to a multi-machine system. Simulation results are also provided to investigate the performance of the proposed technique.**

I. INTRODUCTION

Electric machines have been extensively used in industry for years. Among different types of electric machines, Induction Machines (IMs) are more common due to their ruggedness and less maintenance requirements. However, control of induction machines is not trivial. Different methods including scalar control, Field Oriented Control (FOC) and Direct Torque Control (DTC) have been proposed for the control of IMs. Among them, the DTC offers advantages such as, simplicity, faster torque response, accurate torque control, no need to coordinate transformation, elimination of voltage modulation blocks, absence of current control loop and robustness. In many applications, one motor is controlled by one converter. These systems are called Single-Machine Single-Converter systems. However, in high power applications such as traction systems, conveyer lines and steel processing, two or more machines are fed by one converter. This topology results in a light, more compact and less costly system [1], [2]. These systems are called Multi-Machine Single-Converter systems. Control of multi-machine single-converter systems is the subject of this paper. The paper first reviews existing control strategies for such systems. Then, a new algorithm based on DTC is proposed. This is the first time that the DTC algorithm is used for the control of a multi machine system. The performance of the proposed method is investigated using computer simulations.

Several control methods have been proposed for the control of multi-machine single-converter systems. These methods were primarily based on scalar techniques with slow responses. With the invention of vector control techniques, more effective

methods appeared. These methods improved the response time and accuracy of motor control. But these methods were complex and their implementation was difficult.

Most of the existing control techniques can be divided into two main categories; Master-Slave and Mean Control. In a Master-Slave control system, one motor, which is selected as the master, is directly controlled. In this method, the behavior of slave motors is not considered [3], [4]. Consequently, in some conditions, the performance of slave motors may not be acceptable. In a mean control technique, depending on the control parameter, different averaging systems can be realized. One method is to choose the average of currents of all the machines as the main control signal. The control system is basically similar to that of a single machine. In this method, machine internal parameters such as flux, do not show desirable behavior especially during transients [3], [4]. Another method is to take an average over the parameters of the equivalent circuit of all the parallel machines at steady state [5], [6], [7]. As a result of this averaging, a new equivalent circuit is derived. This method does not yield satisfactory results in case machine parameters are not similar. Another method is described in [5]. In this method, to each machine, a single-machine control technique is applied, and then, a reference voltage is obtained for each machine. Since, an inverter can only provide one reference voltage vector, therefore, a vector average is taken over the machine reference voltages, and the result is generated by the inverter.

There are other methods which cannot be included in the above mentioned divisions. However, they also have shortcomings such as complexity [2].

II. CONTROL PROBLEMS OF MULTI MACHINE SYSTEMS

Before introducing the proposed idea, some issues must be explained regarding parallel induction motors. As a result of applying one voltage vector, stator flux vector of all the parallel induction motors will vary instantaneously in the same direction. Therefore:

$$\frac{d}{dt}(\overline{\psi}_s) = \overline{V}_s - R_s\overline{I}_s \qquad (1)$$

978-1-4244-1871-8/07 $25.00 © 2007 IEEE

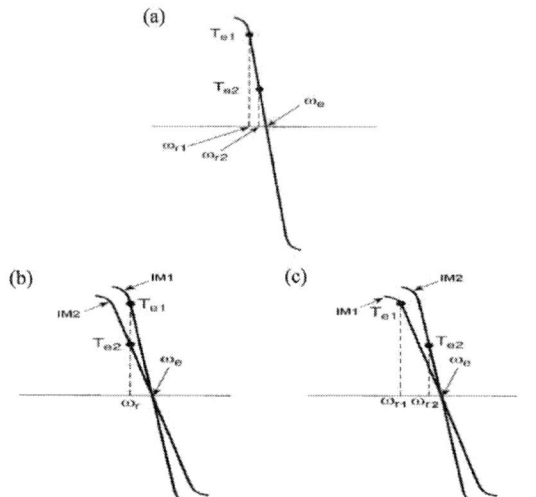

Fig. 1. (a) matched Torque-Speed characteristics of two motor (b),(c) mismatched Torque-Speed characteristics of two motor

Where $\overline{V}_s$, $\overline{I}_s$ and $\overline{\psi}_s$ are stator voltage vector, stator current vector and stator flux vector respectively. R_s is the stator resistance. These Vectors are calculated using abc/dq transformation.

If the applied voltage causes an instantaneous decrease/increase in the stator flux of one motor, stator flux of other motors also decrease/increase, However, the amount of decrease/increase in all motors may not be equal.

Another issue in the control of multi-machine systems is speed control. In speed control applications, it is not always possible to achieve desired operating points for all motors. In practice, there may be some amount of mismatch between machine parameters. Also the motors may have different load torque. Assume that all motors have the same parameters and their characteristics are matched. If load is different, then their speeds will not be equal [1]. This is shown in Fig. 1(a). In this case, the torque of motor 1 is more than that of motor 2 and its speed is less. The difference between the speeds increases as the speed reference decreases. It is due to the fact that the difference between the rotor fluxes increases as the speed reference decreases.

Now, if the parameters of the two motors have some tolerances and their characteristics are mismatched, and the same as before, the torque of motor 1 is higher, then the problem may get better or worse. In Fig. 1(b), the two motors have the same speed, however their torque is different. But in Fig. 1(c) the speed difference between the motors is more.

In multi-machine single-converter systems, usually all the motors are the same and have identical characteristics. With induction motors connected in parallel, the load distribution is influenced only by the correct selection of the torque-speed mechanical characteristic. For squirrel-cage induction motors, no economical method for the adjustment of mechanical characteristic of the ready-made motors exists. For a slip-ring induction motor, the mechanical characteristic can be adjusted

TABLE I

PROPOSED TABLE IN TORQUE CONTROL LOOP

		Motor 2		
	H_{T_e}	-1	0	1
Motor 1	1	-1	-1	0
	0	-1	0	1
	-1	0	1	1

by including rotor resistors [8]. Normally, in such systems, the mechanical load is common between all motors and the control system must keep proper load sharing among them.

III. PROPOSED CONTROL TECHNIQUE

The proposed method is based on the conventional DTC technique of induction motors. In the DTC method, the flux and electromagnetic torque are controlled by adjusting the magnitude and position of the stator flux respectively. This principle is used in the proposed method. The method is explained for a two-machine system and can be extended to multi-machine systems. In the proposed method, to each motor, a conventional single-machine DTC strategy is independently applied. Then, some changes are made in order to make it applicable in a multi-machine system. In the conventional single-machine DTC method, there are two control loops; one for the stator flux control and one for the electromagnetic torque. The proposed method also employs these two control loops but with different procedure in each loop. The procedure of each control loop will be explained hereafter.

A. Electromagnetic torque control loop

The new idea in the torque control loop is to consider motors torque requirement and determine system overall needs before selecting a voltage vector. This is done based on a new look-up table. The table is for the case in which a three-level comparator is used in the torque control loop. The procedure is explained below and shown in Table I. In this table, -1, 0, 1 are outputs of torque error comparator.

• If both motors require a reduction in torque, a vector is applied to decrease torque.

• If no motor requires a torque change, then a vector is applied such that torque is kept constant.

• If both motors require an increase in torque, then a vector is applied to increase torque.

• If one motor requires a decrease in torque but the other one requires no change, then a vector is applied to decrease torque.

• If one motor requires an increase in torque and the other one requires no change, then a vector is applied to increase torque.

• If one motor requires a decrease in torque but the other one requires an increase, then a vector is applied such that torque is kept constant.

The proposed idea is better understood using Fig. 2. In this figure, different possible kinds of torque errors in torque control loop are presented. The black points are the typical position for the electromagnetic torque of the two motors. For

978-1-4244-1871-8/07 $25.00 © 2007 IEEE   1102

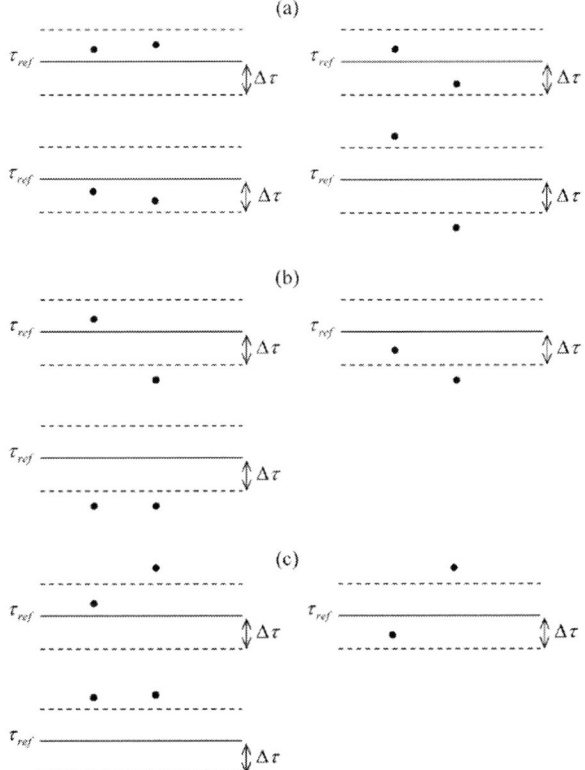

Fig. 2. Different possible cases of torque errors (a) the torque should be kept constant (b) the torque should be increased (c) the torque should be decreased

the conditions depicted in Fig. 2(a), a voltage vector is applied such that the torque does not vary. For the conditions shown in Fig. 2(b), a voltage vector is applied to increase the torque. Also, for the conditions presented in Fig. 2(c), a voltage vector is applied to decrease torque.

Finally, using the output of this table and the output of the stator flux control loop, the appropriate voltage vector is selected based on the conventional DTC switching look-up table.

B. Stator flux control loop

In the stator flux control loop, one should know that the flux of no machine can go beyond its rated value. According to (1), the stator flux of each induction motor sorely depends on the applied voltage. In cases where parameters of the motors are different, or motors load are not the same, stator fluxes will be different. From (1), it can be seen that the stator flux vector only depends on stator resistance among all other parameters. Therefore, when stator resistances are the same, one expects to see the same flux for both machines. This is valid only at steady state, and during transients, the difference between the fluxes may be observed. This difference will also increase as motor speed decreases. For this reason, in cases where motor loads or stator resistances are different, speed reference cannot go below a certain value for speed control applications.

Because, as the stator flux of one motor decreases, its torque generation capability will also decrease.

In these conditions, the mean control strategy cannot be used. Because the flux of one machine can be saturated while its average value is equal to the reference value. Therefore, master-slave control technique can be used for the stator flux control loop. In this way, only the stator flux of one motor is controlled. But the motor with the bigger stator flux magnitude has to be selected as the master, and its stator flux is set to the reference value. To prevent flux saturation at different situations, the master motor may change. Therefore, in the proposed method, switchable master-slave technique is employed for stator flux control.

To accurately select the master motor and prevent flux saturation, an index is needed. The product of stator resistance and electromagnetic torque, i.e. $R_s\tau_e$, is selected as the index. For each motor, this product is calculated. The motor with the smallest $R_s\tau_e$ is chosen as the master motor. In cases where motor parameters are equal, the motor with the lower torque is selected as the master. The duration and amount of difference between indices are important in order to prevent frequent change variation of master motor during transients.

In the conventional DTC, the final step is the selection of voltage vector using a look-up table. The voltage is selected with respect to the section of the stator flux. The proposed method uses the stator flux of the master motor for flux sector selection. The block diagram of the proposed method is shown in Fig. 3.

IV. SIMULATION

The proposed method is simulated for a system with two induction motors. Induction motors parameters are listed in Table II. First both motors are assumed to have the same parameters as IMA in Table II. Simulations are performed using MATLAB-Simulink. The performance of the proposed method is compared with that of the mean control strategy with the conventional DTC technique. In the mean control, speed, electromagnetic torque and the stator flux of each motor are estimated. Then, the average of the estimated values are applied to a DTC method.

TABLE II

INDUCTION MOTOR PARAMETERS

	IMA	IMB
P	4	4
R_s (Ω)	0.3831	0.4597
L_s (H)	0.03334	0.04
R_r' (Ω)	0.2367	0.284
L_r' (H)	0.03334	0.04
L_m (H)	0.03211	0.03853
J (Kgm^2)	0.3	0.3

First, the speed, electromagnetic torque and stator flux magnitude of two machines with different loads are studied at start up and at steady state. The reference speed is $1400rpm$ which is increased in a ramp fashion. The load of motor 1 is $200N.m$ and that of motor 2 is $150N.m$. Fig. 4 and

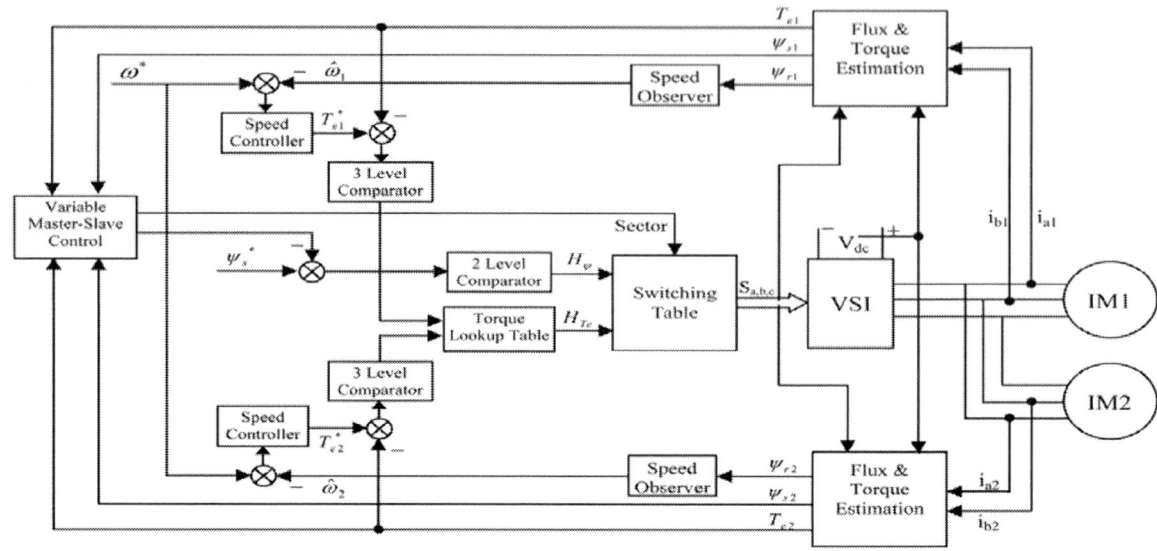

Fig. 3. Block diagram of the proposed control method

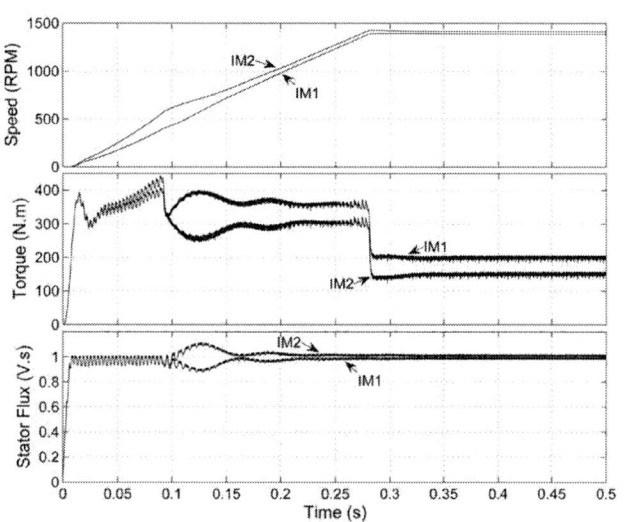

Fig. 4. Speed, electromagnetic torque and stator flux magnitude of two machines with the mean control method at start up

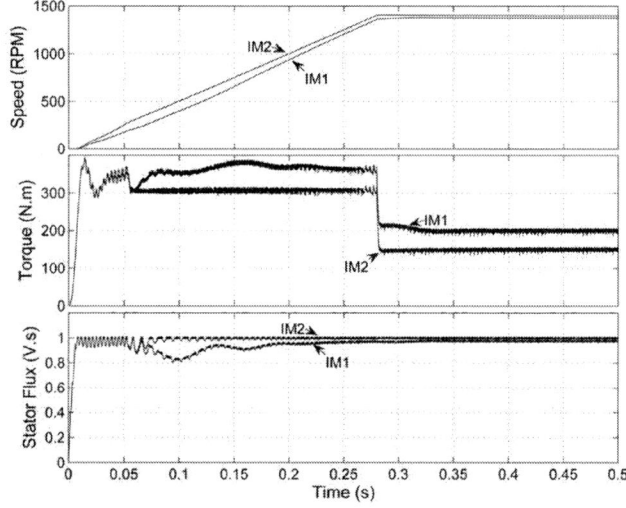

Fig. 5. Speed, electromagnetic torque and stator flux magnitude of two machines with the proposed control method at start up

Fig. 5 show the results of the mean control and the proposed control method respectively. The results indicate that the speed difference between motors is lower at start up when the proposed method is employed. It can also be seen that torque variation is higher in the case of the mean control strategy, while flux of motor 2 also goes over its rated value.

To evaluate the performance of the proposed method when load change occurs, loads of motor 1 and 2 are changed to $200N.m$ and $100N.m$ respectively. Both motors are started with the same load of $150N.m$ and the reference speed of $1400rpm$, and the load change occurs at $t = 0.35s$. The results are depicted in Fig. 6 for the mean control and Fig. 7 for the proposed control method. These figures show that the stator

flux and speed of motor 2 go beyond their rated values when mean control is used. In cases where the reference speed is near the rated speed, the speed can enter into flux weakening region while its stator flux increases.

Finally, the case with different motor parameters is simulated. Parameters of motor 2 are the same as those of IMB in Table II. Again, the reference speed is $1400rpm$ and applied with a ramp fashion. The load of motor 1 is $200N.m$ and that of motor 2 is $150N.m$. The results of the mean control and the proposed control method are shown in Fig. 8 and Fig. 9 respectively. It can be seen that the torque response of mean control has more oscillations as compared to the proposed method.

978-1-4244-1871-8/07 $25.00 © 2007 IEEE

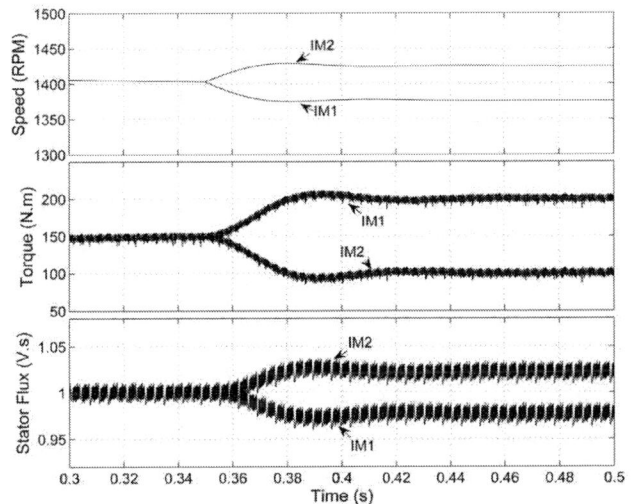

Fig. 6. Speed, electromagnetic torque and stator flux magnitude of two machines with mean control method during load change

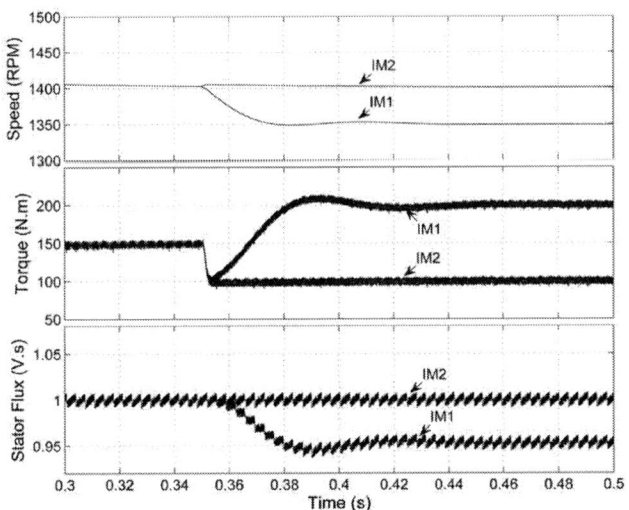

Fig. 7. Speed, electromagnetic torque and stator flux magnitude of two machines with proposed control method during load change

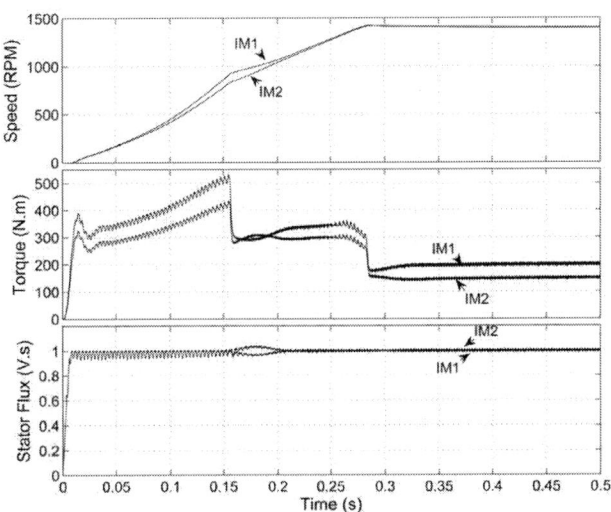

Fig. 8. Speed, electromagnetic torque and stator flux magnitude of two machines with different parameters and mean control method during starting

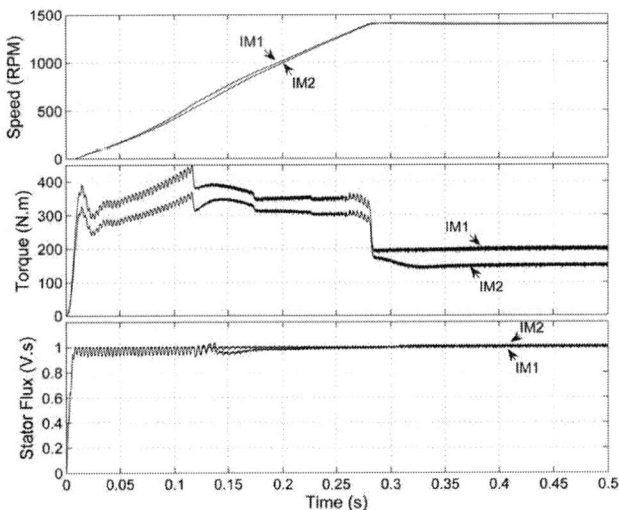

Fig. 9. Speed, electromagnetic torque and stator flux magnitude of two machines with different parameters and proposed control method during starting

A load exchange at $t = 0.5s$ is also applied for this case. The responses are depicted in Fig. 10 and Fig. 11 for the mean control and the proposed control method respectively. The results of these two figures indicate that the proposed method offers better steady state response.

V. CONCLUSION

In this paper, a new method is introduced for a two-machine system which can be easily extended to a multi-machine system. The proposed method is based on conventional DTC of induction motors. This method is very simple and has the advantages of single motor DTC. The principles of both flux and torque control loops are explained. A two-machine system is simulated. Simulation results indicate that the performance of the proposed method in control of two machines

is favorable. Using the proposed method, the behavior of two machines are close even during transients. Furthermore, the proposed method prevents flux saturation of both machines and offers good response at steady state.

REFERENCES

[1] B. K. Bose, *Modern Power Electronics and AC Drives*. Prentice Hall, 1999.
[2] K. Matsuse, H. Kawai, Y. Kouno, and J. Oikawa, "Characteristics of speed sensorless vector controlled dual induction motor drive connected in parallel fed by a single inverter," *IEEE Transactions on Industry Applications*, vol. 40, no. 1, January/February 2004.
[3] P. Escane, M. David, and B. de Fornel, "Optimization of a railway traction system drive control vs slip perturbation," vol. 3. Industry Applications Conference, October 2000, pp. 1909–1916.

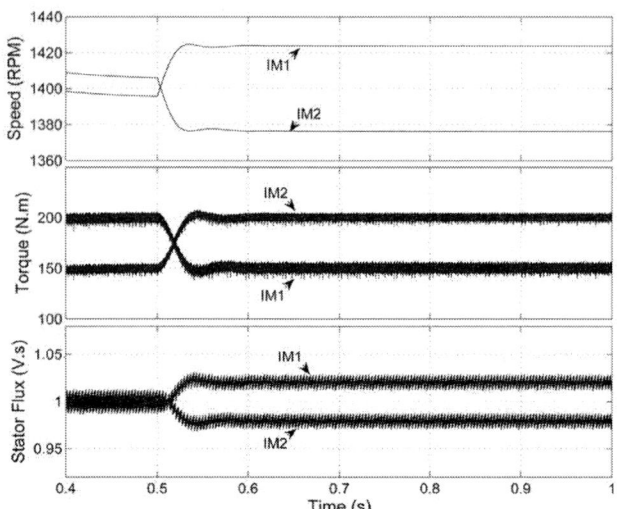

Fig. 10. Speed, electromagnetic torque and stator flux magnitude of two machines with different parameters and mean control method during load exchange

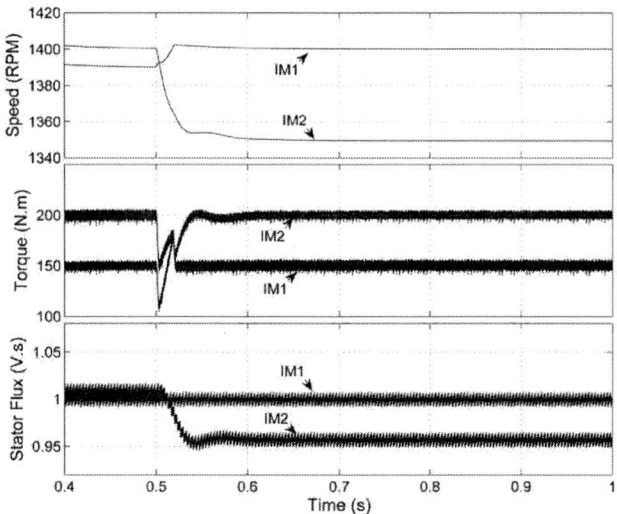

Fig. 11. Speed, electromagnetic torque and stator flux magnitude of two machines with different parameters and proposed control method during load exchange

[4] P. Escane, C. Lochot, M. David, and B. de Fornel, "Electromechanical interactions in a high speed railway traction system comparison between two drive control structures." EPE conference, 1999.

[5] J. Wang, Y. Wang, Z. Wang, J. Yang, Y. Pei, and Q. Dong, "Comparative study of vector control schemes for parallel-connected induction motors," vol. 36. Power Electronics Specialists Conference, 2005, pp. 1264–1270.

[6] Y. Matsumoto, C. Osawa, T. Mizukami, and S. Ozaki, "A stator flux-based vector control method for parallel-connected multiple induction motors fed by a single inverter," in *Proceedings of IEEE Applied Power Electronics Conference and Exposition*, vol. 2, February 1998, pp. 575–580.

[7] Y. Matsumoto, S. Ozaki, and A. Kawamura, "A novel vector control of single-inverter multiple-induction-motors drives for shinkansen traction system," in *Proceedings of Applied Power Electronics Conference and Exposition*, vol. 1, 2001, pp. 608–614.

[8] B. Jeftenic, M. Bebic, and S. statkic, "Controlled multi-motor drives,"

in *International Symposium on Power Electronics, Electrical Drives, Automation and Motion (SPEEDAM)*, 2006, pp. 1392–1398.

New Current Controller for Inverter Fed Medium Voltage Drives with LC Filter

Tomasz Laczynski
University of Hannover
Institute for Drive Systems and Power Electronics
Welfengarten 1, 30167 Hannover, Germany
Email: laczynski@ial.uni-hannover.de

Axel Mertens
University of Hannover
Institute for Drive Systems and Power Electronics
Welfengarten 1, 30167 Hannover, Germany
Email: mertens@ial.uni-hannover.de

Abstract—**In high power medium voltage drives, the switching frequency of power semiconductor devices is restricted because of high switching losses. Such drives often include an LC output filter, introducing a resonant circuit that may be excited for instance by fast control transients. One way to avoid such oscillations is using damping control schemes, which in turn usually require switching frequencies well above the filter resonance. This paper proposes a new predictive stator current controller that avoids the excitation of the filter resonance, performs active damping and allows fast current control while maintaining low switching frequency. Simulation results of a 2.4 kV induction motor drive and experimental results obtained from a 55 kW prototype a.c. drive demonstrate feasibility and good dynamic performance of the proposed control approach.**

I.   INTRODUCTION

Medium voltage high power inverters usually operate at a switching frequency lower than 500 Hz [1-4] in order to minimize the switching losses of the power semiconductors. They are often applied to retrofit of existing fixed speed induction motors with medium voltage variable speed drives, i.e. to achieve energy savings. Because existing motors usually are not designed for inverter supply, the use of LC output filters becomes necessary in order to avoid isolation problems and bearing currents [5]. The filter creates sinusoidal output voltages and thereupon conditions for the machine similar to operation from the grid. The inverter switching frequency that is acceptable for thermal reasons lies only slightly above the resonance frequency of the LC filter. At the same time, a suitable motor current control is necessary that does not excite the filter resonance. Otherwise, sudden changes of the control reference would create considerable weakly damped transient currents [6].

This paper proposes a new predictive stator current control scheme which takes the nonlinear nature of the inverter into account. The controller computes the stator current trajectories for every possible inverter switching state and chooses the one which minimizes a performance index. The approach makes fast motor current control possible while keeping switching frequency at a low level of around 500 Hz. By means of the controller, excitation of filter resonance is avoided and active damping of filter oscillations is achieved.

This paper presents simulation results of the proposed predictive current control method for a 2.4 kV medium voltage drive fed by a Neutral Point Clamped (NPC) three

level inverter. In addition, measurement results of a prototype 55 kW induction motor drive system fed by a two level inverter are presented.

II.   CURRENT CONTROL

A.   System Description

The signal flow graph of the proposed controlling system for the inverter fed induction machine with LC filter is shown in Fig. 1.

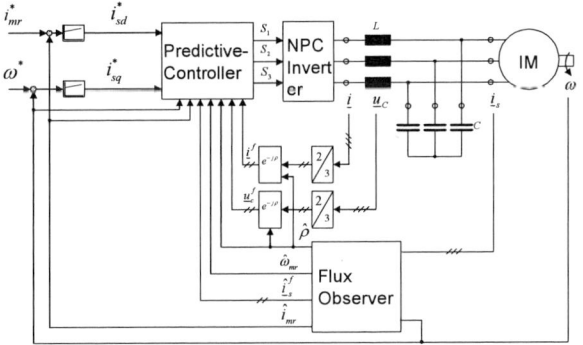

Fig. 1: Signal flow graph of the proposed control structure

The predictive current controller receives its reference value of the complex stator current vector from the flux and speed PI controllers, and computes the switching functions for the inverter using the measured or estimated instantaneous state variables of the motor and filter. The controller makes use of the values of the stator current, the choke current, the capacitor voltage and the rotational speed. Additionally, a flux observer provides values for the magnetizing current, the angle and the angular velocity of the rotor flux. An NPC three level inverter generates 19 different voltage space vectors (Fig. 2). The machine stator currents are the controlled variables. The controller predicts stator current trajectories which come with each of the voltage space vectors by solving a discrete-time state space model of the filter and motor. By means of a weighting function taking into consideration the deviation of the current trajectory from the stator reference current, a performance index is computed. For every sample interval, the switching state having the minimal performance index is identified, and is then generated by the inverter.

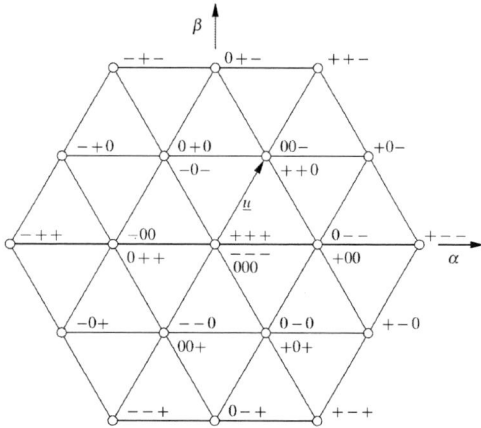

Fig. 2: Voltage space vectors of the NPC three level inverter

B. System Model

The dynamic behavior of an induction machine can be described in field coordinates by the following equations [7]:

$$u_{sd} = R_s i_{sd} + \sigma L_s \frac{di_{sd}}{dt} - \sigma L_s \omega_{mr} i_{sq} + (1-\sigma)L_s \frac{di_{mr}}{dt}, \tag{1}$$

$$u_{sq} = R_s i_{sq} + \sigma L_s \frac{di_{sq}}{dt} + \sigma L_s \omega_{mr} i_{sd} + (1-\sigma)L_s i_{mr} \omega_{mr}, \tag{2}$$

$$i_{sd} = T_r \frac{di_{mr}}{dt} + i_{mr}, \tag{3}$$

$$\omega_{mr} = \frac{i_{sq}}{T_r i_{mr}} + p\omega. \tag{4}$$

The nomenclature used is shown in Tab. 1. The equation of motion is:

$$J\frac{d\omega}{dt} = \frac{3}{2}p(1-\sigma)L_s i_{mr} i_{sq} - m_l. \tag{5}$$

In the above equations an amplitude invariant space vector transformation is used:

$$x_\alpha + x_\beta = \frac{2}{3}\begin{bmatrix} 1 & e^{j\frac{2}{3}\pi} & e^{j\frac{4}{3}\pi} \end{bmatrix}\begin{bmatrix} x_1 \\ x_2 \\ x_3 \end{bmatrix}. \tag{6}$$

The differential equations of the LC filter inductor current space vector are:

$$\frac{di_d}{dt} = \frac{1}{L}u_d - \frac{1}{L}u_{cd} + \omega_{mr}i_q - \frac{R_l + R_c}{L}i_d + \frac{R_c}{L}i_{sd} \tag{7}$$

and

$$\frac{di_q}{dt} = \frac{1}{L}u_q - \frac{1}{L}u_{cq} - \omega_{mr}i_d - \frac{R_l + R_c}{L}i_q + \frac{R_c}{L}i_{sq}. \tag{8}$$

The capacitor voltages are:

$$\frac{du_{cd}}{dt} = \omega_{mr}u_{cq} + \frac{1}{C}i_d - \frac{1}{C}i_{sd} \tag{9}$$

and

$$\frac{du_{cq}}{dt} = -\omega_{mr}u_{cd} + \frac{1}{C}i_q - \frac{1}{C}i_{sq}. \tag{10}$$

Using the above equations a continuous state space system can be formed (11), which allows the computation of the stator current trajectories. The inputs of the model are the inverter output voltages u_d, u_q and the magnetizing current i_{mr}. The prediction is carried out based on the assumption that the motor speed ω, the rotor flux velocity ω_{mr} and magnetising current i_{mr} are constant during the prediction period.

$$\frac{d}{dt}\begin{bmatrix} i_{sd} \\ i_{sq} \\ i_d \\ i_q \\ u_{cd} \\ u_{cq} \end{bmatrix} = \begin{bmatrix} a_{11} & a_{12} & a_{13} & 0 & a_{15} & 0 \\ -a_{12} & a_{11} & 0 & a_{13} & 0 & a_{15} \\ a_{31} & 0 & a_{33} & a_{12} & a_{35} & 0 \\ 0 & a_{31} & -a_{12} & a_{33} & 0 & -a_{35} \\ a_{51} & 0 & -a_{51} & 0 & 0 & a_{12} \\ 0 & a_{51} & 0 & -a_{51} & -a_{12} & 0 \end{bmatrix}\begin{bmatrix} i_{sd} \\ i_{sq} \\ i_d \\ i_q \\ u_{cd} \\ u_{cq} \end{bmatrix}$$
$$+ \begin{bmatrix} 0 & 0 & b_{13} \\ 0 & 0 & b_{23} \\ b_{31} & 0 & 0 \\ 0 & b_{31} & 0 \\ 0 & 0 & 0 \\ 0 & 0 & 0 \end{bmatrix}\begin{bmatrix} u_d \\ u_q \\ i_{mr} \end{bmatrix} = \underline{Ax} + \underline{Bu} \tag{11}$$

The matrix elements of the above state space model are:

$a_{11} = -R_c/(\sigma L_s) - 1/(T_s) - (1-\sigma)/(\sigma T_r)$

$a_{12} = \omega_{mr}$

$a_{13} = -R_c/(\sigma L_s)$

$a_{15} = -1/(\sigma L_s)$

$a_{31} = -R_c/L$

$a_{33} = -(R_l + R_c)/L$

$a_{35} = -1/L$

$a_{51} = -1/C$

$b_{13} = (1-\sigma)/(\sigma T_r)$

$b_{23} = -(1-\sigma)/\sigma\, p\omega$

$b_{31} = -a_{35}$

978-1-4244-1871-8/07 $25.00 © 2007 IEEE   1108

R_s, R_r	stator and rotor resistance
L_s, L_r	stator and rotor self inductance
L_h	magnetizing inductance
T_s	stator time constant, $T_s = L_s/R_s$
T_r	rotor time constant, $T_r = L_r/R_r$
σ	leakage coefficient, $\sigma = 1 - L_h^2/(L_s L_r)$
ω_{mr}	angular velocity of the rotor field
ω	rotational speed
p	number of pole pairs
J	total moment of inertia
m_l	load torque
R_l, R_c	filter inductor and filter capacitance resistance
L, C	filter inductance and capacitance

Table 1: Nomenclature

In order to reduce the computational demand, the current trajectories are predicted by means of a time discrete state space model:

$$\underline{x}(k+1) = \underline{\Phi}\,\underline{x}(k) + \underline{\Gamma}\,\underline{u}(k). \tag{12}$$

The discrete state space model is obtained through discretization of the state space model (11) using the following series:

$$\underline{\Phi}(T) = \sum_{i=0}^{\infty} \frac{\underline{A}^i T_{Step}^i}{i!}, \tag{13}$$

$$\underline{\Gamma}(T) = \sum_{i=1}^{\infty} \frac{\underline{A}^{i-1} T_{Step}^i}{i!} \underline{B}. \tag{14}$$

T_{Step} is the sample period of the discrete prediction model.

C. Predictive stator current controller

An NPC three level inverter generates 19 different voltage space vectors ($z = 1,\dots,19$). Every voltage space vector z causes a specific stator current trajectory. At the beginning of every sample interval $t(i) = T*i$ ($i = 0,1,2,3,\dots$) possible stator current trajectories are predicted by solving (12). Measured instant values of the stator current, the filter inductor current and the filter capacitor voltage provide the initial values. Computation of possible current trajectories $i_{sd}(z,k)$ and $i_{sq}(z,k)$ is carried out at discrete instants $t_{step}(k) = T_{step}*k$ ($k = 0,1,2,3,\dots,k_{max}$). The parameter k_{max} defines the prediction length $T_{pred} = T_{step}*k_{max}$ which has a significant effect on the controller performance.

The resulting trajectory errors are defined as:

$$e(z, k+1) = \sqrt{\left(f_d\left(i_{sd}^*(k) - i_{sd}(z,k+1)\right)\right)^2 + \left(i_{sq}^*(k) - i_{sq}(z,k+1)\right)^2} \tag{15}$$

and are computed for the entire prediction length. The parameter f_d is a weighting factor. The average area between the error function and the abscissa results from:

$$e_A(z, k+1) = \frac{T_{step}}{2}\left(e(z, i+k) + e(z, i+k-1)\right). \tag{16}$$

Summation of the above values yields the performance index:

$$e_{A,tot}(z) = \sum_{k=0}^{k_{max}} e_A(z, i+k). \tag{17}$$

Following the inverter realizes a switching state z with the minimal performance index $e_{A,tot}(z)$ (17).

III. SIMULATION RESULTS

Simulation results for a 2.2 MW medium voltage drive system with an LC filter were obtained using Matlab/Simulink. The drive was fed by an NPC three level inverter. The power semiconductors were modelled as ideal switches and an ideal dc-link with a constant voltage was assumed. The parameters of the simulated model are given in the appendix in Table 2.

A sampling frequency of $f = 1/T = 5$ kHz was used and the prediction length T_{pred} chosen to 800 µs. Fig. 3 shows the magnetising current, speed, motor torque, direct and quadrature components of stator current during start-up and under load with half the rated torque.

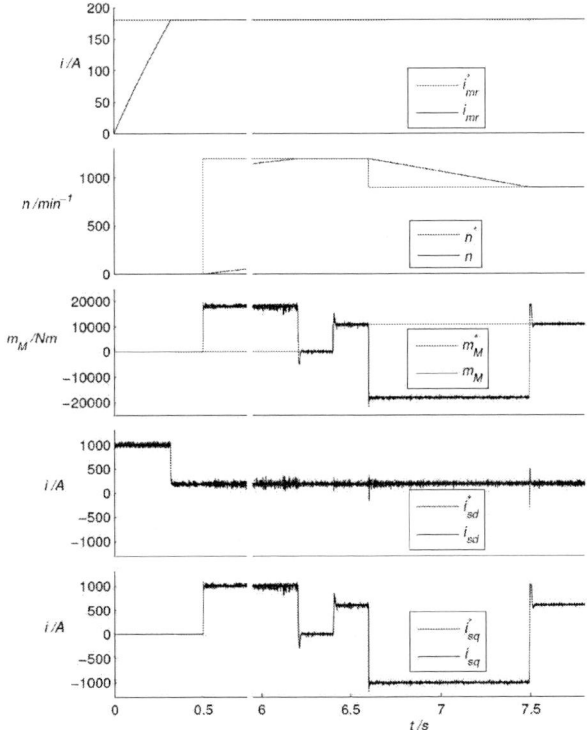

Fig. 3: Magnetising current, speed, torque, direct and quadrature components of stator current (simulation)

978-1-4244-1871-8/07 $25.00 © 2007 IEEE

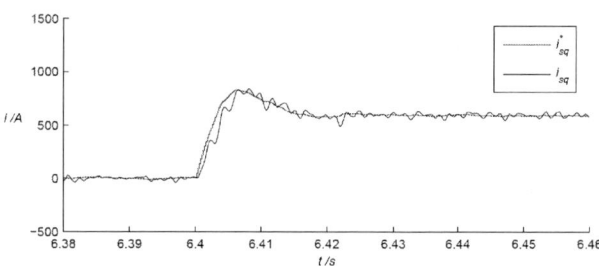

Fig. 4: Quadrature current at load step (simulation)

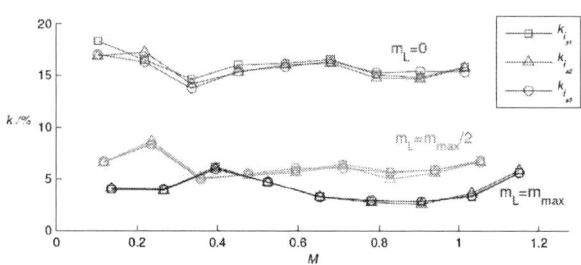

Fig. 5: Total harmonic distortion of the stator currents (simulation)

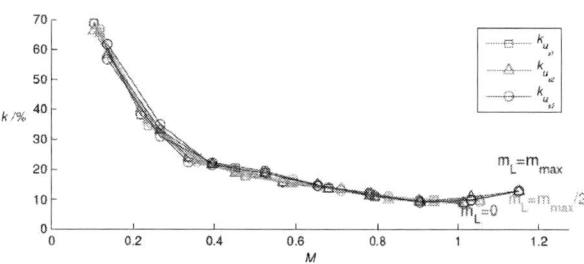

Fig. 6: Total harmonic distortion of the motor phase voltages (simulation)

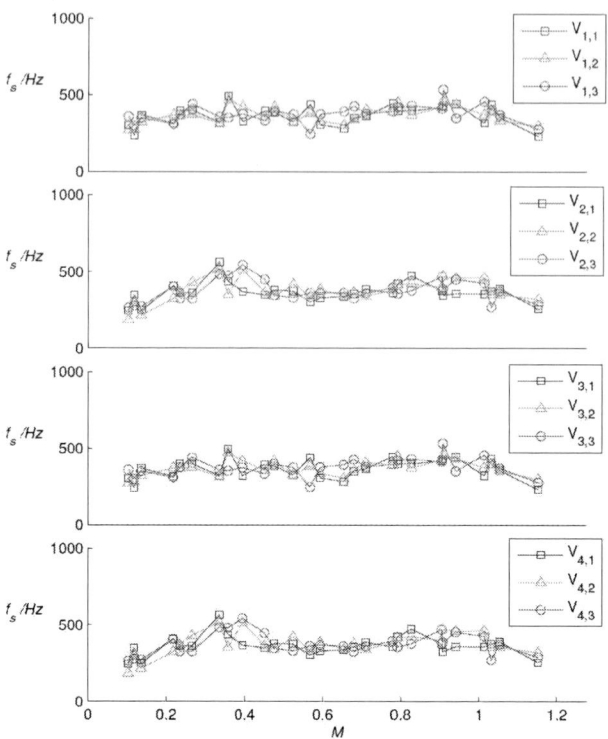

Fig. 7: Switching frequency f_s of the IGBTs (simulation). The first index denotes IGBT number (from top to bottom) and the second index the inverter half bridge (from left to right).

The quadrature current at load step at 6.4 s reaches its maximum value within 6 ms (Fig. 4).

The total harmonic distortion (THD) of the stator currents k_{is123} at rated load torque is about 5% in the interval 0.1,..., 1.15 of the modulation index (Fig. 5). The modulation index is defined as $M = u_1/(U_{dc}/2)$ where u_1 is the fundamental motor phase voltage and U_{dc} the dc-link voltage. The THD of the motor phase voltages k_{us123} is shown in Fig. 6.

In the entire modulation range the maximum switching frequency of all twelve switches is approximately $f_s = 500$ Hz (Fig. 7).

IV. EXPERIMENTAL RESULTS

In order to experimentally verify the performance of the proposed approach, the controller was implemented in a dSPACE DS1103 PPC controller board. The sampling frequency was sat at $f = 5$ kHz. The prediction length was $T_{pred} = 1$ ms and the prediction sampling period $T_{step} = 1/(2f) = 100$ μs. A 55 kW induction machine drive with LC filter was fed by a two level inverter with 540V dc-link voltage. The drive is speed controlled with a d.c. motor delivering the mechanical load. The parameters of the experimental setup are given in the appendix in Table 3.

The dynamic performance of the drive was tested using a reference speed step. Before applying the step, the drive was operated at 20% speed, rated magnetizing current and 10% load torque. The computed motor torque (5) is shown Fig. 8. After applying the reference speed step at 1.496 s, the motor generates a maximal torque of 190 Nm within 1.5 ms. The limitation of the maximum torque to 53 % of the rated torque results from using an inverter with a 540 V dc-link voltage (400 V mains supply) instead of a 932 V dc-link voltage (690 V mains supply).

The quadrature current i_{sq} and its reference value during the speed reference step are shown in Fig. 9. The corresponding stator current is depicted in Fig. 10. The results show that a fast current control is possible without excitation of the LC filter resonance.

The stationary motor current and motor torque during operation at rated speed are presented in Fig. 11 and Fig. 12. The average value of the switching frequency was $f_s = 552$ Hz.

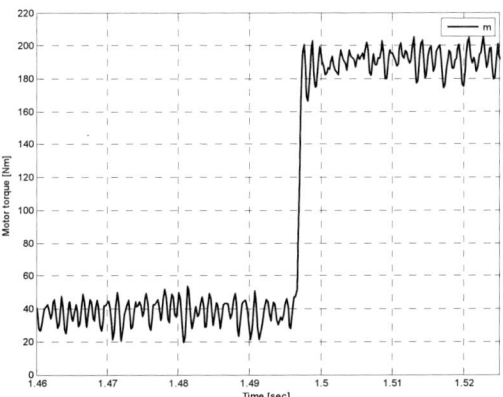

Fig. 8: Motor torque (reference speed step)

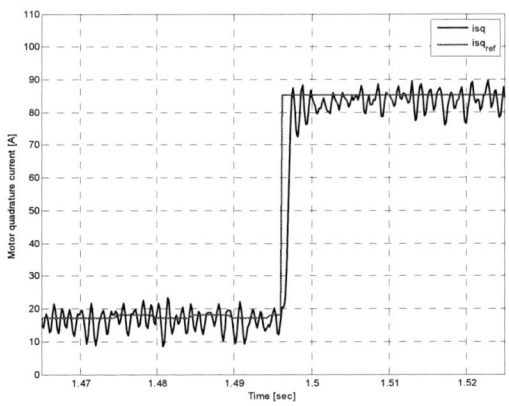

Fig. 9: Quadrature current i_{sq} (reference speed step)

V.   CONCLUSIONS

The proposed predictive current controller enables fast dynamic operation of drive systems with an LC output filter while keeping the switching frequency at a low level (around 500 Hz). Fast stator control is possible, filter resonance is not excited and any filter oscillations due to disturbances are dampened. The proper operation principle of the controller was verified by simulation of a 2.4 kV medium voltage drive and by measurements on a prototype 55 kW drive system.

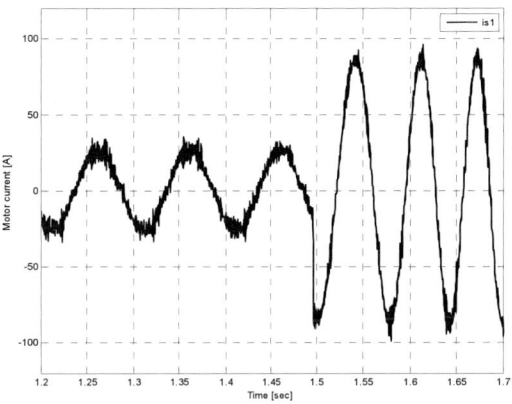

Fig. 10: Stator current (reference speed step)

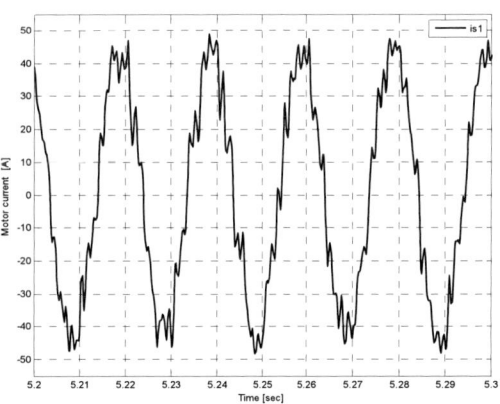

Fig. 11: Stator current (rated speed, maximal torque)

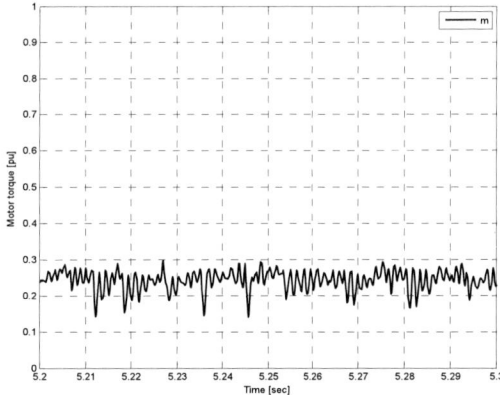

Fig. 12: Motor torque (pu)

978-1-4244-1871-8/07 $25.00 © 2007 IEEE

REFERENCES

[1] A. Nabae, I. Takahashi, H, Akagi, "A New Neutral Point Clamped PWM Inverter", IEEE Trans. Ind. Appl., vol. IA-17, pp. 518-522, 1981.

[2] R. Sommer, A. Mertens, M. Griggs, H.-J. Conraths, M. Bruckmann, T. Greif, "New Medium Voltage Drive Systems Using Three-Level Neutral Point Clamped Inverter With High Voltage IGBT", IEEE IAS 1999, Phoenix, AZ.

[3] P.N. Enjeti, R. Jakkli, " Optimal Power Control Strategies for Neutral Point Clamped (NPC) Inverter Topology", IEEE Trans. Ind. Appl., vol. 28, no. 3, pp. 558-566, 1992.

[4] J. Holtz, "Pulsewidth Modulation for Electronic Power Converters", IEEE Trans. Ind. Appl., vol. 31, no. 5, pp. 1110-1120, 1995.

[5] B.P. Schmitt, R. Sommer, "Retrofit of Fixed Speed Induction Motors With Medium Voltage Drive Converters Using NPC Three-Level Inverter High-Voltage IGBT Based Topology", IEEE ISIE 2001, Pusan, Korea.

[6] J. Holtz, B. Beyer, "Fast Current Trajectory Control Based on Synchronous Optimal Pulsewidth Modulation", IEEE Trans. Ind. Appl., vol. 31, no. 5, pp. 1110-1120, 1995.

[7] W. Leonhard, "Control of Electrical Drives", Springer-Verlag Berlin, 1996.

APPENDIX

The parameters of the used drive systems are listed in Table 2 and Table 3.

U	2.4 kV
I	600 A
n	1190 min^{-1}
f	60 Hz

Table 2: Parameters of the 2.4 kV drive

U	660 V
I	62 A
n	1475 min^{-1}
f	50 Hz
m	357 Nm
J	0.55 kg m^2
P	55 kW
$cos\ \varphi$	0.84
p	2
R_s	114 mΩ
R_r	100 mΩ
L_s	46.932 mH
L_r	56.036 mH
L_h	46 mH
L	3.13 mH
C	39.8 μF
R_l	60 mΩ
U_{dc}	540 V

Table 3: Parameters of the motor and the LC filter (55 kW drive)

Low frequency stability study of a three-phase induction motor

M. Bashir Uddin
Dhaka University of Engg. & Technology
Dept. of Electrical & Electronic Engg.
Gazipur, Dhaka, Bangladesh.
E-mail: bashircse@yahoo.co.uk

Md. Nuruzzaman Pramanik
Dhaka University of Engg. & Technology
Dept. of Electrical & Electronic Engg.
Gazipur, Dhaka, Bangladesh.
E-mail: nzp_bteb@yahoo.com

Sheikh Abu Reza
Dhaka University of Engg. & Technology
Dept. of Electrical & Electronic Engg.
Gazipur, Dhaka, Bangladesh.
E-mail: ecc_iist@yahoo.com

Abstract--A suitable volt/Hz control strategy is proposed for stability improvement in low frequency operation of induction motors fed from variable voltage and variable frequency supply. The analysis is based on small scale perturbation method applied to the linearized model of state-space equations. Stability analyses are carried out through evaluation of eigen values of the linear model characteristic equation. Stability investigation shows that the simultaneous adjustment of stator voltage and frequency provides a convenient means for maintaining stability of induction motor. The results are verified by standard transient analysis method and demonstrate a very close agreement.

I. INTRODUCTION

Technological advancements in the field of variable speed drive have widened the field of application of induction motors. Use of the variable speed drives [1,2] improves the performance of induction motor drives and adds new dimensions to the stability study with other performance criteria of induction motors. There are many problems associated with variable frequency drives that are still being investigated. The stability problem [3,4] is one of the concerns that require investigation of variable voltage and variable frequency induction motor modeling and solution of models by fast computer aided numerical techniques. It has been found that during at low frequency operation [5] an induction motor is lightly damped which can cause sustained oscillation or lead to pull out of step. At lower frequency range there is a tendency towards greater instability, which might be equivalent to the operation at rated frequency with added armature resistance. One approach to improve the stability of an induction motor is to introduce positive damping by simultaneously adjustment of stator voltage and frequency. This approach stems from the logical inference that in volt/Hz modes of operation of induction motor, maximum torque limit can be enhanced by increasing the amplitude of the stator voltage as well as by decreasing frequency at the expense of lowering the speed. Therefore, simultaneous increase in voltage and decrease in frequency can enhance the maximum torque limit, which should be higher than that caused by either voltage change or frequency change individually.

II. MOTOR MODEL

The Park's model is based on transformation of stator variables of induction machine to a reference frame fixed in the rotor. As a result, the time varying inductances from the voltage equations are disappearing. The Park's model [6,7,8,9] of induction motor is applied to the stability analysis. The equivalent circuit of an induction machine can be shown as follows:

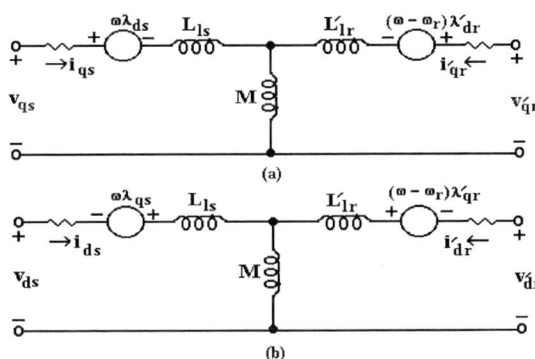

Fig.1: Equivalent circuits of an induction machine Park's Model
(a)   Quadrature-axis equivalent circuit;
(b)   Direct-axis equivalent circuit.

The voltage equations of induction motor in terms of current can be expressed in matrix form as:

$$
\begin{bmatrix} v_{qs} \\ v_{ds} \\ v_{0s} \\ v'_{qr} \\ v'_{dr} \\ v'_{0r} \end{bmatrix} =
\begin{bmatrix}
r_s + \frac{p}{\omega_b}X_{ss} & \frac{\omega}{\omega_b}X_{ss} & 0 & \frac{p}{\omega_b}X_M & \frac{\omega}{\omega_b}X_M & 0 \\
-\frac{\omega}{\omega_b}X_{ss} & r_s + \frac{p}{\omega_b}X_{ss} & 0 & -\frac{\omega}{\omega_b}X_M & \frac{p}{\omega_b}X_M & 0 \\
0 & 0 & r_s + \frac{p}{\omega_b}X_{ls} & 0 & 0 & 0 \\
\frac{p}{\omega_b}X_M & (\frac{\omega-\omega_r}{\omega_b})X_M & 0 & r'_r + \frac{p}{\omega_b}X'_{rr} & (\frac{\omega-\omega_r}{\omega_b})X'_{rr} & 0 \\
-(\frac{\omega-\omega_r}{\omega_b})X_M & \frac{p}{\omega_b}X_M & 0 & -(\frac{\omega-\omega_r}{\omega_b})X'_{rr} & r'_r + \frac{p}{\omega_b}X'_{rr} & 0 \\
0 & 0 & 0 & 0 & 0 & r'_r + \frac{p}{\omega_b}X'_{lr}
\end{bmatrix}
\begin{bmatrix} i_{qs} \\ i_{ds} \\ i_{0s} \\ i'_{qr} \\ i'_{dr} \\ i'_{0r} \end{bmatrix}
\quad (1)
$$

The voltage equations of induction motor in terms of flux linkage can be expressed in matrix form as:

$$
\begin{bmatrix} v_{qs} \\ v_{ds} \\ v_{0s} \\ v'_{qr} \\ v'_{dr} \\ v'_{0r} \end{bmatrix} =
\begin{bmatrix}
\frac{r_s X'_{rr}}{D} + \frac{p}{\omega_b} & \frac{\omega}{\omega_b} & 0 & -\frac{r_s X_M}{D} & 0 & 0 \\
-\frac{\omega}{\omega_b} & \frac{r_s X'_{rr}}{D} + \frac{p}{\omega_b} & 0 & 0 & -\frac{r_s X_M}{D} & 0 \\
0 & 0 & \frac{r_s}{X_{ls}} + \frac{p}{\omega_b} & 0 & 0 & 0 \\
-\frac{r'_r X_M}{D} & 0 & 0 & \frac{r'_r X_{ss}}{D} + \frac{p}{\omega_b} & \frac{\omega-\omega_r}{\omega_b} & 0 \\
0 & -\frac{r'_r X_M}{D} & 0 & -\frac{\omega-\omega_r}{\omega_b} & \frac{r'_r X_{ss}}{D} + \frac{p}{\omega_b} & 0 \\
0 & 0 & 0 & 0 & 0 & \frac{r'_r}{X'_{lr}} + \frac{p}{\omega_b}
\end{bmatrix}
\begin{bmatrix} \psi_{qs} \\ \psi_{ds} \\ \psi_{0s} \\ \psi'_{qr} \\ \psi'_{dr} \\ \psi'_{0r} \end{bmatrix}
\quad (2)
$$

The electromagnetic torque can be expressed as:

$$
T_e = \left(\frac{3}{2}\right)\left(\frac{P}{2}\right)\left(\frac{1}{\omega_b}\right)\left(\psi'_{qr}i'_{dr} - \psi'_{dr}i'_{qr}\right) \quad (3)
$$

III. LINEARIZATION

Linearizing by using small scale perturbation, the differential equations of induction motor become,

$$
\begin{bmatrix}
\Delta v'_{qs} \\[4pt]
\Delta v'_{ds} \\[4pt]
\Delta v'^{e}_{qr} \\[4pt]
\Delta v'^{e}_{dr} \\[4pt]
\Delta T_L
\end{bmatrix}
=
\begin{bmatrix}
r_s + \frac{p}{\omega_b}X_{ss} & \frac{\omega_e}{\omega_b}X_{ss} & \frac{p}{\omega_b}X_M & \frac{\omega_e}{\omega_b}X_M & 0 \\[4pt]
-\frac{\omega_e}{\omega_b}X_{ss} & r_s + \frac{p}{\omega_b}X_{ss} & -\frac{\omega_e}{\omega_b}X_M & \frac{p}{\omega_b}X_M & 0 \\[4pt]
\frac{p}{\omega_b}X_M & s_o\frac{\omega}{\omega_b}X_M & r'_r + \frac{p}{\omega_b}X'_{rr} & s_o\frac{\omega}{\omega_b}X'_{rr} & -X_M i'^{e}_{dso}-X'_{rr}i'^{e}_{dro} \\[4pt]
-s_o\frac{\omega}{\omega_b}X_M & \frac{p}{\omega_b}X_M & -s_o\frac{\omega}{\omega_b}X'_{rr} & r'_r + \frac{p}{\omega_b}X'_{rr} & X_M i'^{e}_{qso}+X'_{rr}i'^{e}_{qro} \\[4pt]
X_M i^{e}_{dro} & -X_M i^{e}_{qro} & -X_M i^{e}_{dso} & X_M i^{e}_{qso} & -2Hp
\end{bmatrix}
\begin{bmatrix}
i^{e}_{qs} \\[4pt]
i^{e}_{ds} \\[4pt]
i'^{e}_{qr} \\[4pt]
i'^{e}_{dr} \\[4pt]
\frac{\Delta\omega_r}{\omega_b}
\end{bmatrix}
\tag{4}
$$

It is convenient to separate out the derivative terms and write equations (4) in the form

$$Epx = Fx + u \tag{5}$$

where

$$(x)^T = \begin{bmatrix} \Delta i^{e}_{qs} & \Delta i^{e}_{ds} & \Delta i'^{e}_{qr} & \Delta i'^{e}_{dr} & \dfrac{\Delta\omega_r}{\omega_b} \end{bmatrix} \tag{6}$$

$$(u)^T = \begin{bmatrix} \Delta v^{e}_{qs} & \Delta v^{e}_{ds} & \Delta v'^{e}_{qr} & \Delta v'^{e}_{dr} & \Delta T_L \end{bmatrix} \tag{7}$$

$$
E = \frac{1}{\omega_b}
\begin{bmatrix}
X_{ss} & 0 & X_M & 0 & 0 \\
0 & X_{ss} & 0 & X_M & 0 \\
X_M & 0 & X'_{rr} & 0 & 0 \\
0 & X_M & 0 & X'_{rr} & 0 \\
0 & 0 & 0 & 0 & -2Hp
\end{bmatrix}
\tag{8}
$$

$$
F = -
\begin{bmatrix}
r_s & \frac{\omega_e}{\omega_b}X_{ss} & 0 & \frac{\omega_e}{\omega_b}X_M & 0 \\[4pt]
-\frac{\omega_e}{\omega_b}X_{ss} & r_s & -\frac{\omega_e}{\omega_b}X_M & 0 & 0 \\[4pt]
0 & s_o\frac{\omega_e}{\omega_b}X_M & r'_r & s_o\frac{\omega_e}{\omega_b}X'_{rr} & -X_M i^{e}_{dso}-X'_{rr}i'^{e}_{dro} \\[4pt]
-s_o\frac{\omega_e}{\omega_b}X_M & 0 & -s_o\frac{\omega_e}{\omega_b}X'_{rr} & r'_r & X_M i^{e}_{qso}+X'_{rr}i'^{e}_{qro} \\[4pt]
X_M i'^{e}_{dro} & -X_M i'^{e}_{qro} & -X_M i^{e}_{dso} & X_M i^{e}_{qso} & 0
\end{bmatrix}
\tag{9}
$$

In linear system analysis it is convenient to express the linear differential equations in the form

$$px = Ax + Bu \tag{10}$$

Equation (10) may be written as,

$$px = (E)^{-1}Fx + (E)^{-1}u \tag{11}$$

which is in the form of (10) with

$$A = (E)^{-1}F \tag{12}$$
$$B = (E)^{-1} \tag{13}$$

It is to be noted that $(E)^{-1}$ is the inverse of E matrix of co-efficient.

IV. STABILITY ANALYSIS BY EIGENVALUES

The range of stability as well as oscillating frequency can be predicted by eigen values determined from the linearized equations of an induction motor. The **u** vector of the linear model equations (10) represents the forcing function of the motor. If **u** is set to zero, the general solution of the force free linear equation becomes,

$$\mathbf{x} = \mathbf{K}\, e^{\mathbf{A}t} \tag{14}$$

Where, **K** vector is determined from the initial conditions. The exponential $e^{\mathbf{A}t}$ represents the unforced response of the system, which is known as state transition matrix. Stability is assumed asymmetrical if all elements of the transition matrix approach zero. The asymmetric behavior of all elements of the matrix occurs, whenever all of the roots of the characteristic equation of **A** is defined by

$$\det(\mathbf{A} - \lambda\mathbf{I}) = 0 \tag{15}$$

In equation (16) **I** is the identity matrix and λ are the roots of the characteristic equation of **A** referred to as eigenvalues.

All the machine parameters and the terminal constraints used in stability studies of this work are in per unit values, given in TABLE 2. The value of each of the motor terminal quantities are varied in a number of steps. The fifth-order equations yield five eigenvalues in each step of calculation. Since the real parts of the eigenvalues are the determining criteria for stability, only the real parts are presented in the graphical form. The straight line parallel to the x-axis passing through zero value of the ordinate forms the boundary between the stable and unstable region of operation. The lower part of the boundary represents the stable region and the upper part signifies the unstable region of operation. The eigenvalues are obtained by using a standard eigenvalue computer routine. A cross check of the computed eigenvalues has also been carried out by the popular software MATLAB [10].

The eigenvalues are obtained by calculating the roots of the characteristic equation given by equations (12). The induction motor operation is described by five state variables and hence five eigenvalues. Two eigenvalues have complex conjugate pairs and one is real value. The real parts of the eigenvalues may be negative or positive. Negative real parts of eigenvalue oscillate and exponentially decay with time indicating stable operation. Positive real parts of eigenvalue oscillate and exponentially increase with time indicates unstable operation.

A. Results

The excursion of eigen values for the rated operating condition of induction motor are depicted in Fig. 2 which indicate that operation becomes unstable at lower frequency. In each step of computation the applied voltage is decreased linearly with frequency. The eigen values are found to cross the boundary at low frequency of 0.116 p. u. (7 Hz) while the corresponding stator voltage is also 0.116 p. u. indicating unstable operation below this frequency.

The effect of amplitude of the stator voltage on stability is illustrated in Fig. 3 by drawing two additional curves. These curves are obtained by decreasing the stator voltage and frequency keeping volt/Hz ratio constant. The reduction of voltage and frequency simultaneously is indicating the induction motor trends to be unstable at lower frequency.

The investigation shows that the stability of the induction motor can be improved by the three modes of operations, termed as, (a) increasing stator voltage at fixed frequency (0.1 p. u.), (b) decreasing frequency at fixed stator voltage and (c) increasing stator voltage and decreasing frequency simultaneously.

Enhanced volt/Hz ratio by increasing voltage at fixed frequency mode:

In Fig. 4, the stator voltage is increased at fixed frequency (0.1 p. u.). So, the induction motor is stable at volt/Hz ratio of 2.5 and the operation drifts to unstable regions below the ratio of 2.5. If we increase the ratio from 2.5 to more, the motor operation becomes more stable.

Enhancing volt/Hz ratio by decreasing frequency at fixed voltage mode:

In this mode, improvement of the stability is achieved by increasing volt/Hz ratio through decreasing frequency at fixed stator voltage. Fig. 5 indicates the induction motor operation is stable at volt/Hz ratio of 2.5 p.u. with reduced frequency (0.04 p.u.) at fixed stator voltage.

Enhancing volt/Hz ratio by increasing voltage & decreasing frequency mode:

In this mode, improvement of the stability is achieved by increasing volt/Hz ratio through enhancing stator voltage and decreasing frequency simultaneously which is evident in Fig. 6. In this figure, the stability is pronounced at a volt/Hz ratio of 2.5 with stator voltage of 0.16 p. u. and at the frequency of 0.06 p. u.(4 Hz).

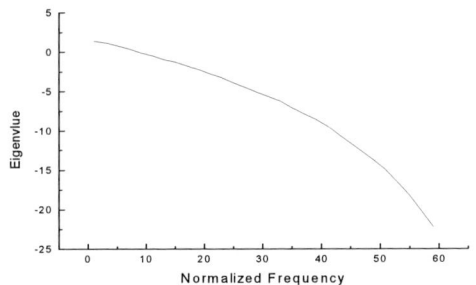

Fig. 2: Illustration of induction motor stability at rated volt/Hz ratio.

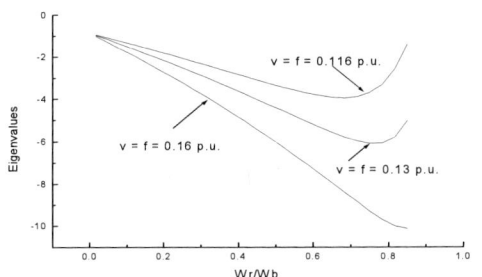

Fig. 3: Effect on stability for decreasing stator voltage and frequency simultaneously with fixed ratio (v/f = 1.0).

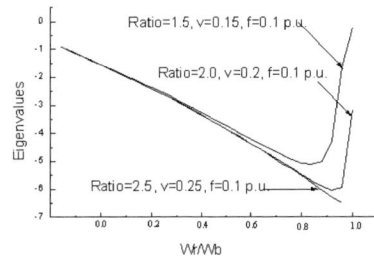

Fig. 4: Increasing stator voltage with fixed frequency (f = 0.1 p. u.) to improved stability.

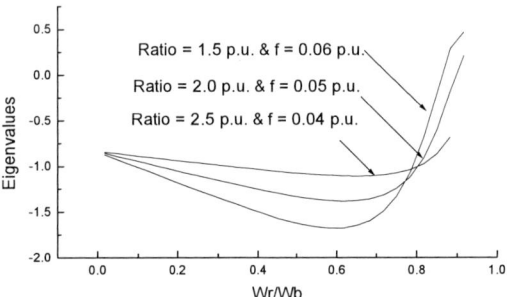

Fig. 5: Decreasing frequency with fixed voltage (v = 0.1 p. u.) to improved stability.

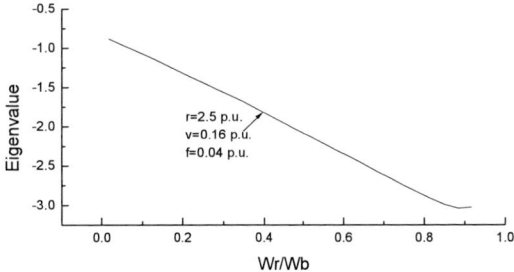

Fig. 6: Increasing stator voltage and decreasing frequency simultaneously to improved stability.

V.   STABILITY STUDY BY TRANSIENT ANALYSIS

The transient analysis of an induction motor expresses the behavioral characteristics of the machine during starting from steady state condition. For stable operation of the induction motor, the oscillation of the developed torque would cease after a certain time. If the oscillation of the torque does not diminish, rather increases with time, the motor operations becomes unstable. So, the transient analysis offers a convenient method for finding the stability behavior of the induction motor. The stability of an induction motor can be studied by changing voltage and frequency simultaneously with fixed volt/Hz ratio.

A. State Space model of an induction machine

The non-linear differential equations (2) in terms of flux linkage as state variables are shown in matrix forms as:

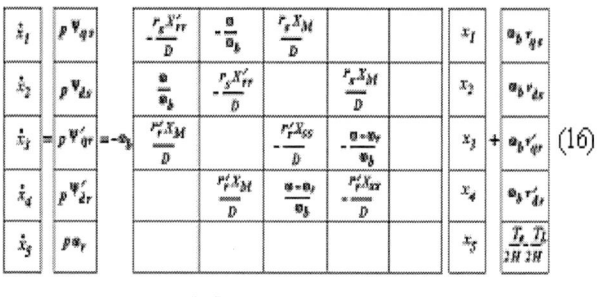

$$[\dot{x}] = \omega_b[A][x] + [B]$$

B. Results

Effect of variable voltage and variable frequency as applied to the stator of the motor causes the operation to be unstable at low frequency range as evident in Fig. 7–10 which demonstrate the operation of the motor with variable voltage and frequency at fixed ratio.

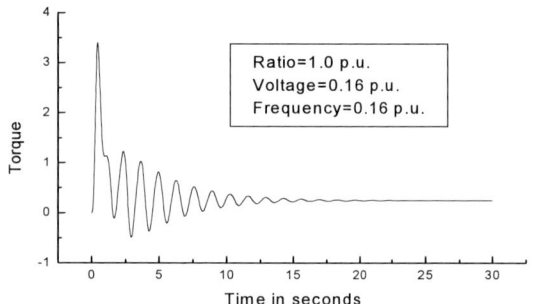

Fig. 7: Decreasing stator voltage and frequency with fixed ratio for stable operation.

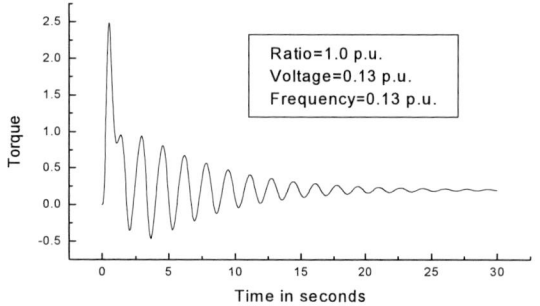

Fig. 8: Decreasing stator voltage and frequency with fixed ratio for stable operation.

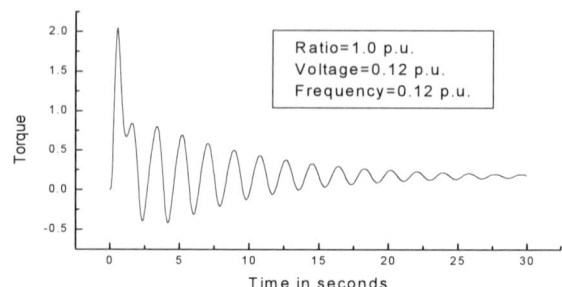

Fig. 9: Decreasing stator voltage and frequency with fixed ratio for motor goes to be unstable.

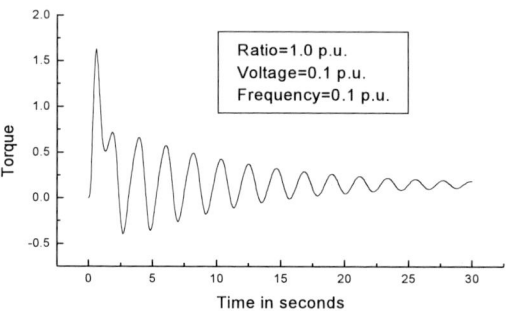

Fig. 10: Decreasing stator voltage and frequency with fixed ratio for motor is unstable.

In similar way as described earlier, the results of stability analysis by transient method may be carried out in three modes, termed as (a) increasing volt/Hz ratio by increasing voltage at fixed frequency, (b) increasing volt/Hz ratio by decreasing frequency at fixed voltage and (c) increasing volt/Hz ratio by increasing voltage & decreasing frequency simultaneously.

Enhanced volt/Hz ratio by increasing voltage at fixed frequency mode:

In this mode, the motor operation tends to be stable. The stability is improved by increasing volt/Hz ratio through increasing stator voltage at fixed frequency. These are shown in Fig. 11–14. Fig. 11-13 indicating stable operation achieved by increasing the voltage from 0.1 to 0.25 p. u. Fig. 14 indicates that higher stator voltage of the motor causes more stable operation at volt/Hz ratio of 3.0.

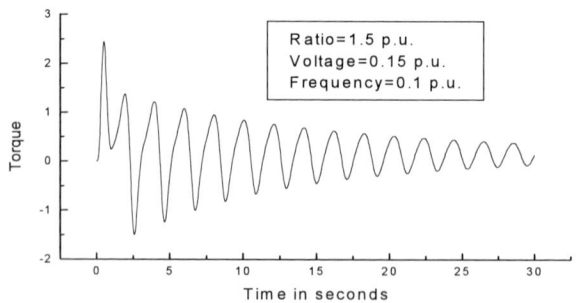

Fig.11: Increasing stator voltage at fixed frequency to improved stability.

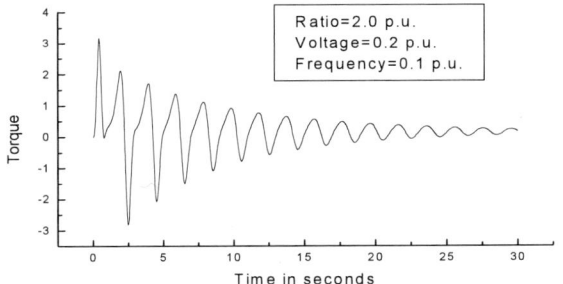

Fig. 12: Increasing stator voltage at fixed frequency to improve stability.

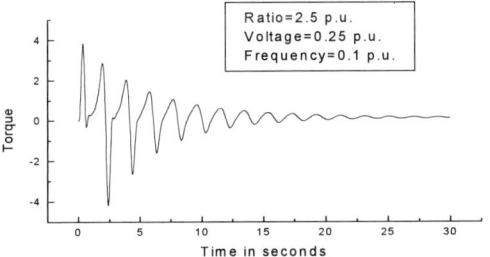

Fig. 13: Increasing stator voltage with fixed frequency achieved stability (v/f=2.5).

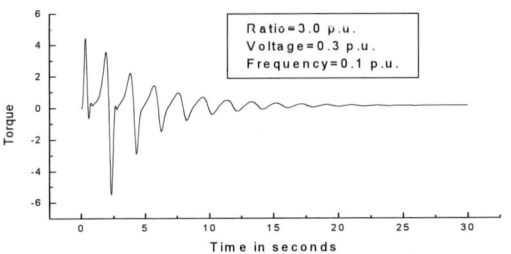

Fig. 14: Motor is more stable with more stator voltage (v/f=3.0).

Enhancing volt/Hz ratio by decreasing frequency at fixed voltage mode:

In this mode, the motor operation tends to be stable by increasing volt/Hzratio. The stability can be improved by increasing volt/Hzratio through decreasing frequency at fixed stator voltage. These are shown in Fig. 15–17. Fig. 15-16 indicates that the induction motor goes to stable operating region for decreasing frequency from 0.06 to 0.05 p. u. at fixed stator voltage. Fig. 17 indicate that in lower range of frequency motor becomes more stable in operation at volt/Hz ratio of 3.0 and frequency of 0.03 p. u.

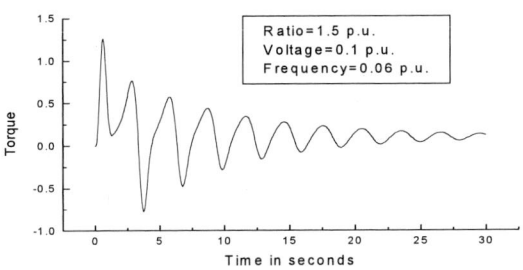

Fig. 15: Decreasing frequency at fixed stator voltage to improved stability.

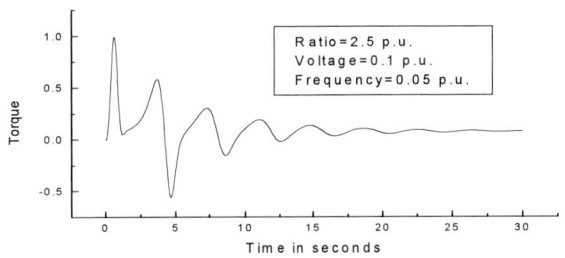

Fig. 16: Decreasing frequency at fixed stator voltage to improved stability.

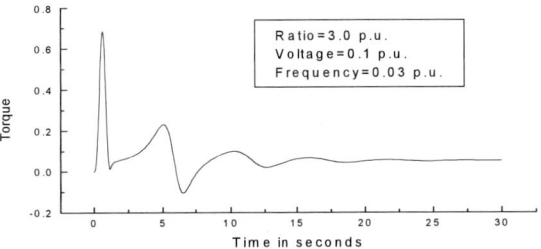

Fig. 17: Motor is more stable with more decreasing frequency (v/f=3.0).

Enhancing volt/Hz ratio by increasing voltage & decreasing frequency mode:

In this mode, the motor operation tends to be stable by increasing volt/Hz ratio. The stability thus improved by increasing volt/Hzrati o through increasing stator voltage and decreasing frequency simultaneously. These are shown in Fig. 18–19 Fig. 1 8 indicates the induction motor operation is stable by increasing stator voltage from 0.1 to 0.16 p. u. and decreasing frequency from 0.1 to 0.06 p. u. simultaneously. Fig. 19 indicates the induction motor operation becomes also stable by increasing stator voltage from 0.1 to 0.2 p. u. and decreasing frequency from 0.1 to 0.08 p. u. simultaneously.

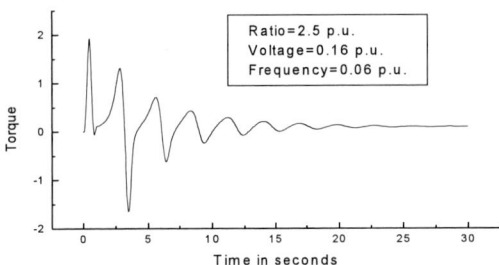

Fig. 18: Stability achieved by increasing stator voltage and decreasing frequency simultaneously.

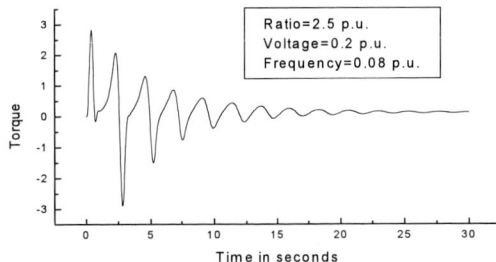

Fig. 19: Stability achieved by increasing stator voltage and decreasing frequency simultaneously.

So, it is evident, the higher stability of the induction motor is achieved at low frequency by enhancing volt/Hz ratio in different ways which are also shown in the TABLE 1.

TABLE 1
Comparison of the Result by Standard Eigen Value & Transient Analysis

Modes of Operations	Result by Eigen Value	Result by Transient Analysis	Remarks
Rated voltage & Frequency	Stable	Stable	
Decrease voltage & frequency at v/f=1.0, v=0.1 & f=0.1	Unstable	Unstable	
Increase volt/hertz ratio from 1.0 to 2.0 by increasing voltage (0.1 to 0.2) at fixed frequency (0.1)	Unstable	Unstable	
Increase volt/hertz ratio from 2.0 to 2.5 by increasing voltage (0.2 to 0.25) at fixed frequency (0.1)	Stable	Stable	The obtained result by standard Eigen value method & Transient Analysis method both are same. So, the result is satisfied.
Increase volt/hertz ratio from 1.0 to 2.0 by decreasing frequency (0.1 to 0.05) at fixed voltage (0.1)	Unstable	Unstable	
Increase volt/hertz ratio from 2.0 to 2.5 by decreasing frequency (0.05 to 0.04) at fixed voltage (0.1)	Stable	Stable	
Increase volt/hertz ratio from 1.0 to 2.5 by increasing voltage (0.1 to 0.16) & decreasing frequency (0.1 to 0.06) simultaneously.	Stable	Stable	

TABLE 2
Induction Machine Parameters

Machines Parameters	Value	Per Unit Value
Horse Power (Hp)	50 hp	-
Voltage (V_L)	460 V	-
Frequency (Hz)	60 Hz	-
Stator Resistance (r_s)	0.087 Ω	0.015336
Stator Reactance (X_{ls})	0.302 Ω	0.053235
Mutual Reactance (X_M)	13.08 Ω	2.30569
Equivalent Rotor Resistance (r'_r)	0.302 Ω	0.040191
Equivalent Rotor Reactance (X'_{lr})	0.228 Ω	0.053235
Moment of Inertia (J)	1.662 Ω	-

VI. CONCLUSION

The stability analysis of variable voltage and variable frequency induction motor have been carried out with an aimed at exploring the advantages of simultaneous adjustment of applied voltage and frequency to improve the stability of the induction motor. A control strategy has been suggested to improve the stability of induction motor by changing the volt/Hz ratio. It has been found that in volt/Hz mode of operation of an induction motor the maximum torque limit increases with increased stator voltage or decreased frequency. Simultaneous increase in stator voltage and decrease in frequency significantly enhance the maximum torque limit. This technique of controlling the maximum torque limit has established the fact that the effective damping may be added in case of sustained oscillations or even in the unstable region of motor operation. The results of the investigation also indicate that this method of control strategy inherits the features of practical importance that may, in some applications, offer a convenient means of stabilizing induction motors at low speed operation. To justify the results of stability analysis of induction motor obtained by standard eigen values is cross checked by transient analysis method. The results in the both the cases demonstrate a very close agreement as evident from the analysis. The effects of non-linearity due to core saturation and eddy currents are not considered in this analysis. Therefore, the analysis and conclusion illustrated of this work are valid for the linear range of operation of the induction motor.

REFERENCES

[1] H. Le-Huy and L.A. Dessaint, "An adaptive current control scheme for PWM motor drives: analysis and simulation:, IEEE Trans. on Power Electronics, vol. 4, no. 4, pp. 486-495 October 1989.

[2] P.M. Hart and W.J. Bonwick, "Harmonic Modeling of induction machines , IEE proc. vol. 135, pt. B no.2, pp. 52-58, March 1988.

[3] R.A. Turton and G.R. Slemon, "Stability of induction motors supplied from current source inverters , IEEE Power Engg. Society, Summer Meeting, Los Angles, Paper no. F78-471-1.

[4] A. Blondel, Transactions of the St. Louis Electrical Congress, 1904.

[5] F. Fallside and A.T. Wortley, "Steady-State Stability of Variable-Frequency Induction Machine Drives , Electronic Letters, vol. 3, No. 7, pp. 334-335, July 1967.

[6] C.A. Nickle and C.A. Pierce, "Stability of induction machine-Effect of Armature Circuit Resistance, AIEE Trans. vol. 149, pp. 338-350, January 1930.

[7] R.H. Park, "Two-reaction theory of induction machines-I , Trans. A.I.E.E., vol. 48, no. 2, pp. 716-227, July 1929.

[8] R.H. Park, "Two-reaction theory of induction machines-II , Trans. A.I.E.E., vol. 52, p. 52, 1933.

[9] Krause P. C., Analysis of Electric Machinery, McGraw Hill, New York, 1986.

[10] Matlab software, Mathwork Inc. U.S.A.

Modeling of Battery Charging Wind Turbines

Justin Reed, Giri Venkataramanan, Jonathan Rose
University of Wisconsin
Department of Electrical and Computer Engineering
1415 Engineering Drive, Madison, WI 53706
Email: jkreed@ieee.org

Abstract—**Small scale wind turbines are typically used in battery charging systems in off-grid and remote applications to provide electric power to small load locations such as homes, cabins and instrumentation stations. This paper is devoted to presenting the results from the modeling case-study of such a turbine. Analytical models, computer simulations and experimental results are presented.**

I. INTRODUCTION

Environmental concerns including the growing consensus on global climate change has prompted a spurt of growth in the application of large scale wind power generation applied in wind-farms across the world [1, 2]. The engineering issues accompanying this growth continue to be addressed by the international technical community [3-5]. On the other hand, small scale wind turbines used for battery charging systems are often applied stand-alone in off-grid remote and rural locations [6].

Large-scale and small-scale wind turbines are topologically quite different from each other. Hence the analysis and design principles of the common large-scale turbines do not directly scale down to smaller scale machines. For example, the large-scale turbines often utilize squirrel cage or doubly-fed induction machines operated typically in conjunction with a gearbox. A number of small-scale electric machines, however, typically use axial flux permanent magnet designs with small radii applied without a gearbox. Sensitivity to cost, complexity and reliability generally mean that the mechanical design and architecture, control systems and power electronics systems, tend to be passive. There are very few definitive technical publications that have focused on the design and performance of small scale wind turbines [7, 8]. This paper is aimed at presenting the modeling results of the electrical system of such a small scale wind turbine suitable for battery charging systems.

The prototype turbine that is presented in this study is of a design promoted by Scoraig Wind Electric from Scotland, which consists of a direct-drive, sub-kW battery-charging wind turbine system [9]. The design of the turbine is simple and permits its construction by relatively untrained personnel working as a team within a few days. A photograph of the prototype turbine is presented in Fig. 1. The prototype was constructed by a team of freshman students in engineering at the University of Wisconsin-Madison in a semester long introductory course in engineering. The components of the prototype turbine may be divided into four major categories:

(a) a tower tall enough to provide unrestricted access to wind; (b) a propeller to harvest the kinetic energy from the wind and convert it into rotational energy; (c) a tail-furling system that directs the propeller into the wind at low speeds to maximize energy harvest and directs the propeller away from the wind at high speeds to limit turbine velocity; and (d) an electrical system consisting of the alternator and a rectifier to convert the rotational energy into electrical energy stored in batteries.

The tower is typically of tilt-up type, consisting of steel pipes hinged to a foundation and held erect using guy-wires of suitable strength anchored to the ground. The propeller consists of three equally spaced blades hand-carved from knot-free soft-wood that roughly follows a tapering width, tapering thickness, NACA 4412 aerofoil [10]. The passive tail-furling system operates by balancing the thrust acting on the propeller against the drag on the tail-vane and the weight of the tail-boom. While the modeling details of the tower, propeller, and the furling system are beyond the scope of this paper, a detailed modeling study of the electrical system is presented in the following sections.

In the following section a brief review of the wind turbine power flow components are presented. In Sections III and IV, the construction details of the alternator and its electrical circuit model are developed. The circuit solution model for the battery charging system is developed in Section V. Selected results from laboratory experiments are presented in Section VI, followed by a brief concluding section.

Fig. 1: Prototype Scoraig wind turbine

978-1-4244-1871-8/07 $25.00 © 2007 IEEE

II. WIND TURBINE POWER FLOW

Wind turbines operate on power flow principles and obey the conservation principles. An abstract sketch of the power flow components are illustrated in Fig. 2. In the turbine, a fraction of the power carried by the wind P_{wind} is harvested as P_{out}.

The power carried by the wind may be derived from the kinetic energy of moving air as shown in (1). The power developed by the propeller $P_{propeller}$ may be determined from the torque generated by the blade sections shown in (2), representing the mechanical power at the turbine shaft. P_{loss} represents the electrical circuit losses, which consist of resistive losses and diode junction losses represented by the two terms in (3). The power fed into the battery is determined by the battery terminal voltage V_{dc} and output current I_{dc}.

$$P_{wind} = \frac{\rho_{air}}{2} \cdot A \cdot v_{wind}^3 \tag{1}$$

$$P_{propeller} = n_{blades} \cdot \omega_r \cdot \int_{r_{in}}^{r_{out}} T(r) \cdot dr \tag{2}$$

$$P_{loss} = \sum I_{ph}^2 R_{ph} + \sum V_d I_{dc} \tag{3}$$

$$P_{out} = V_{dc} \cdot I_{dc} \tag{4}$$

Where

ρ_{air}:	1.225 kg/m³	Density of air
A:	4.52 m²	Swept area of the wind turbine
I_{dc}:	10 A max.	Rectifier output current
I_{ph}:	5 A max.	Electrical phase current
r_{in}:	0.2 m	Inner radius of the blade airfoil shape
r_{out}:	1.2 m	Outer radius of the blade airfoil shape
R_{ph}:	0.42 Ω	Electrical phase resistance
v_{wind}:	5.5 m/s max.	Wind speed
V_d:	0.775 Ω	Diode forward voltage
V_{dc}:	24.4 V min.	Battery terminal voltage

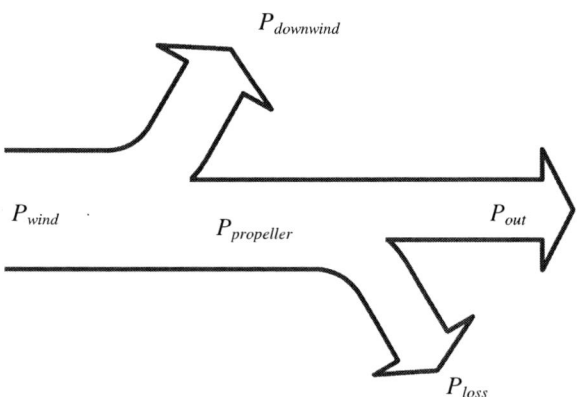

Fig. 2: Energy flow diagram illustrating various power components in a wind turbine electric generator

III. ALTERNATOR CONSTRUCTION

A photograph of the alternator is shown in Fig. 3(a). The machine comprises one stator between two rotors. The alternator consists of five phases and each phase consists of two parallel paths realized using ten rectangular concentrated coils, spaced 36 mechanical degrees apart from each other. The stator windings are embedded in a fiberglass epoxy resin composite as shown in the photograph on Fig. 3(b). The 12 pole rotor consists of 24 NdFeB magnets glued onto the opposite faces of two steel discs that provide structural support and also provide a path for the magnetic circuit. Together, the rotor consists of 6 pairs of poles, with each pole pair consists of four half-segments of magnets. The magnetic face of the rotor discs are also embedded in the fiberglass epoxy resin composite to provide structural protection and weather proofing for the magnets. A photograph of one the rotor plates is shown in Fig. 3(c). The complete absence of stator iron and the large air-gap eliminates all cogging torque and iron-loss induced drag in the machine.

(a)

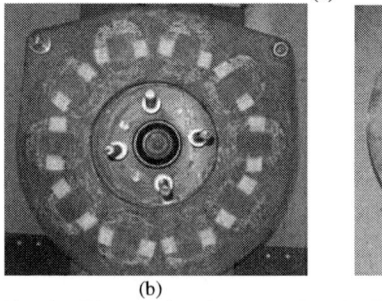

(b)

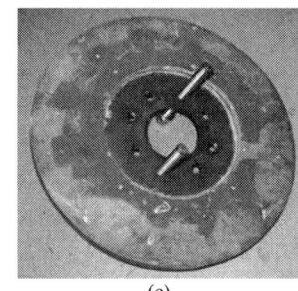

(c)

Fig. 3: Photographs of construction details of the alternator (a) overall assembly on the tower (top rotor partially removed for clarity); (b) permanent magnets on steel rotor discs embedded in fiberglass epoxy composite; and (c) phase coils embedded in fiberglass epoxy composite stator assembly.

IV. ALTERNATOR ELECTRICAL MODEL

A. Magnetic circuit

A two-dimensional sketch of the magnetic circuit of a pole pair is illustrated in Fig. 4. As may be observed from the figure, the nonferrous stator appears a large air-gap to the magnetic circuit. In order to develop a reasonable flux density, the large air-gap requires a large magnetizing force and thus thick NdFeB magnets. A two dimensional flux-plot of the magnetization of a magnetic half-section obtain using a finite element analysis is shown in Fig. 5 [11]. It may be observed that there is negligible flux fringing and hence a first-order model may be used to determine the airgap flux density B_{gap}.

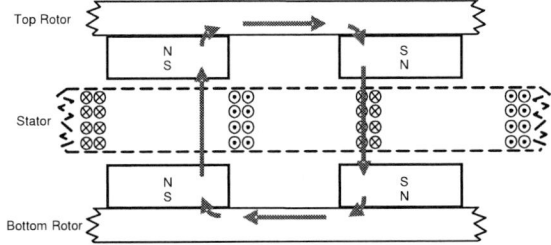

Fig. 4: Magnetic circuit of prototype electric machine, illustrating a flux path.

Fig. 5: FEM results of alternator flux behavior

Neglecting MMF drops due to the steel plate, Ampere's Law is applied around the path illustrated in Fig. 4,

$$4H_c l_m + 2H_g l_g = 0 , \qquad (5)$$

where

H_c:	-955 kA/m	Magnet coercivity,
H_g:	868 kA/m	Gap magnetic field intensity,
l_g:	22 mm	Length of gap,
l_m:	10 mm	Length of magnet.

Using the properties of the magnet, the air-gap flux density may further be determined as [12],

$$B_m\left(H_m\right) = B_r - \frac{B_r}{H_c} \cdot H_m \qquad (6)$$

$$B_{gap} = \frac{2\mu_0 \cdot B_r \cdot H_c \cdot l_m}{2\mu_0 \cdot H_c \cdot l_m - B_r \cdot l_g} = 0.587 \, \text{T} , \qquad (7)$$

where

B_r:	1.27 T	Magnet coercivity,
μ_0:	$4\pi \cdot 10^{-7}$ H/m	Free-space permeability.

B. EMF Generation

Once the air-flux density is known, the rotational EMF generated by the machine may be determined using Faraday's law as:

$$V_{ph} = k_e \cdot \omega_r \cdot \cos\left(P_p \cdot \omega_r \cdot t - \frac{2k\pi}{5} \right), \qquad (8)$$

$$k_e = \frac{\sqrt{2}N_s \cdot N_t \cdot a_m \cdot B_{gap}}{\pi} , \qquad (9)$$

where

k:	0,1...4	Index of electrical phase,
ω_r:	(variable)	Rotor speed,
N_s:	1	Number of series coils,
N_t:	150	Number of turns/coil,
P_p:	6	Number of pole-pairs.

C. Electrical parameters

Based on the geometric parameters of the phase windings, the phase resistance R_{ph} and inductance L_{ph} may be determined analytically as,

$$R_{ph} = \frac{\rho_{cu}}{\pi \cdot r_{wire}^2} \cdot l_{wire} = 0.50 \, \Omega , \qquad (10)$$

$$L_{ph} = 10^{-7} \cdot N_t^2 \cdot r_{coil} \cdot P' = 1.4 \, \text{mH} , \qquad (11)$$

where

ρ_{cu}:	17 nΩ-m	Resistivity of copper,
l_{wire}:	32 m	Wire length,
r_{wire}:	0.59 mm	Wire radius,
r_{coil}:	31.7 mm	Mean radius of coil,
P':	20.11 H/m-turn^2	Geometric parameter.

The L_{ph} estimate in (11) is obtained by approximating the rectangular coil shape as an equivalent circular shape based on equal inner and annular areas. The P' parameter is found in lookup tables of [13] and is dependent upon coil dimensional ratios.

V. BATTERY CHARGING SYSTEM

A simplified schematic of the battery charging system is shown in Fig. 6. An isolated neutral point is formed by connecting the return terminal of each of the coils together. The remaining phase terminal of each of the coils is terminated in a diode half bridge feeding the battery.

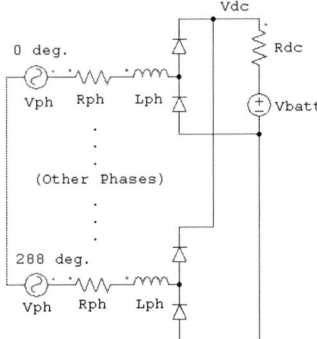

Fig. 6: Simplified schematic of charging system illustrating the interconnection of the alternator rectifier and the battery.

The battery model includes the internal battery resistance, typically about 1 ohm for the deep cycle lead-acid batteries often used in these applications. All system modeling in this paper assumes a battery resistance of 1.2Ω. The behavior of passive diode bridge rectifiers, particularly at variable operating frequencies have generally been difficult to model [14]. The simplifying approximations applied in [14] are inapplicable in this system. Therefore a computer simulation model for the system is used to develop an insight into the operation, and an empirical analytical model is developed on the basis of the simulations.

A. Computer simulations

A computer simulation model for the system was developed using PSIM [15]. Fig. 7 illustrates the phase current and dc terminal voltage waveforms obtained using the simulation under three different conditions: (a) assuming negligible phase impedance, (b) negligible phase reactance (c) actual phase impedance.

As may be observed from Fig 7(a), at any instant of time, the phases generating the largest positive and negative voltages force themselves to conduct current and charge the battery under zero phase impedance conditions. Furthermore, it may be observed from Fig. 7(b), that in the presence of phase resistance in the windings, during transition intervals when the current conduction is handed over from one phase to another phase, a linear fall and rise in currents are observed, and a corresponding dip in the output voltage is observed. In Fig. 7(c), the effect of phase reactance is observed. The commutation interval becomes even more pronounced to the inductive effect thereby delaying the current extinction in the outgoing phase and current inception in the incoming phase. These effects will be used as a basis for developing the average output voltage and charging current in the following subsection.

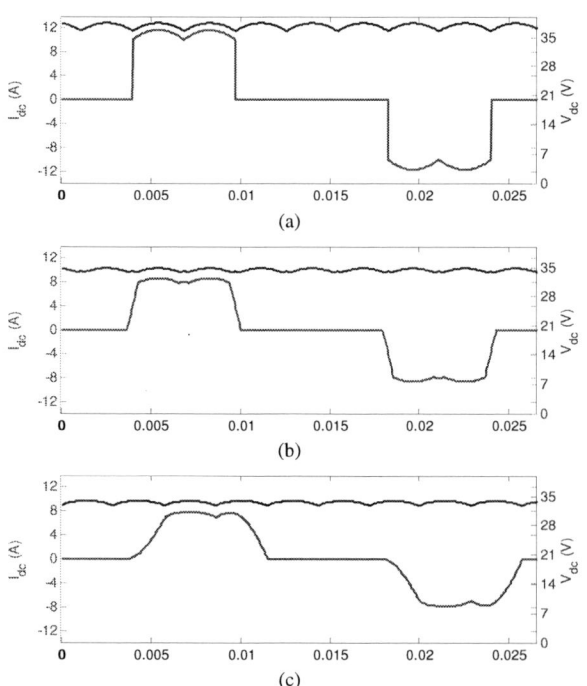

Fig. 7: Battery terminal voltage and phase current waveforms with (a) neglected phase impedance, (b) neglected phase reactance (c) actual phase impedance.

B. Analytical solutions

The battery terminal voltage as a function of the rotor speed due the rectification of phase voltages corresponding to Fig. 7(a) may be determined through a piecewise integration of appropriate segments of the ac voltage as:

$$V_{ave} = \frac{5k_e\omega_r\sqrt{10+2\sqrt{5}}}{4\pi} \cdot \left[\int_0^{\pi/5}\cos\left(\theta - \frac{\pi}{10}\right)d\theta + \int_{\pi/5}^{2\pi/5}\cos\left(\theta - \frac{3\pi}{10}\right)d\theta\right] \quad (12)$$

The effect of resistive commutation and inductive commutation may be modeled as correction terms that account for lost volts from (12). These correction terms follow from determination of length of commutation intervals and the voltage at the bridge terminals during the commutation intervals [16]. Once an accurate enough value for the average battery terminal voltage is obtained, the charging current is determined using Fig. 8 as follows:

$$I_{dc} \approx 2\frac{V_{ave}(\omega_r) - 2V_d - V_{dc}}{3.4R_s + R_{dc} + 2\frac{5}{\pi}L_sP_p\omega_r} \quad (13)$$

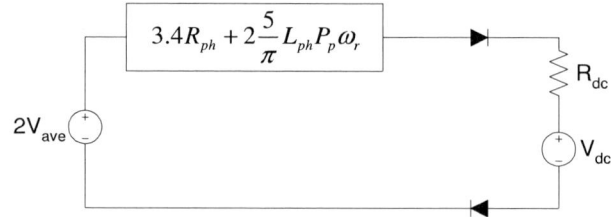

Fig. 8: Simplified schematic of battery charging loop.

VI. LABORATORY TEST RESULTS

In order to verify the analytical and simulation results developed in the previous section, a laboratory test-bed was constructed to obtain experimental measurements. Fig. 9 illustrates the photograph of the experimental test-bed. The alternator shaft may be spun at various speeds using the three phase drive motor fed by a variable frequency inverter. Extensive measurements were carried out to verify the operation of the battery charging system. The experimental test setup uses known R_{dc}=1.2 Ω and V_{batt}=24.4 V. Selected results of measured parameters from the test system are shown in Table I.

Fig. 9: Photograph of the laboratory test setup

Fig. 10 illustrates one half cycle of typical phase current waveform obtained from the experimental measurement that may be compared with the simulation result in Fig. 7(c). Fig. 11 illustrates the variation of the output power as a function of rotor speed obtained from the analytical model, compared against results from the computer simulations and experimental measurements.

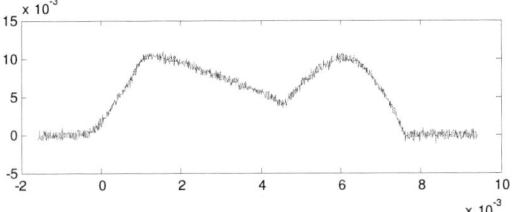

Fig. 10: Phase current waveforms from the laboratory experimental set-up

TABLE I. ANALYTICAL PREDICTION AND EXPERIMENTAL MEASUREMENT OF MACHINE PARAMETERS

Parameter	Analytical Prediction	Measurement
B_{gap}	0.587 T	0.57 T
k_e	0.656 V/(rad/sec)	0.574 V/(rad/sec)
R_{ph}	0.50 Ω	0.42 Ω
L_{ph}	1.44 mH	1.3 mH

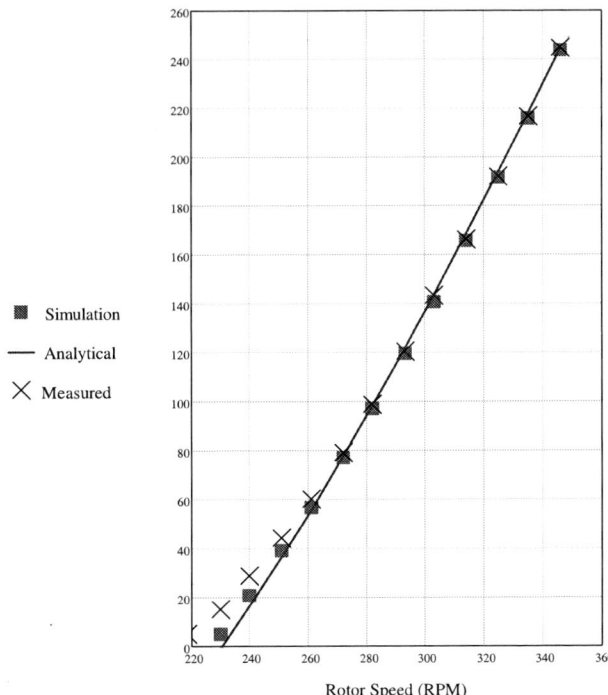

Fig. 11: Variation of output power (W) as a function of alternator speed.

VII. CONCLUSIONS

This paper has presented the electrical modeling features of a small scale wind turbine used for battery charging applications. The geometric construction features of the alternator used with the wind turbine are used to develop an electrical performance model. The analytical results are correlated with experimental measurements and computer simulations. The model may be used to predict the electrical

output, mechanical input and electrical losses in the machine and integrated with the aerodynamic model of the propeller and the tail-furling system to develop an overall model for the turbine system. This is the subject of ongoing investigations and the results will be reported in the future.

VIII. ACKNOWLEDGEMENTS

Prototype turbines were constructed during academic travels supported in part through a sabbatical grant from the University of Wisconsin-Madison, and research and education grants from the Brazilian National Council for Scientific and Technological Development (CNPq), and The Scientific and Technical Research Council of Turkey, (TÜBİTAK). The authors would like to acknowledge the support from the sponsors of the Wisconsin Electric Machines and Power Electronics Consortium for the laboratory investigations. The prototype design assistance from Hugh Piggot, at Scoraig Wind Electric, Scotland, is gratefully acknowledged. Prototype turbines were built at the University of California-Berkeley, USA, at the Federal University of Minas Gerais, Belo Horizonte, Brazil and at Gazi University, Ankara, Turkey. Many thanks are due to the Professor Dan Kammen of Renewable and Alternative Energy Laboratory, UC-Berkeley, Professors Alessandro Moreira, Renato Lyra and José Carlos Rodrigues de Oliveira of UFMG-Belo Horizonte, and Professor Mehmet Timur Aydemir of Gazi University, Olga Yunak of Berkeley, for facilitating different wind turbine construction projects, and numerous volunteer project participants at Scoraig, Berkeley, Belo Horizonte and Ankara.

IX. REFERENCES

[1] Ryan Wiser and Matt Bollinger, *Annual Report on U.S. Wind Power Installation, Cost, and Performance Trends: 2006*, US Department of Energy, May 2007.

[2] G Angelika Pullen and Meera Ghani Eneland, (Eds), *GLOBAL WIND 2006 REPORT*, Global Wind Energy Council, Belgium,

[3] D. Xiang, L. Ran, P.Tavner, S. Yang, "Control of a Doubly Fed Induction Generator in a Wind Turbine During Grid Fault Ride-Through", IEEE Trans. on Energy Conversion, vol. 21, no. 3, Sept. 2006.

[4] A. Petersson, "Analysis Modeling and Control of Doubly-Fed Induction Generators for Wind Turbines", PhD. Thesis, Chalmers University of Technology, Goteborg, Sweden, 2005.

[5] P. Flannery, G. Venkataramanan, "A Grid Fault Tolerant Doubly Fed Induction Generator Wind Turbine via Series Connected Grid Side Converter" WINDPOWER 2006 Conference, June 4-7, 2006.

[6] "Small Wind Electric Systems - A U.S. Consumer's Guide", US Department of Energy, 2007.

[7] Simon Dunnett,Smail Khennas and Hugh Piggott, SMALL WIND SYSTEMS FOR BATTERY CHARGING A guide for development worker, Department for International Development, UK, July 2001.

[8] B. Milkovska, G. Venkataramanan, H. Nehrir and V. Gerez, "Variable Speed Operation of Permanent Magnet Alternator Wind Turbines Using a Single Switch Power Converter", Transactions of the ASME, v 118, n 4, Nov, 1996, p 235-238.

[9] Hugh Piggot, *How to build a wind turbine*, Scoraig Wind Electric, Scotland, 2005.

[10] Teodoro Sanchez, Sunith Fernando and Hugh Piggott, Wind rotor blade construction, Small Wind Systems for Battery Charging, Department for International Development, UK, Oct 2001.

[11] Gemini User Manual, 1995, Infolytica Corporation.

[12] N40 Material Datasheet, 2007, PowerMagnetStore.com.

[13] Frederick W. Grover, "Inductance Calculations, Working Formulas and Tables", Van Nostrand, 1946; Dover, 1962.
[14] Caliskan, V.; Perreault, D.J.; Jahns, T.M.; Kassakian, J.G., "Analysis of three-phase rectifiers with constant-voltage loads", IEEE Transactions on Circuits and Systems I: Regular Papers, Volume 50, Issue 9, Sept. 2003 Page(s):1220 - 1225
[15] Mourant, PSIM User's Manual, Micro Simulation, Boston, MA.
[16] N. Mohan, et al., Power Electronics – Converters, Applications, and Design, John Wiley, 2002.

X. BIOGRAPHIES

Justin Reed (S'04) received the B.S. degree in electrical engineering from the University of Washington, Seattle in 2005, where he worked on the high voltage, undersea NEPTUNE project. He is currently working towards the M.S. degree in electrical engineering at the Wisconsin Electric Machines and Power Electronics Consortium (WEMPEC). He enjoys power engineering for its diverse and plentiful applications, having industrial experience in both high- and low-voltage power conversion. His interests include power electronics, drives, and motors.

Giri Venkataramanan (M'92–SM'06) received the B.E. degree in electrical engineering from the Government College of Technology, Coimbatore, India, the M.S. degree from the California Institute of Technology, Pasadena, and the Ph.D. degree from the University of Wisconsin, Madison in the years 1986, 1987 and 1992 respectively. After teaching electrical engineering at Montana State University, Bozeman, he returned to University of Wisconsin, Madison, as a faculty member in 1999, where he continues to direct research in various areas of electronic power conversion as an Associate Director of the Wisconsin Electric Machines and Power Electronics Consortium (WEMPEC). He holds several U.S. patents and has coauthored more than a hundred technical publications.

A biography of **Jonathan Rose** was unavailable at the time of publication.

Development of Simulator for DFIG-based Wind Turbine

Young-Ger Seo, Young-Chan Kim,
Jong-Sun Ko, Soon-Chan Hong
Dankook University
Department of Electrical Engineering
San 44-1, Jookjun-dong, Suji-gu, Yongin, 448-161
Email: syger01@hanmail.net,
weist14g@paran.com, jsko@dku.edu,
schong@dku.edu

Byung-Moon Han
Myongji University
Department of Electrical Engineering
San 38-2, Nam-dong, Cheoin-gu, Yongin, 449-728
Email: erichan@hju.ac.kr

Abstract—The main goal of this paper is to simulate a Doubly-Fed Induction Generator (DFIG), which is similar to a real system. Wind velocity data is applied to a 2D Lookup table as a speed reference for an induction motor instead of the turbine. A real electric machine's parameters are put in the simulator to get some results of the real system. The Matlab have been generally used to simulate DFIG, but it has some differences from the real system and is difficult to implement. A Simplorer simulator, however, simplifies DFIG simulation. A Space Vector Modulation (SVM) is accepted to control a driver of an induction motor, which plays a part of the turbine. It is possible to retrieve important information, like a generated power and wind quality etc., from the simulator without a huge wind turbine.

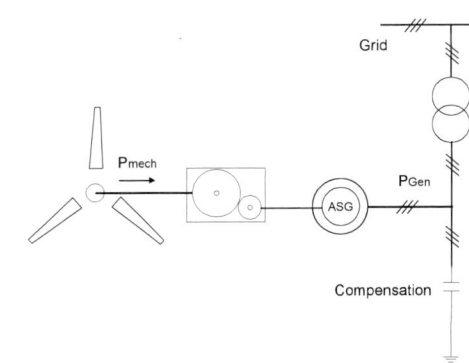

Fig. 1: Fixed speed Danish concept

I. INTRODUCTION

Wind energy is one of the most important and promising sources of renewable energy in the world, mainly because it is considered to be nonpolluting and economically viable. Many wind turbines built to-date were constructed according to the so called 'Danish concept' which is presented in Fig. 1. Wind energy is converted into electrical energy using a simple squirrel cage induction machine directly connected to a three-phase power grid. This Danish turbine basically has to operate at constant speed at any given operating point [1]. The variable speed doubly-fed induction generator wind turbine is the most commonly used wind turbine concept today. The variable speed operation and a grid connection through power electronic interface are clearly assessed in the main trend of modern wind turbines, especially using the Doubly-Fed Induction Generator (DFIG). The DFIG concept uses a partial scale frequency converter in the rotor circuit. This converter controls the rotor voltage and thus performs independent control of active and reactive power at a speed range of +/-30% around synchronous speed. Since only a fraction of the produced power is transferred via the power converter size, costs and losses can be reduced in comparison to a full scale power converter [2].

Modern high-power wind turbines are capable of variable speed operation. Advantages of variable generators compared to the fixed speed generators are an effective cost, a simple pitch control, a reduction of mechanical stresses, an improved power quality and improved system efficiency. Fig. 2 represents typical output speed-power curves as a function of turbine and wind speed [3]. The power is controlled in order to follow a pre-defined speed-power characteristic, named

tracking characteristic. The actual speed of the turbine is measured and the corresponding mechanical power of the tracking characteristic is used as the reference power for the power control loop [4].

The primary goal of this paper is to simulate DFIG, which is similar with a real system. Wind velocity data is applied to a 2D Lookup table as a speed reference for a turbine model of the Simplorer simulator. It is possible to check the generated voltage and wind quality from the simulation result. In other words, it is feasible to penetrate an economical efficiency with this accurate simulator.

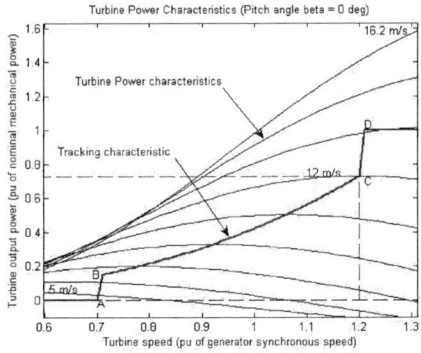

Fig. 2: Turbine power characteristics and tracking characteristic

II. PRINCIPLES OF A DFIG

Doubly fed induction generator can be excited by not only a stator current but also a rotor current, like its name. It is possible to get an output power from the stator and the rotor

978-1-4244-1871-8/07 $25.00 © 2007 IEEE

according to the rotating speed. There are two modes of operation of DFIG which are subsynchronous regeneration and supersynchronous regeneration [5][6]. Fig. 3(a) presents a power flow of DFIG in subsynchronous speed range that rotor rotating speed is slower than stator's synchronous speed. The slip s is positive and the air gap power P_{ag} is negative. Correspondingly, negative slip power sP_{ag} is fed to the rotor from the converter so that the total air gap power is constant.

$$P_1 = P_m + P_2 - P_{cr} - P_{cs} \qquad (1)$$

The stator output power P_1 is supplied from the mechanical power P_m and the slip power sP_{ag}, where P_2 denotes the rotor power, P_{cr} is rotor winding loss, P_{cs} is stator winding loss. The DFIG system requires that the slip power in the rotor flow in the reverse direction. With reverse slip-power flow at subsynchronous speed, the power corresponding to shaft input mechanical power can be pumped out of the stator.

Fig. 3(b) presents a power flow of DFIG in supersynchronous speed range that rotor rotating speed is faster than stator's synchronous speed. In this operation, the stator output power P_1 remains constant, but the additional mechanical input power is reflected as slip power output. The converter phase sequence is reversed so that the rotor field rotates in the opposite direction.

$$P_1 = P_m - P_2 - P_{cr} - P_{cs} \qquad (2)$$
$$P_{ag} = P_m - sP_{ag} \qquad (3)$$
$$sP_{ag} = P_2 \pm P_{cr} \qquad (4)$$

As a result, the stator output power P_1 is always constant at both subsynchronous speed and supersynchronous speed [5][6].

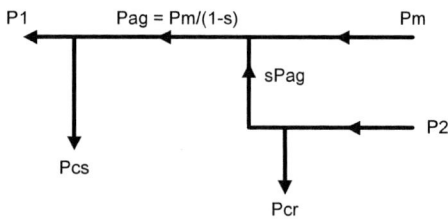

(a) Subsynchronous operation

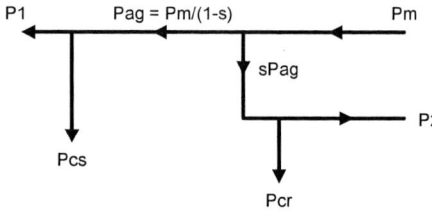

(b) Supersynchronous operation

Fig. 3: Power flow of a DFIG

Power distribution as a function of slip in subsynchronous and supersynchronous speed range is summarized for all modes in Fig. 4, where f_2 is the rotor electrical frequency [7].

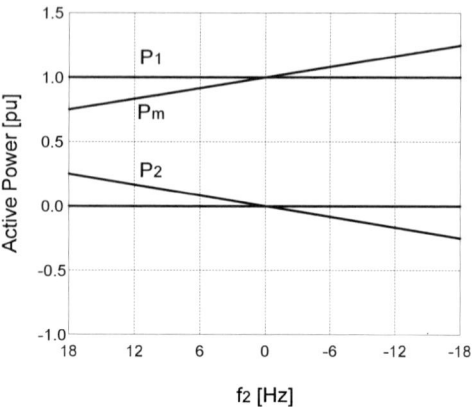

Fig. 4: Power characteristics of ideal DFIG

III. CONFIGURATION OF A DFIG

The DFIG is basically a wound rotor induction machine combined with a voltage source converter in the rotor circuit, which serves to apply a variable voltage to the rotor slip rings. In the case of fixed speed generators, in order to get a constant output power, only proper operating wind speed is valid but other speed is accepted as a slip. However, the variable speed generator, DFIG, can produce power 20% higher than fixed speed generator because of the maximum power point tracking system [2].

Fig. 5 illustrates a configuration of the DFIG wind power generator. Wind energy goes through the blade of the turbine and is converted to mechanical energy, which rotates a rotor of the DFIG. There are two different cases to excite the rotor. One is that the power for excitation flow from the grid to the rotor and the other is from the rotor to the grid. It is exactly the same with the mentioned above.

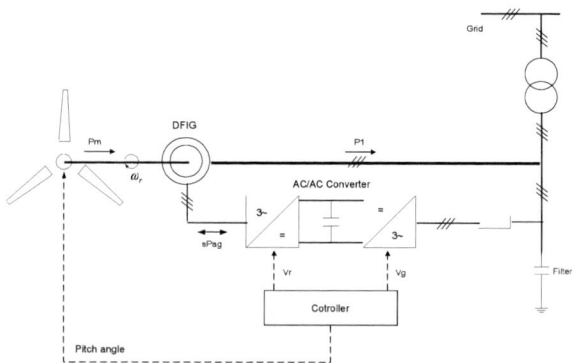

Fig. 5: Configuration of a DFIG wind power generator

The former is operating in the subsynchronous speed; the machine side converter is operated as an inverter and the grid side converter is operated as a rectifier, while the latter is operating in the supersynchronous speed; the machine side converter is operated as a rectifier and the grid side converter is operated as an inverter. The above converter has to be bidirectional to flow in and out the power. A power IGBT,

978-1-4244-1871-8/07 $25.00 © 2007 IEEE

matrix converter and cycloconverter are possible to implement the bidirectional device. The inverter has to make a sinusoidal low frequency rotor current because a stator current is affected by the harmonics of the rotor current. In order to get an optimal power factor of the stator, a driver has to control an excitation frequency and maintain a constant power and speed. A control of excitation voltage is needed to produce the rated field current of the rotor [8][9].

IV. SIMULATION USING SIMPLORER SIMULATOR

Fig. 6 illustrates a configuration of a motor simulator that is a substitute for a real wind turbine. The motor is a squirrel type induction machine, and it has a shaft that is able to mechanically connect with a secondary part. A Simplorer simulator makes the connector possible. The six step inverter is operated by using a Space Vector Modulation (SVM).

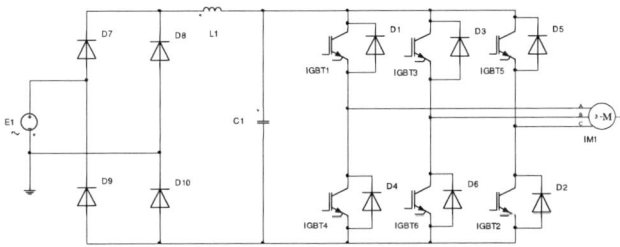

Fig. 6: Configuration of a motor simulator

Fig. 7 represents a gearbox for a speed amplifier, which can amplify the motor speed to the generator speed. The gearbox is necessary because mechanical rotating speed of the turbine blades is slower than the electrical rotating speed. Therefore, the mechanical time constant is more dominant than the electrical time constant. Unless there is a gearbox, operating in supersynchronous speed cannot be satisfied as it is in effect.

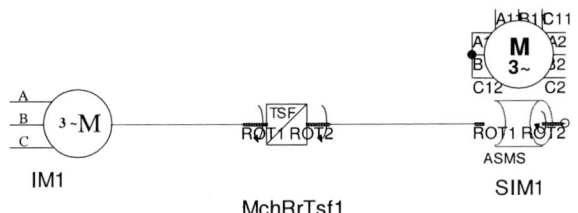

Fig. 7: Gearbox for a speed amplifier

Fig. 8 illustrates the configuration of a slip ring induction generator that will be performed as a DFIG. A three phase voltage source is substituted for the grid. The voltage source excites a stator of the slip ring induction generator. A generated voltage (VM2) and source voltage (VM3) are measured by checking the voltage drop of an external resistor (R4) of the stator. The direction of the generated power flow is confirmed by measuring the current of R4 (AM2). The external resistors (R1, R2 and R3) of the rotor will be replaced

with the matrix converter.

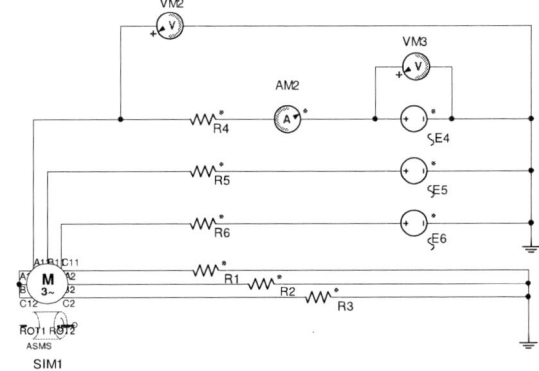

Fig. 8: Configuration of a DFIG wind power generator

Fig. 9 illustrates a configuration of motor-generator set. The squirrel type induction motor is connected parallel to the slip ring induction generator with a gearbox. That means, positive direction of motor is exactly the same with positive direction of generator. When the torque reference is applied to the motor, the direction of torque has to be the same with the direction of rotating field of grid. If the blade rotates in supersynchronous speed, the over generated power is restricted by a negative torque of generator. It is a process that the slip-power flows through the converter from the rotor to grid.

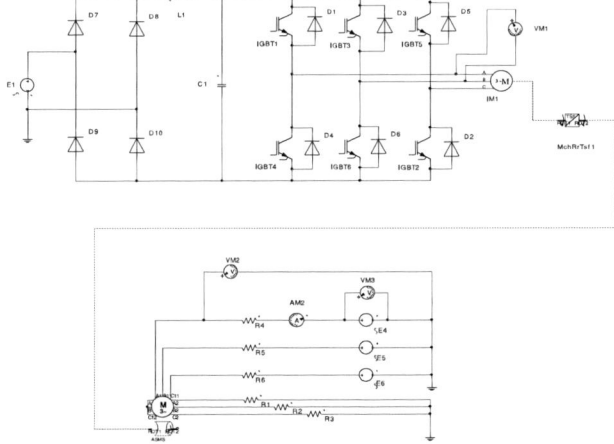

Fig. 9: Configuration of motor-generator set

Fig. 10 is a block diagram of a torque controller using SVM. An angle of synchronously rotating rotor flux is generated by indirect vector control (feedforward manner) and a coupling effect of d, q-axis current is decoupled by decoupling current controller. The SVM method is advanced and possibly the best among all the PWM techniques for variable -frequency drive applications. By using the SVM method, the time intervals T_1 and T_2 satisfy the command voltage, but time T_0 fills up the remaining gap with the zero vectors.

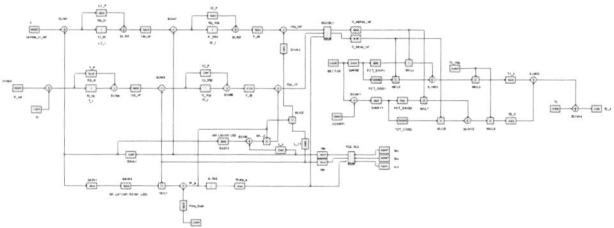

Fig. 10: Block diagram of torque controller using SVM

Fig. 11 is a state diagram for selecting sectors. The sector can be obtained by comparing a trigonometric function with the ratio of a two-phase stationary frame V_α and V_β. The sector information is needed to calculate the time intervals T_1, T_2 and T_0. A command voltage vector is combined with the adjacent two voltage vectors V_1 and V_2, which are in the selected sector.

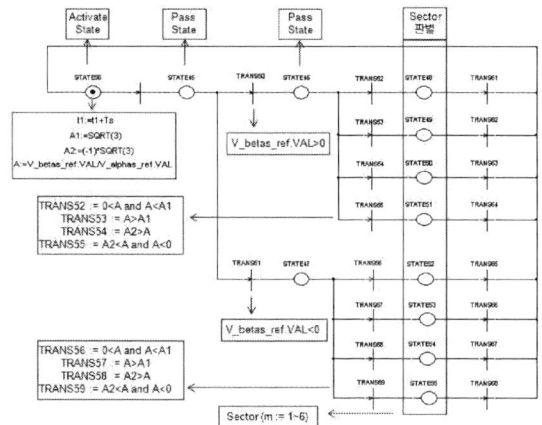

Fig. 11: Sector selection using the state diagram

Fig. 12 illustrates a switching table using the state diagram. By using the SVM method, the time intervals T_1, T_2 and T_0 are obtained as mentioned above. The switching table makes signals to fire pulses at the gate of the switches. Switching sequences are constructed by a double-sided switching pattern. If a switching pattern is settled by a symmetric form during two sampling periods, unnecessary switching sequences can be reduced. That means the harmonics of switching, which creates high frequency of current, is nearly removed.

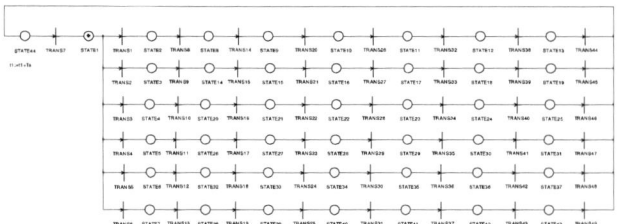

Fig. 12: Switching table using the state diagram

Fig. 13 is an equation block for protecting the over-modulation. When the sum of T_1 and T_2 is over T_s, which is a switching period, it is an over-modulation. In other words, a commanded voltage vector cannot be generated within a sampling period. An inverter is saturated when it is in over-modulation or nonlinear region. It has to maintain a direction of commanded voltage vector and restrict the only magnitude of the vector in that region. Output voltage, which is similar with the reference voltage, can then be obtained.

```
EQU
    T_sum:=T1_d.VAL+T2_d.VA

    IF(T_sum>Tc)
    {
    T1:=T1_d.VAL*Tc/(T1_d.VAL+T2_d.VAL
    T2:=T2_d.VAL*Tc/(T1_d.VAL+T2_d.VAL
    T0:=0;
    }
    ELSE
    {
    T1:=T1_d.VAL;
    T2:=T2_d.VAL;
    T0:=T0_d.VAL;
    }
```

Fig. 13: Equation block for protecting the over-modulation

Fig. 14 represents a turbine model of a Simplorer simulator. A secondary part of the turbine is mechanically connected with inertia load that plays the role of a real generator. Radius and moment of inertia of an actual size blade are inputted into the turbine model, so the moment of inertia of actual generator has to be inputted into the inertia load. Wind velocity and power coefficient (C_p) are needed to operate the turbine model [10]. It is possible to input the wind data and the power coefficient into the turbine model with CONST block and 3D lookup table. A tip-speed ratio and a pitch angle are preprogrammed in the 3D lookup table. Accordingly, the power coefficient can be obtained as an output from the 3D lookup table.

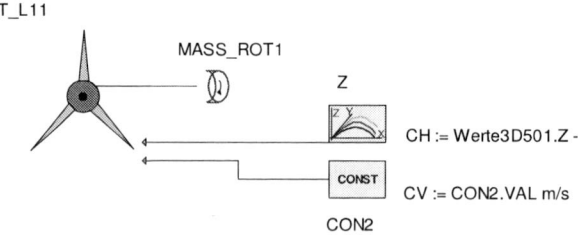

Fig. 14: Turbine model of Simplorer simulator

Fig. 15 points out some results of simulation. The rotor shaft torque of the turbine is applied to a motor as a reference. The torque of the motor can follow the reference torque well. The angular velocity of blade is naturally increased by the wind data and the angular velocity of the motor is increased up to the synchronous speed, and the amplified angular velocity of the generator is provided from the gearbox. The generator's speed can also be increased up to the synchronous speed, because the grid always supplies constant frequency.

978-1-4244-1871-8/07 $25.00 © 2007 IEEE   1128

However, when it surpasses the synchronous speed, the instantaneous power becomes negative. Therefore, it is possible to regenerate in the supersynchronous speed region.

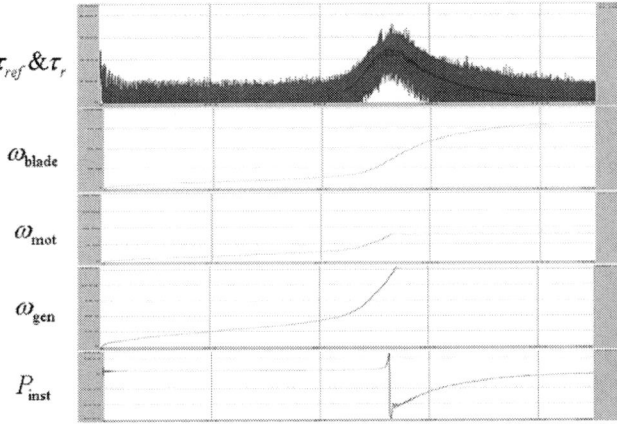

Fig. 15: Results of simulation

V. CONCLUSION

In this paper, a simulation model is developed in the simulation tool Simplorer. A turbine rotates the DFIG and the turbine is simulated by the Simplorer's unique turbine. This turbine is serves to provide characteristics of a real turbine. The rotor shaft torque, which is transferred to the DFIG, is provided from the current wind velocity using the simulated turbine and is then applied to the motor as an input. The motor is operated by a SVM control. The possibility of generating is confirmed by checking the output power. It is possible to get some important information, like generated power and wind quality etc., from the simulator without a huge wind turbine. Hereafter, a matrix converter will be inserted to control the excitation voltage and the rotor frequency. The doubly fed will be realized literally with the matrix converter and the grid connection. As a result, the economical efficiency will be estimated by checking the generated power.

TABLE 1
Squirrel type induction motor parameters

Power	7.5 [kW]
Rated Speed	1750 [rpm]
Voltage	380/220 [V], 3φ, 60 [Hz]
Current	16.5/28.6 [A]
Poles	4 poles
Stator Resistance	0.528 [Ω]
Rotor Resistance	0.519 [Ω]
Stator Inductance	0.0032 [H]
Rotor Inductance	0.0063 [H]
Mutual Inductance	0.099 [H]

TABLE 2
DFIG parameters

Power	3.7 [kW]
Rated Speed	1127 [rpm]
Poles	6 poles
Stator Resistance	4.47 [Ω]
Rotor Resistance	1.6 [Ω]
Stator Inductance	0.762 [H]
Rotor Inductance	0.8 [H]
Mutual Inductance	0.8 [H]
Moment of Inertia	0.038 [kgm^2]

ACKNOWLEDGMENT

This work has been supported by KESRI(R-2005-B-136), which is funded by MOCIE (Ministry of commerce, industry and energy).

REFERENCES

[1] S. Muller, M. Deicke, "Adjustable speed generator for wind turbine based on doubly-fed induction machines and 4-quadrant IGBT converters linked to the rotor", *IAS 2000 Conference Record*, Vol. 4, pp. 2249-2254, 2000

[2] Gabriele Gail, Anca D. Hansen, "Controller design and analysis of a variable speed wind turbine with doubly-fed induction generator"

[3] Documentation of SimPowerSystems, "Wind turbine doubly-fed induction generator(Phasor Type)"

[4] R.Pena, J. C. Clare, G. M. Asher, "Doubly fed induction generator using back-to-back PWM converters and its application to variable-speed wind-energy generation", *IEE Proc.-Electr, Power Appl.*, Vol. 143, No. 3, pp. 231-241, May 1996

[5] Woo-Seok Lee, "A constant power and optimal power factor drive of doubly fed induction generator", *Journal of the Korea Institute of Illuminating and Electrical Installation Engineers*, Vol. 14, No. 4, Page(s): 31-38., July 2000.

[6] Chul-Ho Kim, Young-Taek Seo, Chul-Soo Oh, "A power analysis for DFIG according to rotor excitation", *Journal of the Korea Institute of Electrical Engineers*, Vol. 52B, No. 7, July 2003

[7] B. K. Bose, *Modern Power Electronics and AC Drives*. Upper Saddle River, NJ: Prentice-Hall, 2002.

[8] Byoung-Chang Jeong, Tae-Hwa Kwon, Seung-Ho Song, Eel-Hwan Kim, "Grid connection algorithm for doubly-fed induction generator using rotor side PWM inverter-converter", *Journal of the Korea Institute of Electrical Engineers*, Vol. 52B, No. 10, October 2003

[9] Byoung-Chang Jeong, Seung-Ho Song, Dong-Joon Sim, " A wind turbine simulator for doubly-fed induction-type generator with automatic operation mode change during wind speed variation", *Journal of the Korea Institute of Power Electronics*, Vol. 11, No. 4, October 2006

[10] Byoung-Chang Jeong, Se-Jong Jeong, Seung-Ho Song, "Control algorithm for wind turbine simulator with variable inertia emulation", *Journal of the Korea Institute of Power Electronics*, Vol. 8, No. 3, June 2003

[11] Young-Ger Seo, Jong-Sun Ko, " A Study on a Simplorer Simulator of DFIG for Wind Power Generation", *Korea Institute of Power Electronics*, pp. 231-233, June 2006

[12] Hyeung-Gyun Kim, Dong-Choon Lee, and Jul-Ki Seok, "Maximum Output Power Control for Stand-Alone Wind Power Generation System Using Cage-Type Induction Generators", *Journal of the Korea Institute of Electrical Engineers*, Vol. 9, No. 1, pp. 73-80, February 2004

[13] Arantxa Tapia, Gerardo Tapia, J. Xabier Ostoaza, and Jose Ramon saenz, "Modeling and Control of a Wind Turbine Driven Doubly Fed Induction Generator, *IEEE Transactions on Energy conversion*, Vol. 18, No. 2, pp. 194-204, June 2003

978-1-4244-1871-8/07 $25.00 © 2007 IEEE

[14] Badrul H. Chowdhury, Srinivas Chellapilla, "Double-fed induction g enerator control for variable speed wind power generation", *Electric Power Systems Research*, Vol. 76, pp. 786-800, 2006

[15] Florin Iov, Anca Daniela Hansen, Clemens Jauch, Poul Sorensen and Frede Blaabjerg, "Advanced Tools for Modeling, Design and Optimi zation of Wind Turbine Systems", *Journal of Power Electronics*, Vo l. 5, No. 2, pp. 83-98, April 2005

[16] Stefan Marko, Ivan Darula and Stanislav Vlcek, "Development of wi nd farm models for power system studies", *Journal of Electrical Eng ineering*, Vol. 56, No. 5-6, pp. 165-168, 2005

LCL Filter design for grid-connected NPC inverters in offshore wind turbines

Samuel Vasconcelos Araújo, Alfred Engler,
Benjamin Sahan
ISET e.V. Universität Kassel
Power Electronics Division
Königstör 59, 34119 Kassel, Germany
Email: sva@ieee.org

Fernando Luiz Marcelo Antunes
Universidade Federal do Ceará
Department of Electrical Engineering, GPEC
Campus do Pici – Caixa Postal 6001
Email: fantunes@dee.ufc.br

Abstract— This paper deals with the analysis, design and optimization of a LCL filter topology to connect a 7MW NPC inverter to the grid. Following the requirements based on the IEEE 519-1992 recommendation and the German Guideline VDEW, simulation results were evaluated in order to access the performance of the proposed filter and the quality of the current injected into the grid.

I. INTRODUCTION

The trend towards multi MW wind turbine units has called up for new concepts in the design of wind energy conversion systems. Economic viability of offshore wind turbines clearly scales with power and efficiency of generators and power conversion systems. Within this trend, power electronic multilevel converters have been seen as an appropriate technology for the wind energy conversion system because they can operate at high power and high voltage [1].

Among several proposed multilevel topologies the three-level diode-clamped, or simply called Neutral-Point Clamped - NPC inverter was the first widely implemented by the industry and it continues to be extensively used in high voltage and high power applications.

A NPC inverter leg is shown in Fig. 1. It is composed by four switches S_1 to S_4 and the anti-parallel diodes D_1 to D_4.

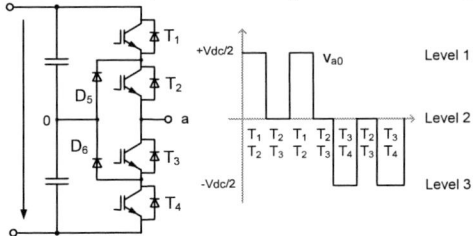

Fig. 1: NPC inverter leg and AC waveform.

Voltages across the switches are only half of the DC input voltage because diodes D5 and D6 are connected to the neutral point [2].

The three-phase NPC inverter presents a line-to-line voltage with five voltage levels which leads to a lower Total Harmonic Distortion (THD) and dv/dt; also resulting in lower Electromagnetic Interference EMI when compared to the conventional two-level inverter. On the other hand, as the number of levels of the inverter increases the THD of the output voltage waveform presents little improvement [3].

This paper will focus the filter design aiming to minimize the harmonic content of the current injected by the grid side

NPC inverter at the Point of Common Coupling (PCC), as depicted in Fig. 2. As a remark, this is a full converter solution in order to avoid dealing with slip rings and possible Low Voltage Ride Through difficulties.

Fig. 2: System diagram.

Several requirements can be considered when analyzing the power quality and grid-compatibility of wind turbines connected to the grid. According to [4] and [5], the following topics are important when studying the grid compatibility of a device: average and maximum produced power, reactive power levels, coupling procedure to the grid, grid short-circuit current (weak or stiff grid conditions), voltage fluctuations (under normal and transient operation), flicker and harmonics.

Concerning wind turbines, almost all of the above listed requirements are influenced either by the behavior of the wind profile or by the dynamic of the mechanic system (including rotor and generator characteristics). However, for this analysis the inverter DC link voltage is strong enough to avoid flicker or voltage fluctuation caused by wind speed variation.

II. STANDARDS REGARDING THE QUALITY OF THE POWER INSERTED INTO THE GRID

The usual diagnosis of harmonic distortion is done by calculating each harmonic component present in the produced current and then determining the distortion on the voltage due to the grid impedance at the considered harmonic frequency level. Knowing that at the design phase, no further information is available about the grid connection and impedances, the analysis will be limited to the harmonic components of the produced current.

For such purpose, standards that presented limits for the currents will be shortly analyzed. Those limits were proposed as guidance for designers that do not have information about the grid characteristics.

A. IEEE Std. 519-1992: Recommended Practices and Requirements for Harmonic Control in Electrical Power Systems [4]

The limits for currents at this recommendation are

978-1-4244-1871-8/07 $25.00 © 2007 IEEE

presented in three tables, depending on the system voltage level and on the ratio between the grid short-circuit current capability and equipment maximal fundamental current. For voltage levels below 69 kV, the established limits are presented in Table 1:

TABLE 1

Current Distortion Limits for General Dist. Systems (120 V – 69 000 V)

	Maximum Harmonic Current Distortion in Percent of I_L Individual Harmonic Order (Odd Harmonics)					
I_{SC}/I_L	<11	11≤h<17	17≤h<23	23≤h<25	35≤h	TDD
<20*	4.0	2.0	1.5	0.6	0.3	5.0
20<50	7.0	3.5	2.5	1.0	0.5	8.0
50<100	10.0	4.5	4.0	1.5	0.7	12.0
100<1000	12.0	5.5	5.0	2.0	1.0	15.0
>1000	15.0	7.0	6.0	2.5	1.4	20

Even harmonics are limited to 25% of the odd harmonic limits above.

Current distortions that result in a dc offset, e.g., half-wave converters, are not allowed

* All power generation equipment is limited to these values current distortion, regardless of actual I_{SC}/I_L

As stated on the table, for generators the limits do not actually depend on the currents ratio. As a remark, Table 1 has been developed for 6 pulse rectifiers. Therefore, the limits may be increased in the case of converters operating with a higher number of pulses per period, which is the case of an inverter operating at a few kHz.

Though clearly stated that the voltage harmonic analysis should be done up to the 50th order, no order limit is actually proposed on this standard for harmonic current components.

B. VDEW - Guideline for connection and parallel operation of generators at the medium voltage grid [6].

The maximum allowed values for harmonic current components ($I_{v,\mu,allowed}$) can be obtained by multiplying the constants of Table 2 ($i_{v,\mu,allowed}$) by the short-circuit power at the connecting point (S_{kv}), as shown in (1):

$$I_{v,\mu,allowed} = i_{v,\mu,zul} \cdot S_{kv} \tag{1}$$

Limits for other voltage levels can be directly calculated with the given values, knowing that those are inversely proportional to the voltage value.

Triple harmonic components and harmonics up to 25th order have their limits based on the ones of next given order (for example, the 9th harmonic order limit is equal to the 11th one), with the condition that produced zero-sequence currents are not inserted on the grid.

TABLE 2

Allowed rated Harmonic Currents

Harmonic Order v, μ	Allowed rated Harmonic Currents $I_{v,\mu,zul}$ in A/MVA	
	10-kV-Grid	20-kV-Grid
5	0.115	0.058
7	0.082	0.041
11	0.052	0.026
13	0.038	0.019
17	0.022	0.011
19	0.018	0.009
23	0.012	0.006
25	0.010	0.005
>25 or Pairs	0.06/v	0.03/v
μ < 40	0.06/μ	0.03/μ
μ > 40*	0.18/μ	0.09/μ

* Integer and not integer within a Bandwidth of 200 Hz

The values presented on Table 2 were chosen in such a way to be valid with a grid inductive impedance under high frequencies; for example in overhead lines. Nevertheless, in grids with noticeable underground cable share, the grid impedance under higher frequencies (over 2000 Hz) is generally smaller in such a way that high frequency components can be neglected. A requirement for such a procedure is the calculation of either the voltage distortion or grid impedances along such harmonic orders at the connection point; keeping in mind that a limit of 0.2% for harmonic voltage components between 2000 and 9000 Hz may not be surpassed.

Since the frequency band from 2000 Hz to 9000 Hz was considered as being important to the characterization of frequency converters by this standard, it was interpreted that the harmonic analysis of the current should be made until this level of frequency.

C. EN 61400-21 Measurement and evaluation of the grid compatibility of grid-connected wind energy installations [7]

Similar to what was proposed by [4], voltage harmonic analysis is done up to the 50th order for integer harmonics and up to 2.5 kHz for inter-harmonics.

Nevertheless, the current standard considers that wind turbines connected to the grid via power converters operating with a switching frequency on the range of kHz will most likely produce harmonic components greater than the 50th order, but this range is still being analyzed by the IEC committee.

The limits assumed by this standard are in accordance with the standard IEC 61000-3-6 (Electromagnetic Compatibility – Limits Assessment of emission limits for distorting loads in MV and HV power systems – basic EMC publication)

In order to perform the harmonic analysis, values of the produced harmonic current components are used to calculate the percentage in relation with the rated current (that means, under nominal power, voltage and frequency conditions). Components with a value under 0.1% may not be considered.

The measurements and stated limits may be valid for both sides of the transformer (primary and secondary) – this means that if the system was approved for a certain level of voltage, it may be also in accordance if a transformer with another turns ratio is used.

Finally, another recommendation of this standard is that the short-circuit power may be at least 50 times greater than the maximum active power produced by the equipment.

III. REVISION ON FILTER TOPOLOGIES

In order to choose an optimal filter topology considering a NPC inverter for offshore wind turbines, parameters like efficiency, weight and volume have to be considered.

Regarding efficiency, filter topologies with reduced losses are required, though those are relatively small when compared to losses in the inverter. Weight and volume are considered as critical characteristics at offshore applications due to difficulties with transportation, installation and maintenance. The filter cost depends basically on the amount of components and materials used, for example the magnetic material for the core of inductors. Last, but certainly not least, the filter shall be able to perform its task within a certain degree of independence of the grid parameters, like resonance susceptibility and dynamic performance are of major importance.

As proposed in [8], filters connected to the output of an inverter have basically the following four-pole circuit configuration as seen in Fig. 3.

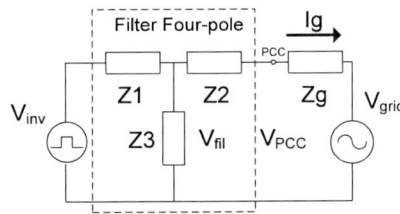

Fig. 3: Generic diagram of a three element filter.

A. L Filter

This topology (Z1 is finite, Z3 is infinite and Z2=0) consists on just an inductive filter connected in series with the converter. Although being the topology with the fewer number of components the system dynamics is poor due to the voltage drop across the inductor causing long time responses.

B. LC Filter

This topology (Z1 is finite, Z3 is finite and Z2=0) has Z3 as a result of association of a capacitor and an inductor. With higher values of capacitance, the inductance can be reduced, leading to reduction of losses and cost. Nevertheless, very high capacitance values are not recommended, since problems may arise with inrush currents, high capacitance current at the fundamental frequency, grid side resonance and dependence of the filter on grid impedance for overall harmonic attenuation.

C. LCL Filter

When compared with the previous topology, the LCL filter has the advantage of providing a better decoupling between filter and grid impedance (reducing the dependence on grid parameters) and a lower ripple current stress across the grid inductor (since the current ripple is reduced by the capacitor, the impedance at the grid side suffers less stress when compared with the LC topology).

Like the LC filter, increasing the capacitor value reduces filter cost and weight but with similar drawbacks. The split factor between the inductances at the inverter and grid side offers a further design flexibility.

D. Tuned Filter with LC Filter

An alternative to the above described topologies is the LC filter with a secondary branch tuned to the switching frequency and connected in parallel. An advantage is that the tuned filter needs to be set up to the harmonic current

components only.

IV. LCL FILTER DESIGN

Though the LCL filter can sometimes cost more than other more simple topologies, its small dependence on the grid parameters is of major importance at high power applications, in order to guarantee a stable power quality level. Furthermore, it provides better attenuation than other filters with the same size and by having an inductive output; it is capable of limiting current inrush problems. This topology is, therefore, the one proposed for the NPC grid side inverter, and analyzed in this paper.

Considering that the future generation of offshore wind energy conversion system points towards turbines in the range of 7 to 10MW [9] the filter will be designed taking into account the following parameters for the grid and the inverter:

TABLE 3
Design Parameters

Grid Line Voltage	$V_n = 1380$ V
Grid Phase Voltage	$V_{ph} = 796.73$ V
NPC DC-Link Voltage	$V_{dc} = 2200$ V
Output Power of the Inverter	$P_n = 7$ MW
Grid Frequency	$f = 50$ Hz
NPC Switching Frequency	$f_{sw} = 2000$ Hz

The nomenclature for the components is based on the schematic of Fig. 4:

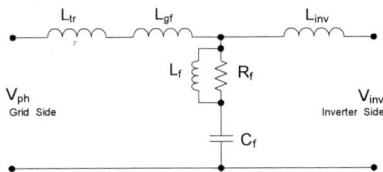

Fig. 4. LCL filter circuit with components nomenclature.

The first step is to design an inductor capable of limiting the ripple at the output current to 10% of the rated amplitude value. The ripple value of a periodic waveform refers to the difference between the instantaneous value of the waveform and its fundamental frequency. However, considering the switching nature of the inverter, it is still necessary to find an appropriate equation to calculate the filter inductance for this particular inverter topology.

For the filter inductance design, the NPC inverter can be modelled at the switching frequency as one half wave buck converter with the condition that the NPC neutral point is connected to the Y (star) point of the grid, and the grid is modelled as a half sinusoidal waveform voltage, as shown in Fig. 6. If the neutral point is not connected to the star point, the common-mode voltage needs to be taken into account and the calculation becomes more complex.

For simplicity reasons, (2), [11], shall be used for both cases:

978-1-4244-1871-8/07 $25.00 © 2007 IEEE

$$L_{inv} = \frac{\frac{V_{dc}}{2} - V_{ph}}{2 \cdot \Delta I_L} \cdot \frac{D}{f_{sw}} \qquad (2)$$

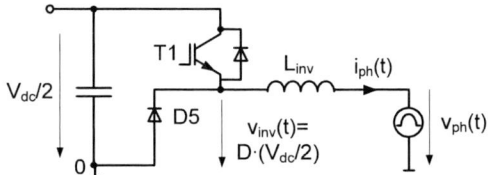

Fig. 5. NPC simplified circuit.

The phase current ripple as a function of time is depicted in Fig. 6, providing a duty-cycle function $D(t) \approx M \cos(\omega t)$, with a modulation index M=1.

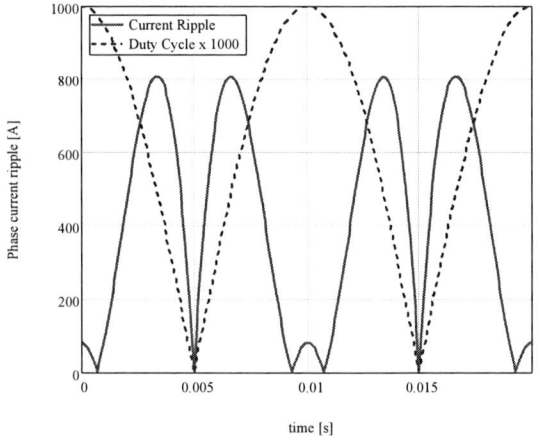

Fig. 6. Phase current ripple (peak-to-peak) for L_{inv}=166uH, f_{sw}= 2kHz, V_{ph}=796V, P=7MW and corresponded duty-cycle value multiplied by 1000.

The current ripple at the output of a dc-dc converter is the largest for a duty cycle of 50%. Therefore:

$$L_{inv} = \frac{\frac{V_{dc}}{2} - \frac{V_{dc}}{4}}{2 \cdot \Delta I_{L\max}} \cdot \frac{1}{2 \cdot f_{sw}} = \frac{V_{dc}}{16 \cdot f_{sw} \cdot \Delta I_{L\max}} \qquad (3)$$

A 10% ripple of the rated current for the design parameters presented in Table 3 is given by (4).

$$\Delta I_{L\max} = 0.10 \cdot \frac{P_n \cdot \sqrt{2}}{3 \cdot V_{ph}} = 414.165 A \qquad (4)$$

The value of the inverter side inductance (L_{inv}) using (3) and (4) is 166 µH or 0.192 pu.

For the design of the filter capacitance, it is considered that the maximum power factor variation seen by the grid is 5%, as it is multiplied by the value of base impedance of the system in (5).

$$C_f = 0.05 \cdot C_b = 585 \mu F = 20 pu \qquad (5)$$

It is important to notice that factors higher than 5% can be used, since they will compensate the inductive reactance of the inductors on the filter and therefore the influence at the power factor of the system will be lesser than expected. In addition, the greater the capacitance, the smaller is the

inductor [9]. Nevertheless, if too large capacitors are used, the ripple on the inductor current will tend to increase [10].

As stated in [14], the main objective of this LCL filter design is in fact to reduce the expected 10% current ripple limit to 20% of its own value, resulting in a ripple value of 2% of the output current. In order to calculate the ripple reduction, the LCL filter equivalent circuit is firstly analyzed considering the inverter as a current source for each harmonic frequency, as seen in Fig. 7.

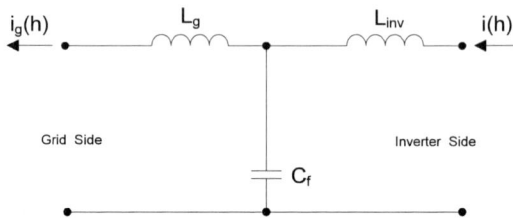

Fig. 7. Simplified LCL Filter scheme at hth order harmonic component.

Equation (6) gives the relation between the harmonic current generated by the inverter and the one injected in the grid (respectively i(h) and ig(h)).

$$\frac{i_g(h)}{i(h_{sw})} \approx \frac{z^2_{LC}}{|w^2_{res} - w^2_{sw}|} \qquad (6)$$

Equation (6) is a ratio between the filter impedance and the difference between resonant frequency and switching frequency (that will be later calculated). Simplifying this equation using the already developed relations for the inductance at the inverter side, results in (7) that represents the ripple attenuation factor:

$$\frac{i_g(h)}{i(h_{sw})} = \frac{1}{|1 + r \cdot [1 - (L \cdot Cb \cdot w^2_{sw}) \cdot x]|} \qquad (7)$$

The constant r is defined as the relation between the inductance at the inverter side and the one at the grid side:

$$L_g = r \cdot L_{inv} \qquad (8)$$

The value of r for a desired ripple attenuation can be obtained from the Fig. 8.

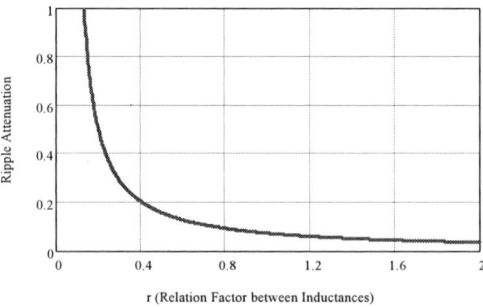

Fig. 8. Ripple attenuation as a function of the relation factor between inductances.

The calculated grid side inductance value is 69.8µH. Since the inverter will be connected to the grid via a power transformer, the transformer leakage inductance shall be considered as part of L_g.

978-1-4244-1871-8/07 $25.00 © 2007 IEEE   1134

A. Design Optimization Potential

By providing different ripple attenuation factors between the inductances at the inverter and grid side, it is possible to optimize the size of the total inductance. This is particularly interesting in order to reduce the voltage drop across the inductors and therefore avoid the use of high values for the modulation index. In Fig. 9 is presented the plot of the total necessary inductance as a function of the admitted ripple at the inverter side inductor. Nevertheless, the optimization was not carried out due to the necessity of physical design considerations, which are out of the scope of this publication.

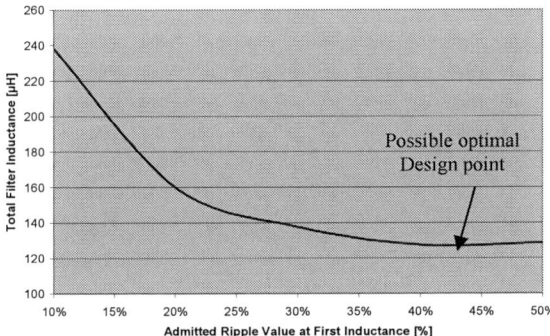

Fig. 9: Total inductance of the filter as a function of the admitted current ripple at first inductance.

B. Damping Branch

The attenuation mentioned above is only effective after considering a damping factor. This is necessary since the transfer function of the filter presents some peaks that may in fact increase the ripple at particular resonant frequencies with the grid impedance.

Instead of using active damping, which would make the control of the inverter more complex, the choice was made for passive damping. A resistance added in series with the capacitor attenuates part of the ripple on the switching frequency in order to avoid the resonance. The value of this resistor should be one third of the impedance of the filter capacitor at the resonant frequency (calculated by (10)).

$$w_{res} = \sqrt{\frac{L_{inv} + L_{gf}}{L_{inv} \cdot L_g \cdot C_f}} = 5144 \, rad/s \qquad (9)$$

The resistor in series with the filter capacitance is:

$$R_f = \frac{1}{3 \cdot w_{res} \cdot C_f} = 0.111\Omega = 0.408\,pu \qquad (10)$$

The resistor losses, rated at 9.4 kW for this given system design (or 0.134% for the inverter operating at nominal power), can be reduced if the resistor value is increased, but with the drawback of reducing the damping effectiveness.

A technique to reduce the filter losses and keeping the damping performance of the filter is proposed in [9]. It consists of adding an inductor in parallel with the damping resistor. The inductor is designed to have an impedance value smaller than the resistor impedance for operation below the resonant frequency (f_{res}). The value of the inductance in parallel with the resistor is determined by (11).

$$L_f = \frac{Rf}{2 \cdot \pi \cdot f_{res}} \approx 22\mu H = 0.025\,pu \qquad (11)$$

A value of 25μH is adopted considering the possible presence of further inductive components caused by the cables that may increase the value of the resonant frequency. The resistance of the windings was neglected. With this inductance, the expected losses on the resistances of the three phases were reduced to 1.9 kW or 0.027% of the inverter rated power, without noticeable prejudice to the damping effectiveness of the filter.

V. SIMULATION RESULTS

Fig. 10 and 11 depict the simulated output current waveforms of the NPC filter with a L filter and with the designed LCL filter.

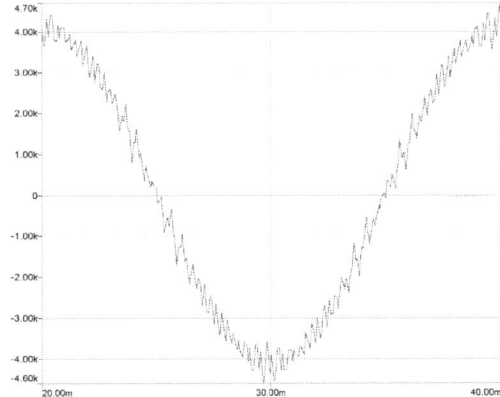

Fig. 10. Output current waveform with L filter [A/ms].

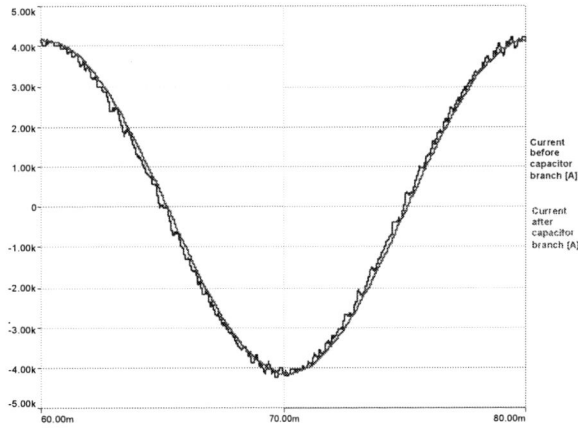

Fig. 11. Output current waveform with LCL filter before and after filter capacitor branch [A/ms].

The performance of the LCL filter design was analyzed using an Excel application that, by receiving the FFT of the output current calculated by Simplorer®, generated the recommended limits based on the standards [1] and [3] for specific system parameters, as illustrated in Fig. 12 and 13.

978-1-4244-1871-8/07 $25.00 © 2007 IEEE

As a remark, since the frequency band from 2000 Hz to 9000 Hz was considered as being important to the characterization of frequency converters [3], it was interpreted that the harmonic analysis of the current should be made up to this level of frequency, though it is of common sense that the voltage distortions shall be analyzed up to the 50th order. Additionally, according to the recommendations of [11], harmonic components smaller than 0.1% of the fundamental were neglected.

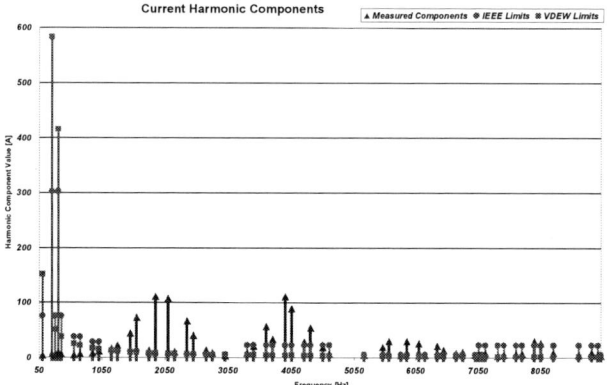

Fig. 12. Harmonic spectrum of the output current and corresponding recommended limits for the simulations with L filter.

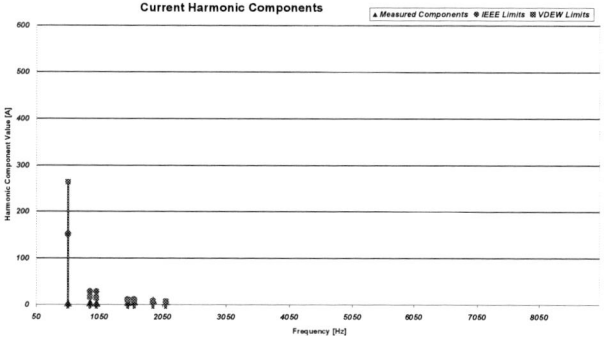

Fig. 13. Harmonic spectrum of the output current and correspondent recommended limits for the simulations with filter.

It was noticeable that the produced harmonics were mostly concentrated around the so-called "characteristic harmonic orders" [1]. After the application of the filter, most of the components were eliminated, though further improvement can be achieved by optimizing control parameters.

VI. CONCLUSIONS

This paper proposed the design and application of a LCL filter aiming to enhance the grid compatibility of a high power Wind Turbine connected to the utility grid via a NPC inverter. In order to evaluate the filter effectiveness, the power quality of the inverter was analyzed through the output currents, using procedures in accordance with the VDEW Standard and IEEE Recommendation.

At the end, the LCL filter topology showed as main advantages: the design flexibility, which allows further optimization; reduced size in comparison to other topologies and finally good capability of operation in a wide range of frequency and voltage.

ACKNOWLEDGEMENT

The authors would like to thank the European Commission and the work package partners for their support in the project "Upwind" (Contract No. SES6-019945). It should be noted that this paper reflects only the author's views and the European Commission is not liable for any use that may be made of the information contained therein. Fernando Antunes would like to thank also UFC and CNPq in Brazil and DAAD in Germany for the support received during his postdoctoral period at ISET.

REFERENCES

[1] J. Pou, "Modulation and control of three-phase PWM multilevel converters", PhD Dissertation, Universidat Politecnica de Catalunya, 2002.

[2] D. Holmes, G. Lipo, A. Thomas A. "Pulse Width Modulation for Power Converters – Principle and practice. IEEE Press, 2003.

[3] Skavarenina, L. Timothy, The Power Electronics Handbook – "Cap. 6 by Corzine, Keith" CRC Press, 2002.

[4] IEEE Std. 519-1992 - IEEE Recommended Practices and Requirement for Harmonic Control in Electrical Power Systems – IEEE Industry Applications Society/ Power Engineering Society.

[5] Netzverträglichkeit von Windenergienalagen in Deutschland und Dänemark/Schweden – DEWI Magazin Nr. 16 Februar 2000.

[6] VDEW Eigenerzeugungsanlagen am Mittelspannungsnetz - Richtlinie für Anschluss und Parallelbetrieb von Eigenerzeugungsanlagen am Mittelspannungsnetz.

[7] DIN EN 61400-21 - Windenergieanlagen - Teil 21 : Messung und Bewertung von netzgekoppelten Windenergieanlagen –VDE Verband der Elektrotechnik Elektronik Informationstechnik e.V.

[8] V. Pradeep, A. Kolwalar, R. Teichmann, "Optimized Filter Design for IEEE 519 compliant grid connected inverters".

[9] B. Sahan, F. L. M. Antunes, A. Engler, "Design And Benchmark Of A Multilevel Converter For Large-Scale Wind Power Systems", European Wind Energy Conference & Exhibition, 2007.

[10] N. Mohan, et al, Power Electronics: Converters, Applications and Design, Second Edition, John Wiley & Sons Inc., 1995.

[11] R. W. Erickson, D. Maksimovic, Fundamentals of Power Electronics, Second Edition, Kluwer Academic Publishers, 2001.

[12] C. T. Wang, Z. Ye, G. Sinha, X. Yuan, "Output Filter Design for a Grid-Interconnected Three-Phase Inverter", Power Electronics Specialist Conference, 2003.

[13] H. R. Karschenas, H. Saghafi, "Performance Investigation of LCL Filter in Grid Connected Converters", 2006 IEEE PES Transmission and Distribution Conference and Exposition Latin America, Venezuela.

[14] M. Liserre, F. Blaabjerg, S. Hansen, "Design and Control of an LCL-Based Three-Phase Active Rectifier", IEEE Transactions on Industry Applications, VOL. 41, NO. 5, Septem-ber/October 2005.

Decoupled Voltage and Frequency Controller for an Isolated Pico Hydro System Feeding Dynamic Loads

Bhim Singh, *Senior Member, IEEE*
Department of Electrical Engineering
Indian institute of Technology, New Delhi India
Email: bhimsinghr@gmail.com

Gaurav Kumar Kasal
Department of Electrical Engineering
Indian Institute of Technology, New Delhi, India
Email: gauravkasal@gmail.com

Abstract— This paper deals with a voltage and frequency controller for an isolated power generation system based on asynchronous generator (AG) driven by pico hydro turbine and supplying 3-phase dynamic (asynchronous motor) loads. The proposed controller is a combination of a static compensator (STATCOM) and an electronic load controller (ELC) for independent control of the reactive and active powers of the system to control the voltage and frequency respectively. The proposed generating system along with its controller is modeled in MATLAB using SIMULINK and PSB (Power System Block Sets) toolboxes. The performance of the controller is verified with direct on line starting of a 3-phase asynchronous motor.

I. INTRODUCTION

Asynchronous machines (AMs) are robust, inexpensive compared with DC and wound rotor synchronous machines, require little maintenance, and have high power weight ratio [1-3]. Despite these favorable features, commercialization of an asynchronous machine as an isolated asynchronous generator is still a bottleneck because of its unsatisfactory voltage and frequency regulation, even when driven under constant speed and feeding varying loads. In view of this, since last two decades number of research publications have been reported on modeling, steady state and transient analysis [1,2] of capacitor excited asynchronous generator (CEAG) along-with its controller [3,4].

In this paper, a new controller is proposed which is having capability of controlling the voltage and frequency in decoupled manner. For controlling the voltage, a static compensator (STATCOM) is used as a reactive power compensator along with harmonic eliminator and a load balancer while for controlling the frequency; an electronic load controller (ELC) is used to regulate the total active power at the terminals of generator. The STATCOM is realized using IGBTs (Insulated gate bipolar junction transistors) based voltage source converter (VSC), and a capacitor as an energy storage element at its DC link, while an ELC consists of a diode bridge rectifier, a chopper switch and an auxilliary load resistance.

II. SYSTEM CONFIGURATION

Fig. 1 shows the system configuration of CEAG, DVFC (consisting 3 leg IGBT based VSC and diode bridge rectifier based ELC) and the consumer load. The delta connected 3-phase capacitor bank is used for the generator excitation nd value of an excitation capacitor is selected to generate the rated voltage at no load. The CEAG generates constant power and when consumer load power changes, the DC chopper of an ELC absorbs the difference in power (generated-consumed) into an auxilliary load, while STATCOM is used to regulate the voltage due to load changes. Thus generated voltage and frequency are not affected and remain constant during the changes in consumer loads.

The DVFC is an arrangement of a STATCOM with an ELC. STATCOM consists of IGBT based current controlled 3-leg VSC, DC bus capacitor and AC inductors. The output of the VSC is connected through the AC filtering inductor to the CEAG terminals. The DC bus capacitor is used to filter voltage ripples and provides self supporting DC bus. A DC chopper in an ELC is used to control the extra power in controller auxilliary load due to change in consumer loads, so that generated power at the generator remains constant.

Fig.2 shows the control technique of decoupled voltage and frequency controller (DVFC) for providing single point operation (constant voltage and frequency along with constant excitation capacitor) of CEAG. The technique for STATCOM control is based on the generation of reference source currents while ELC is controlled for regulating the constant generated power.

III. CONTROL STRATEGY

Fig.2 shows the control strategy of the proposed controller for CEAG. The control technique of STATCOM to regulate the terminal voltage of the CEAG which is based on the generation of source currents have two components, in-phase and quadrature, with AC voltage. The in-phase unit vectors (u_a, u_b and u_c) are three-phase sinusoidal functions, computed by dividing the AC voltages v_a, v_b and v_c by their amplitude V_t. Another set of quadrature unit vectors (w_a, w_b and w_c) are sinusoidal function obtained from in-phase vectors (u_a, u_b and u_c). To regulate AC terminal voltage (V_t), it is sensed and compared with the reference voltage. The voltage error is processed in the PI controller. The output of the PI controller ($I_{smq}{}^*$) for AC voltage control loop decides the amplitude of reactive current to be generated by the STATCOM. Multiplication of quadrature unit vectors (w_a, w_b and w_c) with the output of PI based AC voltage controller ($I_{smq}{}^*$) yields the quadrature component of the reference source currents ($i_{saq}{}^*$, $i_{sbq}{}^*$ and $i_{scq}{}^*$). To provide a self-supporting DC bus of STATCOM, its DC bus voltage is sensed and compared with

978-1-4244-1871-8/07 $25.00 © 2007 IEEE

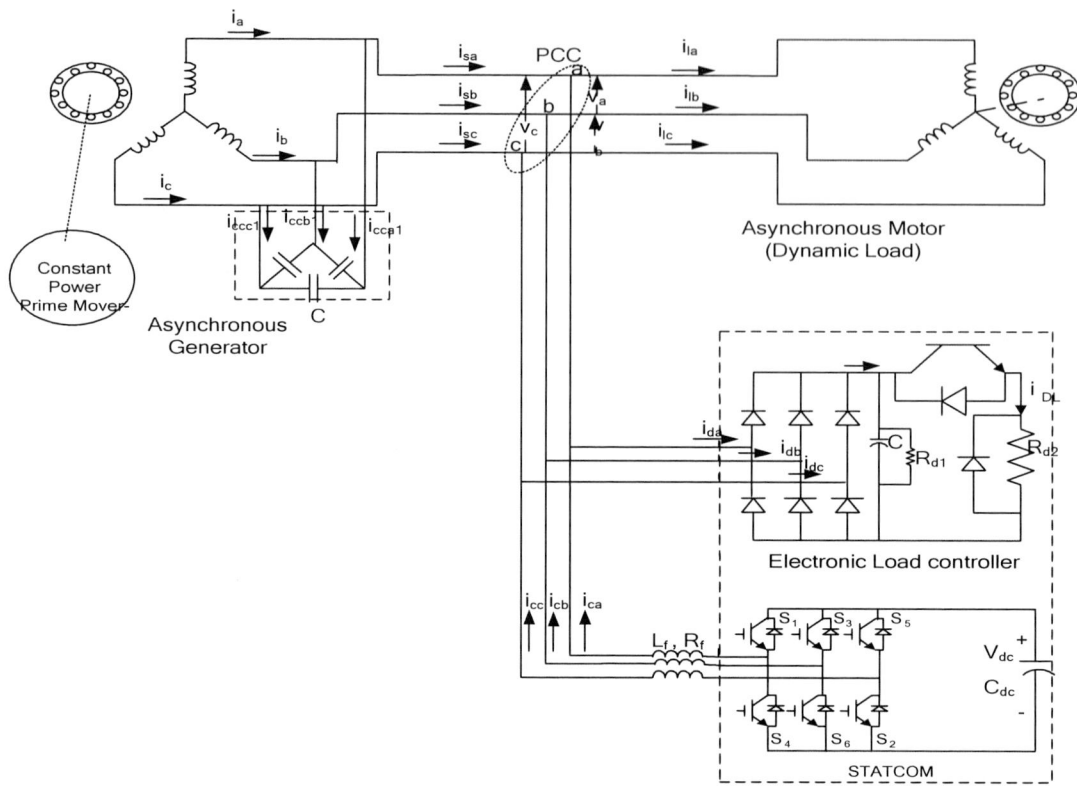

Fig. 1: Schematic diagram of a proposed controller for an isolated power generation and supplying asynchronous motor load.

DC reference voltage. The error voltage is processed in another PI controller. The output of the PI controller (I_{smd}^*) decides the amplitude of active current. Multiplication of in-phase unit vectors (u_a, u_b and u_c) with output of PI controller (I_{smd}^*) yields the in-phase component of the reference source currents (i_{sad}^*, i_{sbd}^* and i_{scd}^*). The instantaneous sum of quadrature and in-phase components gives the reference source currents (i_{sa}^*, i_{sb}^* and i_{sc}^*), which are compared with the sensed source current (i_{sa}, i_{sb} and i_{sc}). These current error signals are amplified and compared with the triangular carrier wave to generate the gating signals.

For controlling the chopper of an electronic load controller, measured power (P_{gen}) is compared with rated power (P_r) of the generator and then output of a power PI controller is compared with saw-tooth carrier wave resulting in PWM output of varying duty cycle for switching of the IGBT of an ELC chopper.

IV. CONTROL ALGORITHM

Basic equations of control scheme of a proposed decoupled controller (DVFC) for CEAG is given here. The control scheme is divided into two sections. Section 'A' describes the equations of controlling the "STATCOM" while section 'B' deals with the chopper control of an electronic load controller (ELC).

A. Control algorithm for STATCOM

Different components of CEAG-DVFC system shown in Fig. 1 are modeled as follows.

Three-phase voltages at the CEAG terminals (v_a, v_b and v_c) are considered sinusoidal and hence their amplitude is computed as:

$$V_t = \sqrt{(2/3)\,(v_a^2 + v_b^2 + v_c^2)} \qquad (1)$$

The unit vector in phase with v_a, v_b and v_c are derived as:

$$u_a = v_a/V_t; \quad u_b = v_b/V_t; \quad u_c = v_c/V_t \qquad (2)$$

The unit vectors in quadrature with v_a, v_b and v_c may be derived using a quadrature transformation of the in-phase unit vectors u_a, u_b and u_c as:

$$w_a = -u_b/\sqrt{3} + u_c/\sqrt{3} \qquad (3)$$

$$w_b = \sqrt{3}\,u_a/2 + (u_b - u_c)/2\sqrt{3} \qquad (4)$$

$$w_c = -\sqrt{3}\,u_a/2 + (u_b - u_c)/2\sqrt{3} \qquad (5)$$

1. Quadrature Component of Reference Source Currents

The AC voltage error $V_{er(n)}$ at the nth sampling instant is:

$$V_{er(n)} = V_{tref(n)} - V_{t(n)} \qquad (6)$$

where $V_{tref(n)}$ is the amplitude of reference AC terminal voltage and $V_{t(n)}$ is the amplitude of the sensed three-phase AC voltage at the CEAG terminals at n^{th} instant.

The output of the PI controller ($I_{smq(n)}^*$) for maintaining AC terminal voltage constant at the nth sampling instant is expressed as:

$$I_{smq(n)}^* = I_{smq(n-1)}^* + K_{pa}\{V_{er(n)} - V_{er(n-1)}\} + K_{ia}\,V_{er(n)} \qquad (7)$$

where K_{pa} and K_{ia} are the proportional and integral gain constants of the proportional integral (PI) controller. $V_{er(n)}$ and $V_{er(n-1)}$ are the voltage errors in n^{th} and $(n-1)^{th}$ instant and $I_{smq(n-1)}^*$ is the amplitude of quadrature component of the reference source current at $(n-1)^{th}$ instant.

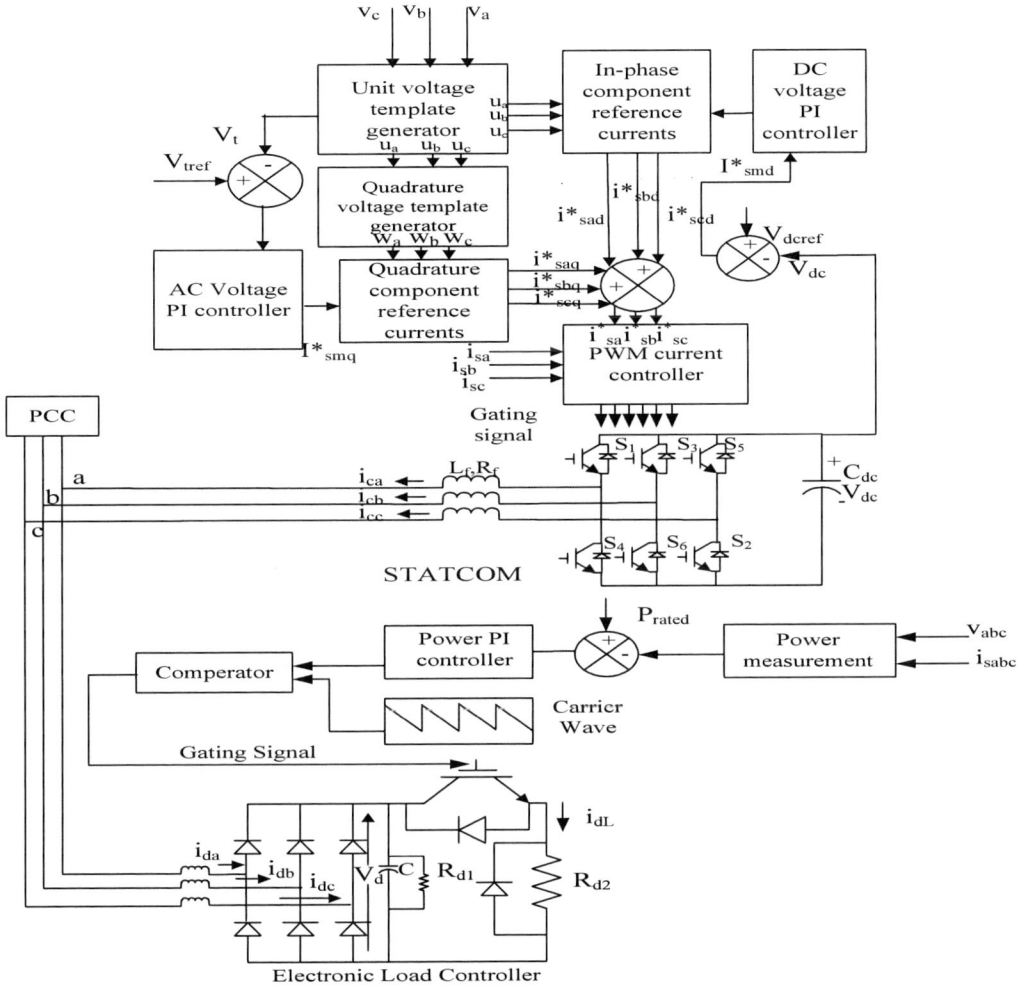

Fig. 2: Control scheme for the proposed voltage and frequency controller.

The quadrature components of the reference source currents are computed as:

$$i^*_{saq} = I^*_{smq} W_a; \ i^*_{sbq} = I^*_{smq} W_b; \ i^*_{scq} = I^*_{smq} W_c \quad (8)$$

2. In-Phase Component of Reference Source Currents

The error in DC bus voltage of STATCOM ($V_{dcer(n)}$) at n^{th} sampling instant is:

$$V_{dcer(n)} = V_{dcref(n)} - V_{dc(n)} \quad (9)$$

where $V_{dcref(n)}$ is the reference DC voltage and $V_{dc(n)}$ is the sensed DC link voltage of the STATCOM. The output of the PI controller for maintaining DC bus voltage of the STATCOM at the nth sampling instant is expressed as:

$$I^*_{smd(n)} = I^*_{smd(n-1)} + K_{pd} \{ V_{dcer(n)} - V_{dcer(n-1)}\} + K_{id} V_{dcer(n)} \quad (10)$$

$I^*_{smd(n)}$ is considered as the amplitude of active source current. K_{pd} and K_{id} are the proportional and integral gain constants of the DC bus PI voltage controller.

In-phase components of reference source currents are computed as:

$$i^*_{sad} = I^*_{smd} u_a; \ i^*_{sbd} = I^*_{smd} u_b; \ i^*_{scd} = I^*_{smd} u_c \quad (11)$$

3. Reference Source Currents

Total reference source currents are sum of in-phase and quadrature components of the reference source currents as:

$$i^*_{sa} = i^*_{saq} + i^*_{sad} \quad (12)$$
$$i^*_{sb} = i^*_{sbq} + i^*_{sbd} \quad (13)$$
$$i^*_{sc} = i^*_{scq} + i^*_{scd} \quad (14)$$

4. PWM Current Controller

The total reference currents (i^*_{sa}, i^*_{sb} and i^*_{sc}) are compared with the sensed source currents (i_{sa}, i_{sb} and i_{sc}). The ON/OFF switching patterns of the gate drive signals to the IGBTs are generated from the PWM current controller. The current errors are computed as:

$$i_{saerr} = i^*_{sa} - i_{sa} \quad (15)$$
$$i_{sberr} = i^*_{sb} - i_{sb} \quad (16)$$
$$i_{scerr} = i^*_{sc} - i_{sc} \quad (17)$$

These current error signals are amplified and then compared with the triangular carrier wave. If the amplified current error signal is greater than the triangular wave signal switch S_4 (lower device) is ON and switch S_1 (upper device) is OFF. If the amplified current error signal corresponding to i_{saerr} is less than the triangular wave, the signal switch S_1 is ON and switch S_4 is OFF. Similar logic applies to other two phases.

B. Control Algorithm for ELC

To maintain the generated power constant at the generator terminals, the measured power (P_{gen}) is compred with generator rated power (P_r) and power error $P_{er(n)}$ at n^{th} sampling instant ia calculated as:

$$P_{er(n)} = P_{r(n)} - P_{gen}(n) \qquad (18)$$

where $P_{r(n)}$ is the reference or rated power $P_{g(n)}$ is the measured power at the nth sampling instant and estimated as:

$$p_{gen} = \frac{1}{\sqrt{3}}\left(v_a i_a + v_b i_b + v_c i_c\right) \qquad (19)$$

where v_a, v_b and v_c are line voltages and the output of the PI power controller at the nth sampling instant, is expressed as:

$$P^*_{con(n)} = P^*_{con(n-1)} + K_{pp}\{ P_{er(n)} - P_{er(n-1)}\} + K_{pi} P_{er(n)} \quad (20)$$

where K_{pp} and K_{pi} are the proportional and integral gain constants of the power controller. The PI controller output ($P^*_{con(n)}$) is compared with the triangular carrier (P_{tri}) waveform and output is fed to the gate of the chopper switch (IGBT)is used in an ELC of DVFC.

When $P^*_{con(n)} > P_{tri}$,   SD = 1 and
When $P^*_{con(n)} < P_{tri}$,   SD = 0 $\qquad (21)$

The SD is the switching function used for generating the gating pulse of IGBT of the ELC chopper.

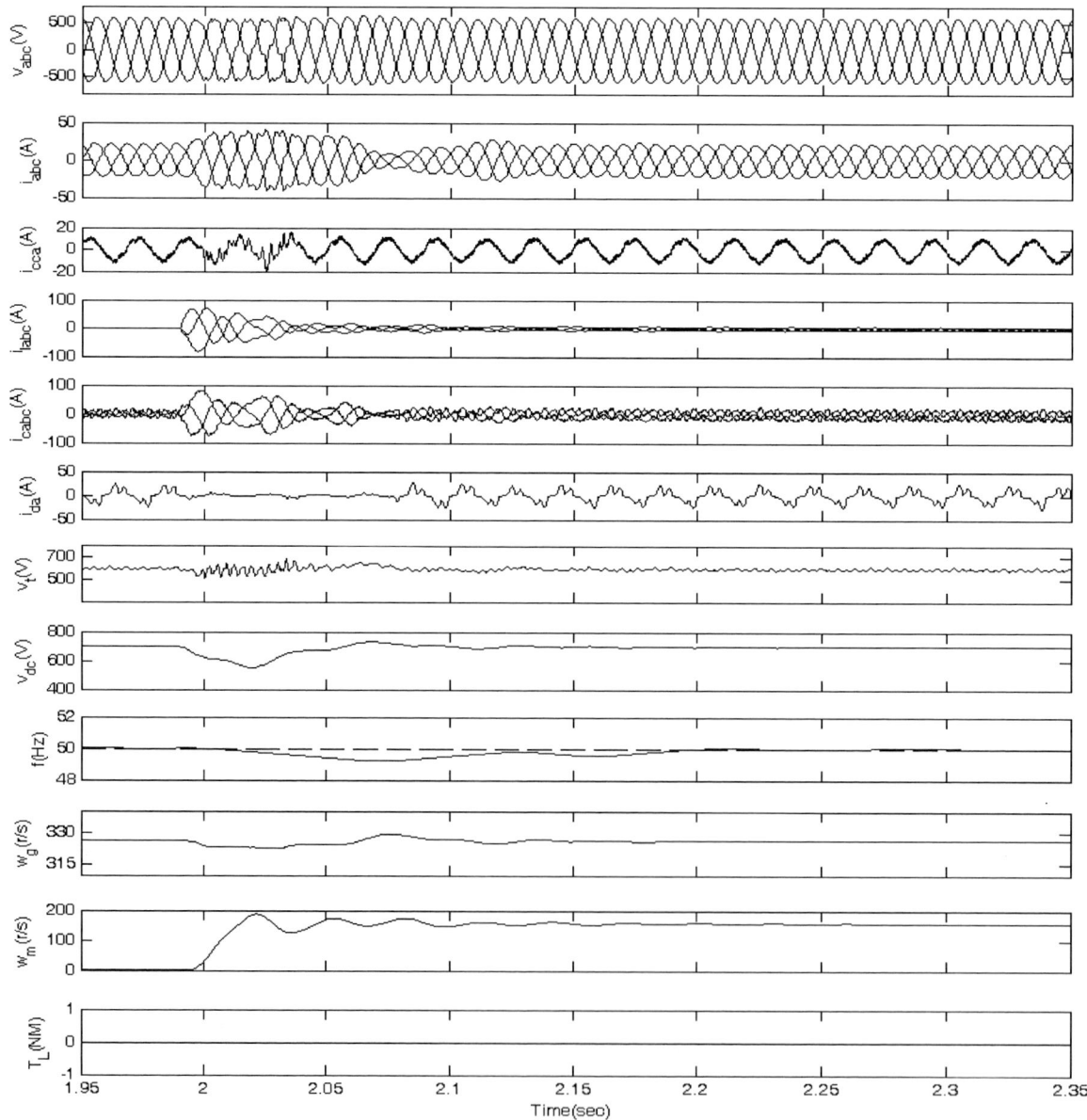

Fig. 3: Transient waveforms during direct online starting of an asynchronous motor load in proposed electrical system

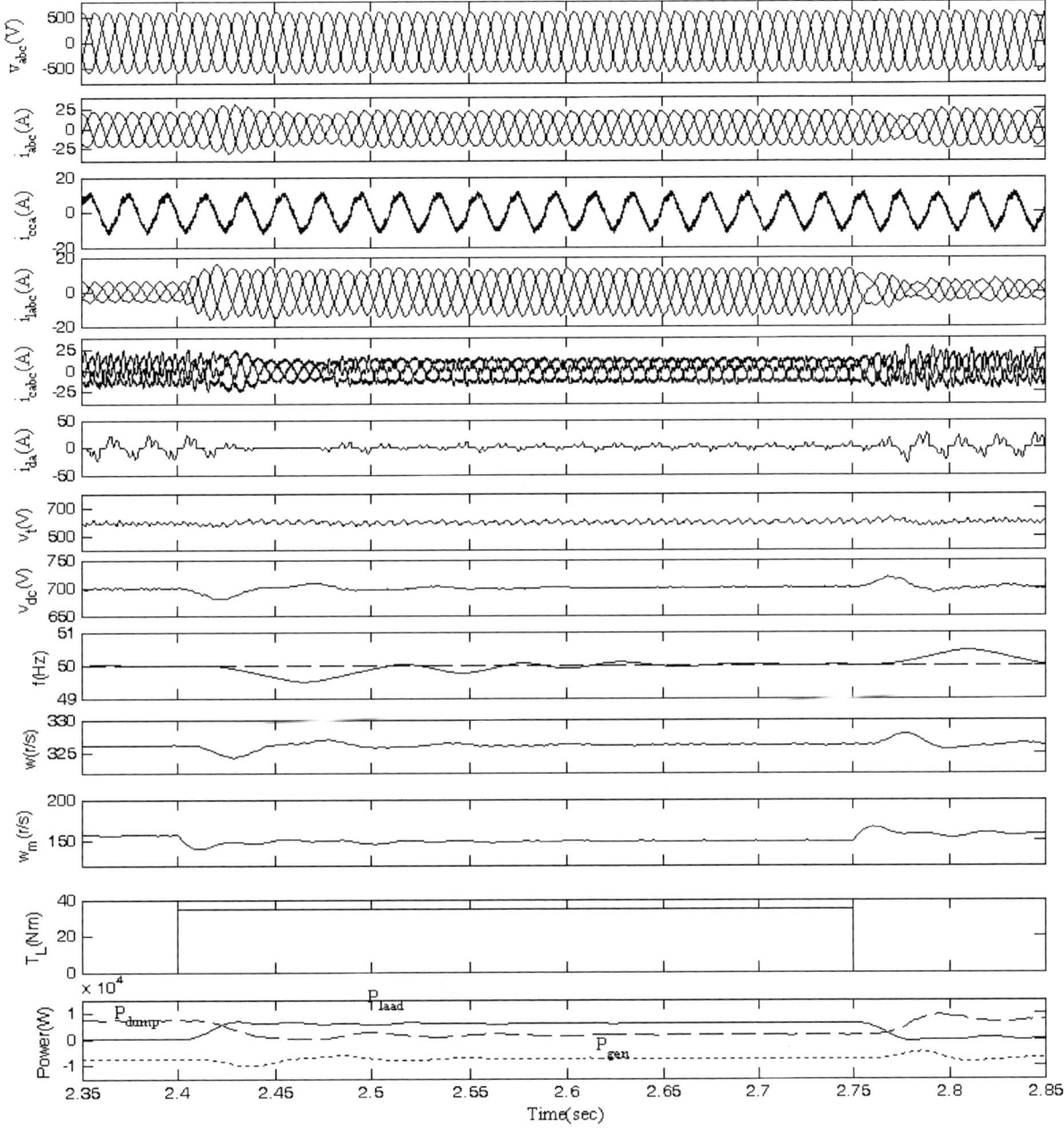

Fig. 4: Transient waveforms during the load torque application and removal of motor load.

V. MATLAB BASED MODELING

MATLAB model of the DVFC asynchronous generator system consists of asynchronous machine with capacitor bank and DVFC are realized in version 7.3. The modeling of CEAG is carried out using 7.5 kW, 415V, 50Hz, Y-connected induction and 5kVAR delta-connected excitation capacitor banks. The controller is realized with 3-leg voltage source converter and a diode bridge rectifier based ELC with DC chopper and auxilliary load. A dynamic load (Asynchronous motor) is considered here to demonstrate the capability of controller. Simulation is carried out in discrete mode at 10e-6 step size with ode23tb (stiff/ TR-BDF-2) solver.

VI. RESULTS AND DISCUSSION

Figs. 3 and 4 show the performance of the proposed system with direct on-line starting of an asynchronous motor and application/removal of a load torque on the running motor respectivaly. Different transient waveforms of generator voltage (v_{abc}) and generator currents (i_{abc}), excitation capacitor currents (i_{cca}), load currents (i_{labc}), STATCOM currents (i_{cabc}), ELC current (i_{da}), amplitude of terminal voltage (v_t), the DC link voltage (V_{dc}), frequency (f), generator and motor speed (ω_g, ω_m) and motor applied load torque (T_L) power (P_{gen}, P_{load} and P_{dump}) etc are shown.

Fig 3 shows the controller performance with direct on line

starting of a 3-phase asynchronous motor. At 2s it is observed that due to sudden starting of an asynchronous motor, the current flowing through load controller is sudden reduced to maintain the power at the generator terminal constant.

At 2.4 s the load torque is applied on the motor as shown in Fig 4 then the current in the stator winding of the motor is increased and the current drawn by the auxiliary load is reduced due to action of the controller while voltage and current at the generator terminal remain constant. At 2.75 s when the load torque on the motor is removed the current drawn by the motor is reduced and additional generated power is absorbed by the auxiliary load. In this way the controller functions as voltage and frequency controller along-with a load balancer.

VII. CONCLUSION

The performance of the proposed voltage and frequency controller has been demonstrated for an isolated power generation system and feeding dynamic loads. Here it is observed that the controller responds in desired manner and regulates the system voltage and frequency under direct on line starting of asynchronous motor and application /removal of load torque.

VIII. APPENDIX

A. The Parameters of 7.5kW, 415V, 50Hz, Y-Connected, 4-pole Asynchronous Machine used as AG are given below

$R_s = 1\Omega$, $R_r = 0.77\Omega$, $X_{lr} = X_{ls} = 1.5\Omega$, $J = 0.1384$ kg-m^2

$L_m = 0.134$ ($I_m < 3.16$)

$L_m = 9e\text{-}5I_m^2 - 0.0087I_m + 0.1643$ ($3.16 < I_m < 12.72$)

$L_m = 0.068$ ($I_m > 12.72$)

B. The Parameter of 4kW, 415V, 50Hz, Asynchronous Machine as a Motor Load

$R_s = 1.4\Omega$, $R_r = 1.39\Omega$, $X_{lr} = X_{ls} = 1.8\Omega$, $J = 0.0131$ kg-m^2.

REFERENCES

[1] R.C. Bansal, "Three phase self excited induction generators: an overview," *IEEE Trans. on Energy Conversion,* vol 20, no. 2, pp. 292-299, June 2005.

[2] S.N. Bhadra, D. Kastha, S.Banerjee, "Wind Electrical Systems" 1st Ed. Oxford University Press, New Delhi 2004.

[3] M. Godoy Simoes, Felix A. Farret, "Renewable Energy Systems," 1st Ed. CRC Press Florida, 2004.

[4] M. H. Salama and P. G. Holmes, "Transient and steady state load performance of a stand alone self excited induction generator," *IEE Proc.- Electr. Power Appl.,* Vol. 143, No. 1, pp. 50-58, January 1996.

[5] S. H. Kim and S. Y. Hahn, "Analysis and design of a induction generator with a superconducting bulk magnet rotor," *IEEE Trans. on Applied Superconductivity,* Vol. 10, No. 1, pp. 931-934, March 2000.

[6] B. Singh, S.S. Murthy and Sushma Gupta, "Transient analysis of self excited induction generator with electronic load controller supplying static and dynamic loads" *IEEE Trans. on Industry Applications,* vol. 41, no. 5, pp.1194-1204, Sept 2005.

[7] O. Chtchetinine, "Voltage stabilization system for induction generator in stand alone mode," *IEEE Trans. on Energy Conversion,* Vol. EC-12, No. 3, pp. 298-303, September 1999.

[8] E. G. Marra and J. A. Pomilio, "Self excited induction generator controlled by a VS-PWM bi-directional converter for rural application," IEEE Trans. on Industry Applications, vol. 35, no. 4, pp. 877-883, July/August 1999.

[9] B.Singh, S.S. Murthy and Sushma Gupta, "A voltage and frequency controller for self-excited induction generators" Electrical Power Components and Systems, vol., 34, pp 141-157, 2006.

AUTHORS' BIOGRAPHY

Bhim Singh (SM'99) was born in Rahampur, India, in 1956. He received the B.E (Electrical) degree from the University of Roorkee, Roorkee, India, in 1977 and the M.Tech and Ph.D. degree from the Indian Institute of Technology (IIT) Delhi, New Delhi, India, In 1979 and 1983, respectively.

In 1983, he joined the Department of Electrical Engineering, University of Roorkee, as a lecturer, and in 1988 became a Reader. In December 1990, he joined the Department of Electrical Engineering, IIT Delhi, as an Assistant Professor. He became an Associate Professor in 1994 and Professor in 1997. His area of interest includes power electronics, electrical machines and drives, active filters, FACTS, HVDC and power quality.

Dr. Singh is a fellow of Indian National Academy of Engineering (INAE), the Institution of Engineers (India) (IE(1)), and the Institution of Electronics and Telecommunication Engineers (IETE), a life Member of the Indian Society for Technical Education(ISTE), the System Society of India (SSI), and the National Institution of Quality and Reliability (NIQR) and Senior Member of Institute of Electrical and Electronics Engineers (IEEE).

Gaurav Kumar Kasal was born in Bhopal, India, in Nov, 1978. He received the B.E (Electrical) and M.Tech degree from the National Institute of Technology (NIT) Allahabad and National Institute of Technology (NIT) Bhopal, India respectively in 2002 and 2004. Since Dec 2004, he has been pursuing the Ph. D. degree with the Department of Electrical Engineering, Indian Institute of Technology (IIT) Delhi, New Delhi, India. His field of interest includes power electronics and drives, renewable energy generation and applications, FACTS and electrical machines.

Power Electronic Conversion Systems for Deepwater Applications

Jacobo Aguillon-Garcia
Korean Advance Institute of
Science and Technology
Dept. of Electrical Engineering
373-1 Guseong-dong, Yuseong-gu
Daejeon, 305-701
jacobo@angel.kaist.ac.kr

Andreas Kristoffersen
Schlumberger S.P.A.

Via Terra Vergine, 14
65129 Pescara, Italia

akristoffersen@pescara.oilfield.slb.com

Tore Undeland
Norwegian University of
Science and Technology
Inst. for Elkraftteknikk
O.S. Bragstads plass 2e,
Gloshaugen, N-7491 Trondheim
tore.undeland@elkraft.ntnu.no

Abstract—This paper describes the research program devoted to the study of electronic components and its function in the power conversion systems to be used on deepwater applications. Frequent approaches for operation of electronic equipment include: Hard shell enclosures maintained at normal air pressure, soft shell with a pressure compensating fluid and free-flooded equipment. This article discusses the last two techniques taking in account the overall cost. In those viewpoints the characteristics and the influence of high hydrostatic pressure on the components and incompatibility of the components with the immersion fluid (if needed), will be presented. Guidelines necessary to alleviate anticipated and known problem areas are proposed.

I. INTRODUCTION

A. Background

Located 120 kilometers from the Norwegian coast, Ormen Lange (Fig. 1) is Norway's second-largest offshore gas field. Gas reserves are estimated at 397 billion m³. Geographical location, weather and sub sea environment make unaffordable a permanent offshore platform.

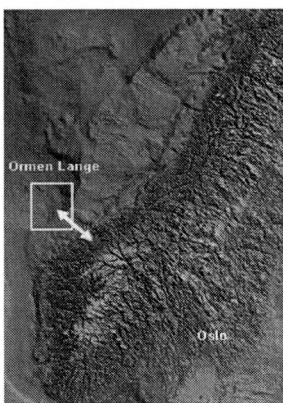

Fig. 1: Ormen Lange location.

If the technology were available for design and development of circuits and equipments that would operate reliably when immersed in deep water locations, most of the problems mentioned above could effectively be addressed.

The need for electronics which will operate in an environment that is pressure equalized to the ocean depths could be understood by a brief review of some of the considerations that stem from the oil industry's interest in the deeper parts of the ocean.

Therefore, some of the objectives of this article contemplate:

- Classification and characterization of materials components, technology, and system concepts as to their potential usefulness in submergence application.
- The development of knowledge and technology to allow the utilization of as many electronic devices and components as practically possible.
- The development of a design and testing procedure which can be used by designers.

B. Basic Approach

The proposed solution of the problem is to classify and separate the various observed effects into physical, physicochemical, shorthand and long-term reliability problem areas. It is expected that further research in various areas will permit successful synthesis and a more rapid, comprehensive solution to the overall problem:

- Develop a model of the electronic material, components, circuits and equipments immersed in a dielectric fluid or free-flooded.
  - ◆ Consistent with reasonable overall constraints on the electronics such as environment, operating conditions, reliable performance, and cost.
  - ◆ That contains several physical and chemical interfaces between materials, components, circuits, dielectric fluid, coating for free-flooded circuitry and housing that require analysis and experimentation.
  - ◆ Provide insight into the proper selection of dielectric fluid, housing, passivating, potting compounds and covering for free-flooded circuitry.
- Identify specific failure mechanisms and develop techniques to harden against such failure; i.e., try to identify a universal pressure-hardening technique, universal fluid, universal coating and contaminant hardening techniques and systematically investigate them.
- Plan test programs and implement them to determine long-term effects including pressure, fluid, contamination, temperature, and cycling.
- Implement representative and feasibility systems to demonstrate usefulness of the results.

II. MATERIALS AND METHODS

The 99% of the known ocean is less than 6,000mts. deep, corresponding to a pressure of around 620bar, a pressure equal to or greater than 1.5 times this pressure would allow operation almost everyplace in the ocean with a reasonable safety factor. Hence, a practical pressure selected for the first series of tests is approximately 1,000bar. It should be

978-1-4244-1871-8/07 $25.00 © 2007 IEEE   1143

noted that 1,000bar is equivalent to 10,287mts. and that the deepest known point in the ocean is a little over 11,000mts. [1]. Therefore, testing to 1,000bar will qualify components to go virtually anywhere in the ocean (excluding any safety factor). Obviously, some components may be well suited to low but not deep operations. Thus, a wider selection of components might be available for shallow operations, with a narrowing selection as depth increases.

A. Pressure Test

Due to different effects of pressure on diverse electronic form factors, diverse test procedures are necessary. This could be a single exposure to pressure, various excursions to pressure, a long-term exposure, a short-term pressure, or a combination of previous mentioned. In this paper, all the samples were immersed or filled with a dielectric fluid [2-4].

Generally, electronic components are physically classified in hermetic packages (with voids), and plastic/ceramic devices (void-less).

The procedure used in metal hermetic devices is simply to pressurize them. The devices are not energized and no measurements are taken under pressure. The most obvious harm on these devices is the structural damage.

By other hand, the test procedure for plastic/ceramic devices (void-less), includes the following procedures:

• Subject samples to cyclic pressure and long-term at defined pressure.
• Energize samples on testing.
• Implement a circuit and realize measurements under pressure.
• Perform failure analysis on fault devices.

The effect on this type of devices reveals three classes of damages: structural, physical and chemical.

The pressurization procedure is divided into series utilizing a facility similar to [5]. One series is defined as 100 cycles of test pressure, afterward a 60-day soak at test pressure. A cycle (0bar to test pressure and backwards) takes about 6 minutes. The devices may be subjected to one or more of these series. The test pressure is 1,000bar unless otherwise stated.

B. Effects on Pressurized Components

For metal packaged devices, the most common form of structural damage is package crushing (Fig. 2). As might be expected, the severity of crushing is directly proportional to pressure and package size.

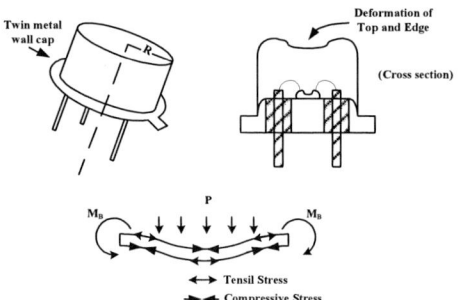

Fig. 2: Effects of pressure on TO's packages.

At any given pressure, a small package will be crushed less than a large package. This failure mode depends on the geometry, dimensions, and material properties. A technique available for hardening are potting with a structurally supportive material (this is helpful in free-flooding approach). Another form of structural damage is cracking of the lead pin to header seal (Fig. 3). This type of damage was observed on TO-72 packages cycled 10 times to 1,000bar. The package is small and strong enough not to crush; however, the relatively weak lead to header glass seals was cracked.

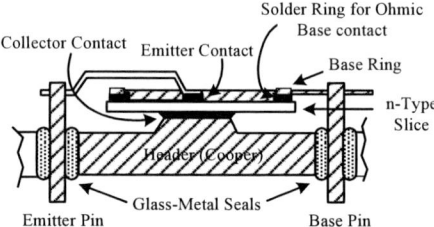

Fig. 3: Typical construction of power transistor.

Similar damage, (for ceramic packages), is the development of leaks in the package. These leaks may either be fine or gross (as defined in MIL-STD 883, Method 1014). For ceramic DIP, leaks began to occur at 55bar (1 leak in 24 samples), proportional to pressure and number of cycles (e.g., 2 leaks in 25 samples at 1,000bar, 36 leaks in 39 samples at 1,000bar).

Two other types of structural damage have been observed, one type for TO-105 and TO-106 packages (ceramic header, epoxy seal), and the other for plastic packages with a large amount of silicone rubber over the chip. On the TO-105 and TO-106 packages, the epoxy seal over the top of the chip had broken off. This is believed to be caused by liquid being forced into the area of the epoxy to header bond during long-term exposure to pressure. The above effect can lead to the second category of damage: physical damage.

The types of physical damage that have been observed to result from pressurization of active devices are broken leads, lifted metallization and cracking of the chip. Broken leads and lifted metallization were the result of package damage and for manufacturing defects, and the cracked chips (accounted for all failures of power transistors), were caused by voids in the chip-to-header bond (Fig. 4).

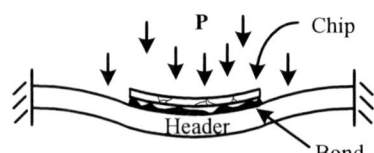

Fig. 4: Cracked chip due to voids on bonding.

Another case of physical harm is the electrical shorting of a device due to the deformation of some part of the device, for example, shorting between the contacts posts in a transistor. Other failure mode is stressing of materials beyond their elastic limits to produce a fracture and bonding of two materials with different compressibility to produce stress. In the case of silicon die, mechanical stress or strain

978-1-4244-1871-8/07 $25.00 © 2007 IEEE

changes the lattice spacing in materials also destroys the crystal symmetry.

Methods of elasticity theory [6] are used in facing the general problem of determining the stress distribution s(x,y,z) in model structures which resemble the component packages most generally used. These structures are composed of laminar slab elements. Acting on these elements are the applied pressure (P), reaction forces (R), and bending moments (M_B), as sketched in figure 5.

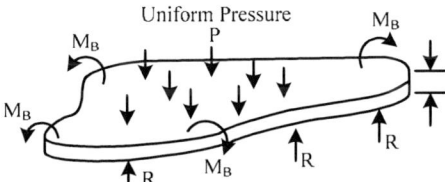

Fig. 5: Forces and moments applied to a laminar slab.

It is often this symmetry change that induces the more pronounced variations in electrical properties. When the lattice spacing and symmetries are altered, in general, cause changes in almost all the material parameters which contribute to the electrical properties. In addition to purely dimensional changes, one finds changes in such important parameters as the amplitude of vibration of the atoms about their normal lattice sites, the energy band structures, the inter and intra atomic interaction, and the impurity lattice interaction.

Consider the effects of stress on resistors. This is one of the easier devices to characterize due to its simplicity of material, phenomena, and configuration. Here the primary concern is with the electrical conductivity (piezoresistance and elastoresistance) [7]. Fortunately, hydrostatic pressure has little effect on the crystals and homogeneous materials used.

Practically no work has been done on the experimental characterization of dielectric materials under pressure. There has been some work on ferroelectrics such as $BaTiO_3$ and $SrTiO_3$ [8, 9] in which large changes in the dielectric constants have been observed with pressure. The dielectric constant of these materials decreases with pressure while the loss tangent first increases and then decreases as pressure is increased. Some of the single cubic crystals of the alkali halides have also been investigated [10]. The optical properties of materials are closely related to the dielectric properties, and some insight has been gained into the dielectric properties of materials by studying their optical properties under pressure.

Most of the above types of damage are directly attributable to a void in the package or in the chip-to-header bond.

The third category of damages is dependent on the pressurization medium [2-4], and it's dissolved or emulsified impurities being forced into the area of the silicon chip. This category is chemical damage. Chemical damage may take many forms. Among the most common are formation of ion chains and inversion layers, and the many reactions that can occur in the presence of moisture. The subject is commented in next subsection; however, a few comments should be made. The extent and speed of chemical damage will depend on the liquid used and its impurities. It will generally be a slow process visible only after long-term testing and results in a slow degradation of the device, rather than a quick, clean failure.

C. Dielectric Fluids

An extra failure mode results from the fluid environment, which is the transport medium for contaminants. These may be caused by the fracture of some encapsulation material, deliberate flooding, or simply the lack of the use of any passivation scheme for device elements. These problems are not easy to classify; however, there are at least three categories:

- Filling of voids.
- Electronic interaction.
- Chemical interaction.

The filling of voids by the immersion fluid can cause the device to explode when pressure is suddenly released.

The second category involving electronic interaction is as broad as the types of elements which will be used, e.g. the fluid or a contaminant transported by the fluid may collect on the surface of an insulating or semi-insulating element and influence electronic charge distribution in the element. Thin film resistors would be very susceptible to this failure mode.

Another example is the diffusion of ions, such as sodium, into critical elements or device parts resulting in degradation of the device functions, e.g., a shift in MOS transistor gate voltage with any sodium ions which could find their way into the oxide or the oxide silicon interface. The bonding of the passivated silicon chip to a gold-plated header presents a pathway for moisture and ionic contaminants along the passivant-gold interface since the adherence of plastics to gold, kovar, nickel alloys, solder, solder plating, and cadmium is poor.

Chemical interaction of the fluid or fluid contaminants is a serious problem; e.g. the degradation of a curing agent in a rigid phenolic outer case of a micro diode produced ammonia which permeated a silicone barrier coating between the phenolic and the semiconductor surface [11]. The ammonia caused an increase in the leakage current of the diode. Chloride ions, from a source like sea water, corrode aluminum leads which result in an open circuit or a high resistance [12].

In order to select the best fluid, it was necessary to establish a scheme. The requirements of such a fluid, relative to the environment, were grouped into three categories, namely chemical, dielectric, and physical. The choice was then based on achieving superior qualifications in all three categories.

The fluids will have to withstand exposure to ambient temperatures ranging from -2° to 150°C, cycling pressures to 1,000bar, and chemically corrosive environments. The fluid will be expected to shield solid-state components from the environment and prevent electronic failures due to energy transfer and contamination, not emulsify water and simultaneously retain a homogeneous fluid phase which would offer maximum dielectric properties [2-4, 13]. Pressure-compensating fluid for electric and electronic components must have good dielectric properties and be inert to the action of electrical equipment. The dielectric quality of a fluid is determined by the dielectric breakdown voltage, dissipation factor, and electrical resistivity. The dielectric properties of a fluid result from its chemical composition and additives.

The dielectric breakdown voltages were measured according to ASTM procedure D877-02e1 with the exception of a 1.27mm electrode gap [14].

The requirements that fluids inhibit corrosion and lubricate dictate the utilization of polar and water-soluble additives. These polar materials not only lower the dielectric breakdown voltage and resistivity, but also increase the dissipation of electrical energy as heat. The ideal dielectric fluid would have a dissipation factor of 0%. A relative measure of acceptable dissipation factor for dielectric fluids has been fixed at 2.0%. The dissipation factors [13, 15] were determined from the measurement of the loss angle of a fluid-filled cell on a capacitance bridge.

Maximum protection against failures of electrical and electronic components states that the dielectric fluid exhibit a minimum resistivity of 3×10^{11} Ω-cm, which is the tentative measure of acceptable resistivity.

D. Device Hardening Materials and Techniques

Since true hydrostatic pressure has little effect on most elements of electronic components, the mayor hardening technique against pressure will be to modify housings and encapsulants to allow the hydrostatic pressure to be transmitted directly to the device elements i.e. free-flooded. In cases where voids either cannot be filled because of the manner in which the device is constructed or because the device operation requires a void (e.g. crystals, distributed gap ferrites) the component must be potted or hardened mechanically against the pressure. The third type of hardening technique involves the passivation or protection of critical device areas against the fluid environment; none of those are free of serious problems.

In potting materials for pressure protection, there are several requirements and tradeoffs to be consider. It must be strong enough to withstand the desired pressure, adhere to all types of materials and particularly metal wires and conducting paths, have a reasonable thermal conductivity, have good insulating properties, possess good chemical resistance to both the fluid and the component materials, and have as little shrinkage during curing as possible. The most important practical problems of the above list are the adhesion and shrinkage during curing; e.g. consider the case where it is desired to fill the inside (void) of housing with a hard material epoxy. If the epoxy shrinks during curing, it almost certainly will separate from the housing. The very opposite type problem results from potting a component in which shrinkage results in very large internal stresses applied to the device [16-19]. Simple problems often result, such as the breaking off of leads during curing.

A synopsis of types of materials for potting electronics (table 1), reveals that low-density epoxy bonded syntactic foam adheres best and shrinks the least on curing.

TABLE 1

Type*	Typical Properties
Low-density Epoxy-bonded syntactic foam	Low coefficient of thermal expansion, low elastic modulus, high percentage of voids, limited strength.
Dense, highly filled, low expansion epoxies	Thermal expansion coefficients closely match component lead materials, strong rigid.
Unfilled silicones and urethanes	High thermal expansion coefficient, low elastic modulus.

*Specific potting materials under consideration include: various types of epoxies, ceramics, and silicones in addition to glass and diallyl-phthalate.

In the case of a passivating or protection coating, the requirements may not be as severe. Chemical inertness is the key for these materials. They must protect against all types of impurities, however, the two most prevalent types are:

- Water vapor.
- Chloride and sodium ions, as well as heavy metals.

This type of passivation can be classified as a coating. All types of semiconductor devices utilize some type of passivation scheme. Integrated circuits have passivation and protective layers formed on them during the device fabrication processing steps. Hence, a good passivant by definition is a good adherent. It is likely that a good passivant against moisture will also be a good passivant against all other impurities. These surfaces will in many instances be the first to which additional coatings will be applied.

III. TEST RESULTS

Table 2 provides a general understanding of the short-term effects of high hydrostatic pressure cycling (0bar - 1,000bar), on electrical and electronic components.

A. Resistors

Resistors exhibit a decrease in resistance under pressure. Carbon types, which are the most pressure sensitive, vary 20% or more, while the metal-film and wire wound resistors vary less than 10% [20].

By other hand, data taken during the test procedure indicated that thick-film resistors are not satisfactory for use in a high-pressure environment. Changes in resistor value vs. pressure were large and inconsistent. Some resistance values increased with pressure, others decreased. Resistance changes as much as 43% were recorded 1,000bar.

Thin-film resistors, on the other hand, showed both small and consistent changes in resistance as a function of pressure. After the measured data was corrected for temperature change, the value for resistance variation as a function of pressure was found to be <5% at 1,000bar.

B. Capacitors

The table includes data from the failure analyses; when the original data were taken, a number of capacitors

978-1-4244-1871-8/07 $25.00 © 2007 IEEE

appeared to fail. During the failure analysis, it was determined that the capacitors actually were not failing but that the substrate was cracking (or separating), giving the appearance of a device failure.

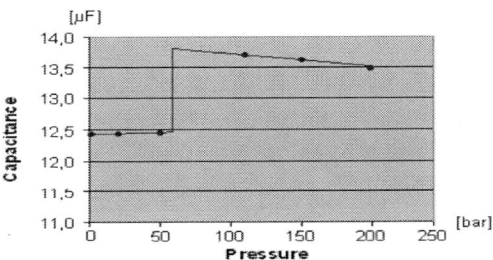

Fig. 6: Capacitance vs. pressure in an electrolytic capacitor.

Dry-type capacitors, namely glass and ceramic, only suffer slight changes with pressure (less than 5%). The wet capacitors, tantalum, paper, and especially electrolytic, are characterized by a considerable increase in capacitance and subsequent case collapse and functional breakdown (Fig. 6).

C. Magnetics

Inductors display the greatest sensitivity to high pressure due to the susceptibility to magnetostrictive effect [21-22]. Pressure has the effect of altering the magnetic permeability of materials [23]. Not only does the saturation magnetic moment change with pressure but the curie-point can also change. In fact, 30% nickel-iron alloys become nonmagnetic at room temperature under a pressure of 12.5kbar [24].

Iron-core types are characterized by large and often erratic variations with some permanent change in inductances and Q. As with resistor materials, granular or powdery materials which are polycrystalline will result in magnetic devices which are very susceptible to pressure-induced changes in their electrical properties. Any time there are voids in materials, the magnetic properties are likely to experience drastic changes with pressure. Core materials for magnetic devices vary greatly in the sensitivity of magnetic parameters, such as permeability, to pressure. Polycrystalline ferrite materials are known to be stress sensitive [25]. If any of these classes of devices are to be operated immersed at high pressure, a careful assessment of their behavior should be made. Iron cores have been found to increase their permeability with pressure, i.e. molybdenum permalloy dust cores, test were performed using Magnetics Designer software [26].

Air-core inductors are less sensitive to pressure. With either type, maximum change occurs during pressure cycling and appears to stabilize at a lower value after soaking.

D. Semiconductors

The influence of mechanical stress on semiconductors, particularly silicon and germanium, received much attention in previous years [24]. As already mentioned, it is the stress-induced changes in the energy band structure of semiconductors that are reflected in property changes. Both the majority and minority carriers are affected. Silicon and germanium exhibit large piezoresistance effects as compared to metals [27]. Hydrostatic pressure effects on the resistance of semiconductors are, in general, much smaller than shear and uniaxial stress effects. This is due to the fact that hydrostatic pressure does not break up the crystal symmetry.

Minority carriers and their physical properties are stressed sensitive, i.e. large uniaxial stresses can cause many orders of magnitude increases in the number of minority carries [28-30]. The effects for purely hydrostatic pressures are much lower than for anisotropic stress effects. Many of the standard semiconductor devices have been studied and their stress dependence characterized [28-32].

Most short-term failures are caused by case deformation or collapse resulting in open circuits and in some instances short circuits. Many samples could be operated by puncturing the case prior to pressurization so that oil could enter the housing. Characteristics of diodes, SCR's, BJT's and switching devices in general, which did not experience catastrophic failure, varied from less than 10% to as much as 50% [30, 33].

E. Circuit Implementation

Having in account the previous considerations on electronic components, the logical step is to create a power conversion circuit to validate the previous mentioned results.

Utilizing specific software [26], and pressure tested parts a small converter was created. Up to now, circuit has not presented different results as in an ordinary pressure.

IV. CONCLUSIONS

In this article is demonstrated that power electronics systems are feasible for deepwater applications. Independently of component's new properties achieved by exposure on high pressure, mostly all components off-the-shelf are reliable under such characteristics. The need of buffer fluid to prevent damage on components appears to be mandatory in this type of environments. The viability to implement high scale circuit blocks under this environment makes them cost effective in the long run.

REFERENCES

[1] "Physical Features of the Ocean" The Museum of Science, Boston1998 <http://www.mos.org/oceans/planet/features.html >.

[2] "Fluorinert Electronic Liquids", *Data sheet*, 3M, Boston 2006 http://products3.3m.com/catalog/us/en001/oil_gas/speciality_materials/.

[3] "Tellus T Oils", *Data sheet*, Shell oils, http://www.shell.com.

[4] "Silicon Dielectric Fluids", *Data sheet*, Dow Corning http://dowcorning.com/content/etronics/.

[5] J. E. Holzschuh, "An Automated Deep Sea Simulation Facility for Testing Electronic Components", *IEEE Eng. In Ocean Env. Conf.*, Vol. 3, Sep 1971, pp.12-15.

[6] S. Timoshenko, J. N. Goodier, "Theory of Elasticity", 3rd. Ed., New York, McGraw-Hill 1970.

[7] P. W. Bridgman, "The Physics of High Pressure", Dover Publications Inc., London, 1971.

[8] W. J. Merz, "The Effect of Hydrostatic Pressure on the Curie Point of Barium Titanate Single Crystals", *Physical Review*, Vol. 78, Apr 1950, pp. 52-54.

[9] G. A. Samara, A. A. Giardini, "Pressure Dependence of the Dielectric Constant of Strontium Titanate" *Physical Review*, Vol. 140, Issue 3A, Nov 1965, pp. A954-A957.

[10] S. Mayburg, "Effect of Pressure on the Low Frequency Dielectric Constant of Ionic Crystals" *Physical Review*, Vol. 79, Issue 2, Jul 1950, pp. 375-582.

[11] J. J. Licari, G. V. Browning, *Electronics*, Vol. 17 Apr 1967, pp. 101-108.

[12] Snow, E. H., et al, "Ion Transport Phenomena in Insulating Film," *Applied Physics*, Vol. 36, No. 5. 1965, pp. 1664.

[13] J. L. March, E. R. Jerome, "Design Guidelines for Selection and Use of Metallic Materials in Sea Water Applications", General Dynamics Corp., Groton, Conn., Dec 1966.

[14] ASTM D877-02e1 Standard Test Method for Dielectric Breakdown Voltage of Insulating Liquids Using Disk Electrodes.

[15] W. M. Otto, "Wet Electronics", *North American Aviation Inc.* Rept. T6-1151/020, Ocean systems Operations, Aug 1966.

[16] J. J. Licari, "Plastic Coatings for Electronics", Krieger Pub. Co., 1981.

[17] L. I. Johnson, R. J. Ryan, "Encapsulated Component Stress Testing," *Proc. Elec. Insulation Conf.*, 13-16 Sep 1965, pp. 11-15.

[18] F. L. Howland, C. H. Zierdt, "Stress Analysis of Semiconductor Device Structures", *Integrated Device and Connection Technology*, Vol. III, Englewood Cliffs, N. J., Prentice-Hall, 1971, pp. 383-396.

[19] A. T. Tweedie, et al, "Final Sealing and Encapsulation", Integrated Circuit Technology, New York, McGraw-Hill, 1967, pp. 139-146.

[20] R. Beaman, "Electronic Component Pressure Testing Program", *NURDC* Rept. IV, Jan 1971.

[21] H. Nagaoka, K. Honda, *Phil. Mag.*, Series 5, Vol. 46, 1898, pp. 261-290.

[22] K. Azumi, J. E. Goldman, "Volume Magnetostriction in Nickel and the Bethe-Slater Interaction Curve", *Physical Review*, Vol. 93, Issue 3, pp. 630-631, Feb 1954.

[23] Chi-Sun, Yeh, *Proc. Am. Acad. Arts Sci.*, Vol. 60, 1925, pp. 503-533.

[24] R. L. Steinburger, *Physics*, Vol. 4, 1933, pp. 153-161.

[25] A. P. Greifer, "Ferrite Memory Materials", *IEEE Trans. on Magnetics*, Vol. 4, MAG-5, 1969, pp. 774-811.

[26] Magnetics Designer v.4.1.0, Intusoft <http://www.intusoft.com/>

[27] C.S. Smith, "Piezoresistance Effect in Germanium and Silicon", *Physical Review*, Vol. 94, Issue 1, Apr 1954, pp. 42-49.

[28] J. J. Wortman, J. R. Hauser, R. M. Burger, "Effect of Mechanical Stress on p-n Junction Device Characteristics", *Journal of Applied Physics*, Vol. 35, 1964, pp. 2122.

[29] J. R. Hauser, J. J. Wortman, "Some Effects of Mechanical Stress on the Breakdown Voltage of p-n Junctions", *Journal of Applied Physics*, Vol. 37, 1966, pp. 3884.

[30] J.J. Wortman, J. R. Hauser, "Effect of Mechanical Stress on p-n Junction Device Characteristics", *Journal of Applied Physics*, Vol. 37, 1966, pp. 3527.

[31] H. Kressel, A. Elsea, *Solid State Electronics*, Vol. 10, 1967, p. 213.

[32] J. J. Wortman, L. K. Monteith, "Effects of dynamic stress on p-n Junctions," *Bull. Am. Phys. Soc.*, vol. 13, p. 708, 1968.

[33] R. Beaman, "Electronic Component Pressure Testing Program", *NURDC* Rept. VI, Apr 1971.

TABLE 2

Components	Limit of Values Tested	Type Tested	Change in Value	Failure Mode	Estimated Performance
RESISTORS					
Carbon Composition	<1mΩ <2W	-	Negative, >20% to Failure	Case Deformation	Poor
Deposited Carbon	<100 kΩ <1/2W	-	Negative, >10%	N/A	Fair
Metal Film	<1 mΩ <1W	-	<5%	N/A	Good
Wire wound	<100 kΩ <10W	-	Negative, >10%	N/A	Good
CAPACITORS					
Dry					
Glass	<0.1 pF 500VDC	-	<2%	N/A	Good
Ceramics	<0.5 pF <50VDC	-	<5%	N/A	Good
Wet					
Tantalum	<100 pF <45VDC	-	Positive, >5% to Failure	Case Deformation	Poor
Paper	<8 pF <600VDC	-	Positive, >10% to Failure	Case Deformation	Poor
Electrolytic	<85 pF <50VDC	-	Positive, >15% to Failure	Case Deformation	Poor
INDUCTORS					
Air Core	<100μH	-	Negative, <10%	N/A	Good
Iron core	<2μH	-	>10%	N/A	Fair
Transformer	Low power	-	10%-50%	N/A	Poor
SEMICONDUCTORS					
Diode & SCR					
Glass	-	Low Power	<10%	N/A	Good
Epoxy	-	Low Power	>10% to Failure	Case Deformation	Poor
Metal	-	High Power	>10% to Failure	Case Deformation	Poor
TRANSISTORS					
Epoxy	-	Low Power	>10% to Failure	Case Deformation	Poor
Metal	-	High Power	Failure	Case Deformation	Poor
IC, MOS, MSI, LSI, Thin and Thick Film	-	-	-	-	Unknown (Probably Poor)

Failure = Crack, crush, break or change in excess 50% conventional value.

Accurate Mixed Electrical and Electromagnetic Model of a 6,5kV IGBT Module

E. Batista[1], J. M. Dienot[2], M. Mermet-Guyennet[1],
[1]Power Electronics Associated Research Laboratory
ALSTOM Transportation 65600 Séméac, France
[2]LabCem, University Institute of Technology,
1 rue Lautreamont 65000 Tarbes, France
email: emmanuel.batista@transport.alstom.com

A. Castellazzi, M. Ciappa, W. Fichtner,
Integrated Systems Laboratory
Swiss Federal Institute of Technology
8092 Zurich, Switzerland
email: castellazzi@iis.ee.ethz.ch

Abstract — The compact model development of a 6,5kV field-stop IGBT module is presented. In particular, the model considers the realistic interconnection of IGBTs and anti-parallel diodes found in commercial modules, providing, next to semiconductor physics, an accurate description of electro-magnetic (EM) phenomena associated with the package and layout. A selection of simulation examples demonstrates the usefulness of the proposed solution.

I. INTRODUCTION

IGBT-modules are relatively complex objects, consisting of many parallel devices, interconnected by means of bond-wires, printed tracks, metal bars and custom designed solutions. This inevitably shows an impact on the power system operation and performance and it is very important to take it into account in the design and analysis phase. However, in-depth experimental characterization is impractical and effects inside the modules are not easily accessible. So, the availability of accurate simulation models is very important.

In this paper, we present the development of a comprehensive electrical compact model for a commercial IGBT module. The device is of planar-gate field stop technology [1]; it is rated at 6,5kV-600A and represents the state-of-the-art for high-voltage power applications. Our focus is on a detailed electromagnetic approach, packaging and layout modeling strategy.

II. MODEL DEVELOPMENT

The module consists of six substrates, connected pair-wise by metal bars, which also implement the user *collector* and *emitter* terminals. Each substrate allocates four transistors and two diodes (the latter acting as freewheeling element for the load current in inverter operation [2]), interconnected by means of bond-wires and printed tracks (images of a substrate are proposed in figure 3 and figure 5). The substrates are mounted on a common base-plate and provide a thermal path towards cooling system, while preserving electrical insulation of the silicon chips. The assembly is covered with an insulating gel and enclosed in a plastic case from which only the metal connections to the device terminals emerge. These are connected together in the application by means of a bus-bar, custom designed by the user. So, a substrate-pair, including the metal-bar

connections, is taken as the basis for the envisaged modeling purposes.

A. Silicon Devices

For the sake of brevity, only the IGBT model is discussed; the diode model can be inferred for analogy. Figure 1 proposes a schematic cross-sectional view of the devices, highlighting the intrinsic components of interest. In building the model we follow the mixed physical and behavioral approach of [3], since, in our opinion, it offers an optimum compromise between accuracy of description and computational effort and is therefore the ideal choice for the simulation of multi-chip assemblies. The original model of [3] needs to be modified and extended to account for the influence of the field-stop layer. The *MOS* section, i_{nC} in figure 1 is described by a behavioral model. The parameters that need to be specified are the channel doping, saturation velocity and surface mobility. The carriers mobility is described with simplified expressions introducing different temperature dependencies for the bulk and channel regions [4].

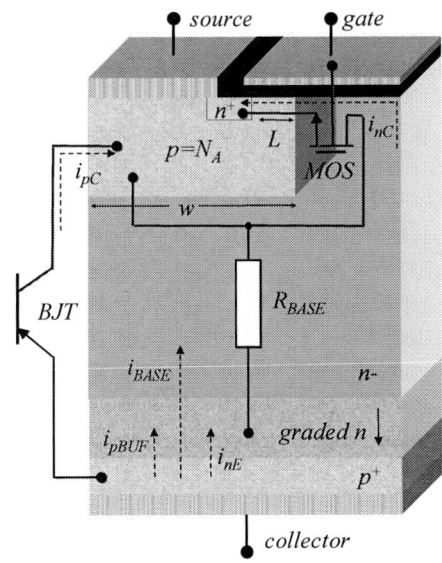

Fig. 1: Schematic cross-sectional view of a planar-gate field-stop IGBT.

The model capacitance is set at negligible values and the various components are added externally, for a more accurate

978-1-4244-1871-8/07 $25.00 © 2007 IEEE

description. The *BJT* section includes the collector current, i_{pC} in figure 1, described after [3]. On the p^+ emitter side, in view of the reduced charge-carrier base lifetime of punch-through-like transistors, this current is modeled with an ambipolar recombination component, i_{base}, plus a contribution due to the diffusion of holes in the buffer layer and to the diffusion of electrons in the p^+ emitter, i_{pBUF} and i_{nE}, respectively. For these components the expressions for low-level injection conditions are used. The base resistance model, charge dependent, follows [5]. The position dependent base doping, $N_B(x)$, is described by means of an exponential profile for the buffer layer:

$$N_B(x) = N_{SUB} + N_{BUF} \cdot \exp\left[-\frac{w_0 - x}{\chi \cdot L_F} \right] \quad (1)$$

The term $\chi.L_F$ describes the decay of the buffer layer doping profile into the base (N_{SUB} indicates the background substrate doping). It holds

$$L_F = \frac{w_{BUF}}{\ln\left(\dfrac{N_{BUF}}{N_{SUB}} \right)} \quad (2)$$

while χ is a fitting parameter. The charge-density in the depletion region is calculated with (1), additionally taking into account the hole-current contribution. The equivalent circuit schematic of the IGBT is presented in figure 2: the device terminals are indicated as G for *gate*, A for *anode* and K for *cathode*, the last two corresponding to the IGBT collector and emitter, respectively; x_J indicates the width of the depletion layer.

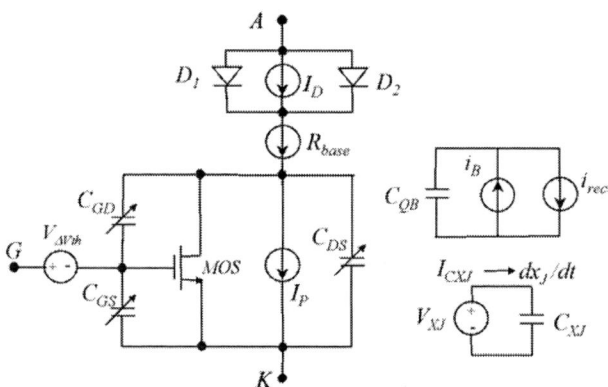

Fig. 2: Schematic Model of the IGBT

It is worth pointing out that temperature dependency is also included in the model equations, so that the ambient temperature value can be varied between different simulation runs.

Furthermore, the model is fully scalable to be representative of a single chip or of a whole module.

B. Packaging and Interconnections

The modeling of layout and interconnection parasitic effects is based on the numerical electromagnetic (EM) analysis of an accurate three dimensional structure file. The electromagnetic properties of the various materials constituting the module are duly introduced (i.e. copper layers, aluminum base-plate, ceramic substrates, dielectric gel, silicon dies and bond-wires) and properly meshed (represented in figure 3). For saving computation time, some geometrical simplifications are added to the structure file, particularly fillet and wire bonding.

Using a combination of the finite element method (FEM) and the method of moment (MoM) [6] [8], magnetic and electrical fields are computed solving the Maxwell's equations, in particular the quasi-static approximation. Layout impact and near-fields couplings are taken into account, described by a behavioral model constituted by resistance, inductance and capacitance matrices.

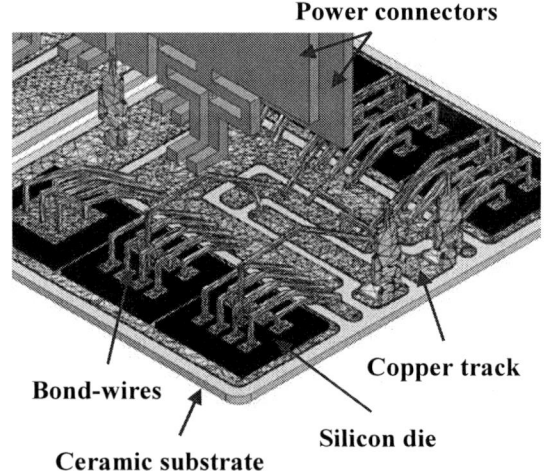

Fig. 3: Example of materials taken into account in the three dimensional model with mesh representation

On the three dimensional structure, conduction path are identified and numbered (i), all matrices exported are i-by-i matrices. In contrast with Partial – Element – Equivalent – Circuit (PEEC) method [8] [11], this numerical analysis doesn't need a decomposition of any conductor into partial elements [9]. The proposed solution gives an easily understandable compact model for integrity signal and near-field couplings modeling of hybrid technology modules [10]. An example of equivalent compact model is depicted in figure 4.

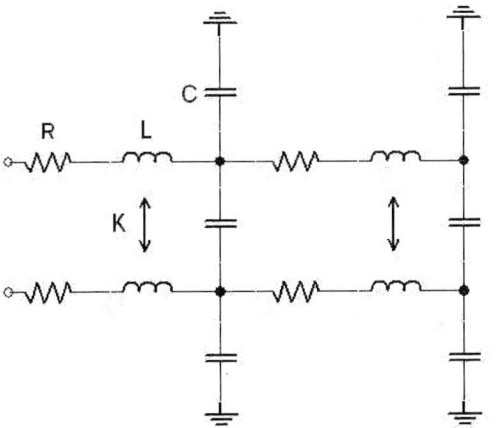

Fig. 4: Indication of two conduction paths, magnetic and electrical couplings equivalent compact model

First, to extract the resistance and inductance matrices, a FEM method for solving current density is computed for each source terminal, as

$$\vec{J} = -\sigma . \nabla \phi \qquad (3)$$

Where σ and ϕ indicates, respectively, the conductivity and the electrical potential. Resistance matrix is given by:

$$R_{ij} = \int_V \vec{J}_i . \frac{\vec{J}_j}{\sigma} dV \qquad (4)$$

In the sequel, the inductance matrix is computed using the relationship

$$L_{ij} = \int_V \vec{A}_i . \vec{J}_j dV \qquad (5)$$

where A is the magnetic vector potential ($\vec{B} = \nabla \times \vec{A}$).

For the extraction of the capacitance matrix, the charge distribution is written

$$\rho = \sum_{j=1}^{N} \alpha_j P_j(x, y, z) \qquad (6)$$

with P_j and α_j indicating, respectively, a unitary charge pulse and an unknown charge coefficient for the j_{th} element, ρ the spatial charge density and N the total number of triangles in the mesh.

Furthermore, some realistic frequency effects are taken into account. In this mode, all currents are surface currents and skin depth in conductors is calculated (skin effect) [6] [12]. These induced currents affect the computation of inductance and resistance matrices. To take this into account,

the surface magnetic field H is computed and the surface current density K is introduced in the inductance computation.

The inductance relationship became

$$L_{ij} = \int_V \vec{A}_i . K_j dV \qquad (7)$$

Figure 5 a) depicts the simulated current density distribution in diode on-state case; figure 5 b) displays the current density distribution in IGBTs on-state case.

The last one highlight density current mismatches of the silicon die positioning and design path lengths impact (c.f. paragraph IV).

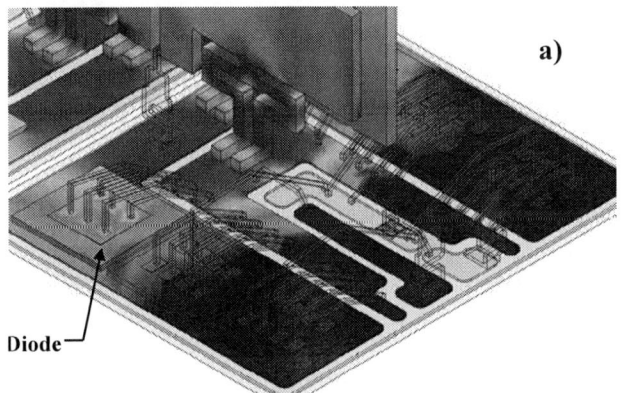

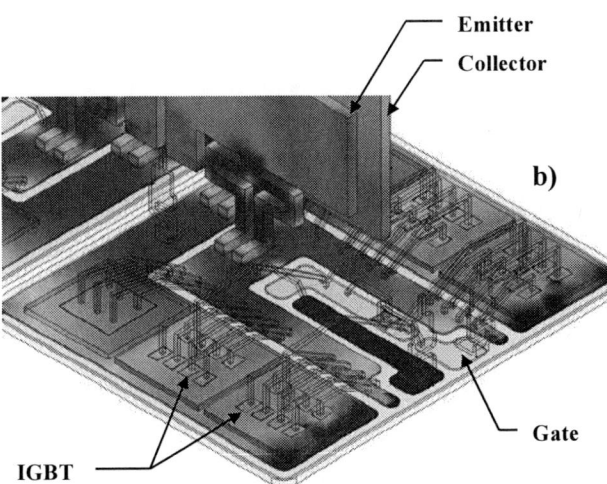

Fig. 5: Simulated substrate current density distribution, a) diode on-state (IGBTs "OFF"), b) IGBTs on-state (diode "OFF").

III. MODEL VALIDATION

So as to ensure the general validity of the IGBT and diode model, final fitting was performed against data-sheet information, which reports "typical" data. For validation, a comparison of data-sheet and simulated output-characteristics is provided in figures 6 and 7 for the IGBT and diode, respectively. In both cases, two temperature values are considered, 25°C and 125°C. In view of the simplified nature of the model, the agreement is deemed very satisfactory.

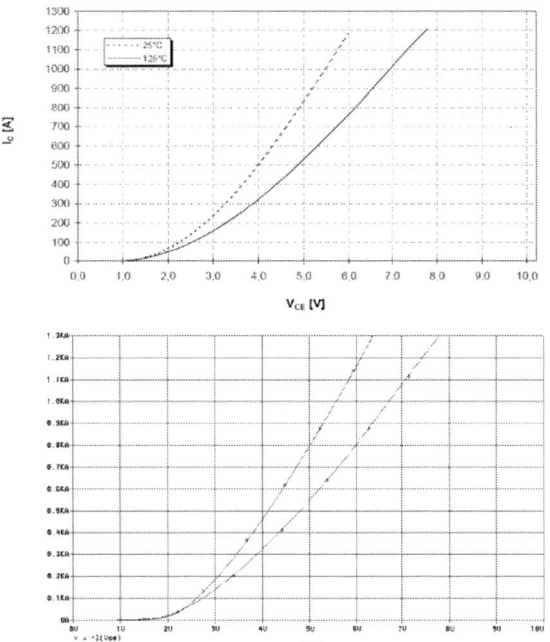

Fig. 6: data-sheet, top, and simulated, bottom, IGBT output characteristics at two different temperature values.

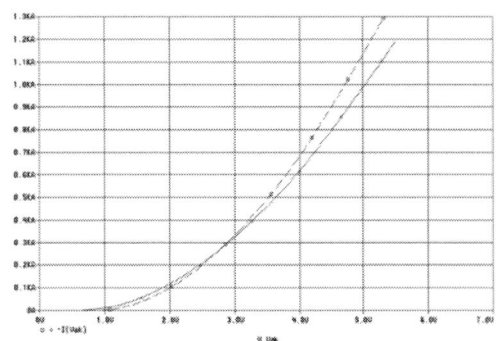

Fig. 7: Data-sheet, top, and simulated, bottom, diode output characteristics at two different temperature values.

IV. SIMULATION RESULTS

The following examples are derived from railway applications. The load, typically a 3-phase ac-motor, is of inductive nature and requires a sinusoidal input current of variable amplitude and frequency. Inverters are used to create the necessary waveform from a dc input voltage [2]. The circuit of figure 8 is used to reproduce the behavior of one inverter phase.

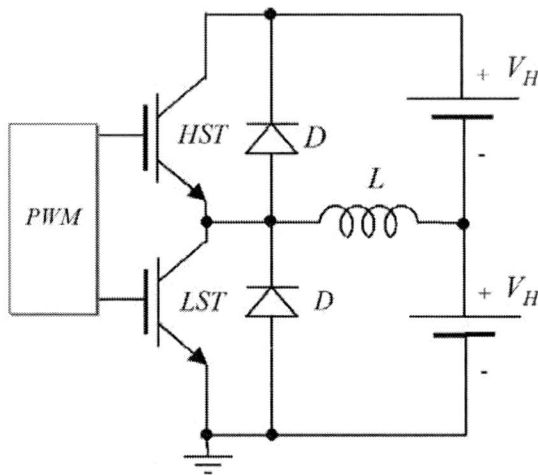

Fig. 8: Circuit schematic of the setup used to simulate one inverter phase.

The high-side switch corresponds to two substrates as per figure 5 (8 IGBTs and 4 diodes); the low-side switch is a single device, properly scaled. The pulse-width-modulation (PWM) command sequence is obtained with a basic sinus-triangle modulation: the intersection points between the two waveforms in figure 9 (top) determine a transition in the driving signal, from low to high or vice versa. The figure 9 (bottom) reports the resulting PWM driving signal and the resulting inductor current for one period of operation.

Figure 10 proposes a detail of the current distribution between two IGBTs of the substrate pair at turn-on: the inequalities are due solely to mismatches of the drive and power path lengths (detailed in table 1). These unbalances can become significant in some operational modes, such as *short-circuit* for example, during which the transistors are also exposed to consistent power dissipation and temperature increase.

Finally, Table 1 reports resistance and inductance of the collector track for each IGBT die, two cases are presented: Z_{DC} represents the static impedance values computed with (4) and (5) and Z_{AC} gives the dynamic values taking into account skin effect and eddy current (computed at frequency=10MHz). Dissimilarities are underlined between the four IGBT positioning; figure 11 displays this different conduction path.

978-1-4244-1871-8/07 $25.00 © 2007 IEEE

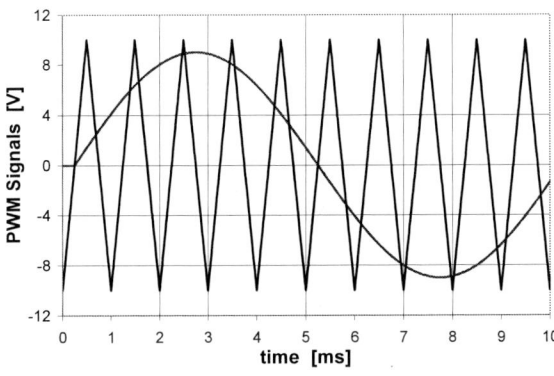

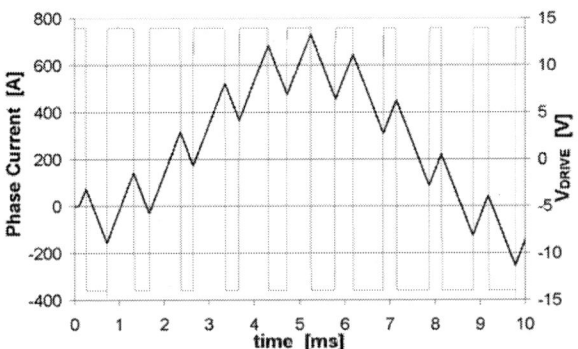

Fig. 9: PWM command Sinus-Triangle modulation waveforms (top); PWM drive signal and corresponding load current for one period of operation (bottom).

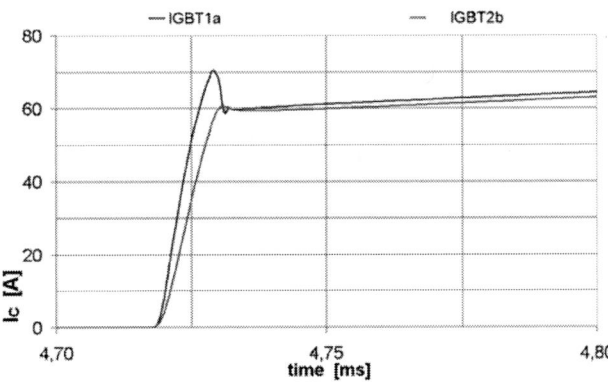

Fig. 10: Simulated turn-on current waveforms for two IGBTs in a substrate pair.

	IGBT 1a	IGBT 2a	IGBT 3a	IGBT 4a
R_{DC} (mΩ)	0.243	0.249	0.365	0.359
R_{AC} (mΩ)	5.434	5.473	7.112	7.075
L_{DC} (nH)	47.802187	48.084676	55.908690	55.647710
L_{AC} (nH)	41.379566	41.705575	45.733878	45.457812

Tab. 1: Resistive and Inductive values of the power path length for each IGBT die.

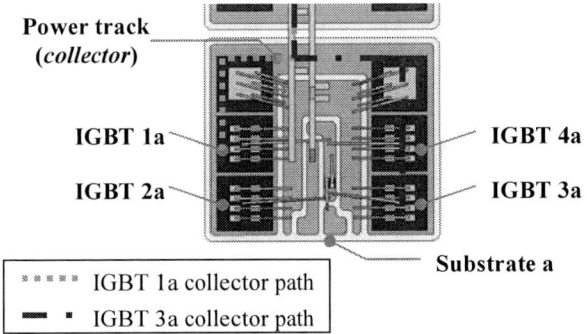

Fig. 11: Simulated turn-on current waveforms for two IGBTs in a substrate pair.

V. CONCLUSIONS

We have proposed a mixed electrical and electromagnetic compact model development of a 6.5 kV IGBT module. The model contains all main electromagnetic effects associated with the layout and packaging. Simulation examples presented in this paper highlight interconnection parasitic effects are non negligible and have demonstrated the validity of the proposed solutions, which qualifies as a powerful tool for diverse lines of investigation, ranging from reliability analysis to overall system design.

REFERENCES

[1] T. Laska, M.Munzer, F. Pfirsch, C. Shaeffer and T. Schmidt, *"The field stop IGBT (FS IGBT) – A new power device concept with a great improvement potential"*. *Proc. ISPD 00*, IEEE International Symposium on Power Semiconductor Devices and ICs, Toulouse, France, May 22-25, 2000.

[2] N. Mohan, T.M. Undeland, W.P. Robbins, *Power Electronics: Converters, Applications and Design*, Second Edition, J. Wiley & Sons, Inc., 1995.

[3] R. Kraus, P. Türkes, J. Sigg, *"Physics-Based Models of Power Semiconductor Devices for the Circuit Simulator SPICE"*, *Proc. PESC 98*, IEEE Power Electronics Specialists Conference, Fukuoka, Japan, 1998.

[4] V. Benda, J. Gowar, D.A. Grant, *Power Semiconductor Devices – Theory and Applications*, John Wiley & Sons, Chichester, England, 1999.

[5] M. Redding, R. Kraus, *"The influence of the base resistance modulation on switching losses in IGBTs"* Proc. IAS 96, *IEEE Industry Applications Conference*, San Diego, California, 1996.

[6] G. Lourdel, J.M. Dienot, *"EMC Investigation Method Using Numerical Tools for Power Hybrid Circuits Design Optimization"*, EMC'Compo'04, March 2004, Angers, France.

[7] Ansoft Corporation. Q3D Extractor - Technical Notes. http://ansoft.com/products/si/q3d_extractor/

[8] T. H. Hubing, *"Survey of Numerical Electromagnetic Modeling Techniques"*, University Missouri-Rolla, EMC Laboratory, Intel Corporation, Technical report TR91-1-001.3, Sept. 1991.

[9] A. E. Ruelhi, *"Equivalent Circuit Models for Three-Dimensional Multiconductor Systems"*, *IEEE Trans. On Microwave Theory and Techniques*, vol. MTT-22, no. 3, March 1974.

[10] J.M. Dienot, *"Power PCB and Packaging modeling"*, Part 5 of *"EMC for Integrated Circuits"*, First Edition, Springer, 2005.

[11] G. Antonini, *"Overview of the Partial Element Equivalent Circuit Method"*, *IEEE International Symposium on Electromagnetic Compatibility*, July 2007.

[12] T. H. Hubing, *"Non-ideal Behavior of Components"*, Global EMC University, *IEEE International Symposium on Electromagnetic Compatibility*, July 2007.

978-1-4244-1871-8/07 $25.00 © 2007 IEEE

Research On A Flexible Waveform Power Amplifier Adopting Switch-Linear Hybrid(SLH) Scheme

Xiaodong Liu,Sucheng Liu, and Lusheng Ge
Key Laboratory of Power Electronics and Motion Control
Anhui University of Technology,Hudong Rd. No.59,Ma'anshan,China,243002
Email: lxdong168@sina.com

Abstract—A novel power amplifier adopting Switch-Linear Hybrid Scheme (SLH-FWPA) is introduced to accomplish flexible waveforms power amplification in a high bandwidth at high output power rating. The final output stage of SLH-FWPA is derived from linear power amplifier architecture to maintain excellent output quality and well dynamic performance; and meanwhile using tracking power supplies for the linear output stage to acquire total system efficiency above 80% generally. Theoretic analysis for SLH-FWPA is given and experimental results verify the implementation of SLH-FWPA.

I. INTRODUCTION

Switch-mode power amplifiers can produce high output voltage with high efficiency but characterized by their low fidelity due to their nonlinear nature of switching operation. Some digital distortion correction technique has been proposed in low power applications such as digital audio devices [1], [2], [3].In high power industrial applications, however, multi-level power conversion is preferred to achieve desired low total harmonic distortion(THD) of output waveforms, but the complexity and cost of overall system are also increased in the same time [4], [5]. On the other hand, Conventional linear power amplifiers have high fidelity and well dynamics but confined by their low efficiency because of their static operation region. For the considerations of cost and synthetic performance, increasing the efficiency of linear power amplifiers has advantages over increasing the fidelity of switch-mode power amplifiers.

Switch-Linear hybrid power conversion (SLH) is such a technique that can boost the efficiency of linear power amplifiers with little increasing the THD figure of linear power amplifiers [6], [7], [8], [9],[10] replacing the constant power supplies of linear power amplifier by dynamic tracking power supplies, the voltage drop cross the output transistors can be decreased to a very low value compared to conventional power amplifiers.

Considering that SLH converters have accomplished single frequency sinusoidal power conversion successfully in low frequency electrical drive application, a flexible waveform power amplifier based on switch-linear hybrid scheme is constructed to achieve flexible waveforms power conversion such as sinusoidal waves, triangular waves and also rectangular waves in a wide frequency band. The

---

Project supported by national nature science foundation of China (50407017) and the international cooperation project of Anhui (06088021)

SLH-FWPA with source follower configuration has very low output impedance, as result, synthetic performance of high efficiency, good dynamics and high output quality can be expected without complex control strategy. In previous paper, small signal analysis of SLH-FWPA was done and simulation results verified the validity of SLH-FWPA [11].

In this paper, the basic principle of switch-linear hybrid scheme is introduced in section II and the topology of SLH-FWPA is also derived in this section. Key problems of efficiency analysis and phase shifting with SLH-FWPA are investigated in Section III. Experimental results are given to verify the implementation of SLH-FWPA in section IV and section V draws the conclusion finally.

II. BASIC SLH SCHEME AND TOPOLOGY OF SLH-FWPA

A. Basic Scheme of Switch-Linear Hybrid Power Conversion

As shown in Fig.1, the unified structure of SLH power converter is mainly composed of three parts: Voltage -gain stage, Switch-Filter unit and linear unit [6].Unlike to conventional linear power amplifier with DC supply, the linear units of SLH power converter are fed by variable power supplies according to output voltage waveforms, so that the output transistors in linear unit operate in quasi-linear region, the efficiency of SLH converters can be increased due to that the voltage drop across the output transistors is reduced to a small value. Besides high efficiency and low THD, robustness of SLH converters was observed from foretime research achievements besides high efficiency and high output quality [6], [7].

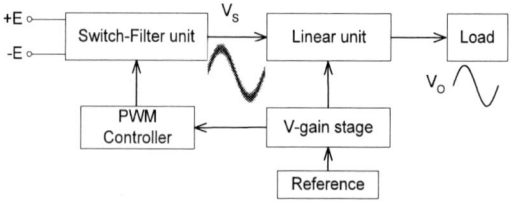

Fig.1. Unified Structure of SLH Converters

B. Topology of SLH-FWPA

Based on SLH scheme, a SLH-FWPA is produced to perform flexible waveforms amplification. Hence, SLH-FWPA will possess same good synthetic performance as SLH converters. Fig.2 shows the topology of SLH-FWPA, which is also mainly composed of three stages: PWM

978-1-4244-1871-8/07 $25.00 © 2007 IEEE

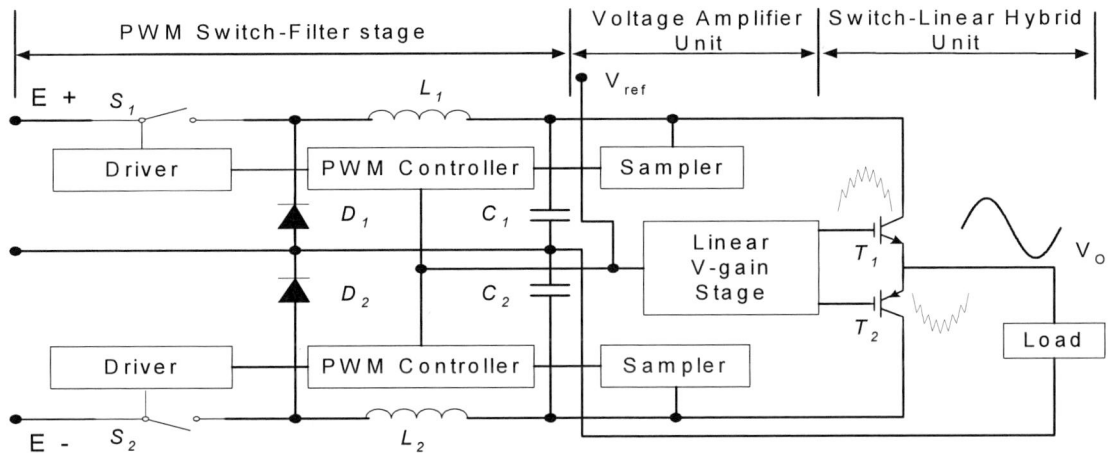

Fig.2. Topology of SLH-FWPA

Switch-Filter stage, voltage amplifier unit and Switch-Linear hybrid unit, where the switch-filter unit acts as dynamic power supplies with the power transistors in the switch-linear hybrid unit which is derived from a source follower power stage. Fed by ripple supply generated from the switch unit by simple voltage feedback control strategy, the power transistors in the source follower power stage operate in a special condition, which neither in their ohmic region nor saturation region but in between them, and the output load waveforms track the gate voltage in wave shape, amplitude and phase. Considering that the robustness is an important system performance feature of power amplifier besides high efficiency and high fidelity in power conversion scope, emitter follower configuration which is formed by two complementary N-channel IGBT and a artificially P-channel IGBT is adopted in SLH-FWPA scheme, where the P-channel IGBT is hybrid by a P-channel MOSFET and a NPN power transistor. Because emitter followers have well stability and excellent tracking performance besides their very low output resistances, so SLH-FWPA can obtain excellent dynamic performance in addition that of high efficiency, high fidelity.

III. EFFICIENCY ANALYSIS AND PHASE SHIFTING

A. Efficiency Calculation and Analysis

For the following efficiency calculations and analysis we take sinusoidal output waveforms for illustration for the sake of simplicity.Fig.3 (a) shows Class B emitter follower output structure; T_1 and T_2 are complementary power transistors. To calculate the transistor losses, constant DC power supplies and tracking power supplies are investigated and their waveforms are illustrated in Fig. 3 (b), (c) respectively. We assume that the output voltage of emitter follower

$$v_o = V_P \sin \omega t \qquad (1)$$

resulting output current flows through resistor load

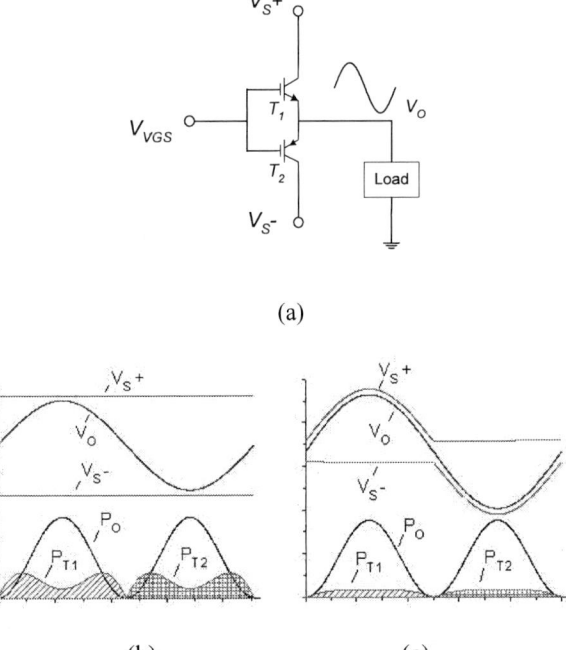

(a)

(b)   (c)

Fig.3. (a)Class B Emitter follower structure. (b)Output voltage, output power and instantaneous power loss of transistor T1 and T2 for constant power supply. (c)Output voltage, output power and instantaneous power loss of transistor T1 and T2 for dynamic tracking power supply.

$$i_o = \frac{V_P}{R_L} \sin \omega t \qquad (2)$$

And then the output power of the emitter follower is given by

$$P_o = \frac{V_P^2}{2R_L} \qquad (3)$$

Considering an emitter follower operating in class-B mode for a constant power supply, in Fig.3 (b), the loss consumed by each transistor T_1 or T_2 is

978-1-4244-1871-8/07 $25.00 © 2007 IEEE

$$P_{T1} = \frac{1}{2\pi} \int_0^\pi (V_S - V_P \sin \omega t) \frac{V_P}{R_L} \sin \omega t d\omega t$$

$$= \frac{V_P}{\pi R_L}(V_S - \frac{\pi}{4}V_P) \tag{4}$$

In case the emitter follower power stage is fed by the dynamic tracking power supply we can also calculate the power loss for each transistor T_1 or T_2, in Fig.3 (c), as

$$P_{T1} = \frac{1}{2\pi} \int_0^\pi V_D \frac{V_P}{R_L} \sin \omega t d\omega t$$

$$= \frac{V_D V_P}{\pi R_L} \tag{5}$$

where V_D is the DC offset voltage for the tracking power supply which is also the voltage drop across the conducting power transistor.

From the above analysis and calculations we can draw the efficiency curves of SLH-FWPA and class B power amplifier to see the efficiency boost of SLH-FWPA compared with conventional class B linear power amplifier apparently.Fig.4 shows that the efficiency of class B power amplifier is significantly improved by SLH based power amplifier.

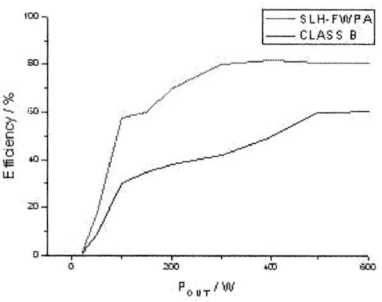

Fig.4. Compared efficiency curves of SLH-FWPA and class B amplifier

B. Phase Shifting and Compensating Network

As seen forms Fig.2, there are LC filters in the switch-filter unit in which some phase shift is introduced to the signal at the output of the filter, particularly at high frequency band. However, the voltage gain stage adds a very little or no phase shift to the signals at its output. Therefore, a phase shifting compensating network has to be added to the voltage gain stage, so that all the signals going to the emitter follower stage are in phase. To solve this problem, we placed a first order RC phase shifting network before the voltage gain stage to compensate the phase discrepancy between switch filter unit and voltage gain stage.

Fig.5 (a) and (b) shows the basic LC filter and RC network respectively, the transfer function of each network can be written with the frequency expressed in radians

$$\frac{U_o(\omega)}{U_i(\omega)} = \frac{1}{1 - LC \cdot \omega^2 + j\omega \cdot L/R} \tag{6}$$

$$\frac{U_o(\omega)}{U_i(\omega)} = \frac{1}{1 + j\omega \cdot RC} \tag{7}$$

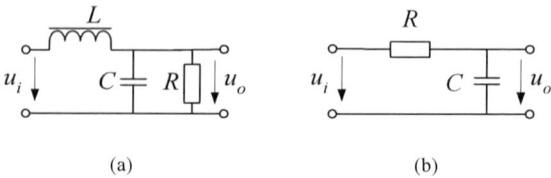

(a)   (b)

Fig.5. (a) Basic LC filter. (b) First-order RC network.

According to (6) and (7), the phase curves of LC filter and RC network versus frequency are easily drawn as in Fig.6.As shown in the figure, the blue curve in the above indicates phase curve of a first order RC circuit, it can compensate the phase shift at the output of switch-filter unit caused by LC filter under the condition that output frequency less than 10kHz.

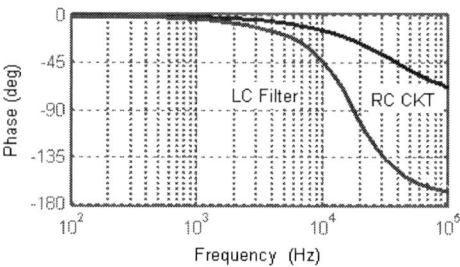

Fig.6. Phase curve of LC filter and RC circuit. The design parameters are: L=300 μ H, C=0.22 μ C for LC filter; R=4k Ω ,C=1nF for RC compensating network.

IV.   EXPERIMENTAL RESULTS

For verify the theoretical considerations a 500W laboratory prototype of SLH-FWPA has been built to perform flexible waveforms amplification for three typical reference signal: sinusoidal wave, triangular wave and rectangular wave; the experimental waveforms for output frequencies at 20Hz and 1kHz are shown in Fig.7. The operating parameters of the prototype are: the switching frequency f_s=200kHz, the input voltage for switch-filter unit E=60V, the load resistor RL =15 Ω ,. The main components employed in the SLH-FWPA are listed in Table 1 except T_2 which is artificially hybrid by a P-channel MOSFET MTP2P50E and a NPN power transistor BU508E.

TABLE 1
List of Power Components of the SLH-FWPA

Name	Denomination	Type
CoolMOS	S_1, S_2	IXKH24N605C
Power Diode	D_1, D_2	RHRG30120
Film Capacitor	C_1, C_2	0.22 μ F/630V
Inductor	L_1, L_2	300 μ H/15A
IGBT	T_1	1MBH60-120D

In Fig.7, only the output voltage waveforms and the negative dynamic tracking power supplies are given because that only one two-channel oscilloscope is used. As shown, the negative

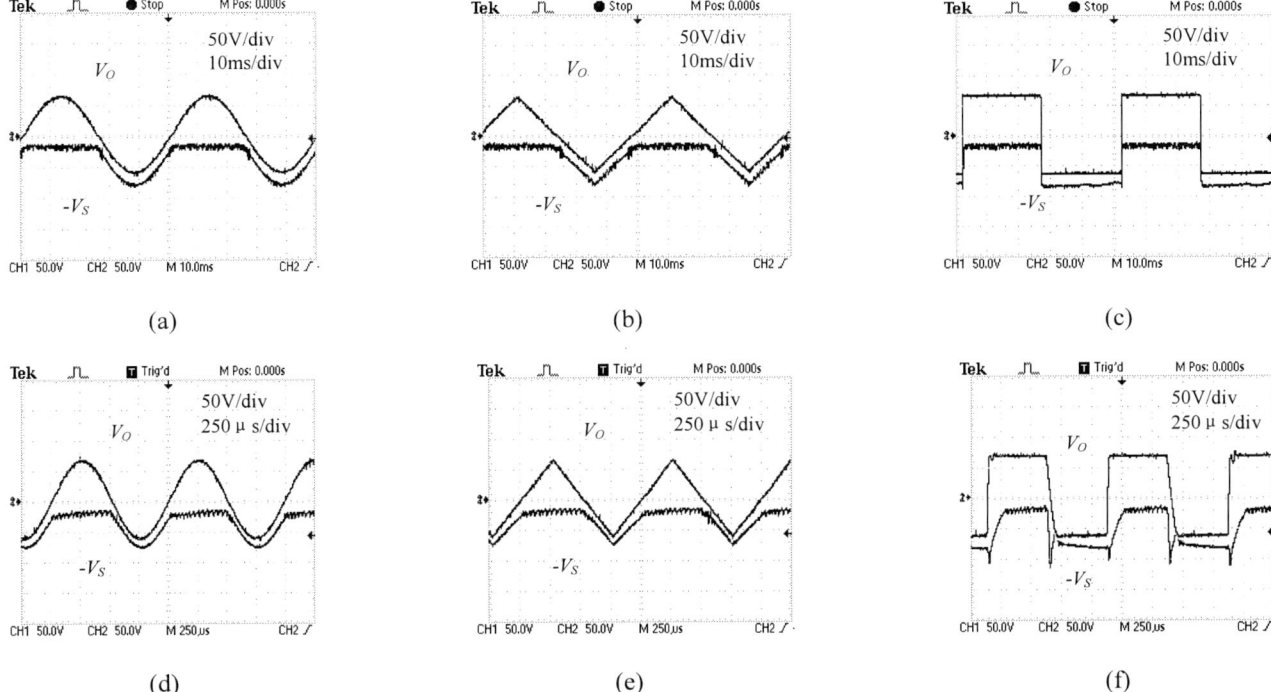

Fig.7. Experimental results for three different reference signals at output frequency 20Hz and 1kHz: (a)120V_{P-P}, 20Hz sinusoidal output. (b)120V_{P-P}, 20Hz triangular output. (c)120V_{P-P}, 20Hz rectangular output. (d)120V_{P-P}, 1kHz sinusoidal output. (e)120V_{P-P}, 1kHz triangular output. (f)120V_{P-P}, 20Hz rectangular output.

dynamic power supplies have good tracking performance, and in consequence the power losses in the output power transistors is lowered while maintaining optimal output waveforms . There are some overshoots and oscillations for rectangular waveforms at high frequency, which will not have negative effect on the output quality if the oscillations can be confined to a desired value.

V.   CONCLUSIONS

The SLH based power amplifier SLH-FWPA can perform flexible waveforms amplification with low THD figure, high efficiency and high power. In essence, the SLH-FWPA can be considered as some kind of special high efficiency linear power amplifier due to the power transistors in the emitter follower operate in the quasi-linear region. The system performance is also optimized by its extremely low output impedance. In addition, the implementation of SLH-FWPA is simple and do not need complex control strategy. Therefore the SLH-FWPA provides effective way for power amplifiers and it is a promising solution to high power amplifiers for industrial applications.

REFERENCES

[1]   Deron K. Jackson and Steven B. Leeb,   *"A Digitally Controlled Amplifier With Ripple Cancellation,"* IEEE Trans. Power Electronics, vol. 18, pp. 486–494, Jan. 2003.

[2]   Pallab Midya, Bill Roeckner and Steve Bergstedt, *"Digital Correction of PWM Switching Amplifiers,"* IEEE Power Electronics Letters, vol. 2, pp. 68–72, Jun. 2004.

[3]   Tripath. Class-T Digital Audio Amplifier Technology Overview. Available:www.tripath.com/downloads/an1.pdf

[4]   S. Mariethoz and A.Rufer. An Overview of the Technique of Asymmetrical Multilevel Inverters. Available: leiwww.epfl.ch/publications/mariethoz_rufer_anae_05.pdf

[5]   Isao Takahashi and Kazuki Iwaya, "100kHz, 10kW Switching Type Power Amplifier Using Multilevel Inverter," in *Conf. Rec. 2001 IEEE Int. Conf. Power Electronics and Drive Systems*, pp. 286–291.

[6]   Qianzhi Zhou, Wenhua Hu, Bin Wu and Mouzhi Dong, "Switch-Linear Power Conversion(I)—The topology based on Source Follower," in *IEEE Conf. Rec. 2006 Industrial Electronics and Applications*, pp. 1118–1124.

[7]   Qianzhi Zhou and Luseng Ge, "Switch-linear hybrid power conversion (SLH) with low output resistance," in *Conf. Proc. 2004 Int. Conf. Power Electronics and Motion Control*, pp. 96–98.

[8]   Kashiwagi. S, *"A High Efficiency Audio Power Amplifier Using A Self-oscillating Switching Regulator,"* IEEE Trans. Industry Applications, Vol. 21, pp.906-911, 1985.

[9]   Guanghai Gong, Hans Ertl and Johann W. Kolar, "High-Frequency Isolated DC/DC Converter for Input Voltage Conditioning of a Linear Power Amplifier," in *Conf. Proc. PESC'2003*, pp. 1929–1934.

[10]   Jae Hoon Jeong, et al. "A High Efficiency Class A Amplifier Accompanied By Class D Switching Amplifier," in *Conf. Proc. PESC'1997*, pp.1210–1216.

[11]   Xiaodong Liu, Sucheng Liu, Mengzeng Cheng and Kekang Jiang. "A Flexible Waveform Power Amplifier Based on Switch-Linear Hybrid Scheme," in *Conf. Proc. ICIEA'2006*, pp.1569–1572.

978-1-4244-1871-8/07 $25.00 © 2007 IEEE

Switching Transient Shaping of RF Power MOSFETs for a 2.5 MHz, Three-Phase PFC

Michael HARTMANN, Andreas MÜSING and Johann W. KOLAR
Power Electronic System Laboratory
ETH Zurich
CH-8092 Zurich, Switzerland
Email: hartmann@lem.ee.ethz.ch

Abstract—To increase the power density of active rectifiers, the switching frequency and switching-speed have to be raised considerably. However, the very fast switching transients induce a strong voltage and current ringing. In this paper, a novel magnetically coupled damping layer is introduced for attenuating these unwanted oscillations. The proposed damping layer can be implemented using standard materials and printed circuit board manufacturing processes. The system behavior is analyzed in detail and design guidelines are given. The effectiveness of the introduced layer is determined by layout parasitics, which are calculated with the Partial Element Equivalent Circuit method and compared to impedance measurements. The performance of the damping layer is demonstrated by simulations and verified via measurements on a laboratory prototype.

I. INTRODUCTION

Traditional requirements on 3-phase active rectifiers such as unity power factor and sinusoidal input currents have been extended over the last few years to include high compactness and high efficiency [1]. Therefore, the power density of active rectifiers, such as the three-level Vienna Rectifier (VR) [2], has to be increased substantially. Whereas the size of the active components (switches and diodes) can be minimized by using a low profile power module including all active components [3], the size of the passive components (EMI-filter, boost inductor etc.) has to be minimized by increasing the switching frequency.

An optimization of the size of the EMI-filter shows that a maximal power density of 24 kW/liter for a system with either CoolMOS or RF MOSFETs, water cooling and a switching frequency of 2.1 MHz could be achieved [4]. For realizing a switching frequency in the MHz range, a very fast switching transition is required in order to limit the switching losses.

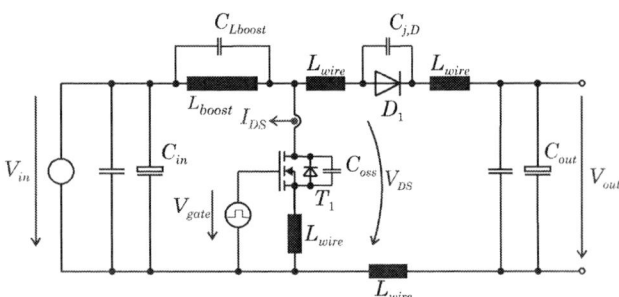

Fig. 1: Schematic of the boost circuit including some parasitic elements.

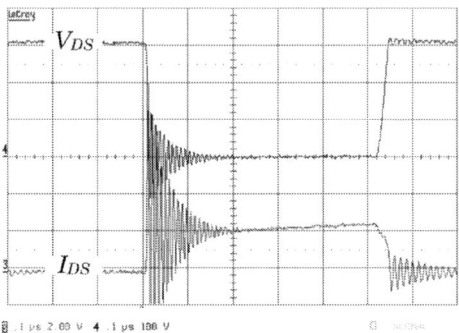

Fig. 2: Current and voltage wave shapes at turn-on of transistor T_1 at 20 A, V_{DS} (100 V/div), I_{DS} (20 A/div), time scale 50 ns/div.

Therefore, voltage slopes of 30 kV/µs and current slopes up to 2 kA/µs have to be realized.

In order to analyze the switching behavior of CoolMOS and RF MOSFET devices in combination with SiC-diodes for the target VR system with regard to switching frequencies in the MHz range in detail, a boost-type test circuit (cf. **Fig. 1**) has been designed. In Fig. 1, the parasitic capacitances C_{oss} of the MOSFET and $C_{j,D}$ of the diode are shown as well as the wiring inductances L_{wire} of the commutation path and the parasitic capacitance C_{Lboost} of the boost inductor. These parasitic inductances and capacitances cause current and voltage oscillations excited by the very fast switching transients (cf. **Fig. 2**). The oscillations increase the voltage stress of the devices and generate EMI noise. Basically, the ringing can be reduced by application of an RC snubber, however, the losses of such snubber circuit would be far too high for a switching frequency in the MHz range, and hence another solution has to be found for reducing the unwanted oscillations.

Some interconnection concepts with low-pass characteristics, based on high-ε-materials, have been reported in literature [5, 6]. However, these approaches are applicable only in case of longer interconnection distances to achieve the desired damping behavior.

In this paper, a novel magnetically coupled damping layer is introduced to realize the required damping. As shown in **Fig. 3** the additional layer is inserted between the two wiring layers. If the copper path of the damping layer now is terminated by an appropriately designed RC network, the currents induced in the damping layer will significantly reduce the parasitic oscillations in the wiring layers. An optimized termination network results in a very good damping

978-1-4244-1871-8/07 $25.00 © 2007 IEEE

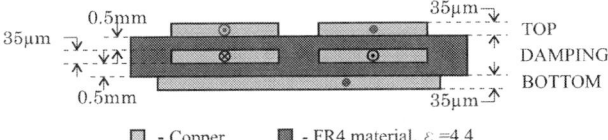

Fig. 3: Layer stack of the system with damping layer.

behavior and much lower losses compared to a classical snubber circuit. Furthermore, the damping network uses standard materials (FR4, copper) and no extra manufacturing processes are needed, which leads to lower realization costs.

In **Section II** the proposed damping concept is analyzed in detail using a simple circuit model. Furthermore, the design and optimization of the termination network is given. Subsequently, a laboratory prototype of the mentioned boost converter employing the proposed damping concept is analyzed in detail in **Section III**. The parasitic wiring elements are calculated using the Partial Element Equivalent Circuit (PEEC) method and compared to measurements of the realized prototype leading finally to a precise model of the system. The resulting voltage and current wave shapes demonstrate the very good performance of the optimized damping.

II. ANALYSIS OF THE PROPOSED DAMPING LAYER

The origin of the undesired switching transient oscillations shall be described briefly in the following. At turn-on of the MOSFET T_1 the converter's input current has to commutate from diode D_1 to MOSFET T_1. Since SiC-Schottky diodes are used, there is no reverse recovery current I_{rr}. However, a displacement current is charging the voltage dependent junction capacitance $C_{j,D}$ of D_1. This capacitance in connection with the wiring inductance L_{wire} forms a series resonant circuit which is damped only by the on-state resistance $R_{DS,on}$ of T_1 and by the high frequency resistance of the wiring.

At turn-off of T_1 the input current has to commutate from MOSFET T_1 to diode D_1 and now the voltage dependent output capacitance C_{oss} of the MOSFET forms a weakly damped series resonant circuit with the wiring inductance L_{wire} of the PCB. In [7], the turn-off behavior of the MOSFET is analyzed and an analytical expression for the turn-off switching transient overvoltage is given. In the following only the behavior of the circuit at turn-on of the MOSFET shall be analyzed for the sake of brevity since the turn off behavior can be treated in a similar way.

The natural frequency f_0 and characteristic impedance Z_0 of the LC-tank at turn-on are given in (1). For SiC-diodes with some 100 pF junction capacitance and some 10 nH stray inductance f_0 lies in the 100 MHz range. To achieve a proper damping, the value of the damping resistor in series to the LC-tank has to be in the range of Z_0

$$f_0 = \frac{1}{2\pi\sqrt{L_{wire}C_{j,D}}}, \quad Z_0 = \sqrt{\frac{L_{wire}}{C_{j,D}}}. \tag{1}$$

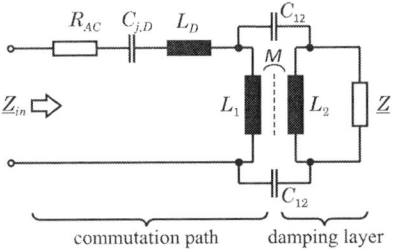

Fig. 4: Model of the circuit at turn-on of the MOSFET with damping layer.

With (1), the stray inductance of the commutation path can easily be determined by measuring the frequency of the voltage or current oscillation and by using the junction capacitance $C_{j,D}$ of the diode, specified in the datasheet.

In **Fig. 4** a simple model for the turn-on behavior of the MOSFET including the damping layer is shown, where the following symbols are used:

R_{AC} HF-resistance of the wiring in the commutation path;
$C_{j,D}$ Junction capacitance of the SiC-Schottky diode;
L_D Parasitic inductance of diode and MOSFET;
L_1 Effective inductance of the commutation path;
L_2 Inductance of the damping layer;
M Mutual inductance between commutation path and damping layer;
C_{12} Coupling capacitance between wiring layer and damping layer;
$\underline{Z}$ Termination network for damping layer.

Although the inductance of the commutation path can be minimized using a proper layout, a small residual value will remain. An excellent PCB layout offers stray inductances in the range of the parasitic inductances of the diode and the MOSFET. Hence the component parasitics are added to the model. Due to skin effect, the current of the high frequency oscillations flows only in a thin layer at the surface of the copper layer and so the HF-resistance R_{AC} has to be considered. Additionally, the skin effect reduces the effective area of the copper "wires" and so the high frequency coupling capacitance C_{12} is reduced to small values so that it can be neglected for sake of an easier modeling. Furthermore, the input inductors parasitic capacitance C_{Lboost} can be neglected because the analyzed oscillations at turn-on of T_1 are not directly affected by C_{Lboost}.

A. Equivalent circuits

For a better understanding of the proposed damping system two equivalent circuit diagrams for the model of Fig. 4 are given in **Fig. 5**. There, the two magnetically coupled layers are modeled by a leakage inductance L_{lki}, a magnetizing inductance L_{mi} and an ideal transformer with the ratio $u_i{:}1$. In Fig. 5(a) u_1 is chosen such, that the full leakage inductance is considered on the primary side:

$$L_{lk1} = L_1 - \frac{M^2}{L_2}, \quad u_1 = \frac{M}{L_2}, \quad L_{m1} = \frac{M^2}{L_2}, \tag{2}$$

978-1-4244-1871-8/07 $25.00 © 2007 IEEE

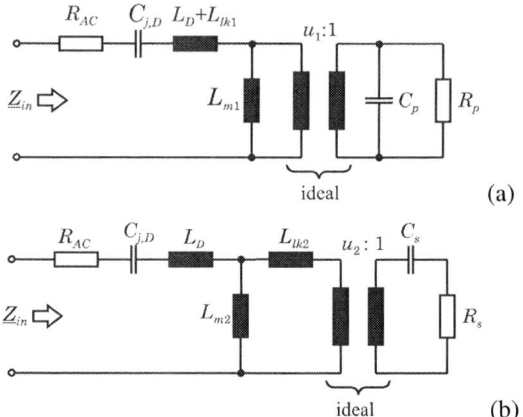

Fig. 5: Equivalent circuits for the model of Fig. 4 for turn-on of the MOSFET with (a) leakage inductance transferred to the primary side and (b) leakage inductance transferred to the secondary side.

where in Fig. 5(b) the transformer conversion ratio u_2 is chosen such that the full leakage inductance is considered on the secondary side:

$$L_{lk2} = \frac{L_1^2 L_2}{M^2} - L_1, \quad u_2 = \frac{L_1}{M}, \quad L_{m2} = L_1. \quad (3)$$

B. Approaches for the termination network

As is well known from coreless transformer designs [8], the magnetizing inductance L_{mi} is quite small as compared to the leakage inductance L_{lki} because of the limited coupling of the layers. If only the resistor R_s (cf. Fig. 5(b)) is used as termination (e.g., realized by resistive materials in the damping layer itself) the major part of the input current is flowing through the magnetizing inductance because the magnetizing reactance $X_{m2} = j\omega L_{m2}$ is much smaller than the impedance of the series connection formed by R'_s (primary side referenced value) and by $X_{lk2} = j\omega L_{lk2}$. The inductance L_{mi} (and hence the coupling of the two layers) could be increased using a ferrite core or a magnetic layer as done in [9]. However this also raises the inductance of the commutation path and, therefore, is no option in the case at hand.

The unwanted leakage inductance can be cancelled if a well designed capacitor C_s is placed in series to the resistor R_s. This capacitance in connection with the leakage inductance L_{lk2} forms a series resonant circuit, which has to be tuned to a resonant frequency being equal to the oscillations to be damped. However, simulations of the proposed system showed that the coupling capacitance C_{12} can not be neglected in this case and that even a very small coupling capacitance of a few pF overrides the positive effect of canceling the leakage inductance. The best results are achieved by using a parallel connection of R and C as termination network which will be discussed in the following.

C. Damping using a parallel R-C connection

As can be seen in Fig. 5(a) the series resonant circuit, formed by $C_{j,D}$ and the sum of the inductances L_{lk1} and L_D, is connected in series with a well damped parallel resonant circuit, formed by the termination network and the inductance

L_{m1}. The two elements C_p and R_p have to be transferred to the primary side considering u_1:

$$C'_p = \frac{C_p}{u_1^2}, \quad R'_p = R_p \cdot u_1^2. \quad (4)$$

If the natural frequency f_{par} of the parallel resonant circuit is chosen according to

$$f_{par} = \frac{1}{2\pi\sqrt{L_{m1} \cdot C'_p}} = f_{ser} = \frac{1}{2\pi\sqrt{(L_D + L_{lk1}) \cdot C_{j,D}}}, \quad (5)$$

leading to

$$C_p = \frac{L_D + L_{lk1}}{L_{m1}} C_{j,D} \cdot u_1^2, \quad (6)$$

and the damping resistor is chosen to

$$R_p = \frac{R'_p}{u_1^2} \approx Z_1 = \sqrt{\frac{L_D + L_{lk1}}{C_{j,D}}}, \quad (7)$$

the damping of the resulting system can be increased significantly. A Bode plot of the input impedance

$$\underline{Z}_{in} = \underline{Z}_{ser} + \underline{Z}_{par} \quad \text{with} \quad (8)$$

$$\underline{Z}_{ser} = \frac{1 + sR_{AC}C_{j,D} + s^2 C_{j,D}(L_D + L_{lk1})}{sC_{j,D}} \quad \text{and} \quad (9)$$

$$\underline{Z}_{par} = \frac{sL_{m1}}{1 + s\dfrac{L_{m1}}{R_p} + s^2 L_{m1} C_p} \quad (10)$$

is given in Fig. 6. The damping of the system is increased when the magnitude of the input impedance is increased at the resonant frequency. Since (7) is only an approximation, the optimal values for a maximal damping can be found by application of the following optimization function:

$$Z_{opt} = \left| \underline{Z}_{in}(f, C_p, R_p) \right| \Big|_{\arg(\underline{Z}_{in}(f))=0°} \rightarrow \max . \quad (11)$$

The result of the optimization for an assumed system with

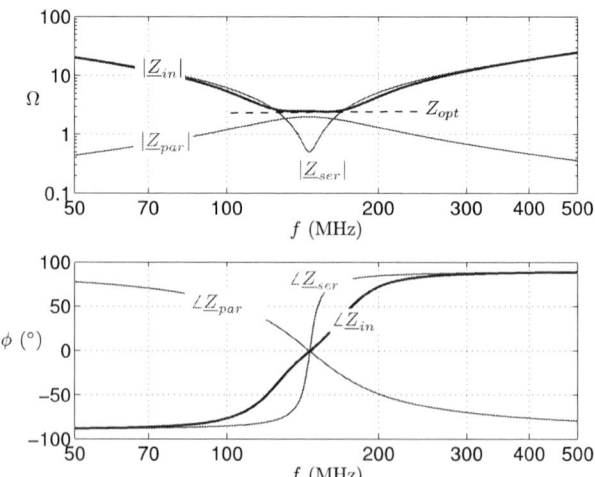

Fig. 6: Calculated Bode diagram of impedances $\underline{Z}_{ser}$, $\underline{Z}_{par}$ and $\underline{Z}_{in}$ for the system: L_1=10 nH, L_2 = 10 nH, M = 5 nH, L_D = 6 nH and $C_{j,D}$ = 136 pF.

978-1-4244-1871-8/07 $25.00 © 2007 IEEE

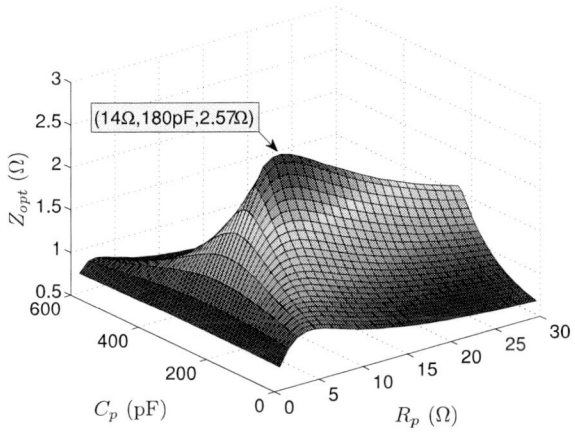

Fig. 7: Calculated magnitude of impedance Z_{opt} as a function of the termination network parameters R_p and C_p.

the parameters $L_1 = 10$ nH, $L_2 = 10$ nH, $M = 5$ nH, $L_D = 6$ nH and $C_{j,D} = 136$ pF is depicted in **Fig. 7** and the results are summarized in **TABLE 1**.

TABLE 1

CALCULATED VALUES FOR THE TERMINATION NETWORK ACCORDING TO (6) AND (7), AND RESULTS OF THE OPTIMIZATION ACCORDING TO (11).

	R_p	C_p
Calculated values	9 Ω	186 pF
Result of optimization	14 Ω	180 pF

III. DESIGN OF A DAMPING LAYER

A. Design of the realized prototype

The design of the damping layer is performed using a boost converter as an example (cf. Fig. 1). This circuit is also used to study the switching behavior of different MOSFET devices. To achieve very fast switching transients, a low impedance gate driver DEIC420 featuring a peak current capability of over 20 A is used. For the boost diode D_1 a 600 V SiC-Schottky diode CSD20060D is used; output capacitance C_{out} and also input capacitance C_{in} are partly realized by several 220 nF / 630 V ceramic type SMD capacitors, providing high current capability and low inductance.

In [10] it is shown that the switching losses are dominated by the intrinsic capacitance $C_{oss}(V_{DS})$ of the MOSFET when very fast switching is realized. Because of the low $R_{DS,on}$ of the CoolMOS SPW47N60C3 and its low output capacitance C_{oss} at $V_{DS} = 300$ V, a superjunction power MOSFET seems to be an ideal choice for high speed switching. Unfortunately, as shown in [11], C_{oss} rises from a few 100 pF to over 10 nF for very low blocking voltage levels. This results in long delay times at turn-off of the device, finally leading to significant current distortions for lower current values in single phase and three phase active rectifier applications [4]. To overcome this drawback the RF switch-mode power MOSFET DE475-501N44 from IXYS-RF is used. This device shows a much less pronounced $C_{oss}(V_{DS})$-characteristic than

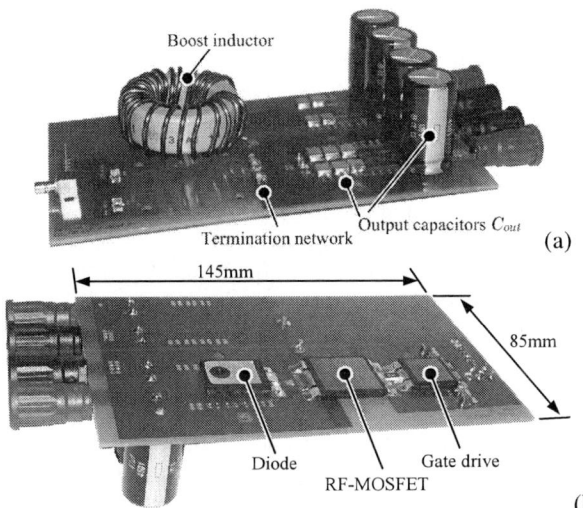

Fig. 8: Realized prototype of the boost circuit with PCB board wiring including a damping layer with optimized termination network. (a) TOP view and (b) BOTTOM view of the prototype which can directly be mounted on a heatsink.

superjunction MOSFETs. Additionally, the RF MOSFET utilizes a DE475 package which is optimized for high speed, high frequency, high power applications (**Fig. 8**). Due to the symmetrical package design, where the two source terminals lie on either side of the drain terminal, the parasitic inductance of the device can be reduced to less than 5 nH [12]. Due to the two source terminals two commutation paths have to be also considered. The layout of the boost circuit is optimized for very low wiring inductances in the commutation path. To form the proposed damping structure a copper loop (terminated by the proposed RC-network) is routed in a layer that is between the two wiring layers (cf. Fig. 3).

B. Current measurement

The MOSFET current I_{DS} is measured using a self-made AC-current probe (**Fig. 9**). The sensor has been designed for a sensitivity of 100 mV/A using a minimum number of turns to achieve a high bandwidth. The performance of this sensor has been compared to a 2 GHz shunt of T&M research [13] and shows very good results for frequencies beyond 100 MHz. Due to its small size the current transformer can be plugged directly on the connection leads of a TO220- or TO247-package. Unfortunately the DE475- package has wide connection leads and, therefore a short wire has to be used to insert the current sensor into the circuit, resulting in an additional inductance of approximately 8 nH for the commutation path which has to be considered in the model of the damping circuit

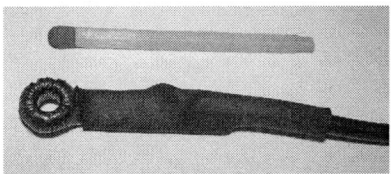

Fig. 9: Self-made AC-current transformer with a sensitivity of 100 mV/A to measure the Drain-Source current I_{DS} of the MOSFET.

978-1-4244-1871-8/07 $25.00 © 2007 IEEE

C. Design of the damping layer using PEEC method

For designing an optimal termination network of the damping layer, the parasitic elements of the system have to be considered. These elements could be determined by impedance measurements on the realized hardware using an adequate impedance analyzer. However, due to the fact that such measurements in the pF / nH-range are rather difficult, an alternative method to determine the parameters in an early state of system design based on simulation without building a dedicated prototype shall be used.

For a simulation including layout parasitics, the PEEC method emerged as a computational effective and accurate technique [14, 15]. Here, the PCB layout is discretized into a large number of individual elements as shown in the 3D-model of the prototype layout (cf. **Fig. 10**). The PEEC method creates matrices of partial elements representing the magnetic and electric field couplings and the resulting equations subsequently are solved in a Spice-like circuit simulator. Recent development makes PEEC an integrated full wave method, which can handle non-orthogonal elements [16] as well as dielectrics [17]. In **TABLE 2** both simulation results of the PEEC method and measurement results, using the impedance analyzer HP4294A, are summarized. The difference between the measured and calculated layout inductances L_{1i}, L_{2i} and M_i has its origin in the measurement technique, which poses a challenge due to the compact geometry of an optimized layout.

Since all parameters of the prototype are identified, a simplified model of the boost circuit can be drawn (cf. **Fig. 11**). Using the optimization of (11) results in $R_{p,opt}$= 22 Ω and $C_{p,opt}$= 184 pF for the termination network. The optimized components of the termination network are added to the PEEC model.

The capacitive coupling C_{12} between the proposed damping layer and the circuit layout, which is neglected for the sake of easy modeling in **Section II**, lowers the damping performance. Due to the distributed nature of the parasitic capacitance and inductance, it is not possible to distinguish clearly between the opposing effects of magnetic and capacitive coupling between the layers. Likewise, the assumption of a frequency independent lumped inductance and capacitance is not an accurate representation. However, the solver [18] takes all

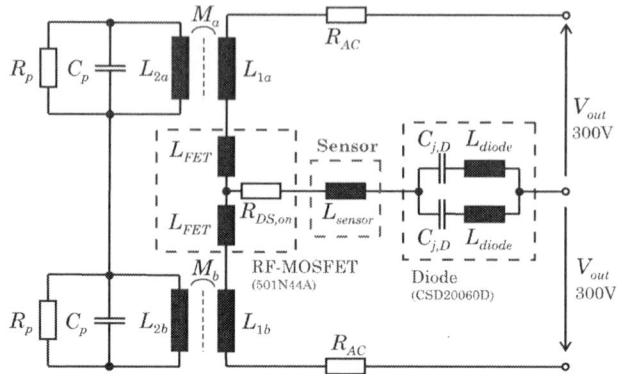

Fig. 11: Simplified model of the realized prototype with damping layer including parasitic elements of the RF-MOSFET, current sensor and the SiC-diode.

these effects into account.

The simulated total impedance of the commutation path with optimized damping network, including the dielectric of the FR4 material (ε_r=4.4), is depicted in **Fig. 12**. There is also an impedance measurement included, taken from the realized prototype, as well as the calculated impedance of the simplified model of Fig. 11. The measurements could only be performed up to 110MHz due to the bandwidth limitation of the used impedance analyzer HP4294A. The damping of the resonance at 100 MHz is apparent. The similarity of simulation and measurement is very good but also the simplified model with lumped elements shows very good results.

TABLE 2
MEASURED AND SIMULATED PARASITICS OF THE REALIZED PROTOTYPE.

	Measurements (HP4294A)	Simulation (PEEC method)
L_{diode}	10.9 nH (per diode)	-
$C_{j,D}$	55 pF	-
L_{FET}	4 nH (per lead)	-
L_{sensor}	8 nH	-
R_{AC}	500mΩ	-
L_{1a} / L_{1b}	8.2 nH / 7.7 nH	11.3 nH / 10.5 nH
L_{2a} / L_{2b}	9.9 nH / 8.1 nH	11.5 nH / 10.6 nH
M_a / M_b	5 nH / 5.25 nH	6.4 nH

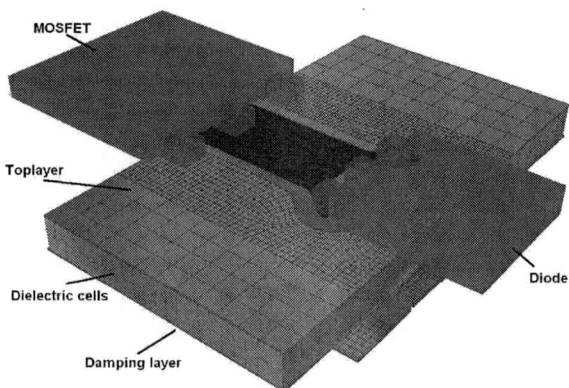

Fig. 10: PEEC model of the realized prototype with damping layer.

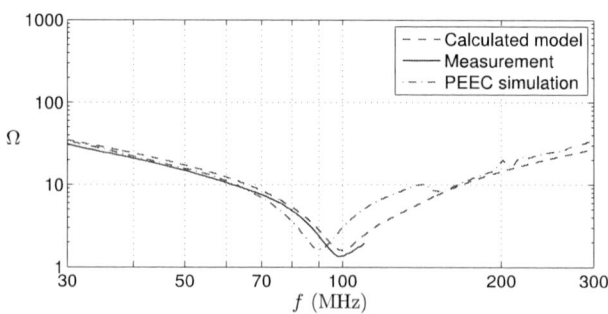

Fig. 12: Impedance measurement, calculated impedance using (8) – (10) and corresponding PEEC simulation of the commutation path, including the effect of the damping layer.

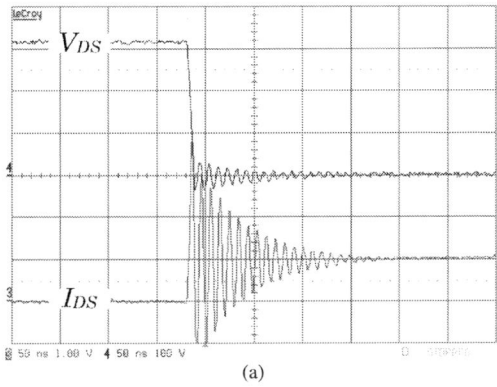 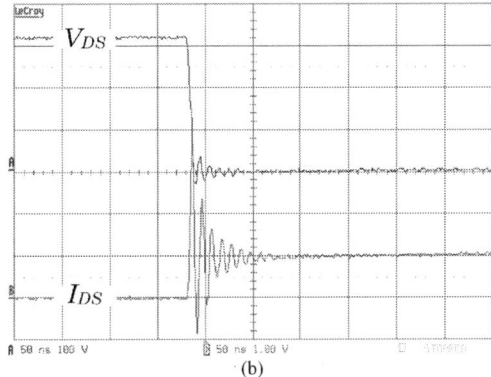

(a)   (b)

Fig. 13: Measurement results taken from the realized prototype (a) without damping layer and (b) with damping layer terminated with an optimized RC-network. U_{DS} (300 V/div), I_{DS} (10A/div), time scale 50 ns/div.

D. Experimental Results

The realized prototype with optimized damping network has been tested for a boost converter output voltage of 300 V and for current levels up to 20 A. The results for a current of 10 A are given in **Fig. 13** (for Fig. 13(a) the damping layer of the PCB has been left open). It can be seen that the switching transient oscillations are reduced significantly using the damping layer. Although the first voltage/current peak is still present, the decay of the ringing is enhanced noticeably.

The time domain response of the commutation path exited by a step function is calculated based on the simulated impedance curve. In **Fig. 14** the numerical result of this transformation is compared to a current measurement taken from the realized prototype for a current of 10 A (the amplitude of the input step of the numerical system is scaled to obtain an equal final value). The oscillation frequency of the simulated system is slightly smaller compared to the measured. This can also be seen at the lower resonance point in Fig. 12. However, the simulated maximum peak current and especially the damping effect are in good agreement with the measurement results. Therefore a PEEC simulation can be used to directly analyze the performance of the designed damping system in an early development state.

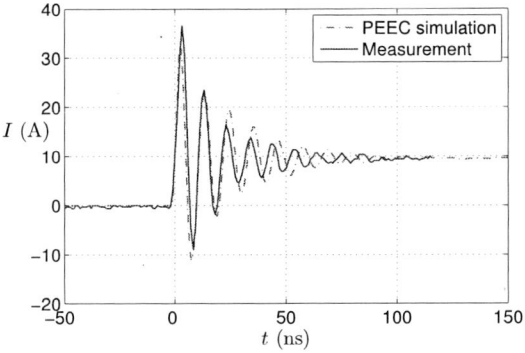

Fig. 14: Measured and calculated current shape obtained from a numeric Laplace transform of the impedance curve of **Fig. 12**.

IV. CONCLUSIONS

In this paper a new passive damping layer is introduced for damping the undesired voltage/current ringing appearing at the switching instants of hard-switched power electronic converters. An additional copper layer, terminated by an optimally designed RC-network, is inserted between the PCB wiring layers of the converter. A very simple model describing the overall system behavior, as well as design guidelines for the damping layer and the RC-termination network are given. The design and performance of the proposed damping layer is shown for a realized boost-converter circuit. For the arrangement and the optimization of the termination network, impedance measurements of the realized PCB would be necessary. It is shown that by application of the PEEC simulation method the system can be designed and analyzed in an early development state without an existing hardware prototype. Measurements, taken from the realized prototype show, that the switching transient oscillations can be reduced significantly but it is also shown that the damping capability of this purely passive approach is limited.

An active control of the switching transient would be an enhanced method for reducing the oscillations. This active control (to be implemented into the gate drive stage of the MOSFET), however, would have to have a control bandwidth in the frequency region of the oscillations to be damped (e.g., typ. 100 MHz for state-of-the-art power MOSFETs). Such active damping seems to be impossible using discrete components, but could be realizable if the gate drive, the active control of the switch and the switch itself are closely integrated.

The switching-transient oscillations at turn-on of the MOSFET are mainly determined by the relatively high junction capacitance of the SiC-diode. The performance of the proposed passive damping layer could be increased considerably if this capacitance is reduced. This can be achieved in future by application of other semiconductor technologies.

978-1-4244-1871-8/07 $25.00 © 2007 IEEE   1163

REFERENCES

[1] A. Siebert, A. Troedson, S. Ebner, "AC to DC power conversion now and in the future," *IEEE Trans. Industrial Applications*, vol.38, no. 4, July-Aug. 2002, pp. 934-940.

[2] J.W. Kolar and F.C. Zach, "A Novel Three-Phase Utility Interface Minimizing Line Current Harmonics of High-Power Telecommunications Rectifier Modules," *IEEE Transactions on Industrial Electronics*, vol. 44, no.4, August 1997, pp. 456-467.

[3] S. Bontemps, A. Calmels, S.D. Round, J.W. Kolar, "Low profile power module combined with state of the art MOSFET switches and SiC diodes allows high frequency and very compact three-phase sinusoidal input rectifiers." *Proc. of the Conf. for Power for Power Electronics, Intelligent Motion and Power Quality (PCIM'07)*, Nuremberg (Germany), 2007.

[4] S.D. Round, P. Karutz, M.L. Heldwein and J.W. Kolar, "Towards a 30kW/liter, Three-Phase Unity Power Factor Rectifier." *Proc. of the 4th Power Conversion Conference (PCC'07)*, Nagoya (Japan), April 2-5, 2007.

[5] J.D. van Wyk Jr., W.A. Cronje, J.D. van Wyk, C.K. Campbell, P.J. Wolmarans, "Power Electronic Interconnects: Skin- and Proximity Effect-Based Frequency Selective Multipath Propagation," *IEEE Transactions on Power Electronics*, vol. 20, no.3, May 2005, pp. 600-609.

[6] K. de Jager, L. Dalessandro, I.W. Hofsajer, W.G. Odendaal, "Wave Analysis of Multilayer Absorptive Low-Pass Interconnects," *Proc. of the 38th IEEE Power Electronics Specialists Conference (PESC'07)*, Orlando (USA), June 17-21, pp. 2121-2127.

[7] W. Teulings, J.L. Schanen, J.Roudet, "MOSFET Switching Behaviour Under Influence of PCB Stray Inductance," *Proc. of the 31st IAS Annual Meeting*, San Diego (USA), 6-10 October 1996, pp.449-1453.

[8] S.C. Tang, S.Y. Hui, H.S. Chung, "Coreless Planar Printed-Circuit-Board (PCB) Transformers – A Fundamental Concept for Signal and Energy Transfer," *IEEE Transactions on Power Electronics*, vol. 15, no. 5, Sept. 2000, pp. 931-941.

[9] J. Biela, A. Wirthmueller, R. Waespe, M.L. Heldwein, J.W. Kolar, E. Waffenschmidt, "Passive and Active Hybrid Integrated EMI Filters," *Proc. of the 21st IEEE Applied Power Electronics Conf. (APEC'06)*, Dallas (USA), vol. 2, March 19-23, 2006, pp. 1174-1180.

[10] K. Adachi, K. Takao, Y. Hayashi, H. Ohashi, "Study on Intrinsic Loss of Unipola Power Device due to Main Junction Capacitance," *IEEJ Transactions on Industry Applications*, vol. 126, no. 7, 2006, pp. 941-945.

[11] L. Lorenz, G. Deboy, I. Zverev, "Matched Pair of CoolMOS Transistor With SiC-Schottky Diode – Advantages in Application," *IEEE Transactions on Industry Applications*, vol. 40, no. 5, Sept./Oct. 2005, pp. 1265-1272.

[12] IXYS-RF, "DE-Series Fast Power MOSFETs – An Introduction," Application Note published at http://www.ixysrf.com/ (2007).

[13] T&M Research, "Series SDN-414 Current Viewing Resistors – Description and Specification," published at http://www.tandmresearch.com (2007).

[14] P. Musznicki, J.L. Schanen, B. Allard and P. Chrzan, "Accurate Modeling of Layout Parasitics to Forecast EMI Emitted from a DC-DC Converter," *35th Annual IEEE Power Electronics Specialists Conference (PESC'04)*, vol. 1, Aachen (Germany), 2004, pp. 278-283.

[15] A. Musing, M.L. Heldwein, T. Friedli, J.W. Kolar, "Steps Towards Prediction of Conducted Emission Levels of an RB-IGBT Indirect Matrix Converter," *Power Conversion Conference (PCC '07)*, Nagoya (Japan), 2-5 April 2007, pp. 1181-1188.

[16] A. Ruehli, G. Antonini, J. Esch, J. Ekman, A. Mayo, A. Orlandi, "Non-Orthogonal PEEC Formulation for Time and Frequency Domain EM and Circuit Modeling," *IEEE Transactions on Electromagnetic Compatibility*, vol. 45, no.2, May 2003, pp. 167-176.

[17] A.E. Ruehli, H. Heeb, "Circuit models for three-dimensional geometries including dielectrics," *IEEE Transactions on Microwave Theory and Techniques*, vol. 40, no. 7, July 1992, pp. 1507-1516.

[18] Luleå University of Technology and University of L'Aquila, "Partial Element Equivalent Circuit (PEEC) Solver Page," http://www.sm.luth.se/~jekman/PEEC/Program/ (2007).

Controlling Voltage Profile in Loop Distribution System with Distributed Generation Using Series Type BTB Converter

Rejeki Simanjorang
Osaka University
Email: rejeki@pe.eei.eng.osaka-u.ac.jp

Yushi Miura
Osaka University
Email: miura@eei.eng.osaka-u.ac.jp

Toshifumi Ise
Osaka University
Email: ise@eei.eng.osaka-u.ac.jp

Abstract—Connecting Distributed Generations (DGs) into distribution system has presented a challenge of controlling voltage profile in distribution system. In the presence of DGs, voltage profile of distribution system becomes non-uniform and may result in large voltage fluctuation at loads. Therefore, there is need for strict limitation on power injected by DGs. This paper proposes an effective method to overcome the problem of increase/decrease of voltages in distribution system due to the presence of DGs, and it also results in higher permissible power injected by DGs. The proposed method solves the problems by firstly, changing radial to loop distribution configuration. Then secondly, by using series type Back-To-Back (BTB) converter to control voltage profile of the resulting loop distribution system. Two control nodes in the loop distribution system are used to control the voltage profile. The series type BTB converter is installed at the sending end voltage node of loop feeder. The effectiveness of the proposed method is shown by the improvement of voltage profile and the small size of series type BTB converter. Theoretical analysis is discussed and simulation results are shown to verify the proposed method.

I. INTRODUCTION

Utilization of distributed generations (DGs) has increased rapidly in recent years as one of the solutions to reduce CO_2 emission. The DGs such as Photovoltaic (PV), wind turbine, micro-gas turbine, etc. have a low adverse environmental impact and they are installed at customers' side.

As a result of the increase in DGs' utilization, the challenges of connecting them to the distribution system have arisen. However, power injected by connecting DGs brings some problems such as increase of load side voltage in the distribution system and unpredictable voltage profile in the feeder.

To overcome such problems, some research on connecting DG to the distribution systems have been carried out. We can categorize these research into two methods of controlling voltage profile. The first one is the use of mechanical switch (tap changer of transformer), while the second one is the use of power electronic apparatus.

The first method can be realized by adjusting tap changer of main transformer of distribution system or tap changer of step voltage regulator (SVR). The limit of power injected by DGs on the distribution system with line drop compensator (LDC) of distribution transformer was evaluated [1]. However, controlling voltage by tap changer of transformer and SVR is step-wise and need delay time for operation of tap

changer. As a result; continuous, precise and fast voltage control can not be realized. In order to overcome this problem, cooperative method using SVR and Static Var Compensator (SVC) was proposed [2]. The purpose of this method is to realize fast controlling voltage and reduce the capacity size of SVC. The limit of power injected by DGs on the distribution system with SVR and SVC was also evaluated [3].

More advanced apparatuses such as Static Synchronous Compensator (STATCOM), Static Synchronous Series Compensator (SSSC), Unified Power Flow Controller (UPFC) and BTB converter were studied [4][5]. The UPFC was evaluated as the most effective voltage control device in the distribution system [4]. The Back-To-Back (BTB) converter connected at the end of the distribution system was studied in order to improve voltage profile and reduce line losses as Loop Power Flow Controller (LPC) [5].

In this paper, a novel method for controlling distribution system voltage profile by smaller series type BTB converter is proposed.

Advantage of this method is easiness of installation in the system because the power converter needs smaller space than LPC and is installed in the area of substation. The proposed method has the same configuration as the Loop Power Balancer introduced by the same authors [6][7].

II. SERIES TYPE BTB CONVERTER FOR CONTROLLING VOLTAGE IN DISTRIBUTION SYSTEM

A. Configuration of the system

One example of a distribution system with multi-feeders is shown in Fig. 1. This distribution system has many radial feeders and connected with a busbar. Power of busbar is supplied through an on-load tap changer power transformer TR.

In the distribution system, it is assumed that two feeders k and l, are in the lightly and heavily loaded condition, respectively, and a bulk DG (clustered PVs and it only generates active power) is installed in the lightly loaded feeder (feeder k). Maximum power of DG is larger than total load of feeder k.

With respect to condition above, two problems of high power losses in heavily loaded feeder and excess power in lightly loaded feeder occur in the distribution system. High

978-1-4244-1871-8/07 $25.00 © 2007 IEEE

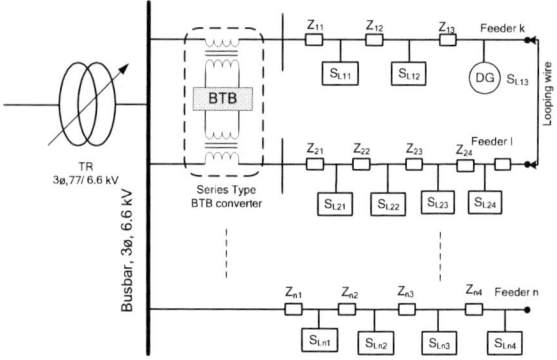

Fig. 1 A model of distribution system with multi-feeders

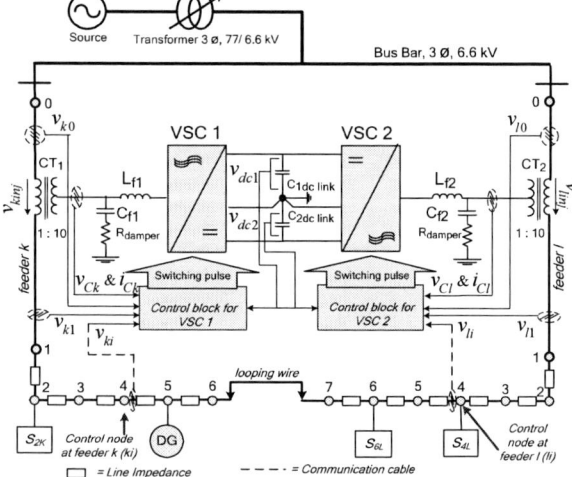

Fig. 2 Detail circuit of the proposed method

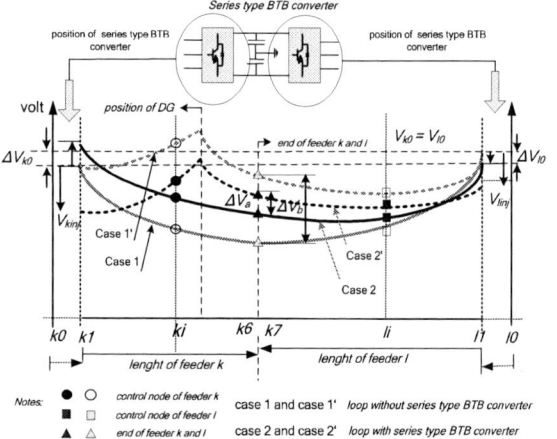

Fig. 3. Controlling voltage profile of distribution system by series type BTB converter

power losses lead to a high voltage drop and excess power lead to voltage increase of load side.

To overcome these problems, this paper proposes the method that the feeders k and l be looped by tying their receiving ends and then a series type BTB converter is installed close to sending end. The resultant loop distribution

system is shown in Figs. 1 and 2. Feeder k has seven nodes $k0$ to $k6$ and feeder l has eight nodes $l0$ to $l7$, respectively.

By changing to loop distribution system, the high voltage drop in heavily loaded feeder can be reduced due to the fact that some of the excess current can now flow in the lightly loaded feeder.

If the DG generates large power, a large voltage fluctuation and over-voltage may occur in the loop distribution system. To avoid these problems, series type BTB converter provides function of controlling voltage profile in loop.

B. Controlling voltage profile of distribution system by series type BTB converter

Series type BTB converter controls voltage profile by controlling voltage at two nodes. These nodes, which are named the control nodes, are located at feeders k and l (ki and li) as shown in Fig. 2. Controlling voltage at each control node is effected by injecting voltage (V_{kinj} and V_{linj}) at the sending end voltage node of each feeder, which is supplied by the series type BTB converter.

In Fig. 2, desired voltages at the control nodes of feeders k and l (V_{ki}^* and V_{li}^*, respectively) can be equal or different. The determination of desired voltages is based on the assumption that these voltages can subsequently control the voltage profile to be in permissible range. Position of control node is around quarter of total length feeders k and l. The reference voltage of control nodes can be obtained by simulation study.

Voltage profiles of the loop distribution system with/without series type BTB converter are assumed as shown in Fig. 3. Cases 1 and 1' are voltage profiles of the loop distribution system without the series type BTB converter. Cases 2 and 2' are voltage profiles of the loop distribution system with the series type BTB converter. Cases 1 and 2 are a result of the DG not injecting power to loop distribution system whereas cases 1' and 2' are a result of the DG injecting power to loop distribuition system. Voltage of busbar will slightly increase by ΔV_{k0} (which is also equal to ΔV_{l0}) while DG injects power to the loop distribution system.

ΔV_a is voltage fluctuation at junction nodes of $k6$ and $l7$ between cases 1 and 1'. ΔV_b is voltage fluctuation at junction nodes of $k6$ and $l7$ between cases 2 and 2'.

In the case of Fig. 3, ΔV_a is higher than ΔV_b. This result shows that voltage fluctuation was mitigated with series type BTB converter. Voltage increase of case 1' at DG voltage node is higher than case 2'. The lower voltage fluctuation in cases 2 and 2' is effected by V_{kinj} and V_{linj} for feeders k and l, respectively.

The series type BTB converter increases or reduces the voltage in feeders k and l as shown by cases 2 and 2' in Fig. 3. Amount of reduced voltage depends on voltage reference at control nodes of feeders k and l.

III.   CONTROL SYSTEM DESIGN FOR SERIES TYPE BTB CONVERTER

The equivalent circuit of the distribution system with DG is shown in Fig. 4. V_{k0} is voltage of busbar, V_{ki} is voltage at

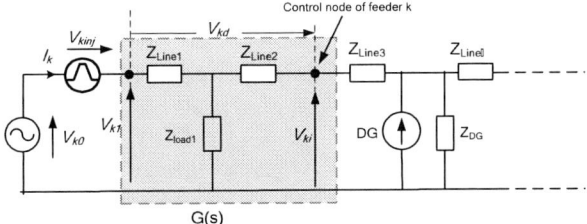

Fig. 4 Equivalent circuit of distribution system at feeder k with series type BTB converter and DG

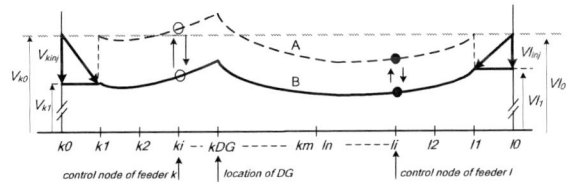

Fig. 5. Concept of controlling voltage profile in loop distribution system

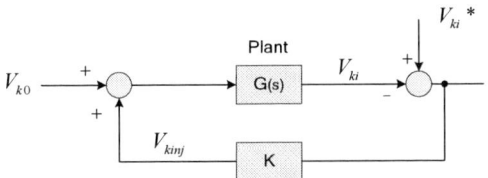

Fig. 6. Feedback control system to control voltage at V_{ki}

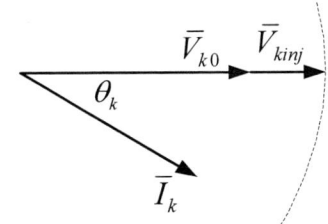

Fig. 7 In-phase injected voltage scheme

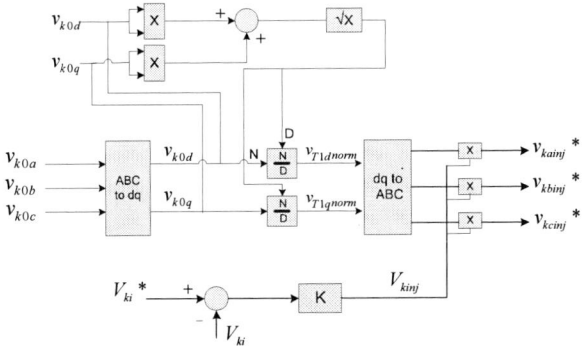

Fig. 8 Determination of injected voltage reference

control node of feeder k and V_{kd} is voltage drop between V_{k0} and V_{ki}. V_{kd} is determined by line impedances Z_{line1} and Z_{line2}, load impedance Z_{load1} and current I_k. The current I_k is also influenced by the DG current.

From Fig. 4, G(s) is a plant of control system. Objective of the control system is to control V_{ki} by injecting voltage V_{kinj}.

A. Concept of Controller

Fig. 5 shows the proposed concept of controlling voltage in the loop distribution system, which is proposed in this paper. In Fig. 5, the series type BTB converter has two outputs V_{kinj} and V_{linj} into the loop distribution system. Controlling voltage profile is done by injecting voltage at sending end feeders, between $k0$ to $k1$ and $l0$ to $l1$, to achieve the desired voltages at ki and li, respectively. V_{kinj} and V_{linj} can be positive or negative.

Feedback control system of controlling voltage profiles is depicted as Fig. 6. The transfer function of the system can be written as

$$\left[\left(V_{ki}{}^* - V_{ki}\right)K + V_{k0}\right]G(s) = V_{ki} \qquad (1)$$

where $V_{ki}{}^*$ is reference voltage at control node of feeder k, V_{ki} is rms of actual voltage at control node of feeder k. V_{k0} is rms of voltage at node 0 of feeder k, and K is proportional gain. According to Fig. 4, G(s) can be written as

$$G(s) = \frac{V_{ki}}{V_{k1}} = \frac{V_{ki}}{V_{k0} + V_{kinj}} \qquad (2)$$

From Fig. 6, we can derive the equation for the injected voltage as

$$V_{kinj} = K\left(V_{ki}{}^* - V_{ki}\right) \qquad (3)$$

where V_{kinj} is reference voltage rms for series type BTB converter.

B. In-phase injected voltage scheme

From (3), V_{kinj} is a scalar value. The injected voltage is done by in-phase voltage scheme as shown in Fig. 7.

According to Fig. 7, active power of in-phase injected voltage scheme can be formulated as

$$P_{kinj} = V_{kinj}I_k \cos\theta_k \qquad (4)$$
$$P_{linj} = V_{linj}I_l \cos\theta_l \qquad (5)$$

where P_{kinj} and P_{linj} are active power injected to feeders k and l respectively; V_{linj} is voltage injected to feeder l; I_k and I_l are currents at feeders k and l respectively; θ_k and θ_l are phase angles between voltage and current at feeders k and l respectively.

The values of voltage, current and phase angle in (4) and (5) are not equal to each other. As a result, active power injected to feeders k and l might be different when using in-phase injected voltage scheme for both sides of BTB converter. However, this does not matter. These active powers are needed to charge capacitor in the dc-link in order to achieve its reference voltage.

C. Control block for injecting voltage

Based on the explanation in sections III.A and III.B, control block for injecting voltage to the distribution system is designed. Both sides of the series type BTB converter use the same model control scheme (in-phase injected voltage scheme). The first one is control block for determining injected voltage as shown in Fig. 8. In this control block, in-phase injected voltage scheme is implemented by generating voltages $v_{kainj}{}^*$, $v_{kbinj}{}^*$ and $v_{kcinj}{}^*$. These voltages

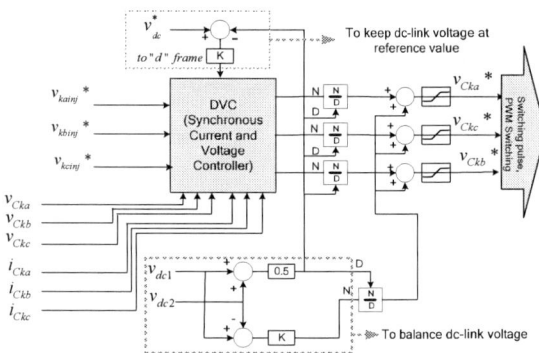

Fig. 9 DVC and dc link voltage controller

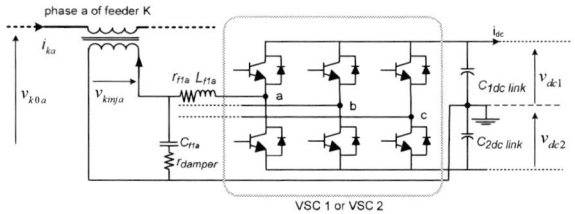

Fig. 10 Series type BTB converter at feeder k

are used as inputs for Double Vector Control (DVC) of Fig. 9. DVC is voltage controller using d-q transformation with ac currents feed back.

In Fig. 9, controller for maintaining dc-link voltage at reference value is also shown. Both sides of the series converter have the controller for maintaining dc-link voltage at the reference value. As a result, the robustness of controlling dc-link voltage is achieved.

The series type BTB converter in this study uses two three-phase four-wire bidirectional inverter, which are connected back to back through a dc-link. Converter side at feeder k is shown in Fig. 10.

IV. SIMULATION MODEL

To investigate operational performance of the proposed method, a simulation, which represents model of Fig. 2 has been carried out by PSCAD/EMTDC software.

Parameters used in the simulation are shown in Tables 1 and 2. It is assumed that feeders k and l have lengths of 5 km and 6 km, respectively. Permissible range of line voltage regulation is +/- 5 % of line voltage.

The DG is represented by clustered PVs. To reduce the simulation time in software, generating power of DG is shortened to a few seconds. For example, in a real situation, PV can generate power starting hours from 0600 hrs to 1800 hrs and the peak power can be reached at 1200 hrs. For the purpose of simulation, the simulation time has been compressed from hours to seconds. We assume the DG produce power from 0.6 sec. to 1.8 sec as representative time from 0600 hrs to 1800 hrs and the peak power can be produced at 1.2 sec. And then, to show the effectiveness of the proposed method, loads of feeders k and l are also

TABLE 1.

PARAMETER OF FEEDERS

Length of feeder	Total Length of feeder k = 5 km Distance of two nodes sequence from nodes 1 to 6 is 1 km
	Total Length of feeder l = 6 km Distance of two nodes sequence from nodes 1 to 7 is 1 km
Distributed Generation	Clustered PV Total capacity 7 MW
On line tap changer transformer (TR).	20 MVA, Δ/Y 77/6.6 kV (X= 6 %) Iron loss = 1.0 %; Copper Loss = 2.5 %
Permissible line voltage regulation	6.3 kV ≤ V ≤ 6.9kV

* Line impedance: R = 0.228 Ω/km and X = 0.395 Ω/km (80 mm²)

TABLE 2.

PARAMETER OF BTB CONVERTER

BTB Converter Capacity	300 kVA
Series Transformer/ Current Transformer (CT) Ratio	3 single phase series Transformer 1 : 10 (X = 2%) Iron loss 1 %; Cooper Loss = 2 %
LC Filter	L_f = 9.2 mH (2.8 %) ; C_f = 5.31μF ; R_f = 0.025 Ω R_{damper} = 10 Ω
$C_{1dc\ link}$	400 μF
$C_{2dc\ link}$	400 μF
v_{dc1}* (Reference)	5000 V
v_{dc2}* (Reference)	5000 V

assumed as constant loads and they are periodic (from 0 sec to 2.4 sec).

Total length of the feeder is 11 km. We positioned two control nodes at $k4$ and $l4$, because both distances from sending end are 3 km, which is nearly quarter of total length of the feeder.

Load Curves

Load curves of feeders k and l are shown in Figs. 11a and 11b, respectively. It can be seen that from 0.0 sec to 2.4 sec, feeder k is lightly loaded feeder and feeder l is heavily loaded feeder. Load at feeder k is placed at node 2 and two loads at feeder l are placed at node 4 and node 6 as shown in Figs. 11a and 11b, respectively.

Active Power of Distributed Generation and Transformer

Output active powers of the DG and transformer (TR) are shown in Fig. 11c.The DG is installed at node 5 of feeder k. The DG starts to generate power from 0.6 sec to 1.8 sec and repeats again from 3.0 sec to 4.2 sec periodically. Maximum power output of the DG is 7 MW.

In Fig. 11c, power injection by the DG reduces input power from TR to the distribution system. While the power

978-1-4244-1871-8/07 $25.00 © 2007 IEEE   1168

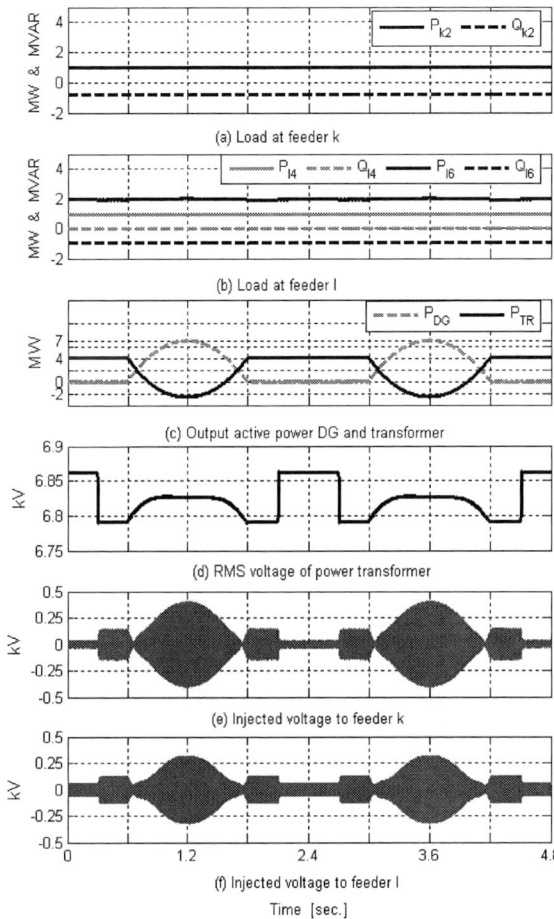

Fig. 11 Condition of load, power and voltage at feeder k and l of loop distribution system

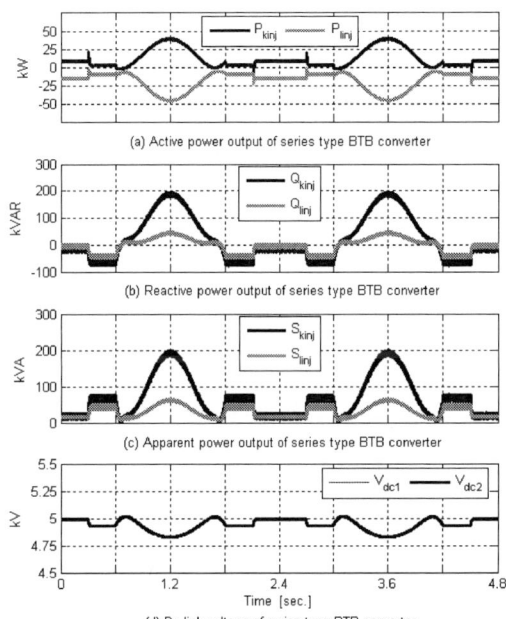

Fig. 12. Output power and voltage of series type BTB converter

injected by the DG is larger than power consumed in the distribution system, reverse power flow to TR occurs at maximum output power of DG around 1.2 sec and 3.6 sec.

Voltage of Transformer

Since the busbar has many feeders, tap changer of distribution transformer may change due to voltage changes in the feeder not in the loop distribution system. To verify the robustness of proposed control system, position of tap changer is changed from 0 sec to 0.3 sec, 2.1 sec to 2.7 sec and 4.5 sec to 4.8 sec as shown in Fig. 12d. In these time durations, ratio of transformer is altered from 77/6.947 kV to 77/6.875 kV.

In Fig. 11d, the actual transformer voltage is slightly different from the ideal transformer voltage (6.947 kV and 6.875 kV, respectively) due to losses inside transformer.

Fig. 11d also shows that transformer voltage increases slightly while the DG injects power to feeder.

Injected Voltage

To control voltage profile in the loop, the series type BTB converter injects voltage to feeders l and k. Voltages injected are shown in Figs.11e and 11f. These injected voltages can follow the change of voltage in the loop smoothly. Smooth

voltage injection can be realized by series type converter such as proposed BTB converter.

Injected Power

Fig. 12 shows the power injected by the series type BTB converter to feeders k and l. Fig. 12a shows that active powers injected for both sides of converter are not equal as explained in section III.B. Active power is drawn from feeder l and small amount of active power is injected to feeder k. The difference of active power is caused by unequal current flow in both feeders. This results in different value of power losses of their switching filters.

Reactive power injected by series type BTB converter is shown in Fig. 12b. When the DG is injecting power to the distribution system, reactive power output of series type BTB converter is increased. This means the series type BTB converter injects capacitive reactive power. When the DG is not supplying power and voltage of transformer decreases, the series type BTB converter injects inductive reactive power to distribution system.

From Figs. 12a and 12b, reactive power has main role of boosting voltage at control nodes.

Capacity of converter is represented by Fig. 12c. Maximum capacity of each side converter is about 200 kVA and 70 kVA, respectively. These values are sufficiently small compared with total power in the loop distribution system.

Dc-link voltage

Fig. 12(d) shows the dc-link voltage during operation of the series type BTB converter. The reference of the dc-link voltage is 5 kV. Fluctuation of dc-link voltage is around 200 V, which is only 4 % of reference dc-link voltage.

Voltage at control nodes

In this simulation, we determined voltage references at control nodes of feeders k and l as 6.6 kV and 6.5 kV, respectively. The series type BTB converter is used to

978-1-4244-1871-8/07 $25.00 © 2007 IEEE

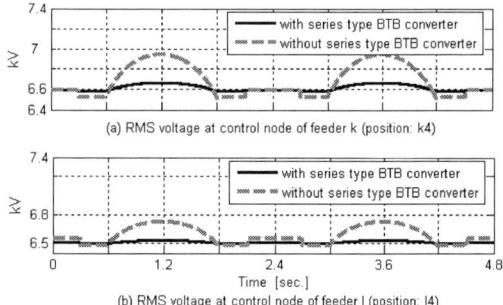

(a) RMS voltage at control node of feeder k (position: k4)

(b) RMS voltage at control node of feeder l (position: l4)

Fig. 13. RMS voltage at control nodes

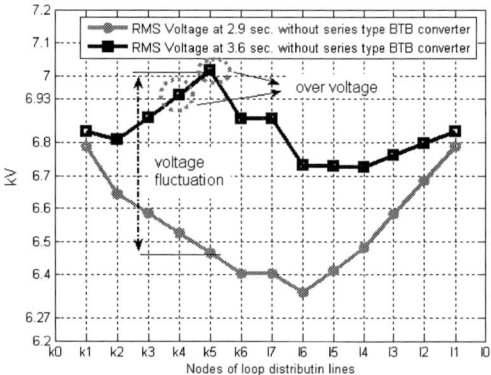

(a). Without series type BTB converter

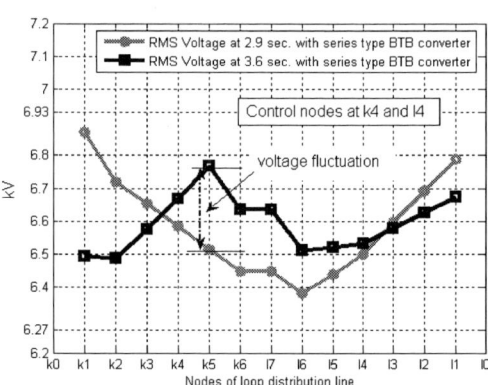

(b) With series type BTB converter

Fig. 14 Voltage profile in loop distribution system

regulate voltage difference at control nodes as shown in Figs. 13a and 13b. In these figures, voltage increases while the DG injects power.

The system without series type BTB converter has a larger voltage fluctuation than the system with series type BTB converter. From Fig. 13a, the maximum fluctuation voltage for the system without the series type BTB converter is around·400 V compared with around 100 V for the system with the series type BTB converter.

Voltage profile in loop distribution system

Voltage at each node of loop distribution system is shown in Fig. 14. Figs. 14a and 14b show voltage profiles in the loop with/without the series type BTB converter. In each figure, two profiles are shown: one is the profile at 2.9 sec when DG

is supplying no power, and the other is at 3.6 sec when DG is supplying the maximum power to the distribution system.

From Figs. 14a and 14b, the largest value of voltage fluctuation occurs at node *k5*. At this node, there is the DG as shown in simulation model. In Fig. 14a, over-voltage (outside permissible range: 6.3 kV ≤ V ≤ 6.9kV) occurs at 3.6 sec when DG injects maximum power to the loop distribution system. Comparing voltage fluctuations between Figs. 14a and 14b, the loop distribution system with series type BTB converter has lower voltage fluctuation. For the system with the series type BTB converter, the voltage fluctuation is around 250 V as compared to voltage fluctuation of around 500 V for the system without the series type BTB converter. From these results, the effectiveness of installing the series type BTB converter due to integration of the DG to reduce voltage fluctuation and to suppress the over voltage is verified.

V.   CONCLUSION

We proposed a method to improve voltage profile in the distribution system with DG. The proposed method is realized by changing radial to loop feeder of distribution system and installing a small capacity of series type BTB converter close to substation.

Since the proposed method is achieved by injecting voltage at the sending end voltage of two looped feeders, this method is suitable for the distribution system with a short distance. For longer line case, distributed series type converter is under investigation.

REFERENCES

[1] T E Kim, J E Kim, "A Method for Determining the Introduction limit of Distributed Generation System in Distribution System", IEEE Power Engineering Society Summer Meeting 2001, Vol. 1, pp 456 – 461

[2] Y. Kubota, T. Genji, K. Miyazato, N. Hayashi, H. Tokuda and Y. Fukuyama: "Verification of cooperation control method for voltage equipment on distribution network simulatior considering interconnection fo wind power generators", IEEE Trans. and Dist. Conference and Exhibition 2002, Vol. 2, pp 1151 – 1156

[3] Y. Kubota, T.Genji, S. Takayama and Y Fukuyama: "Influence of Distribution Voltage Control Method to Maximum Capacity of Distributed Generators", T. IEE Japan, Vol. 124-B, No. 1, pp 7 – 14 (2004 - 1) (in Japanese)

[4] T. Kondo, J. Baba and A. Yokoyama: "Voltage Control of Distribution Network with a Large Photovoltaic Generations using FACTS Devices", T. IEE Japan, Vol. 126-B, No. 3, pp 347-358, (2006-4) (in Japanese)

[5] N Okada and K Kurokawal: "Control of Loop Power Flow Controller using Local Information and Determination of Control Coefficient", T. IEE Japan, Vol. 125-B, No. 4, pp 381-389, (2005-4) (in Japanese)

[6] R. Simanjorang, Y. Miura, T. Ise, S. Shigeyuki and H. Fujita: "Loop Power Balancer System between Distribution Power Transformers Using a Small Capacity Series Type Back to Back Converter", Proceeding of the Joint Technical Meeting on Power Engineering and Power Systems Engineering, IEE Japan, PE-06-073, PSE-06-073, (2006 - 8)

[7] R. Simanjorang, Y. Miura, T. Ise, S. Shigeyuki and H. Fujita: "Application of Series Type Back to Back Converter for Minimizing Circulation Current and Balancing Power Transformers in Loop Distribution Lines", The Fourth Power Conversion Conference (PCC), IAS-IEEE, pp. 997-1004, Nagoya-Japan, (2007-4)

978-1-4244-1871-8/07 $25.00 © 2007 IEEE

A New Direct Current Internal Resistance and State of Charge Relationship for the Li-Ion Battery Pulse Power Estimation

Jong Hoon Kim, Seong Jun Lee, Jae Moon Lee, Bo Hyung Cho
Seoul National University
Department of Electrical Engineering
San 56-1, Sillim-dong, Gwanak-gu, Seoul, 151-742
Email: qwzxas@hanmail.net, jun2u@pesl.snu.ac.kr, zzugumi@pesl.snu.ac.kr, bhcho@snu.ac.kr

Abstract—The conventional test to obtain the direct current internal resistance (DCIR) has only experimented with a duration time of 5 seconds in the discharge region[3]~[5]. To obtain the DCIR, the duration time, $\triangle t$ and the region condition are important for the hybrid electric vehicle (HEV). In this paper, a new measurement method to obtain a direct current internal resistance (DCIR) is proposed. The proposed approach is performed during 10 seconds in the charge and discharge regions in order to obtain the new relationship between the DCIR and the state of charge (SOC). Thus, this obtained data can be used to estimate the battery pulse power using the previous SOC algorithm, extended kalman filter (EKF)[6], which includes the DCIR-SOC relationship. The experiments are achieved using a fresh 1.3Ah 18650 type Li-ion battery at 25℃.

I. INTRODUCTION

The Li-ion battery has the most comprehensive characteristics of all types of the batteries due to high voltage, light mass, low self discharge, more cycle life and other advantages. The Li-ion battery has thus been developed and is widely used in many fields for energy saving systems and for protecting the global environment. Recently, electric vehicle (EV) and hybrid electric vehicle (HEV) applications mostly use the Li-ion battery due to are the most used for weight reduction, downsizing, improvement of input and output performance as well as the discharging and charging efficiency.

To determine the performance of EV and HEV applications, some factors should be well considered. Among these, the power is an important than other factors. Therefore, it is required to obtain precise battery power information. The power is defined as 10 seconds pulse power capability.

There are several ways to obtain the pulse power of an Li-ion battery, of which the Hybrid Pulse Power Characterization and DCIR tests are major methods. The HPPC test[1]~[2] is a representative method for obtaining the pulse power. This test is intended to determine dynamic pulse power capability using a test profile that incorporates both discharge and charge pulse. It is introduced and used under the US Department of Energy's (DOE) Advanced Technology Development (ATD) Program, the Idaho National Engineering and Environmental Laboratory (INEEL), the Freedom Cooperative Automotive Research (FreedomCAR), etc.

The DCIR test is another method used to obtain pulse power together with the HPPC test. This was introduced and used at the Advanced Battery Development Center and Hitachi Research Laboratory. The test to obtain the DCIR is similar with the HPPC test. The two tests are derived from Ohm's law, and thus during the discharging and charging, the voltage of the battery decreases and increases by the battery currents. Then a lumped resistance can be defined as the $\triangle V/I$. Thus, the values of the lumped resistance are approximately the same for the two tests.

To precisely obtain the DCIR, the duration time and region condition in the discharge and charge regions should be well considered. In particular, the length of the pulse and the region applied to the battery are such conditions The conventional DCIR test was experimented with a duration time of 5 seconds and only in the discharge region.

In this paper, a new measurement method to obtain the DCIR is proposed. It is performed during 10 seconds in the charge and discharge regions in order to obtain a new relationship between the DCIR and the SOC. This obtained data can be used to estimate the battery pulse power using the SOC algorithm, the EKF, which includes the DCIR-SOC relationship.

This proposed method is verified through simulations and experiments using a fresh 1.3Ah 18650 type Li-ion at 25℃

II. DIRECT CURRENT INTERNAL RESISTANCE

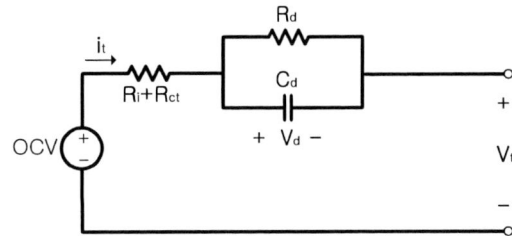

Fig. 1: Lumped parameter battery model

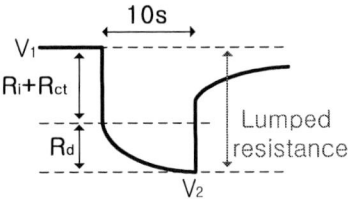

Fig. 2: Lumped resistance

978-1-4244-1871-8/07 $25.00 © 2007 IEEE

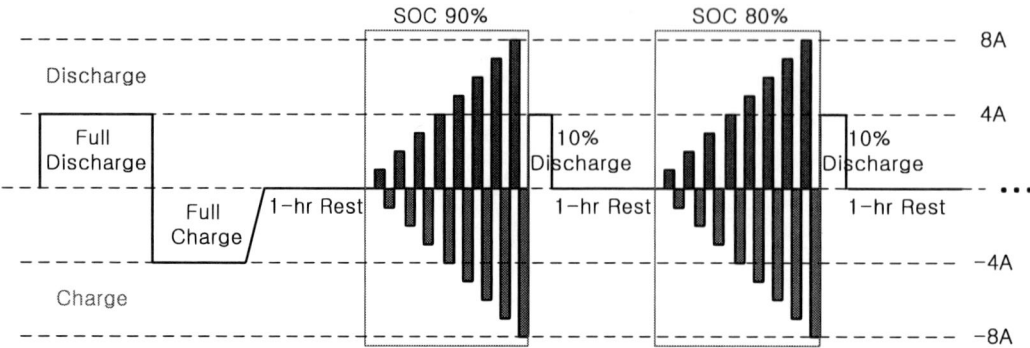

Fig. 3: DCIR test profile

The DCIR is defined as the magnitude of the lumped parameter. In general, the lumped parameter battery model consists of two resistance, namely the Ri+Rct(series resistance, charge transfer resistance), the Rd(diffusion resistance), and one capacitance, Cd(diffusion capacitance) as shown in Fig 1[7].

The lumped parameter of the battery can be modeled based on the OCV. Then, the Rct can be abbreviated due to fast dynamic [5]~[6]. By Ohm's law, the lumped resistance is derived as (1). V_1 and V_2 represent the terminal voltage measured under the pulse current, I. By Ohm's law, the lumped parameter of the battery model can be linearized battery behavior at given point in life based on a repeatable test cycle.

$$DCIR = lumped\ resistance = (V_1 + V_2)/I \approx Ri + Rd \quad (1)$$

Like the HPPC test[1], the objective of the DCIR test is to establish the pulse power estimation of a battery. The profile of the DCIR test is shown in Fig 3. It is made up of single repetitions of the profile, separated by a 10% SOC constant current discharge segment. During a period of 1 hour, the interval between before one profile and the next profile, the rest period, is applied in order to return to the electrochemical and thermal equilibrium condition. The capacity of a battery is measured through a full discharge period. When the battery is fully charged, the DCIR test begins and terminates after completing the final profile at 10% SOC. A pulse current is applied to the battery in the discharge and charge regions and the terminal voltage is measured in the respective region by turns after the duration time of 10 seconds.

The lumped resistance is varied with a duration time that is applied to the battery using a pulse current. Besides, the lumped resistance varies with region conditions. A few published researches only used the DCIR test profile in the discharge region with a duration time of 5 seconds[3]~[5]. In this case, it is difficult to obtain a precise DCIR. As a result, the objective of a longer duration time of 10 seconds is to demonstrate the power at various stages for the Minimum and Maximum Power Assist goals[1]. Thus, in this research,

a duration time of 10 seconds is applied to the battery at the discharge and charge regions

Each OCV during the rest periods is measured. In turn, a rest period of 5 minutes is applied to the battery for both the discharge and charge regions.

Fig 4 shows the DCIR test profile at a certain SOC. This test is repeatedly conducted at the increased current value. Then, the I-$\triangle$V curve is drawn and the resulting slope of the curves is the DCIR

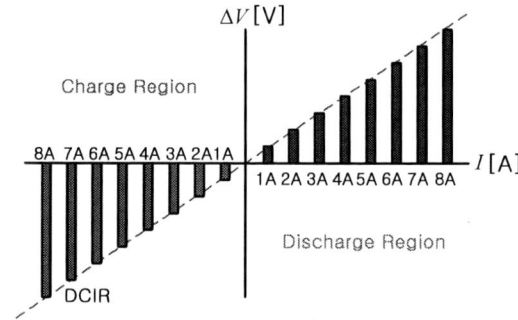

Fig. 4: DCIR test profile at a certain SOC

Under these conditions, the pulse power is obtained from the equation (2). The cutoff voltages in the discharge and charge regions are set as 4.2V and 2.8V, respectively.

$$Po = Io \times cutoff = \frac{OCV - cutoff}{DCIR} \times cutoff \quad (2)$$

III.   EXPERIMENTAL RESULT

In this paper, a fresh Li-ion fresh battery is used the DCIR experimental result. The capacity of the experimental battery is 1.28Ah. This experiment is carried out at room temperature, 25℃. A constant current of 4A is used to fully discharge and charge the battery and is used to discharge at each 10% SOC

978-1-4244-1871-8/07 $25.00 © 2007 IEEE

intervals. Besides, the pulse currents from 1A through 8A, are applied to the battery for obtaining the DCIR at a certain SOC. The current profile and each voltage drop/rise at SOC 30% are shown in Fig 5 and Fig 6. As previously mentioned, the lumped resistance is varied a duration time and region condition. Thus, four test cases (10sec-both, 10sec-dis, 5sec-both, 5sec-dis) are used to identify.

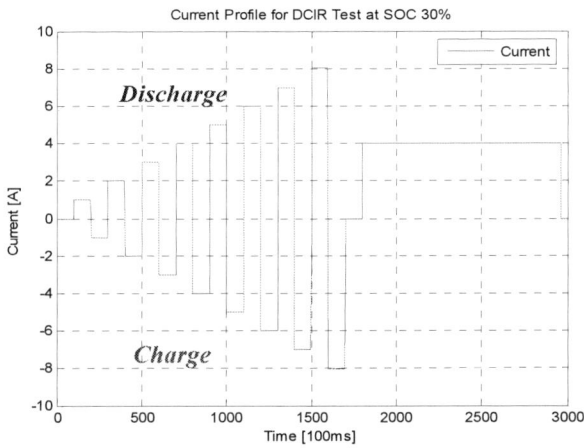

Fig. 5: Current profile for DCIR test at SOC 30% (10sec, both regions)

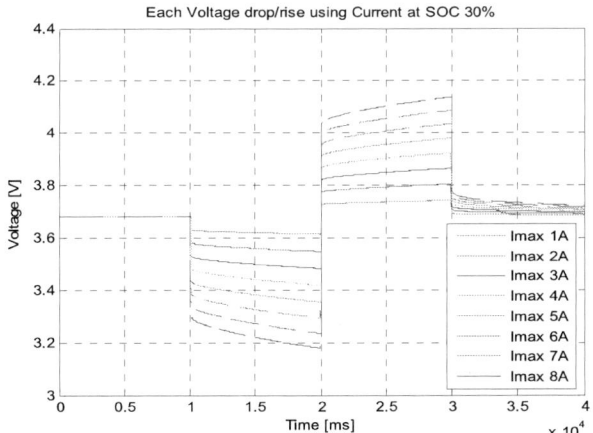

Fig. 6: Each voltage drop/rise at SOC 30% (10sec, both regions)

TABLE 1
Ri+Rct, Rd, average values of DCIR and Power

SOC	Ri+Rct[Ω]	Rd[Ω]	DCIR[Ω]	Power[W]
90%	0.03501	0.02969	0.0647	55.684
80%	0.03581	0.02968	0.0655	53.166
70%	0.03616	0.03073	0.0669	49.352
60%	0.03652	0.03028	0.0668	45.694
50%	0.03671	0.02488	0.0616	47.125
40%	0.03678	0.02632	0.0631	43.131
30%	0.03723	0.02727	0.0645	38.052
20%	0.03701	0.02878	0.0658	33.137
10%	0.03684	0.03285	0.0697	26.833

Table 1 shows the experimental results of the lumped parameters (Ri+Rct, Rd, DCIR) and power. The DCIR is obtained in Fig 4 and the power is obtained from equation (2), respectively. Although Ri+Rct is consistent despite the

differences of the SOC, Rd, DCIR and pulse power are not. Thus, Rd is a major factor in determining the DCIR and the pulse power. As mentioned above, DCIR and power are not consistent due to the differences of the SOC, but they are nearly consistent with respect to pulse current variation. Thus, it can be used for the average DCIR at a certain SOC.

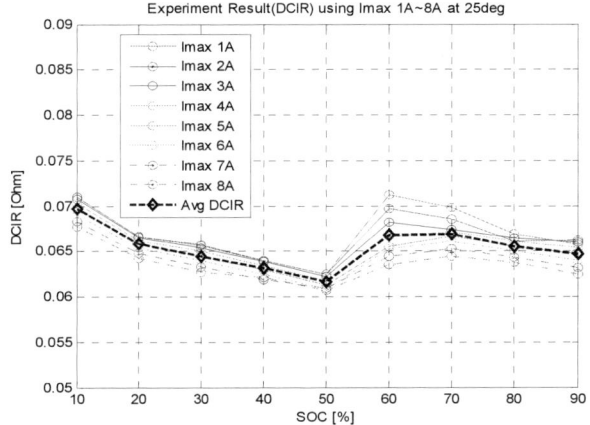

Fig. 8: DCIR experimental result using profile.

TABLE 2
DCIR using the pulse currents 1A~8A [Ω]
(SOC 20, 40, 60, 80%)

Current	SOC 20%	SOC 40%	SOC 60%	SOC 80%
1A	0.06665	0.06400	0.07120	0.06695
2A	0.06669	0.06394	0.06972	0.06607
3A	0.06658	0.06382	0.06813	0.06642
4A	0.06624	0.06337	0.06670	0.06609
5A	0.06584	0.06293	0.06549	0.06555
6A	0.06537	0.06244	0.06532	0.06498
7A	0.06479	0.06188	0.06443	0.06434
8A	0.06418	0.06214	0.06350	0.06369

Fig 8 shows the DCIR experimental result using the pulse currents from 1A through 8A for the entire SOC ranges except for SOC 0% and 100%. Table 2 shows the DCIR experimental results at SOC 20, 40, 60 and 80%. To obtain the pulse power, the OCV can be measured at a certain SOC as shown in Fig 9. Using DCIR and OCV, the experimental result of the pulse power is shown in Fig 10. In addition, like the DCIR experimental result, is the results are shown at SOC 20, 40, 60, 80% in Table 3. The average values of the DCIR and pulse power are shown by the dotted black line in Fig 8 and 10, respectively.

TABLE 3
Pulse power using the pulse currents 1A~8A [W]
(SOC 20, 40, 60, 80%)

Current	SOC 20%	SOC 40%	SOC 60%	SOC 80%
1A	32.7052	42.4944	42.8180	52.0102
2A	32.6856	42.5343	43.7269	52.7029
3A	32.7396	42.6142	44.7474	52.4252
4A	32.9076	42.9168	45.7067	52.6869
5A	33.1075	43.2169	46.5512	53.1210
6A	33.3456	43.5561	46.6724	53.5869
7A	33.6441	43.9502	47.3171	54.1200
8A	33.9639	43.7663	48.0101	54.6723

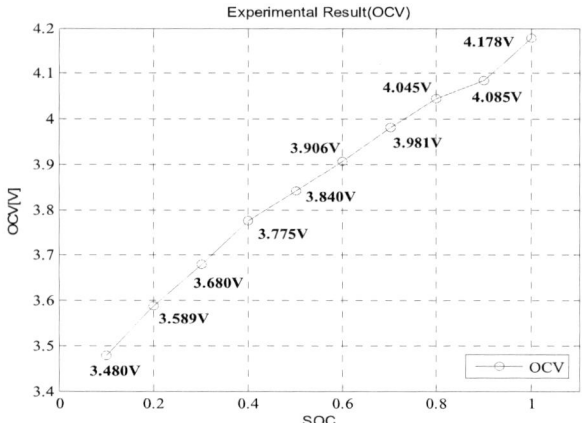

Fig. 9: OCV experimental result

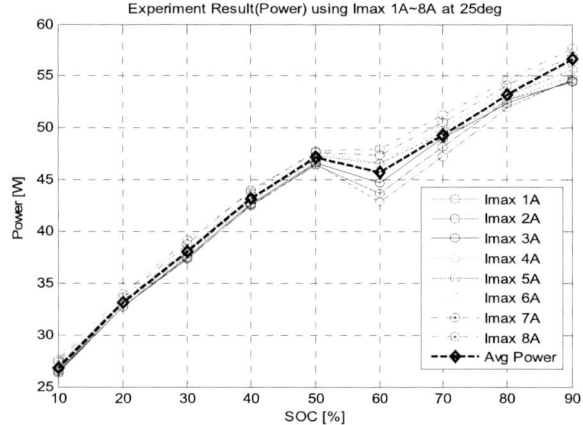

Fig. 10: Power experimental result using DCIR, OCV

The patterns of the DCIR and pulse power are almost the same with respect to SOC variation. Using these results, the DCIR-SOC relationship can be derived. The rule of the DCIR-SOC relationship is the same as the relation of OCV-SOC. Due to small differences among current variation, the average value of the DCIR is used to determining the DCIR-SOC relationship.

As previously mentioned, the lumped resistance is varied with duration time and region condition. Moreover, it is important to determine a precise DCIR-SOC relationship. Therefore, four test cases (10sec-both, 10sec-dis, 5sec-both,

5sec-dis) were used.

Table 4 shows the average values of the DCIR and pulse power using four test cases and the HPPC result in the entire SOC range except for SOC 0% and SOC 100%. The DCIR and pulse power experimental results using a duration time of 10 seconds are more similar with the HPPC experimental results in comparison with the DCIR results from a duration time of 5 seconds.

In addition, the standard deviations of the DCIR and pulse power in both regions are smaller than that in the discharge region. The results are shown in Fig 11, 12. These results are important because the more the standard deviation increases, the less precise the DCIR-SOC relationship.

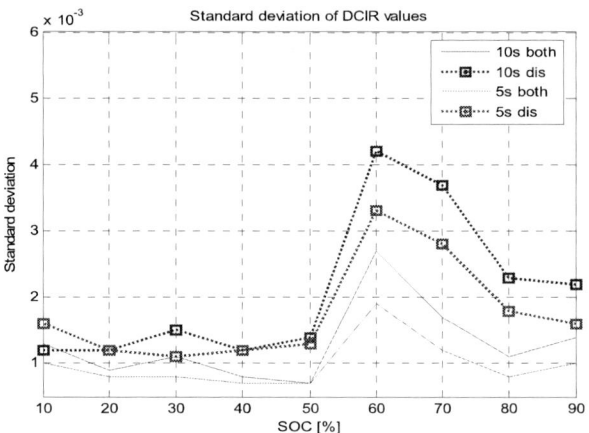

Fig. 11: Standard deviation of four test cases (DCIR)

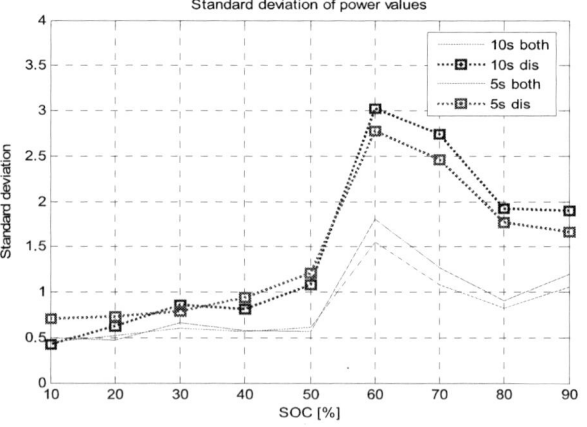

Fig. 12: Standard deviation of four test cases (Pulse Power)

TABLE 4

Average values of DCIR, Power using four test cases and HPPC result at fully SOC range

Lumped resistance[Ω]		SOC90%	SOC80%	SOC70%	SOC60%	SOC50%	SOC40%	SOC30%	SOC20%	SOC10%
DCIR Test	**10s, both**	**0.06470**	**0.06550**	**0.06690**	**0.06680**	**0.06160**	**0.06310**	**0.06450**	**0.06580**	**0.06970**
	10s, dis	0.06420	0.06520	0.06650	0.06380	0.06170	0.06370	0.06510	0.06640	0.07100
	5s, both	0.05970	0.06030	0.06130	0.06150	0.05700	0.05790	0.05920	0.06040	0.06460
	5s, dis	0.05930	0.06010	0.06110	0.05960	0.05680	0.05820	0.05960	0.06080	0.06530
HPPC Test		0.06230	0.06320	0.06370	0.06050	0.06070	0.06300	0.06340	0.06510	0.06720
Power[w]		SOC90%	SOC80%	SOC70%	SOC60%	SOC50%	SOC40%	SOC30%	SOC20%	SOC10%
Power Test	**10s, both**	**55.6838**	**53.1657**	**49.3524**	**45.6937**	**47.1253**	**43.1311**	**38.0522**	**33.1374**	**26.8330**
	10s, dis	56.1866	53.4532	49.7633	47.9312	47.0625	42.6929	37.7494	32.8304	26.3486
	5s, both	60.3520	57.7324	53.8168	49.6310	50.9272	47.0097	41.4596	36.0795	28.9714
	5s, dis	60.7508	57.9551	54.0767	51.3006	51.0610	46.7204	41.1999	35.8516	28.6779
HPPC Test		57.7791	55.0744	51.4339	50.3213	47.4511	42.3155	37.7359	32.5414	26.8590

978-1-4244-1871-8/07 $25.00 © 2007 IEEE

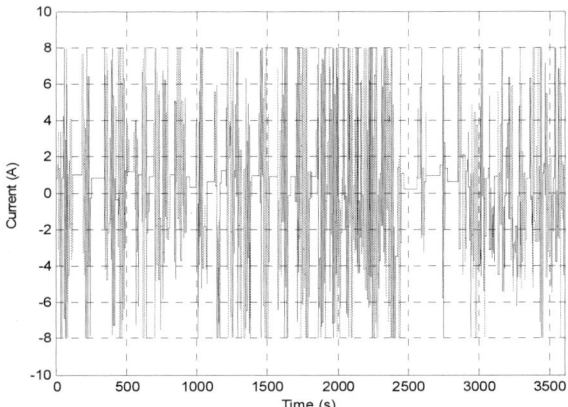

Fig. 13: Current profile

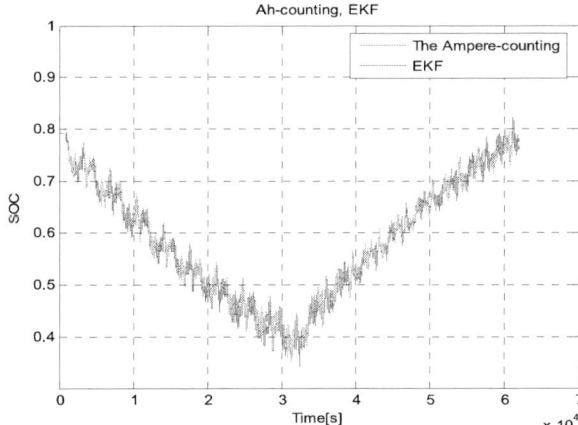

Fig. 14: Ampere-Counting, EKF results

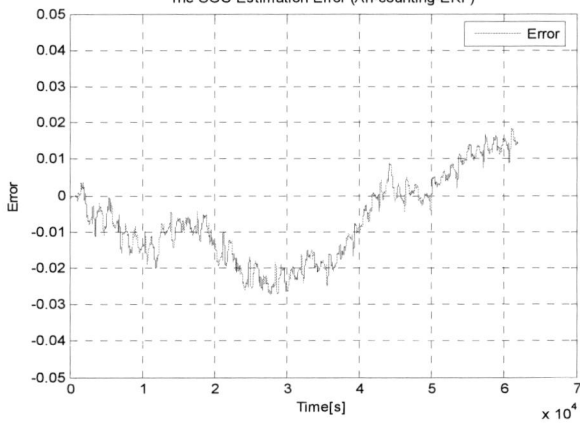

Fig. 15: SOC estimation error (Ampere-Counting – EKF)

The EKF is a good method to estimate the SOC. It has been known to be the optimal adaptive method based on recursive estimation using probability. The EKF [6] for the SOC algorithm is used to estimate pulse power including the DCIR-SOC relationship.

To verify the estimation of the SOC using an experimental battery, the EKF result is compared with the Ah-counting result in Fig 14. Furthermore, the SOC estimation error is shown to be under 2% in Fig 15.

Using the SOC algorithm including the DCIR-SOC relationship, the DCIR and pulse power results are shown in Fig 15 and 16. The results are shown at SOC 40~80%. The pulse power is obtained from equation (2).

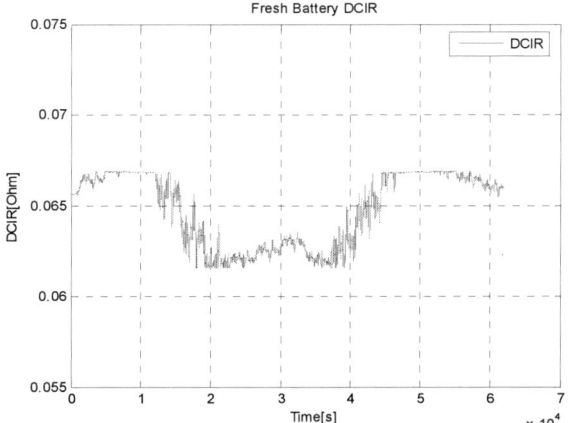

Fig. 16: DCIR result using EKF

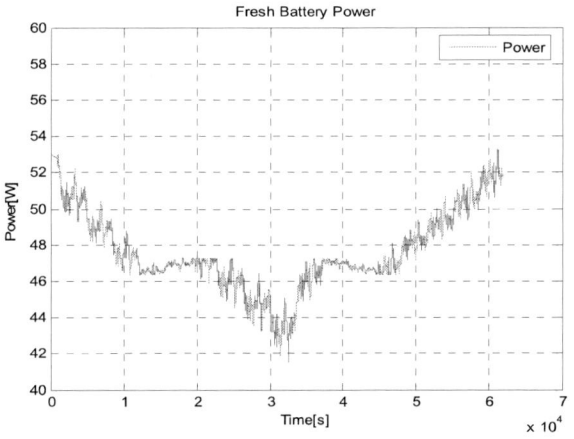

Fig. 17: Pulse Power result using EKF

IV. CONCLUSION

To obtain the DCIR and pulse power, the duration time $\triangle t$ and region condition are important for HEV applications. In this paper, a duration time of 10 seconds and discharge and charge regions are applied to the battery to obtain the DCIR-SOC relationship. For the verification of the proposed method, four test cases and HPPC test results are shown in this paper.

The pulse power is estimated using the SOC algorithm that includes the DCIR-SOC relationship. The pulse power is shown at SOC 40~80%.

978-1-4244-1871-8/07 $25.00 © 2007 IEEE

REFERENCES

[1] FreedomCAR 42V Battery Test Manual, DOE/ID-11070, Feb, 2003

[2] PNGV Battery Test Manual, Rev. 3, DOE/ID-10597, Feb, 2001

[3] Tetsuro Okoshi, Keizo Yamada, Tokiyoshi Hirasawa, Akihiko Emori, "Battery condition monitoring (BCM) technology about lead-acid batteries", J. Power Source 158 (2006) 874-878

[4] T. Horiba, T. Maeshima, T. Matsumura, M. Koseki, J. Arai, Y. Muranaka, "Applications of high power density lithium ion batteries", J. Power Source 146 (2005) 107-110

[5] J. Arai, T. Yamaki, S. Yamauchi, T. Yuasa, T. Maeshima, T. Sakai, M. Koseki, T. Horiba, "Development of a high power lithium secondary battery for hybrid electric vehicles", J. Power Source 146 (2005) 788-792

[6] Oanyong Nam, Jaemoon Lee, Jaeho Lee, Jonghun, Bo H. Cho, "Li-Ion Battery SOC Estimation Method based on the Reduced Order Extended Kalman Filtering", IECEC, 2006.

[7] Jae-Seung Koo, Sun-Soon Park, Kil-Young Youn, Chul-Soo Kim, "Development of SOC Estimation Logic using the Steady State DC-IR for SHEV", 18th Electric Vehicle Symposium, Berlin, 2001

A New Single-Stage PFC AC/DC Converter with Voltage-Doubler Rectified Asymmetric Half-Bridge Converter

Byoung-Hee Lee, Chong-Eun Kim, Ki-Bum Park and Gun-Woo Moon
Korea Advanced Institutes of Science and Technology (KAIST)
Department of Electrical Engineering
373-1 Guseong-Dong, Yuseong-Gu, Daejeon, 305-701, Republic of Korea
Phone:+82-42-869-8075, Fax:+82-42-869-8520
Email: gwmoon@ee.kaist.ac.kr

Abstract— Conventional Single-Stage Power-Factor-Correction (PFC) AC/DC converter has a link capacitor voltage problem at high line input and low load condition. In this paper, this problem is analyzed by using voltage conversion ratio of the DC/DC conversion cell. By applying this analysis, A new Single-Stage PFC AC/DC converter with boost PFC cell is integrated with Voltage-Doubler Rectified Asymmetrical Half-Bridge (VDRAHB) is proposed. The proposed converter features good power factor correction, low current harmonic distortions, tight output regulations and low voltage of link capacitor. An 85W prototype was implemented to show that it meets the harmonic requirements and standards satisfactorily with nearly unity power factor and high efficiency over universal input.

I. INTRODUCTION

Conventional off-line power converters with diode-capacitor rectifier front-end have distorted input current waveform with high harmonic contents. Since these converters have a low power factor, they cannot meet harmonic regulations and standards such as European line-current harmonic regulations defined in the IEC 61000-3-2 document and the Japanese input-harmonic current specifications. To comply with these standards, a number of single-stage and two-stage PFC circuits have been developed and reported in the literature. In the two-stage approach, it is customary to add a power-factor corrector ahead of the dc/dc converter to provide a regulated and isolated dc output. This approach has good characteristics of power-factor correction and fast output regulations, but the power-factor corrector increases the size and cost of overall system. Therefore, the two-stage approach is not desirable in low power applications [1-2].

Recently, many single-stage approaches have been suggested to achieve both power-factor correction and power conversion from the ac line to a desired dc output [3]. Since these topologies have a PFC cell is integrated with a DC/DC conversion cell and both cells share active switch and controller, single-stage approach is a better choice in cost point of view. Unfortunately, most of the proposed converters have one or more of the following disadvantages: large low-frequency output voltage ripple, low efficiency due to the switching loss and the rectification loss, variable switching frequency and high-voltage stress of switch.

In this paper, specifically, link capacitor voltage is

analyzed by using the voltage conversion ratio of the DC/DC conversion cell of single-stage PFC AC/DC converter. Moreover, by adapting this analysis, the new single-stage PFC AC/DC converter with boost for PFC cell is integrated with VDRABH for DC/DC conversion cell is proposed. To confirm the validity of the proposed converter, the operation principle and the experimental results of an 85W prototype will be presented.

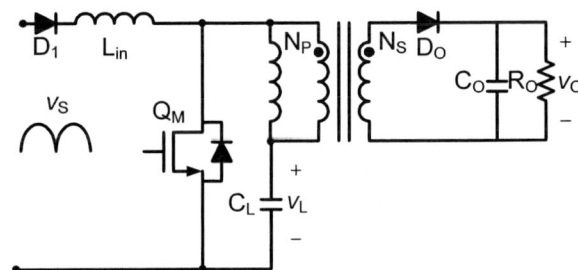

Fig. 1 : Conventional Single-Stage PFC AC/DC Converter

II. ANALYSIS OF LINK CAPACITOR VOLTAGE

Conventional Single-Stage PFC AC/DC converter with PFC cell is operated in Discontinuous Conduction Mode (DCM) and DC/DC cell is operated in Continuous Conduction Mode (CCM) is shown in Fig. 1 [5]. Since DC/DC cell is operated in CCM, duty ratio is not changed with load variation as shown in Fig. 2. Moreover, DC conversion ratio of the PFC cell increases according to decrease of load as shown in Fig. 3. Therefore, link capacitor voltage increases as load decreases. Since the high voltage stress and the impossibility of using commercial capacitor, increase of link capacitor voltage is one major problem of Single-Stage PFC AC/DC converter. To overcome this problem, there were many approaches, for example, Bi-flyback PFC converter [1], DC bus-voltage feedback [3], Charge-pump PFC converter [4] and Variable Frequency Modulation.

The characteristic of this link capacitor voltage can be explained by using voltage conversion ratio of the DC/DC conversion cell of the single-stage PFC AC/DC converter. The relation between link capacitor voltage, v_L, and output voltage, v_O can be expressed as a function of duty ratio, D.

$$\frac{V_O}{V_L} = f(D) \tag{1}$$

978-1-4244-1871-8/07 $25.00 © 2007 IEEE

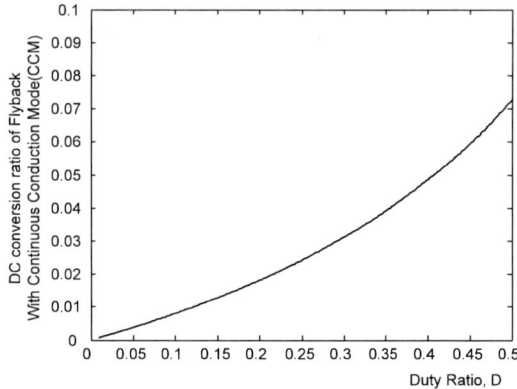

Fig. 2 : DC conversion ratio of Flyback with CCM

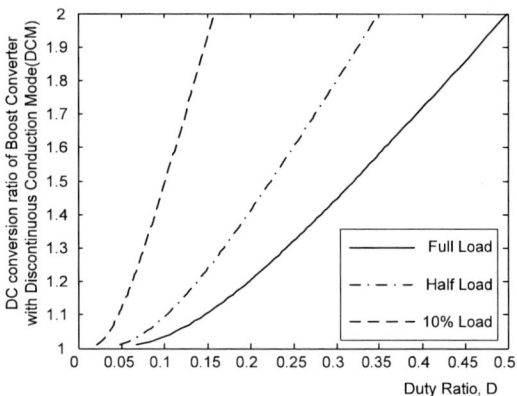

Fig. 3 : DC conversion ratio of Boost with DCM

In general single-stage PFC AC/DC converter, f(D) is highly dependent on duty ratio, D, as shown in Fig. 4(a). Since the voltage conversion ratio, f(D), varies with load changes, the link capacitor voltage, v_L have to vary as load becomes different to regulate the output voltage, v_O. On this account, the link capacitor voltage increases as load decreases.

In case of f(D) is hardly dependent on duty ratio, D, as shown in Fig. 4(b), the link capacitor voltage increase problem can be resolved. Since the voltage conversion ratio, f(D), hardly ever varies with load variations, the link capacitor voltage, v_L only slightly vary to regulate the output voltage, v_O. For this reason, in the event that one converter has the voltage conversion ratio of DC/DC conversion cell as shown in Fig. 4(b) can be applied to single-stage PFC AC/DC Converter without link capacitor voltage problem.

The voltage conversion ratio of the Voltage Doubler Rectified Asymmetric Half-Bridge(VDRAHB) converter is plotted as a function of duty ration, D, as shown in Fig. 5. Since f(D) of the VDRAHB is hardly dependent on duty ratio, D, the new single-stage PFC AC/DC converter that VDRAHB converter is applied to DC/DC conversion cell with boost PFC cell is proposed as shown in Fig. 6.

III.   OPERATIONAL PRINCIPLES

The circuit diagram of the proposed single-stage PFC

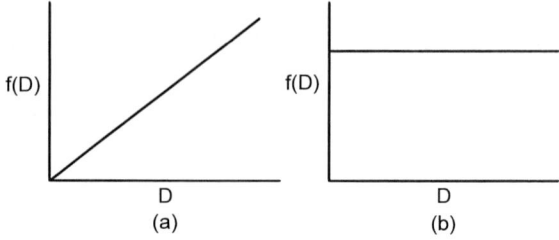

Fig. 4 : (a) Rough voltage conversion ratio of DC/DC conversion cell in general single-stage PFC converter
(b) Roughly desired voltage conversion ratio of DC/DC conversion cell in proposed single-stage PFC converter

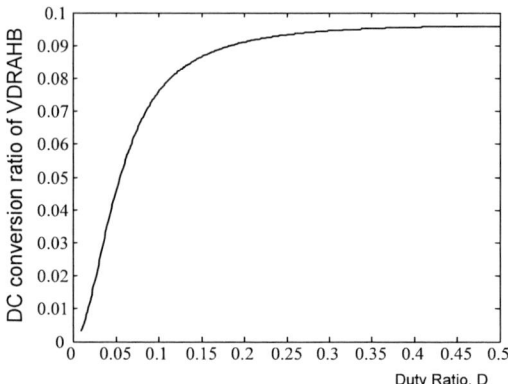

Fig. 5 : DC conversion ratio of VDRAHB

AC/DC converter is shown in Fig. 6. The proposed circuit is composed of a boost for PFC cell is integrated with a VDRABH for DC/DC conversion cell.

A.   Mode Analysis

The operation of the proposed converter can be classified into the three modes ($t_0 \sim t_{0'}$) and the corresponding waveforms are presented in Fig. 7.

Mode 1 (M_1, $t_0 \sim t_1$) : Mode 1 begins at t_0 when Q_M is turned-on. The input voltage, v_S, is applied to the inductor L_{in}, i_{Lin} starts to build up from zero. At the same time, (v_L-v_H) is applied to the primary winding of transformer T_1, i_{Lk} increases. The slopes of these currents, i_{Lin}, i_{Lk} are given by

$$\frac{d}{dt} i_{Lin}(t) = \frac{V_S}{L_{in}} \qquad (2)$$

$$\frac{d}{dt} i_{Lk}(t) = \frac{1}{L_K}\left[V_L - V_H - \frac{N_P}{N_S}(V_o - V_{SB})\right] \qquad (3)$$

Mode 2 (M_2, $t_1 \sim t_2$) : Mode 2 begins at t_1 when Q_M is turned-off and Q_A is turned-on, respectively. (v_S-v_L) is applied to the inductor L_{in}, i_{Lin} starts to flow through Q_A, and decreases to zero. -v_H is applied to the primary winding of T_1, i_{Lk} decreases. The slopes of these currents, i_{Lin}, i_{Lk} are determined by

978-1-4244-1871-8/07 $25.00 © 2007 IEEE   1178

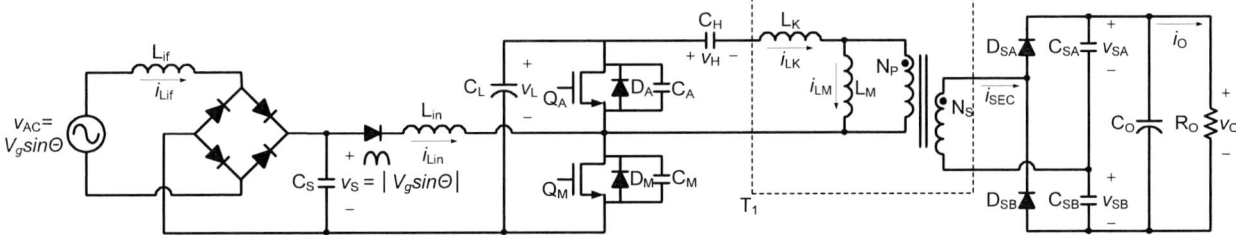

Fig. 6 : Circuit Diagram of the proposed converter

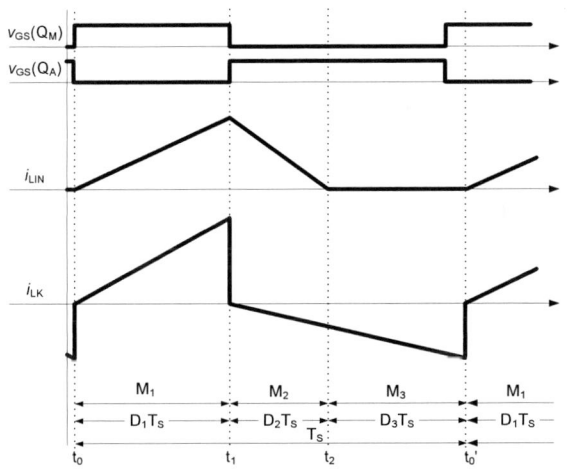

Fig. 7 : Key Waveforms of the proposed converter

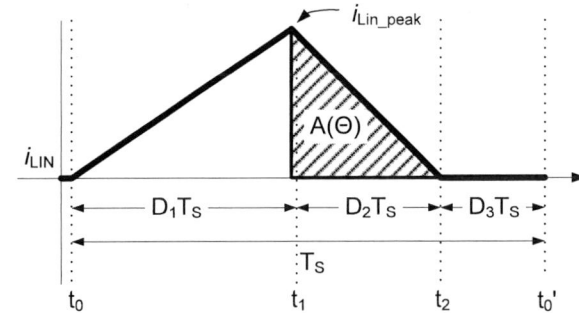

Fig. 8 : Waveform of the boost inductor current

$$M_{boost} = \frac{V_L}{V_g} \tag{7}$$

$$M_{VDRAHB} = \frac{V_O}{V_L} \tag{8}$$

The waveform of the boost inductor current is shown in Fig. 8. By applying the voltage-second balance of L_{in} during one switching cycle, Ts, the following equations are derived.

$$\frac{V_s}{L_{in}} D_1 T_s = \frac{V_L - V_s}{L_{in}} D_2 T_s \tag{9}$$

$$D_2 = \frac{\sin\theta}{M_{boost} - \sin\theta} \times D_1 \tag{10}$$

The peak current of L_{in} is determined by following equation.

$$i_{Lin\_peak} = \frac{V_s}{L_{in}} \times D_1 T_s \tag{11}$$

A(Θ), is representing the charging energy into C_L during the one switching cycle, T_S, is derived as follows by using (7) and (8).

$$A(\theta) = \frac{1}{2} \times i_{Lin\_peak} \times D_2 T_s$$

$$= \frac{D_1^2 \times V_g \sin^2\theta}{2 \times L_{in} \times (M_{boost} - \sin\theta) \times F_s^2} \tag{12}$$

$$\frac{d}{dt} i_{Lin}(t) = \frac{1}{L_{in}}\left(V_s - V_L\right) \tag{4}$$

$$\frac{d}{dt} i_{Lk}(t) = \frac{1}{L_K}\left[-V_H + \frac{N_P}{N_S}(V_o - V_{SA})\right] \tag{5}$$

Mode 3 (M₂, t₂~ t₀·) : Mode 3 begins at t₂ when i_{Lin} decreased to zero and i_{Lk} decreases with same slope during mode 2. Mode 3 ends when Q_M is turned-on and starts another switching cycle at t₀·.
The slope of current, i_{Lk} is determined by following equation.

$$\frac{d}{dt} i_{Lk}(t) = \frac{1}{L_K}\left[-V_H + \frac{N_P}{N_S}(V_o - V_{SA})\right] \tag{6}$$

B. Input-Output Voltage Conversion Ratio

The voltage conversion ratio of the proposed converter can be divided into two conversion ratio M_{boost}, is the voltage conversion ratio of the boost converter between v_g and v_L, and M_{VDRAHB}, is the voltage conversion ratio of the VDRAHB converter between v_L and v_O. Each conversion ratio can be expressed are :

978-1-4244-1871-8/07 $25.00 © 2007 IEEE   1179

,where F_S is the switching frequency of the converter.

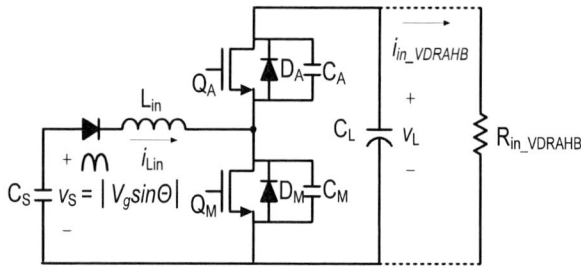

Fig. 9 : Circuit Diagram with equivalent resistance

Since the average of the B(Θ), which is average of the A(Θ) during T_S, during the 120Hz is equal to the value of input current of VDRAHB as shown in Fig. 9, the voltage conversion ratio of the boost converter, M_{boost} can be obtained as follows.

$$\frac{V_L}{R_{in\_VDRAHB}} = \frac{1}{\pi}\int_0^\pi B(\theta)d\theta = \frac{1}{\pi}\int_0^\pi \frac{A(\theta)}{T_s}d\theta$$
$$= \frac{1}{\pi}\int_0^\pi \frac{D_1^2 \times V_g \sin^2\theta}{2\times L_{in}\times F_S\times \pi\times(M_{boost}-\sin\theta)}d\theta \qquad (13)$$

,where R_{in_VDRAHB} is the equivalent input resistance of Voltage-Doubler Rectified Asymmetric Half-Bridge Converter. R_{in_VDRAHB} can be expressed as follows.

$$R_{in\_VDRAHB} = (\frac{V_L}{V_O})^2 \times R_o = \frac{R_o}{M_{VDRAHB}^2} \qquad (14)$$

By arranging the (13), M_{boost}, can be expressed as :

$$M_{boost} = \frac{V_L}{V_g} = \frac{D_1^2 R_{in\_VDRAHB}}{2\times L_{in}\times F_S\times \pi}\int_0^\pi \frac{\sin^2\theta}{M_{boost}-\sin\theta}d\theta \qquad (15)$$

The voltage conversion ratio of the VDRAHB has been described in reference [6] as follows.

Table1. Design Specification

Specification	Value
Input voltage, V_S	90V ~ 270V_AC RMS
Output voltage, V_O	18.5V
Max. output power, $P_{O,max}$	85W
Switching frequency, f_S	100kHz

Table2. Circuit Parameters

Parameter		Value
Transformer	$N_P: N_S$	51 : 5
	L_M	1.37mH
	L_K	12uH
C_H		2.2uF / 630V
C_{SA} & C_{SB}		22μF / 25V, 2EA

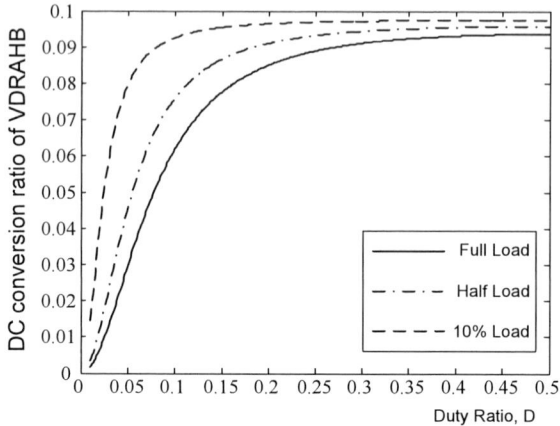

Fig. 10 : DC conversion ratio of VDRAHB

$$M_{VDRAHB} = \frac{V_O}{V_L} = \frac{\frac{N_S}{N_P}D_1^2(1-D_1)^2}{D_1^2(1-D_1)^2 + \frac{2L_kF_S}{R_O}\left(\frac{N_S}{N_P}\right)^2\left[D_1^2+(1-D_1)^2\right]} \qquad (16)$$

,where D_1 is duty ratio of Q_M and F_S is switching frequency.

DC conversion ratio of the M_{boost} and M_{VDRAHB} are plotted as a function of duty ratio in Fig. 3 and Fig. 10 with load variation, respectively, based on the design specifications and the circuit parameters for the 85W prototype in Table 1 and Table 2, respectively.

IV. EXPERIMENTAL RESULTS

To verify the operational principles and the feasibility of the proposed single-stage PFC AC/DC converter, the 85W prototype of the proposed converter has been implemented.

The design specifications and circuit parameters of this prototype are same as presented in Table 1 and Table 2, respectively. Fig. 11(a) and Fig. 11(b) show the Boost PFC

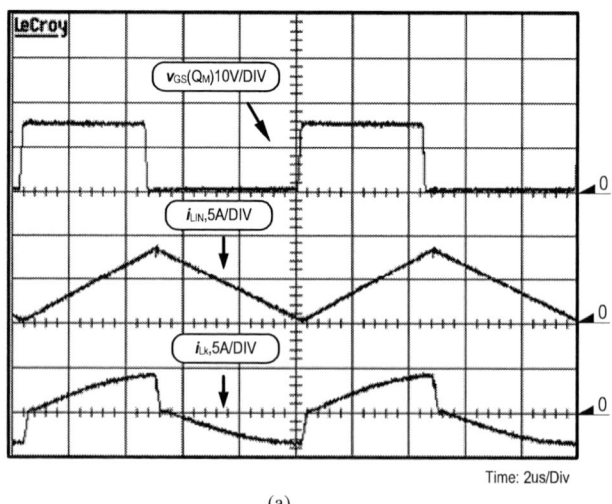

(a)

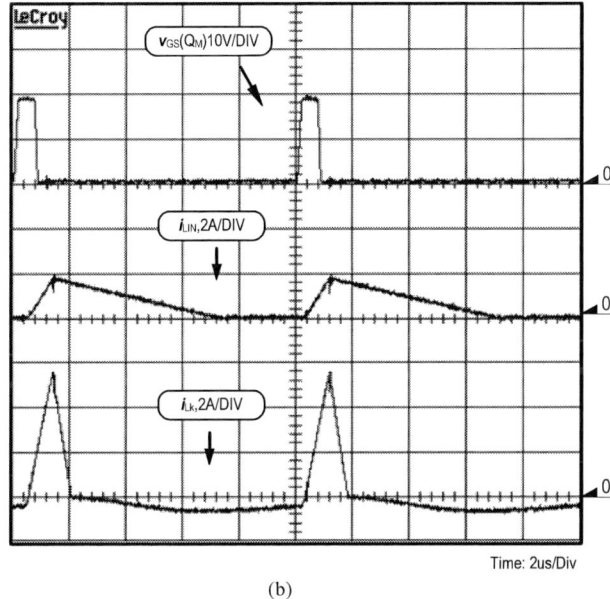

(b)

Fig. 11 : Experimental Key waveforms
(a) V_{AC_RMS} = 90V (b) V_{AC_RMS} = 270V

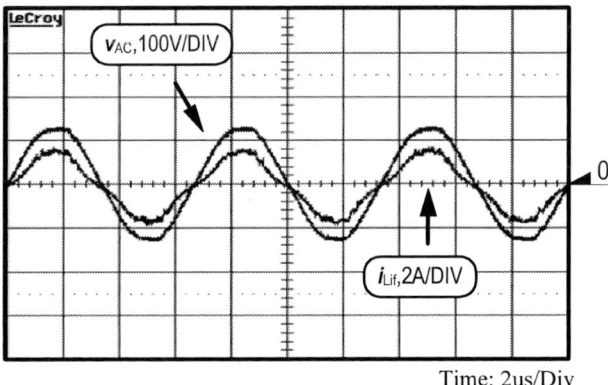

(a)

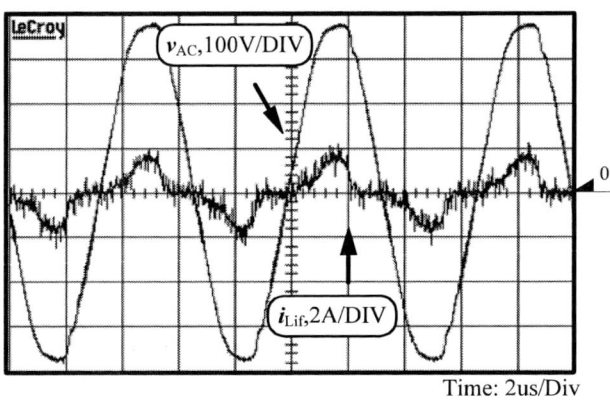

(b)

Fig. 12 : Line Input Voltage and Current Waveforms
(a) V_{AC_RMS} = 90V (b) V_{AC_RMS} = 270V

cell is operated in Boundary Conduction Mode(BCM) at 90 $V_{AC,RMS}$ and 270$V_{AC,RMS}$, respectively. The line input current follow the shape of the line input voltage as shown in Fig. 12(a) and Fig. 12(b) at $90V_{AC,RMS}$, and $270V_{AC,RMS}$ respectively. The power factors and the efficiencies of the proposed converter are measured as presented in Fig. 13(a) and Fig. 13(b), respectively. The harmonic line currents of the proposed converter are analyzed as shown in Fig. 13(c). The proposed converter meets the harmonic requirements of IEC 61000-3-2 Standards. The voltage of link capacitor is sustained below 405V at line and load variations as shown in Fig. 13(d). Therefore, the commercial capacitor can be used in the proposed converter.

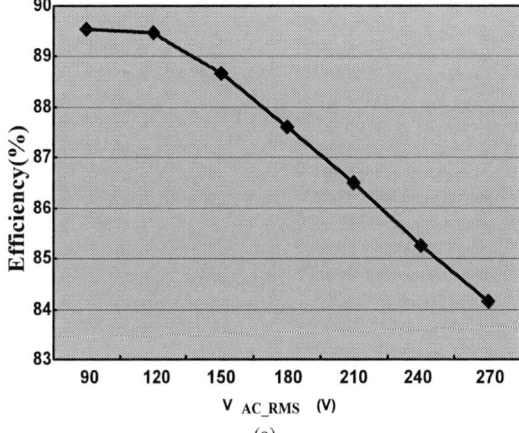

(a)

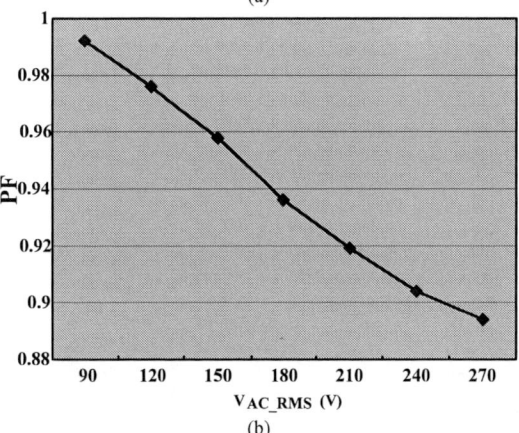

(b)

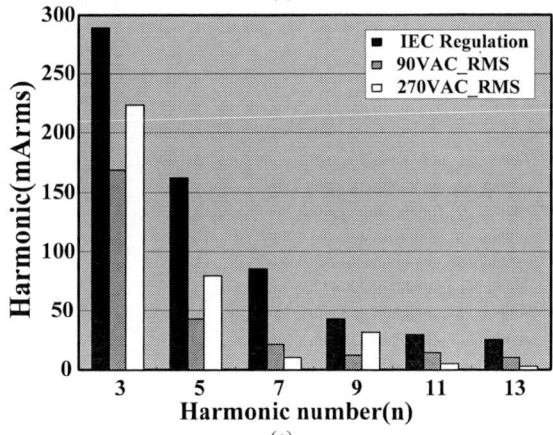

(c)

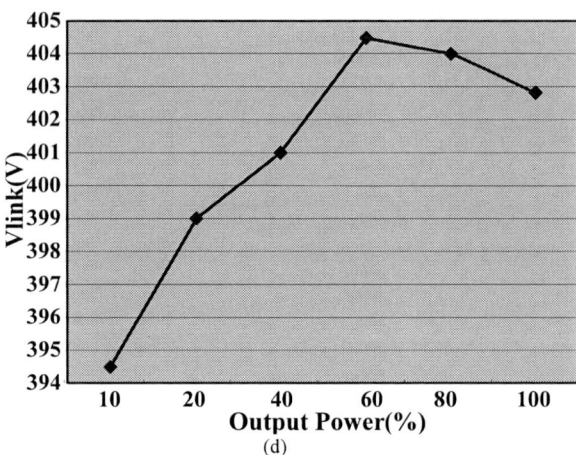

(d)

Fig. 13 : (a)Measured Efficiency, (b)Measured Power Factor,
(c) Measured Harmonic Components, (d)Measured Link Capacitor Voltage

V.   CONCLUSION

In this paper, the link capacitor voltage problem in single-stage PFC AC/DC converter is analyzed in voltage conversion ratio of DC/DC conversion cell point of view and the new single-stage PFC AC/DC converter is proposed. By experimental results of an 85W converter, the operation principle and main features of the proposed converter are confirmed. Since link capacitor voltage of the proposed converter could be sustained below 405V in all load and line input conditions, the commercial capacitor can be used. Moreover, the proposed converter met the harmonic regulations and had high efficiency. Therefore, the proposed converter can be expected to be widely used for low power applications such as laptop adapter.

REFERENCES

[1]   M. H. Kheruluwala, R. L. Steingerwald, and R. Gurumoorthy, "Fast response high power factor converter with a single power stage," in *IEEE Proc. PESC*, 1991, pp. 769-779.

[2]   Jindong Zhang, Milan M. Jovanovic, and Fred C. Lee, " Comparison Between CCM Single-Stage And Two-Stage Boost PFC Converters," in *IEEE Proc. APEC*, 1999, pp. 335-341.

[3]   Redl, R.; Balogh, L.; Sokal, N.O., "A new family of single-stage isolated power-factor correctors with fast regulation of the output voltage", in *IEEE Proc. PESC*, 1994, pp. 1137-1144.

[4]   Qiao, Chongming; Smedley, K.M., "A topology survey of single-stage power factor corrector with a boost type input-current-shaper", in *IEEE Proc. APEC*, 2000, pp. 460-467.

[5]   M. Madigan, R. Efickson, E. Ismail, "Integrated high-quality rectifier-regulators," in *IEEE Proc. PESC*, 1992, pp. 1043-1051.

[6]   C.E. Kim, S.K. Han and G.W. Moon, "A New High Efficiency and Low Profile On-Board DC/DC Converter for Digital Car Audio Amplifier," in *Proc. Journal of Power Electronics*, Vol. 6, No. 1, pp. 83-93 Oct. 2005.

Direct High Frequency Soft Switching PWM Cyclo-Converter-Fed AC-DC Converter without DC Link for Consumer Magnetron Drive

Hisayuki Sugimura[1], Bishwajit Saha[1], Sang Pil Mun [1], Eiji Hiraki[2], Hideki Omori[3], Mutsuo Nakaoka[1, 4]

[1] Kyungnam University, Masan, KOREA
[2] Yamaguchi University, Yamaguchi, JAPAN
[3] Matsushita Electric Industrial Co. Ltd., Osaka, JAPAN
[4] Industrial College of Technology-University, Hyogo, JAPAN
Email: bsaha@ieee.org

Abstract – The oscillation voltage (E_{bm}) is an important parameter of the characteristic constant of a high power microwave generator. When E_{bm} could be made high, it serves so as to improve the power conversion efficiency of a magnetron. Thus, the newly developed magnetron with 7% higher oscillation cut-off voltage than the present conventional one is manufactured by Matsushita Electric Industrial Co. Ltd, In addition to this, the utility AC power line current reduced harmonic type direct inverter AC-DC power converter suitable and acceptable for driving this magnetron is proposed in this paper. This direct high frequency inverter type soft switching PWM AC-DC converter has the voltage-boost function based on active clamp scheme with soft switching pulse modulation strategy topology. This paper describes a novel type active clamp AC-DC power converter circuit defined as direct high frequency inverter and the switching operation of this power converter is described and discussed on the basis of simulation and experimental results. The operating characteristics of this soft switching direct high frequency inverter type AC-DC power converter using IGBTs that incorporates the harmonic current improvement as well as power factor correction are evaluated and discussed from a practical point of view as compared with the conventional AC-DC power converter.

Keywords—AC-DC power conversion, Consumer power electronics, Magnetrons, Direct high frequency inverters.

I. INTRODUCTION

The high frequency inverter type high voltage DC power supply to drive consumer magnetron is based upon the power electronics circuits technology for the microwave power generator, which handles relatively high power conversion process ranging from 1.0[kw] to 1.4[kw]. The high frequency inverter type AC-DC power converter for the microwave oven as well as high frequency inverter type AC-AC converter for induction heating type cooker and hot water producer has attracted special interest as one of consumer power appliances from global environmental conservation and reproduction viewpoints. In recent years, research and development have been actively carried out from a practical point of view for the purposes of realizing miniaturization, downsizing based on high efficiency, lowered electromagnetic noises, high performances related waveform quality and responsibility, lower cost due to minimum circuit components, higher utilizations, and new multi-functions. Of these, the high power microwave output can be required for rapid cooking processing as one of the basic operating performances of a consumer microwave oven. Most of consumer power electronic appliances such as microwave

oven, induction heating cooker and steamer, are connected to the circuit to 100V/200V utility AC power supply, which use the full-wave rectified DC output. As for the rectified smoothing output, making the most of non-smoothing operation can minimize the smoothing filter capacitance of a rectification smoothing circuit. In addition, because of non-smoothing filter design in this power converter, the harmonic current components in the utility AC power line grid side could be minimized so as to be sinewave shaping under a condition of unity power factor. Since the commercial power supply line current can be brought close to sinusoidal wave, the harmonics reduction can be performed by using the AC-DC converter design. But, because of non-linearity of i-v characteristic of the magnetron, high frequency inverter output voltage or the output voltage of voltage doubler could not raise up to magnetron oscillation operation voltage in the lower portion of the utility AC power supply instantaneous voltage. As a result, the magnetron device stops stable oscillation operation and the utility AC line current distortion causes because the utility AC input line current stops. Thus, in order to implement the high power microwave output, the new type magnetron with a high cut-off voltage design has just developed by Matsushita Electric Industrial Co. Ltd, Japan. We developed new high frequency switching AC-DC power conversion processing equipment that takes out the DC high voltage through a high frequency transformer and a voltage doubler rectifier in addition to the control implementation scheme for executing total power factor improvement and harmonic current reduction in the utility AC commercial frequency alternate current line side. The newly-developed direct soft switching high frequency inverter is directly changed into high frequency alternate current from a utility AC commercial alternate current power supply to drive this type of magnetron efficiently with low cost and downsizing. Finally, Proposed are hybrid modulation pattern selection method of the partially-modulated operating frequency in addition to its duty factor control strategy in order to improve harmonic current components and make total harmonic power factor unity under a practical condition of the peak limitation of the magnetron anode current.

II. LOAD CHARACTERISTICS OF THE MAGNETRON FOR MICROWAVE OVEN

A. Structure of Magnetron

978-1-4244-1871-8/07 $25.00 © 2007 IEEE

Figure 1 shows the exterior appearance (see Fig.1 (a)) of magnetron structure in addition to its internal cross section (see Fig.1(b)). The internal construction includes cylindrical cathode and an anode on the concentric circle, and the permanent magnet with ferrite core gives the magnetic field toward axial direction as depicted Fig.2. The big difference between the new model magnetron and conventional magnetron indicates that the magnetic field intensity of the permanent magnet with 2512 [G] is larger as compared with 1862 [G] of the conventional one. The behaviour of the electron in action space is determined by this strong magnetic field intensity, and a magnetron comes to start an oscillation by the high cut-off voltage, so that magnetic field intensity becomes strong. In this case, the cut-off voltage designed for about 4.6 [kV] that reaches an oscillation mode is designed for higher value than that of the conventional magnetron designed for about 3.6 [kV]. Moreover, the conversion efficiency of the new type magnetron is also 78%. As a result, it is higher than conventional one; 75%.

B. i-v Characteristics and Electrical Equivalent Circuit Model of Magnetron

Figure 3 represents the terminal voltage vs. current characteristics of the new magnetron. Concerning to the magnetron v-i characteristics, the magnetron has non-linear v-i characteristics, what is called, piecewise linear characteristics which include high resistance area in non-oscillating range for the magnetron and low resistance area in oscillating range for the magnetron. When the cut-off voltage between anode and cathode exceeds about 4.6[kV], the magnetron anode current begins to flow from anode toward cathode terminal. On the other hand, when the voltage between anode and cathode is lower than a cut-off voltage, the anode current of magnetron does not flow. The static model of the electrical equivalent circuit of the magnetron with these characteristics can simply represent by using two pure piecewise resistances, ideal diode to determine one quadrant operation and ideal battery voltage corresponding to the cut-off voltage Vz as depicted in Fig.4. As illustrated in Fig.4, the load type with two power switches; Load1 in non-oscillation mode and Load2 in oscillation mode is to be automatically selected whether the voltage between anode and cathode is higher than the cut-off voltage Vz 4.6[kV] or not.

III. HIGH FREQUENCY INVERTER WITH BOOST FUNCTION

A. System Consideration
The high efficient magnetron introduced newly has a

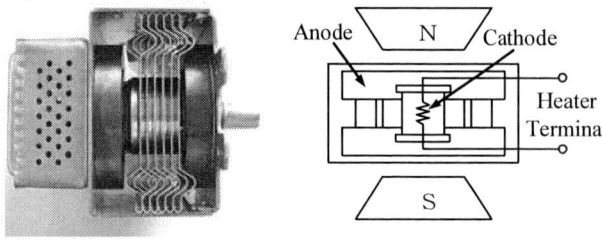

(a) Appearance (b) Internal construction
Fig.1. Construction of Magnetron.

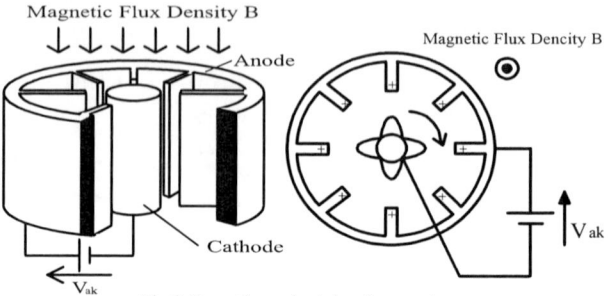

Fig.2 Operation principle of magnetron

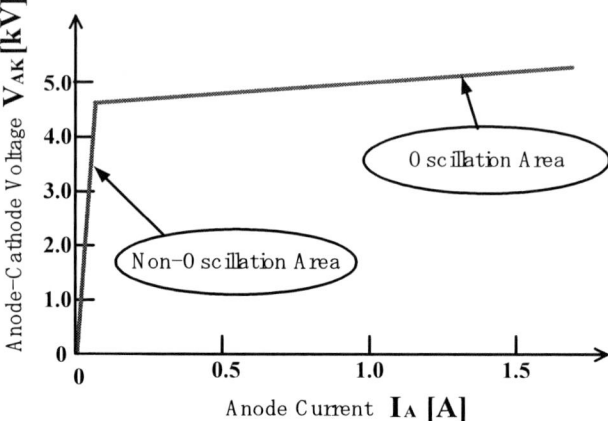

Fig.3. Current vs. voltage characteristics model of magnetron

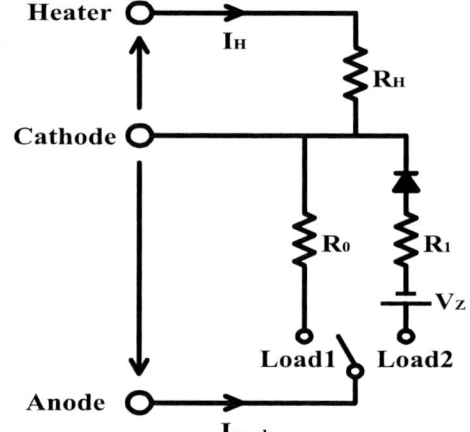

Fig.4 Electrical equivalent circuit modelling of magnetron

specialized feature that the cut-off voltage of the magnetron is higher than the conventional one. Therefore, it is considered that the non-oscillating period of the magnetron increase near the zero valley point area of the utility AC input voltage. As a result, the power factor of the utility input current will be worse than conventional one. Especially, in the latest development of household microwave oven, because the utility input current is actually limited by the indoor wiring capacity, a deterioration of power factor is concerned with decrease of utility input power. As a result, the required microwave output could not provide. Then, a magnetron is needed to drive under a condition of high power factor by reducing the non-operating period of a magnetron and the distortion of the utility AC input line current waveform by maintaining the applied voltage across a high

978-1-4244-1871-8/07 $25.00 © 2007 IEEE   1184

frequency transformer near zero voltage around valley of the utility AC input voltage by adding a particular voltage boost function. But, the power conversion efficiency actually decreases to use the some power conversion processing stages and the increase of circuit components results in the large-sizing volume. Thus, a direct high frequency inverter with the multi function of one stage power conversion that has three operating abilities of rectification, boost and frequency conversion is proposed by the authors. Consequently, this proposed direct inverter type AC-DC power converter has boost function without reducing the power conversion efficiency and increasing circuit components and power semiconductor switching devices.

B. Power Circuit and System Topology

The main power circuit and system topology of the direct high frequency inverter type AC-DC power converter for a microwave oven is shown in Fig.5 in which the utility AC power supply has three winding high frequency transformer, and the design specifications and circuit parameters are indicated in Table I. The multiple power conversion circuit stages consist of the utility power supply, non-smoothing filter, high frequency soft switching inverter, three winding high frequency boost transformer, full-wave voltage doubler rectifier circuit, magnetron circuit represented by electrical equivalent circuit with i-v characteristic as zener diode characteristics, and magnetron heater circuit. The commercial utility AC power source voltage supplies the electric power to a high frequency inverter through the low path filter as non-smoothing filter that consists of an inductor and a capacitor. This low path filter is inserted not to leak the noise by the high frequency inverter to a commercial utility power source side. Moreover, since the capacitance is designed for the value that can maintain a certain constant voltage in the operating frequency area of high frequency inverter, the inductance of the filter inductor is set to 100[μH] and the capacitance of the filter capacitor is set to 6.0[μF]. Therefore, a commercial utility AC power source does not design so as to have smoothing ability for commercial AC frequency and can deliver to the output stage of direct high frequency inverter from 0[V] to 141[V] in non-smoothing filter operation. A high frequency transformer is connected to the primary winding at the middle point of the circuit of a

filter capacitor in series with a power switching semiconductor device, and a reverse conducting diode is connected to a semiconductor switching device, respectively. The capacitor C2 is connected between collector-emitter of the semiconductor-switching device Q1, and works as lossless capacitive snubber at the time at a turn-off of the switching device Q1 and Q2. This capacitor performs a soft switching operation commutation. The capacitors C3 and C4 are connected in parallel with the diodes D1 and D2, respectively, and these capacitors have the clamp action which maintains the voltage applied for a high frequency transformer by charging the energy stored into the high frequency transformer through the reverse conducting diodes D3 and D4. Moreover, the secondary winding of a high frequency transformer is connected to the double voltage rectification circuit that consists of the capacitors C5 and C6 and the diodes D5 and D6.

C. High Frequency Three Winding Transformer

The appearance of a high frequency three winding transformer is shown in Fig.6. Although the independent power supply which is not influenced upon the inverter operation is actually adopted for the heater power supply in an industrial magnetron drive as a plasma generator, the magnetron heater power is to be supplied from installing the 3rd winding in a high frequency high-voltage transformer, from a viewpoint of small size, lightweight and low cost, for the household electric appliances.

D. Gate Pulse Control Implementation

Figure 7 gives the asymmetrical PWM signal pulse timing sequences. These gate pulse voltages are respectively supplied to the power semiconductor switching blocks; the main switching block Q_1 (SW_1/D_4) and the subsidiary switching block Q_2 (SW_2/D_3). In the Fig.8, the constant frequency duty factor (or duty cycle) control scheme is implemented for the continuous power regulation strategy of this high frequency quasi-resonant inverter using IGBTs. The duty factor is defined as the conduction time Ton1 including a dead time Td of the main active power switch SW_1 during one period T. The control variable range of this duty factor

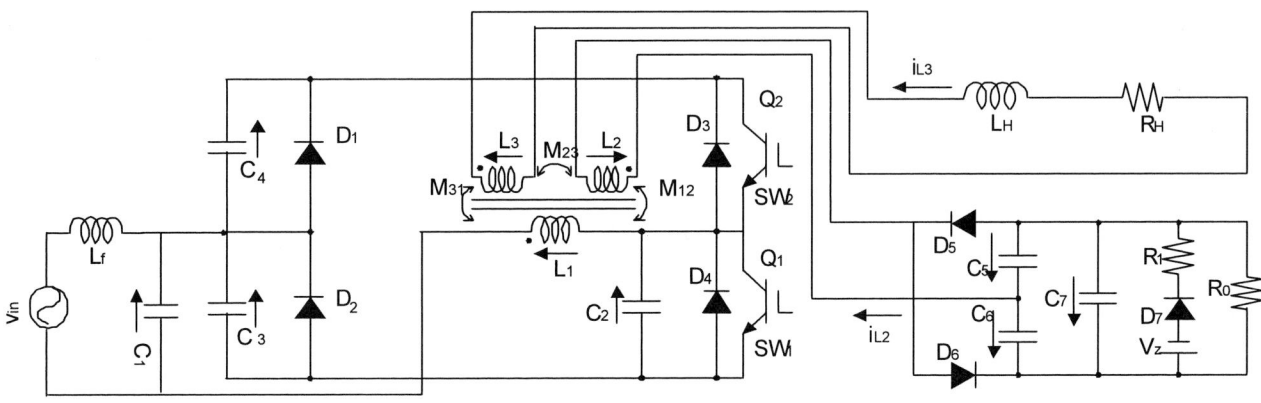

Fig.5 Circuit diagram of proposed converter.

978-1-4244-1871-8/07 $25.00 © 2007 IEEE   1185

is basically from 0 to 1. The output power of this high frequency inverter is controlled smoothly by varying this duty factor defined as the ON time of the gate voltage pulse signal for driving main active power switch SW_1 as a control variable.

IV. EXPERIMENTAL RESULTS AND DISCUSSIONS

A. Steady State Operating Waveforms of the Switches

Figure 8 (a) shows the appearance of newly developed high frequency inverter type AC-DC power converter, which is composed of the direct AC-HFAC power converter, high frequency transformer and full wave voltage doubler and the magnetron. Figure 8 (b) shows the IGBT (40[A]/600[V]) produced by Fairchild Semiconductor Co. Ltd for this high frequency inverter type AC-DC converter. Figure 9 (a) and (b) illustrate the measured voltage and current waveforms of the switching blocks; Q1 and Q2, and Figure 10 shows the simulating waveforms of these power switches. As understood clearly from these operating waveforms, it is noted that the entire switching block in this inverter type AC-DC power converter can achieve the soft switching.

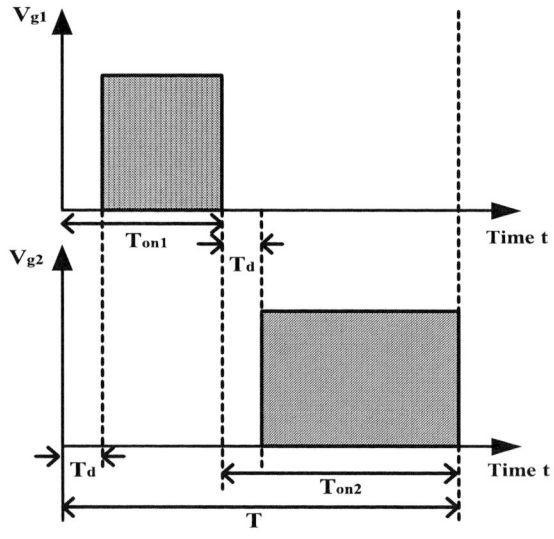

Fig.6. High frequency three winding transformer

Fig.7. Gate voltage pulse signal timing sequence

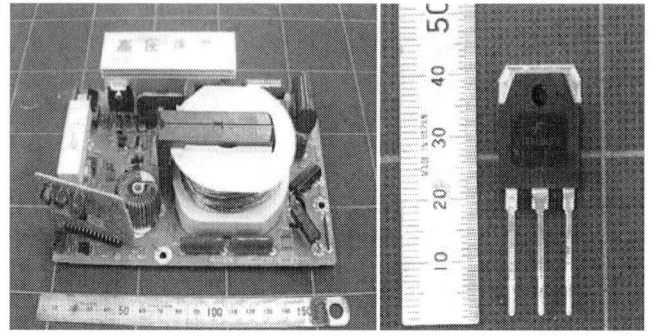

(a) Appearance of AC-DC power converter   (b) IGBT

Fig.8 The appearance of newly developed high frequency inverter type AC-DC power converter and IGBT

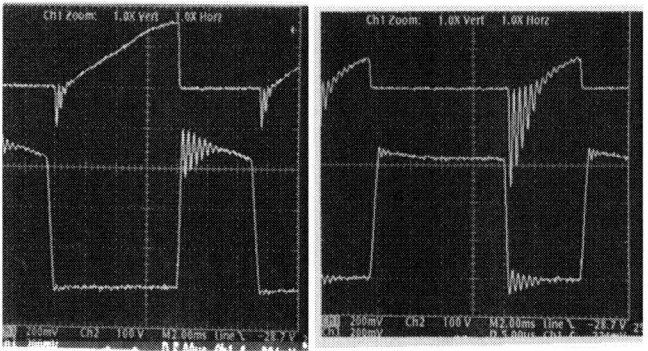

(a) Switching block Q1   (b) Switching block Q2
Current axis; 40A/DIV   Voltage axis; 100V/DIV   Time axis; 5µs/DIV

Fig.9. Experimental waveforms of current and voltage of Q1 and Q2

TABLE I
DESIGN SPECIFICATIONS

Item		Symbol	Value
High frequency inverter		C_1	0.6[µF]
		C_2	0.1[F]
		C_3	4.5[µF]
		C_4	4.5[µF]
		L_f	100[µH]
HF-3 winding transformer	inductance	L_1	60.0[µH]
		L_2	24.0[mH]
		L_3	0.60[µH]
	coupling coefficient	K_{12}	0.81
		K_{23}	0.66
		K_{31}	0.55
Full-wave voltage doubler rectifier		C_5	3600[pF]
		C_6	3600[pF]
Penetration capacitor		C_4	500[pF]
Magnetron	cut-off voltage	Vz	4600[V]
	non-oscillation Load	R_0	500[kΩ]
	oscillation load	R_1	300[Ω]

978-1-4244-1871-8/07 $25.00 © 2007 IEEE

Moreover, the simulated operating waveforms have good agreement of the observed operating waveforms and it is clear that the validity of this simulation method is also more effective. It is confirmed that the soft switching commutation in all the operating area during one cycle period of the utility AC voltage can be achieved for this high frequency inverter type AC-DC power converter.

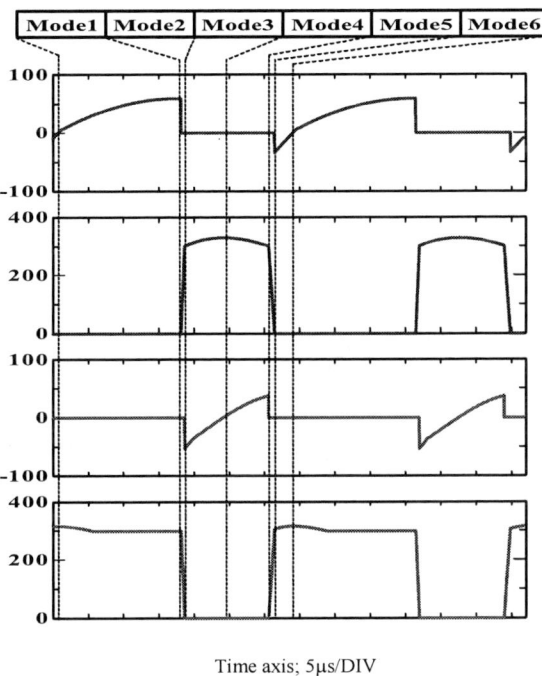

Time axis; 5μs/DIV

Fig.10. Simulation waveforms of current and voltage of Q1 and Q2

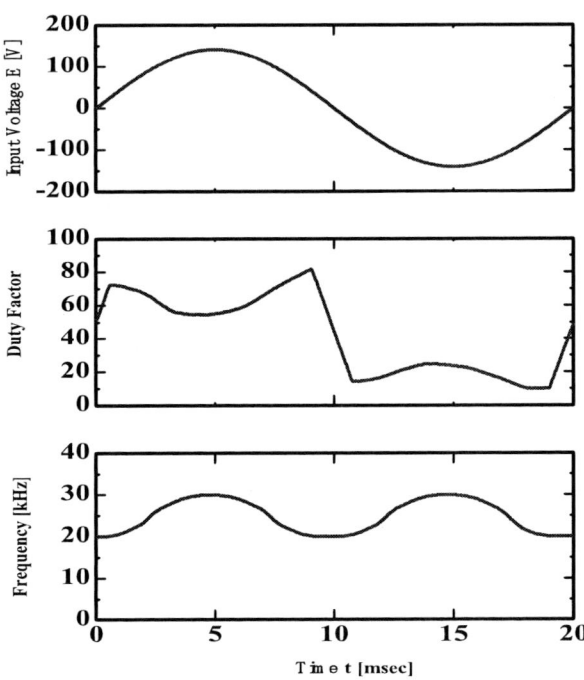

Fig.11. Pulse width and pulse frequency mixed pattern modulation

B. Partially Pulse Modulation

Figure 11 demonstrates how to select modulation patterns in respect of the partially pulse modulated pattern scheme due to inverter operating frequency as well as its duty factor-based asymmetrical pulse width modulation in order to improve line current harmonic distortion and total power factor in the utility AC power grid side. This principle is based on the characteristics of this direct high frequency inverter; when its operation frequency is so as to be high, the output voltage decreases. On the other hand, when duty factor is so as to be high, the output voltage increases. This pulse modulation scheme is based on pulse modulation on the duty factor, which is changed so as to be a low value near peak value of utility input sinewave AC voltage. On the other hand, it is changed so as to be a high value near transient build-up and transient build-down of the utility input sinewave AC voltage. Moreover, this pulse modulation scheme is based on pulse modulation on the operation frequency, which is changed so as to be a high value near peak value of utility input sinewave AC voltage; on the other hand, it is changed so as to be a low value near transient build-up and transient build-down of utility input sinewave AC voltage. A conduction angle of an input current makes wider for the utility sinewave AC input voltage by introducing this modulation strategy. The oscillation period of magnetron increases in accordance with the conduction angle of input line current for positive and negative utility AC voltage half cycle. In addition to this, the peak value of anode current is held with a low value near the peak absolute value of the utility AC input voltage. So this pulse modulation is more effective modulation method for excellent oscillation range of the magnetron and harmonic current reduction. Furthermore, the main control requirement relating to the operating life problem of the magnetron must hold down the peak value of anode current below to 1.2 [A] is fully met from an application point of view.

C. Evaluations of Input Line Current Waveforms and Comparative Studies

Figure 12 (a) shows the utility AC input line current waveform of the modulated high frequency inverter type AC-DC power converter during one cycle period of utility sinewave AC voltage, Figure 12 (b) shows utility AC input line current waveform of the conventional high frequency inverter. The utility AC input line current of newly developed high frequency inverter type AC-DC power converter shown in Fig.12 (a) becomes continuous near zero voltage of the power supply voltage, and these waveforms is good similar to a sinusoidal current wave. This high frequency inverter type AC-DC power converter can control voltage across transformer with boost operation function. The results on FFT analysis of the utility AC input line current of this high frequency inverter type AC-DC power converter and conventional inverter type AC-DC power converter is shown in Fig.13. The utility AC input line current of the developed inverter type AC-DC power converter to this standard value is settled within the standard value in all the current harmonic orders. It is noted that the good input line current waveform can be obtained for this AC-DC power converter. Furthermore, it is noted that harmonics current in utility AC line for newly inverter type AC-DC power converter is less than the conventional high frequency inverter type AC-DC power converter. This result is considered because the total power factor in the utility AC side of this high frequency inverter type AC-DC power converter can be improved as compared with that of conventional inverter type AC-DC power converter. The total power factor and total system efficiency of this high frequency inverter and conventional.

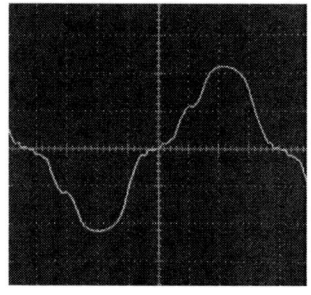 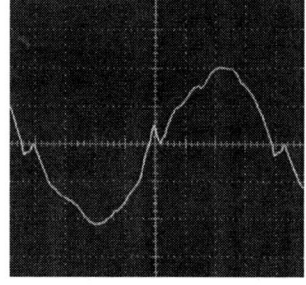

(a) Developed inverter   (b) Conventional inverter
Current axis; 10A/DIV   Time axis; 2ms/DIV
Fig.12 Input side current waveforms

TABLE.II
COMPARATIVE EFFICIENCY AND POWER FACTOR OF CONVENTIONAL
AND PROPOSED POWER CONVERTER SYSTEM

	This Developed Inverter	Conventional Inverter
Power Factor	0.99	0.96
Efficiency	92%	90%

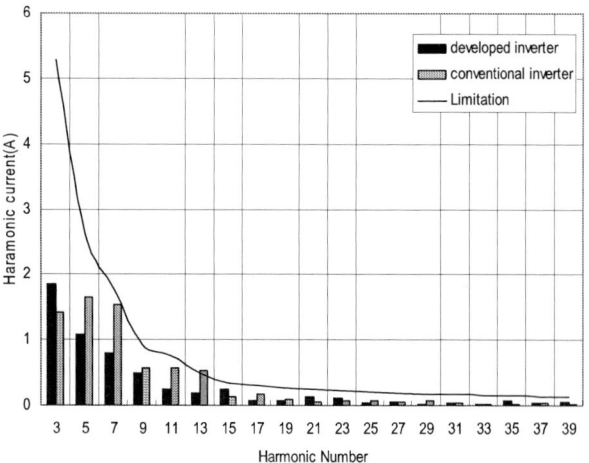

Fig.13. FFT spectrum of input line current in utility grid

V. CONCLUSIONS

The new model has high efficient high-power magnetron with high cut-off DC voltage so as to provide harmonic current reduction in the utility AC grid power side and total power factor improvement in utility AC power grid side in addition to the newly proposed high frequency inverter with boost and active voltage clamp function. Moreover, the AC-DC power converter including high frequency inverter ability, boost function, voltage clamp and a rectification function was integrated as one power converter. The high frequency inverter type AC-DC power converter treated here was more suitable for driving the new type magnetron. Finally, the harmonic components to the 40th order of the distorted harmonics input line current in the commercial sinusoidal wave AC side has been sufficiently suppressed below each regulation value specified by EN 61000-3-2 on harmonic current order. As a result, and AC-DC his power converter system was cost-effective for driving consumer high power magnetron for microwave oven.

REFERENCES

[1] K. Yasui, D. Bessyo, H. Omori, H. Terai, M. Nakaoka" The Inverter Circuit Skills to Realize low-cost, compact-size Power Supply for Microwave Oven, and the Advantages of Improved Defrosting" International Appliance Technical Conference 2000

[2] K. Yasui, D. Bessyo, H. Omori, M. Nakaoka, "The Active-Clamp Quasi-resonant Inverter Power Supply for Magnetron Drive" International Power Electronics Conference 2000 Tokyo

[3] E. Miyata, M. Nakaoka, K. Yasui, D. Bessyo, H. Omori, "Performance Evaluations on an Active Voltage-Clampling Self-Excited Zero Voltage Soft Switching DC-DC Converters for Consumer Magnetron Drive." Energy Engineering in Electronics and Communications, Vol. 100 No.303, 304, pp. 33-40, 2000.

[4] T. Matsushige, M. Ishitobi, M. Nakaoka, D. Bessyo, H. Yamashita, H. Omori, H. Terai, "Pulse Width and Pulse Frequency Modulated Soft Commutation Inverter type AC-DC Power Converter with Lowered Utility 200V AC Grid Side Harmonic Current Components", Proceedings of 2001 International Conference on Power Electronics (ICPE), pp.484-488, October, 2001.

[5] T. Matsushige, S. Chandhaket, S. Moisseev, E. Hiraki, M. Nakaoka, H. Terai, D. Bessyo, K. Yasui, H. Omori, "Cost Effective PWM Soft Switching High Voltage Converter with Utility AC Side Harmonic Current Reduction Strategy for Microwave Oven", Proceedings of International Symposium on Industrial Electronics (ISIE), Vol.1, pp.118-123, December, 2000.

An Estimation Method on the Initial Pole Position of a Z-Axis PMLSM

Jin-Woo Lee
Doowon Technical College
Department of Mechatronics
678, Jangwon-ri, Juksan-myun, Ansung-shi, Kyonggi-do, Korea, 456-718
Email: jinwoo@doowon.ac.kr

Abstract — This paper presents an estimation method on the initial pole position of a z-axis permanent magnet linear synchronous motor (PMLSM) without a magnetic pole sensor such as a Hall sensor. It takes account of the gravitational force of a z-axis motor and also the variation of load mass. First it approximately estimates the value of a given initial pole position by monitoring movements due to test current at predefined different test q-axes. Then it starts the successive estimation by the correction term from the initial estimate of the first step and continues until the estimate converges. The correction term is made of three values corresponding to force due to the test current at three test q-axes in order to avoid the effect of load mass. Experimental results on the test rig show that the proposed algorithm accurately estimates the initial pole position within a small moving range irrespective of load mass.

I. INTRODUCTION

Permanent magnet linear synchronous motors (PMLSM) are widely used in high performance servo applications such as semiconductor and plat panel display industries. In order to accurately control a PMLSM it is necessary to know the exact information of a magnetic pole position, which is usually available from the information of a magnetic pole sensor such as a Hall sensor and the position information of an encoder. But from the viewpoint of cost, size and reliability it is desirable to remove a magnetic pole sensor and also it is very difficult to attach a Hall sensor in case of tubular linear motors. Therefore recently some papers have proposed initial pole position estimation algorithms of a PMLSM with an encoder only [1] – [6].

The simple dc current excitation method and the dither commutation initialization method are adopted in [1]. Applying two currents with electrical phase difference of 90° needs a speed controller by using the concept that force is the same at the same speed [2]. The method of using the steady state position response needs a position controller [3]. A method of applying the same test current to two different q-axes located at ±45° from the estimated q-axis has good estimation accuracy [4]. The method using the secant algorithm shows good estimation accuracy with short moving range [5]. The method based on agreement of two reference frames shows good estimation accuracy [6]. However, these methods do not explicitly describe the vertical application case like a z-axis application.

This paper deals with the initial pole position estimation of a z-axis PMLSM with an encoder only. It analyzes conditions of a z-axis motor and suggests an estimation algorithm on the initial pole position, which aims at good estimation accuracy, short estimation time and short moving range irrespective of load mass. And some experimental results on the test rig are presented to verify the validity of the proposed method.

II. INITIAL POLE POSITION ESTIMATION

The proposed estimation method on the initial pole position of a z-axis PMLSM consists of two steps. The first step approximately estimates the value of a given initial pole position and then the second step successively estimates the initial pole position from the initial estimate until the estimate converges.

A. Model of a permanent magnet linear synchronous motor

The stator voltage and thrust force equations of a PMLSM can be described in the synchronous reference frame as follows [4].

$$v_{ds}^e = r_s i_{ds}^e + L_s \frac{di_{ds}^e}{dt} - \frac{\pi}{\tau} v L_s i_{qs}^e \tag{1}$$

$$v_{qs}^e = r_s i_{qs}^e + L_s \frac{di_{qs}^e}{dt} + \frac{\pi}{\tau} v L_s i_{ds}^e + \frac{\pi}{\tau} v \lambda_f \tag{2}$$

$$F_e = \frac{3}{2} \frac{\pi}{\tau} \lambda_f i_{qs}^e \tag{3}$$

where r_s is stator resistance, L_s is stator inductance, v is mover speed, λ_f is flux linkage produced by the permanent magnet, τ is pole pitch, v_{ds}^e & v_{qs}^e are dq-axes stator voltages, i_{ds}^e & i_{qs}^e are dq-axes stator currents, and F_e is thrust force, respectively.

The mechanical dynamic equation of a z-axis PMLSM is as follows.

$$F_e = m \frac{dv}{dt} + Bv + F_g + F_d \tag{4}$$

where m is mover mass, B is a viscous friction coefficient, F_g is gravitational force, and F_d is disturbance force including load force, respectively.

978-1-4244-1871-8/07 $25.00 © 2007 IEEE 1189

B. First estimation step

Fig. 1 shows the initial mover position of a z-axis PMLSM and the gravitational force (F_g) acting always upon the mover that consists of load and the moving part of a motor. In case of moving upward, the positive thrust force should be larger than the gravitational force.

Fig. 2 shows the stationary reference frame of DQ-axes and the synchronous reference frames of the real dq-axes and the initial control dq-axes. The angle of the real synchronous reference frame in Fig. 2 is θ_e, the angle of a given initial magnetic pole position is ϕ_i and the feedback angle by an encoder is θ_{enc}, respectively. The angle of the control synchronous reference frame is $\theta_{control}$. The initial value of control angle ($\theta_{control}$) is θ_{enc}. But when the estimated angle ($\hat{\phi}_i$) of a given initial pole position is available, the control angle is calculated as follows.

$$\theta_{control} = \theta_{enc} + \hat{\phi}_i \qquad (5)$$

Fig. 3 shows the thrust force of a standstill PMLSM according to the magnetic pole position at the same control q-axis current. It also shows the initial control q-axis of q_i-axis for a given initial pole position of ϕ_i and the positions of real dq-axes. Positions of the real dq-axes correspond to $\phi = -\pi/2$ [rad] and $\phi = 0$ [rad] or $\phi = 3\pi/2$ [rad] and $\phi = 2\pi$ [rad] respectively on the horizontal axis in Fig. 3. The initial pole position of a z-axis PMLSM shown in Fig. 1 depends on both the motor and the mechanical attachment structure and therefore can be located at anywhere between 0 and 2π radian in Fig. 3. The mover shown in Fig. 1 can move upward when the initial pole position is only between ϕ_a and ϕ_b in Fig. 3 for a given motor current. And the force (f_a) at ϕ_a equals the force (f_b) at ϕ_b. In addition, values of ϕ_a and ϕ_b for a given motor current vary with the load mass because the gravitational force is proportional to the mover mass. Angles of ϕ_i, ϕ_a and ϕ_b are also defined with respect to the initial control q-axis and the following relations are satisfied.

$$\phi_i = (\phi_a + \phi_b)/2 \qquad (6)$$

$$\alpha = (\phi_b - \phi_a)/2, \quad \phi_a = \phi_i - \alpha, \quad \phi_b = \phi_i + \alpha \qquad (7)$$

If the initial pole position of ϕ_i is correctly estimated, then the magnetic pole position is always available through the position information of an encoder and the pole pitch of a PMLSM. Consequently the thrust force of a PMLSM without a magnetic pole sensor can be accurately controlled.

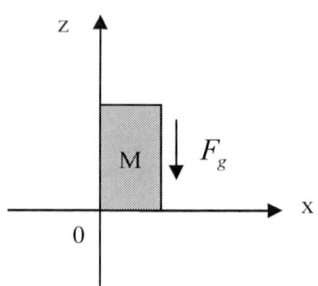

Fig. 1: Initial mover position of a z-axis PMLSM

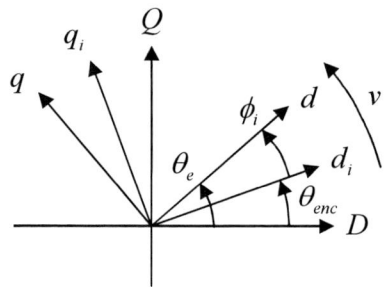

Fig. 2: Reference frames and angle definitions

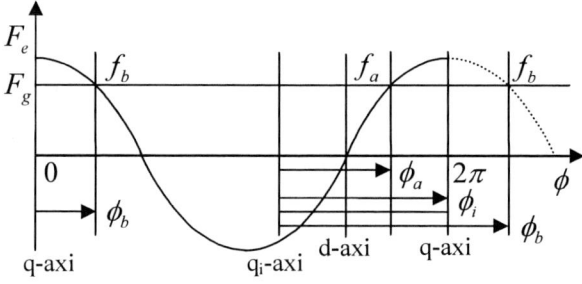

Fig. 3: Thrust force of a standstill PMLSM with respect to the magnetic pole position at the same control q-axis current and the initial control q-axis of q_i-axis for a given initial pole position

Let's develop a new estimation method on the initial pole position of ϕ_i on the basis of Fig. 3 and the above observations.

First, this paper approximately estimates two magnetic pole positions of ϕ_a and ϕ_b shown in Fig. 3 only by monitoring position signal of an encoder while applying test current at the assumed q-axes spaced at predefined angles between 0 and 2π radian. Then the first estimate of a given initial magnetic pole position can be obtained according to (6) from two approximate estimates of $\hat{\phi}_a$ and $\hat{\phi}_b$ as follows.

$$\hat{\phi}_i = (\hat{\phi}_a + \hat{\phi}_b)/2 \qquad (8)$$

978-1-4244-1871-8/07 $25.00 © 2007 IEEE

Additionally, the following relation is always satisfied because this method needs at least a detectable position signal of an encoder.

$$\phi_a < \hat{\phi}_a < \hat{\phi}_b < \phi_b \qquad (9)$$

C. Second estimation step

The second estimation step successively estimates the initial pole position from the first estimate until the estimate converges.

Fig. 4 shows relations among the real dq-axes, the initial control q_i-axis, three test q-axes, and the estimated control $\hat{q}_i$-axis in the dq coordinates.

The second step measures three developed thrust forces of f_a, f_b and f_c due to the same test current at three test $\overline{q}$-axes of $\overline{q}_a$, $\overline{q}_b$ and $\overline{q}_c$ as shown in Fig. 4. The successive estimate based on the relation between the magnetic pole position and the thrust force shown in Fig. 4 can be obtained as follows.

$$\hat{\phi}_i(k+1) = \hat{\phi}_i(k) + \frac{\delta}{2}\frac{f_b - f_a}{f_c - f_a} \qquad (10)$$

where $k = 0, 1, 2, \dots$, $\hat{\phi}_i(0) = \hat{\phi}_i$, $\beta = (\hat{\phi}_b - \hat{\phi}_a)/2$, δ is a known small angle.

Assume f_c equals f_b in (10) and in Fig. 4. First, the next estimate by (10) gives (11).

$$\hat{\phi}_i(k+1) = \hat{\phi}_i(k) + \frac{\delta}{2} \qquad (11)$$

The real initial magnetic pole position in Fig. 4 according to (6) gives (12), which is the same as (11).

$$\begin{aligned}\phi_i &= \frac{(\hat{\phi}_i(k) + \beta + \hat{\phi}_i(k) - \beta + \delta)}{2}\\ &= \hat{\phi}_i(k) + \frac{\delta}{2}\end{aligned} \qquad (12)$$

And also (10) gives the real initial pole position when f_a equals f_b. In other cases (10) gives the better updated estimate. Therefore (10) successively estimates the initial pole position of ϕ_i until the estimate converges.

The successive estimation algorithm needs the developed force of a PMLSM, but it can not be known without the information of the initial pole position. Therefore another method to evaluate (10) has to be studied.

Fortunately, (10) can be indirectly evaluated by using any

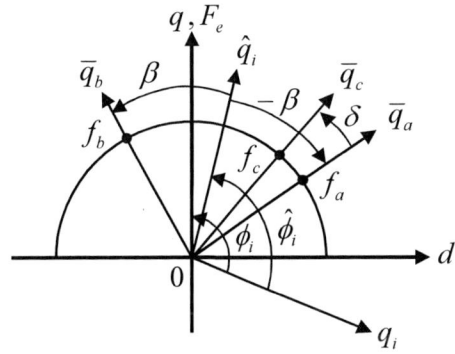

Fig. 4: Real, initial and estimated control, and test q-axes in the synchronous frame and thrust forces due to the same test current at test q-axes

corresponding variable to the thrust force because $(f_b - f_a)/(f_c - f_a)$ of (10) has no physical dimension. Therefore let's find some physical variables corresponding to the thrust force.

Equation (4) includes the thrust force so that mover speed and position are a function of the thrust force and so correspond to the thrust force. Therefore (10) can be evaluated by using the speed or the position corresponding to the thrust force instead of the thrust force itself.

First assume that the system parameters and the disturbance force in (4) are constant and also the viscous friction coefficient is negligible. And in order to have a short moving range and a smooth motion, this paper applies a ramp test current to the test q-axis until the mover gets to the target position. The thrust force due to the ramp test current is as follows.

$$i^e_{qs\_control} = K_I t \qquad (13)$$

$$i^e_{qs} = K_{\Delta\phi} i^e_{qs\_control} \qquad (14)$$

$$F_e = \frac{3}{2}\frac{\pi}{\tau}\lambda_f i^e_{qs} = K K_{\Delta\phi} t \qquad (15)$$

where $K = \frac{3}{2}\frac{\pi}{\tau}\lambda_f K_I$, $t \geq 0$, K_I is a given current constant, and $K_{\Delta\phi}$ is the cosine function of the angle between the real and test q-axes and increases as the test q-axis moves to the real q-axis.

In case of a z-axis PMLSM, the solution of (4) due to (15) under the above assumptions and the zero initial speed is given as follows.

$$v = \frac{1}{m}K K_{\Delta\phi} t_x \frac{t_x}{2} \qquad (16)$$

where $t_x = t - t_0$, $t \geq t_0$, $t_0 = (F_g + F_d)/(K K_{\Delta\phi})$.

The position under the zero initial position is the integral of (16) as follows.

$$S = \frac{1}{m} KK_{\Delta\phi} t_x \frac{t_x^2}{6} \qquad (17)$$

The speed of (16) and the position of (17) include the thrust force of (15) so that speed and position are variables corresponding to the thrust force under the above assumptions.

For the same test current, the thrust force is proportional to $K_{\Delta\phi}$ according to (13) and (15). Therefore the estimate of (10) can be rewritten by using (15) as follows.

$$\hat{\phi}_i(k+1) = \hat{\phi}_i(k) + \frac{\delta}{2} \frac{(K_{\Delta\phi b} - K_{\Delta\phi a})}{(K_{\Delta\phi c} - K_{\Delta\phi a})} \qquad (18)$$

where $K_{\Delta\phi a}$, $K_{\Delta\phi b}$, and $K_{\Delta\phi c}$ are the corresponding values of $K_{\Delta\phi}$ at three test q-axes.

In addition, we can obtain the following term from (17), which is directly proportional to $K_{\Delta\phi}$, as follows.

$$\frac{1}{6m} KK_{\Delta\phi} = \frac{S}{t_x^3} \qquad (19)$$

Consequently the estimate of (18) can be evaluated by using (19), which needs only measurements of the real position and the elapsed time when the mover gets to the target position. Additionally even though the position of (17) is dependent on the load mass, the estimate of (18) does not depend on the load mass because $K/(6m)$ in (19) is cancelled in (18).

III. EXPERIMENTAL RESULTS AND DISCUSSIONS

A. Experimental system

The permanent magnet linear synchronous motor shown in table 1 is tested to verify the validity of the proposed algorithm. The motor has a linear scale with resolution of 1[μm] and also has a Hall sensor as a magnetic pole sensor. The TMS320VC33 DSP-based controller has a 14-bit A/D

TABLE 1
SPECIFICATIONS OF A PMLSM

Rated Thrust Force	84.5[N]	Stator Resistance	3.79[Ω]
Rated Stator Current	2[Arms]	Stator Inductance	13.45[mH]
Number of Poles	8	Mover Mass	2.66[kg]
Thrust Force Constant	42.25[N/Arms]	Pole Pitch	12[mm]

converter, a 12-bit D/A converter and a digital output display.

The switching frequency of the IGBT inverter is 10[kHz] and the sampling period of the current controller is 50[μs]. The measured magnetic pole position is used only for comparison with the estimated value.

B. Experimental results and discussions

Fig. 5 shows experimental waveforms at no-load motor according to the proposed algorithm for two different initial pole positions of 47.25° and -4.55°. Fig. 6 shows the experimental results to the motor with load mass of 2.7[kg] for two different initial pole positions of 46.25° and -3.95°. The estimation errors are 4.75° and 6.5° at no-load and are 4.85° and 4.65° at load, respectively. The angle of δ is 9° and the target position is 7[μm] in the second estimation step. The top trace shows the initial pole position at start and shows the period of the first estimation step with zero value

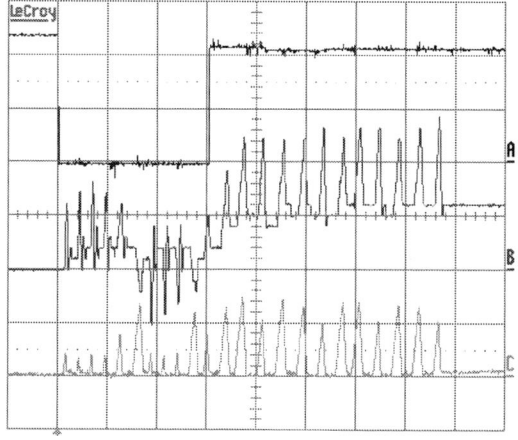

(a) Initial pole position of 47.25°

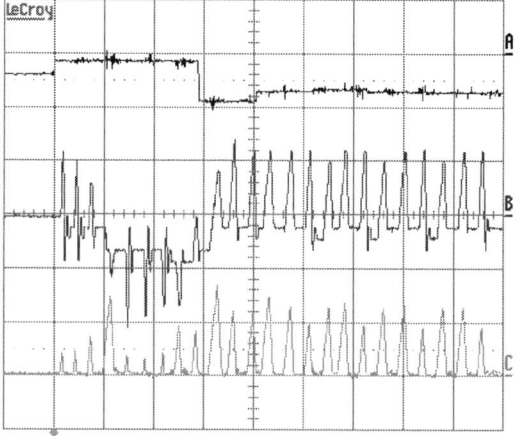

(b) Initial pole position of -4.55°

Fig. 5: Initial pole position estimation at no-load (A: magnetic pole position [20°/div], B: mover position [5μm/div], C: test current at the control q-axis [1A/div], time [0.1s/div])

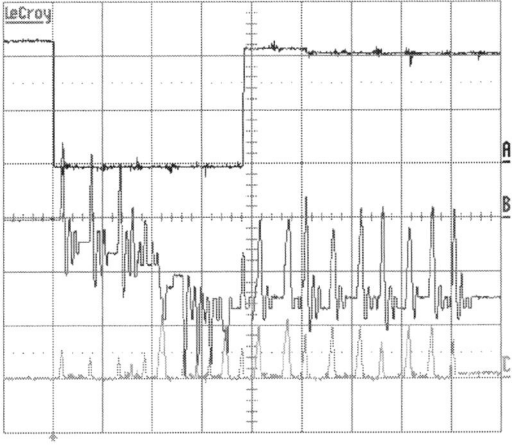

(a) Initial pole position of 46.25°

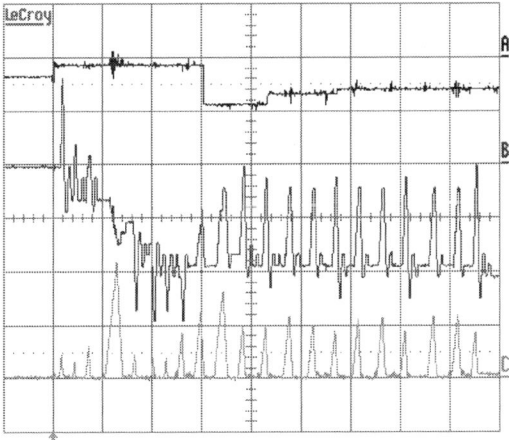

(b) Initial pole position of -3.95°

Fig. 6: Initial pole position estimation at load (A: magnetic pole position [20°/div], B: mover position [5μm/div], C: test current of the control q-axis [1A/div], time [0.1s/div])

of magnetic pole position and then shows the period of the second estimation step with the successive estimated pole position. The middle trace shows the mover position during the estimation process and shows that the moving range is about ± 15[μm]. This trace also shows that the end position is different from the starting position and the negative motion is occurred. This is due to the mechanical damper. The bottom trace shows the ramp test current at the control q-axis.

Consequently the experimental results show that the proposed algorithm to estimate the initial pole position of a z-axis PMLSM has high estimation accuracy and small moving range irrespective of load mass.

IV. CONCLUSIONS

This paper suggests an estimation method on the initial pole position of a z-axis PMLSM with an encoder only. It describes conditions of a z-axis PMLSM. The first estimation step approximately estimates the initial pole position by

monitoring movements due to the test current at predefined control q-axes. The second step successively estimates the initial pole position only by using the mover positions and the test current applied time and also does not depends on the load mass. The experimental results on the test rig show that the proposed algorithm has good estimation accuracy and small moving range irrespective of load mass.

REFERENCES

[1] *Installation & Operational Manual: Omega Series - Digital PWM Brushless Servo Amplifiers*, Glentek Inc. pp.65-68, 2002.

[2] T. W. Kim, J. Watanabe, S. Sonoda, and J. Hirai, " Initial Pole Position Estimation of Surface PM-LSM", Journal of Power Electronics, Vol. 1, No. 1, pp.1-8, 2001, April.

[3] D. H. Jung and I. J. Ha, "An Efficient Method for Identifying the Initial Position of a PMSM with an Incremental Encoder", IEEE Trans. on Industrial Electronics, Vol. 45, No. 4, pp.682-685, 1998, Aug.

[4] Choi J.-W., Yun W.-E. and Kim H.-G., "Initial pole-position estimation of linear motor", IEE Proceedings of Electric Power Applications, Vol. 152, No. 4, pp. 997-1002, 2005, July

[5] J. W. Lee, "A Novel Method of Estimating an Initial Magnetic Pole Position of a PMSM", IEEE Proceedings of PESC, pp. 702-707, 2006, June.

[6] K. Ide, H. S. Song, M. Takaki, S. Morimoto, and S. K. Sul, "Fast Initial Pole-Position Estimation for Non-Salient PM-LSM based on Agreement of Two Reference Frames", IEEE Proceedings of PESC, pp. 1497-1503, 2006, June.

Direct Torque Control Permanent Magnet Synchronous Motor Drive with Asymmetrical Multilevel Inverter Supply

M. N. Abdul Kadir
Malaya University
Department of Electrical Engineering
50603 Kuala Lumpur, Malaysia
Email: makadr@yahoo.com

Saad Mekhilef
Malaya University
Department of Electrical Engineering
50603 Kuala Lumpur, Malaysia
Email: saad@um.edu.my

Hew Wooi Ping
Malaya University
Department of Electrical Engineering
50603 Kuala Lumpur, Malaysia
Email: wphew@um.edu.my

Abstract-This paper demonstrates the Direct Torque Control Permanent Magnet Synchronous Motor (DTC-PMSM) drive fed by asymmetrical multilevel inverter. The cascaded H-bridges topology with ratio-3 dc voltage sourced cells inverter has been employed. The DTC controller receives a reference torque and flux signals, the torque controller output determines the position of the desired flux vector. The amplitude of this vector is determined by the reference flux amplitude. The reference voltage vector is calculated as a continuous value using a technique developed originally for the SVM controlled drive. In this study, however, the reference voltage vector has been approximated to the nearest voltage vector achievable by one switching state of the multilevel inverter. The performance of two- and three-cells per arm inverter drives has been compared to that of the SVM-inverter drive at two values of the switching intervals. The results reflect the low switching frequency and torque ripple advantages of the proposed scheme.

I. INTRODUCTION

The direct torque control (DTC) concept has been originally developed for induction motor drives and it has been well accepted by manufacturers and attracted researchers' attention [1]. It is known that the high torque ripple problem associated with the basic DTC scheme can be eased by effectively increasing the inverter's output resolution. Many developments have been introduced to achieve this such as zero state modulation [2], space vector modulation (SVM) [3] and the use multilevel inverter [4]. In general these modified schemes provide the desired torque smoothing and the price is added switching losses or power circuit complexity.

The DTC technique has been extended to the permanent magnet synchronous motor (PMSM) drives [5, 6]. The advantages of the DTC-PMSM drive compared to other PMSM drives are eliminating the current control and axis transformation besides the capability to operate without position sensor [7]. The torque ripple problem associated with the hysteresis comparators and switching table controller has been treated by introducing the PI controller and the SVM inverter [8, 9]. The SVM-DTC drive combines the basic DTC advantages with the torque and flux ripples reduction provided by the SVM technique, but it requires more powerful processor and operate at higher switching frequency.

The application of modified converter circuits has been considered recently in the PMSM–DTC drives to obtain improved characteristics. The three-level inverter has been used with four-level flux and torque hysteresis controllers and consequently modified switching table [10]. The comparison with the basic two-level system has shown the

harmonic and switching frequency reduction advantages. Another study replaced the rectifier-inverter supply by a matrix converter [11]. The matrix converter based DTC-PMSM drive has been operated in two modes; Basic DTC scheme and SVM-DTC. It has been shown that the matrix converter can maintain unity power at the supply side.

Among the multilevel inverter topologies, the cascaded H-bridge inverter has the advantages of modular structure and simple design. This topology is well suited for asymmetrical multilevel inverter implementation achieved by supplying the cascaded cells by different values of dc voltages and results increased number of levels for a given inverter circuit [12]. The maximum number of levels can be achieved by an inverter with"c" H-bridge cells per arm is 3^c and this occurs when the dc supplies voltages of the cascaded cells are related by ratio 3, i.e. V_s, $3V_s$, $9V_s$, ..etc. [13]

In this paper, ratio –three sourced asymmetrical multilevel inverter is considered for the PMSM-DTC drive. Due to the large number of voltage vectors obtainable, it is not practical to consider the extension of the switching table based-DTC. Therefore the control concept is extracted from the SVM-DTC drive presented in [8] and [9]. This paper compares the performance of two- and three-cells inverter to that of the SVM-inverter fed drives. Up to our knowledge, this is the first report of the application of the ratio-three based multilevel inverter in a DTC drive.

Section II describes the asymmetrical multilevel inverter. Section III presents the continuous voltage reference DTC strategy and the related equations. In section IV the Multilevel inverter PMSM-DTC model is presented. Simulation results are presented in section V.

II. RATIO-THREE BASED SOURCED CASCADED H-BRIDGES MULTILEVEL INVERTER

A. Circuit Topology, Switching States and Voltage Vectors

A three-cell per arm cascaded H-bridge inverter is shown in Fig. 1a. As seen the three cells of each arm are fed by dc voltages of Vs, 3Vs and 9Vs respectively. Each cell can be controlled to produce zero, the positive, or negative of its input. Therefore the arm voltage has a maximum of 13Vs, i.e. the sum of the three dc voltages, and a minimum of −13Vs and 27 levels as seen in Fig 1b.

The inverter seen in Fig 1 is three-cell or 27-steps inverter it can produce 2107 voltage vectors, where the *c*-cells per arm inverter has 3^c levels and the *l*-level three-phase inverter has *v* of distinct voltage vectors given in (1).

$$v = (l)^3 - (l - 1)^3 \qquad (1)$$

978-1-4244-1871-8/07 $25.00 © 2007 IEEE

The switching state of the inverter to be represented by three numbers $[x_a, x_b, x_c]$, where x_a, x_b and x_c represents the switching state of arms a,b anc c respectively. For c-cells per arm inverter: $x_{a,b,c}$ take any integer within the range

$$-\frac{3^c - 1}{2} \le x_{a,b,c} \le +\frac{3^c - 1}{2} \tag{2}$$

With respect to the common point of the three inverter arms, N in Fig. 1a, the line voltage can be represented by:

$$\begin{bmatrix} v_{AN} \\ v_{BN} \\ v_{CN} \end{bmatrix} = V_s \begin{bmatrix} 1 & 0 & 0 \\ 0 & 1 & 0 \\ 0 & 0 & 1 \end{bmatrix} \begin{bmatrix} x_a \\ x_b \\ x_c \end{bmatrix} \tag{3}$$

The load line-to-line voltages and the balanced load phase voltages expressions are given in (4) and (5) respectively

$$\begin{bmatrix} v_{AB} \\ v_{BC} \\ v_{CA} \end{bmatrix} = V_s \begin{bmatrix} 1 & -1 & 0 \\ 0 & 1 & -1 \\ -1 & 0 & 1 \end{bmatrix} \begin{bmatrix} x_a \\ x_b \\ x_c \end{bmatrix} \tag{4}$$

$$\begin{bmatrix} v_{An} \\ v_{Bn} \\ v_{Cn} \end{bmatrix} = \frac{V_s}{3} \begin{bmatrix} 2 & -1 & -1 \\ -1 & 2 & -1 \\ -1 & -1 & 2 \end{bmatrix} \begin{bmatrix} x_a \\ x_b \\ x_c \end{bmatrix} \tag{5}$$

Where n is the load neutral point, not shown in the figures. The output voltage vector corresponding to an inverter state is give by:

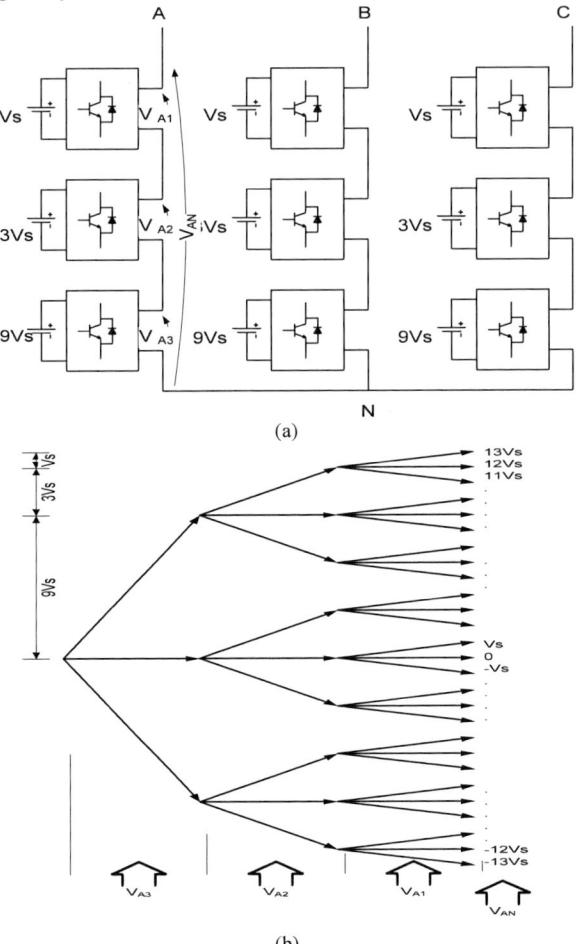

(a)

(b)

Fig. 1. (a) 3-cell, 27-level asymmetrical cascaded H-bridge inverter (b) the voltage levels corresponding to one inverter arm.

$$\overset{\rightarrow}{v_o} = v_d + jv_q = \left[\left(x_a - \frac{x_b}{2} - \frac{x_c}{2} \right) + j \left(\frac{\sqrt{3}x_b}{2} - \frac{\sqrt{3}x_c}{2} \right) \right] V_s \tag{6}$$

B. Voltage Vector Control

While the advantage of the based-3 ratio sourcing is to maximize the number of levels for a given number of cascaded cells, PWM-control is not suitable with this condition. The reason is that certain state transitions between adjacent voltage levels involve the high voltage cell switching results in high switching frequecny of these cells [14]. Therefore, voltage vector control of this inverter has been redefined as operating the inverter in the state that produces the voltage vector nearest to the reference vector rather than constructing the reference vector by adjusting the duty ratio of three inverter vectors.

The best one state vector approximation of a given reference voltage vector ($\overset{\rightarrow}{v_o}^* = v_d^* + jv_q^*$) can be realized by first analysing the reference vector to its reference phase voltage components:

$$\begin{bmatrix} v_a^* \\ v_b^* \\ v_c^* \end{bmatrix} = \frac{1}{3} \begin{bmatrix} 2 & 0 \\ -1 & \sqrt{3} \\ -1 & -\sqrt{3} \end{bmatrix} \begin{bmatrix} v_d^* \\ v_q^* \end{bmatrix} \tag{7}$$

Then approximating the normalized phase reference voltages to the nearest integers, results the inverter switching state:

$$x_{a,b,c} = round \left(\frac{v_{a,b,c}^*}{V_s} \right) \tag{8}$$

where *round* represents the rounding towards the nearest interger. The phase switching state ($x_{a,b,c}$) is converted to the corresponding phase cascaded cells desired gains (X_{c1}, X_{c2} and X_{c3}) using a look up table as shown in Fig. 2 for the three-cell inverter.

III. DTC-PMSM DRIVE WITH CONTINUOUS REFERENCE VOLTAGE

A. The motor equations

The stator side equations of the PMSM motor in the rotor reference frame are:

$$\begin{bmatrix} v_d \\ v_q \end{bmatrix} = \begin{bmatrix} R_s + pL_d & -\omega L_q \\ \omega L_d & R_s + pL_q \end{bmatrix} \begin{bmatrix} i_d \\ i_q \end{bmatrix} + \begin{bmatrix} 0 \\ \omega \varphi_f \end{bmatrix} \tag{9}$$

where $v_{d,q}$ and $i_{d,q}$ are the motor voltage and current components, related to the actual quantities by the inverse transformation of equation (7), R_s, L_d and L_q are the stator circuit resistance and direct and quadrature inductances, ω is the electrical synchronous speed and φ_f is the rotor flux.And the motor torque expression is given by

$$T_M = \frac{3p_p \varphi_s}{4L_d L_q} \left[2\varphi_f L_q sin\delta + \varphi_s (L_d - L_q) sin2\delta \right] \tag{10}$$

where p_p is the number of pole pairs and δ is the angle between the stator and rotor flux vectors referred to as the torque angle. The motor torque is proportional to $\sin \delta$ and can be controlled by changing this angle

B. Reference voltage calculation

The DTC controller used has PI torque controller its output signal $\Delta\delta^*$ represents the desirable change in the torque angle over the next switching signal. The new reference stator flux has an angle equals to the current angle of the stator flux plus $\Delta\delta^*$. The flux amplitude is taken as a constant this amplitude must abide to the stability condition derived in [5] and given by:

$$\varphi_s < \frac{L_q}{L_q - L_d}\varphi_f \qquad (11)$$

The controller determine the stator flux that is required to be build up during the next switching state as shown in Fig. 3, where

$$\Delta\varphi_s^* = \varphi_s^* - \varphi_s \qquad (12)$$

Where φ_s, φ_s^* and $\Delta\varphi_s^*$ are the estimated, next reference, and desired change of the stator flux respectively.

The stationary dq-model stator side voltage equation is given by

$$\vec{v_s} = \vec{\iota_s}R_s + p\varphi_s \qquad (13)$$

Therefore the next reference stator voltage vector is defined as:

$$\vec{v_s}(k)^* = \vec{\iota_s}(k-1)R_s + \frac{\overrightarrow{\Delta\varphi_s^*}}{T_s} \qquad (14)$$

where T_s is the switching interval. This voltage vector is realized using the method described in section II.B

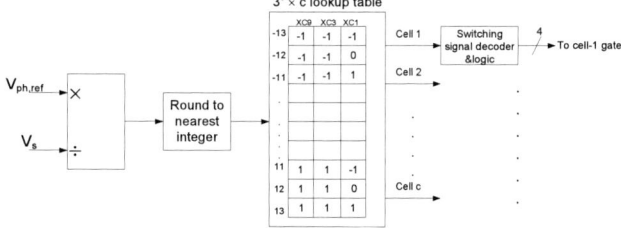

Fig. 2. Multilevel inverter control by the following phase reference voltage (the table shown inside the lookup table block for 3-cell per arm inverter)

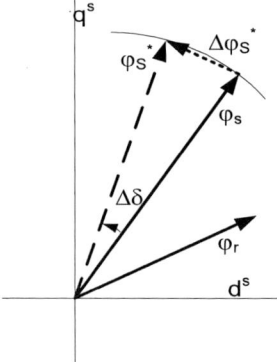

Fig. 3: Stator reference flux and flux difference calculation

IV. DRIVE MODEL DESCRIPTION

The drive controller is constructed based on (9) - (14) and the multilevel inverter as shown in Fig. 4. The parameters of the PMSM are given in Table I. The DTC converter is operated at a sampling time of 100 and 50μsec and the PI torque controller gain is set to K_p=1250 and K_i=0.9×10^6, the output of the controller is multiplied by a constant equivalent to the switching time, not shown in Fig. 4. The summation of the dc voltages for one inverter arm is 195V. The load is modeled by a pure friction torque of 0.033441 N.m.sec. For comparison, a different model has been constructed for the SVM DTC drive with the same motor and controller parameters and SVM controller with voltage control time step equals to 1% of the torque loop time interval, T_s.

In the following section, the reference torque has been set to a constant value of 3.6 N.m. and the reference stator flux amplitude is set to a constant of 0.7Wb. The results compare the performance of the three flowing system:
1) SVM –inverter controlled DTC-PMSM drive
2) 2-cell per arm asymmetrical cascaded H-bridge inverter fed DTC-PMSM drive
3) 3-cell per arm asymmetrical cascaded H-bridge inverter fed DTC-PMSM drive

V. SIMULATION RESULTS AND DISCUSSION

With the model parameters described in section IV, the DTC -PMSM drive has been simulated from starting over 0.4 sec. During the switching interval, T_s, the multilevel inverter holds the state corresponding to the voltage vector nearest to the reference voltage vector. On the other hand, the SVM drive produces the reference voltage vector by adjusting the duty ratios of two nonzero adjacent inverter states and the zero states during this interval.

A. Torque ripple

The simulation results are plotted in Fig. 5 and Fig. 6 for switching intervals of 100 and 50μsec respectively. It can be seen from Fig. 5 that the 3-cell multilevel inverter drive has a peak-to-peak torque ripple of about 20% of that of the SVM drive. At this sampling time, the 2-cell per arm inverter, or the nine-level inverter has a torque ripple slightly less than that of the SVM inverter. At switching interval of 50μsec the SVM drive inverter ripple decreases considerably, however, the torque ripple of the multilevel inverter drives is almost not affected as seen in Fig. 6.

The unaffected torque level with the time interval variation is due to the fact that this limit is set by the inverter resolution. By comparing the torque levels at the two switching intervals, it seems that we can define an optimum sampling time as the longest sampling time beyond which the torque ripple reduction is negligible.

It has to be indicated here that the SVM voltage controller operates at switching time equals to 1% of T_s This implies that the SVM drive requires much faster processor and faster switching devices than the proposed multilevel scheme. However, the proposed multilevel inverter operation cannot be further improved by PWM technique, due to the fact that the high voltage cells will be subjected to high frequency operation which eliminates the hybrid structure advantage.

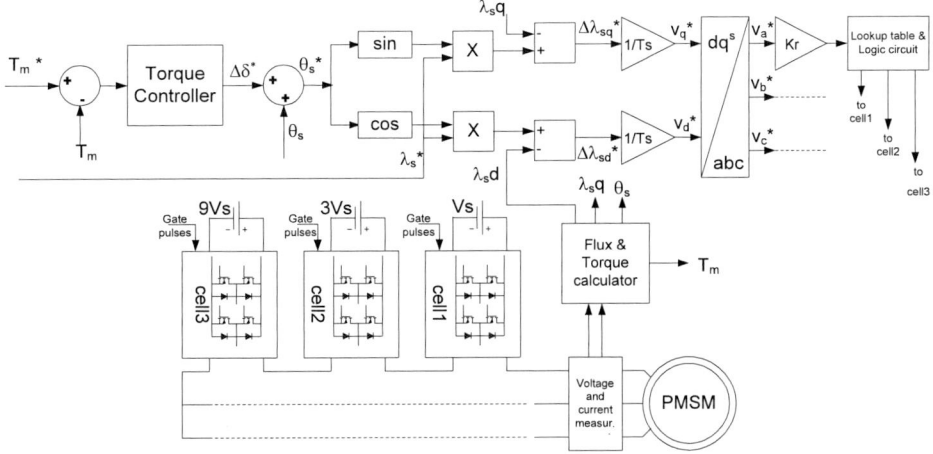

Fig. 4 The proposed multilevel inverter DTC-PMSM drive system

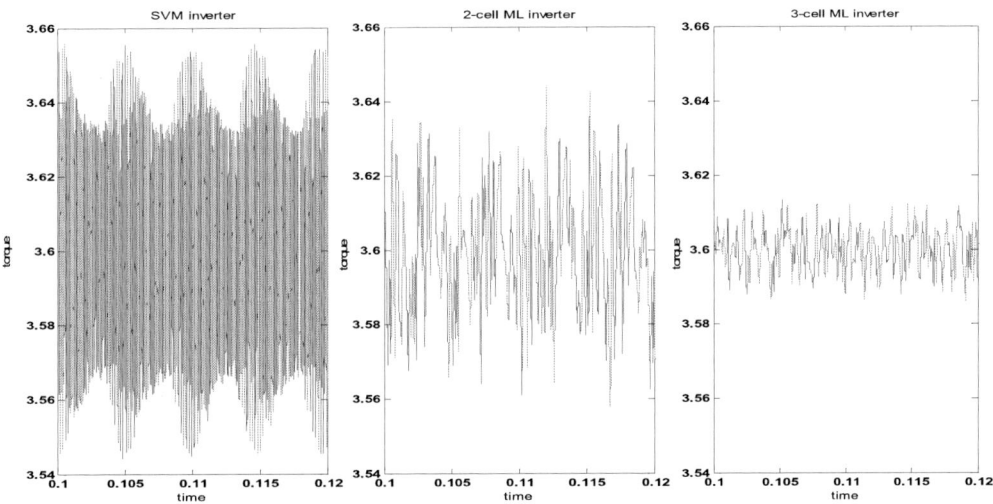

Fig. 5. The torque ripple comparison of the DTC-PMSM drive with, from left, SVM controlled inverter, 2-cell per arm and 3-cell per arm asymmetrical multilevel inverters with switching interval of 100μsec.

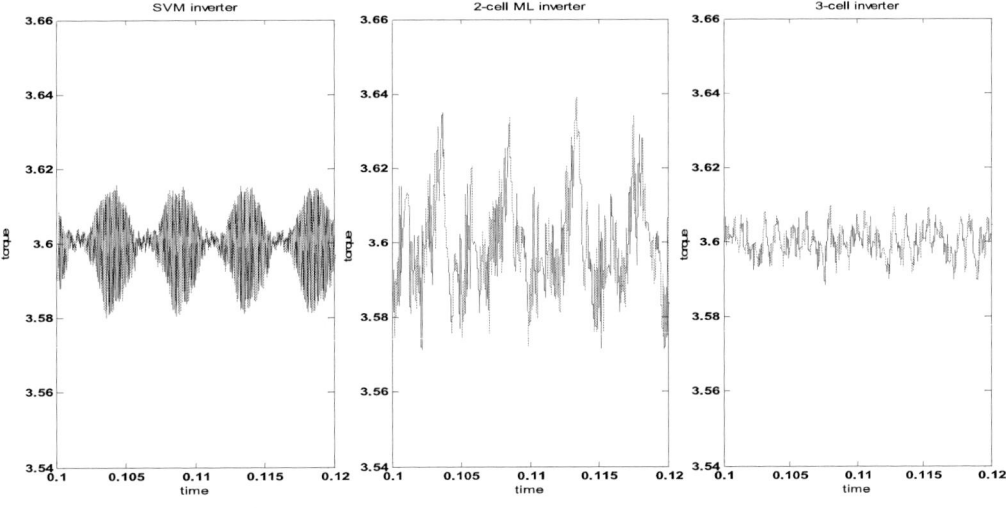

Fig. 6. The torque ripple comparison of the DTC-PMSM drive with, from left, SVM controlled inverter, 2-cell per arm and 3-cell per arm asymmetrical multilevel inverters with switching interval of 50μsec.

TABLE I
PARAMETERS OF THE IPMSM USED IN THE COMPUTER
SIMULATION

Rated output power (W)	1000
Rated phase voltage (V)	138.56
Magnetic flux linkage (Wb)	0.533
Number of poles	4
Rated torque (N.m)	6
Maximum speed (rpm)	1500
Stator resistance (Ω)	5.8
d-axis inductance (mH)	44.8
q-axis inductance (mH)	102.7
Moment of inertia of the motor and load (Kg-m^2)	0.001

B. . Switching losses

As an indication of the switching losses, the sum of switching pulses for the three switching signals has been plotted over the simulation period. Also the multilevel inverter pulses has been counted for the individual H-bridge legs. The results are given in Fig 7 and Fig. 8 for 100µsec and 50µsec switching intervals respectively. It can be seen that the SVM inverter switch operates at an average switching frequency close to5kHz when the sampling time is 100µsec and to 10kHz with sampling time of 50µsec. This rate consists with the basic SVM concept where the switching period is twice the sampling time. The multilevel inverter switching frequency for the lowest voltage switch is only about a quarter of this value and for the higher voltage cell is less by few orders. The total number of switching pulses obtained by adding the pulses for the cascaded cells is also shown. The total number of switching pulses forms about one third of the number of the SVM inverter for the three-cell inverter and about one quarter for the 2-cell inverter. This implies a considerable saving in the switching losses.

Another related advantage of the multilevel inverter that should be mentioned is related to the switching pattern. Whereas, the switching frequency (1/100 or 1/50µsec) represents the maximum switching frequency for the multilevel inverter switching devices, it forms twice the averages frequency for the SVM inverter. The difference is that the average switching frequency gives indication about the switching losses, but the maximum switching frequency is required to determine the switching device speed. If the SVM criteria of 100 steps per switching interval is followed, it implies that the switching device is subject to operate at a switching speed of 2µsec or 1µsec for the two tested switching intervals. In other words the required switching frequency rating of the fastest (low voltage's cell) multilevel inverter switching device is 50 times less than that of the SVM controlled inverter.

VI. CONCLUSION

The cascaded H-bridge cells asymmetrical multilevel inverter has been proposed to supply the PMSM-DTC drive. Two and three cells per arm inverters have been considered.

The DTC control strategy is similar to the one used for the SVM DTC-PMSM drive. This strategy produces a continuous reference voltage vector. The inverter controller approximates this reference vector to the nearest vector that can be produced by one inverter switching state, and the inverter holds this state during the following switching interval.

The proposed drive system has been modeled and, for the purpose of performance evaluation, it has been compared with the SVM drive of the same switching interval. The results obtained demonstrate that the 2-cell drive has a torque ripple level comparable to that of the SVM drive, and the 3-cell drive has a torque ripple level much lower than the SVM drive at switching interval of 100µsec. However, the results show that the SVM torque ripple reduces by decreasing the switching time interval while the torque ripple of the multilevel inverter, with the proposed controller, has a minimum ripple level after which increasing the controller frequency does not reduce it. By comparing the number of switching pulses per arm it has been shown that the multilevel drive has much lower switching losses compared to the SVM drive.

Besides the advantages of the torque ripple and the switching losses reduction, the proposed drive inherits the features of the asymmetrical multilevel inverter drive, such as low voltage stress and the highest power stage operates at the lowest switching frequency.

The drive has two main drawbacks; first the higher number of switching devices and second the requirement for a number of dc power supplies equal to the number of H-bridge cells applied. However it sensible to suggest the system for battery powered drives or for high power or high voltage drives

Other aspects may be further investigated, issues like: modifying the inverter topology to accommodate SVM technique to reduce the ripple and developing a dedicated DTC concepts for this type of inverters.

REFERENCES

[1] M.P Kazmierkowski and G. Buja, "Review of direct torque control methods for voltage source inverter-fed induction motors"; Industrial Electronics Society, 2003. IECON '03. The 29th Annual Conference of the IEEE, Volume 1, pp 981 - 991,2-6 Nov. 2003

[2] A. Arias, J.L. Romeral, E. Aldabas, and M.G. Jayne, "Fuzzy logic direct torque control", Industrial Electronics, 2000. ISIE 2000. Proceedings of the 2000 IEEE International Symposium on Volume 1, pp 253 - 258, 4-8 Dec. 2000

[3] F. Profumo, T.G. Habetler, M. Pastorelli and L.M. Tolbert, "Direct torque control of induction machines using space vector modulation",Industry Applications, IEEE Transactions on, Volume 28, Issue 5, pp1045 – 1053, Sept.-Oct. 1992

[4] X. Toro, S. Calls, M.G. Jayne, P.A. Witting, A. Arias and J.L. Romeral, "Direct torque control of an induction motor using a three-level inverter and fuzzy logic",Industrial Electronics, 2004 IEEE International Symposium on, Volume 2, pp 923 – 927, 4-7 May 2004

[5] L. Zhong, M.F. Rahman and W.Y. Hu,"Analysis of direct torque control in permanent magnet synchronous motor drives," IEEE Trans. Power electron., vol.12, pp528-536, May1997.

[6] M.R. Zolghadri, E.M. Olasagasti, and D. Roye, "Steady State Torque Correction of A Direct Torque Controlled PM Synchronous Machine", Proceedings of IEEE International Electrical Machines and Drives Conference, IEMDC'97, Wisconsin, USA, pp: MC3/4.1-MC3/4.3, May 18-21 1997.

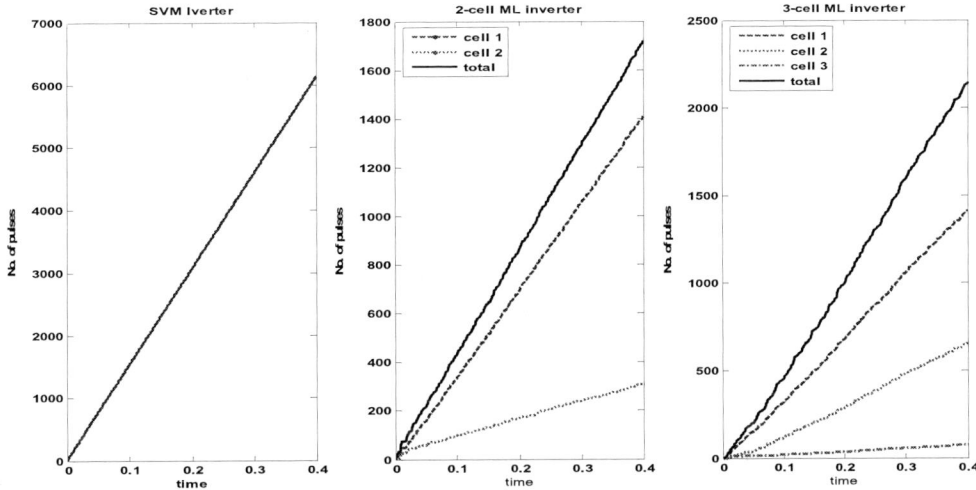

Fig. 7. The number of switching pulses of the three switching signals of DTC-PMSM drive with, from left, SVM controlled inverter, 2-cell per arm and 3-cell per arm asymmetrical multilevel inverters with switching interval of 100μsec.

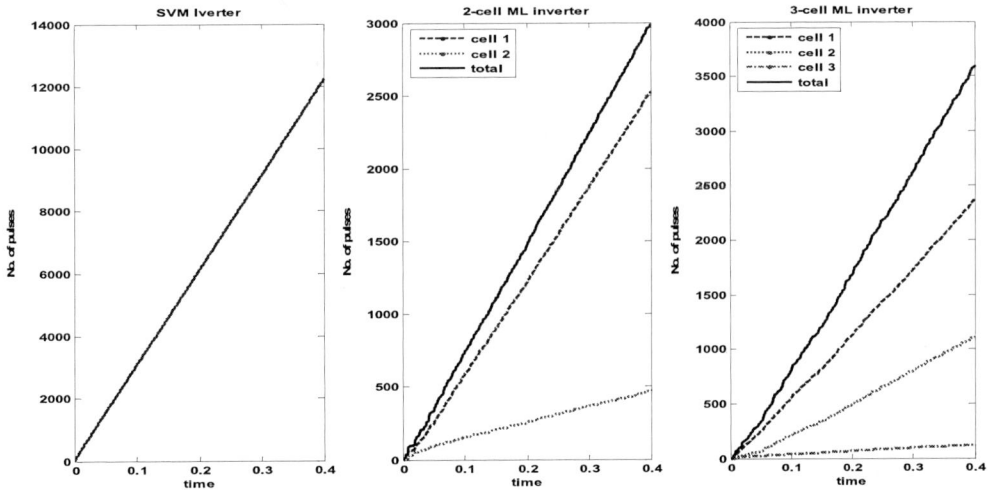

Fig. 8. The number of switching pulses of the three switching signals of the DTC-PMSM drive with, from left, SVM controlled inverter, 2-cell per arm and 3-cell per arm asymmetrical multilevel inverters with switching interval of 50μsec.

[7] M. F. Rahman, L. Zhong, and M. E. Haque, "A Direct Torque-Controlled Interior Permanent-Magnet Synchronous Motor Drive Without a Speed Sensor", IEEE Transactions on Energy Conversion, Vol. 18, No 1, March 2003, pp 17-22

[8] L. Tang, L. Zhong, M.F. Rahman, and Y. Hu, " A novel direct torque controlled interior permanent magnet synchronous machine drive with low ripple in flux and torque and fixed switching frequency", IEEE Trans. Power electron., vol. 19, no. 2, pp 346-354, March 2004.

[9] L. Tang, L. Zhong, M.F. Rahman, and Y. Hu; "A novel direct torque control for interior permanent magnet synchronous machine drive with low ripple in torque and flux-A speed-sensorless approach", IEEE Trans. On Industry Applications, Vol. 39, No. 6, pp. 1748-1756, November 2003

[10] K.E.B. Quindere, F. Ruppert and M.E. de Oliveira, "Direct torque control of permanent magnet synchronous motor drive with a three-level inverter", Power Electronics Specialists Conference, 2006. PESC '06. 37th IEEE, pp:1 – 6,18-22 June 2006

[11] D. Xiao and M.F. Rahman, "Direct Torque Control of an Interior Permanent Magnet Synchronous Machine fed by a Direct AC-AC Converter", Power Electronics and Motion Control Conference, 2006. IPEMC '06. CES/IEEE 5th International Volume 3, pp:1 – 6, Aug. 2006.

[12] D. Lipo, et al., "Hybrid topology for multilevel power conversion," U.S. Patent 6 005 788, Dec.1999

[13] M.N. Abdul Kadir, and Z. F. Hussien, " Asymmetrical Multilevel Inverter: Maximum Resolution for H-Bridge Topology", PEDS-05 The Sixth International Conference on Power Electronics and Drive Systems, Kuala Lumpur, 28 Nov.-1-Dec 2005.

[14] S. Mariethoz,, and A. Rufer, " An Overview of the Technique of Asymmetrical Multilevel Inverters", *In* ANAE : Associazione Nazionale Azionamenti Elettrici, 16o Seminario Interattivo, Azionamenti elettrici : Evoluzione Tecnologica e Problematiche Emergenti,

978-1-4244-1871-8/07 $25.00 © 2007 IEEE

Robust Measurement Disturbance Observer Design for AC Motor Drive Systems with Current Measurement Errors

Kyung-Rae Cho
YeungNam University
Department of Electrical Engineering
214-1, Dae-dong, KyungSan, KyungBuk, 712-749
Email: jkl2000@ynu.ac.kr

Jul-Ki Seok
YeungNam University
Department of Electrical Engineering
214-1, Dae-dong, KyungSan,KyungBuk, 712-749
Email: doljk@ynu.ac.kr

Abstract—In the vector control, current measurement errors deteriorate the motor drive system since the vector control is based on the current control. In this paper, in order to obtain precise torque control and eliminate the speed ripple, an online measurement disturbance observer (MDO) is proposed. Based on parameter sensitivity analysis, a practical method of designing robust MDO gains is proposed. The proposed method is applied to a commercial 1.1-kW permanent magnet synchronous motor to show its effectiveness and applicability.

I. INTRODUCTION

Recently, the vector control has applied to various ac motor drive systems. In the vector control, a precise current measurement is necessary to achieve a high static and dynamic performance since the vector control is based on the current control [1-2]. Currents are measured through current sensors, Op-Amps, measurement resistors and A/D converters. However, the nonlinearity of current sensors and Op-Amps, thermal drift of measurement resistors, the quantization error and nonlinearity of A/D converters result in DC offsets and scaling errors. It causes fluctuations of the motor torque and generates speed oscillations of fundamental and double fundamental frequency [1-2]. This results in the deterioration of the performance of the motor drive system because they act as a disturbance load.

Several papers report on the successful compensation of current measurement errors [2-6]. The methods of [2], [3] utilize the inverse system model of the speed controller to calculate the torque ripple. However, the compensator demands a detailed knowledge of the mechanical parameters. An inaccurate knowledge may cause an instability problem. Another scheme in [4] and [5] requires a torque sensor. Unfortunately, this may be unacceptable in many application fields. The approach of [6] utilizes an integral output of the d-axis current controller to compensate for undesirable periodic speed oscillations. The performance heavily depends on the accuracy of the feed-forward back-EMF voltage in the current controller. In addition, the convergence rate is quite disappointing.

This paper develops an online correction method of current measurement errors for the compensation of periodic motor torque oscillations. The speed ripple caused by the torque ripple is eliminated according to the compensation of the torque ripple. A measurement disturbance observer (MDO) based approach is taken as one possible solution of eliminating the DC offset and the scaling error, particularly the negative sequence component of unbalanced gains in the synchronous reference frame. The proposed MDO approach ensures a robust performance to other system nonlinearity and disturbance since the proposed MDO will suppress the higher frequency content that contains the dead-time and the motor slotting effect. Based on the parameter sensitivity analysis, a practical method of designing robust MDO gains is proposed. The developed algorithm has been implemented on a 1.1-kW permanent magnet synchronous motor (PMSM) drive to confirm the effectiveness and applicability of the proposed scheme.

II. EFFECTS OF CURRENT MEASUREMENT ERRORS

A. Effect of the Offset Error

The main causes of offset currents are drift phenomena or the residual current of current sensors and the offset of Op-Amps in measuring circuits. In this situation, the measured d- and q-axis currents in the synchronous frame are

$$i_{ds\_AD}^e = i_{ds}^e + \Delta I_{as} \cos \theta + \frac{1}{\sqrt{3}} \left(\Delta I_{as} + 2\Delta I_{bs} \right) \sin \theta \qquad (1)$$

$$i_{qs\_AD}^e = i_{qs}^e - \Delta I_{as} \sin \theta + \frac{1}{\sqrt{3}} \left(\Delta I_{as} + 2\Delta I_{bs} \right) \cos \theta \qquad (2)$$

where i_{ds}^e and i_{qs}^e are the actual d- and q-axis motor currents, respectively. ΔI_{as} and ΔI_{bs} indicate the A- and B-phase offset current, and θ represents the electrical angle.

Equation (1), (2) show that the offset current causes the torque of the motor to oscillated at the frequency ω. Since the torque ripple acts as the disturbance load to a speed controller, the speed of the motor also oscillates at the same frequency.

B. Effect of the Scaling Error

The scaling error originates from the nonlinearity and inaccuracy of the current sensors, the nonideal behavior of Op-Amps, inaccuracy of passive devices in measuring circuits, and the quantization error of an A/D converter [7]. If there are inaccurate gains on the A- and B-phase measurement channel, the measured phase currents are

$$i_{as\_AD} = k_a i_{as} = I \sin(\theta - \varphi)$$
$$i_{bs\_AD} = k_b i_{bs} = I \sin(\theta - \varphi - \frac{2\pi}{3}) \qquad (3)$$

where I represents an instantaneous current amplitude and

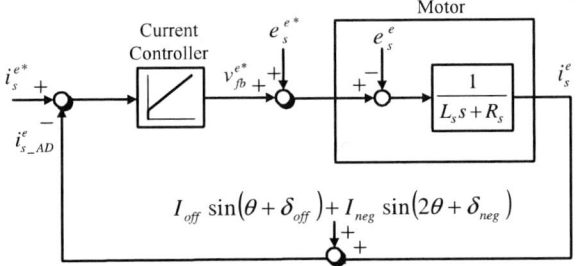

Fig. 1. The Block diagram of estimation strategy on ripple components.

$$\varphi = \tan^{-1}\left(\frac{i^e_{qs\_AD}}{i^e_{ds\_AD}}\right). \quad k_a \text{ and } k_b \text{ are the A- and B-phase gain}$$

ratios, respectively. Then, the synchronous d- and q-axis current errors due to the measurement gain errors are given by

$$\Delta i^e_{ds} = G_b i^e_{ds\_AD} - \frac{1}{\sqrt{3}} I (G_a - G_b) \sin\left(\varphi - \frac{\pi}{6}\right)$$
$$+ \frac{1}{\sqrt{3}} I (G_a - G_b) \sin\left(2\theta - \frac{\pi}{6} - \varphi\right) \quad (4)$$

$$\Delta i^e_{qs} = G_b i^e_{qs\_AD} - \frac{1}{\sqrt{3}} I (G_a - G_b) \sin\left(\varphi + \frac{\pi}{3}\right)$$
$$+ \frac{1}{\sqrt{3}} I (G_a - G_b) \sin\left(2\theta + \frac{\pi}{3} - \varphi\right) \quad (5)$$

where

$$G_a = 1 - \frac{1}{k_a}$$
$$G_b = 1 - \frac{1}{k_b} \quad (6)$$

It is noticed from (4) and (5) that the error component is composed of the DC component (the positive sequence of the scaling errors) and the repetitive ripple with double synchronous frequency (the negative sequence of the scaling errors). In AC drive systems, the control performance less affected by d-axis current error in the synchronous reference frame and the q-axis DC current error in the synchronous reference frame is compensated by the speed controller. However, the 2ω torque ripple results from the q-axis ac current error in the synchronous reference frame and deteriorate the performance of the motor control system, because they act as a disturbance load.

III.   PROPOSED MDO DESIGN

The voltage model is then completed by considering ripple components as shown in Fig. 1 and given by

$$i^e_{s\_AD} = \frac{v^{e*}_{fb}}{L_s s + R_s} + I_{off} \sin(\theta - \delta_{off}) + I_{neg} \sin(2\theta - \delta_{neg})$$
$$(7)$$

where $i^e_{s\_AD}$ and v^{e*}_{fb} are the controlled motor current and the voltage command in an arbitrary synchronous axis. R_s and L_s represent the stator resistance and inductance,

respectively. I_{off} and I_{neg} mean the ripple amplitude that is caused by the offset and the negative sequence component of the scaling error. δ_{off} and δ_{neg} indicate the arbitrary phase angles, respectively. To correct these effects, ripple components can be dynamically estimated by using an observer. They are then injected to the feedback current so as to cancel periodic oscillations in a feed-forward manner. For the purpose of designing a disturbance observer, a state-space representation is first required. From (7), the states are described as

$$x_1 = i^e_{s\_AD}, x_2 = I_{off} \sin(\theta - \delta_{off}), x_3 = I_{off} \cos(\theta - \delta_{off}),$$
$$x_4 = I_{neg} \sin(2\theta - \delta_{neg}), x_5 = I_{neg} \cos(2\theta - \delta_{neg}) \quad (8)$$

Then, the overall state equation can be obtained with a straight-forward extension:

$$\dot{\mathbf{x}} = \mathbf{A}\mathbf{x} + \mathbf{B}v^{e*}_{fb}$$
$$y = \mathbf{C}\mathbf{x} \quad (9a)$$
$$i_{ripple} = x_2 + x_4$$

where, i_{ripple} means the current ripple of fundamental and double fundamental frequency

$$\text{and} \quad \mathbf{A} = \begin{bmatrix} -\dfrac{R_s}{L_s} & \dfrac{R_s}{L_s} & \omega & \dfrac{R_s}{L_s} & 2\omega \\ 0 & 0 & \omega & 0 & 0 \\ 0 & -\omega & 0 & 0 & 0 \\ 0 & 0 & 0 & 0 & 2\omega \\ 0 & 0 & 0 & -2\omega & 0 \end{bmatrix}, \mathbf{B} = \begin{bmatrix} \dfrac{1}{L_s} \\ 0 \\ 0 \\ 0 \\ 0 \end{bmatrix},$$

$$\mathbf{C} = \begin{bmatrix} 1 & 0 & 0 & 0 & 0 \end{bmatrix}$$
$$(9b)$$

where ω is the measured or estimated rotor angular velocity. Here, we choose the linear MDO which is designed as (4).

$$\dot{\hat{\mathbf{x}}} = \mathbf{A}\hat{\mathbf{x}} + \mathbf{B}v^{e*}_{fb} + \mathbf{L}(y - \hat{y})$$
$$\hat{y} = \mathbf{C}\hat{\mathbf{x}} \quad (10)$$
$$\hat{i}_{rpple} = \hat{x}_2 + \hat{x}_4$$

where $\mathbf{L}=[l_1 \ l_2 \ l_3 \ l_4 \ l_5]^T$ is a gain matrix. Here, l_3 and l_5 are just needed to construct the observer and it can be called a dummy gain. The observer design method leads to the advantage of easy implementation of the proposed method. This form of MDO would have a desired property of extracting the $\omega + 2\omega$ component from the measured current and voltage command.

IV.   PARAMETER SENSITIVITY ANALYSIS OF THE PROPOSED MDO

One can see that, in practice, the identification of the MDO model (3) is a nontrivial task due to the presence of the possibly unknown/time-varying motor parameters such as R_s and L_s. For real time implementation, it is important to investigate the parameter sensitivity characteristics because the parameter uncertainty may noticeably alter the overall estimated results. In this section, our development begins with

978-1-4244-1871-8/07 $25.00 © 2007 IEEE

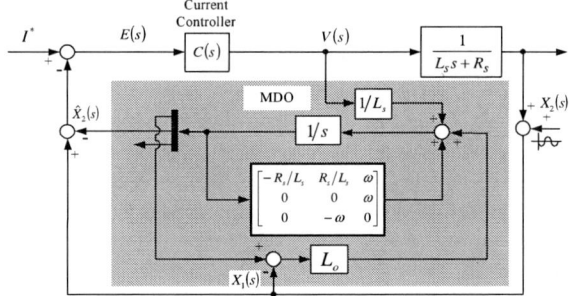

Fig. 2. Compensated feedback system with MDO considering the offset error.

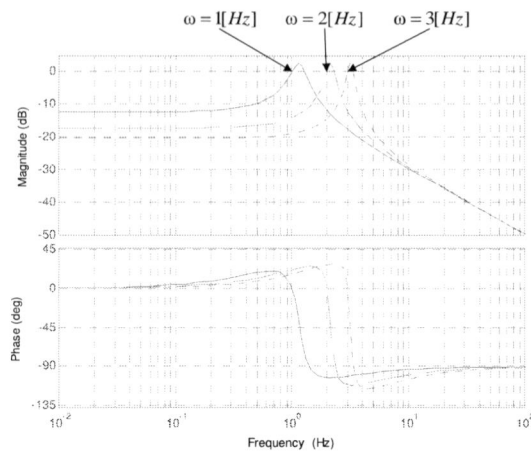

Fig. 3. Frequency response of typical MDO with exact parameters under $l_1 = l_2 = l_3 = 2$.

the MDO system without motor parameter uncertainty. Then, we take a more rigorous look at the parameter sensitivity characteristics of the proposed structure.

A. Analysis of MDO with Exact Motor Parameter

The model in (10) can be revised by the laplace transform as following:

$$s\hat{X}(s) = A\hat{X}(s) + BV(s) + L(X_1(s) - \hat{X}_1(s)) \quad (11)$$

To simplify the MDO analysis, the offset component $\hat{x}_2$ and $\hat{x}_3$ are considered. Then, (11) can be rewritten as

$$s\hat{X}(s) = A_o\hat{X}(s) + B_oV(s) + L_oX_1(s) \quad (12a)$$

where

$$A_o = \begin{bmatrix} -\dfrac{R_s}{L_s} - l_1 & \dfrac{R_s}{L_s} & \omega \\ -l_2 & 0 & \omega \\ -l_3 & -\omega & 0 \end{bmatrix}, \quad B_o = \begin{bmatrix} \dfrac{1}{L_s} \\ 0 \\ 0 \end{bmatrix}, \quad (12b)$$

$$L_o = \begin{bmatrix} l_1 & l_2 & l_3 \end{bmatrix}$$

Fig. 2 shows the compensated feedback system by $\hat{X}_2$ to make the actual plant with the current measurement disturbance behave like a given reference model. Here, I^* represents a DC current command, $C(s)$ is the transfer function of the current controller and $E(s)$ indicates the tracking error. The MDO output $\hat{X}_2$ can be expressed as

$$\hat{X}_2 = D_o\hat{X} \quad (13a)$$

where $D_o = \begin{bmatrix} 0 & 1 & 0 \end{bmatrix}$. (13b)

Combining (12) and (13) gives

$$\hat{X}_2(s) = D_o(sI - A_o)^{-1}B_oV(s) + D_o(sI - A_o)^{-1}L_oX_1(s) \quad (14a)$$
$$= H(s)\left[(R_s + L_s s)X_1(s) - V(s)\right]$$

where $H(s) = \dfrac{l_2 s + l_3\omega}{L_s s^3 + (l_1 L_s + R_s)s^2 + (l_2 R_s + \omega^2 L_s + l_3\omega L_s)s}$
$$+ (R_s + l_1 L_s - l_2 L_s)\omega^2 + l_3\omega R_s$$

(14b)

From Fig. 2, $V(s)$ can be written as

$$V(s) = (R_s + L_s s)(X_1(s) - X_2(s)) \quad (15)$$

Substituting (15) back into (14a) yields

$$\hat{X}_2(s) = H(s)(R_s + L_s s)X_2(s) = G(s)X_2(s) \quad (16)$$

where $G(s)$ indicates the transfer function of a typical

MDO with exact parameters. Analyzing $G(s)$ by plotting its frequency response shows in Fig. 3 shows that the proposed MDO has the unity gain at a specified frequency to extract the concerned ripple component. This feature leaves the estimated current measurement errors unaffected by the dead-time effect (6ω harmonic) during test.

B. Estimation Error Analysis in the Presence of Motor Parameter Errors

When there exists the stator resistance error ΔR_s in the model, $V(s)$ can be expressed as

$$V(s) = (R_s + \Delta R_s + L_s s)(X_1(s) - X_2(s)) \quad (17)$$

Then, the MDO estimate $\hat{X}_2$ in (16) can be reformulated as

$$\hat{X}_2(s) = H(s)\left[(R_s + L_s s)X_1(s) - (R_s + \Delta R_s + L_s s)(X_1(s) - X_2(s))\right]$$
$$= H(s)(R_s + L_s s)X_2(s) - H(s)\Delta R_s(X_1(s) - X_2(s))$$

(18)

From Fig2, If the current control bandwidth is sufficiently large for a good tracking, it is possible to approximate the current error as zero, i.e. $E(s) \cong 0$. This implies that

$$X_1(s) = I^* + \hat{X}_2(s) \quad (19)$$

After substituting (19) back into (18), the MDO output $\hat{X}_2$ becomes

$$\hat{X}_2(s) = \dfrac{(R_s + \Delta R_s + L_s s)H(s)}{1 + \Delta R_s H(s)}X_2(s) - \dfrac{\Delta R_s H(s)}{1 + \Delta R_s H(s)}I^* \quad (20)$$
$$= G_{\Delta R}(s)X_2(s) - G_{DCR}(s)I^*$$

where $G_{\Delta R}(s)$ means the transfer function of the ac ripple component and $G_{DCR}(s)$ represents the transfer function on the DC input, which implies that $\hat{X}_2$ has the residual DC component under the stator resistance mismatch. Thus, the DC term of $\hat{X}_2$ can be computed at $s = 0$, that is

$$\hat{X}_2(0) = -\dfrac{l_3\Delta R_s}{(l_1 L_s + R_s - l_2 L_s)\omega + l_3(R_s + \Delta R_s)}I^* \quad (21)$$

This means that the variation of R_s during the operation

978-1-4244-1871-8/07 $25.00 © 2007 IEEE

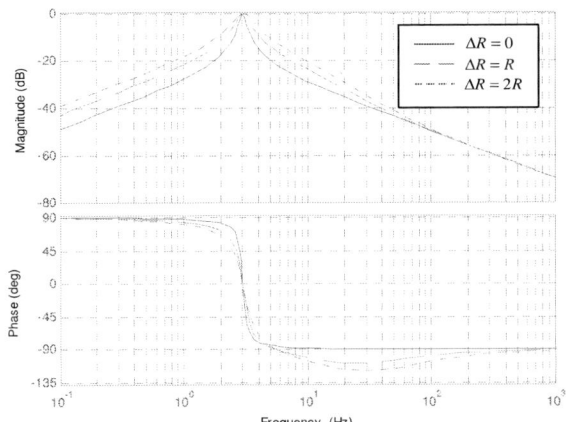

Fig. 4. Bode plot of $G_{\Delta R}(s)\big|_{l_3=0}$ under various resistance errors.

TABLE I
RATING AND NOMINAL PARAMETERS OF PMSM UNDER TEST

Rating and Parameter	Value	Unit
Rated power output	1.1	kW
Rated current	8.8	A_{peak}
Rated speed	2000	r/min
Number of pole pairs	4	poles
Stator resistance	0.38	Ω
Stator inductance	3.88	mH
PM flux linkage	0.1	Wb

will cause a DC estimation error of the MDO.

Similar approach on the stator inductance error ΔL_s gives

$$\hat{X}_2(s) = \frac{(R_s + (L_s + \Delta L_s)s)H(s)}{1 + s\Delta L_s H(s)} X_2(s) - \frac{\Delta L_s H(s)}{1 + s\Delta L_s H(s)} sI^* \quad (22)$$

$$= G_{\Delta L}(s)X_2(s) - G_{DCL}(s)I^*$$

where $G_{\Delta L}(s)$ means the transfer function of the ac ripple component and $G_{DCL}(s)$ represents the transfer function on the DC input under the stator inductance mismatch. Note that the DC gain $G_{DCL}(0)$ is zero. This means that the variation of L_s will not bring any estimation error.

C. Gain Selection of the Proposed MDO

One limitation with the proposed MDO scheme that has been identified from the analysis is that the stator resistance uncertainty provides a DC estimation error which causes a periodic torque oscillation. Our purpose here is to select the adequate MDO gain eliminating the DC estimation error without affecting the ripple component estimation performance. By choosing $l_1 = l_2 = g$ and $l_3 = 0$, the DC term $\hat{X}_2(0)$ in (21) becomes zero even in the presence of the stator resistance error. The constant g can be selected to set the convergence speed in a way similar to pole assignment in linear systems. In order to compare the estimation performance on the ripple component, an overlay Bode plot of $G_{\Delta R}(s)\big|_{l_3=0}$ under various resistance errors is plotted in Fig. 4. The introduction of resistance errors alters the Bode plot of $G_{\Delta R}(s)\big|_{l_3=0}$ but has little effect around the synchronous frequency of the concerned ripple component. This means that the proposed gain selection is effective to eliminate the DC error and gives little negative impact on the ripple component estimation in practice.

V. EXPERIMENTAL RESULTS

The proposed algorithm is implemented on a commercial 1.1-kW PMSM and the motor parameters are listed in Table I.

The PWM inverter consists of 10 kHz switching IGBT modules with a dead-time of 3.5 μs. The classical dead-time compensation scheme relying on the accurate current polarity [8-9] may cause secondary upsets in the proposed scheme. Thus, we implement a state-of-art dead-time compensation scheme [10] to minimize the interaction between in proposed correction action and the dead-time effect, which is based on the support vector regression theory [11-12] and robust to current measurement errors. Two-phase currents are sampled with a rate of 50 μs and the current regulator is tuned for a 6000 rad/s bandwidth. The LA 55-P current sensor of LEM with a maximum level of ±70 A and ±0.65% accuracy is installed in the drive. The online MDO is performed every 50 μs and the rotary encoder with 3000 p/r resolution is attached to the motor for the closed-loop velocity control. The constant gain g is fixed at 5 and dummy gains are set as $l_3 = l_5 = 0$.

Fig. 5 shows the waveforms for the measured q-axis current in the synchronous reference frame and the proposed compensator of the DC offset and the negative sequence component at 100 r/min speed command. An intentional +5% offset error of the actual current is added to the respective channel of an A/D converter for actual phase current signals. The scaling ratio error is given as $k_a = 1.05$ and $k_b = 0.95$. Correct estimation and compensation become effective at 0.6 s when the compensation flag is on. Almost complete suppression of the torque current ripple was realized, i.e., after the online estimation/compensation, at the constant torque current, as seen in Fig 5(a) while the observer estimates the $\omega + 2\omega$ component [see Fig. 5(b) and (c)] in the synchronous reference frame.

Fig. 6 shows the real phase current and the MDO compensation performance of the DC offset and the negative sequence component at 15 r/min. An intentional +4.4% offset error of the actual current is added to the respective channel of an A/D converter for actual phase current signals. The gain ratio error is given as $k_a = 1.1$ and $k_b = 0.9$. Correct estimation and compensation become effective at 5 s after starting. As soon as the online estimation/compensation starts, the controlled phase current in Fig. 6(b) rapidly converges to its real value within two cycles while the observer estimates

978-1-4244-1871-8/07 $25.00 © 2007 IEEE

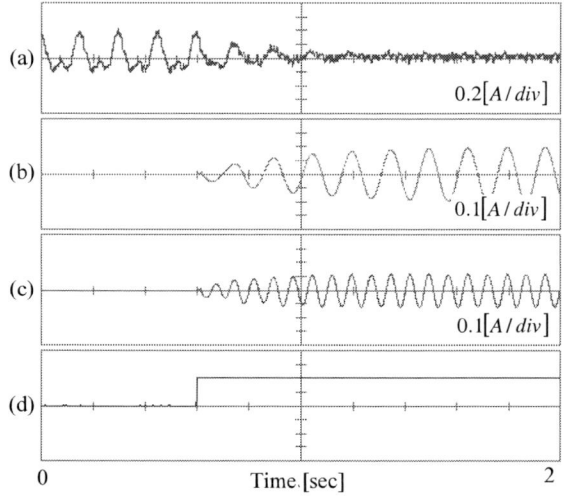

Fig. 5. Dynamic compensation results of the DC offset and the negative sequence component. (a) measured q-axis current. (b) estimated offset error in the synchronous reference frame. (c) estimated negative sequence error in the synchronous reference frame. (d) compensation flag.

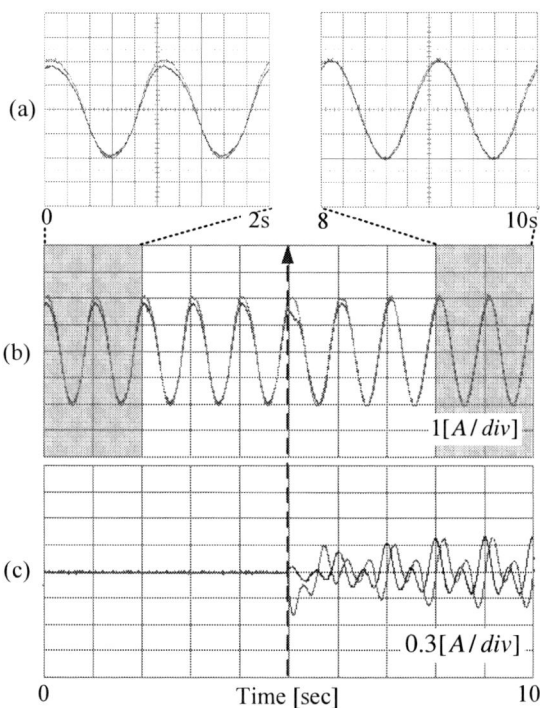

Fig. 6. Dynamic compensation results of the DC offset and the negative sequence component. (a) enlarged responses of the shaded part in (b). (b) controlled and real phase current. (c) d-q axis estimated ripple component in the synchronous reference frame.

the $\omega + 2\omega$ component [Fig. 6(c)] in the synchronous reference frame. The proposed MDO does not interfere with other higher frequency contents such as the dead-time, the motor slotting effect, as shown in Fig. 5 and 6.

To illustrate the improvement in a speed response, we compare a motor speed for the closed-loop velocity control with and without the proposed MDO. Fig. 7 and 8 show the

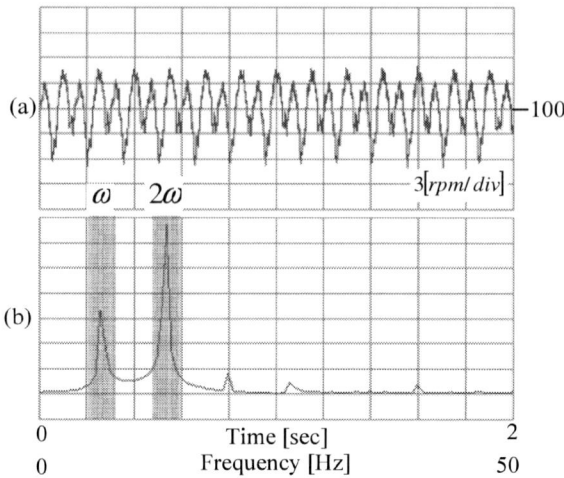

Fig. 7. Steady-state speed responses before the MDO compensation under the DC offset and negative sequence component. (a) motor speed. (b) FFT analysis of the speed.

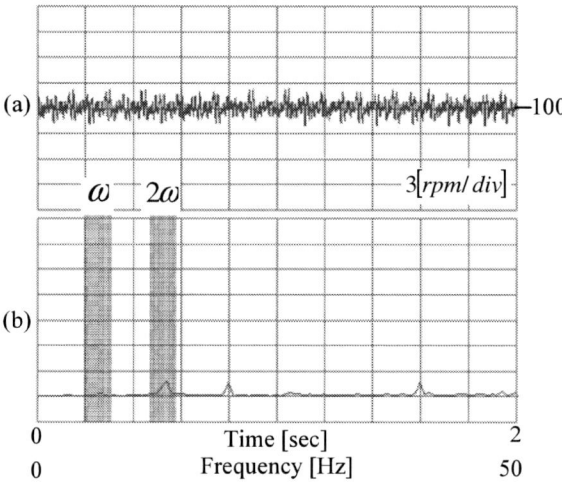

Fig. 8. Steady-state speed responses after the MDO compensation under the DC offset and negative sequence component. (a) motor speed. (b) FFT analysis of the speed.

steady-state characteristics of the motor speed before and after the compensation at 100 r/min. An intentional +4.4% offset error of the actual current is added to the respective channel of an A/D converter for actual phase current signals. The gain ratio error is given as $k_a = 1.1$ and $k_b = 0.9$. Unfortunately, before the compensation of current measurement errors, there exist undesirable ω and 2ω speed ripple as shown in Fig. 7. The reason for this is that current measurement errors act as a disturbance load for the closed-loop velocity control. In Fig. 8, after the compensation of current measurement errors, compared with responses shown in Fig. 7, they exhibit a significant reduction of the speed ripple. It is worth pointing out that the MDO compensates the torque ripple almost completely.

VI. CONCLUSIONS

In this paper, we describe an online correction approach to

compensate current measurement errors for the compensation of the torque ripple in AC motor drive systems. An MDO-based estimation is proposed as one possible solution of eliminating the DC offset and the negative sequence component of unbalanced gains in the synchronous reference frame. Based on the parameter sensitivity analysis, it is confirmed that the proposed MDO design is completely robust to the motor parameter variation by introducing zero dummy gains. From experimental results, it is concluded that the proposed method can cope well with the velocity/torque fluctuation problem in the presence of current measurement errors and motor parameter uncertainties.

ACKNOWLEDGMENT

This work was supported by KESRI (R-2005-7-067), which is funded by MOCIE (Ministry of Commerce, Industry, and Energy).

REFERENCES

[1] K. R. Cho, and J. K. Seok, "Correction on current measurement errors for accurate flux estimation of AC drives at low stator frequency," *IEEE Trans. Ind. Appl.,* to be published.

[2] D. W. Chung and S. K. Sul, "Analysis and compensation of current measurement error in vector-controlled AC motor drives," *IEEE Trans. Ind. Appl.,* vol. 34, pp.340-345, Mar./Apr. 1998.

[3] B. H. Lam, S. K. Panda, J. X. Xu and K. W. Lim, "Torque ripple minimization in PM synchronous motor using iterative learning control," in proceedings of the 1999 Industrial Electronics Society Annual Conference of the IEEE, vol. 3, pp. 1458-1463, 1999.

[4] Robert Barro and Ping Hsu, "Torque ripple compensation of induction motors under field oriented control," *IEEE APEC Conf.,* vol. 1, pp. 527-533, 1997.

[5] Weizhe Qian, S. K. Panda, and J. X. Xu, "Reduction of periodic torque ripple in PM synchronous motors using learning variable structure control," *IEEE IECON Conf.,* vol. 2, pp. 1032-1037, 2002.

[6] H. S. Jung, S. H. Hwang, J. M. Kim, C. U. Kim, and C. Choi, "Diminution of current measurement error for vector-controlled AC motor drives," *IEEE Trans. Ind Appl.,* vol. 42, pp. 1049-1256, Sep./Oct. 2006.

[7] K. R. Cho, and J. K. Seok, "Pure integration-based stator flux acquisition of AC drives at low stator frequency," *IEEE IEMDC Conf.,* pp. 1148-1153, 2007.

[8] J. W. Choi and S. K. Sul, "Inverter output voltage synthesis using novel dead time compensation," *IEEE Trans. Power Electron.,* vol. 11, pp. 221-227, Mar. 1996.

[9] N. Urasaki, T. Senjyu, T. Kinjo, T. Funabashi, and H. Sekine, "Dead-time compensation strategy for permanent magnet synchronous motor drive taking zero-current clamp and parasitic capacitance effects into account," *IEE Proc. Electr. Power Appl.,* vol. 152, no. 4, pp. 845-853, July 2005.

[10] C. H. Choi, K. R. Cho, and J. K. Seok, "Inverter nonlinearity compensation in the presence of current measurement errors and switching device parameter uncertainties," *IEEE Trans. Power Electron.,* vol. 22, No. 2, pp. 576-583, Mar. 2007.

[11] V. Vapnik, *The Nature of Statistical Learning Theory 2nd ed.* Springer, 1999.

[12] H. Drucker *et al.,* "Support vector regression machines," *Neural Information Processing Systems.* Cambridge, MA:MIT Press, vol. 9, 1997.

New Flux Weakening Control for Surface Mounted Permanent Magnet Synchronous Machine Using Gradient Descent Method

Young-Doo Yoon
Seoul National University
Department of Electrical Engineering
San 56-1, Sillim-dong, Gwanak-gu, Seoul, 151-742
Email: birdy003@eepel.snu.ac.kr

Seung-Ki Sul
Seoul National University
Department of Electrical Engineering
San 56-1, Sillim-dong, Gwanak-gu, Seoul, 151-742
Email: sulsk@plaza.snu.ac.kr

Abstract—**This paper presents a new flux weakening algorithm for a surface mounted permanent magnet synchronous machine. A current reference is modified by a controller in order to achieve a flux weakening. The current reference modification is determined by the d-q currents and the output voltage using the gradient descent method. The proposed method consists of two parts, the determination of the flux weakening region and the current reference modification. The flux weakening region is determined by the angle between the contstant torque direction and the voltage decreasing direction, and the current reference is modified by the voltage magnitude and the flux weakening direction according to the flux weakening region. This method is possible to be applied to the infinite speed drive system as well as the finite speed drive system.**

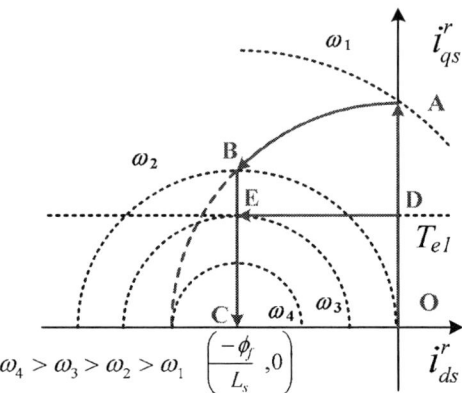

Fig. 1: Flux weakening of SMPMSM in the infinite speed drive system

I. INTRODUCTION

Due to high efficiency and high torque density, a Surface Mounted Permanent Magnet Synchronous Machine (SMPMSM) has been used in many electrical drive systems. Though the capability of the flux weakening is limited, there are many researches to extend the speed range of a SMPMSM as much as possible under the given voltage and current limitations.

Some papers have been presented about the operation region of a SMPMSM [1]-[3]. It is determined by the voltage and current limitations, especially the center of voltage limitation circle, $I_{ch} \equiv \phi_f / L_s$, which is described by the electrical machine parameter. According to the relation between I_{ch} and the maximum current, $I_{s,\max}$, a SMPMSM can be classified into a finite speed drive system and an infinite speed drive system [3]. In the case of the finite speed drive system, where I_{ch} is larger than $I_{s,\max}$, there is the available maximum speed which is restricted by the electrical parameters. On the other hand, in the case of the infinite speed drive system, where I_{ch} is smaller than $I_{s,\max}$, there is no maximum speed restricted by the electrical parameters.

The algorithms in [4]-[8] had used the steady-state voltage equations in order to determine the current reference. Because an available current reference was calculated considering the rotor speed and DC link voltage, these algorithms was sensitive to the parameter uncertainty. The transient performance was not analyzed, either. In [9]-[11], close-loop

control algorithms were proposed. These methods are robust to the parameter error because these didn't use the parameters and voltage equations. However, it is hard to set the gain of the close-loop controller which deals with voltage error. The algorithm in [12] utilized the voltage difference between the input and the output of the over-modulation block for the flux weakening. This method is effective for the maximum output torque operation since it efficiently utilized the dc-link voltage.

While many researches about the flux weakening of SPMSM have been presented, most of the researches were concerned about the finite speed drive system and the study about the infinite speed drive system is insufficient. In this paper, a novel flux weakening algorithm for a SMPMSM is proposed using the gradient descent method [13]. This method is possible to operate in infinite speed drive system as well as the finite speed drive system.

II. CURRENT TRAJECTORY OF SMPMSM

The trajectory of the d-q axes currents under the flux weakening control in the infinite speed drive system is shown in Fig. 1. In this case, I_{ch} is smaller than $I_{s,\max}$. In the low speed range, the operating points are on the Maximum Torque Per Ampere (MTPA) line ($\overline{OA}$) where the d-axis current reference is set to 0. As the speed is increasing, the stator voltage is also increasing. When the stator voltage is over the limit of the voltage from the inverter, the operating point is

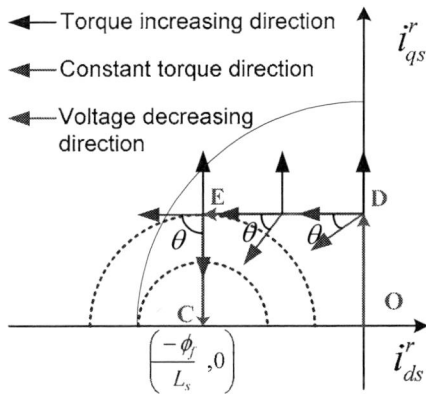

Fig. 2: Flux weakening of SMPMSM

moved along the constant torque line (Flux Weakening Region 1, FWR1) between the MTPA line and the Maximum Torque Per Voltage (MTPV) line ($\overline{BC}$). The constant torque line is parallel with i_{ds}^r-axis. In the higher speed range, the operating point is moved along the MTPV line (Flux Weakening Region 2, FWR2) which is parallel with i_{qs}^r-axis. With a given torque reference, T_{e1}, the operating point is moved from D to C via E as the speed is increasing.

On the other hand, in the case of the finite speed drive system, I_{ch} is larger than $I_{s,\max}$. Therefore, the MTPV line is not appeared in the current trajectory under the flux weakening control. After the operating point on the FWR1 meets the current limitation circle as the speed is increasing, the operating point is moved along the circle of the current limitation.

III. PROPOSED FLUX WEAKENING ALGORITHM

In the case of a SMPMSM, the direction of the current reference trajectory under the flux weakening control is simply determined according to the flux weakening region. The direction on the FWR1 is $-i_{ds}^r$ direction and the direction on the FWR2 is $-i_{qs}^r$ direction. In order to achieve the flux weakening up to the FWR2, the determination of the flux weakening region is necessary. And then, the current reference could be modified using the gradient descent method.

Therefore, the proposed method consists of two parts, the determination of the flux weakening region and the current reference modification.

A. Determination of the flux weakening region

Three directions are shown in Fig. 2. The flux weakening region is possible to be determined using the angle, θ, between the constant torque direction and the voltage decreasing direction as shown in Fig. 2.

On the FWR1, θ is less than $90°$. And, on the FWR2, θ

is $90°$. Therefore, the determination of the flux weakening region at the operating point is possible using θ or $\cos(\theta)$.

The constant torque direction is simply $-i_{ds}^r$ direction. This direction, (X_d, X_q), can be expressed as (-1,0) using the normalized vector form.

The voltage decreasing direction can be described using the gradient descent method [13]. In order to decrease the magnitude of the output voltage, the cost function can be defined as

$$J \equiv \frac{1}{2} \cdot \left[(V_{ds}^{r*})^2 + (V_{qs}^{r*})^2 \right]. \tag{1}$$

The output voltages can be simplified as (2) by neglecting the voltage drops due to the stator resistance.

$$
\begin{aligned}
V_{ds}^{r*} &\approx -L_s \cdot i_{qs}^r \cdot \omega_r, \\
V_{qs}^{r*} &\approx (L_s \cdot i_{ds}^r + \lambda_f) \cdot \omega_r.
\end{aligned} \tag{2}
$$

The direction of ∇J is the orientation in which the directional derivative has the largest value and $|\nabla J|$ is the value of that directional derivative. Since the voltage decreasing direction is required, the normalized vector, (V_d, V_q), which means the direction of the voltage decreasing can be expressed as

$$
\begin{aligned}
-\nabla J &= -\left(\partial J / \partial i_{ds}^r, \partial J / \partial i_{qs}^r \right) \\
&= \left(-L_s \cdot \omega_r \cdot V_{qs}^{r*}, L_s \cdot \omega_r \cdot V_{ds}^{r*} \right) \\
&\equiv (V_1, V_2), \\
V &\equiv \sqrt{(V_1)^2 + (V_2)^2}, \\
(V_d, V_q) &\equiv (V_1 / V, V_2 / V)
\end{aligned} \tag{3}
$$

where V means the magnitude of the vector, (V_1, V_2).

Using (X_d, X_q) and (V_d, V_q), $\cos(\theta)$ can be calculated as (4) using the inner product.

$$\cos(\theta) = (X_d, X_q) \cdot (V_d, V_q) = -V_d. \tag{4}$$

B. Current reference modification

The modification by the gradient descent method requires the direction and the magnitude of a change. The direction is derived from the torque and voltage equations of a SMPMSM. It is simply determined as mentioned above. The magnitude is derived from the output voltage error. Fig. 3 shows the block diagram of the proposed algorithm.

To limit the magnitude of the output voltage, the current reference is modified as

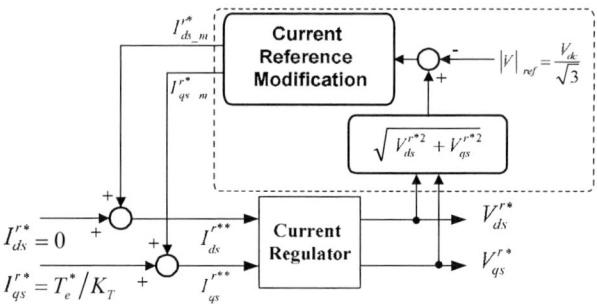

Fig. 3: Block diagram of the proposed algorithm

$$I_{ds}^{r**} = I_{ds}^{r*} + I_{ds\_m}^{r*},$$
$$I_{qs}^{r**} = I_{qs}^{r*} + I_{qs\_m}^{r*}$$

(5)

where I_{ds}^{r*} and I_{qs}^{r*} are the current references directly generated from T_e^*. Usually, I_{ds}^{r*} is set to 0. And, $I_{ds\_m}^{r*}$ and $I_{qs\_m}^{r*}$ are the current reference modification terms, and I_{ds}^{r**} and I_{qs}^{r**} are the modified current references as shown in Fig. 3.

Using the gradient descent method, $I_{ds\_m}^{r*}$ and $I_{qs\_m}^{r*}$ can be calculated as

$$\partial I_{ds\_m}^{r*}/\partial t = M_d,$$
$$\partial I_{qs\_m}^{r*}/\partial t = M_q$$

(6)

where M_d and M_q can be calculated by the direction of current reference modification and the magnitude of output voltage error according to the flux weakening region.

In order to limit the magnitude of the output voltage up to the desired magnitude of the output voltage, the error of output voltage magnitude can be expressed as

$$\Delta V = V_{mag} - V_{ref} = V_{mag} - V_{dc}/\sqrt{3},$$
$$V_{mag} = \sqrt{\left(V_{ds}^{r*}\right)^2 + \left(V_{qs}^{r*}\right)^2}.$$

(7)

where V_{ref} means the desired magnitude of the output voltage. Usually, V_{ref} can be set to $V_{dc}/\sqrt{3}$. And, V_{mag} means the magnitude of the output voltage.

On the FWR1, the current reference is modified along the constant torque direction, $-i_{ds}^r$ direction. On the FWR2, the current reference is modified along the MTPV direction, $-i_{qs}^r$ direction. Therefore, using ΔV, M_d and M_q can be calculated as follows.

TABLE 1
Nominal Machine Parameter

Parameter	Value
Rated power	11 [kW]
Rated speed	1500 [r/min]
Stator resistance	58.6 [mΩ]
Stator inductance	1.8 [mH]
Magnet flux linkage	0.1492 [V·s]
Pole pairs	4
DC link voltage	300 [V]
Rated current	58.6 [A$_{rms}$]
Maximum current	120 [A$_{peak}$]

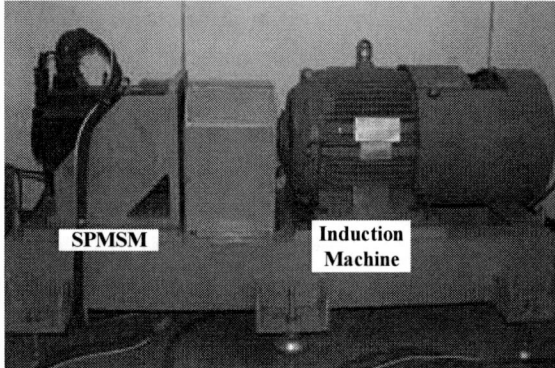

Fig. 4: Experimental setup

$$M_d = \begin{cases} -\alpha_1 \cdot \Delta V, & on\ the\ FWR1 \\ 0, & on\ the\ FWR2 \end{cases},$$
$$M_q = \begin{cases} 0, & on\ the\ FWR1 \\ -\alpha_2 \cdot \Delta V, & on\ the\ FWR2 \end{cases}$$

(8)

where α_1 and α_2 are the reference modification gains.

When the output voltage is not saturated, this algorithm should not be operated in order to prohibit the malfunction of the proposed method. Therefore, if $I_{ds\_m}^{r*}$ is larger than 0, $I_{ds\_m}^{r*}$ should be set to 0.

IV. EXPERIMENTAL RESULTS

Experiments have been performed with a SMPMSM. The nominal SMPMSM parameters of the tested motor are listed in the Table. I. Fig. 4 shows the experimental setup. 11 [kW] Induction Machine (IM) and the SMPMSM are directly coupled. The IM and SMPMSM were driven in the speed and torque control mode, respectively.

The tested machine is a general purpose SMPMSM. In order to verify the feasibility of the proposed method up to FWR2, V_{ref} of these experiments is set to 50 [V] though V_{ref} can generally be set to $V_{dc}/\sqrt{3}$. And the maximum peak current of the SMPMSM is set to 120 [A] to operate the

978-1-4244-1871-8/07 $25.00 © 2007 IEEE

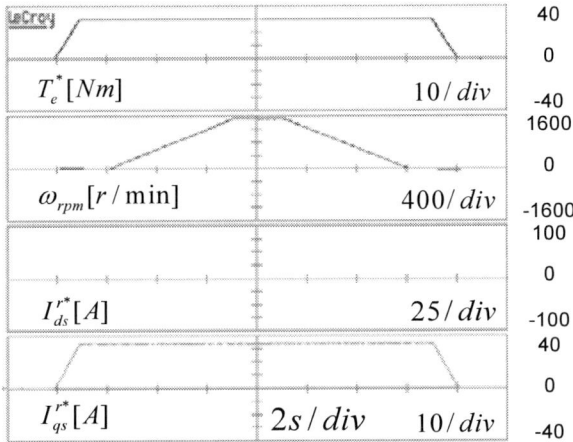

Fig. 5: Experimental conditions

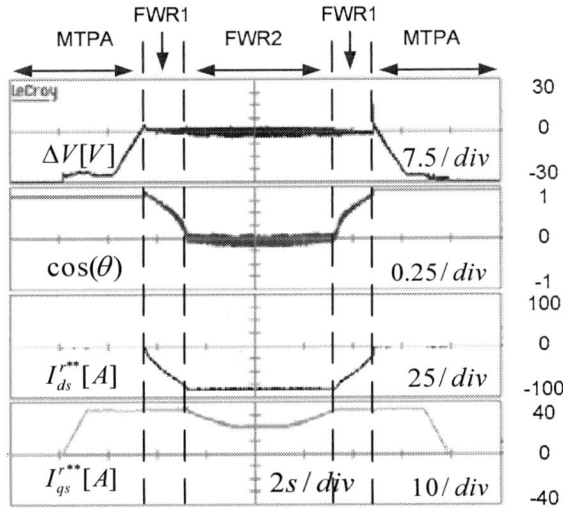

Fig. 6: Operation of the proposed algorithm

machine up to the FWR2. If not, the IM and the SMPMSM would be operated over the maximum speed to verify the proposed method on the FWR2.

Fig. 5 shows an example of the desired T_e^* and ω_{rpm}. T_e^* is set to 30 [Nm]. According to T_e^*, I_{qs}^{r*} are generated as T_e^*/K_T and I_{ds}^{r*} is set to 0. ω_{rpm} is increasing and decreasing from 0 [r/min] to 1500 [r/min] and vice versa. Although ω_{rpm} is increasing, I_{ds}^{r*} and I_{qs}^{r*} are not changing because they are independent of ω_{rpm} and the flux weakening operation is not considered.

Fig. 6 shows the operation of the proposed flux weakening algorithm in the same experimental condition as Fig. 5. While the operating point is on the MTPA line, ΔV is less than 0. The output voltage is not saturated. During ω_{rpm} is increasing, ΔV is increasing. When ΔV becomes 0, the operating point of the SMPMSM is changed from MTPA to FWR1.

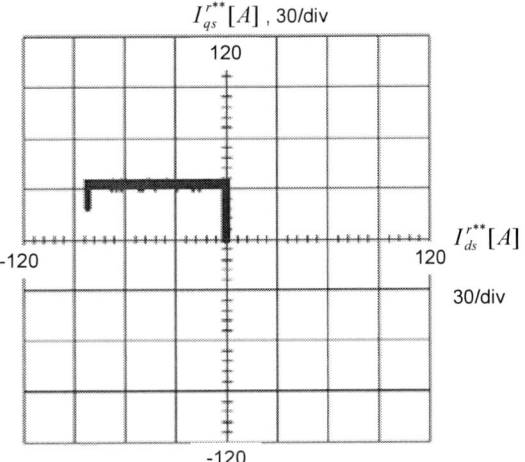

Fig. 7: Current trajectory in the x-y plot

On the FWR1, ΔV is limited to 0 and $\cos(\theta)$ is larger than 0. During ω_{rpm} is increasing, $\cos(\theta)$ is decreasing. The current reference is modified along to the constant torque line to achieve the flux weakening. I_{ds}^{r**} is decreasing and I_{qs}^{r**} is maintained according to the given torque reference. When $\cos(\theta)$ becomes 0, the operating point is changed from the FWR1 to the FWR2.

On the FWR2, $\cos(\theta)$ is limited to 0. During ω_{rpm} is increasing, the current reference is modified along to the MTPV line to achieve the flux weakening. I_{ds}^{r**} is maintained and I_{qs}^{r**} is decreasing.

During ω_{rpm} is decreasing, the proposed method is operated in the reverse order. The operating point is changed from FWR2 to MTPA via FWR1.

Fig. 7 shows the trajectory of the modified current reference, I_{ds}^{r**} and I_{qs}^{r**} in the x-y plot. As shown in Fig. 6 and Fig. 7, the determination of the flux weakening region is possible using $\cos(\theta)$. And I_{ds}^{r**} and I_{qs}^{r**} are modified well from I_{ds}^{r*} and I_{qs}^{r*} up to FWR2 by the proposed algorithm.

V. CONCLUSION

In this paper, a new flux weakening algorithm for a SMPMSM has been proposed. Using the gradient descent method, it modifies the current reference up to the flux weakening region 2. Using the angle between the constant torque direction and the voltage decreasing direction, the determination of the flux weakening region is possible. The current reference is modified using the magnitude of the output voltage and the flux weakening direction according to the flux weakening region. This method could be applied to the infinite speed drive system as well as to the finite speed drive system. The feasibility of the proposed algorithm was verified with the experimental results.

REFERENCES

[1] R. F. Schiferl and T. A. Lipo, "Power capability of salient pole permanent magnet synchronous motors in variable speed drive applications," *IEEE Trans. Ind. Applicat.*, vol. 26, pp. 115-123, Jan./Feb., 1990.

[2] A. K. Adnanes, "Torque analysis of permanent magnet synchronous motors," in *Conf. Rec. IEEE-PESC*, 1991, pp. 695-701.

[3] W. L. Soong and T. J. E. Miller, "Field-weakening performance of brushless synchronous AC motor drives," *IEE Proc.-B*, vol. 141, pp. 331-340, Nov., 1994.

[4] B. Sneyers, D. W. Novotny, and T. A. Lipo, "Field weakening in buried permanent magnet AC motor drives," *IEEE Trans. Ind. Applicat.*, vol. IA-21, pp. 398-407, Mar./Apr., 1985.

[5] S. Morimoto, Y. Takeda, T. Hirasa and K. Taniguchi, "Expansion of operating limits for permanent magnet motor by current vector control considering inverter capacity," *IEEE Trans. Ind. Applicat.*, vol. 26, pp. 866-871, Sept./Oct., 1990.

[6] S. Morimoto, M. Sanada and Y. Takeda, "Wide-speed operation of interior permanent magnet synchronous motors with high-performance current regulator," *IEEE Trans. Ind. Applicat.*, vol. 30, pp. 920-926, July/Aug., 1994.

[7] J. –H. Song and S. –K. Sul, "Torque maximizing control of permanent magnet synchronous motor under voltage and current limitations of PWM inverter," in *Conf. Rec. IEEE-APEC*, vol. 2, 1996, pp. 758-763.

[8] J. –M. Kim, K. Park, S. –J. Kang, S. –K. Sul, and J. –L. Kwon, "Improved dynamic performance of interior permanent magnet synchronous motor drive in flux-weakening operation," in *Conf. Rec. IEEE-PESC*, vol. 2, 1996, pp. 1562-1567.

[9] J. –H. Song, J. –M. Kim, and S. –K. Sul, "A new robust SPMSM control to parameter variations in flux weakening region," in *Proc. IEEE-IECON*, vol. 2, 1996, pp. 1193-1198.

[10] D. S. Maric, S. Hiti, C. C. Stancu, and J. M. Nagashima, "Two improved flux weakening schemes for surface mounted permanent magnet synchronous machine drives employing space vector modulation," in *Proc. IEEE-IECON*, vol. 1, 1998, pp. 508-512.

[11] J. –M. Kim and S. –K. Sul, "Speed control of interior permanent magnet synchronous motor drive for flux weakening operation," *IEEE Trans. Ind. Applicat.*, vol. 33, pp. 43-48, Jan./Feb., 1997.

[12] T. Kwon and S. Sul, "Novel Antiwindup of a Current Regulator of a Surface-Mounted Permanent-Magnet Motor for Flux-Weakening Control," *IEEE Trans. Ind. Applicat.*, Vol. 42, No. 5, pp. 1293-1300, September/October 2006.

[13] P. A. Ioannou and J. Sun, *Robust Adaptive Control*, Upper Saddle River, NJ: Prentice-Hall, 1996, pp. 785-786.

HVDC Control Development for Isolated Small AC System
with Real Time Digital Simulator

Teruo Yoshino
Toshiba Mitsubishi-Electric Industrial Systems Corporation
Power Electronics Systems Division
1, Toshiba-cho, Fuchu, Tokyo, Japan 183-8511
Email: YOSHINO.teruo@tmeic.co.jp

Jin Lee, Chul Kyun Lee, Sung Woo Park
LS Industrial Systems Co., Ltd.
Electrotechnology R&D Center, 1, Songjeong-dong,
Heungdeok-gu, Cheongju-si, Chungcheongbuk-do 361-720
Email: jinlee@ LSIS.BIZ, ckleea@LSIS.BIZ, swpark1@LSIS.BIZ

Abstract—This paper reports the control strategy and its simulator studies for HVDC transmission system feeding power to an isolated weak AC network. Authors investigated the control strategy in which the DC current control is done at the receiving-end. The combination of the receiving-end current control and the weak AC system is one of fields not yet studied in detail. The authors proposed a prototype control and evaluated it through studies with the real time digital simulator. The prototype control showed instabilities in dynamic simulation cases. Then, authors discussed and investigated the instability through the simulation and found the causes of the instability. The first point to be improved was the coordination between the current order and the actual current. The second point was the gamma control considering the large voltage phase shift at the power change in the weak AC system. The improved control was installed again in the real time digital simulator and finally verified to show good performances at step response and at the recovery from AC faults.

I. INTRODUCTION

An international team performed this study. The Japanese member proposed a prototype control for the HVDC for an isolated small AC system. The Korean members investigated its stability with RTDS™ (Real Time Digital Simulator, the trademark of RTDS Inc.)) and showed their expertise to improve the control strategy.

A. Purpose and process of the study

In most of HVDC transmission systems in the world, the sending-end controls the DC current and controls the DC power. However, the isolated small AC system usually receives the DC power and, at the same time, requires the frequency control by the DC power. With the conventional control, the power demand is sent from the receiving-end to the sending-end via fast communication systems. In this paper, the authors propose an alternative control strategy without the fast communication systems. The proposed strategy is hoped to enhance the HVDC performance connected to the weak AC system even where reliable communication system is difficult to be realized. The proposed strategy is tested, investigated and improved through static and dynamic simulation cases with the RTDS.

B. Technical background

The HVDC performance connected to the weak AC system has been investigated by CIGRE and IEEE working groups and others in decades from 1980 to 1990 [1] [2] [3] [4] [5]. However, the all investigations were for the conventional HVDC control scheme where the sending-end controls the DC current and the receiving-end controls the extinction angle, Gamma. Authors studied rather new field of the HVDC performance where the receiving-end has ACR (Automatic Current Regulation) and controls the DC power to the weak AC system.

A reference [6] suggested the control concept without the fast communication link although it studied the multi-terminal HVDC system. The reference also implied the control concept of the receiving-end ACR in degraded conditions. However it did not deals with the connection to the weak AC system. The reference inspired the authors to apply the control to the two terminal HVDC systems. Then, the authors started to study the combination of the receiving-end ACR and the weak AC network.

II. HVDC SYSTEM UNDER STUDY

The isolated small AC system is assumed at the receiving-end to have the SCR (Short Circuit Ratio) less than 2. The AC system is modeled with the infinite bus and the equivalent impedance. The HVDC converter stations are equipped with the harmonics filters and the Var compensation devices. Furthermore, at the receiving-end station, a TCR type SVC is also connected. The circuits are shown in Fig. 1. The major parameters are listed in Table 1.

The per-unit values are based on the HVDC capacity except for impedances of the transformer and the DC reactor. The impedance of the transformer is based on its capacity.

TABLE 1
Major Parameters Of The Circuit Model

Items	Sending-end	Receiving-end
AC system	0.0263 + j0.1861 (pu)	0.1536 + j0.6468(pu)
Filters	6 banks 0.06875 (pu)/bank	6 banks 0.06875 (pu)/bank
Var compensation devices	Capacitor 3 banks 0.06875 (pu)/bank Reactor 1 bank 0.06875 (pu)/bank	Capacitor 2 banks 0.06875 (pu)/bank
SVC	-	+/-0.15 (pu)
Transformer	Capacity 0.625(pu)/pole Impedande 0.13(pu)	Capacity 0.625(pu)/pole Impedande 0.13(pu)
DC reactor	0.2262 (pu)	0.2262 (pu)

978-1-4244-1871-8/07 $25.00 © 2007 IEEE

The impedance of the DC reactor is defined by equation (1) in this paper.

$$x(pu) = \frac{\omega L}{Vdc/Idc} \qquad (1)$$

where, ω is the angular frequency of the AC system, Vdc is the rated DC voltage and the Idc is the rated DC current.

III. PROTOTYPE CONTROL BLOCKS

The prototype of the proposed control is shown in Fig. 2. One of the features is the proportional feedback control. Actually, the blocks are made from the first order delay functions or the lead-lag functions. The second is the theoretical extinction angle control as shown in Fig. 2 (b). In Japan, this type of HVDC control has been successfully applied since this control offers fast and stable characteristics. One of references is found in [7].

The significant difference of the proposed control from the conventional control is that the receiving-end controls the DC current. The sending-end controls the DC voltage instead. With this configuration, the receiving-end can control the DC power without any communication system.

The sending-end control model in Fig.2 (a) consists of the DC voltage regulator AVR (Automatic Voltage Regulator), the DC current regulator ACR and voltage dependent current order limiter VDCOL (Voltage Dependent Current Limiter). The AVR keeps the DC voltage constant. The ACR limits the DC current when faults happen in the DC circuit or in the receiving-end AC system while the VDCOL reduces the current order according to the pre-determined pattern.

The receiving-end control model in Fig.2 (b) consists of the AVR, the ACR and the AGR (Automatic Gamma Regulator). In the receiving-end, the ACR is usually selected and keeps the DC current as ordered. The AGR advances the control angle when the extinction angle becomes lower than the order value to avoid the commutation failure. In this model, the open-loop type AGR was applied in the prototype. In order to keep the extinction angle to be the order, the open-loop AGR calculates the control angle from the DC current, the AC voltage and the transformer impedance by the theoretical equation.

The DC power control APR (Automatic Power Regulator) also consists of the proportional type function block, which

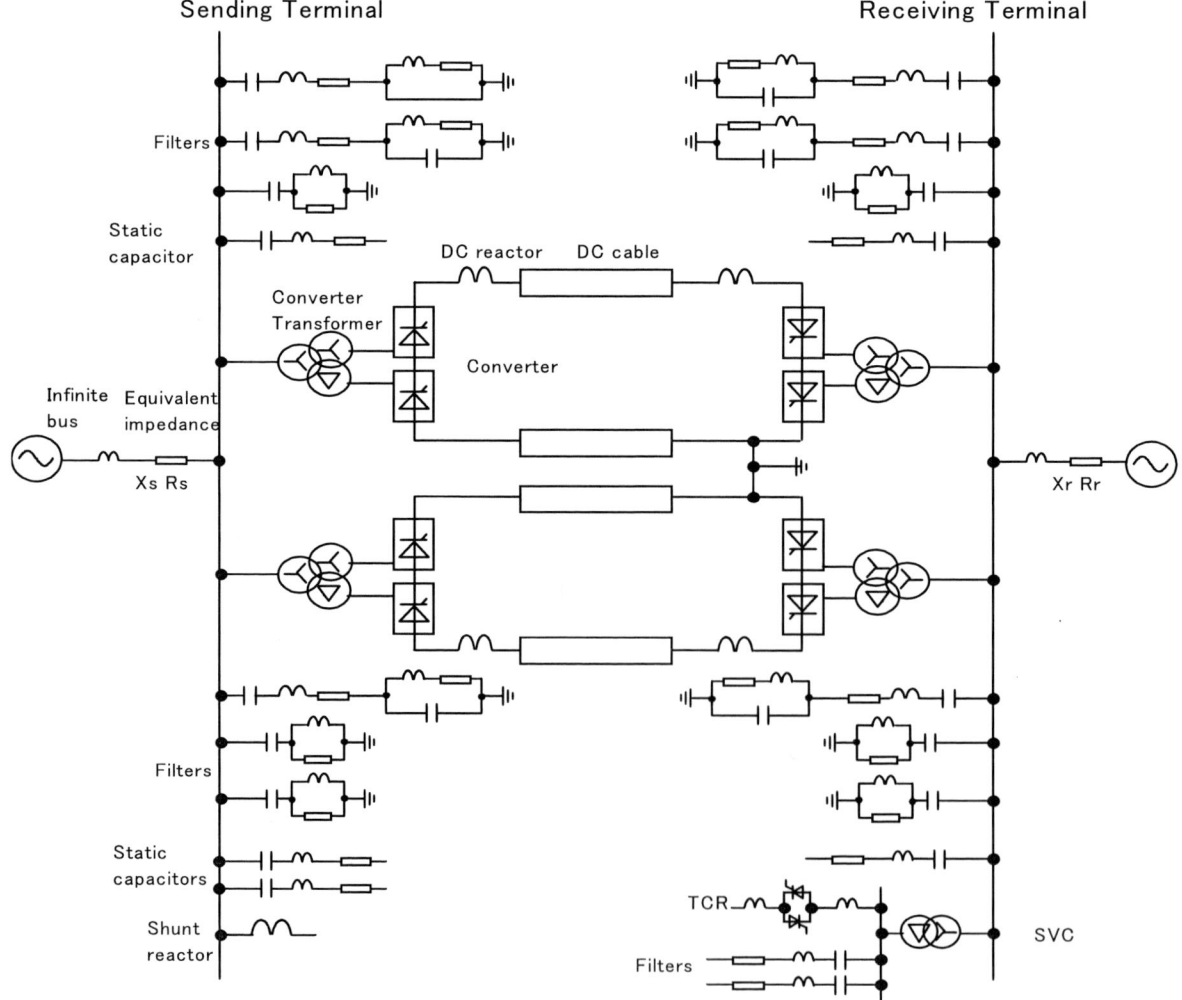

Fig.1: HVDC main circuit model

amplifies the error signal between the power order and the actual power and generate the DC current order to the ACR. This APR compensates only small error since it is based on the assumption that usually the DC voltage is kept at the rated value.

Because the control functions are based on proportional control, the control blocks are expected to show rather faster

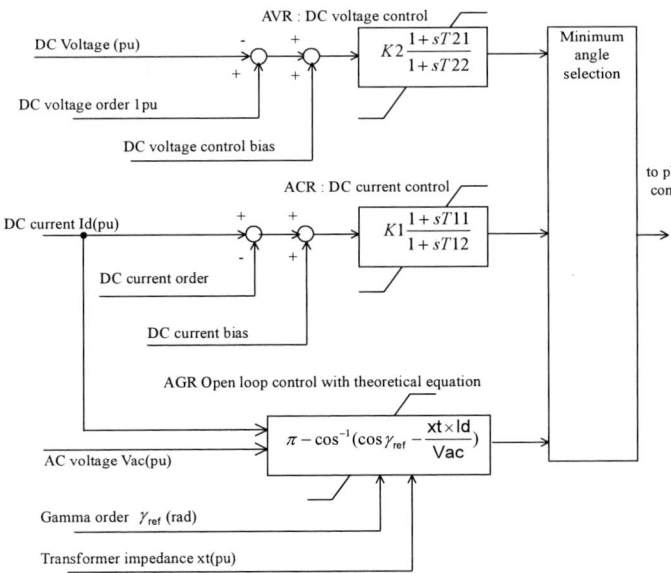

(a)   Sending-end control block

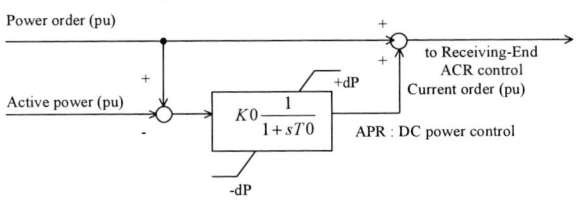

(b)   Receiving-end control block

(c)   DC power control block
Fig. 2 Prototype control blocks

and more damped response than the blocks including the integral function blocks.

IV.   SIMULATOR STUDY WITH PROTOTYPE CONTROL

The prototype control is installed in the RTDS processor cards through RSCAD program. First, the steady state performance is investigated to determine the Var control strategy including the SVC at the receiving-end. With the proposed control, the HVDC is confirmed to send the DC power up to its rated capacity in stable manner.

In addition to the steady state stability, the HVDC is required the capability for fast power increase. The capability is necessary to compensate the power unbalance due to the generator trip for example. In order to recover the balance between load demand and generation, the HVDC should response fast.

To confirm the capability, the power order step changes are performed. However, in very large step case, from 0.5pu to 1pu, the instability is found to appear as shown in Fig. 3(a). The AC voltage decreases deep and a commutation failure takes place when the DC current reaches around 1pu.

V.   DISCUSSION ON THE INSTABILITY

From the detailed waveforms including the control angles, voltage and current, the causes of the instability are analyzed by drawing the operation point locus from Fig. 3 as shown in Fig. 4.

A.   Operation point locus analysis

The HVDC operation point is considered to move from point (1) to (5) as follows;

(1)   Initial steady state before change... The DC voltage is 1pu and the DC current is 0.5pu.

(2)   Control angle advance ... The control error between the current order and the actual current drives the ACR to advance the control angle.

(3)   AC voltage decrease ... The control angle advance results in smaller power factor. Then, the reactive power from the converter increases and the receiving-end AC voltage decreases. The AC voltage continuously decreases along with the increase of the DC current since the reactive power is in proportion to the DC current.

(4)   AGR control selected ... The actual current reaches at 1pu. The current control error becomes small and the ACR retards the control angle. However, at this point, the AGR takes the control since the extinction angle decreases due to the AC voltage decrease.

(5)   Positive feedback ... Once the AGR is selected, the receiving-end DC voltage decreases because the AGR advances the control angle. The voltage difference between the sending-end and receiving-end increases and, thus, the DC current increases. The behaivior makes the AGR further

978-1-4244-1871-8/07 $25.00 © 2007 IEEE   1213

advances the control angle since the extinction angle decreases when the DC current increases. During this positive feedback process, the DC voltage collapses and the DC current increases steeply. Finally, the VDCOL and the ACR at the sending-end activate to limit the DC current.

B. Control angle exessive advance

From the operation point locus analysis, the rapid and continuous control angle advance is considered to be the first cause of the instability. Then, the root cause of the control angle advance is discussed.

The HVDC main circuit current response is represented by

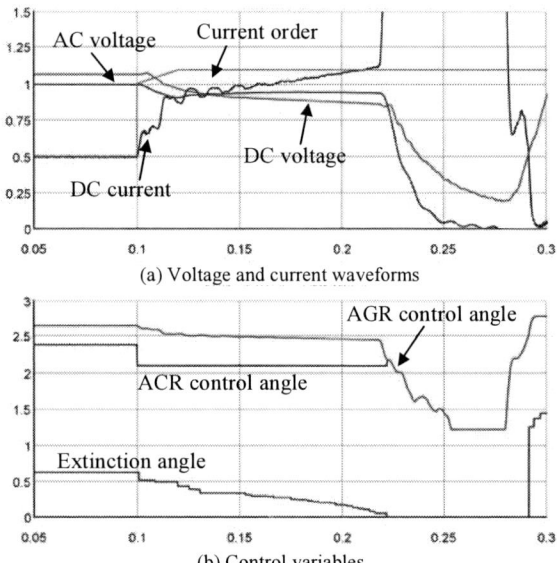

(a) Voltage and current waveforms

(b) Control variables

Fig. 3 Fast power step change with the proto-type control

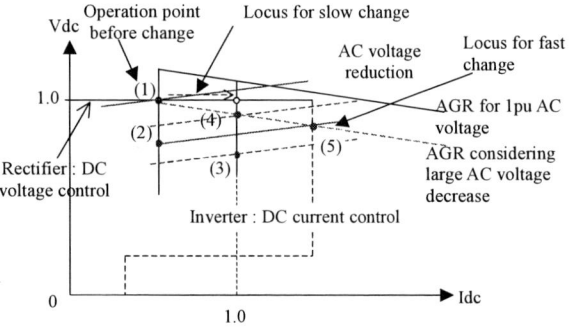

Fig. 4 Operation point locus of the step change

the simlified circuit as shown in Fig. 5. The DC current is determined by the voltage difference between two terminals and the resistance in the circuit including the equivalent resistance corresponding to the converter commutation[7]. In addition to the resistance, in the circuit, the DC reactor and another inductances limit the rate of the current increase. From Fig. 5, the current response to the step voltage change is expressed with a first order delay function.

The equation for the dc current can be written with Laplace expression as follows according to the equivalent circuit shown in Fig. 5.

$$(R + sL)I_d = V_{r0} - V_{i0} \tag{2}$$

where, $R = R_{eqr} + R_{tl} + R_{eqi}$, $L = 2L_{dc}$. V_{r0} and V_{i0} are the sending-end (rectifier) and receiving-end (inverter) no-load dc voltages, respectively.

From the equation (2), the dc current is given as follows;

$$I_d = \frac{1}{R}\frac{1}{1 + sT_{dc}}\left(V_{r0} - V_{i0}\right) \tag{3}$$

where, $T_{dc} = \dfrac{L}{R}$.

From the DC current waveform in Fig. 3, the time constant T_{dc} is estimated around 20 to 30ms in this simulation model. On the other hand, the current order changes stepwise to 1pu. Then, the current error continues to be large and the ACR continues to advance the control angle at its lower limit.

In order to avoid these phenomena, the current order or the power order should be slower than the main circuit current response time constant.

C. Fast voltage phase change

From Fig.3, the AGR advances the control angle when the extinction angle reaches almost zero. This is the secondary point to be discussed. In order to investigate these phenomena, the phase change between the initial state and the final state is calculated as shown in Table 2. The calculation shows that the phase changes wide, almost

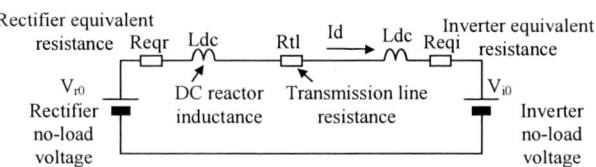

Fig. 5 Equivalent circuit for HVDC

TABLE 2

Phase Change Calculation

Item	DC power 0.5pu	DC power 1pu
Equivalent reactance	0.6468pu based on the HVDC capacity	
AC voltage at the station	1pu	0.85pu
Phase displacement	$\sin\delta = \dfrac{0.5 \times 0.6468}{1 \times 1} = 0.3234$ $\delta = 18.9\deg(0.330rad)$	$\sin\delta' = \dfrac{1 \times 0.6468}{1 \times 0.85} = 0.761$ $\delta' = 49.5\deg(0.864rad)$

Note : $P = \dfrac{Vs \bullet Vl}{X}\sin\delta$, $\sin\delta = \dfrac{P \bullet X}{Vs \bullet Vl}$ where X is the equivalent reactance, Vs is the infinite bus voltage and Vl is the converter station voltage.

30degree, in very short time.

The phase detection and synchronization function in the controller usually tuned assuming moderate phase change. Then, in this case, the relative phase error between the actual voltage phase and the phase signal in the controller becomes large. This means that the actual firing angle of the converter delays than the open-loop AGR calculation. The firing angle delay results in the extinction angle shortage.

VI.   SIMULATOR STUDY WITH IMPROVED CONTROL

A.   Control block modification

Based on the analysis and discussions of the instability causes, the prototype control is modified considering the DC current response, the extinction angle control at fast phase change.

All the modifications are made only in the receiving-end control blocks as shown in Fig. 6. A first order delay function is added to the DC power control block to make the power order and the current order slower than the main circuit response. With this delay function, the DC current is expected to follow the current order with small error and the

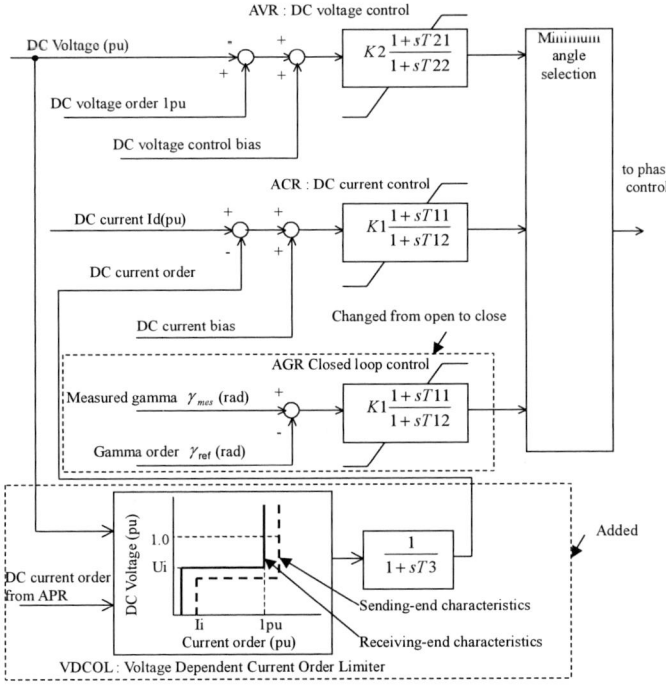

(a)   Modified receiving-end control block

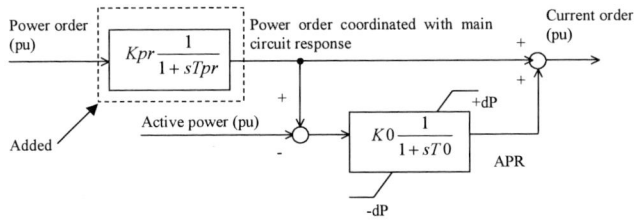

(b)   Modified DC power control block

Fig. 6 Modified control blocks

ACR will not advance the control angle so much. The AGR function is realized by the feedback type control. This type of AGR has been also used in actual fields [8].

The VDCOL function is added to the receiving-end control also considering the recovery performance after the AC system faults. The characteristics of the receiving-end VDCOL are coordinated with that of the sending-end. The current order signal output from the receiving-end VDCOL is always smaller than that of the sending-end.

B.   Simulator study with modified control block

The improved control is then installed in the RTDS and the step response from 0.5pu to 1pu is performed and shows stable behavior. The DC power increased smoothly without overshoot as shown in Fig.7. The power increase is done within around 0.2 sec.

In Fig. 7, the current order and the DC current are seen almost overlapped and the current error is kept small. The control angle of the ACR shows only small advance and fully controls the DC current. The receiving-end AC voltage decrease is suppressed because of the small control angle advance and small power factor decrease. After the DC current reaches 1pu, the AGR take the control in place of the ACR. However, since the AC voltage reduction is not so large, the receiving-end DC voltage is kept high enough to avoid the positive feedback. The operation locus of this stable operation is summarized in Fig. 8 for comparison with Fig. 4.

In Fig.3 (b), the fed-back extinction angle and the calculated extinction angle are plotted for comparison. As clearly seen, the calculation value shows large error at the transient period while the DC power is increasing and the voltage phase angle is changing.

After the power step change test, the AC system disturbance test is also performed. After another investigation and improvement, the VDCOL is added in the receiving-end control as already shown in Fig. 6(a). With the modified control, the HVDC showed successful and fast recovery after the AC system disturbance as shown in Fig 9.

VII.   CONCLUSIONS

The control strategy was proposed to realize the stable and robust operation of the HVDC independent of fast communication systems between two terminals. The prototype control was installed in the RTDS and tested. Based on the test results, the control strategy was improved for better performance. The improved control showed stable performances in step response tests and AC system disturbance tests.

In process of the development, the followings were found to be the key factors for success in the receiving-end current control strategy feeding to the weak AC system.

-   The current and current order is better kept almost

equal for realizing high power factor operation in the receiving-end converter.

- The extinction angle control should be designed considering the large phase changes expected at large DC power change in the weak AC system.

We hope that this control strategy will be applied in an actual HVDC system to enhance the system stability and operation reliability.

ACKNOWLEDGMENT

The authors thank to colleagues to give us good opportunity to perform these simulator studies. Through the works of the international team, we think that we could make small contribution to the technology exchange between Korea and Japan in the field of power electronics for high voltage and high capacity utility application.

REFERENCES

[1] A. Gavlilovic et.al., "INTERACTION BETWEEN AC AND DC SYTEMS", CIGRE Symposium 09-87, 200-20 (1987)

[2] C. Hatziadoniu, G.D. Galanos, "INTERACTION BETWEEN THE AC VOLTAGE AND DC CURRENT IN WEAK AC/DC INTERCONNECTIONS", IEEE Trans. on PD, Vol. 3, No. 4, Oct. (1988), pp.2092-2102.

[3] B. Franken, G. Andersson, "ANALYSIS OF HVDC CONVERTERS CONNECTED TO WEAK AC SYSTEMS", IEEE Trans. on PS, Vol.5, No.1, Feb. (1990), pp. 235-242.

[4] CIGRE WG B14.07, IEEE WG15.05.05, "Guide For Planning DC Links Terminating AC System Locations Having Low Short Circuit Capacities Part I AD/DC System Interaction Phenomena", ELECTRA No. 142 (1992) pp.119-123.

[5] CIGRE WG B14.07, IEEE WG15.05.05, "Guide For Planning DC Links Terminating AC System Locations Having Low Short Circuit Capacities Part II Planning Guidlines", ELECTRA No. 173 (1997) pp.149-152.

[6] T. Sakurai, T. Sakai, et.al, "A NEW CONTROL METHOD FOR MULTITERMINAL HVDC TRANSMISSION WITHOUT FAST COMMUNICATION SYSTEMS", IEEE Trans on PAS, Vol. PAS-102, No.5, May (1983), pp.1140-1150

[7] K. Yanagihashi, J. Arai, T. Yoshino, T. Komukai, "Consideration on stability of current control with constant extinction angle control in HVDC system," (in Japanese) T. IEE Japan, Vol. 115-B, No. 6 (1995), pp.610 -616

[8] EPRI, "HVDC HANDBOOK, Chapter 5 HVDC SYSTEM OPERATION", (1994)

[9] T. Yoshino, M. Sato et.al: "HVDC converter control for fast power recovery after AC system fault", IEEE Trans. on PD, Vol.12 (1997) pp.1955-1950

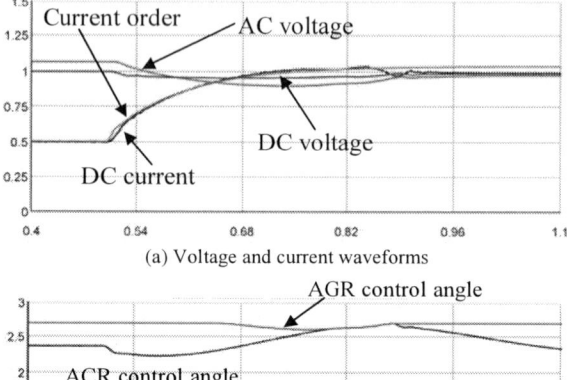

(a) Voltage and current waveforms

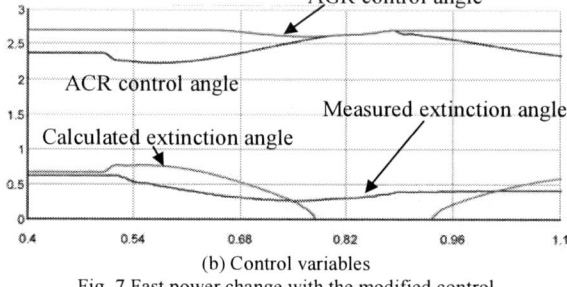

(b) Control variables

Fig. 7 Fast power change with the modified control

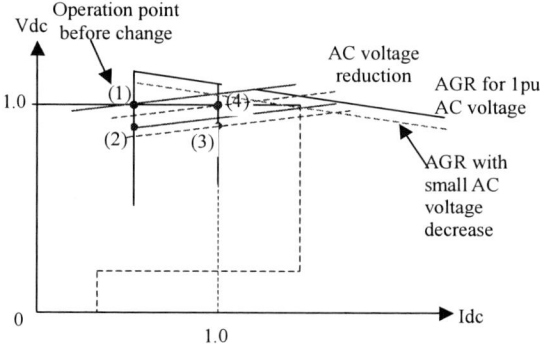

Fig. 8 Operation point locus of the stable response

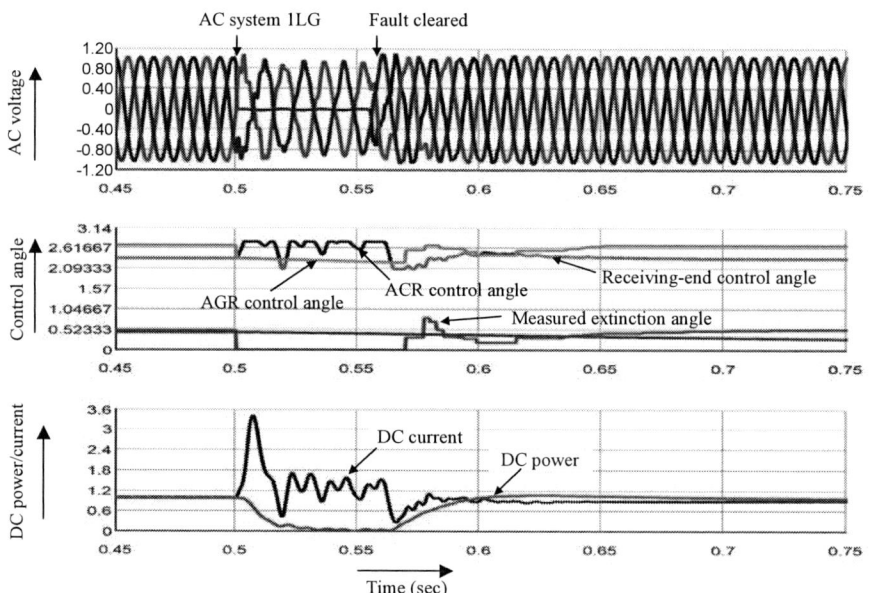

Fig. 9 Recovery from receiving-end 1LG fault

978-1-4244-1871-8/07 $25.00 © 2007 IEEE   1216

Using a Cascaded H-Bridge STATCOM for Rebalancing Unbalanced Voltages

R.E. Betz
School of Electrical Engineering and Computer Science
University of Newcastle, Australia, 2308.
Email: Robert.Betz@newcastle.edu.au

T.J. Summers
School of Electrical Engineering and Computer Science
University of Newcastle, Australia, 2308.
Email: Terry.Summers@newcastle.edu.au

Abstract—This paper considers specific issues related to using the cascade (also known as the H-bridge) multilevel STATCOM with unbalanced voltages and currents at the STATCOM-AC line connection terminals. Under this condition the average power into the individual phase legs of the converter, in general, is not zero, although the total three phase power is zero. The resultant phase cluster power unbalance can result in increasing or decreasing capacitor voltages within a phase leg. This paper will present a technique for zeroing the phase cluster average power under unbalanced line conditions when the control strategy is to rebalance unbalanced voltages.

I. Introduction

The Cascaded H-bridge multilevel Converter (CHC) is currently being studied as a candidate for use in Static Compensator (STATCOM) applications. The motivation for this study is due to some unique advantages over other multilevel converter topologies, namely:

- The number of components in CHC scales linearly with the number of converter levels, whereas other converter types scale quadratically [1].
- Capacitor voltage balancing can be solved for this converter, even with high numbers of output levels [2].
- The individual modules of the CHC are identical and can be made completely modular making construction and future maintenance simple. This feature also makes redundancy simple to implement.
- The Var rating of a CHC scales linearly with increased numbers of modules of the same voltage and current rating.
- The high level number attainable with the CHC means that direct grid connection is possible up to medium voltage levels. This eliminates the traditional coupling transformer from the STATCOM.
- The CHC is ideally suited to the STATCOM application where it does not have to handle real power.

One of the unique features of the CHC is that the capacitor energy storage is isolated to each phase cluster. Compare this to a standard three phase inverter structure, or a neutral point clamped or flying capacitor multilevel structure, where the main capacitor energy storage is common to the three phases. In all STATCOM Var compensation applications (regardless of the topology) the sum of the STATCOM input powers for the three phases are instantaneously zero when operating under balanced conditions. Therefore, under ideal conditions, the net power flowing into a common capacitor present in the standard topologies is zero. This has the additional benefit of low current stress on the common capacitor.

If a CHC based STATCOM is used for Var compensation control, where the line voltages are balanced and the STATCOM is only supplying/absorbing positive sequence currents to/from the AC line, the phase average real power flowing into each individual phase leg of the STATCOM is zero. Under this condition (ideally) the sum of the capacitor voltages in each individual phase leg will neither increase or decrease. However, under practical conditions, there are differences in losses and other non-ideal conditions between the phase legs which can lead to non-zero average power flowing into the phase legs. To compensate for this non-ideality explicit control must be applied to ensure not only individual H-bridge capacitor voltage balancing, but also balancing of each of the phase leg capacitor cluster voltages [3].

When the voltages at the grid-STATCOM connection are unbalanced, or when the STATCOM is supplying negative sequence currents for an unbalanced load, the average real power in the individual phase legs can be non-zero, even though the total three phase average phase leg power is still zero. The average power difference between the phase legs can be significant, resulting in the capacitor cluster voltages either rising or falling in the phase legs depending on whether the leg is absorbing or supplying average real power. Clearly if this condition is left then the voltage rise on the leg could lead to catastrophic failure.

The concept of using STATCOMs in unbalanced situations is well established [4]–[11]. However the issue of phase leg power balancing in CHCs has not been extensively addressed in the literature. Most previously published papers addressed rebalancing control issues for other converter types.

The emphasis in this paper is on the control of the phase leg average powers when operating with a supply voltage rebalancing control strategy. The paper will develop a novel algorithm based on using zero sequence injection to control the average real power in the phase legs of a CHC. The technique can also be used under balanced conditions to compensate for the previously mentioned loss differences.

II. Mathematical Preliminaries

Prior to developing the algorithm, we shall investigate the influence of unbalance on the operation of a STATCOM. This material provides the context for the development of the subsequent algorithms.

978-1-4244-1871-8/07 $25.00 © 2007 IEEE

Using the normal definitions of space vectors [12] one can write the following well known expression for the real power in a three phase system (using power variant transformations):

$$p = \frac{3}{2}\Re\{\underline{v}\,\underline{i}^*\} + 3v_0 i_0 \quad (1)$$

where $\underline{x}$ denotes a space vector and x_0 denotes a zero sequence instantaneous value.

We shall assume that we have positive and negative sequence current feeding into a three port network with positive and negative sequence terminal voltages. This models the operation of a STATCOM under unbalanced operation. We shall assume that sinusoidal steady state conditions exist, and the positive and negative sequence voltages and currents can be written as[1]:

$$v_a(t) = \sqrt{2}\hat{V}_{\mathrm{p}}\cos(\omega t + \alpha_1) + \sqrt{2}\hat{V}_{\mathrm{n}}\cos(\omega t + \alpha_2) \quad (2)$$

$$v_b(t) = \sqrt{2}\hat{V}_{\mathrm{p}}\cos(\omega t + \alpha_1 - \frac{2\pi}{3}) +$$
$$\cdots \sqrt{2}\hat{V}_{\mathrm{n}}\cos(\omega t + \alpha_2 + \frac{2\pi}{3}) \quad (3)$$

$$v_c(t) = \sqrt{2}\hat{V}_{\mathrm{p}}\cos(\omega t + \alpha_1 + \frac{2\pi}{3}) +$$
$$\cdots \sqrt{2}\hat{V}_{\mathrm{n}}\cos(\omega t + \alpha_2 - \frac{2\pi}{3}) \quad (4)$$

and

$$i_a(t) = \sqrt{2}\hat{I}_{\mathrm{p}}\cos(\omega t + \alpha_1 + \phi_1) + \sqrt{2}\hat{I}_{\mathrm{n}}\cos(\omega t + \alpha_2 + \phi_2) \quad (5)$$

$$i_b(t) = \sqrt{2}\hat{I}_{\mathrm{p}}\cos(\omega t + \alpha_1 + \phi_1 - \frac{2\pi}{3}) +$$
$$\cdots \sqrt{2}\hat{I}_{\mathrm{n}}\cos(\omega t + \alpha_2 + \phi_2 + \frac{2\pi}{3}) \quad (6)$$

$$i_c(t) = \sqrt{2}\hat{I}_{\mathrm{p}}\cos(\omega t + \alpha_1 + \phi_1 + \frac{2\pi}{3}) +$$
$$\cdots \sqrt{2}\hat{I}_{\mathrm{n}}\cos(\omega t + \alpha_2 + \phi_2 - \frac{2\pi}{3}) \quad (7)$$

Substituting (2)–(7) into the standard definitions of a space vector [12] allows one to write:

$$\underline{v} = \sqrt{2}\hat{V}_{\mathrm{p}}e^{j(\omega t + \alpha_1)} + \sqrt{2}\hat{V}_{\mathrm{n}}e^{-j(\omega t + \alpha_2)} = \underline{v}_p + \underline{v}_n^* \quad (8)$$

$$\underline{i} = \sqrt{2}\hat{I}_{\mathrm{p}}e^{j(\omega t + \alpha_1 + \phi_1)} + \sqrt{2}\hat{I}_{\mathrm{n}}e^{-j(\omega t + \alpha_2 + \phi_2)} = \underline{i}_p + \underline{i}_n^* \quad (9)$$

Substituting (8) and (9) into (1) gives the space vector version of the real power expression:

$$p = \frac{3}{2}\Re\{\underline{v}\,\underline{i}^*\} = \frac{3}{2}\Re\{(\underline{v}_p + \underline{v}_n^*)(\underline{i}_p^* + \underline{i}_n)\} \quad (10)$$

$$= \frac{3}{2}\;[\underbrace{\Re\{\underline{v}_p \underline{i}_p^* + \underline{v}_n^* \underline{i}_n\}}_{\text{+ve \& -ve sequence real power}} + \underbrace{\Re\{\underline{v}_p \underline{i}_n + \underline{v}_n^* \underline{i}_p^*\}}_{\text{+ve \& -ve sequence reactive power}}]$$
$$(11)$$

Remark 1: One can see from (11) that the first term is a conventional average real power term for a three phase system. Notice that there are positive and negative sequence contributions. In the case of a STATCOM which is controlling Vars the $\underline{v}_p\,\underline{i}_p^*$ term would be approximately zero. The second set of terms, which contain cross terms of the positive

[1]We are assuming that no zero sequence currents can flow in the lines.

and negative sequence components are unfamiliar, and due to the unbalance. We shall examine what they mean in more detail below. ◆

Using (8) and (9) one can write (11) as:

$$p = 3[(\hat{V}_{\mathrm{p}}\hat{I}_{\mathrm{p}}\cos\phi_1 + \hat{V}_{\mathrm{n}}\hat{I}_{\mathrm{n}}\cos\phi_2) +$$
$$(\hat{V}_{\mathrm{p}}\hat{I}_{\mathrm{n}}\cos(2\omega t + \alpha_1 + \alpha_2 + \phi_2) +$$
$$\hat{V}_{\mathrm{n}}\hat{I}_{\mathrm{p}}\cos(2\omega t + \alpha_1 + \alpha_2 + \phi_1))] \quad (12)$$

Remark 2: The second group of terms in (12) are formed from the so-called reactive power terms of (11), and are **three phase reactive power** terms. They have no average power. This is a different type of reactive power, in that it is three phase power flowing into and out of the STATCOM/grid connection at twice the supply frequency. This means that all types of STATCOMs under unbalanced conditions have to cope with these power flows. This clearly has implications on the design of the STATCOM. ◆

Thus-far we have been considering the three phase power situation. This is all that is required for STATCOM topologies with common capacitors in their DC link, but in the case of the CHC based STATCOM there is no common DC link. Therefore, the individual phase powers are important because of the effect that these have on the charge in the individual leg capacitors.

In order to simplify the expression, we shall assume that $\underline{v}_n = 0$ (this is valid under steady state if rebalancing control is active). If one calculates $p_a = v_a i_a$, $p_b = v_b i_b$ and $p_c = v_c i_c$ using the approximation that $\hat{V}_{\mathrm{n}} = 0$ and $\alpha_1 = \alpha_2 = 0$ and $\phi_1 = \frac{\pi}{2}$, then one can, after some tedious but straightforward manipulation, write:

$$p_a = \hat{V}_{\mathrm{p}}\hat{I}_{\mathrm{n}}[\cos(2\omega t + \frac{\pi}{2}) + \cos(2\omega t + \phi_2) + \cos\phi_2] \quad (13)$$

$$p_b = \hat{V}_{\mathrm{p}}\hat{I}_{\mathrm{n}}[\cos(2\omega t + \frac{\pi}{2} - \frac{4\pi}{3}) +$$
$$\cdots \cos(2\omega t + \phi_2) + \cos(\phi_2 + \frac{4\pi}{3})] \quad (14)$$

$$p_c = \hat{V}_{\mathrm{p}}\hat{I}_{\mathrm{n}}[\cos(2\omega t + \frac{\pi}{2} + \frac{4\pi}{3}) +$$
$$\cdots \cos(2\omega t + \phi_2) + \cos(\phi_2 - \frac{4\pi}{3})] \quad (15)$$

Remark 3: One can see from (13)–(15) that the power expressions contain terms that either have an average value of zero when added together (e.g. $3\hat{V}_{\mathrm{p}}\hat{I}_{\mathrm{p}}\cos(2\omega t + \phi_2)$) or all add instantaneously together to be zero. One can see the correspondence of between (13)–(15) and (12), taking into account the fact the $\hat{V}_{\mathrm{n}} = 0$ in the former. One can conclude from this situation that the total average three phase power to or from the STATCOM is zero, although as noted previously there is a "reactive" three phase power flow. ◆

Remark 4: The most significant result from (13)–(15) is that each phase, on a phase basis, can be generating or absorbing real power. This can be deduced from the $\cos\phi_2$ terms that appear in this expression. **This has important implications on the control of the CHC based STATCOM in circumstances where the STATCOM supports negative sequence currents.** Strategies have to be developed that allow the real power flow to be controlled not only on a three phase basis, but also on a phase basis, since each phase leg of a CHC STATCOM cannot generate or absorb average real power. ♦

III. ALGORITHM DEVELOPMENT

This section will develop algorithms that use zero sequence injection to give the extra degree of freedom required to control the phase powers individually. The injection of zero sequences is an appropriate technique since zero sequence injection is decoupled from other STATCOM control strategies. The approach is similar, but more general than that in [13]. The following development will be for a wye connected STATCOM, since the zero sequence injection expressions explicitly involve PCC voltage magnitudes. In the unbalance load case in [13] it was assumed that the voltages at the PCC did not contain negative sequences. The presence of negative sequence voltages cannot be ignored with the wye connection, and hence the following development will concentrate on this case.

Consider the definitions of positive, negative and zero sequence voltage and current phasors in Table I. The standard power expressions for the phase powers with a wye connection, no neutral (i.e. no zero sequence current), and using symmetrical components [14] are:

$$\vec{S}_a = \vec{V}_a \vec{I}_a^* = (\vec{V}_{a0} + \vec{V}_{ap} + \vec{V}_{an})(\vec{I}_{ap}^* + \vec{I}_{an}^*) \quad (16)$$

$$\vec{S}_b = \vec{V}_b \vec{I}_b^* = (\vec{V}_{a0} + a^2 \vec{V}_{ap} + a\vec{V}_{an})((a^2)^* \vec{I}_{ap}^* + a^* \vec{I}_{an}^*) \quad (17)$$

$$\vec{S}_c = \vec{V}_c \vec{I}_c^* = (\vec{V}_{a0} + a\vec{V}_{ap} + a^2 \vec{V}_{an})(a^* \vec{I}_{ap}^* + (a^2)^* \vec{I}_{an}^*) \quad (18)$$

where $a = e^{j\frac{2\pi}{3}}$, and $\vec{S}$ is the complex power.

Equations (16)–(18) can be expanded, and the real part extracted. The expressions for P_a and P_b are:

$$P_a = \Re\{\vec{S}_a\} = \hat{V}_0 \hat{I}_p \cos(\alpha_0 - \theta_{pi}) + \hat{V}_0 \hat{I}_n \cos(\alpha_0 - \theta_{ni}) + \\ \cdots \hat{V}_p \hat{I}_n \cos(\theta_{pv} - \theta_{ni}) + \hat{V}_n \hat{I}_p \cos(\theta_{nv} - \theta_{pi}) + \\ \cdots \hat{V}_p \hat{I}_p \cos(\theta_{pv} - \theta_{pi}) + \hat{V}_n \hat{I}_n \cos(\theta_{nv} - \theta_{ni}) \quad (19)$$

$$P_b = \Re\{\vec{S}_b\} = \hat{V}_0 \hat{I}_p \cos(\alpha_0 - \theta_{pi} - \frac{4\pi}{3}) + \\ \cdots \hat{V}_0 \hat{I}_n \cos(\alpha_0 - \theta_{ni} - \frac{2\pi}{3}) + \hat{V}_p \hat{I}_n \cos(\theta_{pv} - \theta_{ni} + \frac{2\pi}{3}) + \\ \cdots \hat{V}_n \hat{I}_p \cos(\theta_{nv} - \theta_{pi} - \frac{2\pi}{3}) + \hat{V}_p \hat{I}_p \cos(\theta_{pv} - \theta_{pi}) + \\ \cdots \hat{V}_n \hat{I}_n \cos(\theta_{nv} - \theta_{ni}) \quad (20)$$

which form two independent equations.[2] The zero sequence voltage in these expressions is normally zero, but is included as this is the voltage that is to be injected by CHC STATCOM. We need to determine $\vec{V}_0$ so that the negative sequence/positive sequence cross components of the real power are zero – i.e. we want:

$$P_a - [\hat{V}_p \hat{I}_p \cos(\theta_{pv} - \theta_{pi}) + \hat{V}_n \hat{I}_n \cos(\theta_{nv} - \theta_{ni})] = 0 \quad (21)$$

$$P_b - [\hat{V}_p \hat{I}_p \cos(\theta_{pv} - \theta_{pi}) + \hat{V}_n \hat{I}_n \cos(\theta_{nv} - \theta_{ni})] = 0 \quad (22)$$

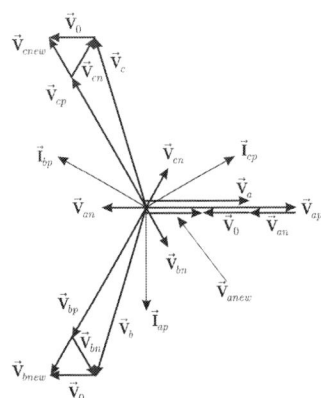

Fig. 1. Vector diagram of zero sequence injection to balance phase voltages.

After some tedious manipulations it is possible to get the following equations:

$$\hat{V}_0 k_1 \cos \alpha_0 + \hat{V}_0 k_2 \sin \alpha_0 = -k_3 \quad (23)$$

$$\hat{V}_0 k_4 \cos \alpha_0 + \hat{V}_0 k_5 \sin \alpha_0 = -k_6 \quad (24)$$

where $k_1 = \hat{I}_p \cos \theta_{pi} + \hat{I}_n \cos \theta_{ni}$; $k_2 = \hat{I}_p \sin \theta_{pi} + \hat{I}_n \sin \theta_{ni}$; $k_3 = \hat{V}_p \hat{I}_n \cos(\theta_{pv} - \theta_{ni}) + \hat{V}_n \hat{I}_p \cos(\theta_{nv} - \theta_{pi})$; $k_4 = \hat{I}_p \cos(\theta_{pi} + \frac{4\pi}{3}) + \hat{I}_n \cos(\theta_{ni} + \frac{2\pi}{3})$; $k_5 = \hat{I}_p \sin(\theta_{pi} + \frac{4\pi}{3}) + \hat{I}_n \sin(\theta_{ni} + \frac{2\pi}{3})$; and $k_6 = \hat{V}_p \hat{I}_n \cos(\theta_{pv} - \theta_{ni} + \frac{2\pi}{3}) + \hat{V}_n \hat{I}_p \cos(\theta_{nv} - \theta_{pi} - \frac{2\pi}{3})$.

The solutions to equations (23) and (24) are:

$$\tan \alpha_0 = \frac{k_6 k_1 - k_3 k_4}{k_3 k_5 - k_6 k_2} \quad (25)$$

$$\hat{V}_0 = \frac{-k_3}{k_1 \cos \alpha_0 + k_2 \sin \alpha_0} \quad (26)$$

or

$$\hat{V}_0 = \frac{-k_6}{k_4 \cos \alpha_0 + k_5 \sin \alpha_0} \quad (27)$$

The choice of equation (26) or (27) for the $\hat{V}_0$ expression depends on the size of the denominator. The expression with the largest denominator should be chosen, so that the calculation is better conditioned. To find α_0 a four quadrant $\tan^{-1}$ function should be used.

Example 1: Let us apply these equations to a simple example. If the a-phase sequence voltages at the terminals of the STATCOM are: $\vec{V}_{ap} = 93.333\angle 0$; $\vec{V}_{an} = 6.6667\angle\pi$ and $\vec{V}_0 = 0$. These values corresponds to an a-phase source voltage amplitude of 80 VRMS while the other two phases are 100 VRMS. It is assumed that the STATCOM is sinking only positive sequence currents of the form $\vec{I}_{ap} = 30.0\angle(\theta_{pv} - \frac{\pi}{2})$.[3]

The real phase powers under these conditions are $P_a = 0$, $P_b = -173.21$ and $P_c = 173.21$. As can be seen the individual phase powers are not zero even though the total three phase power is. If equations (25) and (26) are solved then one gets a zero sequence voltage of:

$$\vec{V}_0 = 6.6667\angle\pi \quad (28)$$

---

[2]The third equation is not independent, therefore only two equations are required for the two unknowns.

[3]The $\pi/2$ phase angle means that the STATCOM is only supplying positive sequence VArs.

978-1-4244-1871-8/07 $25.00 © 2007 IEEE

Phasor		+ve Seq	-ve Seq	0 Seq
Voltage		$\vec{V}_{ap} = \hat{V}_p \angle \theta_{pv}$	$\vec{V}_{an} = \hat{V}_n \angle \theta_{nv}$	$\vec{V}_{a0} = \hat{V}_0 \angle \alpha_0$
		$\vec{V}_{bp} = \hat{V}_p \angle \left(\theta_{pv} - \frac{2\pi}{3}\right)$	$\vec{V}_{bn} = \hat{V}_n \angle \left(\theta_{nv} + \frac{2\pi}{3}\right)$	$\vec{V}_{b0} = \hat{V}_0 \angle \alpha_0$
		$\vec{V}_{cp} = \hat{V}_p \angle \left(\theta_{pv} - \frac{4\pi}{3}\right)$	$\vec{V}_{cn} = \hat{V}_n \angle \left(\theta_{nv} + \frac{4\pi}{3}\right)$	$\vec{V}_{b0} = \hat{V}_0 \angle \alpha_0$
Current		$\vec{I}_{ap} = \hat{I}_p \angle \theta_{pi}$	$\vec{I}_{an} = \hat{I}_n \angle \theta_{ni}$	0
		$\vec{I}_{bp} = \hat{I}_p \angle \left(\theta_{pi} - \frac{2\pi}{3}\right)$	$\vec{I}_{bn} = \hat{I}_n \angle \left(\theta_{ni} + \frac{2\pi}{3}\right)$	0
		$\vec{I}_{cp} = \hat{I}_p \angle \left(\theta_{pi} - \frac{4\pi}{3}\right)$	$\vec{I}_{cn} = \hat{I}_n \angle \left(\theta_{ni} + \frac{4\pi}{3}\right)$	0

TABLE I
TABLE OF POSITIVE, NEGATIVE AND ZERO SEQUENCE PHASOR DEFINITIONS.

If one of the current sources in the STATCOM is replaced by a voltage source to allow the injection of the zero sequence voltage then the real powers in the three phases then become zero.

Fig. 1 shows diagrammatically what the zero sequence injection is doing. Notice that originally the $\vec{V}_b$ and $\vec{V}_c$ phasors are not orthogonal to the phase currents, hence there is real power in these phases. The $\vec{V}_0$ injection however creates resultant phase voltages that are all orthogonal to their respective phase currents.

Similar results can be obtained for arbitrary positive and negative sequence currents. ♦

As eluded to previously, the delta connected STATCOM is slightly simpler with respect to the zero sequence current injection, since the same equations as derived in [13] can be used if it is assumed that the unbalance is due to voltage magnitude difference. These are simply be stated below for convenient reference, but without proof. The interested reader is referred to [13] for more details.

If one has a delta connected STATCOM that is *absorbing negative sequence currents*, a zero sequence current can be injected around the delta to balance the phase powers. In the case where the following expressions are used with unbalanced load under balance voltage conditions the STATCOM *supplies* the loads negative sequence current requirements – i.e. it is a source of negative sequence currents [13]. The zero sequence current magnitude and angle are determined as follows:

$$\gamma = \arctan\left(\frac{z - y}{\sqrt{3}(y + z)}\right) \quad (29)$$

$$\hat{I}_0' = -\frac{y + z}{\cos\gamma} \quad (30)$$

where:

$$\gamma = \theta_{ab} - \alpha_0 \quad (31)$$

$$y = \frac{1}{\sqrt{3}}\cos(\theta_{ab} - \theta_{ni} - \frac{\pi}{6}) \quad (32)$$

$$z = \frac{1}{\sqrt{3}}\cos(\theta_{ab} - \theta_{ni} + \frac{\pi}{2}) \quad (33)$$

$$\hat{I}_0 = \hat{I}_0'\hat{I}_n \quad (34)$$

$$\text{and } \vec{I}_0 = \hat{I}_0 \angle \alpha_0 \quad (35)$$

where θ_{ab} is the phase angle of $\vec{V}_{ab} = \hat{V}_{ab} \angle \theta_{ab}$, and the zero sequence current phasor is denoted as $\vec{I}_0$.

Remark 5: As noted previously, these expressions are valid under conditions of *voltage magnitude unbalance only*. If the unbalance involves phase angle variations from the

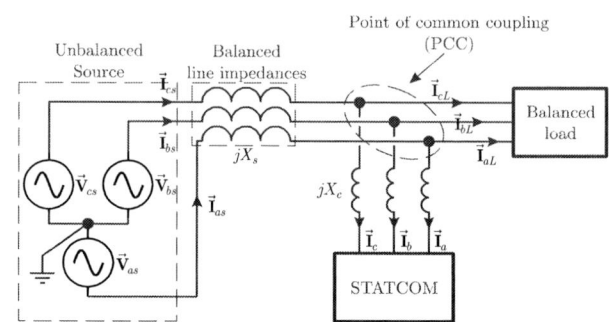

Fig. 2. Block diagram of a STATCOM connection when there is an unbalanced source.

nominal $2\pi/3$ radians then the expressions are more complex. ♦

IV. REBALANCING UNBALANCED VOLTAGES

The use of zero sequence injection techniques for controlling phase power in a CHC STATCOM when it is providing negative sequence currents to rebalance an unbalanced load has already been published [13]. However, in [13] the source voltages, and the voltages at the point of common coupling (PCC) contained only positive sequence components. In this section the delta based algorithm of section III will be used to control the CHC STATCOM phase leg voltages whilst the overall control strategy for the STATCOM is to eliminate the negative sequence voltage at the PCC – i.e. the STATCOM is attempting to make sure that other loads connected to the PCC are only seeing positive sequence voltages. It is well known that STATCOMs can eliminate negative sequences, but there is a price to pay in terms of STATCOM component rating for this [11]. This strategy would normally be a complementary control to a positive sequence control that supports the positive sequence voltage level, so that overall voltage dip support is provided [10].

We shall consider a simple unbalance situation where the source voltage magnitudes are unbalanced, but still at 120° phase difference, and the load is balanced, and the line impedances are assumed to be the same. This situation is depicted in Fig. 2. These conditions mean that unbalance currents flowing to the load will be due solely to the unbalance in the supply voltages. This is important with respect to one of the algorithms being tested, since it is a current controlled algorithm that works by eliminating the negative sequence current components flowing to the load. Clearly, if these negative sequence currents are due to only to the unbalanced source voltages, then their elimination also implies the elimination of the unbalanced voltage at the PCC.

978-1-4244-1871-8/07 $25.00 © 2007 IEEE

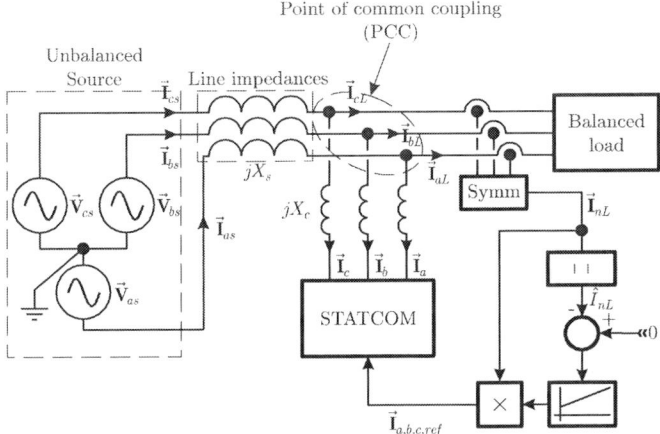

Fig. 3. Block diagram showing the STATCOM and a rebalance control based on feeding back the negative sequence load currents.

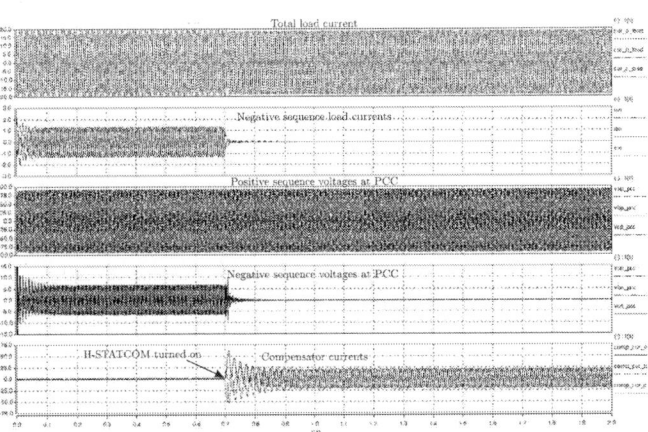

Fig. 4. From the top plot: load currents, negative sequence load current, positive and negative sequence voltages, STATCOM compensator currents with a balanced resistive load and a 20% different voltage in phase 'a' (14.3% unbalance). STATCOM re-balancing occurs from 0.7 seconds.

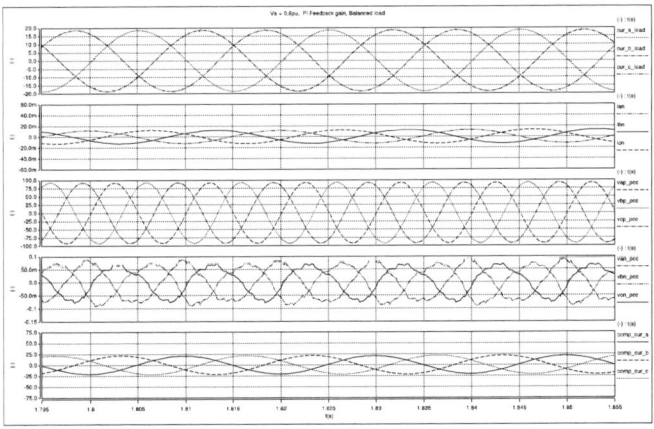

Fig. 5. Close-up of Fig. 4. Time from 1.795→1.855 secs. From the top plot: load currents (vert axis -20.0→20.0 Amp); negative sequence load current (vert axis -0.06→0.06 Amp); positive sequence voltages (vert axis -100.0→100.0 Volts); negative sequence voltages (vert axis -0.1→0.1 Volts); STATCOM compensator currents (vert axis -75.0→75.0 Amp). Conditions: balanced resistive load and a 20% less voltage magnitude in phase 'a' (14.3% unbalance).

The focus of this paper is on the issue of the power balance in the CHC STATCOM legs, therefore the rebalancing algorithms simulated are simple, and are being used to generate the conditions in which to test the phase leg power balance algorithm. The STATCOM used in the simulations is delta connected, and the unbalance is limited to unbalanced voltage magnitude. The control strategy is similar to that in [11], which is the current feed form of that in [10]. The STATCOM operates as a voltage controlled current source (VCCS), since this leads to the lowest rating for the STATCOM as compared to operating as a voltage controlled voltage source (VCVS) [11].

V. SIMULATION STUDIES

A conceptual diagram of a rebalancing control strategy that uses the negative sequence load current as the feedback is shown in Fig. 3. A simulation was developed using the Saber® simulation package. The STATCOM was idealised, and modelled as ideal voltage controlled current sources. The load is a balanced three phase load consisting of 5Ω resistors. The unbalance in the source voltage is created through a dip of 20% in magnitude of phase 'a'. This is representative of the common brown out phenomena caused by a single-line-to-ground fault somewhere on the network. Measurements of the voltages and currents were made using a PLL technique based on [15], [16]. Symmetrical components were generated using traditional phasor based expressions [14].

Figs. 4–6 show the results of a simulation of the rebalancing algorithm shown in Fig. 3. It is very difficult to read the anotation on the axes of these graphs, but their meaning is described in the captions. The specific values are not that important, but the general trends can be seen.

In the simulation of Figs. 4–6 the rebalance compensation algorithm is switched on at 0.7 seconds. The STATCOM then executes the rebalancing algorithm *without* zero sequence power balancing until $t = 1.3$ secs.

Fig. 4 shows the key variables under this control. The second plot from the bottom shows the negative sequence voltage at the PCC. One can see that the control has the desired effect of driving this voltage to zero, which in turns

means that only the positive sequence voltage remains. The third plot from the bottom shows the positive sequence voltage – it is unaltered by the rebalancing at 0.7 secs, and the zero sequence injection control which begins at $t = 1.3$ secs. The negative sequence current flowing to the load decreases to almost zero (second plot from the top). The other important plot in this figure is that for the compensator currents (the bottom plot). The load current is approximately 18 Amp peak, and the compensator current is approximately 24 Amp magnitude. Therefore the compensator, in this case, has to have a current rating higher than the load current. If Vars were to be supplied by the compensator to control the positive sequence voltage then the compensator rating would increase even more. Unseen in this plot is the zero sequence current of $\hat{I}_0 = 4$ Amp.

Fig. 5 is a zoomed in view of Fig. 4. One can see that the negative sequence PCC voltages are very small relative to the positive sequence voltage amplitude, which indicates that the strategy is very effective at rebalancing this degree

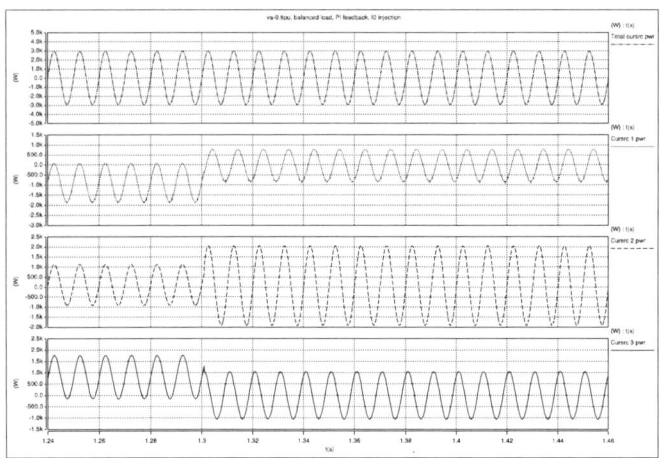

Fig. 6.   Total three phase power and the individual current source powers under the conditions of figures 4 and 5. From the top the plots are: the total three phase power in the three current sources (vert axis -5000→5000 Watts), the powers in each individual current source in the STATCOM (vert axes top to bottom: -3000→ 3000 Watts; -2000→2500 Watts; -1500→2500 Watts).

of unbalance in the supply.

Fig. 6 shows the total power into the STATCOM, and the individual powers in each of the current sources in the STATCOM from the time of 1.24 to 1.45 seconds in the same simulation run as the results shown in Figs. 4 and 5. At time $t = 1.3$ secs the zero sequence injection is turned on.

The top plot in Fig. 6 shows the total three phase power flowing into the STATCOM. As shown in (12) the power due to the interaction of the negative sequence voltages and currents with the positive sequence voltage and currents leads to a double frequency *three phase* "reactive" power – i.e energy flows in a three phase sense into and out of the STATCOM. On the left hand side of the individual current source power plots, one can see that their powers can have a non-zero average value.

The step jump in the bottom three plots is when the zero sequence injection occurs. One can see that the non-zero average powers in the currents sources then move to have a zero average, which means that none of the current sources are receiving or delivering average power. The total three phase power plot is unaffected by this injection, indicating that the zero sequence injection does not affect the external terminal conditions.

One can alter the simulation so that the compensator current is related to the magnitude of the negative sequence voltage at the PCC [10], [11]. The advantage of this approach is that it does not have to make any assumptions about the load. In the balanced load case it is equivalent to the scheme reported above, and simulation studies give similar results to those above.

VI. CONCLUSIONS

This paper has investigated the use of the CHC based STATCOM for rebalancing an unbalanced voltage at the PCC. The main issue with the use of the CHC STATCOM in this type of application is that the phase leg average powers are not zero. If not addressed this can lead to the destruction

of the converter leg/s due to H-bridge capacitor over-voltage. The paper develops expressions for injected zero sequence voltages and zero sequence currents for the wye and delta connected STATCOM respectively, which force the phase leg average powers to zero when the control is attempting to rebalance an unbalanced supply.

Preliminary idealised simulation results are presented which show the viability of the technique. The technique is able to zero the phase leg average powers without any affect on the external performance of the STATCOM. There are a number of additonal issues to investigate with respect to this approach, but the preliminary results are very encouraging.

REFERENCES

[1] J. Rodriguez, J.-S. Lai, and F. Z. Peng, "Multilevel inverters: a survey of topologies, controls, and applications," *IEEE Transactions on Industrial Electronics*, vol. 49, no. 4, pp. 724–738, 2002.

[2] D. Gerry, P. Wheeler, and J. Clare, "A voltage-balance strategy for multi-level, multi-cellular converters," in *CDROM Proceedings of the 10th EPE*, September 2003. ISBN 90-75815-07-7.

[3] T. Yoshii, S. Inoue, and H. Akagi, "Control and performance of a medium-voltage transformerless cascade PWM STATCOM with star-configuration," in *The 2006 IEEE Industry Applications Conference Forty-First IAS Annual Meeting, Conference Record of*, vol. 4, (Tampa, FL), pp. 1716–1723, Oct. 2006.

[4] B. Blazic and I. Papic, "Improved d-statcom control for operation with unbalanced currents and voltages," *IEEE Transactions on Power Delivery*, vol. 21, pp. 225–233, Jan. 2006.

[5] Y. Liu and F. L. Luo, "Trinary hybrid multilevel inverter used in STATCOM with unbalanced voltages," *IEE Proceedings- Electric Power Applications*, vol. 152, pp. 1203–1222, Sept. 9, 2005.

[6] L. Xu, B. R. Andersen, and P. Cartwright, "VSC transmission operating under unbalanced AC conditions - analysis and control design," *IEEE Transactions on Power Delivery*, vol. 20, pp. 427–434, Jan. 2005.

[7] G. Escobar, A. M. Stankovic, and P. Mattavelli, "An adaptive controller in stationary reference frame for d-statcom in unbalanced operation," *IEEE Transactions on Industrial Electronics*, vol. 51, pp. 401–409, Apr. 2004.

[8] C. A. C. Cavaliere, E. H. Watanabe, and M. Aredes, "Multi-pulse STATCOM operation under unbalanced voltages," in *Power Engineering Society Winter Meeting, 2002. IEEE*, vol. 1, pp. 567–572, Jan. 2002.

[9] N. I. A. Wahab, N. Mariun, A. Mohamed, and M. Mohamad, "Response of d-STATCOM under unbalanced voltage condition caused by SLG fault," in *Research and Development, 2003. SCORED 2003. Proceedings. Student Conference on*, pp. 395–400, Aug. 2003.

[10] C. Hochgraf and R. H. Lasseter, "Statcom controls for operation with unbalanced voltages," *IEEE Transactions on Power Delivery*, vol. 13, pp. 538–544, Apr. 1998.

[11] K. Li, J. Liu, B. Wei, and Z. Wang, "Comparison of two control approaches of static VAr generators for compensating source voltage unbalance in three-phase three-wire systems," in *Power Electronics and Motion Control Conference, 2004. IPEMC 2004. The 4th International*, vol. 3, pp. 1213–1218, Aug. 2004.

[12] P. Vas, *Electrical Machines and Drives - A Space Vector Theory Approach*. Monographs in Electrical and Electronic Engineering, Oxford Univeristy Press, 1992. ISBN: 0 19 859378 3.

[13] R. E. Betz, T. Summers, and T. Furney, "Symmetry compensation using a h-bridge multilevel STATCOM with zero sequence injection," in *The 2006 IEEE Industry Applications Conference Forty-First IAS Annual Meeting, Conference Record of*, vol. 4, (Tampa, FL), pp. 1724–1731, Oct. 2006.

[14] W. D. Stevenson, *Elements of Power System Analysis*. McGraw-Hill, 1975. ISBN: 0-07-061285-4.

[15] M. Karimi-Ghartemani and M. R. Iravani, "A new phase-locked loop (PLL) system," in *Circuits and Systems, 2001. MWSCAS 2001. Proceedings of the 44th IEEE 2001 Midwest Symposium on*, vol. 1, (Dayton, OH), pp. 421–424, Aug. 2001.

[16] M. Karimi-Ghartemani, "A novel three-phase magnitude-phase-locked loop system," *Circuits and Systems I: Regular Papers, IEEE Transactions on [see also Circuits and Systems I: Fundamental Theory and Applications, IEEE Transactions on]*, vol. 53, pp. 1792–1802, Aug. 2006.

Distributed Power Supply with Power Factor Correction: a Solution to Feed All Modules in Power Electronic Transformers

M. Saghaleini
University of Tehran
Department of Electrical and
Computer Engineering
Kargar St., Tehran, Iran
Email: saghaleini@ut.ac.ir

S. Farhangi
University of Tehran
Department of Electrical and
Computer Engineering
Kargar St., Tehran, Iran
Email: farhangi@ut.ac.ir

Abstract- The idea of power electronic transformer (PET), which works with solid-state switches, has been considered noticeably recently. Every power electronic transformer has several parts which each of them needs its own reliable supply voltage. So the appropriate performance of a PET depends on the existence of various proper supply voltages. The distributed power supply (DPS), which can produce several different supply voltages simultaneously with various powers and high reliability, may be the optimum choice available for the supply of different modules of a PET. In this paper the idea of a distributed power supply with proper topology for a PET has been introduced and its performance and effects on the network has been demonstrated by simulation.

I. INTRODUCTION

Distribution transformers are the fundamental components in power distribution systems. They are relatively inexpensive, highly reliable, and fairly efficient. However they have some disadvantages such as heavy weight, large size, sensitivity to harmonics, voltage drop under load, (required) protection from system disruptions and overload, and environmental concerns regarding mineral oil. These disadvantages are becoming increasingly important as power quality becomes more of a concern. In this case, power electronic based transformer is a good option for solving above problems [1-3].

The plan to use power electronic converters to change the voltage level with a frequency much more than the network frequency results in creation of power electronic transformers, which have much less copper and iron weight relative to the classic transformers, and probably causes lower cost and more ease of use [4,5].

Because power electronic transformers make use of various converters, they need power supplies to feed their different modules. Therefore, several power supplies with different power and voltage levels are needed. But, using a power supply with distributed structure to feed all the modules have many advantages which some of them are pointed here: less complexity of the control circuit, modularity capability, higher reliability, higher expandability and maintainability. [6, 7].

In this paper, a distributed power supply is presented for a distribution power electronic transformer and the design steps are expressed in order with corresponding justifications. Meanwhile the effects of this DPS on the network are considered and analyzed.

II. PROPOSED TOPOLOGY

In fig. 1 two general structures of the distributed power supply have been shown. The structure of Fig. 1.a uses a bulky ac/dc converter at the first stage and provides the DC voltage and power of the common bus. But the structure of Fig. 1.b uses some smaller paralleled converters for its first stage. This increases the reliability and simplifies the DPS maintenance but has a more difficult control relative to the previous structure, has more number of switches, and is more costly. Moreover in the first structure we only need one PFC module but in the second it isn't so.

According to the mentioned reasons we choose the structure of fig 1.a. In order to prevent harmonic current pollution to the line source, more stringent regulations from IEC 1000-3-2 have been imposed on electronic equipment. To meet the requirement, it is customary to add a power factor corrector (PFC) in front of a DC regulator. Also a two stage approach is relatively mature and viable to applications with a wide power range, it is not an optimal design, and it may suffer from the drawbacks of low-conversion efficiency, high cost, and great design complexity. In an effort to improve the conversion efficiency and also to reduce the component count and cost, a PFC and a fast regulator are integrated to form a singe-stage converter by sharing their active switches [8,9]. The converter can be used as the front-end stage of the DPS. This single-stage converter is introduced in [10, 11] and is shown in Fig. 2. The proposed single-stage converter is a combination of two semi stages, namely buck-boost semi stage and flyback semi stage. In fact, as can be seen in Fig. 2., diodes D5 and D6 in conjunction with Cc, Lpf and M1 form the first semi stage (buck-boost power factor corrector) and diodes D6 and D7 in conjunction with Cc, M1, TX1, Co and Ro form the second semi stage (flyback regulator). So by this topology we have a good power factor in addition to fast and proper regulation.

III. SPECIAL CONSIDERATIONS

The voltage available from the network for the input to the power supply is 20 kV which can't be directly connected to the input of the power supply and should be reduced and brought to a lower level (say 400 V). To accomplish this, a line frequency transformer can be used. It increases the volume of the power supply greatly. Another way is to use a capacitive divider. It has the following features:

978-1-4244-1871-8/07 $25.00 © 2007 IEEE

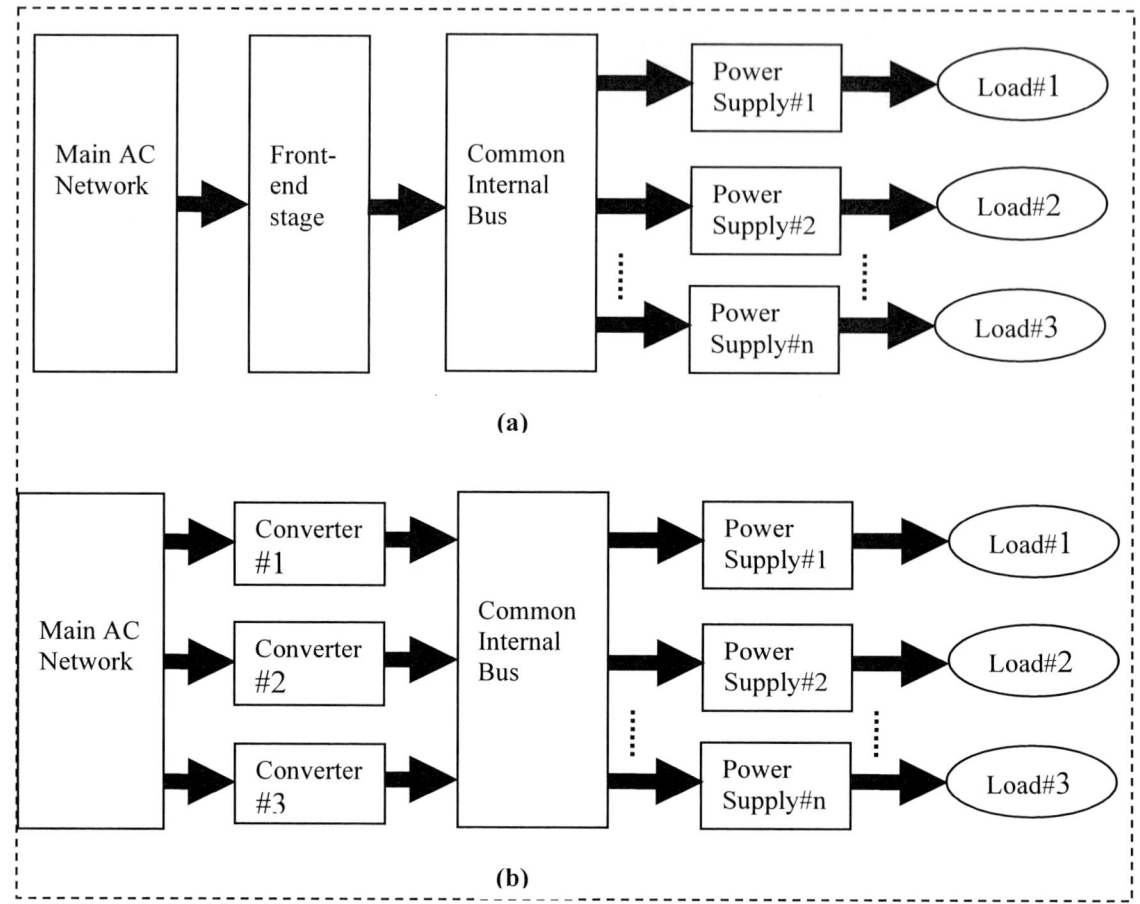

Fig. 1.Two main structures of DPS: a) bulky and b) distributed front-end

First, its volume and cost are considerably lower than a line frequency transformer. Second, it can provide a better voltage regulation. Besides, the repair and maintenance of a capacitive divider is easier and less expensive relative to a conventional transformer, especially at such a low power level [12].

We may consider both two situations. If we want to use a line frequency transformer, because of the PFC capability of the mentioned topology for the front-end stage of the DPS, we can have a unity power factor; and if we want to use a capacitive divider instead of the transformer, because of the existence of two series capacitors, the input current of the DPS is capacitive and does not need to be modified.

However, we can change the shape (phase) of the input current by changing the duty cycle of the switch M1 in the proposed circuit. We show the results of simulation of all

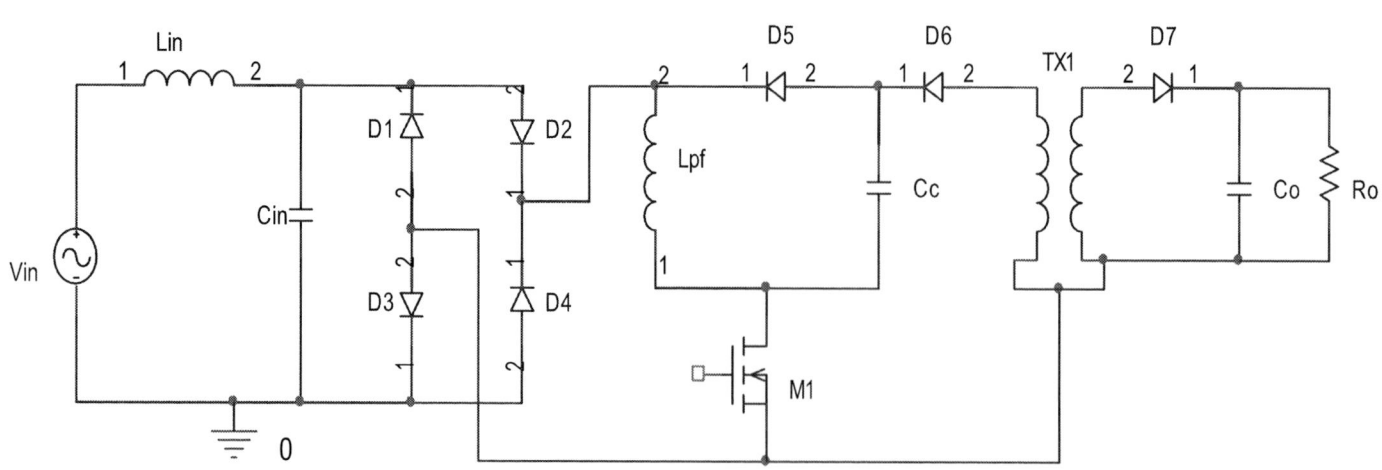

Fig. 2. Selected topology for the front-end stage of the DPS

978-1-4244-1871-8/07 $25.00 © 2007 IEEE   1224

situations in the next section.

IV. SIMULATION RESULTS

In this section we consider the introduced topology of Fig. 2. as the front-end stage of the DPS and simulate the circuit to prove the claimed specifications.

It should be noted that the circuit is just one phase of the main three-phase system and so we consider a lower power to be delivered to the loads. The power needed for all parts of a distribution power electronic transformer is estimated about 1 kW. So the components of the topology are designed so that a 300 W power can be delivered to the loads through a 100 V common DC bus. The values for the circuit components are given in Table I.

To see the power factor correction ability of the introduced topology, we consider that the input voltage is reduced to a lower one (say 200 V) by a line frequency transformer. Then, this reduced voltage is applied to the circuit of Fig. 2. We want to show that we can change the input current phase angle by changing the duty cycle of the switch M1. For this purpose, we choose the duty cycle so that the input current of the circuit is in-phase with the input voltage. The results of the simulation are illustrated in fig. 3.

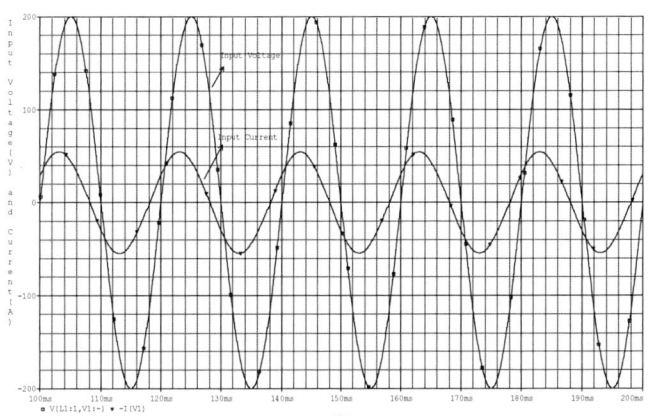

Fig. 4. Input current and voltage of the front-end stage (not in-phase)

As can be seen, the input current is completely in-phase with the input voltage and has a good sinusoidal shape. The switch operational frequency in this simulation was 20 kHz. We can adopt the switch duty cycle to make the input current of the front-end stage even capacitive. The result of an example (for duty cycle equal to 45%) is shown in Fig. 4. By using a lower duty cycle relative to the previous situation, we can cause the input current to have a positive phase angle or be capacitive. As we know, a capacitive sinusoidal current does not have bad effects on the network and even has a role of a compensator in the network.

After justifying the PFC ability incase of application a line frequency transformer, it would be good to add a capacitive divider instead of the transformer and see the performance of the circuit.

In this case, because of the existence of some spikes in the input current, we should add an inductor (Ls) before the capacitive divider stage to smooth the current and suppress the spikes. This inductor is placed in the high-pressure part of the circuit and so is very costly and makes the system more complex. Besides, for the starting time of the system, we should a resistive divider in parallel with the capacitive divider. The overall circuit is illustrated in Fig. 5. The capacitive divider component values are given in Table II.

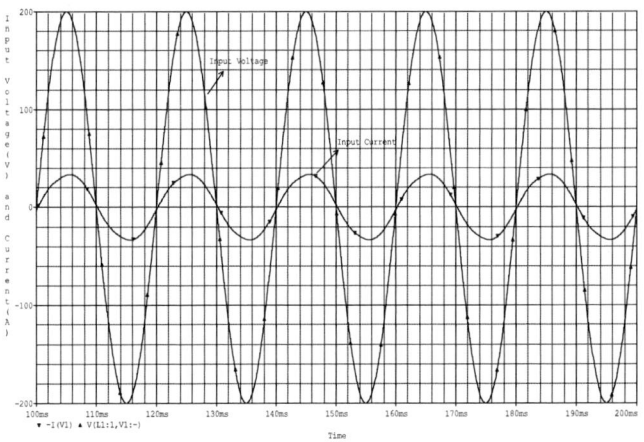

Fig. 3. Input current and voltage of the front-end stage (in-phase)

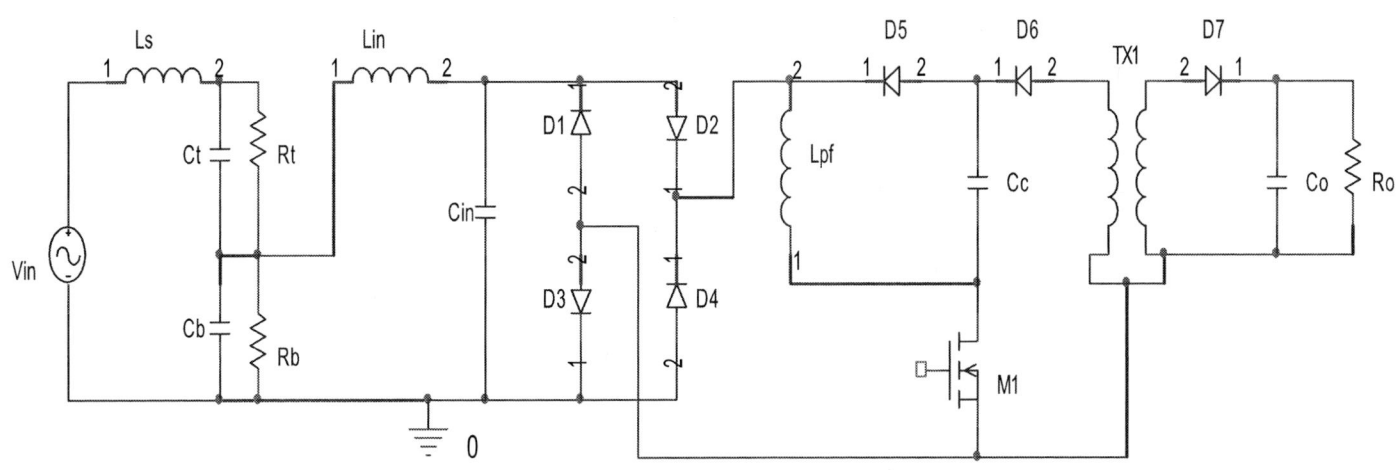

Fig. 5. Front-end stage with capacitive divider

978-1-4244-1871-8/07 $25.00 © 2007 IEEE

Table I
Circuit Values

Vin (V)	fin (Hz)	Lin (mH)	Cin (uF)	Diodes	Lpf (mH)	Cc (uF)	M1	fsw (kHz)	TX1 turns ratio	Co (uF)	Ro (Ω)
3000	50	6	230	BYT30P-600	8	250	IRFP460	20	1:12	600	33

The input current is completely capacitive in this case. So it does not need to be modified. The input current and voltage are shown in Fig. 6. As can be seen from these waveforms, the input current has a phase angle of about 90° leading relative to the voltage and also is in proper sinusoidal shape. So our goal to have a good input current is satisfied. The output voltage of this simulation (common DC bus voltage) is shown in Fig. 7.

Table II
Capacitive Divider Values

Ct (uF)	Rt (MegΩ)	Cb (uF)	Rb (MegΩ)
50	9	450	1

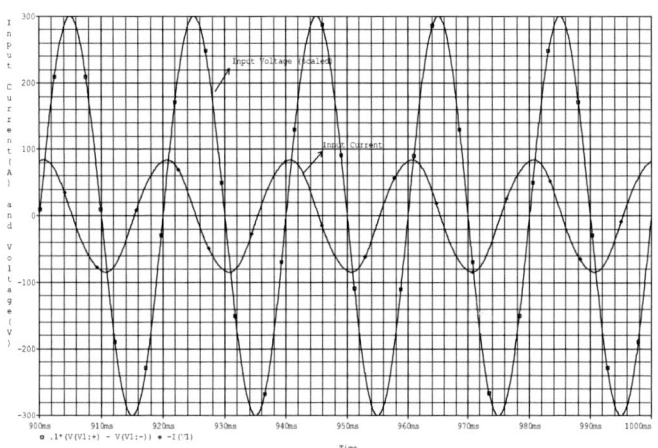

Fig. 6. Input current and voltage waveforms with capacitive divider

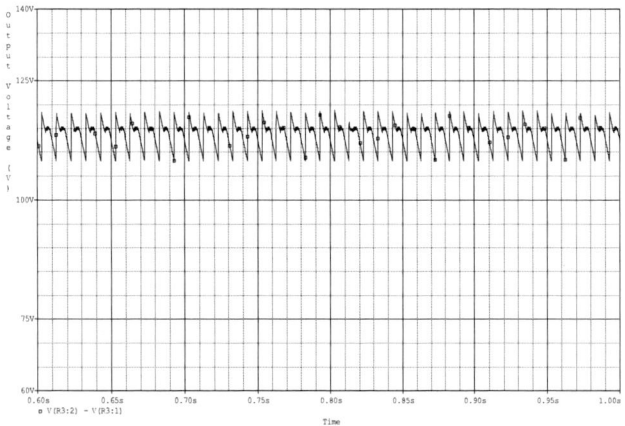

Fig. 7. Output voltage (common DC bus voltage)

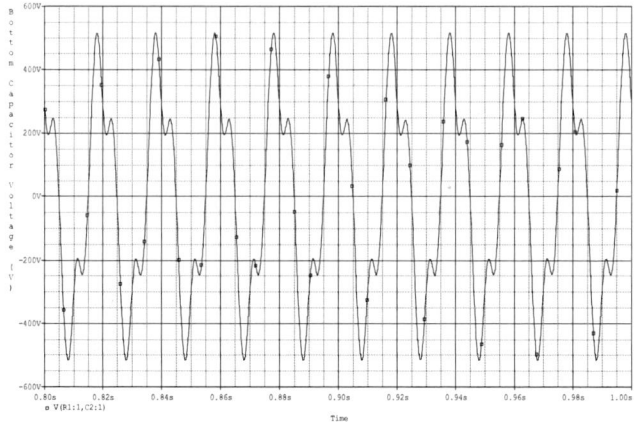

Fig. 8. Capacitive divider voltage

Although we design the circuit to generate a 100 V DC bus with good regulation, but the output voltage has an average of about 110 V and a ripple of approximately 10 Vp-p. However, we can ignore this problem by controlling the power supplies (converters) after the common DC bus.

The voltage of the capacitive divider is shown in Fig. 8. The capacitors are selected so that this voltage magnitude is about 400 V. This voltage does not have a sinusoidal shape because of paralleling capacitors with resistors and nonlinearity effect of the next stages. Anyway, the voltage shape and specially its level are acceptable.

V. CONCLUSION

In this paper the idea of the power electronic transformer and its advantages relative to the conventional transformers were presented. Regarding to the structure of this transformer type and its need for various power supplies, the necessity of distributed power supply and its advantages for PET were discussed. Considering different aspects, the proper topology of a DPS for a PET was chosen. To reach an acceptable voltage level at the input stage of the DPS a capacitive divider was introduced, calculated and well simulated. According to the simulations done by the SPICE program, the voltages generated by this power supply are quite acceptable and its features are promising.

REFERENCES

[1] E. R. Ronan, S. D. Sudhoff, S. F. Glover, and D. L. Galloway, "A power electronic-based distribution transformer," IEEE Trans. Power Delivery., pp. 537-543, Apr. 2002.
[2] L. Heinemann, G. Mauthe, "The universal power electronics based distribution transformer, a unified approach", in Proc. IEEE PESC Conf., pp. 504-509, Jun. 2001.

[3] M. Kang, P. N. Enjeti and I. J. Pitel, "Analysis and design of electronic transformers for electric power distribution system", IEEE Trans. Power Electronics., pp. 1133-1141, Nov. 1999.

[4] M. D. Manjrekar, R. Kieferndorf and G. Venkataramanan, "Power electronic transformers for utility applications", in Proc. IEEE IPC Conf., pp. 2496-2502, Oct. 2000.

[5] H. Iman-Eini, S. Farhangi, "Analysis and design of power electronic transformer for medium voltage levels", in Proc. IEEE PESC Conf., pp. 843-847, Jun. 2006.

[6] F. C. Lee, P. Barbosa, et al, "topologies and design considerations for distributed power system applications", in Proc. IEEE, pp. 939-950, Jun. 2001.

[7] G. Zhu, et al, "Large-signal simulation of a distributed power supply system with power factor correction ", in Proc. IEEE ISCAS Conf., pp. 258-261, 1999.

[8] L. Huber and M. M. Jovanovic, "Single-stage, single-switch, isolated power supply technique with input-current shaping and fast output voltage regulation for universal input-voltage-range application," in *Proc. Applied Power Electronics Conf.*, pp. 272–280, Feb. 1997.

[9] P. Kornetzky, H. Wei, and I. Batarseh, "A novel one-stage power factor correction converter," in *Proc. Applied Power Electronics Conf.*, pp. 251–258, Feb. 1997.

[10] T. F. Wu, T. H. Yu, and Y. H. Chang, "Generation of power converter with graft technique", in proc. 15th Symp. Electrical Power Engineering, pp. 370-376, Nov. 1995.

[11] T. F. Wu, S. A. Liang, and Y. K. Chen, "High-power-factor single-stage converter with robust controller for universal off-line applications", IEEE Trans. Power Electronics, pp. 1078-1085, Nov. 1999.

[12] T. V. Raphalalani, N. M. Ijumba, A. A. Jimoh, "Capacitive divider system for feeding a distribution network from an EHV line", in proc. IEEE ICPS, pp. 299-304, Dec. 2000.

978-1-4244-1871-8/07 $25.00 © 2007 IEEE

Development on 31.5MVA STATCOM and Digital Evaluation Tool for Voltage Flicker Compensation

Koichi Hidese, Tomotsugu Ishizuka, Fumio Aoyama, Satoru Ota, Kouji Ogushi
Toshiba Mitsubishi-Electric Industrial Systems Corporation
Power Electronics Department
1,Toshiba-cho Fuchu-shi, Tokyo, Japan 183-8511
mail: HIDESE.koichi@tmeic.co.jp

Abstract— A large capacity self-commutated static var compensator rated at 31.5MVA has been developed for the voltage flicker compensation. The operation experiences in an actual field and the successful performance of the voltage flicker suppression are introduced in this paper. The principle of the voltage flicker evaluation and the real time digital evaluation tool are also described. This tool can calculate simultaneously the voltage flicker value with and without the STATCOM. It can also analyze the voltage flicker from the arc furnaces which will be compensated by the STATCOM.

I. INTRODUCTION

Theses days, the demand for the iron steel is rapidly increasing. The trend then accelerates the installation of large arc furnaces in the world. To response this market needs, large capacity STATCOM rated at 31.5MVA has been developed with new generation of high-voltage and high-capacity semiconductor device, IEGT (Injection Enhanced insulated Gate Transistor). The developed STATCOM has been installed in steel companies with arc furnaces and showing good performance for the voltage flicker compensation and also helping the iron production by maintaining the mains AC voltage.

Furthermore, a tool has also been developed for real time evaluation of the voltage flicker compensation. The tool can calculate simultaneously the voltage flicker value with and without the STATCOM. It can also analyze the voltage flicker from the arc furnaces which will be compensated by the STATCOM. The tool includes the flicker meter function to measure the flicker index delta-V10 which is popular in Japan. The meter function is also realized by digital calculation.

II. CONFIGURATION OF DEVELOPED STATCOM

The specifications and ratings of the developed STATCOM are shown in Table 1. The circuit configuration is shown in Fig. 1. This equipment applies high-voltage and high-capacity IEGT device developed for the high-power converters. The developed circuit utilizes the features of the IEGT device and is very simple compared with the GTO converter once dominated in the large capacity power electronics technology. The dimensions and the loss become almost a half of the GTO equipment.

10.5MVA unit consists of six IEGT single-phase bridges. Two single-phase converters are connected in series through the transformer primary winding. For the 31.5MVA STATCOM, 3 sets of 10.5MVA units are used. In the 31.5MVA STATCOM, the transformer has six series multiplex windings for improvement of control performance and reduction of harmonic waves.

The pulses to the bridges are displaced each other to make equivalent carrier frequency near 5 kHz. This makes the STATCOM response fast enough to compensate not only the fundamental reactive power and also the harmonics components. This feature results in significant difference from the conventional SVC which only compensates the fundamental components.

The IEGT single phase bridge is shown in Fig. 2. The IEGT converter and capacitor cubicle for 10.5MVA and the STATCOM control cubicle are shown in Fig. 3. The 10.5MVA unit is 2900mm wide, 2800mm high and 2200mm deep.

TABLE1 Developed STATCOM Ratings

Items	Specifications/Ratings
Capacity	31.5MVA
Rated DC voltage	2500V
Equivalent carrier frequency	4680Hz

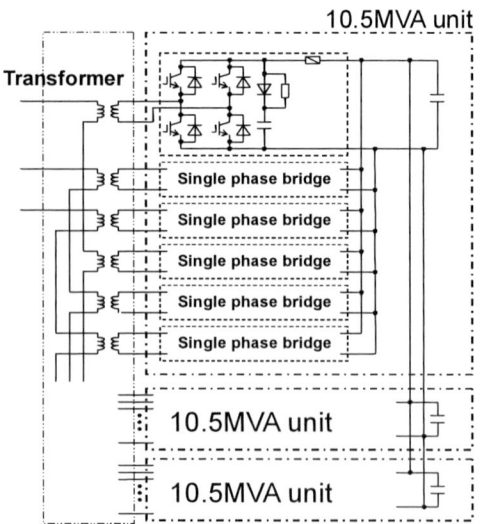

978-1-4244-1871-8/07 $25.00 © 2007 IEEE

Fig. 1. Configuration of the developed STATCOM

Fig. 2: IEGT single-phase bridge unit

Fig. 3: IEGT converter for 10.5MVA and control cubicle

The controller of the STATCOM is made compact. The main part of the controller is the high-performance microprocessor

Fig. 4: Card for the STATCOM controller

and has contributed to improve the flicker suppression performance. Fig.4 shows the main digital control card. The power-electronics oriented digital processor PP7 is mounted on it and carries out ultra-high speed processing. Therefore, the STATCOM can compensate the rapid and large voltage flickers from the load. This control cubicle controls the cooling equipment as well as generates the gate pulses for the IEGT converters. The information required for operation is displayed on the liquid crystal touch panel. All operations including the STATCOM start/stop is possible on this touch panel.

The control block diagram for the voltage flicker suppression is shown in Fig.5. The control cubicle inputs AC voltages and current signals including the arc furnace load currents. The arc furnace current signals are converted to the voltage signals and sent to the main card. It extracts the factors of voltage flicker based on the instantaneous d-q theory and calculates the STATCOM output current reference. The IEGT gate pulses are generated by the pulse width modulation (PWM) cards and sent to the IEGT modules through the optical fibers.

The DC magnetization of the transformer by asymmetrical currents is a serious problem in the STATCOM for the voltage flicker compensation. The DC suppression control block suppresses the DC magnetization by the individual phase current control.

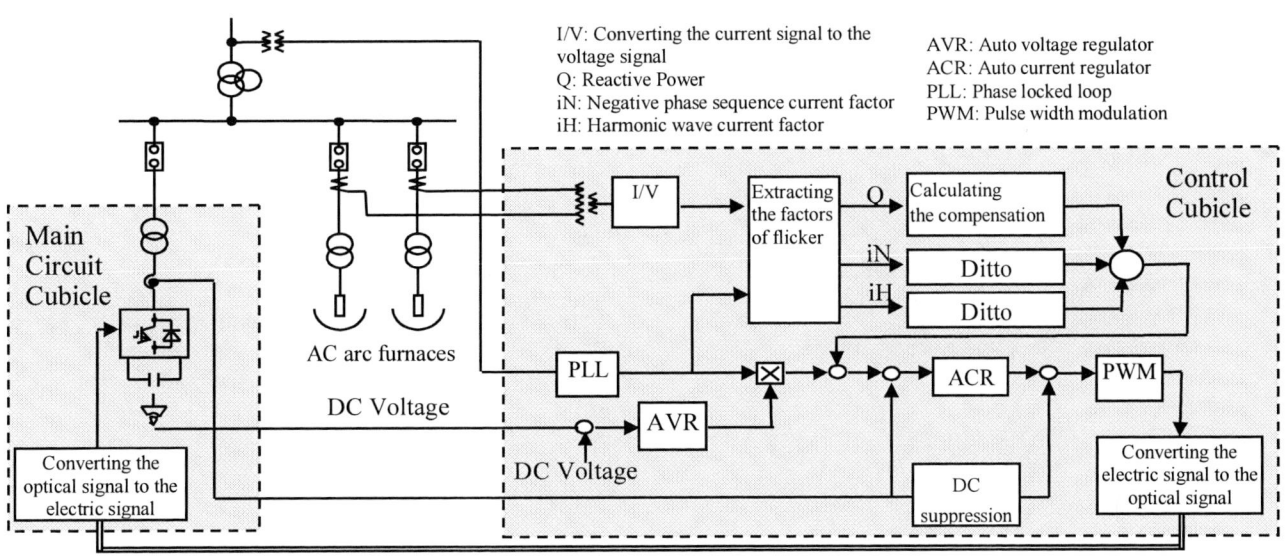

Fig.5: Control block diagram for STATCOM

978-1-4244-1871-8/07 $25.00 © 2007 IEEE   1229

III. OPERATION WAVEFORMS

This type of STATCOM's has been installed in systems with the AC arc furnaces. Fig.6 shows an example of such systems.

The mains AC voltage is 36kV and there are two arc furnaces, which are compensation targets of the STATCOM, and filters.

Fig.7 shows some actual field waveforms in the STATCOM operation. The reactive power of the STATCOM output is inversely proportional to the load reactive power. These waveforms prove that the STATCOM compensates the change of reactive power and stabilize the AC voltage at the common coupling point.

Fig.8 shows the delta-V10 value with and without the STATCOM. The load current changed fast because of the arc furnaces operation and the AC voltage variation also is large without the STATCOM. In this case, the improvement effect of the voltage flicker suppression is over 70% and it proves the STATCOM performance for the voltage flicker compensation.

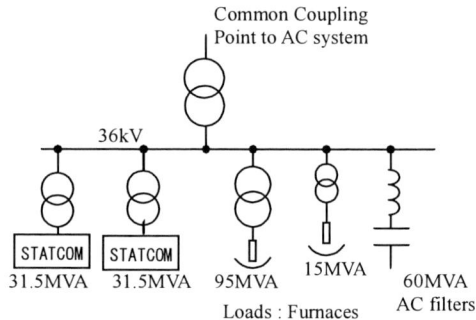

Fig. 6: An actual STATCOM installation configuration

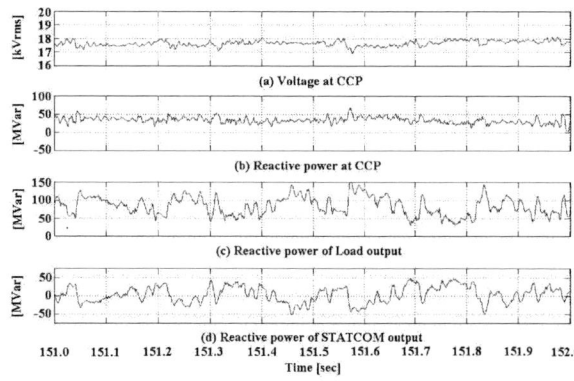

Fig. 7: Actual field waveforms of the STATCOM

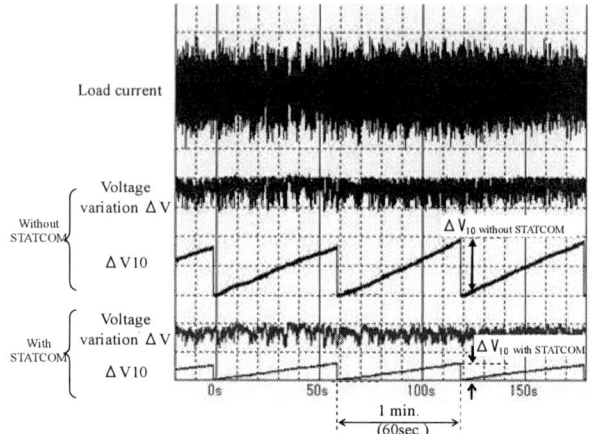

Fig. 8: Function diagram of the voltage flicker evaluation tool

IV. VOLTAGE FLICKER EVALUATION TOOL

The voltage flickers at random due to the arc furnaces operation and the phenomena do never repeat in the same waveform. Therefore, we can not compare simultaneously the voltage flicker values with and without the STATOM at the actual field operation.

The voltage digital flicker evaluation tool has been developed. It can calculate simultaneously the equivalent voltage flicker values with and without the STATCOM and analyze the factor of voltage flicker from the signal of fluctuating load current.

The evaluation tool consists of some fast DSP cards and analog input/output cards. Fig.9 shows the function diagram of the voltage flicker evaluation tool.

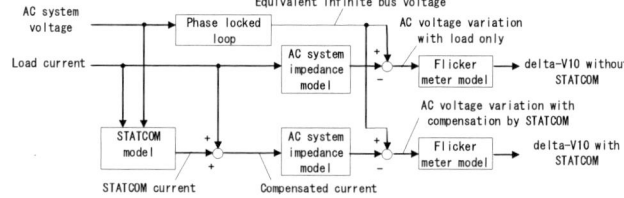

Fig. 9: Function diagram of the voltage flicker evaluation tool

It inputs the mains AC voltage and the load current. It also simulates the STATCOM output at the same time. Then, the voltage variations with and without the STATCOM can be estimated on line at the actual field operation. Then, the signals corresponding to the voltage variations are introduced to the voltage flicker meter models also in the DSP card. These models output the voltage flicker index delta-V10. The photograph of the evaluation tool is shown in Fig. 10 and Fig. 11.

978-1-4244-1871-8/07 $25.00 © 2007 IEEE

Fig. 10: Developed voltage flicker evaluation tool for the STATCOM

The software of the DSP are developed with

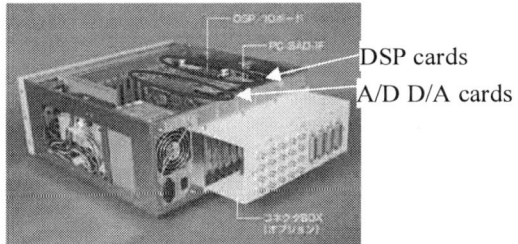

DSP cards
A/D D/A cards

Fig. 11: Developed voltage flicker evaluation tool cover opened

MATLAB/SIMLINK. In the SIMLINK, the AC system impedance, the STATCOM control and the meter models are made off-line. Then, the execution modules are compiled by the C language code generator included in MATLAB/SIMLINK library and installed in the DSP cards for real time calculation. Fig. 12 shows the development flow of the digital evaluation tool software.

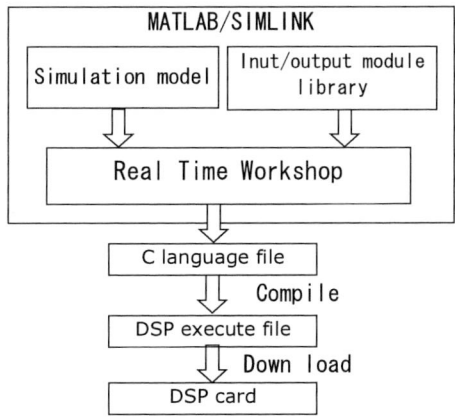

Fig. 12: Development flow of the digital evaluation software

The characteristics of the digital voltage flicker meter model are compared with one of the conventional analog voltage

flicker meters already available in market. The digital meter model characteristics are evaluated to be equivalent the analog meter as shown in Fig. 13. The characteristics are based on the Japanese standard and show the human sensitivity to the visual flicker of the electric light bulb for AC110V system.

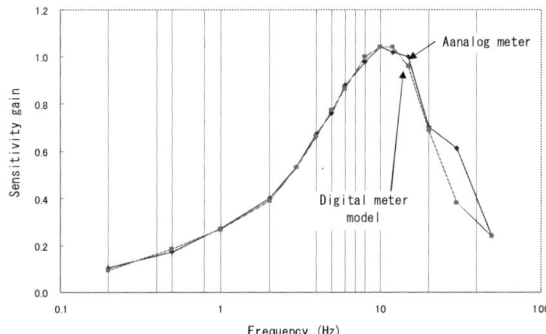

Fig. 13: Sensitivity characteristics installed in the flicker meter

The digital tool is compact and installed just in a small box. This makes the measurement and adjustment in the field very easy. The conventional analog tools are also shown in Fig. 14 for comparison. The analog tool takes long preparation hours, heavy to handle.

The tool can be used for pre-evaluation of the STATCOM installation. Using the model of STATCOM output, the voltage flicker can be simulated on line the actual AC system voltage and the load current at the expected STATCOM installation point. Then, the improvement effect of the STATCOM can be estimated before the installation.

Fig. 14: Conventional analog voltage flicker evaluation tool

978-1-4244-1871-8/07 $25.00 © 2007 IEEE

V. CONCLUSIONS

The 31.5MVA STATCOM and the voltage flicker evaluation tool have been successfully developed at the same time. The authors hope this development helps the industrial power quality improvement, especially the voltage flicker from the fluctuating loads. Furthermore, we also hope that the development contributes the industrial development by offering high quality electronic power leading to the high quality products or the high efficiency of the factory operation.

REFERENCES

[1] M.Tobita and S.Ota: "New Self-Commutated Static Var Compensator using IEGT for flicker suppression" IEEJ -IES conference03, Tokyo ,pp.1-S3-4(2003)

[2] M.Tobita : "Development of the 2nd generation SVCS using IEGT" ,PCC-Osaka, O-27-2 (2002) H. Poor, *An Introduction to Signal Detection and Estimation.* New York: Springer-Verlag, 1985, ch. 4.

[3] S.Ota et.al, :"New Self-Commutated SVC and SFC usin g IEGT" IPEC-Niigata, S71-2, pp.2177 (2003)

AUTHOR INDEX

Abass, A. 659
Abe, S. 252
Adnani, M. 816
Aguillon-Garcia, J. 1143
Ahn, H. 59
Ahn, I-R. 591
Ahn, J. 505
Ahn, J-W. 613, 649, 911
Ahn, J-Y. 90
Akagi, H. 1
Alahuhtala, J. 741
Alizadeh, A. 1101
Alizadeh, O. 748
Alkoshairy, A. 659
An, Y-J. 649
Andersen, M. 759
Andronescu, G. 351
Antunes, F. 510, 1131
Aoyama, F. 1228
Apte, N. 671
Araujo, S. 510, 1131
Bac, N. 1054
Badstuebner, U. 23
Bae, C-H. 800
Bae, H. 9
Baek, H-L. 527
Baek, J. 988
Baek, S-T. 439
Baik, W-S. 639, 723
Baljinnyam, Z. 169
Bapat, V. 671
Batista, E. 1149
Betz, R. 1217
Biela, J. 23
Borle, L. 955
Cabral Ferreira, A. 737

Caixue, C. 213
Canesin, C. 419
Castellazi, A. 1149
Cha, H. 894, 1090
Cha, I-H. 553
Cha, I-S. 521
Cha, J-S. 524
Chang, L. 597, 601
Chang, S-C. 217
Chang, S-K. 771, 800
Chang, Y-T. 624, 1085
Chen, C-Y. 309
Chen, J. 377, 430, 965
Chen, L. 1090
Chen, P-Y. 225
Chen, X. 921
Chen, Z. 601
Cho, B. 9, 988, 1171
Cho, C. 505
Cho, D. 709
Cho, G. 53
Cho, I. 502
Cho, J-H. 705
Cho, J-R. 243
Cho, K-H. 71, 203
Cho, K-M. 709, 882
Cho, K-R. 1200
Cho, M-T. 122, 217, 221
Cho, Y-O. 527
Choe, G-H. 66, 119, 169, 410
Choi, C-H. 357, 387, 723
Choi, C-L. 333
Choi, E-S. 705
Choi, G-S. 122, 217
Choi, H-J. 521, 524
Choi, I. 450

AUTHOR INDEX

Choi, J. 137, 164, 229, 404, 541

Choi, J-D. 315

Choi, J-S. 367, 372, 521, 524

Choi, J-Y. 450, 502

Choi, K-H. 221, 639

Choi, N-S. 337, 1036

Choi, S. 894, 982

Choi, W. 113

Choi, Y. 66, 71, 229

Chu, C-C. 225

Chun, T-W. 159, 337, 591

Chung, C. 541, 547

Chung, D-H. 367, 372

Chung, G-B. 164

Chung, J-M. 439

Chung, Y-H. 435

Ciappa, M. 1149

Cong, T. 580

Dagdag, S. 346

Dai, Z. 1064

Dasgupta, A. 1048

Dash, S. 413

Datta, M. 444, 828

De Freitas, L. 419

De Haan, S. 85, 192

Deng, L. 690

Dienot, J. 1149

Doi, N. 186

Du, S. 601

Duchesne, C. 346

Dugarjav, B. 169

Dutarde, E. 346

Edington, M. 921

Elgendy, O. 659

Engler, A. 1131

Eshtehardiha, S. 153, 246

Fang, Y. 921

Farhangi, S. 425, 180, 748, 1223

Fathi, S. 586

Fernandes, B. 1005

Fitchner, W. 1149

Freere, P. 899

Fukushima, K. 1064

Funabashi, T. 444, 828, 915

Funato, H. 822

Ge, L. 1154

Geetha, E. 805

Gharehpetian, G. 586

Gil, C-H. 243

Gomez, D. 935

Goto, H. 634

Griva, G. 147

Grzesiak, L. 777, 1042

Guang, Z. 300

Gun, S-D. 521

Guo, H-J. 634

Ha, K-M. 232

Ha, S-J. 271

Halasz, S. 1096

Hamada, T. 398

Hamaogi, M. 1085

Han, B-M. 1036, 1125

Han, D. 59

Han, J-H. 454

Han, M. 295

Han, M-K. 71

Han, S-B. 329

Han, S-T. 686

Han, Y. 541, 547

Haneyoshi, T. 143

Harada, Y. 1064

Hartmann, M. 1158

AUTHOR INDEX

Heising, C. 865
Heo, H-S. 66
Hidesi, K. 1228
Hiraki, E. 1183
Hirokawa, M. 252
Ho, S-K. 225
Holmes, D. 899
Hong, C. 295
Hong, S-C. 197, 713, 1125
Hosseiniaghdam, M. 586
Hou, Q. 870
Hsieh, J-C. 309
Hu, J. 850, 944
Huang, J. 53, 232
Huang, T-C. 309
Huang, X-H. 971
Hwang, H-M. 450
Hwang, L-H. 122, 217, 221
Hwang, T-S. 387
Hyun, D-S. 132, 553, 562, 567, 699, 732, 940, 950
Ichinokura, O. 634
Ion, Y-S. 435
Ise, T. 1165
Ishirobi, M. 1075
Ishizuka, T. 1228
Ito, T. 74
Jakobsen, L. 759
Jang, D-U. 800
Jang, S-H. 482
Jang, S-J. 203, 789, 976
Jang, T. 709
Jeon, H-J. 410, 454, 530
Jeon, J. 505
Jeong, B-C. 497
Jeong, C-I. 404
Jeong, H-G. 329

Jianlin, Z. 213
Jog, A. 671
Joung, S-E. 553, 940
Joung, W-T. 387
Jung, B-M. 329
Jung, H-C. 562, 950
Jung, H-J. 439
Jung, H-M. 271
Jung, H-S. 319
Jung, M. 53
Jung, S-M. 810, 1060
Jung, Y-C. 319
Jung, Y-G. 466, 471, 629
Jung, Y-S. 502
Kadar, I. 1096
Kadir, M. 1194
Kang, E-G. 62
Kang, G-I. 90
Kang, H-H. 439
Kasal, G. 1137
Kawamura, A. 575
Kennel, R. 737
Khai, N. 629
Kim, B-C. 278
Kim, B-H. 517
Kim, C-E. 274, 783, 794, 994, 1001, 1177
Kim, C-H. 783, 794
Kim, C-S. 266
Kim, D. 53, 541
Kim, D-E. 765
Kim, D-H. 90, 639
Kim, D-Y. 994
Kim, E-C. 315, 323
Kim, G. 289, 342
Kim, G-D. 203
Kim, H. 229

AUTHOR INDEX

Kim, H-G. 337, 482, 591

Kim, H-J. 435

Kim, H-S. 66, 122, 221

Kim, H-W. 810, 1060

Kim, I-D. 337, 613

Kim, J. 257, 477, 505, 547

Kim, J-B. 127

Kim, J-C. 66

Kim, J-H. 127, 271, 283, 471, 477, 1171

Kim, J-K. 243

Kim, J-S. 79

Kim, J-U. 319

Kim, J-Y. 976

Kim, K-H. 90

Kim, K-S. 295, 435

Kim, M-H. 357, 639, 723

Kim, N-H. 357, 639, 723

Kim, S. 505, 547

Kim, S-C. 232

Kim, S-H. 119, 278, 283, 463

Kim, S-J. 79

Kim, S-K. 460, 463

Kim, T. 257

Kim, T-J. 132, 477

Kim, T-S. 732, 961

Kim, T-W. 404

Kim, W-R. 530

Kim, W-S. 497

Kim, Y-C. 197, 1125

Kim, Y-D. 274

Kim, Y-G. 800

Kim, Y-R. 381, 460, 463, 929

Kim, Y-S. 169, 217, 387

Kim, Y-W. 107

Kimura, N. 398

Kisck, D. 351

Kisck, M. 351

Kitano, T. 571

Ko, J. 197, 367, 372, 713, 1036, 1125

Koczara, W. 43

Koh, K. 571

Kohama, T. 238

Kolar, J. 23, 888, 1011, 1158

Kong, T-W. 860

Kong, W. 899

Koo, C-H. 315

Koo, J-C. 315, 323

Kristoffersen, A. 1143

Kucuk, F. 634

Kumar, S. 665

Kurokawa, K. 487, 834

Kurokawa, F. 261

Kusuhara, Y. 619

Kwak, D-K. 266

Kwak, S. 333

Kwon, B-H. 79, 127

Kwon, B-K. 530

Kwon, J-M. 728

Kwon, T-S. 686, 694

Laczynski, T. 1107

Lai, J-H. 309

Lai, Y-S. 624

Lauttamus, R. 535

Lebey, T. 346

Lee, B-H. 1177

Lee, B-K. 203, 789, 976

Lee, B-S. 197, 266

Lee, C. 976, 1211

Lee, D-C. 482, 765

Lee, D-H. 649, 911

Lee, D-M. 699

Lee, D-S. 333

AUTHOR INDEX

Lee, H. 174, 289, 342

Lee, H-C. 1001

Lee, H-H..............................159, 361, 580, 728

Lee, H-M. 789

Lee, H-S. 62

Lee, J. 113, 137, 164, 229, 257, 1211

Lee, J-B. 694

Lee, J-C. 132

Lee, J-D. 132, 732

Lee, J-H...............119, 381, 387, 439, 783, 794, 860, 929

Lee, J-I. 524

Lee, J-M. 1171

Lee, J-P. 169, 477

Lee, J-S. 62

Lee, J-W..305, 1189

Lee, J-Y. 713

Lee, K. 487, 834

Lee, K-B. 497

Lee, K-J...567

Lee, K-O...502

Lee, K-S. 435

Lee, K-W...882

Lee, S. 243, 955

Lee, S-D...295

Lee, S-H. 221, 723

Lee, S-I. 410

Lee, S-J. 1171

Lee, W-C...381, 860, 929

Lee, W-J...225, 994

Lee, Y-H...613

Lei, X...492

Li, R...430

Li, Y...1036

Liang, J. 649

Liang, L...690

Lijuan, Z. 300

Lim, C-J. 530

Lim, J. 71, 771

Lim, S-B. 713

Lim, S-H. 771

Lim, Y-C. 90, 208, 305, 393, 466, 471, 629

Lin, J...1031

Lin, Y-S. 309

Liu, G. 96

Liu, H. 96

Liu, J...............................33, 102, 676, 850, 870, 944

Liu, S. 1154

Liu, X. 1154

Lorenz, L...682

Mahkhtari, H. 1101

Mahmoud, A. 659

Mannam, R. 607

Martens, A. 1107

Martinez, F. 935

Matsui, M. 571, 840, 846

Matsumoto, Y. 575

Meenakshi, J. 413

Mekhilef, S. 1194

Meng, Y. 850, 944

Mermet-Guyennet, M. 346, 1149

Miller, J. 16

Min, B. 257

Min, B-D...477

Min, B-H. 613

Miura, Y. 1165

Mok, H-S. 119, 278, 283

Mok, J-K. 800

Mokhtari, H. 643

Moon, G-W. 274, 705, 709, 882, 961, 982, 994, 1001, 1177

Moon, J...229, 460

Moon, S-H. 387

Moore, D. 1005

AUTHOR INDEX

Morizane, T. 398
Muesing, A. 1158
Mun, S. 1075, 1183
Mutschler, P. 1025
Na, S-K. 217, 221
Nakagawa, S. 619
Nakaoka, M. 1075, 1183
Nakayama, A. 619
Namnabat, M. 153, 246
Narikiyo, T. 107
Navrapescu, Y. 351
Nayar, C. 955
Nghia, D. 361
Nguyen, H. 159
Nguyen, N-T. 728
Nho, E-C. 337, 591, 613, 1036
Nho, N. 174, 361, 580, 1054
Ninomiya, T. 238, 252, 619, 1064, 1081
Nishida, Y. 137
Noguchi, Y. 915
Nokali, M. 59
Norigoe, I. 1064
Nussbaumer, T. 1011
Ogasawa, S. 822
Ogushi, K. 1228
Oh, H-S. 122, 217, 221
Oh, J. 794
Oh, M. 229
Oh, S. 289
Oh, S-B. 517
Oh, W-S. 243, 709, 882
Oh, Y. 278, 357
Oikawa, R. 575
Okamatsu, M. 261
Okyere, P. 1019
Oleschuk, V. 147

Omori, H. 1183
Ota, S. 1228
Oum, J-H. 466
Paek, S-H. 694
Pan, C-T. 225
Park, B-G. 553, 732
Park, B-S. 372
Park, C-H. 789
Park, D-H. 127
Park, G. 164
Park, H-G. 482
Park, H-S. 208, 367, 783, 794, 1001
Park, J. 988
Park, J-C. 169
Park, J-M. 527
Park, J-S. 810, 961, 1060
Park, K-B. 274, 1001, 1177
Park, K-T. 367, 372
Park, N-C. 283
Park, N-J. 567
Park, S. 79, 329, 342, 1211
Park, S-D. 337
Park, S-H. 381, 929, 961
Park, S-J. 521, 524
Pauta, S. 557
Peng, F. 1090
Peng, Q. 921
Peng, Y. 690
Pettersson, S. 654
Phuc, T. 1054
Piao, Z-G. 527
Ping, H. 1194
Pinggang, S. 597
Pong, C-Y. 875
Poodeh, M. 153, 246
Popescu, M. 351

AUTHOR INDEX

Pramanik, M. 1113

Premrudeepreechacham, S. 557

Profumo, F. 147

Qing, F. 492

Quanhai, W. 300

Ra, B-H. 460, 463

Raggl, K. 1011

Rahman, M. 905, 1069

Reed, J. 1119

Ren, J. 377, 965

Reza, S. 1113

Rose, J. 1119

Ryu, J-S. 732

Ryu, K. 477

Ryu, S. 59, 278

Ryu, Y-H. 381

Ryul, K. 257

Saghaleini, M. 1223

Saha, B. 1075, 1183

Sahoo, A. 413

Saito, T. 186

Salo, M. 654

Samadi, A. 180

Sasaki, N. 74

Sayeef, S. 905

Sekine, H. 444, 828, 915

Senjyu, T. 444, 828, 915

Sensama, P. 665

Sensarma, P. 1048

Seo, H. 229

Seo, I-D. 454

Seo, Y-G. 197, 1125

Seok, J-K. 1200

Shen, Y. 96

Shim, J-S. 266

Shimizu, T. 856

Shin, D-S. 393, 629

Shin, H-H. 62

Shin, Y-C. 460

Shon, J-G. 454, 530

Shoyama, M. 1081

Simanjorang, R. 1165

Singh, B. 1137

So, J-H. 450, 502

Soares, J. 419

Son, Y-H. 463

Song, B. 257

Song, E. 329, 477

Song, H. 221, 229

Song, J-H. 686, 694

Song, K-B. 517

Song, S H. 497, 502, 517

Staudt, V. 865

Steimel, A. 865

Subrahmanyam, V. 805

Sugimura, H. 1075, 1183

Suh, I-Y. 439

Suh, J-S. 521, 524

Sul, S-K. 1206

Summers, T. 1217

Sun, H-G. 393

Sun, X-D. 840, 846

Suzuki, S. 74, 143

Tai, C-C. 309

Takahashi, F. 1085

Takeshita, T. 753

Tanabe, T. 74

Tang, Q. 1090

Taniguchi, K. 398

Tara, E. 643

Tenconi, A. 147

Thornton, R. 921

AUTHOR INDEX

Thyagarajan, T. 413, 805

Toliyar, H. 357, 723

Tomasik, J. 777, 1042

Torrical Bascope, T. 510

Tran, Q-V. 591

Tripathy, P. 1048

Tsukakoshi, K. 1064

Tu, C-C. 309

Tumbelaka, H. 955

Tuusa, H. 535, 654

Tuyen, N. 174

Uddin, M. 1113

Ueda, Y. 74

Undeland, T. 1143

Van Zwam, A. 85, 192

Vangala, N. 607

Venkataramanan, G. 1119

Wada, K. 856

Waffler, S. 888

Wang, C-T. 309

Wang, H. 387, 870, 911

Wang, J. 430

Wang, R. 102, 870

Wang, X. 33, 850, 944

Wang, Y. 85, 192

Wang, Z. 33, 676

Wheeler, P. 404

Won, C-Y. 203, 381, 789, 860, 929, 976

Wu, L. 676, 875

Wu, W. 1031

Xiao, H. 122, 217

Xiao, L-Y. 971

Xiaogao, C. 492

Yabg, O. 357

Yamaguchi, K. 487, 834

Yamamoto, Y. 822

Yanagimura, K. 846

Yang, B. 709

Yang, C-S. 295

Yang, H-Y. 393

Yang, S-H. 208, 305

Yanmin, S. 300

Yazdanian, M. 425

Yeh, C-A. 1085

Yi, K-H. 982

Yim, J. 229

Yisheng, Y. 597

Yokoyama, R. 74

Yokoyama, T. 186

Yona, A. 444, 828, 915

Yong, S-I. 686, 694

Yoo, D. 257, 271, 477

Yoo, H. 547

Yoo, J. 257

Yoo, J-G. 410

Yoo, J-H. 122, 221

Yoo, M-H. 271

Yoon, H-K. 705

Yoon, J. 521, 524, 541

Yoon, M-J. 961

Yoon, Y-D. 1206

Yoshino, T. 1211

Yoshioka, Y. 575

You, K. 1069

Yougui, G. 213

Youn, M-J. 705, 810, 1060

Yousfi, D. 816

Yrasaki, N. 915

Yu, B. 450, 502, 571,840

Yu, G-J. 450, 502

Yu, J-S. 381, 860

Yu, Y. 690

AUTHOR INDEX

Yuan, C. 33, 944

Yun, D-J. 119

Zacharias, P. 510

Zhang, H. 377, 965

Zhi, Y-M. 122, 217

Zhong, Y. 377, 965

Zhou, G-X. 911

Zhou, Q. 1031

Zhuo, F. 676

Zoubek, H. 717

CURRAN ASSOCIATES INC.
proceedings
.com

9781424418718

2009 IEEE 22nd International Conference on Micro Electro Mechanical Systems (MEMS)

Sorrento, Italy
25-29 January 2009

IEEE Catalog Number: CFP09MEM-POD
ISBN: 978-1-42442-977-6